D1339101

ALL THE HELP, RESOURCES, AND PERSONAL SUPPORT YOU AND YOUR STUDENTS NEED!

2-Minute Tutorials and all of the resources you & your students need to get started
www.wileyplus.com/firstday

Student support from an experienced student user
Ask your local representative for details!

Collaborate with your colleagues, find a mentor, attend virtual and live events, and view resources
www.WhereFacultyConnect.com

Pre-loaded, ready-to-use assignments and presentations
www.wiley.com/college/quickstart

Technical Support 24/7
FAQs, online chat, and phone support
www.wileyplus.com/support

Your *WileyPLUS* Account Manager
Training and implementation support
www.wileyplus.com/accountmanage

MAKE IT YOURS!

Introduction to Electric Circuits

EIGHTH EDITION

Richard C. Dorf
University of California

James A. Svoboda
Clarkson University

International Student Version

John Wiley & Sons, Inc.

ISBN: 978-0-470-55302-2

Printed in Asia

10 9 8 7 6 5 4 3 2 1

The scientific nature of the ordinary man
Is to go on out and do the best he can.
—John Prine

But, Captain, I cannot change the laws of physics.
—Lt. Cmdr. Montogomery Scott (Scotty), USS *Enterprise*

Dedicated to our grandchildren:

Ian Christopher Boilard, Kyle Everett Schafer, and Graham Henry Schafer
and
Heather Lynn Svoboda, James Hugh Svoboda, Jacob Arthur Leis,
Maxwell Andrew Leis, and Jack Mandlin Leffler

About the Authors

Richard C. Dorf, professor of electrical and computer engineering at the University of California, Davis, teaches graduate and undergraduate courses in electrical engineering in the fields of circuits and control systems. He earned a PhD in electrical engineering from the U.S. Naval Postgraduate School, an MS from the University of Colorado, and a BS from Clarkson University. Highly concerned with the discipline of electrical engineering and its wide value to social and economic needs, he has written and lectured internationally on the contributions and advances in electrical engineering.

Professor Dorf has extensive experience with education and industry and is professionally active in the fields of robotics, automation, electric circuits, and communications. He has served as a visiting professor at the University of Edinburgh, Scotland, the Massachusetts Institute of Technology, Stanford University, and the University of California at Berkeley.

A Fellow of the Institute of Electrical and Electronic Engineers and the American Society for Engineering Education, Dr. Dorf is widely known to the profession for his *Modern Control Systems*, eleventh edition (Prentice Hall, 2008) and *The International Encyclopedia of Robotics* (Wiley, 1988). Dr. Dorf is also the coauthor of *Circuits, Devices and Systems* (with Ralph Smith), fifth edition (Wiley, 1992). Dr. Dorf edited the widely used *Electrical Engineering Handbook*, third edition (CRC Press and IEEE press), published in 2008. His latest work is Technology Ventures, third edition (McGraw-Hill 2010).

James A. Svoboda is an associate professor of electrical and computer engineering at Clarkson University,where he teaches courses on topics such as circuits, electronics, and computer programming. He earned a PhD in electrical engineering from the University of Wisconsin at Madison, an MS from the University of Colorado, and a BS from General Motors Institute.

Sophomore Circuits is one of Professor Svoboda's favorite courses. He has taught this course to 5,500 undergraduates at Clarkson University over the past 30 years. In 1986, he received Clarkson University's Distinguished Teaching Award.

Professor Svoboda has written several research papers describing the advantages of using nullors to model electric circuits for computer analysis. He is interested in the way technology affects engineering education and has developed several software packages for use in Sophomore Circuits.

Preface

The central theme of *Introduction to Electric Circuits* is the concept that electric circuits are part of the basic fabric of modern technology. Given this theme, we endeavor to show how the analysis and design of electric circuits are inseparably intertwined with the ability of the engineer to design complex electronic, communication, computer, and control systems as well as consumer products.

APPROACH & ORGANIZATION

This book is designed for a one- to three-term course in electric circuits or linear circuit analysis and is structured for maximum *flexibility*. The flowchart in Figure 1 demonstrates alternative chapter organizations that can accommodate different course outlines without disrupting continuity.

The presentation is geared to readers who are being exposed to the basic concepts of electric circuits for the first time, and the scope of the work is broad. Students should come to the course with the basic knowledge of differential and integral calculus.

This book endeavors to prepare the reader to solve realistic problems involving electric circuits. Thus, circuits are shown to be the results of real inventions and the answers to real needs in industry, the office, and the home. Although the tools of electric circuit analysis may be partially abstract, electric circuits are the building blocks of modern society. The analysis and design of electric circuits are critical skills for all engineers.

WHAT'S NEW IN THE 8TH EDITION

Increased use of PSpice® and MATLAB®

Significantly more attention has been given to using PSpice and MATLAB to solve circuits problems. It starts with two new appendixes, one introducing PSpice and the other introducing MATLAB. These appendixes briefly describe the capabilities of the programs and illustrate the steps needed to get started using them. Next, PSpice and MATLAB are used throughout the text to solve various circuit analysis and design problems. For example, PSpice is used in Chapter 5 to find a Thévenin equivalent circuit and in Chapter 15 to represent circuit inputs and outputs as Fourier series. MATLAB is frequently used to obtain plots of circuit inputs and outputs that help us see what our equations are telling us. MATLAB also helps us with some long and tedious arithmetic. For example, in Chapter 10, MATLAB helps us do the

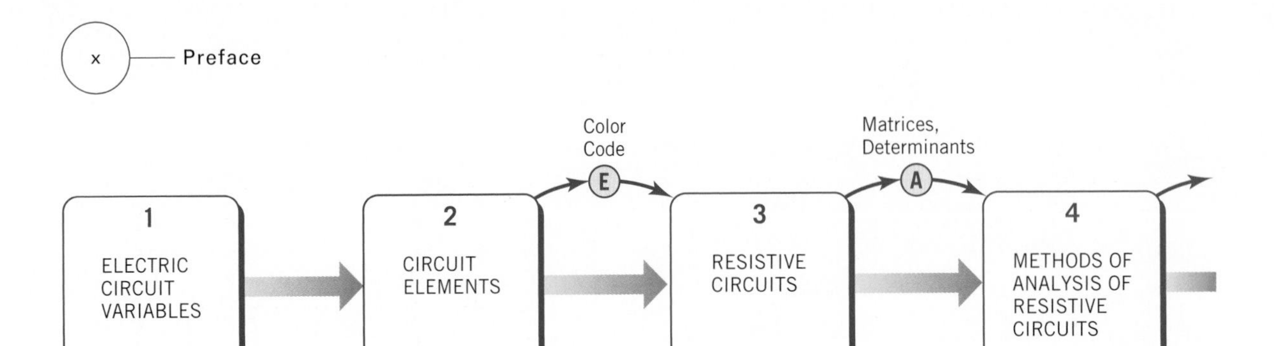

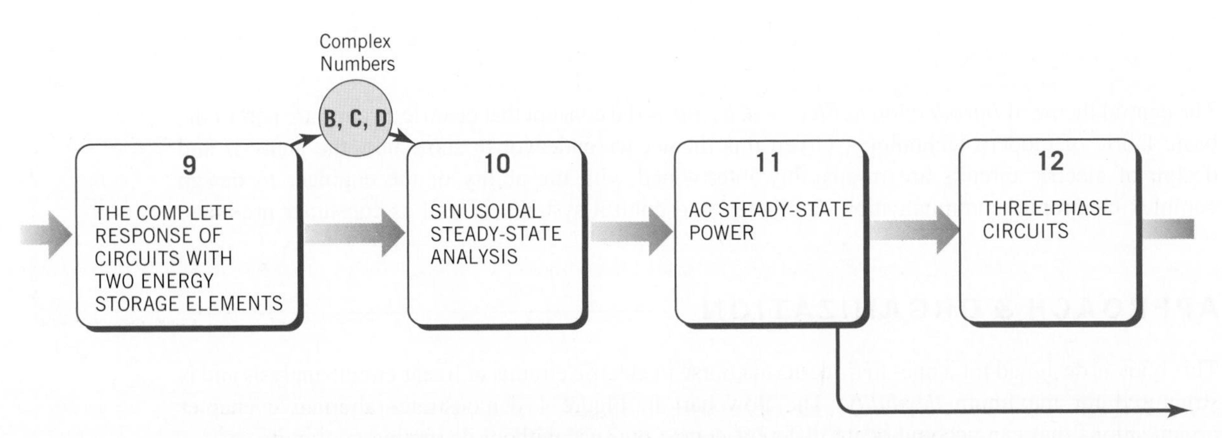

FIGURE 1 Flow chart showing alternative paths through the topics in this textbook.

complex arithmetic to analyze ac circuits and, in Chapter 14, MATLAB helps with the partial fraction required to find inverse Laplace transforms.

Of course, there's more to using PSpice and MATLAB than simply running the programs. We pay particular attention to interpreting the output of these computer programs and checking it to make sure it is correct. Frequently, this is done in the section called, ''How Can We Check . . . '' included in every chapter. For example, Section 8.9 shows how to interpret and check a PSpice transient response, and Section 13.7 shows how to interpret and check a frequency response produced using MATLAB or PSpice.

Revisions to Improve Clarity

Chapter 15 covering the Laplace transform and the Fourier series and transform, Chapters 14 and 15, have been largely rewritten, both to improve clarity of exposition and to significantly increase coverage of MATLAB and PSpice. In addition, revisions have been made throughout the text to improve clarity. Sometimes these revisions are small, involving sentences or paragraphs. Other, larger revisions involve pages or even entire sections.

More Problems

The 8th edition contains 120 new problems, bringing the total number of problems to more than 1,350. This edition uses a variety of problem types and they range in difficulty from simple to challenging, including:

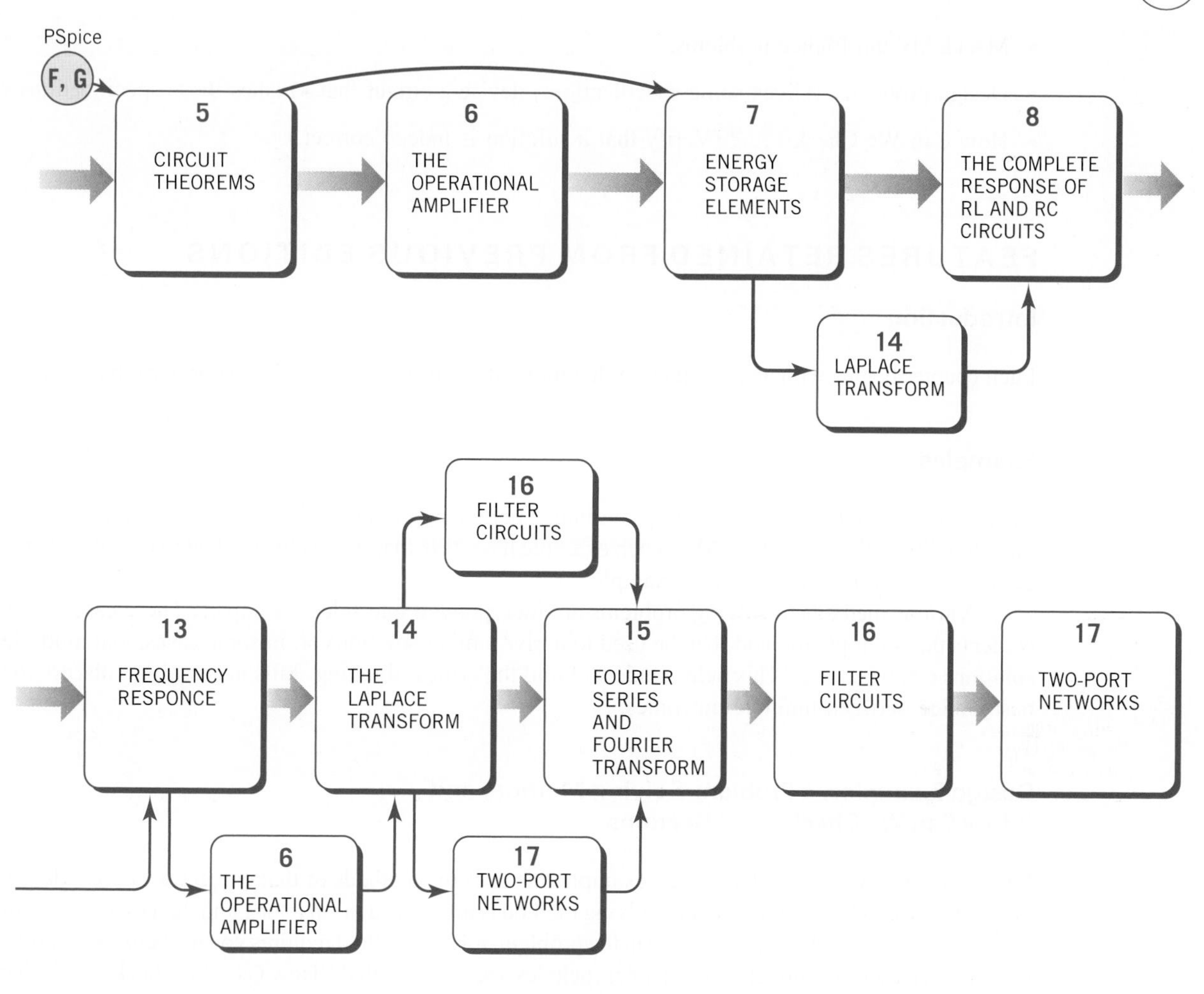

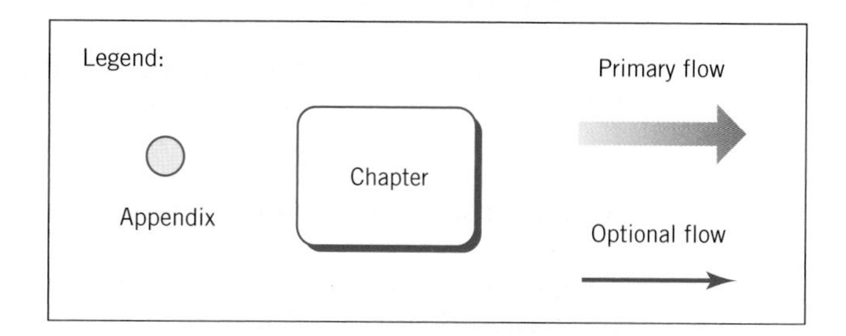

- Straightforward analysis problems.
- Analysis of complicated circuits.
- Simple design problems. (For example, given a circuit and the specified response, determine the required *RLC* values.)
- Compare and contrast, multipart problems that draw attention to similarities or differences between two situations.

- MATLAB and PSpice problems.
- Design problems. (Given some specifications, devise a circuit that satisfies those specifications.)
- How Can We Check . . . ? (Verify that a solution is indeed correct.)

FEATURES RETAINED FROM PREVIOUS EDITIONS

Introduction

Each chapter begins with an introduction that motivates consideration of the material of that chapter.

Examples

Because this book is oriented toward providing expertise in problem solving, we have included more than 260 illustrative examples. Also, each example has a title that directs the student to exactly what is being illustrated in that particular example.

Various methods of solving problems are incorporated into select examples. These cases show students that multiple methods can be used to derive similar solutions or, in some cases, that multiple solutions can be correct. This helps students build the critical thinking skills necessary to discern the best choice between multiple outcomes.

Design Examples, a Problem-Solving Method, and "How Can We Check . . . " Sections

Each chapter concludes with a design example that uses the methods of that chapter to solve a design problem. A formal, five-step problem-solving method is introduced in Chapter 1 and then used in each of the design examples. An important step in the problem-solving method requires you to check your results to verify that they are correct. Each chapter includes a section entitled ''How Can We Check . . . '' that illustrates how the kind of results obtained in that chapter can be checked to ensure correctness.

Key Equations and Formulas

You will find that key equations, formulas, and important notes have been called out in a shaded box to help you pinpoint critical information.

Summarizing Tables and Figures

The procedures and methods developed in this text have been summarized in certain key tables and figures. Students will find these to be an important problem-solving resource.

- Table 1.5-1. The passive convention.
- Figure 2.7-1 and Table 2.7-1. Dependent sources.
- Table 3.10-1. Series and parallel sources.
- Table 3.10-1. Series and parallel elements. Voltage and current division.
- Figure 4.2-3. Node voltages versus element currents and voltages.

- Figure 4.5-4. Mesh currents versus element currents and voltages.
- Figures 5.4-3 and 5.4-4. Thévenin equivalent circuits.
- Figure 6.3-1. The ideal op amp.
- Figure 6.5-1. A catalog of popular op amp circuits.
- Table 7.8-1. Capacitors and inductors.
- Table 7.13-2. Series and parallel capacitors and inductors.
- Table 8.11-1. First-order circuits.
- Tables 9.13-1, 2, and 3. Second-order circuits.
- Table 10.6-1. AC circuits in the frequency domain (phasors and impedances).
- Table 10.8-1. Voltage and current division for AC circuits.
- Table 11.5-1. Power formulas for AC circuits.
- Tables 11.13-1 and 11.13-2. Coupled inductors and ideal transformers.
- Table 13.4-1. Resonant circuits.
- Tables 14.2-1 and 14.2-2. Laplace transform tables.
- Table 14.7-1. s-domain models of circuit elements.
- Table 15.4-1. Fourier series of selected periodic waveforms.

Introduction to Signal Processing

Signal processing is an important application of electric circuits. This book introduces signal processing in two ways. First, two sections (Sections 6.6 and 7.9) describe methods to design electric circuits that implement algebraic and differential equations. Second, numerous examples and problems throughout this book illustrate signal processing. The input and output signals of an electric circuit are explicitly identified in each of these examples and problems. These examples and problems investigate the relationship between the input and output signals that is imposed by the circuit.

Interactive Examples and Exercises

Numerous examples throughout this book are labeled as interactive examples. This label indicates that computerized versions of that example are available at the textbook's companion site, www.wiley.com/dorf. Figure 2 illustrates the relationship between the textbook example and the computerized example available on the Web-Site. Figure 2*a* shows an example from Chapter 3. The problem presented by the interactive example shown in Figure 2*b* is similar to the textbook example but different in several ways:

- The values of the circuit parameters have been randomized.
- The independent and dependent sources may be reversed.
- The reference direction of the measured voltage may be reversed.
- A different question is asked. Here, the student is asked to work the textbook problem backward, using the measured voltage to determine the value of a circuit parameter.

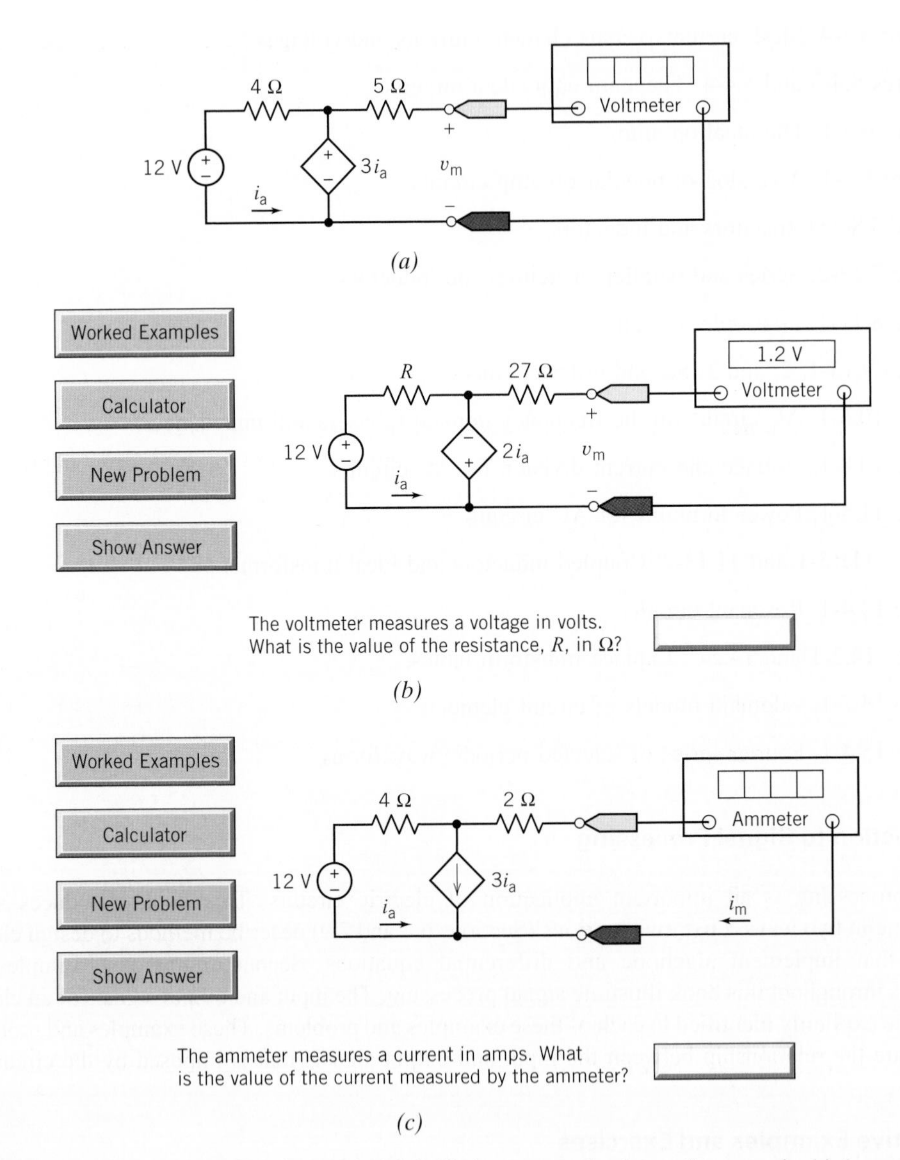

FIGURE 2 (*a*) The circuit considered Example 3.2-5. (*b*) A corresponding interactive example. (*c*) A corresponding interactive exercise.

The interactive example poses a problem and then accepts and checks the user's answer. Students are provided with immediate feedback regarding the correctness of their work. The interactive example chooses parameter values somewhat randomly, providing a seemingly endless supply of problems. This pairing of a solution to a particular problem with an endless supply of similar problems is an effective aid for learning about electric circuits.

The interactive exercise shown in Figure 2*c* considers a similar, but different, circuit. Like the interactive example, the interactive exercise poses a problem and then accepts and checks the user's answer. Student learning is further supported by extensive help in the form of worked example problems, available from within the interactive exercise, using the Worked Example button.

Variations of this problem are obtained using the New Problem button. We can peek at the answer, using the Show Answer button. The interactive examples and exercises provide hundreds of additional practice problems with countless variations, all with answers that are checked immediately by the computer.

SUPPLEMENTS AND WEB-SITE MATERIAL

The almost ubiquitous use of computers and the Web have provided an exciting opportunity to rethink supplementary material. The supplements available have been greatly enhanced.

Book Companion Site

Additional student and instructor resources can be found on the John Wiley & Sons textbook companion site at www.wiley.com/go/global/dorf.

Student

- **Interactive Examples** The interactive examples and exercises are powerful support resources for students. They were created as tools to assist students in mastering skills and building their confidence. The examples selected from the text and included on the Web give students options for navigating through the problem. They can immediately request to see the solution or select a more gradual approach to help. Then they can try their hand at a similar problem by simply electing to change the values in the problem. By the time students attempt the homework, they have built the confidence and skills to complete their assignments successfully. It's a virtual homework helper.
- **MATLAB Tutorial**, by Gary Ybarra and Michael Gustafson of Duke University, builds upon the MATLAB examples in the text. By providing these additional examples, the authors show how this powerful tool is easily used in appropriate areas of introductory circuit analysis. Ten example problems are created in HTML. M-files for the computer-based examples are available for download on the Student Companion site.
- PowerPoints for note taking
- Historical information
- *PSpice for Linear Circuits*, available for purchase
- *WileyPLUS* option

Instructor

- Solutions manual
- PowerPoint slides
- *WileyPLUS* option

WileyPLUS

Pspice for Linear Circuits is a student supplement available for purchase. The *PSpice for Linear Circuits* manual describes in careful detail how to incorporate this valuable tool in solving problems. This manual emphasizes the need to verify the correctness of computer output. No example is finished until the simulation results have been checked to ensure that they are correct.

ACKNOWLEDGMENTS AND COMMITMENT TO ACCURACY

We are grateful to many people whose efforts have gone into the making of this textbook. We are especially grateful to our Associate Publisher Daniel Sayre, Executive Marketing Manager Chris Ruel and Marketing Assistant Diana Smith for their support and enthusiasm. We are grateful to Janet Foxman and Dorothy Sinclair of Wiley and Heather Johnson of Elm Street Publishing Services for their efforts in producing this textbook. We wish to thank Lauren Sapira, Carolyn Weisman, and Andre Legaspi for their significant contributions to this project.

We are particularly grateful to the team of reviewerswho checked the problems and solutions to ensure their accuracy:

Accuracy Checkers

Khalid Al-Olimat, Ohio Northern University
Lisa Anneberg, Lawrence Technological University
Horace Gordon, University of South Florida
Lisimachos Kondi, SUNY, Buffalo
Michael Polis, Oakland University
Sannasi Ramanan, Rochester Institute of Technology
William Robbins, University of Minnesota
James Rowland, University of Kansas
Mike Shen, Duke University
Thyagarajan Srinivasan, Wilkes University
Aaron Still, U.S. Naval Academy
Howard Weinert, Johns Hopkins University
Xiao-Bang Xu, Clemson University
Jiann Shiun Yuan, University of Central Florida

Reviewers

Rehab Abdel-Kader, Georgia Southern University
Said Ahmed-Zaid, Boise State University
Farzan Aminian, Trinity University
Constantin Apostoaia, Purdue University Calumet
Jonathon Bagby, Florida Atlantic University
Carlotta Berry, Tennessee State University
Kiron Bordoloi, University of Louisville
Mauro Caputi, Hofstra University
Edward Collins, Clemson University
Glen Dudevoir, U.S. Miliary Academy
Malik Elbuluk, University of Akron
Prasad Enjeti, Texas A&M University
Ali Eydgahi, University of Maryland Eastern Shore
Carlos Figueroa, Cabrillo College
Walid Hubbi, New Jersey Institute of Technology
Brian Huggins, Bradley University
Chris Ianello, University of Central Florida
Simone Jarzabek, ITT Technical Institute
James Kawamoto, Mission College
Rasool Kenarangui, University of Texas Arlington
Jumoke Ladeji-Osias, Morgan State University
Mark Lau, Universidad del Turabo
Seyed Mousavinezhad, Western Michigan University
Philip Munro, Youngstown State University
Ahmad Nafisi, California Polytechnic State University
Arnost Neugroschel, University of Florida
Tokunbo Ogunfunmi, Santa Clara University
Gary Perks, California Polytechnic State University, San Luis Obispo
Owe Petersen, Milwaukee School of Engineering
Ron Pieper, University of Texas, Tyler
Teodoro Robles, Milwaukee School of Engineering
Pedda Sannuti, Rutgers University
Marcelo Simoes, Colorado School of Mines
Ralph Tanner, Western Michigan University
Tristan Tayag, Texas Christian University
Jean-Claude Thomassian, Central Michigan University
John Ventura, Christian Brothers University
Annette von Jouanne, Oregon State University
Ravi Warrier, Kettering University
Gerald Woelfl, Milwaukee School of Engineering
Hewlon Zimmer, U.S. Merchant Marine Academy

Contents

CHAPTER 1
Electric Circuit Variables 1
1.1 Introduction 1
1.2 Electric Circuits and Current 1
1.3 Systems of Units 5
1.4 Voltage 7
1.5 Power and Energy 7
1.6 Circuit Analysis and Design 11
1.7 How Can We Check . . . ? 13
1.8 Design Example—Jet Valve Controller 14
1.9 Summary 15
Problems 15
Design Problems 19

CHAPTER 2
Circuit Elements 20
2.1 Introduction 20
2.2 Engineering and Linear Models 20
2.3 Active and Passive Circuit Elements 24
2.4 Resistors 25
2.5 Independent Sources 28
2.6 Voltmeters and Ammeters 31
2.7 Dependent Sources 33
2.8 Transducers 37
2.9 Switches 39
2.10 How Can We Check . . . ? 41
2.11 Design Example—Temperature Sensor 42
2.12 Summary 44
Problems 44
Design Problems 52

CHAPTER 3
Resistive Circuits **53**
3.1 Introduction 53
3.2 Kirchhoff's Laws 53
3.3 Series Resistors and Voltage Division 61
3.4 Parallel Resistors and Current Division 66
3.5 Series Voltage Sources and Parallel Current Sources 72
3.6 Circuit Analysis 73
3.7 Analyzing Resistive Circuits Using MATLAB 78
3.8 How Can We Check . . . ? 82
3.9 Design Example—Adjustable Voltage Source 84
3.10 Summary 87
Problems 88
Design Problems 106

CHAPTER 4
Methods of Analysis of Resistive Circuits **108**
4.1 Introduction 108
4.2 Node Voltage Analysis of Circuits with Current Sources 109
4.3 Node Voltage Analysis of Circuits with Current and Voltage Sources 115
4.4 Node Voltage Analysis with Dependent Sources 120
4.5 Mesh Current Analysis with Independent Voltage Sources 122
4.6 Mesh Current Analysis with Current and Voltage Sources 127
4.7 Mesh Current Analysis with Dependent Sources 131
4.8 The Node Voltage Method and Mesh Current Method Compared 134
4.9 Mesh Current Analysis Using MATLAB 136
4.10 Using PSpice to Determine Node Voltages and Mesh Currents 138
4.11 How Can We Check . . . ? 140
4.12 Design Example—Potentiometer Angle Display 143
4.13 Summary 146
Problems 147
PSpice Problems 160
Design Problems 160

CHAPTER 5
Circuit Theorems **162**
5.1 Introduction 162
5.2 Source Transformations 162
5.3 Superposition 167
5.4 Thévenin's Theorem 171
5.5 Norton's Equivalent Circuit 175
5.6 Maximum Power Transfer 179
5.7 Using MATLAB to Determine the Thévenin Equivalent Circuit 182
5.8 Using PSpice to Determine the Thévenin Equivalent Circuit 185
5.9 How Can We Check . . . ? 188
5.10 Design Example—Strain Gauge Bridge 189
5.11 Summary 192
Problems 192
PSpice Problems 205
Design Problems 206

CHAPTER 6
The Operational Amplifier ... 208
6.1 Introduction ... 208
6.2 The Operational Amplifier ... 208
6.3 The Ideal Operational Amplifier ... 210
6.4 Nodal Analysis of Circuits Containing Ideal Operational Amplifiers ... 212
6.5 Design Using Operational Amplifiers ... 217
6.6 Operational Amplifier Circuits and Linear Algebraic Equations ... 222
6.7 Characteristics of Practical Operational Amplifiers ... 227
6.8 Analysis of Op Amp Circuits Using MATLAB ... 234
6.9 Using PSpice to Analyze Op Amp Circuits ... 236
6.10 How Can We Check . . . ? ... 237
6.11 Design Example—Transducer Interface Circuit ... 239
6.12 Summary ... 241
Problems ... 242
PSpice Problems ... 255
Design Problems ... 256

CHAPTER 7
Energy Storage Elements ... 257
7.1 Introduction ... 257
7.2 Capacitors ... 258
7.3 Energy Storage in a Capacitor ... 264
7.4 Series and Parallel Capacitors ... 267
7.5 Inductors ... 269
7.6 Energy Storage in an Inductor ... 274
7.7 Series and Parallel Inductors ... 276
7.8 Initial Conditions of Switched Circuits ... 277
7.9 Operational Amplifier Circuits and Linear Differential Equations ... 281
7.10 Using MATLAB to Plot Capacitor or Inductor Voltage and Current ... 287
7.11 How Can We Check . . . ? ... 289
7.12 Design Example—Integrator and Switch ... 290
7.13 Summary ... 293
Problems ... 294
Design Problems ... 309

CHAPTER 8
The Complete Response of *RL* And *RC* Circuits ... 311
8.1 Introduction ... 311
8.2 First-Order Circuits ... 311
8.3 The Response of a First-Order Circuit to a Constant Input ... 314
8.4 Sequential Switching ... 327
8.5 Stability of First-Order Circuits ... 329
8.6 The Unit Step Source ... 331
8.7 The Response of a First-Order Circuit to a Nonconstant Source ... 335
8.8 Differential Operators ... 340
8.9 Using PSpice to Analyze First-Order Circuits ... 342
8.10 How Can We Check . . . ? ... 345
8.11 Design Example—A Computer and Printer ... 349

8.12 Summary ... 352
Problems ... 353
PSpice Problems ... 366
Design Problems ... 367

CHAPTER 9
The Complete Response of Circuits With Two Energy Storage Elements ... 368
9.1 Introduction ... 368
9.2 Differential Equation for Circuits with Two Energy Storage Elements ... 369
9.3 Solution of the Second-Order Differential Equation—The Natural Response ... 373
9.4 Natural Response of the Unforced Parallel *RLC* Circuit ... 376
9.5 Natural Response of the Critically Damped Unforced Parallel *RLC* Circuit ... 379
9.6 Natural Response of an Underdamped Unforced Parallel *RLC* Circuit ... 380
9.7 Forced Response of an *RLC* Circuit ... 382
9.8 Complete Response of an *RLC* Circuit ... 386
9.9 State Variable Approach to Circuit Analysis ... 389
9.10 Roots in the Complex Plane ... 393
9.11 How Can We Check . . . ? ... 394
9.12 Design Example—Auto Airbag Igniter ... 397
9.13 Summary ... 399
Problems ... 401
PSpice Problems ... 412
Design Problems ... 413

CHAPTER 10
Sinusoidal Steady-State Analysis ... 415
10.1 Introduction ... 415
10.2 Sinusoidal Sources ... 416
10.3 Steady-State Response of an *RL* Circuit for a Sinusoidal Forcing Function ... 421
10.4 Complex Exponential Forcing Function ... 422
10.5 The Phasor ... 426
10.6 Phasor Relationships for *R*, *L*, and *C* Elements ... 430
10.7 Impedance and Admittance ... 434
10.8 Kirchhoff 's Laws Using Phasors ... 438
10.9 Node Voltage and Mesh Current Analysis Using Phasors ... 443
10.10 Superposition, Thévenin and Norton Equivalents, and Source Transformations ... 449
10.11 Phasor Diagrams ... 454
10.12 Phasor Circuits and the Operational Amplifier ... 455
10.13 The Complete Response ... 457
10.14 Using MATLAB for Analysis of Steady-State Circuits with Sinusoidal Inputs ... 464
10.15 Using PSpice to Analyze AC Circuits ... 466
10.16 How Can We Check . . . ? ... 469
10.17 Design Example—Op Amp Circuit ... 471
10.18 Summary ... 474
Problems ... 474
PSpice Problems ... 493
Design Problems ... 494

CHAPTER 11

AC Steady-State Power **496**

11.1 Introduction 496
11.2 Electric Power 496
11.3 Instantaneous Power and Average Power 497
11.4 Effective Value of a Periodic Waveform 501
11.5 Complex Power 503
11.6 Power Factor 511
11.7 The Power Superposition Principle 519
11.8 The Maximum Power Transfer Theorem 522
11.9 Coupled Inductors 523
11.10 The Ideal Transformer 531
11.11 How Can We Check . . . ? 536
11.12 Design Example—Maximum Power Transfer 538
11.13 Summary 540
Problems 542
PSpice Problems 556
Design Problems 556

CHAPTER 12

Three-Phase Circuits **558**

12.1 Introduction 558
12.2 Three-Phase Voltages 559
12.3 The Y-to-Y Circuit 562
12.4 The Δ-Connected Source and Load 571
12.5 The Y-to-Δ Circuit 573
12.6 Balanced Three-Phase Circuits 576
12.7 Instantaneous and Average Power in a Balanced Three-Phase Load 578
12.8 Two-Wattmeter Power Measurement 581
12.9 How Can We Check . . . ? 584
12.10 Design Example—Power Factor Correction 587
12.11 Summary 588
Problems 589
PSpice Problems 593
Design Problems 593

CHAPTER 13

Frequency Response **594**

13.1 Introduction 594
13.2 Gain, Phase Shift, and the Network Function 594
13.3 Bode Plots 606
13.4 Resonant Circuits 623
13.5 Frequency Response of Op Amp Circuits 630
13.6 Plotting Bode Plots Using MATLAB 632
13.7 Using PSpice to Plot a Frequency Response 634
13.8 How Can We Check . . . ? 636

13.9 Design Example—Radio Tuner 640
13.10 Summary 642
Problems 643
PSpice Problems 656
Design Problems 658

CHAPTER 14
The Laplace Transform 660
14.1 Introduction 660
14.2 Laplace Transform 661
14.3 Pulse Inputs 667
14.4 Inverse Laplace Transform 671
14.5 Initial and Final Value Theorems 677
14.6 Solution of Differential Equations Describing a Circuit 680
14.7 Circuit Analysis Using Impedance and Initial Conditions 681
14.8 Transfer Function and Impedance 692
14.9 Convolution 695
14.10 Stability 699
14.11 Partial Fraction Expansion Using MATLAB 702
14.12 How Can We Check . . . ? 707
14.13 Design Example—Space Shuttle Cargo Door 710
14.14 Summary 713
Problems 714
PSpice Problems 728
Design Problems 729

CHAPTER 15
Fourier Series and Fourier Transform 730
15.1 Introduction 730
15.2 The Fourier Series 731
15.3 Symmetry of the Function $f(t)$ 739
15.4 Fourier Series of Selected Waveforms 744
15.5 Exponential Form of the Fourier Series 746
15.6 The Fourier Spectrum 754
15.7 Circuits and Fourier Series 758
15.8 Using PSpice to Determine the Fourier Series 761
15.9 The Fourier Transform 766
15.10 Fourier Transform Properties 769
15.11 The Spectrum of Signals 773
15.12 Convolution and Circuit Response 774
15.13 The Fourier Transform and the Laplace Transform 777
15.14 How Can We Check . . . ? 779
15.15 Design Example—DC Power Supply 781
15.16 Summary 784
Problems 785
PSpice Problems 791
Design Problems 791

CHAPTER 16
Filter Circuits 793
16.1 Introduction 793
16.2 The Electric Filter 793
16.3 Filters 794
16.4 Second-Order Filters 797
16.5 High-Order Filters 805
16.6 Simulating Filter Circuits Using PSpice 811
16.7 How Can We Check . . . ? 815
16.8 Design Example—Anti-Aliasing Filter 817
16.9 Summary 820
Problems 820
PSpice Problems 825
Design Problems 828

CHAPTER 17
Two-Port and Three-Port Networks 829
17.1 Introduction 829
17.2 T-to-Π Transformation and Two-Port Three-Terminal Networks 830
17.3 Equations of Two-Port Networks 832
17.4 *Z* and *Y* Parameters for a Circuit with Dependent Sources 835
17.5 Hybrid and Transmission Parameters 837
17.6 Relationships Between Two-Port Parameters 839
17.7 Interconnection of Two-Port Networks 841
17.8 How Can We Check . . . ? 844
17.9 Design Example—Transistor Amplifier 846
17.10 Summary 848
Problems 848
Design Problems 852

APPENDIX A
Getting Started with PSpice 853

APPENDIX B
MATLAB, Matricies and Complex Arithmetic 860

APPENDIX C
Mathematical Formulas 871

APPENDIX D
Standard Resistor Color Code 874

References 876

Index 879

CHAPTER 1

Electric Circuit Variables

IN THIS CHAPTER

1.1 Introduction
1.2 Electric Circuits and Current
1.3 Systems of Units
1.4 Voltage
1.5 Power and Energy
1.6 Circuit Analysis and Design
1.7 How Can We Check . . . ?
1.8 **DESIGN EXAMPLE**—Jet Valve Controller
1.9 Summary
Problems
Design Problems

1.1 INTRODUCTION

A circuit consists of electrical elements connected together. Engineers use electric circuits to solve problems that are important to modern society. In particular:

1. Electric circuits are used in the generation, transmission, and consumption of electric power and energy.
2. Electric circuits are used in the encoding, decoding, storage, retrieval, transmission, and processing of information.

In this chapter, we will do the following:

- Represent the current and voltage of an electric circuit element, paying particular attention to the reference direction of the current and to the reference direction or polarity of the voltage
- Calculate the power and energy supplied or received by a circuit element
- Use the passive convention to determine whether the product of the current and voltage of a circuit element is the power supplied by that element or the power received by the element
- Use scientific notation to represent electrical quantities with a wide range of magnitudes

1.2 ELECTRIC CIRCUITS AND CURRENT

The outstanding characteristics of electricity when compared with other power sources are its mobility and flexibility. Electrical energy can be moved to any point along a couple of wires and, depending on the user's requirements, converted to light, heat, or motion.

An **electric circuit** or electric network is an interconnection of electrical elements linked together in a closed path so that an electric current may flow continuously.

Consider a simple circuit consisting of two well-known electrical elements, a battery and a resistor, as shown in Figure 1.2-1. Each element is represented by the two-terminal element shown in Figure 1.2-2. Elements are sometimes called devices, and terminals are sometimes called nodes.

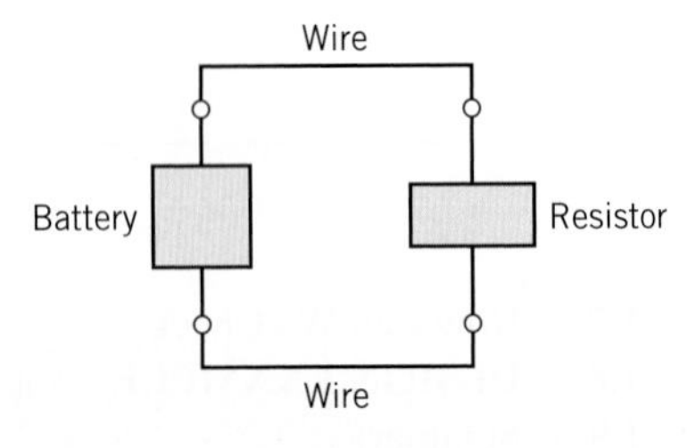

FIGURE 1.2-1 A simple circuit.

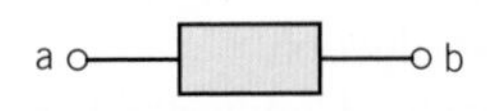

FIGURE 1.2-2 A general two-terminal electrical element with terminals a and b.

Charge may flow in an electric circuit. *Current is the time rate of change of charge past a given point.* Charge is the intrinsic property of matter responsible for electric phenomena. The quantity of charge q can be expressed in terms of the charge on one electron, which is -1.602×10^{-19} coulombs. Thus, -1 coulomb is the charge on 6.24×10^{18} electrons. The current through a specified area is defined by the electric charge passing through the area per unit of time. Thus, q is defined as the charge expressed in coulombs (C).

Charge is the quantity of electricity responsible for electric phenomena.

Then we can express current as

$$i = \frac{dq}{dt} \tag{1.2-1}$$

The unit of current is the ampere (A); an ampere is 1 coulomb per second.

Current is the time rate of flow of electric charge past a given point.

Note that throughout this chapter we use a lowercase letter, such as q, to denote a variable that is a function of time, $q(t)$. We use an uppercase letter, such as Q, to represent a constant.

The flow of current is conventionally represented as a flow of positive charges. This convention was initiated by Benjamin Franklin, the first great American electrical scientist. Of course, we now know that charge flow in metal conductors results from electrons with a negative charge. Nevertheless, we will conceive of current as the flow of positive charge, according to accepted convention.

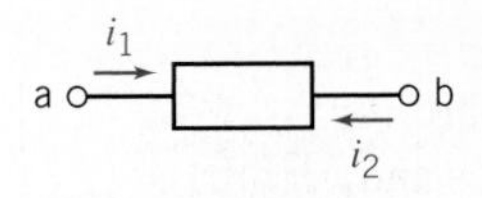

FIGURE 1.2-3 Current in a circuit element.

Figure 1.2-3 shows the notation that we use to describe a current. There are two parts to this notation: a value (perhaps represented by a variable name) and an assigned direction. As a matter of vocabulary, we say that a current exists *in* or *through* an element. Figure 1.2-3 shows that there are two ways to assign the direction of the current through an element. The current i_1 is the rate of flow of electric charge from terminal a to terminal b. On the other hand, the current i_2 is the flow of electric charge from terminal b to terminal a. The currents i_1 and i_2 are

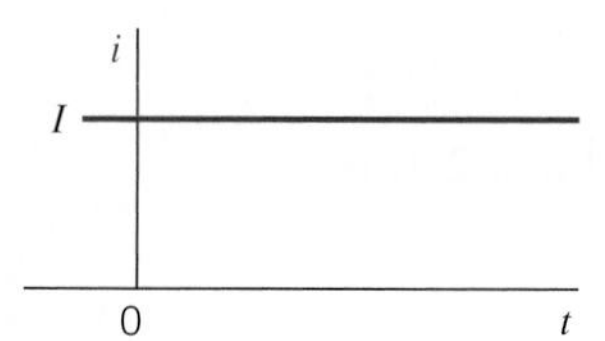

FIGURE 1.2-4 A direct current of magnitude I.

similar but different. They are the same size but have different directions. Therefore, i_2 is the negative of i_1 and

$$i_1 = -i_2$$

We always associate an arrow with a current to denote its direction. A complete description of current requires both a value (which can be positive or negative) and a direction (indicated by an arrow).

If the current flowing through an element is constant, we represent it by the constant I, as shown in Figure 1.2-4. A constant current is called a *direct current* (dc).

A **direct current** (dc) is a current of constant magnitude.

A time-varying current $i(t)$ can take many forms, such as a ramp, a sinusoid, or an exponential, as shown in Figure 1.2-5. The sinusoidal current is called an *alternating current* (ac).

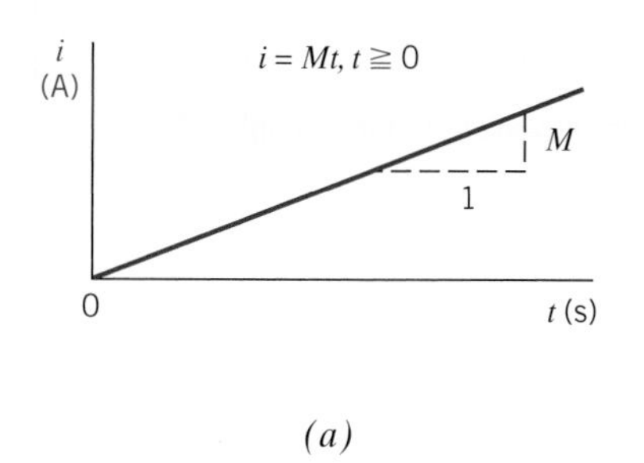

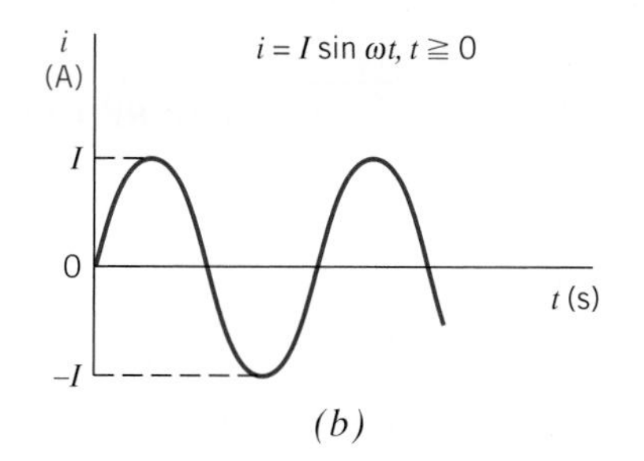

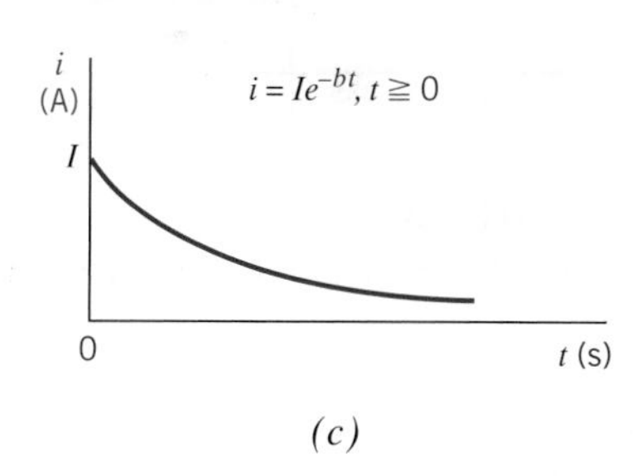

FIGURE 1.2-5 (*a*) A ramp with a slope M. (*b*) A sinusoid. (*c*) An exponential. I is a constant. The current i is zero for $t < 0$.

If the charge q is known, the current i is readily found using Eq. 1.2-1. Alternatively, if the current i is known, the charge q is readily calculated. Note that from Eq. 1.2-1, we obtain

$$q = \int_{-\infty}^{t} i\, d\tau = \int_{0}^{t} i\, d\tau + q(0) \tag{1.2-2}$$

where $q(0)$ is the charge at $t = 0$.

EXAMPLE 1.2-1 Current from Charge

Find the current in an element when the charge entering the element is

$$q = 12t \text{ C}$$

where t is the time in seconds.

Solution

Recall that the unit of charge is coulombs, C. Then the current, from Eq. 1.2-1, is

$$i = \frac{dq}{dt} = 12\ \text{A}$$

where the unit of current is amperes, A.

EXAMPLE 1.2-2 Charge from Current

Find the charge that has entered the terminal of an element from $t = 0$ s to $t = 3$ s when the current entering the element is as shown in Figure 1.2-6.

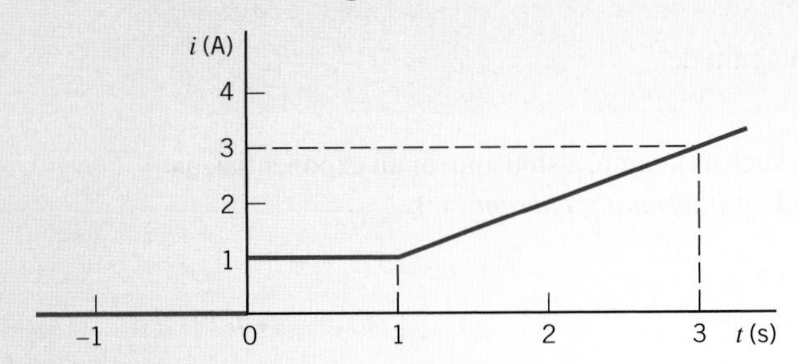

FIGURE 1.2-6 Current waveform for Example 1.2-2.

Solution

From Figure 1.2-6, we can describe $i(t)$ as

$$i(t) = \begin{cases} 0 & t < 0 \\ 1 & 0 < t \leq 1 \\ t & t > 1 \end{cases}$$

Using Eq. 1.2-2, we have

$$\begin{aligned} q(3) - q(0) &= \int_0^3 i(t)dt = \int_0^1 1\,dt + \int_1^3 t\,dt \\ &= t\Big|_0^1 + \frac{t^2}{2}\Big|_1^3 = 1 + \frac{1}{2}(9 - 1) = 5\ \text{C} \end{aligned}$$

Alternatively, we note that integration of $i(t)$ from $t = 0$ to $t = 3$ s simply requires the calculation of the area under the curve shown in Figure 1.2-6. Then, we have

$$q = 1 + 2 \times 2 = 5\ \text{C}$$

EXERCISE 1.2-1 Find the charge that has entered an element by time t when $i = 8t^2 - 4t$ A, $t \geq 0$. Assume $q(t) = 0$ for $t < 0$.

Answer: $q(t) = \frac{8}{3}t^3 - 2t^2$ C

EXERCISE 1.2-2 The total charge that has entered a circuit element is $q(t) = 4 \sin 3t$ C when $t \geq 0$, and $q(t) = 0$ when $t < 0$. Determine the current in this circuit element for $t > 0$.

Answer: $i(t) = \frac{d}{dt} 4 \sin 3t = 12 \cos 3t$ A

1.3 SYSTEMS OF UNITS

In representing a circuit and its elements, we must define a consistent system of units for the quantities occurring in the circuit. At the 1960 meeting of the General Conference of Weights and Measures, the representatives modernized the metric system and created the Système International d'Unites, commonly called SI units.

SI is *Systéme International d'Unités* or the International System of Units.

The fundamental, or base, units of SI are shown in Table 1.3-1. Symbols for units that represent proper (persons') names are capitalized; the others are not. Periods are not used after the symbols, and the symbols do not take on plural forms. The derived units for other physical quantities are obtained by combining the fundamental units. Table 1.3-2 shows the more common derived units along with their formulas in terms of the fundamental units or preceding derived units. Symbols are shown for the units that have them.

Table 1.3-1 SI Base Units

	SI UNIT	
QUANTITY	NAME	SYMBOL
Length	meter	m
Mass	kilogram	kg
Time	second	s
Electric current	ampere	A
Thermodynamic temperature	kelvin	K
Amount of substance	mole	mol
Luminous intensity	candela	cd

Table 1.3-2 Derived Units in SI

QUANTITY	UNIT NAME	FORMULA	SYMBOL
Acceleration — linear	meter per second per second	m/s^2	
Velocity — linear	meter per second	m/s	
Frequency	hertz	s^{-1}	Hz
Force	newton	$kg \cdot m/s^2$	N
Pressure or stress	pascal	N/m^2	Pa
Density	kilogram per cubic meter	kg/m^3	
Energy or work	joule	$N \cdot m$	J
Power	watt	J/s	W
Electric charge	coulomb	$A \cdot s$	C
Electric potential	volt	W/A	V
Electric resistance	ohm	V/A	Ω
Electric conductance	siemens	A/V	S
Electric capacitance	farad	C/V	F
Magnetic flux	weber	$V \cdot s$	Wb
Inductance	henry	Wb/A	H

Table 1.3-3 SI Prefixes

MULTIPLE	PREFIX	SYMBOL
10^{12}	tera	T
10^{9}	giga	G
10^{6}	mega	M
10^{3}	kilo	k
10^{-2}	centi	c
10^{-3}	milli	m
10^{-6}	micro	μ
10^{-9}	nano	n
10^{-12}	pico	p
10^{-15}	femto	f

The basic units such as length in meters (m), time in seconds (s), and current in amperes (A) can be used to obtain the derived units. Then, for example, we have the unit for charge (C) derived from the product of current and time (A · s). The fundamental unit for energy is the joule (J), which is force times distance or N · m.

The great advantage of the SI system is that it incorporates a decimal system for relating larger or smaller quantities to the basic unit. The powers of 10 are represented by standard prefixes given in Table 1.3-3. An example of the common use of a prefix is the centimeter (cm), which is 0.01 meter.

The decimal multiplier must always accompany the appropriate units and is never written by itself. Thus, we may write 2500 W as 2.5 kW. Similarly, we write 0.012 A as 12 mA.

EXAMPLE 1.3-1 SI Units

A mass of 150 grams experiences a force of 100 newtons. Find the energy or work expended if the mass moves 10 centimeters. Also, find the power if the mass completes its move in 1 millisecond.

Solution

The energy is found as

$$\text{energy} = \text{force} \times \text{distance} = 100 \times 0.1 = 10\ \text{J}$$

Note that we used the distance in units of meters. The power is found from

$$\text{power} = \frac{\text{energy}}{\text{time period}}$$

where the time period is 10^{-3} s. Thus,

$$\text{power} = \frac{10}{10^{-3}} = 10^4\ \text{W} = 10\ \text{kW}$$

EXERCISE 1.3-1 Which of the three currents, $i_1 = 45\ \mu\text{A}$, $i_2 = 0.03$ mA, and $i_3 = 25 \times 10^{-4}$ A, is largest?

Answer: i_3 is largest.

1.4 VOLTAGE

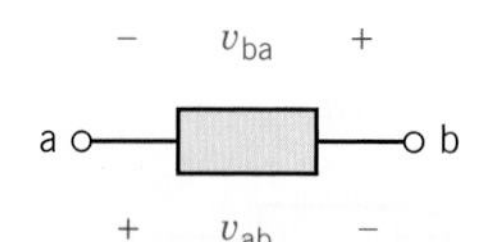

FIGURE 1.4-1 Voltage across a circuit element.

The basic variables in an electrical circuit are current and voltage. These variables describe the flow of charge through the elements of a circuit and the energy required to cause charge to flow. Figure 1.4-1 shows the notation we use to describe a voltage. There are two parts to this notation: a value (perhaps represented by a variable name) and an assigned direction. The value of a voltage may be positive or negative. The direction of a voltage is given by its polarities (+, −). As a matter of vocabulary, we say that a voltage exists *across* an element. Figure 1.4-1 shows that there are two ways to label the voltage across an element. The voltage v_{ba} is proportional to the work required to move a positive charge from terminal a to terminal b. On the other hand, the voltage v_{ab} is proportional to the work required to move a positive charge from terminal b to terminal a. We sometimes read v_{ba} as ''the voltage at terminal b with respect to terminal a.'' Similarly, v_{ab} can be read as ''the voltage at terminal a with respect to terminal b.'' Alternatively, we sometimes say that v_{ba} is the voltage drop from terminal a to terminal b. The voltages v_{ab} and v_{ba} are similar but different. They have the same magnitude but different polarities. This means that

$$v_{ab} = -v_{ba}$$

When considering v_{ba}, terminal b is called the ''+ terminal'' and terminal a is called the ''− terminal.'' On the other hand, when talking about v_{ab}, terminal a is called the ''+ terminal'' and terminal b is called the ''− terminal.''

The **voltage** across an element is the work (energy) required to move a unit positive charge from the − terminal to the + terminal. The unit of voltage is the volt, V.

The equation for the voltage across the element is

$$v = \frac{dw}{dq} \tag{1.4-1}$$

where v is voltage, w is energy (or work), and q is charge. A charge of 1 coulomb delivers an energy of 1 joule as it moves through a voltage of 1 volt.

1.5 POWER AND ENERGY

The power and energy delivered to an element are of great importance. For example, the useful output of an electric lightbulb can be expressed in terms of power. We know that a 300-watt bulb delivers more light than a 100-watt bulb.

Power is the time rate of expending or absorbing energy.

Thus, we have the equation

$$p = \frac{dw}{dt} \tag{1.5-1}$$

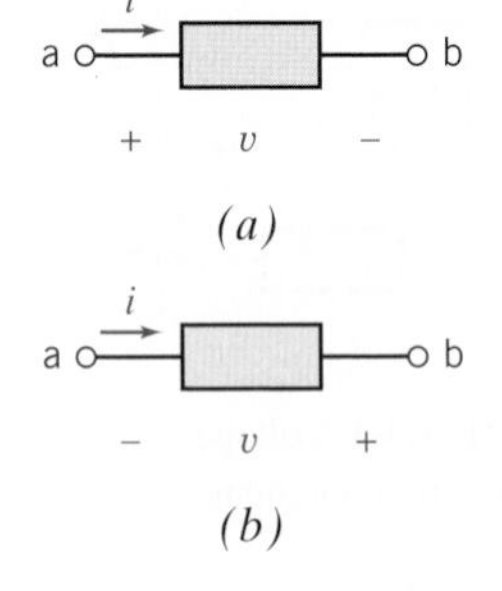

FIGURE 1.5-1 (*a*) The passive convention is used for element voltage and current. (*b*) The passive convention is not used.

where p is power in watts, w is energy in joules, and t is time in seconds. The power associated with the charge flow through an element is

$$p = \frac{dw}{dt} = \frac{dw}{dq} \cdot \frac{dq}{dt} = v \cdot i \tag{1.5-2}$$

From Eq. 1.5-2, we see that the power is simply the product of the voltage across an element times the current through the element. The power has units of watts.

Two circuit variables are assigned to each element of a circuit: a voltage and a current. Figure 1.5-1 shows that there are two different ways to arrange the direction of the current and the polarity of the voltage. In Figure 1.5-1*a*, the current enters the circuit element at the + terminal of the voltage and exits at the − terminal. In contrast, in Figure 1.5-1*b*, the current enters the circuit element at the − terminal of the voltage and exits at the + terminal.

First, consider Figure 1.5-1*a*. When the current enters the circuit element at the + terminal of the voltage and exits at the − terminal, the voltage and current are said to "adhere to the passive convention." In the passive convention, the voltage pushes a positive charge in the direction indicated by the current. Accordingly, the power calculated by multiplying the element voltage by the element current

$$p = vi$$

is the power **absorbed** by the element. (This power is also called "the power received by the element" and "the power dissipated by the element.") The power absorbed by an element can be either positive or negative. This will depend on the values of the element voltage and current.

Next, consider Figure 1.5-1*b*. Here the passive convention has not been used. Instead, the current enters the circuit element at the − terminal of the voltage and exits at the + terminal. In this case, the voltage pushes a positive charge in the direction opposite to the direction indicated by the current. Accordingly, when the element voltage and current do not adhere to the passive convention, the power calculated by multiplying the element voltage by the element current is the power **supplied** by the element. The power supplied by an element can be either positive or negative, depending on the values of the element voltage and current.

The power absorbed by an element and the power supplied by that same element are related by

power absorbed = −power supplied

The rules for the passive convention are summarized in Table 1.5-1. When the element voltage and current adhere to the passive convention, the energy absorbed by an element can be determined from

Table 1.5-1 Power Absorbed or Supplied by an Element

POWER ABSORBED BY AN ELEMENT	POWER SUPPLIED BY AN ELEMENT
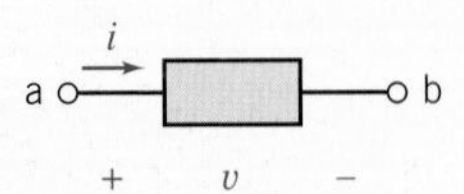	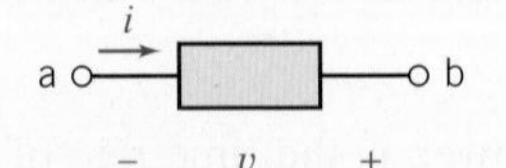
Because the reference directions of v and i adhere to the passive convention, the power $p = vi$ is the power absorbed by the element.	Because the reference directions of v and i do not adhere to the passive convention, the power $p = vi$ is the power supplied by the element.

Eq. 1.5-1 by rewriting it as

$$dw = p\,dt \tag{1.5-3}$$

On integrating, we have

$$w = \int_{-\infty}^{t} p\,d\tau \tag{1.5-4}$$

If the element receives power only for $t \geq t_0$ and we let $t_0 = 0$, then we have

$$w = \int_{0}^{t} p\,d\tau \tag{1.5-5}$$

EXAMPLE 1.5-1 Electrical Power and Energy

Let us consider the element shown in Figure 1.5-1*a* when $v = 4$ V and $i = 10$ A. Find the power absorbed by the element and the energy absorbed over a 10-s interval.

Solution

The power absorbed by the element is

$$p = vi = 4 \cdot 10 = 40 \text{ W}$$

The energy absorbed by the element is

$$w = \int_0^{10} p\,dt = \int_0^{10} 40\,dt = 40 \cdot 10 = 400 \text{ J}$$

EXAMPLE 1.5-2 Electrical Power and the Passive Convention

Consider the element shown in Figure 1.5-2. The current i and voltage v_{ab} adhere to the passive convention, so the power *absorbed* by this element is

$$\text{power absorbed} = i \cdot v_{ab} = 2 \cdot (-4) = -8 \text{ W}$$

The current i and voltage v_{ba} do not adhere to the passive convention, so the power *supplied* by this element is

$$\text{power supplied} = i \cdot v_{ba} = 2 \cdot (4) = 8 \text{ W}$$

As expected

$$\text{power absorbed} = -\text{power supplied}$$

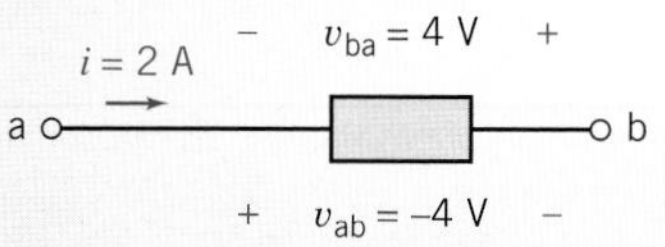

FIGURE 1.5-2 The element considered in Example 1.5-2.

Now let us consider an example when the passive convention is not used. Then $p = vi$ is the power supplied by the element.

EXAMPLE 1.5-3 Power, Energy, and the Passive Convention

Consider the circuit shown in Figure 1.5-3 with $v = 8e^{-t}$ V and $i = 20e^{-t}$ A for $t \geq 0$. Find the power supplied by this element and the energy supplied by the element over the first second of operation. We assume that v and i are zero for $t < 0$.

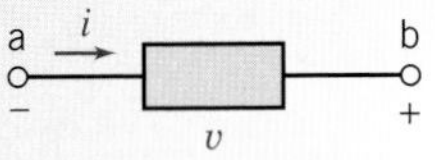

FIGURE 1.5-3 An element with the current flowing into the terminal with a negative voltage sign.

Solution

The power supplied is

$$p = vi = (8e^{-t})(20e^{-t}) = 160e^{-2t} \text{ W}$$

This element is providing energy to the charge flowing through it.

The energy supplied during the first second is

$$w = \int_0^1 p\,dt = \int_0^1 (160e^{-2t})dt$$

$$= 160\frac{e^{-2t}}{-2}\bigg|_0^1 = \frac{160}{-2}(e^{-2} - 1) = 80(1 - e^{-2}) = 69.2 \text{ J}$$

EXAMPLE 1.5-4 Energy in a Thunderbolt

The average current in a typical lightning thunderbolt is 2×10^4 A, and its typical duration is 0.1 s (Williams, 1988). The voltage between the clouds and the ground is 5×10^8 V. Determine the total charge transmitted to the earth and the energy released.

Solution

The total charge is

$$Q = \int_0^{0.1} i(t)\,dt = \int_0^{0.1} 2 \times 10^4\,dt = 2 \times 10^3 \text{ C}$$

The total energy released is

$$w = \int_0^{0.1} i(t) \times v(t)\,dt = \int_0^{0.1} (2 \times 10^4)(5 \times 10^8)\,dt = 10^{12} \text{ J} = 1 \text{ TJ}$$

EXERCISE 1.5-1 Figure E 1.5-1 shows four circuit elements identified by the letters A, B, C, and D.

(a) Which of the devices supply 12 W?

(b) Which of the devices absorb 12 W?

(f) What is the value of the power received by device B?

(g) What is the value of the power delivered by device B?

(h) What is the value of the power delivered by device D?

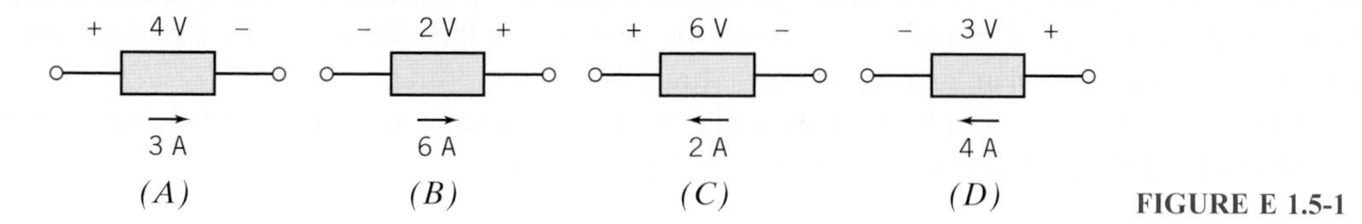

FIGURE E 1.5-1

Answers: **(a)** B and C, **(b)** A and D, **(c)** -12 W, **(d)** 12 W, **(e)** -12 W

1.6 CIRCUIT ANALYSIS AND DESIGN

The analysis and design of electric circuits are the primary activities described in this book and are key skills for an electrical engineer. The *analysis* of a circuit is concerned with the methodical study of a given circuit designed to obtain the magnitude and direction of one or more circuit variables, such as a current or voltage.

The analysis process begins with a statement of the problem and usually includes a given circuit model. The goal is to determine the magnitude and direction of one or more circuit variables, and the final task is to verify that the proposed solution is indeed correct. Usually, the engineer first identifies what is known and the principles that will be used to determine the unknown variable.

The problem-solving method that will be used throughout this book is shown in Figure 1.6-1. Generally, the problem statement is given. The analysis process then moves sequentially through the five steps shown in Figure 1.6-1. First, we describe the situation and the assumptions. We also record or review the circuit model that is provided. Second, we state the goals and requirements, and we

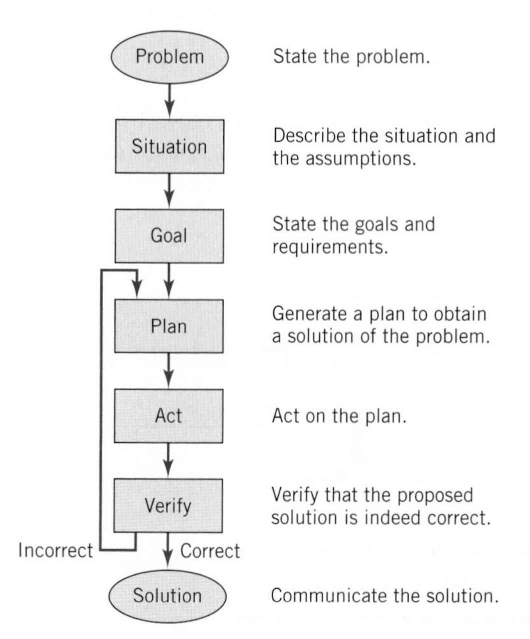

FIGURE 1.6-1 The problem-solving method.

normally record the required circuit variable to be determined. The third step is to create a plan that will help obtain the solution of the problem. Typically, we record the principles and techniques that pertain to this problem. The fourth step is to act on the plan and carry out the steps described in the plan. The final step is to verify that the proposed solution is indeed correct. If it is correct, we communicate this solution by recording it in writing or by presenting it verbally. If the verification step indicates that the proposed solution is incorrect or inadequate, then we return to the plan steps, reformulate an improved plan, and repeat steps 4 and 5.

To illustrate this analytical method, we will consider an example. In Example 1.6-1, we use the steps described in the problem-solving method of Figure 1.6-1.

EXAMPLE 1.6-1 The Formal Problem-Solving Method

An experimenter in a lab assumes that an element is absorbing power and uses a voltmeter and ammeter to measure the voltage and current as shown in Figure 1.6-2. The measurements indicate that the voltage is $v = +12$ V and the current is $i = -2$ A. Determine whether the experimenter's assumption is correct.

Describe the Situation and the Assumptions: Strictly speaking, the element *is* absorbing power. The value of the power absorbed by the element may be positive or zero or negative. When we say that someone "assumes that an element is absorbing power," we mean that someone assumes that the power absorbed by the element is positive.

The meters are ideal. These meters have been connected to the element in such a way as to measure the voltage labeled v and the current labeled i. The values of the voltage and current are given by the meter readings.

State the Goals: Calculate the power absorbed by the element to determine whether the value of the power absorbed is positive.

Generate a Plan: Verify that the element voltage and current adhere to the passive convention. If so, the power absorbed by the device is $p = vi$. If not, the power absorbed by the device is $p = -vi$.

Act on the Plan: Referring to Table 1.5-1, we see that the element voltage and current do adhere to the passive convention. Therefore, power absorbed by the element is

$$p = vi = 12 \cdot (-2) = -24 \text{ W}$$

The value of the power absorbed is not positive.

Verify the Proposed Solution: Let's reverse the ammeter probes as shown in Figure 1.6-3. Now the ammeter measures the current i_1 rather than the current i, so $i_1 = 2$ A and $v = 12$ V. Because i_1 and v do not adhere to the passive convention, $p = i_1 \cdot v = 24$ W is the power supplied by the element. Supplying 24 W is equivalent to absorbing -24 W, thus verifying the proposed solution.

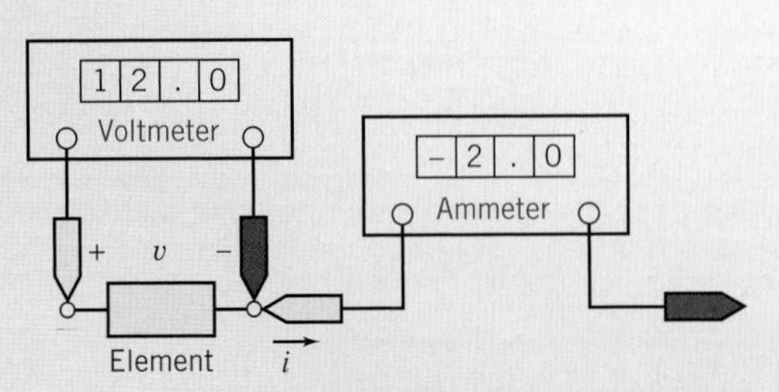

FIGURE 1.6-2 An element with a voltmeter and ammeter.

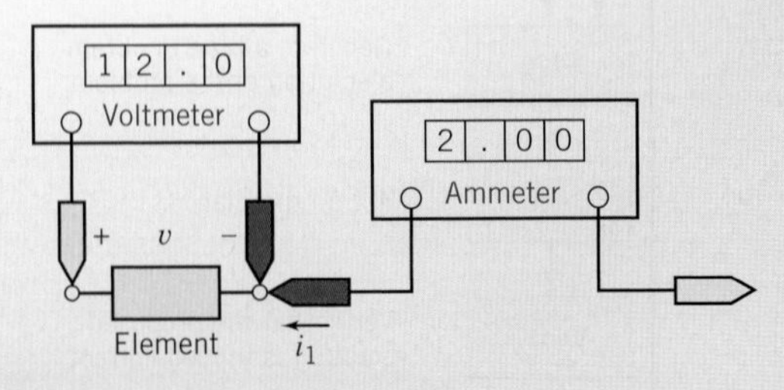

FIGURE 1.6-3 The circuit from Figure 1.6-2 with the ammeter probes reversed.

Design is a purposeful activity in which a designer visualizes a desired outcome. It is the process of originating circuits and predicting how these circuits will fulfill objectives. Engineering design is the process of producing a set of descriptions of a circuit that satisfy a set of performance requirements and constraints.

The design process may incorporate three phases: analysis, synthesis, and evaluation. The first task is to diagnose, define, and prepare—that is, to understand the problem and produce an explicit statement of goals; the second task involves finding plausible solutions; the third concerns judging the validity of solutions relative to the goals and selecting among alternatives. A cycle is implied in which the solution is revised and improved by reexamining the analysis. These three phases are part of a framework for planning, organizing, and evolving design projects.

Design is the process of creating a circuit to satisfy a set of goals.

The problem-solving process shown in Figure 1.6-1 is used in certain Design Examples included in each chapter.

1.7 HOW CAN WE CHECK . . . ?

Engineers are frequently called upon to check that a solution to a problem is indeed correct. For example, proposed solutions to design problems must be checked to confirm that all of the specifications have been satisfied. In addition, computer output must be reviewed to guard against data-entry errors, and claims made by vendors must be examined critically.

Engineering students are also asked to check the correctness of their work. For example, occasionally just a little time remains at the end of an exam. It is useful to be able quickly to identify those solutions that need more work.

This text includes some examples that illustrate techniques useful for checking the solutions of the particular problems discussed in that chapter. At the end of each chapter, some problems are presented that provide an opportunity to practice these techniques.

EXAMPLE 1.7-1 How Can We Check Power and the Passive Convention?

A laboratory report states that the measured values of v and i for the circuit element shown in Figure 1.7-1 are -5 V and 2 A, respectively. The report also states that the power absorbed by the element is 10 W. **How can we check** the reported value of the power absorbed by this element?

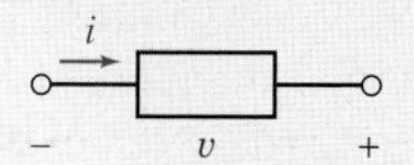

FIGURE 1.7-1 A circuit element with measured voltage and current.

Solution

Does the circuit element absorb -10 W or $+10$ W? The voltage and current shown in Figure 1.7-1 do not adhere to the passive sign convention. Referring to Table 1.5-1, we see that the product of this voltage and current is the power supplied by the element rather than the power absorbed by the element.

Then the power supplied by the element is

$$p = vi = (-5)(2) = -10 \text{ W}$$

The power absorbed and the power supplied by an element have the same magnitude but the opposite sign. Thus, we have verified that the circuit element is indeed absorbing 10 W.

1.8 DESIGN EXAMPLE

JET VALVE CONTROLLER

A small, experimental space rocket uses a two-element circuit, as shown in Figure 1.8-1, to control a jet valve from point of liftoff at $t = 0$ until expiration of the rocket after one minute. The energy that must be supplied by element 1 for the one-minute period is 40 mJ. Element 1 is a battery to be selected.

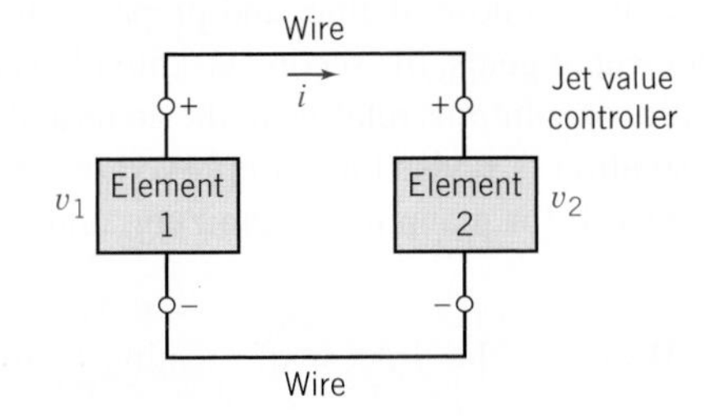

FIGURE 1.8-1 The circuit to control a jet valve for a space rocket.

It is known that $i(t) = De^{-t/60}$ mA for $t \geq 0$, and the voltage across the second element is $v_2(t) = Be^{-t/60}$ V for $t \geq 0$. The maximum magnitude of the current, D, is limited to 1 mA. Determine the required constants D and B and describe the required battery.

Describe the Situation and the Assumptions

1. The current enters the plus terminal of the second element.
2. The current leaves the plus terminal of the first element.
3. The wires are perfect and have no effect on the circuit (they do not absorb energy).
4. The model of the circuit, as shown in Figure 1.8-1, assumes that the voltage across the two elements is equal; that is, $v_1 = v_2$.
5. The battery voltage v_1 is $v_1 = Be^{-t/60}$ V where B is the initial voltage of the battery that will discharge exponentially as it supplies energy to the valve.
6. The circuit operates from $t = 0$ to $t = 60$ s.
7. The current is limited, so $D \leq 1$ mA.

State the Goal

Determine the energy supplied by the first element for the one-minute period and then select the constants D and B. Describe the battery selected.

Generate a Plan

First, find $v_1(t)$ and $i(t)$ and then obtain the power, $p_1(t)$, supplied by the first element. Next, using $p_1(t)$, find the energy supplied for the first 60 s.

GOAL	EQUATION	NEED	INFORMATION
The energy w_1 for the first 60 s	$w_1 = \int_0^{60} p_1(t)\, dt$	$p_1(t)$	v_1 and i known except for constants D and B

Act on the Plan

First, we need $p_1(t)$, so we first calculate

$$\begin{aligned} p_1(t) &= iv_1 = \left(De^{-t/60} \times 10^{-3}\text{ A}\right)\left(Be^{-t/60}\text{ V}\right) \\ &= DBe^{-t/30} \times 10^{-3}\text{ W} = DBe^{-t/30}\text{ mW} \end{aligned}$$

Second, we need to find w_1 for the first 60 s as

$$w_1 = \int_0^{60} \left(DBe^{-t/30} \times 10^{-3}\right) dt = \frac{DB \times 10^{-3} e^{-t/30}}{-1/30}\Bigg|_0^{60}$$
$$= -30DB \times 10^{-3}(e^{-2} - 1) = 25.9DB \times 10^{-3} \text{ J}$$

Because we require $w_1 \geq 40$ mJ,

$$40 \leq 25.9DB$$

Next, select the limiting value, $D = 1$, to get

$$B \geq \frac{40}{(25,.9)(1)} = 1.54 \text{ V}$$

Thus, we select a 2-V battery so that the magnitude of the current is less than 1 mA.

Verify the Proposed Solution

We must verify that at least 40 mJ is supplied using the 2-V battery. Because $i = e^{-t/60}$ mA and $v_2 = 2e^{-t/60}$ V, the energy supplied by the battery is

$$w = \int_0^{60} \left(2e^{-t/60}\right)\left(e^{-t/60} \times 10^{-3}\right) dt = \int_0^{60} 2e^{-t/30} \times 10^{-3}\, dt = 51.8 \text{ mJ}$$

Thus, we have verified the solution, and we communicate it by recording the requirement for a 2-V battery.

1.9 SUMMARY

- Charge is the intrinsic property of matter responsible for electric phenomena. The current in a circuit element is the rate of movement of charge through the element. The voltage across an element indicates the energy available to cause charge to move through the element.
- Given the current, i, and voltage, v, of a circuit element, the power, p, and energy, w, are given by

$$p = v \cdot i \quad \text{and} \quad w = \int_0^t p d\tau$$

- Table 1.5-1 summarizes the use of the passive convention when calculating the power supplied or received by a circuit element.
- The SI units (Table 1.3-1) are used by today's engineers and scientists. Using decimal prefixes (Table 1.3-3), we may simply express electrical quantities with a wide range of magnitudes.

PROBLEMS

Section 1.2 Electric Circuits and Current

P 1.2-1 The total charge that has entered a circuit element is $q(t) = 0.30(1 - e^{-5t})$ when $t \geq 0$ and $q(t) = 0$ when $t < 0$. Determine the current in this circuit element for $t \geq 0$.

Answer: $i(t) = 1.5e^{-5t}$ A

P 1.2-2 The current in a circuit element is $i(t) = 6(1 - e^{-5t})$ A when $t \geq 0$ and $i(t) = 0$ when $t < 0$. Determine the total charge that has entered a circuit element for $t \geq 0$.

Hint: $q(0) = \int_{-\infty}^{0} i(\tau)\, d\tau = \int_{-\infty}^{0} 0\, d\tau = 0$

Answer: $q(t) = 6t + 1.2e^{-5t} - 1.2$ C for $t \geq 0$

P 1.2-3 The current in a circuit element is $i(t) = 5 \sin 6t$ when $t \geq 0$ and $i(t) = 0$ when $t < 0$. Determine the total charge that has entered a circuit element for $t \geq 0$.

Hint: $q(0) = \int_{-\infty}^{0} i(\tau)\, d\tau = \int_{-\infty}^{0} 0\, d\tau = 0$

P 1.2-4 The current in a circuit element is

$$i(t) = \begin{cases} 0 & t < 2 \\ 2 & 2 < t < 4 \\ -1 & 4 < t < 8 \\ 0 & 8 < t \end{cases}$$

where the units of current are A and the units of time are s. Determine the total charge that has entered a circuit element for $t \geq 0$.

Answer:

$$q(t) = \begin{cases} 0 & t < 2 \\ 2t - 4 & 2 < t < 4 \\ 8 - t & 4 < t < 8 \\ 0 & 8 < t \end{cases}$$ where the units of charge are C.

P 1.2-5 The total charge $q(t)$, in coulombs, that enters the terminal of an element is

$$q(t) = \begin{cases} 0 & t < 0 \\ 3t & 0 \leq t \leq 2 \\ 5 + e^{-2(t-2)} & t > 2 \end{cases}$$

Find the current $i(t)$ and sketch its waveform for $t \geq 0$.

P 1.2-6 An electroplating bath, as shown in Figure P 1.2-6, is used to plate silver uniformly onto objects such as kitchenware and plates. A current of 460 A flows for 30 minutes, and each coulomb transports 1.120 mg of silver. What is the weight of silver deposited in grams?

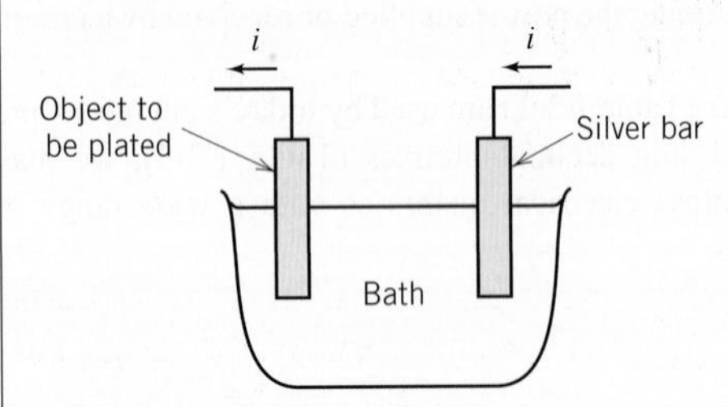

Figure P 1.2-6 An electroplating bath.

P 1.2-7 Find the charge, $q(t)$, and sketch its waveform when the current entering a terminal of an element is as shown in Figure P 1.2-7. Assume that $q(t) = 0$ for $t < 0$.

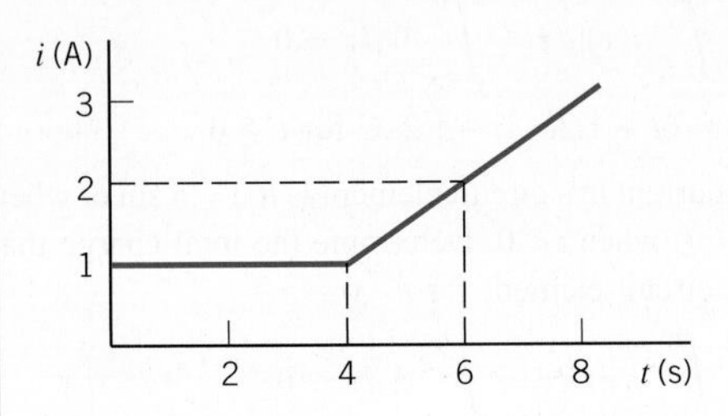

Figure P 1.2-7

Section 1.3 Systems of Units

P 1.3-1 A constant current of 3.5 μA flows through an element. What is the charge that has passed through the element in the first millisecond?

Answer: 3.5 nC

P 1.3-2 A charge of 50 nC passes through a circuit element during a particular interval of time that is 8 ms in duration. Determine the average current in this circuit element during that interval of time.

Answer: $i = 6.25\ \mu$A

P 1.3-3 Twenty billion electrons per second pass through a particular circuit element. What is the average current in that circuit element?

Answer: $i = 3.204$ nA

P 1.3-4 The charge flowing in a wire is plotted in Figure P 1.3-4. Sketch the corresponding current.

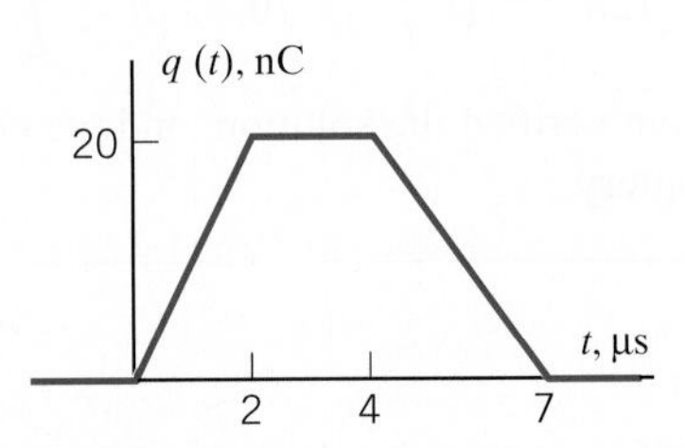

Figure P 1.3-4

P 1.3-5 The current in a circuit element is plotted in Figure P 1.3-5. Sketch the corresponding charge flowing through the element for $t > 0$.

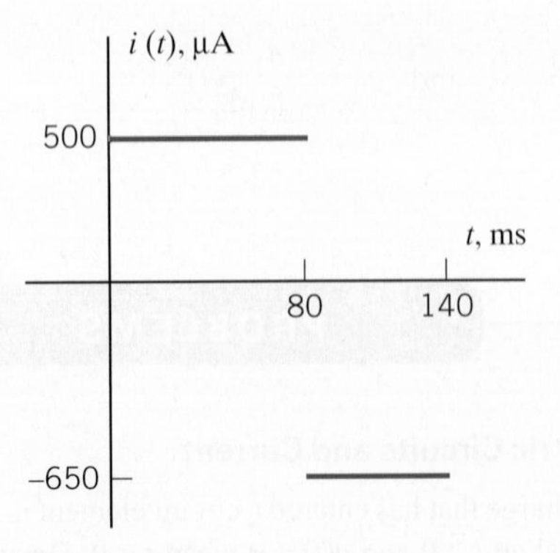

Figure P 1.3-5

P 1.3-6 The current in a circuit element is plotted in Figure P 1.3-6. Determine the total charge that flows through the circuit element between 300 and 1500 μs.

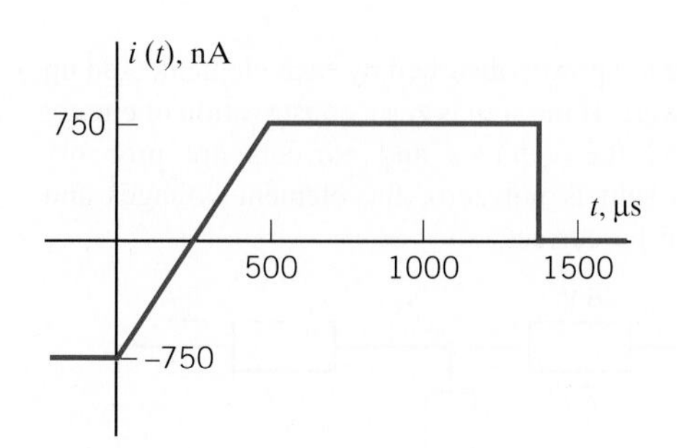

Figure P 1.3-6

Section 1.5 Power and Energy

P 1.5-1 Figure P 1.5-1 shows four circuit elements identified by the letters *A*, *B*, *C*, and *D*.

(a) Which of the devices supply 60 mW?
(b) Which of the devices absorb 0.06 W?
(c) What is the value of the power received by device *B*?
(d) What is the value of the power delivered by device *B*?
(e) What is the value of the power delivered by device *C*?

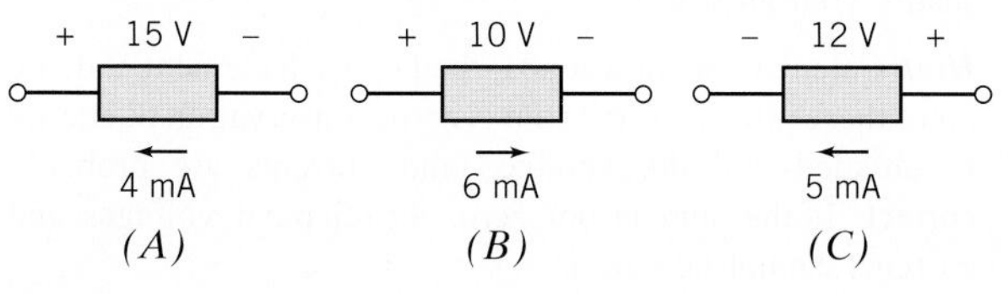

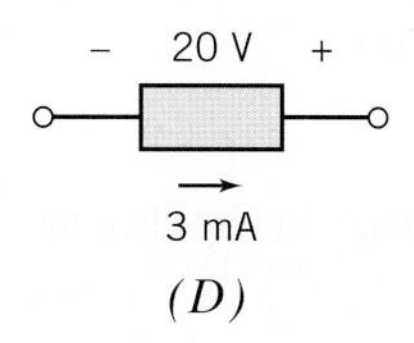

Figure P 1.5-1

P 1.5-2 An electric range has a constant current of 10 A entering the positive voltage terminal with a voltage of 110 V. The range is operated for two hours. (a) Find the charge in coulombs that passes through the range. (b) Find the power absorbed by the range. (c) If electric energy costs 12 cents per kilowatt-hour, determine the cost of operating the range for two hours.

P 1.5-3 A walker's cassette tape player uses four AA batteries in series to provide 8 V to the player circuit. The four alkaline battery cells store a total of 220 watt-seconds of energy. If the cassette player is drawing a constant 10 mA from the battery pack, how long will the cassette operate at normal power?

P 1.5-4 The current through and voltage across an element vary with time as shown in Figure P 1.5-4. Sketch the power delivered to the element for $t > 0$. What is the total energy delivered to the element between $t = 0$ and $t = 25$ s? The element voltage and current adhere to the passive convention.

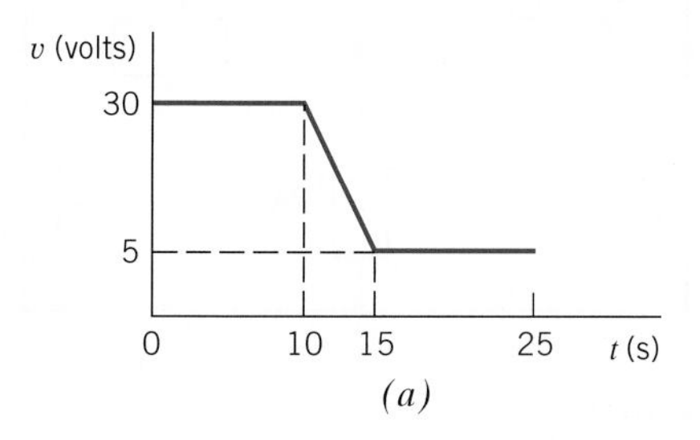

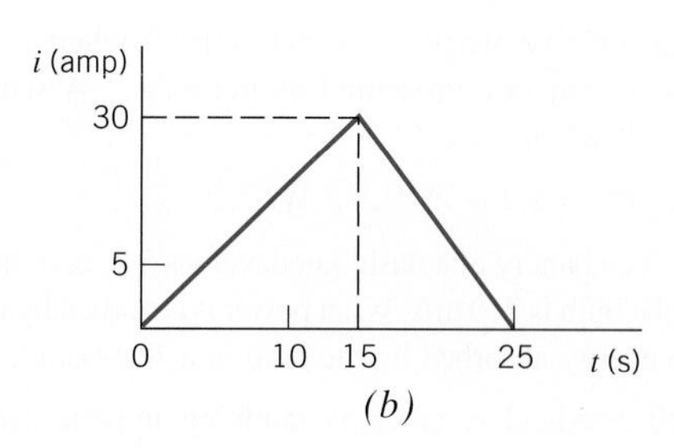

Figure P 1.5-4 (*a*) Voltage $v(t)$ and (*b*) current $i(t)$ for an element.

P 1.5-5 An automobile battery is charged with a constant current of 3 A for six hours. The terminal voltage of the battery is $v = 11 + 0.5t$ V for $t > 0$, where t is in hours. (a) Find the energy delivered to the battery during the six hours. (b) If electric energy costs 15 cents/kWh, find the cost of charging the battery for six hours.

Answer: (b) 2.76 cents

P 1.5-6 Find the power, $p(t)$, supplied by the element shown in Figure P 1.5-6 when $v(t) = 4 \cos 3t$ V and $i(t) = \dfrac{\sin 3t}{12}$ A. Evaluate $p(t)$ at $t = 0.5$ s and at $t = 1$ s. Observe that the power supplied by this element has a positive value at some times and a negative value at other times.

Hint: $(\sin at)(\cos bt) = \dfrac{1}{2}(\sin(a + b)t + \sin(a - b)t)$

Answer:

$$p(t) = \frac{1}{6}\sin 6t \text{ W}, \quad p(0.5) = 0.0235 \text{ W}, \quad p(1) = -0.0466 \text{ W}$$

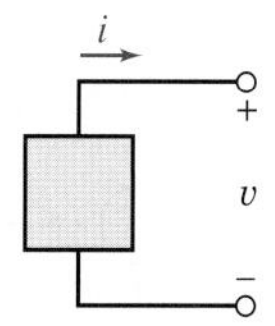

Figure P 1.5-6 An element.

P 1.5-7 Find the power, $p(t)$, supplied by the element shown in Figure P 1.5-6 when $v(t) = 8 \sin 3t$ V and $i(t) = 2 \sin 3t$ A.

Hint: $(\sin at)(\sin bt) = \dfrac{1}{2}(\cos(a - b)t - \cos(a + b)t)$

Answer: $p(t) = 8 - 8\cos 6t$ W

P 1.5-8 Find the power, $p(t)$, supplied by the element shown in Figure P 1.5-6. The element voltage is represented as

$v(t) = 4(1 - e^{-2t})$V when $t \geq 0$ and $v(t) = 0$ when $t < 0$. The element current is represented as $i(t) = 2e^{-2t}$ A when $t \geq 0$ and $i(t) = 0$ when $t < 0$.

Answer: $p(t) = 8(1 - e^{-2t})e^{-2t}$ W

P 1.5-9 The battery of a flashlight develops 5 V, and the current through the bulb is 210 mA. What power is absorbed by the bulb? Find the energy absorbed by the bulb in a five-minute period.

P 1.5-10 Medical researchers studying hypertension often use a technique called ''2D gel electrophoresis'' to analyze the protein content of a tissue sample. An image of a typical ''gel'' is shown in Figure P1.5-10*a*.

The procedure for preparing the gel uses the electric circuit illustrated in Figure 1.5-10*b*. The sample consists of a gel and a filter paper containing ionized proteins. A voltage source causes a large, constant voltage 500 V, across the sample. The large, constant voltage moves the ionized proteins from the filter paper to the gel. The current in the sample is given by

$$i(t) = 2 + 30e^{-at} \text{ mA}$$

where t is the time elapsed since the beginning of the procedure and the value of the constant a is

$$a = 0.85 \frac{1}{\text{hr}}$$

Determine the energy supplied by the voltage source when the gel preparation procedure lasts 3 hours.

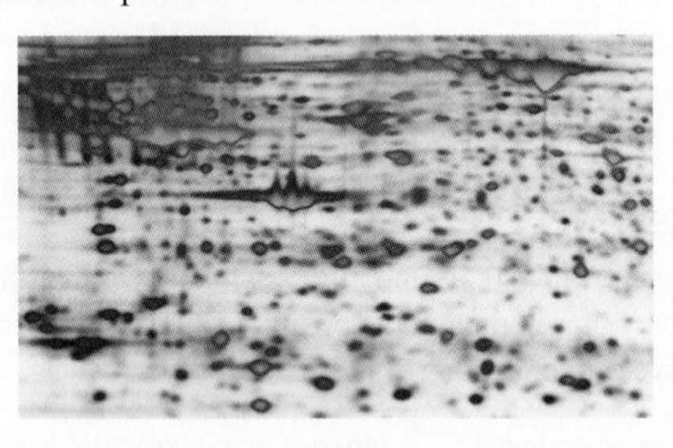

(*a*)

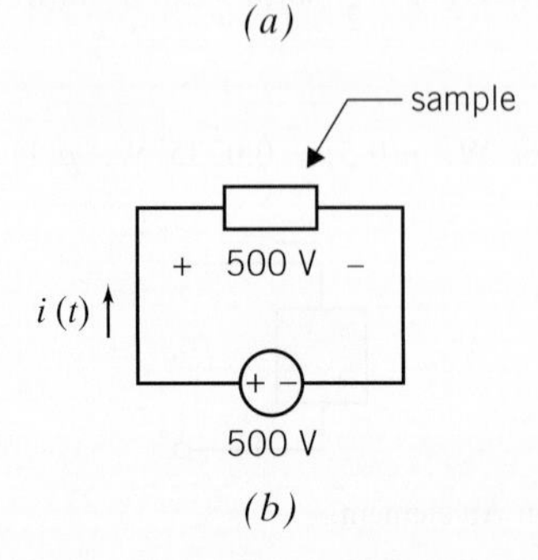

(*b*)

Figure P 1.5-10 (*a*) An image of a gel and (*b*) the electric circuit used to prepare gel.

Section 1.7 How Can We Check . . . ?

P 1.7-1 Conservation of energy requires that the sum of the power absorbed by all of the elements in a circuit be zero. Figure P 1.7-1 shows a circuit. All of the element voltages and currents are specified. Are these voltage and currents correct? Justify your answer.

Hint: Calculate the power absorbed by each element. Add up all of these powers. If the sum is zero, conservation of energy is satisfied and the voltages and currents are probably correct. If the sum is not zero, the element voltages and currents cannot be correct.

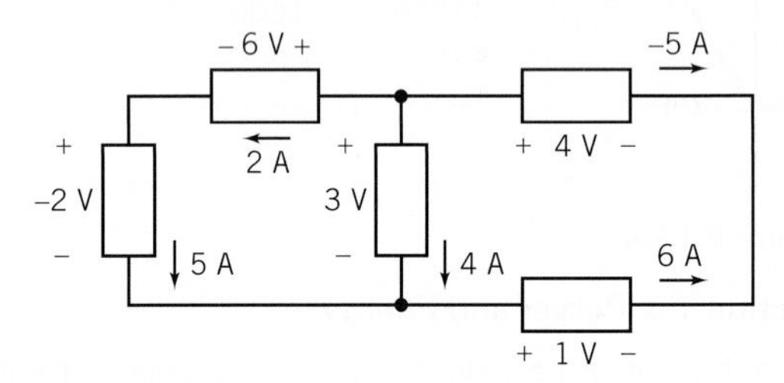

Figure P 1.7-1

P 1.7-2 Conservation of energy requires that the sum of the power absorbed by all of the elements in a circuit be zero. Figure P 1.7-2 shows a circuit. All of the element voltages and currents are specified. Are these voltage and currents correct? Justify your answer.

Hint: Calculate the power absorbed by each element. Add up all of these powers. If the sum is zero, conservation of energy is satisfied and the voltages and currents are probably correct. If the sum is not zero, the element voltages and currents cannot be correct.

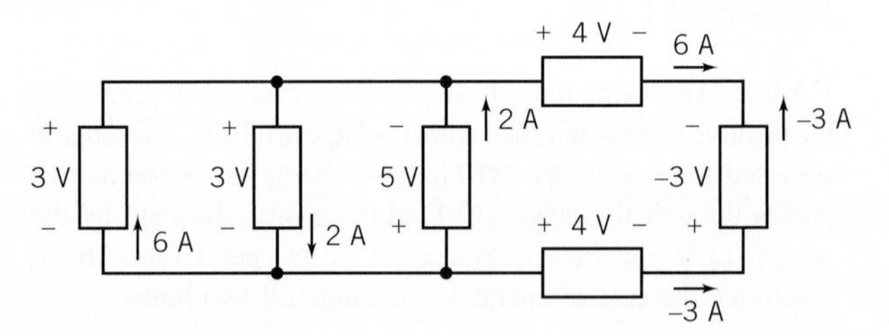

Figure P 1.7-2

P 1.7-3 The element currents and voltages shown in Figure P 1.7-3 are correct with one exception: the reference direction of exactly one of the element currents is reversed. Determine which reference direction has been reversed.

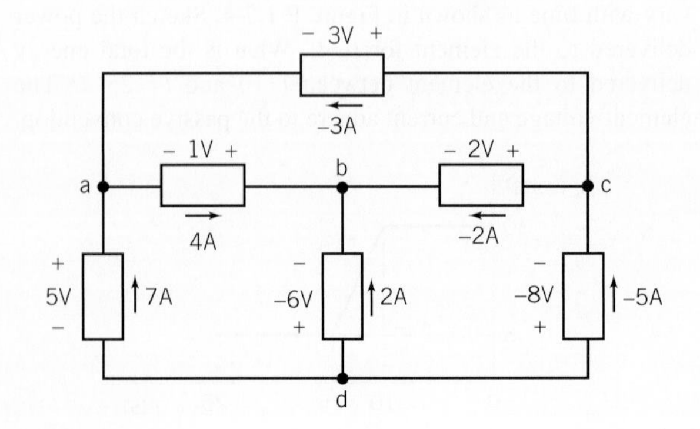

Figure P 1.7-3

Design Problems

DP 1-1 A particular circuit element is available in three grades. Grade A guarantees that the element can safely absorb 1/2 W continuously. Similarly, Grade B guarantees that 1/4 W can be absorbed safely, and Grade C guarantees that 1/8 W can be absorbed safely. As a rule, elements that can safely absorb more power are also more expensive and bulkier.

The voltage across an element is expected to be about 20 V, and the current in the element is expected to be about 8 mA. Both estimates are accurate to within 25 percent. The voltage and current reference adhere to the passive convention.

Specify the grade of this element. Safety is the most important consideration, but don't specify an element that is more expensive than necessary.

DP 1-2 The voltage across a circuit element is $v(t) = 20\,(1-e^{-8t})$ V when $t \geq 0$ and $v(t) = 0$ when $t < 0$. The current in this element is $i(t) = 30e^{-8t}$ mA when $t \geq 0$ and $i(t) = 0$ when $t < 0$. The element current and voltage adhere to the passive convention. Specify the power that this device must be able to absorb safely.

Hint: Use MATLAB, or a similar program, to plot the power.

Circuit Elements

IN THIS CHAPTER

2.1 Introduction
2.2 Engineering and Linear Models
2.3 Active and Passive Circuit Elements
2.4 Resistors
2.5 Independent Sources
2.6 Voltmeters and Ammeters
2.7 Dependent Sources
2.8 Transducers
2.9 Switches
2.10 How Can We Check . . . ?
2.11 **DESIGN EXAMPLE**—Temperature Sensor
2.12 Summary
Problems
Design Problems

2.1 INTRODUCTION

Not surprisingly, the behavior of an electric circuit depends on the behaviors of the individual circuit elements that comprise the circuit. Of course, different types of circuit elements behave differently. The equations that describe the behaviors of the various types of circuit elements are called the constitutive equations. Frequently, the constitutive equations describe a relationship between the current and voltage of the element. Ohm's law is a well-known example of a constitutive equation.

In this chapter, we will investigate the behavior of several common types of circuit element:

- Resistors
- Independent voltage and current sources
- Open circuits and short circuits
- Voltmeters and ammeters
- Dependent sources
- Transducers
- Switches

2.2 ENGINEERING AND LINEAR MODELS

The art of engineering is to take a bright idea and, using money, materials, knowledgeable people, and a regard for the environment, produce something the buyer wants at an affordable price.

Engineers use *models* to represent the elements of an electric circuit. A model is a description of those properties of a device that we think are important. Frequently, the model will consist of an equation relating the element voltage and current. Though the model is different from the electric device, the model can be used in pencil-and-paper calculations that will predict how a circuit composed of actual devices will operate. Engineers frequently face a trade-off when selecting a model for a device. Simple models are easy to work with but may not be accurate. Accurate models are usually more complicated and harder to use. The conventional wisdom suggests that simple models be used first. The results obtained using the models must be checked to verify that use of these simple models is appropriate. More accurate models are used when necessary.

The idealized models of electric devices are precisely defined. It is important to distinguish between actual devices and their idealized models, which we call circuit elements. The goal of circuit analysis is to predict the quantitative electrical behavior of physical circuits. Its aim is to predict and to explain the terminal voltages and terminal currents of the circuit elements and thus the overall operation of the circuit.

Models of circuit elements can be categorized in a variety of ways. For example, it is important to distinguish linear models from nonlinear models because circuits that consist entirely of linear circuit elements are easier to analyze than circuits that contain some nonlinear elements.

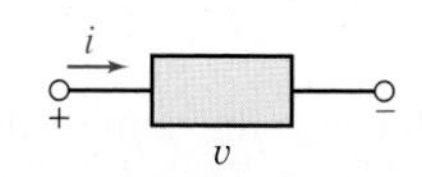

FIGURE 2.2-1 An element with an excitation current i and a response v.

An element or circuit is *linear* if the element's excitation and response satisfy certain properties. Consider the element shown in Figure 2.2-1. Suppose that the excitation is the current i and the response is the voltage v. When the element is subjected to a current i_1, it provides a response v_1. Furthermore, when the element is subjected to a current i_2, it provides a response v_2. For a linear element, it is necessary that the excitation $i_1 + i_2$ result in a response $v_1 + v_2$. This is usually called the *principle of superposition.*

Also, multiplying the input of a linear device by a constant must have the consequence of multiplying the output by the same constant. For example, doubling the size of the input causes the size of the output to double. This is called the *property of homogeneity.* An element is linear if, and only if, the properties of superposition and homogeneity are satisfied for all excitations and responses.

A **linear element** satisfies the properties of both superposition and homogeneity.

Let us restate mathematically the two required properties of a linear circuit, using the arrow notation to imply the transition from excitation to response:

$$i \to v$$

Then we may state the two properties required as follows.

Superposition:

$$i_1 \to v_1$$

$$i_2 \to v_2$$

then

$$i_1 + i_2 \to v_1 + v_2 \tag{2.2-1}$$

Homogeneity:

$$i \to v$$

then

$$ki \to kv \tag{2.2-2}$$

A device that does not satisfy either the superposition or the homogeneity principle is said to be nonlinear.

EXAMPLE 2.2-1 A Linear Device

Consider the element represented by the relationship between current and voltage as

$$v = Ri$$

Determine whether this device is linear.

Solution

The response to a current i_1 is

$$v_1 = Ri_1$$

The response to a current i_2 is

$$v_2 = Ri_2$$

The sum of these responses is

$$v_1 + v_2 = Ri_1 + Ri_2 = R(i_1 + i_2)$$

Because the sum of the responses to i_1 and i_2 is equal to the response to $i_1 + i_2$, the principle of superposition is satisfied. Next, consider the principle of homogeneity. Because

$$v_1 = Ri_1$$

we have for an excitation $i_2 = ki_1$

$$v_2 = Ri_2 = Rki_1$$

Therefore,

$$v_2 = kv_1$$

satisfies the principle of homogeneity. Because the element satisfies the properties of both superposition and homogeneity, it is linear.

EXAMPLE 2.2-2 A Nonlinear Device

Now let us consider an element represented by the relationship between current and voltage:

$$v = i^2$$

Determine whether this device is linear.

Solution

The response to a current i_1 is

$$v_1 = i_1^2$$

The response to a current i_2 is

$$v_2 = i_1^2$$

The sum of these responses is

$$v_1 + v_2 = i_1^2 + i_1^2$$

The response to $i_1 + i_2$ is

$$(i_1 + i_2)^2 = i_1^2 + 2i_1 i_2 + i_1^2$$

Because

$$i_1^2 + i_1^2 \neq (i_1 + i_2)^2$$

the principle of superposition is not satisfied. Therefore, the device is nonlinear.

EXAMPLE 2.2-3 A Model of a Linear Device

A linear element has voltage v and current i as shown in Figure 2.2-2*a*. Values of the current i and corresponding voltage v have been tabulated as shown in Figure 2.2-2*b*. Represent the element by an equation that expresses v as a function of i. This equation is a model of the element. Use the model to predict the value of v corresponding to a current of $i = 100$ mA and the value of i corresponding to a voltage of $v = 18$ V.

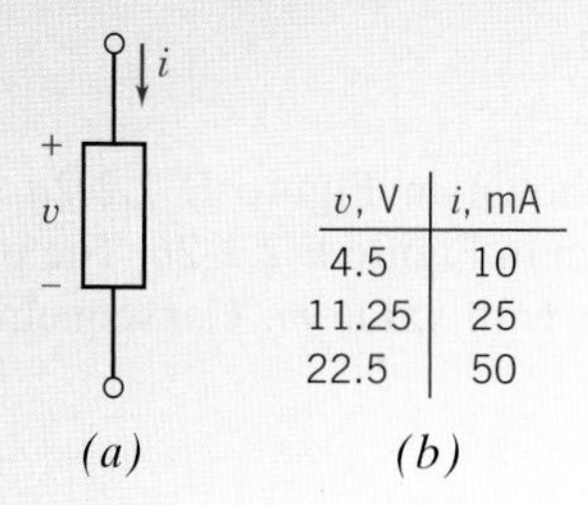

v, V	i, mA
4.5	10
11.25	25
22.5	50

FIGURE 2.2-2 (*a*) A linear circuit element and (*b*) a tabulation of corresponding values of its voltage and current.

FIGURE 2.2-3 A plot of voltage versus current for the linear element from Figure 2.2-2.

Solution

Figure 2.2-3 is a plot of the voltage v versus the current i. The points marked by dots represent corresponding values of v and i from the rows of the table in Figure 2.2-2*b*. Because the circuit element is linear, we expect these points to lie on a straight line, and indeed they do. We can represent the straight line by the equation

$$v = mi + b$$

where m is the slope and b is the v-intercept. Noticing that the straight line passes through the origin, $v = 0$ when $i = 0$, we see that $b = 0$. We are left with

$$v = mi$$

The slope m can be calculated from the data in any two rows of the table in Figure 2.2-2*b*. For example:

$$\frac{11.25 - 4.5}{25 - 10} = 0.45\ \frac{\text{V}}{\text{mA}},\ \frac{22.5 - 11.25}{50 - 25} = 0.45\ \frac{\text{V}}{\text{mA}},\ \text{and}\ \frac{22.5 - 4.5}{50 - 10} = 0.45\ \frac{\text{V}}{\text{mA}}$$

Consequently,

$$m = 0.45\ \frac{\text{V}}{\text{mA}} = 450\ \frac{\text{V}}{\text{A}}$$

and

$$v = 450i$$

This equation is a model of the linear element. It predicts that the voltage $v = 450(0.1) = 45$ V corresponds to the current $i = 100\text{ mA} = 0.1$ A and that the current $i = 18/450 = 0.04\text{ A} = 40$ mA corresponds to the voltage $v = 18$ V.

EXERCISE 2.2-1 Consider the circuit element shown in Figure E 2.2-1*a*. A plot of the element voltage, v, versus the element current, i, is shown in Figure E 2.2-1*b*. The plot is a straight line that passes through the origin and has a slope with value m. Consequently, v and i are related by

$$v = mi$$

Show that this device is linear.

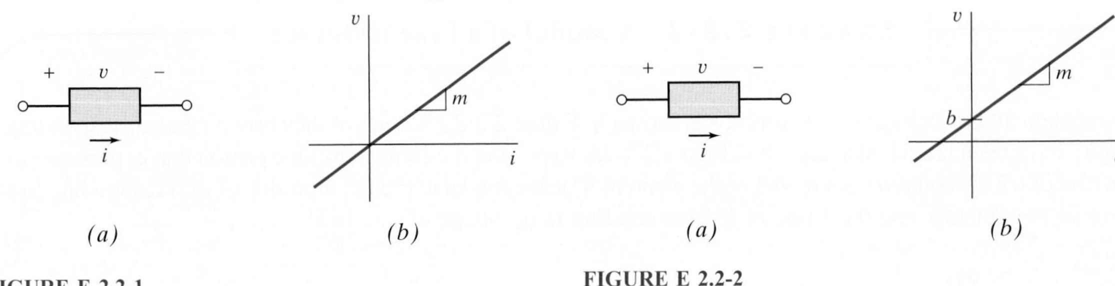

FIGURE E 2.2-1

FIGURE E 2.2-2

EXERCISE 2.2-2 Consider the circuit element shown in Figure E 2.2-2*a*. A plot of the element voltage, v, versus the element current, i, is shown in Figure E 2.2-2*b*. The plot is a straight line that has a y-intercept with value b and has a slope with value m. Consequently, v and i are related by

$$v = mi + b$$

Show that this device is not linear.

2.3 ACTIVE AND PASSIVE CIRCUIT ELEMENTS

We may classify circuit elements in two categories, *passive* and *active*, by determining whether they absorb energy or supply energy. An element is said to be passive if the total energy delivered to it from the rest of the circuit is always nonnegative (zero or positive). Then for a passive element, with the current flowing into the + terminal as shown in Figure 2.3-1*a*, this means that

$$w = \int_{-\infty}^{t} vi \, d\tau \geq 0 \tag{2.3-1}$$

for all values of t.

A **passive element** absorbs energy.

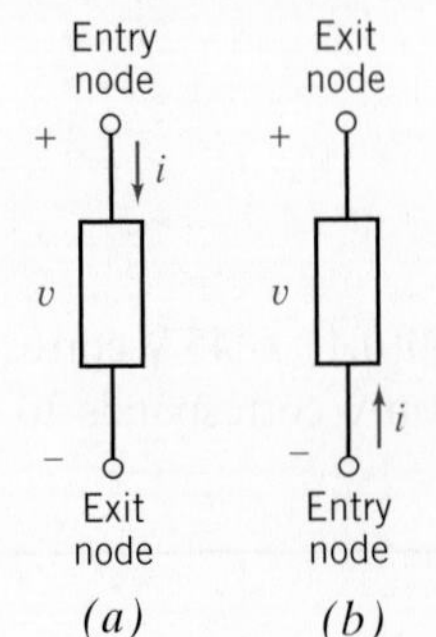

FIGURE 2.3-1 (*a*) The entry node of the current i is the positive node of the voltage v; (*b*) the entry node of the current i is the negative node of the voltage v. The current flows from the entry node to the exit node.

An element is said to be *active* if it is capable of delivering energy. Thus, an active element violates Eq. 2.3-1 when it is represented by Figure 2.3-1*a*. In other words, an active element is one that is capable of generating energy. Active elements are potential sources of energy, whereas passive elements are sinks or absorbers of energy. Examples of active elements include batteries and generators. Consider the element shown in Figure 2.3-1*b*. Note that the current flows into the negative terminal and out of the positive terminal. This element is said to be active if

$$w = \int_{-\infty}^{t} vi\,d\tau \geq 0 \tag{2.3-2}$$

for at least one value of t.

An **active element** is capable of supplying energy.

EXAMPLE 2.3-1 An Active Circuit Element

A circuit has an element represented by Figure 2.3-1b where the current is a constant 5 A and the voltage is a constant 6 V. Find the energy supplied over the time interval 0 to T.

Solution

Because the current enters the negative terminal, the energy *supplied* by the element is given by

$$w = \int_{0}^{T} (6)(5)d\tau = 30T \text{ J}$$

Thus, the device is a generator or an active element, in this case a dc battery.

2.4 RESISTORS

The ability of a material to resist the flow of charge is called its *resistivity*, ρ. Materials that are good electrical insulators have a high value of resistivity. Materials that are good conductors of electric current have low values of resistivity. Resistivity values for selected materials are given in Table 2.4-1. Copper is commonly used for wires because it permits current to flow relatively unimpeded. Silicon is commonly used to provide resistance in semiconductor electric circuits. Polystyrene is used as an insulator.

Resistance is the physical property of an element or device that impedes the flow of current; it is represented by the symbol R.

Georg Simon Ohm was able to show that the current in a circuit composed of a battery and a conducting wire of uniform cross-section could be expressed as

$$i = \frac{Av}{\rho L} \tag{2.4-1}$$

Table 2.4-1 **Resistivities of Selected Materials**

MATERIAL	RESISTIVITY ρ (OHM.CM)
Polystyrene	1×10^{18}
Silicon	2.3×10^{5}
Carbon	4×10^{-3}
Aluminum	2.7×10^{-6}
Copper	1.7×10^{-6}

FIGURE 2.4-1 Georg Simon Ohm (1787–1854), who determined Ohm's law in 1827. The ohm was chosen as the unit of electrical resistance in his honor.

where A is the cross-sectional area, ρ the resistivity, L the length, and v the voltage across the wire element. Ohm, who is shown in Figure 2.4-1, defined the constant resistance R as

$$R = \frac{\rho L}{A} \tag{2.4-2}$$

Ohm's law, which related the voltage and current, was published in 1827 as

$$v = Ri \tag{2.4-3}$$

The unit of resistance R was named the ohm in honor of Ohm and is usually abbreviated by the Ω (capital omega) symbol, where $1\ \Omega = 1$ V/A. The resistance of a 10-m length of common TV cable is 2 mΩ.

An element that has a resistance R is called a *resistor*. A resistor is represented by the two-terminal symbol shown in Figure 2.4-2. Ohm's law, Eq. 2.4-3, requires that the i-versus-v relationship be linear. As shown in Figure 2.4-3, a resistor may become nonlinear outside its normal rated range of operation. We will assume that a resistor is linear unless stated otherwise. Thus, we will use a linear model of the resistor as represented by Ohm's law.

In Figure 2.4-4, the element current and element voltage of a resistor are labeled. The relationship between the directions of this current and voltage is important. The voltage direction marks one resistor terminal + and the other −. The current i_a flows from the terminal marked + to the terminal marked −. This relationship between the current and voltage reference directions is a convention called the passive convention. Ohm's law states that when the element voltage and the element current adhere to the passive convention, then

$$v = Ri_a \tag{2.4-4}$$

Consider Figure 2.4-4. The element currents i_a and i_b are the same except for the assigned direction, so

$$i_a = -i_b$$

The element current i_a and the element voltage v adhere to the passive convention,

$$v = Ri_a$$

Replacing i_a by $-i_b$ gives

$$v = -Ri_b$$

There is a minus sign in this equation because the element current i_b and the element voltage v do not adhere to the passive convention. We must pay attention to the current direction so that we don't overlook this minus sign.

Ohm's law, Eq. 2.4-3, can also be written as

$$i = Gv \tag{2.4-5}$$

where G denotes the *conductance* in siemens (S) and is the reciprocal of R; that is, $G = 1/R$. Many engineers denote the units of conductance as mhos with the ℧ symbol, which is an inverted omega (mho is *ohm* spelled backward). However, we will use SI units and retain siemens as the units for conductance.

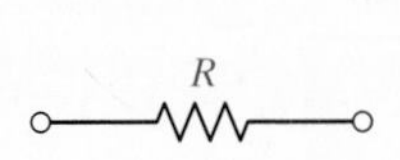

FIGURE 2.4-2 Symbol for a resistor having a resistance of R ohms.

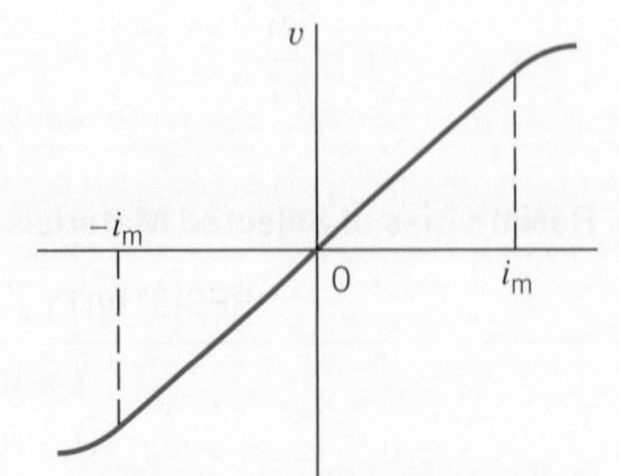

FIGURE 2.4-3 A resistor operating within its specified current range, $\pm i_m$, can be modeled by Ohm's law.

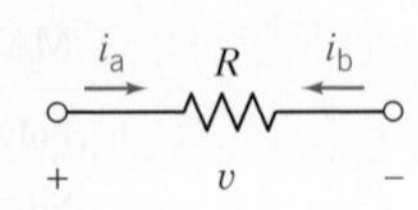

FIGURE 2.4-4 A resistor with element current and element voltage.

FIGURE 2.4-5 (*a*) Wirewound resistor with an adjustable center tap. (*b*) Wirewound resistor with a fixed tap. Courtesy of Dale Electronics.

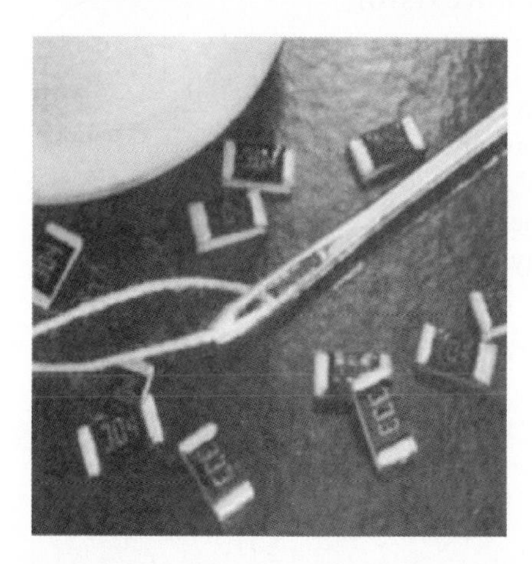

FIGURE 2.4-6 Small thick-film resistor chips used for miniaturized circuits. Courtesy of Corning Electronics.

FIGURE 2.4-7 A 1/4-watt metal film resistor. The body of the resistor is 6 mm long. Courtesy of Dale Electronics.

Most discrete resistors fall into one of four basic categories: carbon composition, carbon film, metal film, or wirewound. Carbon composition resistors have been in use for nearly 100 years and are still popular. Carbon film resistors have supplanted carbon composition resistors for many general-purpose uses because of their lower cost and better tolerances. Two wirewound resistors are shown in Figure 2.4-5.

Thick-film resistors, as shown in Figure 2.4-6, are used in circuits because of their low cost and small size. General-purpose resistors are available in standard values for tolerances of 2, 5, 10, and 20 percent. Carbon composition resistors and some wirewounds have a color code with three to five bands. A color code is a system of standard colors adopted for identification of the resistance of resistors. Figure 2.4-7 shows a metal film resistor with its color bands. This is a 1/4-watt resistor, implying that it should be operated at or below 1/4 watt of power delivered to it. The normal range of resistors is from less than 1 ohm to 10 megohms. Typical values of some commercially available resistors are given in Appendix D.

The power delivered to a resistor (when the passive convention is used) is

$$p = vi = v\left(\frac{v}{R}\right) = \frac{v^2}{R} \tag{2.4-6}$$

Alternatively, because $v = iR$, we can write the equation for power as

$$p = vi = (iR)i = i^2R \tag{2.4-7}$$

Thus, the power is expressed as a nonlinear function of the current i through the resistor or of the voltage v across it.

Recall the definition of a passive element as one for which the energy absorbed is always nonnegative. The equation for energy delivered to a resistor is

$$w = \int_{-\infty}^{t} p\,d\tau = \int_{-\infty}^{t} i^2R\,d\tau \tag{2.4-8}$$

Because i^2 is always positive, the energy is always positive and the resistor is a passive element.

Resistance is a measure of an element's ability to dissipate power irreversibly.

EXAMPLE 2.4-1 Power Dissipated by a Resistor

Let us devise a model for a car battery when the lights are left on and the engine is off. We have all experienced or seen a car parked with its lights on. If we leave the car for a period, the battery will run down or go dead. An auto battery is a 12-V constant-voltage source, and the lightbulb can be modeled by a resistor of 6 ohms. The circuit is shown in Figure 2.4-8. Let us find the current i, the power p, and the energy supplied by the battery for a four-hour period.

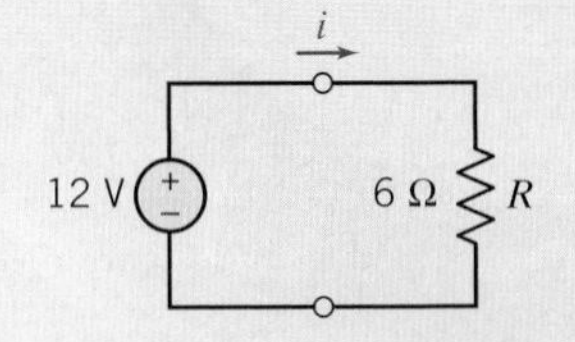

FIGURE 2.4-8 Model of a car battery and the headlight lamp.

Solution

According to Ohm's law, Eq. 2.4-3, we have

$$v = Ri$$

Because $v = 12$ V and $R = 6\ \Omega$, we have $i = 2$ A.

To find the power delivered by the battery, we use

$$p = vi = 12(2) = 24 \text{ W}$$

Finally, the energy delivered in the four-hour period is

$$w = \int_0^t p\,d\tau = 24t = 24(60 \times 60 \times 4) = 3.46 \times 10^5 \text{ J}$$

Because the battery has a finite amount of stored energy, it will deliver this energy and eventually be unable to deliver further energy without recharging. We then say the battery is run down or dead until recharged. A typical auto battery may store 10^6 J in a fully charged condition.

EXERCISE 2.4-1 Find the power absorbed by a 100-ohm resistor when it is connected directly across a constant 10-V source.

Answer: 1-W

EXERCISE 2.4-2 A voltage source $v = 10 \cos t$ V is connected across a resistor of 10 ohms. Find the power delivered to the resistor.

Answer: $10 \cos^2 t$ W

2.5 INDEPENDENT SOURCES

Some devices are intended to supply energy to a circuit. These devices are called sources. Sources are categorized as being one of two types: voltage sources and current sources. Figure 2.5-1a shows the symbol that is used to represent a voltage source. The voltage of a voltage source is specified, but the current is determined by the rest of the circuit. A voltage source is described by specifying the function $v(t)$, for example,

$$v(t) = 12 \cos 1000t \quad \text{or} \quad v(t) = 9 \quad \text{or} \quad v(t) = 12 - 2t$$

An active two-terminal element that supplies energy to a circuit is a *source* of energy. An *independent voltage source* provides a specified voltage independent of the current through it and is independent of any other circuit variable.

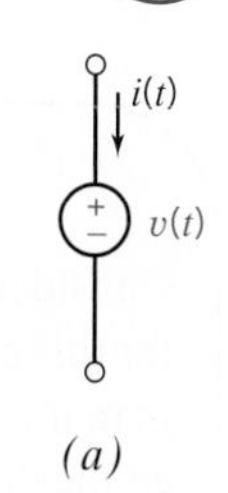

(a)

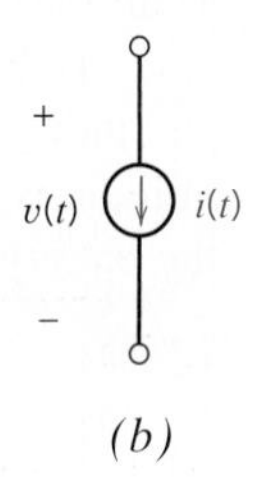

(b)

FIGURE 2.5-1 (*a*) Voltage source. (*b*) Current source.

A **source** is a voltage or current generator capable of supplying energy to a circuit.

An *independent current source* provides a current independent of the voltage across the source element and is independent of any other circuit variable. Thus, when we say a source is independent, we mean it is independent of any other voltage or current in the circuit.

An **independent source** is a voltage or current generator not dependent on other circuit variables.

Suppose the voltage source is a battery and

$$v(t) = 9 \text{ volts}$$

The voltage of this battery is known to be 9 volts regardless of the circuit in which the battery is used. In contrast, the current of the voltage source is not known and depends on the circuit in which the source is used. The current could be 6 amps when the voltage source is connected to one circuit and 6 milliamps when the voltage source is connected to another circuit.

Figure 2.5-1*b* shows the symbol that is used to represent a current source. The current of a current source is specified, but the voltage is determined by the rest of the circuit. A current source is described by specifying the function $i(t)$, for example,

$$i(t) = 6 \sin 500t \quad \text{or} \quad i(t) = -0.25 \quad \text{or} \quad i(t) = t + 8$$

A current source specified by $i(t) = -0.25$ milliamps will have a current of -0.25 milliamps in any circuit in which it is used. The voltage across this current source will depend on the particular circuit.

The preceding paragraphs have ignored some complexities to give a simple description of the way sources work. The voltage across a 9-volt battery may not actually be 9 volts. This voltage depends on the age of the battery, the temperature, variations in manufacturing, and the battery current. It is useful to make a distinction between real sources, such as batteries, and the simple voltage and current sources described in the preceding paragraphs. It would be *ideal* if the real sources worked like these simple sources. Indeed, the word *ideal* is used to make this distinction. The simple sources described in the previous paragraph are called the *ideal voltage source* and the *ideal current source*.

The voltage of an **ideal voltage source** is given to be a specified function, say $v(t)$. The current is determined by the rest of the circuit.

The current of an **ideal current source** is given to be a specified function, say $i(t)$. The voltage is determined by the rest of the circuit.

An **ideal source** is a voltage or a current generator independent of the current through the voltage source or the voltage across the current source.

EXAMPLE 2.5-1 A Battery Modeled as a Voltage Source

Consider the plight of the engineer who needs to analyze a circuit containing a 9-volt battery. Is it really necessary for this engineer to include the dependence of battery voltage on the age of the battery, the temperature, variations in manufacturing, and the battery current in this analysis? Hopefully not. We expect the battery to act enough like an ideal 9-volt voltage source that the differences can be ignored. In this case, it is said that the battery is *modeled* as an ideal voltage source.

To be specific, consider a battery specified by the plot of voltage versus current shown in Figure 2.5-2*a*. This plot indicates that the battery voltage will be $v = 9$ volts when $i \leq 10$ milliamps. As the current increases above 10 milliamps, the voltage decreases from 9 volts. When $i \leq 10$ milliamps, the dependence of the battery voltage on the battery current can be ignored and the battery can be modeled as an ideal voltage source.

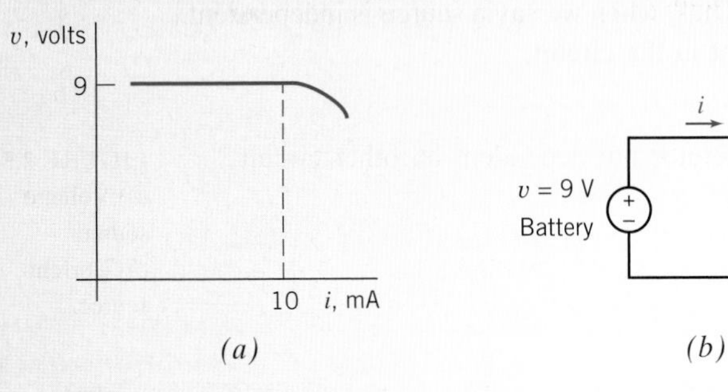

FIGURE 2.5-2 (*a*) A plot of battery voltage versus battery current. (*b*) The battery is modeled as an independent voltage source.

Suppose a resistor is connected across the terminals of the battery as shown in Figure 2.5-2*b*. The battery current will be

$$i = \frac{v}{R} \tag{2.5-1}$$

The relationship between v and i shown in Figure 2.5-2*a* complicates this equation. This complication can be safely ignored when $i \leq 10$ milliamps. When the battery is modeled as an ideal 9-volt voltage source, the voltage source current is given by

$$i = \frac{9}{R} \tag{2.5-2}$$

The distinction between these two equations is important. Eq. 2.5-1, involving the v–i relationship shown in Figure 2.5-2*a*, is more accurate but also more complicated. Equation 2.5-2 is simpler but may be inaccurate.

Suppose that $R = 1000$ ohms. Equation 2.5-2 gives the current of the ideal voltage source:

$$i = \frac{9}{1000} = 9 \text{ mA} \tag{2.5-3}$$

Because this current is less than 10 milliamps, the ideal voltage source is a good model for the battery, and it is reasonable to expect that the battery current is 9 milliamps.

Suppose, instead, that $R = 600$ ohms. Once again, Eq. 2.5-2 gives the current of the ideal voltage source:

$$i = \frac{9}{600} = 15 \text{ mA} \tag{2.5-4}$$

Because this current is greater than 10 milliamps, the ideal voltage source is not a good model for the battery. In this case, it is reasonable to expect that the battery current is different from the current for the ideal voltage source.

Engineers frequently face a trade-off when selecting a model for a device. Simple models are easy to work with but may not be accurate. Accurate models are usually more complicated and harder to use. The conventional wisdom suggests that simple models be used first. The results obtained using the models must be checked to verify that use of these simple models is appropriate. More accurate models are used when necessary.

The short circuit and open circuit are special cases of ideal sources. A *short circuit* is an ideal voltage source having $v(t) = 0$. The current in a short circuit is determined by the rest of the circuit. An *open circuit* is an ideal current source having $i(t) = 0$. The voltage across an open circuit is determined by the rest of the circuit. Figure 2.5-3 shows the symbols used to represent the short circuit and the open circuit. Notice that the power absorbed by each of these devices is zero.

Open and short circuits can be added to a circuit without disturbing the branch currents and voltages of all the other devices in the circuit. Figure 2.6-3 shows how this can be done. Figure 2.6-3*a* shows an example circuit. In Figure 2.6-3*b* an open circuit and a short circuit have been added to this example circuit. The open circuit was connected between two nodes of the original circuit. In contrast, the short circuit was added by cutting a wire and inserting the short circuit. Adding open circuits and short circuits to a network in this way does not change the network.

Open circuits and short circuits can also be described as special cases of resistors. A resistor with resistance $R = 0$ $(G = \infty)$ is a short circuit. A resistor with conductance $G = 0$ $(R = \infty)$ is an open circuit.

$i(t) = 0$, $v(t)$ (*a*)

$v(t) = 0$, $i(t)$ (*b*)

FIGURE 2.5-3 (*a*) Open circuit. (*b*) Short circuit.

2.6 VOLTMETERS AND AMMETERS

Measurements of dc current and voltage are made with direct-reading (analog) or digital meters, as shown in Figure 2.6-1. A direct-reading meter has an indicating pointer whose angular deflection depends on the magnitude of the variable it is measuring. A digital meter displays a set of digits indicating the measured variable value.

To measure a voltage or current, a meter is connected to a circuit, using terminals called probes. These probes are color coded to indicate the reference direction of the variable being measured. Frequently, meter probes are colored red and black. An ideal voltmeter measures the voltage from the red to the black probe. The red terminal is the positive terminal, and the black terminal is the negative terminal (see Figure 2.6-2*b*).

An ideal ammeter measures the current flowing through its terminals, as shown in Figure 2.6-2*a* and has zero voltage, v_m, across its terminals. An ideal voltmeter measures the voltage across its terminals, as shown in Figure 2.6-2*b*, and has terminal current, i_m, equal to zero. Practical measuring instruments only approximate the ideal conditions. For a practical ammeter, the voltage across its terminals is usually negligibly small. Similarly, the current into a voltmeter is usually negligible.

Ideal voltmeters act like open circuits, and ideal ammeters act like short circuits. In other words, the model of an ideal voltmeter is an open circuit, and the model of an ideal ammeter is a short circuit. Consider the circuit of Figure 2.6-3*a* and then add an open circuit with a voltage v and a short circuit with a current i as shown in Figure 2.6-3*b*. In Figure 2.6-3*c*, the open circuit has been replaced by a voltmeter, and the short circuit has been replaced by an ammeter. The voltmeter will measure the voltage labeled v in Figure 2.6-3*b* whereas the ammeter will measure the current labeled i. Notice that Figure 2.6-3*c* could be obtained from Figure 2.6-3*a* by adding a voltmeter

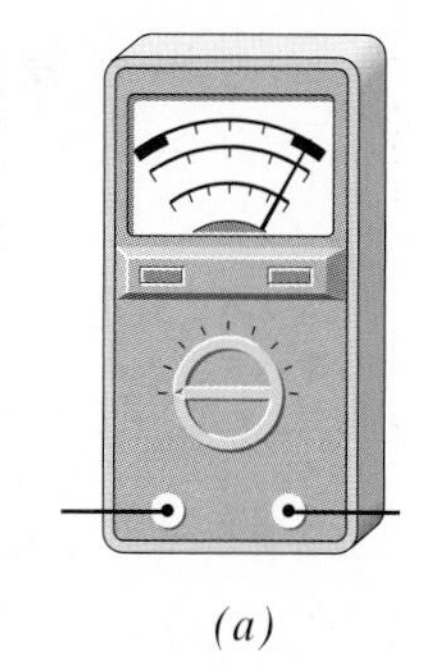

(*a*)

2.9

(*b*)

FIGURE 2.6-1 (*a*) A direct-reading (analog) meter. (*b*) A digital meter.

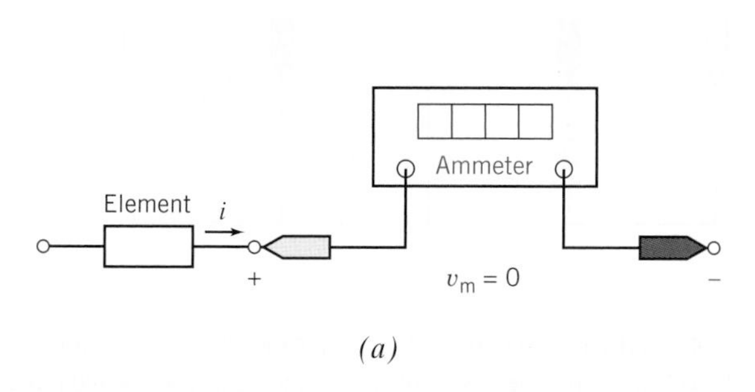

(*a*)

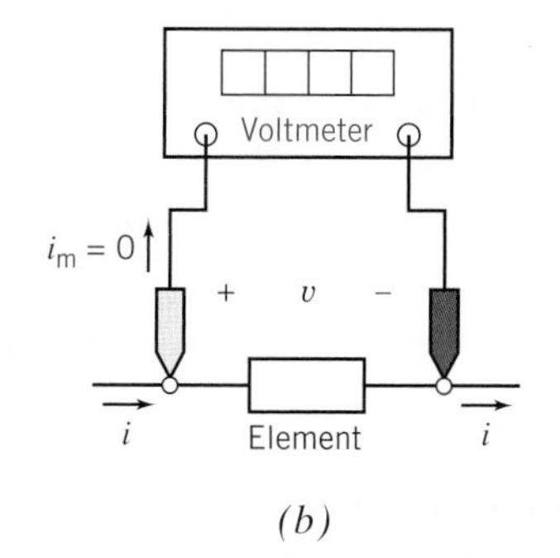

(*b*)

FIGURE 2.6-2 (*a*) Ideal ammeter. (*b*) Ideal voltmeter.

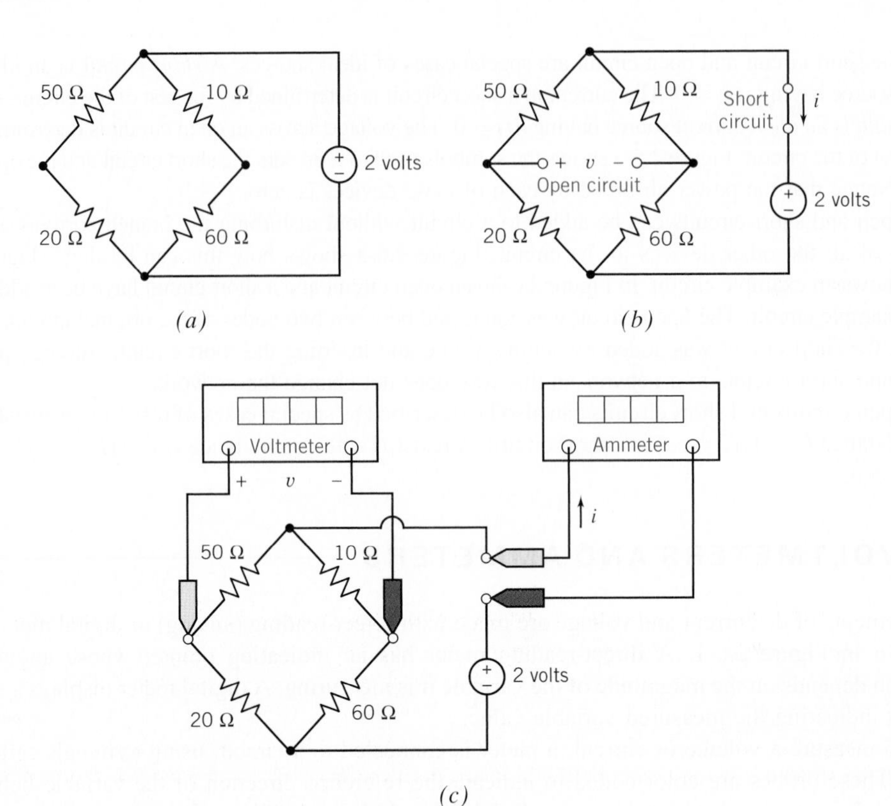

FIGURE 2.6-3 (*a*) An example circuit, (*b*) plus an open circuit and a short circuit. (*c*) The open circuit is replaced by a voltmeter, and the short circuit is replaced by an ammeter.

and an ammeter. Ideally, adding the voltmeter and ammeter in this way does not disturb the circuit. One more interpretation of Figure 2.6-3 is useful. Figure 2.6-3*b* could be formed from Figure 2.6-3*c* by replacing the voltmeter and the ammeter by their (ideal) models.

The reference direction is an important part of an element voltage or element current. Figures 2.6-4 and 2.6-5 show that attention must be paid to reference directions when measuring an element voltage or element current. Figure 2.6-4*a* shows a voltmeter. Voltmeters have two color-coded probes. This color coding indicates the reference direction of the voltage being measured. In Figures 2.6-4*b* and Figure 2.6-4*c* the voltmeter is used to measure the voltage across the 6-kΩ resistor. When the voltmeter is connected to the circuit as shown in Figure 2.6-4*b*, the voltmeter measures v_a, with + on

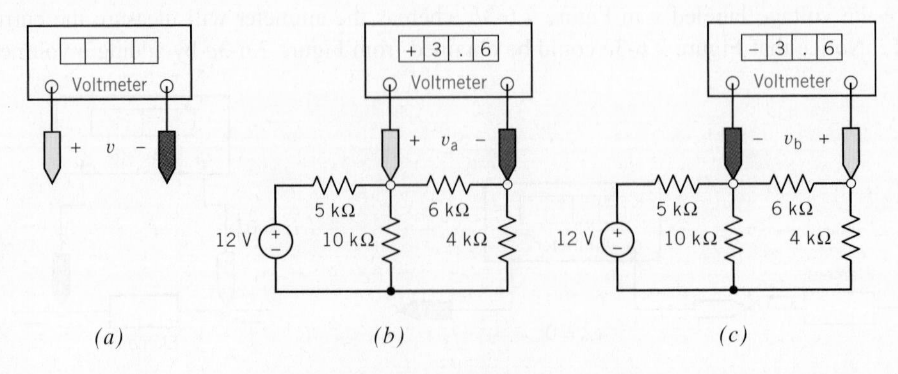

FIGURE 2.6-4 (*a*) The correspondence between the color-coded probes of the voltmeter and the reference direction of the measured voltage. In (*b*), the + sign of v_a is on the left, whereas in (*c*), the + sign of v_b is on the right. The colored probe is shown here in blue. In the laboratory this probe will be red. We will refer to the colored probe as the ''red probe.''

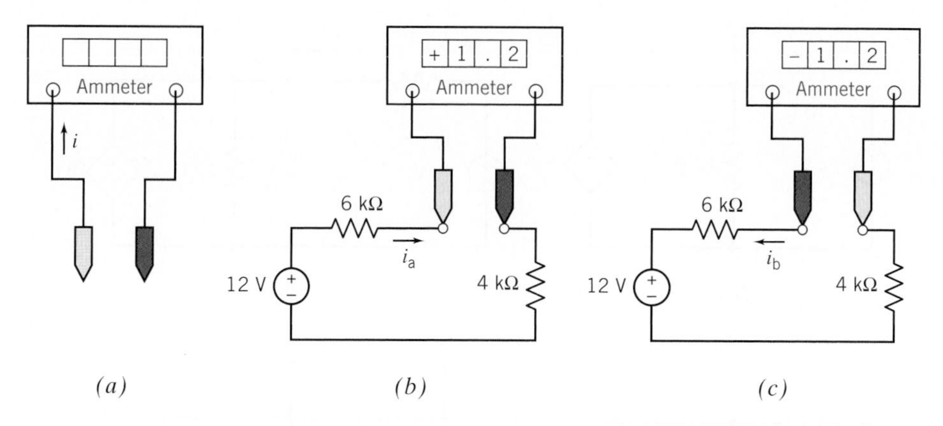

FIGURE 2.6-5 (*a*) The correspondence between the color-coded probes of the ammeter and the reference direction of the measured current. In (*b*) the current i_a is directed to the right, while in (*c*) the current i_b is directed to the left. The colored probe is shown here in blue. In the laboratory this probe will be red. We will refer to the colored probe as the "red probe."

the left, at the red probe. When the voltmeter probes are interchanged as shown in Figure 2.6-4*c*, the voltmeter measures v_b, with + on the right, again at the red probe. Note $v_b = -v_a$.

Figure 2.6-5*a* shows an ammeter. Ammeters have two color-coded probes. This color coding indicates the reference direction of the current being measured. In Figures 2.6-5*b* and *c*, the ammeter is used to measure the current in the 6-kΩ resistor. When the ammeter is connected to the circuit as shown in Figure 2.6-5*b*, the ammeter measures i_a, directed from the red probe toward the black probe. When the ammeter probes are interchanged as shown in Figure 2.6-5*c*, the ammeter measures i_b, again directed from the red probe toward the black probe. Note $i_b = -i_a$.

2.7 DEPENDENT SOURCES

Dependent sources model the situation in which the voltage or current of one circuit element is proportional to the voltage or current of the second circuit element. (In contrast, a resistor is a circuit element in which the voltage of the element is proportional to the current in the *same* element.) Dependent sources are used to model electronic devices such as transistors and amplifiers. For example, the output voltage of an amplifier is proportional to the input voltage of that amplifier, so an amplifier can be modeled as a dependent source.

Figure 2.7-1*a* shows a circuit that includes a dependent source. The diamond symbol represents a dependent source. The plus and minus signs inside the diamond identify the dependent source as a voltage source and indicate the reference polarity of the element voltage. The label "5*i*" represents the voltage of this dependent source. This voltage is a product of two factors, 5 and *i*. The second factor, *i*, indicates that the voltage of this dependent source is controlled by the current, *i*, in the 18-Ω resistor. The first factor, 5, is the gain of this dependent source. The gain of this dependent source is the ratio of the controlled voltage, 5*i*, to the controlling current, *i*. This gain has units of V/A or Ω. Because this dependent source is a voltage source and because a current controls the voltage, the dependent source is called a current-controlled voltage source (CCVS).

Figure 2.7-1*b* shows the circuit from 2.7-1*a*, using a different point of view. In Figure 2.7-1*b*, a short circuit has been inserted in series with the 18-Ω resistor. Now we think of the controlling current *i* as the current in a short circuit rather than the current in the 18-Ω resistor itself. In this way, we can always treat the controlling current of a dependent source as the current in a short circuit. We will use this second point of view to categorize dependent sources in this section.

Figure 2.7-1*c* shows a circuit that includes a dependent source, represented by the diamond symbol. The arrow inside the diamond identifies the dependent source as a current source and indicates the reference direction of the element current. The label "0.2*v*" represents the current of this

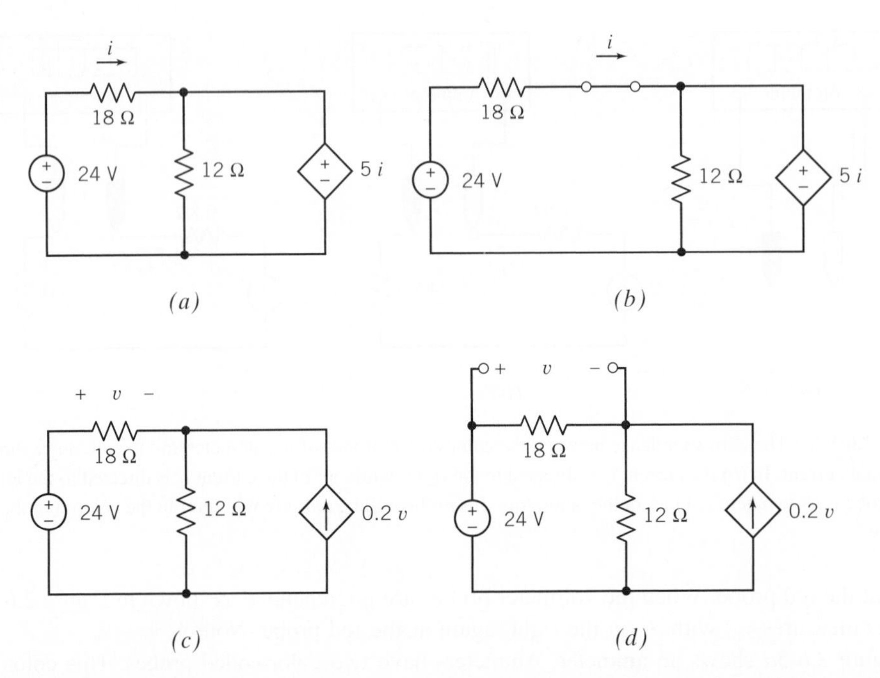

FIGURE 2.7-1 The controlling current of a dependent source shown as (*a*) the current in an element and as (*b*) the current in a short circuit in series with that element. The controlling voltage of a dependent source shown as (*c*) the voltage across an element and as (*d*) the voltage across an open circuit in parallel with that element.

dependent source. This current is a product of two factors, 0.2 and v. The second factor, v, indicates that the current of this dependent source is controlled by the voltage, v, across the 18-Ω resistor. The first factor, 0.2, is the gain of this dependent source. The gain of this dependent source is the ratio of the controlled current, $0.2v$, to the controlling voltage, v. This gain has units of A/V. Because this dependent source is a current source and because a voltage controls the current, the dependent source is called a voltage-controlled current source (VCCS).

Figure 2.7-1*d* shows the circuit from Figure 2.7-1*c*, using a different point of view. In Figure 2.7-1*d*, an open circuit has been added in parallel with the 18-Ω resistor. Now we think of the controlling voltage v as the voltage across an open circuit Figure 2.7-1, rather than the voltage across the 18-Ω resistor itself. In this way, we can always treat the controlling voltage of a dependent source as the voltage across an open circuit.

We are now ready to categorize dependent source. Each dependent source consists of two parts: the controlling part and the controlled part. The controlling part is either an open circuit or a short circuit. The controlled part is either a voltage source or a current source. There are four types of dependent source that correspond to the four ways of choosing a controlling part and a controlled part. These four dependent sources are called the voltage-controlled voltage source (VCVS), current-controlled voltage source (CCVS), voltage-controlled current source (VCCS), and current-controlled current source (CCCS). The symbols that represent dependent sources are shown in Table 2.7-1.

Consider the CCVS shown in Table 2.7-1. The controlling element is a short circuit. The element current and voltage of the controlling element are denoted as i_c and v_c. The voltage across a short circuit is zero, so $v_c = 0$. The short-circuit current, i_c, is the controlling signal of this dependent source. The controlled element is a voltage source. The element current and voltage of the controlled element are denoted as i_d and v_d. The voltage v_d is controlled by i_c:

$$v_d = ri_c$$

The constant r is called the gain of the CCVS. The current i_d, like the current in any voltage source, is determined by the rest of the circuit.

Table 2.7-1 Dependent Sources

DESCRIPTION	SYMBOL
Current-Controlled Voltage Source (CCVS) r is the gain of the CCVS. r has units of volts/ampere.	$v_c = 0$, i_c; i_d, $v_d = ri_c$
Voltage-Controlled Voltage Source (VCVS) b is the gain of the VCVS. b has units of volts/volt.	$i_c = 0$, v_c; i_d, $v_d = bv_c$
Voltage-Controlled Current Source (VCCS) g is the gain of the VCCS. g has units of amperes/volt.	$i_c = 0$, v_c; v_d, $i_d = gv_c$
Current-Controlled Current Source (CCCS) d is the gain of the CCCS. d has units of amperes/ampere.	$v_c = 0$, i_c; v_d, $i_d = di_c$

Next, consider the VCVS shown in Table 2.7-1. The controlling element is an open circuit. The current in an open circuit is zero, so $i_c = 0$. The open-circuit voltage, v_c, is the controlling signal of this dependent source. The controlled element is a voltage source. The voltage v_d is controlled by v_c:

$$v_d = bv_c$$

The constant b is called the gain of the VCVS. The current i_d is determined by the rest of the circuit.

The controlling element of the VCCS shown in Table 2.7-1 is an open circuit. The current in this open circuit is $i_c = 0$. The open-circuit voltage, v_c, is the controlling signal of this dependent source. The controlled element is a current source. The current i_d is controlled by v_c:

$$i_d = gv_c$$

The constant g is called the gain of the VCCS. The voltage v_d, like the voltage across any current source, is determined by the rest of the circuit.

The controlling element of the CCCS shown in Table 2.7-1 is a short circuit. The voltage across this open circuit is $v_c = 0$. The short-circuit current, i_c, is the controlling signal of this dependent source. The controlled element is a current source. The current i_d is controlled by i_c:

$$i_d = di_c$$

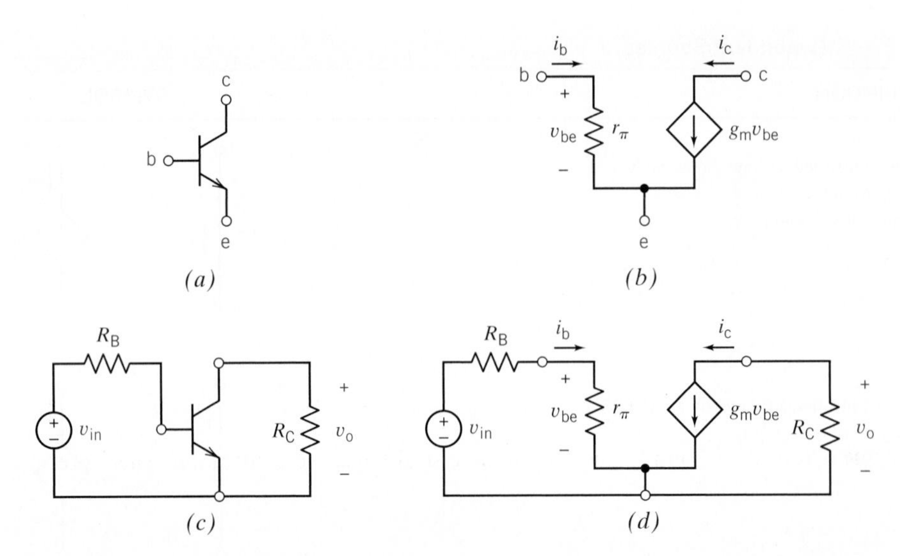

FIGURE 2.7-2 (*a*) A symbol for a transistor. (*b*) A model of the transistor. (*c*) A transistor amplifier. (*d*) A model of the transistor amplifier.

The constant d is called the gain of the CCCS. The voltage v_d, like the voltage across any current source, is determined by the rest of the circuit.

Figure 2.7-2 illustrates the use of dependent sources to model electronic devices. In certain circumstances, the behavior of the transistor shown in Figure 2.7-2*a* can be represented using the model shown in Figure 2.7-2*b*. This model consists of a dependent source and a resistor. The controlling element of the dependent source is an open circuit connected across the resistor. The controlling voltage is v_{be}. The gain of the dependent source is g_m. The dependent source is used in this model to represent a property of the transistor, namely, that the current i_c is proportional to the voltage v_{be}, that is,

$$i_c = g_m v_{be}$$

where g_m has units of amperes/volt. Figures 2.7-2*c* and *d* illustrate the utility of this model. Figure 2.7-2*d* is obtained from Figure 2.7-2*c* by replacing the transistor by the transistor model.

EXAMPLE 2.7-1 Power and Dependent Sources

Determine the power absorbed by the VCVS in Figure 2.7-3.

Solution

The VCVS consists of an open circuit and a controlled-voltage source. There is no current in the open circuit, so no power is absorbed by the open circuit.

The voltage, v_c, across the open circuit is the controlling signal of the VCVS. The voltmeter measures v_c to be

$$v_c = 2\text{ V}$$

The voltage of the controlled voltage source is

$$v_d = 2\,v_c = 4\text{ V}$$

The ammeter measures the current in the controlled voltage source to be

$$i_d = 1.5\text{ A}$$

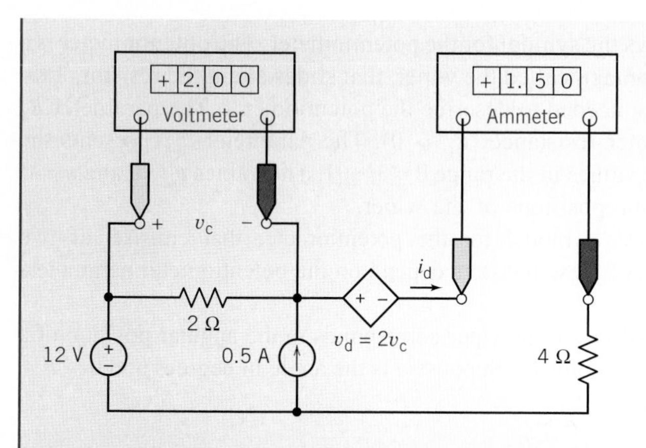

FIGURE 2.7-3 A circuit containing a VCVS. The meters indicate that the voltage of the controlling element is $v_c = 2.0$ volts and that the current of the controlled element is $i_d = 1.5$ amperes.

The element current, i_d, and voltage, v_d, adhere to the passive convention. Therefore,

$$p = i_d v_d = (1.5)(4) = 6 \text{ W}$$

is the power absorbed by the VCVS.

EXERCISE 2.7-1 Find the power absorbed by the CCCS in Figure E 2.7-1.

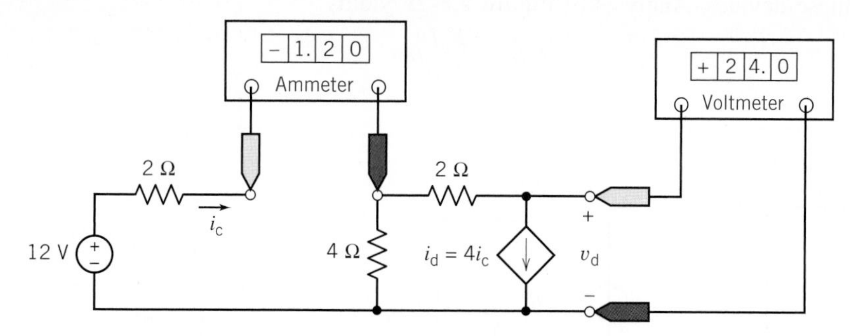

FIGURE E 2.7-1 A circuit containing a CCCS. The meters indicate that the current of the controlling element is $i_c = -1.2$ amperes and that the voltage of the controlled element is $v_d = 24$ volts.

Hint: The controlling element of this dependent source is a short circuit. The voltage across a short circuit is zero. Hence, the power absorbed by the controlling element is zero. How much power is absorbed by the controlled element?

Answer: -115.2 watts are absorbed by the CCCS. (The CCCS delivers $+115.2$ watts to the rest of the circuit.)

2.8 TRANSDUCERS

Transducers are devices that convert physical quantities to electrical quantities. This section describes two transducers: potentiometers and temperature sensors. Potentiometers convert position to resistance, and temperature sensors convert temperature to current.

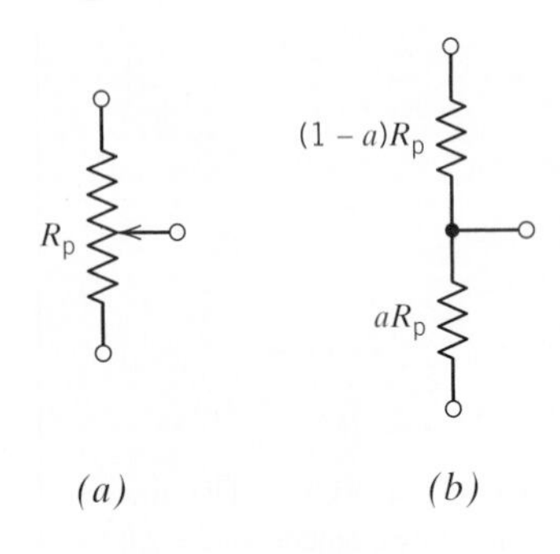

FIGURE 2.8-1 (*a*) The symbol and (*b*) a model for the potentiometer.

Figure 2.8-1*a* shows the symbol for the potentiometer. The potentiometer is a resistor having a third contact, called the wiper, that slides along the resistor. Two parameters, R_p and a, are needed to describe the potentiometer. The parameter R_p specifies the potentiometer resistance ($R_p > 0$). The parameter a represents the wiper position and takes values in the range $0 \le a \le 1$. The values $a = 0$ and $a = 1$ correspond to the extreme positions of the wiper.

Figure 2.8-1*b* shows a model for the potentiometer that consists of two resistors. The resistances of these resistors depend on the potentiometer parameters R_p and a.

Frequently, the position of the wiper corresponds to the angular position of a shaft connected to the potentiometer. Suppose θ is the angle in degrees and $0 \le \theta \le 360$. Then,

$$a = \frac{\theta}{360}$$

EXAMPLE 2.8-1 Potentiometer Circuit

Figure 2.8-2*a* shows a circuit in which the voltage measured by the meter gives an indication of the angular position of the shaft. In Figure 2.8-2*b*, the current source, the potentiometer, and the voltmeter have been replaced by models of these devices. Analysis of Figure 2.8-2*b* yields

$$v_m = R_p I a = \frac{R_p I}{360}\theta$$

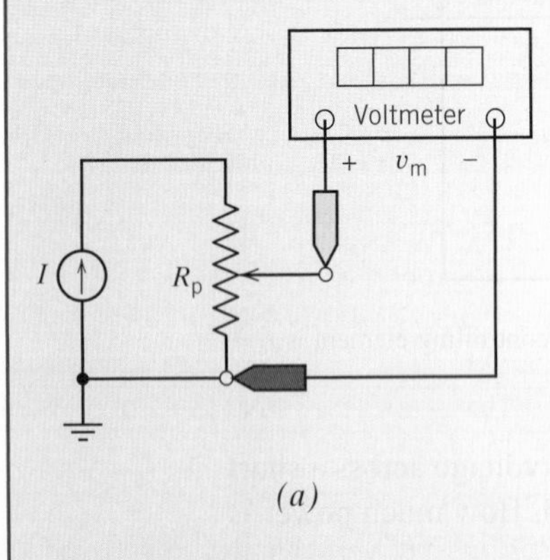

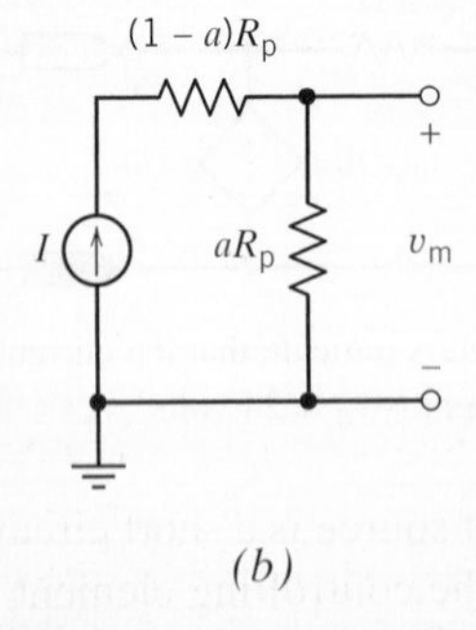

FIGURE 2.8-2 (*a*) A circuit containing a potentiometer. (*b*) An equivalent circuit containing a model of the potentiometer.

Solving for the angle gives

$$\theta = \frac{360}{R_p I} v_m$$

Suppose $R_p = 10\ \text{k}\Omega$ and $I = 1$ mA. An angle of 163° would cause an output of $v_m = 4.53$ V. A meter reading of 7.83 V would indicate that $\theta = 282°$.

Temperature sensors, such as the AD590 manufactured by Analog Devices, are current sources having current proportional to absolute temperature. Figure 2.8-3*a* shows the symbol used to represent the temperature sensor. Figure 2.8-3*b* shows the circuit model of the temperature sensor. For the temperature sensor to operate properly, the branch voltage v must satisfy the condition

$$4 \text{ volts} \leq v \leq 30 \text{ volts}$$

When this condition is satisfied, the current, i, in microamps, is numerically equal to the temperature T, in degrees Kelvin. The phrase *numerically equal* indicates that the current and temperature have the same value but different units. This relationship can be expressed as

$$i = k \cdot T$$

where $k = 1 \dfrac{\mu\text{A}}{°\text{K}}$, a constant associated with the sensor.

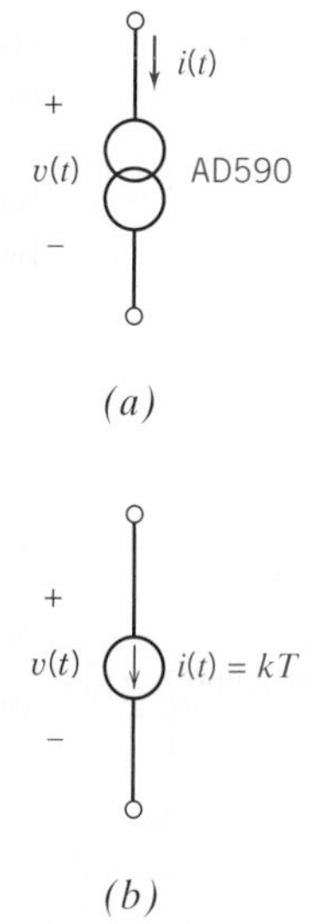

+ $v(t)$ − $i(t)$ AD590

(*a*)

+ $v(t)$ − $i(t) = kT$

(*b*)

FIGURE 2.8-3 (*a*) The symbol and (*b*) a model for the

EXERCISE 2.8-1 For the potentiometer circuit of Figure 2.8-2, calculate the meter voltage, v_{m}, when $\theta = 45°$, $R_{\text{p}} = 20\ \text{k}\Omega$, and $I = 2$ mA.

Answer: $v_{\text{m}} = 5$ V

EXERCISE 2.8-2 The voltage and current of an AD590 temperature sensor of Figure 2.8-3 are 10 V and 280 μA, respectively. Determine the measured temperature.

Answer: $T = 280°$K, or approximately 6.85°C

2.9 SWITCHES

Switches have two distinct states: open and closed. Ideally, a switch acts as a short circuit when it is closed and as an open circuit when it is open.

Figures 2.9-1 and 2.9-2 show several types of switches. In each case, the time when the switch changes state is indicated. Consider first the single-pole, single-throw (SPST) switches shown in Figure 2.9-1. The switch in Figure 2.9-1*a* is initially open. This switch changes state, becoming closed, at time $t = 0$ s. When this switch is modeled as an ideal switch, it is treated like an open circuit when $t < 0$ s and like a short circuit when $t > 0$ s. The ideal switch changes state instantaneously. The switch in Figure 2.9-1*b* is initially closed. This switch changes state, becoming open, at time $t = 0$ s.

Next, consider the single-pole, double-throw (SPDT) switch shown in Figure 2.9-1*a*. This SPDT switch acts like two SPST switches, one between terminals c and a, another between terminals c and b. Before $t = 0$ s, the switch between c and a is closed and the switch between c and b is open. At $t = 0$ s, both switches change state; that is, the switch between a and c opens, and the switch between c and b closes. Once again, the ideal switches are modeled as open circuits when they are open and as short circuits when they are closed.

In some applications, it makes a difference whether the switch between c and b closes before, or after, the switch between c and a opens. Different symbols are used to represent these two types of

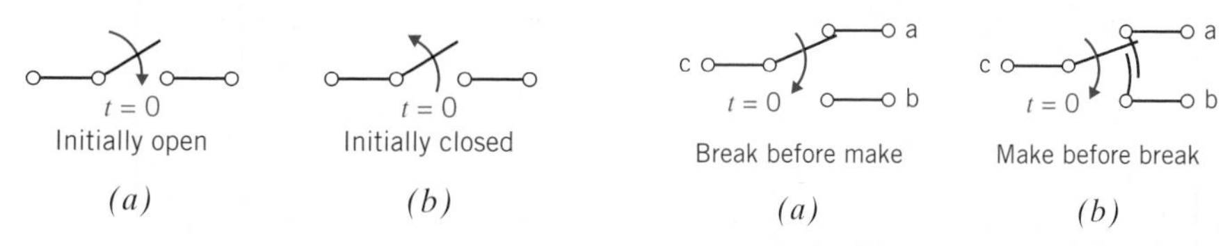

FIGURE 2.9-1 SPST switches. (*a*) Initially open and (*b*) initially closed.

FIGURE 2.9-2 SPDT switches. (*a*) Break before make and (*b*) make before break.

single-pole, double-throw switch. The break-before-make switch is manufactured so that the switch between c and b closes after the switch between c and a opens. The symbol for the break-before-make switch is shown in Figure 2.9-2*a*. The make-before-break switch is manufactured so that the switch between c and b closes before the switch between c and a opens. The symbol for the make-before-break switch is shown in Figure 2.9-2*b*. Remember: the switch transition from terminal a to terminal b is assumed to take place instantaneously. This instantaneous transition is an accurate model when the actual make-before-break transition is very fast compared to the circuit time response.

EXAMPLE 2.9-1 Switches

Figure 2.9-3 illustrates the use of open and short circuits for modeling ideal switches. In Figure 2.9-3*a*, a circuit containing three switches is shown. In Figure 2.9-3*b*, the circuit is shown as it would be modeled before $t = 0$ s. The two single-pole, single-throw switches change state at time $t = 0$ s. Figure 2.9-3*c* shows the circuit as it would be modeled when the time is between 0 s and 2 s. The single-pole, double-throw switch changes state at time $t = 2$ s. Figure 2.9-3*d* shows the circuit as it would be modeled after 2 s.

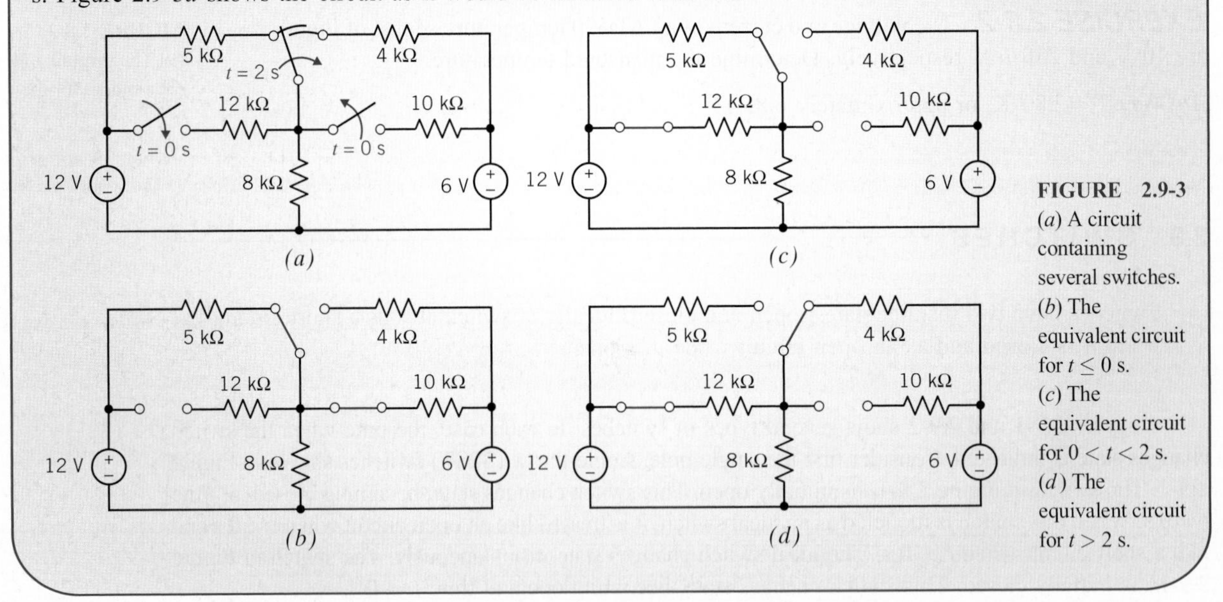

FIGURE 2.9-3 (*a*) A circuit containing several switches. (*b*) The equivalent circuit for $t \leq 0$ s. (*c*) The equivalent circuit for $0 < t < 2$ s. (*d*) The equivalent circuit for $t > 2$ s.

EXERCISE 2.9-1 What is the value of the current i in Figure E 2.9-1 at time $t = 4$ s?

Answer: $i = 0$ amperes at $t = 4$ s (both switches are open).

EXERCISE 2.9-2 What is the value of the voltage v in Figure E 2.9-2 at time $t = 4$ s? At $t = 6$ s?

Answer: $v = 6$ volts at $t = 4$ s, and $v = 0$ volts at $t = 6$ s.

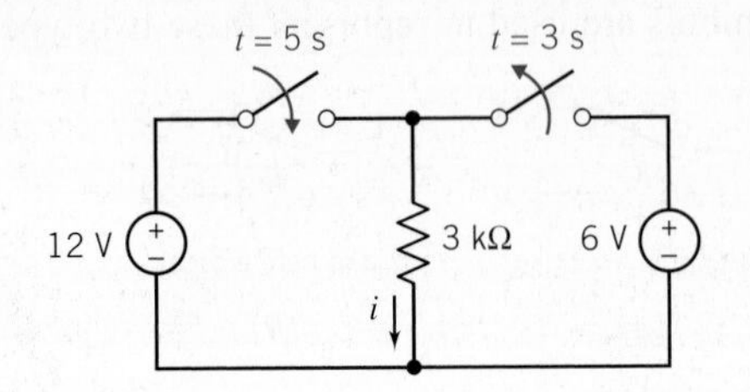

FIGURE E 2.9-1 A circuit with two SPST switches.

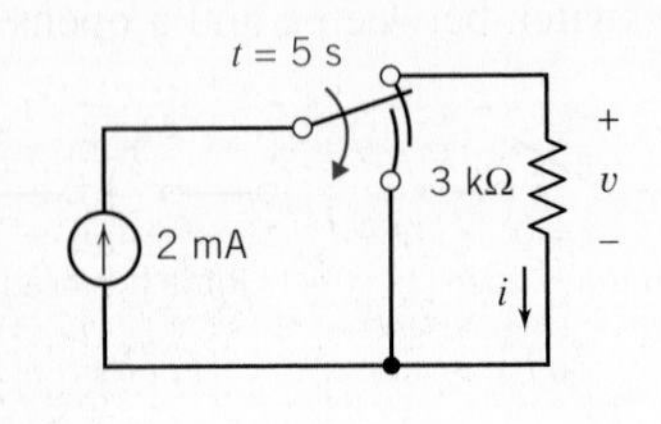

FIGURE E 2.9-2 A circuit with a make-before-break SPDT switch.

2.10 HOW CAN WE CHECK . . . ?

Engineers are frequently called upon to check that a solution to a problem is indeed correct. For example, proposed solutions to design problems must be checked to confirm that all of the specifications have been satisfied. In addition, computer output must be reviewed to guard against data-entry errors, and claims made by vendors must be examined critically.

Engineering students are also asked to check the correctness of their work. For example, occasionally just a little time remains at the end of an exam. It is useful to be able quickly to identify those solutions that need more work.

The following example illustrates techniques useful for checking the solutions of the sort of problem discussed in this chapter.

EXAMPLE 2.10-1 How Can We Check Voltage and Current Values?

The meters in the circuit of Figure 2.10-1 indicate that $v_1 = -4$ V, $v_2 = 8$ V and that $i = 1$ A. **How can we check** that the values of v_1, v_2, and i have been measured correctly? Let's check the values of v_1, v_2, and i in two ways:

(a) Verify that the given values satisfy Ohm's law for both resistors.

(b) Verify that the power supplied by the voltage source is equal to the power absorbed by the resistors.

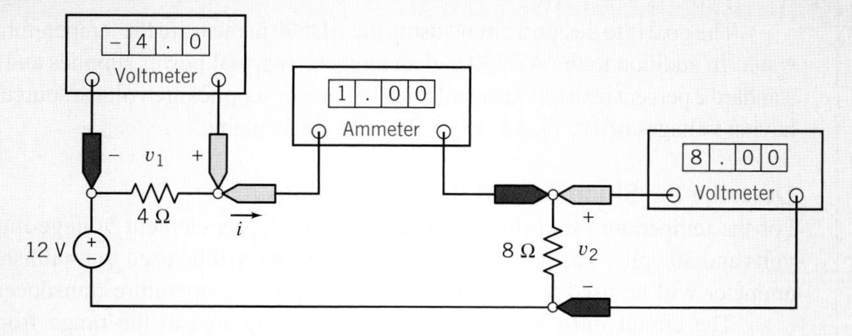

FIGURE 2.10-1 A circuit with meters.

Solution

(a) Consider the 8-Ω resistor. The current i flows through this resistor from top to bottom. Thus, the current i and the voltage v_2 adhere to the passive convention. Therefore, Ohm's law requires that $v_2 = 8i$. The values $v_2 = 8$ V and $i = 1$ A satisfy this equation.

Next, consider the 4-Ω resistor. The current i flows through this resistor from left to right. Thus, the current i and the voltage v_1 do not adhere to the passive convention. Therefore, Ohm's law requires that $v_1 = 4(-i)$. The values $v_1 = -4$ V and $i = 1$ A satisfy this equation.

Thus, Ohm's law is satisfied.

(b) The current i flows through the voltage source from bottom to top. Thus the current i and the voltage 12 V do not adhere to the passive convention. Therefore, $12i = 12(1) = 12$ W is the power supplied by the voltage source. The power absorbed by the 4-Ω resistor is $4i^2 = 4(1^2) = 4$ W, and the power absorbed by the 8-Ω resistor is $8i^2 = 8(1^2) = 8$ W. The power supplied by the voltage source is indeed equal to the power absorbed by the resistors.

2.11 DESIGN EXAMPLE

TEMPERATURE SENSOR

Currents can be measured easily, using ammeters. A temperature sensor, such as Analog Devices' AD590, can be used to measure temperature by converting temperature to current. Figure 2.11-1 shows a symbol used to represent a temperature sensor. For this sensor to operate properly, the voltage v must satisfy the condition

$$4 \text{ volts} \leq v \leq 30 \text{ volts}$$

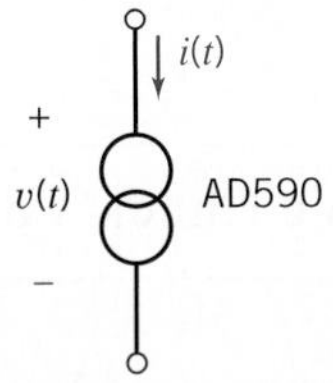

$i(t)$ + $v(t)$ – AD590

FIGURE 2.11-1
A temperature sensor.

When this condition is satisfied, the current i, in μA, is numerically equal to the temperature T, in °K. The phrase *numerically equal* indicates that the two variables have the same value but different units.

$$i = k \cdot T \quad \text{where} \quad k = 1\frac{\mu\text{A}}{°\text{K}}$$

The goal is to design a circuit using the AD590 to measure the temperature of a container of water. In addition to the AD590 and an ammeter, several power supplies and an assortment of standard 2 percent resistors are available. The power supplies are voltage sources. Power supplies having voltages of 10, 12, 15, 18, or 24 volts are available.

Describe the Situation and the Assumptions

For the temperature transducer to operate properly, its element voltage must be between 4 volts and 30 volts. The power supplies and resistors will be used to establish this voltage. An ammeter will be used to measure the current in the temperature transducer.

The circuit must be able to measure temperatures in the range from 0°C to 100°C because water is a liquid at these temperatures. Recall that the temperature in °C is equal to the temperature in °K minus 273°.

State the Goal

Use the power supplies and resistors to cause the voltage, v, of the temperature transducer to be between 4 volts and 30 volts.

Use an ammeter to measure the current, i, in the temperature transducer.

Generate a Plan

Model the power supply as an ideal voltage source and the temperature transducer as an ideal current source. The circuit shown in Figure 2.11-2*a* causes the voltage across the temperature transducer to be equal to the power supply voltage. Because all of the available power supplies have voltages between 4 volts and 30 volts, any one of the power supplies can be used. Notice that the resistors are not needed.

In Figure 2.11-2*b*, a short circuit has been added in a way that does not disturb the network. In Figure 2.11-2*c*, this short circuit has been replaced with an (ideal) ammeter. Because the ammeter will measure the current in the temperature transducer, the ammeter reading will be numerically equal to the temperature in °K.

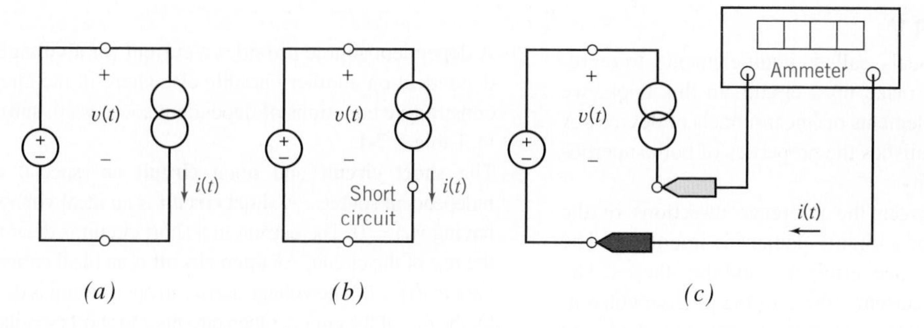

FIGURE 2.11-2 (*a*) Measuring temperature with a temperature sensor. (*b*) Adding a short circuit. (*c*) Replacing the short circuit by an ammeter.

Although any of the available power supplies is adequate to meet the specifications, there may still be an advantage to choosing a particular power supply. For example, it is reasonable to choose the power supply that causes the transducer to absorb as little power as possible.

Act on the Plan

The power absorbed by the transducer is

$$p = v \cdot i$$

where v is the power supply voltage. Choosing v as small as possible, 10 volts in this case, makes the power absorbed by the temperature transducer as small as possible. Figure 2.11-3*a* shows the final design. Figure 2.11-3*b* shows a graph that can be used to find the temperature corresponding to any ammeter current.

Verify the Proposed Solution

Let's try an example. Suppose the temperature of the water is 80.6°F. This temperature is equal to 27°C or 300°K. The current in the temperature sensor will be

$$i = \left(1 \frac{\mu\text{A}}{°\text{K}}\right) 300°\text{K} = 300\ \mu\text{A}$$

Next, suppose that the ammeter in Figure 2.11-3*a* reads 300 μA. A sensor current of 300 μA corresponds to a temperature of

$$T = \frac{300\ \mu\text{A}}{1 \frac{\mu\text{A}}{°\text{K}}} = 300°\text{K} = 27°\text{C} = 80.6°\text{F}$$

The graph in Figure 2.11-3*b* indicates that a sensor current of 300 μA does correspond to a temperature of 27°C.

This example shows that the circuit is working properly.

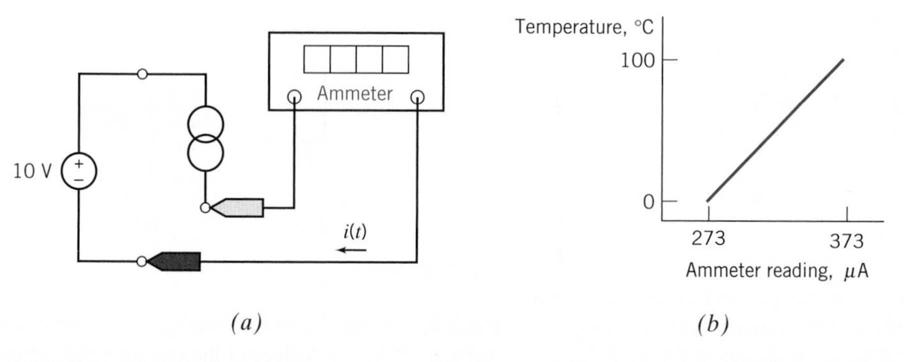

FIGURE 2.11-3 (*a*) Final design of a circuit that measures temperature with a temperature sensor. (*b*) Graph of temperature versus ammeter current.

2.12 SUMMARY

- The engineer uses models, called circuit elements, to represent the devices that make up a circuit. In this book, we consider only linear elements or linear models of devices. A device is linear if it satisfies the properties of both superposition and homogeneity.
- The relationship between the reference directions of the current and voltage of a circuit element is important. The voltage polarity marks one terminal + and the other −. The element voltage and current adhere to the passive convention if the current is directed from the terminal marked + to the terminal marked −.
- Resistors are widely used as circuit elements. When the resistor voltage and current adhere to the passive convention, resistors obey Ohm's law; the voltage across the terminals of the resistor is related to the current into the positive terminal as $v = Ri$. The power delivered to a resistance is $p = i^2R = v^2/R$ watts.
- An independent source provides a current or a voltage independent of other circuit variables. The voltage of an independent voltage source is specified, but the current is not. Conversely, the current of an independent current source is specified whereas the voltage is not. The voltages of independent voltage sources and currents of independent current sources are frequently used as the inputs to electric circuits.
- A dependent source provides a current (or a voltage) that is dependent on another variable elsewhere in the circuit. The constitutive equations of dependent sources are summarized in Table 2.7-1.
- The **short circuit** and **open circuit** are special cases of independent sources. A **short circuit** is an ideal voltage source having $v(t) = 0$. The current in a short circuit is determined by the rest of the circuit. An **open circuit** is an ideal current source having $i(t) = 0$. The voltage across an open circuit is determined by the rest of the circuit. Open circuits and short circuits can also be described as special cases of resistors. A resistor with resistance $R = 0$ ($G = \infty$) is a short circuit. A resistor with conductance $G = 0$ ($R = \infty$) is an open circuit.
- An ideal ammeter measures the current flowing through its terminals and has zero voltage across its terminals. An ideal voltmeter measures the voltage across its terminals and has terminal current equal to zero. Ideal voltmeters act like open circuits, and ideal ammeters act like short circuits.
- Transducers are devices that convert physical quantities, such as rotational position, to an electrical quantity such as voltage. In this chapter, we describe two transducers: potentiometers and temperature sensors.
- Switches are widely used in circuits to connect and disconnect elements and circuits. They can also be used to create discontinuous voltages or currents.

PROBLEMS

Section 2.2 Engineering and Linear Models

P 2.2-1 An element has voltage v and current i as shown in Figure P 2.2-1*a*. Values of the current i and corresponding voltage v have been tabulated as shown in Figure P 2.2-1*b*. Determine whether the element is linear.

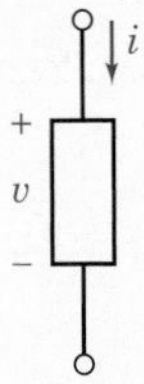

(*a*)

v, V	i, A
−5	−5
−6	−4
0	0
10	4
34	6
50	5

(*b*)

Figure P 2.2-1

P 2.2-2 A linear element has voltage v and current i as shown in Figure P 2.2-2*a*. Values of the current i and corresponding voltage v have been tabulated as shown in Figure P 2.2-2*b*. Represent the element by an equation that expresses v as a function of i. This equation is a model of the element. (a) Verify that the model is linear. (b) Use the model to predict the value of v corresponding to a current of $i = 50$ mA. (c) Use the model to predict the value of i corresponding to a voltage of $v = 6$ V.

Hint: Plot the data. We expect the data points to lie on a straight line. Obtain a linear model of the element by representing that straight line by an equation.

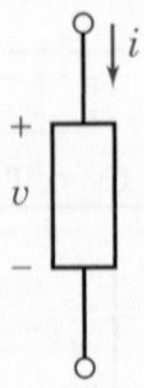

(*a*)

v, V	i, A
4.0	35
2.8	25
6.4	55

(*b*)

Figure P 2.2-2

P 2.2-3 A linear element has voltage v and current i as shown in Figure P 2.2-3*a*. Values of the current i and corresponding voltage v have been tabulated as shown in Figure P 2.2-3*b*. Represent the element by an equation that expresses v as a function of i. This equation is a model of the element. (a) Verify that the model is linear. (b) Use the model to predict the value of v corresponding to a current of $i = 5$ mA. (c) Use the model to predict the value of i corresponding to a voltage of $v = 15$ V.

Hint: Plot the data. We expect the data points to lie on a straight line. Obtain a linear model of the element by representing that straight line by an equation.

(a)

v, V	i, mA
4.078	17
6.13	25
13.8	55

(b)

Figure P 2.2-3

P 2.2-4 An element is represented by the relation between current and voltage as

$$v = 6i + 10$$

Determine whether the element is linear.

P 2.2-5 The circuit shown in Figure P 2.2-5 consists of a current source, a resistor, and element A. Consider three cases.

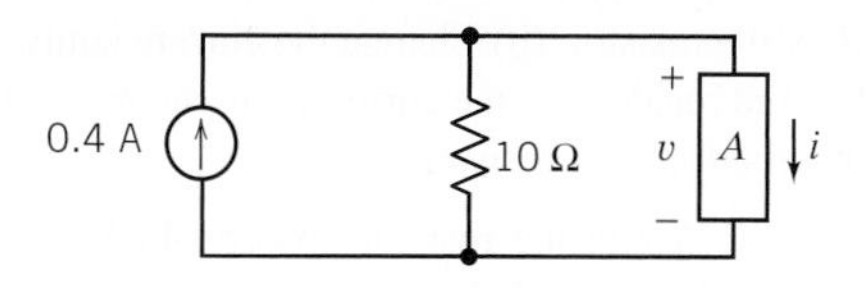

Figure P 2.2-5

(a) When element A is a 40-Ω resistor, described by $i = v/40$, then the circuit is represented by

$$0.4 = \frac{v}{10} + \frac{v}{40}$$

Determine the values of v and i. Notice that the above equation has a unique solution.

(b) When element A is a nonlinear resistor described by $i = v^2/2$, then the circuit is represented by

$$0.4 = \frac{v}{10} + \frac{v^2}{2}$$

Determine the values of v and i. In this case, there are two solutions of the above equation. Nonlinear circuits exhibit more complicated behavior than linear circuits.

(c) When element A is a nonlinear resistor described by $i = 0.8 + \frac{v^2}{2}$, then the circuit is described by

$$0.4 = \frac{v}{10} + 0.8 + \frac{v^2}{2}$$

Show that this equation has no solution. This result usually indicates a modeling problem. At least one of the three elements in the circuit has not been modeled accurately.

Section 2.4 Resistors

P 2.4-1 A current source and a resistor are connected in series in the circuit shown in Figure P 2.4-1. Elements connected in series have the same current, so $i = i_s$ in this circuit. Suppose that $i_s = 5$ A and $R = 10\ \Omega$. Calculate the voltage v across the resistor and the power absorbed by the resistor.

Answer: $v = 50$ V and the resistor absorbs 250 W.

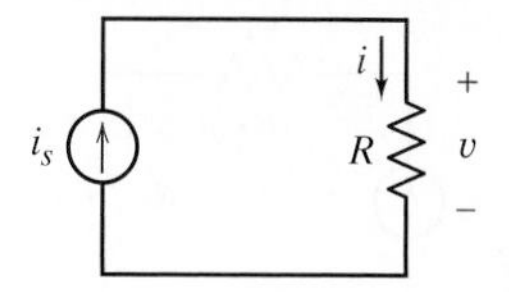

Figure P 2.4-1

P 2.4-2 A current source and a resistor are connected in series in the circuit shown in Figure P 2.4-1. Elements connected in series have the same current, so $i = i_s$ in this circuit. Suppose that $i = 5$ mA and $v = 50$ V. Calculate the resistance R and the power absorbed by the resistor.

P 2.4-3 A voltage source and a resistor are connected in parallel in the circuit shown in Figure P 2.4-3. Elements connected in parallel have the same voltage, so $v = v_s$ in this circuit. Suppose that $v_s = 15$ V and $R = 3\ \Omega$. Calculate the current i in the resistor and the power absorbed by the resistor.

Answer: $i = 5$ A and the resistor absorbs 75 W.

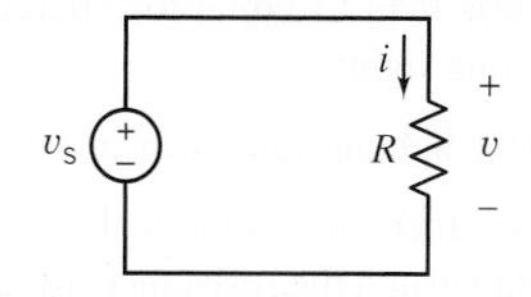

Figure P 2.4-3

P 2.4-4 A voltage source and a resistor are connected in parallel in the circuit shown in Figure P 2.4-3. Elements connected in parallel have the same voltage, so $v = v_s$ in this circuit. Suppose that $v_s = 25$ V and $i = 5$ A. Calculate the resistance R and the power absorbed by the resistor.

P 2.4-5 A voltage source and two resistors are connected in parallel in the circuit shown in Figure P 2.4-5. Elements connected in parallel have the same voltage, so $v_1 = v_s$ and $v_2 = v_s$ in this circuit. Suppose that $v_s = 200$ V, $R_1 = 50\ \Omega$, and $R_2 = 25\ \Omega$. Calculate the current in each resistor and the power absorbed by each resistor.

Hint: Notice the reference directions of the resistor currents.

Answer: $i_1 = 4$ A and $i_2 = -8$ A. R_1 absorbs 800 W and R_2 absorbs 900 W.

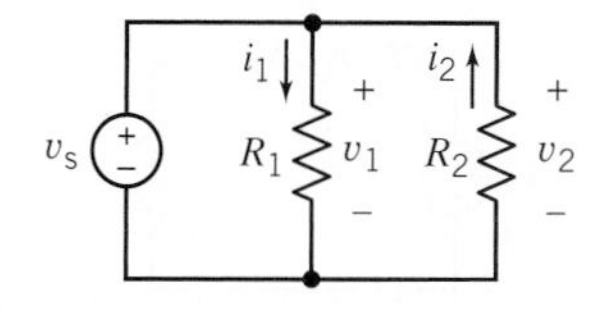

Figure P 2.4-5

P 2.4-6 A current source and two resistors are connected in series in the circuit shown in Figure P 2.4-6. Elements connected in series have the same current, so $i_1 = i_s$ and $i_2 = i_s$ in this circuit. Suppose that $i_s = 50$ mA, $R_1 = 8\ \Omega$, and $R_2 = 16\ \Omega$. Calculate the voltage across each resistor and the power absorbed by each resistor.

Hint: Notice the reference directions of the resistor voltages.

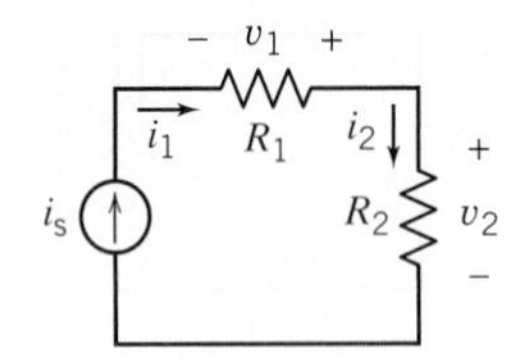

Figure P 2.4-6

P 2.4-7 An electric heater is connected to a constant 200-V source and absorbs 1000 W. Subsequently, this heater is connected to a constant 220-V source. What power does it absorb from the 220-V source? What is the resistance of the heater?

Hint: Model the electric heater as a resistor.

P 2.4-8 The portable lighting equipment for a mine is located 100 meters from its dc supply source. The mine lights use a total of 5 kW and operate at 120 V dc. Determine the required cross-sectional area of the copper wires used to connect the source to the mine lights if we require that the power lost in the copper wires be less than or equal to 5 percent of the power required by the mine lights.

Hint: Model both the lighting equipment and the wire as resistors.

P 2.4-9 The resistance of a practical resistor depends on the nominal resistance and the resistance tolerance as follows:

$$R_{nom}\left(1 - \frac{t}{100}\right) \le R \le R_{nom}\left(1 + \frac{t}{100}\right)$$

where R_{nom} is the nominal resistance and t is the resistance tolerance expressed as a percentage. For example, a 100-Ω, 2 percent resistor will have a resistance given by

$$98\ \Omega \le R \le 102\ \Omega$$

The circuit shown in Figure P 2.4-9 has one input, v_s, and one output, v_o. The gain of this circuit is given by

$$\text{gain} = \frac{v_o}{v_s} = \frac{R_2}{R_1 + R_2}$$

Determine the range of possible values of the gain when R_1 is the resistance of a 100-Ω, 2 percent resistor and R_2 is the resistance of a 400-Ω, 5 percent resistor. Express the gain in terms of a nominal gain and a gain tolerance.

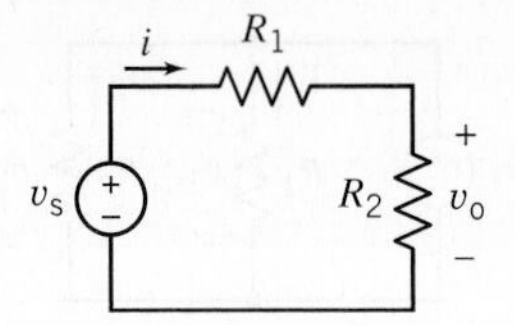

Figure P 2.4-9

P 2.4-10 The voltage source shown in Figure P 2.4-10 is an adjustable dc voltage source. In other words, the voltage v_s is a constant voltage, but the value of that constant can be adjusted. The tabulated data were collected as follows. The voltage, v_s, was set to some value, and the voltages across the resistor, v_a and v_b, were measured and recorded. Next, the value of v_s was changed, and the voltages across the resistors were measured again and recorded. This procedure was repeated several times. (The values of v_s were not recorded.) Determine the value of the resistance, R.

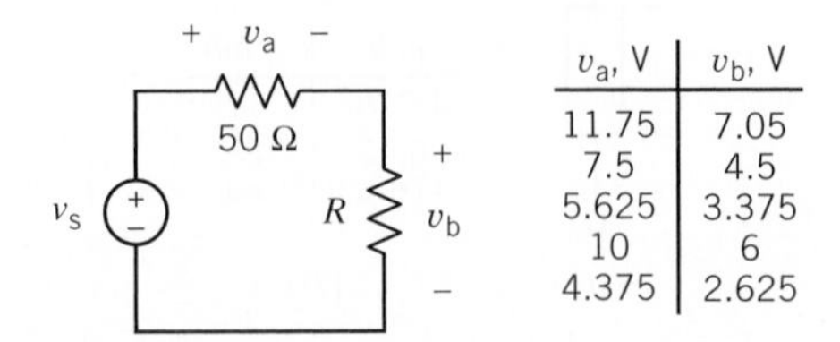

v_a, V	v_b, V
11.75	7.05
7.5	4.5
5.625	3.375
10	6
4.375	2.625

Figure P 2.4-10

Section 2.5 Independent Sources

P 2.5-1 A current source and a voltage source are connected in parallel with a resistor as shown in Figure P 2.5-1. All of the elements connected in parallel have the same voltage, v_s in this circuit. Suppose that $v_s = 10$ V, $i_s = 2$ A, and $R = 5\ \Omega$. (a) Calculate the current i in the resistor and the power absorbed by the resistor. (b) Change the current source current to $i_s = 4$ A and recalculate the current, i, in the resistor and the power absorbed by the resistor.

Answer: $i = 2$ A and the resistor absorbs 45 W both when $i_s = 2$ A and when $i_s = 4$ A.

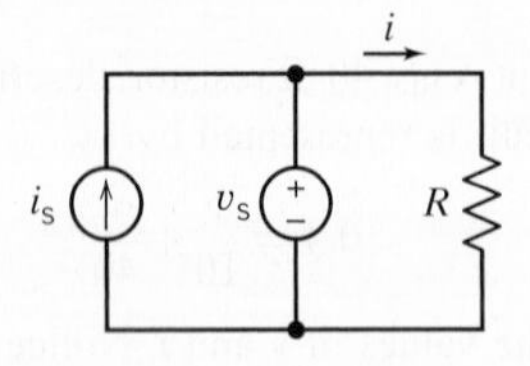

Figure P 2.5-1

P 2.5-2 A current source and a voltage source are connected in series with a resistor as shown in Figure P 2.5-2. All of the elements connected in series have the same current, i_s, in this circuit. Suppose that $v_s = 15$ V, $i_s = 5$ A, and $R = 5\ \Omega$. (a) Calculate the voltage v across the resistor and the power absorbed by the resistor. (b) Change the voltage source voltage to $v_s = 10$ V and recalculate the voltage, v, across the resistor and the power absorbed by the resistor.

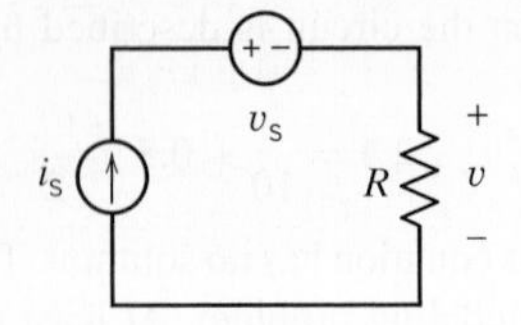

Figure P 2.5-2

P 2.5-3 The current source and voltage source in the circuit shown in Figure P 2.5-3 are connected in parallel so that they both have the same voltage, v_s. The current source and voltage source are also connected in series so that they both have the same current, i_s. Suppose that $v_s = 10$ V and $i_s = 5$ A. Calculate the power supplied by each source.

Answer: The voltage source supplies 50 W, and the current source supplies 50 W.

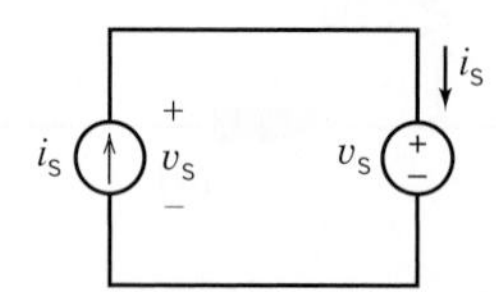

Figure P 2.5-3

P 2.5-4 The current source and voltage source in the circuit shown in Figure P 2.5-4 are connected in parallel so that they both have the same voltage, v_s. The current source and voltage source are also connected in series so that they both have the same current, i_s. Suppose that $v_s = 10$ V and $i_s = 3$ A. Calculate the power supplied by each source.

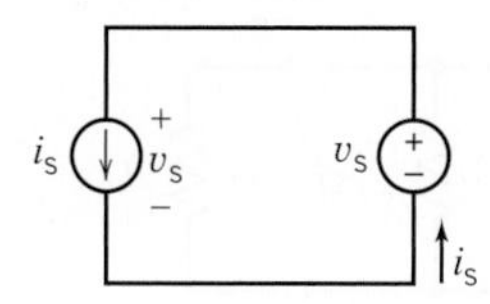

Figure P 2.5-4

P 2.5-5

(a) Find the power supplied by the voltage source shown in Figure P 2.5-5 when for $t \geq 0$ we have

$$v = 5 \cos t \text{ V}$$

and

$$i = 20 \cos t \text{ mA}$$

(b) Determine the energy supplied by this voltage source for the period $0 \leq t \leq 1$ s.

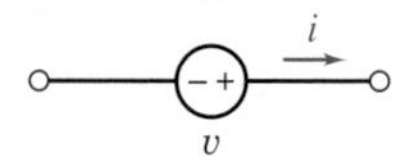

Figure P 2.5-5

P 2.5-6 Figure P 2.5.6 shows a battery connected to a load. The load in Figure P 2.5.6 might represent automobile headlights, a digital camera, or a cell phone. The energy supplied by the battery to load is given by

$$w = \int_{t_1}^{t_2} vi \, dt$$

When the battery voltage is constant and the load resistance is fixed, then the battery current will be constant and

$$w = vi(t_2 - t_1)$$

The capacity of a battery is the product of the battery current and time required to discharge the battery. Consequently, the energy stored in a battery is equal to the product of the battery voltage and the battery capacity. The capacity is usually given with the units of Ampere-hours (Ah). A new 12-V battery having a capacity of 800 mAh is connected to a load that draws a current of 25 mA. (a) How long will it take for the load to discharge the battery? (b) How much energy will be supplied to the load during the time required to discharge the battery?

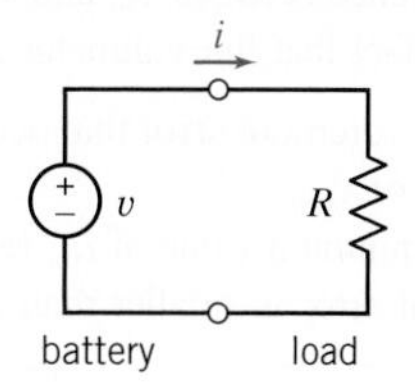

Figure P 2.5-6

Section 2.6 Voltmeters and Ammeters

P 2.6-1 For the circuit of Figure P 2.6-1:

(a) What is the value of the resistance R?

(b) How much power is delivered by the voltage source?

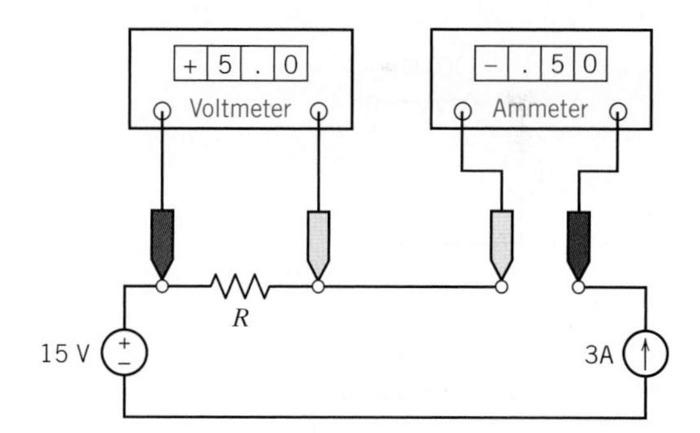

Figure P 2.6-1

P 2.6-2 The current source in Figure P 2.6-2 supplies 40 W. What values do the meters in Figure P 2.6-2 read?

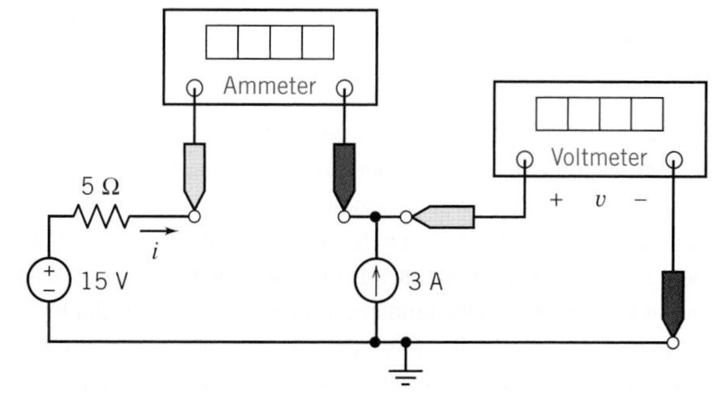

Figure P 2.6-2

P 2.6-3 An ideal voltmeter is modeled as an open circuit. A more realistic model of a voltmeter is a large resistance. Figure P 2.6-3*a* shows a circuit with a voltmeter that measures the voltage v_m. In Figure P 2.6-3*b*, the voltmeter is replaced by the model of an ideal voltmeter, an open circuit. Ideally, there is no current in the 100-Ω resistor, and the voltmeter measures $v_{mi} = 12$ V, the ideal value of v_m. In Figure P 2.6-3*c*, the voltmeter is modeled by the resistance R_m. Now the voltage measured by the voltmeter is

$$v_m = \left(\frac{R_m}{R_m + 100}\right)12$$

Because $R_m \to \infty$, the voltmeter becomes an ideal voltmeter, and $v_m \to v_{mi} = 12$ V . When $R_m < \infty$, the voltmeter is not ideal, and $v_m < v_{mi}$. The difference between v_m and v_{mi} is a measurement error caused by the fact that the voltmeter is not ideal.

(a) Express the measurement error that occurs when $R_m = 900$ Ω as a percent of v_{mi}.

(b) Determine the minimum value of R_m required to ensure that the measurement error is smaller than 2 percent of v_{mi}.

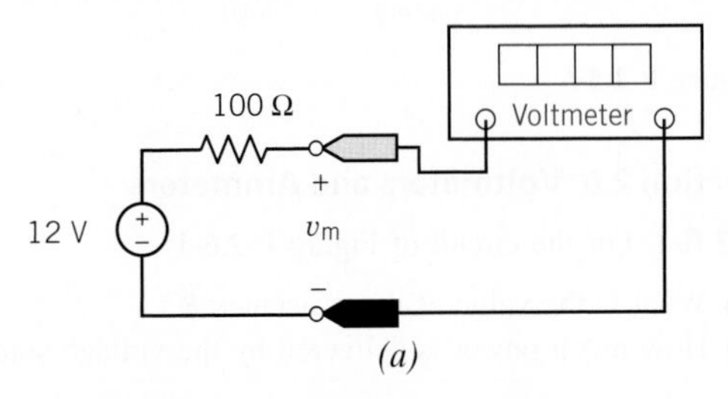

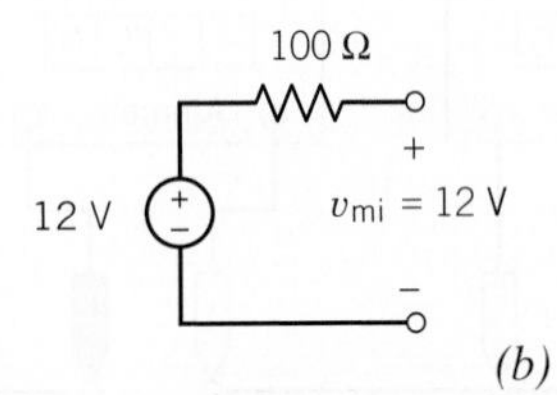

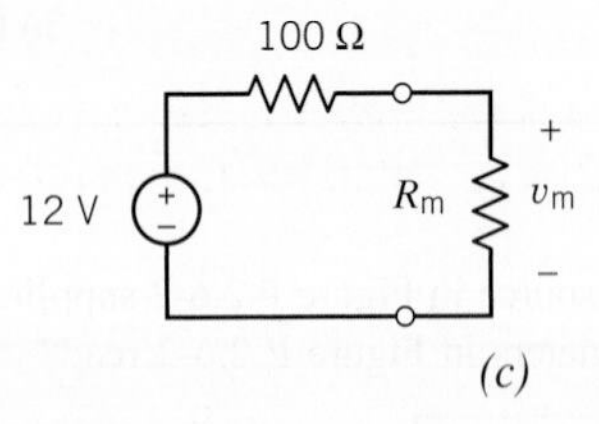

Figure P 2.6-3

P 2.6-4 An ideal ammeter is modeled as a short circuit. A more realistic model of an ammeter is a small resistance. Figure P 2.6-4*a* shows a circuit with an ammeter that measures the current i_m. In Figure P 2.6-4*b*, the ammeter is replaced by the model of an ideal ammeter, a short circuit. Ideally, there is no voltage across the 1-kΩ resistor, and the ammeter measures $i_{mi} = 2$ A, the ideal value of i_m. In Figure P 2.6-4*c*, the ammeter is modeled by the resistance R_m. Now the current measured by the ammeter is

$$i_m = \left(\frac{1000}{1000 + R_m}\right)2$$

As $R_m \to 0$, the ammeter becomes an ideal ammeter, and $i_m \to i_{mi} = 2$ A. When $R_m > 0$, the ammeter is not ideal, and $i_m < i_{mi}$. The difference between i_m and i_{mi} is a measurement error caused by the fact that the ammeter is not ideal.

(a) Express the measurement error that occurs when $R_m = 20$ Ω as a percent of i_{mi}.

(b) Determine the maximum value of R_m required to ensure that the measurement error is smaller than 5 percent.

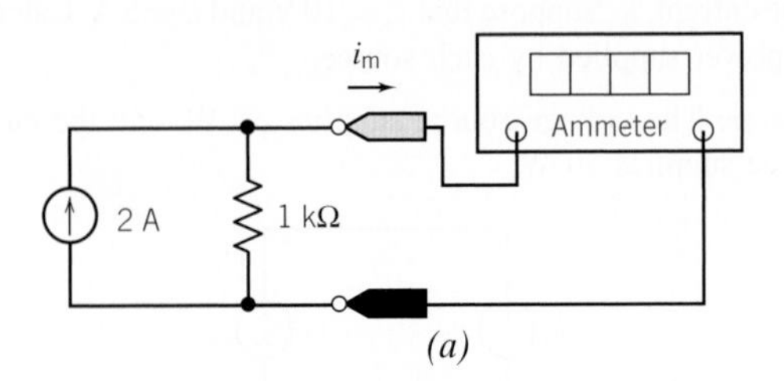

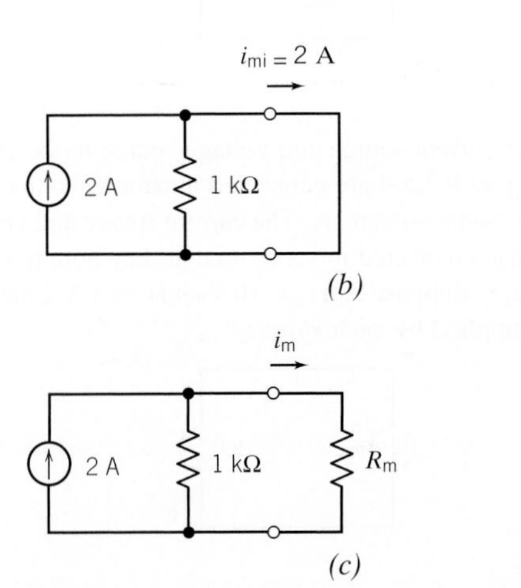

Figure P 2.6-4

P 2.6-5 The voltmeter in Figure P 2.6-5*a* measures the voltage across the current source. Figure P 2.6-5*b* shows the circuit after removing the voltmeter and labeling the voltage measured by the voltmeter as v_m. Also, the other element voltages and currents are labeled in Figure P 2.6-5*b*.

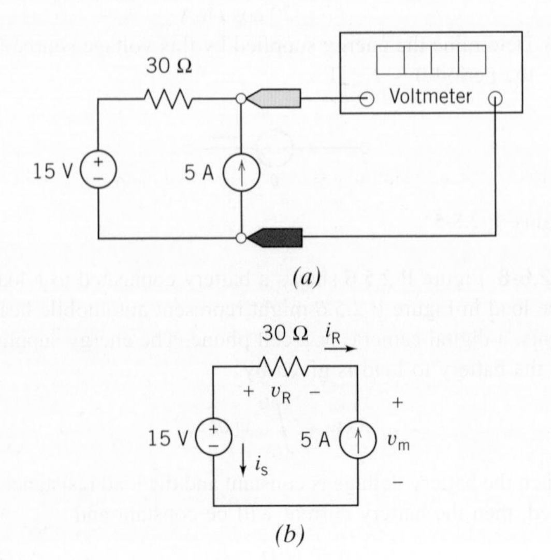

Figure P 2.6-5

Given that

$$15 = v_R + v_m \text{ and } -i_R = i_s = 5\text{ A}$$

and

$$v_R = 30 i_R$$

(a) Determine the value of the voltage measured by the meter.
(b) Determine the power supplied by each element.

P 2.6-6 The ammeter in Figure P 2.6-6*a* measures the current in the voltage source. Figure P 2.6-6*b* shows the circuit after removing the ammeter and labeling the current measured by the ammeter as i_m. Also, the other element voltages and currents are labeled in Figure P 2.6-6*b*.
Given that

$$2 + i_m = i_R \text{ and } v_R = v_s = 15\text{ V}$$

and

$$v_R = 30 i_R$$

(a) Determine the value of the current measured by the meter.
(b) Determine the power supplied by each element.

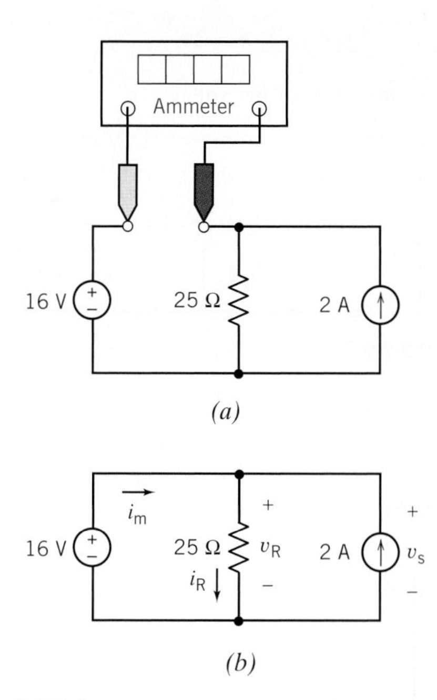

Figure P 2.6-6

Section 2.7 Dependent Sources

P 2.7-1 The ammeter in the circuit shown in Figure P 2.7-1 indicates that $i_a = 5$ A, and the voltmeter indicates that $v_b = 10$ V. Determine the value of r, the gain of the CCVS.

Answer: $r = 2\ \Omega$

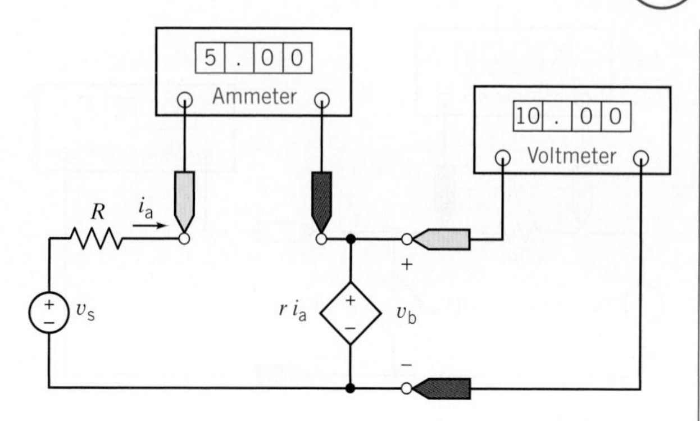

Figure P 2.7-1

P 2.7-2 The ammeter in the circuit shown in Figure P 2.7-2 indicates that $i_a = 5$ A, and the voltmeter indicates that $v_b = 10$ V. Determine the value of g, the gain of the VCCS.

Answer: $g = 0.5$ A/V

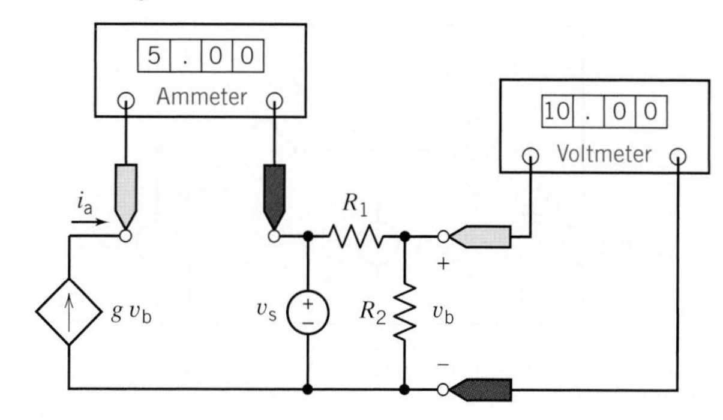

Figure P 2.7-2

P 2.7-3 The ammeters in the circuit shown in Figure P 2.7-3 indicate that $i_a = 40$ A and $i_b = 10$ A. Determine the value of d, the gain of the CCCS.

Answer: $d = 4$ A/A

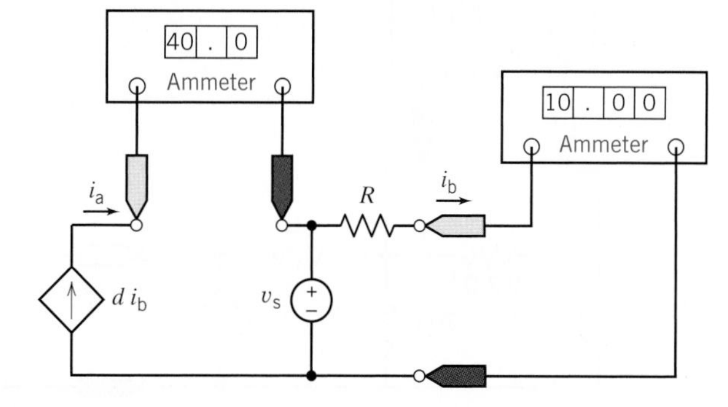

Figure P 2.7-3

P 2.7-4 The voltmeters in the circuit shown in Figure P 2.7-4 indicate that $v_a = 3$ V and $v_b = 9$ V. Determine the value of b, the gain of the VCVS.

Answer: $b = 3$ V/V

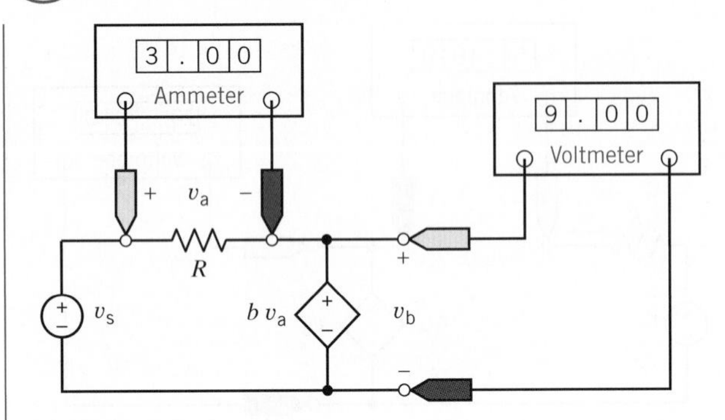

Figure P 2.7-4

P 2.7-5 The values of the current and voltage of each circuit element are shown in Figure P 2.7-5.

Determine the values of the resistance, R, and of the gain of the dependent source, A.

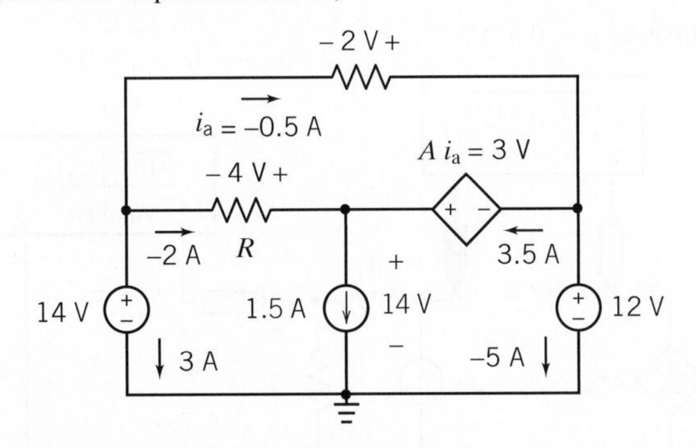

Figure P 2.7-5

P 2.7-6 Find the power supplied by the VCCS in Figure P 2.7-6.

Answer: 40 watts are supplied by the VCCS. (−40 watts are absorbed by the VCCS.)

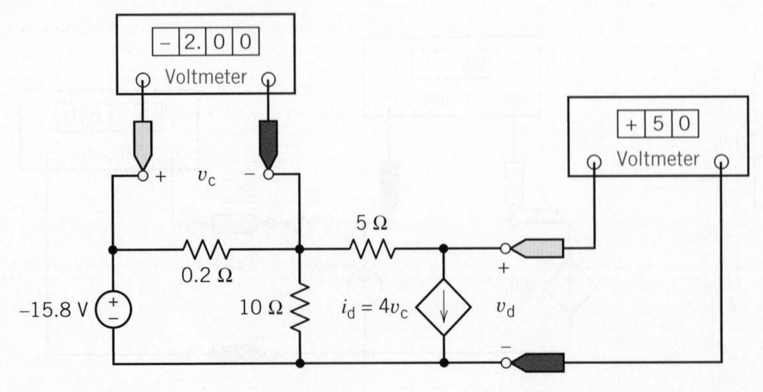

Figure P 2.7-6

P 2.7-7 The circuit shown in Figure P 2.7.7 contains a dependent source. Determine the value of the gain k of that dependent source.

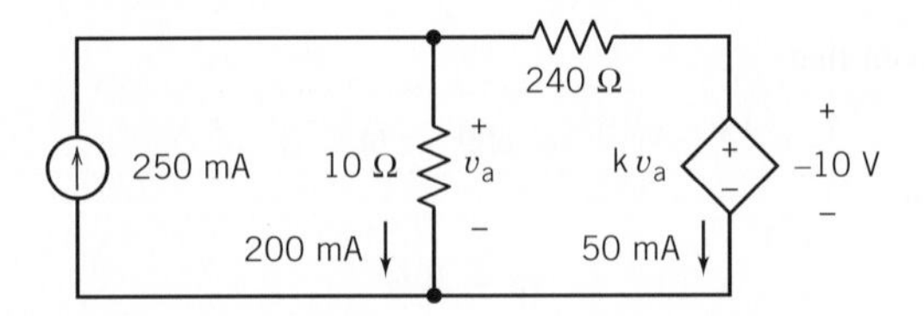

Figure P 2.7-7

P 2.7-8 The circuit shown in Figure P 2.7-8 contains a dependent source. Determine the value of the gain k of that dependent source.

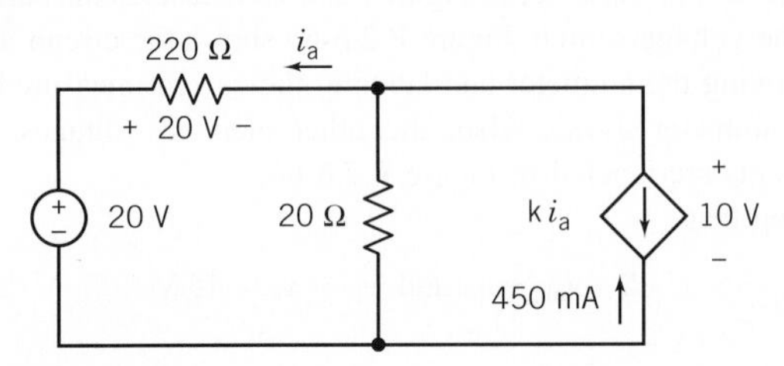

Figure P 2.7-8

P 2.7-9 The circuit shown in Figure P 2.7-9 contains a dependent source. The gain of that dependent source is

$$k = 25\frac{\text{V}}{\text{A}}$$

Determine the value of the voltage v_b.

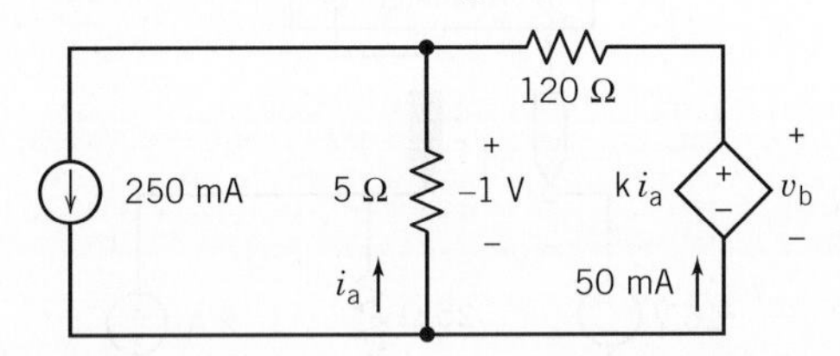

Figure P 2.7-9

P 2.7-10 The circuit shown in Figure P 2.7-10 contains a dependent source. The gain of that dependent source is

$$k = 100\frac{\text{mA}}{\text{V}} = .1\frac{\text{A}}{\text{V}}$$

Determine the value of the current i_b.

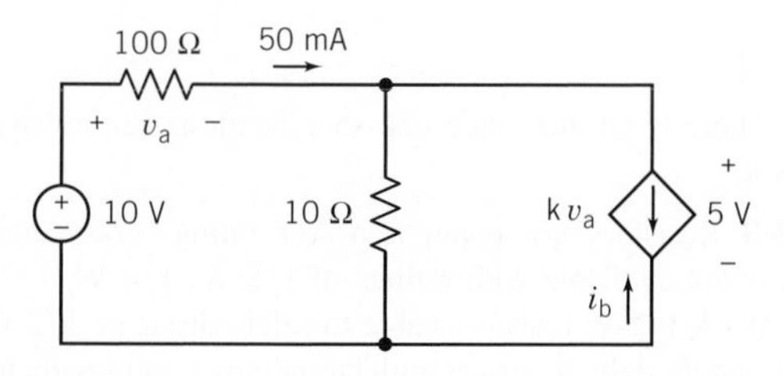

Figure P 2.7-10

Section 2.8 Transducers

P 2.8-1 For the potentiometer circuit of Figure 2.8-2, the current source current and potentiometer resistance are 2 mA and 110 kΩ, respectively. Calculate the required angle, θ, so that the measured voltage is 25 V.

P 2.8-2 An AD590 sensor has an associated constant $k = 1\frac{\mu A}{^\circ K}$. The sensor has a voltage $v = 20$ V; and the measured current, $i(t)$, as shown in Figure 2.8-3, is $4\ \mu A < i < 13\ \mu A$ in a laboratory setting. Find the range of measured temperature.

Section 2.9 Switches

P 2.9-1 Determine the current, i, at $t = 1$ s and at $t = 4$ s for the circuit of Figure P 2.9-1.

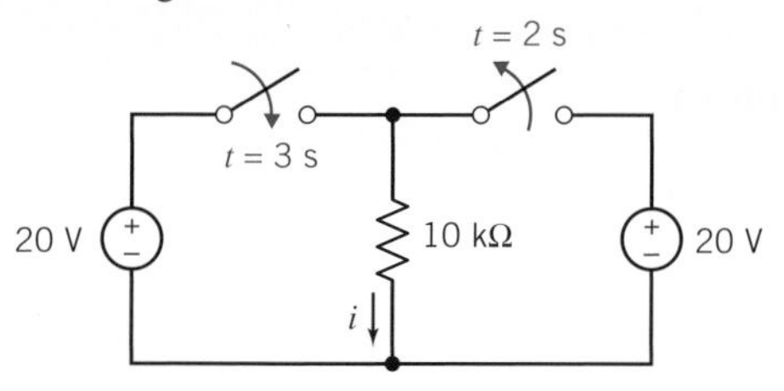

Figure P 2.9-1

P 2.9-2 Determine the voltage, v, at $t = 1$ s and at $t = 4$ s for the circuit shown in Figure P 2.9-2.

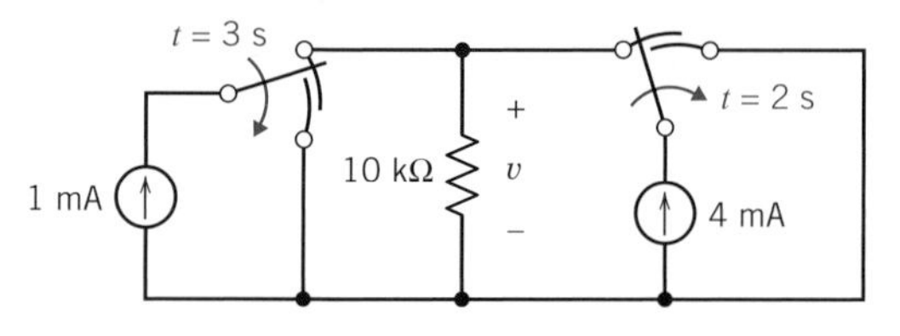

Figure P 2.9-2

P 2.9-3 Ideally, an open switch is modeled as an open circuit and a closed switch is modeled as a closed circuit. More realistically, an open switch is modeled as a large resistance, and a closed switch is modeled as a small resistance.

Figure P 2.9-3*a* shows a circuit with a switch. In Figure P 2.9-3*b*, the switch has been replaced with a resistance. In Figure P 2.9-3*b*, the voltage v is given by

$$v = \left(\frac{100}{R_s + 100}\right)12$$

Determine the value of v for each of the following cases.

(a) The switch is closed and $R_s = 0$ (a short circuit).
(b) The switch is closed and $R_s = 5\ \Omega$.
(c) The switch is open and $R_s = \infty$ (an open circuit).
(d) The switch is open and $R_s = 10\ \text{k}\Omega$.

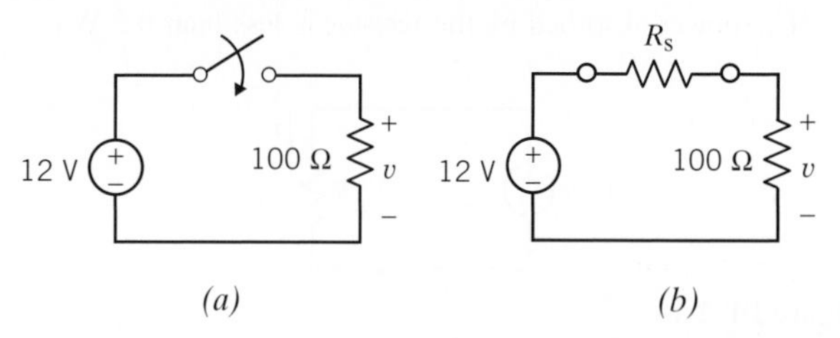

Figure P 2.9-3

Section 2-10 How Can We Check . . . ?

P 2.10-1 The circuit shown in Figure P 2.10-1 is used to test the CCVS. Your lab partner claims that this measurement shows that the gain of the CCVS is −20 V/A instead of +20 V/A. Do you agree? Justify your answer.

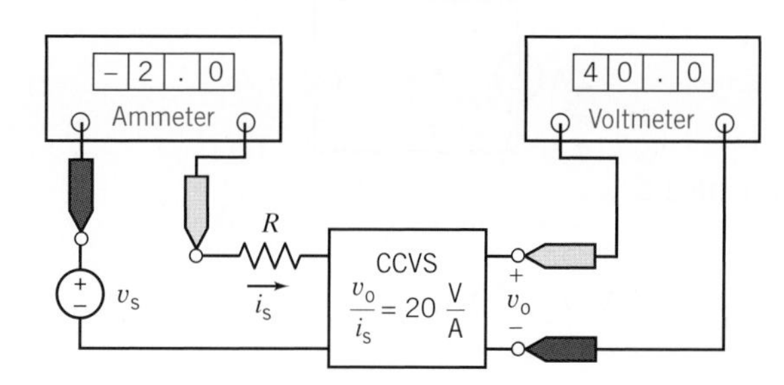

Figure P 2.10-1

P 2.10-2 The circuit of Figure P 2.10-2 is used to measure the current in the resistor. Once this current is known, the resistance can be calculated as $R = \frac{v_s}{i}$. The circuit is constructed using a voltage source with $v_s = 12$ V and a 25-Ω, 1/2-W resistor. After a puff of smoke and an unpleasant smell, the ammeter indicates that $i = 0$ A. The resistor must be bad. You have more 25-Ω, 1/2-W resistors. Should you try another resistor? Justify your answer.

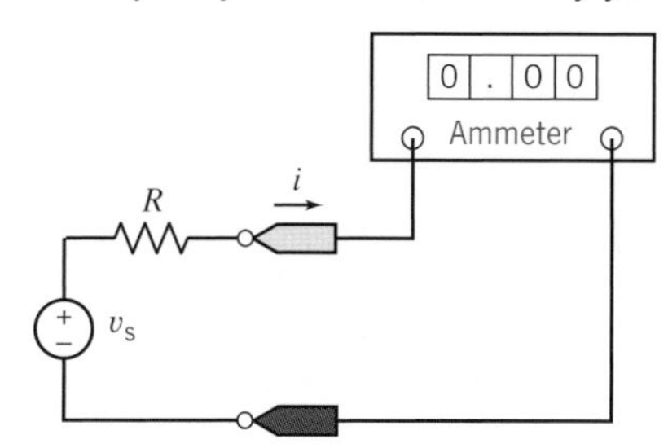

Figure P 2.10-2

Hint: 1/2-W resistors are able to safely dissipate one 1/2 W of power. These resistors may fail if required to dissipate more than 1/2 watt of power.

Design Problems

DP 2-1 Specify the resistance R in Figure DP 2-1 so that both of the following conditions are satisfied:

1. $i > 40$ mA.
2. The power absorbed by the resistor is less than 0.5 W.

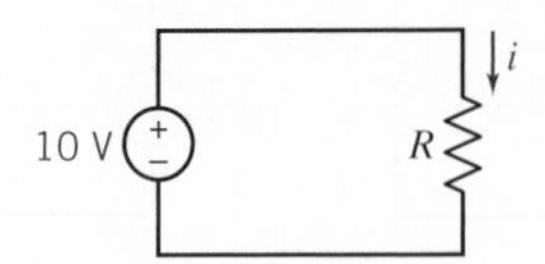

Figure DP 2-1

DP 2-2 Specify the resistance R in Figure DP 2-2 so that both of the following conditions are satisfied:

1. $v > 40$ V.
2. The power absorbed by the resistor is less than 15 W.

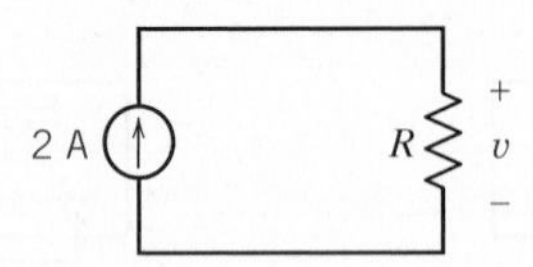

Figure DP 2-2

Hint: There is no guarantee that specifications can always be satisfied.

DP 2-3 Resistors are given a power rating. For example, resistors are available with ratings of 1/8 W, 1/4 W, 1/2 W, and 1 W. A 1/2-W resistor is able to safely dissipate 1/2 W of power, indefinitely. Resistors with larger power ratings are more expensive and bulkier than resistors with lower power ratings. Good engineering practice requires that resistor power ratings be specified to be as large as, but not larger than, necessary.

Consider the circuit shown in Figure DP 2-3. The values of the resistances are

$$R_1 = 1000\ \Omega,\ R_2 = 2000\ \Omega,\ \text{and}\ R_3 = 4000\ \Omega$$

The value of the current source current is

$$i_s = 30\ \text{mA}$$

Specify the power rating for each resistor.

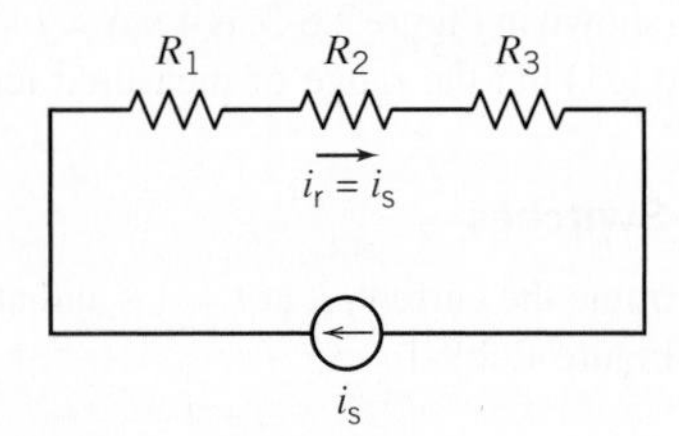

Figure DP 2-3

Resistive Circuits

IN THIS CHAPTER

3.1 Introduction
3.2 Kirchhoff's Laws
3.3 Series Resistors and Voltage Division
3.4 Parallel Resistors and Current Division
3.5 Series Voltage Sources and Parallel Current Sources
3.6 Circuit Analysis
3.7 Analyzing Resistive Circuits Using MATLAB
3.8 How Can We Check . . . ?
3.9 **DESIGN EXAMPLE**—Adjustable Voltage Source
3.10 Summary
Problems
Design Problems

3.1 INTRODUCTION

In this chapter, we will do the following:

- Write equations using Kirchhoff's laws.
  Not surprisingly, the behavior of an electric circuit is determined both by the types of elements that comprise the circuit and by the way those elements are connected together. The constitutive equations describe the elements themselves, and Kirchhoff's laws describe the way the elements are connected to each other to form the circuit.
- Analyze simple electric circuits, using only Kirchhoff's laws and the constitutive equations of the circuit elements.
- Analyze two very common circuit configurations: series resistors and parallel resistors.
  We will see that series resistors act like a "voltage divider," and parallel resistors act like a "current divider." Also, series resistors and parallel resistors provide our first examples of an "equivalent circuit." Figure 3.1-1 illustrates this important concept. Here, a circuit has been partitioned into two parts, A and B. Replacing B by an equivalent circuit, B_{eq}, does not change the current or voltage of any circuit element in part A. It is in this sense that B_{eq} is equivalent to B. We will see how to obtain an equivalent circuit when part B consists either of series resistors or of parallel resistors.
- Determine equivalent circuits for series voltage sources and parallel current sources.
- Determine the equivalent resistance of a resistive circuit.

Often, circuits consisting entirely of resistors can be reduced to a single equivalent resistor by repeatedly replacing series and/or parallel resistors by equivalent resistors.

3.2 KIRCHHOFF'S LAWS

An electric circuit consists of circuit elements that are connected together. The places where the elements are connected to each other are called nodes. Figure 3.2-1a shows an electric circuit that consists of six elements connected together at four nodes. It is common practice to draw electric

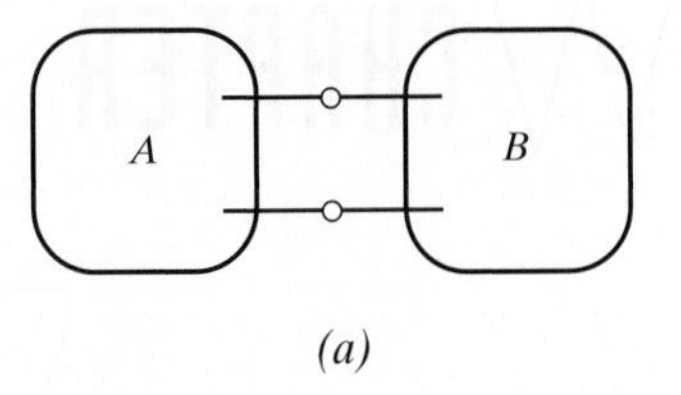

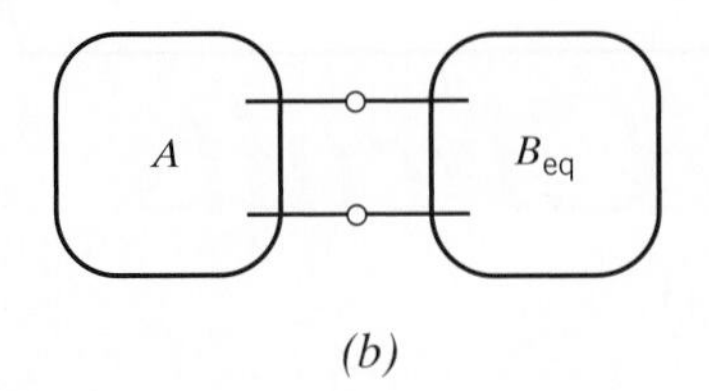

FIGURE 3.1-1 Replacing B by an equivalent circuit, B_{eq}, does not change the current or voltage of any circuit element in A.

circuits using straight lines and to position the elements horizontally or vertically as shown in Figure 3.2-1*b*.

The circuit is shown again in Figure 3.2-1*c*, this time emphasizing the nodes. Notice that redrawing the circuit, using straight lines and horizontal and vertical elements, has changed the way that the nodes are represented. In Figure 3.2-1*a*, nodes are represented as points. In Figures 3.2-1*b*,*c*, nodes are represented using both points and straight-line segments.

The same circuit can be drawn in several ways. One drawing of a circuit might look much different from another drawing of the same circuit. How can we tell when two circuit drawings represent the same circuit? Informally, we say that two circuit drawings represent the same circuit if

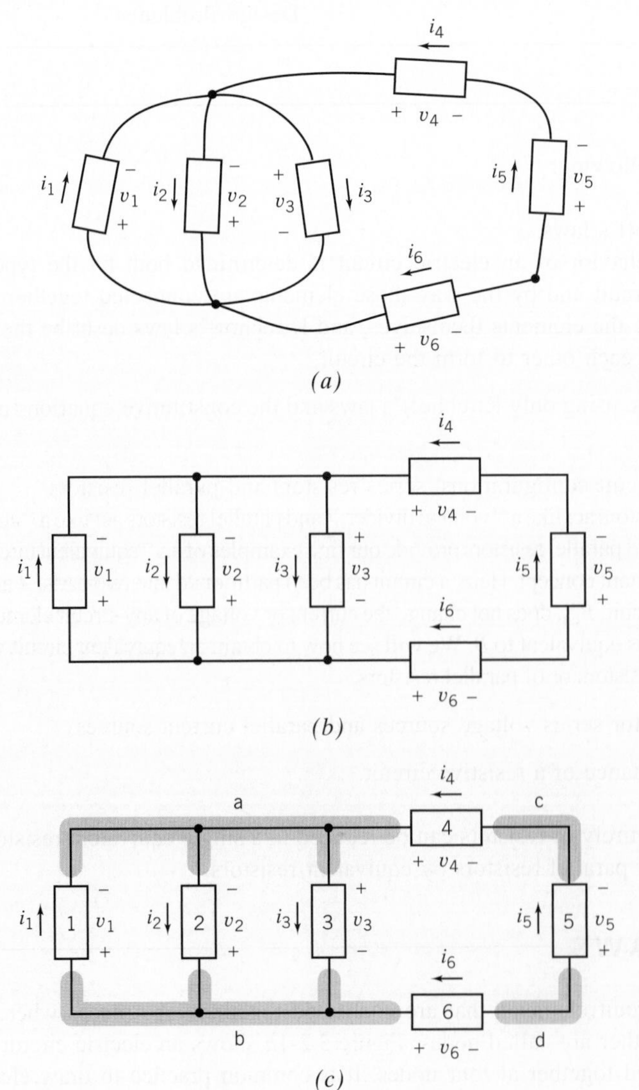

FIGURE 3.2-1 (*a*) An electric circuit. (*b*) The same circuit, redrawn using straight lines and horizontal and vertical elements. (*c*) The circuit after labeling the nodes and elements.

corresponding elements are connected to corresponding nodes. More formally, we say that circuit drawings A and B represent the same circuit when the following three conditions are met.

1. There is a one-to-one correspondence between the nodes of drawing A and the nodes of drawing B. (A one-to-one correspondence is a matching. In this one-to-one correspondence, each node in drawing A is matched to exactly one node of drawing B and vice versa. The position of the nodes is not important.)
2. There is a one-to-one correspondence between the elements of drawing A and the elements of drawing B.
3. Corresponding elements are connected to corresponding nodes.

EXAMPLE 3.2-1 Different Drawings of the Same Circuit

Figure 3.2-2 shows four circuit drawings. Which of these drawings, if any, represent the same circuit as the circuit drawing in Figure 3.2-1*c*?

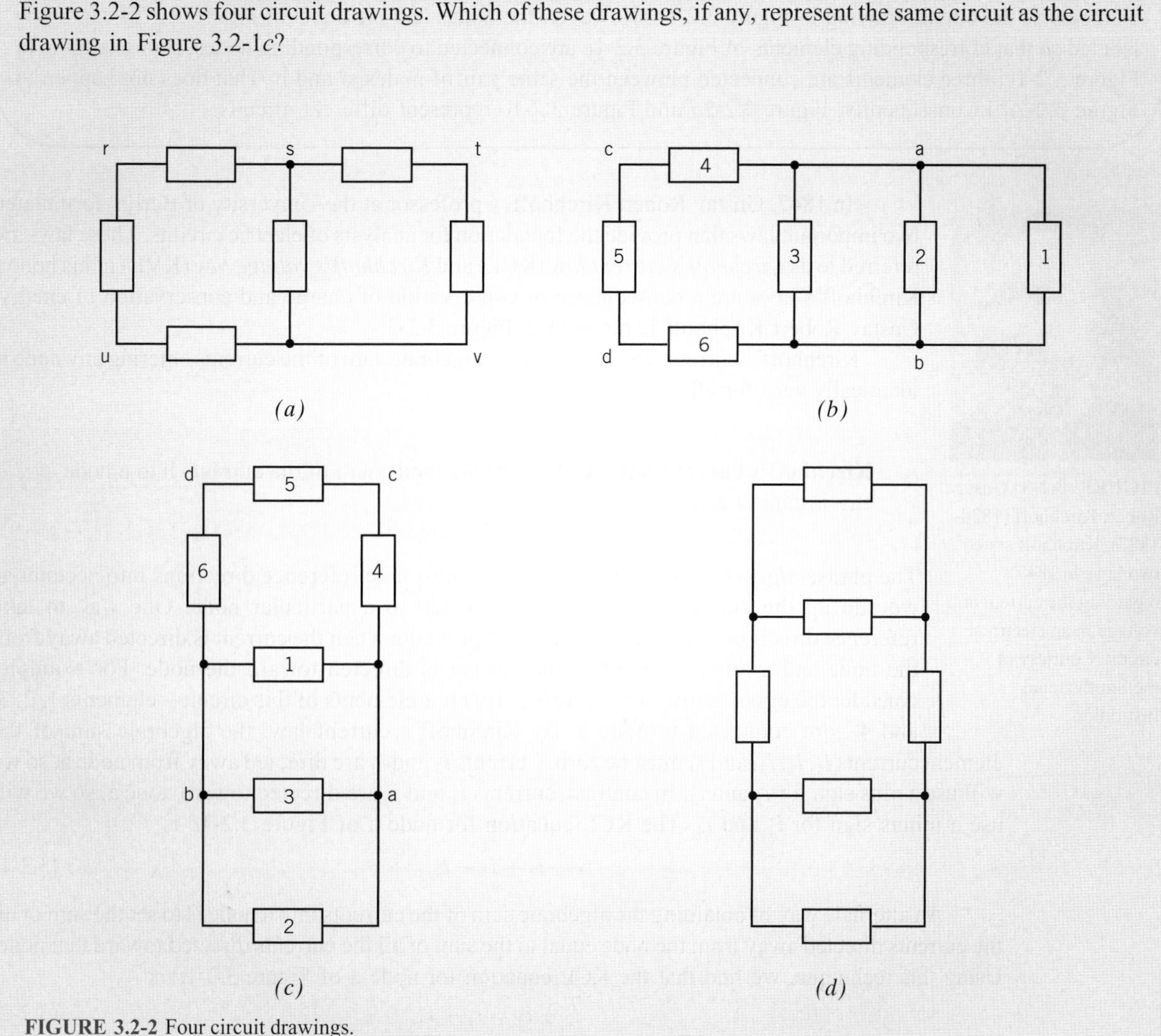

FIGURE 3.2-2 Four circuit drawings.

Solution

The circuit drawing shown in Figure 3.2-2*a* has five nodes, labeled r, s, t, u, and v. The circuit drawing in Figure 3.2-1*c* has four nodes. Because the two drawings have different numbers of nodes, there cannot be a one-to-one correspondence between the nodes of the two drawings. Hence, these drawings represent different circuits.

The circuit drawing shown in Figure 3.2-2*b* has four nodes and six elements, the same numbers of nodes and elements as the circuit drawing in Figure 3.2-1*c*. The nodes in Figure 3.2-2*b* have been labeled in the same way as the corresponding nodes in Figure 3.2-1*c*. For example, node c in Figure 3.2-2*b* corresponds to node c in Figure 3.2-1*c*. The elements in Figure 3.2-2*b* have been labeled in the same way as the corresponding elements in Figure 3.2-1*c*. For example, element 5 in Figure 3.2-2*b* corresponds to element 5 in Figure 3.2-1*c*. Corresponding elements are indeed connected to corresponding nodes. For example, element 2 is connected to nodes a and b, in both Figure 3.2-2*b* and in Figure 3.2-1*c*. Consequently, Figure 3.2-2*b* and Figure 3.2-1*c* represent the same circuit.

The circuit drawing shown in Figure 3.2-2*c* has four nodes and six elements, the same number of nodes and elements as the circuit drawing in Figure 3.2-1*c*. The nodes and elements in Figure 3.2-2*c* have been labeled in the same way as the corresponding nodes and elements in Figure 3.2-1*c*. Corresponding elements are indeed connected to corresponding nodes. Therefore, Figure 3.2-2*c* and Figure 3.2-1*c* represent the same circuit.

The circuit drawing shown in Figure 3.2-2*d* has four nodes and six elements, the same numbers of nodes and elements as the circuit drawing in Figure 3.2-1*c*. However, the nodes and elements of Figure 3.2-2*d* cannot be labeled so that corresponding elements of Figure 3.2-1*c* are connected to corresponding nodes. (For example, in Figure 3.2-1*c*, three elements are connected between the same pair of nodes, a and b. That does not happen in Figure 3.2-2*d*.) Consequently, Figure 3.2-2*d* and Figure 3.2-1*c* represent different circuits.

FIGURE 3.2-3 Gustav Robert Kirchhoff (1824-1887). Kirchhoff stated two laws in 1847 regarding the current and voltage in an electrical circuit. Courtesy of the Smithsonian Institution.

In 1847, Gustav Robert Kirchhoff, a professor at the University of Berlin, formulated two important laws that provide the foundation for analysis of electric circuits. These laws are referred to as *Kirchhoff's current law* (KCL) and *Kirchhoff's voltage law* (KVL) in his honor. Kirchhoff's laws are a consequence of conservation of charge and conservation of energy. Gustav Robert Kirchhoff is pictured in Figure 3.2-3.

Kirchhoff's current law states that the algebraic sum of the currents entering any node is identically zero for all instants of time.

Kirchhoff's current law (KCL): The algebraic sum of the currents into a node at any instant is zero.

The phrase *algebraic sum* indicates that we must take reference directions into account as we add up the currents of elements connected to a particular node. One way to take reference directions into account is to use a plus sign when the current is directed away from the node and a minus sign when the current is directed toward the node. For example, consider the circuit shown in Figure 3.2-1*c*. Four elements of this circuit—elements 1, 2, 3, and 4—are connected to node a. By Kirchhoff's current law, the algebraic sum of the element currents i_1, i_2, i_3, and i_4 must be zero. Currents i_2 and i_3 are directed away from node a, so we will use a plus sign for i_2 and i_3. In contrast, currents i_1 and i_4 are directed toward node a, so we will use a minus sign for i_1 and i_4. The KCL equation for node a of Figure 3.2-1*c* is

$$-i_1 + i_2 + i_3 - i_4 = 0 \tag{3.2-1}$$

An alternate way of obtaining the algebraic sum of the currents into a node is to set the sum of all the currents directed away from the node equal to the sum of all the currents directed toward that node. Using this technique, we find that the KCL equation for node a of Figure 3.2-1*c* is

$$i_2 + i_3 = i_1 + i_4 \tag{3.2-2}$$

Clearly, Eqs. 3.2-1 and 3.2-2 are equivalent.

Similarly, the Kirchhoff's current law equation for node b of Figure 3.2-1*c* is

$$i_1 = i_2 + i_3 + i_6$$

Before we can state Kirchhoff's voltage law, we need the definition of a loop. A *loop* is a closed path through a circuit that does not encounter any intermediate node more than once. For example, starting at node a in Figure 3.2-1*c*, we can move through element 4 to node c, then through element 5 to node d, through element 6 to node b, and finally through element 3 back to node a. We have a closed path, and we did not encounter any of the intermediate nodes—b, c, or d—more than once. Consequently, elements 3, 4, 5, and 6 comprise a loop. Similarly, elements 1, 4, 5, and 6 comprise a loop of the circuit shown in Figure 3.2-1*c*. Elements 1 and 3 comprise yet another loop of this circuit. The circuit has three other loops: elements 1 and 2, elements 2 and 3, and elements 2, 4, 5, and 6.

We are now ready to state Kirchhoff's voltage law.

Kirchhoff's voltage law (KVL): The algebraic sum of the voltages around any loop in a circuit is identically zero for all time.

The phrase *algebraic sum* indicates that we must take polarity into account as we add up the voltages of elements that comprise a loop. One way to take polarity into account is to move around the loop in the clockwise direction while observing the polarities of the element voltages. We write the voltage with a plus sign when we encounter the + of the voltage polarity before the −. In contrast, we write the voltage with a minus sign when we encounter the − of the voltage polarity before the +. For example, consider the circuit shown in Figure 3.2-1*c*. Elements 3, 4, 5, and 6 comprise a loop of the circuit. By Kirchhoff's voltage law, the algebraic sum of the element voltages v_3, v_4, v_5, and v_6 must be zero. As we move around the loop in the clockwise direction, we encounter the + of v_4 before the −, the − of v_5 before the +, the − of v_6 before the +, and the − of v_3 before the +. Consequently, we use a minus sign for v_3, v_5, and v_6 and a plus sign for v_4. The KCL equation for this loop of Figure 3.2-1*c* is

$$v_4 - v_5 - v_6 - v_3 = 0$$

Similarly, the Kirchhoff's voltage law equation for the loop consisting of elements 1, 4, 5, and 6 is

$$v_4 - v_5 - v_6 + v_1 = 0$$

The Kirchhoff's voltage law equation for the loop consisting of elements 1 and 2 is

$$-v_2 + v_1 = 0$$

EXAMPLE 3.2-2 Kirchhoff's Laws

INTERACTIVE EXAMPLE

Consider the circuit shown in Figure 3.2-4*a*. Determine the power supplied by element *C* and the power received by element *D*.

Solution

Figure 3.2-4*a* provides a value for the current in element *C* but not for the voltage, *v*, across element *C*. The voltage and current of element *C* given in Figure 3.2-4*a* adhere to the passive convention, so the product of this voltage and current is the power *received* by element *C*. Figure 3.2-4*a* provides a value for the voltage across element *D* but not for the current, *i*, in element *D*. The voltage and current of element *D* given in Figure 3.2-4*a* do not adhere to the passive convention, so the product of this voltage and current is the power *supplied* by element *D*.

We need to determine the voltage, *v*, across element *C* and the current, *i*, in element *D*. We will use Kirchhoff's laws to determine values of *v* and *i*. First, we identify and label the nodes of the circuit as shown in Figure 3.2-4*b*.

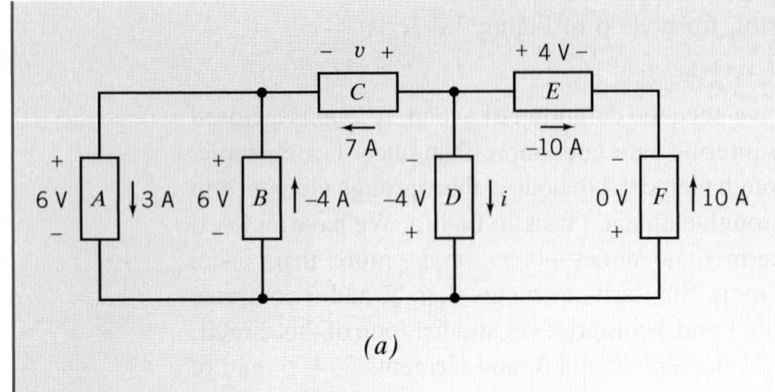

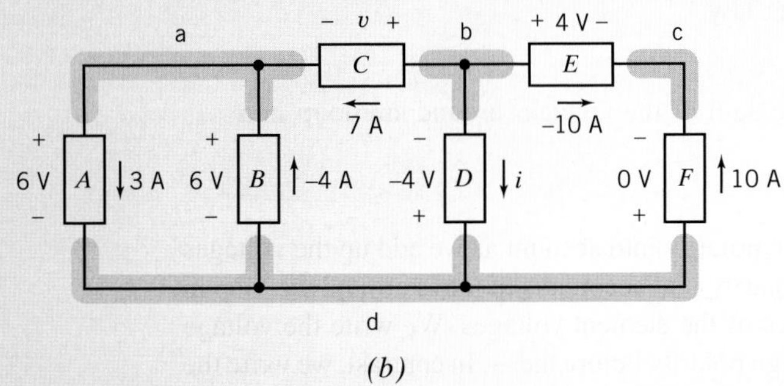

FIGURE 3.2-4 (*a*) The circuit considered in Example 3.2-2 and (*b*) the circuit redrawn to emphasize the nodes.

Apply Kirchhoff's voltage law (KVL) to the loop consisting of elements *C*, *D*, and *B* to get

$$-v - (-4) - 6 = 0 \Rightarrow v = -2\text{ V}$$

The value of the current in element *C* in Figure 3.2-4*b* is 7 A. The voltage and current of element *C* given in Figure 3.2-4*b* adhere to the passive convention, so

$$p_C = v(7) = (-2)(7) = -14\text{ W}$$

is the power *received* by element *C*. Therefore, element *C supplies* 14 W.

Next, apply Kirchhoff's current law (KCL) at node b to get

$$7 + (-10) + i = 0 \Rightarrow i = 3\text{ A}$$

The value of the voltage across element *D* in Figure 3.2-4*b* is –4 V. The voltage and current of element *D* given in Figure 3.2-4*b* do not adhere to the passive convention, so the power *supplied* by element *D* is given by

$$p_D = (-4)i = (-4)(3) = -12\text{ W}$$

Therefore, element *D receives* 12 W.

EXAMPLE 3.2-3 Ohm's and Kirchhoff's Laws

Consider the circuit shown in Figure 3.2-5. Notice that the passive convention was used to assign reference directions to the resistor voltages and currents. This anticipates using Ohm's law. Find each current and each voltage when $R_1 = 8\ \Omega$, $v_2 = -10$ V, $i_3 = 2$ A, and $R_3 = 1\ \Omega$. Also, determine the resistance R_2.

Solution

The sum of the currents entering node a is

$$i_1 - i_2 - i_3 = 0$$

Using Ohm's law for R_3, we find that

$$v_3 = R_3 i_3 = 1(2) = 2\text{ V}$$

Kirchhoff's voltage law for the bottom loop incorporating v_1, v_3, and the 10-V source is

$$-10 + v_1 + v_3 = 0$$

Therefore,

$$v_1 = 10 - v_3 = 8\text{ V}$$

Ohm's law for the resistor R_1 is

$$v_1 = R_1 i_1$$

or

$$i_1 = v_1/R_1 = 8/8 = 1\text{ A}$$

Next, apply Kirchhoff's current law at node a to get

$$i_2 = i_1 - i_3 = 1 - 2 = -1\text{ A}$$

We can now find the resistance R_2 from

$$v_2 = R_2 i_2$$

or

$$R_2 = v_2/i_2 = -10/-1 = 10\ \Omega$$

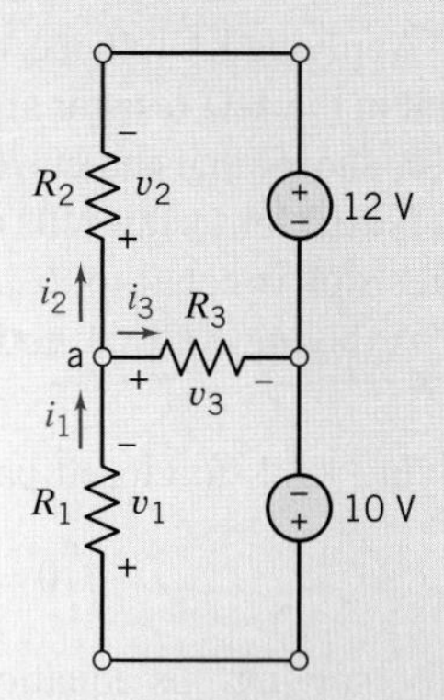

FIGURE 3.2-5 Circuit with two constant-voltage sources.

EXAMPLE 3.2-4 Ohm's and Kirchhoff's Laws

INTERACTIVE EXAMPLE

Determine the value of the current, in amps, measured by the ammeter in Figure 3.2-6*a*.

Solution

An ideal ammeter is equivalent to a short circuit. The current measured by the ammeter is the current in the short circuit. Figure 3.2-6*b* shows the circuit after replacing the ammeter by the equivalent short circuit.

The circuit has been redrawn in Figure 3.2-7 to label the nodes of the circuit. This circuit consists of a voltage source, a dependent current source, two resistors, and two short circuits. One of the short circuits is the controlling element of the CCCS, and the other short circuit is a model of the ammeter.

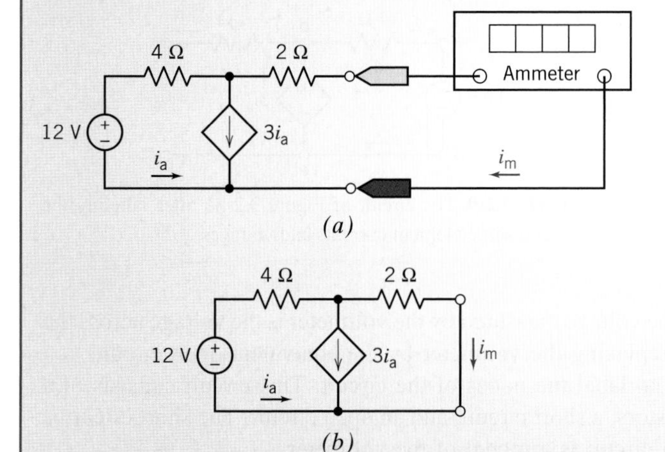

FIGURE 3.2-6 (*a*) A circuit with dependent source and an ammeter. (*b*) The equivalent circuit after replacing the ammeter by a short circuit.

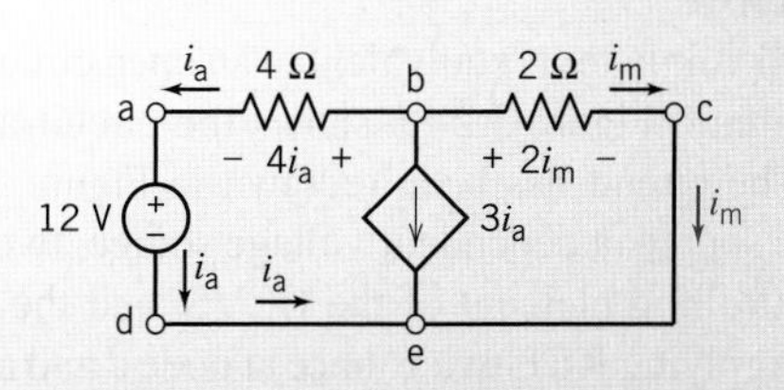

FIGURE 3.2-7 The circuit of Figure 3.2-6 after labeling the nodes and some element currents and voltages.

Applying KCL twice, once at node d and again at node a, shows that the current in the voltage source and the current in the 4-Ω resistor are both equal to i_a. These currents are labeled in Figure 3.2-7. Applying KCL again, at node c, shows that the current in the 2-Ω resistor is equal to i_m. This current is labeled in Figure 3.2-7.

Next, Ohm's law tells us that the voltage across the 4-Ω resistor is equal to $4i_a$ and that the voltage across the 2-Ω resistor is equal to $2i_m$. Both of these voltages are labeled in Figure 3.2-7.

Applying KCL at node b gives

$$-i_a - 3i_a - i_m = 0$$

Applying KVL to closed path a-b-c-e-d-a gives

$$0 = -4i_a + 2i_m - 12 = -4\left(-\frac{1}{4}i_m\right) + 2i_m - 12 = 3i_m - 12$$

Finally, solving this equation gives

$$i_m = 4\text{ A}$$

EXAMPLE 3.2-5 Ohm's and Kirchhoff's Laws

INTERACTIVE EXAMPLE

Determine the value of the voltage, in volts, measured by the voltmeter in Figure 3.2-8*a*.

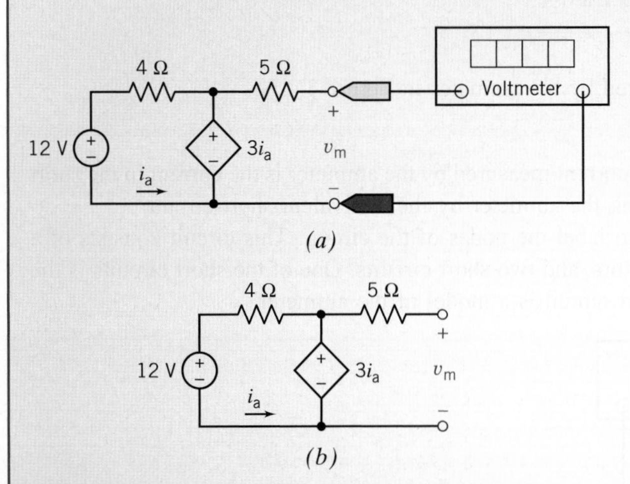

FIGURE 3.2-8 (*a*) A circuit with dependent source and a voltmeter. (*b*) The equivalent circuit after replacing the voltmeter by an open circuit.

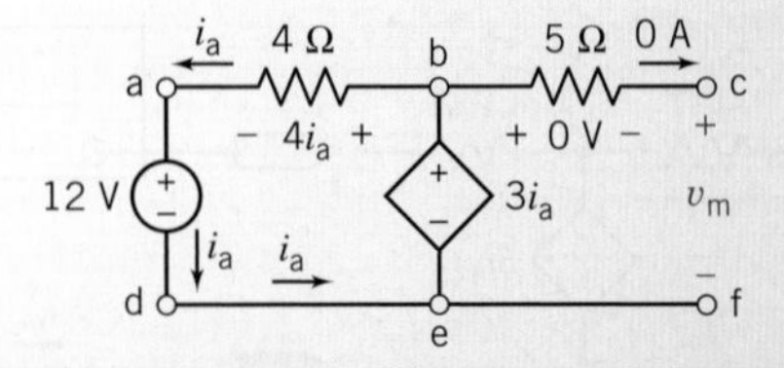

FIGURE 3.2-9 The circuit of Figure 3.2-8*b* after labeling the nodes and some element currents and voltages.

Solution

An ideal voltmeter is equivalent to an open circuit. The voltage measured by the voltmeter is the voltage across the open circuit. Figure 3.2-8*b* shows the circuit after replacing the voltmeter by the equivalent open circuit.

The circuit has been redrawn in Figure 3.2-9 to label the nodes of the circuit. This circuit consists of a voltage source, a dependent voltage source, two resistors, a short circuit, and an open circuit. The short circuit is the controlling element of the CCVS, and the open circuit is a model of the voltmeter.

Applying KCL twice, once at node d and again at node a, shows that the current in the voltage source and the current in the 4-Ω resistor are both equal to i_a. These currents are labeled in Figure 3.2-9. Applying KCL again, at node c, shows that the current in the 5-Ω resistor is equal to the current in the open circuit, that is, zero. This current is labeled in Figure 3.2-9. Ohm's law tells us that the voltage across the 5-Ω resistor is also equal to zero. Next, applying KVL to the closed path b-c-f-e-b gives $v_m = 3i_a$.

Applying KVL to the closed path a-b-e-d-a gives

$$-4i_a + 3i_a - 12 = 0$$

so

$$i_a = -12 \text{ A}$$

Finally

$$v_m = 3i_a = 3(-12) = -36 \text{ V}$$

EXERCISE 3.2-1 Determine the values of i_3, i_4, i_6, v_2, v_4, and v_6 in Figure E 3.2-1.

Answer: $i_3 = -3$ A, $i_4 = 3$ A, $i_6 = 4$ A, $v_2 = -3$ V, $v_4 = -6$ V, $v_6 = 6$ V

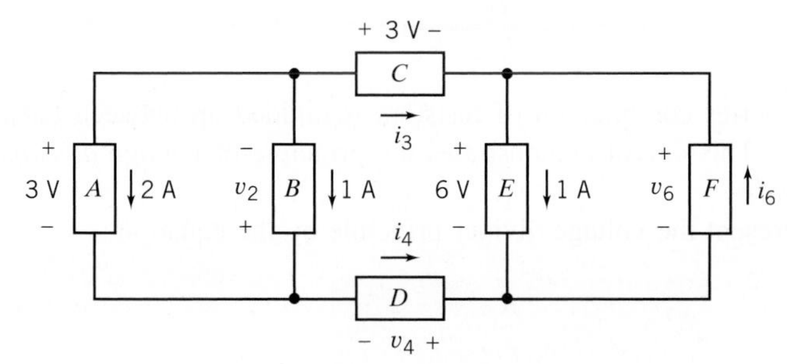

FIGURE E 3.2-1

3.3 SERIES RESISTORS AND VOLTAGE DIVISION

Let us consider a single-loop circuit, as shown in Figure 3.3-1. In anticipation of using Ohm's law, the passive convention has been used to assign reference directions to resistor voltages and currents.

The connection of resistors in Figure 3.3-1 is said to be a *series* connection because all the elements carry the same current. To identify a pair of series elements, we look for two elements connected to a single node that has no other elements connected to it. Notice, for example, that resistors R_1 and R_2 are both connected to node b and that no other circuit elements are connected to node b. Consequently, $i_1 = i_2$, so both resistors have the same current. A similar argument shows that resistors R_2 and R_3 are also connected in series. Noticing that R_2 is connected in series with both R_1 and R_3, we say that all three resistors are connected in series. The order of series resistors is not important. For example, the voltages and currents of the three resistors in Figure 3.3-1 will not change if we interchange the positions R_2 and R_3.

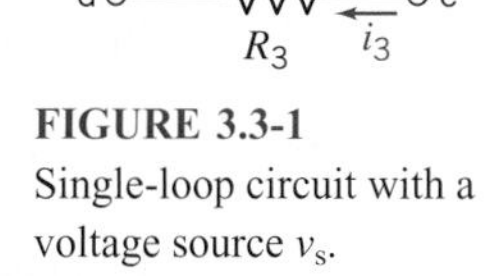

FIGURE 3.3-1 Single-loop circuit with a voltage source v_s.

Using KCL at each node of the circuit in Figure 3.3-1, we obtain

$$\begin{aligned} \text{a:}\ i_s &= i_1 \\ \text{b:}\ i_1 &= i_2 \\ \text{c:}\ i_2 &= i_3 \\ \text{d:}\ i_3 &= i_s \end{aligned}$$

Consequently,

$$i_s = i_1 = i_2 = i_3$$

To determine i_1, we use KVL around the loop to obtain

$$v_1 + v_2 + v_3 - v_s = 0$$

where, for example, v_1 is the voltage across the resistor R_1. Using Ohm's law for each resistor,

$$R_1 i_1 + R_2 i_2 + R_2 i_3 - v_s = 0 \Rightarrow R_1 i_1 + R_2 i_1 + R_2 i_1 = v_s$$

Solving for i_1, we have

$$i_1 = \frac{v_s}{R_1 + R_2 + R_3}$$

Thus, the voltage across the nth resistor R_n is v_n and can be obtained as

$$v_n = i_1 R_n = \frac{v_s R_n}{R_1 + R_2 + R_3}$$

For example, the voltage across resistor R_2 is

$$v_2 = \frac{R_2}{R_1 + R_2 + R_3} v_s$$

Thus, the voltage across the series combination of resistors is divided up between the individual resistors in a predictable way. This circuit demonstrates the principle of *voltage division*, and the circuit is called a *voltage divider*.

In general, we may represent the voltage divider principle by the equation

$$v_n = \frac{R_n}{R_1 + R_2 + \cdots + R_N} v_s$$

where v_n is the voltage across the nth resistor of N resistors connected in series.

We can replace series resistors by an equivalent resistor. This is illustrated in Figure 3.3-2. The series resistors R_1, R_2, and R_3 in Figure 3.3-2*a* are replaced by a single, equivalent resistor R_s in Figure 3.3-2*b*. R_s is said to be equivalent to the series resistors R_1, R_2, and R_3 when replacing R_1, R_2, and R_3 by R_s does not change the current or voltage of any other element of the circuit. In this case, there is only one other element in the circuit, the voltage source. We must choose the value of the resistance R_s so that replacing R_1, R_2, and R_3 by R_s will not change the current of the voltage source. In Figure 3.3-2*a*, we have

$$i_s = \frac{v_s}{R_1 + R_2 + R_3}$$

In Figure 3.3-2*b*, we have

$$i_s = \frac{v_s}{R_s}$$

Because the voltage source current must be the same in both circuits, we require that

$$R_s = R_1 + R_2 + R_3$$

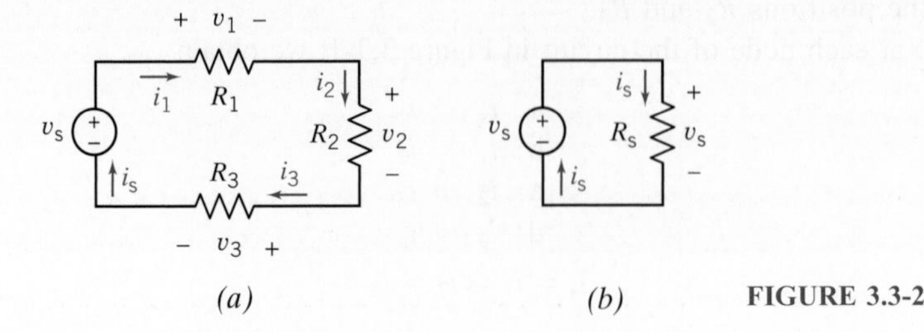

(*a*) (*b*) **FIGURE 3.3-2**

In general, the series connection of N resistors having resistances $R_1, R_2 \ldots R_N$ is equivalent to the single resistor having resistance

$$R_s = R_1 + R_2 + \cdots + R_N$$

Replacing series resistors by an equivalent resistor does not change the current or voltage of any other element of the circuit.

Next, let's calculate the power absorbed by the series resistors in Figure 3.3-2*a*:

$$p = i_s^2 R_1 + i_s^2 R_2 + i_s^2 R_3$$

Doing a little algebra gives

$$p = i_s^2 (R_1 + R_2 + R_3) = i_s^2 R_s$$

which is equal to the power absorbed by the equivalent resistor in Figure 3.3-2b. We conclude that the power absorbed by series resistors is equal to the power absorbed by the equivalent resistor.

EXAMPLE 3.3-1 Voltage Divider

Let us consider the circuit shown in Figure 3.3-3 and determine the resistance R_2 required so that the voltage across R_2 will be 1/4 of the source voltage when $R_1 = 9\ \Omega$. Determine the current i when $v_s = 12$ V.

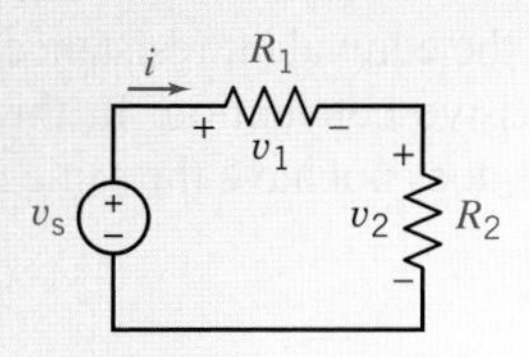

FIGURE 3.3-3 Voltage divider circuit with $R_1 = 9\ \Omega$.

Solution

The voltage across resistor R_2 will be

$$v_2 = \frac{R_2}{R_1 + R_2} v_s$$

Because we desire $v_2 / v_s = 1/4$, we have

$$\frac{R_2}{R_1 + R_2} = \frac{1}{4}$$

or

$$R_1 = 3R_2$$

Because $R_1 = 9\ \Omega$, we require that $R_2 = 3\ \Omega$. Using KVL around the loop, we have

$$-v_s + v_1 + v_2 = 0$$

or

$$v_s = iR_1 + iR_2$$

Therefore,

$$i = \frac{v_s}{R_1 + R_2} = \frac{12}{9 + 3} = 1 \text{ A} \tag{3.3-1}$$

EXAMPLE 3.3-2 Series Resistors

For the circuit of Figure 3.3-4*a*, find the current measured by the ammeter. Then show that the power absorbed by the two resistors is equal to that supplied by the source.

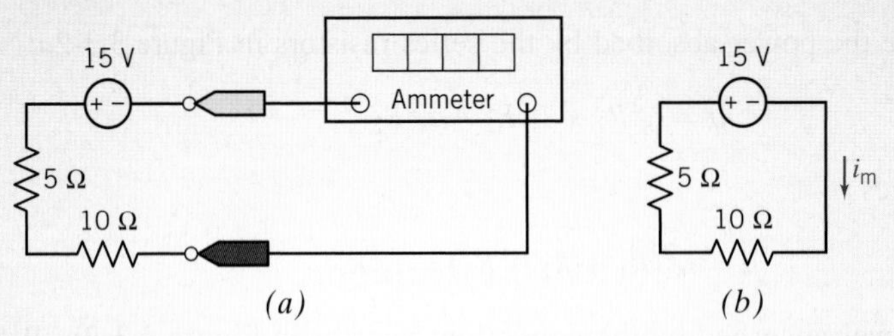

FIGURE 3.3-4 (*a*) A circuit containing series resistors. (*b*) The circuit after the ideal ammeter has been replaced by the equivalent short circuit, and a label has been added to indicate the current measured by the ammeter, i_m.

Solution

Figure 3.3-4*b* shows the circuit after the ideal ammeter has been replaced by the equivalent short circuit and a label has been added to indicate the current measured by the ammeter, i_m. Applying KVL gives

$$15 + 5i_m + 10i_m = 0$$

The current measured by the ammeter is

$$i_m = -\frac{15}{5+10} = -1\text{ A}$$

(Why is i_m negative? Why can't we just divide the source voltage by the equivalent resistance? Recall that when we use Ohm's law, the voltage and current must adhere to the passive convention. In this case, the current calculated by dividing the source voltage by the equivalent resistance does not have the same reference direction as i_m, so we need a minus sign.)

The total power absorbed by the two resistors is

$$p_R = 5{i_m}^2 + 10{i_m}^2 = 15(1^2) = 15\text{ W}$$

The power supplied by the source is

$$p_s = -v_s\, i_m = -15(-1) = 15\text{ W}$$

Thus, the power supplied by the source is equal to that absorbed by the series connection of resistors.

EXAMPLE 3.3-3 Voltage Divider Design

The input to the voltage divider in Figure 3.3-5 is the voltage, v_s, of the voltage source. The output is the voltage, v_o, measured by the voltmeter. Design the voltage divider; that is, specify values of the resistances, R_1 and R_2, to satisfy both of these specifications.

Specification 1: The input and output voltages are related by $v_o = 0.8\ v_s$.

Specification 2: The voltage source is required to supply no more than 1 mW of power when the input to the voltage divider is $v_s = 20$ V.

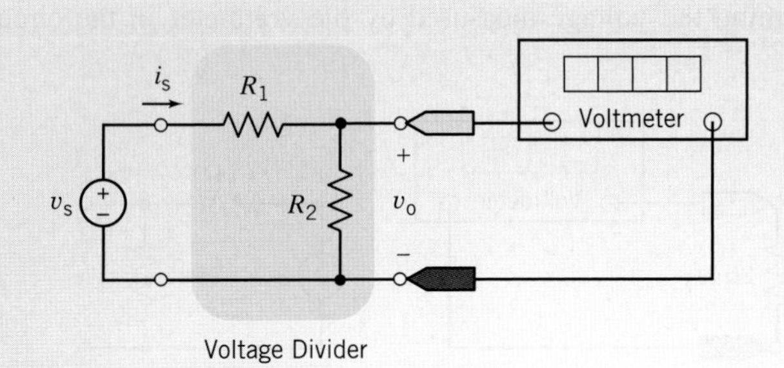

FIGURE 3.3-5 A voltage divider.

Solution

We'll examine each specification to see what it tells us about the resistor values.

Specification 1: The input and output voltages of the voltage divider are related by

$$v_o = \frac{R_2}{R_1 + R_2} v_s$$

So specification 1 requires

$$\frac{R_2}{R_1 + R_2} = 0.8 \quad \Rightarrow \quad R_2 = 4R_1$$

Specification 2: The power supplied by the voltage source is given by

$$p_s = i_s v_s = \left(\frac{v_s}{R_1 + R_2}\right) v_s = \frac{{v_s}^2}{R_1 + R_2}$$

So specification 2 requires

$$0.001 \geq \frac{20^2}{R_1 + R_2} \quad \Rightarrow \quad R_1 + R_2 \geq 400 \times 10^3 = 400\ \text{k}\Omega$$

Combining these results gives

$$5R_1 \geq 400\ \text{k}\Omega$$

The solution is not unique. One solution is

$$R_1 = 100\ \text{k}\Omega \text{ and } R_2 = 400\ \text{k}\Omega$$

EXERCISE 3.3-1 Determine the voltage measured by the voltmeter in the circuit shown in Figure E 3.3-1*a*.

Hint: Figure E 3.3-1*b* shows the circuit after the ideal voltmeter has been replaced by the equivalent open circuit and a label has been added to indicate the voltage measured by the voltmeter, v_m.

Answer: $v_m = 2$ V

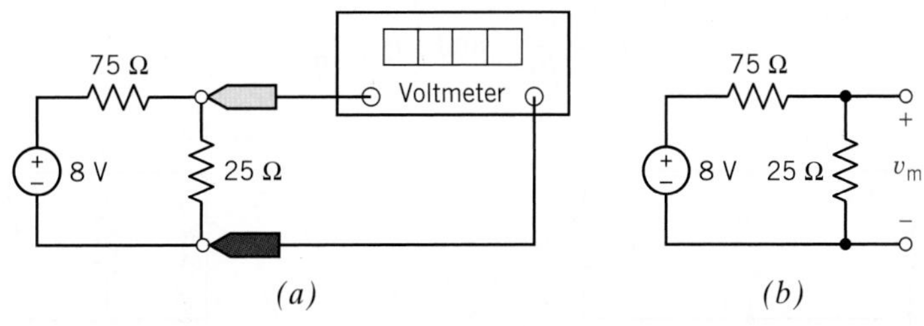

FIGURE E 3.3-1 (*a*) A voltage divider. (*b*) The voltage divider after the ideal voltmeter has been replaced by the equivalent open circuit and a label has been added to indicate the voltage measured by the voltmeter, v_m.

EXERCISE 3.3-2 Determine the voltage measured by the voltmeter in the circuit shown in Figure E 3.3-2*a*.

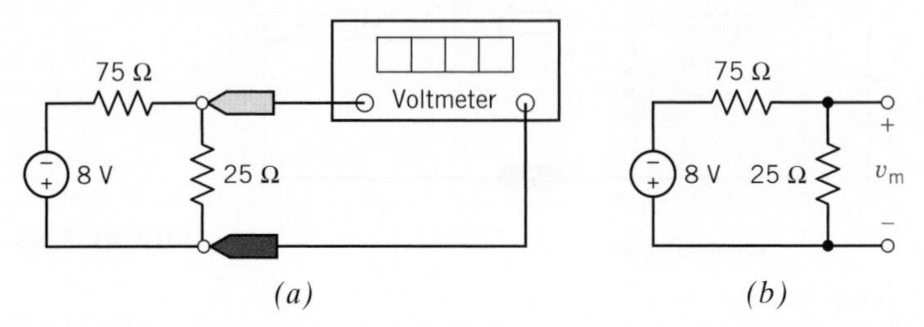

FIGURE E 3.3-2 (*a*) A voltage divider. (*b*) The voltage divider after the ideal voltmeter has been replaced by the equivalent open circuit and a label has been added to indicate the voltage measured by the voltmeter, v_m.

Hint: Figure E 3.3-2*b* shows the circuit after the ideal voltmeter has been replaced by the equivalent open circuit and a label has been added to indicate the voltage measured by the voltmeter, v_m.

Answer: $v_m = -2$ V

3.4 PARALLEL RESISTORS AND CURRENT DIVISION

Circuit elements, such as resistors, are connected in *parallel* when the voltage across each element is identical. The resistors in Figure 3.4-1 are connected in *parallel*. Notice, for example, that resistors R_1 and R_2 are each connected to both node a and node b. Consequently, $v_1 = v_2$, so both resistors have the same voltage. A similar argument shows that resistors R_2 and R_3 are also connected in parallel. Noticing that R_2 is connected in parallel with both R_1 and R_3, we say that all three resistors are connected in parallel. The order of parallel resistors is not important. For example, the voltages and currents of the three resistors in Figure 3.4-1 will not change if we interchange the positions R_2 and R_3.

The defining characteristic of parallel elements is that they have the same voltage. To identify a pair of parallel elements, we look for two elements connected between the same pair of nodes.

Consider the circuit with two resistors and a current source shown in Figure 3.4-2. Note that both resistors are connected to terminals a and b and that the voltage v appears across each parallel element. In anticipation of using Ohm's law, the passive convention is used to assign reference directions to the resistor voltages and currents. We may write KCL at node a (or at node b) to obtain

$$i_s - i_1 - i_2 = 0$$

or

$$i_s = i_1 + i_2$$

However, from Ohm's law

$$i_1 = \frac{v}{R_1} \quad \text{and} \quad i_2 = \frac{v}{R_2}$$

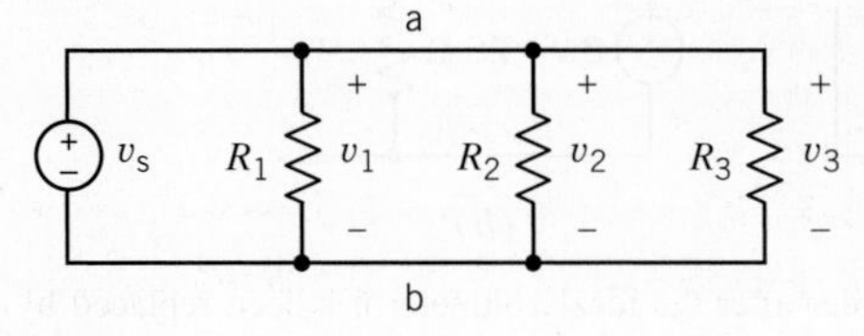

FIGURE 3.4-1 A circuit with parallel resistors.

FIGURE 3.4-2 Parallel circuit with a current source.

Then

$$i_s = \frac{v}{R_1} + \frac{v}{R_2} \tag{3.4-1}$$

Recall that we defined conductance G as the inverse of resistance R. We may therefore rewrite Eq. 3.4-1 as

$$i_s = G_1 v + G_2 v = (G_1 + G_2)v \tag{3.4-2}$$

Thus, the equivalent circuit for this parallel circuit is a conductance G_p, as shown in Figure 3.4-3, where

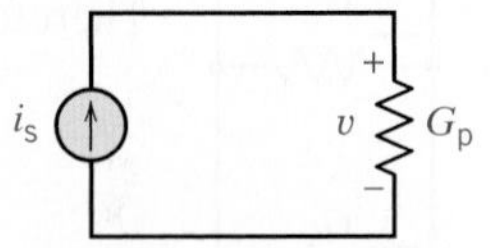

$$G_p = G_1 + G_2$$

FIGURE 3.4-3 Equivalent circuit for a parallel circuit.

The equivalent resistance for the two-resistor circuit is found from

$$G_p = \frac{1}{R_1} + \frac{1}{R_2}$$

Because $G_p = 1/R_p$, we have

$$\frac{1}{R_p} = \frac{1}{R_1} + \frac{1}{R_2}$$

or

$$R_p = \frac{R_1 R_2}{R_1 + R_2} \tag{3.4-3}$$

Note that the total conductance, G_p, increases as additional parallel elements are added and that the total resistance, R_p, declines as each resistor is added.

The circuit shown in Figure 3.4-2 is called a *current divider* circuit because it divides the source current. Note that

$$i_1 = G_1 v \tag{3.4-4}$$

Also, because $i_s = (G_1 + G_2)v$, we solve for v, obtaining

$$v = \frac{i_s}{G_1 + G_2} \tag{3.4-5}$$

Substituting v from Eq. 3.4-5 into Eq. 3.4-4, we obtain

$$i_1 = \frac{G_1 i_s}{G_1 + G_2} \tag{3.4-6}$$

Similarly,

$$i_2 = \frac{G_2 i_s}{G_1 + G_2}$$

Note that we may use $G_2 = 1/R_2$ and $G_1 = 1/R_1$ to obtain the current i_2 in terms of two resistances as follows:

$$i_2 = \frac{R_1 i_s}{R_1 + R_2}$$

The current of the source divides between conductances G_1 and G_2 in proportion to their conductance values.

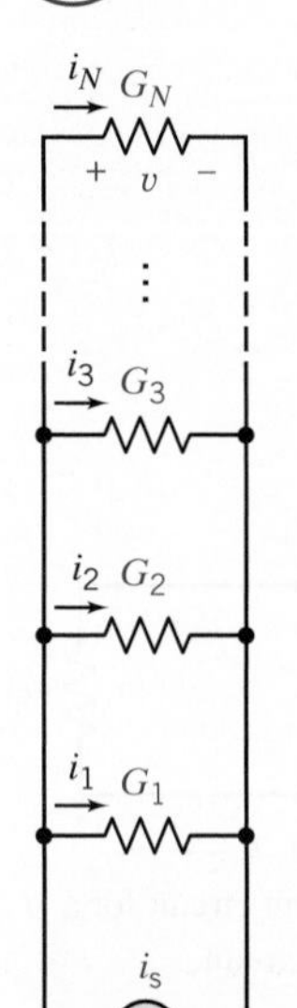

FIGURE 3.4-4 Set of N parallel conductances with a current source i_s.

Let us consider the more general case of current division with a set of N parallel conductors as shown in Figure 3.4-4. The KCL gives

$$i_s = i_1 + i_2 + i_3 + \cdots + i_N \tag{3.4-7}$$

for which

$$i_n = G_n v \tag{3.4-8}$$

for $n = 1, \ldots, N$. We may write Eq. 3.4-7 as

$$i_s = (G_1 + G_2 + G_3 + \cdots + G_N)v \tag{3.4-9}$$

Therefore,

$$i_s = v \sum_{n=1}^{N} G_n \tag{3.4-10}$$

Because $i_n = G_n v$, we may obtain v from Eq. 3.4-10 and substitute it in Eq. 3.4-8, obtaining

$$i_n = \frac{G_n i_s}{\sum_{n=1}^{N} G_n} \tag{3.4-11}$$

Recall that the equivalent circuit, Figure 3.4-12, has an equivalent conductance G_p such that

$$G_p = \sum_{n=1}^{N} G_n \tag{3.4-12}$$

Therefore,

$$i_n = \frac{G_n i_s}{G_p} \tag{3.4-13}$$

which is the basic equation for the current divider with N conductances. Of course, Eq. 3.4-12 can be rewritten as

$$\frac{1}{R_p} = \sum_{n=1}^{N} \frac{1}{R_n} \tag{3.4-14}$$

EXAMPLE 3.4-1 Parallel Resistors

For the circuit in Figure 3.4-5, find (a) the current in each branch, (b) the equivalent circuit, and (c) the voltage v. The resistors are

$$R_1 = \frac{1}{2}\ \Omega, \quad R_2 = \frac{1}{4}\ \Omega, \quad R_3 = \frac{1}{8}\ \Omega$$

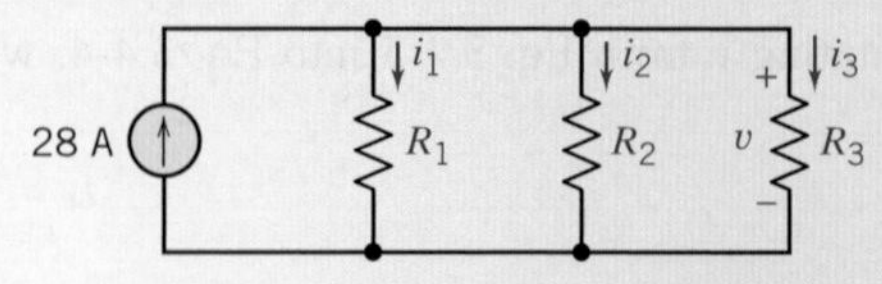

FIGURE 3.4-5 Parallel circuit for Example 3.3-2.

Solution

The current divider follows the equation

$$i_n = \frac{G_n i_s}{G_p}$$

so it is wise to find the equivalent circuit, as shown in Figure 3.4-6, with its equivalent conductance G_p. We have

$$G_p = \sum_{n=1}^{N} G_n = G_1 + G_2 + G_3 = 2 + 4 + 8 = 14\ \text{S}$$

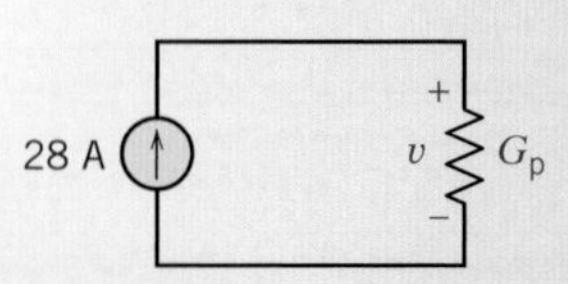

FIGURE 3.4-6 Equivalent circuit for the parallel circuit of Figure 3.4-5.

Recall that the units for conductance are siemens (S). Then

$$i_1 = \frac{G_1 i_s}{G_p} = \frac{2}{14}(28) = 4\text{ A}$$

Similarly,

$$i_2 = \frac{G_2 i_s}{G_p} = \frac{4(28)}{14} = 8\text{ A}$$

and

$$i_3 = \frac{G_3 i_s}{G_p} = 16\text{ A}$$

Because $i_n = G_n v$, we have

$$v = \frac{i_1}{G_1} = \frac{4}{2} = 2\text{ V}$$

EXAMPLE 3.4-2 Parallel Resistors

INTERACTIVE EXAMPLE

For the circuit of Figure 3.4-7*a*, find the voltage measured by the voltmeter. Then show that the power absorbed by the two resistors is equal to that supplied by the source.

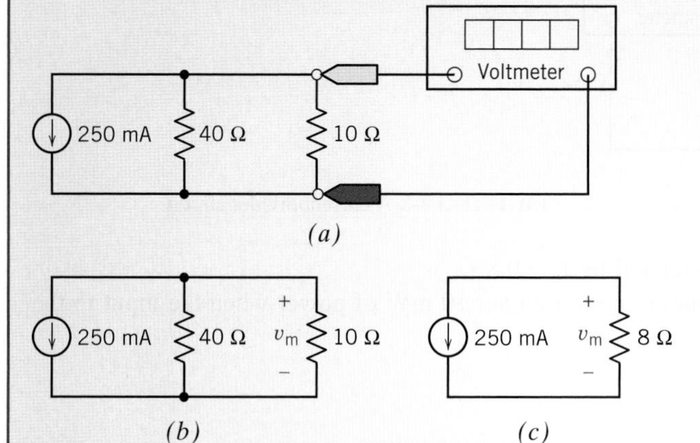

FIGURE 3.4-7 (*a*) A circuit containing parallel resistors. (*b*) The circuit after the ideal voltmeter has been replaced by the equivalent open circuit and a label has been added to indicate the voltage measured by the voltmeter, v_m. (*c*) The circuit after the parallel resistors have been replaced by an equivalent resistance.

Solution

Figure 3.4-7*b* shows the circuit after the ideal voltmeter has been replaced by the equivalent open circuit, and a label has been added to indicate the voltage measured by the voltmeter, v_m. The two resistors are connected in parallel and can be replaced with a single equivalent resistor. The resistance of this equivalent resistor is calculated as

$$\frac{40 \cdot 10}{40 + 10} = 8\ \Omega$$

Figure 3.4-7*c* shows the circuit after the parallel resistors have been replaced by the equivalent resistor. The current in the equivalent resistor is 250 mA, directed upward. This current and the voltage v_m do not adhere to the passive convention. The current in the equivalent resistance can also be expressed as −250 mA, directed downward. This current and the voltage v_m do adhere to the passive convention. Ohm's law gives

$$v_m = 8(-0.25) = -2\text{ V}$$

The voltage v_m in Figure 3.4-7*b* is equal to the voltage v_m in Figure 3.4-7*c*. This is a consequence of the equivalence of the 8-Ω resistor to the parallel combination of the 40-Ω and 10-Ω resistors. Looking at Figure 3.4-7*b*, we see that the power absorbed by the resistors is

$$p_R = \frac{{v_m}^2}{40} + \frac{{v_m}^2}{10} = \frac{2^2}{40} + \frac{2^2}{10} = 0.1 + 0.4 = 0.5\text{ W}$$

The voltage v_m and the current of the current source adhere to the passive convention, so

$$p_s = v_m(0.25) = (-2)(0.25) = -0.5\text{ W}$$

is the power received by the current source. The current source supplies 0.5 W.

Thus, the power absorbed by the two resistors is equal to that supplied by the source.

EXAMPLE 3.4-3 Current Divider Design

The input to the current divider in Figure 3.4-8 is the current, i_s, of the current source. The output is the current, i_o, measured by the ammeter. Specify values of the resistances, R_1 and R_2, to satisfy both of these specifications:

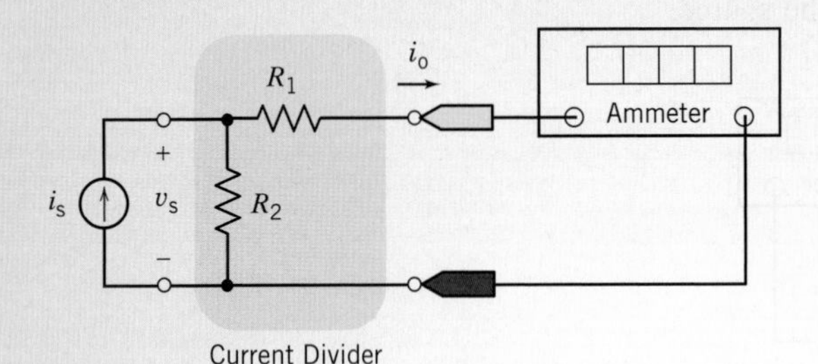

FIGURE 3.4-8 A current divider circuit.

Specification 1: The input and output currents are related by $i_o = 0.8\ i_s$.
Specification 2: The current source is required to supply no more than 10 mW of power when the input to the current divider is $i_s = 2$ mA.

Solution

We'll examine each specification to see what it tells us about the resistor values.
Specification 1: The input and output currents of the current divider are related by

$$i_o = \frac{R_2}{R_1 + R_2} i_s$$

So specification 1 requires

$$\frac{R_2}{R_1 + R_2} = 0.8 \quad \Rightarrow \quad R_2 = 4R_1$$

Specification 2: The power supplied by the current source is given by

$$p_s = i_s v_s = i_s\left(i_s\left(\frac{R_1 R_2}{R_1 + R_2}\right)\right) = {i_s}^2\left(\frac{R_1 R_2}{R_1 + R_2}\right)$$

So specification 2 requires

$$0.01 \geq (0.002)^2 \left(\frac{R_1 R_2}{R_1 + R_2} \right) \quad \Rightarrow \quad \frac{R_1 R_2}{R_1 + R_2} \leq 2500$$

Combining these results gives

$$\frac{R_1(4R_2)}{R_1 + 4R_2} \leq 2500 \quad \Rightarrow \quad \frac{4}{5} R_1 \leq 2500 \quad \Rightarrow \quad R_1 \leq 3125\ \Omega$$

The solution is not unique. One solution is

$$R_1 = 3\ \text{k}\Omega \quad \text{and} \quad R_2 = 12\ \text{k}\Omega$$

EXERCISE 3.4-1 A resistor network consisting of parallel resistors is shown in a package used for printed circuit board electronics in Figure E 3.4-1*a*. This package is only 2 cm × 0.7 cm, and each resistor is 1 kΩ. The circuit is connected to use four resistors as shown in Figure E 3.4-1*b*. Find the equivalent circuit for this network. Determine the current in each resistor when $i_s = 1$ mA.

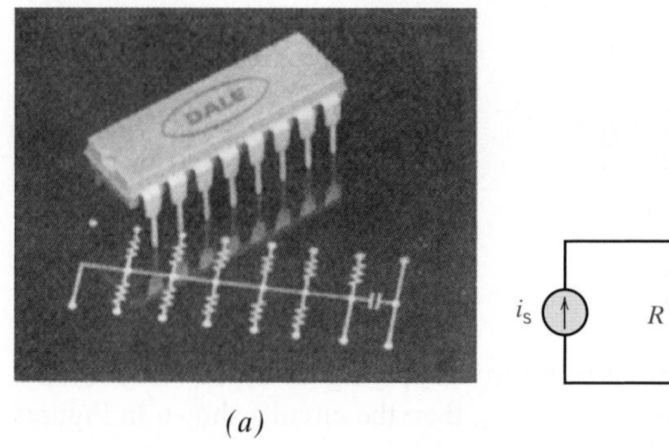

(*a*)

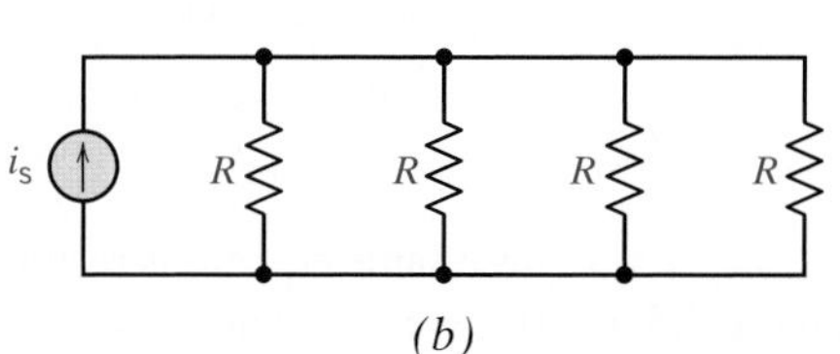

(*b*)

FIGURE E 3.4-1 (*a*) A parallel resistor network. Courtesy of Dale Electronics. (*b*) The connected circuit uses four resistors where $R =$ 1 kΩ.

Answer: $R_p = 250\ \Omega$

EXERCISE 3.4-2 Determine the current measured by the ammeter in the circuit shown in Figure E 3.4-2*a*.

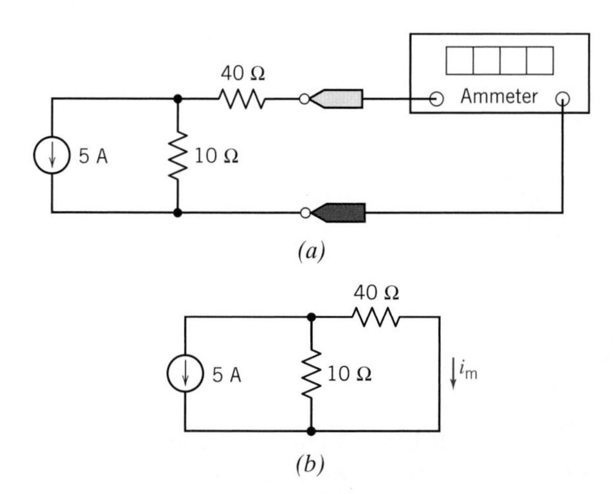

FIGURE E 3.4-2 (*a*) A current divider. (*b*) The current divider after the ideal ammeter has been replaced by the equivalent short circuit and a label has been added to indicate the current measured by the ammeter, i_m.

Hint: Figure E 3.4-2*b* shows the circuit after the ideal ammeter has been replaced by the equivalent short circuit, and a label has been added to indicate the current measured by the ammeter, i_m.

Answer: $i_m = -1$ A

3.5 SERIES VOLTAGE SOURCES AND PARALLEL CURRENT SOURCES

Voltage sources connected in series are equivalent to a single voltage source. The voltage of the equivalent voltage source is equal to the algebraic sum of voltages of the series voltage sources.

Consider the circuit shown in Figure 3.5-1*a*. Notice that the currents of both voltage sources are equal. Accordingly, define the current, i_s, to be

$$i_s = i_a = i_b \tag{3.5-1}$$

Next, define the voltage, v_s, to be

$$v_s = v_a + v_b \tag{3.5-2}$$

Using KCL, KVL, and Ohm's law, we can represent the circuit in Figure 3.5-1*a* by the equations

$$i_c = \frac{v_1}{R_1} + i_s \tag{3.5-3}$$

$$i_s = \frac{v_2}{R_2} + i_3 \tag{3.5-4}$$

$$v_c = v_1 \tag{3.5-5}$$

$$v_1 = v_s + v_2 \tag{3.5-6}$$

$$v_2 = i_3 R_3 \tag{3.5-7}$$

where $i_s = i_a = i_b$ and $v_s = v_a + v_b$. These same equations result from applying KCL, KVL, and Ohm's law to the circuit in Figure 3.5-1*b*. If $i_s = i_a = i_b$ and $v_s = v_a + v_b$, then the circuits shown in Figures 3.5-1*a* and 3.5-1*b* are equivalent because they are both represented by the same equations.

For example, suppose that $i_c = 4$ A, $R_1 = 2\ \Omega$, $R_2 = 6\ \Omega$, $R_3 = 3\ \Omega$, $v_a = 1$ V, and $v_b = 3$ V. The equations describing the circuit in Figure 3.5-1*a* become

$$4 = \frac{v_1}{2} + i_s \tag{3.5-8}$$

$$i_s = \frac{v_2}{6} + i_3 \tag{3.5-9}$$

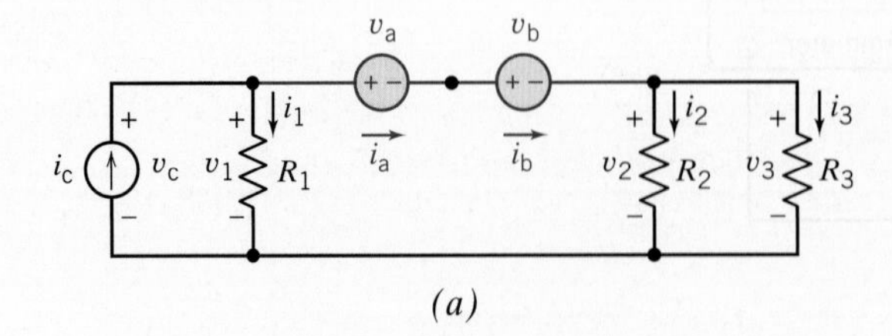

(*a*)

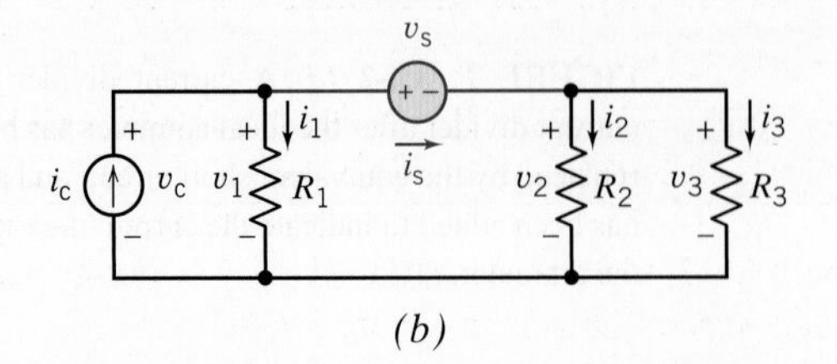

(*b*)

FIGURE 3.5-1 (*a*) A circuit containing voltage sources connected in series and (*b*) an equivalent circuit.

Table 3.5-1 Parallel and Series Voltage and Current Sources

CIRCUIT	EQUIVALENT CIRCUIT	CIRCUIT	EQUIVALENT CIRCUIT
v_a (+ −) v_b (+ −)	$v_a + v_b$ (+ −)	i_a (↑) i_b (↑)	$i_a + i_b$ (↑)
v_a (+ −) v_b (− +)	$v_a - v_b$ (+ −)	i_a (↑) i_b (↓)	$i_a - i_b$ (↑)
i_a (→) i_b (→)	Not allowed	v_a (+ −) v_b (+ −)	Not allowed

$$v_c = v_1 \tag{3.5-10}$$

$$v_1 = 4 + v_2 \tag{3.5-11}$$

$$v_2 = 3i_3 \tag{3.5-12}$$

The solution to this set of equations is $v_1 = 6$ V, $i_s = 1$ A, $i_3 = 0.66$ A, $v_2 = 2$ V, and $v_c = 6$ V. Eqs. 3.5-8 to 3.5-12 also describe the circuit in Figure 3.5-1*b*. Thus, $v_1 = 6$ V, $i_s = 1$ A, $i_3 = 0.66$ A, $v_2 = 2$ V, and $v_c = 6$ V in both circuits. Replacing series voltage sources by a single, equivalent voltage source does not change the voltage or current of other elements of the circuit.

Figure 3.5-2*a* shows a circuit containing parallel current sources. The circuit in Figure 3.5-2*b* is obtained by replacing these parallel current sources by a single, equivalent current source. The current of the equivalent current source is equal to the algebraic sum of the currents of the parallel current sources.

We are not allowed to connect independent current sources in series. Series elements have the same current. This restriction prevents series current sources from being independent. Similarly, we are not allowed to connect independent voltage sources in parallel.

Table 3.5-1 summarizes the parallel and series connections of current and voltage sources.

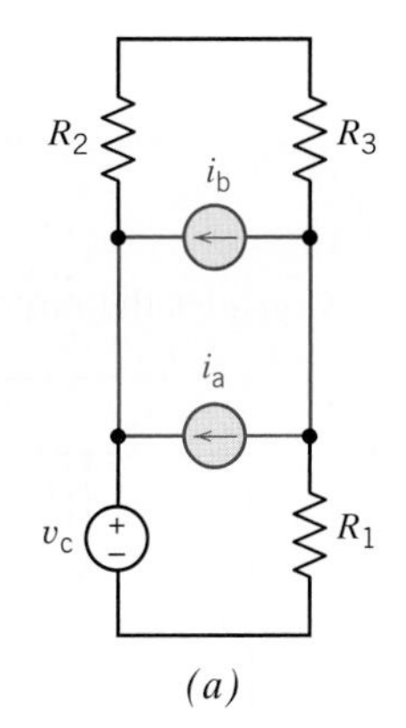

(*a*)

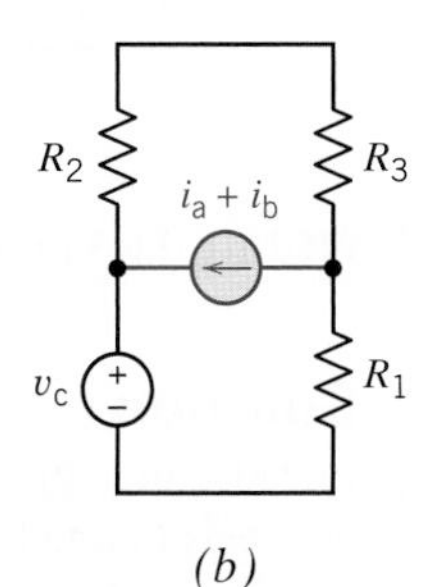

(*b*)

FIGURE 3.5-2 (*a*) A circuit containing parallel current sources and (*b*) an equivalent circuit.

3.6 CIRCUIT ANALYSIS

In this section, we consider the analysis of a circuit by replacing a set of resistors with an equivalent resistance, thus reducing the network to a form easily analyzed.

Consider the circuit shown in Figure 3.6-1. Note that it includes a set of resistors that is in series and another set of resistors that is in parallel. It is desired to find the output voltage v_o, so we wish to reduce the circuit to the equivalent circuit shown in Figure 3.6-2.

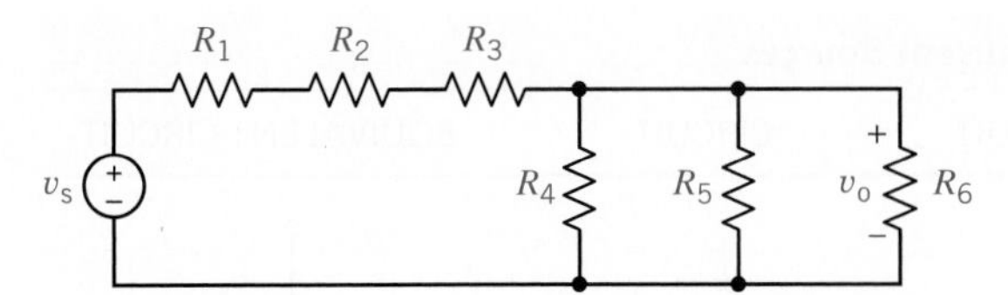

FIGURE 3.6-1 Circuit with a set of series resistors and a set of parallel resistors.

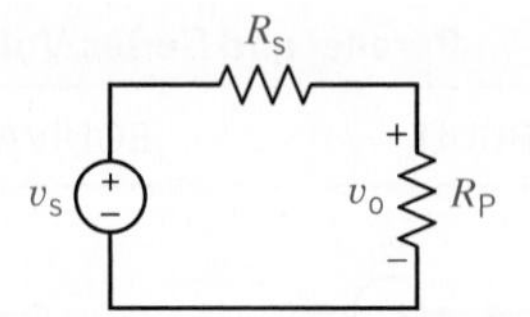

FIGURE 3.6-2 Equivalent circuit for the circuit of Figure 3.6-2.

We note that the equivalent series resistance is

$$R_s = R_1 + R_2 + R_3$$

and the equivalent parallel resistance is

$$R_p = \frac{1}{G_p}$$

where

$$G_p = G_4 + G_5 + G_6$$

Then, using the voltage divider principle, with Figure 3.6-2, we have

$$v_o = \frac{R_p}{R_s + R_p} v_s$$

Replacing the series resistors by the equivalent resistor R_s did not change the current or voltage of any other circuit element. In particular, the voltage v_o did not change. Also, the voltage v_o across the equivalent resistor R_p is equal to the voltage across each of the parallel resistors. Consequently, the voltage v_o in Figure 3.6-2 is equal to the voltage v_o in Figure 3.6-1. We can analyze the simple circuit in Figure 3.6-2 to find the value of the voltage v_o and know that the voltage v_o in the more complicated circuit shown in Figure 3.6-1 has the same value.

EXAMPLE 3.6-1 Series and Parallel Resistors

Consider the circuit shown in Figure 3.6-3. Find the current i_1 when

$$R_4 = 2\ \Omega \quad \text{and} \quad R_2 = R_3 = 8\ \Omega$$

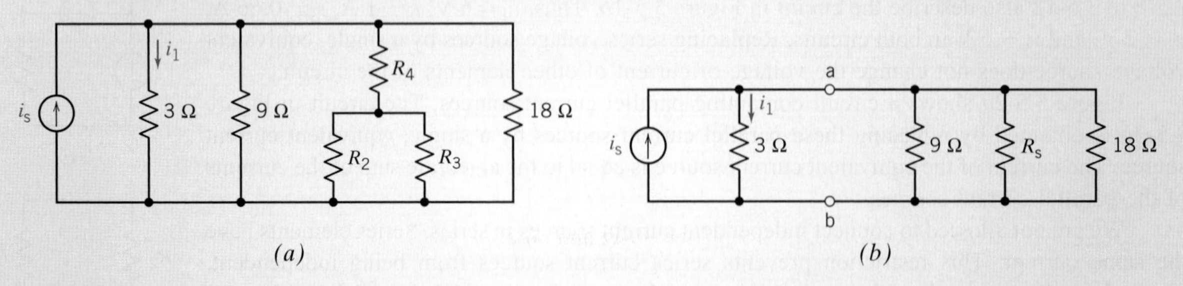

FIGURE 3.6-3 (*a*) Circuit for Example 3.6-1. (*b*) Partially reduced circuit for Example 3.6-1.

Solution

Because the objective is to find i_1, we will attempt to reduce the circuit so that the 3-Ω resistor is in parallel with one resistor and the current source i_s. Then we can use the current divider principle to obtain i_1. Because R_2 and R_3 are in parallel, we find an equivalent resistance as

$$R_{p1} = \frac{R_2 R_3}{R_2 + R_3} = 4\ \Omega$$

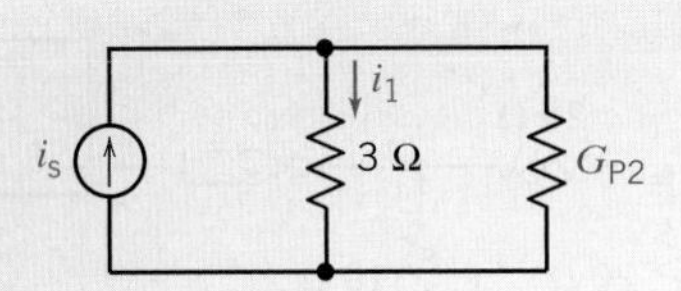

FIGURE 3.6-4 Equivalent circuit for Figure 3.6-3.

This equivalent resistor is connected in series with R_4. Then adding R_{p1} to R_4, we have a series equivalent resistor

$$R_s = R_4 + R_{p1} = 2 + 4 = 6\ \Omega$$

Now the R_s resistor is in parallel with three resistors as shown in Figure 3.6-3*b*. However, we wish to obtain the equivalent circuit as shown in Figure 3.6-4 so that we can find i_1. Therefore, we combine the 9-Ω resistor, the 18-Ω resistor, and R_s shown to the right of terminals a-b in Figure 3.6-3*b* into one parallel equivalent conductance G_{p2}. Thus, we find

$$G_{p2} = \frac{1}{9} + \frac{1}{18} + \frac{1}{R_s} = \frac{1}{9} + \frac{1}{18} + \frac{1}{6} = \frac{1}{3}\ \mathrm{S} \quad \Rightarrow \quad R_{p2} = \frac{1}{G_{p2}} = 3\ \Omega$$

Then, using the current divider principle,

$$i_1 = \frac{G_1 i_s}{G_p}$$

where

$$G_p = G_1 + G_{p2} = \frac{1}{3} + \frac{1}{3} = \frac{2}{3}$$

Therefore,

$$i_1 = \frac{1/3}{2/3} i_s = \frac{1}{2} i_s$$

EXAMPLE 3.6-2 Equivalent Resistance

The circuit in Figure 3.6-5*a* contains an ohmmeter. An ohmmeter is an instrument that measures resistance in ohms. The ohmmeter will measure the equivalent resistance of the resistor circuit connected to its terminals. Determine the resistance measured by the ohmmeter in Figure 3.6-5*a*.

Solution

Working from left to right, the 30-Ω resistor is parallel to the 60-Ω resistor. The equivalent resistance is

$$\frac{60 \cdot 30}{60 + 30} = 20\ \Omega$$

In Figure 3.6-5*b*, the parallel combination of the 30-Ω and 60-Ω resistors has been replaced with the equivalent 20-Ω resistor. Now the two 20-Ω resistors are in series.

The equivalent resistance is

$$20 + 20 = 40\ \Omega$$

In Figure 3.6-5*c*, the series combination of the two 20-Ω resistors has been replaced with the equivalent 40-Ω resistor. Now the 40-Ω resistor is parallel to the 10-Ω resistor. The equivalent resistance is

$$\frac{40 \cdot 10}{40 + 10} = 8\ \Omega$$

In Figure 3.6-5*d* the parallel combination of the 40-Ω and 10-Ω resistors has been replaced with the equivalent 8-Ω resistor. Thus, the ohmmeter measures a resistance equal to 8 Ω.

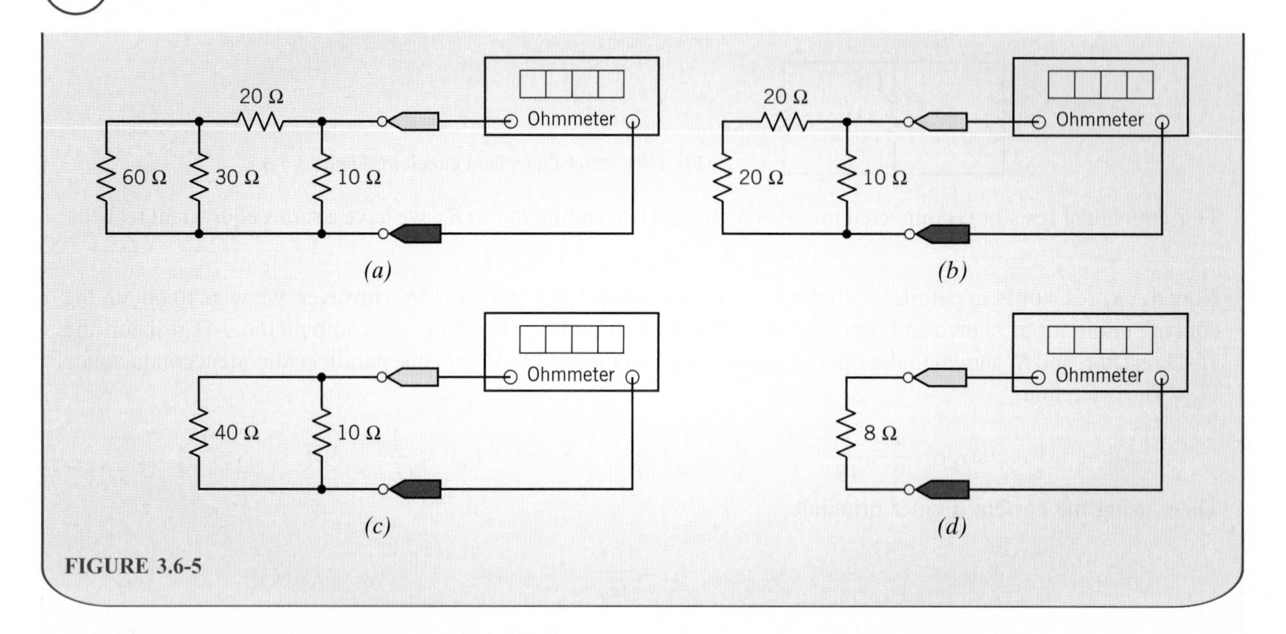

FIGURE 3.6-5

EXAMPLE 3.6-3 Circuit Analysis Using Equivalent Resistances

Determine the values of i_3, v_4, i_5, and v_6 in circuit shown in Figure 3.6-6.

Solution

The circuit shown in Figure 3.6-7 has been obtained from the circuit shown in Figure 3.6-6 by replacing series and parallel combinations of resistances by equivalent resistances. We can use this equivalent circuit to solve this problem in three steps:

1. Determine the values of the resistances R_1, R_2, and R_3 in Figure 3.6-7 that make the circuit in Figure 3.6-7 equivalent to the circuit in Figure 3.6-6.

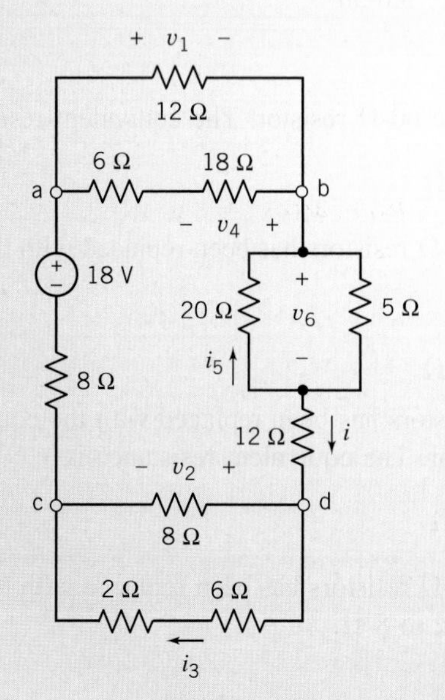

FIGURE 3.6-6 The circuit considered in Example 3.6-3.

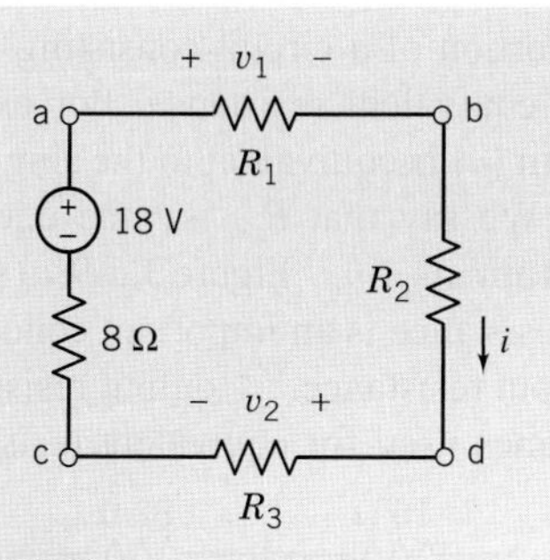

FIGURE 3.6-7 An equivalent circuit for the circuit in Figure 3.3-6.

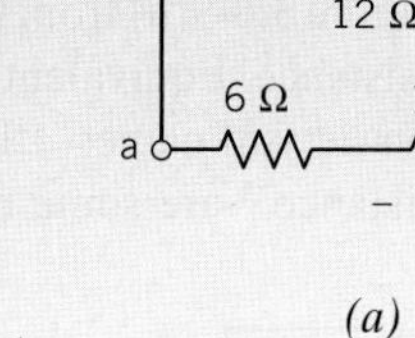

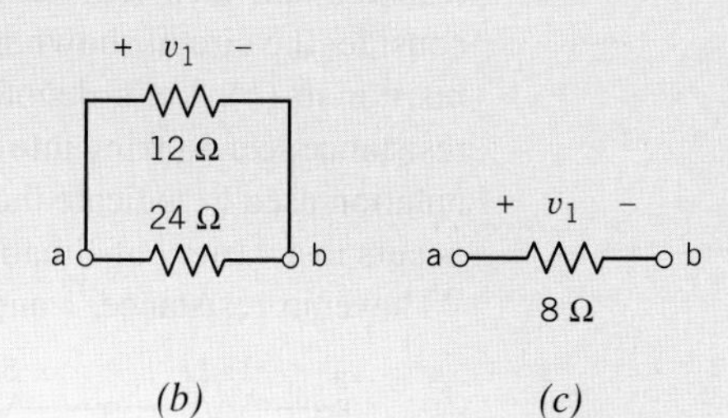

(a) (b) (c)

FIGURE 3.6-8

2. Determine the values of v_1, v_2, and i in Figure 3.6-7.
3. Because the circuits are equivalent, the values of v_1, v_2, and i in Figure 3.6-6 are equal to the values of v_1, v_2, and i in Figure 3.6-7. Use voltage and current division to determine the values of i_3, v_4, i_5, and v_6 in Figure 3.6-6.

Step 1: Figure 3.6-8*a* shows the three resistors at the top of the circuit in Figure 3.6-6. We see that the 6-Ω resistor is connected in series with the 18-Ω resistor. In Figure 3.6-8*b*, these series resistors have been replaced by the equivalent 24-Ω resistor. Now the 24-Ω resistor is connected in parallel with the 12-Ω resistor. Replacing series resistors by an equivalent resistance does not change the voltage or current in any other element of the circuit. In particular, v_1, the voltage across the 12-Ω resistor, does not change when the series resistors are replaced by the equivalent resistor. In contrast, v_4 is not an element voltage of the circuit shown in Figure 3.6-8*b*.

In Figure 3.6-8*c*, the parallel resistors have been replaced by the equivalent 8-Ω resistor. The voltage across the equivalent resistor is equal to the voltage across each of the parallel resistors, v_1 in this case. In summary, the resistance R_1 in Figure 3.6-7 is given by

$$R_1 = 12 \parallel (6 + 18) = 8\,\Omega$$

Similarly, the resistances R_2 and R_3 in Figure 3.6-7 are given by

$$R_2 = 12 + (20 \parallel 5) = 16\,\Omega$$

$$R_3 = 8 \parallel (2 + 6) = 4\,\Omega$$

Step 2: Apply KVL to the circuit of Figure 3.6-7 to get

$$R_1 i + R_2 i + R_3 i + 8i - 18 = 0 \quad \Rightarrow \quad i = \frac{18}{R_1 + R_2 + R_3 + 8} = \frac{18}{8 + 16 + 4 + 8} = 0.5\text{ A}$$

Next, Ohm's law gives

$$v_1 = R_1 i = 8(0.5) = 4\text{ V} \quad \text{and} \quad v_2 = R_3 i = 4(0.5) = 2\text{ V}$$

Step 3: The values of v_1, v_2, and i in Figure 3.6-6 are equal to the values of v_1, v_2, and i in Figure 3.6-7. Returning our attention to Figure 3.6-6, and paying attention to reference directions, we can determine the values of i_3, v_4, i_5, and v_6 using voltage division, current division, and Ohm's law:

$$\begin{aligned}
i_3 &= \frac{8}{8 + (2 + 6)} i = \frac{1}{2}(0.5) = 0.25\text{ A} \\
v_4 &= -\frac{18}{6 + 18} v_1 = -\frac{3}{4}(4) = -3\text{ V} \\
i_5 &= -\frac{5}{20 + 5} i = -\left(\frac{1}{5}\right)(0.5) = -0.1\text{ A} \\
v_6 &= (20 \parallel 5) i = 4(0.5) = 2\text{ V}
\end{aligned}$$

In general, we may find the equivalent resistance for a portion of a circuit consisting only of resistors and then replace that portion of the circuit with the equivalent resistance. For example, consider the circuit shown in Figure 3.6-9. The resistive circuit in (*a*) is equivalent to the single 56 Ω resistor in (*b*). Let's denote the equivalent resistance as R_{eq}. We say that R_{eq} is "the equivalent resistance seen looking into the circuit of Figure 3.6-9(*a*) from terminals *a*-*b*." Figure 3.6-9(*c*) shows a notation used to indicate the equivalent resistance. Equivalent resistance is an important concept that occurs in a variety of situations and has a variety of names. "Input resistance," "output resistance," "Thevenin resistance," and "Norton resistance" are some names used for equivalent resistance.

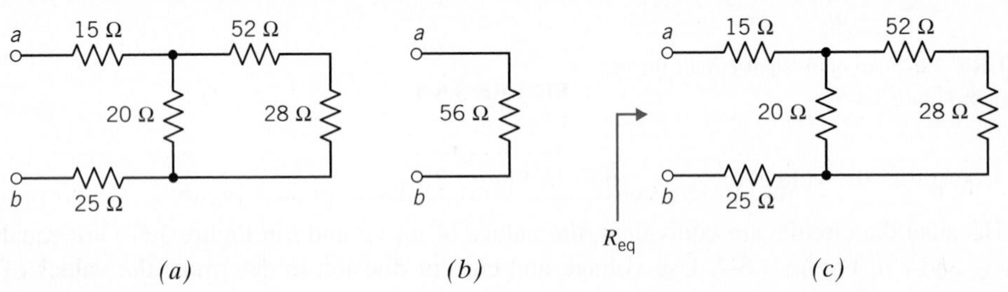

FIGURE 3.6-9 The resistive circuit in (*a*) is equivalent to the single resistor in (*b*). The notation used to indicate the equivalent resistance is shown in (*c*).

EXERCISE 3.6-1 Determine the resistance measured by the ohmmeter in Figure E 3.6-1.

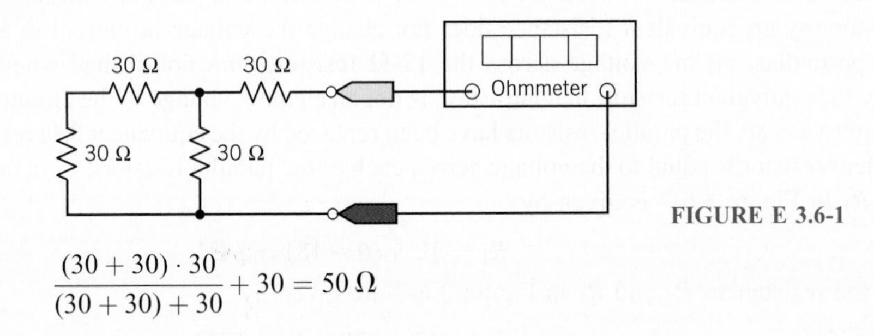

FIGURE E 3.6-1

Answer: $\dfrac{(30+30)\cdot 30}{(30+30)+30} + 30 = 50\ \Omega$

3.7 ANALYZING RESISTIVE CIRCUITS USING MATLAB

We can analyze simple circuits by writing and solving a set of equations. We use Kirchhoff's law and the element equations, for instance, Ohm's law, to write these equations. As the following example illustrates, MATLAB provides a convenient way to solve the equations describing an electric circuit.

EXAMPLE 3.7-1 MATLAB for Simple Circuits

Determine the values of the resistor voltages and currents for the circuit shown in Figure 3.7-1.

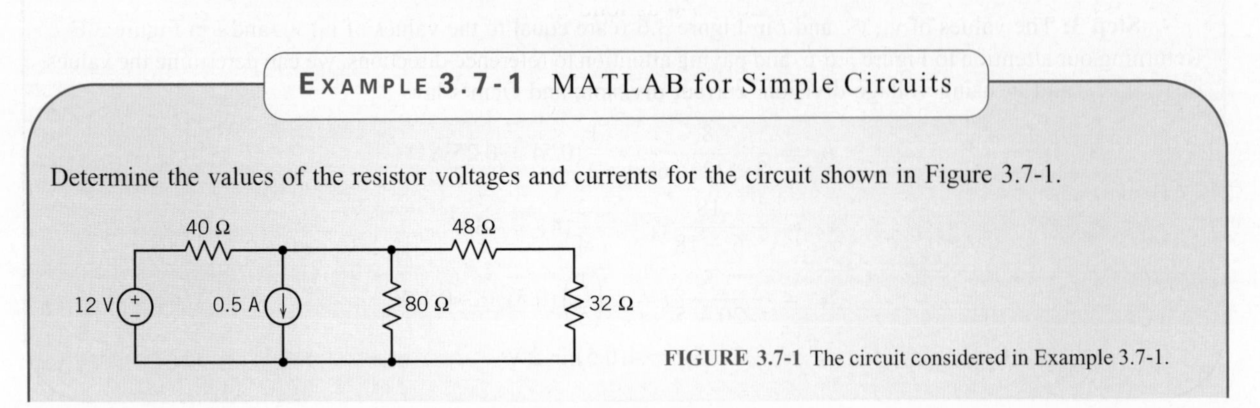

FIGURE 3.7-1 The circuit considered in Example 3.7-1.

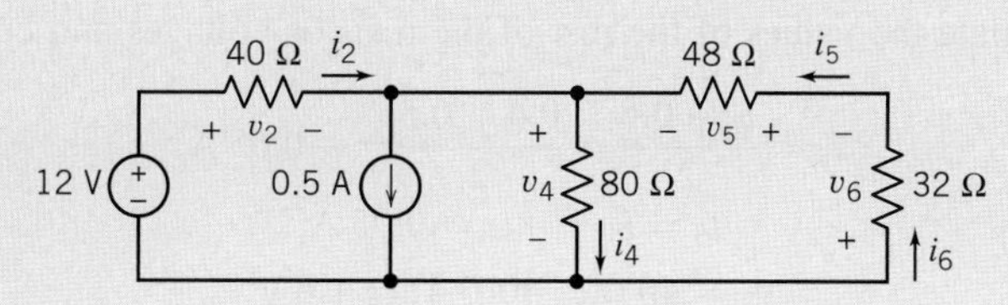

FIGURE 3.7-2 The circuit from Figure 3.7-1 after labeling the resistor voltages and currents.

Solution

Let's label the resistor voltages and currents. In anticipation of using Ohm's law, we will label the voltage and current of each resistor to adhere to the passive convention. (Pick one of the variables—the resistor current or the resistor voltage—and label the reference direction however you like. Label the reference direction of the other variable to adhere to the passive convention with the first variable.) Figure 3.7-2 shows the labeled circuit.

Next, we will use Kirchhoff's laws. First, apply KCL to the node at which the current source and the 40-Ω, 48-Ω, and 80-Ω resistors are connected together to write

$$i_2 + i_5 = 0.5 + i_4 \tag{3.7-1}$$

Next, apply KCL to the node at which the 48-Ω and 32-Ω resistors are connected together to write

$$i_5 = i_6 \tag{3.7-2}$$

Apply KVL to the loop consisting of the voltage source and the 40-Ω and 80-Ω resistors to write

$$12 = v_2 + v_4 \tag{3.7-3}$$

Apply KVL to the loop consisting of the 48-Ω, 32-Ω, and 80-Ω resistors to write

$$v_4 + v_5 + v_6 = 0 \tag{3.7-4}$$

Apply Ohm's law to the resistors.

$$v_2 = 40\,i_2,\ v_4 = 80\,i_4,\ v_5 = 48\,i_5,\ v_6 = 32\,i_6 \tag{3.7-5}$$

We can use the Ohm's law equations to eliminate the variables representing resistor voltages. Doing so enables us to rewrite Eq. 3.7-3 as:

$$12 = 40\,i_2 + 80\,i_4 \tag{3.7-6}$$

Similarly, we can rewrite Eq. 3.7-4 as

$$80\,i_4 + 48\,i_5 + 32\,i_6 = 0 \tag{3.7-7}$$

Next, use Eq. 3.7-2 to eliminate i_6 from Eq. 3.7-6 as follows

$$80\,i_4 + 48\,i_5 + 32\,i_5 = 0 \quad \Rightarrow \quad 80\,i_4 + 80\,i_5 = 0 \quad \Rightarrow \quad i_4 = -i_5 \tag{3.7-8}$$

Use Eq. 3.7-8 to eliminate i_5 from Eq. 3.7-1.

$$i_2 - i_4 = 0.5 + i_4 \quad \Rightarrow \quad i_2 = 0.5 + 2\,i_4 \tag{3.7-9}$$

Use Eq. 3.7-9 to eliminate i_4 from Eq. 3.7-6. Solve the resulting equation to determine the value of i_2.

$$12 = 40\,i_2 + 80\left(\frac{i_2 - 0.5}{2}\right) = 80\,i_2 - 20 \quad \Rightarrow \quad i_2 = \frac{12 + 20}{80} = 0.4\text{ A} \tag{3.7-10}$$

Now we are ready to calculate the values of the rest of the resistor voltages and currents as follows:

$$i_4 = \frac{i_2 - 0.5}{2} = \frac{0.4 - 0.5}{2} = -0.05\text{ A},$$
$$i_6 = i_5 = -i_4 = 0.05\text{ A},$$
$$v_2 = 40\,i_2 = 40(0.4) = 1.6\text{ V},$$
$$v_4 = 80\,i_4 = 80(-0.05) = -4\text{ V},$$
$$v_5 = 48\,i_5 = 48(0.05) = 2.4\text{ V},$$

and

$$v_6 = 32\,i_6 = 32(0.05) = 1.6\text{ V}.$$

MATLAB Solution 1

The preceding algebra shows that this circuit can be represented by these equations:

$$12 = 80\,i_2 - 20, i_4 = \frac{i_2 - 0.5}{2}, i_6 = i_5 = -i_4, v_2 = 40\,i_2,\ v_4 = 80\,i_4,$$
$$v_5 = 48\,i_5, \text{ and } v_6 = 32\,i_6$$

These equations can be solved consecutively, using MATLAB as shown in Figure 3.7-3.

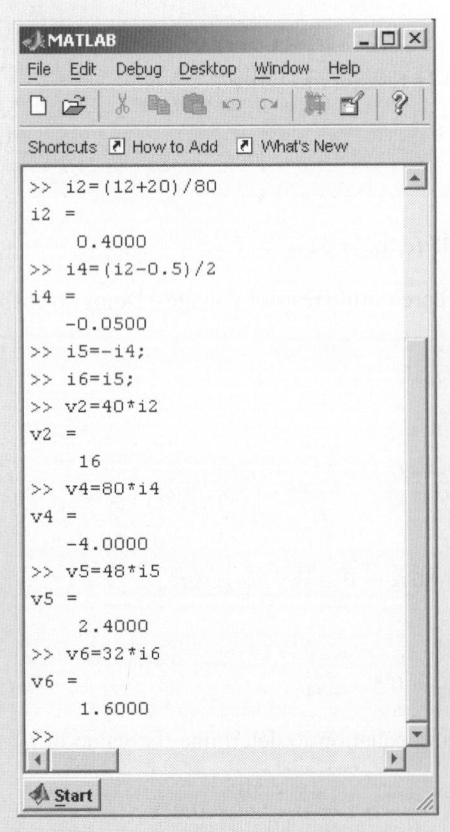

FIGURE 3.7-3 Consecutive equations.

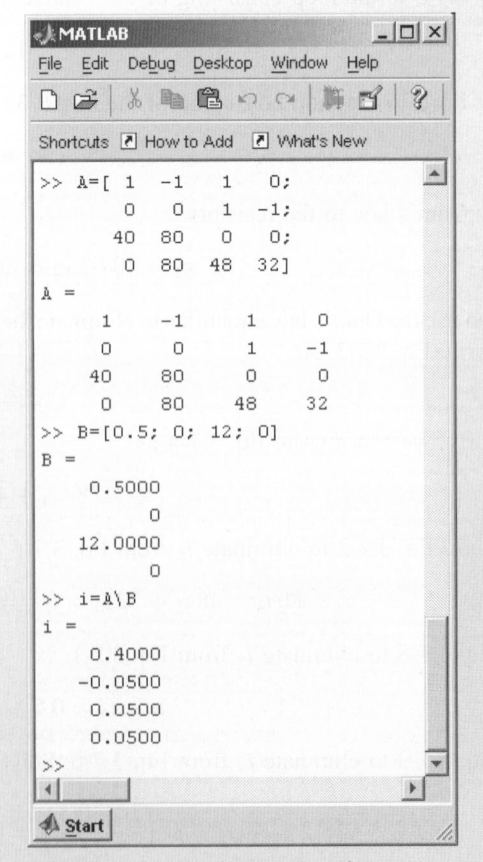

FIGURE 3.7-4 Simultaneous equations.

MATLAB Solution 2

We can avoid some algebra if we are willing to solve simultaneous equations.

After applying Kirchhoff's laws and then using the Ohm's law equations to eliminate the variables representing resistor voltages, we have Eqs 3.7-1, 2, 6, and 7:

$$i_2 + i_5 = 0.5 + i_4,\ i_5 = i_6, 12 = 40\, i_2 + 80\, i_4,$$

and

$$80\, i_4 + 48\, i_5 + 32\, i_6 = 0$$

This set of four simultaneous equations in i_2, i_4, i_5, and i_6 can be written as a single matrix equation.

$$\begin{bmatrix} 1 & -1 & 1 & 0 \\ 0 & 0 & 1 & -1 \\ 40 & 80 & 0 & 0 \\ 0 & 80 & 48 & 32 \end{bmatrix} \begin{bmatrix} i_2 \\ i_4 \\ i_5 \\ i_6 \end{bmatrix} = \begin{bmatrix} 0.5 \\ 0 \\ 12 \\ 0 \end{bmatrix} \qquad (3.7\text{-}11)$$

We can write this equation as

$$Ai = B \qquad (3.7\text{-}12)$$

where

$$A = \begin{bmatrix} 1 & -1 & 1 & 0 \\ 0 & 0 & 1 & -1 \\ 40 & 80 & 0 & 0 \\ 0 & 80 & 48 & 32 \end{bmatrix},\ i = \begin{bmatrix} i_2 \\ i_4 \\ i_5 \\ i_6 \end{bmatrix} \text{ and } B = \begin{bmatrix} 0.5 \\ 0 \\ 12 \\ 0 \end{bmatrix}$$

This matrix equation can be solved using MATLAB as shown in Figure 3.7-4. After entering matrices A and B, the statement

i = A/B

tells MATLAB to calculate i by solving Eq 3.7-12.

A circuit consisting of n elements has n currents and n voltages. A set of equations representing that circuit could have as many as $2n$ unknowns. We can reduce the number of unknowns by labeling the currents and voltages carefully. For example, suppose two of the circuit elements are connected in series. We can choose the reference directions for the currents in those elements so that they are equal and use one variable to represent both currents. Table 3.7-1 presents some guidelines that will help us reduce the number of unknowns in the set of equations describing a given circuit.

Table 3.7-1 Guidelines for Labeling Circuit Variables

CIRCUIT FEATURE	GUIDELINE
Resistors	Label the voltage and current of each resistor to adhere to the passive convention. Use Ohm's law to eliminate either the current or voltage variable.
Series elements	Label the reference directions for series elements so that their currents are equal. Use one variable to represent the currents of series elements.
Parallel elements	Label the reference directions for parallel elements so that their voltages are equal. Use one variable to represent the voltages of parallel elements.
Ideal Voltmeter	Replace each (ideal) voltmeter by an open circuit. Label the voltage across the open circuit to be equal to the voltmeter voltage.
Ideal Ammeter	Replace each (ideal) ammeter by a short circuit. Label the current in the short circuit to be equal to the ammeter current.

3.8 HOW CAN WE CHECK . . . ?

Engineers are frequently called upon to check that a solution to a problem is indeed correct. For example, proposed solutions to design problems must be checked to confirm that all of the specifications have been satisfied. In addition, computer output must be reviewed to guard against data-entry errors, and claims made by vendors must be examined critically.

Engineering students are also asked to check the correctness of their work. For example, occasionally just a little time remains at the end of an exam. It is useful to be able to quickly identify those solutions that need more work.

The following example illustrates techniques useful for checking the solutions of the sort of problem discussed in this chapter.

EXAMPLE 3.8-1 How Can We Check Voltage and Current Values?

The circuit shown in Figure 3.8-1*a* was analyzed by writing and solving a set of simultaneous equations:

$$12 = v_2 + 4i_3, i_4 = \frac{v_2}{5} + i_3, v_5 = 4i_3, \text{ and } \frac{v_5}{2} = i_4 + 5i_4$$

The computer Mathcad (*Mathcad User's Guide*, 1991) was used to solve the equations as shown in Figure 3.8-1*b*. It was determined that

$$v_2 = -60\text{ V}, i_3 = 18\text{ A}, i_4 = 6\text{ A}, \text{ and } v_5 = 72\text{ V}.$$

How can we check that these currents and voltages are correct?

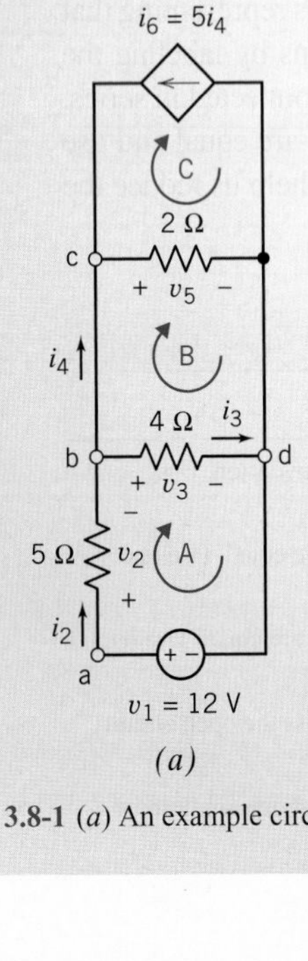

```
v2 := 0   i3 := 0   i4 := 0   v5 := 0
Given
12 ≈ v2 + 4 · i3              Apply KVL to loop A.
      v2
i4 ≈ ---- + i3                Apply KCL at node b.
      5
v5 ≈ 4 · i3                   Apply KVL to loop B.
v5
--- ≈ i4 + 5 · i4             Apply KCL at node c.
2
                              [-60]
                              [ 18]
Find (v2,i3,i4,v5) =          [  6]
                              [ 72]
```

(*a*) (*b*)

FIGURE 3.8-1 (*a*) An example circuit and (*b*) computer analysis using Mathcad.

Solution

The current i_2 can be calculated from v_2, i_3, i_4, and v_5 in a couple of different ways. First, Ohm's law gives

$$i_2 = \frac{v_2}{5} = \frac{-60}{5} = -12\ \text{A}$$

Next, applying KCL at node b gives

$$i_2 = i_3 + i_4 = 18 + 6 = 24\ \text{A}$$

Clearly, i_2 cannot be both -12 and 24 A, so the values calculated for v_2, i_3, i_4, and v_5 cannot be correct. Checking the equations used to calculate v_2, i_3, i_4, and v_5, we find a sign error in the KCL equation corresponding to node b. This equation should be

$$i_4 = \frac{v_2}{5} - i_3$$

After making this correction, v_2, i_3, i_4, and v_5 are calculated to be

$$v_2 = 7.5\ \text{V},\ i_3 = 1.125\ \text{A},\ i_4 = 0.375\ \text{A},\ v_5 = 4.5\ \text{V}$$

Now

$$i_2 = \frac{v_2}{5} = \frac{7.5}{5} = 1.5\ \text{A}$$

and

$$i_2 = i_3 + i_4 = 1.125 + 0.375 = 1.5\ \text{A}$$

This checks as we expected.

As an additional check, consider v_3. First, Ohm's law gives

$$v_3 = 4i_3 = 4(1.125) = 4.5\ \text{V}$$

Next, applying KVL to the loop consisting of the voltage source and the 4-Ω and 5-Ω resistors gives

$$v_3 = 12 - v_2 = 12 - 7.5 = 4.5\ \text{V}$$

Finally, applying KVL to the loop consisting of the 2-Ω and 4-Ω resistors gives

$$v_3 = v_5 = 4.5\ \text{V}$$

The results of these calculations agree with each other, indicating that

$$v_2 = 7.5\ \text{V},\ i_3 = 1.125\ \text{A},\ i_4 = 0.375\ \text{A},\ v_5 = 4.5\ \text{V}$$

are the correct values.

3.9 DESIGN EXAMPLE

ADJUSTABLE VOLTAGE SOURCE

A circuit is required to provide an adjustable voltage. The specifications for this circuit are that:

1. It should be possible to adjust the voltage to any value between −5 V and +5 V. It should not be possible accidentally to obtain a voltage outside this range.
2. The load current will be negligible.
3. The circuit should use as little power as possible.

The available components are:

1. Potentiometers: resistance values of 10 kΩ, 20 kΩ, and 50 kΩ are in stock
2. A large assortment of standard 2 percent resistors having values between 10 Ω and 1 MΩ (see Appendix D)
3. Two power supplies (voltage sources): one 12-V supply and one −12-V supply, both rated at 100 mA (maximum)

Describe the Situation and the Assumptions

Figure 3.9-1 shows the situation. The voltage v is the adjustable voltage. The circuit that uses the output of the circuit being designed is frequently called the load. In this case, the load current is negligible, so $i = 0$.

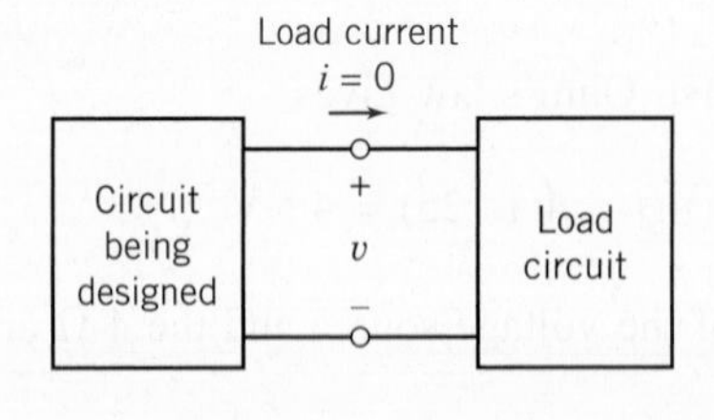

FIGURE 3.9-1 The circuit being designed provides an adjustable voltage, v, to the load circuit.

State the Goal

A circuit providing the adjustable voltage

$$-5\text{V} \leq v \leq +5\text{V}$$

must be designed using the available components.

Generate a Plan

Make the following observations.

1. The adjustability of a potentiometer can be used to obtain an adjustable voltage v.
2. Both power supplies must be used so that the adjustable voltage can have both positive and negative values.
3. The terminals of the potentiometer cannot be connected directly to the power supplies because the voltage v is not allowed to be as large as 12 V or −12 V.

These observations suggest the circuit shown in Figure 3.9-2*a*. The circuit in Figure 3.9-2*b* is obtained by using the simplest model for each component in Figure 3.9-2*a*.

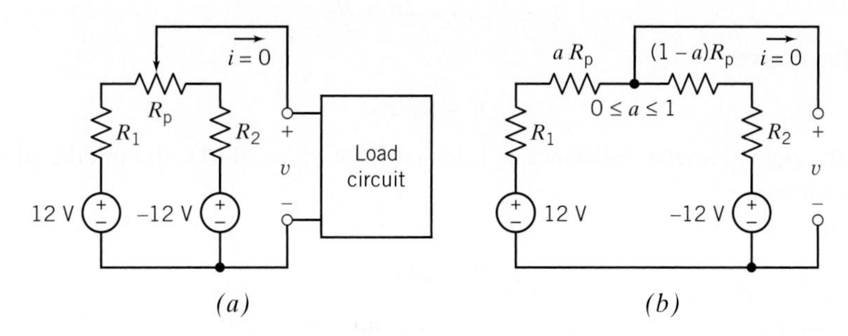

FIGURE 3.9-2 (*a*) A proposed circuit for producing the variable voltage, v, and (*b*) the equivalent circuit after the potentiometer is modeled.

To complete the design, values need to be specified for R_1, R_2, and R_p. Then several results need to be checked and adjustments made, if necessary.

1. Can the voltage v be adjusted to any value in the range -5 V to $+5$V?
2. Are the voltage source currents less than 100 mA? This condition must be satisfied if the power supplies are to be modeled as ideal voltage sources.
3. Is it possible to reduce the power absorbed by R_1, R_2, and R_p?

Act on the Plan

It seems likely that R_1 and R_2 will have the same value, so let $R_1 = R_2 = R$. Then it is convenient to redraw Figure 3.9-2*b* as shown in Figure 3.9-3.

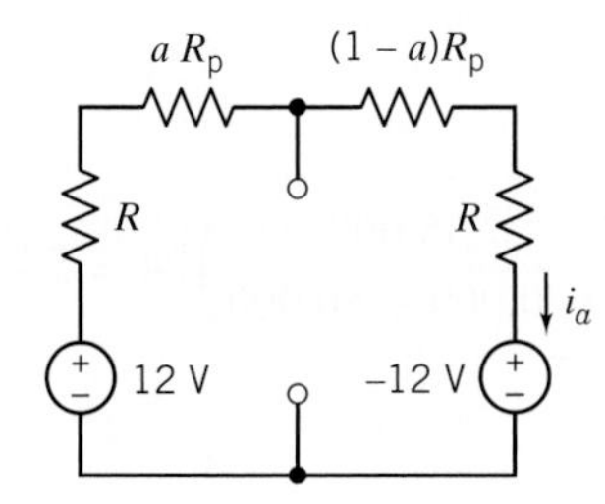

FIGURE 3.9-3 The circuit after setting $R_1 = R_2 = R$.

Applying KVL to the outside loop yields

$$-12 + Ri_a + aR_p\, i_a + (1 - a)R_p\, i_a + Ri_a - 12 = 0$$

so

$$i_a = \frac{24}{2R + R_p}$$

Next, applying KVL to the left loop gives

$$v = 12 - (R + aR_p)i_a$$

Substituting for i_a gives

$$v = 12 - \frac{24(R + aR_p)}{2R + R_p}$$

When $a=0$, v must be 5 V, so

$$5 = 12 - \frac{24R}{2R + R_p}$$

Solving for R gives

$$R = 0.7R_p$$

Suppose the potentiometer resistance is selected to be $R_p = 20$ kΩ, the middle of the three available values.

Then,

$$R = 14 \text{ k}\Omega$$

Verify the Proposed Solution

As a check, notice that when $a=1$,

$$v = 12 - \left(\frac{14{,}000 + 20{,}000}{28{,}000\text{ k} + 20{,}000}\right)24 = -5$$

as required. The specification that

$$-5\text{ V} \le v \le 5\text{ V}$$

has been satisfied. The power absorbed by the three resistances is

$$p = {i_a}^2(2R + R_p) = \frac{24^2}{2R + R_p}$$

so

$$p = 12\text{ mW}$$

Notice that this power can be reduced by choosing R_p to be as large as possible, 50 kΩ in this case. Changing R_p to 50 kΩ requires a new value of R:

$$R = 0.7 \times R_p = 35\text{ k}\Omega$$

Because

$$-5\text{ V} = 12 - \left(\frac{35{,}000 + 50{,}000}{70{,}000 + 50{,}000}\right)24 \le v \le 12 - \left(\frac{35{,}000}{70{,}000 + 50{,}000}\right)24 = 5\text{ V}$$

the specification that

$$-5\text{ V} \le v \le 5\text{ V}$$

has been satisfied. The power absorbed by the three resistances is now

$$p = \frac{24^2}{50{,}000 + 70{,}000} = 5\text{ mW}$$

Finally, the power supply current is

$$i_a = \frac{24}{50{,}000 + 70{,}000} = 0.2\text{ mA}$$

which is well below the 100 mA that the voltage sources are able to supply. The design is complete.

3.10 SUMMARY

- Kirchhoff's current law (KCL) states that the algebraic sum of the currents entering a node is zero. Kirchhoff's voltage law (KVL) states that the algebraic sum of the voltages around a closed path (loop) is zero.
- Simple electric circuits can be analyzed using only Kirchhoff's laws and the constitutive equations of the circuit elements.
- Series resistors act like a ''voltage divider,'' and parallel resistors act like a ''current divider.'' The first two rows of Table 3.10-1 summarize the relevant equations.
- Series resistors are equivalent to a single ''equivalent resistor.'' Similarly, parallel resistors are equivalent to a single ''equivalent resistor.'' The first two rows of Table 3.10-1 summarize the relevant equations.
- Series voltage sources are equivalent to a single ''equivalent voltage source.'' Similarly, parallel current sources are equivalent to a single ''equivalent current.'' The last two rows of Table 3.10-1 summarize the relevant equations.
- Often circuits consisting entirely of resistors can be reduced to a single equivalent resistor by repeatedly replacing series and/or parallel resistors by equivalent resistors.

Table 3.10-1 Equivalent Circuits for Series and Parallel Elements

Series resistors

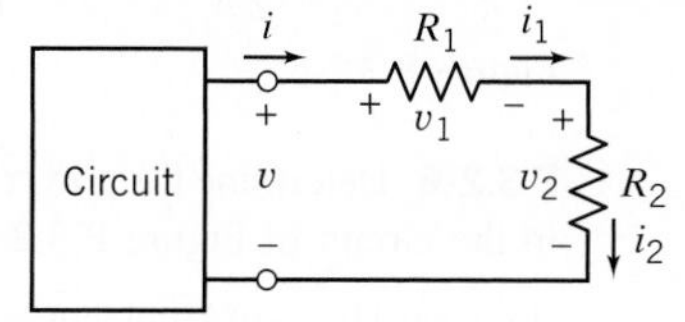

$i = i_1 = i_2$, $v_1 = \dfrac{R_1}{R_1 + R_2} v$, and $v_2 = \dfrac{R_2}{R_1 + R_2} v$

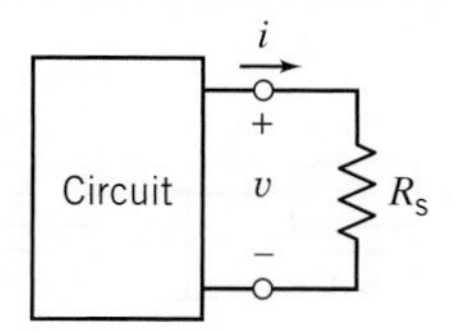

$R_s = R_1 + R_2$ and $v = R_s i$

Parallel resistors

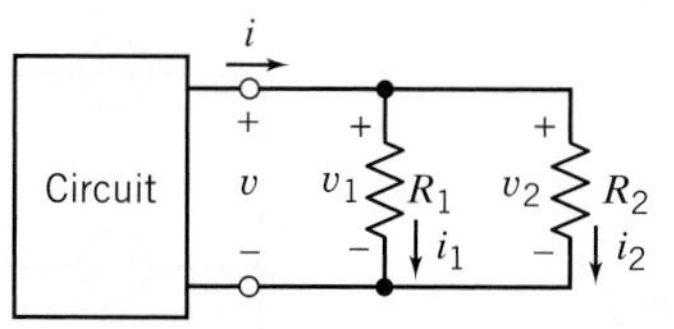

$v = v_1 = v_2$, $i_1 = \dfrac{R_2}{R_1 + R_2} i$, and $i_2 = \dfrac{R_1}{R_1 + R_2} i$

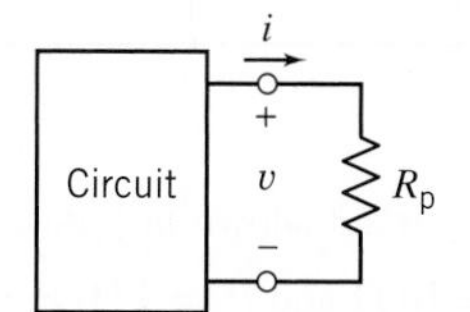

$R_p = \dfrac{R_1 R_2}{R_1 + R_2}$ and $v = R_p i$

Series voltage sources

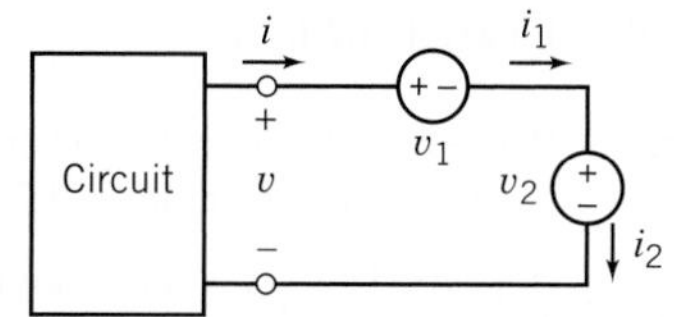

$i = i_1 = i_2$ and $v = v_1 + v_2$

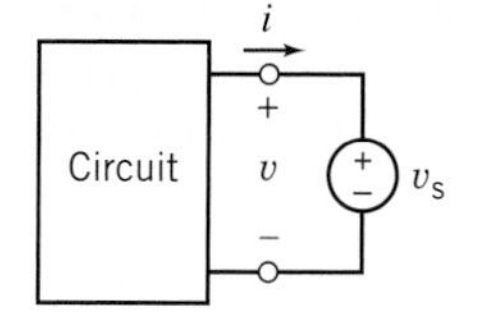

$v_s = v_1 + v_2$

Parallel current sources

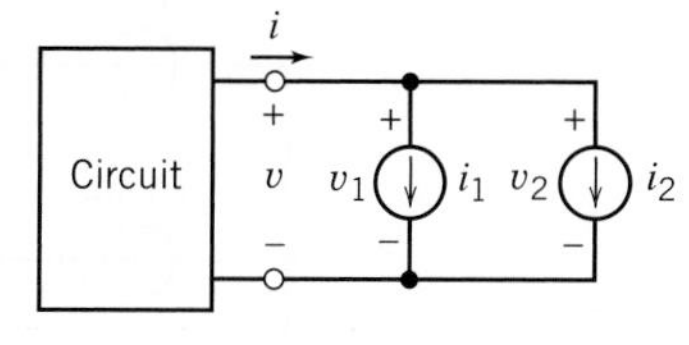

$v = v_1 = v_2$ and $i = i_1 + i_2$

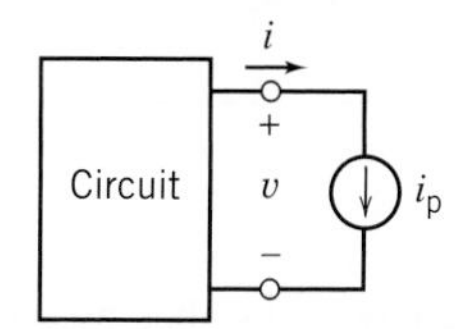

$i_p = i_1 + i_2$

PROBLEMS

Section 3.2 Kirchhoff's Laws

P 3.2-1 Consider the circuit shown in Figure P 3.2-1. Determine the values of the power supplied by branch B and the power supplied by branch F.

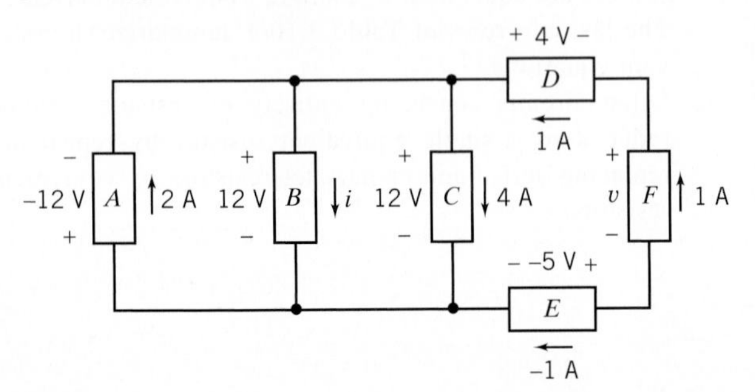

Figure P 3.2-1

P 3.2-2 Determine the values of i_2, i_4, v_2, v_3, and v_6 in Figure P 3.2-2.

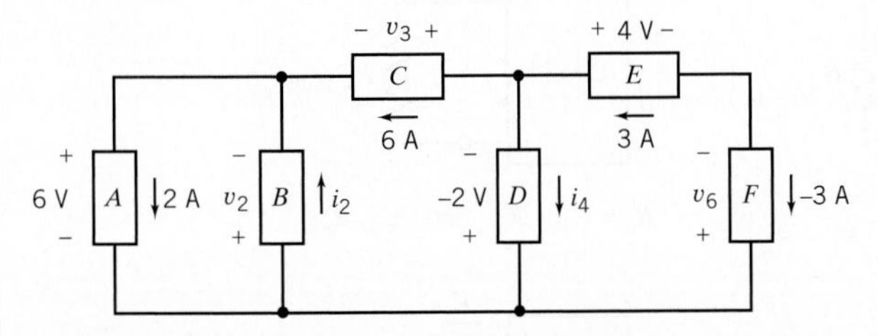

Figure P 3.2-2

P 3.2-3 Consider the circuit shown in Figure P 3.2-3.

(a) Suppose that $R_1 = 10\ \Omega$ and $R_2 = 5\ \Omega$. Find the current i and the voltage v.

(b) Suppose, instead, that $i = 2.75$ A and $v = 45$ V. Determine the resistances R_1 and R_2.

(c) Suppose, instead, that the voltage source supplies 25 W of power and that the current source supplies 10 W of power. Determine the current i, the voltage v, and the resistances R_1 and R_2.

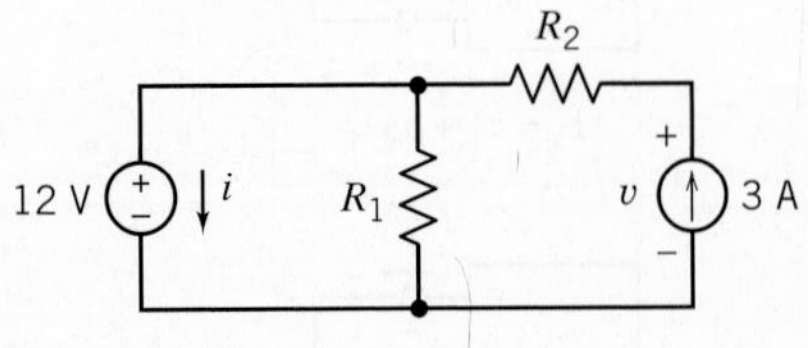

Figure P 3.2-3

P 3.2-4 Determine the power absorbed by each of the resistors in the circuit shown in Figure P 3.2-4.

Answer: The 5-Ω resistor absorbs 80 kW, the 6-Ω resistor absorbs 24 W, and the 9-Ω resistor absorbs 81 W.

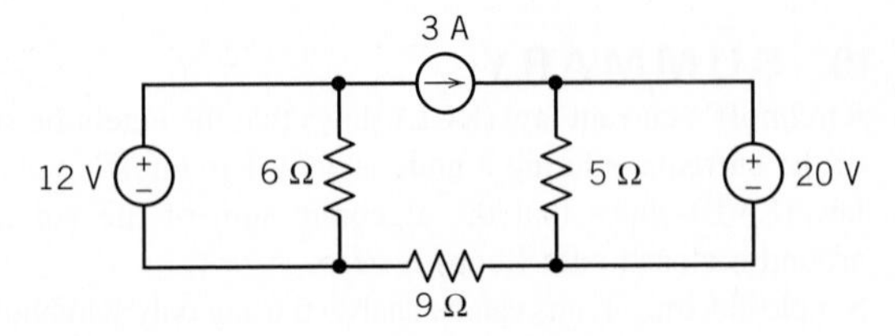

Figure P 3.2-4

P 3.2-5 Determine the power absorbed by each of the resistors in the circuit shown in Figure P 3.2-5.

Answer: The 4-Ω resistor absorbs 16 W, the 6-Ω resistor absorbs 24 W, and the 8-Ω resistor absorbs 8 W.

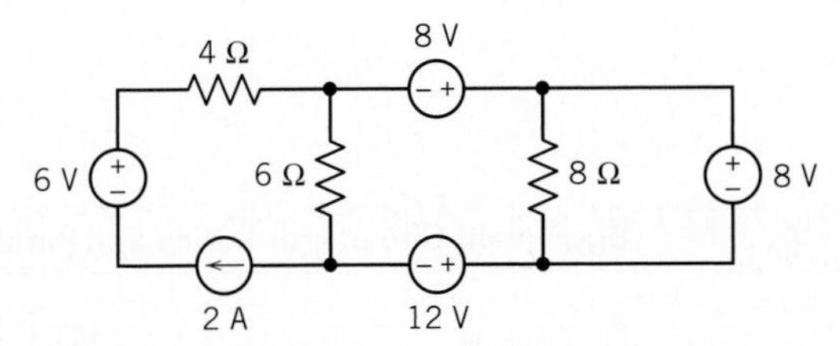

Figure P 3.2-5

P 3.2-6 Determine the power supplied by each current source in the circuit of Figure P 3.2-6.

Answer: The 4-mA current source supplies 12 mW, and the 2-mA current source supplies −14 mW.

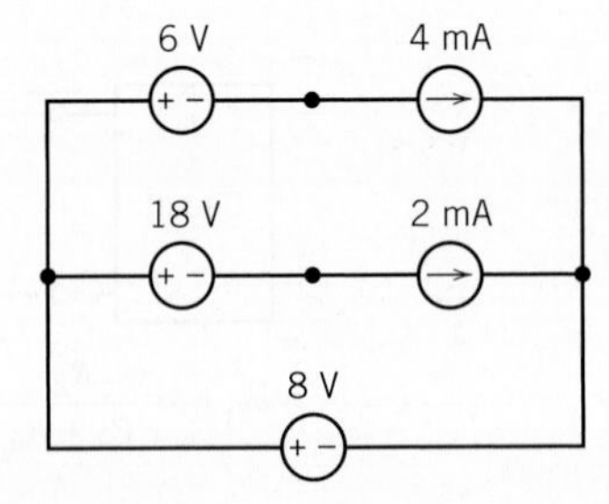

Figure P 3.2-6

P 3.2-7 Determine the power supplied by each voltage source in the circuit of Figure P 3.2-7.

Answer: The 4-V voltage source supplies 4 mW and the 6-V voltage source supplies −12 mW.

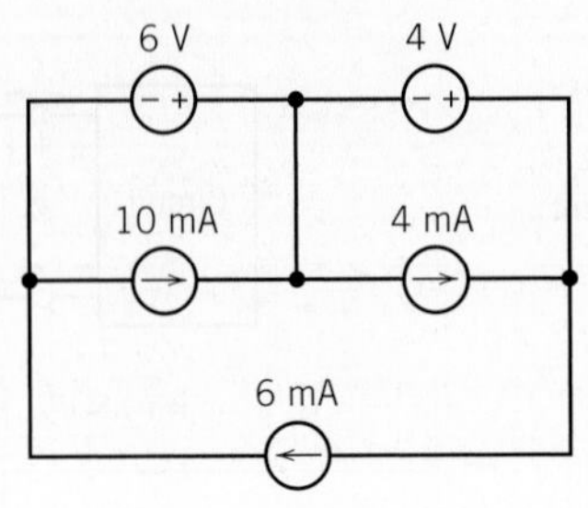

Figure P 3.2-7

P 3.2-8 What is the value of the resistance R in Figure P 3.2-8.

Hint: Assume an ideal ammeter. An ideal ammeter is equivalent to a short circuit.

Answer: $R = 4\ \Omega$

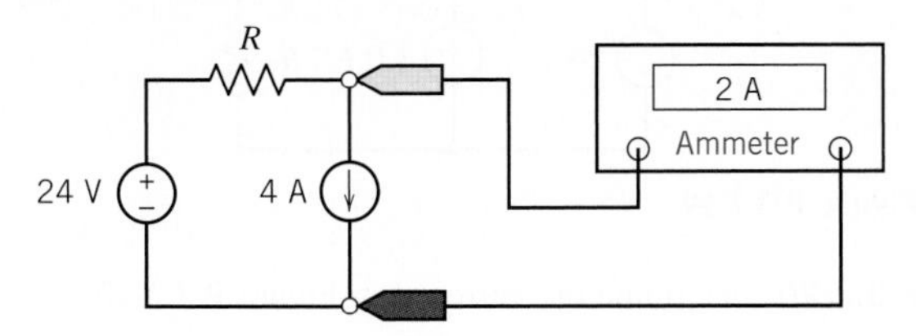

Figure P 3.2-8

P 3.2-9 The voltmeter in Figure P 3.2-9 measures the value of the voltage across the current source to be 56 V. What is the value of the resistance R?

Hint: Assume an ideal voltmeter. An ideal voltmeter is equivalent to an open circuit.

Answer: $R = 7.3\ \Omega$

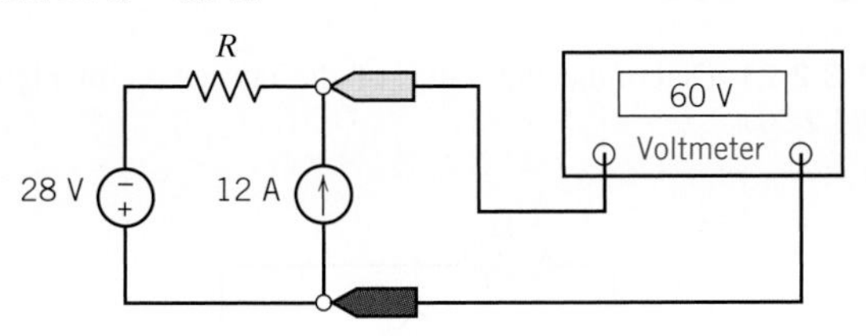

Figure P 3.2-9

P 3.2-10 Determine the values of the resistances R_1 and R_2 in Figure P 3.2-10.

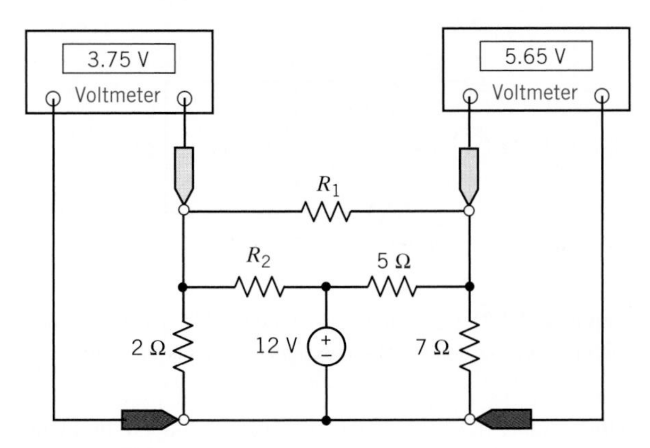

Figure P 3.2-10

P 3.2-11 The circuit shown in Figure P 3.2-11 consists of five voltage sources and four current sources. Express the power supplied by each source in terms of the voltage source voltages and the current source currents.

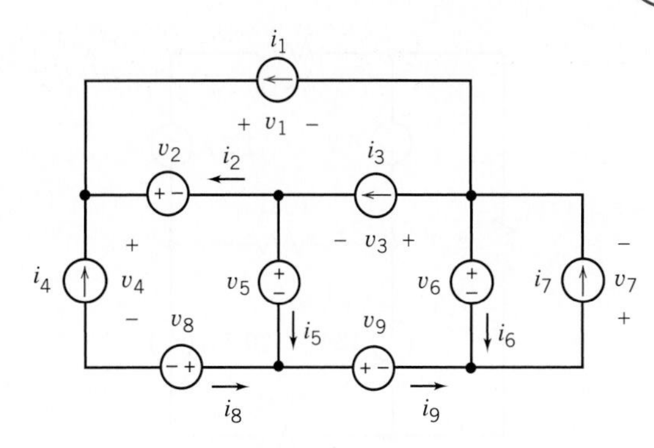

Figure P 3.2-11

P 3.2-12 Determine the power received by each of the resistors in the circuit shown in Figure P 3.2-12.

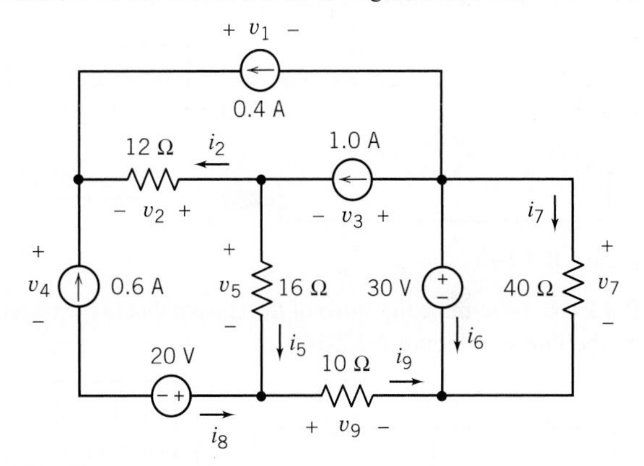

Figure P 3.2-12

P 3.2-13 Determine the voltage and current of each of the circuit elements in the circuit shown in Figure P 3.2-13.

Hint: You'll need to specify reference directions for the element voltages and currents. There is more than one way to do that, and your answers will depend on the reference directions that you choose.

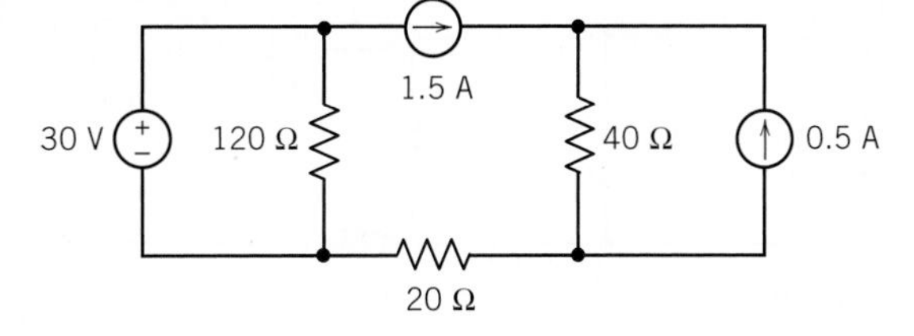

Figure P 3.2-13

P 3.2-14 Determine the voltage and current of each of the circuit elements in the circuit shown in Figure P 3.2-14.

Hint: You'll need to specify reference directions for the element voltages and currents. There is more than one way to do that, and your answers will depend on the reference directions that you choose.

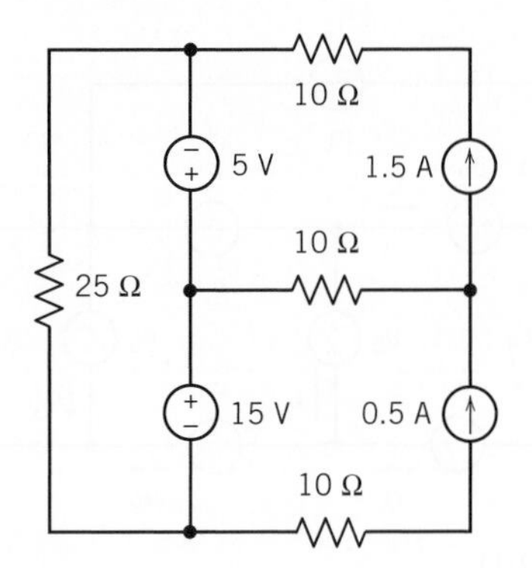

Figure P 3.2-14

P 3.2-15 Determine the value of the current that is measured by the meter in Figure P 3.2-15.

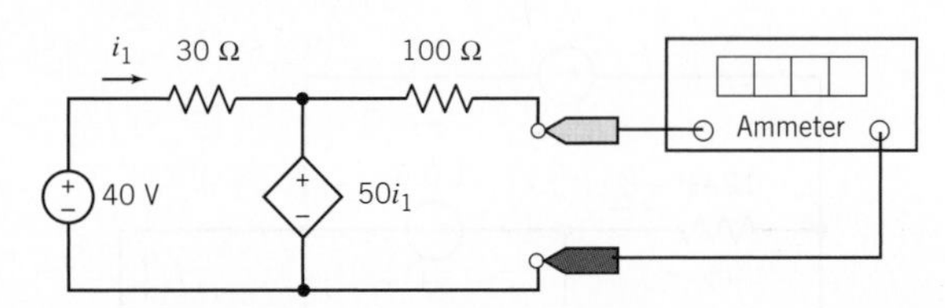

Figure P 3.2-15

P 3.2-16 Determine the value of the current that is measured by the meter in Figure P 3.2-16.

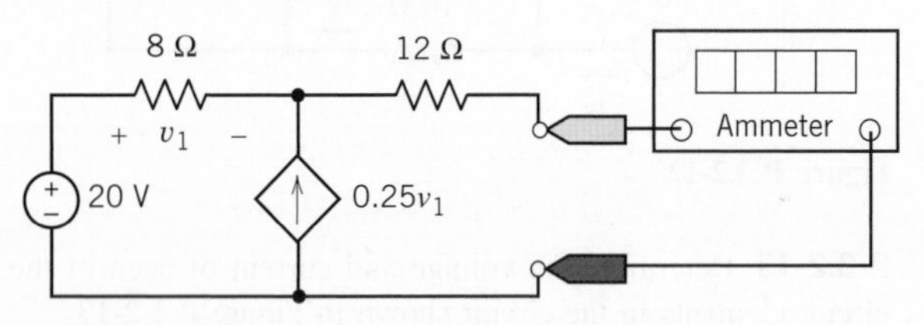

Figure P 3.2-16

P 3.2-17 Determine the value of the voltage that is measured by the meter in Figure P 3.2-17.

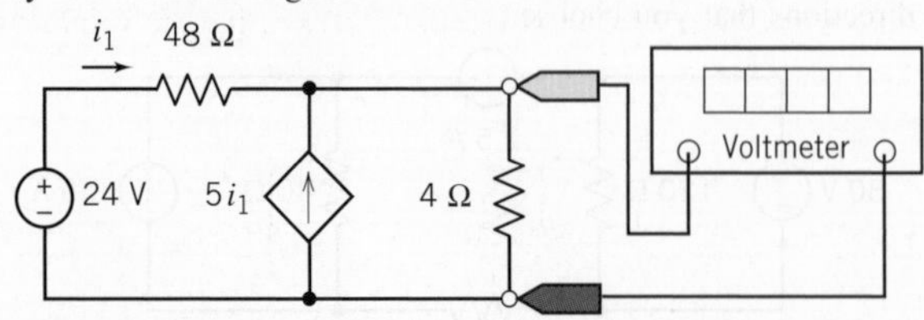

Figure P 3.2-17

P 3.2-18 Determine the value of the voltage that is measured by the meter in Figure P 3.2-18.

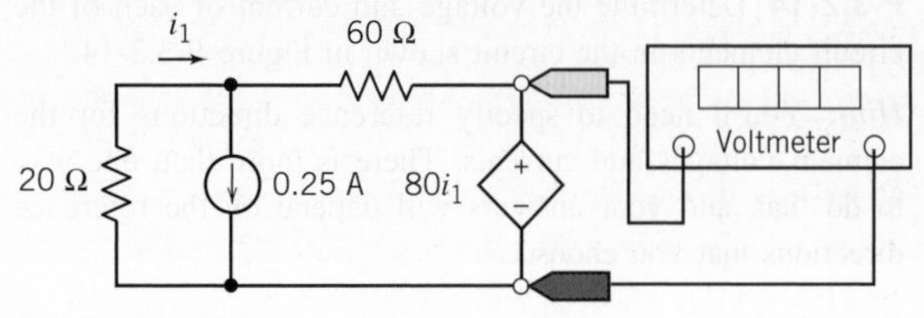

Figure P 3.2-18

P 3.2-19 The voltage source in Figure P 3.2-19 supplies 3.6 W of power. The current source supplies 5.0 W. Determine the values of the resistances, R_1 and R_2.

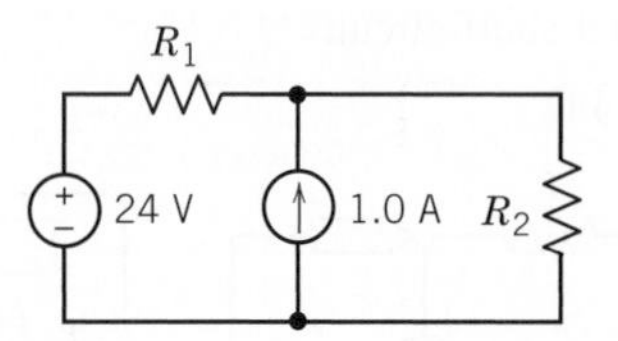

Figure P 3.2-19

P 3.2-20 Determine the current i in Figure P 3.2-20.

Answer: $i = 4$ A

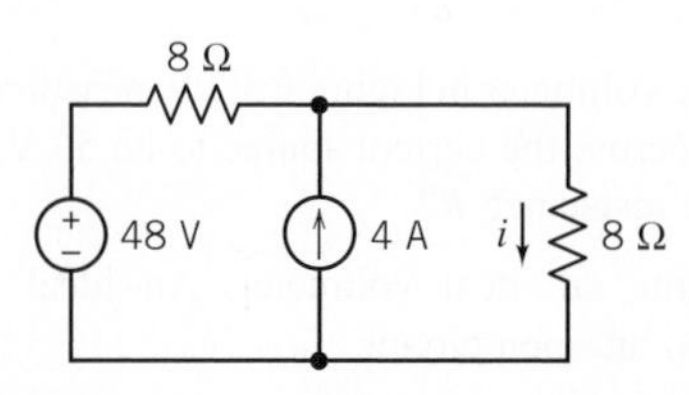

Figure P 3.2-20

P 3.2-21 Determine the value of the current i_m in Figure P 3.2-21*a*.

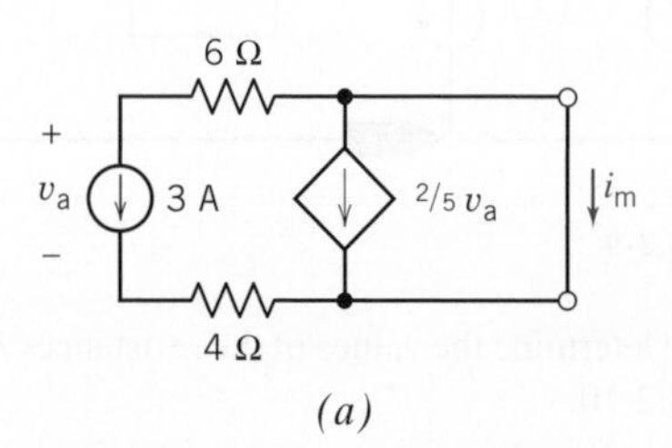

(*a*)

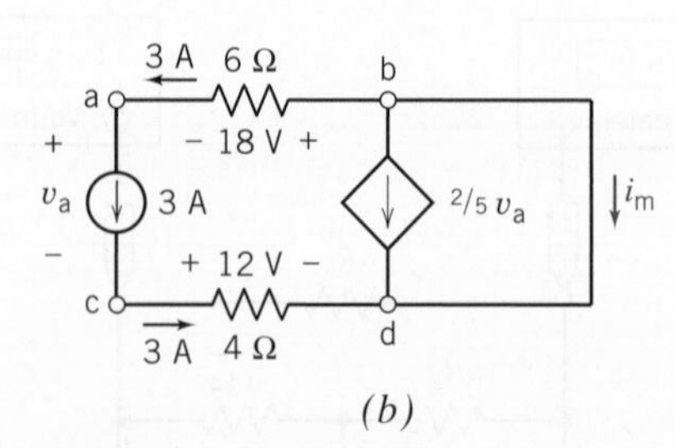

(*b*)

Figure P 3.2-21 (*a*) A circuit containing a VCCS. (*b*) The circuit after labeling the nodes and some element currents and voltages.

Hint: Apply KVL to the closed path a-b-d-c-a in Figure P 3.2-21*b* to determine v_a. Then apply KCL at node b to find i_m.

Answer: $i_m = 9$ A

P 3.2-22 Determine the value of the voltage v_m in Figure P 3.2-22*a*.

Hint: Apply KVL to the closed path a-b-d-c-a in Figure P 3.2-22*b* to determine v_a.

Answer: $v_m = 41.2$ V

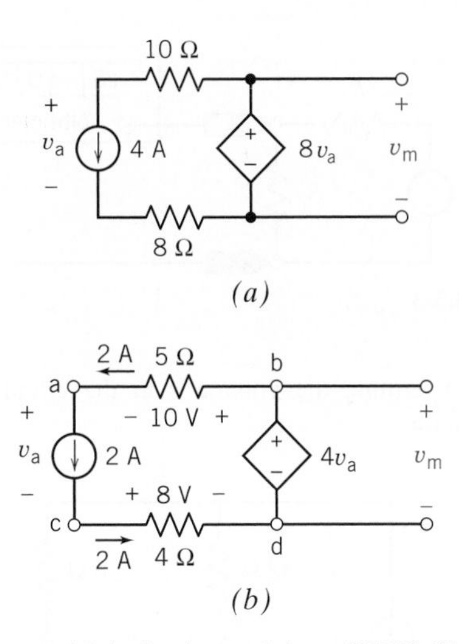

Figure P 3.2-22 (*a*) A circuit containing a VCVS. (*b*) The circuit after labeling the nodes and some element currents and voltages.

P 3.2-23 Determine the value of the voltage v_6 for the circuit shown in Figure P 3.2-23.

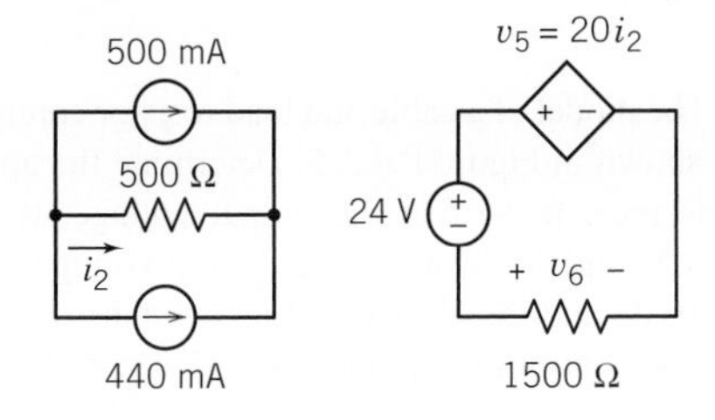

Figure P 3.2-23

P 3.2-24 Determine the value of the voltage v_6 for the circuit shown in Figure P 3.2-24.

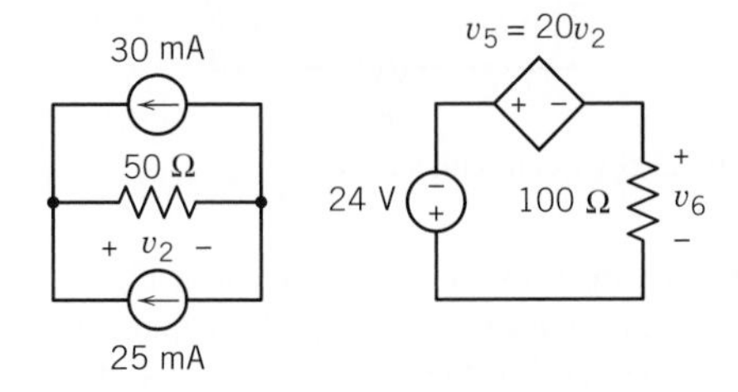

Figure P 3.2-24

P 3.2-25 Determine the value of the voltage v_5 for the circuit shown in Figure P 3.2-25.

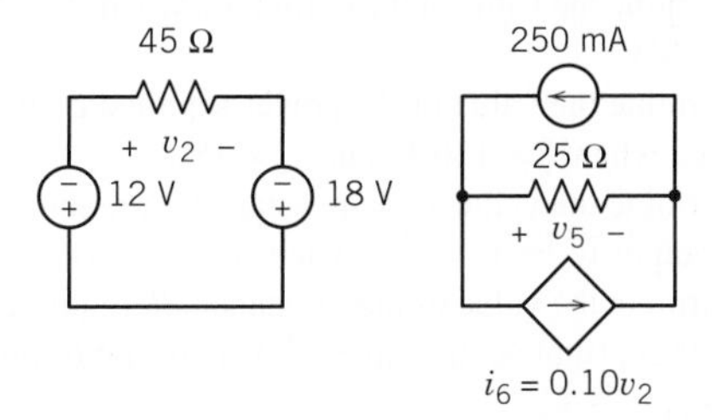

Figure P 3.2-25

P 3.2-26 Determine the value of the voltage v_5 for the circuit shown in Figure P 3.2-26.

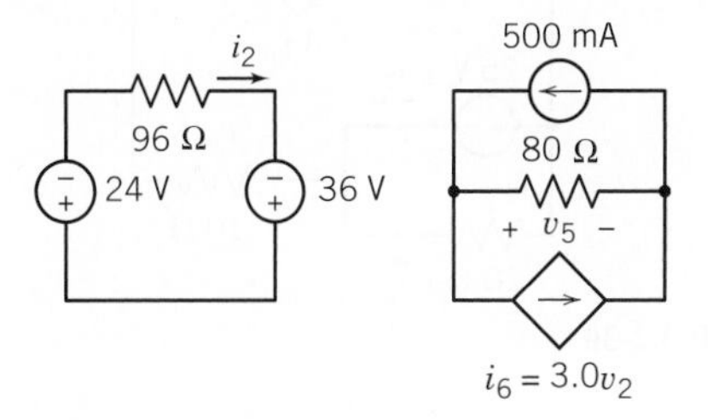

Figure P 3.2-26

P 3.2-27 Determine the value of the voltage v_6 for the circuit shown in Figure P 3.2-27.

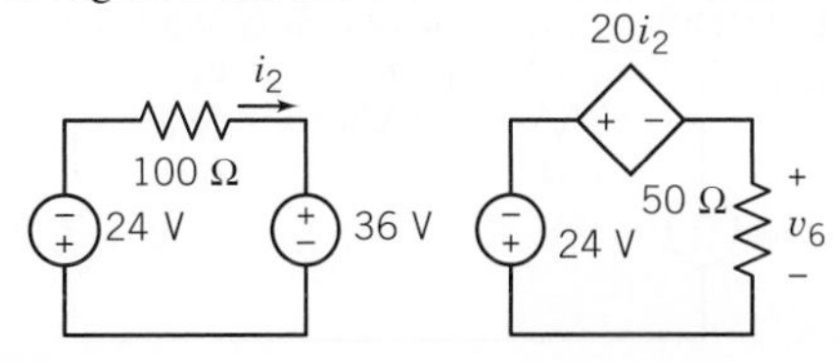

Figure P 3.2-27

P 3.2-28 Determine the value of the voltage v_5 for the circuit shown in Figure P 3.2-28.

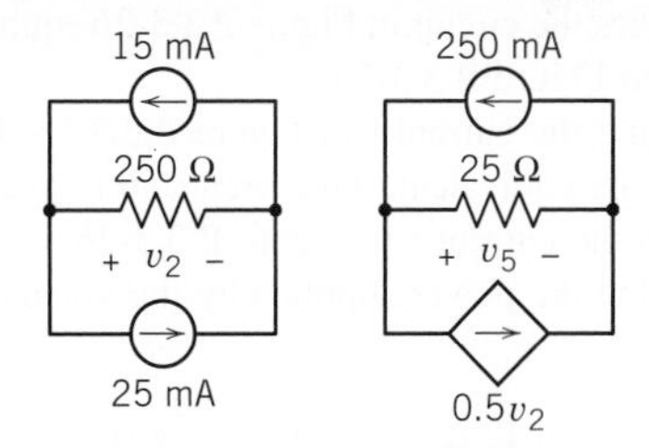

Figure P 3.2-28

P 3.2-29 The voltage source in the circuit shown in Figure P 3.2-29 supplies 2 W of power. The value of the voltage across the 25-Ω resistor is $v_2 = 4$ V. Determine the values of the resistance R_1 and of the gain, G, of the VCCS.

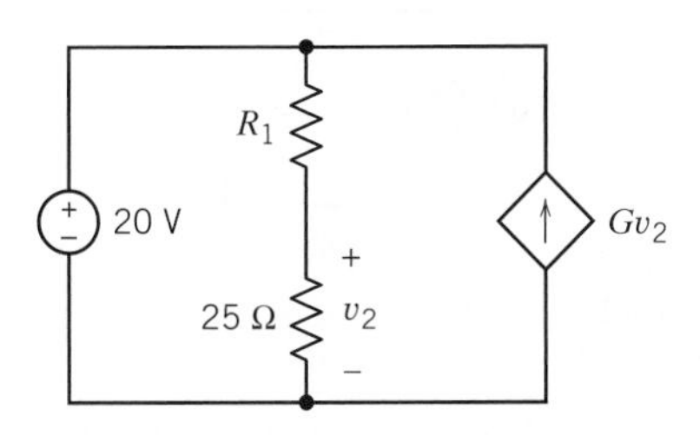

Figure P 3.2-29

P 3.2-30 Consider the circuit shown in Figure P 3.2-30. Determine the values of

(a) The current i_a in the 20-Ω resistor.
(b) The voltage v_b across the 10-Ω resistor.
(c) The current i_c in the independent voltage source.

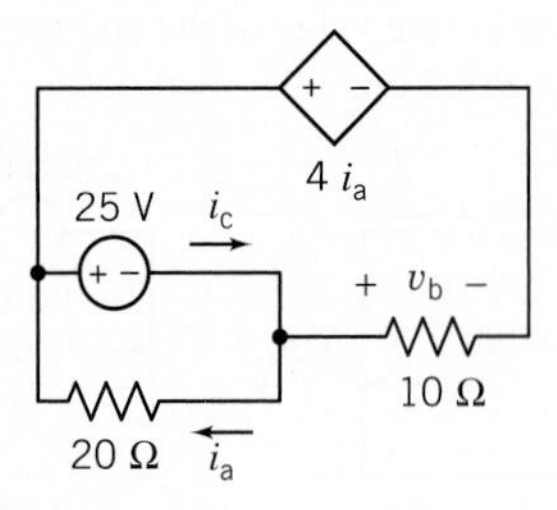

Figure P 3.2-30

Section 3.3 Series Resistors and Voltage Division

P 3.3-1 Use voltage division to determine the voltages v_1, v_2, v_3, and v_4 in the circuit shown in Figure P 3.3-1.

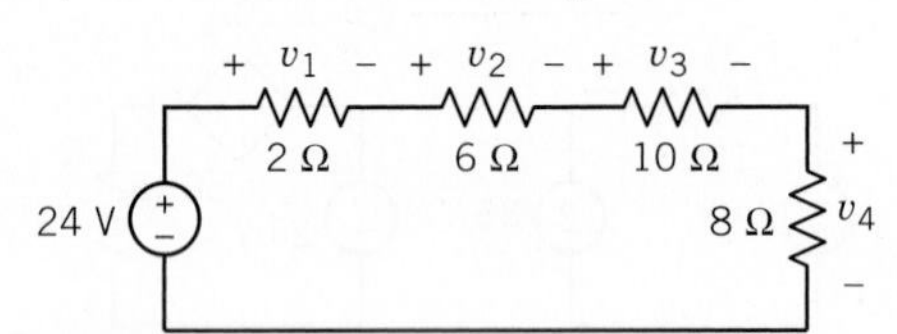

Figure P 3.3-1

P 3.3-2 Consider the circuits shown in Figure P 3.3-2.

(a) Determine the value of the resistance R in Figure P 3.3-2*b* that makes the circuit in Figure P 3.3-2*b* equivalent to the circuit in Figure P 3.3-2*a*.

(b) Determine the current i in Figure P 3.3-2*b*. Because the circuits are equivalent, the current i in Figure P 3.3-2*a* is equal to the current i in Figure P 3.3-2*b*.

(c) Determine the power supplied by the voltage source.

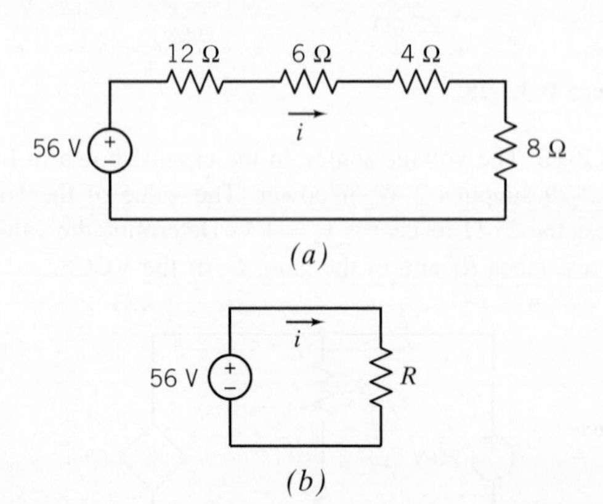

Figure P 3.3-2

P 3.3-3 The ideal voltmeter in the circuit shown in Figure P 3.3-3 measures the voltage v.

(a) Suppose $R_2 = 50\ \Omega$. Determine the value of R_1.

(b) Suppose, instead, $R_1 = 50\ \Omega$. Determine the value of R_2.

(c) Suppose, instead, that the voltage source supplies 1.2 W of power. Determine the values of both R_1 and R_2.

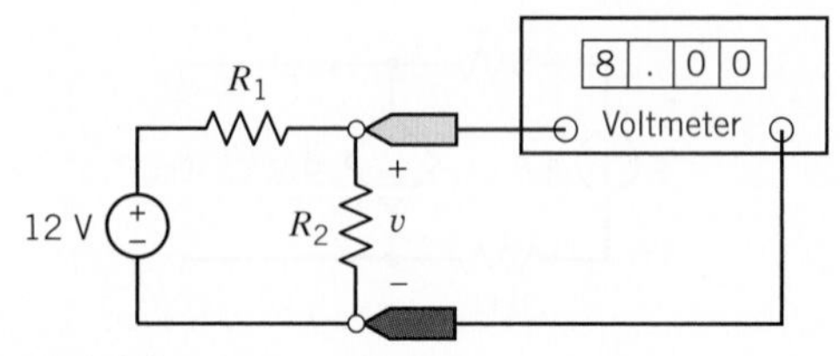

Figure P 3.3-3

P 3.3-4 Determine the voltage v in the circuit shown in Figure P 3.3-4.

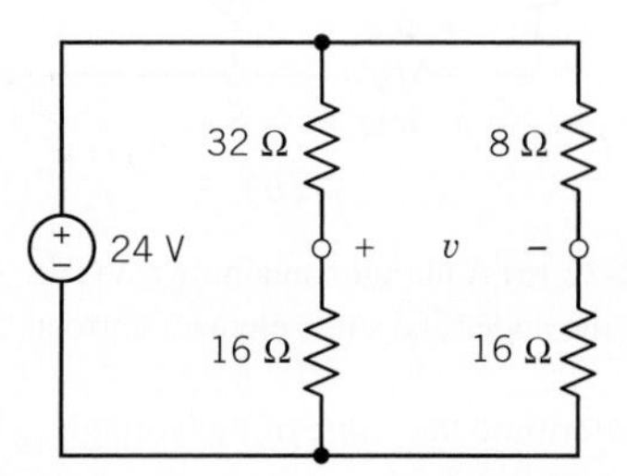

Figure P 3.3-4

P 3.3-5 The model of a cable and load resistor connected to a source is shown in Figure P 3.3-5. Determine the appropriate cable resistance, R, so that the output voltage, v_o, remains between 8 V and 14 V when the source voltage, v_s, varies between 19 V and 29 V. The cable resistance can assume integer values only in the range $20 < R < 100\ \Omega$.

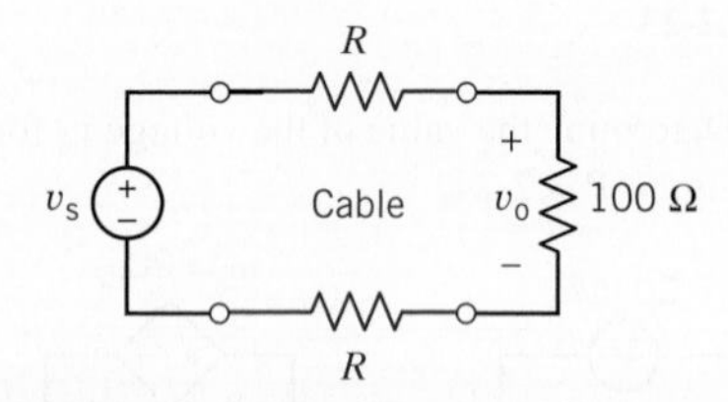

Figure P 3.3-5 Circuit with a cable.

P 3.3-6 The input to the circuit shown in Figure P 3.3-6 is the voltage of the voltage source, v_a. The output of this circuit is the voltage measured by the voltmeter, v_b. This circuit produces an output that is proportional to the input, that is,

$$v_b = k\, v_a$$

where k is the constant of proportionality.

(a) Determine the value of the output, v_b, when $R = 180\ \Omega$ and $v_a = 18$ V.

(b) Determine the value of the power supplied by the voltage source when $R = 180\ \Omega$ and $v_a = 18$ V.

(c) Determine the value of the resistance, R, required to cause the output to be $v_b = 2$ V when the input is $v_a = 18$ V.

(d) Determine the value of the resistance, R, required to cause $v_b = 0.2 v_a$ (that is, the value of the constant of proportionality is $k = \frac{2}{10}$).

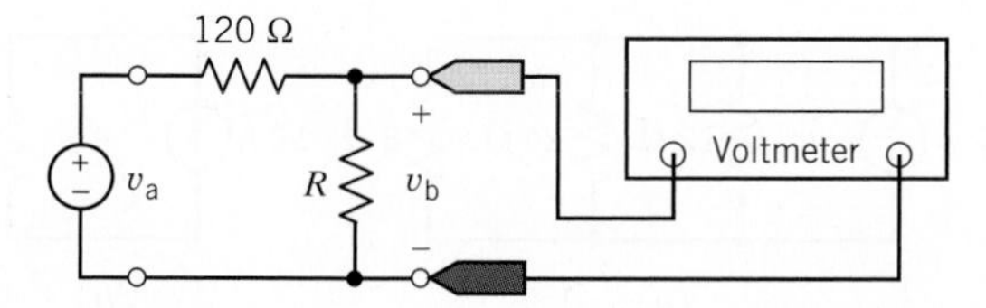

Figure P 3.3-6

P 3.3-7 Determine the value of voltage v in the circuit shown in Figure P 3.3-7.

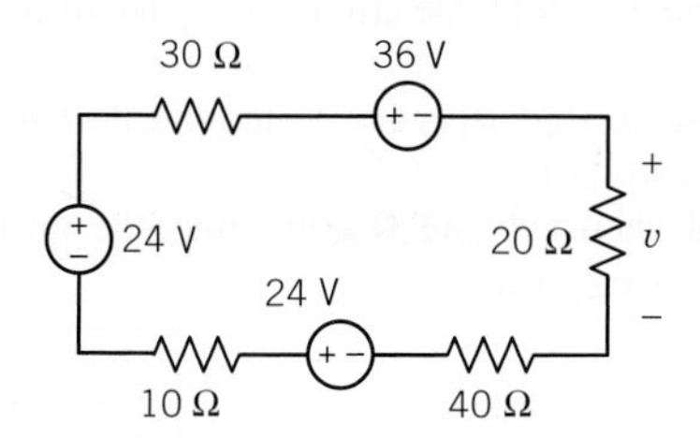

Figure P 3.3-7

P 3.3-8 Determine the power supplied by the dependent source in the circuit shown in Figure P 3.3-8.

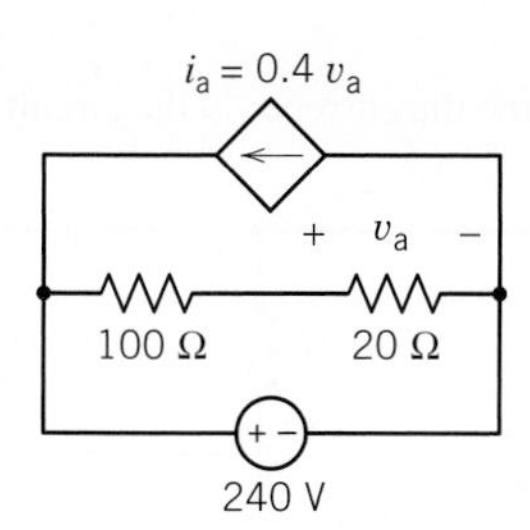

Figure P 3.3-8

P 3.3-9 A potentiometer can be used as a transducer to convert the rotational position of a dial to an electrical quantity. Figure P 3.3-9 illustrates this situation. Figure P 3.3-9*a* shows a potentiometer having resistance R_p connected to a voltage source. The potentiometer has three terminals, one at each end and one connected to a sliding contact called a wiper. A voltmeter measures the voltage between the wiper and one end of the potentiometer.

Figure P 3.3-9*b* shows the circuit after the potentiometer is replaced by a model of the potentiometer that consists of two resistors. The parameter a depends on the angle, θ, of the dial. Here $a = \frac{\theta}{360°}$, and θ is given in degrees. Also, in Figure P 3.3-9*b*, the voltmeter has been replaced by an open circuit, and the voltage measured by the voltmeter, v_m, has been labeled. The input to the circuit is the angle θ, and the output is the voltage measured by the meter, v_m.

(a) Show that the output is proportional to the input.

(b) Let $R_p = 1$ kΩ and $v_s = 24$ V. Express the output as a function of the input. What is the value of the output when $\theta = 45°$? What is the angle when $v_m = 10$ V?

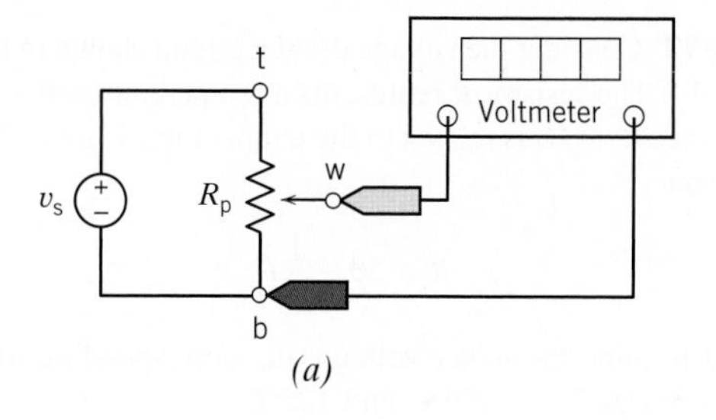

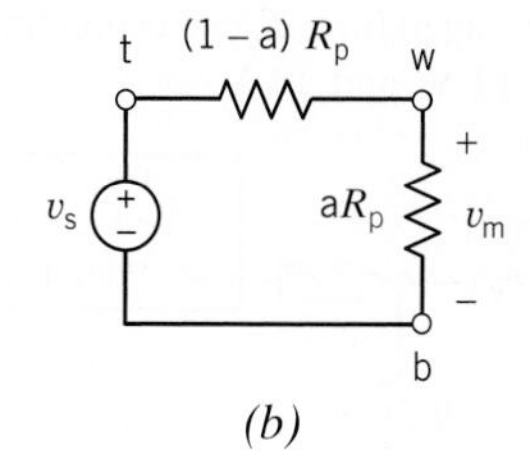

Figure P 3.3-9

P 3.3-10 Determine the value of the voltage measured by the meter in Figure P 3.3-10.

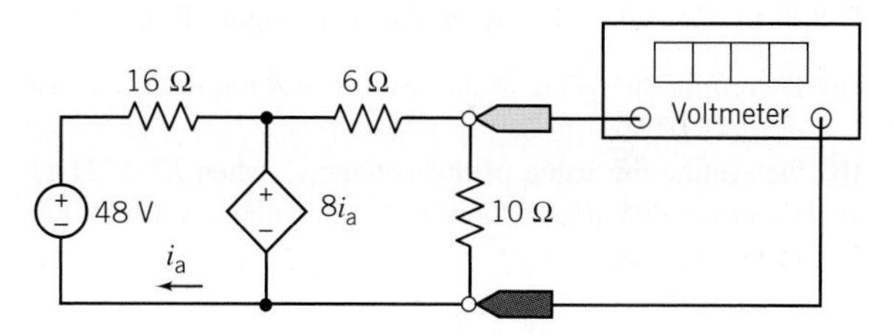

Figure P 3.3-10

P 3.3-11 For the circuit of Figure P 3.3-11, find the voltage v_3 and the current i and show that the power delivered to the three resistors is equal to that supplied by the source.

Answer: $v_3 = 6$ V, $i = 1$ A

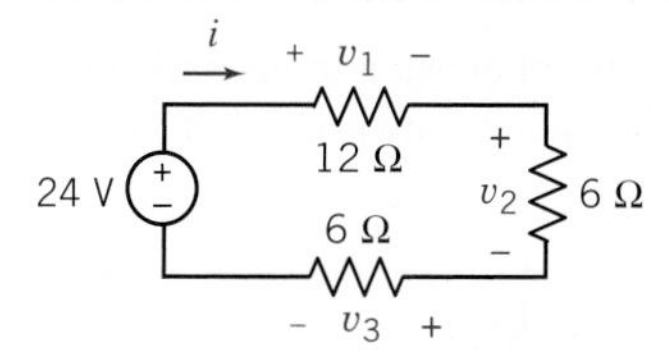

Figure P 3.3-11

P 3.3-12 Consider the voltage divider shown in Figure P 3.3-12 when $R_1 = 16$ Ω. It is desired that the output power absorbed by R_1 be 9 W. Find the voltage v_o and the required source v_s.

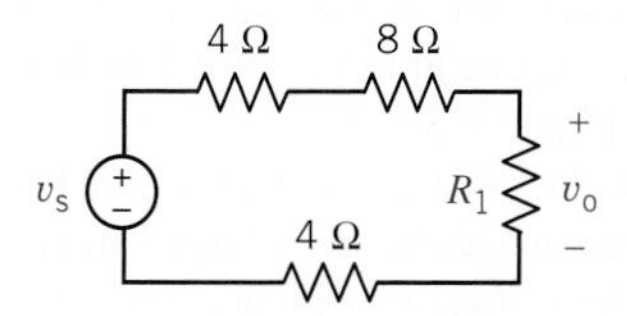

Figure P 3.3-12

P 3.3-13 Consider the voltage divider circuit shown in Figure P 3.3-13. The resistor R represents a temperature sensor. The resistance R, in Ω, is related to the temperature T, in °C, by the equation

$$R = 50 + \frac{1}{2}T$$

(a) Determine the meter voltage, v_m, corresponding to temperatures 25°C, 100°C and 125°C.
(b) Determine the temperature, T, corresponding to the meter voltages 10 V, 12 V, and 14 V.

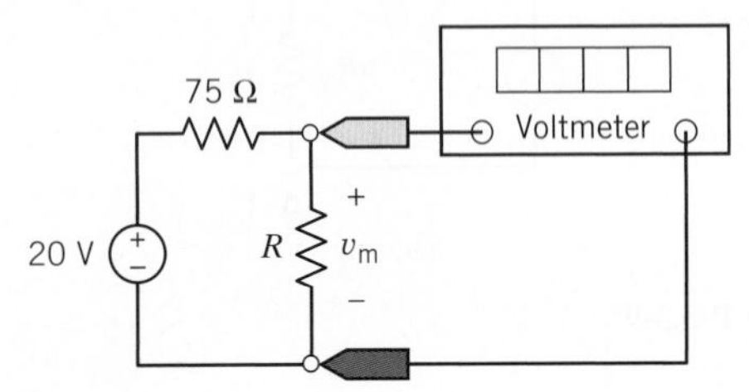

Figure P 3.3-13

P 3.3-14 Consider the circuit shown in Figure P 3.3-14.

(a) Determine the value of the resistance R required to cause $v_o = 17.07$ V.
(b) Determine the value of the voltage v_o when $R = 21\ \Omega$.
(c) Determine the power supplied by the voltage source when $v_o = 14.22$ V.

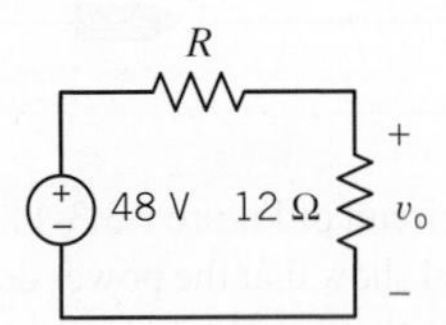

Figure P 3.3-14

Section 3.4 Parallel Resistors and Current Division

P 3.4-1 Use current division to determine the currents i_1, i_2, i_3, and i_4 in the circuit shown in Figure P 3.4-1.

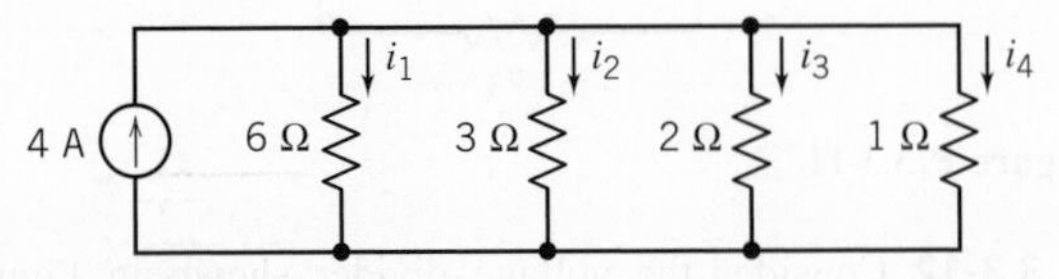

Figure P 3.4-1

P 3.4-2 Consider the circuits shown in Figure P 3.4-2.

(a) Determine the value of the resistance R in Figure P 3.4-2b that makes the circuit in Figure P 3.4-2b equivalent to the circuit in Figure P 3.4-2a.
(b) Determine the voltage v in Figure P 3.4-2b. Because the circuits are equivalent, the voltage v in Figure P 3.4-2a is equal to the voltage v in Figure P 3.4-2b.
(c) Determine the power supplied by the current source.

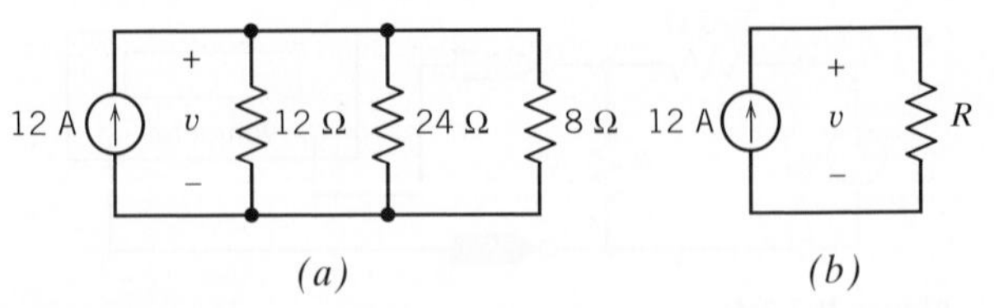

Figure P 3.4-2

P 3.4-3 The ideal voltmeter in the circuit shown in Figure P 3.4-3 measures the voltage v.

(a) Suppose $R_2 = 6\ \Omega$. Determine the value of R_1 and of the current i.
(b) Suppose, instead, $R_1 = 6\ \Omega$. Determine the value of R_2 and of the current i.
(c) Instead, choose R_1 and R_2 to minimize the power absorbed by any one resistor.

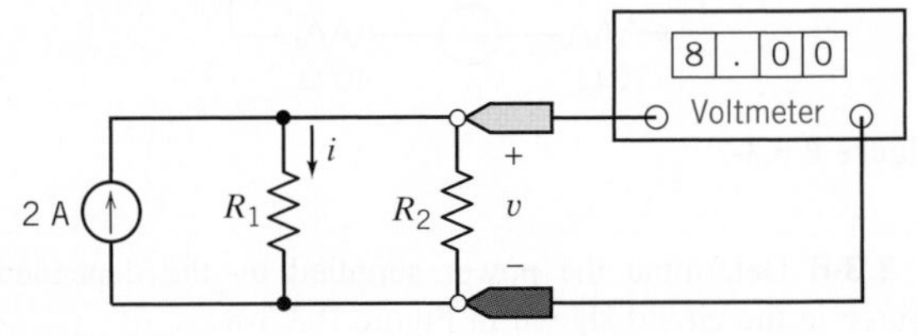

Figure P 3.4-3

P 3.4-4 Determine the current i in the circuit shown in Figure P 3.4-4.

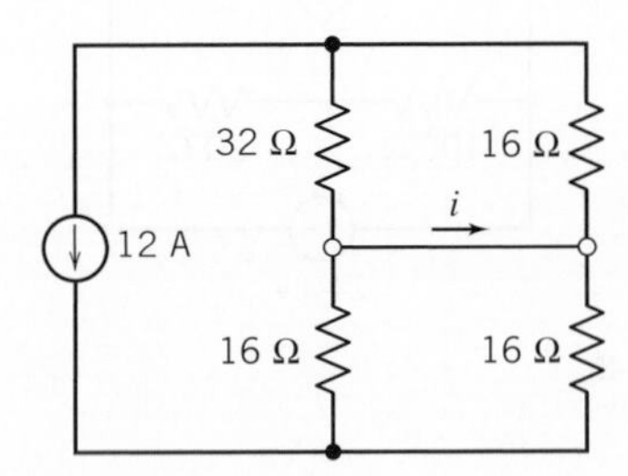

Figure P 3.4-4

P 3.4-5 Consider the circuit shown in Figure P 3.4-5 when $4\ \Omega \le R_1 \le 6\ \Omega$ and $R_2 = 12\ \Omega$. Select the source i_s so that v_o remains between 8 V and 14 V.

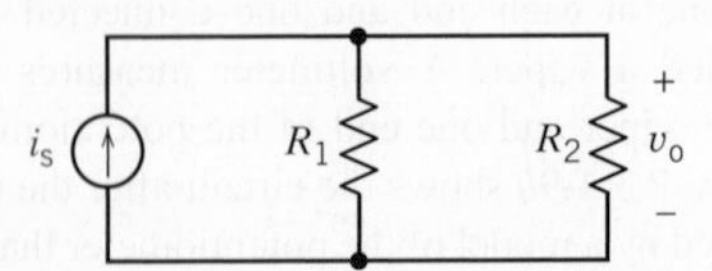

Figure P 3.4-5

P 3.4-6 The input to the circuit shown in Figure P 3.4-6 is the current of the current source, i_a. The output of this circuit is the current measured by the ammeter, i_b. This circuit produces an output that is proportional to the input, that is,

$$i_b = k\ i_a$$

where k is the constant of proportionality.
(a) Determine the value of the output, i_b, when $R = 24\ \Omega$ and $i_a = 2.1$ A.

(b) Determine the value of the resistance, R, required to cause the output to be $i_b = 1.5$ A when the input is $i_a = 2$ A.

(c) Determine the value of the resistance, R, required to cause $i_b = 0.4\, i_a$ (that is, the value of the constant of proportionality is $k = \frac{4}{10}$).

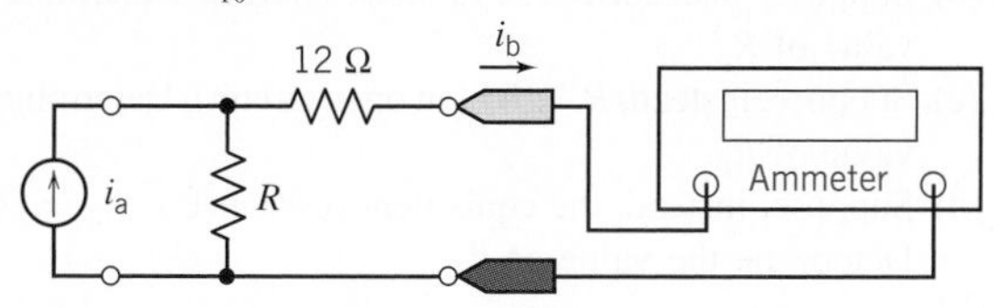

Figure P 3.4-6

P 3.4-7 Figure P 3.4-7 shows a transistor amplifier. The values of R_1 and R_2 are to be selected. Resistances R_1 and R_2 are used to bias the transistor, that is, to create useful operating conditions. In this problem, we want to select R_1 and R_2 so that $v_b = 5$ V. We expect the value of i_b to be approximately 10 μA. When $i_1 \leq 10 i_b$, it is customary to treat i_b as negligible, that is, to assume $i_b = 0$. In that case, R_1 and R_2 comprise a voltage divider.

(a) Select values for R_1 and R_2 so that $v_b = 5$ V, and the total power absorbed by R_1 and R_2 is no more than 5 mW.

(b) An inferior transistor could cause i_b to be larger than expected. Using the values of R_1 and R_2 from part (a), determine the value of v_b that would result from $i_b = 15$ μA.

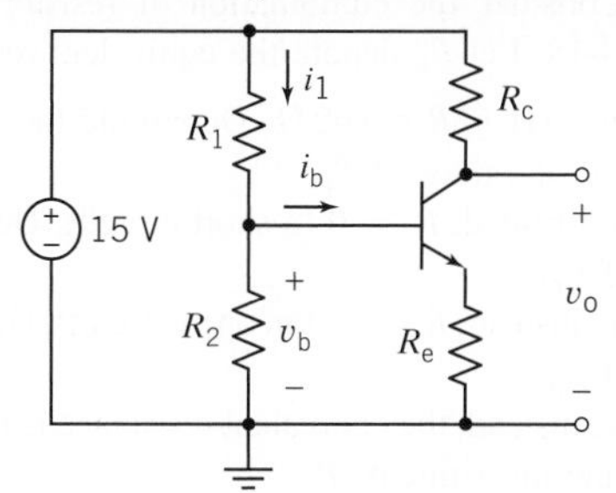

Figure P 3.4-7

P 3.4-8 Determine the value of the current i in the circuit shown in Figure P 3.4-8.

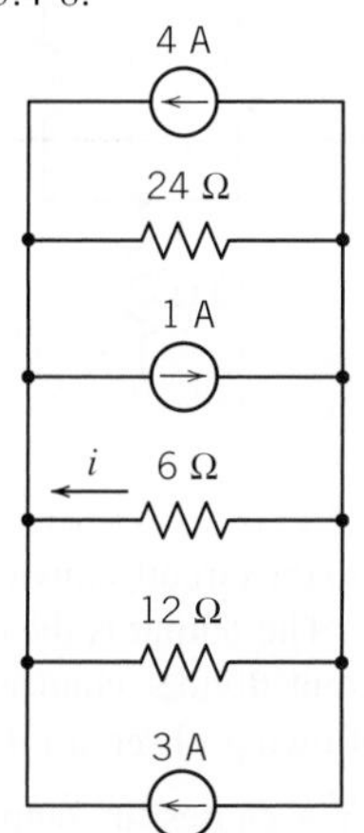

Figure P 3.4-8

P 3.4-9 Determine the value of the voltage v in Figure P 3.4-9.

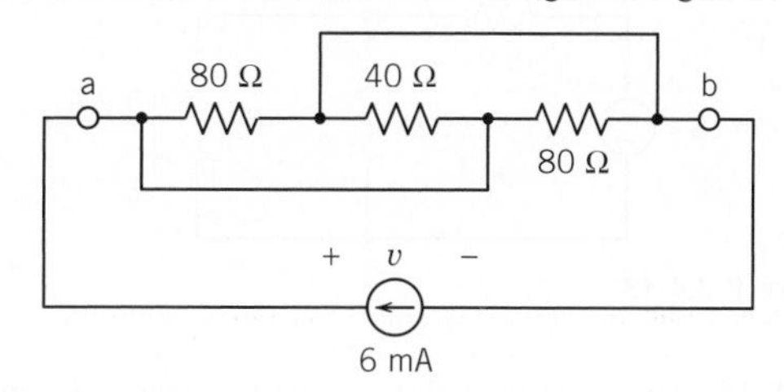

Figure P 3.4-9

P 3.4-10 A solar photovoltaic panel may be represented by the circuit model shown in Figure P 3.4-10, where R_L is the load resistor. Determine the values of the resistances R_1 and R_L.

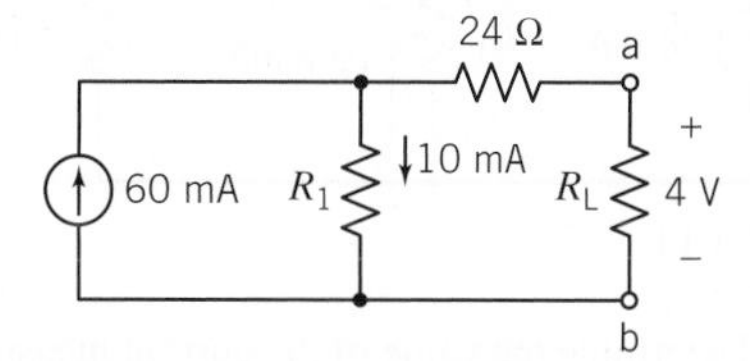

Figure P 3.4-10

P 3.4-11 Determine the power supplied by the dependent source in Figure P 3.4-11.

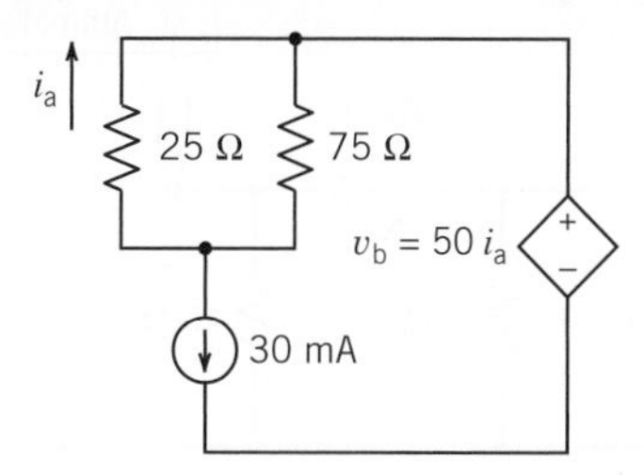

Figure P 3.4-11

P 3.4-12 The voltmeter in Figure P 3.4-12 measures the value of the voltage v_m.

(a) Determine the value of the resistance R.

(b) Determine the value of the power supplied by the current source.

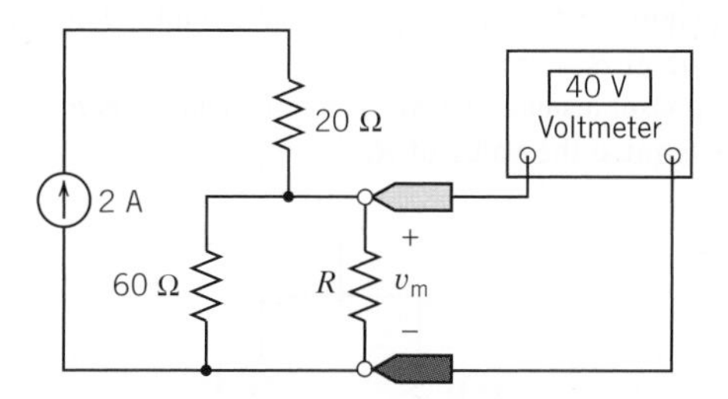

Figure P 3.4-12

P 3.4-13 Determine the values of the resistances R_1 and R_2 for the circuit shown in Figure P 3.4-13.

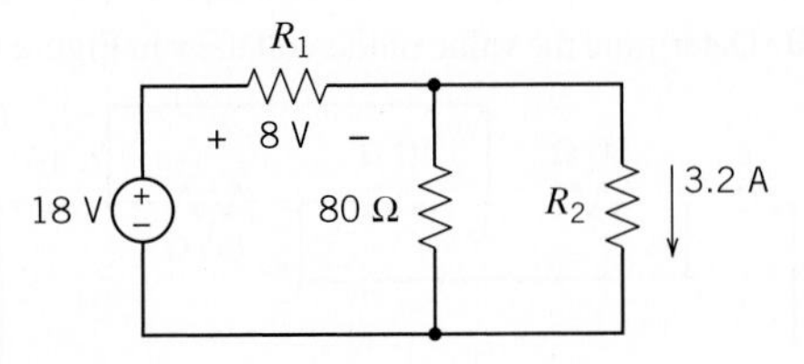

Figure P 3.4-13

P 3.4-14 Determine the values of the resistances R_1 and R_2 for the circuit shown in Figure P 3.4-14.

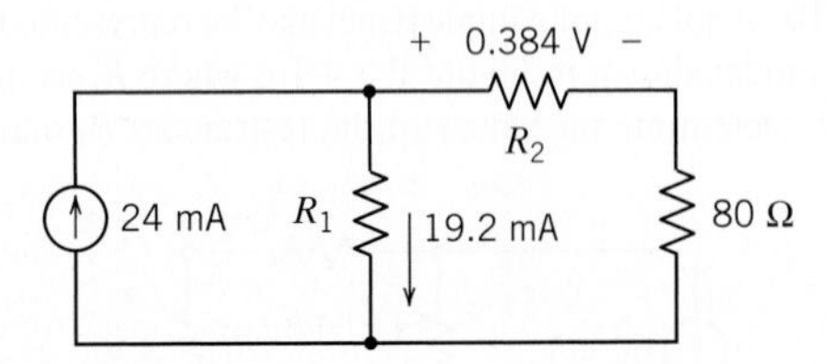

Figure P 3.4-14

P 3.4-15 Determine the value of the current measured by the meter in Figure P 3.4-15.

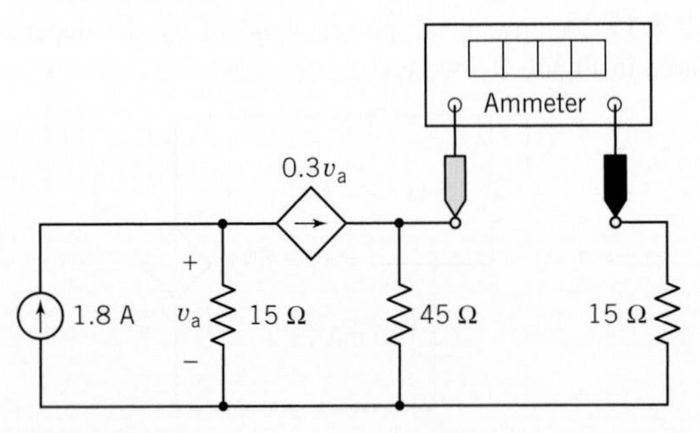

Figure P 3.4-15

P 3.4-16 Consider the combination of resistors shown in Figure P 3.4-16. Let R_p denote the equivalent resistance.

(a) Suppose $20\ \Omega \leq R \leq 320\ \Omega$. Determine the corresponding range of values of R_p.
(b) Suppose, instead, $R = 0$ (a short circuit). Determine the value of R_p.
(c) Suppose, instead, $R = \infty$ (an open circuit). Determine the value of R_p.
(d) Suppose, instead, the equivalent resistance is $R_p = 40\ \Omega$. Determine the value of R.

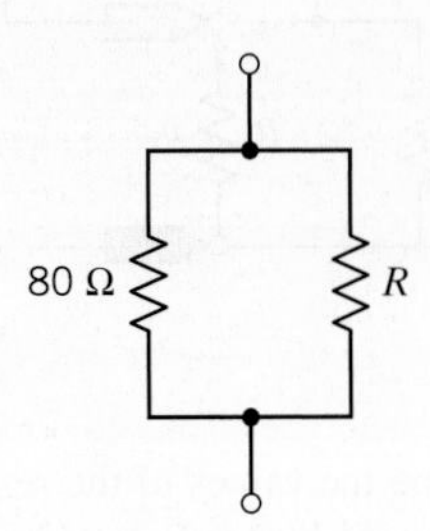

Figure P 3.4-16

P 3.4-17 Consider the combination of resistors shown in Figure P 3.4-17. Let R_p denote the equivalent resistance.

(a) Suppose $40\ \Omega \leq R \leq 400\ \Omega$. Determine the corresponding range of values of R_p.
(b) Suppose, instead, $R = 0$ (a short circuit). Determine the value of R_p.
(c) Suppose, instead, $R = \infty$ (an open circuit). Determine the value of R_p.
(d) Suppose, instead, the equivalent resistance is $R_p = 80\ \Omega$. Determine the value of R.

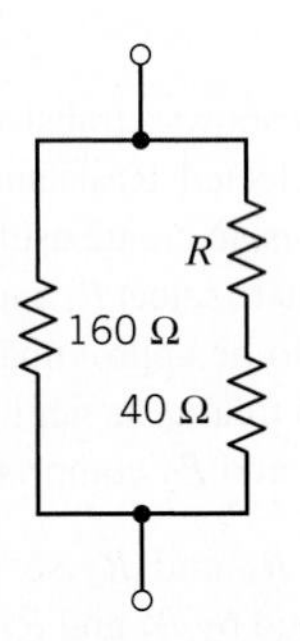

Figure P 3.4-17

P 3.4-18 Consider the combination of resistors shown in Figure P 3.4-18. Let R_p denote the equivalent resistance.

(a) Suppose $50\ \Omega \leq R \leq 800\ \Omega$. Determine the corresponding range of values of R_p.
(b) Suppose, instead, $R = 0$ (a short circuit). Determine the value of R_p.
(c) Suppose, instead, $R = \infty$ (an open circuit). Determine the value of R_p.
(d) Suppose, instead, the equivalent resistance is $R_p = 150\ \Omega$. Determine the value of R.

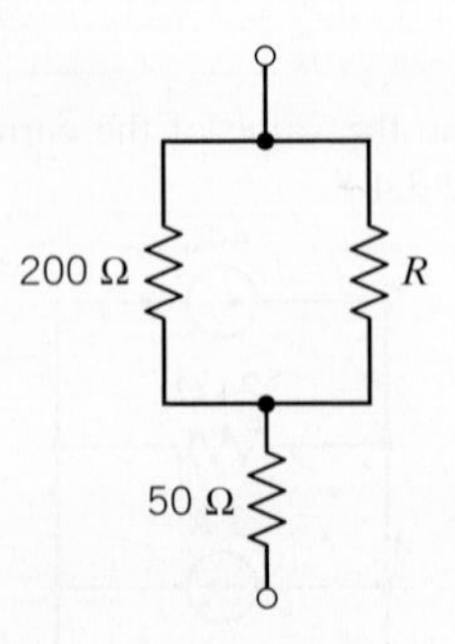

Figure P 3.4-18

P 3.4-19 The input to the circuit shown in Figure P 3.4-19 is the source current, i_s. The output is the current measured by the meter, i_o. A current divider connects the source to the meter. Given the following observations:

(a) The input $i_s = 5$ A causes the output to be $i_o = 2$ A.
(b) When $i_s = 2$ A, the source supplies 48 W.

Determine the values of the resistances R_1 and R_2.

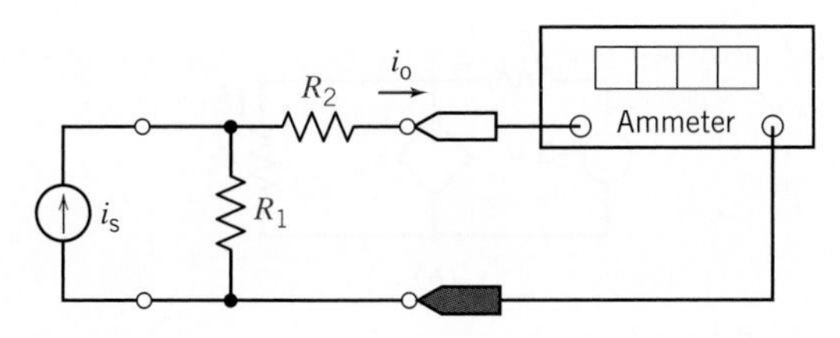

Figure P 3.4-19

Section 3.5 Series Voltage Sources and Parallel Current Sources

P 3.5-1 Determine the power supplied by each source in the circuit shown in Figure P 3.5-1.

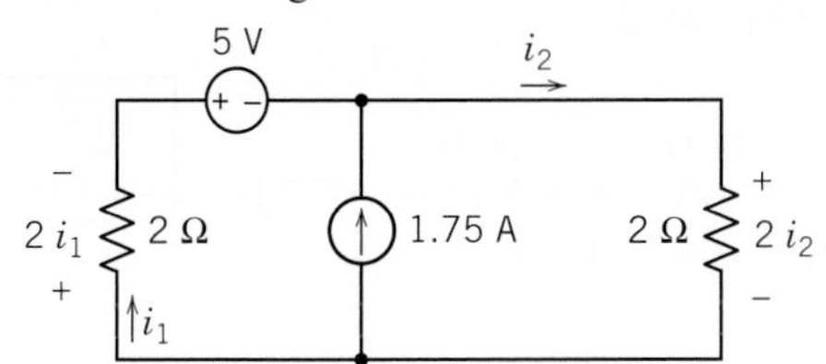

Figure P 3.5-1

P 3.5-2 Determine the power supplied by each source in the circuit shown in Figure P 3.5-2.

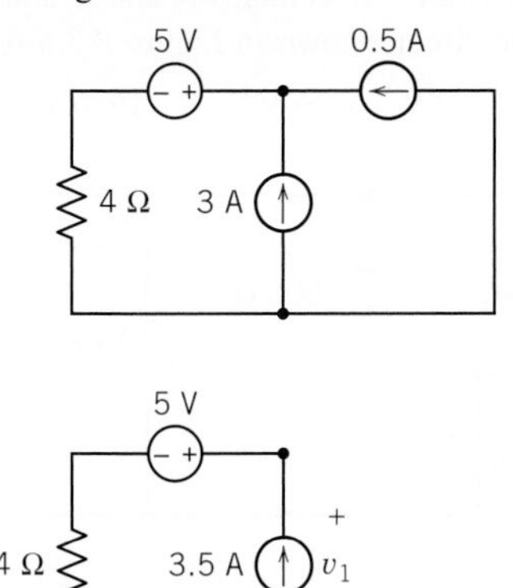

Figure P 3.5-2

P 3.5-3 Determine the power received by each resistor in the circuit shown in Figure P 3.5-3.

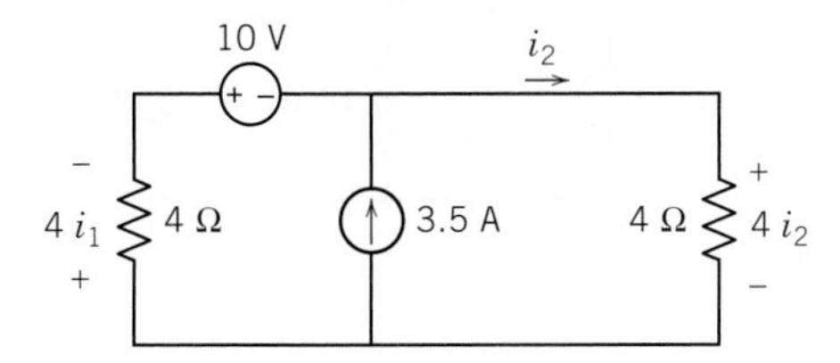

Figure P 3.5-3

Section 3.6 Circuit Analysis

P 3.6-1 The circuit shown in Figure P 3.6-1*a* has been divided into two parts. In Figure P 3.6-1*b*, the right-hand part has been replaced with an equivalent circuit. The left-hand part of the circuit has not been changed.

(a) Determine the value of the resistance R in Figure P 3.6-1*b* that makes the circuit in Figure P 3.6-1*b* equivalent to the circuit in Figure P 3.6-1*a*.

(b) Find the current i and the voltage v shown in Figure P 3.6-1*b*. Because of the equivalence, the current i and the voltage v shown in Figure P 3.6-1*a* are equal to the current i and the voltage v shown in Figure P 3.6-1*b*.

(c) Find the current i_2, shown in Figure P 3.6-1*a*, using current division.

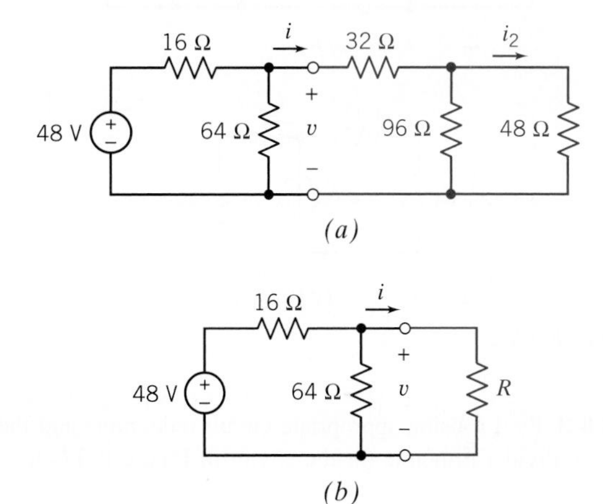

Figure P 3.6-1

P 3.6-2 The circuit shown in Figure P 3.6-2*a* has been divided into three parts. In Figure P 3.6-2*b*, the rightmost part has been replaced with an equivalent circuit. The rest of the circuit has not been changed. The circuit is simplified further in Figure 3.6-2*c*. Now the middle and rightmost parts have been replaced by a single equivalent resistance. The leftmost part of the circuit is still unchanged.

(a) Determine the value of the resistance R_1 in Figure P 3.6-2*b* that makes the circuit in Figure P 3.6-2*b* equivalent to the circuit in Figure P 3.6-2*a*.

(b) Determine the value of the resistance R_2 in Figure P 3.6-2*c* that makes the circuit in Figure P 3.6-2*c* equivalent to the circuit in Figure P 3.6-2*b*.

(c) Find the current i_1 and the voltage v_1 shown in Figure P 3.6-2*c*. Because of the equivalence, the current i_1 and the voltage v_1 shown in Figure P 3.6-2*b* are equal to the current i_1 and the voltage v_1 shown in Figure P 3.6-2*c*.

Hint: $24 = 6(i_1 - 2) + i_1 R_2$

(d) Find the current i_2 and the voltage v_2 shown in Figure P 3.6-2*b*. Because of the equivalence, the current i_2 and the voltage v_2 shown in Figure P 3.6-2*a* are equal to the current i_2 and the voltage v_2 shown in Figure P 3.6-2*b*.

Hint: Use current division to calculate i_2 from i_1.

(e) Determine the power absorbed by the 3-Ω resistance shown at the right of Figure P 3.6-2*a*.

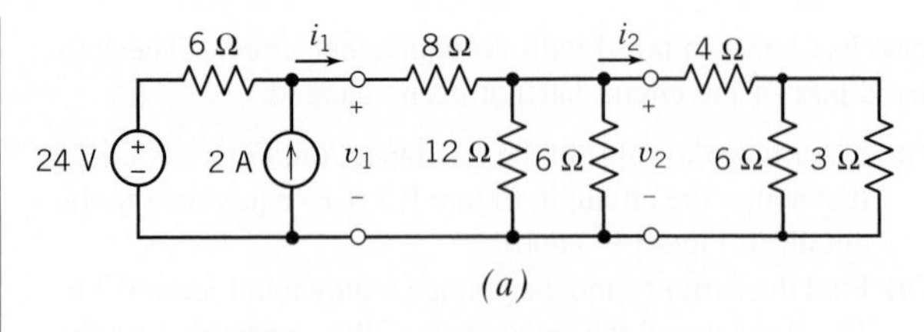

(a)

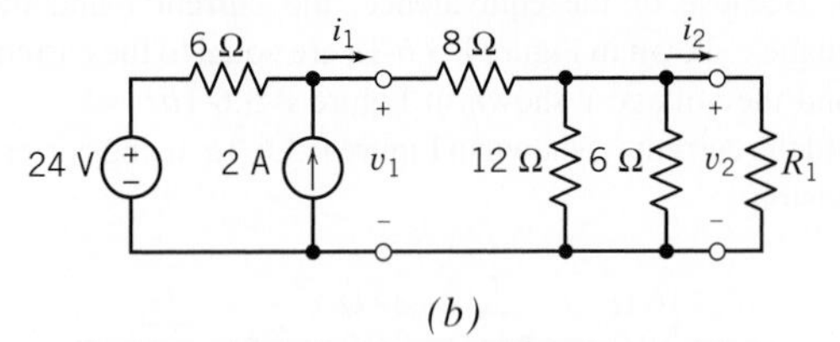

(b)

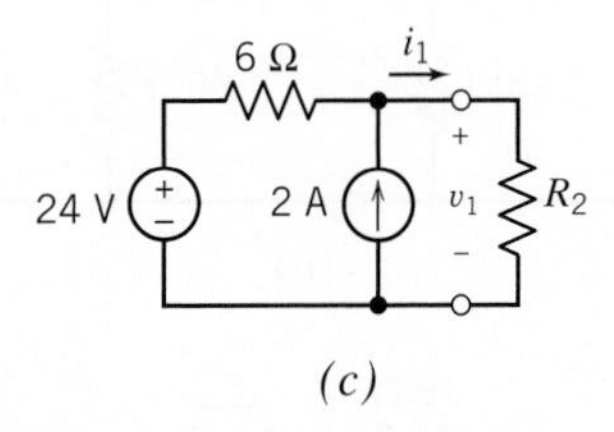

(c)

Figure P 3.6-2

P 3.6-3 Find i, using appropriate circuit reductions and the current divider principle for the circuit of Figure P 3.6-3.

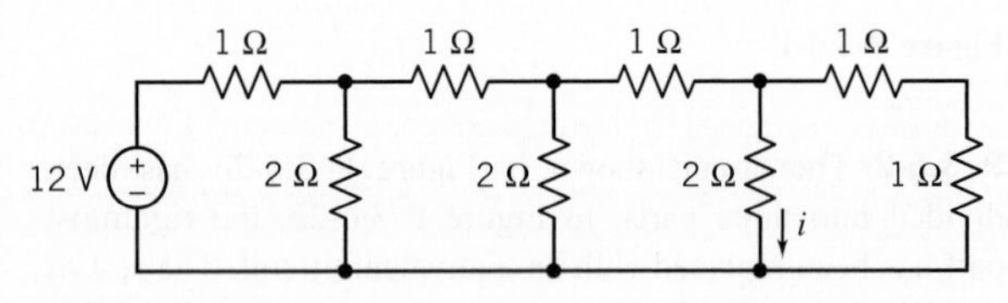

Figure P 3.6-3

P 3.6-4

(a) Determine values of R_1 and R_2 in Figure P 3.6-4*b* that make the circuit in Figure P 3.6-4*b* equivalent to the circuit in Figure P 3.6-4*a*.

(b) Analyze the circuit in Figure P 3.6-4*b* to determine the values of the currents i_a and i_b.

(c) Because the circuits are equivalent, the currents i_a and i_b shown in Figure P 3.6-4*b* are equal to the currents i_a and i_b shown in Figure P 3.6-4*a*. Use this fact to determine values of the voltage v_1 and current i_2 shown in Figure P 3.6-4*a*.

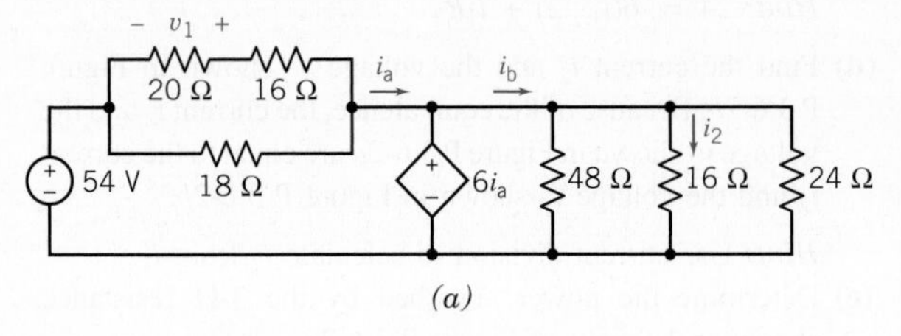

(a)

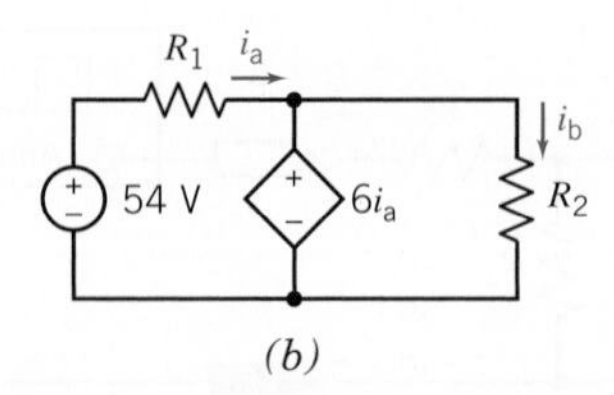

(b)

Figure P 3.6-4

P 3.6-5 The voltmeter in the circuit shown in Figure P 3.6-5 shows that the voltage across the 30-Ω resistor is 6 volts. Determine the value of the resistance R_1.

Hint: Use the voltage division twice.

Answer: $R_1 = 80\ \Omega$

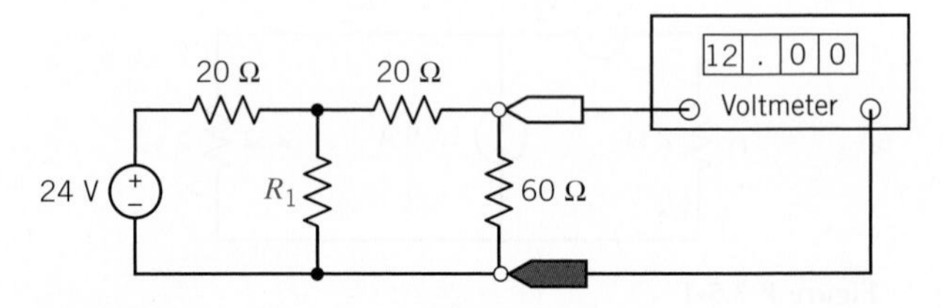

Figure P 3.6-5

P 3.6-6 Determine the voltages v_a and v_c and the currents i_b and i_d for the circuit shown in Figure P 3.6-6.

Answer: $v_a = -2$ V, $v_c = 6$ V, $i_b = -16$ mA, and $i_d = 2$ mA

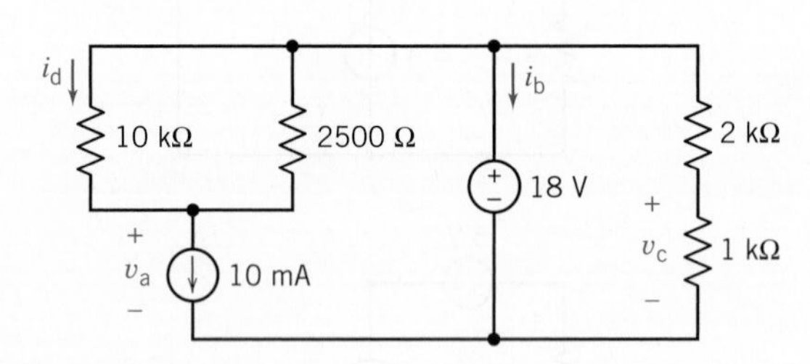

Figure P 3.6-6

P 3.6-7 Determine the value of the resistance R in Figure P 3.6-7.

Answer: $R = 56$ kΩ

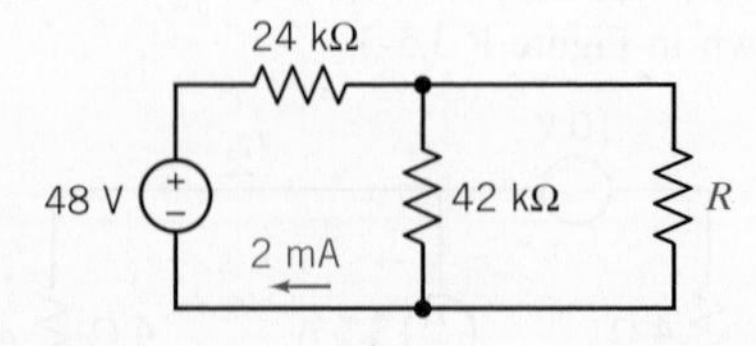

Figure P 3.6-7

P 3.6-8 Most of us are familiar with the effects of a mild electric shock. The effects of a severe shock can be devastating and often fatal. Shock results when current is passed through the body. A person can be modeled as a network of resistances. Consider the model circuit shown in Figure P 3.6-8. Determine the voltage developed across the heart and the current flowing through the heart of the person when he or she firmly grasps one end of a voltage source whose other end is connected to the floor. The heart is represented by R_h. The floor has resistance to current flow equal to R_f, and the person is standing barefoot on the floor. This type of accident might occur at a swimming pool or boat dock. The upper-body resistance R_u and lower-body resistance R_L vary from person to person.

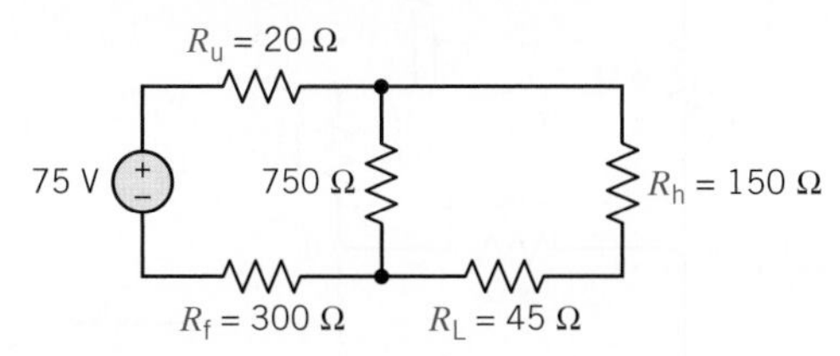

Figure P 3.6-8

P 3.6-9 Determine the value of the current i in Figure 3.6-9.

Answer: $i = 0.5$ mA

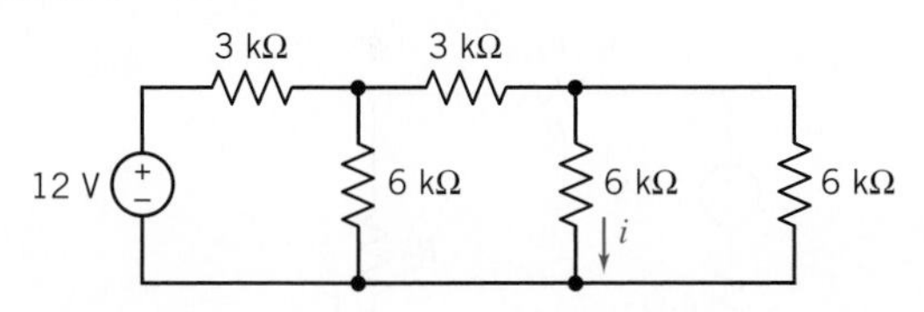

Figure P 3.6-9

P 3.6-10 Determine the values of i_a, i_b, and v_c in Figure P 3.6-10.

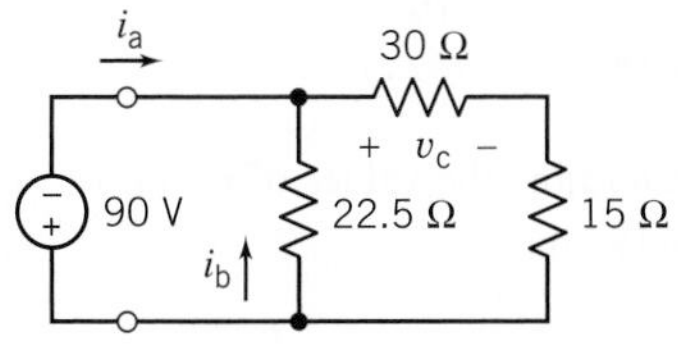

Figure P 3.6-10

P 3.6-11 Find i and $R_{eq\ a\text{-}b}$ if $v_{ab} = 40$ V in the circuit of Figure P 3.6-11.

Answer: $R_{eq\ a-b} = 8\ \Omega$, $i = 5/6$ A

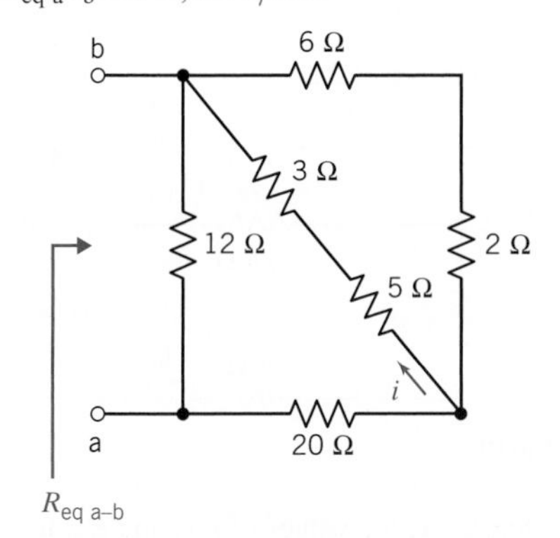

Figure P 3.6-11

P 3.6-12 The ohmmeter in Figure P 3.6-12 measures the equivalent resistance, R_{eq}, of the resistor circuit. The value of the equivalent resistance, R_{eq}, depends on the value of the resistance R.

(a) Determine the value of the equivalent resistance, R_{eq}, when $R = 9\ \Omega$.

(b) Determine the value of the resistance R required to cause the equivalent resistance to be $R_{eq} = 12\ \Omega$.

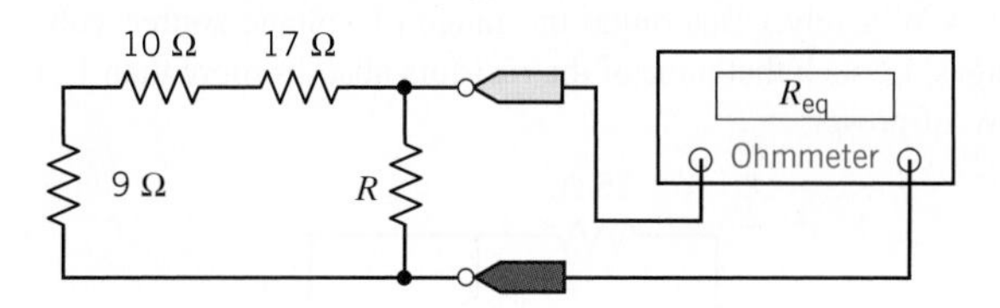

Figure P 3.6-12

P 3.6-13 Find the R_{eq} at terminals a-b in Figure P 3.6-13. Also determine i, i_1, and i_2.

Answer: $R_{eq} = 12\ \Omega$, $i = 5$ A, $i_1 = \frac{5}{3}$ A, $i_2 = \frac{5}{2}$ A

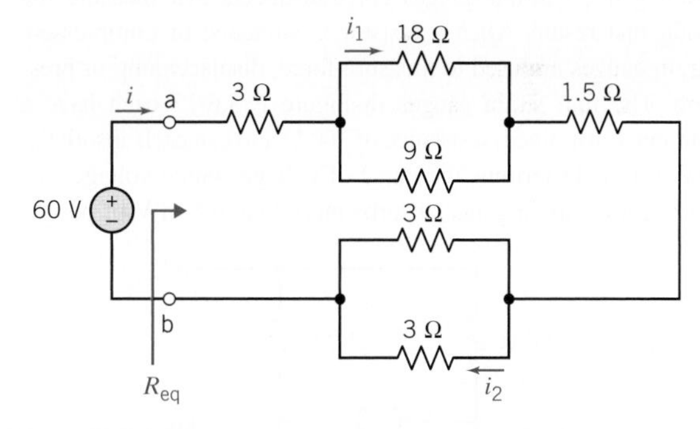

Figure P 3.6-13

P 3.6-14 All of the resistances in the circuit shown in Figure P 3.6-14 are multiples of R. Determine the value of R.

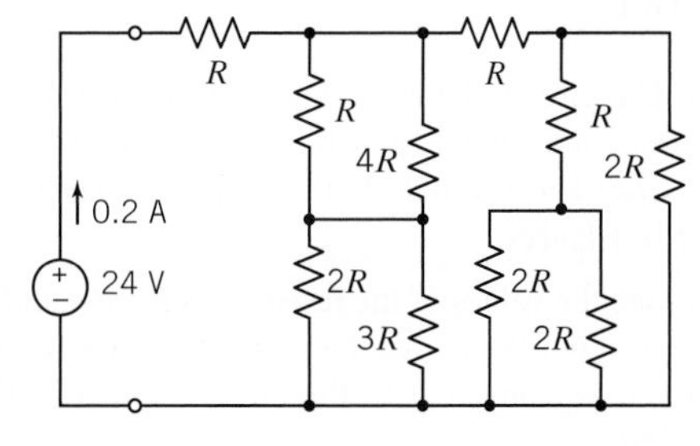

Figure P 3.6-14

P 3.6-15 The circuit shown in Figure P 3.6-15 contains seven resistors, each having resistance R. The input to this circuit is the voltage source voltage, v_s. The circuit has two outputs, v_a and v_b. Express each output as a function of the input.

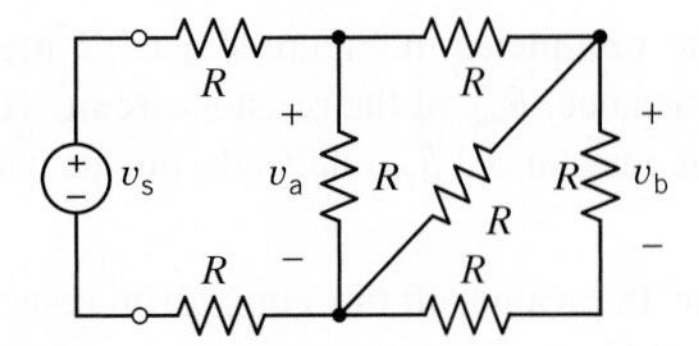

Figure P 3.6-15

P 3.6-16 The circuit shown in Figure P 3.6-16 contains three 15-Ω, 1/4-W resistors. (Quarter-watt resistors can dissipate 1/4 W safely.) Determine the range of voltage source voltages, v_s, such that none of the resistors absorbs more than 1/4 W of power.

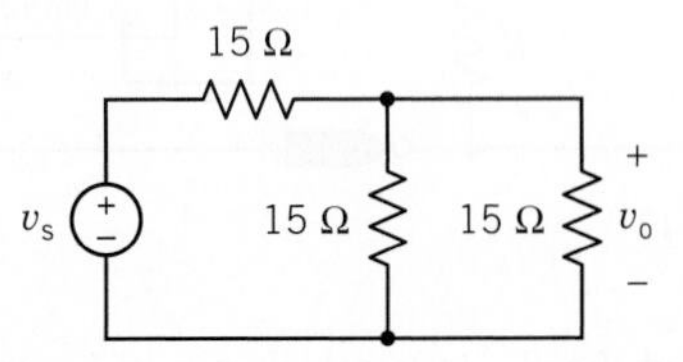

Figure P 3.6-16

P 3.6-17 The four resistors shown in Figure P 3.6-17 represent strain gauges. Strain gauges are transducers that measure the strain that results when a resistor is stretched or compressed. Strain gauges are used to measure force, displacement, or pressure. The four strain gauges in Figure P 3.6-17 each have a nominal (unstrained) resistance of 200 Ω and can each absorb 0.5 mW safely. Determine the range of voltage source voltages, v_s, such that no strain gauge absorbs more than 0.5 mW of power.

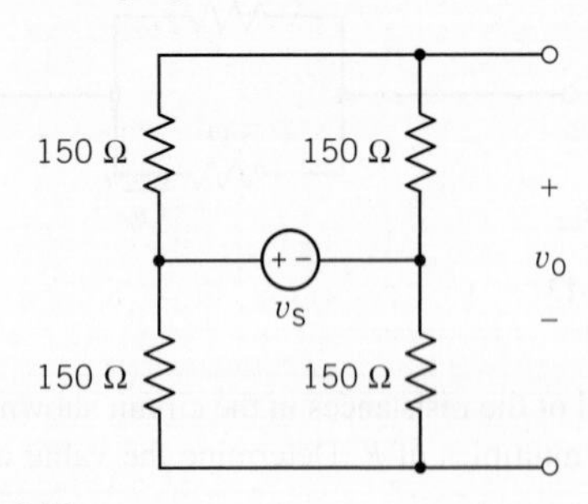

Figure P 3.6-17

P 3.6-18 The circuit shown in Figure P 3.6-18*b* has been obtained from the circuit shown in Figure P 3.6-18*a* by replacing series and parallel combinations of resistances by equivalent resistances.

(a) Determine the values of the resistances R_1, R_2, and R_3 in Figure P 3.6-18*b* so that the circuit shown in Figure P 3.6-18*b* is equivalent to the circuit shown in Figure P 3.6-18*a*.

(b) Determine the values of v_1, v_2, and i in Figure P 3.6-18*b*.

(c) Because the circuits are equivalent, the values of v_1, v_2, and i in Figure P 3.6-18*a* are equal to the values of v_1, v_2, and i in Figure P 3.6-18*b*. Determine the values of v_4, i_5, i_6, and v_7 in Figure P 3.6-18*a*.

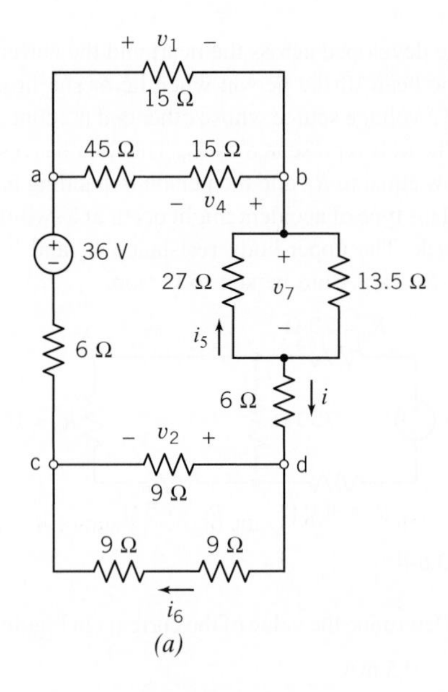

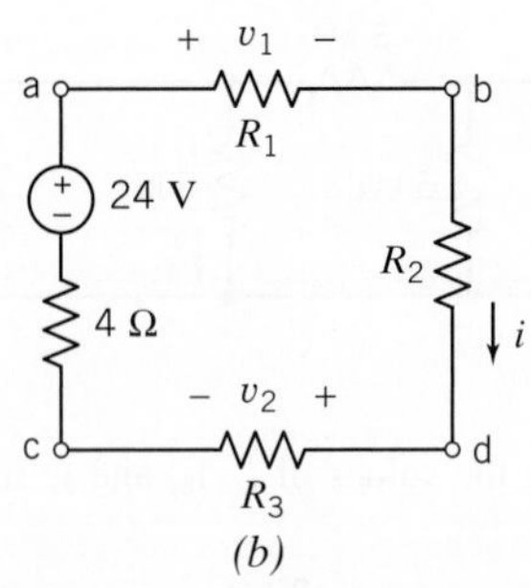

Figure P 3.6-18

P 3.6-19 Determine the values of v_1, v_2, i_3, v_4, v_5, and i_6 in Figure P 3.6-19.

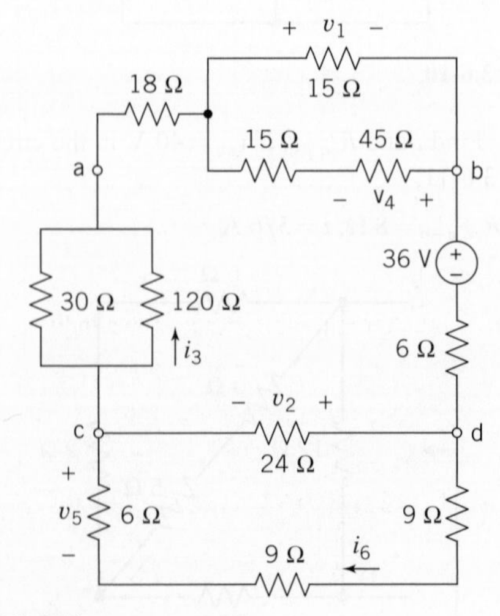

Figure P 3.6-19

P 3.6-20 Determine the values of i, v, and R_{eq} for the circuit shown in Figure P 3.6-20, given that $v_{ab} = 27$ V.

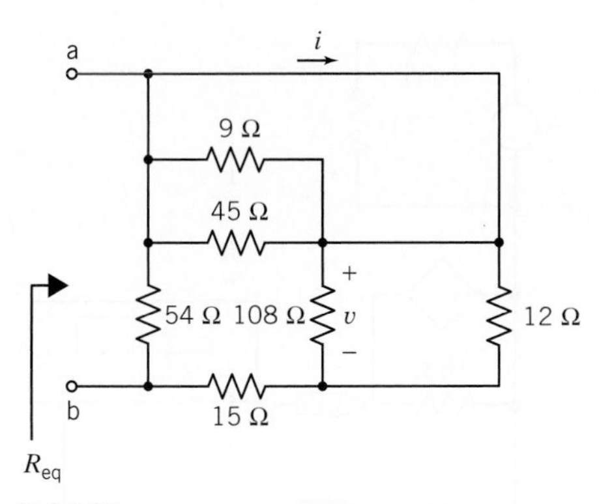

Figure P 3.6-20

P 3.6-21 Determine the value of the resistance R in the circuit shown in Figure P 3.6-21, given that $R_{eq} = 18\ \Omega$.

Answer: $R = 30\ \Omega$

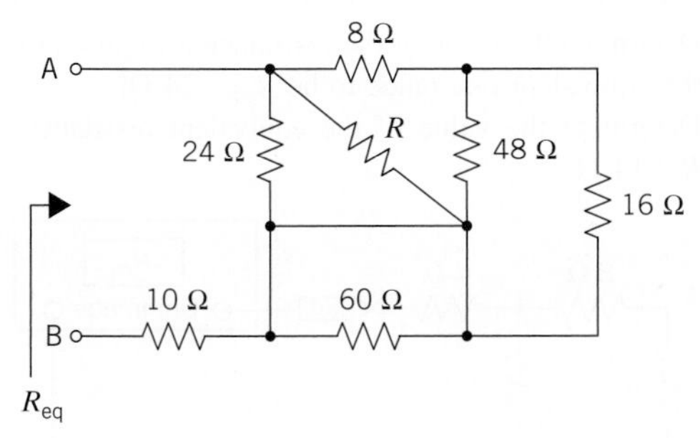

Figure P 3.6-21

P 3.6-22 Determine the value of the resistance R in the circuit shown in Figure P 3.6-22, given that $R_{eq} = 40\ \Omega$.

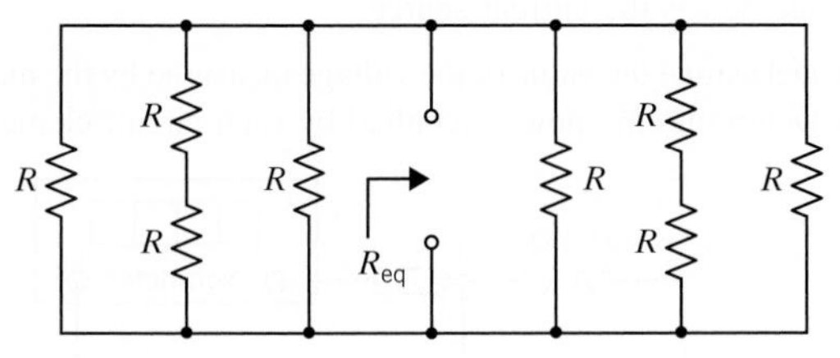

Figure P 3.6-22

P 3.6-23 Determine the values of r, the gain of the CCVS, and g, the gain of the VCCS, for the circuit shown in Figure P 3.6-23.

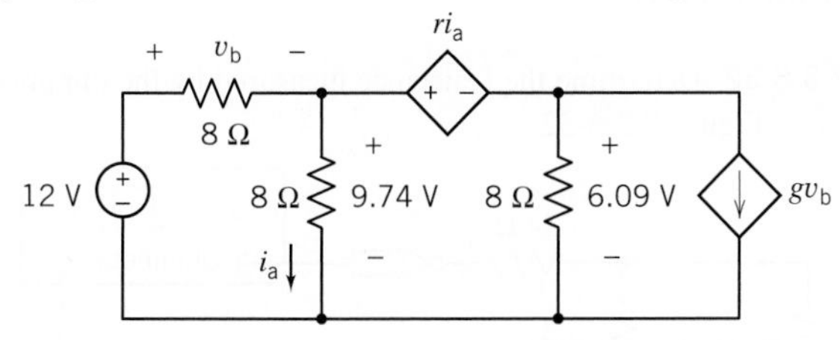

Figure P 3.6-23

P 3.6-24 The input to the circuit in Figure P 3.6-24 is the voltage of the voltage source, v_s. The output is the voltage measured by the meter, v_o. Show that the output of this circuit is proportional to the input. Determine the value of the constant of proportionality.

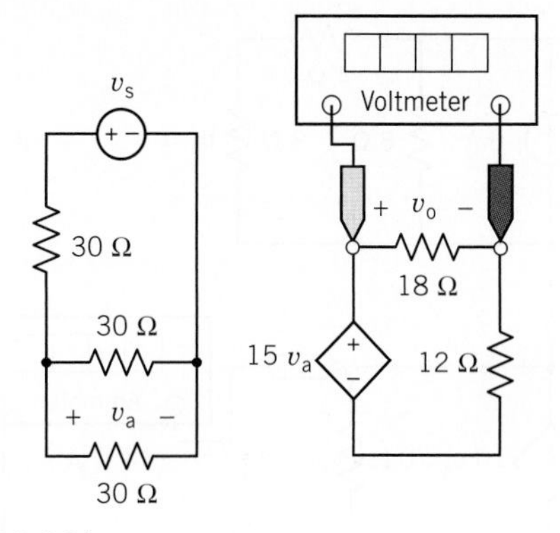

Figure P 3.6-24

P 3.6-25 The input to the circuit in Figure P 3.6-25 is the voltage of the voltage source, v_s. The output is the current measured by the meter, i_o. Show that the output of this circuit is proportional to the input. Determine the value of the constant of proportionality.

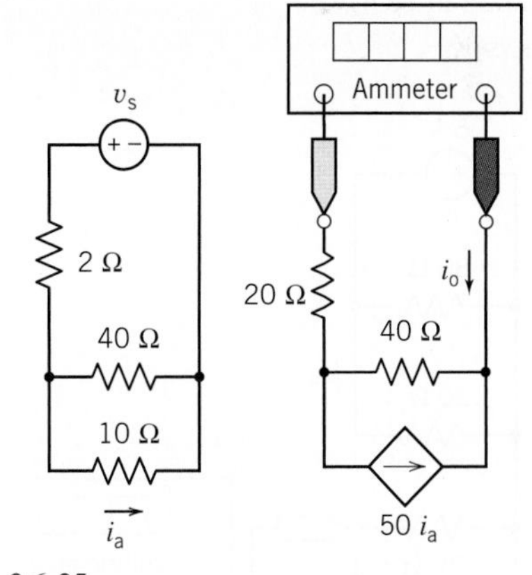

Figure P 3.6-25

P 3.6-26 Determine the voltage measured by the voltmeter in the circuit shown in Figure P 3.6-26.

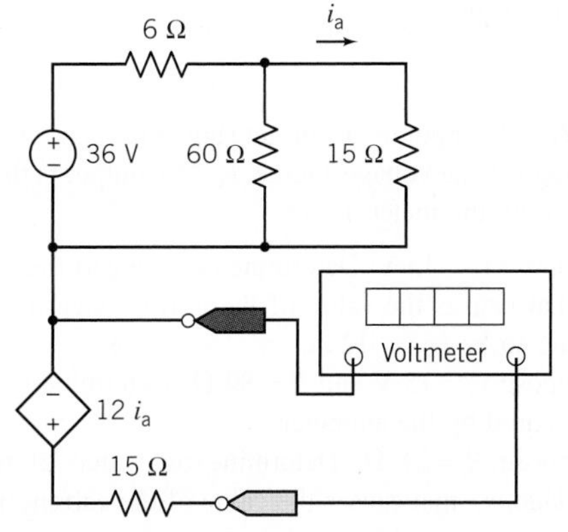

Figure P 3.6-26

P 3.6-27 Determine the current measured by the ammeter in the circuit shown in Figure P 3.6-27.

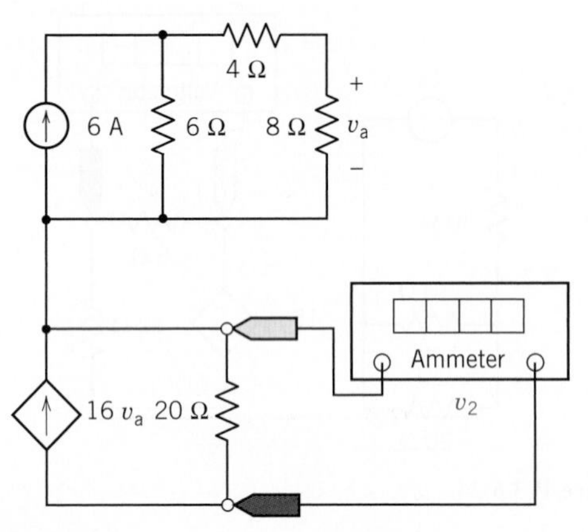

Figure P 3.6-27

P 3.6-28 Determine the value of the resistance R that causes the voltage measured by the voltmeter in the circuit shown in Figure P 3.6-28 to be 6 V.

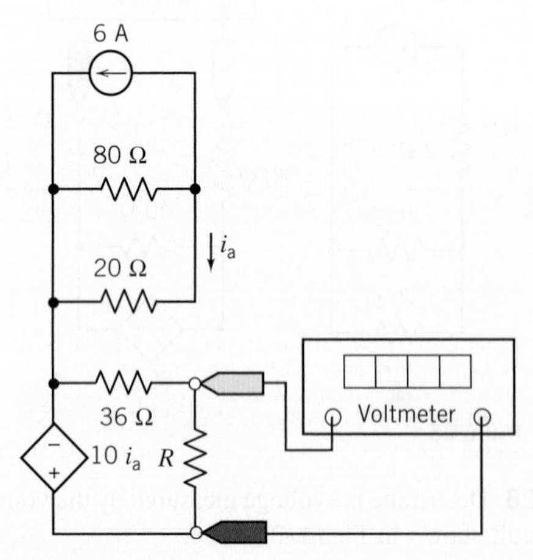

Figure P 3.6-28

P 3.6-29 The input to the circuit shown in Figure P 3.6-29 is the voltage of the voltage source, v_s. The output is the current measured by the meter, i_m.

(a) Suppose $v_s = 15$ V. Determine the value of the resistance R that causes the value of the current measured by the meter to be $i_m = 12$ A.

(b) Suppose $v_s = 15$ V and $R = 80\ \Omega$. Determine the current measured by the ammeter.

(c) Suppose $R = 24\ \Omega$. Determine the value of the input voltage, v_s, that causes the value of the current measured by the meter to be $i_m = 3$ A.

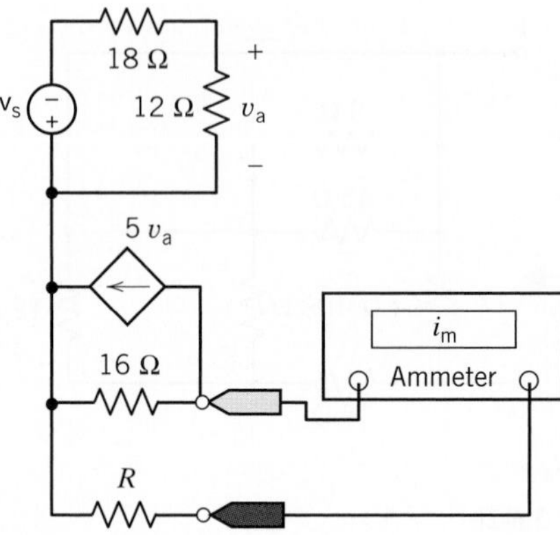

Figure P 3.6-29

P 3.6-30 The ohmmeter in Figure P 3.6-30 measures the equivalent resistance of the resistor circuit connected to the meter probes.

(a) Determine the value of the resistance R required to cause the equivalent resistance to be $R_{eq} = 24\ \Omega$.

(b) Determine the value of the equivalent resistance when $R = 14\ \Omega$.

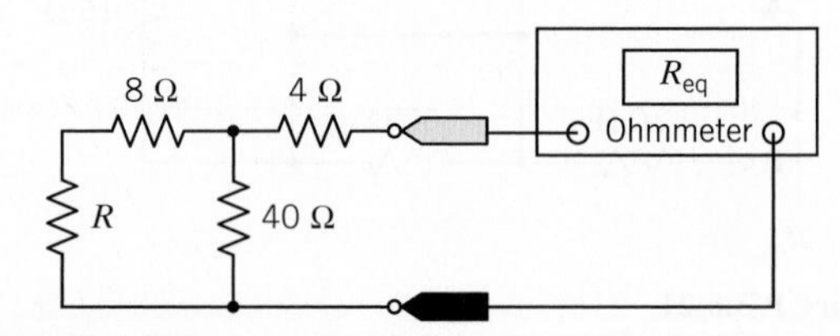

Figure P 3.6-30

P 3.6-31 The voltmeter in Figure P 3.6-31 measures the voltage across the current source.

(a) Determine the value of the voltage measured by the meter.

(b) Determine the power supplied by each circuit element.

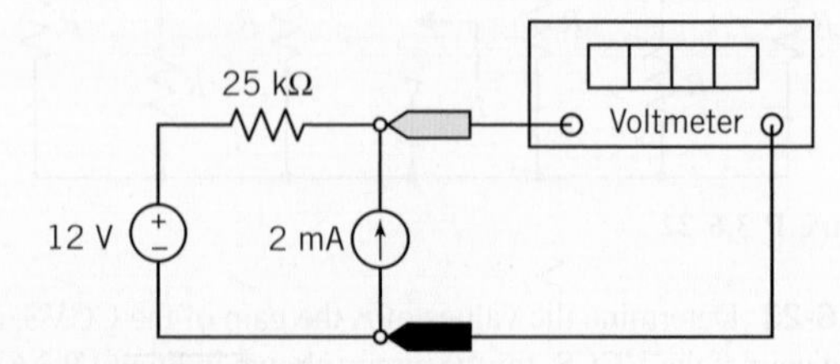

Figure P 3.6-31

P 3.6-32 Determine the resistance measured by the ohmmeter in Figure P 3.6-32.

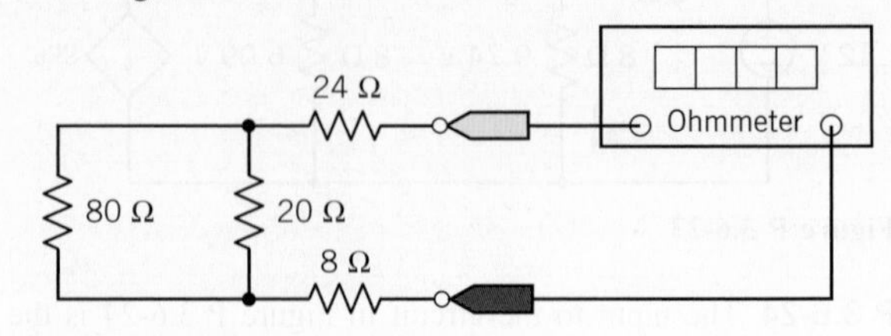

Figure P 3.6-32

P 3.6-33 Determine the resistance measured by the ohmmeter in Figure P 3.6-33.

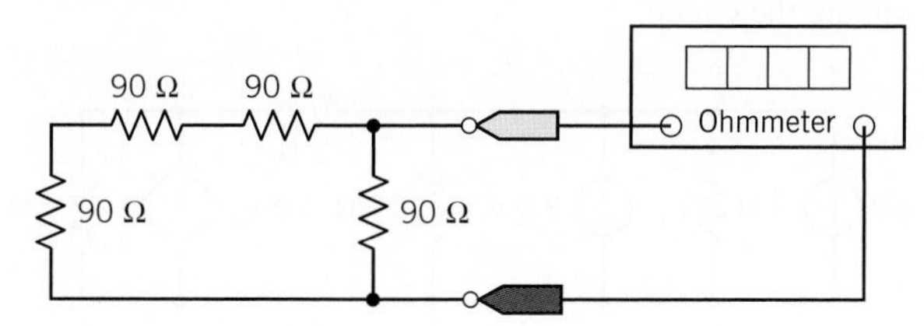

Figure P 3.6-33

P 3.6-34 Consider the circuit shown in Figure P 3.6-34. Given the values of the following currents and voltages:

$$i_1 = 0.625\text{ A},\ v_2 = -25\text{ V},\ i_3 = -1.25\text{ A},\ \text{and } v_4 = -18.75\text{ V},$$

determine the values of R_1, R_2, R_3, and R_4.

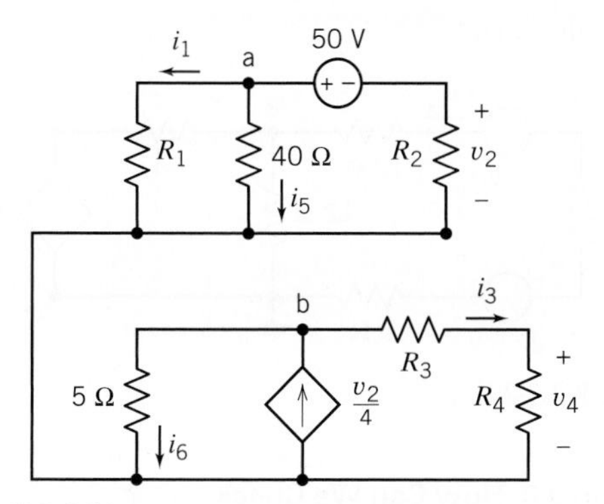

Figure P 3.6-34

P 3.6-35 Consider the circuits shown in Figure P 3.6-35. The equivalent circuit is obtained from the original circuit by replacing series and parallel combinations of resistors with equivalent resistors. The value of the current in the equivalent circuit is $i_s = 0.8$ A. Determine the values of R_1, R_2, R_5, v_2, and i_3.

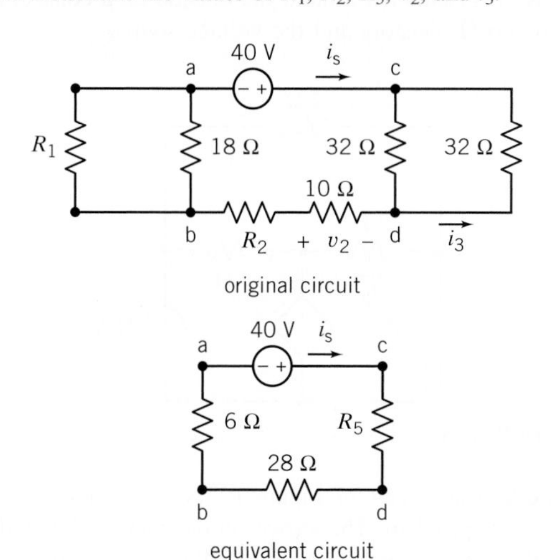

Figure P 3.6-35

P 3.6-36 Consider the circuit shown in Figure P 3.6-36. Given

$$v_2 = \frac{2}{3}v_s,\ i_3 = \frac{1}{5}i_1,\ \text{and } v_4 = \frac{3}{8}v_2,$$

determine the values of R_1, R_2, and R_4.

Hint: Interpret $v_2 = \frac{2}{3}v_s$, $i_3 = \frac{1}{5}i_1$, and $v_4 = \frac{3}{8}v_2$ as current and voltage division.

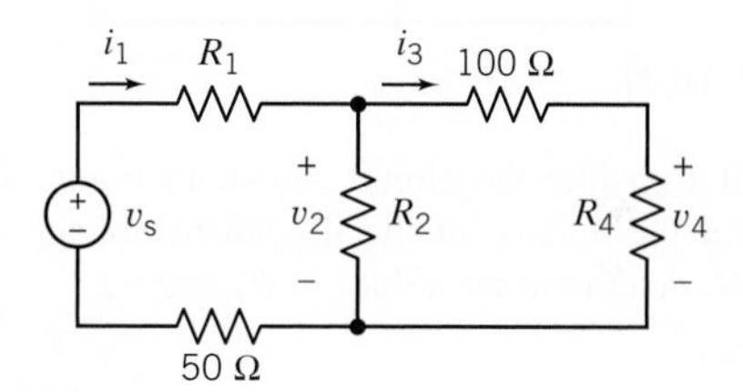

Figure P 3.6-36

P 3.6-37 Consider the circuit shown in Figure P 3.6-37. Given

$$i_2 = \frac{2}{5}i_s,\ v_3 = \frac{2}{3}v_1,\ \text{and } i_4 = \frac{4}{5}i_2,$$

determine the values of R_1, R_2, and R_4.

Hint: Interpret $i_2 = \frac{2}{5}i_s$, $v_3 = \frac{2}{3}v_1$, and $i_4 = \frac{4}{5}i_2$ as current and voltage division.

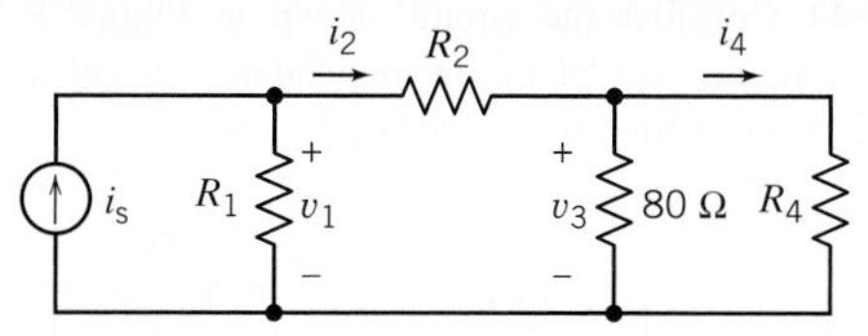

Figure P 3.6-37

P 3.6-38 Consider the circuit shown in Figure P 3.6-38.

(a) Suppose $i_3 = \frac{1}{3}i_1$. What is the value of the resistance R?
(b) Suppose, instead, $v_2 = 4.8$ V. What is the value of the equivalent resistance of the parallel resistors?
(c) Suppose, instead, $R = 20\ \Omega$. What is the value of the current in the 60-Ω resistor?

Hint: Interpret $i_3 = \frac{1}{3}i_1$ as current division.

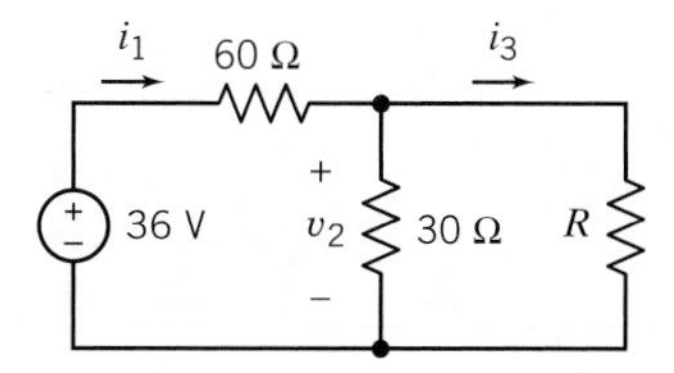

Figure P 3.6-38

P 3.6-39 Consider the circuit shown in Figure P 3.6-39.

(a) Suppose $v_3 = \frac{1}{4}v_1$. What is the value of the resistance R?
(b) Suppose $i_2 = 1.2$ A. What is the value of the resistance R?
(c) Suppose $R = 70\ \Omega$. What is the voltage across the 20-Ω resistor?

(d) Suppose $R = 30\ \Omega$. What is the value of the current in this 30-Ω resistor?

Hint: Interpret $v_3 = \frac{1}{4} v_1$ as voltage division.

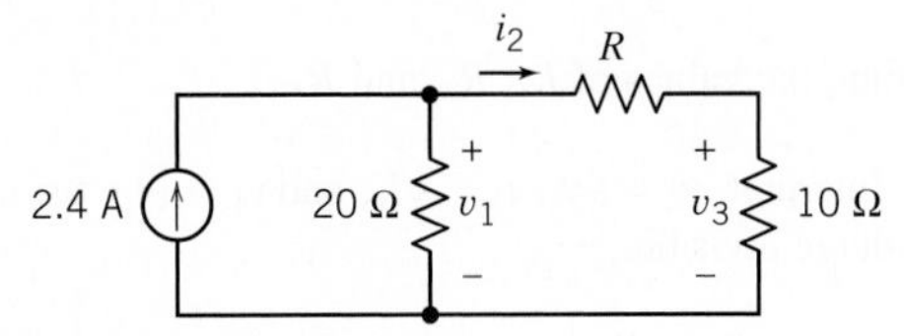

Figure P 3.6-39

P 3.6-40 Consider the circuit shown in Figure P 3.6-40. Given that the voltage of the dependent voltage source is $v_a = 8$ V, determine the values of R_1 and v_o.

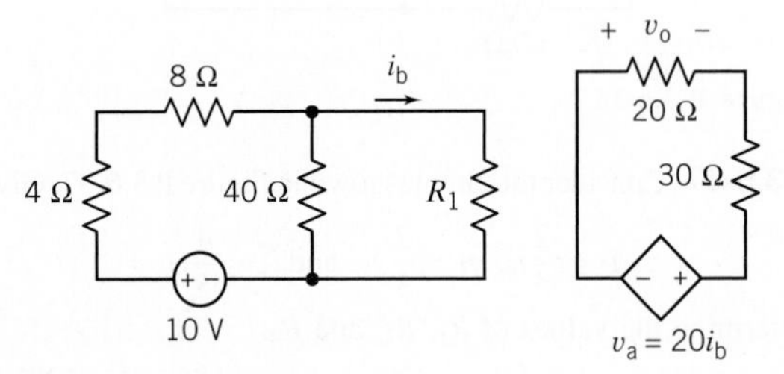

Figure P 3.6-40

P 3.6-41 Consider the circuit shown in Figure P 3.6-41. Given that the current of the dependent current source is $i_a = 2$ A, determine the values of R_1 and i_o.

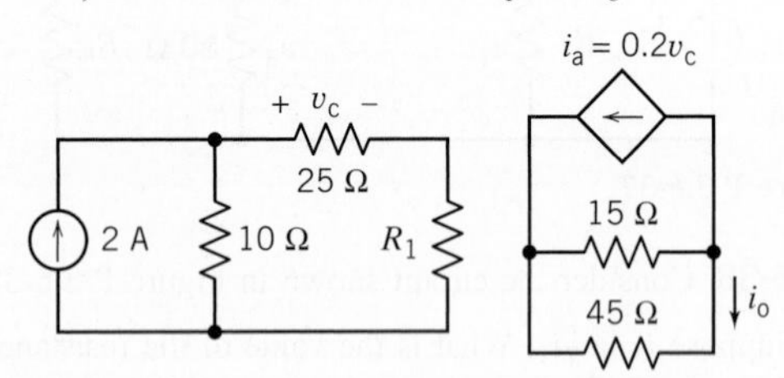

Figure P 3.6-41

P 3.6-42 Determine the values of i_a, i_b, i_2, and v_1 in the circuit shown in Figure P 3.6-42.

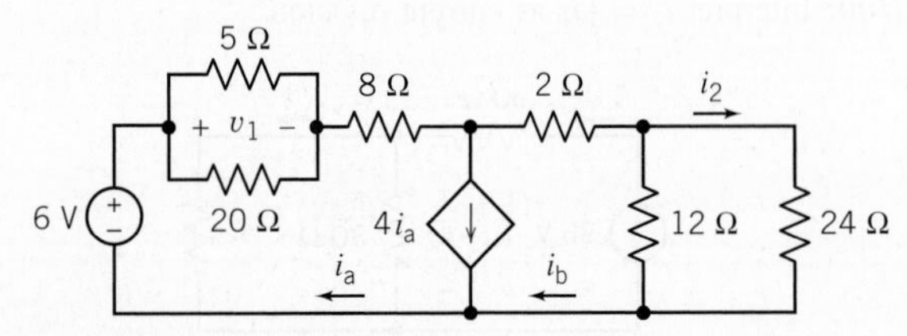

Figure P 3.6-42

Section 3.7 Analyzing Resistive Circuits Using MATLAB

P 3.7-1 Determine the power supplied by each of the sources, independent and dependent, in the circuit shown in Figure P 3.7-1.

Hint: Use the guidelines given in Section 3.7 to label the circuit diagram. Use MATLAB to solve the equations representing the circuit.

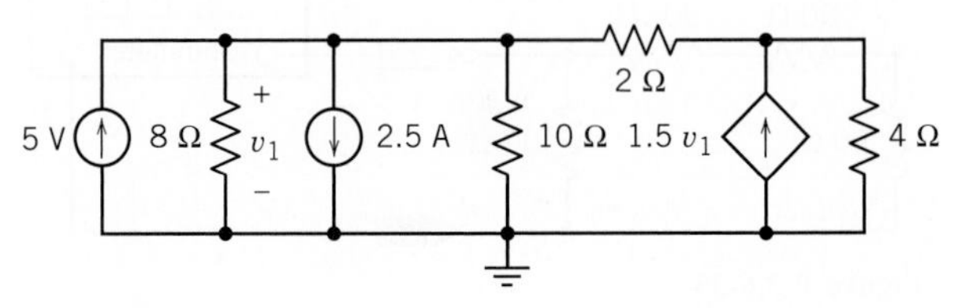

Figure P 3.7-1

P 3.7-2 Determine the power supplied by each of the sources, independent and dependent, in the circuit shown in Figure P 3.7-2.

Hint: Use the guidelines given in Section 3.7 to label the circuit diagram. Use MATLAB to solve the equations representing the circuit.

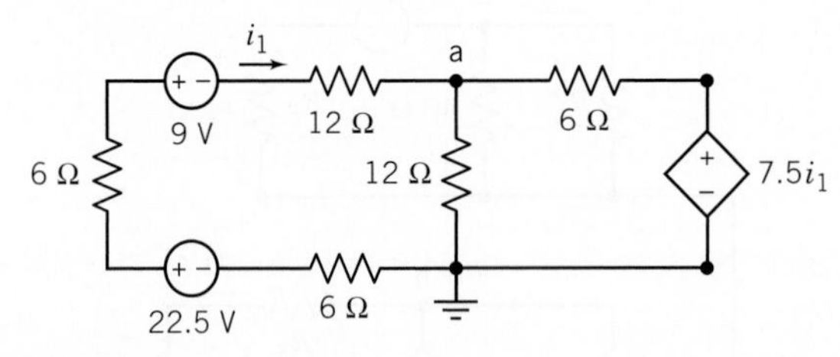

Figure P 3.7-2

Section 3.8 How Can We Check . . . ?

P 3.8-1 A computer analysis program, used for the circuit of Figure P 3.8-1, provides the following branch currents and voltages: $i_1 = -0.833$ A, $i_2 = -0.333$ A, $i_3 = -1.167$ A, and $v = -2.0$ V. Are these answers correct?

Hint: Verify that KCL is satisfied at the center node and that KVL is satisfied around the outside loop consisting of the two 6-Ω resistors and the voltage source.

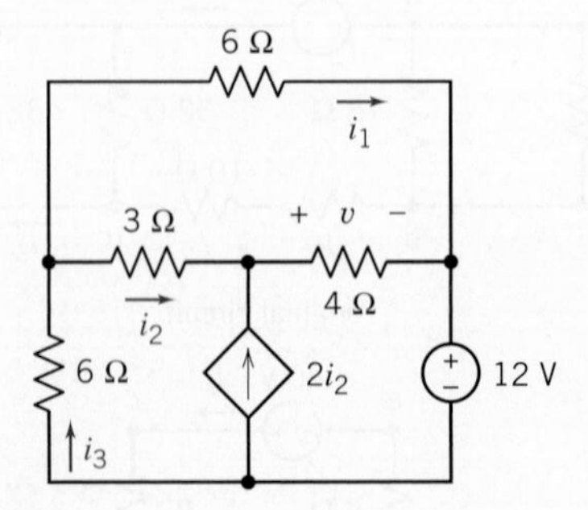

Figure P 3.8-1

P 3.8-2 The circuit of Figure P 3.8-2 was assigned as a homework problem. The answer in the back of the textbook says the current, i, is 1.25 A. Verify this answer, using current division.

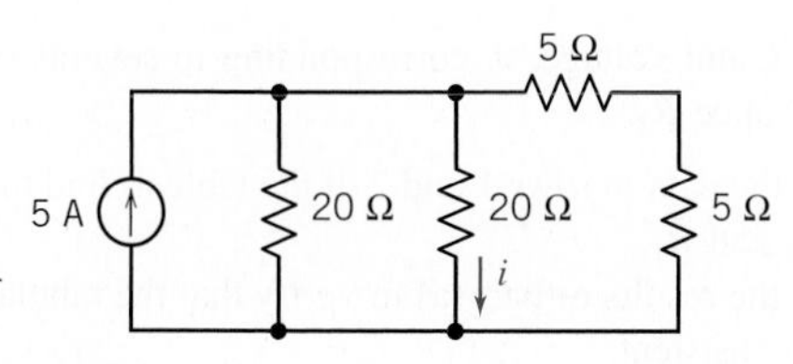

Figure P 3.8-2

P 3.8-3 The circuit of Figure P 3.8-3 was built in the lab, and v_o was measured to be 6.25 V. Verify this measurement, using the voltage divider principle.

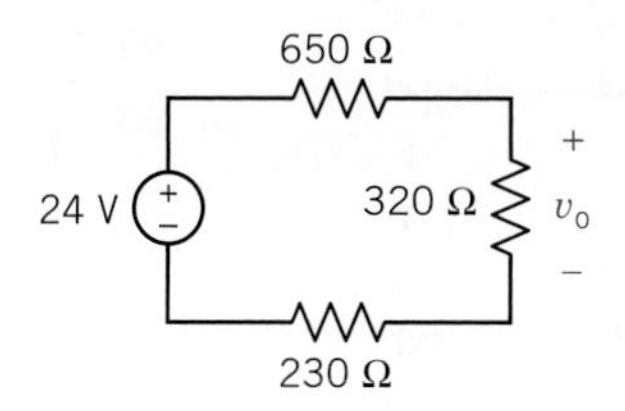

Figure P 3.8-3

P 3.8-4 The circuit of Figure P 3.8-4 represents an auto's electrical system. A report states that $i_H = 9$ A, $i_B = -9$ A, and $i_A = 19.1$ A. Verify that this result is correct.

Hint: Verify that KCL is satisfied at each node and that KVL is satisfied around each loop.

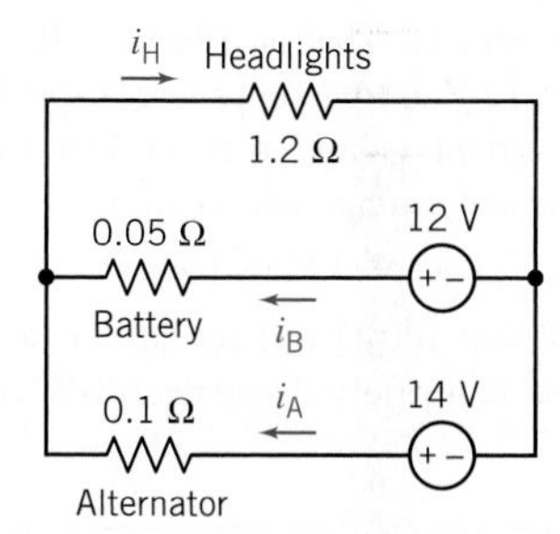

Figure P 3.8-4 Electric circuit model of an automobile's electrical system.

P 3.8-5 Computer analysis of the circuit in Figure P 3.8-5 shows that $i_a = -0.5$ mA, and $i_b = -2$ mA. Was the computer analysis done correctly?

Hint: Verify that the KVL equations for all three meshes are satisfied when $i_a = -0.5$ mA, and $i_b = -2$ mA.

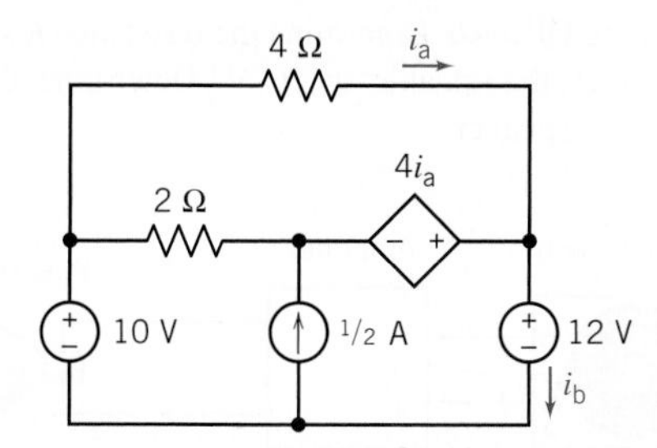

Figure P 3.8-5

P 3.8-6 Computer analysis of the circuit in Figure P 3.8-6. shows that $i_a = 0.5$ mA and $i_b = 4.5$ mA. Was the computer analysis done correctly?

Hint: First, verify that the KCL equations for all five nodes are satisfied when $i_a = 0.5$ mA, and $i_b = 4.5$ mA. Next, verify that the KVL equation for the lower left mesh (a-e-d-a) is satisfied. (The KVL equations for the other meshes aren't useful because each involves an unknown voltage.)

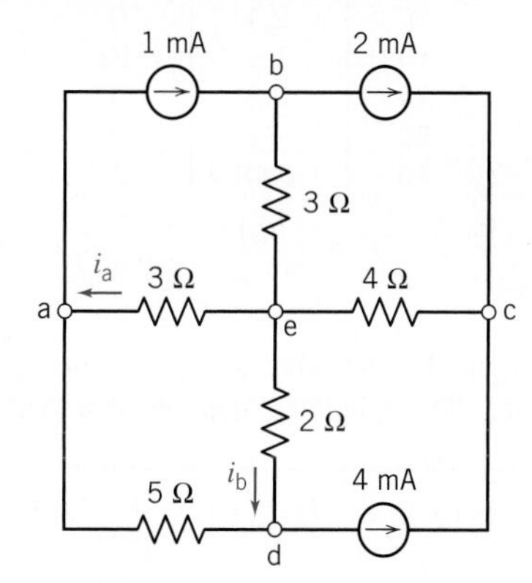

Figure P 3.8-6

P 3.8-7 Verify that the element currents and voltages shown in Figure P 3.8-7 satisfy Kirchhoff's laws:

(a) Verify that the given currents satisfy the KCL equations corresponding to nodes a, b, and c.

(b) Verify that the given voltages satisfy the KVL equations corresponding to loops a-b-d-c-a and a-b-c-d-a.

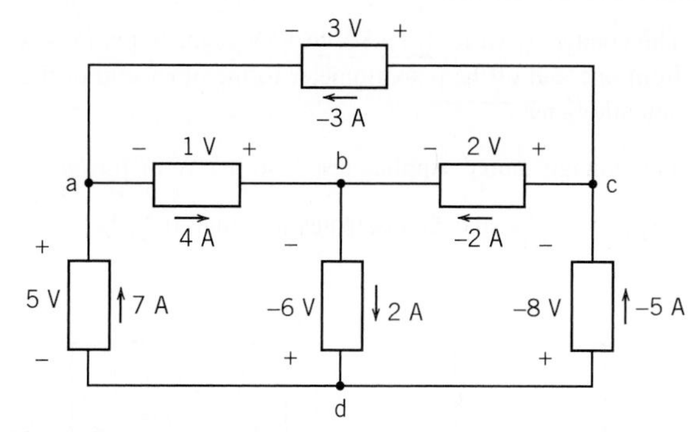

Figure P 3.8-7

***P 3.8-8** Figure P 3.8-8 shows a circuit and some corresponding data. The tabulated data provides values of the current, i, and voltage, v, corresponding to several values of the resistance R_2.

(a) Use the data in rows 1 and 2 of the table to find the values of v_s and R_1.

(b) Use the results of part (a) to verify that the tabulated data are consistent.

(c) Fill in the missing entries in the table.

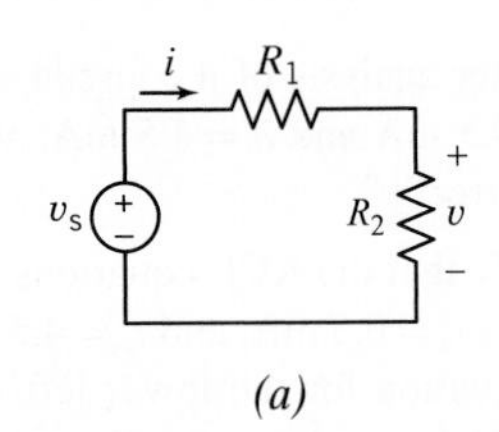

(a)

R_2, Ω	i, A	v, V
0	2.4	0
10	1.2	12
20	0.8	16
30	?	18
40	0.48	?

(b)

Figure P 3.8-8

***P 3.8-9** Figure P 3.8-9 shows a circuit and some corresponding data. The tabulated data provide values of the current, i, and voltage, v, corresponding to several values of the resistance R_2.

(a) Use the data in rows 1 and 2 of the table to find the values of i_s and R_1.

(b) Use the results of part (a) to verify that the tabulated data are consistent.

(c) Fill in the missing entries in the table.

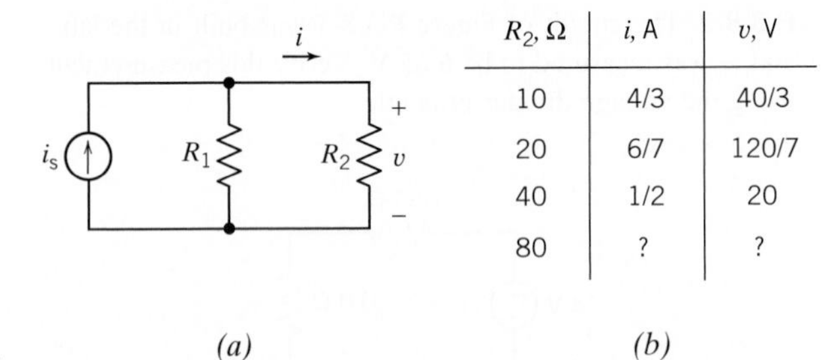

R_2, Ω	i, A	v, V
10	4/3	40/3
20	6/7	120/7
40	1/2	20
80	?	?

(a) (b)

Figure P 3.8-9

Design Problems

DP 3-1 The circuit shown in Figure DP 3-1 uses a potentiometer to produce a variable voltage. The voltage v_m varies as a knob connected to the wiper of the potentiometer is turned. Specify the resistances R_1 and R_2 so that the following three requirements are satisfied:

1. The voltage v_m varies from 8 V to 12 V as the wiper moves from one end of the potentiometer to the other end of the potentiometer.
2. The voltage source supplies less than 0.5 W of power.
3. Each of R_1, R_2, and R_P dissipates less than 0.25 W.

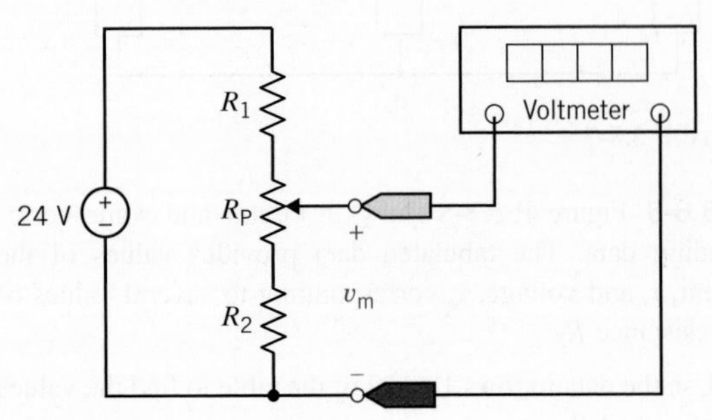

Figure DP 3-1

DP 3-2 The resistance R_L in Figure DP 3-2 is the equivalent resistance of a pressure transducer. This resistance is specified to be 200 Ω ± 5 percent. That is, 190 Ω ≤ R_L ≤ 210 Ω. The voltage source is a 12 V ± 1 percent source capable of supplying 5 W. Design this circuit, using 5 percent, 1/8-watt resistors for R_1 and R_2, so that the voltage across R_L is

$$v_o = 4\text{ V} \pm 10\%$$

(A 5 percent, 1/8-watt 100-Ω resistor has a resistance between 95 and 105 Ω and can safely dissipate 1/8-W continuously.)

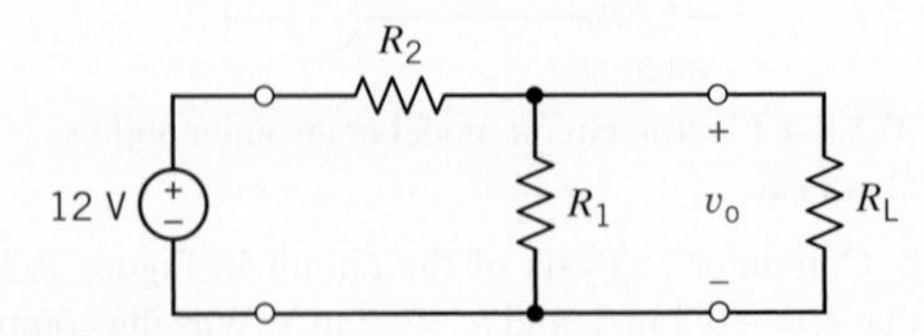

Figure DP 3-2

DP 3-3 A phonograph pickup, stereo amplifier, and speaker are shown in Figure DP 3-3a and redrawn as a circuit model as shown in Figure DP 3-3b. Determine the resistance R so that the voltage v across the speaker is 16 V. Determine the power delivered to the speaker.

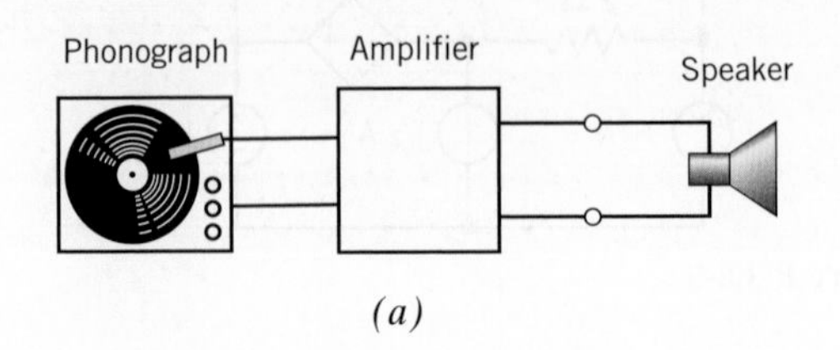

(a)

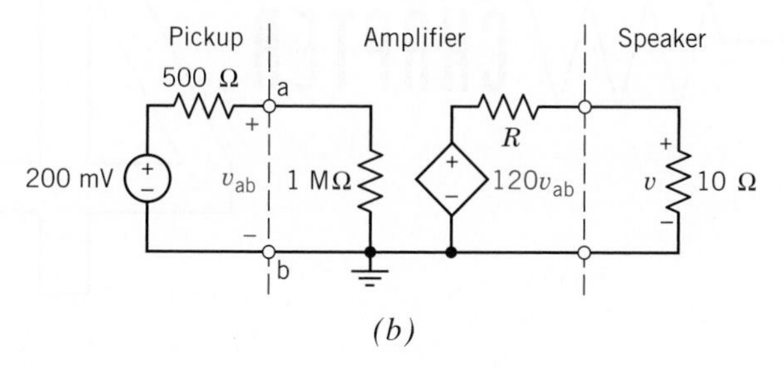

Figure DP 3-3 A phonograph stereo system.

DP 3-4 A Christmas tree light set is required that will operate from a 6-V battery on a tree in a city park. The heavy-duty battery can provide 9 A for the four-hour period of operation each night. Design a parallel set of lights (select the maximum number of lights) when the resistance of each bulb is 12 Ω.

DP 3-5 The input to the circuit shown in Figure DP 3-5 is the voltage source voltage, v_s. The output is the voltage v_o. The output is related to the input by

$$v_o = \frac{R_2}{R_1 + R_2} v_s = g v_s$$

The output of the voltage divider is proportional to the input. The constant of proportionality, g, is called the gain of the voltage divider and is given by

$$g = \frac{R_2}{R_1 + R_2}$$

The power supplied by the voltage source is

$$p = v_s i_s = v_s \left(\frac{v_s}{R_1 + R_2} \right) = \frac{v_s^2}{R_1 + R_2} = \frac{v_s^2}{R_{in}}$$

where

$$R_{in} = R_1 + R_2$$

is called the input resistance of the voltage divider.

(a) Design a voltage divider to have a gain, $g = 0.65$.
(b) Design a voltage divider to have a gain, $g = 0.65$, and an input resistance, $R_{in} = 2500\ \Omega$.

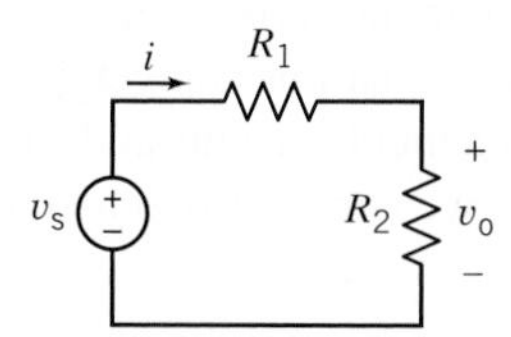

Figure DP 3-5

DP 3-6 The input to the circuit shown in Figure DP 3-6 is the current source current, i_s. The output is the current i_o. The output is related to the input by

$$i_o = \frac{R_1}{R_1 + R_2} i_s = g i_s$$

The output of the current divider is proportional to the input. The constant of proportionality, g, is called the gain of the current divider and is given by

$$g = \frac{R_1}{R_1 + R_2}$$

The power supplied by the current source is

$$p = v_s i_s = \left[i_s \left(\frac{R_1 R_2}{R_1 + R_2} \right) \right] i_s = \frac{R_1 R_2}{R_1 + R_2} i_s^2 = R_{in} i_s^2$$

where

$$R_{in} = \frac{R_1 R_2}{R_1 + R_2}$$

is called the input resistance of the current divider.

(a) Design a current divider to have a gain, $g = 0.65$.
(b) Design a current divider to have a gain, $g = 0.65$, and an input resistance, $R_{in} = 10000\ \Omega$.

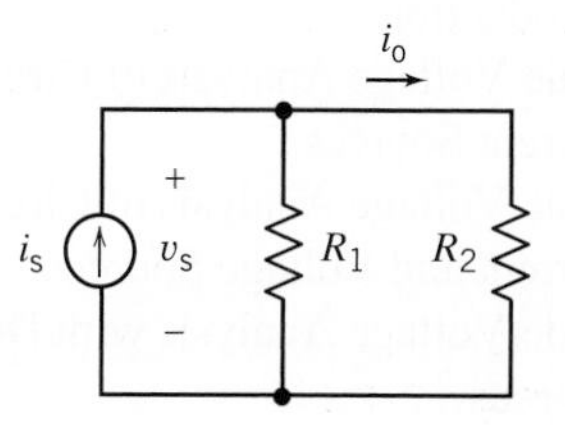

Figure DP 3-6

DP 3-7 Design the circuit shown in Figure DP 3-7 to have an output $v_o = 8.5$ V when the input is $v_s = 12$ V. The circuit should require no more than 1 mW from the voltage source.

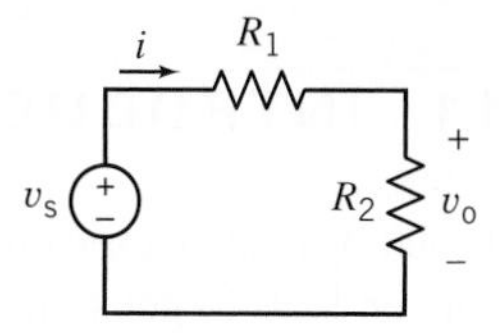

Figure DP 3-7

DP 3-8 Design the circuit shown in Figure DP 3-8 to have an output $i_o = 1.8$ mA when the input is $i_s = 5$ mA. The circuit should require no more than 1 mW from the current source.

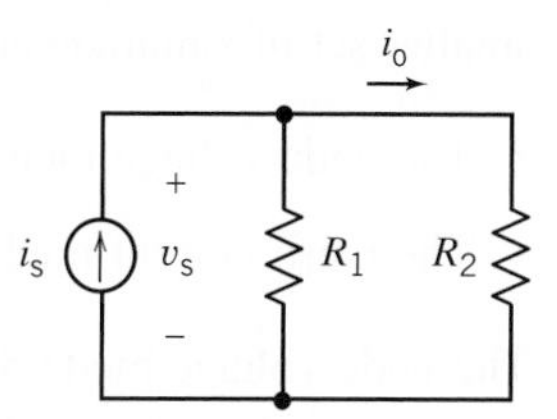

Figure DP 3-8

CHAPTER 4

Methods of Analysis of Resistive Circuits

IN THIS CHAPTER

4.1 Introduction
4.2 Node Voltage Analysis of Circuits with Current Sources
4.3 Node Voltage Analysis of Circuits with Current and Voltage Sources
4.4 Node Voltage Analysis with Dependent Sources
4.5 Mesh Current Analysis with Independent Voltage Sources
4.6 Mesh Current Analysis with Current and Voltage Sources
4.7 Mesh Current Analysis with Dependent Sources
4.8 The Node Voltage Method and Mesh Current Method Compared
4.9 Mesh Current Analysis Using MATLAB
4.10 Using PSpice to Determine Node Voltages and Mesh Currents
4.11 How Can We Check . . . ?
4.12 **DESIGN EXAMPLE**—Potentiometer Angle Display
4.13 Summary
Problems
PSpice Problems
Design Problems

4.1 INTRODUCTION

To analyze an electric circuit, we write and solve a set of equations. We apply Kirchhoff's current and voltage laws to get some of the equations. The constitutive equations of the circuit elements, such as Ohm's law, provide the remaining equations. The unknown variables are element currents and voltages. Solving the equations provides the values of the element current and voltages.

This method works well for small circuits, but the set of equations can get quite large for even moderate-sized circuits. A circuit with only 6 elements has 6 element currents and 6 element voltages. We could have 12 equations in 12 unknowns. In this chapter, we consider two methods for writing a smaller set of simultaneous equations:

- The node voltage method
- The mesh current method

The node voltage method uses a new type of variable called the node voltage. The ''node voltage equations'' or, more simply, the ''node equations,'' are a set of simultaneous equations that represent a given electric circuit. The unknown variables of the node voltage equations are the node voltages. After solving the node voltage equations, we determine the values of the element currents and voltages from the values of the node voltages.

It's easier to write node voltage equations for some types of circuit than for others. Starting with the easiest case, we will learn how to write node voltage equations for circuits that consist of:

- Resistors and independent current sources
- Resistors and independent current and voltage sources
- Resistors and independent and dependent voltage and current sources

The mesh current method uses a new type of variable called the mesh current. The ''mesh current equations'' or, more simply, the ''mesh equations,'' are a set of simultaneous equations that represent a given electric circuit. The unknown variables of the mesh current equations are the mesh currents. After solving the mesh current equations, we determine the values of the element currents and voltages from the values of the mesh currents.

It's easier to write mesh current equations for some types of circuit than for others. Starting with the easiest case, we will learn how to write mesh current equations for circuits that consist of:

- Resistors and independent voltage sources
- Resistors and independent current and voltage sources
- Resistors and independent and dependent voltage and current sources

4.2 NODE VOLTAGE ANALYSIS OF CIRCUITS WITH CURRENT SOURCES

Consider the circuit shown in Figure 4.2-1*a*. This circuit contains four elements: three resistors and a current source. The *nodes* of a circuit are the places where the elements are connected together. The circuit shown in Figure 4.2-1*a* has three nodes. It is customary to draw the elements horizontally or vertically and to connect these elements by horizontal and vertical lines that represent wires. In other words, nodes are drawn as points or are drawn using horizontal or vertical lines. Figure 4.2-1*b* shows the same circuit, redrawn so that all three nodes are drawn as points rather than lines. In Figure 4.2-1*b*, the nodes are labeled as node a, node b, and node c.

Analyzing a connected circuit containing n. nodes will require $n-1$ KCL equations. One way to obtain these equations is to apply KCL at each node of the circuit except for one. The node at which

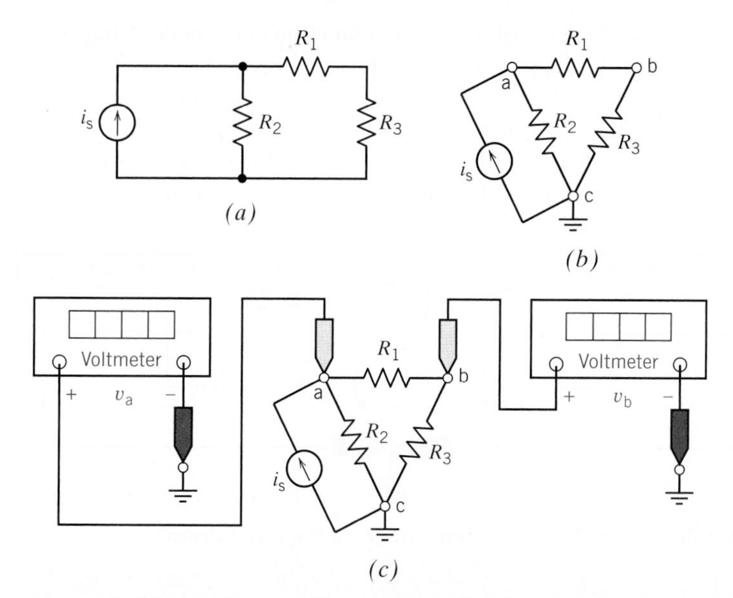

FIGURE 4.2-1 (*a*) A circuit with three nodes. (*b*) The circuit after the nodes have been labeled and a reference node has been selected and marked. (*c*) Using voltmeters to measure the node voltages.

KCL is not applied is called the reference node. Any node of the circuit can be selected to be the reference node. We will often choose the node at the bottom of the circuit to be the reference node. (When the circuit contains a grounded power supply, the ground node of the power supply is usually selected as the reference node.) In Figure 4.2-1*b*, node c is selected as the reference node and marked with the symbol used to identify the reference node.

The voltage at any node of the circuit, relative to the reference node, is called a **node voltage**. In Figure 4.2-1*b*, there, are two node voltages: the voltage at node a with respect to the reference node, node c, and the voltage at node b, again with respect to the reference node, node c. In Figure 4.2-1*c*, voltmeters are added to measure the node voltages. To measure node voltage at node a, connect the red probe of the voltmeter at node a and connect the black probe at the reference node, node c. To measure node voltage at node b, connect the red probe of the voltmeter at node b and connect the black probe at the reference node, node c.

The node voltages in Figure 4.2-1*c* can be represented as v_{ac} and v_{bc}, but it is conventional to drop the subscript c and refer to these as v_a and v_b. Notice that the node voltage at the reference node is $v_{cc} = v_c = 0$ V because a voltmeter measuring the node voltage at the reference node would have both probes connected to the same point.

One of the standard methods for analyzing an electric circuit is to write and solve a set of simultaneous equations called the node equations. The unknown variables in the node equations are the node voltages of the circuit. We determine the values of the node voltages by solving the node equations.

To write a set of node equations, we do two things:

1. Express element current as functions of the node voltages.
2. Apply Kirchhoff's current law (KCL) at each of the nodes of the circuit except for the reference node.

Consider the problem of expressing element currents as functions of the node voltages. Although our goal is to express element *currents* as functions of the node voltages, we begin by expressing element *voltages* as functions of the node voltages. Figure 4.2-2 shows how this is done. The voltmeters in Figure 4.2-2 measure the node voltages, v_1 and v_2, at the nodes of the circuit element. The element voltage has been labeled as v_a. Applying Kirchhoff's voltage law to the loop shown in Figure 4.2-2 gives

$$v_a = v_1 - v_2$$

This equation expresses the element voltage, v_a, as a function of the node voltages, v_1 and v_2. (There is an easy way to remember this equation. Notice the reference polarity of the element voltage, v_a. The element voltage is equal to the node voltage at the node near the + of the reference polarity minus the node voltage at the node near the – of the reference polarity.)

Now consider Figure 4.2-3. In Figure 4.2-3*a*, we use what we have learned to express the voltage of a circuit element as a function of node voltages. The circuit element in Figure 4.2-3*a* could be

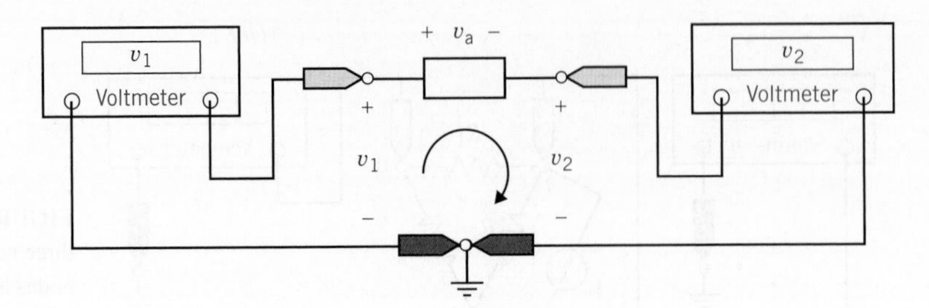

FIGURE 4.2-2 Node voltages, v_1 and v_2, and element voltage, v_a, of a circuit element.

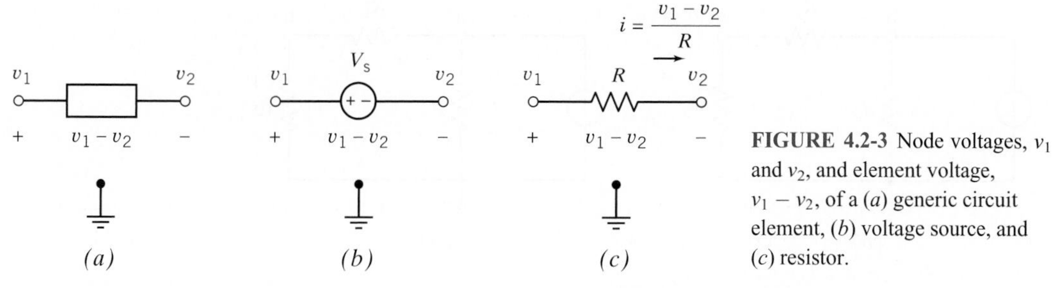

FIGURE 4.2-3 Node voltages, v_1 and v_2, and element voltage, $v_1 - v_2$, of a (*a*) generic circuit element, (*b*) voltage source, and (*c*) resistor.

anything: a resistor, a current source, a dependent voltage source, and so on. In Figures 4.2-3*b* and *c*, we consider specific types of circuit element. In Figure 4.2-3*b*, the circuit element is a voltage source. The element voltage has been represented twice, once as the voltage source voltage, V_s, and once as a function of the node voltages, $v_1 - v_2$. Noticing that the reference polarities for V_s and $v_1 - v_2$ are the same (both + on the left), we write

$$V_s = v_1 - v_2$$

This is an important result. Whenever we have a voltage source connected between two nodes of a circuit, we can express the voltage source voltage, V_s, as a function of the node voltages, v_1 and v_2.

Frequently, we know the value of the voltage source voltage. For example, suppose that $V_s = 12$ V. Then

$$12 = v_1 - v_2$$

This equation relates the values of two of the node voltages.

Next, consider Figure 4.2-3*c*. In Figure 4.2-3*c*, the circuit element is a resistor. We will use Ohm's law to express the resistor current, i, as a function of the node voltages. First, we express the resistor voltage as a function of the node voltages, $v_1 - v_2$. Noticing that the resistor voltage, $v_1 - v_2$, and the current, i, adhere to the passive convention, we use Ohm's law to write

$$i = \frac{v_1 - v_2}{R}$$

Frequently, we know the value of the resistance. For example, when $R = 8\ \Omega$, this equation becomes

$$i = \frac{v_1 - v_2}{8}$$

This equation expresses the resistor current, i, as a function of the node voltages, v_1 and v_2.

Next, let's write node equations to represent the circuit shown in Figure 4.2-4*a*. The input to this circuit is the current source current, i_s. To write node equations, we will first express the resistor currents as functions of the node voltages and then apply Kirchhoff's current law at nodes a and b. The resistor voltages are expressed as functions of the node voltages in Figure 4.2-4*b*, and then the resistor currents are expressed as functions of the node voltages in Figure 4.2-4*c*.

The node equations representing the circuit in Figure 4.2-4 are obtained by applying Kirchhoff's current law at nodes a and b. Using KCL at node a gives

$$i_s = \frac{v_a}{R_2} + \frac{v_a - v_b}{R_1} \tag{4.2-1}$$

Similarly, the KCL equation at node b is

$$\frac{v_a - v_b}{R_1} = \frac{v_b}{R_3} \tag{4.2-2}$$

If $R_1 = 1\ \Omega$, $R_2 = R_3 = 0.5\ \Omega$, and $i_s = 4$ A, and Eqs. 4.2-1 and 4.2-2 may be rewritten as

$$4 = \frac{v_a - v_b}{1} + \frac{v_a}{0.5} \tag{4.2-3}$$

$$\frac{v_a - v_b}{1} = \frac{v_b}{0.5} \tag{4.2-4}$$

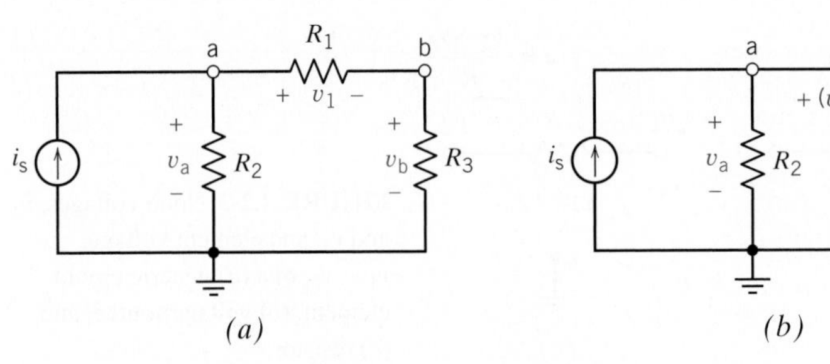

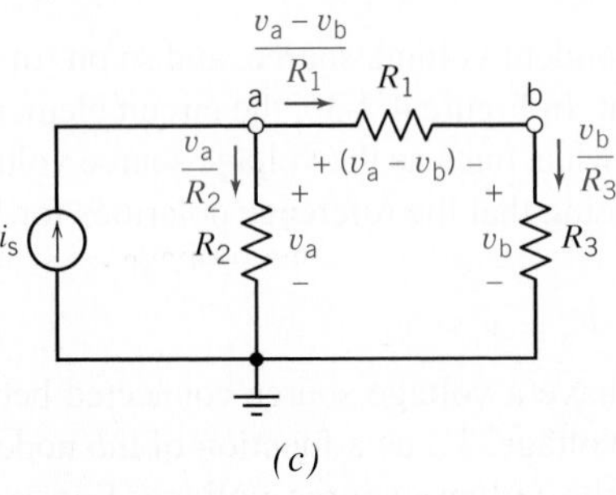

FIGURE 4.2-4 (*a*) A circuit with three resistors. (*b*) The resistor voltages expressed as functions of the node voltages. (*c*) The resistor currents expressed as functions of the node voltages.

Solving Eq. 4.2-4 for v_b gives

$$v_b = \frac{v_a}{3} \tag{4.2-5}$$

Substituting Eq. 4.2-5 into Eq. 4.2-3 gives

$$4 = v_a - \frac{v_a}{3} + 2v_a = \frac{8}{3}v_a \tag{4.2-6}$$

Solving Eq. 4.2-6 for v_a gives

$$v_a = \frac{3}{2}\ \mathrm{V}$$

Finally, Eq. 4.2-5 gives

$$v_b = \frac{1}{2}\ \mathrm{V}$$

Thus, the node voltages of this circuit are

$$v_a = \frac{3}{2}\ \mathrm{V} \quad \text{and} \quad v_b = \frac{1}{2}\ \mathrm{V}$$

EXAMPLE 4.2-1 Node Equations

Determine the value of the resistance R in the circuit shown in Figure 4.2-5*a*.

Solution

Let v_a denote the node voltage at node a and v_b denote the node voltage at node b. The voltmeter in Figure 4.2-5 measures the value of the node voltage at node b, v_b. In Figure 4.2-5*b*, the resistor currents are expressed as functions of the node voltages. Apply KCL at node a to obtain

$$4 + \frac{v_a}{10} + \frac{v_a - v_b}{5} = 0$$

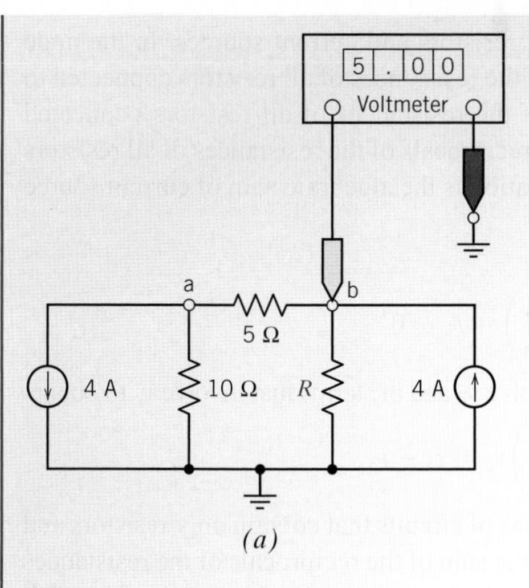

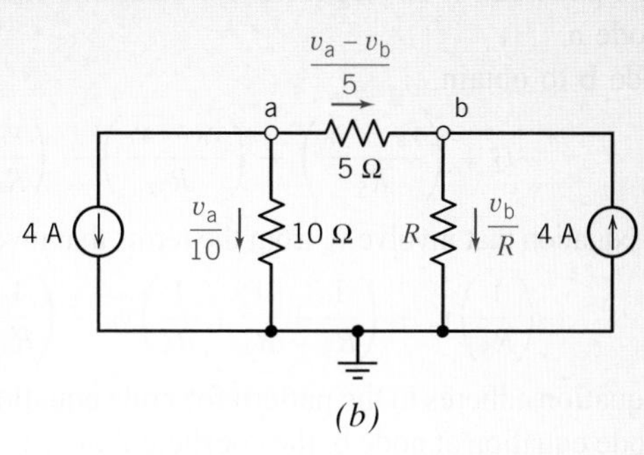

FIGURE 4.2-5 (*a*) The circuit for Example 4.2-1. (*b*) The circuit after the resistor currents are expressed as functions of the node voltages.

Using $v_b = 5$ V gives

$$4 + \frac{v_a}{10} + \frac{v_a - 5}{5} = 0$$

Solving for v_a, we get

$$v_a = -10\text{ V}$$

Next, apply KCL at node b to obtain

$$-\left(\frac{v_a - v_b}{5}\right) + \frac{v_b}{R} - 4 = 0$$

Using $v_a = -10$ V and $v_b = 5$ V gives

$$-\left(\frac{-10 - 5}{5}\right) + \frac{5}{R} - 4 = 0$$

Finally, solving for R gives

$$R = 5\,\Omega$$

Example 4.2-2 Node Equations

Obtain the node equations for the circuit in Figure 4.2-6.

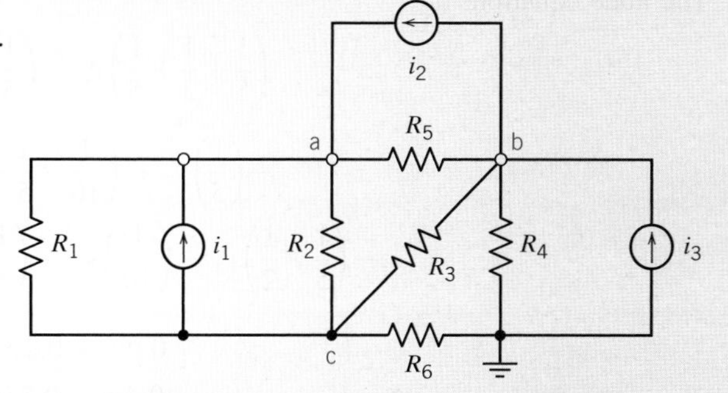

FIGURE 4.2-6 The circuit for Example 4.2-2

Solution

Let v_a denote the node voltage at node a, v_b denote the node voltage at node b, and v_c denote the node voltage at node c. Apply KCL at node a to obtain

$$-\left(\frac{v_a - v_c}{R_1}\right) + i_1 - \left(\frac{v_a - v_c}{R_2}\right) + i_2 - \left(\frac{v_a - v_b}{R_5}\right) = 0$$

Separate the terms of this equation that involve v_a from the terms that involve v_b and the terms that involve v_c to obtain.

$$\left(\frac{1}{R_1} + \frac{1}{R_2} + \frac{1}{R_5}\right)v_a - \left(\frac{1}{R_5}\right)v_b - \left(\frac{1}{R_1} + \frac{1}{R_2}\right)v_c = i_1 + i_2$$

There is a pattern in the node equations of circuits that contain only resistors and current sources. In the node equation at node a, the coefficient of v_a is the sum of the reciprocals of the resistances of all resistors connected to node a. The coefficient of v_b is minus the sum of the reciprocals of the resistances of all resistors connected between node b and node a. The coefficient v_c is minus the sum of the reciprocals of the resistances of all resistors connected between node c and node a. The right-hand side of this equation is the algebraic sum of current source currents directed into node a.

Apply KCL at node b to obtain

$$-i_2 + \left(\frac{v_a - v_b}{R_5}\right) - \left(\frac{v_b - v_c}{R_3}\right) - \left(\frac{v_b}{R_4}\right) + i_3 = 0$$

Separate the terms of this equation that involve v_a from the terms that involve v_b and the terms that involve v_c to obtain

$$-\left(\frac{1}{R_5}\right)v_a + \left(\frac{1}{R_3} + \frac{1}{R_4} + \frac{1}{R_5}\right)v_b - \left(\frac{1}{R_3}\right)v_c = i_3 - i_2$$

As expected, this node equation adheres to the pattern for node equations of circuits that contain only resistors and current sources. In the node equation at node b, the coefficient of v_b is the sum of the reciprocals of the resistances of all resistors connected to node b. The coefficient of v_a is minus the sum of the reciprocals of the resistances of all resistors connected between node a and node b. The coefficient of v_c is minus the sum of the reciprocals of the resistances of all resistors connected between node c and node b. The right-hand side of this equation is the algebraic sum of current source currents directed into node b.

Finally, use the pattern for the node equations of circuits that contain only resistors and current sources to obtain the node equation at node c:

$$-\left(\frac{1}{R_1} + \frac{1}{R_2}\right)v_a - \left(\frac{1}{R_3}\right)v_b + \left(\frac{1}{R_1} + \frac{1}{R_2} + \frac{1}{R_3} + \frac{1}{R_6}\right)v_c = i_1$$

Example 4.2-3 Node Equations

Determine the node voltages for the circuit in Figure 4.2-6 when $i_1 = 1$ A, $i_2 = 2$ A, $i_3 = 3$ A, $R_1 = 5\ \Omega$, $R_2 = 2\ \Omega$, $R_3 = 10\ \Omega$, $R_4 = 4\ \Omega$, $R_5 = 5\ \Omega$, and $R_6 = 2\ \Omega$.

Solution

The node equations are

$$\left(\frac{1}{5} + \frac{1}{2} + \frac{1}{5}\right)v_a - \left(\frac{1}{5}\right)v_b - \left(\frac{1}{5} + \frac{1}{2}\right)v_c = 1 + 2$$

$$-\left(\frac{1}{5}\right)v_a + \left(\frac{1}{10} + \frac{1}{5} + \frac{1}{4}\right)v_b - \left(\frac{1}{10}\right)v_c = -2 + 3$$

$$-\left(\frac{1}{5} + \frac{1}{2}\right)v_a - \left(\frac{1}{10}\right)v_b + \left(\frac{1}{5} + \frac{1}{2} + \frac{1}{10} + \frac{1}{2}\right)v_c = -1$$

$$0.9v_a - 0.2v_b - 0.7v_c = 3$$

$$-0.2v_a + 0.55v_b - 0.1v_c = 1$$

$$-0.7v_a - 0.1v_b + 1.3v_c = -1$$

The node equations can be written using matrices as

$$A\,v = b$$

where

$$A = \begin{bmatrix} 0.9 & -0.2 & -0.7 \\ -0.2 & 0.55 & -0.1 \\ -0.7 & 0.1 & 1.3 \end{bmatrix}, \; b = \begin{bmatrix} 3 \\ 1 \\ -1 \end{bmatrix} \text{ and, } v = \begin{bmatrix} v_a \\ v_b \\ v_c \end{bmatrix}$$

This matrix equation is solved using MATLAB in Figure 4.2-7.

$$v = \begin{bmatrix} v_\mathrm{a} \\ v_\mathrm{b} \\ v_\mathrm{c} \end{bmatrix} = \begin{bmatrix} 7.1579 \\ 5.0526 \\ 3.4737 \end{bmatrix}$$

Consequently, $v_\mathrm{a} = 7.1579$ V, $v_\mathrm{b} = 5.0526$ V, and $v_\mathrm{c} = 3.4737$ V

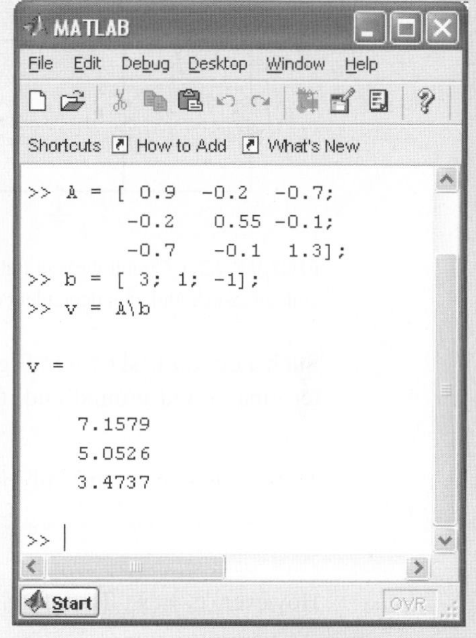

FIGURE 4.2-7 Using MATLAB to solve the node equation in Example 4.2-3.

EXERCISE 4.2-1 Determine the node voltages, v_a and v_b, for the circuit of Figure E 4.2-1.

Answer: $v_\mathrm{a} = 3$ V and $v_\mathrm{b} = 11$ V

EXERCISE 4.2-2 Determine the node voltages, v_a and v_b, for the circuit of Figure E 4.2-2.

Answer: $v_\mathrm{a} = -4/3$ V and $v_\mathrm{b} = 4$ V

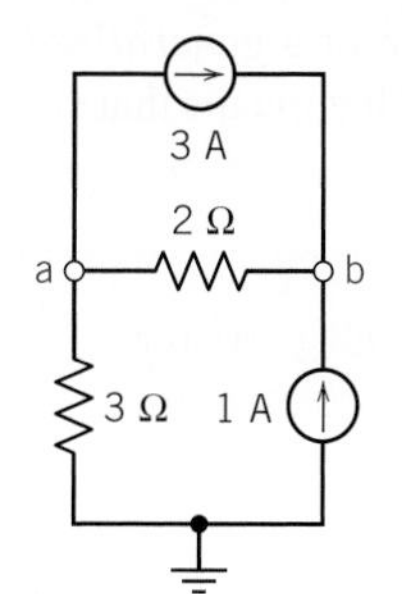

FIGURE E 4.2-1

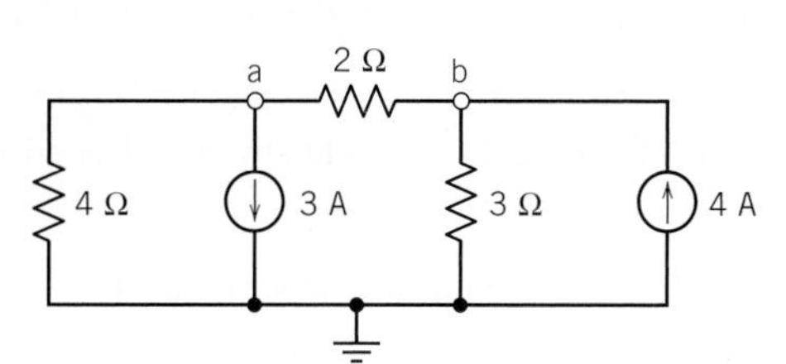

FIGURE E 4.2-2

4.3 NODE VOLTAGE ANALYSIS OF CIRCUITS WITH CURRENT AND VOLTAGE SOURCES

In the preceding section, we determined the node voltages of circuits with independent current sources only. In this section, we consider circuits with both independent current and voltage sources.

First we consider the circuit with a voltage source between ground and one of the other nodes. Because we are free to select the reference node, this particular arrangement is easily achieved.

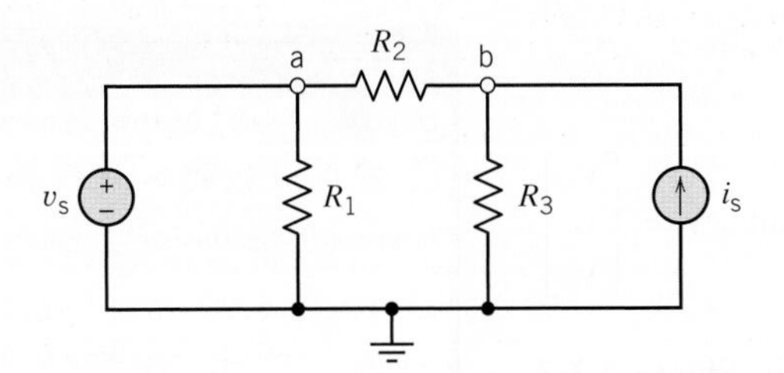

FIGURE 4.3-1 Circuit with an independent voltage source and an independent current source.

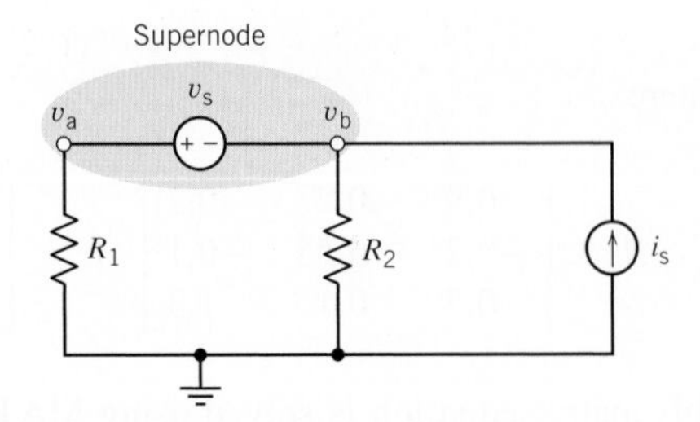

FIGURE 4.3-2 Circuit with a supernode that incorporates v_a and v_b.

Such a circuit is shown in Figure 4.3-1. We immediately note that the source is connected between terminal a and ground and, therefore,

$$v_a = v_s$$

Thus, v_a is known and only v_b is unknown. We write the KCL equation at node b to obtain

$$i_s = \frac{v_b}{R_3} + \frac{v_b - v_a}{R_2}$$

However, $v_a = v_s$. Therefore,

$$i_s = \frac{v_b}{R_3} + \frac{v_b - v_s}{R_2}$$

Then, solving for the unknown node voltage v_b, we get

$$v_b = \frac{R_2 R_3 i_s + R_3 v_s}{R_2 + R_3}$$

Next, let us consider the circuit of Figure 4.3-2, which includes a voltage source between two nodes. Because the source voltage is known, use KVL to obtain

$$v_a - v_b = v_s$$

or

$$v_a - v_s = v_b$$

To account for the fact that the source voltage is known, we consider both node a and node b as part of one larger node represented by the shaded ellipse shown in Figure 4.3-2. We require a larger node because v_a and v_b are dependent. This larger node is often called a *supernode* or a *generalized node*. KCL says that the algebraic sum of the currents entering a supernode is zero. That means that we apply KCL to a supernode in the same way that we apply KCL to a node.

A **supernode** consists of two nodes connected by an independent or a dependent voltage source.

We then can write the KCL equation at the supernode as

$$\frac{v_a}{R_1} + \frac{v_b}{R_2} = i_s$$

However, because $v_a = v_s + v_b$, we have

$$\frac{v_s + v_b}{R_1} + \frac{v_b}{R_2} = i_s$$

Then, solving for the unknown node voltage v_b, we get

$$v_b = \frac{R_1 R_2 i_s - R_2 v_s}{R_1 + R_2}$$

We can now compile a summary of both methods of dealing with independent voltage sources in a circuit we wish to solve by node voltage methods, as recorded in Table 4.3-1.

Table 4.3-1 Node Voltage Analysis Method with a Voltage Source

CASE	METHOD
1. The voltage source connects node q and the reference node (ground).	Set v_q equal to the source voltage accounting for the polarities and proceed to write the KCL at the remaining nodes.
2. The voltage source lies between two nodes, a and b.	Create a supernode that incorporates a and b and equate the sum of all the currents into the supernode to zero.

EXAMPLE 4.3-1 Node Equations for a Circuit Containing Voltage Sources

Determine the node voltages for the circuit shown in Figure 4.3-3.

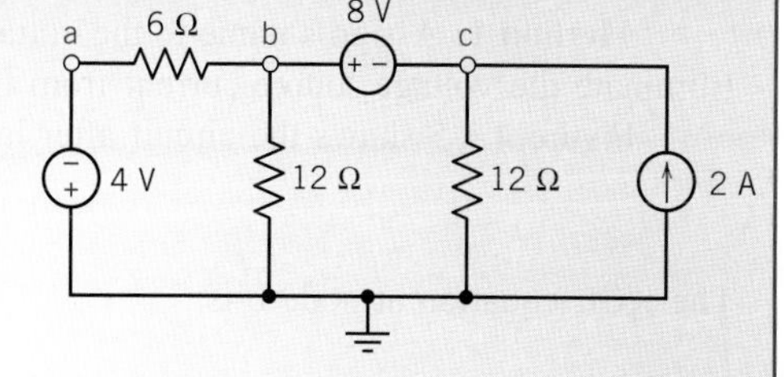

FIGURE 4.3-3 A circuit containing two voltage sources, only one of which is connected to the reference node.

Solution

The methods summarized in Table 4.3-1 are exemplified in this solution. The 4-V voltage source connected to node a exemplifies method 1. The 8-V source between nodes b and c exemplifies method 2.

Using method 1 for the 4-V source, we note that

$$v_a = -4 \text{ V}$$

Using method 2 for the 8-V source, we have a supernode at nodes b and c. The node voltages at nodes b and c are related by

$$v_b = v_c + 8$$

Writing a KCL equation for the supenode, we have

$$\frac{v_b - v_a}{6} + \frac{v_b}{12} + \frac{v_c}{12} = 2$$

or

$$3\,v_b + v_c = 24 + 2\,v_a$$

Using $v_a = -4$ V and $v_b = v_c + 8$ to eliminate v_a and v_b, we have

$$3(v_c + 8) + v_s = 24 + 2(-4)$$

Solving this equation for v_c, we get

$$v_c = -2 \text{ V}$$

Now we calculate v_b to be

$$v_b = v_c + 8 = -2 + 8 = 6 \text{ V}$$

EXAMPLE 4.3-2 Supernodes

Determine the values of the node voltages, v_a and v_b, for the circuit shown in Figure 4.3-4.

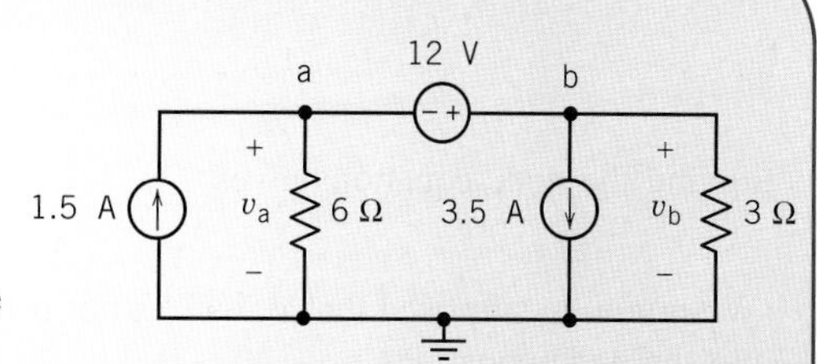

FIGURE 4.3-4 The circuit for Example 4.3-2.

Solution

We can write the first node equation by considering the voltage source. The voltage source voltage is related to the node voltages by

$$v_b - v_a = 12 \Rightarrow v_b = v_a + 12$$

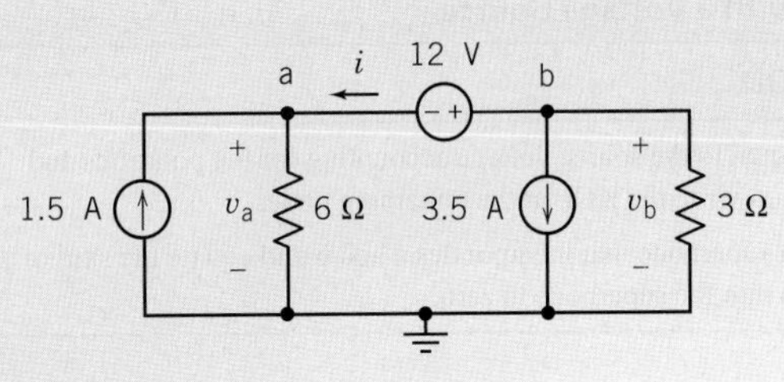

FIGURE 4.3-5 Method 1 For Example 4.3-2.

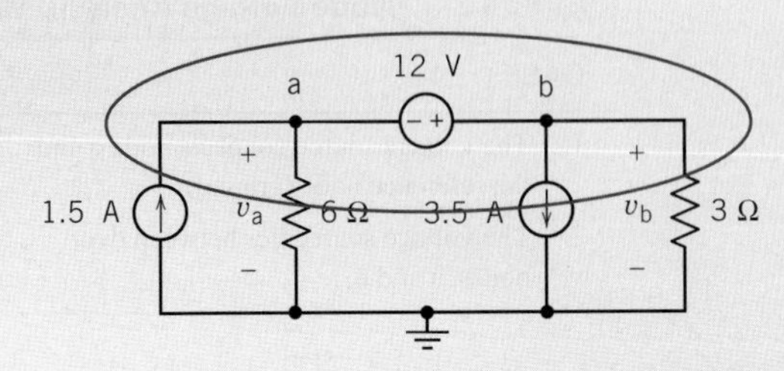

FIGURE 4.3-6 Method 2 for Example 4.3-2.

To write the second node equation, we must decide what to do about the voltage source current. (Notice that there is no easy way to express the voltage source current in terms of the node voltages.) In this example, we illustrate two methods of writing the second node equation.

Method 1: Assign a name to the voltage source current. Apply KCL at both of the voltage source nodes. Eliminate the voltage source current from the KCL equations.

Figure 4.3-5 shows the circuit after labeling the voltage source current. The KCL equation at node a is

$$1.5 + i = \frac{v_a}{6}$$

The KCL equation at node b is

$$i + 3.5 + \frac{v_b}{3} = 0$$

Combining these two equations gives

$$1.5 - \left(3.5 + \frac{v_b}{3}\right) = \frac{v_a}{6} \quad \Rightarrow \quad -2.0 = \frac{v_a}{6} + \frac{v_b}{3}$$

Method 2: Apply KCL to the supernode corresponding to the voltage source. Shown in Figure 4.3-6, this supernode separates the voltage source and its nodes from the rest of the circuit. (In this small circuit, the rest of the circuit is just the reference node.)

Apply KCL to the supernode to get

$$1.5 = \frac{v_a}{6} + 3.5 + \frac{v_b}{3} \quad \Rightarrow \quad -2.0 = \frac{v_a}{6} + \frac{v_b}{3}$$

This is the same equation that was obtained using method 1. Applying KCL to the supernode is a shortcut for doings three things:

1. Labeling the voltage source current as i
2. Applying KCL at both nodes of the voltage source
3. Eliminating i from the KCL equations

In summary, the node equations are

$$v_b - v_a = 12$$

and

$$\frac{v_a}{6} + \frac{v_b}{3} = -2.0$$

Solving the node equations gives

$$v_a = -12\text{ V}, \text{ and } v_b = 0\text{ V}$$

(We might be surprised that v_b is 0 V, but it is easy to check that these values are correct by substituting them into the node equations.)

EXAMPLE 4.3-3 Node Equations for a Circuit Containing Voltage Sources

Determine the node voltages for the circuit shown in Figure 4.3-7.

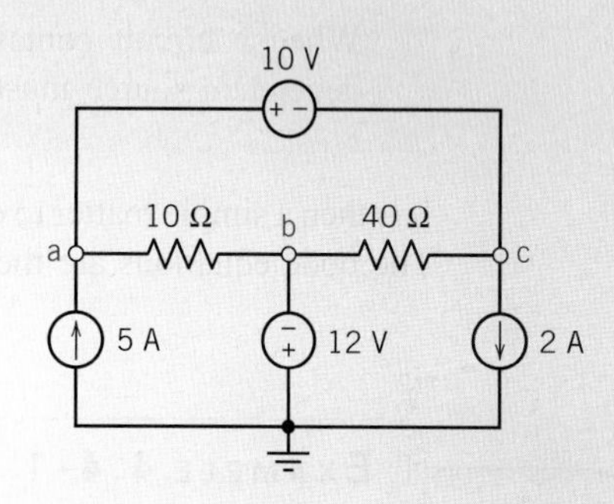

FIGURE 4.3-7 The circuit for Example 4.3-3.

Solution

We will calculate the node voltages of this circuit by writing a KCL equation for the supernode corresponding to the 10-V voltage source. First notice that

$$v_b = -12 \text{ V}$$

and that

$$v_a = v_c + 10$$

Writing a KCL equation for the supernode, we have

$$\frac{v_a - v_b}{10} + 2 + \frac{v_c - v_b}{40} = 5$$

or

$$4\,v_a + v_c - 5\,v_b = 120$$

Using $v_a = v_c + 10$ and $v_b = -12$ to eliminate v_a and v_b, we have

$$4(v_c + 10) + v_c - 5(-12) = 120$$

Solving this equation for v_c, we get

$$v_c = 4 \text{ V}$$

EXERCISE 4.3-1 Find the node voltages for the circuit of Figure E 4.3-1.

Hint: Write a KCL equation for the supernode corresponding to the 10-V voltage source.

Answer: $2 + \frac{v_b + 10}{20} + \frac{v_b}{30} = 5 \Rightarrow v_b = 30 \text{ V} \text{ and } v_a = 40 \text{ V}$

EXERCISE 4.3-2 Find the voltages v_a and v_b for the circuit of Figure E 4.3-2.

Answer: $\frac{(v_b + 8) - (-12)}{10} + \frac{v_b}{40} = 3 \Rightarrow v_b = 8 \text{ V} \text{ and } v_a = 16 \text{ V}$

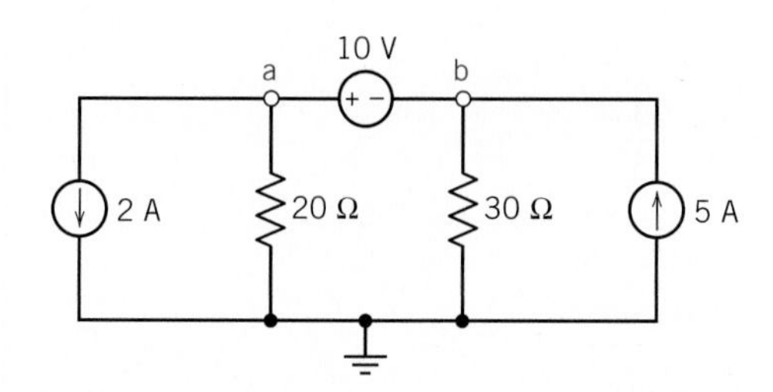

FIGURE E 4.3-1

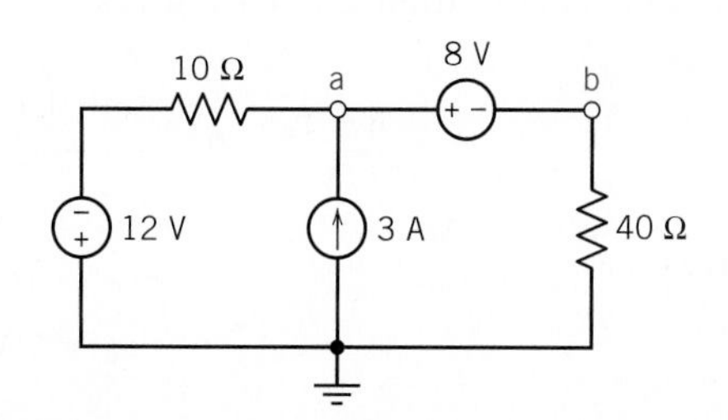

FIGURE E 4.3-2

4.4 NODE VOLTAGE ANALYSIS WITH DEPENDENT SOURCES

When a circuit contains a dependent source the controlling current or voltage of that dependent source must be expressed as a function of the node voltages.

It is then a simple matter to express the controlled current or voltage as a function of the node voltages. The node equations are then obtained using the techniques described in the previous two sections.

EXAMPLE 4.4-1 Node Equations for a Circuit Containing a Dependent Source

Determine the node voltages for the circuit shown in Figure 4.4-1.

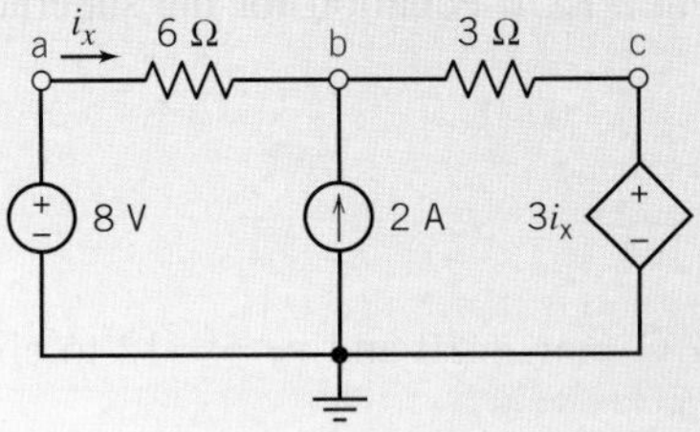

FIGURE 4.4-1 A circuit with a CCVS.

Solution

The controlling current of the dependent source is i_x. Our first task is to express this current as a function of the node voltages:

$$i_x = \frac{v_a - v_b}{6}$$

The value of the node voltage at node a is set by the 8-V voltage source to be

$$v_a = 8\ \text{V}$$

So

$$i_x = \frac{8 - v_b}{6}$$

The node voltage at node c is equal to the voltage of the dependent source, so

$$v_c = 3i_x = 3\left(\frac{8 - v_b}{6}\right) = 4 - \frac{v_b}{2} \tag{4.4-1}$$

Next, apply KCL at node b to get

$$\frac{8 - v_b}{6} + 2 = \frac{v_b - v_c}{3} \tag{4.4-2}$$

Using Eq. 4.4-1 to eliminate v_c from Eq. 4.4-2 gives

$$\frac{8 - v_b}{6} + 2 = \frac{v_b - \left(4 - \frac{v_b}{2}\right)}{3} = \frac{v_b}{2} - \frac{4}{3}$$

Solving for v_b gives

$$v_b = 7\ \text{V}$$

Then,

$$v_c = 4 - \frac{v_b}{2} = \frac{1}{2}\ \text{V}$$

EXAMPLE 4.4-2

Determine the node voltages for the circuit shown in Figure 4.4-2.

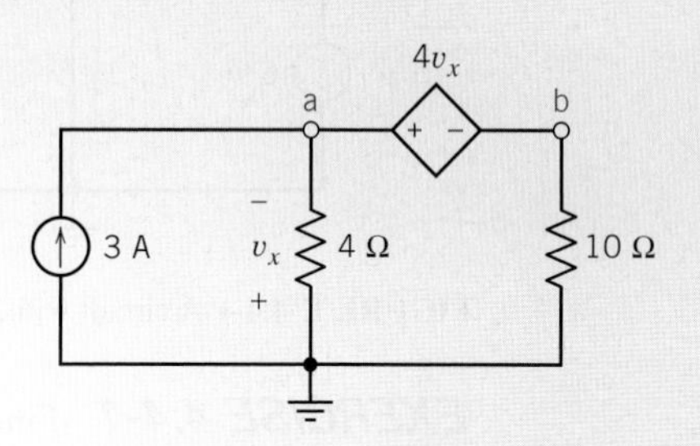

FIGURE 4.4-2 A circuit with a VCVS.

Solution

The controlling voltage of the dependent source is v_x. Our first task is to express this voltage as a function of the node voltages:

$$v_x = -v_a$$

The difference between the node voltages at nodes a and b is set by voltage of the dependent source:

$$v_a - v_b = 4\,v_x = 4(-v_a) = -4\,v_a$$

Simplifying this equation gives

$$v_b = 5\,v_a \tag{4.4-3}$$

Applying KCL to the supernode corresponding to the dependent voltage source gives

$$3 = \frac{v_a}{4} + \frac{v_b}{10} \tag{4.4-4}$$

Using Eq. 4.4-3 to eliminate v_b from Eq. 4.4-4 gives

$$3 = \frac{v_a}{4} + \frac{5v_a}{10} = \frac{3}{4}v_a$$

Solving for v_a, we get

$$v_a = 4\text{ V}$$

Finally,

$$v_b = 5\,v_a = 20\text{ V}$$

EXAMPLE 4.4-3

Determine the node voltages corresponding to nodes a and b for the circuit shown in Figure 4.4-3.

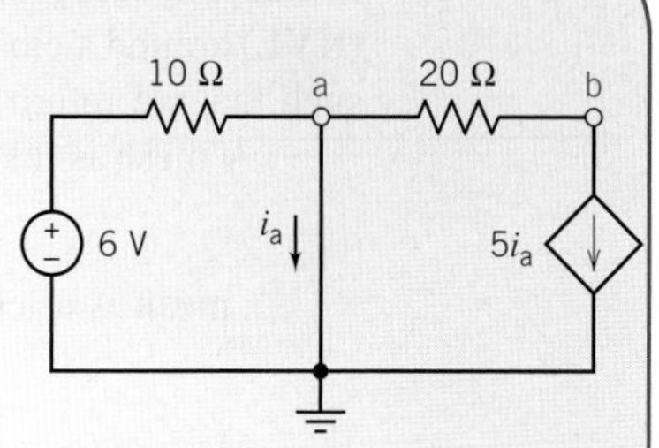

FIGURE 4.4-3 A circuit with a CCCS.

Solution

The controlling current of the dependent source is i_a. Our first task is to express this current as a function of the node voltages. Apply KCL at node a to get

$$\frac{6 - v_a}{10} = i_a + \frac{v_a - v_b}{20}$$

Node a is connected to the reference node by a short circuit, so $v_a = 0$ V. Substituting this value of v_a into the preceding equation and simplifying gives

$$i_a = \frac{12 + v_b}{20} \tag{4.4-5}$$

Next, apply KCL at node b to get

$$\frac{0 - v_b}{20} = 5\,i_a \tag{4.4-6}$$

Using Eq. 4.4-5 to eliminate i_a from Eq. 4.4-6 gives

$$\frac{0 - v_b}{20} = 5\left(\frac{12 + v_b}{20}\right)$$

Solving for v_b gives

$$v_b = -10\text{ V}$$

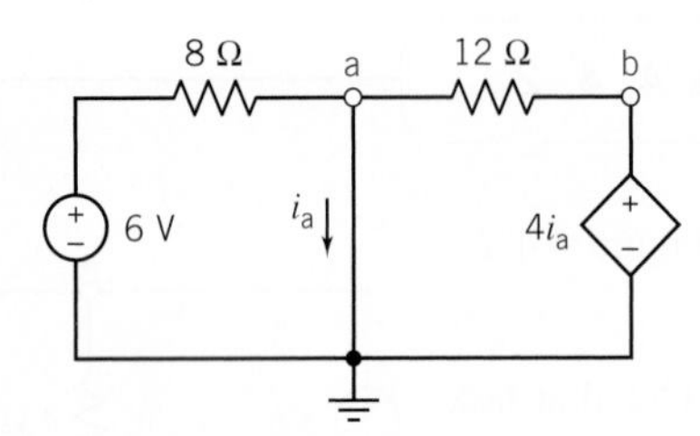

FIGURE E 4.4-1 A circuit with a CCVS.

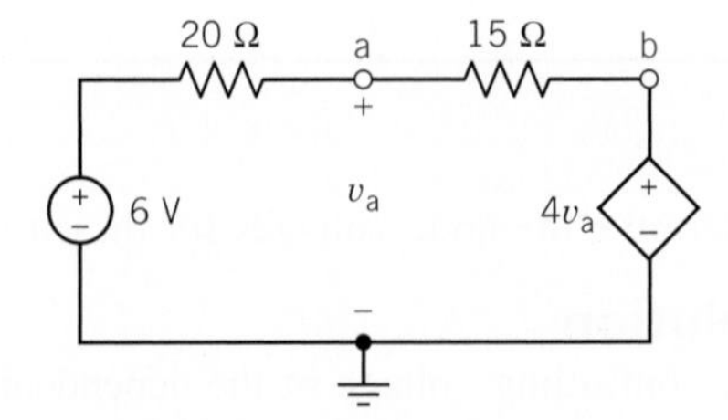

FIGURE E 4.4-2 A circuit with a VCVS.

EXERCISE 4.4-1 Find the node voltage v_b for the circuit shown in Figure E 4.4-1.

Hint: Apply KCL at node a to express i_a as a function of the node voltages. Substitute the result into $v_b = 4i_a$ and solve for v_b.

Answer: $-\frac{6}{8} + \frac{v_b}{4} - \frac{v_b}{12} = 0 \Rightarrow v_b = 4.5\text{ V}$

EXERCISE 4.4-2 Find the node voltages for the circuit shown in Figure E 4.4-2.

Hint: The controlling voltage of the dependent source is a node voltage, so it is already expressed as a function of the node voltages. Apply KCL at node a.

Answer: $\frac{v_a - 6}{20} + \frac{v_a - 4v_a}{15} = 0 \Rightarrow v_a = -2\text{ V}$

4.5 MESH CURRENT ANALYSIS WITH INDEPENDENT VOLTAGE SOURCES

In this and succeeding sections, we consider the analysis of circuits using Kirchhoff's voltage law (KVL) around a closed path. A *closed path* or a *loop* is drawn by starting at a node and tracing a path such that we return to the original node without passing an intermediate node more than once.

A mesh is a special case of a loop.

A **mesh** is a loop that does not contain any other loops within it.

Mesh current analysis is applicable only to planar networks. A planar circuit is one that can be drawn on a plane, without crossovers. An example of a nonplanar circuit is shown in Figure 4.5-1, in which the crossover is identified and cannot be removed by redrawing the circuit. For planar networks, the meshes in the network look like windows. There are four meshes in the circuit shown in Figure 4.5-2.

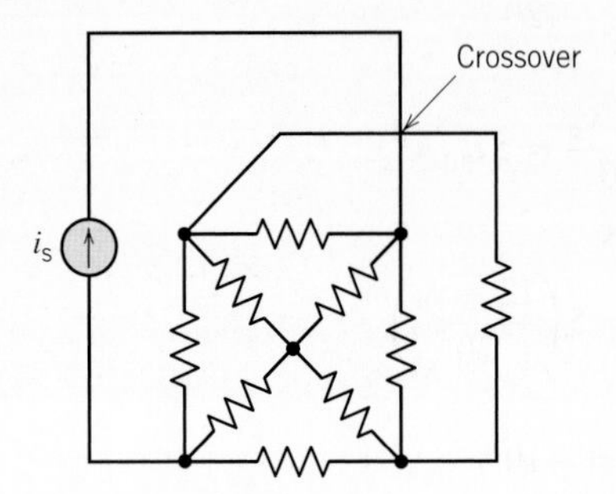

FIGURE 4.5-1 Nonplanar circuit with a crossover.

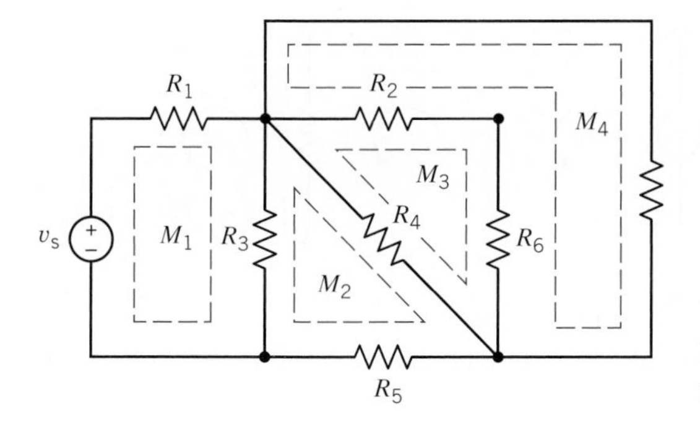

FIGURE 4.5-2 Circuit with four meshes. Each mesh is identified by dashed lines.

They are identified as M_i. Mesh 2 contains the elements R_3, R_4, and R_5. Note that the resistor R_3 is common to both mesh 1 and mesh 2.

We define a mesh current as the current that flows through the elements constituting the mesh. Figure 4.5-3*a* shows a circuit having two meshes with the mesh currents labeled as i_1 and i_2. We will use the convention of a mesh current flowing clockwise as shown in Figure 4.5-3*a*. In Figure 4.5-3*b*, ammeters have been inserted into the meshes to measure the mesh currents.

One of the standard methods for analyzing an electric circuit is to write and solve a set of simultaneous equations called the mesh equations. The unknown variables in the mesh equations are the mesh currents of the circuit. We determine the values of the mesh currents by solving the mesh equations.

To write a set of mesh equations, we do two things:

1. Express element voltages as functions of the mesh currents
2. Apply Kirchhoff's voltage law (KVL) to each of the meshes of the circuit

Consider the problem of expressing element voltages as functions of the mesh currents. Although our goal is to express element *voltages* as functions of the mesh currents, we begin by expressing element *currents* as functions of the mesh currents. Figure 4.5-3*b* shows how this is done. The ammeters in Figure 4.5-3*b* measure the mesh currents, i_1 and i_2. Elements C and E are in the right mesh but not in the left mesh. Apply Kirchhoff's current law at node c and then at node f to see that the currents in elements C and

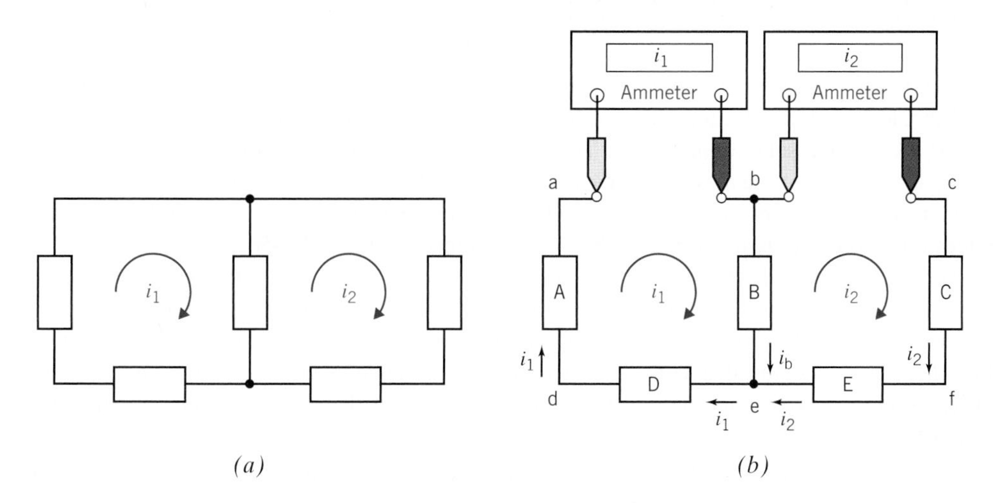

FIGURE 4.5-3 (*a*) A circuit with two meshes. (*b*) Inserting ammeters to measure the mesh currents.

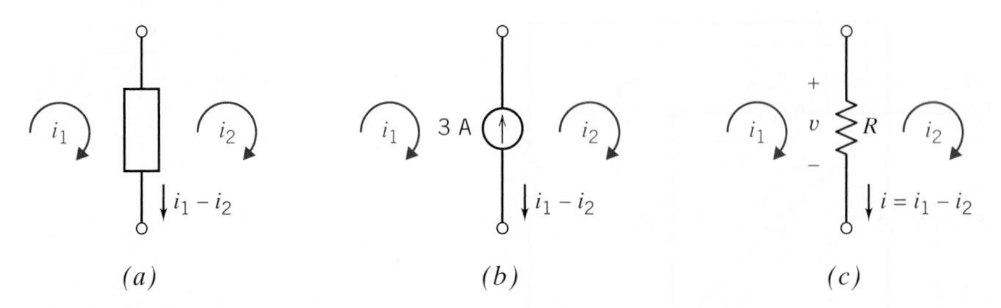

FIGURE 4.5-4 Mesh currents, i_1 and i_2, and element current, $i_1 - i_2$, of a (*a*) generic circuit element, (*b*) current source, and (*c*) resistor.

E are equal to the mesh current of the right mesh, i_2, as shown in Figure 4.5-3*b*. Similarly, elements A and D are only in the left mesh. The currents in elements A and D are equal to the mesh current of the left mesh, i_1, as shown in Figure 4.5-3*b*.

Element B is in both meshes. The current of element B has been labeled as i_b. Applying Kirchhoff's current law at node b in Figure 4.5-3*b* gives

$$i_b = i_1 - i_2$$

This equation expresses the element current, i_b, as a function of the mesh currents, i_1 and i_2.

Figure 4.5-4*a* shows a circuit element that is in two meshes. The current of the circuit element is expressed as a function of the mesh currents of the two meshes. The circuit element in Figure 4.5-4*a* could be anything: a resistor, a current source, a dependent voltage source, and so on. In Figures 4.5-4*b* and *c*, we consider specific types of circuit element. In Figure 4.5-4*b*, the circuit element is a current source. The element current has been represented twice, once as the current source current, 3 A, and once as a function of the mesh currents, $i_1 - i_2$. Noticing that the reference directions for 3 A and $i_1 - i_2$ are different (one points up, the other points down), we write

$$-3 = i_1 - i_2$$

This equation relates the values of two of the mesh currents.

Next consider Figure 4.5-4*c*. In Figure 4.5-4*c*, the circuit element is a resistor. We will use Ohm's law to express the resistor voltage, v, as functions of the mesh currents. First, we express the resistor current as a function of the mesh currents, $i_1 - i_2$. Noticing that the resistor current, $i_1 - i_2$, and the voltage, v, adhere to the passive convention, we use Ohm's law to write

$$v = R(i_1 - i_2)$$

Frequently, we know the value of the resistance. For example, when $R = 8\ \Omega$, this equation becomes

$$v = 8(i_1 - i_2)$$

This equation expresses the resistor voltage, v, as a function of the mesh currents, i_1 and i_2.

Next, let's write mesh equations to represent the circuit shown in Figure 4.5-5*a*. The input to this circuit is the voltage source voltage, v_s. To write mesh equations, we will first express the resistor voltages as functions of the mesh currents and then apply Kirchhoff's voltage law to the meshes. The resistor currents are expressed as functions of the mesh currents in Figure 4.5-5*b*, and then the resistor voltages are expressed as functions of the mesh currents in Figure 4.5-5*c*.

We may use Kirchhoff's voltage law around each mesh. We will use the following convention for obtaining the algebraic sum of voltages around a mesh. We will move around the mesh in the clockwise direction. If we encounter the + sign of the voltage reference polarity of an element voltage before the − sign, we add that voltage. Conversely, if we encounter the − of the voltage reference polarity of an element voltage before the + sign, we subtract that voltage. Thus, for the circuit of Figure 4.5-5*c*, we have

$$\text{mesh 1:}\quad -v_s + R_1 i_1 + R_3(i_1 - i_2) = 0 \tag{4.5-1}$$

$$\text{mesh 2:}\quad -R_3(i_1 - i_2) + R_2 i_2 = 0 \tag{4.5-2}$$

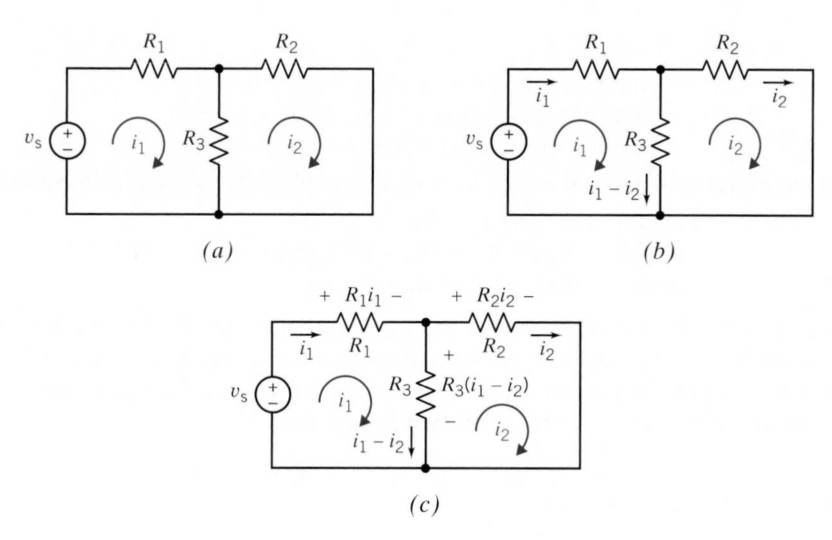

FIGURE 4.5-5 (*a*) A circuit. (*b*) The resistor currents expressed as functions of the mesh currents. (*c*) The resistor voltages expressed as functions of the mesh currents.

Note that the voltage across R_3 in mesh 1 is determined from Ohm's law, where

$$v = R_3 i_a = R_3(i_1 - i_2)$$

where i_a is the actual element current flowing downward through R_3.

Equations 4.5-1 and 4.5-2 will enable us to determine the two mesh currents, i_1 and i_2. Rewriting the two equations, we have

$$i_1(R_1 + R_3) - i_2 R_3 = v_s$$

and

$$-i_1 R_3 + i_2(R_3 + R_2) = 0$$

If $R_1 = R_2 = R_3 = 1\ \Omega$, we have

$$2i_1 - i_2 = v_s$$

and

$$-i_1 + 2i_2 = 0$$

Add twice the first equation to the second equation, obtaining $3i_1 = 2v_s$. Then we have

$$i_1 = \frac{2v_s}{3} \text{ and } i_2 = \frac{v_s}{3}$$

Thus, we have obtained two independent mesh current equations that are readily solved for the two unknowns. If we have N meshes and write N mesh equations in terms of N mesh currents, we can obtain N independent mesh equations. This set of N equations is independent and thus guarantees a solution for the N mesh currents.

A circuit that contains only independent voltage sources and resistors results in a specific format of equations that can readily be obtained. Consider a circuit with three meshes, as shown in Figure 4.5-6. Assign the clockwise direction to all of the mesh currents. Using KVL, we obtain the three mesh

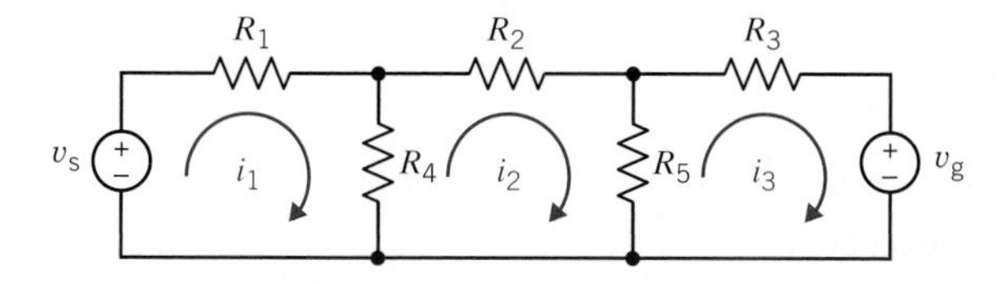

FIGURE 4.5-6 Circuit with three mesh currents and two voltage sources.

equations

$$\begin{aligned}
&\text{mesh 1: } -v_s + R_1 i_1 + R_4(i_1 - i_2) = 0\\
&\text{mesh 2: } R_2 i_2 + R_5(i_2 - i_3) + R_4(i_2 - i_1) = 0\\
&\text{mesh 3: } R_5(i_3 - i_2) + R_3 i_3 + v_g = 0
\end{aligned}$$

These three mesh equations can be rewritten by collecting coefficients for each mesh current as

$$\begin{aligned}
&\text{mesh 1: } (R_1 + R_4)i_1 - R_4 i_2 = v_s\\
&\text{mesh 2: } -R_4 i_1 + R_5 + (R_4 + R_2 + R_5)i_2 - R_5 i_3 = 0\\
&\text{mesh 3: } -R_5 i_2 + (R_3 + R_5)i_3 = -v_g
\end{aligned}$$

Hence, we note that the coefficient of the mesh current i_1 for the first mesh is the sum of resistances in mesh 1, and the coefficient of the second mesh current is the negative of the resistance common to meshes 1 and 2. In general, we state that for mesh current i_n, the equation for the nth mesh with independent voltage sources only is obtained as follows:

$$-\sum_{q=1}^{Q} R_k i_q + \sum_{j=1}^{P} R_j i_n = -\sum_{n=1}^{N} v_{sn} \tag{4.5-3}$$

That is, for mesh n we multiply i_n by the sum of all resistances R_j around the mesh. Then we add the terms due to the resistances in common with another mesh as the negative of the connecting resistance R_k, multiplied by the mesh current in the adjacent mesh i_q for all Q adjacent meshes. Finally, the independent voltage sources around the loop appear on the right side of the equation as the negative of the voltage sources encountered as we traverse the loop in the direction of the mesh current. Remember that the preceding result is obtained assuming all mesh currents flow clockwise.

The general matrix equation for the mesh current analysis for independent voltage sources present in a circuit is

$$\mathbf{R}\,\mathbf{i} = \mathbf{v}_s \tag{4.5-4}$$

where $\mathbf{R}$ is a symmetric matrix with a diagonal consisting of the sum of resistances in each mesh and the off-diagonal elements are the negative of the sum of the resistances common to two meshes. The matrix $\mathbf{i}$ consists of the mesh current as

$$\mathbf{i} = \begin{bmatrix} i_1 \\ i_2 \\ \cdot \\ \cdot \\ \cdot \\ i_N \end{bmatrix}$$

For N mesh currents, the source matrix $\mathbf{v_s}$ is

$$\mathbf{v_s} = \begin{bmatrix} v_{s1} \\ v_{s2} \\ \cdot \\ \cdot \\ \cdot \\ v_{sN} \end{bmatrix}$$

where v_{sj} is the algebraic sum of the voltages of the voltage sources in the jth mesh with the appropriate sign assigned to each voltage.

For the circuit of Figure 4.5-6 and the matrix Eq. 4.5-4, we have

$$\mathbf{R} = \begin{bmatrix} (R_1 + R_4) & -R_4 & 0 \\ -R_4 & (R_2 + R_4 + R_5) & -R_5 \\ 0 & -R_5 & (R_3 + R_5) \end{bmatrix}$$

Note that $\mathbf{R}$ is a symmetric matrix, as we expected.

EXERCISE 4.5-1 Determine the value of the voltage measured by the voltmeter in Figure E 4.5-1.

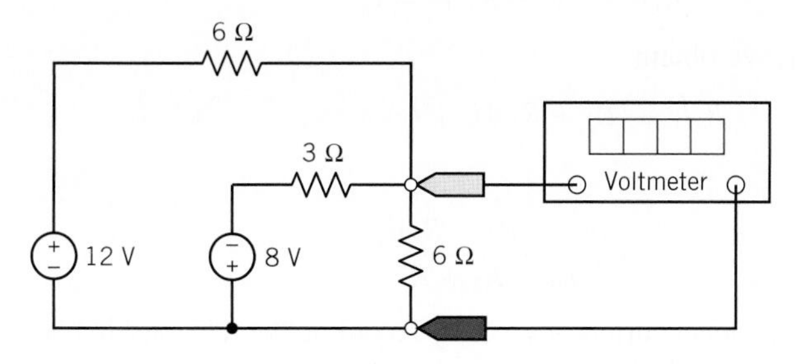

FIGURE E 4.5-1

Answer: −1 V

4.6 MESH CURRENT ANALYSIS WITH CURRENT AND VOLTAGE SOURCES

Heretofore, we have considered only circuits with independent voltage sources for analysis by the mesh current method. If the circuit has an independent current source, as shown in Figure 4.6-1, we recognize that the second mesh current is equal to the negative of the current source current. We can then write

$$i_2 = -i_s$$

and we need only determine the first mesh current i_1. Writing KVL for the first mesh, we obtain

$$(R_1 + R_2)i_1 - R_2 i_2 = v_s$$

Because $i_2 = -i_s$, we have

$$i_1 = \frac{v_s - R_2 i_s}{R_1 + R_2} \tag{4.6-1}$$

where i_s and v_s are sources of known magnitude.

If we encounter a circuit as shown in Figure 4.6-2, we have a current source i_s that has an unknown voltage v_{ab} across its terminals. We can readily note that

$$i_2 - i_1 = i_s \tag{4.6-2}$$

by writing KCL at node a. The two mesh equations are

$$\text{mesh 1:}\ R_1 i_1 + v_{ab} = v_s \tag{4.6-3}$$

$$\text{mesh 2:}\ (R_2 + R_3)i_2 - v_{ab} = 0 \tag{4.6-4}$$

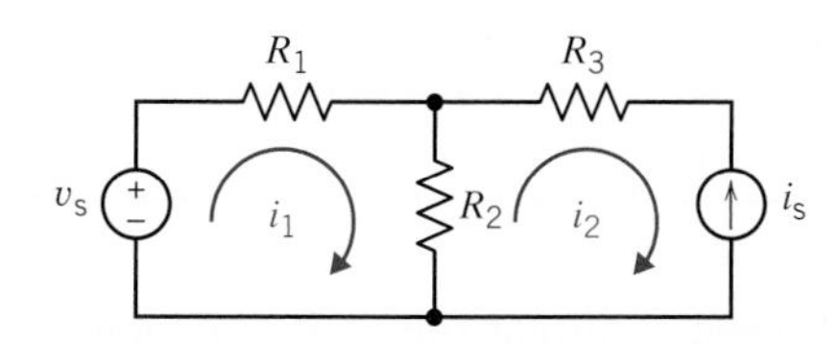

FIGURE 4.6-1 Circuit with an independent voltage source and an independent current source.

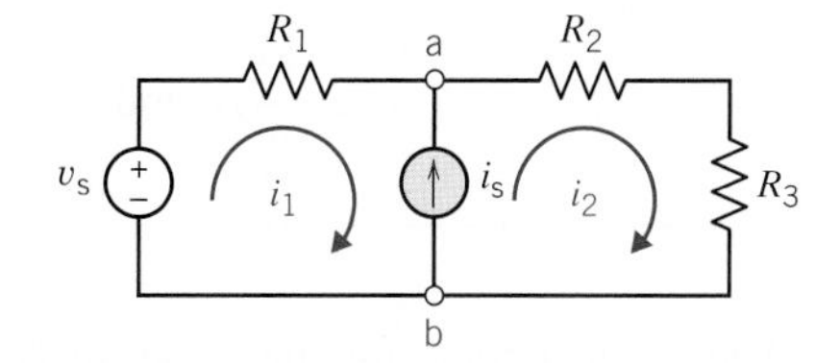

FIGURE 4.6-2 Circuit with an independent current source common to both meshes.

We note that if we add Eqs. 4.6-3 and 4.6-4, we eliminate v_{ab}, obtaining

$$R_1 i_1 + (R_2 + R_3) i_2 = v_s$$

However, because $i_2 = i_s + i_1$, we obtain

$$R_1 i_1 + (R_2 + R_3)(i_s + i_1) = v_s$$

or

$$i_1 = \frac{v_s - (R_2 + R_3) i_s}{R_1 + R_2 + R_3} \tag{4.6-5}$$

Thus, we account for independent current sources by recording the relationship between the mesh currents and the current source current. If the current source influences *only one* mesh current, we write the equation that relates that mesh current to the current source current and write the KVL equations for the remaining meshes. If the current source influences two mesh currents, we write the KVL equation for both meshes, assuming a voltage v_{ab} across the terminals of the current source. Then, adding these two mesh equations, we obtain an equation independent of v_{ab}.

EXAMPLE 4.6-1 Mesh Equations

Consider the circuit of Figure 4.6-3 where $R_1 = R_2 = 1\ \Omega$ and $R_3 = 2\ \Omega$. Find the three mesh currents.

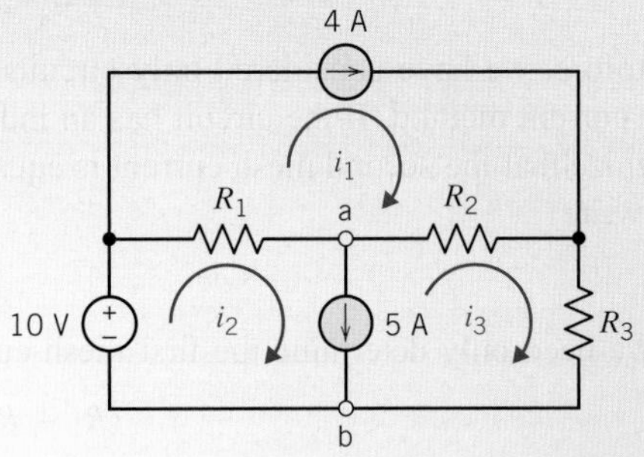

FIGURE 4.6-3 Circuit with two independent current sources.

Solution

Because the 4-A source is in mesh 1 only, we note that

$$i_1 = 4$$

For the 5-A source, we have

$$i_2 - i_3 = 5 \tag{4.6-6}$$

Writing KVL for mesh 2 and mesh 3, we obtain

$$\text{mesh 2: } R_1(i_2 - i_1) + v_{ab} = 10 \tag{4.6-7}$$

$$\text{mesh 3: } R_2(i_3 - i_1) + R_3 i_3 - v_{ab} = 0 \tag{4.6-8}$$

We substitute $i_1 = 4$ and add Eqs. 4.6-7 and 4.6-8 to obtain

$$R_1(i_2 - 4) + R_2(i_3 - 4) + R_3 i_3 = 10 \tag{4.6-9}$$

From Eq. 4.6-6, $i_2 = 5 + i_3$, substituting into Eq. 4.6-9, we have

$$R_1(5 + i_3 - 4) + R_2(i_3 - 4) + R_3 i_3 = 10$$

Using the values for the resistors, we obtain

$$i_3 = \frac{13}{4}\text{A and } i_2 = 5 + i_3 = \frac{33}{4}\text{A}$$

Another technique for the mesh analysis method when a current source is common to two meshes involves the concept of a supermesh. A *supermesh* is one mesh created from two meshes that have a current source in common, as shown in Figure 4.6-4.

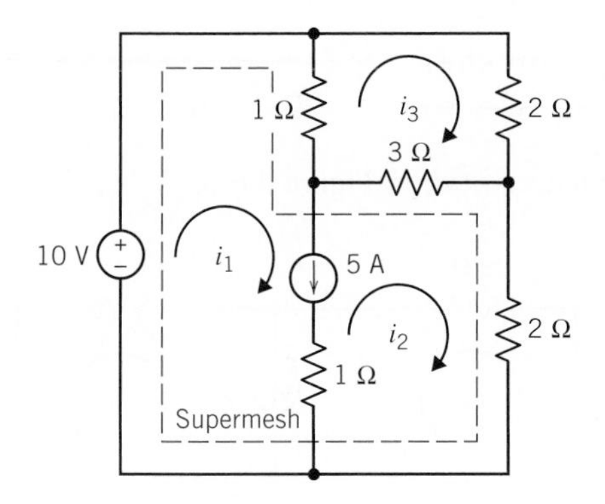

FIGURE 4.6-4 Circuit with a supermesh that incorporates mesh 1 and mesh 2. The supermesh is indicated by the dashed line.

A **supermesh** is one larger mesh created from two meshes that have an independent or dependent current source in common.

For example, consider the circuit of Figure 4.6-4. The 5-A current source is common to mesh I and mesh 2. The supermesh consists of the interior of mesh 1 and mesh 2. Writing KVL around the periphery of the supermesh shown by the dashed lines, we obtain

$$-10 + 1(i_1 - i_3) + 3(i_2 - i_3) + 2i_2 = 0$$

For mesh 3, we have

$$1(i_3 - i_1) + 2i_3 + 3(i_3 - i_2) = 0$$

Finally, the equation that relates the current source current to the mesh currents is

$$i_1 - i_2 = 5$$

Then the three equations may be reduced to

$$\begin{aligned}
&\text{supermesh:} && 1i_1 + 5i_2 - 4i_3 && = 10\\
&\text{mesh 3:} && -1i_1 - 3i_2 + 6i_3 && = 0\\
&\text{current source:} && 1i_1 - 1i_2 && = 5
\end{aligned}$$

Therefore, solving the three equations simultaneously, we find that $i_2 = 2.5\text{A}, i_1 = 7.5$ A, and $i_3 = 2.5\text{A}$.

The methods of mesh current analysis used when a current source is present are summarized in Table 4.6-1.

Table 4.6-1 Mesh Current Analysis Methods with a Current Source

CASE	METHOD
1. A current source appears on the periphery of only one mesh, n.	Equate the mesh current i_n to the current source current, accounting for the direction of the current source.
2. A current source is common to two meshes.	A. Assume a voltage v_{ab} across the terminals of the current source, write the KVL equations for the two meshes, and add them to eliminate v_{ab}, or, B. create a supermesh as the periphery of the two meshes and write one KVL equation around the periphery of the supermesh. In addition, write the constraining equation for the two mesh currents in terms of the current source.

EXAMPLE 4.6-2 Supermeshes

Determine the values of the mesh currents, i_1 and i_2, for the circuit shown in Figure 4.6-5.

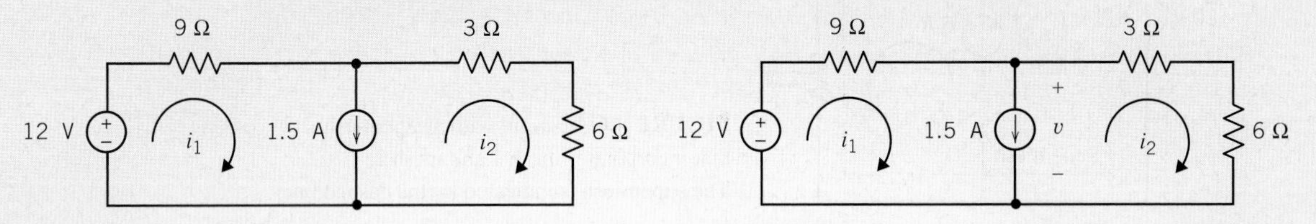

FIGURE 4.6-5 The circuit for Example 4.6-2.

FIGURE 4.6-6 Method 1 of Example 4.6-2.

Solution

We can write the first mesh equation by considering the current source. The current source current is related to the mesh currents by

$$i_1 - i_2 = 1.5 \quad \Rightarrow \quad i_1 = i_2 + 1.5$$

To write the second mesh equation, we must decide what to do about the current source voltage. (Notice that there is no easy way to express the current source voltage in terms of the mesh currents.) In this example, we illustrate two methods of writing the second mesh equation.

Method 1: Assign a name to the current source voltage. Apply KVL to both of the meshes. Eliminate the current source voltage from the KVL equations.

Figure 4.6-6 shows the circuit after labeling the current source voltage. The KVL equation for mesh 1 is

$$9i_1 + v - 12 = 0$$

The KVL equation for mesh 2 is

$$3i_2 + 6i_2 - v = 0$$

Combining these two equations gives

$$9i_1 + (3i_2 + 6i_2) - 12 = 0 \quad \Rightarrow \quad 9i_1 + 9i_2 = 12$$

Method 2: Apply KVL to the supermesh corresponding to the current source. Shown in Figure 4.6-7, this supermesh is the perimeter of the two meshes that each contain the current source. Apply KVL to the supermesh to get

$$9i_1 + 3i_2 + 6i_2 - 12 = 0 \quad \Rightarrow \quad 9i_1 + 9i_2 = 12$$

This is the same equation that was obtained using method 1. Applying KVL to the supermesh is a shortcut for doing three things:

1. Labeling the current source voltage as v
2. Applying KVL to both meshes that contain the current source
3. Eliminating v from the KVL equations

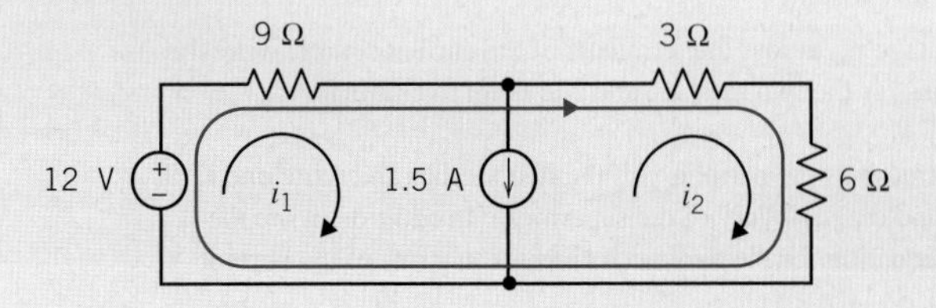

FIGURE 4.6-7 Method 2 of Example 4.6-2.

In summary, the mesh equations are

$$i_1 = i_2 + 1.5$$

and

$$9i_1 + 9i_2 = 12$$

Solving the node equations gives

$$i_1 = 1.4167\text{A} \quad \text{and} \quad i_2 = -83.3\text{ mA}$$

EXERCISE 4.6-1 Determine the value of the voltage measured by the voltmeter in Figure E 4.6-1.

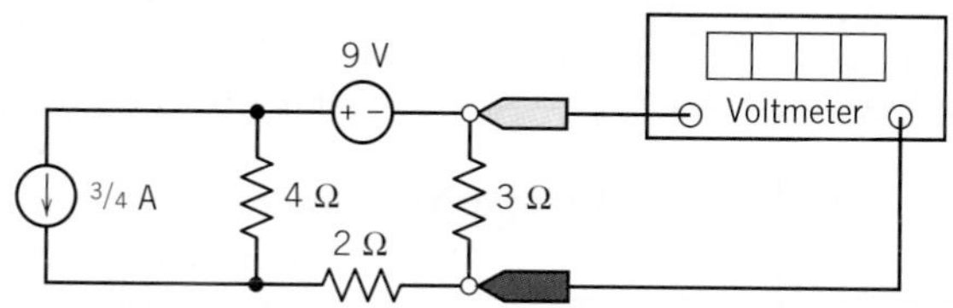

FIGURE E 4.6-1

Hint: Write and solve a single mesh equation to determine the current in the 3 Ω resistor.

Answer: −4 V

EXERCISE 4.6-2 Determine the value of the current measured by the ammeter in Figure E 4.6-2.

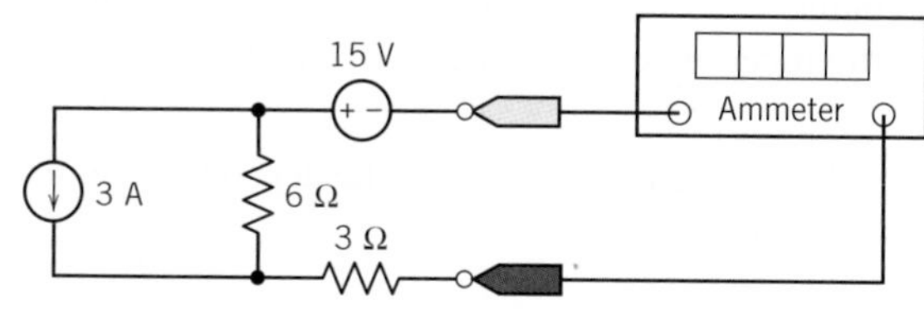

FIGURE E 4.6-2

Hint: Write and solve a single mesh equation.

Answer: −3.67 A

4.7 MESH CURRENT ANALYSIS WITH DEPENDENT SOURCES

When a circuit contains a dependent source the controlling current or voltage of that dependent source must be expressed as a function of the mesh currents.

It is then a simple matter to express the controlled current or voltage as a function of the mesh currents. The mesh equations can then be obtained by applying Kirchhoff s voltage law to the meshes of the circuit.

Example 4.7-1 Mesh Equations and Dependent Sources

INTERACTIVE EXAMPLE

Consider the circuit shown in Figure 4.7-1*a*. Find the value of the voltage measured by the voltmeter.

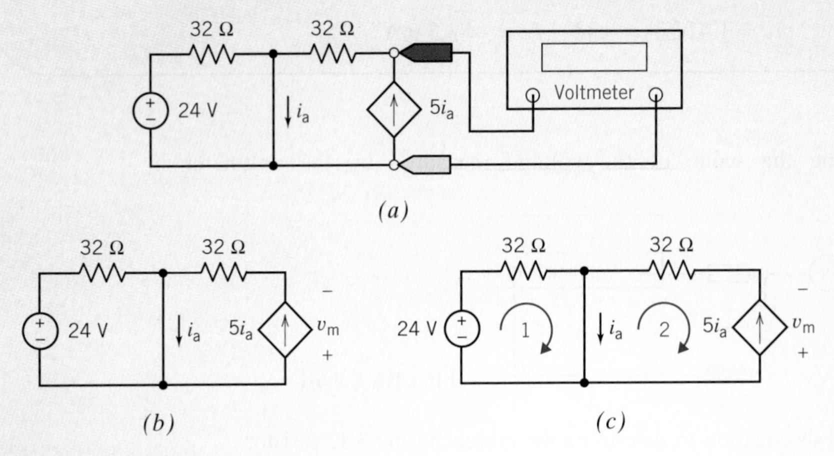

FIGURE 4.7-1 (*a*) The circuit considered in Example 4.7-1. (*b*) The circuit after replacing the voltmeter by an open circuit. (*c*) The circuit after labeling the meshes.

Solution

Figure 4.7-1*b* shows the circuit after replacing the voltmeter by an equivalent open circuit and labeling the voltage, v_m, measured by the voltmeter. Figure 4.7-1*c* shows the circuit after numbering the meshes. Let i_1 and i_2 denote the mesh currents in meshes 1 and 2, respectively.

The controlling current of the dependent source, i_a, is the current in a short circuit. This short circuit is common to meshes 1 and 2. The short-circuit current can be expressed in terms of the mesh currents as

$$i_a = i_1 - i_2$$

The dependent source is in only one mesh, mesh 2. The reference direction of the dependent source current does not agree with the reference direction of i_2. Consequently,

$$5i_a = -i_2$$

Solving for i_2 gives

$$i_2 = -5i_a = -5(i_1 - i_2)$$

Therefore,

$$-4i_2 = -5i_1 \Rightarrow i_2 = \frac{5}{4}i_1$$

Apply KVL to mesh 1 to get

$$32i_1 - 24 = 0 \Rightarrow i_1 = \frac{3}{4}\,\text{A}$$

Consequently, the value of i_2 is

$$i_2 = \frac{5}{4}\left(\frac{3}{4}\right) = \frac{15}{16}\,\text{A}$$

Apply KVL to mesh 2 to get

$$32i_2 - v_m = 0 \Rightarrow v_m = 32i_2$$

Finally,

$$v_m = 32\left(\frac{15}{16}\right) = 30\ \text{V}$$

EXAMPLE 4.7-2 Mesh Equations and Dependent Sources

INTERACTIVE EXAMPLE

Consider the circuit shown in Figure 4.7-2*a*. Find the value of the gain, *A*, of the CCVS.

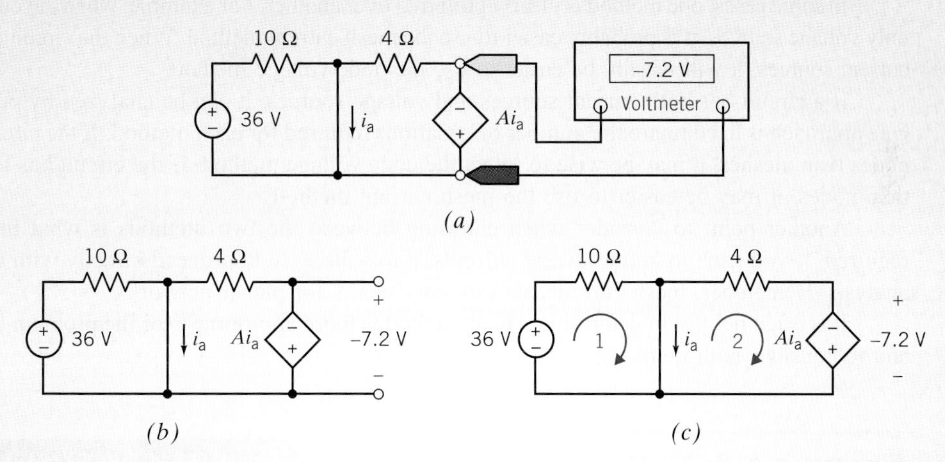

FIGURE 4.7-2 (*a*) The circuit considered in Example 4.7-2. (*b*) The circuit after replacing the voltmeter by an open circuit. (*c*) The circuit after labeling the meshes.

Solution

Figure 4 7-2*b* shows the circuit after replacing the voltmeter by an equivalent open circuit and labeling the voltage measured by the voltmeter. Figure 4.7-2*c* shows the circuit after numbering the meshes. Let i_1 and i_2 denote the mesh currents in meshes 1 and 2, respectively.

The voltage across the dependent source is represented in two ways. It is Ai_a with the + of reference direction at the bottom and −7.2 V with the + at the top. Consequently,

$$Ai_a = -(-7.2) = 7.2 \text{ V}$$

The controlling current of the dependent source, i_a, is the current in a short circuit. This short circuit is common to meshes 1 and 2. The short-circuit current can be expressed in terms of the mesh currents as

$$i_a = i_1 - i_2$$

Apply KVL to mesh 1 to get

$$10i_1 - 36 = 0 \Rightarrow i_1 = 3.6 \text{ A}$$

Apply KVL to mesh 2 to get

$$4i_2 + (-7.2) = 0 \Rightarrow i_2 = 1.8 \text{ A}$$

Finally,

$$A = \frac{Ai_a}{i_a} = \frac{Ai_a}{i_1 - i_2} = \frac{7.2}{3.6 - 1.8} = 4 \text{ V/A}$$

4.8 THE NODE VOLTAGE METHOD AND MESH CURRENT METHOD COMPARED

The analysis of a complex circuit can usually be accomplished by either the node voltage or the mesh current method. The advantage of using these methods is the systematic procedures provided for obtaining the simultaneous equations.

In some cases, one method is clearly preferred over another. For example, when the circuit contains only voltage sources, it is probably easier to use the mesh current method. When the circuit contains only current sources, it will usually be easier to use the node voltage method.

If a circuit has both current sources and voltage sources, it can be analyzed by either method. One approach is to compare the number of equations required for each method. If the circuit has fewer nodes than meshes, it may be wise to select the node voltage method. If the circuit has fewer meshes than nodes, it may be easier to use the mesh current method.

Another point to consider when choosing between the two methods is what information is required. If you need to know several currents, it may be wise to proceed directly with mesh current analysis. Remember, mesh current analysis only works for planar networks.

It is often helpful to determine which method is more appropriate for the problem requirements and to consider both methods.

EXAMPLE 4.8-1 Mesh Equations

INTERACTIVE EXAMPLE

Consider the circuit shown in Figure 4.8-1. Find the value of the resistance, R.

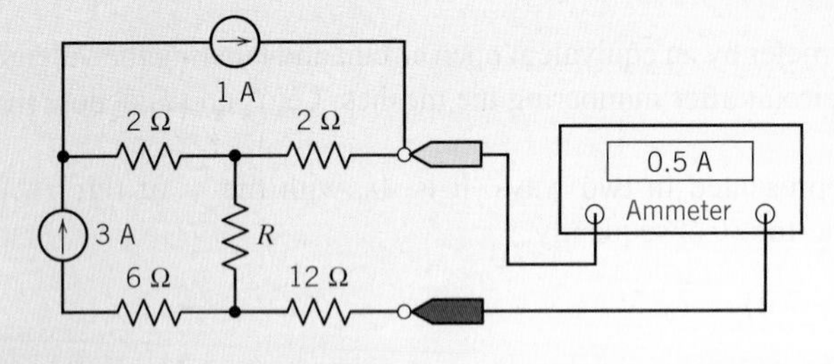

FIGURE 4.8-1 The circuit considered in Example 4.8-1.

Solution

Figure 4.8-2*a* shows the circuit from Figure 4.8-1 after replacing the ammeter by an equivalent short circuit and labeling the current measured by the ammeter. This circuit can be analyzed using mesh equations or using node equations. To decide which will be easier, we first count the nodes and meshes. This circuit has five nodes. Selecting a

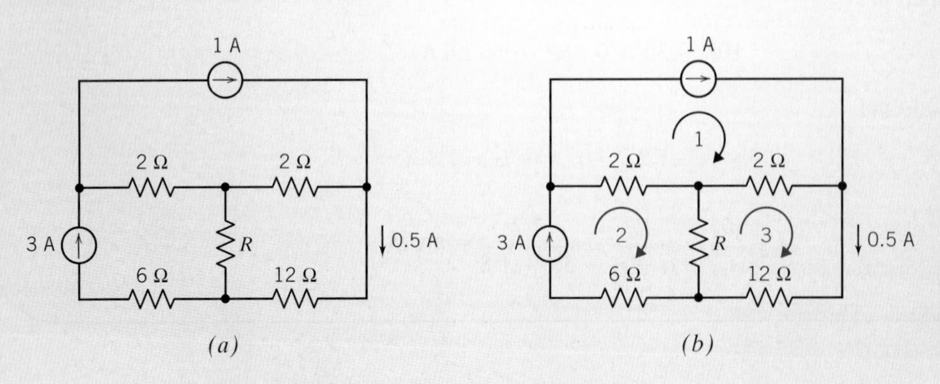

FIGURE 4.8-2 (*a*) The circuit from Figure 4.8-1 after replacing the ammeter by a short circuit. (*b*) The circuit after labeling the meshes.

reference node and then applying KCL at the other four nodes will produce a set of four node equations. The circuit has three meshes. Applying KVL to these three meshes will produce a set of three mesh equations. Hence, analyzing this circuit using mesh equations instead of node equations will produce a smaller set of equations. Further, notice that two of the three mesh currents can be determined directly from the current source currents. This makes the mesh equations easier to solve. We will analyze this circuit by writing and solving mesh equations.

Figure 4.8-2*b* shows the circuit after numbering the meshes. Let i_1, i_2, and i_3 denote the mesh currents in meshes 1, 2, and 3, respectively. The mesh current i_1 is equal to the current in the 1-A current source, so

$$i_1 = 1\text{ A}$$

The mesh current i_2 is equal to the current in the 3-A current source, so

$$i_2 = 3\text{ A}$$

The mesh current i_3 is equal to the current in the short circuit that replaced the ammeter, so

$$i_3 = 0.5\text{ A}$$

Apply KVL to mesh 3 to get

$$2(i_3 - i_1) + 12(i_3) + R(i_3 - i_2) = 0$$

Substituting the values of the mesh currents gives

$$2(0.5 - 1) + 12(0.5) + R(0.5 - 3) = 0 \quad \Rightarrow \quad R = 2\ \Omega$$

EXAMPLE 4.8-2 Node Equations

INTERACTIVE EXAMPLE

Consider the circuit shown in Figure 4.8-3. Find the value of the resistance, R.

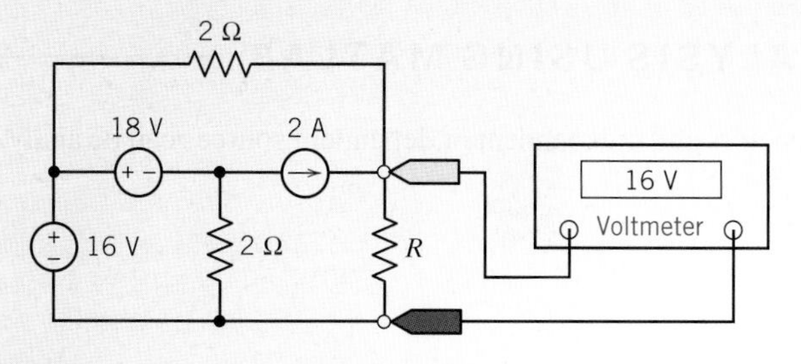

FIGURE 4.8-3 The circuit considered in Example 4.8-2.

Solution

Figure 4.8-4*a* shows the circuit from Figure 4.8-3 after replacing the voltmeter by an equivalent open circuit and labeling the voltage measured by the voltmeter. This circuit can be analyzed using mesh equations or node equations. To decide which will be easier, we first count the nodes and meshes. This circuit has four nodes. Selecting a reference node and then applying KCL at the other three nodes will produce a set of three node equations. The circuit has three meshes. Applying KVL to these three meshes will produce a set of three mesh equations. Analyzing this circuit using mesh equations requires the same number of equations as are required to analyze the circuit using node equations. Notice that one of the three mesh currents can be determined directly from the current source current, but two of the three node voltages can be determined directly from the voltage source voltages. This makes the node equations easier to solve. We will analyze this circuit by writing and solving node equations.

Figure 4.8-4*b* shows the circuit after selecting a reference node and numbering the other nodes. Let v_1, v_2, and v_3 denote the node voltages at nodes 1, 2, and 3, respectively. The voltage of the 16-V voltage source can be expressed in terms of the node voltages as

$$16 = v_1 - 0 \quad \Rightarrow \quad v_1 = 16\text{ V}$$

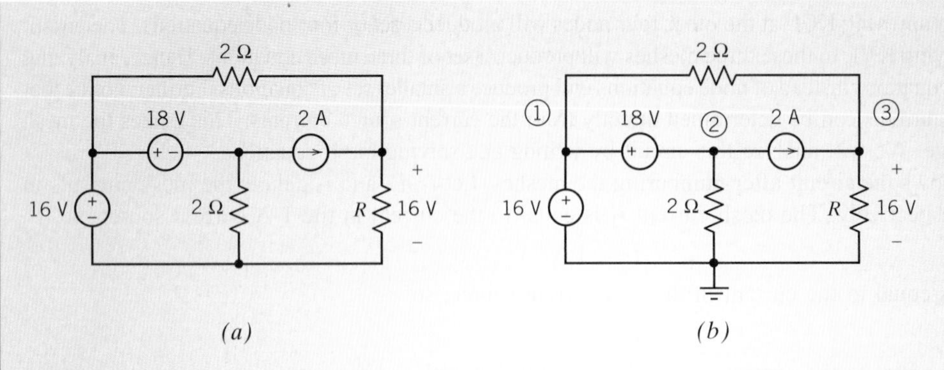

FIGURE 4.8-4 (*a*) The circuit from Figure 4.8-3 after replacing the voltmeter by an open circuit. (*b*) The circuit after labeling the nodes.

The voltage of the 18-V voltage source can be expressed in terms of the node voltages as

$$18 = v_1 - v_2 \quad \Rightarrow \quad 18 = 16 - v_2 \quad \Rightarrow \quad v_2 = -2\text{ V}$$

The voltmeter measures the node voltage at node 3, so

$$v_3 = 16\text{ V}$$

Applying KCL at node 3 to get

$$\frac{v_1 - v_3}{2} + 2 = \frac{v_3}{R}$$

Substituting the values of the node voltages gives

$$\frac{16 - 16}{2} + 2 = \frac{16}{R} \quad \Rightarrow \quad R = 8\,\Omega$$

4.9 MESH CURRENT ANALYSIS USING MATLAB

We have seen that circuits that contain resistors and independent or dependent sources can be analyzed in the following way:

1. Writing a set of node or mesh equations
2. Solving those equations simultaneously

In this section, we will use the MATLAB computer program to solve the equations.

Consider the circuit shown in Figure 4.9-1*a*. This circuit contains a potentiometer. In Figure 4.9-1*b*, the potentiometer has been replaced by a model of a potentiometer. R_p is the resistance of

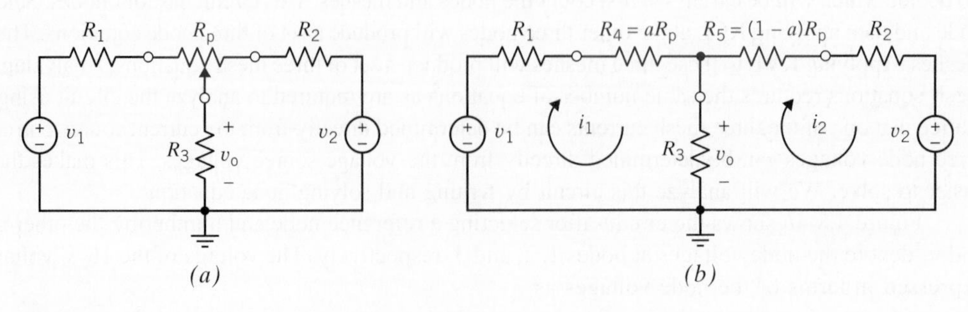

FIGURE 4.9-1 (*a*) A circuit that contains a potentiometer and (*b*) an equivalent circuit formed by replacing the potentiometer with a model of a potentiometer ($0 < a < 1$).

the potentiometer. The parameter a varies from 0 to 1 as the wiper of the potentiometer is moved from one end of the potentiometer to the other. The resistances R_4 and R_5 are described by the equations

$$R_4 = aR_p \tag{4.9-1}$$

and

$$R_5 = (1 - a)R_p \tag{4.9-2}$$

Our objective is to analyze this circuit to determine how the output voltage changes as the position of the potentiometer wiper is changed.

```
% mesh.m solves mesh equations
%------------------------------------------------------------
%  Enter values of the parameters that describe the circuit.
%------------------------------------------------------------
                        % circuit parameters
R1=1000;                % ohms
R2=1000;                % ohms
R3=5000;                % ohms
V1= 15;                 % volts
V2=-15;                 % volts

                        % potentiometer parameters
Rp=20e3;                % ohms
%------------------------------------------------------------
%  the parameter a varies from 0 to 1 in 0.05 increments.
%------------------------------------------------------------

a=0:0.05:1;             % dimensionless

for k=1:length(a)
    %--------------------------------------------------------
    % Here is the mesh equation, RI=V:
    %--------------------------------------------------------

    R = [R1+a(k)*Rp+R3        -R3;                   % ------
              -R3       (1-a(k))*Rp+R2+R3];          % eqn.
    V = [ V1;                                        % 4.9-6
        -V2];                                        % ------

    %--------------------------------------------------------
    % Tell MATLAB to solve the mesh equation:
    %--------------------------------------------------------
    I = V'/R;

    %--------------------------------------------------------
    % Calculate the output voltage from the mesh currents.
    %--------------------------------------------------------

    Vo(k) = R3*(I(1)-I(2)); % eqn. 4.9-7

end

%------------------------------------------------------------
%  Plot Vo versus a
%------------------------------------------------------------
plot(a, Vo)
axis([0 1 -15 15])
xlabel('a, dimensionless')
ylabel('Vo, V')
```

FIGURE 4.9-2 MATLAB input file used to analyze the circuit shown in Figure 4.9-1.

The circuit in Figure 4.9-1b can be represented by mesh equations as

$$\begin{aligned} R_1 i_1 + R_4 i_1 + R_3(i_1 - i_2) - v_1 &= 0 \\ R_5 i_2 + R_2 i_2 + [v_2 - R_3(i_1 - i_2)] &= 0 \end{aligned} \tag{4.9-3}$$

These mesh equations can be rearranged as

$$\begin{aligned} (R_1 + R_4 + R_3)i_1 - R_3 i_2 &= v_1 \\ -R_3 i_1 + (R_5 + R_2 + R_3)i_2 &= -v_2 \end{aligned} \tag{4.9-4}$$

Substituting Eqs. 4.9-1 and 4.9-2 into Eq. 4.9-4 gives

$$\begin{aligned} (R_1 + aR_p + R_3)i_1 - R_3 i_2 &= v_1 \\ -R_3 i_1 + \left[(1-a)R_p + R_2 + R_3\right]i_2 &= -v_2 \end{aligned} \tag{4.9-5}$$

Equation 4.9-5 can be written using matrices as

$$\begin{bmatrix} R_1 + aR_P + R_3 & -R_3 \\ -R_3 & (1-a)R_P + R_2 + R_3 \end{bmatrix} \begin{bmatrix} i_1 \\ i_2 \end{bmatrix} = \begin{bmatrix} v_1 \\ -v_2 \end{bmatrix} \tag{4.9-6}$$

Next, i_1 and i_2 are calculated by using MATLAB to solve the mesh equation, Eq. 4.9-6. Then the output voltage is calculated as

$$v_o = R_3(i_1 - i_2) \tag{4.9-7}$$

Figure 4.9-2 shows the MATLAB input file. The parameter a varies from 0 to 1 in increments of 0.05. At each value of a, MATLAB solves Eq. 4.9-6 and then uses Eq. 4.9-7 to calculate the output voltage. Finally, MATLAB produces the plot of v_o versus a that is shown in Figure 4.9-3.

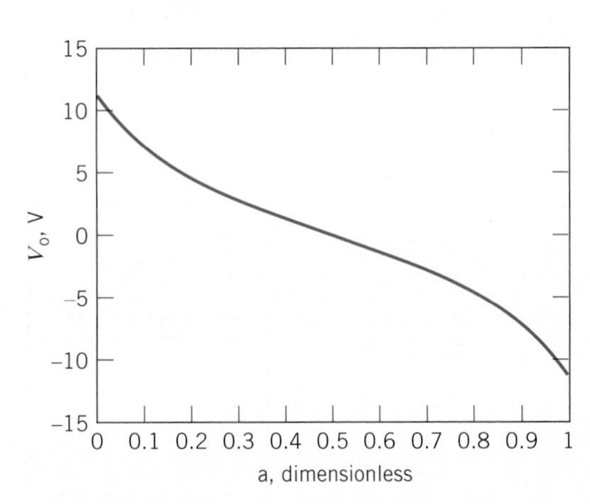

FIGURE 4.9-3 Plot of v_o versus a for the circuit shown in Figure 4.9-1.

4.10 USING PSPICE TO DETERMINE NODE VOLTAGES AND MESH CURRENTS

To determine the node voltages of a dc circuit using PSpice, we

1. Draw the circuit in the OrCAD Capture workspace
2. Specify a 'Bias Point' simulation
3. Run the simulation

PSpice will label the nodes with the values of the node voltages.

An extra step is needed to use PSpice to determine the mesh currents. PSpice does not label the values of the mesh currents, but it does provide the value of the current in each voltage source. Recall that a 0-V voltage source is equivalent to a short circuit. Consequently, we can insert 0-V current sources into the circuit without altering the values of the mesh currents. We will insert those sources into the circuit in such a way that their currents are also the mesh currents. To determine the mesh currents of a dc circuit using PSpice, we

1. Draw the circuit in the OrCAD Capture workspace.
2. Add 0-V voltage sources to measure the mesh currents.
3. Specify a Bias Point simulation.
4. Run the simulation.

PSpice will write the voltage source currents in the output file.

EXAMPLE 4.10-1 Using PSpice to Find Node Voltages and Mesh Currents

Use PSpice to determine the values of the node voltages and mesh currents for the circuit shown in Figure 4.10-1.

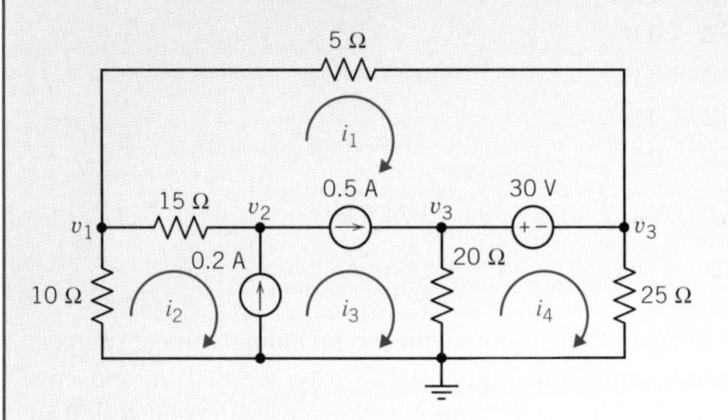

FIGURE 4.10-1 A circuit having node voltages v_1, v_2, v_3, and v_4 and mesh currents i_1, i_2, i_3, and i_4.

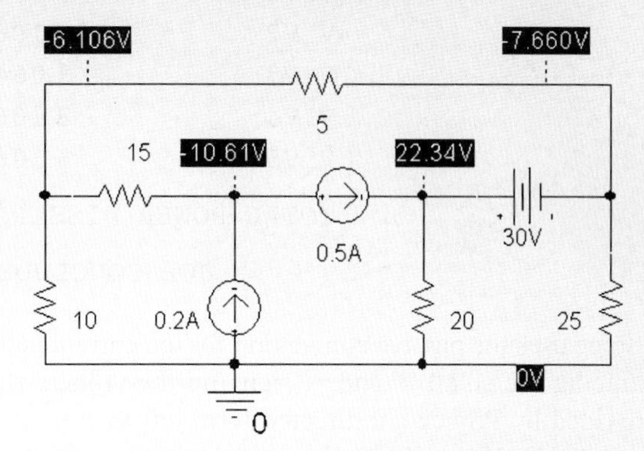

FIGURE 4.10-2 The circuit from Figure 4.10-1 drawn in the OrCAD workspace. The white numbers shown on black backgrounds are the values of the node voltages.

Solution

Figure 4.10-2 shows the result of drawing the circuit in the OrCAD workspace (see Appendix A) and performing a Bias Point simulation. (Select PSpice\New Simulation Profile from the OrCAD Capture menu bar; then choose Bias Point from the Analysis Type drop-down list in the Simulation Settings dialog box to specify a bias point simulation. Select PSpice\Run Simulation Profile from the OrCAD Capture menu bar to run the simulation.) PSpice labels the nodes with the values of the node voltages using white numbers shown on black backgrounds. Comparing Figures 4.10-1 and 4.10-2, we see that the node voltages are

$$v_1 = -6.106\text{ V}, v_2 = -10.61\text{ V}, v_3 = 22.34\text{ V}, \text{ and } v_4 = -7.660\text{ V}.$$

Figure 4.10-3 shows the circuit from Figure 4.10-2 after inserting a 0-V current source on the outside of each mesh. The currents in these 0-V sources will be the mesh currents shown in Figure 4.10-1. In particular, source V2 measures mesh current i_1, source V3 measures mesh current i_2, source V4 measures mesh current i_3, and source V5 measures mesh current i_4.

After we rerun the simulation (Select PSpice\Run from the OrCAD Capture menu bar), OrCAD Capture will open a Schematics window. Select View\Output File from the menu bar in the Schematics window. Scroll

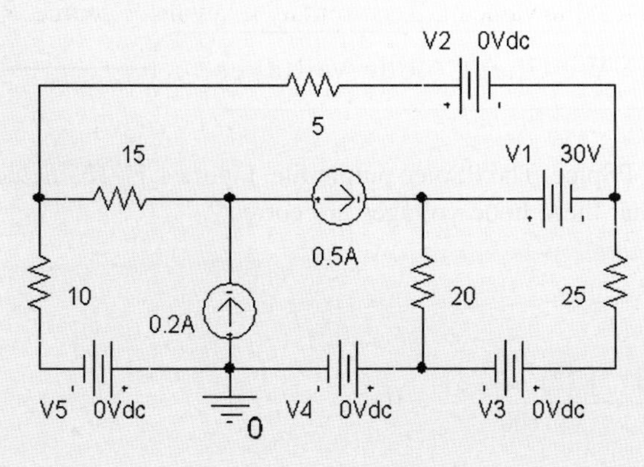

FIGURE 4.10-3 The circuit from Figure 4.10-1 drawn in the OrCAD workspace with 0-V voltage sources added to measure the mesh currents.

down through the output file to find the currents in the voltage sources:

```
VOLTAGE SOURCE CURRENTS
NAME            CURRENT

V_V1          - 6.170E - 01
V_V2            3.106E - 01
V_V3          - 3.064E - 01
V_V4            8.106E - 01
V_V5            6.106E - 01

TOTAL POWER DISSIPATION  1.85E + 01  WATTS

          JOB CONCLUDED
```

PSpice uses the passive convention for the current and voltage of all circuit elements, including voltage sources. Noticing the small + and – signs on the voltage source symbols in Figure 4.10-3, we see that the currents provided by PSpice are directed form left to right in sources VI and V2 and are directed from right to left in sources V3, V4, and V5. In particular, the mesh currents are

$$i_1 = 0.3106\text{ A}, i_2 = 0.6106\text{ A}, i_3 = 0.8106\text{ A}, \text{ and } i_4 = -0.3064\text{ A}.$$

4.11 HOW CAN WE CHECK . . . ?

Engineers are frequently called upon to check that a solution to a problem is indeed correct. For example, proposed solutions to design problems must be checked to confirm that all of the specifications have been satisfied. In addition, computer output must be reviewed to guard against data-entry errors, and claims made by vendors must be examined critically.

Engineering students are also asked to check the correctness of their work. For example, occasionally just a little time remains at the end of an exam. It is useful to be able quickly to identify those solutions that need more work.

The following examples illustrate techniques useful for checking the solutions of the sort of problem discussed in this chapter.

EXAMPLE 4.11-1 How Can We Check Node Voltages?

The circuit shown in Figure 4.11-1*a* was analyzed using PSpice. The PSpice output file, Figure 4.11-1*b*, includes the node voltages of the circuit. **How can we check** that these node voltages are correct?

Solution

The node equation corresponding to node 2 is

$$\frac{V(2) - V(1)}{100} + \frac{V(2)}{200} + \frac{V(2) - V(3)}{100} = 0$$

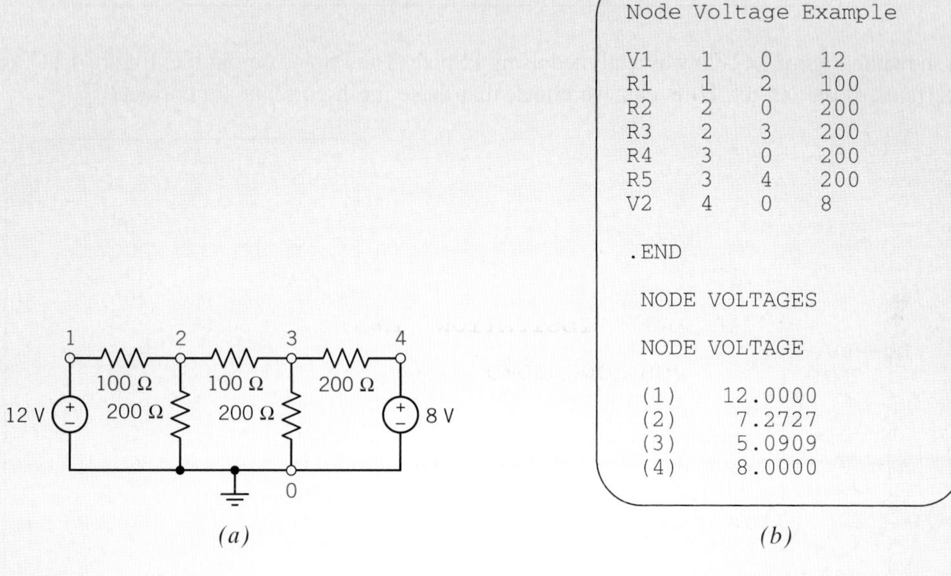

FIGURE 4.11-1 (*a*) A circuit and (*b*) the node voltages calculated using PSpice. The bottom node has been chosen as the reference node, which is indicated by the ground symbol and the node number 0. The voltages and resistors have units of voltages and ohms, respectively.

where, for example, $V(2)$ is the node voltage at node 2. When the node voltages from Figure 4.11-1*b* are substituted into the left-hand side of this equation, the result is

$$\frac{7.2727 - 12}{100} + \frac{7.2727}{200} + \frac{7.2727 - 5.0909}{100} = 0.011$$

The right-hand side of this equation should be 0 instead of 0.011. It looks like something is wrong. Is a current of only 0.011 negligible? Probably not in this case. If the node voltages were correct, then the currents of the 100-Ω resistors would be 0.047 A and 0.022 A, respectively. The current of 0.011 A does not seem negligible when compared to currents of 0.047 A and 0.022 A.

Is it possible that PSpice would calculate the node voltages incorrectly? Probably not, but the PSpice input file could easily contain errors. In this case, the value of the resistance connected between nodes 2 and 3 has been mistakenly specified to be 200 Ω. After changing this resistance to 100 Ω, PSpice calculates the node voltages to be

$$V(1) = 12.0, \; V(2) = 7.0, \; V(3) = 5.5, \; V(4) = 8.0$$

Substituting these voltages into the node equation gives

$$\frac{7.0 - 12.0}{100} + \frac{7.0}{200} + \frac{7.0 - 5.5}{100} = 0.0$$

so these node voltages do satisfy the node equation corresponding to node 2.

EXAMPLE 4.11-2 How Can We Check Mesh Currents?

The circuit shown in Figure 4.11-2*a* was analyzed using PSpice. The PSpice output file, Figure 4.11-2*b*, includes the mesh currents of the circuit. **How can we check** that these mesh currents are correct?

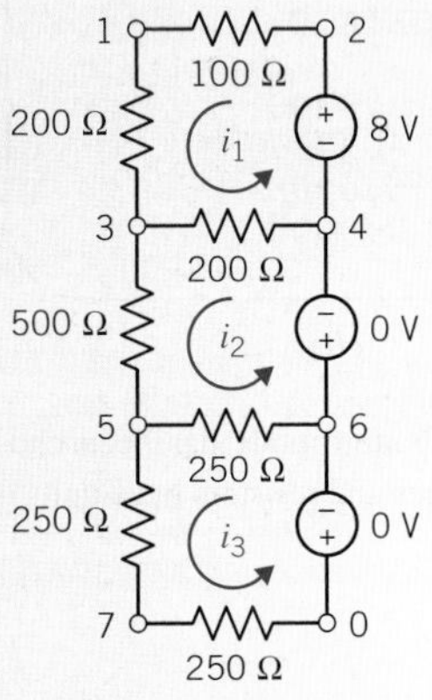

(*a*)

```
Mesh Current Example

R1   1   2   100
R2   1   3   200
V1   2   4   8
R3   3   4   200
R5   3   5   500
V2   4   6   0
R6   5   6   250
R7   5   7   250
V3   6   0   0
R8   7   0   250

.END

 MESH CURRENTS

 NAME    CURRENT

 I1     1.763E-02
 I2    -4.068E-03
 I3    -1.356E-03
```

(*b*)

FIGURE 4.11–2 (*a*) A circuit and (*b*) the mesh currents calculated using PSpice. The voltages and resistances are given in volts and ohms, respectively.

(The PSpice output file will include the currents through the voltage sources. Recall that PSpice uses the passive convention, so the current in the 8-V source will be $-i_1$ instead of i_1. The two 0-V sources have been added to include mesh currents i_2 and i_3 in the PSpice output file.)

Solution

The mesh equation corresponding to mesh 2 is

$$200(i_2 - i_1) + 500 i_2 + 250(i_2 - i_3) = 0$$

When the mesh currents from Figure 4.11-2*b* are substituted into the left-hand side of this equation, the result is

$$200(-0.004068 - 0.01763) + 500(-0.004068) + 250(-0.004068 - (-0.001356)) = 1.629$$

The right-hand side of this equation should be 0 instead of 1.629. It looks like something is wrong. Most likely, the PSpice input file contains an error. This is indeed the case. The nodes of both 0-V voltage sources have been entered in the wrong order. Recall that the first node should be the positive node of the voltage source. After correcting this error, PSpice gives

$$i_1 = 0.01763, \quad i_2 = 0.004068, \quad i_3 = 0.001356$$

Using these values in the mesh equation gives

$$200(0.004068 - 0.01763) + 500(0.004068) + 250(0.004068 - 0.001356) = 0.0$$

These mesh currents do indeed satisfy the mesh equation corresponding to mesh 2.

4.12 DESIGN EXAMPLE

POTENTIOMETER ANGLE DISPLAY

A circuit is needed to measure and display the angular position of a potentiometer shaft. The angular position, θ, will vary from -180° to 180°.

Figure 4.12-1 illustrates a circuit that could do the job. The +15-V and –15-V power supplies, the potentiometer, and resistors R_1 and R_2 are used to obtain a voltage, v_i, that is proportional to θ. The amplifier is used to change the constant of proportionality to obtain a simple relationship between θ and the voltage, v_o, displayed by the voltmeter. In this example, the amplifier will be used to obtain the relationship

$$v_o = k \cdot \theta \text{ where } k = 0.1 \frac{\text{volt}}{\text{degree}} \tag{4.12-1}$$

so that θ can be determined by multiplying the meter reading by 10. For example, a meter reading of -7.32 V indicates that $\theta = -73.2^\circ$.

Describe the Situation and the Assumptions

The circuit diagram in Figure 4.12-2 is obtained by modeling the power supplies as ideal voltage sources, the voltmeter as an open circuit, and the potentiometer by two resistors. The parameter, a, in the model of the potentiometer varies from 0 to 1 as θ varies from -180° to 180°. That means

$$a = \frac{\theta}{360^\circ} + \frac{1}{2} \tag{4.12-2}$$

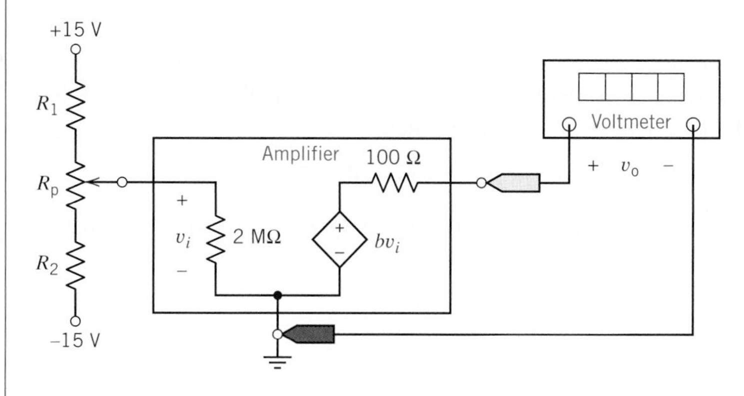

FIGURE 4.12-1 Proposed circuit for measuring and displaying the angular position of the potentiometer shaft.

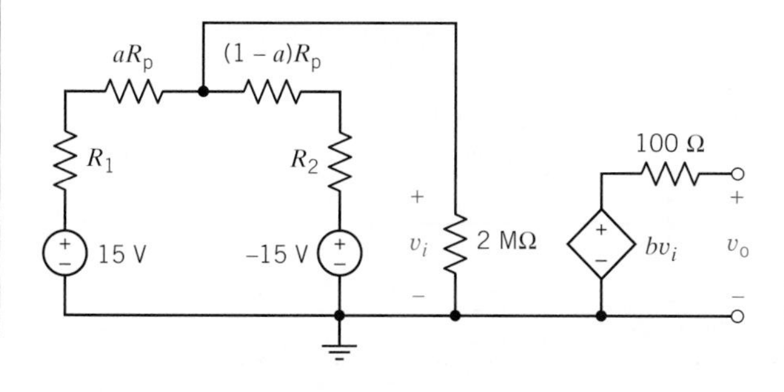

FIGURE 4.12-2 Circuit diagram containing models of the power supplies, voltmeter, and potentiometer.

Solving for θ gives

$$\theta = \left(a - \frac{1}{2}\right) \cdot 360^\circ \tag{4.12-3}$$

State the Goal

Specify values of resistors R_1 and R_2, the potentiometer resistance R_P, and the amplifier gain b that will cause the meter voltage, v_o, to be related to the angle θ by Eq. 4.12-1.

Generate a Plan

Analyze the circuit shown in Figure 4.12-2 to determine the relationship between v_i and θ. Select values of R_1, R_2, and R_p. Use these values to simplify the relationship between v_i and θ. If possible, calculate the value of b that will cause the meter voltage, v_o, to be related to the angle θ by Eq. 4.12-1. If this isn't possible, adjust the values of R_1, R_2, and R_p and try again.

Act on the Plan

The circuit has been redrawn in Figure 4.12-3. A single node equation will provide the relationship between between v_i and θ:

$$\frac{v_i}{2\,\text{M}\Omega} + \frac{v_i - 15}{R_1 + aR_p} + \frac{v_i - (-15)}{R_2 + (1-a)R_p} = 0$$

Solving for v_i *gives*

$$v_i = \frac{2\,\text{M}\Omega\left(R_p(2a-1) + R_1 - R_2\right)15}{\left(R_1 + aR_p\right)\left(R_2 + (1-a)R_p\right) + 2\,\text{M}\Omega\left(R_1 + R_2 + R_p\right)} \tag{4.12-4}$$

This equation is quite complicated. Let's put some restrictions on R_1, R_2, and R_p that will make it possible to simplify this equation. First, let $R_1 = R_2 = R$. Second, require that both R and R_p be much smaller than 2 MΩ (for example, $R < 20\,\text{k}\Omega$). Then,

$$\left(R + aR_p\right)\left(R + (1-a)R_p\right) \ll 2\,\text{M}\Omega\left(2R + R_p\right)$$

That is, the first term in the denominator of the left side of Eq. 4.12-4 is negligible compared to the second term. Equation 4.12-4 can be simplified to

$$v_i = \frac{R_p(2a-1)15}{2R + R_p}$$

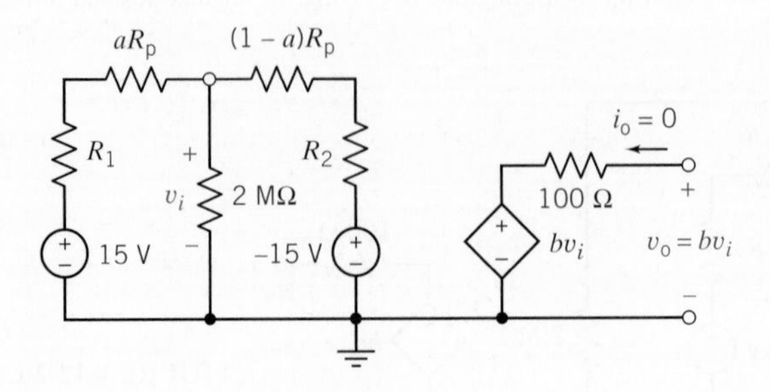

FIGURE 4.12-3 The redrawn circuit showing the mode v_i.

Next, using Eq. 4.12-3,

$$v_i = \left(\frac{R_p}{2R + R_p}\right)\left(\frac{15\text{ V}}{180^\circ}\right)\theta$$

It is time to pick values for R and R_p. Let $R = 5\text{ k}\Omega$ and $R_p = 10\text{ k}\Omega$; then

$$v_i = \left(\frac{7.5\text{ V}}{180^\circ}\right)$$

Referring to Figure 4.12-2, the amplifier output is given by

$$v_o = bv_i \tag{4.12-5}$$

so

$$v_o = b\left(\frac{7.5\text{ V}}{180^\circ}\right)\theta$$

Comparing this equation to Eq. 4.12-1 gives

$$b\left(\frac{7.5\text{ V}}{180^\circ}\right) = 0.1\frac{\text{volt}}{\text{degree}}$$

or

$$b = \frac{180}{7.5}(0.1) = 2.4$$

The final circuit is shown in Figure 4.12-4.

Verify the Proposed Solution

As a check, suppose $\theta = 150^\circ$. From Eq. 4.12-2, we see that

$$a = \frac{150^\circ}{360^\circ} + \frac{1}{2} = 0.9167$$

Using Eq. 4.12-4, we calculate

$$v_i = \frac{2\text{ M}\Omega(10\text{ k}\Omega(2 \times 0.9167 - 1))15}{(5\text{ k}\Omega + 0.9167 \times 10\text{ k}\Omega)(5\text{ k}\Omega + (1 - 0.9167)10\text{ k}\Omega) + 2\text{ M}\Omega(2 \times 5\text{ k}\Omega + 10\text{ k}\Omega)} = 6.24$$

Finally, Eq. 4.12-5 indicates that the meter voltage will be

$$v_o \times 2.4 \cdot 6.24 = 14.98$$

This voltage will be interpreted to mean that the angle was

$$\theta = 10 \cdot v_o = 149.8^\circ$$

which is correct to three significant digits.

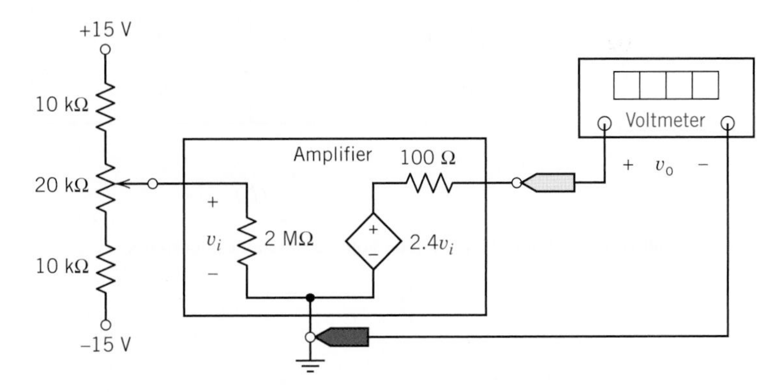

FIGURE 4.12-4 The final designed circuit.

4.13 SUMMARY

- The node voltage method of circuit analysis identifies the nodes of a circuit where two or more elements are connected. When the circuit consists of only resistors and current sources, the following procedure is used to obtain the node equations.
  1. We choose one node as to the reference node. Label the node voltages at the other nodes.
  2. Express element currents as functions of the node voltages. Figure 4.13-1*a* illustrates the relationship between the current in a resistor and the voltages at the nodes of the resistor.
  3. Apply KCL at all nodes except for the reference node. Solution of the simultaneous equations results in knowledge of the node voltages. All the voltages and currents in the circuit can be determined when the node voltages are known.
- When a circuit has voltage sources as well as current sources, we can still use the node voltage method by using the concept of a supernode. A supernode is a large node that includes two nodes connected by a known voltage source. If the voltage source is directly connected between a node q and the reference node, we may set $v_q = v_s$ and write the KCL equations at the remaining nodes.
- If the circuit contains a dependent source, we first express the controlling voltage or current of the dependent source as a function of the node voltages. Next, we express the controlled voltage or current as a function of the node voltages. Finally, we apply KCL to nodes and supernodes.
- Mesh current analysis is accomplished by applying KVL to the meshes of a planar circuit. When the circuit consists of only resistors and voltage sources, the following procedure is used to obtain the mesh equations.
  1. Label the mesh currents.
  2. Express element voltages as functions of the mesh currents. Figure 4.13-1*b* illustrates the relationship between the voltage across a resistor and the currents of the meshes that include the resistor.
  3. Apply KVL to all meshes.
  Solution of the simultaneous equations results in knowledge of the mesh currents. All the voltages and currents in the circuit can be determined when the mesh currents are known.
- If a current source is common to two adjoining meshes, we define the interior of the two meshes as a supermesh. We then write the mesh current equation around the periphery of the supermesh. If a current source appears at the periphery of only one mesh, we may define that mesh current as equal to the current of the source, accounting for the direction of the current source.
- If the circuit contains a dependent source, we first express the controlling voltage or current of the dependent source as a function of the mesh currents. Next, we express the controlled voltage or current as a function of the mesh currents. Finally, we apply KVL to meshes and supermeshes.
- In general, either node voltage or mesh current analysis can be used to obtain the currents or voltages in a circuit. However, a circuit with fewer node equations than mesh current equations may require that we select the node voltage method. Conversely, mesh current analysis is readily applicable for a circuit with fewer mesh current equations than node voltage equations.
- MATLAB greatly reduces the drudgery of solving node or mesh equations.

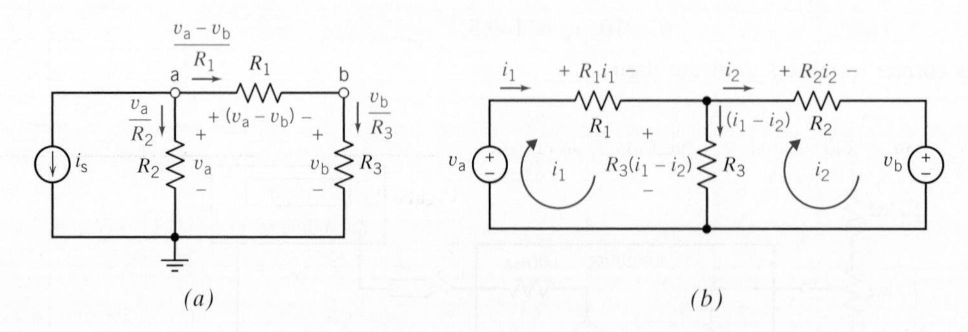

FIGURE 4.13-1 Expressing resistor currents and voltages in terms of (*a*) node voltage or (*b*) mesh currents.

PROBLEMS

Section 4.2 Node Voltage Analysis of Circuits with Current Sources

P 4.2-1 The node voltages in the circuit of Figure P 4.2-1 are $v_1 = -8$ V and $v_2 = 4$ V. Determine i, the current of the current source.

Answer: $i = 1.5$ A

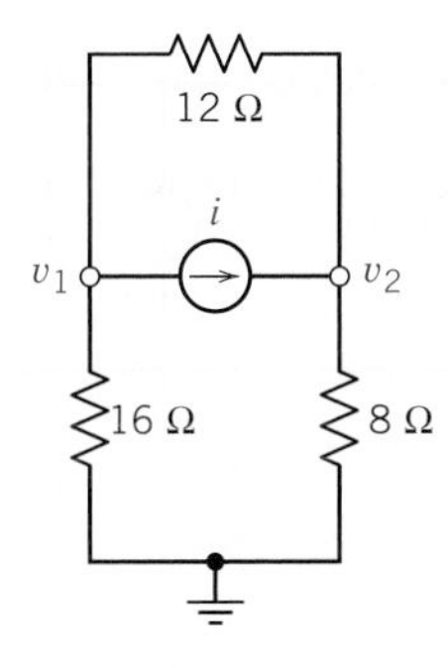

Figure P 4.2-1

P 4.2-2 Determine the node voltages for the circuit of Figure P 4.2-2.

Answer: $v_1 = 4$ V, $v_2 = 60$ V, and $v_3 = 48$ V

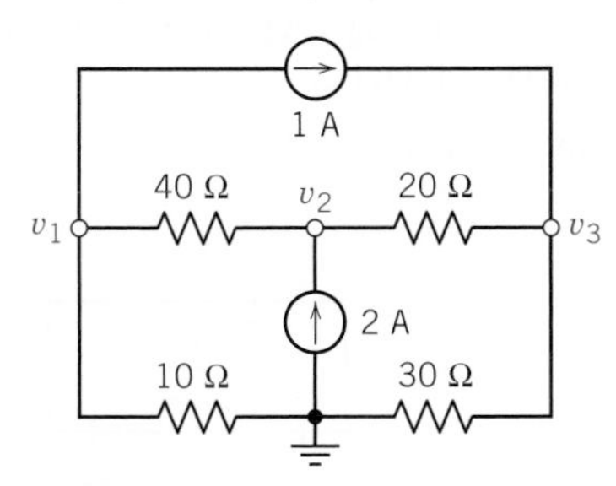

Figure P 4.2-2

P 4.2-3 The node voltages in the circuit of Figure P 4.2-3 are $v_1 = 8$ V, $v_2 = 30$ V, and $v_3 = 36$ V. Determine i_1 and i_2, the currents of the current sources.

Answer: $i_1 = -2$ A and $i_2 = 2$ A

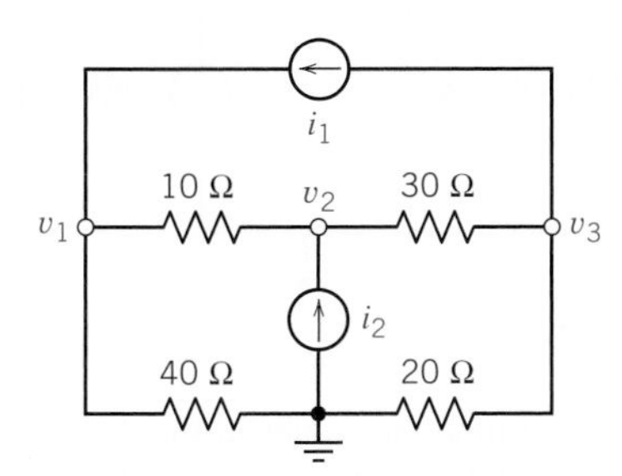

Figure P 4.2-3

P 4.2-4 Consider the circuit shown in Figure P 4.2-4. Find values of the resistances R_1 and R_2 that cause the voltages v_1 and v_2 to be $v_1 = 2$V and $v_2 = 4$ V.

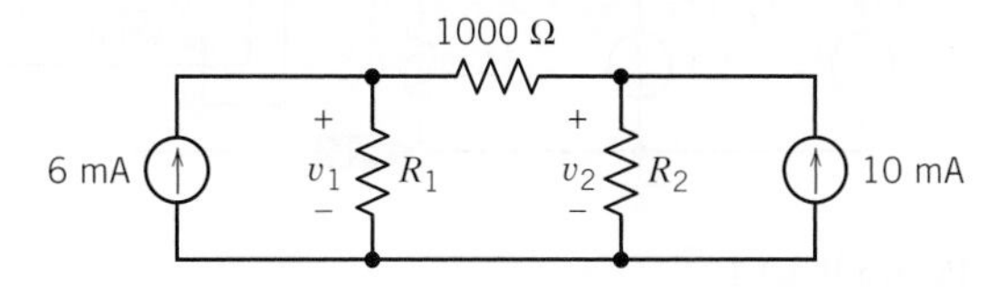

Figure P 4.2-4

P 4.2-5 Find the voltage v for the circuit shown in Figure P 4.2-5.

Answer: $v = 32$ mV

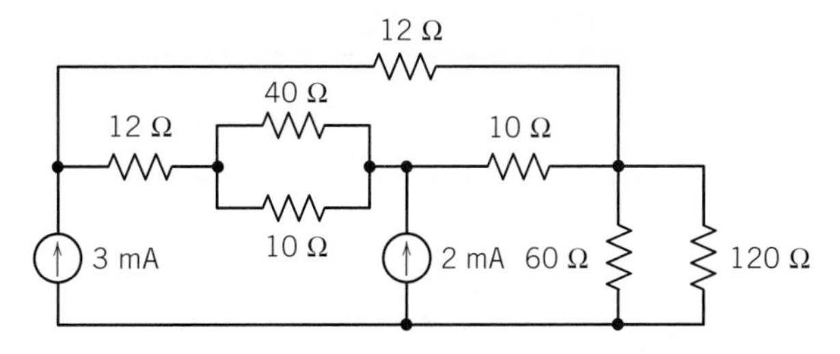

Figure P 4.2-5

P 4.2-6 Simplify the circuit shown in Figure P 4.2-6 by replacing series and parallel resistors with equivalent resistors; then analyze the simplified circuit by writing and solving node equations. (*a*) Determine the power supplied by each current source. (*b*) Determine the power received by the 12-Ω resistor.

Figure P 4.2-6

P 4.2-7 The node voltages in the circuit shown in Figure P 4.2-7 are $v_a = 14$ V and $v_b = 20$ V. Determine values of the current source current, i_s, and the resistance, R.

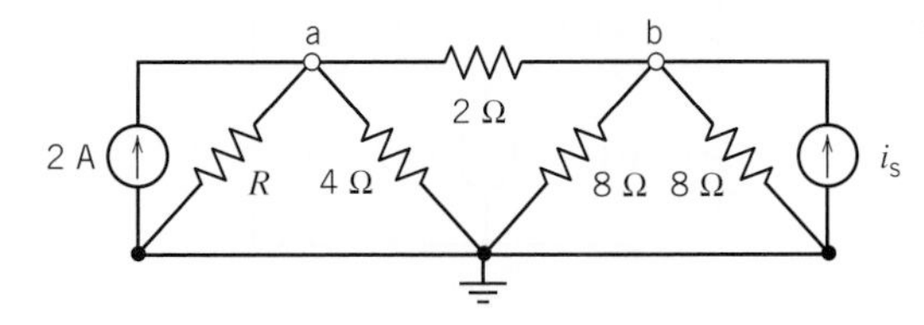

Figure P 4.2-7

Section 4.3 Node Voltage Analysis of Circuits with Current and Voltage Sources

P 4.3-1 The voltmeter in Figure P 4.3-1 measures v_c, the node voltage at node c. Determine the value of v_c.

Answer: $v_c = 12$ V

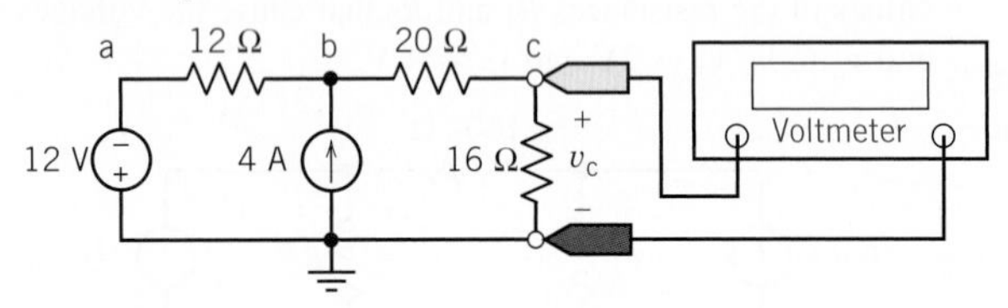

Figure P 4.3-1

P 4.3-2 The voltages v_a, v_b, v_c, and v_d in Figure P 4.3-2 are the node voltages corresponding to nodes a, b, c, and d. The current i is the current in a short circuit connected between nodes b and c. Determine the values of v_a, v_b, v_c, and v_d and of i.

Answer:
$v_a = -18$ V, $v_b = v_c = 10.5$ V, $v_d = -1.5$ V, $i = 1.75$ mA

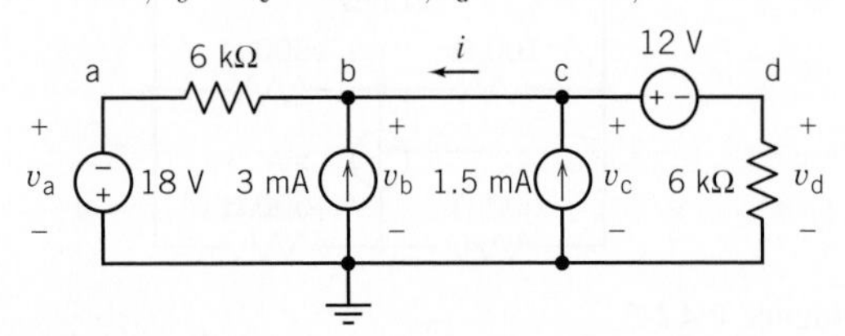

Figure P 4.3-2

P 4.3-3 Determine the node voltage v_a for the circuit of Figure P 4.3-3.

Answer: $v_a = 11.25$ V

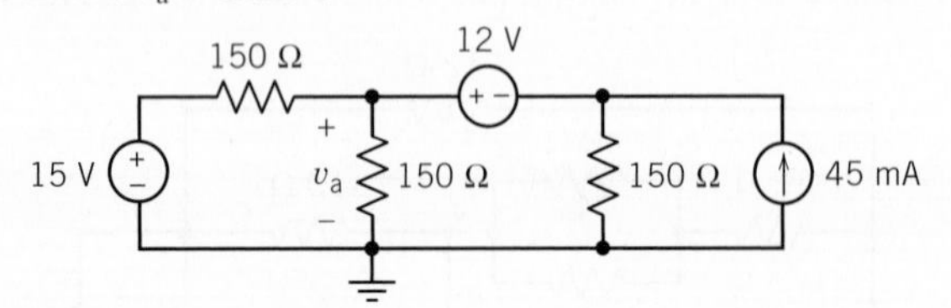

Figure P 4.3-3

P 4.3-4 Determine the node voltage v_a for the circuit of Figure P 4.3-4.

Answer: $v_a = 4$ V

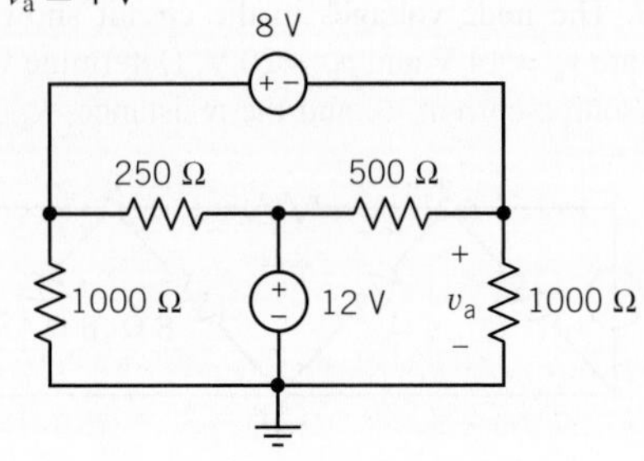

Figure P 4.3-4

P 4.3-5 The voltages v_a, v_b, and v_c in Figure P 4.3-5 are the node voltages corresponding to nodes a, b, and c. The values of these voltages are:

$$v_a = 24 \text{ V}, v_b = 19.76 \text{ V, and } v_c = 10.58 \text{ V}$$

Determine the power supplied by the voltage source.

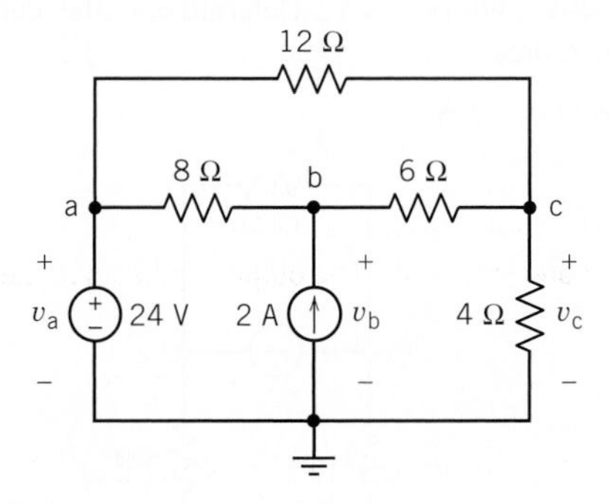

Figure P 4.3-5

P 4.3-6 The voltmeter in the circuit of Figure P 4.3-6 measures a node voltage. The value of that node voltage depends on the value of the resistance R.

(a) Determine the value of the resistance R that will cause the voltage measured by the voltmeter to be 4 V.

(b) Determine the voltage measured by the voltmeter when $R = 1.2$ kΩ $= 1200$ Ω.

Answers: **(a)** 6 kΩ **(b)** 1.7 V

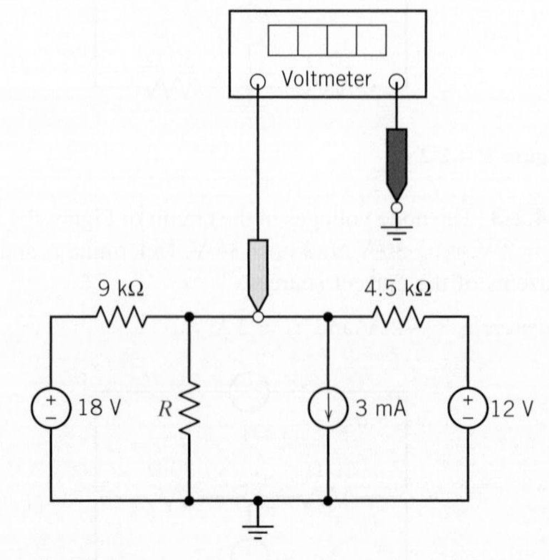

Figure P 4.3-6

P 4.3-7 Determine the values of the node voltages, v_1 and v_2, in Figure P 4.3-7. Determine the values of the currents i_a and i_b.

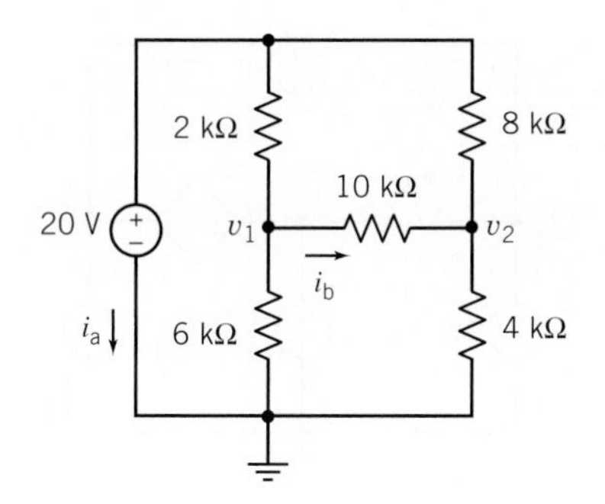

Figure P 4.3-7

P 4.3-8 The circuit shown in Figure P 4.3-8 has two inputs, v_1 and v_2, and one output, v_o. The output is related to the input by the equation

$$v_o = av_1 + bv_2$$

where a and b are constants that depend on R_1, R_2, and R_3.

(a) Determine the values of the coefficients a and b when $R_1 = 10\ \Omega, R_2 = 40\ \Omega,$ and $R_3 = 8\ \Omega$.
(b) Determine the values of the coefficients a and b when $R_1 = R_2$ and $R_3 = R_1 || R_2$.

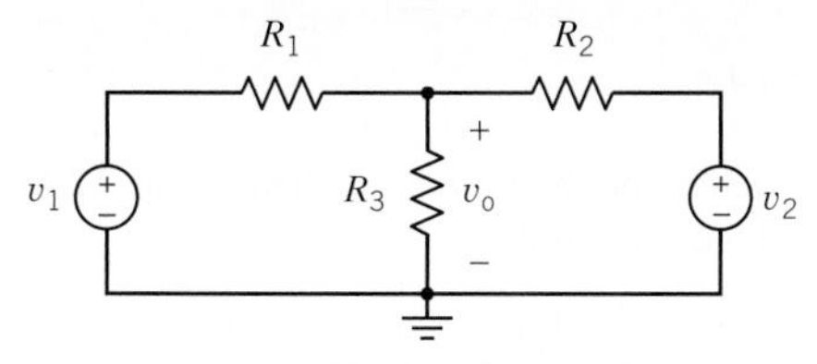

Figure P 4.3-8

P 4.3-9 Determine the values of the node voltages of the circuit shown in Figure P 4.3-9.

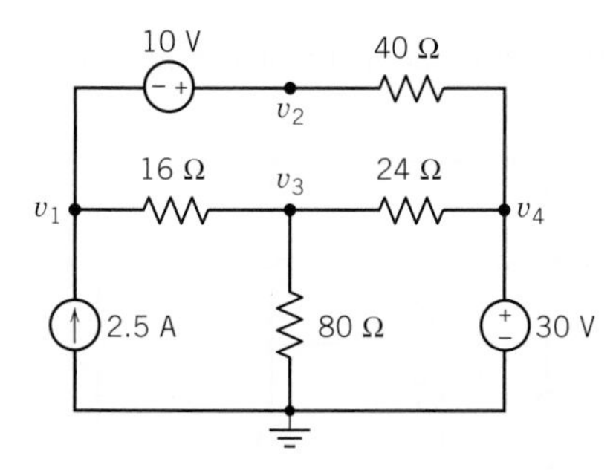

Figure P 4.3-9

P 4.3-10 Figure P 4.3-10 shows a measurement made in the laboratory. Your lab partner forgot to record the values of R_1, R_2, and R_3. He thinks that the two resistors were 10-kΩ resistors and the other was a 5-kΩ resistor. Is this possible? Which resistor is the 5-kΩ resistor?

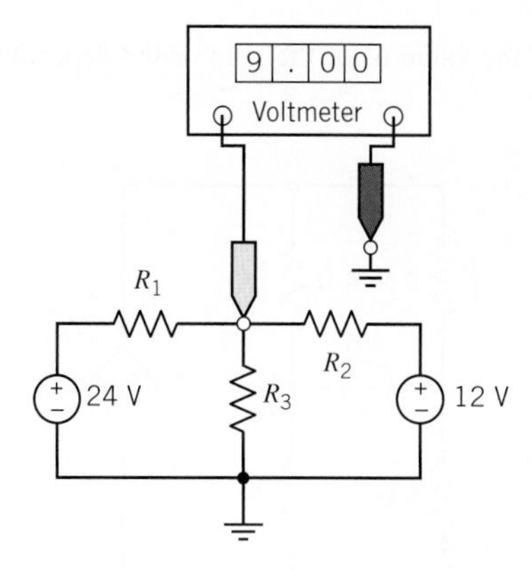

Figure P 4.3-10

***P 4.3-11** Determine the values of the node voltages of the circuit shown in Figure P 4.3-11.

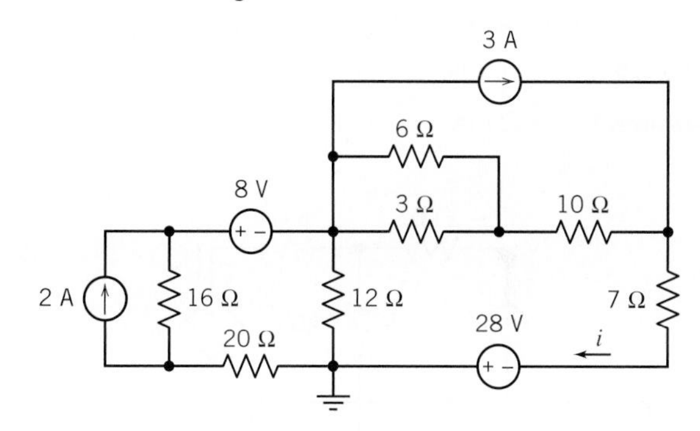

Figure P 4.3-11

P 4.3-12 Determine the values of the node voltages of the circuit shown in Figure P 4.3-12.

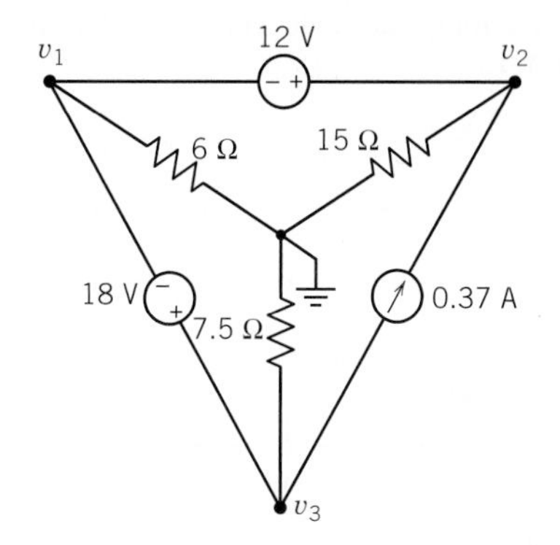

Figure P 4.3-12

Section 4.4 Node Voltage Analysis with Dependent Sources

P 4.4-1 The voltages v_a, v_b, and v_c in Figure P 4.4-1 are the node voltages corresponding to nodes a, b, and c. The values of these voltages are:

$$v_a = 8.667\text{ V}, v_b = 2\text{ V}, \text{ and } v_c = 10\text{ V}$$

Determine the value of A, the gain of the dependent source.

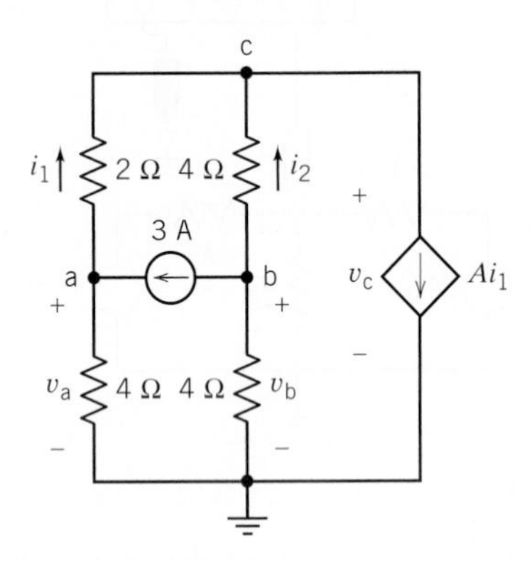

Figure P 4.4-1

P 4.4-2 Find i_b for the circuit shown in Figure P 4.4-2.

Answer: $i_b = -12$ mA

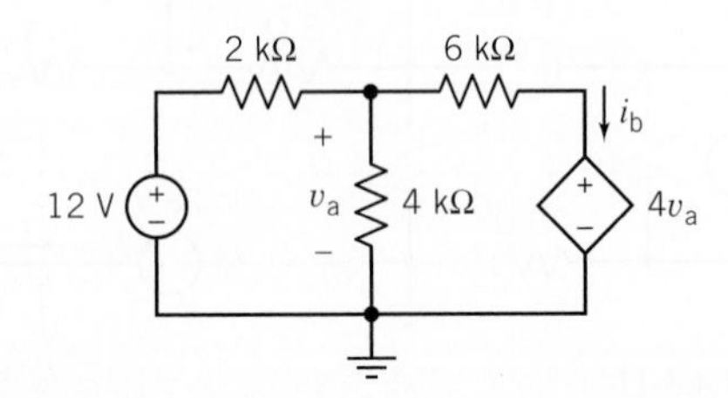

Figure P 4.4-2

P 4.4-3 Determine the node voltage v_b for the circuit of Figure P 4.4-3.

Answer: $v_b = 3$ V

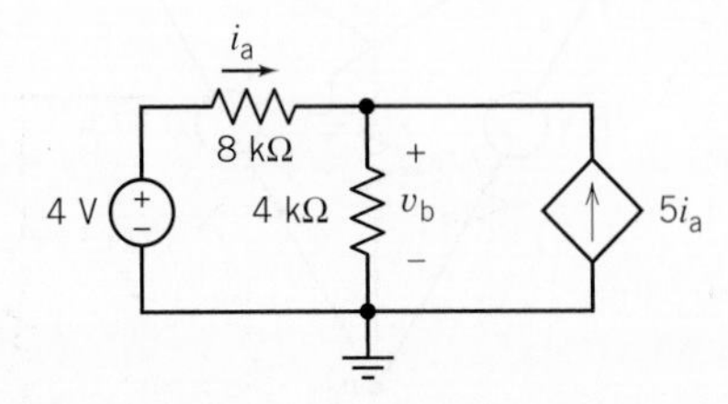

Figure P 4.4-3

P 4.4-4 The circled numbers in Figure P 4.4-4 are node numbers. The node voltages of this circuit are $v_1 = 15$ V, $v_2 = 21$ V, and $v_3 = 18$ V.

(a) Determine the value of the current i_b.
(b) Determine the value of r, the gain of the CCVS.

Answers: **(a)** -1.75 A **(b)** 6 V/A

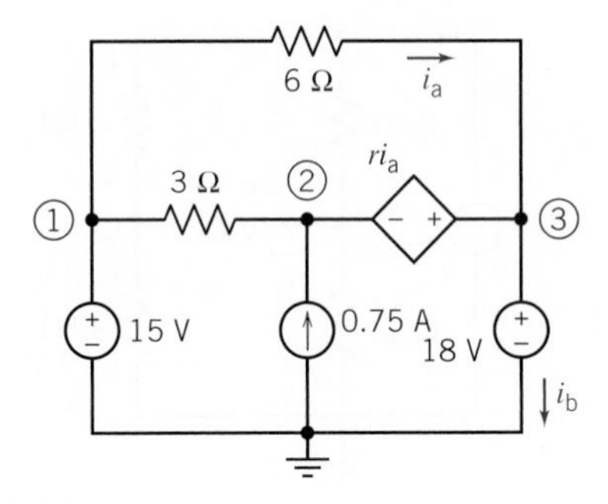

Figure P 4.4-4

P 4.4-5 Determine the value of the current i_x in the circuit of Figure P 4.4-5.

Answer: $i_x = 3.43$ A

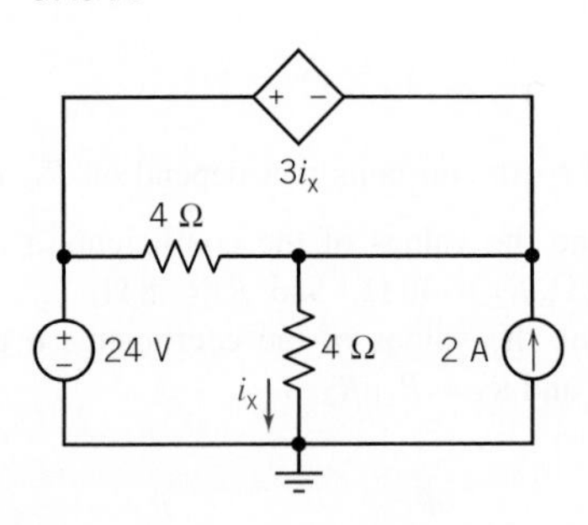

Figure P 4.4-5

P 4.4-6 Determine the power supplied by the 12-V voltage source in Figure P 4.4-6.

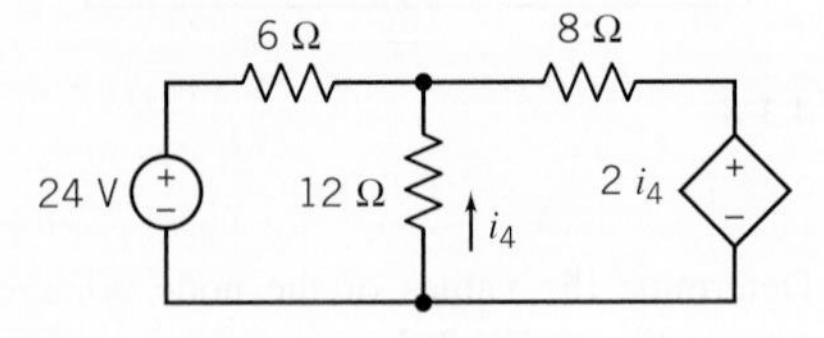

Figure P 4.4-6

P 4.4-7 Determine the value of the current i_c in Figure P 4.4-7.

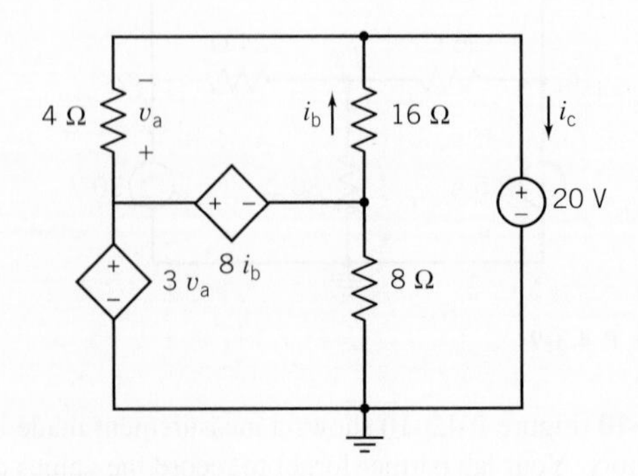

Figure P 4.4-7

P 4.4-8 Determine the value of the power supplied by the dependent source in Figure P 4.4-8

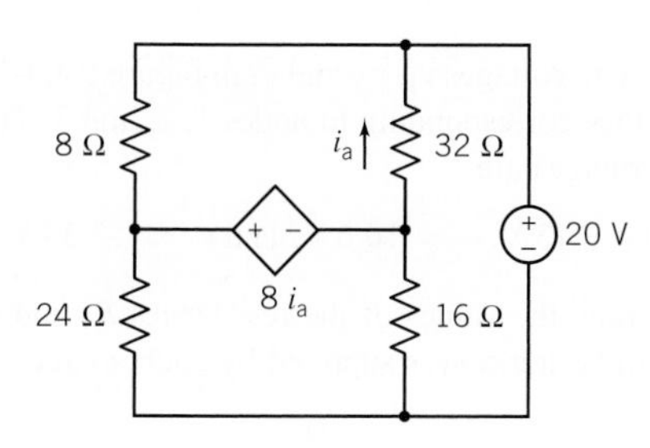

Figure P 4.4-8

P 4.4-9 The node voltages in the circuit shown in Figure P 4.4-9 are

$$v_1 = 8\text{ V}, v_2 = 0\text{ V, and } v_3 = -12\text{ V}$$

Determine the values of the resistance, R, and of the gain, b, of the CCCS.

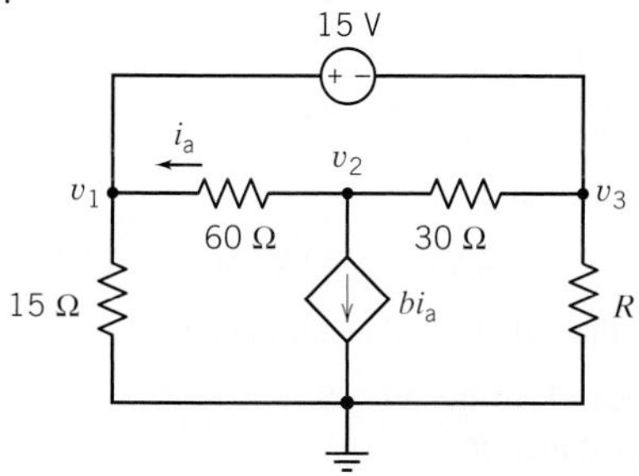

Figure P 4.4-9

P 4.4-10 The value of the node voltage at node b in the circuit shown in Figure P 4.4-10 is $v_b = 18$ V.

(a) Determine the value of A, the gain of the dependent source.
(b) Determine the power supplied by the dependent source.

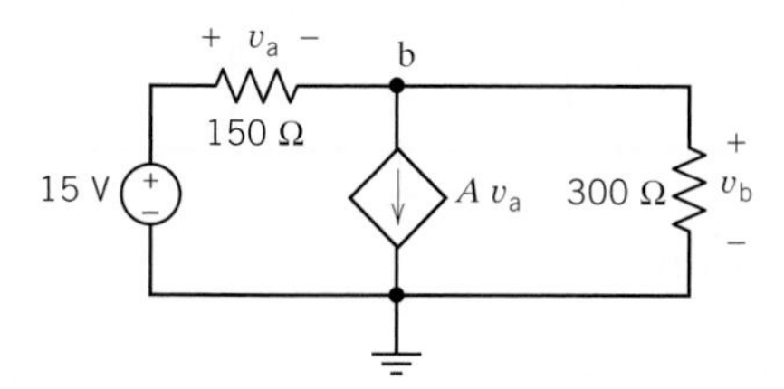

Figure P 4.4-10

***P 4.4-11** Determine the power supplied by the dependent source in the circuit shown in Figure P 4.4-11.

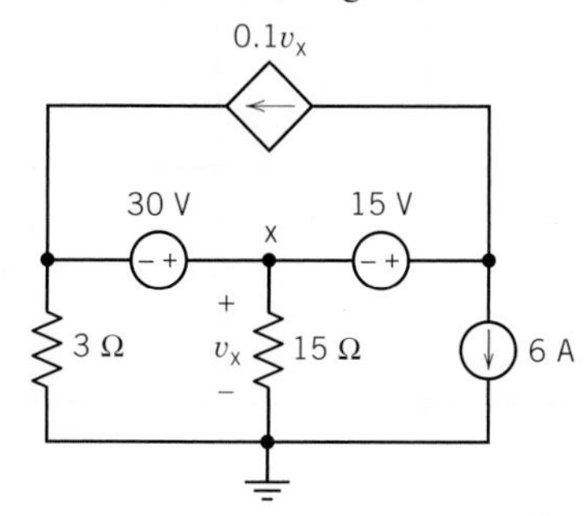

Figure P 4.4-11

***P 4.4-12** Determine values of the node voltages, v_1, v_2, v_3, v_4, and v_5 in the circuit shown in Figure P 4.4-12.

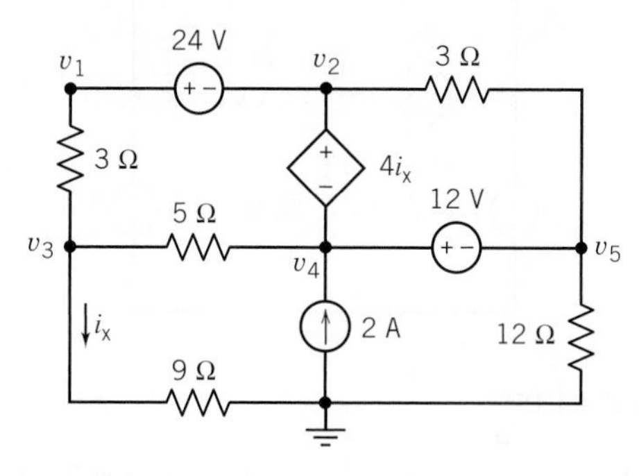

Figure P 4.4-12

***P 4.4-13** Determine values of the node voltages, v_1, v_2, v_3, v_4, and v_5 in the circuit shown in Figure P 4.4-13.

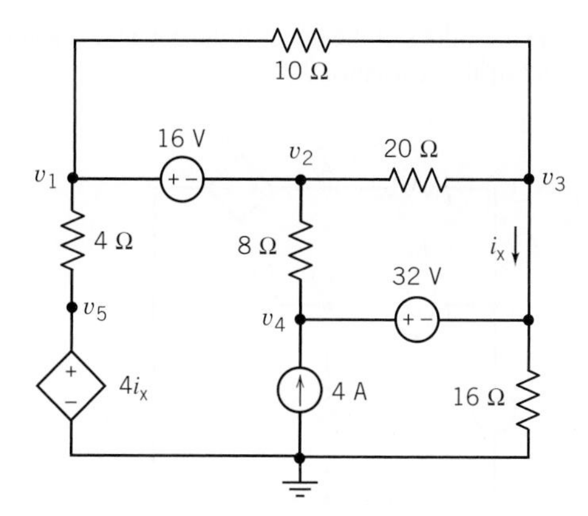

Figure P 4.4-13

***P 4.4-14** Determine values of the node voltages, v_1, v_2, v_3, v_4, and v_5 in the circuit shown in Figure P 4.4-14.

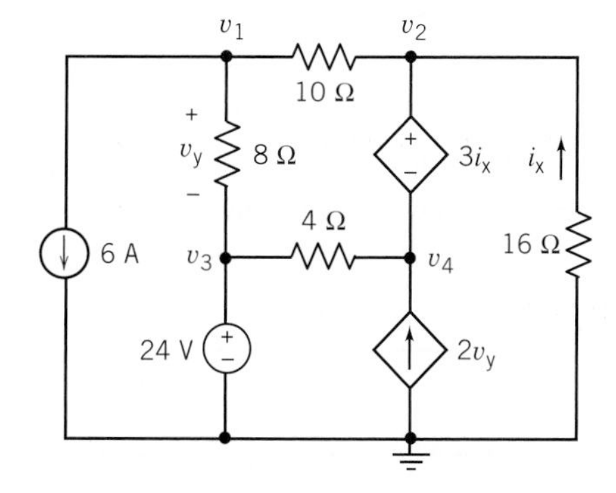

Figure P 4.4-14

P 4.4-15 The voltages v_1, v_2, v_3, and v_4 are the node voltages corresponding to nodes 1, 2, 3, and 4 in Figure P 4.4-15. Determine the values of these node voltages.

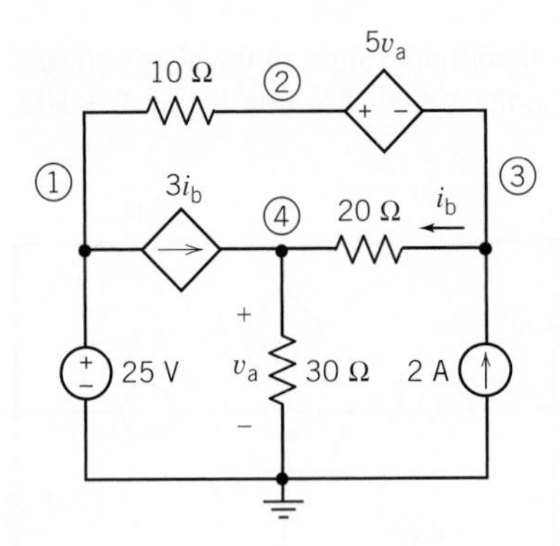

Figure P 4.4-15

P 4.4-16 The voltages v_1, v_2, v_3, and v_4 in Figure P 4.4-16 are the node voltages corresponding to nodes 1, 2, 3, and 4. The values of these voltages are

$$v_1 = 10\text{ V}, v_2 = 75\text{ V}, v_3 = -15\text{ V}, \text{ and } v_4 = 22.5\text{ V}$$

Determine the values of the gains of the dependent sources, A and B, and of the resistance R_1.

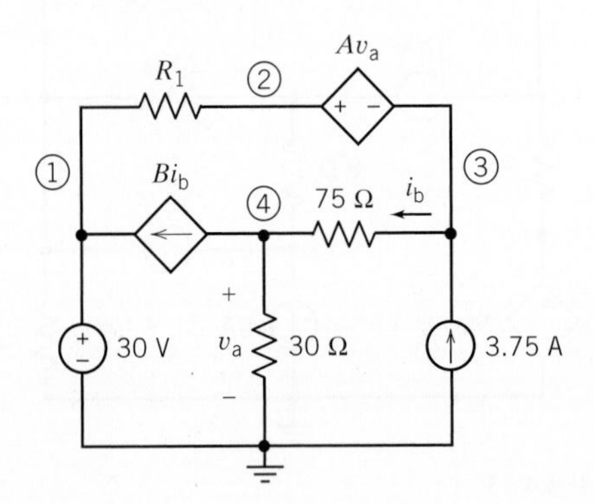

Figure P 4.4-16

P 4.4-17 The voltages v_1, v_2, and v_3 in Figure P 4.4-17 are the node voltages corresponding to nodes 1, 2, and 3. The values of these voltages are

$$v_1 = 12\text{ V}, v_2 = 21\text{ V}, \text{ and } v_3 = -3\text{ V}$$

(a) Determine the values of the resistances R_1 and R_2.
(b) Determine the power supplied by each source.

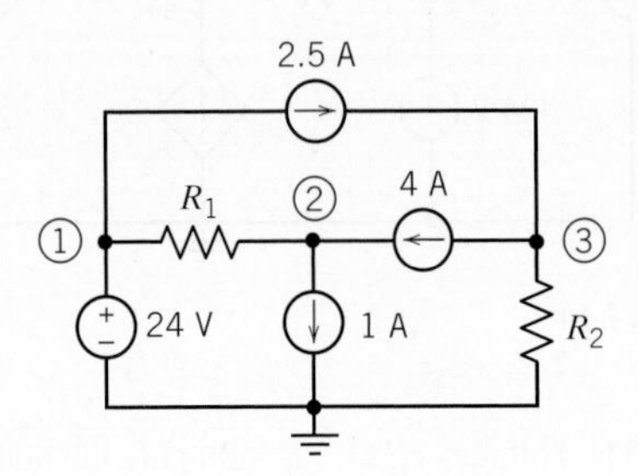

Figure P 4.4-17

P 4.4-18 The voltages v_1, v_2, and v_3 in Figure P 4.4-18 are the node voltages corresponding to nodes 1, 2, and 3. The values of these voltages are

$$v_1 = 13\text{ V}, v_2 = 10.6\text{ V}, \text{ and } v_3 = -2.33\text{ V}$$

(a) Determine the values of the resistances R_1 and R_2.
(b) Determine the power supplied by each source.

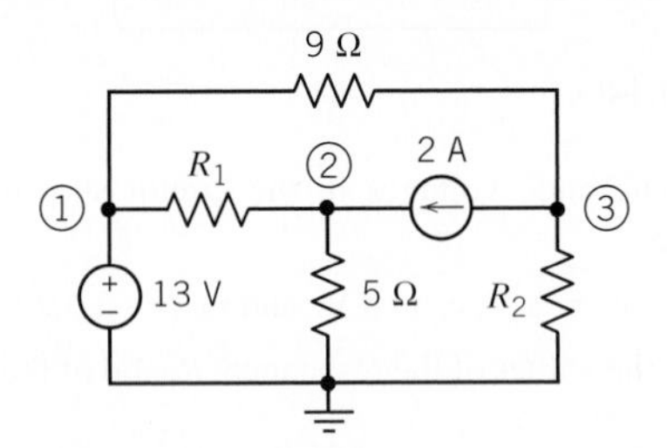

Figure P 4.4-18

P 4.4-19 The voltages v_2, v_3, and v_4 for the circuit shown in Figure P 4.4-19 are:

$$v_2 = 24\text{ V}, v_3 = 12\text{ V}, \text{ and } v_4 = 9\text{ V}$$

Determine the values of the following:

(a) The gain, A, of the VCVS
(b) The resistance R_5
(c) The currents i_b and i_c
(d) The power received by resistor R_4

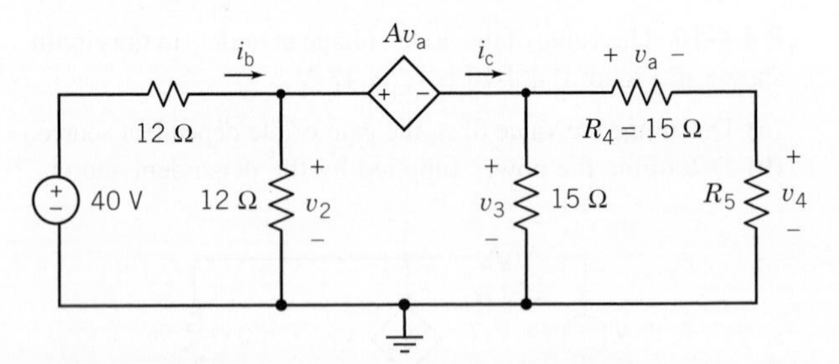

Figure P 4.4-19

P 4.4-20 Determine the values of the node voltages v_1 and v_2 for the circuit shown in Figure P 4.4-20.

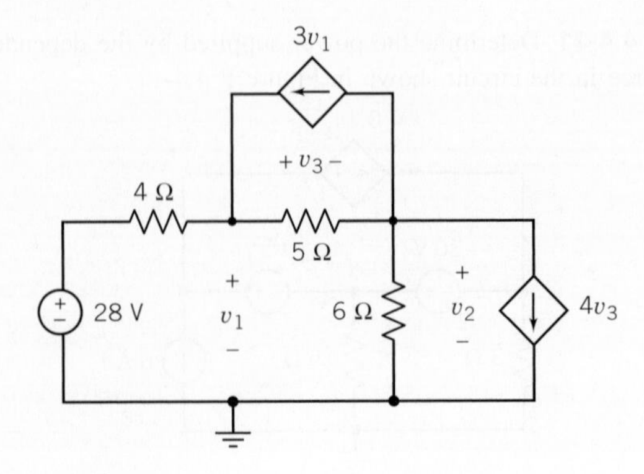

Figure P 4.4-20

P 4.4-21 The encircled numbers in Figure P 4.4-21 are node numbers. Determine the values of v_1, v_2, and v_3, the node voltages corresponding to nodes 1, 2, and 3.

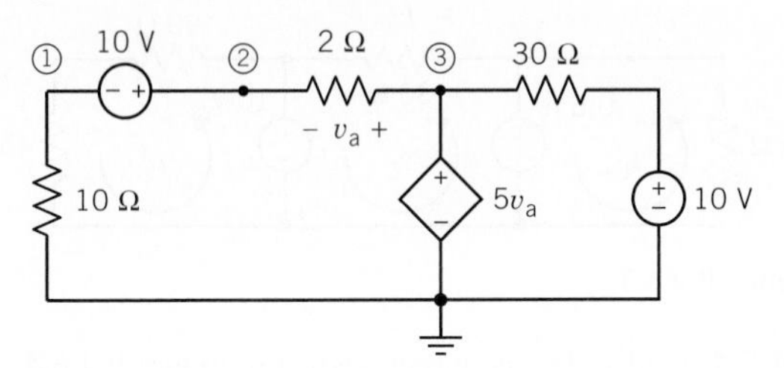

Figure P 4.4-21

P 4.4-22 Determine the values of the node voltages v_1, v_2, and v_3 for the circuit shown in Figure P 4.4-22.

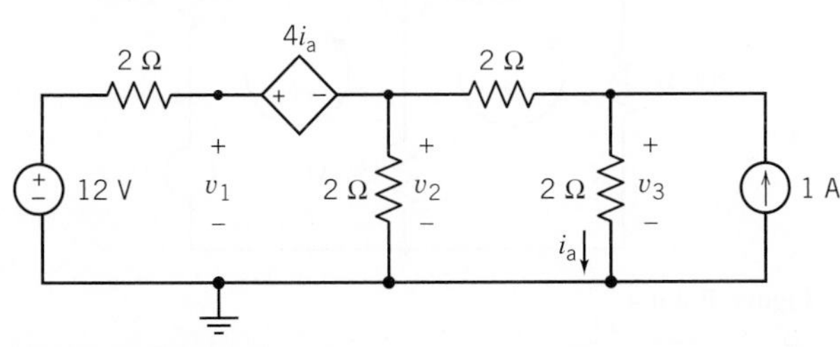

Figure P 4.4-22

P 4.4-23 Determine the values of the node voltages v_1, v_2, and v_3 for the circuit shown in Figure P 4.4-23.

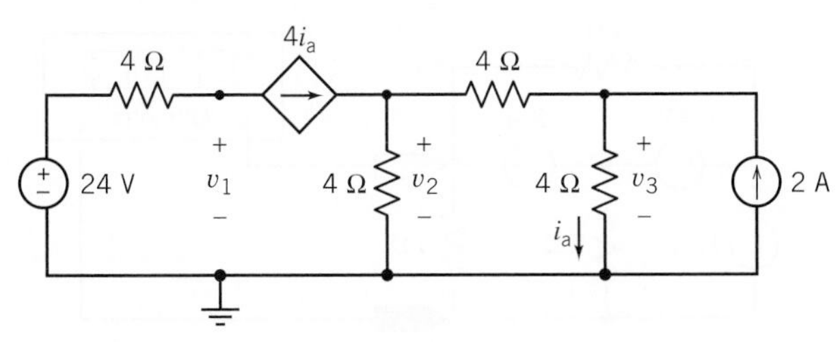

Figure P 4.4-23

Section 4.5 Mesh Current Analysis with Independent Voltage Sources

P 4.5-1 Determine the mesh currents, i_1, i_2 and i_3 for the circuit shown in Figure P 4.5-1.

Answers: $i_1 = 3$ A, $i_2 = 2$ A, and $i_3 = 4$ A

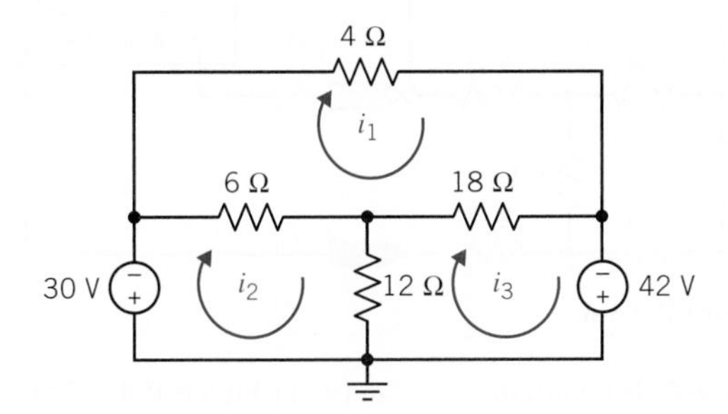

Figure P 4.5-1

P 4.5-2 The values of the mesh currents in the circuit shown in Figure P 4.5-2 are $i_1 = 4$ A, $i_2 = 6$ A, and $i_3 = 8$ A. Determine the values of the resistance R and of the voltages v_1 and v_2 of the voltage sources.

Answers: $R = 24\ \Omega$, $v_1 = -16$ V, and $v_2 = -112$ V

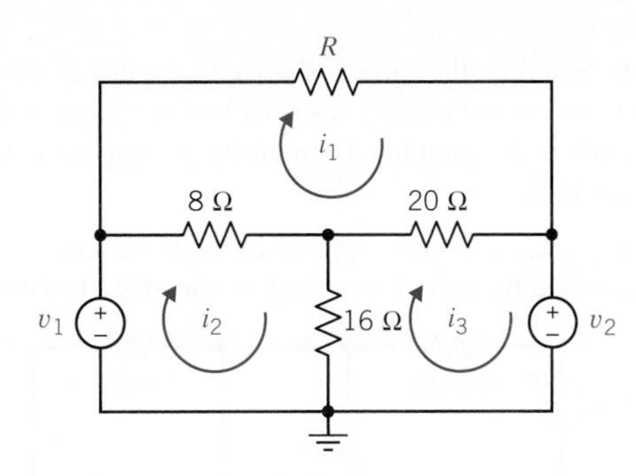

Figure P 4.5-2

P 4.5-3 The currents i_1 and i_2 in Figure P 4.5-3 are the mesh currents. Determine the value of the resistance R required to cause $v_a = -6$ V.

Answer: 15.7 Ω

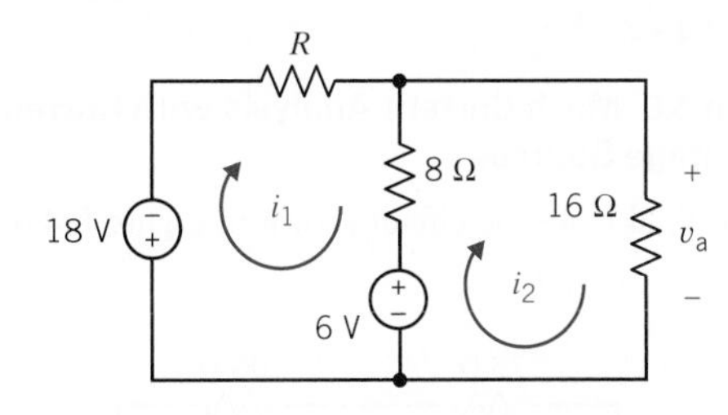

Figure P 4.5-3

P 4.5-4 Determine the mesh currents i_a and i_b in the circuit shown in Figure P 4.5-4.

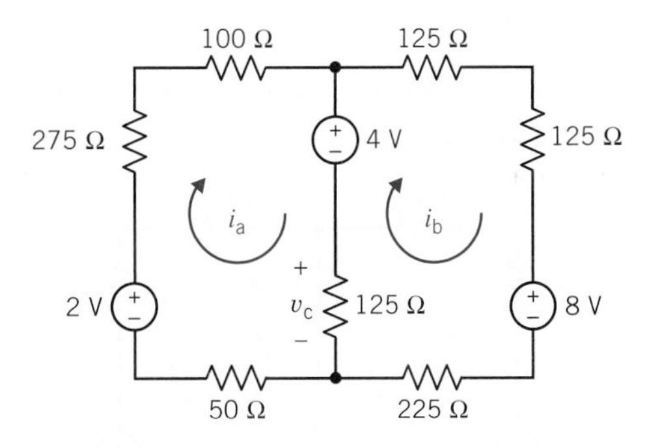

Figure P 4.5-4

P 4.5-5 Find the current i for the circuit of Figure P 4.5-5.

Hint: A short circuit can be treated as a 0-V voltage source.

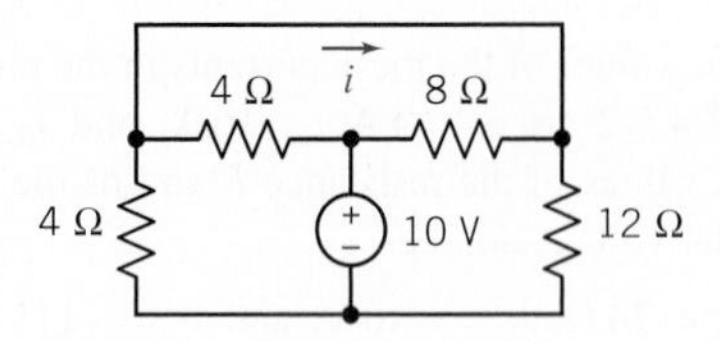

Figure P 4.5-5

P 4.5-6 Simplify the circuit shown in Figure P 4.5-6 by replacing series and parallel resistors by equivalent resistors. Next, analyze the simplified circuit by writing and solving mesh equations.

(a) Determine the power supplied by each source,
(b) Determine the power absorbed by the 30-Ω resistor.

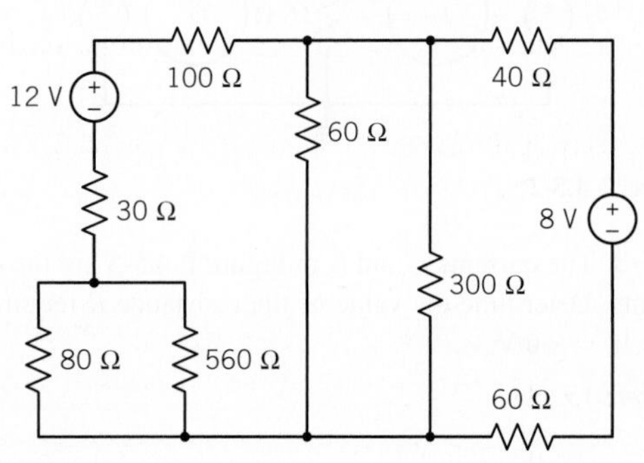

Figure P 4.5-6

Section 4.6 Mesh Current Analysis with Current and Voltage Sources

P 4.6-1 Find i_b for the circuit shown in Figure P 4.6-1.

Answer: $i_b = 0.57$ A

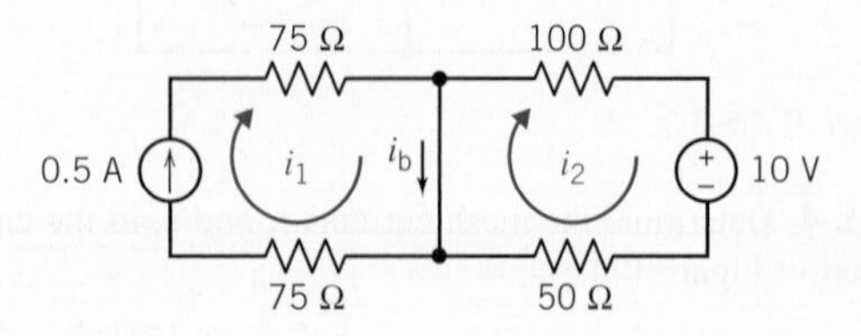

Figure P 4.6-1

P 4.6-2 Find v_c for the circuit shown in Figure P 4.6-2.

Answer: $v_c = 18.7$ V

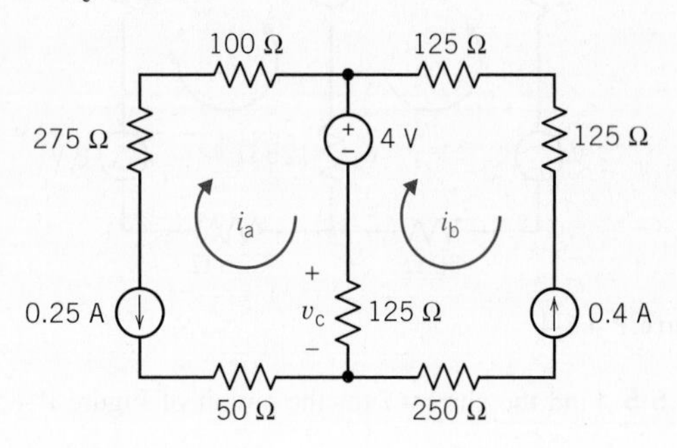

Figure P 4.6-2

P 4.6-3 Find v_2 for the circuit shown in Figure P 4.6-3.

Answer: $v_2 = 3.2$ V

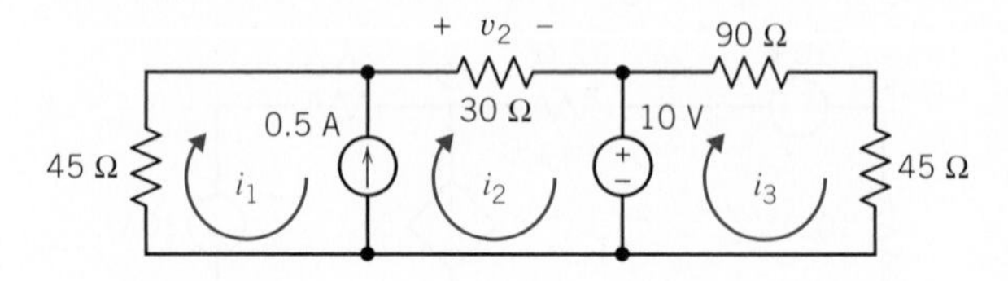

Figure P 4.6-3

P 4.6-4 Find v_c for the circuit shown in Figure P 4.6-4.

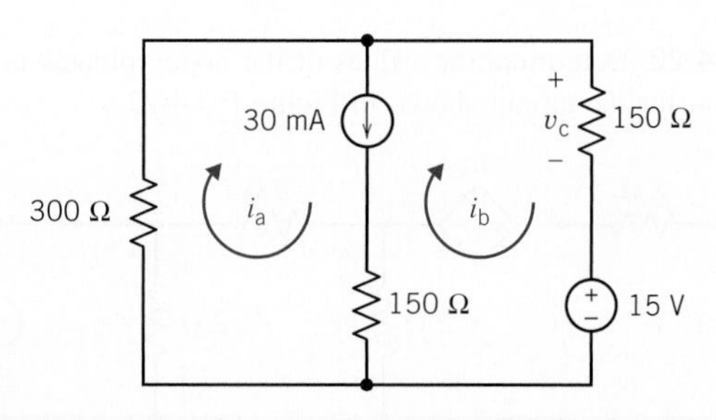

Figure P 4.6-4

P 4.6-5 Determine the value of the voltage measured by the voltmeter in Figure P 4.6-5.

Answer: 8 V

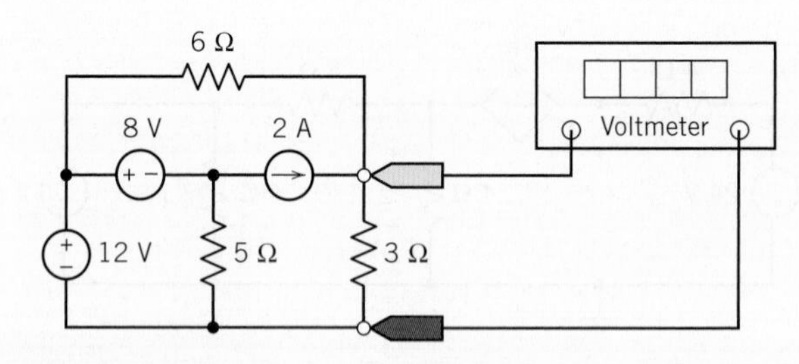

Figure P 4.6-5

P 4.6-6 Determine the value of the current measured by the ammeter in Figure P 4.6-6.

Hint: Write and solve a single mesh equation.

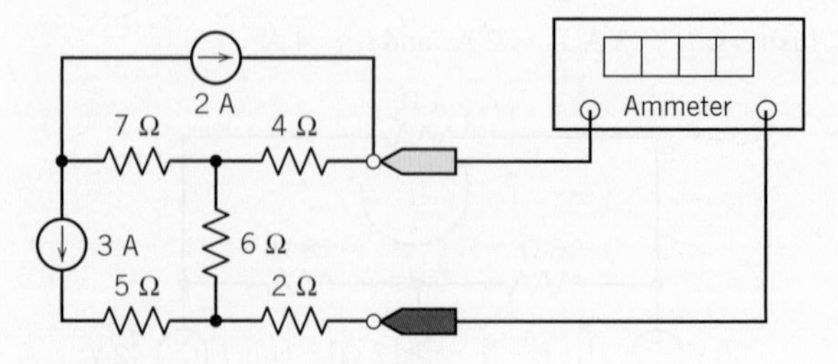

Figure P 4.6-6

P 4.6-7 The currents i_1, i_2 and i_3 in Figure P 4.6-7 are the mesh currents. Determine the value of the resistance R.

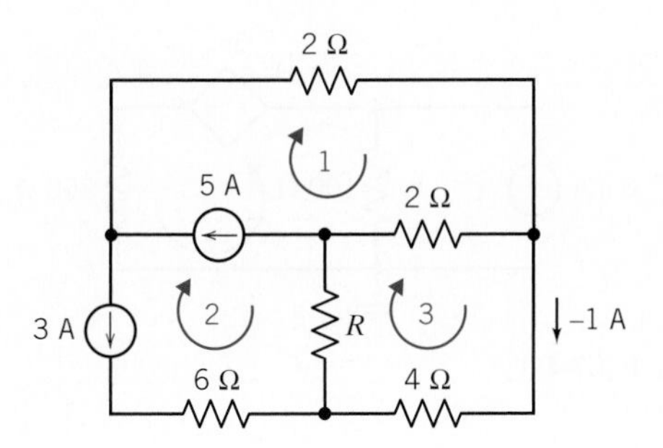

Figure P 4.6-7

P 4.6-8 Determine values of the mesh currents, i_1, i_2 and i_3 in the circuit shown in Figure P 4.6-8.

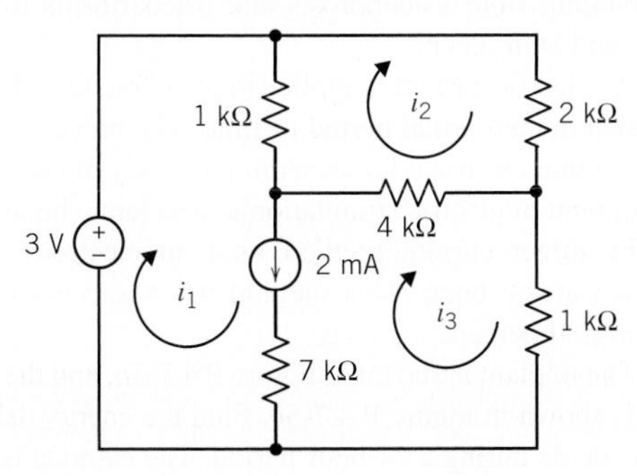

Figure P 4.6-8

*__P 4.6-9__ The circuit shown in Figure P 4.6-9 has three inputs: i_x, i_y, and v_z. The output of the circuit is i_o. The output is related to the inputs by

$$i_o = a\,i_x + b\,i_y + c\,v_z$$

where a, b, and c are constants. Determine the values of a, b, and c.

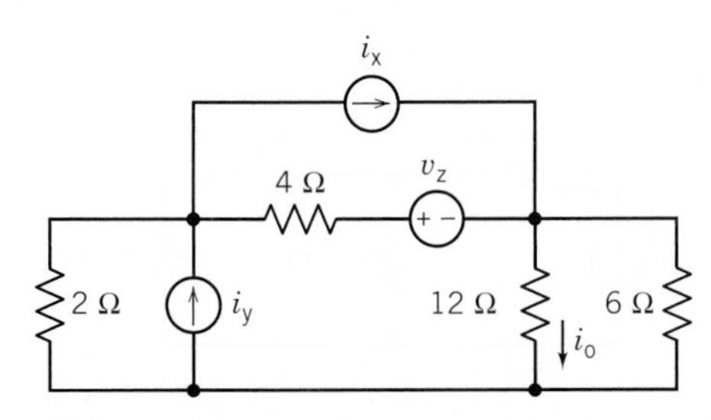

Figure P 4.6-9

P 4.6-10 The mesh currents in the circuit shown in Figure P 4.6-10 are

$$i_1 = -2.2213\text{ A}, i_2 = 0.7787\text{ A, and } i_3 = 0.0770\text{ A}$$

(a) Determine the values of the resistances R_1 and R_3.
(b) Determine the value of the power supplied by the current source.

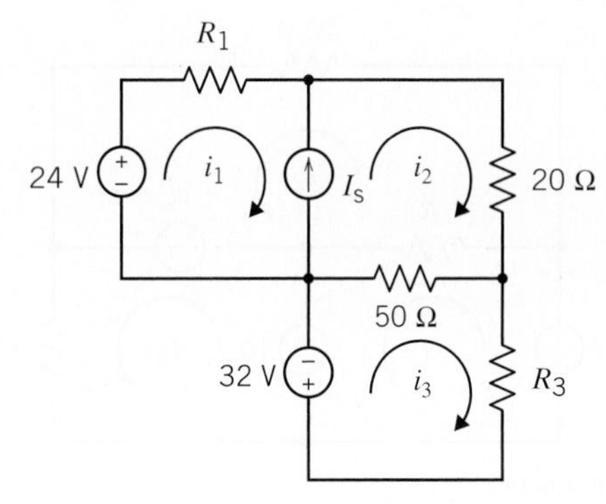

Figure P 4.6-10

P 4.6-11 Determine the value of the voltage measured by the voltmeter in Figure P 4.6-11.

Hint: Apply KVL to a supermesh to determine the current in the 2-Ω resistor.

Answer: 4/3 V

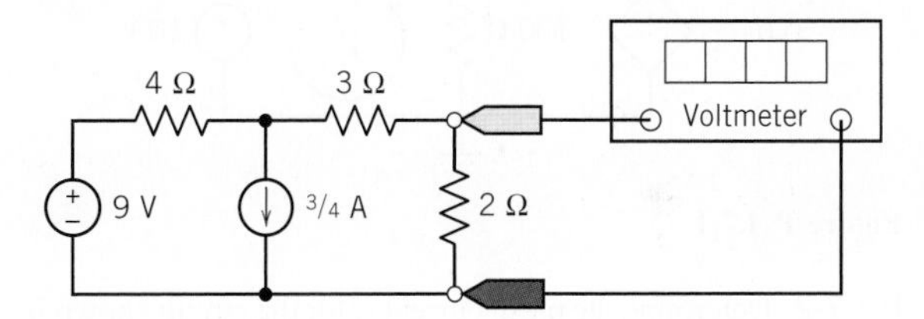

Figure P 4.6-11

P 4.6-12 Determine the value of the current measured by the ammeter in Figure P 4.6-12.

Hint: Apply KVL to a supermesh.

Answer: −0.333 A

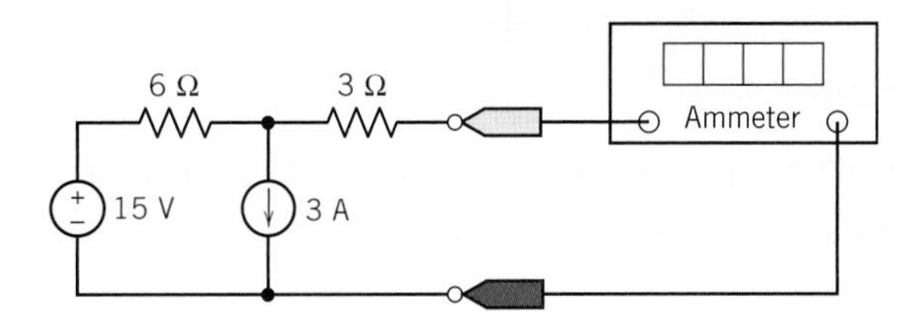

Figure P 4.6-12

P 4.6-13 The values of the mesh currents in the circuit shown in Figure P 4.6-13 are

$$i_1 = 0.2\text{ A}, i_2 = 0.7\text{ A, and } i_3 = 0.8\text{ A}$$

Determine the values of the following:

(a) The power supplied by each voltage source
(b) The resistance R
(c) The current source current
(d) The voltage v_s across the current source

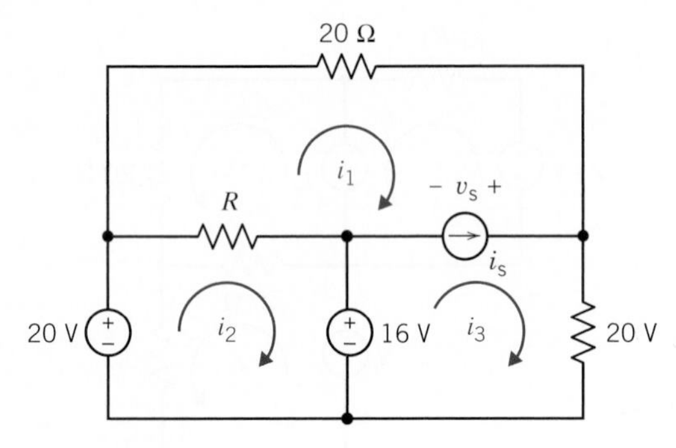

Figure P 4.6-13

Section 4.7 Mesh Current Analysis with Dependent Sources

P 4.7-1 Find v_2 for the circuit shown in Figure P 4.7-1.

Answer: $v_2 = 10$ V

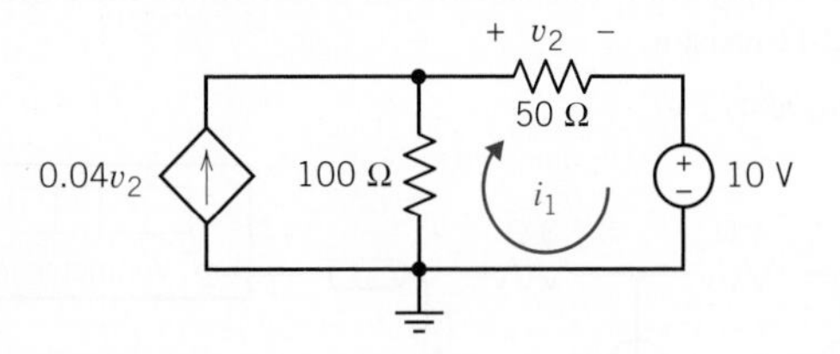

Figure P 4.7-1

P 4.7-2 Determine the mesh current i_a for the circuit shown in Figure P 4.7-2.

Answer: $i_a = -48$ mA

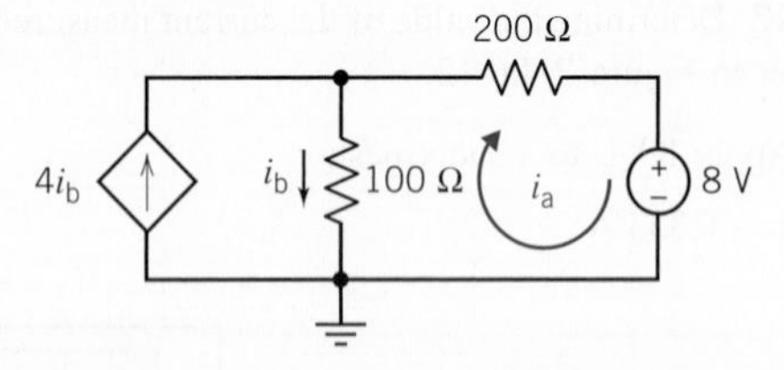

Figure P 4.7-2

P 4.7-3 Find v_o for the circuit shown in Figure P 4.7-3.

Answer: $v_o = 2.5$ V

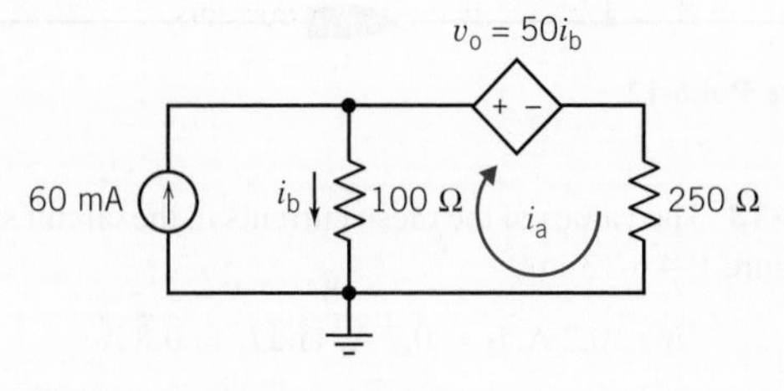

Figure P 4.7-3

P 4.7-4 Determine the mesh current i_a for the circuit shown in Figure P 4.7-4.

Answer: $i_a = -24$ mA

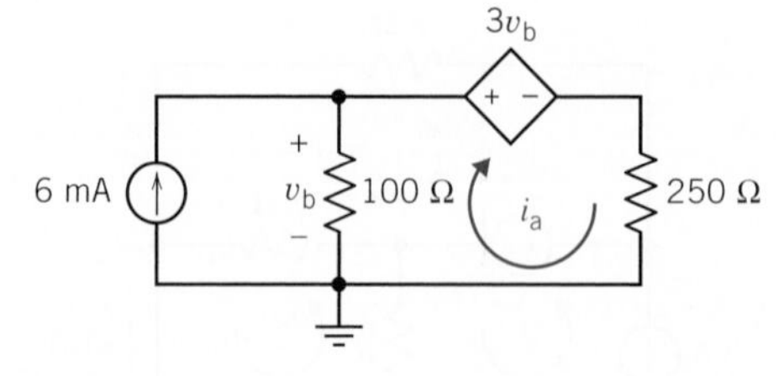

Figure P 4.7-4

P 4.7-5 Although scientists continue to debate exactly why and how it works, the process of using electricity to aid in the repair and growth of bones—which has been used mainly with fractures—may soon be extended to an array of other problems, ranging from osteoporosis and osteoarthritis to spinal fusions and skin ulcers.

An electric current is applied to bone fractures that have not healed in the normal period of time. The process seeks to imitate natural electrical forces within the body. It takes only a small amount of electric stimulation to accelerate bone recovery. The direct current method uses an electrode that is implanted at the bone. This method has a success rate approaching 80 percent.

The implant is shown in Figure P 4.7-5*a*, and the circuit model is shown in Figure P 4.7-5*b*. Find the energy delivered to the cathode during a 24-hour period. The cathode is represented by the dependent voltage source and the 100-kΩ resistor.

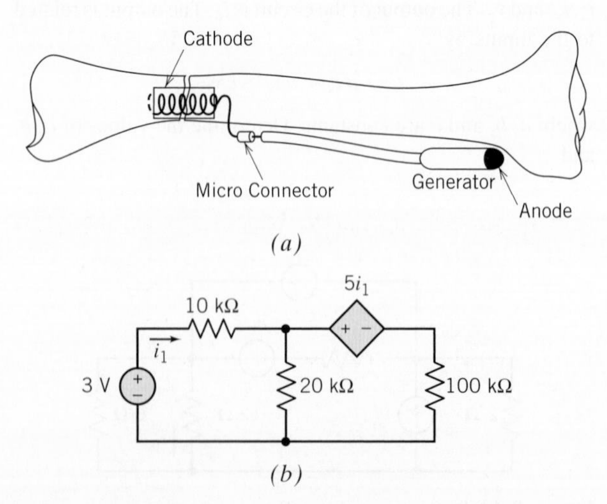

Figure P 4.7-5 (*a*) Electric aid to bone repair. (*b*) Circuit model.

P 4.7-6 The model of a bipolar junction transistor (BJT) amplifier is shown in Figure P 4.7-6.

(a) Determine the gain v_o/v_i.

(b) Calculate the required value of g to obtain a gain $v_o/v_i = -170$ when $R_L = 5$ kΩ, $R_1 = 100$ Ω, and $R_2 = 1$ kΩ.

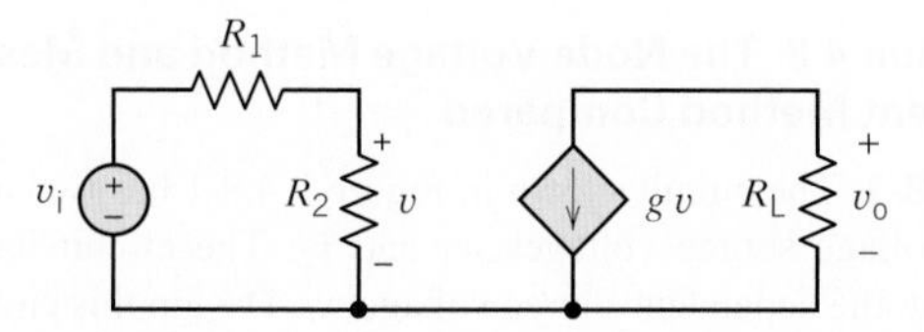

Figure P 4.7-6

P 4.7-7 The currents i_1, i_2, and i_3, are the mesh currents of the circuit shown in Figure P 4.7-7. Determine the values of i_1, i_2, and i_3.

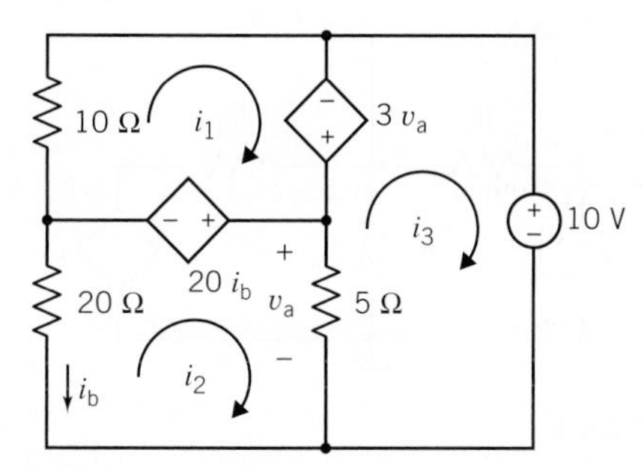

Figure P 4.7-7

P 4.7-8 Determine the value of the power supplied by the dependent source in Figure P 4.7-8.

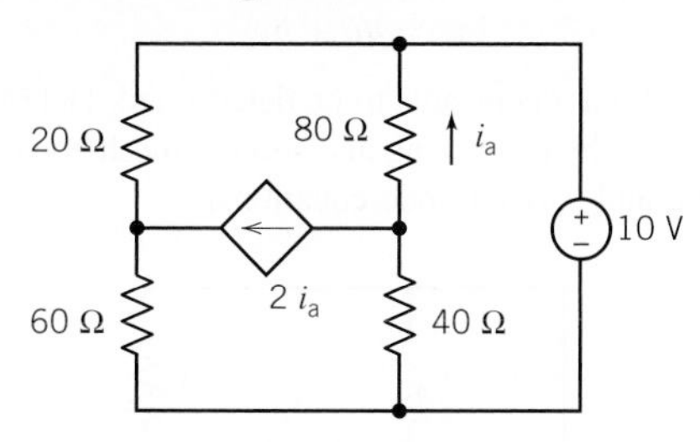

Figure P 4.7-8

P 4.7-9 Determine the value of the resistance R in the circuit shown in Figure P 4.7-9.

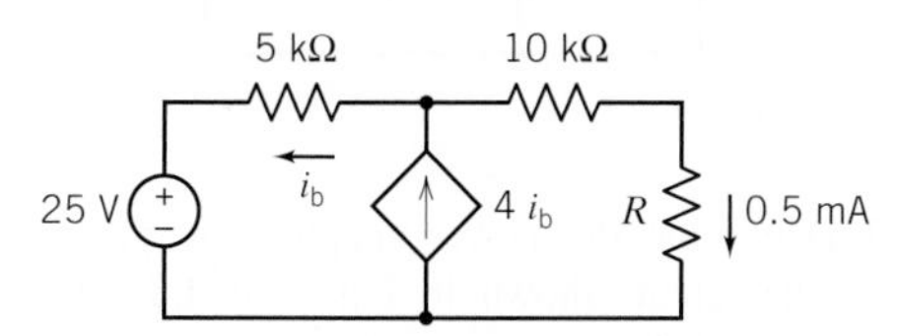

Figure P 4.7-9

P 4.7-10 The circuit shown in Figure P 4.7-10 is the small signal model of an amplifier. The input to the amplifier is the voltage source voltage, v_s. The output of the amplifier is the voltage v_o.

(a) The ratio of the output to the input, v_o/v_s, is called the gain of the amplifier. Determine the gain of the amplifier.

(b) The ratio of the current of the input source to the input voltage, age i_b/v_s, is called the input resistance of the amplifier. Determine the input resistance.

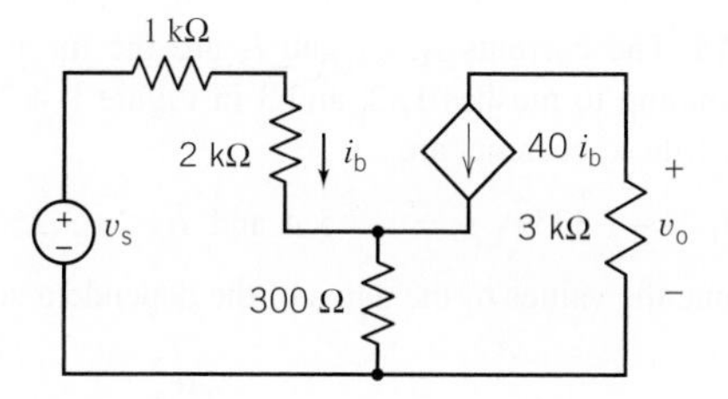

Figure P 4.7-10

P 4.7-11 Determine values of the mesh currents if, i_1, i_2, i_3, and i_4 in the circuit shown in Figure P 4.7-11.

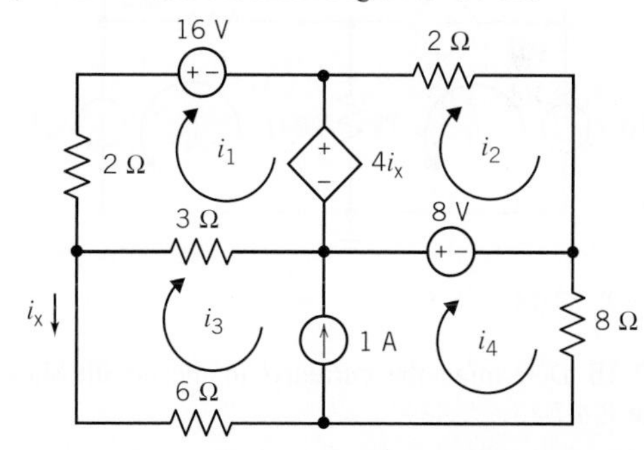

Figure P 4.7-11

P 4.7-12 Determine the values of the mesh currents of the circuit shown in Figure P 4.7-12.

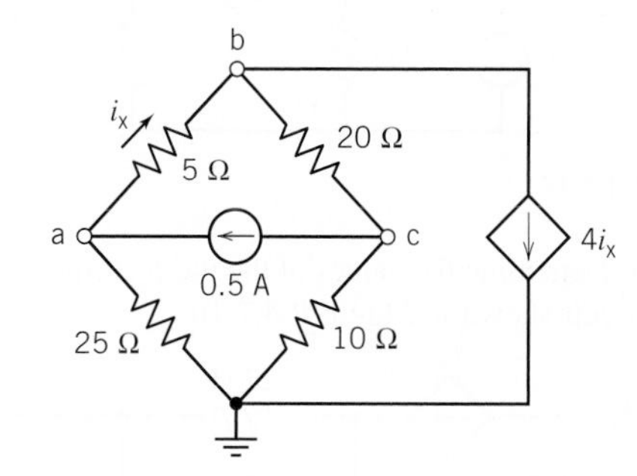

Figure P 4.7-12

P 4.7-13 The currents i_1, i_2, and i_3 are the mesh currents corresponding to meshes 1, 2, and 3 in Figure P 4.7-13. Determine the values of these mesh currents.

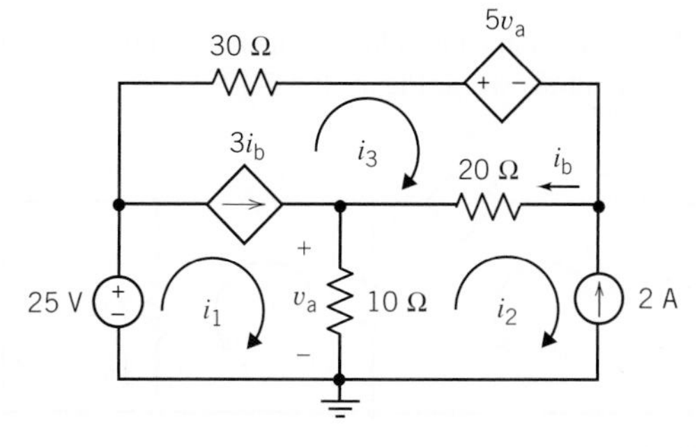

Figure P 4.7-13

P 4.7-14 The currents i_1, i_2, and i_3 are the mesh currents corresponding to meshes 1, 2, and 3 in Figure P 4.7-14. The values of these currents are

$$i_1 = -1.375\text{ A}, i_2 = -2.5\text{ A and } i_3 = -3.25\text{ A}$$

Determine the values of the gains of the dependent sources, A and B.

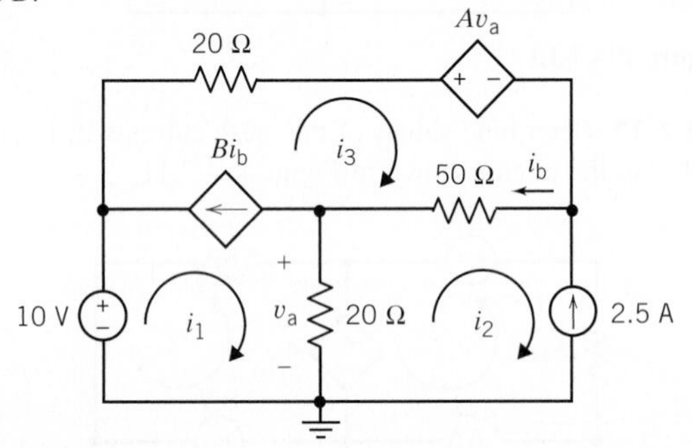

Figure P 4.7-14

P 4.7-15 Determine the current i in the circuit shown in Figure P 4.7-15.

Answer: $i = 3$ A

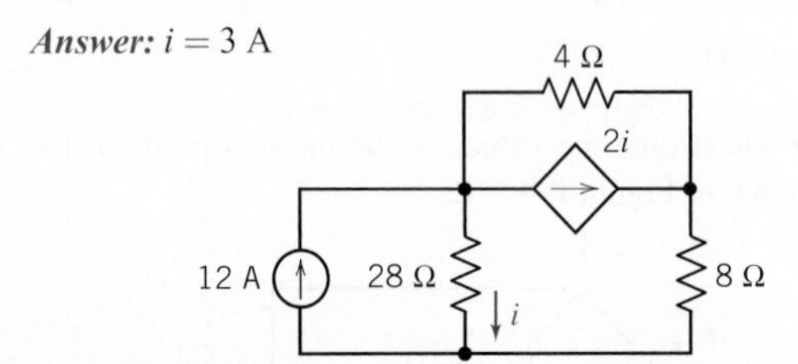

Figure P 4.7-15

P 4.7-16 Determine the values of the mesh currents i_1 and i_2 for the circuit shown in Figure P 4.7-16

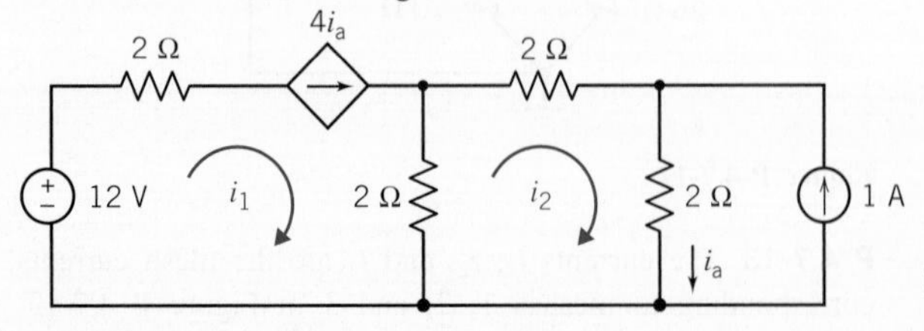

Figure P 4.7-16

P 4.7-17 Determine the values of the mesh currents i_1 and i_2 for the circuit shown in Figure P 4.7-17

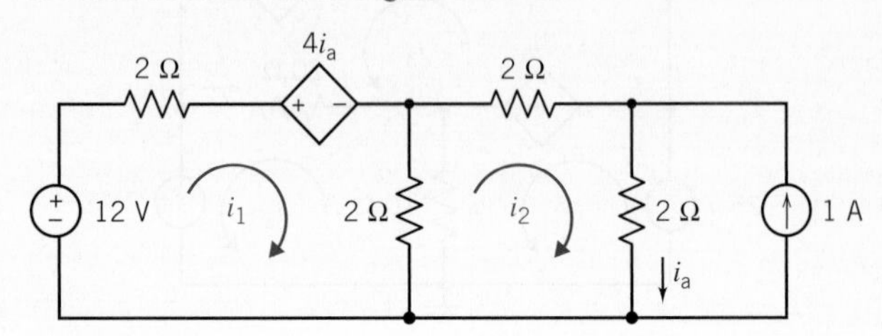

Figure P 4.7-17

Section 4.8 The Node Voltage Method and Mesh Current Method Compared

*__P 4.8-1__ The circuit shown in Figure P 4.8-1 has two inputs, the voltage source voltages, v_1 and v_2. The circuit has one output, the dependent source voltage, v_o. Design this circuit so that the output is related to the inputs by

$$v_o = 2v_1 + 0.5v_2$$

Hint: Determine the required values of A, R_1, R_2, R_3, and R_4.

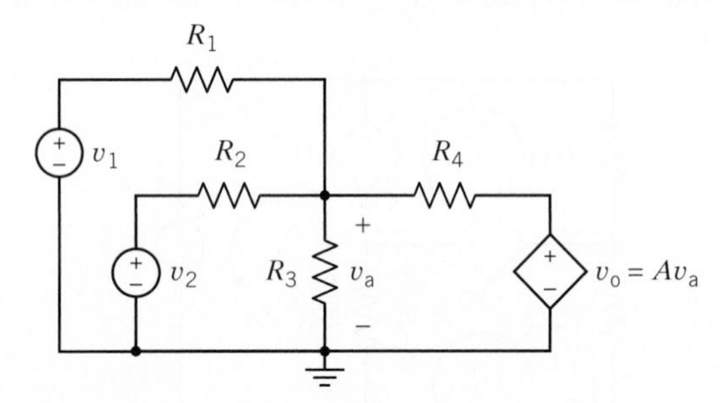

Figure P 4.8-1

P 4.8-2 The circuit shown in Figure P 4.8-2 has two inputs, v_s and i_s and one output v_o. The output is related to the inputs by the equation

$$v_o = ai_s + bv_s$$

where a and b are constants to be determined. Determine the values a and b by (a) writing and solving mesh equations and (b) writing and solving node equations.

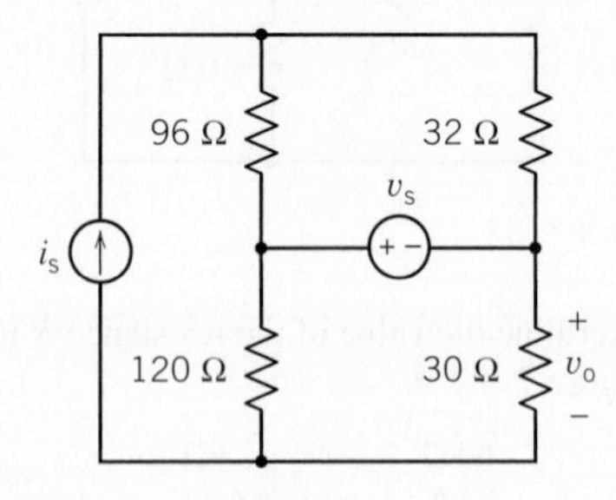

Figure P 4.8-2

P 4.8-3 Determine the power supplied by the dependent source in the circuit shown in Figure P 4.8-3 by writing and solving (a) node equations and (b) mesh equations.

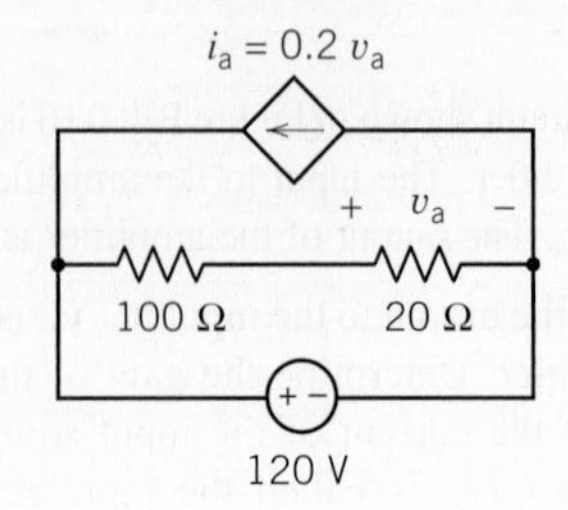

Figure P 4.8-3

Section 4.11 How Can We Check . . . ?

P 4.11-1 Computer analysis of the circuit shown in Figure P 4.11-1 indicates that the node voltages are $v_a = 5.2$ V, $v_b = -4.8$ V, and $v_c = 3.0$ V. Is this analysis correct?

Hint: Use the node voltages to calculate all the element currents. Check to see that KCL is satisfied at each node.

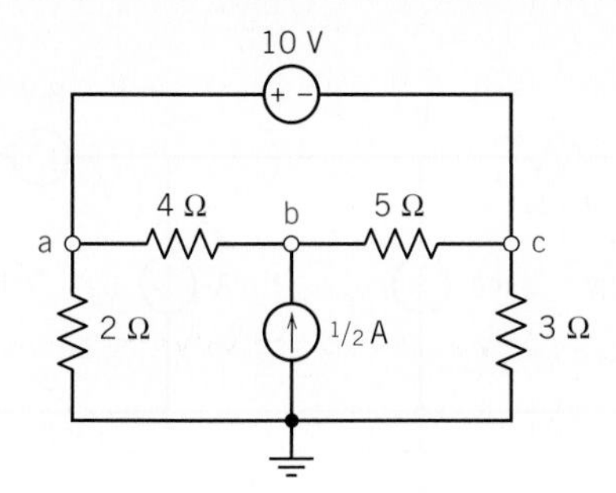

Figure P 4.11-1

P 4.11-2 An old lab report asserts that the node voltages of the circuit of Figure P 4.11-2 are $v_a = 4$ V, $v_b = 20$ V, and $v_c = 12$ V. Are these correct?

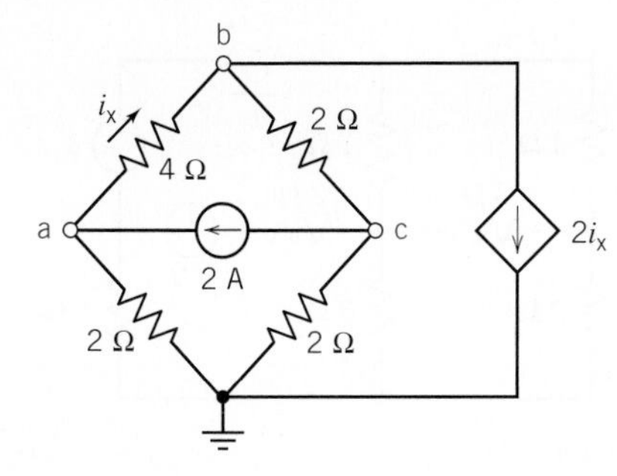

Figure P 4.11-2

P 4.11-3 Your lab partner forgot to record the values of R_1, R_2, and R_3. He thinks that two of the resistors in Figure P 4.11-3 had values of 10 kΩ and that the other had a value of 5 kΩ. Is this possible? Which resistor is the 5-kΩ resistor?

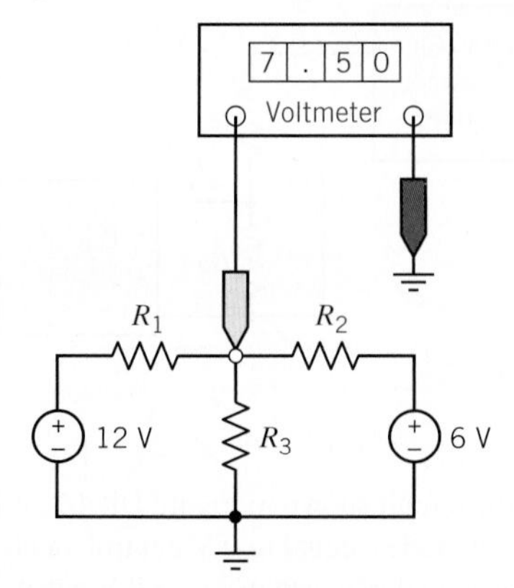

Figure P 4.11-3

P 4.11-4 Computer analysis of the circuit shown in Figure P 4.11-4 indicates that the node voltages are $v_1 = -8$ V, $v_2 = -20$ V, and $v_3 = -6$ V. Verify that this analysis is correct.

Hint: Use the node voltages to calculate the element currents. Verify that KCL is satisfied at each node.

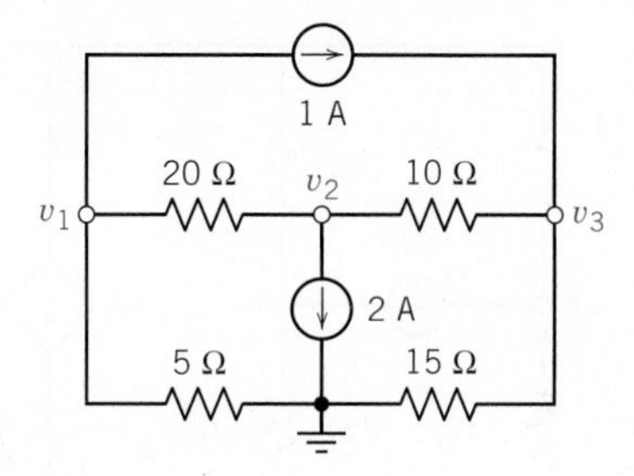

Figure P 4.11-4

P 4.11-5 Computer analysis of the circuit shown in Figure P 4.11-5 indicates that the mesh currents are $i_1 = 2$ A, $i_2 = 4$ A, and $i_3 = 3$ A. Verify that this analysis is correct.

Hint: Use the mesh currents to calculate the element voltages. Verify that KVL is satisfied for each mesh.

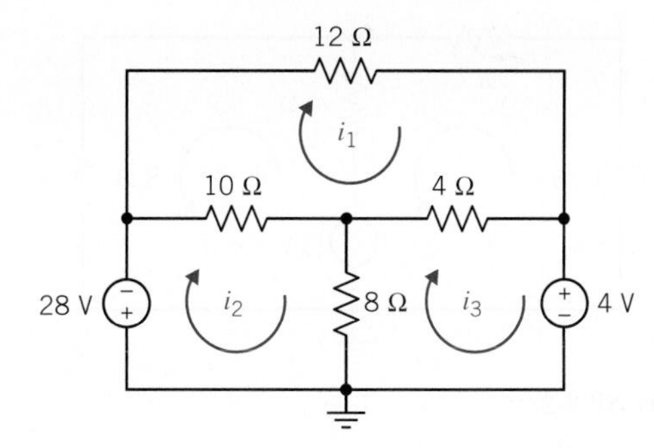

Figure P 4.11-5

PSpice Problems

SP 4-1 Use PSpice to determine the node voltages of the circuit shown in Figure SP 4-1.

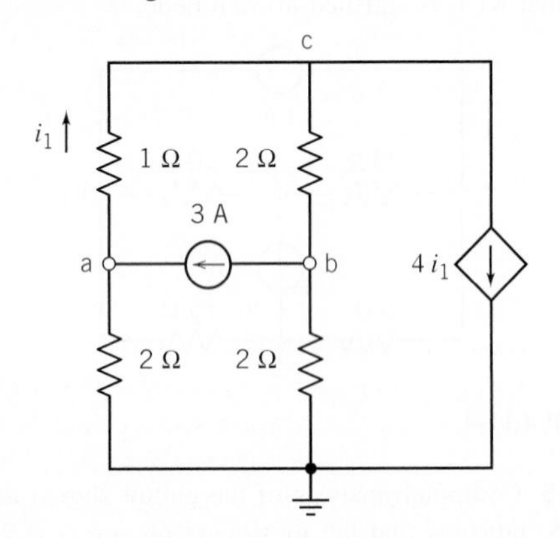

Figure SP 4-1

SP 4-2 Use PSpice to determine the mesh currents of the circuit shown in Figure SP 4-2.

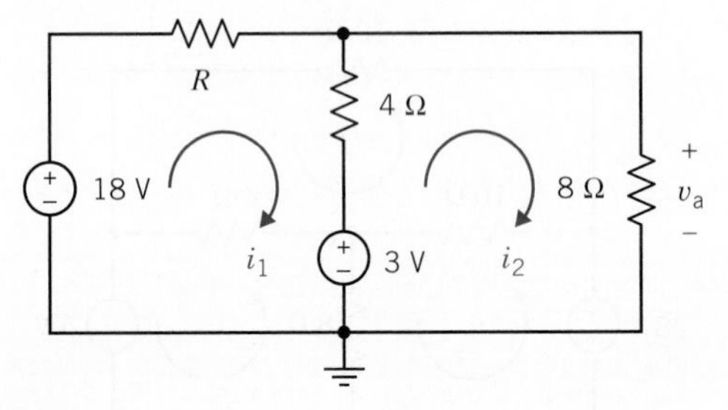

Figure SP 4-2

SP 4-3 The voltages v_a, v_b, v_c, and v_d in Figure SP 4-3 are the node voltages corresponding to nodes a, b, c and d. The current i is the current in a short circuit connected between nodes b and c. Use PSpice to determine the values of v_a, v_b, v_c, and v_d and of i.

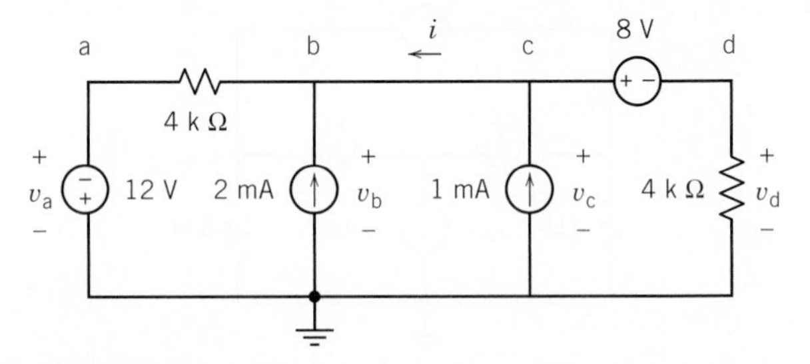

Figure SP 4-3

SP 4-4 Determine the current, i, shown in Figure SP 4-4.

Answer: $i = 0.56$ A

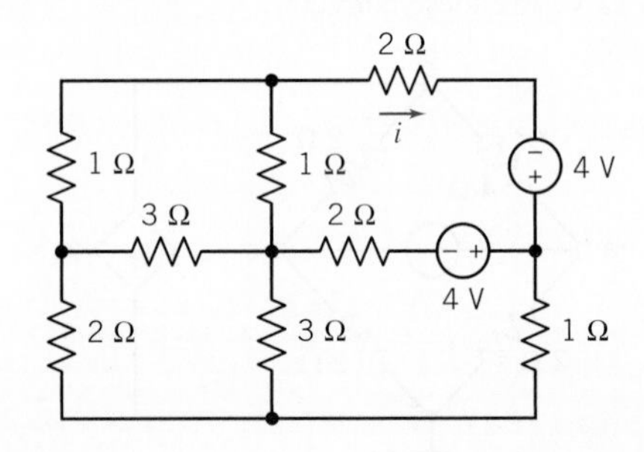

Figure SP 4-4

Design Problems

DP 4-1 An electronic instrument incorporates a 15-V power supply. A digital display is added that requires a 5-V power supply. Unfortunately, the project is over budget, and you are instructed to use the existing power supply. Using a voltage divider, as shown in Figure DP 4-1, you are able to obtain 5 V. The specification sheet for the digital display shows that the display will operate properly over a supply voltage range of 4.8 V to 5.4 V. Furthermore, the display will draw 300 mA (I) when the display is active and 100 mA when quiescent (no activity).

(a) Select values of R_1 and R_2 so that the display will be supplied with 4.8 V to 5.4 V under all conditions of current I.

(b) Calculate the maximum power dissipated by each resistor, R_1 and R_2, and the maximum current drawn from the 15-V supply.

(c) Is the use of the voltage divider a good engineering solution? If not, why? What problems might arise?

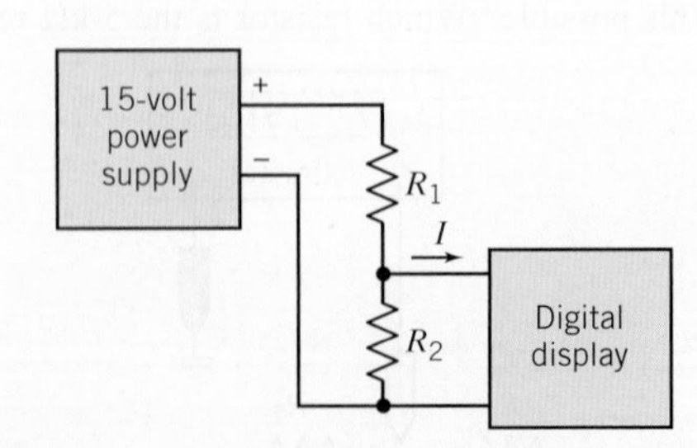

Figure DP 4-1

DP 4-2 For the circuit shown in Figure DP 4-2, it is desired to set the voltage at node a equal to 0 V control an electric motor. Select voltages v_1 and v_2 to achieve $v_a = 0$ V when v_1 and v_2 are less than 20 V and greater than zero and $R = 2\ \Omega$.

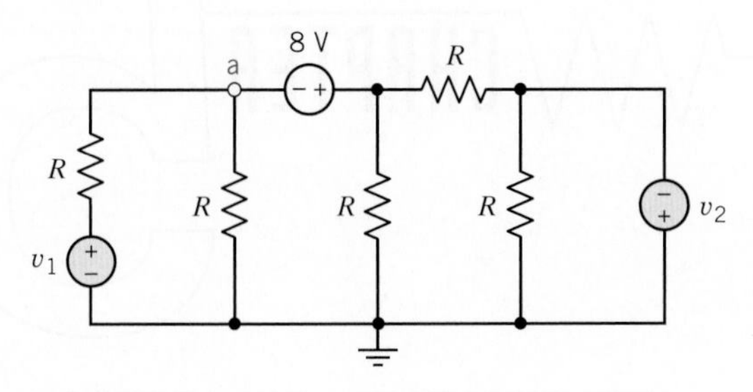

Figure DP 4-2

DP 4-3 A wiring circuit for a special lamp in a home is shown in Figure DP 4-3. The lamp has a resistance of 2 Ω, and the designer selects $R = 100\ \Omega$. The lamp will light when $I \geq 50$ mA but will burn out when $I > 75$ mA.

(a) Determine the current in the lamp and whether it will light for $R = 100\ \Omega$.

(b) Select R so that the lamp will light but will not burn out if R changes by ± 10 percent because of temperature changes in the home.

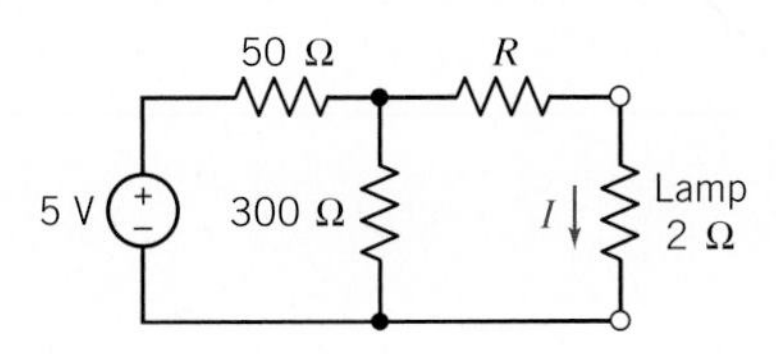

Figure DP 4-3 A lamp circuit.

D P 4-4 To control a device using the circuit shown in Figure DP 4-4, it is necessary that $v_{ab} = 10$ V. Select the resistors when it is required that all resistors be greater than 1 Ω and $R_3 + R_4 = 20\ \Omega$.

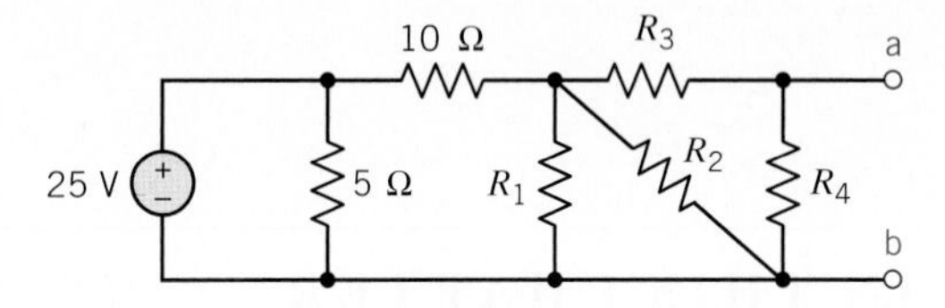

Figure DP 4-4

DP 4-5 The current i shown in the circuit of Figure DP 4-5 is used to measure the stress between two sides of an earth fault line. Voltage v_1 is obtained from one side of the fault, and v_2 is obtained from the other side of the fault. Select the resistances R_1, R_2, and R_3 so that the magnitude of the current i will remain in the range between 0.5 mA and 2 mA when v_1 and v_2 may each vary independently between +1 V and +2 V ($1\ \text{V} \leq v_n \leq 2\ \text{V}$).

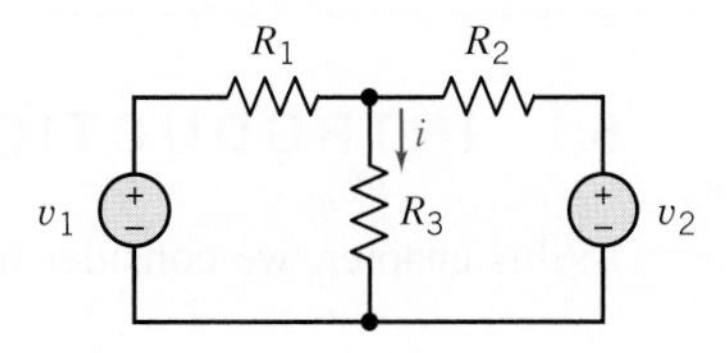

Figure DP 4-5 A circuit for earth fault-line stress measurement.

Circuit Theorems

IN THIS CHAPTER

5.1 Introduction
5.2 Source Transformations
5.3 Superposition
5.4 Thévenin's Theorem
5.5 Norton's Equivalent Circuit
5.6 Maximum Power Transfer
5.7 Using MATLAB to Determine the Thévenin Equivalent Circuit
5.8 Using PSpice to Determine the Thévenin Equivalent Circuit
5.9 How Can We Check . . . ?
5.10 **DESIGN EXAMPLE**—Strain Gauge Bridge
5.11 Summary
Problems
PSpice Problems
Design Problems

5.1 INTRODUCTION

In this chapter, we consider five circuit theorems:

- A **source transformation** allows us to replace a voltage source and series resistor by a current source and parallel resistor. Doing so does not change the element current or voltage of any other element of the circuit.
- **Superposition** says that the response of a linear circuit to several inputs working together is equal to the sum of the responses to each of the inputs working separately.
- **Thévenin's theorem** allows us to replace part of a circuit by a voltage source and series resistor. Doing so does not change the element current or voltage of any other element of the circuit.
- **Norton's theorem** allows us to replace part of a circuit by a current source and parallel resistor. Doing so does not change the element current or voltage of any other element of the circuit.
- The **maximum power transfer theorem** describes the condition under which one circuit transfers as much power as possible to another circuit.

Each of these circuit theorems can be thought of as a shortcut, a way to reduce the complexity of an electric circuit so that it can be analyzed more easily. More important, these theorems provide insight into the nature of linear electric circuits.

5.2 SOURCE TRANSFORMATIONS

The ideal voltage source is the simplest model of a voltage source, but occasionally we need a more accurate model. Figure 5.2-1*a* shows a more accurate but more complicated model of a voltage source. The circuit shown in Figure 5.2-1 is sometimes called a nonideal voltage source. (The voltage of a practical voltage source decreases as the voltage source supplies more power. The nonideal voltage source models this behavior, whereas the ideal voltage source does not. The

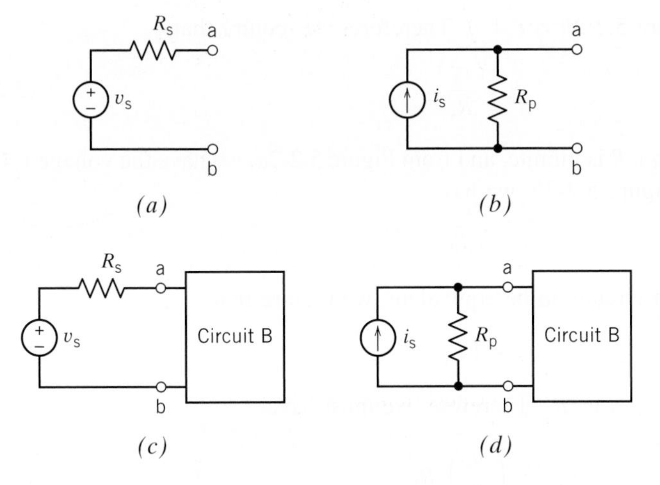

FIGURE 5.2-1 (*a*) A nonideal voltage source. (*b*) A nonideal current source. (*c*) Circuit B connected to the nonideal voltage source. (*d*) Circuit B connected to the nonideal current source.

nonideal voltage source is a more accurate model of a practical voltage source than the ideal voltage source, but it is also more complicated. We will usually use ideal voltage sources to model practical voltage sources but will occasionally need to use a nonideal voltage source.) Figure 5.2-1*b* shows a nonideal current source. It is a more accurate but more complicated model of a practical current source.

Under certain conditions ($R_p = R_s$ and $v_s = R_s i_s$), the nonideal voltage source and the nonideal current source are equivalent to each other. Figure 5.2-1 illustrates the meaning of "equivalent." In Figure 5.2-1*c*, a nonideal voltage source is connected to circuit B. In Figure 5.2-1*d*, a nonideal current source is connected to that same circuit B. Perhaps Figure 5.2-1*d* was obtained from Figure 5.2-1*c*, by replacing the nonideal voltage source with a nonideal current source. Replacing the nonideal voltage source by the *equivalent* nonideal current source does not change the voltage or current of any element in circuit B. That means that if you looked at a list of the values of the currents and voltages of all the circuit elements in circuit B, you could not tell whether circuit B was connected to a nonideal voltage source or to an equivalent nonideal current source. Similarly, we can imagine that Figure 5.2-1*c* was obtained from Figure 5.2-1*d* by replacing the nonideal current source with a nonideal voltage source. Replacing the nonideal current source by the *equivalent* nonideal voltage source does not change the voltage or current of any element in circuit B. The process of transforming Figure 5.2-1*c* into Figure 5.2-1*d*, or vice versa, is called a source transformation.

We want the circuit of Figure 5.2-1*a* to transform into that of Figure 5.2-1*b*. We then require that both circuits have the same characteristic for all values of an external resistor R connected between terminals a–b (Figures 5.2-2*a,b*). We will try the two extreme values $R = 0$ and $R = \infty$.

When the external resistance $R = 0$, we have a short circuit across terminals a–b. First, we require the short-circuit current to be the same for each circuit. The short-circuit current for Figure 5.2-2*a* is

$$i = \frac{v_s}{R_s} \tag{5.2-1}$$

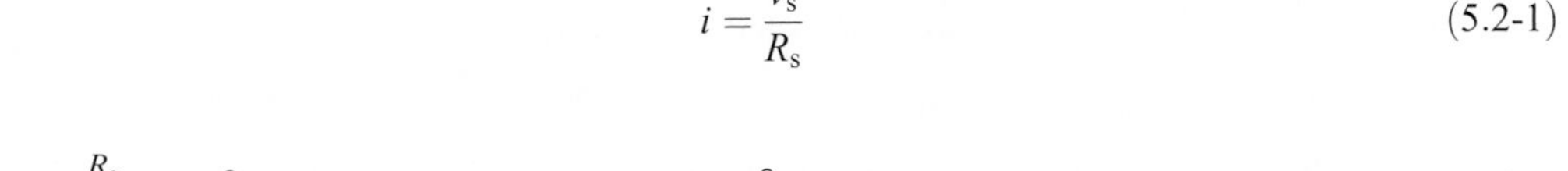

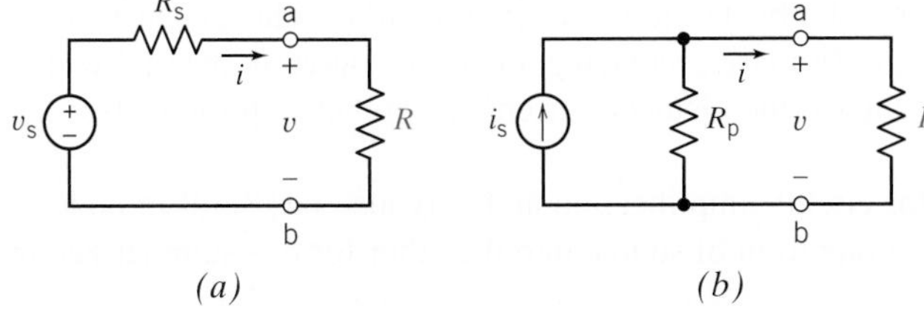

FIGURE 5.2-2 (*a*) Voltage source with an external resistor R. (*b*) Current source with an external resistance R.

The short-circuit current for Figure 5.2-2*b* is $i = i_s$. Therefore, we require that

$$i_s = \frac{v_s}{R_s} \tag{5.2-2}$$

For the open-circuit condition, R is infinite, and from Figure 5.2-2*a*, we have the voltage $v = v_s$. For the open-circuit voltage of Figure 5.2-2*b*, we have

$$v = i_s R_p$$

Because v must be equal for both circuits to be equivalent, we require that

$$v_s = i_s R_p \tag{5.2-3}$$

Also, from Eq. 5.2-2, we require $i_s = v_s/R_s$. Therefore, we must have

$$v_s = \left(\frac{v_s}{R_s}\right) R_p$$

and, therefore, we require that

$$R_s = R_p \tag{5.2-4}$$

Equations 5.2-2 and 5.2-4 must be true simultaneously for the two nonideal sources to be equivalent. Of course, we have proved that the two sources are equivalent at two values ($R = 0$ and $R = \infty$). We have not proved that the circuits are equal for all R, but we assert that the equality relationship holds for all R for these two circuits as we show below.

For the circuit of Figure 5.2-2*a*, we use KVL to obtain

$$v_s = iR_s + v$$

Dividing by R_s gives

$$\frac{v_s}{R_s} = i + \frac{v}{R_s}$$

If we use KCL for the circuit of Figure 5.2-2*b*, we have

$$i_s = i + \frac{v}{R_p}$$

Thus, the two circuits are equivalent when $i_s = v_s/R_s$ and $R_s = R_p$.

A voltage source v_s connected in series with a resistor R_s and a current source i_s connected in parallel with a resistor R_p are equivalent circuits provided that

$$R_p = R_s \quad \text{and} \quad v_s = R_s i_s$$

Replacing a voltage source in series with a resistor by its equivalent circuit will not change the element currents or voltages in the rest of the circuit. Similarly, replacing a current source in parallel with a resistor by its equivalent circuit will not change the element currents or voltages in the rest of the circuit.

Source transformations are useful for circuit simplification and may also be useful in node or mesh analysis. The method of transforming one form of source into the other form is summarized in Figure 5.2-3.

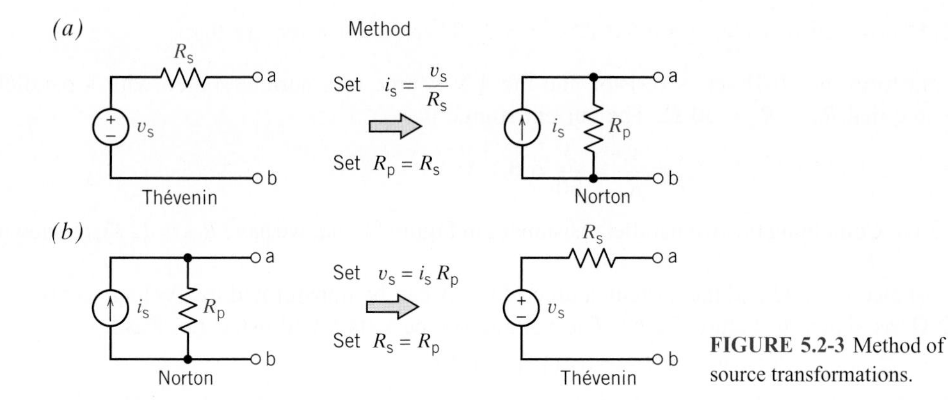

FIGURE 5.2-3 Method of source transformations.

EXAMPLE 5.2-1 Source Transformations

Find the source transformation for the circuits shown in Figures 5.2-4*a*,*b*.

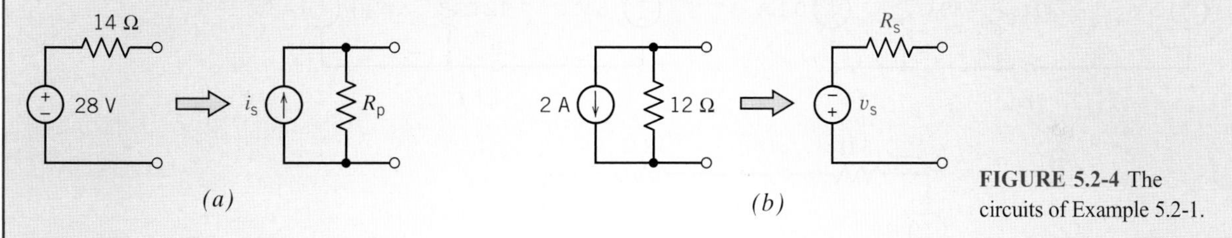

FIGURE 5.2-4 The circuits of Example 5.2-1.

Solution

Using the method summarized in Figure 5.2-3, we note that the voltage source of Figure 5.2-4*a* can be transformed to a current source with $R_p = R_s = 14\ \Omega$. The current source is

$$i_s = \frac{v_s}{R_s} = \frac{28}{14} = 2\ \text{A}$$

The resulting transformed source is shown on the right side of Figure 5.2-4*a*.

Starting with the current source of Figure 5.2-4*b*, we have $R_s = R_p = 12\ \Omega$. The voltage source is

$$v_s = i_s R_p = 2(12) = 24\ \text{V}$$

The resulting transformed source is shown on the right side of Figure 5.2-4*b*. Note that the positive sign of the voltage source v_s appears on the lower terminal because the current source arrow points downward.

EXAMPLE 5.2-2 Source Transformations

A circuit is shown in Figure 5.2-5. Find the current i by reducing the circuit to the right of terminals a–b to its simplest form, using source transformations.

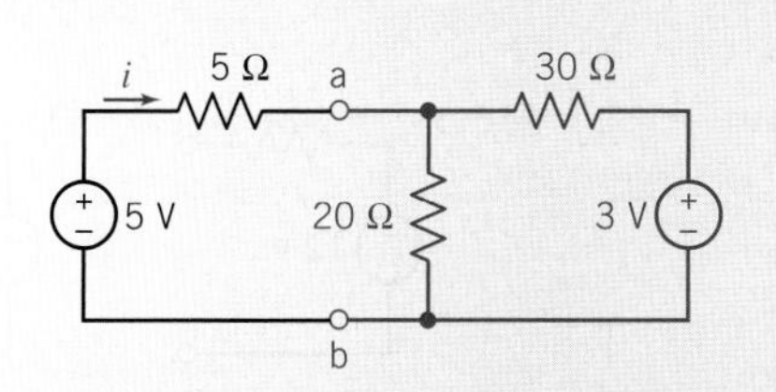

FIGURE 5.2-5 The circuit of Example 5.2-2.

Solution

The first step is to transform the 30-Ω series resistor and the 3-V source to a current source with a parallel resistance. First, we note that $R_p = R_s = 30\ \Omega$. The current source is

$$i_s = \frac{v_s}{R_p} = \frac{3}{30} = 0.1\ \text{A}$$

as shown in Figure 5.2-6*a*. Combining the two parallel resistances in Figure 5.2-6*a*, we have $R_{p2} = 12\ \Omega$, as shown in Figure 5.2-6*b*.

The parallel resistance of 12 Ω and the current source of 0.1 A can be transformed to a voltage source in series with $R_{s2} = 12\ \Omega$, as shown in Figure 5.2-6*c*. The voltage source v_s is found using Eq. 5.2-3:

$$v_s = i_s R_{s2} = 0.1(12) = 1.2\ \text{V}$$

Source transformations do not disturb the currents and voltages in the rest of the circuit. Therefore, the current i in Figure 5.2-5 is equal to the current i in Figure 5.2-6*c*. The current i is found by using KVL around the loop of Figure 5.2-6*c*, yielding $i = 3.8/17 = 0.224$ A.

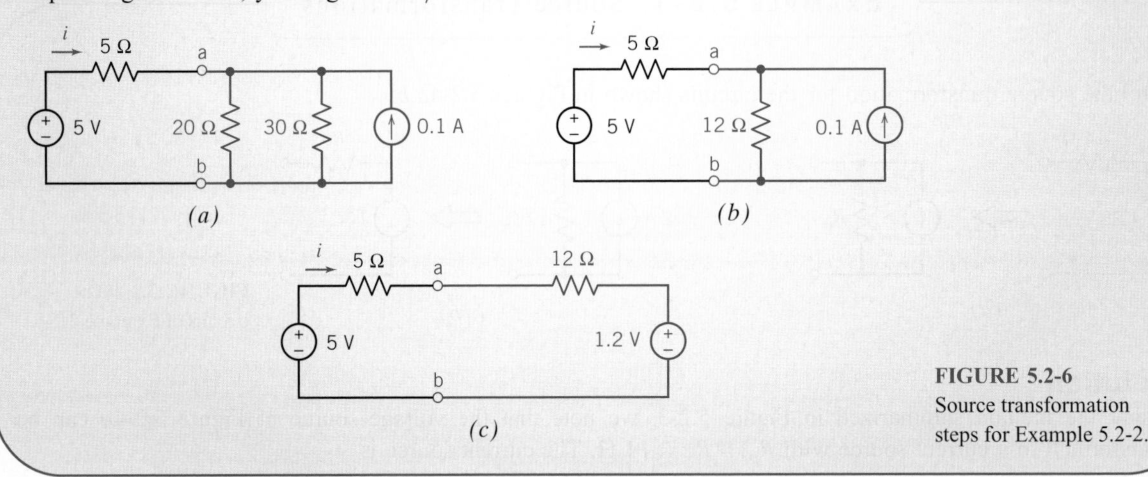

FIGURE 5.2-6 Source transformation steps for Example 5.2-2.

EXERCISE 5.2-1 Determine values of R and i_s so that the circuits shown in Figures E 5.2-1*a*,*b* are equivalent to each other due to a source transformation.
Answer: $R = 10\ \Omega$ and $i_s = 1.2$ A

EXERCISE 5.2-2 Determine values of R and i_s so that the circuits shown in Figures E 5.2-2*a*,*b* are equivalent to each other due to a source transformation.

Hint: Notice that the polarity of the voltage source in Figure E 5.2-2*a* is not the same as in Figure E 5.2-1*a*.

Answer: $R = 10\ \Omega$ and $i_s = -1.2$ A

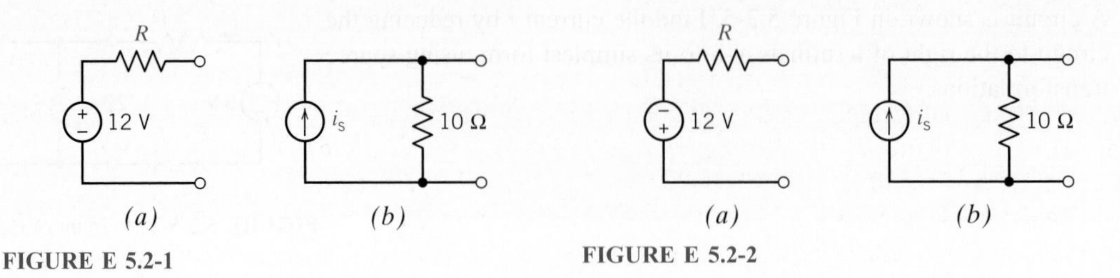

FIGURE E 5.2-1

FIGURE E 5.2-2

EXERCISE 5.2-3 Determine values of R and v_s so that the circuits shown in Figures E 5.2-3*a*, *b* are equivalent to each other due to a source transformation.

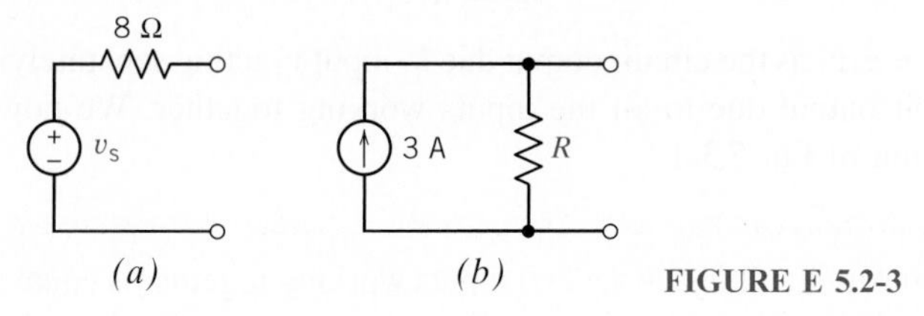

FIGURE E 5.2-3

Answer: $R = 8\ \Omega$ and $v_s = 24$ V

EXERCISE 5.2-4 Determine values of R and v_s so that the circuits shown in Figures E 5.2-4*a*, *b* are equivalent to each other due to a source transformation.

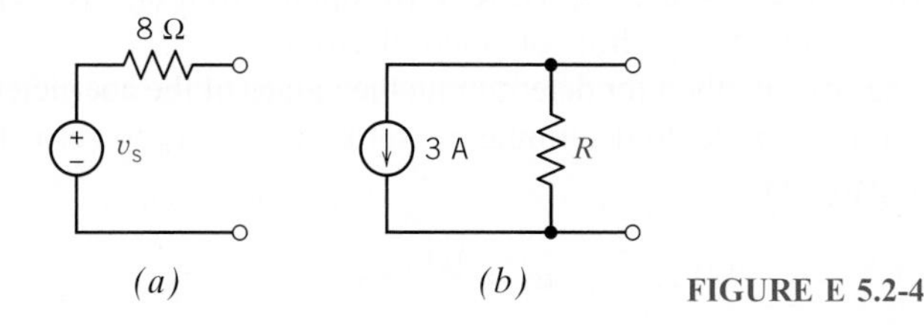

FIGURE E 5.2-4

Hint: Notice that the reference direction of the current source in Figure E 5.2-4*b* is not the same as in Figure E 5.2-3*b*.

Answer: $R = 8\ \Omega$ and $v_s = -24$ V

5.3 SUPERPOSITION

The output of a linear circuit can be expressed as a linear combination of its inputs. For example, consider any circuit having the following three properties:

1. The circuit consists entirely of resistors and dependent and independent sources.
2. The circuit inputs are the voltages of all the independent voltage sources and the currents of all the independent current sources.
3. The output is the voltage or current of any element of the circuit.

Such a circuit is a linear circuit. Consequently, the circuit output can be expressed as a linear combination of the circuit input. For example,

$$v_o = a_1 v_1 + a_2 v_2 + \cdots + a_n v_n \tag{5.3-1}$$

where v_o is the output of the circuit (it could be a current instead of a voltage) and $v_1, v_2, \ldots, v_n$ are the inputs to the circuit (any or all the inputs could be currents instead of voltages). The coefficients $a_1, a_2, \ldots, a_n$ of the linear combination are real constants called gains.

Next, consider what would happen if we set all but one input to zero. Let v_{oi} denote output when all inputs except the ith input have been set to zero. For example, suppose we set $v_2, v_3, \ldots, v_n$ to zero.

Then

$$v_{o1} = a_1 v_1 \tag{5.3-2}$$

We can interpret $v_{o1} = a_1 v_1$ as the circuit output due to input v_1 acting separately. In contrast, the v_o in Eq 5.3-1 is the circuit output due to all the inputs working together. We now have the following important interpretation of Eq. 5.3-1:

> The output of a linear circuit due to several inputs working together is equal to the sum of the outputs due to each input working separately.

The inputs to our circuit are voltages of independent voltage sources and the currents of independent current sources. When we set all but one input to zero the other inputs become 0-V voltage sources and 0-A current sources. Because 0-V voltage sources are equivalent to short circuits and 0-A current sources are equivalent to open circuits, we replace the sources corresponding to the other inputs by short or open circuits.

Equation 5.3-2 suggests a method for determining the values of the coefficients $a_1, a_2, \ldots, a_n$ of the linear combination. For example, to determine a_1, set $v_2, v_3, \ldots, v_n$ to zero. Then, dividing both sides of Eq. 5.5-2 by v_1, we get

$$a_1 = \frac{v_{o1}}{v_1}$$

The other gains are determined similarly.

EXAMPLE 5.3-1 Superposition

The circuit shown in Figure 5.3-1 has one output, v_o, and three inputs, v_1, i_2, and v_3. (As expected, the inputs are voltages of independent voltage sources and the currents of independent current sources.) Express the output as a linear combination of the inputs.

Solution

Let's analyze the circuit using node equations. Label the node voltage at the top node of the current source and identify the supernode corresponding to the horizontal voltage source as shown in Figure 5.3-2. Apply KCL to the supernode to get

$$\frac{v_1 - (v_3 + v_o)}{40} + i_2 = \frac{v_o}{10}$$

Multiply both sides of this equation by 40 to eliminate the fractions. Then we have

$$v_1 - (v_3 + v_o) + 40 i_2 = 4 v_o \quad \Rightarrow \quad v_1 + 40 i_2 - v_3 = 5 v_o$$

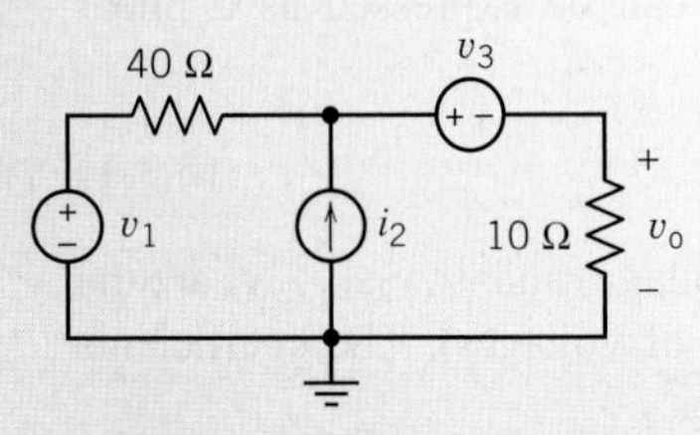

FIGURE 5.3-1 The linear circuit for Example 5.3-1.

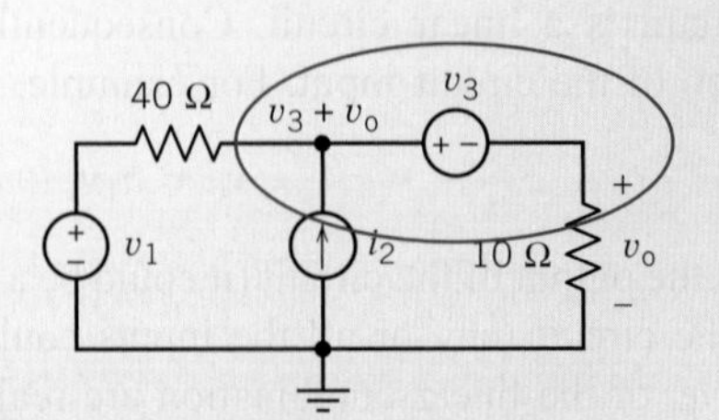

FIGURE 5.3-2 A supernode.

Dividing both sides by 5 expresses the output as a linear combination of the inputs:

$$v_o = \frac{v_1}{5} + 8i_2 - \frac{v_3}{5}$$

Also, the coefficients of the linear combination can now be determined to be

$$a_1 = \frac{v_{o1}}{v_1} = \frac{1}{5}\text{V/V}, a_2 = \frac{v_{o2}}{i_2} = 8\text{V/A, and } a_3 = \frac{v_{o3}}{v_3} = -\frac{1}{5}\text{V/V}$$

Alternate Solution

Figure 5.3-3 shows the circuit from Figure 5.3-1 when $i_2 = 0$ A and $v_3 = 0$ V. (A zero current source is equivalent to an open circuit, and a zero voltage source is equivalent to a short circuit.)

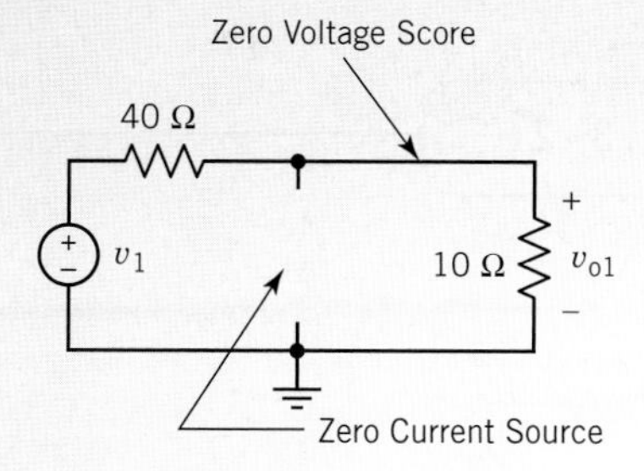

FIGURE 5.3-3 Output due to the first input.

Using voltage division

$$v_{o1} = \frac{10}{40+10}v_1 = \frac{1}{5}v_1$$

In other words,

$$a_1 = \frac{v_{o1}}{v_1} = \frac{1}{5}\text{ V/V}$$

Next, Figure 5.3-4 shows the circuit when $v_1 = 0$ V and $v_3 = 0$ V. The resistors are connected in parallel. Applying Ohm's law to the equivalent resistance gives

$$v_{o2} = \frac{40 \times 10}{40+10}i_2 = 8i_2$$

In other words,

$$a_2 = \frac{v_{o2}}{i_2} = 8\text{V/A}$$

Finally, Figure 5.3-5 shows the circuit when $v_1 = 0$ V and $i_2 = 0$ A. Using voltage division,

$$v_{o3} = \frac{10}{40+10}(-v_3) = -\frac{1}{5}v_3$$

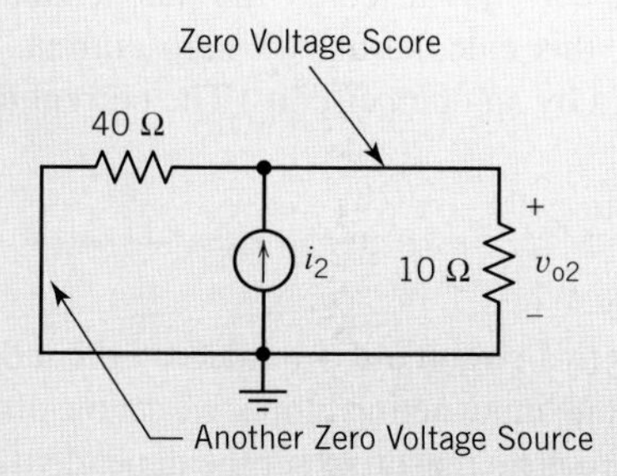

FIGURE 5.3-4 Output due to the second input.

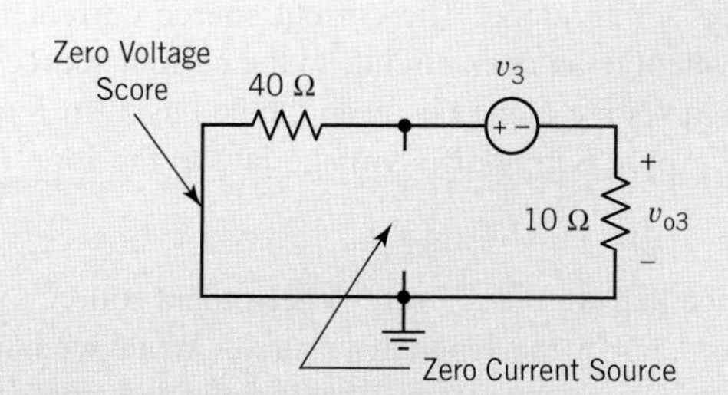

FIGURE 5.3-5 Output due to the third input.

In other words,

$$a_3 = \frac{v_{o3}}{v_3} = -\frac{1}{5}\text{ V/V}$$

Now the output can be expressed as a linear combination of the inputs

$$v_o = a_1 v_1 + a_2 i_2 + a_3 v_3 = \frac{1}{5} v_1 + 8 i_2 + \left(-\frac{1}{5}\right) v_3$$

as before.

EXAMPLE 5.3-2

Find the current i for the circuit of Figure 5.3-6a.

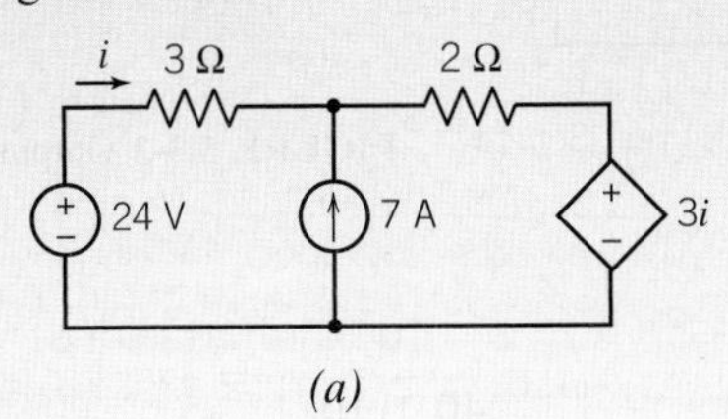

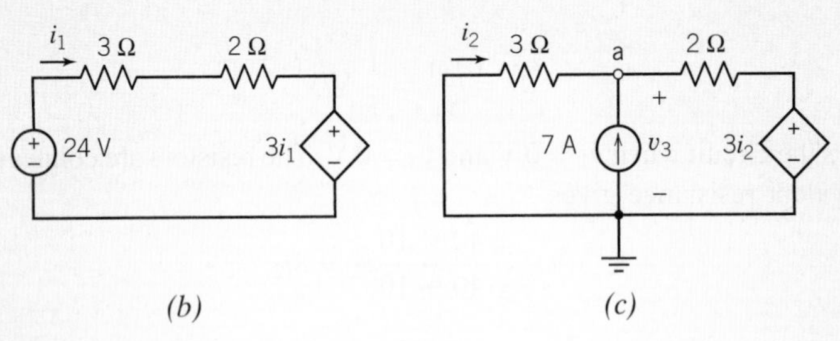

FIGURE 5.3-6 (a) The circuit for Example 5.3-2. (b) The independent voltage source acting alone. (c) The independent current source acting alone.

Solution

Independent sources provide the inputs to a circuit. The circuit in Figure 5.3-6a has two inputs: the voltage of the independent voltage source and the current of the independent current source. The current, i, caused by the two sources acting together is equal to the sum of the currents caused by each independent source acting separately.

Step 1: Figure 5.3-6b shows the circuit used to calculate the current caused by the independent voltage source acting alone. The current source current is set to zero for this calculation. (A zero current source is equivalent to an open circuit, so the current source has been replaced by an open circuit.) The current due to the voltage source alone has been labeled as i_1 in Figure 5.3-6b.

Apply Kirchhoff's voltage law to the loop in Figure 5.3-6b to get

$$-24 + (3+2)i_1 + 3i_1 + 0 \quad \Rightarrow \quad i_1 = 3\text{ A}$$

(Notice that we did not set the dependent source to zero. The inputs to a circuit are provided by the independent sources, not by the dependent sources. When we find the response to one input acting alone, we set the other inputs to zero. Hence, we set the other independent sources to zero, but there is no reason to set the dependent source to zero.)

Step 2: Figure 5.3-6c shows the circuit used to calculate the current caused by the current source acting alone. The voltage of the independent voltage is set to zero for this calculation. (A zero voltage source is equivalent to a short circuit, so the independent voltage source has been replaced by a short circuit.) The current due to the voltage source alone has been labeled as i_2 in Figure 5.3-6c.

First, express the controlling current of the dependent source in terms of the node voltage, v_a, using Ohm's law:

$$i_2 = -\frac{v_a}{3} \quad \Rightarrow \quad v_a = -3i_2$$

Next, apply Kirchhoff's current law at node a to get

$$i_2 + 7 = \frac{v_a - 3i_2}{2} \quad \Rightarrow \quad i_2 + 7 = \frac{-3i_2 - 3i_2}{2} \quad \Rightarrow \quad i_2 = -\frac{7}{4}\text{ A}$$

Step 3: The current, i, caused by the two independent sources acting together is equal to the sum of the currents, i_1 and i_2, caused by each source acting separately:

$$i = i_1 + i_2 = 3 - \frac{7}{4} = \frac{5}{4}\text{ A}$$

5.4 THÉVENIN'S THEOREM

In this section, we introduce the Thévenin equivalent circuit, based on a theorem developed by M. L. Thévenin, a French engineer, who first published the principle in 1883. Thévenin, who is credited with the theorem, probably based his work on earlier work by Hermann von Helmholtz (see Figure 5.4-1).

FIGURE 5.4-1 Hermann von Helmholtz (1821–1894), who is often credited with the basic work leading to Thévenin's theorem. Courtesy of the New York Public Library.

Figure 5.4-2 illustrates the use of the Thévenin equivalent circuit. In Figure 5.4-2a, a circuit is partitioned into two parts—circuit A and circuit B—that are connected at a single pair of terminals. (This is the only connection between circuits A and B. In particular, if the overall circuit contains a dependent source, then either both parts of that dependent source must be in circuit A or both parts must be in circuit B.) In Figure 5.4-2b, circuit A is replaced by its Thévenin equivalent circuit, which consists of an ideal voltage source in series with a resistor. Replacing circuit A by its Thévenin equivalent circuit does not change the voltage or current of any element in circuit B. This means that if you looked at a list of the values of the currents and voltages of all the circuit elements in circuit B, you could not tell whether circuit B was connected to circuit A or connected to its Thévenin equivalent circuit.

Finding the Thévenin equivalent circuit of circuit A involves three parameters: the open-circuit voltage, v_{oc}, the short-circuit current, i_{sc}, and the Thévenin resistance, R_t. Figure 5.4-3 illustrates the meaning of these three parameters. In Figure 5.4-3a, an open circuit is connected across the terminals of circuit A. The voltage across that open circuit is

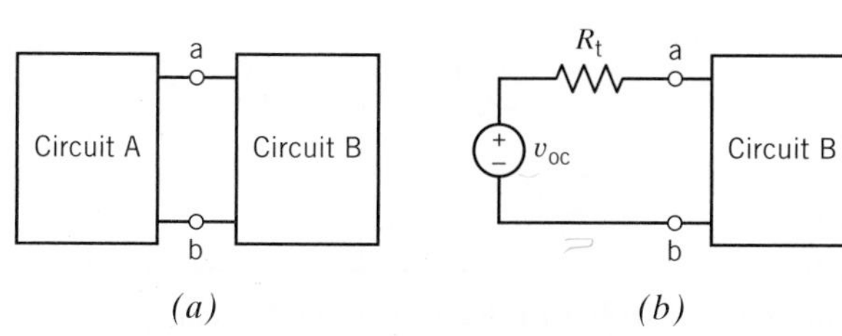

FIGURE 5.4-2 (a) A circuit partitioned into two parts: circuit A and circuit B. (b) Replacing circuit A by its Thévenin equivalent circuit.

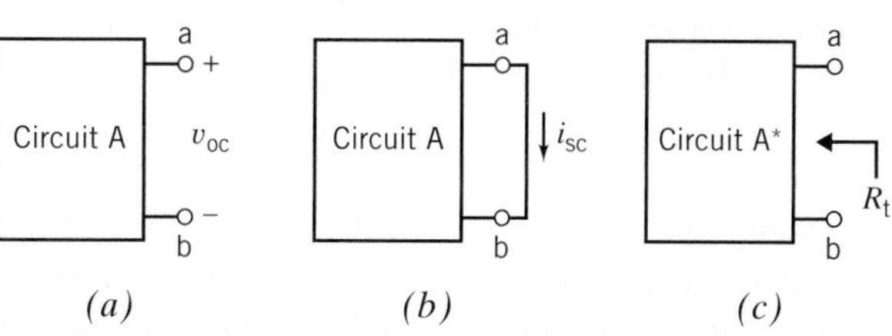

FIGURE 5.4-3 The Thévenin equivalent circuit involves three parameters: (a) the open-circuit voltage, v_{oc}, (b) the short-circuit current, i_{sc}, and (c) the Thévenin resistance, R_t.

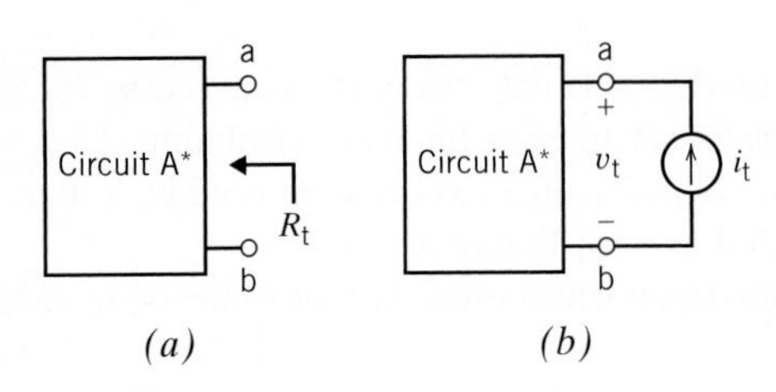

FIGURE 5.4-4 (*a*) The Thévenin resistance, R_t, and (*b*) a method for measuring or calculating the Thévenin resistance, R_t.

the open-circuit voltage, v_{oc}. In Figure 5.4-3*b*, a short circuit is connected across the terminals of circuit A. The current in that short circuit is the short-circuit current, i_{sc}.

Figure 5.4-3*c* indicates that the Thévenin resistance, R_t, is the equivalent resistance of circuit A*. Circuit A* is formed from circuit A by replacing all the *independent* voltage sources by short circuits and replacing all the *independent* current sources by open circuits. (*Dependent* current and voltage sources are not replaced with open circuits or short circuits.) Frequently, the Thévenin resistance, R_t, can be determined by repeatedly replacing series or parallel resistors by equivalent resistors. Sometimes, a more formal method is required. Figure 5.4-4 illustrates a formal method for determining the value of the Thévenin resistance. A current source having current i_t is connected across the terminals of circuit A*. The voltage, v_t, across the current source is calculated or measured. The Thévenin resistance is determined from the values of i_t and v_t, using

$$R_t = \frac{v_t}{i_t}$$

The open-circuit voltage, v_{oc}, the short-circuit current, i_{sc}, and the Thévenin resistance, R_t, are related by the equation

$$v_{oc} = R_t i_{sc}$$

Consequently, the Thévenin resistance can be calculated from the open-circuit voltage and the short-circuit current.

In summary, the Thévenin equivalent circuit for circuit A consists of an ideal voltage source, having voltage v_{oc}, in series with a resistor, having resistance R_t. Replacing circuit A by its Thévenin equivalent circuit does not change the voltage or current of any element in circuit B.

EXAMPLE 5.4-1 Thévenin Equivalent Circuit

Using Thévenin's theorem, find the current i through the resistor R in the circuit of Figure 5.4-5.

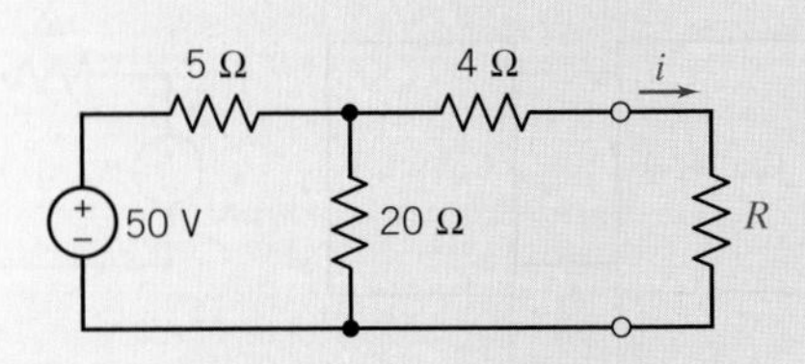

FIGURE 5.4-5 Circuit for Example 5.4-1.

Solution

Because we are interested in the current i, we identify the resistor R as circuit B. Then circuit A is as shown in Figure 5.4-6*a*. The Thévenin resistance R_t is found from Figure 5.4-6*b*, where we have set the voltage source voltage to zero and then replaced the 0-V source by a short circuit. We calculate the equivalent resistance looking into the terminals, obtaining $R_t = 8\ \Omega$.

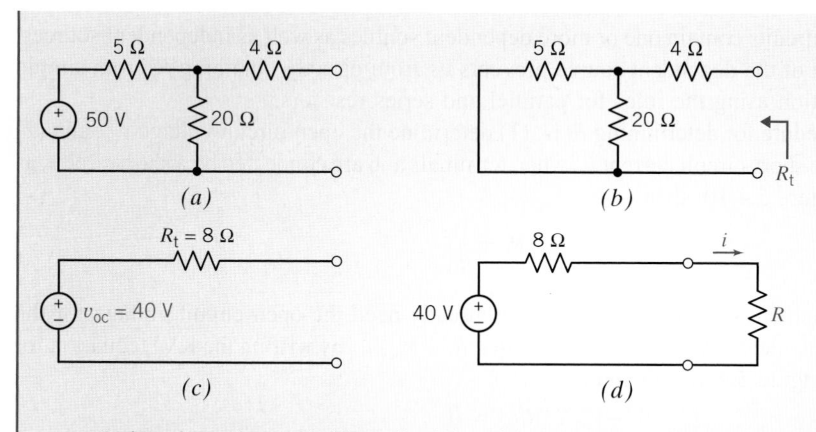

FIGURE 5.4-6 Steps for determining the Thévenin equivalent circuit for the circuit left of the terminals of Figure 5.4-5.

Using the voltage divider principle with the circuit of Figure 5.4-6*a*, we find $v_{oc} = 40$ V. Reconnecting circuit B to the Thévenin equivalent circuit as shown in Figure 5.4-6*d*, we obtain

$$i = \frac{40}{R + 8} \text{ A}$$

EXAMPLE 5.4-2 Thévenin Equivalent Circuit

Find the Thévenin equivalent circuit for the circuit shown in Figure 5.4-7.

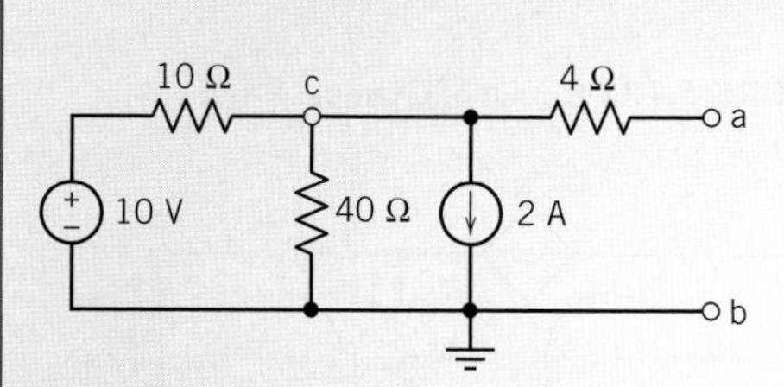

FIGURE 5.4-7 Circuit for Example 5.4-2.

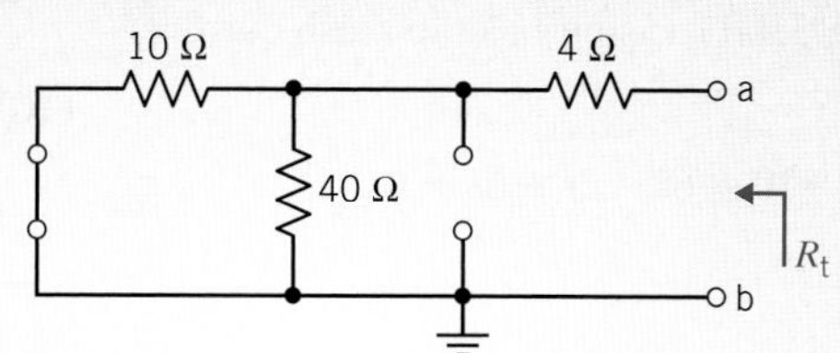

FIGURE 5.4-8 Circuit of Figure 5.4-7 with all the sources deactivated.

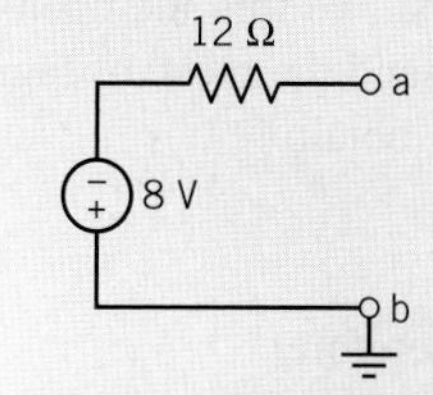

FIGURE 5.4-9 Thévenin equivalent circuit for the circuit of Figure 5.4-7.

Solution

One approach is to find the open-circuit voltage and the circuit's Thévenin equivalent resistance R_t. First, let us find the resistance R_t. Figure 5.4-8 shows the circuit after replacing the voltage source by a short circuit and replacing the current source by an open circuit. Look into the circuit at terminals a–b to find R_t. The 10-Ω resistor in parallel with the 40-Ω resistor results in an equivalent resistance of 8 Ω. Adding 8 Ω to 4 Ω in series, we obtain

$$R_t = 12\,\Omega$$

Next, we wish to determine the open-circuit voltage at terminals a–b. Because no current flows through the 4-Ω resistor, the open-circuit voltage is identical to the voltage across the 40-Ω resistor, v_c. Using the bottom node as the reference, we write KCL at node c of Figure 5.4-7 to obtain

$$\frac{v_c - 10}{10} + \frac{v_c}{40} + 2 = 0$$

Solving for v_c yields

$$v_c = -8 \text{ V}$$

Therefore, the Thévenin equivalent circuit is as shown in Figure 5.4-9.

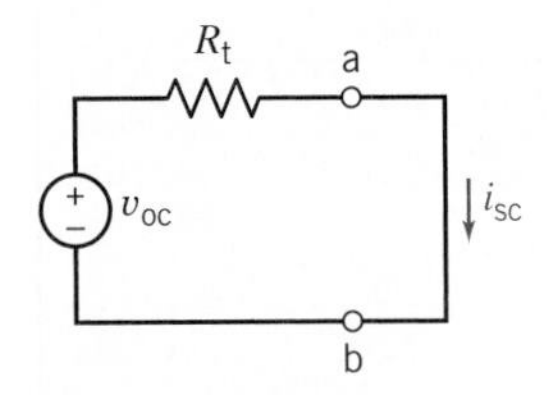

FIGURE 5.4-10 Thévenin circuit with a short circuit at terminals a–b.

Some circuits contain one or more dependent sources as well as independent sources. The presence of the dependent source prevents us from directly obtaining R_t from simple circuit reduction using the rules for parallel and series resistors.

A procedure for determining R_t is: (1) determine the open-circuit voltage v_{oc}, and (2) determine the short-circuit current i_{sc} when terminals a–b are connected by a short circuit, as shown in Figure 5.4-10; then

$$R_t = \frac{v_{oc}}{i_{sc}}$$

This method is attractive because we already need the open-circuit voltage for the Thévenin equivalent circuit. We can show that $R_t = v_{oc}/i_{sc}$ by writing the KVL equation for the loop of Figure 5.4-10, obtaining

$$-v_{oc} + R_t i_{sc} = 0$$

Clearly, $R_t = v_{oc}/i_{sc}$.

EXAMPLE 5.4-3 Thévenin Equivalent Circuits and Dependent Sources

Find the Thévenin equivalent circuit for the circuit shown in Figure 5.4-11, which includes a dependent source.

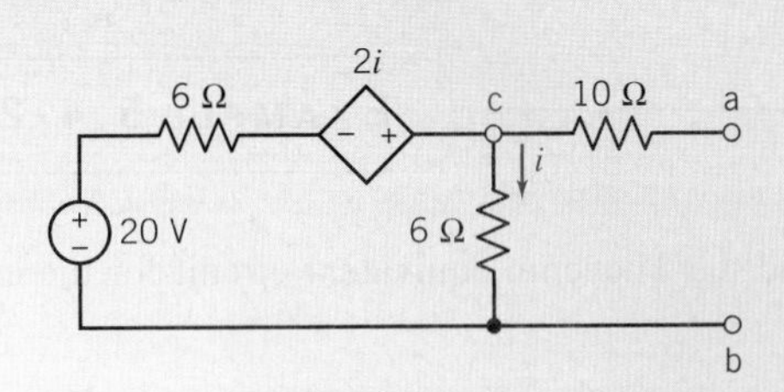

FIGURE 5.4-11 Circuit of Example 5.4-3.

Solution

First, we find the open-circuit voltage $v_{oc} = v_{ab}$. Writing KVL around the mesh of Figure 5.4-11 (using i as the mesh current), we obtain

$$-20 + 6i - 2i + 6i = 0$$

Therefore,

$$i = 2\text{ A}$$

Because no current is flowing through the 10-Ω resistor, the open-circuit voltage is identical to the voltage across the resistor between terminals c and b. Therefore,

$$v_{oc} = 6i = 12\text{ V}$$

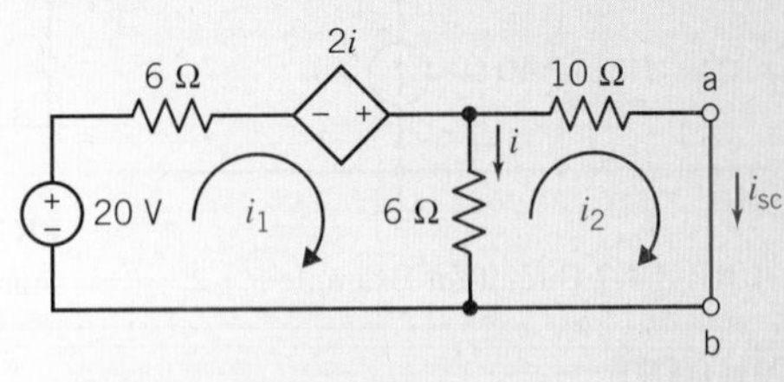

FIGURE 5.4-12 Circuit of Figure 5.4-11 with output terminals a–b short-circuited.

The next step is to determine the short-circuit current for the circuit of Figure 5.4-12. Using the two mesh currents indicated, we have

$$-20 + 6i_1 - 2i + 6(i_1 - i_2) = 0$$

and

$$6(i_2 - i_1) + 10i_2 = 0$$

Substitute $i = i_1 - i_2$ and rearrange the two equations to obtain

$$10i_1 - 4i_2 = 20$$

and

$$-6i_1 + 16i_2 = 0$$

Therefore, we find that $i_2 = i_{sc} = 120/136$ A. The Thévenin resistance is

$$R_t = \frac{v_{oc}}{i_{sc}} = \frac{12}{120/136} = 13.6\ \Omega$$

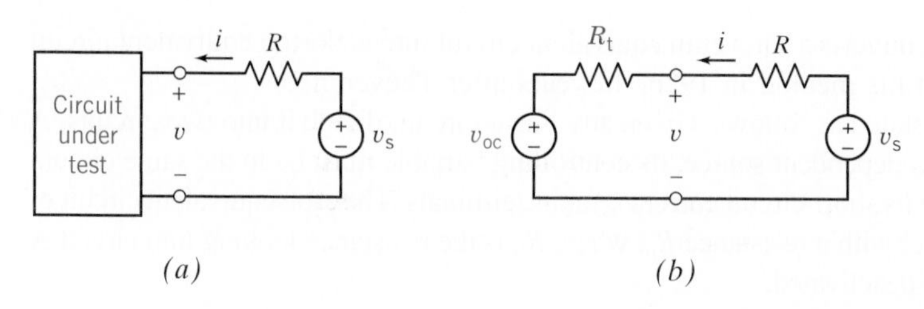

FIGURE 5.4-13 (*a*) Circuit under test with laboratory source v_s and resistor R. (*b*) Circuit of (*a*) with Thévenin equivalent circuit replacing the test circuit.

A laboratory procedure for determining the Thévenin equivalent of a black box circuit (see Figure 5.4.13*a*) is to measure i and v for two or more values of v_s and a fixed value of R. For the circuit of Figure 5.4.13*b*, we replace the test circuit with its Thévenin equivalent, obtaining

$$v = v_{oc} + iR_t \tag{5.4-1}$$

The procedure is to measure v and i for a fixed R and several values of v_s. For example, let $R = 10\ \Omega$ and consider the two measurement results

$$(1)\quad v_s = 49\ \text{V}:\ i = 0.5\ \text{A},\ v = 44\ \text{V}$$

and

$$(2)\quad v_s = 76\ \text{V}:\ i = 2\ \text{A},\ v = 56\ \text{V}$$

Then we have two simultaneous equations (using Eq. 5.51):

$$\begin{aligned} 44 &= v_{oc} + 0.5R_t \\ 56 &= v_{oc} + 2R_t \end{aligned}$$

Solving these simultaneous equations, we get $R_t = 8\ \Omega$ and $v_{oc} = 40$ V, thus obtaining the Thévenin equivalent of the black box circuit.

EXERCISE 5.4-1 Determine values of R_t and v_{oc} that cause the circuit shown in Figure E 5.4-1*b* to be the Thévenin equivalent circuit of the circuit in Figure E 5.4-1*a*.

Answer: $R_t = 8\ \Omega$ and $v_{oc} = 2$ V

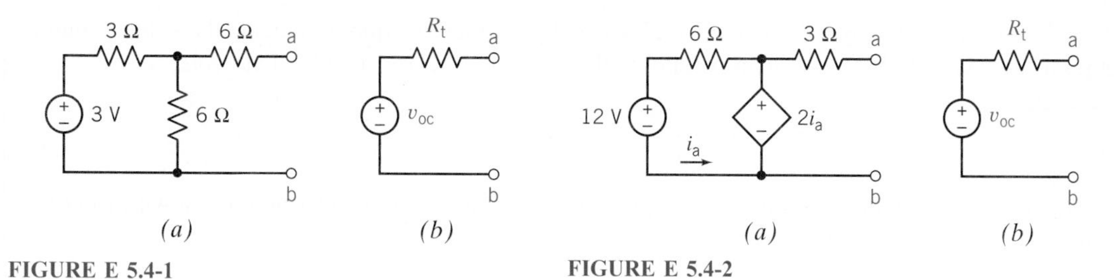

FIGURE E 5.4-1

FIGURE E 5.4-2

EXERCISE 5.4-2 Determine values of R_t and v_{oc} that cause the circuit shown in Figure E 5.4-2*b* to be the Thévenin equivalent circuit of the circuit in Figure E 5.4-2*a*.

Answer: $R_t = 3\ \Omega$ and $v_{oc} = -6$ V

5.5 NORTON'S EQUIVALENT CIRCUIT

An American engineer, E. L. Norton at Bell Telephone Laboratories, proposed an equivalent circuit for circuit A of Figure 5.4-2, using a current source and an equivalent resistance. The Norton equivalent circuit is related to the Thévenin equivalent circuit by a source transformation. In other words, a source transformation converts a Thévenin equivalent circuit into a Norton equivalent circuit or vice versa. Norton published his method in 1926, 43 years after Thévenin.

Norton's theorem may be stated as follows: Given any linear circuit, divide it into two circuits, A and B. If either A or B contains a dependent source, its controlling variable must be in the same circuit. Consider circuit A and determine its short-circuit current i_{sc} at its terminals. Then the equivalent circuit of A is a current source i_{sc} in parallel with a resistance R_n, where R_n is the resistance looking into circuit A with all its independent sources deactivated.

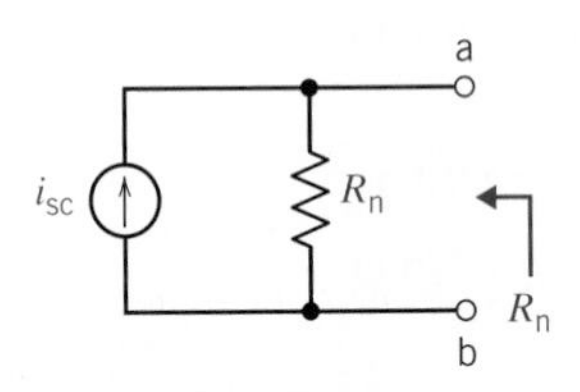

FIGURE 5.5-1 Norton equivalent circuit for a linear circuit A.

Norton's theorem requires that, for any circuit of resistance elements and energy sources with an identified terminal pair, the circuit can be replaced by a parallel combination of an ideal current source i_{sc} and a conductance G_n, where i_{sc} is the short-circuit current at the two terminals and G_n is the ratio of the short-circuit current to the open-circuit voltage at the terminal pair.

We therefore have the Norton circuit for circuit A as shown in Figure 5.5-1. Finding the Thévenin equivalent circuit of the circuit in Figure 5.5-1 shows that $R_n = R_t$ and $v_{oc} = R_t i_{sc}$. The Norton equivalent is simply the source transformation of the Thévenin equivalent.

EXAMPLE 5.5-1 Norton Equivalent Circuit

Find the Norton equivalent circuit for the circuit of Figure 5.5-2.

Solution

We can replace the voltage source by a short circuit and find R_n by circuit reduction. Replacing the voltage source by a short circuit, we have a 6-kΩ resistor in parallel with (8 kΩ + 4 kΩ) = 12 kΩ. Therefore,

$$R_n = \frac{6 \times 12}{6 + 12} = 4\,\text{k}\Omega$$

To determine i_{sc}, we short-circuit the output terminals with the voltage source activated as shown in Figure 5.5-3. Writing KCL at node a, we have

$$-\frac{15\text{ V}}{12\text{ k}\Omega} + i_{sc} = 0$$

or

$$i_{sc} = 1.25\text{ mA}$$

Thus, the Norton equivalent (Figure 5.5-1) has $R_n = 4$ kΩ and $i_{sc} = 1.25$ mA.

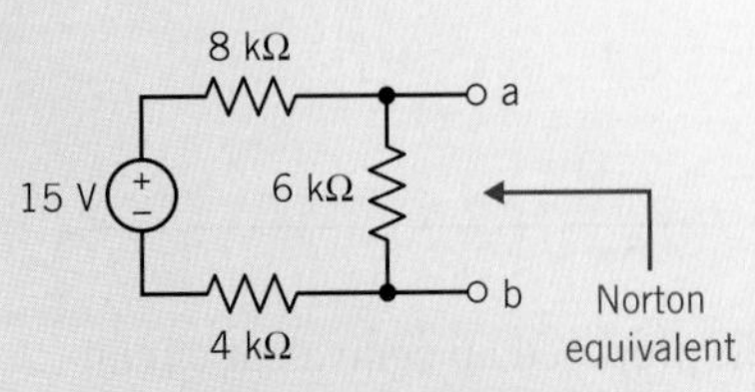

FIGURE 5.5-2 Circuit of Example 5.5-1.

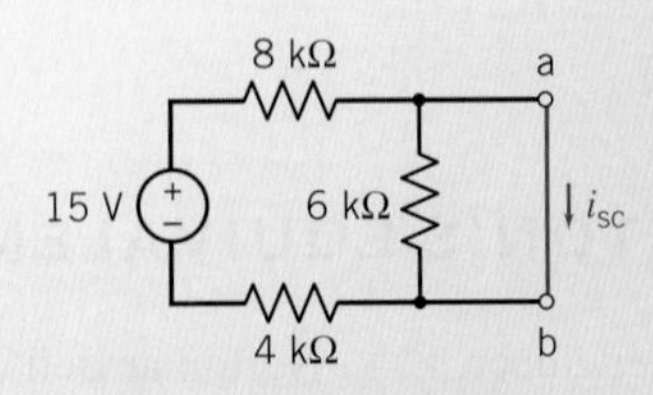

FIGURE 5.5-3 Short circuit connected to output terminals.

EXAMPLE 5.5-2 Norton Equivalent Circuit

Find the Norton equivalent circuit for the circuit of Figure 5.5-4.

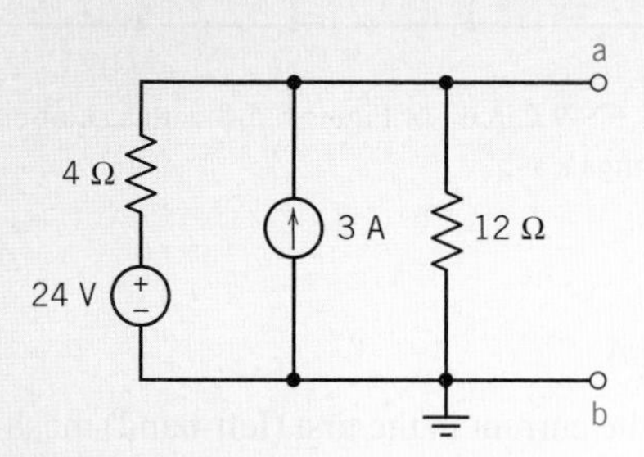

FIGURE 5.5-4 Circuit of Example 5.5-2. Resistances in ohms.

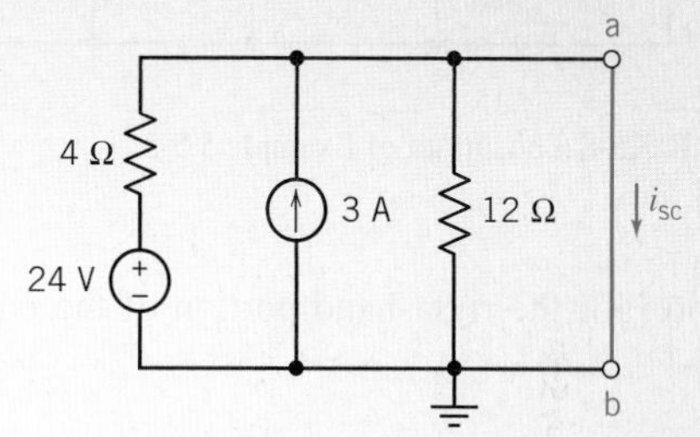

FIGURE 5.5-5 Short circuit connected to terminals a–b of the circuit of Figure 5.5-4. Resistances in ohms.

Solution

First, determine the current i_{sc} for the short-circuit condition shown in Figure 5.5-5. Writing KCL at a, we obtain

$$-\frac{24}{4} - 3 + i_{sc} = 0$$

Note that no current flows in the 12-Ω resistor because it is in parallel with a short circuit. Also, because of the short circuit, the 24-V source causes 24 V to appear across the 4-Ω resistor. Therefore,

$$i_{sc} = \frac{24}{4} + 3 = 9 \text{ A}$$

Now determine the equivalent resistance $R_n = R_t$. Figure 5.5-6 shows the circuit after replacing the voltage source by a short circuit and replacing the current source by an open circuit. Clearly, $R_n = 3\ \Omega$. Thus, we obtain the Norton equivalent circuit as shown in Figure 5.5-7.

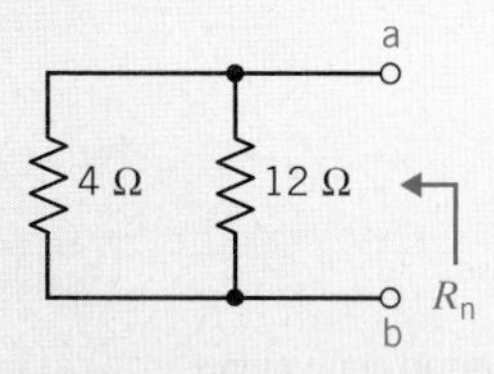

FIGURE 5.5-6 Circuit of Figure 5.5-4 with its sources deactivated. The voltage source becomes a short circuit, and the current source is replaced by an open circuit.

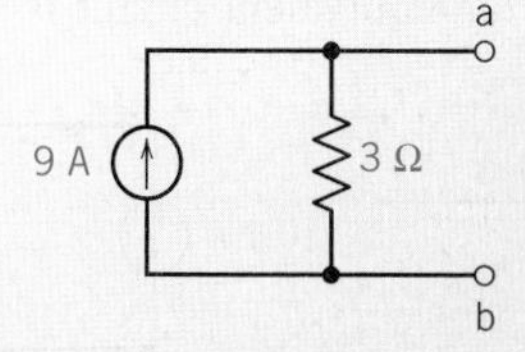

FIGURE 5.5-7 Norton equivalent of the circuit of Figure 5.5-4.

EXAMPLE 5.5-3 Norton Equivalent Circuits and Dependent Sources

Find the Norton equivalent to the left of terminals a–b for the circuit of Figure 5.5-8.

Solution

First, we need to determine the short-circuit current i_{sc}, using Figure 5.5-9. Note that $v_{ab} = 0$ when the terminals are short circuited. Then,

$$i = 5/500 = 10 \text{ mA}$$

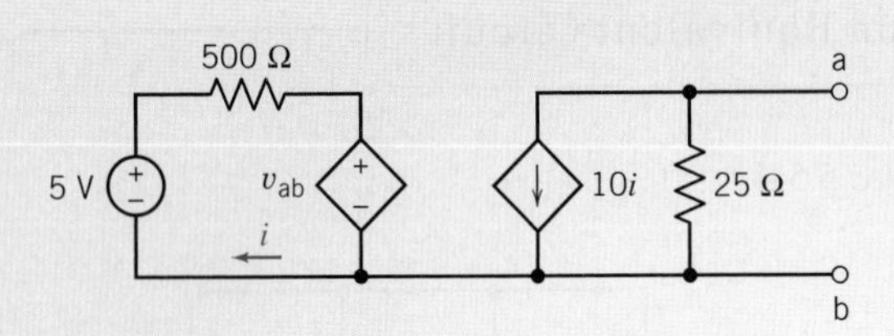

FIGURE 5.5-8 The circuit of Example 5.5-3.

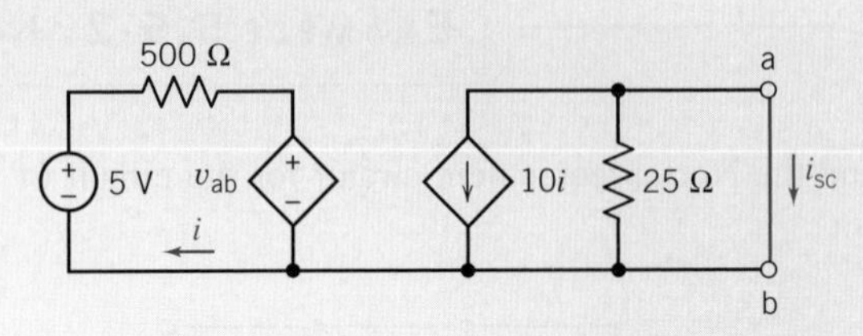

FIGURE 5.5-9 Circuit of Figure 5.5-8 with a short circuit at the terminals a–b.

Therefore, for the right-hand portion of the circuit,

$$i_{sc} = -10i = -100\text{ mA}$$

Now, to obtain R_t, we need $v_{oc} = v_{ab}$ from Figure 5.5-8, where i is the current in the first (left-hand) mesh. Writing the mesh current equation, we have

$$-5 + 500i + v_{ab} = 0$$

Also, for the right-hand mesh of Figure 5.5-8, we note that

$$v_{ab} = -25(10i) = -250i$$

Therefore,

$$i = \frac{-v_{ab}}{250}$$

Substituting i into the first mesh equation, we obtain

$$500\left(\frac{-v_{ab}}{250}\right) + v_{ab} = 5$$

Therefore,

$$v_{ab} = -5\text{ V}$$

and

$$R_t = \frac{v_{ab}}{i_{sc}} = \frac{-5}{-0.1} = 50\ \Omega$$

The Norton equivalent circuit is shown in Figure 5.5-10.

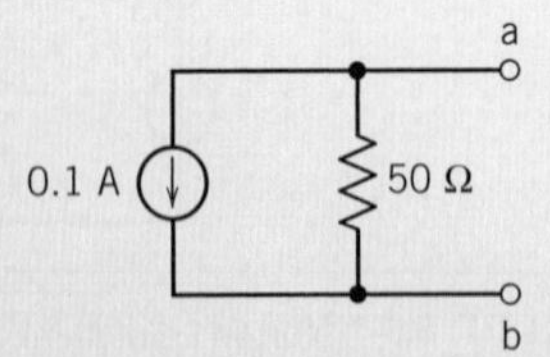

FIGURE 5.5-10 The Norton equivalent circuit for Example 5.5-3.

EXERCISE 5.5-1 Determine values of R_t and i_{sc} that cause the circuit shown in Figure E 5.5-1*b* to be the Norton equivalent circuit of the circuit in Figure E 5.5-1*a*.

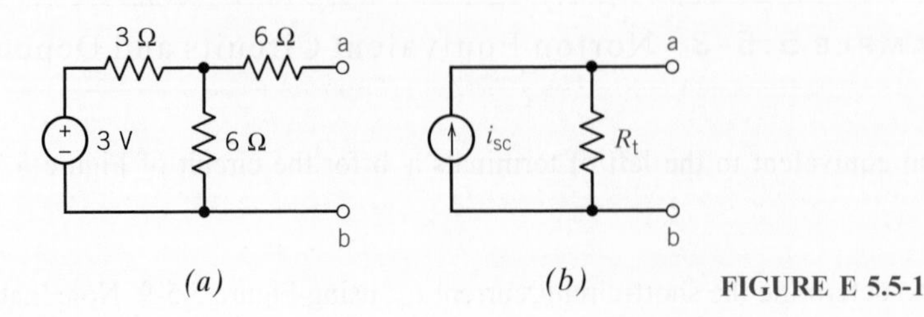

FIGURE E 5.5-1

Answer: $R_t = 8\ \Omega$ and $i_{sc} = 0.25$ A

5.6 MAXIMUM POWER TRANSFER

Many applications of circuits require the maximum power available from a source to be transferred to a load resistor R_L. Consider the circuit A shown in Figure 5.6-1, terminated with a load R_L. As demonstrated in Section 5.4, circuit A can be reduced to its Thévenin equivalent, as shown in Figure 5.6-2.

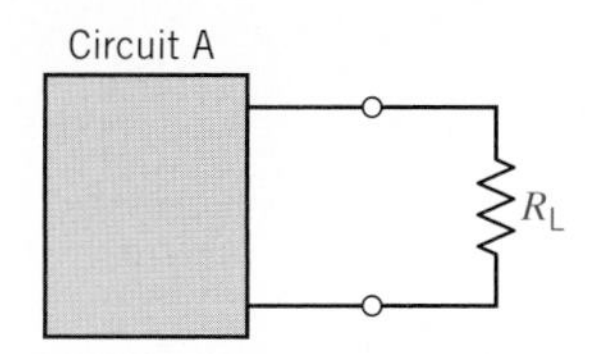

FIGURE 5.6-1 Circuit A contains resistors and independent and dependent sources. The load is the resistor R_L.

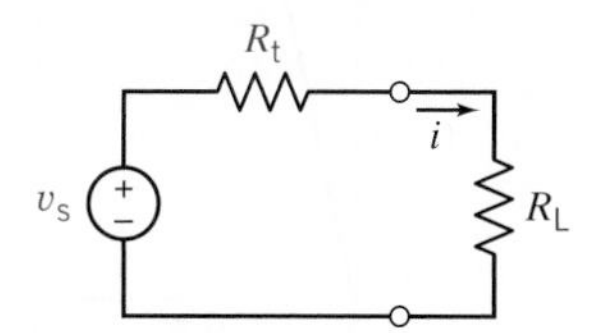

FIGURE 5.6-2 The Thévenin equivalent is substituted for circuit A. Here we use v_s for the Thévenin source voltage.

The general problem of power transfer can be discussed in terms of efficiency and effectiveness. Power utility systems are designed to transport the power to the load with the greatest efficiency by reducing the losses on the power lines. Thus, the effort is concentrated on reducing R_t, which would represent the resistance of the source plus the line resistance. Clearly, the idea of using superconducting lines that would exhibit no line resistance is exciting to power engineers.

In the case of signal transmission, as in the electronics and communications industries, the problem is to attain the maximum signal strength at the load. Consider the signal received at the antenna of an FM radio receiver from a distant station. It is the engineer's goal to design a receiver circuit so that the maximum power ultimately ends up at the output of the amplifier circuit connected to the antenna of your FM radio. Thus, we may represent the FM antenna and amplifier by the Thévenin equivalent circuit shown in Figure 5.6-2.

Let us consider the general circuit of Figure 5.6-2. We wish to find the value of the load resistance R_L such that maximum power is delivered to it. First, we need to find the power from

$$p = i^2 R_L$$

Because the current i is

$$i = \frac{v_s}{R_L + R_t}$$

we find that the power is

$$p = \left(\frac{v_s}{R_L + R_t}\right)^2 R_L \tag{5.6-1}$$

Assuming that v_s and R_t are fixed for a given source, the maximum power is a function of R_L. To find the value of R_L that maximizes the power, we use the differential calculus to find where the derivative dp/dR_L equals zero. Taking the derivative, we obtain

$$\frac{dp}{dR_L} = {v_s}^2 \frac{(R_t + R_L)^2 - 2(R_t + R_L)R_L}{(R_L + R_t)^4}$$

The derivative is zero when

$$(R_t + R_L)^2 - 2(R_t + R_L)R_L = 0 \tag{5.6-2}$$

or

$$(R_t + R_L)(R_t + R_L - 2R_L) = 0 \tag{5.6-3}$$

Solving Eq. 5.6-3, we obtain

$$R_L = R_t \tag{5.6-4}$$

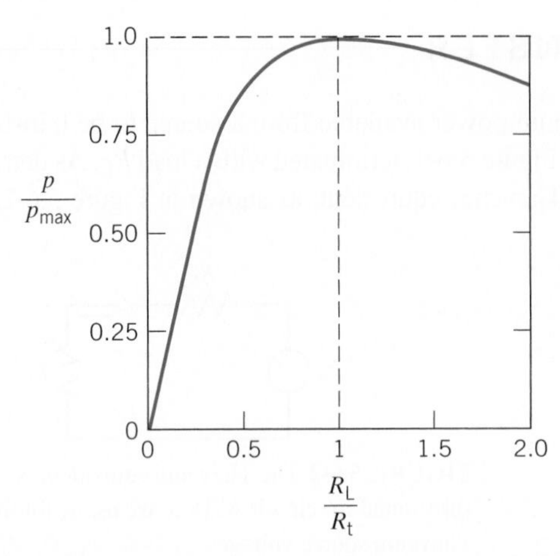

FIGURE 5.6-3 Power actually attained as R_L varies in relation to R_t.

To confirm that Eq. 5.6-4 corresponds to a maximum, it should be shown that $d^2p/dR_L{}^2 < 0$. Therefore, the maximum power is transferred to the load when R_L is equal to the Thévenin equivalent resistance R_t.

The maximum power, when $R_L = R_t$, is then obtained by substituting $R_L = R_t$ in Eq. 5.6-1 to yield

$$p_{max} = \frac{v_s{}^2 R_t}{(2R_t)^2} = \frac{v_s{}^2}{4R_t}$$

The power delivered to the load will differ from the maximum attainable as the load resistance R_L departs from $R_L = R_t$. The power attained as R_L varies from R_t is portrayed in Figure 5.6-3.

The **maximum power transfer** theorem states that the maximum power delivered to a load by a source is attained when the load resistance, R_L, is equal to the Thévenin resistance, R_t, of the source.

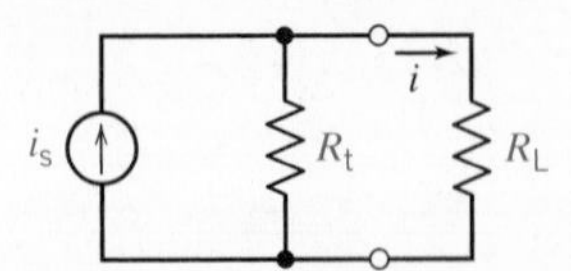

FIGURE 5.6-4 Norton's equivalent circuit representing the source circuit and a load resistor R_L. Here we use i_s as the Norton source current.

We may also use Norton's equivalent circuit to represent circuit A in Figure 5.6.1. We then have a circuit with a load resistor R_L as shown in Figure 5.6-4. The current i may be obtained from the current divider principle to yield

$$i = \frac{R_t}{R_t + R_L} i_s$$

Therefore, the power p is

$$p = i^2 R_L = \frac{i_s{}^2 R_t{}^2 R_L}{(R_t + R_L)^2} \tag{5.6-5}$$

Using calculus, we can show that the maximum power occurs when

$$R_L = R_t \tag{5.6-6}$$

Then the maximum power delivered to the load is

$$p_{max} = \frac{R_t i_s{}^2}{4} \tag{5.6-7}$$

EXAMPLE 5.6-1 Maximum Power Transfer

Find the load resistance R_L that will result in maximum power delivered to the load for the circuit of Figure 5.6-5. Also, determine the maximum power delivered to the load resistor.

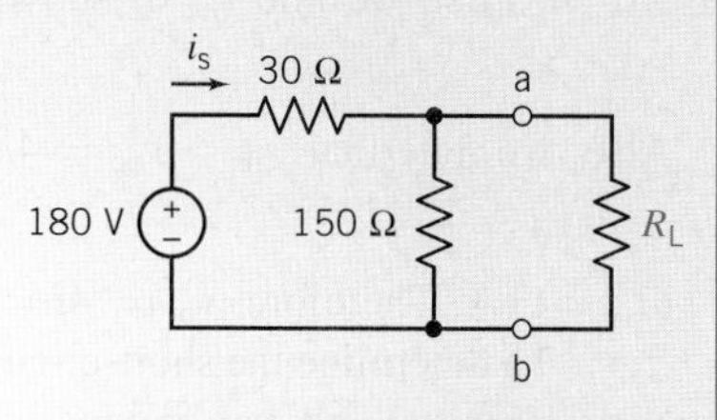

FIGURE 5.6-5 Circuit for Example 5.6-1. Resistances in ohms.

Solution

First, we determine the Thévenin equivalent circuit for the circuit to the left of terminals a–b. Disconnect the load resistor. The Thévenin voltage source v_t is

$$v_t = \frac{150}{180} \times 180 = 150 \text{ V}$$

The Thévenin resistance R_t is

$$R_t = \frac{30 \times 150}{30 + 150} = 25 \ \Omega$$

The Thévenin circuit connected to the load resistor is shown in Figure 5.6-6. Maximum power transfer is obtained when $R_L = R_t = 25 \ \Omega$.

Then the maximum power is

$$p_{max} = \frac{v_s^2}{4R_L} = \frac{(150)^2}{4 \times 25} = 225 \text{ W}$$

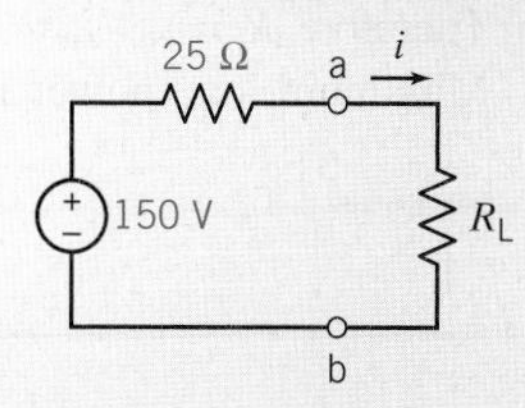

FIGURE 5.6-6 Thévenin equivalent circuit connected to R_L for Example 5.6-1.

EXAMPLE 5.6-2 Maximum Power Transfer

Find the load R_L that will result in maximum power delivered to the load of the circuit of Figure 5.6-7*a*. Also, determine p_{max} delivered.

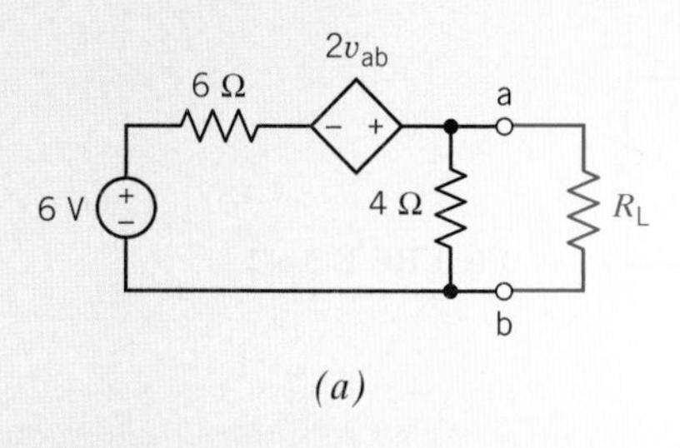

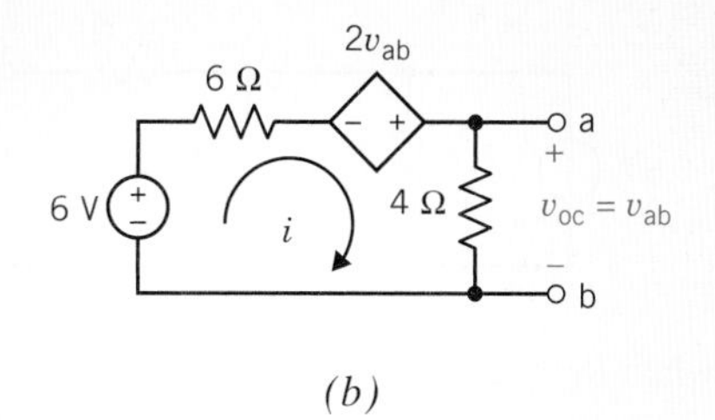

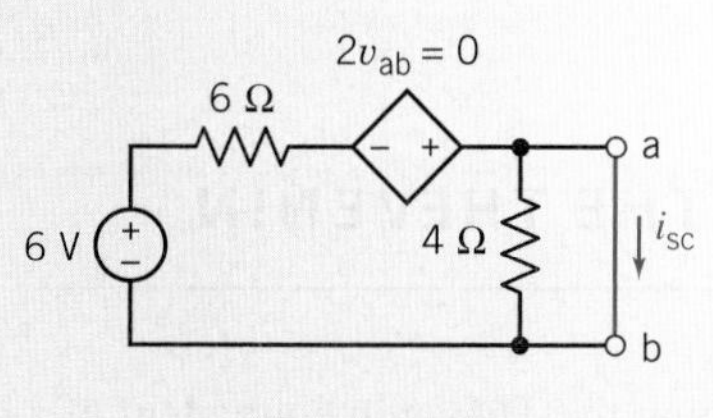

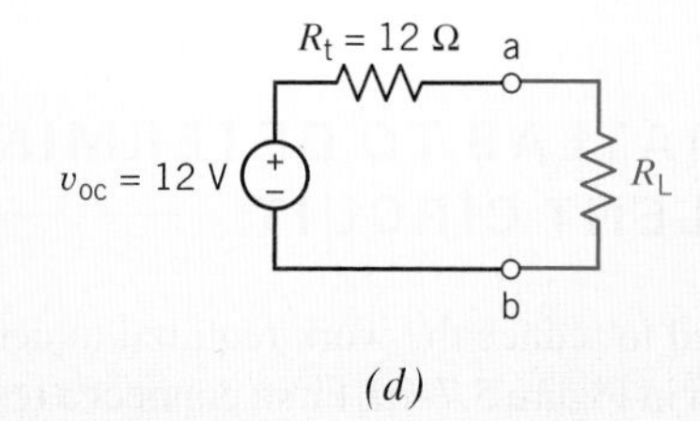

FIGURE 5.6-7 Determination of maximum power transfer to a load R_L.

Solution

We will obtain the Thévenin equivalent circuit for the part of the circuit to the left of terminals a,b in Figure 5.6-7*a*. First, we find v_{oc} as shown in Figure 5.6-7*b*. The KVL gives

$$-6 + 10i - 2v_{ab} = 0$$

Also, we note that $v_{ab} = v_{oc} = 4i$. Therefore,

$$10i - 8i = 6$$

or $i = 3$ A. Therefore, $v_{oc} = 4i = 12$ V.

To determine the short-circuit current, we add a short circuit as shown in Figure 5.6-7*c*. The 4-Ω resistor is short circuited and can be ignored. Writing KVL, we have

$$-6 + 6i_{sc} = 0$$

Hence, $i_{sc} = 1$ A.

Therefore, $R_t = v_{oc}/i_{sc} = 12\ \Omega$. The Thévenin equivalent circuit is shown in Figure 5.6-7*d* with the load resistor. Maximum load power is achieved when $R_L = R_t = 12\ \Omega$. Then,

$$p_{max} = \frac{v_{oc}^2}{4R_L} = \frac{12^2}{4(12)} = 3\ \text{W}$$

EXERCISE 5.6-1 Find the maximum power that can be delivered to R_L for the circuit of Figure E 5.6-1, using a Thévenin equivalent circuit.

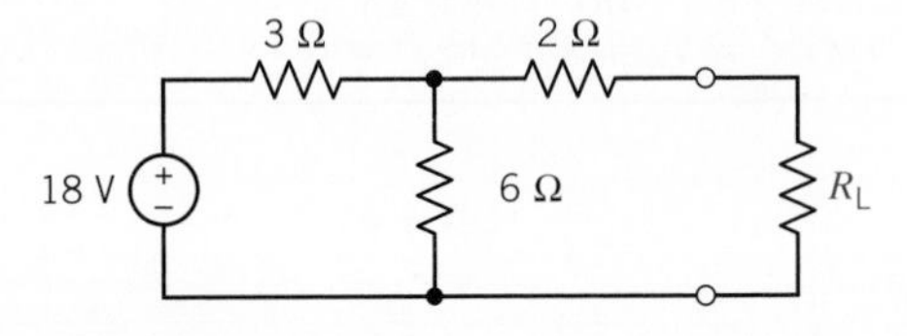

FIGURE E 5.6-1

Answer: 9 W when $R_L = 4\ \Omega$

EXERCISE 5.6-2 Find the maximum power delivered to R_L for the circuit of Figure E 5.6-2, using a Norton equivalent circuit.

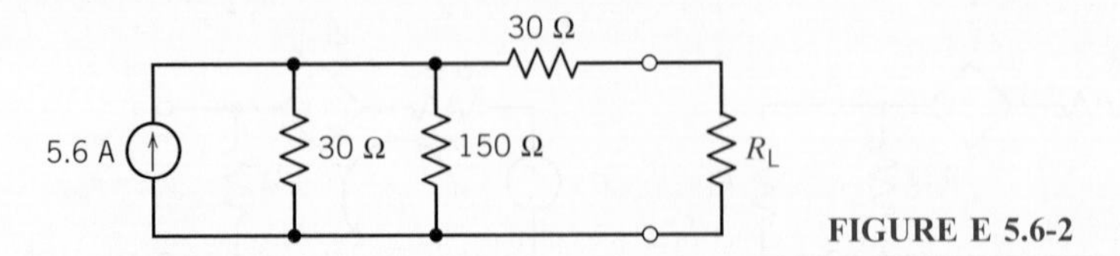

FIGURE E 5.6-2

Answer: 175 W when $R_L = 28\ \Omega$

5.7 USING MATLAB TO DETERMINE THE THÉVENIN EQUIVALENT CIRCUIT

MATLAB can be used to reduce the work required to determine the Thévenin equivalent of a circuit such as the one shown in Figure 5.7-1*a*. First, connect a resistor, R, across the terminals of the network, as shown in Figure 5.7-1*b*. Next, write node or mesh equations to describe the circuit with the resistor

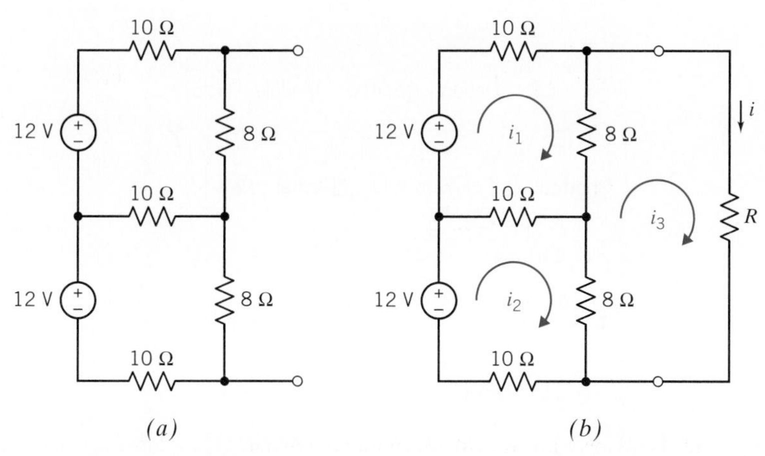

FIGURE 5.7-1 The circuit in (*b*) is obtained by connecting a resistor, *R*, across the terminals of the circuit in (*a*).

connected across its terminals. In this case, the circuit in Figure 5.7-1*b* is represented by the mesh equations

$$\begin{aligned} 12 &= 28i_1 - 10i_2 - 8i_3 \\ 12 &= -10i_1 + 28i_2 - 8i_3 \\ 0 &= -8i_1 - 8i_2 + (16 + R)i_3 \end{aligned} \tag{5.7-1}$$

The current i in the resistor R is equal to the mesh current in the third mesh, that is,

$$i = i_3 \tag{5.7-2}$$

The mesh equations can be written using matrices such as

$$\begin{bmatrix} 28 & -10 & -8 \\ -10 & 28 & -8 \\ -8 & -8 & 16 + R \end{bmatrix} \begin{bmatrix} i_1 \\ i_2 \\ i_3 \end{bmatrix} = \begin{bmatrix} 12 \\ 12 \\ 0 \end{bmatrix} \tag{5.7-3}$$

Notice that $i = i_3$ in Figure 5.7-1*b*.

Figure 5.7-2*a* shows a MATLAB file named ch5ex.m that solves Eq. 5.7-1. Figure 5.7-3 illustrates the use of this MATLAB file and shows that when $R = 6\ \Omega$, then $i = 0.7164$ A, and that when $R = 12$ W, then $i = 0.5106$ A.

```
% ch5ex.m  -  MATLAB input file for Section 5-7

Z  =  [ 28     -10      -8;        %
       -10      28      -8;        %  Mesh Equation
        -8                         %
                -8    16+R];       %  Equation 5.7-3
V  =  [ 12;                        %
        12;                        %
         0];                       %

Im  = Z\V;          %  Calculate the mesh currents.

I   = Im(3)         %  Equation 5.7-2
```

FIGURE 5.7-2 MATLAB file used to solve the mesh equation representing the circuit shown in Figure 5.7-1*b*.

FIGURE 5.7-3 Computer screen showing the use of MATLAB to analyze the circuit shown in Figure 5.7-1.

Next, consider Figure 5.7-4, which shows a resistor R connected across the terminals of a Thévenin equivalent circuit. The circuit in Figure 5.7-4 is represented by the mesh equation

$$V_t = R_t i + R i \tag{5.7-4}$$

As a matter of notation, let $i = i_a$ when $R = R_a$. Similarly, let $i = i_b$ when $R = R_b$. Equation 5.7-4 indicates that

$$\begin{aligned} V_t &= R_t i_a + R_a i_a \\ V_t &= R_t i_b + R_b i_b \end{aligned} \tag{5.7-5}$$

Equation 5.7-5 can be written using matrices as

$$\begin{bmatrix} R_a i_a \\ R_b i_b \end{bmatrix} = \begin{bmatrix} 1 & -i_a \\ 1 & -i_b \end{bmatrix} \begin{bmatrix} V_t \\ R_t \end{bmatrix} \tag{5.7-6}$$

Given i_a, R_a, i_b, and R_b, this matrix equation can be solved for V_t and R_t, the parameters of the Thévenin equivalent circuit. Figure 5.7-5 shows a MATLAB file that solves Eq. 5.7-6, using the values $i_b = 0.7164$ A, $R_b = 6\ \Omega$, $i_a = 0.5106$ A, and $R_a = 12\ \Omega$. The resulting values of V_t and R_t are

$$V_t = 10.664\ \text{V} \quad \text{and} \quad R_t = 8.8863\ \Omega$$

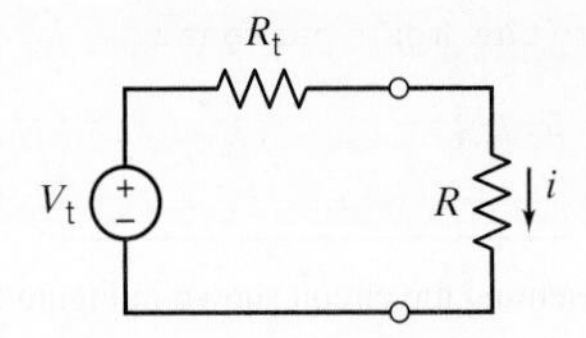

FIGURE 5.7-4 The circuit obtained by connecting a resistor, R, across the terminals of a Thévenin equivalent circuit.

```
% Find the Thevenin equivalent of the circuit
% connected to the resister R.

Ra = 12;  ia = 0.5106;      % When R=Ra then i=ia

Rb = 6;   ib = 0.7164;      % When R=Rb then i=ib

A = [1 -ia;   %
     1 -ib];  %
              % Eqn 5.7-6
b = [Ra*ia;   %
     Rb*ib];  %

X = A\b;

Vt = X(1)  % Open-Circuit Voltage

Rt = X(2)  % Thevenin Resistance
```

FIGURE 5.7-5 MATLAB file used to calculate the open-circuit voltage and Thévenin resistance.

5.8 USING PSPICE TO DETERMINE THE THÉVENIN EQUIVALENT CIRCUIT

We can use the computer program PSpice to find the Thévenin or Norton equivalent circuit for circuits even though they are quite complicated. Figure 5.8-1 illustrates this method. We calculate the Thévenin equivalent of the circuit shown in Figure 5.8-1*a* by calculating its open-circuit voltage, v_{oc}, and its short-circuit current, i_{sc}. To do so, we connect a resistor across its terminals as shown in Figure 5.8-1*b*. When the resistance of this resistor is infinite, the resistor voltage will be equal to the open-circuit voltage, v_{oc}, as shown m Figure 5.8-1*b*. On the other hand, when the resistance of this resistor is zero, the resistor current will be equal to the short-circuit current, i_{sc}, as shown in Figure 5.8-1*c*.

We can't use either infinite or zero resistances in PSpice, so we will approximate the infinite resistance by a resistance that is several orders of magnitude larger than the largest resistance in circuit A. We can check whether our resistance is large enough by doubling it and rerunning the PSpice simulation. If the computed value of v_{oc} does not change, our large resistance is effectively infinite. Similarly, we can approximate a zero resistance by a resistance that is several orders of magnitude smaller than the smallest resistance in circuit A. Our small resistance is effectively zero when halving it does not change the computed value of i_{sc}.

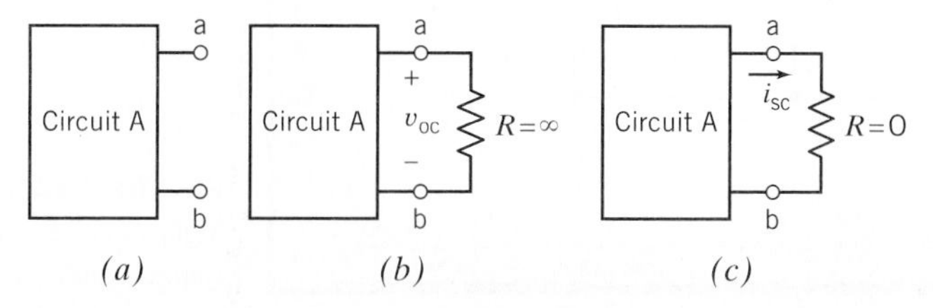

FIGURE 5.8-1 A method for computing the values of v_{oc} and i_{sc}, using PSpice.

EXAMPLE 5.8-1 Using-PSpice to find a Thévenin Equivalent Circuit

Use PSpice to determine the values of the open-circuit voltage, v_{oc}, and the short-circuit current, i_{sc}, for the circuit shown in Figure 5.8-2.

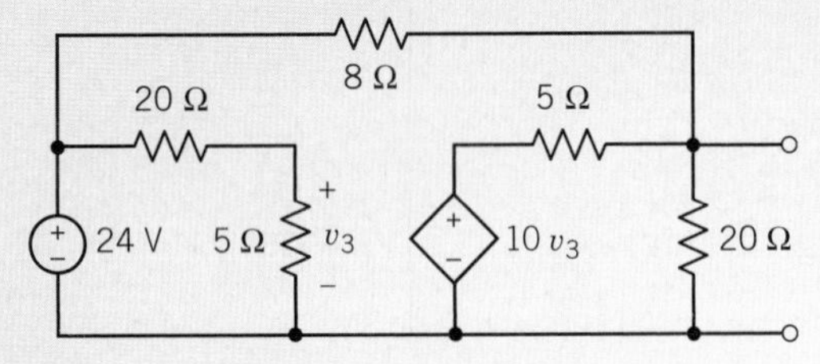

FIGURE 5.8-2 The circuit considered in Example 5.8-1.

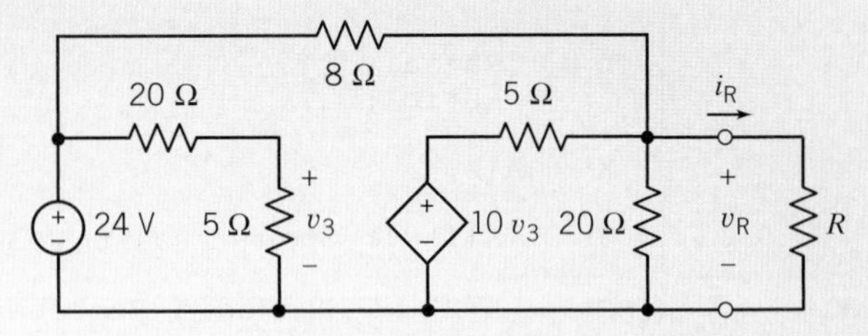

FIGURE 5.8-3 The circuit from Figure 5.8-2 after adding a resistor across its terminals.

Solution

Following our method, we add a resistor across the terminals of the circuit as shown in Figure 5.8-3. Noticing that the largest resistance in our circuit is 20 Ω and the smallest is 5 Ω, we will determine v_{oc} and i_{sc}, using

$$v_{oc} \approx v_R \quad \text{when} \quad R \gg 20\ \Omega$$

and

$$v_{sc} \approx i_R = \frac{v_R}{R} \quad \text{when} \quad R \ll 5\ \Omega$$

Using PSpice begins with drawing the circuit in the OrCAD Capture workspace as shown in Figure 5.8-4 (see Appendix A). The VCVS in Figure 5.8-3 is represented by a PSpice ''Part E'' in Figure 5.8-4. Figure 5.8-5 illustrates the correspondence between the VCVS and the PSpice ''Part E.''

To determine the open circuit voltage, we set the resistance R to a very large value and perform a 'Bias Point' simulation (see Appendix A). Figure 5.8-6 shows the simulation results when $R = 20\ \text{M}\Omega$. The voltage across the resistor R is 33.6 V, so $v_{oc} = 33.6$ V. (Doubling the value of R and rerunning the simulation did not change the value of the voltage across R, so we are confident that $v_{oc} = 33.6$ V.)

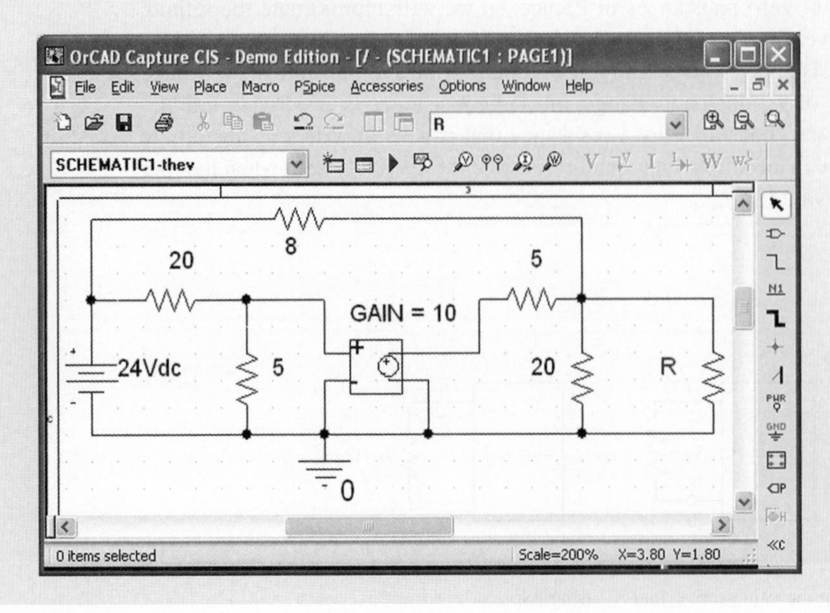

FIGURE 5.8-4 The circuit from Figure 5.8-3 drawn in the OrCAD Capture workspace.

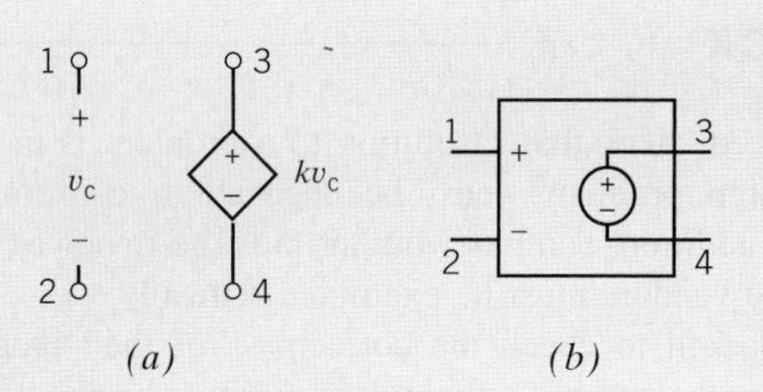

FIGURE 5.8-5 A VCVS (a) and the corresponding PSpice ''Part E'' (b).

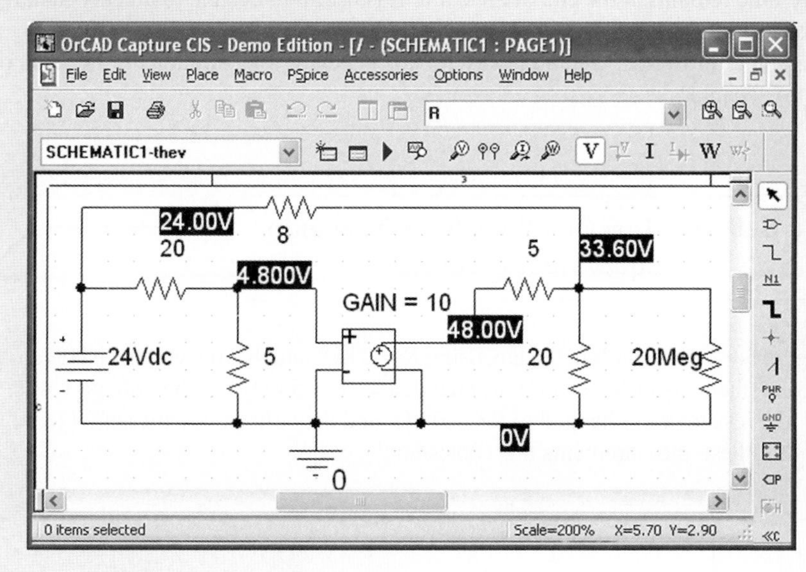

FIGURE 5.8-6 Simulation results for $R = 20\ \text{M}\Omega$.

To determine the short-circuit current, we set the resistance R to a very small value and perform a 'Bias Point' simulation (see Appendix A). Figure 5.8-7 shows the simulation results when $R = 1\ \text{m}\Omega$. The voltage across the resistor R is 12.6 mV. Using Ohm's law, the value of the short-circuit current is

$$i_{sc} = \frac{12.6 \times 10^{-3}}{1 \times 10^{-3}} = 12.6\ \text{A}$$

(Halving the value of R and rerunning the simulation did not change the value of the voltage across R, so we are confident that $i_{sc} = 12.6$ A.)

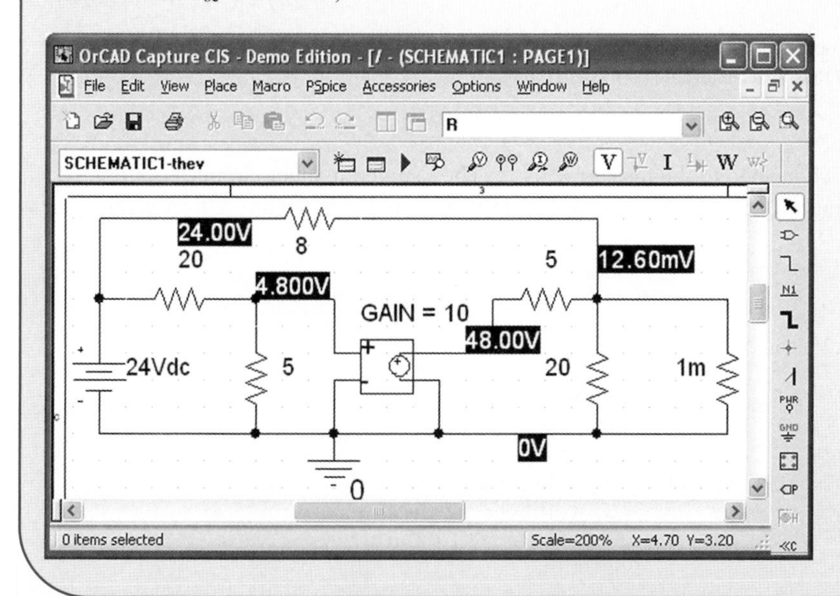

FIGURE 5.8-7 Simulation results for $R = 1\ \text{M}\Omega = 0.001\ \Omega$.

5.9 HOW CAN WE CHECK . . . ?

Engineers are frequently called upon to check that a solution to a problem is indeed correct. For example, proposed solutions to design problems must be checked to confirm that all of the specifications have been satisfied. In addition, computer output must be reviewed to guard against data-entry errors, and claims made by vendors must be examined critically.

Engineering students are also asked to check the correctness of their work. For example, occasionally just a little time remains at the end of an exam. It is useful to be able to quickly identify those solutions that need more work.

The following example illustrates techniques useful for checking the solutions of the sort of problem discussed in this chapter.

EXAMPLE 5.9-1 How Can We Check Thévenin Equivalent Circuits?

Suppose that the circuit shown in Figure 5.9-1*a* was built in the lab, using $R = 2\ \text{k}\Omega$, and that the voltage labeled v was measured to be $v = -1.87$ V. Next, the resistor labeled R was changed to $R = 5\ \text{k}\Omega$, and the voltage v was measured to be $v = -3.0$ V. Finally, the resistor was changed to $R = 10\ \text{k}\Omega$, and the voltage was measured to be $v = -3.75$ V. **How can we check** that these measurements are consistent?

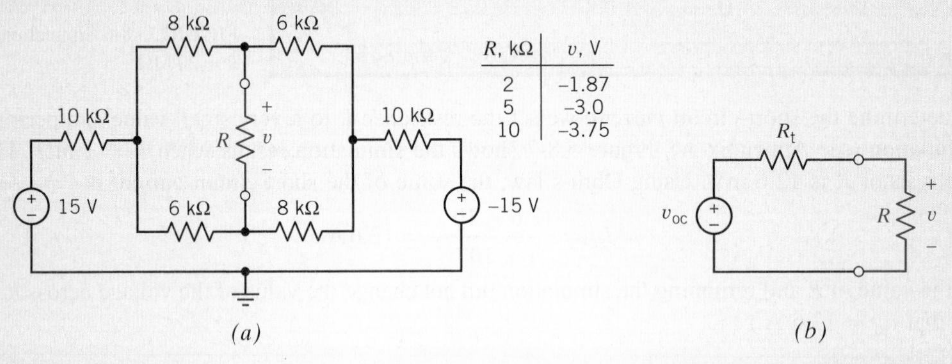

R, kΩ	v, V
2	-1.87
5	-3.0
10	-3.75

FIGURE 5.9-1 (*a*) A circuit with data obtained by measuring the voltage across the resistor R, and (*b*) the circuit obtained by replacing the part of the circuit connected to R by its Thévenin equivalent circuit.

Solution

Let's replace the part of the circuit connected to the resistor R by its Thévenin equivalent circuit. Figure 5.9-1*b* shows the resulting circuit. Applying the voltage division principle to the circuit in Figure 5.9-1*b* gives

$$v = \frac{R}{R + R_t} v_{oc} \tag{5.9-1}$$

When $R = 2\ \text{k}\Omega$, then $v = -1.87$ V, and Eq. 5.9-1 becomes

$$-1.87 = \frac{2000}{2000 + R_t} v_{oc} \tag{5.9-2}$$

Similarly, when $R = 5\ \text{k}\Omega$, then $v = -3.0$ V, and Eq. 5.9-1 becomes

$$-3.0 = \frac{5000}{5000 + R_t} v_{oc} \tag{5.9-3}$$

Equations 5.9-2 and 5.9-3 constitute a set of two equations in two unknowns, v_{oc} and R_t. Solving these equations gives $v_{oc} = -5$ V and $R_t = 3333\ \Omega$. Substituting these values into Eq. 5.9-1 gives

$$v = \frac{R}{R + 3333}(-5) \qquad (5.9\text{-}4)$$

Equation 5.9-4 can be used to predict the voltage that would be measured if $R = 10\ \text{k}\Omega$. If the value of v obtained using Eq. 5.9-4 agrees with the measured value of v, then the measured data are consistent. Letting $R = 10\ \text{k}\Omega$ in Eq. 5.9-4 gives

$$v = \frac{10{,}000}{10{,}000 + 3333}(-5) = -3.75\ \text{V} \qquad (5.9\text{-}5)$$

Because this value agrees with the measured value of v, the measured data are indeed consistent.

5.10 DESIGN EXAMPLE

STRAIN GAUGE BRIDGE

Strain gauges are transducers that measure mechanical strain. Electrically, the strain gauges are resistors. The strain causes a change in resistance that is proportional to the strain.

Figure 5.10-1 shows four strain gauges connected in a configuration called a bridge. Strain gauge bridges measure force or pressure (Doebelin, 1966).

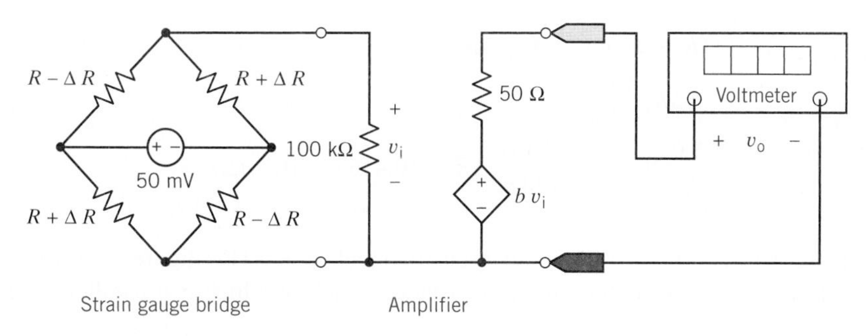

FIGURE 5.10-1 Design problem involving a strain gauge bridge.

The bridge output is usually a small voltage. In Figure 5.10-1, an amplifier multiplies the bridge output, v_i, by a gain to obtain a larger voltage, v_o, which is displayed by the voltmeter.

Describe the Situation and the Assumptions

A strain gauge bridge is used to measure force. The strain gauges have been positioned so that the force will increase the resistance of two of the strain gauges while, at the same time, decreasing the resistance of the other two strain gauges.

The strain gauges used in the bridge have nominal resistances of $R = 120\ \Omega$. (The nominal resistance is the resistance when the strain is zero.) This resistance is expected to increase or decrease by no more than 2 Ω due to strain. This means that

$$-2\ \Omega \le \Delta R \le 2\ \Omega \qquad (5.10\text{-}1)$$

The output voltage, v_o, is required to vary from -10 V to $+10$ V as ΔR varies from $-2\,\Omega$ to $2\,\Omega$.

State the Goal

Determine the amplifier gain, b, needed to cause v_o to be related to ΔR by

$$v_o = 5\frac{\text{volt}}{\text{ohm}} \cdot \Delta R \tag{5.10-2}$$

Generate a Plan

Use Thévenin's theorem to analyze the circuit shown in Figure 5.10-1 to determine the relationship between v_i and ΔR. Calculate the amplifier gain needed to satisfy Eq. 5.10-2.

Act on the Plan

We begin by finding the Thévenin equivalent of the strain gauge bridge. This requires two calculations: one to find the open-circuit voltage, v_t, and the other to find the Thévenin resistance R_t. Figure 5.10-2*a* shows the circuit used to calculate v_t. Begin by finding the currents i_1 and i_2.

$$i_1 = \frac{50\text{ mV}}{(R - \Delta R) + (R + \Delta R)} = \frac{50\text{ mV}}{2R}$$

Similarly

$$i_2 = \frac{50\text{ mV}}{(R + \Delta R) + (R - \Delta R)} = \frac{50\text{ mV}}{2R}$$

Then

$$\begin{aligned} v_t &= (R + \Delta R)i_1 - (R - \Delta R)i_2 \\ &= (2\Delta R)\frac{50\text{ mV}}{2R} \\ &= \frac{\Delta R}{R}50\text{ mV} = \frac{50\text{ mV}}{120\,\Omega}\Delta R = (0.4167 \times 10^{-3})\Delta R \end{aligned} \tag{5.10-3}$$

Figure 5.10-2*b* shows the circuit used to calculate R_t. This figure shows that R_t is composed of a series connection of two resistances, each of which is a parallel connection

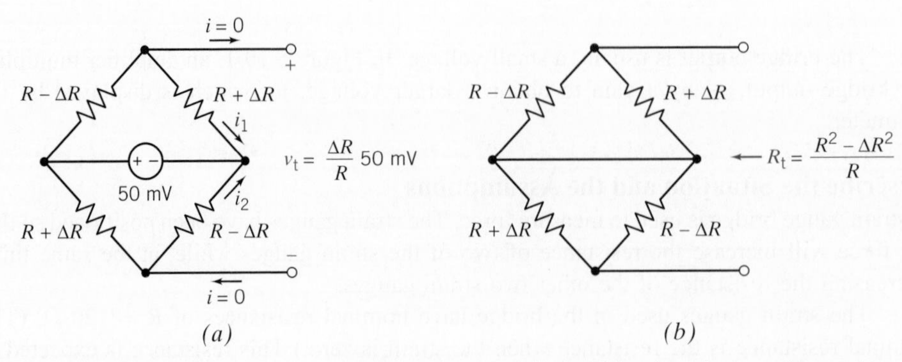

FIGURE 5.10-2 Calculating (*a*) the open-circuit voltage, and (*b*) the Thévenin resistance of the strain gauge bridge.

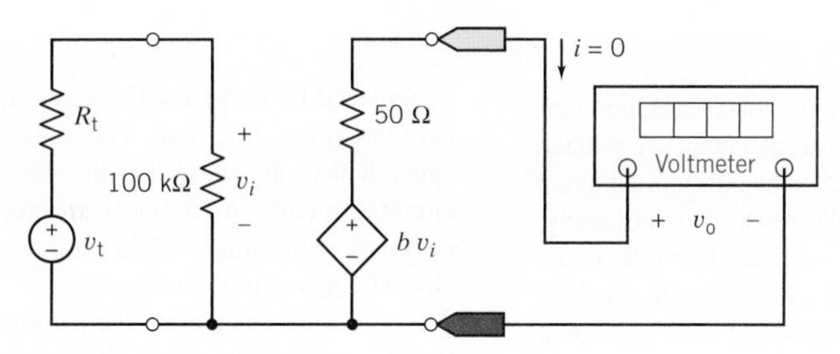

FIGURE 5.10-3 Solution to the design problem.

of two strain gauge resistances

$$R_t = \frac{(R - \Delta R)(R + \Delta R)}{(R - \Delta R) + (R + \Delta R)} + \frac{(R + \Delta R)(R - \Delta R)}{(R + \Delta R) + (R - \Delta R)} = 2\frac{R^2 - \Delta R^2}{2R}$$

Because R is much larger than ΔR, this equation can be simplified to

$$R_t = R$$

In Figure 5.10-3 the strain gauge bridge has been replaced by its Thévenin equivalent circuit. This simplification allows us to calculate v_i using voltage division

$$v_i = \frac{100\text{ k}\Omega}{100\text{ k}\Omega + R_t} v_t = 0.9988 v_t = \left(0.4162 \times 10^{-3}\right)\Delta R \qquad (5.10\text{-}4)$$

Model the voltmeter as an ideal voltmeter. Then the voltmeter current is $i = 0$ as shown in Figure 5.10-3. Applying KVL to the right-hand mesh gives

$$v_o + 50(0) - b v_i = 0$$

or

$$v_o = b v_i = b\left(0.4162 \times 10^{-3}\right)\Delta R \qquad (5.10\text{-}5)$$

Comparing Eq. 5.10-5 to Eq. 5.10-2 shows that the amplifier gain, b, must satisfy

$$b\left(0.4162 \times 10^{-3}\right) = 5$$

Hence, the amplifier gain is

$$b = 12{,}013$$

Verify the Proposed Solution

Substituting $b = 12{,}013$ into Eq. 5.10-5 gives

$$v_o = (12{,}013)\left(0.4162 \times 10^{-3}\right)\Delta R = 4.9998\,\Delta R \qquad (5.10\text{-}6)$$

which agrees with Eq. 5.10-2.

5.11 SUMMARY

- Source transformations, summarized in Table 5.11-1, are used to transform a circuit into an equivalent circuit. A voltage source v_{oc} in series with a resistor R_t can be transformed into a current source $i_{sc} = v_{oc}/R_t$ and a parallel resistor R_t. Conversely, a current source i_{sc} in parallel with a resistor R_t can be transformed into a voltage source $v_{oc} = R_t i_{sc}$ in series with a resistor R_t. The circuits in Table 5.11-1 are equivalent in the sense that the voltage and current of all circuit elements in circuit B are unchanged by the source transformation.
- The superposition theorem permits us to determine the total response of a linear circuit to several independent sources by finding the response to each independent source separately and then adding the separate responses algebraically.
- Thévenin and Norton equivalent circuits, summarized in Table 5.11-2, are used to transform a circuit into a smaller, yet equivalent, circuit. First the circuit is separated into two parts, circuit A and circuit B, in Table 5.11-2. Circuit A can be replaced by either its Thévenin equivalent circuit or its Norton equivalent circuit. The circuits in Table 5.11-2 are equivalent in the sense that the voltage and current of all circuit elements in circuit B are unchanged by replacing circuit A with either its Thévenin equivalent circuit or its Norton equivalent circuit.
- Procedures for calculating the parameters v_{oc}, i_{sc}, and R_t of the Thévenin and Norton equivalent circuits are summarized in Figures 5.4-3 and 5.4-4.
- The goal of many electronic and communications circuits is to deliver maximum power to a load resistor R_L. Maximum power is attained when R_L is set equal to the Thévenin resistance, R_t, of the circuit connected to R_L. This results in maximum power at the load when the series resistance R_t cannot be reduced.
- The computer program MATLAB can be used to reduce the computational burden of calculating the parameters v_{oc}, i_{sc}, and R_t of the Thévenin and Norton equivalent circuits.

Table 5.11-1 Source Transformations

THÉVENIN CIRCUIT	NORTON CIRCUIT

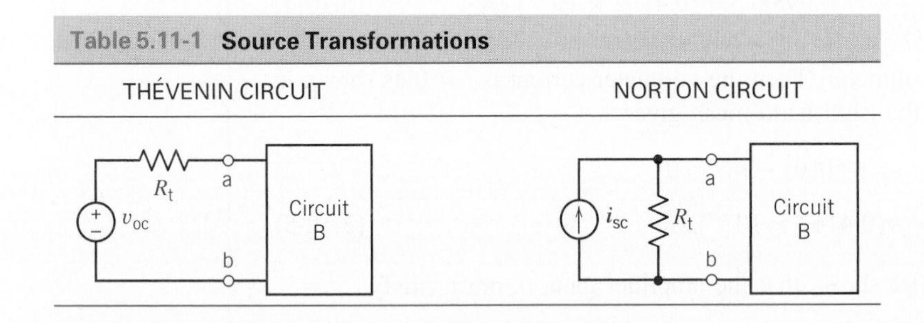

Table 5.11-2 Thévenin and Norton Equivalent Circuits

ORIGINAL CIRCUIT	THÉVENIN CIRCUIT	NORTON EQUIVALENT CIRCUIT

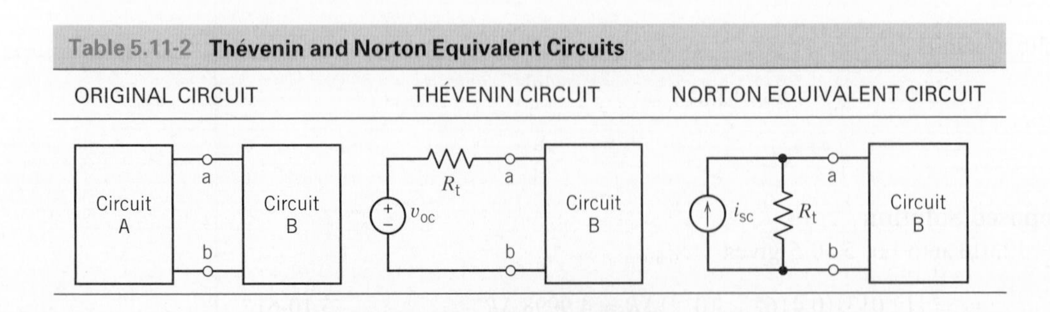

PROBLEMS

Section 5.2 Source Transformations

P 5.2-1 The circuit shown in Figure P 5.2-1*a* has been divided into two parts. The circuit shown in Figure P 5.2-1*b* was obtained by simplifying the part to the right of the terminals using source transformations. The part of the circuit to the left of the terminals was not changed.

(a) Determine the values of R_t and v_t in Figure P 5.2-1*b*.

(b) Determine the values of the current i and the voltage v in Figure P 5.2-1b. The circuit in Figure P 5.2-1b is equivalent to the circuit in Figure P 5.2-1a. Consequently, the current i and the voltage v in Figure P 5.2-1a have the same values as do the current i and the voltage v in Figure P 5.2-1b.

(c) Determine the value of the current i_a in Figure P 5.2-1a.

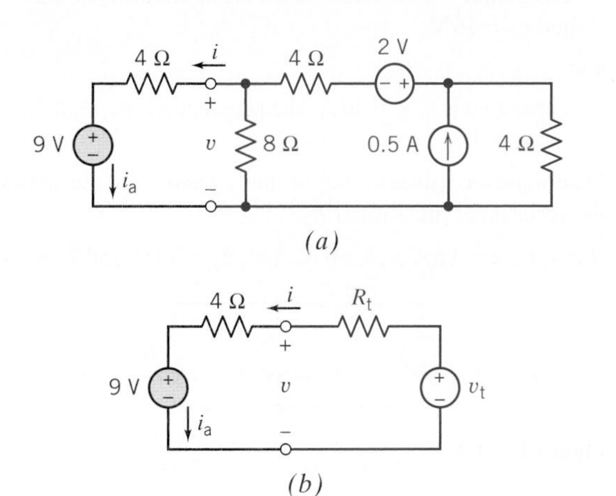

Figure P 5.2-1

P 5.2-2 Consider the circuit of Figure P 5.2-2. Find i_a by simplifying the circuit (using source transformations) to a single-loop circuit so that you need to write only one KVL equation to find i_a.

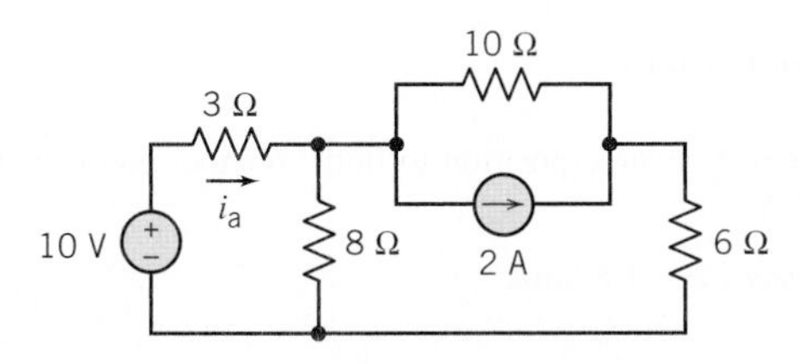

Figure P 5.2-2

P 5.2-3 Find v_o using source transformations if $i = 5/2$ A in the circuit shown in Figure P 5.2-3.

Hint: Reduce the circuit to a single mesh that contains the voltage source labeled v_o.

Answer: $v_o = 28$ V

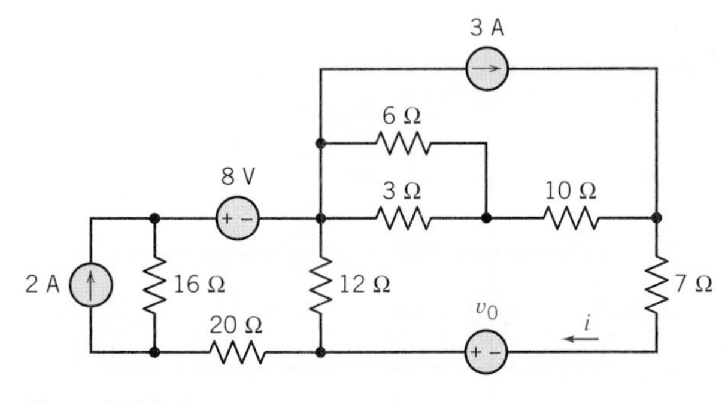

Figure P 5.2-3

P 5.2-4 Determine the value of the current i_a in the circuit shown in Figure P 5.2-4.

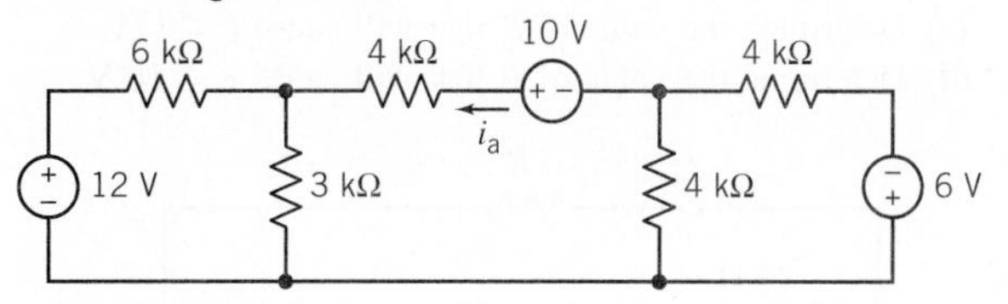

Figure P 5.2-4

P 5.2-5 Use source transformations to find the current i_a in the circuit shown in Figure P 5.2-5.

Answer: $i_a = 1.2$ A

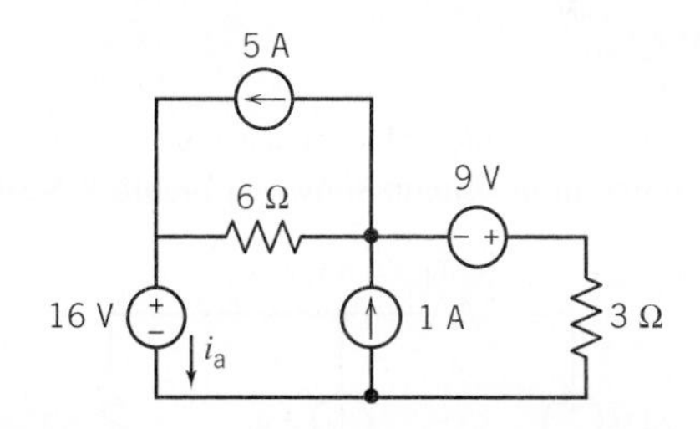

Figure P 5.2-5

P 5.2-6 Use source transformations to find the value of the voltage v_a in Figure P 5.2-6.

Answer: $v_a = 6.6$ V

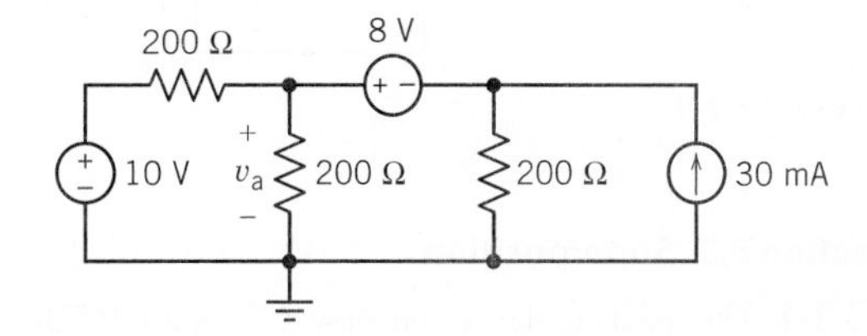

Figure P 5.2-6

***P 5.2-7** Determine the power supplied by each of the sources in the circuit shown in Figure P 5.2-7.

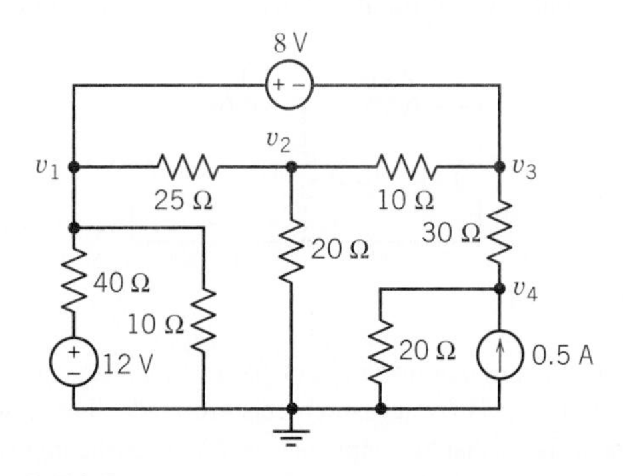

Figure P 5.2-7

P 5.2-8 The circuit shown in Figure P 5.2-8 contains an unspecified resistance R.

(a) Determine the value of the current i when $R = 4\,\Omega$.
(b) Determine the value of the voltage v when $R = 8\,\Omega$.
(c) Determine the value of R that will cause $i = 1$ A.
(d) Determine the value of R that will cause $v = 16$ V.

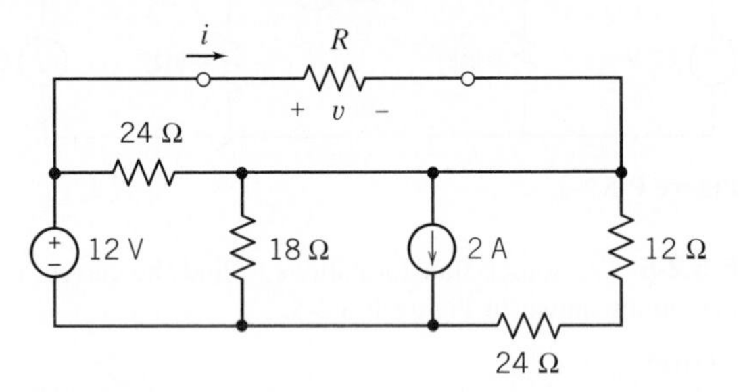

Figure P 5.2-8

P 5.2-9 Determine the value of the power supplied by the current source in the circuit shown in Figure P 5.2-9.

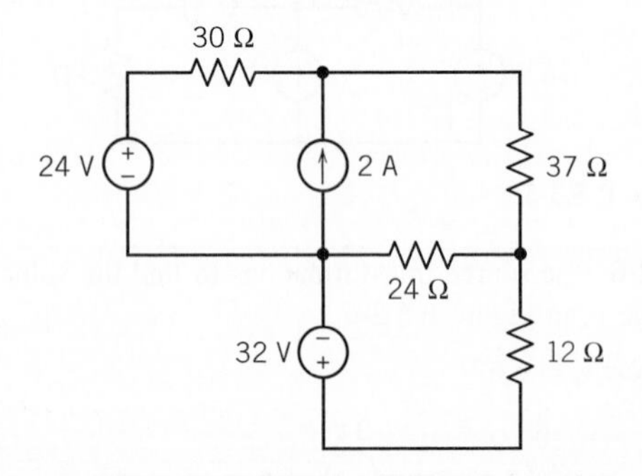

Figure P 5.2-9

Section 5.3 Superposition

P 5.3-1 The inputs to the circuit shown in Figure P 5.3-1 are the voltage source voltages v_1 and v_2. The output of the circuit is the voltage v_o. The output is related to the inputs by

$$v_o = av_1 + bv_2$$

where a and b are constants. Determine the values of a and b.

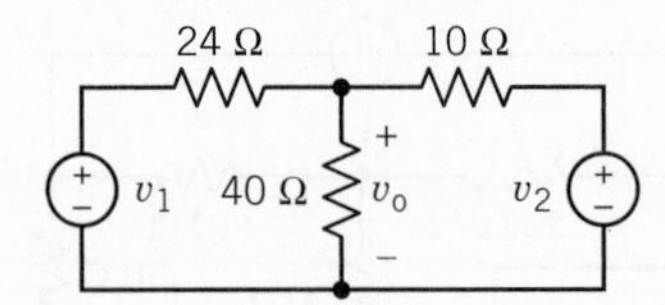

Figure P 5.3-1

P 5.3-2 A particular linear circuit has two inputs, v_1 and v_2, and one output, v_o. Three measurements are made. The first measurement shows that the output is $v_o = 2$ V when the inputs are $v_1 = 4$ V and $v_2 = 0$. The second measurement shows that the output is $v_o = 5$ V when the inputs are $v_1 = 0$ and $v_2 = -5$ V. In the third measurement, the inputs are $v_1 = 6$ V and $v_2 = 6$ V. What is the value of the output in the third measurement?

P 5.3-3 The circuit shown in Figure P 5.3-3 has two inputs, v_s and i_s, and one output i_o. The output is related to the inputs by the equation

$$i_o = ai_s + bv_s$$

Given the following two facts:

The output is $i_o = 0.45$ A when the inputs are $i_s = 0.25$ A and $v_s = 15$ V

and

The output is $i_o = 0.30$ A when the inputs are $i_s = 0.50$ A and $v_s = 0$ V

Determine the values of the constants a and b and the values of the resistances are R_1 and R_2.

Answers: $a = 0.6$ A/A, $b = 0.02$ A/V, $R_1 = 30\,\Omega$, and $R_2 = 20\,\Omega$.

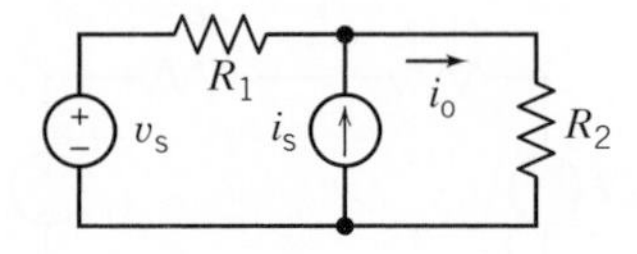

Figure P 5.3-3

P 5.3-4 Use superposition to find v for the circuit of Figure P 5.3-4.

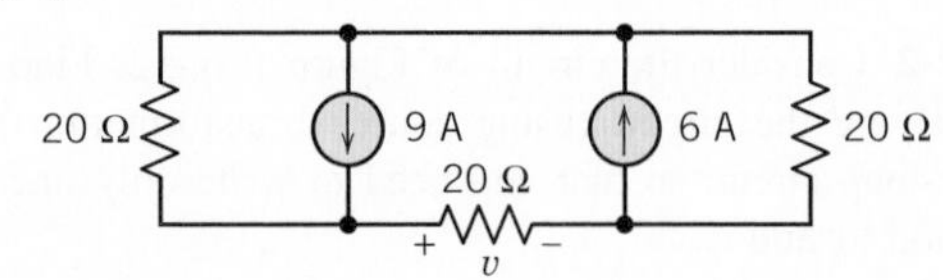

Figure P 5.3-4

P 5.3-5 Use superposition to find i for the circuit of Figure P 5.3-5.

Answer: $i = -1.57$ mA

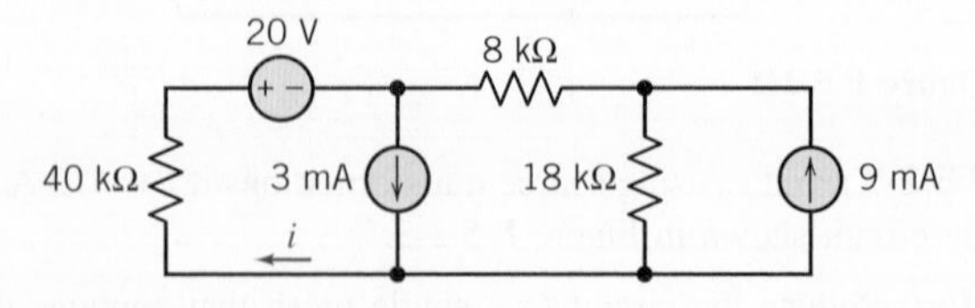

Figure P 5.3-5

P 5.3-6 Use superposition to find i for the circuit of Figure P5.3-6.

Answer: $i = 3.5$ mA

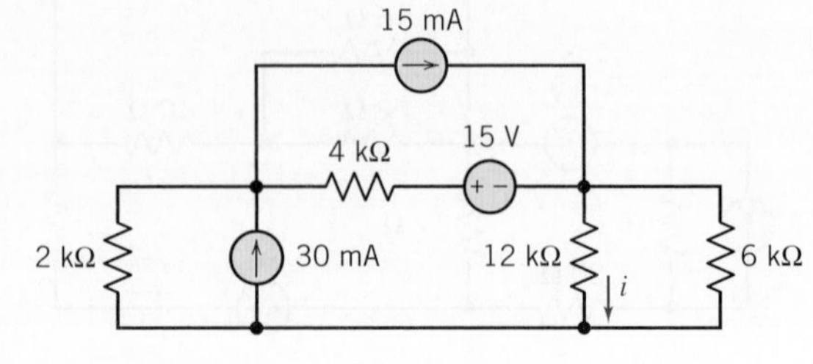

Figure P 5.3-6

P 5.3-7 Use superposition to find the value of the voltage v_a in Figure P 5.3-7.

Answer: $v_a = 7$ V

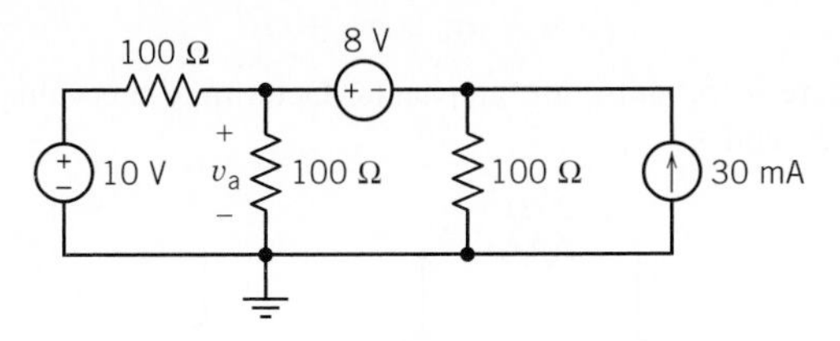

Figure P 5.3-7

P 5.3-8 Use superposition to find the value of the current i_x in Figure P 5.3-8.

Answer: $i_x = 1/6$ A

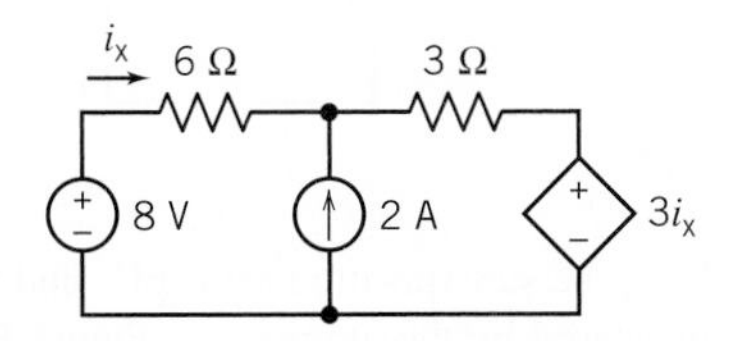

Figure P 5.3-8

P 5.3-9 The input to the circuit shown in Figure P 5.3-9*a* is the voltage source voltage v_s. The output is the voltage v_o. The current source current, i_a, is used to adjust the relationship between the input and output. The plot shown in Figure P 5.3-9*b* specifies a relationship between the input and output of the circuit. Design the circuit shown in Figure P 5.3-9*a* to satisfy the specification shown in Figure P 5.3-9*b*.

Hint: Use superposition to express the output as $v_o = cv_s + di_a$ where c and d are constants that depend on R_1, R_2, and A. Specify values of R_1, R_2, and A to cause the required value of c.

Finally, specify a value of i_a to cause the required value of di_a.

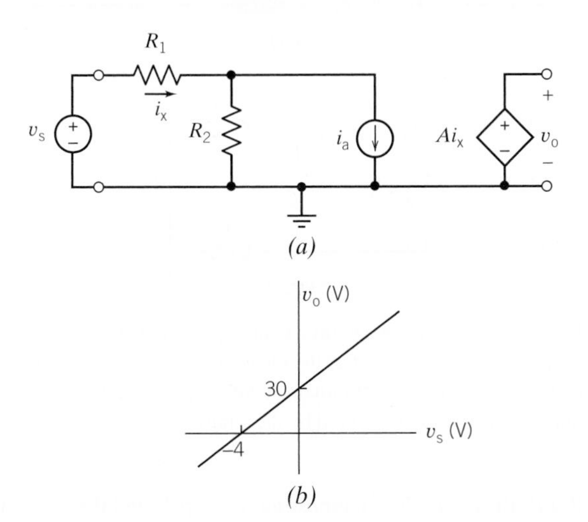

Figure P 5.3-9

***P 5.3-10** The input to the circuit shown in Figure P 5.3-10 is the voltage source voltage, v_s. The output is the voltage v_o. The current source current, i_a, is used to adjust the relationship between the input and output. Design the circuit so that input and output are related by the equation $v_o = 2v_s + 9$.

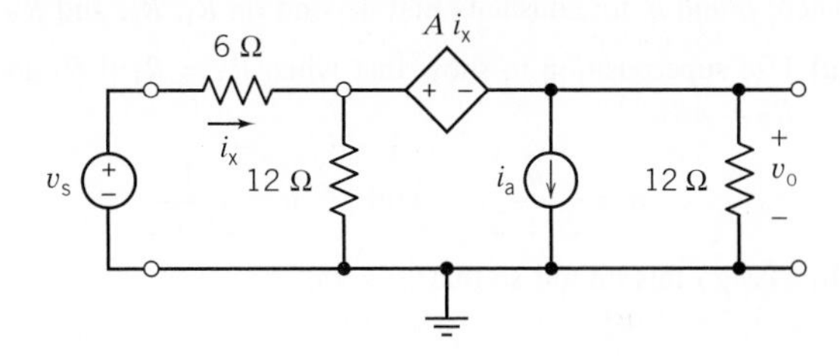

Figure P 5.3-10

Hint: Determine the required values of A and i_a.

P 5.3-11 The circuit shown in Figure P 5.3-11 has three inputs: v_1, v_2, and i_3. The output of the circuit is v_o. The output is related to the inputs by

$$v_o = av_1 + bv_2 + ci_3$$

where a, b, and c are constants. Determine the values of a, b, and c.

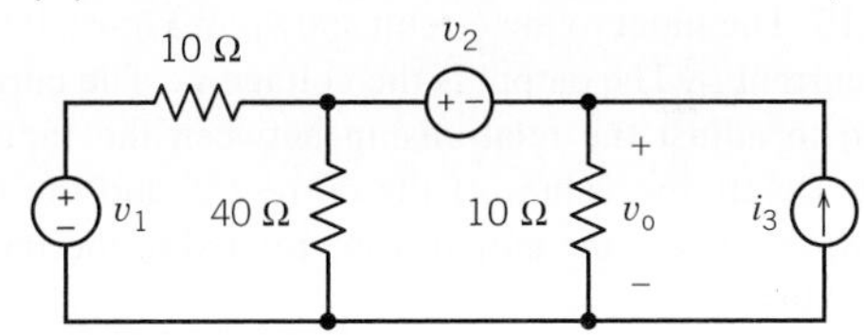

Figure P 5.3-11

P 5.3-12 Determine the voltage $v_o(t)$ for the circuit shown in Figure P 5.3-12.

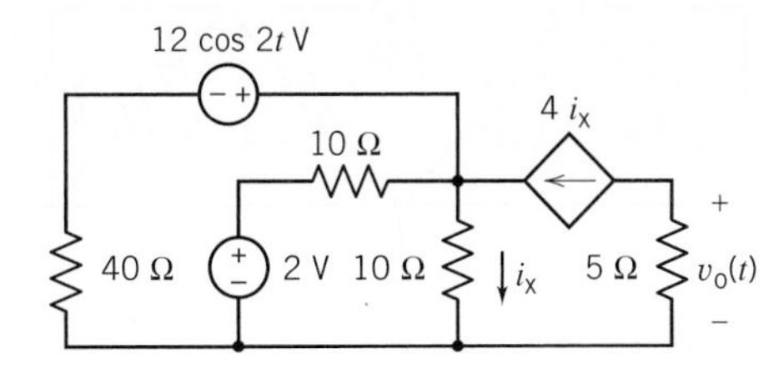

Figure P 5.3-12

P 5.3-13 Determine the value of the voltage v_o in the circuit shown in Figure P 5.3-13.

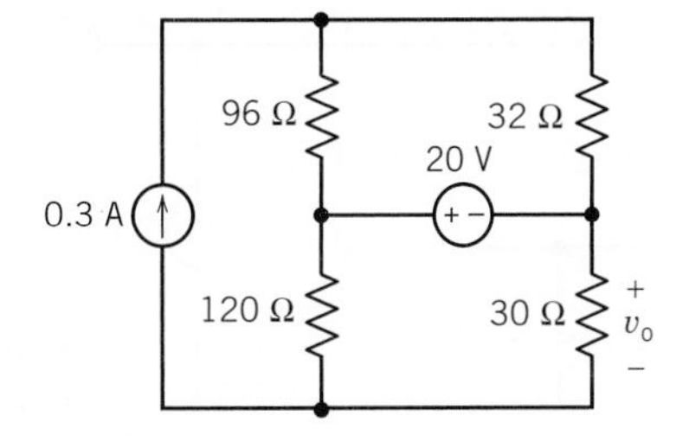

Figure P 5.3-13

***P 5.3-14** The circuit shown in Figure P 5.3-14 has two inputs, v_1 and v_2, and one output, v_o. The output is related to the input by the equation

$$v_o = av_1 + bv_2$$

where a and b are constants that depend on R_1, R_2, and R_3.

(a) Use superposition to show that when $R_3 = R_1 \parallel R_2$ and $R_2 = nR_1$,

$$a = \frac{n}{2n+2} \quad \text{and} \quad b = \frac{1}{2n+2}$$

(b) Design this circuit so that $a = 4b$.

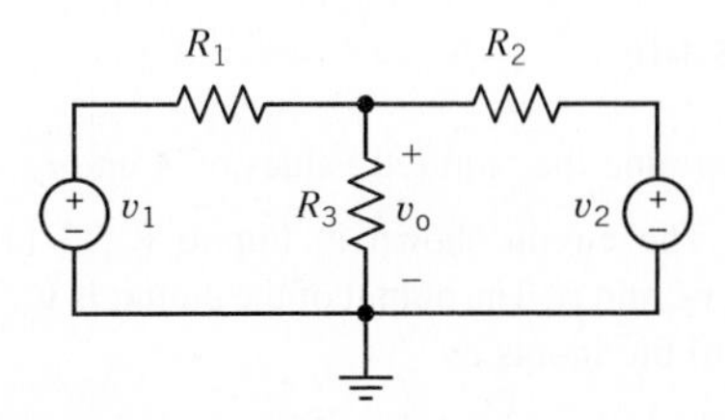

Figure P 5.3-14

P 5.3-15 The input to the circuit shown in Figure P 5.3-15 is the current i_1. The output is the voltage v_o. The current i_2 is used to adjust the relationship between the input and output. Determine values of the current i_2 and the resistance, R, that cause the output to be related to the input by the equation

$$v_o = -0.5i_1 + 4$$

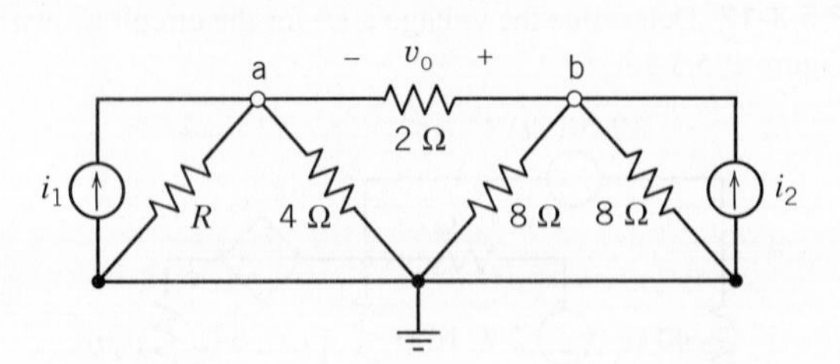

Figure P 5.3-15

P 5.3-16 Determine values of the current, i_a, and the resistance, R, for the circuit shown in Figure P 5.3-16.

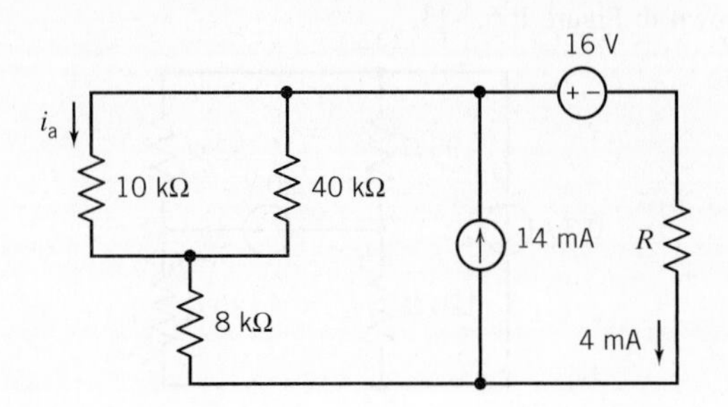

Figure P 5.3-16

P 5.3-17 The circuit shown in Figure P 5.3-17 has three inputs: v_1, i_2, and v_3. The output of the circuit is the current i_o. The output of the circuit is related to the inputs by

$$i_1 = av_o + bv_2 + ci_3$$

where a, b, and c are constants. Determine the values of a, b, and c.

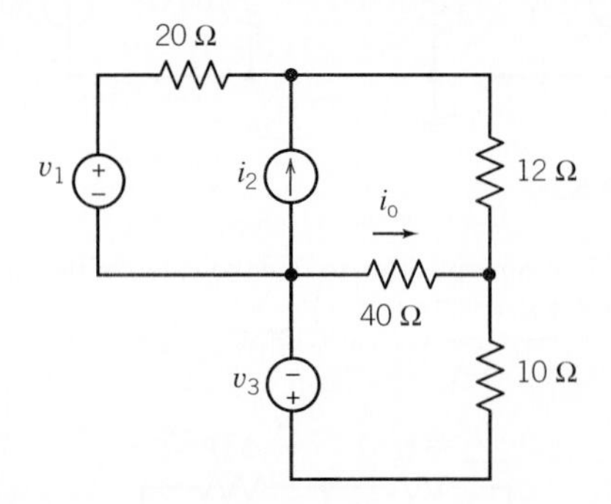

Figure P 5.3-17

P 5.3-18 Using the superposition principle, find the value of the current measured by the ammeter in Figure P 5.3-18*a*.

Hint: Figure P 5.3-18*b* shows the circuit after the ideal ammeter has been replaced by the equivalent short circuit and a label has been added to indicate the current measured by the ammeter, i_m.

Answer: $i_m = \frac{25}{3+2} - \frac{3}{2+3}5 = 5 - 3 = 2\text{ A}$

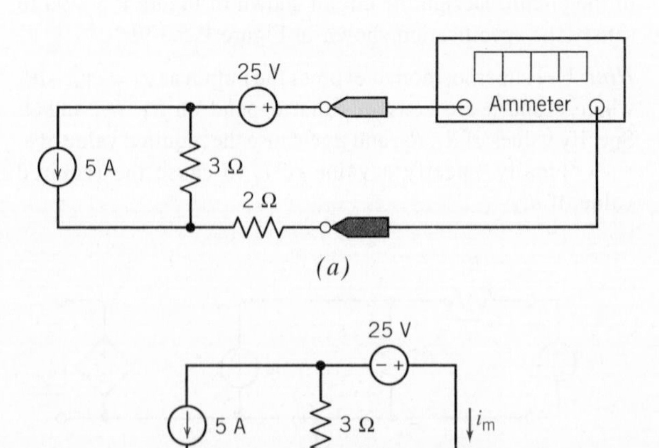

Figure P 5.3-18 (*a*) A circuit containing two independent sources. (*b*) The circuit after the ideal ammeter has been replaced by the equivalent short circuit and a label has been added to indicate the current measured by the ammeter, i_m.

P 5.3-19 Using the superposition principle, find the value of the voltage measured by the voltmeter in Figure P 5.3-19*a*.

Hint: Figure P 5.3-19*b* shows the circuit after the ideal voltmeter has been replaced by the equivalent open circuit and a label has been added to indicate the voltage measured by the voltmeter, v_m.

Answer: $v_m = 3\left(\frac{3}{3+(3+3)}5\right) - \frac{3}{3+(3+3)}18 = 5 - 6 = -1\text{ V}$

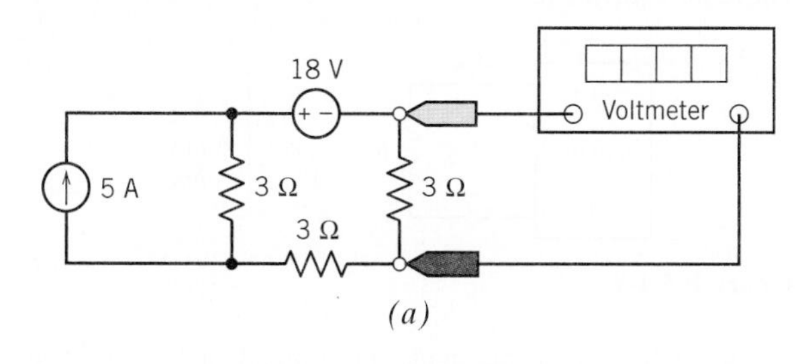

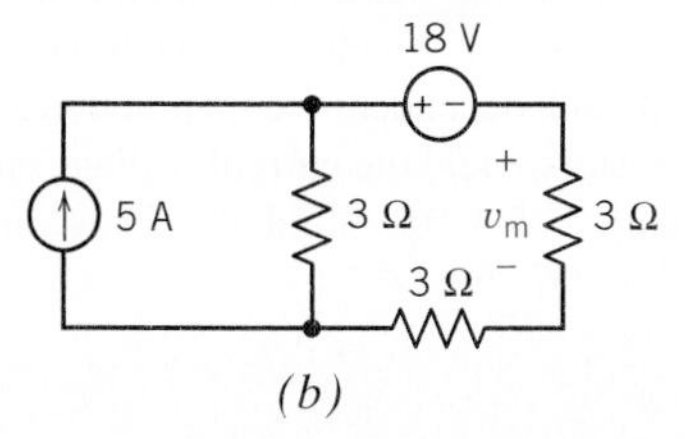

Figure P 5.3-19 (*a*) A circuit containing two independent sources. (*b*) The circuit after the ideal voltmeter has been replaced by the equivalent open circuit and a label has been added to indicate the voltage measured by the voltmeter, v_m.

Section 5.4 Thévenin's Theorem

P 5.4-1 Determine values of R_t and v_{oc} that cause the circuit shown in Figure P 5.4-1*b* to be the Thévenin equivalent circuit of the circuit in Figure P 5.4-1*a*.

Hint: Use source transformations and equivalent resistances to reduce the circuit in Figure P 5.4-1*a* until it is the circuit in Figure P 5.4-1*b*.

Answer: $R_t = 8\ \Omega$ and $v_{oc} = 6$ V

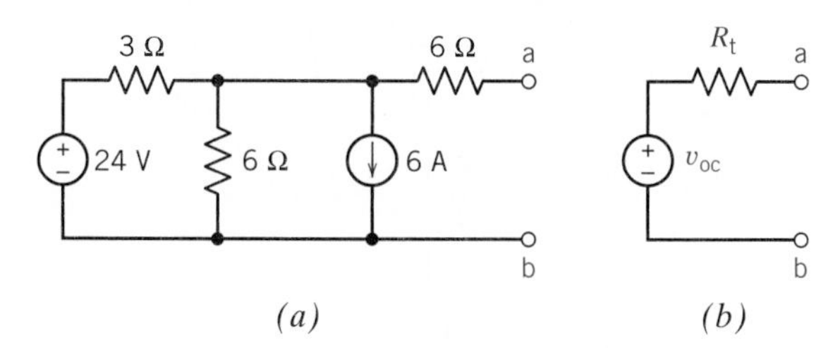

Figure P 5.4-1

P 5.4-2 The circuit shown in Figure P 5.4-2*b* is the Thévenin equivalent circuit of the circuit shown in Figure P 5.4-2*a*. Find the value of the open-circuit voltage, v_{oc}, and Thévenin resistance, R_t.

Answer: $v_{oc} = -16$ V and $R_t = 16\ \Omega$

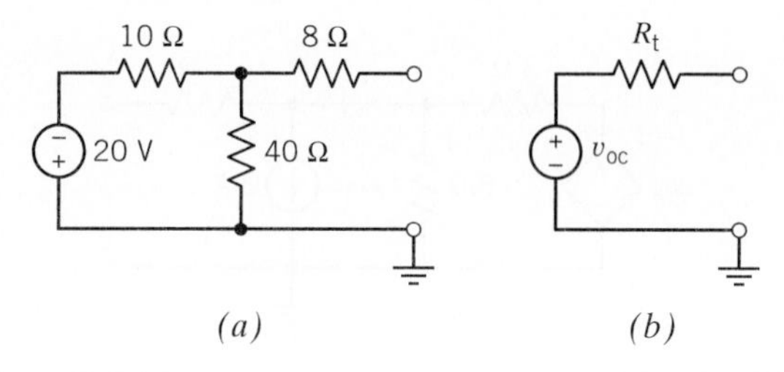

Figure P 5.4-2

P 5.4-3 The circuit shown in Figure P 5.4-3*b* is the Thévenin equivalent circuit of the circuit shown in Figure P 5.4-3*a*. Find the value of the open-circuit voltage, v_{oc}, and Thévenin resistance, R_t.

Answer: $v_{oc} = 2$ V and $R_t = 4\ \Omega$

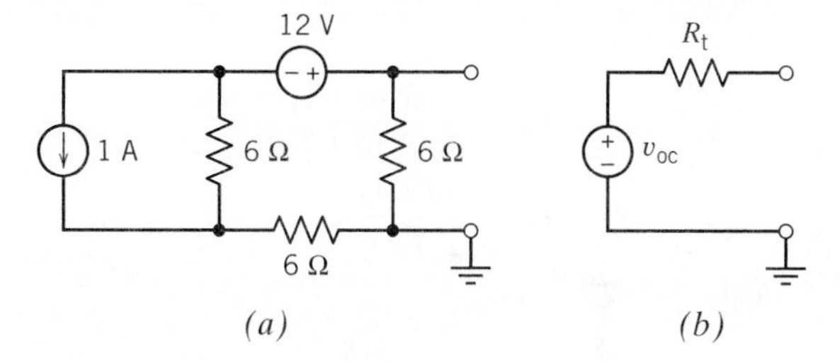

Figure P 5.4-3

P 5.4-4 Find the Thévenin equivalent circuit for the circuit shown in Figure P 5.4-4.

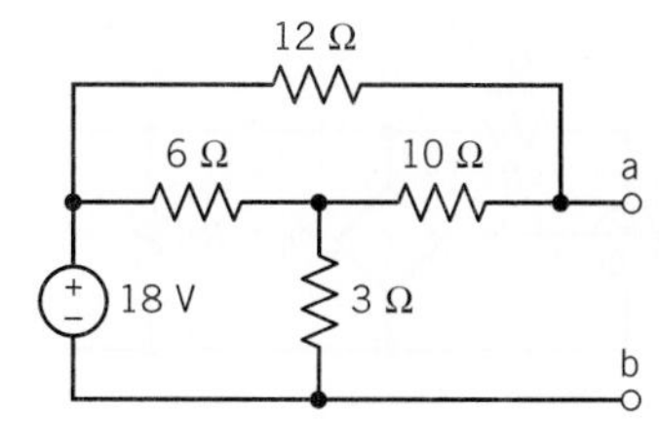

Figure P 5.4-4

P 5.4-5 Find the Thévenin equivalent circuit for the circuit shown in Figure P 5.4-5.

Answer: $v_{oc} = -2$ V and $R_t = -8/3\ \Omega$

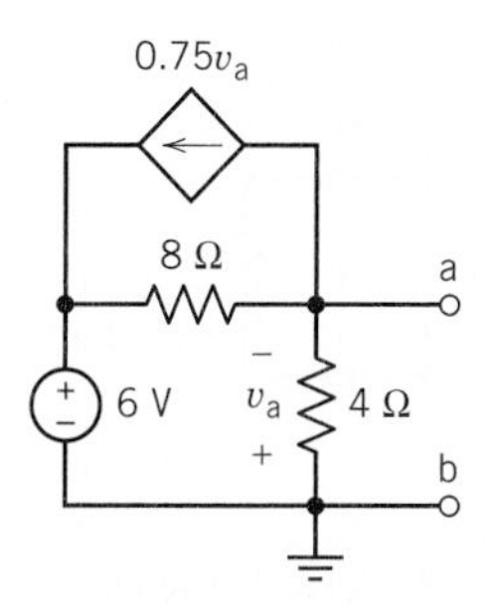

Figure P 5.4-5

P 5.4-6 Find the Thévenin equivalent circuit for the circuit shown in Figure P 5.4-6.

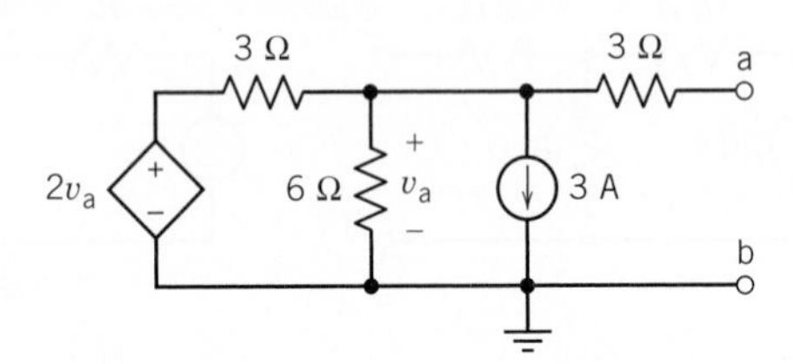

Figure P 5.4-6

P 5.4-7 The circuit shown in Figure P 5.4-7 has four unspecified circuit parameters: v_s, R_1, R_2, and d, where d is the gain of the CCCS.

(a) Show that the open-circuit voltage, v_{oc}, the short-circuit current, i_{sc}, and the Thévenin resistance, R_t, of this circuit are given by

$$v_{oc} = \frac{R_2(d+1)}{R_1 + (d+1)R_2} v_s$$

$$i_{sc} = \frac{(d+1)}{R_1} v_s$$

and

$$R_t = \frac{R_1 R_2}{R_1 + (d+1)R_2}$$

(b) Let $R_1 = R_2 = 1\ \text{k}\Omega$. Determine the values of v_s and d required to cause $v_{oc} = 5$ V and $R_t = 625\ \Omega$.

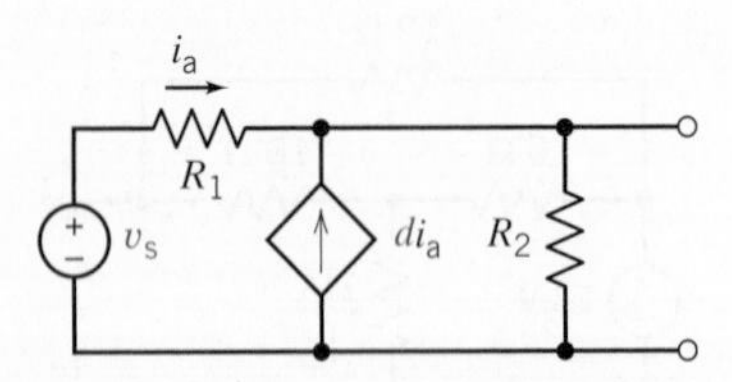

Figure P 5.4-7

P 5.4-8 A resistor, R, was connected to a circuit box as shown in Figure P 5.4-8. The voltage, v, was measured. The resistance was changed, and the voltage was measured again. The results are shown in the table. Determine the Thévenin equivalent of the circuit within the box and predict the voltage, v, when $R = 8\ \text{k}\Omega$.

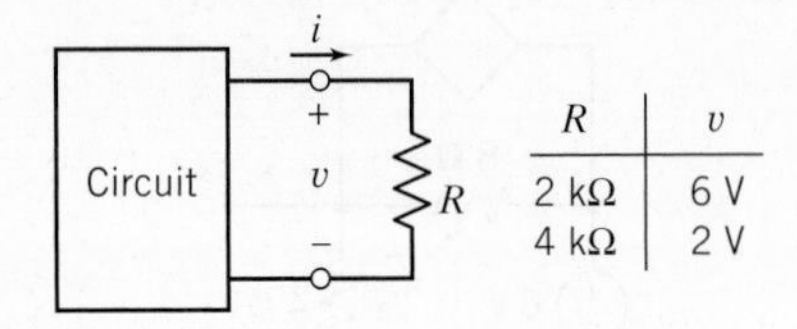

R	v
2 kΩ	6 V
4 kΩ	2 V

Figure P 5.4-8

P 5.4-9 A resistor, R, was connected to a circuit box as shown in Figure P 5.4-9. The current, i, was measured. The resistance was changed, and the current was measured again. The results are shown in the table.

(a) Specify the value of R required to cause $i = 2$ mA.
(b) Given that $R > 0$, determine the maximum possible value of the current i.

Hint: Use the data in the table to represent the circuit by a Thévenin equivalent.

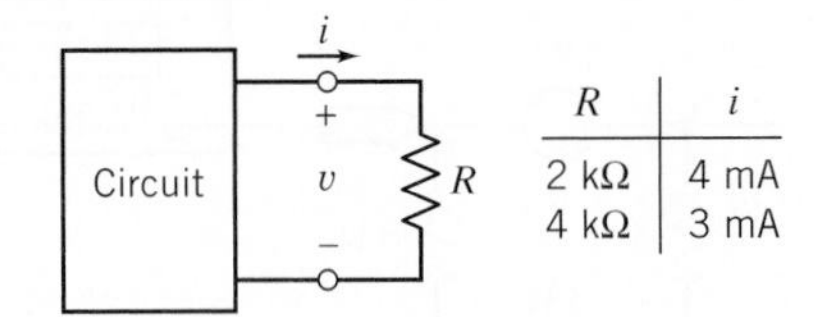

R	i
2 kΩ	4 mA
4 kΩ	3 mA

Figure P 5.4-9

P 5.4-10 Measurements made on terminals a–b of a linear circuit, Figure P 5.4-10*a*, which is known to be made up only of independent and dependent voltage sources and current sources and resistors, yield the current–voltage characteristics shown in Figure P 5.4-10*b*. Find the Thévenin equivalent circuit.

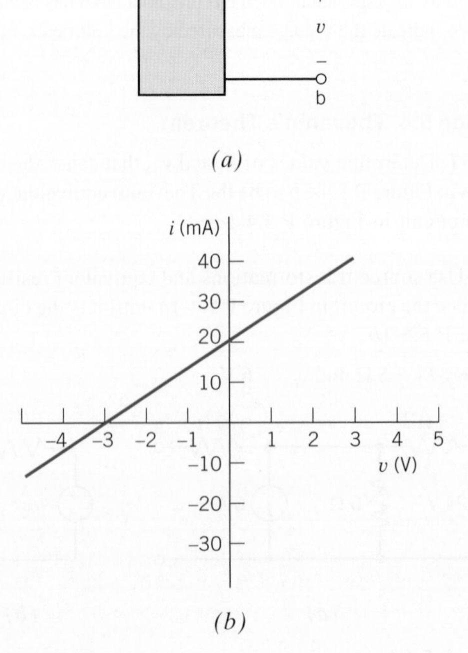

Figure P 5.4-10

P 5.4-11 For the circuit of Figure P 5.4-11, specify the resistance R that will cause current i_b to be 2 mA. The current i_a has units of amps.

Hint: Find the Thévenin equivalent circuit of the circuit connected to R.

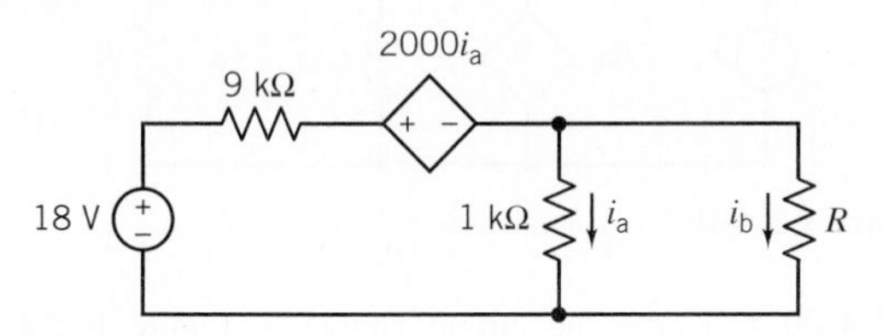

Figure P 5.4-11

P 5.4-12 For the circuit of Figure P 5.4-12, specify the value of the resistance R_L that will cause current i_L to be -2 A.

Answer: $R_L = 12\,\Omega$

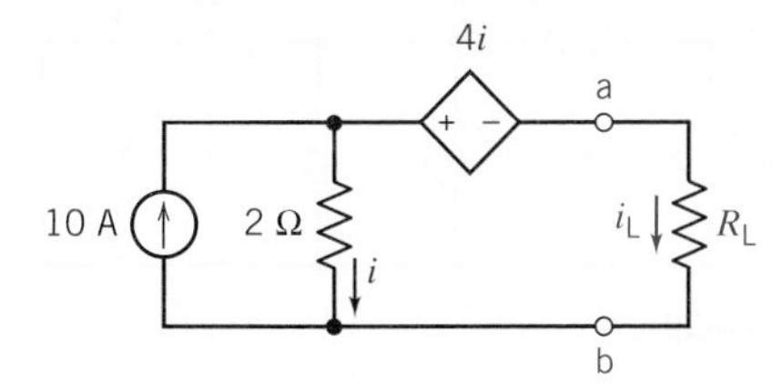

Figure P 5.4-12

P 5.4-13 The circuit shown in Figure P 5.4-13 contains an adjustable resistor. The resistance R can be set to any value in the range $0 \leq R \leq 100$ kΩ.

(a) Determine the maximum value of the current i_a that can be obtained by adjusting R. Determine the corresponding value of R.

(b) Determine the maximum value of the voltage v_a that can be obtained by adjusting R. Determine the corresponding value of R.

(c) Determine the maximum value of the power supplied to the adjustable resistor that can be obtained by adjusting R. Determine the corresponding value of R.

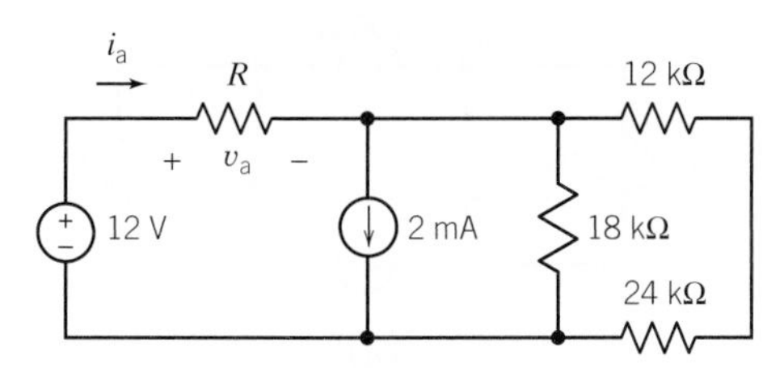

Figure P 5.4-13

P 5.4-14 The circuit shown in Figure P 5.4-14 consists of two parts, the source (to the left of the terminals) and the load. The load consists of a single adjustable resistor having resistance $0 \leq R_L \leq 20\,\Omega$. The resistance R is fixed but unspecified. When $R_L = 4\,\Omega$, the load current is measured to be $i_o = 0.375$ A. When $R_L = 8\,\Omega$, the value of the load current is $i_o = 0.300$ A.

(a) Determine the value of the load current when $R_L = 10\,\Omega$.

(b) Determine the value of R.

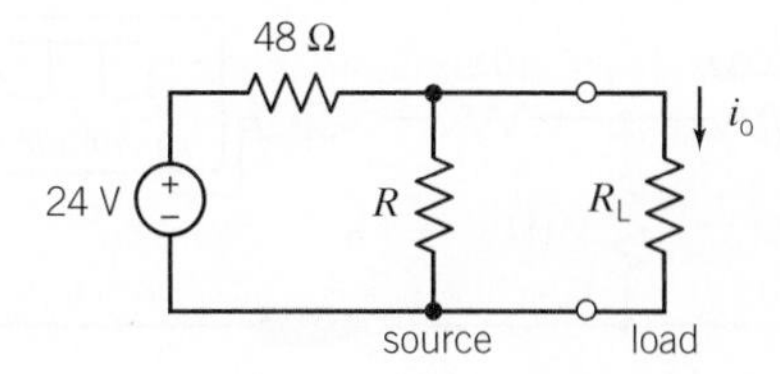

Figure P 5.4-14

P 5.4-15 The circuit shown in Figure P 5.4-15 contains an unspecified resistance, R. Determine the value of R in each of the following two ways.

(a) Write and solve mesh equations.

(b) Replace the part of the circuit connected to the resistor R by a Thévenin equivalent circuit. Analyze the resulting circuit.

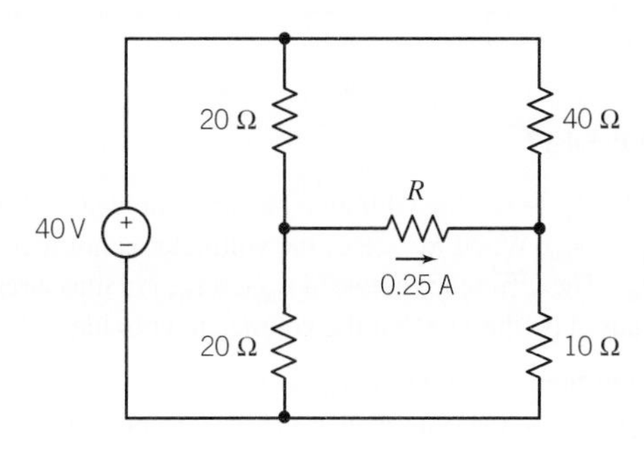

Figure P 5.4-15

P 5.4-16 Consider the circuit shown in Figure P 5.4-16. Replace the part of the circuit to the left of terminals a–b by its Thévenin equivalent circuit. Determine the value of the current i_o.

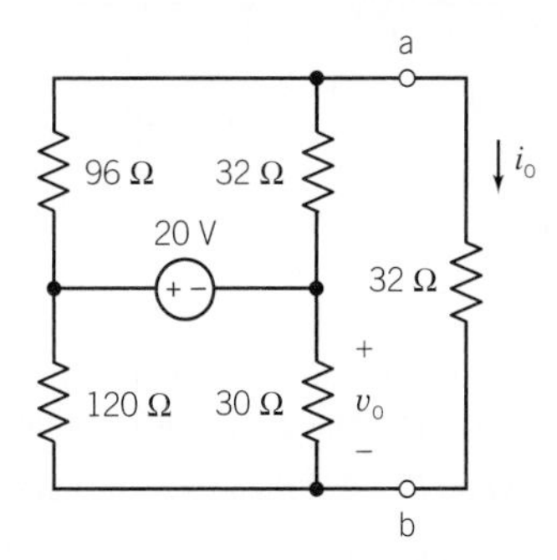

Figure P 5.4-16

P 5.4-17 An ideal voltmeter is modeled as an open circuit. A more realistic model of a voltmeter is a large resistance. Figure P 5.4-17*a* shows a circuit with a voltmeter that measures the voltage v_m. In Figure P 5.4-17*b*, the voltmeter is replaced by the model of an ideal voltmeter, an open circuit. The voltmeter measures v_{mi}, the ideal value of v_m.

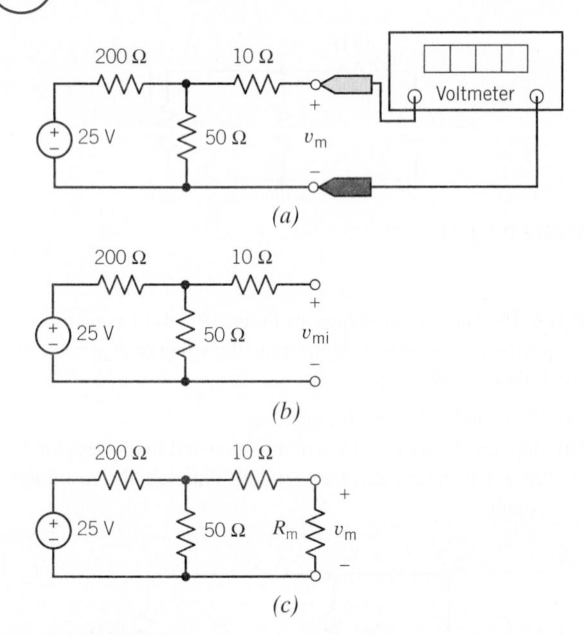

Figure P 5.4-17

As $R_m \to \infty$, the voltmeter becomes an ideal voltmeter and $v_m \to v_{mi}$. When $R_m < \infty$, the voltmeter is not ideal and $v_m > v_{mi}$. The difference between v_m and v_{mi} is a measurement error caused by the fact that the voltmeter is not ideal.

(a) Determine the value of v_{mi}.
(b) Express the measurement error that occurs when $R_m = 1000\,\Omega$ as a percentage of v_{mi}.
(c) Determine the minimum value of R_m required to ensure that the measurement error is smaller than 2 percent of v_{mi}.

P 5.4-18 Determine the Thévenin equivalent circuit for the circuit shown in Figure P 5.4-18.

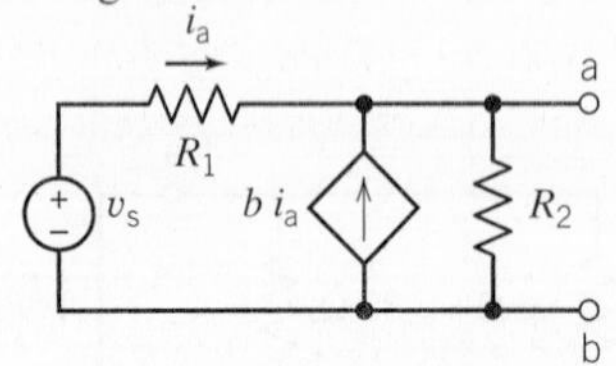

Figure P 5.4-18

P 5.4-19 Given that $0 \le R \le \infty$ in the circuit shown in Figure P 5.4-19, consider these two observations:

Observation 1: When $R = 2\,\Omega$ then $v_R = 4$ V and $i_R = 2$ A.

Observation 1: When R = $6\,\Omega$ then $v_R = 6$ V and $i_R = 1$ A.

Determine the following:

(a) The maximum value of i_R and the value of R that causes i_R to be maximal.
(b) The maximum value of v_R and the value of R that causes v_R to be maximal.
(c) The maximum value of $p_R = i_R v_R$ and the value of R that causes p_R to be maximal.

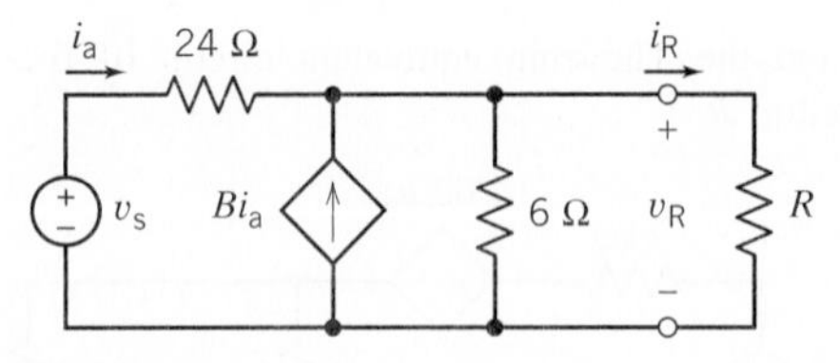

Figure P 5.4-19

P 5.4-20 Consider the circuit shown in Figure P 5.4-20. Determine

(a) The value of v_R that occurs when $R = 9\,\Omega$.
(b) The value of R that causes $v_R = 5.4$ V.
(c) The value of R that causes $i_R = 300$ mA.

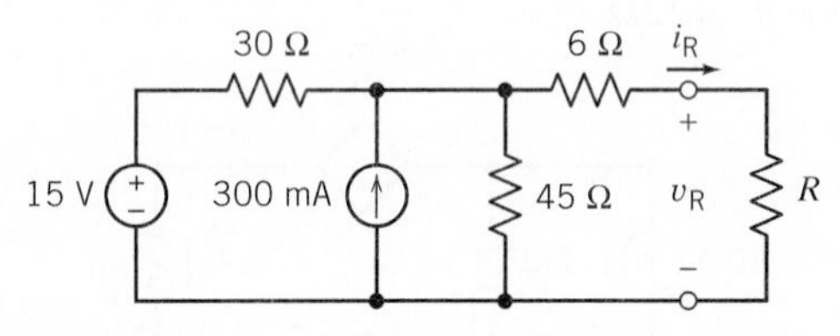

Figure P 5.4-20

Section 5.5 Norton's Equivalent Circuit

P 5.5-1 The part of the circuit shown in Figure P 5.5-1*a* to the left of the terminals can be reduced to its Norton equivalent circuit using source transformations and equivalent resistance. The resulting Norton equivalent circuit, shown in Figure P 5.5-1*b*, will be characterized by the parameters:

$$i_{sc} = 0.5\text{ A and } R_t = 20\,\Omega$$

(a) Determine the values of v_S and R_1.
(b) Given that $0 \le R_2 \le \infty$, determine the maximum values of the voltage, v, and of the power, $p = vi$.

Answers: $v_s = 50$ V, $R_1 = 25\,\Omega$, max $v = 10$ V and max $p = 1.25$ W

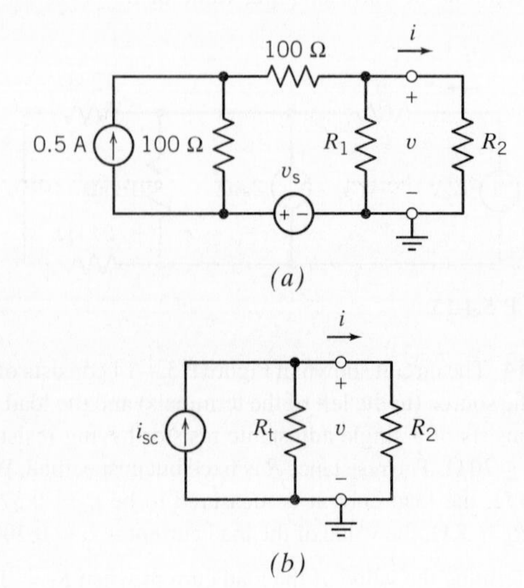

Figure P 5.5-1

P 5.5-2 Two black boxes are shown in Figure P 5.5-2. Box A contains the Thévenin equivalent of some linear circuit, and box B contains the Norton equivalent of the same circuit. With access to just the outsides of the boxes and their terminals, how can you determine which is which, using only one shorting wire?

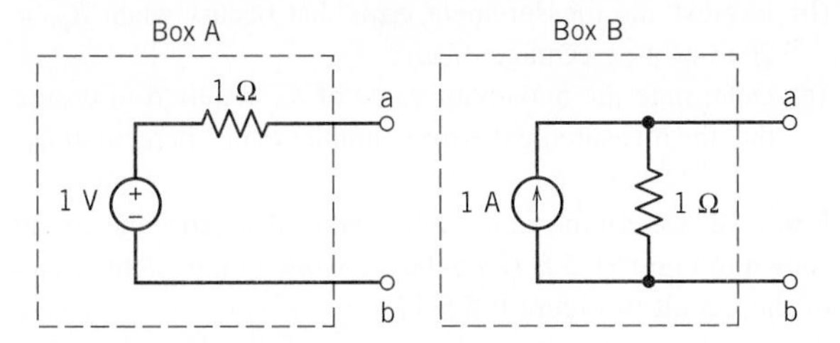

Figure P 5.5-2 Black boxes problem.

P 5.5-3 Find the Norton equivalent circuit for the circuit shown in Figure P 5.5-3.

Answer: $R_t = 2\,\Omega$ and $i_{sc} = -7.5$ A

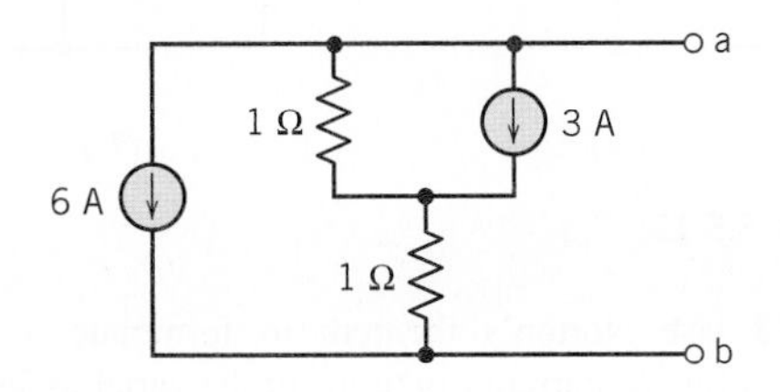

Figure P 5.5-3

P 5.5-4 Find the Norton equivalent circuit for the circuit shown in Figure P 5.5-4.

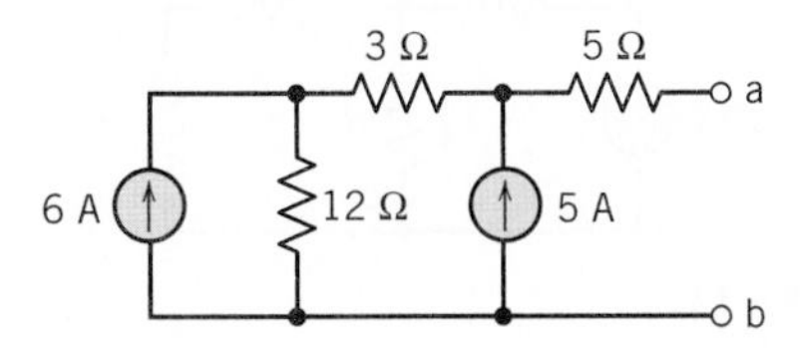

Figure P 5.5-4

P 5.5-5 The circuit shown in Figure P 5.5-5*b* is the Norton equivalent circuit of the circuit shown in Figure P 5.5-5*a*. Find the value of the short-circuit current, i_{sc}, and Thévenin resistance, R_t.

Answer: $i_{sc} = 1.13$ A and $R_t = 7.57\,\Omega$

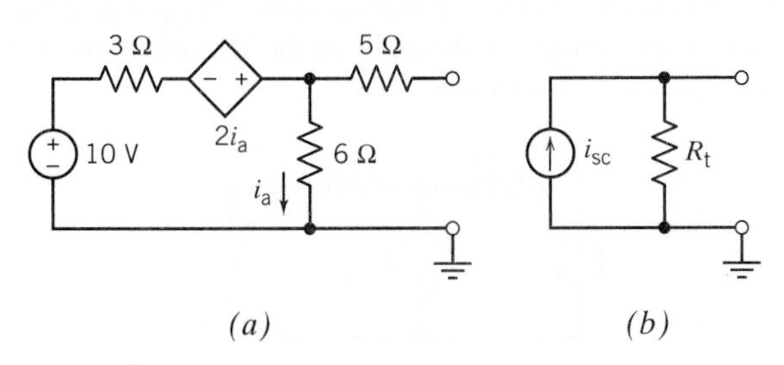

Figure P 5.5-5

P 5.5-6 The circuit shown in Figure P 5.5-6*b* is the Norton equivalent circuit of the circuit shown in Figure P 5.5-6*a*. Find the value of the short-circuit current, i_{sc}, and Thévenin resistance, R_t.

Answer: $i_{sc} = -24$ A and $R_t = -3\,\Omega$

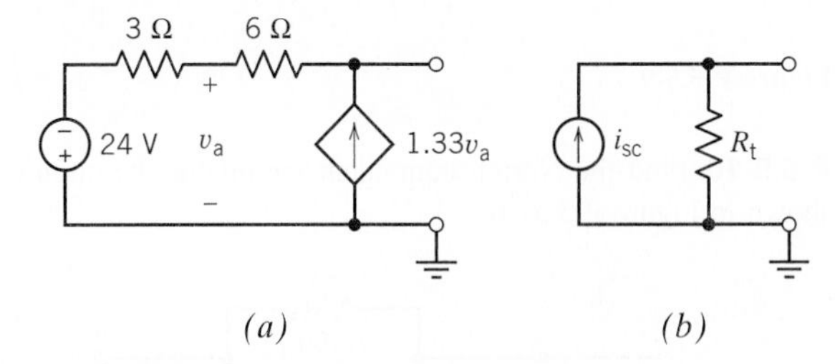

Figure P 5.5-6

P 5.5-7 Determine the value of the resistance R in the circuit shown in Figure P 5.5-7 by each of the following methods:

(a) Replace the part of the circuit to the left of terminals a–b by its Norton equivalent circuit. Use current division to determine the value of R.

(b) Analyze the circuit shown Figure P 5.5-7 using mesh equations. Solve the mesh equations to determine the value of R.

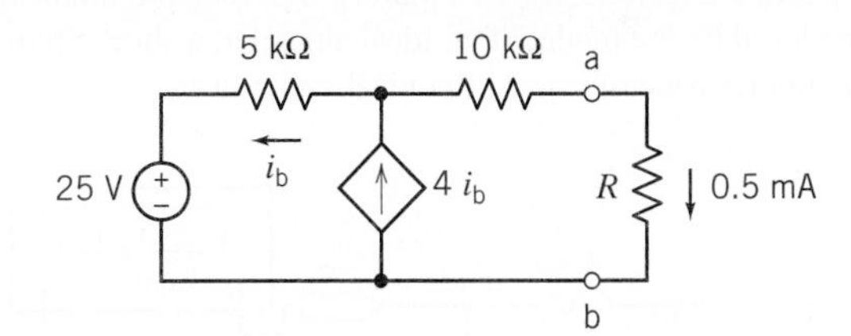

Figure P 5.5-7

***P 5.5-8** The device to the right of terminals a–b in Figure P 5.5-8 is a nonlinear resistor characterized by

$$i = \frac{v^2}{2}$$

Determine the values of i and v.

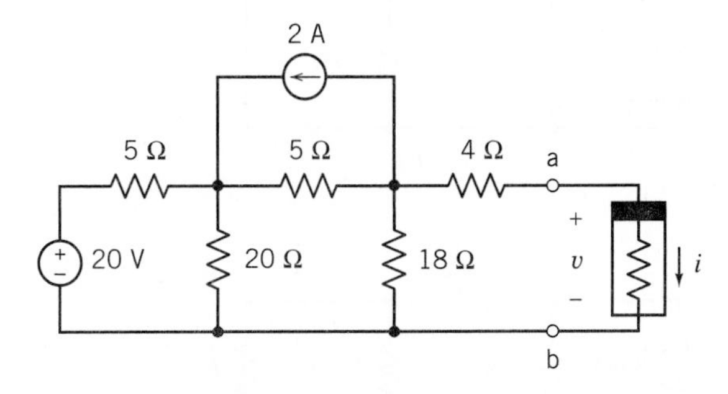

Figure P 5.5-8

P 5.5-9 Find the Norton equivalent circuit for the circuit shown in Figure P 5.5-9.

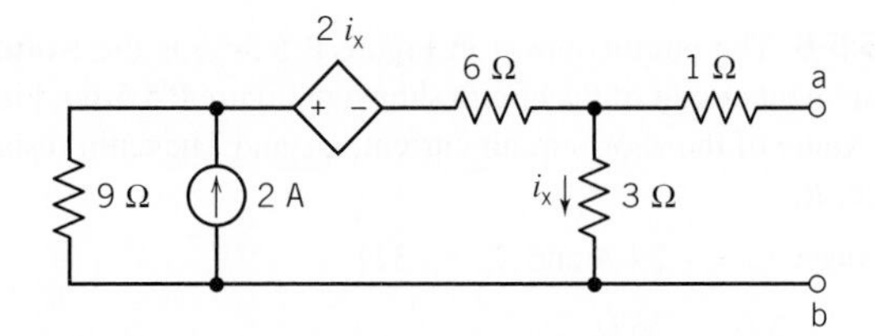

Figure P 5.5-9

P 5.5-10 Find the Norton equivalent circuit for the circuit shown in Figure P 5.5-10.

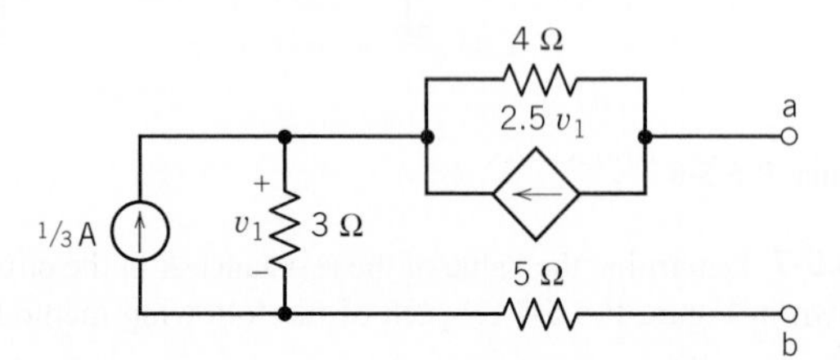

Figure P 5.5-10

P 5.5-11 An ideal ammeter is modeled as a short circuit. A more realistic model of an ammeter is a small resistance. Figure P 5.5-11*a* shows a circuit with an ammeter that measures the current i_m. In Figure P 5.5-10*b*, the ammeter is replaced by the model of an ideal ammeter, a short circuit. The ammeter measures i_{mi}, the ideal value of i_m.

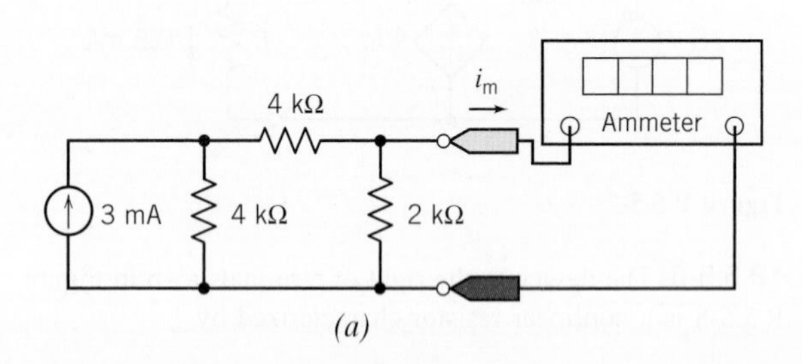

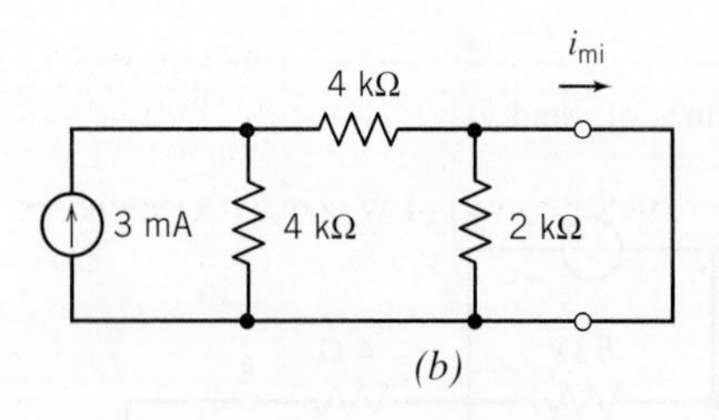

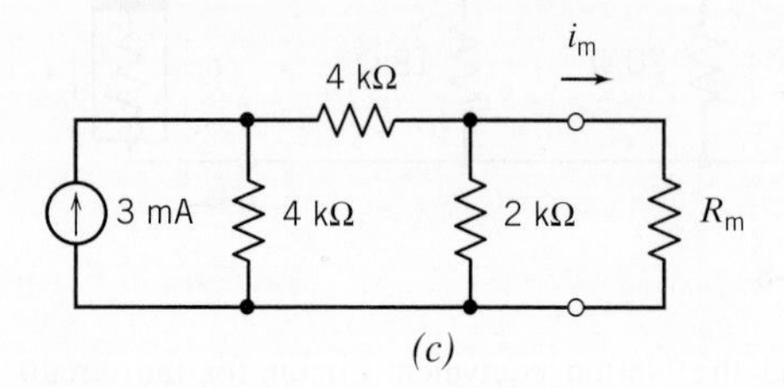

Figure P 5.5-11

As $R_m \to 0$, the ammeter becomes an ideal ammeter and $i_m \to i_{mi}$. When $R_m > 0$, the ammeter is not ideal and $i_m < i_{mi}$. The difference between i_m and i_{mi} is a measurement error caused by the fact that the ammeter is not ideal.

(a) Determine the value of i_{mi}.
(b) Express the measurement error that occurs when $R_m = 20\ \Omega$ as a percentage of i_{mi}.
(c) Determine the maximum value of R_m required to ensure that the measurement error is smaller than 2 percent of i_{mi}.

P 5.5-12 Determine values of R_t and i_{sc} that cause the circuit shown in Figure P 5.5-12*b* to be the Norton equivalent circuit of the circuit in Figure P 5.5-12*a*.

Answer: $R_t = 3\ \Omega$ and $i_{sc} = -1.6$ A

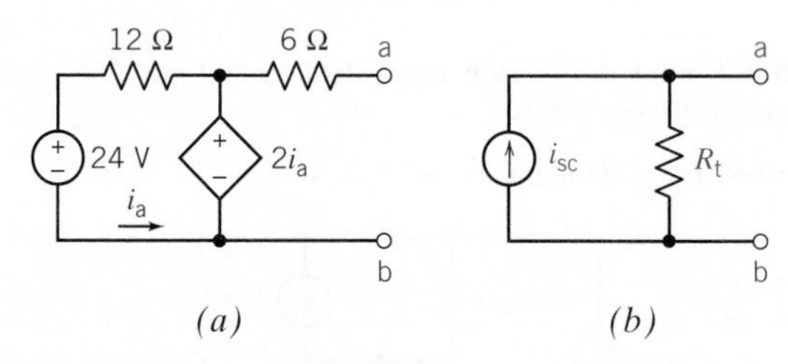

Figure P 5.5-12

P 5.5-13 Use Norton's theorem to formulate a general expression for the current i in terms of the variable resistance R shown in Figure P 5.5-13.

Answer: $i = 20/(8 + R)$ A

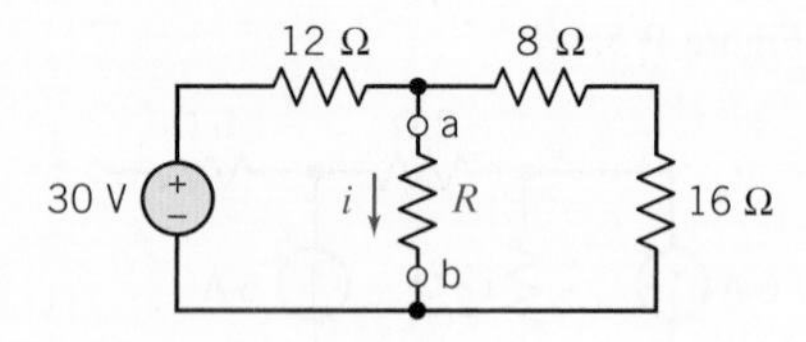

Figure P 5.5-13

Section 5.6 Maximum Power Transfer

P 5.6-1 The circuit shown in Figure P 5.6-1 consists of two parts separated by a pair of terminals. Consider the part of the circuit to the left of the terminals. The open circuit voltage is $v_{oc} = 8$ V, and short-circuit current is $i_{sc} = 2$ A. Determine the values of (a) the voltage source voltage, v_s, and the resistance R_2, and (b) the resistance R that maximizes the power delivered to the resistor to the right of the terminals, and the corresponding maximum power.

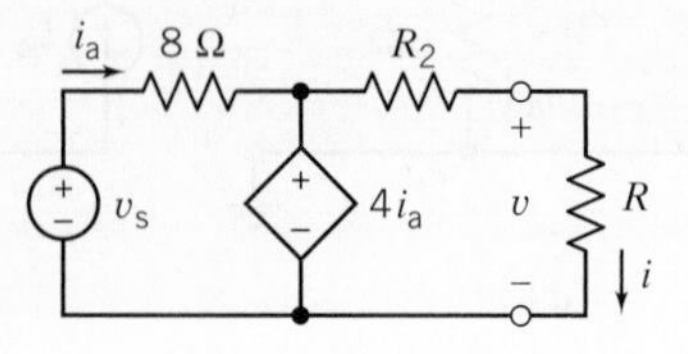

Figure P 5.6-1

P 5.6-2 The circuit model for a photovoltaic cell is given in Figure P 5.6-2 (Edelson, 1992). The current i_s is proportional to the solar insolation (kW/m^2).

(a) Find the load resistance, R_L, for maximum power transfer.
(b) Find the maximum power transferred when $i_s = 1$ A.

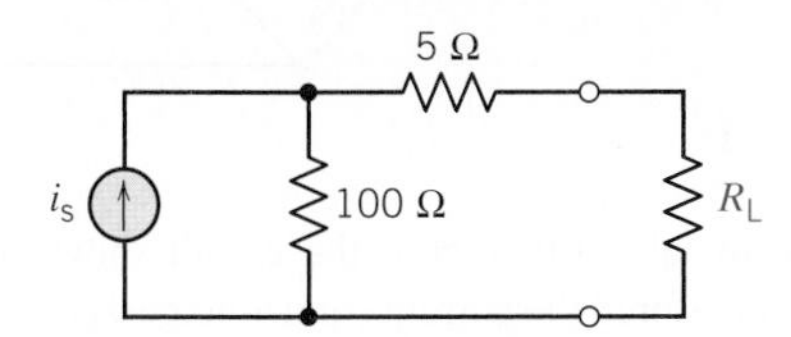

Figure P 5.6-2 Circuit model of a photovoltaic cell.

P 5.6-3 For the circuit in Figure P 5.6-3, (a) find R such that maximum power is dissipated in R, and (b) calculate the value of maximum power.

Answer: $R = 120\ \Omega$ and $P_{max} = 108$ mW

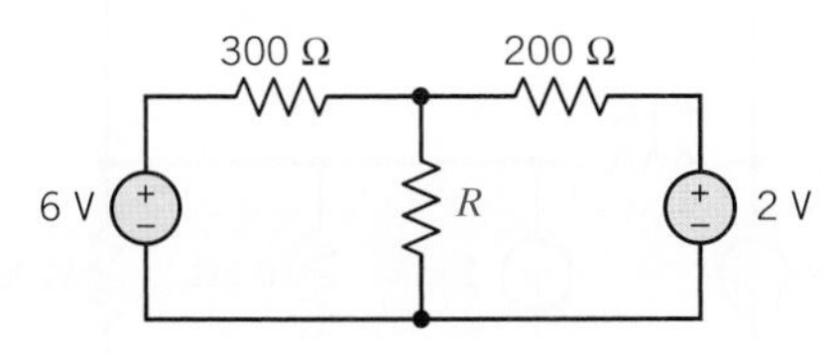

Figure P 5.6-3

P 5.6-4 For the circuit in Figure P 5.6-4, prove that for R_s variable and R_L fixed, the power dissipated in R_L is maximum when $R_s = 0$.

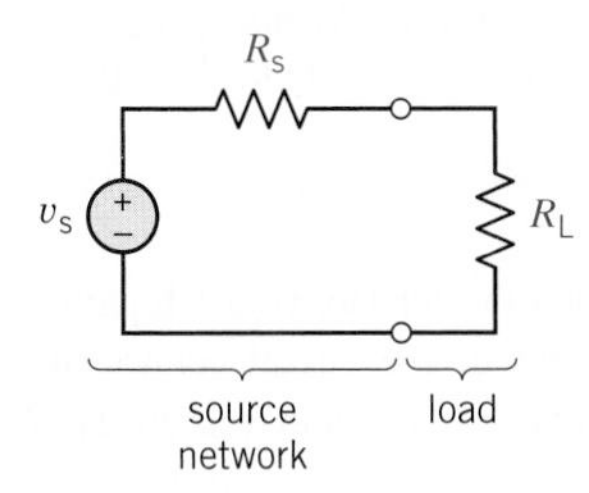

Figure P 5.6-4

P 5.6-5 Find the maximum power to the load R_L if the maximum power transfer condition is met for the circuit of Figure P 5.6-5.

Answer: max $p_L = 0.75$ W

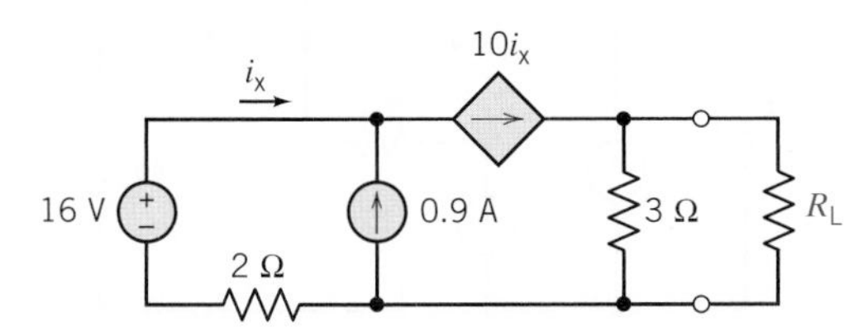

Figure P 5.6-5

P 5.6-6 Determine the maximum power that can be absorbed by a resistor, R, connected to terminals a–b of the circuit shown in Figure P 5.6-6. Specify the required value of R.

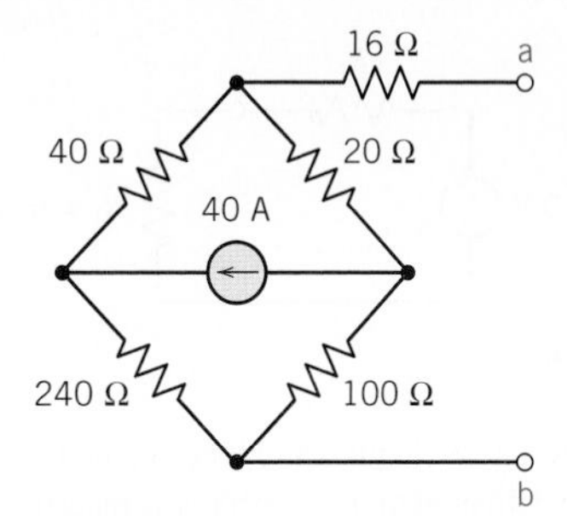

Figure P 5.6-6 Bridge circuit.

P 5.6-7 Figure P 5.6-7 shows a source connected to a load through an amplifier. The load can safely receive up to 15 W of power. Consider three cases:

(a) $A = 20$ V/V and $R_o = 10\ \Omega$. Determine the value of R_L that maximizes the power delivered to the load and the corresponding maximum load power.
(b) $A = 20$ V/V and $R_L = 8\ \Omega$. Determine the value of R_o that maximizes the power delivered to the load and the corresponding maximum load power.
(c) $R_o = 10\ \Omega$ and $R_L = 8\ \Omega$. Determine the value of A that maximizes the power delivered to the load and the corresponding maximum load power.

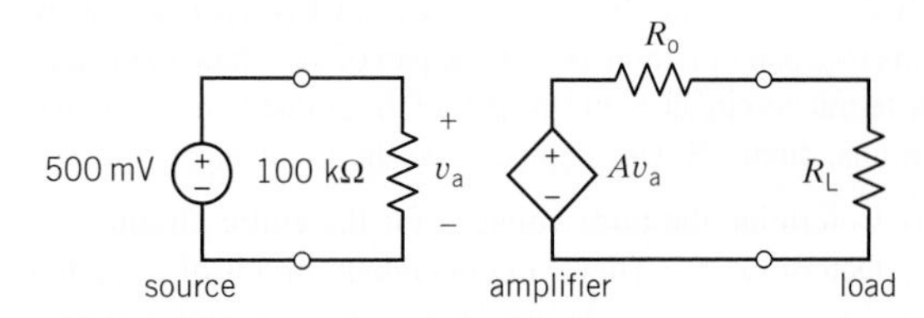

Figure P 5.6-7

P 5.6-8 The circuit in Figure P 5.6-8 contains a variable resistance, R, implemented using a potentiometer. The resistance of the variable resistor varies over the range $0 \le R \le 1000\ \Omega$. The variable resistor can safely receive 1/4 W power. Determine the maximum power received by the variable resistor. Is the circuit safe?

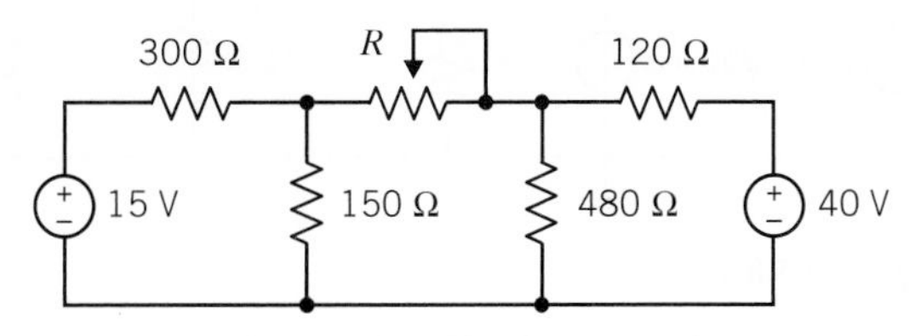

Figure P 5.6-8

P 5.6-9 For the circuit of Figure P 5.6-9, find the power delivered to the load when R_L is fixed and R_t may be varied between 1 Ω and 5 Ω. Select R_t so that maximum power is delivered to R_L.

Answer: 13.9 W

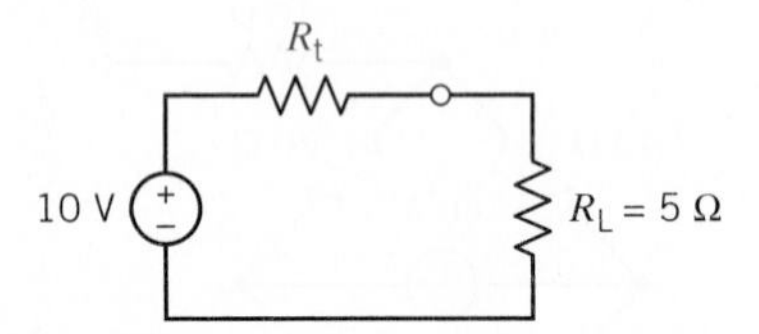

Figure P 5.6-9

P 5.6-10 A resistive circuit was connected to a variable resistor, and the power delivered to the resistor was measured as shown in Figure P 5.6-10. Determine the Thévenin equivalent circuit.

Answer: $R_t = 20\ \Omega$ and $v_{oc} = 20$ V

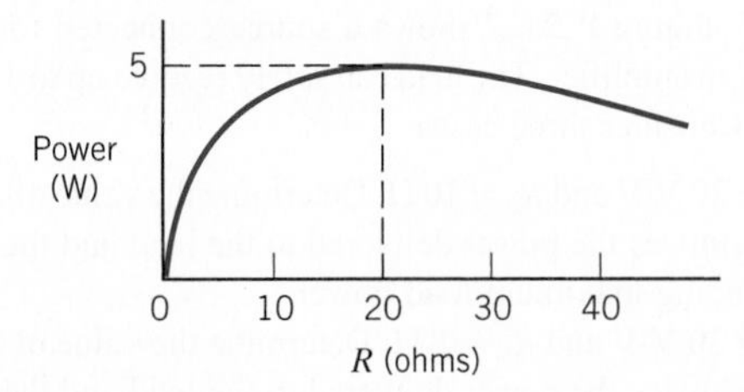

Figure P 5.6-10

Section 5.8 Using PSpice to Determine the Thévenin Equivalent Circuit

P 5.8-1 The circuit shown in Figure P 5.8-1 is separated into two parts by a pair of terminals. Call the part of the circuit to the left of the terminals circuit A and the part of the circuit to the right of the terminal circuit B. Use PSpice to do the following:

(a) Determine the node voltages for the entire circuit.
(b) Determine the Thévenin equivalent circuit of circuit A.
(c) Replace circuit A by its Thévenin equivalent and determine the node voltages of the modified circuit.
(d) Compare the node voltages of circuit B before and after replacing circuit A by its Thévenin equivalent.

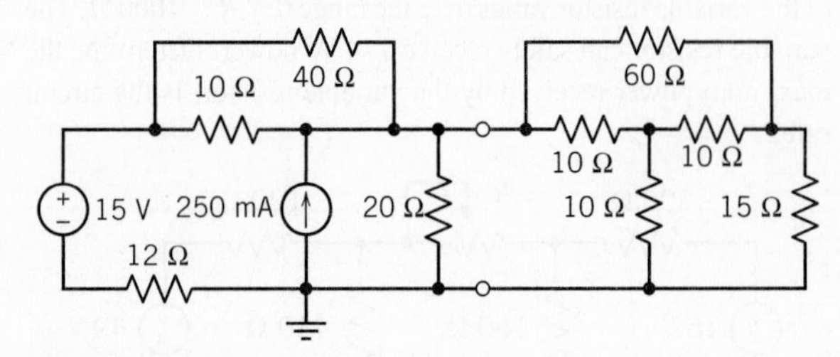

Figure P 5.8-1

Section 5.9 How Can We Check . . . ?

P 5.9-1 For the circuit of Figure P 5.9-1, the current i has been measured for three different values of R and is listed in the table. Are the data consistent?

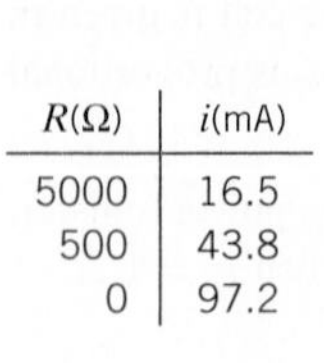

R(Ω)	i(mA)
5000	16.5
500	43.8
0	97.2

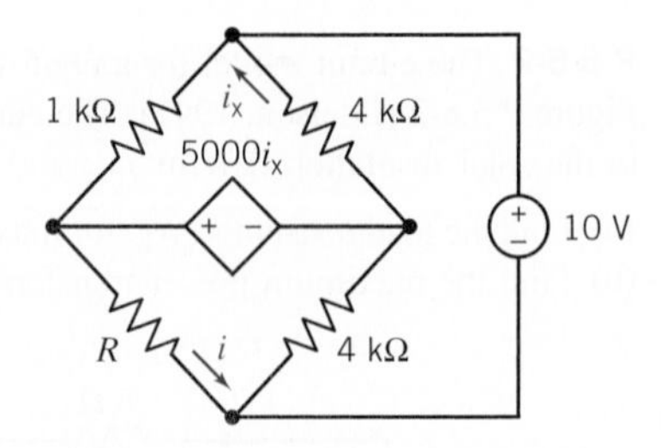

Figure P 5.9-1

P 5.9-2 Your lab partner built the circuit shown in Figure P 5.9-2 and measured the current i and voltage v corresponding to several values of the resistance R. The results are shown in the table in Figure P 5.9-2. Your lab partner says that $R_L = 8000\ \Omega$ is required to cause $i = 1$ mA. Do you agree? Justify your answer.

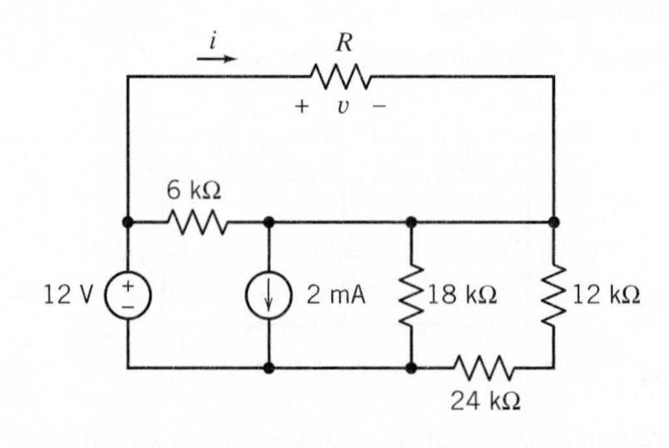

R	i	v
open	0 mA	12 V
10 kΩ	0.857 mA	8.57 V
short	3 mA	0 V

Figure P 5.9-2

P 5.9-3 In preparation for lab, your lab partner determined the Thévenin equivalent of the circuit connected to R_L in Figure P 5.9-3. She says that the Thévenin resistance is $R_t = \frac{6}{11} R$ and the open-circuit voltage is $v_{oc} = \frac{60}{11}$ V. In lab, you built the circuit using $R = 110\ \Omega$ and $R_L = 40\ \Omega$ and measured that $i = 54.5$ mA. Is this measurement consistent with the prelab calculations? Justify your answers.

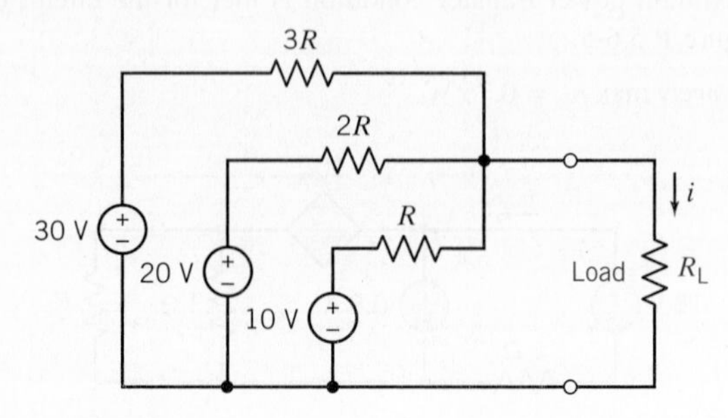

Figure P 5.9-3

P 5.9-4 Your lab partner claims that the current i in Figure P 5.9-4 will be no greater than 10 mA, regardless of the value of the resistance R. Do you agree?

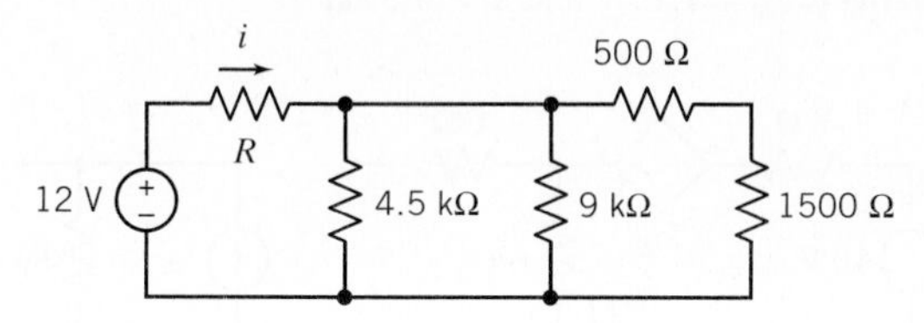

Figure P 5.9-4

P 5.9-5 Figure P 5.9-5 shows a circuit and some corresponding data. Two resistances, R_1 and R, and the current source current are unspecified. The tabulated data provide values of the current, i, and voltage, v, corresponding to several values of the resistance R.

(a) Consider replacing the part of the circuit connected to the resistor R by a Thévenin equivalent circuit. Use the data in rows 2 and 3 of the table to find the values of R_t and v_{oc}, the Thévenin resistance, and the open-circuit voltage.

(b) Use the results of part (a) to verify that the tabulated data are consistent.

(c) Fill in the blanks in the table.

(d) Determine the values of R_1 and i_s.

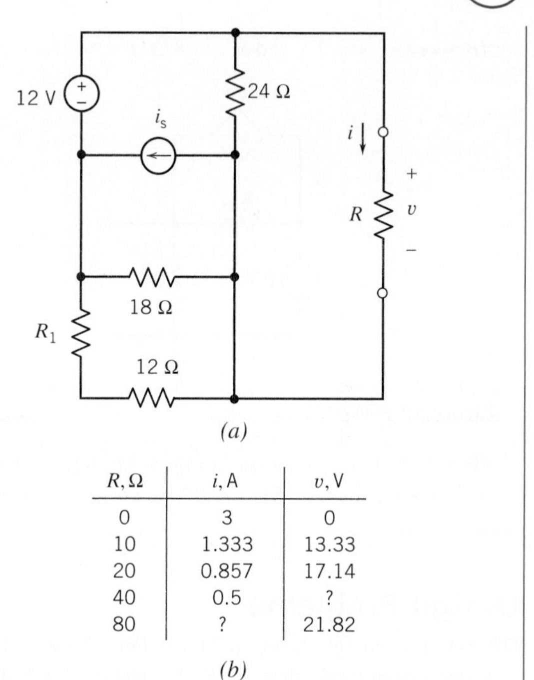

R, Ω	i, A	v, V
0	3	0
10	1.333	13.33
20	0.857	17.14
40	0.5	?
80	?	21.82

(b)

Figure P 5.9-5

PSpice Problems

SP 5-1 The circuit in Figure SP 5-1 has three inputs: v_1, v_2, and i_3. The circuit has one output, v_o. The equation

$$v_o = a\, v_1 + b\, v_2 + c\, i_3$$

expresses the output as a function of the inputs. The coefficients a, b, and c are real constants.

(a) Use PSpice and the principle of superposition to determine the values of a, b, and c.

(b) Suppose $v_1 = 10$ V and $v_2 = 8$ V, and we want the output to be $v_o = 7$ V. What is the required value of i_3?

Hint: The output is given by $v_o = a$ when $v_1 = 1$ V, $v_2 = 0$ V, and $i_3 = 0$ A.

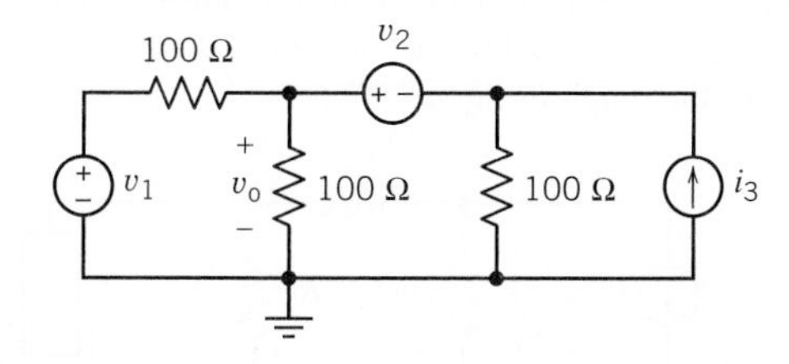

Figure SP 5-1

Answer: (a) $v_o = 0.3333 v_1 + 0.3333 v_2 + 33.33 i_3$, (b) $i_3 = 30$ mA

SP 5-2 The pair of terminals a–b partitions the circuit in Figure SP 5-2 into two parts. Denote the node voltages at nodes 1 and 2 as v_1 and v_2. Use PSpice to demonstrate that performing a source transformation on the part of the circuit to the left of the terminal does not change anything to the right of the terminals. In particular, show that the current, i_o, and the node voltages, v_1 and v_2, have the same values after the source transformation as before the source transformation.

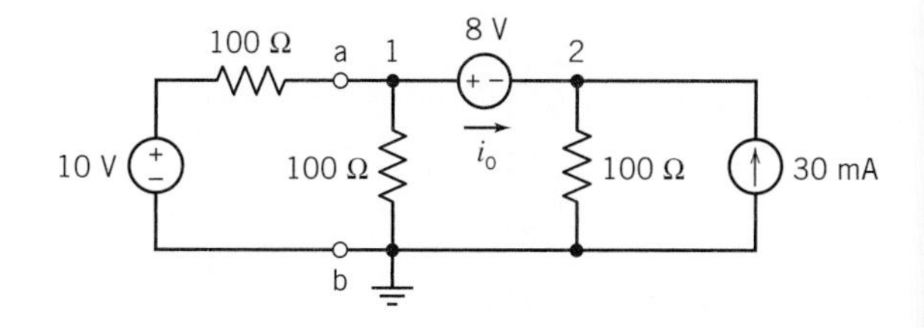

Figure SP 5-2

SP 5-3 Use PSpice to find the Thévenin equivalent circuit for the circuit shown in Figure SP 5-3.

Answer: $v_{oc} = -2$ V and $R_t = -8/3\ \Omega$

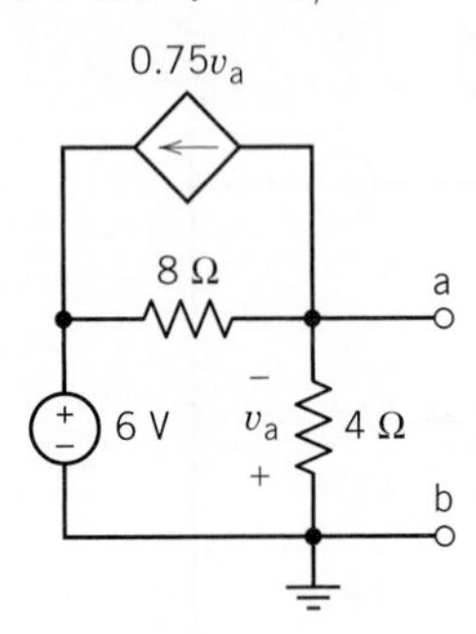

Figure SP 5-3

SP 5-4 The circuit shown in Figure SP 5-4*b* is the Norton equivalent circuit of the circuit shown in Figure SP 5-4*a*. Find the value of the short-circuit current, i_{sc}, and Thévenin resistance, R_t.

Answer: $i_{sc} = 1.13$ V and $R_t = 7.57\ \Omega$

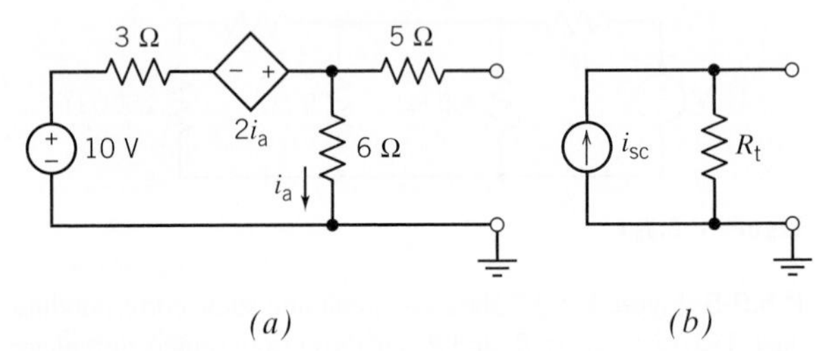

Figure SP 5-4

Design Problems

DP 5-1 The circuit shown in Figure DP 5-1*a* has four unspecified circuit parameters: v_s, R_1, R_2, and R_3. To design this circuit, we must specify the values of these four parameters. The graph shown in Figure DP 5-1*b* describes a relationship between the current i and the voltage v.

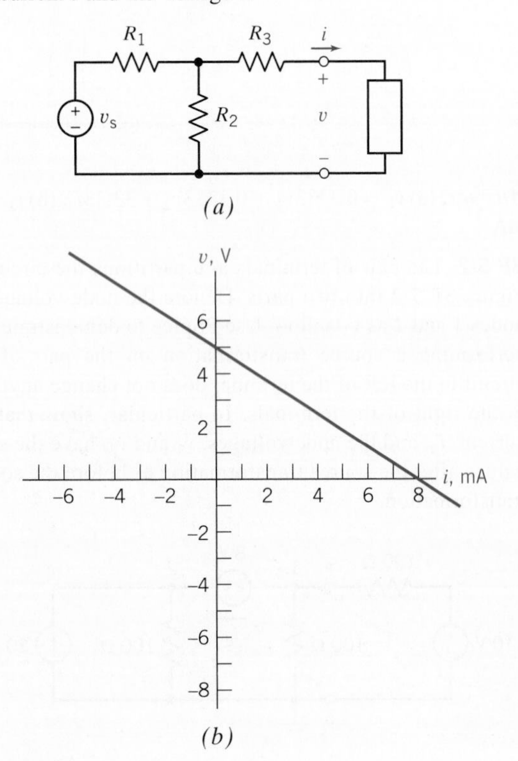

Figure DP 5-1

Specify values of v_s, R_1, R_2, and R_3 that cause the current i and the voltage v in Figure DP 5-1*a* to satisfy the relationship described by the graph in Figure DP 5-1*b*.

First Hint: The equation representing the straight line in Figure DP 5-1*b* is

$$v = -R_t i + v_{oc}$$

That is, the slope of the line is equal to -1 times the Thévenin resistance, and the v-intercept is equal to the open-circuit voltage.

Second Hint: There is more than one correct answer to this problem. Try setting $R_1 = R_2$.

DP 5-2 The circuit shown in Figure DP 5-2*a* has four unspecified circuit parameters: i_s, R_1, R_2, and R_3. To design this circuit, we must specify the values of these four parameters. The graph shown in Figure DP 5-2*b* describes a relationship between the current i and the voltage v.

Specify values of i_s, R_1, R_2, and R_3 that cause the current i and the voltage v in Figure DP 5-2*a* to satisfy the relationship described by the graph in Figure DP 5-2*b*.

First Hint: Calculate the open-circuit voltage, v_{oc}, and the Thévenin resistance, R_t, of the part of the circuit to the left of the terminals in Figure DP 5-2*a*.

Second Hint: The equation representing the straight line in Figure DP 5-2*b* is

$$v = -R_t i + v_{oc}$$

That is, the slope of the line is equal to -1 times the Thévenin resistance, and the v-intercept is equal to the open-circuit voltage.

Third Hint: There is more than one correct answer to this problem. Try setting both R_3 and $R_1 + R_2$ equal to twice the slope of the graph in Figure DP 5-2*b*.

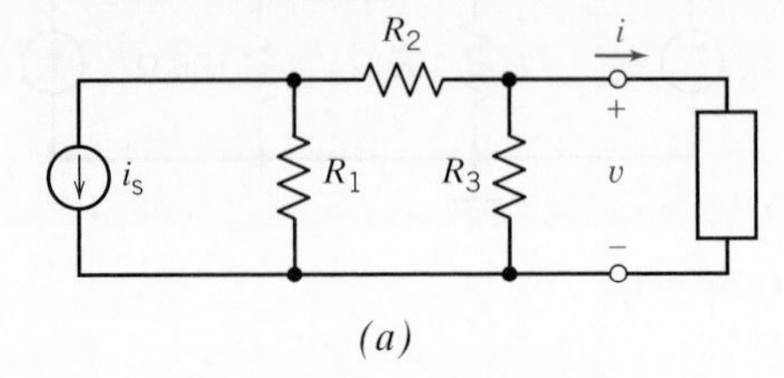

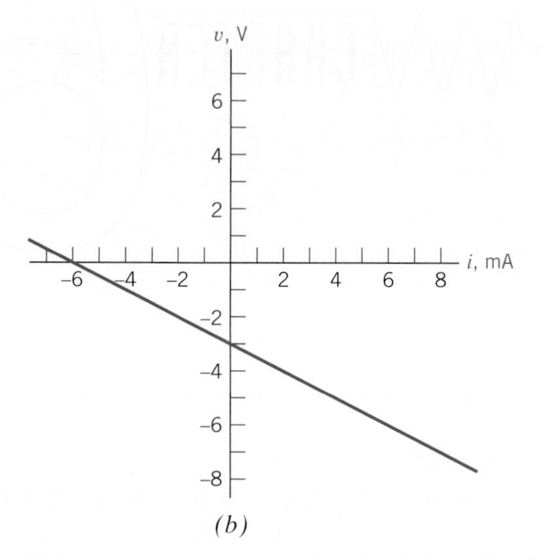

Figure DP 5-2

DP 5-3 The circuit shown in Figure DP 5-3*a* has four unspecified circuit parameters: v_s, R_1, R_2, and R_3. To design this circuit, we must specify the values of these four parameters. The graph shown in Figure DP 5-3*b* describes a relationship between the current i and the voltage v.

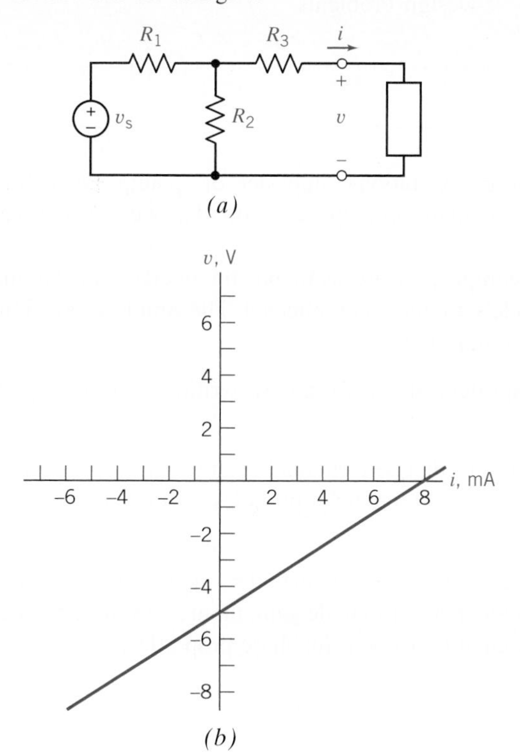

Figure DP 5-3

Is it possible to specify values of v_s, R_1, R_2, and R_3 that cause the current i and the voltage v in Figure DP 5-1*a* to satisfy the relationship described by the graph in Figure DP 5-3*b*? Justify your answer.

DP 5-4 The circuit shown in Figure DP 5-4*a* has four unspecified circuit parameters: v_s, R_1, R_2, and d, where d is the gain of the CCCS. To design this circuit, we must specify the values of these four parameters. The graph shown in Figure DP 5-4*b* describes a relationship between the current i and the voltage v.

Specify values of v_s, R_1, R_2, and d that cause the current i and the voltage v in Figure DP 5-4*a* to satisfy the relationship described by the graph in Figure DP 5-4*b*.

First Hint: The equation representing the straight line in Figure DP 5-4*b* is

$$v = -R_t i + v_{oc}$$

That is, the slope of the line is equal to -1 times the Thévenin resistance and the v-intercept is equal to the open-circuit voltage.

Second Hint: There is more than one correct answer to this problem. Try setting $R_1 = R_2$.

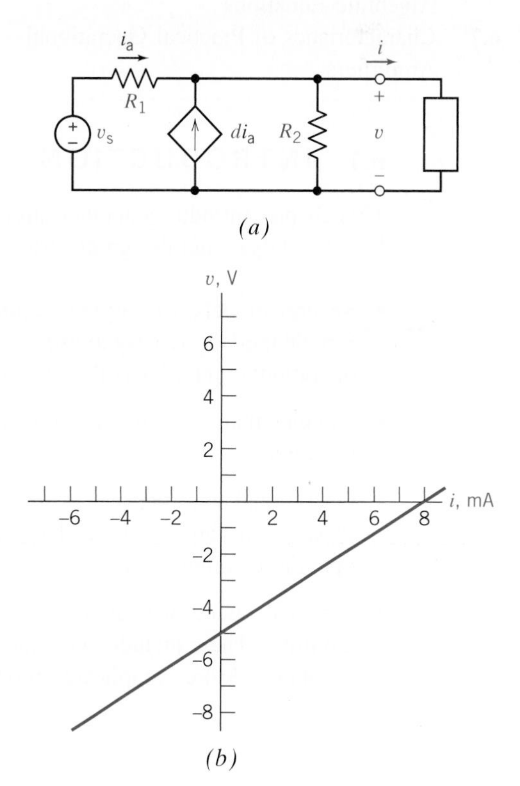

Figure DP 5-4

The Operational Amplifier

IN THIS CHAPTER

6.1 Introduction
6.2 The Operational Amplifier
6.3 The Ideal Operational Amplifier
6.4 Nodal Analysis of Circuits Containing Ideal Operational Amplifiers
6.5 Design Using Operational Amplifiers
6.6 Operational Amplifier Circuits and Linear Algebraic Equations
6.7 Characteristics of Practical Operational Amplifiers
6.8 Analysis of Op Amp Circuits Using MATLAB
6.9 Using PSpice to Analyze Op Amp Circuits
6.10 How Can We Check . . . ?
6.11 **DESIGN EXAMPLE**—Transducer Interface Circuit
6.12 Summary
Problems
PSpice Problems
Design Problems

6.1 INTRODUCTION

This chapter introduces another circuit element, the operational amplifier, or op amp. We will learn how to analyze and design electric circuits that contain op amps. In particular, we will see that:

- Several models, of varying accuracy and complexity, are available for operational amplifiers. Simple models are easy to use. Accurate models are more complicated. The simplest model of the operational amplifier is the ideal operational amplifier.
- Circuits that contain ideal operational amplifiers are analyzed by writing and solving node equations.
- Operational amplifiers can be used to build circuits that perform mathematical operations. Many of these circuits are widely used and have been named. Figure 6.5-1 provides a catalog of some useful operational amplifier circuits.
- Practical operational amplifiers have properties that are not included in the ideal operational amplifier. These include the input offset voltage, bias current, dc gain, input resistance, and output resistance. More complicated models are needed to account for these properties.

6.2 THE OPERATIONAL AMPLIFIER

The *operational amplifier* is an electronic circuit element designed to be used with other circuit elements to perform a specified signal-processing operation. The μA741 operational amplifier is shown in Figure 6.2-1a. It has eight pin connections, whose functions are indicated in Figure 6.2-1b.

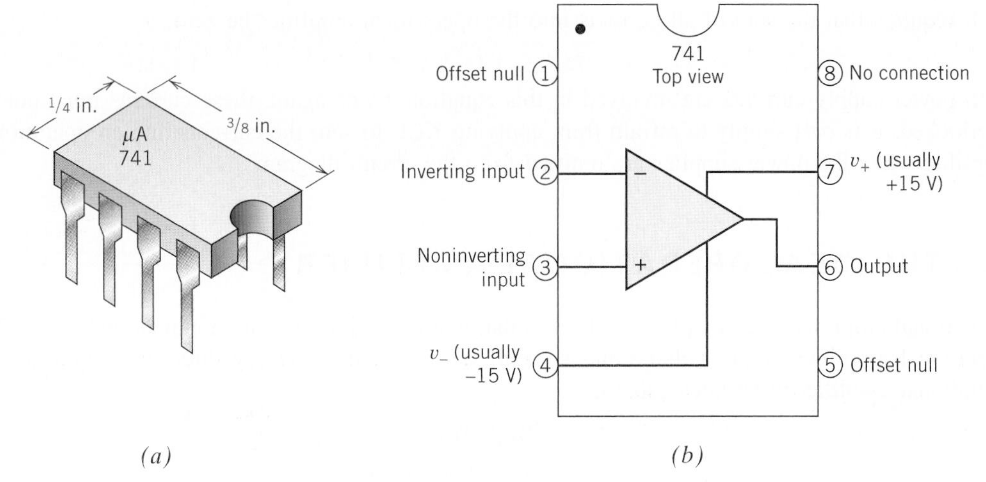

FIGURE 6.2-1 (*a*) A μA741 integrated circuit has eight connecting pins. (*b*) The correspondence between the circled pin numbers of the integrated circuit and the nodes of the operational amplifier.

The operational amplifier shown in Figure 6.2-2 has five terminals. The names of these terminals are shown in both Figure 6.2-1*b* and Figure 6.2-2. Notice the plus and minus signs in the triangular part of the symbol of the operational amplifier. The plus sign identifies the noninverting input, and the minus sign identifies the inverting input.

The power supplies are used to bias the operational amplifier. In other words, the power supplies cause certain conditions that are required for the operational amplifier to function properly. It is inconvenient to include the power supplies in drawings of operational amplifier circuits. These power supplies tend to clutter drawings of operational amplifier circuits, making them harder to read. Consequently, the power supplies are frequently omitted from drawings that accompany explanations of the function of operational amplifier circuits, such as the drawings found in textbooks. It is understood that power supplies are part of the circuit even though they are not shown. (Schematics, the drawings used to describe how to assemble a circuit, are a different matter.) The power supply voltages are shown in Figure 6.2-2, denoted as v_+ and v_-.

Because the power supplies are frequently omitted from the drawing of an operational amplifier circuit, it is easy to overlook the power supply currents. This mistake is avoided by careful application of Kirchhoff's current law (KCL). As a general rule, it is not helpful to apply KCL in a way that involves any power supply current. Two specific cases are of particular importance. First, the ground node in Figure 6.2-2 is a terminal of both power supplies. Both power supply currents would be involved if KCL were applied to the ground node. These currents must not be overlooked. It is best simply to refrain from applying KCL at the ground node of an operational amplifier circuit. Second,

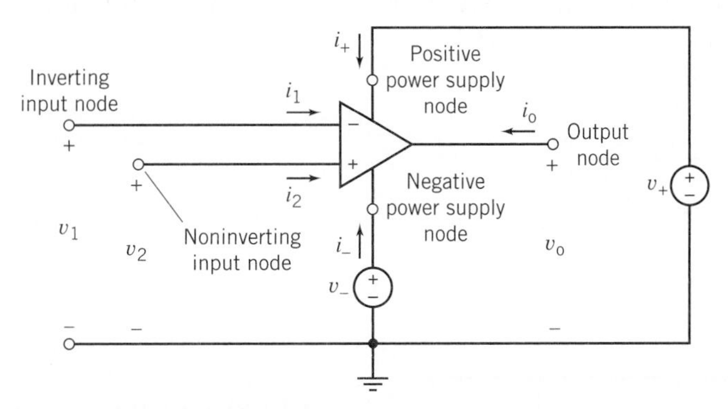

FIGURE 6.2-2 An op amp, including power supplies v_+ and v_-.

KCL requires that the sum of all currents into the operational amplifier be zero:

$$i_1 + i_2 + i_o + i_+ + i_- = 0$$

Both power supply currents are involved in this equation. Once again, these currents must not be overlooked. It is best simply to refrain from applying KCL to sum the currents into an operational amplifier when the power supplies are omitted from the circuit diagram.

6.3 THE IDEAL OPERATIONAL AMPLIFIER

Operational amplifiers are complicated devices that exhibit both linear and nonlinear behavior. The operational amplifier output voltage and current, v_o and i_o, must satisfy three conditions for an operational amplifier to be linear, that is:

$$|v_o| \le v_{sat}$$
$$|i_o| \le i_{sat}$$
$$\left|\frac{dv_o(t)}{dt}\right| \le SR \tag{6.3-1}$$

The saturation voltage, v_{sat}, the saturation current, i_{sat}, and the slew rate limit, SR, are all parameters of an operational amplifier. For example, if a μA741 operational amplifier is biased using +15-V and −15-V power supplies, then

$$v_{sat} = 14\text{ V}, \quad i_{sat} = 2\text{ mA}, \quad \text{and} \quad SR = 500{,}000\,\frac{\text{V}}{\text{s}} \tag{6.3-2}$$

These restrictions reflect the fact that operational amplifiers cannot produce arbitrarily large voltages or arbitrarily large currents or change output voltage arbitrarily quickly.

Figure 6.3-1 describes the *ideal operational amplifier*. The ideal operational amplifier is a simple model of an operational amplifier that is linear. The ideal operational amplifier is characterized by restrictions on its input currents and voltages. The currents into the input terminals of an ideal operational amplifier are zero. Consequently, in Figure 6.3-1,

$$i_1 = 0 \quad \text{and} \quad i_2 = 0$$

The node voltages at the input nodes of an ideal operational amplifier are equal. Consequently, in Figure 6.3-1,

$$v_2 = v_1$$

The ideal operational amplifier is a model of a linear operational amplifier, so the operational amplifier output current and voltage must satisfy the restrictions in Eq. 6.3-1. If they do not, then the ideal operational amplifier is not an appropriate model of the real operational amplifier. The output current and voltage depend on the circuit in which the operational amplifier is used. The ideal op amp conditions are summarized in Table 6.3-1.

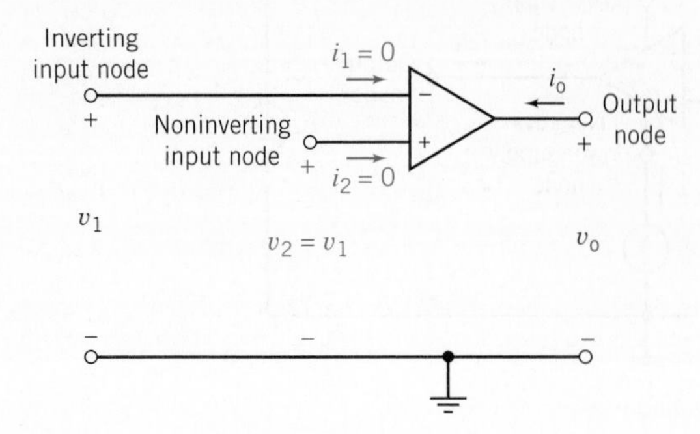

FIGURE 6.3-1 The ideal operational amplifier.

Table 6.3-1 Operating Conditions for an Ideal Operational Amplifier

VARIABLE	IDEAL CONDITION
Inverting node input current	$i_1 = 0$
Noninverting node input current	$i_2 = 0$
Voltage difference between inverting node voltage v_1 and noninverting node voltage v_2	$v_2 - v_1 = 0$

EXAMPLE 6.3-1 Ideal Operational Amplifier

Consider the circuit shown in Figure 6.3.2*a*. Suppose the operational amplifier is a μA741 operational amplifier. Model the operational amplifier as an ideal operational amplifier. Determine how the output voltage, v_o, is related to the input voltage, v_s.

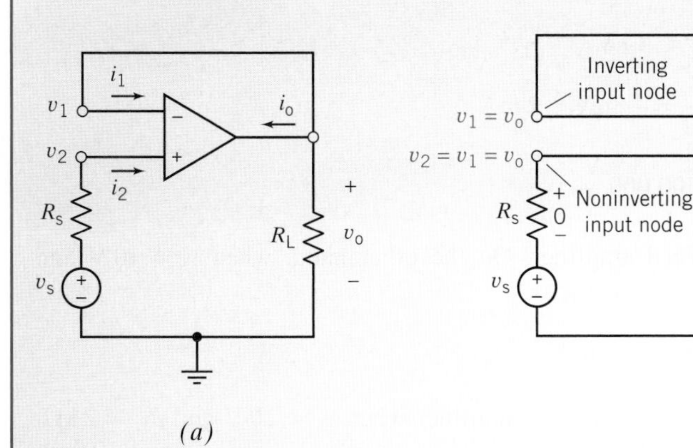

FIGURE 6.3-2 (*a*) The operational amplifier circuit for Example 6.3-1 and (*b*) an equivalent circuit showing the consequences of modeling the operational amplifier as an ideal operational amplifier. The voltages v_1, v_2, and v_o are node voltages.

Solution

Figure 6.3-2*b* shows the circuit when the operational amplifier of Figure 6.3-2*a* is modeled as an ideal operational amplifier.

1. The inverting input node and output node of the operational amplifier are connected by a short circuit, so the node voltages at these nodes are equal:

$$v_1 = v_o$$

2. The voltages at the inverting and noninverting nodes of an ideal op amp are equal:

$$v_2 = v_1 = v_o$$

3. The currents into the inverting and noninverting nodes of an operational amplifier are zero, so

$$i_1 = 0 \quad \text{and} \quad i_2 = 0$$

4. The current in resistor R_s is $i_2 = 0$, so the voltage across R_s is 0 V. The voltage across R_s is $v_s - v_2 = v_s - v_o$; hence,

$$v_s - v_o = 0$$

or

$$v_s = v_o$$

Does this solution satisfy the requirements of Eqs. 6.3-1 and 6.3-2? The output current of the operational amplifier must be calculated. Apply KCL at the output node of the operational amplifier to get

$$i_1 + i_o + \frac{v_o}{R_L} = 0$$

Because $i_1 = 0$,

$$i_o = -\frac{v_o}{R_L}$$

Now Eqs. 6.3-1 and 6.3-2 require

$$|v_s| \leq 14\text{ V}$$

$$\left|\frac{v_s}{R_L}\right| \leq 2\text{ mA}$$

$$\left|\frac{d}{dt}v_s\right| \leq 500{,}000\,\frac{\text{V}}{\text{s}}$$

For example, when $v_s = 10$ V and $R_L = 20$ kΩ, then

$$|v_s| = 10\text{ V} < 14\text{ V}$$

$$\left|\frac{v_s}{R_L}\right| = \frac{10\text{ V}}{20\text{ k}\Omega} = \frac{1}{2}\text{ mA} < 2\text{ mA}$$

$$\left|\frac{d}{dt}v_s\right| = 0 < 500{,}000\,\frac{\text{V}}{\text{s}}$$

This is consistent with the use of the ideal operational amplifier. On the other hand, when $v_s = 10$ V and $R_L = 2$ kΩ, then

$$\frac{v_s}{R_L} = 5\text{ mA} > 2\text{ mA}$$

so it is not appropriate to model the μA741 as an ideal operational amplifier when $v_s = 10$ V and $R_L = 2$ kΩ. When $v_s = 10$ V, we require $R_L > 5$ kΩ to satisfy Eq. 6.3-1.

6.4 NODAL ANALYSIS OF CIRCUITS CONTAINING IDEAL OPERATIONAL AMPLIFIERS

It is convenient to use node equations to analyze circuits containing ideal operational amplifiers.

There are three things to remember.

1. The node voltages at the input nodes of ideal operational amplifiers are equal. Thus, one of these two node voltages can be eliminated from the node equations. For example, in Figure 6.4-1, the voltages at the input nodes of the ideal operational amplifier are v_1 and v_2. Because

$$v_1 = v_2$$

v_2 can be eliminated from the node equations.

2. The currents in the input leads of an ideal operational amplifier are zero. These currents are involved in the KCL equations at the input nodes of the operational amplifier.

3. The output current of the operational amplifier is not zero. This current is involved in the KCL equations at the output node of the operational amplifier. Applying KCL at this node adds another unknown to the node equations. If the output current of the operational amplifier is not to be determined, then it is not necessary to apply KCL at the output node of the operational amplifier.

EXAMPLE 6.4-1 Difference Amplifier

The circuit shown in Figure 6.4-1 is called a difference amplifier. The operational amplifier has been modeled as an ideal operational amplifier. Use node equations to analyze this circuit and determine v_o in terms of the two source voltages, v_a and v_b.

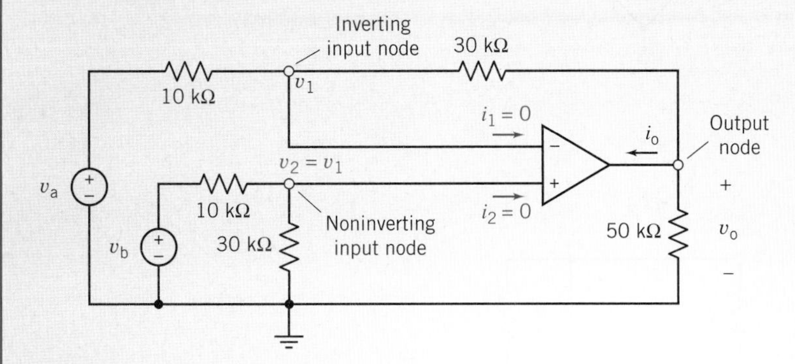

FIGURE 6.4-1 Circuit of Example 6.4-1.

Solution

The node equation at the noninverting node of the ideal operational amplifier is

$$\frac{v_2}{30{,}000} + \frac{v_2 - v_b}{10{,}000} + i_2 = 0$$

Because $v_2 = v_1$ and $i_2 = 0$, this equation becomes

$$\frac{v_1}{30{,}000} + \frac{v_1 - v_b}{10{,}000} = 0$$

Solving for v_1, we have

$$v_1 = 0.75 \cdot v_b$$

The node equation at the inverting node of the ideal operational amplifier is

$$\frac{v_1 - v_a}{10{,}000} + \frac{v_1 - v_o}{30{,}000} + i_1 = 0$$

Because $v_1 = 0.75 v_b$ and $i_1 = 0$, this equation becomes

$$\frac{0.75 \cdot v_b - v_a}{10{,}000} + \frac{0.75 \cdot v_b - v_o}{30{,}000} = 0$$

Solving for v_o, we have

$$v_o = 3(v_b - v_a)$$

The difference amplifier takes its name from the fact that the output voltage, v_o, is a function of the difference, $v_b - v_a$, of the input voltages.

EXAMPLE 6.4-2 Analysis of a Bridge Amplifier

Next, consider the circuit shown in Figure 6.4-2*a*. This circuit is called a bridge amplifier. The part of the circuit that is called a bridge is shown in Figure 6.4-2*b*. The operational amplifier and resistors, R_5 and R_6, are used to amplify the output of the bridge. The operational amplifier in Figure 6.4-2*a* has been modeled as an ideal operational amplifier. As a consequence, $v_1 = 0$ and $i_1 = 0$, as shown. Determine the output voltage, v_o, in terms of the source voltage, v_s.

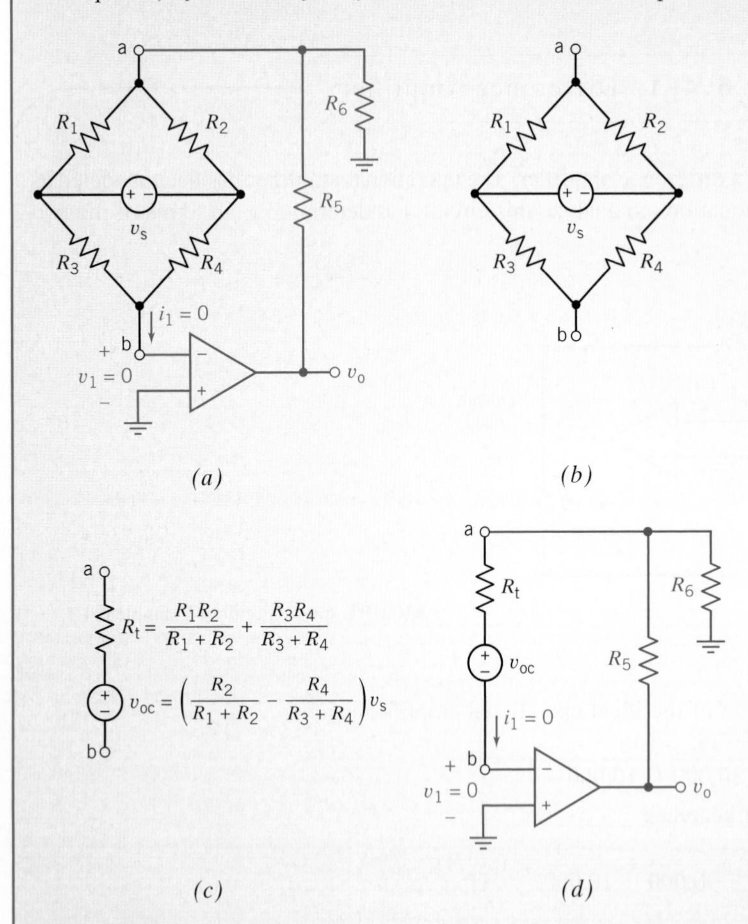

FIGURE 6.4-2 (*a*) A bridge amplifier, including the bridge circuit. (*b*) The bridge circuit and (*c*) its Thévenin equivalent circuit. (*d*) The bridge amplifier, including the Thévenin equivalent of the bridge.

Solution

Here is an opportunity to use Thévenin's theorem. Figure 6.4-2*c* shows the Thévenin equivalent of the bridge circuit. Figure 6.4-2*d* shows the bridge amplifier after the bridge has been replaced by its Thévenin equivalent. Figure 6.4-2*d* is simpler than Figure 6.4-2*a*. It is easier to write and solve the node equations representing Figure 6.4-2*d* than it is to write and solve the node equations representing Figure 6.4-2*a*. Thévenin's theorem assures us that the voltage v_o in Figure 6.4-2*d* is the same as the voltage v_o in Figure 6.4-2*a*.

Let us write node equations representing the circuit in Figure 6.4-2*d*. First, notice that the node voltage v_a is given by (using KVL)

$$v_a = v_1 + v_{oc} + R_t i_1$$

Because $v_1 = 0$ and $i_1 = 0$,

$$v_a = v_{oc}$$

Now, writing the node equation at node a

$$i_1 + \frac{v_a - v_o}{R_5} + \frac{v_a}{R_6} = 0$$

Because $v_a = v_{oc}$ and $i_1 = 0$,

$$\frac{v_{oc} - v_o}{R_5} + \frac{v_{oc}}{R_6} = 0$$

Solving for v_o, we have

$$v_o = \left(1 + \frac{R_5}{R_6}\right) v_{oc} = \left(1 + \frac{R_5}{R_6}\right)\left(\frac{R_2}{R_1 + R_2} - \frac{R_4}{R_3 + R_4}\right) v_s$$

EXAMPLE 6.4-3 Analysis of an Op Amp Circuit Using Node Equations

Consider the circuit shown in Figure 6.4-3. Find the value of the voltage measured by the voltmeter.

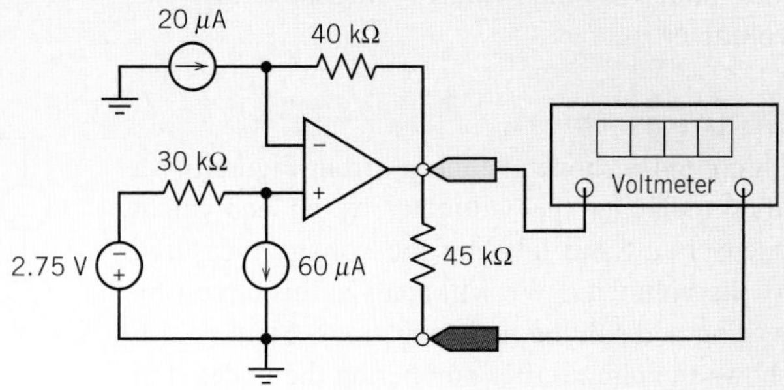

FIGURE 6.4-3 The circuit considered in Example 6.4-3.

Solution

Figure 6.4-4 shows the circuit from Figure 6.4-3 after replacing the voltmeter by an equivalent open circuit and labeling the voltage measured by the voltmeter. We will analyze this circuit by writing and solving node equations. The nodes of the circuit are numbered in Figure 6.4-4. Let v_1, v_2, v_3, and v_4 denote the node voltages at nodes 1, 2, 3, and 4, respectively.

The output of this circuit is the voltage measured by the voltmeter. The output voltage is related to the node voltages by

$$v_m = v_4 - 0 = v_4$$

The inputs to this circuit are the voltage of the voltage source and the currents of the current sources. The voltage of the voltage source is related to the node voltages at the nodes of the voltage source by

$$0 - v_3 = 2.75 \quad \Rightarrow \quad v_3 = -2.75 \text{ V}$$

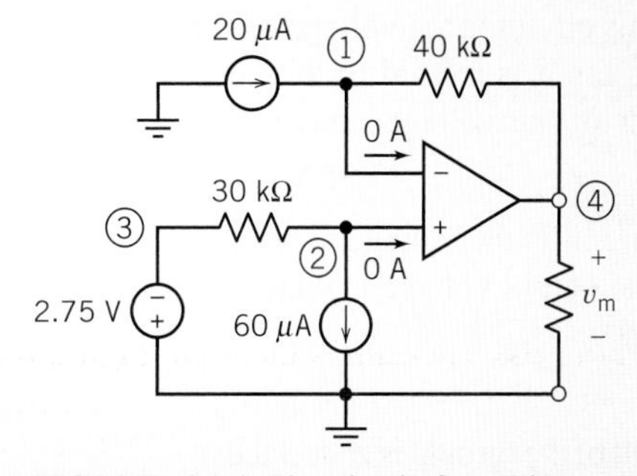

FIGURE 6.4-4 The circuit from Figure 6.4-3 after replacing the voltmeter by an open circuit and labeling the nodes. (Circled numbers are node numbers.)

Apply KCL to node 2 to get

$$\frac{v_3 - v_2}{30{,}000} = 0 + 60 \times 10^{-6} \quad \Rightarrow \quad v_3 - v_2 = 1.8 \text{ V}$$

Using $v_3 = -2.75$ V gives

$$v_2 = -4.55 \text{ V}$$

The noninverting input of the op amp is connected to node 2. The node voltage at the inverting input of an ideal op amp is equal to the node voltage at the noninverting input. The inverting input of the op amp is connected to node 1. Consequently,

$$v_1 = v_2 = -4.55 \text{ V}$$

Apply KCL to node 1 to get

$$20 \times 10^{-6} = 0 + \frac{v_1 - v_4}{40{,}000} \quad \Rightarrow \quad v_1 - v_4 = 0.8 \text{ V}$$

Using $v_m = v_4$ and $v_1 = -4.55$ V gives the value of the voltage measured by the voltmeter to be

$$v_m = -4.55 - 0.8 = -5.35 \text{ V}$$

Example 6.4-4 Analysis of an Op Amp Circuit

Consider the circuit shown in Figure 6.4-5. Find the value of the voltage measured by the voltmeter.

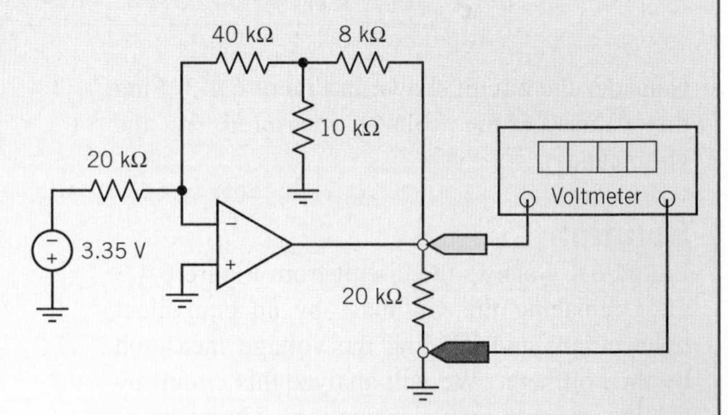

FIGURE 6.4-5 The circuit considered in Example 6.4-4.

Solution

Figure 6.4-6 shows the circuit from Figure 6.4-5 after replacing the voltmeter by an equivalent open circuit and labeling the voltage measured by the voltmeter. We will analyze this circuit by writing and solving node equations. Figure 6.4-6 shows the circuit after numbering the nodes. Let v_1, v_2, v_3, and v_4 denote the node voltages at nodes 1, 2, 3, and 4, respectively.

The input to this circuit is the voltage of the voltage source. This input is related to the node voltages at the nodes of the voltage source by

$$0 - v_1 = 3.35 \quad \Rightarrow \quad v_1 = -3.35 \text{ V}$$

The output of this circuit is the voltage measured by the voltmeter. The output voltage is related to the node voltages by

$$v_m = v_4 - 0 = v_4$$

The noninverting input of the op amp is connected to the reference node. The node voltage at the inverting input of an ideal op amp is equal to the node voltage at the noninverting input. The inverting input of the op amp is connected to node 2. Consequently,

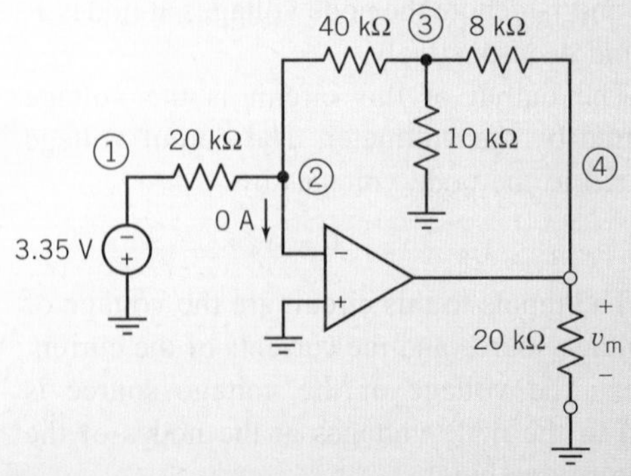

FIGURE 6.4-6 The circuit from Figure 6.4-5 after replacing the voltmeter by an open circuit and labeling the nodes. (Circled numbers are node numbers.)

$$v_2 = 0 \text{ V}$$

Apply KCL to node 2 to get

$$\frac{v_1 - v_2}{20{,}000} = 0 + \frac{v_2 - v_3}{40{,}000} \quad \Rightarrow \quad v_3 = -2v_1 + 3v_2 = -2v_1$$

Apply KCL to node 3 to get

$$\frac{v_2 - v_3}{40{,}000} = \frac{v_3}{10{,}000} + \frac{v_3 - v_4}{8000} \quad \Rightarrow \quad 5v_4 = -v_2 + 10v_3 = 10v_3$$

Combining these equations gives

$$v_4 = 2v_3 = -4v_1$$

Using $v_m = v_4$ and $v_1 = -3.35$ V gives the value of the voltage measured by the voltmeter to be

$$v_m = -4(-3.35) = 13.4 \text{ V}$$

6.5 DESIGN USING OPERATIONAL AMPLIFIERS

One of the early applications of operational amplifiers was to build circuits that performed mathematical operations. Indeed, the operational amplifier takes its name from this important application. Many of the operational amplifier circuits that perform mathematical operations are used so often that they have been given names. These names are part of an electrical engineer's vocabulary. Figure 6.5-1 shows several standard operational amplifier circuits. The next several examples show how to use Figure 6.5-1 to design simple operational amplifier circuits.

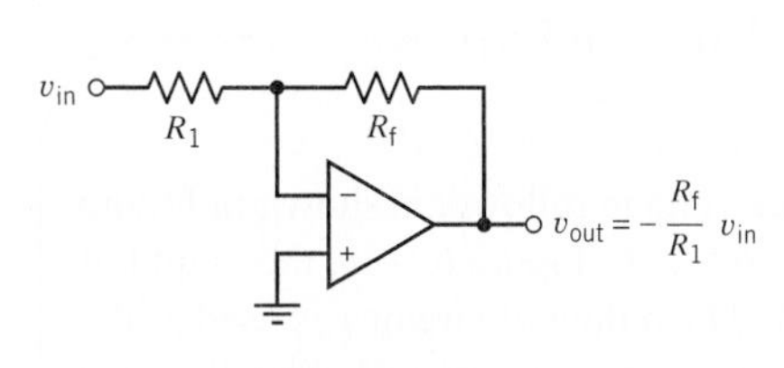

(a) Inverting amplifier

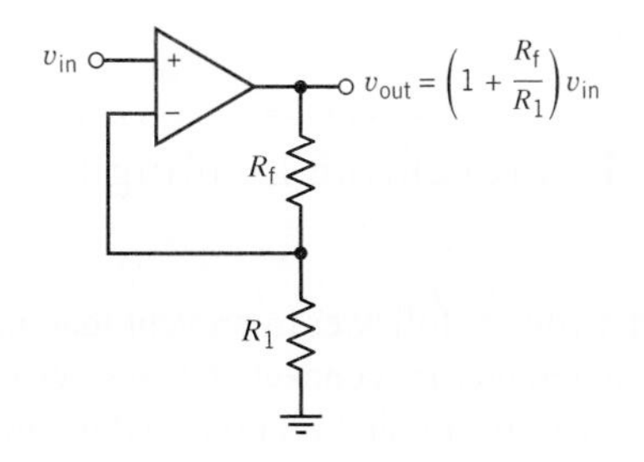

(b) Noninverting amplifier

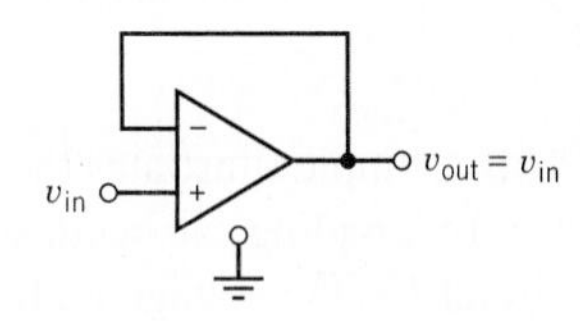

(c) Voltage follower (buffer amplifier)

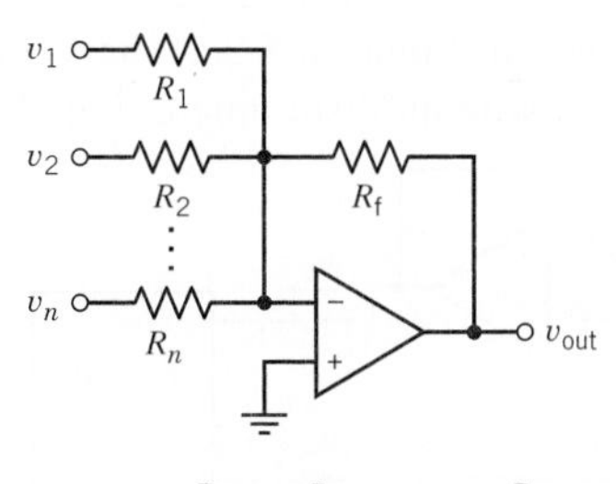

(d) Summing amplifier

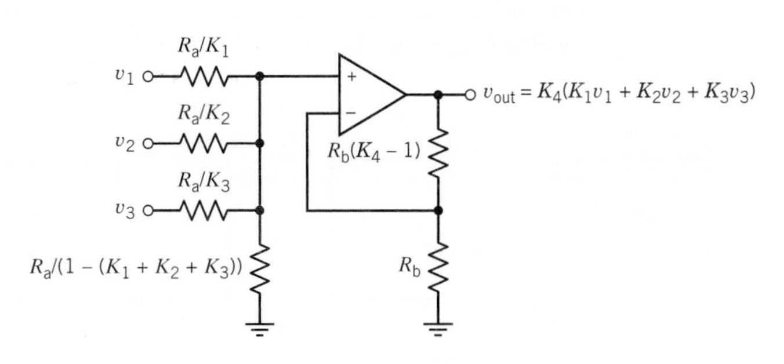

(e) Noninverting summing amplifier

FIGURE 6.5-1 A brief catalog of operational amplifier circuits. Note that all node voltages are referenced to the ground node.

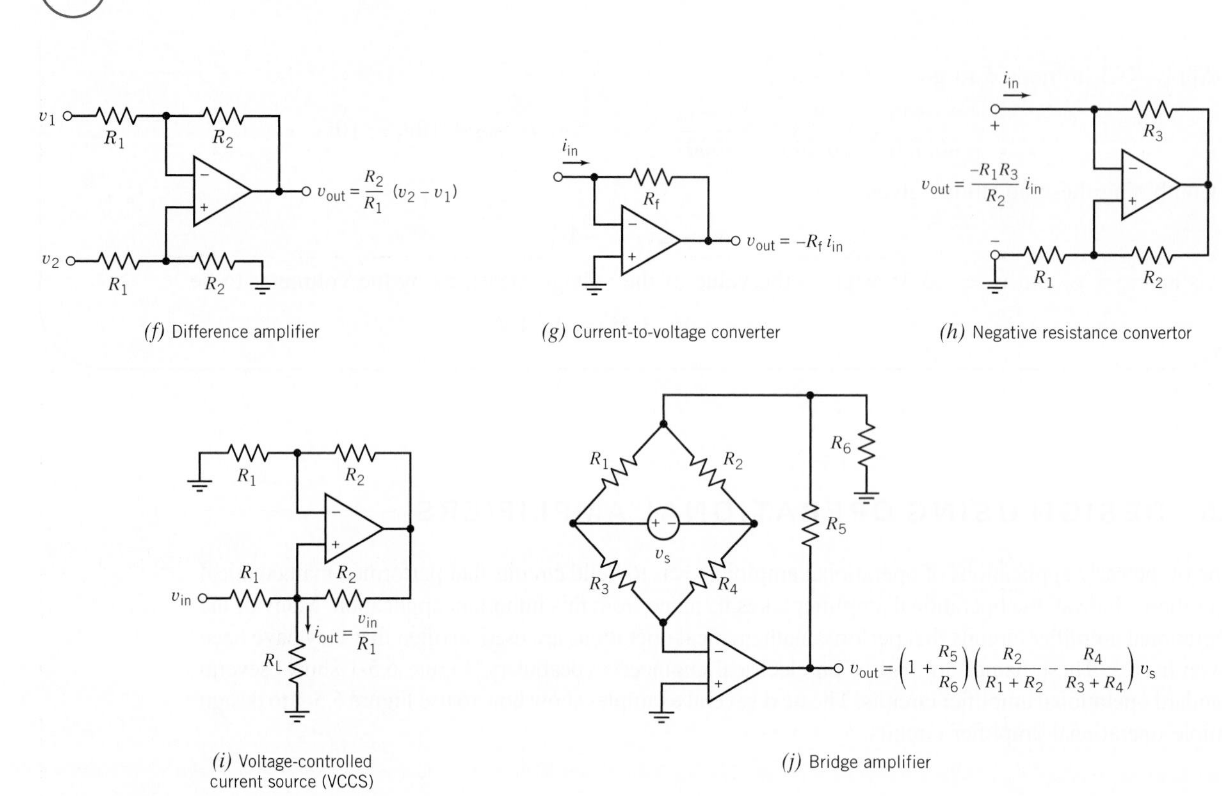

(*f*) Difference amplifier

(*g*) Current-to-voltage converter

(*h*) Negative resistance convertor

(*i*) Voltage-controlled current source (VCCS)

(*j*) Bridge amplifier

FIGURE 6.5-1 (*Continued*)

Example 6.5-1 Preventing Loading Using a Voltage Follower

This example illustrates the use of a voltage follower to prevent loading. The voltage follower is shown in Figure 6.5-1*c*. Loading can occur when two circuits are connected. Consider Figure 6.5-2. In Figure 6.5-2*a*, the output of circuit 1 is the voltage v_a. In Figure 6.5-2*b*, circuit 2 is connected to circuit 1. The output of circuit 1 is used as the input to circuit 2. Unfortunately, connecting circuit 2 to circuit 1 can change the output of circuit 1. This is called *loading*. Referring again to Figure 6.5-2, circuit 2 is said to load circuit 1 if $v_b \neq v_a$. The current i_b is called the load current. Circuit 1 is required to provide this current in Figure 6.5-2*b* but not in Figure 6.5-2*a*. This is the cause of the loading. The load current can be eliminated using a voltage follower as shown in Figure 6.5-2*c*. The voltage follower copies voltage v_a from the output of circuit 1 to the input of circuit 2 without disturbing circuit 1.

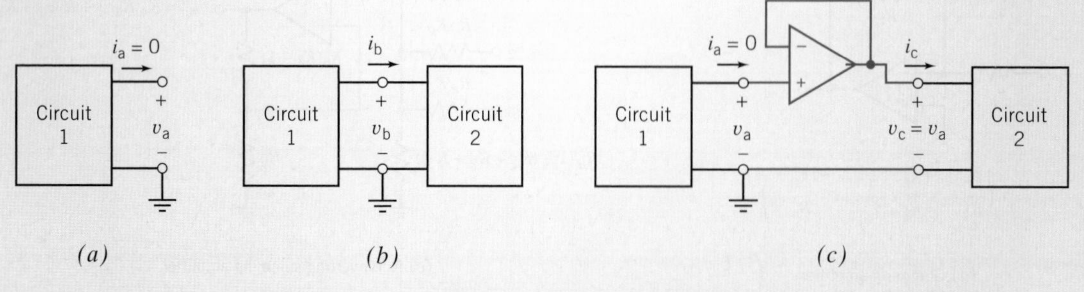

FIGURE 6.5-2 Circuit 1 (*a*) before and (*b*) after circuit 2 is connected. (*c*) Preventing loading, using a voltage follower.

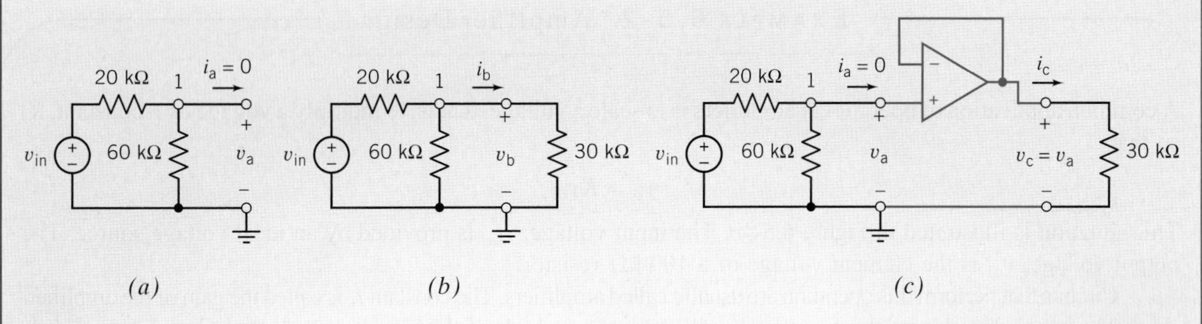

FIGURE 6.5-3 A voltage divider (*a*) before and (*b*) after a 30-kΩ resistor is added. (*c*) A voltage follower is added to prevent loading.

Solution

As a specific example, consider Figure 6.5-3. The voltage divider shown in Figure 6.5-3*a* can be analyzed by writing a node equation at node 1:

$$\frac{v_a - v_{in}}{20{,}000} + \frac{v_a}{60{,}000} = 0$$

Solving for v_a, we have

$$v_a = \frac{3}{4} v_{in}$$

In Figure 6.5-3*b*, a resistor is connected across the output of the voltage divider. This circuit can be analyzed by writing a node equation at node 1:

$$\frac{v_b - v_{in}}{20{,}000} + \frac{v_b}{60{,}000} + \frac{v_b}{30{,}000} = 0$$

Solving for v_b, we have

$$v_b = \frac{1}{2} v_{in}$$

Because $v_b \neq v_a$, connecting the resistor directly to the voltage divider loads the voltage divider. This loading is caused by the current required by the 30-kΩ resistor. Without the voltage follower, the voltage divider must provide this current.

In Figure 6.5-3*c*, a voltage follower is used to connect the 30-kΩ resistor to the output of the voltage divider. Once again, the circuit can be analyzed by writing a node equation at node 1:

$$\frac{v_c - v_{in}}{20{,}000} + \frac{v_c}{60{,}000} = 0$$

Solving for v_c, we have

$$v_c = \frac{3}{4} v_{in}$$

Because $v_c = v_a$, loading is avoided when the voltage follower is used to connect the resistor to the voltage divider. The voltage follower, not the voltage divider, provides the current required by the 30-kΩ resistor.

EXAMPLE 6.5-2 Amplifier Design

A common application of operational amplifiers is to scale a voltage, that is, to multiply a voltage by a constant, K, so that

$$v_o = K v_{in}$$

This situation is illustrated in Figure 6.5-4*a*. The input voltage, v_{in}, is provided by an ideal voltage source. The output voltage, v_o, is the element voltage of a 100-kΩ resistor.

Circuits that perform this operation are usually called amplifiers. The constant K is called the gain of the amplifier.

The required value of the constant K will determine which of the circuits is selected from Figure 6.5-1. There are four cases to consider: $K < 0$, $K > 1$, $K = 1$, and $0 < K < 1$.

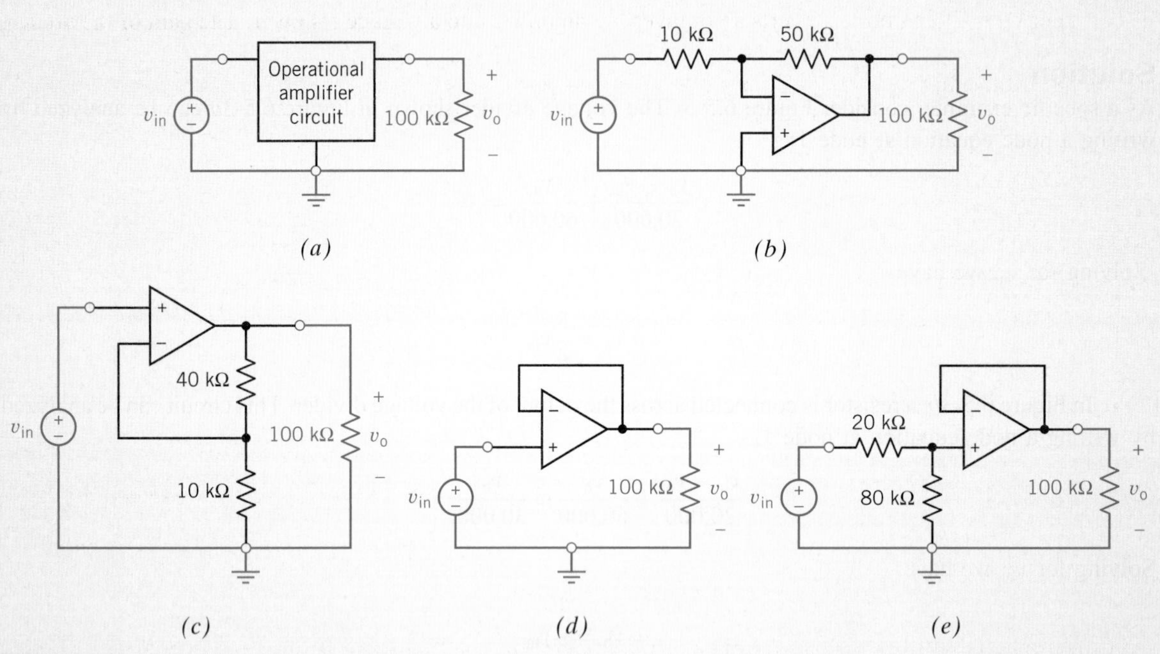

FIGURE 6.5-4 (*a*) An amplifier is required to make $v_o = K v_{in}$. The choice of amplifier circuit depends on the value of the gain K. Four cases are shown: (*b*) $K = -5$, (*c*) $K = 5$, (*d*) $K = 1$, and (*e*) $K = 0.8$.

Solution

Because resistor values are positive, the gain of the inverting amplifier, shown in Figure 6.5-1*a*, is negative. Accordingly, when $K < 0$ is required, an inverting amplifier is used. For example, suppose we require $K = -5$. From Figure 6.5-1*a*,

$$-5 = -\frac{R_f}{R_1}$$

so

$$R_f = 5R_1$$

As a rule of thumb, it is a good idea to choose resistors in operational amplifier circuits that have values between 5 kΩ and 500 kΩ when possible. Choosing

$$R_1 = 10\ \text{k}\Omega$$

gives

$$R_f = 50\ \text{k}\Omega$$

The resulting circuit is shown in Figure 6.5-4*b*.

Next, suppose we require $K = 5$. The noninverting amplifier, shown in Figure 6.5-1*b*, is used to obtain gains greater than 1. From Figure 6.5-1*b*

$$5 = 1 + \frac{R_f}{R_1}$$

so

$$R_f = 4R_1$$

Choosing $R_1 = 10\,\text{k}\Omega$ gives $R_f = 40\,\text{k}\Omega$. The resulting circuit is shown in Figure 6.5-4*c*.

Consider using the noninverting amplifier of Figure 6.5-1*b* to obtain a gain $K = 1$. From Figure 6.5-1*b*,

$$1 = 1 + \frac{R_f}{R_1}$$

so

$$\frac{R_f}{R_1} = 0$$

This can be accomplished by replacing R_f by a short circuit ($R_f = 0$) or by replacing R_1 by an open circuit ($R_1 = \infty$) or both. Doing both converts a noninverting amplifier into a voltage follower. The gain of the voltage follower is 1. In Figure 6.5-4*d*, a voltage follower is used for the case $K = 1$.

There is no amplifier in Figure 6.5-1 that has a gain between 0 and 1. Such a circuit can be obtained using a voltage divider together with a voltage follower. Suppose we require $K = 0.8$. First, design a voltage divider to have an attenuation equal to K:

$$0.8 = \frac{R_2}{R_1 + R_2}$$

so

$$R_2 = 4 \cdot R_1$$

Choosing $R_1 = 20\,\text{k}\Omega$ gives $R_2 = 80\,\text{k}\Omega$. Adding a voltage follower gives the circuit shown in Figure 6.5-4*e*.

EXAMPLE 6.5-3 Designing a Noninverting Summing Amplifier

Design a circuit having one output, v_o, and three inputs, v_1, v_2, and v_3. The output must be related to the inputs by

$$v_o = 2v_1 + 3v_2 + 4v_3$$

In addition, the inputs are restricted to having values between -1 V and 1 V, that is,

$$|v_i| \leq 1\,\text{V} \quad i = 1, 2, 3$$

Consider using an operational amplifier having $i_{sat} = 2$ mA and $v_{sat} = 15$ V and design the circuit.

Solution

The required circuit must multiply each input by a separate positive number and add the results. The noninverting summer shown in Figure 6.5-1*e* can do these operations. This circuit is represented by six parameters: K_1, K_2, K_3, K_4, R_a, and R_b. Designing the noninverting summer amounts to choosing values for these six parameters. Notice that $K_1 + K_2 + K_3 < 1$ is required to ensure that all of the resistors have positive values. Pick $K_4 = 10$ (a convenient value that is just a little larger than $2 + 3 + 4 = 9$). Then,

$$v_o = 2v_1 + 3v_2 + 4v_3 = 10(0.2v_1 + 0.3v_2 + 0.4v_3)$$

That is, $K_4 = 10$, $K_1 = 0.2$, $K_2 = 0.3$, and $K_3 = 0.4$. Figure 6.5-1*e* does not provide much guidance in picking values of R_a and R_b. Try $R_a = R_b = 100\ \Omega$. Then,

$$(K_4 - 1)R_b = (10 - 1)100 = 900\,\Omega$$

Figure 6.5-5 shows the resulting circuit. It is necessary to check this circuit to ensure that it satisfies the specifications. Writing node equations

$$\frac{v_a - v_1}{500} + \frac{v_a - v_2}{333} + \frac{v_a - v_3}{250} + \frac{v_a}{1000} = 0$$

$$-\frac{v_o - v_a}{900} + \frac{v_a}{100} = 0$$

and solving these equations yield

$$v_o = 2v_1 + 3v_2 + 4v_3 \quad \text{and} \quad v_a = \frac{v_o}{10}$$

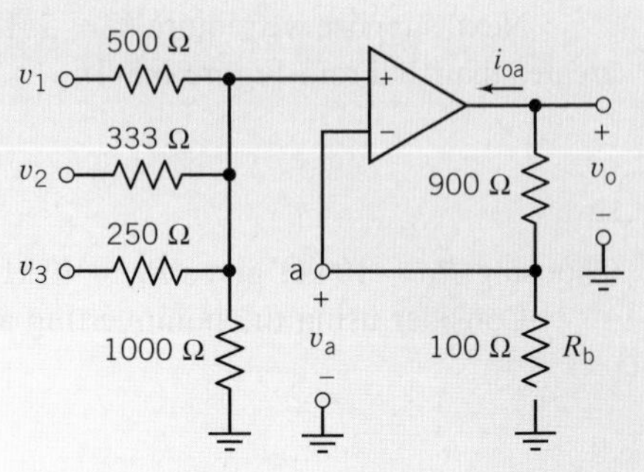

FIGURE 6.5-5 The proposed noninverting summing amplifier.

The output current of the operational amplifier is given by

$$i_{oa} = \frac{v_a - v_o}{900} = -\frac{v_o}{1000} \tag{6.5-1}$$

How large can the output voltage be? We know that

$$|v_o| = |2v_1 + 3v_2 + 4v_3|$$

so

$$|v_o| \leq 2|v_1| + 3|v_2| + 4|v_3| \leq 9\text{ V}$$

The operational amplifier output voltage will always be less than v_{sat}. That's good. Now what about the output current? Notice that $|v_o| \leq 9$ V. From Eq. 6.5-1,

$$|i_{oa}| = \left|\frac{-v_o}{1000\ \Omega}\right| \leq \left|\frac{-9\text{ V}}{1000\ \Omega}\right| = 9\text{ mA}$$

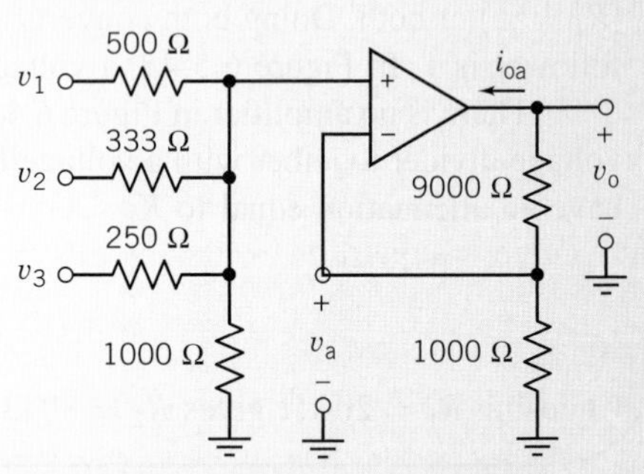

FIGURE 6.5-6 The final design of the noninverting summing amplifier.

The operational amplifier output current exceeds $i_{sat} = 2$ mA. This is not allowed. Increasing R_b will reduce i_o. Try $R_b = 1000\ \Omega$. Then,

$$(K_4 - 1)R_b = (10 - 1)1000 = 9000\ \Omega$$

This produces the circuit shown in Figure 6.5-6. Increasing R_a and R_b does not change the operational amplifier output voltage. As before,

$$v_o = 2v_1 + 3v_2 + 4v_3$$

and

$$|v_o| \leq 2|v_1| + 3|v_2| + 4|v_3| \leq 9\text{ V}$$

Increasing R_b does reduce the operational amplifier output current. Now,

$$|i_{oa}| \leq \left|\frac{-9\text{ V}}{10{,}000\ \Omega}\right| = 0.9\text{ mA}$$

so $|i_{oa}| < 2$ mA and $|v_o| < 15$ V, as required.

6.6 OPERATIONAL AMPLIFIER CIRCUITS AND LINEAR ALGEBRAIC EQUATIONS

This section describes a procedure for designing operational amplifier circuits to implement linear algebraic equations. Some of the node voltages of the operational amplifier circuit will represent the variables in the algebraic equation. For example, the equation

$$z = 4x - 5y + 2 \tag{6.6-1}$$

will be represented by an operational amplifier circuit that has node voltages v_x, v_y, and v_z that are related by the equation

$$v_z = 4v_x - 5v_y + 2 \tag{6.6-2}$$

A voltage or current that is used to represent something is called a signal.

That 'something' could be a temperature or a position or a force or something else. In this case, v_x, v_y, and v_z are signals representing the variables x, y, and z.

Equation 6.6-1 shows how the value of z can be obtained from values of x and y. Similarly, Eq. 6.6-2 shows how the value of v_z can be obtained from values of v_x and v_y. The operational amplifier circuit will have one output, v_z, and two inputs, v_x and v_y.

The design procedure has two steps. First, we represent the equation by a diagram called a block diagram. Second, we implement each block of the block diagram as an operational amplifier circuit.

We will start with the algebraic equation. Equation 6.6-1 indicates that the value of variable z can be calculated from the values of the variables x and y using the operations of addition, subtraction, and multiplication by a constant multiplier. Equation 6.6-1 can be rewritten as

$$z = 4x + (-5)y + 2 \tag{6.6-3}$$

Equation 6.6-3 indicates that z can be obtained from x and y using only addition and multiplication, though one of the multipliers is now negative.

Figure 6.6-1 shows symbolic representations of the operations of addition and multiplication by a constant. In Figure 6.6-1*a*, the operation of multiplication by a constant multiplier is represented by a rectangle together with two arrows, one pointing toward and one pointing away from the rectangle. The arrow pointing toward the rectangle is labeled by a variable representing the input to the operation, that is, the variable that is to be multiplied by the constant. Similarly, the arrow pointing away from the rectangle is labeled by a variable representing the output, or result, of the operation. The rectangle itself is labeled with the value of the multiplier. The symbol shown in Figure 6.6-1*b* represents the operation of addition. The rectangle is labeled with a plus sign. The arrows that point toward the rectangle are labeled by the variables that are to be added. There are as many of these arrows as there are variables to be added. One arrow points away from the rectangle. This arrow is labeled by the variable representing the sum.

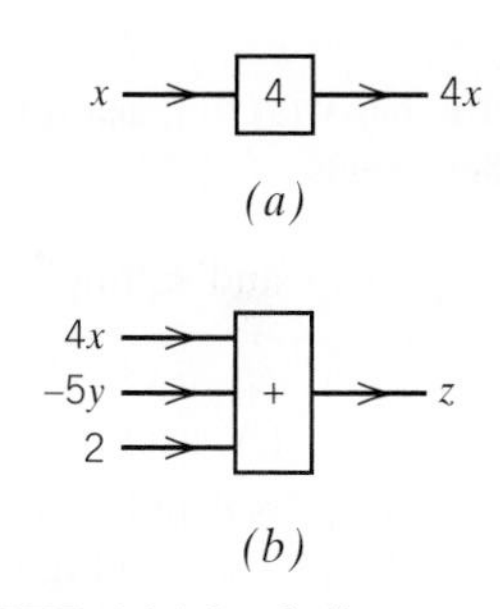

FIGURE 6.6-1 Symbolic representations of (*a*) multiplication by a constant and (*b*) addition.

The rectangles that represent addition and multiplication by a constant are called blocks. A diagram composed of such blocks is called a block diagram. Figure 6.6-2 represents Eq. 6.6-3 as a block diagram. Each block in the block diagram corresponds to an operation in the equation. Notice, in particular, that the product $4x$ has two roles in Eq. 6.6-3. The product $4x$ is both the output of one operation, multiplying x by the constant 4, and one of the inputs to another operation, adding $4x$ to $-5y$ and 2 to obtain z. This observation is used to construct the block diagram. The product $4x$ is the output of one block and the input to another. Indeed, this observation explains why the output of the block that multiplies x by 4 is connected to an input of the block that adds $4x$ to $-5y$ and 2.

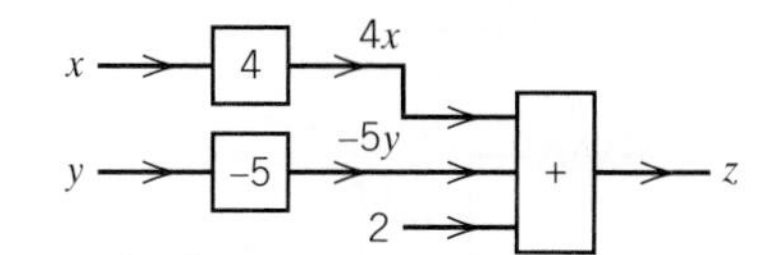

FIGURE 6.6-2 A block diagram representing Eq. 6.6-3.

Next, consider designing an operational amplifier circuit to implement the block diagram in Figure 6.6-2. The blocks representing multiplication by a constant multiplier can be implemented using either inverting or noninverting amplifiers, depending on the sign of the multiplier. To do so, design the amplifier to have a gain that is equal to the multiplier of the corresponding block. (Noninverting amplifiers can be used when the constant is both positive and greater than 1. Example 6.5-2 shows that a circuit consisting of a voltage divider and voltage follower can be used when the constant is positive and less than 1.)

Figures 6.6-3*b*,*d*,*f* show operational amplifier circuits that implement the blocks shown in Figures 6.6-3*a*,*c*,*e*, respectively. The block in Figure 6.6-3*a* requires multiplication by a positive constant, 4. Figure 6.6-3*b* shows the corresponding operational amplifier circuit, a noninverting amplifier having a gain equal to 4. This noninverting amplifier is designed by referring to Figure 6.5-1*b*

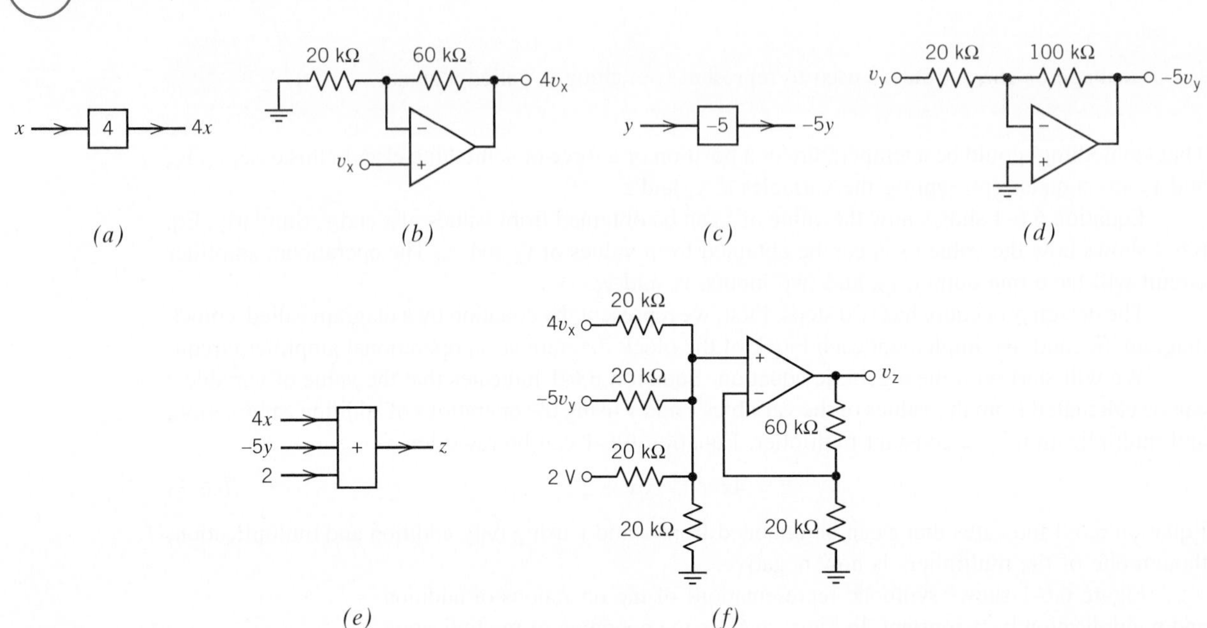

FIGURE 6.6-3 (*a*), (*c*), and (*e*) show the blocks from Figure 6.6-2, whereas (*b*), (*d*), and (*f*) show the corresponding operational amplifier circuits.

and setting

$$R_1 = 20\ \text{k}\Omega \quad \text{and} \quad R_f = 3R_1 = 60\ \text{k}\Omega$$

(A useful rule of thumb suggests selecting resistors for operational amplifier circuits to have resistances in the range 5 kΩ to 500 kΩ.)

In Figure 6.6-3*b*, the notation $v_x = x$ indicates that v_x is a voltage that represents x. A voltage or current that is used to represent something else is called a signal, so v_x is the signal representing x.

The block in Figure 6.6-3*c* requires multiplication by a negative constant, −5. Figure 6.6-3*d* shows the corresponding operational amplifier circuit, an inverting amplifier having a gain equal to −5. Design this inverting amplifier by referring to Figure 6.5-1*a* and setting

$$R_1 = 20\ \text{k}\Omega \quad \text{and} \quad R_f = 5\,R_1 = 100\ \text{k}\Omega$$

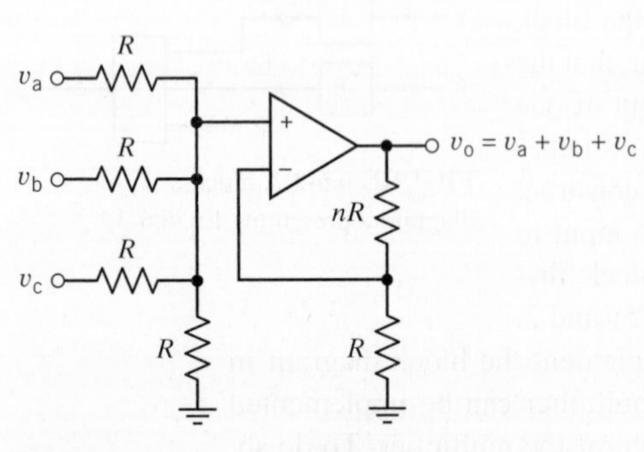

FIGURE 6.6-4 The noninverting summer. The integer n indicates the number of inputs to the circuit.

The block in Figure 6.6-3*e* requires adding three terms. Figure 6.6-3*f* shows the corresponding operational amplifier circuit, a noninverting summer. Design the noninverting summer by referring to Figure 6.6-4 and setting

$$R_1 = 20\ \text{k}\Omega, \quad n = 3, \quad \text{and} \quad nR = 3(20{,}000) = 60\ \text{k}\Omega$$

(The noninverting summer is a special case of the noninverting-summing amplifier shown in Figure 6.5-1*e*. Take $K_1 = K_2 = K_3 = 1/(n+1)$, $K_4 = n$, $R_b = R$, and $R_a = R/(n+1)$ in Figure 6.5-1*e* to get the circuit shown in Figure 6.6-4.)

Figure 6.6-5 shows the circuit obtained by replacing each block in Figure 6.6-2 by the corresponding operational amplifier circuit from Figure 6.6-3. The circuit in Figure 6.6-5 does indeed implement Eq. 6.6-3, but it's possible to improve this circuit.

The constant input to the summer has been implemented using a 2-V voltage source. Although correct, this may be more expensive than necessary. Voltage sources are relatively expensive devices, considerably more expensive than resistors or operational amplifiers. We can reduce the cost of this circuit by using a voltage source we already have instead of getting a new one. Recall that we need power supplies to bias the operational amplifier. Suppose that ±15-V voltage sources

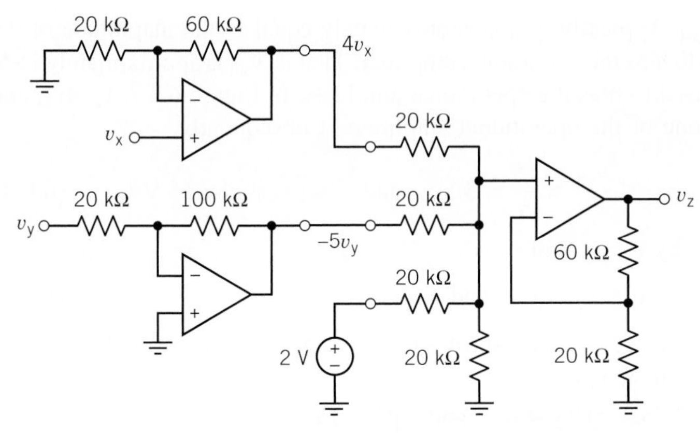

FIGURE 6.6-5 An operational amplifier circuit that implements Eq. 6.6-2.

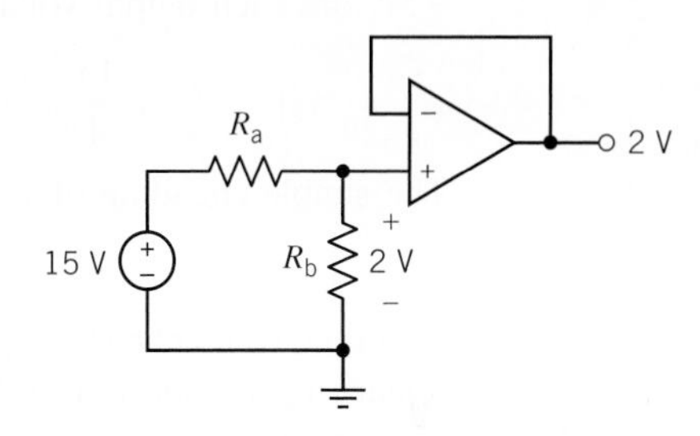

FIGURE 6.6-6 Using the operational amplifier power supply to obtain a 2-V signal.

are used to bias the operational amplifier. We can reduce costs by using the ±15-V voltage source together with a voltage divider and a voltage follower to obtain the 2-V input for the summer. Figure 6.6-6 illustrates the situation. The voltage divider produces a constant voltage equal to 2 V. The voltage follower prevents loading (see Example 6.5-1).

Applying the voltage division rule in Figure 6.6-6 requires that

$$\frac{R_b}{R_a + R_b} = \frac{2}{15} = 0.133 \quad \Rightarrow \quad R_a = 6.5\, R_b$$

The solution to this equation is not unique. One solution is $R_a = 130$ kΩ and $R_b = 20$ kΩ. Figure 6.6-7 shows the improved operational amplifier circuit. We can verify, perhaps by writing node equations, that

$$v_z = 4v_x - 5v_y + 2$$

Voltage saturation of the operational amplifiers should be considered when defining the relationship between the signals v_x, v_y, and v_z and the variables x, y, and z. The output voltage of an operational

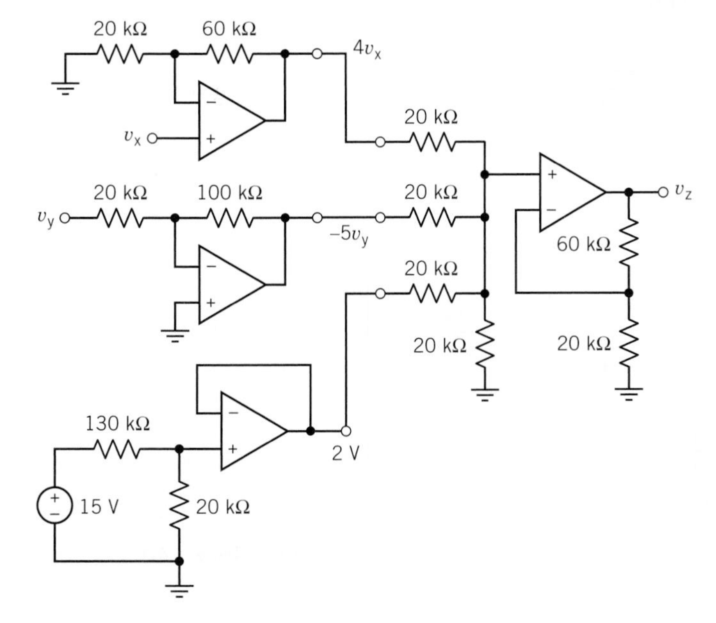

FIGURE 6.6-7 An improved operational amplifier circuit that implements Eq. 6.6-2.

amplifier is restricted by $|v_o| \le v_{sat}$. Typically, v_{sat} is approximately equal to the magnitude of the voltages of the power supplies used to bias the operational amplifier. That is, v_{sat} is approximately 15 V when ±15-V voltage sources are used to bias the operational amplifier. In Figure 6.6.7, v_z, $4v_x$, and $-5v_y$ are each output voltages of one of the operational amplifiers. Consequently,

$$|v_x| \le \frac{v_{sat}}{4} \approx \frac{15}{4} = 3.75 \text{ V}, \quad |v_y| \le \frac{v_{sat}}{5} \approx \frac{15}{5} = 3 \text{ V}, \quad \text{and} \quad |v_z| \le v_{sat} \approx 15 \text{ V} \tag{6.6-4}$$

The simple encoding of x, y, and z by v_x, v_y, and v_z is

$$v_x = x, \quad v_y = y, \quad \text{and} \quad v_z = z \tag{6.6-5}$$

This is convenient because, for example, $v_z = 4.5$ V indicates that $z = 4.5$. However, using Eq. 6.6-3 to replace v_x, v_y, and v_z in Eq. 6.6-4 with x, y, and z gives

$$|x| \le 3.75, \quad |y| \le 3.0, \quad \text{and} \quad |z| \le 15$$

Should these conditions be too restrictive, consider defining the relationship between the signals v_x, v_y, and v_z and the variables, x, y, and z differently. For example, suppose

$$v_x = \frac{x}{10}, \quad v_y = \frac{y}{10}, \quad \text{and} \quad v_z = \frac{z}{10} \tag{6.6-6}$$

Now we need to multiply the value of v_z by 10 to get the value of z. For example, $v_z = 4.5$ V indicates that $z = 45$. On the other hand, the circuit can accommodate larger values of x, y, and z. Equations 6.6-4 and 6.6-6 imply that

$$|x| \le 37.5, \quad |y| \le 30.0, \quad \text{and} \quad |z| \le 150.0$$

EXERCISE 6.6-1 Specify the values of R_1 and R_2 in Figure E 6.6-1 that are required to cause v_3 to be related to v_1 and v_2 by the equation $v_3 = (4)v_1 - \left(\frac{1}{5}\right)v_2$.

Answer: $R_1 = 10\text{ k}\Omega$ and $R_2 = 2.5\text{ k}\Omega$

EXERCISE 6.6-2 Specify the values of R_1 and R_2 in Figure E 6.6-1 that are required to cause v_3 to be related to v_1 and v_2 by the equation $v_3 = (6)v_1 - \left(\frac{4}{5}\right)v_2$.

Answer: $R_1 = 20\text{ k}\Omega$ and $R_2 = 40\text{ k}\Omega$

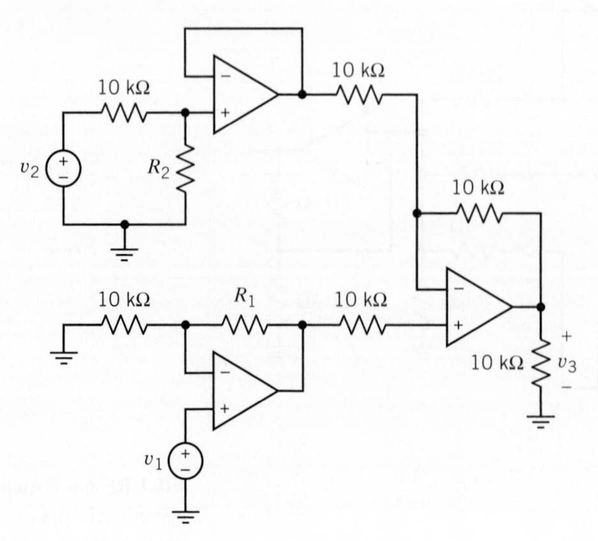

FIGURE E 6.6-1

6.7 CHARACTERISTICS OF PRACTICAL OPERATIONAL AMPLIFIERS

The ideal operational amplifier is the simplest model of an operational amplifier. This simplicity is obtained by ignoring some imperfections of practical operational amplifiers. This section considers some of these imperfections and provides alternate operational amplifier models to account for these imperfections.

Consider the operational amplifier shown in Figure 6.7-1*a*. If this operational amplifier is ideal, then

$$i_1 = 0, \quad i_2 = 0, \quad \text{and} \quad v_1 - v_2 = 0 \tag{6.7-1}$$

In contrast, the operational amplifier model shown in Figure 6.7-1*d* accounts for several nonideal parameters of practical operational amplifiers, namely:

- Nonzero bias currents
- Nonzero input offset voltage
- Finite input resistance
- Nonzero output resistance
- Finite voltage gain

This model more accurately describes practical operational amplifiers than does the ideal operational amplifier. Unfortunately, the more accurate model of Figure 6.7-1*d* is much more complicated and

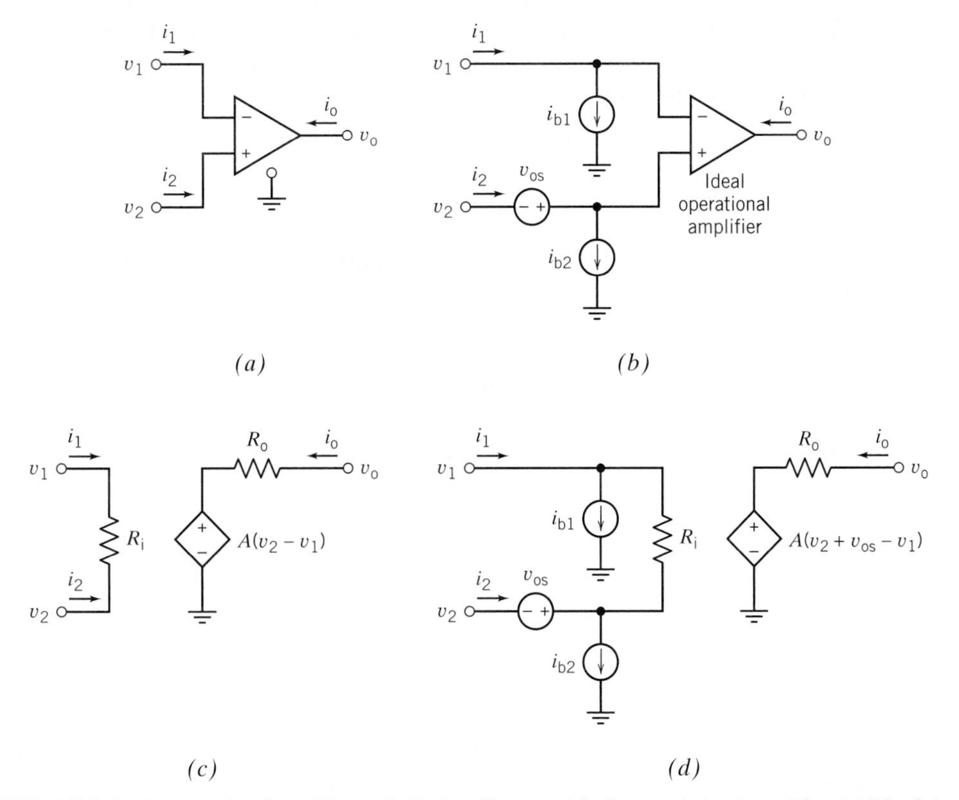

FIGURE 6.7-1 (*a*) An operational amplifier and (*b*) the offsets model of an operational amplifier. (*c*) The finite gain model of an operational amplifier. (*d*) The offsets and finite gain model of an operational amplifier.

much more difficult to use than the ideal operational amplifier. The models in Figures 6.7-1*b* and 6.7-1*c* provide a compromise. These models are more accurate than the ideal operational amplifier but easier to use than the model in Figure 6.7-1*d*. It will be convenient to have names for these models. The model in Figure 6.7-1*b* will be called the offsets model of the operational amplifier. Similarly, the model in Figure 6.7-1*c* will be called the finite gain model of the operational amplifier, and the model in Figure 6.7-1*d* will be called the offsets and finite gain model of the operational amplifier.

The operational amplifier model shown in Figure 6.7-1*b* accounts for the nonzero bias current and nonzero input offset voltage of practical operational amplifiers but not the finite input resistance, the nonzero output resistance, or the finite voltage gain. This model consists of three independent sources and an ideal operational amplifier. In contrast to the ideal operational amplifier, the operational amplifier model that accounts for offsets is represented by the equations

$$i_1 = i_{b1}, \quad i_2 = i_{b2}, \quad \text{and} \quad v_1 - v_2 = v_{os} \tag{6.7-2}$$

The voltage v_{os} is a small, constant voltage called the input offset voltage. The currents i_{b1} and i_{b2} are called the bias currents of the operational amplifier. They are small, constant currents. The difference between the bias currents is called the input offset current, i_{os}, of the amplifier:

$$i_{os} = i_{b1} - i_{b2}$$

Notice that when the bias currents and input offset voltage are all zero, Eq. 6.7-2 is the same as Eq. 6.7-1. In other words, the offsets model reverts to the ideal operational amplifier when the bias currents and input offset voltage are zero.

Frequently, the bias currents and input offset voltage can be ignored because they are very small. However, when the input signal to a circuit is itself small, the bias currents and input voltage can become important.

Manufacturers specify a maximum value for the bias currents, the input offset current, and the input offset voltage. For the μA741, the maximum bias current is specified to be 500 nA, the maximum input offset current is specified to be 200 nA, and the maximum input offset voltage is specified to be 5 mV. These specifications guarantee that

$$|i_{b1}| \leq 500\text{ nA} \quad \text{and} \quad |i_{b2}| \leq 500\text{ nA}$$
$$|i_{b1} - i_{b2}| \leq 200\text{ nA}$$
$$|v_{os}| \leq 5\text{ mV}$$

Table 6.7-1 shows the bias currents, offset current, and input offset voltage *typical* of several types of operational amplifier.

Table 6.7-1 Selected Parameters of Typical Operational Amplifiers

PARAMETER	UNITS	μA741	LF351	TL051C	OPA101 AM	OP-07E
Saturation voltage, v_{sat}	V	13	13.5	13.2	13	13
Saturation current, i_{sat}	mA	2	15	6	30	6
Slew rate, SR	V/μS	0.5	13	23.7	6.5	0.17
Bias current, i_b	nA	80	0.05	0.03	0.012	1.2
Offset current, i_{os}	nA	20	0.025	0.025	0.003	0.5
Input offset voltage, v_{os}	mV	1	5	0.59	0.1	0.03
Input resistance, R_i	MΩ	2	10^6	10^6	10^6	50
Output resistance, R_o	Ω	75	1000	250	500	60
Differential gain, A	V/mV	200	100	105	178	5000
Common mode rejection ratio, CMRR	V/mv	31.6	100	44	178	1413
Gain bandwidth product, B	MHz	1	4	3.1	20	0.6

EXAMPLE 6.7-1 Offset Voltage and Bias Currents

The inverting amplifier shown in Figure 6.7-2*a* contains a μA741 operational amplifier. This inverting amplifier designed in Example 6.5-2 has a gain of -5, that is,

$$v_o = -5 \cdot v_{in}$$

The design of the inverting amplifier is based on the ideal model of an operational amplifier and so did not account for the bias currents and input offset voltage of the μA741 operational amplifier. In this example, the offsets model of an operational amplifier will be used to analyze the circuit. This analysis will tell us what effect the bias currents and input offset voltage have on the performance of this circuit.

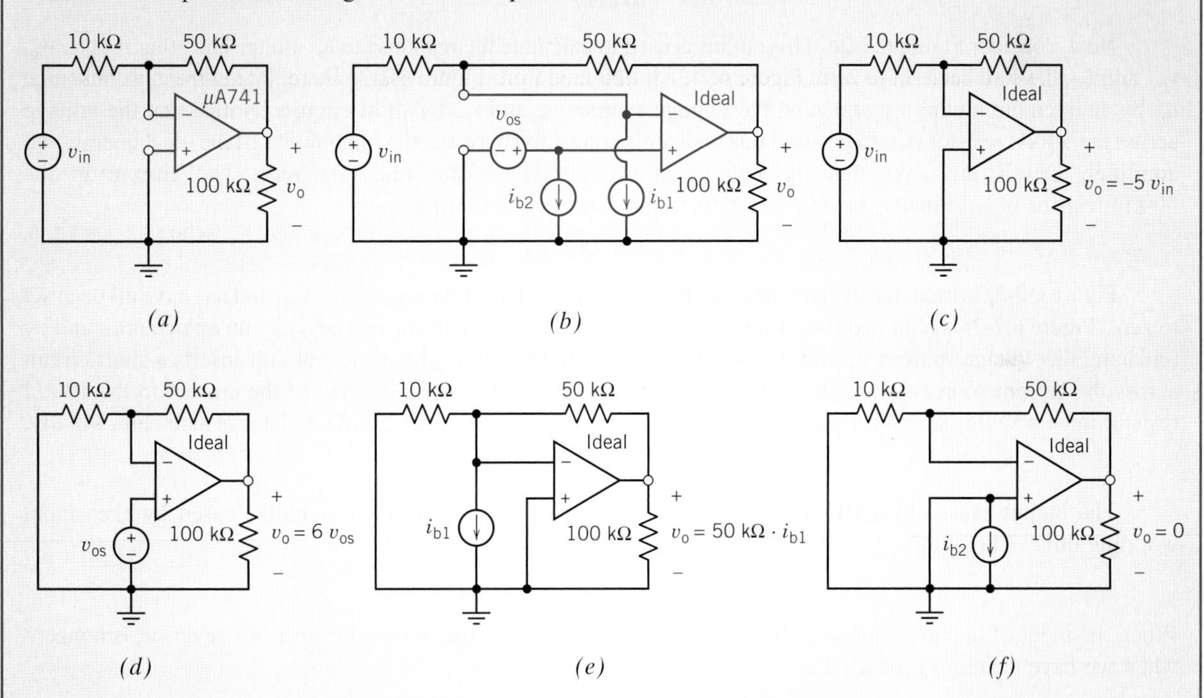

FIGURE 6.7-2 (*a*) An inverting amplifier and (*b*) an equivalent circuit that accounts for the input offset voltage and bias currents of the operational amplifier. (*c*)–(*f*) Analysis using superposition.

Solution

In Figure 6.7-2*b*, the operational amplifier has been replaced by the offsets model of an operational amplifier. Notice that the operational amplifier in Figure 6.7-2*b* is the ideal operational amplifier that is part of the model of the operational amplifier used to account for the offsets. The circuit in Figure 6.7-2*b* contains four inputs that correspond to the four independent sources, v_{in}, i_{b1}, i_{b2}, and v_{os}. (The input v_{in} is obtained by connecting a voltage source to the circuit. In contrast, the "inputs" i_{b1}, i_{b2}, and v_{os} are the results of imperfections of the operational amplifier. These inputs are part of the operational amplifier model and do not need to be added to the circuit.) Superposition can be used to good advantage in analyzing this circuit. Figures 6.7-2*c*–6.7-2*f* illustrate this process. In each of these figures, all but one input has been set to zero, and the output due to that one input has been calculated.

Figure 6.7-2*c* shows the circuit used to calculate the response to v_{in} alone. The other inputs, i_{b1}, i_{b2}, and v_{os}, have all been set to zero. Recall that zero current sources act like open circuits and zero voltage sources act like short circuits. Figure 6.7-2*c* is obtained from Figure 6.7-2*b* by replacing the current sources i_{b1}, i_{b2} by open circuits and by replacing the voltage source v_{os} by a short circuit. The operational amplifier in Figure

6.7-2*c* is the ideal operational amplifier that is part of the offsets model. Analysis of the inverting amplifier in Figure 6.7-2*c* gives

$$v_o = -5 \cdot v_{in}$$

Next, consider Figure 6.7-2*d*. This circuit is used to calculate the response to v_{os} alone. The other inputs, v_{in}, i_{b1}, and i_{b2}, have all been set to zero. Figure 6.7-2*d* is obtained from Figure 6.7-2*b* by replacing the current sources i_{b1} and i_{b2} by open circuits and by replacing the voltage source v_{in} by a short circuit. Again, the operational amplifier is the ideal operational amplifier from the offsets model. The circuit in Figure 6.7-2*d* is one we have seen before; it is the noninverting amplifier (Figure 6.5-1*b*). Analysis of this noninverting amplifier gives

$$v_o = \left(1 + \frac{50\ \text{k}\Omega}{10\ \text{k}\Omega}\right) \cdot v_{os} = 6\ v_{os}$$

Next, consider Figure 6.7-2*e*. This circuit is used to calculate the response to i_{b1} alone. The other inputs, v_{in}, v_{os}, and i_{b2}, have all been set to zero. Figure 6.7-2*e* is obtained from Figure 6.7-2*b* by replacing the current source i_{b2} by an open circuit and replacing the voltage sources v_{in} and v_{os} by short circuits. Notice that the voltage across the 10-kΩ resistor is zero because this resistor is connected between the input nodes of the ideal operational amplifier. Ohm's law says that the current in the 10-kΩ resistor must be zero. The current in the 50-kΩ resistor is i_{b1}. Finally, paying attention to the reference directions,

$$v_o = 50\ \text{k}\Omega \cdot i_{b1}$$

Figure 6.7-2*f* is used to calculate the response to i_{b2} alone. The other inputs, v_{in}, v_{os}, and i_{b1}, have all been set to zero. Figure 6.7-2*f* is obtained from Figure 6.7-2*b* by replacing the current source i_{b1} by an open circuit and by replacing the voltage sources v_{in} and v_{os} by short circuits. Replacing v_{os} by a short circuit inserts a short circuit across the current source i_{b2}. Again, the voltage across the 10-kΩ resistor is zero, so the current in the 10-kΩ resistor must be zero. Kirchhoff's current law shows that the current in the 50-kΩ resistor is also zero. Finally,

$$v_o = 0$$

The output caused by all four inputs working together is the sum of the outputs caused by each input working alone. Therefore,

$$v_o = -5 \cdot v_{in} + 6 \cdot v_{os} + (50\ \text{k}\Omega)i_{b1}$$

When the input of the inverting amplifier, v_{in}, is zero, the output v_o also should be zero. However, v_o is nonzero when we have a finite v_{os} or i_{b1}. Let

$$\text{output offset voltage} = 6 \cdot v_{os} + (50\ \text{k}\Omega)i_{b1}$$

Then

$$v_o = -5 \cdot v_{in} + \text{output offset voltage}$$

Recall that when the operational amplifier is modeled as an ideal operational amplifier, analysis of this inverting amplifier gives

$$v_o = -5 \cdot v_{in}$$

Comparing these last two equations shows that bias currents and input offset voltage cause the output offset voltage. Modeling the operational amplifier as an ideal operational amplifier amounts to assuming that the output offset voltage is not important and thus ignoring it. Using the operational amplifier model that accounts for offsets is more accurate but also more complicated.

How large is the output offset voltage of this inverting amplifier? The input offset voltage of a μA741 operational amplifier will be at most 5 mV, and the bias current will be at most 500 nA, so

$$\text{output offset voltage} \leq 6 \cdot 5\ \text{mV} + (50\ \text{k}\Omega)\ 500\ \text{nA} = 55\ \text{mV}$$

We note that we can ignore the effect of the offset voltage only when $|5\ v_{in}| > 500$ mV or $|v_{in}| > 100$ mV. The output offset error can be reduced by using a better operational amplifier, that is, one that guarantees smaller bias currents and input offset voltage.

Now, let us turn our attention to different parameters of practical operational amplifiers. The operational amplifier model shown in Figure 6.7-1*c* accounts for the finite input resistance, the nonzero output resistance, and the finite voltage gain of practical operational amplifiers but not the nonzero bias current and nonzero input offset voltage. This model consists of two resistors and a VCVS.

The finite gain model reverts to an ideal operational amplifier when the gain, A, becomes infinite. To see that this is so, notice that in Figure 6.7-1*c*

$$v_o = A(v_2 - v_1) + R_o i_o$$

so

$$v_2 - v_1 = \frac{v_o - R_o i_o}{A}$$

The models in Figure 6.7-1, as well as the model of the ideal operational amplifier, are valid only when v_o and i_o satisfy Eq. 6.3-1. Therefore,

$$|v_o| \le v_{sat} \quad \text{and} \quad |i_o| \le i_{sat}$$

Then

$$|v_2 - v_1| \le \frac{v_{sat} + R_o i_{sat}}{A}$$

Therefore,

$$\lim_{A \to \infty} (v_2 - v_1) = 0$$

Next, because

$$i_1 = -\frac{v_2 - v_1}{R_i} \quad \text{and} \quad i_2 = \frac{v_2 - v_1}{R_i}$$

we conclude that

$$\lim_{A \to \infty} i_1 = 0 \quad \text{and} \quad \lim_{A \to \infty} i_2 = 0$$

Thus, i_1, i_2, and $v_2 - v_1$ satisfy Eq. 6.7-1. In other words, the finite gain model of the operational amplifier reverts to the ideal operational amplifier as the gain becomes infinite. The gain for practical op amps ranges from 100,000 to 10^7.

EXAMPLE 6.7-2 Finite Gain

In Figure 6.7-3, a voltage follower is used as a buffer amplifier. Analysis based on the ideal operational amplifier shows that the gain of the buffer amplifier is

$$\frac{v_o}{v_s} = 1$$

What effects will the input resistance, output resistance, and finite voltage gain of a practical operational amplifier have on the performance of this circuit? To answer this question, replace the operational amplifier by the operational amplifier model that accounts for finite voltage gain. This gives the circuit shown in Figure 6.7-3*b*.

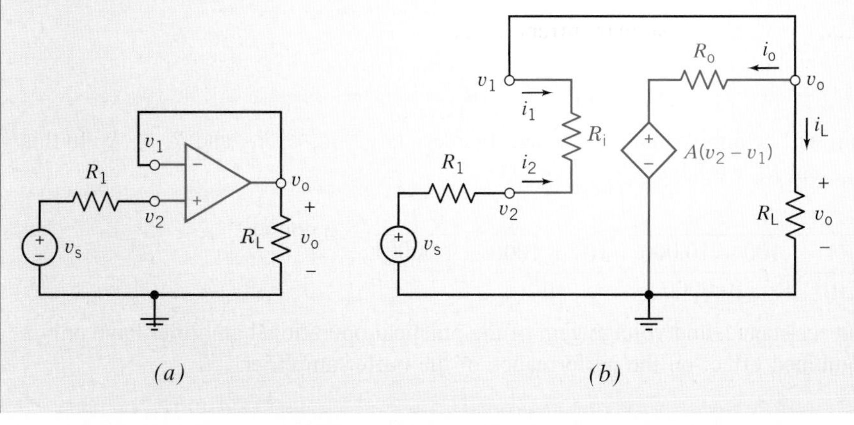

FIGURE 6.7-3 (*a*) A voltage follower used as a buffer amplifier and (*b*) an equivalent circuit with the operational amplifier model that accounts for finite voltage gain.

Solution

To be specific, suppose $R_1 = 1\,\text{k}\Omega$; $R_L = 10\,\text{k}\Omega$; and the parameters of the practical operational amplifier are $R_i = 100\,\text{k}\Omega$, $R_o = 100\Omega$, and $A = 10^5\,\text{V/V}$.

Suppose that $v_o = 10$ V. We can find the current, i_L in the output resistor as

$$i_L = \frac{v_o}{R_L} = \frac{10\text{ V}}{10^4\ \Omega} = 10^{-3}\text{ A}$$

Apply KCL at the top node of R_L to get

$$i_1 + i_o + i_L = 0$$

It will turn out that i_1 will be much smaller than both i_o and i_L. It is useful to make the approximation that $i_1 = 0$. (We will check this assumption later in this example.) Then,

$$i_o = -i_L$$

Next, apply KVL to the mesh consisting of the VCVS, R_o, and R_L to get

$$-A(v_2 - v_1) - i_o R_o + i_L R_L = 0$$

Combining the last two equations and solving for $(v_2 - v_1)$ gives

$$v_2 - v_1 = \frac{i_L(R_o + R_L)}{A} = \frac{10^{-3}(100 + 10{,}000)}{10^5} = 1.01 \times 10^{-4}\text{ V}$$

Now i_1 can be calculated using Ohm's law:

$$i_1 = \frac{v_1 - v_2}{R_i} = \frac{-1.01 \times 10^{-4}\text{ V}}{100\text{ k}\Omega} = -1.01 \times 10^{-9}\text{ A}$$

This justifies our earlier assumption that i_1 is negligible compared with i_o and i_L.

Applying KVL to the outside loop gives

$$-v_s - i_1 R_1 - i_1 R_i + v_o = 0$$

Now, let us do some algebra to determine v_s:

$$\begin{aligned} v_s &= v_o - i_1(R_1 + R_i) = v_o + i_2(R_1 + R_i) \\ &= v_o + \frac{v_2 - v_1}{R_i} \times (R_1 + R_i) \\ &= v_o + \frac{i_L(R_o + R_L)}{A} \times \frac{(R_1 + R_i)}{R_i} \\ &= v_o + \frac{v_o}{R_L} \times \frac{(R_o + R_L)}{A} \times \frac{(R_1 + R_i)}{R_i} \end{aligned}$$

The gain of this circuit is

$$\frac{v_o}{v_s} = \frac{1}{1 + \dfrac{1}{A} \times \dfrac{R_o + R_L}{R_L} \times \dfrac{R_i + R_1}{R_i}}$$

This equation shows that the gain will be approximately 1 when A is very large, $R_o \ll R_L$, and $R_1 \ll R_i$. In this example, for the specified A, R_o, and R_i, we have

$$\frac{v_o}{v_s} = \frac{1}{1 \times \dfrac{1}{10^5} \times \dfrac{100 + 10{,}000}{10{,}000} \times \dfrac{10^5 + 1000}{10^5}} = \frac{1}{1.00001} = 0.99999$$

Thus, the input resistance, output resistance, and voltage gain of the practical operational amplifier have only a small, essentially negligible, combined effect on the performance of the buffer amplifier.

Table 6.7-1 lists two other parameters of practical operational amplifiers that have not yet been mentioned. They are the *common mode rejection ratio* (*CMRR*) and the gain bandwidth product. Consider first the common mode rejection ratio. In the finite gain model, the voltage of the dependent source is

$$A(v_2 - v_1)$$

In practice, we find that dependent source voltage is more accurately expressed as

$$A(v_2 - v_1) + A_{\mathrm{cm}}\left(\frac{v_1 + v_2}{2}\right)$$

where $v_2 - v_1$ is called the differential input voltage,

$\frac{v_1 + v_2}{2}$ is called the common mode input voltage,

and A_{cm} is called the common mode gain.

The gain A is sometimes called the differential gain to distinguish it from A_{cm}. The common mode rejection ratio is defined to be the ratio of A to A_{cm}

$$\mathrm{CMRR} = \frac{A}{A_{\mathrm{cm}}}$$

The dependent source voltage can be expressed using A and CMRR as

$$\begin{aligned} A(v_2 - v_1) + A_{\mathrm{cm}}\frac{v_1 + v_2}{2} &= A(v_2 - v_1) + \frac{A}{\mathrm{CMRR}}\frac{v_1 + v_2}{2} \\ &= A\left[\left(1 + \frac{1}{2\,\mathrm{CMRR}}\right)v_2 - \left(1 - \frac{1}{2\,\mathrm{CMRR}}\right)v_1\right] \end{aligned}$$

CMRR can be added to the finite gain model by changing the voltage of the dependent source. The appropriate change is

replace $A(v_2 - v_1)$ by $A\left[\left(1 + \frac{1}{2\,\mathrm{CMRR}}\right)v_2 - \left(1 - \frac{1}{2\,\mathrm{CMRR}}\right)v_1\right]$

This change will make the model more accurate but also more complicated. Table 6.7-1 shows that CMRR is typically very large. For example, a typical LF351 operational amplifier has $A = 100\mathrm{V/mV}$ and $\mathrm{CMRR} = 100\ \mathrm{V/mV}$. This means that

$$A\left[\left(1 + \frac{1}{2\,\mathrm{CMRR}}\right)v_2 - \left(1 - \frac{1}{2\,\mathrm{CMRR}}\right)v_1\right] = 100,000.5v_2 - 99,999.5v_1$$

compared to $\mathrm{A}(v_2 - v_1) = 100{,}000v_2 - 100{,}000v_1$

In most cases, negligible error is caused by ignoring the CMRR of the operational amplifier. The CMRR does not need to be considered unless accurate measurements of very small differential voltages must be made in the presence of very large common mode voltages.

Next, we consider the gain bandwidth product of the operational amplifier. The finite gain model indicates that the gain, A, of the operational amplifier is a constant. Suppose

$$v_1 = 0 \quad \text{and} \quad v_2 = M \sin \omega t$$

so that $v_2 - v_1 = M \sin \omega t$

The voltage of the dependent source in the finite gain model will be

$$A(v_2 - v_1) = A \cdot M \sin \omega t$$

The amplitude, $A \cdot M$, of this sinusoidal voltage does not depend on the frequency, ω. Practical operational amplifiers do not work this way. The gain of a practical amplifier is a function of frequency, say $A(\omega)$. For many practical amplifiers, $A(\omega)$ can be adequately represented as

$$A(\omega) = \frac{B}{j\omega}$$

It is not necessary to know now how this function behaves. Functions of this sort will be discussed in Chapter 13. For now, it is enough to realize that the parameter B is used to describe the dependence of the operational amplifier gain on frequency. The parameter B is called the gain bandwidth product of the operational amplifier.

EXERCISE 6.7-1 The input offset voltage of a *typical* μA741 operational amplifier is 1 mV, and the bias current is 80 nA. Suppose the operational amplifier in Figure 6.7-2*a* is a typical μA741. Show that the output offset voltage of the inverting amplifier will be at most 10 mV.

EXERCISE 6.7-2 Suppose the 10-kΩ resistor in Figure 6.7-2*a* is changed to 2 kΩ and the 50-kΩ resistor is changed to 10 kΩ. (These changes will not change the gain of the inverting amplifier. It will still be -5.) Show that the *maximum* output offset voltage is reduced to 35 mV. (Use $i_b = 500$ nA and $v_{os} = 5$ mV to calculate the *maximum* output offset voltage that could be caused by the μA741 amplifier.)

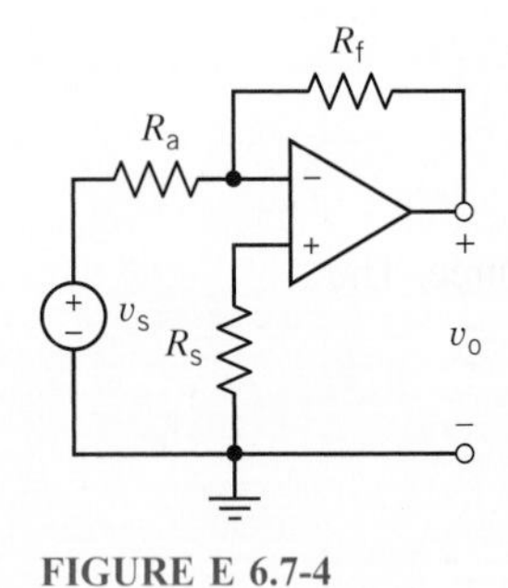

FIGURE E 6.7-4

EXERCISE 6.7-3 Suppose the μA741 operational amplifier in Figure 6.7-2*a* is replaced with a *typical* OPA101AM operational amplifier. Show that the output offset voltage of the inverting amplifier will be at most 0.6 mV.

EXERCISE 6.7-4

a. Determine the voltage ratio v_o/v_s for the op amp circuit shown in Figure E 6.7-4.

b. Calculate v_o/v_s for a practical op amp with $A = 10^5$, $R_o = 100\ \Omega$, and $R_i = 500$ kΩ. The circuit resistors are $R_s = 10$ kΩ, $R_f = 50$ kΩ, and $R_a = 25$ kΩ.

Answer: (b) $v_o/v_s = -2$

6.8 ANALYSIS OF OP AMP CIRCUITS USING MATLAB

Figure 6.8-1 shows an inverting amplifier. Model the operational amplifier as an ideal op amp. Then the output voltage of the inverting amplifier is related to the input voltage by

$$v_o(t) = -\frac{R_2}{R_1} v_s(t) \tag{6.8-1}$$

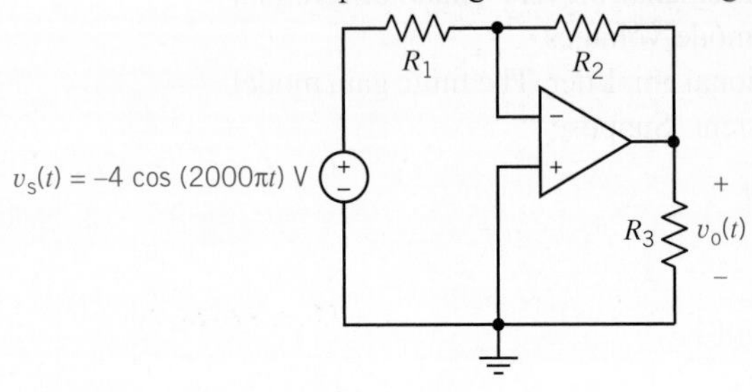

FIGURE 6.8-1 An inverting amplifier.

Suppose that $R_1 = 2$ kΩ, $R_2 = 50$ kΩ, and $v_s = -4$ cos $(2000\,\pi t)$ V. Using these values in Eq. 6.8-1 gives $v_o(t) = 100$ cos$(2000\,\pi t)$ V. This is not a practical answer. It's likely that the operational amplifier saturates, and, therefore, the ideal op amp is not an appropriate model of the operational amplifier. When voltage saturation is included in the model of the operational amplifier, the inverting amplifier is described by

$$v_o(t) = \begin{cases} v_{sat} & \text{when } -\dfrac{R_2}{R_1}v_s(t) > v_{sat} \\ -\dfrac{R_2}{R_1}v_s(t) & \text{when } -v_{sat} < -\dfrac{R_2}{R_1}v_s(t) < v_{sat} \\ -v_{sat} & \text{when } -\dfrac{R_2}{R_1}v_s(t) < -v_{sat} \end{cases} \tag{6.8-2}$$

where v_{sat} denotes the saturation voltage of the operational amplifier. Equation 6.8-2 is a more accurate, but more complicated, model of the inverting amplifier than Eq. 6.8-1. Of course, we prefer the simpler model, and we use the more complicated model only when we have reason to believe that answers based on the simpler model are not accurate.

Figures 6.8-2 and 6.8-3 illustrate the use of MATLAB to analyze the inverting amplifier when the operational amplifier model includes voltage saturation. Figure 6.8-2 shows the MATLAB input file, and Figure 6.8-3 shows the resulting plot of the input and output voltages of the inverting amplifier.

```
% Saturate.m simulates op amp voltage saturation
%------------------------------------------------------------
% Enter values of the parameters that describe the circuit.
%------------------------------------------------------------
                            % circuit parameters
R1=2e3;                     % resistance, ohms
R2=50e3;                    % resistance, ohms
R3=20e3;                    % resistance, ohms

                            % op amp parameter
vsat=15;                    % saturation voltage, V

                            % source parameters
M=4;                        % amplitude, V
f=1000;                     % frequency, Hz
w=2*pi*f;                   % frequency, rad/s
theta=(pi/180)*180;         % phase angle, rad

%------------------------------------------------------------
% Divide the time interval (0, tf) into N increments
%------------------------------------------------------------
tf=2/f;                     % final time
N=200;                      % number of incerments
t=0:tf/N:tf;                % time, s

%------------------------------------------------------------
% at each time t=k*(tf/N), calculate vo from vs
%------------------------------------------------------------
vs = M*cos(w*t+theta);      % input voltage

for k=1:length(vs)

      if      (-(R2/R1)*vs(k) < -vsat)  vo(k) = -vsat; % ------
      elseif  (-(R2/R1)*vs(k) >  vsat)  vo(k) =  vsat; % eqn.
      else    vo(k) = -(R2/R1)*vs(k);                  % 6.8-2
      end                                              % ------

end
%------------------------------------------------------------
% Plot Vo and vs versus t
%------------------------------------------------------------
plot(t, vo, t, vs)          % plot the transfer characteristic
axis([0 tf -20 20])
xlabel('time, s')
ylabel('vo(t), V')
```

FIGURE 6.8-2 MATLAB input file corresponding to the circuit shown in Figure 6.8-1.

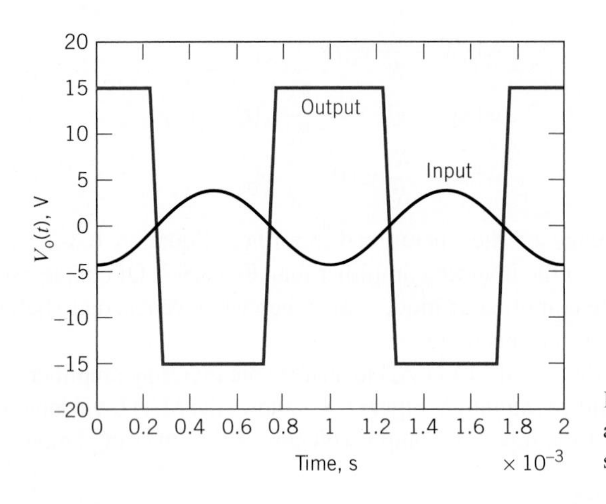

FIGURE 6.8-3 Plots of the input and output voltages of the circuit shown in Figure 6.8-1.

6.9 USING PSPICE TO ANALYZE OP AMP CIRCUITS

Consider an op amp circuit having one input, v_i, and one output, v_o. Let's plot the output voltage as a function of the input voltage using PSpice. We need to do the following:

1. Draw the circuit in the OrCAD Capture workspace.
2. Specify a DC Sweep simulation.
3. Run the simulation.
4. Plot the simulation results.

The DC Sweep simulation provides a way to vary the input of a circuit and then plot the output as a function of the input.

EXAMPLE 6.9-1 Using PSpice to Analyze an Op Amp Circuit

The input to the circuit shown in Figure 6.9-1 is the voltage source voltage, v_i. The response is the voltage, v_o. Use PSpice to plot the output voltage as a function of the input voltage.

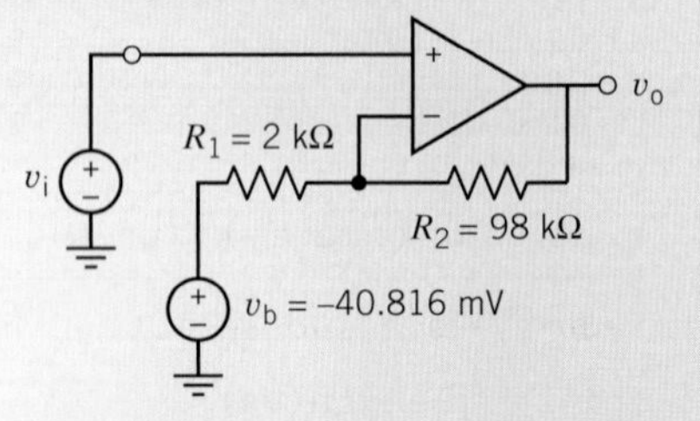

FIGURE 6.9-1 The circuit considered in Example 6.9-1.

Solution

We begin by drawing the circuit in the OrCAD workspace as shown in Figure 6.9-2 (see Appendix A). The op amp in Figure 6.9-2 is represented by the PSpice part named OPAMP from the ANALOG library. The circuit output is a node voltage. It's convenient to give the output voltage a PSpice name. In Figure 6.9-2, a PSpice part called an off-page connector is used to label the output node as ''o.'' Labeling the output node in this way gives the circuit output the PSpice name, V(o).

We will perform a DC Sweep simulation. (Select PSpice\New Simulation Profile from the OrCAD Capture menu bar, then DC Sweep from the Analysis Type drop-down list. Specify the Sweep variable to be the input voltage by selecting Voltage Source and identifying the voltage source as Vi. Specify a linear sweep and the desired range of input voltages.) Select PSpice\Run Simulation Profile from the OrCAD Capture menu bar to ran the simulation.

After a successful DC Sweep simulation, OrCAD Capture will automatically open a Schematics window. Select Trace/Add Trace from the Schematics menus to pop up the Add Traces dialog box. Select V(o) from the

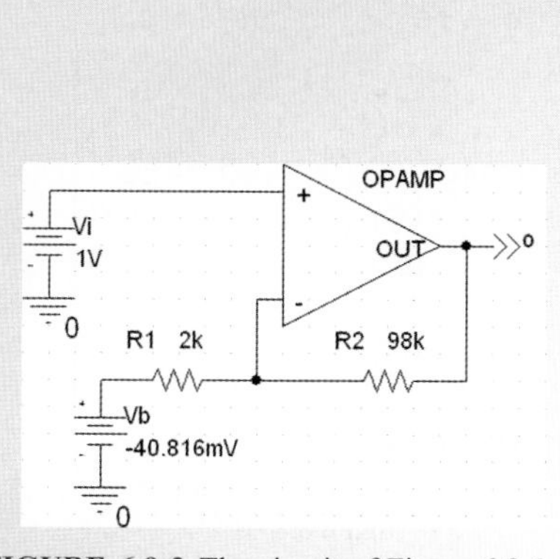

FIGURE 6.9-2 The circuit of Figure 6.9-1 as drawn in the OrCAD workspace.

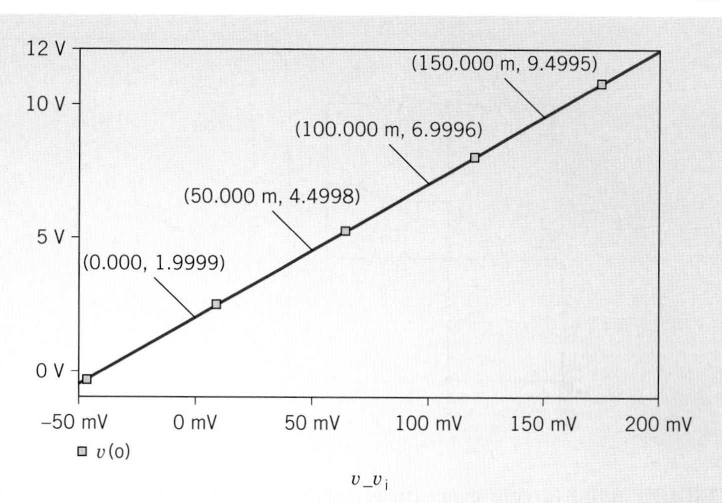

FIGURE 6.9-3 The plot of the output voltage as a function of the input voltage.

Simulation Output Variables list. Close the Add Traces dialog box. Figure 6.9-3 shows the resulting plot after removing the grid and labeling some points. The plot is a straight line. Consequently, the circuit output is related to the circuit input by an equation of the form

$$v_o = mv_i + b$$

where the values of the slope m and intercept b can be determined from the points labeled in Figure 6.9-3. In particular,

$$m = \frac{6.9996 - 4.4998}{0.100 - 0.050} = 49.996 \approx 50 \, \frac{\text{V}}{\text{V}}$$

and

$$1.9999 = 59.996(0) + b \quad \Rightarrow \quad b = 1.9999 \approx 2 \text{ V}$$

The circuit output is related to the circuit input by the equation

$$v_o = 50v_i + 2$$

6.10 HOW CAN WE CHECK . . . ?

Engineers are frequently called upon to check that a solution to a problem is indeed correct. For example, proposed solutions to design problems must be checked to confirm that all of the specifications have been satisfied. In addition, computer output must be reviewed to guard against data-entry errors, and claims made by vendors must be examined critically.

Engineering students are also asked to check the correctness of their work. For example, occasionally just a little time remains at the end of an exam. It is useful to be able to quickly identify those solutions that need more work.

The following example illustrates techniques useful for checking the solutions of the sort of problems discussed in this chapter.

EXAMPLE 6.10-1 How Can We Check Op Amp Circuits?

The circuit in Figure 6.10-1a was analyzed by writing and solving the following set of simultaneous equations

$$\frac{v_6}{10} + i_5 = 0$$

$$10i_5 = v_4$$

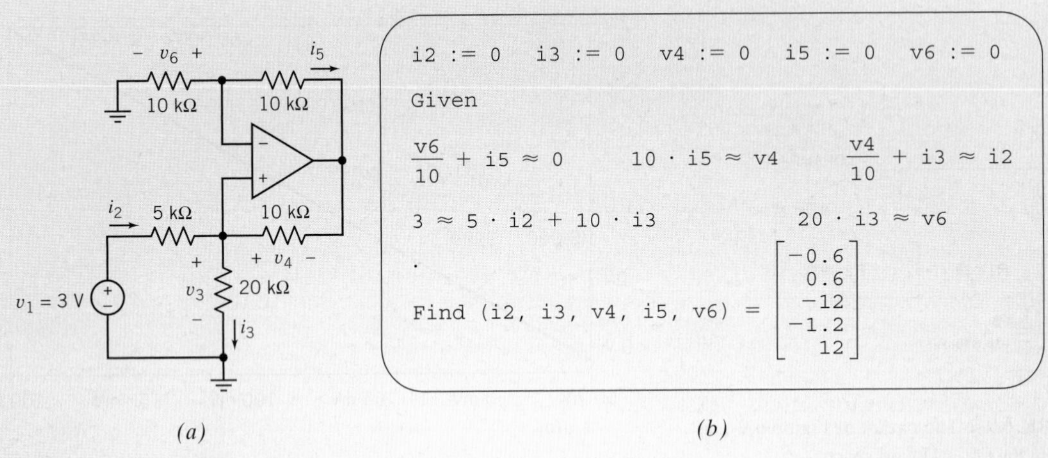

FIGURE 6.10-1 (*a*) An example circuit and (*b*) computer analysis using Mathcad.

$$\frac{v_4}{10} + i_3 = i_2$$

$$3 = 5i_2 + 10i_3$$

$$20i_3 = v_6$$

(These equations use units of volts, milliamps, and kohms.) A computer and the program Mathcad were used to solve these equations as shown in Figure 6.10-1*b*. The solution of these equations indicates that

$$i_2 = -0.6 \text{ mA}, \quad i_3 = 0.6 \text{ mA}, \quad v_4 = -12 \text{ V},$$
$$i_5 = -1.2 \text{ mA}, \quad \text{and} \quad v_6 = 12 \text{ V}$$

How can we check that these voltage and current values are correct?

Solution

Consider the voltage v_3. Using Ohm's law,

$$v_3 = 20i_3 = 20(0.6) = 12 \text{ V}$$

Remember that resistances are in kΩ and currents in milliamps. Applying KVL to the mesh consisting of the voltage source and the 5-kΩ and 20-kΩ resistors gives

$$v_3 = 3 - 5i_2 = 3 - 5(-0.6) = 6 \text{ V}$$

Clearly, v_3 cannot be both 12 and 6, so the values obtained for i_2, i_3, v_4, i_5, and v_6 cannot all be correct. Checking the simultaneous equations, we find that a resistor value has been entered incorrectly. The KVL equation corresponding to the mesh consisting of the voltage source and the 5-kΩ and 20-kΩ resistors should be

$$3 = 5i_2 + 20i_3$$

Note that $10i_3$ was incorrectly used in the fourth line of the Mathcad program of Figure 6.10-1. After making this correction, i_2, i_3, v_4, i_5, and v_6 are calculated to be

$$i_2 = -0.2 \text{ mA}, \quad i_3 = 0.2 \text{ mA}, \quad v_4 = -4 \text{ V},$$
$$i_5 = 0.4 \text{ mA}, \quad \text{and} \quad v_6 = 4 \text{ V}$$

Now

$$v_3 = 20i_3 = 20(0.2) = 4$$

and

$$v_3 = 3 - 5i_2 = 3 - 5(-0.2) = 4$$

This agreement suggests that the new values of i_2, i_3, v_4, i_5, and v_6 are correct. As an additional check, consider v_5. First, Ohm's law gives

$$v_5 = 10i_5 = 10(-0.4) = -4$$

Next, applying KVL to the loop consisting of the two 10-kΩ resistors and the input of the operational amplifier gives

$$v_5 = 0 + v_4 = 0 + (-4) = -4$$

This increases our confidence that the new values of i_2, i_3, v_4, i_5, and v_6 are correct.

6.11 DESIGN EXAMPLE

TRANSDUCER INTERFACE CIRCUIT

A customer wants to automate a pressure measurement, which requires converting the output of the pressure transducer to a computer input. This conversion can be done using a standard integrated circuit called an analog-to-digital converter (ADC). The ADC requires an input voltage between 0 V and 10 V, whereas the pressure transducer output varies between −250 mV and 250 mV. Design a circuit to interface the pressure transducer with the ADC. That is, design a circuit that translates the range −250 mV to 250 mV to the range 0 V to 10 V.

Describe the Situation and the Assumptions

The situation is shown in Figure 6.11-1.

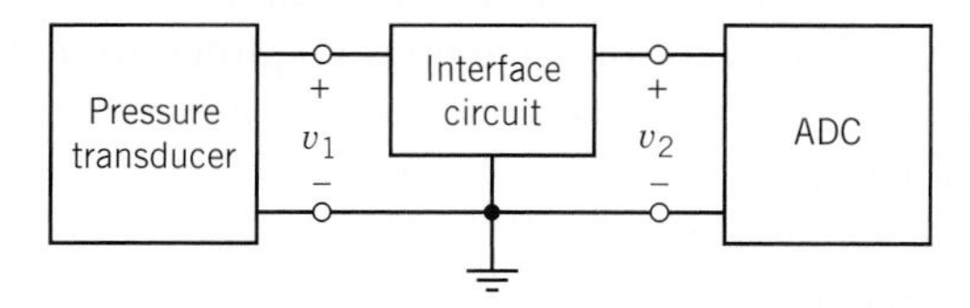

FIGURE 6.11-1 Interfacing a pressure transducer with an analog-to-digital converter (ADC).

The specifications state that

$$-250\text{ mV} \le v_1 \le 250\text{ mV}$$
$$0\text{ V} \le v_2 \le 10\text{ V}$$

A simple relationship between v_2 and v_1 is needed so that information about the pressure is not obscured. Consider

$$v_2 = a \cdot v_1 + b$$

The coefficients, a and b, can be calculated by requiring that $v_2 = 0$ when $v_1 = -250$ mV and that $v_2 = 10$ V when $v_1 = 250$ mV, that is,

$$0\text{ V} = a\,(-250\text{ mV}) + b$$
$$10\text{ V} = a\,(250\text{ mV}) + b$$

Solving these simultaneous equations gives $a = 20$ V/V and $b = 5$ V.

State the Goal

Design a circuit having input voltage v_1 and output voltage v_2. These voltages should be related by

$$v_2 = 20\,v_1 + 5\text{ V} \tag{6.11-1}$$

Generate a Plan

Figure 6.11-2 shows a plan (or a structure) for designing the interface circuit. The operational amplifiers are biased using +15-V and −15-V power supplies. The constant 5-V input is generated from the 15-V power supply by multiplying by a gain of 1/3. The input voltage, v_1, is multiplied by a gain of 20. The summer (adder) adds the outputs of the two amplifiers to obtain v_2.

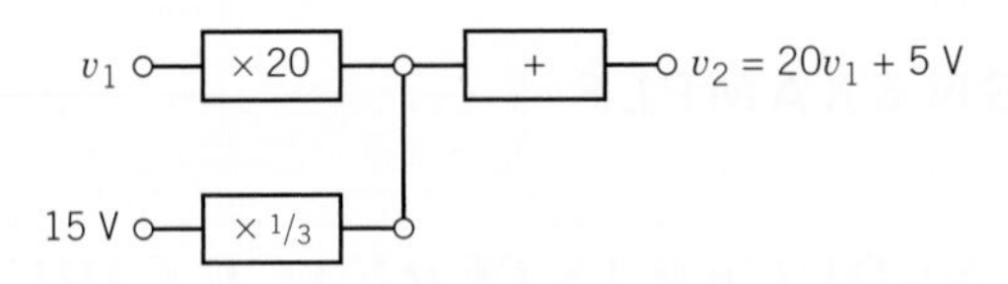

FIGURE 6.11-2 A structure (or plan) for the interface circuit.

Each block in Figure 6.11-2 will be implemented using an operational amplifier circuit.

Act on the Plan

Figure 6.11-3 shows one proposed interface circuit. Some adjustments have been made to the plan. The summer is implemented using the inverting summing amplifier from Figure 6.5-1*d*. The inputs to this inverting summing amplifier must be $-20v_i$ and -5 V instead of $20v_i$ and 5 V. Consequently, an inverting amplifier is used to multiply v_1 by -20. A voltage follower prevents the summing amplifier from loading the voltage divider. To make the signs work out correctly, the -15-V power supply provides the input to the voltage divider.

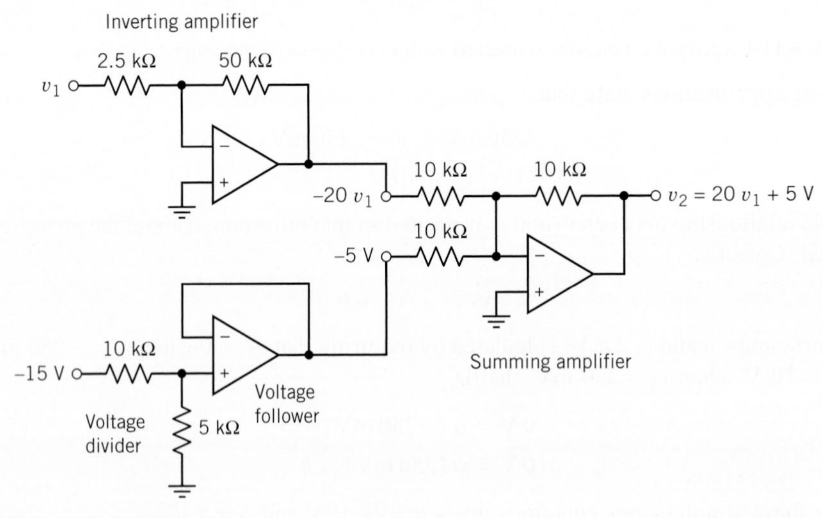

FIGURE 6.11-3 One implementation of the interface circuit.

The circuit shown in Figure 6.11-3 is not the only circuit that solves this design challenge. There are several circuits that implement

$$v_2 = 20v_1 + 5 \text{ V}$$

We will be satisfied with having found one circuit that does the job.

Verify the Proposed Solution

The circuit shown in Figure 6.11-3 was simulated using PSpice. The result of this simulation is the plot of the v_2 versus v_1 shown in Figure 6.11-4. Because this plot shows a straight line, v_2 is related to v_1 by the equation of a straight line

$$v_2 = mv_1 + b$$

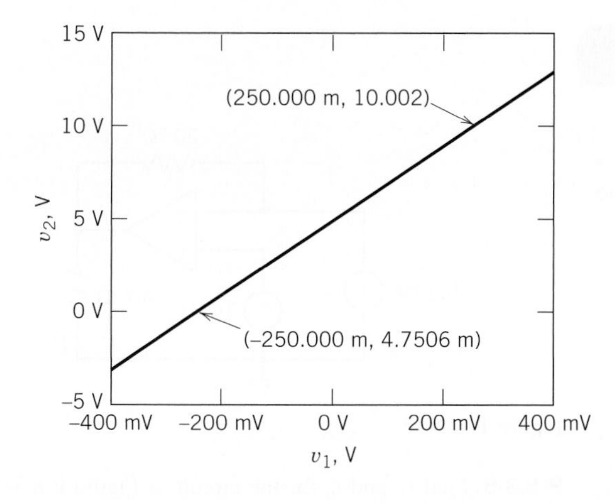

FIGURE 6.11-4 PSpice simulation of the circuit shown in Figure 6.11-3.

where m is the slope of the line and b is the intercept of the line with the vertical axis. Two points on the line have been labeled to show that $v_2 = 10.002$ V when $v_1 = 0.250$ V and that $v_2 = 0.0047506$ V when $v_1 = -0.250$ V. The slope, m, and intercept, b, can be calculated from these points. The slope is given by

$$m = \frac{10.002 - (0.0047506)}{0.250 - (-0.250)} = 19.994$$

The intercept is given by

$$b = 10.002 - 19.994 \times 0.0250 = 5.003$$

Thus,

$$v_2 = 19.994 v_1 + 5.003 \qquad (6.11\text{-}2)$$

Comparing Eqs. 6.11-1 and 6.11-2 verifies that the proposed solution is indeed correct.

6.12 SUMMARY

- Several models are available for operational amplifiers. Simple models are easy to use. Accurate models are more complicated. The simplest model of the operational amplifier is the ideal operational amplifier.
- The currents into the input terminals of an ideal operational amplifier are zero, and the voltages at the input nodes of an ideal operational amplifier are equal.
- It is convenient to use node equations to analyze circuits that contain ideal operational amplifiers.
- Operational amplifiers are used to build circuits that perform mathematical operations. Many of these circuits have been used so often that they have been given names. The inverting amplifier gives a response of the form $v_o = -Kv_i$ where K is a positive constant. The noninverting amplifier gives a response of the form $v_o = Kv_i$ where K is a positive constant. Another useful operational amplifier circuit is the noninverting amplifier with a gain of $K = 1$, often called a voltage follower or buffer. The output of the voltage follower faithfully follows the input voltage. The voltage follower reduces loading by isolating its output terminal from its input terminal.
- Figure 6.5-1 is a catalog of some frequently used operational amplifier circuits.
- Practical operational amplifiers have properties that are not included in the ideal operational amplifier. These include the input offset voltage, bias current, dc gain, input resistance, and output resistance. More complicated models are needed to account for these properties.
- PSpice can be used to reduce the drudgery of analyzing operational amplifier circuits with complicated models.

PROBLEMS

Section 6.3 The Ideal Operational Amplifier

P 6.3-1 Determine the value of voltage measured by the voltmeter in Figure P 6.3-1.

Answer: −4 V

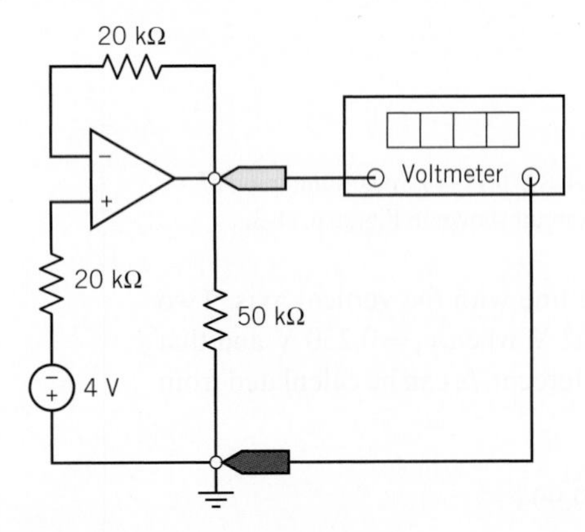

Figure P 6.3-1

P 6.3-2 Find v_o and i_o for the circuit of Figure P 6.3-2.

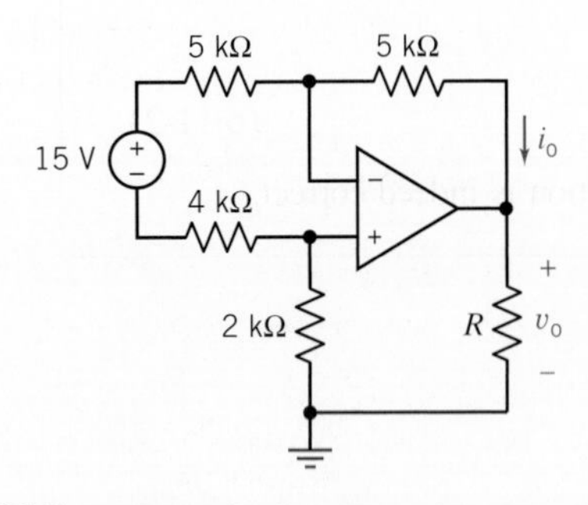

Figure P 6.3-2

P 6.3-3 Find v_o and i_o for the circuit of Figure P 6.3-3.

Answer: $v_o = -30$ V and $i_o = 3.5$ mA

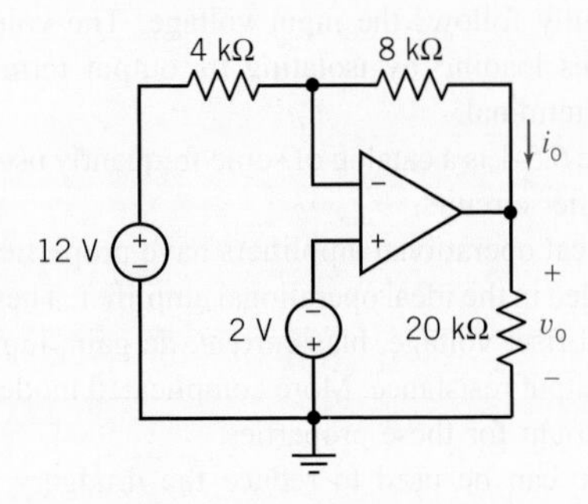

Figure P 6.3-3

P 6.3-4 Find v and i for the circuit of Figure P 6.3-4.

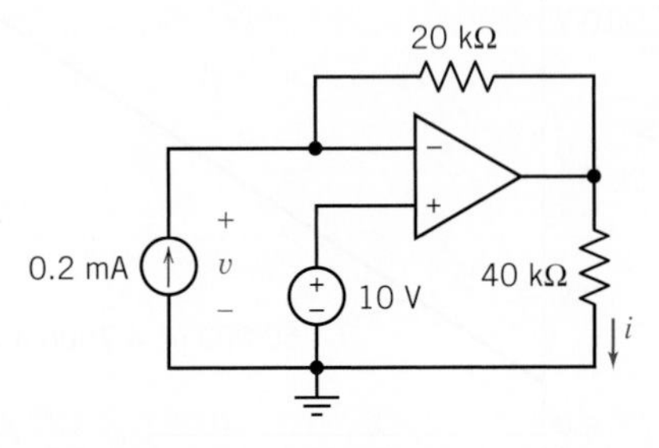

Figure P 6.3-4

P 6.3-5 Find v_o and i_o for the circuit of Figure P 6.3-5.

Answer: $v_o = -42$ V and $i_o = 10.5$ mA

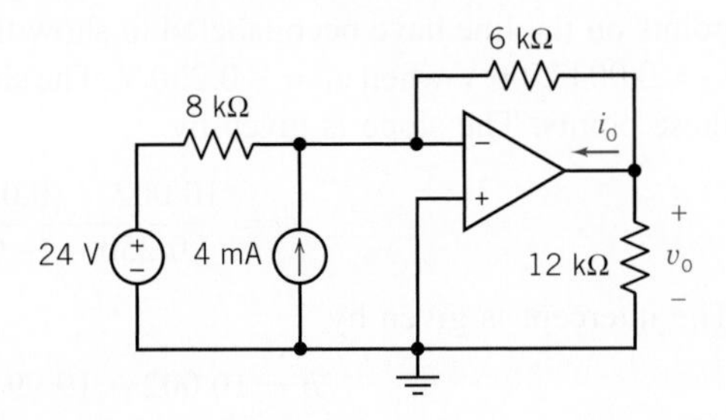

Figure P 6.3-5

P 6.3-6 Determine the value of voltage measured by the voltmeter in Figure P 6.3-6.

Answer: 9.9 V

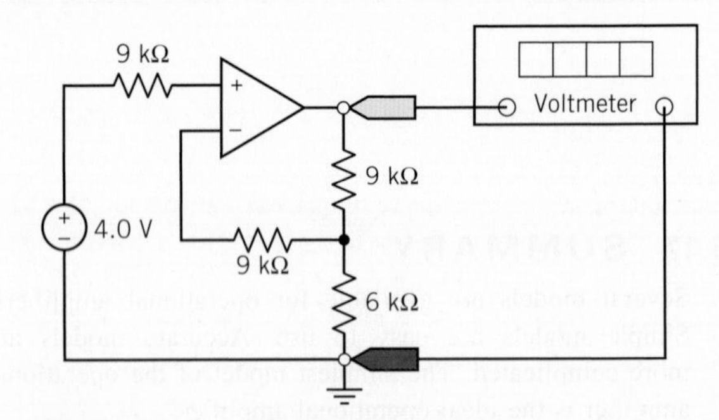

Figure P 6.3-6

P 6.3-7 Find v_o and i_o for the circuit of Figure P 6.3-7.

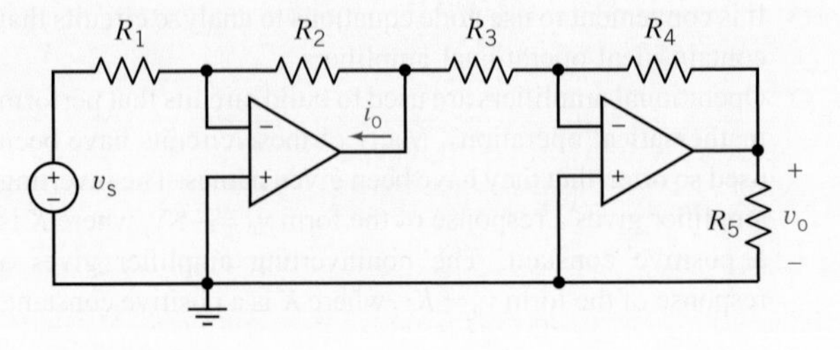

Figure P 6.3-7

P 6.3-8 Determine the current i_o for the circuit shown in Figure P 6.3-8.

Answer: $i_o = 2.5$ mA

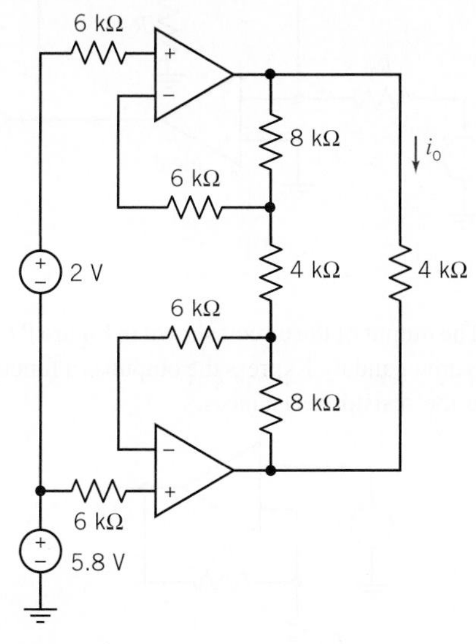

Figure P 6.3-8

P 6.3-9 Determine the voltage v_o for the circuit shown in Figure P 6.3-9.

Answer: $v_o = -12$ V

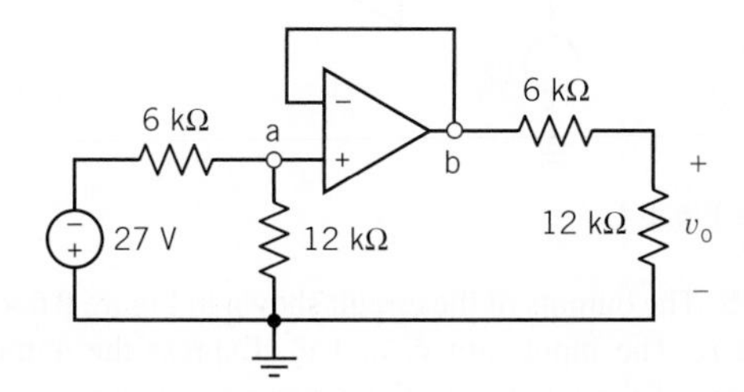

Figure P 6.3-9

P 6.3-10 The circuit shown in Figure P 6.3-10 has one input, i_s, and one output, v_o. Show that the output is proportional to the input. Design the circuit so that the gain is $\frac{v_o}{i_s} = 20\ \frac{\text{V}}{\text{mA}}$.

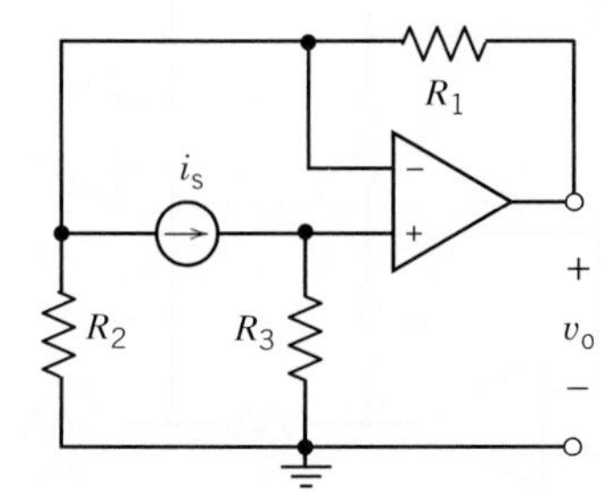

Figure P 6.3-10

P 6.3-11 The circuit shown in Figure P 6.3-11 has one input, v_s, and one output, v_o. Show that the output is proportional to the input. Design the circuit so that $v_o = 5\ v_s$.

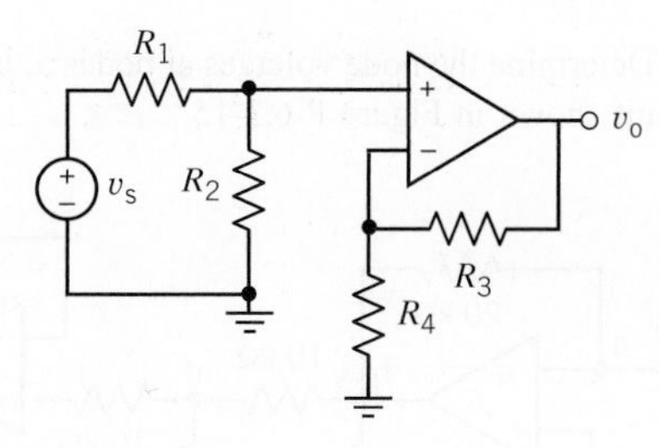

Figure P 6.3-11

P 6.3-12 The input to the circuit shown in Figure P 6.3-12 is the voltage v_s. The output is the voltage v_o. The output is related to the input by the equation $v_o = mv_s + b$ where m and b are constants. Determine the values of m and b.

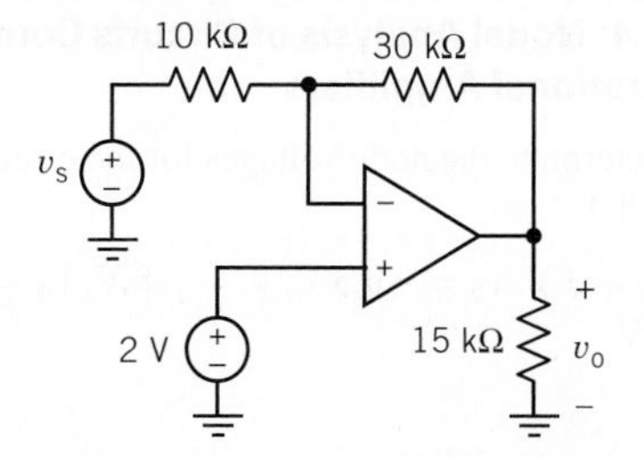

Figure P 6.3-12

P 6.3-13 The output of the circuit shown in Figure P 6.3-13 is $v_o = 3.5$ V. Determine the value of (a) the resistance R, (b) the power supplied be each independent source, and (c) the power, $p_{oa} = i_{oa} \times v_o$ supplied by the op amp.

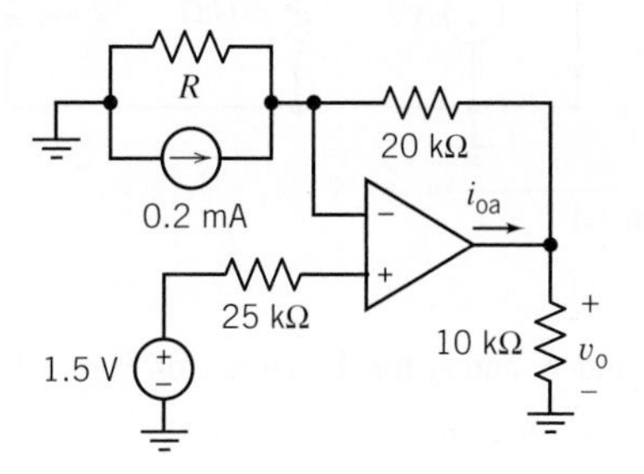

Figure P 6.3-13

P 6.3-14 Determine the node voltages at nodes a, b, c, and d of the circuit shown in Figure P 6.3-14.

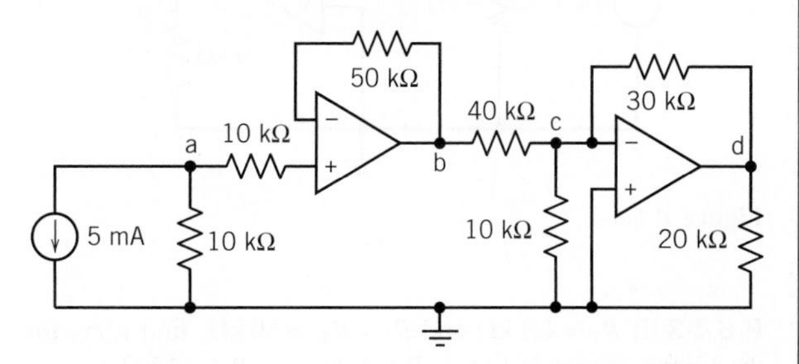

Figure P 6.3-14

P 6.3-15 Determine the node voltages at nodes a, b, c, and d of the circuit shown in Figure P 6.3-15.

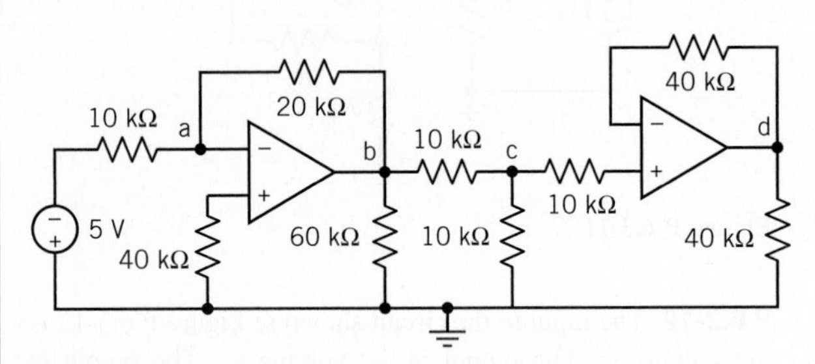

Figure P 6.3-15

Section 6.4 Nodal Analysis of Circuits Containing Ideal Operational Amplifiers

P 6.4-1 Determine the node voltages for the circuit shown in Figure P 6.4-1.

Answer: $v_a = 3$ V, $v_b = -0.5$ V, $v_c = -8$ V, $v_d = -5$ V, and $v_e = -0.5$ V

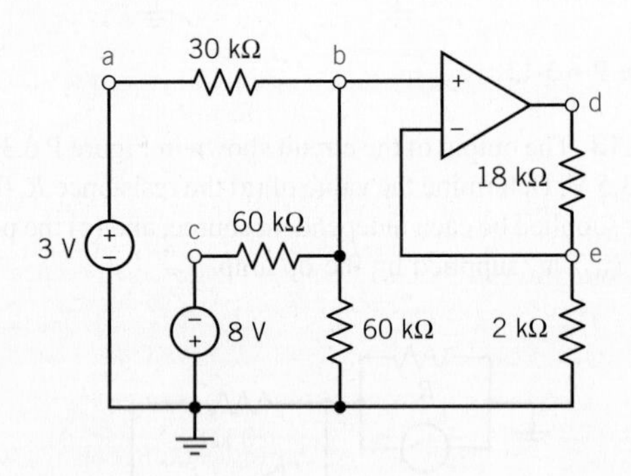

Figure P 6.4-1

P 6.4-2 Find v_o and i_o for the circuit of Figure P 6.4-2.

Answer: $v_o = -6$ V and $i_o = 1.33$ mA

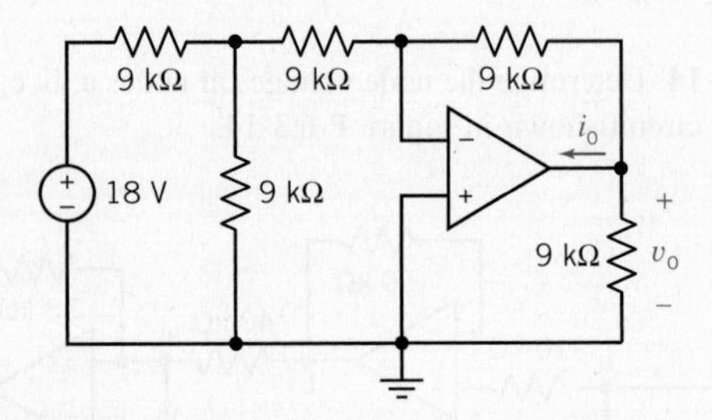

Figure P 6.4-2

P 6.4-3 If $R_1 = 4.8$ kΩ and $R_2 = R_4 = 30$ kΩ, find v_o/v_s for the circuit shown in Figure P 6.4-3 when $R_3 = 1$ kΩ.

Answer: $v_o/v_s = -200$

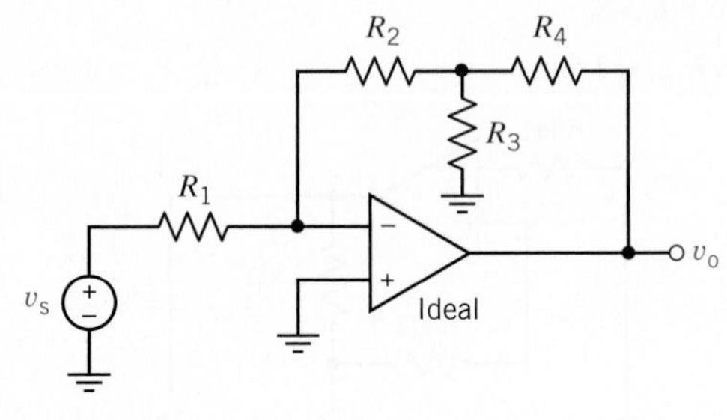

Figure P 6.4-3

P 6.4-4 The output of the circuit shown in Figure P 6.4-4 is v_o. The inputs are v_1 and v_2. Express the output as a function of the inputs and the resistor resistances.

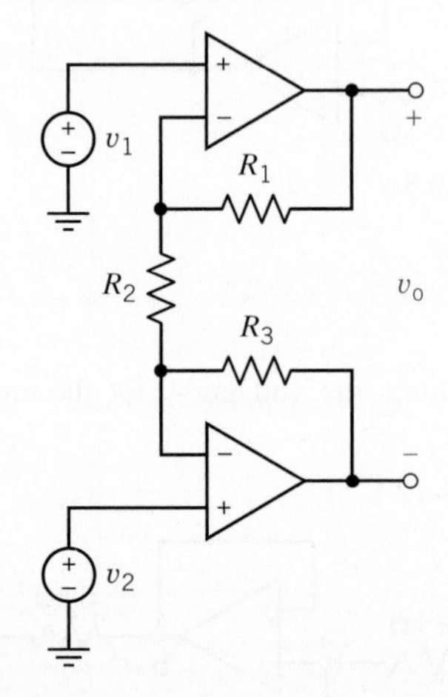

Figure P 6.4-4

P 6.4-5 The outputs of the circuit shown in Figure P 6.4-5 are v_o and i_o. The inputs are v_1 and v_2. Express the outputs as functions of the inputs and the resistor resistances.

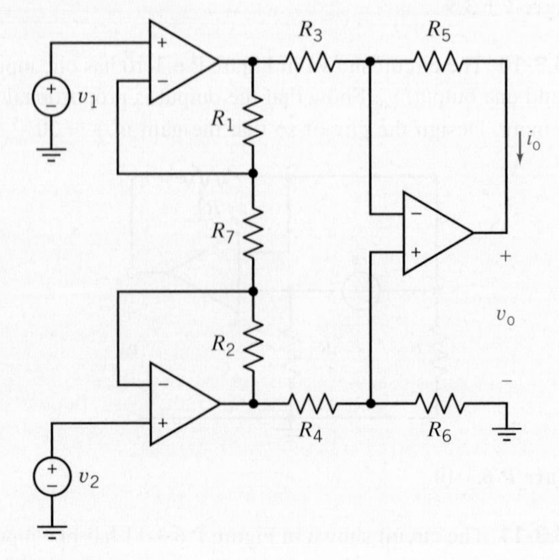

Figure P 6.4-5

P 6.4-6 Determine the node voltages for the circuit shown in Figure P 6.4-6.

Answer: $v_a = -1.85$ V, $v_b = 0$ V, and $v_c = 2.31$ V

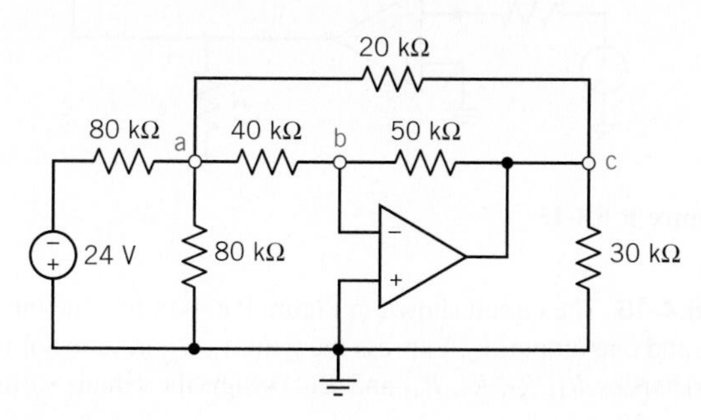

Figure P 6.4-6

P 6.4-7 Find v_o and i_o for the circuit shown in Figure P 6.4-7.

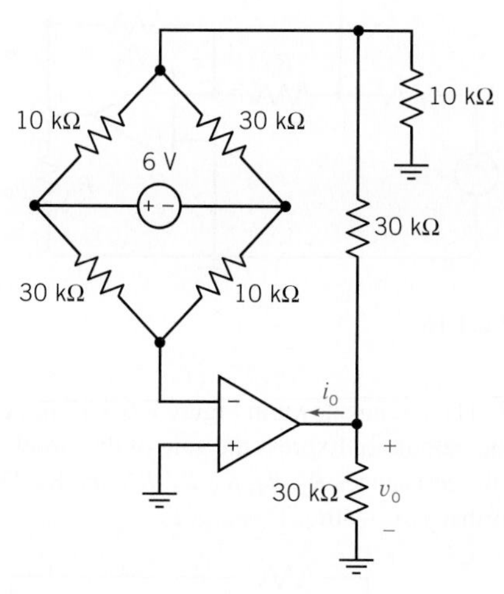

Figure P 6.4-7

P 6.4-8 Find v_o and i_o for the circuit shown in Figure P 6.4-8.

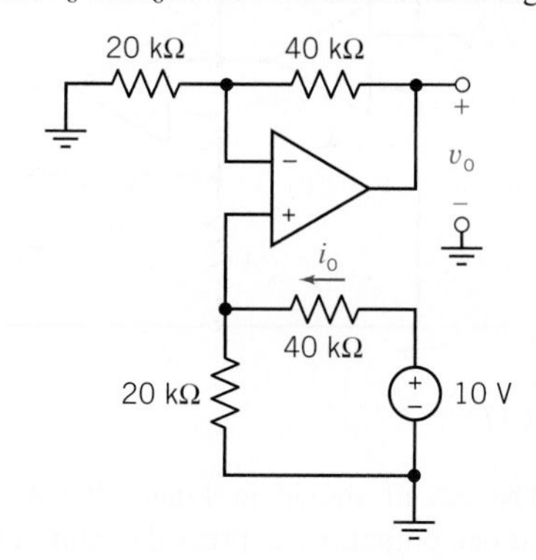

Figure P 6.4-8

P 6.4-9 Determine the node voltages for the circuit shown in Figure P 6.4-9.

Answer: $v_a = -12$ V, $v_b = -4$ V, $v_c = -4$ V, $v_d = -4$ V, $v_e = -3.2$ V, $v_f = -4.8$ V, and $v_g = -3.2$ V

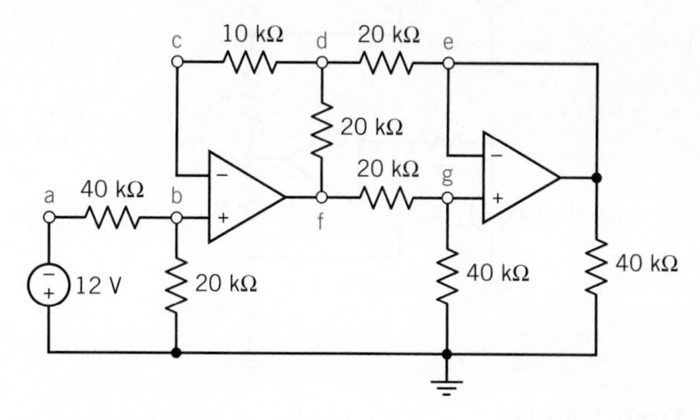

Figure P 6.4-9

P 6.4-10 The circuit shown in Figure P 6.4-10 includes a simple strain gauge. The resistor R changes its value by ΔR when it is twisted or bent. Derive a relation for the voltage gain v_o/v_s and show that it is proportional to the fractional change in R, namely, $\Delta R/R_o$.

Answer: $v_o = \dfrac{R_o}{R_o + R_1} \dfrac{\Delta R}{R_o}$

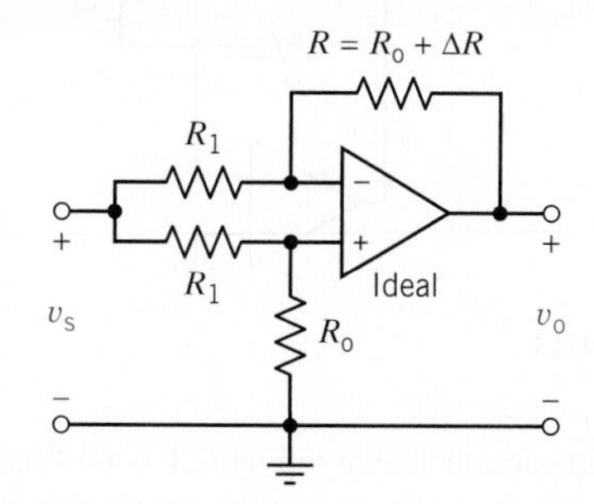

Figure P 6.4-10 A strain gauge circuit.

P 6.4-11 Find v_o for the circuit shown in Figure P 6.4-11.

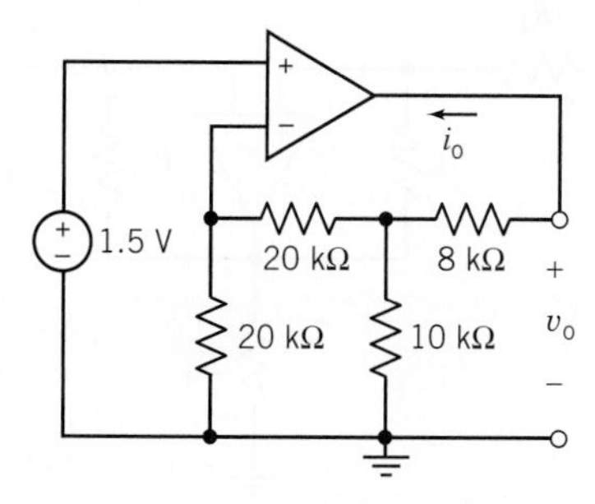

Figure P 6.4-11

P 6.4-12 The circuit shown in Figure P 6.4-12 has one output, v_o, and two inputs, v_1 and v_2. Show that when $\frac{R_3}{R_4} = \frac{R_6}{R_5}$, the output is proportional to the difference of the inputs, $v_1 - v_2$. Specify resistance values to cause $v_o = 5\,(v_1 - v_2)$.

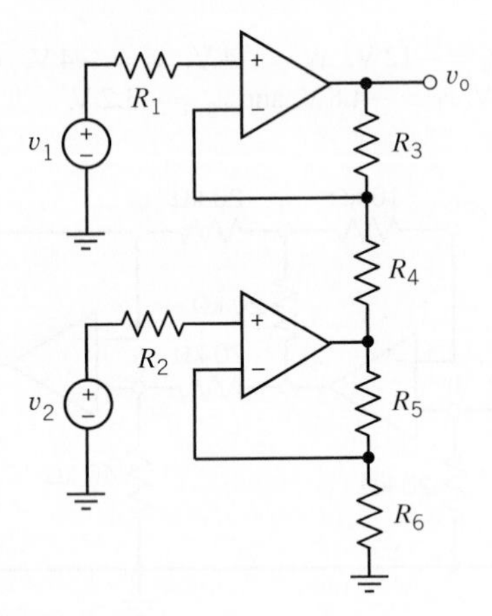

Figure P 6.4-12

P 6.4-13 The circuit shown in Figure P 6.4-13 has one output, v_o, and one input, v_i. Show that the output is proportional to the input. Specify resistance values to cause $v_o = 20v_i$.

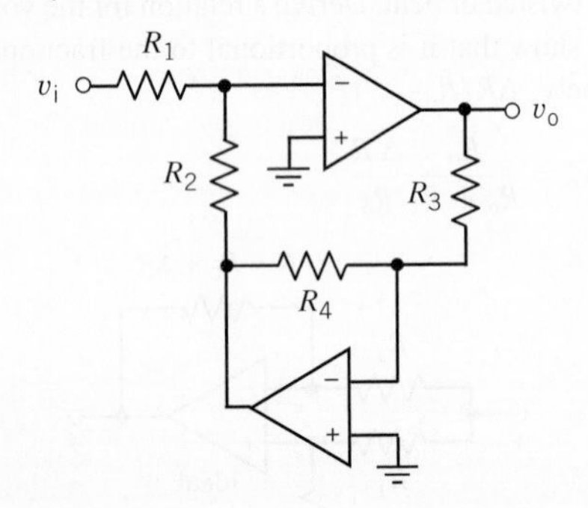

Figure P 6.4-13

P 6.4-14 The circuit shown in Figure P 6.4-14 has one input, v_s, and one output, v_o. Show that the output is proportional to the input. Design the circuit so that $v_o = 20v_s$.

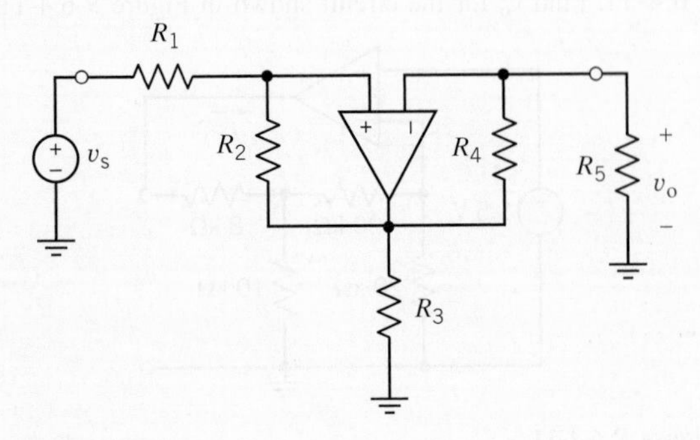

Figure P 6.4-14

P 6.4-15 The circuit shown in Figure P 6.4-15 has one input, v_s, and one output, v_o. The circuit contains seven resistors having equal resistance, R. Express the gain of the circuit, v_o/v_s, in terms of the resistance R.

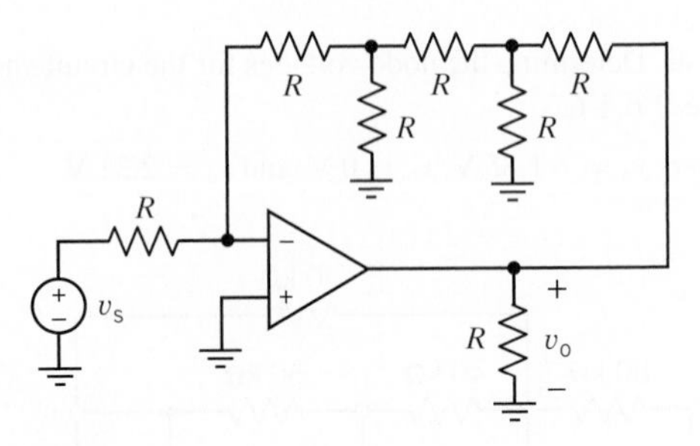

Figure P 6.4-15

P 6.4-16 The circuit shown in Figure P 6.4-16 has one input, v_s, and one output, v_o. Express the gain, v_o/v_s, in terms of the resistances R_1, R_2, R_3, R_4, and R_5. Design the circuit so that $v_o = -30\ v_s$.

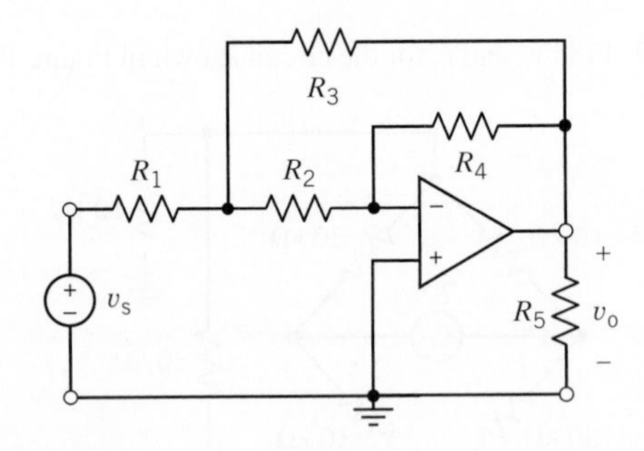

Figure P 6.4-16

P 6.4-17 The circuit shown in Figure P 6.4-17 has one input, v_s, and one output, v_o. Express the gain of the circuit, v_o/v_s, in terms of the resistances R_1, R_2, R_3, R_4, R_5, and R_6. Design the circuit so that $v_o = -40v_s$.

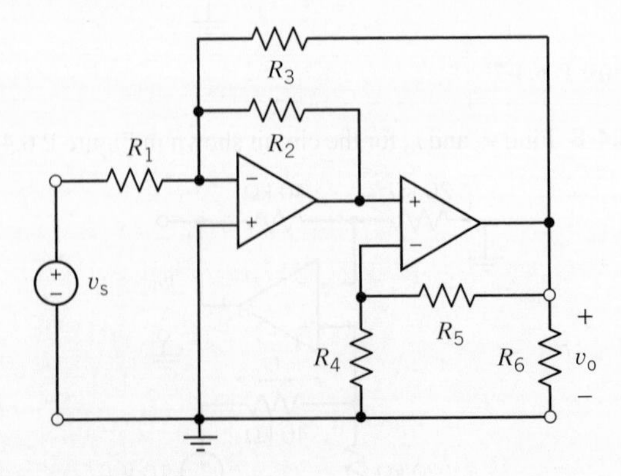

Figure P 6.4-17

P 6.4-18 The circuit shown in Figure P 6.4-18 has one input, v_s, and one output, i_o. Express the gain of the circuit, i_o/v_s, in terms of the resistances R_1, R_2, R_3, and R_o. (This circuit contains a pair of resistors having resistance R_1 and another pair having resistance R_2.) Design the circuit so that $i_o = 0.02v_s$.

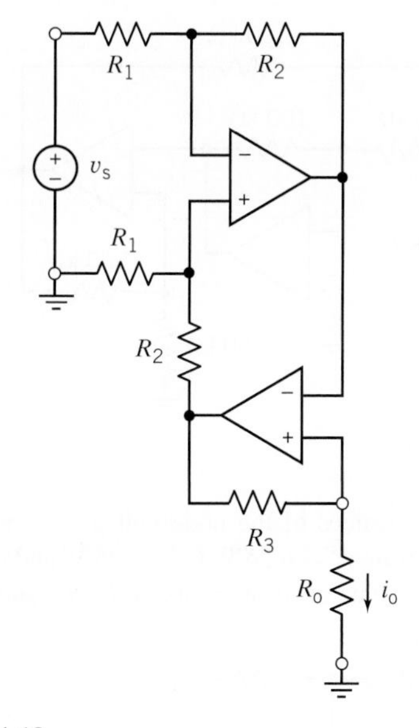

Figure P 6.4-18

P 6.4-19 The circuit shown in Figure P 6.4-19 has one input, v_s, and one output, v_o. The circuit contains one unspecified resistance, R.

(a) Express the gain of the circuit, v_o/v_s, in terms of the resistance R.

(b) Determine the range of values of the gain that can be obtained by specifying a value for the resistance R.

(c) Design the circuit so that $v_o = -2v_s$.

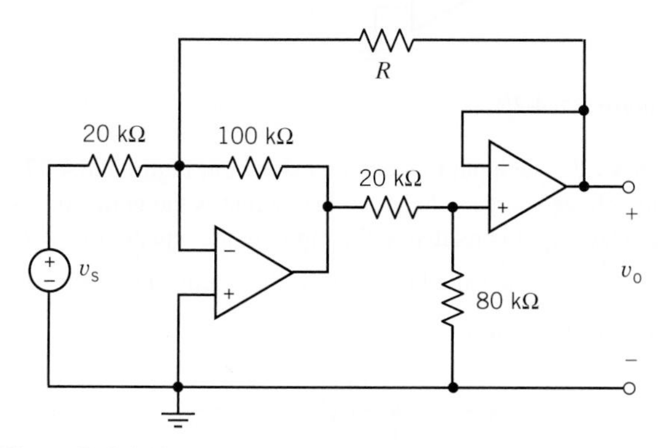

Figure P 6.4-19

P 6.4-20 The circuit shown in Figure P 6.4-20 has one input, v_s, and one output, v_o. The circuit contains one unspecified resistance, R.

(a) Express the gain of the circuit, v_o/v_s, in terms of the resistance R.

(b) Determine the range of values of the gain that can be obtained by specifying a value for the resistance R.

(c) Design the circuit so that $v_o = -5v_s$.

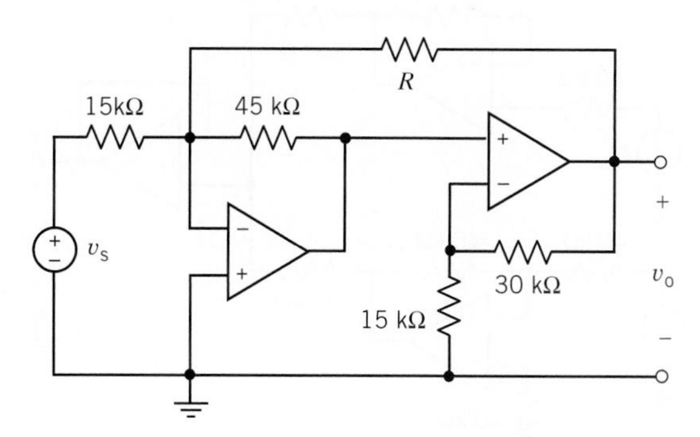

Figure P 6.4-20

P 6.4-21 The circuit shown in Figure P 6.4-21 has three inputs: v_1, v_2, and v_3. The output of the circuit is v_o. The output is related to the inputs by

$$v_o = av_1 + bv_2 + cv_3$$

where a, b, and c are constants. Determine the values of a, b, and c.

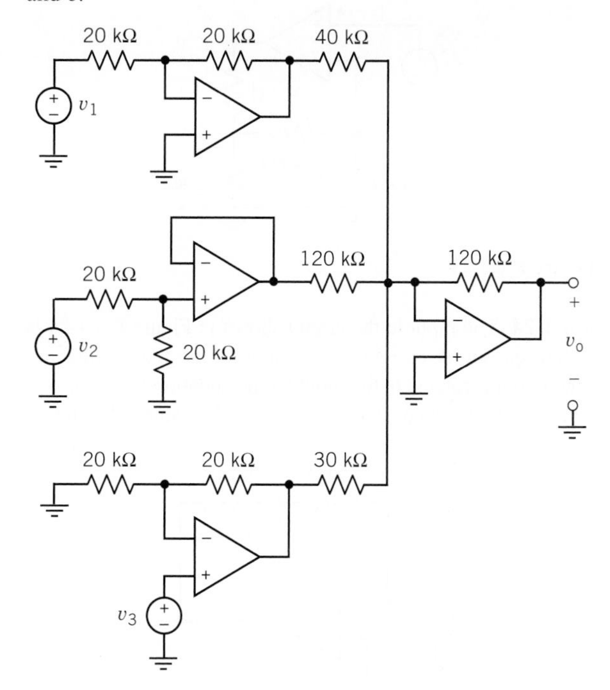

Figure P 6.4-21

P 6.4-22 The circuit shown in Figure P 6.4-22 has two inputs: v_1 and v_2. The output of the circuit is v_o. The output is related to the inputs by

$$v_o = av_1 + bv_2$$

where a and b are constants. Determine the values of a and b.

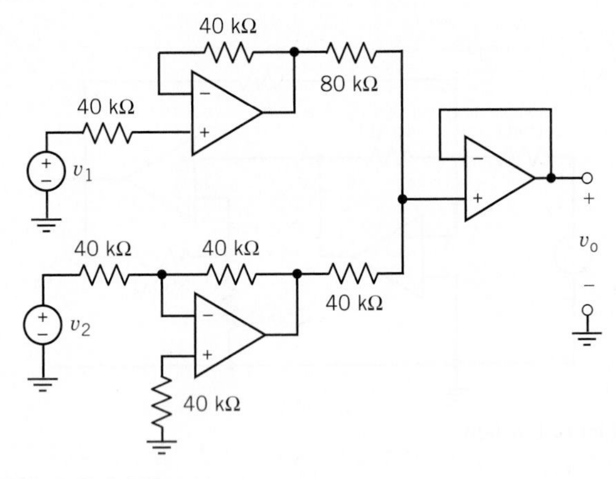

Figure P 6.4-22

P 6.4-23 The input to the circuit shown in Figure P 6.4-23 is the voltage source voltage v_s. The output is the node voltage v_o The output is related to the input by the equation $v_o = kv_s$ where $k = \frac{v_o}{v_s}$ is called the gain of the circuit. Determine the value of the gain k.

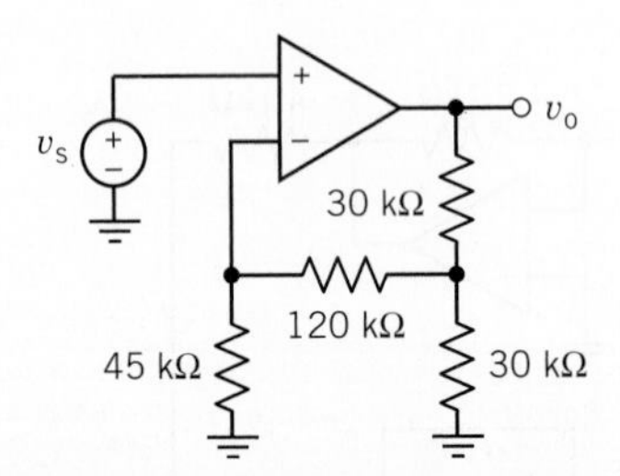

Figure P 6.4-23

P 6.4-24 The input to the circuit shown in Figure P 6.4-24 is the current source current i_s. The output is the node voltage v_o. The output is related to the input by the equation $v_o = mi_s + b$ where m and b are constants. Determine the values of m and b.

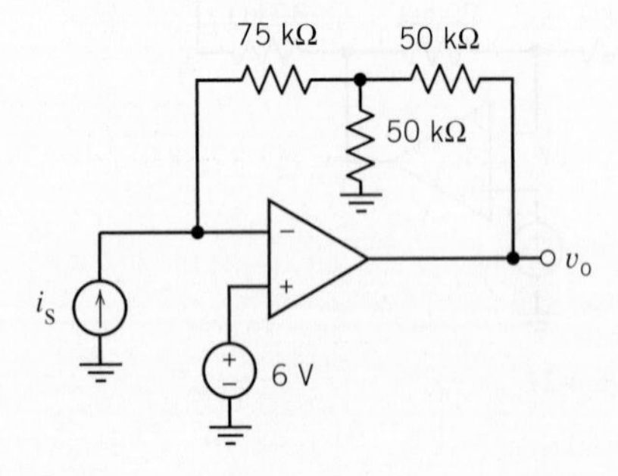

Figure P 6.4-24

P 6.4-25 The input to the circuit shown in Figure P 6.4-25 is the node voltage v_s. The output is the node voltage v_o. The output is related to the input by the equation $v_o = kv_s$ where $k = \frac{v_o}{v_s}$ is called the gain of the circuit. Determine the value of the gain k.

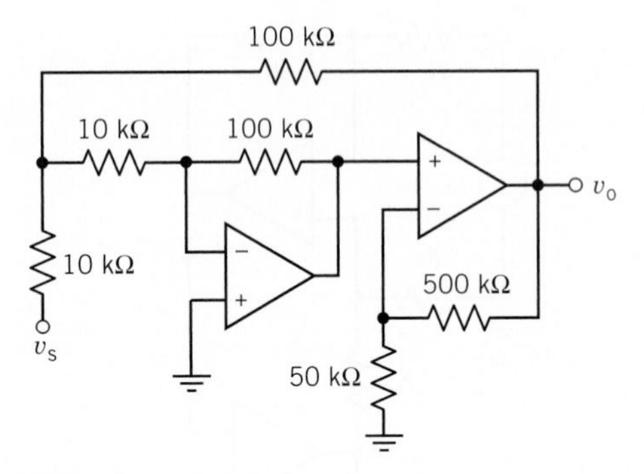

Figure P 6.4-25

P 6.4-26 The values of the node voltages v_1, v_2, and v_o in Figure P 6.4-26 are $v_1 = 6.25$ V, $v_2 = 3.75$ V, and $v_o = -15$ V. Determine the value of the resistances R_1, R_2, and R_3.

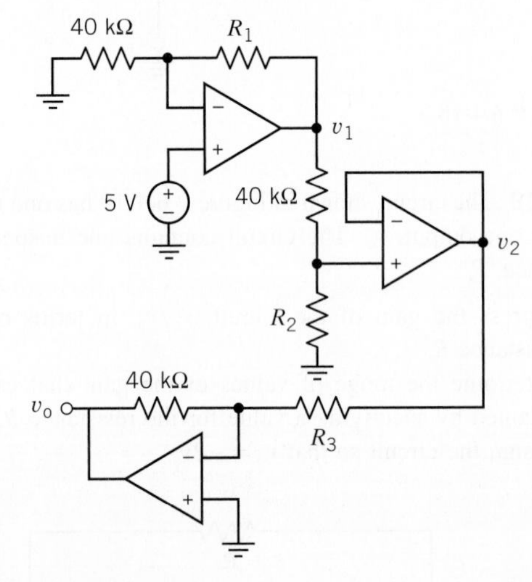

Figure P 6.4-26

P 6.4-27 The input to the circuit shown in Figure P 6.4-27 is the voltage source voltage, v_i, The output is the node voltage, v_o. The output is related to the input by the equation $v_o = kv_i$ where $k = \frac{v_o}{v_i}$ is called the gain of the circuit. Determine the value of the gain k.

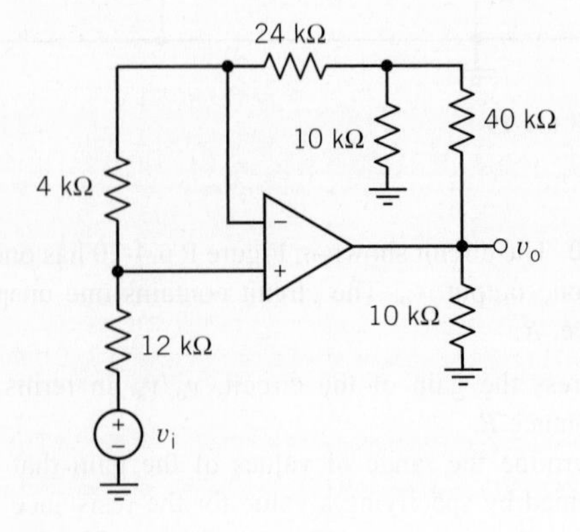

Figure P 6.4-27

Section 6.5 Design Using Operational Amplifiers

P 6.5-1 Design the operational amplifier circuit in Figure P 6.5-1 so that

$$v_{out} = r \cdot i_{in}$$

where

$$r = 20 \frac{V}{mA}$$

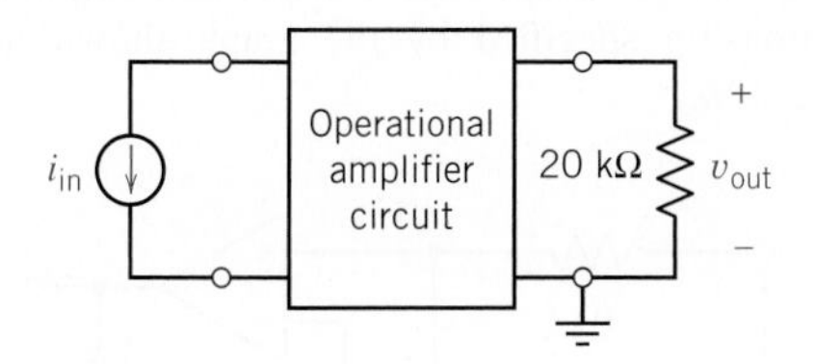

Figure P 6.5-1

P 6.5-2 Design the operational amplifier circuit in Figure P 6.5-2 so that

$$i_{out} = g \cdot v_{in}$$

where

$$g = 2 \frac{mA}{V}$$

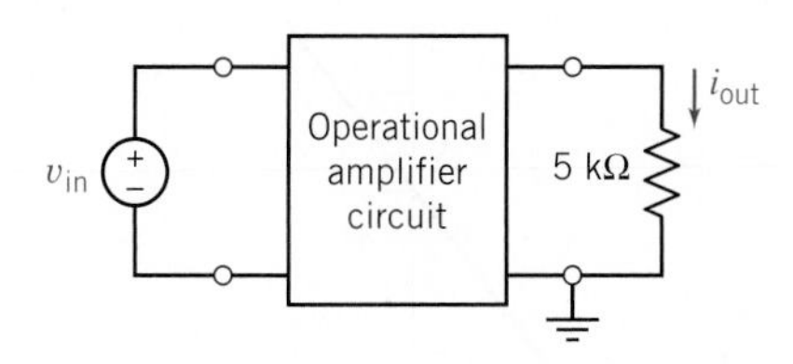

Figure P 6.5-2

P 6.5-3 Design the operational amplifier circuit in Figure P 6.5-3 so that

$$v_{out} = 5 \cdot v_1 + 2 \cdot v_2$$

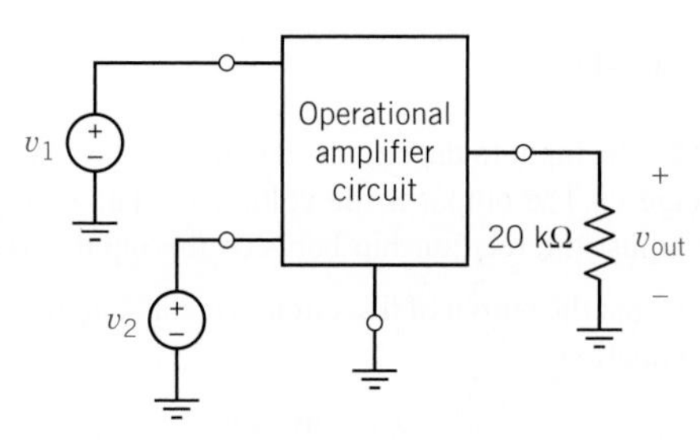

Figure P 6.5-3

P 6.5-4 Design the operational amplifier circuit in Figure P 6.5-3 so that

$$v_{out} = 5 \cdot (v_1 - v_2)$$

P 6.5-5 Design the operational amplifier circuit in Figure P 6.5-3 so that

$$v_{out} = 5 \cdot v_1 - 2 \cdot v_2$$

P 6.5-6 The voltage divider shown in Figure P 6.5-6 has a gain of

$$\frac{v_{out}}{v_{in}} = \frac{-10\ k\Omega}{5\ k\Omega + (-10\ k\Omega)} = 2$$

Design an operational amplifier circuit to implement the -10-kΩ resistor.

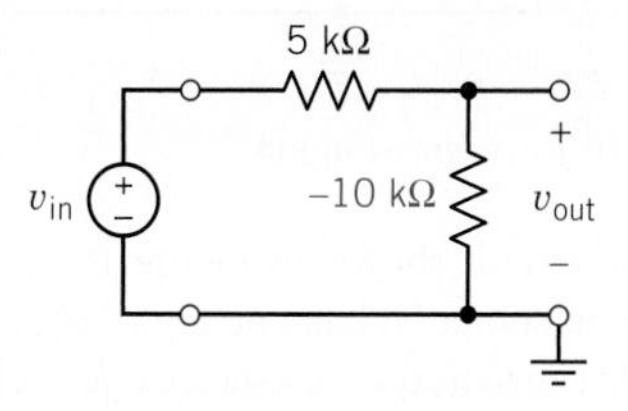

Figure P 6.5-6 A circuit with a negative resistor.

P 6.5-7 Design the operational amplifier circuit in Figure P 6.5-7 so that

$$i_{in} = 0 \quad \text{and} \quad v_{out} = 3 \cdot v_{in}$$

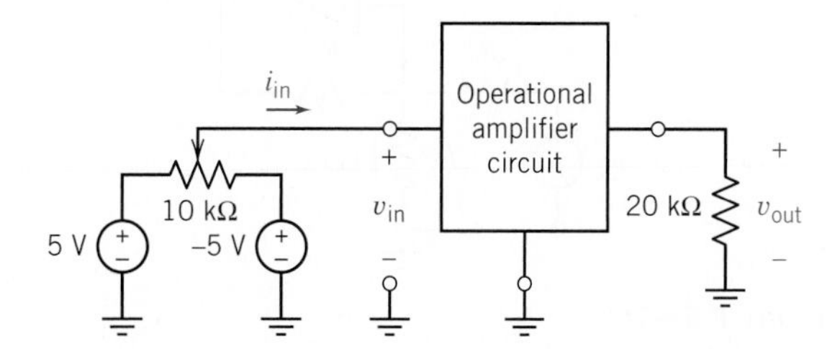

Figure P 6.5-7

P 6.5-8 Design an operational amplifier circuit with output $v_o = 6\ v_1 + 2\ v_2$, where v_1 and v_2 are input voltages.

P 6.5-9 Determine the voltage v_o for the circuit shown in Figure P 6.5-9.

Hint: Use superposition.

Answer: $v_o = (-3)(3) + (4)(-4) + (4)(8) = 7$ V

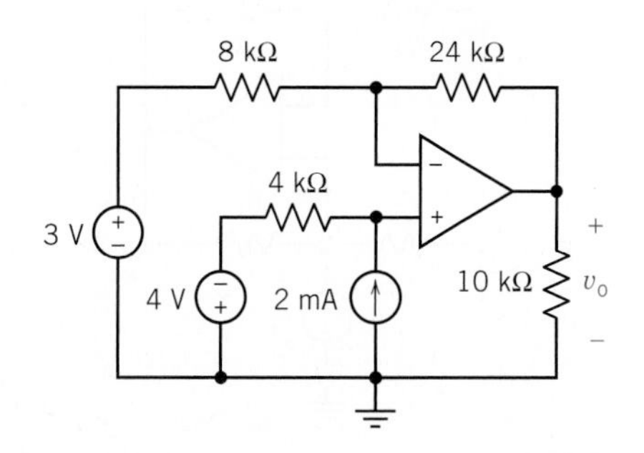

Figure P 6.5-9

P 6.5-10 For the op amp circuit shown in Figure P 6.5-10, find and list all the possible voltage gains that can be achieved by connecting the resistor terminals to either the input or the output voltage terminals.

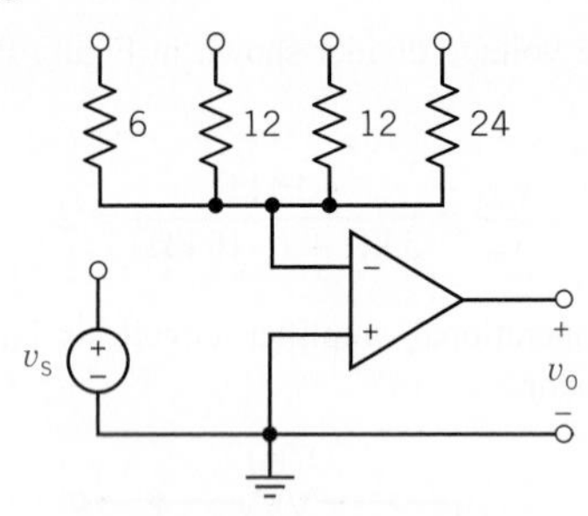

Figure P 6.5-10 Resistances in kΩ.

P 6.5-11 The circuit shown in Figure P 6.5-11 is called a Howland current source. It has one input, v_{in}, and one output, i_{out}. Show that when the resistances are chosen so that $R_2R_3 = R_1R_4$, the output is related to the input by the equation

$$i_{out} = \frac{v_{in}}{R_1}$$

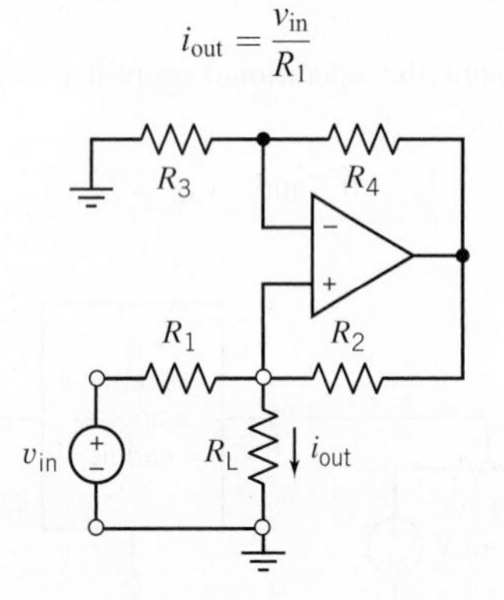

Figure P 6.5-11

P 6.5-12 The circuit shown in Figure P 6.5-12 is used to calculate the output resistance of the Howland current source. It has one input, i_t, and one output, v_t. The output resistance, R_o, is given by

$$R_o = \frac{v_t}{i_t}$$

Express the output resistance of the Howland current source in terms of the resistances R_1, R_2, R_3, and R_4.

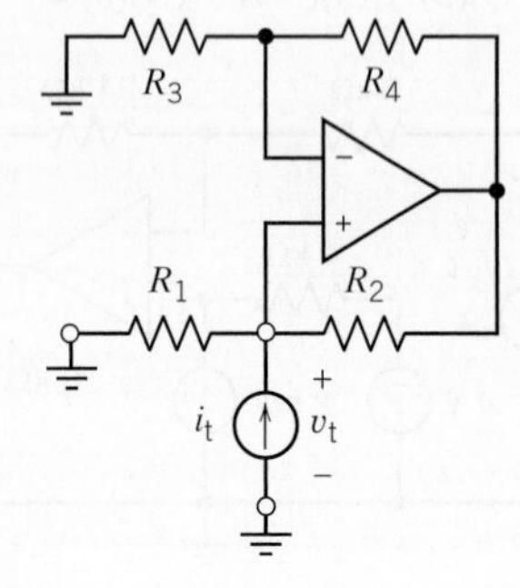

Figure P 6.5-12

P 6.5-13 The input to the circuit shown in Figure P 6.5-13*a* is the voltage v_s. The output is the voltage v_o. The voltage v_b is used to adjust the relationship between the input and output.

(a) Show that the output of this circuit is related to the input by the equation

$$v_o = av_s + b$$

where a and b are constants that depend on R_1, R_2, R_3, R_4, R_5, and v_b.

(b) Design the circuit so that its input and output have the relationship specified by the graph shown in Figure P 6.5-13*b*.

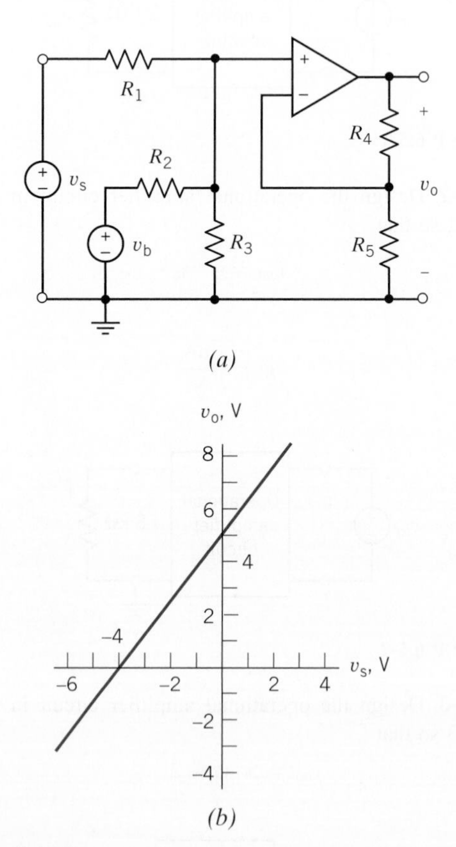

Figure P 6.5-13

P 6.5-14 The input to the circuit shown in Figure P 6.5-14*a* is the voltage v_s. The output is the voltage v_o. The voltage v_b is used to adjust the relationship between the input and output.

(a) Show that the output of this circuit is related to the input by the equation

$$v_o = av_s + b$$

where a and b are constants that depend on R_1, R_2, R_3, R_4, and v_b.

(b) Design the circuit so that its input and output have the relationship specified by the graph shown in Figure P 6.5-14*b*.

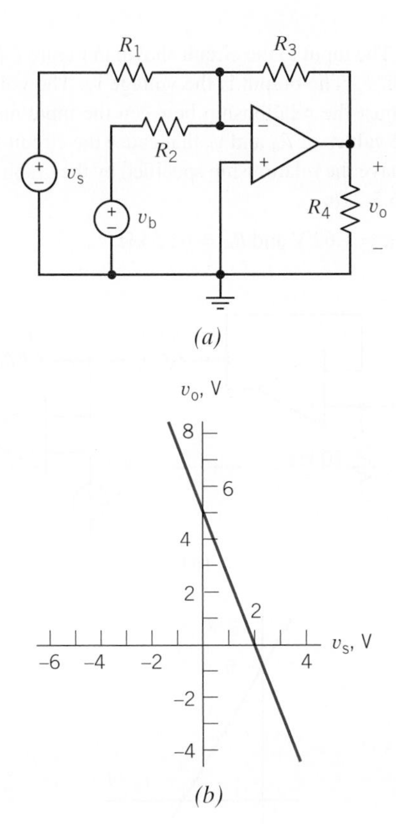

Figure P 6.5-14

***P 6.5-15** The circuit shown in Figure P 6.5-15 contains both an op amp and a potentiometer. This circuit is called an active potentiometer (Graeme, 1982) because the equivalent resistance, R_{eq}, takes both positive and negative values as the position of the potentiometer wiper varies. R_p is the potentiometer resistance. The expressions aR_p and $(1-a)R_p$ indicate the resistances that appear between potentiometer terminals y–w and x–w, respectively. Express the equivalent resistance of the active potentiometer source in terms of R, R_p, and a.

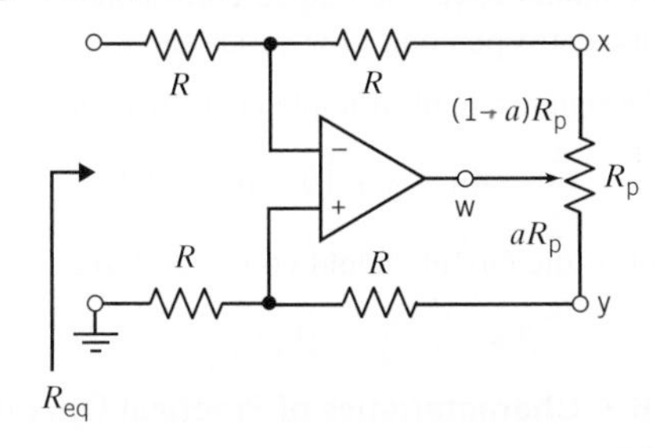

Figure P 6.5-15

***P 6.5-16** The circuit shown in Figure P 6.5-16 contains both op amps and a potentiometer. This circuit has an adjustable gain, v_o/v_1, that takes both positive and negative values as the position of the potentiometer wiper varies (Albean, 1997). R_p is the potentiometer resistance. The expression aR_p indicates the part of R_p that appears between potentiometer terminals y–w.

(a) Express the gain in terms of the resistor resistances, R_p and a.

(b) Set $R_1 = R_3 = R_4 = \frac{1}{2}R_p$. Design the circuit so that the gain varies from −10 V to 10 V as the position of the potentiometer wiper is varied through its full range.

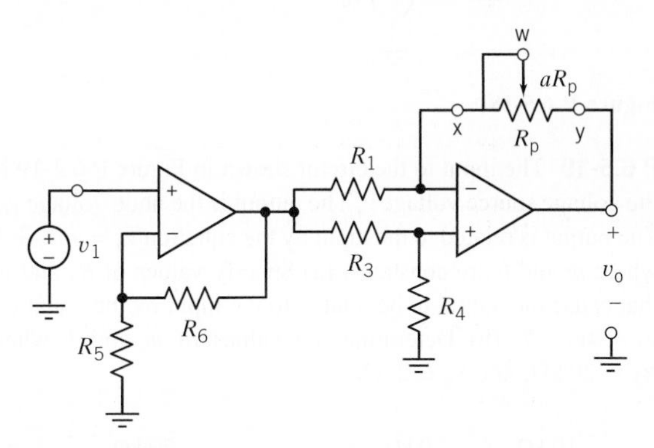

Figure P 6.5-16

P 6.5-17 The input to the circuit shown in Figure P 6.5-17 is the voltage source voltage v_s. The output is the node voltage v_o. The output is related to the input by the equation $v_o = kv_s$ where $k = \frac{v_o}{v_s}$ is called the gain of the circuit. (In Figure P 6.5-17, a and b are positive real constants, so the resistance aR and bR are a and b times as large as the resistances R). Derive an equation that shows how to pick values of a and b that cause the circuit to have a given gain k. Use this equation to design the circuit to have a gain $k =$ 8 V/V using $R = 20\,\text{k}\Omega$.

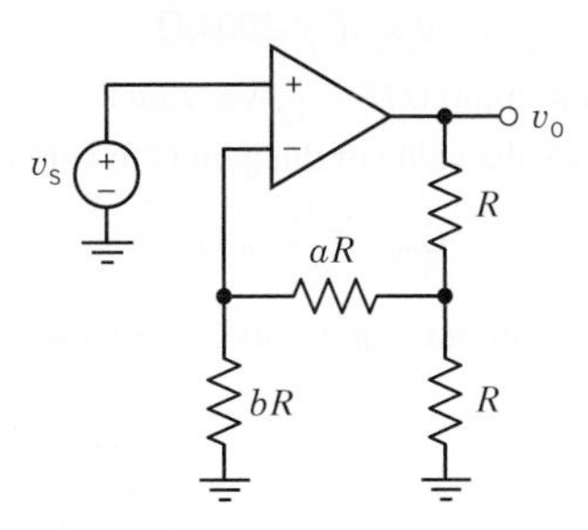

Figure P 6.5-17

P 6.5-18 The input to the circuit shown in Figure P 6.5-18 is the current source current i_s. The output is the node voltage v_o. The output is related to the input by the equation $v_o = mi_s + b$ where m and b are constants. (In Figure P 6.5-18, c and d are positive real constants, so the resistance cR and dR are c and d times as large as the resistance R.) Derive an equation that shows how to pick values of c and d that cause the circuit to have given values of m and b. Use this equation to design the circuit to have $m =$ −125 V/mA and $b = 12$ V when $R = 25\,\text{k}\Omega$.

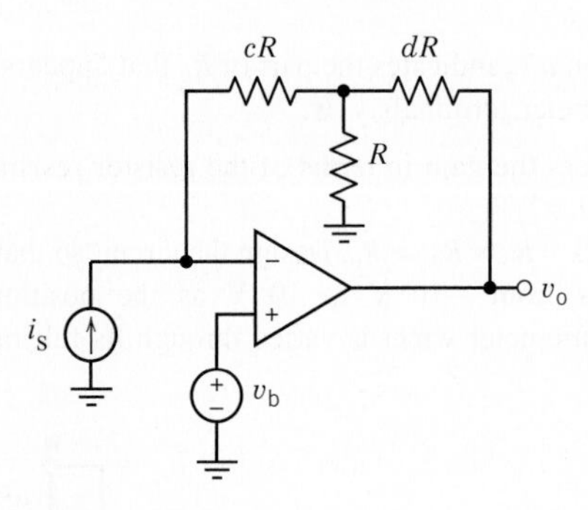

Figure P 6.5-18

P 6.5-19 The input to the circuit shown in Figure P 6.4-19 is the voltage source voltage v_s The output is the node voltage v_o. The output is related to the input by the equation $v_o = mv_s + b$ where m and b are constants. (a) Specify values of R_3 and v_a that cause the output to be related to the input by the equation $v_o = 4v_s + 7$. (b) Determine the values of m and b when $R_3 = 20\ \text{k}\Omega$, and $v_a = 2.5$ V.

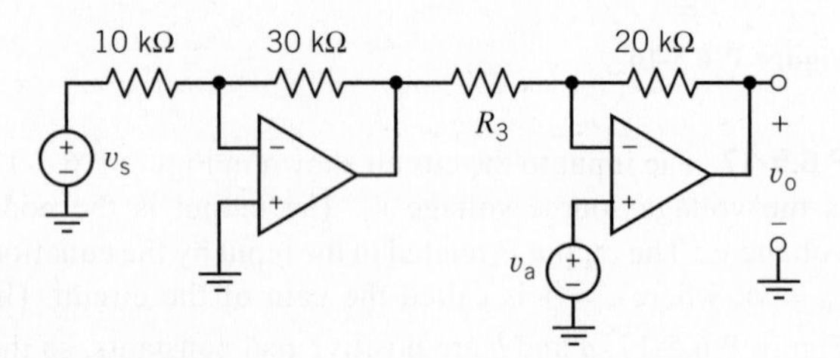

Figure P 6.5-19

P 6.5-20 The circuit shown in Figure P 6.5-20 uses a potentiometer to implement a variable resistor having a resistance R that varies over the range

$$0 < R < 200\ \text{k}\Omega$$

The gain of this circuit is $G = \frac{v_o}{v_s}$. Varying the resistance R over it's range causes the value of the gain G to vary over the range

$$G_{\text{min}} \leq \frac{v_o}{v_s} \leq G_{\text{max}}$$

Determine the minimum and maximum values of the gains, G_{min} and G_{max}.

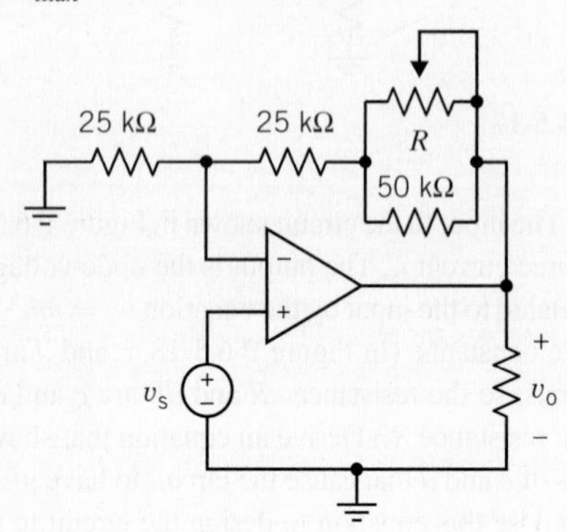

Figure P 6.5-20

P 6.5-21 The input to the circuit shown in Figure P 6.5-21*a* is the voltage, v_s. The output is the voltage v_o. The voltage v_b is used to adjust the relationship between the input and output. Determine values of R_4 and v_b that cause the circuit input and output to have the relationship specified by the graph shown in Figure P 6.5-21*b*.

Answer: $v_b = 1.62$ V and $R_4 = 62.5\ \text{k}\Omega$

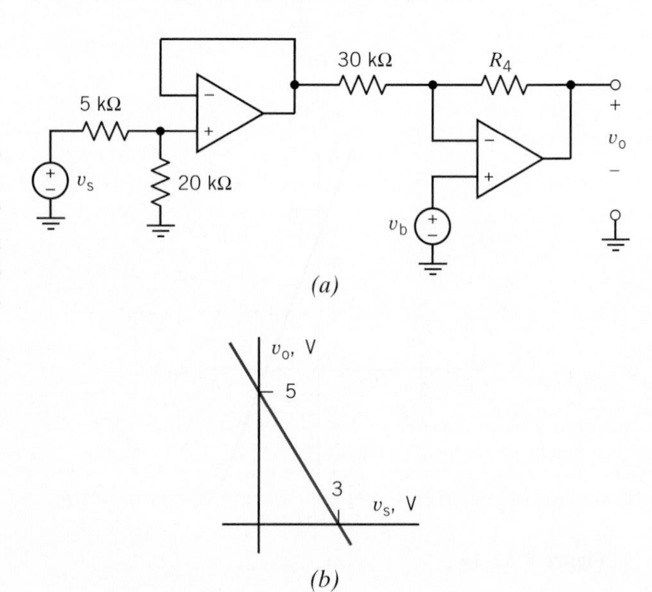

Figure P 6.5-21

Section 6.6 Operational Amplifier Circuits and Linear Algebraic Equations

P 6.6-1 Design a circuit to implement the equation

$$z = 4w + \frac{x}{4} - 3y$$

The circuit should have one output, corresponding to z, and three inputs, corresponding to w, x, and y.

P 6.6-2 Design a circuit to implement the equation

$$0 = 4w + x + 10 - (6y + 2z)$$

The output of the circuit should correspond to z.

Section 6.7 Characteristics of Practical Operational Amplifiers

P 6.7-1 Consider the inverting amplifier shown in Figure P 6.7-1. The operational amplifier is a typical OP-07E (Table 6.7-1). Use the offsets model of the operational amplifier to calculate the output offset voltage. (Recall that the input, v_{in}, is set to zero when calculating the output offset voltage.)

Answer: 0.45 mV

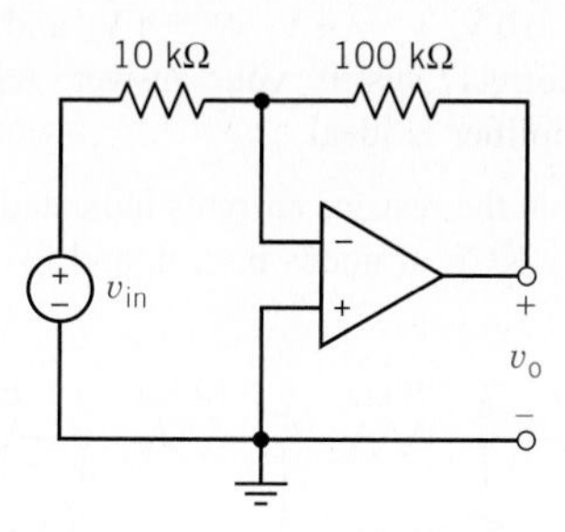

Figure P 6.7-1

P 6.7-2 Consider the noninverting amplifier shown in Figure P 6.7-2. The operational amplifier is a typical LF351 (Table 6.7-1). Use the offsets model of the operational amplifier to calculate the output offset voltage. (Recall that the input, v_{in}, is set to zero when calculating the output offset voltage.)

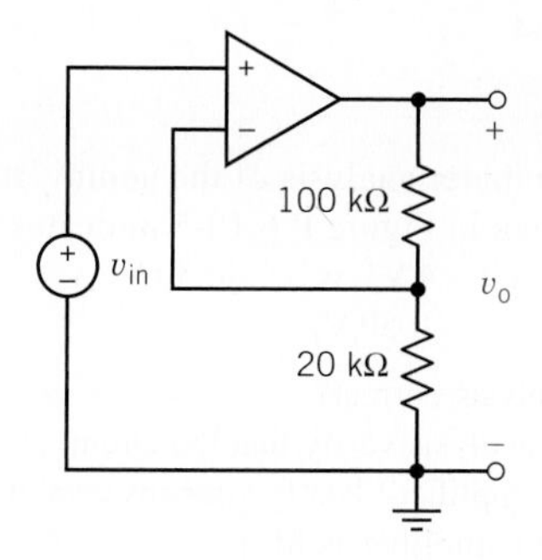

Figure P 6.7-2

P 6.7-3 Consider the inverting amplifier shown in Figure P 6.7-3. Use the finite gain model of the operational amplifier (Figure 6.7-1*c*) to calculate the gain of the inverting amplifier. Show that

$$\frac{v_o}{v_{in}} = \frac{R_{in}(R_o - AR_2)}{(R_1 + R_{in})(R_o + R_2) + R_1 R_{in}(1 + A)}$$

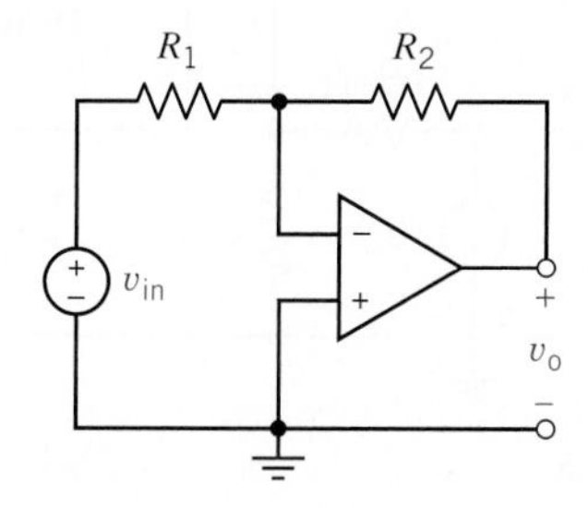

Figure P 6.7-3

P 6.7-4 Consider the inverting amplifier shown in Figure P 6.7-3. Suppose the operational amplifier is ideal, $R_1 = 5\,k\Omega$, and $R_2 = 50\,k\Omega$. The gain of the inverting amplifier will be

$$\frac{v_o}{v_{in}} = -10$$

Use the results of Problem P 6.7-3 to find the gain of the inverting amplifier in each of the following cases:

(a) The operational amplifier is ideal, but 2 percent resistors are used and $R_1 = 5.6\ k\Omega$ and $R_2 = 54\ k\Omega$.
(b) The operational amplifier is represented using the finite gain model with $A = 400{,}000$, $R_i = 4\ M\Omega$, and $R_o = 150\ \Omega$; $R_1 = 10\ k\Omega$ and $R_2 = 100\ k\Omega$.
(c) The operational amplifier is represented using the finite gain model with $A = 400{,}000$, $R_i = 4\ M\Omega$, and $R_o = 150\ \Omega$; $R_1 = 10.2\ k\Omega$ and $R_2 = 98\ k\Omega$.

P 6.7-5 The circuit in Figure P 6.7-5 is called a difference amplifier and is used for instrumentation circuits. The output of a measuring element is represented by the common mode signal v_{cm} and the differential signal $(v_n + v_p)$. Using an ideal operational amplifier, show that

$$v_o = -\frac{R_4}{R_1}(v_n + v_p)$$

when

$$\frac{R_4}{R_1} = \frac{R_3}{R_2}$$

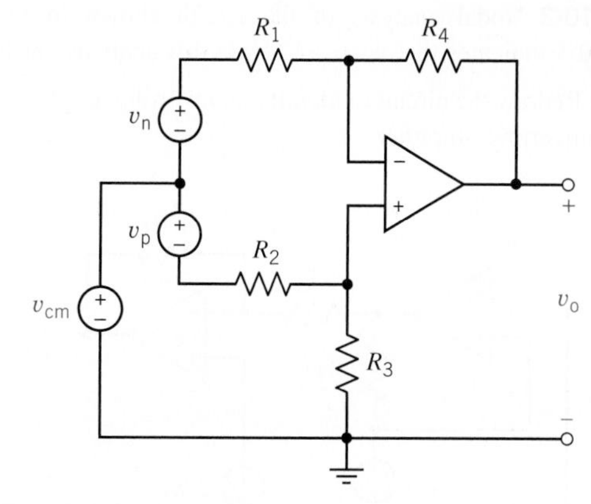

Figure P 6.7-5

Section 6.10 How Can We Check . . . ?

P 6.10-1 Analysis of the circuit in Figure P 6.10-1 shows that $i_o = -1$ mA and $v_o = 7$ V. Is this analysis correct?

Hint: Is KCL satisfied at the output node of the op amp?

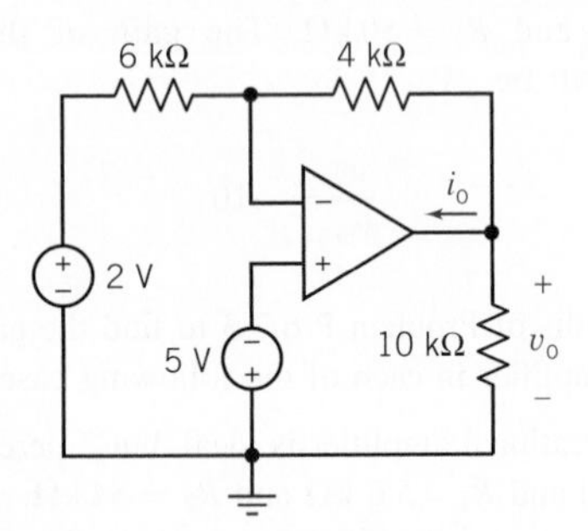

Figure P 6.10-1

P 6.10-2 Your lab partner measured the output voltage of the circuit shown in Figure P 6.10-2 to be $v_o = 9.6$ V. Is this the correct output voltage for this circuit?

Hint: Ask your lab partner to check the polarity of the voltage that he or she measured.

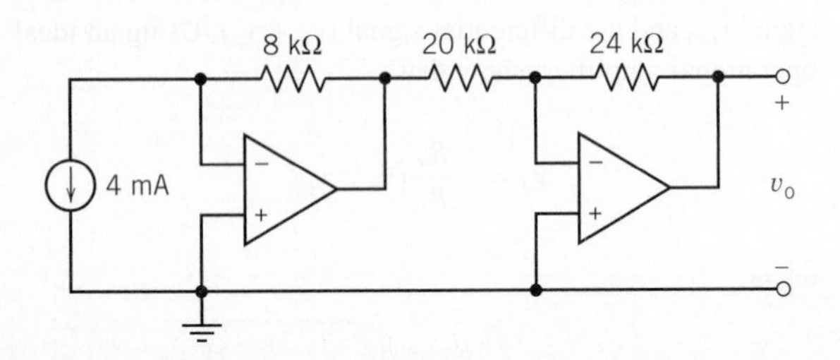

Figure P 6.10-2

P 6.10-3 Nodal analysis of the circuit shown in Figure P 6.10-3 indicates that $v_o = -12$ V. Is this analysis correct?

Hint: Redraw the circuit to identify an inverting amplifier and a noninverting amplifier.

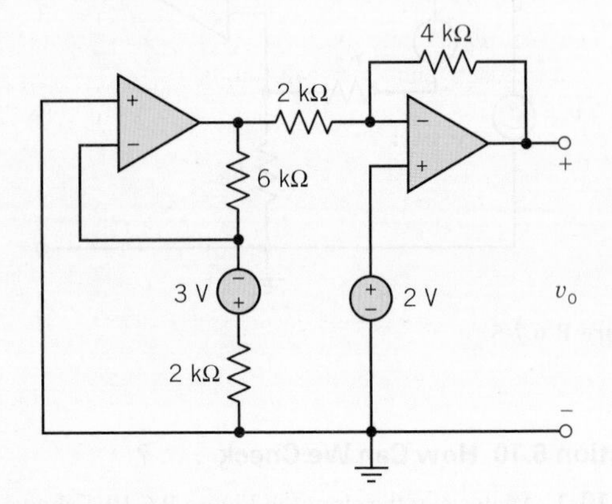

Figure P 6.10-3

P 6.10-4 Computer analysis of the circuit in Figure P 6.10-4 indicates that the node voltages are $v_a = -10$ V, $v_b = 0$ V, $v_c = 4$ V, $v_d = 10$ V, $v_e = 4$ V, $v_f = 4$ V, and $v_g = 22$ V. Is this analysis correct? Justify your answer. Assume that the operational amplifier is ideal.

Hint: Verify that the resistor currents indicated by these node voltages satisfy KCL at nodes b, c, d, and f.

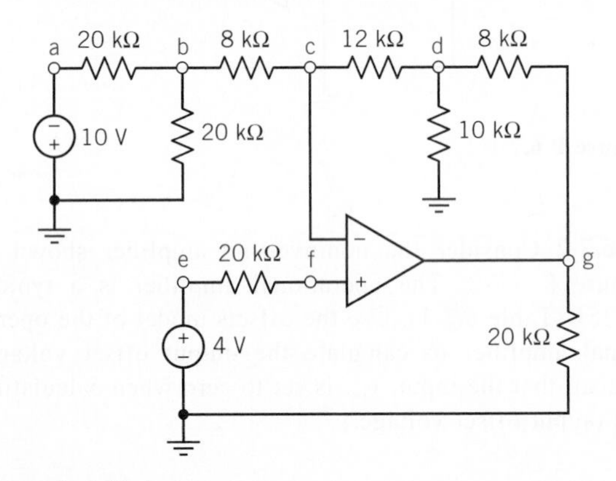

Figure P 6.10-4

P 6.10-5 Computer analysis of the noninverting summing amplifier shown in Figure P 6.10-5 indicates that the node voltages are $v_a = 4$ V, $v_b = -0.50$ V, $v_c = -10$ V, $v_d = -5.0$ V, and $v_e = -0.50$ V.

(a) Is this analysis correct?

(b) Does this analysis verify that the circuit is a noninverting summing amplifier? Justify your answers. Assume that the operational amplifier is ideal.

1st Hint: Verify that the resistor currents indicated by these node voltages satisfy KCL at nodes b and e.

2nd Hint: Compare to Figure 6.5-1e to see that $R_a = 10$ kΩ and $R_b = 1$ kΩ. Determine K_1, K_2, and K_4 from the resistance values. Verify that $v_d = K_4(K_1 v_a + K_2 v_c)$.

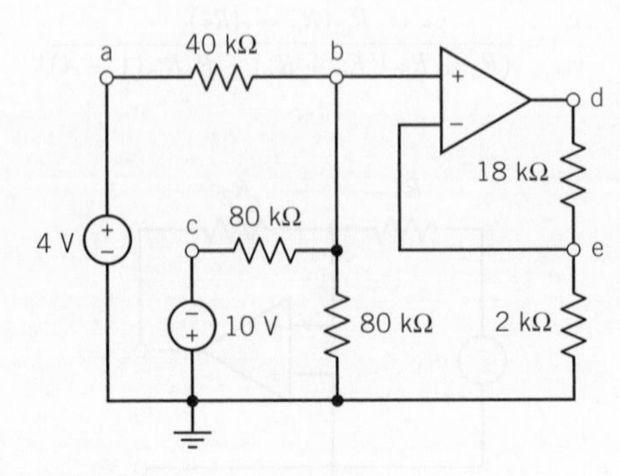

Figure P 6.10-5

PSpice Problems

SP 6-1 The circuit in Figure SP 6-1 has three inputs: v_w, v_x, and v_y. The circuit has one output, v_z. The equation

$$v_z = av_w + bv_x + cv_y$$

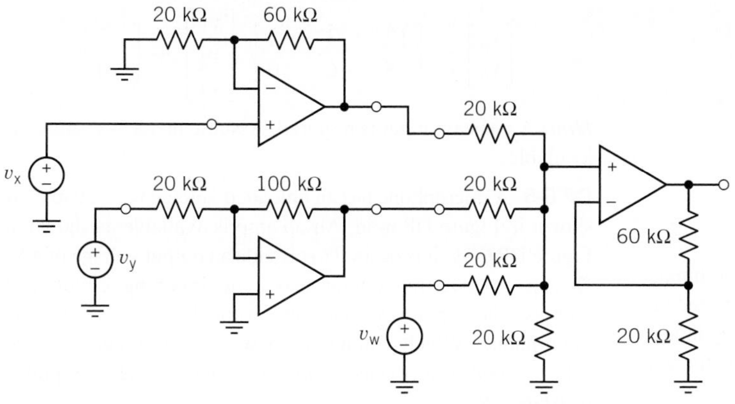

Figure SP 6-1

expresses the output as a function of the inputs. The coefficients a, b, and c are real constants.

(a) Use PSpice and the principle of superposition to determine the values of a, b, and c.

(b) Suppose $v_w = 2$ V, $v_x = x$, $v_y = y$ and we want the output to be $v_z = z$. Express z as a function of x and y.

Hint: The output is given by $v_z = a$ when $v_w = 1$ V, $v_x = 0$ V, and $v_y = 0$ V.

Answer: (a) $v_z = v_w + 4\,v_x - 5\,v_y$ (b) $z = 4\,x - 5\,y + 2$

SP 6-2 The input to the circuit in Figure SP 6-2 is v_s, and the output is v_o. (a) Use superposition to express v_o as a function of v_s. (b) Use the DC Sweep feature of PSpice to plot v_o as a function of v_s. (c) Verify that the results of parts (a) and (b) agree with each other.

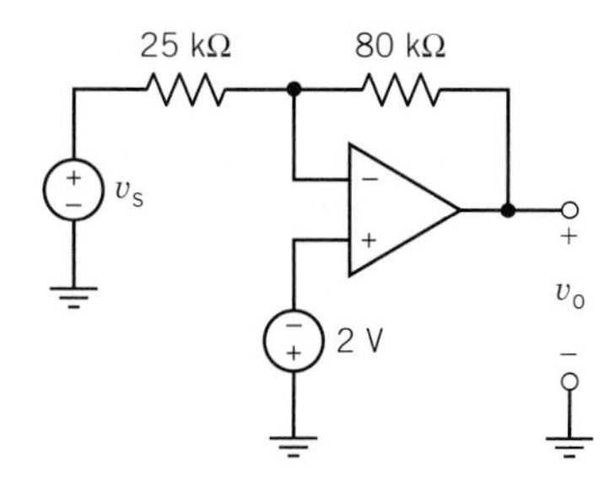

Figure SP 6-2

SP 6-3 A circuit with its nodes identified is shown in Figure SP 6-3. Determine v_{34}, v_{23}, v_{50}, and i_o.

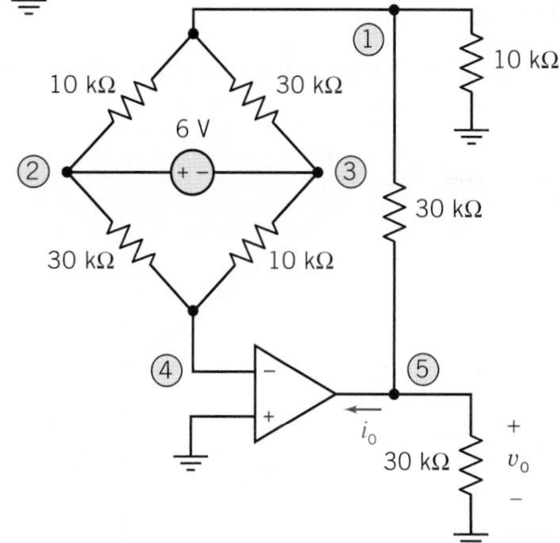

Figure SP 6-3 Bridge circuit.

SP 6-4 Use PSpice to analyze the VCCS shown in Figure SP 6-4. Consider two cases:

(a) The operational amplifier is ideal.

(b) The operational amplifier is a typical μA741 represented by the offsets and finite gain model.

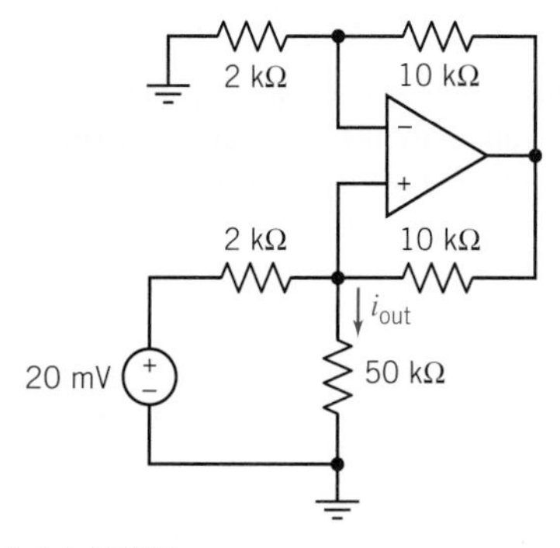

Figure SP 6-4 A VCCS.

Design Problems

DP 6-1 Design the operational amplifier circuit in Figure DP 6-1 so that

$$i_{\text{out}} = \frac{1}{4} \cdot i_{\text{in}}$$

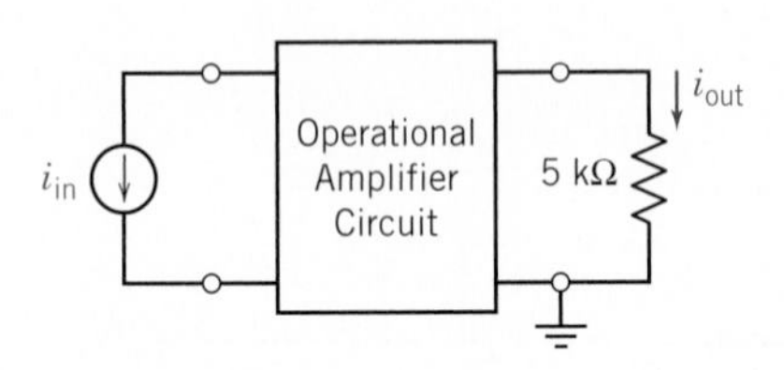

Figure DP 6-1

DP 6-2 Figure DP 6-2*a* shows a circuit that has one input, v_i, and one output, v_o. Figure DP 6-2*b* shows a graph that specifies a relationship between v_o and v_i. Design a circuit having input, v_i, and output, v_o, that have the relationship specified by the graph in Figure DP 6-2*b*.

Hint: A constant input is required. Assume that a 5-V source is available.

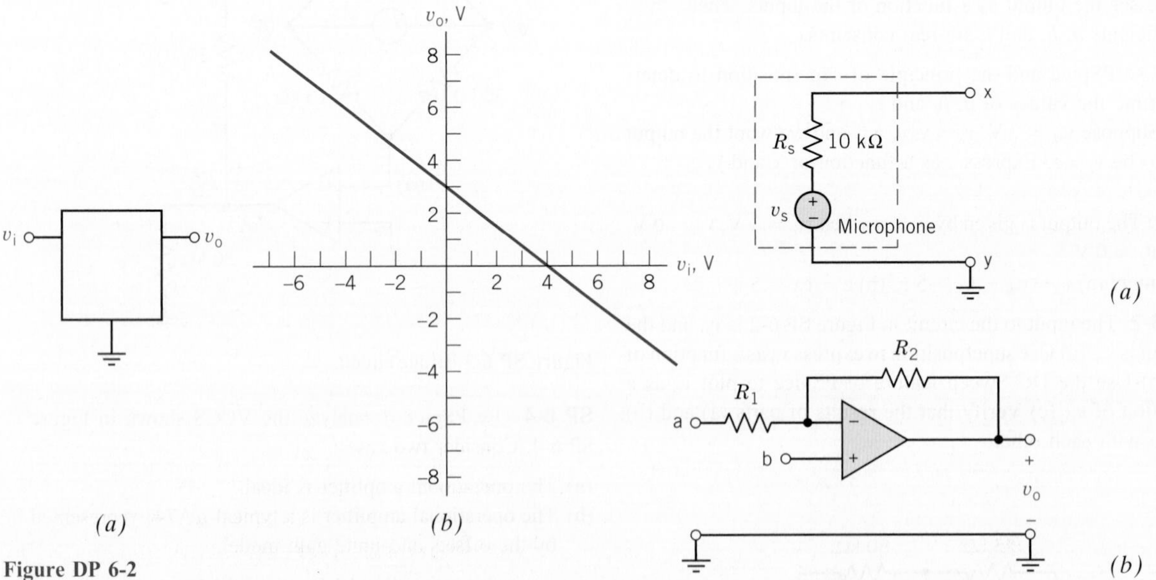

Figure DP 6-2

DP 6-3 Design a circuit having input, v_i, and output, v_o, that are related by the equations (a) $v_o = 12v_i + 6$, (b) $v_o = 12v_i - 6$, (c) $v_o = -12v_i + 6$, and (d) $v_o = -12v_i - 6$.

Hint: A constant input is required. Assume that a 5-V source is available.

DP 6-4 Design a circuit having three inputs, v_1, v_2, v_3, and two outputs, v_a, v_b, that are related by the equation

$$\begin{bmatrix} v_a \\ v_b \end{bmatrix} = \begin{bmatrix} 12 & 3 & -2 \\ 8 & -6 & 0 \end{bmatrix} \begin{bmatrix} v_1 \\ v_2 \\ v_3 \end{bmatrix} + \begin{bmatrix} 2 \\ -4 \end{bmatrix}$$

Hint: A constant input is required. Assume that a 5-V source is available.

DP 6-5 A microphone has an unloaded voltage $v_s = 20$ mV, as shown in Figure DP 6-5*a*. An op amp is available as shown in Figure DP 6-5*b*. It is desired to provide an output voltage of 4 V. Design an inverting circuit and a noninverting circuit and contrast the input resistance at terminals x–y seen by the microphone. Which configuration would you recommend to achieve good performance in spite of changes in the microphone resistance R_s?

Hint: We plan to connect terminal a to terminal x and terminal b to terminal y or vice versa.

Figure DP 6-5 Microphone and op amp circuit.

CHAPTER 7

Energy Storage Elements

IN THIS CHAPTER

7.1 Introduction
7.2 Capacitors
7.3 Energy Storage in a Capacitor
7.4 Series and Parallel Capacitors
7.5 Inductors
7.6 Energy Storage in an Inductor
7.7 Series and Parallel Inductors
7.8 Initial Conditions of Switched Circuits
7.9 Operational Amplifier Circuits and Linear Differential Equations
7.10 Using MATLAB to Plot Capacitor or Inductor Voltage and Current
7.11 How Can We Check . . . ?
7.12 **DESIGN EXAMPLE**—Integrator and Switch
7.13 Summary
Problems
Design Problem

7.1 INTRODUCTION

This chapter introduces two more circuit elements, the capacitor and the inductor. The constitutive equations for the devices involve either integration or differentiation. Consequently:

- Electric circuits that contain capacitors and/or inductors are represented by differential equations. Circuits that do not contain capacitors or inductors are represented by algebraic equations. We say that circuits containing capacitors and/or inductors are **dynamic** circuits, whereas circuits that do not contain capacitors or inductors are **static** circuits.
- Circuits that contain capacitors and/or inductors are able to store energy.
- Circuits that contain capacitors and/or inductors have memory. The voltages and currents at a particular time depend not only on other voltages at currents at that same instant of time but also on previous values of those currents and voltages.

In addition, we will see that:

- In the absence of unbounded currents or voltages, capacitor voltages and inductor currents are continuous functions of time.
- In a dc circuit, capacitors act like open circuits, and inductors act like short circuits.
- A set of series or parallel capacitors can be reduced to an equivalent capacitor. A set of series or parallel inductors can be reduced to an equivalent inductor. Doing so does not change the element current or voltage of any other circuit element.

- An op amp and a capacitor can be used to make circuits that perform the mathematical operations of integration or differentiation. Appropriately, these important circuits are called the integrator and the differentiator.
- The element voltages and currents in a circuit containing capacitors and inductors can be complicated functions of time. MATLAB is useful for plotting these functions.

7.2 CAPACITORS

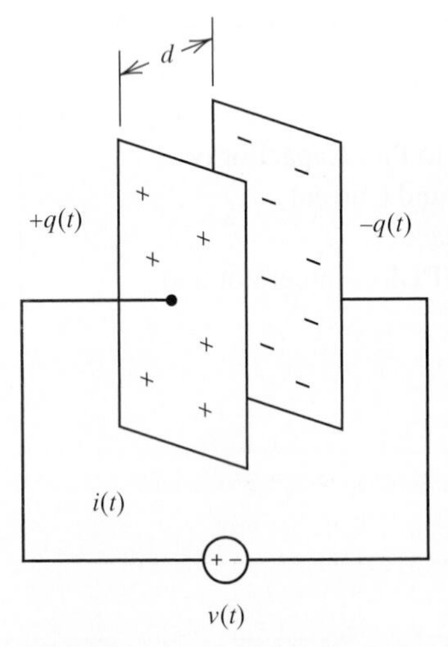

FIGURE 7.2-1 A capacitor connected to a voltage source.

A capacitor is a circuit element that stores energy in an electric field. A capacitor can be constructed by using two parallel conducting plates separated by distance d as shown in Figure 7.2-1. Electric charge is stored on the plates, and a uniform electric field exists between the conducting plates whenever there is a voltage across the capacitor. The space between the plates is filled with a dielectric material. Some capacitors use impregnated paper for a dielectric, whereas others use mica sheets, ceramics, metal films, or just air. A property of the dielectric material, called the dielectric constant, describes the relationship between the electric field strength and the capacitor voltage. Capacitors are represented by a parameter called the *capacitance*. The capacitance of a capacitor is proportional to the dielectric constant and surface area of the plates and is inversely proportional to the distance between the plates. In other words, the capacitance C of a capacitor is given by

$$C = \frac{\in A}{d}$$

where $\in$ is the dielectric constant, A the area of the plates, and d the distance between plates. The unit of capacitance is coulomb per volt and is called farad (F) in honor of Michael Faraday.

A capacitor voltage $v(t)$ deposits a charge $+q(t)$ on one plate and a charge $-q(t)$ on the other plate. We say that the charge $q(t)$ is stored by the capacitor. The charge stored by a capacitor is proportional to the capacitor voltage, $v(t)$. Thus, we write

$$q(t) = Cv(t) \tag{7.2-1}$$

where the constant of proportionality, C, is the capacitance of the capacitor.

Capacitance is a measure of the ability of a device to store energy in the form of a separated charge or an electric field.

In general, the capacitor voltage $v(t)$ varies as a function of time. Consequently, $q(t)$, the charge stored by the capacitor, also varies as a function of time. The variation of the capacitor charge with respect to time implies a capacitor current, $i(t)$, given by

$$i(t) = \frac{d}{dt}q(t)$$

We differentiate Eq. 7.2-1 to obtain

$$i(t) = C\frac{d}{dt}v(t) \tag{7.2-2}$$

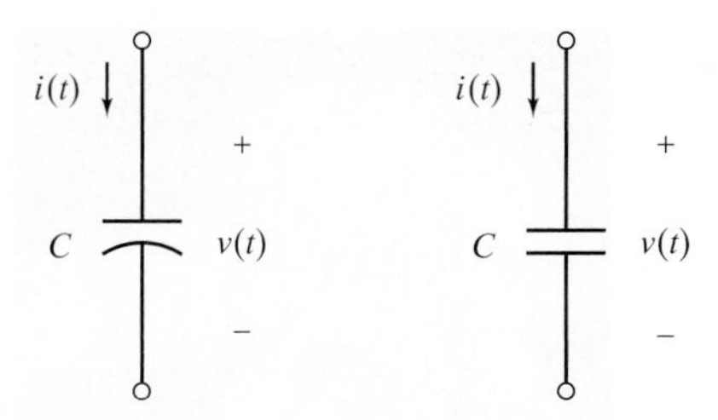

FIGURE 7.2-2 Circuit symbols of a capacitor.

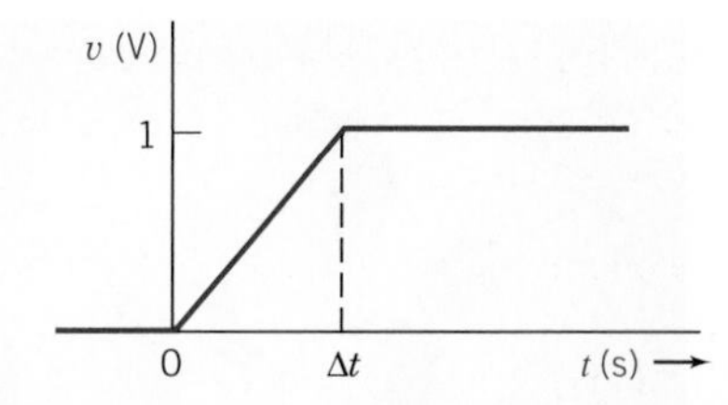

FIGURE 7.2-3 Voltage waveform in which the change in voltage occurs over an increment of time, Δt.

Equation 7.2-2 is the current–voltage relationship of a capacitor. The current and voltage in Eq. 7.7-2 adhere to the passive convention. Figure 7.2-2 shows two alternative symbols to represent capacitors in circuit diagrams. In both Figure 7.2-2(*a*) and (*b*), the capacitor current and voltage adhere to the passive sign convention and are related by Eq. 7.2-2.

Now consider the waveform shown in Figure 7.2-3, in which the voltage changes from a constant voltage of zero to another constant voltage of 1 over an increment of time, Δt. Using Eq. 7.2-2, we obtain

$$i(t) = \begin{cases} 0 & t < 0 \\ \dfrac{C}{\Delta t} & 0 < t < \Delta t \\ 0 & t > \Delta t \end{cases}$$

Thus, we obtain a pulse of height equal to $C/\Delta t$. As Δt decreases, the current will increase. Clearly, Δt cannot decline to zero or we would experience an infinite current. An infinite current is an impossibility because it would require infinite power. Thus, an instantaneous $(\Delta t = 0)$ change of voltage across the capacitor is not possible. In other words, we cannot have a discontinuity in $v(t)$.

The voltage across a **capacitor** cannot change instantaneously.

Now, let us find the voltage $v(t)$ in terms of the current $i(t)$ by integrating both sides of Eq. 7.2-2. We obtain

$$v(t) = \frac{1}{C}\int_{-\infty}^{t} i(\tau)d\tau \qquad (7.2\text{-}3)$$

This equation says that the capacitor voltage $v(t)$ can be found by integrating the capacitor current from time $-\infty$ until time t. To do so requires that we know the value of the capacitor current from time $\tau = -\infty$ until time $\tau = t$. Often, we don't know the value of the current all the way back to $\tau = -\infty$. Instead, we break the integral up into two parts:

$$v(t) = \frac{1}{C}\int_{t_0}^{t} i(\tau)d\tau + \frac{1}{C}\int_{-\infty}^{t_0} i(\tau)d\tau = \frac{1}{C}\int_{t_0}^{t} i(\tau)d\tau + v(t_0) \qquad (7.2\text{-}4)$$

This equation says that the capacitor voltage $v(t)$ can be found by integrating the capacitor current from some convenient time $\tau = t_0$ until time $\tau = t$, provided that we also know the capacitor voltage at time t_0. Now we are required to know only the capacitor current from time $\tau = t_0$ until time $\tau = t$. The time t_0 is called the **initial time**, and the capacitor voltage $v(t_0)$ is called the **initial condition**. Frequently, it is convenient to select $t_0 = 0$ as the initial time.

Capacitors are commercially available in a variety of types and capacitance values. Capacitor types are described in terms of the dielectric material and the construction technique. Miniature metal film capacitors are shown in Figure 7.2-4. Miniature hermetically sealed polycarbonate capacitors are

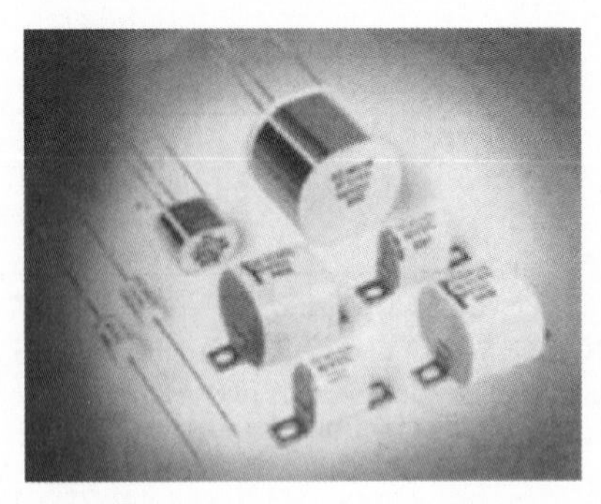

FIGURE 7.2-4 Miniature metal film capacitors ranging from 1 mF to 50 mF. Courtesy of Electronic Concepts Inc.

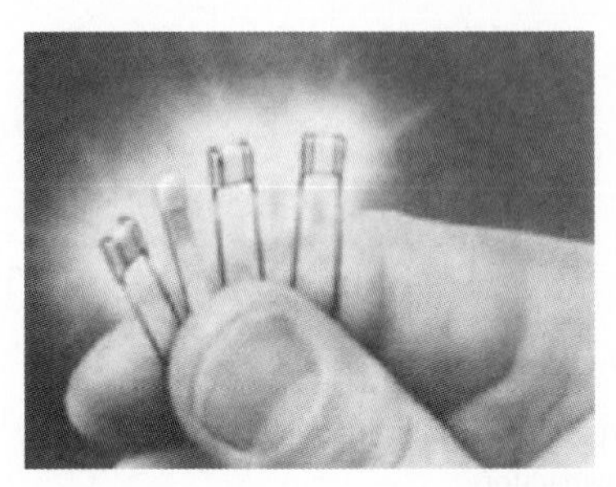

FIGURE 7.2-5 Miniature hermetically sealed polycarbonate capacitors ranging from 1 μF to 50 μF. Courtesy of Electronic Concepts Inc.

shown in Figure 7.2-5. Capacitance values typically range from picofarads (pF) to microfarads (μF). Two pieces of insulated wire about an inch long when twisted together will have a capacitance of about 1 pF. On the other hand, a power supply capacitor about an inch in diameter and a few inches long may have a capacitance of 0.01 F.

Actual capacitors have some resistance associated with them. Fortunately, it is easy to include approximate resistive effects in the circuit models. In capacitors, the dielectric material between the plates is not a perfect insulator and has some small conductivity. This can be represented by a very high resistance in parallel with the capacitor. Ordinary capacitors can hold a charge for hours, and the parallel resistance is then hundreds of megaohms. For this reason, the resistance associated with a capacitor is usually ignored.

EXAMPLE 7.2-1 Capacitor Current and Voltage

Find the current for a capacitor $C = 1$ mF when the voltage across the capacitor is represented by the signal shown in Figure 7.2-6.

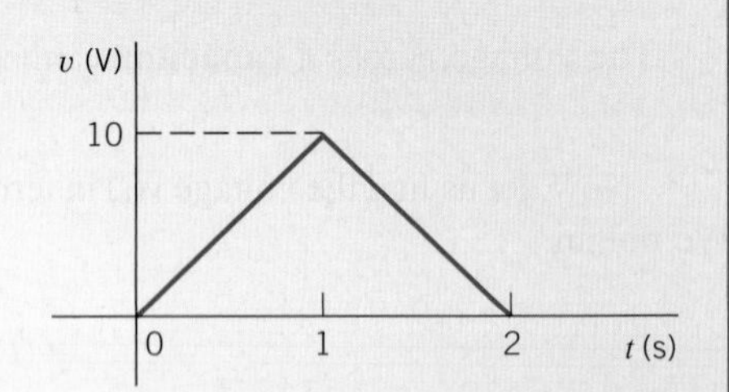

FIGURE 7.2-6 Waveform of the voltage across a capacitor for Example 7.2-1. The units are volts and seconds.

Solution

The voltage (with units of volts) is given by

$$v(t) = \begin{cases} 0 & t \le 0 \\ 10t & 0 \le t \le 1 \\ 20 - 10t & 1 \le t \le 2 \\ 0 & t \ge 2 \end{cases}$$

Then, because $i = C\,dv/dt$, where $C = 10^{-3}$ F, we obtain

$$i(t) = \begin{cases} 0 & t < 0 \\ 10^{-2} & 0 < t < 1 \\ -10^{-2} & 1 < t < 2 \\ 0 & t > 2 \end{cases}$$

Therefore, the resulting current is a series of two pulses of magnitudes 10^{-2} A and -10^{-2} A, respectively, as shown in Figure 7.2-7.

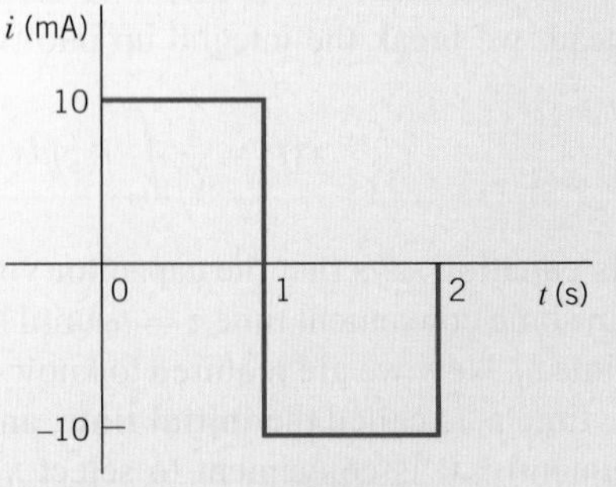

FIGURE 7.2-7 Current for Example 7.2-1.

Example 7.2-2 Capacitor Current and Voltage

Find the voltage $v(t)$ for a capacitor $C = 1/2$ F when the current is as shown in Figure 7.2-8 and $v(0) = 0$.

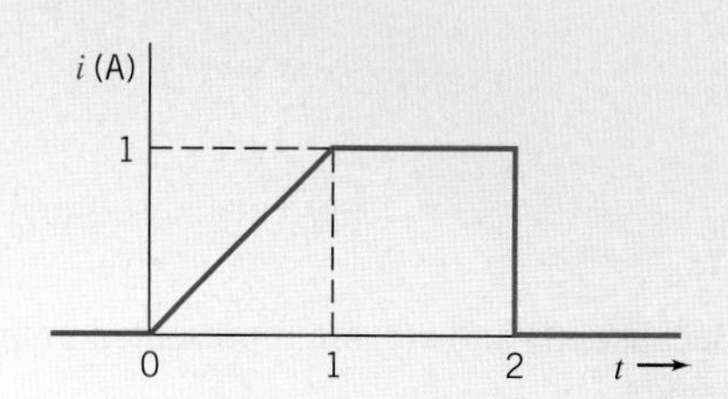

FIGURE 7.2-8 Circuit waveform for Example 7.2-2. The units are in amperes and seconds.

Solution

First, we write the equation for $i(t)$ as

$$i(t) = \begin{cases} 0 & t \le 0 \\ t & 0 \le t \le 1 \\ 1 & 1 \le t \le 2 \\ 0 & 2 < t \end{cases}$$

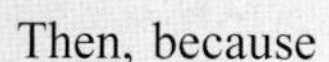

Then, because

$$v(t) = \frac{1}{C}\int_0^t i(\tau)d\tau$$

and $C = 1/2$, we have

$$v(t) = \begin{cases} 0 & t \le 0 \\ 2\int_0^t \tau d\tau & 0 \le t \le 1 \\ 2\int_1^t (1)d\tau + v(1) & 1 \le t \le 2 \\ v(2) & 2 \le t \end{cases}$$

with units of volts. Therefore, for $0 < t \le 1$, we have

$$v(t) = t^2$$

For the period $1 \le t \le 2$, we note that $v(1) = 1$ and, therefore, we have

$$v(t) = 2(t-1) + 1 = (2t-1)\text{ V}$$

The resulting voltage waveform is shown in Figure 7.2-9. The voltage changes with t^2 during the first 1 s, changes linearly with t during the period from 1 to 2 s, and stays constant equal to 3 V after $t = 2$ s.

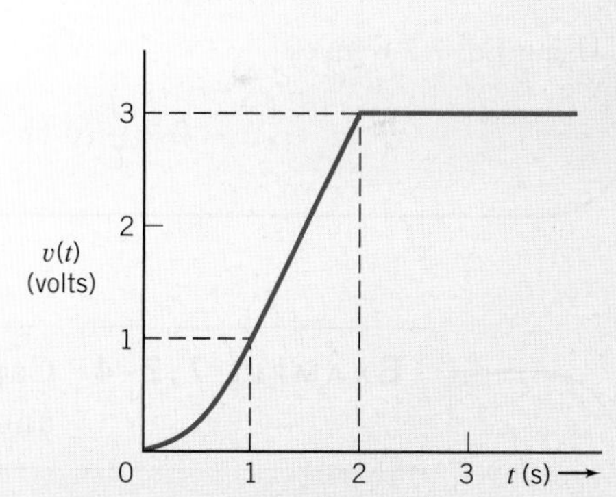

FIGURE 7.2-9 Voltage waveform for Example 7.2-2.

Example 7.2-3 Capacitor Current and Voltage

Figure 7.2-10 shows a circuit together with two plots. The plots represent the current and voltage of the capacitor in the circuit. Determine the value of the capacitance of the capacitor.

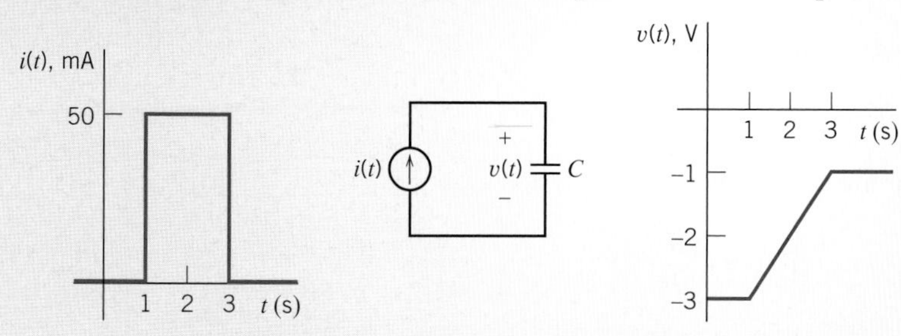

FIGURE 7.2-10 The circuit and plots considered in Example 7.2-3.

Solution

The current and voltage of the capacitor are related by

$$v(t) = \frac{1}{C}\int_{t_0}^{t} i(\tau)\, d\tau + v(t_0) \tag{7.2-5}$$

or

$$v(t) - v(t_0) = \frac{1}{C}\int_{t_0}^{t} i(\tau)\, d\tau \tag{7.2-6}$$

Because $i(t)$ and $v(t)$ are represented graphically by plots rather than equations, it is useful to interpret Eq. 7.2-6 using

$$v(t) - v(t_0) = \text{the difference between the values of voltage at times } t \text{ and } t_0$$

and

$$\int_{t_0}^{t} i(\tau)\, d\tau = \text{ the area under the plot of } i(t) \text{ versus } t \text{ for times between } t \text{ and } t_0$$

Pick convenient values t and t_0, for example, $t_0 = 1$ s and $t = 3$ s. Then,

$$v(t) - v(t_0) = -1 - (-3) = 2\text{ V}$$

and

$$\int_{t_0}^{t} i(\tau)\, d\tau = \int_{1}^{3} 0.05\, d\tau = (0.05)(3-1) = 0.1\text{ A}\cdot\text{s}$$

Using Eq. 7.2-6 gives

$$2 = \frac{1}{C}(0.1) \quad \Rightarrow \quad C = 0.05\,\frac{\text{A}\cdot\text{s}}{\text{V}} = 0.05\text{ F} = 50\text{ mF}$$

EXAMPLE 7.2-4 Capacitor Current and Voltage

INTERACTIVE EXAMPLE

Figure 7.2-11 shows a circuit together with two plots. The plots represent the current and voltage of the capacitor in the circuit. Determine the values of the constants, a and b, used to label the plot of the capacitor current.

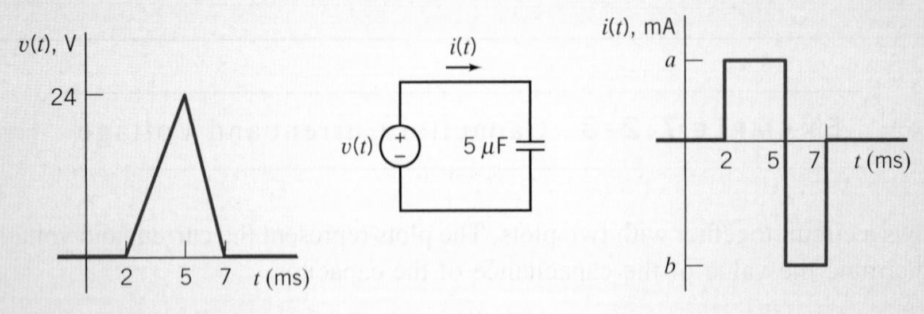

FIGURE 7.2-11 The circuit and plots considered in Example 7.2-4.

Solution

The current and voltage of the capacitor are related by

$$i(t) = C\frac{d}{dt}v(t) \tag{7.2-7}$$

Beacause $i(t)$ and $v(t)$ are represented graphically, by plots rather than equations, it is useful to interpret Eq. 7.2-7 as

$$\text{the value of } i(t) = C \times \text{the slope of } v(t)$$

To determine the value of a, pick a time when $i(t) = a$ and the slope of $v(t)$ is easily determined. For example, at time $t = 3$ ms,

$$\frac{d}{dt}v(0.003) = \frac{0-24}{0.002-0.005} = 8000\,\frac{\text{V}}{\text{s}}$$

(The notation $\frac{d}{dt}v(0.003)$ indicates that the derivative $\frac{d}{dt}v(t)$ is evaluated at time $t = 0.003$ s.) Using Eq. 7.2-7 gives

$$a = \left(5 \times 10^{-6}\right)(8000) = 40\ \text{mA}$$

To determine the value of b, pick $t = 6$ ms;

$$\frac{d}{dt}v\,(0.006) = \frac{24-0}{0.005-0.007} = 12 \times 10^3\,\frac{\text{V}}{\text{s}}$$

Using Eq. 7.2-7 gives

$$b = \left(5 \times 10^{-6}\right)\left(12 \times 10^3\right) = 60\ \text{mA}$$

EXAMPLE 7.2-5 Capacitor Current and Voltage

The input to the circuit shown in Figure 7.2-12 is the current

$$i(t) = 3.75e^{-1.2t}\text{A} \quad \text{for } t > 0$$

The output is the capacitor voltage

$$v(t) = 4 - 1.25e^{-1.2t}\ \text{V} \quad \text{for } t > 0$$

Find the value of the capacitance, C.

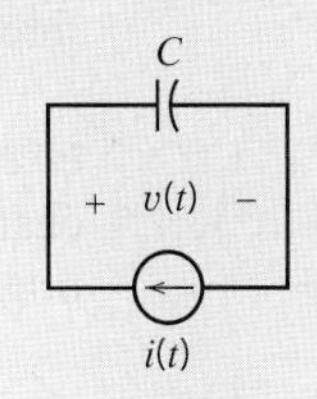

FIGURE 7.2-12 The circuit considered in Example 7.2-5.

Solution

The capacitor voltage is related to the capacitor current by

$$v(t) = \frac{1}{C}\int_0^t i(\tau)d\tau + v(0)$$

That is,

$$4 - 1.25e^{-1.2t} = \frac{1}{C}\int_0^t 3.75e^{-1.2\tau}d\tau + v(0) = \frac{3.75}{C(-1.2)}e^{-1.2\tau}\Big|_0^t + v(0) = \frac{-3.125}{C}\left(e^{-1.2t} - 1\right) + v(0)$$

Equating the coefficients of $e^{-1.2t}$ gives

$$1.25 = \frac{3.125}{C} \quad \Rightarrow \quad C = \frac{3.125}{1.25} = 2.5\ \text{F}$$

EXERCISE 7.2-1 Determine the current $i(t)$ for $t > 0$ for the circuit of Figure E 7.2-1b when $v_s(t)$ is the voltage shown in Figure E 7.2-1a.

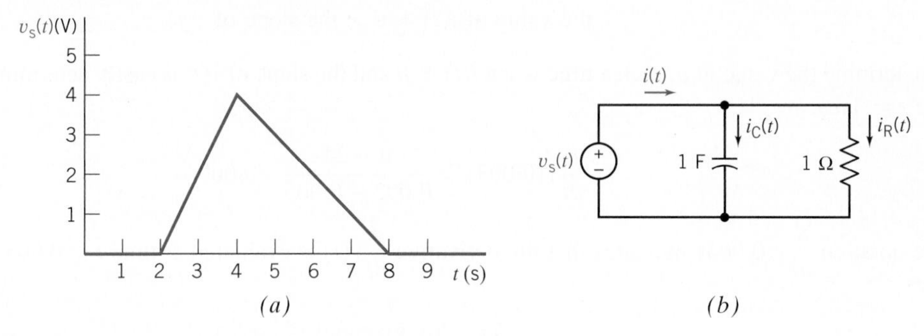

FIGURE E 7.2-1 (a) The voltage source voltage. (b) The circuit.

Hint: Determine $i_C(t)$ and $i_R(t)$ separately, then use KCL.

Answer: $v(t) = \begin{cases} 2t-2 & 2 < t < 4 \\ 7-t & 4 < t < 8 \\ 0 & \text{otherwise} \end{cases}$

7.3 ENERGY STORAGE IN A CAPACITOR

Consider a capacitor that has been connected to a battery of voltage v. A current flows and a charge is stored on the plates of the capacitor, as shown in Figure 7.3-1. Eventually, the voltage across the capacitor is a constant, and the current through the capacitor is zero. The capacitor has stored energy by virtue of the separation of charges between the capacitor plates. These charges have an electrical force acting on them.

The forces acting on the charges stored in a capacitor are said to result from an electric field. An *electric field* is defined as the force acting on a unit positive charge in a specified region. Because the charges have a force acting on them along a direction x, we recognize that the energy required originally to separate the charges is now stored by the capacitor in the electric field.

The energy stored in a capacitor is

$$w_c(t) = \int_{-\infty}^{t} vi\, d\tau$$

Remember that v and i are both functions of time and could be written as $v(t)$ and $i(t)$. Because

$$i = C\frac{dv}{dt}$$

we have

$$w_c = \int_{-\infty}^{t} vC\frac{dv}{dt}\, d\tau = C\int_{v(-\infty)}^{v(t)} v\, dv = \frac{1}{2}Cv^2\bigg|_{v(-\infty)}^{v(t)}$$

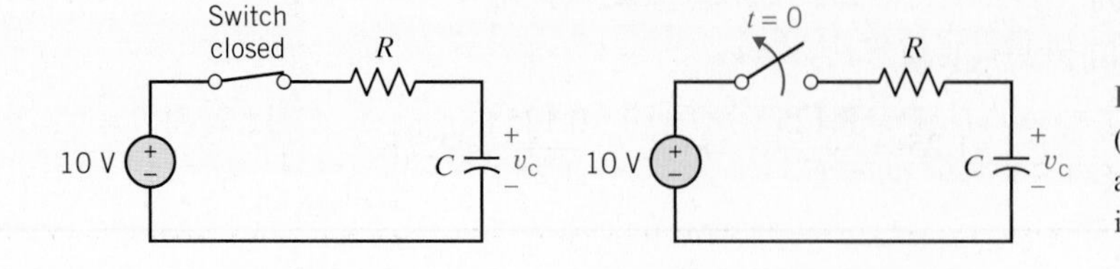

FIGURE 7.3-1 A circuit (a) where the capacitor is charged and $v_c = 10$ V and (b) the switch is opened at $t = 0$.

Because the capacitor was uncharged at $t = -\infty$, set $v(-\infty) = 0$. Therefore,

$$w_c(t) = \frac{1}{2}Cv^2(t) \text{ J} \tag{7.3-1}$$

Therefore, as a capacitor is being charged and $v(t)$ is changing, the energy stored, w_c, is changing. Note that $w_c(t) > 0$ for all $v(t)$, so the element is said to be passive.

Because $q = Cv$, we may rewrite Eq. 7.3-1 as

$$w_c = \frac{1}{2C}q^2(t) \text{ J} \tag{7.3-2}$$

The capacitor is a storage element that stores but does not dissipate energy. For example, consider a 100-mF capacitor that has a voltage of 100 V across it. The energy stored is

$$w_c = \frac{1}{2}Cv^2 = \frac{1}{2}(0.1)(100)^2 = 500 \text{ J}$$

As long as the capacitor is not connected to any other element, the energy of 500 J remains stored. Now if we connect the capacitor to the terminals of a resistor, we expect a current to flow until all the energy is dissipated as heat by the resistor. After all the energy dissipates, the current is zero and the voltage across the capacitor is zero.

As noted in the previous section, the requirement of conservation of charge implies that the voltage on a capacitor is continuous. Thus, the *voltage and charge on a capacitor cannot change instantaneously*. This statement is summarized by the equation

$$v(0^+) = v(0^-)$$

where the time just prior to $t = 0$ is called $t = 0^-$ and the time immediately after $t = 0$ is called $t = 0^+$. The time between $t = 0^-$ and $t = 0^+$ is infinitely small. Nevertheless, the voltage will not change abruptly.

To illustrate the continuity of voltage for a capacitor, consider the circuit shown in Figure 7.3-1. For the circuit shown in Figure 7.3-1*a*, the switch has been closed for a long time and the capacitor voltage has become $v_c = 10$ V. At time $t = 0$, we open the switch, as shown in Figure 7.3-1*b*. Because the voltage on the capacitor is continuous,

$$v_c(0^+) = v_c(0^-) = 10 \text{ V}$$

EXAMPLE 7.3-1 Energy Stored by a Capacitor

A 10-mF capacitor is charged to 100 V, as shown in the circuit of Figure 7.3-2. Find the energy stored by the capacitor and the voltage of the capacitor at $t = 0^+$ after the switch is opened.

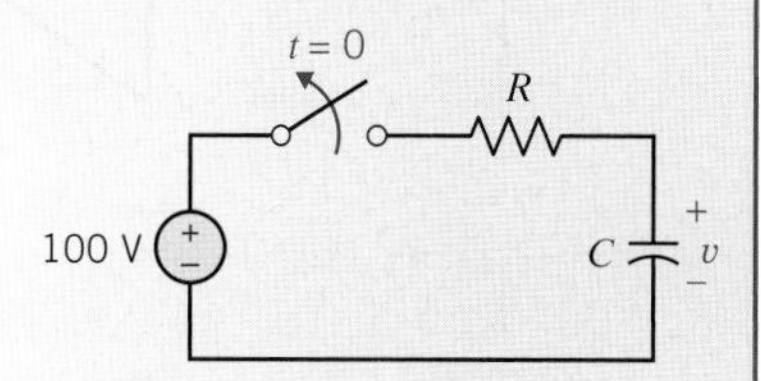

FIGURE 7.3-2 Circuit of Example 7.3-1 with $C = 10$ mF.

Solution

The voltage of the capacitor is $v = 100$ V at $t = 0^-$. Because the voltage at $t = 0^+$ cannot change from the voltage at $t = 0^-$, we have

$$v(0^+) = v(0^-) = 100 \text{ V}$$

The energy stored by the capacitor at $t = 0^+$ is

$$w_c = \frac{1}{2}Cv^2 = \frac{1}{2}(10^{-2})(100)^2 = 50 \text{ J}$$

EXAMPLE 7.3-2 Power and Energy for a Capacitor

The voltage across a 5-mF capacitor varies as shown in Figure 7.3-3. Determine and plot the capacitor current, power, and energy.

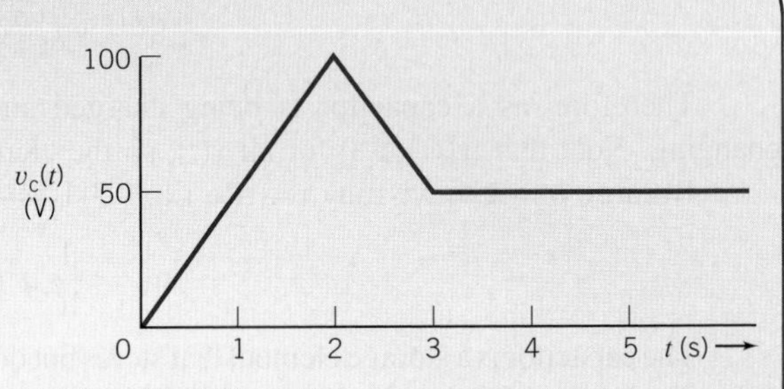

FIGURE 7.3-3 The voltage across a capacitor.

Solution

The current is determined from $i_c = C\,dv/dt$ and is shown in Figure 7.3-4*a*. The power is $v(t)i(t)$—the product of the current curve (Figure 7.3-4*a*) and the voltage curve (Figure 7.3-3)—and is shown in Figure 7.3-4*b*. The capacitor receives energy during the first two seconds and then delivers energy for the period $2 < t < 3$.

The energy is $\omega = \int p\,dt$ and can be found as the area under the $p(t)$ curve. The curve for the energy is shown in Figure 7.3-4*c*. Note that the capacitor increasingly stores energy from $t = 0$ s to $t = 2$ s, reaching a maximum energy of 25 J. Then the capacitor delivers a total energy of 18.75 J to the external circuit from $t = 2$ s to $t = 3$ s. Finally, the capacitor holds a constant energy of 6.25 J after $t = 3$ s.

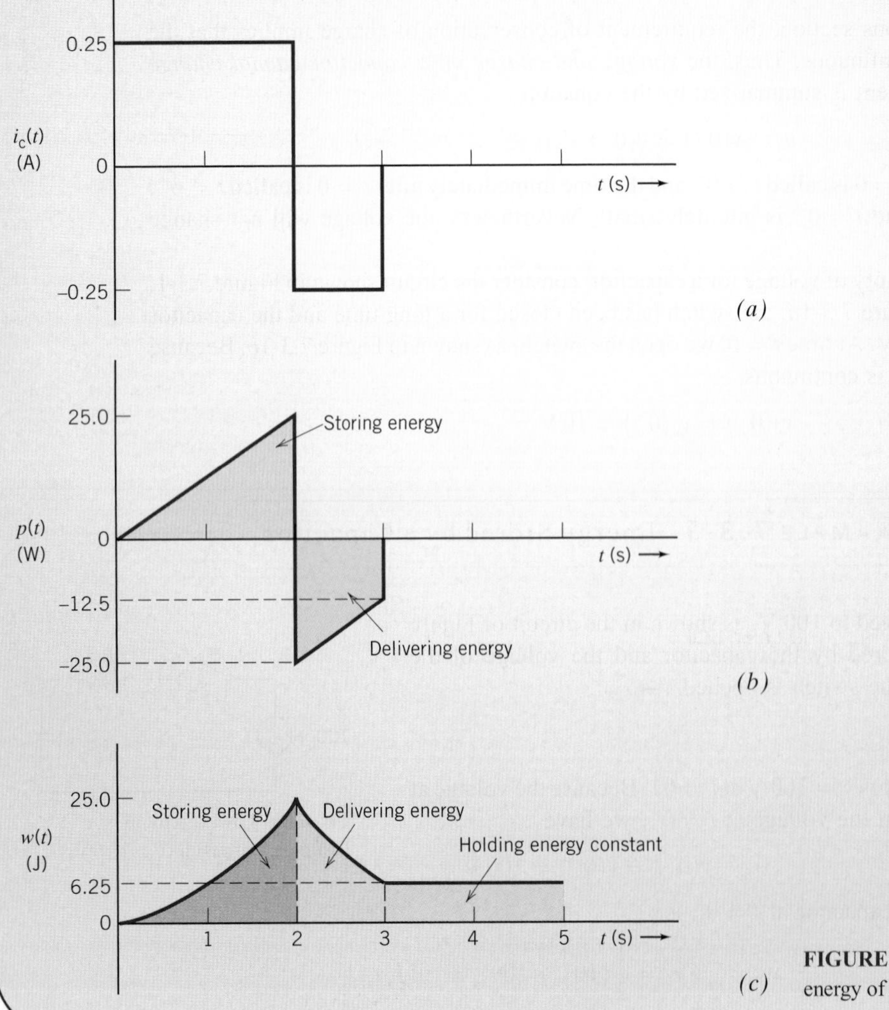

FIGURE 7.3-4 The current, power, and energy of the capacitor of Example 7.3-2.

EXERCISE 7.3-1 A 200-μF capacitor has been charged to 100 V. Find the energy stored by the capacitor. Find the capacitor voltage at $t = 0^+$ if $v(0^-) = 100$ V.

Answer: $w(1) = 1$ J and $v(0^+) = 100$ V

EXERCISE 7.3-2 A constant current $i = 2$ A flows into a capacitor of 100μF after a switch is closed at $t = 0$. The voltage of the capacitor was equal to zero at $t = 0^-$. Find the energy stored at (a) $t = 1$ s and (b) $t = 100$ s.

Answer: $w(1) = 20$ kJ and $w(100) = 200$ MJ

7.4 SERIES AND PARALLEL CAPACITORS

First, let us consider the parallel connection of N capacitors as shown in Figure 7.4-1. We wish to determine the equivalent circuit for the N parallel capacitors as shown in Figure 7.4-2.

Using KCL, we have

$$i = i_1 + i_2 + i_3 + \cdots + i_N$$

Because
$$i_n = C_n \frac{dv}{dt}$$

and v appears across each capacitor, we obtain

$$\begin{aligned} i &= C_1\frac{dv}{dt} + C_2\frac{dv}{dt} + C_3\frac{dv}{dt} + \cdots + C_N\frac{dv}{dt} \\ &= (C_1 + C_2 + C_3 + \cdots + C_N)\frac{dv}{dt} \\ &= \left(\sum_{n=1}^{N} C_n\right)\frac{dv}{dt} \end{aligned} \tag{7.4-1}$$

For the equivalent circuit shown in Figure 7.4-2,

$$i = C_p \frac{dv}{dt} \tag{7.4-2}$$

Comparing Eqs. 7.4-1 and 7.4-2, it is clear that

$$C_p = C_1 + C_2 + C_3 + \cdots + C_N = \sum_{n=1}^{N} C_n$$

Thus, the equivalent capacitance of a set of N parallel capacitors is simply the sum of the individual capacitances. It must be noted that all the parallel capacitors will have the same initial condition, $v(0)$.

Now let us determine the equivalent capacitance C_s of a set of N series-connected capacitances, as shown in Figure 7.4-3. The equivalent circuit for the series of capacitors is shown in Figure 7.4-4.

Using KVL for the loop of Figure 7.4-3, we have

$$v = v_1 + v_2 + v_3 + \cdots + v_N \tag{7.4-3}$$

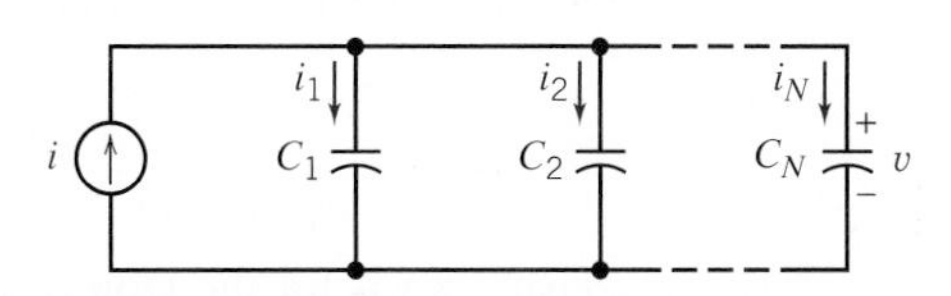

FIGURE 7.4-1 Parallel connection of N capacitors.

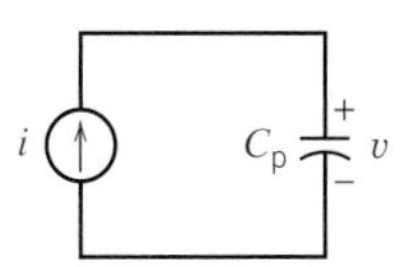

FIGURE 7.4-2 Equivalent circuit for N parallel capacitors.

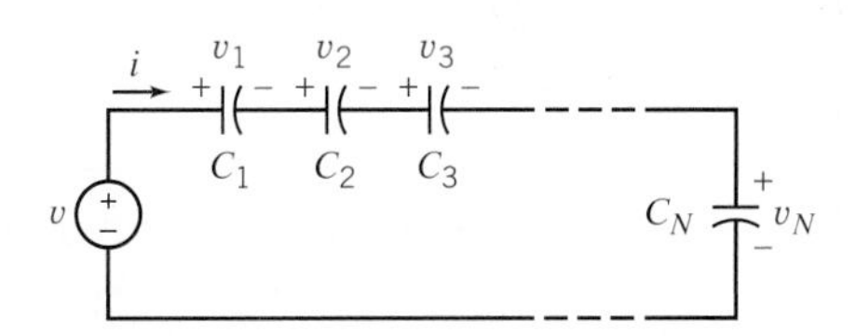

FIGURE 7.4-3 Series connection of N capacitors.

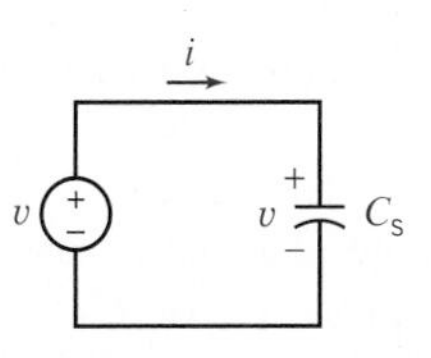

FIGURE 7.4-4 Equivalent circuit for N series capacitors.

Because, in general,

$$v_n(t) = \frac{1}{C_n}\int_{t_0}^{t} i\,d\tau + v_n(t_0)$$

where i is common to all capacitors, we obtain

$$\begin{aligned} v &= \frac{1}{C_1}\int_{t_0}^{t} i\,d\tau + v_1(t_0) + \cdots + \frac{1}{C_N}\int_{t_0}^{t} i\,d\tau + v_N(t_0) \\ &= \left(\frac{1}{C_1} + \frac{1}{C_2} + \cdots + \frac{1}{C_N}\right)\int_{t_0}^{t} i\,d\tau + \sum_{n=1}^{N} v_n(t_0) \\ &= \sum_{n=1}^{N}\frac{1}{C_n}\int_{t_0}^{t} i\,d\tau + \sum_{n=1}^{N} v_n(t_0) \end{aligned} \tag{7.4-4}$$

From Eq. 7.4-3, we note that at $t = t_0$,

$$v(t_0) = v_1(t_0) + v_2(t_0) + \cdots + v_N(t_0) = \sum_{n=1}^{N} v_n(t_0) \tag{7.4-5}$$

Substituting Eq. 7.4-5 into Eq. 7.4-4, we obtain

$$v = \left(\sum_{n=1}^{N}\frac{1}{C_n}\right)\int_{t_0}^{t} i\,d\tau + v(t_0) \tag{7.4-6}$$

Using KVL for the loop of the equivalent circuit of Figure 7.4-4 yields

$$v = \frac{1}{C_s}\int_{t_0}^{t} i\,d\tau + v(t_0) \tag{7.4-7}$$

Comparing Eqs. 7.4-6 and 7.4-7, we find that

$$\frac{1}{C_s} = \sum_{n=1}^{N}\frac{1}{C_n} \tag{7.4-8}$$

For the case of two series capacitors, Eq. 7.4-8 becomes

$$\frac{1}{C_s} = \frac{1}{C_1} + \frac{1}{C_2}$$

or

$$C_s = \frac{C_1 C_2}{C_1 + C_2} \tag{7.4-9}$$

EXAMPLE 7.4-1 Parallel and Series Capacitors

Find the equivalent capacitance for the circuit of Figure 7.4-5 when $C_1 = C_2 = C_3 = 2$ mF, $v_1(0) = 10$ V, and $v_2(0) = v_3(0) = 20$ V.

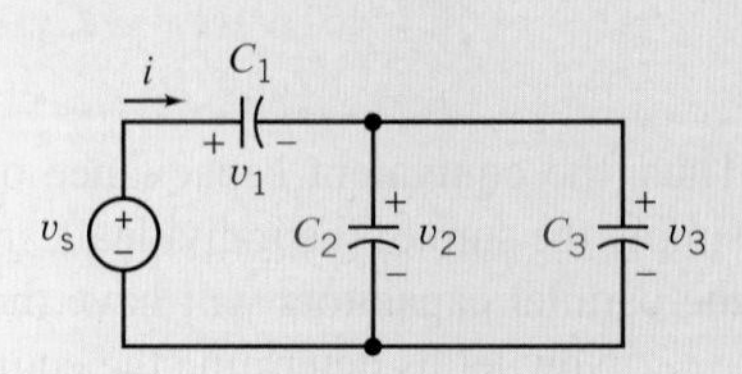

FIGURE 7.4-5 Circuit for Example 7.4-1.

Solution

Because C_2 and C_3 are in parallel, we replace them with C_p, where

$$C_p = C_2 + C_3 = 4\text{ mF}$$

The voltage at $t = 0$ across the equivalent capacitance C_p is equal to the voltage across C_2 or C_3, which is $v_2(0) = v_3(0) = 20$ V. As a result of replacing C_2 and C_3 with C_p, we obtain the circuit shown in Figure 7.4-6.

We now want to replace the series of two capacitors C_1 and C_p with one equivalent capacitor. Using the relationship of Eq. 7.4-9, we obtain

$$C_s = \frac{C_1 C_p}{C_1 + C_p} = \frac{(2 \times 10^{-3})(4 \times 10^{-3})}{(2 \times 10^{-3}) + (4 \times 10^{-3})} = \frac{8}{6} \text{ mF}$$

The voltage at $t = 0$ across C_s is

$$v(0) = v_1(0) + v_p(0)$$

where $v_p(0) = 20$ V, the voltage across the capacitance C_p at $t = 0$. Therefore, we obtain

$$v(0) = 10 + 20 = 30 \text{ V}$$

Thus, we obtain the equivalent circuit shown in Figure 7.4-7.

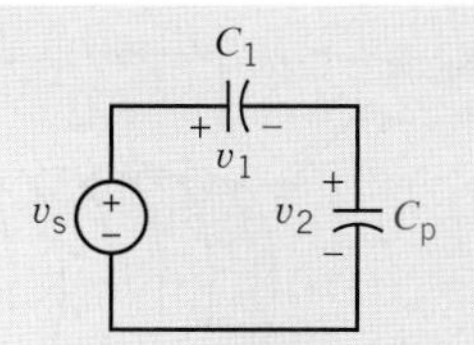

FIGURE 7.4-6 Circuit resulting from Figure 7.4-5 by replacing C_2 and C_3 with C_p.

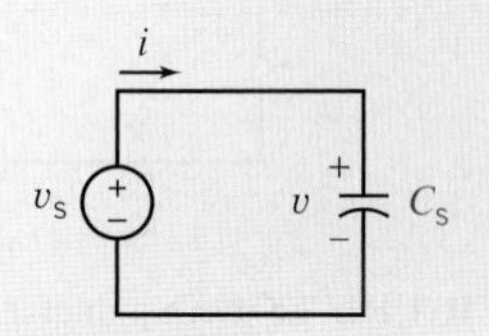

FIGURE 7.4-7 Equivalent circuit for the circuit of Example 7.4-1.

EXERCISE 7.4-1 Find the equivalent capacitance for the circuit of Figure E 7.4-1

Answer: $C_{eq} = 4$ mF

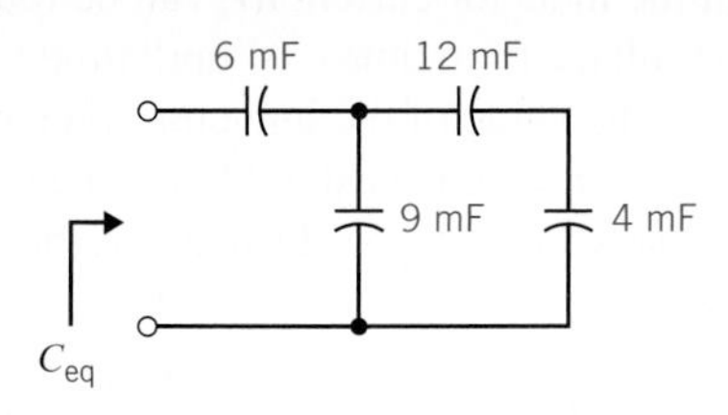

FIGURE E 7.4-1

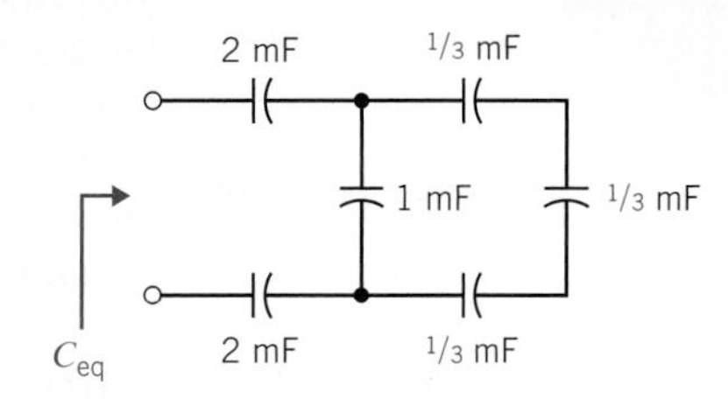

FIGURE E 7.4-2

EXERCISE 7.4-2 Determine the equivalent capacitance C_{eq} for the circuit shown in Figure E 7.4-2.

Answer: 10/19 mF

7.5 INDUCTORS

An inductor is a circuit element that stores energy in a magnetic field. An inductor can be constructed by winding a coil of wire around a magnetic core as shown in Figure 7.5-1. Inductors are represented by a parameter called the *inductance*. The inductance of an inductor depends on its size, materials, and method of construction. For example, the inductance of the inductor shown in Figure 7.5-1 is given by

$$L = \frac{\mu N^2 A}{l}$$

where N is the number of turns—that is, the number of times that the wire is wound around the core—A is the cross-sectional area of the core in square meters; l the length of the winding in meters; and μ is a property of the magnetic core known as the permeability. The unit of inductance

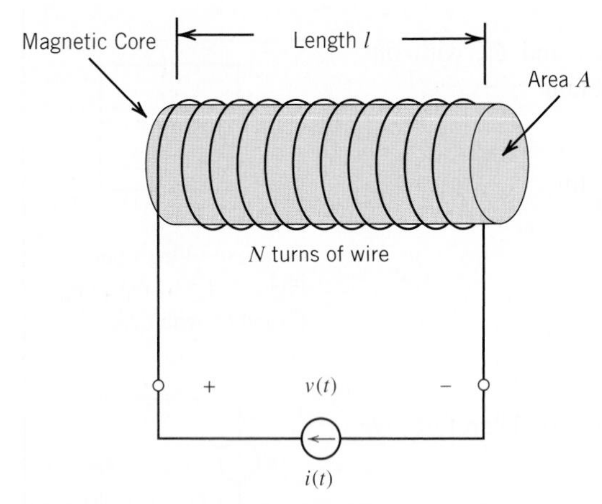

FIGURE 7.5-1 An inductor connected to a current source.

FIGURE 7.5-2 Coil with a large inductance. Courtesy of MuRata Company.

FIGURE 7.5-3 Elements with inductances arranged in various forms of coils. Courtesy of Dale Electronic Inc.

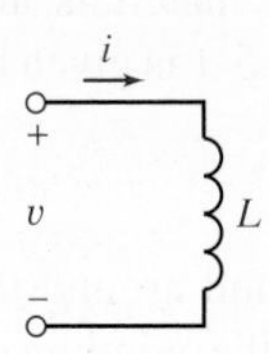

FIGURE 7.5-4 Circuit symbol for an inductor.

is called henry (H) in honor of the American physicist Joseph Henry. Practical inductors have inductances ranging from 1 μH to 10 H. An example of a coil with a large inductance is shown in Figure 7.5-2. Inductors are wound in various forms, as shown in Figure 7.5-3.

> **Inductance** is a measure of the ability of a device to store energy in the form of a magnetic field.

In Figure 7.5-1, a current source is used to cause a coil current $i(t)$. We find that the voltage $v(t)$ across the coil is proportional to the rate of change of the coil current. That is,

$$v(t) = L\frac{d}{dt}i(t) \tag{7.5-1}$$

where the constant of proportionality is L, the inductance of the inductor.

Integrating both sides of Eq. 7.5-1, we obtain

$$i(t) = \frac{1}{L}\int_{-\infty}^{t} v(\tau)d\tau \tag{7.5-2}$$

This equation says that the inductor current $i(t)$ can be found by integrating the inductor voltage from time $-\infty$ until time t. To do so requires that we know the value of the inductor voltage from time $\tau = -\infty$ until time $\tau = t$. Often, we don't know the value of the voltage all the way back to $\tau = -\infty$. Instead, we break the integral up into two parts:

$$i(t) = \frac{1}{L}\int_{-\infty}^{t_0} v(\tau)d\tau + \frac{1}{L}\int_{t_0}^{t} v(\tau)d\tau = i(t_0) + \frac{1}{L}\int_{t_0}^{t} v(\tau)d\tau \tag{7.5-3}$$

This equation says that the inductor current $i(t)$ can be found by integrating the inductor voltage from some convenient time $\tau = t_0$ until time $\tau = t$, provided that we also know the inductor current at time t_0. Now we are required to know only the inductor voltage from time $\tau = t_0$ until time $\tau = t$. The time t_0 is called the **initial time**, and the inductor current $i(t_0)$ is called the **initial condition**. Frequently, it is convenient to select $t_0 = 0$ as the initial time.

Equations 7.5-1 and 7.5-3 describe the current–voltage relationship of an inductor. The current and voltage in these equations adhere to the passive convention. The circuit symbol for an inductor is shown in Figure 7.5-4. The inductor current and voltage in Figure 7.5-4 adhere to the passive sign convention and are related by Eqs. 7.5-1 and 7.5-3.

Consider the voltage of an inductor when the current changes at $t = 0$ from zero to a constantly increasing current and eventually levels off as shown in Figure 7.5-5. Let us determine the voltage of the inductor. We may describe the

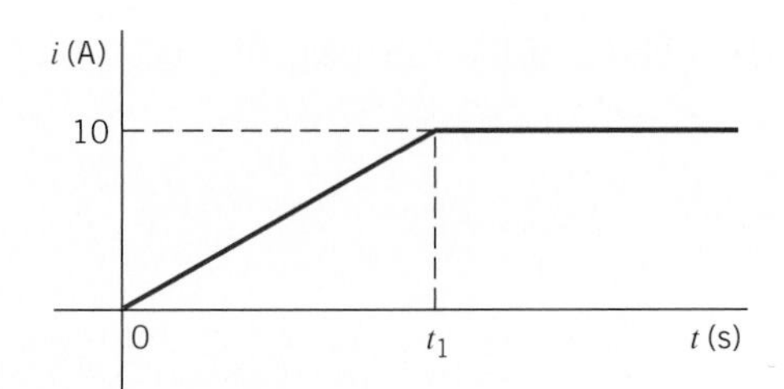

FIGURE 7.5-5 A current waveform. The current is in amperes.

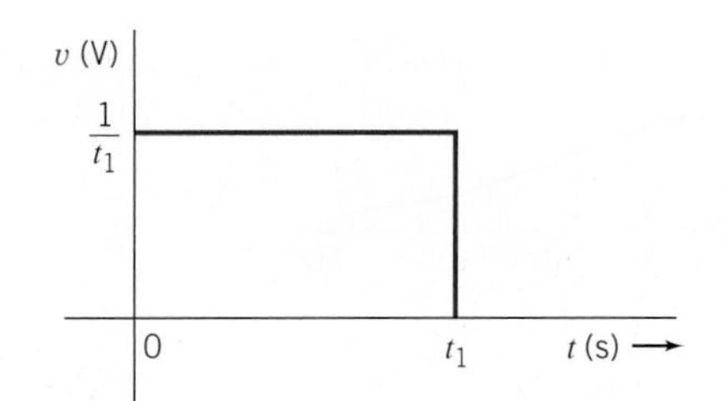

FIGURE 7.5-6 Voltage response for the current waveform of Figure 7.5-7 when $L = 0.1$ H.

current (in amperes) by

$$i(t) = \begin{cases} 0 & t \le 0 \\ \dfrac{10t}{t_1} & 0 \le t \le t_1 \\ 10 & t \ge t_1 \end{cases}$$

Let us consider a 0.1-H inductor and find the voltage waveform. Because $v = L(di/dt)$, we have (in volts)

$$v(t) = \begin{cases} 0 & t < 0 \\ \dfrac{1}{t_1} & 0 < t < t_1 \\ 0 & t > t_1 \end{cases}$$

The resulting voltage pulse waveform is shown in Figure 7.5-6. Note that as t_1 decreases, the magnitude of the voltage increases. Clearly, we cannot let $t_1 = 0$ because the voltage required would then become infinite, and we would require infinite power at the terminals of the inductor. Thus, instantaneous changes in the current through an inductor are not possible.

The current in an inductance cannot change instantaneously.

An ideal inductor is a coil wound with resistanceless wire. Practical inductors include the actual resistance of the copper wire used in the coil. For this reason, practical inductors are far from ideal elements and are typically modeled by an ideal inductance in series with a small resistance.

EXAMPLE 7.5-1 Inductor Current and Voltage

Find the voltage across an inductor, $L = 0.1$ H, when the current in the inductor is

$$i(t) = 20te^{-2t} \text{ A}$$

for $t > 0$ and $i(0) = 0$.

Solution

The voltage for $t < 0$ is

$$v(t) = L\frac{di}{dt} = (0.1)\frac{d}{dt}\left(20te^{-2t}\right) = 2\left(-2te^{-2t} + e^{-2t}\right) = 2e^{-2t}(1 - 2t) \text{ V}$$

The voltage is equal to 2 V when $t = 0$, as shown in Figure 7.5-7*b*. The current waveform is shown in Figure 7.5-7*a*. Note that the current reaches a maximum value, and the voltage is zero at $t = 0.5$ s.

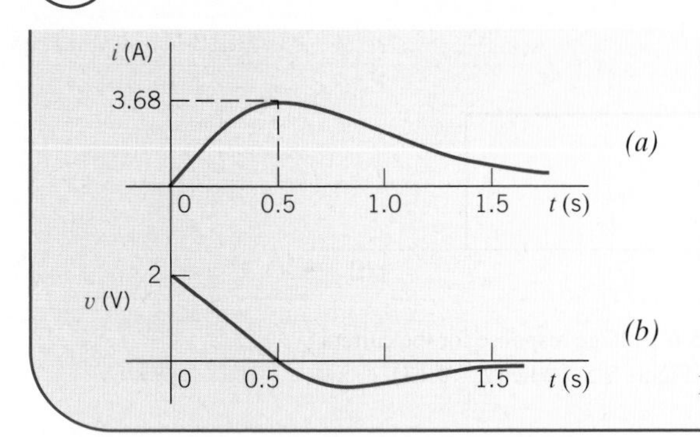

FIGURE 7.5-7 Voltage and current waveforms for Example 7.5-1.

EXAMPLE 7.5-2 Inductor Current and Voltage

INTERACTIVE EXAMPLE

Figure 7.5-8 shows a circuit together with two plots. The plots represent the current and voltage of the inductor in the circuit. Determine the value of the inductance of the inductor.

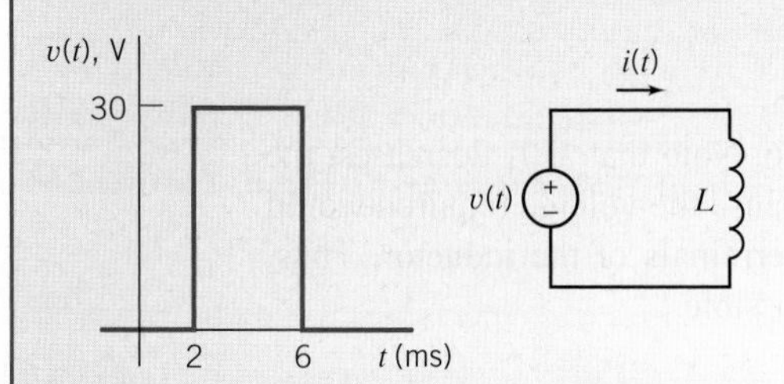

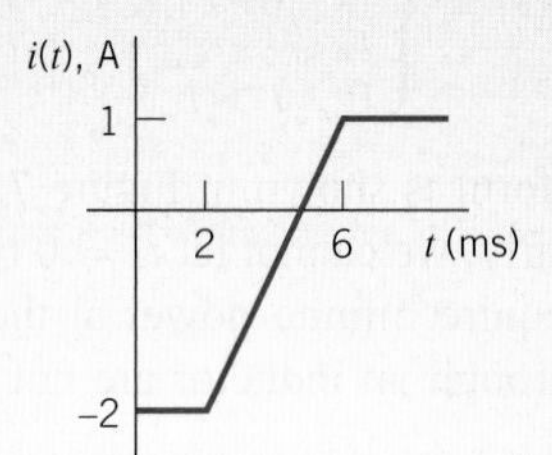

i(t), A
1
2
6
t (ms)
−2

FIGURE 7.5-8 The circuit and plots considered in Example 7.5-2.

Solution

The current and voltage of the inductor are related by

$$i(t) = \frac{1}{L}\int_{t_0}^{t} v(\tau)\, d\tau + i(t_0) \tag{7.5-4}$$

or

$$i(t) - i(t_0) = \frac{1}{L}\int_{t_0}^{t} v(\tau)\, d\tau \tag{7.5-5}$$

Because $i(t)$ and $v(t)$ are represented graphically, by plots rather than equations, it is useful to interpret Eq. 7.5-5 using

$$i(t) - i(t_0) = \text{the difference between the values of current at times } t \text{ and } t_0$$

and

$$\int_{t_0}^{t} v(\tau) d\tau = \text{the area under the plot of } v(t) \text{ versus } t \text{ for times between } t \text{ and } t_0$$

Pick convenient values t and t_0, for example, $t_0 = 2$ ms and $t = 6$ ms. Then,

$$i(t) - i(t_0) = 1 - (-2) = 3\text{ A}$$

and

$$\int_{t_0}^{t} v(\tau)\, d\tau = \int_{0.002}^{0.006} 30\, d\tau = (30)(0.006 - 0.002) = 0.12\text{ V}\cdot\text{s}$$

Using Eq. 7.5-5 gives

$$3 = \frac{1}{L}(0.12) \quad \Rightarrow \quad L = 0.040\ \frac{\text{V}\cdot\text{s}}{\text{A}} = 0.040\text{ H} = 40\text{ mH}$$

EXAMPLE 7.5-3 Inductor Current and Voltage

The input to the circuit shown in Figure 7.5-9 is the voltage

$$v(t) = 4e^{-20t}\ \text{V} \quad \text{for } t > 0$$

The output is the current

$$i(t) = -1.2e^{-20t} - 1.5\ \text{A} \quad \text{for } t > 0$$

The initial inductor current is $i_L(0) = -3.5$ A. Determine the values of the inductance, L, and resistance, R.

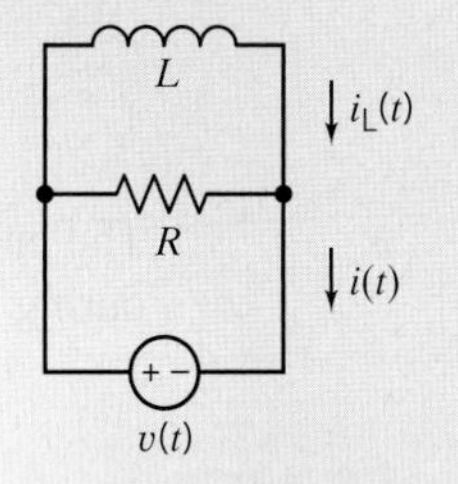

FIGURE 7.5-9 The circuit considered in Example 7.5-3.

Solution

Apply KCL at either node to get

$$i(t) = \frac{v(t)}{R} + i_L(t) = \frac{v(t)}{R} + \left[\frac{1}{L}\int_0^t v(\tau)d\tau + i(0)\right]$$

That is

$$-1.2e^{-20t} - 1.5 = \frac{4e^{-20t}}{R} + \frac{1}{L}\int_0^t 4e^{-20t}d\tau - 3.5 = \frac{4e^{-20t}}{R} + \frac{4}{L(-20)}(e^{-20t} - 1) - 3.5$$

$$= \left(\frac{4}{R} - \frac{1}{5L}\right)e^{-20t} + \frac{1}{5L} - 3.5$$

Equating coefficients gives

$$-1.5 = \frac{1}{5L} - 3.5 \Rightarrow L = 0.1\ \text{H}$$

and

$$-1.2 = \frac{4}{R} - \frac{1}{5L} = \frac{4}{R} - \frac{1}{5(0.1)} = \frac{4}{R} - 2 \Rightarrow R = 5\ \Omega$$

EXERCISE 7.5-1 Determine the voltage $v(t)$ for $t > 0$ for the circuit of Figure E 7.5-1*b* when $i_s(t)$ is the current shown in Figure E 7.5-1*a*.

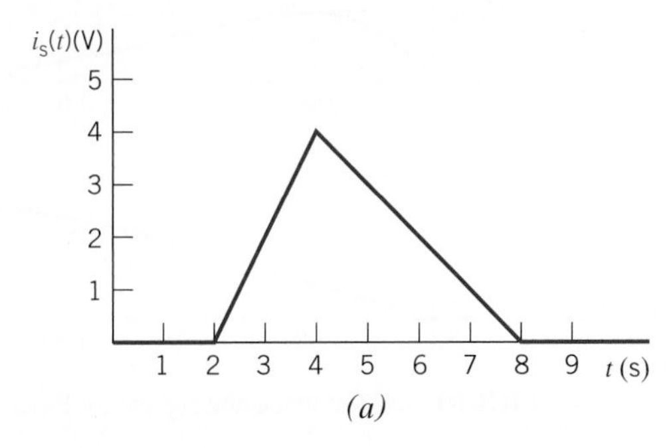

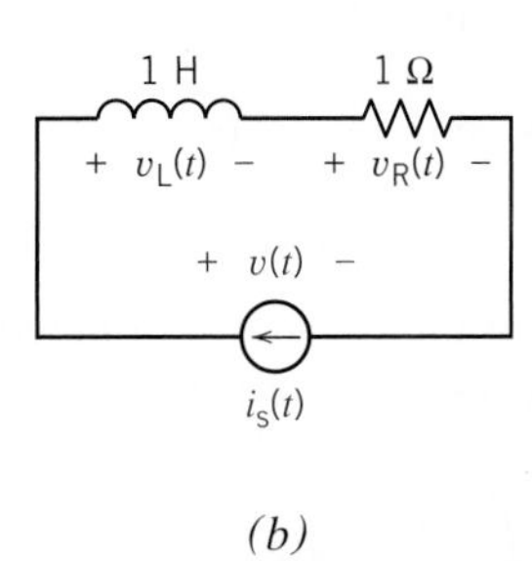

FIGURE E 7.5-1 (*a*) The current source current. (*b*) The circuit.

Hint: Determine $v_L(t)$ and $v_R(t)$ separately, then use KVL.

Answer: $v(t) = \begin{cases} 2t - 2 & 2 < t < 4 \\ 7 - t & 4 < t < 8 \\ 0 & \text{otherwise} \end{cases}$

7.6 ENERGY STORAGE IN AN INDUCTOR

The power in an inductor is

$$p = vi = \left(L\frac{di}{dt}\right)i \tag{7.6-1}$$

The energy stored in the inductor is stored in its magnetic field. The energy stored in the inductor during the interval t_0 to t is given by

$$w = \int_{t_0}^{t} p\,d\tau = L\int_{i(t_0)}^{i(t)} i\,di$$

Integrating the current between $i(t_0)$ and $i(t)$, we obtain

$$w = \frac{L}{2}\left[i^2(t)\right]_{i(t_0)}^{i(t)} = \frac{L}{2}i^2(t) - \frac{L}{2}i^2(t_0) \tag{7.6-2}$$

Usually, we select $t_0 = -\infty$ for the inductor and then the current $i(-\infty) = 0$. Then we have

$$w = \frac{1}{2}Li^2 \tag{7.6-3}$$

Note that $w(t) \geq 0$ for all $i(t)$, so the inductor is a passive element. The inductor does not generate or dissipate energy but only stores energy. It is important to note that inductors and capacitors are fundamentally different from other devices considered in earlier chapters in that they have memory.

EXAMPLE 7.6-1 Inductor Voltage and Current

Find the current in an inductor, $L = 0.1$ H, when the voltage across the inductor is

$$v = 10te^{-5t}\ \text{V}$$

Assume that the current is zero for $t \leq 0$.

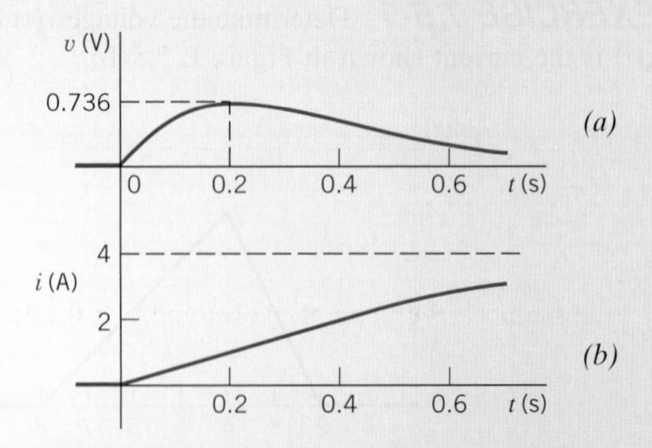

FIGURE 7.6-1 Voltage and current for Example 7.6-1.

Solution

The voltage as a function of time is shown in Figure 7.6-1*a*. Note that the voltage reaches a maximum at $t = 0.2$ s. The current is

$$i = \frac{1}{L}\int_0^t v\,d\tau + i(t_0)$$

Because the voltage is zero for $t < 0$, the current in the inductor at $t = 0$ is $i(0) = 0$. Then we have

$$i = 10\int_0^t 10\,\tau e^{-5\tau}\,d\tau = 100\left[\frac{-e^{-5\tau}}{25}(1 + 5\tau)\right]_0^t = 4\left(1 - e^{-5t}(1 + 5t)\right)\ \text{A}$$

The current as a function of time is shown in Figure 7.6-1*b*.

EXAMPLE 7.6-2 Power and Energy for an Inductor

Find the power and energy for an inductor of 0.1 H when the current and voltage are as shown in Figures 7.6-2*a*,*b*.

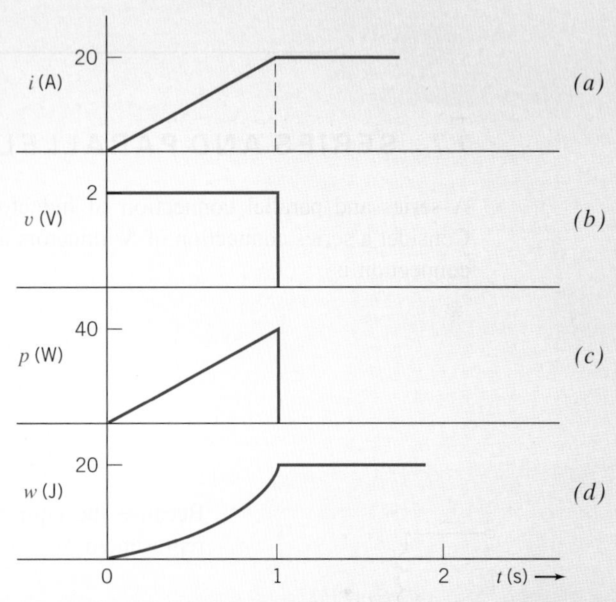

FIGURE 7.6-2 Current, voltage, power, and energy for Example 7.6-2.

Solution

First, we write the expression for the current and the voltage. The current is

$$
\begin{aligned}
i &= 0 \quad t < 0 \\
&= 20t \quad 0 \le t \le 1 \\
&= 20 \quad 1 \le t
\end{aligned}
$$

The voltage is expressed as

$$
\begin{aligned}
v &= 0 \quad t < 0 \\
&= 2 \quad 0 < t < 1 \\
&= 0 \quad 1 < t
\end{aligned}
$$

You can verify the voltage by using $v = L(di/dt)$. Then the power is

$$p = vi = 40t \text{ W}$$

for $0 \le t < 1$ and zero for all other time.

The energy, in joules, is then

$$
\begin{aligned}
w &= \frac{1}{2}Li^2 \\
&= 0.05(20t)^2 \quad 0 \le t \le 1 \\
&= 0.05(20)^2 \quad 1 < t
\end{aligned}
$$

and zero for all $t < 0$.

The power and energy are shown in Figures 7.6-2*c*,*d*.

EXAMPLE 7.6-3 Power and Energy for an Inductor

Find the power and the energy stored in a 0.1-H inductor when $i = 20te^{-2t}$ A and $v = 2e^{-2t}(1 - 2t)$ V for $t \ge 0$ and $i = 0$ for $t < 0$. (See Example 7.5-1.)

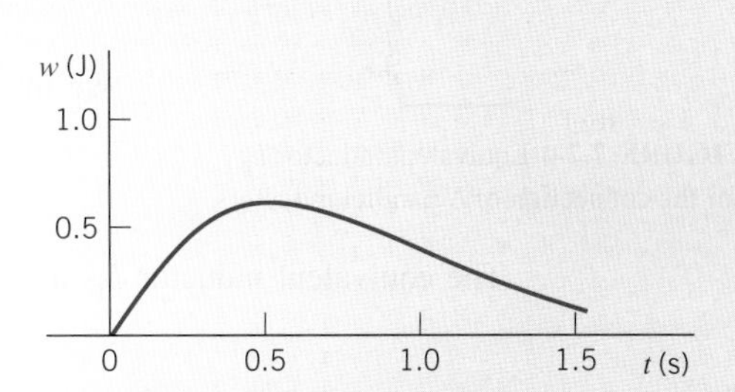

FIGURE 7.6-3 Energy stored in the inductor of Example 7.6-3.

Solution

The power is

$$p = iv = (20te^{-2t})[2e^{-2t}(1 - 2t)] = 40te^{-4t}(1 - 2t) \text{ W} \quad t > 0$$

The energy is then

$$w = \frac{1}{2}Li^2 = 0.05\left(20te^{-2t}\right)^2 = 20t^2e^{-4t} \text{ J} \quad t > 0$$

Note that w is positive for all values of $t > 0$. The energy stored in the inductor is shown in Figure 7.6-3.

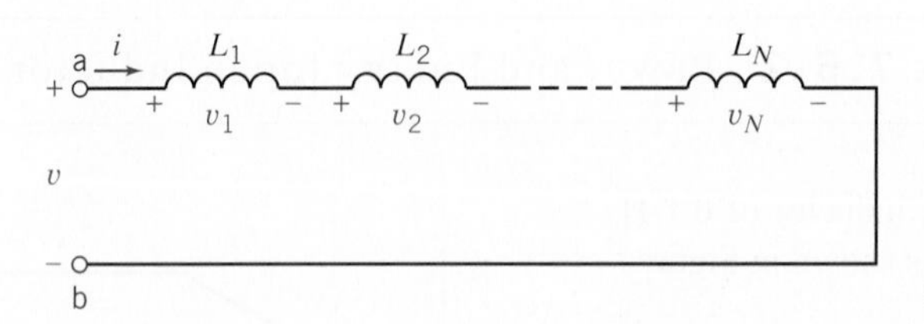

FIGURE 7.7-1 Series of N inductors.

7.7 SERIES AND PARALLEL INDUCTORS

A series and parallel connection of inductors can be reduced to an equivalent simple inductor. Consider a series connection of N inductors as shown in Figure 7.7-1. The voltage across the series connection is

$$\begin{aligned} v &= v_1 + v_2 + \cdots + v_N \\ &= L_1 \frac{di}{dt} + L_2 \frac{di}{dt} + \cdots + L_N \frac{di}{dt} \\ &= \left(\sum_{n=1}^{N} L_N \right) \frac{di}{dt} \end{aligned}$$

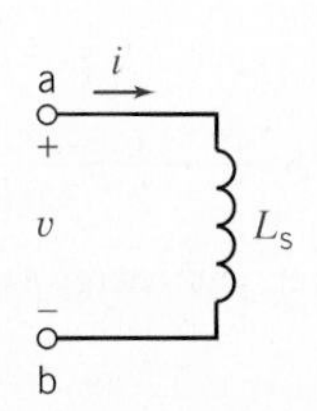

FIGURE 7.7-2 Equivalent inductor L_s for N series inductors.

Because the equivalent series inductor L_s, as shown in Figure 7.7-2, is represented by

$$v = L_s \frac{di}{dt}$$

we require that

$$L_s = \sum_{n=1}^{N} L_n \tag{7.7-1}$$

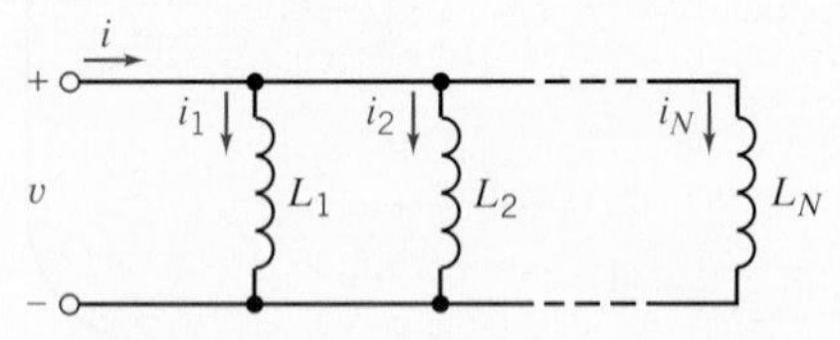

FIGURE 7.7-3 Connection of N parallel inductors.

Thus, an equivalent inductor for a series of inductors is the sum of the N inductors.

Now, consider the set of N inductors in parallel, as shown in Figure 7.7-3. The current i is equal to the sum of the currents in the N inductors:

$$i = \sum_{n=1}^{N} i_n$$

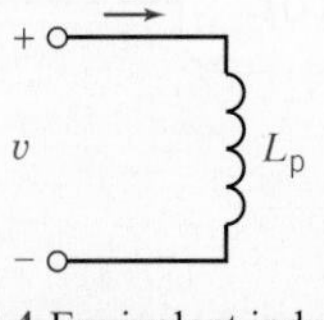

However, because

$$i_n = \frac{1}{L_n} \int_{t_0}^{t} v \, d\tau + i_n(t_0)$$

we may obtain the expression

FIGURE 7.7-4 Equivalent inductor L_p for the connection of N parallel inductors.

$$i = \sum_{n=1}^{N} \frac{1}{L_n} \int_{t_0}^{t} v \, d\tau + \sum_{n=1}^{N} i_n(t_0) \tag{7.7-2}$$

The equivalent inductor L_p, as shown in Figure 7.7-4, is represented by the equation

$$i = \frac{1}{L_p} \int_{t_0}^{t} v \, d\tau + i(t_0) \tag{7.7-3}$$

When Eqs. 7.7-2 and 7.7-3 are set equal to each other, we have

$$\frac{1}{L_p} = \sum_{n=1}^{N} \frac{1}{L_n} \tag{7.7-4}$$

and
$$i(t_0) = \sum_{n=1}^{N} i_n(t_0) \tag{7.7-5}$$

EXAMPLE 7.7-1 Series and Parallel Inductors

Find the equivalent inductance for the circuit of Figure 7.7-5. All the inductor currents are zero at t_0.

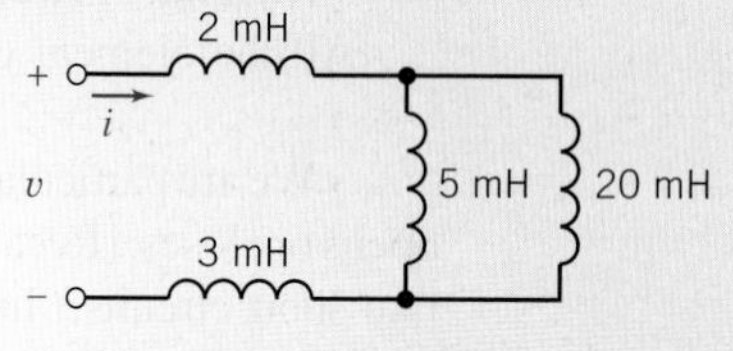

FIGURE 7.7-5 The circuit of Example 7.7-1.

Solution

First, we find the equivalent inductance for the 5-mH and 20-mH inductors in parallel.

From Eq. 7.7-4, we obtain

$$\frac{1}{L_p} = \frac{1}{L_1} + \frac{1}{L_2}$$

or
$$L_p = \frac{L_1 L_2}{L_1 + L_2} = \frac{5 \times 20}{5 + 20} = 4 \text{ mH}$$

This equivalent inductor is in series with the 2-mH and 3-mH inductors. Therefore, using Eq. 7.7-1, we obtain

$$L_{eq} = \sum_{n=1}^{N} L_n = 2 + 3 + 4 = 9 \text{ mH}$$

EXERCISE 7.7-1 Find the equivalent inductance of the circuit of Figure E 7.7-1.

Answer: $L_{eq} = 14$ mH

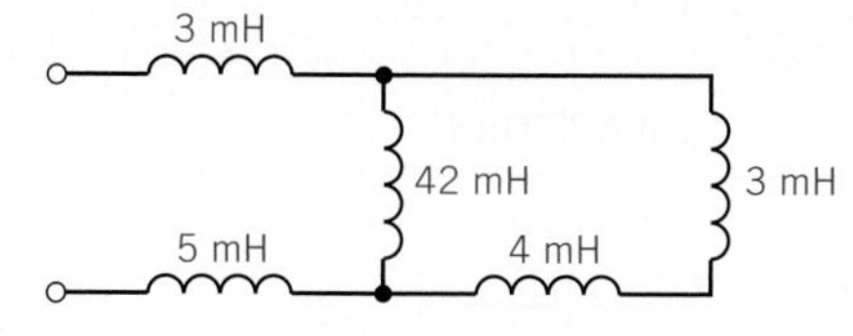

FIGURE E 7.7-1

FIGURE E 7.7-2

EXERCISE 7.7-2 Find the equivalent inductance of the circuit of Figure E 7.7-2.

Answer: $L_{eq} = 4$ mH

7.8 INITIAL CONDITIONS OF SWITCHED CIRCUITS

In this section, we consider switched circuits. These circuits have the following characteristics:

1. All of the circuit inputs, that is, the independent voltage source voltages and independent current source currents, are constant functions of time.

2. The circuit includes one or more switches that open or close at time t_0. We denote the time immediately before the switch opens or closes as t_0^- and the time immediately after the switch opens or closes as t_0^+. Often, we will assume that $t_0 = 0$.
3. The circuit includes at least one capacitor or inductor.
4. We will assume that the switches in a circuit have been in position for a long time at $t = t_0$, the switching time. We will say that such a circuit is at *steady state* immediately before the time of switching. A circuit that contains only constant sources and is at steady state is called a *dc circuit*. All the element currents and voltages in a dc circuit are constant functions of time.

We are particularly interested in the current and voltage of energy storage elements after the switch opens or closes. (Recall from Section 2.9 that open switches act like open circuits and closed switches act like short circuits.) In Table 7.8-1, we summarize the important characteristics of the behavior of an inductor and a capacitor. In particular, notice that neither a capacitor voltage nor an inductor current can change instantaneously. (Recall from Sections 7.2 and 7.5 that such changes would require infinite power, something that is not physically possible.) However, instantaneous changes to an inductor voltage or a capacitor current are quite possible.

Suppose that a dc circuit contains an inductor. The inductor current, like every other voltage and current in the dc circuit, will be a constant function of time. The inductor voltage is proportional to the derivative of the inductor current, $v = L(di/dt)$, so the inductor voltage is zero. Consequently, the inductor acts like a short circuit.

An **inductor** in a **dc circuit** behaves as a **short circuit**.

Similarly, the voltage of a capacitor in a dc circuit will be a constant function of time. The capacitor current is proportional to the derivative of the capacitor voltage, $i = C(dv/dt)$, so the capacitor current is zero. Consequently, the capacitor acts like a open circuit.

Table 7.8-1 Characteristics of Energy Storage Elements

VARIABLE	INDUCTORS	CAPACITORS
Passive sign convention	i → L, + v −	i → C, + v −
Voltage	$v = L\dfrac{di}{dt}$	$v = \dfrac{1}{C}\int_0^t i d\tau + v(0)$
Current	$i = \dfrac{1}{L}\int_0^t v d\tau + i(0)$	$i = C\dfrac{dv}{dt}$
Power	$i = Li\dfrac{di}{dt}$	$p = Cv\dfrac{dv}{dt}$
Energy	$w = \dfrac{1}{2}Li^2$	$w = \dfrac{1}{2}Cv^2$
An instantaneous change is not permitted for the element's	Current	Voltage
Will permit an instantaneous change in the element's	Voltage	Current
This element acts as a (see note below)	Short circuit to a constant current into its terminals	Open circuit to a constant voltage across its terminals

Note: Assumes that the element is in a circuit with steady-state condition.

A **capacitor** in a **dc circuit** behaves as an **open circuit**.

Our plan to analyze switched circuits has two steps:

1. Analyze the dc circuit that exists before time t_0 to determine the capacitor voltages and inductor currents. In doing this analysis, we will take advantage of the fact that capacitors behave as open circuits and inductors behave as short circuits when they are in dc circuits.
2. Recognize that capacitor voltages and inductor currents cannot change instantaneously, so the capacitor voltages and inductor currents at time t_0^+ have the same values that they had at time t_0^-.

The following examples illustrate this plan.

EXAMPLE 7.8-1 Initial Conditions in a Switched Circuit

Consider the circuit Figure 7.8-1. Prior to $t = 0$, the switch has been closed for a long time. Determine the values of the capacitor voltage and inductor current immediately after the switch opens at time $t = 0$.

Solution

1. To find $v_c(0^-)$ and $i_L(0^-)$, we consider the circuit before the switch opens, that is for $t < 0$. The circuit input, the voltage source voltage, is constant. Also, before the switch opens, the circuit is at steady state. Because the circuit is a dc circuit, the capacitor will act like an open circuit, and the inductor will act like a short circuit. In Figure 7.8-2, we replace the capacitor by an open circuit having voltage $v_c(0^-)$ and the inductor by a short circuit having current $i_L(0^-)$. First, we notice that

$$i_L(0^-) = \frac{10}{5} = 2\text{ A}$$

Next, using the voltage divider principle, we see that

$$v_c(0^-) = \left(\frac{3}{5}\right)10 = 6\text{ V}$$

2. The capacitor voltage and inductor current cannot change instantaneously, so

$$v_c(0^+) = v_c(0^-) = 6\text{ V}$$

and

$$i_L(0^+) = i_L(0^-) = 2\text{ A}$$

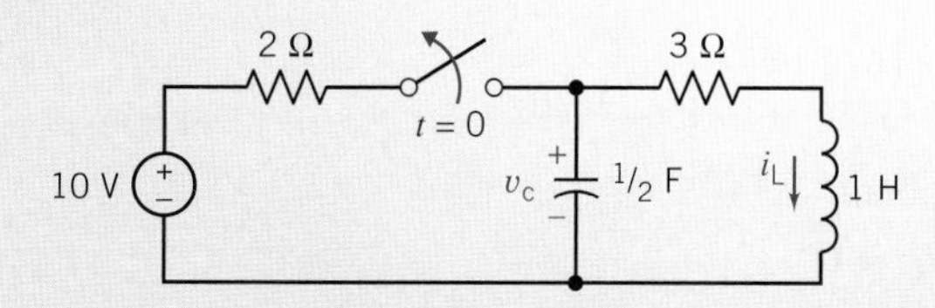

FIGURE 7.8-1 Circuit with an inductor and a capacitor. The switch is closed for a long time prior to opening at $t = 0$.

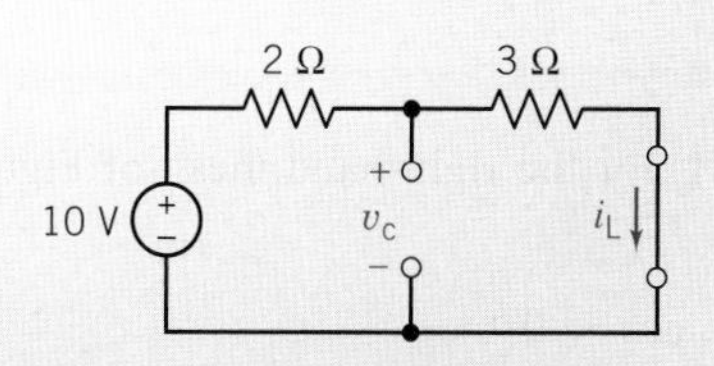

FIGURE 7.8-2 Circuit of Figure 7.8-1 for $t < 0$.

EXAMPLE 7.8-2 Initial Conditions in a Switched Circuit

Find $i_L(0^+)$, $v_c(0^+)$, $dv_c(0^+)/dt$, and $di_L(0^+)/dt$ for the circuit of Figure 7.8-3. We will use $dv_c(0^+)/dt$ to denote $dv_c(t)/dt|_{t=0^+}$.

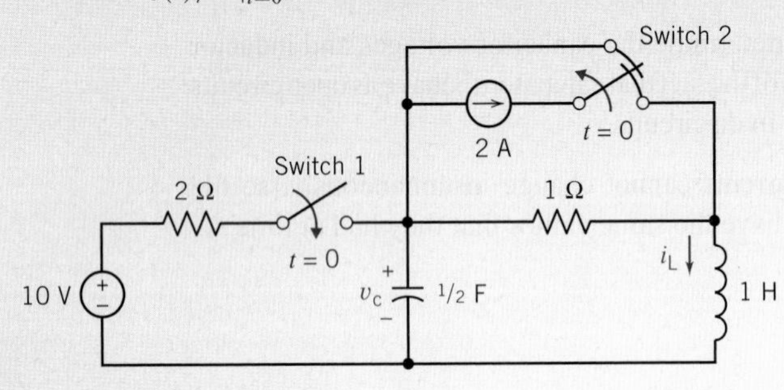

FIGURE 7.8-3 Circuit for Example 7.8-2. Switch 1 closes at $t = 0$ and switch 2 opens at $t = 0$.

Assume that switch 1 has been open and switch 2 has been closed for a long time and steady-state conditions prevail at $t = 0^-$.

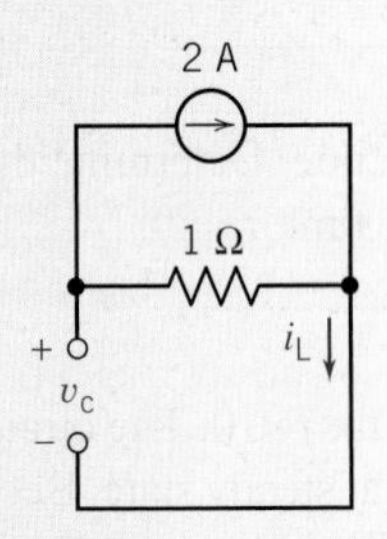

FIGURE 7.8-4 Circuit of Figure 7.8-3 at $t = 0^-$.

Solution

First, we redraw the circuit for $t = 0^-$ by replacing the inductor with a short circuit and the capacitor with an open circuit, as shown in Figure 7.8-4. Then we note that

$$i_L(0^-) = 0$$

and

$$v_c(0^-) = -2\text{ V}$$

Therefore, we have

$$i_L(0^+) = i_L(0^-) = 0$$

and

$$v_c(0^+) = v_c(0^-) = -2\text{ V}$$

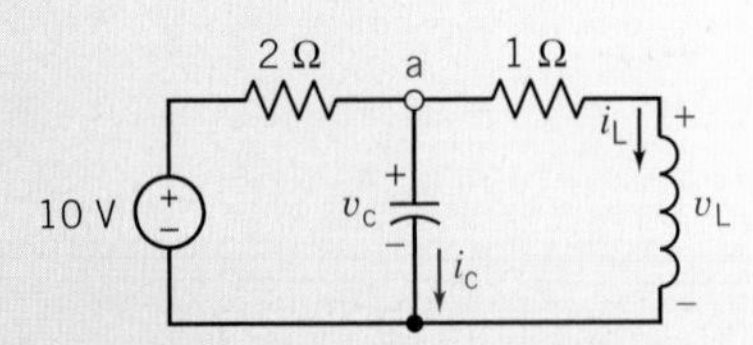

FIGURE 7.8-5 Circuit of Figure 7.8-3 at $t = 0^+$ with the switch closed and the current source disconnected.

To find $dv_c(0^+)/dt$ and $di_L(0^+)/dt$, we throw the switch at $t = 0$ and redraw the circuit of Figure 7.8-3, as shown in Figure 7.8-5. (We did not draw the current source because its switch is open.)

Because we wish to find $dv_c(0^+)/dt$, we recall that

$$i_c = C\frac{dv_c}{dt}$$

so

$$\frac{dv_c(0^+)}{dt} = \frac{i_c(0^+)}{C}$$

Similarly, because for the inductor

$$v_L = L\frac{di_L}{dt}$$

we may obtain $di_L(0^+)/dt$ as

$$\frac{di_L(0^+)}{dt} = \frac{v_L(0^+)}{L}$$

Using KVL for the right-hand mesh of Figure 7.8-5, we obtain

$$v_L - v_c + 1i_L = 0$$

Therefore, at $t = 0^+$,

$$v_L(0^+) = v_c(0^+) - i_L(0^+) = -2 - 0 = -2\text{ V}$$

Hence, we obtain

$$\frac{di_L(0^+)}{dt} = -2 \text{ A/s}$$

Similarly, to find i_c, we write KCL at node a to obtain

$$i_c + i_L + \frac{v_c - 10}{2} = 0$$

Consequently, at $t = 0^+$,

$$i_c(0^+) = \frac{10 - v_c(0^+)}{2} - i_L(0^+) = 6 - 0 = 6 \text{ A}$$

Accordingly,

$$\frac{dv_c(0^+)}{dt} = \frac{i_c(0^+)}{C} = \frac{6}{1/2} = 12 \text{ V/s}$$

Thus, we found that at the switching time $t = 0$, the current in the inductor and the voltage of the capacitor remained constant. However, the inductor voltage did change instantaneously from $v_L(0^-) = 0$ to $v_L(0^+) = -2$ V, and we determined that $di_L(0^+)/dt = -2$ A/s. Also, the capacitor current changed instantaneously from $i_c(0^-) = 0$ to $i_c(0^+) = 6$ A, and we found that $dv_c(0^+)/dt = 12$ V/s.

7.9 OPERATIONAL AMPLIFIER CIRCUITS AND LINEAR DIFFERENTIAL EQUATIONS

This section describes a procedure for designing operational amplifier circuits that implement linear differential equations such as

$$2\frac{d^3}{dt^3}y(t) + 5\frac{d^2}{dt^2}y(t) + 4\frac{d}{dt}y(t) + 3y(t) = 6x(t) \tag{7.9-1}$$

The solution of this equation is a function, $y(t)$, that depends both on the function $x(t)$ and on a set of initial conditions. It is convenient to use the initial conditions:

$$\frac{d^2}{dt^2}y(t) = 0, \quad \frac{d}{dt}y(t) = 0, \quad \text{and} \quad y(t) = 0 \tag{7.9-2}$$

Having specified these initial conditions, we expect a unique function $y(t)$ to correspond to any given function $x(t)$. Consequently, we will treat $x(t)$ as the input to the differential equation and $y(t)$ as the output.

Section 6.6 introduced the notion of diagramming operations as blocks and equations as block diagrams. Section 6.6 also introduced blocks to represent addition and multiplication by a constant. Figure 7.9-1 illustrates two additional blocks, representing integration and differentiation.

Suppose that we were somehow to obtain $\frac{d^3}{dt^3}y(t)$. We could then integrate three times to obtain $\frac{d^2}{dt^2}y(t)$, $\frac{d}{dt}y(t)$, and $y(t)$, as illustrated in Figure 7.9-2.

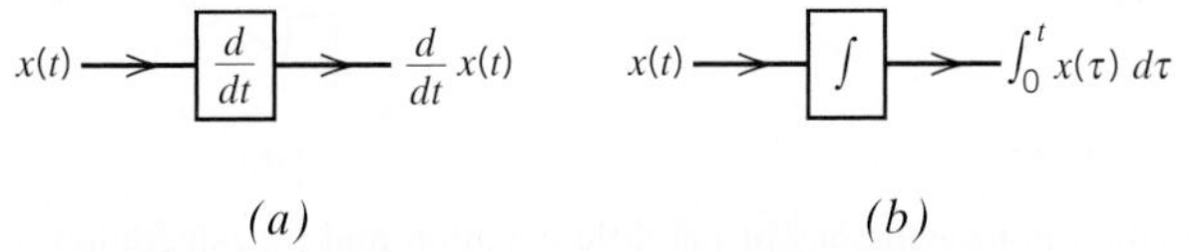

FIGURE 7.9-1 Block diagram representations of (*a*) differentiation and (*b*) integration.

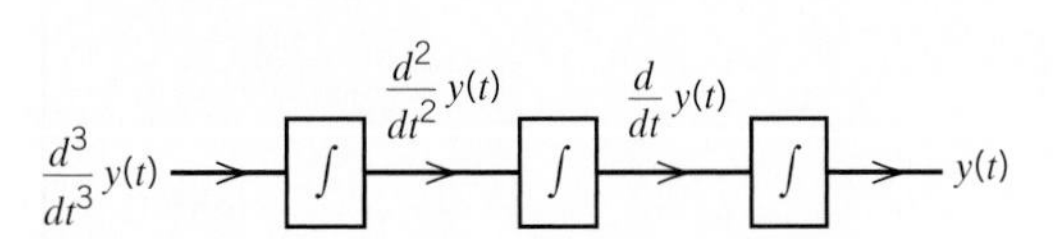

FIGURE 7.9-2 The first partial block diagram.

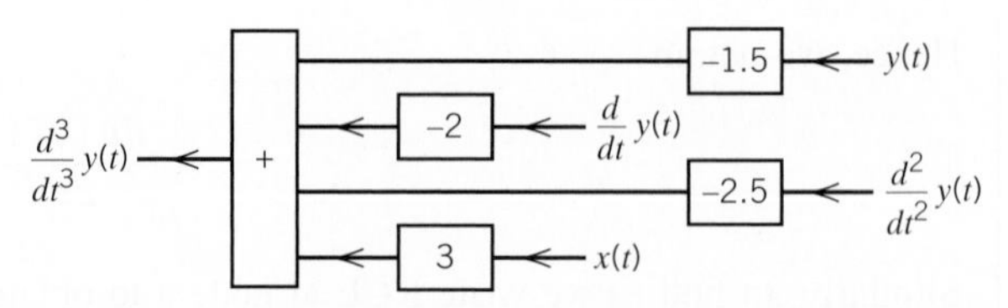

FIGURE 7.9-3 A block diagram that represents Eq. 7.9-3.

Now we must obtain $\frac{d^3}{dt^3}y(t)$. To do so, solve Eq. 7.9-1 for $\frac{d^3}{dt^3}y(t)$ to get

$$\frac{d^3}{dt^3}y(t) = 3x(t) - \left[2.5\frac{d^2}{dt^2}y(t) + 2\frac{d}{dt}y(t) + 1.5y(t)\right] \tag{7.9-3}$$

Next, represent Eq. 7.9-3 by a block diagram such as the diagram shown in Figure 7.9-3. Finally, the block diagrams in Figures 7.9-2 and 7.9-3 can be combined as shown in Figure 7.9-4 to obtain the block diagram of Eq. 7.9-1.

Our next task is to implement the block diagram as an operational amplifier circuit. Figure 7.9-5 provides operational amplifier circuits to implement both differentiation and integration. To see how the integrator works, consider Figure 7.9-6. The nodes of the integrator in Figure 7.9-6 have been labeled in anticipation of writing node equations. Let v_1, v_2, and v_3 denote the node voltages at nodes 1, 2, and 3, respectively.

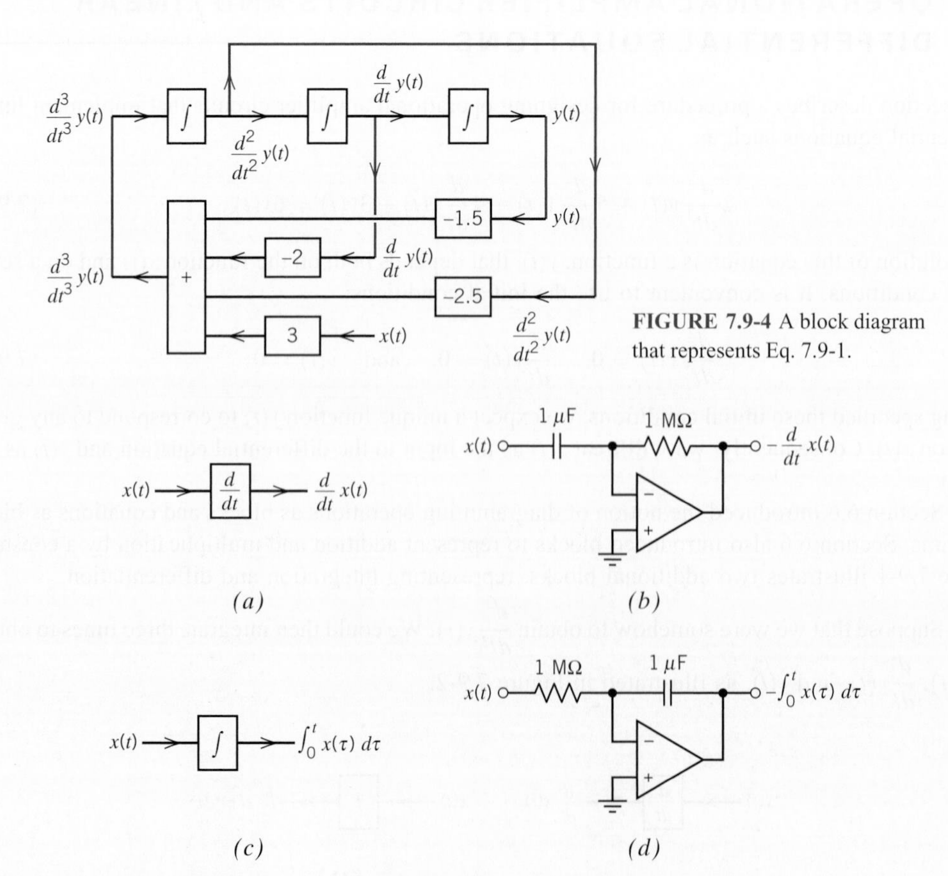

FIGURE 7.9-4 A block diagram that represents Eq. 7.9-1.

FIGURE 7.9-5 Block diagram representations of (*a*) differentiation and (*c*) integration. Corresponding operational amplifier circuits that (*b*) differentiate and (*d*) integrate.

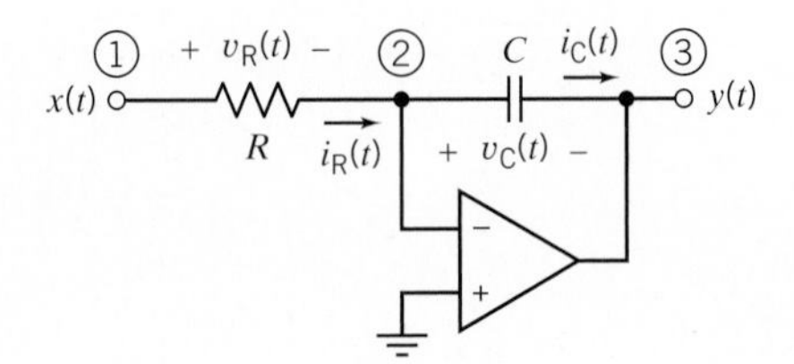

FIGURE 7.9-6 The integrator.

The input to the integrator is $x(t)$, the node voltage at node 1. Thus, $v_1 = x(t)$. The output of the integrator is $y(t)$, the node voltage at node 3. Thus, $v_3 = y(t)$. The noninverting input of the ideal operational amplifier is attached to the reference node, and the inverting input is connected to node 2. The node voltages at these two nodes must be equal, so $v_2 = 0$.

The voltage across the resistor is related to the node voltages at the resistor nodes by

$$v_R(t) = v_1(t) - v_2(t) = x(t) - 0 = x(t)$$

The resistor current is calculated, using Ohm's law, to be

$$i_R(t) = \frac{v_R(t)}{R} = \frac{x(t)}{R}$$

The value of the current flowing into an input of an ideal operational amplifier is zero, so applying KCL at node 2 gives

$$i_C(t) = i_R(t) = \frac{x(t)}{R}$$

The voltage across the capacitor is related to the node voltages at the capacitor nodes by

$$v_C(t) = v_2(t) - v_3(t) = 0 - y(t) = -y(t) \tag{7.9-4}$$

The capacitor voltage is related to the capacitor current by

$$v_C(t) = \frac{1}{C}\int_0^t i_C(\tau)\,d\tau + v_C(0)$$

Recall that $y(0) = 0$. Thus, $v_C(0) = 0$, and

$$v_C(t) = \frac{1}{C}\int_0^t i_C(\tau)\,d\tau = \frac{1}{C}\int_0^t \frac{x(\tau)}{R}\,d\tau = \frac{1}{RC}\int_0^t x(\tau)\,d\tau$$

Finally, using Eq. 7.9-4 gives

$$y(t) = -\frac{1}{RC}\int_0^t x(\tau)\,d\tau = -k\int_0^t x(\tau)\,d\tau \tag{7.9-5}$$

where $k = \frac{1}{RC}$.

Equation 7.9-5 indicates that the integrator does two things. First, the input is integrated. Second, the integral is multiplied by a negative constant, k. In Figure 7.9-5*d*, values of R and C have been selected to make $k = -1$. Multiplying a function by -1 reflects the graph of the function across the time axis. This reflection is called an inversion, and the circuit is said to be an inverting circuit. Consequently, the integrator shown in Figure 7.9-5*d* is sometimes called an inverting integrator. We will call this circuit an integrator unless we want to call attention to the inversion, in which case, we will call the circuit an inverting integrator.

Analysis of the summing integrator shown in Figure 7.9-7 is similar to the analysis of the integrator. The inputs to the summing integrator are $x_1(t)$, the node voltage at node 1, and $x_2(t)$, the

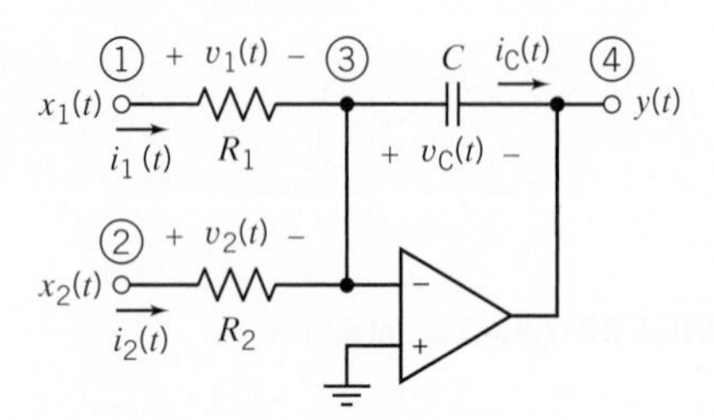

FIGURE 7.9-7 The summing integrator.

node voltage at node 2. The output of the integrator is $y(t)$, the node voltage at node 4. The ideal operational amplifier causes the voltage at node 3 to be zero. Hence,

$$v_1(t) = x_1(t), \quad v_2(t) = x_2(t), \quad v_3(t) = 0, \quad \text{and} \quad v_4(t) = y(t)$$

Using Ohm's law shows the currents in the resistors to be

$$i_1(t) = \frac{v_1(t)}{R_1} = \frac{x_1(t)}{R_1} \quad \text{and} \quad i_2(t) = \frac{v_2(t)}{R_2} = \frac{x_2(t)}{R_2}$$

The value of the current flowing into an input of an ideal operational amplifier is zero, so applying KCL at node 3 gives

$$i_C(t) = i_1(t) + i_2(t) = \frac{x_1(t)}{R_1} + \frac{x_2(t)}{R_2}$$

The voltage across the capacitor is related to the node voltages at the capacitor nodes by

$$v_C(t) = v_3(t) - v_4(t) = 0 - y(t) = -y(t) \tag{7.9-6}$$

The capacitor voltage is related to the capacitor current by

$$v_C(t) = \frac{1}{C}\int_0^t i_C(\tau)\,d\tau + v_C(0)$$

Recall that $y(0) = 0$. Thus, $v_C(0) = 0$, and

$$v_C(t) = \frac{1}{C}\int_0^t i_C(\tau)\,d\tau = \frac{1}{C}\int_0^t \left(\frac{x_1(\tau)}{R_1} + \frac{x_2(\tau)}{R_2}\right)d\tau = \int_0^t \left(\frac{x_1(\tau)}{R_1 C} + \frac{x_2(\tau)}{R_2 C}\right)d\tau$$

Finally, using Eq. 7.9-6 gives

$$y(t) = -\int_0^t \left(\frac{x_1(\tau)}{R_1 C} + \frac{x_2(\tau)}{R_2 C}\right)d\tau = -\int_0^t (k_1 x_1(\tau) + k_2 x_2(\tau))\,d\tau \tag{7.9-7}$$

where $k_1 = \dfrac{1}{R_1 C}$ and $k_2 = \dfrac{1}{R_2 C}$.

Equation 7.9-7 indicates that the summing integrator does four things. First, each input is multiplied by a separate constant: x_1 is multiplied by k_1, and x_2 is multiplied by k_2. Second, the products are summed. Third, the sum is integrated. Fourth, the integral is multiplied by -1. (Like the inverting integrator, this circuit inverts its output. It is sometimes called an inverting summing integrator. Fortunately, we don't need to use that long name very often.)

The summing amplifier in Figure 7.9-7 accommodates two inputs. To accommodate additional inputs, we add more input resistors, each connected between an input node and the inverting input node of the operational amplifier. (The operational amplifier circuit that implements Eq. 7.9-1 will require a four-input summing integrator.)

We will design an operational amplifier circuit to implement Eq. 7.9-1 by replacing the blocks in the block diagram of Eq. 7.9-1 by operational amplifier circuits. This process will be easier if we first modify the block diagram to accommodate *inverting* integrators. Figures 7.9-8 and 7.9-9 show

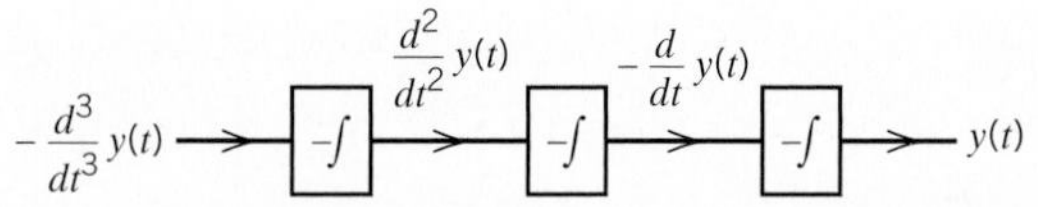

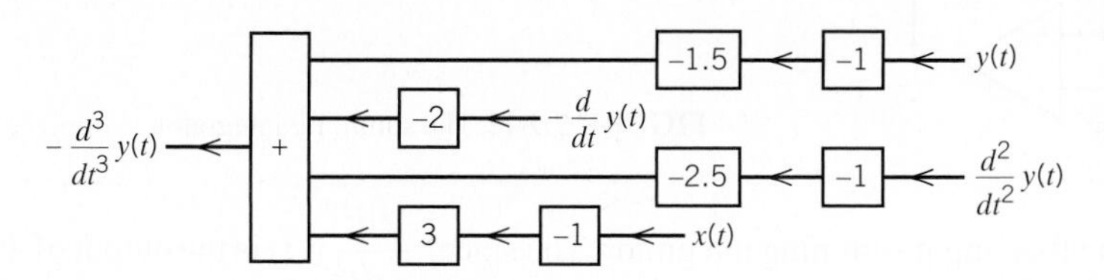

FIGURE 7.9-8 The block diagram from Figure 7.9-2, adjusted to accommodate inverting integrators.

FIGURE 7.9-9 The block diagram from Figure 7.9-3, adjusted to accommodate the consequences of using inverting integrators.

modified versions of the block diagrams from Figures 7.9-2 and 7.9-3. Replace all the integrators in Figure 7.9-2 by inverting integrators to get Figure 7.9-8. It's necessary to set the input equal to $-\frac{d^3}{dt^3}y(t)$ instead of $\frac{d^3}{dt^3}y(t)$ to cause the output to be equal to $y(t)$ instead of $-y(t)$.

The block diagram in Figure 7.9-9 produces $-\frac{d^3}{dt^3}y(t)$ from $\frac{d^2}{dt^2}y(t)$, $-\frac{d}{dt}y(t)$, and $y(t)$. The block diagrams in Figures 7.9-8 and 7.9-9 can be combined as shown in Figure 7.9-10 to obtain the block diagram of Eq. 7.9-1.

A summing integrator can multiply each of its inputs by a separate constant, add the products, and integrate the sum. The block diagram shown in Figure 7.9-11 emphasizes the blocks that can be implemented by a single four-input summing integrator.

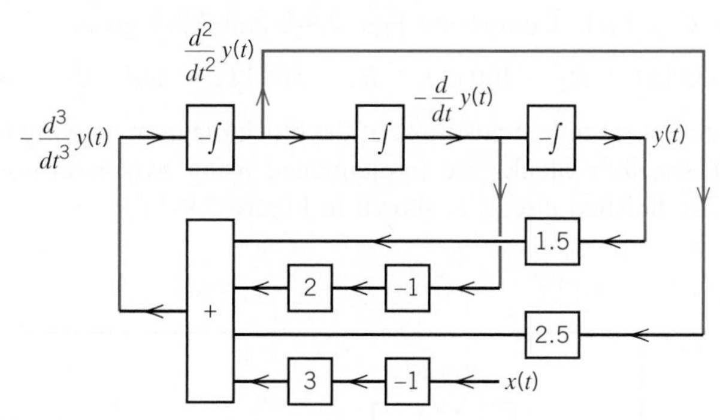

FIGURE 7.9-10 The block diagram representing Eq. 7.9-1, adjusted to accommodate inverting integrators.

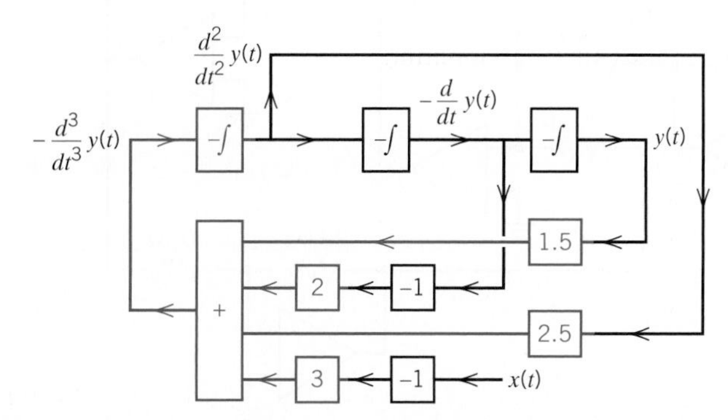

FIGURE 7.9-11 The block diagram representing Eq. 7.9-1, emphasizing the part implemented by the summing integrator.

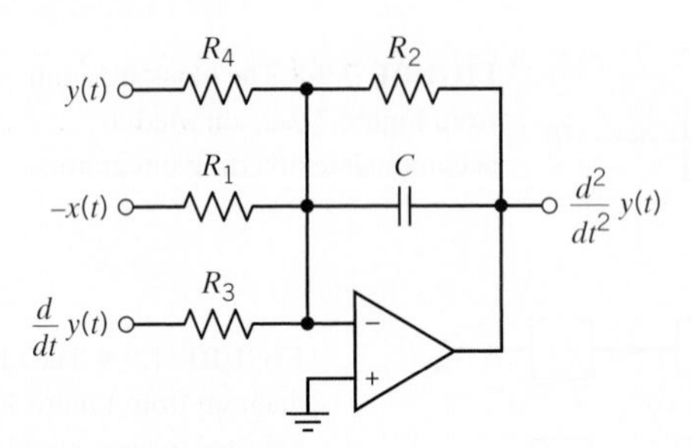

FIGURE 7.9-12 The summing integrator.

Figure 7.9-12 shows the four-input summing integrator. The signal $-\frac{d^2}{dt^2}y(t)$ is the output of this circuit and is also one of the inputs to the circuit. The resistor R_2 is connected between this input and the node connected to the inverting input of the operational amplifier. The summing integrator is represented by the equation

$$\frac{d^2}{dt^2}y(t) = -\int_0^t \left(\frac{1}{R_1 C}[-x(t)] + \frac{1}{R_2 C}\frac{d^2}{dt^2}y(t) + \frac{1}{R_3 C}\left[\frac{d}{dt}y(t)\right] + \frac{1}{R_4 C}y(t) \right) d\tau \qquad (7.9\text{-}8)$$

Integrating both sides of Eq. 7.9-3 gives

$$\frac{d^2}{dt^2}y(t) = -\int_0^t \left(3[-x(t)] + 2.5\frac{d^2}{dt^2}y(t) + 2\left[\frac{d}{dt}y(t)\right] + 1.5\,y(t) \right) d\tau \qquad (7.9\text{-}9)$$

For convenience, pick $C = 1\,\mu\text{F}$. Comparing Eqs. 7.9-8 and 7.9-9 gives

$$R_1 = 333\,\text{k}\Omega, \quad R_2 = 400\,\text{k}\Omega, \quad R_3 = 500\,\text{k}\Omega, \quad \text{and} \quad R_4 = 667\,\text{k}\Omega$$

The summing integrator implements most of the block diagram, leaving only four other blocks to be implemented. Those four blocks are implemented using two inverting integrators and two inverting amplifiers. The finished circuit is shown in Figure 7.9-13.

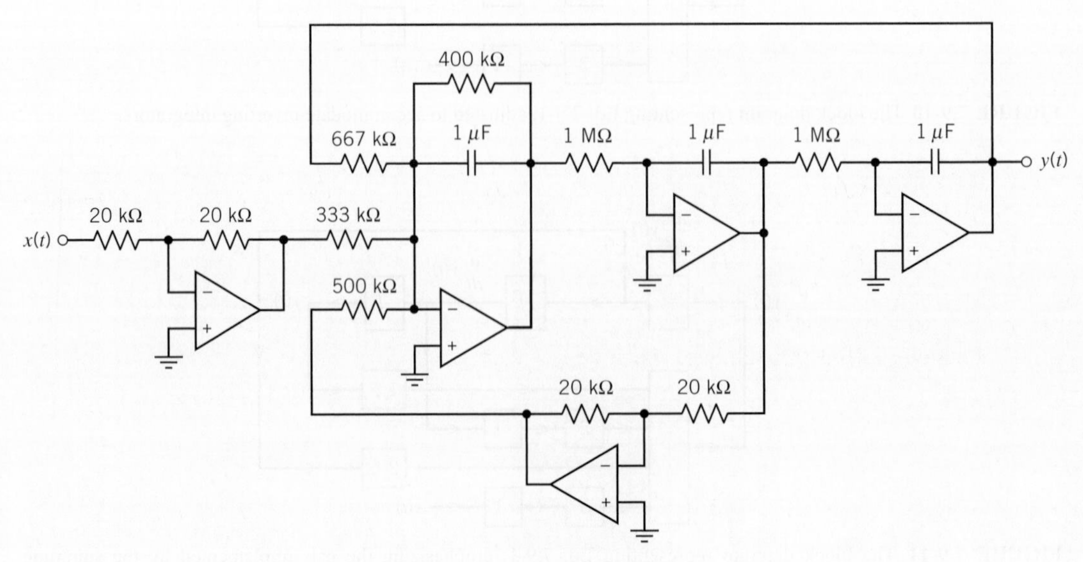

FIGURE 7.9-13 An operational amplifier circuit that implements Eq. 7.9-1.

7.10 USING MATLAB TO PLOT CAPACITOR OR INDUCTOR VOLTAGE AND CURRENT

Suppose that the current in a 2-F capacitor is

$$i(t) = \begin{cases} 4 & t \leq 2 \\ t+2 & 2 \leq t \leq 6 \\ 20-2t & 6 \leq t \leq 14 \\ -8 & t \geq 4 \end{cases} \tag{7.10-1}$$

where the units of current are A and the units of time are s. When the initial capacitor voltage is $v(0) = -5$ V, the capacitor voltage can be calculated using

$$v(t) = \frac{1}{2}\int_0^t i(\tau)d\tau - 5 \tag{7.10-2}$$

Equation 7.10-1 indicates that $i(t) = 4$ A, whereas $t < 2$ s. Using this current in Eq. 7.10-2 gives

$$v(t) = \frac{1}{2}\int_0^t 4d\tau - 5 = 2t - 5 \tag{7.10-3}$$

when $t < 2$ s. Next, Eq. 7.10-1 indicates that $i(t) = t + 2$ A, whereas $2 < t < 6$ s. Using this current in Eq. 7.10-2 gives

$$v(t) = \frac{1}{2}\left(\int_2^t (t+2)d\tau + \int_0^2 4\,d\tau\right) - 5 = \frac{1}{2}\int_2^t (t+2)d\tau - 1 = \frac{t^2}{4} + t - 4 \tag{7.10-4}$$

when $2 < t < 6$ s. Continuing in this way, we calculate

$$\begin{aligned} v(t) &= \frac{1}{2}\left(\int_6^t (20-2t)\,d\tau + \int_2^6 (t+2)\,d\tau + \int_0^2 4\,d\tau\right) - 5 \\ &= \frac{1}{2}\int_6^t (20-2t)\,d\tau + 11 = -\frac{t^2}{2} + 10t - 31 \end{aligned} \tag{7.10-5}$$

when $6 < t < 14$ s, and

$$\begin{aligned} v(t) &= \frac{1}{2}\left(\int_{14}^t -8\,d\tau + \int_6^{14} (20-2t)\,d\tau + \int_2^6 (t+2)\,d\tau + \int_0^2 4\,d\tau\right) - 5 \\ &= \frac{1}{2}\int_{14}^t -8\,d\tau + 11 = 67 - 4t \end{aligned} \tag{7.10-6}$$

when $t > 14$ s.

Equations 7.10-3 through 7.10-6 can be summarized as

$$v(t) = \begin{cases} 2t-5 & t \leq 2 \\ \dfrac{t^2}{4} + t - 4 & 2 \leq t \leq 6 \\ -\dfrac{t^2}{2} + 10t - 31 & 6 \leq t \leq 14 \\ 67 - 4t & t \geq 14 \end{cases} \tag{7.10-7}$$

```
function i = CapCur(t)
  if t<2
     i=4;
  elseif t<6
     i=t+2;
  elseif t<14
     i=20 - 2*t;
  else
     i = - 8;
  end
```

(a)

```
function v = CapVol(t)
  if t<2
     v = 2*t - 5;
  elseif t<6
     v = 0.25*t*t + t - 4;
  elseif t<14
     v = -.5*t*t + 10*t - 31;
  else
     v = 67 - 4*t;
  end
```

(b)

```
t = 0:1:20;
for k = 1:1:length(t)
   i(k) = CapCur(k - 1);
   v(k) = CapVol(k - 1);
end
plot(t,i,t,v)
text(12,10, 'v(t), V')
text(10, -5, 'i(t), A')
title('Capacitor Voltage and Current')
xlabel('time, s')
```

(c)

FIGURE 7.10-1 MATLAB input files representing (*a*) the capacitor current and (*b*) the capacitor voltage; (*c*) the MATLAB input file used to plot the capacitor current and voltage.

Equations 7.10-1 and 7.10-7 provide an analytic representation of the capacitor current and voltage. MATLAB provides a convenient way to obtain graphical representation of these functions. Figures 7.10-1*a*,*b* show MATLAB input files that represent the capacitor current and voltage. Notice that the MATLAB input file representing the current, Figure 7.10-1*a*, is very similar to Eq. 7.10-1, whereas the MATLAB input file representing the voltage, Figure 7.10-1*b*, is very similar to Eq. 7.10-7. Figure 7.10-1*c* shows the MATLAB input file used to plot the capacitor current and voltage. Figure 7.10-2 shows the resulting plots of the capacitor current and voltage.

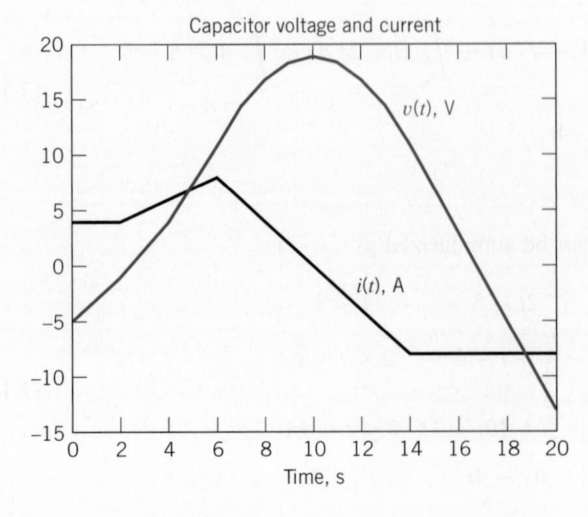

FIGURE 7.10-2 A plot of the voltage and current of a capacitor.

7.11 HOW CAN WE CHECK . . . ?

Engineers are frequently called upon to check that a solution to a problem is indeed correct. For example, proposed solutions to design problems must be checked to confirm that all of the specifications have been satisfied. In addition, computer output must be reviewed to guard against data-entry errors, and claims made by vendors must be examined critically.

Engineering students are also asked to check the correctness of their work. For example, occasionally just a little time remains at the end of an exam. It is useful to be able to quickly identify those solutions that need more work.

The following example illustrates techniques useful for checking the solutions of the sort of problems discussed in this chapter.

EXAMPLE 7.11-1 How Can We Check the Voltage and Current of a Capacitor?

A homework solution indicates that the current and voltage of a 2-F capacitor are

$$i(t) = \begin{cases} 4 & t < 2 \\ t + 2 & 2 < t < 6 \\ 20 - 2t & 6 < t < 14 \\ -8 & t > 14 \end{cases} \tag{7.11-1}$$

and

$$v(t) = \begin{cases} 2t - 5 & t < 2 \\ \frac{t^2}{4} + t - 4 & 2 < t < 6 \\ -\frac{t^2}{2} + 10t - 21 & 6 < t < 14 \\ 67 - 4t & t > 14 \end{cases} \tag{7.11-2}$$

where the units of current are A, the units of voltage are V, and the units of time are s. **How can we check** this homework solution to see whether it is correct?

Solution

The capacitor voltage cannot change instantaneously. The capacitor voltage is given by

$$v(t) = 2t - 5 \tag{7.11-3}$$

when $t < 2$ s and by

$$v(t) = \frac{t^2}{4} + t - 4 \tag{7.11-4}$$

when $2 < t < 6$ s. Because the capacitor voltage cannot change instantaneously, Eqs. 7.11-3 and 7.11-4 must both give the same value for $v(2)$, the capacitor voltage at time $t = 2$ s. Solving Eq. 7.11-3 gives

$$v(2) = 2(2) - 5 = -1 \text{ V}$$

Also, solving Eq. 7.11-4 gives

$$v(2) = \frac{2^2}{4} + 2 - 4 = -1 \text{ V}$$

These values agree, so we haven't found an error. Next, let's check $v(6)$, the capacitor voltage at time $t = 6$ s. The capacitor voltage is given by

$$v(t) = -\frac{t^2}{2} + 10t - 21 \tag{7.11-5}$$

when $6 < t < 14$ s. Equations 7.11-4 and 7.11-5 must both give the same value for $v(6)$. Solving Eq. 7.11-4 gives

$$v(6) = \frac{6^2}{4} + 6 - 4 = 11 \text{ V}$$

whereas solving Eq. 7.11-5 gives

$$v(6) = -\frac{6^2}{2} + 10(6) - 21 = 21 \text{ V}$$

These values don't agree. That means that $v(t)$ changes instantaneously at $t = 6$ s, so $v(t)$ cannot be the voltage across the capacitor. The homework solution is not correct.

7.12 DESIGN EXAMPLE

INTEGRATOR AND SWITCH

This design challenge involves an integrator and a voltage-controlled switch.

An integrator is a circuit that performs the mathematical operation of integration. The output of an integrator, say $v_o(t)$, is related to the input of the integrator, say $v_s(t)$, by the equation

$$v_o(t_2) = K \cdot \int_{t_1}^{t_2} v_s(t)dt + v_o(t_1) \tag{7.12-1}$$

The constant K is called the gain of the integrator.

Integrators have many applications. One application of an integrator is to measure an interval of time. Suppose $v_s(t)$ is a constant voltage, V_s. Then,

$$v_o(t_2) = K \cdot (t_2 - t_1) \cdot V_s + v_o(t_1) \tag{7.12-2}$$

This equation indicates that the output of the integrator at time t_2 is a measure of the time interval $t_2 - t_1$.

Switches can be controlled electronically. Figure 7.12-1 illustrates an electronically controlled SPST switch. The symbol shown in Figure 7.12-1*a* is sometimes used to emphasize that a switch is controlled electronically. The node voltage $v_c(t)$ is called the control voltage. Figure 7.12-1*b* shows a typical control voltage. This voltage-controlled switch is closed when $v_c(t) = v_H$ and open when $v_c(t) = v_L$. The switch shown in Figure 7.12-1 is open before time t_1. It closes at time t_1 and stays closed until time t_2. The switch opens at time t_2 and remains open.

Consider Figure 7.12-2. The voltage $v_c(t)$ controls the switch. The integrator converts the time interval $t_2 - t_1$ to a voltage that is displayed using the voltmeter. The time

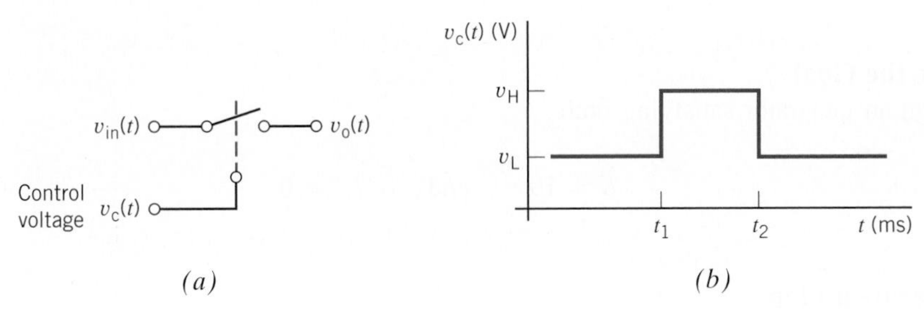

FIGURE 7.12-1 The voltage-controlled switch. (*a*) Switch symbol. (*b*) Typical control voltage.

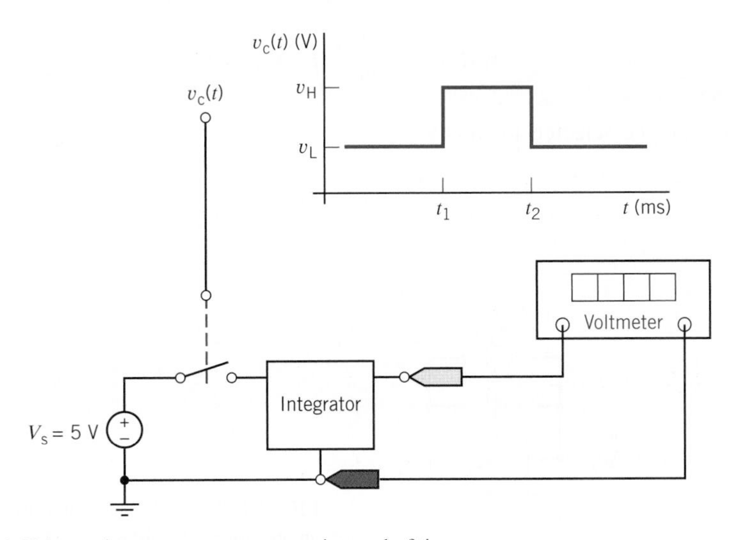

FIGURE 7.12-2 Using an integrator to measure an interval of time.

interval to be measured could be as small as 5 ms or as large as 200 ms. The challenge is to design the integrator. The available components include:

- Standard 2 percent resistors (see Appendix D)
- 1-μF, 0.2-μF, and 0.1-μF capacitors
- Operational amplifiers
- +15-V and − 15-V power supplies
- 1-kΩ, 10-kΩ, and 100-kΩ potentiometers
- Voltage-controlled SPST switches

Describe the Situation and the Assumptions

It is convenient to set the integrator output to zero at time t_1. The relationship between the integrator output voltage and the time interval should be simple. Accordingly, let

$$v_o(t_2) = \frac{10\ \text{V}}{200\ \text{ms}} \cdot (t_2 - t_1) \tag{7.12-3}$$

Figure 7.12-2 indicates that $V_s = 5$ V. Comparing Eqs. 7.12-2 and 7.12-3 yields

$$K \cdot V_s = \frac{10\ \text{V}}{200\ \text{ms}} \quad \text{and, therefore,} \quad K = 10\ \frac{1}{\text{s}} \tag{7.12-4}$$

State the Goal

Design an integrator satisfying both

$$K = 10\ \frac{1}{\text{s}} \quad \text{and} \quad v_o(t_1) = 0 \tag{7.12-5}$$

Generate a Plan

Let us use the integrator described in Section 7.9. Adding a switch as shown in Figure 7.12-3 satisfies the condition $v_o(t_1) = 0$. The analysis performed in Section 7.9 showed that

$$v_o(t_2) = -\frac{1}{RC} \cdot \int_{t_1}^{t_2} v_s(t)\, dt \tag{7.12-6}$$

so R and C must be selected to satisfy

$$\frac{1}{RC} = K = 10\ \frac{1}{\text{s}} \tag{7.12-7}$$

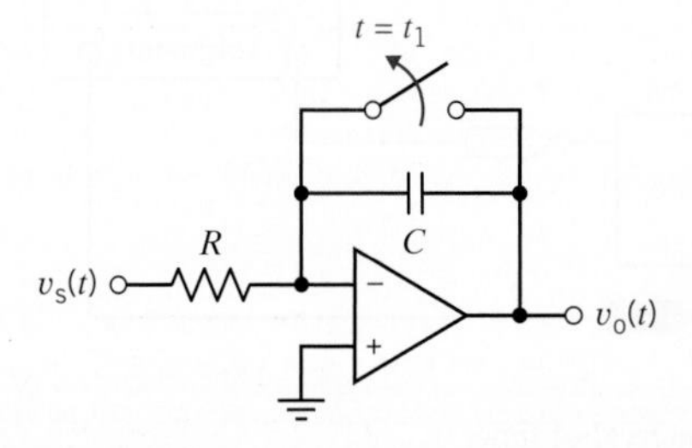

FIGURE 7.12-3 An integrator using an operational amplifier.

Act on the Plan

Any of the available capacitors would work. Select $C = 1\ \mu\text{F}$. Then,

$$R = \frac{1}{10 \frac{\text{V}}{\text{s}} \cdot 1\ \mu\text{F}} = 100\ \text{k}\Omega \tag{7.12-8}$$

The final design is shown in Figure 7.12-4.

Verify the Proposed Solution

The output voltage of the integrator is given by

$$v_o(t) = -\frac{1}{RC} \int_{t_1}^{t} v_s(\tau)\, d\tau + v_o(0) = \frac{-1}{(100 \cdot 10^3)(10^{-6})} \int_{t_1}^{t} 5\, d\tau = -50\,(t - t_1)$$

where the units of voltage are V and the units of time are s. The interval of time can be calculated from the output voltage, using

$$-(t - t_1) = \frac{v_o(t)}{50}$$

For example, an output voltage of -4 V indicates a time interval of $\frac{4}{50}$ s = 80 ms.

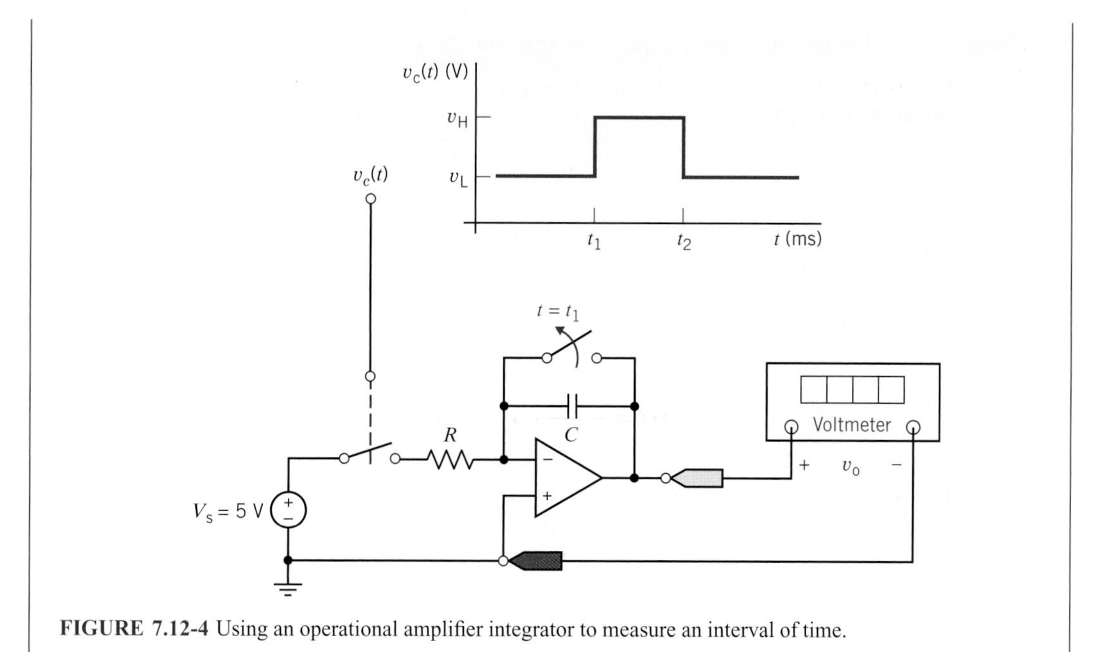

FIGURE 7.12-4 Using an operational amplifier integrator to measure an interval of time.

7.13 SUMMARY

- Table 7.13-1 summarizes the element equations for capacitors and inductors. (Notice that the voltage and current referred to in these equations adhere to the passive convention.) Unlike the circuit elements we encountered in previous chapters, the element equations for capacitors and inductors involve derivatives and integrals.
- Circuits that contain capacitors and/or inductors are able to store energy. The energy stored in the electric field of a capacitor is equal to $\frac{1}{2}Cv^2(t)$, where $v(t)$ is the voltage across the capacitor. The energy stored in the magnetic field of a inductor is equal to $\frac{1}{2}Li^2(t)$, where $i(t)$ is the current in the inductor.
- Circuits that contain capacitors and/or inductors have memory. The voltages and currents in that circuit at a particular time depend not only on other voltages and currents at that same instant of time but also on previous values of those currents and voltages. For example, the voltage across a capacitor at time t_1 depends on the voltage across that capacitor at an earlier time t_0 and on the value of the capacitor current between t_0 and t_1.
- A set of series or parallel capacitors can be reduced to an equivalent capacitor. A set of series or parallel inductors can readily be reduced to an equivalent inductor. Table 7.13-2 summarizes the equations required to do so.

Table 7.13-1 Element Equations for Capacitors and Inductors

CAPACITOR	INDUCTOR
+ $v(t)$ − ; $i(t)$ → ; C	+ $v(t)$ − ; $i(t)$ → ; L
$i(t) = C\frac{d}{dt}v(t)$	$i(t) = \frac{1}{L}\int_{t_0}^{t} v(\tau)d\tau + i(t_0)$
$v(t) = \frac{1}{C}\int_{t_0}^{t} i(\tau)d\tau + v(t_0)$	$v(t) = L\frac{d}{dt}i(t)$

Table 7.13-2 Parallel and Series Capacitors and Inductors

SERIES OR PARALLEL CIRCUIT	EQUIVALENT CIRCUIT	EQUATION
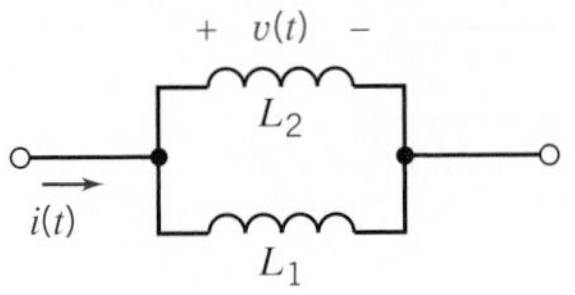 + $v(t)$ −; L_2; L_1; $i(t)$	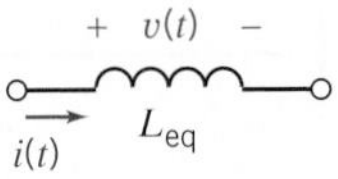 + $v(t)$ −; L_{eq}; $i(t)$	$L_{eq} = \dfrac{1}{\dfrac{1}{L_1} + \dfrac{1}{L_2}}$
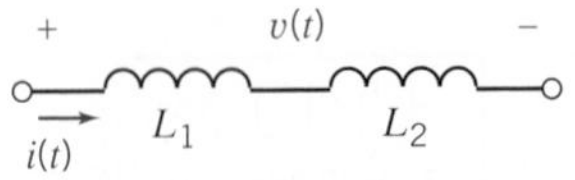 + $v(t)$ −; L_1; L_2; $i(t)$	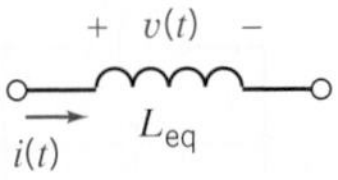 + $v(t)$ −; L_{eq}; $i(t)$	$L_{eq} = L_1 + L_2$
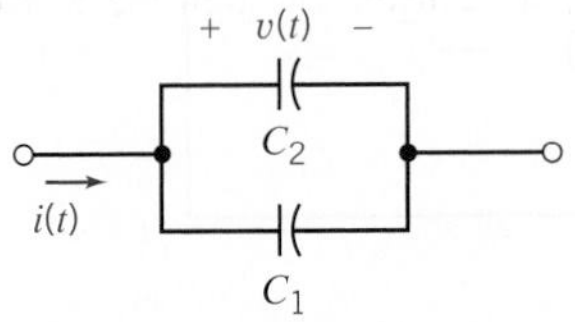 + $v(t)$ −; C_2; C_1; $i(t)$	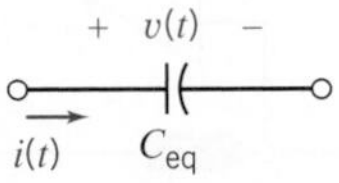 + $v(t)$ −; C_{eq}; $i(t)$	$C_{eq} = C_1 + C_2$
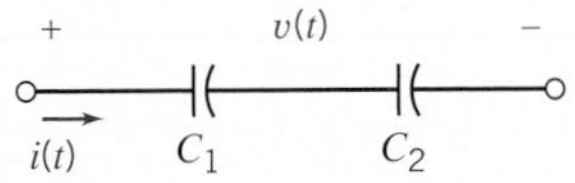 + $v(t)$ −; C_1; C_2; $i(t)$	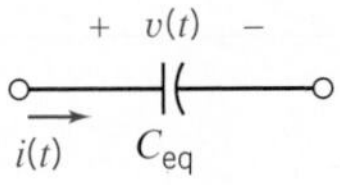 + $v(t)$ −; C_{eq}; $i(t)$	$C_{eq} = \dfrac{1}{\dfrac{1}{C_1} + \dfrac{1}{C_2}}$

- In the absence of unbounded currents, the voltage across a capacitor cannot change instantaneously. Similarly, in the absence of unbounded voltages, the current in an inductor cannot change instantaneously. In contrast, the current in a capacitor and voltage across an inductor are both able to change instantaneously.
- We sometimes consider circuits that contain capacitors and inductors and have only constant inputs. (The voltages of the independent voltage sources and currents of the independent current sources are all constant.) When such a circuit is at steady state, all the currents and voltages in that circuit will be constant. In particular, the voltage across any capacitor will be constant. The current in that capacitor will be zero due to the derivative in the equation for the capacitor current. Similarly, the current through any inductor will be constant and the voltage across any inductor will be zero. Consequently, the capacitors will act like open circuits and the inductors will act like short circuits. Notice that this situation occurs only when all of the inputs to the circuit are constant.
- An op amp and a capacitor can be used to make circuits that perform the mathematical operations of integration and differentiation. Appropriately, these important circuits are called the integrator and the differentiator.
- The element voltages and currents in a circuit containing capacitors and inductors can be complicated functions of time. MATLAB is useful for plotting these functions.

PROBLEMS

Section 7.2 Capacitors

P 7.2-1 A 20-μF capacitor has a voltage of 5 V across it at $t = 0$. If a constant current of 30 mA flows through the capacitor, how long will it take for the capacitor to charge up to 180 μC?

Answer: $t = 2.7$ ms

P 7.2-2 The voltage, $v(t)$, across a capacitor and current, $i(t)$, in that capacitor adhere to the passive convention. Determine the current, $i(t)$, when the capacitance is $C = 0.125$ F, and the voltage is $v(t) = 12\cos(2t + 30°)$ V.

Hint:
$$\begin{aligned}\frac{d}{dt}A\cos(\omega t + \theta) &= -A\sin(\omega t + \theta)\cdot\frac{d}{dt}(\omega t + \theta)\\ &= -A\omega\sin(\omega t + \theta)\\ &= A\omega\cos\left(\omega t + \left(\theta + \frac{\pi}{2}\right)\right)\end{aligned}$$

Answer: $i(t) = 3\cos(2t + 120°)$ A

P 7.2-3 The voltage, $v(t)$, across a capacitor and current, $i(t)$, in that capacitor adhere to the passive convention. Determine the capacitance when the voltage is $v(t) = 12\cos(500t - 45°)$ V and the current is $i(t) = 3\cos(500t + 45°)$ mA.

Answer: $C = 0.5\ \mu$F

P 7.2-4 Determine $v(t)$ for the circuit shown in Figure P 7.2-4$a(t)$ when the $i_s(t)$ is as shown in Figure P 7.2-4b and $v_o(0^-) = -1$ mV.

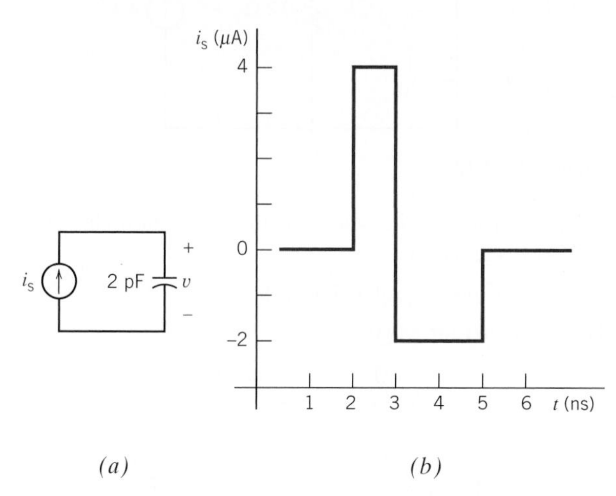

Figure P 7.2-4 (*a*) Circuit and (*b*) waveform of current source.

P 7.2-5 The voltage, $v(t)$, and current, $i(t)$, of a 1-F capacitor adhere to the passive convention. Also, $v(0) = 0$ V and $i(0) = 0$ A. (a) Determine $v(t)$ when $i(t) = x(t)$, where $x(t)$ is shown in Figure P 7.2-5 and $i(t)$ has units of A. (b) Determine $i(t)$ when $v(t) = x(t)$, where $x(t)$ is shown in Figure P 7.2-5 and $v(t)$ has units of V.

Hint: $x(t) = 4t - 4$ when $1 < t < 2$, and $x(t) = -4t + 12$ when $2 < t < 3$.

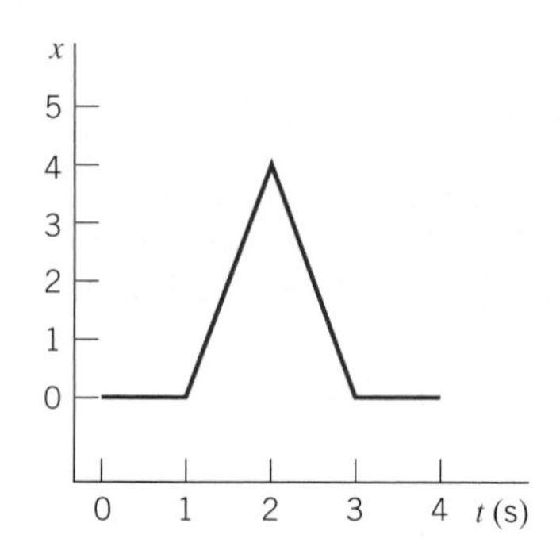

Figure P 7.2-5

P 7.2-6 The voltage, $v(t)$, and current, $i(t)$, of a 0.5-F capacitor adhere to the passive convention. Also, $v(0) = 0$ V and $i(0) = 0$ A. (a) Determine $v(t)$ when $i(t) = x(t)$, where $x(t)$ is shown in Figure P 7.2-6 and $i(t)$ has units of A. (b) Determine $i(t)$ when $v(t) = x(t)$, where $x(t)$ is shown in Figure P 7.2-6 and $v(t)$ has units of V.

Hint: $x(t) = 0.2t - 0.4$ when $2 < t < 6$.

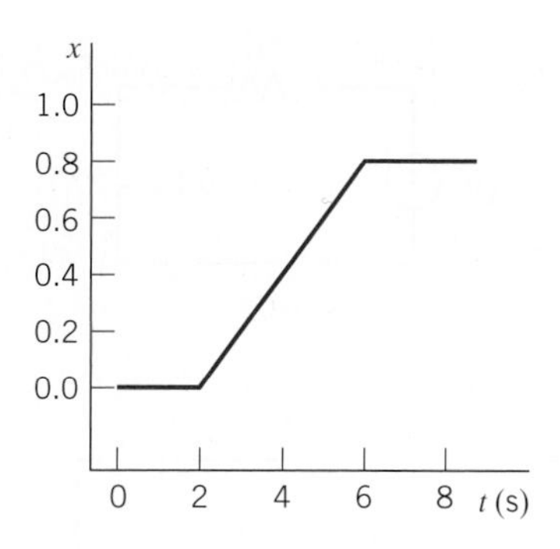

Figure P 7.2-6

P 7.2-7 The voltage across a 40-μF capacitor is 30 V at $t_0 = 0$. If the current through the capacitor as a function of time is given by $i(t) = 6e^{-6t}$ mA for $t < 0$, find $v(t)$ for $t > 0$.

Answer: $v(t) = 55 - 25e^{-6t}$ V

P 7.2-8 Find i for the circuit of Figure P 7.2-8 if $v = 5(1 - 2e^{-2t})$ V.

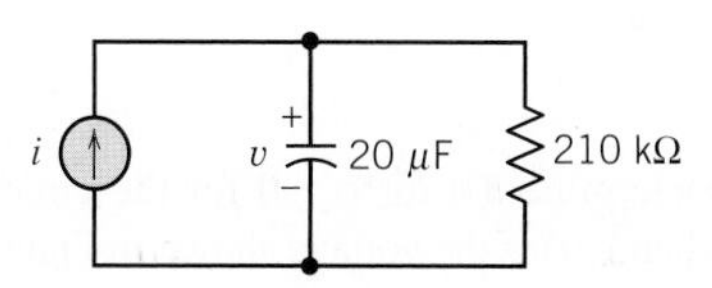

Figure P 7.2-8

P 7.2-9 Determine $v(t)$ for $t \geq 0$ for the circuit of Figure P 7.2-9a when $i_s(t)$ is the current shown in Figure P 7.2-9b and $v(0) = 1$ V.

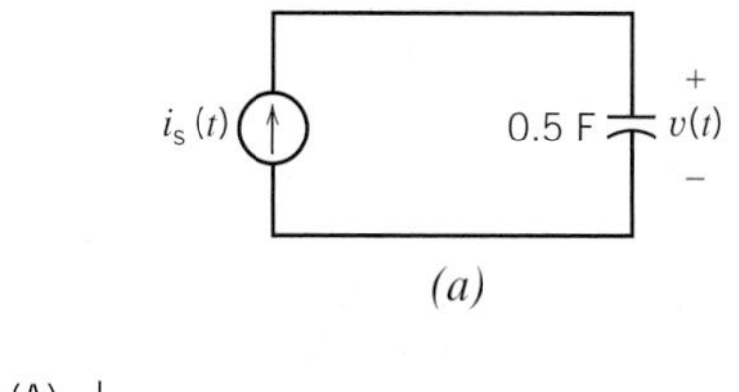

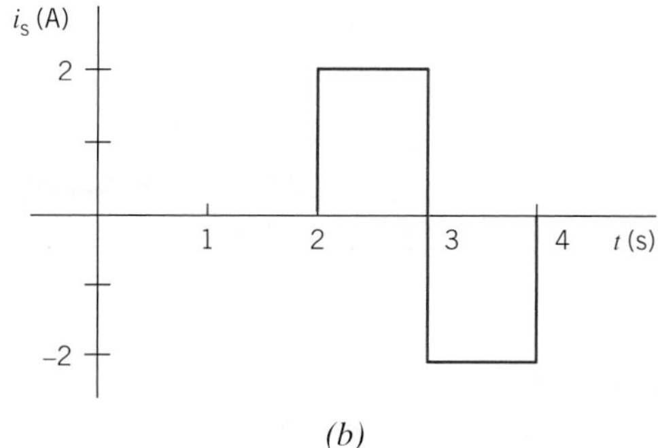

Figure P 7.2-9

P 7.2-10 Determine $v(t)$ for $t \geq 0$ for the circuit of Figure P 7.2-10a when $v(0) = -4$ V and $i_s(t)$ is the current shown in Figure P 7.2-10b.

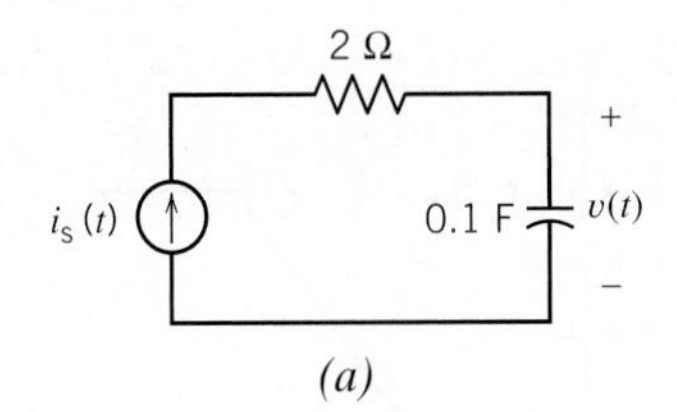

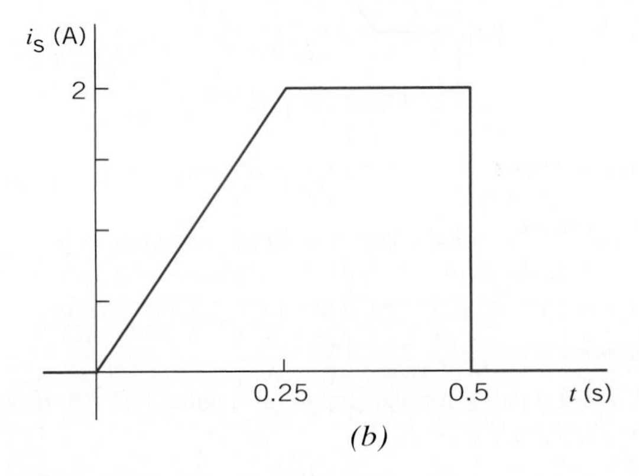

Figure P 7.2-10

P 7.2-11 Determine $i(t)$ for $t \geq 0$ for the circuit of Figure P 7.2-11*a* when $v_s(t)$ is the voltage shown in Figure P 7.2-11*b*.

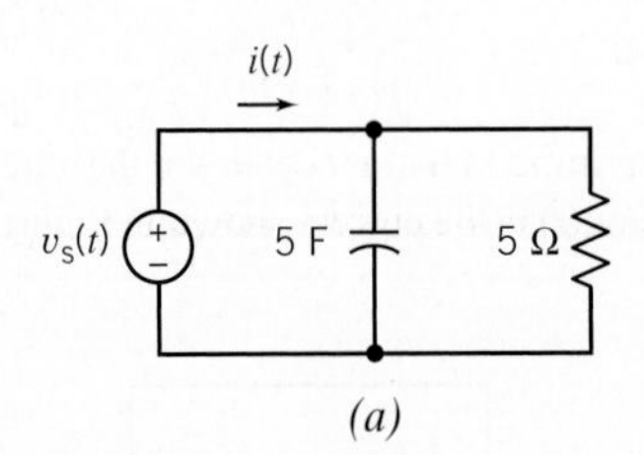

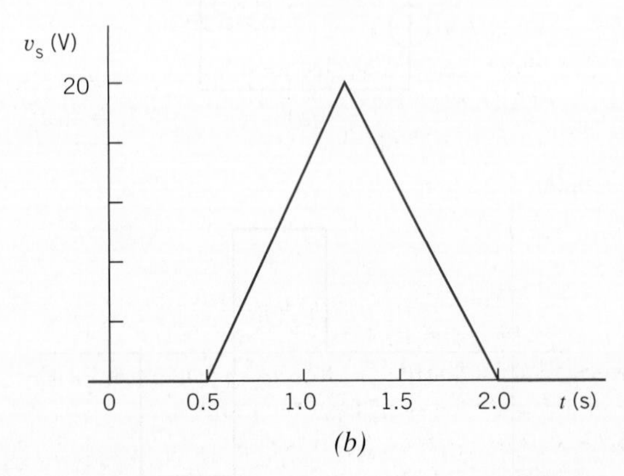

Figure P 7.2-11

P 7.2-12 The capacitor voltage in the circuit shown in Figure P 7.2-12 is given by

$$v(t) = 12 - 10e^{-2t}\text{ V} \quad \text{for } t \geq 0$$

Determine $i(t)$ for $t > 0$.

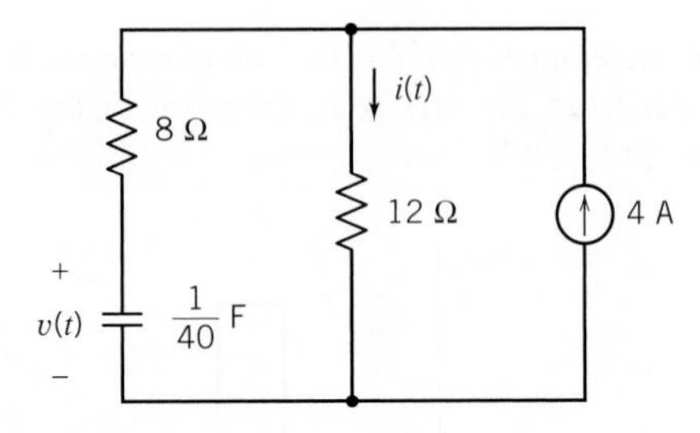

Figure P 7.2-12

P 7.2-13 The capacitor voltage in the circuit shown in Figure P 7.2-13 is given by

$$v(t) = 2.4 + 5.6e^{-5t}\text{ V} \quad \text{for } t \geq 0$$

Determine $i(t)$ for $t > 0$.

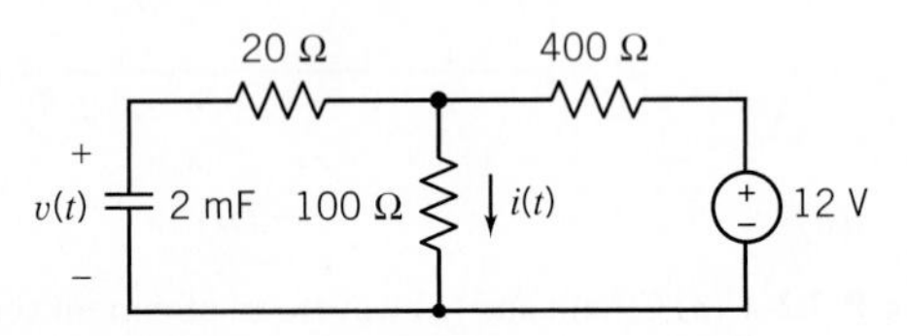

Figure P 7.2-13

P 7.2-14 The capacitor voltage in the circuit shown in Figure P 7.2-14 is given by

$$v(t) = 10 - 8e^{-5t}\text{ V} \quad \text{for } t \geq 0$$

Determine $i(t)$ for $t > 0$.

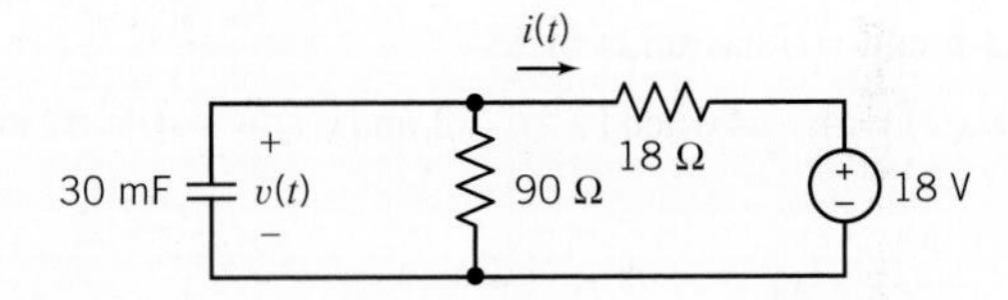

Figure P 7.2-14

P 7.2-15 Determine the voltage $v(t)$ for $t > 0$ for the circuit of Figure P 7.2-15*b* when $i_s(t)$ is the current shown in Figure P 7.2-15*a*. The capacitor voltage at time $t = 0$ is $v(0) = -12$ V.

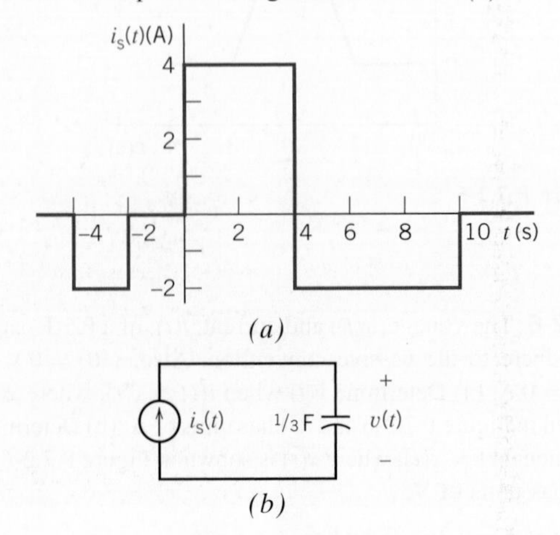

Figure P 7.2-15 (*a*) The voltage source voltage. (*b*) The circuit.

P 7.2-16 The input to the circuit shown in Figure P 7.2-16 is the current

$$i(t) = 3.75e^{-1.2t}\text{ A} \quad \text{for } t > 0$$

The output is the capacitor voltage

$$v(t) = 4 - 1.25e^{-1.2t}\text{ V} \quad \text{for } t > 0$$

Find the value of the capacitance, C.

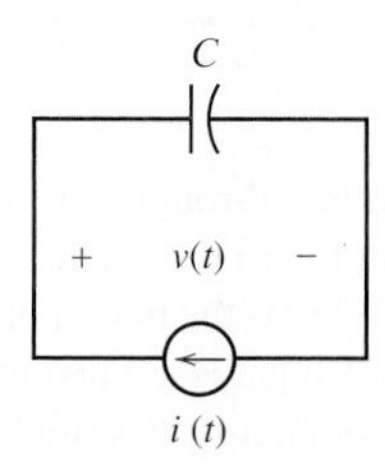

Figure P 7.2-16

P 7.2-17 The input to the circuit shown in Figure P 7.2-17 is the current

$$i(t) = 3e^{-25t}\text{ A} \quad \text{for } t > 0$$

The initial capacitor voltage is $v_C(0) = -2$ V. Determine the current source voltage, $v(t)$, for $t > 0$.

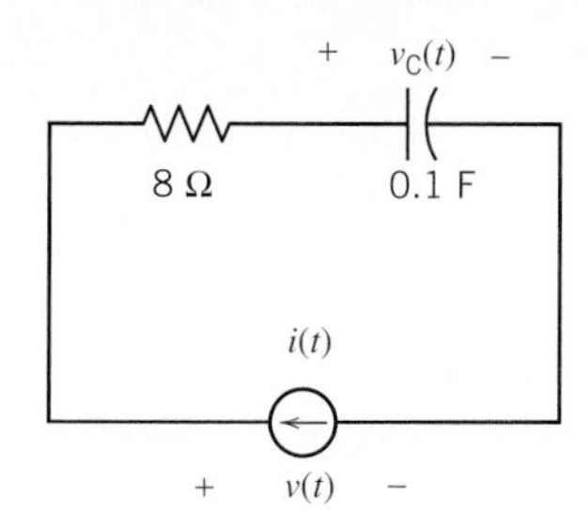

Figure P 7.2-17

P 7.2-18 The input to the circuit shown in Figure P 7.2-18 is the current

$$i(t) = 5e^{-25t}\text{ A} \quad \text{for } t > 0$$

The output is the voltage

$$v(t) = 9.8e^{-5t} + 0.6\text{ V} \quad \text{for } t > 0$$

The initial capacitor voltage is $v_C(0) = -2$ V. Determine the values of the capacitance, C, and resistance, R.

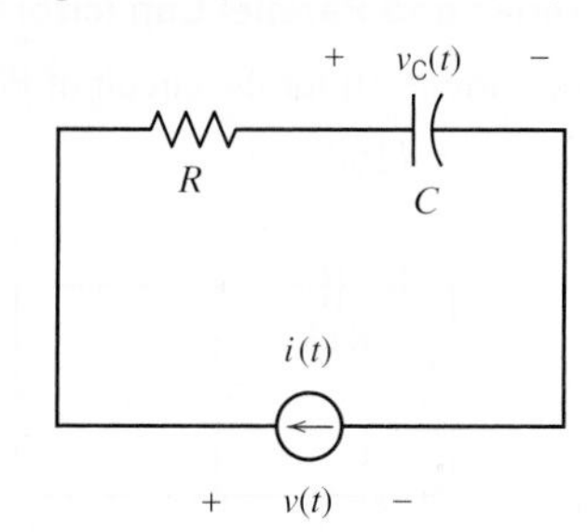

Figure P 7.2-18

P 7.2-19 The input to the circuit shown in Figure P 7.2-19 is the voltage

$$v(t) = 8 + 5e^{-10t}\text{ V} \quad \text{for } t > 0$$

Determine the current, $i(t)$ for $t > 0$.

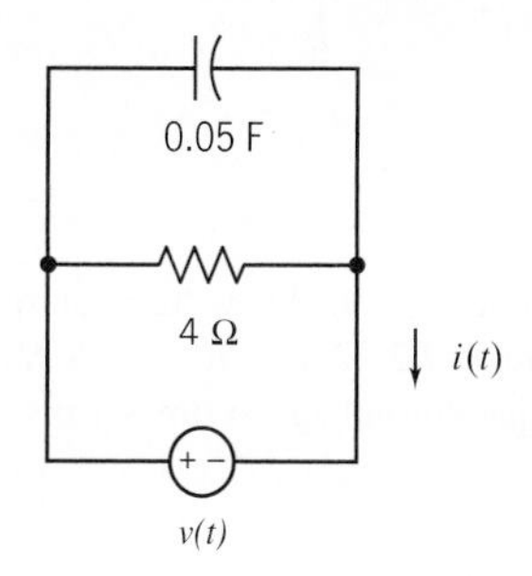

Figure P 7.2-19

P 7.2-20 The input to the circuit shown in Figure P 7.2-20 is the voltage:

$$v(t) = 3 + 4e^{-2t}\text{ A} \quad \text{for } t > 0$$

The output is the current, $i(t) = 0.3 - 1.6e^{-2t}$ V for $t > 0$

Determine the values of the resistance and capacitance.

Answers: $R = 10\ \Omega$ and $C = 0.25$ F

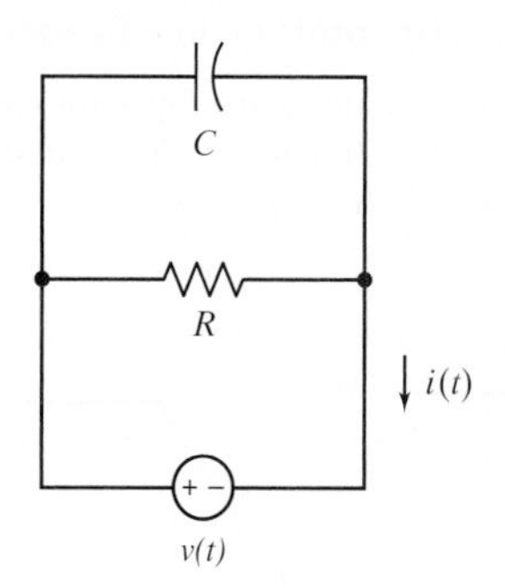

Figure P 7.2-20

P 7.2-21 Consider the capacitor shown in Figure P 7.2-21. The current and voltage are given by

$$i(t) = \begin{cases} 0.5 & 0 < t < 0.5 \\ 2 & 0.5 < t < 1.5 \\ 0 & t > 1.5 \end{cases}$$

and

$$v(t) = \begin{cases} 2t + 8.6 & 0 \le t \le 0.5 \\ at + b & 0.5 \le t \le 1.5 \\ c & t \ge 1.5 \end{cases}$$

where a, b, and c are real constants. (The current is given in Amps, the voltage in Volts, and the time in seconds.) Determine the values of a, b, and c.

Answers: $a = 8$ V/s, $b = 5.6$ V, and $c = 17.6$ V

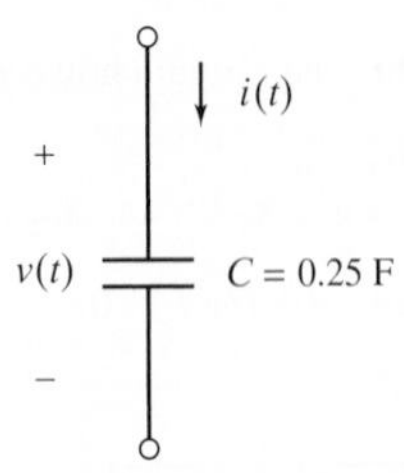

Figure P 7.2-21

P 7.2-22 At time $t = 0$, the voltage across the capacitor shown in Figure P 7.2-22 is $v(0) = -20$ V. Determine the values of the capacitor voltage at times 1 ms, 3 ms, and 7 ms.

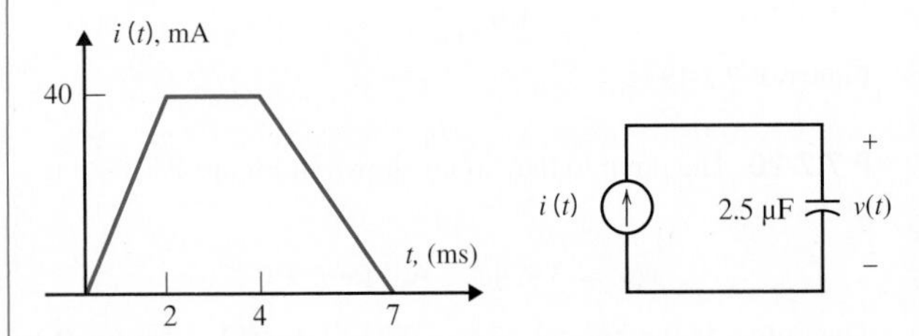

Figure P 7.2-22

Section 7.3 Energy Storage in a Capacitor

P 7.3-1 The current, i, through a capacitor is shown in Figure P 7.3-1. When $v(0) = 0$ and $C = 0.5$ F, determine and plot $v(t)$, $p(t)$, and $w(t)$ for $0\text{ s} < t < 6$ s.

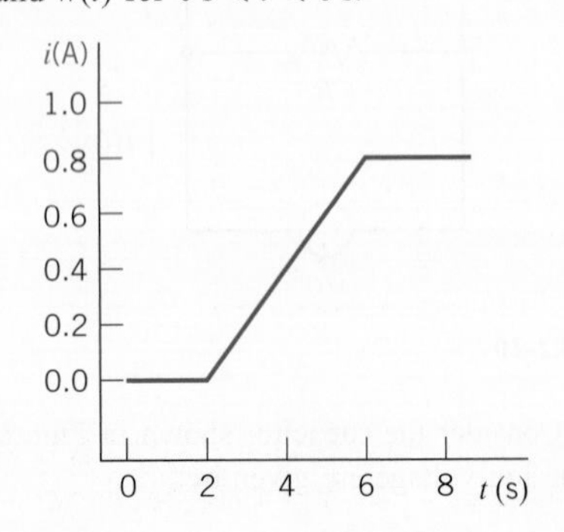

Figure P 7.3-1

P 7.3-2 In a pulse power circuit, the voltage of a 10-μF capacitor is zero for $t < 0$ and

$$v = 5\left(1 - e^{-4000t}\right)\text{ V} \quad t \geq 0$$

Determine the capacitor current and the energy stored in the capacitor at $t = 0$ ms and $t = 10$ ms.

P 7.3-3 If $v_c(t)$ is given by the waveform shown in Figure P 7.3-3, sketch the capacitor current for $-1\text{ s} < t < 2$ s. Sketch the power and the energy for the capacitor over the same time interval when $C = 1$ mF.

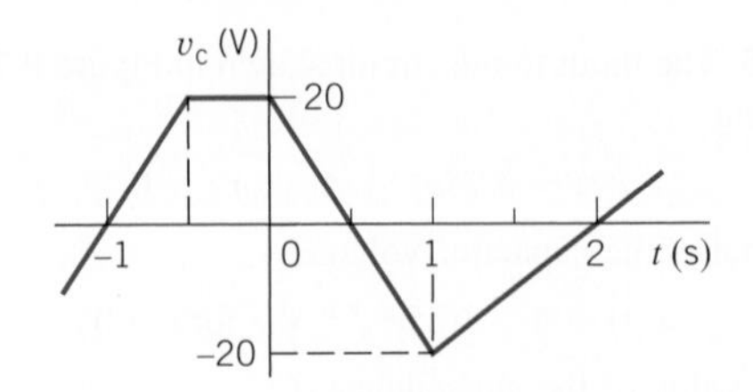

Figure P 7.3-3

P 7.3-4 The current through a 4-μF capacitor is $60\cos(10t + \pi/6)$ μA for all time. The average voltage across the capacitor is zero. What is the maximum value of the energy stored in the capacitor? What is the first nonnegative value of t at which the maximum energy is stored?

P 7.3-5 A capacitor is used in the electronic flash unit of a camera. A small battery with a constant voltage of 9 V is used to charge a capacitor with a constant current of 15 μA. How long does it take to charge the capacitor when $C = 15\ \mu$F? What is the stored energy?

P 7.3-6 The initial capacitor voltage of the circuit shown in Figure P 7.3-6 is $v_c(0^-) = 3$ V. Determine (a) the voltage $v(t)$ and (b) the energy stored in the capacitor at $t = 0.2$ s and $t = 0.8$ s when

$$i(t) = \begin{cases} 3e^{5t}\text{ A} & 0 < t < 1 \\ 0 & t \geq 1\text{ s} \end{cases}$$

Answers:

(a) $18e^{5t}$ V, $0 \leq t < 1$
(b) $w(0.2) = 6.65$ J and $w(0.8) = 2.68$ kJ

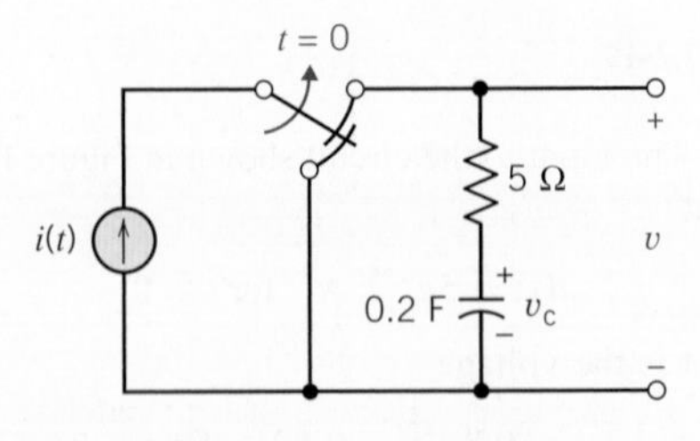

Figure P 7.3-6

Section 7.4 Series and Parallel Capacitors

P 7.4-1 Find the current $i(t)$ for the circuit of Figure P 7.4-1.

Answer: $i(t) = -3.24\sin 100t$ mA

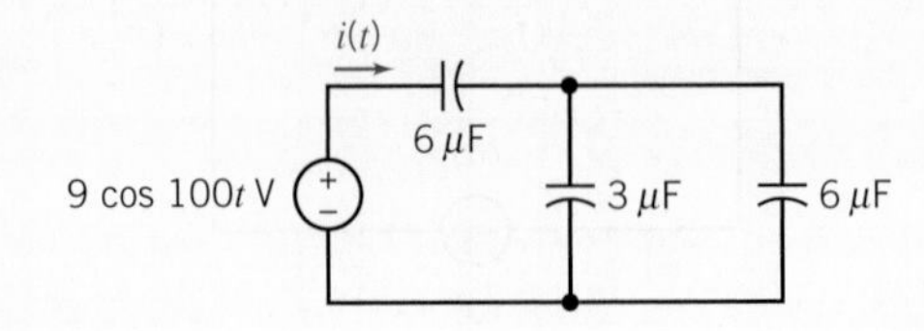

Figure P 7.4-1

P 7.4-2 Find the current $i(t)$ for the circuit of Figure P 7.4-2.

Answer: $i(t) = -1.5e^{-250t}$ mA

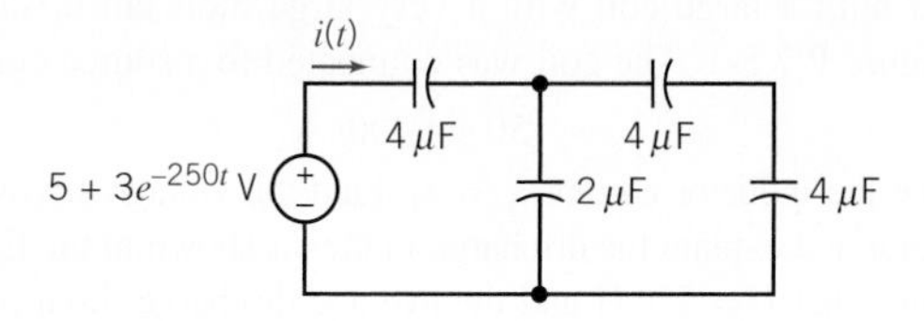

Figure P 7.4-2

P 7.4-3 The circuit of Figure P 7.4-3 contains five identical capacitors. Find the value of the capacitance C.

Answer: $C = 10\ \mu$F

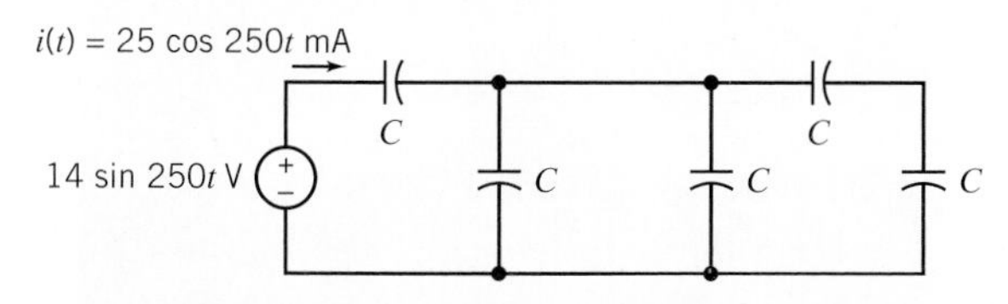

Figure P 7.4-3

P 7.4-4 The circuit shown in Figure P 7.4-4 contains seven capacitors, each having capacitance C. The source voltage is given by

$$v(t) = 4\cos(3t)\ \text{V}$$

Find the current $i(t)$ when $C = 1$ F.

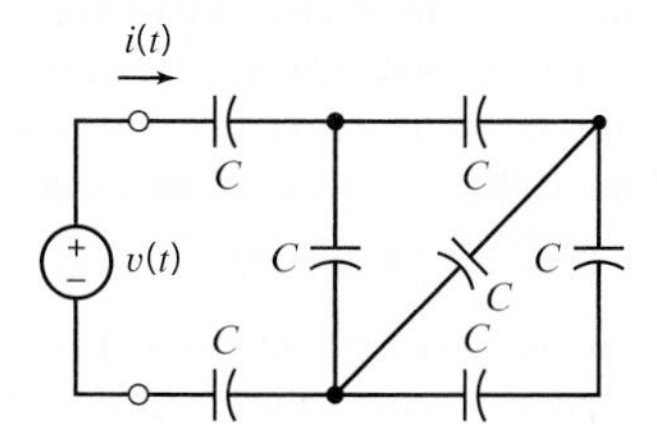

Figure P 7.4-4

P 7.4-5 Determine the value of the capacitance C in the circuit shown in Figure P 7.4-5, given that $C_{eq} = 8$ F.

Answer: $C = 20$ F

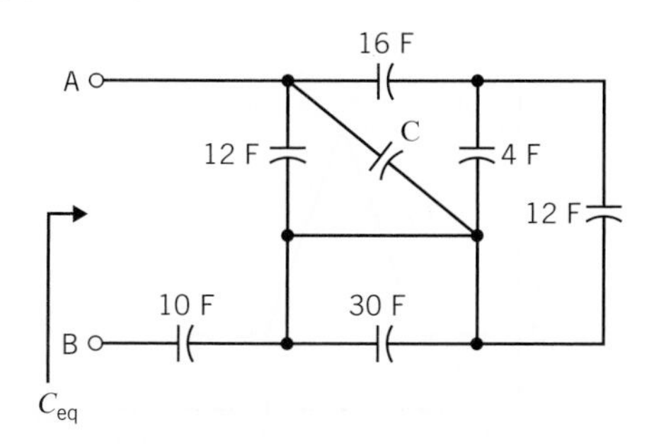

Figure P 7.4-5

P 7.4-6 Determine the value of the equivalent capacitance, C_{eq}, in the circuit shown in Figure P 7.4-6.

Answer: $C_{eq} = 14.3$ F

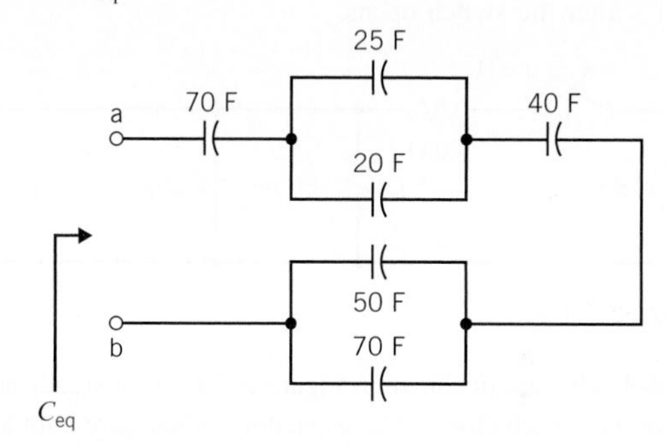

Figure P 7.4-6

P 7.4-7 The circuit shown in Figure P 7.4-7 consists of nine capacitors having equal capacitance, C. Determine the value of the capacitance C, given that $C_{eq} = 50$ mF.

Answer: $C = 90$ mF

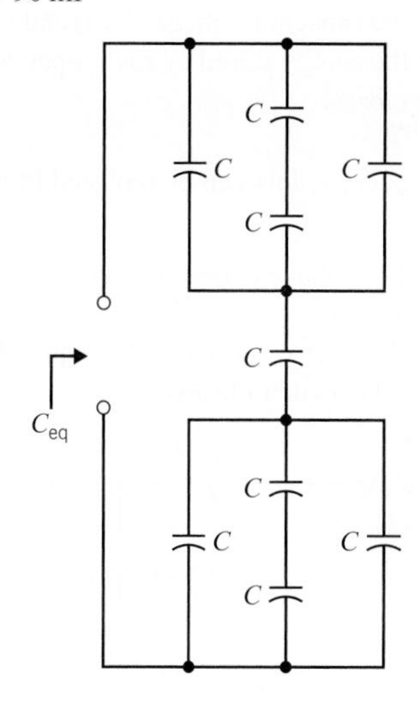

Figure P 7.4-7

P 7.4-8 The circuit shown in Figure P 7.4-8 is at steady state before the switch opens at time $t = 0$. The voltage $v(t)$ is given by

$$v(t) = \begin{cases} 3.6\ \text{V} & \text{for } t \leq 0 \\ 3.6e^{-2.5t}\ \text{V} & \text{for } t \geq 0 \end{cases}$$

(a) Determine the energy stored by each capacitor before the switch opens.

(b) Determine the energy stored by each capacitor 1 s after the switch opens.

The parallel capacitors can be replaced by an equivalent capacitor.

(c) Determine the energy stored by the equivalent capacitor before the switch opens.

(d) Determine the energy stored by the equivalent capacitor 1 s after the switch opens.

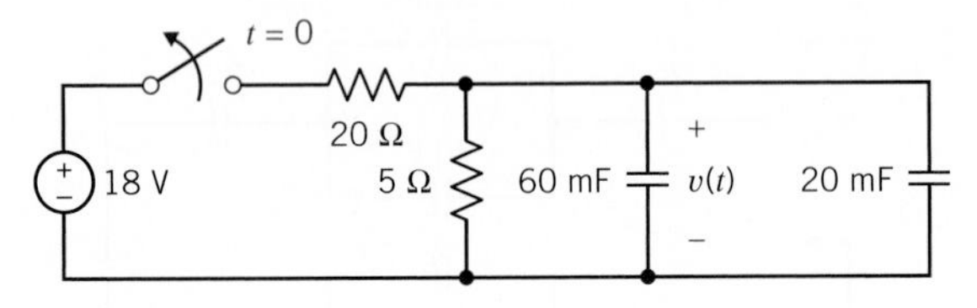

Figure P 7.4-8

P 7.4-9 The circuit shown in Figure P 7.4-9 is at steady state before the switch closes. The capacitor voltages are both zero before the switch closes ($v_1(0) = v_2(0) = 0$). The current $i(t)$ is given by

$$i(t) = \begin{cases} 0\ \text{A} & \text{for } t < 0 \\ 2.4e^{-30t}\ \text{A} & \text{for } t > 0 \end{cases}$$

(a) Determine the capacitor voltages, $v_1(t)$ and $v_2(t)$, for $t \geq 0$.

(b) Determine the energy stored by each capacitor 20 ms after the switch closes.

The series capacitors can be replaced by an equivalent capacitor.

(c) Determine the voltage across the equivalent capacitor, + on top, for $t \geq 0$.

(d) Determine the energy stored by the equivalent capacitor 20 ms after the switch closes.

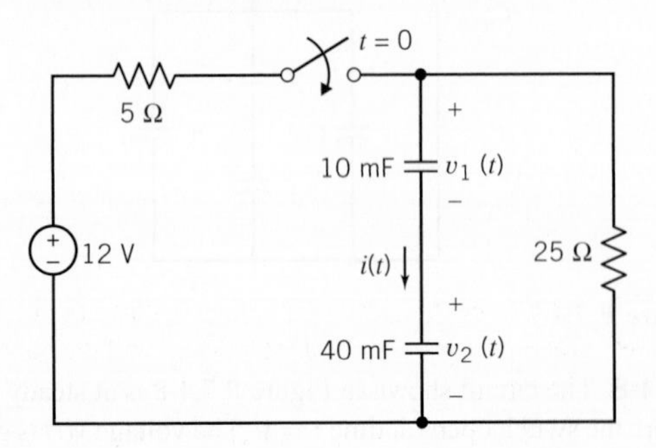

Figure P 7.4-9

P 7.4-10 Find the relationship for the division of current between two parallel capacitors as shown in Figure P 7.4-10.

Answer: $i_n = iC_n/(C_1 + C_2)$, $n = 1, 2$

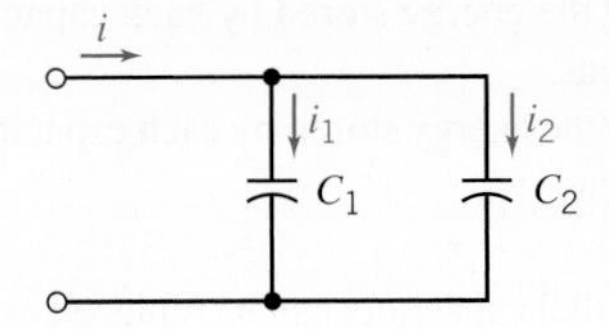

Figure P 7.4-10

Section 7.5 Inductors

P 7.5-1 Nikola Tesla (1857–1943) was an American electrical engineer who experimented with electric induction. Tesla built a large coil with a very large inductance, shown in Figure P 7.5-1. The coil was connected to a source current

$$i_s = 150 \sin 400t\ \text{A}$$

so that the inductor current $i_L = i_s$. Find the voltage across the inductor and explain the discharge in the air shown in the figure. Assume that $L = 250$ H and the average discharge distance is 3 m. Note that the dielectric strength of air is 4×10^6 V/m.

Figure P 7.5-1 Nikola Tesla sits impassively as alternating current induction coils discharge millions of volts with a roar audible 10 miles away (about 1910). Courtesy of Burndy Library.

P 7.5-2 The model of an electric motor consists of a series combination of a resistor and inductor. A current $i(t) = 4te^{-t}$ A flows through the series combination of a 20-Ω resistor and 0.2-H inductor. Find the voltage across the combination.

Answer: $v(t) = 0.8e^{-t} + 79.2te^{-t}$ V

P 7.5-3 The voltage, $v(t)$, and current, $i(t)$, of a 1-H inductor adhere to the passive convention. Also, $v(0) = 0$ V and $i(0) = 0$ A.

(a) Determine $v(t)$ when $i(t) = x(t)$, where $x(t)$ is shown in Figure P 7.5-3 and $i(t)$ has units of A.

(b) Determine $i(t)$ when $v(t) = x(t)$, where $x(t)$ is shown in Figure P 7.5-3, and $v(t)$ has units of V.

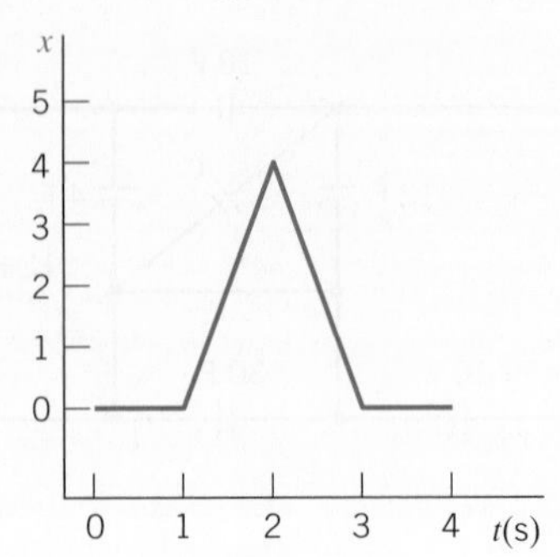

Figure P 7.5-3

Hint: $x(t) = 4t - 4$ when $1 < t < 2$, and $x(t) = -4t + 12$ when $2 < t < 3$.

P 7.5-4 The voltage, $v(t)$, across an inductor and current, $i(t)$, in that inductor adhere to the passive convention. Determine the voltage, $v(t)$, when the inductance is $L = 300$ mH, and the current is $i(t) = 150 \sin(500t - 30°)$ mA.

Hint:
$$\begin{aligned}\frac{d}{dt}A\sin(\omega t+\theta) &= A\cos(\omega t+\theta)\cdot\frac{d}{dt}(\omega t+\theta)\\ &= A\omega\cos(\omega t+\theta)\\ &= A\omega\sin\left(\omega t+\left(\theta+\frac{\pi}{2}\right)\right)\end{aligned}$$

Answer: $v(t) = 22.5\cos(500t - 30°)$ V

P 7.5-5 Determine $i_L(t)$ for $t > 0$ when $i_L(0) = -2\ \mu$A for the circuit of Figure P 7.5-5*a* when $v_s(t)$ is as shown in Figure P 7.5-5*b*.

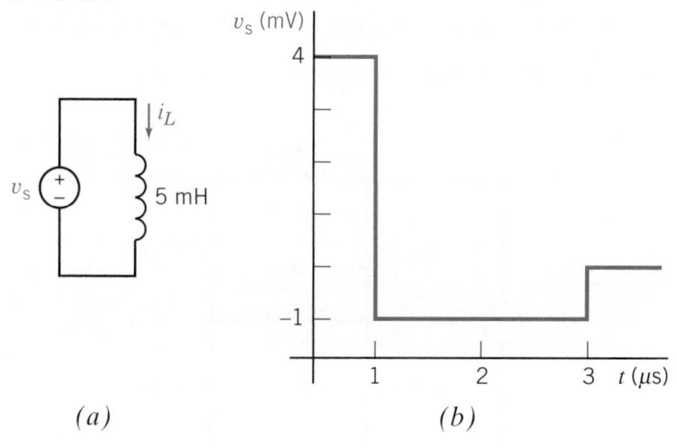

Figure P 7.5-5

P 7.5-6 Determine $v(t)$ for $t > 0$ for the circuit of Figure P 7.5-6*a* when $i_L(0) = 0$ and i_s is as shown in Figure P 7.5-6*b*.

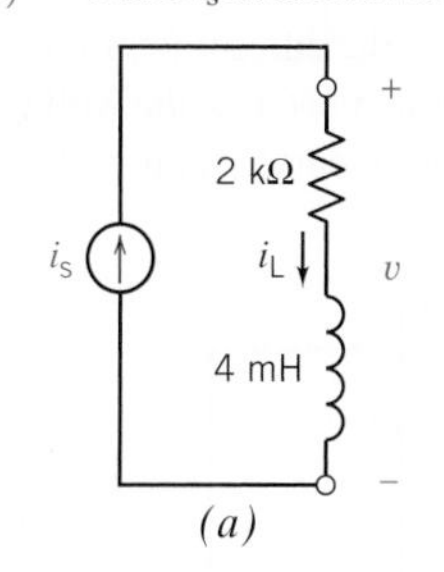

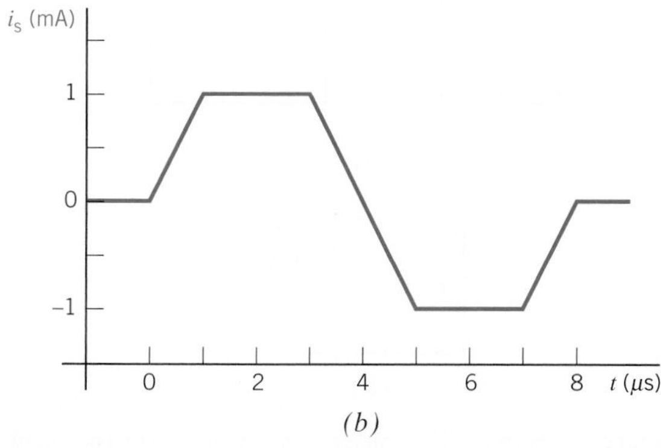

Figure P 7.5-6

P 7.5-7 The voltage, $v(t)$, and current, $i(t)$, of a 0.5-H inductor adhere to the passive convention. Also, $v(0) = 0$ V, and $i(0) = 0$ A.

(a) Determine $v(t)$ when $i(t) = x(t)$, where $x(t)$ is shown in Figure P 7.5-7 and $i(t)$ has units of A.

(b) Determine $i(t)$ when $v(t) = x(t)$, where $x(t)$ is shown in Figure P 7.5-7 and $v(t)$ has units of V.

Hint: $x(t) = 0.2t - 0.4$ when $2 < t < 6$.

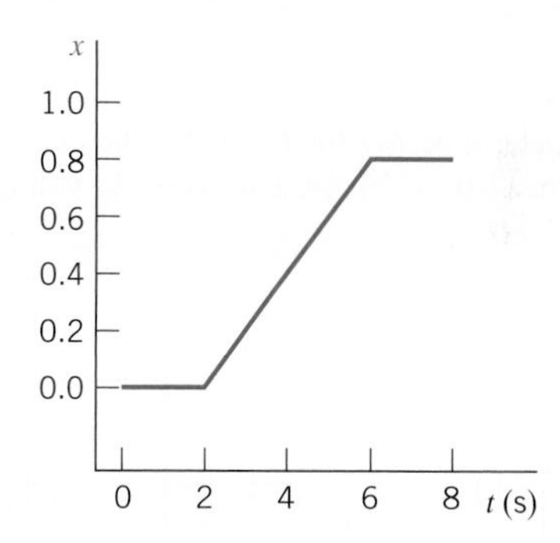

Figure P 7.5-7

P 7.5-8 Determine $i(t)$ for $t \geq 0$ for the current of Figure P 7.5-8*a* when $i(0) = 30$ mA and $v_s(t)$ is the voltage shown in Figure P 7.5-8*b*.

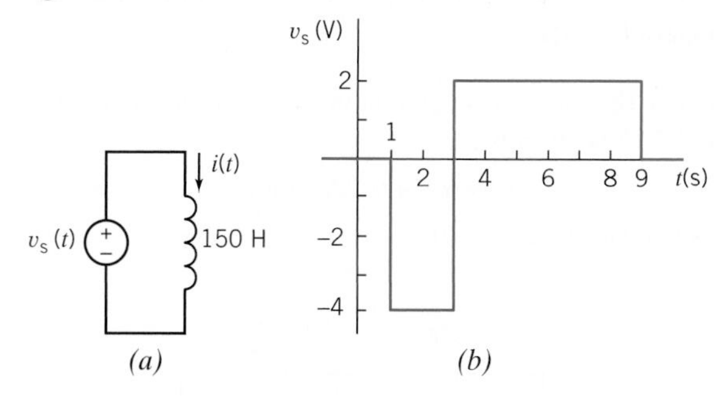

Figure P 7.5-8

P 7.5-9 Determine $i(t)$ for $t \geq 0$ for the current of Figure P 7.5-9*a* when $i(0) = -2$ A and $v_s(t)$ is the voltage shown in Figure P 7.5-9*b*.

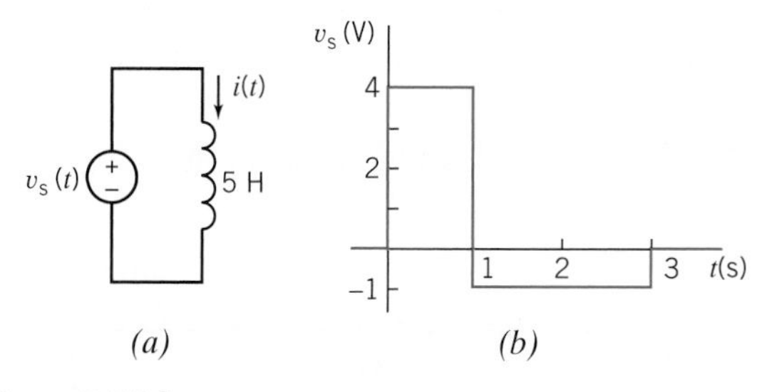

Figure P 7.5-9

P 7.5-10 Determine $i(t)$ for $t \geq 0$ for the current of Figure P 7.5-10*a* when $i(0) = 1$ A and $v_s(t)$ is the voltage shown in Figure P 7.5-10*b*.

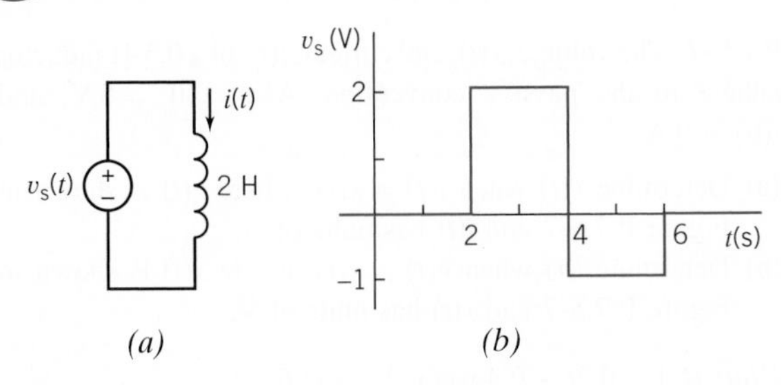

Figure P 7.5-10

P 7.5-11 Determine $i(t)$ for $t \geq 0$ for the circuit of Figure P 7.5-11*a* when $i(0) = 25$ mA and $v_s(t)$ is the voltage shown in Figure P 7.5-11*b*.

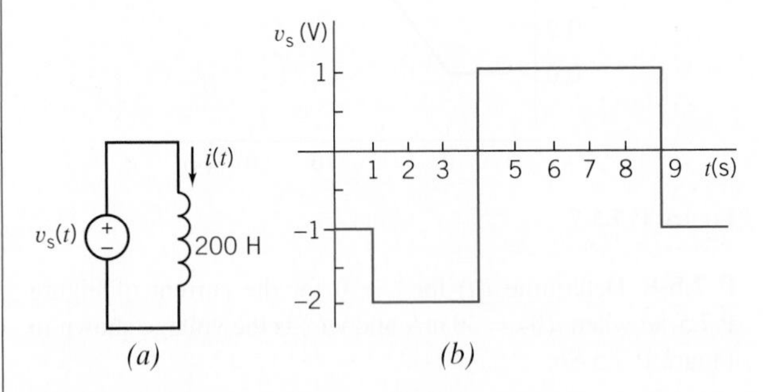

Figure P 7.5-11

P 7.5-12 The inductor current in the circuit shown in Figure P 7.5-12 is given by

$$i(t) = 6 + 4e^{-8t} \text{ A} \quad \text{for } t \geq 0$$

Determine $v(t)$ for $t > 0$.

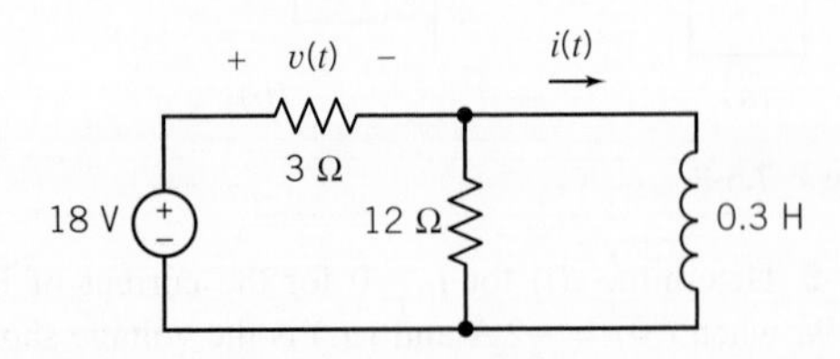

Figure P 7.5-12

P 7.5-13 The inductor current in the circuit shown in Figure P 7.5-13 is given by

$$i(t) = 5 - 3e^{-4t} \text{ A} \quad \text{for } t \geq 0$$

Determine $v(t)$ for $t > 0$.

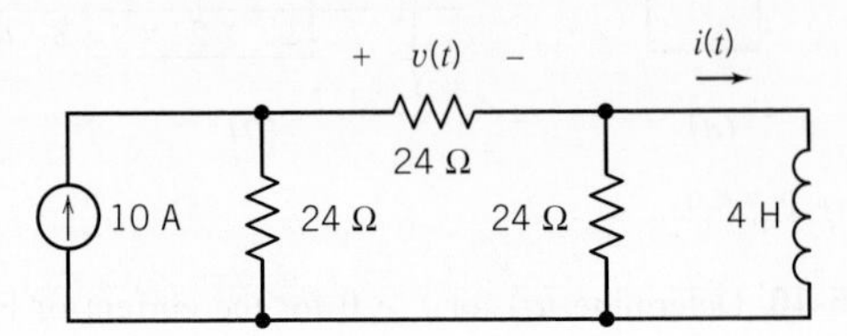

Figure P 7.5-13

P 7.5-14 The inductor current in the circuit shown in Figure P 7.5-14 is given by

$$i(t) = 3 + 2e^{-3t} \text{ A} \quad \text{for } t \geq 0$$

Determine $v(t)$ for $t > 0$.

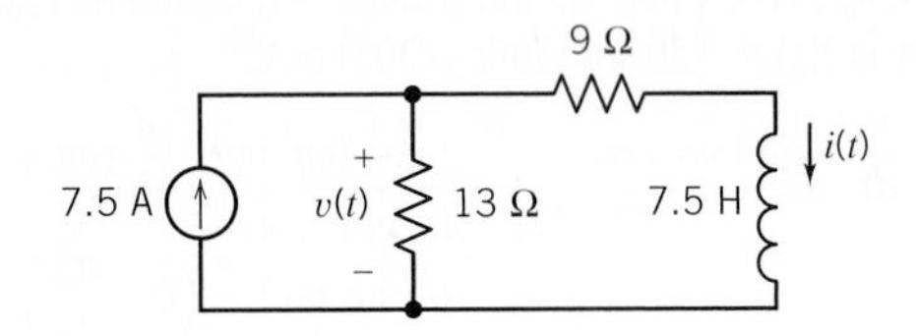

Figure P 7.5-14

***P 7.5-15** The inductor current in the circuit shown in Figure P 7.5-15 is given by

$$i(t) = 240 + 193e^{-6.25t} \cos(9.27t - 102°) \text{ mA} \quad \text{for } t \geq 0$$

Determine the capacitor voltage, $v(t)$, for $t > 0$.

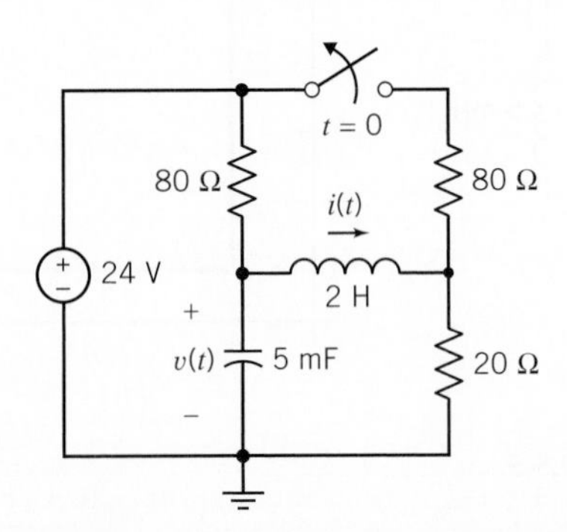

Figure P 7.5-15

P 7.5-16 Determine the current $i(t)$ for $t > 0$ for the circuit of Figure P 7.5-16*b* when $v_s(t)$ is the voltage shown in Figure P 7.5-16*a*. The inductor current at time $t = 0$ is $i(0) = -12$ A.

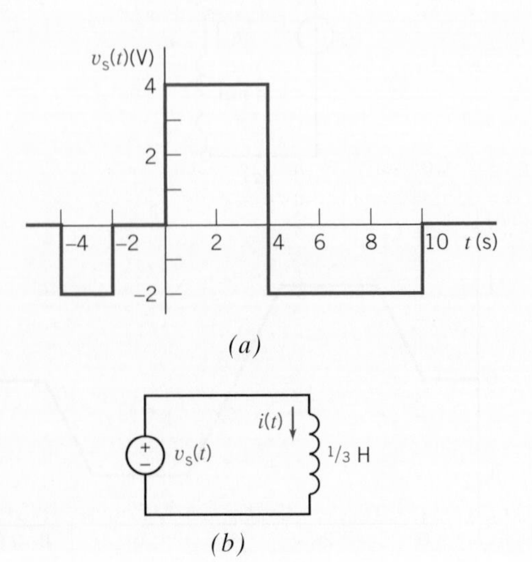

Figure P 7.5-16 (*a*) The voltage source voltage. (*b*) The circuit.

P 7.5-17 The input to the circuit shown in Figure P 7.5-17 is the voltage

$$v(t) = 15e^{-4t}\ \text{V} \quad \text{for } t > 0$$

The initial current in the inductor is $i(0) = 4$ A. Determine the inductor current, $i(t)$, for $t > 0$.

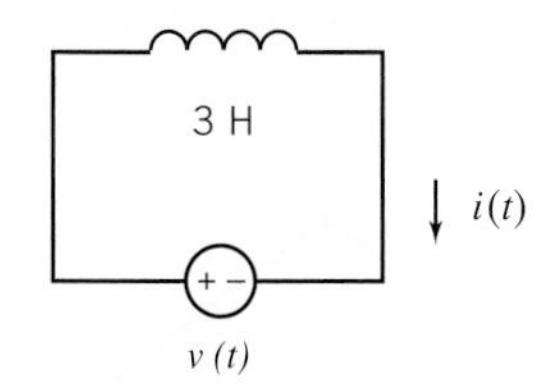

Figure P 7.5-17

P 7.5-18 The input to the circuit shown in Figure P 7.5-18 is the voltage

$$v(t) = 4e^{-20t}\ \text{V} \quad \text{for } t > 0$$

The output is the current

$$i(t) = -1.2e^{-20t} - 1.5\ \text{A} \quad \text{for } t > 0$$

The initial inductor current is $i_L(0) = -3.5$ A. Determine the values of the inductance, L, and resistance, R.

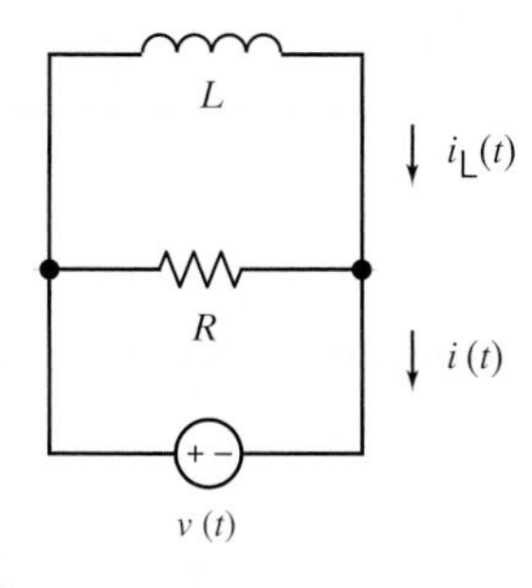

Figure P 7.5-18

P 7.5-19 Consider the inductor shown in Figure P 7.5-19. The current and voltage are given by

$$i(t) = \begin{cases} 5t - 4.6 & 0 \le t \le 0.2 \\ at + b & 0.2 \le t \le 0.5 \\ c & t \ge 0.5 \end{cases}$$

$$\text{and} \quad v(t) = \begin{cases} 12.5 & 0 < t < 0.2 \\ 25 & 0.2 < t < 0.5 \\ 0 & t > 0.5 \end{cases}$$

where a, b, and c are real constants. (The current is given in Amps, the voltage in Volts, and the time in seconds.) Determine the values of a, b, and c.

Answers: $a = 10$ A/s, $b = -5.6$ A, and $c = -0.6$ A

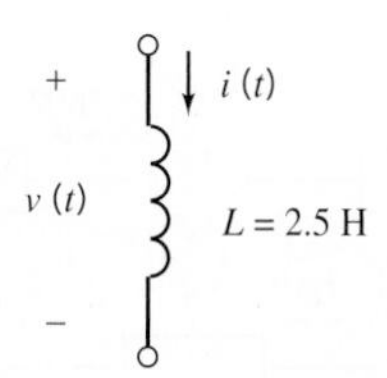

Figure P 7.5-19

P 7.5-20 At time $t = 0$, the current in the inductor shown in Figure P 7.5-20 is $i(0) = 45$ mA. Determine the values of the inductor current at times 1 ms, 4 ms, and 6 ms.

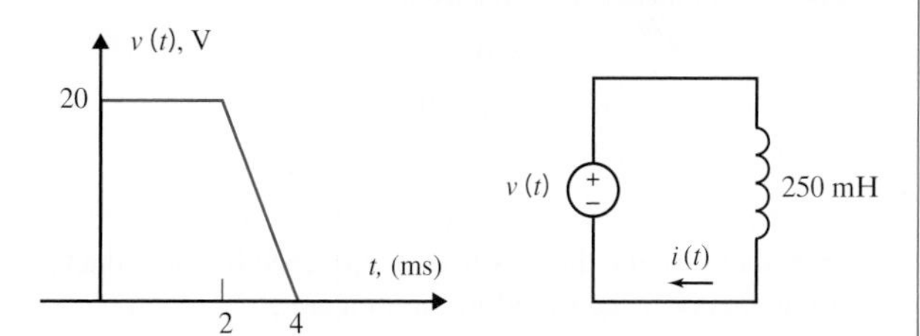

Figure P 7.5-20

P 7.5-21 One of the three elements shown in Figure P 7.5-21 is a resistor, one is a capacitor, and one is an inductor. Given

$$i(t) = 0.3\cos(2t)\ \text{A},$$

and $v_a(t) = -10\sin(2t)$ V, $v_b(t) = 10\sin(2t)$ V, and $v_c(t) = 10\cos(2t)$ V, determine the resistance of the resistor, the capacitance of the capacitor, and the inductance of the inductor. (We require positive values of resistance, capacitance, and inductance.)

Answers: resistance = 33.3 Ω, capacitance = 0.015 F, and inductance = 16.7 H

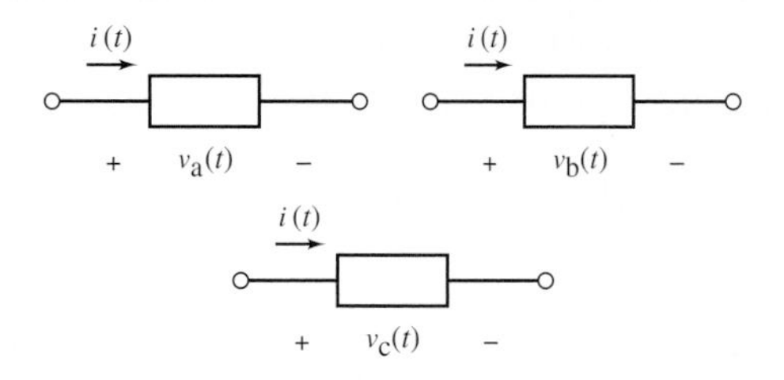

Figure P 7.5-21

P 7.5-22 One of the three elements shown in Figure P 7.5-22 is a resistor, one is a capacitor, and one is an inductor. Given

$$v(t) = 24\cos(5t)\ \text{V},$$

and $i_a(t) = 3\cos(5t)$ A, $i_b(t) = 12\sin(5t)$ A and $i_c(t) = -1.8\sin(5t)$ A, determine the resistance of the resistor, the capacitance of the capacitor, and the inductance of the inductor. (We require positive values of resistance, capacitance, and inductance.)

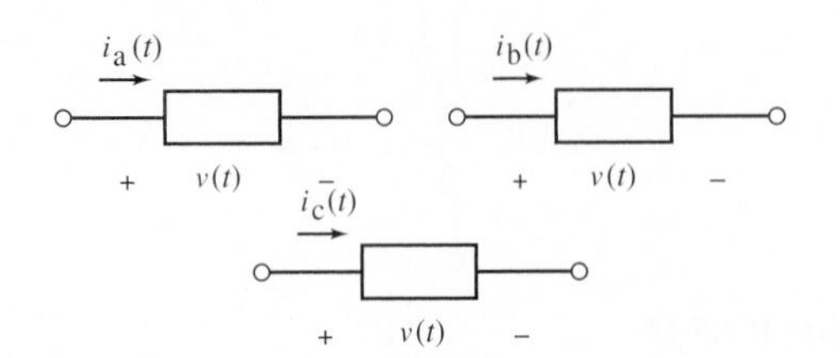

Figure P 7.5-22

Section 7.6 Energy Storage in an Inductor

P 7.6-1 The current, $i(t)$, in a 100-mH inductor connected in a telephone circuit changes according to

$$i(t) = \begin{cases} 0 & t \le 0 \\ 4t & 0 \le t \le 1 \\ 4 & t \ge 1 \end{cases}$$

where the units of time are seconds and the units of current are amperes. Determine the power, $p(t)$, absorbed by the inductor and the energy, $w(t)$, stored in the inductor.

Answers: $p(t) = \begin{cases} 0 & t \le 0 \\ 1.6t & 0 < t < 1 \\ 0 & t \ge 1 \end{cases}$ and

$$w(t) = \begin{cases} 0 & t \le 0 \\ 0.8t^2 & 0 < t < 1 \\ 0.8 & t \ge 1 \end{cases}$$

The units of $p(t)$ are W and the units of $w(t)$ are J.

P 7.6-2 The current, $i(t)$, in a 5-H inductor is

$$i(t) = \begin{cases} 0 & t \le 0 \\ 4 \sin 2t & t \ge 0 \end{cases}$$

where the units of time are s and the units of current are A. Determine the power, $p(t)$, absorbed by the inductor and the energy, $w(t)$, stored in the inductor.

Hint: $2\,(\cos A)(\sin B) = \sin(A + B) + \sin(A - B)$

P 7.6-3 The voltage, $v(t)$, across a 25-mH inductor used in a fusion power experiment is

$$v(t) = \begin{cases} 0 & t \le 0 \\ 6 \cos 100t & t \ge 0 \end{cases}$$

where the units of time are s and the units of voltage are V. The current in this inductor is zero before the voltage changes at $t = 0$. Determine the power, $p(t)$, absorbed by the inductor and the energy, $w(t)$, stored in the inductor.

Hint: $2(\cos A)(\sin B) = \sin(A + B) + \sin(A - B)$

Answer: $p(t) = 7.2 \sin 200t$ W and $w(t) = 3.6[1 - \cos 200t]$ mJ

P 7.6-4 The current in an inductor, $L = 1/4$ H, is $i = 4te^{-t}$ A for $t \ge 0$ and $i = 0$ for $t < 0$. Find the voltage, power, and energy in this inductor.

Partial Answer: $w = 2t^2e^{-2t}$ J

P 7.6-5 The current through the inductor of a television tube deflection circuit is shown in Figure P 7.6-5 when $L = 1/2$ H. Find the voltage, power, and energy in the inductor.

Partial Answer:

$$\begin{aligned} p &= 2t \text{ for } 0 \le t < 1 \\ &= 2(t - 2) \text{ for } 1 < t < 2 \\ &= 0 \text{ for other } t \end{aligned}$$

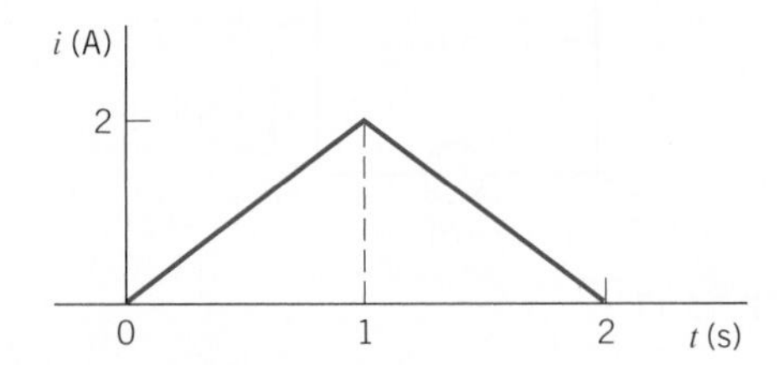

Figure P 7.6-5

Section 7.7 Series and Parallel Inductors

P 7.7-1 Find the current $i(t)$ for the circuit of Figure P 7.7-1.

Answer: $i(t) = 7.5 \sin 100t$ mA

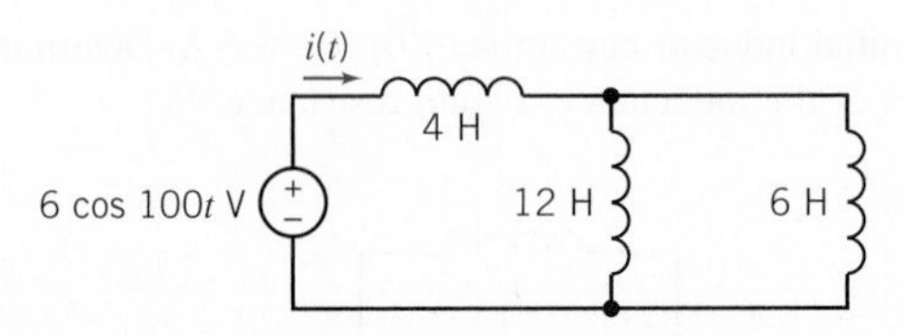

Figure P 7.7-1

P 7.7-2 Find the voltage $v(t)$ for the circuit of Figure P 7.7-2.

Answer: $v(t) = -9e^{-250t}$ mV

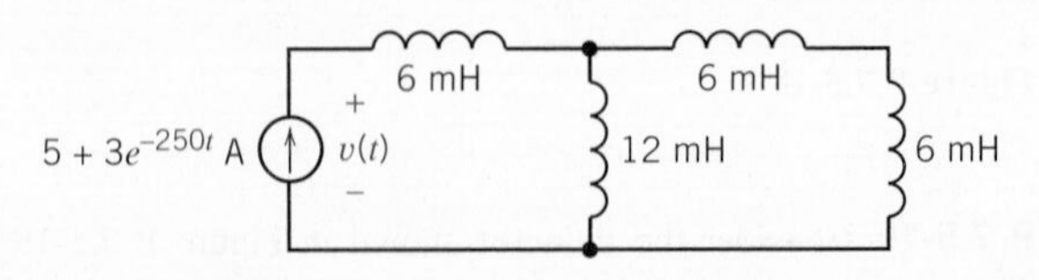

Figure P 7.7-2

P 7.7-3 The circuit of Figure P 7.7-3 contains four identical inductors. Find the value of the inductance L.

Answer: $L = 2.86$ H

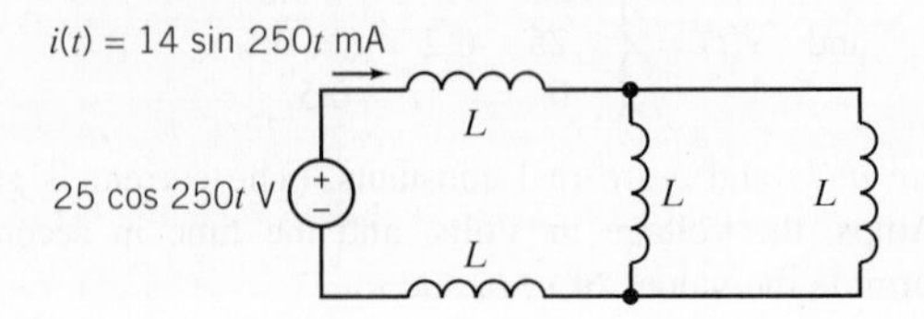

Figure P 7.7-3

P 7.7-4 The circuit shown in Figure P 7.7-4 contains seven inductors, each having inductance L. The source voltage is given by

$$v(t) = 4\cos(3t)\ \text{V}$$

Find the current $i(t)$ when $L = 4$ H.

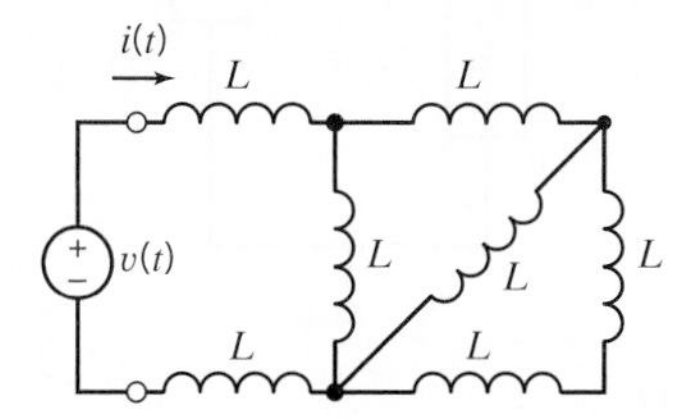

Figure P 7.7-4

P 7.7-5 Determine the value of the inductance L in the circuit shown in Figure P 7.7-5, given that $L_{eq} = 28$ H.

Answer: $L = 13.88$ H

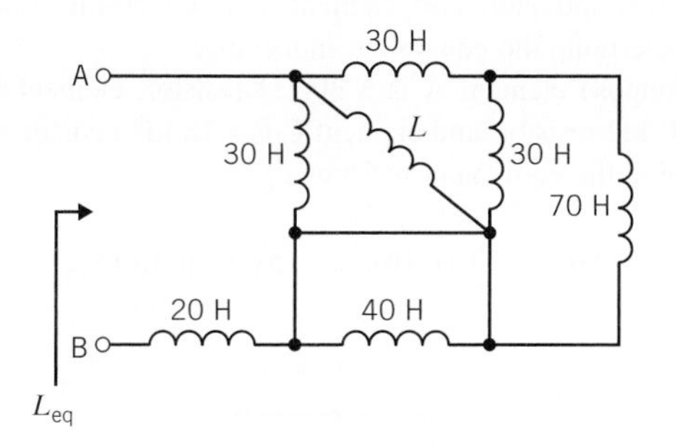

Figure P 7.7-5

P 7.7-6 Determine the value of the equivalent inductance, L_{eq}, for the circuit shown in Figure P 7.7-6.

Answer: $L_{eq} = 120$ H

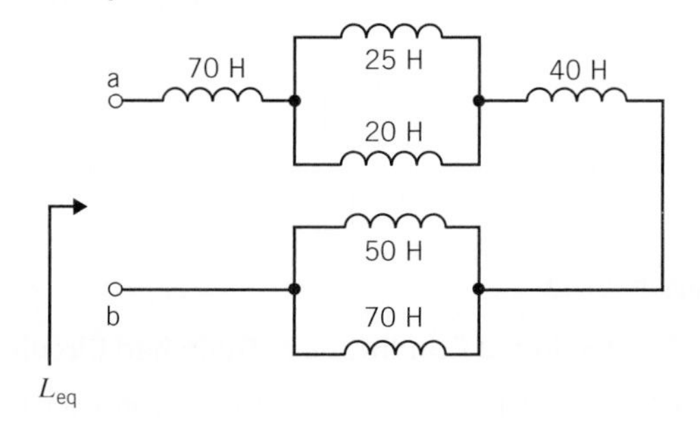

Figure P 7.7-6

P 7.7-7 The circuit shown in Figure P 7.7-7 consists of 10 inductors having equal inductance, L. Determine the value of the inductance L, given that $L_{eq} = 12$ mH.

Answer: $L = 35$ mH

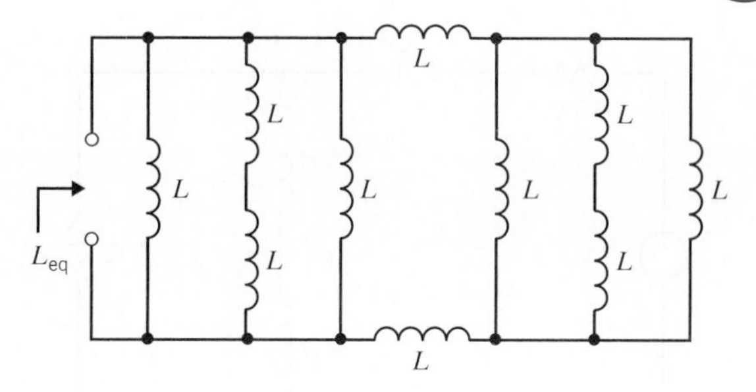

Figure P 7.7-7

P 7.7-8 The circuit shown in Figure P 7.7-8 is at steady state before the switch closes. The inductor currents are both zero before the switch closes ($i_1(0) = i_2(0) = 0$).

The voltage $v(t)$ is given by

$$v(t) = \begin{cases} 0 \ \text{V} & \text{for } t < 0 \\ 4e^{-5t} \ \text{V} & \text{for } t > 0 \end{cases}$$

(a) Determine the inductor currents, $i_1(t)$ and $i_2(t)$, for $t \geq 0$.
(b) Determine the energy stored by each inductor 200 ms after the switch closes.

The parallel inductors can be replaced by an equivalent inductor.

(c) Determine the current in the equivalent inductor, directed downward, for $t \geq 0$.
(d) Determine the energy stored by the equivalent inductor 200 ms after the switch closes.

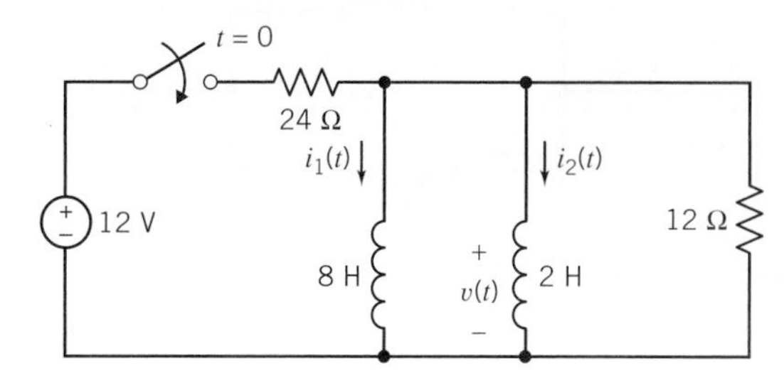

Figure P 7.7-8

P 7.7-9 The circuit shown in Figure P 7.7-9 is at steady state before the switch opens at time $t = 0$. The current $i(t)$ is given by

$$i(t) = \begin{cases} 0.8 \ \text{A} & \text{for } t \leq 0 \\ 0.8e^{-2t} \ \text{A} & \text{for } t \geq 0 \end{cases}$$

(a) Determine the energy stored by each inductor before the switch opens.
(b) Determine the energy stored by each inductor 200 ms after the switch opens.

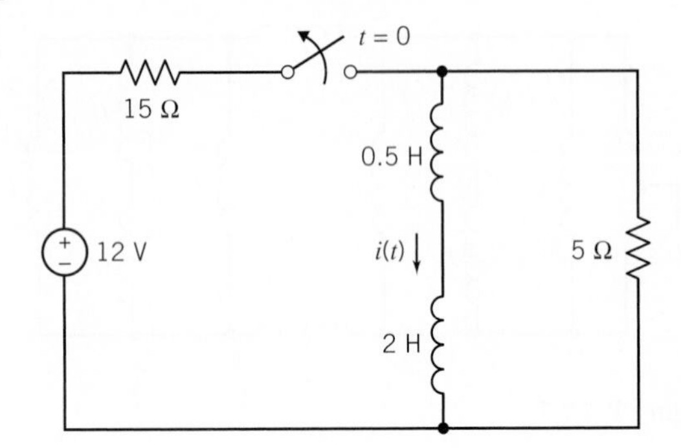

Figure P 7.7-9

The series inductors can be replaced by an equivalent inductor.

(c) Determine the energy stored by the equivalent inductor before the switch opens.

(d) Determine the energy stored by the equivalent inductor 200 ms after the switch opens.

P 7.7-10 Determine the current ratio i_1/i for the circuit shown in Figure P 7.7-10. Assume that the initial currents are zero at t_0.

Answer: $\frac{i_1}{i} = \frac{L_1}{L_1 + L_2}$

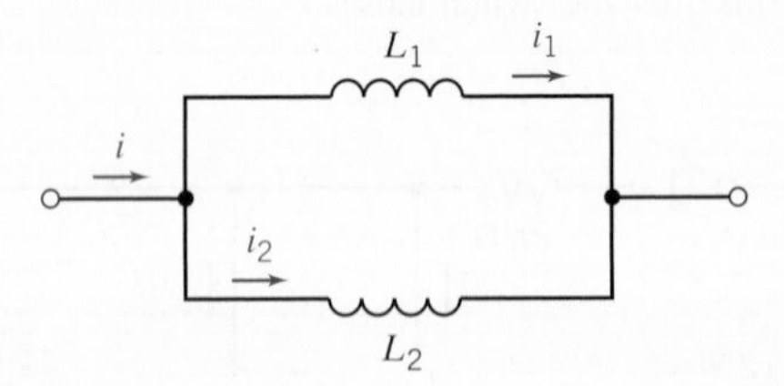

Figure P 7.7-10

P 7.7-11 Consider the combination of circuit elements shown in Figure P 7.7-11.

(a) Suppose element A is a 30-μF capacitor, element B is a 10-μF capacitor, and element C is a 30-μF capacitor. Determine the equivalent capacitance.

(b) Suppose element A is a 60-mH inductor, element B is a 40-mH inductor, and element C is a 30-mH inductor. Determine the equivalent inductance.

(c) Suppose element A is a 10-kΩ resistor, element B is a 8-kΩ resistor and element C is a 10-kΩ resistor. Determine the equivalent resistance.

Answers: **(a)** C_{eq}=37.5 μF, **(b)** L_{eq}=23.08 mH, and **(c)** R_{eq}=6.4 kΩ

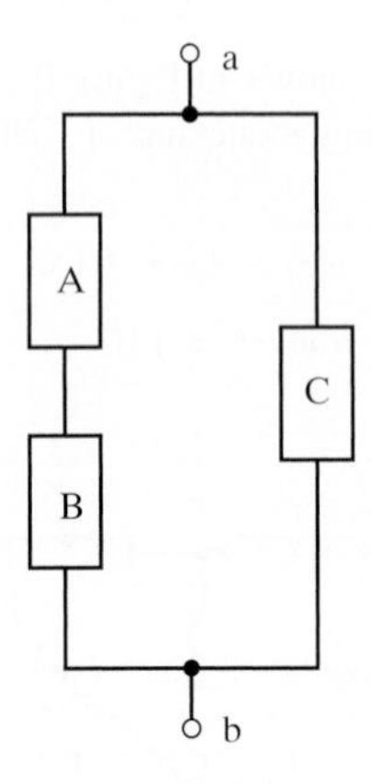

Figure P 7.7-11

P 7.7-12 Consider the combination of circuit elements shown in Figure P 7.7-12.

(a) Suppose element A is a 10-μF capacitor, element B is a 20-μF capacitor, and element C is a 15-μF capacitor. Determine the equivalent capacitance.

(b) Suppose element A is a 30-mH inductor, element B is a 6-mH inductor, and element C is an 10-mH inductor. Determine the equivalent inductance.

(c) Suppose element A is a 30-kΩ resistor, element B is a 40-kΩ resistor, and element C is a 16-kΩ resistor. Determine the equivalent resistance.

Answers: **(a)** C_{eq}=10 μF, **(b)** L_{eq}=15 mH, and **(c)** R_{eq}=33.14 kΩ

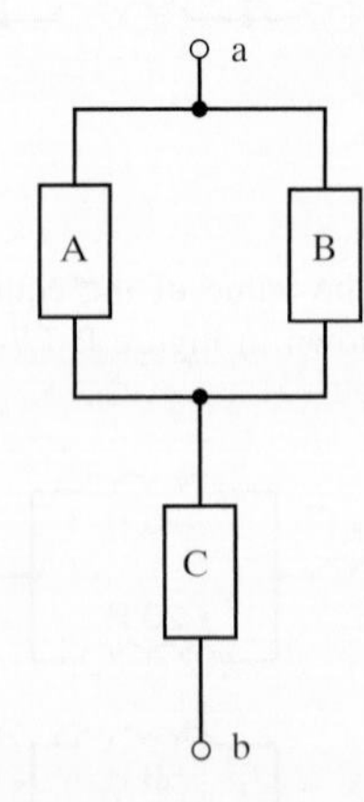

Figure P 7.7-12

Section 7.8 Initial Conditions of Switched Circuits

P 7.8-1 The switch in Figure P 7.8-1 has been open for a long time before closing at time $t = 0$. Find $v_c(0^+)$ and $i_L(0^+)$, the values of the capacitor voltage and inductor current immediately after the switch closes. Let $v_c(\infty)$ and $i_L(\infty)$ denote the values of the capacitor voltage and inductor current after the switch has been closed for a long time. Find $v_c(\infty)$ and $i_L(\infty)$.

Answers: $v_c(0^+) = 12$ V, $i_L(0^+) = 0$, $v_c(\infty) = 4$ V, and $i_L(\infty) = 1$ mA

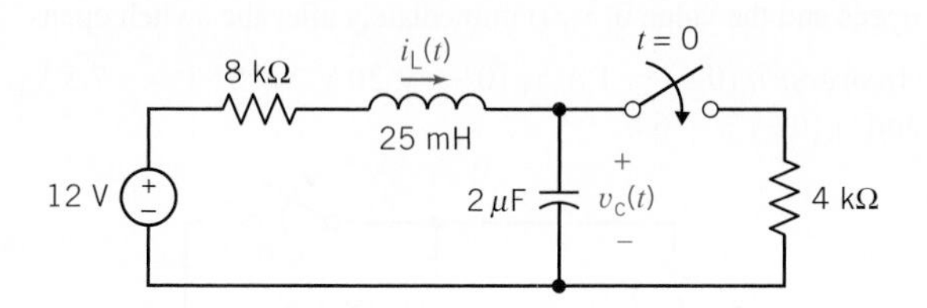

Figure P 7.8-1

P 7.8-2 The switch in Figure P 7.8-2 has been open for a long time before closing at time $t = 0$. Find $v_c(0^+)$ and $i_L(0^+)$, the values of the capacitor voltage and inductor current immediately after the switch closes. Let $v_c(\infty)$ and $i_L(\infty)$ denote the values of the capacitor voltage and inductor current after the switch has been closed for a long time. Find $v_c(\infty)$ and $i_L(\infty)$.

Answer: $v_c(0^+) = 9$ V, $i_L(0^+) = 1$ mA, $v_c(\infty) = 4.5$ V, and $i_L(\infty) = 1.5$ mA

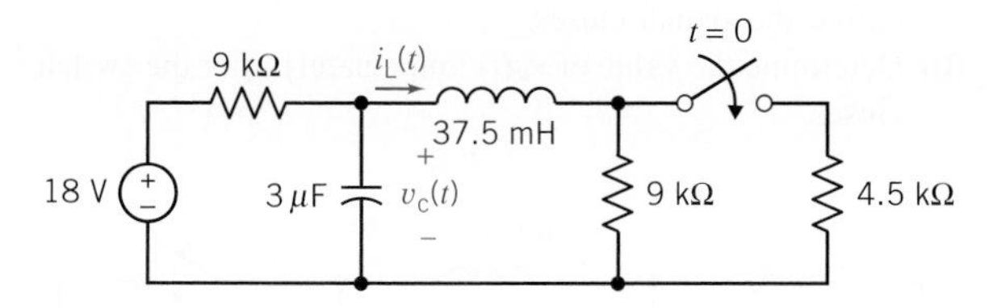

Figure P 7.8-2

P 7.8-3 The switch in Figure P 7.8-3 has been open for a long time before closing at time $t = 0$. Find $v_c(0^+)$ and $i_L(0^+)$, the values of the capacitor voltage and inductor current immediately after the switch closes. Let $v_c(\infty)$ and $i_L(\infty)$ denote the values of the capacitor voltage and inductor current after the switch has been closed for a long time. Find $v_c(\infty)$ and $i_L(\infty)$.

Answers: $v_c(0^+) = 0$ V, $i_L(0^+) = 0$, $v_c(\infty) = 8$ V, and $i_L(\infty) = 0.5$ mA

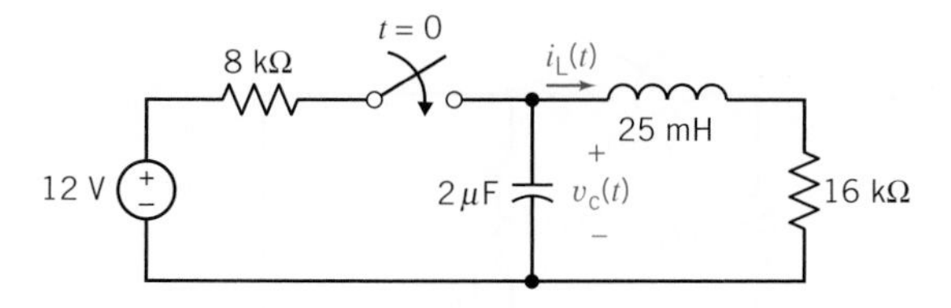

Figure P 7.8-3

P 7.8-4 Find $v_c(0^+)$ and $dv_c(0^+)/dt$ if $v(0^-) = 22.5$ V for the circuit of Figure P 7.8-4.

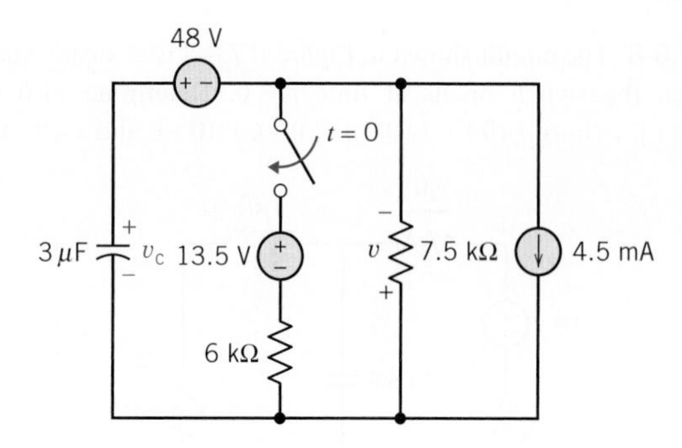

Figure P 7.8-4

P 7.8-5 For the circuit shown in Figure P 7.8-5, find $dv_c(0^+)/dt$, $di_L(0^+)/dt$, and $i(0^+)$ if $v(0^-) = 16$ V. Assume that the switch was closed for a long time prior to $t = 0$.

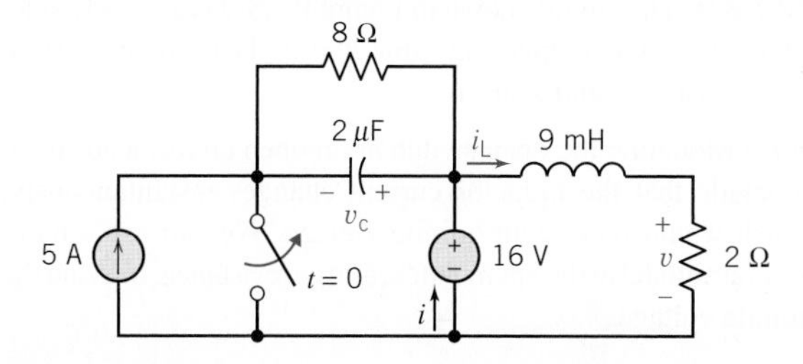

Figure P 7.8-5

P 7.8-6 For the circuit of Figure P 7.8-6, determine the current and voltage of each passive element at $t = 0^-$ and $t = 0^+$. The current source is $i_s = 0$ for $t < 0$ and $i_s = 4$ A for $t \geq 0$.

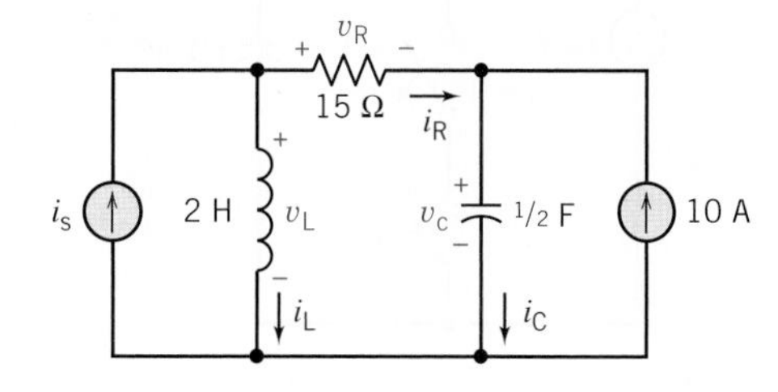

Figure P 7.8-6

P 7.8-7 The circuit shown in Figure P 7.8-7 is at steady state when the switch closes at time $t = 0$. Determine $v_1(0-)$, $v_1(0+)$, $i_2(0-)$, and $i_2(0+)$.

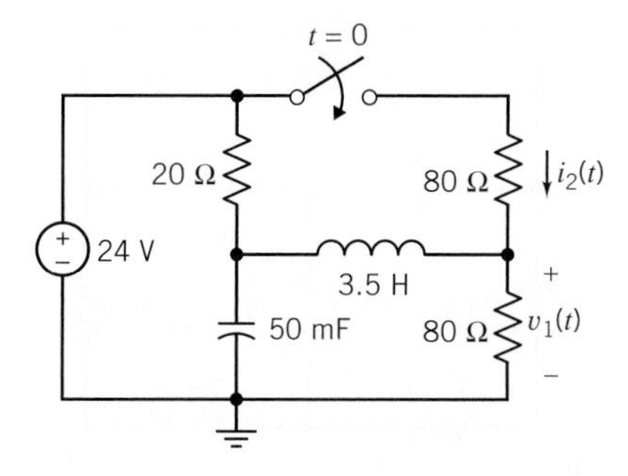

Figure P 7.8-7

P 7.8-8 The circuit shown in Figure P 7.8-8 is at steady state when the switch opens at time $t = 0$. Determine $v_1(0-)$, $v_1(0+)$, $i_2(0-)$, $i_2(0+)$, $i_3(0-)$, $i_3(0+)$, $v_4(0-)$, and $v_4(0+)$.

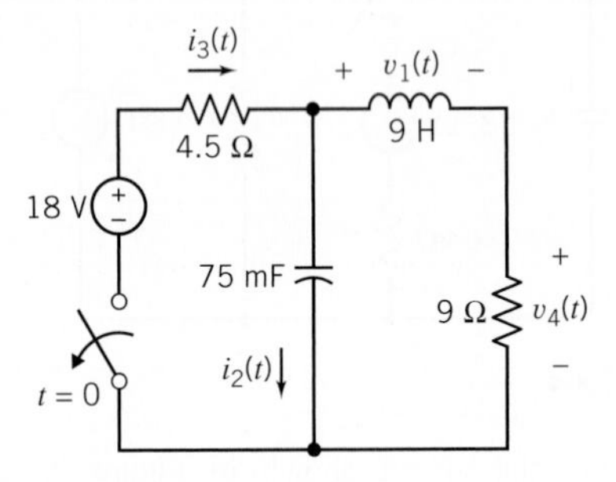

Figure P 7.8-8

***P 7.8-9** The circuit shown in Figure P 7.8-9 is at steady state when the switch opens at time $t = 0$. Determine $v_1(0-)$, $v_1(0+)$, $i_2(0-)$, and $i_2(0+)$.

Hint: Modeling the open switch as an open circuit leads us to conclude that the inductor current changes instantaneously, which would require an infinite voltage. We can use a more accurate model of the open switch, a large resistance, to avoid the infinite voltage.

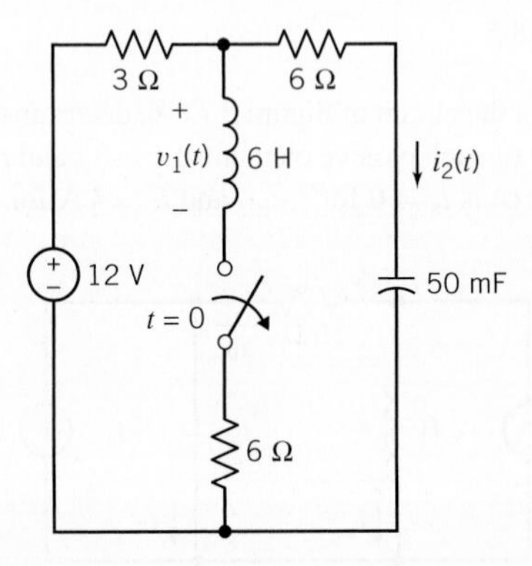

Figure P 7.8-9

P 7.8-10 The circuit shown in Figure P 7.8-10 is at steady state when the switch closes at time $t = 0$. Determine $v_1(0-)$, $v_1(0+)$, $i_2(0-)$, and $i_2(0+)$.

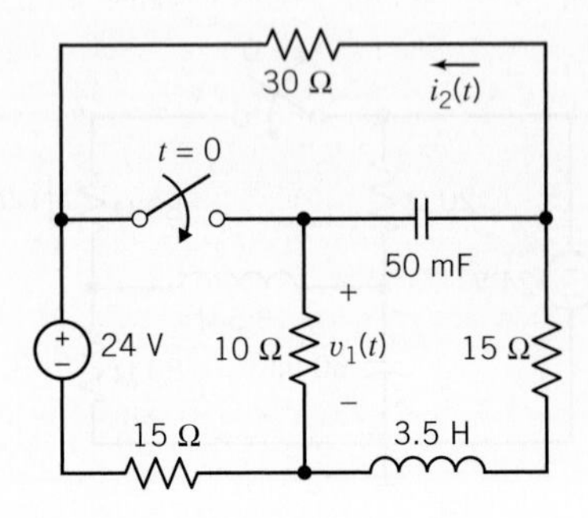

Figure P 7.8-10

P 7.8-11 The circuit shown in Figure P 7.8-11 has reached steady state before the switch opens at time $t = 0$. Determine the values of $i_L(t)$, $v_C(t)$, and $v_R(t)$ immediately before the switch opens and the value of $v_R(t)$ immediately after the switch opens.

Answers: $i_L(0-) = 1$ A, $v_C(0-) = 30$ V, $v_R(0-) = -7.5$ V, and $v_R(0+) = -6$ V

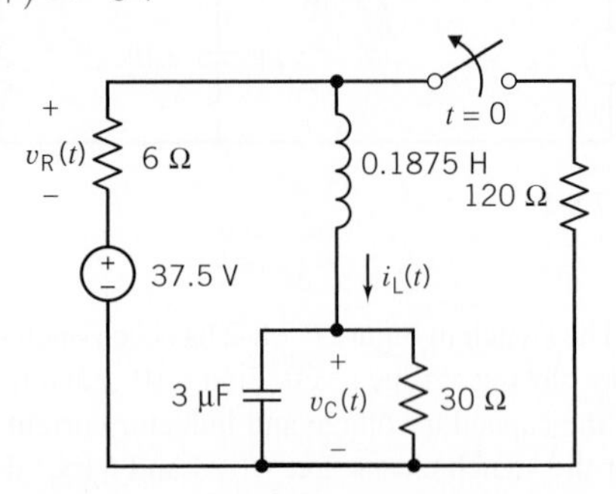

Figure P 7.8-11

P 7.8-12 The circuit shown in Figure P 7.8-12 has reached steady state before the switch closes at time $t = 0$.

(a) Determine the values of $i_L(t)$, $v_C(t)$, and $v_R(t)$ immediately before the switch closes.

(b) Determine the value of $v_R(t)$ immediately after the switch closes.

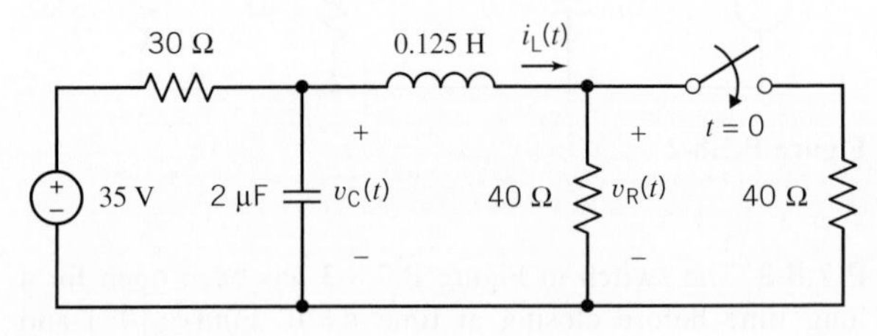

Figure P 7.8-12

P 7.8-13 The circuit shown in Figure P 7.8-13 has reached steady state before the switch opens at time $t = 0$. Determine the values of $i_L(t)$, $v_C(t)$, and $v_R(t)$ immediately before the switch opens and the value of $v_R(t)$ immediately after the switch opens.

Answers: $i_L(0-) = 0.4$ A, $v_C(0-) = 16$ V, $v_R(0-) = 0$ V, and $v_R(0+) = -12$ V

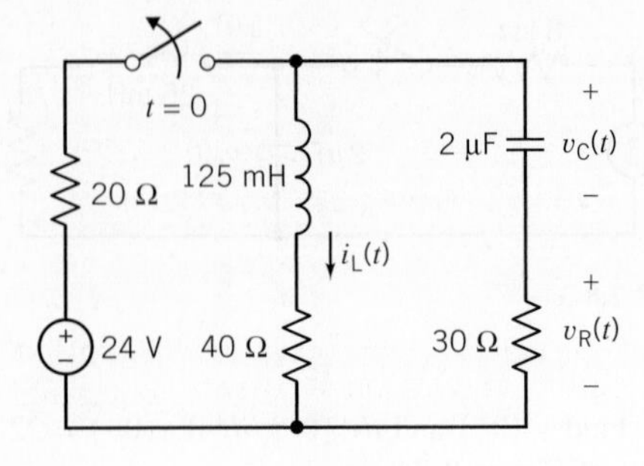

Figure P 7.8-13

Section 7.9 Operational Amplifier Circuits and Linear Differential Equations

P 7.9-1 Design a circuit with one input, $x(t)$, and one output, $y(t)$, that are related by this differential equation:

$$\frac{1}{2}\frac{d^2}{dt^2}y(t) + 4\frac{d}{dt}y(t) + y(t) = \frac{5}{2}x(t)$$

P 7.9-2 Design a circuit with one input, $x(t)$, and one output, $y(t)$, that are related by this differential equation:

$$\frac{1}{2}\frac{d^2}{dt^2}y(t) + y(t) = -\frac{5}{2}x(t)$$

P 7.9-3 Design a circuit with one input, $x(t)$, and one output, $y(t)$, that are related by this differential equation:

$$2\frac{d^3}{dt^3}y(t) + 16\frac{d^2}{dt^2}y(t) + 8\frac{d}{dt}y(t) + 10y(t) = -4x(t)$$

P 7.9-4 Design a circuit with one input, $x(t)$, and one output, $y(t)$, that are related by this differential equation:

$$\frac{d^3}{dt^3}y(t) + 16\frac{d^2}{dt^2}y(t) + 8\frac{d}{dt}y(t) + 10y(t) = 4x(t)$$

Section 7.11 How Can We Check . . . ?

P 7.11-1 A homework solution indicates that the current and voltage of a 100-H inductor are

$$i(t) = \begin{cases} 0.05 & t < 1 \\ -\frac{t}{50} + 0.13 & 1 < t < 3 \\ \frac{t}{100} - 0.23 & 3 < t < 9 \\ 0.13 & t < 9 \end{cases}$$

and

$$v(t) = \begin{cases} 0 & t < 1 \\ -4 & 1 < t < 3 \\ 2 & 3 < t < 9 \\ 0 & t > 9 \end{cases}$$

where the units of current are A, the units of voltage are V, and the units of time are s. Verify that the inductor current does not change instantaneously.

P 7.11-2 A homework solution indicates that the current and voltage of a 100-H inductor are

$$i(t) = \begin{cases} -\frac{t}{300} + 0.0375 & t < 1 \\ -\frac{t}{150} + 0.045 & 1 < t < 4 \\ \frac{t}{150} - 0.045 & 4 < t < 9 \\ 0.0225 & t < 9 \end{cases}$$

and

$$v(t) = \begin{cases} -1 & t < 1 \\ -2 & 1 < t < 4 \\ 1 & 4 < t < 9 \\ 0 & t > 9 \end{cases}$$

where the units of current are A, the units of voltage are V, and the units of time are s. Is this homework solution correct? Justify your answer.

Design Problems

DP 7-1 Consider a single-circuit element, that is, a single resistor, capacitor, or inductor. The voltage, $v(t)$, and current, $i(t)$, of the circuit element adhere to the passive convention. Consider the following cases:

(a) $v(t) = 6 + 3e^{-4.5t}$ V and $i(t) = -4.5e^{-4.5t}$ A for $t > 0$
(b) $v(t) = -4.5e^{-4.5t}$ V and $i(t) = 6 + 3e^{-4.5t}$ A for $t > 0$
(c) $v(t) = 6 + 3e^{-4.5t}$ V and $i(t) = 3 + 1.5e^{-4.5t}$ A for $t > 0$

For each case, specify the circuit element to be a capacitor, resistor, or inductor and give the value of its capacitance, resistance, or inductance.

DP 7-2 Figure DP 7-2 shows a voltage source and unspecified circuit elements. Each circuit element is a single resistor, capacitor, or inductor. Consider the following cases:

(a) $i(t) = 1.131\cos(2t + 45°)$ A
(b) $i(t) = 1.131\cos(2t - 45°)$ A

For each case, specify each circuit element to be a capacitor, resistor, or inductor and give the value of its capacitance, resistance, or inductance.

Hint: $\cos(\theta + \phi) = \cos\theta\cos\phi - \sin\theta\sin\phi$

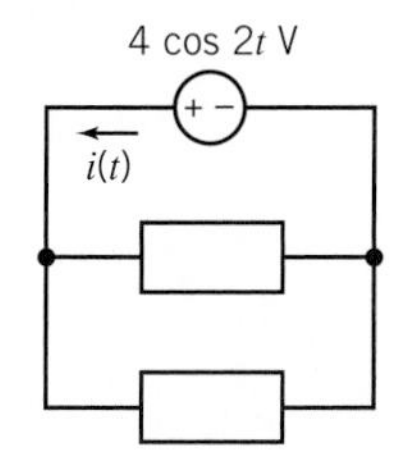

Figure DP 7-2

DP 7-3 Figure DP 7-3 shows a voltage source and unspecified circuit elements. Each circuit element is a single resistor, capacitor, or inductor. Consider the following cases:

(a) $v(t) = 11.31 \cos(2t + 45°)$ V
(b) $v(t) = 11.31 \cos(2t - 45°)$ V

For each case, specify each circuit element to be a capacitor, resistor, or inductor and give the value of its capacitance, resistance, or inductance.

Hint: $\cos(\theta + \phi) = \cos\theta\cos\phi - \sin\theta\sin\phi$

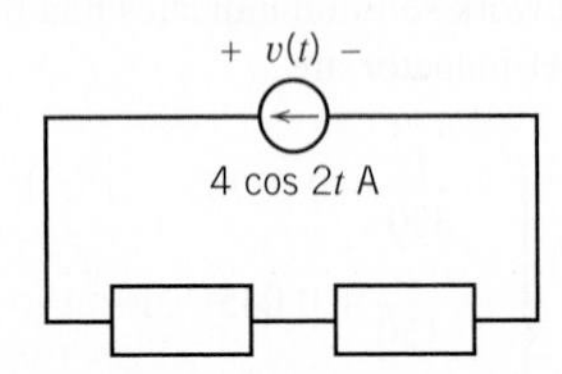

Figure DP 7-3

DP 7-4 A high-speed flash unit for sports photography requires a flash voltage $v(0^+) = 3$ V and

$$\left.\frac{dv(t)}{dt}\right|_{t=0} = 24 \text{ V/s}$$

The flash unit uses the circuit shown in Figure DP 7-4. Switch 1 has been closed a long time, and switch 2 has been open a long time at $t = 0$. Actually, the long time in this case is 3 s. Determine the required battery voltage, V_B, when $C = 1/8$ F.

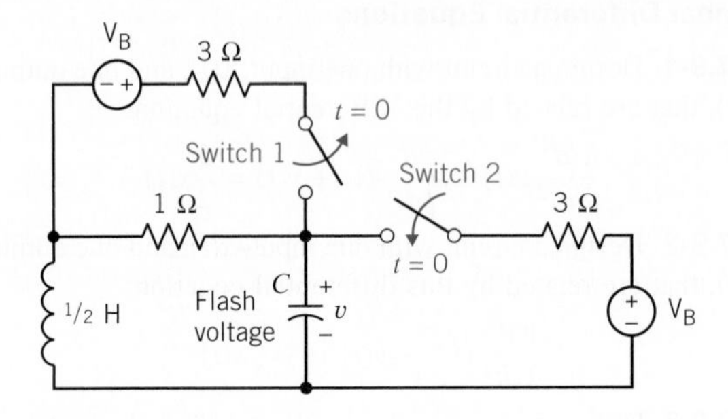

Figure DP 7-4

DP 7-5 For the circuit shown in Figure DP 7-5, select a value of R so that the energy stored in the inductor is equal to the energy stored in the capacitor at steady state.

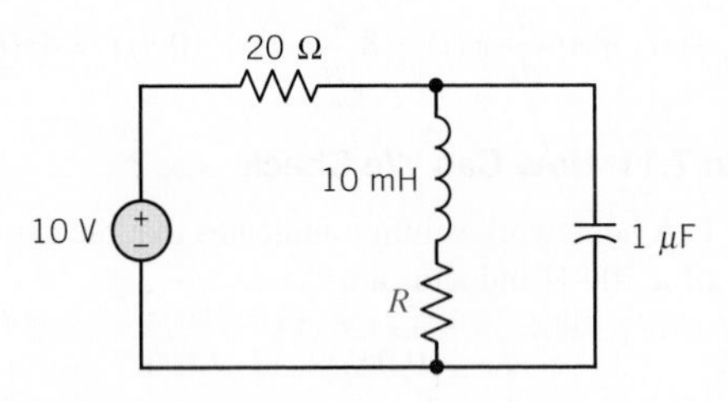

Figure DP 7-5

CHAPTER 8

The Complete Response of RL and RC Circuits

IN THIS CHAPTER

8.1 Introduction
8.2 First-Order Circuits
8.3 The Response of a First-Order Circuit to a Constant Input
8.4 Sequential Switching
8.5 Stability of First-Order Circuits
8.6 The Unit Step Source
8.7 The Response of a First-Order Circuit to a Nonconstant Source
8.8 Differential Operators
8.9 Using PSpice to Analyze First-Order Circuits
8.10 How Can We Check . . . ?
8.11 **DESIGN EXAMPLE**—A Computer and Printer
8.12 Summary
Problems
PSpice Problems
Design Problems

8.1 INTRODUCTION

In this chapter, we consider the response of *RL* and *RC* circuits to abrupt changes. The abrupt change might be a change to the circuit, as when a switch opens or closes. Alternately, the abrupt change might be a change to the input to the circuit, as when the voltage of a voltage source is a discontinuous function of time.

RL and *RC* circuits are called first-order circuits. In this chapter, we will do the following:

- Develop vocabulary that will help us talk about the response of a first-order circuit.
- Analyze first-order circuits with inputs that are constant after some particular time, t_0.
- Introduce the notion of a stable circuit and use it to identify stable first-order circuits.
- Analyze first-order circuits that experience more than one abrupt change.
- Introduce the step function and use it to determine the step response of a first-order circuit.
- Analyze first-order circuits with inputs that are not constant.

8.2 FIRST-ORDER CIRCUITS

Circuits that contain capacitors and inductors can be represented by differential equations. The order of the differential equation is usually equal to the number of capacitors plus the number of inductors in the circuit.

Circuits that contain only one inductor and no capacitors or only one capacitor and no inductors can be represented by a first-order differential equation. These circuits are called **first-order circuits**.

Thévenin and Norton equivalent circuits simplify the analysis of first-order circuits by showing that all first-order circuits are equivalent to one of two simple first-order circuits. Figure 8.2-1 shows how this is accomplished. In Figure 8.2-1*a*, a first-order circuit is partitioned into two parts. One part is the single capacitor or inductor that we expect to find in a first-order circuit. The other part is the rest of the circuit—everything except that capacitor or inductor. The next step, shown in Figure 8.2-1*b*, depends on whether the energy storage element is a capacitor or an inductor. If it is a capacitor, then the rest of the circuit is replaced by its Thévenin equivalent circuit. The result is a simple first-order circuit—a series circuit consisting of a voltage source, a resistor, and a capacitor. On the other hand, if the energy storage element is an inductor, then the rest of the circuit is replaced by its Norton equivalent circuit. The result is another simple first-order circuit—a parallel circuit consisting of a current source, a resistor, and an inductor. Indeed, all first-order circuits are equivalent to one of these two simple first-order circuits.

Consider the first-order circuit shown in Figure 8.2-2*a*. The input to this circuit is the voltage $v_s(t)$. The output, or response, of this circuit is the voltage across the capacitor. This circuit is at steady state before the switch is closed at time $t = 0$. Closing the switch disturbs this circuit. Eventually, the

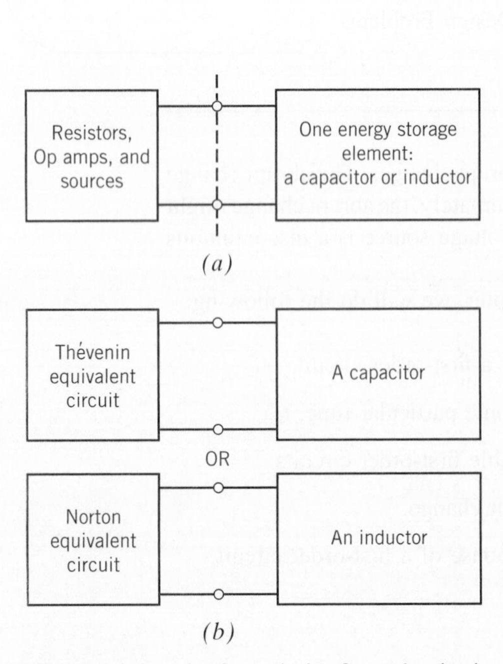

FIGURE 8.2-1 A plan for analyzing first-order circuits. (*a*) First, separate the energy storage element from the rest of the circuit. (*b*) Next, replace the circuit connected to a capacitor by its Thévenin equivalent circuit or replace the circuit connected to an inductor by its Norton equivalent circuit.

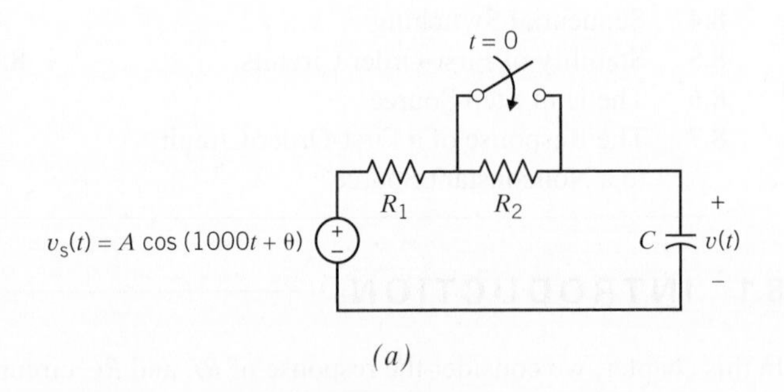

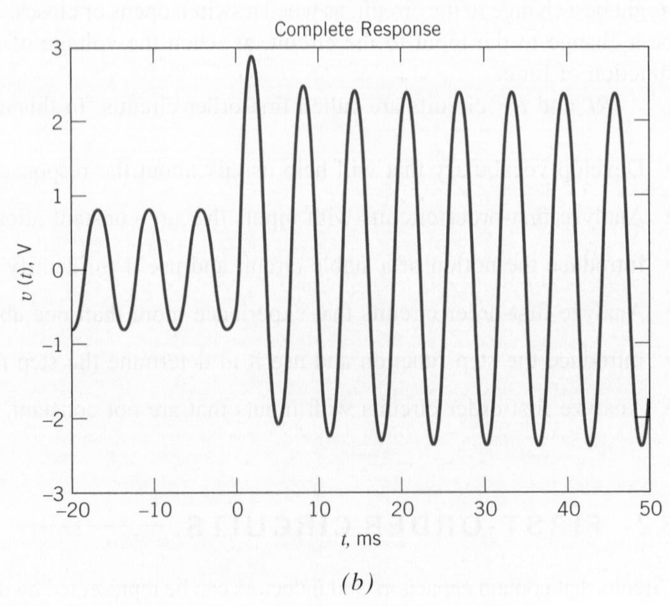

FIGURE 8.2-2 (*a*) A circuit and (*b*) its complete response.

disturbance dies out and the circuit is again at steady state. The steady-state condition with the switch closed will probably be different from the steady-state condition with the switch open. Figure 8.2-2*b* shows a plot of the capacitor voltage versus time.

When the input to a circuit is sinusoidal, the steady-state response is also sinusoidal. Furthermore, the frequency of the response sinusoid must be the same as the frequency of the input sinusoid. The circuit shown in Figure 8.2-2*a* is at steady state before the switch is closed. The steady-state capacitor voltage will be

$$v(t) = B\cos(1000t + \phi), \; t < 0 \tag{8.2-1}$$

The switch closes at time $t = 0$. The value of the capacitor voltage at the time the switch closes is

$$v(0) = B\cos(\phi), \; t = 0 \tag{8.2-2}$$

After the switch closes, the response will consist of two parts: a transient part that eventually dies out and a steady-state part. The steady-state part of the response will be sinusoidal and will have the frequency of the input. For a first-order circuit, the transient part of the response is exponential. Indeed, we consider first-order circuits separately to take advantage of the simple form of the transient response of these circuits. After the switch is closed, the capacitor voltage is

$$v(t) = Ke^{-t/\tau} + M\cos(1000t + \delta) \tag{8.2-3}$$

Notice that $Ke^{-t/\tau}$ goes to zero as t becomes large. This is the transient part of the response, which dies out, leaving the steady-state response, $M\cos(1000t + \delta)$.

As a matter of vocabulary, the ''transient part of the response'' is frequently shortened to the **transient response**, and the ''steady-state part of the response'' is shortened to the ''steady-state response.'' The response, $v(t)$, given by Eq. 8.2-3, is called the **complete response** to contrast it with the transient and steady-state responses.

complete response = transient response + steady-state response

(The term *transient response* is used in two different ways by electrical engineers. Sometimes it refers to the ''transient part of the complete response,'' and at other times, it refers to a complete response, which includes a transient part. In particular, PSpice uses the term *transient response* to refer to the complete response. This can be confusing, so the term *transient response* must be used carefully.)

In general, the complete response of a first-order circuit can be represented as the sum of two parts, the **natural response** and the **forced response**:

complete response = natural response + forced response

The natural response is the general solution of the differential equation representing the first-order circuit, when the input is set to zero. The forced response is a particular solution of the differential equation representing the circuit.

The complete response of a first-order circuit will depend on an initial condition, usually a capacitor voltage or an inductor current at a particular time. Let t_0 denote the time at which the initial condition is given. The natural response of a first-order circuit will be of the form

$$\text{natural response} = Ke^{-(t-t_0)/\tau}$$

When $t_0 = 0$, then

$$\text{natural response} = Ke^{-t/\tau}$$

The constant K in the natural response depends on the initial condition, for example, the capacitor voltage at time t_0.

In this chapter, we will consider three cases. In these cases, the input to the circuit after the disturbance will be (1) a constant, for example,

$$v_s(t) = V_0$$

or (2) an exponential, for example,

$$v_s(t) = V_0 e^{-t/\tau}$$

or (3) a sinusoid, for example,

$$v_s(t) = V_0 \cos(\omega t + \theta)$$

These three cases are special because the forced response will have the same form as the input. For example, in Figure 8.2-2, both the forced response and the input are sinusoidal, and the frequency of the forced response is the same as the frequency of the input. For other inputs, the forced response may not have the same form as the input. For example, when the input is a square wave, the forced response is not a square wave.

When the input is a constant or a sinusoid, the forced response is also called the steady-state response, and the natural response is called the transient response.

Here is our plan for finding the complete response of first-order circuits:

Step 1: Find the forced response before the disturbance. Evaluate this response at time $t = t_0$ to obtain the initial condition of the energy storage element.

Step 2: Find the forced response after the disturbance.

Step 3: Add the natural response $= Ke^{-t/\tau}$ to the forced response to get the complete response. Use the initial condition to evaluate the constant K.

8.3 THE RESPONSE OF A FIRST-ORDER CIRCUIT TO A CONSTANT INPUT

In this section, we find the complete response of a first-order circuit when the input to the circuit is constant after time t_0. Figure 8.3-1 illustrates this situation. In Figure 8.3-1*a*, we find a first-order circuit that contains a single capacitor and no inductors. This circuit is at steady state before the switch closes, disturbing the steady state. The time at which steady state is disturbed is denoted as t_0. In Figure 8.3-1*a*, $t_0 = 0$. Closing the switch removes the resistor R_1 from the circuit. (A closed switch is modeled by a short circuit. A short circuit in parallel with a resistor is equivalent to a short circuit.) After the switch closes, the circuit can be represented as shown in Figure 8.3-1*b*. In Figure 8.3-1*b*, the part of the circuit that is connected to the capacitor has been replaced by its Thévenin equivalent circuit. Therefore,

$$V_{oc} = \frac{R_3}{R_2 + R_3} V_s \quad \text{and} \quad R_t = \frac{R_2 R_3}{R_2 + R_3}$$

Let's represent the circuit in Figure 8.3-1*b* by a differential equation. The capacitor current is given by

$$i(t) = C \frac{d}{dt} v(t)$$

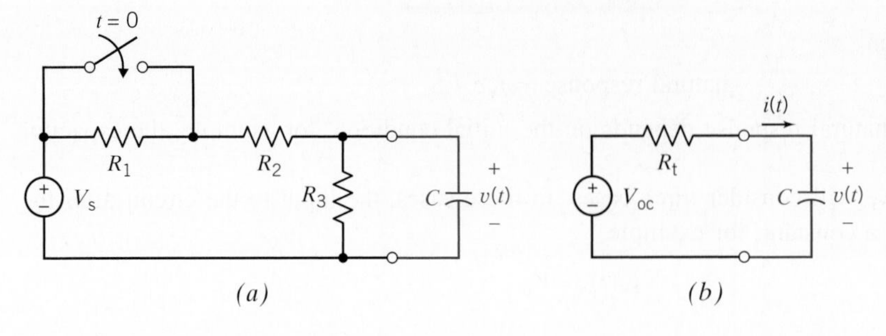

FIGURE 8.3-1 (*a*) A first-order circuit containing a capacitor. (*b*) After the switch closes, the circuit connected to the capacitor is replaced by its Thévenin equivalent circuit.

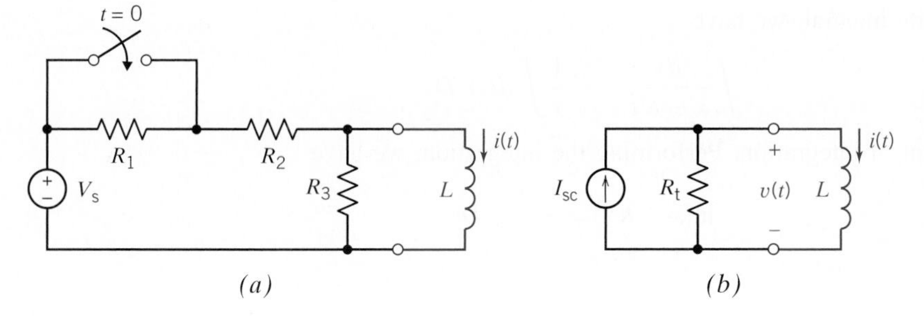

FIGURE 8.3-2 (*a*) A first-order circuit containing an inductor. (*b*) After the switch closes, the circuit connected to the inductor is replaced by its Norton equivalent circuit.

The same current, $i(t)$, passes through the resistor. Apply KVL to Figure 8.3-1*b* to get

$$V_{oc} = R_t i(t) + v(t) = R_t\left(C\frac{d}{dt}v(t)\right) + v(t)$$

Therefore,

$$\frac{d}{dt}v(t) + \frac{v(t)}{R_t C} = \frac{V_{oc}}{R_t C} \tag{8.3-1}$$

The highest-order derivative in this equation is first order, so this is a first-order differential equation.

Next, let's turn our attention to the circuit shown in Figure 8.3-2*a*. This circuit contains a single inductor and no capacitors. This circuit is at steady state before the switch closes at time $t_0 = 0$, disturbing the steady state. After the switch closes, the circuit can be represented as shown in Figure 8.3-2*b*. In Figure 8.3-2*b*, the part of the circuit that is connected to the inductor has been replaced by its Norton equivalent circuit. We calculate

$$I_{sc} = \frac{V_s}{R_2} \quad \text{and} \quad R_t = \frac{R_2 R_3}{R_2 + R_3}$$

Let's represent the circuit in Figure 8.3-2*b* by a differential equation. The inductor voltage is given by

$$v(t) = L\frac{d}{dt}i(t)$$

The voltage, $v(t)$, appears across the resistor. Apply KCL to the top node in Figure 8.3-2*b* to get

$$I_{sc} = \frac{v(t)}{R_t} + i(t) = \frac{L\frac{d}{dt}i(t)}{R_t} + i(t)$$

Therefore,

$$\frac{d}{dt}i(t) + \frac{R_t}{L}i(t) = \frac{R_t}{L}I_{sc} \tag{8.3-2}$$

As before, this is a first-order differential equation.

Equations 8.3-1 and 8.3-2 have the same form. That is,

$$\frac{d}{dt}x(t) + \frac{x(t)}{\tau} = K \tag{8.3-3}$$

The parameter τ is called the time constant. We will solve this differential equation by separating the variables and integrating. Then we will use the solution of Eq. 8.3-3 to obtain solutions of Eqs. 8.3-1 and 8.3-2.

We may rewrite Eq. 8.3-3 as

$$\frac{dx}{dt} = \frac{K\tau - x}{\tau}$$

or, separating the variables,

$$\frac{dx}{x - K\tau} = -\frac{dt}{\tau}$$

Forming the indefinite integral, we have

$$\int \frac{dx}{x - K\tau} = -\frac{1}{\tau}\int dt + D$$

where D is a constant of integration. Performing the integration, we have

$$\ln(x - K\tau) = -\frac{t}{\tau} + D$$

Solving for x gives

$$x(t) = K\tau + Ae^{-t/\tau}$$

where $A = e^D$, which is determined from the initial condition, $x(0)$. To find A, let $t = 0$. Then

$$x(0) = K\tau + Ae^{-0/\tau} = K\tau + A$$

or

$$A = x(0) - K\tau$$

Therefore, we obtain

$$x(t) = K\tau + [x(0) - K\tau]e^{-t/\tau} \tag{8.3-4}$$

Because

$$x(\infty) = \lim_{t\to\infty} x(t) = K\tau$$

Equation 8.3-4 can be written as

$$x(t) = x(\infty) + [x(0) - x(\infty)]e^{-t/\tau}$$

Taking the derivative of $x(t)$ with respect to t leads to a procedure for measuring or calculating the time constant:

$$\frac{d}{dt}x(t) = -\frac{1}{\tau}[x(0) - x(\infty)]e^{-t/\tau}$$

Now let $t = 0$ to get

$$\left.\frac{d}{dt}x(t)\right|_{t=0} = -\frac{1}{\tau}[x(0) - x(\infty)]$$

or

$$\tau = \frac{x(\infty) - x(0)}{\left.\frac{d}{dt}x(t)\right|_{t=0}} \tag{8.3-5}$$

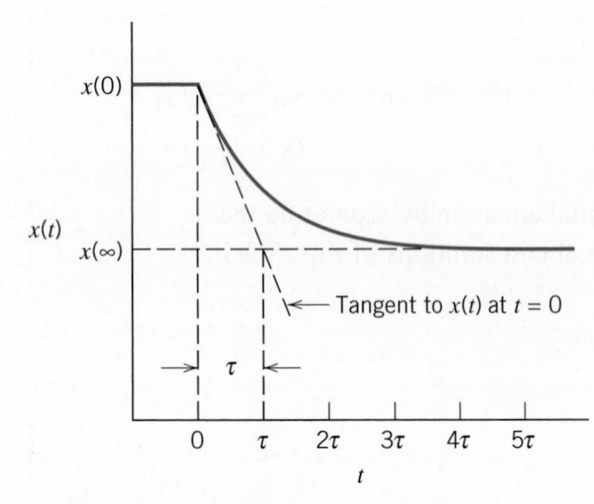

FIGURE 8.3-3 A graphical technique for measuring the time constant of a first-order circuit.

Figure 8.3-3 shows a plot of $x(t)$ versus t. We can determine the values of (1) the slope of the plot at time $t = 0$, (2) the initial value of $x(t)$, and (3) the final value of $x(t)$ from this plot. Equation 8.3-5 can be used to calculate the time constant from these values. Equivalently, Figure 8.3-3 shows how to measure the time constant from a plot of $x(t)$ versus t.

Next, we apply these results to the *RC* circuit in Figure 8.3-1. Comparing Eqs. 8.3-1 and 8.3-3, we see that

$$x(t) = v(t),\ \tau = R_t C,\ \text{and}\ K = \frac{V_{oc}}{R_t C}$$

Making these substitutions in Eq. 8.3-4 gives

$$v(t) = V_{oc} + (v(0) - V_{oc})e^{-t/(R_t C)} \tag{8.3-6}$$

The second term on the right-hand side of Eq. 8.3-6 dies out as t increases. This is the transient or natural response. At $t = 0$, $e^{-0} = 1$. Letting $t = 0$ in Eq. 8.3-6 gives $v(0) = v(0)$, as required. When $t = 5\tau$, $e^{-5} = 0.0067 \approx 0$, so at time $t = 5\tau$, the capacitor voltage will be

$$v(5\tau) = 0.9933\ V_{oc} + 0.0067\ v(0) \approx V_{oc}$$

This is the steady-state or forced response. The forced response is of the same form, a constant, as the input to the circuit. The sum of the natural and forced responses is the complete response:

$$\text{complete response} = v(t), \quad \text{forced response} = V_{oc}$$

and

$$\text{natural response} = (v(0) - V_{oc})e^{-t/(R_t C)}$$

Next, compare Eqs. 8.3-2 and 8.3-3 to find the solution of the RL circuit in Figure 8.3-2. We see that

$$x(t) = i(t), \quad \tau = \frac{L}{R_t}, \quad \text{and } K = \frac{L}{R_t} I_{sc}$$

Making these substitutions in Eq. 8.3-4 gives

$$i(t) = I_{sc} + (i(0) - I_{sc})e^{-(R_t/L)t} \tag{8.3-7}$$

Again, the complete response is the sum of the forced (steady-state) response and the transient (natural) response:

$$\text{complete response} = i(t), \quad \text{forced response} = I_{sc}$$

and

$$\text{natural response} = (i(0) - I_{sc})e^{-(R_t/L)t}$$

EXAMPLE 8.3-1 First-Order Circuit with a Capacitor

Find the capacitor voltage after the switch opens in the circuit shown in Figure 8.3-4*a*. What is the value of the capacitor voltage 50 ms after the switch opens?

Solution

The 2-volt voltage source forces the capacitor voltage to be 2 volts until the switch opens. Because the capacitor voltage cannot change instantaneously, the capacitor voltage will be 2 volts immediately after the switch opens. Therefore, the initial condition is

$$v(0) = 2\text{ V}$$

Figure 8.3-4*b* shows the circuit after the switch opens. Comparing this circuit to the RC circuit in Figure 8.3-1*b*, we see that

$$R_t = 10\text{ k}\Omega \quad \text{and} \quad V_{oc} = 8\text{ V}$$

The time constant for this first-order circuit containing a capacitor is

$$\tau = R_t C = \left(10 \times 10^3\right)\left(2 \times 10^{-6}\right) = 20 \times 10^{-3} = 20\text{ ms}$$

Substituting these values into Eq. 8.3-6 gives

$$v(t) = 8 - 6e^{-t/20}\text{ V} \tag{8.3-8}$$

where t has units of ms. To find the voltage 50 ms after the switch opens, let $t = 50$. Then,

$$v(50) = 8 - 6e^{-50/20} = 7.51\text{ V}$$

Figure 8.3-4c shows a plot of the capacitor voltage as a function of time.

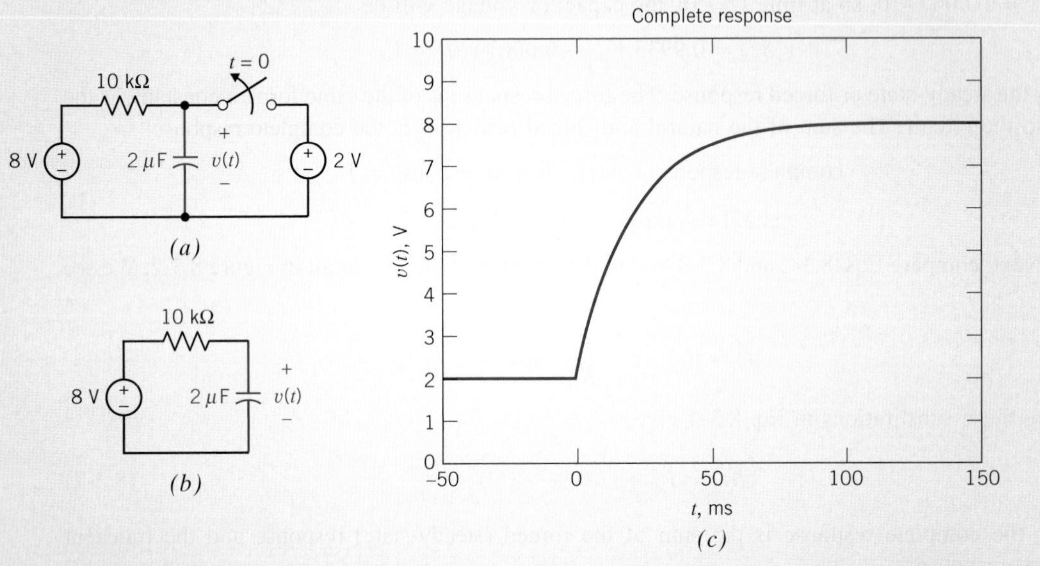

FIGURE 8.3-4 (a) A first-order circuit and (b) an equivalent circuit that is valid after the switch opens. (c) A plot of the complete response, $v(t)$, given in Eq. 8.3-8.

EXAMPLE 8.3-2 First-Order Circuit with an Inductor

Find the inductor current after the switch closes in the circuit shown in Figure 8.3-5a. How long will it take for the inductor current to reach 2 mA?

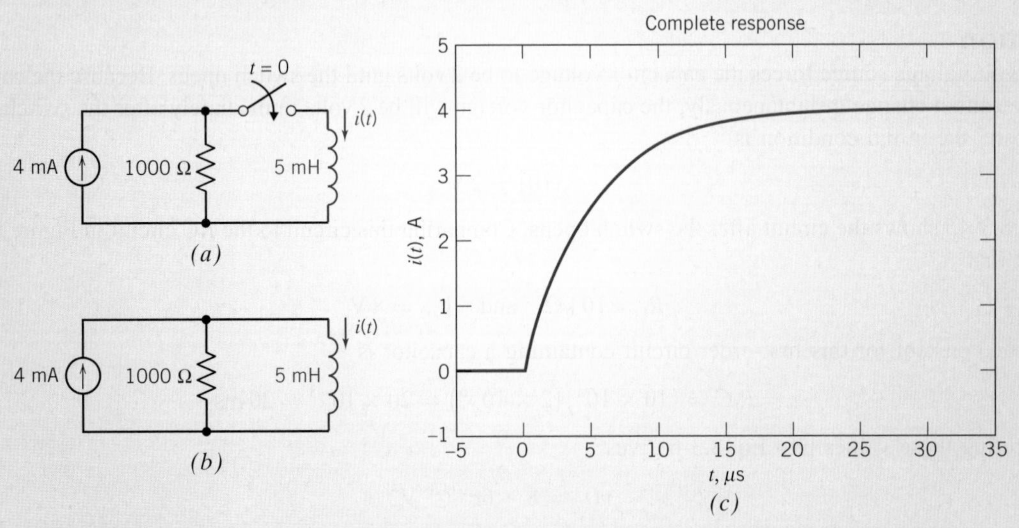

FIGURE 8.3-5 (a) A first-order circuit and (b) an equivalent circuit that is valid after the switch closes. (c) A plot of the complete response, $i(t)$, given by Eq. 8.3-9.

Solution

The inductor current will be 0 A until the switch closes. Because the inductor current cannot change instantaneously, it will be 0 A immediately after the switch closes. Therefore, the initial condition is

$$i(0) = 0$$

Figure 8.3-5*b* shows the circuit after the switch closes. Comparing this circuit to the *RL* circuit in Figure 8.3-2*b*, we see that

$$R_t = 1000\ \Omega \quad \text{and} \quad I_{sc} = 4\ \text{mA}$$

The time constant for this first-order circuit containing an inductor is

$$\tau = \frac{L}{R_t} = \frac{5 \times 10^{-3}}{1000} = 5 \times 10^{-6} = 5\ \mu\text{s}$$

Substituting these values into Eq. 8.3-7 gives

$$i(t) = 4 - 4e^{-t/5}\ \text{mA} \tag{8.3-9}$$

where t has units of microseconds. To find the time when the current reaches 2 mA, substitute $i(t) = 2$ mA. Then

$$2 = 4 - 4e^{-t/5}\ \text{mA}$$

Solving for t gives

$$t = -5 \times \ln\left(\frac{2-4}{-4}\right) = 3.47\ \mu\text{s}$$

Figure 8.3-5*c* shows a plot of the inductor current as a function of time.

EXAMPLE 8.3-3 First-Order Circuit

INTERACTIVE EXAMPLE

The switch in Figure 8.3-6*a* has been open for a long time, and the circuit has reached steady state before the switch closes at time $t = 0$. Find the capacitor voltage for $t \geq 0$.

Solution

The switch has been open for a long time before it closes at time $t = 0$. The circuit will have reached steady state before the switch closes. Because the input to this circuit is a constant, all the element currents and voltages will be constant when the circuit is at steady state. In particular, the capacitor voltage will be constant. The capacitor current will be

$$i(t) = C\frac{d}{dt}v(t) = C\frac{d}{dt}(\text{a constant}) = 0$$

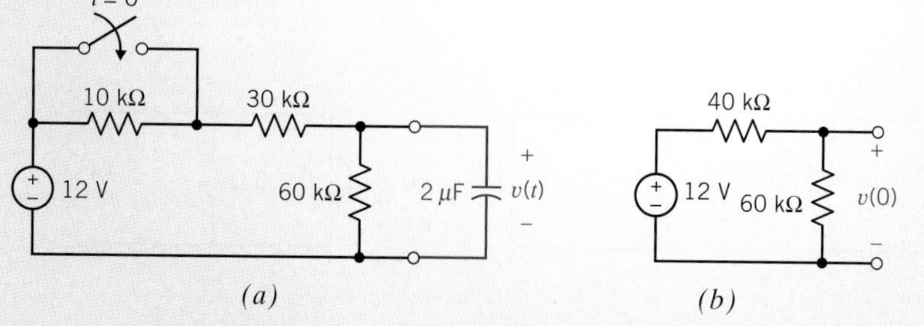

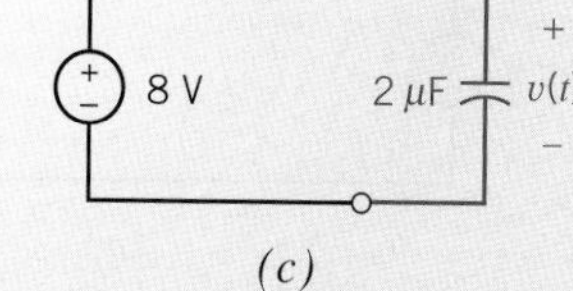

FIGURE 8.3-6 (*a*) A first-order circuit. The equivalent circuit for (*b*) $t < 0$ and (*c*) $t > 0$.

The capacitor voltage is unknown, but the capacitor current is zero. In other words, the capacitor acts like an open circuit when the input is constant and the circuit is at steady state. (By a similar argument, inductors act like short circuits when the input is constant and the circuit is at steady state.)

Figure 8.3-6*b* shows the appropriate equivalent circuit while the switch is open. An open switch acts like an open circuit; thus, the 10-kΩ and 30-kΩ resistors are in series. They have been replaced by an equivalent 40-kΩ resistor. The input to the circuit is a constant (12 volts), and the circuit is at steady state; therefore, the capacitor acts like an open circuit. The voltage across this open circuit is the capacitor voltage. Because we are interested in the initial condition, the capacitor voltage has been labeled as $v(0)$. Analyzing the circuit in Figure 8.3-6*b* using voltage division gives

$$v(0) = \frac{60 \times 10^3}{40 \times 10^3 + 60 \times 10^3} 12 = 7.2 \text{ V}$$

Figure 8.3-6*c* shows the appropriate equivalent circuit after the switch closes. Closing the switch shorts out the 10-kΩ resistor, removing it from the circuit. (A short circuit in parallel with any resistor is equivalent to a short circuit.) The part of the circuit that is connected to the capacitor has been replaced by its Thévenin equivalent circuit. After the switch is closed,

$$V_{oc} = \frac{60 \times 10^3}{30 \times 10^3 + 60 \times 10^3} 12 = 8 \text{ V}$$

and

$$R_t = \frac{30 \times 10^3 \times 60 \times 10^3}{30 \times 10^3 + 60 \times 10^3} = 20 \times 10^3 = 20 \text{ k}\Omega$$

and the time constant is

$$\tau = R_t \times C = \left(20 \times 10^3\right) \times \left(2 \times 10^{-6}\right) = 40 \times 10^{-3} = 40 \text{ ms}$$

Substituting these values into Eq. 8.3-6 gives

$$v(t) = 8 - 0.8e^{-t/40} \text{ V}$$

where t has units of ms.

EXAMPLE 8.3-4 First-Order Circuit

INTERACTIVE EXAMPLE

The switch in Figure 8.3-7*a* has been open for a long time, and the circuit has reached steady state before the switch closes at time $t = 0$. Find the inductor current for $t \geq 0$.

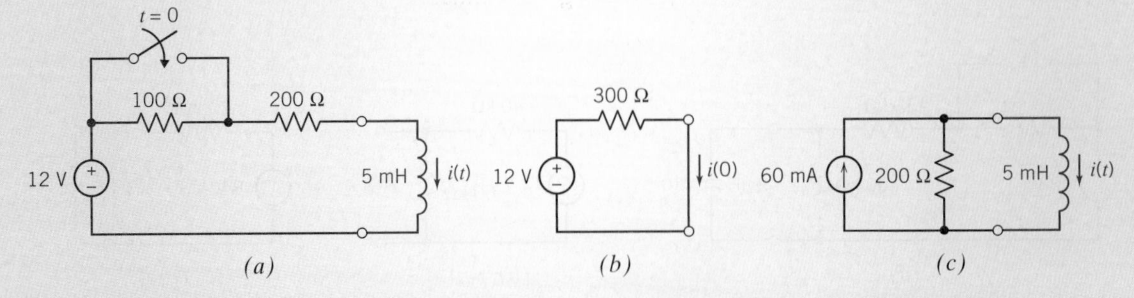

FIGURE 8.3-7 (*a*) A first-order circuit. The equivalent circuit for (*b*) $t < 0$ and (*c*) $t > 0$.

Solution

Figure 8.3-7*b* shows the appropriate equivalent circuit while the switch is open. The 100-Ω and 200-Ω resistors are in series and have been replaced by an equivalent 300-Ω resistor. The input to the circuit is a constant (12 volts), and the circuit is at steady state; therefore, the inductor acts like a short circuit. The current in this short circuit is the inductor current. Because we are interested in the initial condition, the initial inductor current has been labeled as $i(0)$. This current can be calculated using Ohm's law:

$$i(0) = \frac{12}{300} = 40 \text{ mA}$$

Figure 8.3-7*c* shows the appropriate equivalent circuit after the switch closes. Closing the switch shorts out the 100-Ω resistor, removing it from the circuit. The part of the circuit that is connected to the inductor has been replaced by its Norton equivalent circuit. After the switch is closed,

$$I_{sc} = \frac{12}{200} = 60 \text{ mA} \quad \text{and} \quad R_t = 200 \, \Omega$$

and the time constant is

$$\tau = \frac{L}{R_t} = \frac{5 \times 10^{-3}}{200} = 25 \times 10^{-6} = 25 \, \mu\text{s}$$

Substituting these values into Eq. 8.3-7 gives

$$i(t) = 60 - 20e^{-t/25} \text{ mA}$$

where t has units of microseconds.

EXAMPLE 8.3-5 First-Order Circuit

The circuit in Figure 8.3-8*a* is at steady state before the switch opens. Find the current $i(t)$ for $t > 0$.

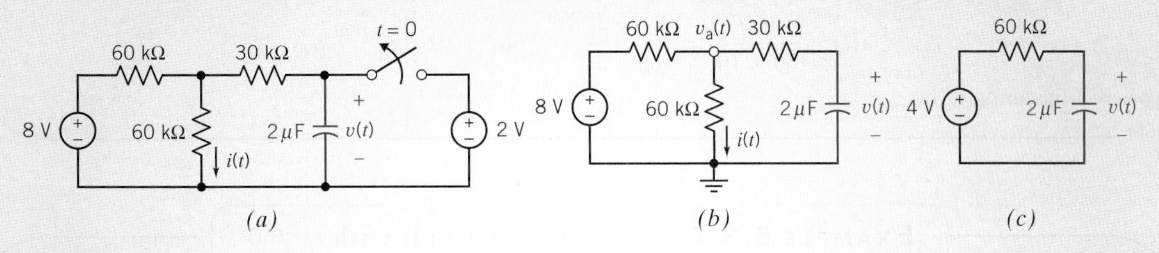

FIGURE 8.3-8 (*a*) A first-order circuit, (*b*) the circuit after the switch opens, and (*c*) the equivalent circuit after the switch opens.

Solution

The response or output of a circuit can be any element current or voltage. Frequently, the response is not the capacitor voltage or inductor current. In Figure 8.3-8*a*, the response is the current $i(t)$ in a resistor rather than the capacitor voltage. In this case, two steps are required to solve the problem. First, find the capacitor voltage using the methods already described in this chapter. Once the capacitor voltage is known, write node or mesh equations to express the response in terms of the input and the capacitor voltage.

First we find the capacitor voltage. Before the switch opens, the capacitor voltage is equal to the voltage of the 2-volt source. The initial condition is

$$v(0) = 2 \text{ V}$$

Figure 8.3-8*b* shows the circuit as it will be after the switch is opened. The part of the circuit connected to the capacitor has been replaced by its Thévenin equivalent circuit in Figure 8.3-8*c*. The parameters of the Thévenin equivalent circuit are

$$V_{oc} = \frac{60 \times 10^3}{60 \times 10^3 + 60 \times 10^3} 8 = 4\ \text{V}$$

and $$R_t = 30 \times 10^3 + \frac{60 \times 10^3 \times 60 \times 10^3}{60 \times 10^3 + 60 \times 10^3} = 60 \times 10^3 = 60\ \text{k}\Omega$$

The time constant is

$$\tau = R_t \times C = (60 \times 10^3) \times (2 \times 10^{-6}) = 120 \times 10^{-3} = 120\ \text{ms}$$

Substituting these values into Eq. 8.3-6 gives

$$v(t) = 4 - 2e^{-t/120}\ \text{V}$$

where t has units of ms.

Now that the capacitor voltage is known, we return to the circuit in Figure 8.3-8*b*. Notice that the node voltage at the middle node at the top of the circuit has been labeled as $v_a(t)$. The node equation corresponding to this node is

$$\frac{v_a(t) - 8}{60 \times 10^3} + \frac{v_a(t)}{60 \times 10^3} + \frac{v_a(t) - v(t)}{30 \times 10^3} = 0$$

Substituting the expression for the capacitor voltage gives

$$\frac{v_a(t) - 8}{60 \times 10^3} + \frac{v_a(t)}{60 \times 10^3} + \frac{v_a(t) - \left(4 - 2e^{-t/120}\right)}{30 \times 10^3} = 0$$

or $$v_a(t) - 8 + v_a(t) + 2\left[v_a(t) - \left(4 - 2e^{-t/120}\right)\right] = 0$$

Solving for $v_a(t)$, we get

$$v_a(t) = \frac{8 + 2(4 - 2e^{-t/120})}{4} = 4 - e^{-t/120}\ \text{V}$$

Finally, we calculate $i(t)$ using Ohm's law:

$$i(t) = \frac{v_a(t)}{60 \times 10^3} = \frac{4 - e^{-t/120}}{60 \times 10^3} = 66.7 - 16.7e^{-t/120}\ \mu\text{A}$$

where t has units of ms.

EXAMPLE 8.3-6 First-Order Circuit with $t_0 \neq 0$

Find the capacitor voltage after the switch opens in the circuit shown in Figure 8.3-9*a*. What is the value of the capacitor voltage 50 ms after the switch opens?

Solution

This example is similar to Example 8.3-1. The difference between the two examples is the time at which the switch opens. The switch opens at time $t = 0$ in Example 8.3-1 and at time $t = 50$ ms $= 0.05$ s in this example.

The 2-volt voltage source forces the capacitor voltage to be 2 volts until the switch opens. Consequently,

$$v(t) = 2\ \text{V} \quad \text{for} \quad t \leq 0.05\ \text{s}$$

In particular, the initial condition is

$$v(0.05) = 2\ \text{V}$$

Figure 8.3-9*b* shows the circuit after the switch opens. Comparing this circuit to the *RC* circuit in Figure 8.3-1*b*,

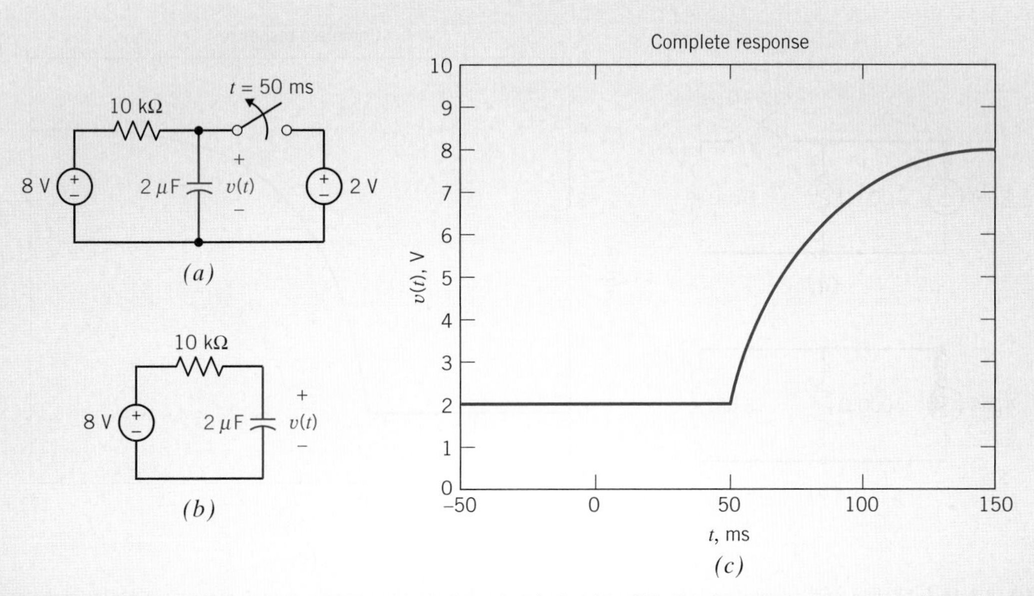

FIGURE 8.3-9 (*a*) A first-order circuit and (*b*) an equivalent circuit that is valid after the switch opens. (*c*) A plot of the complete response, $v(t)$, given by Eq. 8.3-10.

we see that

$$R_t = 10\,\text{k}\Omega \quad \text{and} \quad V_{oc} = 8\,\text{V}$$

The time constant for this first-order circuit containing a capacitor is

$$\tau = R_t C = 0.020\,\text{s}$$

A plot of the capacitor voltage in this example will have the same shape as did the plot of the capacitor voltage in Example 8.3-1, but the capacitor voltage in this example will be delayed by 50 ms because the switch opened 50 ms later. To account for this delay, we replace t by $t - 50$ ms in the equation that represents the capacitor voltage. Consequently, the voltage of the capacitor in this example is given by

$$v(t) = 8 - 6e^{-(t-50)/20}\,\text{V} \tag{8.3-10}$$

where t has units of ms. (Compare Eq. 8.3-8 and 8.3-10.) To find the voltage 50 ms after the switch opens, let $t =$ 100 ms. Then,

$$v(100) = 8 - 6e^{-(100-50)/20} = 7.51\,\text{V}$$

The value of the capacitor voltage 50 ms after the switch opens is the same here as it was in Example 8.3-1. Figure 8.3-9*c* shows a plot of the capacitor voltage as a function of time. As expected, this plot is a delayed copy of the plot shown in Figure 8.3-4*c*.

EXAMPLE 8.3-7 First-Order Circuit with $t_0 \neq 0$

Find the inductor current after the switch closes in the circuit shown in Figure 8.3-10*a*. How long will it take for the inductor current to reach 2 mA?

Solution

This example is similar to Example 8.3-2. The difference between the two examples is the time at which the switch closes. The switch closes at time $t = 0$ in Example 8.3-2 and at time $t = 10$ μs in this example.

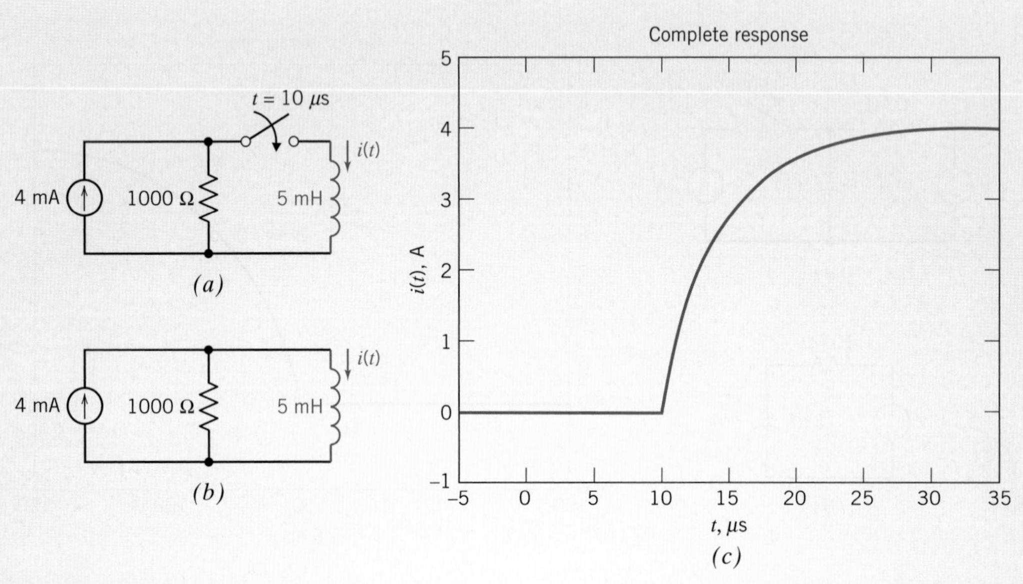

FIGURE 8.3-10 (*a*) A first-order circuit and (*b*) an equivalent circuit that is valid after the switch closes. (*c*) A plot of the complete response, $i(t)$, given by Eq. 8.3-11.

The inductor current will be 0 A until the switch closes. Because the inductor current cannot change instantaneously, it will be 0 A immediately after the switch closes. Therefore, the initial condition is

$$i(10\ \mu\text{s}) = 0\ \text{A}$$

Figure 8.3-10*b* shows the circuit after the switch closes. Comparing this circuit to the *RL* circuit in Figure 8.3-2*b*, we see that

$$R_t = 1000\ \Omega \quad \text{and} \quad I_{sc} = 4\ \text{mA}$$

The time constant for this first-order circuit containing an inductor is

$$\tau = \frac{L}{R_t} = \frac{5 \times 10^{-3}}{1000} = 5 \times 10^{-6} = 5\ \mu\text{s}$$

A plot of the inductor current in this example will have the same shape as did the plot of the inductor current in Example 8.3-2, but the inductor current in this example will be delayed by 10 μs because the switch closed 10 μs later. To account for this delay, we replace t by $t-10\ \mu$s in the equation that represents the inductor current. Consequently, the current of the inductor in this example is given by

$$i(t) = 4 - 4e^{-(t-10)/5}\ \text{mA} \tag{8.3-11}$$

where t has units of microseconds. (Compare Eq. 8.3-9 and 8.3-11.) To find the time when the current reaches 2 mA, substitute $i(t) = 2$ mA. Then

$$2 = 4 - 4e^{-(t-10)/5}\ \text{mA}$$

Solving for t gives

$$t = -5 \times \ln\left(\frac{2-4}{-4}\right) + 10 = 13.47\ \mu\text{s}$$

Because the switch closes at time 10 μs, an additional time of 3.47 μs after the switch closes is required for the value of the current to reach 2 mA. Figure 8.3-10*c* shows a plot of the inductor current as a function of time. As expected, this plot is a delayed copy of the plot shown in Figure 8.3-5*c*.

EXAMPLE 8.3-8 Exponential Response of a First-Order Circuit

Figure 8.3-11a shows a plot of the voltage across the inductor in Figure 8.3-11b.

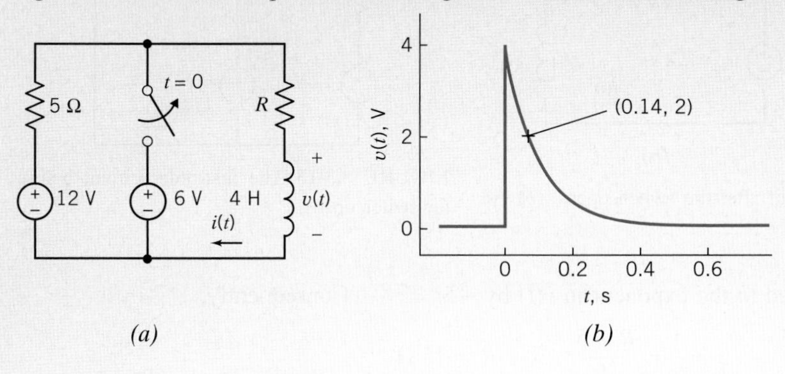

FIGURE 8.3-11 (a) A first-order circuit and (b) a plot of the inductor voltage.

(a) Determine the equation that represents the inductor voltage as a function of time.
(b) Determine the value of the resistance R.
(c) Determine the equation that represents the inductor current as a function of time.

Solution

(a) The inductor voltage is represented by an equation of the form

$$v(t) = \begin{cases} D & \text{for } t < 0 \\ E + F\,e^{-at} & \text{for } t \geq 0 \end{cases}$$

where D, E, F, and a are unknown constants. The constants D, E, and F are described by

$$D = v(t) \quad \text{when } t < 0, \quad E = \lim_{t \to \infty} v(t), \quad \text{and } E + F = \lim_{t \to 0+} v(t)$$

From the plot, we see that

$$D = 0,\ E = 0, \quad \text{and} \quad E + F = 4\text{ V}$$

Consequently,

$$v(t) = \begin{cases} 0 & \text{for } t < 0 \\ 4e^{-at} & \text{for } t \geq 0 \end{cases}$$

To determine the value of a, we pick a time when the circuit is not at steady state. One such point is labeled on the plot in Figure 8.3-11. We see $v\,(0.14) = 2$ V; that is, the value of the voltage is 2 volts at time 0.14 seconds. Substituting these into the equation for $v(t)$ gives

$$2 = 4e^{-a(0.14)} \quad \Rightarrow \quad a = \frac{\ln(0.5)}{-0.14} = 5$$

Consequently,

$$v(t) = \begin{cases} 0 & \text{for } t < 0 \\ 4e^{-5t} & \text{for } t \geq 0 \end{cases}$$

(b) Figure 8.3-12a shows the circuit immediately after the switch opens. In Figure 8.3-12b, the part of the circuit connected to the inductor has been replaced by its Thévenin equivalent circuit.
The time constant of the circuit is given by

$$\tau = \frac{L}{R_t} = \frac{4}{R + 5}$$

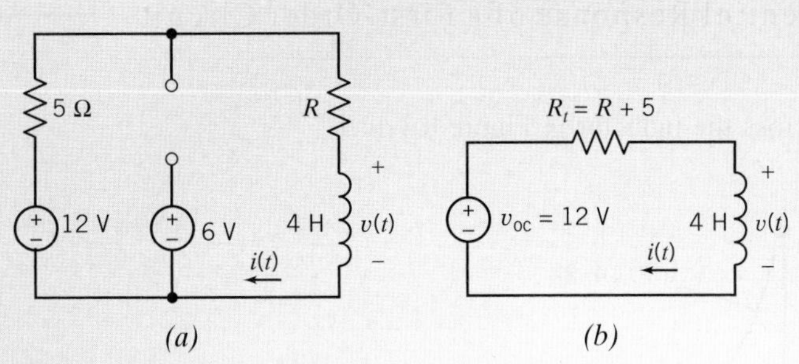

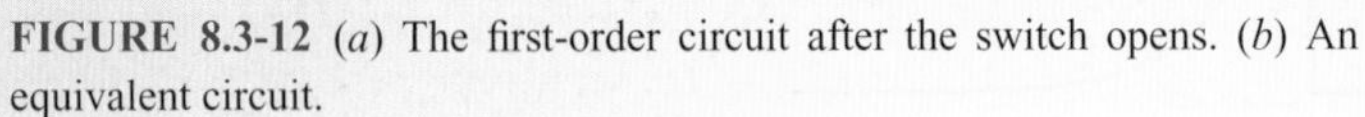
FIGURE 8.3-12 (*a*) The first-order circuit after the switch opens. (*b*) An equivalent circuit.

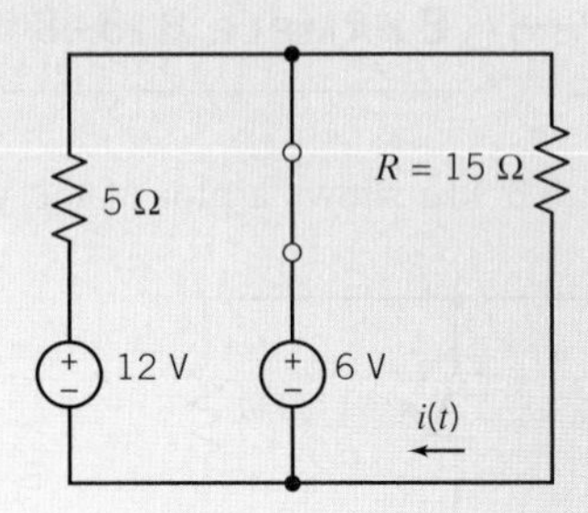

FIGURE 8.3-13 The first-order circuit before the switch opens.

Also, the time constant is related to the exponent in $v(t)$ by $-5t = -\frac{t}{\tau}$. Consequently,

$$5 = \frac{1}{\tau} = \frac{R+5}{4} \quad \Rightarrow \quad R = 15\,\Omega$$

(c) The inductor current is related to the inductor voltage by

$$i(t) = \frac{1}{L}\int_0^t v(\tau)d\tau + i(0)$$

Figure 8.3-13 shows the circuit before the switch opens. The closed switch is represented by a short circuit. The circuit is at steady state, and the voltage sources have constant voltages, so the inductor acts like a short circuit. The inductor current is given by

$$i(t) = \frac{6}{15} = 0.4\text{ A}$$

In particular, $i(0-) = 0.4$ A. The current in an inductor is continuous, so $i(0+) = i(0-)$. Consequently,

$$i(0) = 0.4\text{ A}$$

Returning to the equation for the inductor current, after the switch opens, we have

$$i(t) = \frac{1}{4}\int_0^t 4e^{-5\tau}d\tau + 0.4 = \frac{1}{-5}\left(e^{-5t} - 1\right) + 0.4 = 0.6 - 0.2e^{-5t}$$

In summary,

$$i(t) = \begin{cases} 0.4 & \text{for } t < 0 \\ 0.6 - 0.2e^{-5t} & \text{for } t \geq 0 \end{cases}$$

EXERCISE 8.3-1 The circuit shown in Figure E 8.3-1 is at steady state before the switch closes at time $t = 0$. Determine the capacitor voltage, $v(t)$, for $t \geq 0$.

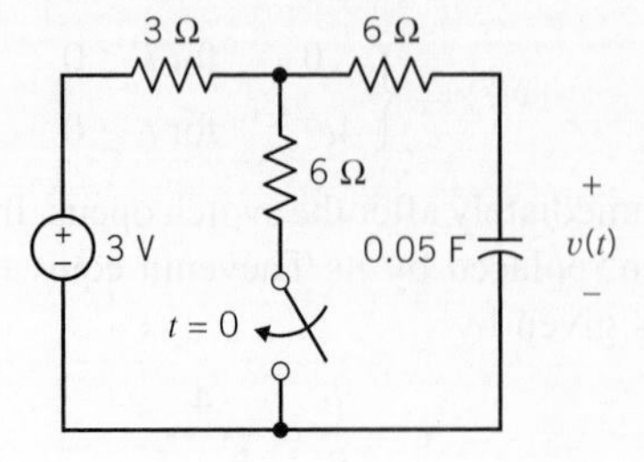

FIGURE E 8.3-1

Answer: $v(t) = 2 + e^{-2.5t}$ V for $t > 0$

EXERCISE 8.3-2 The circuit shown in Figure E 8.3-2 is at steady state before the switch closes at time $t = 0$. Determine the inductor current, $i(t)$, for $t > 0$.

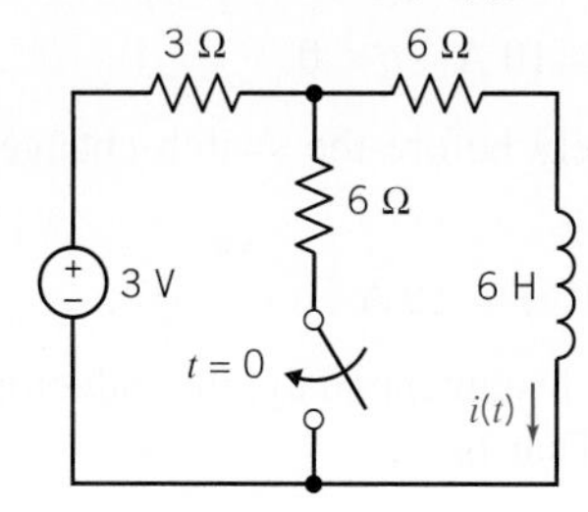

FIGURE E 8.3-2

Answer: $i(t) = \frac{1}{4} + \frac{1}{12}e^{-1.33t}$ A for $t > 0$

8.4 SEQUENTIAL SWITCHING

Often, circuits contain several switches that are not switched at the same time. For example, a circuit may have two switches where the first switch changes state at time $t = 0$ and the second switch closes at $t = 1$ ms.

Sequential switching occurs when a circuit contains two or more switches that change state at different instants.

Circuits with sequential switching can be solved using the methods described in the previous sections, based on the fact that inductor currents and capacitor voltages do not change instantaneously.

As an example of sequential switching, consider the circuit shown in Figure 8.4-1*a*. This circuit contains two switches—one that changes state at time $t = 0$ and a second that closes at $t = 1$ ms. Suppose this circuit has reached steady state before the switch changes state at time $t = 0$. Figure 8.4-1*b* shows the equivalent circuit that is appropriate for $t < 0$. Because the circuit is at steady state and the input is constant, the inductor acts like a short circuit and the current in this short circuit is the

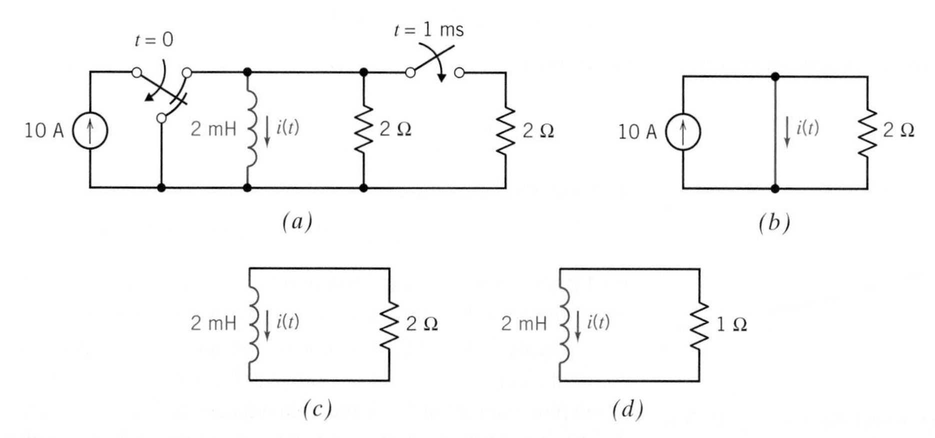

FIGURE 8.4-1 (*a*) A circuit with sequential switching. (*b*) The equivalent circuit before $t = 0$. (*c*) The equivalent circuit for $0 < t < 1$ ms. (*d*) The equivalent circuit after $t = 1$ ms.

inductor current. The short circuit forces the voltage across the resistor to be zero, so the current in the resistor is also zero. As a result, all of the source current flows in the short circuit and

$$i(t) = 10\text{ A} \quad t < 0$$

The inductor current will be 10 A immediately before the switch changes state at time $t = 0$. We express this as

$$i(0^-) = 10\text{ A}$$

Because the inductor current does not change instantaneously, the inductor current will also be 10 A immediately after the switch changes state. That is,

$$i(0^+) = 10\text{ A}$$

This is the initial condition that is used to calculate the inductor current after $t = 0$. Figure 8.4-1*c* shows the equivalent circuit that is appropriate after one switch changes state at time $t = 0$ and before the other switch closes at time $t = 1$ ms. We see that the Norton equivalent of the part of the circuit connected to the inductor has the parameters

$$I_{sc} = 0\text{ A} \quad \text{and} \quad R_t = 2\ \Omega$$

The time constant of this first-order circuit is

$$\tau = \frac{L}{R_t} = \frac{2 \times 10^{-3}}{2} = 1 \times 10^{-3} = 1\text{ ms}$$

The inductor current is

$$i(t) = i(0)e^{-t/\tau} = 10e^{-t}\text{ A}$$

for $0 < t < 1$ ms. Notice that t has units of ms. Immediately before the other switch closes at time $t = 1$ ms, the inductor current will be

$$i(1^-) = 10e^{-1} = 3.68\text{ A}$$

Because the inductor current does not change instantaneously, the inductor current will also be 3.68 A immediately after the switch changes state. That is,

$$i(1^+) = 3.68\text{ A}$$

This is the initial condition that is used to calculate the inductor current after the switch closes at time $t = 1$ ms. Figure 8.4-1*d* shows the appropriate equivalent circuit. We see that the Norton equivalent of the part of the circuit connected to the inductor has the parameters

$$I_{sc} = 0\text{ A} \quad \text{and} \quad R_t = 1\ \Omega$$

The time constant of this first-order circuit is

$$\tau = \frac{L}{R_t} = \frac{2 \times 10^{-3}}{1} = 2 \times 10^{-3} = 2\text{ ms}$$

The inductor current is

$$i(t) = i(t_0)e^{-(t-t_0)/\tau} = 3.68e^{-(t-1)/2}\text{ A}$$

for 1 ms $< t$. Once again, t has units of ms. Also, t_0 denotes the time when the switch changes state—1 ms in this example.

Figure 8.4-2 shows a plot of the inductor current. The time constant changes when the second switch closes. As a result, the slope of the plot changes at $t = 1$ ms. Immediately before the switch closes, the slope is -3.68 A/ms. Immediately after the switch closes, the slope becomes $-3.68/2$ A/ms.

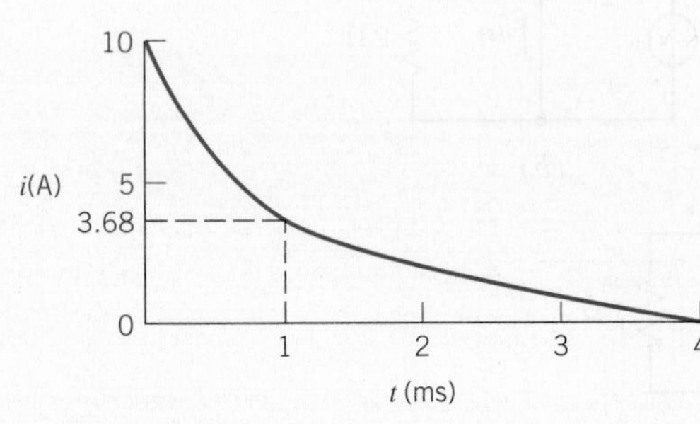

FIGURE 8.4-2 Current waveform for $t \geq 0$. The exponential has a different time constant for $0 \leq t < t_1$ and for $t \geq t_1$ where $t_1 = 1$ ms.

8.5 STABILITY OF FIRST-ORDER CIRCUITS

We have shown that the natural response of a first-order circuit is

$$x_n(t) = Ke^{-t/\tau}$$

and that the complete response is the sum of the natural and forced responses:

$$x(t) = x_n(t) + x_f(t)$$

When $\tau > 0$, the natural response vanishes as $t \to 0$, leaving the forced response. In this case, the circuit is said to be *stable*. When $\tau < 0$, the natural response grows without bound as $t \to 0$. The forced response becomes negligible, compared to the natural response. The circuit is said to be *unstable*. When a circuit is stable, the forced response depends on the input to the circuit. That means that the forced response contains information about the input. When the circuit is unstable, the forced response is negligible, and this information is lost. In practice, the natural response of an unstable circuit is not unbounded. This response will grow until something happens to change the circuit. Perhaps that change will be saturation of an op amp or of a dependent source. Perhaps that change will be the destruction of a circuit element. In most applications, the behavior of unstable circuits is undesirable and is to be avoided.

How can we design first-order circuits to be stable? Recalling that $\tau = R_t C$ or $\tau = L/R_t$, we see that

$R_t > 0$ is required to make a first-order circuit stable.

This condition will always be satisfied whenever the part of the circuit connected to the capacitor or inductor consists of only resistors and independent sources. Such circuits are guaranteed to be stable. In contrast, a first-order circuit that contains op amps or dependent sources may be unstable.

EXAMPLE 8.5-1 Response of an Unstable First-Order Circuit

The first-order circuit shown in Figure 8.5-1*a* is at steady state before the switch closes at $t = 0$. This circuit contains a dependent source and so may be unstable. Find the capacitor voltage, $v(t)$, for $t > 0$.

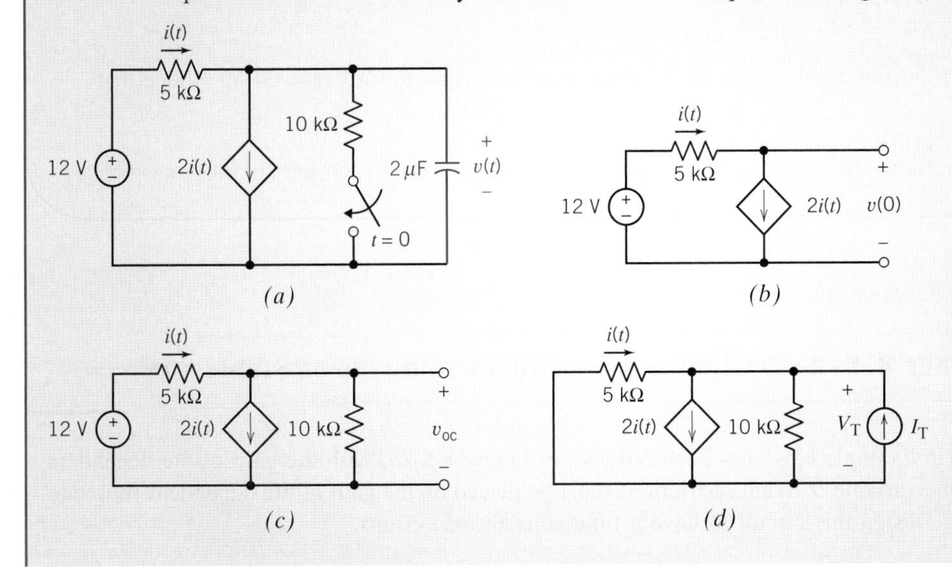

FIGURE 8.5-1 (*a*) A first-order circuit containing a dependent source. (*b*) The circuit used to calculate the initial condition. (*c*) The circuit used to calculate V_{oc}. (*d*) The circuit used to calculate R_t.

Solution

The input to the circuit is a constant, so the capacitor acts like an open circuit at steady state. We calculate the initial condition from the circuit in Figure 8.5-2*b*. Applying KCL to the top node of the dependent current source, we get

$$-i + 2i = 0$$

Therefore, $i = 0$. Consequently, there is no voltage drop across the resistor, and

$$v(0) = 12 \text{ V}$$

Next, we determine the Thévenin equivalent circuit for the part of the circuit connected to the capacitor. This requires two calculations. First, calculate the open-circuit voltage, using the circuit in Figure 8.5-1*c*. Writing a KVL equation for the loop consisting of the two resistors and the voltage source, we get

$$12 = \left(5 \times 10^3\right) \times i + \left(10 \times 10^3\right) \times (i - 2i)$$

Solving for the current, we find

$$i = -2.4 \text{ mA}$$

Applying Ohm's law to the 10-kΩ resistor, we get

$$V_{oc} = \left(10 \times 10^3\right) \times (i - 2i) = 24 \text{ V}$$

Now calculate the Thévenin resistance using the circuit shown in Figure 8.5-1*d*. Apply KVL to the loop consisting of the two resistors to get

$$0 = \left(5 \times 10^3\right) \times i + \left(10 \times 10^3\right) \times (I_T + i - 2i)$$

Solving for the current,

$$i = 2I_T$$

Applying Ohm's law to the 10-kΩ resistor, we get

$$V_T = 10 \times 10^3 \times (I_T + i - 2i) = -10 \times 10^3 \times I_T$$

The Thévenin resistance is given by

$$R_t = \frac{V_T}{I_T} = -10 \text{ k}\Omega$$

The time constant is

$$\tau = R_t C = -20 \text{ ms}$$

This circuit is unstable. The complete response is

$$v(t) = 24 - 12\, e^{t/20}$$

The capacitor voltage *decreases* from $v(0) = 12$ V rather than *increasing* toward $v_f = 24$ V. Notice that

$$v(\infty) = \lim_{t \to \infty} v(t) = -\infty$$

It's not appropriate to refer to the forced response as a steady-state response when the circuit is unstable.

EXAMPLE 8.5-2 Designing First-Order Circuits to be Stable

The circuit considered in Example 8.5-1 has been redrawn in Figure 8.5-2*a*, with the gain of the dependent source represented by the variable *B*. What restrictions must be placed on the gain of the dependent source to ensure that it is stable? Design this circuit to have a time constant of +20 ms.

Solution

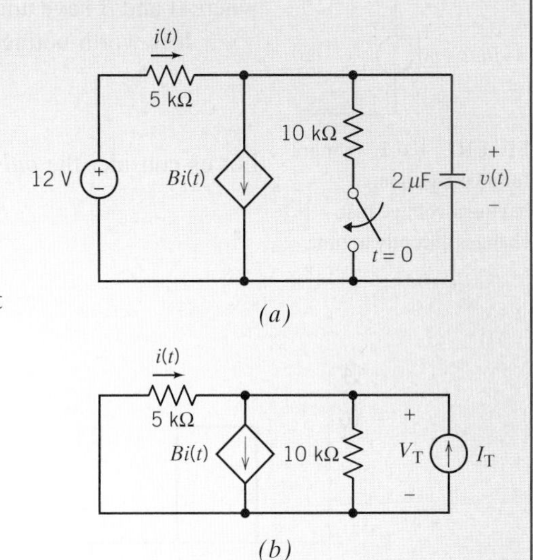

FIGURE 8.5-2 (*a*) A first-order circuit containing a dependent source. (*b*) The circuit used to calculate the Thévenin resistance of the part of the circuit connected to the capacitor.

Figure 8.5-2*b* shows the circuit used to calculate R_t. Applying KVL to the loop consisting of the two resistors,

$$5 \times 10^3 \times i + V_T = 0$$

Solving for the current gives

$$i = -\frac{V_T}{5 \times 10^3}$$

Applying KCL to the top node of the dependent source, we get

$$-i + Bi + \frac{V_T}{10 \times 10^3} - I_T = 0$$

Combining these equations, we get

$$\left(\frac{1 - B}{5 \times 10^3} + \frac{1}{10 \times 10^3}\right) V_T - I_T = 0$$

The Thévenin resistance is given by

$$R_t = \frac{V_T}{I_T} = -\frac{10 \times 10^3}{2B - 3}$$

The condition $B < 3/2$ is required to ensure that R_t is positive and the circuit is stable.

To obtain a time constant of +20 ms requires

$$R_t = \frac{\tau}{C} = \frac{20 \times 10^{-3}}{2 \times 10^{-6}} = 10 \times 10^3 = 10\,\text{k}\Omega$$

which in turn requires

$$10 \times 10^3 = -\frac{10 \times 10^3}{2B - 3}$$

Therefore $B = 1$. This suggests that we can fix the unstable circuit by decreasing the gain of the dependent source from 2 A/A to 1 A/A.

8.6 THE UNIT STEP SOURCE

The unit step function provides a convenient way to represent an abrupt change in a voltage or current.

We define the *unit step function* as a function of time that is zero for $t < t_0$ and unity for $t > t_0$. At $t = t_0$, the value changes from zero to one. We represent the unit step function by $u(t - t_0)$, where

$$u(t - t_0) = \begin{cases} 0 & t < t_0 \\ 1 & t > t_0 \end{cases} \tag{8.6-1}$$

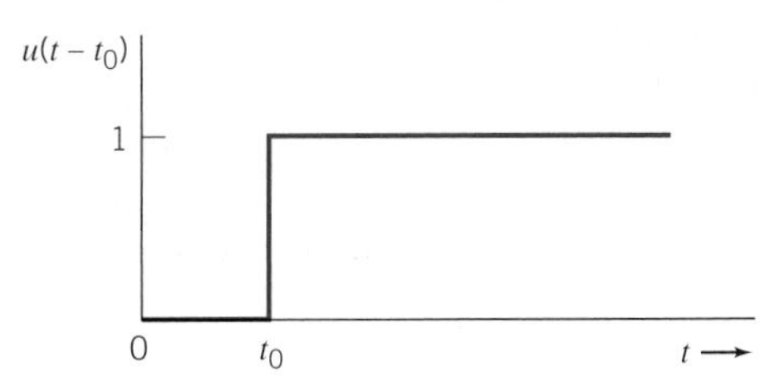

FIGURE 8.6-1 Unit step forcing function, $u(t - t_0)$.

The value of $u(t - t_0)$ is not defined at $t = t_0$, where it switches instantaneously from a value of zero to one. The unit step function is shown in Figure 8.6-1. We will often consider $t_0 = 0$.

The unit step function is dimensionless. To represent a voltage that changes abruptly from one constant value to another constant value at time $t = t_0$, we can write

$$v(t) = A + B\,u(t - t_0)$$

which indicates that

$$v(t) = \begin{cases} A & t < t_0 \\ A + B & t > t_0 \end{cases}$$

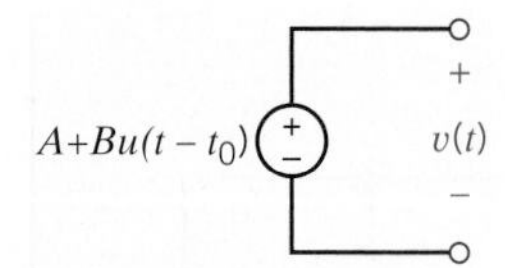

FIGURE 8.6-2 Symbol for a voltage source having a voltage that changes abruptly at time $t = t_0$

where A and B have units of Volt. Figure 8.6-2 shows a voltage source having this voltage.

It is worth noting that $u(-t)$ indicates that we have a value of 1 for $t < 0$, so that

$$u(-t) = \begin{cases} 1 & t < 0 \\ 0 & t > 0 \end{cases}$$

Let us consider the *pulse* source

$$v(t) = \begin{cases} 0 & t < t_0 \\ V_0 & t_0 < t < t_1 \\ 0 & t_1 < t \end{cases}$$

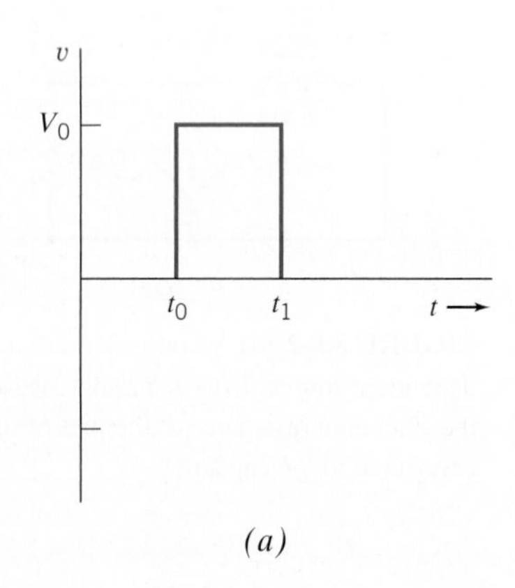

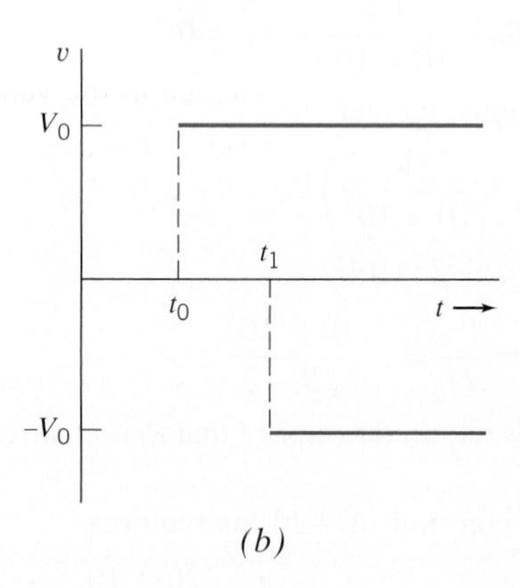

FIGURE 8.6-3 (*a*) Rectangular voltage pulse. (*b*) Two-step voltage waveforms that yield the voltage pulse.

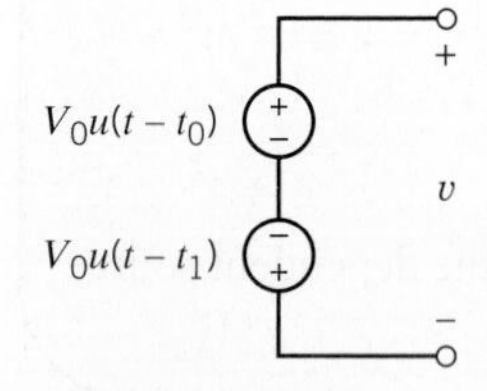

FIGURE 8.6-4 Two-step voltage sources that yield a rectangular voltage pulse, $v(t)$, with a magnitude of V_0 and a duration of $(t_1 - t_0)$ where $t_0 < t_1$.

which is shown in Figure 8.6-3*a*. As shown in Figure 8.6-3*b*, the pulse can be obtained from two-step voltage sources, the first of value V_0 occurring at $t = t_0$ and the second equal to $-V_0$ occurring at $t = t_1$. Thus, the two-step sources of magnitude V_0 shown in Figure 8.6-4 will yield the desired pulse. We have $v(t) = V_0u(t - t_0) - V_0u(t - t_1)$ to provide the pulse. Notice how easy it is to use two-step function symbols to represent this pulse source. The pulse is said to have a duration of $(t_1 - t_0)$ s.

A **puise signal** has a constant nonzero value for a time duration of $\Delta_t = t_1 - t_0$.

We recognize that the unit step function is an ideal model. No real element can switch instantaneously. However, if the switching time is very short compared to the time constant of the circuit, we can approximate the switching as instantaneous.

EXAMPLE 8.6-1 First-Order Circuit

INTERACTIVE EXAMPLE

Figure 8.6-5 shows a first-order circuit. The input to the circuit is the voltage of the voltage source, $v_s(t)$. The output is the current of the inductor, $i_o(t)$. Determine the output of this circuit when the input is $v_s(t) = 4 - 8u(t)$ V.

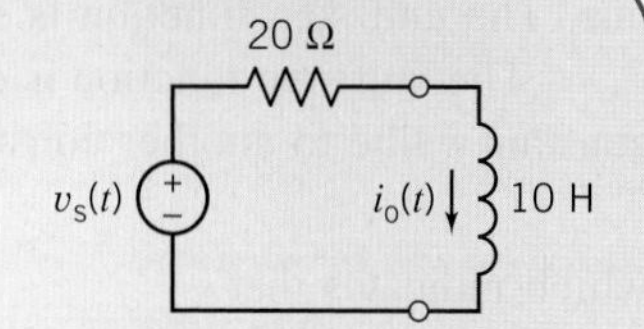

FIGURE 8.6-5 The circuit considered in Example 8.6-1.

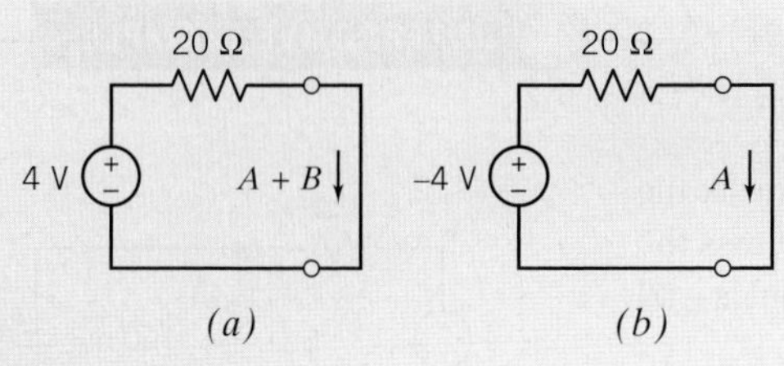

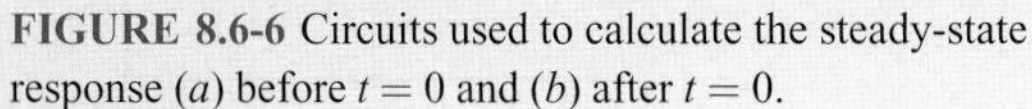

FIGURE 8.6-6 Circuits used to calculate the steady-state response (a) before $t = 0$ and (b) after $t = 0$.

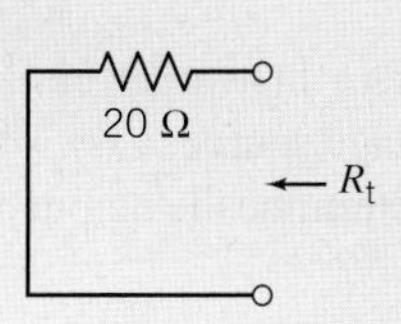

FIGURE 8.6-7 The circuit used to calculate R_t.

Solution

The value of the input is one constant, 4 V, before time $t = 0$ and a different constant, −4 V, after time $t = 0$. The response of the first-order circuit to the change in the value of the input will be

$$i_o(t) = A + Be^{-at} \quad \text{for} \quad t > 0 \tag{8.6-2}$$

where the values of the three constants A, B, and a are to be determined.

The values of A and B are determined from the steady-state responses of this circuit before and after the input changes value. Figures 8.6-6a,b show the circuits used to calculate those steady-state responses. Figures 8.6-6a,b require some explanation.

Inductors act like short circuits when the input is constant and the circuit is at steady state. Consequently, the inductor is replaced by a short circuit in Figure 8.6-6a and in Figure 8.6-6b.

The value of the inductor current at time $t = 0$ will be equal to the steady-state inductor current before the input changes. At time $t = 0$, the output current is

$$i_o(0) = A + Be^{-a(0)} = A + B$$

Consequently, the inductor current is labeled as $A + B$ in Figure 8.6-6a.

The value of the inductor current at time $t = \infty$ will be equal to the steady-state inductor current after the input changes. At time $t = \infty$, the output current is

$$i_o(\infty) = A + Be^{-a(\infty)} = A$$

Consequently, the inductor current is labeled as A in Figure 8.6-6b.

Analysis of the circuit in Figure 8.6-6a gives

$$A + B = 0.2 \text{ A}$$

Analysis of the circuit in Figure 8.6-6b gives

$$A = -0.2 \text{ A}$$

Therefore,

$$B = 0.4 \text{ A}$$

The value of the constant a in Eq. 8.6-2 is determined from the time constant, τ, which in turn is calculated from the values of the inductance L and of the Thévenin resistance, R_t, of the circuit connected to the inductor.

$$\frac{1}{a} = \tau = \frac{L}{R_t}$$

Figure 8.6-7 shows the circuit used to calculate R_t. It is seen from Figure 8.6-7 that

$$R_t = 20\ \Omega$$

Therefore,

$$a = \frac{20}{10} = 2\frac{1}{\text{s}}$$

(The time constant is $\tau = 10/20 = 0.5$ s.) Substituting the values of A, B, and a into Eq. 8.6-2 gives

$$i_o(t) = \begin{cases} 0.2 \text{ A} & \text{for } t \le 0 \\ -0.2 + 0.4\,e^{-2t} \text{ A} & \text{for } t \ge 0 \end{cases}$$

EXAMPLE 8.6-2 First-Order Circuit

INTERACTIVE EXAMPLE

Figure 8.6-8 shows a first-order circuit. The input to the circuit is the voltage of the voltage source, $v_s(t)$. The output is the voltage across the capacitor, $v_o(t)$. Determine the output of this circuit when the input is $v_s(t) = 7 - 14u(t)$ V.

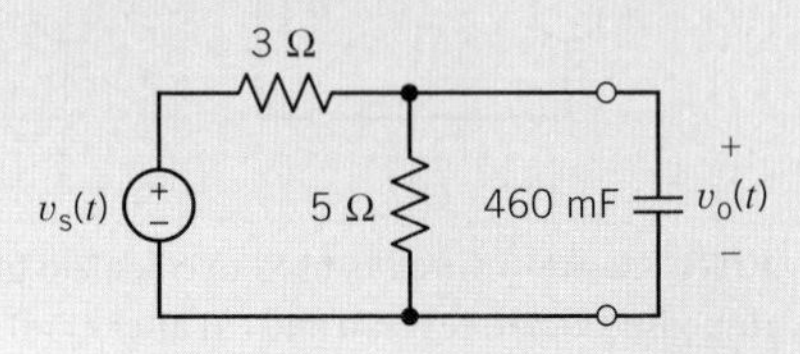

FIGURE 8.6-8 The circuit considered in Example 8.6-2.

Solution

The value of the input is one constant, 7 V, before time $t = 0$ and a different constant, -7 V, after time $t = 0$. The response of the first-order circuit to the change in the value of the input will be

$$v_o(t) = A + Be^{-at} \quad \text{for} \quad t > 0 \tag{8.6-3}$$

where the values of the three constants A, B, and a are to be determined.

The values of A and B are determined from the steady-state responses of this circuit before and after the input changes value. Figures 8.6-9*a*, *b* show the circuits used to calculate those steady-state responses. Figures 8.6-9*a*, *b* require some explanation.

Capacitors act like open circuits when the input is constant and the circuit is at steady state. Consequently, the capacitor is replaced by an open circuit in Figure 8.6-9*a* and in Figure 8.6-9*b*.

The value of the capacitor voltage at time $t = 0$ will be equal to the steady-state capacitor voltage before the input changes. At time $t = 0$, the output voltage is

$$v_o(0) = A + Be^{-a(0)} = A + B$$

Consequently, the capacitor voltage is labeled as $A + B$ in Figure 8.6-9*a*.

The value of the capacitor voltage at time $t = \infty$ will be equal to the steady-state capacitor voltage after the input changes. At time $t = \infty$, the output voltage is

$$v_o(\infty) = A + Be^{-a(\infty)} = A$$

Consequently, the capacitor voltage is labeled as A in Figure 8.6-9*b*.

Apply the voltage division rule to the circuit in Figure 8.6-9*a* to get

$$A + B = \frac{5}{3+5} \times 7 = 4.38 \text{ V}$$

Apply the voltage division rule to the circuit in Figure 8.6-9*b* to get

$$A = \frac{5}{3+5} \times (-7) = -4.38 \text{ V}$$

Therefore,

$$B = 8.76 \text{ V}$$

The value of the constant a in Eq. 8.6-3 is determined from the time constant, τ, which in turn is calculated from the values of the capacitance C and of the Thévenin resistance, R_t, of the circuit connected to

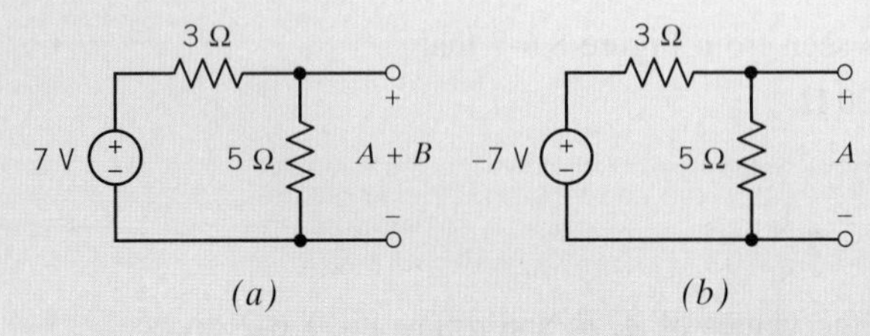

FIGURE 8.6-9 Circuits used to calculate the steady-state response (*a*) before $t = 0$ and (*b*) after $t = 0$.

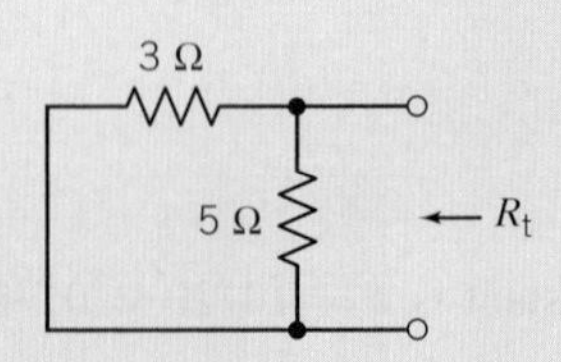

FIGURE 8.6-10 The circuit used to calculate R_t.

the capacitor:

$$\frac{1}{a} = \tau = R_t C$$

Figure 8.6-10 shows the circuit used to calculate R_t. It is seen from Figure 8.6-10 that

$$R_t = \frac{(5)(3)}{5+3} = 1.875\ \Omega$$

Therefore,

$$a = \frac{1}{(1.875)(460 \times 10^{-3})} = 1.16 \frac{1}{\text{s}}$$

(The time constant is $\tau = (1.875)(460 \times 10^{-3}) = 0.86$ s.) Substituting the values of A, B, and a into Eq. 8.6-3 gives

$$v_o(t) = \begin{cases} -4.38\ \text{V} & \text{for } t \le 0 \\ -4.38 + 8.76\, e^{-1.16\, t}\ \text{V} & \text{for } t \ge 0 \end{cases}$$

8.7 THE RESPONSE OF A FIRST-ORDER CIRCUIT TO A NONCONSTANT SOURCE

In the previous sections, we wisely used the fact that the forced response to a constant source will be a constant itself. It now remains to determine what the response will be when the forcing function is not a constant.

The differential equation described by an *RL* or *RC* circuit is represented by the general form

$$\frac{dx(t)}{dt} + ax(t) = y(t) \tag{8.7-1}$$

where $y(t)$ is a constant only when we have a constant-current or constant-voltage source and where $a = 1/\tau$ is the reciprocal of the time constant.

In this section, we introduce the *integrating factor method*, which consists of multiplying Eq. 8.7-1 by a factor that makes the left-hand side a perfect derivative, and then integrating both sides.

Consider the derivative of a product of two terms such that

$$\frac{d}{dt}(xe^{at}) = \frac{dx}{dt}e^{at} + axe^{at} = \left(\frac{dx}{dt} + ax\right)e^{at} \tag{8.7-2}$$

The term within the parentheses on the right-hand side of Eq. 8.7-2 is exactly the form on the left-hand side of Eq. 8.7-1.

Therefore, if we multiply both sides of Eq. 8.7-1 by e^{at}, the left-hand side of the equation can be represented by the perfect derivative, $d(xe^{at})/dt$. Carrying out these steps, we show that

$$\left(\frac{dx}{dt} + ax\right)e^{at} = ye^{at}$$

or

$$\frac{d}{dt}(xe^{at}) = ye^{at}$$

Integrating both sides of the second equation, we have

$$xe^{at} = \int ye^{at}\,dt + K$$

where K is a constant of integration. Therefore, solving for $x(t)$, we multiply by e^{-at} to obtain

$$x = e^{-at}\int ye^{at}\,dt + Ke^{-at} \tag{8.7-3}$$

When the source is a constant so that $y(t) = M$, we have

$$x = e^{-at}M \int e^{at}dt + Ke^{-at} = \frac{M}{a} + Ke^{-at} = x_f + x_n$$

where the natural response is $x_n = Ke^{-at}$ and the forced response is $x_f = M/a$, a constant.

Now consider the case in which $y(t)$, the forcing function, is not a constant. Considering Eq. 8.7-3, we see that the natural response remains $x_n = Ke^{-at}$. However, the forced response is

$$x_f = e^{-at} \int y(t)e^{at}\, dt$$

Thus, the forced response will be dictated by the form of $y(t)$. Let us consider the case in which $y(t)$ is an exponential function so that $y(t) = e^{bt}$. We assume that $(a + b)$ is not equal to zero. Then we have

$$x_f = e^{-at} \int e^{bt}e^{at}dt = e^{-at} \int e^{(a+b)t}dt = \frac{1}{a+b}e^{-at}e^{(a+b)} = \frac{e^{bt}}{a+b} \tag{8.7-4}$$

Therefore, the forced response of an *RL* or *RC* circuit to an exponential forcing function is of the same form as the forcing function itself. When $a + b$ is not equal to zero, we assume that the forced response will be of the same form as the forcing function itself, and we try to obtain the relationship that will be satisfied under those conditions.

EXAMPLE 8.7-1 First-Order Circuit with Nonconstant Source

Find the current i for the circuit of Figure 8.7-1*a* for $t > 0$ when

$$v_s = 10e^{-2t}u(t)\text{ V}$$

Assume the circuit is in steady state at $t = 0^-$.

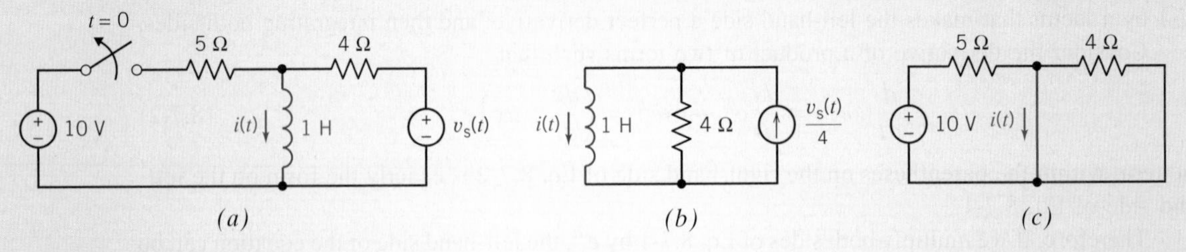

FIGURE 8.7-1 (*a*) A circuit with a nonconstant source, (*b*) the appropriate equivalent circuit after the switch opens, and (*c*) the appropriate equivalent circuit before the switch opens.

Solution

Because the forcing function is an exponential, we expect an exponential for the forced response, i_f. Therefore, we expect i_f to be

$$i_f = Be^{-2t}$$

for $t \geq 0$. Writing KVL around the right-hand mesh, we have

$$L\frac{di}{dt} + Ri = v_s$$

or

$$\frac{di}{dt} + 4i = 10e^{-2t}$$

for $t > 0$. Substituting $i_f = Be^{-2t}$, we have

$$-2Be^{-2t} + 4Be^{-2t} = 10e^{-2t}$$

or

$$(-2B + 4B)e^{-2t} = 10e^{-2t}$$

Hence, $B = 5$ and

$$i_f = 5e^{-2t}$$

The natural response can be obtained by considering the circuit shown in Figure 8.7-1*b*. This is the equivalent circuit that is appropriate after the switch has opened. The part of the circuit that is connected to the inductor has been replaced by its Norton equivalent circuit. The natural response is

$$i_n = Ae^{-(R_t/L)t} = Ae^{-4t}$$

The complete response is

$$i = i_n + i_f = Ae^{-4t} + 5e^{-2t}$$

The constant A can be determined from the value of the inductor current at time $t = 0$. The initial inductor current, $i(0)$, can be obtained by considering the circuit shown in Figure 8.7-1*c*. This is the equivalent circuit that is appropriate before the switch opens. Because $v_s(t) = 0$ for $t < 0$ and a zero voltage source is a short circuit, the voltage source at the right side of the circuit has been replaced by a short circuit. Also, because the circuit is at steady state before the switch opens and the only input is the constant 10-volt source, the inductor acts like a short circuit. The current in the short circuit that replaces the inductor is the initial condition, $i(0)$. From Figure 8.7-1*c*,

$$i(0) = \frac{10}{5} = 2\text{ A}$$

Therefore, at $t = 0$,

$$i(0) = Ae^{-4\times 0} + 5e^{-2\times 0} = A + 5$$

or

$$2 = A + 5$$

or $A = -3$. Therefore,

$$i = \left(-3e^{-4t} + 5e^{-2t}\right)\text{ A} \quad t > 0$$

The voltage source of Example 8.7-1 is a decaying exponential of the form

$$v_s = 10e^{-2t}u(t)\text{ V}$$

This source is said to be *aperiodic* (nonperiodic). A periodic source is one that repeats itself exactly after a fixed length of time. Thus, the signal $f(t)$ is *periodic* if there is a number T such that for all t

$$f(t + T) = f(t) \tag{8.7-5}$$

The smallest positive number T that satisfies Eq. 8.7-5 is called the *period*. The period defines the duration of one complete cycle of $f(t)$. Thus, any source for which there is no value of T satisfying Eq. 8.7-5 is said to be aperiodic. An example of a periodic source is 10 sin $2t$, which we consider in Example 8.7-2. The period of this sinusoidal source is π s.

EXAMPLE 8.7-2 First-Order Circuit with Nonconstant Source

Find the response $v(t)$ for $t > 0$ for the circuit of Figure 8.7-2*a*. The initial voltage $v(0) = 0$, and the current source is $i_s = (10 \sin 2t)u(t)$ A.

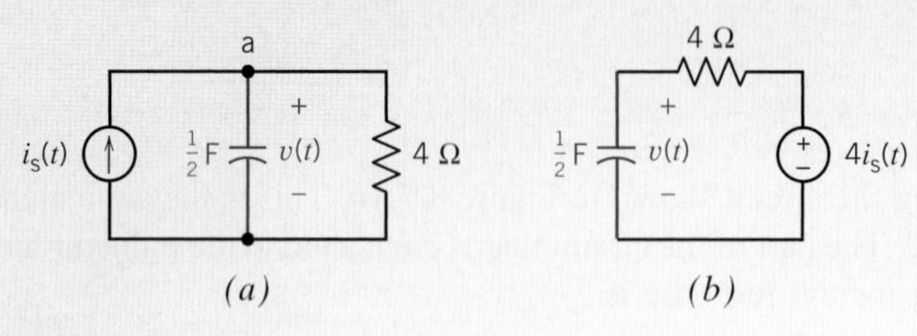

FIGURE 8.7-2 (*a*) A circuit with a nonconstant source. (*b*) The equivalent circuit for $t > 0$.

Solution

Because the forcing function is a sinusoidal function, we expect that v_f is of the same form. Writing KCL at node a, we obtain

$$C\frac{dv}{dt} + \frac{v}{R} = i_s$$

or

$$0.5\frac{dv}{dt} + \frac{v}{4} = 10 \sin 2t \tag{8.7-6}$$

for $t > 0$. We assume that v_f will consist of the sinusoidal function $\sin 2t$ and its derivatives.

Examining Eq. 8.7-6, $v_f/4$ plus $0.5\, dv_f/dt$ must equal $10 \sin 2t$. However, $d(\sin 2t)/dt = 2 \cos 2t$. Therefore, the trial v_f needs to contain both $\sin 2t$ and $\cos 2t$ terms. Thus, we try the proposed solution

$$v_f = A \sin 2t + B \cos 2t$$

The derivative of v_f is then

$$\frac{dv_f}{dt} = 2A \cos 2t - 2B \sin 2t$$

Substituting v_f and dv_f/dt into Eq. 8.7-6, we obtain

$$(A \cos 2t - B \sin 2t) + \frac{1}{4}(A \sin 2t + B \cos 2t) = 10 \sin 2t$$

Therefore, equating $\sin 2t$ terms and $\cos 2t$ terms, we obtain

$$\left(\frac{A}{4} - B\right) = 10 \quad \text{and} \quad \left(A + \frac{B}{4}\right) = 0$$

Solving for A and B, we obtain

$$A = \frac{40}{17} \quad \text{and} \quad B = \frac{-160}{17}$$

Consequently,

$$v_f = \frac{40}{17} \sin 2t - \frac{160}{17} \cos 2t$$

It is necessary that v_f be made up of $\sin 2t$ and $\cos 2t$ because the solution has to satisfy the differential equation. Of course, the derivative of $\sin \omega t$ is $\omega \cos \omega t$.

The natural response can be obtained by considering the circuit shown in Figure 8.7-2*b*. This is the equivalent circuit that is appropriate for $t > 0$. The part of the circuit connected to the capacitor has been replaced by its Thévenin equivalent circuit. The natural response is

$$v_n = De^{-t/(R_t C)} = De^{-t/2}$$

The complete response is then

$$v = v_n + v_f = De^{-t/2} + \frac{40}{17}\sin 2t - \frac{160}{17}\cos 2t$$

Because $v(0) = 0$, we obtain at $t = 0$

$$0 = D - \frac{160}{17}$$

or

$$D = \frac{160}{17}$$

Then the complete response is

$$v = \left(\frac{160}{17}e^{-t/2} + \frac{40}{17}\sin 2t - \frac{160}{17}\cos 2t\right)\text{V}$$

Table 8.7-1 Forced Response to a Forcing Function

FORCING FUNCTION, $y(t)$	FORCED RESPONSE, $x_f(t)$
1. Constant	
$y(t) = M$	$x_f = N$, a constant
2. Exponential	
$y(t) = Me^{-bt}$	$x_f = Ne^{-bt}$
3. Sinusoid	
$y(t) = M\sin(\omega t + \theta)$	$x_f = A\sin\omega t + B\cos\omega t$

A special case for the forced response of a circuit may occur when the forcing function is a damped exponential when we have $y(t) = e^{-bt}$. Referring back to Eq. 8.7-4, we can show that

$$x_f = \frac{e^{-bt}}{a - b}$$

when $y(t) = e^{-bt}$. Note that here we have e^{-bt} whereas we used e^{bt} for Eq. 8.7-4. For the special case when $a = b$, we have $a - b = 0$, and this form of the response is indeterminate. For the special case, we must use $x_f = te^{-bt}$ as the forced response. The solution, x_f, for the forced response when $a = b$ will satisfy the original differential Eq. (8.7-1). Thus, when the natural response already contains a term of the same form as the forcing function, we need to multiply the assumed form of the forced response by t.

The forced response to selected forcing functions is summarized in Table 8.7-1. We note that if a circuit is linear, at steady state, and excited by a single sinusoidal source having frequency ω, then all the element currents and voltages are sinusoids having frequency ω.

EXERCISE 8.7-1 The electrical power plant for the orbiting space station shown in Figure E 8.7-1*a* uses photovoltaic cells to store energy in batteries. The charging circuit is modeled by the circuit shown in Figure E 8.7-1*b*, where $v_s = 10\sin 20t$ V. If $v(0^-) = 0$, find $v(t)$ for $t > 0$.

(*a*)

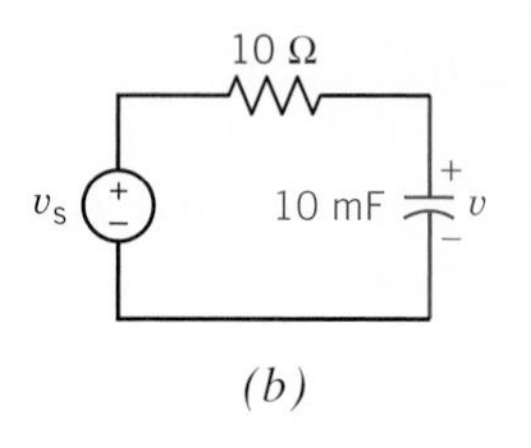

(*b*)

FIGURE E 8.7-1 (*a*) The NASA space station design shows the longer habitable modules that would house an orbiting scientific laboratory. (*b*) The circuit for energy storage for the laboratories. Photograph courtesy of the National Aeronautics and Space Administration.

Answer: $v = 4\,e^{-10t} - 4\cos 20t + 2\sin 20t$ V

8.8 DIFFERENTIAL OPERATORS

In this section, we introduce the differential operator, s.

An **operator** is a symbol that represents a mathematical operation. We can define a differential *operator* s such that

$$sx = \frac{dx}{dt} \quad \text{and} \quad s^2x = \frac{d^2x}{dt^2}$$

Thus, the operator s denotes differentiation of the variable with respect to time. The utility of the operator s is that it can be treated as an algebraic quantity. This permits the replacement of differential equations with algebraic equations, which are easily handled.

Use of the s operator is particularly attractive when higher-order differential equations are involved. Then we use the s operator, so that

$$s^n x = \frac{d^n x}{dt^n} \quad \text{for } n \geq 0$$

We assume that $n = 0$ represents no differentiation, so that

$$s^0 = 1$$

which implies $s^0 x = x$.

Because integration is the inverse of differentiation, we define

$$\frac{1}{s}x = \int_{-\infty}^{t} x\,d\tau \tag{8.8-1}$$

The operator $1/s$ must be shown to satisfy the usual rules of algebraic manipulations. Of these rules, the commutative multiplication property presents the only difficulty. Thus, we require

$$s \cdot \frac{1}{s} = \frac{1}{s} \cdot s = 1 \tag{8.8-2}$$

Is this true for the operator s? First, we examine Eq. 8.8-1. Multiplying Eq. 8.8-1 by s yields

$$s \cdot \frac{1}{s}x = \frac{d}{dt}\int_{-\infty}^{t} x\,d\tau$$

or

$$x = x$$

as required. Now we try the reverse order by multiplying sx by the integration operator to obtain

$$\frac{1}{s}sx = \int_{-\infty}^{t} \frac{dx}{d\tau}\,d\tau = x(t) - x(-\infty)$$

Therefore,

$$\frac{1}{s}sx = x$$

only when $x(-\infty) = 0$. From a physical point of view, we require that all capacitor voltages and inductor currents be zero at $t = -\infty$. Then the operator $1/s$ can be said to satisfy Eq. 8.8-2 and can be manipulated as an ordinary algebraic quantity.

Differential operators can be used to find the natural solution of a differential equation. For example, consider the first-order differential equation

$$\frac{d}{dt}x(t) + ax(t) = by(t) \tag{8.8-3}$$

The natural solution of this differential equation is

$$x_n(t) = Ke^{st} \tag{8.8-4}$$

The homogeneous form of a differential equation is obtained by setting the forcing function equal to zero. The forcing function in Eq. 8.8-3 is $y(t)$. The homogeneous form of this equation is

$$\frac{d}{dt}x(t) + ax(t) = 0 \tag{8.8-5}$$

To see that $x_n(t)$ is a solution of the homogeneous form of the differential equation, we substitute Eq. 8.8-4 into Eq. 8.8-5.

$$\frac{d}{dt}(Ke^{st}) + a(Ke^{st}) = sKe^{st} + aKe^{st} = 0$$

To obtain the parameter s in Eq. 8.8-4, replace d/dt in Eq. 8.8-5 by the differential operator s. This results in

$$sx + ax = (s + a)x = 0 \tag{8.8-6}$$

This equation has two solutions: $x = 0$ and $s = -a$. The solution $x = 0$ isn't useful, so we use the solution $s = -a$. Substituting this solution into Eq. 8.8-4 gives

$$x_n(t) = Ke^{-at}$$

This is the same expression for the natural response that we obtained earlier in this chapter by other methods. That's reassuring but not new. Differential operators will be quite useful when we analyze circuits that are represented by second- and higher-order differential equations.

As a second application of differential operators, consider using the computer program MATLAB to find the complete response of a first-order circuit. Differential operators are used to describe differential equations to MATLAB. As an example, consider the circuit shown in Figure 8.8-1*a*. To represent this circuit by a differential equation, apply KVL to get

$$10 \times 10^3 \left(1 \times 10^{-6} \frac{d}{dt}v(t)\right) + v(t) - 4\cos(100t) = 0$$

or

$$0.01\frac{d}{dt}v(t) + v(t) = 4\cos(100t) \tag{8.8-7}$$

In the syntax used by MATLAB, the differential operator is represented by D instead of s. Replace d/dt in Eq. 8.8-7 by the differential operator D to get

$$0.01\,Dv + v = 4\cos(100t)$$

Entering the MATLAB commands

```
v = dsolve('0.01*Dv + v = 4*cos(100*t)', 'v(0) = -8')
ezplot(v, [0, 2])
```

tells MATLAB to solve the differential equation using the initial condition $v(0) = -8$ volts and then plot the result. (The function named dsolve determines the symbolic solution of ordinary differential equations. This function is provided with the student edition of version 4 of MATLAB.) MATLAB

10 kΩ

4 cos (100t) 1 μF $v(t)$

(a)

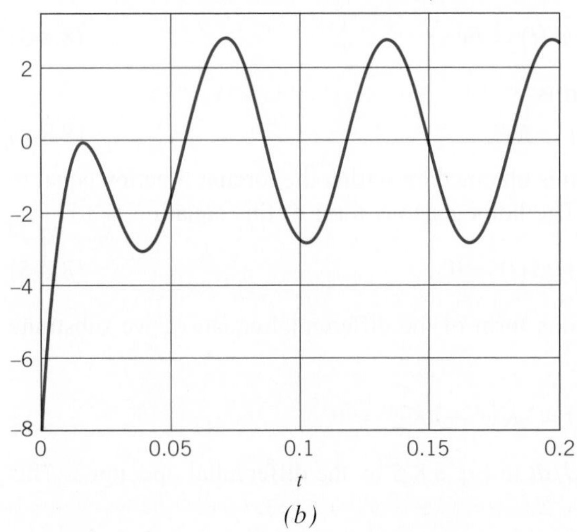

(b)

FIGURE 8.8-1 (a) A first-order circuit with a sinusoidal input and (b) a plot of its complete response produced using MATLAB.

responds by providing the complete solution of the differential equation

$$v = 2.^{*}\cos(100^{*}t) + 2.^{*}\sin(100^{*}t) - 10.^{*}\exp(-100.^{*}t)$$

and the plot of $v(t)$ versus t shown in Figure 8.8-1b.

8.9 USING PSPICE TO ANALYZE FIRST-ORDER CIRCUITS

To use PSpice to analyze a first-order circuit, we do the following:

1. Draw the circuit in the OrCAD Capture workspace
2. Specify a Time Domain (Transient) simulation
3. Run the simulation
4. Plot the simulation results

Time domain analysis is most interesting for circuits that contain capacitors or inductors or both. PSpice provides parts representing capacitors and inductors in the ANALOG parts library. The part name for the capacitor is C. The part properties that are of the most interest are the capacitance and the initial condition, both of which are specified using the OrCAD Capture property editor. (The initial condition of a capacitor is the value of the capacitor voltage at time $t = 0$.) The part name for the inductor is L. The inductance and the initial condition of the inductor are specified using the property editor. (The initial condition of an inductor is the value of the inductor current at time $t = 0$.)

The voltage and current sources that represent time-varying inputs are provided in the SOURCE parts library. Table 8.9-1 summarizes these voltage sources. The voltage waveform describes the

Table 8.9-1 PSpice Voltage Sources for Transient Response Simulations

NAME	SYMBOL	VOLTAGE WAVEFORM
VEXP	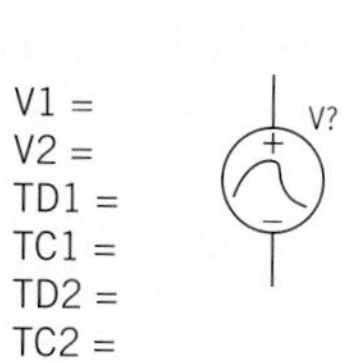	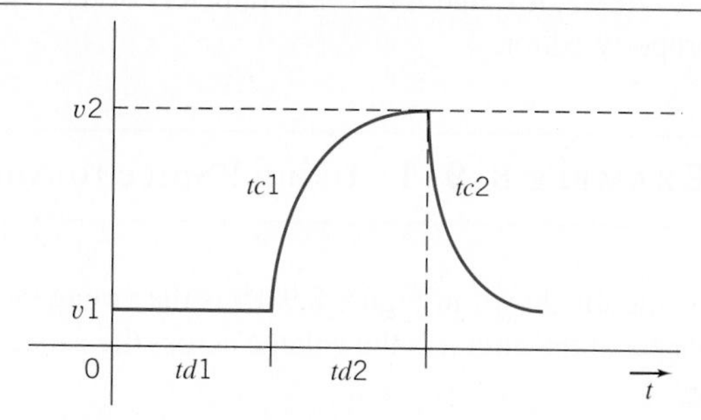
VPULSE	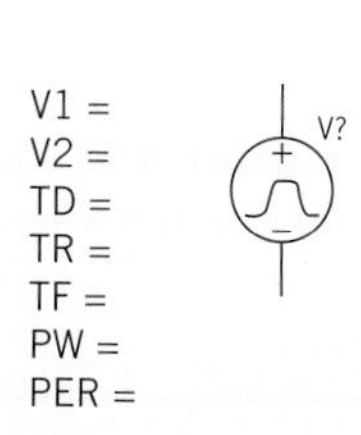	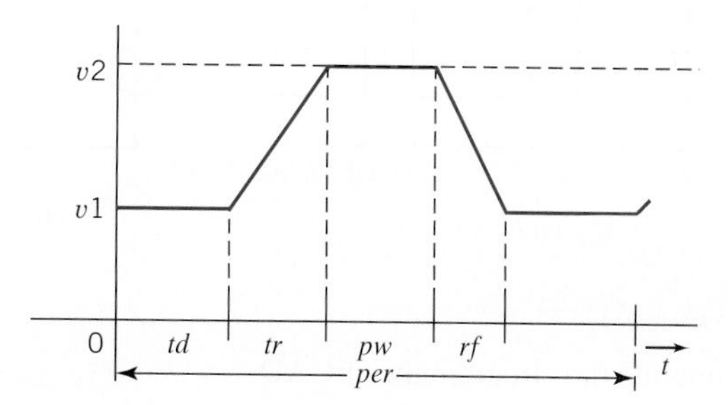
VPWL	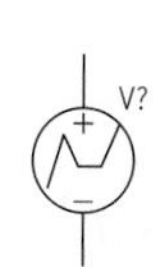	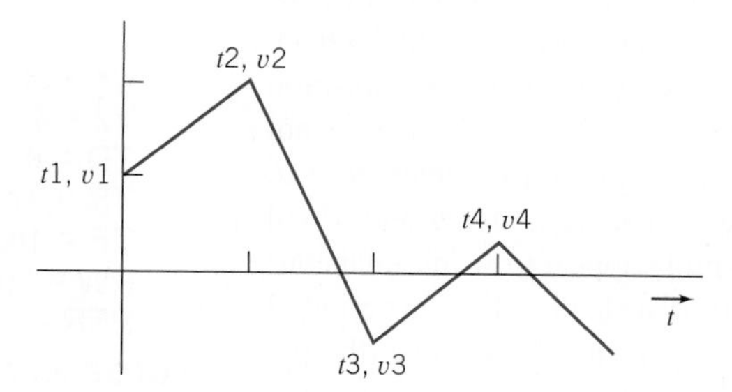
VSIN	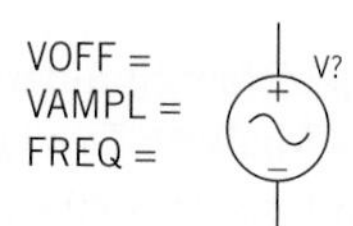	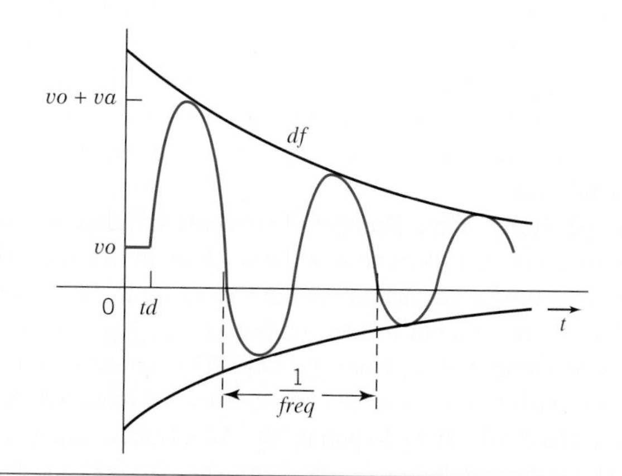

shape of the voltage source voltage as a function of time. Each voltage waveform is described using a series of parameters. For example, the voltage of an exponential source, VEXP, is described using *v*l, *v*2, *td*l, *td*2, *tc*1, and *tc*2. The parameters of the voltage sources in Table 8.9-1 are specified using the property editor.

EXAMPLE 8.9-1 Using PSpice to Analyze First-Order Circuits

The input to the circuit shown in Figure 8.9-1*a* is the voltage source voltage, $v_i(t)$, shown in Figure 8.9-l*a*. The output, or response, of the circuit is the voltage across the capacitor, $v_o(t)$. Use PSpice to plot the response of this circuit.

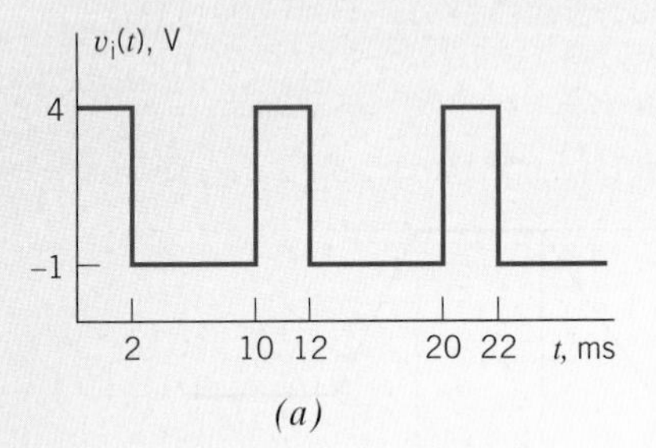

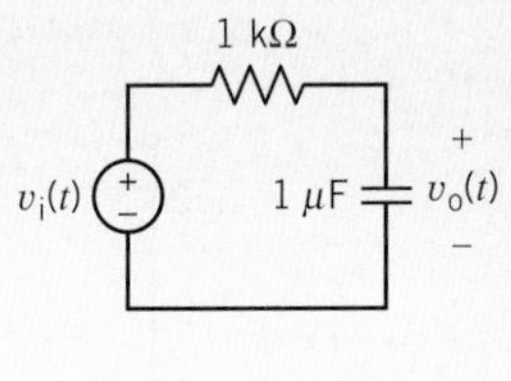

FIGURE 8.9-1 An *RC* circuit (*b*) with a pulse input (*a*).

Solution

We begin by drawing the circuit in the OrCAD workspace as shown in Figure 8.9-2 (see Appendix A). The voltage source is a VPULSE part (see the second row of Table 8.9-1). Figure 8.9-l*a* shows $v_i(t)$ making the transition from −1 V to 4 V instantaneously. Zero is not an acceptable value for the parameters *tr* or *tf*. Choosing a very small value for *tr* and *tf* will make the transitions appear to be instantaneous when using a time scale that shows a period of the input waveform. In this example, the period of the input waveform is 10 ms, so 1 ns is a reasonable choice for the values of *tr* and *tf*.

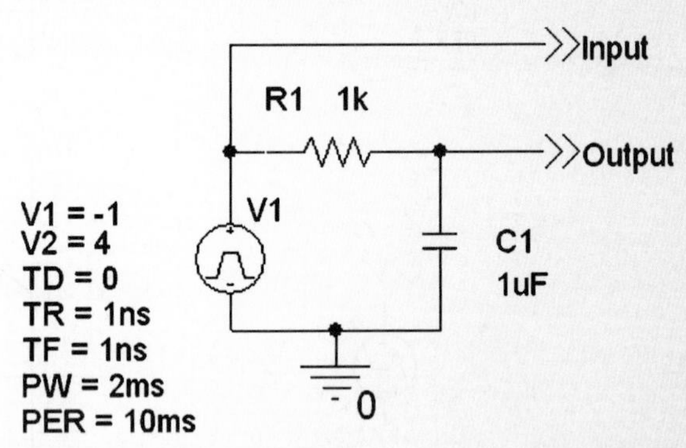

FIGURE 8.9-2 The circuit of Figure 8.9-1 as drawn in the OrCAD workspace.

It's convenient to set *td*, the delay before the periodic part of the waveform, to zero. Then the values of *v*l and *v*2 are −1 and 4, respectively. The value of *pw* is the length of time that $v_i(t) = v2 = 4$ V, so $pw = 2$ ms in this example. The pulse input is a periodic function of time. The value of *per* is the period of the pulse function, 10 ms.

The circuit shown in Figure 8.9-1*b* does not have a ground node. PSpice requires that all circuits have a ground node, so it is necessary to select a ground node. Figure 8.9-2 shows that the bottom node has been selected to be the ground node.

We will perform a Time Domain (Transient) simulation. (Select PSpice\New Simulation Profile from the OrCAD Capture menu bar; then choose Time Domain (Transient) from the Analysis Type drop-down list. The simulation starts at time zero and ends at the Run to Time. Specify the Run to Time as 20 ms to run the simulation for two full periods of the input waveform. Select the Skip The Initial Transient Bias Point Calculation (SKJPBP) check box.) Select PSpice\Run from the OrCAD Capture menu bar to run the simulation.

After a successful Time Domain (Transient) simulation, OrCAD Capture will automatically open a Schematics window. Select Trace/Add Trace to pop up the Add Traces dialog box. Add the traces V(OUTPUT) and V(INPUT). Figure 8.9-3 shows the resulting plot after removing the grid and labeling some points.

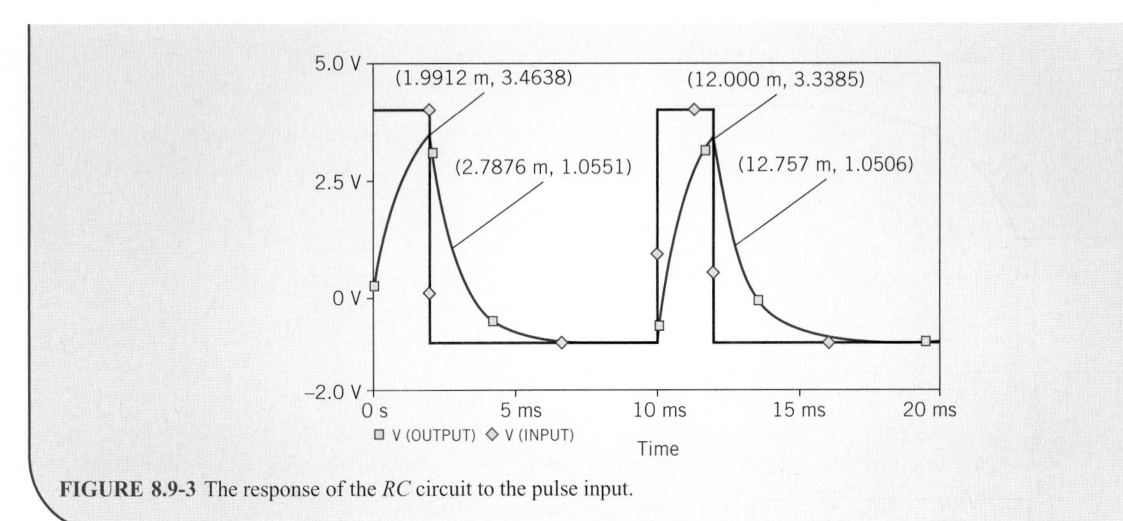

FIGURE 8.9-3 The response of the *RC* circuit to the pulse input.

8.10 HOW CAN WE CHECK . . . ?

Engineers are frequently called upon to check that a solution to a problem is indeed correct. For example, proposed solutions to design problems must be checked to confirm that all of the specifications have been satisfied. In addition, computer output must be reviewed to guard against data-entry errors, and claims made by vendors must be examined critically.

Engineering students are also asked to check the correctness of their work. For example, occasionally just a little time remains at the end of an exam. It is useful to be able to quickly identify those solutions that need more work.

The following examples illustrate techniques useful for checking the solutions of the sort of problems discussed in this chapter.

EXAMPLE 8.10-1 How Can We Check the Response of a First-Order Circuit?

Consider the circuit and corresponding transient response shown in Figure 8.10-1. **How can we check** whether the transient response is correct? Three things need to be verified: the initial voltage, $v_o(t_0)$; the final voltage, $v_o(\infty)$; and the time constant, τ.

Solution

Consider first the initial voltage, $v_o(t_0)$. (In this example, $t_0 = 10\ \mu s$.) Before time $t_0 = 10\ \mu s$, the switch is closed and has been closed long enough for the circuit to reach steady state, that is, for any transients to have died out. To calculate $v_o(t_0)$, we simplify the circuit in two ways. First, replace the switch with a short circuit because the switch is closed. Second, replace the inductor with a short circuit because inductors act like short circuits when all the inputs are constants and the circuit is at steady state. The resulting circuit is shown in Figure 8.10-2*a*. After replacing the parallel 300-Ω and 600-Ω resistors by the equivalent 200-Ω resistor, the initial voltage is calculated using voltage division as

$$v_o(t_0) = \frac{200}{200 + 200} 8 = 4\ \text{V}$$

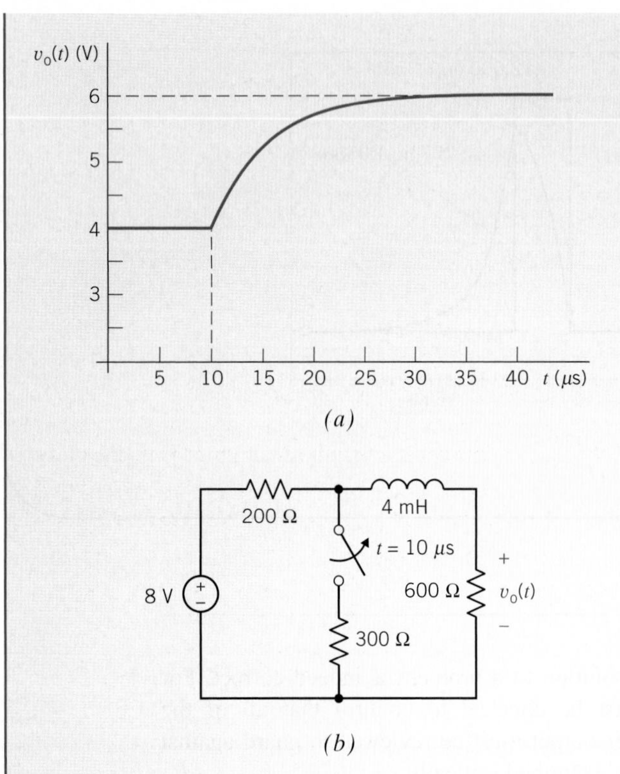

FIGURE 8.10-1 (*a*) A transient response and (*b*) the corresponding circuit.

Next consider the final voltage, $v_o(\infty)$. In this case, the switch is open and the circuit has reached steady state. Again, the circuit is simplified in two ways. The switch is replaced with an open circuit because the switch is open. The inductor is replaced by a short circuit because inductors act like short circuits when all the inputs are constants and the circuit is at steady state. The simplified circuit is shown in Figure 8.10-2*b*. The final voltage is calculated using voltage division as

$$v_o(\infty) = \frac{600}{200 + 600} 8 = 6 \text{ V}$$

The time constant is calculated from the circuit shown in Figure 8.10-2*c*. This circuit has been simplified by setting the input to zero (a zero voltage source acts like a short circuit) and replacing the switch by an open circuit. The time constant is

$$\tau = \frac{L}{R_t} = \frac{4 \times 10^{-3}}{200 + 600} = 5 \times 10^{-6} = 5 \ \mu\text{s}$$

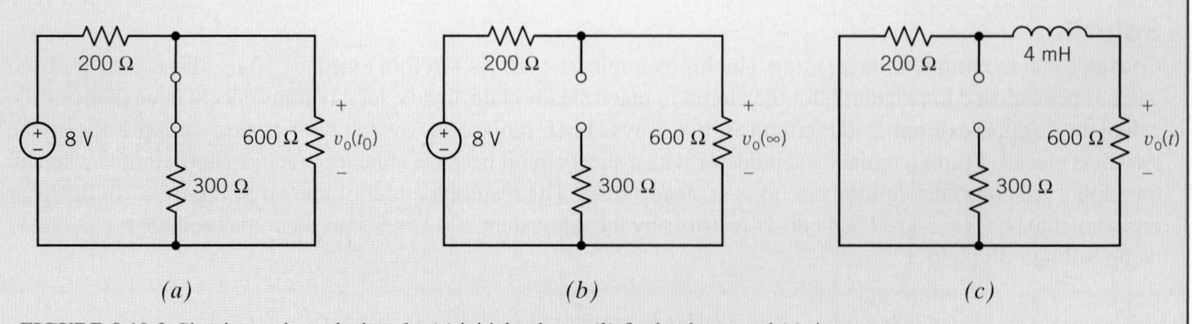

FIGURE 8.10-2 Circuits used to calculate the (*a*) initial voltage, (*b*) final voltage, and (*c*) time constant.

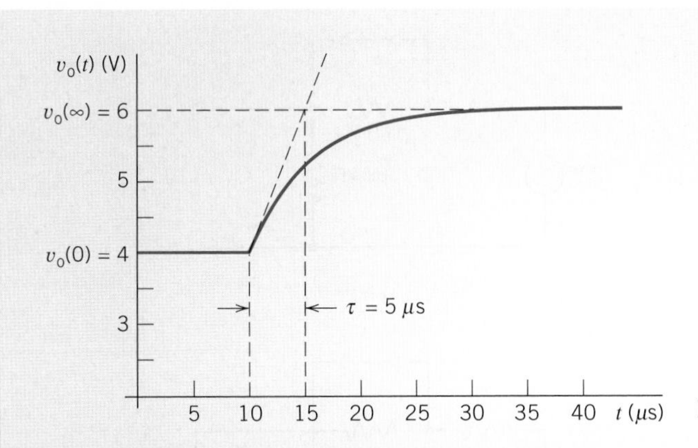

FIGURE 8.10-3 Interpretation of the transient response.

Figure 8.10-3 shows how the initial voltage, final voltage, and time constant can be determined from the plot of the transient response. (Recall that a procedure for determining the time constant graphically was illustrated in Figure 8.3-3.) Because the values of $v_o(t_0)$, $v_o(\infty)$, and τ obtained from the transient response are the same as the values obtained by analyzing the circuit, we conclude that the transient response is indeed correct.

EXAMPLE 8.10-2 How Can We Check the Response of a First-Order Circuit?

Consider the circuit and corresponding transient response shown in Figure 8.10-4. **How can we check** whether the transient response is correct? Four things need to be verified: the steady-state capacitor voltage when the switch is open, the steady-state capacitor voltage when the switch is closed, the time constant when the switch is open, and the time constant when the switch is closed.

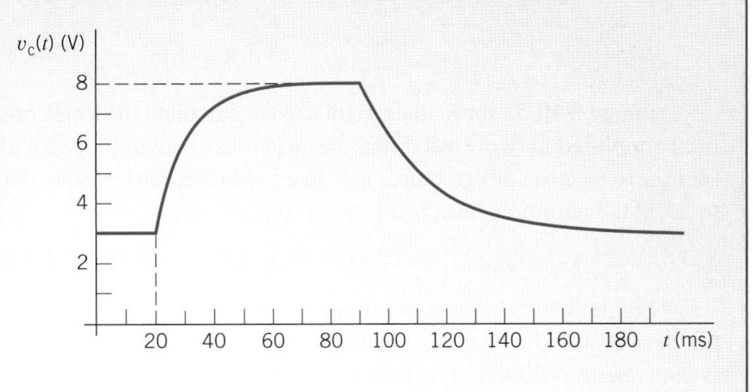

(*a*)

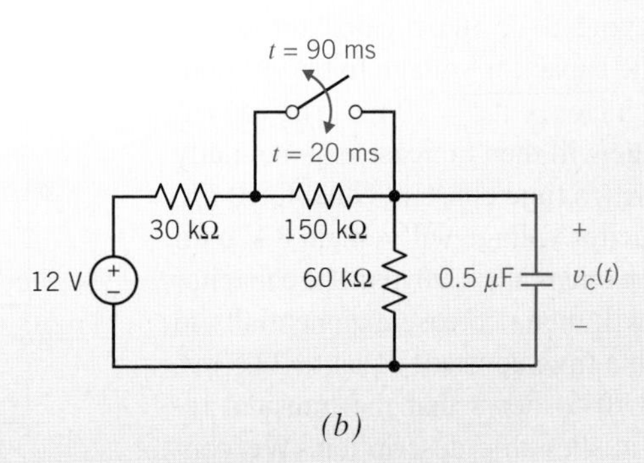

(*b*)

FIGURE 8.10-4 (*a*) A transient response and (*b*) the corresponding circuit.

Solution

Figure 8.10-5*a* shows the circuit used to calculate the steady-state capacitor voltage when the switch is open. The circuit has been simplified in two ways. First, the switch has been replaced with an open circuit. Second, the capacitor has been replaced with an open circuit because capacitors act like open circuits when all the inputs are constants and the circuit is at steady state. The steady-state capacitor voltage is calculated using voltage division as

$$v_c(\infty) = \frac{60}{60 + 30 + 150} 12 = 3\ \text{V}$$

Figure 8.10-5*b* shows the circuit used to calculate the steady-state capacitor voltage

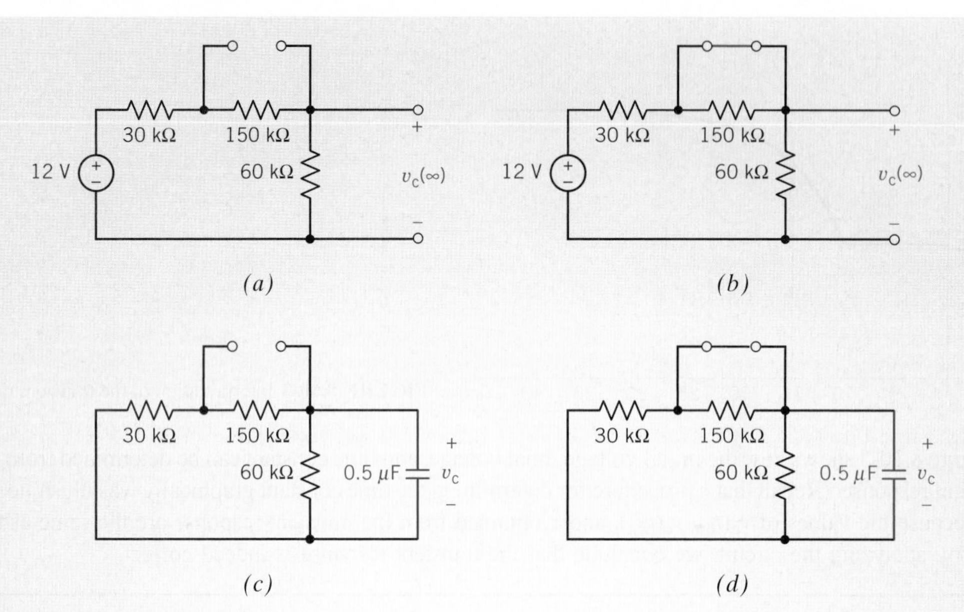

FIGURE 8.10-5 Circuits used to calculate (*a*) the steady-state voltage when the switch is open, (*b*) the steady-state voltage when the switch is closed, (*c*) the time constant when the switch is open, and (*d*) the time constant when the switch is closed.

when the switch is closed. Again, this circuit has been simplified in two ways. First, the switch has been replaced with a short circuit. Second, the capacitor has been replaced with an open circuit. The steady-state capacitor voltage is calculated using voltage division as

$$v_c(\infty) = \frac{60}{60+30}12 = 8\text{ V}$$

Figure 8.10-5*c* shows the circuit used to calculate the time constant when the switch is open. This circuit has been simplified in two ways. First, the switch has been replaced with an open circuit. Second, the input has been set to zero (a zero voltage source acts like a short circuit). Notice that 180 kΩ in parallel with 60 kΩ is equivalent to 45 kΩ. The time constant is

$$\tau = (45 \times 10^3) \cdot (0.5 \times 10^{-6}) = 22.5 \times 10^{-3} = 22.5\text{ ms}$$

Figure 8.10-5*d* shows the circuit used to calculate the time constant when the switch is closed. The switch has been replaced with a short circuit, and the input has been set to zero. Notice that 30 kΩ in parallel with 60 kΩ is equivalent to 20 kΩ. The time constant is

$$\tau = (20 \times 10^3) \cdot (0.5 \times 10^{-6}) = 10^{-2} = 10\text{ ms}$$

Having done these calculations, we expect the capacitor voltage to be 3 V until the switch closes at $t = 20$ ms. The capacitor voltage will then increase exponentially to 8 V, with a time constant equal to 10 ms. The capacitor voltage will remain 8 V until the switch opens at $t = 90$ ms. The capacitor voltage will then decrease exponentially to 3 V, with a time constant equal to 22.5 ms. Figure 8.10-6 shows that the transient response satisfies this description. We conclude that the transient response is correct.

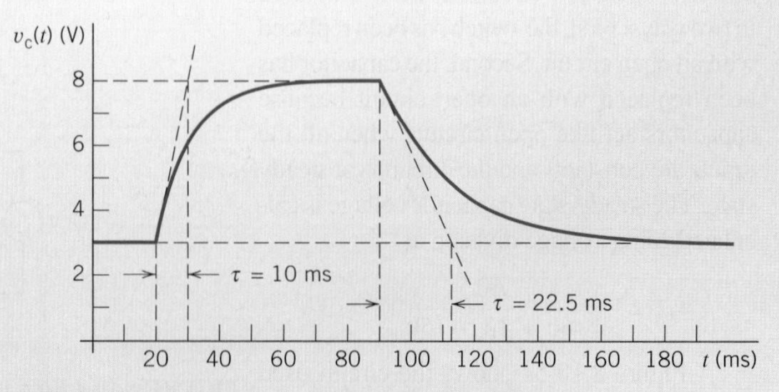

FIGURE 8.10-6 Interpretation of the transient response.

8.11 DESIGN EXAMPLE

A COMPUTER AND PRINTER

It is frequently necessary to connect two pieces of electronic equipment together so that the output from one device can be used as the input to another device. For example, this situation occurs when a printer is connected to a computer, as shown in Figure 8.11-1*a*. This situation is represented more generally by the circuit shown in Figure 8.11-1*b*. The driver sends a signal through the cable to the receiver. Let us replace the driver, cable, and receiver with simple models. Model the driver as a voltage source, the cable as an *RC* circuit, and the receiver as an open circuit. The values of resistance and capacitance used to model the cable will depend on the length of the cable. For example, when RG58 coaxial cable is used,

$$R = r \cdot \ell \text{ where } r = 0.54\ \frac{\Omega}{\text{m}}$$

and

$$C = c \cdot \ell \text{ where } c = 88\ \frac{\text{pF}}{\text{m}}$$

and ℓ is the length of the cable in meters, Figure 8.11-1*c* shows the equivalent circuit.

Suppose that the circuits connected by the cable are digital circuits. The driver will send 1's and 0's to the receiver. These 1's and 0's will be represented by voltages. The output of the driver will be one voltage, V_{OH}, to represent logic 1 and another voltage, V_{OL}, to represent a logic 0. For example, one popular type of logic, called TTL logic, uses $V_{OH} = 2.4$ V and $V_{OL} = 0.4$ V. (TTL stands for transistor–transistor logic.) The receiver uses two different voltages, V_{IH} and V_{IL}, to represent 1's and 0's. (This is done to provide noise immunity, but that is another story.) The receiver will interpret its input, v_b, to be a logic 1 whenever $v_b > V_{IH}$ and to be a logic 0 whenever $v_b < V_{IL}$. (Voltages between V_{IH} and V_{IL} will occur only during transitions between logic 1 and logic 0. These voltages will sometimes be interpreted as logic 1 and other times as logic 0.) TTL logic uses $V_{IH} = 2.0$ V and $V_{IL} = 0.8$ V.

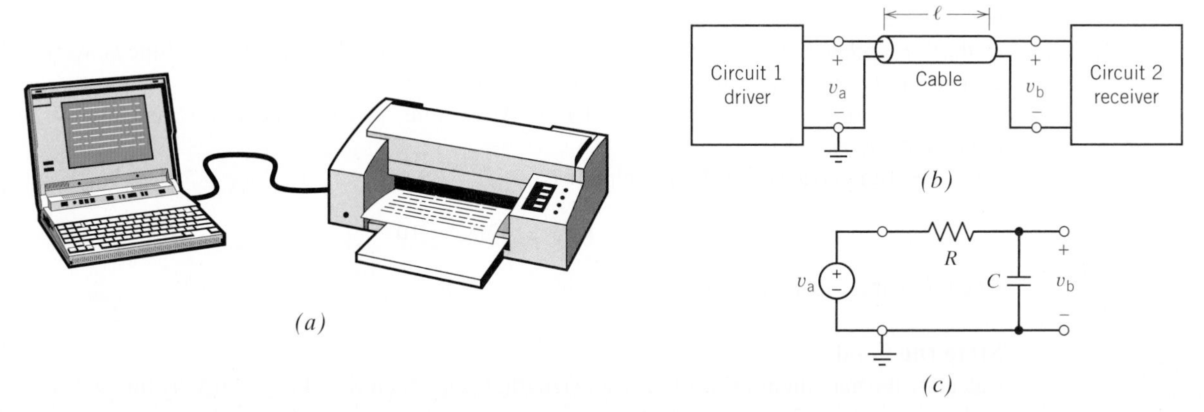

FIGURE 8.11-1 (*a*) A printer connected to a laptop computer. (*b*) Two circuits connected by a cable. (*c*) An equivalent circuit.

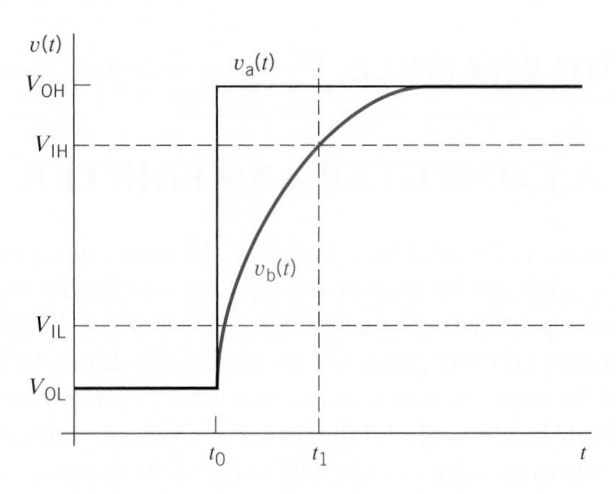

FIGURE 8.11-2 Voltages that occur during a transition from a logic 0 to a logic 1.

Figure 8.11-2 shows what happens when the driver output changes from logic 0 to logic 1. Before time t_0,

$$v_a = V_{OL} \quad \text{and} \quad v_b < V_{IL} \quad \text{for} \quad t < t_0$$

In words, a logic 0 is sent and received. The driver output switches to V_{OH} at time t_0. The receiver input, v_b, makes this transition more slowly. Not until time t_1 does the receiver input become large enough to be interpreted as a logic 1. That is,

$$v_b > V_{IH} \quad \text{for} \quad t > t_1$$

The time that it takes for the receiver to recognize the transition from logic 0 to logic 1

$$\Delta t = t_1 - t_0$$

is called the delay. This delay is important because it puts a limit on how fast 1's and 0's can be sent from the driver to the receiver. To ensure that the 1's and 0's are received reliably, each 1 and each 0 must last at least Δt. The rate at which 1's and 0's are sent from the driver to the receiver is inversely proportional to the delay.

Suppose two TTL circuits are connected using RG58 coaxial cable. What restriction must be placed on the length of the cable to ensure that the delay, Δt, is less than 2 ns?

Describe the Situation and the Assumptions

The voltage $v_b(t)$ is the capacitor voltage of an *RC* circuit. The *RC* circuit is at steady state just before time t_0.

The input to the *RC* circuit is $v_a(t)$. Before time t_0, $v_a(t) = V_{OL} = 0.4$ V. At time t_0, $v_a(t)$ changes abruptly. After time t_0, $v_a(t) = V_{OH} = 2.4$ V.

Before time t_0, $v_b(t) = V_{OL} = 0.4$ V. After time t_0, $v_b(t)$ increases exponentially. Eventually, $v_b(t) = V_{OH} = 2.4$ V.

The time constant of the *RC* circuit is

$$\tau = R \cdot C = rc\ell^2 = 47.52 \times 10^{-2} \cdot \ell^2$$

where ℓ is the cable length in meters.

State the Goal

Calculate the maximum value of the cable length, ℓ, for which $v_b > V_{IH} = 2.0$ V by time $t = t_0 + \Delta t$, where $\Delta t = 2$ ns.

Generate a Plan

Calculate the voltage $v_b(t)$ in Figure 8.11-1*b*. The voltage $v_b(t)$ will depend on the length of the cable, ℓ, because the time constant of the *RC* circuit is a function of ℓ. Set $v_b = V_{IH}$ at time $t = t_0 + \Delta t$. Solve the resulting equation for the length of the cable.

Act on the Plan

Using the notation introduced in this chapter,

$$v_b(0) = V_{OL} = 0.4\text{ V}$$
$$v_b(\infty) = V_{OH} = 2.4\text{ V}$$

and

$$\tau = 47.52 \times 10^{-12} \cdot \ell^2$$

Using Eq. 8.3-6, we express the voltage $v_b(t)$ as

$$v_b(t) = V_{OH} + (V_{OL} - V_{OH})e^{-(t-t_0)/\tau}$$

The capacitor voltage, v_b, will be equal to V_{IH} at time $t_1 = t_0 + \Delta t$, so

$$V_{IH} = V_{OH} + (V_{OL} - V_{OH})e^{-\Delta t/\tau}$$

Solving for the delay, Δt, gives

$$\Delta t = -\tau \ln\left[\frac{V_{IH} - V_{OH}}{V_{OL} - V_{OH}}\right] = -47.52 \times 10^{-12} \cdot \ell^2 \cdot \ln\left[\frac{V_{IH} - V_{OH}}{V_{OL} - V_{OH}}\right]$$

In this case,

$$\ell = \sqrt{\frac{-\Delta t}{47.52 \times 10^{-12} \cdot \ln\left[\dfrac{V_{IH} - V_{OH}}{V_{OL} - V_{OH}}\right]}}$$

and, therefore,

$$\ell = \sqrt{\frac{-2 \cdot 10^{-9}}{47.52 \times 10^{-12} \cdot \ln\left[\dfrac{2.0 - 2.4}{0.4 - 2.4}\right]}} = 5.11\text{ m} = 16.8\text{ ft}$$

Verify the Proposed Solution

When $\ell = 5.11$ m, then

$$R = 0.54 \times 5.11 = 2.76\ \Omega$$

and

$$C = \left(88 \times 10^{-12}\right) \times 5.11 = 450\text{ pF}$$

so

$$\tau = 2.76 \times \left(450 \times 10^{-12}\right) = 1.24\text{ ns}$$

Finally,

$$\Delta t = -1.24 \times 10^{-9} \times \ln\left[\frac{2.0 - 2.4}{0.4 - 2.4}\right] = 1.995\text{ ns}$$

Because $\Delta t < 2$ ns, the specifications have been satisfied but with no margin for error.

8.12 SUMMARY

- Voltages and currents can be used to encode, store, and process information. When a voltage or current is used to represent information, that voltage or current is called a signal. Electric circuits that process that information are called signal-processing circuits.
- Circuits that contain energy-storing elements, that is, capacitors and inductors, are represented by differential equations rather than by algebraic equations. Analysis of these circuits requires the solution of differential equations.
- In this chapter, we restricted our attention to first-order circuits. First-order circuits contain one energy storage element and are represented by first-order differential equations, which are reasonably easy to solve. We solved first-order differential equations, using the method called separation of variables.
- The *complete response* of a circuit is the sum of the *natural response* and *the forced response*. The natural response is the general solution of the differential equation that represents the circuit when the input is set to zero. The forced response is the particular solution of the differential equation representing the circuit.
- The complete response can be separated into the *transient response* and the *steady-state response*. The transient response vanishes with time, leaving the steady-state response. When the input to the circuit is either a constant or a sinusoid, the steady-state response can be used as the forced response.
- The term *transient response* sometimes refers to the ''transient part of the complete response'' and other times to a complete response that includes a transient part. In particular, PSpice uses the term *transient response* to refer to the complete response. Because this can be confusing, the term must be used carefully.
- The *step response* of a circuit is the response when the input is equal to a unit step function and all the initial conditions of the circuit are equal to zero.
- We used Thévenin and Norton equivalent circuits to reduce the problem of analyzing any first-order circuit to the problem of analyzing one of two simple first-order circuits. One of the simple first-order circuits is a series circuit consisting of a voltage source, a resistor, and a capacitor. The other is a parallel circuit consisting of a current source, a resistor, and an inductor. Table 8.12-1 summarizes the equations used to determine the complete response of a first-order circuit.
- The parameter τ in the first-order differential equation $\frac{d}{dt}x(t) + \frac{x(t)}{\tau} = K$ is called the time constant. The time constant τ is the time for the response of a first-order circuit to complete 63 percent of the transition from initial value to final value.
- Stability is a property of well-behaved circuits. It is easy to tell whether a first-order circuit is stable. A first-order circuit is stable if, and only if, its time constant is not negative, that is, $\tau \geq 0$.

Table 8.12-1 Summary of First-Order Circuits

FIRST-ORDER CIRCUIT CONTAINING A CAPACITOR	FIRST-ORDER CIRCUIT CONTAINING AN INDUCTOR
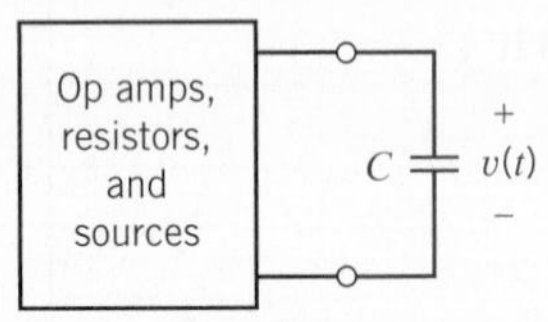	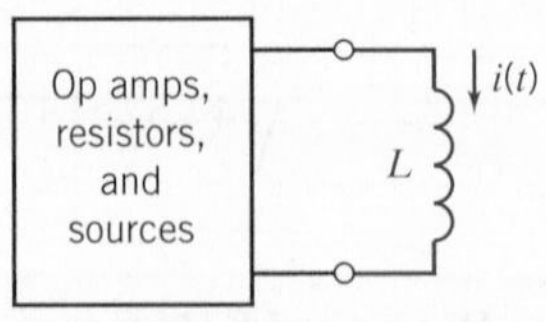
Replace the circuit consisting of op amps, resistors, and sources by its Thévenin equivalent circuit:	Replace the circuit consisting of op amps, resistors, and sources by its Norton equivalent circuit:
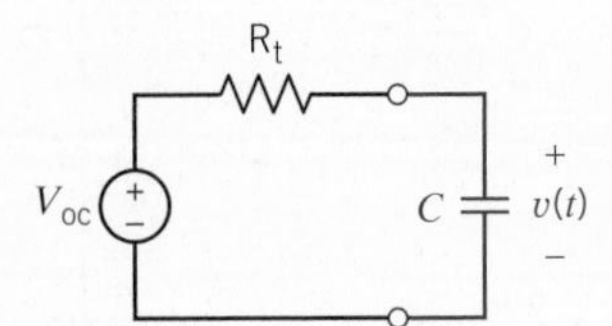	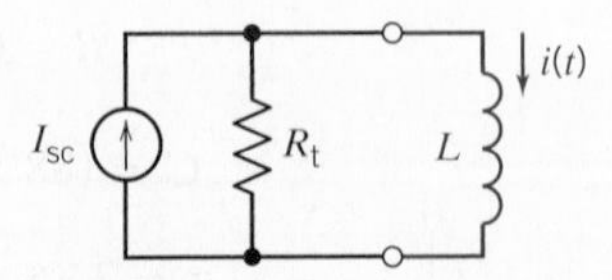
The capacitor voltage is: $v(t) = V_{oc} + (v(0) - V_{oc})e^{-t/\tau}$ where the time constant, τ, is $\tau = R_t C$ and the initial condition, $v(0)$, is the capacitor voltage at time $t = 0$.	The inductor current is $i(t) = I_{sc} + (i(0) - I_{sc})e^{-t/\tau}$ where the time constant, τ, is $\tau = \frac{L}{R_t}$ and the initial condition, $i(0)$, is the inductor current at time $t = 0$.

PROBLEMS

Section 8.3 The Response of a First-Order Circuit to a Constant Input

P 8.3-1 The circuit shown in Figure P 8.3-1 is at steady state before the switch closes at time $t = 0$. The input to the circuit is the voltage of the voltage source, 18 V. The output of this circuit is the voltage across the capacitor, $v(t)$. Determine $v(t)$ for $t > 0$.

Answer: $v(t) = 9 - 3e^{-0.89t}$ V for $t > 0$

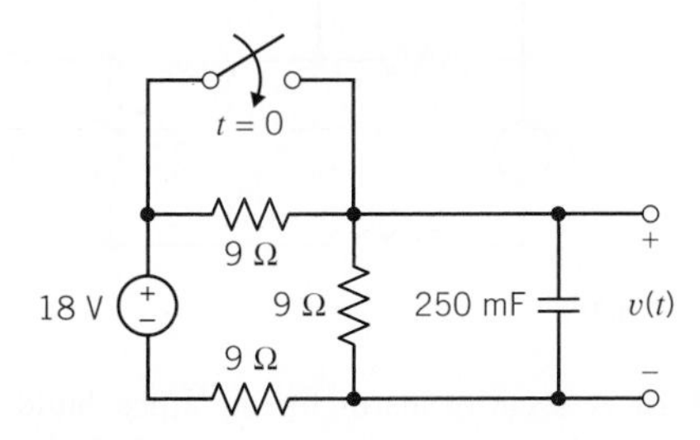

Figure P 8.3-1

P 8.3-2 The circuit shown in Figure P 8.3-2 is at steady state before the switch opens at time $t = 0$. The input to the circuit is the voltage of the voltage source, 12 V. The output of this circuit is the current in the inductor, $i(t)$. Determine $i(t)$ for $t > 0$.

Answer: $i(t) = 1.5 + 1.5e^{-0.6t}$A for $t > 0$

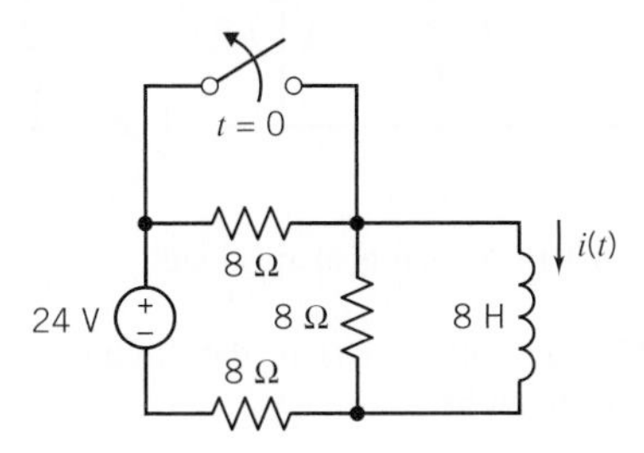

Figure P 8.3-2

P 8.3-3 The circuit shown in Figure P 8.3-3 is at steady state before the switch closes at time $t = 0$. Determine the capacitor voltage, $v(t)$, for $t > 0$.

Answer: $v(t) = -6 + 18e^{-6.67t}$ V for $t > 0$

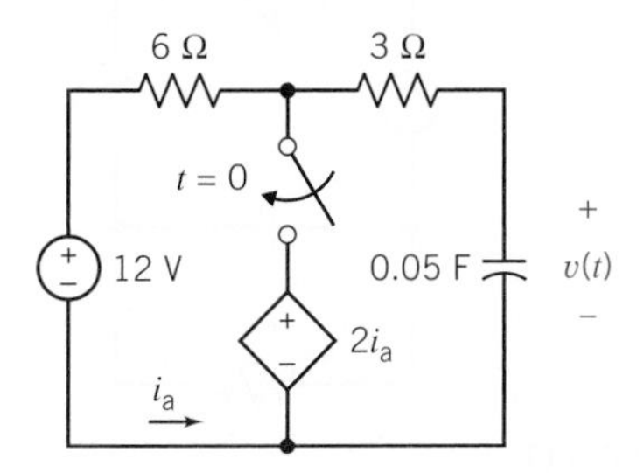

Figure P 8.3-3

P 8.3-4 The circuit shown in Figure P 8.3-4 is at steady state before the switch closes at time $t = 0$. Determine the inductor current, $i(t)$, for $t > 0$.

Answer: $i(t) = -2 + \frac{10}{3}e^{-0.5t}$ A for $t > 0$

Figure P 8.3-4

P 8.3-5 The circuit shown in Figure P 8.3-5 is at steady state before the switch opens at time $t = 0$. Determine the voltage, $v_o(t)$, for $t > 0$.

Answer: $v_o(t) = 10 - 5e^{-3.1t}$ V for $t > 0$

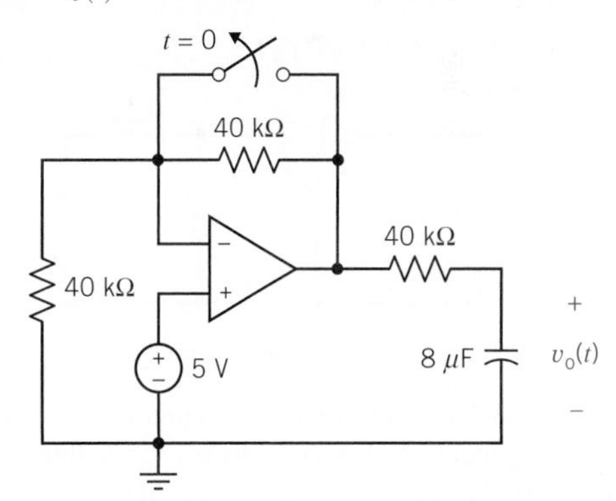

Figure P 8.3-5

P 8.3-6 The circuit shown in Figure P 8.3-6 is at steady state before the switch opens at time $t = 0$. Determine the voltage, $v_o(t)$, for $t > 0$.

Answer: $v_o(t) = 5e^{-4000t}$ V for $t > 0$

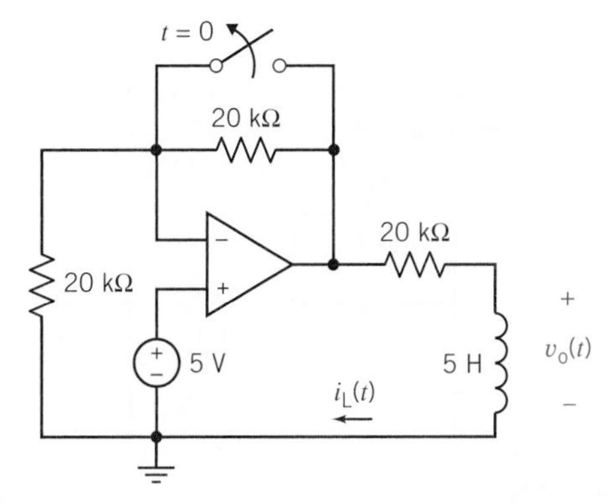

Figure P 8.3-6

P 8.3-7 Figure P 8.3-7*a* shows astronaut Dale Gardner using the manned maneuvering unit to dock with the spinning *Westar VI* satellite on November 14, 1984. Gardner used a large tool called the apogee capture device (ACD) to stabilize the satellite and capture it for recovery, as shown in Figure P 8.3-7*a*. The ACD can be modeled by the circuit of Figure P 8.3-7*b*. Find the inductor current i_L for $t > 0$.

Answer: $i_L(t) = 6e^{-20t}$ A

(*a*)

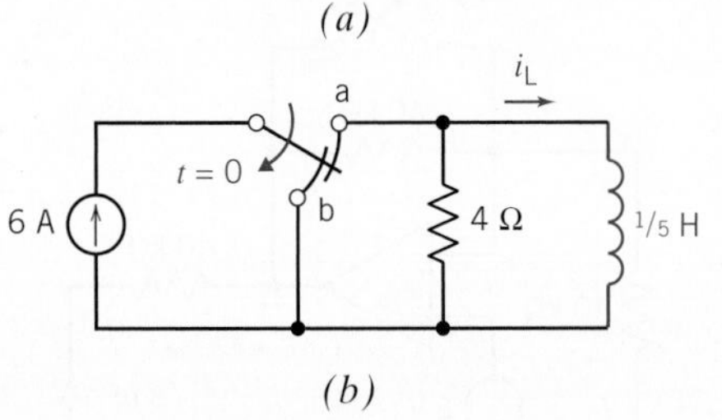

(*b*)

Figure P 8.3-7 (*a*) Astronaut Dale Gardner using the manned maneuvering unit to dock with the *Westar VI* satellite. Courtesy of NASA. (*b*) Model of the apogee capture device. Assume that the switch has been in position for a long time at $t = 0^-$.

P 8.3-8 The circuit shown in Figure P 8.3-8 is at steady state before the switch opens at time $t = 0$. The input to the circuit is the voltage of the voltage source, V_s. This voltage source is a dc voltage source; that is, V_s is a constant. The output of this circuit is the voltage across the capacitor, $v_o(t)$. The output voltage is given by

$$v_o(t) = 2 + 8e^{-0.5t} \text{ V for } t > 0$$

Determine the values of the input voltage, V_s, the capacitance, C, and the resistance, R.

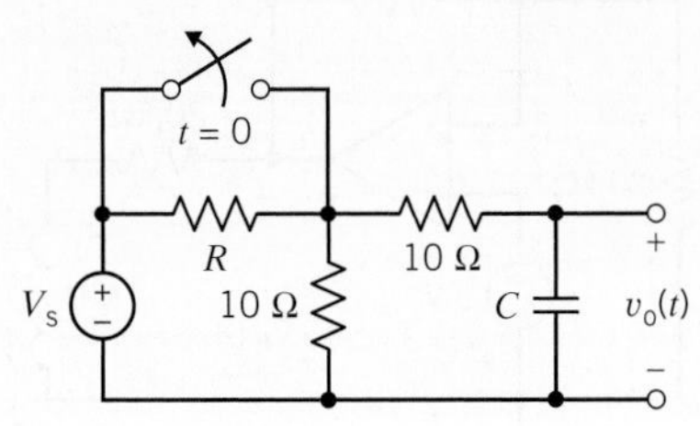

Figure P 8.3-8

P 8.3-9 The circuit shown in Figure P 8.3-9 is at steady state before the switch closes at time $t = 0$. The input to the circuit is the voltage of the voltage source, 24 V. The output of this circuit, the voltage across the 3-Ω resistor, is given by

$$v_o(t) = 6 - 3e^{-0.35t} \text{ V when } t > 0$$

Determine the value of the inductance, L, and of the resistances, R_1 and R_2.

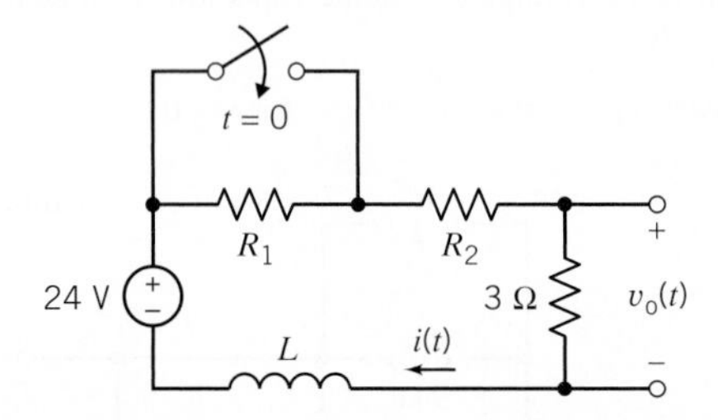

Figure P 8.3-9

P 8.3-10 A security alarm for an office building door is modeled by the circuit of Figure P 8.3-10. The switch represents the door interlock, and v is the alarm indicator voltage. Find $v(t)$ for $t > 0$ for the circuit of Figure P 8.3-10. The switch has been closed for a long time at $t = 0^-$.

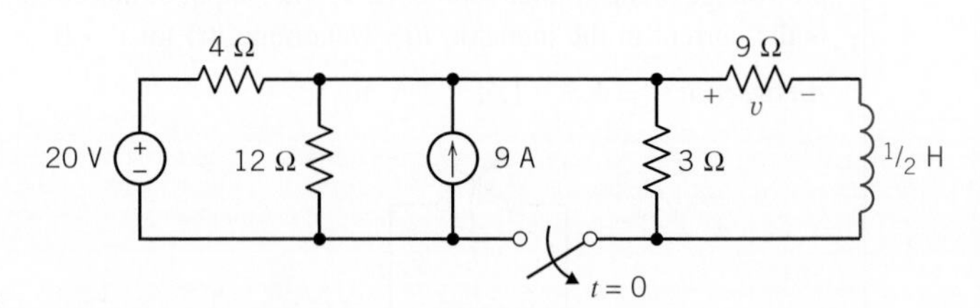

Figure P 8.3-10 A security alarm circuit.

P 8.3-11 The voltage $v(t)$ in the circuit shown in Figure P 8.3-11 is given by

$$v(t) = 8 + 4e^{-2t} \text{ V for } t > 0$$

Determine the values of R_1, R_2, and C.

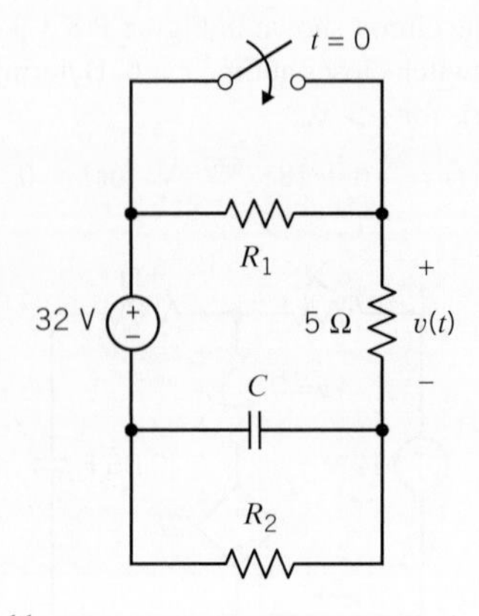

Figure P 8.3-11

P 8.3-12 The circuit shown in Figure P 8.3-12 is at steady state when the switch opens at time $t = 0$. Determine $i(t)$ for $t \geq 0$.

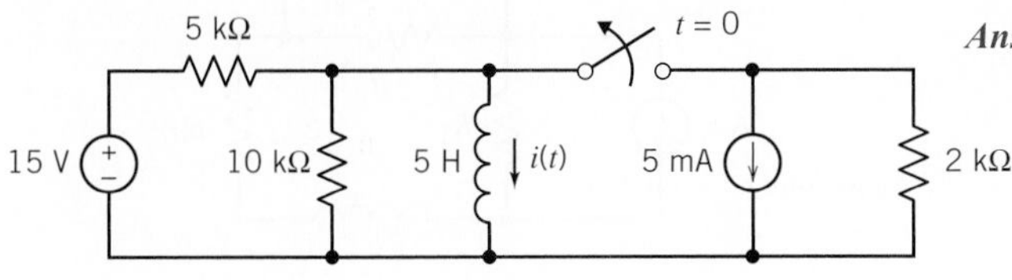

Figure P 8.3-12

P 8.3-13 The circuit shown in Figure P 8.3-13 is at steady state when the switch opens at time $t = 0$. Determine $v(t)$ for $t \geq 0$.

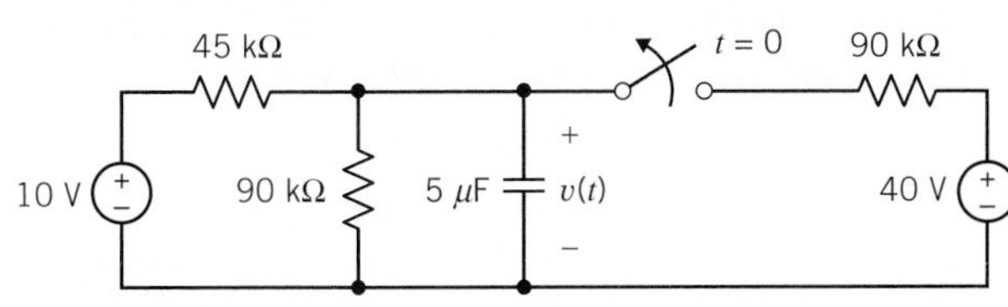

Figure P 8.3-13

P 8.3-14 The circuit shown in Figure P 8.3-14 is at steady state when the switch closes at time $t = 0$. Determine $i(t)$ for $t \geq 0$.

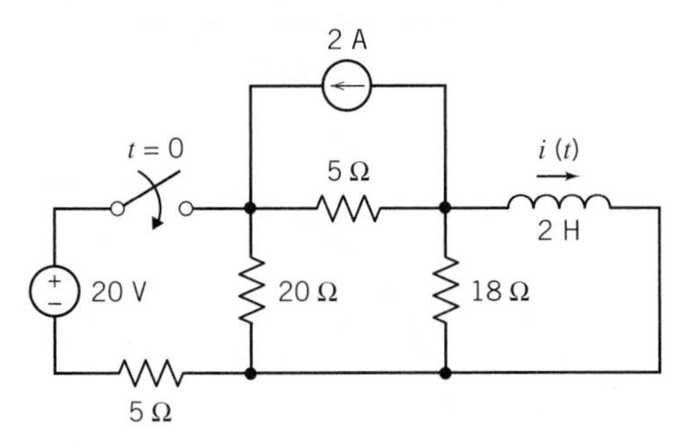

Figure P 8.3-14

P 8.3-15 The circuit in Figure P 8.3-15 is at steady state before the switch closes. Find the inductor current after the switch closes.

Hint: $i(0) = 0.1$ A, $I_{sc} = 0.3$ A, $R_t = 40\ \Omega$

Answer: $i(t) = 0.3 - 0.2e^{-2t}$ A $\quad t \geq 0$

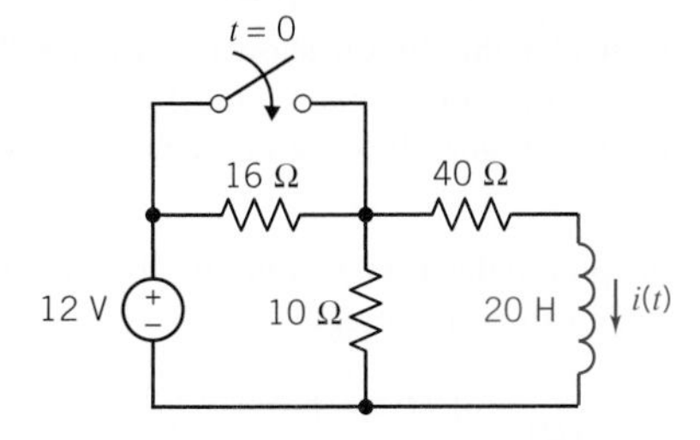

Figure P 8.3-15

P 8.3-16 The circuit in Figure P 8.3-16 is at steady state before the switch closes. Find the capacitor voltage for $t \geq 0$.

Hint: $v(0) = 12$ V, $V_{oc} = 12$ V

Answer: $v(t) = 12.0$ V

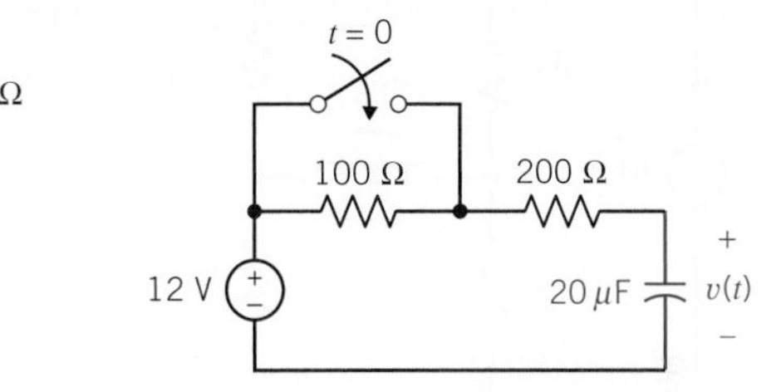

Figure P 8.3-16

P 8.3-17 The circuit shown in Figure P 8.3-17 is at steady state before the switch closes. The response of the circuit is the voltage $v(t)$. Find $v(t)$ for $t > 0$.

Hint: After the switch closes, the inductor current is $i(t) = 0.2\,(1 - e^{-1.8t})$ A

Answer: $v(t) = 8 + e^{-1.8t}$ V

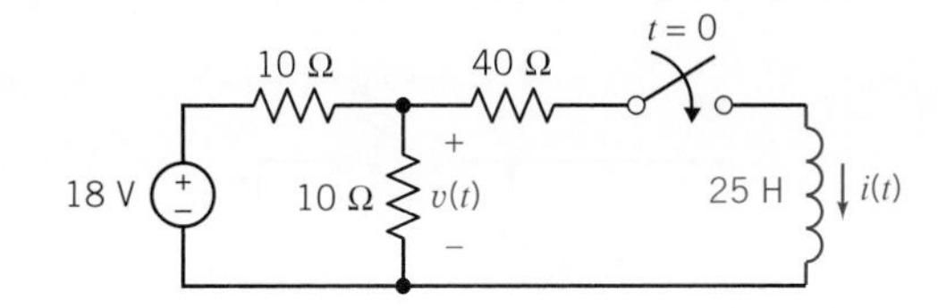

Figure P 8.3-17

P 8.3-18 The circuit shown in Figure P 8.3-18 is at steady state before the switch closes. The response of the circuit is the voltage $v(t)$. Find $v(t)$ for $t > 0$.

Answer: $v(t) = 37.5 - 97.5e^{-6400t}$ V

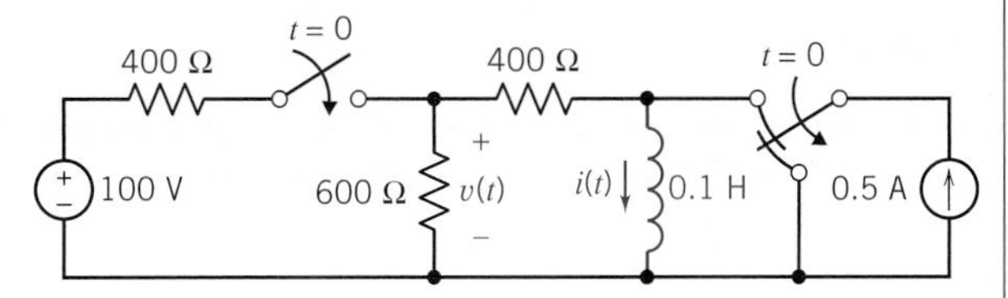

Figure P 8.3-18

P 8.3-19 The circuit shown in Figure P 8.3-19 is at steady state before the switch closes. Find $v(t)$ for $t \geq 0$.

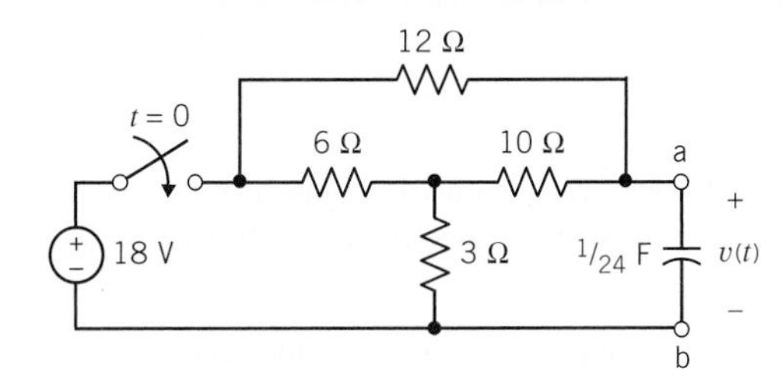

Figure P 8.3-19

P 8.3-20 The circuit shown in Figure P 8.3-20 is at steady state before the switch closes. Determine $i(t)$ for $t \geq 0$.

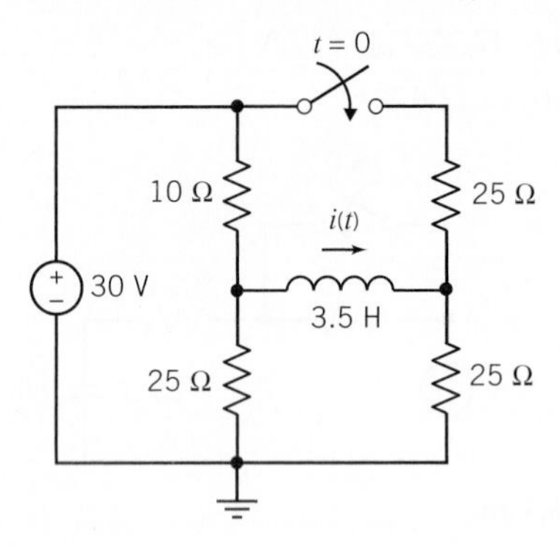

Figure P 8.3-20

***P 8.3-21** The circuit shown in Figure P 8.3-21 is at steady state before the switch closes at time $t = 0$. The current $i(t)$ is given by

$$i(t) = 15 + 53.6e^{-548t} \text{ mA} \quad \text{for } t \geq 0$$

Determine the values of R_1, R_2, and L.

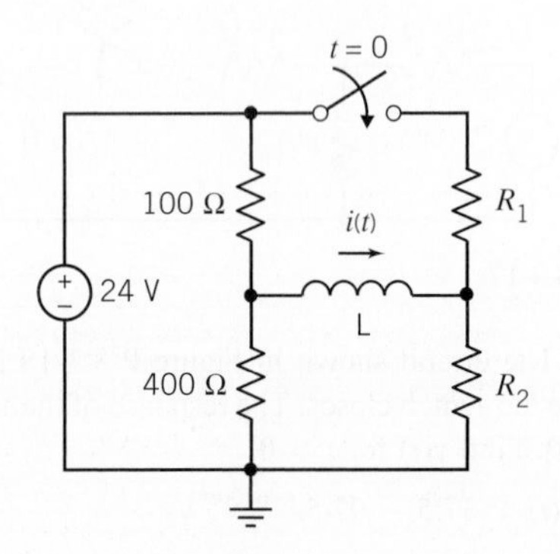

Figure P 8.3-21

P 8.3-22 The circuit shown in Figure P 8.3-22 is at steady state when the switch closes at time $t = 0$. Determine $i(t)$ for $t \geq 0$.

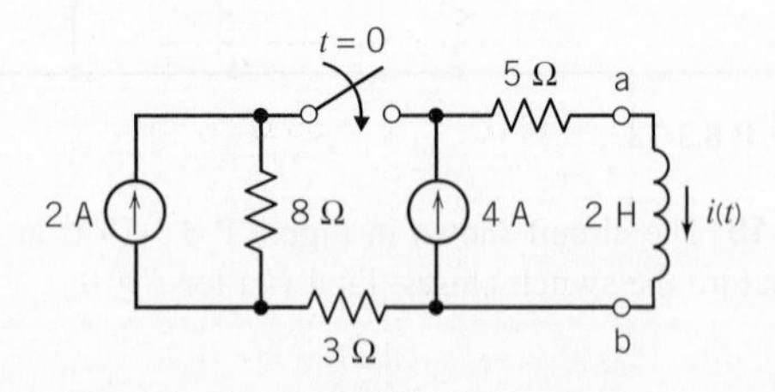

Figure P 8.3-22

P 8.3-23 The circuit shown in Figure P 8.3-23 is at steady state before the switch closes at time $t = 0$. The voltage $v(t)$ is given by

$$v(t) = 4 - 2e^{-3t} \text{ V} \quad \text{for } t > 0$$

Determine the values of R_1, R_2, and L.

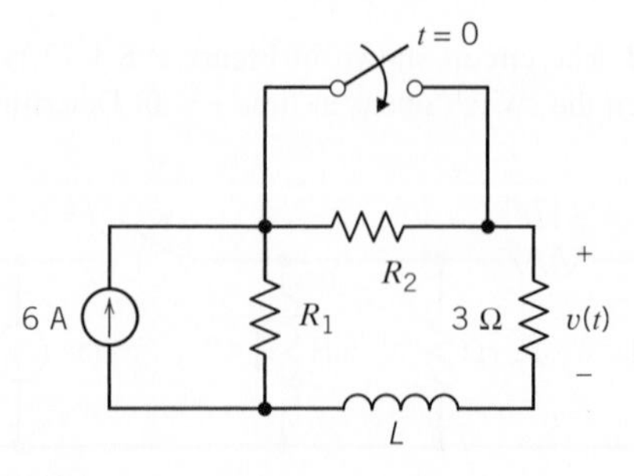

Figure P 8.3-23

P 8.3-24 Consider the circuit shown in Figure P 8.3-24*a* and corresponding plot of the inductor current shown in Figure P 8.3-24*b*. Determine the values of L, R_1, and R_2.

Hint: Use the plot to determine values of D, E, F, and a such that the inductor current can be represented as

$$i(t) = \begin{cases} D & \text{for } t \leq 0 \\ E + Fe^{-at} & \text{for } t \geq 0 \end{cases}$$

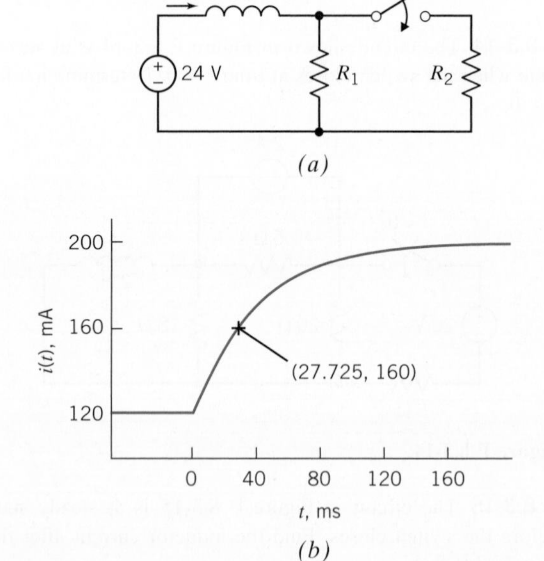

Figure P 8.3-24

Answers: $L = 4.8$ H, $R_1 = 200\ \Omega$, and $R_2 = 300\ \Omega$

P 8.3-25 Consider the circuit shown in Figure P 8.3-25*a* and corresponding plot of the voltage across the 40-Ω resistor shown in Figure P 8.3-25*b*. Determine the values of L and R_2.

Hint: Use the plot to determine values of D, E, F, and a such that the voltage can be represented as

$$v(t) = \begin{cases} D & \text{for } t < 0 \\ E + Fe^{-at} & \text{for } t > 0 \end{cases}$$

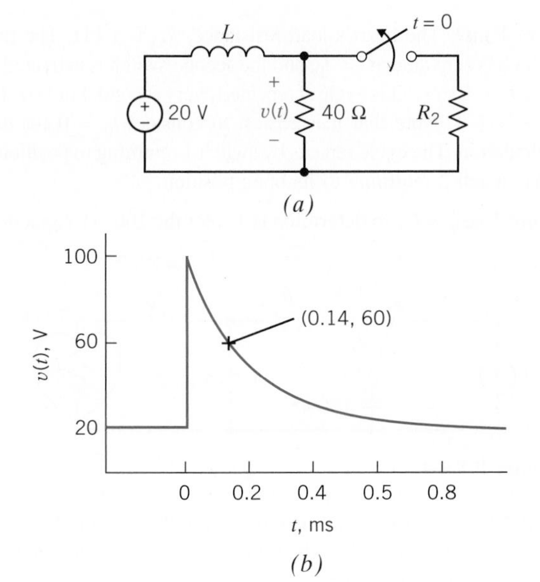

Figure P 8.3-25

Answers: $L = 8$ H and $R_2 = 10\ \Omega$.

P 8.3-26 The voltage shown in Figure P 8.3-26 can be represented by an equation of the form

$$v(t) = \begin{cases} D & \text{for } t < 0 \\ E + Fe^{-at} & \text{for } t > 0 \end{cases}$$

Determine the values of the constants D, E, F, and a.

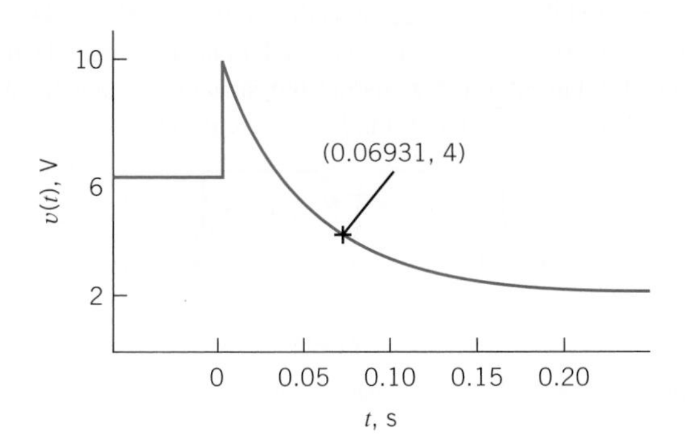

Figure P 8.3-26

P 8.3-27 The circuit shown in Figure P 8.3-27 is at steady state before the switch closes at time $t = 0$. After the switch closes, the inductor current is given by

$$i(t) = 0.9 - 0.3e^{-5t}\ \text{A} \quad \text{for } t \geq 0$$

Determine the values of R_1, R_2, and L.

Answers: $R_1 = 16.6\ \Omega$, $R_2 = 8.4\ \Omega$, and $L = 3.3$ H

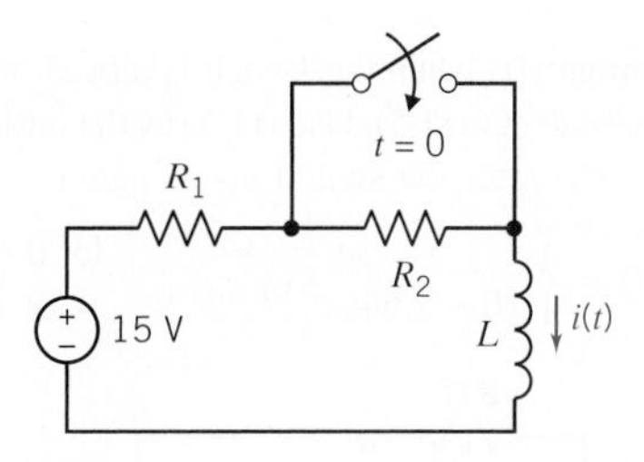

Figure P 8.3-27

P 8.3-28 After time $t = 0$, a given circuit is represented by the circuit diagram shown in Figure P 8.3-28.

(a) Suppose that the inductor current is

$$i(t) = 21.6 + 28.4e^{-4t}\ \text{mA} \quad \text{for } t \geq 0$$

Determine the values of R_1 and R_3.

(b) Suppose instead that $R_1 = 16\ \Omega$, $R_3 = 20\ \Omega$, and the initial condition is $i(0) = 10$ mA.
Determine the inductor current for $t \geq 0$.

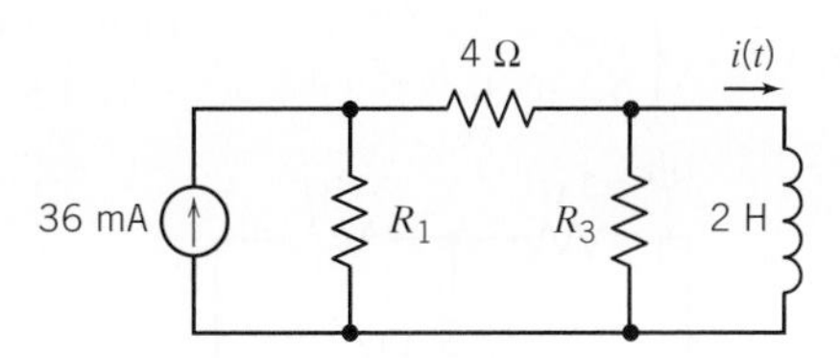

Figure P 8.3-28

P 8.3-29 Consider the circuit shown in Figure P 8.3-29.

(a) Determine the time constant, τ, and the steady state capacitor voltage, $v(\infty)$, when the switch is **open.**

(b) Determine the time constant, τ, and the steady state capacitor voltage, $v(\infty)$, when the switch is **closed.**

Answers: **(a)** $\tau = 3$ s, and $v(\infty) = 24$ V; **(b)** $\tau = 2.25$ s, and $v(\infty) = 12$ V

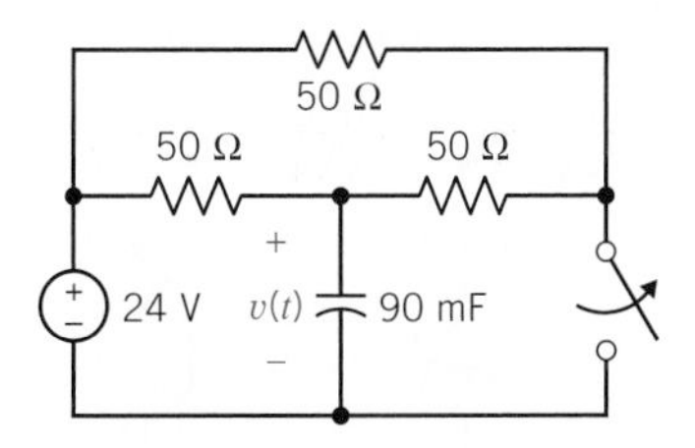

Figure P 8.3-29

Section 8.4 Sequential Switching

P 8.4-1 The circuit shown in Figure P 8.4-1 is at steady state before the switch closes at time $t = 0$. The switch remains closed for 1.5 s and then opens. Determine the capacitor voltage, $v(t)$, for $t > 0$.

Hint: Determine $v(t)$ when the switch is closed. Evaluate $v(t)$ at time $t = 1.5$ s to get $v(1.5)$. Use $v(1.5)$ as the initial condition to determine $v(t)$ after the switch opens again.

Answer: $v(t) = \begin{cases} 5 + 5e^{-0.5t} \text{ V} & \text{for } 0 < t < 1.5 \text{ s} \\ 10 - 2.64e^{-2.5(t-1.5)} \text{ V} & \text{for } 1.5 \text{ s} < t \end{cases}$

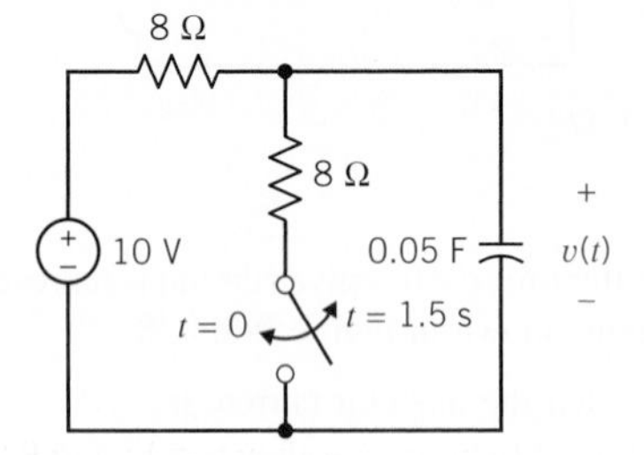

Figure P 8.4-1

P 8.4-2 The circuit shown in Figure P 8.4-2 is at steady state before the switch closes at time $t = 0$. The switch remains closed for 1.5 s and then opens. Determine the inductor current, $i(t)$, for $t > 0$.

Answer: $v(t) = \begin{cases} 2 + e^{-0.5t} \text{ A} & \text{for } 0 < t < 1.5 \text{ s} \\ 3 - 0.53e^{-0.667(t-1.5)} \text{ A} & \text{for } 1.5 \text{ s} < t \end{cases}$

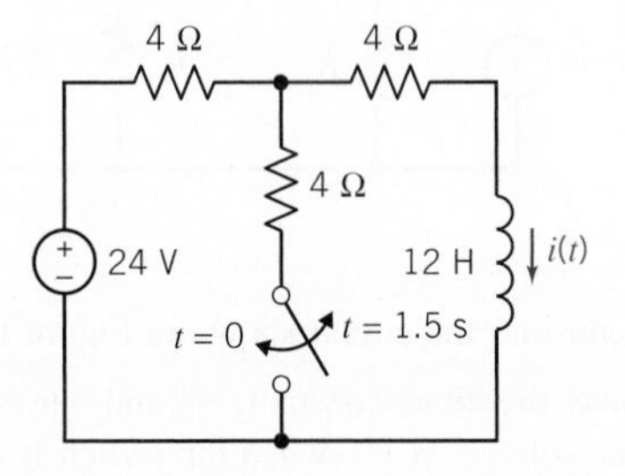

Figure P 8.4-2

P 8.4-3 Find $i(t)$ for $t > 0$ for the circuit shown in Figure P 8.4-3. The circuit is in steady state at $t = 0^-$.

Answer:
$i(t) = 1.4e^{-9t}$ A for $0 \le t \le 51$ ms
$i(t) = 0.8e^{-21(t-0.051)}$ A for $t > 51$ ms

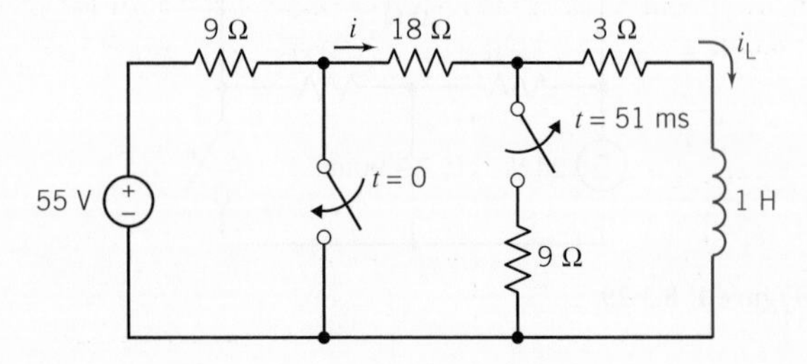

Figure P 8.4-3

P 8.4-4 Cardiac pacemakers are used by people to maintain regular heart rhythm when they have a damaged heart. The circuit of a pacemaker can be represented as shown in Figure P 8.4-4. The resistance of the wires, R, can be neglected because $R < 1$ mΩ. The heart's load resistance, R_L, is 1 kΩ. The first switch is activated at $t = t_0$, and the second switch is activated at $t_1 = t_0 + 10$ ms. This cycle is repeated every second. Find $v(t)$ for $t_0 \le t \le 1$. Note that it is easiest to consider $t_0 = 0$ for this calculation. The cycle repeats by switch 1 returning to position a and switch 2 returning to its open position.

Hint: Use $q = Cv$ to determine $v(0^-)$ for the 100-μF capacitor.

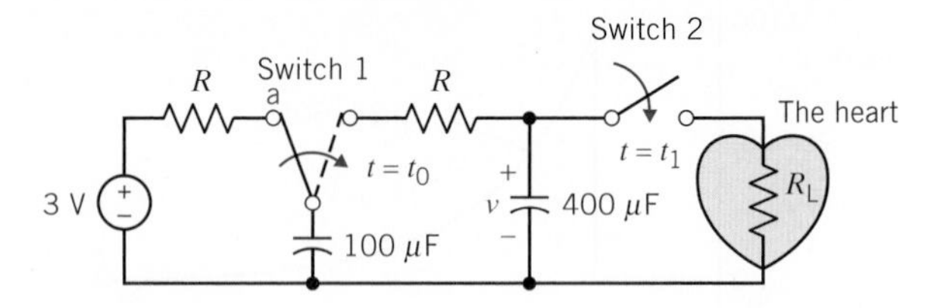

Figure P 8.4-4

P 8.4-5 Determine and sketch $i(t)$ for the circuit shown in Figure P 8.4-5. Calculate the time required for $i(t)$ to reach 99 percent of its final value.

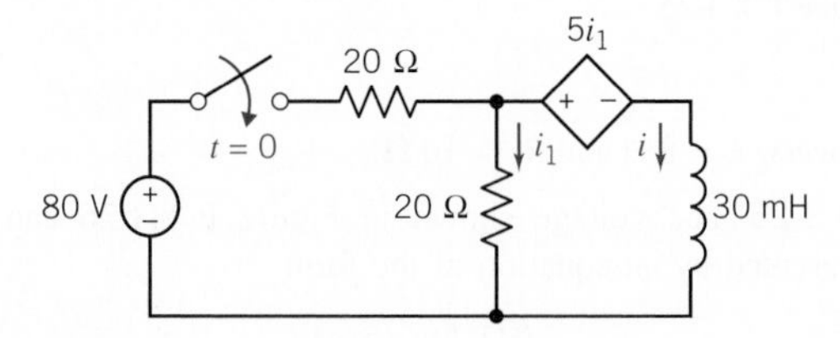

Figure P 8.4-5

P 8.4-6 An electronic flash on a camera uses the circuit shown in Figure P 8.4-6. Harold E. Edgerton invented the electronic flash in 1930. A capacitor builds a steady-state voltage and then discharges it as the shutter switch is pressed. The discharge produces a very brief light discharge. Determine the elapsed time t_1 to reduce the capacitor voltage to one-half of its initial voltage. Find the current, $i(t)$, at $t = t_1$.

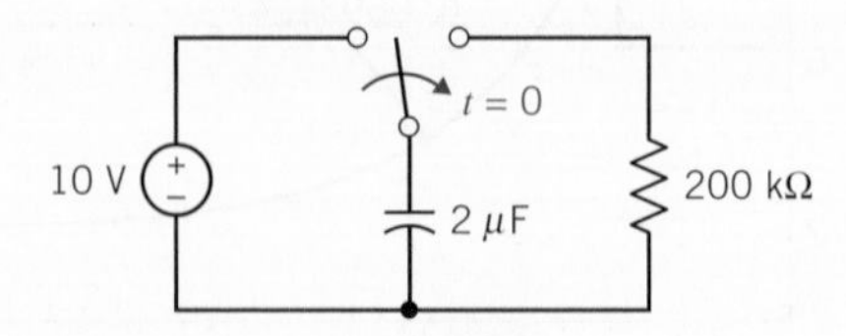

Figure P 8.4-6 Electronic flash circuit.

P 8.4-7 The circuit shown in Figure P 8.4-7 is at steady state before the switch opens at $t = 0$. The switch remains open for 0.5 second and then closes. Determine $v(t)$ for $t \ge 0$.

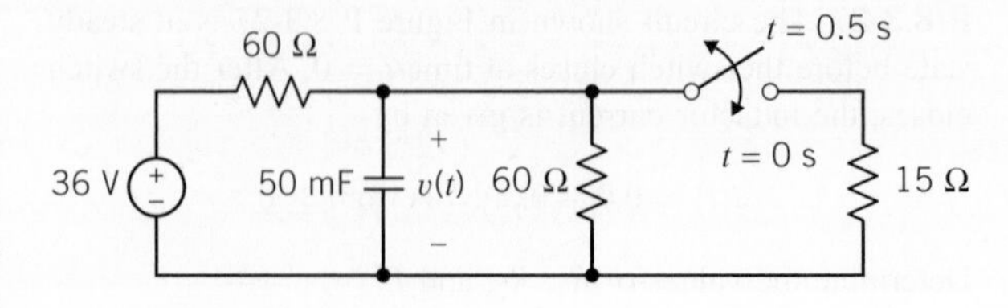

Figure P 8.4-7

Section 8.5 Stability of First-Order Circuits

P 8.5-1 The circuit in Figure P 8.5-1 contains a current controlled voltage source. What restriction must be placed on the gain, R, of this dependent source to guarantee stability?

Answer: $R < 400\ \Omega$

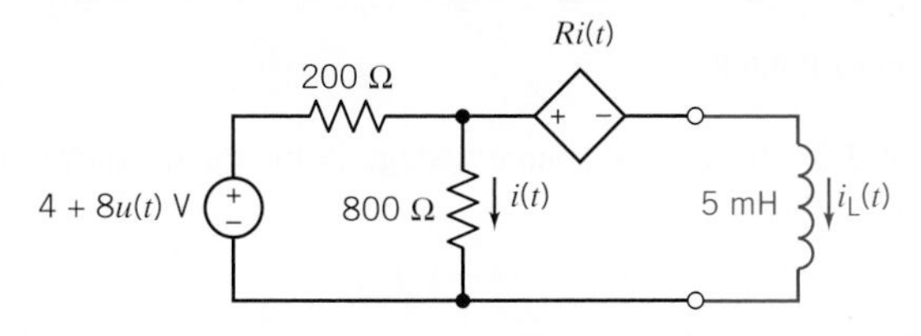

Figure P 8.5-1

P 8.5-2 The circuit in Figure P 8.5-2 contains a voltage-controlled voltage source. What restriction must be placed on the gain, A, of this dependent source to guarantee stability?

Answer: $A < 5$

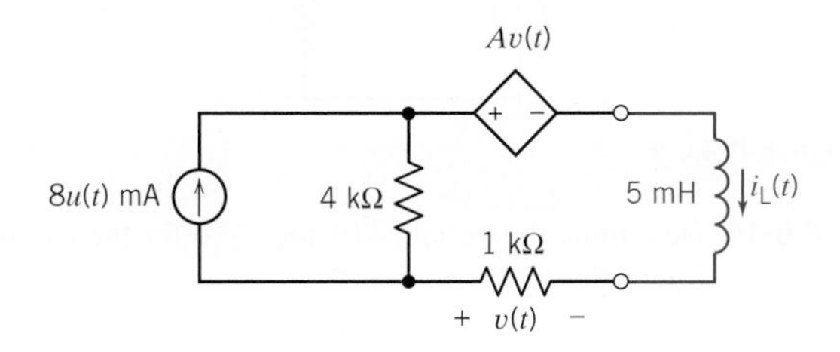

Figure P 8.5-2

P 8.5-3 The circuit in Figure P 8.5-3 contains a current-controlled current source. What restriction must be placed on the gain, B, of this dependent source to guarantee stability?

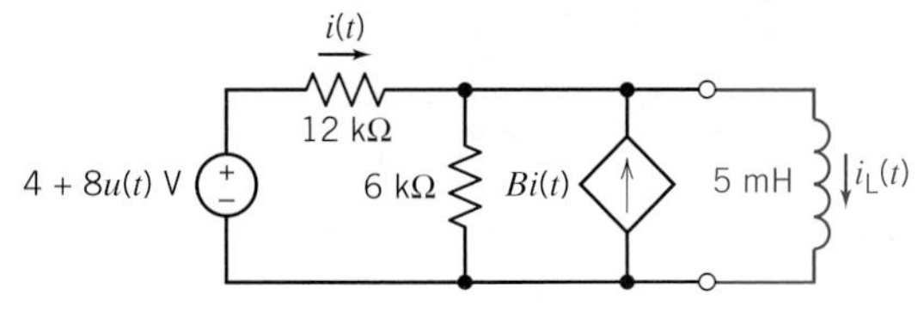

Figure P 8.5-3

P 8.5-4 The circuit in Figure P 8.5-4 contains a voltage-controlled voltage source. What restriction must be placed on the gain, A, of this dependent source to guarantee stability?

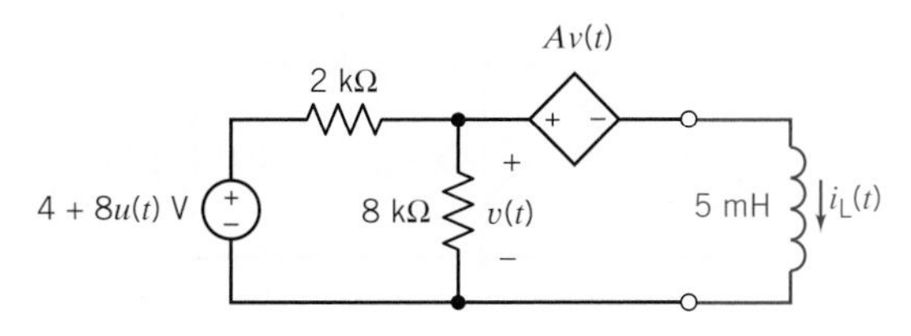

Figure P 8.5-4

Section 8.6 The Unit Step Source

P 8.6-1 The input to the circuit shown in Figure P 8.6-1 is the voltage of the voltage source, $v_s(t)$. The output is the voltage across the capacitor, $v_o(t)$. Determine the output of this circuit when the input is $v_s(t) = 10 - 17\,u(t)$ V.

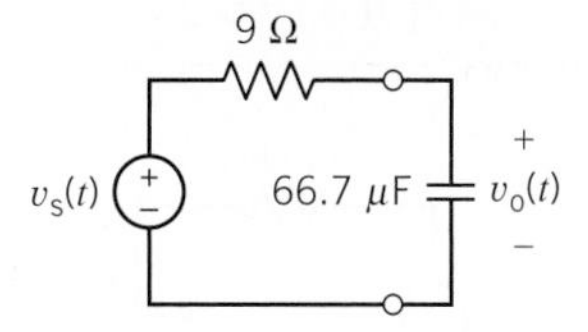

Figure P 8.6-1

P 8.6-2 The input to the circuit shown in Figure P 8.6-2 is the voltage of the voltage source, $v_s(t)$. The output is the voltage across the capacitor, $v_o(t)$. Determine the output of this circuit when the input is $v_s(t) = 3 + 3\,u(t)$ V.

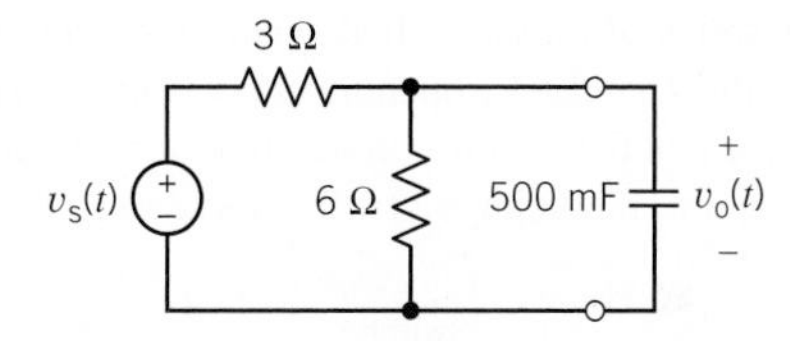

Figure P 8.6-2

P 8.6-3 The input to the circuit shown in Figure P 8.6-3 is the voltage of the voltage source, $v_s(t)$. The output is the current across the inductor, $i_o(t)$. Determine the output of this circuit when the input is $v_s(t) = -7 + 13\,u(t)$ V.

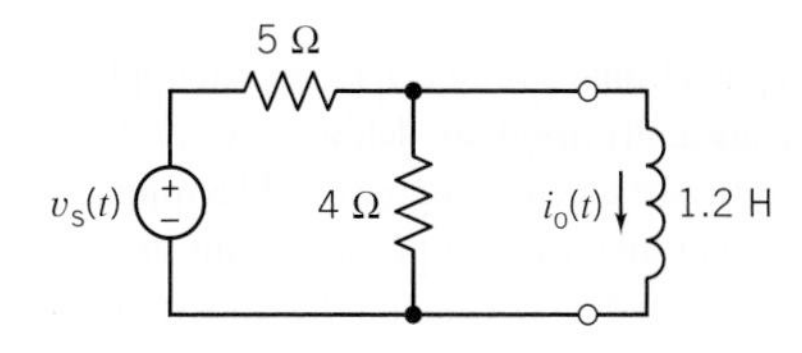

Figure P 8.6-3

P 8.6-4 Use step functions to represent the signal of Figure P 8.6-4.

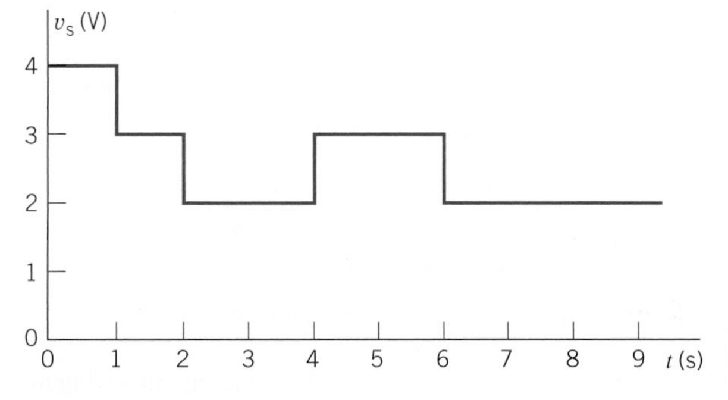

Figure P 8.6-4

P 8.6-5 The initial voltage of the capacitor of the circuit shown in Figure P 8.6-5 is zero. Determine the voltage $v(t)$ when the source is a pulse, described by

$$v_s = \begin{cases} 0 & t < 1\text{ s} \\ 4\text{ V} & 1 < t < 2\text{ s} \\ 0 & t > 2\text{ s} \end{cases}$$

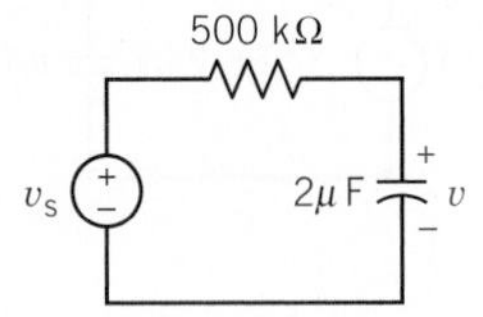

Figure P 8.6-5

P 8.6-6 Studies of an artificial insect are being used to understand the nervous system of animals. A model neuron in the nervous system of the artificial insect is shown in Figure P 8.6-6. A series of pulses, called synapses, is required. The switch generates a pulse by opening at $t = 0$ and closing at $t = 0.5$ s. Assume that the circuit is in steady state and that $v(0^-) = 10$ V. Determine the voltage $v(t)$ for $0 < t < 2$ s.

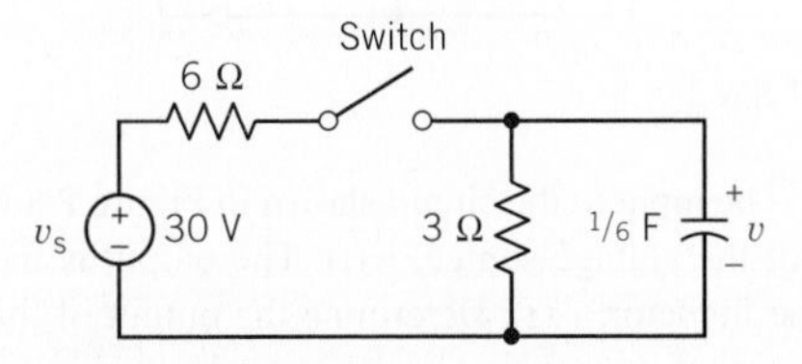

Figure P 8.6-6 Neuron circuit model.

P 8.6-7 An electronic circuit can be used to replace the springs and levers normally used to detonate a shell in a handgun (Jurgen, 1989). The electric trigger would eliminate the clicking sensation, which may cause a person to misaim. The proposed trigger uses a magnet and a solenoid with a trigger switch. The circuit of Figure P 8.6-7 represents the trigger circuit with $i_s(t) = 40\,[u(t) - u(t - t_0)]$ A, where $t_0 = 1$ ms. Determine and plot $v(t)$ for $0 < t < 0.3$ s.

Answer:

$$v = \begin{cases} 480(1 - e^{-1000t}) & 0 < t < 1\text{ ms} \\ 480(1 - e^{-1})e^{-1000(t-t_0)} & t > 1\text{ ms}, t_0 = 1\text{ ms} \end{cases}$$

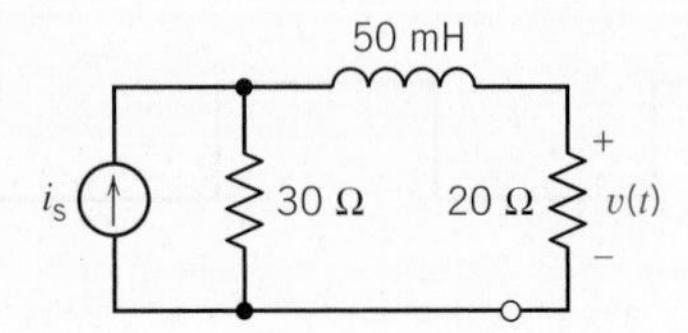

Figure P 8.6-7 Electric trigger circuit for handgun.

P 8.6-8 Determine $v_c(t)$ for $t > 0$ for the circuit of Figure P 8.6-8.

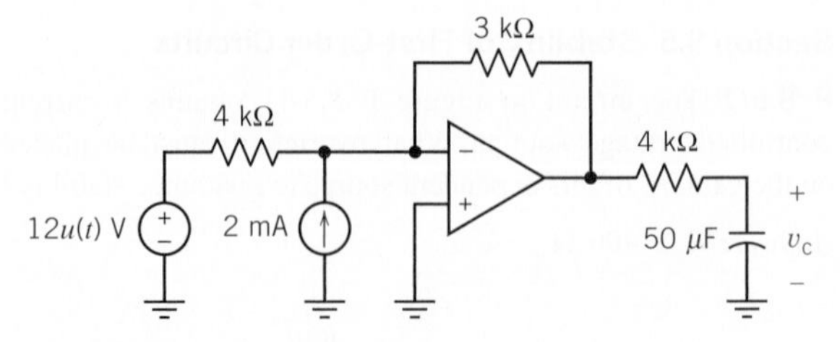

Figure P 8.6-8

P 8.6-9 The voltage source voltage in the circuit shown in Figure P 8.6-9 is

$$v_s(t) = 7 - 14u(t)\ \text{V}$$

Determine $v(t)$ for $t > 0$.

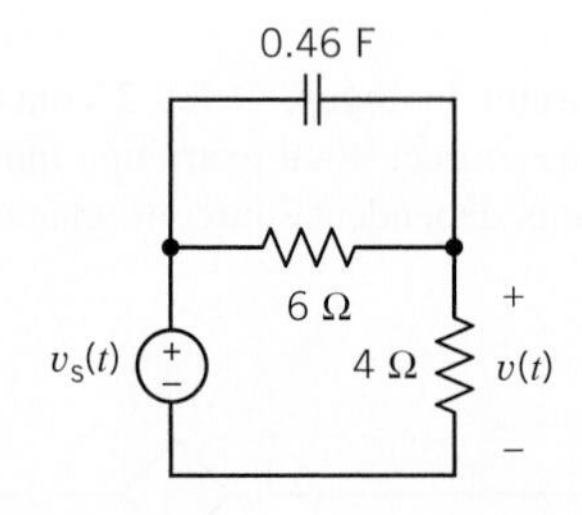

Figure P 8.6-9

P 8.6-10 Determine the voltage $v(t)$ for $t \geq 0$ for the circuit shown in Figure P 8.6-10.

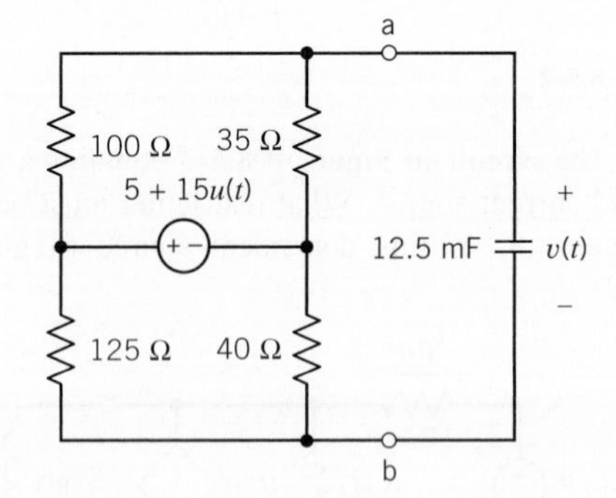

Figure P 8.6-10

P 8.6-11 The voltage source voltage in the circuit shown in Figure P 8.6-11 is

$$v_s(t) = 5 + 20u(t)$$

Determine $i(t)$ for $t \geq 0$.

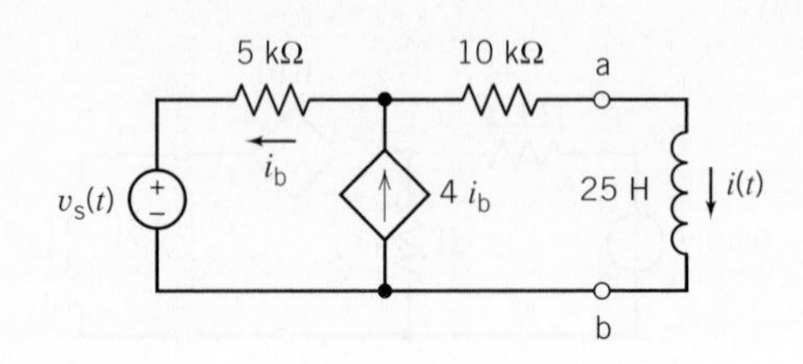

Figure P 8.6-11

P 8.6-12 The voltage source voltage in the circuit shown in Figure P 8.6-12 is

$$v_s(t) = 15 - 9u(t)\text{ V}$$

Determine $v(t)$ for $t \geq 0$.

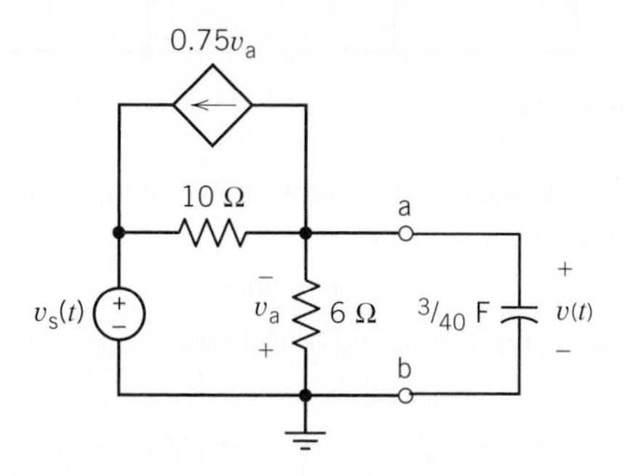

Figure P 8.6-12

P 8.6-13 Determine $i(t)$ for $t \geq 0$ for the circuit shown in Figure P 8.6-13.

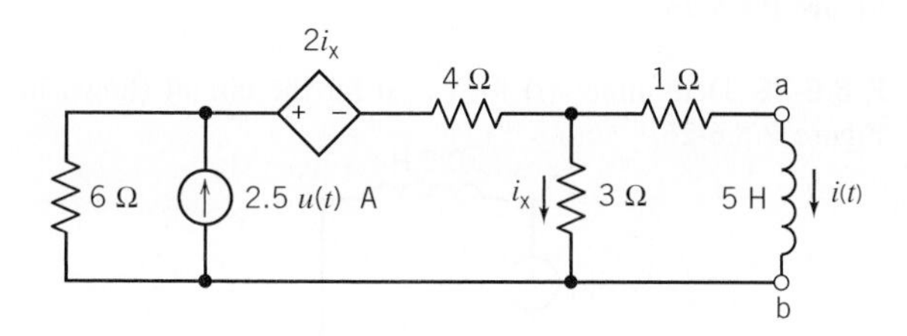

Figure P 8.6-13

P 8.6-14 Determine $i(t)$ for $t \geq 0$ for the circuit shown in Figure P 8.6-14.

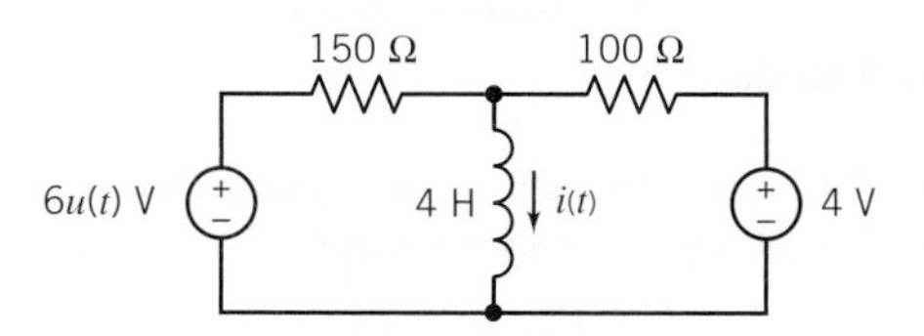

Figure P 8.6-14

P 8.6-15 Determine $v(t)$ for $t \geq 0$ for the circuit shown in Figure P 8.6-15.

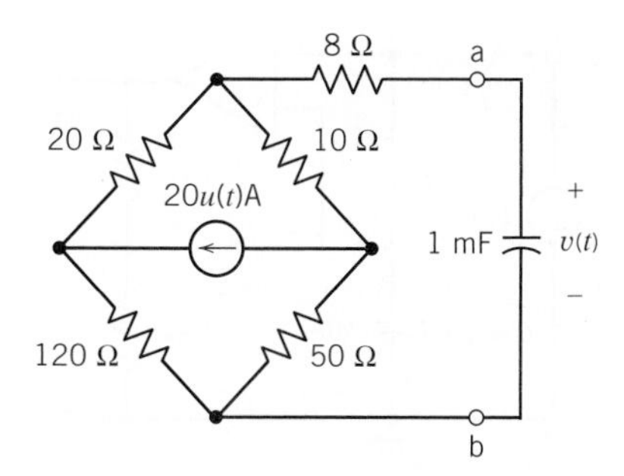

Figure P 8.6-15

P 8.6-16 Determine $v(t)$ for $t \geq 0$ for the circuit shown in Figure P 8.6-16.

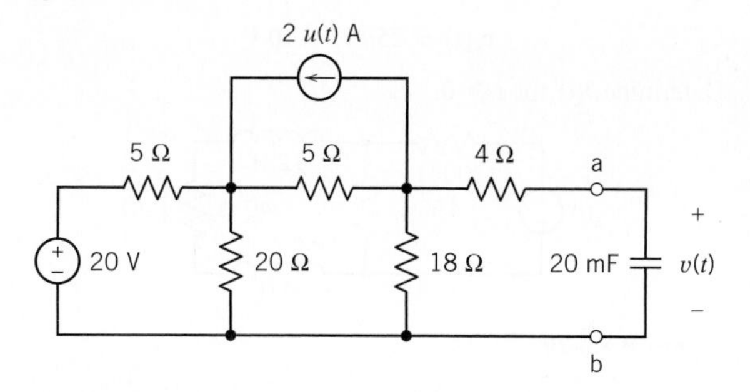

Figure P 8.6-16

P 8.6-17 Determine $i(t)$ for $t \geq 0$ for the circuit shown in Figure P 8.6-17.

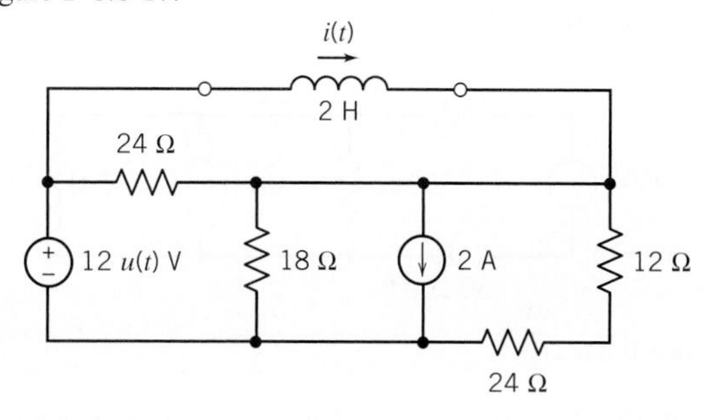

Figure P 8.6-17

P 8.6-18 The voltage source voltage in the circuit shown in Figure P 8.6-18 is

$$v_s(t) = 8 + 12u(t)\text{ V}$$

Determine $v(t)$ for $t \geq 0$.

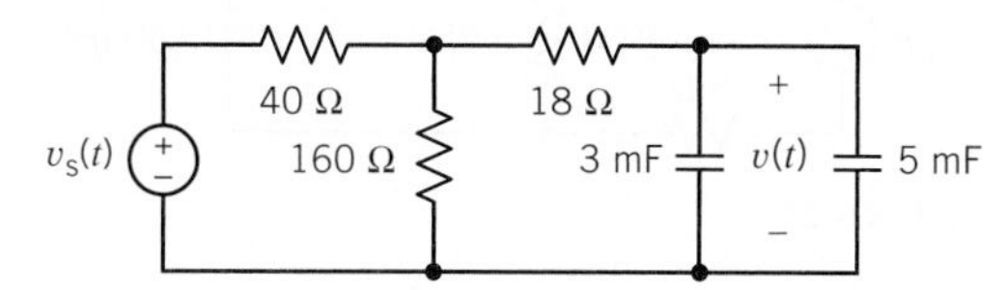

Figure P 8.6-18

P 8.6-19 The circuit shown in Figure P 8.6-19*a* has a current source as shown in Figure P 8.6-19*b*. Determine the current $i(t)$ in the inductor.

Answer: $i(t) = \begin{cases} 5(1 - e^{-10t})\text{ A} & t \leq 0.2\text{ s} \\ 4.32e^{-10\,(t-0.2)}\text{ A} & t \geq 0.2\text{ s} \end{cases}$

$i(t)$ i_s $i_s(t)$ 2 Ω 0.2H 5A 0 0.2 t(s)

(*a*) (*b*)

Figure P 8.6-19

P 8.6-20 The voltage source voltage in the circuit shown in Figure P 8.6-20 is

$$v_s(t) = 25u(t) - 10 \text{ V}$$

Determine $i(t)$ for $t \geq 0$.

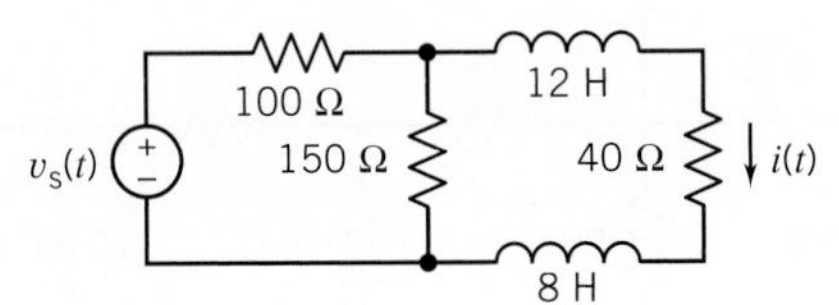

Figure P 8.6-20

P 8.6-21 The voltage source voltage in the circuit shown in Figure P 8.6-21 is

$$v_s(t) = 30 - 24u(t) \text{ V}$$

Determine $i(t)$ for $t \geq 0$.

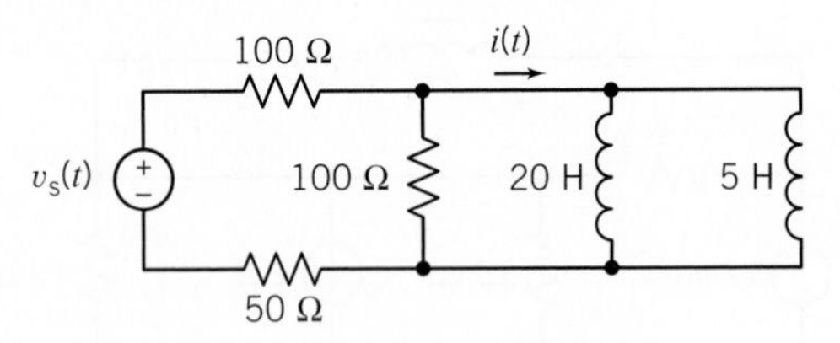

Figure P 8.6-21

P 8.6-22 The voltage source voltage in the circuit shown in Figure P 8.6-22 is

$$v_s(t) = 10 + 40u(t) \text{ V}$$

Determine $v(t)$ for $t \geq 0$.

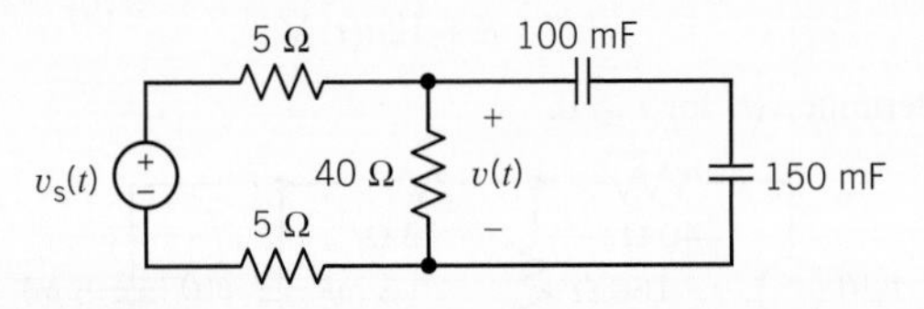

Figure P 8.6-22

P 8.6-23 Determine $v(t)$ for $t > 0$ for the circuit shown in Figure P 8.6-23.

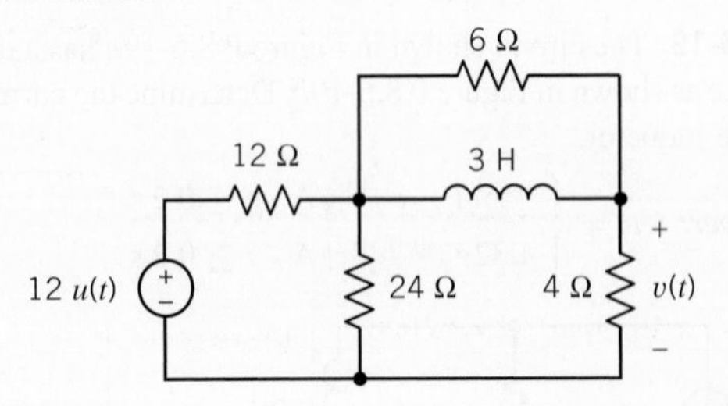

Figure P 8.6-23

P 8.6-24 The input to the circuit shown in Figure P 8.6-24 is the current source current

$$i_s(t) = 4 + 8u(t) \text{ A}$$

The output is the voltage $v(t)$. Determine $v(t)$ for $t > 0$.

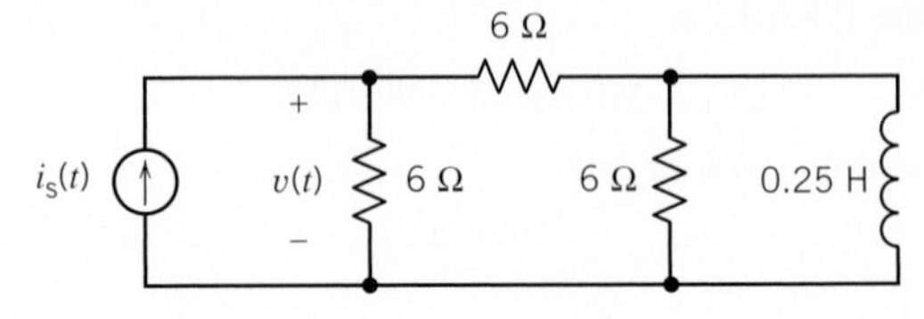

Figure P 8.6-24

P 8.6-25 The input to the circuit shown in Figure P 8.6-25 is the voltage source voltage

$$v_s = 6 + 6u(t)$$

The output is the voltage $v_o(t)$. Determine $v_o(t)$ for $t > 0$.

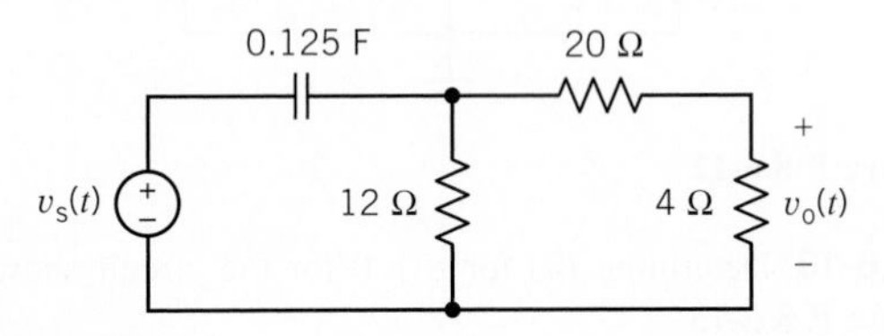

Figure P 8.6-25

P 8.6-26 Determine $v(t)$ for $t > 0$ for the circuit shown in Figure P 8.6-26.

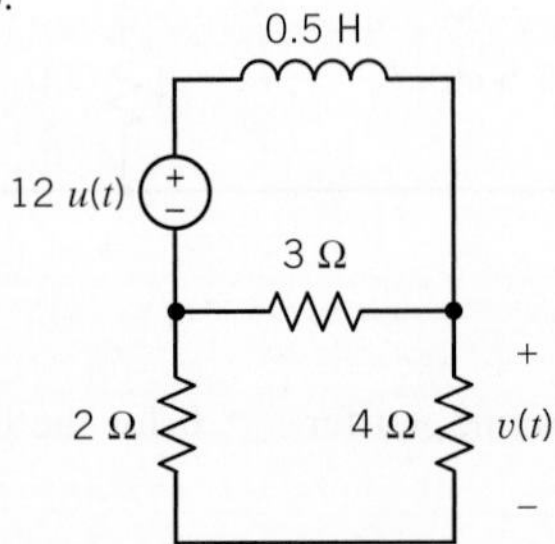

Figure P 8.6-26

***P 8.6-27** When the input to the circuit shown in Figure P 8.6-27 is the voltage source voltage

$$v_s(t) = 3 - u(t) \text{ V}$$

the output is the voltage

$$v_o(t) = 10 + 5\,e^{-50t} \text{ V} \quad \text{for } t \geq 0$$

Determine the values of R_1 and R_2.

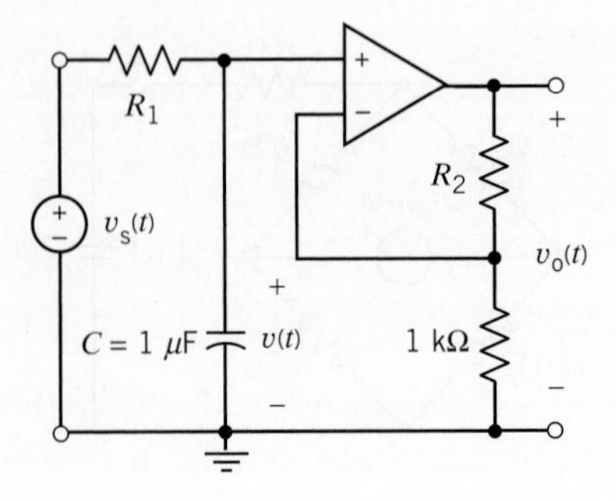

Figure P 8.6-27

P 8.6-28 The time constant of a particular circuit is $\tau = 0.25$ s. In response to a step input, a capacitor voltage changes from -3 V to 5 V. How long did it take for the capacitor voltage to increase from -2.0 V to $+2.0$ V?

Section 8.7 The Response of a First-Order Circuit to a Nonconstant Source

P 8.7-1 Find $v_c(t)$ for $t > 0$ for the circuit shown in Figure P 8.7-1 when $v_1 = 8e^{-5t}u(t)$ V. Assume the circuit is in steady state at $t = 0^-$.

Answer: $v_c(t) = 4e^{-9t} + 18e^{-5t}$ V

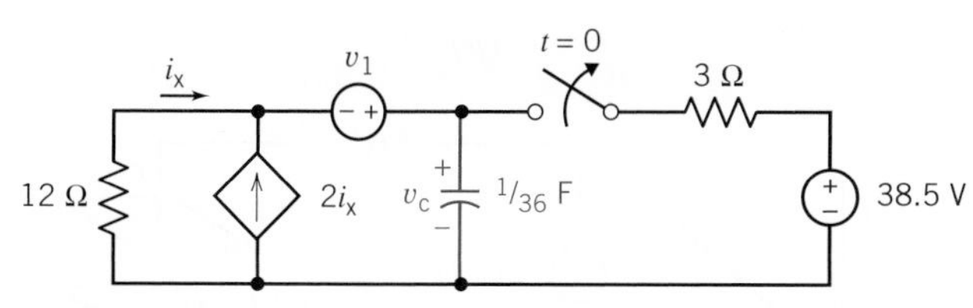

Figure P 8.7-1

P 8.7-2 Find $v(t)$ for $t > 0$ for the circuit shown in Figure P 8.7-2. Assume steady state at $t = 0^-$.

Answer: $v(t) = 20e^{-10t/3} - 12e^{-2t}$ V

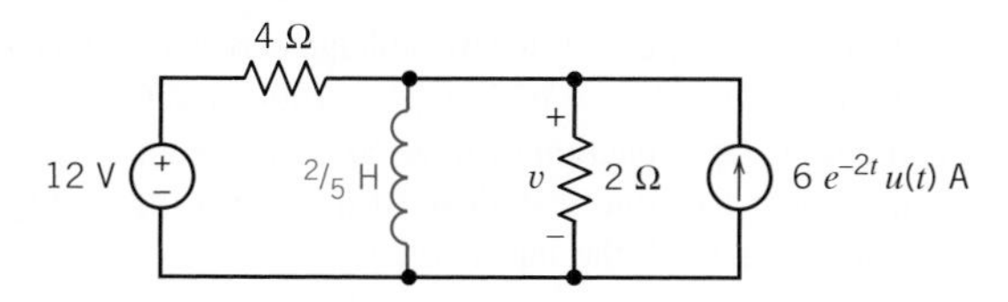

Figure P 8.7-2

P 8.7-3 Find $v(t)$ for $t > 0$ for the circuit shown in Figure P 8.7-3 when $v_1 = (25 \sin 4000t)u(t)$ V. Assume steady state at $t = 0^-$.

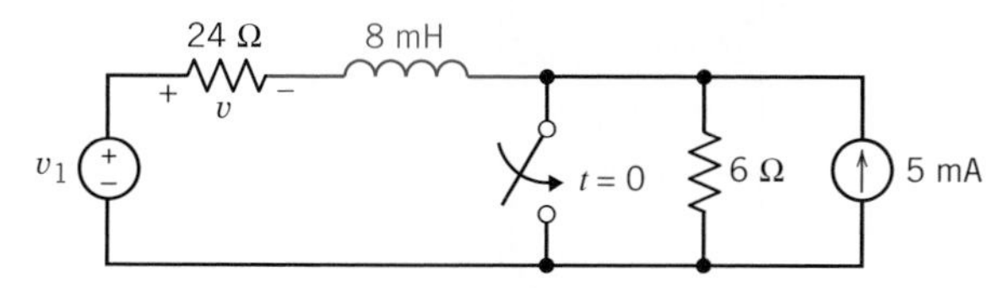

Figure P 8.7-3

P 8.7-4 Find $v_c(t)$ for $t > 0$ for the circuit shown in Figure P 8.7-4 when $i_s = [2 \cos 2t]\ u(t)$ mA.

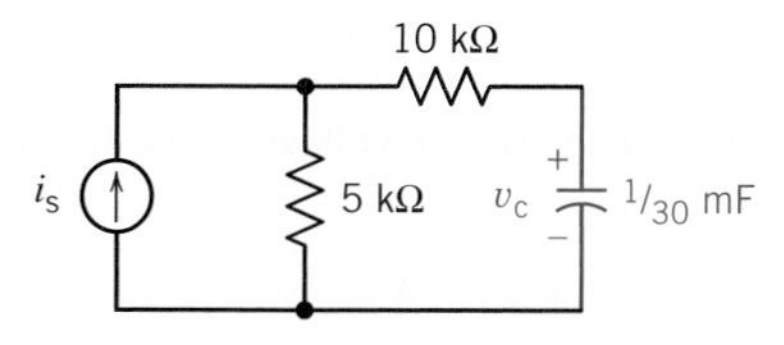

Figure P 8.7-4

P 8.7-5 Many have witnessed the use of an electrical megaphone for amplification of speech to a crowd. A model of a microphone and speaker is shown in Figure P 8.7-5*a*, and the circuit model is shown in Figure P 8.7-5*b*. Find $v(t)$ for $v_s = 10$ (sin $100t)u(t)$, which could represent a person whistling or singing a pure tone.

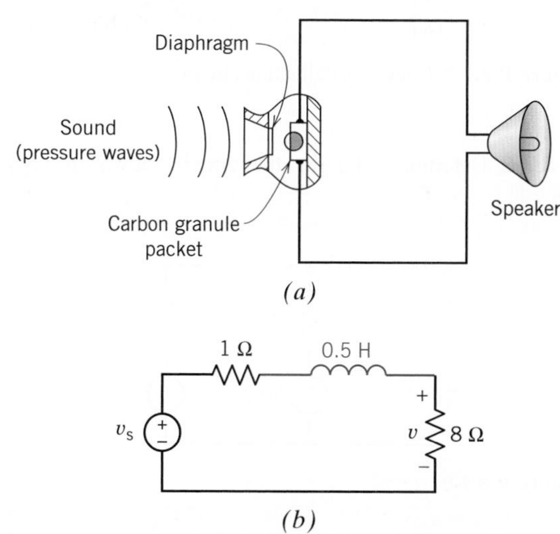

Figure P 8.7-5 Megaphone circuit.

P 8.7-6 A lossy integrator is shown in Figure P 8.7-6. The lossless capacitor of the ideal integrator circuit has been replaced with a model for the lossy capacitor, namely, a lossless capacitor in parallel with a 1-kΩ resistor. If $v_s = 15e^{-2t}u(t)$ V and $v_o(0) = 10$ V, find $v_o(t)$ for $t > 0$. Assume an ideal op amp.

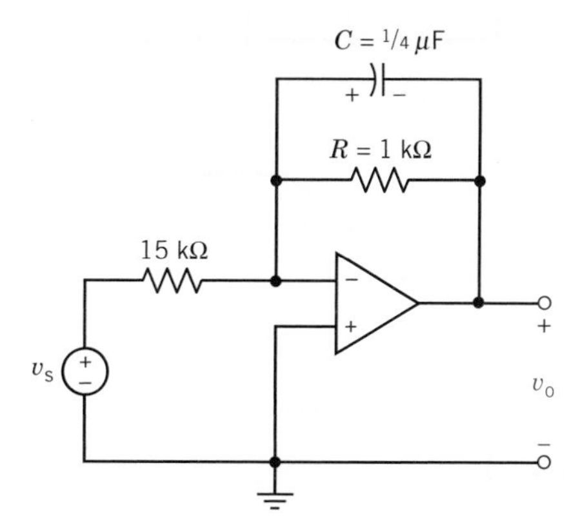

Figure P 8.7-6 Integrator circuit.

P 8.7-7 Most television sets use magnetic deflection in the cathode-ray tube. To move the electron beam across the screen, it is necessary to have a ramp of current, as shown in Figure P 8.7-7*a*, to flow through the deflection coil. The deflection coil circuit is shown in Figure P 8.7-7*b*. Find the waveform v_1 that will generate the current ramp, i_L.

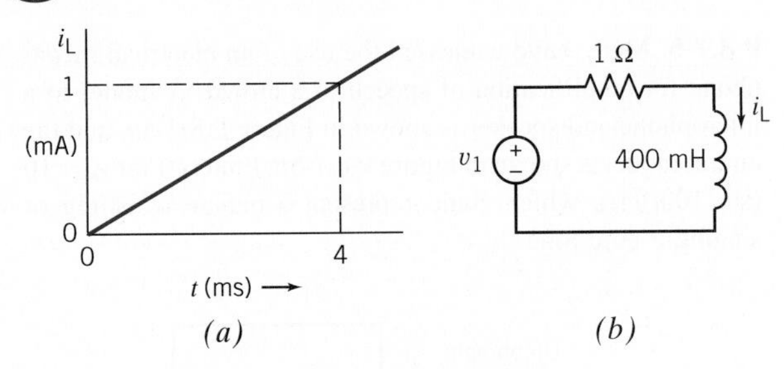

Figure P 8.7-7 Television deflection circuit.

P 8.7-8 Determine $v(t)$ for the circuit shown in Figure P 8.7-8.

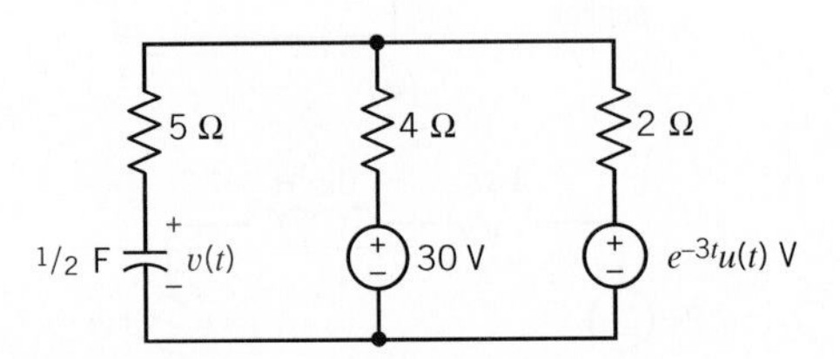

Figure P 8.7-8

P 8.7-9 Determine $v(t)$ for the circuit shown in Figure P 8.7-9a when v_s varies as shown in Figure P 8.7-9b. The initial capacitor voltage is $v_c(0) = 0$.

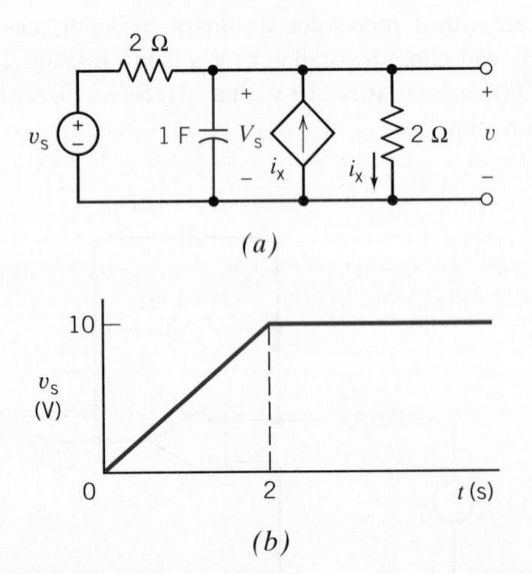

Figure P 8.7-9

P 8.7-10 The electron beam, which is used to draw signals on an oscilloscope, is moved across the face of a cathode-ray tube (CRT) by a force exerted on electrons in the beam. The basic system is shown in Figure P 8.7-10a. The force is created from a time-varying, ramp-type voltage applied across the vertical or the horizontal plates. As an example, consider the simple circuit of Figure P 8.7-10b for horizontal deflection in which the capacitance between the plates is C.

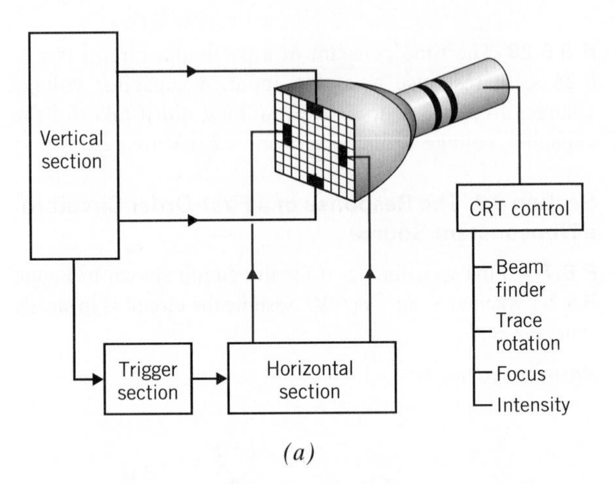

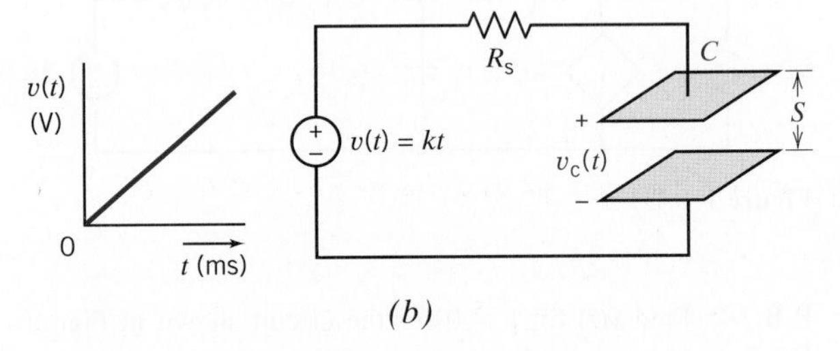

Figure P 8.7-10 Cathode-ray tube beam circuit.

Derive an expression for the voltage across the capacitance. If $v(t) = kt$ and $R_s = 625$ kΩ, $k = 1000$, and $C = 2000$ pF, compute v_c as a function of time. Sketch $v(t)$ and $v_c(t)$ on the same graph for time less than 10 ms. Does the voltage across the plates track the input voltage?

P 8.7-11 Determine the voltage $v(t)$ for $t \geq 0$ for the circuit shown in Figure P 8.7-11.

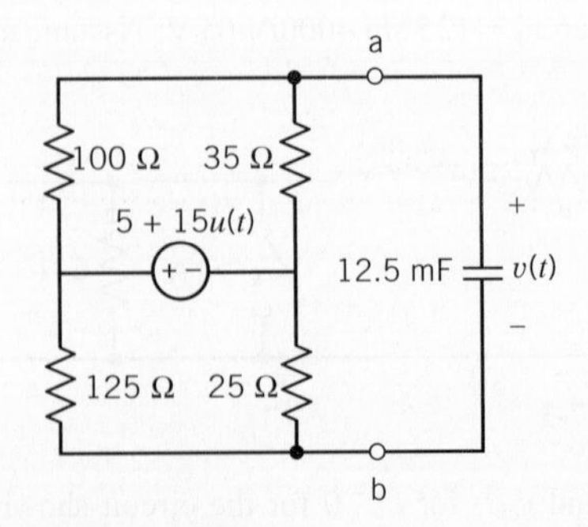

Figure P 8.7-11

P 8.7-12 The voltage source voltage in the circuit shown in Figure P 8.7-12 is

$$v_s(t) = 5 + 20u(t)$$

Determine $i(t)$ for $t \geq 0$.

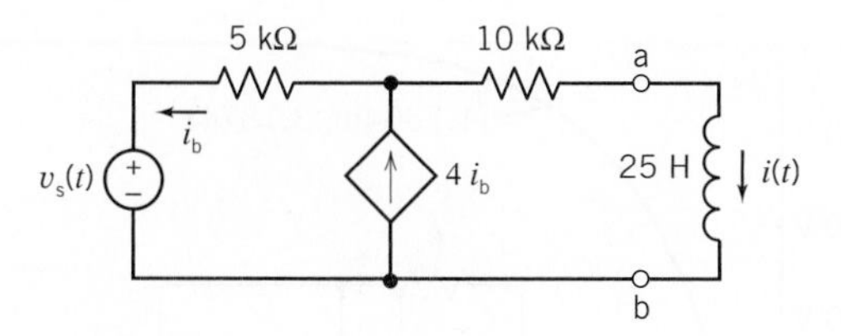

Figure P 8.7-12

P 8.7-13 Find the current i in the circuit of Figure P 8.7-13 for $t > 0$ when $i_s = 10e^{-5t}u(t)$ A and $i(0^-) = 0$.

Answer: $i = 10.53(e^{-5t} - e^{-100t})$ A

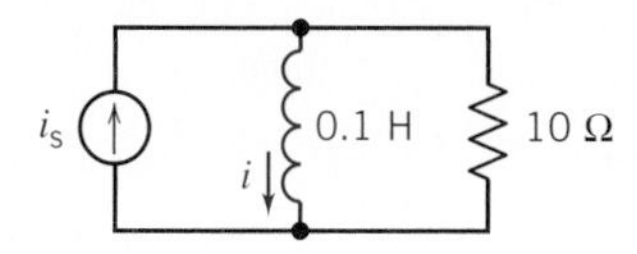

Figure P 8.7-13

P 8.7-14 An experimenter is working in her laboratory with an electromagnet as shown in Figure P 8.7-14. She notices that whenever she turns off the electromagnet, a big spark appears at the switch contacts. Explain the occurrence of the spark. Suggest a way to suppress the spark by adding one element.

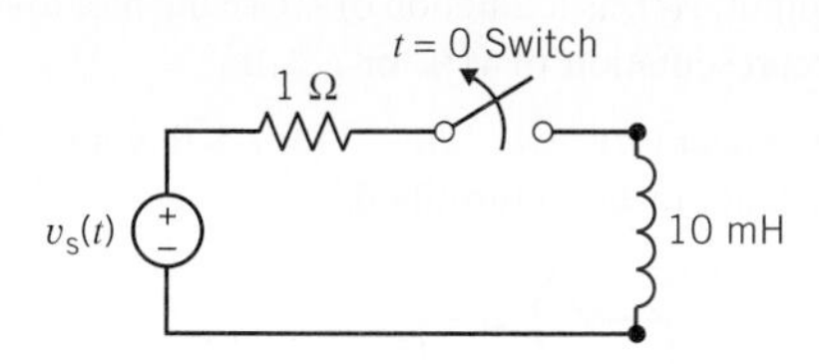

Figure P 8.7-14

Section 8.10 How Can We Check . . . ?

P 8.10-1 Figure P 8.10-1 shows the transient response of a first-order circuit. This transient response was obtained using the computer program, PSpice. A point on this transient response has been labeled. The label indicates a time and the capacitor voltage at that time. Placing the circuit diagram on the plot suggests that the plot corresponds to the circuit. Verify that the plot does indeed represent the voltage of the capacitor in this circuit.

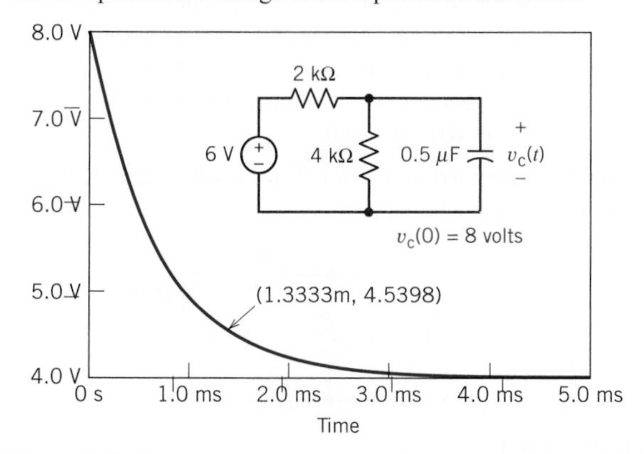

Figure P 8.10-1

P 8.10-2 Figure P 8.10-2 shows the transient response of a first-order circuit. This transient response was obtained using the computer program, PSpice. A point on this transient response has been labeled. The label indicates a time and the inductor current at that time. Placing the circuit diagram on the plot suggests that the plot corresponds to the circuit. Verify that the plot does indeed represent the current of the inductor in this circuit.

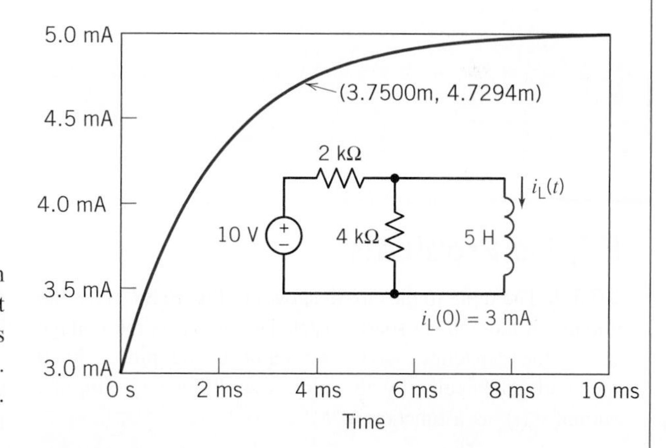

Figure P 8.10-2

P 8.10-3 Figure P 8.10-3 shows the transient response of a first-order circuit. This transient response was obtained using the computer program, PSpice. A point on this transient response has been labeled. The label indicates a time and the inductor current at that time. Placing the circuit diagram on the plot suggests that the plot corresponds to the circuit. Specify that value of the inductance, L, required to cause the current of the inductor in this circuit to be accurately represented by this plot.

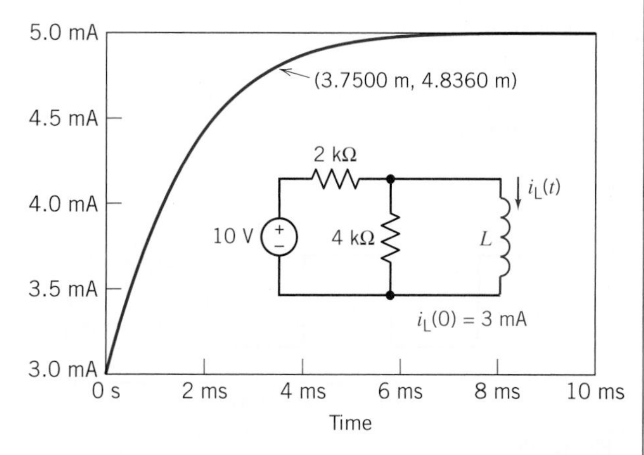

Figure P 8.10-3

P 8.10-4 Figure P 8.10-4 shows the transient response of a first-order circuit. This transient response was obtained using the computer program, PSpice. A point on this transient response has been labeled. The label indicates a time and the capacitor voltage at that time. Assume that this circuit has reached steady state before time $t = 0$. Placing the circuit diagram on the plot suggests that the plot corresponds to the circuit. Specify values of A, B, R_1, R_2, and C that cause the voltage across the capacitor in this circuit to be accurately represented by this plot.

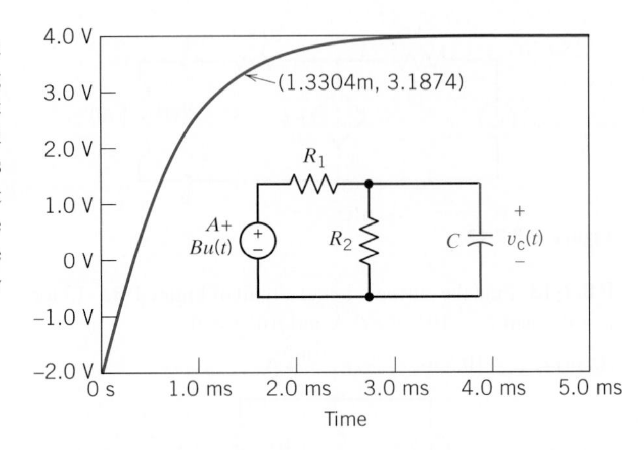

Figure P 8.10-4

PSpice Problems

SP 8-1 The input to the circuit shown in Figure SP 8-1 is the voltage of the voltage source, $v_i(t)$. The output is the voltage across the capacitor, $v_o(t)$. The input is the pulse signal specified graphically by the plot. Use PSpice to plot the output, $v_o(t)$, as a function of t.

Hint: Represent the voltage source, using the PSpice part named VPULSE.

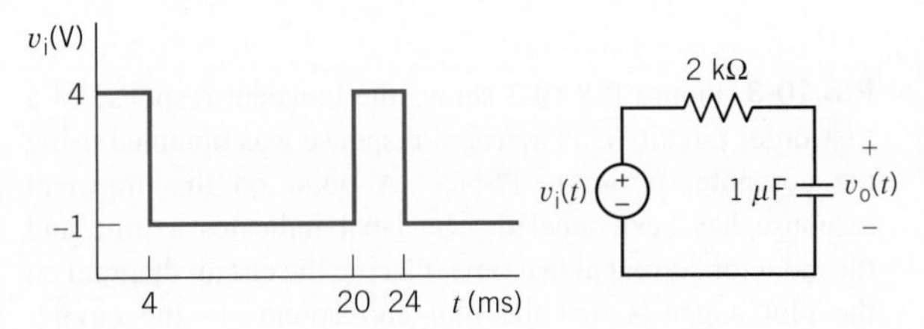

Figure SP 8-1

SP 8-2 The input to the circuit shown in Figure SP 8-2 is the voltage of the voltage source, $v_i(t)$. The output is the current in the inductor, $i_o(t)$. The input is the pulse signal specified graphically by the plot. Use PSpice to plot the output, $i_o(t)$, as a function of t.

Hint: Represent the voltage source, using the PSpice part named VPULSE.

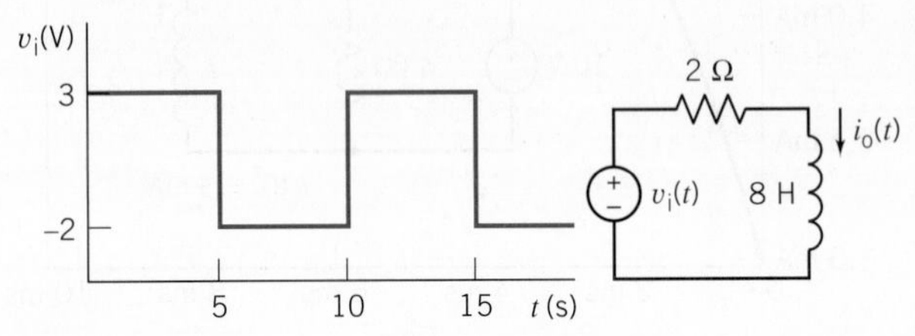

Figure SP 8-2

SP 8-3 The circuit shown in Figure SP 8-3 is at steady state before the switch closes at time $t = 0$. The input to the circuit is the voltage of the voltage source, 12 V. The output of this circuit is the voltage across the capacitor, $v(t)$. Use PSpice to plot the output, $v(t)$, as a function of t. Use the plot to obtain an analytic representation of $v(t)$ for $t > 0$.

Hint: We expect $v(t) = A + Be^{-t/\tau}$ for $t > 0$, where A, B, and τ are constants to be determined.

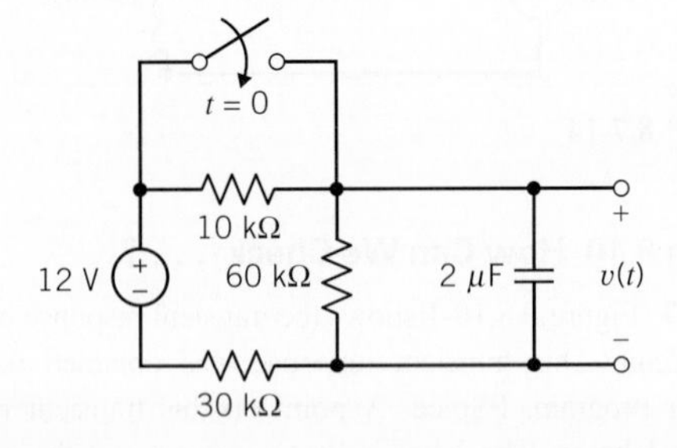

Figure SP 8-3

SP 8-4 The circuit shown in Figure SP 8-4 is at steady state before the switch closes at time $t = 0$. The input to the circuit is the current of the current source, 4 mA. The output of this circuit is the current in the inductor, $i(t)$. Use PSpice to plot the output, $i(t)$, as a function of t. Use the plot to obtain an analytic representation of $i(t)$ for $t > 0$.

Hint: We expect $i(t) = A + Be^{-t/\tau}$ for $t > 0$, where A, B, and τ are constants to be determined.

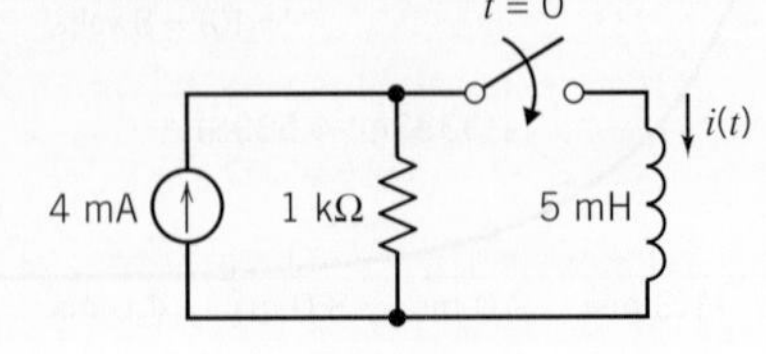

Figure SP 8-4

Design Problems

DP 8-1 Design the circuit in Figure DP 8-1 so that $v(t)$ makes the transition from $v(t) = 6$ V to $v(t) = 10$ V in 10 ms after the switch is closed. Assume that the circuit is at steady state before the switch is closed. Also assume that the transition will be complete after 5 time constants.

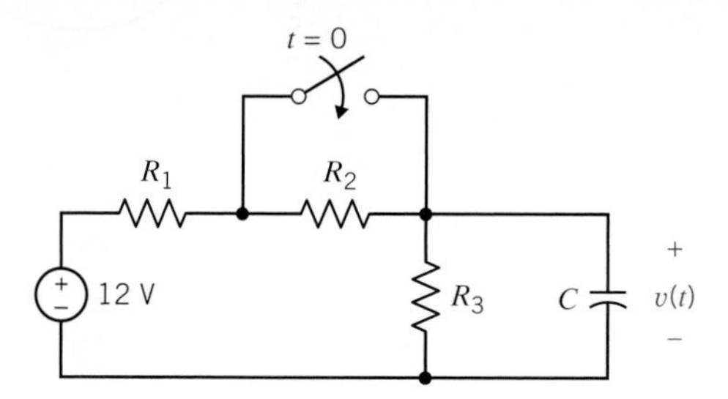

Figure DP 8-1

DP 8-2 Design the circuit in Figure DP 8-2 so that $i(t)$ makes the transition from $i(t) = 1$ mA to $i(t) = 6$ mA in 10 ms after the switch is closed. Assume that the circuit is at steady state before the switch is closed. Also assume that the transition will be complete after 5 time constants.

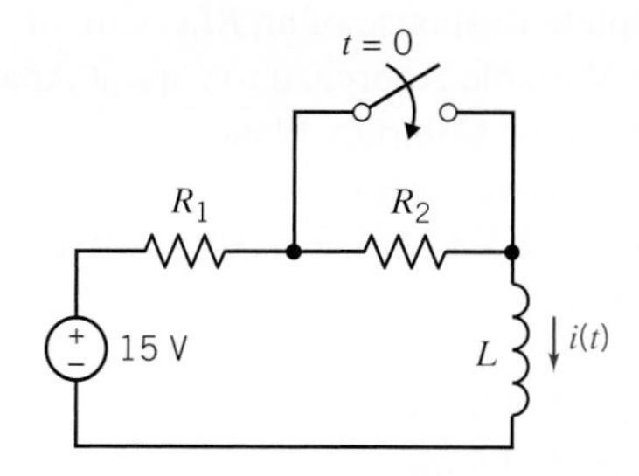

Figure DP 8-2

DP 8-3 The switch in Figure DP 8-3 closes at time 0, $2\Delta t$, $4\Delta t$, $\ldots 2k\Delta t$ and opens at times Δt, $3\Delta t$, $5\Delta t$, $\ldots .(2k+1)\Delta t$. When the switch closes, $v(t)$ makes the transition from $v(t) = 0$ V to $v(t) =$ 5 V. Conversely, when the switch opens, $v(t)$ makes the transition from $v(t) = 5$ V to $v(t) = 0$ V. Suppose we require that $\Delta t = 5\tau$ so that one transition is complete before the next one begins. (a) Determine the value of C required so that $\Delta t = 1\ \mu$s. (b) How large must Δt be when $C = 2\ \mu$F?

Answer: (a) $C = 4$ pF; (b) $\Delta t = 0.5$s

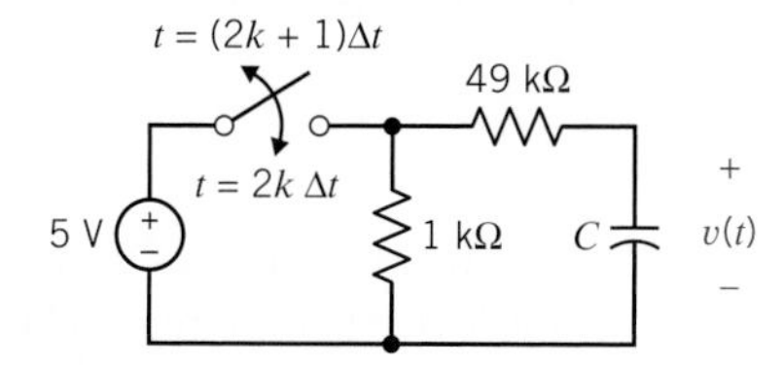

Figure DP 8-3

DP 8-4 The switch in Figure DP 8-3 closes at time 0, $2\Delta t$, $4\Delta t$, $\ldots 2k\Delta t$ and opens at times Δt, $3\Delta t$, $5\Delta t$, $\ldots .\ (2k+1)\Delta t$. When the switch closes, $v(t)$ makes the transition from $v(t) = 0$ V to $v(t) = 5$ V. Conversely, when the switch opens, $v(t)$ makes the transition from $v(t) = 5$ V to $v(t) = 0$ V. Suppose we require that one transition be 95 percent complete before the next one begins. (a) Determine the value of C required so that $\Delta t = 1\ \mu$s. (b) How large must Δt be when $C = 2\ \mu$F?

Hint: Show that $\Delta t = -\tau \ln(1 - k)$ is required for the transition to be 100 k percent complete.

Answer: (a) $C = 6.67$ pF; (b) $\Delta t = 0.3$ s

DP 8-5 A laser trigger circuit is shown in Figure DP 8-5. To trigger the laser, we require 60 mA $< |i| <$ 180 mA for $0 < t <$ 200 μs. Determine a suitable value for R_1 and R_2.

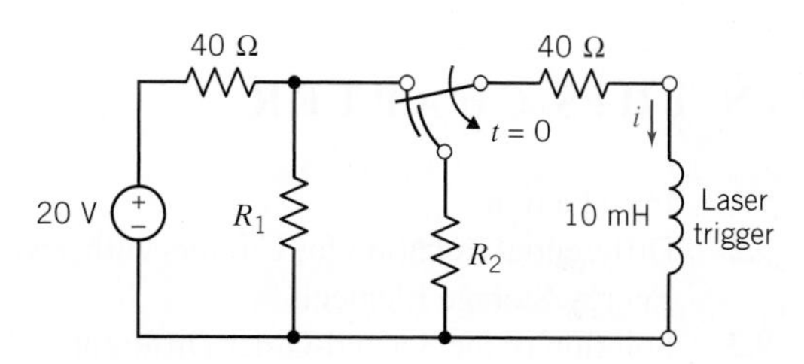

Figure DP 8-5 Laser trigger circuit

DP 8-6 Fuses are used to open a circuit when excessive current flows (Wright, 1990). One fuse is designed to open when the power absorbed by R exceeds 10 W for 0.5 s. Consider the circuit shown in Figure DP 8-6. The input is given by $v_s = A[u(t) - u(t - 0.75)]$ V. Assume that $i_L(0^-) = 0$. Determine the largest value of A that will not cause the fuse to open.

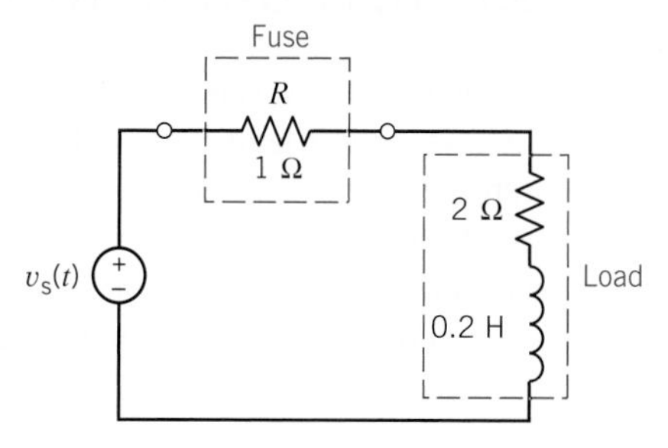

Figure DP 8-6 Fuse circuit.

The Complete Response of Circuits with Two Energy Storage Elements

IN THIS CHAPTER

9.1 Introduction
9.2 Differential Equation for Circuits with Two Energy Storage Elements
9.3 Solution of the Second-Order Differential Equation—The Natural Response
9.4 Natural Response of the Unforced Parallel *RLC* Circuit
9.5 Natural Response of the Critically Damped Unforced Parallel *RLC* Circuit
9.6 Natural Response of an Underdamped Unforced Parallel *RLC* Circuit
9.7 Forced Response of an *RLC* Circuit
9.8 Complete Response of an *RLC* Circuit
9.9 State Variable Approach to Circuit Analysis
9.10 Roots in the Complex Plane
9.11 How Can We Check . . . ?
9.12 **DESIGN EXAMPLE**—Auto Airbag Igniter
9.13 Summary
Problems
PSpice Problems
Design Problems

9.1 INTRODUCTION

In this chapter, we consider second-order circuits. A second-order circuit is a circuit that is represented by a second-order differential equation. As a rule of thumb, the order of the differential equation that represents a circuit is equal to the number of capacitors in the circuit plus the number of inductors. For example, a second-order circuit might contain one capacitor and one inductor, or it might contain two capacitors and no inductors.

For example, a second-order circuit could be represented by the equation

$$\frac{d^2}{dt^2}x(t) + 2\alpha\frac{d}{dt}x(t) + \omega_0^2\,x(t) = f(t)$$

where $x(t)$ is the output of the circuit, and $f(t)$ is the input to the circuit. The output of the circuit, also called the response of the circuit, can be the current or voltage of any device in the circuit. The output is frequently chosen to be the current of an inductor or the voltage of a capacitor. The voltages of independent voltage sources and/or currents of independent current sources provide the input to the circuit. The coefficients of this differential equation have names: α is called the damping coefficient, and ω_0 is called the resonant frequency.

To find the response of the second-order circuit, we:

- Represent the circuit by a second-order differential equation.
- Find the general solution of the homogeneous differential equation. This solution is the natural response, $x_n(t)$. The natural response will contain two unknown constants that will be evaluated later.
- Find a particular solution of the differential equation. This solution is the forced response, $x_f(t)$.
- Represent the response of the second-order circuit as $x(t) = x_n(t) + x_f(t)$.
- Use the initial conditions, for example, the initial values of the currents in inductors and the voltages across capacitors, to evaluate the unknown constants.

9.2 DIFFERENTIAL EQUATION FOR CIRCUITS WITH TWO ENERGY STORAGE ELEMENTS

In Chapter 8, we considered circuits that contained only one energy storage element, and these could be described by a first-order differential equation. In this section, we consider the description of circuits with two irreducible energy storage elements that are described by a second-order differential equation. Later, we will consider circuits with three or more irreducible energy storage elements that are described by a third-order (or higher) differential equation. We use the term *irreducible* to indicate that all parallel or series connections or other reducible combinations of like storage elements have been reduced to their irreducible form. Thus, for example, any parallel capacitors have been reduced to an equivalent capacitor, C_p.

In the following paragraphs, we use two methods to obtain the second-order differential equation for circuits with two energy storage elements. Then, in the next section, we obtain the solution to these second-order differential equations.

First, let us consider the circuit shown in Figure 9.2-1, which consists of a parallel combination of a resistor, an inductor, and a capacitor. Writing the nodal equation at the top node, we have

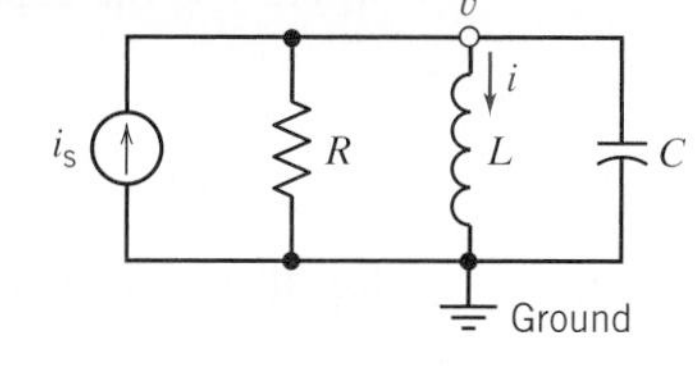

FIGURE 9.2-1 A parallel *RLC* circuit.

$$\frac{v}{R} + i + C\frac{dv}{dt} = i_s \tag{9.2-1}$$

Then we write the equation for the inductor as

$$v = L\frac{di}{dt} \tag{9.2-2}$$

Substitute Eq. 9.2-2 into Eq. 9.2-1, obtaining

$$\frac{L}{R}\frac{di}{dt} + i + CL\frac{d^2 i}{dt^2} = i_s \tag{9.2-3}$$

which is the second-order differential equation we seek. Solve this equation for $i(t)$. If $v(t)$ is required, use Eq. 9.2-2 to obtain it.

This method of obtaining the second-order differential equation may be called the *direct method* and is summarized in Table 9.2-1.

In Table 9.2-1, the circuit variables are called x_1 and x_2. In any example, x_1 and x_2 will be specific element currents or voltages. When we analyzed the circuit of Figure 9.2-1, we used $x_1 = v$ and $x_2 = i$. In contrast, to analyze the circuit of Figure 9.2-2, we will use $x_1 = i$ and $x_2 = v$, where i is the inductor current and v is the capacitor voltage.

Now let us consider the *RLC* series circuit shown in Figure 9.2-2 and use the direct method to obtain the second-order differential equation. We chose $x_1 = i$ and $x_2 = v$. First, we seek an equation

Table 9.2-1 The Direct Method for Obtaining the Second-Order Differential Equation of a Circuit

Step 1	Identify the first and second variables, x_1 and x_2. These variables are capacitor voltages and/or inductor currents.
Step 2	Write one first-order differential equation, obtaining $\frac{dx_1}{dt} = f(x_1, x_2)$.
Step 3	Obtain an additional first-order differential equation in terms of the second variable so that $\frac{dx_2}{dt} = Kx_1$ or $x_1 = \frac{1}{K}\frac{dx_2}{dt}$.
Step 4	Substitute the equation of step 3 into the equation of step 2, thus obtaining a second-order differential equation in terms of x_2.

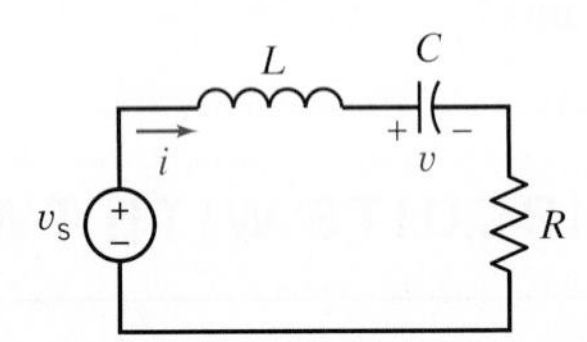

FIGURE 9.2-2 A series *RLC* circuit.

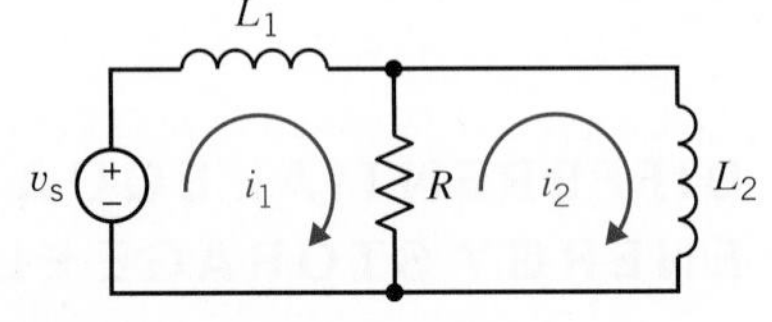

FIGURE 9.2-3 Circuit with two inductors.

for $dx_1/dt = di/dt$. Writing KVL around the loop, we have

$$L\frac{di}{dt} + v + Ri = v_s \tag{9.2-4}$$

where v is the capacitor voltage. This equation may be written as

$$\frac{di}{dt} + \frac{v}{L} + \frac{R}{L}i = \frac{v_s}{L} \tag{9.2-5}$$

Recall $v = x_2$ and obtain an equation in terms of $\frac{dx_2}{dt}$. Because

$$C\frac{dv}{dt} = i \tag{9.2-6}$$

or

$$C\frac{dx_2}{dt} = x_1 \tag{9.2-7}$$

substitute Eq. 9.2-6 into Eq. 9.2-5 to obtain the desired second-order differential equation:

$$C\frac{d^2v}{dt^2} + \frac{v}{L} + \frac{RC}{L}\frac{dv}{dt} = \frac{v_s}{L} \tag{9.2-8}$$

Equation 9.2-8 may be rewritten as

$$\frac{d^2v}{dt^2} + \frac{R}{L}\frac{dv}{dt} + \frac{1}{LC}v = \frac{v_s}{LC} \tag{9.2-9}$$

Another method of obtaining the second-order differential equation describing a circuit is called the *operator method*. First, we obtain differential equations describing node voltages or mesh currents and use operators to obtain the differential equation for the circuit.

As a more complicated example of a circuit with two energy storage elements, consider the circuit shown in Figure 9.2-3. This circuit has two inductors and can be described by the mesh currents as shown in Figure 9.2-3. The mesh equations are

$$L_1\frac{di_1}{dt} + R(i_1 - i_2) = v_s \tag{9.2-10}$$

and

$$R(i_2 - i_1) + L_2\frac{di_2}{dt} = 0 \tag{9.2-11}$$

Now, let us use $R = 1\ \Omega$, $L_1 = 1$ H, and $L_2 = 2$ H. Then we have

$$\frac{di_1}{dt} + i_1 - i_2 = v_s$$

and

$$i_2 - i_1 + 2\frac{di_2}{dt} = 0 \tag{9.2-12}$$

In terms of i_1 and i_2, we may rearrange these equations as

$$\frac{di_1}{dt} + i_1 - i_2 = v_s \tag{9.2-13}$$

and

$$-i_1 + i_2 + 2\frac{di_2}{dt} = 0 \tag{9.2-14}$$

It remains to obtain one second-order differential equation. This is done in the second step of the operator method. The differential operator s, where $s = d/dt$, is used to transform differential equations into algebraic equations. Upon replacing d/dt by s, Eqs. 9.2-13 and 9.2-14 become

$$si_1 + i_1 - i_2 = v_s$$

and

$$-i_1 + i_2 + 2si_2 = 0$$

These two equations may be rewritten as

$$(s+1)i_1 - i_2 = v_s$$

and

$$-i_1 + (2s+1)i_2 = 0$$

We may solve for i_2, obtaining

$$i_2 = \frac{1v_s}{(s+1)(2s+1) - 1} = \frac{v_s}{2s^2 + 3s}$$

Therefore,

$$\left(2s^2 + 3s\right)i_2 = v_s$$

Now, replacing s^2 by $\dfrac{d^2}{dt^2}$ and s by $\dfrac{d}{dt}$, we obtain the differential equation

$$2\frac{d^2 i_2}{dt^2} + 3\frac{di_2}{dt} = v_s \tag{9.2-15}$$

The operator method for obtaining the second-order differential equation is summarized in Table 9.2-2.

Table 9.2-2 Operator Method for Obtaining the Second-Order Differential Equation of a Circuit

Step 1	Identify the variable x_1 for which the solution is desired.
Step 2	Write one differential equation in terms of the desired variable x_1 and a second variable, x_2.
Step 3	Obtain an additional equation in terms of the second variable and the first variable.
Step 4	Use the operator $s = d/dt$ and $1/s = \int dt$ to obtain two algebraic equations in terms of s and the two variables x_1 and x_2.
Step 5	Using Cramer's rule, solve for the desired variable so that $x_1 = f(s, \text{sources}) = P(s)/Q(s)$, where $P(s)$ and $Q(s)$ are polynomials in s.
Step 6	Rearrange the equation of step 5 so that $Q(s)x_1 = P(s)$.
Step 7	Convert the operators back to derivatives for the equation of step 6 to obtain the second-order differential equation.

EXAMPLE 9.2-1 Representing a Circuit by a Differential Equation

Find the differential equation for the current i_2 for the circuit of Figure 9.2-4.

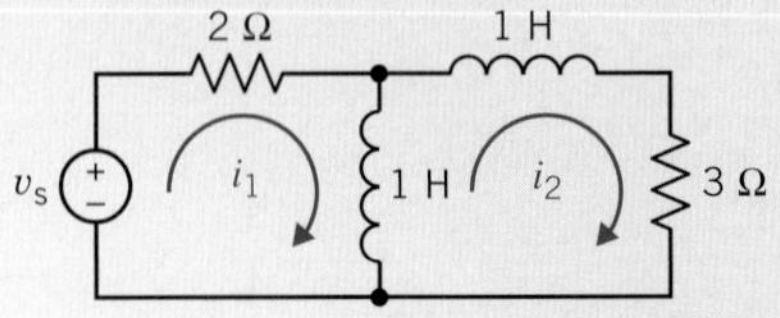

FIGURE 9.2-4 Circuit for Example 9.2-1.

Solution

Write the two mesh equations, using KVL to obtain

$$2i_1 + \frac{di_1}{dt} - \frac{di_2}{dt} = v_s$$

$$-\frac{di_1}{dt} + 3\,i_2 + 2\frac{di_2}{dt} = 0$$

Using the operator $s = d/dt$, we have

$$(2+s)i_1 - si_2 = v_s$$

and

$$-\,si_1 + (3+2s)i_2 = 0$$

Using Cramer's rule to solve for i_2, we obtain

$$i_2 = \frac{sv_s}{(2+s)(3+2s) - s^2} = \frac{sv_s}{s^2 + 7s + 6} \tag{9.2-16}$$

Rearranging Eq. 9.2-16, we obtain

$$\left(s^2 + 7s + 6\right)i_2 = sv_s \tag{9.2-17}$$

Therefore, the differential equation for i_2 is

$$\frac{d^2 i_2}{dt^2} + 7\frac{di_2}{dt} + 6i_2 = \frac{dv_s}{dt} \tag{9.2-18}$$

EXAMPLE 9.2-2 Representing a Circuit by a Differential Equation

Find the differential equation for the voltage v for the circuit of Figure 9.2-5.

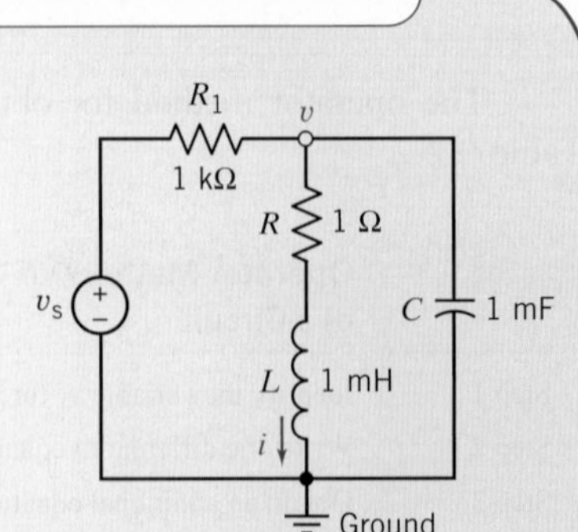

FIGURE 9.2-5 The *RLC* circuit for Example 9.2-2.

Solution

The KCL node equation at the upper node is

$$\frac{v - v_s}{R_1} + i + C\frac{dv}{dt} = 0 \tag{9.2-19}$$

Because we wish to determine the equation in terms of v, we need a second equation in terms of the current i. Write the equation for the current through the branch containing the inductor as

$$Ri + L\frac{di}{dt} = v \tag{9.2-20}$$

Using the operator $s = d/dt$, we have the two equations

$$\frac{v}{R_1} + Csv + i = \frac{v_s}{R_1}$$

and $$-v + Ri + Lsi = 0$$

Substituting the parameter values and rearranging, we have

$$\left(10^{-3} + 10^{-3}s\right)v + i = 10^{-3}v_s$$

and $$-v + \left(10^{-3}s + 1\right)i = 0$$

Using Cramer's rule, solve for v to obtain

$$v = \frac{(s+1000)v_s}{(s+1)(s+1000)+10^6} = \frac{(s+1000)v_s}{s^2 + 1001s + 1001 \times 10^3}$$

Therefore, we have

$$\left(s^2 + 1001s + 1001 \times 10^3\right)v = (s+1000)v_s$$

or the differential equation we seek is

$$\frac{d^2v}{dt^2} + 1001\frac{dv}{dt} + 1001 \times 10^3 v = \frac{dv_s}{dt} + 1000v_s$$

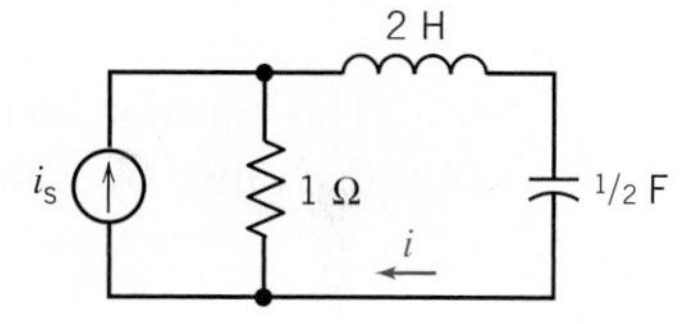

FIGURE E 9.2-1

EXERCISE 9.2-1 Find the second-order differential equation for the circuit shown in Figure E 9.2-1 in terms of i, using the direct method.

Answer: $\dfrac{d^2i}{dt^2} + \dfrac{1}{2}\dfrac{di}{dt} + i = \dfrac{1}{2}\dfrac{di_s}{dt}$

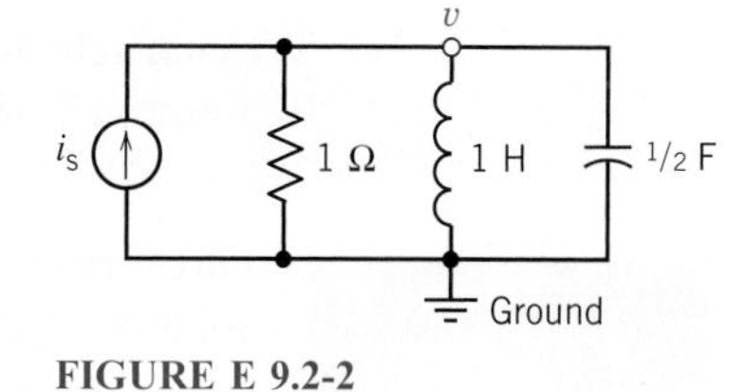

FIGURE E 9.2-2

EXERCISE 9.2-2 Find the second-order differential equation for the circuit shown in Figure E 9.2-2 in terms of v using the operator method.

Answer: $\dfrac{d^2v}{dt^2} + 2\dfrac{dv}{dt} + 2v = 2\dfrac{di_s}{dt}$

9.3 SOLUTION OF THE SECOND-ORDER DIFFERENTIAL EQUATION—THE NATURAL RESPONSE

In the preceding section, we found that a circuit with two irreducible energy storage elements can be represented by a second-order differential equation of the form

$$a_2\frac{d^2x}{dt^2} + a_1\frac{dx}{dt} + a_0x = f(t)$$

where the constants a_2, a_1, a_0 are known and the forcing function $f(t)$ is specified.

The complete response $x(t)$ is given by

$$x = x_n + x_f \tag{9.3-1}$$

where x_n is the natural response and x_f is a forced response. The natural response satisfies the unforced differential equation when $f(t) = 0$. The forced response x_f satisfies the differential equation with the forcing function present.

The natural response of a circuit, x_n, will satisfy the equation

$$a_2 \frac{d^2 x_n}{dt^2} + a_1 \frac{dx_n}{dt} + a_0 x_n = 0 \tag{9.3-2}$$

Because x_n and its derivatives must satisfy the equation, we postulate the exponential solution

$$x_n = Ae^{st} \tag{9.3-3}$$

where A and s are to be determined. The exponential is the only function that is proportional to all of its derivatives and integrals and, therefore, is the natural choice for the solution of a differential equation with constant coefficients. Substituting Eq. 9.3-3 in Eq. 9.3-2 and differentiating where required, we have

$$a_2 A s^2 e^{st} + a_1 A s e^{st} + a_0 A e^{st} = 0 \tag{9.3-4}$$

Because $x_n = Ae^{st}$, we may rewrite Eq. 9.3-4 as

$$a_2 s^2 x_n + a_1 s x_n + a_0 x_n = 0$$

or

$$\left(a_2 s^2 + a_1 s + a_0\right) x_n = 0$$

Because we do not accept the trivial solution, $x_n = 0$, it is required that

$$\left(a_2 s^2 + a_1 s + a_0\right) = 0 \tag{9.3-5}$$

This equation, in terms of s, is called a *characteristic equation*. It is readily obtained by replacing the derivative by s and the second derivative by s^2. Clearly, we have returned to the familiar operator

$$s^n = \frac{d^n}{dt^n}$$

The **characteristic equation** is derived from the governing differential equation for a circuit by setting all independent sources to zero value and assuming an exponential solution.

FIGURE 9.3-1 Oliver Heaviside (1850–1925). Photograph courtesy of the Institution of Electrical Engineers.

Oliver Heaviside (1850–1925), shown in Figure 9.3-1, advanced the theory of operators for the solution of differential equations.

The solution of the quadratic equation (9.3-5) has two roots, s_1 and s_2, where

$$s_1 = \frac{-a_1 + \sqrt{a_1^2 - 4a_2 a_0}}{2a_2} \tag{9.3-6}$$

and

$$s_2 = \frac{-a_1 - \sqrt{a_1^2 - 4a_2 a_0}}{2a_2} \tag{9.3-7}$$

When there are two distinct roots, the natural response is of the form

$$x_n = A_1 e^{s_1 t} + A_2 e^{s_2 t} \tag{9.3-8}$$

where A_1 and A_2 are unknown constants that will be evaluated later. We will delay considering the special case when $s_1 = s_2$.

The **roots** of the characteristic equation contain all the information necessary for determining the character of the natural response.

EXAMPLE 9.3-1 Natural Response of a Second-Order Circuit

Find the natural response of the circuit current i_2 shown in Figure 9.3-2. Use operators to formulate the differential equation and obtain the response in terms of two arbitrary constants.

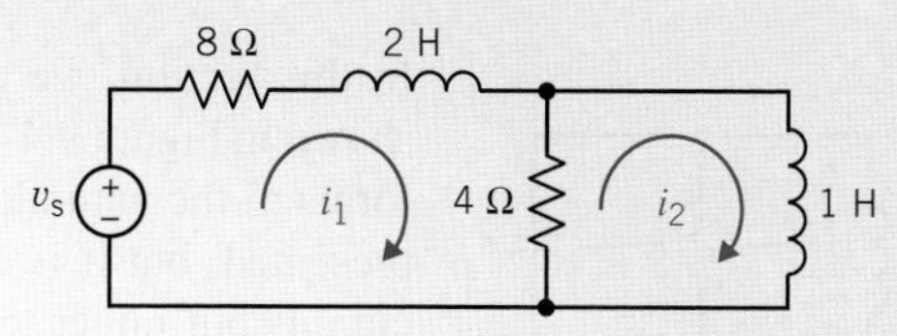

FIGURE 9.3-2 Circuit of Example 9.3-1.

Solution

Writing the two mesh equations, we have

$$12i_1 + 2\frac{di_1}{dt} - 4\,i_2 = v_s$$

and

$$-4\,i_1 + 4\,i_2 + 1\frac{di_2}{dt} = 0$$

Using the operator $s=d/dt$, we obtain

$$(12 + 2s)i_1 - 4i_2 = v_s \tag{9.3-9}$$

$$-4i_1 + (4 + s)i_2 = 0 \tag{9.3-10}$$

Solving for i_2, we have

$$i_2 = \frac{4\,v_s}{(12 + 2s)(4 + s) - 16} = \frac{4\,v_s}{2s^2 + 20s + 32} = \frac{2\,v_s}{s^2 + 10s + 16}$$

Therefore,

$$(s^2 + 10s + 16)i_2 = 2v_s$$

Note that $(s^2 + 10s + 16) = 0$ is the characteristic equation. Thus, the roots of the characteristic equation are $s_1 = -2$ and $s_2 = -8$. Therefore, the natural response is

$$x_n = A_1e^{-2t} + A_2e^{-8t}$$

where $x = i_2$. The roots s_1 and s_2 are the *characteristic roots* and are often called the *natural frequencies*. The reciprocals of the magnitude of the real characteristic roots are the *time constants*. The time constants of this circuit are $1/2$ s and $1/8$ s.

EXERCISE 9.3-1 Find the characteristic equation and the natural frequencies for the circuit shown in Figure E 9.3-1.

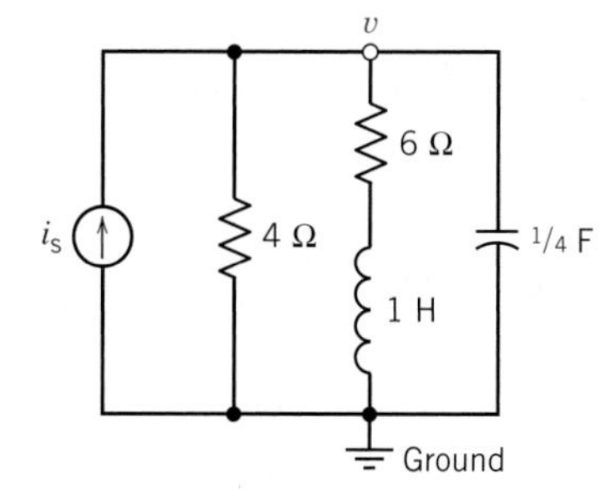

FIGURE E 9.3-1

Answer: $s^2 + 7s + 10 = 0$

$s_1 = -2$

$s_2 = -5$

9.4 NATURAL RESPONSE OF THE UNFORCED PARALLEL *RLC* CIRCUIT

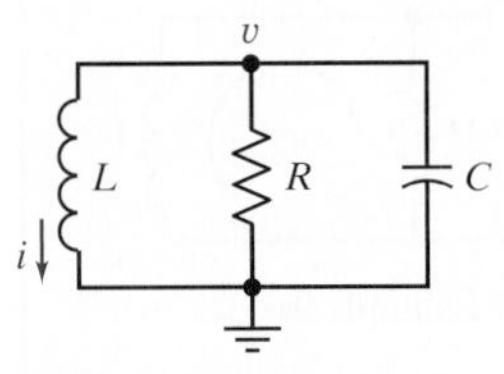

FIGURE 9.4-1 Parallel *RLC* circuit.

In this section, we consider the (unforced) natural response of the parallel *RLC* circuit shown in Figure 9.4-1. We choose to examine the parallel *RLC* circuit to illustrate the three forms of the natural response. An analogous discussion of the series *RLC* circuit could be presented, but it is omitted because the purpose is not to obtain the solution to specific circuits but rather to illustrate the general method.

A circuit that contains one capacitor and one inductor is represented by a second-order differential equation,

$$\frac{d^2}{dt^2}x(t) + 2\alpha\frac{d}{dt}x(t) + \omega_0^2 x(t) = f(t)$$

where $x(t)$ is the output of the circuit, and $f(t)$ is the input to the circuit. The output of the circuit, also called the response of the circuit, can be the current or voltage of any device in the circuit. The output is frequently chosen to be the current of an inductor or the voltage of a capacitor. The voltages of independent voltage sources and/or currents of independent current sources provide the input to the circuit. The coefficients of this differential equation have names: α is called the damping coefficient, and ω_0 is called the resonant frequency.

The circuit shown in Figure 9.4-1 does not contain any independent sources, so the input, $f(t)$, is zero. The differential equation with $f(t) = 0$ is called a homogeneous differential equation. We will take the output to be the voltage, $v(t)$, at the top node of the circuit. Consequently, we will represent the circuit in Figure 9.4-1 by a homogeneous differential equation of the form

$$\frac{d^2}{dt^2}v(t) + 2\alpha\frac{d}{dt}v(t) + \omega_0^2 v(t) = 0$$

Write the KCL at the top node to obtain

$$\frac{v}{R} + \frac{1}{L}\int_0^t v\,d\tau + i(0) + C\frac{dv}{dt} = 0 \tag{9.4-1}$$

Taking the derivative of Eq. 9.4-1, we have

$$C\frac{d^2v}{dt^2} + \frac{1}{R}\frac{dv}{dt} + \frac{1}{L}v = 0 \tag{9.4-2}$$

Dividing both sides of Eq. 9.4-2 by C, we have

$$\frac{d^2v}{dt^2} + \frac{1}{RC}\frac{dv}{dt} + \frac{1}{LC}v = 0 \tag{9.4-3}$$

Using the operator s, we obtain the characteristic equation

$$s^2 + \frac{1}{RC}s + \frac{1}{LC} = 0 \tag{9.4-4}$$

Comparing Eq. 9.4-4 to Eq. 9.4-1, we see

$$\alpha = \frac{1}{2RC} \quad \text{and} \quad \omega_0^2 = \frac{1}{LC} \tag{9.4-5}$$

The two roots of the characteristic equation are

$$s_1 = -\frac{1}{2RC} + \sqrt{\left(\frac{1}{2RC}\right)^2 - \frac{1}{LC}} \quad \text{and} \quad s_2 = -\frac{1}{2RC} - \sqrt{\left(\frac{1}{2RC}\right)^2 - \frac{1}{LC}} \tag{9.4-6}$$

When s_1 is not equal to s_2, the solution to the second-order differential Eq. 9.4-3 for $t > 0$ is

$$v_n = A_1 e^{s_1 t} + A_2 e^{s_2 t} \tag{9.4-7}$$

The roots of the characteristic equation may be rewritten as

$$s_1 = -\alpha + \sqrt{\alpha^2 - \omega_0^2} \quad \text{and} \quad s_2 = -\alpha - \sqrt{\alpha^2 - \omega_0^2} \tag{9.4-8}$$

The damped resonant frequency, ω_d, is defined to be

$$\omega_d = \sqrt{\omega_0^2 - \alpha^2}$$

When $\omega_0 > \alpha$, the roots of the characteristic equation are complex and can be expressed as

$$s_1 = -\alpha + j\omega_d \quad \text{and} \quad s_2 = -\alpha - j\omega_d$$

The roots of the characteristic equation assume three possible conditions:

1. Two real and distinct roots when $\alpha^2 > \omega_0^2$
2. Two real equal roots when $\alpha^2 = \omega_0^2$
3. Two complex roots when $\alpha^2 < \omega_0^2$

When the two roots are real and distinct, the circuit is said to be *overdamped*. When the roots are both real and equal, the circuit is *critically damped*. When the two roots are complex conjugates, the circuit is said to be *underdamped*.

Let us determine the natural response for the overdamped *RLC* circuit of Figure 9.4-1 when the initial conditions are $v(0)$ and $i(0)$ for the capacitor and the inductor, respectively. Notice that because the circuit in Figure 9.4-1 has no input, $v_n(0)$ and $v(0)$ are both names for the same voltage. Then, at $t = 0$ for Eq. 9.4-7, we have

$$v_n(0) = A_1 + A_2 \tag{9.4-9}$$

Because A_1 and A_2 are both unknown, we need one more equation at $t = 0$. Rewriting Eq. 9.4-1 at $t = 0$, we have[1]

$$\frac{v(0)}{R} + i(0) + C\frac{dv(0)}{dt} = 0$$

Because $i(0)$ and $v(0)$ are known, we have

$$\frac{dv(0)}{dt} = -\frac{v(0)}{RC} - \frac{i(0)}{C} \tag{9.4-10}$$

Thus, we now know the initial value of the derivative of v in terms of the initial conditions. Taking the derivative of Eq. 9.4-7 and setting $t = 0$, we obtain

$$\frac{dv_n(0)}{dt} = s_1 A_1 + s_2 A_2 \tag{9.4-11}$$

Using Eqs. 9.4-10 and 9.4-11, we obtain a second equation in terms of the two constants as

$$s_1 A_1 + s_2 A_2 = -\frac{v(0)}{RC} - \frac{i(0)}{C} \tag{9.4-12}$$

Using Eqs. 9.4-9 and 9.4-12, we may obtain A_1 and A_2.

[1] *Note*: $\dfrac{dv(0)}{dt}$ means $\left.\dfrac{dv(t)}{dt}\right|_{t=0}$

EXAMPLE 9.4-1 Natural Response of an Overdamped Second-Order Circuit

Find the natural response of $v(t)$ for $t > 0$ for the parallel *RLC* circuit shown in Figure 9.4-1 when $R = 2/3\ \Omega$, $L = 1$ H, $C = 1/2$ F, $v(0) = 10$ V, and $i(0) = 2$ A.

Solution

Using Eq. 9.4-4, the characteristic equation is

$$s^2 + \frac{1}{RC}s + \frac{1}{LC} = 0$$

or

$$s^2 + 3s + 2 = 0$$

Therefore, the roots of the characteristic equation are

$$s_1 = -1 \quad \text{and} \quad s_2 = -2$$

Then the natural response is

$$v_n = A_1 e^{-t} + A_2 e^{-2t} \tag{9.4-13}$$

The initial capacitor voltage is $v(0) = 10$, so we have

$$v_n(0) = A_1 + A_2$$

or

$$10 = A_1 + A_2 \tag{9.4-14}$$

We use Eq. 9.4-12 to obtain the second equation for the unknown constants. Then

$$s_1 A_1 + s_2 A_2 = -\frac{v(0)}{RC} - \frac{i(0)}{C}$$

or

$$-A_1 - 2A_2 = -\frac{10}{1/3} - \frac{2}{1/2}$$

Therefore, we have

$$-A_1 - 2A_2 = -34 \tag{9.4-15}$$

Solving Eqs. 9.4-14 and 9.4-15 simultaneously, we obtain $A_2 = 24$ and $A_1 = -14$. Therefore, the natural response is

$$v_n = \left(-14e^{-t} + 24e^{-2t}\right) \text{ V}$$

The natural response of the circuit is shown in Figure 9.4-2.

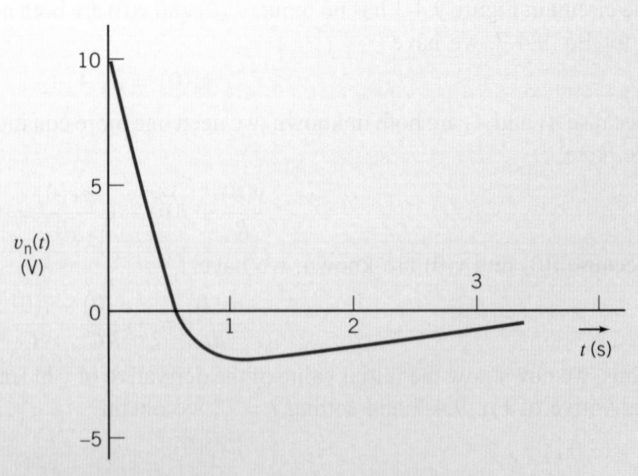

FIGURE 9.4-2 Response of the *RLC* circuit of Example 9.4-1.

EXERCISE 9.4-1 Find the natural response of the *RLC* circuit of Figure 9.4-1 when $R = 6\ \Omega$, $L = 7$ H, and $C = 1/42$ F. The initial conditions are $v(0) = 0$ and $i(0) = 10$ A.

Answer: $v_n(t) = -84(e^{-t} - e^{-6t})$ V

9.5 NATURAL RESPONSE OF THE CRITICALLY DAMPED UNFORCED PARALLEL *RLC* CIRCUIT

Again we consider the parallel *RLC* circuit, and here we will determine the special case when the characteristic equation has two equal real roots. Two real, equal roots occur when $\alpha^2 = \omega_0^2$, where

$$\alpha = \frac{1}{2RC} \quad \text{and} \quad \omega_0^2 = \frac{1}{LC}$$

Let us assume that $s_1 = s_2$ and proceed to find $v_n(t)$. We write the natural response as the sum of two exponentials as

$$v_n = A_1 e^{s_1 t} + A_2 e^{s_1 t} = A_3 e^{s_1 t} \qquad (9.5\text{-}1)$$

where $A_3 = A_1 + A_2$. Because the two roots are equal, we have only one undetermined constant, but we still have two initial conditions to satisfy. Clearly, Eq. 9.5-1 is not the total solution for the natural response of a critically damped circuit. We need the solution that will contain two arbitrary constants, so with some foreknowledge, we try the solution

$$v_n = e^{s_1 t}(A_1 t + A_2) \qquad (9.5\text{-}2)$$

Let us consider a parallel *RLC* circuit in which $L = 1$ H, $R = 1\ \Omega$, $C = 1/4$ F, $v(0) = 5$ V, and $i(0) = -6$ A. The characteristic equation for the circuit is

$$s^2 + \frac{1}{RC}s + \frac{1}{LC} = 0$$

or

$$s^2 + 4s + 4 = 0$$

The two roots are then $s_1 = s_2 = -2$. Using Eq. 9.5-2 for the natural response, we have

$$v_n = e^{-2t}(A_1 t + A_2) \qquad (9.5\text{-}3)$$

Because $v_n(0) = 5$, we have at $t = 0$

$$5 = A_2$$

Now, to obtain A_1, we proceed to find the derivative of v_n and evaluate it at $t = 0$. The derivative of v_n is found by differentiating Eq. 9.5-3 to obtain

$$\frac{dv}{dt} = -2A_1 t e^{-2t} + A_1 e^{-2t} - 2A_2 e^{-2t} \qquad (9.5\text{-}4)$$

Evaluating Eq. 9.5-4 at $t = 0$, we have

$$\frac{dv(0)}{dt} = A_1 - 2A_2$$

Again, we may use Eq. 9.4-10 so that

$$\frac{dv(0)}{dt} = -\frac{v(0)}{RC} - \frac{i(0)}{C}$$

or

$$A_1 - 2A_2 = \frac{-5}{1/4} - \frac{-6}{1/4} = 4$$

Therefore, $A_1 = 14$ and the natural response is

$$v_n = e^{-2t}(14t + 5) \quad \text{V}$$

The critically damped natural response of this *RLC* circuit is shown in Figure 9.5-1.

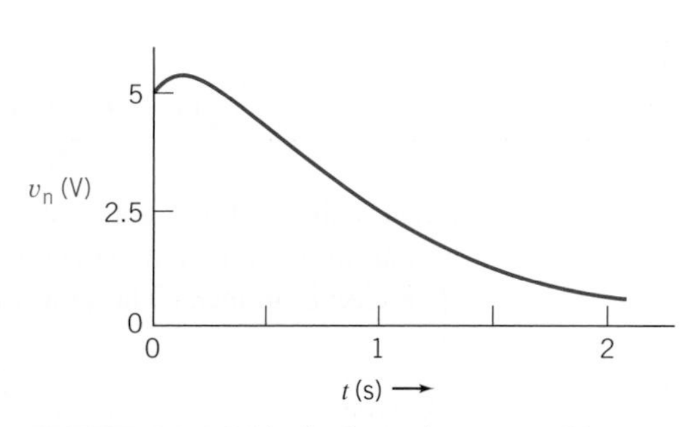

FIGURE 9.5-1 Critically damped response of the parallel *RLC* circuit.

EXERCISE 9.5-1 A parallel *RLC* circuit has $R = 10\ \Omega$, $C = 1$ mF, $L = 0.4$ H, $v(0) = 8$ V, and $i(0) = 0$. Find the natural response $v_n(t)$ for $t < 0$.

Answer: $v_n(t) = e^{-50t}(8 - 400t)$ V

9.6 NATURAL RESPONSE OF AN UNDERDAMPED UNFORCED PARALLEL *RLC* CIRCUIT

The characteristic equation of the parallel *RLC* circuit will have two complex conjugate roots when $\alpha^2 < \omega_0^2$. This condition is met when

$$LC < (2RC)^2$$

or when

$$L < 4R^2C$$

Recall that

$$v_n = A_1 e^{s_1 t} + A_2 e^{s_2 t} \tag{9.6-1}$$

where

$$s_{1,2} = -\alpha \pm \sqrt{\alpha^2 - \omega_0^2}$$

When

$$\omega_0^2 > \alpha^2$$

we have

$$s_{1,2} = -\alpha \pm j\sqrt{\omega_0^2 - \alpha_0^2}$$

where

$$j = \sqrt{-1}$$

See Appendix B for a review of complex numbers.

The complex roots lead to an oscillatory-type response. We define the square root $\sqrt{\omega_0^2 - \alpha^2}$ as ω_d, which we will call the *damped resonant frequency*. The factor α, called the *damping coefficient*, determines how quickly the oscillations subside. Then the roots are

$$s_{1,2} = -\alpha \pm j\omega_d$$

Therefore, the natural response is

$$v_n = A_1 e^{-\alpha t} e^{j\omega_d t} + A_2 e^{-\alpha t} e^{-j\omega_d t}$$

or

$$v_n = e^{-\alpha t}\left(A_1 e^{j\omega_d t} + A_2 e^{-j\omega_d t}\right) \tag{9.6-2}$$

Let us use the Euler identity[2]

$$e^{\pm j\omega t} = \cos \omega t \pm j \sin \omega t \tag{9.6-3}$$

Let $\omega = \omega_d$ in Eq. 9.6-3 and substitute into Eq. 9.6-2 to obtain

$$\begin{aligned} v_n &= e^{-\alpha t}(A_1 \cos \omega_d t + jA_1 \sin \omega_d t + A_2 \cos \omega_d t - jA_2 \sin \omega_d t) \\ &= e^{-\alpha t}[(A_1 + A_2) \cos \omega_d t + j(A_1 - A_2) \sin \omega_d t] \end{aligned} \tag{9.6-4}$$

Because the unknown constants A_1 and A_2 remain arbitrary, we replace $(A_1 + A_2)$ and $j(A_1 - A_2)$ with new arbitrary (yet unknown) constants B_1 and B_2. A_1 and A_2 must be complex conjugates so that B_1 and B_2 are real numbers. Therefore, Eq. 9.6-4 becomes

$$v_n = e^{-\alpha t}(B_1 \cos \omega_d t + B_2 \sin \omega_d t) \tag{9.6-5}$$

[2] See Appendix B for a discussion of Euler's identity.

where B_1 and B_2 will be determined by the initial conditions, $v(0)$ and $i(0)$.

The natural underdamped response is oscillatory with a decaying magnitude. The rapidity of decay depends on α, and the frequency of oscillation depends on ω_d.

Let us find the general form of the solution for B_1 and B_2 in terms of the initial conditions when the circuit is unforced. Then, at $t = 0$, we have

$$v_n(0) = B_1$$

To find B_2, we evaluate the first derivative of v_n and then let $t = 0$. The derivative is

$$\frac{dv_n}{dt} = e^{-\alpha t}[(\omega_d B_2 - \alpha B_1)\cos\omega_d t - (\omega_d B_1 + \alpha B_2)\sin\omega_d t]$$

and, at $t = 0$, we obtain

$$\frac{dv_n(0)}{dt} = \omega_d B_2 - \alpha B_1 \tag{9.6-6}$$

Recall that we found earlier that Eq. 9.4-10 provides $dv(0)/dt$ for the parallel *RLC* circuit as

$$\frac{dv_n(0)}{dt} = -\frac{v(0)}{RC} - \frac{i(0)}{C} \tag{9.6-7}$$

Therefore, we use Eqs. 9.6-6 and 9.6-7 to obtain

$$\omega_d B_2 = \alpha B_1 - \frac{v(0)}{RC} - \frac{i(0)}{C} \tag{9.6-8}$$

EXAMPLE 9.6-1 Natural Response of an Underdamped Second-Order Circuit

Consider the parallel *RLC* circuit when $R = 25/3\ \Omega$, $L = 0.1$ H, $C = 1$ mF, $v(0) = 10$ V, and $i(0) = -0.6$ A. Find the natural response $v_n(t)$ for $t > 0$.

Solution

First, we determine α^2 and ω_0^2 to determine the form of the response. Consequently, we obtain

$$\alpha = \frac{1}{2RC} = 60 \quad \text{and} \quad \omega_0^2 = \frac{1}{LC} = 10^4$$

Therefore, $\omega_0^2 > \alpha^2$, and the natural response is underdamped. We proceed to determine the damped resonant frequency ω_d as

$$\omega_d = \sqrt{\omega_0^2 - \alpha^2} = \sqrt{10^4 - 3.6 \times 10^3} = 80 \text{ rad/s}$$

Hence, the characteristic roots are

$$s_1 = -\alpha + j\omega_d = -60 + j80 \quad \text{and} \quad s_2 = -\alpha - j\omega_d = -60 - j80$$

Consequently, the natural response is obtained from Eq. 9.6-5 as

$$v_n(t) = B_1 e^{-60t}\cos 80t + B_2 e^{-60t}\sin 80t$$

Because $v(0) = 10$, we have

$$B_1 = v(0) = 10$$

We can use Eq. 9.6-8 to obtain B_2 as

$$B_2 = \frac{\alpha}{\omega_d}B_1 - \frac{v(0)}{\omega_d RC} - \frac{i(0)}{\omega_d C}$$
$$= \frac{60 \times 10}{80} - \frac{10}{80 \times 25/3000} - \frac{-0.6}{80 \times 10^{-3}} = 7.5 - 15.0 + 7.5 = 0$$

Therefore, the natural response is

$$v_n(t) = 10e^{-60t}\cos 80\,t \text{ V}$$

A sketch of this response is shown in Figure 9.6-1. Although the response is oscillatory in form because of the cosine function, it is damped by the exponential function, e^{-60t}.

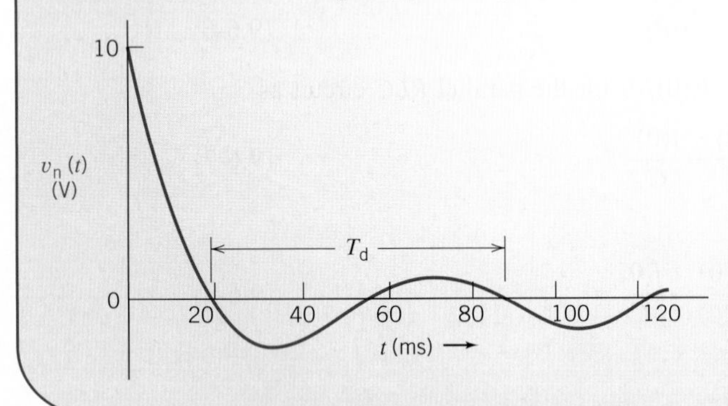

FIGURE 9.6-1 Natural response of the underdamped parallel *RLC* circuit.

The *period of the damped oscillation* is the time interval, denoted as T_d, expressed as

$$T_d = \frac{2\pi}{\omega_d} \quad (9.6\text{-}9)$$

The natural response of an underdamped circuit is not a pure oscillatory response, but it does exhibit the form of an oscillatory response. Thus, we may approximate T_d by the period between the first and third zero-crossings, as shown in Figure 9.6-1. Therefore, the frequency in hertz is

$$f_d = \frac{1}{T_d}$$

The period of the oscillation of the circuit of Example 9.6-1 is

$$T_d = \frac{2\pi}{80} = 79 \text{ ms}$$

EXERCISE 9.6-1 A parallel *RLC* circuit has $R = 62.5\ \Omega$, $L = 10$ mH, $C = 1\ \mu$F, $v(0) = 10$ V, and $i(0) = 80$ mA. Find the natural response $v_n(t)$ for $t > 0$.

Answer: $v_n(t) = e^{-8000t}[10\cos 6000t - 26.7\sin 6000t]$ V

9.7 FORCED RESPONSE OF AN *RLC* CIRCUIT

The forced response of an *RLC* circuit described by a second-order differential equation must satisfy the differential equation and contain no arbitrary constants. As we noted earlier, the response to a forcing function will often be of the same form as the forcing function. Again, we consider the

Table 9.7-1 Forced Responses

FORCING FUNCTION	ASSUMED RESPONSE
K	A
Kt	$At+B$
Kt^2	At^2+Bt+C
$K \sin \omega t$	$A \sin \omega t + B \cos \omega t$
Ke^{-at}	Ae^{-at}

differential equation for the second-order circuit as

$$\frac{d^2x}{dt^2} + a_1 \frac{dx}{dt} + a_0 x = f(t) \tag{9.7-1}$$

The forced response x_f must satisfy Eq. 9.7-1. Therefore, substituting x_f, we have

$$\frac{d^2x_f}{dt^2} + a_1 \frac{dx_f}{dt} + a_0 x_f = f(t) \tag{9.7-2}$$

We need to determine x_f so that x_f and its first and second derivatives all satisfy Eq. 9.7-2.

If the forcing function is a constant, we expect the forced response also to be a constant because the derivatives of a constant are zero. If the forcing function is of the form $f(t) = Be^{-at}$, then the derivatives of $f(t)$ are all exponentials of the form Qe^{-at}, and we expect

$$x_f = De^{-at}$$

If the forcing function is a sinusoidal function, we can expect the forced response to be a sinusoidal function. If $f(t) = A \sin \omega_0 t$, we will try

$$x_f = M \sin \omega_0 t + N \cos \omega_0 t = Q \sin(\omega_0 t + \theta)$$

Table 9.7-1 summarizes selected forcing functions and their associated assumed solutions.

EXAMPLE 9.7-1 Forced Response to an Exponential Input

Find the forced response for the inductor current i_f for the parallel *RLC* circuit shown in Figure 9.7-1 when $i_s = 8e^{-2t}$ A. Let $R = 6\ \Omega$, $L = 7$ H, and $C = 1/42$ F.

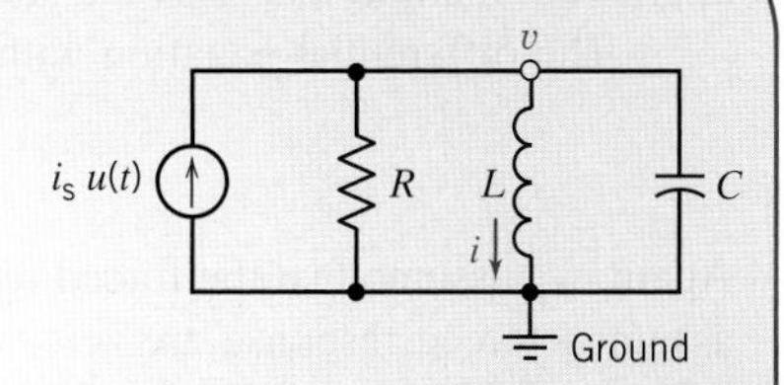

FIGURE 9.7-1 Circuit for Examples 9.7-1 and 9.7-2.

Solution

The source current is applied at $t = 0$ as indicated by the unit step function $u(t)$. After $t = 0$, the KCL equation at the upper node is

$$i + \frac{v}{R} + C\frac{dv}{dt} = i_s \tag{9.7-3}$$

We note that

$$v = L\frac{di}{dt} \tag{9.7-4}$$

so

$$\frac{dv}{dt} = L\frac{d^2i}{dt^2} \tag{9.7-5}$$

Substituting Eqs. 9.7-4 and 9.7-5 into Eq. 9.7-3, we have

$$i + \frac{L}{R}\frac{di}{dt} + CL\frac{di^2}{dt^2} = i_s$$

Then we divide by LC and rearrange to obtain the familiar second-order differential equation

$$\frac{d^2 i}{dt^2} + \frac{1}{RC}\frac{di}{dt} + \frac{1}{LC} i = \frac{i_s}{LC} \qquad (9.7\text{-}6)$$

Substituting the component values and the source i_s, we obtain

$$\frac{d^2 i}{dt^2} + 7\frac{di}{dt} + 6\,i = 48e^{-2t} \qquad (9.7\text{-}7)$$

We wish to obtain the forced response, so we assume that the response will be

$$i_f = Be^{-2t} \qquad (9.7\text{-}8)$$

where B is to be determined. Substituting the assumed solution, Eq. 9.7-8, into the differential equation, we have

$$4Be^{-2t} + 7\left(-2Be^{-2t}\right) + 6\,Be^{-2t} = 48e^{-2t}$$

or

$$(4 - 14 + 6)Be^{-2t} = 48e^{-2t}$$

Therefore, $B = -12$ and

$$i_f = -12e^{-2t}\text{ A}$$

EXAMPLE 9.7-2 Forced Response to a Constant Input

Find the forced response i_f of the circuit of Example 9.7-1 when $i_s = I_0$, where I_0 is a constant.

Solution

Because the source is a constant applied at $t = 0$, we expect the forced response to be a constant also. As a first method, we will use the differential equation to find the forced response. Second, we will demonstrate the alternative method that uses the steady-state behavior of the circuit to find i_f.

The differential equation with the constant source is obtained from Eq. 9.7-6 as

$$\frac{d^2 i}{dt^2} + 7\frac{di}{dt} + 6i = 6I_0$$

Again, we assume that the forced response is $i_f = D$, a constant. Because the first and second derivatives of the assumed forced response are zero, we have

$$6D = 6I_0$$

or

$$D = I_0$$

Therefore,

$$i_f = I_0$$

Another approach is to determine the steady-state response i_f of the circuit of Figure 9.7-1 by drawing the steady-state circuit model. The inductor acts like a short circuit, and the capacitor acts like an open circuit, as shown in Figure 9.7-2. Clearly, because the steady-state model of the inductor is a short circuit, all the source current flows through the inductor at steady state, and

$$i_f = I_0$$

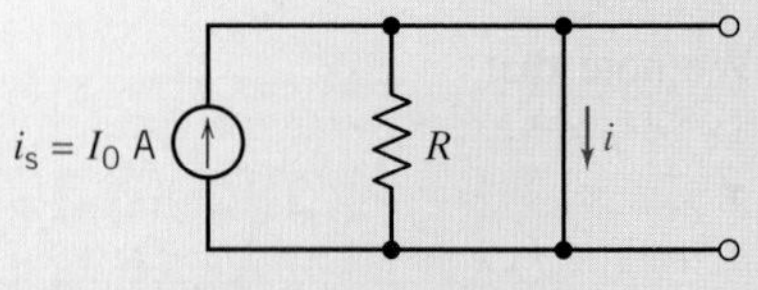

$i_s = I_0$ A R i

FIGURE 9.7-2 Parallel RLC circuit at steady state for a constant input.

The two previous examples showed that it is relatively easy to obtain the response of the circuit to a forcing function. However, we are sometimes confronted with a special case where the form of the forcing function is the same as the form of one of the components of the natural response.

Again, consider the circuit of Examples 9.7-1 and 9.7-2 (Figure 9.7-1) when the differential equation is

$$\frac{d^2 i}{dt^2} + 7\frac{di}{dt} + 6\,i = 6\,i_s \tag{9.7-9}$$

Suppose

$$i_s = 3\,e^{-6t}$$

Substituting this input into Eq. 9.7-9, we have

$$\frac{d^2 i}{dt^2} + 7\frac{di}{di} + 6i = 18\,e^{-6t} \tag{9.7-10}$$

The characteristic equation of the circuit is

$$s^2 + 7s + 6 = 0$$

or

$$(s+1)(s+6) = 0$$

Thus, the natural response is

$$i_n = A_1 e^{-t} + A_2 e^{-6t}$$

Then at first, we, expect the forced response to be

$$i_f = Be^{-6t} \tag{9.7-11}$$

However, the forced response and one component of the natural response would then both have the form De^{-6t}. Will this work? Let's try substituting Eq. 9.7-11 into the differential equation (9.7-10). We then obtain

$$36Be^{-6t} - 42Be^{-6t} + 6Be^{-6t} = 18e^{-6t}$$

or

$$0 = 18e^{-6t}$$

which is an impossible solution. Therefore, we need another form of the forced response when one of the natural response terms has the same form as the forcing function.

Let us try the forced response

$$i_f = Bte^{-6t} \tag{9.7-12}$$

Then, substituting Eq. 9.7-12 into Eq. 9.7-10, we have

$$B\left(-6e^{-6t} - 6e^{-6t} + 36t\,e^{-6t}\right) + 7B\left(e^{-6t} - 6t\,e^{-6t}\right) + 6Bt\,e^{-6t} = 18\,e^{-6t} \tag{9.7-13}$$

Simplifying Eq. 9.7-13, we have

$$B = -\frac{18}{5}$$

Therefore,

$$i_f = -\frac{18}{5}te^{-6t}$$

In general, if the forcing function is of the same form as one of the components of the natural response, x_{n1}, we will use

$$x_f = t^p\,x_{n1}$$

where the integer p is selected so that the x_f is not duplicated in the natural response. Use the lowest power, p, of t that is not duplicated in the natural response.

EXERCISE 9.7-1 A circuit is described for $t > 0$ by the equation

$$\frac{d^2i}{dt^2} + 9\frac{di}{dt} + 20i = 6i_s$$

where $i_s = 6 + 2t$ A. Find the forced response i_f for $t > 0$.

Answer: $i_f = 1.53 + 0.6t$ A

9.8 COMPLETE RESPONSE OF AN *RLC* CIRCUIT

We have succeeded in finding the natural response and the forced response of a circuit described by a second-order differential equation. We wish to proceed to determine the complete response for the circuit.

The *complete response* is the sum of the natural response and the forced response; thus,

$$x = x_n + x_f$$

Let us consider the series *RLC* circuit of Figure 9.2-2 with a differential equation (9.2-8) as

$$LC\frac{d^2v}{dt^2} + RC\frac{dv}{dt} + v = v_s$$

When $L = 1$ H, $C = 1/6$ F, and $R = 5\ \Omega$, we obtain

$$\frac{d^2v}{dt^2} + 5\frac{dv}{dt} + 6v = 6\,v_s \tag{9.8-1}$$

We let $v_s = \dfrac{2e^{-t}}{3}$ V, $v(0) = 10$ V, and $dv(0)/dt = -2$ V/s.

We will first determine the form of the natural response and then determine the forced response. Adding these responses, we have the complete response with two unspecified constants. We will then use the initial conditions to specify these constants to obtain the complete response.

To obtain the natural response, we write the characteristic equation, using operators as

$$s^2 + 5s + 6 = 0$$

or

$$(s+2)(s+3) = 0$$

Therefore, the natural response is

$$v_n = A_1e^{-2t} + A_2e^{-3t}$$

The forced response is obtained by examining the forcing function and noting that its exponential response has a different time constant than the natural response, so we may write

$$v_f = Be^{-t} \tag{9.8-2}$$

We can determine B by substituting Eq. 9.8-2 into Eq. 9.8-1. Then we have

$$Be^{-t} + 5(-Be^{-t}) + 6(Be^{-t}) = 4e^{-t}$$

or

$$B = 2$$

The complete response is then

$$v = v_n + v_f = A_1e^{-2t} + A_2e^{-3t} + 2e^{-t}$$

To find A_1 and A_2, we use the initial conditions. At $t = 0$, we have $v(0) = 10$, so we obtain

$$10 = A_1 + A_2 + 2 \tag{9.8-3}$$

From the fact that $dv/dt = -2$ at $t = 0$, we have

$$-2A_1 - 3A_2 - 2 = -2 \tag{9.8-4}$$

Solving Eqs. 9.8-3 and 9.8-4, we have $A_1 = 24$ and $A_2 = -16$. Therefore,

$$v = 24\,e^{-2t} - 16\,e^{-3t} + 2\,e^{-t}\ \mathrm{V}$$

EXAMPLE 9.8-1 Complete Response of a Second-Order Circuit

Find the complete response $v(t)$ for $t > 0$ for the circuit of Figure 9.8-1. Assume the circuit is at steady state at $t = 0^-$.

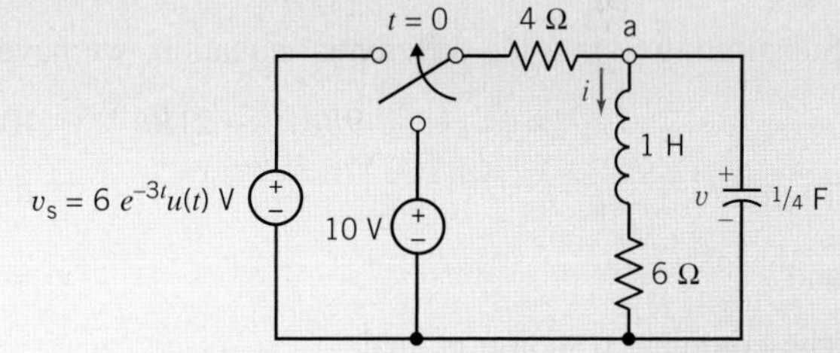

FIGURE 9.8-1 Circuit of Example 9.8-1.

Solution

First, we determine the initial conditions of the circuit. At $t = 0^-$, we have the circuit model shown in Figure 9.8-2, where we replace the capacitor with an open circuit and the inductor with a short circuit. Then the voltage is

$$v(0^-) = 6\ \mathrm{V}$$

and the inductor current is

$$i(0^-) = 1\ \mathrm{A}$$

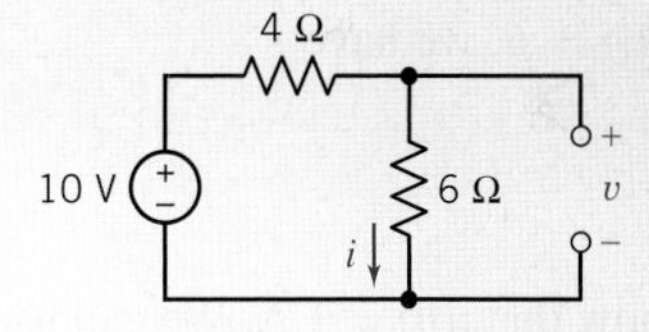

FIGURE 9.8-2 Circuit of Example 9.8-1 at $t = 0^-$.

After the switch is thrown, we can write the KVL for the right-hand mesh of Figure 9.8-1 to obtain

$$-v + \frac{di}{dt} + 6i = 0 \tag{9.8-5}$$

The KCL equation at node a will provide a second equation in terms of v and i as

$$\frac{v - v_s}{4} + i + \frac{1}{4}\frac{dv}{dt} = 0 \tag{9.8-6}$$

Equations 9.8-5 and 9.8-6 may be rearranged as

$$\left(\frac{di}{dt} + 6i\right) - v = 0 \tag{9.8-7}$$

$$i + \left(\frac{v}{4} + \frac{1}{4}\frac{dv}{dt}\right) = \frac{v_s}{4} \tag{9.8-8}$$

We will use operators so that $s = d/dt$, $s^2 = d^2/dt^2$, and $1/s = \int dt$. Then we obtain

$$(s + 6)i - v = 0 \tag{9.8-9}$$

$$i + \frac{1}{4}(s + 1)v = v_s/4 \tag{9.8-10}$$

Solving Eq. 9.8-10 for i and substituting the result into Eq. 9.8-9, we get

$$((s + 6)(s + 1) + 4)v = (s + 6)v_s$$

Or, equivalently,

$$(s^2 + 7s + 10)v = (s + 6)v_s$$

Hence, the second-order differential equation is

$$\frac{d^2v}{dt^2} + 7\frac{dv}{dt} + 10v = \frac{dv_s}{dt} + 6v_s \tag{9.8-11}$$

The characteristic equation is

$$s^2 + 7s + 10 = 0$$

Therefore, the roots of the characteristic equation are

$$s_1 = -2 \quad \text{and} \quad s_2 = -5$$

The natural response v_n is

$$v_n = A_1 e^{-2t} + A_2 e^{-5t}$$

The forced response is assumed to be of the form

$$v_f = Be^{-3t} \tag{9.8-12}$$

Substituting v_f into the differential equation, we have

$$9Be^{-3t} - 21Be^{-3t} + 10Be^{-3t} = -18e^{-3t} + 36e^{-3t}$$

Therefore,

$$B = -9$$

and

$$v_f = -9e^{-3t}$$

The complete response is then

$$v = v_n + v_f = A_1 e^{-2t} + A_2 e^{-5t} - 9e^{-3t} \tag{9.8-13}$$

Because $v(0) = 6$, we have

$$v(0) = 6 = A_1 + A_2 - 9$$

or

$$A_1 + A_2 = 15 \tag{9.8-14}$$

We also know that $i(0) = 1$ A. We can use Eq. 9.8-8 to determine $dv(0)/dt$ and then evaluate the derivative of Eq. 9.8-13 at $t = 0$. Equation 9.8-8 states that

$$\frac{dv}{dt} = -4\,i - v + v_s$$

At $t = 0$, we have

$$\frac{dv(0)}{dt} = -4\,i(0) - v(0) + v_s(0) = -4 - 6 + 6 = -4$$

Let us take the derivative of Eq. 9.8-13 to obtain

$$\frac{dv}{dt} = -2A_1 e^{-2t} - 5A_2 e^{-5t} + 27e^{-3t}$$

At $t = 0$, we obtain

$$\frac{dv(0)}{dt} = -2A_1 - 5A_2 + 27$$

Because $dv(0)/dt = -4$, we have

$$2A_1 + 5A_2 = 31 \tag{9.8-15}$$

Solving Eqs. 9.8-15 and 9.8-14 simultaneously, we obtain

$$A_1 = \frac{44}{3} \quad \text{and} \quad A_2 = \frac{1}{3}$$

Therefore,

$$v = \frac{44}{3}e^{-2t} + \frac{1}{3}e^{-5t} - 9e^{-3t}\ \text{V}$$

Note that we used the capacitor voltage and the inductor current as the unknowns. This is very convenient because you will normally have the initial conditions of these variables. These variables, v_c and i_L, are known as the *state variables*. We will consider this approach more fully in the next section.

9.9 STATE VARIABLE APPROACH TO CIRCUIT ANALYSIS

The *state variables* of a circuit are a set of variables associated with the energy of the energy storage elements of the circuit. Thus, they describe the complete response of a circuit to a forcing function and the circuit's initial conditions. Here the word *state* means ''condition,'' as in *state of the union*. We will choose as the state variables those variables that describe the energy storage of the circuit. Thus, we will use the independent capacitor voltages and the independent inductor currents.

Consider the circuit shown in Figure 9.9-1. The two energy storage elements are C_1 and C_2, and the two capacitors cannot be reduced to one. We expect the circuit to be described by a second-order differential equation. However, let us first obtain the two first-order differential equations that describe the response for $v_1(t)$ and $v_2(t)$, which are the state variables of the circuit. If we know the value of the state variables at one time and the value of the input variables thereafter, we can find the value of any state variable for any subsequent time.

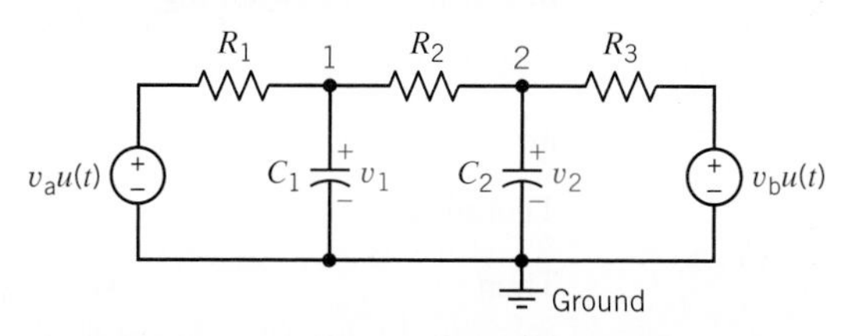

FIGURE 9.9-1 Circuit with two energy storage elements.

Writing the KCL at nodes 1 and 2, we have

$$\text{node 1:} \qquad C_1\frac{dv_1}{dt} = \frac{v_a - v_1}{R_1} + \frac{v_2 - v_1}{R_2} \tag{9.9-1}$$

$$\text{node 2:} \qquad C_2\frac{dv_2}{dt} = \frac{v_b - v_2}{R_3} + \frac{v_1 - v_2}{R_2} \tag{9.9-2}$$

Equations 9.9-1 and 9.9-2 can be rewritten as

$$\frac{dv_1}{dt} + \frac{v_1}{C_1R_1} + \frac{v_1}{C_1R_2} - \frac{v_2}{C_1R_2} = \frac{v_a}{C_1R_1} \tag{9.9-3}$$

$$\frac{dv_2}{dt} + \frac{v_2}{C_2R_3} + \frac{v_2}{C_2R_2} - \frac{v_1}{C_2R_2} = \frac{v_b}{C_2R_3} \tag{9.9-4}$$

Assume that $C_1R_1 = 1$, $C_1R_2 = 1$, $C_2R_3 = 1$, and $C_2R_2 = 1/2$. Then we have

$$\frac{dv_1}{dt} + 2v_1 - v_2 = v_a \tag{9.9-5}$$

and

$$-2v_1 + \frac{dv_2}{dt} + 3v_2 = v_b \tag{9.9-6}$$

Using operators, we have

$$(s+2)v_1 - v_2 = v_a$$
$$-2v_1 + (s+3)v_2 = v_b$$

If we wish to solve for v_1, we use Cramer's rule to obtain

$$v_1 = \frac{(s+3)v_a + v_b}{(s+2)(s+3) - 2} \tag{9.9-7}$$

The characteristic equation is obtained from the denominator and has the form

$$s^2 + 5s + 4 = 0$$

The characteristic roots are $s = -4$ and $s = -1$. The second-order differential equation can be obtained by rewriting Eq. 9.9-7 as

$$(s^2 + 5s + 4)v_1 = (s + 3)v_a + v_b$$

Then the differential equation for v_1 is

$$\frac{d^2 v_1}{dt^2} + 5\frac{dv_1}{dt} + 4v_1 = \frac{dv_a}{dt} + 3v_a + v_b \tag{9.9-8}$$

We now proceed to obtain the natural response

$$v_{1n} = A_1 e^{-t} + A_2 e^{-4t}$$

and the forced response, which depends on the form of the forcing function. For example, if $v_a = 10$ V and $v_b = 6$ V, v_{1f} will be a constant (see Table 9.7-1). We obtain v_{1f} by substituting into Eq. 9.9-8, obtaining

$$4v_{1f} = 3v_a + v_b$$

or

$$4v_{1f} = 30 + 6 = 36$$

Therefore,

$$v_{1f} = 9$$

Then

$$v_1 = v_{1n} + v_{1f} = A_1 e^{-t} + A_2 e^{-4t} + 9 \tag{9.9-9}$$

We will usually know the initial conditions of the energy storage elements. For example, if we know that $v_1(0) = 5$ V and $v_2(0) = 10$ V, we first use $v_1(0) = 5$ along with Eq. 9.9-9 to obtain

$$v_1(0) = A_1 + A_2 + 9$$

and, therefore,

$$A_1 + A_2 = -4 \tag{9.9-10}$$

Now we need the value of dv_1/dt at $t = 0$. Referring back to Eq. 9.9-5, we have

$$\frac{dv_1}{dt} = v_a + v_2 - 2\,v_1$$

Therefore, at $t = 0$, we have

$$\frac{dv_1(0)}{dt} = v_a(0) + v_2(0) - 2v_1(0) = 10 + 10 - 2(5) = 10$$

The derivative of the complete solution, Eq. 9.9-9, at $t = 0$ is

$$\frac{dv_1(0)}{dt} = -A_1 - 4\,A_2$$

Therefore,

$$A_1 + 4\,A_2 = -10 \tag{9.9-11}$$

Solving Eqs. 9.9-10 and 9.9-11, we have

$$A_1 = -2 \quad \text{and} \quad A_2 = -2$$

Therefore,

$$v_1(t) = -2e^{-t} - 2e^{-4t} + 9\text{ V}$$

As you encounter circuits with two or more energy storage elements, you should consider using the state variable method of describing a set of first-order differential equations.

The **state variable method** uses a first-order differential equation for each state variable to determine the complete response of a circuit.

A summary of the state variable method is given in Table 9.9-1. We will use this method in Example 9.9-1.

Table 9.9-1 State Variable Method of Circuit Analysis

1. Identify the state variables as the independent capacitor voltages and inductor currents.
2. Determine the initial conditions at $t = 0$ for the capacitor voltages and the inductor currents.
3. Obtain a first-order differential equation for each state variable, using KCL or KVL.
4. Use the operator s to substitute for d/dt.
5. Obtain the characteristic equation of the circuit by noting that it can be obtained by setting the determinant of Cramer's rule equal to zero.
6. Determine the roots of the characteristic equation, which then determine the form of the natural response.
7. Obtain the second-order (or higher-order) differential equation for the selected variable x by Cramer's rule.
8. Determine the forced response x_f by assuming an appropriate form of x_f and determining the constant by substituting the assumed solution in the second-order differential equation.
9. Obtain the complete solution $x = x_n + x_f$.
10. Use the initial conditions on the state variables along with the set of first-order differential equations (step 3) to obtain $dx(0)/dt$.
11. Using $x(0)$ and $dx(0)/dt$ for each state variable, find the arbitrary constants $A_1, A_2, \ldots A_n$ to obtain the complete solution $x(t)$.

EXAMPLE 9.9-1 Complete Response of a Second-Order Circuit

Find $i(t)$ for $t > 0$ for the circuit shown in Figure 9.9-2 when $R = 3\ \Omega$, $L = 1$ H, $C = 1/2$ F, and $i_s = 2e^{-3t}$ A. Assume steady state at $t = 0^-$.

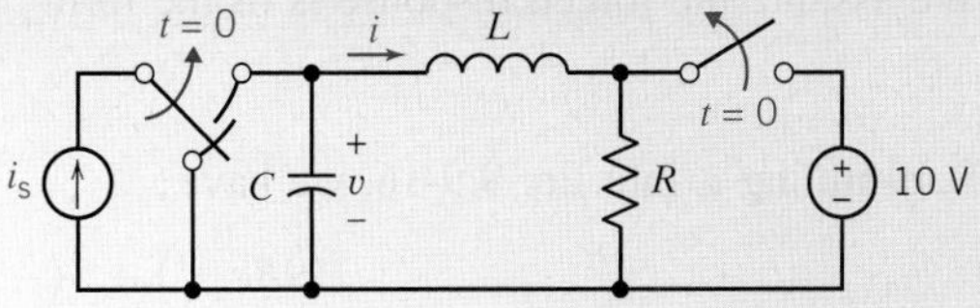

FIGURE 9.9-2 Circuit of Example 9.9-1.

Solution

First, we identify the state variables as i and v. The initial conditions at $t = 0$ are obtained by considering the circuit with the 10-V source connected for a long time at $t = 0^-$. At $t = 0$, the voltage source is disconnected and the current source is connected. Then $v(0) = 10$ V and $i(0) = 0$ A.

Consider the circuit after time $t = 0$. The first differential equation is obtained by using KVL around the *RLC* mesh to obtain

$$L\frac{di}{dt} + Ri = v$$

The second differential equation is obtained by using KCL at the node at the top of the capacitor to get

$$C\frac{dv}{dt} + i = i_s$$

We may rewrite these two first-order differential equations as

$$\frac{di}{dt} + \frac{R}{L}i - \frac{v}{L} = 0$$

and

$$\frac{dv}{dt} + \frac{i}{C} = \frac{i_s}{C}$$

Substituting the component values, we have

$$\frac{di}{dt} + 3i - v = 0 \tag{9.9-12}$$

and
$$\frac{dv}{dt} + 2i = 2i_s \tag{9.9-13}$$

Using the operator $s = d/dt$, we have

$$(s+3)i - v = 0 \tag{9.9-14}$$

$$2i + sv = 2i_s \tag{9.9-15}$$

Therefore, the characteristic equation obtained from the determinant is

$$(s+3)s + 2 = 0$$

or
$$s^2 + 3s + 2 = 0$$

Thus, the roots of the characteristic equation are

$$s_1 = -2 \quad \text{and} \quad s_2 = -1$$

Because we wish to solve for $i(t)$ for $t > 0$, we use Cramer's rule to solve Eqs. 9.9-14 and 9.9-15 for i, obtaining

$$i = \frac{2i_s}{s^2 + 3s + 2}$$

Therefore, the differential equation is

$$\frac{d^2i}{dt^2} + 3\frac{di}{dt} + 2i = 2i_s \tag{9.9-16}$$

The natural response is

$$i_n = A_1 e^{-t} + A_2 e^{-2t}$$

We assume the forced response is of the form

$$i_f = Be^{-3t}$$

Substituting i_f into Eq. 9.9-16, we have

$$(9Be^{-3t}) + 3(-3Be^{-3t}) + 2\,Be^{-3t} = 2(2e^{-3t})$$

or
$$9B - 9B + 2B = 4$$

Therefore, $B = 2$ and

$$i_f = 2e^{-3t}$$

The complete response is

$$i = A_1 e^{-t} + A_2 e^{-2t} + 2e^{-3t}$$

Because $i(0) = 0$,

$$0 = A_1 + A_2 + 2 \tag{9.9-17}$$

We need to obtain $di(0)/dt$ from Eq 9.9-12, which we repeat here as

$$\frac{di}{dt} + 3i - v = 0$$

Therefore, at $t = 0$, we have

$$\frac{di(0)}{dt} = -3i(0) + v(0) = 10$$

The derivative of the complete response at $t = 0$ is

$$\frac{di(0)}{dt} = -A_1 - 2A_2 - 6$$

Because $di(0)/dt = 10$, we have

$$-A_1 - 2A_2 = 16$$

and, repeating Eq. 9.9-17, we have

$$A_1 + A_2 = -2$$

Adding these two equations, we determine that $A_1 = 12$ and $A_2 = -14$. Then we have the complete solution for i as

$$i = 12e^{-t} - 14e^{-2t} + 2e^{-3t} \text{ A}$$

We recognize that the state variable method is particularly powerful for finding the response of energy storage elements in a circuit. This is also true if we encounter higher-order circuits with three or more energy storage elements. For example, consider the circuit shown in Figure 9.9-3. The state variables are v_1, v_2, and i. Two first-order differential equations are obtained by writing the KCL equations at node a and node b. Then a third first-order differential equation is obtained by writing the KVL around the middle mesh containing i. The solution for one or more of these variables can then be obtained by proceeding with the state variable method summarized in Table 9.9-1.

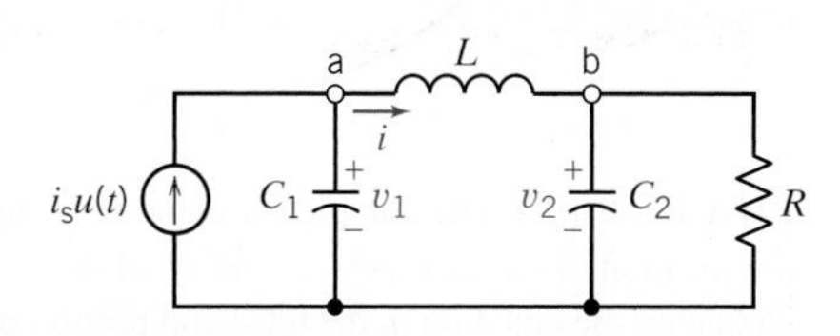

FIGURE 9.9-3 Circuit with three energy storage elements.

EXERCISE 9.9-1 Find $v_2(t)$ for $t > 0$ for the circuit of Figure E 9.9-1. Assume there is no initial stored energy.

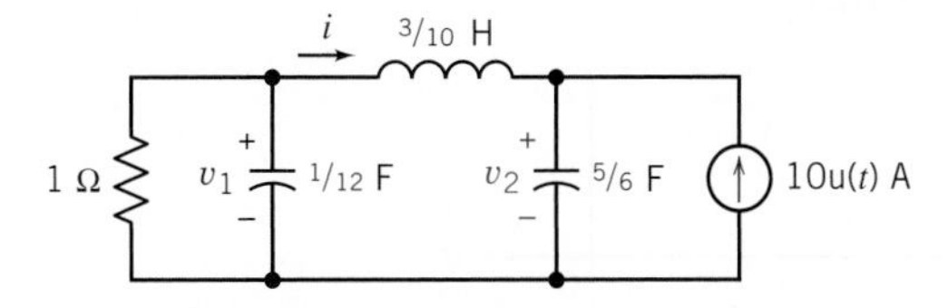

FIGURE E 9.9-1

Answer: $v_2(t) = -15e^{-2t} + 6e^{-4t} - e^{-6t} + 10 \text{ V}$

9.10 ROOTS IN THE COMPLEX PLANE

We have observed that the character of the natural response of a second-order system is determined by the roots of the characteristic equation. Let us consider the roots of a parallel RLC circuit. The characteristic equation (9.4-3) is

$$s^2 + \frac{s}{RC} + \frac{1}{LC} = 0$$

and the roots are given by Eq. 9.4-8 to be

$$s = -\alpha \pm \sqrt{\alpha^2 - \omega_0^2}$$

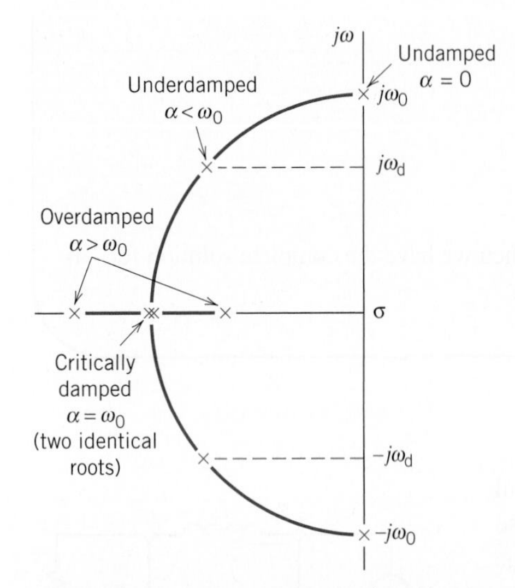

FIGURE 9.10-1 The complete s-plane showing the location of the two roots, s_1 and s_2, of the characteristic equation in the left-hand portion of the s-plane. The roots are designated by the × symbol.

where $\alpha = 1/(2\,RC)$ and $\omega_0^2 = 1/(LC)$. When $\omega_0 > \alpha$, the roots are complex and

$$s = -\alpha \pm j\sqrt{\omega_0^2 - \alpha^2} = -\alpha \pm j\omega_d \qquad (9.10\text{-}1)$$

In general, roots are located in the complex plane, the location being defined by coordinates measured along the real or σ-axis and the imaginary or $j\omega$-axis. This is referred to as the s-plane or, because s has the units of frequency, as the *complex frequency plane*. When the roots are real, negative, and distinct, the response is the sum of two decaying exponentials and is said to be overdamped. When the roots are complex conjugates, the natural response is an exponentially decaying sinusoid and is said to be underdamped or oscillatory.

Now, let us show the location of the roots of the characteristic equation for the four conditions: (a) undamped, $\alpha = 0$; (b) underdamped, $\alpha < \omega_0$; (c) critically damped, $\alpha = \omega_0$; and (d) overdamped, $\alpha > \omega_0$. These four conditions lead to root locations on the s-plane as shown in Figure 9.10-1. When $\alpha = 0$, the two complex roots are $\pm j\omega_0$. When $\alpha < \omega_0$, the roots are $s = -\alpha \pm j\omega_d$. When $\alpha = \omega_0$, there are two roots at $s = -\alpha$. Finally, when $\alpha > \omega_0$, there are two real roots, $s = -\alpha \pm \sqrt{\alpha^2 - \omega_0^2}$.

A summary of the root locations, the type of response, and the form of the response is presented in Table 9.10-1.

EXERCISE 9.10-1 A parallel RLC circuit has $L = 0.1$ H and $C = 100$ mF. Determine the roots of the characteristic equation and plot them on the s-plane when (a) $R = 0.4\ \Omega$ and (b) $R = 1.0\ \Omega$.

Answer: (a) $s = -5, -20$ (Figure E 9.10-1)

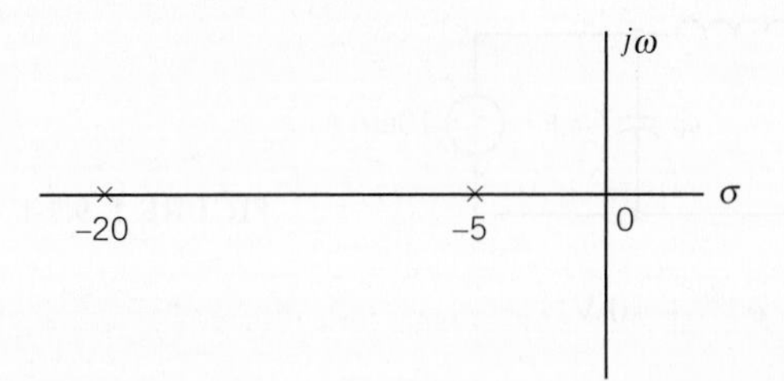

FIGURE E 9.10-1

9.11 HOW CAN WE CHECK . . . ?

Engineers are frequently called upon to check that a solution to a problem is indeed correct. For example, proposed solutions to design problems must be checked to confirm that all of the specifications have been satisfied. In addition, computer output must be reviewed to guard against data-entry errors, and claims made by vendors must be examined critically.

Engineering students are also asked to check the correctness of their work. For example, occasionally just a little time remains at the end of an exam. It is useful to be able to quickly identify those solutions that need more work.

The following example illustrates techniques useful for checking the solutions of the sort of problem discussed in this chapter.

Table 9.10-1 The Natural Response of a Parallel *RLC* Circuit*

TYPE OF RESPONSE	ROOT LOCATION	FORM OF RESPONSE
Overdamped	$j\omega$, σ	$i(t)$, A; 1; t, s
Critically damped	$j\omega$, σ	$i(t)$, A; 1; t, s
Underdamped	$j\omega$, σ	$i(t)$, A; 1; t, s
Undamped	$j\omega$, σ	$i(t)$, A; 1; t, s

*The $i(t)$ is the inductor current in the circuit shown in Figure 9.4-1 for the initial conditions $i(0) = 1$ and $v(0) = 0$.

EXAMPLE 9.11-1 How Can We Check an Underdamped Response?

Figure 9.11-1*b* shows an *RLC* circuit. The voltage, $v_s(t)$, of the voltage source is the square wave shown in Figure 9.11-1*a*. Figure 9.11-2 shows a plot of the inductor current, $i(t)$, which was obtained by simulating this circuit, using PSpice. **How can we check** that the plot of $i(t)$ is correct?

Solution

Several features of the plot can be checked. The plot indicates that steady-state values of the inductor current are $i(\infty) = 0$ and $i(\infty) = 200$ mA and that the circuit is underdamped. In addition, some points on the response have

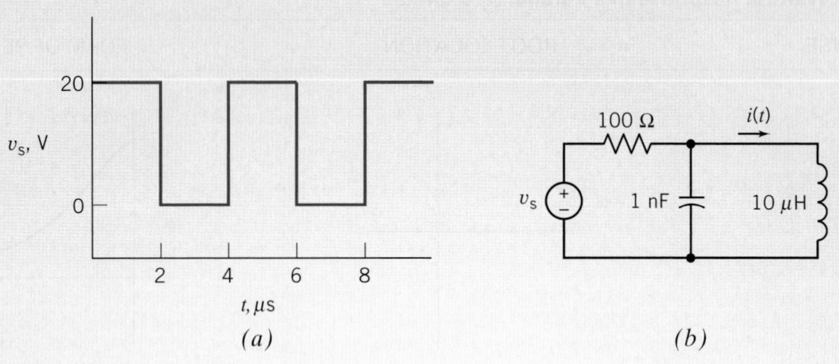

FIGURE 9.11-1 An *RLC* circuit (*b*) excited by a square wave (*a*).

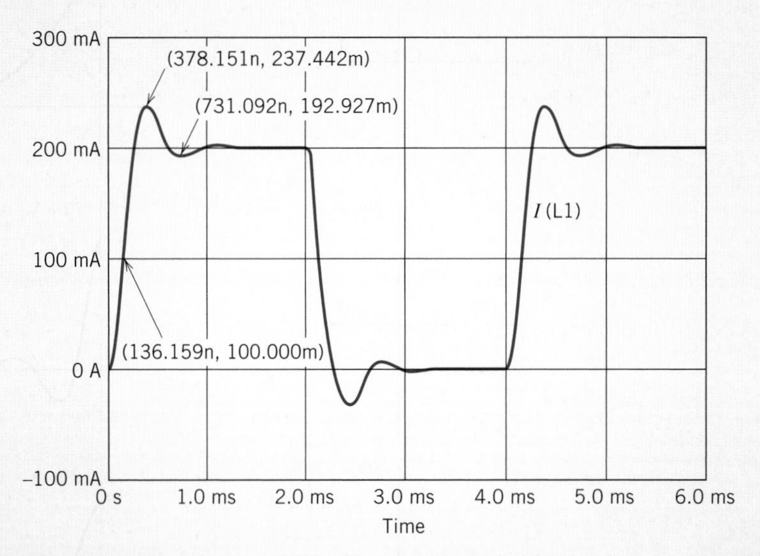

FIGURE 9.11-2 PSpice plot of the inductor current, $i(t)$, for the circuit shown in Figure 9.11-1.

been labeled to give the corresponding values of time and current. These values can be used to check the value of the damped resonant frequency, ω_d.

If the voltage of the voltage source were a constant, $v_s(t) = V_s$, then the steady-state inductor current would be

$$i(t) = \frac{V_s}{100}$$

Thus, we expect the steady-state inductor current to be $i(\infty) = 0$ when $V_s = 0$ V and to be $i(\infty) = 200$ mA when $V_s = 20$ V. The plot in Figure 9.11-2 shows that the steady-state values of the inductor current are indeed $i(\infty) = 0$ and $i(\infty) = 200$ mA.

The plot in Figure 9.11-2 shows an underdamped response. The *RLC* circuit will be underdamped if

$$10^{-5} = L < 4R^2C = 4 \times 100^2 \times 10^{-9}$$

Because this inequality is satisfied, the circuit is indeed underdamped, as indicated by the plot.

The damped resonant frequency, ω_d, is given by

$$\omega_d = \sqrt{\frac{1}{LC} - \left(\frac{1}{2RC}\right)^2} = \sqrt{\frac{1}{10^{-5} \times 10^{-9}} - \left(\frac{1}{2 \times 100 \times 10^{-9}}\right)^2} = 8.66 \times 10^6 \text{ rad/s}$$

The plot indicates that the plot has a maxima at 378 ns and a minima at 731 ns. Therefore, the period of the damped oscillation can be approximated as

$$T_d = 2\left(731 \times 10^{-9} - 378 \times 10^{-9}\right) = 706 \times 10^{-9} \text{ s}$$

The damped resonant frequency, ω_d, is related to T_d by Eq. 9.6-9. Therefore,

$$\omega_d = \frac{2\pi}{T_d} = \frac{2\pi}{706 \times 10^{-9}} = 8.90 \times 10^6 \text{ rad/s}$$

The value of ω_d obtained from the plot agrees with the value obtained from the circuit.
We conclude that the plot is correct.

9.12 DESIGN EXAMPLE

AUTO AIRBAG IGNITER

Airbags are widely used for driver and passenger protection in automobiles. A pendulum is used to switch a charged capacitor to the inflation ignition device, as shown in Figure 9.12-1. The automobile airbag is inflated by an explosive device that is ignited by the energy absorbed by the resistive device represented by R. To inflate, it is required that the energy dissipated in R be at least 1 J. It is required that the ignition device trigger within 0.1 s. Select the L and C that meet the specifications.

Describe the Situation and the Assumptions

1. The switch is changed from position 1 to position 2 at $t = 0$.
2. The switch was connected to position 1 for a long time.
3. A parallel RLC circuit occurs for $t \geq 0$.

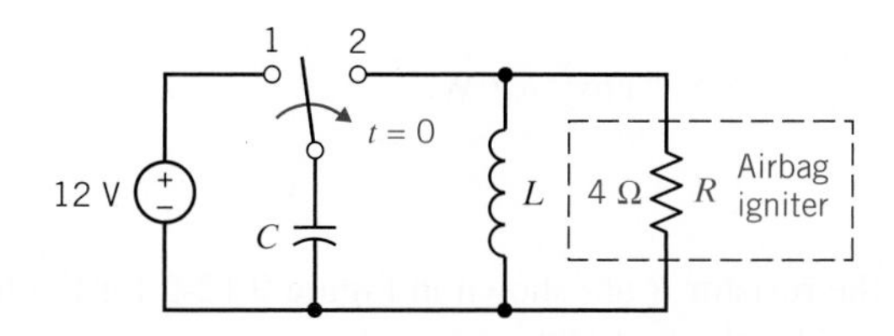

FIGURE 9.12-1 An automobile airbag ignition device.

State the Goal

Select L and C so that the energy stored in the capacitor is quickly delivered to the resistive device R.

Generate a Plan

1. Select L and C so that an underdamped response is obtained with a period of less than or equal to 0.4 s ($T \leq 0.4$ s).
2. Solve for $v(t)$ and $i(t)$ for the resistor R.

Act on the Plan

We assume that the initial capacitor voltage is $v(0) = 12$ V and $i_L(0) = 0$ because the switch is in position 1 for a long time prior to $t = 0$. The response of the parallel RLC circuit for an underdamped response is of the form

$$v(t) = e^{-\alpha t}(B_1 \cos \omega_d t + B_2 \sin \omega_d t) \qquad (9.12\text{-}1)$$

This natural response is obtained when $\alpha^2 < \omega_0^2$ or $L < 4R^2C$. We choose an underdamped response for our design but recognize that an overdamped or critically damped response may satisfy the circuit's design objectives. Furthermore, we recognize that the parameter values selected below represent only one acceptable solution.

Because we want a rapid response, we will select $\alpha = 2$ (a time constant of $1/2$ s) where $\alpha = 1/(2RC)$. Therefore, we have

$$C = \frac{1}{2R\alpha} = \frac{1}{16} \text{ F}$$

Recall that $\omega_0^2 = 1/(LC)$ and it is required that $\alpha^2 < \omega_0^2$. Because we want a rapid response, we select the natural frequency ω_0 so that (recall $T \approx 0.4$ s)

$$\omega_0 = \frac{2\pi}{T} = \frac{2\pi}{0.4} = 5\pi \text{ rad/s}$$

Therefore, we obtain

$$L = \frac{1}{\omega_0^2 C} = \frac{1}{25\pi^2(1/16)} = 0.065 \text{ H}$$

Thus, we will use $C = 1/16$ F and $L = 65$ mH. We then find that $\omega_d = 15.58$ rad/s and, using Eq. 9.6-5, we have

$$v(t) = e^{-2t}(B_1 \cos \omega_d t + B_2 \sin \omega_d t) \qquad (9.12\text{-}2)$$

Then $B_1 = v(0) = 12$ and

$$\omega_d B_2 = \alpha B_1 - \frac{B_1}{RC} = (2 - 4)12 = -24$$

Therefore, $B_2 = -24/15.58 = -1.54$. Because $B_2 \ll B_1$, we can approximate Eq. 9.12-2 as

$$v(t) \cong 12e^{-2t} \cos \omega_d t \text{ V}$$

The power is then

$$p = \frac{v^2}{R} = 36e^{-4t} \cos^2 \omega_d t \text{ W}$$

Verify the Proposed Solution

The actual voltage and current for the resistor R are shown in Figure 9.12-2 for the first 100 ms. If we sketch the product of v and i for the first 100 ms, we obtain a linear approximation

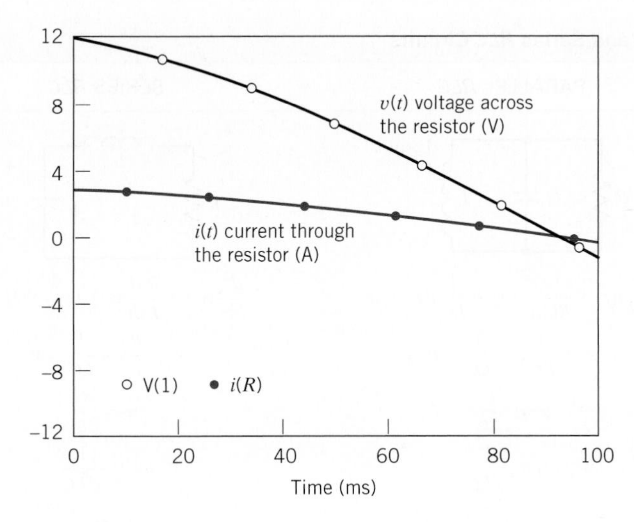

FIGURE 9.12-2 The response of the *RLC* circuit.

declining from 36 W at $t = 0$ to 0 W at $t = 95$ ms. The energy absorbed by the resistor over the first 100 ms is then

$$w \cong \frac{1}{2}(36)(0.1\text{ s}) = 1.8\text{ J}$$

Therefore, the airbag will trigger in less than 0.1 s, and our objective is achieved.

9.13 SUMMARY

- Second-order circuits are circuits that are represented by a second-order differential equation, for example,

$$\frac{d^2}{dt^2}x(t) + 2\alpha\frac{d}{dt}x(t) + \omega_0^2 x(t) = f(t)$$

where $x(t)$ is the output current or voltage of the circuit and $f(t)$ is the input to the circuit. The output of the circuit, also called the response of the circuit, can be the current or voltage of any device in the circuit. The output is frequently chosen to be the current of an inductor or the voltage of a capacitor. The input to the circuit is provided by the voltages of independent voltage sources and/or currents of independent current sources. The coefficients of this differential equation have names: α is called the damping coefficient, and ω_0 is called the resonant frequency.

- Obtaining the differential equation to represent an arbitrary circuit can be challenging. This chapter presents three methods for obtaining that differential equation: the direct method (Section 9.2), the operator method (Section 9.2), and the state variable method (Section 9.10).
- The characteristic equation of a second-order circuit is

$$s^2 + 2\alpha s + \omega_0^2 = 0$$

This second-order equation has two solutions, s_1 and s_2. These solutions are called the natural frequencies of the second-order circuit.

- Second-order circuits are characterized as overdamped, critically damped, or underdamped. A second-order circuit is overdamped when s_1 and s_2 are real and unequal, or, equivalently, $\alpha > \omega_0$. A second-order circuit is critically damped when s_1 and s_2 are real and equal, or, equivalently, $\alpha = \omega_0$. A second-order circuit is underdamped when s_1 and s_2 are real and equal, or, equivalently, $\alpha < \omega_0$.
- Table 9.13-1 describes the natural frequencies of overdamped, underdamped, and critically damped parallel and series *RLC* circuits.
- The complete response of a second-order circuit is the sum of the natural response and the forced response

$$x = x_n + x_f$$

The form of the natural response depends on the natural frequencies of the circuit as summarized in Table 9.13-2. The form of the forced response depends on the input to the circuit as summarized in Table 9.13-3.

Table 9.13-1 Natural Frequencies of Parallel *RLC* and Series *RLC* Circuits

	PARALLEL *RLC*	SERIES *RLC*
Circuit		
Differential equation	$\frac{d^2}{dt^2}i(t)+\frac{1}{RC}\frac{d}{dt}i(t)+\frac{1}{LC}i(t)=0$	$\frac{d^2}{dt^2}v(t)+\frac{R}{L}\frac{d}{dt}v(t)+\frac{1}{LC}v(t)=0$
Characteristic equation	$s^2+\frac{1}{RC}s+\frac{1}{LC}=0$	$s^2+\frac{R}{L}s+\frac{1}{LC}=0$
Damping coefficient, rad/s	$\alpha=\frac{1}{2RC}$	$\alpha=\frac{R}{2L}$
Resonant frequency, rad/s	$\omega_0=\frac{1}{\sqrt{LC}}$	$\omega_0=\frac{1}{\sqrt{LC}}$
Damped resonant frequency, rad/s	$\omega_d=\sqrt{\left(\frac{1}{2RC}\right)^2-\frac{1}{LC}}$	$\omega_d=\sqrt{\left(\frac{R}{2L}\right)^2-\frac{1}{LC}}$
Natural frequencies: overdamped case	$s_1,s_2=-\frac{1}{2RC}\pm\sqrt{\left(\frac{1}{2RC}\right)^2-\frac{1}{LC}}$ when $R<\frac{1}{2}\sqrt{\frac{L}{C}}$	$s_1,s_2=-\frac{R}{2L}\pm\sqrt{\left(\frac{R}{2L}\right)^2-\frac{1}{LC}}$ when $R>2\sqrt{\frac{L}{C}}$
Natural frequencies: critically damped case	$s_1=s_2=-\frac{1}{2RC}$ when $R=\frac{1}{2}\sqrt{\frac{L}{C}}$	$s_1=s_2=-\frac{R}{2L}$ when $R=2\sqrt{\frac{L}{C}}$
Natural frequencies: underdamped case	$s_1,s_2=-\frac{1}{2RC}\pm j\sqrt{\frac{1}{LC}-\left(\frac{1}{2RC}\right)^2}$ when $R>\frac{1}{2}\sqrt{\frac{L}{C}}$	$s_1,s_2=-\frac{R}{2L}\pm j\sqrt{\frac{1}{LC}-\left(\frac{R}{2L}\right)^2}$ when $R<2\sqrt{\frac{L}{C}}$

Table 9.13-2 Natural Response of Second-Order Circuits

CASE	NATURAL FREQUENCIES	NATURAL RESPONSE, x_n
Overdamped	$s_1,s_2=-\alpha\pm\sqrt{\alpha^2-\omega_0^2}$	$A_1e^{s_1t}+A_2e^{s_2t}$
Critically damped	$s_1,s_2=-\alpha$	$(A_1+A_2t)e^{-\alpha t}$
Underdamped	$s_1,s_2=-\alpha\pm j\sqrt{\omega_0^2-\alpha^2}=-\alpha\pm j\omega_d$	$(A_1\cos\omega_d t+A_2\sin\omega_d t)e^{-\alpha t}$

Table 9.13-3 Forced Response of Second-Order Circuits

	INPUT, *f(t)*	FORCED RESPONSE, x_f
Constant	K	A
Ramp	Kt	$A+Bt$
Sinusoid	$K\cos\omega t$, $K\sin\omega t$, or $K\cos(\omega t+\theta)$	$A\cos\omega t+B\sin\omega t$
Exponential	Ke^{-bt}	Ae^{-bt}

PROBLEMS

Section 9.2 Differential Equation for Circuits with Two Energy Storage Elements

P 9.2-1 Find the differential equation for the circuit shown in Figure P 9.2-1 using the direct method.

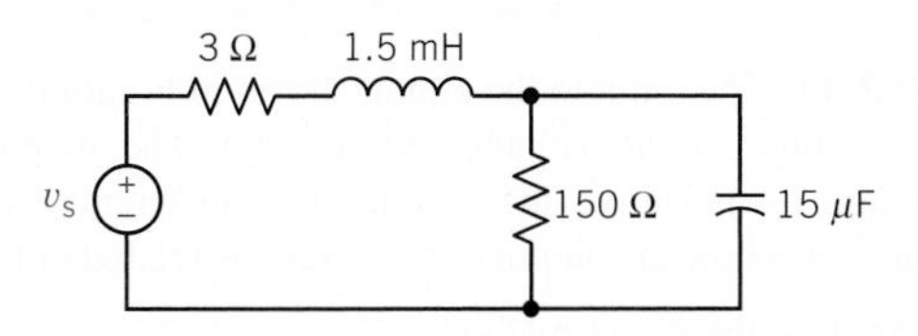

Figure P 9.2-1

P 9.2-2 Find the differential equation for the circuit shown in Figure P 9.2-2 using the operator method.

Answer:

$$\frac{d^2}{dt^2}i_L(t)+10.4\times10^3\frac{d}{dt}i_L(t)+0.48\times10^8 i_L(t)=0.4\times10^8 i_s(t)$$

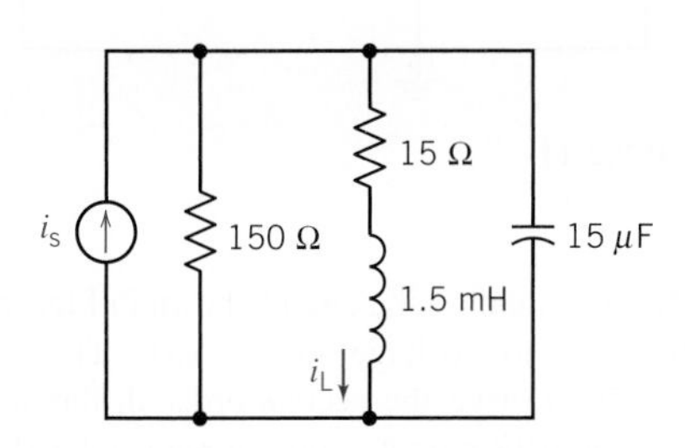

Figure P 9.2-2

P 9.2-3 Find the differential equation for $i_L(t)$ for $t > 0$ for the circuit of Figure P 9.2-3.

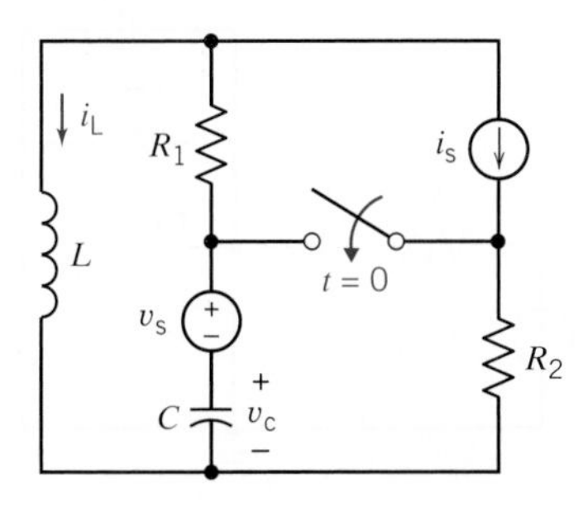

Figure P 9.2-3

P 9.2-4 The input to the circuit shown in Figure P 9.2-4 is the voltage of the voltage source, V_s. The output is the inductor current $i(t)$. Represent the circuit by a second-order differential equation that shows how the output of this circuit is related to the input for $t > 0$.

Hint: Use the direct method.

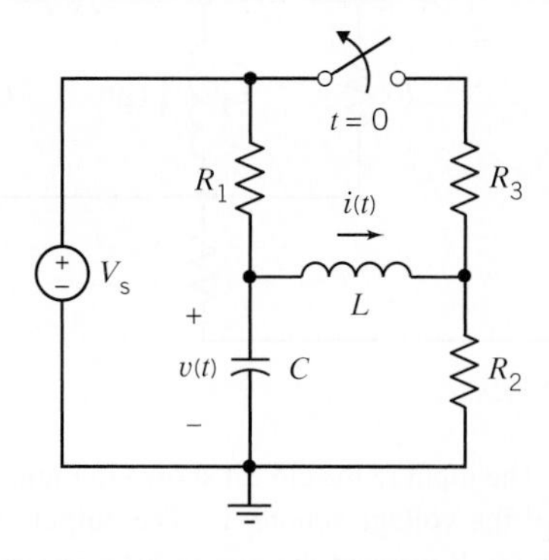

Figure P 9.2-4

P 9.2-5 The input to the circuit shown in Figure P 9.2-5 is the voltage of the voltage source, v_s. The output is the capacitor voltage $v(t)$. Represent the circuit by a second-order differential equation that shows how the output of this circuit is related to the input for $t > 0$.

Hint: Use the direct method.

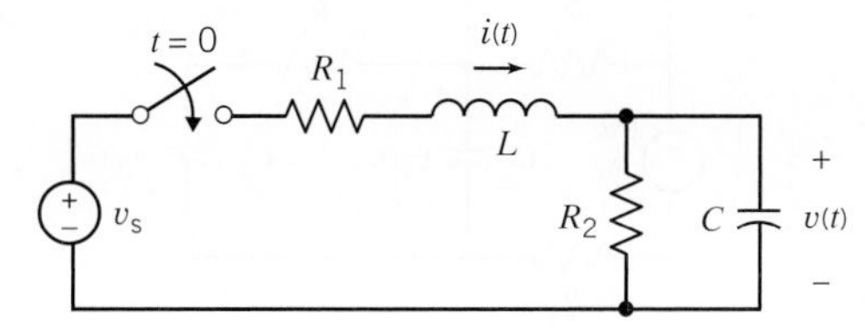

Figure P 9.2-5

P 9.2-6 The input to the circuit shown in Figure P 9.2-6 is the voltage of the voltage source, v_s. The output is the inductor current $i(t)$. Represent the circuit by a second-order differential equation that shows how the output of this circuit is related to the input for $t > 0$.

Hint: Use the direct method.

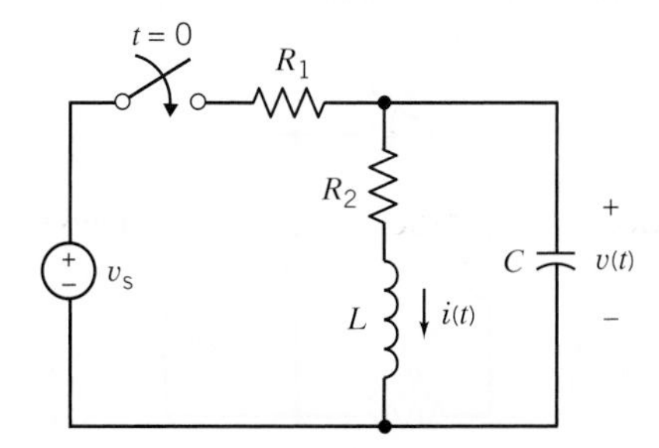

Figure P 9.2-6

P 9.2-7 The input to the circuit shown in Figure P 9.2-7 is the voltage of the voltage source, v_s. The output is the inductor current $i_2(t)$. Represent the circuit by a second-order differential equation that shows how the output of this circuit is related to the input for $t > 0$.

Hint: Use the operator method.

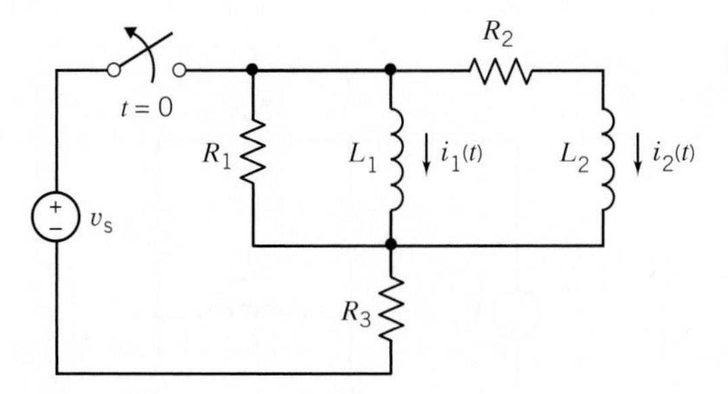

Figure P 9.2-7

P 9.2-8 The input to the circuit shown in Figure P 9.2-8 is the voltage of the voltage source, v_s. The output is the capacitor voltage $v_2(t)$. Represent the circuit by a second-order differential equation that shows how the output of this circuit is related to the input for $t > 0$.

Hint: Use the operator method.

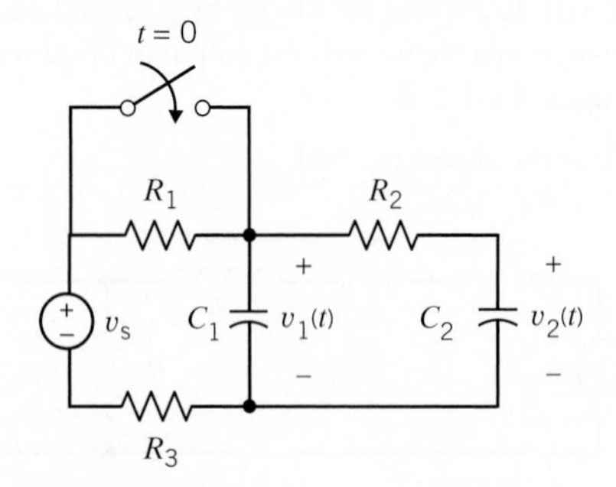

Figure P 9.2-8

P 9.2-9 The input to the circuit shown in Figure P 9.2-9 is the voltage of the voltage source, v_s. The output is the capacitor voltage $v(t)$. Represent the circuit by a second-order differential equation that shows how the output of this circuit is related to the input for $t > 0$.

Hint: Use the direct method.

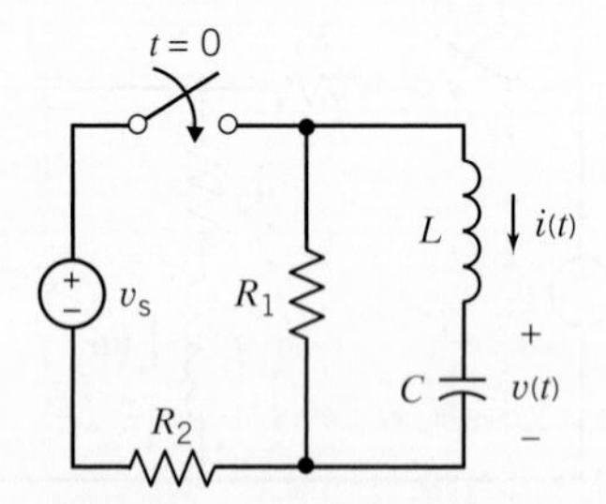

Figure P 9.2-9

P 9.2-10 The input to the circuit shown in Figure P 9.2-10 is the voltage of the voltage source, v_s. The output is the capacitor voltage $v(t)$. Represent the circuit by a second-order differential equation that shows how the output of this circuit is related to the input for $t > 0$.

Hint: Find a Thévenin equivalent circuit.

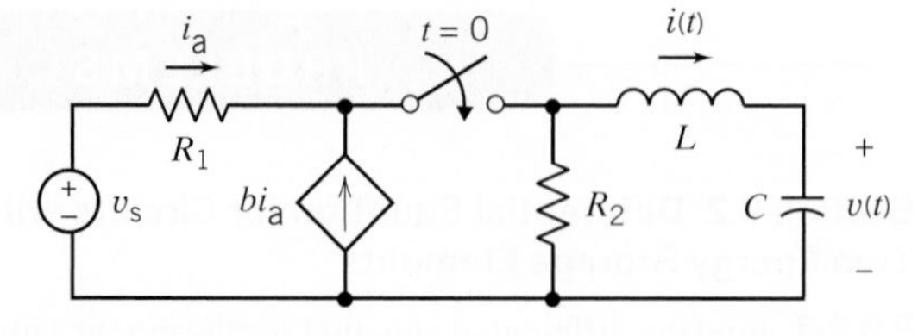

Figure P 9.2-10

P 9.2-11 The input to the circuit shown in Figure P 9.2-11 is the voltage of the voltage source, $v_s(t)$. The output is the voltage $v_2(t)$. Derive the second-order differential equation that shows how the output of this circuit is related to the input.

Hint: Use the direct method.

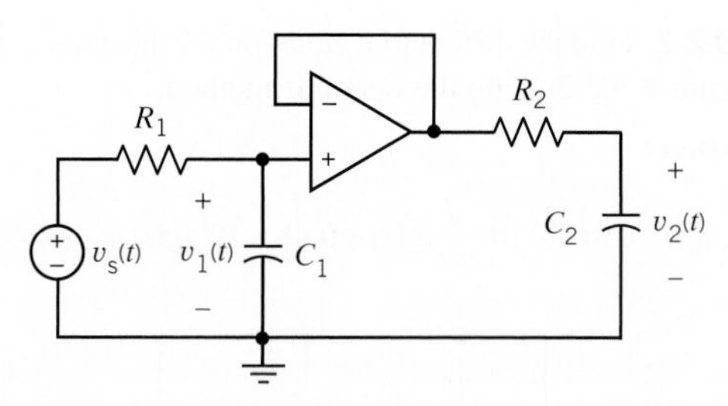

Figure P 9.2-11

P 9.2-12 The input to the circuit shown in Figure P 9.2-12 is the voltage of the voltage source, $v_s(t)$. The output is the voltage $v_o(t)$. Derive the second-order differential equation that shows how the output of this circuit is related to the input.

Hint: Use the operator method.

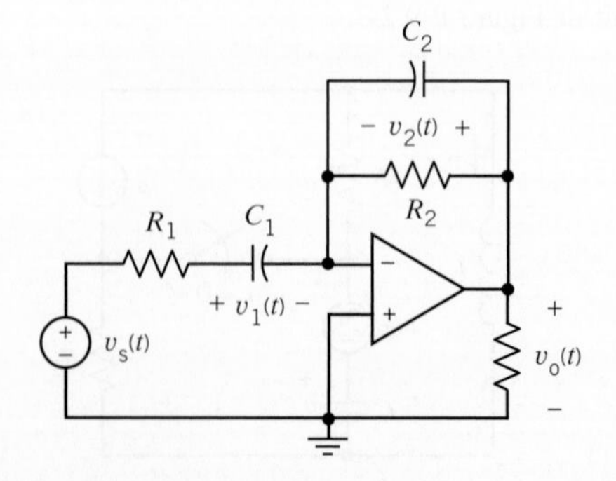

Figure P 9.2-12

P 9.2-13 The input to the circuit shown in Figure P 9.2-13 is the voltage of the voltage source, $v_s(t)$. The output is the voltage $v_o(t)$. Derive the second-order differential equation that shows how the output of this circuit is related to the input.

Hint: Use the direct method.

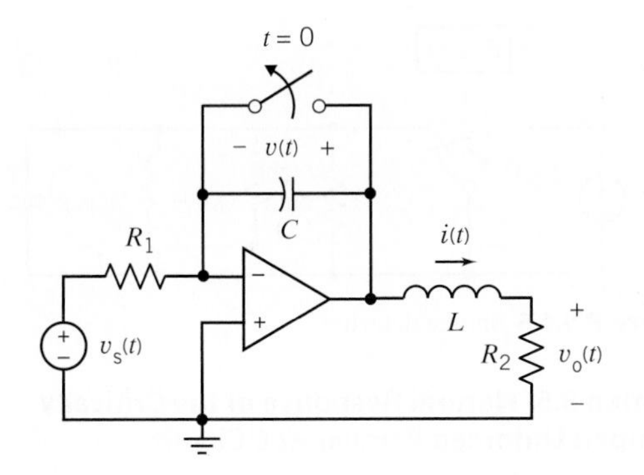

Figure P 9.2-13

P 9.2-14 The input to the circuit shown in Figure P 9.2-14 is the voltage of the voltage source, $v_s(t)$. The output is the voltage $v_2(t)$. Derive the second-order differential equation that shows how the output of this circuit is related to the input.

Hint: Use the direct method.

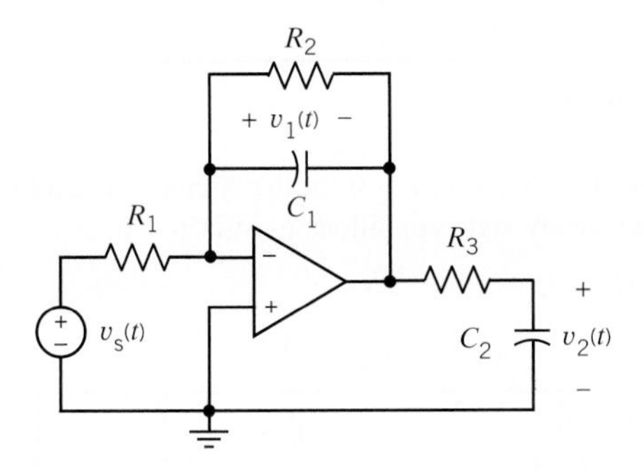

Figure P 9.2-14

P 9.2-15 Find the second-order differential equation for i_2 for the circuit of Figure P 9.2-15 using the operator method. Recall that the operator for the integral is $1/s$.

Answer: $6\dfrac{d^2i_2}{dt^2} + 5\dfrac{di_2}{dt} + i_2 = 2\dfrac{d^2v_s}{dt^2}$

2 Ω 4 Ω vs i1 2 H i2 1 F

Figure P 9.2-15

Section 9.3 Solution of the Second-Order Differential Equation—The Natural Response

P 9.3-1 Find the characteristic equation and its roots for the circuit of Figure P 9.2-2.

P 9.3-2 Find the characteristic equation and its roots for the circuit of Figure P 9.3-2.

Answer: $s^2 + 400s + 13.3 \times 10^3 = 0$
roots: $s = -36.60,\ -363.40$

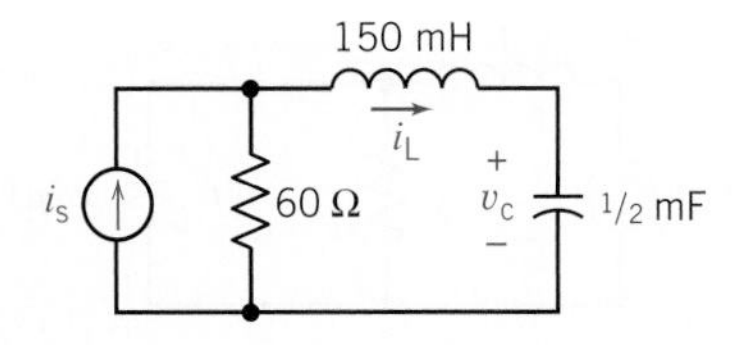

Figure P 9.3-2

P 9.3-3 Find the characteristic equation and its roots for the circuit shown in Figure P 9.3-3.

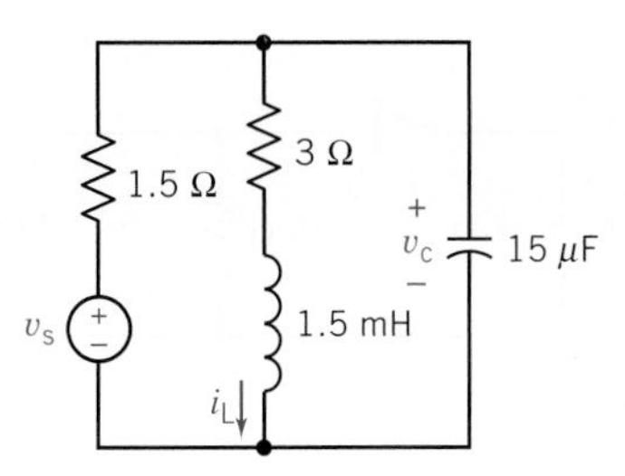

Figure P 9.3-3

P 9.3-4 German automaker Volkswagen, in its bid to make more efficient cars, has come up with an auto whose engine saves energy by shutting itself off at stoplights. The stop–start system springs from a campaign to develop cars in all its world markets that use less fuel and pollute less than vehicles now on the road. The stop–start transmission control has a mechanism that senses when the car does not need fuel: coasting downhill and idling at an intersection. The engine shuts off, but a small starter flywheel keeps turning so that power can be quickly restored when the driver touches the accelerator.

A model of the stop–start circuit is shown in Figure P 9.3-4. Determine the characteristic equation and the natural frequencies for the circuit.

Answer: $s^2 + 20s + 400 = 0$
$s = -10 \pm j17.3$

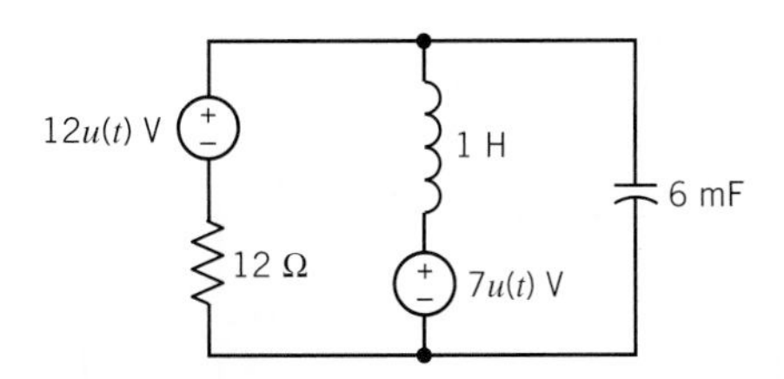

Figure P 9.3-4 Stop–start circuit.

Section 9.4 Natural Response of the Unforced Parallel *RLC* Circuit

P 9.4-1 Determine $v(t)$ for the circuit of Figure P 9.4-1 when $L = 1.5$ H and $v_s = 0$ for $t \geq 0$. The initial conditions are $v(0) = 9$ V and $dv/dt(0) = -4500$ V/s.

Answer: $v(t) = -3.82e^{-71.61t} + 12.82e^{-372.39t}$ V

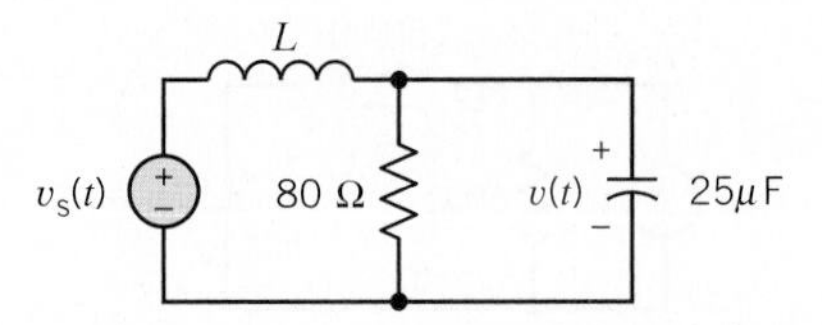

Figure P 9.4-1

P 9.4-2 An *RLC* circuit is shown in Figure P 9.4-2 , in which $v(0) = 2$ V. The switch has been open for a long time before closing at $t = 0$. Determine and plot $v(t)$.

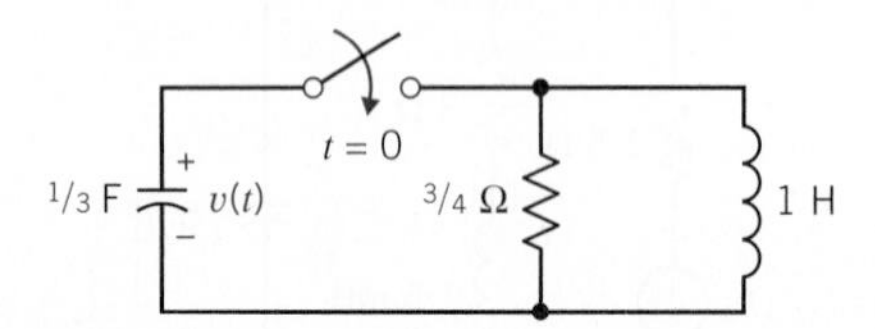

Figure P 9.4-2

P 9.4-3 Determine $i_1(t)$ and $i_2(t)$ for the circuit of Figure P 9.4-3 when $i_1(0) = i_2(0) = 11$ A.

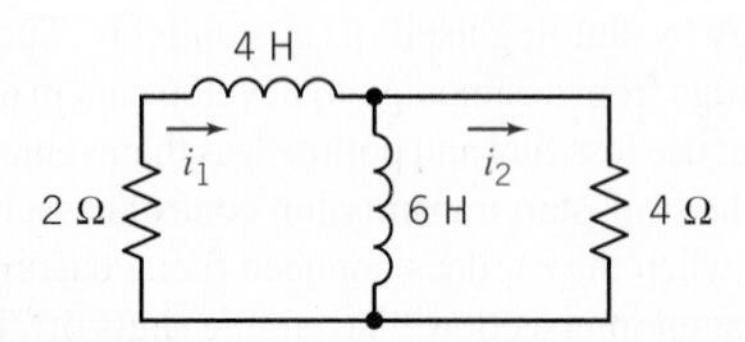

Figure P 9.4-3

P 9.4-4 The circuit shown in Figure P 9.4-4 contains a switch that is sometimes open and sometimes closed. Determine the damping factor, α, the resonant frequency, ω_0, and the damped resonant frequency, ω_d, of the circuit when (a) the switch is open and (b) the switch is closed.

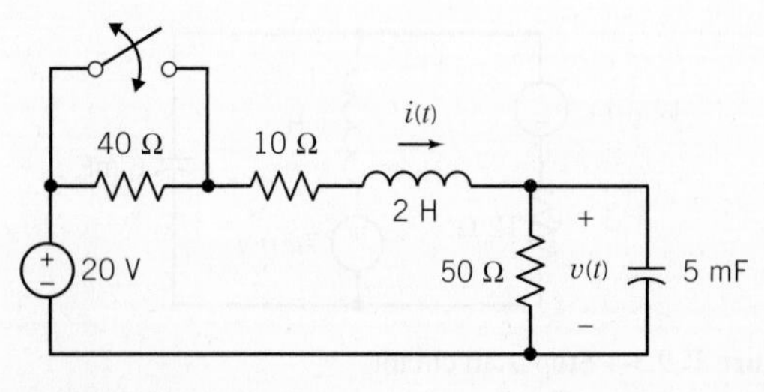

Figure P 9.4-4

P 9.4-5 The circuit shown in Figure P 9.4-5 is used in airplanes to detect smokers, who surreptitiously light up before they can take a single puff. The sensor activates the switch, and the change in the voltage $v(t)$ activates a light at the flight attendant's station. Determine the natural response $v(t)$.

Answer: $v(t) = -2.23e^{-2.1t} + 2.23e^{-37.9t}$ V

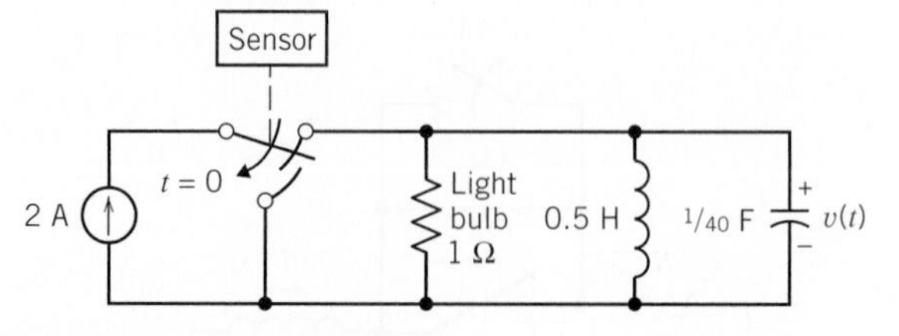

Figure P 9.4-5 Smoke detector.

Section 9.5 Natural Response of the Critically Damped Unforced Parallel *RLC* Circuit

P 9.5-1 Find $v_c(t)$ for $t > 0$ for the circuit shown in Figure P 9.5-1.

Answer: $v_c(t) = 3.3e^{-1667t} + 5498te^{-2000t}$ V

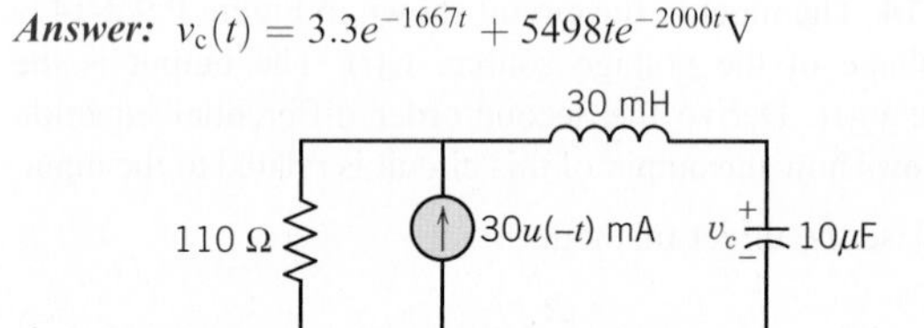

Figure P 9.5-1

P 9.5-2 Find $v_c(t)$ for $t > 0$ for the circuit of Figure P 9.5-2. Assume steady-state conditions exist at $t = 0^-$.

Answer: $v_c(t) = -8te^{-2t}$ V

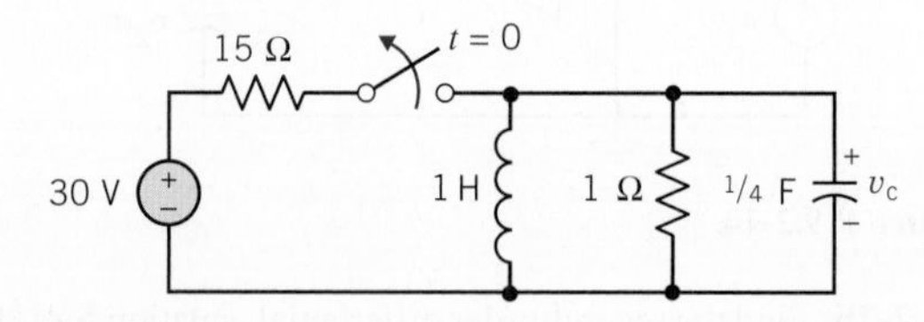

Figure P 9.5-2

P 9.5-3 Police often use stun guns to incapacitate potentially dangerous felons. The handheld device provides a series of high-voltage, low-current pulses. The power of the pulses is far below lethal levels, but it is enough to cause muscles to contract and put the person out of action. The device provides a pulse of up to 50,000 V, and a current of 1 mA flows through an arc. A model of the circuit for one period is shown in Figure P 9.5-3. Find $v(t)$ for $0 < t < 1$ ms. The resistor R represents the spark gap. Select C so that the response is critically damped.

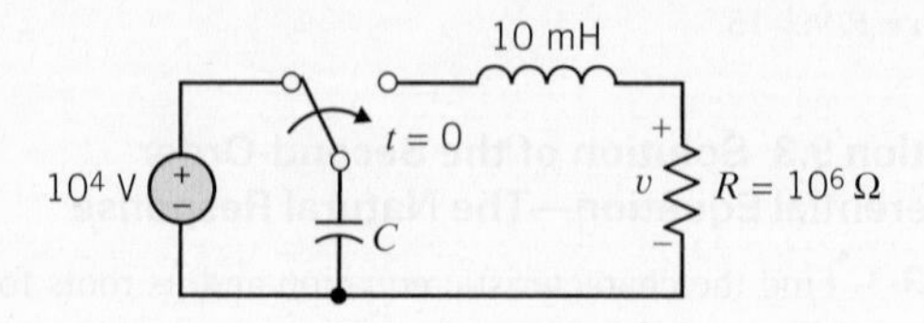

Figure P 9.5-3

P 9.5-4 Reconsider Problem P 9.4-1 when $L = 640$ mH and the other parameters and conditions remain the same.

Answer: $vt = (6 - 1500t)e^{-250t}$ V

P 9.5-5 An automobile ignition uses an electromagnetic trigger. The *RLC* trigger circuit shown in Figure P 9.5-5 has a step input of 6 V, and $v(0) = 2$ V and $i(0) = 0$. The resistance R must be selected from $2\ \Omega < R < 7\ \Omega$ so that the current $i(t)$ exceeds 0.6 A for greater than 0.5 s to activate the trigger. A critically damped response $i(t)$ is required to avoid oscillations in the trigger current. Select R and determine and plot $i(t)$.

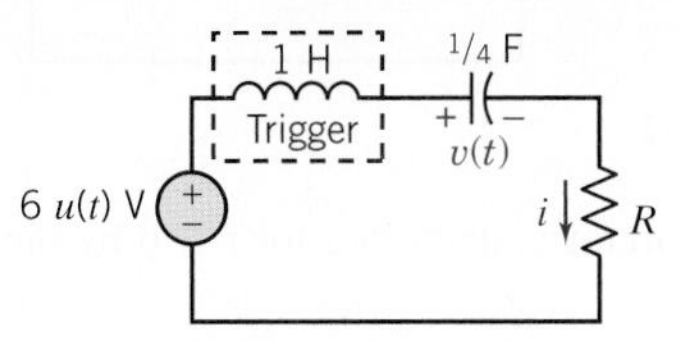

Figure P 9.5-5

Section 9.6 Natural Response of an Underdamped Unforced Parallel *RLC* Circuit

P 9.6-1 A communication system from a space station uses short pulses to control a robot operating in space. The transmitter circuit is modeled in Figure P 9.6-1. Find the output voltage $v_c(t)$ for $t > 0$. Assume steady-state conditions at $t = 0^-$.

Answer: $v_c(t) = e^{-278t}[2.7 \cos 362t + 2.07 \sin 362t]$ V

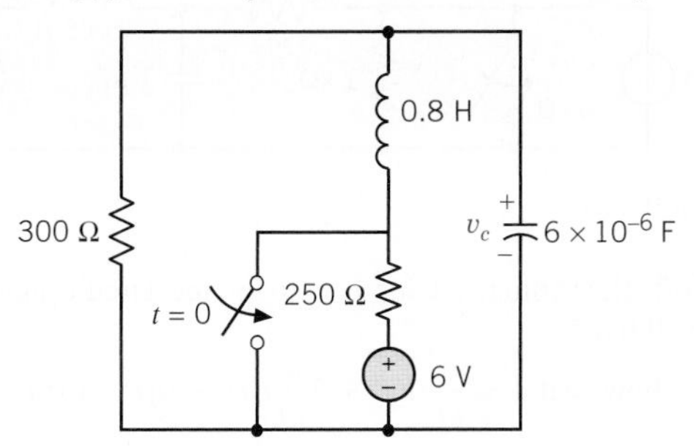

Figure P 9.6-1

P 9.6-2 The switch of the circuit shown in Figure P 9.6-2 is opened at $t = 0$. Determine and plot $v(t)$ when $C = 1/4$ F. Assume steady state at $t = 0^-$.

Answer: $v(t) = -4e^{-2t} \sin 2t$ V

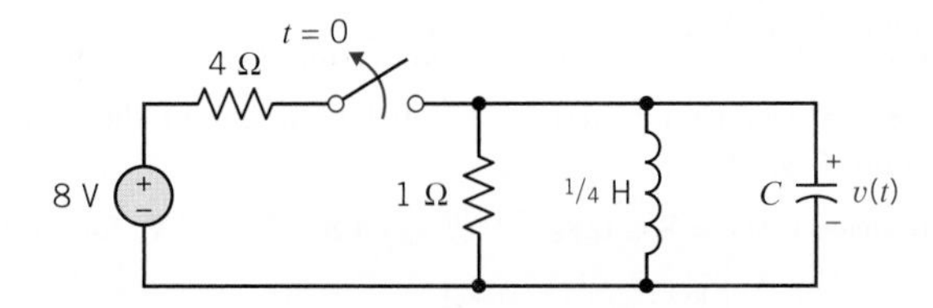

Figure P 9.6-2

P 9.6-3 A 240-W power supply circuit is shown in Figure P 9.6-3*a*. This circuit employs a large inductor and a large capacitor. The model of the circuit is shown in Figure P 9.6-3*b*. Find $i_L(t)$ for $t > 0$ for the circuit of Figure P 9.6-3*b*. Assume steady-state conditions exist at $t = 0^-$.

Answer: $i_L(t) = e^{-1.75t}[-3.2 \cos 0.7t + 8 \sin 0.7t]$ A

(a)

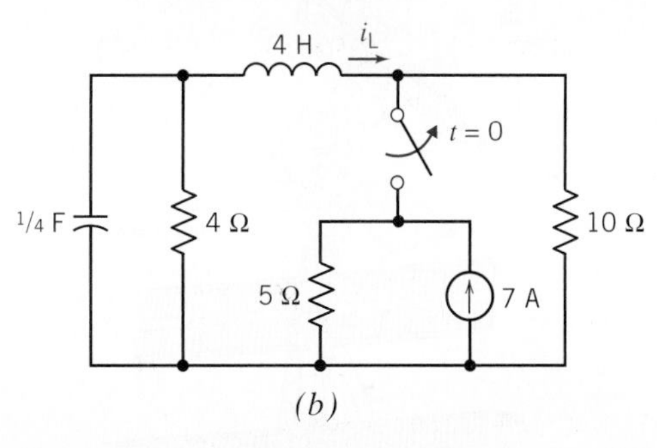

(b)

Figure P 9.6-3 *(a)* A 240-W power supply. Courtesy of Kepco, Inc. *(b)* Model of the power supply circuit.

P 9.6-4 The natural response of a parallel *RLC* circuit is measured and plotted as shown in Figure P 9.6-4. Using this chart, determine an expression for $v(t)$.

Hint: Notice that $v(t) = 260$ mV at $t = 5$ ms and that $v(t) = -200$ mV at $t = 7.5$ ms. Also, notice that the time between the first and third zero-crossings is 5 ms.

Answer: $v(t) = 544e^{-276t} \sin 1257t$ V

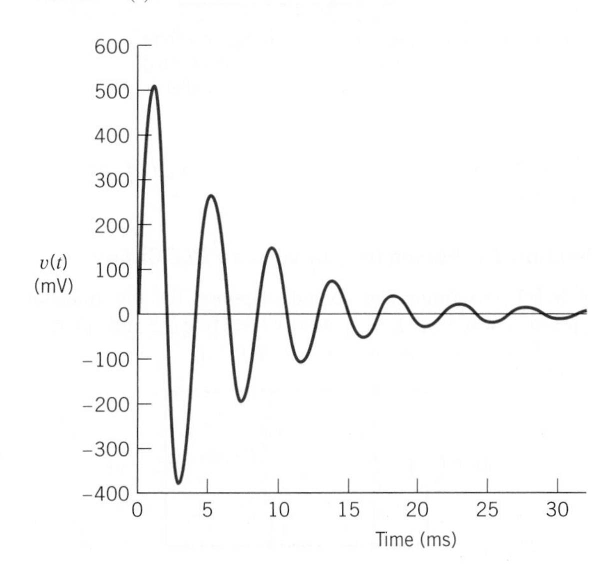

Figure P 9.6-4 The natural response of a parallel *RLC* circuit.

P 9.6-5 The photovoltaic cells of the proposed space station shown in Figure P 9.6-5*a* provide the voltage $v(t)$ of the circuit shown in Figure P 9.6-5*b*. The space station passes behind the shadow of earth (at $t = 0$) with $v(0) = 2$ V and $i(0) = 1/10$ A. Determine and sketch $v(t)$ for $t > 0$.

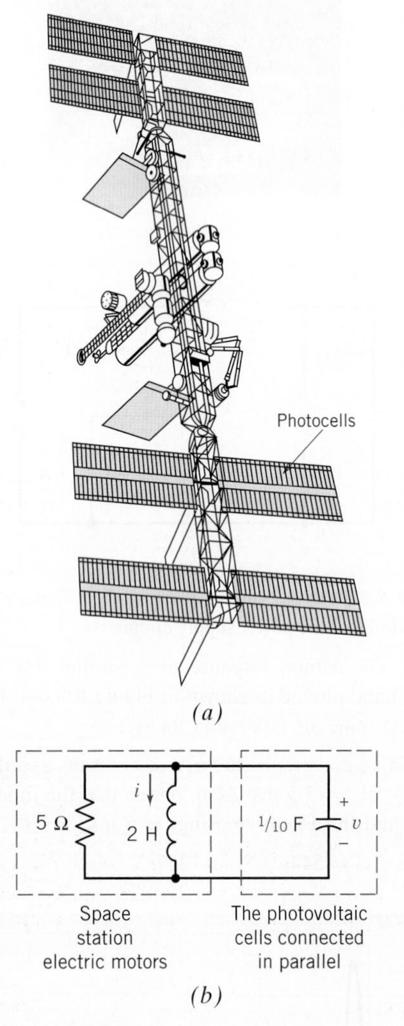

Figure P 9.6-5 (*a*) Photocells on space station. (*b*) Circuit with photocells.

Section 9.7 Forced Response of an *RLC* Circuit

P 9.7-1 Determine the forced response for the inductor current i_f when (a) $i_s = 1$ A, (b) $i_s = 0.5t$ A, and (c) $i_s = 2e^{-250t}$ A for the circuit of Figure P 9.7-1.

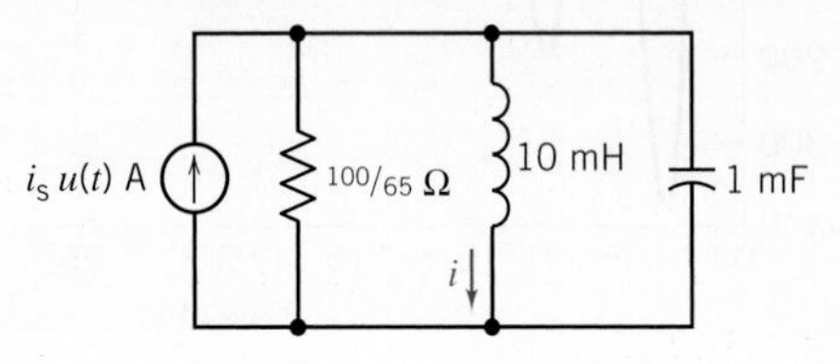

Figure P 9.7-1

P 9.7-2 Determine the forced response for the capacitor voltage, v_f, for the circuit of Figure P 9.7-2 when (a) $v_s = 2$ V, (b) $v_s = 0.2t$ V, and (c) $v_s = 1e^{-30t}$ V.

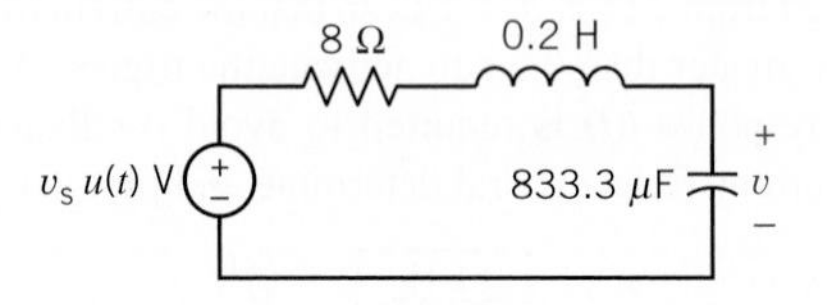

Figure P 9.7-2

P 9.7-3 A circuit is described for $t > 0$ by the equation

$$\frac{d^2v}{dt^2} + 5\frac{dv}{dt} + 6v = v_s$$

Find the forced response v_f for $t > 0$ when (a) $v_s = 8$ V, (b) $v_s = 3e^{-4t}$ V, and (c) $v_s = 2e^{-2t}$ V.

Answer: (a) $v_f = 8/6$ V (b) $v_f = \frac{3}{2}e^{-4t}$ V (c) $v_f = 2te^{-2t}$ V

Section 9.8 Complete Response of an *RLC* Circuit

P 9.8-1 Determine $i(t)$ for $t > 0$ for the circuit shown in Figure P 9.8-1.

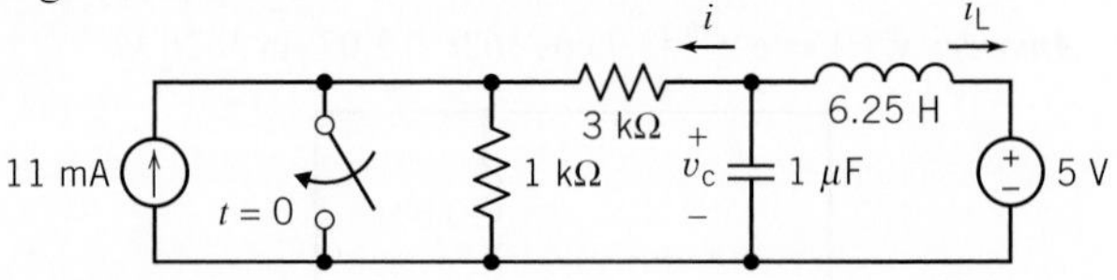

Figure P 9.8-1

P 9.8-2 Determine $i(t)$ for $t > 0$ for the circuit shown in Figure P 9.8-2.

Hint: Show that $1 = \frac{d^2}{dt^2}i(t) + 5\frac{d}{dt}i(t) + 5i(t)$ for $t > 0$

Answer: $i(t) = 0.2 + 0.246\,e^{-3.62t} - 0.646\,e^{-1.38t}$ A for $t > 0$.

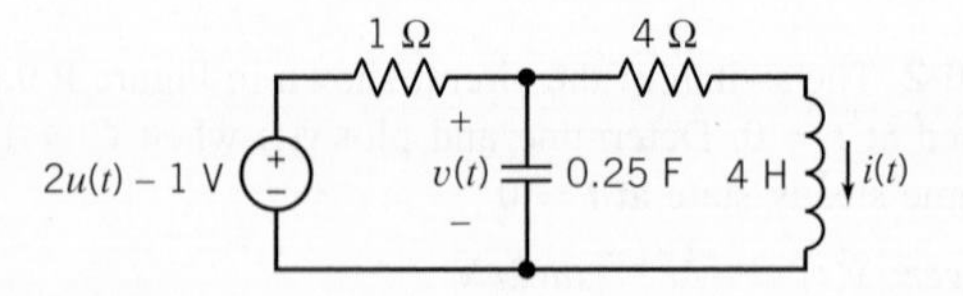

Figure P 9.8-2

P 9.8-3 Determine $v_1(t)$ for $t > 0$ for the circuit shown in Figure P 9.8-3.

Answer: $v_1(t) = 8 + 0.8e^{-3.1\times10^4 t} - 4.8\,e^{-5.2\times10^3 t}$ V for $t > 0$

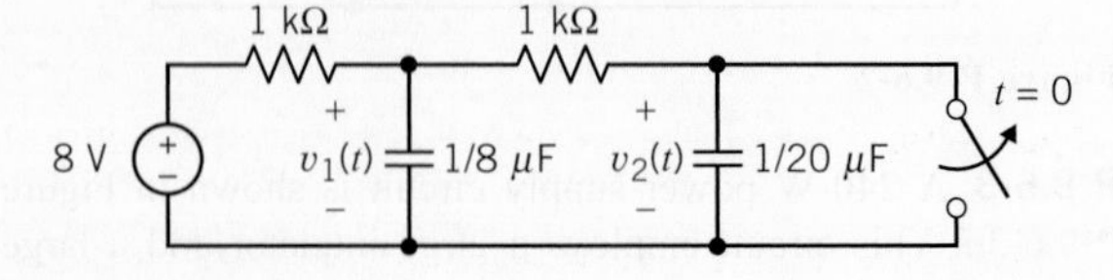

Figure P 9.8-3

P 9.8-4 Find $v(t)$ for $t > 0$ for the circuit shown in Figure P 9.8-4 when $v(0) = 1$ V and $i_L(0) = 0$.

Answer: $v = 25e^{-3t} - \frac{1}{17}[429e^{-4t} - 21\cos t + 33\sin t]$ V

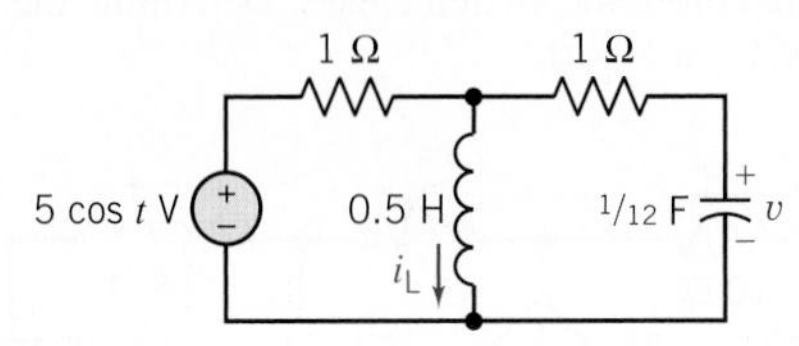

Figure P 9.8-4

P 9.8-5 Find $v(t)$ for $t > 0$ for the circuit of Figure P 9.8-5.

Answer: $v(t) = [16 - 17.6e^{-0.4t} + 17.6e^{-7.6t}]u(t) + [-16 + 17.6e^{-0.4(t-2)} - 17.6e^{-7.6(t-2)}]u(t-2)\ V$

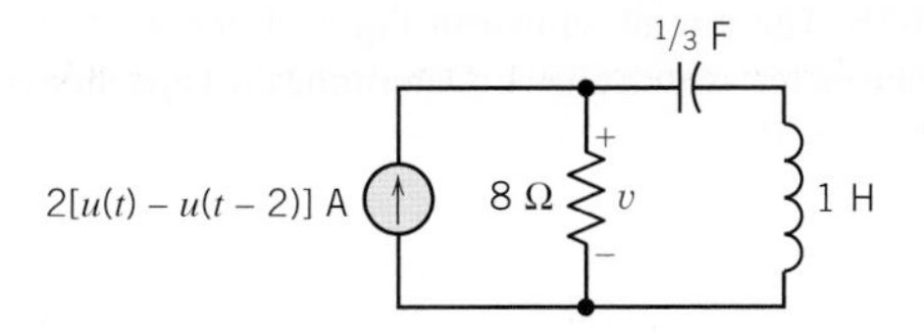

Figure P 9.8-5

P 9.8-6 An experimental space station power supply system is modeled by the circuit shown in Figure P 9.8-6. Find $v(t)$ for $t > 0$. Assume steady-state conditions at $t = 0^-$.

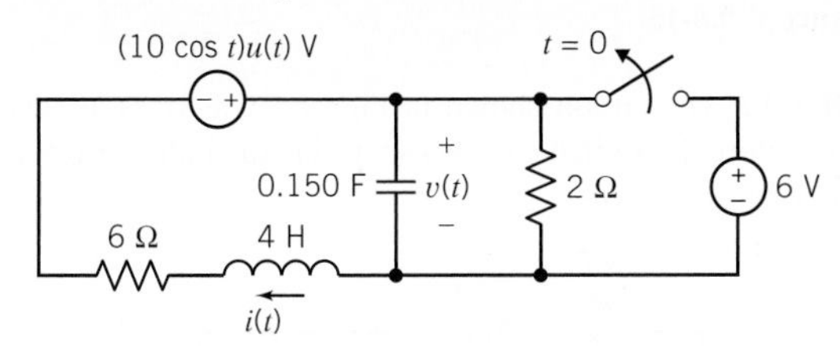

Figure P 9.8-6

P 9.8-7 Find $v_c(t)$ for $t > 0$ in the circuit of Figure P 9.8-7 when (a) $C = 1/18$ F, (b) $C = 1/10$ F, and (c) $C = 1/20$ F.

Answers:

(a) $v_c(t) = 8e^{-3t} + 24te^{-3t} - 8$ V
(b) $v_c(t) = 10e^{-t} - 2e^{-5t} - 8$ V
(c) $v_c(t) = e^{-3t}(8\cos t + 24\sin t) - 8$ V

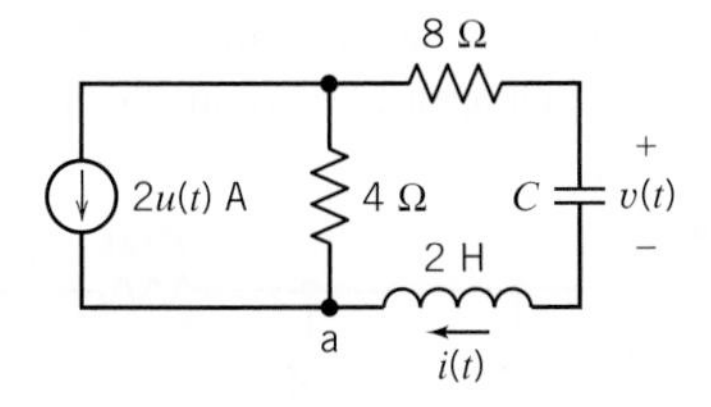

Figure P 9.8-7

P 9.8-8 Find $v_c(t)$ for $t > 0$ for the circuit shown in Figure P 9.8-8.

Hint: $2 = \frac{d^2}{dt^2}v_c(t) + 6\frac{d}{dt}v_c(t) + 2v_c(t)$ for $t > 0$

Answer: $v_c(t) = 0.123e^{-5.65t} + 0.877e^{-0.35t} + 1$ V for $t > 0$.

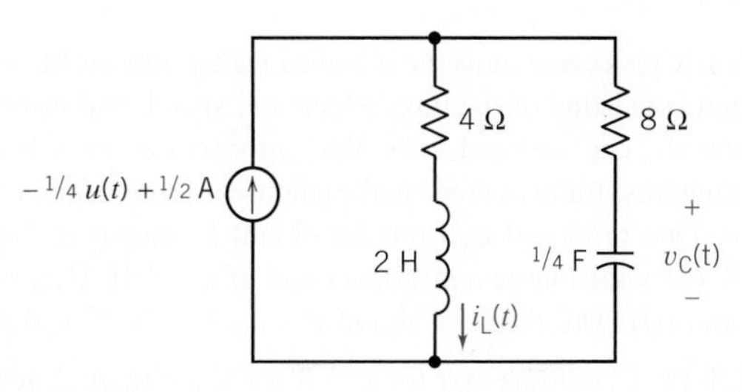

Figure P 9.8-8

P 9.8-9 In Figure P 9.8-9, determine the inductor current $i(t)$ when $i_s = 6u(t)$ A. Assume that $i(0) = 0$, $v_c(0) = 0$.

Answer: $i(t) = 6 + e^{-2t}[-6\cos 5.2t - 2.3\sin 5.2t]$ A

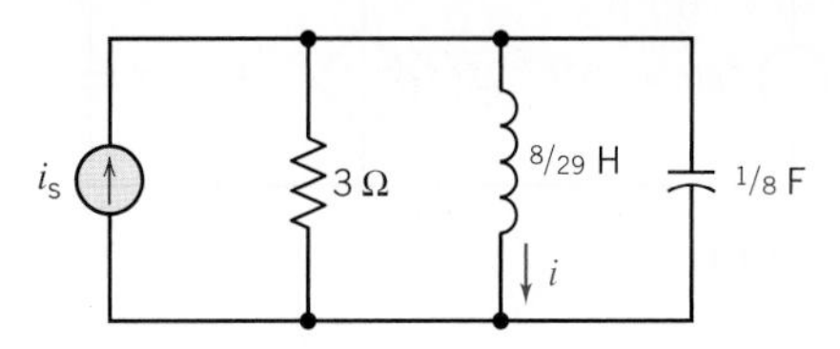

Figure P 9.8-9

P 9.8-10 Railroads widely use automatic identification of railcars. When a train passes a tracking station, a wheel detector activates a radio-frequency module. The module's antenna, as shown in Figure P 9.8-10*a*, transmits and receives a signal that bounces off a transponder on the locomotive. A

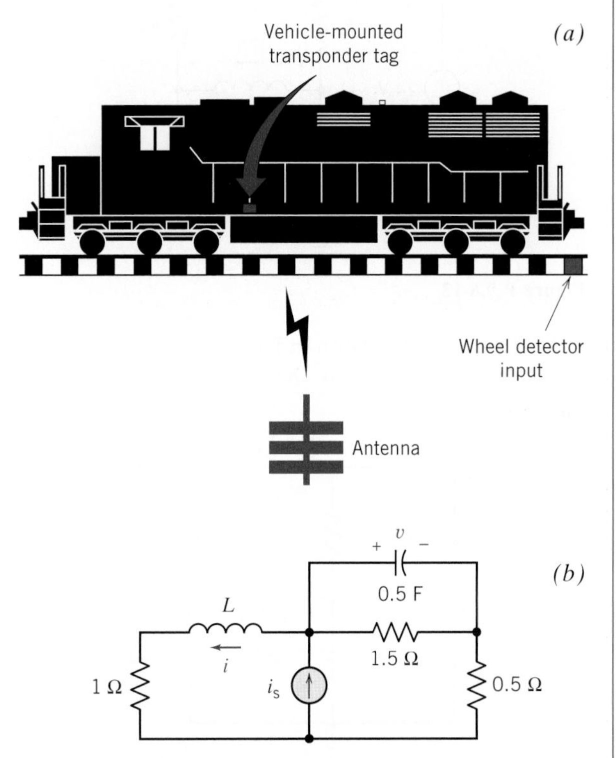

Figure P 9.8-10 (*a*) Railroad identification system. (*b*) Transponder circuit.

trackside processor turns the received signal into useful information consisting of the train's location, speed, and direction of travel. The railroad uses this information to schedule locomotives, trains, crews, and equipment more efficiently.

One proposed transponder circuit is shown in Figure P 9.8-10b with a large transponder coil of $L = 5$ H. Determine $i(t)$ and $v(t)$. The received signal is $i_s = 9 + 3e^{-2t}\,u(t)$ A.

P 9.8-11 Determine $v(t)$ for $t > 0$ for the circuit shown in Figure P 9.8-11.

Answer: $v_c(t) = 0.75\,e^{-4t} - 6.75\,e^{-36t} + 16$ V for $t > 0$

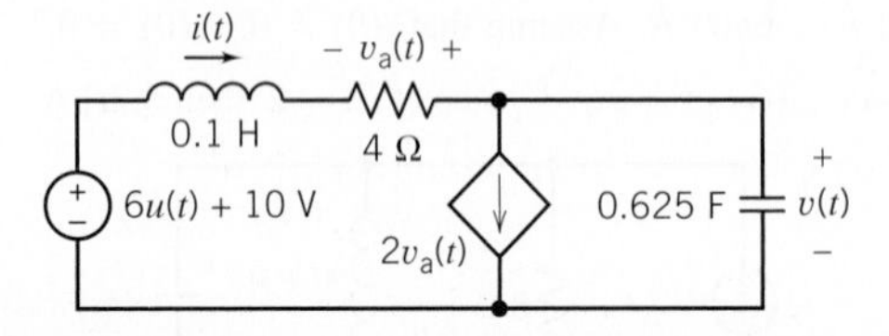

Figure P 9.8-11

P 9.8-12 The circuit shown in Figure P 9.8-12 is at steady state before the switch opens. The inductor current is given to be

$$i(t) = 240 + 193e^{-6.25t}\cos(9.27t - 102°)\text{ mA} \quad \text{for } t \geq 0$$

Determine the values of R_1, R_3, C, and L.

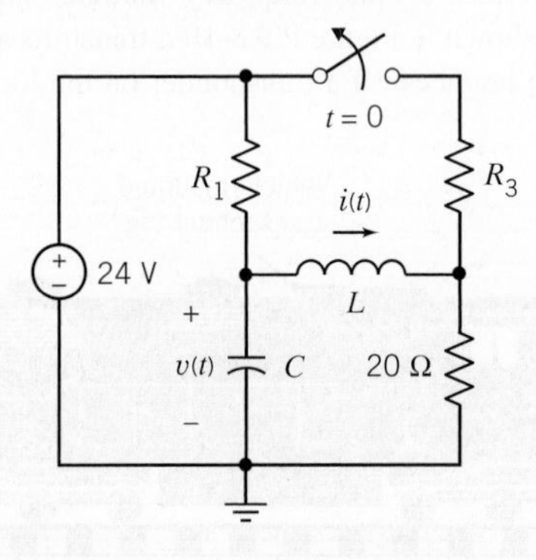

Figure P 9.8-12

P 9.8-13 The circuit shown in Figure P 9.8-13 is at steady state before the switch opens. Determine the inductor current, $i(t)$, for $t > 0$.

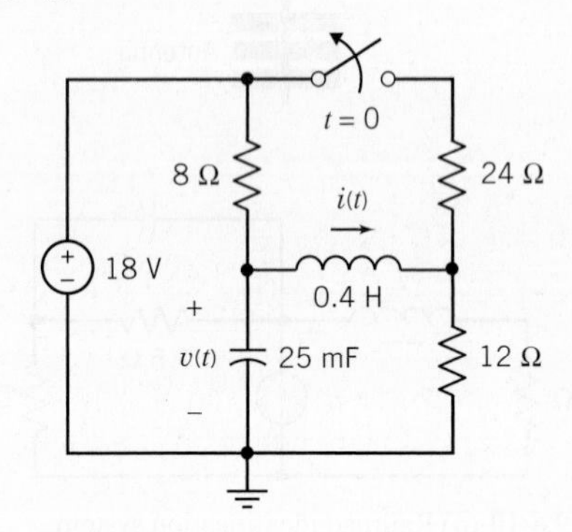

Figure P 9.8-13

***P 9.8-14** The circuit shown in Figure P 9.8-14 is at steady state before the switch closes. Determine the capacitor voltage, $v(t)$, for $t > 0$.

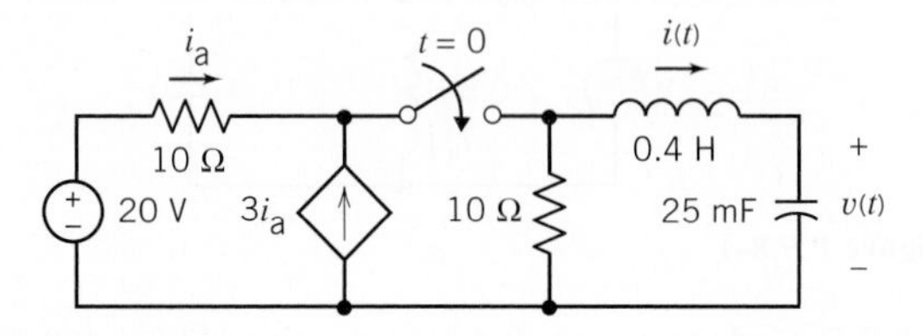

Figure P 9.8-14

P 9.8-15 The circuit shown in Figure P 9.8-15 is at steady state before the switch closes. Determine the capacitor voltage, $v(t)$, for $t > 0$.

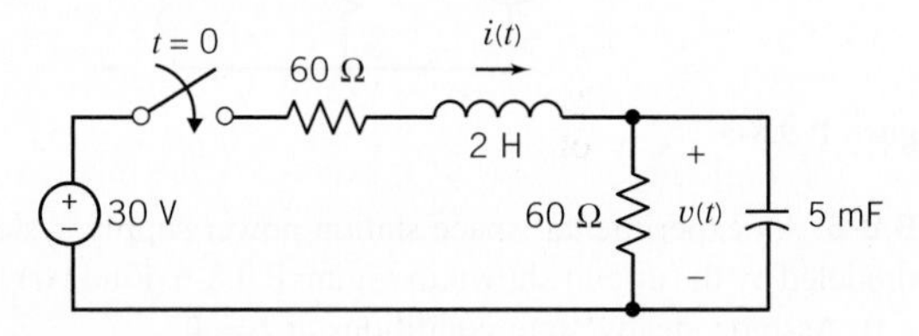

Figure P 9.8-15

P 9.8-16 The circuit shown in Figure P 9.8-16 is at steady state before the switch closes. Determine the inductor current, $i(t)$, for $t > 0$.

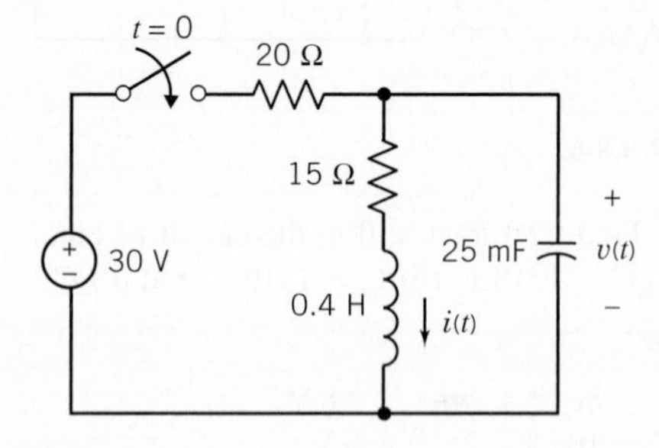

Figure P 9.8-16

P 9.8-17 The circuit shown in Figure P 9.8-17 is at steady state before the switch opens. Determine the inductor current, $i_2(t)$, for $t > 0$.

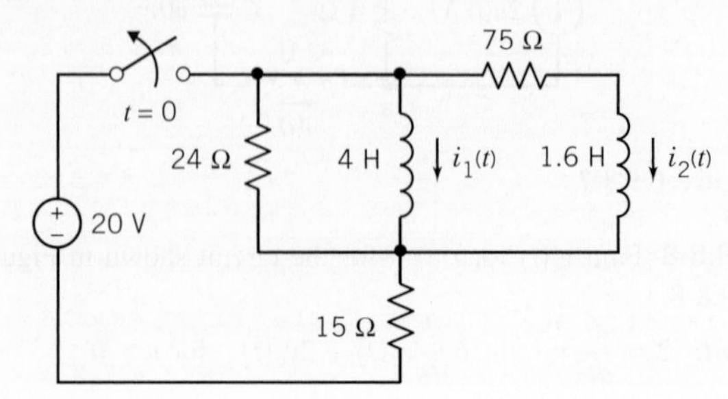

Figure P 9.8-17

P 9.8-18 The circuit shown in Figure P 9.8-18 is at steady state before the switch closes. Determine the capacitor voltage, $v(t)$, for $t > 0$.

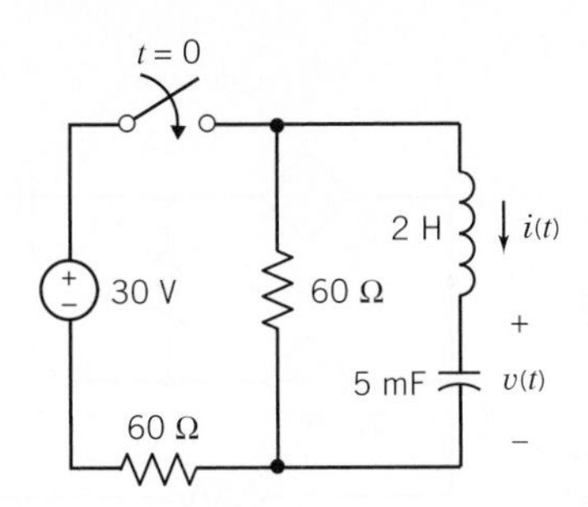

Figure P 9.8-18

P 9.8-19 Find the differential equation for $v_c(t)$ in the circuit of Figure P 9.8-19, using the direct method. Find $v_c(t)$ for time $t > 0$ for each of the following sets of component values:

(a) $C = 1$ F, $L = 0.25$ H, $R_1 = R_2 = 1.309\ \Omega$
(b) $C = 1$ F, $L = 1$ H, $R_1 = 3\ \Omega$, $R_2 = 1\ \Omega$
(c) $C = 0.125$ F, $L = 0.5$ H, $R_1 = 1\ \Omega$, $R_2 = 4\ \Omega$

Answer:

(a) $v_c(t) = \frac{1}{2} - e^{-2t} + \frac{1}{2}e^{-4t}$ V

(b) $v_c(t) = \frac{1}{4} - \left(\frac{1}{4} + \frac{1}{2}t\right) e^{-2t}$ V

(c) $v_c(t) = 0.8 - e^{-2t}(0.8 \cos 4t + 0.4 \sin 4t)$ V

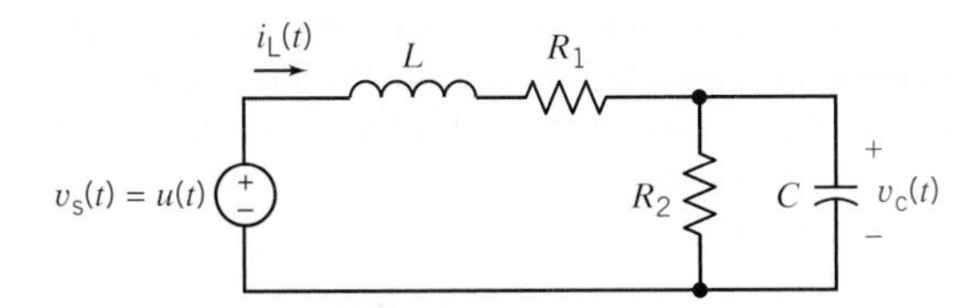

Figure P 9.8-19

P 9.8-20 Find the differential equation for $v_o(t)$ in the circuit of Figure P 9.8-20, using the direct method. Find $v_o(t)$ for time $t > 0$ for each of the following sets of component values:

(a) $C = 1$ F, $L = 0.25$ H, $R_1 = R_2 = 1.309\ \Omega$
(b) $C = 1$ F, $L = 1$ H, $R_1 = 1\ \Omega$, $R_2 = 3\ \Omega$
(c) $C = 0.125$ F, $L = 0.5$ H, $R_1 = 4\ \Omega$, $R_2 = 1\ \Omega$

Answer:

(a) $v_o(t) = \frac{1}{2} - e^{-2t} + \frac{1}{2}e^{-4t}$ V

(b) $v_o\ (t) = \frac{3}{4} - \left(\frac{3}{4} + \frac{3}{2}t\right) e^{-2t}$ V

(c) $v_o(t) = 0.2 - e^{-2t}(0.2 \cos 4t + 0.1 \sin 4t)$ V

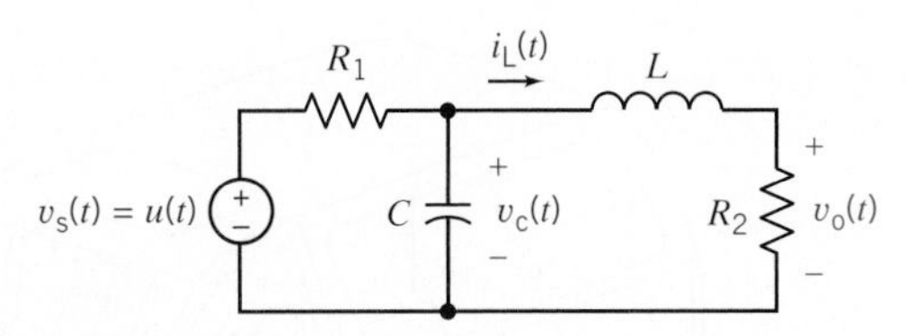

Figure P 9.8-20

Section 9.9 State Variable Approach to Circuit Analysis

P 9.9-1 Find $v(t)$ for $t > 0$, using the state variable method of Section 9.9 when $C = 1/5$ F in the circuit of Figure P 9.9-1. Sketch the response for $v(t)$ for $0 < t < 10$ s.

Answer: $v(t) = -25e^{-t} + e^{-5t} + 24$ V

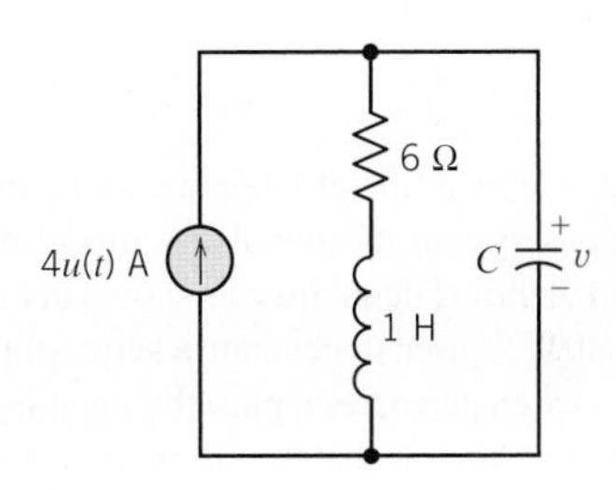

Figure P 9.9-1

P 9.9-2 Repeat Problem P 9.9-1 when $C = 1/10$ F. Sketch the response for $v(t)$ for $0 < t < 3$ s.

Answer: $v(t) = e^{-3t}(-24 \cos t - 32 \sin t) + 24$ V

P 9.9-3 Determine the current $i(t)$ and the voltage $v(t)$ for the circuit of Figure P 9.9-3.

Answer: $i(t) = (3.434e^{-5.6t} - 0.434e^{-44.3t} - 6)$ A

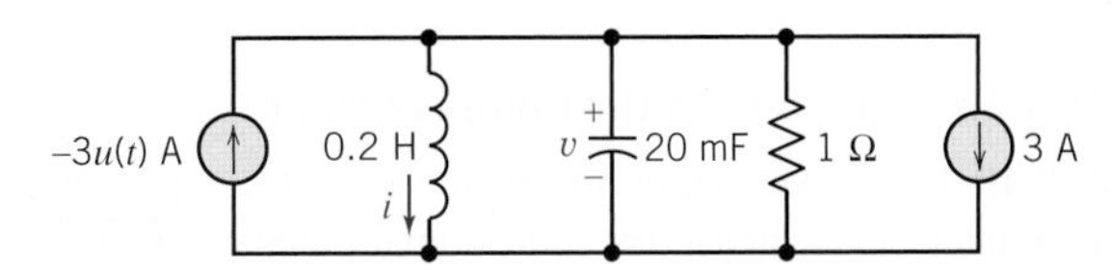

Figure P 9.9-3

P 9.9-4 Clean-air laws are pushing the auto industry toward the development of electric cars. One proposed vehicle using an ac motor is shown in Figure P 9.9-4*a*. The motor-controller circuit is shown in Figure P 9.9-4*b* with $L = 100$ mH and $C = 10$ mF. Using the state equation approach, determine $i(t)$ and $v(t)$ where $i(t)$ is the motor-control current. The initial conditions are $v(0) = 10$ V and $i(0) = 0$.

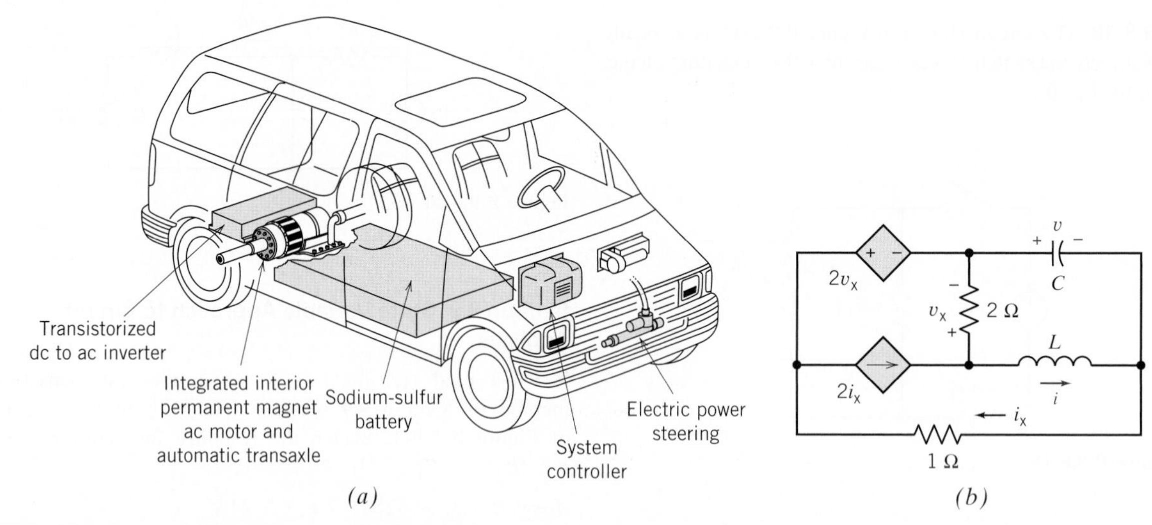

Figure P 9.9-4 (*a*) Electric vehicle. (*b*) Motor-controller circuit.

P 9.9-5 Studies of an artificial insect are being used to understand the nervous system of animals. A model neuron in the nervous system of the artificial insect is shown in Figure P 9.9-5. The input signal, v_s, is used to generate a series of pulses, called synapses. The switch generates a pulse by opening at $t = 0$ and closing at $t = 0.5$ s. Assume that the circuit is at steady state and that $v(0^-) = 10$ V. Determine the voltage $v(t)$ for $0 < t < 2$ s.

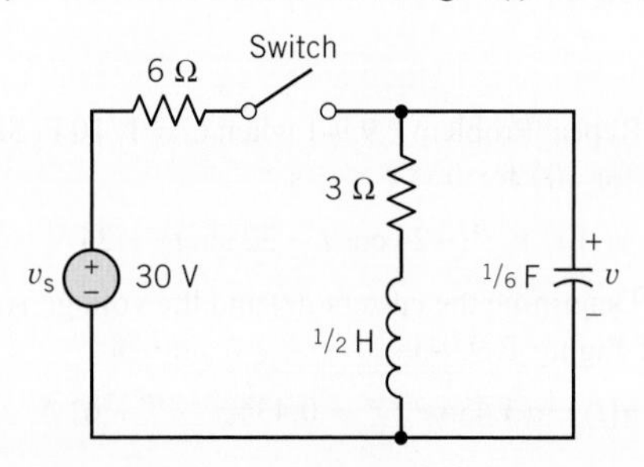

Figure P 9.9-5 Neuron circuit model.

Section 9.10 Roots in the Complex Plane

P 9.10-1 For the circuit of Figure P 9.10-1, determine the roots of the characteristic equation and plot the roots on the s-plane.

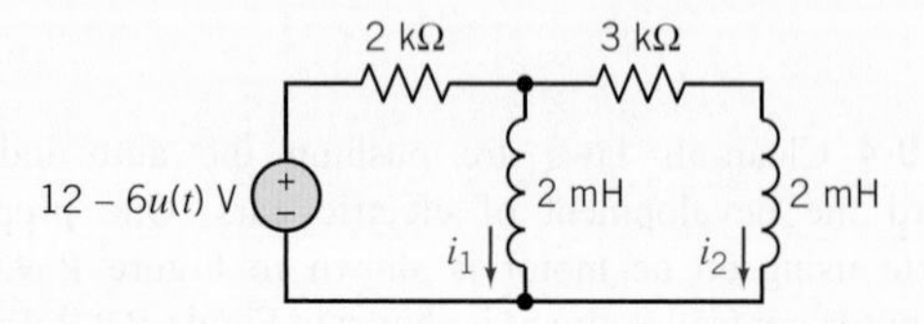

Figure P 9.10-1

P 9.10-2 For the circuit of Figure P 9.6-1, determine the roots of the characteristic equation and plot the roots on the s-plane.

P 9.10-3 For the circuit of Figure P 9.10-3, determine the roots of the characteristic equation and plot the roots on the s-plane.

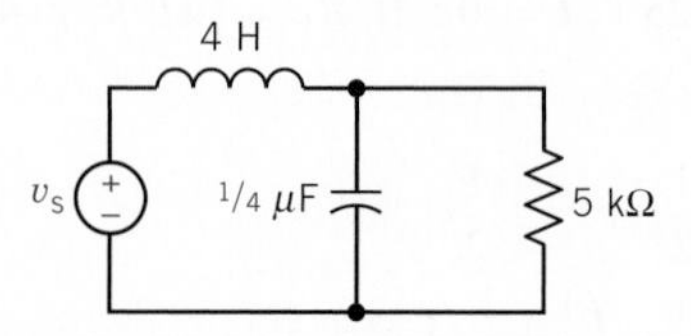

Figure P 9.10-3

P 9.10-4 An *RLC* circuit is shown in Figure P 9.10-4.

(a) Obtain the two-node voltage equations, using operators.
(b) Obtain the characteristic equation for the circuit.
(c) Show the location of the roots of the characteristic equation in the s-plane.
(d) Determine $v(t)$ for $t > 0$.

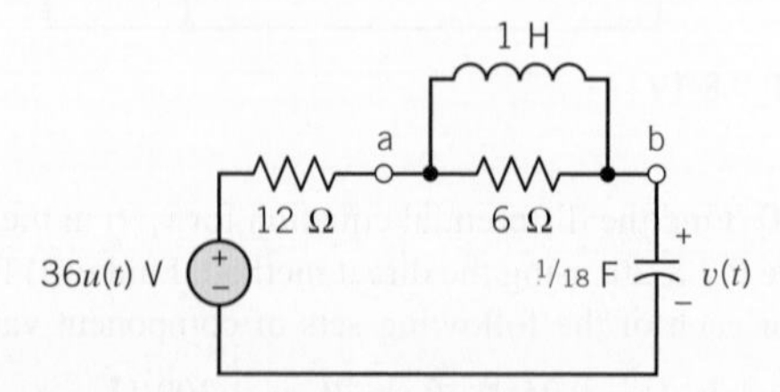

Figure P 9.10-4

Section 9.11 How Can We Check . . . ?

P 9.11-1 Figure P 9.11-1*a* shows an *RLC* circuit. The voltage, $v_s(t)$, of the voltage source is the square wave shown in Figure P 9.11-1*a*. Figure P 9.11-1*c* shows a plot of the inductor current, $i(t)$, which was obtained by simulating this circuit, using PSpice. Verify that the plot of $i(t)$ is correct.

Answer: The plot is correct.

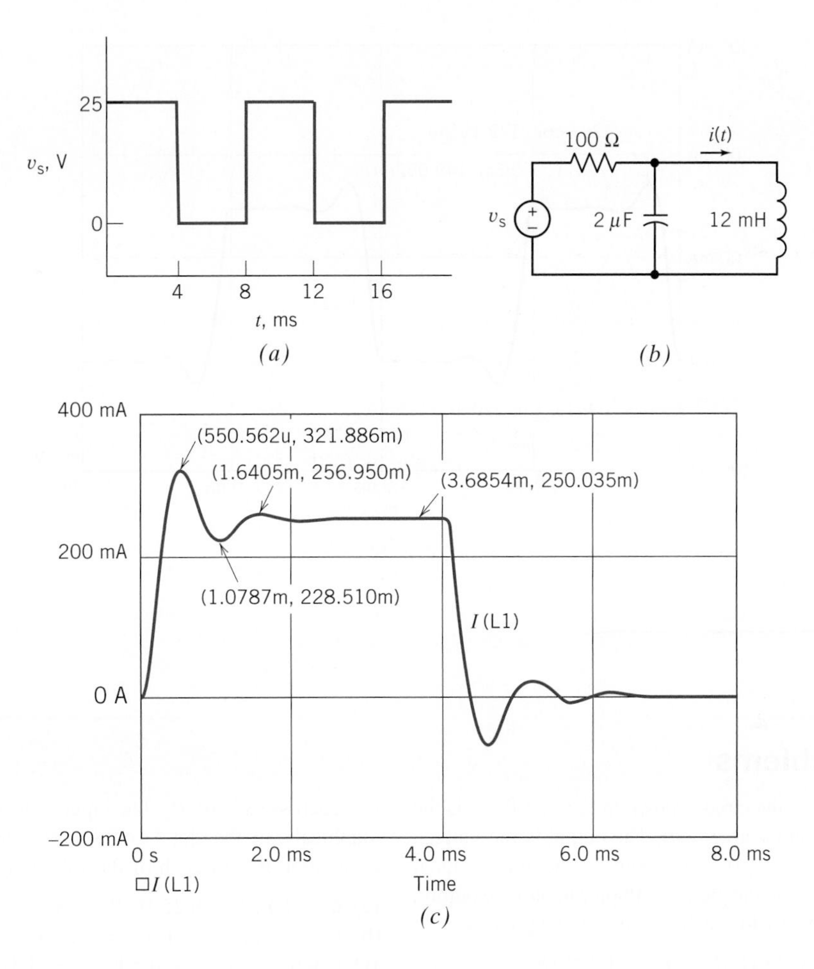

Figure P 9.11-1

P 9.11-2 Figure P 9.11-2*b* shows an *RLC* circuit. The voltage, $v_s(t)$, of the voltage source is the square wave shown in Figure P 9.11-2*a*. Figure P 9.11-2*c* shows a plot of the inductor current, $i(t)$, which was obtained by simulating this circuit, using PSpice. Verify that the plot of $i(t)$ is correct.

Answer: The plot is not correct.

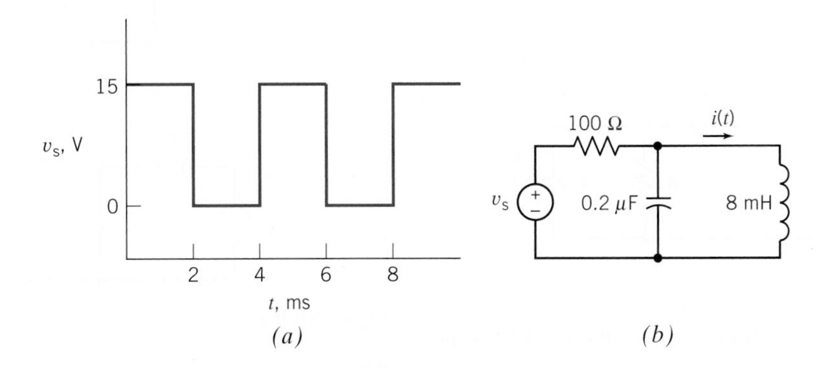

Figure P 9.11-2

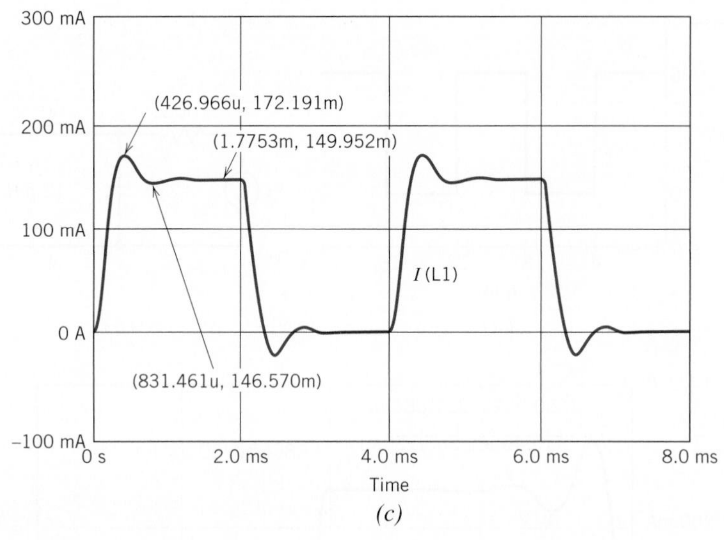

(c)

Figure P 9.11-2 (*Continued*)

PSpice Problems

SP 9-1 The input to the circuit shown in Figure SP 9-1 is the voltage of the voltage source, $v_i(t)$. The output is the voltage across the capacitor, $v_o(t)$. The input is the pulse signal specified graphically by the plot. Use PSpice to plot the output, $v_o(t)$, as a function of t for each of the following cases:

(a) $C = 1$ F, $L = 0.25$ H, $R_1 = R_2 = 1.309\ \Omega$
(b) $C = 1$ F, $L = 1$ H, $R_1 = 3\ \Omega$, $R_2 = 1\ \Omega$
(c) $C = 0.125$ F, $L = 0.5$ H, $R_1 = 1\ \Omega$, $R_2 = 4\ \Omega$

Plot the output for these three cases on the same axis.

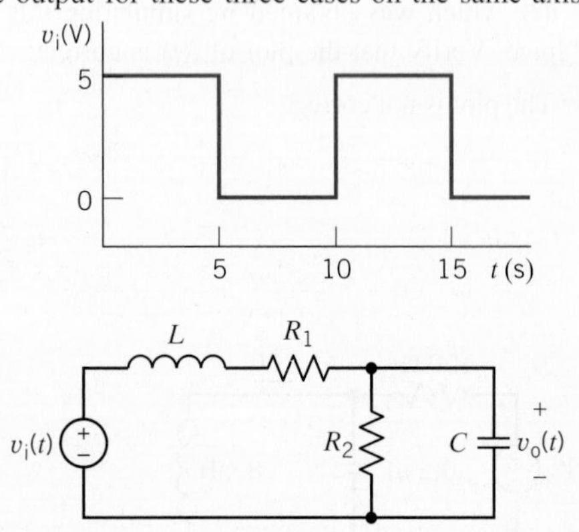

Figure SP 9-1

Hint: Represent the voltage source, using the PSpice part named VPULSE.

SP 9-2 The input to the circuit shown in Figure SP 9-2 is the voltage of the voltage source, $v_i(t)$. The output is the voltage, $v_o(t)$, across resistor, R_2. The input is the pulse signal specified graphically by the plot. Use PSpice to plot the output, $v_o(t)$, as a function of t for each of the following cases:

(a) $C = 1$ F, $L = 0.25$ H, $R_1 = R_2 = 1.309\ \Omega$
(b) $C = 1$ F, $L = 1$ H, $R_1 = 3\ \Omega$, $R_2 = 1\ \Omega$
(c) $C = 0.125$ F, $L = 0.5$ H, $R_1 = 1\ \Omega$, $R_2 = 4\ \Omega$

Plot the output for these three cases on the same axis.

Hint: Represent the voltage source, using the PSpice part named VPULSE.

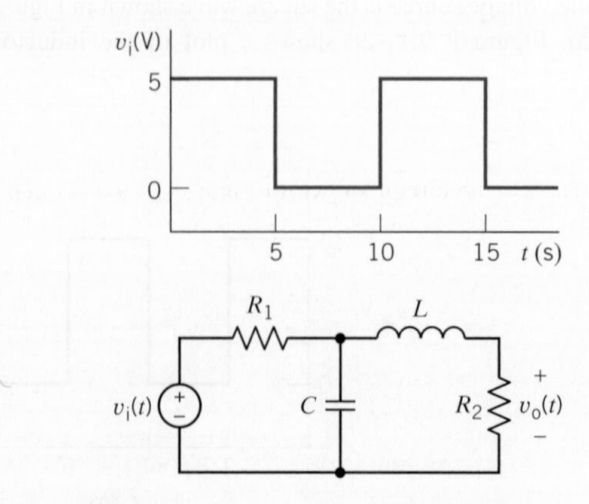

Figure SP 9-2

SP 9-3 Determine and plot the capacitor voltage $v(t)$ for $0 < t < 300\ \mu$s for the circuit shown in Figure SP 9-3*a*. The sources are pulses as shown in Figures SP 9-3*b*,*c*.

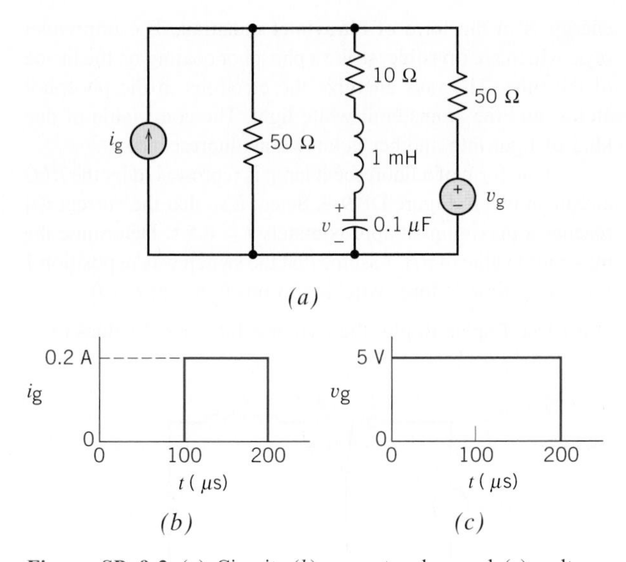

Figure SP 9-3 (*a*) Circuit, (*b*) current pulse, and (*c*) voltage pulse.

SP 9-4 Determine and plot $v(t)$ for the circuit of Figure SP 9-4 when $v_s(t) = 5u(t)$ V. Plot $v(t)$ for $0 < t < 0.25$ s.

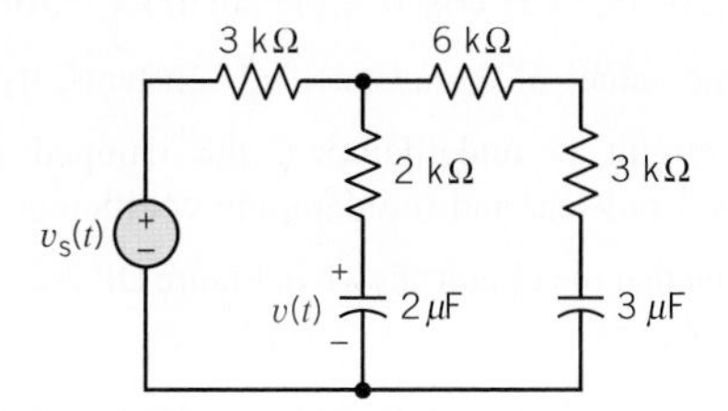

Figure SP 9-4

Design Problems

DP 9-1 Design the circuit shown in Figure DP 9-1 so that

$$v_c(t) = \frac{1}{2} + A_1 e^{-2t} + A_2 e^{-4t} \text{ V} \quad \text{for } t > 0$$

Determine the values of the unspecified constants, A_1 and A_2.

Hint: The circuit is overdamped, and the natural frequencies are 2 and 4 rad/sec.

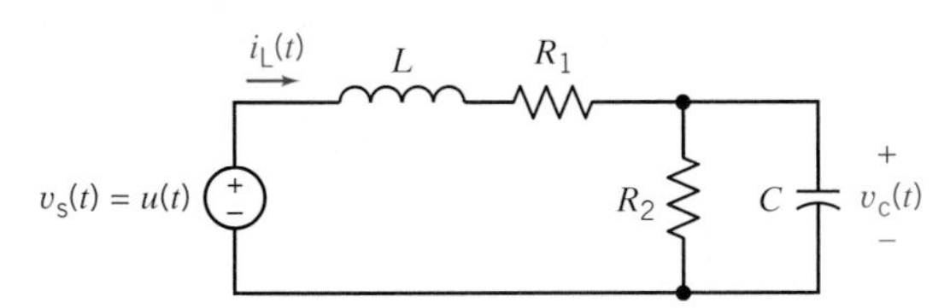

Figure DP 9-1

DP 9-2 Design the circuit shown in Figure DP 9-1 so that

$$v_c(t) = \frac{1}{4} + (A_1 + A_2 t)e^{-2t} \text{ V} \quad \text{for } t > 0$$

Determine the values of the unspecified constants, A_1 and A_2.

Hint: The circuit is critically damped, and the natural frequencies are both 2 rad/sec.

DP 9-3 Design the circuit shown in Figure DP 9-1 so that

$$v_c(t) = 0.8 + e^{-2t}(A_1 \cos 4t + A_2 \sin 4t) \text{ V} \quad \text{for } t > 0$$

Determine the values of the unspecified constants, A_1 and A_2.

Hint: The circuit is underdamped, the damped resonant frequency is 4 rad/sec, and the damping coefficient is 2.

DP 9-4 Show that the circuit shown in Figure DP 9-1 cannot be designed so that

$$v_c(t) = 0.5 + e^{-2t}(A_1 \cos 4t + A_2 \sin 4t) \text{ V} \quad \text{for } t > 0$$

Hint: Show that such a design would require $1/RC + 10RC = 4$ where $R = R_1 = R_2$. Next, show that $1/RC + 10\,RC = 4$ would require the value of RC to be complex.

DP 9-5 Design the circuit shown in Figure DP 9-5 so that

$$v_o(t) = \frac{1}{2} + A_1\, e^{-2t} + A_2\, e^{-4t} \text{ V} \quad \text{for } t > 0$$

Determine the values of the unspecified constants, A_1 and A_2.

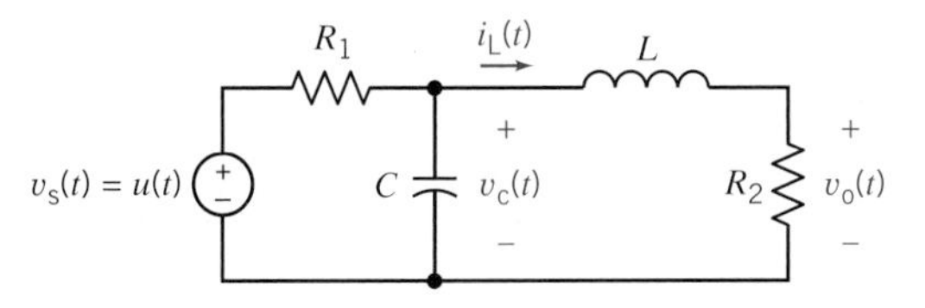

Figure DP 9-5

Hint: The circuit is overdamped, and the natural frequencies are 2 and 4 rad/sec.

DP 9-6 Design the circuit shown in Figure DP 9-5 so that

$$v_o(t) = \frac{3}{4} + (A_1 + A_2 t)e^{-2t} \text{ V} \quad \text{for } t > 0$$

Determine the values of the unspecified constants, A_1 and A_2.

Hint: The circuit is critically damped, and the natural frequencies are both 2 rad/sec.

DP 9-7 Design the circuit shown in Figure DP 9-5 so that

$$v_c(t) = 0.2 + e^{-2t}(A_1 \cos 4t + A_2 \sin 4t) \text{ V} \quad \text{for } t > 0$$

Determine the values of the unspecified constants, A_1 and A_2.

Hint: The circuit is underdamped, the damped resonant frequency is 4 rad/sec, and the damping coefficient is 2.

DP 9-8 Show that the circuit shown in Figure DP 9-5 cannot be designed so that

$$v_c(t) = 0.5 + e^{-2t}(A_1 \cos 4t + A_2 \sin 4t) \text{ V} \quad \text{for } t > 0$$

Hint: Show that such a design would require $1/RC + 10\,RC = 4$ where $R = R_1 = R_2$. Next, show that $1/RC + 10\,RC = 4$ would require the value of RC to be complex.

DP 9-9 A fluorescent light uses cathodes (coiled tungsten filaments coated with an electron-emitting substance) at each end that send current through mercury vapors sealed in the tube. Ultraviolet radiation is produced as electrons from the cathodes knock mercury electrons out of their natural orbits. Some of the displaced electrons settle back into orbit, throwing off the excess energy absorbed in the collision. Almost all of this energy is in the form of ultraviolet radiation. The ultraviolet rays, which are invisible, strike a phosphor coating on the inside of the tube. The rays energize the electrons in the phosphor atoms, and the atoms emit white light. The conversion of one kind of light into another is known as fluorescence.

One form of a fluorescent lamp is represented by the *RLC* circuit shown in Figure DP 9-9. Select L so that the current $i(t)$ reaches a maximum at approximately $t = 0.5$ s. Determine the maximum value of $i(t)$. Assume that the switch was in position 1 for a long time before switching to position 2 at $t = 0$.

Hint: Use PSpice to plot the response for several values of L.

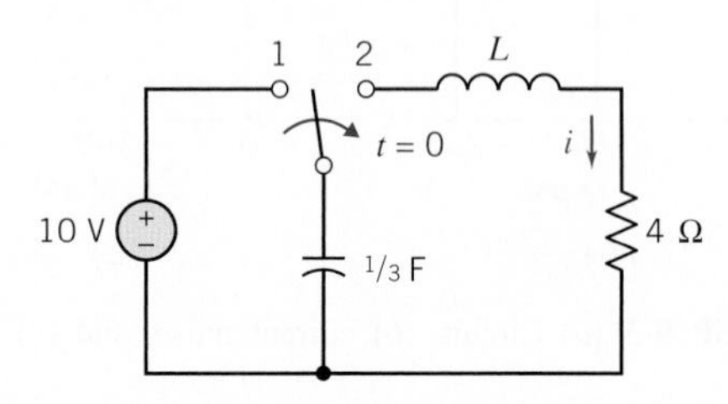

Figure DP 9-9 Flourescent lamp circuit.

CHAPTER 10

Sinusoidal Steady-State Analysis

IN THIS CHAPTER

10.1 Introduction
10.2 Sinusoidal Sources
10.3 Steady-State Response of an *RL* Circuit for a Sinusoidal Forcing Function
10.4 Complex Exponential Forcing Function
10.5 The Phasor
10.6 Phasor Relationships for *R*, *L*, and *C* Elements
10.7 Impedance and Admittance
10.8 Kirchhoff's Laws Using Phasors
10.9 Node Voltage and Mesh Current Analysis Using Phasors
10.10 Superposition, Thévenin and Norton Equivalents, and Source Transformations
10.11 Phasor Diagrams
10.12 Phasor Circuits and the Operational Amplifier
10.13 The Complete Response
10.14 Using MATLAB for Analysis of Steady-State Circuits with Sinusoidal Inputs
10.15 Using PSpice to Analyze AC Circuits
10.16 How Can We Check . . . ?
10.17 **DESIGN EXAMPLE**—Op Amp Circuit
10.18 Summary
Problems
PSpice Problems
Design Problems

10.1 INTRODUCTION

Consider the experiment illustrated in Figure 10.1-1. Here, a function generator provides the input to a linear circuit and the oscilloscope displays the output, or response, of the linear circuit. The linear circuit itself consists of resistors, capacitors, inductors, and perhaps dependent sources and/or op amps. The function generator allows us to choose from several types of input function. These input functions are called waveforms or waves. A typical function generator will provide square waves, pulse waves, triangular waves, and sinusoidal waves.

The output of the circuit will consist of two parts: a transient part that dies out as time increases and a steady-state part that persists. Typically, the transient part dies out quickly, perhaps in a couple of milliseconds. We expect that the oscilloscope in Figure 10.1-1 will display the steady-state response of the linear circuit to the input provided by the function generator.

Suppose we select a sinusoidal input. The function generator permits us to adjust the amplitude, phase angle, and frequency of the input. We notice that no matter what adjustments we make, the (steady-state) response is always a sine wave at the same frequency as the input. The amplitude and phase angle of the output differ from the input, but the frequency is always the same.

Suppose we select a square wave input. The steady-state response is not a square wave. Similarly, the steady-state responses to pulse waves and triangular waves do not have the same shape as the input.

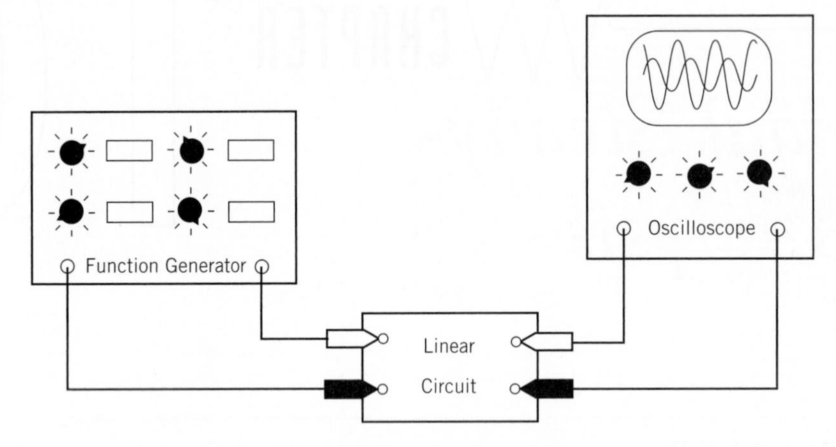

FIGURE 10.1-1 Measuring the input and output of a linear circuit.

Linear circuits with sinusoidal inputs that are at steady state are called ac circuits. The electric power system that provides us with convenient electricity is a very large ac circuit. AC circuits are the subject of this chapter. In particular, we will see that:

- It's useful to associate a complex number with a sinusoid. Doing so allows us to define phasors and impedances.
- Using phasors and impedances, we obtain a new representation of the linear circuit, called the ''frequency-domain representation.''
- We can analyze ac circuits in the frequency domain to determine their steady-state response.

10.2 SINUSOIDAL SOURCES

In electrical engineering, sinusoidal inputs are particularly important because power sources and communication signals are usually transmitted as sinusoids or modified sinusoids. The input causes the forced response, and the natural response is caused by the internal dynamics of the circuit. The natural response will normally decay after some period of time, but the forced, or steady state, response continues indefinitely. Therefore, in this chapter, we are interested primarily in the steady-state response of a circuit to the sinusoidal input.

We consider the input

$$v_s = V_m \sin \omega t \tag{10.2-1}$$

or, in the case of a current source,

$$i_s = I_m \sin \omega t \tag{10.2-2}$$

The amplitude of the sinusoid is V_m, and the radian frequency is ω(rad/s). The sinusoid is a *periodic function* defined by the property

$$x(t + T) = x(t)$$

for all t and where T is the period of oscillation.

The reciprocal of T defines the *frequency* or number of cycles per second, denoted by f, where

$$f = \frac{1}{T}$$

The frequency f is in cycles per second, more commonly referred to as hertz (Hz) in honor of the scientist Heinrich Hertz, shown in Figure 10.2-1.Therefore, the angular (radian) frequency of the sinusoidal function is

$$\omega = 2\pi f = \frac{2\pi}{T}$$

FIGURE 10.2-1 Heinrich R. Hertz (1857–1894). Courtesy of the Institution of Electrical Engineers.

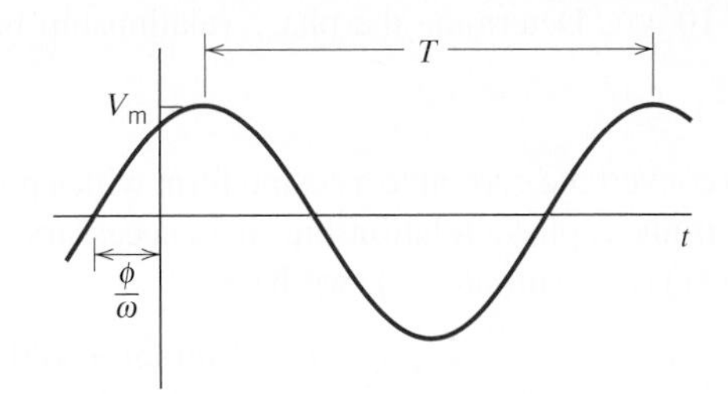

FIGURE 10.2-2 Sinusoidal voltage source $v_s = V_m \sin(\omega t + \phi)$.

The angular frequency ω is in radians per second.

For the voltage source of Eq. 10.2-1, the maximum value is V_m. If the sinusoidal voltage has an associated *phase angle* ϕ, the voltage source is

$$v_s = V_m \sin(\omega t + \phi) \tag{10.2-3}$$

The sinusoidal voltage of Eq. 10.2-3 is represented by Figure 10.2-2.

Because, conventionally, the angle ϕ may be expressed in degrees, you will encounter the notation

$$v_s = V_m \sin(4t + 30^\circ)$$

or, alternatively,

$$v_s = V_m \sin\left(4t + \frac{\pi}{6}\right)$$

where the angle ϕ is expressed in radians. This angular inconsistency will not deter us as long as we recognize that in the actual calculation of sin θ, θ must be in degrees or radians as our calculator requires.

In addition, it is worth noting that

$$V_m \sin(\omega t + 30^\circ) = V_m \cos(\omega t - 60^\circ)$$

This relationship can be deduced using the trigonometric formulas summarized in Appendix C.

If a circuit has a voltage across an element as

$$v = V_m \sin \omega t$$

and a current flows through the element

$$i = I_m \sin(\omega t + \phi)$$

we have the v and the i shown in Figure 10.2-3. We say that the current *leads* the voltage by ϕ radians. Examining Figure 10.2-3, we note that the current reaches its peak value before the voltage and thus is said to lead the voltage. Alternately, we could say that voltage lags the current by ϕ radians.

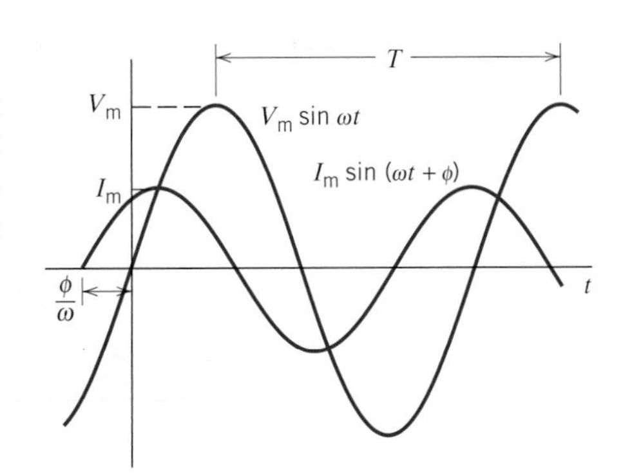

FIGURE 10.2-3 Voltage and current of a circuit element.

Consider a sine waveform with

$$v = 2\sin(3t + 20^\circ)\ \text{V}$$

and the associated current waveform

$$i = 4\sin(3t - 10^\circ)\ \text{A}$$

Clearly, the voltage v leads the current i by 30°, or $\pi/6$ radians.

Example 10.2-1 Phase Angles

The voltage across an element is $v = 3\cos 3t$ V, and the associated current through the element is $i = -2\sin(3t + 10^\circ)$ A. Determine the phase relationship between the voltage and current.

Solution

First, we need to convert the current to a cosine form with a positive magnitude so that it can be contrasted with the voltage. To determine a phase relationship, it is necessary to express both waveforms in a consistent form.

Because $-\sin\omega t = \sin(\omega t + \pi)$, we have

$$i = 2\sin(3t + 180^\circ + 10^\circ) \text{ A}$$

Also, we note that

$$\sin\theta = \cos(\theta - 90^\circ)$$

Therefore,

$$i = 2\cos(3t + 180^\circ + 10^\circ - 90^\circ) = 2\cos(3t + 100^\circ) \text{ A}$$

Recall that $v = 3\cos 3t$. Therefore, the current leads the voltage by 100°.

The sinusoidal function $C\cos(\omega t - \theta)$ can also be represented as $A\cos\omega t + B\sin\omega t$. Occasionally, we will need to convert from one representation to the other. To see how this is accomplished, consider a voltage

$$v(t) = A\cos\omega t + B\sin\omega t \tag{10.2-4}$$

Equation 10.2-4 may also be written as

$$v(t) = \sqrt{A^2 + B^2}\left(\frac{A}{\sqrt{A^2 + B^2}}\cos\omega t + \frac{B}{\sqrt{A^2 + B^2}}\sin\omega t\right)$$

Consider the triangle shown in Figure 10.2-4*a*, which illustrates the situation when $A > 0$, and note that

$$\sin\theta = \frac{B}{\sqrt{A^2 + B^2}}, \quad \cos\theta = \frac{A}{\sqrt{A^2 + B^2}}, \quad \text{and } \tan\theta = \frac{\sin\theta}{\cos\theta} = \frac{B}{A}$$

Then we have for $v(t)$

$$v(t) = C(\cos\theta\cos\omega t + \sin\theta\sin\omega t) \tag{10.2-5}$$

where $C = \sqrt{A^2 + B^2}$. Also, comparing Eqs. 10.2-4 and 10.2-5, we see that $A = C\cos\theta$ and $B = C\sin\theta$. Finally, using a formula from Appendix C, we write Eq. 10.2-5 as

$$v(t) = C\cos(\omega t - \theta) \tag{10.2-6}$$

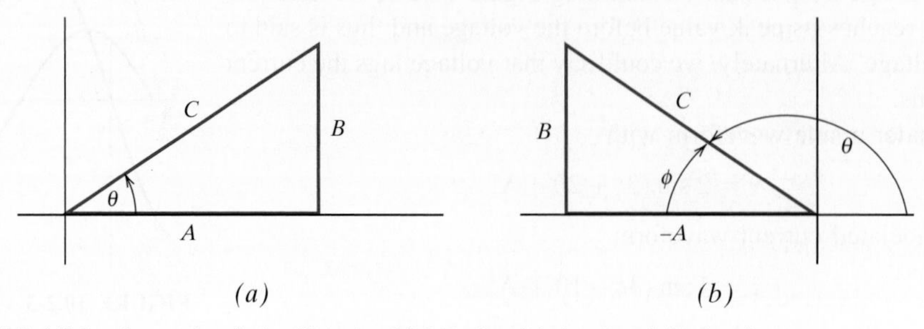

FIGURE 10.2-4 Triangles used to derive Equation 10.2-7 when (*a*) $A > 0$ and (*b*) $A < 0$.

Figure 10.2-4b illustrates the situation when $A < 0$. This case is similar to the previous case except now the phase angle is calculated as

$$\theta = 180^\circ - \phi = 180^\circ - \tan^{-1}\left(\frac{B}{-A}\right) = 180^\circ + \tan^{-1}\left(\frac{B}{A}\right)$$

In summary,

$$A\cos\omega t + B\sin\omega t = C\cos(\omega t - \theta) \tag{10.2-7}$$

where

$$C = \sqrt{A^2 + B^2},\ A = C\cos\theta,\ B = C\sin\theta$$

and

$$\theta = \begin{cases} \tan^{-1}\left(\dfrac{B}{A}\right) & \text{when } A > 0 \\ 180^\circ + \tan^{-1}\left(\dfrac{B}{A}\right) & \text{when } A < 0 \end{cases}$$

EXAMPLE 10.2-2 Magnitude and Phase Angle

A current has the form $i = -6\cos 2t + 8\sin 2t$. Find the current restated in the form of Eq. 10.2-6.

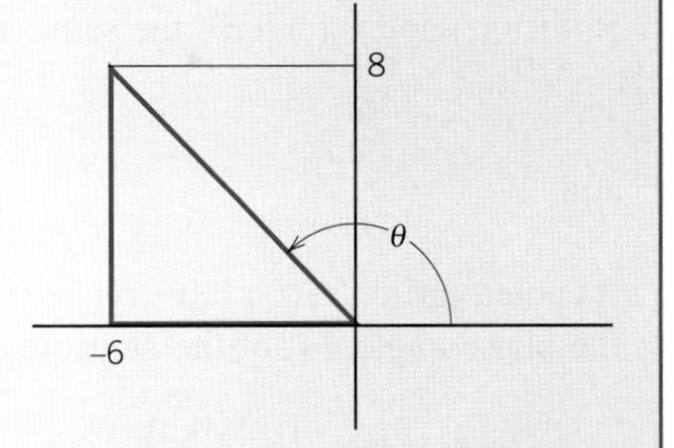

FIGURE 10.2-5 The A–B triangle for Example 10.2-2.

Solution

The triangle for A and B is shown in Figure 10.2-5. Because the coefficient A is equal to -6 and B is $+8$, we have the angle θ shown. Therefore,

$$\theta = 180^\circ + \tan^{-1}\left(\frac{8}{-6}\right) = 180^\circ - 53.1^\circ = 126.9^\circ$$

Hence,

$$i = 10\cos(2t - 126.9^\circ)$$

Next, consider the problem of obtaining an analytic representation, $A\cos(\omega t + \theta)$, of a sinusoid that is given graphically. This problem is frequently encountered by engineers and engineering students in the laboratory. Frequently, an engineer will see a sinusoidal voltage displayed on an oscilloscope and need to represent that voltage using an equation. The analytic representation of the sinusoid is obtained in three steps. The first two are straightforward. The third requires some attention. The procedure is illustrated in Figure 10.2-6, which shows two sinusoidal voltages.

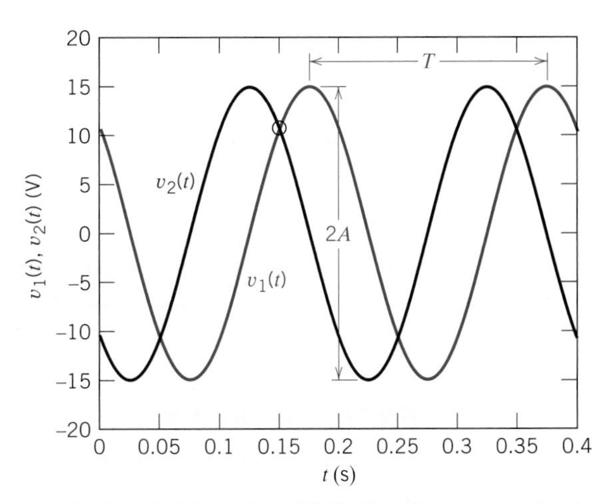

FIGURE 10.2-6 Two sinusoids having the same amplitude and period but different phase angles.

1. Measure the amplitude, A. The location of the time axis may not be obvious when the sinusoidal voltage is displayed on an oscilloscope, so it may be more convenient to measure the peak-to-peak amplitude, $2A$, as shown in Figure 10.2-6.
2. Measure the period, T, in s and calculate the frequency, $\omega = 2\pi/T$, in rad/s.

3. Pick a time and measure the voltage at that time. For example, $t = t_1 = 0.15$ s at the point marked in Figure 10.2-6. Notice that $v_1(t_1) = v_2(t_1) = 10.6066$ V, but $v_1(t_1)$ and $v_2(t_1)$ are clearly not the same sinusoid. The additional information needed to distinguish these two sinusoids is that $v_1(t)$ is increasing (positive slope) at time t_1, whereas $v_2(t)$ is decreasing (negative slope) at time t_1. Finally, calculate the phase angle, θ, of a sinusoidal voltage $v(t)$ as

$$\theta = \begin{cases} -\cos^{-1}\left(\dfrac{v(t_1)}{A}\right) - \omega t_1 & \text{when } v(t) \text{ is increasing at time } t_1 \\ \cos^{-1}\left(\dfrac{v(t_1)}{A}\right) - \omega t_1 & \text{when } v(t) \text{ is decreasing at time } t_1 \end{cases}$$

EXAMPLE 10.2-3 Graphical and Analytic Representation of Sinusoids

Determine the analytic representations of the sinusoidal voltages $v_1(t)$ and $v_2(t)$ shown in Figure 10.2-6.

Solution

Both $v_1(t)$ and $v_2(t)$ have the same amplitude and period:

$$2A = 30 \quad \Rightarrow \quad A = 15 \text{ V}$$

and

$$T = 0.2 \text{ s} \quad \Rightarrow \quad \omega = \frac{2\pi}{0.2} = 10\pi \text{ rad/s}$$

As noted earlier, $v_1(t_1) = v_2(t_1) = 10.6066$ V at $t_1 = 0.15$ s. Because $v_1(t)$ is increasing (positive slope) at time t_1, the phase angle, θ_1, of the sinusoidal voltage $v_1(t)$ is calculated as

$$\theta_1 = -\cos^{-1}\left(\frac{v(t_1)}{A}\right) - \omega t_1 = -\cos^{-1}\left(\frac{10.6066}{15}\right) - (10\pi)(0.15) = -5.498 \text{ rad} = -315° = 45°$$

(Notice that the units of ωt_1 are radians, so $\cos^{-1}\left(\dfrac{v(t_1)}{A}\right)$ must also be calculated in radians so that we can do the subtraction.) Finally, $v_1(t)$ is represented as

$$v_1(t) = 15\cos(10\pi t + 45°) \text{ V}$$

Next, because $v_2(t)$ is decreasing (negative slope) at time t_1, the phase angle, θ_2, of the sinusoidal voltage $v_2(t)$ is calculated as

$$\theta_2 = \cos^{-1}\left(\frac{v(t_1)}{A}\right) - \omega t_1 = \cos^{-1}\left(\frac{10.6066}{15}\right) - (10\pi)(0.15) = -3.927 \text{ rad} = -225° = 135°$$

Finally, $v_2(t)$ is represented as

$$v_2(t) = 15\cos(10\pi t + 135°) \text{ V}$$

EXERCISE 10.2-1 A voltage is $v = 6\cos(4t + 30°)$. (a) Find the period of oscillation. (b) State the phase relation to the associated current $i = 8\cos(4t - 70°)$.

Answers: **(a)** $T = 2\pi/4$

(b) The voltage leads the current by 100°.

EXERCISE 10.2-2 A voltage is $v = 3\cos 4t + 4\sin 4t$. Find the voltage in the form of Eq. 10.2-6.

Answer: $v = 5\cos(4t - 53^\circ)$ V

EXERCISE 10.2-3 A current is $i = 12\sin 5t - 5\cos 5t$. Find the current in the form of Eq. 10.2-6.

Answer: $i = 13\cos(15t - 112.6^\circ)$ A

10.3 STEADY-STATE RESPONSE OF AN *RL* CIRCUIT FOR A SINUSOIDAL FORCING FUNCTION

As an example of the task of determining the steady-state response of a linear circuit to a sinusoidal input, consider the *RL* circuit shown in Figure 10.3-1. The input to this circuit is the voltage of the voltage source

$$v_s(t) = V_m \cos \omega t$$

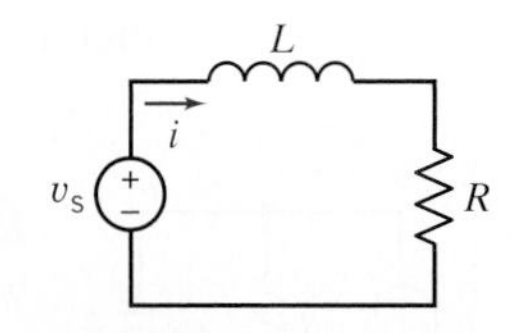

FIGURE 10.3-1 An *RL* circuit.

The response of this circuit is the current i. This response will be of the form

$$i = i_n + i_f = Ke^{-t/\tau} + I_m \cos(\omega t + \phi)$$

The values of the real constants K, τ, I_m, and ϕ are yet to be determined. Also, the value of K depends on the initial condition $i(0)$.

$$\text{As } t \to \infty,\ Ke^{-t/\tau} \to 0 \quad \text{and} \quad i \to i_f = I_m \cos \omega t$$

In other words, as time proceeds, the term $Ke^{-t/\tau}$ dies out, leaving the term $I_m \cos \omega t$. For this reason, $i_f = I_m \cos \omega t$ is called the steady-state response. We expect that the steady-state response of a linear circuit to a sinusoidal input will itself be sinusoidal and will have the same frequency, ω, as the input.

The governing differential equation of the *RL* circuit is given by

$$L\frac{di}{dt} + Ri = V_m \cos \omega t \tag{10.3-1}$$

Following the method of the previous chapter, we assume that

$$i_f = A\cos \omega t + B \sin \omega t \tag{10.3-2}$$

At this point, because we are solving only for the forced response, we drop the subscript *f* notation. Substituting the assumed solution of Eq. 10.3-2 into the differential equation and completing the derivative, we have

$$L(-\omega A \sin \omega t + \omega B \cos \omega t) + R(A\cos \omega t + B \sin \omega t) = V_m \cos \omega t$$

Equating the coefficients of $\cos \omega t$, we obtain

$$\omega LB + RA = V_m$$

Next, equating the coefficients of $\sin \omega t$, we obtain

$$-\omega LA + RB = 0$$

Solving for A and B, we have

$$A = \frac{RV_m}{R^2 + \omega^2 L^2}$$

and

$$B = \frac{\omega L V_m}{R^2 + \omega^2 L^2}$$

The response to the sinusoidal input is then

$$i = A\cos\omega t + B\sin\omega t$$

or

$$i = \frac{V_m}{Z}\cos(\omega t - \beta)$$

where

$$Z = \sqrt{R^2 + \omega^2 L^2}$$

and

$$\beta = \tan^{-1}\frac{\omega L}{R}$$

Thus, the forced (steady-state) response is of the form

$$i = I_m\cos(\omega t + \phi)$$

where

$$I_m = \frac{V_m}{Z}$$

and

$$\phi = -\beta$$

In this case, we have found only the steady-state response of a circuit with one energy storage element. Clearly, this approach can be quite complicated if the circuit has several storage elements.

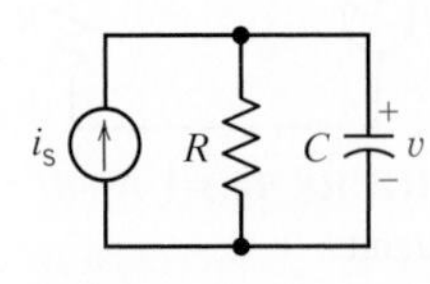

FIGURE E 10.3-1

EXERCISE 10.3-1 Find the forced response v for the RC circuit shown in Figure E 10.3-1 when $i_s = I_m\cos\omega t$.

Answer: $v = (RI_m/P)\cos(\omega t - \theta)$ and $P = \sqrt{1 + \omega^2 R^2 C^2}$, $\theta = \tan^{-1}(\omega RC)$

EXERCISE 10.3-2 Find the forced response $i(t)$ for the RL circuit of Figure 10.3-1 when $R = 2\ \Omega$, $L = 1$ H, and $v_s = 10\cos 3t$ V.

Answer: $i = 2.77\cos(3t - 56.3°)$ A

10.4 COMPLEX EXPONENTIAL FORCING FUNCTION

Upon reviewing the preceding section, we see that the input to the circuit in Figure 10.3-1 was of the form

$$v_s = V_m\cos\omega t$$

and the steady-state response was

$$i = \frac{V_m}{Z}\cos(\omega t - \beta)$$

Thus, the steady-state response to a sinusoidal input is also sinusoidal and has the same frequency as the input but has a different amplitude and phase angle than the original voltage source.

It is useful to consider the exponential signal

$$v_e = V_m e^{j\omega t} \tag{10.4-1}$$

Using Euler's equation, we can relate the exponential signal to a sinusoidal signal

$$v_s = V_m\cos\omega t = \text{Re}\{V_m e^{j\omega t}\} = \text{Re}\{v_e\}$$

The notation $\text{Re}\{a + jb\}$ is read as the real part of the complex number $(a + jb)$. For example,

$$\text{Re}\{a + jb\} = a$$

Let us try the exponential source v_e of Eq. 10.4-1 with the differential equation of the RL circuit shown in Figure 10.4-1

$$L\frac{di_e}{dt} + Ri_e = v_e \tag{10.4-2}$$

FIGURE 10.4-1

where i_e is the response to the exponential input. Because the source is an exponential, we try the solution

$$i_e = Ae^{j\omega t} \tag{10.4-3}$$

and substitute into Eq. 10.4-2 to obtain

$$(j\omega L + R)Ae^{j\omega t} = V_{\rm m}e^{j\omega t}$$

Hence,[1]

$$A = \frac{V_{\rm m}}{R + j\omega L} = \frac{V_{\rm m}}{Z}e^{-j\beta}$$

where

$$\beta = \tan^{-1}\frac{\omega L}{R}$$

and

$$Z = \sqrt{R^2 + \omega^2 L^2}$$

Therefore, substituting for A, we have

$$i_e = \frac{V_{\rm m}}{Z}e^{-j\beta}e^{j\omega t} \tag{10.4-4}$$

Again, noting that the original forcing function was

$$v_{\rm s} = {\rm Re}\{V_{\rm m}e^{j\omega t}\} = V_{\rm m}\cos\omega t$$

we expect that

$$i = {\rm Re}\{i_e\} = {\rm Re}\left\{\frac{V_{\rm m}}{Z}e^{-j\beta}e^{j\omega t}\right\}$$

Accordingly,

$$i = \frac{V_{\rm m}}{Z}{\rm Re}\{e^{-j\beta}e^{j\omega t}\} = \frac{V_{\rm m}}{Z}{\rm Re}\left\{e^{j(\omega t - \beta)}\right\} = \frac{V_{\rm m}}{Z}\cos(\omega t - \beta)$$

In general, we are finding the sinusoidal response

$$i = I_{\rm m}\cos(\omega t - \beta) = {\rm Re}\left\{\frac{V_{\rm m}}{Z}e^{j(\omega t - \beta)}\right\}$$

to the sinusoidal excitation

$$v_{\rm s} = V_{\rm m}\cos\omega t = {\rm Re}\{V_{\rm m}e^{j\omega t}\}$$

We have learned that this response is readily obtained by using the complex exponential excitation, ${\rm Re}\{V_{\rm m}e^{j\omega t}\}$.

As an example, let us find the steady-state response for the RLC circuit shown in Figure 10.4-2. This circuit is represented by the differential equation

$$\frac{d^2i}{dt^2} + \frac{di}{dt} + 12i = 12\cos 3t \tag{10.4-5}$$

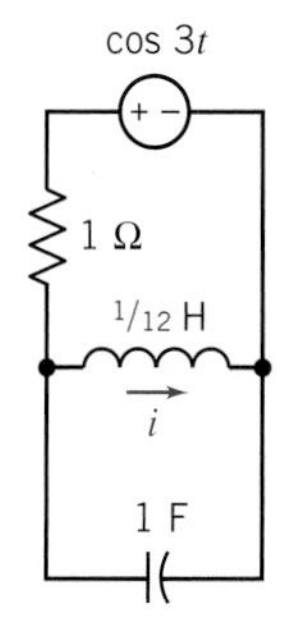

FIGURE 10.4-2

[1] *Note*: See Appendix B for a review of complex numbers.

First, replace the real excitation by the complex exponential excitation

$$v_e = 12e^{j3t}$$

Then we have Eq. 10.4-5 restated as

$$\frac{d^2 i_e}{dt^2} + \frac{di_e}{dt} + 12 i_e = 12 e^{j3t} \tag{10.4-6}$$

We expect the response to the exponential input to be of the form

$$i_e = Ae^{j3t} \tag{10.4-7}$$

The first and second derivatives of the i_e of Eq. 10.4-7 are

$$\frac{di_e}{dt} = j3Ae^{j3t}$$

and

$$\frac{d^2 i_e}{dt^2} = -9\,Ae^{j3t}$$

Substituting into Eq. 10.4-6, we have

$$(-9 + j3 + 12)Ae^{j3t} = 12e^{j3t} \tag{10.4-8}$$

Solving for A, we obtain

$$A = \frac{12}{3+j3} = \frac{12(3-j3)}{(3+j3)(3-j3)} = \frac{12(3-j3)}{18} = 2\sqrt{2}\,\underline{/-45^\circ}$$

Therefore,

$$i_e = Ae^{j3t} = 2\sqrt{2}e^{-j(\pi/4)}e^{j3t} = 2\sqrt{2}e^{j(3t-\pi/4)}$$

Recall that Euler's identity is $e^{j\phi} = \cos\phi + j\sin\phi$. Thus, the desired answer for the steady-state current is[2]

$$i(t) = \mathrm{Re}\{i_e\} = \mathrm{Re}\left\{2\sqrt{2}e^{j(3t-\pi/4)}\right\} = 2\sqrt{2}\cos\ (3t - 45^\circ)$$

Note that we have changed from $\pi/4$ radians to 45°, which are interchangeable and equivalent. Both degree and radian notation are acceptable and interchangeable.

Compare Eqs. 10.4-6 and 10.4-7. Eq. 10.4-6 is a differential equation, which is what we expect for the equation representing a circuit that contains capacitors or inductors. In contrast, Eq. 10.4-8 is an algebraic equation, involving addition and multiplication but not integration or differentiation. The coefficients of Eq. 10.4-8 are complex numbers whereas the coefficients of Eq. 10.4-6 are real numbers. Algebraic equations are easier to solve than differential equations, so we prefer to solve Eq. 10.4-8, even though it contains complex coefficients.

We have developed a straightforward method for determining the steady-state response of a circuit to a sinusoidal excitation. The process is as follows: (1) Instead of applying the actual forcing function, we apply a complex exponential forcing function, and (2) we then obtain the complex response whose real part is the desired response. The process is summarized in Table 10.4-1. Let us use this process in another example.

[2] See Appendix B for a discussion of Euler's equation.

Table 10.4-1 Use of the Complex Exponential Excitation to Determine a Circuit's Steady-State Response to a Sinusoidal Source

1. Write the excitation (forcing function) as a cosine waveform with a phase angle so that $y_s = Y_m \cos(\omega t + \phi)$, where y_s is a source current, i_s, or a source voltage, v_s, in the circuit.
2. Recall Euler's identity, which is

$$e^{j\alpha} = \cos\alpha + j\sin\alpha$$

where $\alpha = \omega t + \phi$ in this case.
3. Introduce the complex excitation so that for a voltage source, for example, we have

$$v_s = \mathrm{Re}\{V_m e^{j(\omega t+\phi)}\}$$

where $V_m e^{j(\omega t+\phi)}$ is a complex exponential excitation.
4. Use the complex excitation and the differential equation along with the assumed response $x_e = Ae^{j(\omega t+\phi)}$, where A is to be determined. Note that A will normally be a complex quantity.
5. Determine the constant $A = Be^{-j\beta}$, so that

$$x_e = Ae^{j(\omega t+\phi)} = Be^{j(\omega t+\phi-\beta)}$$

6. Recognize that the desired response is

$$x(t) = \mathrm{Re}\{x_e\} = B\cos(\omega t + \phi - \beta)$$

Example 10.4-1 Response of an AC Circuit

Find the steady-state response i of the RL circuit of Figure 10.4-1 when $R = 2\ \Omega$, $L = 1$ H, and $v_s = 10 \sin 3t$ V.

Solution

First, we will rewrite the voltage source so that it is expressed as a cosine waveform as follows:

$$v_s = 10\sin 3t = 10\cos(3t - 90°)$$

Using the complex excitation, we have

$$v_e = 10\, e^{j(3t-90°)}$$

Introduce the complex excitation into the circuit's differential equation, which is

$$L\frac{di_e}{dt} + Ri_e = v_e$$

obtaining

$$\frac{di_e}{dt} + 2i_e = 10\, e^{j(3t-90°)}$$

Assume that the response is

$$i_e = Ae^{j(3t-90°)} \qquad (10.4\text{-}9)$$

where A is a complex quantity to be determined. Substituting the assumed solution, Eq. 10.4-9, into the differential equation and taking the derivative, we have

$$j3Ae^{j(3t-90°)} + 2Ae^{j(3t-90°)} = 10e^{j(3t-90°)}$$

Therefore,

$$j3A + 2A = 10$$

or

$$A = \frac{10}{j3+2} = \frac{10}{\sqrt{9+4}}\, e^{-j\beta}$$

where

$$\beta = \tan^{-1}\frac{3}{2} = 56.3^\circ$$

Then the solution is

$$i_e = Ae^{j(3t-90^\circ)} = \frac{10}{\sqrt{13}}\,e^{-j56.3^\circ}\,e^{j(3t-90^\circ)} = \frac{10}{\sqrt{13}}\,e^{j(3t-146.3^\circ)}$$

Consequently, the actual response is

$$i(t) = \text{Re}\{i_e\} = \frac{10}{\sqrt{13}}\cos(3t - 146.3^\circ)\text{ A}$$

Steinmetz observed the process we just used and decided to formulate a method for solving the sinusoidal steady-state response of circuits using, complex number notation. The development of this approach is the subject of the next section.

EXERCISE 10.4-1 Find a and b when

$$\frac{10}{a + jb} = 2.36e^{j45}$$

Answers: $a = 3$ and $b = -3$

EXERCISE 10.4-2 Find A and θ when

$$\left[A\angle\theta\right](-3 + j8) = j32$$

Answers: $A = 3.75$ and $\theta = -20.56^\circ$

10.5 THE PHASOR

A sinusoidal current or voltage at a given frequency is characterized by its amplitude and phase angle. For example, the current response in the *RL* circuit considered in Example 10.4-1 was

$$i(t) = \text{Re}\{I_\text{m}e^{j(\omega t + \phi - \beta)}\} = I_\text{m}\cos(\omega t + \phi - \beta)$$

The magnitude I_m and the phase angle $(\phi - \beta)$, along with knowledge of ω, completely specify the response. Thus, we may write $i(t)$ as

$$i(t) = \text{Re}\{I_\text{m}e^{j(\phi-\beta)}e^{j\omega t}\}$$

However, we note that the complex factor $e^{j\omega t}$ remained unchanged throughout all our previous calculations. Thus, the information we seek is represented by

$$\mathbf{I} = I_\text{m}e^{j(\phi-\beta)} = I_\text{m}\angle\phi - \beta \tag{10.5-1}$$

where **I** is called a *phasor*. A phasor is a complex number that represents the magnitude and phase of a sinusoid. The term *phasor* is used instead of *vector* because the angle is time based rather than space based. A phasor may be written in exponential form, polar form, or rectangular form.

Phasors may be used when the circuit is linear, the steady-state response is sought, and all independent sources are sinusoidal and have the same frequency.

A real sinusoidal current, where $\theta = (\phi - \beta)$, is written as

$$i(t) = I_{\mathrm{m}} \cos(\omega t + \theta)$$

It can be represented by

$$i(t) = \mathrm{Re}\{I_{\mathrm{m}} e^{j(\omega t + \theta)}\}$$

We then decide to drop the notation Re and the redundant $e^{j\omega t}$ to obtain the phasor representation

$$\mathbf{I} = I_{\mathrm{m}} e^{j\theta} = I_{\mathrm{m}} \angle \theta$$

This abbreviated representation is the *phasor notation*. Phasor quantities are complex and thus are printed in boldface in this book. You may choose to use the underline notation as follows:

$$\underline{I} = I_{\mathrm{m}} \angle \theta$$

Although we have dropped or suppressed the complex frequency $e^{j\omega t}$, we continue to note that we are in the complex frequency form and are performing calculations in the *frequency domain*. We have transformed the problem from the time domain to the frequency domain by the use of phasor notation. A transform is a means of encoding to simplify a calculation process. One example of a mathematical transform is the logarithmic transform.

> A **transform** is a change in the mathematical description of a physical variable to facilitate computation.

The actual steps involved in transforming a function in the time domain to the frequency domain are summarized in Table 10.5-1. Because it is easy to move through these steps, we usually jump directly from step 1 to step 4.

For example, let us determine the phasor notation for

$$i = 5 \sin(100t + 120^\circ)$$

We have chosen to use cosine functions as the standard for phasor notation. Thus, we express the current as a cosine waveform:

$$i = 5 \cos(100t + 30^\circ)$$

At this point, it is easy to see that the information we require is the amplitude and the phase. Thus, the phasor is

$$\mathbf{I} = 5 \angle 30^\circ$$

Of course, the reverse process from phasor notation to time notation is exactly the reverse of the steps required to go from the time to the phasor notation. Thus, if we have a voltage in phasor notation:

$$\mathbf{V} = 24 \angle 125^\circ$$

the time-domain notation is

$$v(t) = 24 \cos(\omega t + 125^\circ)$$

Table 10.5-1 Transformation from the Time Domain to the Frequency Domain

1. Write the function in the time domain, $y(t)$, as a cosine waveform with a phase angle ϕ as
$$y(t) = Y_{\mathrm{m}} \cos(\omega t + \phi)$$
2. Express the cosine waveform as the real part of a complex quantity by using Euler's identity so that
$$y(t) = \mathrm{Re}\{Y_{\mathrm{m}} e^{j(\omega t + \theta)}\}$$
3. Drop the real part notation.
4. Suppress the $e^{j\omega t}$ while noting the value of ω for later use, obtaining the phasor
$$\mathbf{Y} = Y_{\mathrm{m}} e^{j\phi} = Y_{\mathrm{m}} \angle \phi$$

Table 10.5-2 Transformation from the Frequency Domain to the Time Domain

1. Write the phasor in exponential form as
$$\mathbf{Y} = Y_{\mathrm{m}} e^{j\beta}$$
2. Reinsert the factor $e^{j\omega t}$ so that you have
$$Y_{\mathrm{m}} e^{j\beta} e^{j\omega t}$$
3. Reinsert the real part operator Re as
$$\mathrm{Re}\{Y_{\mathrm{m}} e^{j\beta} e^{j\omega t}\}$$
4. Use Euler's identity to obtain the time function
$$y(t) = \mathrm{Re}\{Y_{\mathrm{m}} e^{j(\omega t+\beta)}\} = Y_{\mathrm{m}} \cos(\omega t + \beta)$$

where the frequency ω was noted in the original statement of the circuit formulation. This transformation from the frequency domain to the time domain is summarized in Table 10.5-2.

A **phasor** is a transformed version of a sinusoidal voltage or current waveform and consists of the magnitude and phase angle information of the sinusoid.

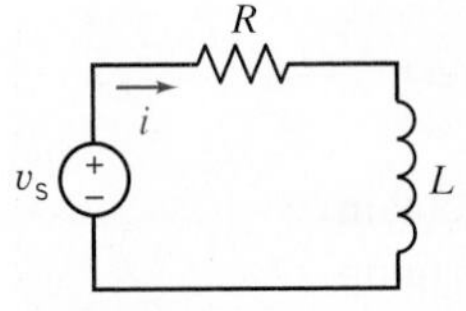

FIGURE 10.5-1
RL circuit.

The phasor method uses the transformation from the time domain to the frequency domain to obtain more easily the sinusoidal steady-state solution of the differential equation. Consider the *RL* circuit of Figure 10.5-1. We wish to find the solution for the steady-state current i when the voltage source is $v_s = V_m \cos \omega t$ V and $\omega = 100$ rad/s. Also, for this circuit, let $R = 200\ \Omega$ and $L = 2$ H. Then we may write the differential equation as

$$L\frac{di}{dt} + Ri = v_s \tag{10.5-2}$$

Because
$$v_s = V_m \cos(\omega t + \phi) = \mathrm{Re}\{V_m e^{j(\omega t+\phi)}\} \tag{10.5-3}$$

we will use the assumed solution

$$i = I_m \cos(\omega t + \beta) = \mathrm{Re}\{I_m e^{j(\omega t+\beta)}\} \tag{10.5-4}$$

Therefore, we may substitute Eqs. 10.5-3 and 10.5-4 into Eq. 10.5-2 and suppress the Re notation to obtain

$$(j\omega L I_m + R I_m)e^{j(\omega t+\beta)} = V_m e^{j(\omega t+\phi)}$$

Suppress the $e^{j\omega t}$ to obtain

$$(j\omega L + R) I_m e^{j\beta} = V_m e^{j\phi}$$

Now we recognize the phasors

$$\mathbf{I} = I_m e^{j\beta}$$

and
$$\mathbf{V}_s = V_m e^{j\phi}$$

Therefore, in phasor notation, we have

$$(j\omega L + R)\mathbf{I} = \mathbf{V}_s$$

Solving for $\mathbf{I}$, we have

$$\mathbf{I} = \frac{\mathbf{V}_s}{j\omega L + R} = \frac{\mathbf{V}_s}{j200 + 200}$$

for $\omega = 100$, $L = 2$, and $R = 200$. Therefore, because

$$\mathbf{V}_s = V_m \underline{/0^\circ}$$

we have

$$\mathbf{I} = \frac{V_m}{283 \underline{/45^\circ}} = \frac{V_m}{283} \underline{/-45^\circ}$$

Using the method of Table 10.5-2, we may transform this result back to the time domain to obtain the steady-state time solution as

$$i(t) = \frac{V_m}{283} \cos(100t - 45^\circ) \text{ A}$$

It is clear that we can use phasors directly to obtain a linear algebraic equation expressed in terms of the phasors and complex numbers and then solve for the phasor variable of interest. After obtaining the phasor we desire, we simply transform it back to the time domain to obtain the steady-state solution.

EXAMPLE 10.5-1 Analysis of AC Circuits Using Phasors

Find the steady-state voltage v for the RC circuit shown in Figure 10.5-2 when $i = 10 \cos \omega t$ A, $R = 1\ \Omega$, $C = 10$ mF, and $\omega = 100$ rad/s.

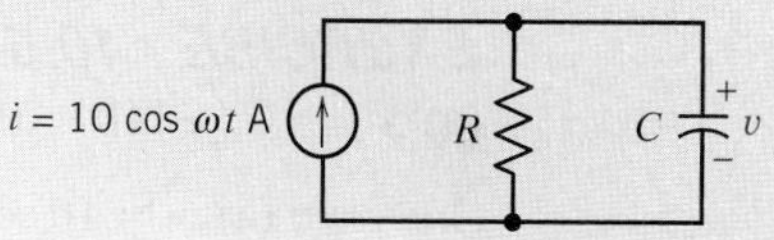

$i = 10 \cos \omega t$ A, R, C, $+ v -$

FIGURE 10.5-2 An RC circuit with a sinusoidal current source.

Solution

First, we find the phasor representation of the source current as

$$\mathbf{I} = I_m \underline{/0^\circ} = 10 \underline{/0^\circ} \tag{10.5-5}$$

We seek to find the voltage v by first obtaining the phasor $\mathbf{V}$.

Write the node voltage differential equation for the circuit to obtain

$$\frac{v}{R} + C\frac{dv}{dt} = i \tag{10.5-6}$$

Because

$$i = 10 \,\text{Re}\{e^{j\omega t}\}$$

and

$$v = V_m \text{Re}\{e^{j(\omega t + \phi)}\}$$

we substitute into Eq. 10.5-6 and suppress the Re notation to obtain

$$\frac{V_m}{R} e^{j(\omega t+\phi)} + j\omega C V_m e^{j(\omega t+\phi)} = 10 e^{j\omega t}$$

We now suppress the $e^{j\omega t}$ and obtain

$$\left(\frac{1}{R} + j\omega C\right) V_m e^{j\phi} = 10 e^{j0^\circ}$$

Recalling the phasor representation of Eq. 10.5-5, we have

$$\left(\frac{1}{R} + j\omega C\right)\mathbf{V} = \mathbf{I}$$

Because $R = 1$, $C = 10^{-2}$, and $\omega = 100$, we have

$$(1 + j1)\mathbf{V} = \mathbf{I}$$

or
$$\mathbf{V} = \frac{\mathbf{I}}{1 + j1}$$

Therefore,
$$\mathbf{V} = \frac{10}{\sqrt{2}\angle 45^\circ} = \frac{10}{\sqrt{2}}\angle -45^\circ$$

Transforming from the phasor notation back to the steady-state time solution, we have
$$v = \frac{10}{\sqrt{2}}\cos(100t - 45^\circ)\ \text{V}$$

EXERCISE 10.5-1 Express the current i as a phasor.
(a) $i = 4\cos(\omega t - 80^\circ)$ **(b)** $i = 10\cos(\omega t + 20^\circ)$ **(c)** $i = 8\sin(\omega t - 20^\circ)$

Answers: **(a)** $4\angle -80^\circ$
(b) $10\angle +20^\circ$
(c) $8\angle -110^\circ$

EXERCISE 10.5-2 Find the steady-state voltage v represented by the phasor
(a) $\mathbf{V} = 10\angle -140^\circ$ **(b)** $\mathbf{V} = 80 + j75$

Answers: **(a)** $v = 10\cos(\omega t - 140^\circ)$
(b) $109.7\cos(\omega t + 43.2^\circ)$

10.6 PHASOR RELATIONSHIPS FOR *R*, *L*, AND *C* ELEMENTS

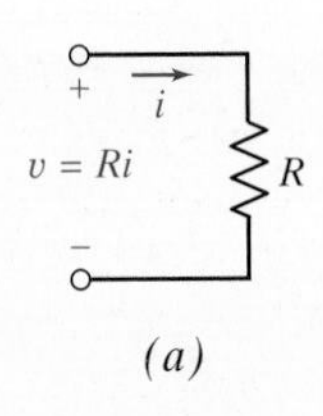

(a)

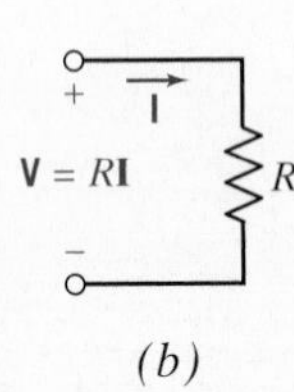

(b)

FIGURE 10.6-1
(*a*) The v–i time-domain relationship for R.
(*b*) The frequency-domain relationship for R.

In the preceding section, we found that the phasor representation is actually a transformation from the time domain into the frequency domain. With this transform, we have converted the solution of a differential equation into the solution of an algebraic equation.

In this section, we determine the relationship between the phasor voltage and the phasor current of the elements: R, L, and C. We use the transformation from the time to the frequency domain and then solve for the phasor relationship for a specified element. We are using the method of Tables 10.5-1 and 10.5-2 as recorded in the last section.

Let us begin with the resistor, as shown in Figure 10.6-1*a*. The voltage–current relationship in the time domain is
$$v = Ri \tag{10.6-1}$$

Now consider the steady-state voltage
$$v = V_m\cos(\omega t + \phi)$$

Then
$$v = \text{Re}\{V_m e^{j(\omega t + \phi)}\} \tag{10.6-2}$$

Assume that the current is of the form
$$i = \text{Re}\{I_m e^{j(\omega t + \beta)}\} \tag{10.6-3}$$

Then substitute Eqs. 10.6-2 and 10.6-3 into Eq. 10.6-1 and suppress the Re notation to obtain

$$V_{\mathrm{m}} e^{j(\omega t+\phi)} = R I_{\mathrm{m}} e^{j(\omega t+\beta)}$$

Suppress $e^{j\omega t}$ to obtain

$$V_{\mathrm{m}} e^{j\phi} = R I_{\mathrm{m}} e^{j\beta}$$

Therefore, we note that $\beta = \phi$, and

$$\mathbf{V} = R\mathbf{I} \tag{10.6-4}$$

Because $\beta = \phi$, the current and voltage waveforms are in phase. This phasor relationship is shown in Figure 10.6-1*b*

For example, if the voltage across a resistor is $v = 10 \cos 10t$, we know that the current will be

$$i = \frac{10}{R} \cos 10t$$

in the time domain.

In the frequency domain, we first note that the voltage is

$$\mathbf{V} = 10 \angle 0^\circ$$

Then, using the phasor relationship of the resistor, Eq. 10.6-4, we have

$$\mathbf{I} = \frac{\mathbf{V}}{R} = \frac{10 \angle 0^\circ}{R}$$

Then, obtaining the time-domain expression for **I**, we have

$$i = \frac{10}{R} \cos 10t$$

Now, consider the inductor as shown in Figure 10.6-2*a*. The time-domain voltage–current relationship is

$$v = L\frac{di}{dt} \tag{10.6-5}$$

Again, we use the complex voltage as

$$v = \mathrm{Re}\{V_{\mathrm{m}} e^{j(\omega t+\phi)}\} \tag{10.6-6}$$

and assume that the current is

$$i = \mathrm{Re}\{I_{\mathrm{m}} e^{j(\omega t+\beta)}\} \tag{10.6-7}$$

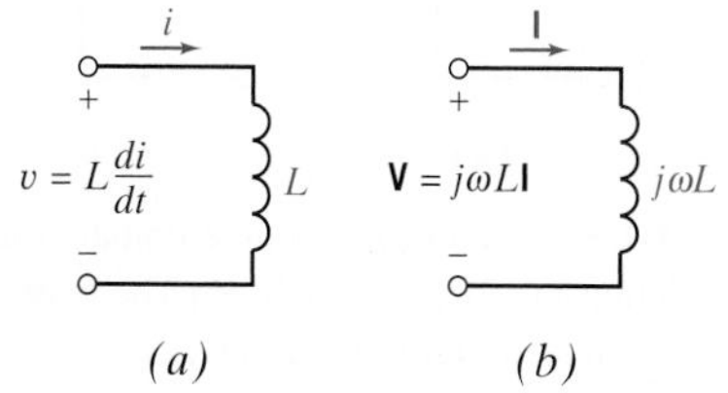

FIGURE 10.6-2 (*a*) The time-domain v–i relationship for an inductor. (*b*) The frequency-domain relationship for an inductor.

Substituting Eqs. 10.6-6 and 10.6-5 into Eq. 10.6-5 and suppressing the Re notation, we have

$$V_{\mathrm{m}} e^{j\phi} e^{j\omega t} = L\frac{d}{dt}\{I_{\mathrm{m}} e^{j\omega t} e^{j\beta}\}$$

Taking the derivative, we have

$$V_{\mathrm{m}} e^{j\phi} e^{j\omega t} = j\omega L I_{\mathrm{m}} e^{j\omega t} e^{j\beta}$$

Now suppressing the $e^{j\omega t}$, we have

$$V_{\mathrm{m}} e^{j\phi} = j\omega L I_{\mathrm{m}} e^{j\beta} \tag{10.6-8}$$

or

$$\mathbf{V} = j\omega L\mathbf{I} \tag{10.6-9}$$

This phasor relationship is shown in Figure 10.6-2*b*. Because $j = e^{j\,90^\circ}$, Eq. 10.6-8 can also be written as

$$V_{\mathrm{m}} e^{j\phi} = \omega L I_{\mathrm{m}} e^{j\,90^\circ} e^{j\beta}$$

Therefore,

$$\phi = \beta + 90^\circ$$

Thus, the voltage leads the current by exactly 90°.

As an illustration, consider an inductor of 2 H with $\omega = 100$ rad/s and with voltage $v = 10\cos(\omega t + 50^\circ)$ V. Then the phasor voltage is

$$\mathbf{V} = 10\,\underline{/50^\circ}$$

and the phasor current is

$$\mathbf{I} = \frac{\mathbf{V}}{j\omega L}$$

Because $\omega L = 200$, we have

$$\mathbf{I} = \frac{\mathbf{V}}{j200} = \frac{10\,\underline{/50^\circ}}{200\,\underline{/90^\circ}} = 0.05\,\underline{/-40^\circ}\ \mathrm{A}$$

Then the current expressed in the time domain is

$$i = 0.05\cos(100t - 40^\circ)\ \mathrm{A}$$

Therefore, the current lags the voltage by 90°.

Finally, let us consider the case of the capacitor, as shown in Figure 10.6-3*a*. The current–voltage relationship is

$$i = C\frac{dv}{dt} \tag{10.6-10}$$

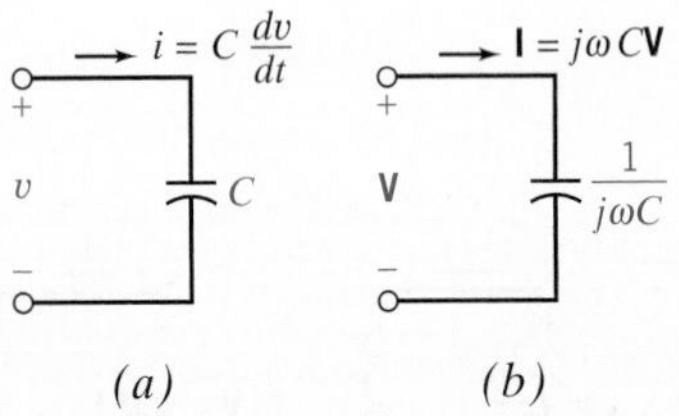

We assume that the voltage is

$$v = V_{\mathrm{m}}\cos(\omega t + \phi) = \mathrm{Re}\{V_{\mathrm{m}} e^{j(\omega t + \phi)}\} \tag{10.6-11}$$

and the current is of the form

$$i = \mathrm{Re}\{I_{\mathrm{m}} e^{j(\omega t + \beta)}\} \tag{10.6-12}$$

FIGURE 10.6-3 (*a*) The time-domain *v*–*i* relationship for a capacitor. (*b*) The frequency-domain relationship for a capacitor.

Suppress the Re notation in Eqs. 10.6-11 and 10.6-12 and substitute them into Eq. 10.6-10 to obtain

$$I_{\mathrm{m}} e^{j(\omega t + \beta)} = C\frac{d}{dt}(V_{\mathrm{m}} e^{j(\omega t + \phi)})$$

Taking the derivative, we have

$$I_{\mathrm{m}} e^{j\omega t} e^{j\beta} = j\omega C V_{\mathrm{m}} e^{j\omega t} e^{j\phi}$$

Suppressing the $e^{j\omega t}$, we obtain

$$I_{\mathrm{m}} e^{j\beta} = j\omega C V_{\mathrm{m}} e^{j\phi}$$

or

$$\mathbf{I} = j\omega C\mathbf{V} \tag{10.6-13}$$

This phasor relationship is shown in Figure 10.6-3*b*. Because $j = e^{j90^\circ}$, the current leads the voltage by 90°. As an example, consider a voltage $v = 100\cos\omega t$ V and let us find the current when $\omega = 1000$ rad/s and $C = 1$ mF. Because

$$\mathbf{V} = 100\,\underline{/0^\circ}$$

we have

$$\mathbf{I} = j\omega C\mathbf{V} = (\omega C e^{j90^\circ})100e^{j0} = (1e^{j90^\circ})100 = 100\,\underline{/90^\circ}$$

Therefore, transforming this phasor into the time domain, we have

$$i = 100 \cos(\omega t + 90°) \text{ A}$$

We can rewrite Eq. 10.6-13 as

$$\mathbf{V} = \frac{1}{j\omega C}\mathbf{I} \tag{10.6-14}$$

Using this form, we summarize the phasor equations for sources and the resistor, inductor, and capacitor in Table 10.6-1, where the phasor voltage is expressed in its relationship to the phasor current.

Table 10.6-1 Time-Domain and Frequency-Domain Relationships

ELEMENT	TIME DOMAIN	FREQUENCY DOMAIN
Corrent Source	 $i(t) = A\cos(\omega t + \theta)$	 $\mathbf{I}(\omega) = Ae^{j\theta}$
Voltage source	 $v(t) = B\cos(\omega t + \phi)$	 $\mathbf{V}(\omega) = Be^{j\phi}$
Resistor	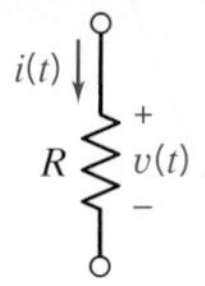 $i(t)$, R, $v(t)$; $v(t) = R\,i(t)$	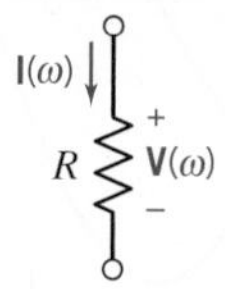 $\mathbf{I}(\omega)$, R, $\mathbf{V}(\omega)$; $\mathbf{V}(\omega) = R\,\mathbf{I}(\omega)$
Capacitor	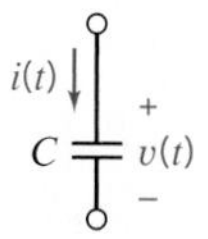 $i(t)$, C, $v(t)$; $v(t) = \frac{1}{C}\int_{-\infty}^{t} i(\tau)\,d\tau$	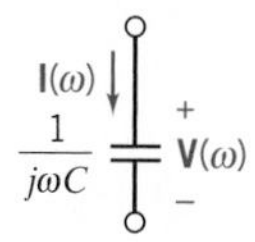 $\mathbf{I}(\omega)$, $\frac{1}{j\omega C}$, $\mathbf{V}(\omega)$; $\mathbf{V}(\omega) = \frac{1}{j\omega C}\mathbf{I}(\omega)$
Inductor	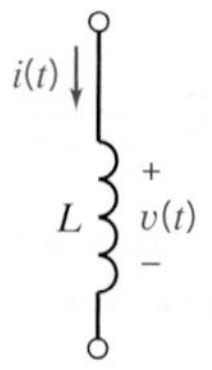 $i(t)$, L, $v(t)$; $v(t) = L\frac{d}{dt}i(t)$	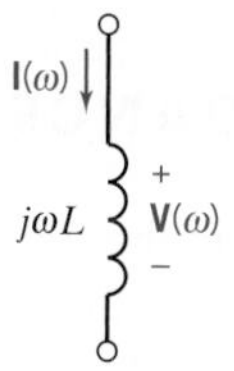 $\mathbf{I}(\omega)$, $j\omega L$, $\mathbf{V}(\omega)$; $\mathbf{V}(\omega) = j\omega L\,\mathbf{I}(\omega)$
CCVS	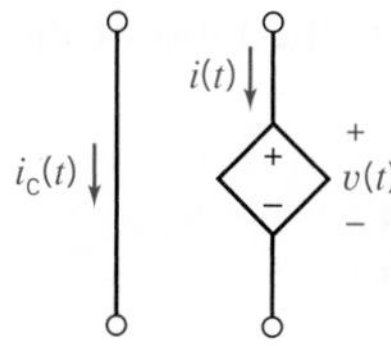 $i_c(t)$, $i(t)$; $v(t) = K\,i_c(t)$	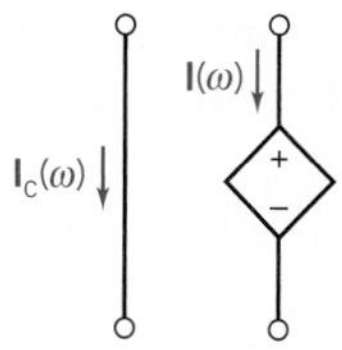 $\mathbf{I}_c(\omega)$, $\mathbf{I}(\omega)$; $\mathbf{V}(\omega) = K\,\mathbf{I}_c(\omega)$
Ideal op amp	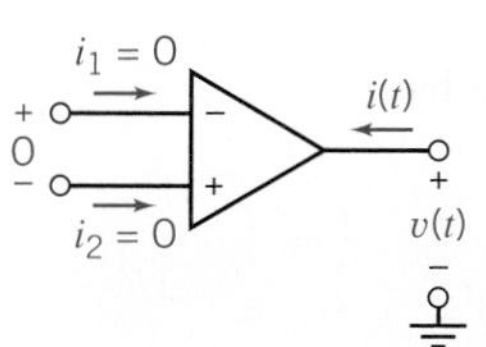 $i_1 = 0$, $i_2 = 0$, 0, $i(t)$, $v(t)$	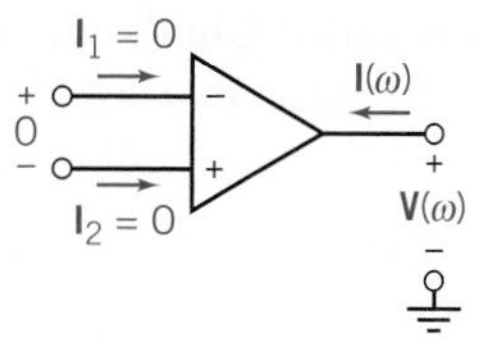 $\mathbf{I}_1 = 0$, $\mathbf{I}_2 = 0$, 0, $\mathbf{I}(\omega)$, $\mathbf{V}(\omega)$

EXERCISE 10.6-1 A current in an element is $i = 5 \cos 100t$ A. Find the steady-state voltage $v(t)$ across the element for (a) a resistor of 10 Ω, (b) an inductor $L = 10$ mH, and (c) a capacitor $C = 1$ mF.

Answers: **(a)** $50 \cos 100t$ V
(b) $5 \cos(100t + 90°)$ V
(c) $50 \cos(100t - 90°)$ V

EXERCISE 10.6-2 A capacitor $C = 10\,\mu$F has a steady-state voltage across it of $v = 100 \cos(500t + 30°)$ V. Find the steady-state current in the capacitor.

Answer: $i = 0.5 \cos(500t - 120°)$ A

EXERCISE 10.6-3 The voltage $v(t)$ and current $i(t)$ for an element are shown in Figure E 10.6-3. Determine whether the element is an inductor or a capacitor.

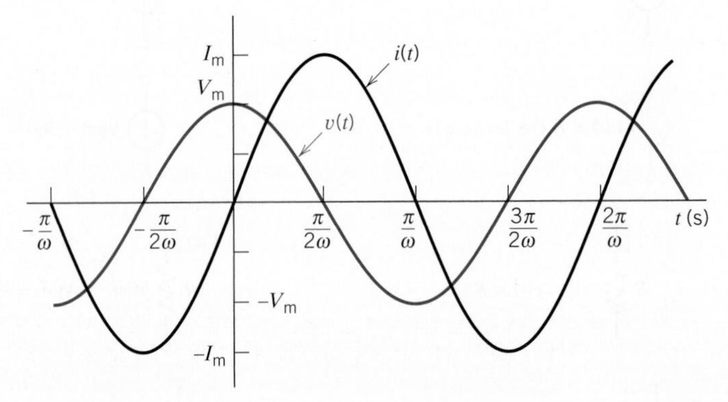

FIGURE E 10.6-3

10.7 IMPEDANCE AND ADMITTANCE

The relationships in the frequency domain for the phasor current and phasor voltage of a capacitor, inductor, and resistor are summarized in Table 10.6-1. These relationships appear to be similar to Ohm's law for resistors.

We will define the *impedance* of an element as the ratio of the phasor voltage to the phasor current, which we denote by **Z**. Therefore,

$$\mathbf{Z} = \frac{\mathbf{V}}{\mathbf{I}} \tag{10.7-1}$$

This is called Ohm's law in phasor notation.

Because $\mathbf{V} = V_m \angle \phi$ and $\mathbf{I} = I_m \angle \beta$, we have

$$\mathbf{Z} = \frac{V_m \angle \phi}{I_m \angle \beta} = \frac{V_m}{I_m} \angle \phi - \beta \tag{10.7-2}$$

Thus, the impedance is said to have a magnitude $|\mathbf{Z}|$ and a phase angle $\theta = \phi - \beta$. Therefore,

$$|\mathbf{Z}| = \frac{V_\mathrm{m}}{I_\mathrm{m}} \tag{10.7-3}$$

and

$$\theta = \phi - \beta \tag{10.7-4}$$

Impedance in ac circuits has a role similar to the role of resistance in dc circuits.

Also, because it is a ratio of volts to amperes, impedance has units of ohms. Impedance is the ratio of two phasors; however, it is not a phasor itself. Impedance is a complex number that relates one phasor **V** to the other phasor **I** as

$$\mathbf{V} = \mathbf{ZI} \tag{10.7-5}$$

The phasors **V** and **I** may be transformed to the time domain to yield the steady-state voltage or current, respectively. Impedance has no meaning in the time domain, however.

With the concept of impedance, we can solve for the behavior of sinusoidally excited circuits, using complex algebra in the same way we solved resistive circuits.

Because the impedance is a complex number, it may be written in several forms, as follows:

$$\begin{aligned} \mathbf{Z} &= |\mathbf{Z}|\angle\theta \quad \rightarrow \text{polar form} \\ &= Ze^{j\theta} \quad \rightarrow \text{exponential form} \\ &= R + jX \quad \rightarrow \text{rectangular form} \end{aligned} \tag{10.7-6}$$

where R is the real part and X is the imaginary part of the complex number **Z**. We introduce the notation, in Eq. 10.7-6, $|\mathbf{Z}| = Z$. Thus, the magnitude of the impedance can be written as Z (not boldface). The $R = \mathrm{Re}\,\mathbf{Z}$ is called the resistive part of the impedance, and $X = \mathrm{Im}\,\mathbf{Z}$ is called the reactive part of the impedance. Both R and X are measured in ohms.

We also note that the magnitude of the impedance is

$$Z = \sqrt{R^2 + X^2} \tag{10.7-7}$$

and the phase angle is

$$\theta = \tan^{-1}\frac{X}{R} \tag{10.7-8}$$

These relationships are summarized graphically, in the complex plane, in Figure 10.7-1. As an example, let us consider

$$\mathbf{Z} = 2 + j2$$

Then,

$$Z = \sqrt{8}$$

and

$$\theta = 45^\circ$$

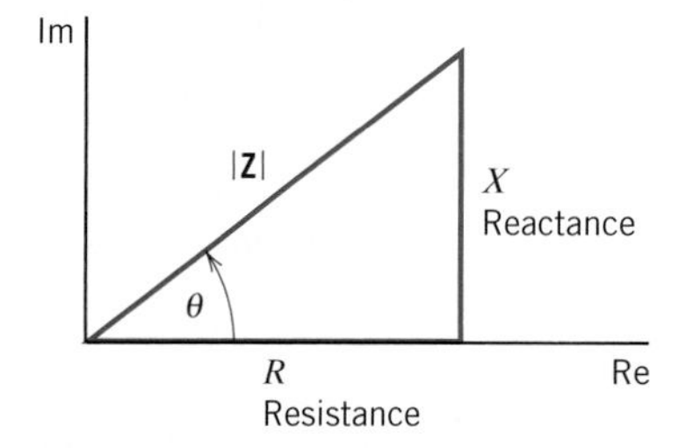

FIGURE 10.7-1 Graphical representation of impedance.

The three elements, R, L, and C, are uniquely represented by an impedance that follows from their **V–I** relationship. For a resistor, we have

$$\mathbf{V} = R\mathbf{I}$$

and, therefore,

$$\mathbf{Z} = R \tag{10.7-9}$$

For the inductor, we have

$$\mathbf{V} = j\omega L\mathbf{I}$$

and, therefore,

$$\mathbf{Z} = j\omega L \tag{10.7-10}$$

Finally, for the capacitor, we have

$$\mathbf{V} = \frac{\mathbf{I}}{j\omega C}$$

so that

$$\mathbf{Z} = \frac{1}{j\omega C} = \frac{-j}{\omega C} \tag{10.7-11}$$

The impedances for R, L, and C are used in Table 10.6-1 to represent resistors, inductors, and capacitors in the frequency domain. The unit for an impedance is ohms.

The reciprocal of impedance is called the *admittance* and is denoted by $\mathbf{Y}$:

$$\mathbf{Y} = \frac{1}{\mathbf{Z}} \tag{10.7-12}$$

Admittance is analogous to conductance for resistive circuits. The units of admittance are siemens, abbreviated as S. Recalling from Eq. 10.7-6 that $\mathbf{Z} = Z\angle\theta$, we have

$$\mathbf{Y} = \frac{1}{|\mathbf{Z}|\angle\theta} = |\mathbf{Y}|\angle-\theta \tag{10.7-13}$$

Therefore, $|\mathbf{Y}| = 1/|\mathbf{Z}|$ and the angle of $\mathbf{Y}$ is $-\theta$. We may also write the magnitude relation as $Y = 1/Z$.

Using the form

$$\mathbf{Z} = R + jX$$

we obtain

$$\mathbf{Y} = \frac{1}{R + jX} = \frac{R - jX}{R^2 + X^2} = G + jB \tag{10.7-14}$$

Note that G is not simply the reciprocal of R, nor is B the reciprocal of X. The real part of admittance, G, is called the *conductance*, and the imaginary part, B, is called the *susceptance*. The units of G and B are siemens.

The impedance of an element is $\mathbf{Z} = R + jX$. The element is inductive if the reactive part X is positive, capacitive if X is negative. Because $\mathbf{Y}$ is the reciprocal of $\mathbf{Z}$ and $\mathbf{Y} = G + jB$, one can also say that if B is positive, the element is capacitive and that a negative B indicates an inductive element.

Let us consider the capacitor $C = 1$ mF and find its impedance and admittance. The impedance of a capacitor is

$$\mathbf{Z} = \frac{1}{j\omega C}$$

Therefore, in addition to the value of $C = 1$ mF, we need the frequency ω. If we consider the case $\omega = 100$ rad/s, we obtain

$$\mathbf{Z} = \frac{1}{j0.1} = \frac{10}{j} = -10j = 10\angle-90^\circ\ \Omega$$

To find the admittance, we note that

$$\mathbf{Y} = \frac{1}{\mathbf{Z}} = j\omega C = j0.1 = 0.1\angle 90^\circ\ \text{S}$$

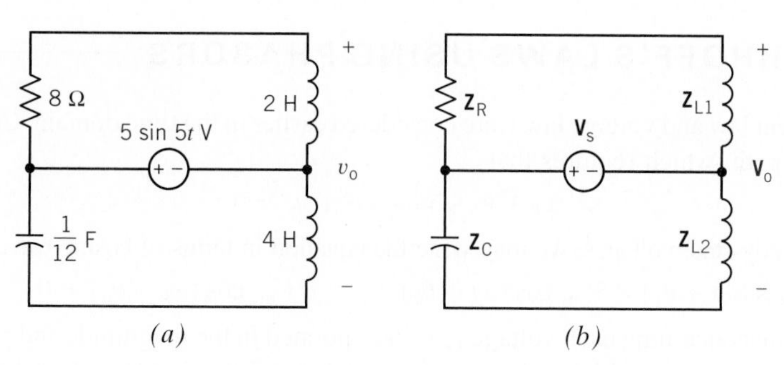

FIGURE E 10.7-1 A circuit represented (*a*) in the time domain and (*b*) in the frequency domain.

EXERCISE 10.7-1 Figure E 10.7-1*a* shows a circuit represented in the time domain. Figure 10.7.1*b* shows the same circuit represented in the frequency domain, using phasors and impedances. $\mathbf{Z}_R$, $\mathbf{Z}_C$, $\mathbf{Z}_{L1}$, and $\mathbf{Z}_{L2}$ are the impedances corresponding to the resistor, capacitor, and two inductors in Figure 10.7-1*a*. $\mathbf{V}_s$ is the phasor corresponding to the voltage of the voltage source. Determine $\mathbf{Z}_R$, $\mathbf{Z}_C$, $\mathbf{Z}_{L1}$, $\mathbf{Z}_{L2}$, and $\mathbf{V}_s$.

Hint: $5 \sin 5t = 5 \cos(5t - 90°)$

Answer: $\mathbf{Z}_R = 8\ \Omega, \mathbf{Z}_C = \dfrac{1}{j5\left(\dfrac{1}{12}\right)} = \dfrac{2.4}{j} = \dfrac{j2.4}{j \times j} = -j2.4\ \Omega, \mathbf{Z}_{L1} = j5(2) = j10\ \Omega,$

$\mathbf{Z}_{L2} = j5(4) = j20\ \Omega$, and $\mathbf{V}_s = 5\angle{-90°}$ V

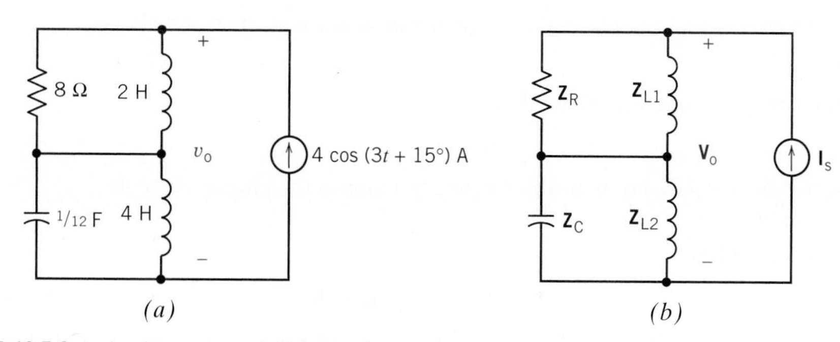

FIGURE E 10.7-2 A circuit represented (*a*) in the time domain and (*b*) in the frequency domain.

EXERCISE 10.7-2 Figure E 10.7-2*a* shows a circuit represented in the time domain. Figure E 10.7-2*b* shows the same circuit represented in the frequency domain, using phasors and impedances. $\mathbf{Z}_R$, $\mathbf{Z}_C$, $\mathbf{Z}_{L1}$, and $\mathbf{Z}_{L2}$ are the impedances corresponding to the resistor, capacitor, and two inductors in Figure E 10.7-2*a*. $\mathbf{I}_s$ is the phasor corresponding to the current of the current source. Determine $\mathbf{Z}_R$, $\mathbf{Z}_C$, $\mathbf{Z}_{L1}$, $\mathbf{Z}_{L2}$, and $\mathbf{I}_s$.

Answer: $\mathbf{Z}_R = 8\ \Omega,\ \mathbf{Z}_C = \dfrac{1}{j3\left(\dfrac{1}{12}\right)} = \dfrac{4}{j} = \dfrac{j4}{j \times j} = -j4\ \Omega,\ \mathbf{Z}_{L1} = j3(2) = j6\ \Omega,$

$\mathbf{Z}_{L2} = j3(4) = j\,12\ \Omega$, and $\mathbf{I}_s = 4\angle{15°}$ A

10.8 KIRCHHOFF'S LAWS USING PHASORS

Kirchhoff's current law and voltage law were considered earlier in the time domain. Consider the KVL around a closed path, which requires that

$$v_1 + v_2 + v_3 + \cdots + v_n = 0 \tag{10.8-1}$$

For sinusoidal steady-state voltages, we may write the equation in terms of cosine waveforms as

$$V_{m_1} \cos(\omega t + \theta_1) + V_{m_2} \cos(\omega t + \theta_2) + \cdots + V_{m_n} \cos(\omega t + \theta_n) = 0 \tag{10.8-2}$$

All the information concerning each voltage v_n is incorporated in the magnitude and phase, V_{m_n} and θ_n (assuming we keep note of ω, which is the same for each term). Equation 10.8-2 can be rewritten, using Euler's identity, as

$$\text{Re}\{V_{m_1} e^{j\theta_1} e^{j\omega t}\} + \cdots + \text{Re}\{V_{m_n} e^{j\theta_n} e^{j\omega t}\} = 0$$

or

$$\text{Re}\{V_{m_1} e^{j\theta_1} e^{j\omega t} + \cdots + V_{m_n} e^{j\theta_n} e^{j\omega t}\} = 0$$

We can factor out the $e^{j\omega t}$ to obtain

$$\text{Re}\{(V_{m_1} e^{j\theta_1} + \cdots + V_{m_n} e^{j\theta_n}) e^{j\omega t}\} = 0$$

Writing $V_{m_p} e^{j\theta_p}$ as $\mathbf{V}_p$, we have

$$\text{Re}(\mathbf{V}_1 + \mathbf{V}_2 + \cdots + \mathbf{V}_n) e^{j\omega t}\} = 0$$

Because $e^{j\omega t}$ cannot equal zero, we require that

$$\mathbf{V}_1 + \mathbf{V}_2 + \cdots + \mathbf{V}_n = 0 \tag{10.8-3}$$

Therefore, we have the important result that the sum of the *phasor voltages* in a closed path is zero. Thus,

> Kirchhoff's voltage law holds in the frequency domain with phasor voltages.

Using a similar process, one can show that

> Kirchhoff's current law holds in the frequency domain for phasor currents.

so that at a node, we have

$$\mathbf{I}_1 + \mathbf{I}_2 + \cdots + \mathbf{I}_n = 0 \tag{10.8-4}$$

Because both the KVL and the KCL hold in the frequency domain, it is easy to conclude that all the techniques of analysis we developed for resistive circuits hold for phasor currents and voltages. For example, we can use the principle of superposition, source transformations, Thévenin and Norton equivalent circuits, and node voltage and mesh current analysis. All these methods apply as long as the circuit is linear.

First, let us consider impedances connected in series, as shown in Figure 10.8-1. The phasor current $\mathbf{I}$ flows through each impedance. Applying KVL, we can write

$$\mathbf{V}_1 + \mathbf{V}_2 + \cdots + \mathbf{V}_n = \mathbf{V}$$

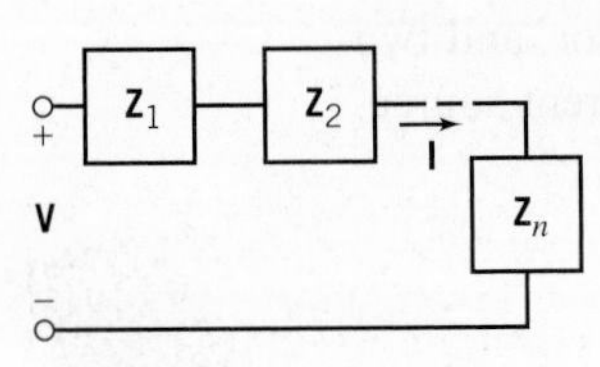

FIGURE 10.8-1 Impedances in series.

Because $\mathbf{V}_j = \mathbf{Z}_j\mathbf{I}$, we have

$$(\mathbf{Z}_1 + \mathbf{Z}_2 + \cdots + \mathbf{Z}_n)\mathbf{I} = \mathbf{V}$$

Therefore, the equivalent impedance seen at the input terminals is

$$\mathbf{Z}_{eq} = \mathbf{Z}_1 + \mathbf{Z}_2 + \cdots + \mathbf{Z}_n \tag{10.8-5}$$

Thus, the equivalent impedance for a series of impedances is the sum of the individual impedances.

Consider the set of parallel admittances shown in Figure 10.8-2. It easily can be shown that the equivalent admittance $\mathbf{Y}_{eq}$ is

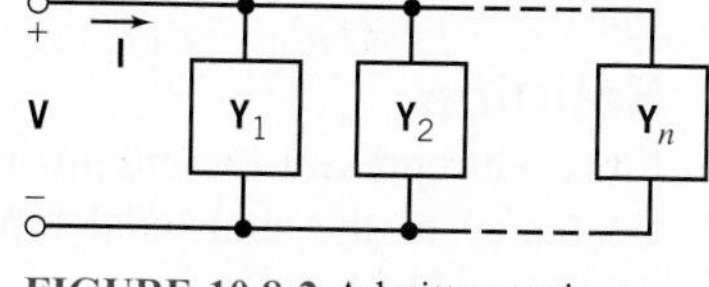

FIGURE 10.8-2 Admittances in parallel.

$$\mathbf{Y}_{eq} = \mathbf{Y}_1 + \mathbf{Y}_2 + \cdots + \mathbf{Y}_n \tag{10.8-6}$$

In the case of two parallel admittances, we have

$$\mathbf{Y}_{eq} = \mathbf{Y}_1 + \mathbf{Y}_2$$

and the corresponding equivalent impedance is

$$\mathbf{Z}_{eq} = \frac{1}{\mathbf{Y}_{eq}} = \frac{1}{\mathbf{Y}_1 + \mathbf{Y}_2} = \frac{\mathbf{Z}_1\mathbf{Z}_2}{\mathbf{Z}_1 + \mathbf{Z}_2} \tag{10.8-7}$$

Similarly, the current divider and voltage divider rules hold for phasor currents and voltages. Table 10.8-1 summarizes the equations for voltage and current division in the frequency domain.

Table 10.8-1 Voltage and Current Division in the Frequency Domain

	CIRCUIT	EQUATIONS
Voltage division	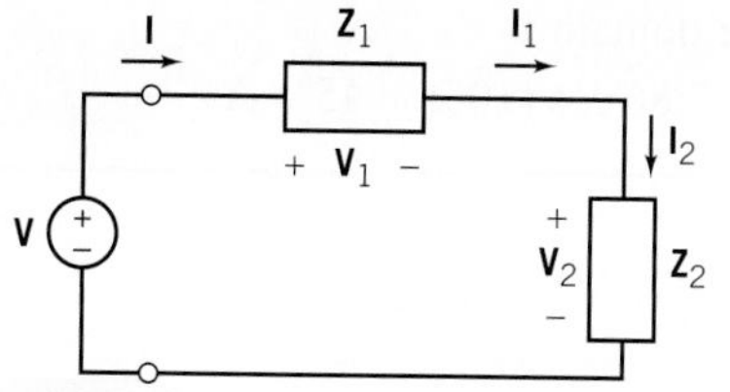	$\mathbf{I}_1 = \mathbf{I}_2 = \mathbf{I}$ $\mathbf{V}_1 = \frac{\mathbf{Z}_1}{\mathbf{Z}_1 + \mathbf{Z}_2}\mathbf{V}$ $\mathbf{V}_2 = \frac{\mathbf{Z}_2}{\mathbf{Z}_1 + \mathbf{Z}_2}\mathbf{V}$
Current division	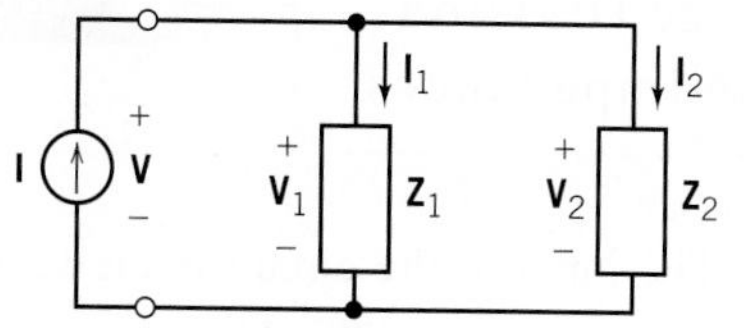	$\mathbf{V}_1 = \mathbf{V}_2 = \mathbf{V}$ $\mathbf{I}_1 = \frac{\mathbf{Z}_2}{\mathbf{Z}_1 + \mathbf{Z}_2}\mathbf{I}$ $\mathbf{I}_2 = \frac{\mathbf{Z}_1}{\mathbf{Z}_1 + \mathbf{Z}_2}\mathbf{I}$

EXAMPLE 10.8-1 Analysis of AC Circuits Using Impedances

Determine the steady-state current $i(t)$ in the *RLC* circuit shown in Figure 10.8-3*a*, using phasors and impedances.

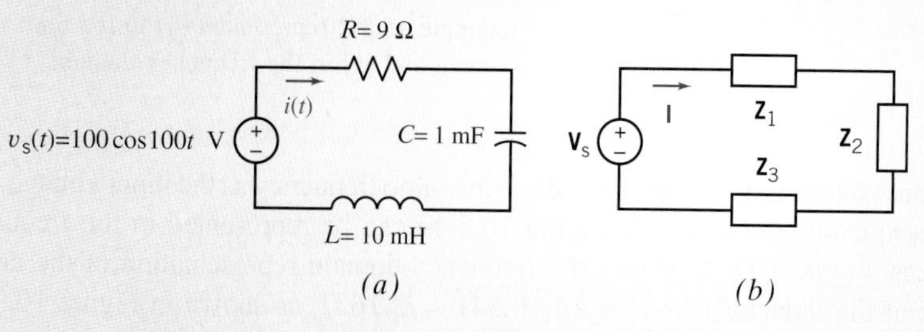

FIGURE 10.8-3 The circuit from Example 10.8-1 represented (*a*) in the time domain and (*b*) in the frequency domain.

Solution

First, we represent the circuit in using phasors and impedances as shown in Figure 10.8-3*b*. Noticing that the frequency of the sinusoidal input in Figure 10.8-3*a* is $\omega = 100$ rad/s, the impedances in Figure 10.8-3*b* are determined to be

$$\mathbf{Z}_1 = R = 9\,\Omega,\ \mathbf{Z}_2 = \frac{1}{j\omega C} = \frac{1}{j(100)(0.001)} = \frac{10}{j} = -j10\,\Omega$$

and

$$\mathbf{Z}_3 = j\omega L = j(100)(0.001) = j1\,\Omega$$

The input phasor in Figure 10.8-3*b* is

$$\mathbf{V}_s = 100\,\angle 0^\circ\ \mathrm{V}$$

Next, we use KVL in Figure 10.8-3*b* to obtain

$$\mathbf{Z}_1\mathbf{I} + \mathbf{Z}_2\mathbf{I} + \mathbf{Z}_3\mathbf{I} = \mathbf{V}_s$$

Substituting for the impedances and the input phasor gives

$$(9 - j10 + j1)\mathbf{I} = 100\,\angle 0^\circ$$

or

$$\mathbf{I} = \frac{100\,\angle 0^\circ}{9 - j9} = \frac{10\,\angle 0^\circ}{9\sqrt{2}\,\angle -45^\circ} = 7.86\,\angle 45^\circ\ \mathrm{A}$$

Therefore, the steady-state current in the time domain is

$$i(t) = 7.86\cos(100t + 45^\circ)\ \mathrm{A}$$

EXAMPLE 10.8-2 Voltage Division Using Impedances

INTERACTIVE EXAMPLE

Consider the circuit shown in Figure 10.8-4*a*. The input to the circuit is the voltage of the voltage source,

$$v_s(t) = 7.28\cos(4t + 77^\circ)\ \mathrm{V}$$

The output is the voltage across the inductor, $v_o(t)$. Determine the steady-state output voltage, $v_o(t)$.

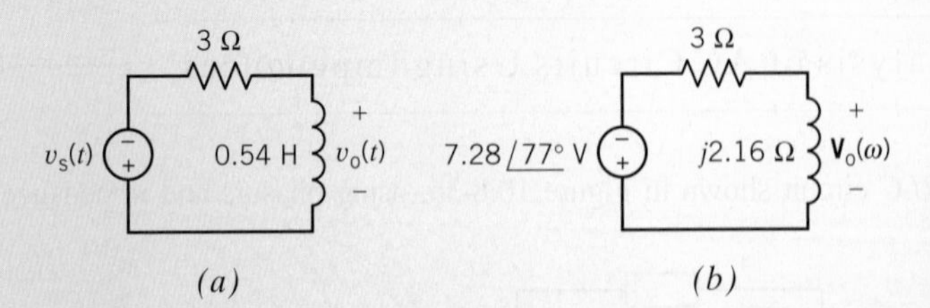

FIGURE 10.8-4 The circuit considered in Example 10.8-2 represented (*a*) in the time domain and (*b*) in the frequency domain.

Solution

The input voltage is sinusoid. The output voltage is also sinusoid and has the same frequency as the input voltage. The circuit has reached steady state. Consequently, the circuit in Figure 10.8-4*a* can be represented in the frequency domain, using phasors and impedances. Figure 10.8-4*b* shows the frequency-domain representation of the circuit from Figure 10.8-4*a*. The impedance of the inductor is $j\omega L = j(4)(0.54) = j2.16\,\Omega$, as shown in Figure 10.8-4*b*.

Apply the voltage divider principle to the circuit in Figure 10.8-4*b* to represent the output voltage in the frequency domain as

$$\mathbf{V}_o(\omega) = \frac{j2.16}{3+j2.16}\left(-7.28\angle 77^\circ\right) = \frac{2.16\angle 90^\circ}{3.70\angle 36^\circ}\left(-7.28\angle 77^\circ\right)$$

$$= \frac{(2.16)(-7.28)}{3.70}\angle (90^\circ + 77^\circ) - 36^\circ$$

$$= -4.25\angle 131^\circ = 4.25\angle 311^\circ$$

In the time domain, the output voltage is represented as

$$v_o(t) = 4.25 \cos(4t + 311^\circ)\ \text{V}$$

Example 10.8-3 AC Circuit Analysis

INTERACTIVE EXAMPLE

Consider the circuit shown in Figure 10.8-5*a*. The input to the circuit is the voltage of the voltage source,

$$v_s(t) = 7.68 \cos(2t + 47^\circ)\ \text{V}$$

The output is the voltage across the resistor,

$$v_o(t) = 1.59 \cos(2t + 125^\circ)\ \text{V}$$

Determine capacitance, *C*, of the capacitor.

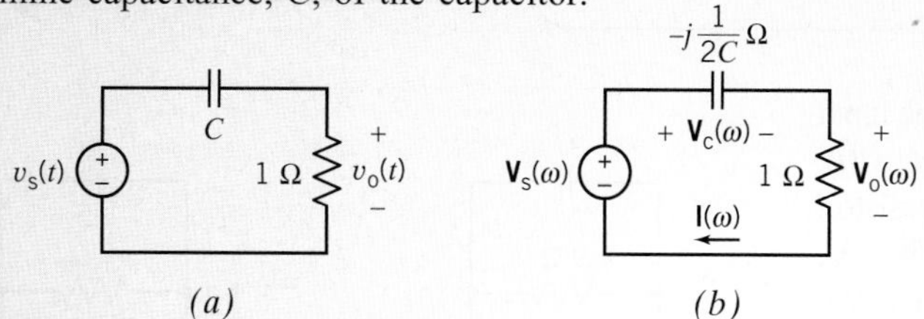

FIGURE 10.8-5 The circuit considered in Example 10.8-3 represented (*a*) in the time domain and (*b*) in the frequency domain.

Solution

The input voltage is sinusoid. The output voltage is also sinusoid and has the same frequency as the input voltage. Apparently, the circuit has reached steady state. Consequently, the circuit in Figure 10.8-5*a* can be represented in the frequency domain, using phasors and impedances. Figure 10.8-5*b* shows the frequency-domain representation of the circuit from Figure 10.8-5*a*. The impedance of the capacitor is

$$\frac{1}{j\omega C} = \frac{j}{j^2\omega C} = -\frac{j}{\omega C} = -\frac{j}{2\,C}$$

The phasors corresponding to the input and output sinusoids are

$$\mathbf{V}_s(\omega) = 7.68\angle 47^\circ\ \text{V}$$

and

$$\mathbf{V}_o(\omega) = 1.59\angle 125^\circ\ \text{V}$$

The current $\mathbf{I}(\omega)$ in Figure 10.8-5*b* is given by

$$\mathbf{I}(\omega) = \frac{\mathbf{V}_o(\omega)}{1} = \frac{1.59\angle 125^\circ}{1\angle 0^\circ} = 1.59\angle 125^\circ\ \text{A}$$

The capacitor voltage, $\mathbf{V}_c(\omega)$, in Figure 10.8-5*b* is given by

$$\begin{aligned}\mathbf{V}_c(\omega) = \mathbf{V}_s(\omega) - \mathbf{V}_o(\omega) &= 7.68\angle 45^\circ - 1.59\angle 125^\circ \\ &= (5.23 + j5.62) - (-0.91 + 1.30) \\ &= (5.23 + 0.91) + j(5.62 - 1.30) \\ &= 6.14 + j4.32 \\ &= 7.51\angle 35^\circ\end{aligned}$$

The impedance of the capacitor is given by

$$-j\frac{1}{2C} = \frac{\mathbf{V}_c(\omega)}{\mathbf{I}(\omega)} = \frac{7.51\angle 35^\circ}{1.59\angle 125^\circ} = 4.72\angle -90^\circ$$

Solving for C gives

$$C = \frac{-j}{2\left(4.72\angle -90^\circ\right)} = \frac{1\angle -90^\circ}{2\left(4.72\angle -90^\circ\right)} = 0.106\text{ F}$$

EXAMPLE 10.8-4 AC Circuit Analysis

INTERACTIVE EXAMPLE

Consider the circuit shown in Figure 10.8-6*a*. The input to the circuit is the voltage of the voltage source, $v_s(t)$, and the output is the voltage across the 4-Ω resistor, $v_o(t)$. When the input is $v_s(t) = 8.93\cos(2t + 54^\circ)$ V, the corresponding output is $v_o(t) = 3.83\cos(2t + 83^\circ)$ V. Determine the voltage across the 9-Ω resistor, $v_a(t)$, and the value of the capacitance, C, of the capacitor.

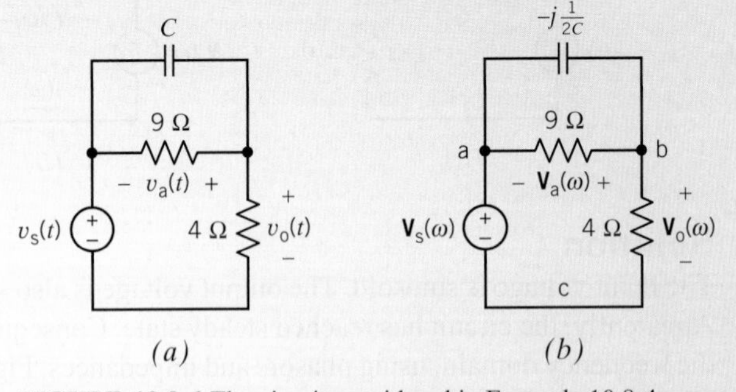

FIGURE 10.8-6 The circuit considered in Example 10.8-4 represented (*a*) in the time domain and (*b*) in the frequency domain.

Solution

The input voltage is a sinusoid. The output voltage is also sinusoid and has the same frequency as the input voltage. Apparently, the circuit has reached steady state. Consequently, the circuit in Figure 10.8-6*a* can be represented in the frequency domain, using phasors and impedances. Figure 10.8-6*b* shows the frequency-domain representation of the circuit from Figure 10.8-6*a*. The voltages $\mathbf{V}_s(\omega)$, $\mathbf{V}_a(\omega)$, and $\mathbf{V}_o(\omega)$ in Figure 10.8-6*b* are the phasors corresponding to $v_s(t)$, $v_a(t)$, and $v_o(t)$ from Figure 10.8-6*a*. The capacitor and the resistors are represented as impedances in Figure 10.8-6*b*. The impedance of the capacitor is $-j1/\omega C = -j1/2C$ where 2 rad/s is the value of the frequency of $v_s(t)$.

The phasors corresponding to the input and output sinusoids are

$$\mathbf{V}_s(\omega) = 8.93\angle 54^\circ\text{ V}$$

and

$$\mathbf{V}_o(\omega) = 3.83\angle 83^\circ\text{ V}$$

First, we calculate the value of $\mathbf{V}_a(\omega)$. Apply KVL to the mesh in Figure 10.8-6*b* that consists of the two resistors and the voltage source to get

$$\begin{aligned}\mathbf{V}_a(\omega) = \mathbf{V}_o(\omega) - \mathbf{V}_s(\omega) &= \left(3.83\angle 83^\circ\right) - \left(8.93\angle 54^\circ\right)\\ &= (0.47 + j3.80) - (5.25 + j7.22)\\ &= -4.78 - j3.42\\ &= 5.88\angle 216^\circ\end{aligned}$$

The voltage across the 9-Ω resistor, $v_a(t)$, is the sinusoid corresponding to this phasor

$$v_a(t) = 5.88\cos(2t + 216^\circ)\text{ V}$$

We can determine the value of the capacitance by applying Kirchhoff's current law (KCL) at node b in Figure 10.8-6*a*:

$$\frac{\mathbf{V}_a(\omega)}{-j\frac{1}{2C}} + \frac{\mathbf{V}_a(\omega)}{9} + \frac{\mathbf{V}_o(\omega)}{4} = 0$$

$$(j2C)\mathbf{V}_a(\omega) + \frac{\mathbf{V}_a(\omega)}{9} + \frac{\mathbf{V}_o(\omega)}{4} = 0$$

Solving this equation for $j2C$ gives

$$j2C = \frac{4\mathbf{V}_a(\omega) + 9\mathbf{V}_o(\omega)}{-36\mathbf{V}_a(\omega)}$$

Substituting the values of the phasors $\mathbf{V}_a(\omega)$ and $\mathbf{V}_o(\omega)$ into this equation gives

$$\begin{aligned}j2C &= \frac{4(-4.78 - j3.42) + 9(0.47 + j3.80)}{-36\left(5.88\angle 216^\circ\right)}\\ &= \frac{-14.89 + j20.52}{-36\left(5.88\angle 216^\circ\right)}\\ &= \frac{25.35\angle 126^\circ}{\left(36\angle -180^\circ\right)\left(5.88\angle 216^\circ\right)}\\ &= \frac{25.35}{(36)(5.88)}\angle 126^\circ - (-180^\circ + 216^\circ)\\ &= 0.120\angle 90^\circ\\ &= j0.120\end{aligned}$$

Therefore, the value of the capacitance is $C = \frac{0.12}{2} = 0.06 = 60\text{ mF}$.

10.9 NODE VOLTAGE AND MESH CURRENT ANALYSIS USING PHASORS

Circuit analysis in the frequency domain follows the same procedure as we used for resistive circuits; however, we use impedances and phasors instead of resistances and time functions. Because Ohm's law can be used in the frequency domain, we use the relationship $\mathbf{V} = \mathbf{ZI}$ for the passive elements and proceed to use the node voltage and mesh current techniques.

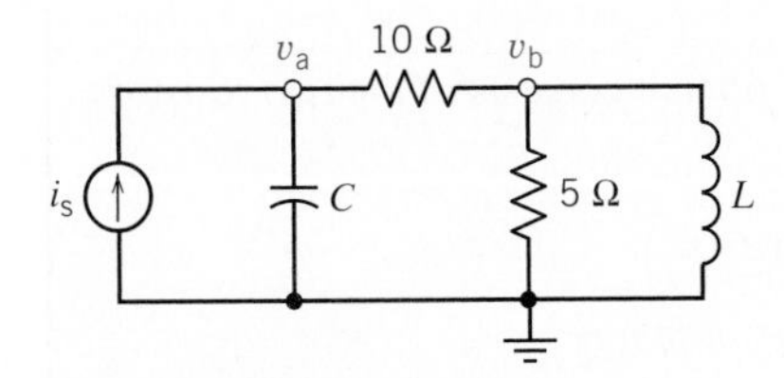

FIGURE 10.9-1 Circuit for which we wish to determine v_a and v_b.

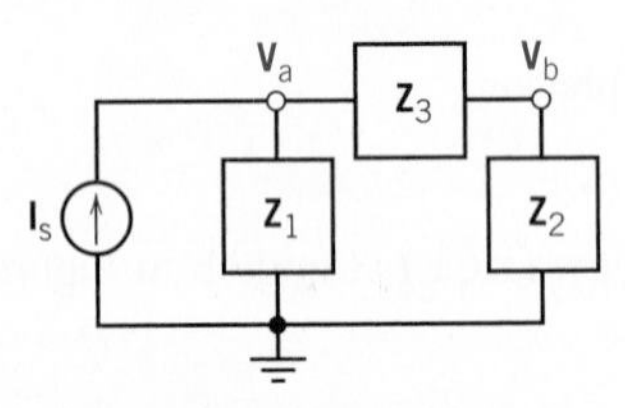

FIGURE 10.9-2 Circuit equivalent to that of Figure 10.9-1 in phasor form.

As an example of the node voltage method using phasors, consider the circuits of Figure 10.9-1 when $i_s = I_m \cos \omega t$. For a specified ω and for specified L and C, we can obtain the impedance for the L and C elements. When $\omega = 1000$ rad/s and $C = 100\ \mu$F, we obtain

$$\mathbf{Z}_1 = \frac{1}{j\omega C} = -j10\ \Omega$$

When $L = 5$ mH for the inductor, we have the impedance

$$\mathbf{Z}_L = j\omega L = j5\ \Omega$$

Then, we may redraw the circuit shown in Figure 10.9-1, using the phasor format shown in Figure 10.9-2. Clearly, $\mathbf{Z}_3 = 10\ \Omega$, and $\mathbf{Z}_2$ is obtained from the parallel combination of the 5-Ω resistor and the inductor's impedance, $\mathbf{Z}_L$. Rather than obtaining $\mathbf{Z}_2$, let us determine $\mathbf{Y}_2$, which is readily found by adding the two parallel admittances as follows:

$$\mathbf{Y}_2 = \frac{1}{5} + \frac{1}{\mathbf{Z}_L} = \frac{1}{5} + \frac{1}{j5} = \frac{1}{5}(1 - j)\ \text{S}$$

Using KCL at node a, we have

$$\frac{\mathbf{V}_a}{\mathbf{Z}_1} + \frac{\mathbf{V}_a - \mathbf{V}_b}{\mathbf{Z}_3} = \mathbf{I}_s \qquad (10.9\text{-}1)$$

At node b, we have

$$\frac{\mathbf{V}_b}{\mathbf{Z}_2} + \frac{\mathbf{V}_b - \mathbf{V}_a}{\mathbf{Z}_3} = 0 \qquad (10.9\text{-}2)$$

Rearranging Eqs. 10.9-1 and 10.9-2, we obtain

$$(\mathbf{Y}_1 + \mathbf{Y}_3)\mathbf{V}_a + (-\mathbf{Y}_3)\mathbf{V}_b = \mathbf{I}_s \qquad (10.9\text{-}3)$$

$$(-\mathbf{Y}_3)\mathbf{V}_a + (\mathbf{Y}_2 + \mathbf{Y}_3)\mathbf{V}_b = 0 \qquad (10.9\text{-}4)$$

where we use the admittance $\mathbf{Y}_n = 1/\mathbf{Z}_n$ and $\mathbf{I}_s = I_m \angle 0^\circ$

We find that Eqs. 10.9-3 and 10.9-4 are similar to the node voltage equations we found in Chapter 4 for resistive circuits. In this case, however, we obtain the node voltage equations in terms of phasor currents, phasor voltages, and complex impedances and admittances.

In general, we may state that for circuits containing only admittances and independent sources, KCL at node k requires that the coefficient of $\mathbf{V}_k$ be the sum of the admittances at node k, and the coefficients of the other terms be the negative of the admittance between those nodes and the kth node.

Let us proceed to solve for $\mathbf{V}_a$ for the circuit shown in Figures 10.9-1 and 10.9-2 when $I_m = 10$ A. Substituting the admittances into Eqs. 10.9-3 and 10.9-4, we have

$$\left(\frac{1}{-j10} + \frac{1}{10}\right)\mathbf{V}_a + \frac{-1}{10}\mathbf{V}_b = 10 \qquad (10.9\text{-}5)$$

$$\frac{-1}{10}\mathbf{V}_a + \left[\frac{1}{5}(1 - j) + \frac{1}{10}\right]\mathbf{V}_b = 0 \qquad (10.9\text{-}6)$$

We then use Cramer's rule to solve for $\mathbf{V}_a$, obtaining

$$\mathbf{V}_a = \frac{100(3 - 2j)}{4 + j} = \frac{100(3 - 2j)(4 - j)}{17} = \frac{100}{17}(10 - 11j) = 87.5 \angle -47.7^\circ$$

Therefore, we have the steady-state voltage v_a:

$$v_a = 87.5 \cos(1000t - 47.7^\circ)\ \text{V}$$

The general nodal analysis methods of Chapter 4 may be used here, where we are careful to note that we use complex impedances and admittances and phasor voltages and currents. After we have determined the desired phasor currents or voltages, we transform them back to the time domain to obtain the steady-state sinusoidal current or voltage desired. We use the concept of a supernode, if necessary, and include the effect of a dependent source, if required.

EXAMPLE 10.9-1 AC Circuit with a Supernode

A circuit is shown in Figure 10.9-3 with $\omega = 10$ rad/s, $L = 0.5$ H, and $C = 10$ mF. Find the node voltage v in its sinusoidal steady-state form when $v_s = 10 \cos \omega t$ V.

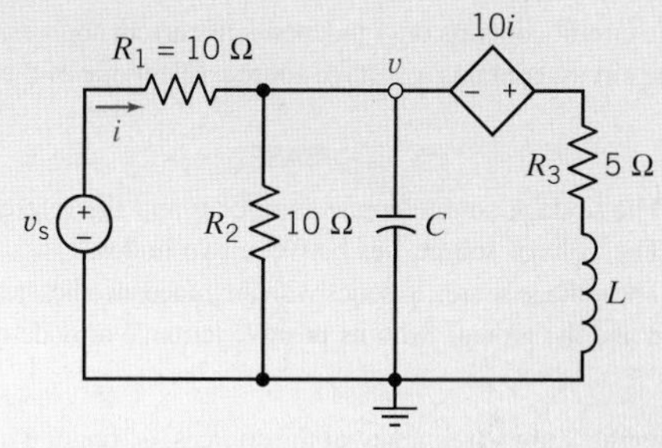

FIGURE 10.9-3 Circuit for Example 10.9-1.

Solution

The circuit has a dependent voltage source between two nodes, so we identify a supernode as shown in Figure 10.9-4, where we also show the impedance for each element. For example, the impedance of the inductor is $\mathbf{Z}_L = j\omega L = j5$. Similarly, the impedance for the capacitor is

$$\mathbf{Z}_c = \frac{1}{j\omega C} = \frac{10}{j} = -j10$$

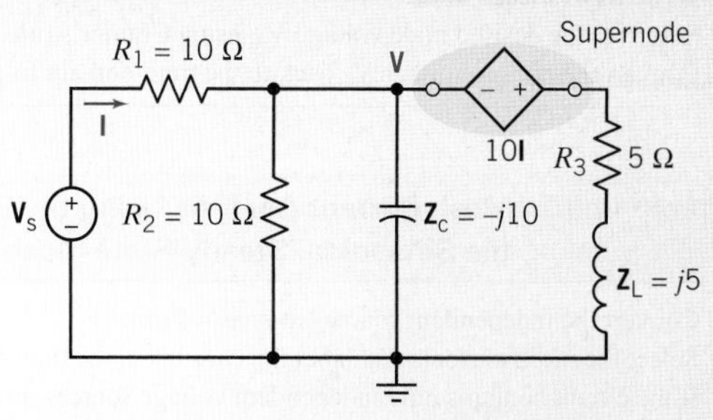

FIGURE 10.9-4 Frequency-domain representation of the circuit for Example 10.9-1.

First, we note that $\mathbf{Y}_1 = 1/R_1 = 1/10$. We now wish to bring together the two parallel admittances for R_2 and C to yield one admittance $\mathbf{Y}_2$ as shown in Figure 10.9-5. We then obtain

$$\mathbf{Y}_2 = \frac{1}{R_2} + \frac{1}{\mathbf{Z}_c} = \frac{1}{10} + \frac{j}{10} = \frac{1}{10}(1 + j)\ \text{S}$$

We may obtain $\mathbf{Y}_3$ for the series resistance and inductance as

$$\mathbf{Y}_3 = \frac{1}{\mathbf{Z}_3}$$

where $\mathbf{Z}_3 = R_3 + \mathbf{Z}_L = 5 + j5\Omega$. Therefore, we have

$$\mathbf{Y}_3 = \frac{1}{5 + j5} = \frac{1}{50}(5 - j5)\ \text{S}$$

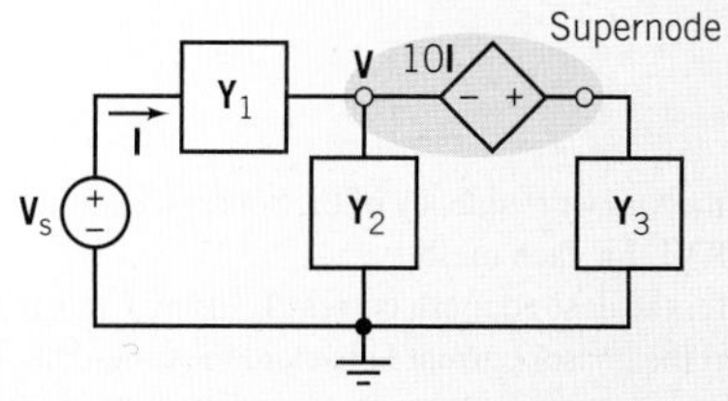

FIGURE 10.9-5 Circuit for Example 10.9-1 with three admittances and the supernode identified.

Writing the KCL at the supernode of Figure 10.9-5, we have

$$\mathbf{Y}_1(\mathbf{V} - \mathbf{V}_s) + \mathbf{Y}_2\mathbf{V} + \mathbf{Y}_3(\mathbf{V} + 10\mathbf{I}) = 0 \tag{10.9-7}$$

Furthermore, we note that

$$\mathbf{I} = \mathbf{Y}_1(\mathbf{V}_s - \mathbf{V}) \tag{10.9-8}$$

Substituting Eq. 10.9-8 into Eq. 10.9-7, we obtain

$$\mathbf{Y}_1(\mathbf{V} - \mathbf{V}_s) + \mathbf{Y}_2\mathbf{V} + \mathbf{Y}_3[\mathbf{V} + 10\mathbf{Y}_1(\mathbf{V}_s - \mathbf{V})] = 0$$

Rearranging, we have

$$(\mathbf{Y}_1 + \mathbf{Y}_2 + \mathbf{Y}_3 - 10\mathbf{Y}_1\mathbf{Y}_3)\mathbf{V} = (\mathbf{Y}_1 - 10\mathbf{Y}_1\mathbf{Y}_3)\mathbf{V}_s$$

Therefore,

$$\mathbf{V} = \frac{(\mathbf{Y}_1 - 10\mathbf{Y}_1\mathbf{Y}_3)\mathbf{V}_s}{\mathbf{Y}_1 + \mathbf{Y}_2 + \mathbf{Y}_3 - 10\mathbf{Y}_1\mathbf{Y}_3}$$

Because $\mathbf{V}_s = 10\underline{/0^\circ}$, we have

$$\mathbf{V} = \frac{\left(\frac{1}{10} - \frac{1}{50}(5 - j5)\right)10}{\frac{1}{10} + \frac{1}{10}(1 + j)} = \frac{1 - (1 - j)}{\frac{1}{10}(2 + j)} = \frac{10j}{2 + j}$$

Therefore, we obtain

$$v = \frac{10}{\sqrt{5}} \cos(10t + 63.4^\circ)\ \text{V}$$

Table 10.9-1 Node Voltage Analysis Using the Phasor Concept to Find the Sinusoidal Steady-State Node Voltages

1. Convert the independent sources to phasor form.
2. Select the nodes and the reference node and label the node voltages in the time domain, v_n, and their corresponding phasor voltages, $\mathbf{V}_n$.
3. If the circuit contains only independent current sources, proceed to step 5; otherwise, proceed to step 4.
4. If the circuit contains a voltage source, select one of the following three cases and the associated method:

CASE	METHOD
a. The voltage source connects node q and the reference node.	Set $\mathbf{V}_q = \mathbf{V}_s$ and proceed.
b. The voltage source lies between two nodes.	Create a supernode including both nodes.
c. The voltage source in series with an impedance lies between node d and the ground, with its positive terminal at node d.	Replace the voltage source and series impedance with a parallel combination of an admittance $\mathbf{Y}_1 = 1/\mathbf{Z}_1$ and a current source $\mathbf{I}_1 = \mathbf{V}_S\mathbf{Y}_1$ entering node d.

5. Using the known frequency of the sources, ω, find the impedance of each element in the circuit.
6. For each branch at a given node, find the equivalent admittance of that branch, $\mathbf{Y}_n$.
7. Write KCL at each node.
8. Solve for the desired node voltage $\mathbf{V}_a$, using Cramer's rule.
9. Convert the phasor voltage $\mathbf{V}_a$ back to the time-domain form.

Table 10.9-2 Mesh Current Analysis Using the Phasor Concept to Find the Sinusoidal Steady-State Mesh Currents

1. Convert the independent sources to phasor form.
2. Select the mesh currents and label the currents in the time domain, i_n, and the corresponding phasor currents, $\mathbf{I}_n$.
3. If the circuit contains only independent voltage sources, proceed to step 5; otherwise, proceed to step 4.
4. If the circuit contains a current source, select one of the following two cases and the associated method:

CASE	METHOD
a. The current source appears as an element of only one mesh, n.	Equate the mesh current $\mathbf{I}_n$ to the current of the current source, accounting for the direction of the source current.
b. b. The current source is common to two meshes.	Create a supermesh as the periphery of the two meshes. In step 6, write one KVL equation around the periphery of the supermesh. Also record the constraining equation incurred by the current source.

5. Using the known frequency of the sources, ω, find the impedance of each element in the circuit.
6. Write KVL for each mesh.
7. Solve for the desired mesh current $\mathbf{I}_n$, using Cramer's rule.
8. Convert the phasor current $\mathbf{I}_n$ back to the time-domain form.

The processes of node voltage and mesh current analysis, using phasors for determining the steady-state sinusoidal response of a circuit, are recorded in Tables 10.9-1 and 10.9-2, respectively.

Mesh current analysis, using the method of Table 10.9-2, is relatively straightforward. When you have the impedance of each element, you may readily write the KVL equations for each mesh.

EXAMPLE 10.9-2 AC Circuit Analysis Using Mesh Equations

Find the steady-state sinusoidal current i_1 for the circuit of Figure 10.9-6 when $v_s = 10\sqrt{2}\cos(\omega t + 45°)$ V and $\omega = 100$ rad/s. Also, $L = 30$ mH and $C = 5$ mF.

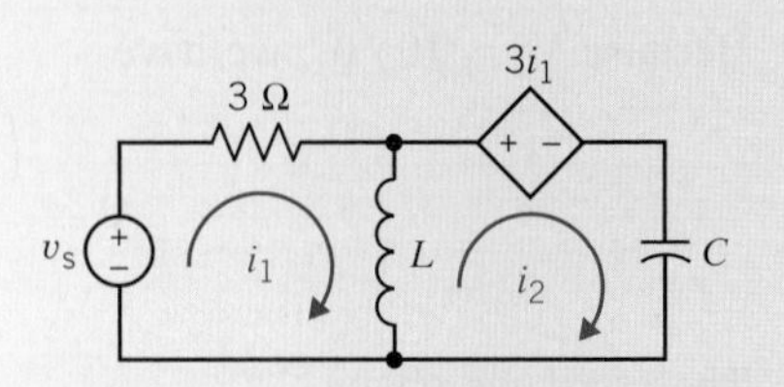

FIGURE 10.9-6 Circuit of Example 10.9-2.

Solution

First, we transform the source voltage to phasor form to obtain

$$\mathbf{V}_s = 10\sqrt{2}\angle 45^\circ = 10 + 10j\text{ V}$$

We then select the two mesh currents as $\mathbf{I}_1$ and $\mathbf{I}_2$, as shown in Figure 10.9-7. Because the frequency of the source is $\omega = 100$, we find that the inductance has an impedance of

$$\mathbf{Z}_L = j\omega L = j3\ \Omega$$

The capacitor has an impedance of

$$\mathbf{Z}_c = \frac{1}{j\omega C} = \frac{1}{j\left(\frac{1}{2}\right)} = -j2\ \Omega$$

We can then summarize the circuit's phasor currents and the impedance of each element by redrawing the circuit in terms of phasors, as shown in Figure 10.9-7. Now we can write the KVL equations for each mesh, obtaining

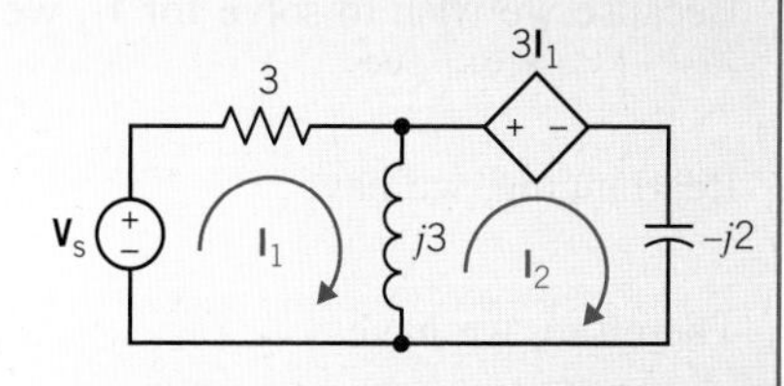

FIGURE 10.9-7 Circuit of Example 10.9-2 with phasors and impedances.

$$\text{mesh 1:}\quad (3 + j3)\mathbf{I}_1 - j3\mathbf{I}_2 = \mathbf{V}_s$$
$$\text{mesh 2:}\quad (3 - j3)\mathbf{I}_1 + (j3 - j2)\mathbf{I}_2 = 0$$

Solving for $\mathbf{I}_1$, using Cramer's rule, we have

$$\mathbf{I}_1 = \frac{(10 + j10)j}{\Delta}$$

where the determinant is

$$\Delta = (3 + j3)(j) + j3(3 - j3) = 6 + 12j$$

Therefore, we have

$$\mathbf{I}_1 = \frac{10j - 10}{6 + 12j}$$

Continuing, we obtain

$$\mathbf{I}_1 = \frac{10(j-1)}{6(1+2j)} = \frac{10\left(\sqrt{2}\angle 135^\circ\right)}{6\left(\sqrt{5}\angle 63.4^\circ\right)} = 1.05\angle 71.6^\circ$$

Thus, the steady-state time response is

$$i_1 = 1.05\cos(100t + 71.6^\circ)\text{ A}$$

EXAMPLE 10.9-3 AC Circuit Analysis Using Impedances

Find the steady-state current i_1 when the voltage source is $v_s = 10\sqrt{2}\cos(\omega t + 45^\circ)$ V and the current source is $i_s = 3\cos\omega t$ A for the frequency-domain circuit of Figure 10.9-8. The circuit of the figure provides the impedance in ohms for each element at the specified ω.

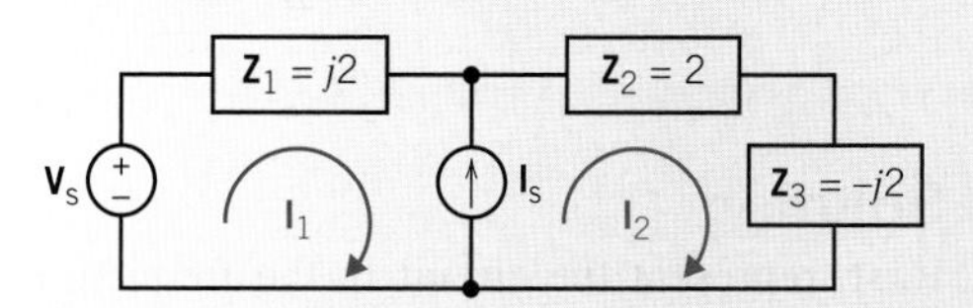

FIGURE 10.9-8 Frequency-domain circuit of Example 10.9-3.

Solution

First, we transform the independent sources into phasor form. The voltage source is

$$\mathbf{V}_s = 10\sqrt{2}\,\underline{/45^\circ} = 10(1 + j)\ \text{V}$$

and the current source is

$$\mathbf{I}_s = 3\,\underline{/0^\circ}\ \text{A}$$

We note that the current source connects the two meshes and provides a constraining equation:

$$\mathbf{I}_2 - \mathbf{I}_1 = \mathbf{I}_s \tag{10.9-9}$$

Creating a supermesh around the periphery of the two meshes, we write one KVL equation, obtaining

$$\mathbf{I}_1\mathbf{Z}_1 + \mathbf{I}_2(\mathbf{Z}_2 + \mathbf{Z}_3) = \mathbf{V}_s \tag{10.9-10}$$

Because we wish to solve for $\mathbf{I}_1$, we will use $\mathbf{I}_2$ from Eq. 10.9-9 and substitute it into Eq. 10.9-10, obtaining

$$\mathbf{I}_1\mathbf{Z}_1 + (\mathbf{I}_s + \mathbf{I}_1)(\mathbf{Z}_2 + \mathbf{Z}_3) = \mathbf{V}_s$$

Rearranging, we have

$$(\mathbf{Z}_1 + \mathbf{Z}_2 + \mathbf{Z}_3)\mathbf{I}_1 = \mathbf{V}_s - (\mathbf{Z}_2 + \mathbf{Z}_3)\mathbf{I}_s$$

Therefore, we have

$$\mathbf{I}_1 = \frac{\mathbf{V}_s - (\mathbf{Z}_2 + \mathbf{Z}_3)\mathbf{I}_s}{\mathbf{Z}_1 + \mathbf{Z}_2 + \mathbf{Z}_3}$$

Substituting the impedances and the sources, we have

$$\mathbf{I}_1 = \frac{(10 + j10) - (2 - j2)3}{2} = 2 + j8 = 8.25\,\underline{/76^\circ}\ \text{A}$$

Thus, we obtain

$$i_1 = 8.25 \cos(\omega t + 76^\circ)\ \text{A}$$

Example 10.9-4 AC Circuit Analysis Using Node Equations

Find the steady-state voltage v for the circuit of Figure 10.9-9*a*

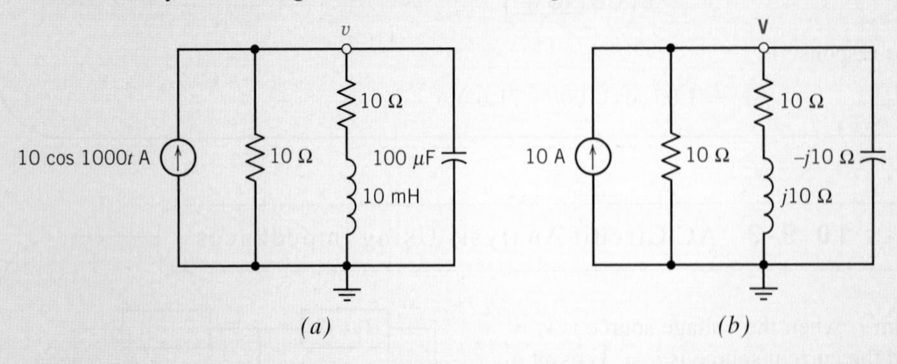

FIGURE 10.9-9 (*a*) Time-domain and (*b*) frequency-domain representation of the circuit for Example 10.9-4.

Solution

First, represent the circuit in the frequency domain, using impedances and phasors. The impedance of the inductor is

$$j\omega L = j1000(10 \times 10^{-3}) = j10\ \Omega$$

The impedance of the capacitor is

$$\frac{1}{j\omega C} = \frac{1}{j1000\left(100 \times 10^{-6}\right)} = \frac{10}{j} = -j10\,\Omega$$

The phasor representation of the input current is

$$10\underline{/0^\circ} = 10\text{ A}$$

Figure 10.9-9*b* shows the frequency-domain representation of the circuit. The phasor voltage **V** can be obtained by applying Kirchhoff's current law at the top node of the circuit in Figure 10.9-9*b* to get

$$\frac{\mathbf{V}}{10} + \frac{\mathbf{V}}{10 + j10} + \frac{\mathbf{V}}{-j10} = 10$$

or

$$\frac{\mathbf{V}}{10} + \frac{\mathbf{V}}{10 + j10}\left(\frac{10 - j10}{10 - j10}\right) + \frac{\mathbf{V}}{-j10} = 0.1\mathbf{V} + (0.05 - j0.05)\mathbf{V} + j0.1\mathbf{V} = 10$$

Solving for **V**, we have

$$\mathbf{V} = \frac{10}{0.158\underline{/-18.4^\circ}} = 63.3\underline{/-18.4^\circ}$$

Therefore, we have the steady-state voltage as

$$v = 63.3\cos\left(1000t - 18.4^\circ\right)\text{ V}.$$

10.10 SUPERPOSITION, THÉVENIN AND NORTON EQUIVALENTS, AND SOURCE TRANSFORMATIONS

Circuits in the frequency domain with phasor currents and voltages and impedances are analogous to the resistive circuits we considered earlier. Because they are linear, we expect that the principle of superposition and the source transformation method will hold. Furthermore, we can define Thévenin and Norton equivalent circuits in terms of impedance or admittance.

First, let us consider the *superposition principle*, which may be restated as follows: For a linear circuit containing two or more independent sources, any circuit voltage or current may be calculated as the algebraic sum of all the individual currents or voltages caused by each independent source acting alone.

If a linear circuit is excited by several sinusoidal sources all having the same frequency, ω, then superposition *may* be used. If a linear circuit is excited by several sources all having different frequencies, then superposition *must* be used.

The superposition principle is particularly useful if a circuit has two or more sources acting at different frequencies. Clearly, the circuit will have one set of impedance values at one frequency and a different set of impedance values at another frequency. We can determine the phasor response at each frequency. Then we find the time response corresponding to each phasor response and add them. Note that superposition, in the case of sources operating at different frequencies, applies to time responses only. We cannot superpose the phasor responses.

EXAMPLE 10.10-1 Superposition

Using the superposition principle, find the steady-state current i for the circuit shown in Figure 10.10-1 when $v_s = 10 \cos 10t$ V, $i_s = 3$ A, $L = 1.5$ H, and $C = 10$ mF.

FIGURE 10.10-1 Circuit of Example 10.10-1.

Solution

The principle of superposition says that the response to the voltage source and current source acting together is equal to the sum of the response to the voltage source acting alone plus the response to the current source acting alone. Let i_1 denote the response to the voltage source acting alone. Figure 10.10-2*a* shows the circuit that is used to calculate i_1. In Figure 10.10-2*b*, this circuit has been represented in the frequency domain using impedances and phasors. Similarly, let i_2 denote the response to the voltage source acting alone. Figure 10.10-3*a* shows the circuit that is used to calculate i_2. In Figure 10.10-3*b*, this circuit has been represented in the frequency domain.

The first step is to convert the independent sources into phasor form, noting that the sources operate at different frequencies. For the voltage source operating at $\omega = 10$, we have

$$\mathbf{V}_s = 10\angle 0^\circ \text{ V}$$

We note that the current source is a direct current, so we can state that $\omega = 0$ for the current source. The phasor form of the current source is then

$$\mathbf{I}_s = 3\angle 0^\circ \text{ A}$$

The second step is to convert the circuit to phasor form with the impedance of each element as shown in Figure 10.10-2*b*.

Now let us determine the phasor current $\mathbf{I}_1$, which is the component of current $\mathbf{I}$ due to the voltage source. We remove the current source, replacing it with an open circuit across the 10-Ω resistor. Then we may find the current $\mathbf{I}_1$ due to the first source as

$$\mathbf{I}_1 = \frac{\mathbf{V}_s}{5 + j\omega L + \mathbf{Z}_p} \tag{10.10-1}$$

where $\mathbf{Z}_p$ is the impedance of the capacitor and the 10-Ω resistance in parallel. Recall that $\omega = 10$ and $C = 10$ mF. Therefore, because $\mathbf{Z}_c = -j10\ \Omega$, we have

$$\mathbf{Z}_p = \frac{\mathbf{Z}_c R}{R + \mathbf{Z}_c} = \frac{(-j10)10}{10 - j10} = 5(1 - j)\ \Omega$$

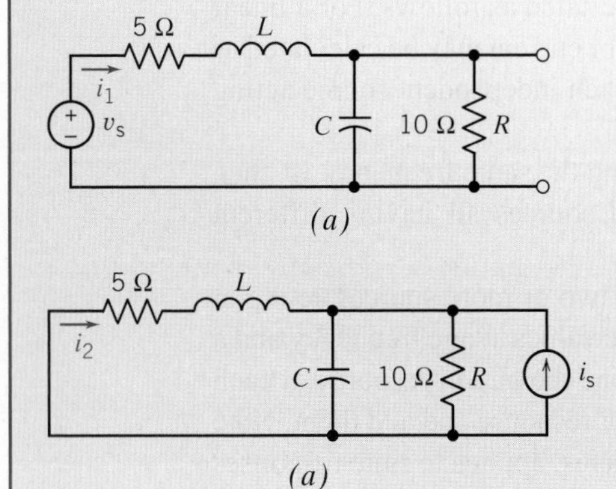

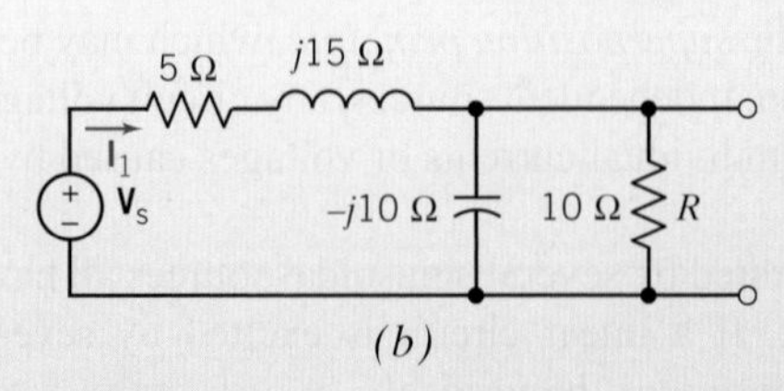

FIGURE 10.10-2 (*a*) Circuit for Example 10.10-1 for the voltage source acting alone. (*b*) Representation in the frequency domain.

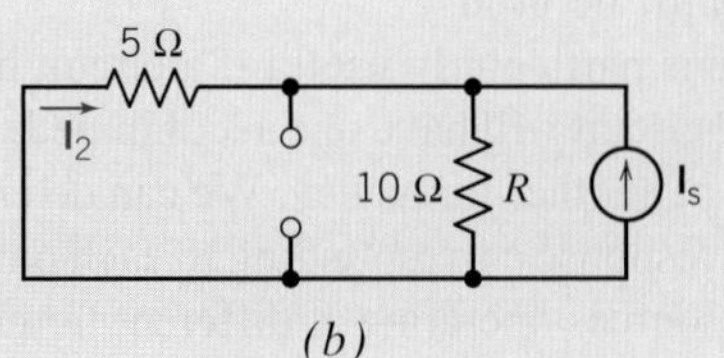

FIGURE 10.10-3 (*a*) Circuit for Example 10.10-1 for the current source acting alone. (*b*) Representation in the frequency domain.

Substituting $\mathbf{Z}_p$ and $\omega L = 15$ into Eq. 10.10-1, we have

$$\mathbf{I}_1 = \frac{10\angle 0^\circ}{5 + j15 + (5 - j5)} = \frac{10}{10 + j10} = \frac{10}{\sqrt{200}}\angle -45^\circ$$

Therefore, the time-domain current resulting from the voltage source is

$$i_1 = 0.71 \cos(10t - 45^\circ) \text{ A}$$

Now let us determine the phasor current $\mathbf{I}_2$ due to the current source. Setting the voltage source to zero results in a short circuit. Because $\omega = 0$ for the dc source, the capacitor impedance becomes an open circuit because $\mathbf{Z}_C = 1/j\omega C = \infty$. The inductor's impedance becomes a short circuit because $\mathbf{Z}_L = j\omega L = 0$. Hence, we obtain the circuit shown in Figure 10.10-3*b*. We see that we have returned to a familiar resistive circuit for a dc source. Then the response due to the current source is

$$\mathbf{I}_2 = -\frac{10}{15}(3) = -2 \text{ A}$$

Therefore, using the principle of superposition, the total steady-state current is $i = i_1 + i_2$ or

$$i = 0.71 \cos(10t - 45^\circ) - 2 \text{ A}$$

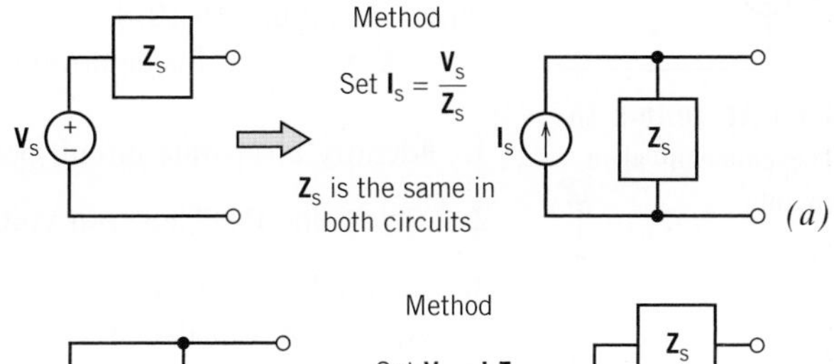

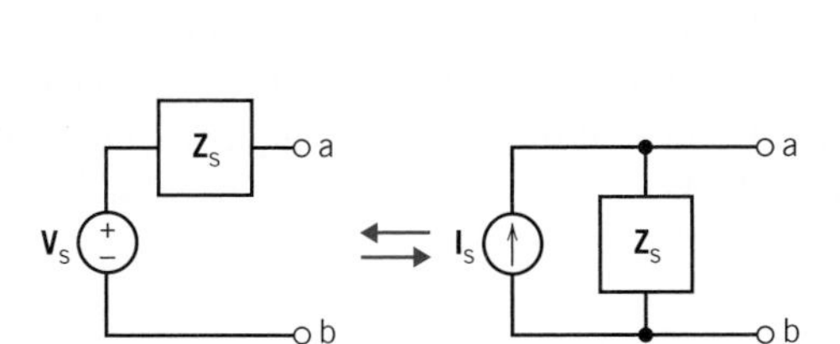

FIGURE 10.10-4 Two equivalent sources when $\mathbf{V}_s = \mathbf{Z}_s\mathbf{I}_s$.

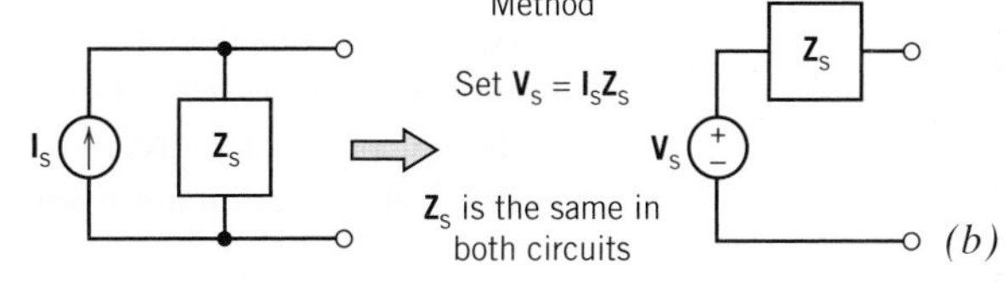

FIGURE 10.10-5 Method of source transformations. (*a*) Converting a voltage source to a current source. (*b*) Converting a current source to a voltage source.

Now let us consider the *source transformations* for frequency-domain (phasor) circuits. The techniques considered for resistive circuits discussed in Chapter 5 can readily be extended. The source transformation is concerned with transforming a voltage source and its associated series impedance to a current source and its associated parallel impedance, or vice versa, as shown in Figure 10.10-4. The method of transforming from one source to another source is summarized in Figure 10.10-5.

EXAMPLE 10.10-2 Source Transformations in AC Circuits

A circuit has a voltage source v_s in series with two elements, as shown in Figure 10.10-6. Determine the phasor equivalent current source form when $v_s = 10\cos(\omega t + 45^\circ)$ V and $\omega = 100$ rad/s.

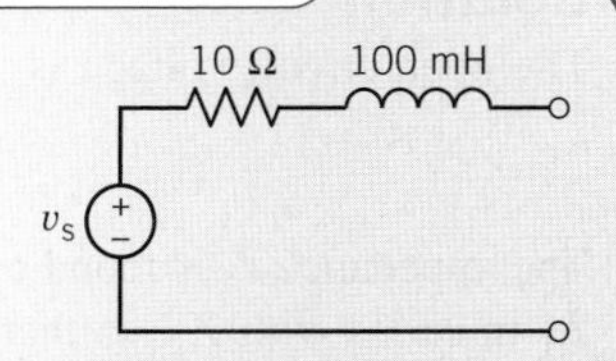

FIGURE 10.10-6 Circuit of Example 10.10-2.

Solution

First, we determine the equivalent current source as

$$\mathbf{I}_s = \frac{\mathbf{V}_s}{\mathbf{Z}_s}$$

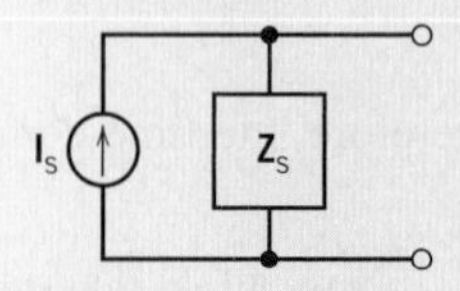

FIGURE 10.10-7 Circuit of Example 10.10-2 transformed to a current source where $\mathbf{Z}_s = 10 + j10\ \Omega$ and $\mathbf{I}_s = 1/\sqrt{2}$ A.

Because $\mathbf{Z}_s = 10 + j10$ and $\mathbf{V}_s = 10\angle 45°$, we obtain

$$\mathbf{I}_s = \frac{10\angle 45°}{\sqrt{200}\angle 45°} = \frac{10}{\sqrt{200}}\angle 0°\ \text{A}$$

The equivalent current source circuit is shown in Figure 10.10-7

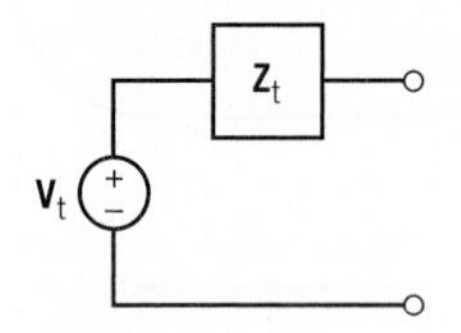

FIGURE 10.10-8 The Thévenin equivalent circuit.

Thévenin's and Norton's theorems apply to phasor current or voltages and impedances in the same way that they do for resistive circuits. The Thévenin theorem is used to obtain an equivalent circuit as discussed in Chapter 5. The Thévenin equivalent circuit is shown in Figure 10.10-8.

A procedure for determining the Thévenin equivalent circuit is as follows:

1. Identify a separate circuit portion of a total circuit.
2. Determine the Thévenin voltage $\mathbf{V}_t = \mathbf{V}_{oc}$, the open-circuit voltage at the terminals.
3. (a) Find $\mathbf{Z}_t$ by deactivating all the independent sources and reducing the circuit to an equivalent impedance; (b) if the circuit has one or more dependent sources, then either short-circuit the terminals and determine $\mathbf{I}_{sc}$ from which $\mathbf{Z}_t = \mathbf{V}_{oc}/\mathbf{I}_{sc}$; or (c) deactivate the independent sources, attach a current source at the terminals, and determine both $\mathbf{V}$ and $\mathbf{I}$ at the terminals from which $\mathbf{Z}_t = \mathbf{V}/\mathbf{I}$.

EXAMPLE 10.10-3 Thévenin Equivalent Circuit

Find the Thévenin equivalent circuit for the circuit shown in Figure 10.10-9 when $\mathbf{Z}_1 = 1 + j\Omega$ and $\mathbf{Z}_2 = -j1\ \Omega$

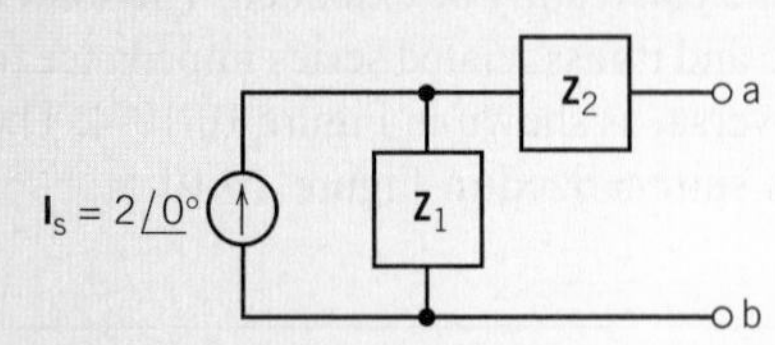

FIGURE 10.10-9 Circuit of Example 10.10-3.

Solution

The open-circuit voltage is

$$\mathbf{V}_{oc} = \mathbf{I}_s\mathbf{Z}_1 = \left(2\angle 0°\right)(1 + j) = 2\sqrt{2}\angle 45°\ \text{V}$$

The impedance $\mathbf{Z}_t$ is found by deactivating the current source by replacing it with an open circuit. Then we have $\mathbf{Z}_1$ in series with $\mathbf{Z}_2$, so that

$$\mathbf{Z}_t = \mathbf{Z}_1 + \mathbf{Z}_2 = (1 + j) - j = 1\ \Omega$$

EXAMPLE 10.10-4 Thévenin Equivalent Circuit

Find the Thévenin equivalent circuit of the frequency-domain circuit shown in Figure 10.10-10 in phasor form.

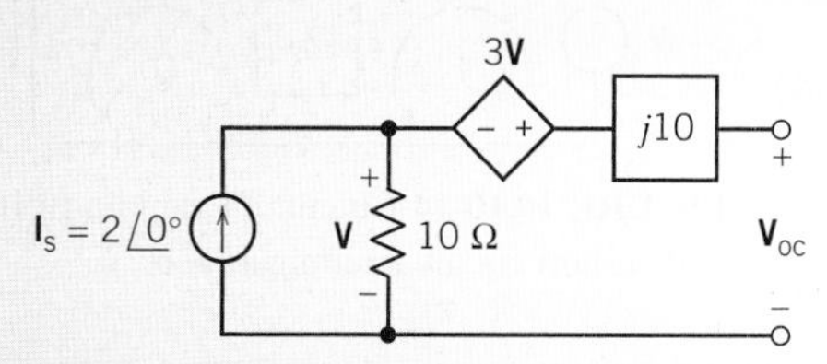

FIGURE 10.10-10 Circuit of Example 10.10-4.

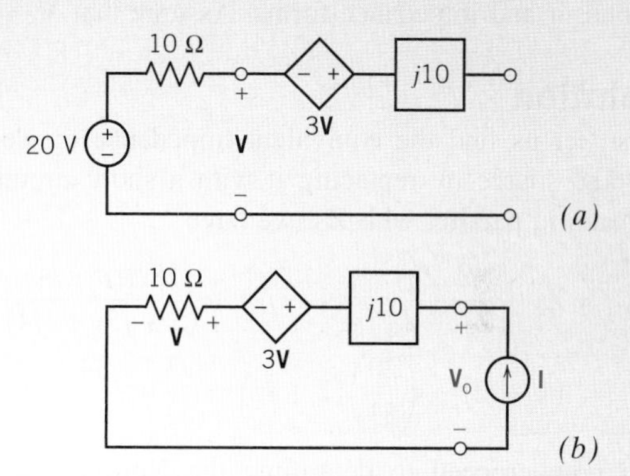

FIGURE 10.10-11 (*a*) Circuit of Example 10.10-4 with an open circuit at the output and the current source transformed to a voltage source. (*b*) Circuit with a test current source connected at the output terminal.

Solution

The Thévenin voltage $\mathbf{V}_t = \mathbf{V}_{oc}$, so we first determine $\mathbf{V}_{oc}$. Note that with the open circuit,

$$\mathbf{V} = 10\,\mathbf{I}_s = 20\,\angle 0°\text{ V}$$

Then, for the mesh on the right, using KVL, we have

$$\mathbf{V}_{oc} = 3\mathbf{V} + \mathbf{V} = 4\mathbf{V} = 80\,\angle 0°\text{ V}$$

Examining the circuit of Figure 10.10-10, we transform the current source and 10-Ω resistance to the voltage source and 10-Ω series resistance as shown in Figure 10.10-11*a*. When the voltage source is deactivated and a current source is connected at the terminals as shown in Figure 10.10-11*b*, KVL gives

$$\mathbf{V}_o = j10\mathbf{I} + 4\mathbf{V} = (j10 + 40)\,\mathbf{I}$$

Therefore,

$$\mathbf{Z}_t = 40 + j10\ \Omega$$

Now let us consider the procedure for finding the Norton equivalent circuit. The steps are similar to those used for the Thévenin equivalent because $\mathbf{Z}_t$ in series with the Thévenin voltage is equal to the Norton impedance in parallel with the Norton current source. The Norton equivalent circuit is shown in Figure 10.10-12.

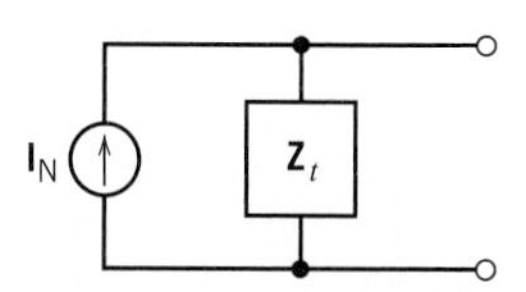

FIGURE 10.10-12 The Norton equivalent circuit expressed in terms of a phasor current and an impedance.

To determine the Norton circuit, we follow this procedure:

1. Identify a separate circuit portion of a total circuit.
2. The Norton current $\mathbf{I}_n$ is the current through a short circuit at the terminals, so $\mathbf{I}_n = \mathbf{I}_{sc}$.
3. Find $\mathbf{Z}_t$ by (a) deactivating all the independent sources and reducing the circuit to an equivalent impedance, *or* (b) if the circuit has one or more dependent sources, find the open-circuit voltage at the terminals, $\mathbf{V}_{oc}$, so that

$$\mathbf{Z}_t = \frac{\mathbf{V}_{oc}}{\mathbf{I}_{sc}}$$

EXAMPLE 10.10-5 Norton Equivalent Circuit

Find the Norton equivalent of the circuit shown in Figure 10.10-13 in phasor and impedance forms. Assume that $\mathbf{V}_s = 100\,\underline{/0^\circ}$ V.

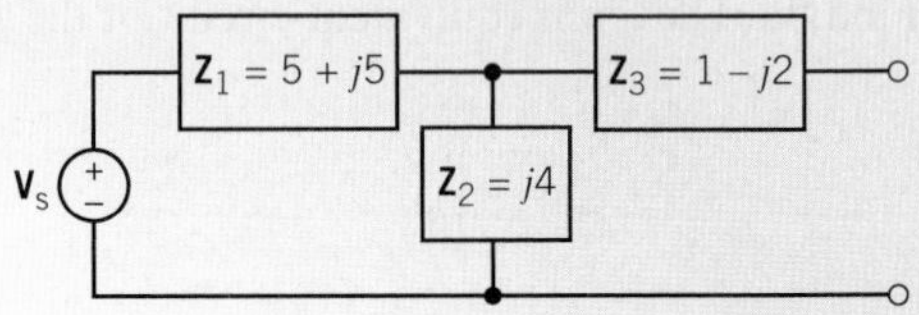

FIGURE 10.10-13 Circuit of Example 10.10-5.

Solution

First, let us find the equivalent impedance by deactivating the voltage source by replacing it with a short circuit. Because $\mathbf{Z}_1$ appears in parallel with $\mathbf{Z}_2$, we have

$$\mathbf{Z}_t = \mathbf{Z}_3 + \frac{\mathbf{Z}_1\mathbf{Z}_2}{\mathbf{Z}_1 + \mathbf{Z}_2} = (1 - j2) + \frac{(5 + j5)(j4)}{(5 + j5) + (j4)}$$

$$= (1 - j2) + \left(\frac{20}{53}\right)(2 + j7) = \frac{93}{53} + j\frac{34}{53} = \left(\frac{1}{53}\right)(93 + j34)$$

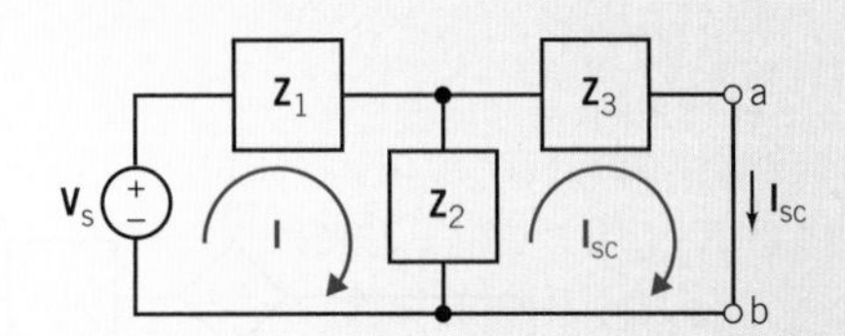

FIGURE 10.10-14 Circuit of Example 10.10-5 with a short circuit at terminals a–b.

We now proceed to determine the Norton equivalent current source by determining the current flowing through a short circuit connected at terminals a–b, as shown in Figure 10.10-14

We will use mesh currents to find $\mathbf{I}_{sc}$ as shown in Figure 10.10-14. The two mesh KVL equations are

$$\text{mesh 1:}\quad (\mathbf{Z}_1 + \mathbf{Z}_2)\mathbf{I} + (-\mathbf{Z}_2)\mathbf{I}_{sc} = \mathbf{V}_s$$

$$\text{mesh 2:}\quad (-\mathbf{Z}_2)\mathbf{I} + (\mathbf{Z}_2 + \mathbf{Z}_3)\mathbf{I}_{sc} = 0$$

Using Cramer's rule, we find $\mathbf{I}_N = \mathbf{I}_{sc}$ as follows:

$$\mathbf{I}_{sc} = \frac{\mathbf{Z}_2\mathbf{V}_s}{(\mathbf{Z}_1 + \mathbf{Z}_2)(\mathbf{Z}_2 + \mathbf{Z}_3) - \mathbf{Z}_2^2} = \frac{(j4)100}{(5 + j9)(1 + j2) - (-16)}$$

$$= \frac{j400}{3 + j19} = \frac{400}{370}(19 + 3j)\ \text{A}$$

10.11 PHASOR DIAGRAMS

Phasors representing the voltage or current of a circuit are time quantities transformed or converted into the frequency domain. Phasors are complex numbers and can be portrayed in a complex plane. The relationship of phasors on a complex plane is called a *phasor diagram*.

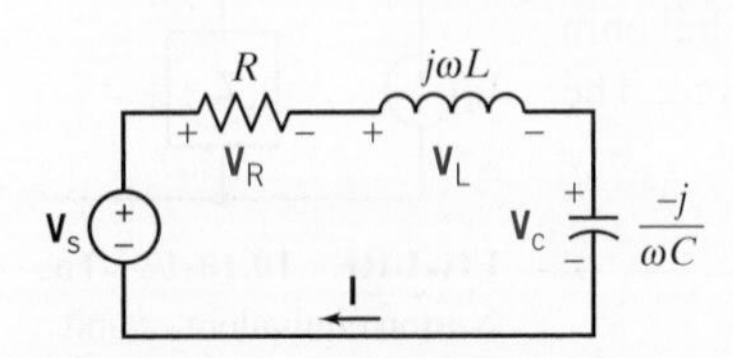

FIGURE 10.11-1 An *RLC* circuit.

Let us consider an *RLC* series circuit as shown in Figure 10.11-1. The impedance of each element is also identified in the diagram. Because the current flows through all elements and is common to all, we take $\mathbf{I}$ as the reference phasor.

$$\mathbf{I} = I\,\underline{/0^\circ}$$

Then the voltage phasors are

$$\mathbf{V}_R = R\mathbf{I} = RI\,\underline{/0^\circ} \tag{10.11-1}$$

$$\mathbf{V}_L = j\omega L\mathbf{I} = \omega LI\,\underline{/90^\circ} \tag{10.11-2}$$

$$\mathbf{V}_c = \frac{-j\mathbf{I}}{\omega C} = \frac{I}{\omega C}\,\underline{/-90^\circ} \tag{10.11-3}$$

These phasors are shown in the phasor diagram of Figure 10.11-2. Note that KVL for this circuit requires that

$$\mathbf{V}_s = \mathbf{V}_R + \mathbf{V}_L + \mathbf{V}_c$$

A **phasor diagram** is a graphical representation of phasors and their relationship on the complex plane.

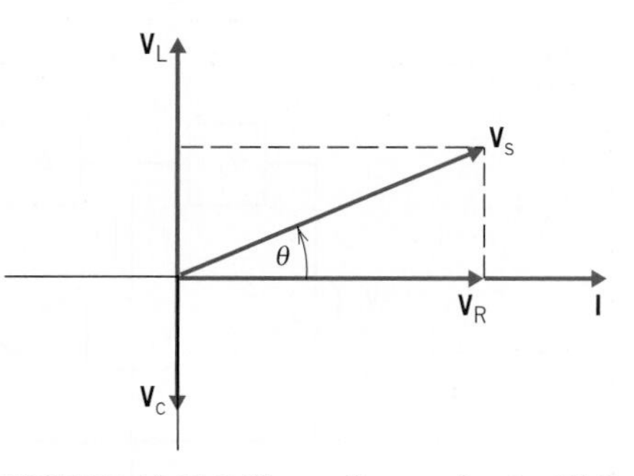

FIGURE 10.11-2 Phasor diagram for the *RLC* circuit of Figure 10.11-1.

The current **I** and the voltage across the resistor are in phase. The inductor voltage leads the current by 90°, and the capacitor voltage lags the current by 90°. For a given L and C, there will be a frequency ω that results in

$$|\mathbf{V}_L| = |\mathbf{V}_c|$$

Referring to Eqs. 10.11-2 and 10.11-3, this equality of voltage magnitudes occurs when

$$\omega L = \frac{1}{\omega C}$$

or

$$\omega^2 = \frac{1}{LC}$$

When $\omega^2 = 1/LC$, the magnitudes of the inductor voltage and capacitor voltage are equal. Because they are out of phase by 180°, they cancel, and the resulting condition is

$$\mathbf{V}_s = \mathbf{V}_R$$

and then $\mathbf{V}_s$ is in phase with **I**. This condition is called *resonance*.

EXERCISE 10.11-1 Consider the *RLC* series circuit of Figure 10.11-1 when $L = 1$ mH and $C = 1$ mF. Find the frequency ω when the current, source voltage, and $\mathbf{V}_R$ are all in phase.

Answer: $\omega = 1000$ rad/s

EXERCISE 10.11-2 Draw the phasor diagram for the circuit of Figure E 10.11-2 when $\mathbf{V} = V \angle 0°$. Show each current on the diagram.

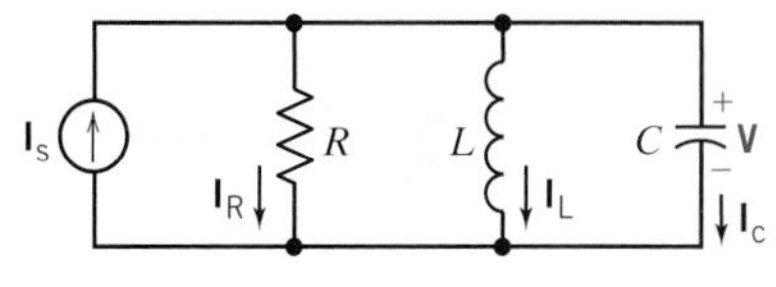

FIGURE E 10.11-2

10.12 PHASOR CIRCUITS AND THE OPERATIONAL AMPLIFIER

The discussion in the prior sections considered the behavior of operational amplifiers and their associated circuits in the time domain. In this section, we consider the behavior of operational amplifiers and associated *RLC* circuits in the frequency domain, using phasors.

Figure 10.12-1 shows two frequently used operational amplifier circuits, the inverting amplifier and the noninverting amplifier. These circuits are represented using impedances and phasors. This representation is appropriate when the input is sinusoidal and the circuit is at steady state. $\mathbf{V}_s$ is the phasor corresponding to a sinusoidal input voltage, and $\mathbf{V}_o$ is the phasor representing the resulting sinusoidal output voltage. Both circuits involve two impedances, $\mathbf{Z}_1$ and $\mathbf{Z}_2$.

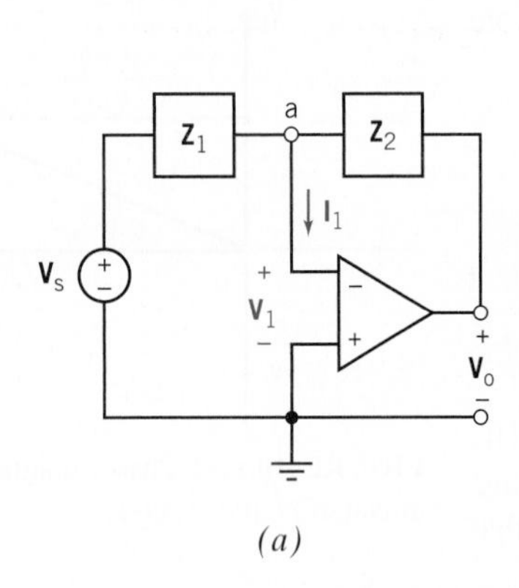

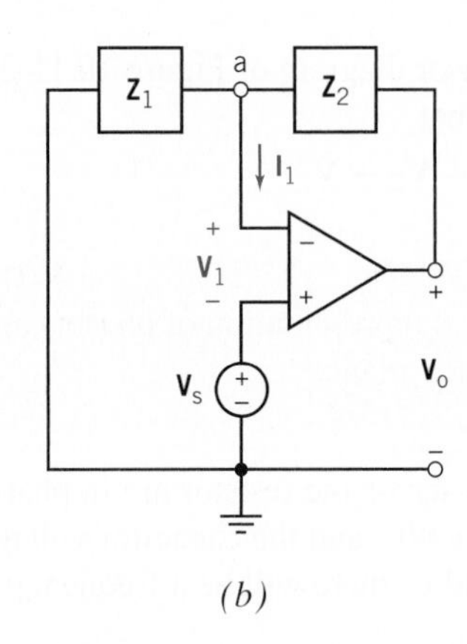

FIGURE 10.12-1 (*a*) An inverting amplifier and (*b*) a noninverting amplifier.

Now let us determine the ratio of output-to-input voltage, $\mathbf{V}_o/\mathbf{V}_s$, for the inverting amplifier shown in Figure 10.12-1*a*. This circuit can be analyzed by writing the node equation at node a as

$$\frac{\mathbf{V}_s - \mathbf{V}_1}{\mathbf{Z}_1} + \frac{\mathbf{V}_o - \mathbf{V}_1}{\mathbf{Z}_2} - \mathbf{I}_1 = 0 \tag{10.12-1}$$

When the operational amplifier is ideal, $\mathbf{V}_1$ and $\mathbf{I}_1$ are both 0. Then,

$$\frac{\mathbf{V}_s}{\mathbf{Z}_1} + \frac{\mathbf{V}_o}{\mathbf{Z}_2} = 0 \tag{10.12-2}$$

Finally,

$$\frac{\mathbf{V}_o}{\mathbf{V}_s} = -\frac{\mathbf{Z}_2}{\mathbf{Z}_1} \tag{10.12-3}$$

Next, we will determine the ratio of output-to-input voltage, $\mathbf{V}_o/\mathbf{V}_s$, for the noninverting amplifier shown in Figure 10.12-1*b*. This circuit can be analyzed by writing the node equation at node a as

$$\frac{(\mathbf{V}_s + \mathbf{V}_1)}{\mathbf{Z}_1} - \frac{\mathbf{V}_o - (\mathbf{V}_s + \mathbf{V}_1)}{\mathbf{Z}_2} + \mathbf{I}_1 = 0 \tag{10.12-4}$$

When the operational amplifier is ideal, $\mathbf{V}_1$ and $\mathbf{I}_1$ are both 0. Then,

$$\frac{\mathbf{V}_s}{\mathbf{Z}_1} - \frac{\mathbf{V}_o - \mathbf{V}_s}{\mathbf{Z}_2} = 0$$

Finally,

$$\frac{\mathbf{V}_o}{\mathbf{V}_s} = \frac{\mathbf{Z}_1 + \mathbf{Z}_2}{\mathbf{Z}_1} \tag{10.12-5}$$

Typically, impedances $\mathbf{Z}_1$ and $\mathbf{Z}_2$ are obtained using only resistors and capacitors. Of course, in theory, we could use inductors, but their cost and size relative to capacitors result in little use of inductors with operational amplifiers.

An example of the inverting amplifier is shown in Figure 10.12-2, The impedance $\mathbf{Z}_n$, where n is equal to 1 or 2, is a parallel R_nC_n impedance so that

$$\mathbf{Z}_n = \frac{R_n \dfrac{1}{j\omega C_n}}{R_n + \dfrac{1}{j\omega C_n}} = \frac{R_n}{1 + j\omega C_n R_n} \tag{10.12-6}$$

Using Eqs. 10.12-3 and 10.12-6, one may obtain the ratio $\mathbf{V}_o/\mathbf{V}_s$.

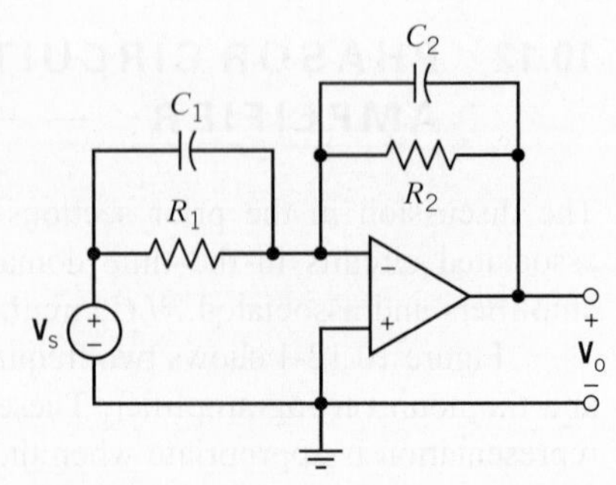

FIGURE 10.12-2 Operational amplifier with two *RC* circuits connected.

EXAMPLE 10.12-1 AC Amplifier

Find the ratio $\mathbf{V}_o/\mathbf{V}_s$ for the circuit of Figure 10.12-2 when $R_1 = 1\ \text{k}\Omega$, $R_2 = 10\ \text{k}\Omega$, $C_1 = 0$, and $C_2 = 0.1\ \mu\text{F}$ for $\omega = 1000$ rad/s.

Solution

The circuit of Figure 10.12-2 is an example of the inverting amplifier shown in Figure 10.12-1*a*. Using Eqs. 10.12-3 and 10.12-6, we obtain

$$\frac{\mathbf{V}_o}{\mathbf{V}_s} = -\frac{\mathbf{Z}_1}{\mathbf{Z}_2} = -\frac{\dfrac{R_2}{1 + j\omega C_2 R_2}}{\dfrac{R_1}{1 + j\omega C_1 R_1}} = -\frac{R_2(1 + j\omega C_1 R_1)}{R_1(1 + j\omega C_2 R_2)}$$

Substituting the given values of R_1, R_2, C_1, C_2, and ω gives

$$\frac{\mathbf{V}_o}{\mathbf{V}_s} = -\frac{10^4(1 + j10^0(0)10^3)}{10^3(1 + j10^3(0.1 \times 10^{-6})10^4)} = -\frac{10}{1 + j} = 7.07\angle 135^\circ$$

EXERCISE 10.12-1 Find the ratio $\mathbf{V}_o/\mathbf{V}_s$ for the circuit shown in Figure 10.12-2 when $R_1 = R_2 = 1\ \text{k}\Omega$, $C_2 = 0$, $C_1 = 1\ \mu\text{F}$, and $\omega = 1000$ rad/s.

Answer: $\mathbf{V}_o/\mathbf{V}_s = -1 - j$

10.13 THE COMPLETE RESPONSE

Next, we consider circuits with sinusoidal inputs that are subject to abrupt changes, as when a switch opens or closes. To find the complete response of such circuits, we:

- Represent the circuit by a differential equation.
- Find the general solution of the homogeneous differential equation. This solution is the natural response, $v_n(t)$. The natural response will contain unknown constants that will be evaluated later.
- Find a particular solution of the differential equation. This solution is the forced response, $v_f(t)$.
- Represent the response of the circuit as $v(t) = v_n(t) + v_f(t)$.
- Use the initial conditions, for example, the initial values of the currents in inductors and the voltages across capacitors to evaluate the unknown constants.

Consider the circuit shown in Figure 10.13-1. Before time $t = 0$, this circuit is at steady state, so all its voltages and currents are sinusoidal with a frequency of 5 rad/s. At time $t = 0$, the switch closes, disturbing the circuit. Immediately after $t = 0$, the currents and voltages are not sinusoidal. Eventually, the disturbance dies out and the circuit is again at steady state (most likely a different steady state). Once again, the currents and voltages are all sinusoidal with a frequency of 5 rad/s.

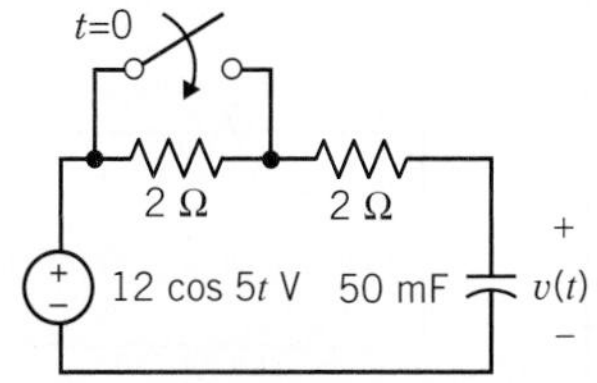

FIGURE 10.13-1 The circuit considered in Example 10.13-1.

Two different steady-state responses are used to find the complete response of this circuit. The steady-state response before the switch closes is used to determine the initial condition. The steady-state response after the switch closes is used as the particular solution of the differential equation representing the circuit.

EXAMPLE 10.13-1 Complete Response

Determine $v(t)$, the voltage across the capacitor in Figure 10.13-1, both before and after the switch closes.

Solution

Step 1: For $t < 0$, the switch is open and the circuit is at steady state.

The open switch acts like an open circuit, so the two 2-Ω resistors are connected in series. Replacing the series resistors with an equivalent resistor produces the circuit shown in Figure 10.13-2*a*. Next, we use impedances and phasors to represent the circuit in the frequency domain as shown in Figure 10.13-2*b*.

Using voltage division in the frequency domain gives

$$\mathbf{V}(\omega) = \left(\frac{-j4}{4-j4}\right)\left(12\angle 0^\circ\right) = \frac{48\angle -90^\circ}{5.66\angle -45^\circ} = 8.485\angle -45^\circ \text{ V}$$

In the time domain,

$$v(t) = 8.485 \cos\,(5t - 45^\circ) \text{ V}$$

Immediately before the switch closes, the capacitor voltage is

$$v(0-) = \lim_{t \to 0-} v(t) = 8.485 \cos\,(0 - 45^\circ) = 6 \text{ V}$$

The capacitor voltage is continuous, so the capacitor voltage immediately after the switch closes is the same as immediately before the switch closes. That is,

$$v(0+) = v(0-) = 6 \text{ V}$$

Step 2: For $t > 0$, the switch is closed. Eventually, the circuit will reach a new steady state.

The closed switch acts like a short circuit. A short circuit in parallel with a resistor is equivalent to a short circuit, so we have the circuit shown in Figure 10.13-3*a*. The steady-state response of the circuit can be obtained by representing the circuit in the frequency domain as shown in Figure 10.13-3*b*.

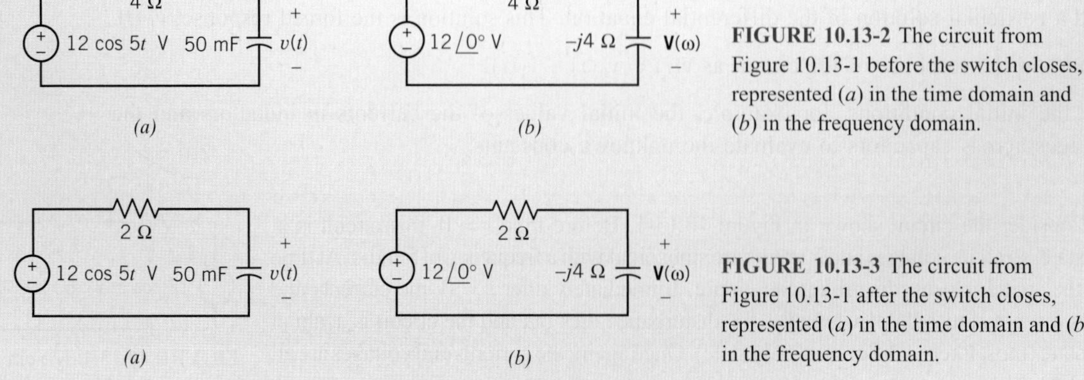

FIGURE 10.13-2 The circuit from Figure 10.13-1 before the switch closes, represented (*a*) in the time domain and (*b*) in the frequency domain.

FIGURE 10.13-3 The circuit from Figure 10.13-1 after the switch closes, represented (*a*) in the time domain and (*b*) in the frequency domain.

Using voltage division in the frequency domain gives

$$\mathbf{V}(\omega) = \left(\frac{-j4}{2-j4}\right)\left(12\,\underline{/0^\circ}\right) = \frac{48\,\underline{/-90^\circ}}{4.47\,\underline{/-63.4^\circ}} = 10.74\,\underline{/-26.6^\circ}\text{ V}$$

In the time domain,

$$v(t) = 10.74\cos(5t - 26.6^\circ)\text{ V}$$

Step 3: Immediately after $t = 0$, the switch is closed but the circuit is not at steady state. We must find the complete response of a first-order circuit.

In Figure 10.13-2*a*, the capacitor is connected to a series voltage source and resistor, that is, a Thévenin equivalent circuit. We can identify R_t and v_{oc} as shown in Figure 10.13-4.

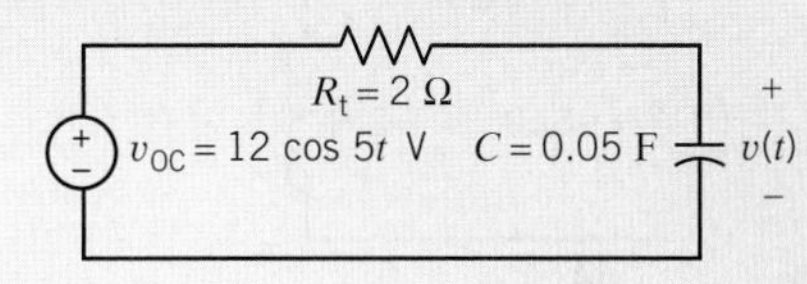

FIGURE 10.13-4 Identifying R_t and v_{oc} in Figure 10.13-2*a*.

Consequently, the time constant of the circuit is

$$\tau = R_t C = 2 \times 0.05 = 0.1 \quad 1/\text{s}$$

The natural response of the circuit is

$$v_n(t) = Ke^{-10t}$$

The steady-state response for $t > 0$ can be used as the forced response, so

$$v_f(t) = 10.74\cos(5t - 26.6^\circ)\text{ V}$$

The complete response is

$$v(t) = v_n(t) + v_f(t) = Ke^{-10t} + 10.74\cos(5t - 26.6^\circ)$$

The constant, K, is evaluated using the initial capacitor voltage, $v(0+)$:

$$6 = v(0+) = Ke^{-0} + 10.74\cos(0 - 26.6^\circ) = K + 9.6$$

Thus, $K = -3.6$ and

$$v(t) = -3.6e^{-10t} + 10.74\cos(5t - 26.6^\circ)\text{ V}$$

Step 4: Summarize the results.

The capacitor voltage is

$$v(t) = \begin{cases} 8.485\cos(5t - 45^\circ)\text{ V} & \text{for } t \le 0 \\ -3.6e^{-10t} + 10.74\cos(5t - 26.6^\circ)\text{ V} & \text{for } t \ge 0 \end{cases}$$

Figure 10.13-5 shows the capacitor voltage as a function of time:

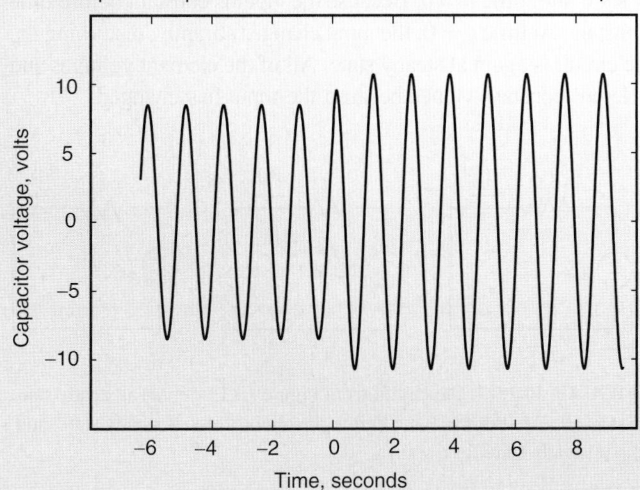

FIGURE 10.13-5 The complete response, plotted using MATLAB.

EXAMPLE 10.13-2 Responses of Various Types of Circuits

The input to each of the circuits shown in Figure 10.13-6 is the voltage source voltage. The output of each circuit is the current $i(t)$. Determine the output of each of the circuits.

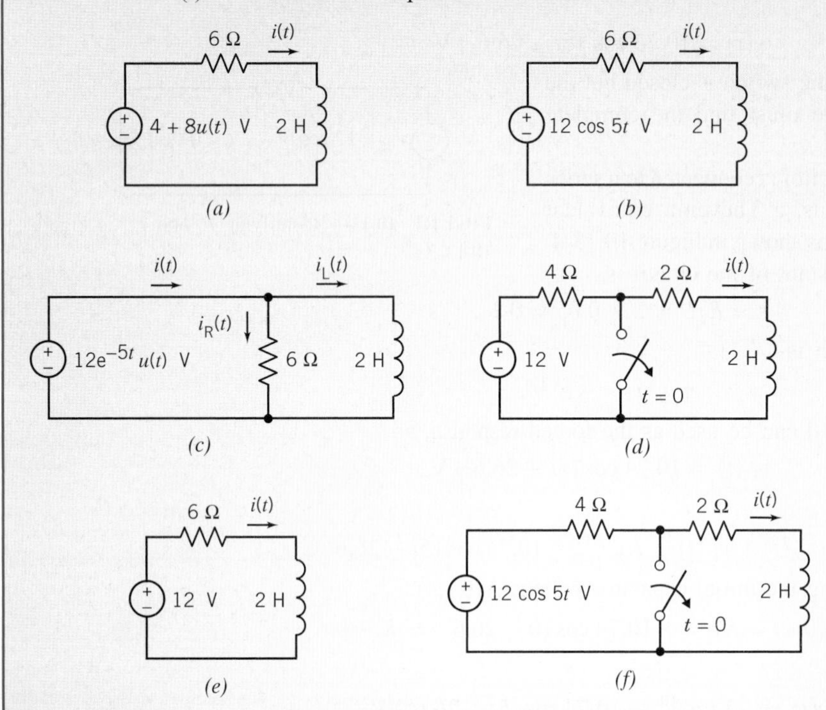

FIGURE 10.13-6 Six circuits considered in Example 10.13-2.

Solution

In this example, we consider similar circuits in contrasting situations. In some cases, the circuit changes abruptly at time $t = 0$. Consequently, the circuit is not at steady state and we seek a complete response—consisting of both a steady-state part and a transient part. In other cases, there is no abrupt change and so no transient part of the response. We seek only the steady-state response. In one case, the input provides the inductor voltage directly, and we can determine the response using the constitutive equation of the inductor.

Case 1: The circuit in Figure 10.13-6*a* will be at steady state until time $t = 0$. Because the input is constant before time $t = 0$, all of the element voltages and currents will be constant. At time $t = 0$, the input changes abruptly, disturbing the steady state. Eventually the disturbance dies out and the circuit is again at steady state. All of the element voltages and currents will again be constant, but they will have different constant values because the input has changed.

The three stages can be illustrated as shown in Figure 10.13-7. Figure 10.13-7*a* represents the circuit for $t < 0$. The source voltage is constant and the circuit is at steady state, so the inductor acts like a short circuit. The inductor current is

$$i(t) = \frac{4}{6} = \frac{2}{3}\ \text{A}$$

In particular, immediately before $t = 0$, $i(0-) = 0.667$ A. The current in an inductor is continuous, so

$$i(0+) = i(0-) = 0.667\ \text{A}$$

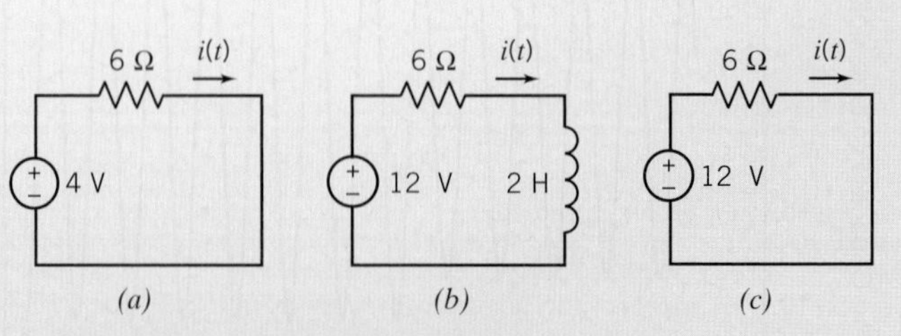

FIGURE 10.13-7 The circuit from Figure 10.13-6*a*, (*a*) at steady state for $t < 0$, (*b*) after $t = 0$ but before the circuit reaches steady state, and (*c*) at steady state for $t > 0$.

Figure 10.13-7*b* represents the circuit immediately after $t = 0$. The input is constant but the circuit is not at steady state, so the inductor does not act like a short circuit. The part of the circuit that is connected to the inductor has the form of a Thévenin equivalent circuit, so we recognize that

$$R_t = 6\,\Omega \quad \text{and} \quad v_{oc} = 12\text{ V}$$

Consequently,

$$i_{sc} = \frac{12}{6} = 2\text{ A}$$

The time constant of the circuit is

$$\tau = \frac{L}{R_t} = \frac{2}{6} = \frac{1}{3}$$

Finally,

$$i(t) = i_{sc} + (i(0+) - i_{sc})e^{-t/\tau} = 2 + (0.667 - 2)e^{-3t} = 2 - 1.33e^{-3t}\text{ A}$$

As t increases, the exponential part of $i(t)$ gets smaller. When $t = 5\tau = 1.667$ s,

$$i(t) = 2 - 1.33e^{-3(1.667)} = 2 - 0.009 \approx 2\text{ A}$$

The exponential part of $i(t)$ has become negligible, so we recognize that the circuit is again at steady state and that the new steady-state current is $i(t) = 2$ A.

Figure 10.13-7*c* represents the circuit after the disturbance has died out and the circuit has reached steady state, that is, when $t > 5\tau$. The source voltage is constant and the circuit is at steady state, so the inductor acts like a short circuit. As expected, the inductor current is 2 A.

Case 2: The circuit in Figure 10.13-6*b* does not contain a switch and the input does not change abruptly, so we expect the circuit to be at steady state. The input is sinusoidal at a frequency of 5 rad/s, so all of the element currents and voltages will be sinusoidal at a frequency of 5 rad/s. We can find the steady-state response by representing the circuit in the frequency domain, using impedances and phasors as shown in Figure 10.13-8.

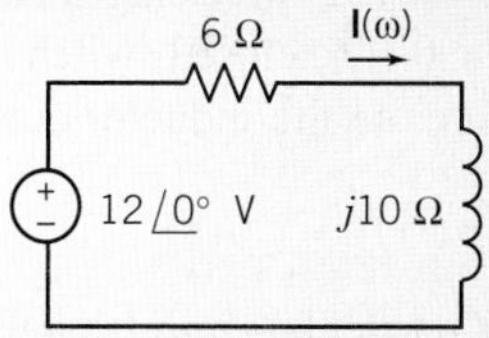

FIGURE 10.13-8 The circuit in Figure 10.13-6*b* is represented in the frequency domain.

Ohm's law gives

$$\mathbf{I}(\omega) = \frac{12\angle 0^\circ}{6 + j10} = \frac{12\angle 0^\circ}{11.66\angle 59^\circ} = 1.03\angle -59^\circ\text{ A}$$

The corresponding current in the time domain is

$$i(t) = 1.03\cos(5t - 59^\circ)\text{ A}$$

Case 3: The voltage source, resistor, and inductor in the circuit in Figure 10.13-6*c* are connected in parallel. The element voltage of the resistor and inductor are each equal to the voltage source voltage. The current in the resistor is given by Ohm's law to be

$$i_R(t) = \frac{12e^{-5t}}{6} = 2e^{-5t}\text{ A}$$

The current in the inductor is

$$\begin{aligned} i_L(t) = \frac{1}{L}\int_0^t v(\tau)\,d\tau + i_L(0) &= \frac{1}{2}\int_0^t 12e^{-5\tau}d\tau + i_L(0) \\ &= \frac{12}{2(-5)}(e^{-5t} - 1) + i_L(0) = -1.2e^{-5t} + 1.2 + i_L(0) \end{aligned}$$

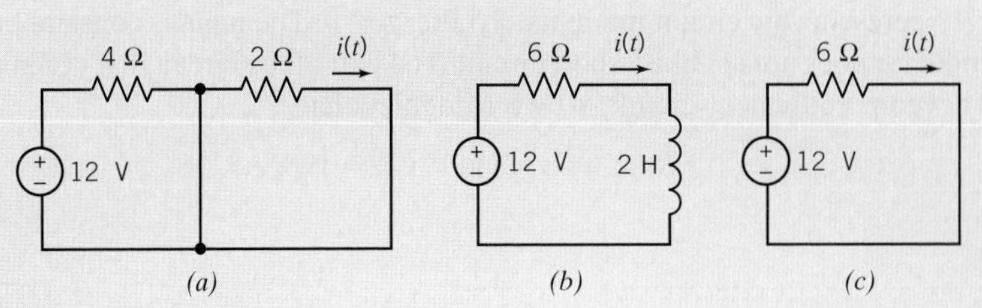

FIGURE 10.13-9 The circuit from Figure 10.13-6*d*, (*a*) steady state for $t < 0$, (*b*) after $t = 0$ but before the circuit reaches steady state, and (*c*) at steady state for $t > 0$.

Finally, using KCL gives

$$i(t) = i_R(t) + i_L(t) = 2e^{-5t} - 1.2e^{-5t} + 1.2 + i_L(0) = 0.8e^{-5t} + 1.2 + i_L(0)$$

Before time $t = 0$, the voltage of source voltage is zero. If the circuit is at steady state, $i_L(0) = 0$. Then

$$i(t) = 0.8e^{-5t} + 1.2\text{ A}$$

Case 4: The circuit in Figure 10.13-6*d* will be at steady state until the switch opens at time $t = 0$. Because the source voltage is constant, all of the element voltages and currents will be constant. At time $t = 0$, the switch opens, disturbing the steady state. Eventually the disturbance dies out and the circuit is again at steady state. All of the element voltages and currents will be constant, but they will have different constant values because the circuit has changed.

The three stages can be illustrated as shown in Figure 10.13-9. Figure 10.13-9*a* represents the circuit for $t < 0$. The closed switch is represented as a short circuit. The source voltage is constant and the circuit is at steady state, so the inductor acts like a short circuit. The inductor current is

$$i(t) = 0\text{ A}$$

In particular, immediately before $t = 0$, $i(0-) = 0$ A. The current in an inductor is continuous, so

$$i(0+) = i(0-) = 0\text{ A}$$

Figure 10.13-9*b* represents the circuit immediately after $t = 0$. The input is constant but the circuit is not at steady state, so the inductor does not act like a short circuit. The part of the circuit that is connected to the inductor has the form of a Thévenin equivalent circuit, so we recognize that

$$R_t = 6\,\Omega \quad \text{and} \quad v_{oc} = 12\text{ V}$$

Consequently,

$$i_{sc} = \frac{12}{6} = 2\text{ A}$$

The time constant of the circuit is

$$\tau = \frac{L}{R_t} = \frac{2}{6} = \frac{1}{3}$$

Finally,

$$i(t) = i_{sc} + (i(0+) - i_{sc})e^{-t/\tau} = 2 + (0 - 2)e^{-3t} = 2 - 2e^{-3t}\text{ A}$$

As t increases, the exponential part of $i(t)$ gets smaller. When $t = 5\tau = 1.667$ s,

$$i(t) = 2 - 2e^{-3(1.667)} = 2 - 0.013 \approx 2\text{ A}$$

The exponential part of $i(t)$ has become negligible, so we recognize that the circuit is again at steady state and that the steady state current is $i(t) = 2$ A.

Figure 10.13-9*c* represents the circuit after the disturbance has died out and the circuit has reached steady state, that is, when $t > 5\tau$. The source voltage is constant and the circuit is at steady state, so the inductor acts like a short circuit. As expected, the inductor current is 2 A.

Case 5: The circuit in Figure 10.13-6*e* does not contain a switch and the input does not change abruptly, so we expect the circuit to be at steady state. Because the source voltage is constant, all of the element voltages and currents will be constant. Because the source voltage is constant and the circuit is at steady state, the inductor acts like a short circuit. (We've encountered this circuit twice before in this example, after the disturbance died out in cases 2 and 4.) The current is given by

$$i(t) = \frac{12}{6} = 2 \text{ A}$$

Case 6: We expect that the circuit in Figure 10.13-6*f* will be at steady state before the switch opens. As before, opening the switch will change the circuit and disturb the steady state. Eventually, the disturbance will die out and the circuit will again be at steady state. We will see that the steady-state current is constant before the switch opens and sinusoidal after the switch opens.

Figure 10.13-10*a* shows the circuit before the switch opens. Applying KVL gives

$$2i(t) + 2\frac{d}{dt}i(t) = 0$$

Consequently, the inductor current is $i(t) = 0$ before the switch opens. The current in an inductor is continuous, so

$$i(0+) = i(0-) = 0 \text{ A}$$

Figure 10.13-10*b* represents the circuit after the switch opens. We can determine the inductor current by adding the natural response to the forced response and then using the initial condition to evaluate the constant in the natural response.

First, we find the natural response. The part of the circuit that is connected to the inductor has the form of the Thévenin equivalent circuit, so we recognize that

$$R_t = 6\ \Omega$$

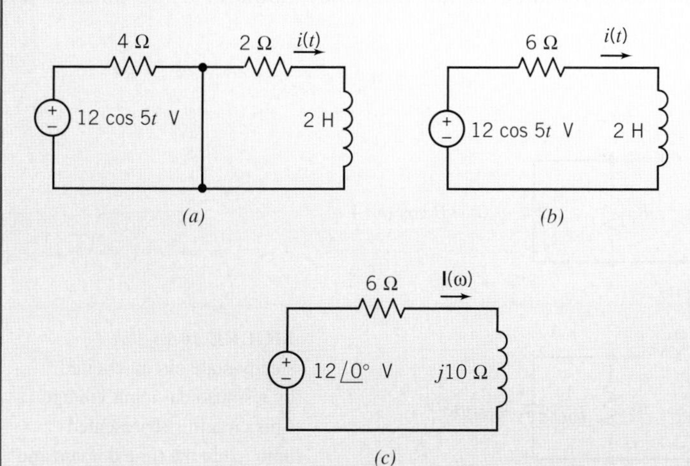

FIGURE 10.13-10 The circuit from Figure 10.13-6*f*, (*a*) before the switch opens, (*b*) after the switch opens, and (*c*) the steady-state circuit for $t > 0$ represented in the frequency domain.

The time constant of the circuit is

$$\tau = \frac{L}{R_t} = \frac{2}{6} = \frac{1}{3}$$

The natural response is

$$i_n(t) = Ke^{-3t} \text{ A}$$

We can use the steady-state response as the forced response. As in case 2, we obtain the steady-state response by representing the circuit in the frequency as shown in Figure 10.13-10*c*. As before, we find $\mathbf{I}(\omega) = 1.03 \angle -59^\circ$ A. The forced response is

$$i_f(t) = 1.03 \cos(5t - 59^\circ) \text{ A}$$

Then,
$$i(t) = i_n(t) + i_f(t) = Ke^{-3t} + 1.03 \cos(5t - 59^\circ) \text{ A}.$$

At $t = 0$,

$$i(0) = Ke^{-0} + 1.03 \cos(-59^\circ) = K + 0.53$$

so
$$i(t) = -0.53e^{-3t} + 1.03 \cos(5t - 59^\circ) \text{ A}$$

10.14 USING MATLAB FOR ANALYSIS OF STEADY-STATE CIRCUITS WITH SINUSOIDAL INPUTS

Analysis of steady-state linear circuits with sinusoidal inputs using phasors and impedances requires complex arithmetic. MATLAB can be used to reduce the effort required to do this complex arithmetic. Consider the circuit shown in Figure 10.14-1*a*. The input to this circuit, $v_s(t)$, is a sinusoidal voltage. At steady state, the output, $v_o(t)$, will also be a sinusoidal voltage as shown in Figure 10.14-1*a*. This circuit can be represented in the frequency domain, using phasors and

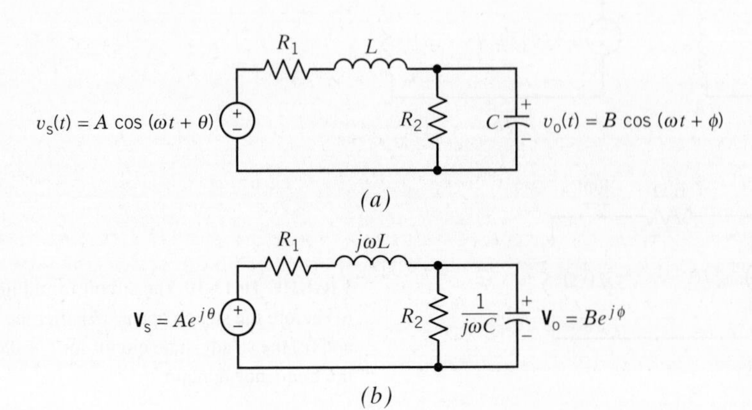

FIGURE 10.14-1 A steady-state circuit excited by a sinusoidal input voltage. This circuit is represented both (*a*) in the time domain and (*b*) in the frequency domain.

impedances as shown in Figure 10.14-1*b*. Analysis of this circuit proceeds as follows. Let $\mathbf{Z}_1$ denote the impedance of the series combination of R_1 and $j\omega L$. That is,

$$\mathbf{Z}_1 = R_1 + j\omega L \tag{10.14-1}$$

Next, let $\mathbf{Y}_2$ denote the admittance of the parallel combination of R_2 and $1/j\omega C$. That is,

$$\mathbf{Y}_2 = \frac{1}{R_2} + j\omega C \tag{10.14-2}$$

Let $\mathbf{Z}_2$ denote the corresponding impedance, that is,

$$\mathbf{Z}_2 = \frac{1}{\mathbf{Y}_2} \tag{10.14-3}$$

```
%-------------------------------------------------
%        Describe the input voltage source.
%-------------------------------------------------
w = 2;
A = 12;
theta = (pi/180)*60;
Vs = A*exp(j*theta)
%-------------------------------------------------
%Describe the resistors, inductor and capacitor.
%-------------------------------------------------
R1 = 6;
L = 4;
R2 = 12;
C = 1/24;
%-------------------------------------------------
% Calculate the equivalent impedances of the
%  series resistor and inductor and of the
%        parallel resistor and capacitor
%-------------------------------------------------
Z1 = R1 + j*w*L               % Eqn 10.14-1
Y2 = 1/R2 + j*w*C;            % Eqn 10.14-2
Z2 = 1 / Y2                   % Eqn 10.14-3
%-------------------------------------------------
%  Calculate the phasor corresponding to the
%                output voltage.
%-------------------------------------------------
Vo = Vs * Z2/(Z1 + Z2) % Eqn 10.14-4
B = abs(Vo);
phi = angle(Vo);
%-------------------------------------------------
%
%-------------------------------------------------
T = 2*pi/w;
tf = 2*T; N = 100; dt = tf/N;
t = 0 : dt : tf;
%-------------------------------------------------
%       Plot the input and output voltages.
%-------------------------------------------------
for k = 1 : 101
    vs(k) = A * cos(w * t(k) + theta);
    vo(k) = B * cos(w * t(k) + phi);
end
plot (t, vs, t, vo)
```

FIGURE 10.14-2 MATLAB input file corresponding to the circuit shown in Figure 10.14-1.

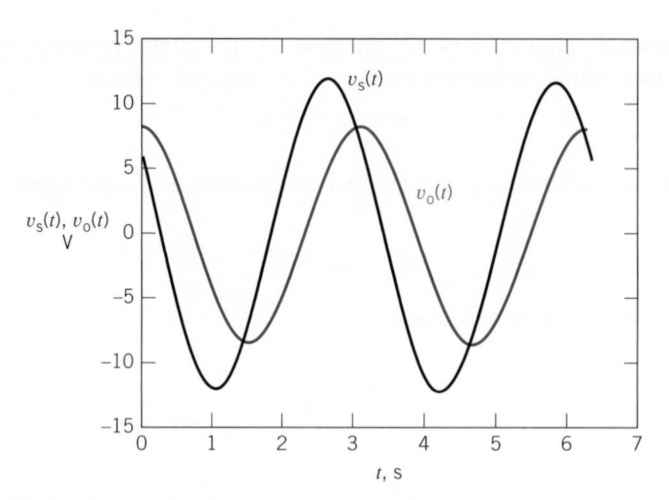

FIGURE 10.14-3 MATLAB plots showing the input and output voltages of the circuit shown in Figure 10.14-1.

Finally, $\mathbf{V}_o$ is calculated from $\mathbf{V}_s$ using voltage division. That is,

$$\mathbf{V}_o = \frac{\mathbf{Z}_2}{\mathbf{Z}_1 + \mathbf{Z}_2}\mathbf{V}_s \tag{10.14-4}$$

Figure 10.14-2 shows a MATLAB input file that uses Eqs. 10.14-1 through 10.14-3 to find the steady-state response of the circuit shown in Figure 10.14-1. Equation 10.14-4 is used to calculate $\mathbf{V}_o$. Next, $B = |\mathbf{V}_o|$ and $\phi = \angle\mathbf{V}_o$ are calculated and used to determine the magnitude and phase angle of the sinusoidal output voltage. Notice that MATLAB, not the user, does the complex arithmetic needed to solve these equations. Finally, MATLAB produces the plot shown in Figure 10.14-3, which displays the sinusoidal input and output voltages in the time domain.

10.15 USING PSPICE TO ANALYZE AC CIRCUITS

To use PSpice to analyze an ac circuit, we do the following:

1. Draw the circuit in the OrCAD Capture workspace.
2. Specify a AC Sweep\Noise simulation.
3. Run the simulation.
4. Open an output file to view the simulation results.

Table 10.15-1 shows some PSpice parts used to analyze ac circuits. When simulating ac circuits, we will represent independent voltage and current sources using the PSpice parts VAC and IAC, respectively. These PSpice parts each have properties named ACMAG and ACPHASE. We will edit the value of these properties to specify the amplitude and phase angle of a sinusoid. (Consequently, ACMAG and ACPHASE also represent the magnitude and phase angle of the phasor corresponding to the sinusoid.)

Table 10.15-1 PSpice Parts for AC Circuits and the Libraries in Which They Are Found

SYMBOL	DESCRIPTION	PSPICE NAME	LIBRARY
V? 1Vac 0Vdc	AC voltage source	VAC	SOURCE
I? 1Aac 0Adc	AC current source	IAC	SOURCE
	Print element voltage	VPRINT2	SPECIAL
	Print node voltage	VPRINT1	SPECIAL
IPRINT	Print element current	IPRINT	SPECIAL

We will add the PSpice parts VPRINT1, VPRINT2, and IPRINT from Table 10.15-1 to specify those current and voltage values that PSpice is to print into the output file. Each of these PSpice parts has properties named AC, REAL, IMAG, MAG, and PHASE. We will edit the value of each of these properties to be y. Then, when we simulate the circuit, PSpice will print the value of the corresponding phasor in both rectangular form and polar form.

EXAMPLE 10.15-1 Using PSpice to Analyze AC Circuits

Consider the ac circuit shown in Figure 10.15-1, in which

$$v_s(t) = 12\cos(100t + 15^\circ)\text{ V} \quad \text{and} \quad i_s(t) = 1.5\cos(100t + 135^\circ)\text{ A}$$

Use PSpice to determine the voltages v_1 and v_3 and the current i_2.

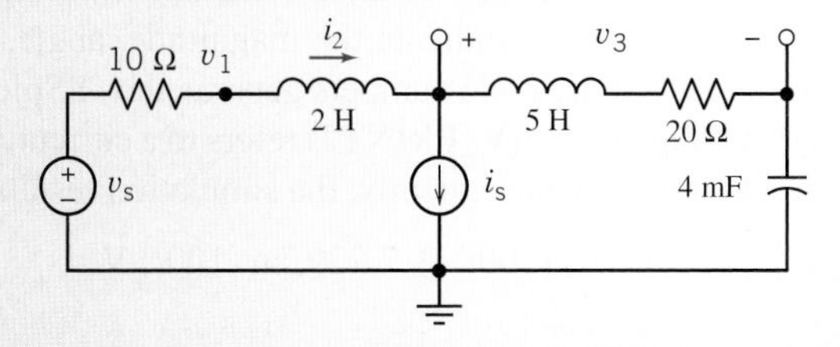

FIGURE 10.15-1 An AC circuit.

Solution

We begin by drawing the circuit in the OrCAD workspace as shown in Figure 10.15-2 (see Appendix A). Notice that we have used the PSpice parts VAC and IAC from Table 10.15-1 to represent the sources. Also, we have

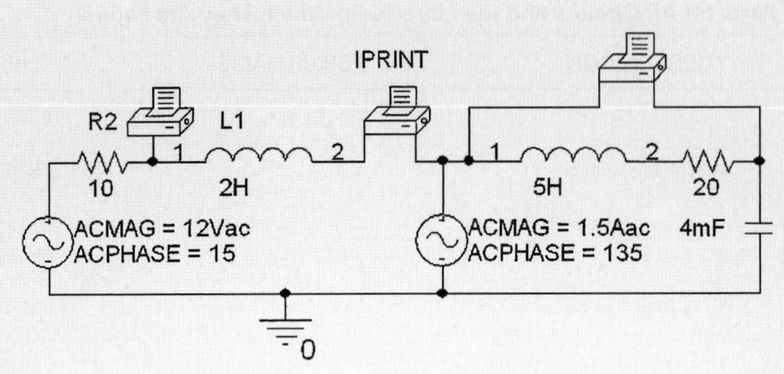

FIGURE 10.15-2 The circuit of Figure 10.15-1 as drawn in the OrCAD workspace.

edited the ACMAG and ACPHASE properties of these sources, setting ACMAG =12 and ACPHASE = 15 for the voltage source and ACMAG = 1.5 and ACPHASE = 135 for the current source.

Figure 10.15-2 also shows that we have added PSpice parts VPRINT1, IPRINT, and VPRINT2 to measure v_1, i_2, and v_3. These printers are connected to the circuit in the same way that ammeters and voltmeters would be connected to measure v_1, i_2, and v_3. Notice the minus sign on the VPRINT2 printer. It indicates the terminal near the minus sign of the polarity of the measured voltage. Similarly, the current measured by the IPRINT printer is the current directed toward the terminal marked by the minus sign. The minus sign on the VPRINT1 printer can be ignored. This printer measures the node voltage at the node to which it is connected.

We will perform a AC Sweep\Noise simulation. (Select Pspice\New Simulation Profile from the OrCAD Capture menu bar; then select AC Sweep\Noise from the Analysis Type drop-down list. Set both the Start Frequency and End Frequency to $100/(2\pi) = 15.92$. Select a Linear Sweep and set the Total Points to 1.) Select PSpice\Run Simulation Profile from the OrCAD Capture menu bar to run the simulation.

After we run the simulation, OrCAD Capture will open a Schematics window. Select View\Output File from the menu bar on the Schematics window. Scroll down through the output file to find the printer voltage and currents:

```
FREQ          VM(N615)          VP(N615)          VR(N615)          VI(N615)
15.92E+00     1.579E+01         -8.112E+00        1.564E+01         -2.229E+00
FREQ          IM(V_PRINT2)      IP(V_PRINT2)      IR(V_PRINT2)      II (V_PRINT2)
15.92E+00     6.694E-01         1.272E+02         -4.045E-01        5.334E-01
FREQ          VM(N256,N761)     VP(N256,N761)     VR(N256,N761)     VI(N256,N761)
15.92E+00     4.533E+01         2.942E+01         3.949E+01         2.227E+01
```

This output requires some interpretation. The labels VM, VP, VR, and VI indicate the magnitude, angle, real part, and imaginary part of a voltage, and the labels IM, IP, IR, and II indicate the magnitude, angle, real part, and imaginary part of a current. The labels N614, N256, and N761 are node numbers generated by PSpice. VM(N615) refers to the voltage at a single node, that is, the node voltage v_1. IM(V_PRINT2) refers to a current, that is, i_2. VM (N256,N761) refers to a voltage between two nodes, that is, v_3. Consequently, the simulation results indicate that

$$v_1(t) = 15.79 \cos(100t - 8.1^\circ) = 15.64 \cos(100t) + 2.229 \sin(100t) \text{ V},$$

$$i_2(t) = 0.6694 \cos(100t + 127.2^\circ) = -0.4045 \cos(100t) - 0.5334 \sin(100t) \text{ V},$$

and

$$v_3(t) = 45.33 \cos(100t + 29.40) = 39.49 \cos(100t) - 22.27 \sin(100t) \text{ V}$$

10.16 HOW CAN WE CHECK . . . ?

Engineers are frequently called upon to check that a solution to a problem is indeed correct. For example, proposed solutions to design problems must be checked to confirm that all of the specifications have been satisfied. In addition, computer output must be reviewed to guard against data-entry errors, and claims made by vendors must be examined critically.

Engineering students are also asked to check the correctness of their work. For example, occasionally just a little time remains at the end of an exam. It is useful to be able to quickly identify those solutions that need more work.

The following examples illustrate techniques useful for checking the solutions of the sort of problem discussed in this chapter.

EXAMPLE 10.16-1 How Can We Check Arithmetic with Complex Numbers?

It is known that

$$\frac{10}{R - j4} = A\angle 53^\circ$$

A computer program states that $A = 2$. **How can we check** this result? (Notice that values are given to only two significant figures.)

Solution

The equation for the angle is

$$-\tan^{-1}\left(\frac{-4}{R}\right) = 53^\circ$$

Then, we have

$$R = \frac{-4}{\tan(-53^\circ)} = 3.014$$

Solving for A in terms of R, we obtain

$$A = \frac{10}{(R^2 + 16)^{1/2}} = 1.997$$

Therefore, $A = 2$ is correct to two significant figures.

EXAMPLE 10.16-2 How Can We Check AC Circuit Analysis?

Consider the circuit shown in Figure 10.16-1. Suppose we know that the capacitor voltages are

$$1.96\cos(100t - 101.3^\circ)\text{ V} \quad \text{and} \quad 4.39\cos(100t - 37.88^\circ)\text{ V}$$

but we do not know which voltage is $v_1(t)$ and which is $v_2(t)$. **How can we check** the capacitor voltages?

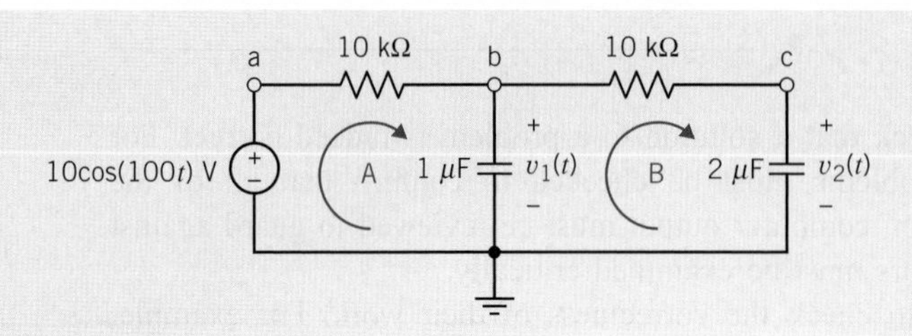

FIGURE 10.16-1 An example circuit.

Solution

Let us guess that

$$v_1(t) = 1.96\cos(100t - 101.3°)$$

and

$$v_2(t) = 4.39\cos(100t - 37.88°)$$

and then check to see whether this choice satisfies the node equations representing the circuit. These node equations are

$$\frac{10 - \mathbf{V}_1}{R_1} = j\omega C_1 \mathbf{V}_1 + \frac{\mathbf{V}_1 - \mathbf{V}_2}{R_2}$$

and

$$j\omega C_2 \mathbf{V}_2 = \frac{\mathbf{V}_1 - \mathbf{V}_2}{R_2}$$

where $\mathbf{V}_1$ and $\mathbf{V}_2$ are the phasors corresponding to $v_1(t)$ and $v_2(t)$. That is,

$$\mathbf{V}_1 = 1.96e^{-j101.3°} \quad \text{and} \quad \mathbf{V}_2 = 4.39e^{-j37.88°}$$

Substituting the phasors $\mathbf{V}_1$ and $\mathbf{V}_2$ into the left-hand side of the first node equation gives

$$\frac{10 - 1.96e^{-j101.3}}{10 \times 10^3} = 0.001 + j1.92 \times 10^{-4}$$

Substituting the phasors $\mathbf{V}_1$ and $\mathbf{V}_2$ into the right-hand side of the first node equation gives

$$j \cdot 100 \times 10^{-6} \cdot 1.96e^{-j101.3} + \frac{1.96e^{-j101.3} - 4.39e^{-j37.88}}{10 \times 10^3}$$
$$= -19.3 \times 10^{-4} + j3.89 \times 10^{-5}$$

Because the right-hand side is not equal to the left-hand side, $\mathbf{V}_1$ and $\mathbf{V}_2$ do not satisfy the node equation. That means that the selected order of $v_1(t)$ and $v_2(t)$ is not correct. Instead, use the reverse order so that

$$v_1(t) = 4.39\cos(100t - 37.88°)$$

and

$$v_2(t) = 1.96\cos(100t - 101.3°)$$

Now the phasors $\mathbf{V}_1$ and $\mathbf{V}_2$ will be

$$\mathbf{V}_1 = 4.39e^{-j37.88°} \quad \text{and} \quad \mathbf{V}_2 = 1.96e^{-j101.3°}$$

Substituting the new values of the phasors $\mathbf{V}_1$ and $\mathbf{V}_2$ into the left-hand side of the first node equation gives

$$\frac{10 - 4.39e^{-j37.88}}{10 \times 10^3} = 6.353 \times 10^{-4} + j2.696 \times 10^{-4}$$

Substituting the new values of the phasors $\mathbf{V}_1$ and $\mathbf{V}_2$ into the right-hand side of the first node equation gives

$$j \cdot 100 \cdot 10^{-6} \cdot 4.39e^{-j37.88} + \frac{4.39e^{-j37.88} - 1.96e^{-j101.3}}{10 \times 10^3}$$
$$= +6.545 \times 10^{-4} + j2.69 \times 10^{-4}$$

Because the right-hand side is very close to equal to the left-hand side, $\mathbf{V}_1$ and $\mathbf{V}_2$ satisfy the first node equation. That means that $v_1(t)$ and $v_2(t)$ are probably correct. To be certain, we will also check the second node equation. Substituting the phasors $\mathbf{V}_1$ and $\mathbf{V}_2$ into the left-hand side of the second node equation gives

$$j \cdot 100 \cdot 2 \times 10^{-6} \cdot 1.96e^{-j101.3} = +3.84 \times 10^{-4} - j7.681 \times 10^{-5}$$

Substituting the phasors $\mathbf{V}_1$ and $\mathbf{V}_2$ into the right-hand side of the second node equation gives

$$\frac{4.39e^{-j37.88} - 1.96e^{-j101.3}}{10 \times 10^3} = 3.85 \times 10^{-4} - j7.735 \times 10^{-5}$$

Because the right-hand side is equal to the left-hand side, $\mathbf{V}_1$ and $\mathbf{V}_2$ satisfy the second node equation. Now we are certain that

$$v_1(t) = 4.39 \cos(100t - 37.88°) \text{ V}$$

and

$$v_2(t) = 1.96 \cos(100t - 101.3°) \text{ V}$$

10.17 DESIGN EXAMPLE

OP AMP CIRCUIT

Figure 10.17-1*a* shows two sinusoidal voltages, one labeled as input and the other labeled as output. We want to design a circuit that will transform the input sinusoid into the output sinusoid. Figure 10.17-1*b* shows a candidate circuit. We must first determine whether this circuit can do the job. Then, if it can, we will design the circuit, that is, specify the required values of R_1, R_2, and C.

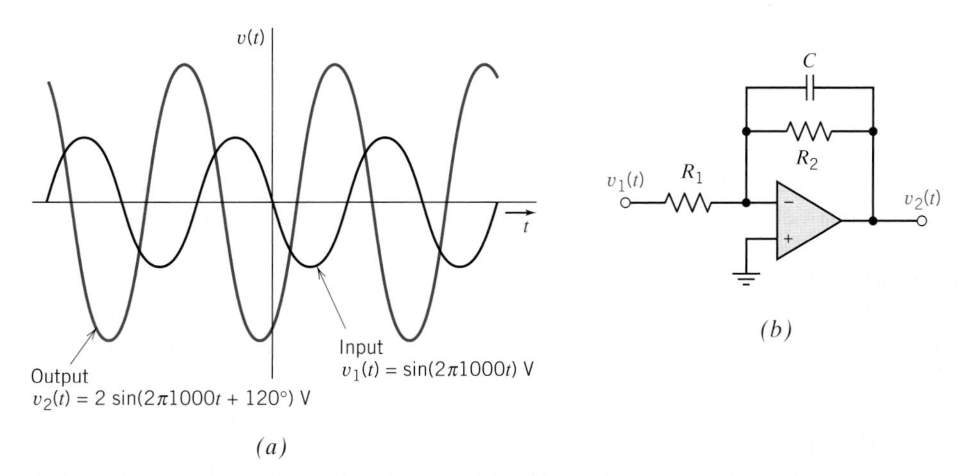

FIGURE 10.17-1 (*a*) Input and output voltages. (*b*) Proposed circuit.

Define the Situation and the Assumptions

The input and output sinusoids have different amplitudes and phase angles but the same frequency:

$$f = 1000\ \text{Hz}$$

or, equivalently,

$$\omega = 2\pi 1000\ \text{rad/s}$$

We now know that this must be the case. When the input to a linear circuit is a sinusoid, the steady-state output will also be a sinusoid having the same frequency.

In this case, the input sinusoid is

$$v_1(t) = \sin(2\pi 1000 t) = \cos(2\pi 1000 t - 90^\circ)\ \text{V}$$

and the corresponding phasor is

$$\mathbf{V}_1 = 1e^{-j90^\circ} = 1\angle{-90^\circ}\ \text{V}$$

The output sinusoid is

$$v_2(t) = 2\sin(2\pi 1000 t + 120^\circ) = 2\cos(2\pi 1000 t + 30^\circ)\ \text{V}$$

and the corresponding phasor is

$$\mathbf{V}_2 = 2e^{j30^\circ}\ \text{V}$$

The ratio of these phasors is

$$\frac{\mathbf{V}_2}{\mathbf{V}_1} = \frac{2e^{j30^\circ}}{1e^{-j90^\circ}} = 2e^{j120^\circ}$$

The magnitude of this ratio, called the gain, G, of the circuit used to transform the input sinusoid into the output sinusoid is

$$G = \left|\frac{\mathbf{V}_2}{\mathbf{V}_1}\right| = 2$$

The angle of this ratio is called the phase shift, θ, of the required circuit:

$$\theta = \angle \frac{\mathbf{V}_2}{\mathbf{V}_1} = 120^\circ$$

Therefore, we need a circuit that has a gain of 2 and a phase shift of 120°.

State the Goal

Determine whether it is possible to design the circuit shown in Figure 10.17-1*b* to have a gain of 2 and a phase shift of 120°. If it is possible, specify the appropriate values of R_1, R_2, and C.

Generate a Plan

Analyze the circuit shown in Figure 10.17-1*b* to determine the ratio of the output phasor to the input phasor, $\mathbf{V}_2/\mathbf{V}_1$. Determine whether this circuit can have a gain of 2 and a phase shift of 120°. If so, determine the required values of R_1, R_2, and C.

Act on the Plan

The circuit in Figure 10.17-1*b* is a special case of the circuit shown in Figure 10.12-1. The impedance $\mathbf{Z}_1$ in Figure 10.12-1 corresponds to the resistor R_1 in Figure 10.17-1*b*, and impedance $\mathbf{Z}_2$ corresponds to the parallel combination of resistor R_2 and capacitor C. That is,

$$\mathbf{Z}_1 = R_1$$

and

$$\mathbf{Z}_2 = \frac{R_2(1/j\omega C)}{R_2 + 1/j\omega C} = \frac{R_2}{1 + j\omega C R_2}$$

Then, using Eq. 10.12-3,

$$\frac{\mathbf{V}_2}{\mathbf{V}_1} = -\frac{\mathbf{Z}_2}{\mathbf{Z}_1} = -\frac{R_2/(1 + j\omega C R_2)}{R_1} = -\frac{R_2/R_1}{1 + j\omega C R_2}$$

The phase shift of the circuit in Figure 10.17-1b is given by

$$\theta = \angle \frac{\mathbf{V}_2}{\mathbf{V}_1} = \angle -\frac{R_2/R_1}{1 + j\omega C R_2} = 180^\circ - \tan^{-1} \omega C R_2 \tag{10.17-1}$$

What values of phase shift are possible? Notice that ω, C, and R_2 are all positive, which means that

$$0^\circ \leq \tan^{-1} \omega C R_2 \leq 90^\circ$$

Therefore, the circuit shown in Figure 10.17-1b can be used to obtain phase shifts between 90° and 180°. Hence, we can use this circuit to produce a phase shift of 120°.

The gain of the circuit in Figure 10.17-1b is given by

$$\begin{aligned} G &= \left|\frac{\mathbf{V}_2}{\mathbf{V}_1}\right| = \left|-\frac{R_2/R_1}{1 + j\omega C R_2}\right| \\ &= \frac{R_2/R_1}{\sqrt{1 + \omega^2 C^2 R_2^2}} = \frac{R_2/R_1}{\sqrt{1 + \tan^2(180^\circ - \theta)}} \end{aligned} \tag{10.17-2}$$

Next, first solve Eq. 10.17-1 for R_2 and then Eq. 10.17-1 for R_1 to get

$$R_2 = \frac{\tan\,(180^\circ - \theta)}{\omega C}$$

and

$$R_1 = \frac{R_2/G}{\sqrt{1 + \tan^2\,(180^\circ - \theta)}}$$

These equations can be used to design the circuit. First, pick a convenient, readily available, and inexpensive value of the capacitor, say,

$$C = 0.02\ \mu\text{F}$$

Next, calculate values of R_1 and R_2 from the values of ω, C, G, and θ. For $\omega = 6283$ rad/s, $C = 0.02\ \mu$F, $G = 2$, and $\theta = 120^\circ$, we calculate

$$R_1 = 3446\ \Omega \quad \text{and} \quad R_2 = 13.78\ \text{k}\Omega$$

and the design is complete.

Verify the Proposed Solution

When $C = 0.02\ \mu F$, $R_1 = 3446\ \Omega$, and $R_2 = 13.78\ k\Omega$, the network function of the circuit is

$$\frac{\mathbf{V}_2}{\mathbf{V}_1} = -\frac{R_2/R_1}{1 + j\omega CR_2} = -\frac{4}{1 + j\omega(0.2756 \times 10^{-3})}$$

In this case, $\omega = 2\pi 1000$, and $\mathbf{V}_1 = 1\angle -90°$, so

$$\frac{\mathbf{V}_2}{\mathbf{V}_1} = -\frac{4}{1 + j(2\pi \times 10^3)(0.2756 \times 10^{-3})} = 2\angle 120°$$

as required by the specifications.

10.18 SUMMARY

- With the pervasive use of ac electric power in the home and industry, it is important for engineers to analyze circuits with sinusoidal independent sources.
- The steady-state response of a linear circuit to a sinusoidal input is itself a sinusoid having the same frequency as the input signal.
- Circuits that contain inductors and capacitors are represented by differential equations. When the input to the circuit is sinusoidal, the phasors and impedances can be used to represent the circuit in the frequency domain. In the frequency domain, the circuit is represented by algebraic equations. The original circuit, represented by a differential equation, is called the time-domain representation of the circuit.
- The steady-state response of a linear circuit with a sinusoidal input is obtained as follows:
  1. Transform the circuit into the frequency domain, using phasors and impedances.
  2. Represent the frequency-domain circuit by algebraic equations, for example, mesh or node equations.
  3. Solve the algebraic equations to obtain the response of the circuit.
  4. Transform the response into the time domain, using phasors.
- Table 10.6-1 summarizes the relationships used to transform a circuit from the time domain to the frequency domain or vice versa.
- When a circuit contains several sinusoidal sources, we distinguish two cases.
  1. When all of the sinusoidal sources have the same frequency, the response will be a sinusoid with that frequency, and the problem can be solved in the same way that it would be if there was only one source.
  2. When the sinusoidal sources have different frequencies, superposition is used to break the time-domain circuit up into several circuits, each with sinusoidal inputs all at the same frequency. Each of the separate circuits is analyzed separately and the responses are summed *in the time domain*.
- MATLAB greatly reduces the computational burden associated with solving mesh or node equations having complex coefficients.

PROBLEMS

Section 10.2 Sinusoidal Sources

P 10.2-1 Express the following summations of sinusoids in the general form $A \sin(\omega t + \theta)$ by using trigonometric identities.

(a) $i(t) = 2\cos(6t + 120°) + 4\sin(6t - 60°)$
(b) $v(t) = 5\sqrt{2}\cos 8t + 10\sin(8t + 45°)$

P 10.2-2 A sinusoidal voltage has a maximum value of 100 V, and the value is 20 V at $t = 0$. The period is $T = 2$ ms. Determine $v(t)$.

P 10.2-3 A sinusoidal current is given as $i = 400\cos(1400\pi t + 70°)$ mA. Determine the frequency f and the value of the current at $t = 2$ ms.

P 10.2-4 Plot a graph of the voltage signal

$$v(t) = 15 \cos(628t + 45°) \text{ mV}$$

P 10.2-5 Figure P 10.2-5 shows a sinusoidal voltage, $v(t)$, plotted as a function of time, t. Represent $v(t)$ by a function of the form $A \cos(\omega t + \theta)$.

Answer: $v(t) = 18 \cos(393t - 27°)$

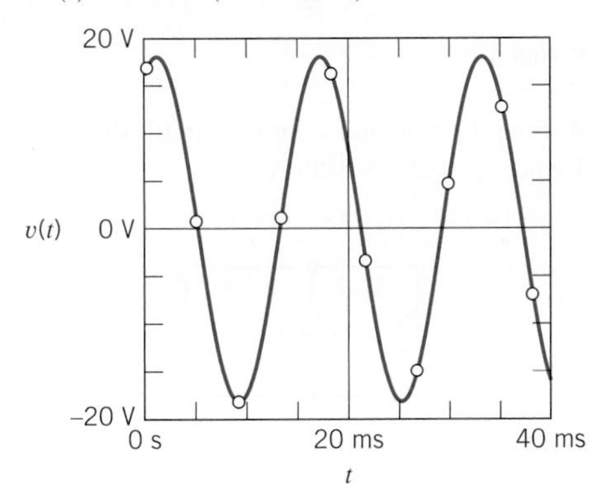

Figure P 10.2-5

P 10.2-6 Figure P 10.2-6 shows a sinusoidal voltage, $v(t)$, plotted as a function of time, t. Represent $v(t)$ by a function of the form $A \cos(\omega t + \theta)$.

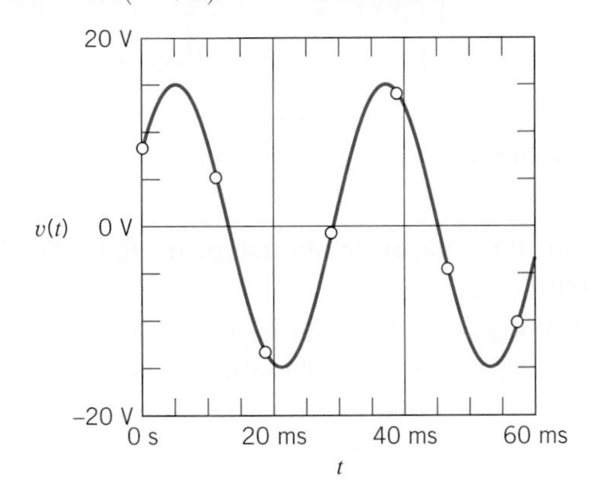

Figure P 10.2-6

Section 10.3 Steady-State Response of an *RL* Circuit for a Sinusoidal Forcing Function

P 10.3-1 Find the forced response i for the circuit of Figure P 10.3-1 when $v_s(t) = 10 \cos(300t)$ V.

Answer: $i(t) = 1.24 \cos(300t - 68°)$ A

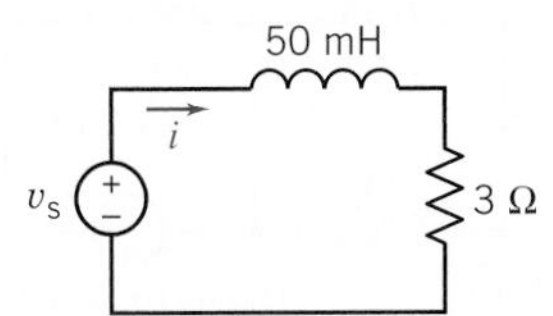

Figure P 10.3-1

P 10.3-2 Find the forced response v for the circuit of Figure P 10.3-2 when $i_s(t) = 0.5 \cos \omega t$ A and $\omega = 1000$ rad/s.

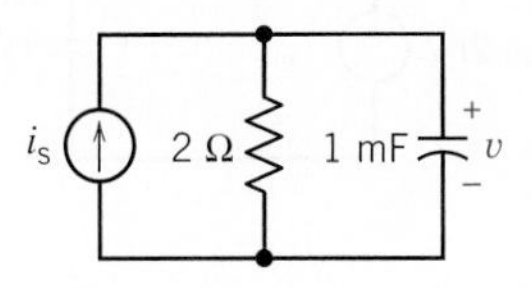

Figure P 10.3-2

P 10.3-3 Find the forced response $i(t)$ for the circuit of Figure P 10.3-3.

Answer: $i(t) = 1.5 \cos(4t + 45°)$ mA

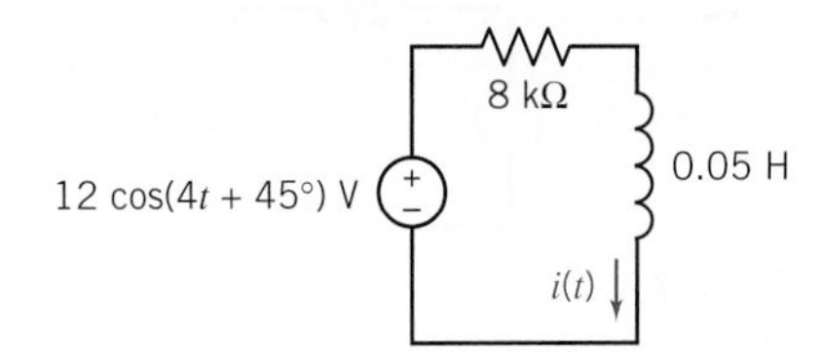

Figure P 10.3-3

Section 10.4 Complex Exponential Forcing Function

P 10.4-1 Determine the polar form of the quantity

$$\frac{(6\angle 36.9°)(10\angle -53.1°)}{(6 + 4j) + (6 - j8)}$$

Answer: $\frac{3}{2}\sqrt{10}\angle 2.23°$

P 10.4-2 Determine the polar and rectangular form of the expression

$$10\angle +81.87° \left(4 - j3 + \frac{3\sqrt{2}\angle -45°}{7 - j1}\right)$$

Answer: $28\angle +45° = 28\sqrt{2} + j28\sqrt{2}$

P 10.4-3 Given $\mathbf{A} = 3 + j7$, $\mathbf{B} = 6\angle 15°$, and $\mathbf{C} = 5e^{j2.3°}$, find $(\mathbf{A}^*\mathbf{C}^*)/\mathbf{B}$.

Answer: $0.65 - j6.32$

P 10.4-4 Determine a and b when (angles in degrees)

$$(6\angle 120°)(-4 + j3 + 2e^{j15}) = a + jb$$

P 10.4-5 Find a, b, A, and θ as required (angles given in degrees).

(a) $Ae^{j120} + jb = -5 + j3$
(b) $6e^{j120}(-4 + jb + 8e^{j\theta}) = 18$
(c) $(a + j4)j2 = 2 + Ae^{j60}$

P 10.4-6 Find the steady-state response, $v(t)$, for the circuit shown in Figure P 10.4-6.

Answer: $v(t) = \frac{5}{\sqrt{29}} \cos(2t - 21.8°)$ V

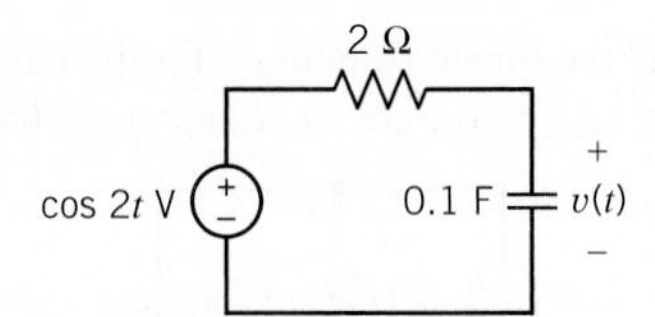

Figure P 10.4-6

P 10.4-7 Find the steady-state response, $v(t)$, for the circuit shown in Figure P 10.4-7.

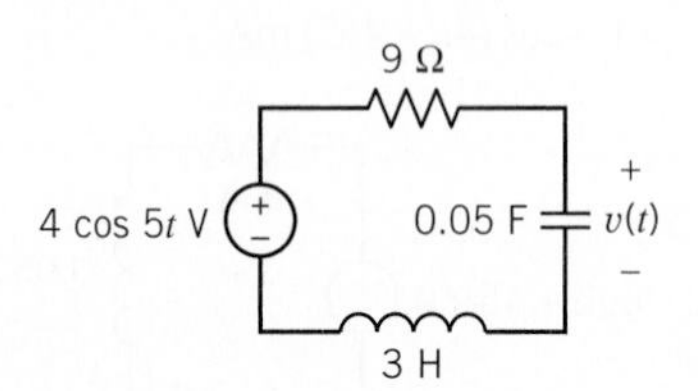

Figure P 10.4-7

Section 10.5 The Phasor

P 10.5-1 Find the steady-state response, $v(t)$, for the circuit shown in Figure P 10.5-1.

Hint: First, show $2\frac{d}{dt}i + 6i = 20 \cos 4t$

Answer: $v(t) = 16 \cos(4t + 37°)$ V

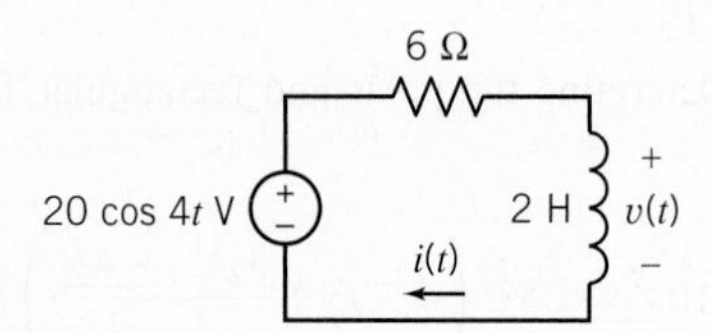

Figure P 10.5-1

P 10.5-2 Find the steady-state response, $i(t)$, for the circuit shown in Figure P 10.5-2.

Answer: $i(t) = 0.398 \cos(2t - 85°)$ A

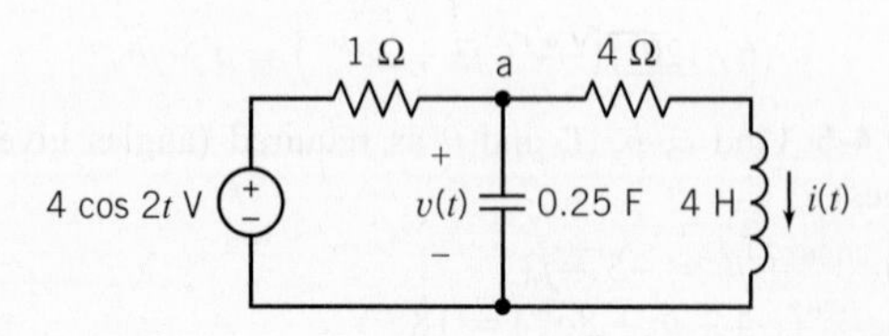

Figure P 10.5-2

P 10.5-3 For the circuit of Figure P 10.5-3, find $v(t)$ when $v_s = 2 \sin 500t$ V.

Answer: $v(t) = 1.25 \cos(500t - 141°)$ V

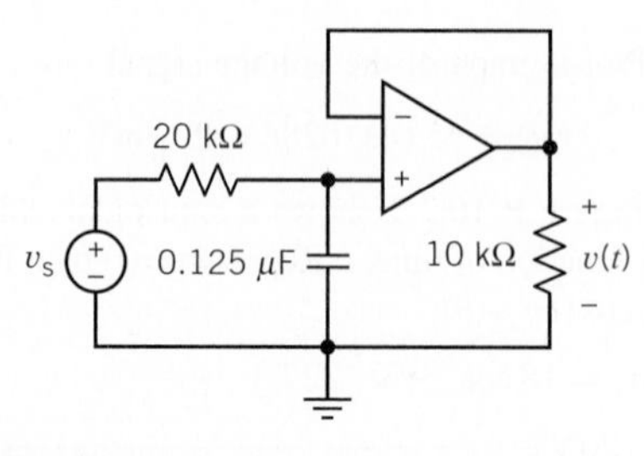

Figure P 10.5-3

P 10.5-4 Find the response v for the circuit shown in Figure P 10.5-4 when $i_s = 20 \cos 100t$ A.

Answer: $v = 14.142 \cos(100t - 45°)$

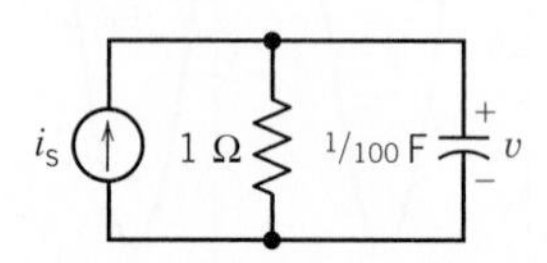

Figure P 10.5-4

P 10.5-5 Find the current $i(t)$ for the *RLC* circuit of Figure P 10.5-5 when $v_s = 4 \cos 100t$ V.

Answer: $i(t) = 2\sqrt{2} \cos(100t + 45°)$ A

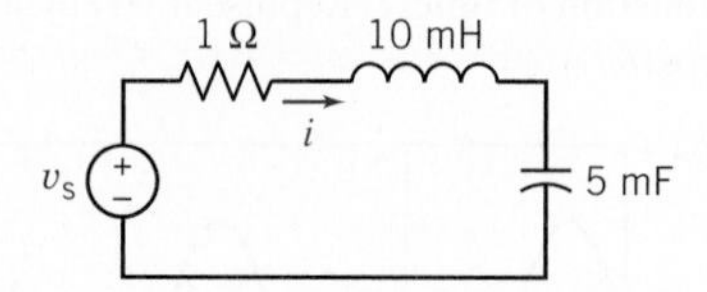

Figure P 10.5-5

Section 10.6 Phasor Relationships for *R, L,* and *C* Elements

P 10.6-1 Represent the circuit shown in Figure P 10.6-1 in the frequency domain, using impedances and phasors.

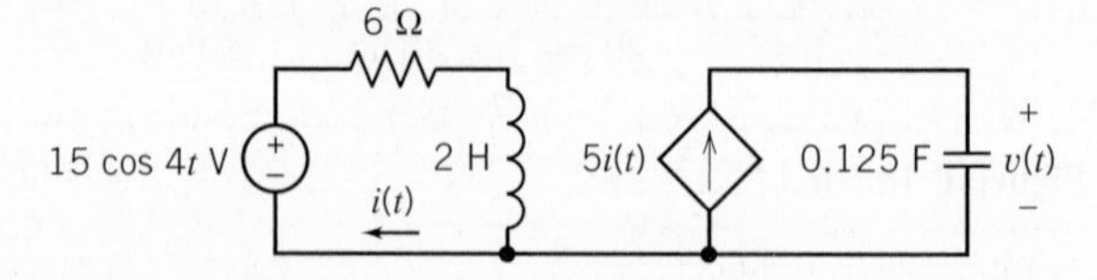

Figure P 10.6-1

P 10.6-2 Represent the circuit shown in Figure P 10.6-2 in the frequency domain, using impedances and phasors.

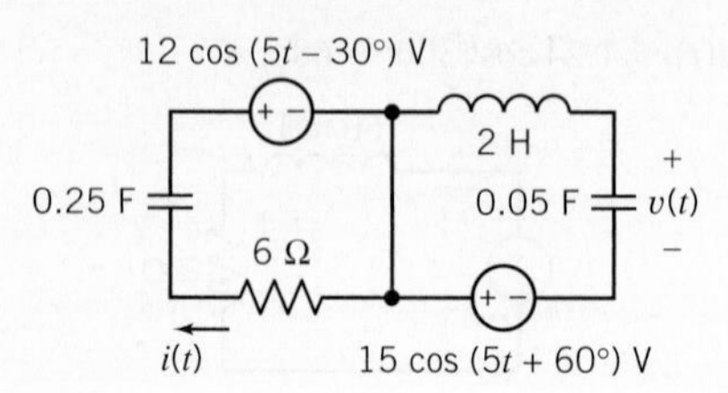

Figure P 10.6-2

P 10.6-3 Represent the circuit shown in Figure P 10.6-3 in the frequency domain, using impedances and phasors.

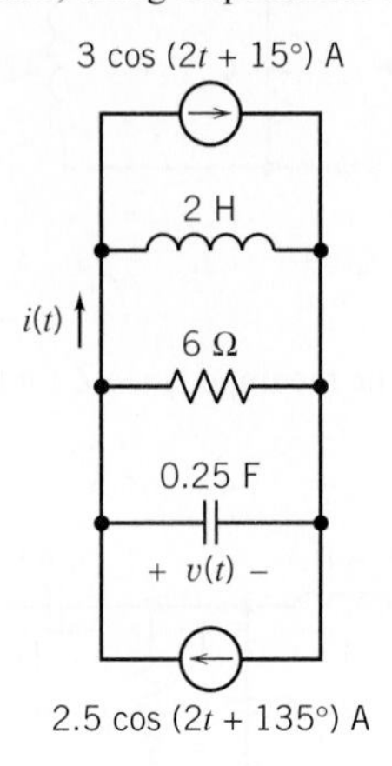

Figure P 10.6-3

P 10.6-4 Represent the circuit shown in Figure P 10.6-4 in the frequency domain, using impedances and phasors.

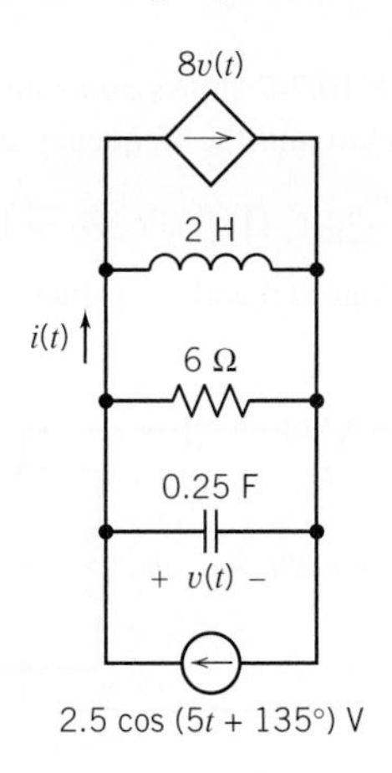

Figure P 10.6-4

P 10.6-5 Each of the following pairs of element voltage and element current adheres to the passive convention. Indicate whether the element is capacitive, inductive, or resistive and find the element value.

(a) $v(t) = 30 \cos(400t + 30°)$; $i = 3 \sin(400t + 30°)$
(b) $v(t) = 8 \sin(900t + 50°)$; $i = 2 \sin(900t + 140°)$
(c) $v(t) = 20 \cos(250t + 60°)$; $i = 10 \sin(250t + 150°)$

Answers: **(a)** $L = 25$ mH
(b) $C = 277.77\ \mu$F
(c) $R = 2\ \Omega$

P 10.6-6 Two circuit elements are connected in series, so $v = v_1 + v_2$. Find $v(t)$ when $v_1(t) = 150 \cos(377t - \pi/6)$ V and $V_2 = 250 \angle{+60°}$ V.

Answer: $v(t) = 292 \cos(377t + 29.1°)$ V

P 10.6-7 The voltage and current for the circuit shown in Figure P 10.6-7 are given by

$$v(t) = 20 \cos(20t + 15°) \text{ V and } i(t) = 2 \cos(20t + 63°) \text{ A}$$

Determine the values of the resistance, R, and capacitance, C

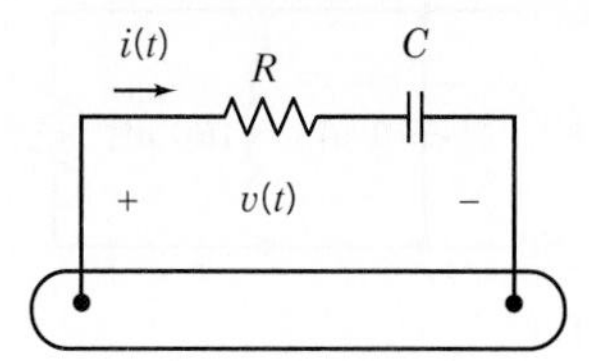

Figure P 10.6-7

P 10.6-8 Given that

$$i_1(t) = 30 \cos(4t + 45°) \text{ mA}$$

and

$$i_2(t) = -40 \cos(4t) \text{ mA}$$

Determine $v(t)$ for the circuit shown in Figure P 10.6-8.

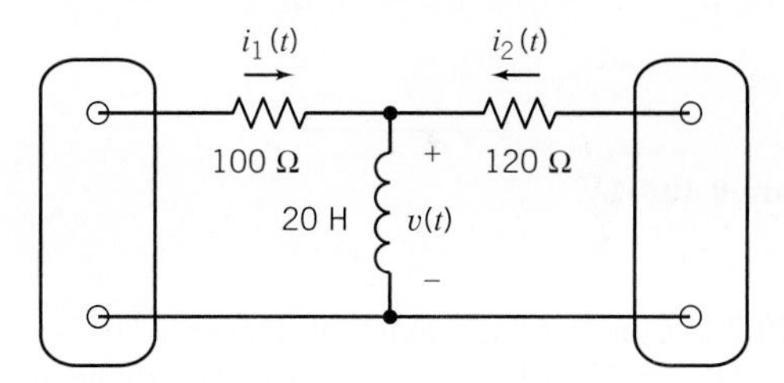

Figure P 10.6-8

P 10.6-9 Figure P 10.6-9 shows an ac circuit represented in both the time domain and the frequency domain. Determine the values of A, B, a, and b.

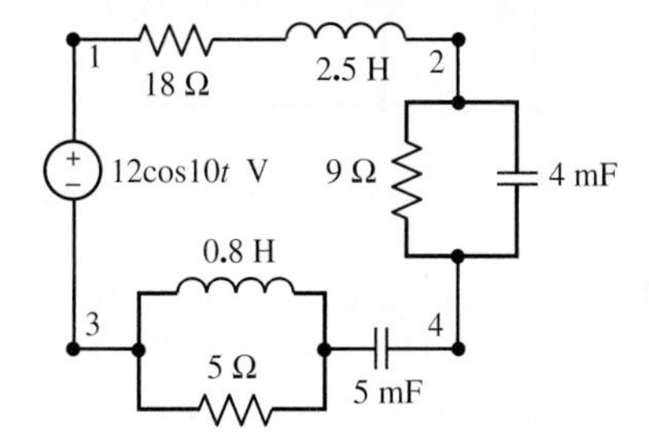

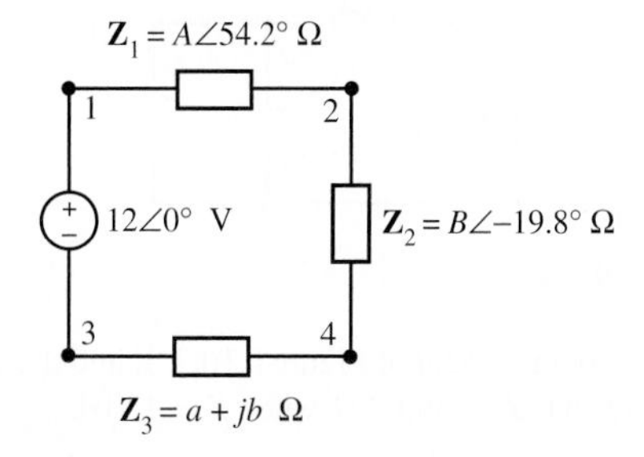

Figure P 10.6-9

Section 10.7 Impedance and Admittance

P 10.7-1 Find **Z** and **Y** for the circuit of Figure P 10.7-1 operating at 10 kHz.

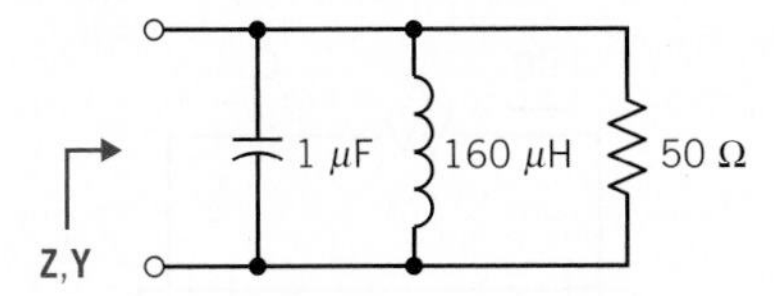

Figure P 10.7-1

P 10.7-2 Find R and L of the circuit of Figure P 10.7-2 when $v(t) = 20\cos(\omega t + 40^\circ)$ V; $i(t) = 2\cos(\omega t + 15^\circ)$ mA, and $\omega = 2 \times 10^6$ rad/s.

Answer: $R = 9.064\ \text{k}\Omega$, $L = 2.113$ mH

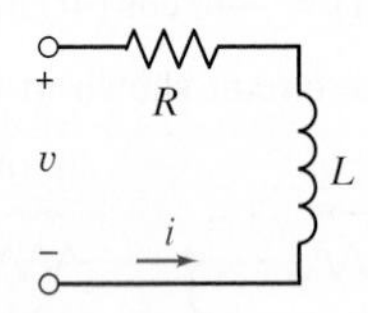

Figure P 10.7-2

P 10.7-3 Consider the circuit of Figure P 10.7-3 when $R = 6\ \Omega$, $L = 27$ mH, and $C = 22\ \mu\text{F}$. Determine the frequency f when the impedance **Z** is purely resistive, and find the input resistance at that frequency.

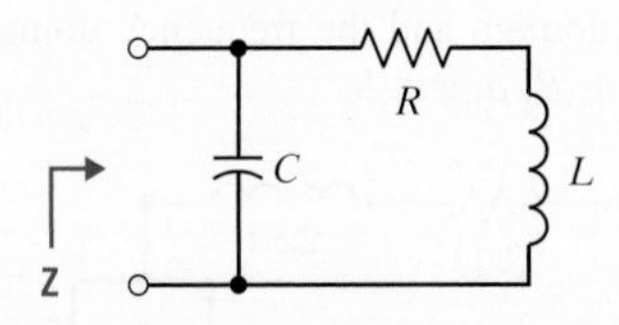

Figure P 10.7-3

P 10.7-4 Consider the circuit of Figure P 10.7-4 when $R = 10\ \text{k}\Omega$ and $f = 1$ kHz. Find L and C so that $\mathbf{Z} = 100 + j0\Omega$.

Answer: $L = 0.1587$ H and $C = 0.158\ \mu\text{F}$

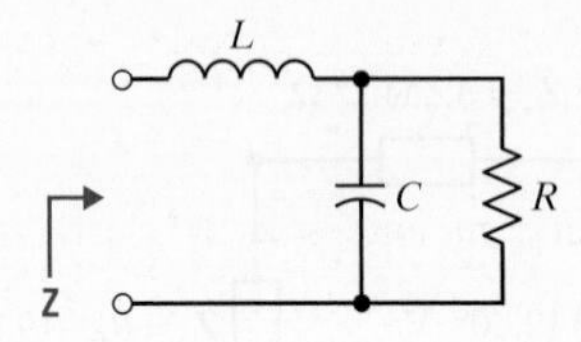

Figure P 10.7-4

P 10.7-5 For the circuit of Figure P 10.7-5, find the value of C required so that $\mathbf{Z} = 590.7\ \Omega$ when $f = 1$ MHz.

Answer: $C = 0.27$ nF

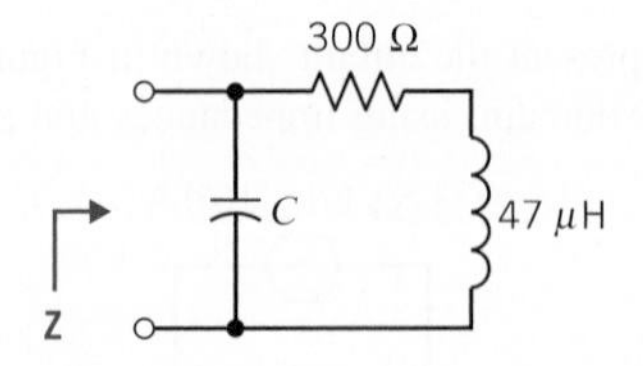

Figure P 10.7-5

P 10.7-6 Determine the impedance **Z** for the circuit shown in Figure P 10.7-6.

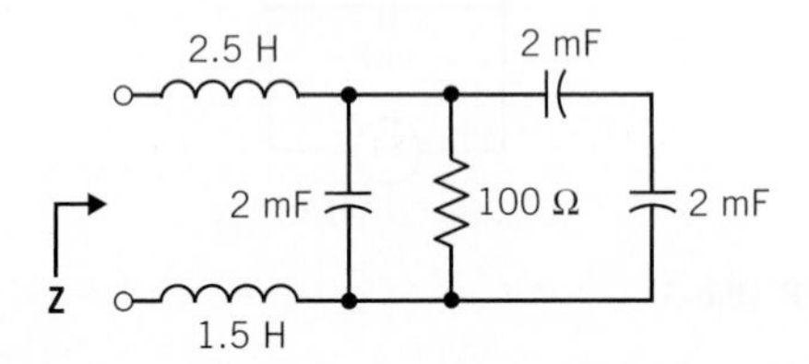

Figure P 10.7-6

P 10.7-7 Figure P 10.7-7 shows an ac circuit represented in both the time domain and the frequency domain. Suppose

$$\mathbf{Z}_1 = 15.3\angle{-24.1^\circ}\ \Omega \quad \text{and} \quad \mathbf{Z}_2 = 14.4\angle{53.1^\circ}\ \Omega$$

Determine the voltage $v(t)$ and the values of R_1, R_2, L, and C.

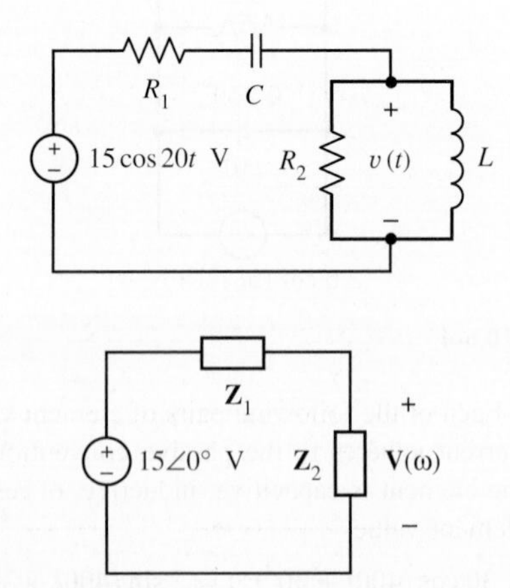

Figure P 10.7-7

Section 10.8 Kirchhoff's Laws Using Phasors

P 10.8-1 For the circuit shown in Figure P 10.8-1, find (a) the impedances $\mathbf{Z}_1$ and $\mathbf{Z}_2$ in polar form, (b) the total combined impedance in polar form, and (c) the steady-state current $i(t)$.

Answers: (a) $\mathbf{Z}_1 = 6.4\angle{38.7^\circ}\ \Omega$; $\mathbf{Z}_2 = 8\sqrt{2}\angle{-45^\circ}\ \Omega$

(b) $\mathbf{Z}_1 + \mathbf{Z}_2 = 13.6\angle{-17.1^\circ}\ \Omega$

(c) $i(t) = (7.35)\cos(1250t + 17.1^\circ)$ A

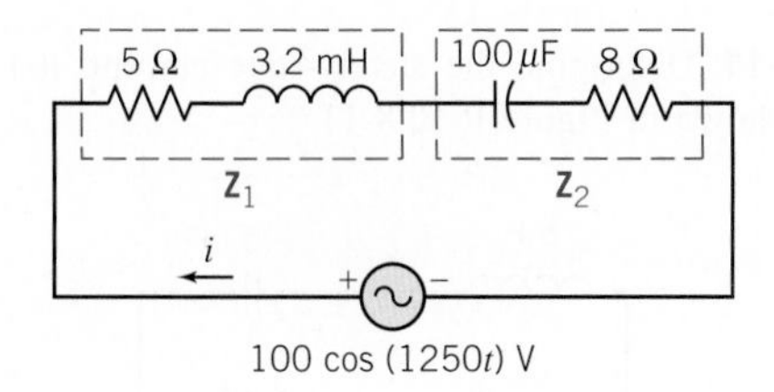

Figure P 10.8-1

P 10.8-2 The circuit shown in Figure P 10.8-2 is at steady state. The voltages $v_s(t)$ and $v_2(t)$ are given by

$$v_s(t) = 7.75 \cos (2t + 47°) \text{ V}$$

and

$$v_2(t) = 1.59 \cos (2t + 125°) \text{ V}$$

Find the steady-state voltage $v_1(t)$

Answer: $v_1(t) = 7.59 \cos (2t + 35.2°)$ V

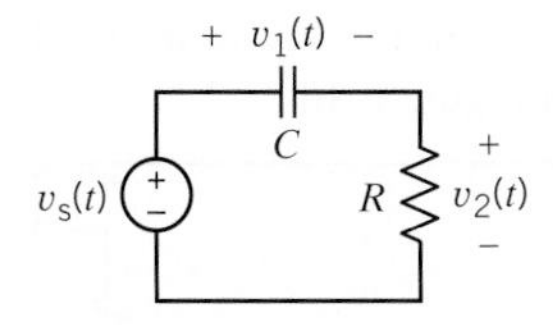

Figure P 10.8-2

P 10.8-3 The circuit shown in Figure P 10.8-3 is at steady state. The currents $i_1(t)$ and $i_2(t)$ are given by

$$i_1(t) = 750 \cos (2t - 118°) \text{ mA}$$

and

$$i_2(t) = 540.5 \cos (2t + 100°) \text{ mA}$$

Find the steady-state current $i(t)$.

Answer: $i(t) = 465 \cos (2t + 196°)$ mA

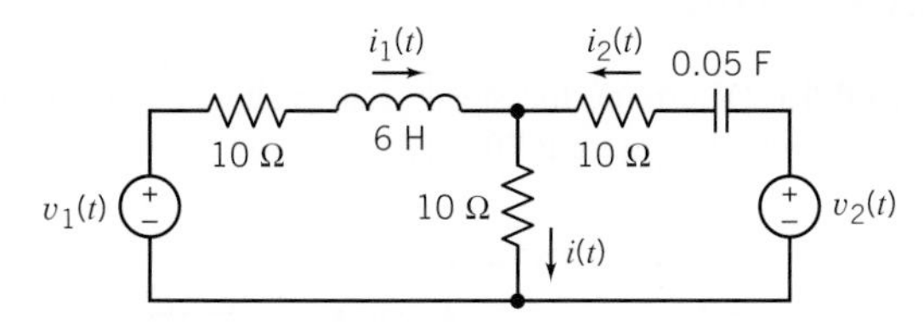

Figure P 10.8-3

P 10.8-4 Determine $i(t)$ of the *RLC* circuit shown in Figure P 10.8-4 when $v_s = 2 \cos (4t + 30°)$ V.

Answer: $i(t) = 0.149 \cos (4t - 12°)$ A

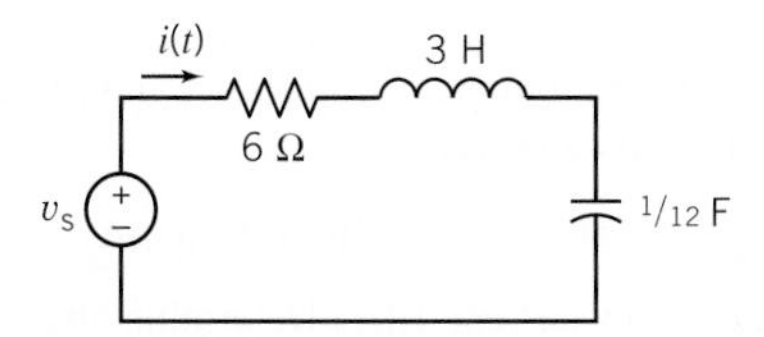

Figure P 10.8-4

P 10.8-5 The big toy from the hit movie *Big* is a child's musical fantasy come true—a sidewalk-sized piano. Like a hopscotch grid, this once-hot Christmas toy invites anyone who passes to jump on, move about, and make music. The developer of the toy piano used a tone synthesizer and stereo speakers as shown in Figure P 10.8-5 (Gardner, 1988). Determine the current $i(t)$ for a tone at 796 Hz when $C = 10\ \mu$F.

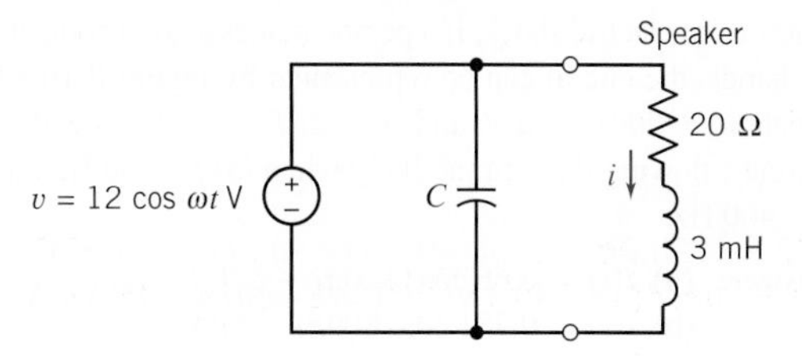

Figure P 10.8-5 Tone synthesizer.

P 10.8-6 Determine B and L for the circuit of Figure P 10.8-6 when $i(t) = B \cos (3t - 51.87°)$ A.

Answer: $B = 1.6$ and $L = 2$ H

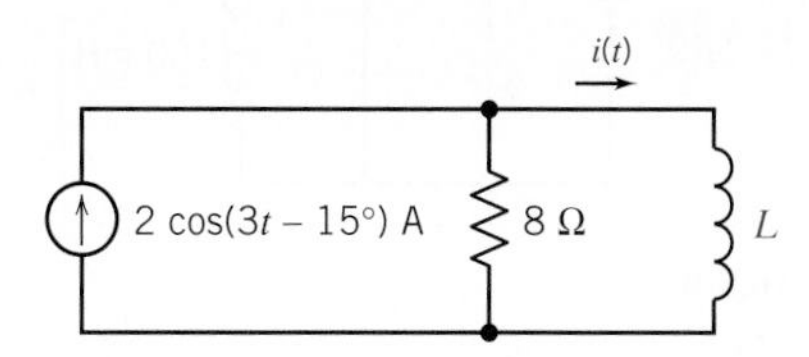

Figure P 10.8-6

P 10.8-7 Determine $i(t)$, $v(t)$, and L for the circuit shown in Figure P 10.8-7.

Answer: $i(t) = 1.34 \cos (2t - 87°)$ A, $v(t) = 7.29 \cos (2t - 24°)$ V, and $L = 4$ H

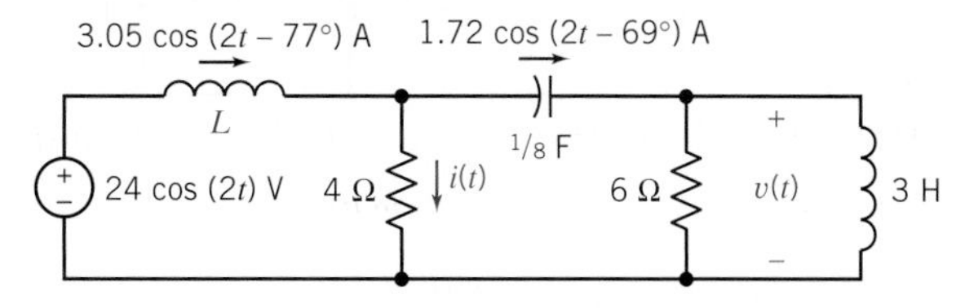

Figure P 10.8-7

P 10.8-8 Spinal cord injuries result in paralysis of the lower body and can cause loss of bladder control. Numerous electrical devices have been proposed to replace the normal nerve pathway stimulus for bladder control. Figure P 10.8-8 shows the model of a bladder control system in which $v_s = 30 \cos \omega t$ V and $\omega = 100$ rad/s. Find the steady-state voltage across the 10-Ω load resistor.

Answer: $v(t) = 15\sqrt{2} \cos (100t + 45°)$ V

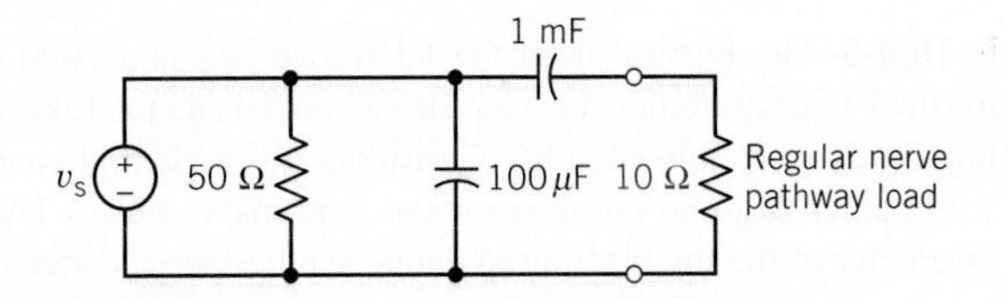

Figure P 10.8-8

P 10.8-9 There are 500 to 1000 deaths each year in the United States from electric shock. If a person makes a good contact with his hands, the circuit can be represented by Figure P 10.8-9, in which $v_s = 200 \cos \omega t$ V and $\omega = 2\pi f$. Find the steady-state current i flowing through the body when (a) $f = 60$ Hz and (b) $f = 400$ Hz.

Answer: (a) $i(t) = 0.66 \cos (120\pi t + 5.9°)$
(b) $i(t) = 0.781 \cos (800\pi t + 59.9°)$ A

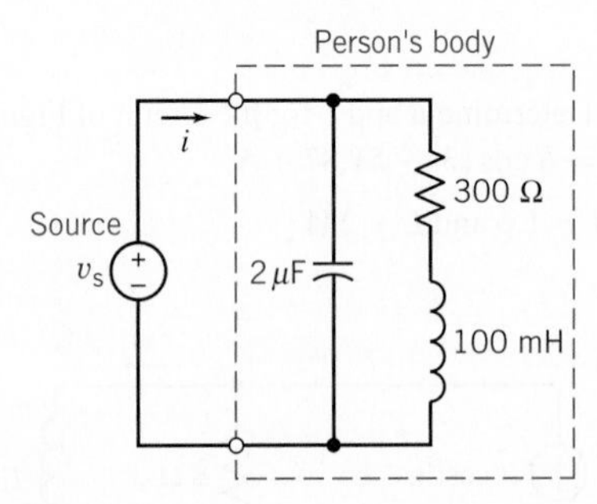

Figure P 10.8-9

P 10.8-10 Determine the steady-state voltage, $v(t)$, and current, $i(t)$, for each of the circuits shown in Figure P 10.8-10.

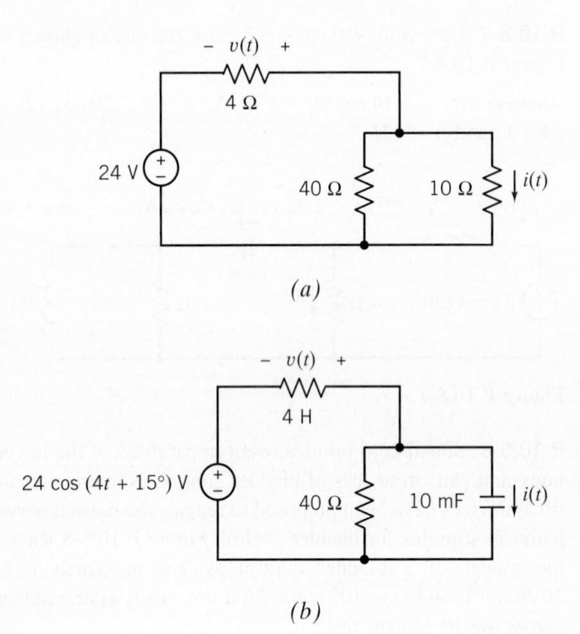

Figure P 10.8-10

P 10.8-11 Determine the steady-state current, $i(t)$, for the circuit shown in Figure P 10.8-11.

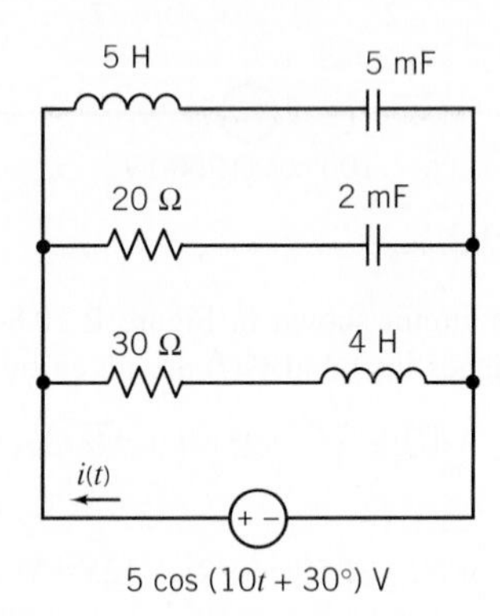

Figure P 10.8-11

P 10.8-12 Determine the steady-state voltage, $v(t)$, for the circuit shown in Figure P 10.8-12.

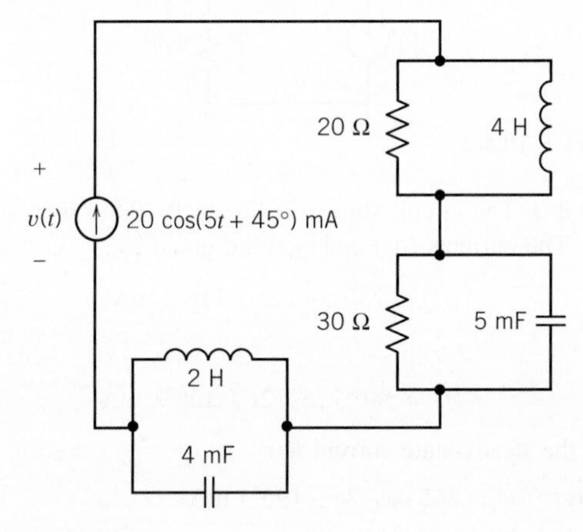

Figure P 10.8-12

P 10.8-13 Determine the steady-state voltage, $v(t)$, for the circuit shown in Figure P 10.8-13.

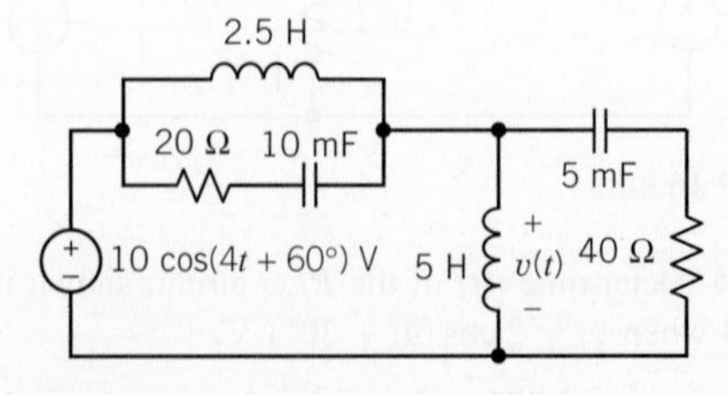

Figure P 10.8-13

P 10.8-14 The input to the circuit shown in Figure P 10.8-14 is the current source current

$$i_s(t) = 25 \cos (10t + 15°) \text{ mA}$$

The output is the current $i_1(t)$. Determine the steady-state response, $i_1(t)$.

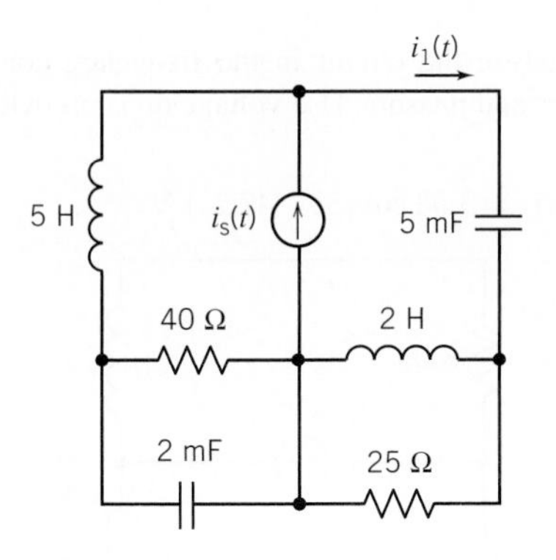

Figure P 10.8-14

P 10.8-15 Determine the steady-state voltage, $v(t)$, and current, $i(t)$, for each of the circuits shown in Figure P 10.8-15.

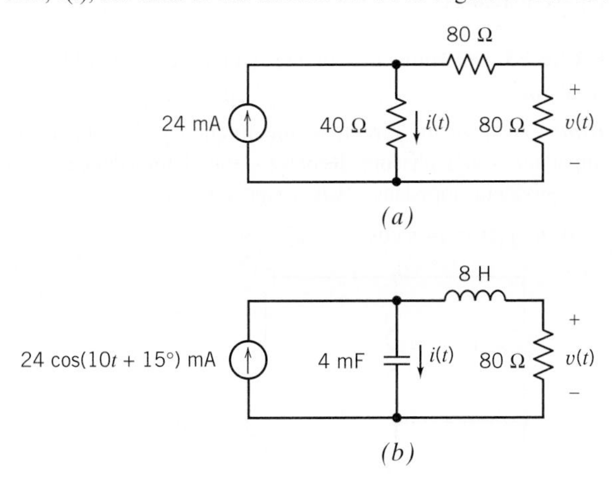

Figure P 10.8-15

P 10.8-16 Determine the steady-state current, $i(t)$, for the circuit in Figure P 10.8-16.

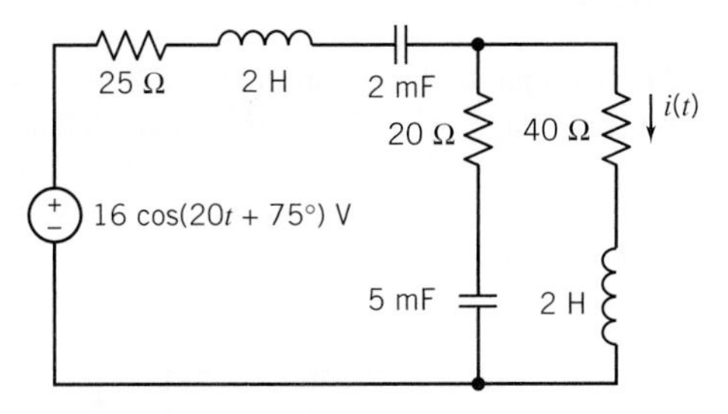

Figure P 10.8-16

P 10.8-17 When the switch in the circuit shown in Figure P 10.8-17 is open and the circuit is at steady state, the capacitor voltage is

$$v(t) = 14.14 \cos(100t - 45^\circ) \text{ V}$$

When the switch is closed and the circuit is at steady state, the capacitor voltage is

$$v(t) = 17.89 \cos(100t - 26.6^\circ) \text{ V}$$

Determine the values of the resistances R_1 and R_2.

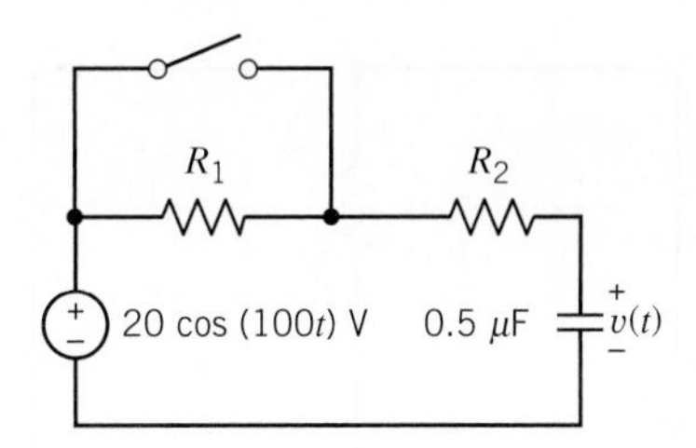

Figure P 10.8-17

P 10.8-18 Determine the steady-state current, $i(t)$, for the circuit shown in Figure P 10.8-18.

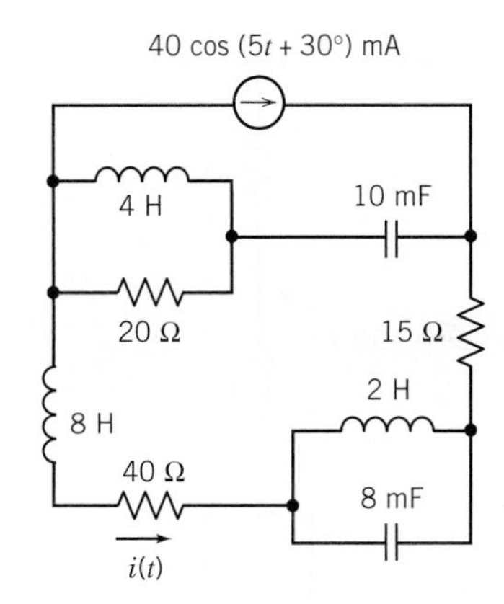

Figure P 10.8-18

P 10.8-19 Determine the steady-state voltage, $v(t)$, and current, $i(t)$, for each of the circuits shown in Figure P 10.8-19.

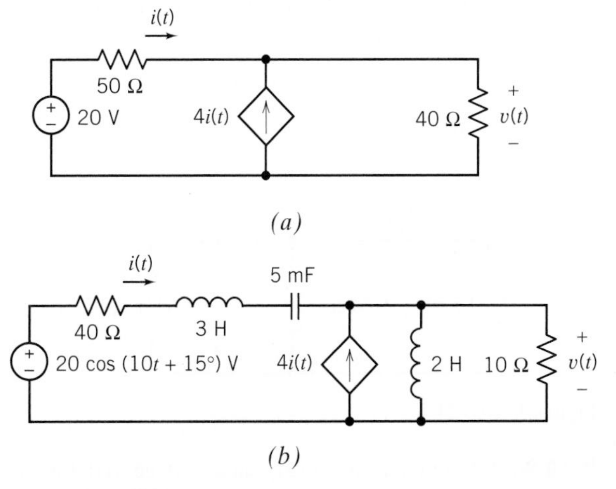

Figure P 10.8-19

P 10.8-20 Determine the steady-state voltage, $v(t)$, for each of the circuits shown in Figure P 10.8-20.

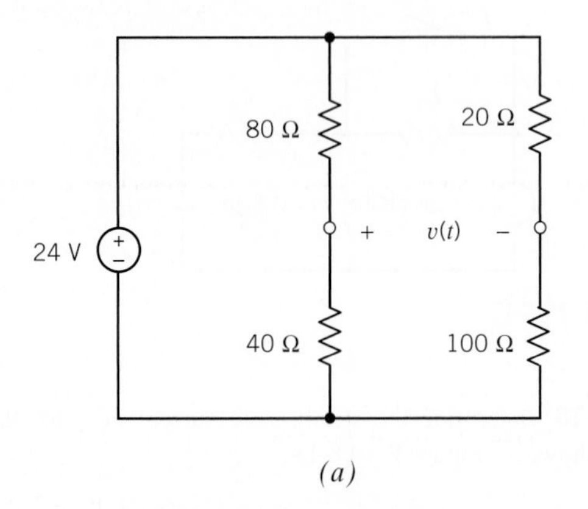

(a)

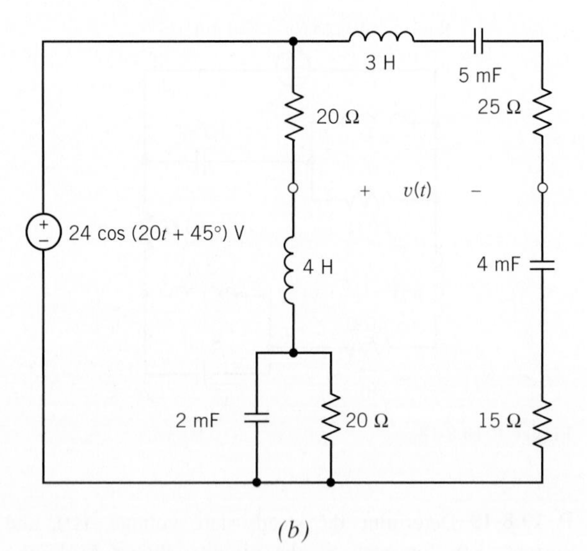

(b)

Figure P 10.8-20

P 10.8-21 The input to the circuit shown in Figure P 10.8-21 is the voltage of the voltage source

$$v_s(t) = 5\cos(2t + 45°)\ \mathrm{V}$$

The output is the inductor voltage, $v(t)$. Determine the steady-state output voltage.

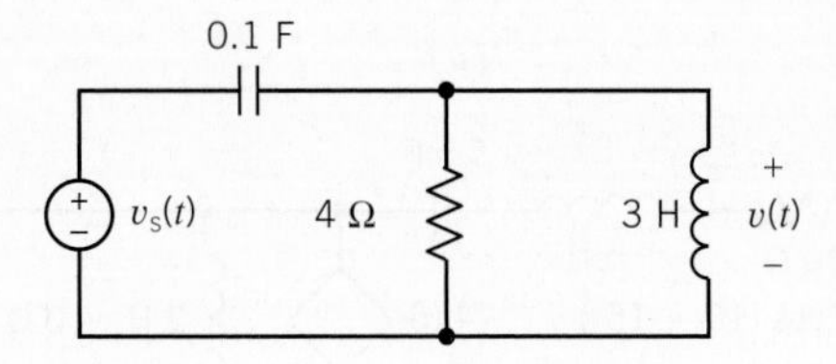

Figure P 10.8-21

P 10.8-22 Determine the steady-state voltage $v(t)$ for the circuit of Figure P 10.8-22.

Hint: Analyze the circuit in the frequency domain, using impedances and phasors. Use voltage division twice. Add the results.

Answer: $v(t) = 3.58\cos(5t + 47.2°)$ V

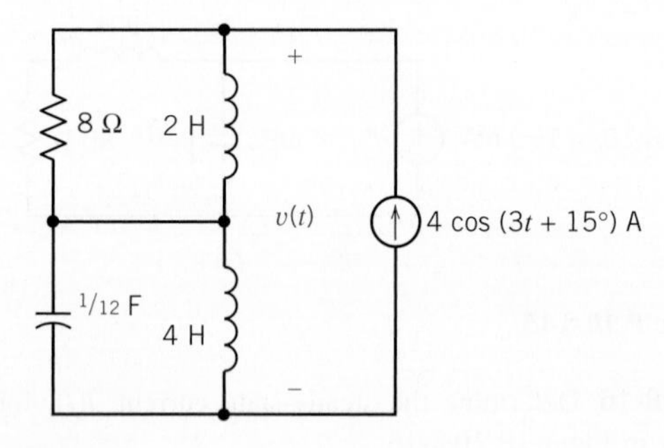

Figure P 10.8-22

P 10.8-23 Determine the voltage $v(t)$ for the circuit of Figure P 10.8-23.

Hint: Analyze the circuit in the frequency domain, using impedances and phasors. Replace parallel impedances with an equivalent impedance twice. Apply KVL.

Answer: $v(t) = 14.4\cos(3t - 22°)$ V

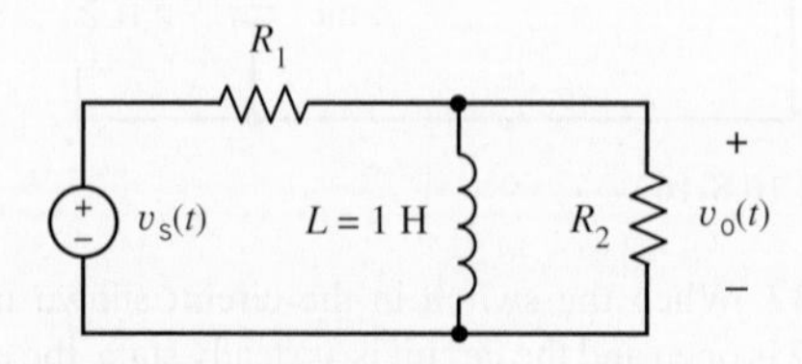

Figure P 10.8-23

P 10.8-24 The input to the circuit in Figure P 10.8-24 is the voltage source voltage, $v_s(t)$. The output is the voltage $v_o(t)$. When the input is $v_s(t) = 8\cos(40t)$ V, the output is $v_o(t) = 2.5\cos(40t + 14°)$ V. Determine the values of the resistances R_1 and R_2.

Figure P 10.8-24

Section 10-9 Node Voltage and Mesh Current Analysis Using Phasors

P 10.9-1 Find the phasor voltage $\mathbf{V}_c$ for the circuit shown in Figure P 10.9-1.

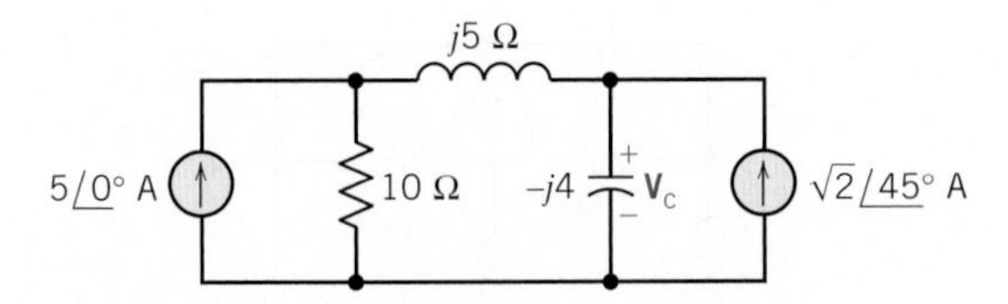

Figure P 10.9-1

P 10.9-2 For the circuit shown in Figure P 10.9-2, determine the phasor currents $\mathbf{I}_s$, $\mathbf{I}_c$, $\mathbf{I}_L$, and $\mathbf{I}_R$ if $\omega = 1000$ rad/s.

Answer: $\mathbf{I}_s = 0.347\angle{-25.5^\circ}$ A

$\mathbf{I}_c = 0.461\angle{112.9^\circ}$ A

$\mathbf{I}_L = 0.720\angle{-67.1^\circ}$ A

$\mathbf{I}_R = 0.230\angle{22.9^\circ}$ A

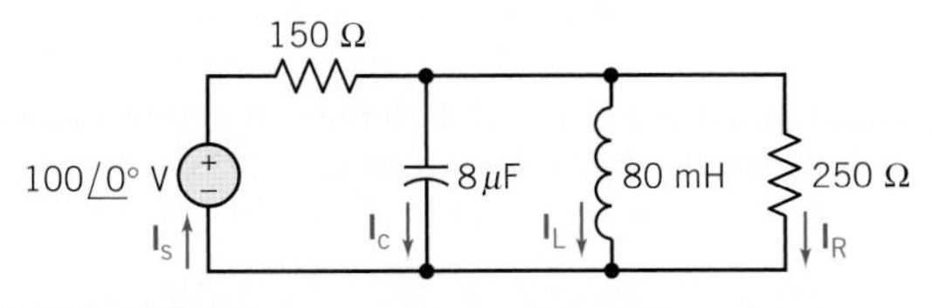

Figure P 10.9-2

P 10.9-3 Find the two node voltages, $v_a(t)$ and $v_b(t)$, for the circuit of Figure P 10.9-3 when $v_s(t) = 1.2\cos 4000t$.

Answer: $v_s(t) = 1.97\cos(4000t - 171^\circ)$ V

$v_b(t) = 2.21\cos(4000t - 144^\circ)$ V

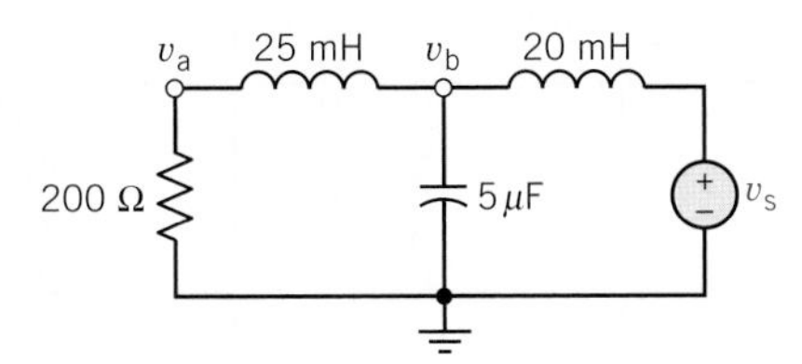

Figure P 10.9-3

P 10.9-4 Determine the voltage v_a for the circuit in Figure P 10.9-4 when $i_s = 20\cos(\omega t + 53.13^\circ)$ A and $\omega = 10^4$rad/s.

Answer: $v_a(t) = 582.1\cos(\omega t + 39.1^\circ)$ V

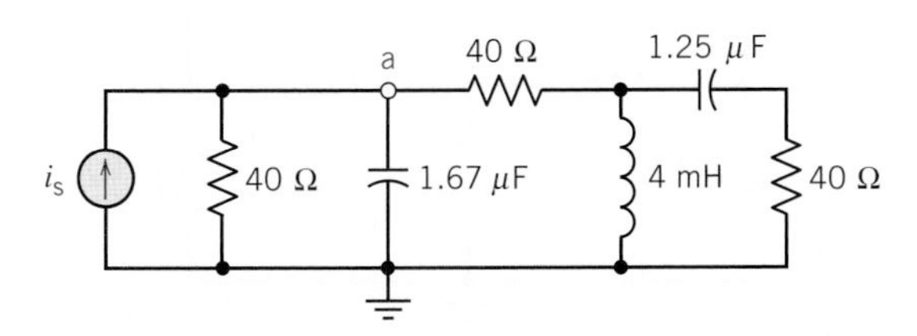

Figure P 10.9-4

P 10.9-5 A commercial airliner has sensing devices to indicate to the cockpit crew that each door and baggage hatch is closed. A device called a search coil magnetometer, also known as a proximity sensor, provides a signal indicative of the proximity of metal or other conducting material to an inductive sense coil. The inductance of the sense coil changes as the metal gets closer to the sense coil. The sense coil inductance is compared to a reference coil inductance with a circuit called a balanced inductance bridge (see Figure P 10.9-5). In the inductance bridge, a signal indicative of proximity is observed between terminals a and b by subtracting the voltage at b, v_b, from the voltage at a, v_a (Lenz, 1990).

The bridge circuit is excited by a sinusoidal voltage source $v_s = \sin(800\pi t)$ V. The two resistors, $R = 100\ \Omega$, are of equal resistance. When the door is open (no metal is present), the sense coil inductance, L_S, is equal to the reference coil inductance, $L_R = 40$ mH. In this case, what is the magnitude of the signal $\mathbf{V}_a - \mathbf{V}_b$?

When the airliner door is completely closed, $L_S = 60$ mH. With the door closed, what is the phasor representation of the signal $\mathbf{V}_a - \mathbf{V}_b$?

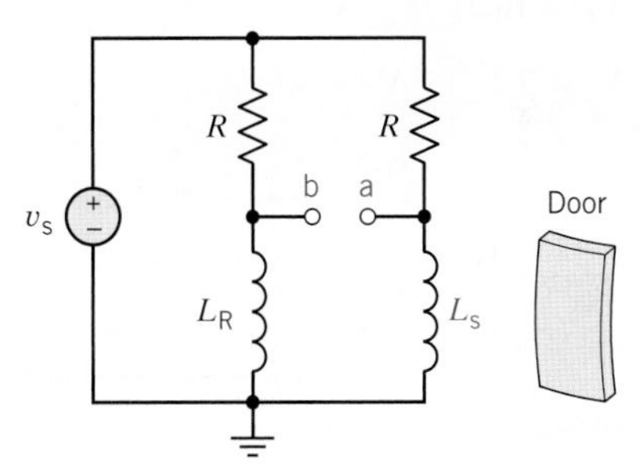

Figure P 10.9-5 Airline door sensing unit.

P 10.9-6 Using a tiny, diamond-studded burr operating at 190,000 rpm, cardiologists can remove life-threatening plaque deposits in coronary arteries. The procedure is fast, uncomplicated, and relatively painless (McCarty, 1991). The Rotablator, an angioplasty system, consists of an advancer/catheter, a guide wire, a console, and a power source. The advancer/catheter contains a tiny turbine that drives the flexible shaft that rotates the catheter burr. The model of the operational and control circuit is shown in Figure P 10.9-6. Determine $v(t)$, the voltage that drives the tip, when $v_s = \sqrt{2}\cos(40t - 135^\circ)$ V.

Answer: $v(t) = \sqrt{2}\cos(40t - 135^\circ)$ V

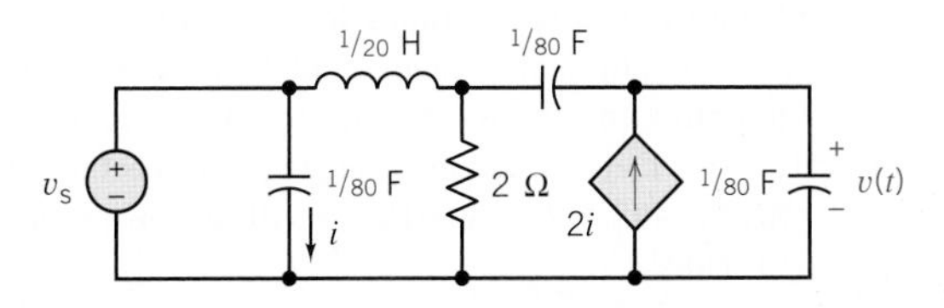

Figure P 10.9-6 Control circuit for Rotablator.

P 10.9-7 For the circuit of Figure P 10.9-7, it is known that

$$v_2(t) = 0.7580 \cos(2t + 66.7^\circ) \text{ V}$$
$$v_3(t) = 0.6064 \cos(2t - 69.8^\circ) \text{ V}$$

Determine $i_1(t)$.

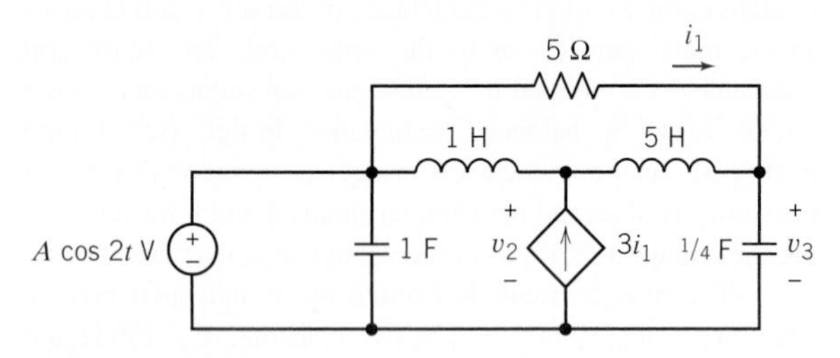

Figure P 10.9-7

P 10.9-8 Determine $\mathbf{I}_1$, $\mathbf{I}_2$, $\mathbf{V}_L$, and $\mathbf{V}_c$ for the circuit of Figure P 10.9-8, using KVL and mesh analysis.

Answer: $\mathbf{I}_1 = 2.5 \angle 29.0^\circ$ A

$\mathbf{I}_2 = 1.8 \angle 105^\circ$ A

$\mathbf{V}_L = 16.3 \angle 78.7^\circ$ V

$\mathbf{V}_c = 7.2 \angle 15^\circ$ V

Figure P 10.9-8

P 10.9-9 Determine the current $i(t)$ for the circuit of Figure P 10.9-9, using mesh currents when $\omega = 1000$ rad/s.

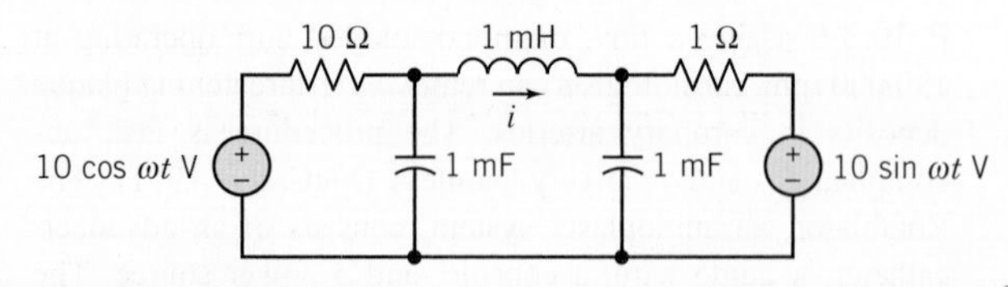

Figure P 10.9-9

P 10.9-10 The idea of using an induction coil in a lamp isn't new, but applying it in a commercially available product is. An induction coil in a bulb induces a high-frequency energy flow in mercury vapor to produce light. The lamp uses about the same amount of energy as a fluorescent bulb but lasts six times longer, with 60 times the life of a conventional incandescent bulb. The circuit model of the bulb and its associated circuit are shown in Figure P 10.9-10. Determine the voltage $v(t)$ across the 2-Ω resistor when $C = 40\ \mu$F, $L = 40\ \mu$H, $v_s = 10 \cos(\omega_0 t + 30^\circ)$, and $\omega_0 = 10^5$ rad/s.

Answer: $v(t) = 6.45 \cos\left(10^5 t + 44^\circ\right)$ V

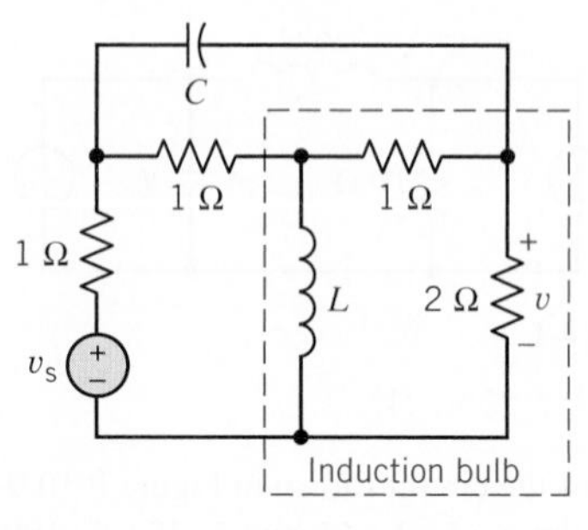

Figure P 10.9-10 Induction bulb circuit.

P 10.9-11 The development of coastal hotels in various parts of the world is a rapidly growing enterprise. The need for environmentally acceptable shark protection is manifest where these developments take place alongside shark-infested waters (Smith, 1991). One concept is to use an electrified line submerged in the water to deter the sharks, as shown in Figure P 10.9-11*a*. The circuit model of the electric fence is shown in Figure P 10.9-11*b*, in which the shark is represented by an equivalent resistance of 100 Ω. Determine the current flowing through the shark's body, $i(t)$, when $v_s = 375 \cos 400t$ V.

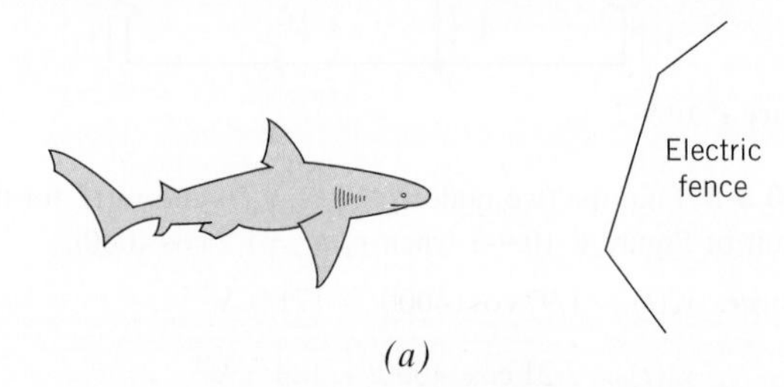

(*a*)

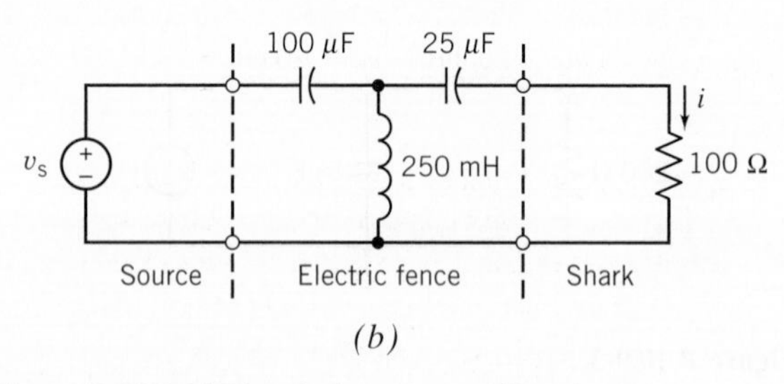

(*b*)

Figure P 10.9-11 Electric fence for repelling sharks.

P 10.9-12 Determine the node voltages at nodes a and b of each of the circuits shown in Figure P 10.9-12.

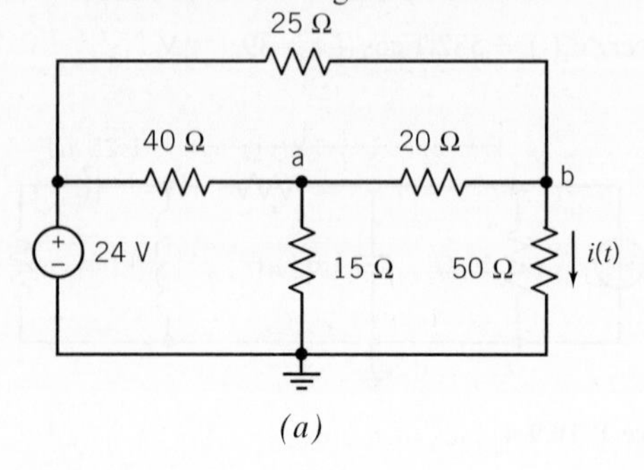

(*a*)

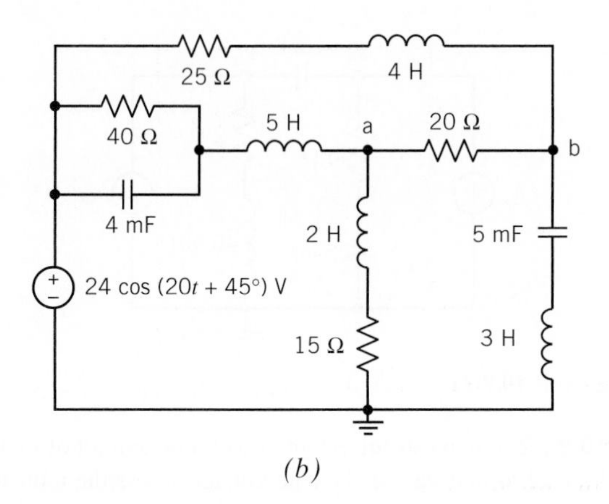

Figure P 10.9-12

P 10.9-13 Determine the steady-state voltage, $v(t)$, for the circuit shown in Figure P 10.9-13.

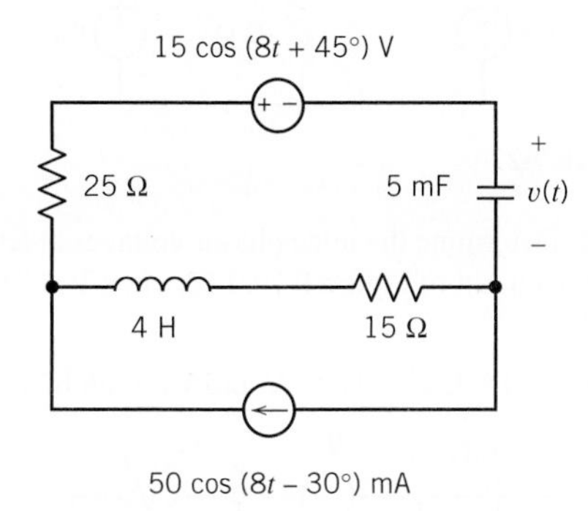

Figure P 10.9-13

P 10.9-14 The input to the circuit shown in Figure P 10.9-14 is the voltage source voltage, $v_s(t)$. The output is the resistor voltage, $v_o(t)$. Determine the output voltage when the circuit is at steady state and the input is

$$v_s(t) = 25 \cos (100t - 15°) \text{ V}$$

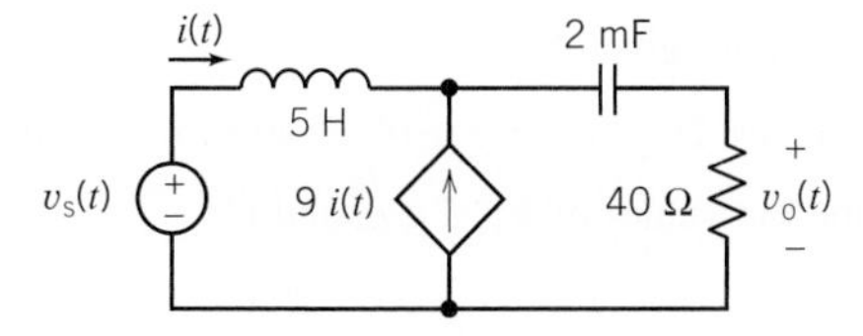

Figure P 10.9-14

P 10.9-15 When the circuit shown in Figure P 10.9-15 is at steady state, the mesh current is

$$i(t) = 0.9 \cos (10t + 138.5°) \text{ A}$$

Determine the values of L and R.

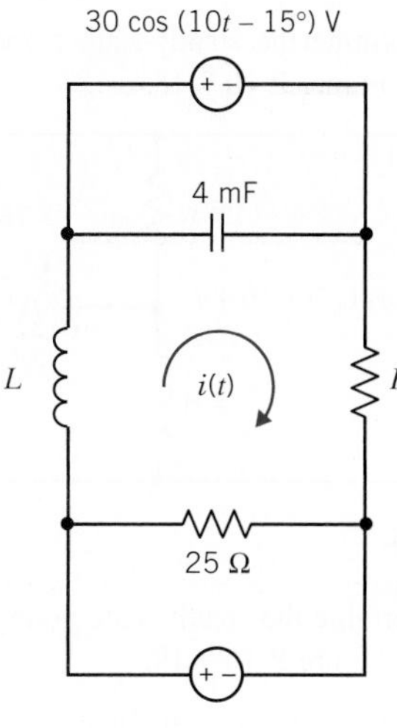

Figure P 10.9-15

P 10.9-16 The circuit shown in Figure P 10.9-16 has two inputs:

$$v_1(t) = 50 \cos (20t - 75°) \text{ V}$$
$$v_2(t) = 35 \cos (20t + 110°) \text{ V}$$

When the circuit is at steady state, the node voltage is

$$v(t) = 21.25 \cos (20t - 168.8°) \text{ V}$$

Determine the values of R and L.

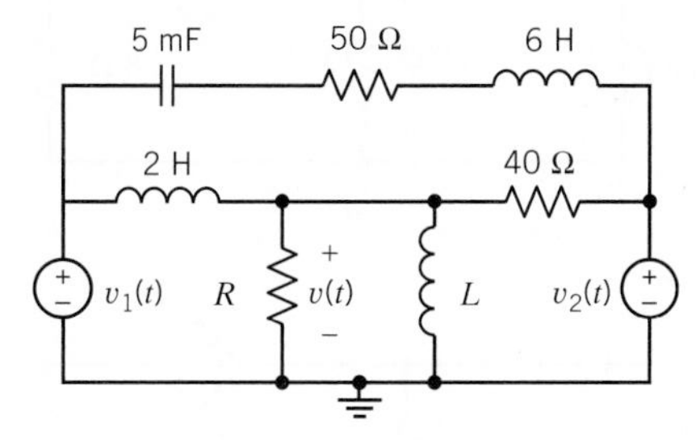

Figure P 10.9-16

P 10.9-17 Determine the steady-state current, $i(t)$, for the circuit shown in Figure P 10.9-17.

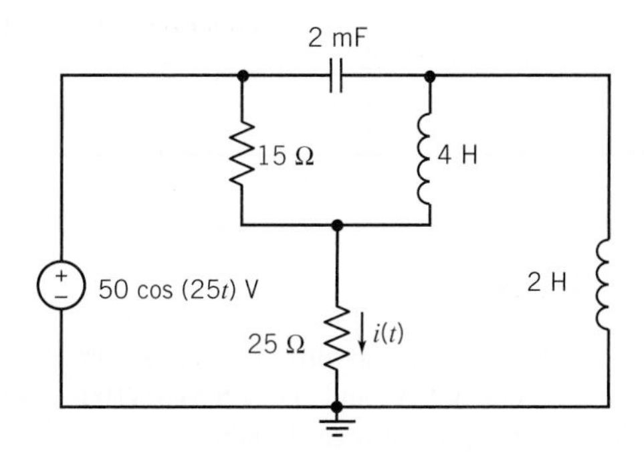

Figure P 10.9-17

P 10.9-18 Determine the steady-state current, $i(t)$, for the circuit shown in Figure P 10.9-18.

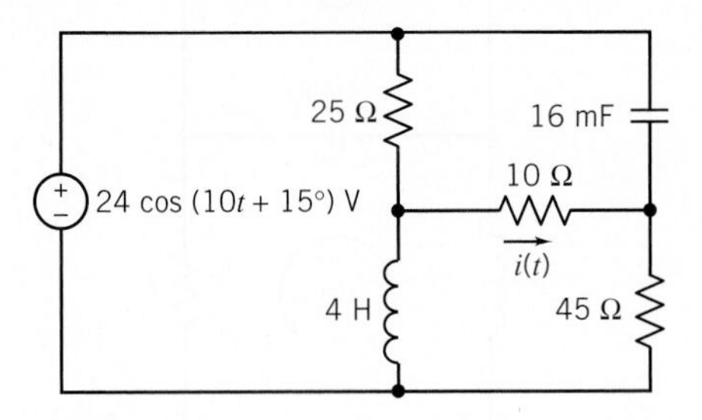

Figure P 10.9-18

P 10.9-19 Determine the steady state voltage, $v_o(t)$, for the circuit shown in Figure P 10.9-19.

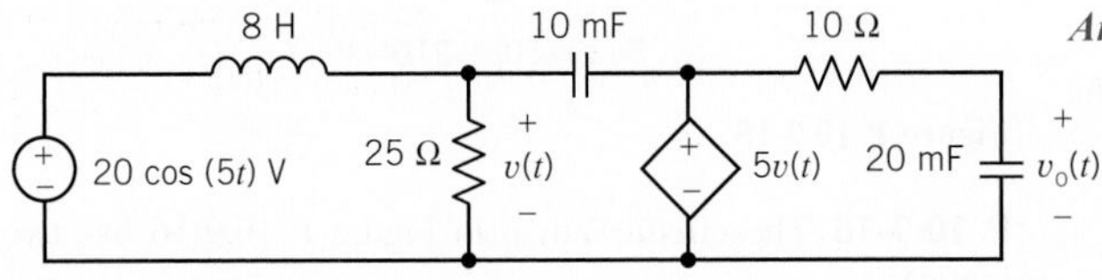

Figure P 10.9-19

P 10.9-20 Determine the steady-state current, $i(t)$, for each of the circuits shown in Figure P 10.9-20.

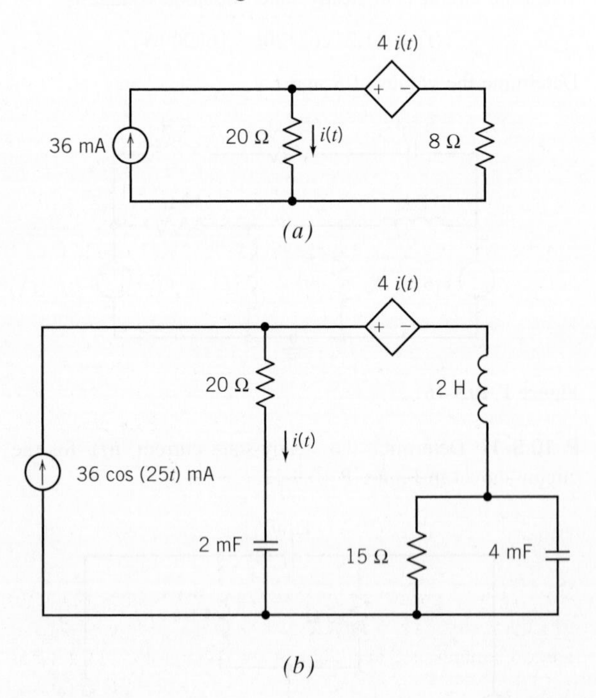

Figure P 10.9-20

P 10.9-21 A circuit has the form shown in Figure P 10.9-21 when $i_{s1} = 1\cos 100t$ A and $i_{s2} = 0.5\cos(100t - 90°)$ A. Find the voltage v_a in the time domain.

Answer: $v_a = \sqrt{5}\cos(100t - 63.5°)$ V

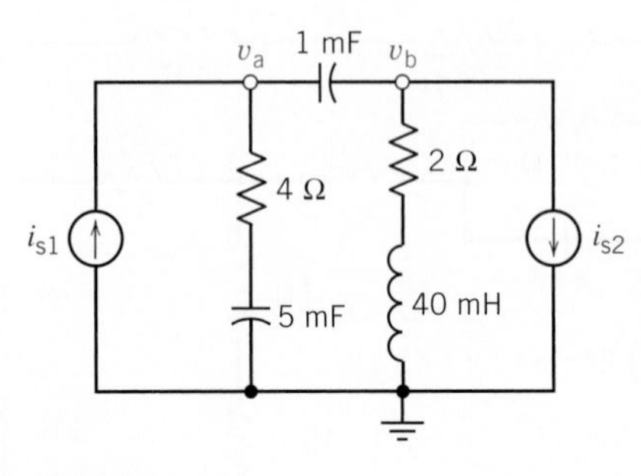

Figure P 10.9-21

P 10.9-22 Use mesh current analysis for the circuit of Figure P 10.9-22 to find the steady-state voltage across the inductor, v_L, when $v_{s1} = 20\cos\omega t$ V, $v_{s2} = 30\cos(\omega t - 90°)$ V, and $\omega = 1000$ rad/s.

Answer: $v_L = 24\sqrt{2}\cos(\omega t + 82°)$ V

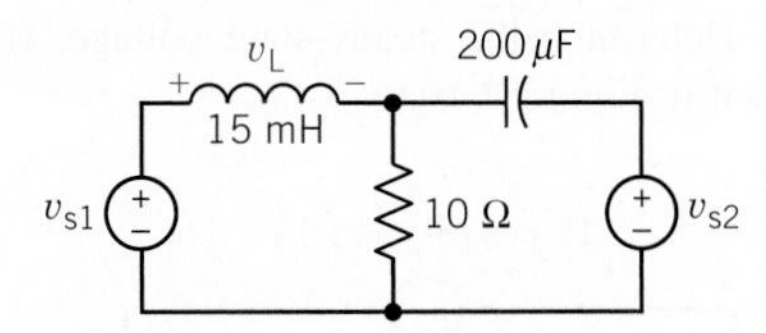

Figure P 10.9-22

P 10.9-23 Determine the node phasor voltages at terminals a and b for the circuit of Figure P 10.9-23 when $\mathbf{V}_s = j50$ V and $\mathbf{V}_1 = j30$ V.

Answer: $\mathbf{V}_a = 14.33\angle{-71.75°}$ V and $\mathbf{V}_b = 36.67\angle{83°}$ V

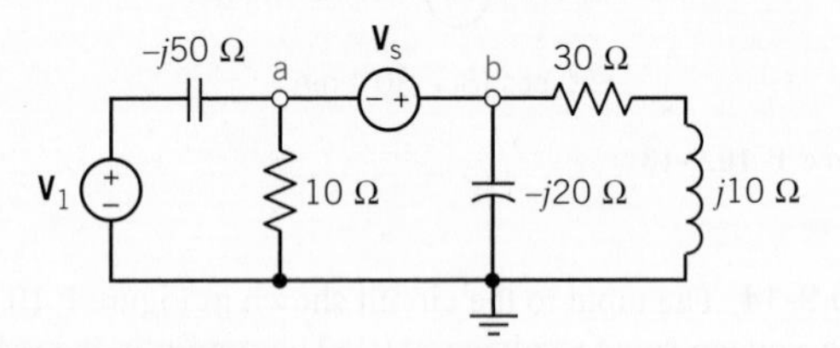

Figure P 10.9-23

P 10.9-24 The circuit shown in Figure P 10.9-24 is at steady state. The voltage source voltages are given by

$$v_1(t) = 12\cos(2t - 90°)\text{ V and } v_2(t) = 5\cos(2t + 90°)\text{ V}$$

The currents are given by

$$i_1(t) = 744\cos(2t - 118°)\text{ mA}, i_2(t) = 540.5\cos(2t + 100°)\text{ mA}$$

Determine the values of R_1, R_2, L, and C

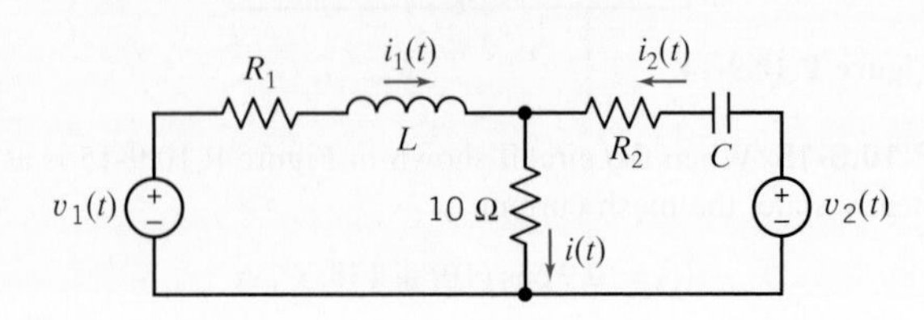

Figure P 10.9-24

Section 10.10 Superposition, Thévenin and Norton Equivalents, and Source Transformations

P 10.10-1 For the circuit of Figure P 10.10-1, find $i(t)$ when $v_1 = 15\cos(4000t + 45°)$ V and $v_2 = 5\cos 3000t$ V.

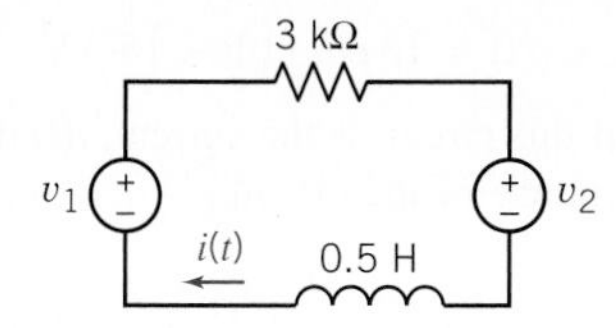

Figure P 10.10-1

P 10.10-2 Determine $i(t)$ of the circuit of Figure P 10.10-2.

Hint: Replace the voltage source by a series combination of a dc voltage and a sinusoidal voltage source.

Answer: $i(t) = 0.166\cos(4t - 135°) + 0.33$ mA

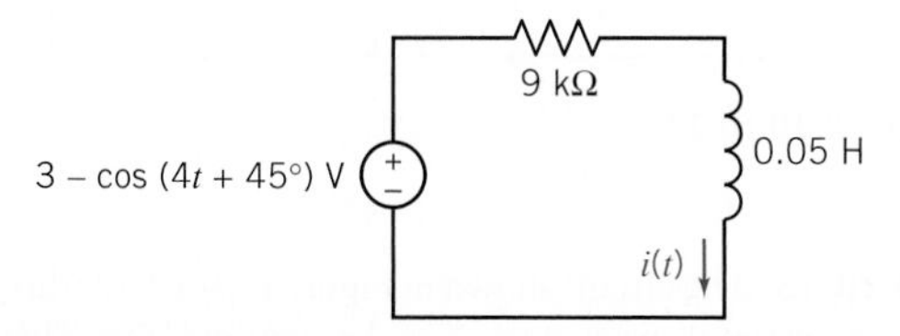

Figure P 10.10-2

P 10.10-3 Determine $i(t)$ for the circuit of Figure P 10.10-3.

Answer: $i(t) = 1.3\cos(4t + 32.5°) - 0.548\cos(3t - 95.4°)$ mA

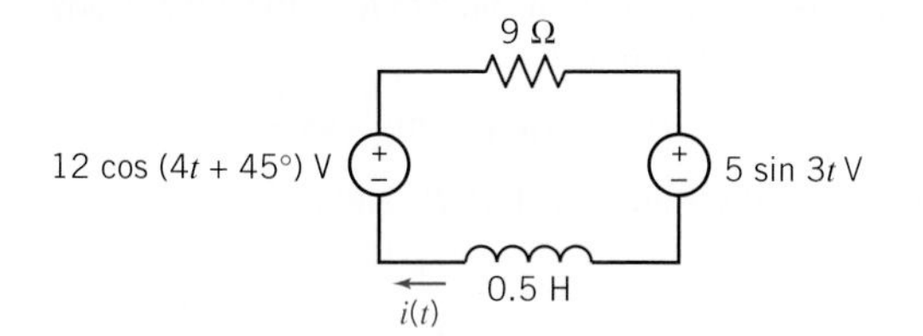

Figure P 10.10-3

P 10.10-4 Determine the Thévenin equivalent circuit for the circuit shown in Figure P 10.10-4 when $v_s = 5\cos(4000t - 30°)$.

Answer: $\mathbf{V}_t = 5.7\angle{-21.9°}$ V

$\mathbf{Z}_t = 23\angle{-81.9°}\ \Omega$

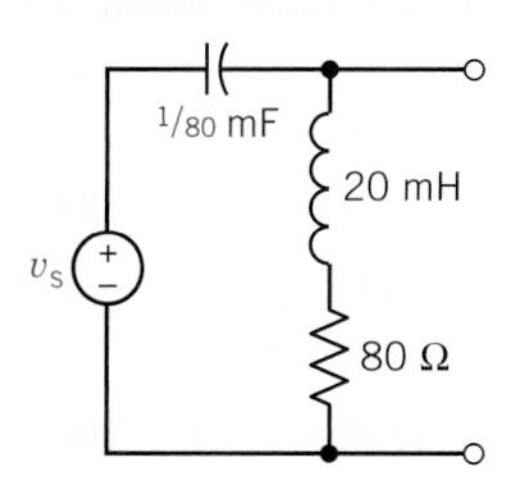

Figure P 10.10-4

P 10.10-5 Find the Thévenin equivalent circuit for the circuit shown in Figure P 10.10-5, using the mesh current method.

Answer: $\mathbf{V}_t = 3.71\angle{-16°}$ V

$\mathbf{Z}_t = 247\angle{-16°}\ \Omega$

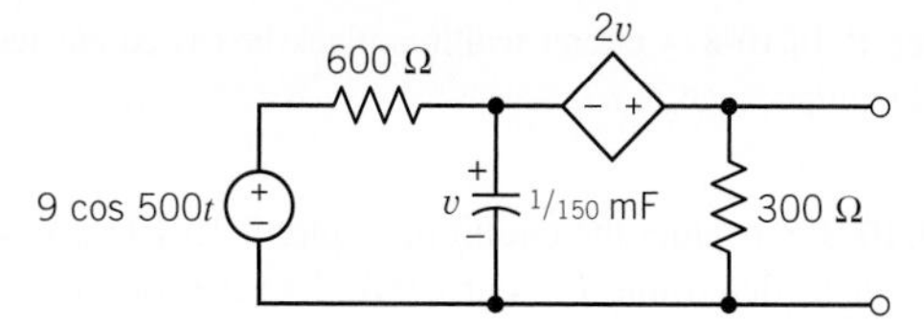

Figure P 10.10-5

P 10.10-6 A pocket-sized minidisc CD player system has an amplifier circuit shown in Figure P 10.10-6 with a signal $v_s = 10\cos(\omega t + 53.1°)$ at $\omega = 10{,}000$ rad/s. Determine the Thévenin equivalent at the output terminals a–b.

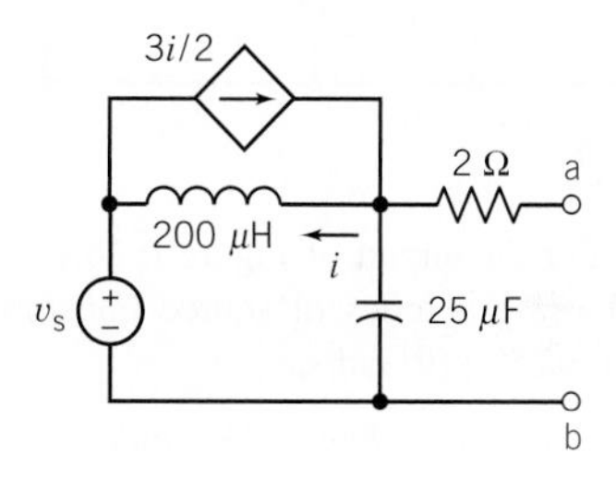

Figure P 10.10-6

P 10.10-7 An AM radio receiver uses the parallel *RLC* circuit shown in Figure P 10.10-7. Determine the frequency, f_0, at which the admittance **Y** is a pure conductance. The AM radio will receive the signal broadcast at the frequency f_0. What is the "number" of this station on the AM radio dial?

Answer: $f_0 = 800$ kHz, which corresponds to 80 on the AM radio dial.

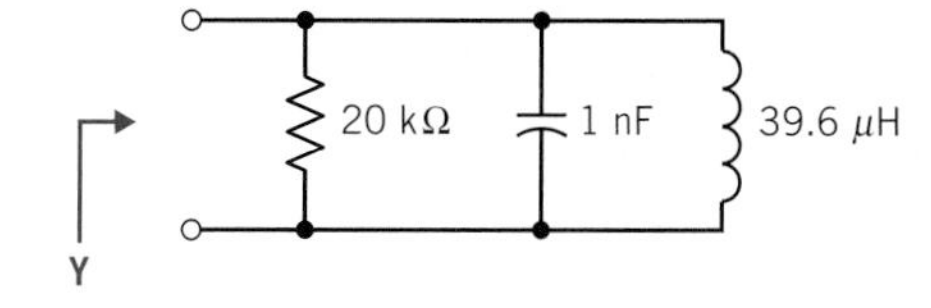

Figure P 10.10-7

P 10.10-8 A linear circuit is placed within a black box with only the terminals a–b available, as shown in Figure P 10.10-8. Three elements are available in the laboratory: (1) a 50-Ω resistor, (2) a 2.5-μF capacitor, and (3) a 50-mH inductor. These three elements are placed across terminals a–b as the load $\mathbf{Z}_L$, and the magnitude of **V** is measured as (1) 25 V, (2) 100 V, and (3) 50 V, respectively. It is known that the sources within the box are sinusoidal with $\omega = 2 \times 10^3$ rad/s. Determine the Thévenin equivalent for the circuit in the box as shown in Figure P 10.10-8.

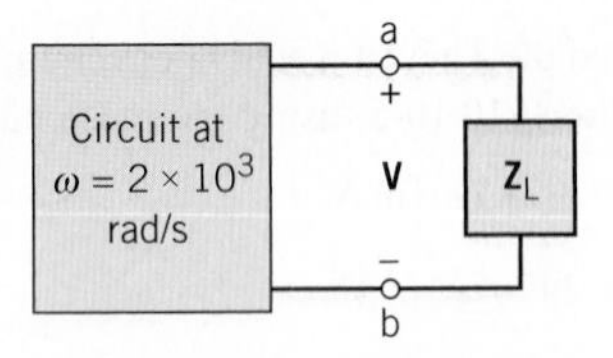

Figure P 10.10-8 A circuit within a black box is connected to a selected impedance $\mathbf{Z}_L$.

P 10.10-9 Consider the circuit of Figure P 10.10-9, of which we wish to determine the current **I**. Use a series of source transformations to reduce the part of the circuit connected to the 2-Ω resistor to a Norton equivalent circuit, and then find the current in the 2-Ω resistor by current division.

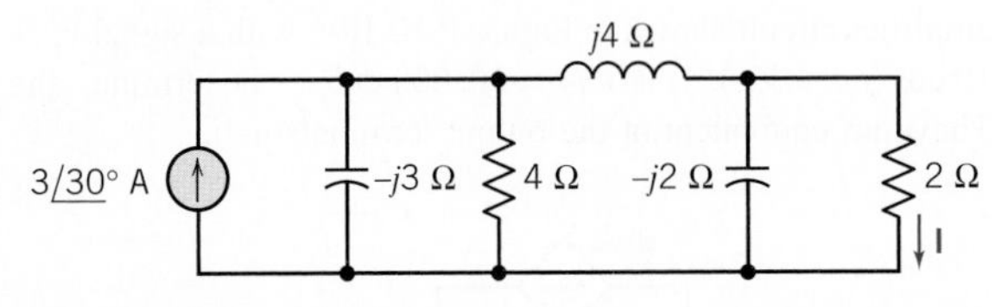

Figure P 10.10-9

P 10.10-10 For the circuit of Figure P 10.10-10, determine the current **I** using a series of source transformations. The source has $\omega = 25 \times 10^3$ rad/s.

Answer: $i(t) = 4 \cos(25{,}000t - 44°)$ mA

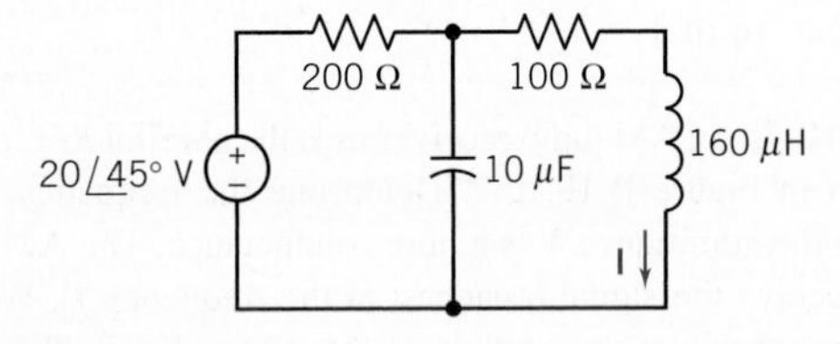

Figure P 10.10-10

P 10.10-11 The input to the circuit shown in Figure P 10.10-11 is the current source current

$$i_s(t) = 36 \cos(25t) + 48 \cos(50t + 45°) \text{ mA}$$

Determine the steady-state current, $i(t)$.

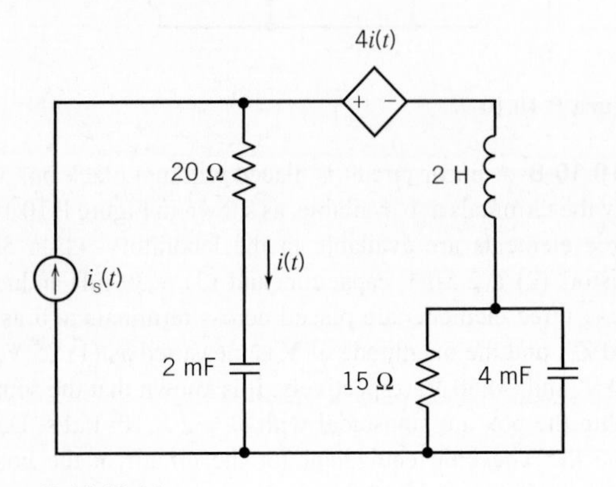

Figure P 10.10-11

P 10.10-12 The inputs to the circuit shown in Figure P 10.10-12 are

$$v_{s1}(t) = 30 \cos(20t + 70°) \text{ V}$$

and

$$v_{s2}(t) = 18 \cos(10t - 15°) \text{ V}$$

The response of this circuit is the current, $i(t)$. Determine the steady-state response of the circuit.

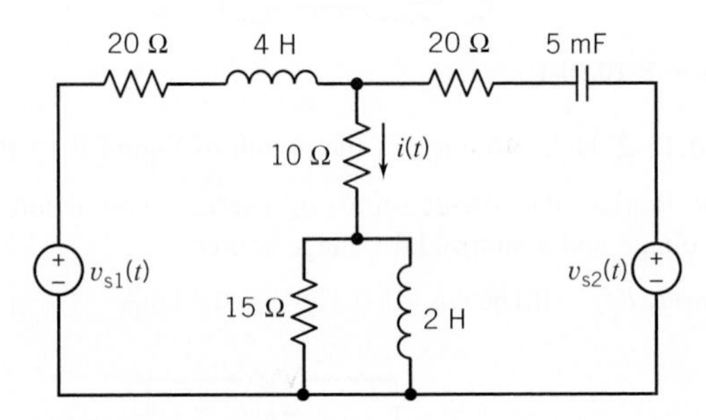

Figure P 10.10-12

P 10.10-13 The circuit shown in Figure P 10.10-13 illustrates an experimental procedure for determining the Thévenin equivalent of an ac circuit. When $R = 20\ \Omega$, the steady-state voltage across terminals a–b is measured to be

$$v(t) = 3.0 \cos(20t - 100.9°) \text{ V}$$

When that resistance is changed to $R = 40\ \Omega$, the steady-state voltage is measured to be

$$v(t) = 4.88 \cos(20t - 95.8°) \text{ V}$$

Determine the values of A, θ, R_t, and L_t.

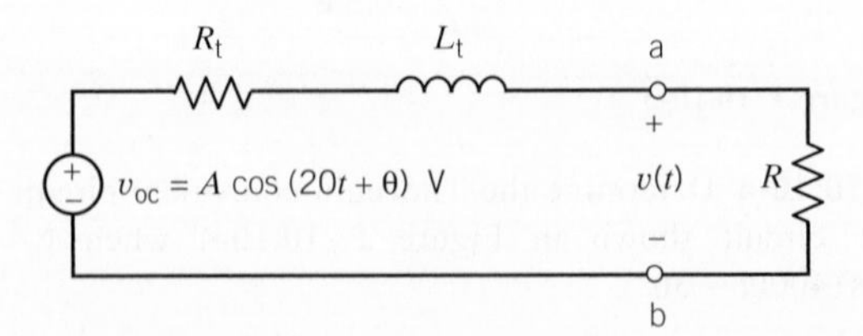

Figure P 10.10-13

P 10.10-14 The circuit in Figure P 10.10-14 illustrates an experimental procedure for determining the Norton equivalent of an ac circuit. When $R = 20\ \Omega$, the steady-state output current is measured to be

$$i(t) = 1.025 \cos(10t - 108.5°) \text{ A}$$

When the resistance is changed to $R = 40\ \Omega$, the steady-state output current is measured to be

$$i(t) = 0.848 \cos(10t - 100.7°) \text{ A}$$

Determine the values of B, θ, R_t, and L_t.

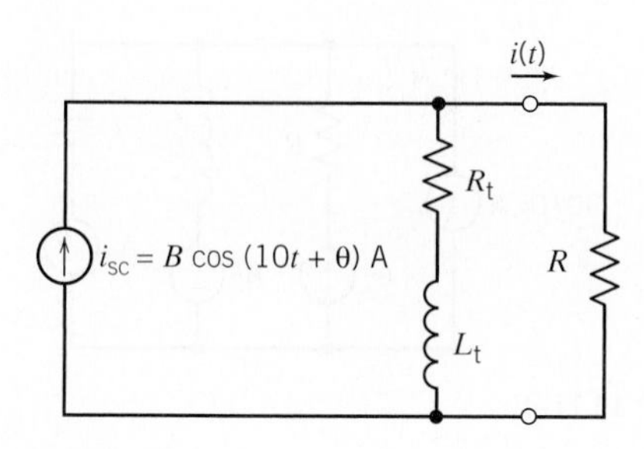

Figure P 10.10-14

***P 10.10-15** The input to the circuit shown in Figure P 10.10-15 is the voltage source voltage

$$v_s(t) = A\cos(25t + \theta)\text{ V}$$

The output is the voltage, $v(t)$. The "*RLC* circuit" consists only of resistors, capacitors, and inductors. Consider the following experiment. We connect a series resistor and inductor between terminals a and b, as shown, and measure the steady-state voltage, $v(t)$. When $R = 10\ \Omega$; and $L = 5$ H, we measure

$$v(t) = 7.063\cos(25t + 50.2^\circ)\text{ V}$$

When we change the series resistance and inductance to $R = 25\ \Omega$ and $L = 10$ H, we measure

$$v(t) = 8.282\cos(25t + 47.8^\circ)\text{ V}$$

Determine the steady-state voltage, $v(t)$, that we will measure after changing the resistance and inductance to $R = 10\ \Omega$ and $L = 8$ H.

Hint: Determine the Thévenin equivalent of the circuit to the left of terminals a–b.

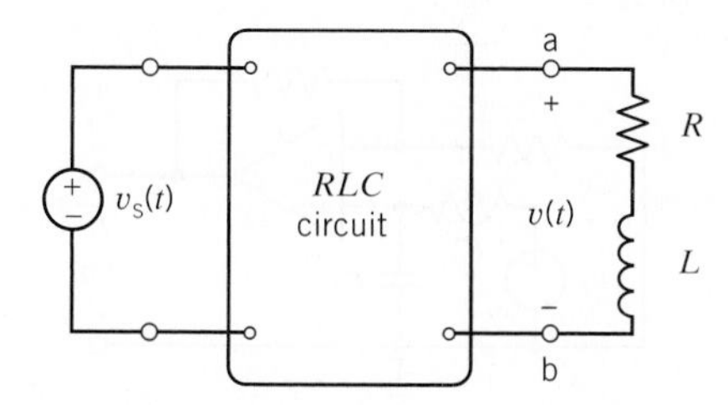

Figure P 10.10-15

***P 10.10-16** The input to the circuit shown in Figure P 10.10-16 is the voltage source voltage

$$v_s(t) = A\cos(15t + \theta)\text{ V}$$

The output is the voltage, $v(t)$. The "*RLC* circuit" consists only of resistors, capacitors, and inductors. Consider the following experiment. We connect a 25-Ω resistor across terminals a–b as shown and measure the steady-state output voltage to be

$$v(t) = 9.77\cos(15t + 31.6^\circ)\text{ V}$$

Next, we replace the 25-Ω resistor with a 4-H inductor and measure the steady-state output voltage to be

$$v(t) = 18.9\cos(15t + 90.9^\circ)\text{ V}$$

Then we replace the 4-H inductor with a capacitor having capacitance C and measure the steady-state output voltage as

$$v(t) = B\cos(15t - 45^\circ)\text{ V}$$

What value of capacitance C is required to cause the phase angle of the output to be -45°?

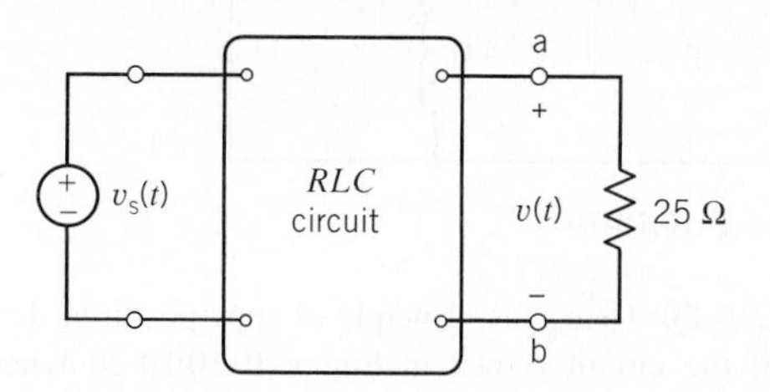

Figure P 10.10-16

P 10.10-17 The input to the circuit shown in Figure P 10.10-17 is the voltage source voltage

$$v_s(t) = 5 + 30\cos(100t)\text{ V}$$

Determine the steady-state current, $i(t)$.

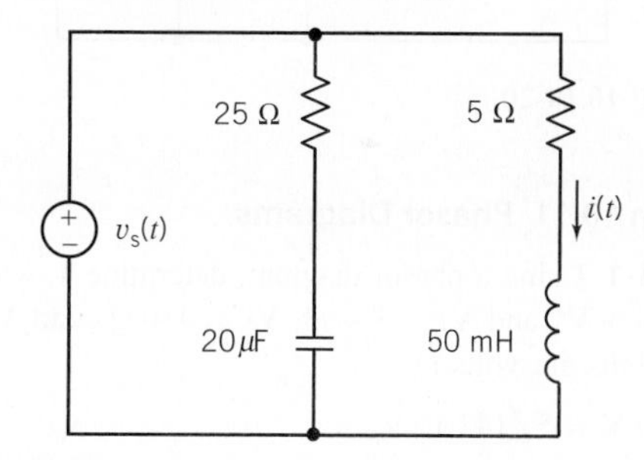

Figure P 10.10-17

P 10.10-18 Determine the value of $\mathbf{V}_t$ and $\mathbf{Z}_t$ such that the circuit shown in Figure P 10.10-18*b* is the Thévenin equivalent circuit of the circuit shown in Figure P 10.10-18*a*.

Answer: $\mathbf{V}_t = 3.58\angle 47^\circ$ and $\mathbf{Z}_t = 4.9 + j1.2\ \Omega$

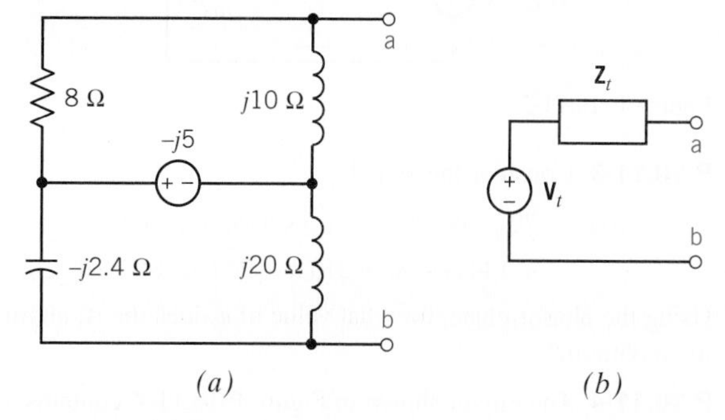

Figure P 10.10-18

P 10.10-19 Determine the voltage $v(t)$ for the circuit of Figure P 10.10-19.

Hint: Use superposition.

Answer: $v(t) = 3.58\cos(5t + 47.2^\circ) + 14.4\cos(3t - 22^\circ)$ V

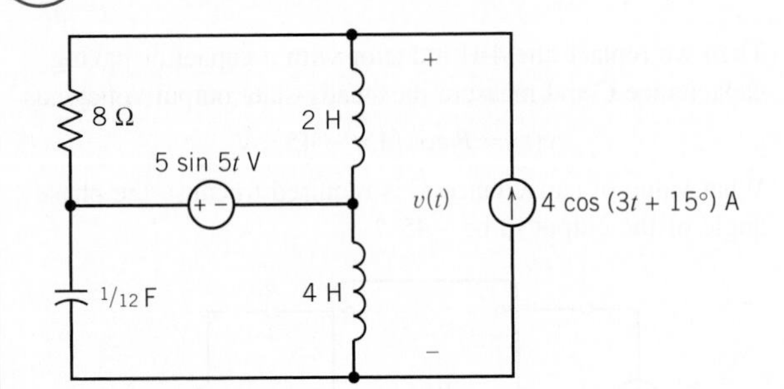

Figure P 10.10-19

P 10.10-20 Using the principle of superposition, determine $i(t)$ of the circuit shown in Figure P 10.10-20 when $v_1 = 10 \cos 10t$ V.

Answer: $i = -2 + 0.71 \cos(10t - 45°)$ A

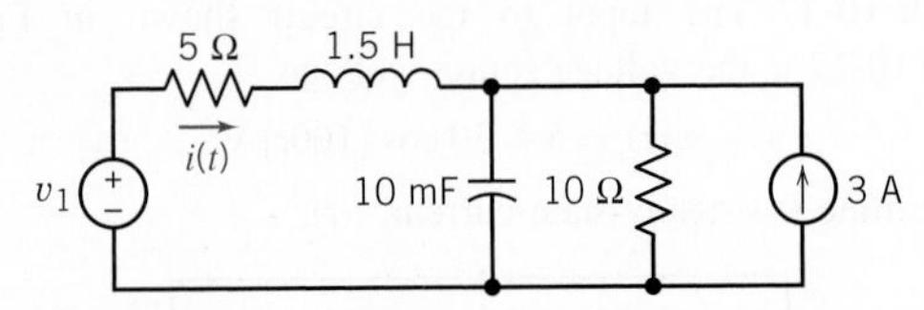

Figure P 10.10-20

Section 10.11 Phasor Diagrams

P 10.11-1 Using a phasor diagram, determine **V** when $\mathbf{V} = \mathbf{V}_1 - \mathbf{V}_2 + \mathbf{V}_3^*$ and $\mathbf{V}_1 = 3 + j3$, $\mathbf{V}_2 = 4 + j2$, and $\mathbf{V}_3 = -3 - j2$. (Units are volts.)

Answer: $\mathbf{V} = 5\angle 143.1°$ V

P 10.11-2 Consider the series *RLC* circuit of Figure P 10.11-2 when $R = 10\ \Omega$, $L = 1$ mH, $C = 100\ \mu$F, and $\omega = 10^3$ rad/s. Find **I** and plot the phasor diagram.

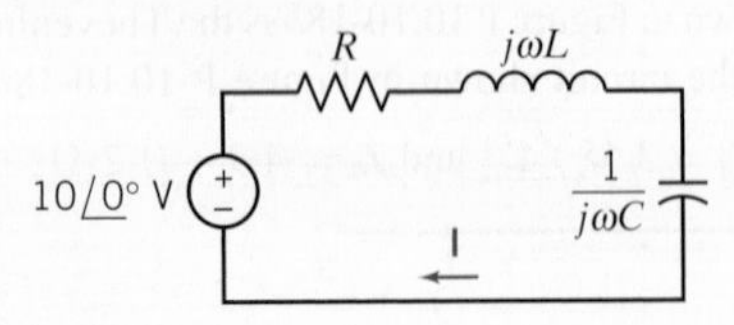

Figure P 10.11-2

P 10.11-3 Consider the signal

$$i(t) = 72\sqrt{3} \cos 8t + 40\sqrt{3} \sin(8t + 140°) + 144 \cos(8t + 210°) + 25 \cos(8t + \phi)$$

Using the phasor plane, for what value of ϕ does the $|\mathbf{I}|$ attain its maximum?

P 10.11-4 The circuit shown in Figure P 10.11-4 contains a sinusoidal current source of $25\ \angle 0°$ A. An ammeter reads the magnitude of the current. Ammeter A_1 reads 15 A, and ammeter A_2 reads 6 A. Find the reading of ammeter A_3.

Hint: The ammeter measures the magnitude of the current through the ammeter.

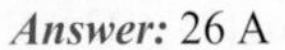

Answer: 26 A

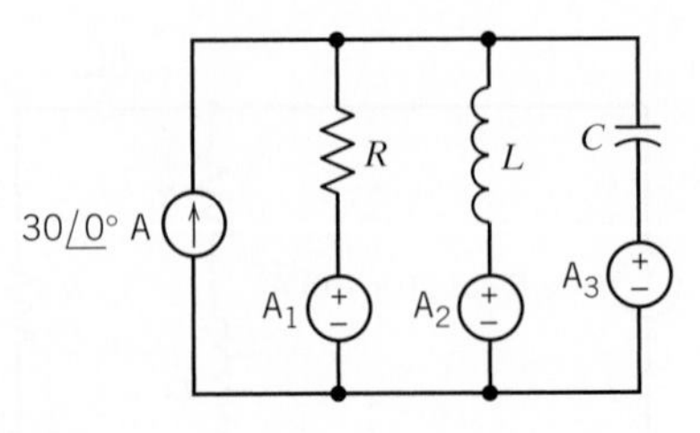

Figure P 10.11-4

Section 10.12 Phasor Circuits and the Operational Amplifier

P 10.12-1 Find the steady-state response $v_o(t)$ if $v_s(t) = \sqrt{2} \cos 1000t$ for the circuit of Figure P 10.12-1.

Answer: $v_o(t) = 10 \cos(1000t - 225°)$

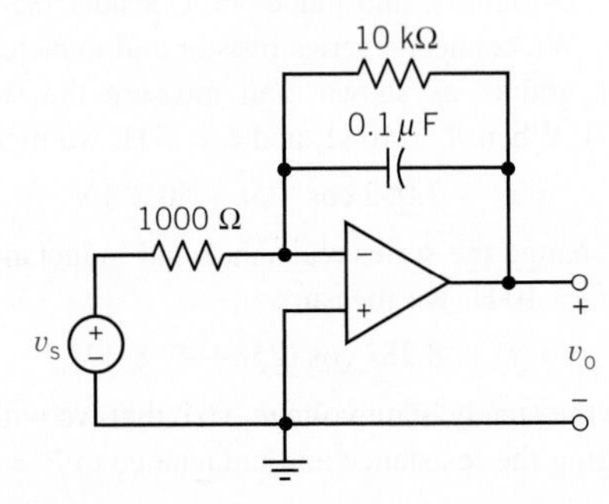

Figure P 10.12-1

P 10.12-2 Determine $\mathbf{V}_o/\mathbf{V}_s$ for the op amp circuit shown in Figure P 10.12-2.

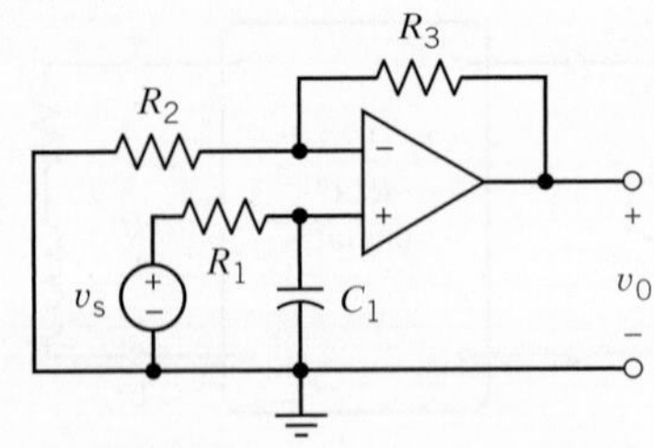

Figure P 10.12-2 Amplifier circuit for disc player.

P 10.12-3 Determine $\mathbf{V}_o/\mathbf{V}_s$ for the op amp circuit shown in Figure P 10.12-3.

Answer: $\dfrac{\mathbf{V}_o}{\mathbf{V}_s} = \dfrac{j\omega R_1 C_1 (1 + R_3/R_2)}{1 + j\omega R_1 C_1}$

Figure P 10.12-3

P 10.12-4 For the circuit of Figure P 10.12-4, determine $v_o(t)$ when $v_s = 20 \cos \omega t$ mV and $f = 10$ kHz.

Answer: $v_o = 0.5 \cos(\omega t - 89.5°)$ mV

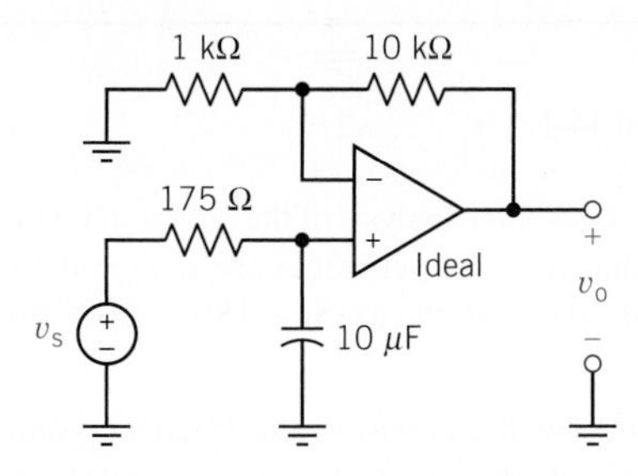

Figure P 10.12-4

P 10.12-5 Determine the ratio $\mathbf{V}_o/\mathbf{V}_s$ for the circuit shown in Figure P 10.12-5.

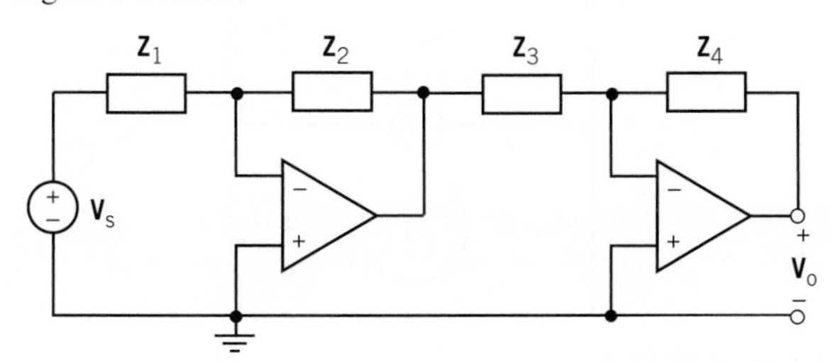

Figure P 10.12-5

P 10.12-6 Determine the ratio $\mathbf{V}_o/\mathbf{V}_s$ for both of the circuits shown in Figure P 10.12-6.

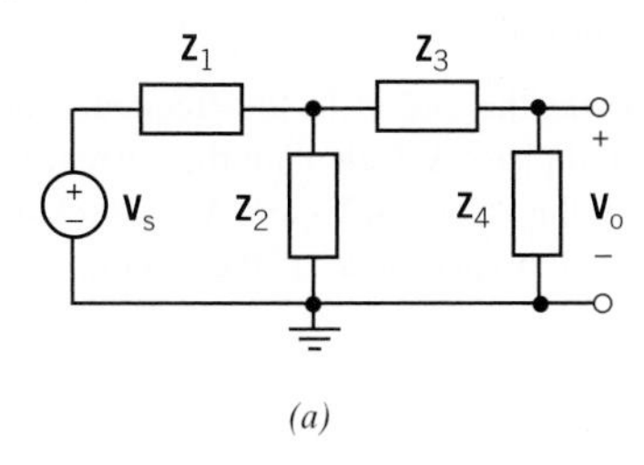

(a)

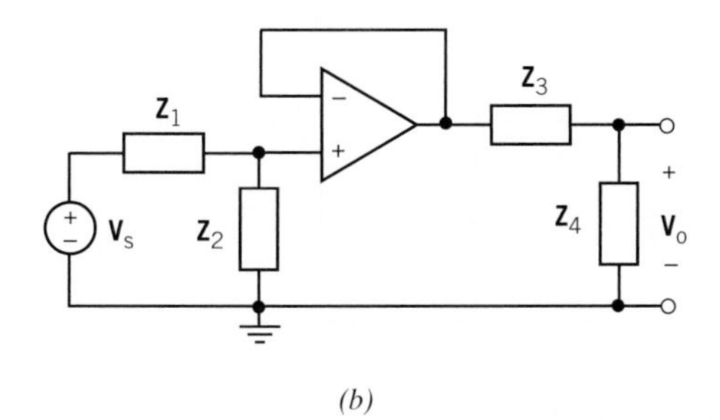

(b)

Figure P 10.12-6

P 10.12-7 Determine the ratio $\mathbf{V}_o/\mathbf{V}_s$ for the circuit shown in Figure P 10.12-7.

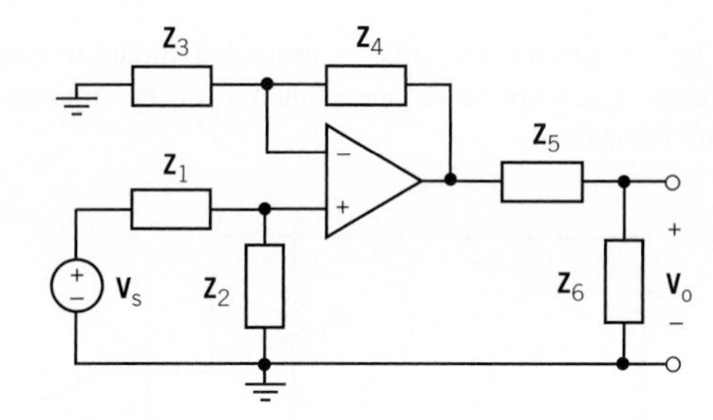

Figure P 10.12-7

P 10.12-8 When the input to the circuit shown in Figure P 10.12-8 is the voltage source voltage

$$v_s(t) = 2 \cos(1000t) \text{ V}$$

the output is the voltage

$$v_o(t) = 6 \cos(1000t - 71.6°) \text{ V}$$

Determine the values of the resistances R_1 and R_2.

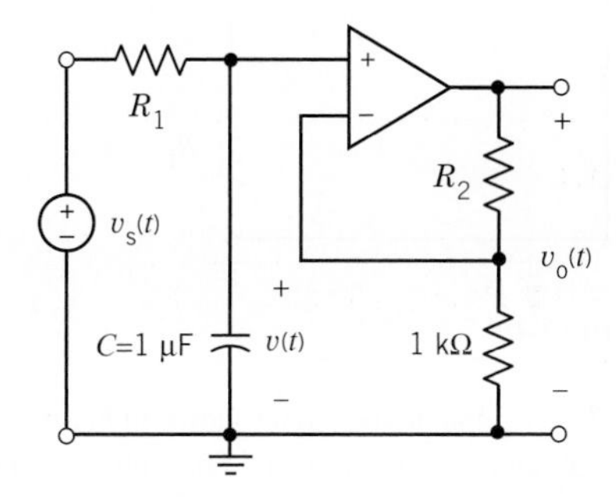

Figure P 10.12-8

P 10.12-9 When the input to the circuit shown in Figure P 10.12-9 is the voltage source voltage

$$v_s(t) = 4 \cos(100t) \text{ V}$$

the output is the voltage

$$v_o(t) = 8 \cos(100t + 135°) \text{ V}$$

Determine the values of C and R

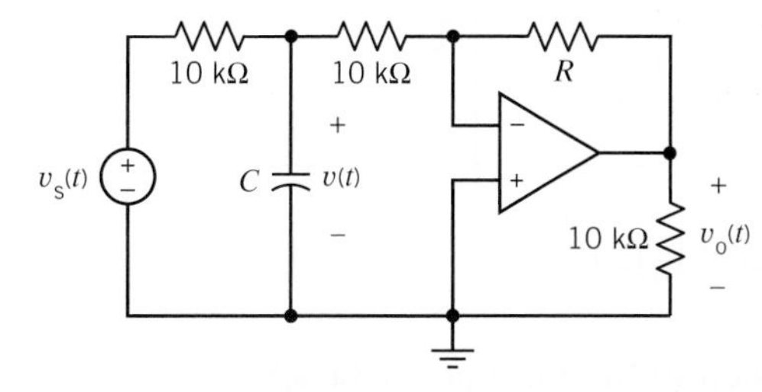

Figure P 10.12-9

P 10.12-10 The circuit shown in Figure P 10.12-10 is called a grounded simulated inductor because its input impedance is given by

$$\mathbf{Z} = j\omega L_{eq}$$

That is, the circuit acts like a grounded inductor having inductance L_{eq}. Express L_{eq} as a function of the capacitance and the resistances.

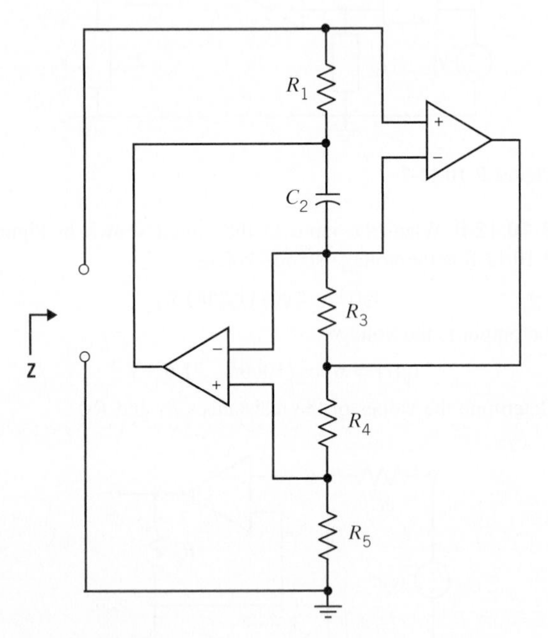

Figure P 10.12-10

P 10.12-11 The input to the circuit shown in Figure P 10.12-11 is the voltage source voltage, $v_s(t)$. The output is the voltage $v_o(t)$. The input $v_s(t) = 2.5\cos(1000t)$ V causes the output to be $v_o(t) = 8\cos(1000t + 104°)$ V. Determine the values of the resistances R_1 and R_2.

Answers: $R_1 = 1515\ \Omega$ and $R_2 = 20\ k\Omega$.

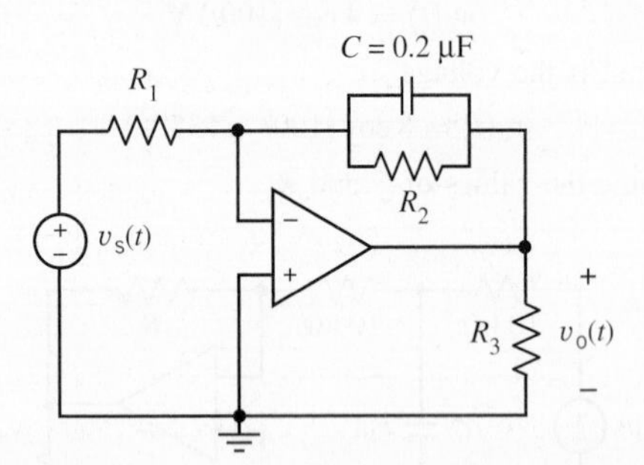

Figure P 10.12-11

Section 10.16 How Can We Check . . . ?

P 10.16-1 Computer analysis of the circuit in Figure P 10.16-1 indicates that the values of the node voltages are $\mathbf{V}_1 = 20\angle{-90°}$ and $\mathbf{V}_2 = 44.7\angle{-63.4°}$. Are the values correct?

Hint: Calculate the current in each circuit element, using the values of $\mathbf{V}_1$ and $\mathbf{V}_2$. Check to see whether KCL is satisfied at each node of the circuit.

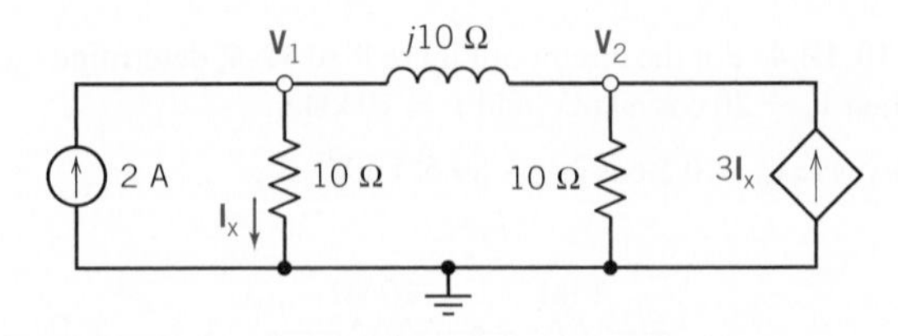

Figure P 10.16-1

P 10.16-2 Computer analysis of the circuit in Figure P 10.16-2 indicates that the mesh currents are $i_1(t) = 0.39\cos(5t + 39°)$ A and $i_2(t) = 0.28\cos(5t + 180°)$ A. Is this analysis correct?

Hint: Represent the circuit in the frequency domain, using impedances and phasors. Calculate the voltage across each circuit element, using the values of $\mathbf{I}_1$ and $\mathbf{I}_2$. Check to see whether KVL is satisfied for each mesh of the circuit.

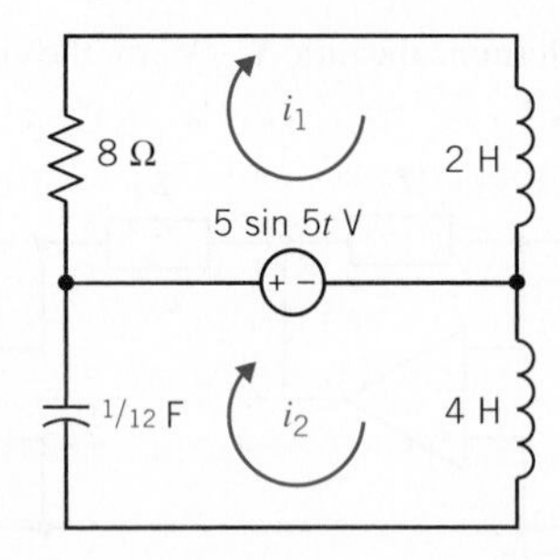

Figure P 10.16-2

P 10.16-3 Computer analysis of the circuit in Figure P 10.16-3 indicates that the values of the node voltages are $v_1(t) = 19.2\cos(3t + 68°)$ V and $v_2(t) = 2.4\cos(3t + 105°)$ V. Is this analysis correct?

Hint: Represent the circuit in the frequency domain, using impedances and phasors. Calculate the current in each circuit element, using the values of $\mathbf{V}_1$ and $\mathbf{V}_2$. Check to see whether KCL is satisfied at each node of the circuit.

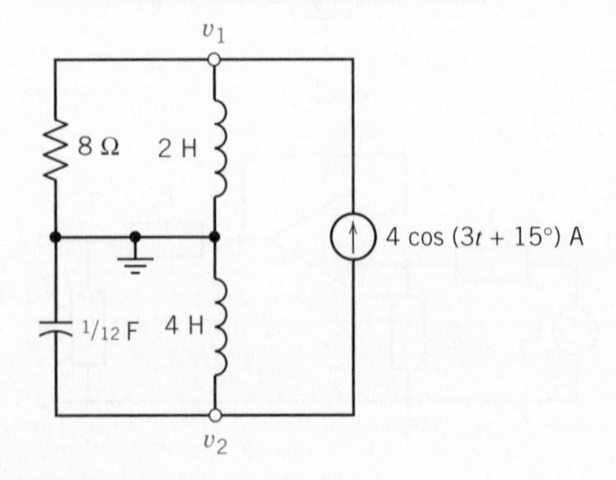

Figure P 10.16-3

P 10.16-4 A computer program reports that the currents of the circuit of Figure P 10.16-4 are $\mathbf{I} = 0.2\angle{53.1°}$ A, $\mathbf{I}_1 = 632\angle{-18.4°}$ mA, and $\mathbf{I}_2 = 190\angle{71.6°}$ mA. Verify this result.

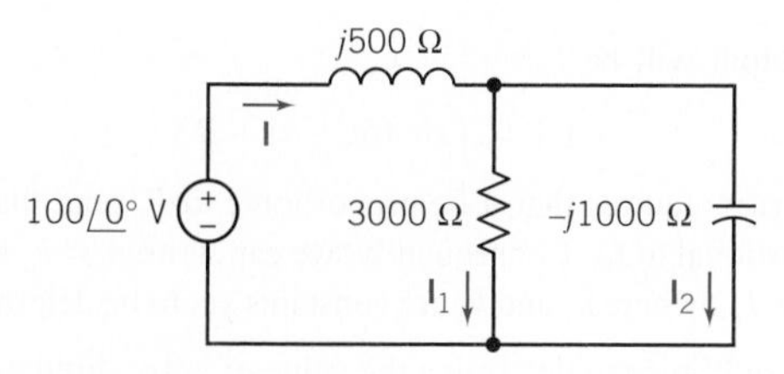

Figure P 10.16-4

P 10.16-5 The circuit shown in Figure P 10.16-5 was built using a 2-percent resistor having a nominal resistance of 500 Ω and a 10-percent capacitor with a nominal capacitance of 5 μF. The steady-state capacitor voltage was measured to be

$$v(t) = 18.3 \cos(200t - 24°) \text{ V}$$

The voltage source represents a signal generator. Suppose that the signal generator was adjusted so carefully that errors in the amplitude, frequency, and angle of the voltage source voltage are all negligible. Is the measured response explained by the component tolerances? That is, could the measured $v(t)$ have been produced by this circuit with a resistance R that is within 2 percent of 500 Ω and a capacitance C that is within 5 percent of 5 μF?

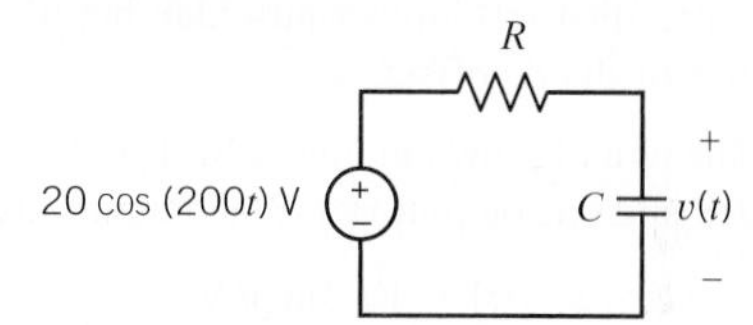

Figure P 10.16-5

PSpice Problems

SP 10-1 The circuit shown in Figure SP 10-1 has two inputs, $v_s(t)$ and $i_s(t)$, and one output, $v(t)$. The inputs are given by

$$v_s(t) = 10 \sin(6t + 45°) \text{ V}$$

and

$$i_s(t) = 2 \sin(6t + 60°) \text{ A}$$

Use PSpice to demonstrate superposition. Simulate three versions of the circuit simultaneously. (Draw the circuit in the PSpice workspace. Cut and paste to make two copies. Edit the part names in the copies to avoid duplicate names. For example, the resistor will be R1 in the original circuit. Change R1 to R2 and R3 in the two copies.) Use the given $v_s(t)$ and $i_s(t)$ in the first version. Set $i_s(t) = 0$ in the second version and $v_s(t) = 0$ in the third version. Plot the capacitor voltage, $v(t)$, for all three versions of the circuit. Show that the capacitor voltage in the first version of the circuit is equal to the sum of the capacitor voltages in the second and third versions.

Hint: Use PSpice parts VSIN and ISIN for the voltage and current source. PSpice uses hertz rather than rad/s as the unit for frequency.

Remark: Notice that $v(t)$ is sinusoidal and has the same frequency as $v_s(t)$ and $i_s(t)$.

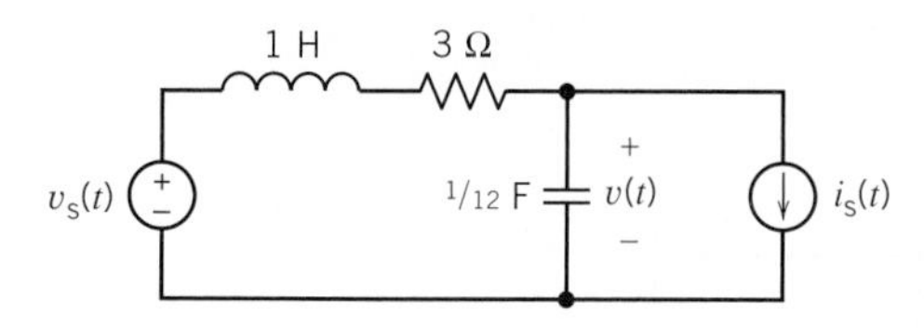

Figure SP 10.1

SP 10-2 The circuit shown in Figure SP 10-1 has two inputs, $v_s(t)$ and $i_s(t)$, and one output, $v(t)$. The inputs are given by

$$v_s(t) = 10 \sin(6t + 45°) \text{ V}$$

and

$$i_s(t) = 2 \sin(18t + 60°) \text{ A}$$

Use PSpice to demonstrate superposition. Simulate three versions of the circuit simultaneously. (Draw the circuit in the PSpice workspace. Cut and paste to make two copies. Edit the part names in the copies to avoid duplicate names. For example, the resistor will be R1 in the original circuit. Change R1 to R2 and R3 in the two copies.) Use the given $v_s(t)$ and $i_s(t)$ in the first version. Set $i_s(t) = 0$ in the second version and $v_s(t) = 0$ in the third version. Plot the capacitor voltage, $v(t)$, for all three versions of the circuit. Show that the capacitor voltage in the first version of the circuit is equal to the sum of the capacitor voltages in the second and third versions.

Hint: Use PSpice parts VSIN and ISIN for the voltage and current source. PSpice uses hertz rather than rad/s as the unit for frequency.
Remark: Notice that $v(t)$ is not sinusoidal.

SP 10-3 The circuit shown in Figure SP 10-1 has two inputs, $v_s(t)$ and $i_s(t)$, and one output, $v(t)$. The inputs are given by

$$v_s(t) = 10 \sin(6t + 45°) \text{ V}$$

and

$$i_s(t) = 0.8 \text{ A}$$

Use PSpice to demonstrate superposition. Simulate three versions of the circuit simultaneously. (Draw the circuit in the PSpice workspace. Cut and paste to make two copies. Edit the part names in the copies to avoid duplicate names. For example, the resistor will be R1 in the original circuit. Change R1 to R2 and R3 in the two copies.) Use the given $v_s(t)$ and $i_s(t)$ in the first version. Set $i_s(t) = 0$ in the second version and

$v_s(t) = 0$ in the third version. Plot the capacitor voltage, $v(t)$, for all three versions of the circuit. Show that the capacitor voltage in the first version of the circuit is equal to the sum of the capacitor voltages in the second and third versions.

Hint: Use PSpice parts VSIN and IDC for the voltage and current source. PSpice uses hertz rather than rad/s as the unit for frequency.

Remark: Notice that $v(t)$ looks sinusoidal, but it's not sinusoidal because of the dc offset.

SP 10-4 The circuit shown in Figure SP 10-1 has two inputs, $v_s(t)$ and $i_s(t)$, and one output, $v(t)$. When inputs are given by

$$v_s(t) = V_m \sin 6t \text{ V}$$

and

$$i_s(t) = I_m \text{ A}$$

the output will be

$$v_o(t) = A \sin(6t + \theta) + B \text{ V}$$

Linearity requires that A be proportional to V_m and that B be proportional to I_m. Consequently, we can write $A = k_1 V_m$ and $B = k_2 I_m$, where k_1 and k_2 are constants yet to be determined.

(a) Use PSpice to determine the value of k_1 by simulating the circuit, using $V_m = 1$ V and $I_m = 0$.

(b) Use PSpice to determine the value of k_2 by simulating the circuit, using $V_m = 0$ V and $I_m = 1$.

(c) Knowing k_1 and k_2, specify the values of V_m and I_m that are required to cause

$$v_o(t) = 5 \sin(6t + \theta) + 5 \text{ V}$$

Simulate the circuit, using PSpice to verify the specified values of V_m and I_m.

Design Problems

DP 10-1 Design the circuit shown in Figure DP 10-1 to produce the specified output voltage $v_o(t)$ when provided with the given input voltage $v_i(t)$.

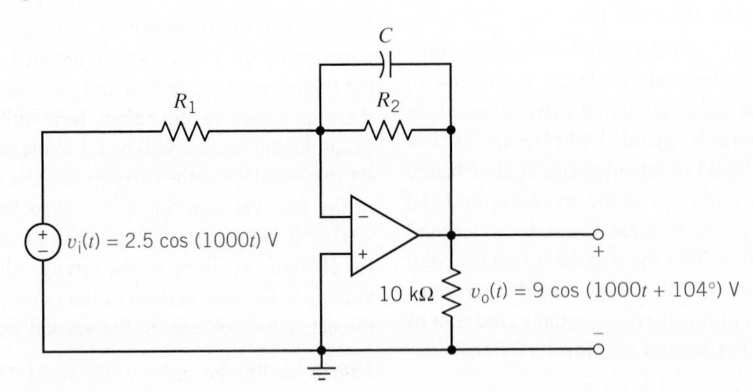

Figure DP 10-1

DP 10-2 Design the circuit shown in Figure DP 10-2 to produce the specified output voltage $v_o(t)$ when provided with the given input voltage $v_i(t)$.

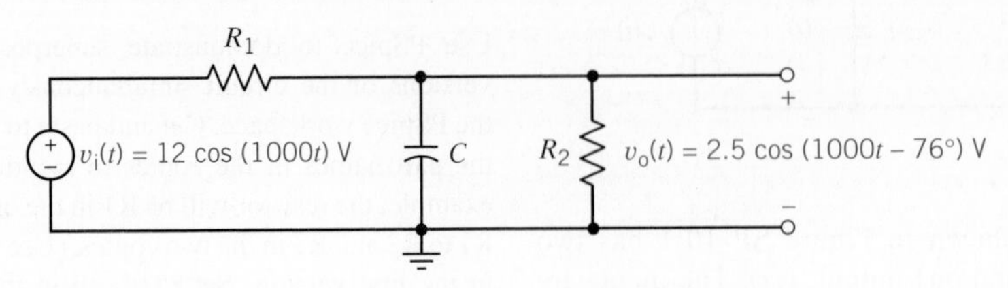

Figure DP 10-2

DP 10-3 Design the circuit shown in Figure DP 10-3 to produce the specified output voltage $v_o(t)$ when provided with the given input voltage $v_i(t)$.

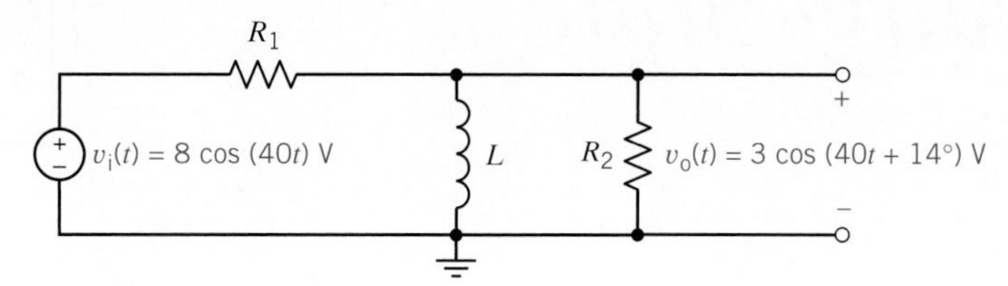

Figure DP 10-3

DP 10-4 Show that it is not possible to design the circuit shown in Figure DP 10-4 to produce the specified output voltage $v_o(t)$ when provided with the given input voltage $v_i(t)$.

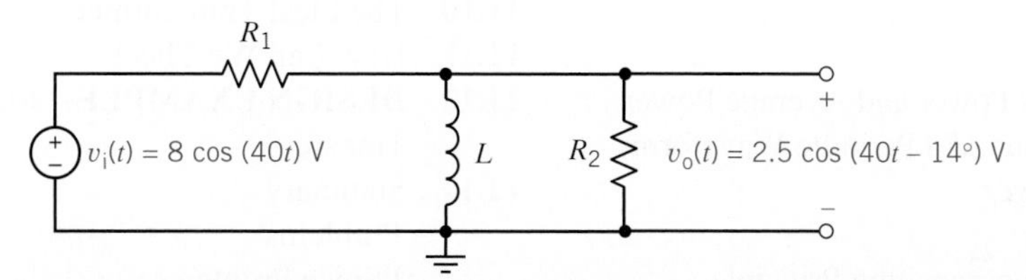

Figure DP 10-4

DP 10-5 A circuit with an unspecified R, L, and C is shown in Figure DP 10-5. The input source is $i_s = 10 \cos 1000t$ A, and the goal is to select the R, L, and C so that the node voltage is $v = 80 \cos 1000t$ V.

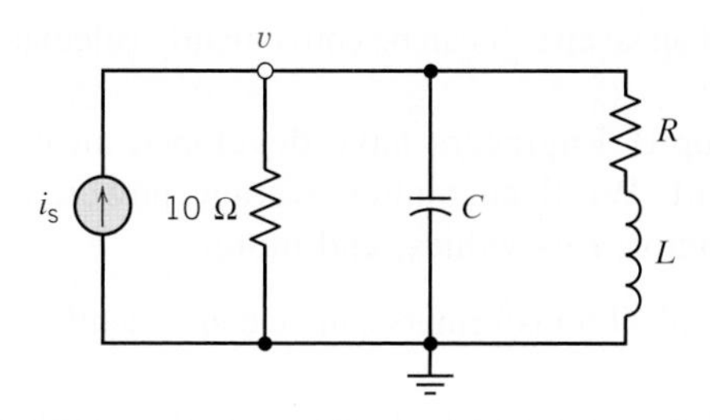

Figure DP 10-5

DP 10-6 The input to the circuit shown in FigureDP 10-6 is the voltage source voltage

$$v_s(t) = 10 \cos (1000t) \text{ V}$$

The output is the steady-state capacitor voltage

$$v_o(t) = A \cos (1000t + \theta) \text{ V}$$

(a) Specify values for R and C such that $\theta = -35°$. Determine the resulting value of A.

(b) Specify values for R and C such that $A = 5$ V. Determine the resulting values of θ.

(c) Is it possible to specify values for R and C such that $A = 4$ and $\theta = -60°$? (If not, justify your answer. If so, specify R and C.)

(d) Is it possible to specify values of R and C such that $A = 7.07$ V and $\theta = -45°$? (If not, justify your answer. If so, specify R and C.)

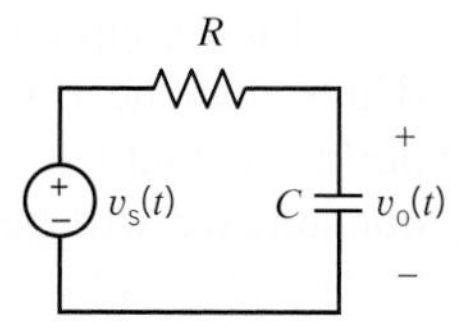

Figure DP 10-6

CHAPTER 11

AC Steady-State Power

IN THIS CHAPTER

11.1 Introduction
11.2 Electric Power
11.3 Instantaneous Power and Average Power
11.4 Effective Value of a Periodic Waveform
11.5 Complex Power
11.6 Power Factor
11.7 The Power Superposition Principle
11.8 The Maximum Power Transfer Theorem
11.9 Coupled Inductors
11.10 The Ideal Transformer
11.11 How Can We Check . . . ?
11.12 **DESIGN EXAMPLE**—Maximum Power Transfer
11.13 Summary
Problems
PSpice Problems
Design Problems

11.1 INTRODUCTION

In this chapter, we continue our study of ac circuits. In particular, we will see the following:

- The power supplied or received by any element of an ac circuit can be conveniently calculated after representing the circuit in the frequency domain.

  Power in ac circuits is an important topic. Engineers have developed an extensive vocabulary to describe power in an ac circuit. We'll encounter average power, real and reactive power, complex power, the power factor, rms values, and more.

- AC circuits that contain coupled inductors and/or ideal transformers can be conveniently analyzed in the frequency domain.

  Both coupled inductors and ideal transformers consist of magnetically coupled coils. (Coils may be tightly coupled or loosely coupled. The coils of an ideal transformer are perfectly coupled.) After representing coupled inductors and transformers in the frequency domain, we will be able to analyze ac circuits containing these devices.

11.2 ELECTRIC POWER

Human civilization's progress has been enhanced by society's ability to control and distribute energy. Electricity serves as a carrier of energy to the user. Energy present in a fossil fuel or a nuclear fuel is converted to electric power to transport and readily distribute it to customers. By means of transmission lines, electric power is transmitted and distributed to essentially all the residences, industries, and commercial buildings in the United States and Canada.

FIGURE 11.2-1 AC power high-voltage transmission lines. Courtesy of Pacific Gas and Electric Company.

FIGURE 11.2-2 A large hydroelectric power plant. Courtesy of Hydro Quebec.

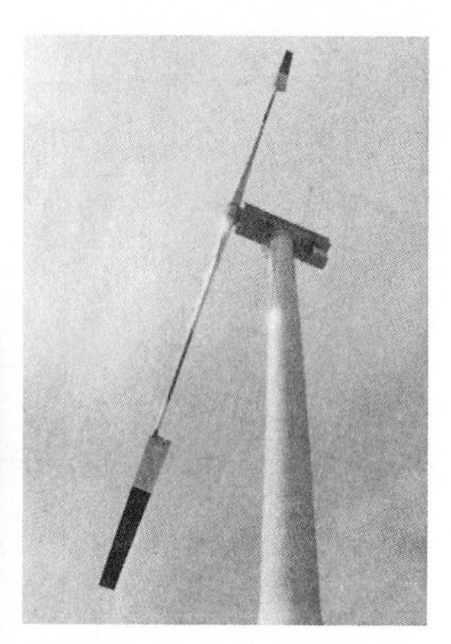

FIGURE 11.2-3 A large wind-power turbine and generator. Courtesy of *EPRI Journal*.

Electric power may be transported readily with low attendant losses, and improved methods for safe handling of electric power have been developed over the past 90 years. Furthermore, methods of converting fossil fuels to electric power are well developed, economical, and safe.

Means of converting solar and nuclear energy to electric power are currently in various stages of development or of proven safety. Geothermal energy, tidal energy, and wind energy may also be converted to electric power. The kinetic energy of falling water may readily be used to generate hydroelectric power.

The necessity of transmitting electrical power over long distances fostered the development of ac high-voltage power lines from power plant to end user. A modern transmission line is shown in Figure 11.2-1.

Electric energy generation uses original sources such as hydropower, coal, and nuclear energy. An example of a large hydroelectric power project is shown in Figure 11.2-2. A typical hydroelectric power plant can generate 1000 MW. On the other hand, many regions are turning to small generators such as the wind-power device shown in Figure 11.2-3. A typical wind-power machine may be capable of generating 75 kW.

A unique element of the American power system is its interconnectedness. Although the power system of the United States consists of many independent companies, it is interconnected by large transmission facilities. An electric utility is often able to save money by buying electricity from another utility and by transmitting the energy over the transmission lines of a third utility.

The power levels for selected electrical devices or phenomena are shown in Figure 11.2-4.

11.3 INSTANTANEOUS POWER AND AVERAGE POWER

We are interested in determining the power generated and absorbed in a circuit or in an element of a circuit. Electrical engineers talk about several types of power, for example, instantaneous power, average power, and complex power. We will start with an examination of the instantaneous power, which is the product of the time-domain voltage and current associated with one or more circuit elements. The instantaneous power is likely to be a complicated function of time. This prompts us to

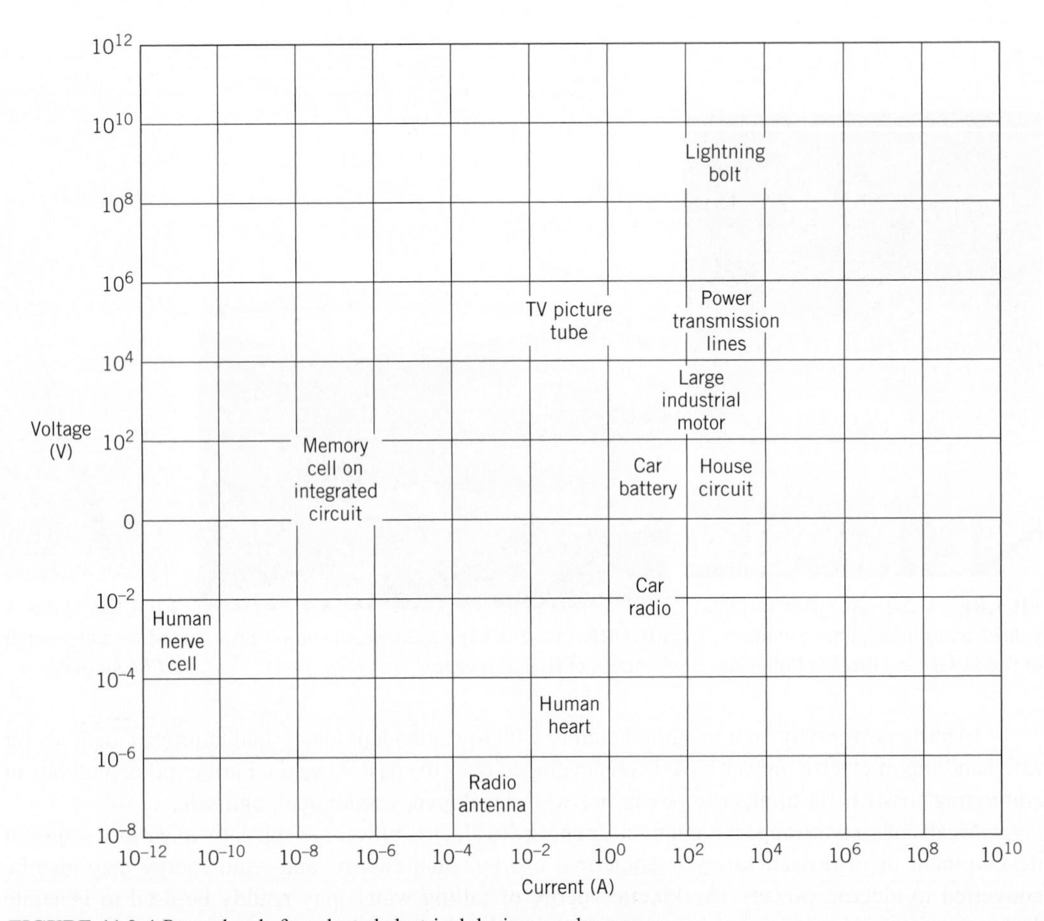

FIGURE 11.2-4 Power levels for selected electrical devices or phenomena.

look for a simpler measure of the power generated and absorbed in a circuit element, such as the average power.

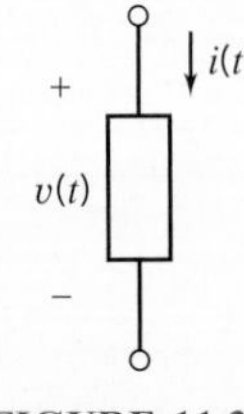

FIGURE 11.3-1 A circuit element.

Consider the circuit element shown in Figure 11.3-1. Notice that the element voltage $v(t)$ and the element current $i(t)$ adhere to the passive convention. The **instantaneous power** delivered to this circuit element is the product of the voltage $v(t)$ and the current $i(t)$, so that

$$p(t) = v(t)\, i(t) \tag{11.3-1}$$

The unit of power is watts (W). We can always calculate the instantaneous power because no restrictions have been placed on either $v(t)$ or $i(t)$. The instantaneous power can be a quite complicated function of t when $v(t)$ or $i(t)$ is itself a complicated function of t.

Suppose that the voltage $v(t)$ is a periodic function having period T. That is,

$$v(t) = v(t + T)$$

because the voltage repeats every T seconds. Then, for a linear circuit, the current will also be a periodic function having the same period, so

$$i(t) = i(t + T)$$

Therefore, the instantaneous power is

$$\begin{aligned} p(t) &= v(t)i(t) \\ &= v(t+T)i(t+T) \end{aligned}$$

The *average value* of a periodic function is the integral of the time function over a complete period, divided by the period. We use a capital P to denote average power and a lowercase p to denote instantaneous power. Therefore, the average power P is given by

$$P = \frac{1}{T}\int_{t_0}^{t_0+T} p(t)dt \tag{11.3-2}$$

where t_0 is an arbitrary starting point in time.

Next, suppose that the voltage $v(t)$ is sinusoidal, that is,

$$v(t) = V_{\text{m}}(\cos \omega t + \theta_{\text{V}})$$

Then, for a linear circuit at steady state, the current will also be sinusoidal and will have the same frequency, so

$$i(t) = I_{\text{m}}(\cos \omega t + \theta_{\text{I}})$$

The period and frequency of $v(t)$ and $i(t)$ are related by

$$\omega = \frac{2\pi}{T}$$

The instantaneous power delivered to the element is

$$p(t) = V_{\text{m}} I_{\text{m}} \cos(\omega t + \theta_{\text{V}}) \cos(\omega t + \theta_{\text{I}})$$

Using the trigonometric identity (see Appendix C) for the product of two cosine functions,

$$p(t) = \frac{V_{\text{m}} I_{\text{m}}}{2}[\cos(\theta_{\text{V}} - \theta_{\text{I}}) + \cos(2\omega t + \theta_{\text{V}} + \theta_{\text{I}})]$$

We see that the instantaneous power has two terms. The first term within the brackets is independent of time, and the second term varies sinusoidally over time at twice the radian frequency of $v(t)$.

The average power delivered to the element is

$$P = \frac{1}{T}\int_0^T \frac{V_{\text{m}} I_{\text{m}}}{2}[\cos(\theta_{\text{V}} - \theta_{\text{I}}) + \cos(2\omega t + \theta_{\text{V}} + \theta_{\text{I}})]\,dt$$

where we have chosen $t_0 = 0$. Then we have

$$\begin{aligned} P &= \frac{1}{T}\int_0^T \frac{V_{\text{m}} I_{\text{m}}}{2}\cos(\theta_{\text{V}} - \theta_{\text{I}})dt + \frac{1}{T}\int_0^T \frac{V_{\text{m}} I_{\text{m}}}{2}\cos(2\omega t + \theta_{\text{V}} + \theta_{\text{I}})dt \\ &= \frac{V_{\text{m}} I_{\text{m}}\cos(\theta_{\text{V}} - \theta_{\text{I}})}{2T}\int_0^T dt + \frac{V_{\text{m}} I_{\text{m}}}{2T}\int_0^T \cos(2\omega t + \theta_{\text{V}} + \theta_{\text{I}})dt \end{aligned}$$

The second integral is zero because the average value of the cosine function over a complete period is zero. Then we have

$$P = \frac{V_{\text{m}} I_{\text{m}}}{2}\cos(\theta_{\text{V}} - \theta_{\text{I}}) \tag{11.3-3}$$

EXAMPLE 11.3-1 Average Power

Find the average power delivered to a resistor R when the current through the resistor is $i(t)$, as shown in Figure 11.3-2.

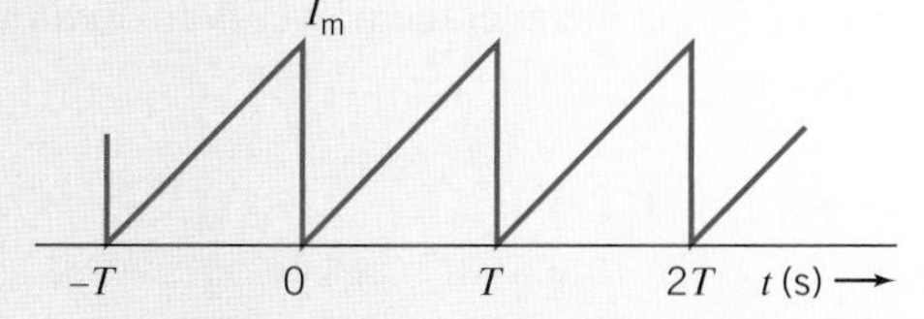

FIGURE 11.3-2 Current through a resistor in Example 11.3-1.

Solution

The current waveform repeats every T seconds and attains a maximum value of I_m. Using the period from $t = 0$ to $t = T$, we have

$$i = \frac{I_m}{T} t \quad 0 \le t < T$$

Then the instantaneous power is

$$p = i^2 R = \frac{I_m^2 t^2}{T^2} R \quad 0 \le t < T$$

It is sufficient to find the average power over $0 < t < T$ because the power is periodic with period T. Then the average power is

$$P = \frac{1}{T} \int_0^T \frac{I_m^2 R}{T^2} t^2 \, dt$$

Integrating, we have

$$P = \frac{I_m^2 R}{T^3} \int_0^T t^2 \, dt = \frac{I_m^2 R}{T^3} \frac{T^3}{3} = \frac{I_m^2 R}{3} \text{ W}$$

EXAMPLE 11.3-2 Average Power

The circuit shown in Figure 11.3-3 is at steady state. The mesh current is

$$i(t) = 721 \cos(100t - 41^\circ) \text{ mA}$$

The element voltages are

$$\begin{aligned} v_s(t) &= 20 \cos(100t - 15^\circ) \text{ V} \\ v_R(t) &= 18 \cos(100t - 41^\circ) \text{ V} \\ v_L(t) &= 8.66 \cos(100t + 49^\circ) \text{ V} \end{aligned}$$

Find the average power *delivered to* each device in this circuit.

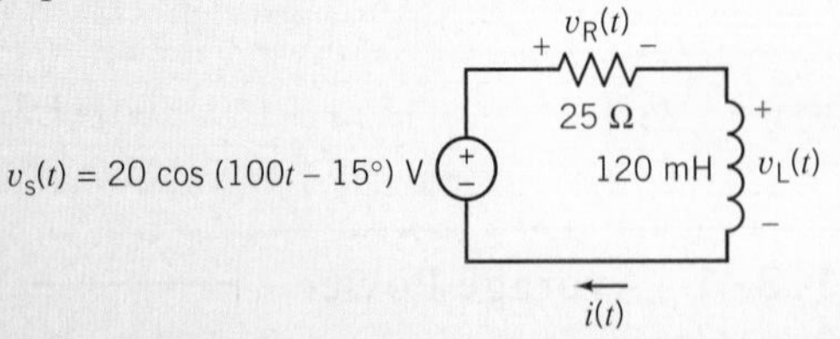

FIGURE 11.3-3 An *RL* circuit with a sinusoidal voltage source.

Solution

Notice that $v_s(t)$ and $i(t)$ don't adhere to the passive convention. Thus, $v_s(t)\, i(t)$ is the power *delivered by* the voltage source. Therefore, the average power calculated using Eq. 11.3-3 is the average power *delivered by* the voltage source. The average power *delivered by* the voltage source is

$$P_s = \frac{(20)(0.721)}{2} \cos(-15^\circ - (-41^\circ)) = 6.5 \text{ W}$$

The average power *delivered to* the voltage source is −6.5 W.

Because $v_R(t)$ and $i(t)$ do adhere to the passive convention, the average power calculated using Eq. 11.3-3 is the average power *delivered to* the resistor. The power *delivered to* the resistor is

$$P_R = \frac{(18)(0.721)}{2}\cos(-41^\circ - (-41^\circ)) = 6.5\ \text{W}$$

The power *delivered to* the inductor is

$$P_L = \frac{(8.66)(0.721)}{2}\cos(49^\circ - (-41^\circ)) = 0\ \text{W}$$

Why is the average power delivered to the inductor equal to zero? The angle of the element voltage is 90° larger than the angle of the element current. Because $\cos(90^\circ) = 0$, the average power delivered to the inductor is zero. The angle of the inductor voltage will always be 90° larger than the angle of the inductor current. Therefore, the average power delivered to any inductor is zero.

EXERCISE 11.3-1 Determine the instantaneous power delivered to an element and sketch $p(t)$ when the element is (a) a resistance R and (b) an inductor L. The voltage across the element is $v(t) = V_m \cos(\omega t + \theta)$ V.

Answers:

(a) $P_R = \frac{V_m^2}{2R}[1 + \cos(2\omega t + 2\theta)]$ W

(b) $P_L = \frac{V_m^2}{2\omega L}\cos(2\omega t + 2\theta - 90^\circ)$ W

11.4 EFFECTIVE VALUE OF A PERIODIC WAVEFORM

The voltage available from a wall plug in a residence is said to be 110 V. Of course, this is not the average value of the sinusoidal voltage because we know that the average would be zero. It is also not the instantaneous value or the maximum value, V_m, of the voltage $v = V_m \cos \omega t$.

The effective value of a voltage is a measure of its effectiveness in delivering power to a load resistor. The concept of an *effective value* comes from a desire to have a sinusoidal voltage (or current) deliver to a load resistor the same average power as an equivalent dc voltage (or current). The goal is to find a dc V_{eff} (or I_{eff}) that will deliver the same average power to the resistor as would be delivered by a periodically varying source, as shown in Figure 11.4-1.

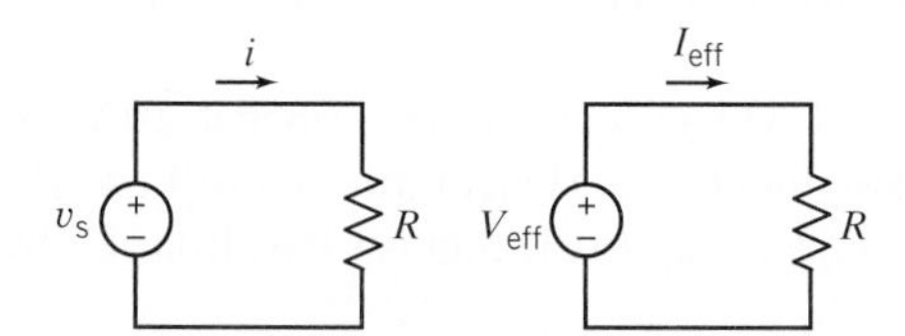

FIGURE 11.4-1 The goal is to find a dc voltage, V_{eff}, for a specified $v_s(t)$ that will deliver the same average power to R as would be delivered by the ac source.

The average power delivered to the resistor R by a periodic current is

$$P = \frac{1}{T}\int_0^T i^2 R\, dt \tag{11.4-1}$$

We select the period, T, of the periodic current as the integration interval.

The power delivered by a direct current is

$$P = I_{\text{eff}}^2 R \tag{11.4-2}$$

where I_{eff} is the dc current that will deliver the same power as the periodically varying current. That is, I_{eff} is defined as the steady (constant) current that is as *effective* in delivering power as the periodically varying current.

We equate Eqs. 11.4-1 and 11.4-2, obtaining

$$I_{\text{eff}}^2 R = \frac{R}{T}\int_0^T i^2\, dt$$

Solving for I_{eff}, we have

$$I_{\text{eff}} = \sqrt{\frac{1}{T}\int_0^T i^2\, dt} \tag{11.4-3}$$

We see that I_{eff} is the square root of the mean of the squared value. Thus, the effective current I_{eff} is commonly called the root-mean-square current I_{rms}.

The **effective value** of a current is the steady current (dc) that transfers the same average power as the given varying current.

Of course, the effective value of the voltage in a circuit is similarly found from the equation

$$V_{\text{eff}}^2 = V_{\text{rms}}^2 = \frac{1}{T}\int_0^T v^2\, dt$$

Thus

$$V_{\text{rms}} = \sqrt{\frac{1}{T}\int_0^T v^2 dt}$$

Now let us find the I_{rms} of a sinusoidally varying current $i = I_{\text{m}} \cos \omega t$. Using Eq. 11.4-3 and a trigonometric formula from Appendix C, we have

$$I_{\text{rms}} = \sqrt{\frac{1}{T}\int_0^T I_{\text{m}}^2 \cos^2 \omega t\, dt} = \sqrt{\frac{I_{\text{m}}^2}{T}\int_0^T \frac{1}{2}(1 + \cos 2\omega t)\, dt} = \frac{I_{\text{m}}}{\sqrt{2}} \tag{11.4-4}$$

because the integral of $\cos 2\omega t$ is zero over the period T. Remember that Eq. 11.4-4 is true only for sinusoidal currents.

In practice, we must be careful to determine whether a sinusoidal voltage is expressed in terms of its effective value or its maximum value I_{m}. In the case of power transmission and use in the home, the voltage is said to be 110 V or 220 V, and it is understood that these values refer to the rms or effective values of the sinusoidal voltage.

In electronics or communications circuits, the voltage could be described as 10 V, and the person is typically indicating the maximum or peak amplitude, V_{m}. Henceforth, we will use V_{m} as the peak value and V_{rms} as the rms value. Sometimes it is necessary to distinguish V_{rms} from V_{m} by the context in which the voltage is given.

Example 11.4-1 Effective Value

Find the effective value of the current for the sawtooth waveform shown in Figure 11.4-2.

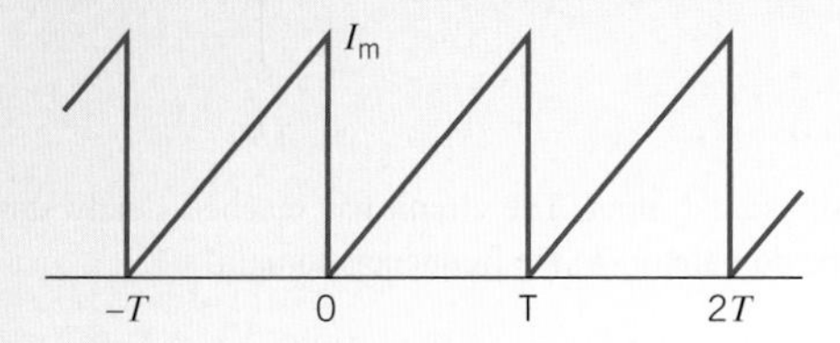

FIGURE 11.4-2 A sawtooth current waveform.

Solution

First, we will express the current waveform over the period $0 \leq t < T$. The current is then

$$i = \frac{I_m}{T} t \quad 0 \leq t < T$$

The effective value is found from

$$I_{eff}^2 = \frac{1}{T}\int_0^T i^2\, dt = \frac{1}{T}\int_0^T \frac{I_m^2}{T^2} t^2\, dt = \frac{I_m^2}{T^3}\left[\frac{t^3}{3}\right]_0^T = \frac{I_m^2}{3}$$

Therefore, solving for I_{eff}, we have

$$I_{eff} = \frac{I_m}{\sqrt{3}}$$

It's worth noticing that the rms value of a sawtooth waveform with amplitude I_m is different than the rms value of a sinusoidal waveform having amplitude I_m.

EXERCISE 11.4-1 Find the effective value of the following currents: (a) $\cos 3t + \cos 3t$; (b) $\sin 3t + \cos(3t + 60°)$; (c) $2 \cos 3t + 3 \cos 5t$

Answer: (a) $\sqrt{2}$ (b) 0.366 (c) 2.55

11.5 COMPLEX POWER

Suppose that a linear circuit with a sinusoidal input is at steady state. All the element voltages and currents will be sinusoidal, with the same frequency as the input. Such a circuit can be analyzed in the frequency domain, using phasors. Indeed, we can calculate the power generated or absorbed in a circuit or in any element of a circuit, in the frequency domain, using phasors.

Figure 11.5-1 represents the voltage and current of an element in both the time domain and the frequency domain. Notice that the element current and voltage adhere to the passive convention. In a previous section, the instantaneous power and the average power were calculated from the time-domain representations of the element current and voltage, $v(t)$ or $i(t)$. In contrast, we now turn our attention to the frequency-domain representations of the element current and voltage

$$\mathbf{I}(\omega) = I_m \angle \theta_I \quad \text{and} \quad \mathbf{V}(\omega) = V_m \angle \theta_V \tag{11.5-1}$$

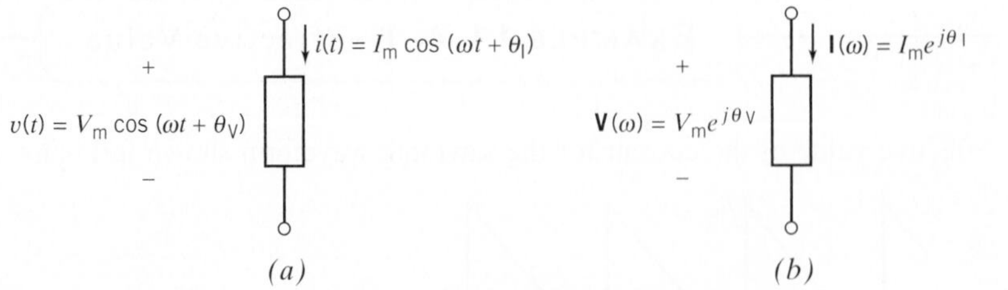

$$v(t) = V_m \cos(\omega t + \theta_V) \qquad i(t) = I_m \cos(\omega t + \theta_I)$$

$$\mathbf{V}(\omega) = V_m e^{j\theta_V} \qquad \mathbf{I}(\omega) = I_m e^{j\theta_I}$$

(a) (b)

FIGURE 11.5-1 A linear circuit is excited by a sinusoidal input. The circuit has reached steady state. The element voltage and current can be represented in (*a*) the time domain or (*b*) the frequency domain.

The **complex power** delivered to the element is defined to be

$$\mathbf{S} = \frac{\mathbf{VI}^*}{2} = \frac{(V_m \angle \theta_V)(I_m \angle -\theta_I)}{2} = \frac{V_m I_m}{2} \angle \theta_V - \theta_I \tag{11.5-2}$$

where $\mathbf{I}^*$ denotes the complex conjugate of **I** (see Appendix B). The magnitude of **S**

$$|\mathbf{S}| = \frac{V_m I_m}{2} \tag{11.5-3}$$

is called the **apparent-power**.

Converting the complex power, **S**, from polar to rectangular form gives

$$\mathbf{S} = \frac{V_m I_m}{2} \cos(\theta_V - \theta_I) + j\frac{V_m I_m}{2} \sin(\theta_V - \theta_I) \tag{11.5-4}$$

The real part of **S** is equal to the average power that we calculated previously in the time domain! (See Eq. 11.3-3.) Recall that the average power was denoted as P. We can represent the complex power as

$$\mathbf{S} = P + jQ \tag{11.5-5}$$

where

$$P = \frac{V_m I_m}{2} \cos(\theta_V - \theta_I) \tag{11.5-6}$$

is the **average power** and

$$Q = \frac{V_m I_m}{2} \sin(\theta_V - \theta_I) \tag{11.5-7}$$

is the **reactive power**. The complex power, average power, and reactive power are all the product of a voltage and a current. Nonetheless, it is conventional to use different units for these three types of power. We have already seen that the units of the average power are watts. The units of complex power are volt-amps (VA), and the units of reactive power are volt-amps reactive (VAR). The formulas used to calculate power in the frequency domain are summarized in Table 11.5-1.

Let's return to Figure 11.5-1*b*. The impedance of the element can be expressed as

$$\mathbf{Z}(\omega) = \frac{\mathbf{V}(\omega)}{\mathbf{I}(\omega)} = \frac{V_m \angle \theta_V}{I_m \angle \theta_I} = \frac{V_m}{I_m} \angle \theta_V - \theta_I \tag{11.5-8}$$

Converting the impedance, **Z**, from polar to rectangular form gives

$$\mathbf{Z}(\omega) = \frac{V_m}{I_m} \cos(\theta_V - \theta_I) + j\frac{V_m}{I_m} \sin(\theta_V - \theta_I) \tag{11.5-9}$$

We can represent the impedance as

$$\mathbf{Z}(\omega) = R + jX$$

Table 11.5-1 Frequency-Domain Power Relationships

QUANTITY	RELATIONSHIP USING PEAK VALUES	RELATIONSHIP USING rms VALUES	UNITS
Element voltage, $v(t)$	$v(t) = V_m \cos(\omega t + \theta_V)$	$v(t) = V_{rms}\sqrt{2}\cos(\omega t + \theta_V)$	V
Element current, $i(t)$	$i(t) = I_m \cos(\omega t + \theta_I)$	$i(t) = I_{rms}\sqrt{2}\cos(\omega t + \theta_I)$	A
Complex power, $\mathbf{S}$	$\mathbf{S} = \frac{V_m I_m}{2}\cos(\theta_V - \theta_I) + j\frac{V_m I_m}{2}\sin(\theta_V - \theta_I)$	$\mathbf{S} = V_{rms}I_{rms}\cos(\theta_V - \theta_I) + jV_{rms}I_{rms}\sin(\theta_V - \theta_I)$	VA
Apparent power, $\lvert\mathbf{S}\rvert$	$\lvert\mathbf{S}\rvert = \frac{V_m I_m}{2}$	$\lvert\mathbf{S}\rvert = V_{rms}I_{rms}$	VA
Average power, P	$P = \frac{V_m I_m}{2}\cos(\theta_V - \theta_I)$	$P = V_{rms}I_{rms}\cos(\theta_V - \theta_I)$	W
Reactive power, Q	$Q = \frac{V_m I_m}{2}\sin(\theta_V - \theta_I)$	$Q = V_{rms}I_{rms}\sin(\theta_V - \theta_I)$	VAR

where $R = \frac{V_m}{I_m}\cos(\theta_V - \theta_I)$ is the resistance and $X = \frac{V_m}{I_m}\sin(\theta_V - \theta_I)$ is the reactance.

The similarity between Eqs. 11.5-4 and 11.5-9 suggests that the complex power can be expressed in terms of the impedance

$$\begin{aligned}\mathbf{S} &= \frac{V_m I_m}{2}\cos(\theta_V - \theta_I) + j\frac{V_m I_m}{2}\sin(\theta_V - \theta_I)\\ &= \left(\frac{I_m^2}{2}\right)\frac{V_m}{I_m}\cos(\theta_V - \theta_I) + j\left(\frac{I_m^2}{2}\right)\frac{V_m}{I_m}\sin(\theta_V - \theta_I)\\ &= \left(\frac{I_m^2}{2}\right)\text{Re}(\mathbf{Z}) + j\left(\frac{I_m^2}{2}\right)I_m(\mathbf{Z})\end{aligned} \tag{11.5-10}$$

In particular, the average power delivered to the element is given by

$$P = \left(\frac{I_m^2}{2}\right)\text{Re}(\mathbf{Z}) \tag{11.5-11}$$

When the element is a resistor, then $\text{Re}(\mathbf{Z}) = R$

$$P_R = \left(\frac{I_m^2}{2}\right)R$$

When the element is a capacitor or an inductor, then $\text{Re}(\mathbf{Z}) = 0$; thus, the average power delivered to a capacitor or an inductor is zero.

Figure 11.5-2 summarizes Eqs. 11.5-4 and 11.5-9, using (a) the impedance triangle and (b) the power triangle.

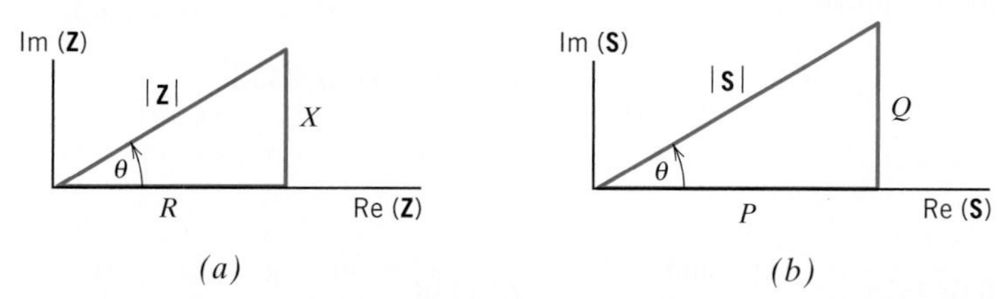

FIGURE 11.5-2 (*a*) The impedance triangle where $\mathbf{Z} = R + jX = Z$. (*b*) The complex power triangle where $\mathbf{S} = P + jQ$.

EXAMPLE 11.5-1 Complex Power

The circuit shown in Figure 11.5-3 consists of a source driving a load. The current source current is

$$i(t) = 1.25 \cos(5t - 15^\circ) \text{ A}$$

(a) What is the value of the complex power delivered by the source to the load when $R = 20\ \Omega$ and $L = 3$ H?

(b) What are the values of the resistance, R, and inductance, L, when the source delivers $11.72 + j11.72$ VA to the load?

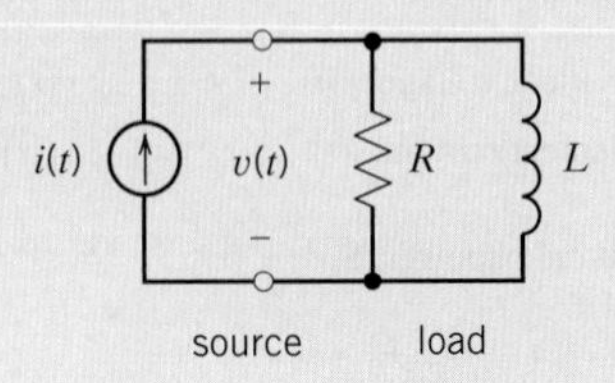

FIGURE 11.5-3 A circuit consisting of a source driving a load.

Solution

Represent the circuit in the frequency domain as shown in Figure 11.5-4, where $\mathbf{I} = 1.25\angle{-15^\circ}$A. The equivalent impedance of the parallel resistor and inductor is

$$\mathbf{Z} = \frac{j\omega LR}{R + j\omega L}$$

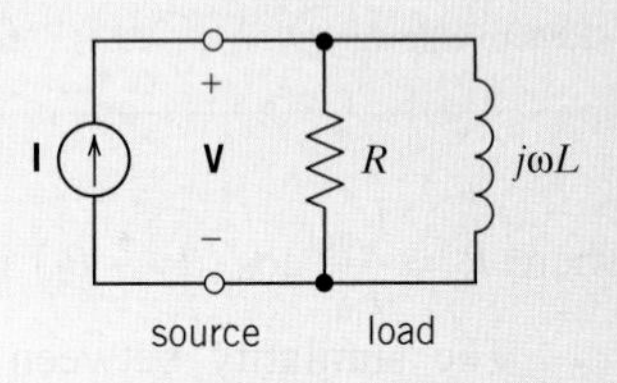

FIGURE 11.5-4 The circuit from Figure 11.5-3, represented in the frequency domain.

(a) When $R = 20\ \Omega$ and $L = 3$ H, the equivalent impedance is

$$\mathbf{Z} = \frac{j300}{20 + j15} = 12\angle{53^\circ}\ \Omega$$

The voltage across this impedance is

$$\mathbf{V} = \mathbf{IZ} = (1.25\angle{-15^\circ})(12\angle{53^\circ}) = 15\angle{38^\circ}\text{ V}$$

The complex power delivered by the source is

$$\mathbf{S} = \frac{\mathbf{VI}^*}{2} = \frac{(15\angle{38^\circ})(1.25 - \angle{15^\circ})^*}{2} = \frac{(15\angle{38^\circ})(1.25\angle{15^\circ})}{2} = 9.375\angle{53^\circ}\text{ VA}$$

(b) The voltage across the equivalent impedance can be calculated from the complex power and the current, using

$$\mathbf{S} = \frac{\mathbf{VI}^*}{2} \Rightarrow \mathbf{V} = \frac{2\mathbf{S}}{\mathbf{I}^*}$$

When $\mathbf{S} = 11.72 + j11.72 = 16.57\angle{45^\circ}$ VA

$$\mathbf{V} = \frac{2\mathbf{S}}{\mathbf{I}^*} = \frac{2(16.57\angle{45^\circ})}{(1.25\angle{-15^\circ})^*} = \frac{2(16.57\angle{45^\circ})}{1.25\angle{15^\circ}} = 26.52\angle{30^\circ}\text{ V}$$

The equivalent impedance is

$$\mathbf{Z} = \frac{\mathbf{V}}{\mathbf{I}} = \frac{26.52\angle{30^\circ}}{1.25\angle{-15^\circ}} = 21.21\angle{45^\circ}\ \Omega$$

It's convenient to take the reciprocal:

$$\frac{1}{R} - j\frac{1}{\omega L} = \frac{1}{21.21\angle{45^\circ}} = 0.033338 - j0.033338$$

Consequently,

$$R = \frac{1}{0.033338} = 30\ \Omega \quad \text{and} \quad 5L = \frac{1}{0.033338} = 30 \quad \Rightarrow \quad L = 6\text{ H}$$

EXAMPLE 11.5-2 Parallel Loads

The circuit shown in Figure 11.5-5 consists of a source driving a load that consists of the parallel connection of two loads. The voltage source voltage is

$$v(t) = 24\cos(5t + 30^\circ)\ \text{V}$$

Load A receives

$$\mathbf{S}_\text{A} = 9.216 + j6.912\ \text{VA}$$

The impedance of load B is

$$\mathbf{Z}_\text{B} = 42.426\angle 45^\circ\ \text{VA}$$

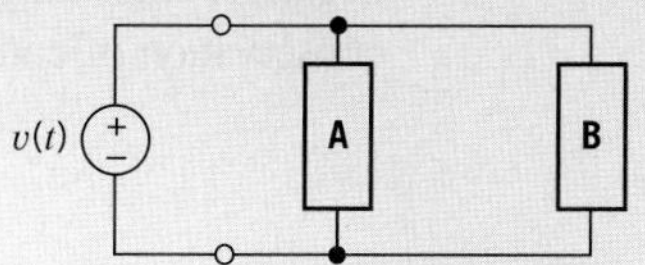

FIGURE 11.5-5 A circuit consisting of a source driving a parallel load.

(a) Determine the value of the complex power delivered by the source to the parallel load.

(b) Determine the value of the equivalent impedance of the parallel load.

Solution

Represent the circuit in the frequency domain as shown in Figure 11.5-6, where $\mathbf{V} = 24\angle 30^\circ$ V. The current in load A can be calculated from the complex power received by load A, using

$$\mathbf{S}_\text{A} = \frac{\mathbf{V}\mathbf{I}_1^*}{2} \quad \Rightarrow \quad \mathbf{I}_1 = \left(\frac{2\mathbf{S}_\text{A}}{\mathbf{V}}\right)^*$$

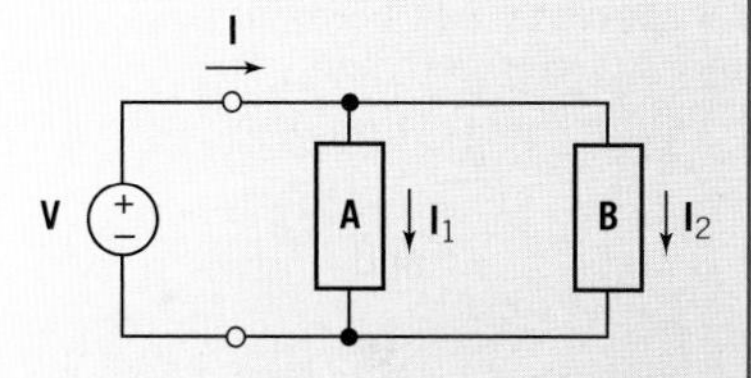

FIGURE 11.5-6 The circuit from Figure 11.5-5, represented in the frequency domain.

When $\mathbf{S}_\text{A} = 9.216 + j6.912 = 11.52\angle 36.9^\circ$ VA

$$\mathbf{I}_1 = \left(\frac{2(11.52\angle 36.9^\circ)}{24\angle 30^\circ}\right)^* = (0.96\angle 7^\circ)^* = 0.96\angle -7^\circ\ \text{A}$$

The current in load B can be calculated as

$$\mathbf{I}_2 = \frac{\mathbf{V}}{\mathbf{Z}_\text{B}} = \frac{24\angle 30^\circ}{42.426\angle 45^\circ} = 0.566\angle -15^\circ\ \text{A}$$

The source current is

$$\mathbf{I} = \mathbf{I}_1 + \mathbf{I}_2 = 1.522\angle -9.9^\circ\ \text{A}$$

(a) The complex power delivered by the source is

$$\mathbf{S} = \frac{\mathbf{V}\mathbf{I}^*}{2} = \frac{(24\angle 30^\circ)(1.522\angle -9.9^\circ)^*}{2} = 18.265\angle 39.9^\circ = 14.02 + j11.71\ \text{VA}$$

(b) The equivalent impedance of the parallel load is

$$\mathbf{Z} = \frac{\mathbf{V}}{\mathbf{I}} = \frac{24\angle 30^\circ}{1.522\angle -9.9^\circ} = 15.768\angle 39.9^\circ\ \Omega$$

Complex power is conserved. The sum of the complex power received by all the elements of a circuit is zero. This fact can be expressed by the equation

$$\sum_{\substack{all \\ elements}} \frac{\mathbf{V}_k \mathbf{I}_k^{\,*}}{2} = 0 \qquad (11.5\text{-}12)$$

where $\mathbf{V}_k$ and $\mathbf{I}_k$ are the phasors corresponding to the element voltage and current of the kth element of the circuit. The phasors $\mathbf{V}_k$ and $\mathbf{I}_k$ must adhere to the passive convention so that $\mathbf{V}_k\mathbf{I}_k^*/2$ is the complex power *received* by the kth branch. The summation in Eq. 11.5-12 adds up the complex powers in all elements of the circuit. When an element of the circuit is a source that is supplying power to the circuit, $\mathbf{V}_k\mathbf{I}_k^*/2$ will be negative, indicating that positive complex power is being supplied rather than received. Sometimes conservation of complex power is expressed as

$$\sum_{sources} \frac{\mathbf{V}_k\mathbf{I}_k^*}{2} = \sum_{\substack{other\\elements}} \frac{\mathbf{V}_k\mathbf{I}_k^*}{2} \tag{11.5-13}$$

where phasors $\mathbf{V}_k$ and $\mathbf{I}_k$ adhere to the passive convention for the "other elements" but do not adhere to the passive convention for the sources. When $\mathbf{V}_k$ and $\mathbf{I}_k$ do not adhere to the passive convention, then $\mathbf{V}_k\mathbf{I}_k^*/2$ is the complex power *supplied* by the kth branch. We read Eq. 11.5-13 to say that the total complex power supplied by the sources is equal to the total complex power received by the other elements of the circuit.

Equation 11.5-12 implies that both

$$\sum_{\substack{all\\Elements}} \mathrm{Re}\left(\frac{\mathbf{V}_k\mathbf{I}_k^*}{2}\right) = 0$$

and

$$\sum_{\substack{all\\Elements}} \mathrm{Im}\left(\frac{\mathbf{V}_k\mathbf{I}_k^*}{2}\right) = 0$$

Therefore,

$$\sum_{\substack{all\\elements}} P_k = 0 \quad \text{and} \quad \sum_{\substack{all\\elements}} Q_k = 0$$

In other words, average power and reactive power are both conserved.

Example 11.5-3 Conservation of Complex Power

Verify that complex power is conserved in the circuit of Figure 11.5-7 when $v_s = 100 \cos 1000t$ V.

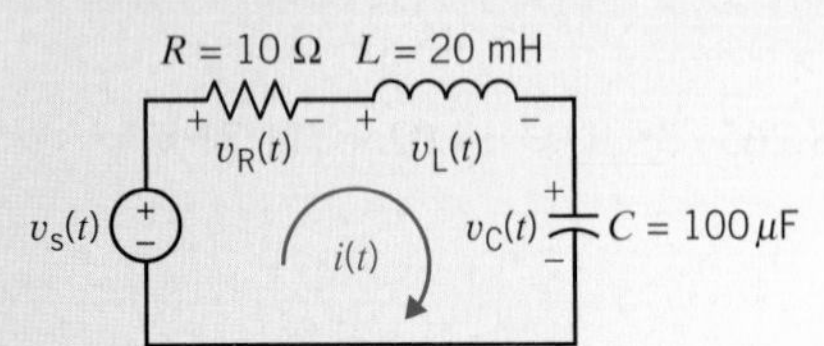

FIGURE 11.5-7 Circuit for Examples 11.5-3 and 11.5-4.

Solution

The phasor corresponding to the source voltage is

$$\mathbf{V}_s(\omega) = 100\angle 0\ \mathrm{V}$$

Writing and solving a mesh equation, we find that the phasor corresponding to the mesh current is

$$\mathbf{I}(\omega) = \frac{\mathbf{V}_s(\omega)}{R + j\omega L - j\dfrac{1}{\omega C}} = \frac{100}{10 + j(1000)(0.02) - j\dfrac{1}{(1000)10^{-4}}} = 7.07\angle -45^\circ\ \mathrm{A}$$

Ohm's law provides the phasors corresponding to the element voltages:

$$\begin{aligned}\mathbf{V}_{\mathrm{R}}(\omega) &= R\,\mathbf{I}(\omega) = 10(7.07\angle{-45^\circ}) = 70.7\angle{-45^\circ}\ \mathrm{V}\\ \mathbf{V}_{\mathrm{L}}(\omega) &= j\omega L\,\mathbf{I}(\omega) = j(1000)(0.02)(7.07\angle{45^\circ})\\ &= (20\angle{90^\circ})(7.07\angle{-45^\circ}) = 141.4\ \angle{45^\circ}\ \mathrm{V}\\ \mathbf{V}_{\mathrm{C}}(\omega) &= -j\frac{1}{\omega C}\mathbf{I}(\omega) = -j\frac{1}{(1000)(10^{-4})}(7.07\angle{-45^\circ})\\ &= (10\angle{90^\circ})(7.07\angle{-45^\circ}) = 70.7\angle{-135^\circ}\ \mathrm{V}\end{aligned}$$

Consider the voltage source. The phasors $\mathbf{V}_s$ and $\mathbf{I}$ do not adhere to the passive convention. The complex power

$$\begin{aligned}\mathbf{S}_{\mathrm{V}} = \frac{\mathbf{V}_s\mathbf{I}^*}{2} = \frac{100(7.07\angle{-45^\circ})^*}{2} &= \frac{100(7.07\ \angle{45^\circ})}{2}\\ &= \frac{100(7.07)}{2}\ \angle{45^\circ} = 353.5\angle{45^\circ}\ \mathrm{VA}\end{aligned}$$

is the complex power supplied by the voltage source.

The phasors $\mathbf{I}$ and $\mathbf{V}_{\mathrm{R}}$ do adhere to the passive convention. The complex power

$$\begin{aligned}\mathbf{S}_{\mathrm{R}} = \frac{\mathbf{V}_{\mathrm{R}}\mathbf{I}^*}{2} &= \frac{(70.7\angle{-45^\circ})(7.07\angle{-45^\circ})^*}{2}\\ &= \frac{(70.7\angle{-45^\circ})(7.07\angle{45^\circ})}{2} = \frac{(70.7)(7.07)}{2}\ \angle{-45^\circ + 45^\circ} = 250\angle{0}\ \mathrm{VA}\end{aligned}$$

is the complex power absorbed by the resistor. Similarly,

$$\mathbf{S}_{\mathrm{L}} = \frac{\mathbf{V}_{\mathrm{L}}\mathbf{I}^*}{2} = \frac{(141.4\angle{45^\circ})(7.07\angle{45^\circ})}{2} = \frac{(141.4)(7.07)}{2}\ \angle{45^\circ + 45^\circ} = 500\angle{90^\circ}\ \mathrm{VA}$$

is the complex power delivered to the inductor, and

$$\begin{aligned}\mathbf{S}_{\mathrm{C}} &= \frac{\mathbf{V}_{\mathrm{C}}\mathbf{I}^*}{2} = \frac{(70.7\angle{-135^\circ})(7.07\angle{45^\circ})}{2} = \frac{(70.7)(7.07)}{2}\ \angle{-135^\circ + 45^\circ}\\ &= 250\angle{-90^\circ}\ \mathrm{VA}\end{aligned}$$

is the complex power delivered to the capacitor.

To verify that complex power has been conserved, we calculate the complex power received by the "other elements" and compare it to the complex power supplied by the source:

$$\begin{aligned}\mathbf{S}_{\mathrm{R}} + \mathbf{S}_{\mathrm{L}} + \mathbf{S}_{\mathrm{C}} &= 250\angle{0^\circ} + 500\angle{90^\circ} + 250\angle{-90^\circ}\\ &= (250 + j0) + (0 + j500) + (0 - j250)\\ &= 250 + j250 = 353.5\angle{45^\circ} = \mathbf{S}_{\mathrm{V}}\end{aligned}$$

As expected, the complex power supplied by the source is equal to the complex power received by the other elements of the circuit.

EXAMPLE 11.5-4 Conservation of Average Power

Verify that average power is conserved in the circuit of Figure 11.5-7 when $v_s = 100 \cos 1000t$ V.

Solution

The phasor corresponding to the source voltage is

$$\mathbf{V}_s(\omega) = 100\angle 0 \text{ V}$$

Writing and solving a mesh equation, we find that the phasor corresponding to the mesh current is

$$\mathbf{I}(\omega) = \frac{\mathbf{V}_s(\omega)}{R + j\omega L - j\dfrac{1}{\omega C}} = \frac{100}{10 + j(1000)(0.02) - j\dfrac{1}{(1000)10^{-4}}} = 7.07\angle -45^\circ \text{ A}$$

The average power absorbed by the resistor, the capacitor, and the inductor can be calculated using

$$P = \left(\frac{I_m^2}{2}\right)\text{Re}(\mathbf{Z})$$

Because $\text{Re}(\mathbf{Z}) = 0$ for the capacitor and for the inductor, the average power absorbed by each of these devices is zero. $\text{Re}(\mathbf{Z}) = R$ for the resistor, so

$$P_R = \left(\frac{I_m^2}{2}\right)R = \frac{(7.07^2)}{2}10 = 250 \text{ W}$$

The average power supplied by the source is

$$P_V = \text{Re}(\mathbf{S}_V) = \text{Re}\left(\frac{\mathbf{V}_s\mathbf{I}^*}{2}\right) = \text{Re}\left(\frac{100(7.07)}{2}\angle 45^\circ\right) = \text{Re}(353.5\angle 45^\circ) = 250 \text{ W}$$

To verify that average power has been conserved, we calculate the average power received by the "other elements" and compare it to the average power supplied by the source:

$$P_R + P_L + P_C = 250 + 0 + 0 = 250 = P_V$$

As expected, the average power supplied by the sources is equal to the average power received by the other elements of the circuit.

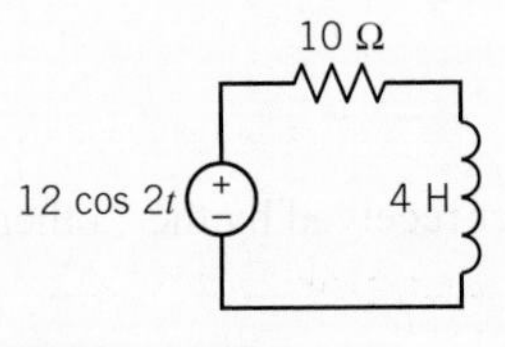

FIGURE E 11.5-1

EXERCISE 11.5-1 Determine the average power delivered to each element of the circuit shown in Figure E 11.5-1. Verify that average power is conserved.

Answer: $4.39 + 0 = 4.39$ W

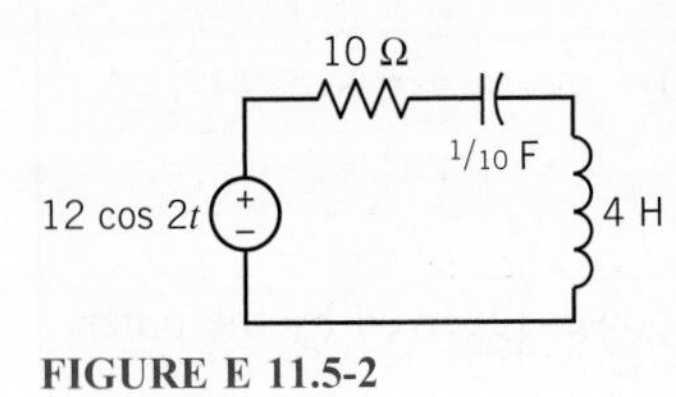

FIGURE E 11.5-2

EXERCISE 11.5-2 Determine the complex power delivered to each element of the circuit shown in Figure E 11.5-2. Verify that complex power is conserved.

Answer: $6.606 + j5.248 - j3.303 = 6.606 + j1.982$ VA

11.6 POWER FACTOR

In this section, as in the previous section, we consider a linear circuit with a sinusoidal input that is at steady state. All the element voltages and currents will be sinusoidal and will have the same frequency as the input. Such a circuit can be analyzed in the frequency domain, using phasors. In particular, we can calculate the power generated or absorbed in a circuit or in any element of a circuit, in the frequency domain, using phasors and impedances.

Recall that in Section 11.5, we showed that the average power absorbed by the element shown in Figure 11.5-1 is

$$P = \frac{V_m I_m}{2} \cos(\theta_V - \theta_I)$$

and that the apparent power is

$$|\mathbf{S}| = \frac{V_m I_m}{2}$$

The ratio of the average power to the apparent power is called the **power factor** (*pf*). The power factor is calculated as

$$pf = \cos(\theta_V - \theta_I)$$

The angle $(\theta_V - \theta_I)$ is often referred to as the *power factor angle*. The average power absorbed by the element shown in Figure 11.5-1 can be expressed as

$$P = \frac{V_m I_m}{2} pf \tag{11.6-1}$$

The cosine is an even function, that is, $\cos(\theta) = \cos(-\theta)$. So

$$pf = \cos(\theta_V - \theta_I) = \cos(\theta_I - \theta_V)$$

This causes a small difficulty. We can't calculate $\theta_V - \theta_I$ from *pf* without some additional information. For example, suppose $pf = 0.8$. We calculate

$$36.87^\circ = \cos^{-1}(0.8)$$

but that's not enough to determine $\theta_V - \theta_I$ uniquely. Because the cosine is even, both $\cos(36.87^\circ) = 0.8$ and $\cos(-36.87^\circ) = 0.8$, so either $\theta_V - \theta_I = 36.87^\circ$ or $\theta_V - \theta_I = -36.87^\circ$. This difficulty is resolved by labeling the power factor as *leading* or *lagging*. When $\theta_V - \theta_I > 0$, the power factor is said to be lagging, and when $\theta_V - \theta_I < 0$, the power factor is said to be leading. If the power factor is specified to be 0.8 leading, then $\theta_V - \theta_I = -36.87^\circ$. On the other hand, if the power factor is specified to be 0.8 lagging, then $\theta_V - \theta_I = 36.87^\circ$.

The significance of the power factor is illustrated by the circuit shown in Figure 11.6-1. This circuit models the transmission of electric power from a power utility company to a customer. The customer's load is connected to the power company's power plant by a transmission line. Typically, the customer requires power at a specified voltage. The power company must supply both the power used by the customer and the power absorbed by the transmission line. The power absorbed by the transmission line is lost; it doesn't do anybody any good, and we want to minimize it.

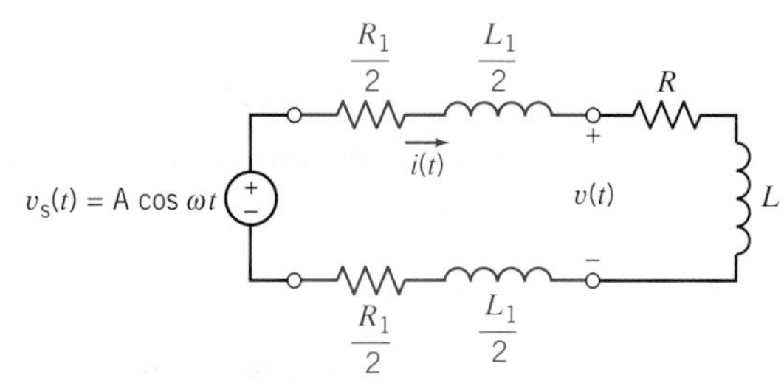

FIGURE 11.6-1 Power plant supplying a customer's electrical load. A transmission line connects the power plant to the customer's terminals.

The circuit in Figure 11.6-2 models the transmission of electric power from a power utility company to a customer in the frequency domain, using impedances and phasors. Our objective is to find a way to reduce the power absorbed by the transmission line. In this situation, it is likely that we cannot change the transmission line, so we can't change R_1 or $j\omega L_1$. Similarly, because the customer requires a specified average power at a specified voltage, we can't change V_m or P. In the following analysis, we leave R_1, L_1, V_m, and P as variables for the sake of generality. We won't need to repeat the analysis later if we encounter a similar situation with a different customer and a different transmission line. We will see that it is possible to adjust the power factor by adding a compensating impedance to the customer's load. We will leave the power factor, *pf*, as a variable in our analysis because we plan to vary the power factor to reduce the power absorbed by the load.

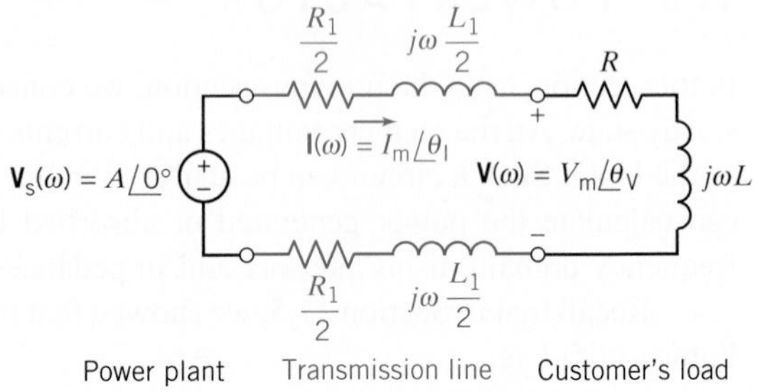

FIGURE 11.6-2 Frequency-domain representation of the power plant supplying a customer's electrical load.

The impedance of the line is

$$\mathbf{Z}_{\text{LINE}}(\omega) = \frac{R_1}{2} + j\omega\frac{L_1}{2} + \frac{R_1}{2} + j\omega\frac{L_1}{2} = R_1 + j\omega L_1$$

The average power absorbed by the line is

$$P_{\text{LINE}} = \frac{I_m^2}{2}\text{Re}(\mathbf{Z}_{\text{LINE}}) = \frac{I_m^2}{2}R_1$$

Because the customer requires power at a specified voltage, we will treat the voltage across the load, V_m, and the average power delivered to the load, P, as known quantities. Recall from Eq. 11.6-1 that

$$P = \frac{V_m I_m}{2} pf$$

Solving for I_m gives

$$I_m = \frac{2P}{V_m\, pf}$$

so

$$P_{\text{LINE}} = 2\left(\frac{P}{V_m\, pf}\right)^2 R_1$$

Increasing *pf* will reduce the power absorbed in the transmission line. The power factor is the cosine of an angle, so its maximum value is 1. Notice that $pf = 1$ occurs when $\theta_V = \theta_I$, that is, when the load appears to be resistive.

In Figure 11.6-3, a compensating impedance has been attached across the terminals of the customer's load. We plan to use this impedance to adjust the power factor of the customer's load. Because it is to the advantage of both the power company and the user to keep the power factor of a load as close to unity as feasible, we say that we are *correcting* the power factor of the load. We will denote the corrected power factor as *pfc* and the corresponding phase angle as θ_C. That is,

$$pfc = \cos\theta_C$$

We can represent the impedance of the load as

$$\mathbf{Z} = R + jX$$

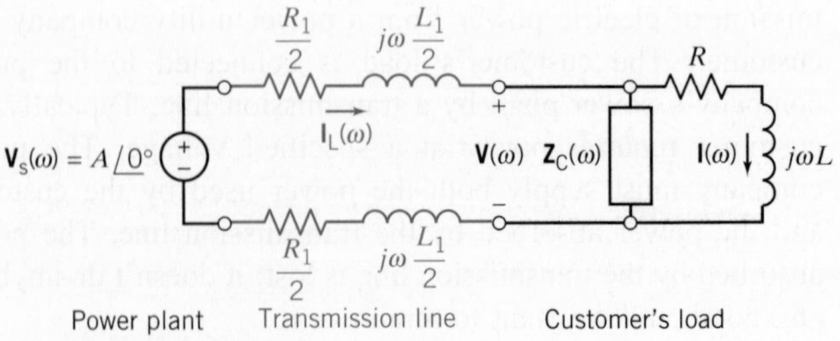

FIGURE 11.6-3 Power plant supplying a customer's electrical load. A compensating impedance has been added to the customer's load to correct the power factor.

Similarly, we can represent the impedance of the compensating impedance as

$$\mathbf{Z}_C = R_C + jX_C$$

Because $\mathbf{Z}$ is connected to draw a current $\mathbf{I}$, the power delivered to $\mathbf{Z}$ will remain P. The benefit of the parallel impedance is that the parallel combination appears as the load to the source, and $\mathbf{I}_L$ is the current that flows through the transmission line. We want $\mathbf{Z}_C$ to absorb no average power. Therefore, we choose a reactive element so that

$$\mathbf{Z}_C = jX_C$$

The impedance of the parallel combination, $\mathbf{Z}_P$, is

$$\mathbf{Z}_P = \frac{\mathbf{Z}\mathbf{Z}_C}{\mathbf{Z} + \mathbf{Z}_C}$$

The parallel impedance may be written as

$$\mathbf{Z}_P = R_P + jX_P = Z_P \angle \theta_P$$

and the power factor of the new combination is

$$pfc = \cos\theta_P = \cos\left(\tan^{-1}\frac{X_P}{R_P}\right) \quad (11.6\text{-}2)$$

where pfc is the corrected power factor, and the corrected phase $\theta_C = \theta_P$. Some algebra is needed to calculate R_P and X_P:

$$\begin{aligned} \mathbf{Z}_P &= \frac{(R + jX)\, jX_C}{R + jX + jX_C} \\ &= \frac{RX_C^2 + j\left[R^2X_C + (X_C + X)X\,X_C\right]}{R^2 + (X + X_C)^2} \\ &= \frac{RX_C^2}{R^2 + (X + X_C)^2} + j\frac{R^2X_C + (X_C + X)X\,X_C}{R^2 + (X + X_C)^2} \end{aligned}$$

Therefore, the ratio of X_P to R_P is

$$\frac{X_P}{R_P} = \frac{R^2 + (X_C + X)X}{RX_C} \quad (11.6\text{-}3)$$

Equation 11.6-2 may be written as

$$\frac{X_P}{R_P} = \tan\left(\cos^{-1} pfc\right) \quad (11.6\text{-}4)$$

Combining Eqs. 11.6-3 and 11.6-4 and solving for X_C, we have

$$X_C = \frac{R^2 + X^2}{R\tan\left(\cos^{-1} pfc\right) - X} \quad (11.6\text{-}5)$$

We note that X_C may be positive or negative, depending on the required pfc and the original R and X of the load. The factor $\tan[\cos^{-1}(pfc)]$ will be positive if pfc is specified as lagging and negative if it is specified as leading.

Typically, we will find that the customer's load is inductive, and we will need a capacitive impedance $\mathbf{Z}_C$. Recall that for a capacitor, we have

$$\mathbf{Z}_C = \frac{-j}{\omega C} = jX_C \quad (11.6\text{-}6)$$

Note that we determine that X_C is typically negative. Combining Eqs. 11.6-5 and 11.6-6 gives

$$\frac{-1}{\omega C} = \frac{R^2 + X^2}{R \tan(\cos^{-1} pfc) - X}$$

Solving for ωC gives

$$\omega C = \frac{X - R\tan(\cos^{-1} pfc)}{R^2 + X^2} = \frac{R}{R^2 + X^2}\left(\frac{X}{R} - \tan\left(\cos^{-1} pfc\right)\right)$$

Let $\theta = \tan^{-1}\left(\frac{X}{R}\right)$. Then

$$\omega C = \frac{R}{R^2 + X^2}(\tan\theta - \tan\theta_C) \tag{11.6-7}$$

where $\theta = \cos^{-1}(pf)$ and $\theta_C = \cos^{-1}(pfc)$.

EXAMPLE 11.6-1 Parallel Loads

A customer's plant has two parallel loads connected to the power utility's distribution lines. The first load consists of 50 kW of heating and is resistive. The second load is a set of motors that operate at 0.86 lagging power factor. The motors' load is 100 kVA. Power is supplied to the plant at 10,000 volts rms. Determine the total current flowing from the utility's lines into the plant and the plant's overall power factor.

Solution

Figure 11.6-4*a* summarizes what is known about this power system.

First, consider the heating load. Because this load is resistive, the reactive power is zero. Therefore,

$$\mathbf{S}_1 = P_1 = 50 \text{ kW}$$

Next, consider the motors. The power factor is lagging, so $\theta_2 > 0°$:

$$\theta_2 = \cos^{-1}(pf_2) = \cos^{-1}(0.86) = 30.7°$$

The complex power absorbed by the motors is

$$\mathbf{S}_2 = |\mathbf{S}_2| \angle \theta_2 = 100 \angle 30.7° \text{ kVA}$$

The average power and reactive power absorbed by the motors is obtained by converting the complex power to rectangular form:

$$\mathbf{S}_2 = |\mathbf{S}_2| \cos\theta_2 + j|\mathbf{S}_2| \sin\theta_2 = 100 \cos 30.7° + j100 \sin 30.7° = 86 + j51 \text{ kVA}$$

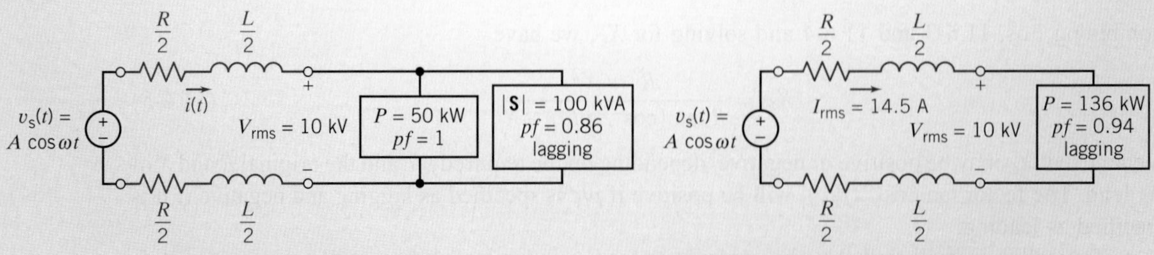

FIGURE 11.6-4 Power system for Example 11.6-1.

Therefore, $P_2 = 86\ \text{kW} \quad \text{and} \quad Q_2 = 51\ \text{kVAR}$

The total complex power **S** delivered to the total load is the sum of the complex power delivered to each load:

$$\mathbf{S} = \mathbf{S}_1 + \mathbf{S}_2 = 50 + (86 + j51) = 136 + j51\ \text{kVA}$$

The average power and reactive power of the customer's load is

$$P = 136\ \text{kW} \quad \text{and} \quad Q = 51\ \text{kVAR}$$

To calculate the power factor of the customer's load, first convert **S** to polar form:

$$\mathbf{S} = 145.2\angle 20.6^\circ\ \text{kVA}$$

Then

$$pf = \cos(20.6^\circ) = 0.94 \text{ lagging}$$

The total current flowing from the utility's lines into the plant can be calculated from the apparent power absorbed by the customer's load and the voltage across the terminals of the customer's load. Recall that

$$|\mathbf{S}| = \frac{V_m I_m}{2} = V_{\text{rms}} I_{\text{rms}}$$

Solving for the current gives

$$I_{\text{rms}} = \frac{|\mathbf{S}|}{V_{\text{rms}}} = \frac{145{,}200}{10^4} = 14.52\ \text{A rms}$$

Figure 11.6-4*b* summarizes the results of this example.

Example 11.6-2 Power Factor Correction

A load as shown in Figure 11.6-5 has an impedance of $\mathbf{Z} = 100 + j100\ \Omega$. Find the parallel capacitance required to correct the power factor to (a) 0.95 lagging and (b) 1.0. Assume that the source is operating at $\omega = 377$ rad/s.

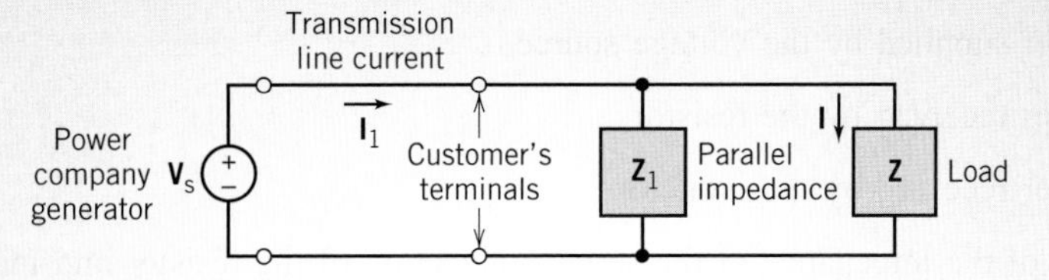

FIGURE 11.6-5 Use of an added parallel impedance $\mathbf{Z}_1$ to correct the customer's power factor.

Solution

The phase angle of the impedance is $\theta = 45^\circ$, so the original load has a lagging power factor with

$$\cos\theta = \cos 45^\circ = 0.707$$

First, we wish to correct the *pf* so that *pfc* = 0.95 lagging. Then, we use Eq. 11.6-5 as follows:

$$X_C = \frac{100^2 + 100^2}{100\tan(\cos^{-1} 0.95) - 100} = -297.9\ \Omega$$

The capacitor required is determined from

$$-\frac{1}{\omega C} = X_C$$

Therefore, because $\omega = 377$ rad/s,

$$C = -\frac{1}{\omega X_C} = \frac{-1}{377(-297.9)} = 8.9\ \mu\text{F}$$

If we wish to correct the load to $pfc = 1$, we have

$$X_C = \frac{2 \times 10^4}{100 \tan(\cos^{-1} 1) - 100} = -200$$

The capacitor required to correct the power factor to 1.0 is determined from

$$C = \frac{-1}{\omega X_C} = \frac{-1}{377(-200)} = 13.3\ \mu\text{F}$$

Because the uncorrected power factor is lagging, we can alternatively use Eq. 11.6-7 to determine C. For example, it follows that $pfc = 1$. Then $\theta_C = 0°$. Therefore,

$$\omega C = \frac{100}{2 \times 10^4}(\tan\theta - \tan\theta_C) = (5 \times 10^{-3})(\tan(45°) - \tan(0°)) = 5 \times 10^{-3}$$

and

$$C = \frac{5 \times 10^{-3}}{377} = 13.3\ \mu\text{F}$$

As expected, this is the same value of capacitance as was calculated using Eq. 11.6-5.

EXAMPLE 11.6-3 Complex Power

INTERACTIVE EXAMPLE

The input to the circuit shown in Figure 11.6-6a is the voltage of the voltage source,

$$v_s(t) = 7.28 \cos(4t + 77°)\ \text{V}$$

The output is the voltage across the inductor,

$$v_o(t) = 4.254 \cos(4t + 311°)\ \text{V}$$

Determine the following:

(a) The average power supplied by the voltage source

(b) The average power received by the resistor

(c) The average power received by the inductor

(d) The power factor of the impedance of the series connection of the resistor and inductor

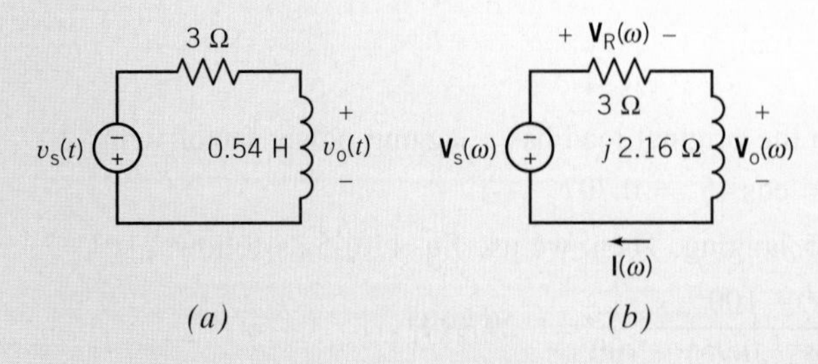

FIGURE 11.6-6 The circuit considered in Example 11.6-3 represented (a) in the time domain and (b) in the frequency domain.

Solution

The input voltage is sinusoid. The output voltage is also sinusoid and has the same frequency as the input voltage. Apparently, the circuit has reached steady state. Consequently, the circuit in Figure 11.6-6a can be represented in the frequency domain, using phasors and impedances. Figure 11.6-6*b* shows the frequency-domain representation of the circuit from Figure 11.6-6*a*. The impedance of the inductor is $j\omega L = j(4)(0.54) = j2.16\ \Omega$, as shown in Figure 11.6-6*b*.

The phasors corresponding to the input and output sinusoids are

$$\mathbf{V}_S(\omega) = 7.28\angle 77^\circ\ \text{V}$$

and

$$\mathbf{V}_o(\omega) = 4.254\angle 311^\circ\ \text{V}$$

The current $\mathbf{I}(\omega)$ in Figure 11.6-6*b* is calculated from $\mathbf{V}_o(\omega)$ and the impedance of the inductor, using Ohm's law:

$$\mathbf{I}(\omega) = \frac{\mathbf{V}_o(\omega)}{j\,2.16} = \frac{4.254\angle 311^\circ}{2.16\angle 90^\circ} = \frac{4.254}{2.16}\angle 311^\circ - 90^\circ = 1.969\angle 221^\circ\text{A}$$

Once we know $\mathbf{I}(\omega)$, we are ready to answer the questions asked in this example.

(a) The average power supplied by the source is calculated from $\mathbf{I}(\omega)$ and $\mathbf{V}_s(\omega)$. The average power of the source is given by

$$\begin{aligned}\frac{|\mathbf{V}_s(\omega)||\mathbf{I}(\omega)|}{2}\cos(\angle\mathbf{V}_s(\omega) - \angle\mathbf{I}(\omega)) &= \frac{(7.28)(1.969)}{2}\cos(77^\circ - 221^\circ) \\ &= 7.167\cos(-144^\circ) = -5.8\ \text{W}\end{aligned} \tag{11.6-8}$$

Notice that $\mathbf{I}(\omega)$ and $\mathbf{V}_s(\omega)$ adhere to the passive convention. Consequently, Eq. 11.6-8 gives the power received by the voltage source rather than the power supplied by the voltage source. The power supplied is the negative of the power received. Therefore, the power supplied by the voltage source is

$$P_s = 5.8\ \text{W}$$

(b) The resistor voltage, $\mathbf{V}_R(\omega)$, in Figure 11.6-6*b* is given by

$$\mathbf{V}_R(\omega) = R\,\mathbf{I}(\omega) = 3(1.969\angle 221^\circ) = 5.907\angle 221^\circ\ \text{V}$$

The average power received by the resistor is calculated from $\mathbf{I}(\omega)$ and $\mathbf{V}_R(\omega)$:

$$\begin{aligned}P_R = \frac{|\mathbf{V}_R(\omega)||\mathbf{I}(\omega)|}{2}(\cos(\angle\mathbf{V}_R(\omega) - \angle\mathbf{I}(\omega))) &= \frac{(5.907)(1.969)}{2}\cos(221^\circ - 221^\circ) \\ &= 5.8\cos(0^\circ) = 5.8\ \text{W}\end{aligned} \tag{11.6-9}$$

Notice that $\mathbf{I}(\omega)$ and $\mathbf{V}_R(\omega)$ adhere to the passive convention. Consequently, P_R is the power received by the resistor, as required.

Alternately, the power received by a resistor can be calculated from the current $\mathbf{I}(\omega)$ and the resistance, R. To see how, first notice that the voltage and current of a resistor are related by

$$\mathbf{V}_R(\omega) = R\mathbf{I}(\omega) \quad\Rightarrow\quad |\mathbf{V}_R(\omega)|\angle\mathbf{V}_R(\omega) = R(|\mathbf{I}(\omega)|\angle\mathbf{I}(\omega)) \quad\Rightarrow\quad \begin{cases}|\mathbf{V}_R(\omega)| = R|\mathbf{I}(\omega)| \\ \angle\mathbf{V}_R(\omega) = \angle\mathbf{I}(\omega)\end{cases}$$

Substituting these expressions for $|\mathbf{V}_R(\omega)|$ and $\angle\mathbf{V}_R(\omega)$ into Eq. 11.6-9 gives

$$\begin{aligned}P_R = \frac{|R\mathbf{I}(\omega)|\,|\mathbf{I}(\omega)|}{2}\cos(\angle\mathbf{I}(\omega) - \angle\mathbf{I}(\omega)) &= \frac{R|\mathbf{I}(\omega)|^2}{2} \\ &= \frac{(3)(1.969)^2}{2} = 5.8\ \text{W}\end{aligned}$$

(c) The average power received by the inductor is calculated from $\mathbf{I}(\omega)$ and $\mathbf{V}_o(\omega)$:

$$\begin{aligned}P_L = \frac{|\mathbf{V}_o(\omega)||\mathbf{I}(\omega)|}{2}\cos(\angle\mathbf{V}_o(\omega) - \angle\mathbf{I}(\omega)) &= \frac{(4.254)(1.969)}{2}\cos(311^\circ - 221^\circ) \\ &= 4.188\cos(90^\circ) = 0\ \text{W}\end{aligned} \tag{11.6-10}$$

The phase angle of the inductor voltage is always 90° greater than the phase angle of the inductor current. Consequently, the value of average power received by any inductor is zero.

(e) The power factor of the impedance of the series connection of the resistor and inductor can be calculated from $\mathbf{I}(\omega)$ and the voltage across the impedance. That voltage is $\mathbf{V}_R(\omega) + \mathbf{V}_o(\omega)$, which is calculated by applying Kirchhoff's voltage law to the circuit in Figure 11.6-6*b*:

$$\mathbf{V}_R(\omega) + \mathbf{V}_o(\omega) + \mathbf{V}_s(\omega) = 0$$

$$\begin{aligned}\mathbf{V}_R(\omega) + \mathbf{V}_o(\omega) = -\mathbf{V}_s(\omega) &= -7.28\,\underline{/77^\circ} \\ &= (1\,\underline{/180^\circ})(7.28\,\underline{/77^\circ}) \\ &= 7.28\,\underline{/257^\circ}\end{aligned}$$

Now the power factor is calculated as

$$pf = \cos\left(\underline{/(\mathbf{V}_R(\omega) + \mathbf{V}_o(\omega))} - \underline{/\mathbf{I}(\omega))}\right) = \cos(257^\circ - 221^\circ) = 0.809$$

The power factor is said to be lagging because $257^\circ - 221^\circ = 36^\circ > 0$.

Average power is conserved. In this example, that means that the average power supplied by the voltage source must be equal to the sum of the average powers received by the resistor and the inductor. This fact provides a check on the accuracy of our calculations.

If the value of $\mathbf{V}_o(\omega)$ had not been given, then $\mathbf{I}(\omega)$ would be calculated by writing and solving a mesh equation. Referring to Figure 11.6-6*b*, the mesh equation is

$$3\mathbf{I}(\omega) + j2.16\,\mathbf{I}(\omega) + 7.28\,\underline{/77^\circ} = 0$$

Solving for $\mathbf{I}(\omega)$ gives

$$\begin{aligned}\mathbf{I}(\omega) &= \frac{-7.28\,\underline{/77^\circ}}{3 + j2.16} = \frac{(1\,\underline{/180^\circ})(7.28\,\underline{/77^\circ})}{3.697\,\underline{/36^\circ}} \\ &= \frac{(1)(7.28)}{3.697}\,\underline{/180 + 77 - 36} = 1.969\,\underline{/221^\circ}\ \text{A}\end{aligned}$$

as before.

EXERCISE 11.6-1 A circuit has a large motor connected to the ac power lines $[\omega = (2\pi)60 = 377$ rad/s]. The model of the motor is a resistor of 100 Ω in series with an inductor of 5 H. Find the power factor of the motor.

Answer: $pf = 0.053$ lagging

EXERCISE 11.6-2 A circuit has a load impedance $Z = 50 + j80\ \Omega$, as shown in Figure 11.6-5. Determine the power factor of the uncorrected circuit. Determine the impedance $\mathbf{Z}_C$ required to obtain a corrected power factor of 1.0.

Answer: $pf = 0.53$ lagging, $\mathbf{Z}_C = -j111.25\ \Omega$

EXERCISE 11.6-3 Determine the power factor for the total plant of Example 11.6-1 when the resistive heating load is decreased to 30 kW. The motor load and the supply voltage remain as described in Example 11.6-1.

Answer: $pf = 0.915$

EXERCISE 11.6-4 A 4-kW, 110-V_{rms} load, as shown in Figure 11.6-5, has a power factor of 0.82 lagging. Find the value of the parallel capacitor that will correct the power factor to 0.95 lagging when $\omega = 377$ rad/s.

Answer: $C = 0.324$ mF

11.7 THE POWER SUPERPOSITION PRINCIPLE

In this section, let us consider the case when the circuit contains two or more sources. For example, consider the circuit shown in Figure 11.7-1*a* with two sinusoidal voltage sources. The principle of superposition states that the response to both sources acting together is equal to the sum of the responses to each voltage source acting alone. The application of the principle of superposition is illustrated in Figure 11.7-1*b*, where i_1 is the response to source 1 acting alone, and the response i_2 is the response to source 2 acting alone. The total response is

$$i = i_1 + i_2 \tag{11.7-1}$$

The instantaneous power is

$$p = i^2 R = R(i_1 + i_2)^2 = R(i_1^2 + i_2^2 + 2i_1 i_2)$$

where R is the resistance of the circuit. Then the average power is

$$\begin{aligned} P &= \frac{1}{T}\int_0^T p\,dt = \frac{R}{T}\int_0^T \left(i_1^2 + i_2^2 + 2i_1 i_2\right) dt \\ &= \frac{R}{T}\int_0^T i_1^2\,dt + \frac{R}{T}\int_0^T i_2^2\,dt + \frac{2R}{T}\int_0^T i_1 i_2\,dt = P_1 + P_2 + \frac{2R}{T}\int_0^T i_1 i_2\,dt \end{aligned} \tag{11.7-2}$$

where P_1 is the average power due to v_1 and P_2 is the average power due to v_2. We will see that when v_1 and v_2 are sinusoids having different frequencies, then

$$\frac{2R}{T}\int_0^T i_1 i_2\,dt = 0 \tag{11.7-3}$$

When Eq. 11.7-3 is satisfied, then Eq. 11.7-2 reduces to

$$P = P_1 + P_2 \tag{11.7-4}$$

This equation states that the average power delivered to the resistor by both sources acting together is equal to the sum of the average power delivered to the resistor by each voltage source acting alone. This is the principle of power superposition. Notice that the principle of power superposition is valid only when Eq. 11.7-3 is satisfied.

Now let us determine under what conditions Eq. 11.7-3 is satisfied. Let the radian frequency for the first source be $m\omega$, and let the radian frequency for the second source be $n\omega$. The currents can be represented by the general form

$$i_1 = I_1 \cos\,(m\omega t + \phi)$$

and

$$i_2 = I_2 \cos\,(n\omega t + \theta)$$

It can be shown that

$$\int_0^T \cos(m\omega t + \phi)\cos(n\omega t + \theta)dt = \begin{cases} 0 & m \neq n \\ \cos(\phi - \theta) & m = n \end{cases}$$

Consequently,

$$\frac{2R}{T}\int_0^T i_1 i_2 dt = \begin{cases} 0 & m \neq n \\ RI_1 I_2 \cos(\phi - \theta) & m = n \end{cases} \tag{11.7-5}$$

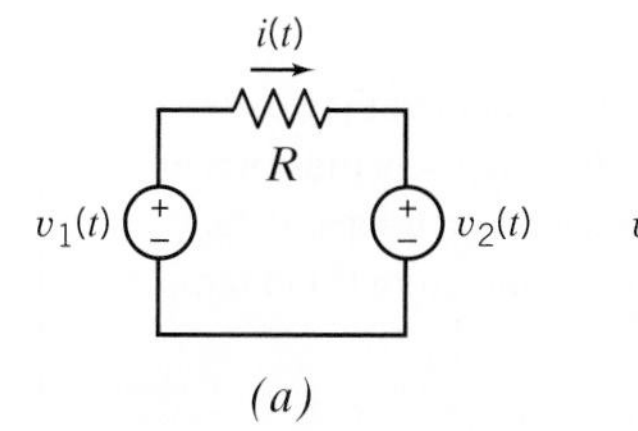

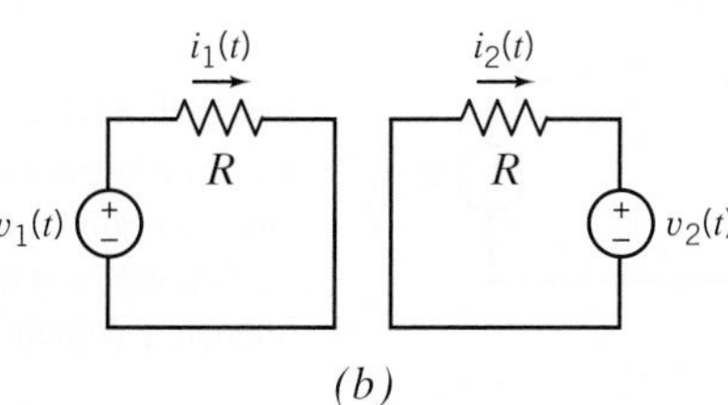

FIGURE 11.7-1 (*a*) A circuit with two sources. (*b*) Using superposition to calculate the resistor current as $i(t) = i_1(t) + i_2(t)$.

Therefore, in summary, the *superposition of average power* states that the average power delivered to a circuit by several sinusoidal sources, acting together, is equal to the sum of the average power delivered to the circuit by each source acting alone, if, and only if, no two of the sources have the same frequency. Similar arguments show that superposition can be used to calculate the reactive power or the complex power delivered to a circuit by several sinusoidal sources, provided again that no two sources have the same frequency.

If two or more sources are operating at the same frequency, the principle of *power* superposition is not valid, but the principle of superposition remains valid. In this case, we use the principle of superposition to find each phasor current and then add the currents to obtain the total phasor current

$$\mathbf{I} = \mathbf{I}_1 + \mathbf{I}_2 + \cdots + \mathbf{I}_N$$

for N sources. Then we have the average power

$$P = \frac{I_\text{m}^2 R}{2} \tag{11.7-6}$$

where $|\mathbf{I}| = I_\text{m}$.

EXAMPLE 11.7-1 Power Superposition

The circuit in Figure 11.7-2 contains two sinusoidal sources. To illustrate power superposition, consider two cases:

(1) $v_\text{A}(t) = 12 \cos 3t$ V and $v_\text{B}(t) = 4 \cos 4t$ V

(2) $v_\text{A}(t) = 12 \cos 4t$ V and $v_\text{B}(t) = 4 \cos 4t$ V

Find the average power absorbed by the 6-Ω resistor.

Solution

The application of the principle of superposition is illustrated in Figure 11.7-2*b*, where i_1 is the response to the voltage source A acting alone, and the response i_2 is the response to the voltage source B acting alone. The total

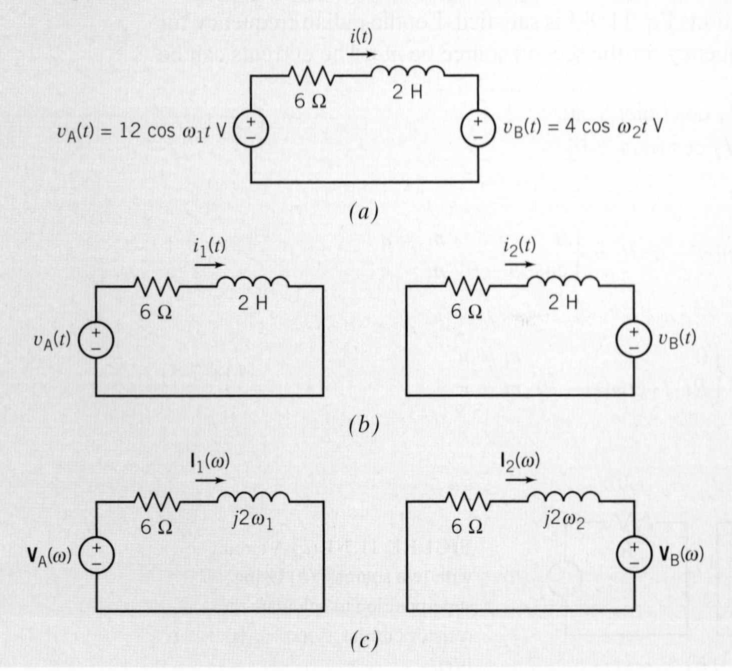

FIGURE 11.7-2 (*a*) A circuit with two sinusoidal sources. (*b*) Using superposition to find the response to each source separately. (*c*) Representing the circuits from (*b*) in the frequency domain.

response is $i = i_1 + i_2$. In Figure 11.7-2*c*, the circuits from Figure 11.7-2*b* are represented in the frequency domain, using impedances and phasors.

Now consider the two cases.

Case 1: Analysis of the circuits in Figure 11.7-2*c* gives

$$\mathbf{I}_1(\omega) = 1.414\angle{-45^\circ}\text{ A} \quad \text{and} \quad \mathbf{I}_2(\omega) = 0.4\angle{127^\circ}\text{ A}$$

These phasors correspond to different frequencies and cannot be added. The corresponding time-domain currents are

$$i_1(t) = 1.414 \cos(3t - 45^\circ)\text{A} \quad \text{and} \quad i_2(t) = 0.4 \cos(4t - 143^\circ)\text{ A}$$

Using superposition, we find that the total current in the resistor is

$$i(t) = 1.414 \cos(3t - 45^\circ) + 0.4 \cos(4t + 127^\circ)\text{ A}$$

The average power could be calculated as

$$P = \frac{R}{T}\int_0^T i^2\,dt = \frac{R}{T}\int_0^T (1.414 \cos(3t - 45^\circ) + 0.4 \cos(4t + 127^\circ))^2\,dt$$

Because the two sinusoidal sources have different frequencies, the average power can be calculated more easily using power superposition:

$$P = P_1 + P_2 = \frac{1.414^2}{2}6 + \frac{0.4^2}{2}6 = 6.48\text{ W}$$

Notice that both superposition and power superposition were used in this case. First, superposition was used to calculate $\mathbf{I}_1(\omega)$ and $\mathbf{I}_2(\omega)$. Next, P_1 was calculated using $\mathbf{I}_1(\omega)$, and P_2 was calculated using $\mathbf{I}_2(\omega)$. Finally, power superposition was used to calculate P from P_1 and P_2.

Case 2: Analysis of the circuits in Figure 11.7-2*c* gives

$$\mathbf{I}_1(\omega) = 1.2\angle{-53^\circ}\text{ A} \quad \text{and} \quad \mathbf{I}_2(\omega) = 0.4\angle{127^\circ}\text{ A}$$

Both of these phasors correspond to the same frequency, $\omega = 4$ rad/s. Therefore, these phasors can be added to obtain the phasor corresponding to $i(t)$.

$$\mathbf{I}(\omega) = \mathbf{I}_1(\omega) + \mathbf{I}_2(\omega) = (1.2\angle{-53^\circ}) + (0.4\angle{127^\circ}) = 0.8\angle{-53^\circ}\text{ A}$$

The sinusoidal current corresponding to this phasor is

$$i(t) = 0.8 \cos(4t - 53^\circ)\text{ A}$$

The average power absorbed by the resistor is

$$P = \frac{0.8^2}{2}6 = 1.92\text{ W}$$

Alternately, the time-domain currents corresponding to $\mathbf{I}_1(\omega)$ and $\mathbf{I}_2(\omega)$ are

$$i_1(t) = 1.2 \cos(4t - 53^\circ)\text{A} \quad \text{and} \quad i_2(t) = 0.4 \cos(4t + 127^\circ)\text{ A}$$

Using superposition, we find that the total current in the resistor is

$$i(t) = 1.2 \cos(4t - 53^\circ) + 0.4 \cos(4t + 127^\circ) = 0.8 \cos(4t - 53^\circ)\text{ A}$$

So $P = 1.92$ W, as before.

Power superposition cannot be used in this case because the two sinusoidal sources have the same frequency.

EXERCISE 11.7-1 Determine the average power absorbed by the resistor in Figure 11.7-2*a* for these two cases:

(a) $v_A(t) = 12 \cos 3t$ V and $v_B(t) = 4 \cos 3t$ V;

(b) $v_A(t) = 12 \cos 4t$ V and $v_B(t) = 4 \cos 3t$ V

Answers: (a) 2.66 W (b) 4.99 W

11.8 THE MAXIMUM POWER TRANSFER THEOREM

In Chapter 5, we proved that for a resistive network, maximum power is transferred from a source to a load when the load resistance is set equal to the Thévenin resistance of the Thévenin equivalent source. Now let us consider a circuit represented by a Thévenin equivalent circuit for a sinusoidal steady-state circuit, as shown in Figure 11.8-1, when the load is $\mathbf{Z}_L$.

We then have

$$\mathbf{Z}_t = R_t + jX_t$$

and

$$\mathbf{Z}_L = R_L + jX_L$$

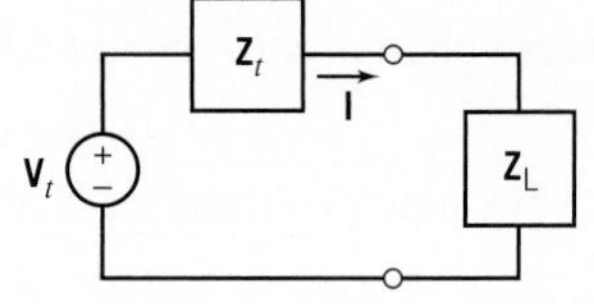

FIGURE 11.8-1 The Thévenin equivalent circuit with a load impedance.

The average power delivered to the load is

$$P = \frac{I_m^2}{2} R_L$$

The phasor current **I** is given by

$$\mathbf{I} = \frac{\mathbf{V}_t}{\mathbf{Z}_t + \mathbf{Z}_L} = \frac{\mathbf{V}_t}{(R_t + jX_t) + (R_L + jX_L)}$$

where we may select the values of R_L and X_L. The average power delivered to the load is

$$P = \frac{I_m^2 R_L}{2} = \frac{|\mathbf{V}_t|^2 R_L / 2}{(R_t + R_L)^2 + (X_t + X_L)^2}$$

and we wish to maximize P. The term $(X_t + X_L)^2$ can be eliminated by setting $X_L = -X_t$. We have

$$P = \frac{|\mathbf{V}_t|^2 R_L}{2(R_t + R_L)^2}$$

The value of R_L that maximizes P is determined by taking the derivative dP/dR_L and setting it equal to zero. Then we find that $dP/dR_L = 0$ when $R_L = R_t$.

Consequently, we have

$$\mathbf{Z}_L = R_t - jX_t$$

Thus, the *maximum power transfer* from a circuit with a Thévenin equivalent circuit with an impedance $\mathbf{Z}_t$ is obtained when $\mathbf{Z}_L$ is set equal to $\mathbf{Z}_t^*$, the complex conjugate of $\mathbf{Z}_t$.

EXAMPLE 11.8-1 Maximum Power Transfer

Find the load impedance that transfers maximum power to the load and determine the maximum power delivered to the load for the circuit shown in Figure 11.8-2.

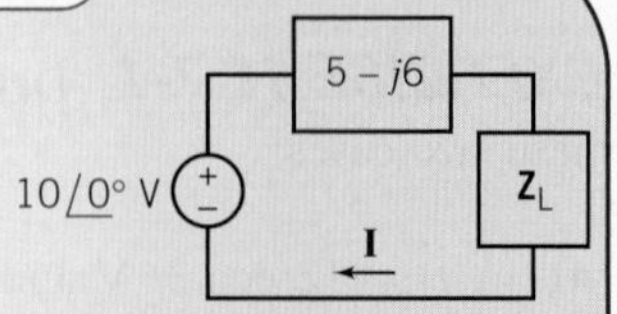

5 − j6
10/0° V
Z_L
I

FIGURE 11.8-2 Circuit for Example 11.8-1. Impedances in ohms.

Solution

We select the load impedance, $\mathbf{Z}_L$, to be the complex conjugate of $\mathbf{Z}_t$ so that

$$\mathbf{Z}_L = \mathbf{Z}_t^* = 5 + j6\ \Omega$$

Then the maximum power transferred can be obtained by noting that

$$\mathbf{I} = \frac{10\,\angle 0^\circ}{5+5} = 1\,\angle 0^\circ\ \mathrm{A}$$

Therefore, the average power transferred to the load is

$$P = \frac{I_m^2}{2} R_L = \frac{(1)^2}{2} 5 = 2.5\ \mathrm{W}$$

EXERCISE 11.8-1 For the circuit of Figure 11.8-1, find $\mathbf{Z}_L$ to obtain the maximum power transferred when the Thévenin equivalent circuit has $\mathbf{V}_t = 100\,\angle 0^\circ\mathrm{V}$ and $\mathbf{Z}_t = 10 + j14\ \Omega$. Also, determine the maximum power transferred to the load.

Answer: $\mathbf{Z}_L = 10 - j14\ \Omega$ and $P = 125$ W

EXERCISE 11.8-2 A television receiver uses a cable to connect the antenna to the TV, as shown in Figure E 11.8-2, with $v_s = 4 \cos \omega t$ mV. The TV station is received at 52 MHz. Determine the average power delivered to each TV set if (a) the load impedance is $\mathbf{Z} = 300\ \Omega$; (b) two identical TV sets are connected in parallel with $\mathbf{Z} = 300\ \Omega$ for each set; (c) two identical sets are connected in parallel and $\mathbf{Z}$ is to be selected so that maximum power is delivered at each set.

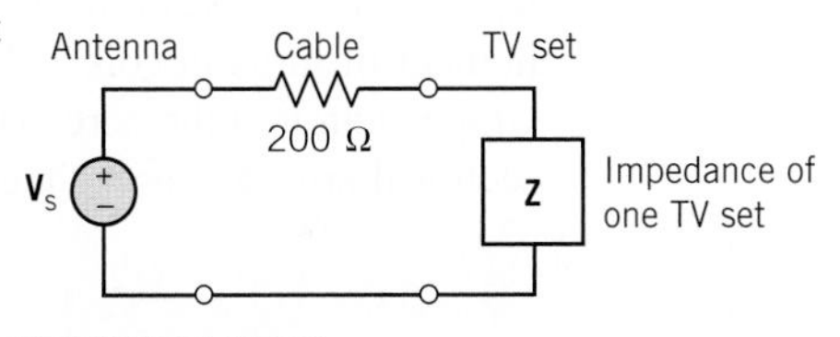

FIGURE E 11.8-2

Answers: (a) 9.6 nW (b) 4.9 nW (c) 5 nW

11.9 COUPLED INDUCTORS

The concept of self-inductance was introduced in Chapter 7. We commonly use the term *inductance* for self-inductance, and we are familiar with circuits that have inductors. In this section, we consider coupled inductors, which are useful in circuits with sinusoidal steady-state (ac) voltages and currents and are also widely used in electronic circuits.

Coupled inductors, or *coupled coils*, are magnetic devices that consist of two or more multiturn coils wound on a common core.

Figure 11.9-1*a* shows two coils of wire wrapped around a magnetic core. These coils are said to be magnetically coupled. A voltage applied to one coil, as shown in Figure 11.9-1*a*, causes a voltage across the second coil. Here's why. The input voltage, $v_1(t)$, causes a current $i_1(t)$ in coil 1. The current and voltage are related by

$$v_1 = L_1 \frac{di_1}{dt} \tag{11.9-1}$$

where L_1 is the self-inductance of coil 1. The current $i_1(t)$ causes a flux in the magnetic core. This flux is related to the current by

$$\phi = c_1 N_1 i_1 \tag{11.9-2}$$

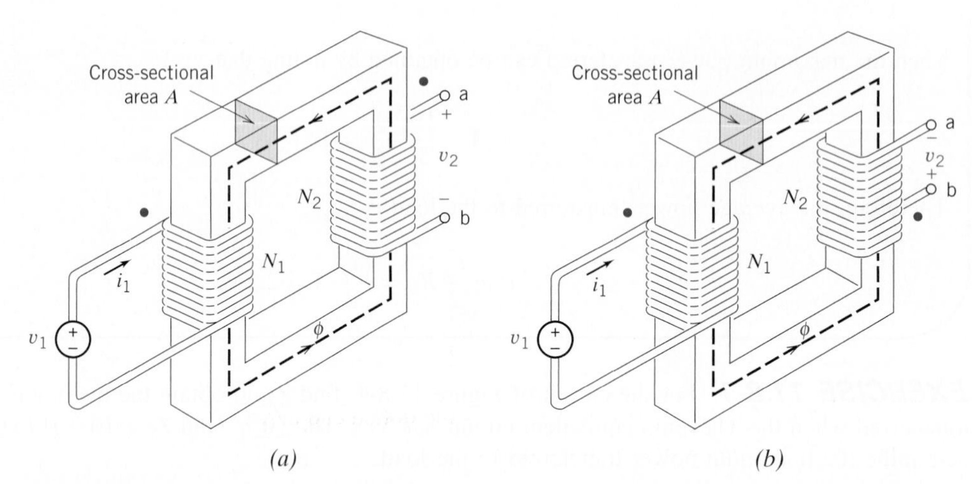

FIGURE 11.9-1 Two magnetically coupled coils mounted on a magnetic material. The flux ϕ is contained within the magnetic core.

where c_1 is a constant that depends on the magnetic properties and geometry of the core, and N_1 is the number of turns in coil 1. The number of turns in a coil indicates the number of times the wire is wrapped around the core. The flux, ϕ, is contained within the magnetic core. The core has a cross-sectional area A. The voltage across the coil 1 is related to the flux by

$$v_1 = N_1 \frac{d\phi}{dt} = N_1 \frac{d}{dt}(c_1 N_1 i_1) = c_1 N_1^2 \frac{di_1}{dt} \tag{11.9-3}$$

Comparing Eqs. 11.9-1 and 11.9-3 shows that

$$L_1 = c_1 N_1^2 \tag{11.9-4}$$

A voltage, v_2, at the terminals of the second coil is induced by ϕ, which flows through the second coil. This voltage is related to the flux by

$$v_2 = N_2 \frac{d\phi}{dt} = c_M \, N_1 N_2 \frac{di_1}{dt} = M \frac{di_1}{dt} \tag{11.9-5}$$

where c_M is a constant that depends on the magnetic properties and geometry of the core, N_2 is the number of turns in the second coil, and $M = c_M N_1 N_2$ is a positive number called the **mutual inductance**. The unit of mutual inductance is the henry, H.

The polarity of the voltage v_2, compared to the polarity of v_1, depends on the way in which the coils are wrapped on the core. There are two distinct cases, and they are shown in Figures 11.9-1*a*,*b*. The difference between these two figures is the direction in which coil 2 is wrapped around the core. A **dot convention** is used to indicate the way the coils have been wrapped on the coil. Notice that one end of each coil is marked with a dot. When the reference direction of the current of one coil enters the dotted end of that coil, the reference polarity of the induced voltage is positive at the dotted end of the other coil. For example, in Figures 11.9-1*a*, *b*, the reference direction of the current i_1 enters the dotted end of the left coil. Consequently, in Figures 11.9-1*a*,*b*, the + sign of the reference polarity of v_2 is located at the dotted end of the right coil.

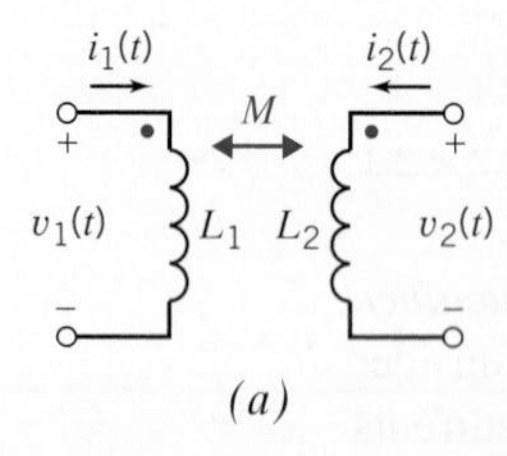

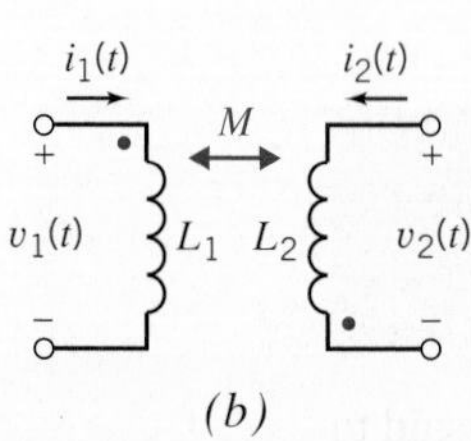

FIGURE 11.9-2 Circuit symbol for coupled inductors. In (*a*), both coil currents enter the dotted ends of the coils. In (*b*), one coil current enters the dotted end of the coil, but the other coil current enters the undotted end.

The circuit symbol that is used to represent coupled inductors is shown in Figure 11.9-2 with the dots shown and the mutual inductance identified as M. Two cases are shown in Figure 11.9-2. In Figure 11.9-2*a*, both coil currents enter the dotted ends of the coils. In Figure 11.9-2*b*, one current, i_1, enters the dotted end of a coil, but the other current, i_2, enters the undotted end on the coil. In both cases, the reference directions of the voltage and current of each coil adhere to the passive convention.

Suppose both coil currents enter the dotted ends of the coils, as in Figure 11.9-1*a*, or both coil currents enter the undotted ends of the coils. The voltage across the first coil, v_1, is related to the coil currents by

$$v_1 = L_1 \frac{di_1}{dt} + M \frac{di_2}{dt} \tag{11.9-6}$$

Similarly, the voltage across the second coil is related to the coil currents by

$$v_2 = L_2 \frac{di_2}{dt} + M \frac{di_1}{dt} \tag{11.9-7}$$

In contrast, suppose one coil current enters the dotted end of a coil while the other coil current enters the undotted end of a coil, as in Figure 11.9-2*b*. The voltage across the first coil, v_1, is related to the coil currents by

$$v_1 = L_1 \frac{di_1}{dt} - M \frac{di_2}{dt} \tag{11.9-8}$$

Similarly, the voltage across the second coil is related to the coil currents by

$$v_2 = L_2 \frac{di_2}{dt} - M \frac{di_1}{dt} \tag{11.9-9}$$

Thus, the mutual inductance can be seen to induce a voltage in a coil due to the current in the other coil.

Coupled inductors can be modeled using inductors (without coupling) and dependent sources. Figure 11.9-3 shows an equivalent circuit for coupled inductors.

The use of coupled inductors is usually limited to non-dc applications because coils behave as short circuits for a steady current.

Suppose that coupled inductors are part of a linear circuit with a sinusoidal input and that the circuit is at steady state. Such a circuit can be analyzed in the frequency domain, using phasors. The coupled inductors shown in Figure 11.9-2*a* are represented by the phasor equations

$$\mathbf{V}_1 = j\omega L_1 \mathbf{I}_1 + j\omega M\,\mathbf{I}_2 \tag{11.9-10}$$

and

$$\mathbf{V}_2 = j\omega L_2 \mathbf{I}_2 + j\omega M\,\mathbf{I}_1 \tag{11.9-11}$$

In contrast, the coupled inductors shown in Figure 11.9-2*b* are represented by the phasor equations

$$\mathbf{V}_1 = j\omega L_1 \mathbf{I}_1 - j\omega M\,\mathbf{I}_2 \tag{11.9-12}$$

and

$$\mathbf{V}_2 = j\omega L_2 \mathbf{I}_2 - j\omega M\,\mathbf{I}_1 \tag{11.9-13}$$

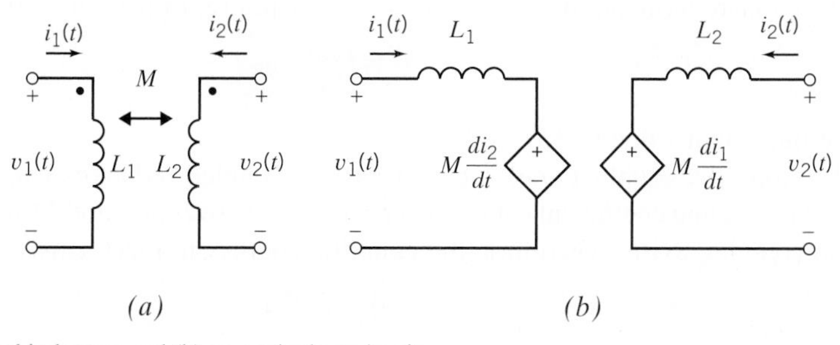

FIGURE 11.9-3 (*a*) Coupled inductors and (b) an equivalent circuit.

The inductances, L_1 and L_2, and mutual inductance, M, each depend on the magnetic properties and geometry of the core and the number of turns in the coils. Referring to Eqs. 11.9-4 and 11.9-5, we can write

$$L_1L_2 = \left(c_1N_1^2\right)\left(c_2N_2^2\right) = c_1c_2(N_1N_2)^2 = \left(\frac{c_MN_1N_2}{k}\right)^2 = \frac{M^2}{k^2} \qquad (11.9\text{-}14)$$

where the constant $k = c_M/\sqrt{c_1c_2}$ is called the **coupling coefficient**. Because the coupling coefficient depends on c_1, c_2, and c_M, it depends on the magnetic properties and geometry of the core. Solving Eq. 11.9-14 for the coupling coefficient gives

$$k = \frac{M}{\sqrt{L_1L_2}} \qquad (11.9\text{-}15)$$

The instantaneous power absorbed by coupled inductors is

$$\begin{aligned} p(t) &= v_1(t)i_1(t) + v_2(t)i_2(t) \\ &= \left(L_1\frac{d}{dt}i_1(t) \pm M\frac{d}{dt}i_2(t)\right)i_1(t) + \left(L_2\frac{d}{dt}i_2(t) \pm M\frac{d}{dt}i_1(t)\right)i_2(t) \\ &= L_1i_1(t)\frac{d}{dt}i_1(t) \pm M\frac{d}{dt}(i_1(t)i_2(t)) + L_2i_2(t)\frac{d}{dt}i_2(t) \end{aligned} \qquad (11.9\text{-}16)$$

where $-M$ is used if one current enters the undotted end of a coil while the other current enters the dotted end; otherwise, $+M$ is used. The energy stored in the coupled inductors is calculated by integrating the power absorbed by the coupled inductors. The energy stored in coupled inductors is

$$w(t) = \int_{-\infty}^{t} p(\tau)d\tau = \frac{1}{2}L_1i_1^2 + \frac{1}{2}L_2i_2^2 \pm Mi_1i_2 \qquad (11.9\text{-}17)$$

where, again, $-M$ is used if one current enters the undotted end of a coil while the other current enters the dotted end; otherwise, $+M$ is used. We can use this equation to find how large a value M can attain in terms of L_1 and L_2. Because coupled inductors are a passive element, the energy stored must be greater than or equal to zero. The limiting quantity for M is obtained when $w = 0$ in Eq. 11.9-17. Then we have

$$\frac{1}{2}L_1i_1^2 + \frac{1}{2}L_2i_2^2 - Mi_1\,i_2 = 0 \qquad (11.9\text{-}18)$$

as the limiting condition for the case in which one current enters the dotted terminal and the other current leaves the dotted terminal. Now add and subtract the term $i_1i_2 = \sqrt{L_1L_2}$ in the equation to generate a term that is a perfect square as follows:

$$\left(\sqrt{\frac{L_1}{2}}i_1 - \sqrt{\frac{L_2}{2}}i_2\right)^2 + i_1i_2\left(\sqrt{L_1L_2} - M\right) = 0$$

The perfect square term can be positive or zero. Therefore, to have $w \geq 0$, we require that

$$\sqrt{L_1L_2} \geq M \qquad (11.9\text{-}19)$$

Thus, the maximum value of M is $\sqrt{L_1L_2}$.

Therefore, the coupling coefficient of passive coupled inductors can be no larger than 1. In addition, the coupling coefficient cannot be negative because L_1, L_2, and M are all nonnegative. When $k = 0$, no coupling exists. Therefore, the coupling coefficient must satisfy

$$0 \leq k \leq 1 \qquad (11.9\text{-}20)$$

Most power system transformers have a k that approaches 1, whereas k is low for radio circuits.

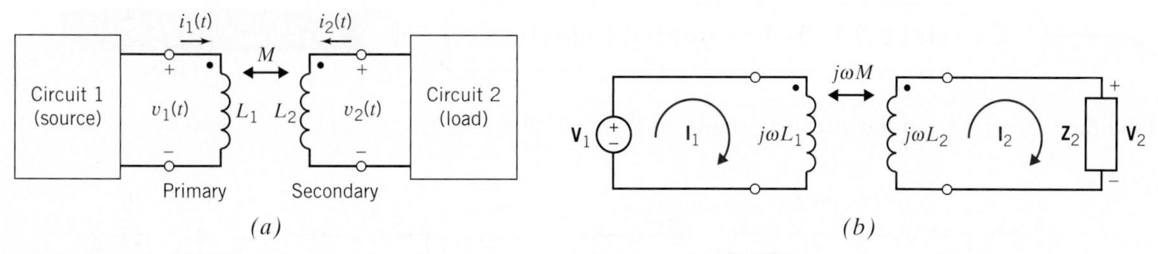

FIGURE 11.9-4 (*a*) Coupled inductors used as a transformer to couple two circuits magnetically and (*b*) a transformer used to couple a voltage source magnetically to an impedance.

Figure 11.9-4*a* shows coupled inductors used as a transformer to connect a source to a load. The coil connected to the source is called the **primary coil**, and the coil connected to the load is called the **secondary coil**. Circuit 2 is connected to circuit 1 through the magnetic coupling of the transformer, but there is no electrical connection between these two circuits. For example, there is no path for current to flow from circuit 1 to circuit 2. In addition, no circuit element is connected between a node of circuit 1 and a node of circuit 2.

Figure 11.9-4*b* shows a specific example of the situation shown in Figure 11.9-4*a*. The source is a single sinusoidal voltage source, and the load is a single impedance. The circuit has been represented in the frequency domain, using phasors and impedances. The circuit in Figure 11.9-4*b* can be analyzed by writing mesh equations. The two mesh equations are

$$j\omega L_1 \mathbf{I}_1 - j\omega M\, \mathbf{I}_2 = \mathbf{V}_1$$
$$-j\omega M\, \mathbf{I}_1 + (j\omega L_2 + \mathbf{Z}_2)\mathbf{I}_2 = 0$$

Solving for $\mathbf{I}_2$ in terms of $\mathbf{V}_1$, we have

$$\mathbf{I}_2 = \left[\frac{j\omega\, M}{\left((j\omega)^2 (L_1 L_2 - M^2) + (j\omega\, L_1 \mathbf{Z}_2)\right)}\right]\mathbf{V}_1 \qquad (11.9\text{-}21)$$

When the coupling coefficient of the coupled inductors is unity, then $M = \sqrt{L_1 L_2}$ and Eq. 11.9-21 reduces to

$$\mathbf{I}_2 = \left[\frac{j\omega\, M}{j\omega\, L_1 \mathbf{Z}_2}\right]\mathbf{V}_1 = \left[\frac{j\omega\sqrt{L_1 L_2}}{j\omega L_1 \mathbf{Z}_2}\right]\mathbf{V}_1 = \frac{\sqrt{L_2}}{\mathbf{Z}_2\sqrt{L_1}}\mathbf{V}_1 \qquad (11.9\text{-}22)$$

The voltage across the impedance is given by

$$\mathbf{V}_2 = \mathbf{Z}_2 \mathbf{I}_2 = \sqrt{\frac{L_2}{L_1}}\mathbf{V}_1 \qquad (11.9\text{-}23)$$

The ratio of the inductances is related to the magnetic properties and geometry of the core and the number of turns in the coils. Referring to Eq. 11.9-4, we can write

$$\frac{L_2}{L_1} = \frac{c_2 N_2^2}{c_1 N_1^2}$$

When both coils are wound symmetrically on the same core, then $c_1 = c_2$. In this case,

$$\frac{L_2}{L_1} = \frac{N_2^2}{N_1^2} = n^2 \qquad (11.9\text{-}24)$$

where n is called the **turns ratio** of the transformer. Combining Eqs. 11.9-23 and 11.9-24 gives

$$\mathbf{V}_2 = n\mathbf{V}_1 \qquad (11.9\text{-}25)$$

where $\mathbf{V}_1$ is the voltage across the primary coil, $\mathbf{V}_2$ is the voltage across the secondary coil, and n is the turns ratio.

EXAMPLE 11.9-1 Coupled Inductors

INTERACTIVE EXAMPLE

Find the voltage $v_2(t)$ in the circuit as shown in Figure 11.9-5*a*.

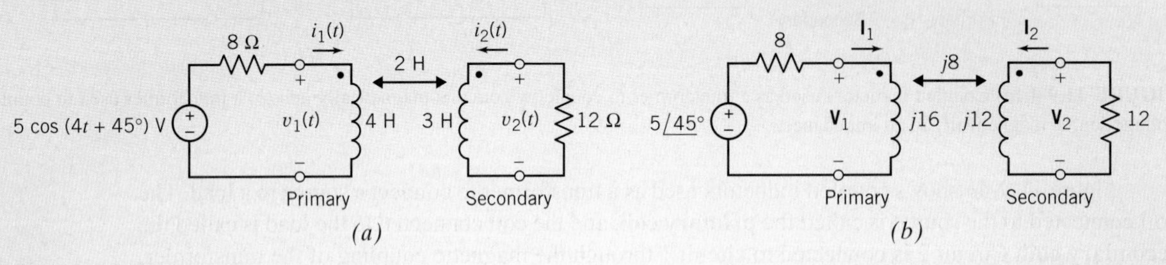

FIGURE 11.9-5 A circuit in which coupled inductors are used as a transformer. The circuit is represented (*a*) in the time domain and (*b*) in the frequency domain, using phasors and impedances.

Solution

First, represent the circuit in the frequency domain, using phasors and impedances, as shown in Figure 11.9-5*b*. Notice that the coil currents, $\mathbf{I}_1$ and $\mathbf{I}_2$, both enter the dotted end of the coils. Express the coil voltages as functions of the coil currents, using the equations that describe the coupled inductors, Eqs. 11.9-10 and 11.9-11.

$$\mathbf{V}_1 = j16\,\mathbf{I}_1 + j8\,\mathbf{I}_2$$
$$\mathbf{V}_2 = j8\,\mathbf{I}_1 + j12\,\mathbf{I}_2$$

Next, write two mesh equations

$$5\angle 45^\circ = 8\,\mathbf{I}_1 + \mathbf{V}_1$$

and

$$\mathbf{V}_2 = -12\,\mathbf{I}_2$$

Substituting the equations for the coil voltages into the mesh equations gives

$$5\angle 45^\circ = 8\,\mathbf{I}_1 + (j16\,\mathbf{I}_1 + j8\,\mathbf{I}_2) = (8 + j16)\mathbf{I}_1 + j8\,\mathbf{I}_2$$

and

$$j8\,\mathbf{I}_1 + j12\,\mathbf{I}_2 = -12\,\mathbf{I}_2$$

Solving for $\mathbf{I}_2$ gives

$$\mathbf{I}_2 = 0.138\angle -141^\circ \text{ A}$$

Next, $\mathbf{V}_2$ is given by

$$\mathbf{V}_2 = -12\,\mathbf{I}_2 = 1.656\angle 39^\circ \text{ V}$$

Returning to the time domain,

$$v_2(t) = 1.656\cos(4t + 39^\circ)\text{ V}$$

EXAMPLE 11.9-2 Coupled Inductors

INTERACTIVE EXAMPLE

The input to the circuit shown in Figure 11.9-6*a* is the voltage of the voltage source,

$$v_s(t) = 5.94\cos(3t + 140^\circ)\ \text{V}$$

The output is the voltage across the right-hand coil, $v_o(t)$. Determine the output voltage, $v_o(t)$.

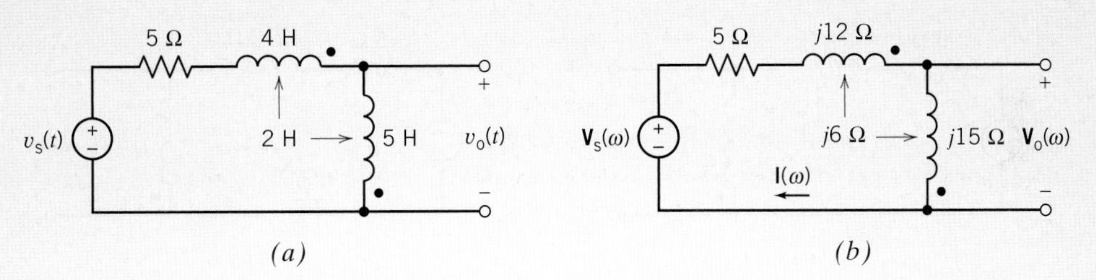

FIGURE 11.9-6 The circuit considered in Example 11.9-2 represented (*a*) in the time domain and (*b*) in the frequency domain.

Solution

The input voltage is a sinusoid. The output voltage is also a sinusoid and has the same frequency as the input voltage. Apparently, the circuit is at steady state. Consequently, the circuit in Figure 11.9-6*a* can be represented in the frequency domain, using phasors and impedances. Figure 11.9-6*b* shows the frequency-domain representation of the circuit from Figure 11.9-6*a*.

The phasor corresponding to the input sinusoids is

$$\mathbf{V}_s(\omega) = 5.94\angle 140^\circ\ \text{V}$$

The circuit in Figure 11.9-6*b* consists of a single mesh. Notice that the mesh current, $\mathbf{I}(\omega)$, enters the undotted ends of both coils. Apply KVL to the mesh to get

$$5\,\mathbf{I}(\omega) + (j12\,\mathbf{I}(\omega) + j6\,\mathbf{I}(\omega)) + (j6\,\mathbf{I}(\omega) + j15\,\mathbf{I}(\omega)) - 5.94\angle 140^\circ = 0$$

$$5\,\mathbf{I}(\omega) + (j12 + j6 + j6 + j15)\mathbf{I}(\omega) - 5.94\angle 140^\circ = 0$$

Solving for $\mathbf{I}(\omega)$ gives

$$\mathbf{I}(\omega) = \frac{5.94\angle 140^\circ}{5 + j(12 + 6 + 6 + 15)} = \frac{5.94\angle 140^\circ}{5 + j39} = \frac{5.94\angle 140^\circ}{39.3\angle 83^\circ} = 0.151\angle 57^\circ\ \text{A}$$

Notice that the voltage, $\mathbf{V}_o(\omega)$, across the right-hand coil and the mesh current, $\mathbf{I}(\omega)$, adhere to the passive convention. The voltage across the right-hand coil is given by

$$\begin{aligned}\mathbf{V}_o(\omega) = j15\,\mathbf{I}(\omega) + j6\,\mathbf{I}(\omega) = j21\,\mathbf{I}(\omega) &= j21(0.151\angle 57^\circ)\\ &= (21\angle 90^\circ)(0.151\angle 57^\circ)\\ &= 3.17\angle 147^\circ\ \text{V}\end{aligned}$$

In the time domain, the output voltage is given by

$$v_o(t) = 3.17\cos(3t + 147^\circ)\ \text{V}$$

EXAMPLE 11.9-3 Coupled Inductors

INTERACTIVE EXAMPLE

The input to the circuit shown in Figure 11.9-7*a* is the voltage of the voltage source,

$$v_s(t) = 5.94 \cos(3t + 140°) \text{ V}$$

The output is the voltage across the right-hand coil, $v_o(t)$. Determine the output voltage, $v_o(t)$.

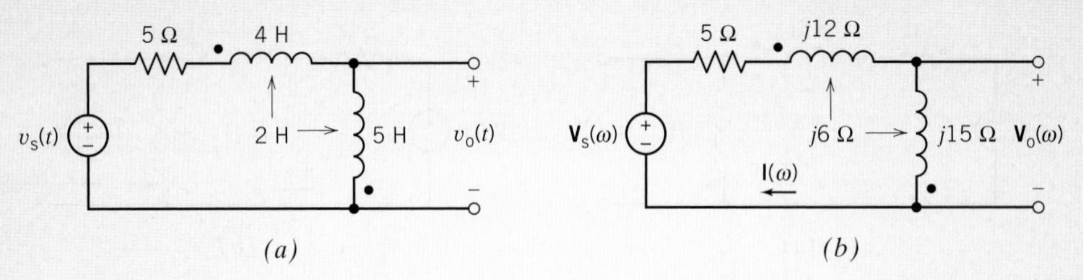

FIGURE 11.9-7 The circuit considered in Example 11.9-3, represented (*a*) in the time domain and (*b*) in the frequency domain.

Solution

The circuit shown in Figure 11.9-7*b* is very similar to the circuit shown in Figure 11.9-6*a*. There is only one difference: the dot of the left-hand coil is located at the right of the coil in Figure 11.9-6*a* and at the left of the coil in Figure 11.9-7*a*. As in Example 11.9-2, our first step is to represent the circuit in the frequency domain, using phasors and impedances. Figure 11.9-7*b* shows the frequency-domain representation of the circuit from Figure 11.9-7*a*.

The phasor corresponding to the input sinusoids is

$$\mathbf{V}_s(\omega) = 5.94 \angle 140° \text{ V}$$

The circuit in Figure 11.9-7 consists of a single mesh. Notice that the mesh current, $\mathbf{I}(\omega)$, enters the dotted end of the left-hand coil and the undotted end of the right-hand coil. Apply KVL to the mesh to get

$$5\,\mathbf{I}(\omega) + (j12\,\mathbf{I}(\omega) - j6\,\mathbf{I}(\omega)) + (-j6\,\mathbf{I}(\omega) + j15\,\mathbf{I}(\omega)) - 5.94\angle 140° = 0$$
$$5\,\mathbf{I}(\omega) + (j12 - j6 - j6 + j15)\,\mathbf{I}(\omega) - 5.94\angle 140° = 0$$

Solving for $\mathbf{I}(\omega)$ gives

$$\mathbf{I}(\omega) = \frac{5.94\angle 140°}{5 + j(12 - 6 - 6 + 15)} = \frac{5.94\angle 140°}{5 + j15} = \frac{5.94\angle 140°}{15.8\angle 71.6} = 0.376\angle 68.4° \text{ A}$$

Notice that the voltage, $\mathbf{V}_o(\omega)$, across the right-hand coil and the mesh current, $\mathbf{I}(\omega)$, adhere to the passive convention. The voltage across the right-hand coil is given by

$$\begin{aligned}\mathbf{V}_o(\omega) = j15\,\mathbf{I}(\omega) - j6\,\mathbf{I}(\omega) = j9\,\mathbf{I}(\omega) &= j9(0.376\angle 68.4°)\\ &= (9\angle 90°)(0.376\angle 68.4°)\\ &= 3.38\angle 158.4° \text{ V}\end{aligned}$$

In the time domain, the output voltage is given by

$$v_o(t) = 3.38 \cos(3t + 158.4°) \text{ V}$$

EXERCISE 11.9-1 Determine the voltage v_o for the circuit of Figure E 11.9-1.

Hint: Write a single mesh equation. The currents in the two coils are equal to each other and equal to the mesh current.

Answer: $v_o = 14 \cos 4t$ V

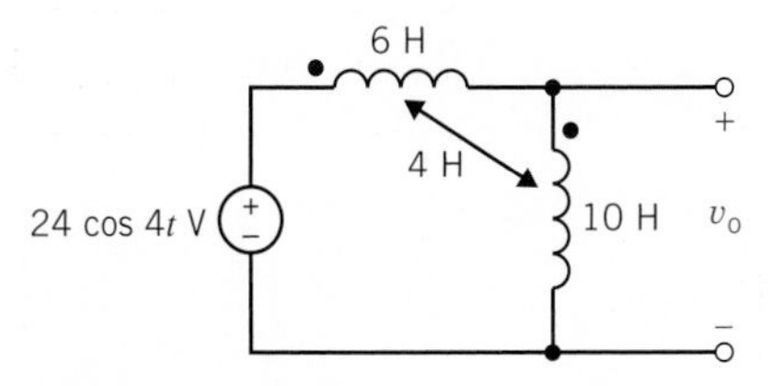

FIGURE E 11.9-1

EXERCISE 11.9-2 Determine the voltage v_o for the circuit of Figure E 11.9-2.

Hint: This exercise is the same as Exercise 11.9-1, except for the position of the dot on the vertical coil.

Answer: $v_o = 18 \cos 4t$ V

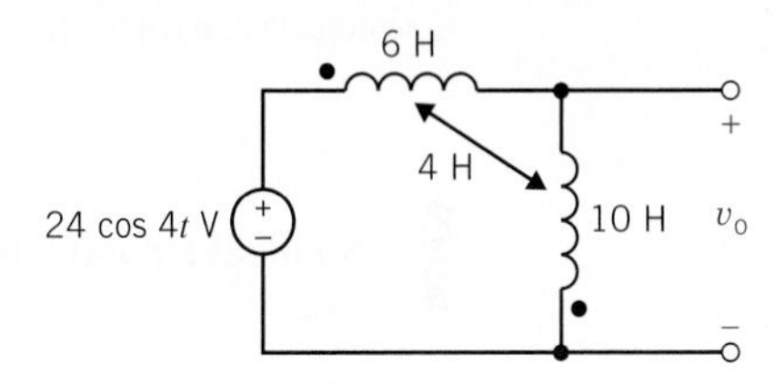

FIGURE E 11.9-2

EXERCISE 11.9-3 Determine the current i_o for the circuit of Figure E 11.9-3.

Hint: The voltage across the vertical coil is zero because of the short circuit. The voltage across the horizontal coil induces a current in the vertical coil. Consequently, the current in the vertical coil is not zero.

Answer: $i_o = 1.909 \cos (4t - 90°)$ A

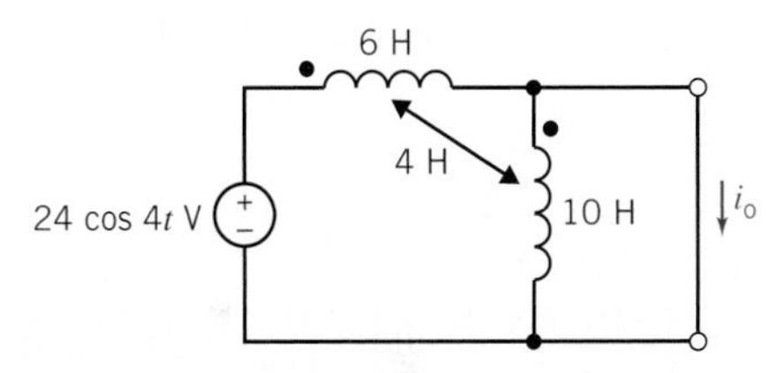

FIGURE E 11.9-3

EXERCISE 11.9-4 Determine the current i_o for the circuit of Figure E 11.9-4.

Hint: This exercise is the same as Exercise 11.9-3, except for the position of the dot on the vertical coil.

Answer: $i_o = 0.818 \cos (4t - 90°)$ A

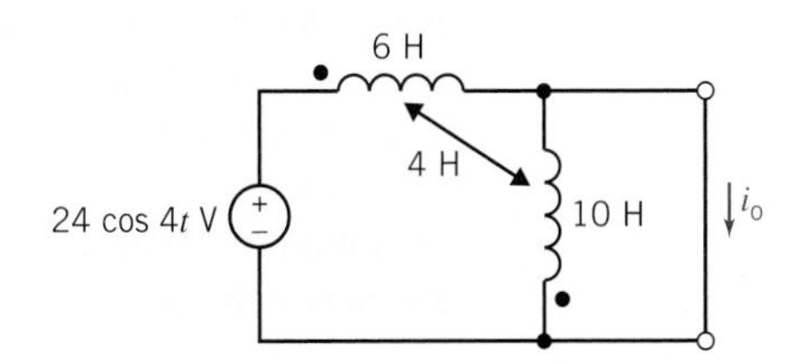

FIGURE E 11.9-4

11.10 THE IDEAL TRANSFORMER

One major use of transformers is in ac power distribution. Transformers possess the ability to step up or step down ac voltages or currents. Transformers are used by power utilities to raise (step up) the voltage from 10 kV at a generating plant to 200 kV or higher for transmission over long distances. Then, at a receiving plant, transformers are used to reduce (step down) the voltage to 220 or 110 V for use by the customer (Coltman, 1988).

In addition to power systems, transformers are commonly used in electronic and communication circuits. They provide the ability to raise or reduce voltages and to isolate one circuit from another.

One of the coils, typically drawn on the left of the diagram of a transformer, is designated as the *primary coil*, and the other is called the *secondary coil* or winding. The primary coil is connected to the energy source, and the secondary coil is connected to the load.

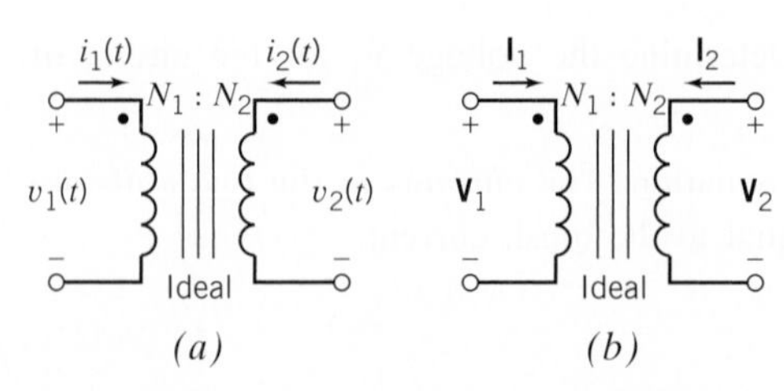

FIGURE 11.10-1 Circuit symbol for an ideal transformer. The ideal transformer has the same representation in (*a*) the time domain and (*b*) the frequency domain.

An **ideal transformer** is a model of a transformer with a coupling coefficient equal to unity.

The symbol for the ideal transformer is shown in Figure 11.10-1, where N_1 and N_2 are the number of turns in the primary and secondary coils. The time-domain representation of the transformer is shown in Figure 11.10-1*a*. In the time domain, the two defining equations for an ideal transformer are

$$v_2(t) = \frac{N_2}{N_1} v_1(t) \tag{11.10-1}$$

and

$$i_2(t) = \frac{N_2}{N_1} i_2(t) \tag{11.10-2}$$

where $N_2/N_1 = n$ is called the *turns ratio* of the transformer. The use of transformers is usually limited to non-dc applications because the primary and secondary windings behave as short circuits for a steady current.

The frequency-domain representation of the transformer is shown in Figure 11.10-1*b*. The operation of the ideal transformer is the same in the time domain as in the frequency domain. In the frequency domain, the two defining equations for an ideal transformer are

$$\mathbf{V}_2 = \frac{N_2}{N_1} \mathbf{V}_1 \tag{11.10-3}$$

and

$$\mathbf{I}_1 = \frac{N_2}{N_1} \mathbf{I}_2 \tag{11.10-4}$$

The vertical bars in Figure 11.10-1 indicate the iron core, and we write *ideal* with the transformer to ensure recognition of the ideal case. An ideal transformer can be modeled using dependent sources, as shown in Figure 11.10-2.

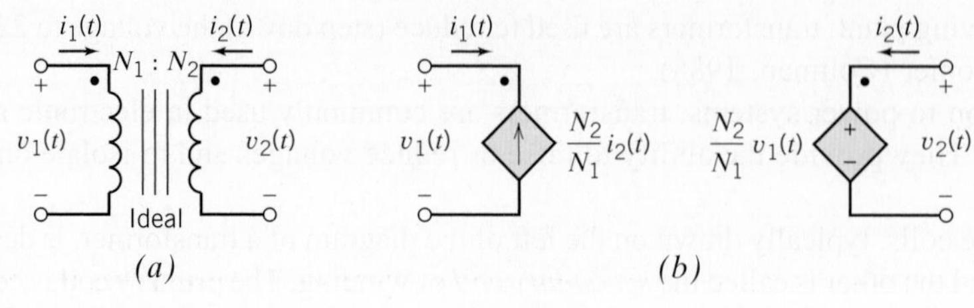

FIGURE 11.10-2 (*a*) Ideal transformer and (*b*) an equivalent circuit.

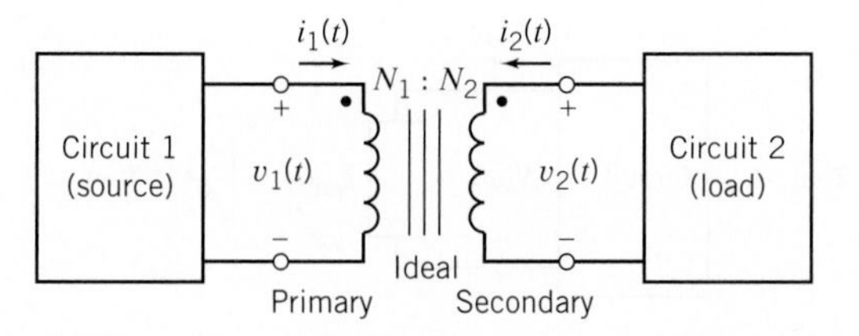

FIGURE 11.10-3 An ideal transformer used to couple two circuits magnetically.

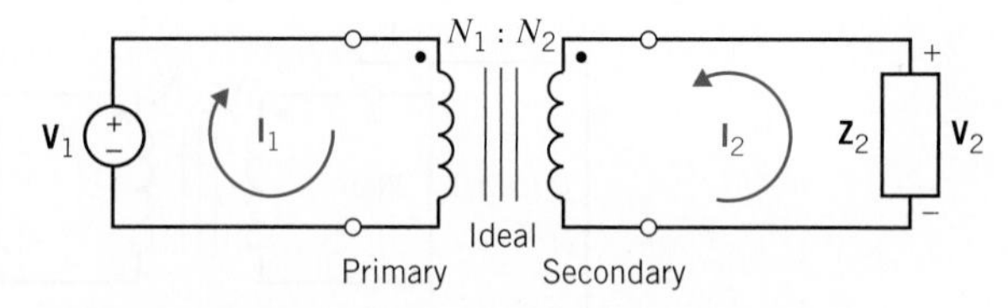

FIGURE 11.10-4 An ideal transformer used to couple an impedance magnetically to a sinusoidal voltage source. This circuit is represented in the frequency domain, using impedances and phasors.

Notice that the voltage and current of both coils of the transformer in Figure 11.10-1 adhere to the passive convention. The instantaneous power absorbed by the ideal transformer is

$$p(t) = v_1(t)i_1(t) + v_2(t)i_2(t) = v_1(t)(-ni_2(t)) + (nv_1(t))i_2(t) = 0 \tag{11.10-5}$$

The ideal transformer is said to be *lossless* because instantaneous power absorbed by it is zero. A similar argument shows that the ideal transformer absorbs zero complex power, zero average power, and zero reactive power.

Figure 11.10-3 shows an ideal transformer that is used to connect a source to a load. The coil connected to the source is called the primary coil, and the coil connected to the load is called the secondary coil. Circuit 2 is connected to circuit 1 through the magnetic coupling of the transformer, but there is no electrical connection between these two circuits. Because the ideal transformer is lossless, all of the power delivered to the ideal transformer by circuit 1 is in turn delivered to circuit 2 by the ideal transformer.

Let us consider the circuit of Figure 11.10-4, which has a load impedance $\mathbf{Z}_2$ magnetically coupled to a voltage source, using an ideal transformer.

The input impedance of the circuit connected to the voltage source is

$$\mathbf{Z}_1 = \frac{\mathbf{V}_1}{\mathbf{I}_1}$$

$\mathbf{Z}_1$ is called the impedance, seen at the primary of the transformer, or the impedance, seen by the voltage source.

The transformer is represented by the equations

$$\mathbf{V}_1 = \mathbf{V}_2/n$$

and

$$\mathbf{I}_1 = -n\mathbf{I}_2$$

where $n = N_2/N_1$ is the turns ratio of the transformer.

The current and voltage of the impedance, $\mathbf{I}_2$ and $\mathbf{V}_2$, do not adhere to the passive convention, so

$$\mathbf{V}_2 = -\mathbf{Z}_2\mathbf{I}_2$$

Therefore, for $\mathbf{Z}_1$, we have

$$\mathbf{Z}_1 = \frac{\mathbf{V}_2/n}{-n\,\mathbf{I}_2} = \frac{1}{n^2}\left(-\frac{\mathbf{V}_2}{\mathbf{I}_2}\right) = \frac{1}{n^2}\mathbf{Z}_2$$

The source experiences the impedance $\mathbf{Z}_1$, which is equal to $\mathbf{Z}_2$ scaled by the factor $1/n^2$. We sometimes say that $\mathbf{Z}_1$ is the impedance $\mathbf{Z}_2$ reflected to the primary of the transformer.

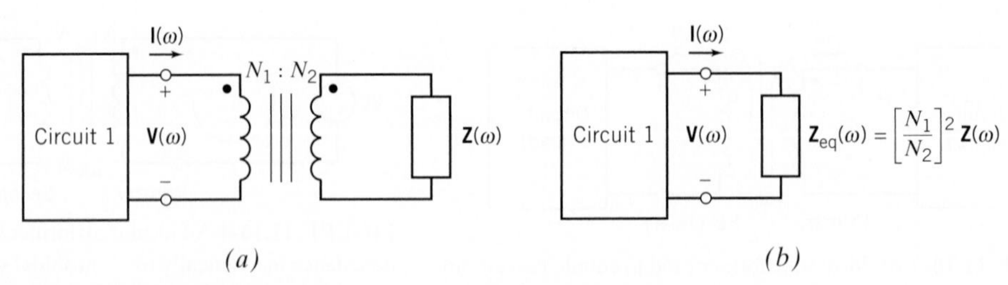

FIGURE 11.10-5 The circuit shown in (*b*) is equivalent to the circuit shown in (*a*).

Suppose we are going to connect a load impedance to a source. If we connect the load impedance directly to the source, then the source sees the load impedance $\mathbf{Z}_2$. In contrast, if we connect the load impedance to the source, using an ideal transformer, the source sees the impedance $\mathbf{Z}_1$. In this context, we say that the transformer has changed the impedance seen by the source from $\mathbf{Z}_2$ to $\mathbf{Z}_1$.

We can formalize this result as the circuit equivalence illustrated in Figure 11.10-5. Figure 11.10-5*a* shows circuit 1 connected to the left-hand coil of an ideal transformer. An impedance, $\mathbf{Z}(\omega)$, is connected in parallel with the right-hand coil of the ideal transformer. In Figure 11.10-5*b*, the ideal transformer and impedance have been replaced by a single equivalent impedance, $\mathbf{Z}_{eq}(\omega)$. The equivalent impedance is related to the original impedance by

$$\mathbf{Z}_{eq}(\omega) = \left(\frac{N_1}{N_2}\right)^2 \mathbf{Z}(\omega) = \frac{1}{n^2}\mathbf{Z}(\omega)$$

The two circuits in Figure 11.10-5 are equivalent. All the currents and voltages of circuit 1, including $\mathbf{I}(\omega)$ and $\mathbf{V}(\omega)$, are the same in Figure 11.10-5*b* as they are in Figure 11.10-5*a*. We can determine the values of $\mathbf{I}(\omega)$ and $\mathbf{V}(\omega)$ in Figure 11.10-5*a* by calculating values of $\mathbf{I}(\omega)$ and $\mathbf{V}(\omega)$ in Figure 11.10-5*b*.

EXAMPLE 11.10-1 Maximum Power Transfer

Often, we can use an ideal transformer to represent a transformer that connects the output of a stereo amplifier, $\mathbf{V}_1$, to a stereo speaker, as shown in Figure 11.10-6. Find the value of the turns ratio n that is required to cause maximum power to be transferred to the load when $R_L = 8\ \Omega$ and $R_s = 48\ \Omega$.

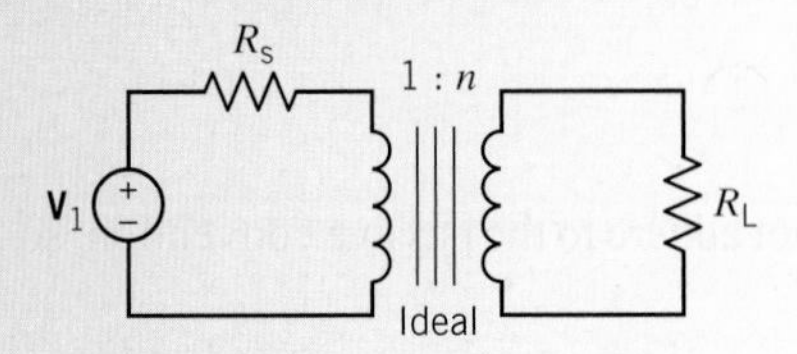

FIGURE 11.10-6 Output of an amplifier connected to a stereo speaker with resistance R_L.

Solution

The impedance seen at the primary due to R_L is

$$\mathbf{Z}_1 = \frac{R_L}{n^2} = \frac{8}{n^2}$$

To achieve maximum power transfer, we require that

$$\mathbf{Z}_1 = R_s$$

Because $R_s = 48\ \Omega$, we require that $\mathbf{Z}_1 = 48\ \Omega$, so

$$n^2 = \frac{8}{48} = \frac{1}{6}$$

and, therefore,

$$\left(\frac{N_2}{N_1}\right)^2 = \frac{1}{6}$$

or

$$N_1 = \sqrt{6}N_2$$

Example 11.10-2 Transformer Circuit

INTERACTIVE EXAMPLE

The input to the circuit shown in Figure 11.10-7 is the voltage of the voltage source, $v_s(t)$. The output is the voltage across the 9-H inductor, $v_o(t)$. Determine the output voltage, $v_o(t)$.

Solution

The input voltage is a sinusoid. The output voltage is also a sinusoid and has the same frequency as the input voltage. Apparently, the circuit is at steady state. Consequently, the circuit in Figure 11.10-7 can be represented in the frequency domain, using phasors and impedances. Figure 11.10-8 shows the frequency-domain representation of the circuit from Figure 11.10-7.

In Figure 11.10-8, the impedance of the inductor is connected in series with the impedance of the 30-Ω resistor. This series impedance is connected in parallel with the right-hand coil of the transformer. Replace the transformer and the series impedance with the equivalent impedance, as shown in Figure 11.10-9. The equivalent impedance is given by

$$\mathbf{Z}_{eq} = \left(\frac{3}{2}\right)^2 (30 + j36) = 67.5 + j81\ \Omega$$

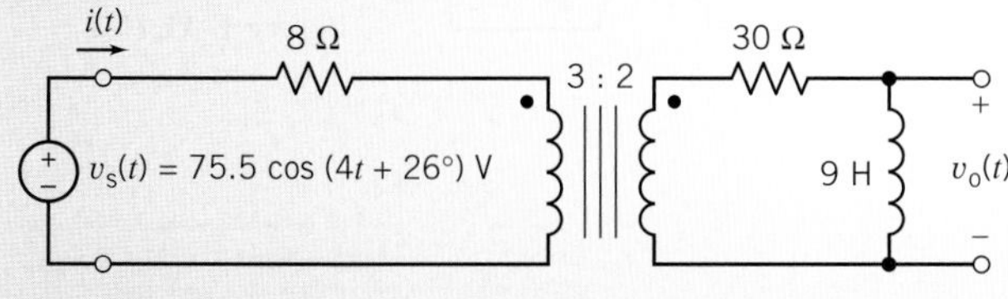

FIGURE 11.10-7 The circuit considered in Example 11.10-2.

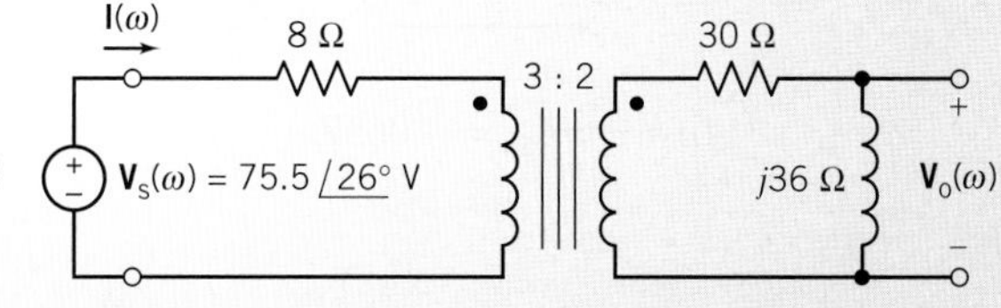

FIGURE 11.10-8 The circuit from Figure 11.10-7, represented in the frequency domain, using impedances and phasors.

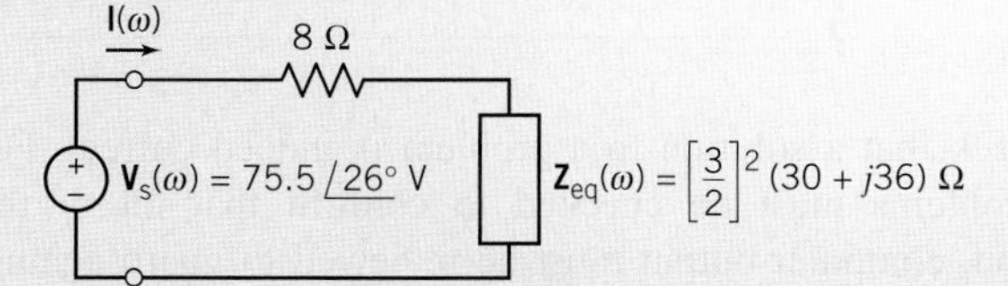

FIGURE 11.10-9 The circuit from Figure 11.10-8, after replacing the transformer and the impedance of the series resistor and inductor with the equivalent impedance.

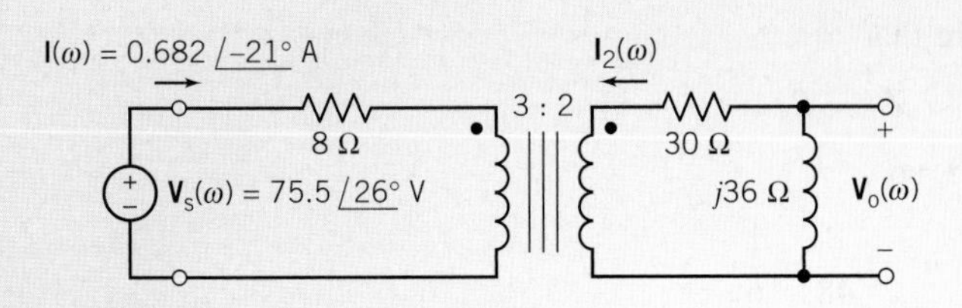

FIGURE 11.10-10 The circuit from Figure 11.10-9 after determining the current $\mathbf{I}(\omega)$.

In Figure 11.10-9, the impedance of the 8-Ω resistor is connected in series with the equivalent impedance, $\mathbf{Z}_{eq}(\omega)$. The current, $\mathbf{I}(\omega)$, is the current in this series impedance, and $\mathbf{V}_s(\omega)$ is the voltage across the series impedance. Applying Ohm's law gives

$$\mathbf{I}(\omega) = \frac{\mathbf{V}_s(\omega)}{8 + \mathbf{Z}_{eq}(\omega)} = \frac{75.5\angle 26^\circ}{8 + 67.5 + j81} = \frac{75.5\angle 26^\circ}{110.73\angle 47^\circ} = 0.682\angle -21^\circ \text{ A} \tag{11.10-6}$$

Because the circuits in Figures 11.10-8 and 11.10-9 are equivalent, the current $\mathbf{I}(\omega)$ in Figure 11.10-10 is also given by Eq. 11.10-6. Figure 11.10-10 shows the circuit from Figure 11.10-8 redrawn with the current $\mathbf{I}(\omega)$ labeled.

Also, the current in the right-hand coil of the transformer has been labeled as $\mathbf{I}_2(\omega)$. Because $\mathbf{I}(\omega)$ and $\mathbf{I}_2(\omega)$ are the currents in the coils of the ideal transformer, they are related by the equations describing the transformer:

$$\mathbf{I}_2(\omega) = -\left(\frac{3}{2}\right)\mathbf{I}(\omega) = -1.023\angle -21^\circ \text{ A}$$

Notice that $\mathbf{I}_2(\omega)$ and $\mathbf{V}_o(\omega)$, the current and voltage of the $j36$-Ω impedance in Figure 11.10-10, do not adhere to the passive convention. Consequently,

$$\mathbf{V}_o(\omega) = -j36\,\mathbf{I}_2(\omega) = (j36)(1.023\angle -21^\circ) = (36\angle 90^\circ)(1.023\angle -21^\circ) = 36.82\angle 69^\circ \text{ V}$$

In the time domain, the output voltage is given by

$$v_o(t) = 36.82\cos(4t + 69^\circ) \text{ V}$$

EXERCISE 11.10-1 Determine the impedance $\mathbf{Z}_{ab}$ for the circuit of Figure E 11.10-1. All the transformers are ideal.

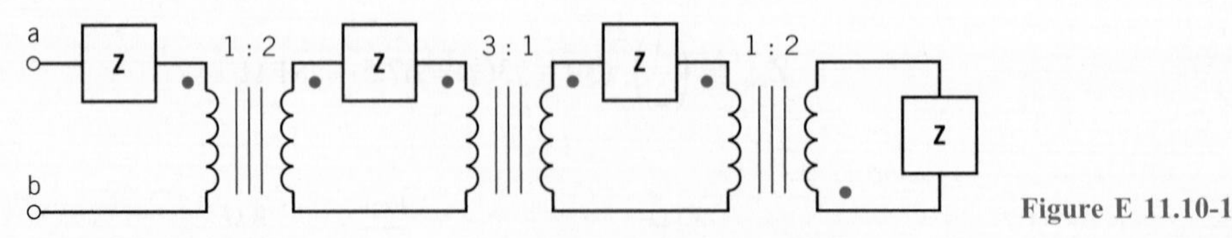

Figure E 11.10-1

Answer: $\mathbf{Z}_{ab} = 4.063\mathbf{Z}$

11.11 HOW CAN WE CHECK . . . ?

Engineers are frequently called upon to check that a solution to a problem is indeed correct. For example, proposed solutions to design problems must be checked to confirm that all of the specifications have been satisfied. In addition, computer output must be reviewed to guard against data-entry errors, and claims made by vendors must be examined critically.

Engineering students are also asked to check the correctness of their work. For example, occasionally just a little time remains at the end of an exam. It is useful to be able to quickly identify those solutions that need more work.

The following example illustrates techniques useful for checking the solutions of the sort of problem discussed in this chapter.

EXAMPLE 11.11-1 How Can We Check Power in AC Circuits?

The circuit shown in Figure 11.11-1*a* has been analyzed using a computer, and the results are tabulated in Figure 11.11-1*b*. The labels Xp and Xs refer to the primary and secondary coils of the transformer. The passive convention is used for all elements, including the voltage sources, which means that

$$\frac{(30)(1.76)}{2}\cos(133^\circ - 0) = -18.00$$

is the average power *absorbed* by the voltage source. The average power *supplied* by the voltage source is $+18.00$ W.

How can we check that the computer analysis of this circuit is indeed correct?

Solution

Several things can be easily checked.

(1) The element current and voltage of each inductor should be 90° out of phase with each other so that the average power delivered to each inductor is zero. The element current and voltage of both L_1 and L_2 satisfy this condition.

(2) An ideal transformer absorbs zero average power. The sum of the average power absorbed by the transformer primary and the secondary is

$$\frac{(5.2)(1.76)}{2}(\cos(9^\circ - (-47^\circ))) + \frac{(7.8)(1.17)}{2}\cos(133^\circ - 9^\circ) = 2.56 + (-2.55) \approx 0\text{ W}$$

so this condition is satisfied.

(3) All of the power delivered to the primary of the transformer is in turn delivered to the load. In this example, the load consists of the inductor L_2 and the resistor R_2. Because the average power delivered to

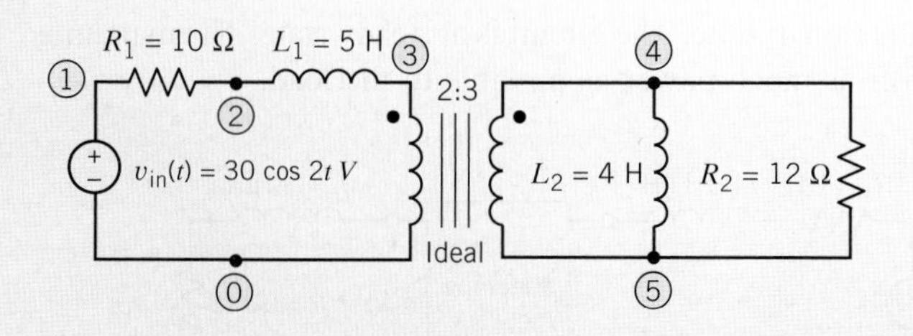

(*a*)

Element	Voltage	Current
Vin 1 0 30 ∠ 0	30 ∠ 0°	1.76 ∠ 133°
R1 1 2 10	17.6 ∠ -47°	1.76 ∠ -47°
L1 2 3 5	17.6 ∠ 43°	1.76 ∠ -47°
Xp 3 0 2	5.2 ∠ 9°	1.76 ∠ -47°
Xs 4 5 3	7.8 ∠ 9°	1.17 ∠ 133°
R2 4 5 12	7.8 ∠ 9°	0.65 ∠ 9°
L2 4 5 4	7.8 ∠ 9°	0.98 ∠ -81°

Steady-state response: $\omega = 2$ rad/s

(*b*)

FIGURE 11.11-1 (*a*) A circuit and (*b*) the results from computer analysis for the circuit.

the inductor is zero, all the power delivered to the transformer primary should be delivered by the secondary to the resistor R_2. The power delivered to the transformer primary is

$$\frac{(5.2)(1.76)}{2}\cos(9^\circ - (-47^\circ)) = 2.56\ \mathrm{W}$$

The power delivered to R_2 is

$$\frac{(7.8)(0.65)}{2}\cos(0) = 2.53\ \mathrm{W}$$

There seems to be some roundoff error in the voltages and currents provided by the computer. Nonetheless, it seems reasonable to conclude that all the power delivered to the transformer primary is delivered by the secondary to the resistor R_2.

(4) The average power supplied by the voltage source should be equal to the average power absorbed by the resistors. We have already calculated that the average power delivered by the voltage source is 18 W. The average power absorbed by the resistors is

$$\frac{(17.6)(1.76)}{2}\cos(0) + \frac{(7.8)(0.65)}{2}\cos(0) = 15.49 + (2.53) = 18.02\ \mathrm{W}$$

so this condition is satisfied.

Because these four conditions are satisfied, we are confident that the computer analysis of the circuit is correct.

11.12 DESIGN EXAMPLE

MAXIMUM POWER TRANSFER

The matching network in Figure 11.12-1 is used to interface the source with the load, which means that the matching network is used to connect the source to the load in a desirable way. In this case, the purpose of the matching network is to transfer as much power as possible to the load. This problem occurs frequently enough that it has been given a name, the maximum power transfer problem.

An important example of the application of maximum power transfer is the connection of a cellular phone or wireless radio transmitter to the cell's antenna. For example, the input impedance of a practical cellular telephone antenna is $\mathbf{Z} = (10 + j6.28)\ \Omega$ (Dorf, 1998).

Describe the Situation and the Assumptions

The input voltage is a sinusoidal function of time. The circuit is at steady state. The matching network is to be designed to deliver as much power as possible to the load.

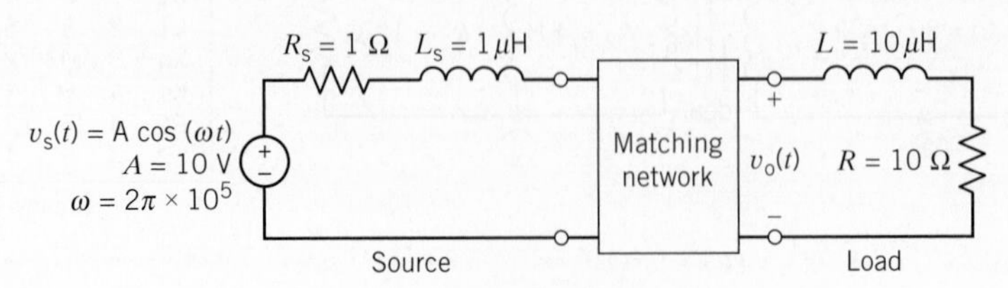

FIGURE 11.12-1 Design the matching network to transfer maximum power to the load where the load is the model of an antenna of a wireless communication system.

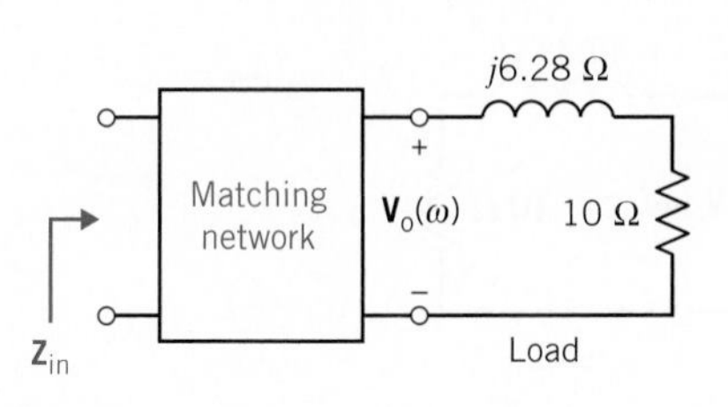

FIGURE 11.12-2 $\mathbf{Z}_{in}$ is the impedance seen looking into the matching network.

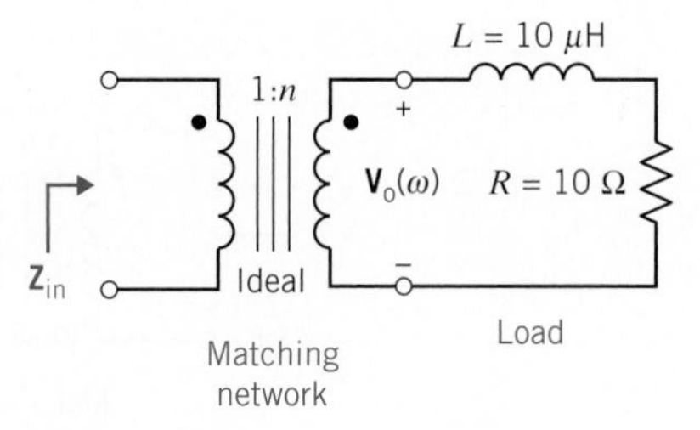

FIGURE 11.12-3 Using an ideal transformer as the matching network.

State the Goal

To achieve maximum power transfer, the matching network should match the load and source impedances. The source impedance is

$$\mathbf{Z}_s = R_s + j\omega L_s = 1 + j(2 \cdot \pi \cdot 10^5)(10^{-6}) = 1 + j0.628\ \Omega$$

For maximum power transfer, the impedance $\mathbf{Z}_{in}$, shown in Figure 11.12-2, must be the complex conjugate of $\mathbf{Z}_s$. That is,

$$\mathbf{Z}_{in} = \mathbf{Z}_s^* = 1 - j\,0.628\ \Omega$$

Generate a Plan

Let us use a transformer for the matching network as shown in Figure 11.12-3. The impedance $\mathbf{Z}_{in}$ will be a function of n, the turns ratio of the transformer. We will set $\mathbf{Z}_{in}$ equal to the complex conjugate of $\mathbf{Z}_s$ and solve the resulting equation to determine the turns ratio, n.

Act on the Plan

$$\mathbf{Z}_{in} = \frac{1}{n^2}(R + j\omega L) = \frac{1}{n^2}(10 + j6.28)$$

We require that

$$\frac{1}{n^2}(10 + j6.28) = 1 - j0.628$$

This requires both

$$\frac{1}{n^2}10 = 1 \qquad (11.12\text{-}1)$$

and

$$\frac{1}{n^2}6.28 = -0.628 \qquad (11.12.2)$$

Selecting

$$n = 3.16$$

(for example, $N_2 = 158$ and $N_1 = 50$) satisfies Eq. 11.12-1 but not Eq. 11.12-2. Indeed, no positive value of n will satisfy Eq. 11.12-2.

We need to modify the matching network to make the imaginary part of $\mathbf{Z}_{in}$ negative. This can be accomplished by adding a capacitor, as shown in Figure 11.12-4. Then,

$$\mathbf{Z}_{in} = \frac{1}{n^2}\left(R + j\omega L - j\frac{1}{\omega C}\right) = \frac{1}{n^2}\left(10 + j6.28 - j\frac{1}{2\pi \cdot 10^5 \cdot C}\right)$$

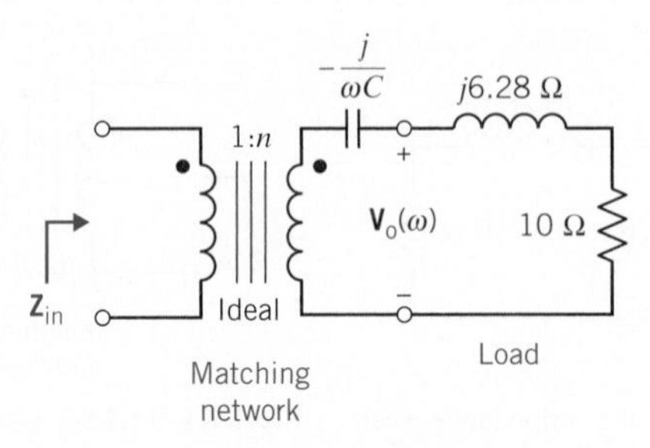

FIGURE 11.12-4 The matching network is modified by adding a capacitor.

We require that

$$\frac{1}{n^2}\left(10 + j6.28 - j\frac{1}{2\pi \cdot 10^5 \cdot C}\right) = 1 - j0.628$$

This requires both

$$\frac{1}{n^2}10 = 1 \qquad (11.12\text{-}3)$$

and

$$\frac{1}{n^2}\left(6.28 - \frac{1}{2\pi \cdot 10^5 \cdot C}\right) = -\ 0.628 \qquad (11.12\text{-}4)$$

First, solving Eq. 11.12-3 gives

$$n = 3.16$$

Next, solving Eq. 11.12-4 gives

$$C = 0.1267\ \mu\text{F}$$

and the design is complete.

Verify the Proposed Solution

When $n = 3.16$ and $C = 0.1267\ \mu$F, the input impedance of the matching network is

$$\begin{aligned}\mathbf{Z}_{\text{in}} &= \frac{1}{n^2}\left(R + j\omega L + \frac{1}{j\omega C}\right)\\ &= \frac{1}{3.16^2}\left(10 + j(2\pi \times 10^5)(10^{-5}) + \frac{1}{j(2\pi \times 10^5)(0.1267 \times 10^{-6})}\right)\\ &= 1 - j0.629\end{aligned}$$

as required.

11.13 SUMMARY

- With the adoption of ac power as the generally used conventional power for industry and the home, engineers became involved in analyzing ac power relationships.
- The instantaneous power delivered to this circuit element is the product of the element voltage and current. Let $v(t)$ and $i(t)$ be the element voltage and current, chosen to adhere to the passive convention. Then $p(t) = v(t)\,i(t)$ is the instantaneous power delivered to this circuit element. Instantaneous power is calculated in the time domain.
- The instantaneous power can be a quite complicated function of t. When the element voltage and current are periodic functions having the same period, T, it is convenient to calculate the average power $P = \frac{1}{T}\int_{t_0}^{t_0+T} i(t)v(t)\,dt$.
- The effective value of a current is the constant (dc) current that delivers the same average power to a 1-Ω resistor as the given varying current. The effective value of a voltage is the

Table 11.13-1 Coupled Inductors

DEVICE SYMBOL (INCLUDING REFERENCE DIRECTIONS OF ELEMENT VOLTAGES AND CURRENTS)	DEVICE EQUATIONS IN THE TIME DOMAIN	DEVICE EQUATIONS IN THE FREQUENCY DOMAIN
$i_1(t)$, $i_2(t)$, M, L_1, L_2, $v_1(t)$, $v_2(t)$	$v_1 = L_1 \frac{di_1}{dt} + M \frac{di_2}{dt}$ $v_2 = L_2 \frac{di_2}{dt} + M \frac{di_1}{dt}$	$\mathbf{V}_1 = j\omega L_1 \mathbf{I}_1 + j\omega M \mathbf{I}_2$ $\mathbf{V}_2 = j\omega L_2 \mathbf{I}_2 + j\omega M \mathbf{I}_1$
$i_1(t)$, $i_2(t)$, M, L_1, L_2, $v_1(t)$, $v_2(t)$	$v_1 = L_1 \frac{di_1}{dt} - M \frac{di_2}{dt}$ $v_2 = L_2 \frac{di_2}{dt} - M \frac{di_1}{dt}$	$\mathbf{V}_1 = j\omega L_1 \mathbf{I}_1 - j\omega M \mathbf{I}_2$ $\mathbf{V}_2 = j\omega L_2 \mathbf{I}_2 - j\omega M \mathbf{I}_1$

constant (dc) voltage that delivers the same average power as the given varying voltage.

- Consider a linear circuit with a sinusoidal input that has reached steady state. All the element voltages and currents will be sinusoidal, with the same frequency as the input. Such a circuit can be analyzed in the frequency domain, using phasors and impedances. Indeed, we can calculate the power generated or absorbed in a circuit or in any element of a circuit, in the frequency domain, using phasors. Table 11.5.1 summarizes the equations used to calculate average power, complex power, or reactive power in the frequency domain.
- Because it is important to keep the current I as small as possible in the transmission lines, engineers strive to achieve a power factor close to 1. The power factor is equal to $\cos\theta$, where θ is the phase angle difference between the sinusoidal steady-state load voltage and current. A purely reactive impedance in parallel with the load is used to correct the power factor.
- Finally, we considered the coupled coils and transformers. Coupled inductors and transformers exhibit mutual inductance, which relates the voltage in one coil to the change in current in another coil. The equations that describe coupled coils and transformers are collected in Tables 11.13-1 and 11.13-2.

Table 11.13-2 Ideal Transformers

DEVICE SYMBOL (INCLUDING REFERENCE DIRECTIONS OF ELEMENT VOLTANGES AND CURRENTS)	DEVICE EQUATIONS IN THE FREQUENCY DOMAIN
$i_1(t)$, $i_2(t)$, $N_1 : N_2$, $v_1(t)$, $v_2(t)$, Ideal	$\mathbf{V}_1 = \frac{N_1}{N_2} \mathbf{V}_2$ $\mathbf{I}_1 = -\frac{N_2}{N_1} \mathbf{I}_2$
$i_1(t)$, $i_2(t)$, $N_1 : N_2$, $v_1(t)$, $v_2(t)$, Ideal	$\mathbf{V}_1 = -\frac{N_1}{N_2} \mathbf{V}_2$ $\mathbf{I}_1 = \frac{N_2}{N_1} \mathbf{I}_2$

PROBLEMS

Section 11.3 Instantaneous Power and Average Power

P 11.3-1 An *RLC* circuit is shown in Figure P 11.3-1. Find the instantaneous power delivered to the inductor when $i_s = 1 \cos \omega t$ A and $\omega = 6283$ rad/s.

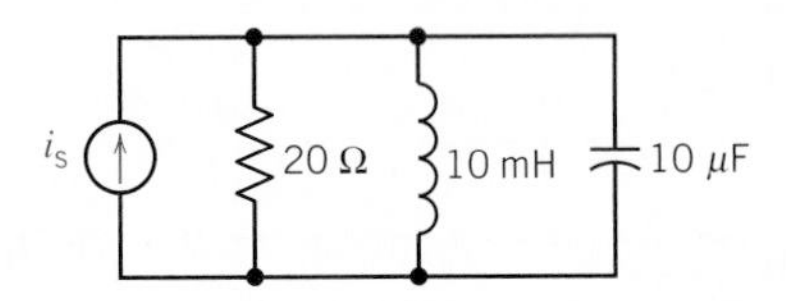

Figure P 11.3-1

P 11.3-2 Find the average power absorbed by the 0.9-kΩ resistor and the average power supplied by the current source for the circuit of Figure P 11.3-2.

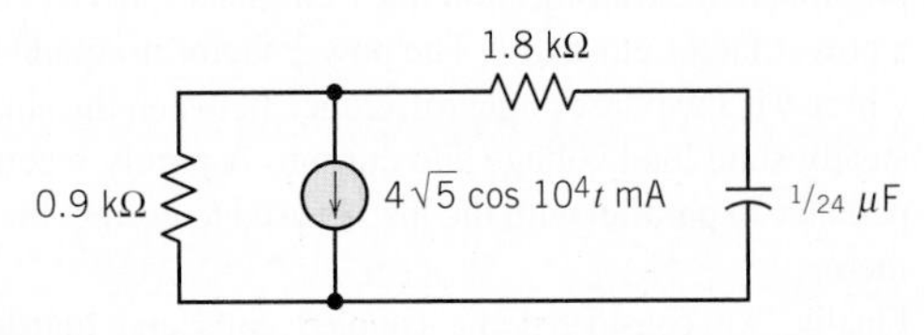

Figure P 11.3-2

P 11.3-3 Use nodal analysis to find the average power absorbed by the 20-Ω resistor in the circuit of Figure P 11.3-3.

Answer: $P = 200$ W

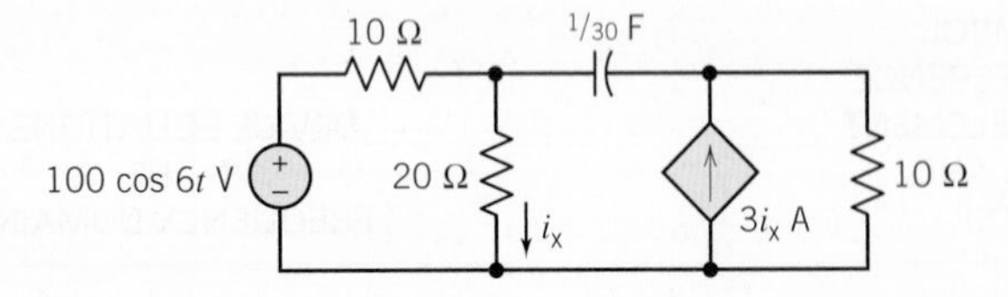

Figure P 11.3-3

P 11.3-4 Nuclear power stations have become very complex to operate, as illustrated by the training simulator for the operating room of the Pilgrim Power Station shown in Figure P 11.3-4*a*. One control circuit has the model shown in Figure P 11.3-4*b*. Find the average power delivered to each element.

Answer: $P_{\text{source current}} = -12.8$ W

$P_{8\Omega} = 6.4$ W

$P_L = 0$ W

$P_{\text{voltage source}} = 6.4$ W

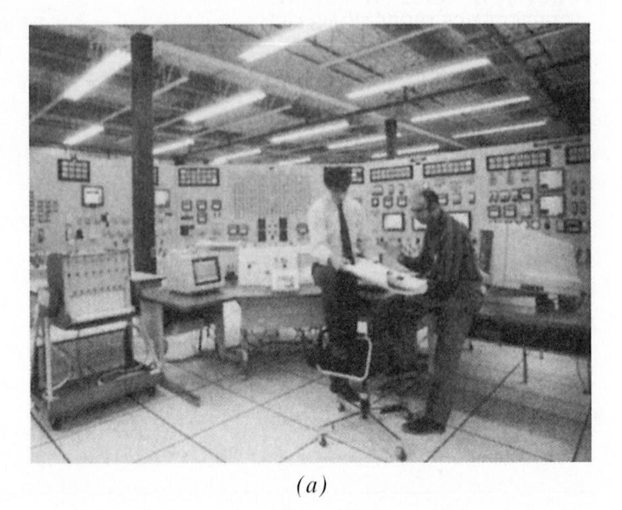

(*a*)

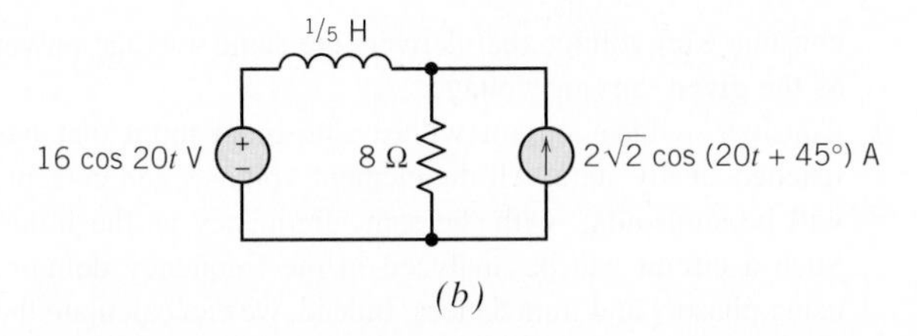

(*b*)

Figure P 11.3-4 (*a*) The simulation training room for the Pilgrim Power Station. The power station is located at Plymouth, Massachusetts, and generates 700 MW. It commenced operation in 1972. Courtesy of Boston Edison. (*b*) One control circuit of the reactor.

P 11.3-5 Find the average power delivered to each element for the circuit of Figure P 11.3-5.

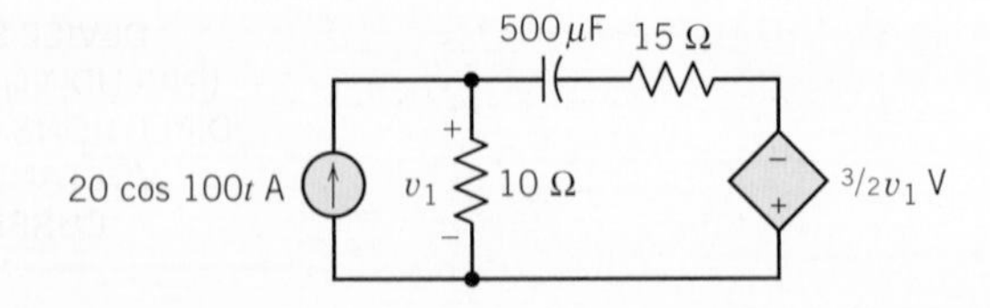

Figure P 11.3-5

P 11.3-6 A student experimenter in the laboratory encounters all types of electrical equipment. Some pieces of test equipment are battery operated or operate at low voltage so that any hazard is minimal. Other types of equipment are isolated from electrical ground so that there is no problem if a grounded object makes contact with the circuit. Some types of test equipment, however, are supplied by voltages that can be hazardous or have dangerous voltage outputs. The standard power supply used in the United States for power and lighting in laboratories is the 150, grounded, 60-Hz sinusoidal supply. This supply provides power for much of the laboratory equipment, so an understanding of its operation is essential in its safe use (Bernstein, 1991).

Consider the case in which the experimenter has one hand on a piece of electrical equipment and the other hand on a ground connection, as shown in the circuit diagram of Figure P 11.3-6*a*.

The hand-to-hand resistance is 200 Ω. Shocks with an energy of 30 J are hazardous to humans. Consider the model shown in Figure P 11.3-6*b*, which represents the human with *R*. Determine the energy delivered to the human in 1 s.

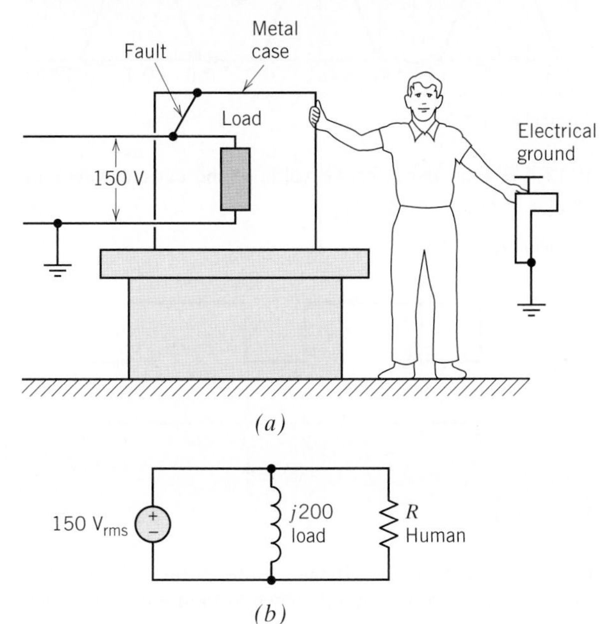

Figure P 11.3-6 Student experimenter touching an electrical device.

P 11.3-7 An *RLC* circuit is shown in Figure P 11.3-7 with a voltage source $v_s = 10 \cos 10t$ V.

(a) Determine the instantaneous power delivered to the circuit by the voltage source.

(b) Find the instantaneous power delivered to the inductor.

Answers:

(a) $p = 15.48 + 31.25 \cos(20t - 60.3°)$ W

(b) $p = 58.6 \cos(20t - 30.6°)$ W

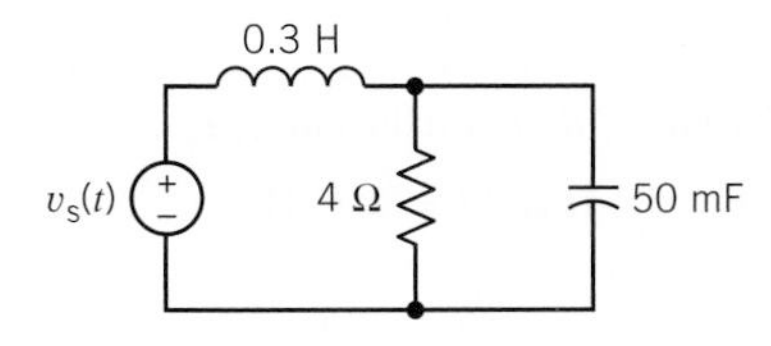

Figure P 11.3-7

P 11.3-8

(a) Find the average power delivered by the source to the circuit shown in Figure P 11.3-8.

(b) Find the power absorbed by resistor R_1.

Answers: **(a)** 67.5 W **(b)** 45 W

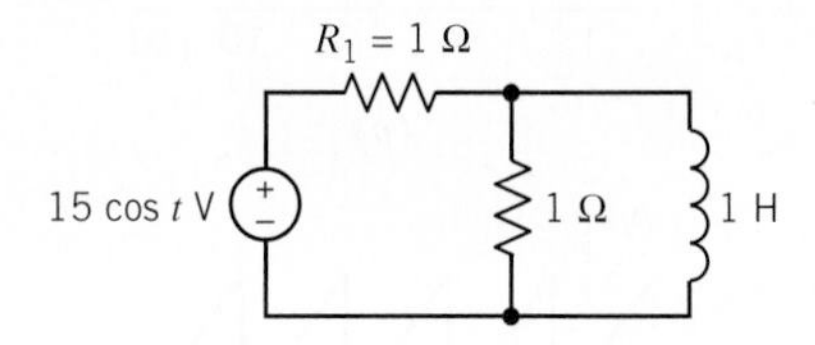

Figure P 11.3-8

Section 11.4 Effective Value of a Periodic Waveform

P 11.4-1 Find the rms value of the current *i* for (a) $i = 2 - 6 \cos 2t$ A, (b) $i = 3 \sin \pi t + \sqrt{2} \cos \pi t$ A, and (c)$i = 2 \cos 2t + 4\sqrt{2} \cos(2t + 45°) + 12 \sin 2t$ A.

Answers: **(a)** 4.7A **(b)** 2.35 A **(c)** $5\sqrt{2}$ A

P 11.4-2 Determine the rms value for each of the waveforms shown in Figure P 11.4-2.

Answers: **(a)** 4.10 V **(b)** 4.81 V **(c)** 4.10

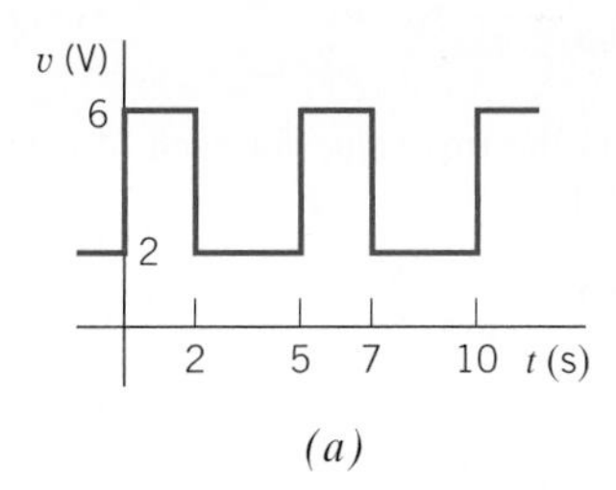

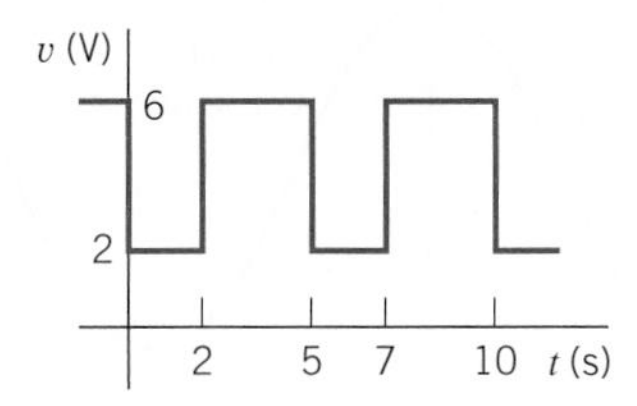

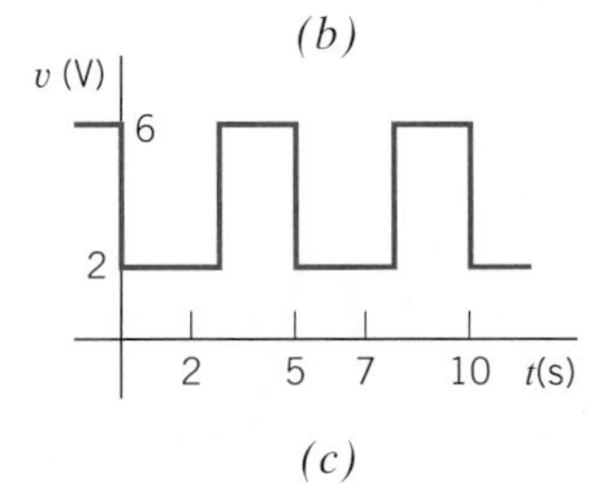

Figure P 11.4-2

P 11.4-3 Determine the rms value for each of the waveforms shown in Figure P 11.4-3.

Answers: **(a)** 4.16 V **(b)** 4.16 V **(c)** 4.16

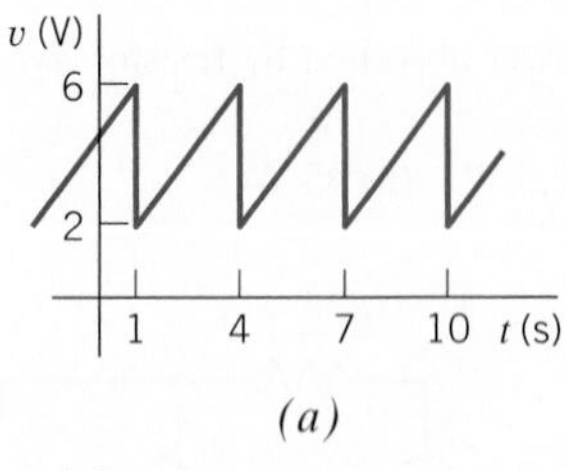

(a)

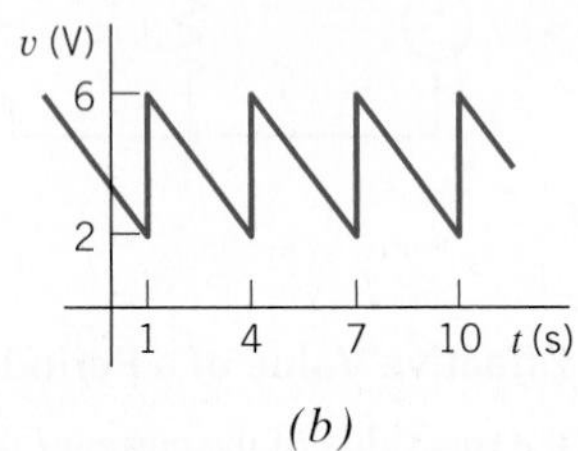

(b)

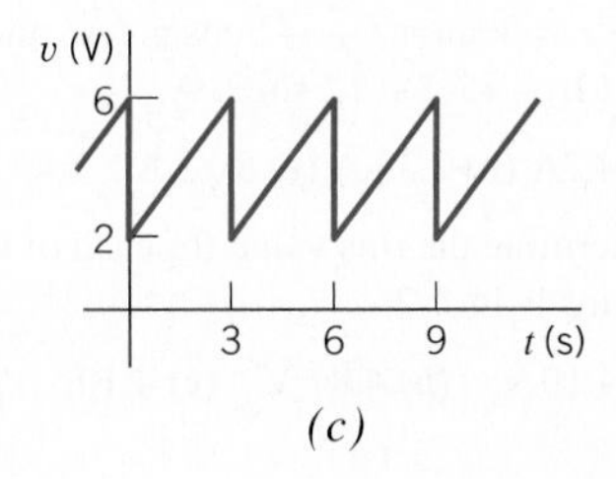

(c)

Figure P 11.4-3

P 11.4-4 Find the rms value for each of the waveforms of Figure P 11.4-4.

Answers: $V_{rms} = 1.225$ V
$I_{rms} = 5$ mA

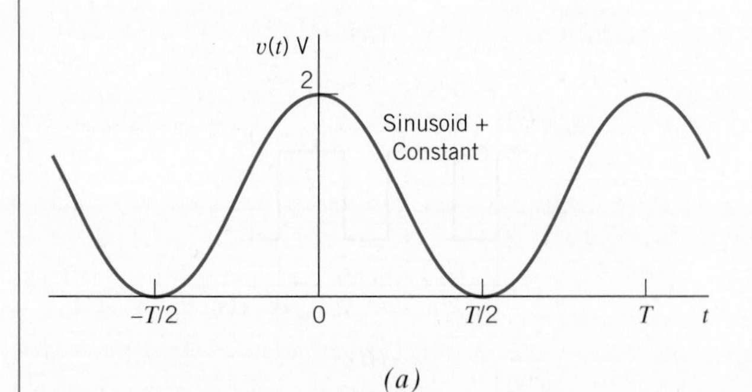

(a)

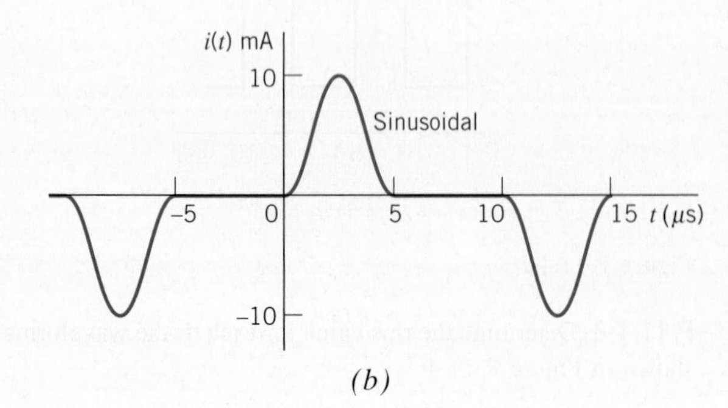

(b)

Figure P 11.4-4

P 11.4-5 Find the rms value of the voltage $v(t)$ shown in Figure P 11.4-5.

Answer: $V_{rms} = 4.24$ V

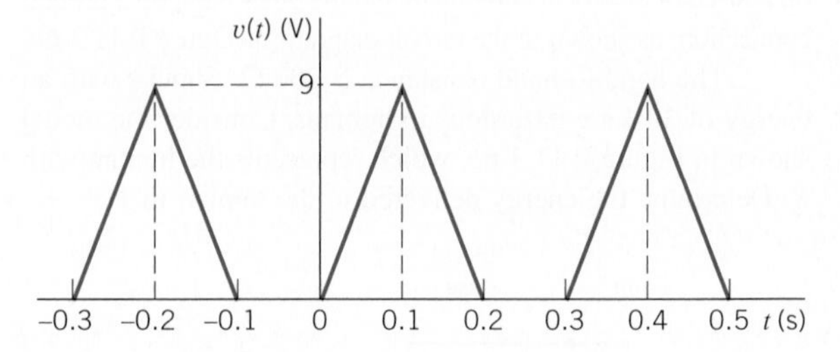

Figure P 11.4-5

P 11.4-6 Find the effective value of the current waveform shown in Figure P 11.4-6.

Answer: $I_{eff} = 8.66$

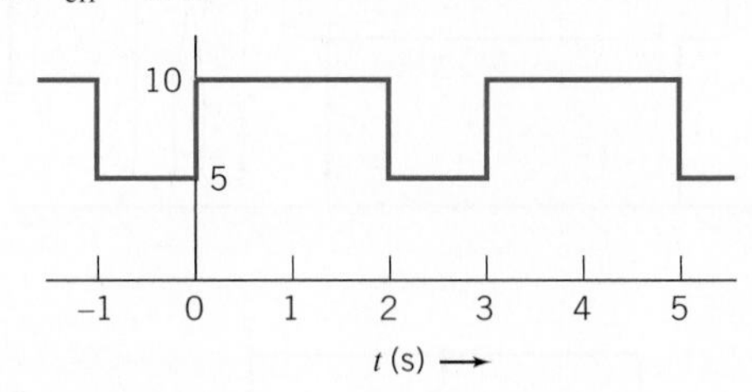

Figure P 11.4-6

P 11.4-7 Calculate the effective value of the voltage across the resistance R of the circuit shown in Figure P 11.4-7 when $\omega = 100$ rad/s.

Hint: Use superposition.

Answer: $V_{eff} = 4.82$ V

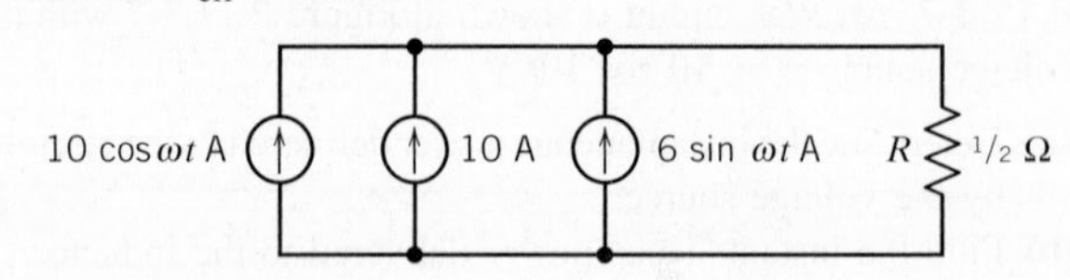

Figure P 11.4-7

Section 11.5 Complex Power

P 11.5-1 The complex power delivered by the voltage source in Figure P 11.5-1 is $\mathbf{S} = 3.6 + j7.2$ V A. Determine the values of the resistance, R, and inductance, L.

Answer: $R = 5.4\ \Omega$ and $L = 2.73$ H

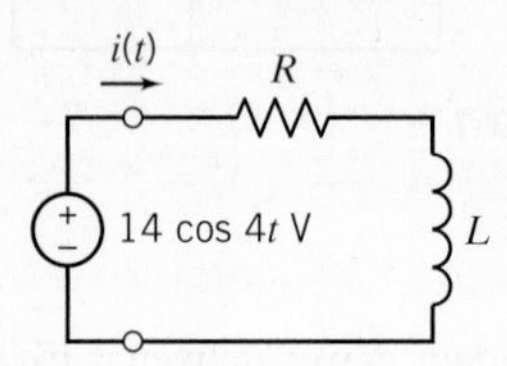

Figure P 11.5-1

P 11.5-2 The complex power delivered by the voltage source in Figure P 11.5-2 is $\mathbf{S} = 18 + j9$ VA. Determine the values of the resistance, R, and inductance, L.

Answers: $R = 4\ \Omega$ and $L = 2$ H

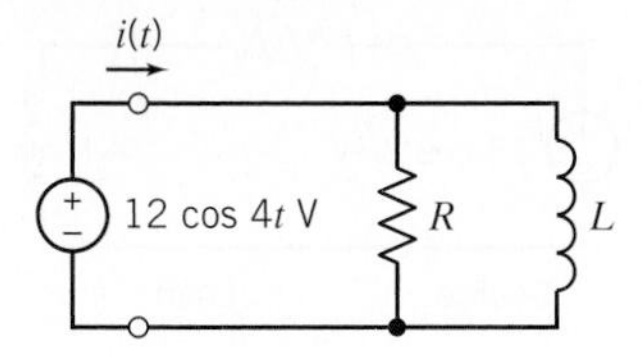

Figure P 11.5-2

P 11.5-3 Determine the complex power delivered by the voltage source in the circuit shown in Figure P 11.5-3.

Answer: $\mathbf{S} = 7.2 + j3.6$ VA

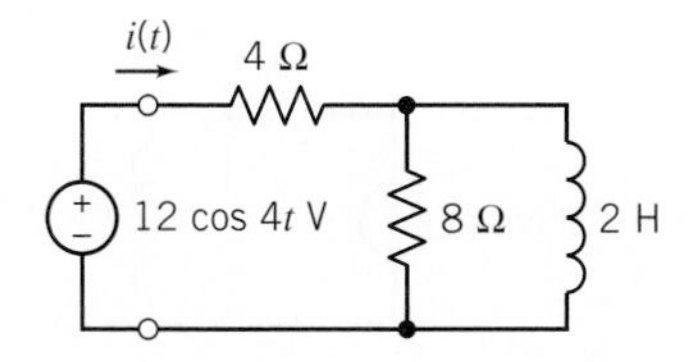

Figure P 11.5-3

P 11.5-4 Many engineers are working to develop photovoltaic power plants that provide ac power. An example of an experimental photovoltaic system is shown in Figure P 11.5-4*a*. A model of one portion of the energy conversion circuit is shown in Figure P 11.5-4*b*. Find the average, reactive, and complex power delivered by the dependent source.

Answer: $\mathbf{S} = +j8/9$ VA

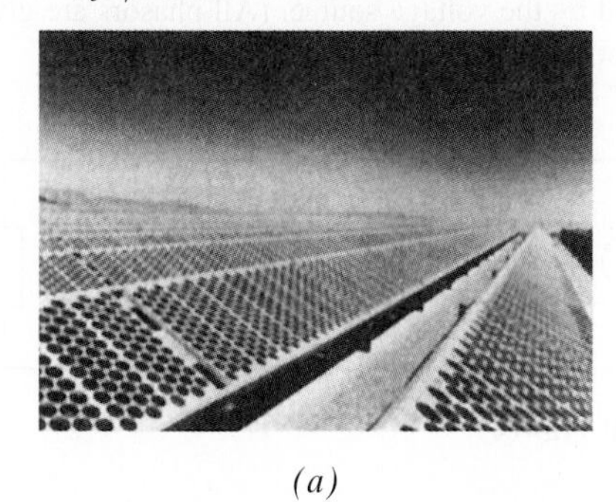

(*a*)

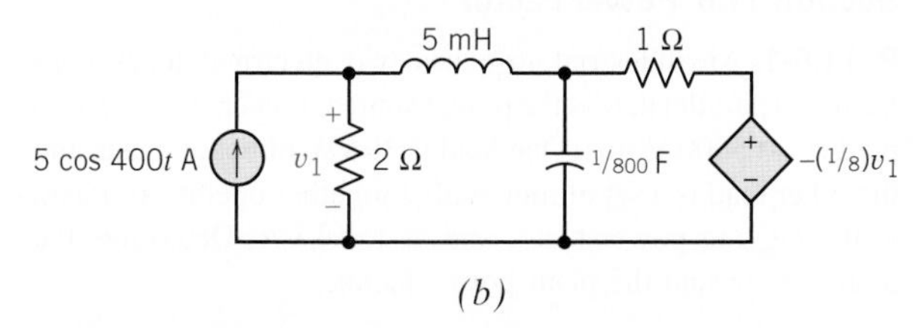

(*b*)

Figure P 11.5-4 (*a*) An experimental photovoltaic power plant. (*b*) Model of part of the energy conversion circuit. Courtesy of *EPRI Journal*.

P 11.5-5 For the circuit shown in Figure P 11.5-5, determine $\mathbf{I}$ and the complex power $\mathbf{S}$ delivered by the source when $\mathbf{V} = 100\angle 120°$ V rms.

Answer: $\mathbf{S} = 400 + j300$ VA

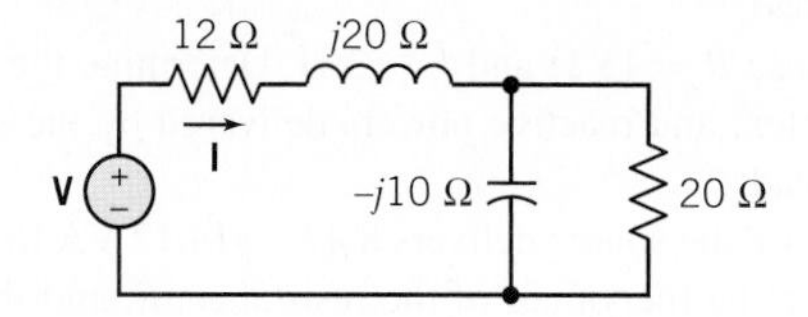

Figure P 11.5-5

P 11.5-6 For the circuit of Figure P 11.5-6, determine the complex power of the R, L, and C elements and show that the complex power delivered by the sources is equal to the complex power absorbed by the R, L, and C elements.

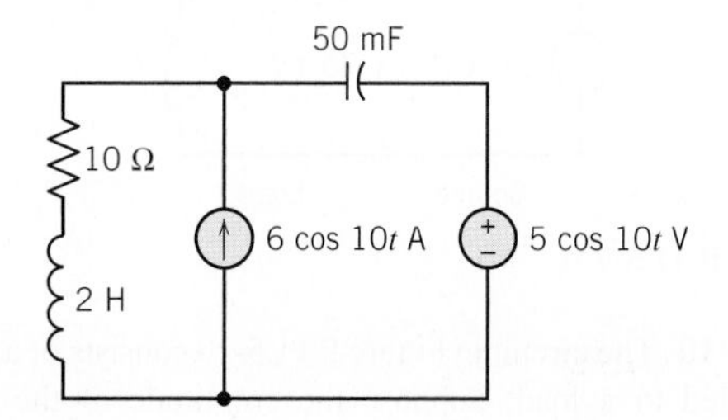

Figure P 11.5-6

P 11.5-7 A circuit is shown in Figure P 11.5-7 with an unknown impedance $\mathbf{Z}$. However, it is known that $v(t) = 100 \cos(100t + 20°)$ V and $i(t) = 50 \cos(100t - 10°)$ A. (a) Find $\mathbf{Z}$. (b) Find the power absorbed by the impedance. (c) Determine the type of element and its magnitude that should be placed across the impedance $\mathbf{Z}$ (connected to terminals a–b) so that the voltage $v(t)$ and the current entering the parallel elements are in phase.

Answers: (a) $2\angle 30°\ \Omega$ (b) 2165 W (c) 2.5 mF

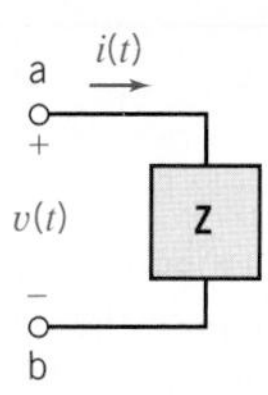

Figure P 11.5-7

P 11.5-8 Find the complex power delivered by the voltage source and the power factor seen by the voltage source for the circuit of Figure P 11.5-8.

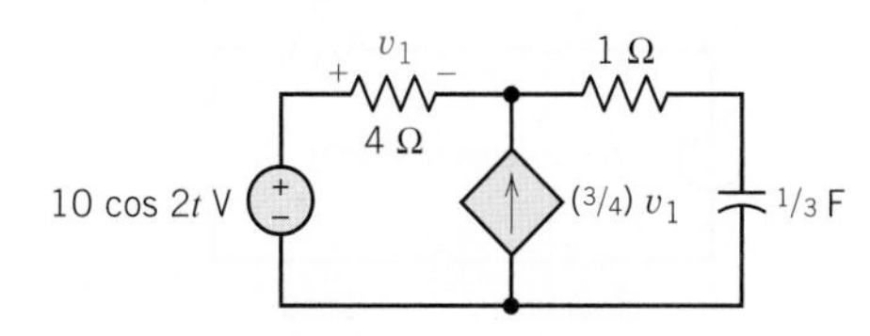

Figure P 11.5-8

P 11.5-9 The circuit in Figure P 11.5-9 consists of a source connected to a load.

(a) Suppose $R = 9\ \Omega$ and $L = 5$ H. Determine the average, complex, and reactive powers delivered by the source to the load.
(b) Suppose $R = 15\ \Omega$ and $L = 3$ H. Determine the average, complex, and reactive powers delivered by the source to the load.
(c) Suppose the source delivers $8.47 + j14.12$ VA to the load. Determine the values of the resistance, R, and the inductance, L.
(d) Suppose the source delivers $14.12 + j8.47$ VA to the load. Determine the values of the resistance, R, and the inductance, L.

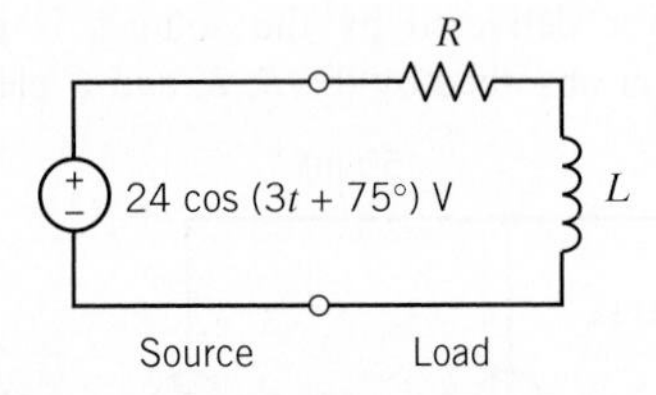

Figure P 11.5-9

P 11.5-10 The circuit in Figure P 11.5-10 consists of a source connected to a load. Suppose the amplitude of the source voltage is doubled so that $v_i(t) = 48 \cos(3t + 75°)$ V. How will each of the following change?

(a) The impedance of the load
(b) The complex power delivered to the load
(c) The load current

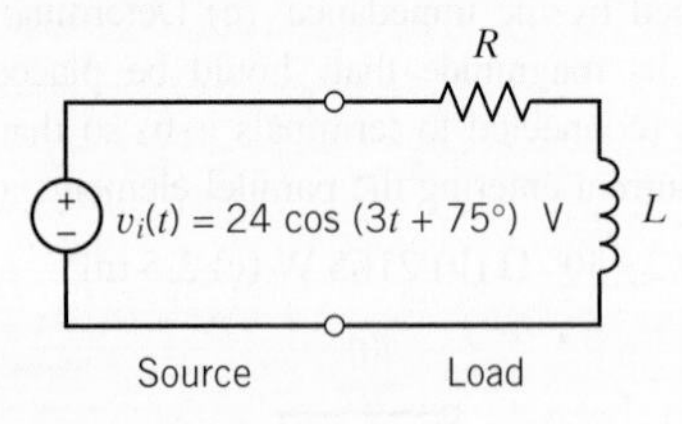

Figure P 11.5-10

P 11.5-11 The circuit in Figure P 11.5-11 consists of a source connected to a load. Suppose the phase angle of the source voltage is doubled so that $v_i(t) = 24 \cos(3t + 150°)$ V. How will the following change?

(a) The impedance of the load
(b) The complex power delivered to the load
(c) The load current

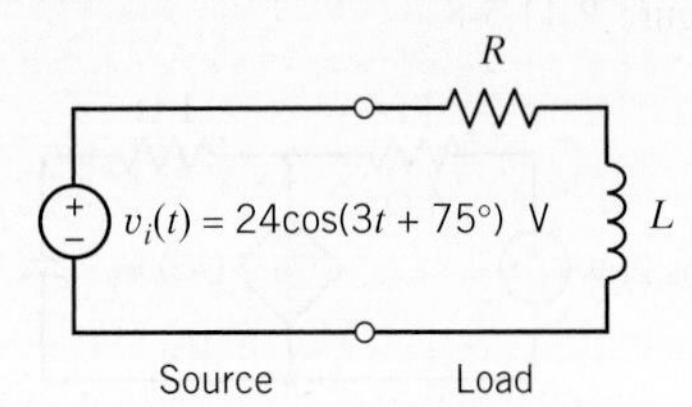

Figure P 11.5-11

P 11.5-12 The circuit in Figure P 11.5-12 consists of a source connected to a load. The complex power delivered by the source to the load is $\mathbf{S} = 6.61 + j1.98$ VA. Determine the values of R and C.

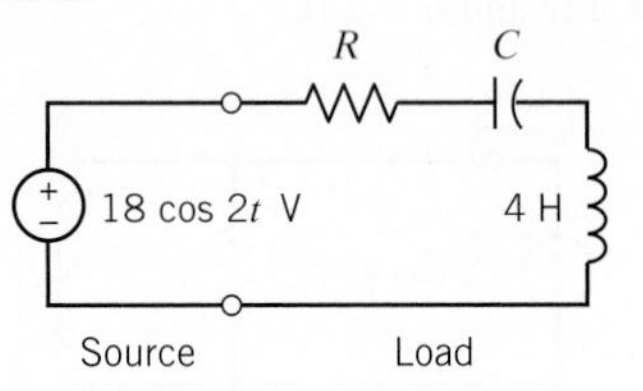

Figure P 11.5-12

P 11.5-13 Design the circuit shown in Figure P 11.5-13, that is, specify values for R and L so that the complex power delivered to the RL circuit is $8 + j6$ VA.

Answer: $R = 5.76\ \Omega$ and $L = 2.16$ H

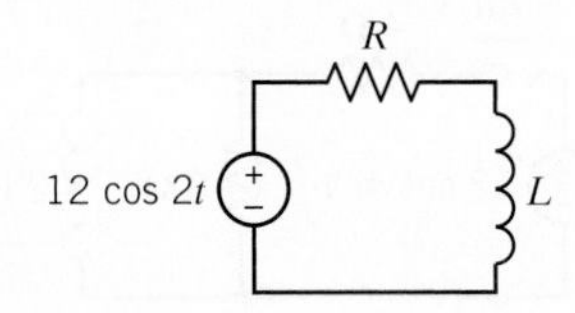

Figure P 11.5-13

P 11.5-14 The source voltage in the circuit shown in Figure P 11.5-14 is $\mathbf{V}_s = 24\angle 30°$ V. Consequently,

$$\mathbf{I}_1 = 2.41\angle -16.1°\text{ A},\ \mathbf{I}_2 = 1.53\angle 11.3°\text{ A} \quad\text{and}\quad \mathbf{V}_4 = 6.86\angle -52.1°\text{ V}$$

Determine (a) the average power absorbed by $\mathbf{Z}_4$, (b) the average power absorbed by $\mathbf{Z}_1$, and (c) the complex power delivered by the voltage source. (All phasors are given using peak, not rms, values.)

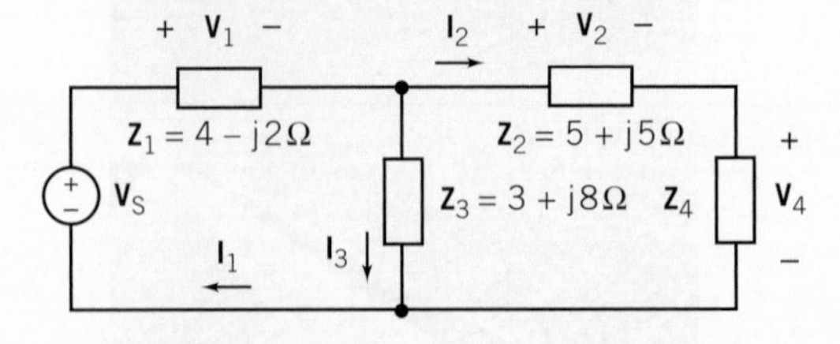

Figure P 11.5-14

Section 11.6 Power Factor

P 11.6-1 An industrial firm has two electrical loads connected in parallel across the power source. Power is supplied to the firm at 5000 V rms. One load is 30 kW of heating use, and the other load is a set of motors that together operate as a load at 0.6 lagging power factor and at 150 kVA. Determine the total current and the plant power factor.

Answer: $I = 34$ A rms and $pf = 1/\sqrt{2}$

P 11.6-2 Two electrical loads are connected in parallel to a 400-V rms, 60-Hz supply. The first load is 12 kVA at 0.7

lagging power factor; the second load is 10 kVA at 0.8 lagging power factor. Find the average power, the apparent power, and the power factor of the two combined loads.

Answer: Total power factor = 0.75 lagging

P 11.6-3 The source of Figure P 11.6-3 delivers 50 VA with a power factor of 0.8 lagging. Find the unknown impedance **Z**.

Answer: $\mathbf{Z} = 1.1\angle 36.2^\circ\ \Omega$

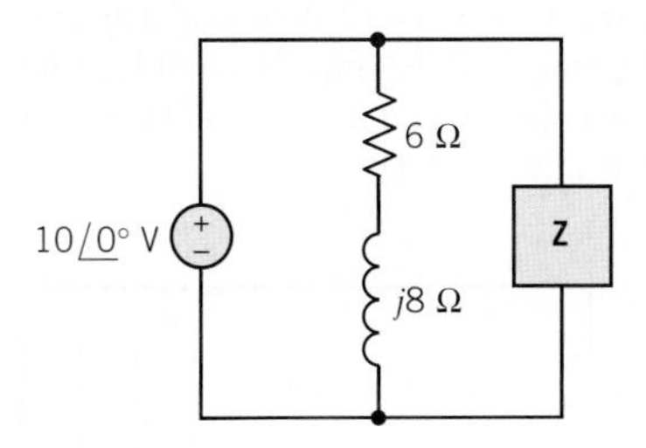

Figure P 11.6-3

P 11.6-4 Manned space stations require several continuously available ac power sources. Also, it is desired to keep the power factor close to 1. Consider the model of one communication circuit, shown in Figure P 11.6-4. If an average power of 500 W is dissipated in the 20-Ω resistor, find (a) V_{rms}, (b) $I_{s\ rms}$, (c) the power factor seen by the source, and (d) $|\mathbf{V}_s|$.

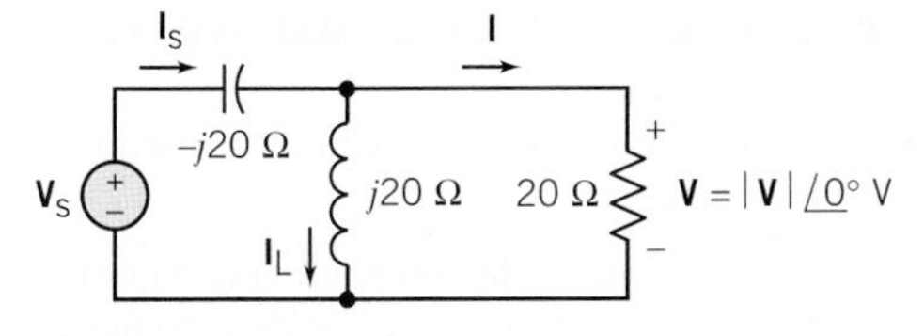

Figure P 11.6-4

P 11.6-5 Two impedances are supplied by $\mathbf{V} = 125\angle 160^\circ\ \mathrm{V_{rms}}$, as shown in Figure P 11.6-5, where $\mathbf{I} = 2\angle 190^\circ$ A rms. The first load draws $P_1 = 23.2$ W, and $Q_1 = 50$ VAR. Calculate $\mathbf{I}_1$, $\mathbf{I}_2$, the power factor of each impedance, and the total power factor of the circuit.

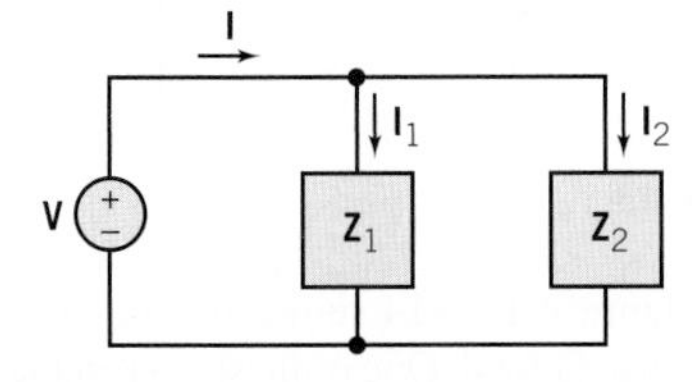

Figure P 11.6-5

P 11.6-6 A residential electric supply three-wire circuit from a transformer is shown in Figure P 11.6-6*a*. The circuit model is shown in Figure P 11.6-6*b*. From its nameplate, the refrigerator motor is known to have a rated current of 8.5 A rms. It is reasonable to assume an inductive impedance angle of 45° for a small motor at rated load. Lamp and range loads are 100 W and 12 kW, respectively.

(a) Calculate the currents in line 1, line 2, and the neutral wire.

(b) Calculate: (i) P_{refrig}, Q_{refrig}, (ii) P_{lamp}, Q_{lamp}, and (iii) P_{total}, Q_{total}, $\mathbf{S}_{total}$, and overall power factor.

(c) The neutral connection resistance increases, because of corrosion and looseness, to 20 Ω. (This must be included as part of the neutral wire.) Use mesh analysis and calculate the voltage across the lamp.

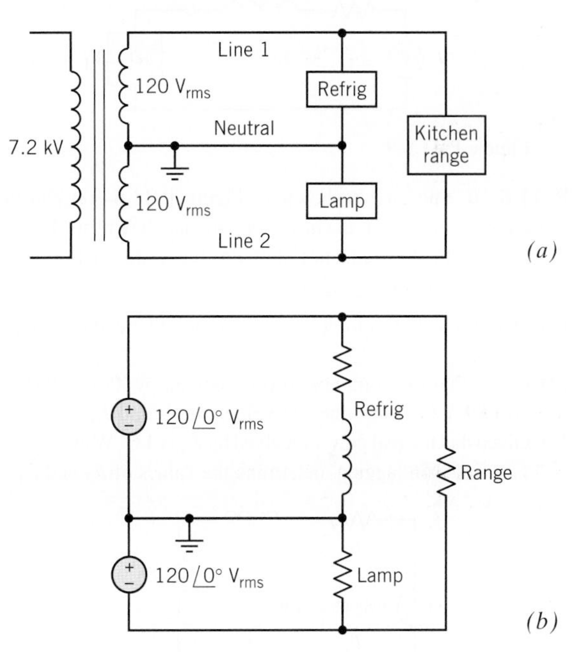

Figure P 11.6-6 Residential circuit with selected loads.

P 11.6-7 A motor connected to a 220-V supply line from the power company has a current of 8 A. Both the current and the voltage are rms values. The average power delivered to the motor is 1317 W.

(a) Find the apparent power, the reactive power, and the power factor when $\omega = 377$ rad/s.

(b) Find the capacitance of a parallel capacitor that will result in a unity power factor of the combination.

(c) Find the current in the utility lines after the capacitor is installed.

Answers: **(a)** $pf = 0.748$ **(b)** $C = 64.1\ \mu$F **(c)** $I = 6.0$ A rms

P 11.6-8 Two loads are connected in parallel across a 1000-V rms, 60-Hz source. One load absorbs 500 kW at 0.6 power factor lagging, and the second load absorbs 400 kW and 600 kVAR. Determine the value of the capacitor that should be added in parallel with the two loads to improve the overall power factor to 0.9 lagging.

Answer: $C = 2.2\ \mu$F

P 11.6-9 A voltage source with a complex internal impedance is connected to a load, as shown in Figure P 11.6-9. The load absorbs 1 kW of average power at 100 V rms with a power factor of 0.80 lagging. The source frequency is 200 rad/s.

(a) Determine the source voltage $\mathbf{V}_1$.
(b) Find the type and value of the element to be placed in parallel with the load so that maximum power is transferred to the load.

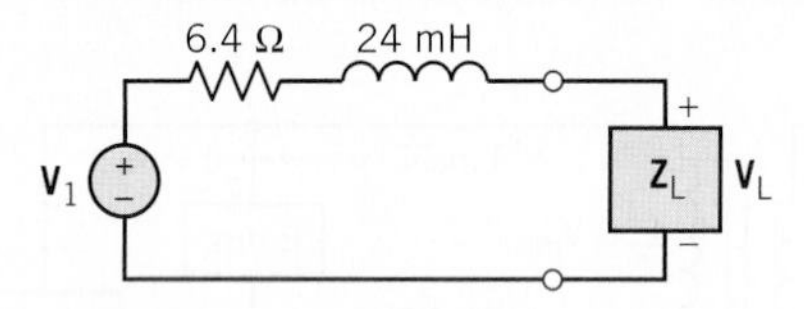

Figure P 11.6-9

P 11.6-10 The circuit shown in Figure P 11.6-10*a* can be represented in the frequency domain as shown in Figure P 11.6-10*b*. In the frequency domain, the value of the mesh current is $\mathbf{I} = 1.076\angle{-8.3}$ A.

(a) Determine the complex power supplied by the voltage source.
(b) Given that the complex power received by $\mathbf{Z}_1$ is $6.945 + j\,13.89$ VA, determine the values of R_1 and L_1.
(c) Given that the real power received by $\mathbf{Z}_3$ is 4.63 W at a power factor of 0.56 lagging, determine the values of R_3 and L_3.

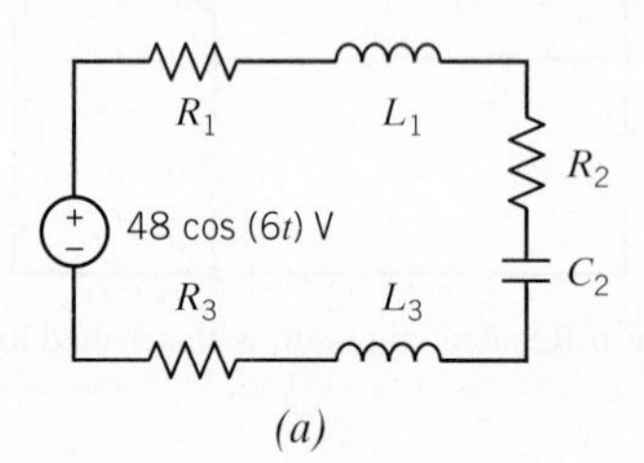

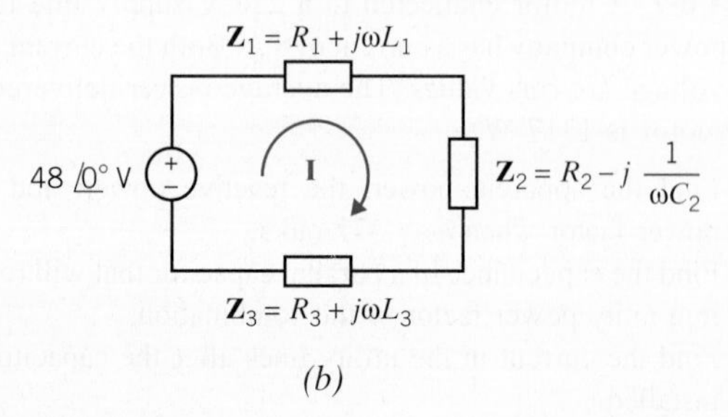

Figure P 11.6-10

P 11.6-11 The circuit in Figure P 11.6-11 consists of a source connected to a load. The source delivers 14.12 W to the load at a power factor of 0.857 lagging. What are the values of the resistance, R, and the inductance, L?

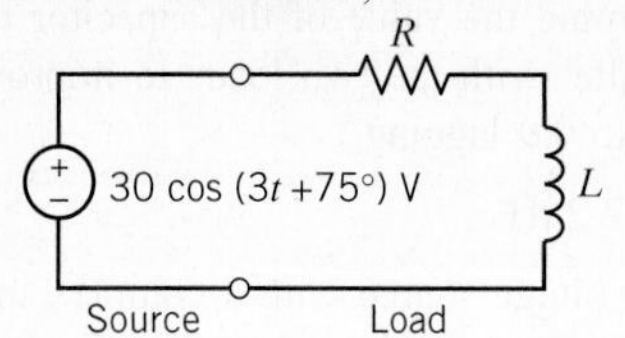

Figure P 11.6-11

P 11.6-12 The circuit in Figure P 11.6-12 consists of a source connected to a load. Determine the impedance of the load and the complex power delivered by the source to the load under each of the following conditions:

(a) The source delivers $14.12 + j8.47$ VA to load A and $8.47 + j14.12$ VA to load B.
(b) The source delivers $8.47 + j14.12$ VA to load A, and the impedance of load B is $15 + j9\ \Omega$.
(c) The source delivers 14.12 W to load A at a power factor of 0.857 lagging, and the impedance of load B is $9 + j15\ \Omega$.
(d) The impedance of load A is $15 + j9\ \Omega$, and the impedance of load B is $9 + j15\ \Omega$.

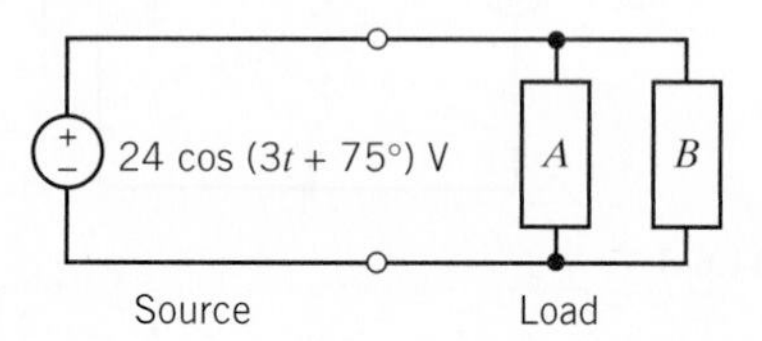

Figure P 11.6-12

P 11.6-13 Figure P 11.6-13 shows two possible representations of an electrical load. One of these representations is used when the power factor of the load is lagging, and the other is used when the power factor is leading. Consider two cases:

(a) At the frequency $\omega = 4$ rad/s, the load has the power factor $pf = 0.8$ lagging.
(b) At the frequency $\omega = 4$ rad/s, the load has the power factor $pf = 0.8$ leading.

In each case, choose one of the two representations of the load. Let $R = 6\ \Omega$ and determine the value of the capacitance, C, or the inductance, L.

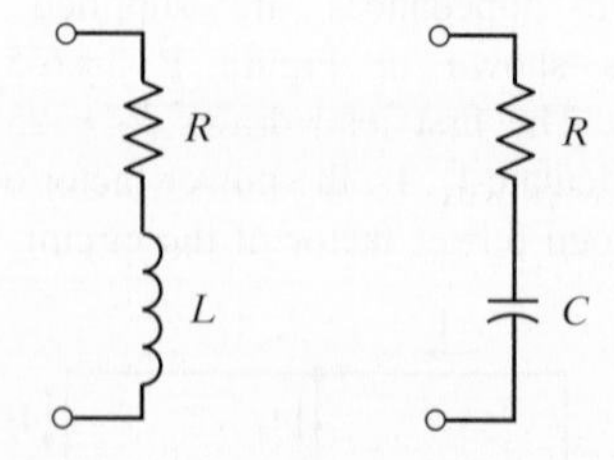

Figure P 11.6-13

P 11.6-14 Figure P 11.6-14 shows two possible representations of an electrical load. One of these representations is used when the power factor of the load is lagging, and the other is used when the power factor is leading. Consider two cases:

(a) At the frequency $\omega = 4$ rad/s, the load has the power factor $pf = 0.8$ lagging.
(b) At the frequency $\omega = 4$ rad/s, the load has the power factor $pf = 0.8$ leading.

In each case, choose one of the two representations of the load. Let $R = 6\ \Omega$ and determine the value of the capacitance, C, or the inductance, L.

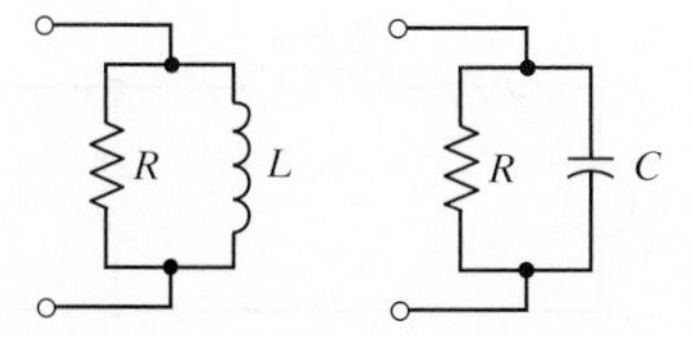

Figure P 11.6-14

P 11.6-15 Figure P 11.6-15 shows two electrical loads. Express the power factor of each load in terms of ω, R, and L.

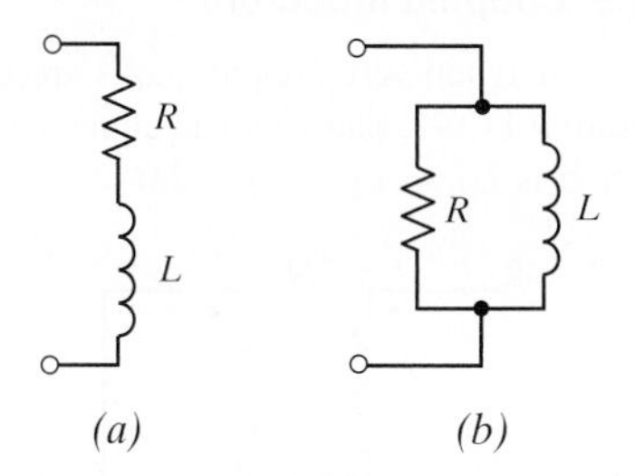

Figure P 11.6-15

P 11.6-16 Figure P 11.6-16 shows two electrical loads. Express the power factor of each load in terms of ω, R, and C.

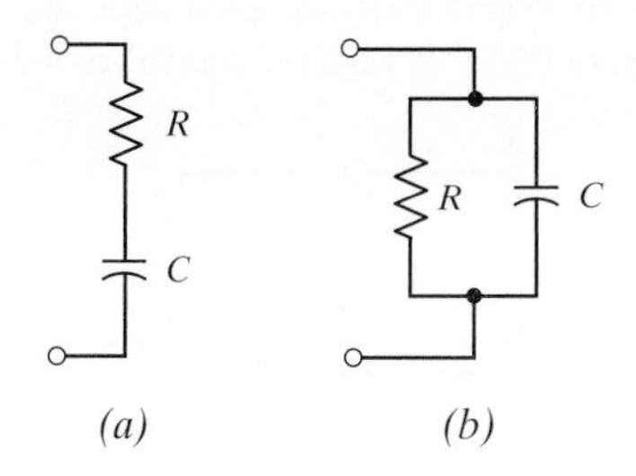

Figure P 11.6-16

P 11.6-17 The source voltage in the circuit shown in Figure P 11.6-17 is $\mathbf{V}_s = 24\,\angle 30^\circ$ V. Consequently,

$$\mathbf{I}_1 = 2.41\,\angle -16.1^\circ \text{ A},\ \mathbf{I}_2 = 1.53\,\angle 11.3^\circ \text{ A and } \mathbf{V}_4 = 6.86\,\angle -52.1^\circ \text{ V}$$

Determine (a) the power factor of $\mathbf{Z}_1$, (b) the power factor of $\mathbf{Z}_3$, and (c) the power factor of $\mathbf{Z}_4$. Include the indication of leading or lagging.

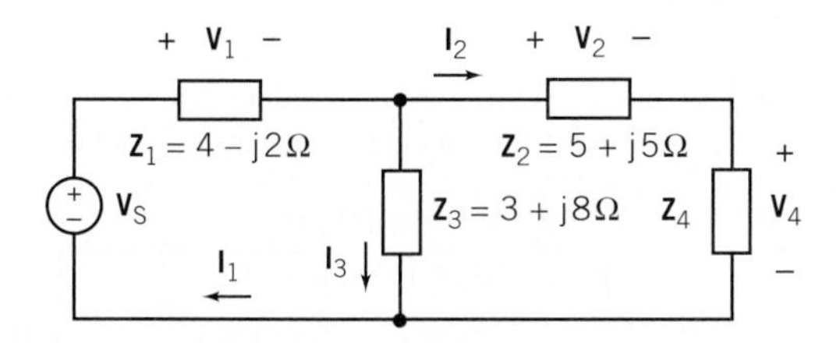

Figure P 11.6-17

Section 11.7 The Power Superposition Principle

P 11.7-1 Find the average power absorbed by the 2-Ω resistor in the circuit of Figure P 11.7-1.

Answer: $P = 413$ W

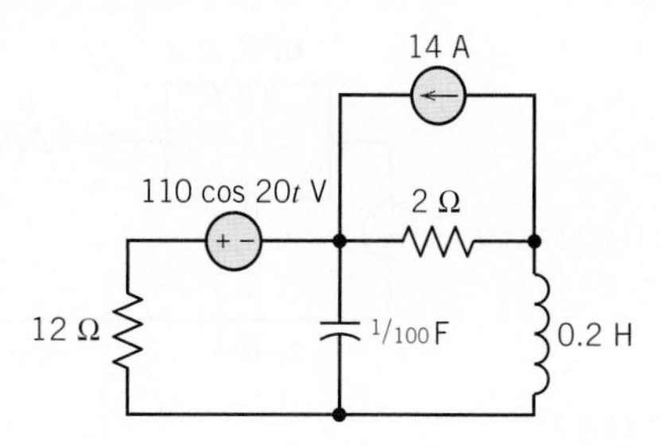

Figure P 11.7-1

P 11.7-2 Find the average power absorbed by the 8-Ω resistor in the circuit of Figure P 11.7-2.

Answer: $P = 22$ W

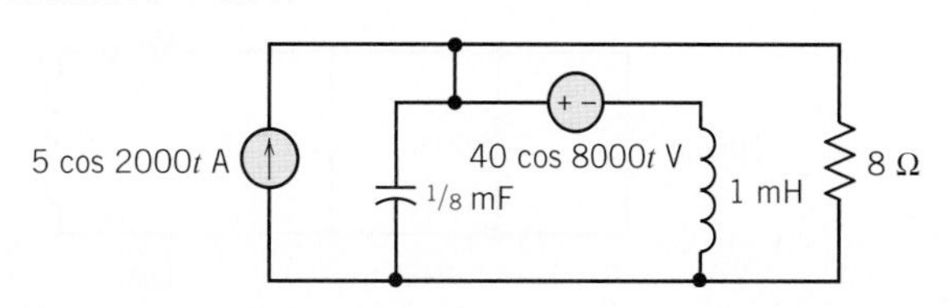

Figure P 11.7-2

P 11.7-3 For the circuit shown in Figure P 11.7-3, determine the average power absorbed by each resistor, R_1 and R_2. The voltage source is $v_s = 10 + 10\cos(5t + 40^\circ)$ V, and the current source is $i_s = 4\cos(5t - 30^\circ)$ A.

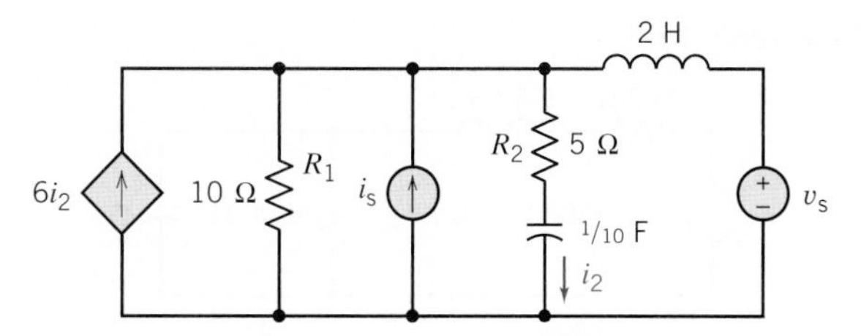

Figure P 11.7-3

P 11.7-4 For the circuit shown in Figure P 11.7-4, determine the effective value of the resistor voltage v_R and the capacitor voltage v_C.

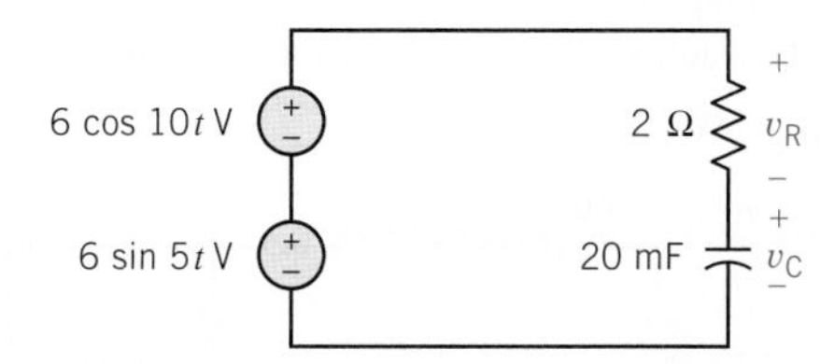

Figure P 11.7-4

Section 11.8 The Maximum Power Transfer Theorem

P 11.8-1 Determine values of R and L for the circuit shown in Figure P 11.8-1 that cause maximum power transfer to the load.

Answer: $R = 1800\ \Omega$ and $L = 0.6$ H

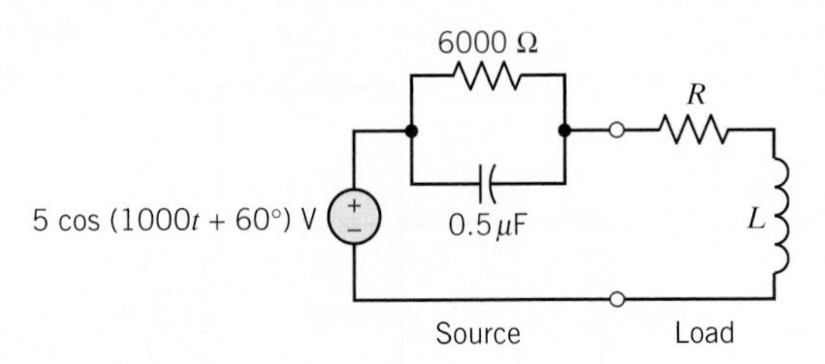

Figure P 11.8-1

P 11.8-2 Is it possible to choose R and L for the circuit shown in Figure P 11.8-2 so that the average power delivered to the load is 12 mW?

Answer: Yes

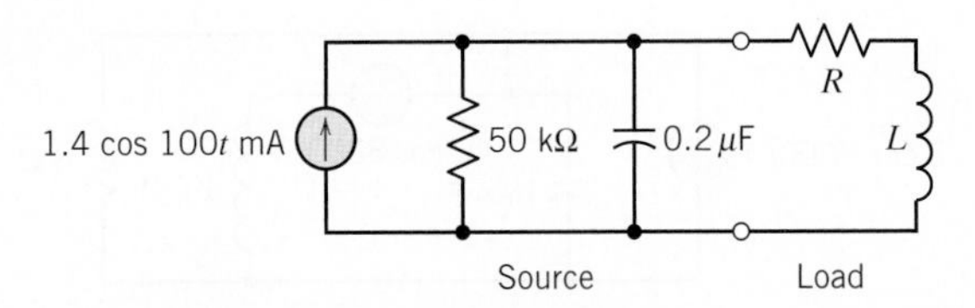

Figure P 11.8-2

P 11.8-3 The capacitor has been added to the load in the circuit shown in Figure P 11.8-3 to maximize the power absorbed by the 4000-Ω resistor. What value of capacitance should be used to accomplish that objective?

Answer: 0.1 μF

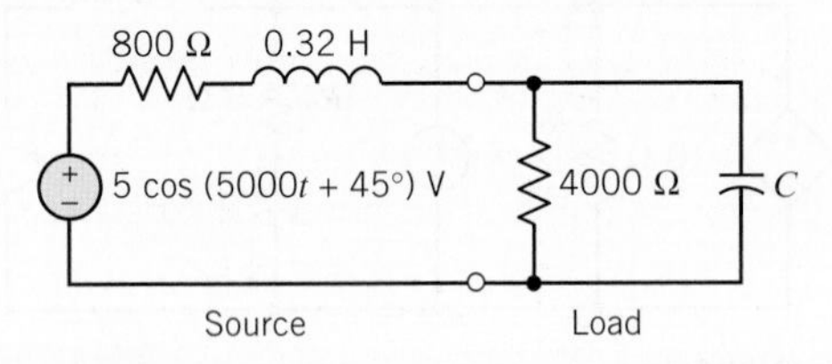

Figure P 11.8-3

P 11.8-4 What is the value of the average power delivered to the 2000-Ω resistor in the circuit shown in Figure P 11.8-4? Can the average power delivered to the 2000-Ω resistor be increased by adjusting the value of the capacitance?

Answers: 8 mW. No.

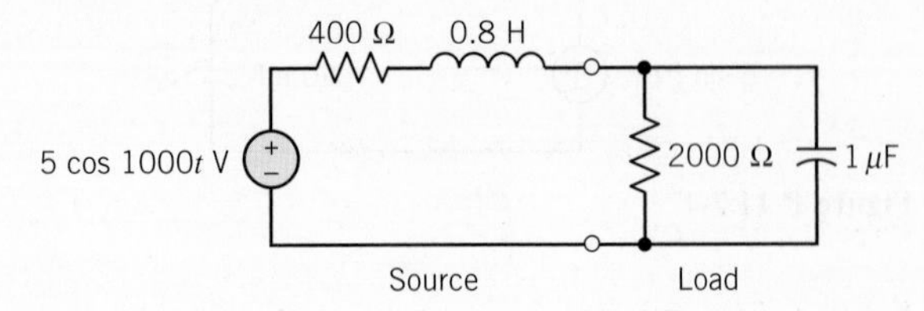

Figure P 11.8-4

P 11.8-5 What is the value of the resistance R in Figure P 11.8-5 that maximizes the average power delivered to the load?

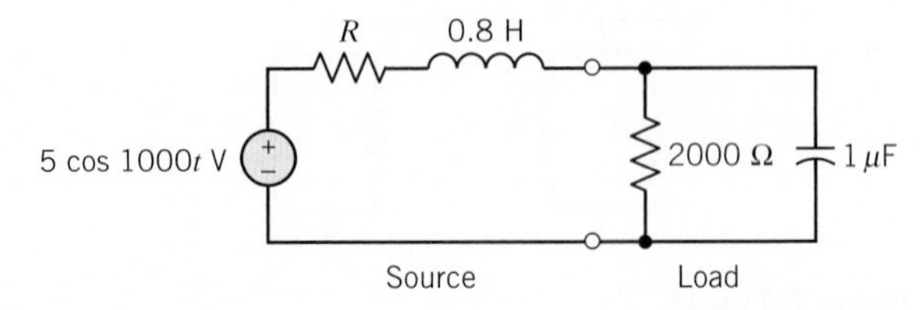

Figure P 11.8-5

Section 11.9 Coupled Inductors

P 11.9-1 Two magnetically coupled coils are connected as shown in Figure P 11.9-1. Show that an equivalent inductance at terminals a–b is $L_{ab} = L_1 + L_2 - 2M$.

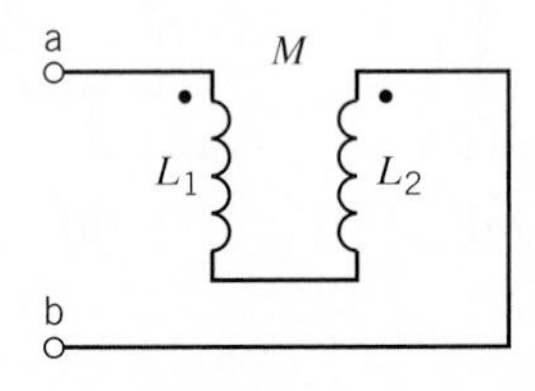

Figure P 11.9-1

P 11.9-2 Two magnetically coupled coils are shown connected in Figure P 11.9-2. Find the equivalent inductance L_{ab}.

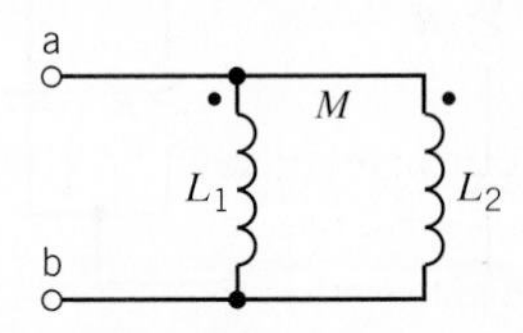

Figure P 11.9-2

P 11.9-3 The source voltage of the circuit shown in Figure P 11.9-3 is $v_s = 142 \cos 100t$ V. Determine $i_1(t)$ and $i_2(t)$.

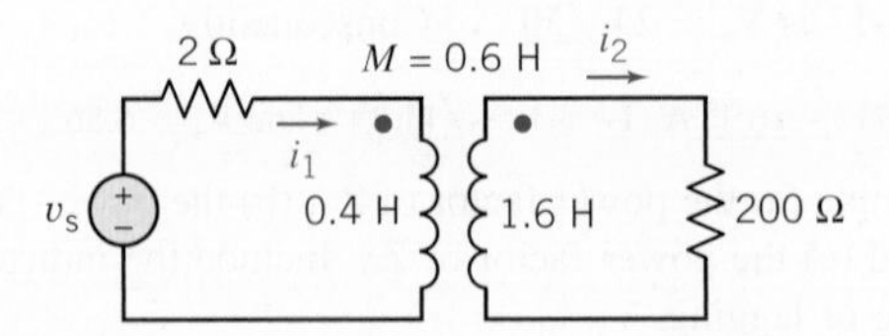

Figure P 11.9-3

P 11.9-4 A circuit with a mutual inductance is shown in Figure P 11.9-4. Find the voltage $\mathbf{V}_2$ when $\omega = 5000$.

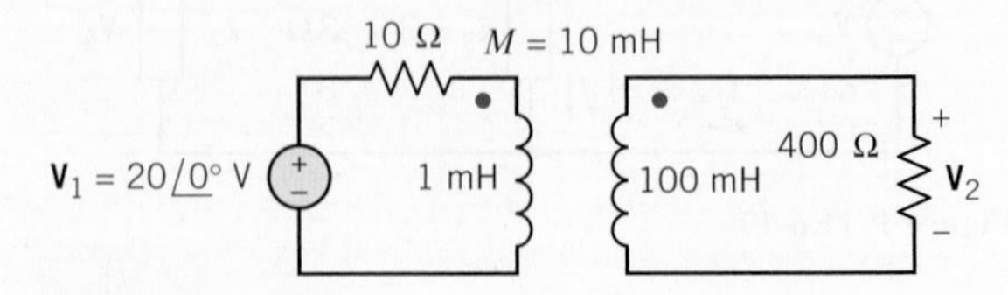

Figure P 11.9-4

P 11.9-5 Determine $v(t)$ for the circuit of Figure P 11.9-5 when $v_s = 30 \cos 30t$ V.

Answer: $v(t) = 68.6 \cos(30t + 10.2°)$V

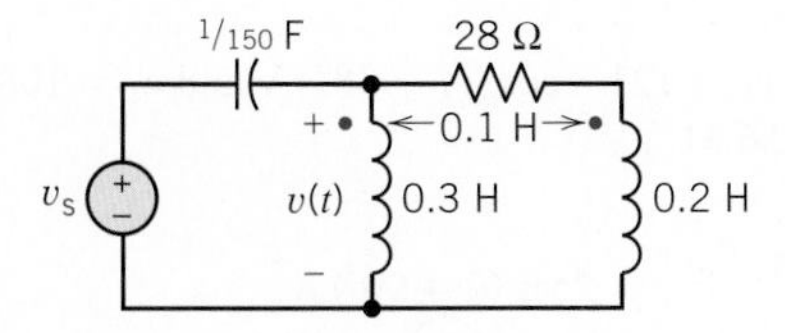

Figure P 11.9-5

P 11.9-6 Find the total energy stored in the circuit shown in Figure P 11.9-6 at $t = 0$ if the secondary winding is (a) open-circuited, (b) short-circuited, (c) connected to the terminals of a 7-Ω resistor.

Answers: (a) 15 J (b) 0 J (c) 5 J

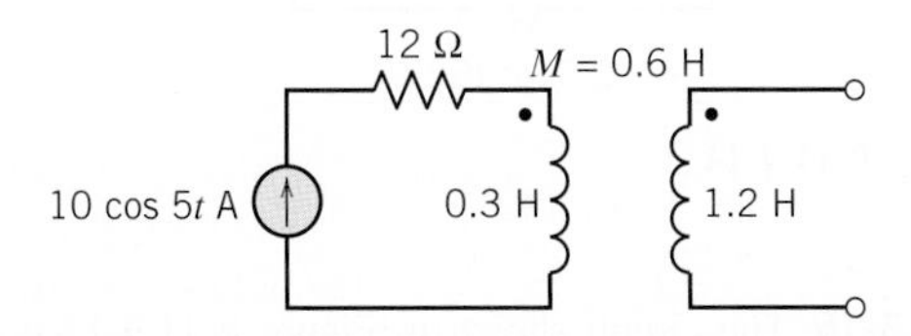

Figure P 11.9-6

P 11.9-7 Find the input impedance, **Z**, of the circuit of Figure P 11.9-7 when $\omega = 1000$ rad/s.

Answer: $\mathbf{Z} = 8.4\angle 14°\ \Omega$

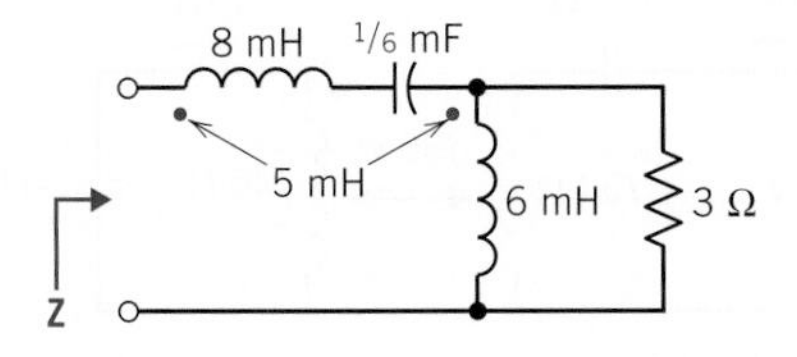

Figure P 11.9-7

P 11.9-8 A circuit with three mutual inductances is shown in Figure P 11.9-8. When $v_s = 20 \cos 2t$ V, $M_1 = 2$ H, and $M_2 = M_3 = 1$ H, determine the capacitor voltage $v(t)$.

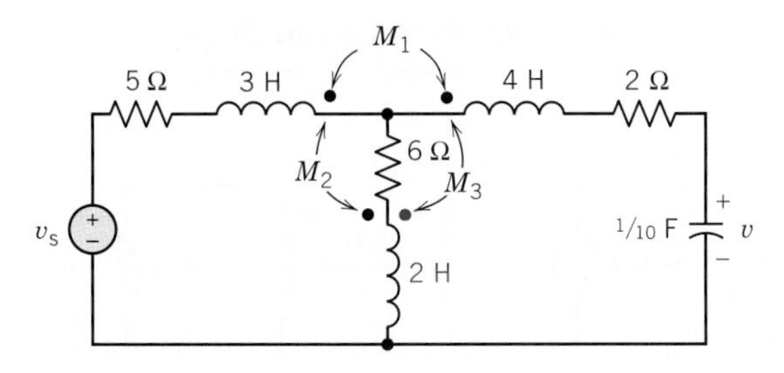

Figure P 11.9-8

P 11.9-9 The currents $i_1(t)$ and $i_2(t)$ in Figure P 11.9-9 are mesh currents. Represent the circuit in the frequency domain and write the mesh equations.

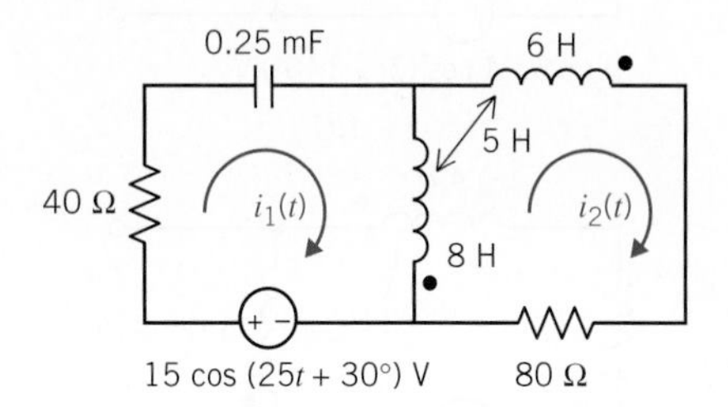

Figure P 11.9-9

P 11.9-10 Determine the mesh currents for the circuit shown in Figure P 11.9-10.

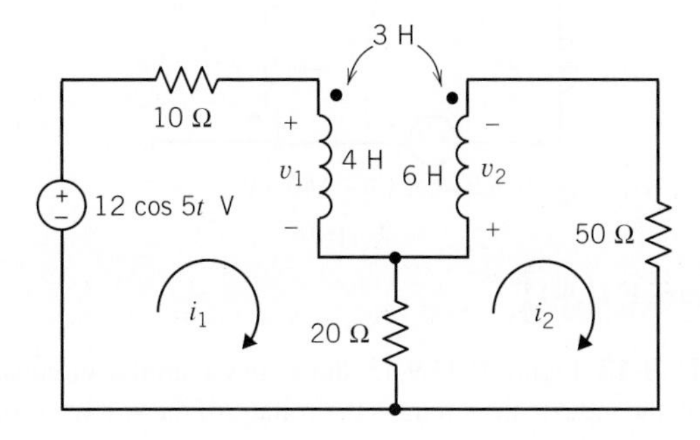

Figure P 11.9-10

P 11.9-11 Determine the coil voltages, v_1, v_2, v_3, and v_4, for the circuit shown in Figure P 11.9-11.

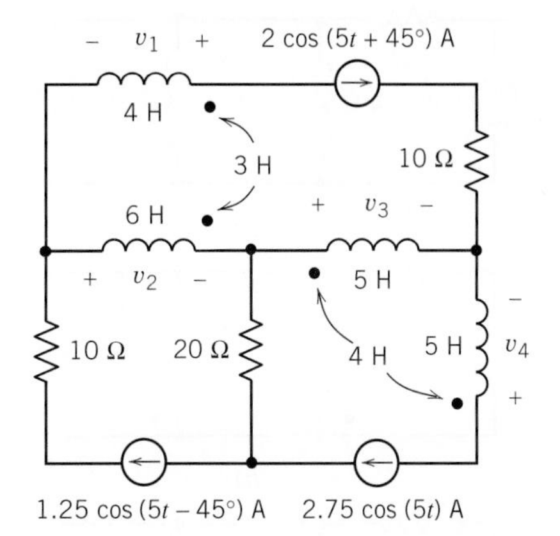

Figure P 11.9-11

P 11.9-12 Figure P 11.9-12 shows three similar circuits. In each, the input to the circuit is the voltage of the voltage source, $v_s(t)$. The output is the voltage across the right-hand coil, $v_o(t)$. Determine the steady-state output voltage, $v_o(t)$, for each of the three circuits.

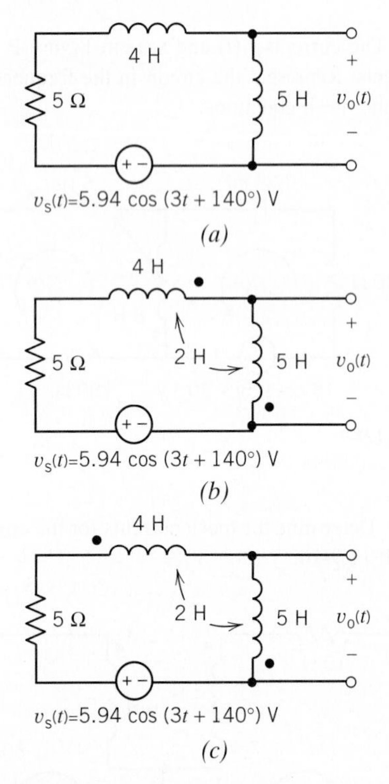

Figure P 11.9-12

P 11.9-13 Figure P 11.9-13 shows three similar circuits. In each, the input to the circuit is the voltage of the voltage source,

$$v_s(t) = 5.7\cos(4t + 158°)\ \text{V}$$

The output in each circuit is the voltage across the right-hand coil, $v_o(t)$. Determine the steady-state output voltage, $v_o(t)$, for each of the three circuits.

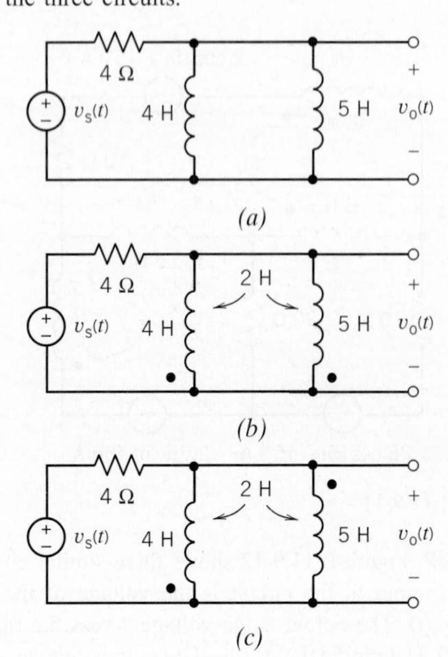

Figure P 11.9-13

P 11.9-14 The circuit shown in Figure P 11.9-14 is represented in the time domain. Determine coil voltages v_1 and v_2.

Answers: $v_1 = 121.8\cos(6t + 38°)$ V and $v_2 = 108.17\cos(6t + 56.31°)$ V

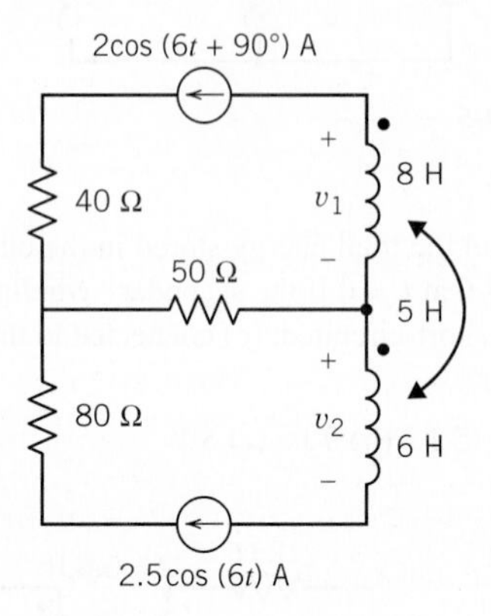

Figure P 11.9-14

P 11.9-15 The circuit shown in Figure P 11.9-15 is represented in the frequency domain. (For example, $-j30\ \Omega$ is the impedance due to the mutual inductance of the coupled coils.) Suppose $\mathbf{V}(\omega) = 70\angle 0°$ V. Then $\mathbf{I}_1(\omega) = B\angle\theta$ A and $\mathbf{I}_2(\omega) = 0.875\angle -90°$ A. Determine the values of B and θ.

Answers: $B = 1.75$ A and $\theta = -90°$

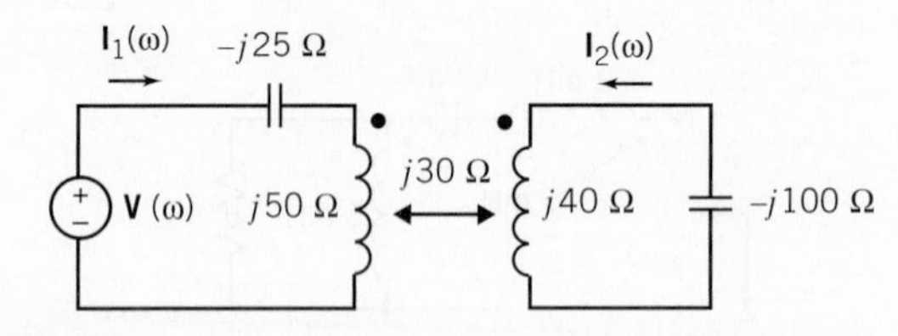

Figure P 11.9-15

P 11.9-16 Determine the values of the inductances L_1 and L_2 in the circuit shown in Figure P 11.9-16, given that

$$i(t) = 0.319\cos(4t - 82.23°)\ \text{A}$$

and

$$v(t) = 0.9285\cos(4t - 62.20°)\ \text{V}.$$

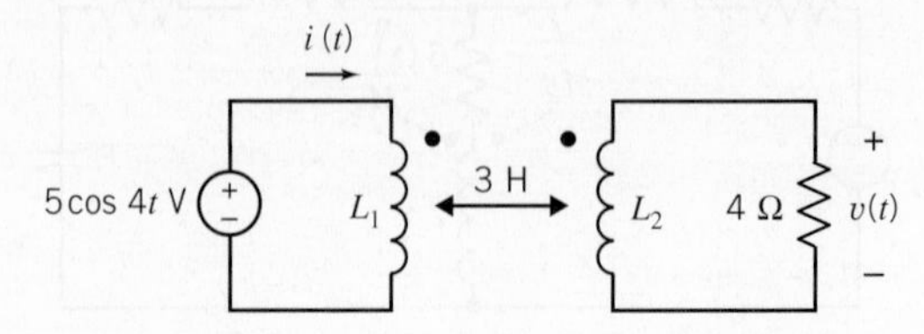

Figure P 11.9-16

P 11.9-17 Determine the complex power supplied by the source in the circuit shown in Figure P 11.9-17.

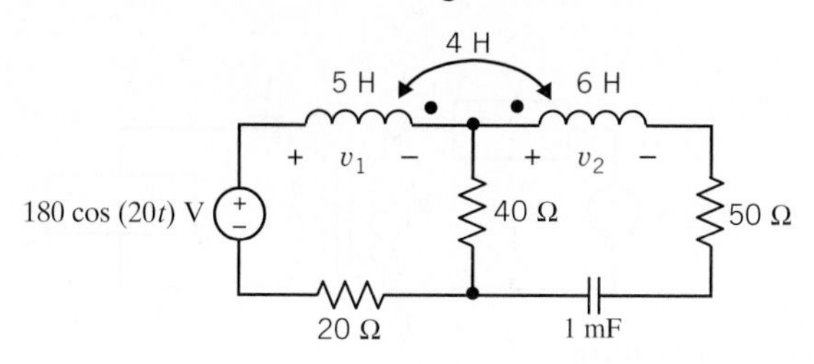

Figure P 11.9-17

Section 11.10 The Ideal Transformer

P 11.10-1 Find $\mathbf{V}_1$, $\mathbf{V}_2$, $\mathbf{I}_1$, and $\mathbf{I}_2$ for the circuit of Figure P 11.10-1 when $n = 5$.

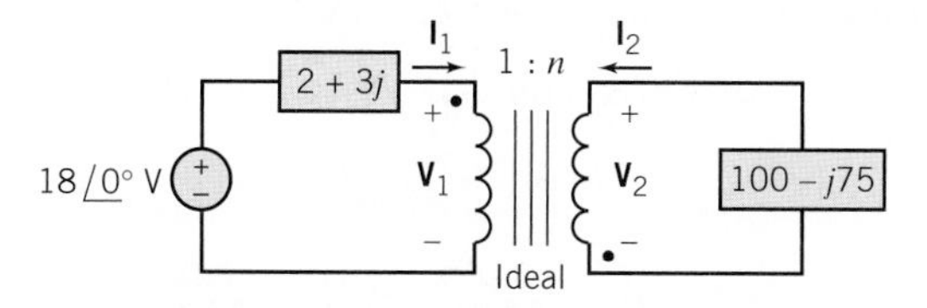

Figure P 11.10-1

P 11.10-2 A circuit with a transformer is shown in Figure P 11.10-2.

(a) Determine the turns ratio, n.
(b) Determine the value of R_{ab}.
(c) Determine the current, i, supplied by the voltage source.

Answers: **(a)** $n = 5$ **(b)** $R_{ab} = 400\ \Omega$

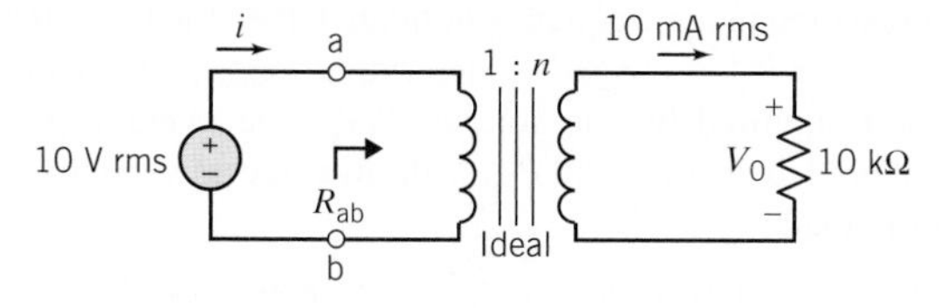

Figure P 11.10-2

P 11.10-3 Find the voltage $\mathbf{V}_c$ in the circuit shown in Figure P 11.10-3 . Assume an ideal transformer. The turns ratio is $n = 1/3$.

Answer: $\mathbf{V}_c = 21.0\angle -105.3^\circ$

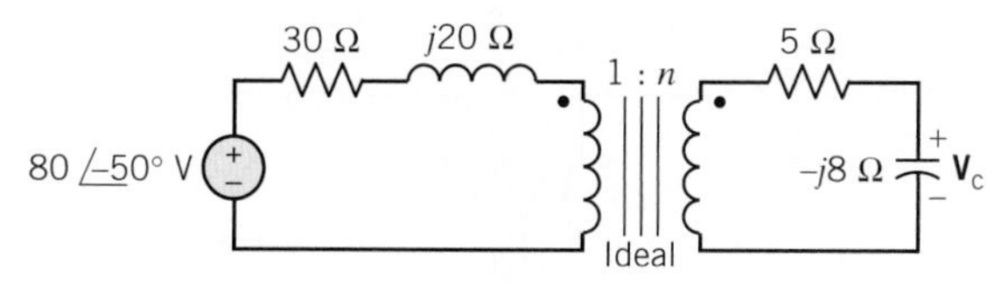

Figure P 11.10-3

P 11.10-4 An ideal transformer is connected in the circuit shown in Figure P 11.10-4, where $v_s = 50 \cos 1000t$ V and $n = N_2/N_1 = 5$. Calculate $\mathbf{V}_1$ and $\mathbf{V}_2$.

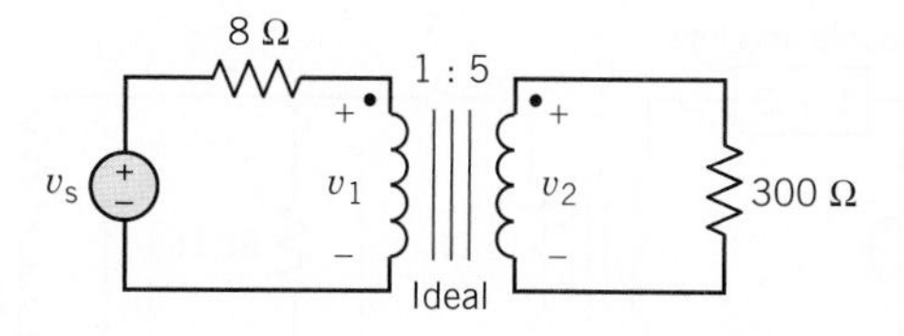

Figure P 11.10-4

P 11.10-5 The circuit of Figure P 11.10-5 is operating at 10^5 rad/s. Determine the inductance L and the turns ratio n to achieve maximum power transfer to the load.

Answer: $n = 2$

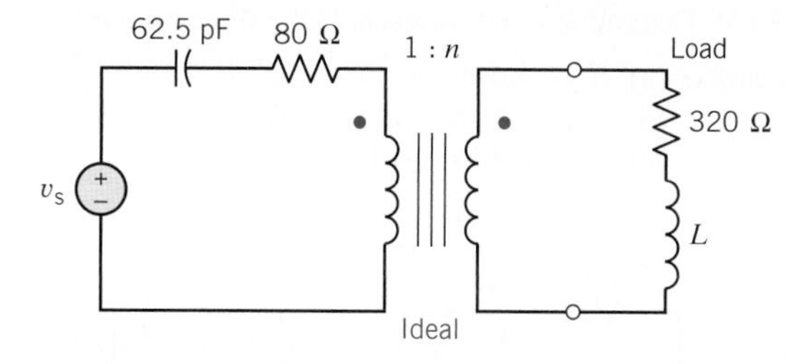

Figure P 11.10-5

P 11.10-6 Find the Thévenin equivalent at terminals a–b for the circuit of Figure P 11.10-6 when $v = 16 \cos 3t$ V.

Answer: $V_{oc} = 12$ and $Z_t = 3.75\ \Omega$

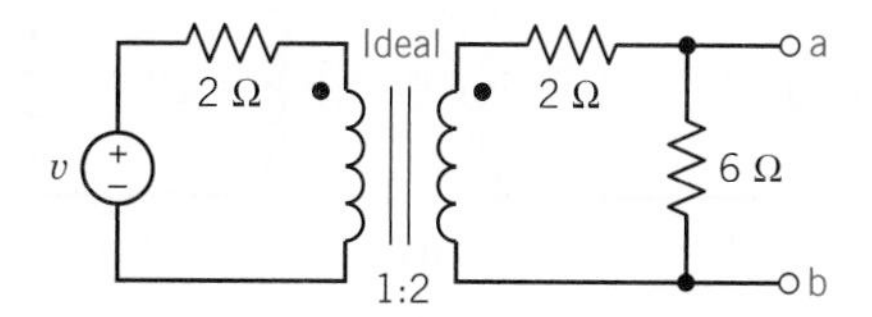

Figure P 11.10-6

P 11.10-7 Find the input impedance $\mathbf{Z}$ for the circuit of Figure P 11.10-7.

Answer: $\mathbf{Z} = 6\ \Omega$

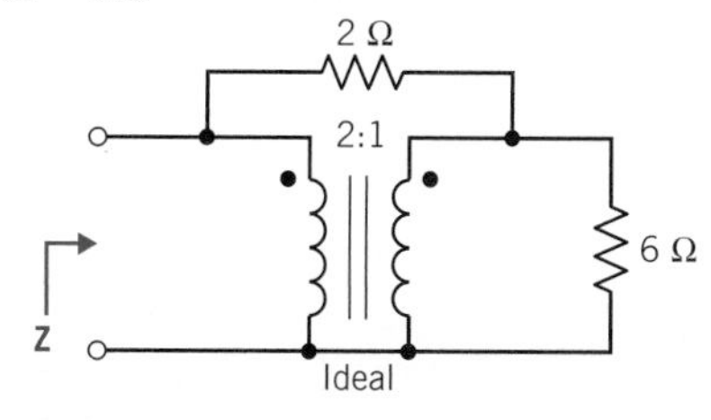

Figure P 11.10-7

P 11.10-8 In less developed regions in mountainous areas, small hydroelectric generators are used to serve several residences (Mackay, 1990). Assume each house uses an electric range and an electric refrigerator, as shown in Figure P 11.10-8. The generator is represented as $\mathbf{V}_s$ operating at 60 Hz and $\mathbf{V}_2 = 230\angle 0^\circ$V. Calculate the power consumed by each home connected to the hydroelectric generator when $n = 5$.

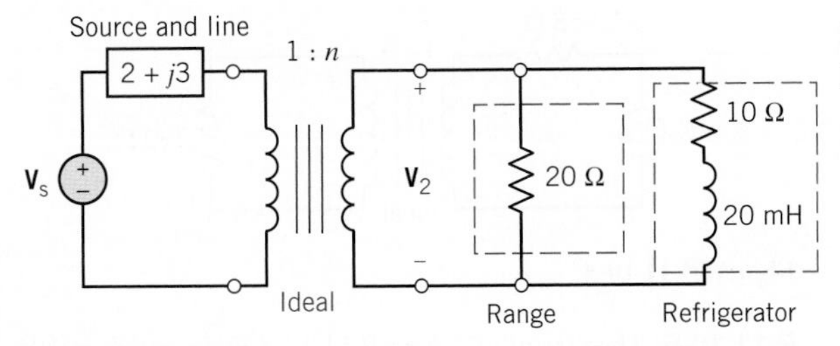

Figure P 11.10-8

P 11.10-9 Three similar circuits are shown in Figure P 11.10-9. In each of these circuits, $v_s(t) = 5\cos(4t + 45°)$ V. Determine $v_2(t)$ for each of the three circuits.

Answers: **(a)** $v_2(t) = 0$ V
(b) $v_2(t) = 1.656\cos(4t + 39°)$ V
(c) $v_2(t) = 2.88\cos(4t + 45°)$ V

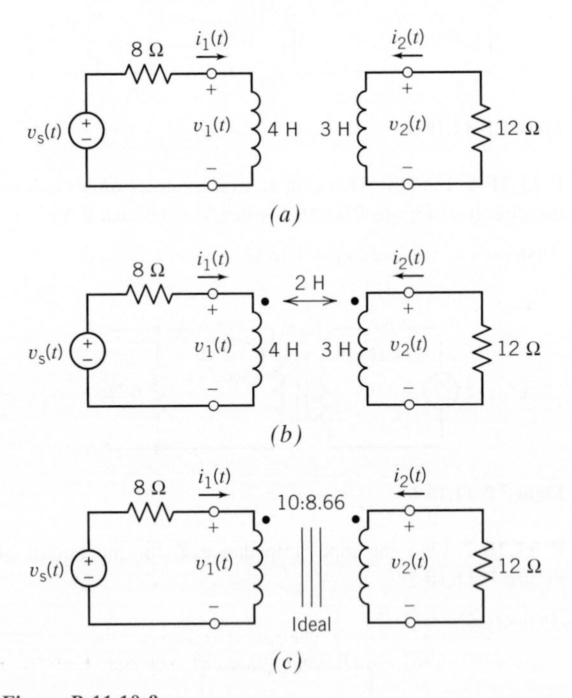

Figure P 11.10-9

P 11.10-10 Find $\mathbf{V}_1$ and $\mathbf{I}_1$ for the circuit of Figure P 11.10-10 when $n = 5$.

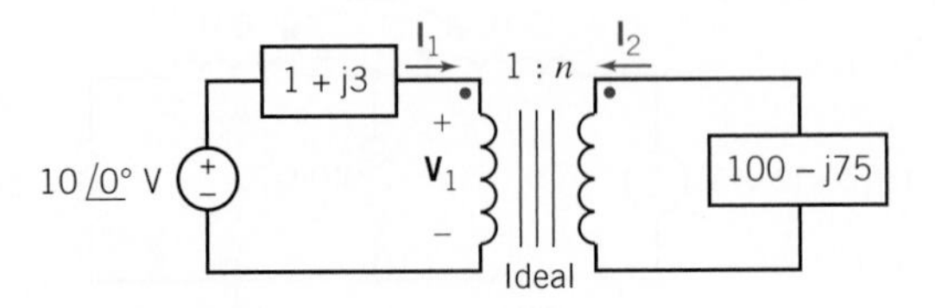

Figure P 11.10-10

P 11.10-11 Determine v_2 and i_2 for the circuit shown in Figure P 11.10-11 when $n = 2$. Note that i_2 does not enter the dotted terminal.

Answers: $v_2 = 6.08\cos(10t + 47.7°)$ V

$i_2 = 3.34\cos(10t + 42°)$ V

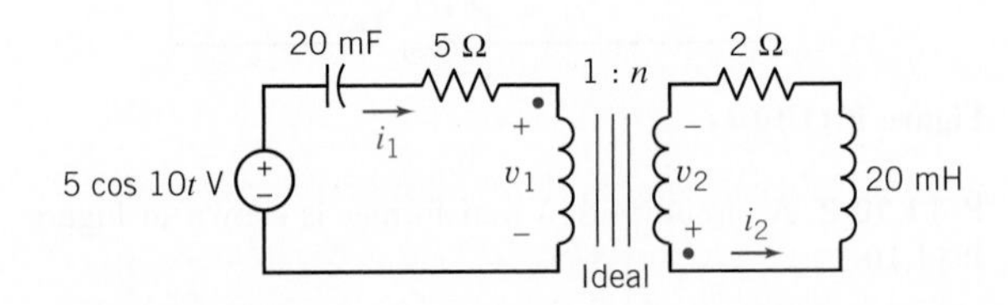

Figure P 11.10-11

P 11.10-12 The circuit shown in Figure P 11.10-12 is represented in the frequency domain. Given the line current $\mathbf{I}_{\text{Line}} = 0.5761\angle{-75.88°}$ A, determine P_{Source}, the average power supplied by the source; P_{Line}, the average power delivered to the line; and P_{Load}, the average power delivered to the load.

Hint: Use conservation of (average) power to check your answers.

Answer: $P_{\text{Source}} = 42.15$ W, $P_{\text{Line}} = 0.6638$ W, and $P_{\text{Laod}} = 41.49$ W

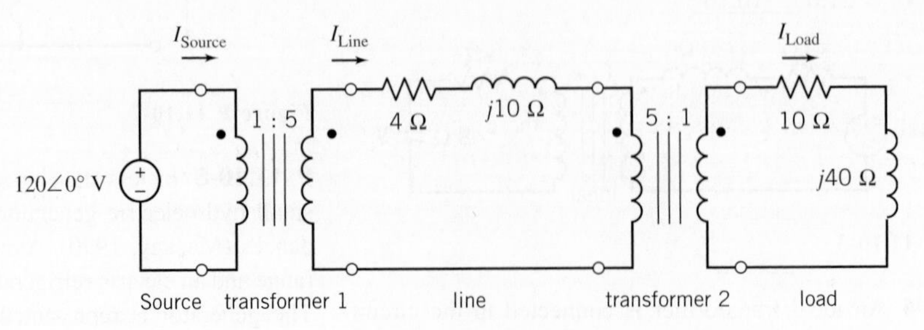

Figure P 11.10-12

P 11.10-13 The circuit shown in Figure P 11.10-13 is represented in the frequency domain. Determine R and X, the real and imaginary parts of the equivalent impedance, $\mathbf{Z}_{eq}$.

Answer: $R = 180\ \Omega$ and $X = 60\ \Omega$

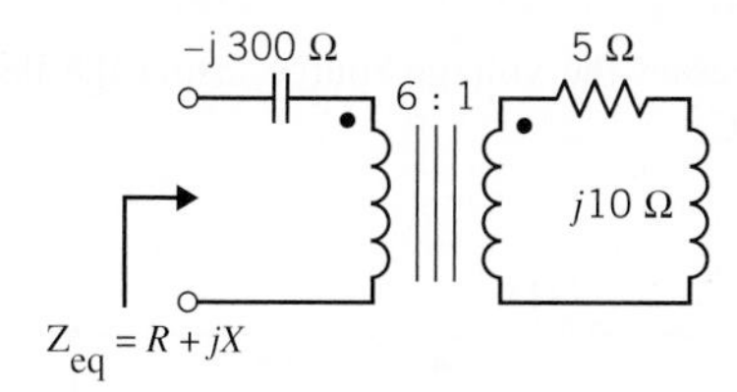

Figure P 11.10-13

Section 11.11 How Can We Check . . . ?

P 11.11-1 Computer analysis of the circuit shown in Figure P 11.11-1 indicates that when

$$v_s(t) = 12 \cos(4t + 30°) \text{ V}$$

the mesh currents are given by

$$i_1(t) = 2.327 \cos(4t - 25.22°) \text{ A}$$

and

$$i_2(t) = 1.129 \cos(4t - 11.19°) \text{ A}$$

Check the results of this analysis by checking that the average power supplied by the voltage source is equal to the sum of the average powers received by the other circuit elements.

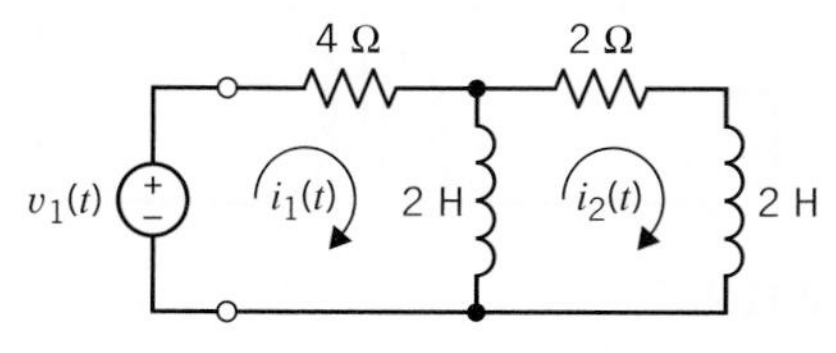

Figure P 11.11-1

P 11.11-2 Computer analysis of the circuit shown in Figure P 11.11-2 indicates that when

$$v_s(t) = 12 \cos(4t + 30°) \text{ V}$$

the mesh currents are given by

$$i_1(t) = 1.647 \cos(4t - 17.92°) \text{ A}$$

and

$$i_2(t) = 1.094 \cos(4t - 13.15°) \text{ A}$$

Check the results of this analysis by checking that the complex power supplied by the voltage source is equal to the sum of the complex powers received by the other circuit elements.

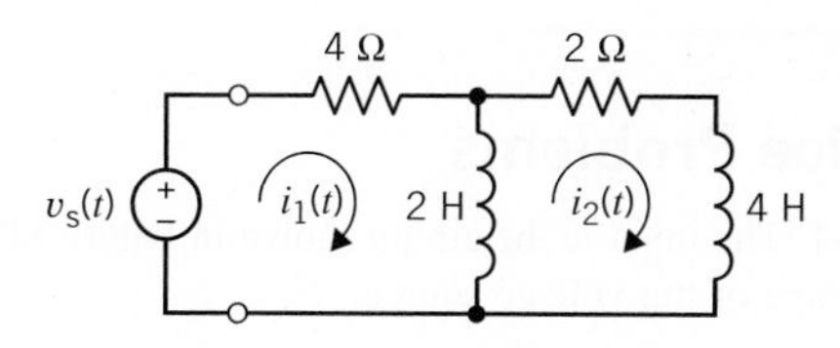

Figure P 11.11-2

P 11.11-3 Computer analysis of the circuit shown in Figure P 11.11-3 indicates that when

$$v_s(t) = 12 \cos(4t + 30°) \text{ V}$$

the mesh currents are given by

$$i_1(t) = 1.001 \cos(4t - 47.01°) \text{ A}$$

and

$$i_2(t) = 0.4243 \cos(4t - 15.00°) \text{ A}$$

Check the results of this analysis by checking that the equations describing currents and voltages of coupled coils are satisfied.

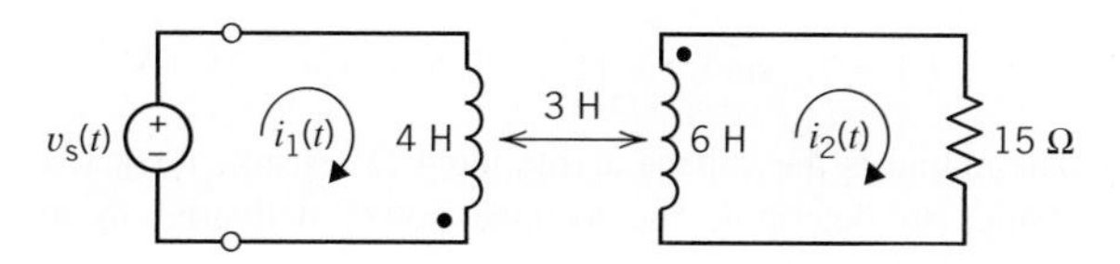

Figure P 11.11-3

P 11.11-4 Computer analysis of the circuit shown in Figure P 11.11-4 indicates that when

$$v_s(t) = 12 \cos(4t + 30°) \text{ V}$$

the mesh currents are given by

$$i_1(t) = 25.6 \cos(4t + 30°) \text{ mA}$$

and

$$i_2(t) = 64 \cos(4t + 30°) \text{ mA}$$

Check the results of this analysis by checking that the equations describing currents and voltages of ideal transformers are satisfied.

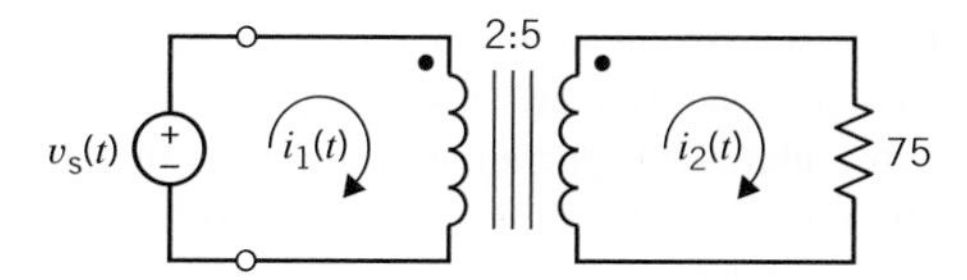

Figure P 11.11-4

PSpice Problems

SP 11-1 The input to the circuit shown in Figure SP 11-1 is the voltage of the voltage source

$$v_s(t) = 7.5 \sin(5t + 15°)\text{ V}$$

The output is the voltage across the 4-Ω resistor, $v_o(t)$. Use PSpice to plot the input and output voltages.

Hint: Represent the voltage source, using the PSpice part called VSIN.

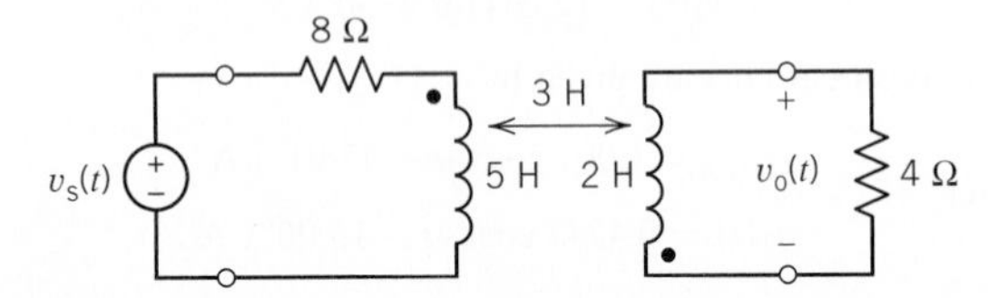

Figure SP 11-1

SP 11-2 The input to the circuit shown in Figure SP 11-1 is the voltage of the voltage source

$$v_s(t) = 7.5 \sin(5t + 15°) = 7.5 \cos(5t - 75°)\text{ V}$$

The output is the voltage across the 4-Ω resistor, $v_o(t)$. Use PSpice to determine the average power delivered to the coupled inductors.

Hint: Represent the voltage source, using the PSpice part called VAC. Use printers (PSpice parts called IPRINT and VPRINT) to measure the ac current and voltage of each coil.

SP 11-3 The input to the circuit shown in Figure SP 11-3 is the voltage of the voltage source,

$$v_s(t) = 48 \cos(4t + 114°)\text{ V}$$

The output is the voltage across the 9-Ω resistor, $v_o(t)$. Use PSpice to determine the average power delivered to the transformer.

Hint: Represent the voltage source, using the PSpice part called VAC.

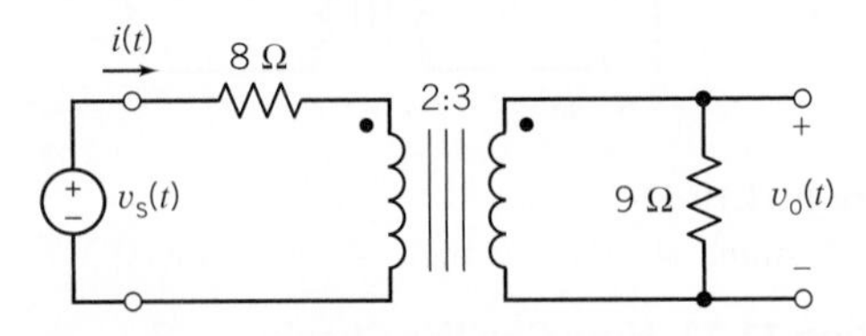

Figure SP 11-3

SP 11-4 Determine the value of the input impedance, $\mathbf{Z}_t$, of the circuit shown in Figure SP 11-4 at the frequency $\omega = 4$ rad/s.

Hint: Connect a current source across the terminals of the circuit. Measure the voltage across the current source. The value of impedance will be equal to the ratio of the voltage to the current.

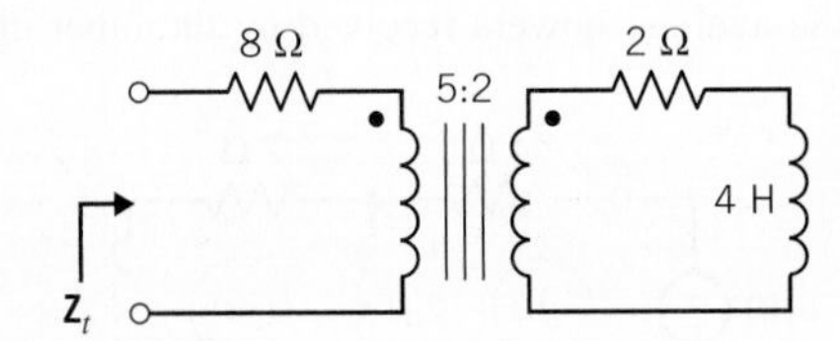

Figure SP 11-4

Design Problems

DP 11-1 A 100-kW induction motor, shown in Figure DP 11-1, is receiving 100 kW at 0.8 power factor lagging. Determine the additional apparent power in kVA that is made available by improving the power factor to (a) 0.95 lagging and (b) 1.0. (c) Find the required reactive power in kVAR provided by a set of parallel capacitors for parts (a) and (b). (d) Determine the ratio of kVA released to the kVAR of capacitors required for parts (a) and (b) alone. Set up a table, recording the results of this problem for the two values of power factor attained.

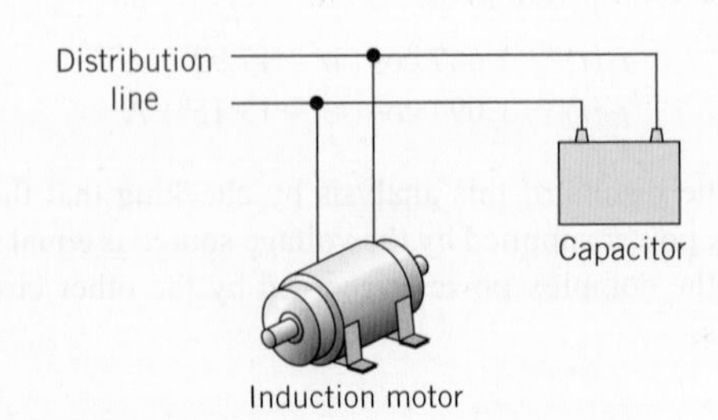

Figure DP 11-1 Induction motor with parallel capacitor.

DP 11-2 Two loads are connected in parallel and supplied from a 7.2-kV rms 60-Hz source. The first load is 50-kVA at 0.9 lagging power factor, and the second load is 45 kW at 0.91 lagging power factor. Determine the kVAR rating and capacitance required to correct the overall power factor to 0.97 lagging.

Answer: $C = 1.01\ \mu\text{F}$

DP 11-3

(a) Determine the load impedance $\mathbf{Z}_{ab}$ that will absorb maximum power if it is connected to terminals a–b of the circuit shown in Figure DP 11-3.
(b) Determine the maximum power absorbed by this load.
(c) Determine a model of the load and indicate the element values.

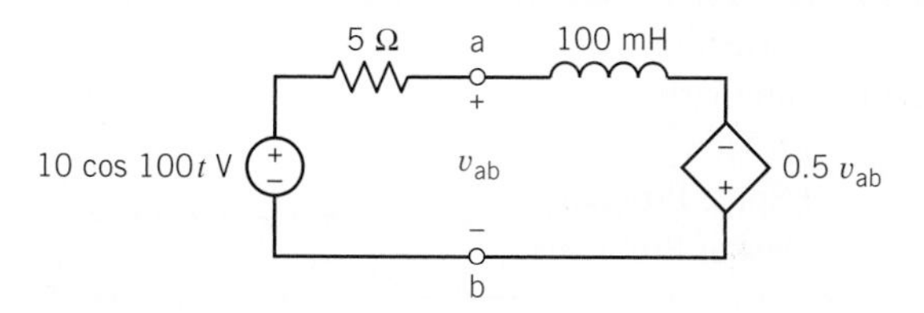

Figure DP 11-3

DP 11-4 Select the turns ratio n necessary to provide maximum power to the resistor R of the circuit shown in Figure DP 11-4. Assume an ideal transformer. Select n when $R = 4$ and 8 Ω.

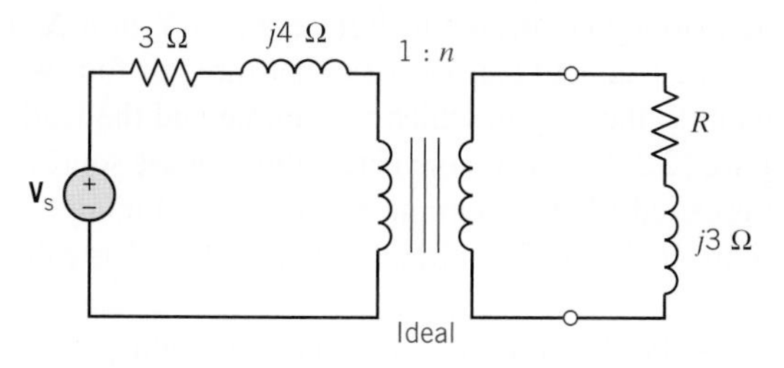

Figure DP 11-4

DP 11-5 An amplifier in a short-wave radio operates at 100 kHz. The load $\mathbf{Z}_2$ is connected to a source through an ideal transformer, as shown in Figure DP 11-5. The load is a series connection of a 10-Ω resistance and a 10-μH inductance. The $\mathbf{Z}_s$ consists of a 1-Ω resistance and a 1-μH inductance.

(a) Select an integer n to maximize the energy delivered to the load. Calculate $\mathbf{I}_2$ and the energy to the load.
(b) Add a capacitance C in series with $\mathbf{Z}_2$ to improve the energy delivered to the load.

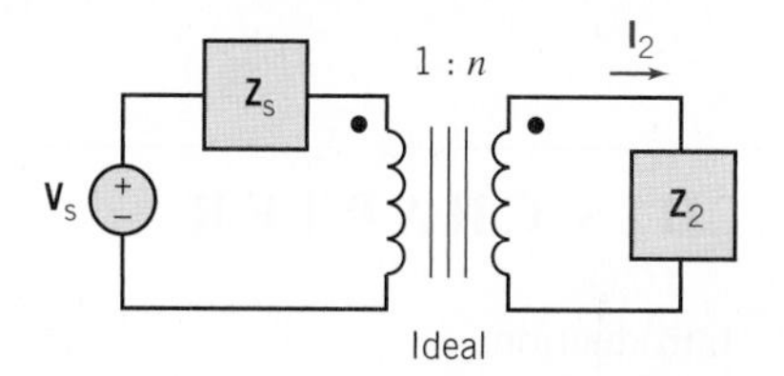

Figure DP 11-5

DP 11-6 A new electronic lamp (e-lamp) has been developed that uses a radio-frequency sinusoidal oscillator and a coil to transmit energy to a surrounding cloud of mercury gas as shown in Figure DP 11-6*a*. The mercury gas emits ultraviolet light that is transmitted to the phosphor coating, which, in turn, emits visible light. A circuit model of the e-lamp is shown in Figure DP 11-6*b*. The capacitance C and the resistance R are dependent on the lamp's spacing design and the type of phosphor. Select R and C so that maximum power is delivered to R, which relates to the phosphor coating (Adler, 1992). The circuit operates at $\omega_0 = 10^7$ rad/s.

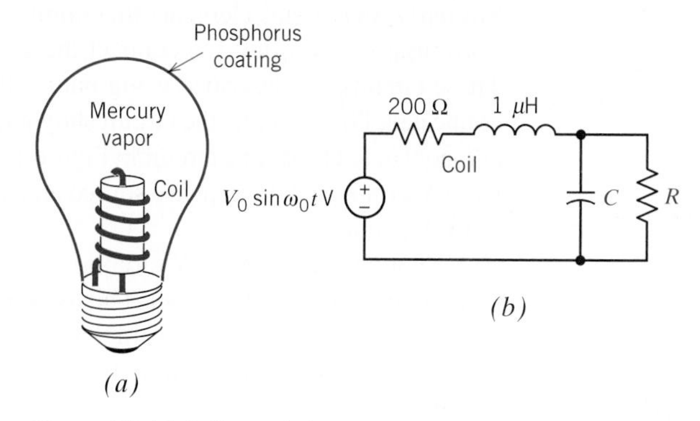

Figure DP 11-6 Electronic lamp.

CHAPTER 12

Three-Phase Circuits

IN THIS CHAPTER

12.1 Introduction
12.2 Three-Phase Voltages
12.3 The Y-to-Y Circuit
12.4 The Δ-Connected Source and Load
12.5 The Y-to-Δ Circuit
12.6 Balanced Three-Phase Circuits
12.7 Instantaneous and Average Power in a Balanced Three-Phase Load
12.8 Two-Wattmeter Power Measurement
12.9 How Can We Check . . . ?
12.10 **DESIGN EXAMPLE**—Power Factor Correction
12.11 Summary
Problems
PSpice Problems
Design Problems

12.1 INTRODUCTION

In this chapter, we will begin to analyze *three-phase circuits*. These circuits consist of three parts: a three-phase source, a three-phase load, and a transmission line. The three-phase source consists of either three Y-connected sinusoidal voltage sources or three Δ-connected sinusoidal voltage sources. Similarly, the circuit elements that comprise the load are connected to form either a Y or a Δ. The transmission line is used to connect the source to the load and consists of either three or four wires. These circuits are described using names that identify the way in which the source and the load are connected. For example, the circuit shown in Figure 12.3-1 has a Y-connected, three-phase source and a Y-connected load. The circuit in Figure 12.3-1 is called a Y-to-Y circuit. The circuit in Figure 12.5-1 has a Y-connected three-phase source and a Δ-connected load. The circuit in Figure 12.5-1 is called a Y-to-Δ circuit.

Notice that the Y-to-Y circuit in Figure 12.3-1 has been represented in the frequency domain, using impedances and phasors. This is appropriate because the three voltage sources that comprise a three-phase source are sinusoidal sources having the *same frequency*. Analysis of three-phase circuits using phasors and impedances will determine the *steady-state response* of the three-phase circuit.

Before beginning our analysis of three-phase circuits, it is helpful to recall why it is advantageous to use phasors to find the steady-state response of linear circuits to sinusoidal inputs. Circuits that contain capacitors or inductors are represented by differential equations in the time domain. We can solve these *differential equations*, but it is a lot of work. Impedances and phasors represent the circuit in the frequency domain. Linear circuits are represented by *algebraic equations* in the frequency domain. These algebraic equations involve complex numbers, but they are still easier to solve than the differential equations. Solving these algebraic equations provides the phasor corresponding to the output voltage or current. We know that the steady-state output voltage or current will be sinusoidal and will have the same frequency as the input sinusoid. The magnitude and phase angle of the phasor corresponding to the output voltage or current provide the magnitude and phase angle of the output sinusoid.

Table 12.1-1 Frequency Domain Power Relationships

QUANTITY	RELATIONSHIP USING PEAK VALUES	RELATIONSHIP USING RMS VALUES	UNITS
Element voltage, $v(t)$	$v(t) = V_m \cos(\omega t + \theta_V)$	$v(t) = V_{rms}\sqrt{2}\cos(\omega t + \theta_V)$	V
Element current, $i(t)$	$i(t) = I_m \cos(\omega t + \theta_I)$	$i(t) = I_{rms}\sqrt{2}\cos(\omega t + \theta_I)$	A
Complex power, **S**	$\mathbf{S} = \frac{V_m I_m}{2}\cos(\theta_V - \theta_I) + j\frac{V_m I_m}{2}\sin(\theta_v - \theta_I)$	$\mathbf{S} = V_{rms} I_{rms}\cos(\theta_V - \theta_I) + jV_{rms} I_{rms}\sin(\theta_V - \theta_I)$	VA
Apparent power, $\lvert\mathbf{S}\rvert$	$\lvert\mathbf{S}\rvert = \frac{V_m I_m}{2}$	$\lvert\mathbf{S}\rvert = V_{rms} I_{rms}$	VA
Average power, P	$P = \frac{V_m I_m}{2}\cos(\theta_V - \theta_I)$	$P = V_{rms} I_{rms}\cos(\theta_V - \theta_I)$	W
Reactive power, Q	$Q = \frac{V_m I_m}{2}\sin(\theta_V - \theta_I)$	$Q = V_{rms} I_{rms}\sin(\theta_V - \theta_I)$	VAR

We will be particularly interested in the power the three-phase source delivers to the three-phase load. Table 12.1-1 summarizes the formulas that can be used to calculate the power delivered to an element when the element voltage and current adhere to the passive convention. Table 12.1-1 also provides the equations for the sinusoidal element current and voltage. In the table, I_m and V_m are the magnitudes of the sinusoidal current and voltage, whereas I_{rms} and V_{rms} are the corresponding effective values of the current and voltage. Notice that the formulas for power in terms of I_{rms} and V_{rms} are simpler than the corresponding formulas in terms of I_m and V_m. In contrast, the equations giving the sinusoidal voltage and current are simpler when I_m and V_m are used. When engineers are interested primarily in power, they are likely to use I_{rms} and V_{rms}. On the other hand, when engineers are interested primarily in the sinusoidal currents and voltages, they are likely to use I_m and V_m. In this chapter, we are interested mainly in power and will use effective values.

12.2 THREE-PHASE VOLTAGES

The generation and transmission of electrical power are more efficient in polyphase systems employing combinations of two, three, or more sinusoidal voltages. In addition, polyphase circuits and machines possess some unique advantages. For example, the power transmitted in a three-phase circuit is constant or independent of time rather than pulsating, as it is in a single-phase circuit. In addition, three-phase motors start and run much better than do single-phase motors. The most common form of polyphase system employs three balanced voltages, equal in magnitude and differing in phase by $360°/3 = 120°$.

An elementary ac generator consists of a rotating magnet and a stationary winding. The turns of the winding are spread along the periphery of the machine. The voltage generated in each turn of the winding is slightly out of phase with the voltage generated in its neighbor because it is cut by maximum magnetic flux density an instant earlier or later. The voltage produced in the first winding is $v_{aa'}$.

If the first winding were continued around the machine, the voltage generated in the last turn would be 180° out of phase with that in the first, and they would cancel, producing no useful effect. For this reason, one winding is commonly spread over no more than one-third of the periphery; the other two-thirds of the periphery can hold two more windings used to generate two other similar voltages. A simplified version of three windings around the periphery of a cylindrical drum is shown in

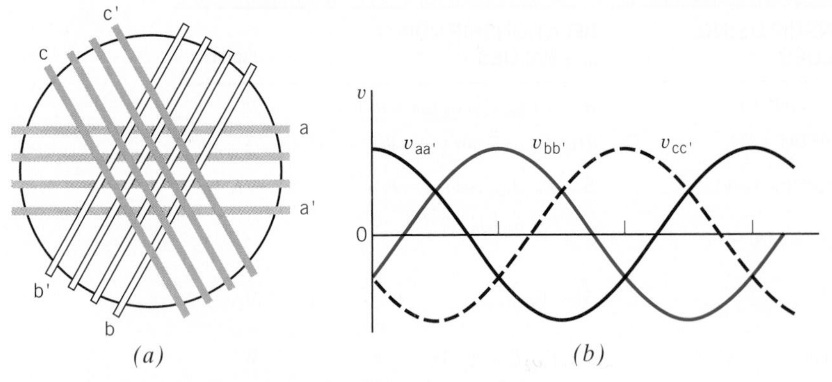

FIGURE 12.2-1 (*a*) The three windings on a cylindrical drum used to obtain three-phase voltages (end view). (*b*) Balanced three-phase voltages.

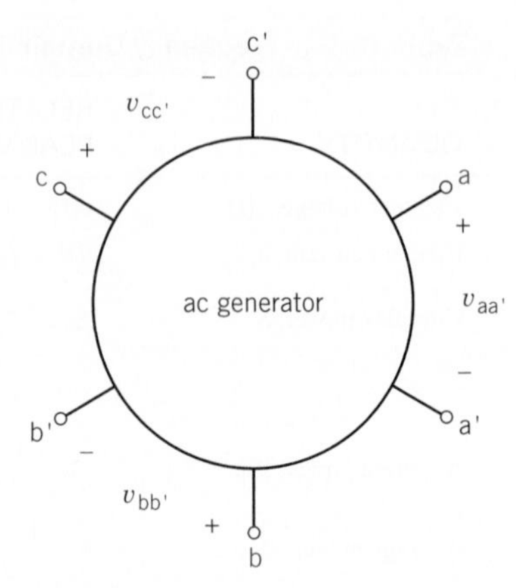

FIGURE 12.2-2 Generator with six terminals.

Figure 12.2-1*a*. The three sinusoids (sinusoids are obtained with a proper winding distribution and magnet shape) generated by the three similar windings are shown in Figure 12.2-1*b*. Defining $v_{aa'}$ as the potential of terminal a with respect to terminal a′, we describe the voltages as

$$\begin{aligned} v_{aa'} &= V_m \cos \omega t \\ v_{bb'} &= V_m \cos (\omega t - 120^\circ) \\ v_{cc'} &= V_m \cos (\omega t - 240^\circ) \end{aligned} \tag{12.2-1}$$

where V_m is the peak value.

A **three-phase circuit** generates, distributes, and uses energy in the form of three voltages equal in magnitude and symmetric in phase.

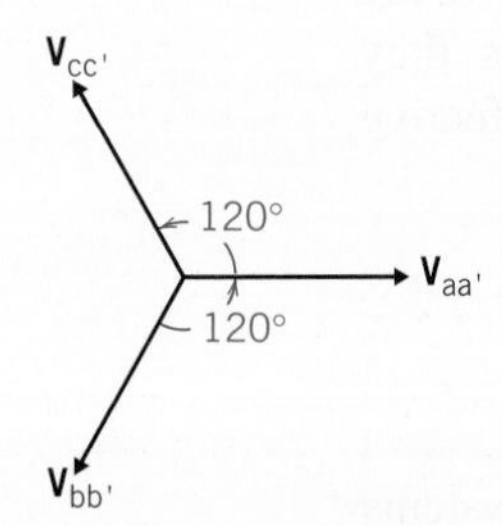

FIGURE 12.2-3 Phasor representation of the positive phase sequence of the balanced three-phase voltages.

The three similar portions of a three-phase system are called *phases*. Because the voltage in phase aa′ reaches its maximum first, followed by that in phase bb′ and then by that in phase cc′, we say the phase rotation is *abc*. This is an arbitrary convention; for any given generator, the phase rotation may be reversed by reversing the direction of rotation. The six-terminal ac generator is shown in Figure 12.2-2.

Using phasor notation, we may write Eq. 12.2-1 as

$$\begin{aligned} \mathbf{V}_{aa'} &= V_m \angle 0^\circ \\ \mathbf{V}_{bb'} &= V_m \angle -120^\circ \\ \mathbf{V}_{cc'} &= V_m \angle -240^\circ = V_m \angle 120^\circ \end{aligned} \tag{12.2-2}$$

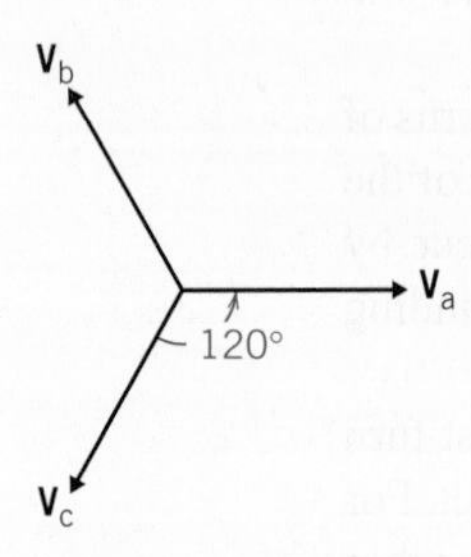

FIGURE 12.2-4 The negative phase sequence acb in the Y connection.

The three voltages are said to be *balanced voltages* because they have identical amplitude, V_m, and frequency, ω, and are out of phase with each other by exactly 120°. The phasor diagram of the balanced three-phase voltages is shown in Figure 12.2-3. Examining Figure 12.2-3, we find

$$\mathbf{V}_{aa'} + \mathbf{V}_{bb'} + \mathbf{V}_{cc'} = 0 \tag{12.2-3}$$

For notational ease, we henceforth use $\mathbf{V}_{aa'} = \mathbf{V}_a$, $\mathbf{V}_{bb'} = \mathbf{V}_b$, and $\mathbf{V}_{cc'} = \mathbf{V}_c$ as the three voltages.

The **positive phase sequence** is *abc*, as shown in Figure 12.2-3. The sequence *acb* is called the negative phase sequence, as shown in Figure 12.2-4.

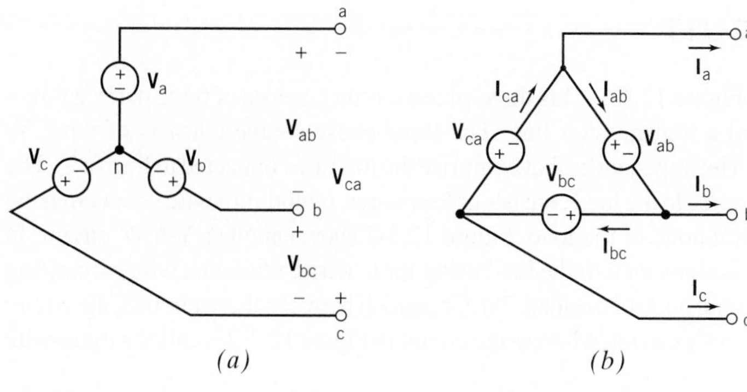

FIGURE 12.2-5 (*a*) Y-connected sources. The voltages $\mathbf{V}_a$, $\mathbf{V}_b$, and $\mathbf{V}_c$ are called phase voltages, and the voltages $\mathbf{V}_{ab}$, $\mathbf{V}_{bc}$, and $\mathbf{V}_{ca}$ are called line-to-line voltages, (*b*) Δ-connected sources. The currents $\mathbf{I}_a$, $\mathbf{I}_b$, and $\mathbf{I}_c$ are called line currents, and the currents $\mathbf{I}_{ab}$, $\mathbf{I}_{bc}$, and $\mathbf{I}_{ca}$ are called phase currents.

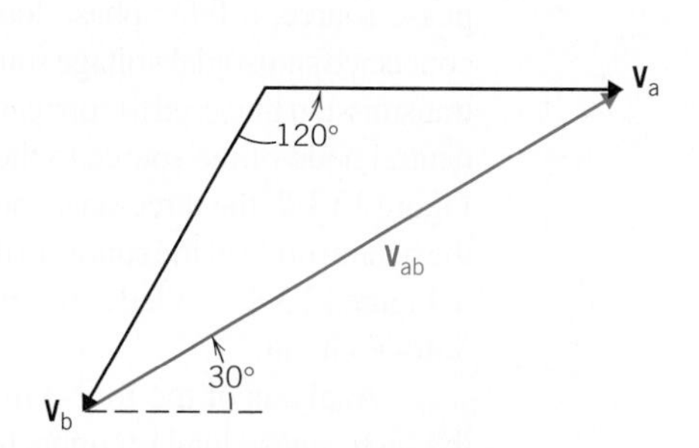

FIGURE 12.2-6 The line-to-line voltage $\mathbf{V}_{ab}$ of the Y-connected source.

Often, the phase voltage in the Y connection is written as

$$\mathbf{V}_a = V_m \angle 0^\circ$$

where V_m is the magnitude of the phase voltage.

Referring to the generator of Figure 12.2-2, there are six terminals and three voltages, v_a, v_b, and v_c. We use phasor notation and assume that each phase winding provides a source voltage in series with a negligible impedance. Under these assumptions, there are two ways of interconnecting the three sources, as shown in Figure 12.2-5. The common terminal of the Y connection is called the *neutral terminal* and is labeled *n*. The neutral terminal may or may not be available for connection. Balanced loads result in no current in a neutral wire, and thus it is often not needed.

The connection shown in Figure 12.2-5*a* is called the Y connection, and the Δ connection is shown in Figure 12.2-5*b*. The Y connection selects terminals a', b', and c' and connects them together as neutral. Then the line-to-line voltage, $\mathbf{V}_{ab}$, of the Y-connected sources is

$$\mathbf{V}_{ab} = \mathbf{V}_a - \mathbf{V}_b \tag{12.2-4}$$

as is evident by examining Figure 12.2-5*a*. Because $\mathbf{V}_a = V_m \angle 0^\circ$ and $\mathbf{V}_b = V_m \angle -120^\circ$, we have

$$\begin{aligned} \mathbf{V}_{ab} &= V_m - V_m(-0.5 - j0.866) \\ &= V_m(1.5 + j0.866) \\ &= \sqrt{3}\, V_m \angle 30^\circ \end{aligned} \tag{12.2-5}$$

This relationship is also demonstrated by the phasor diagram of Figure 12.2-6. Similarly,

$$\mathbf{V}_{bc} = \sqrt{3}\, V_m \angle -90^\circ \tag{12.2-6}$$

and

$$\mathbf{V}_{ca} = \sqrt{3}\, V_m \angle -210^\circ \tag{12.2-7}$$

Therefore, in a Y connection, the line-to-line voltage is $\sqrt{3}$ times the phase voltage and is displaced 30° in phase. The line current is equal to the phase current.

EXERCISE 12.2-1 The Y-connected three-phase voltage source has $\mathbf{V}_c = 120 \angle -240^\circ$ V rms. Find the line-to-line voltage $\mathbf{V}_{bc}$.

Answer: $207.8 \angle -90^\circ$ V rms

12.3 THE Y-TO-Y CIRCUIT

Consider the Y-to-Y circuit shown in Figure 12.3-1. This three-phase circuit consists of three parts: a three-phase source, a three-phase load, and a transmission line. The three-phase source consists of three Y-connected sinusoidal voltage sources. The impedances that comprise the load are connected to form a Y. The transmission line used to connect the source to the load consists of four wires, including a wire connecting the neutral node of the source to the neutral node of the load. Figure 12.3-2 shows another Y-to-Y circuit. In Figure 12.3-2, the three-phase source is connected to the load using three wires, without a wire connecting the neutral node of the source to the neutral node of the load. To distinguish between these circuits, the circuit in Figure 12.3-1 is called a four-wire Y-to-Y circuit, whereas the circuit in Figure 12.3-2 is called a three-wire Y-to-Y circuit.

Analysis of the four-wire Y-to-Y circuit in Figure 12.3-1 is relatively easy. Each impedance of the three-phase load is connected directly across a voltage source of the three-phase source. Therefore, the voltage across the impedance is known, and the line currents are easily calculated as

$$\mathbf{I}_{aA} = \frac{\mathbf{V}_a}{\mathbf{Z}_A}, \quad \mathbf{I}_{bB} = \frac{\mathbf{V}_b}{\mathbf{Z}_B}, \quad \text{and} \quad \mathbf{I}_{cC} = \frac{\mathbf{V}_c}{\mathbf{Z}_C} \tag{12.3-1}$$

The current in the wire connecting the neutral node of the source to the neutral node of the load is

$$\mathbf{I}_{Nn} = \mathbf{I}_{aA} + \mathbf{I}_{bB} + \mathbf{I}_{cC} = \frac{\mathbf{V}_a}{\mathbf{Z}_A} + \frac{\mathbf{V}_b}{\mathbf{Z}_B} + \frac{\mathbf{V}_c}{\mathbf{Z}_C} \tag{12.3-2}$$

The average power delivered by the three-phase source to the three-phase load is calculated by adding up the average power delivered to each impedance of the load.

$$P = P_A + P_B + P_C \tag{12.3-3}$$

where, for example, P_A is the average power absorbed by $\mathbf{Z}_A$. P_A is easily calculated once $\mathbf{I}_{aA}$ is known.

For convenience, let the phase voltages of the Y-connected source be

$$\mathbf{V}_a = V_p \angle 0^\circ \text{ V rms}, \quad \mathbf{V}_b = V_p \angle -120^\circ \text{ V rms}, \quad \text{and} \quad \mathbf{V}_c = V_p \angle 120^\circ \text{ V rms}$$

Notice that we are using effective values because the units of V_p are V rms.

When $\mathbf{Z}_A = \mathbf{Z}_B = \mathbf{Z}_C = \mathbf{Z} = Z \angle \theta$, the load is said to be a *balanced load*. In general, analysis of balanced three-phase circuits is easier than analysis of unbalanced three-phase circuits. The line currents in the balanced, four-wire Y-to-Y circuit are given by

$$\mathbf{I}_{aA} = \frac{\mathbf{V}_a}{\mathbf{Z}} = \frac{V_p \angle 0^\circ}{Z \angle \theta}, \quad \mathbf{I}_{bB} = \frac{\mathbf{V}_b}{\mathbf{Z}} = \frac{V_p - \angle 120^\circ}{Z \angle \theta}, \quad \text{and} \quad \mathbf{I}_{cC} = \frac{\mathbf{V}_c}{\mathbf{Z}} = \frac{V_p \angle 120^\circ}{Z \angle \theta}$$

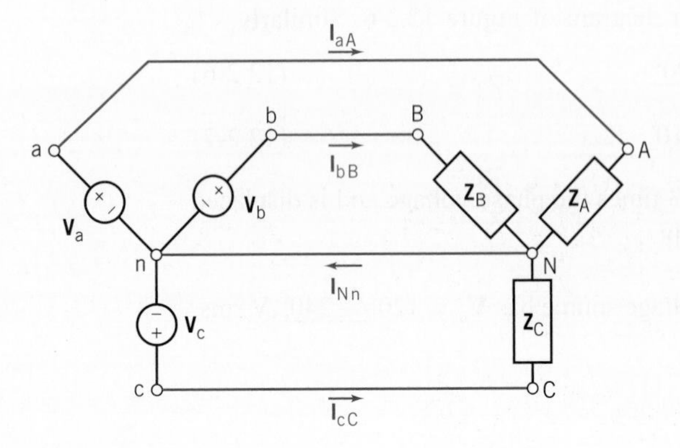

FIGURE 12.3-1 A four-wire Y-to-Y circuit.

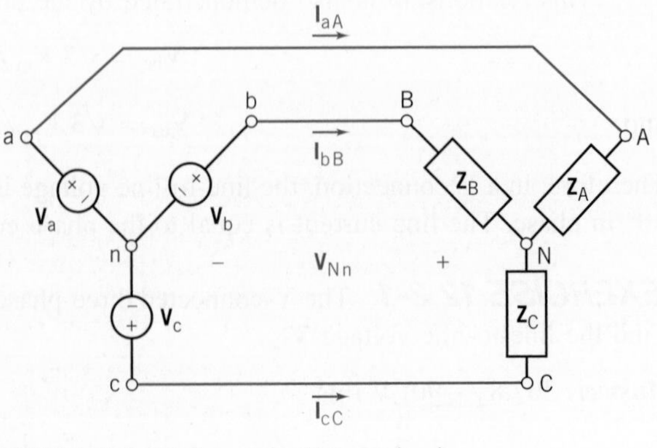

FIGURE 12.3-2 A three-wire Y-to-Y circuit.

Then

$$\mathbf{I}_{aA} = \frac{V_p}{Z}\angle{-\theta}, \quad \mathbf{I}_{bB} = \frac{V_p}{Z}\angle{-\theta - 120°}, \quad \text{and} \quad \mathbf{I}_{cC} = \frac{V_p}{Z}\angle{-\theta + 120°} \tag{12.3-4}$$

The line currents have equal magnitudes and differ in phase by 120°. $\mathbf{I}_{bB}$ and $\mathbf{I}_{cC}$ can be calculated from $\mathbf{I}_{aA}$ by subtracting and adding 120° to the phase angle of $\mathbf{I}_{aA}$.

The current in the wire connecting the neutral node of the source to the neutral node of the load is

$$\mathbf{I}_{Nn} = \mathbf{I}_{aA} + \mathbf{I}_{bB} + \mathbf{I}_{cC} = \frac{V_p}{Z\angle\theta}(\angle{10°} + 1\angle{-120°} + 1\angle{120°})$$

$$\mathbf{I}_{Nn} = 0 \tag{12.3-5}$$

There is no current in the wire connecting the neutral node of the source to the neutral node of the load.

Because effective, or rms, values of the sinusoidal voltages and currents have been used instead of peak values, the appropriate formulas for power are those given in the ''rms values'' column of Table 12.1-1. The average power delivered to the load is

$$P = P_A + P_B + P_C = V_p\frac{V_p}{Z}\cos(-\theta) + V_p\frac{V_p}{Z}\cos(-\theta) + V_p\frac{V_p}{Z}\cos(-\theta)$$

$$P = 3\frac{V_p^2}{Z}\cos(\theta) \tag{12.3-6}$$

where, for example, P_A is the average power absorbed by $\mathbf{Z}_A$. Equal power is absorbed by each impedance of the three-phase load, $\mathbf{Z}_A$, $\mathbf{Z}_B$, and $\mathbf{Z}_C$. It is not necessary to calculate P_A, P_B, and P_C separately. The average power delivered to the load can be determined by calculating P_A and multiplying by 3.

Next, consider the three-wire Y-to-Y circuit shown in Figure 12.3-2. The phase voltages of the Y-connected source are $\mathbf{V}_a = V_p\angle{0°}$ V rms, $\mathbf{V}_b = V_p\angle{-120°}$ V rms, and $\mathbf{V}_c = V_p\angle{120°}$ V rms. The first step in the analysis of this circuit is to calculate $\mathbf{V}_{Nn}$, the voltage at the neutral node of the three-phase load with respect to the voltage at the neutral node of the three-phase source. (This step wasn't needed when the four-wire Y-to-Y circuit was analyzed because the fourth wire forced $\mathbf{V}_{Nn} = 0$.) It is convenient to select node n, the neutral node of the three-phase source, to be the reference node. Then $\mathbf{V}_a$, $\mathbf{V}_b$, $\mathbf{V}_c$, and $\mathbf{V}_{Nn}$ are the node voltages of the circuit. Write a node equation at node N to get

$$\begin{aligned} 0 &= \frac{\mathbf{V}_a - \mathbf{V}_{Nn}}{\mathbf{Z}_A} + \frac{\mathbf{V}_b - \mathbf{V}_{Nn}}{\mathbf{Z}_B} + \frac{\mathbf{V}_c - \mathbf{V}_{Nn}}{\mathbf{Z}_C} \\ &= \frac{(V_p\angle{0°}) - \mathbf{V}_{Nn}}{\mathbf{Z}_A} + \frac{(V_p\angle{-120°}) - \mathbf{V}_{Nn}}{\mathbf{Z}_B} + \frac{(V_p\angle{120°}) - \mathbf{V}_{Nn}}{\mathbf{Z}_C} \end{aligned} \tag{12.3-7}$$

Solving for $\mathbf{V}_{Nn}$ gives

$$\mathbf{V}_{Nn} = \frac{(V_p\angle{-120°})\mathbf{Z}_A\mathbf{Z}_C + (V_p\angle{120°})\mathbf{Z}_A\mathbf{Z}_B + (V_p\angle{0°})\mathbf{Z}_B\mathbf{Z}_C}{\mathbf{Z}_A\mathbf{Z}_C + \mathbf{Z}_A\mathbf{Z}_B + \mathbf{Z}_B\mathbf{Z}_C} \tag{12.3-8}$$

Once $\mathbf{V}_{Nn}$ has been determined, the line currents can be calculated using

$$\mathbf{I}_{aA} = \frac{\mathbf{V}_a - \mathbf{V}_{Nn}}{\mathbf{Z}_A}, \quad \mathbf{I}_{bB} = \frac{\mathbf{V}_b - \mathbf{V}_{Nn}}{\mathbf{Z}_B}, \text{ and } \mathbf{I}_{cC} = \frac{\mathbf{V}_c - \mathbf{V}_{Nn}}{\mathbf{Z}_C} \tag{12.3-9}$$

Analysis of the three-wire Y-to-Y circuit is much simpler when the circuit is balanced, that is, when $\mathbf{Z}_A = \mathbf{Z}_B = \mathbf{Z}_C = \mathbf{Z} = Z\angle\theta$. When the circuit is balanced, Eq. 12.3-8 becomes

$$\mathbf{V}_{Nn} = \frac{(V_P\angle{-120^\circ})\mathbf{ZZ} + (V_P\angle{120^\circ})\mathbf{ZZ} + (V_P\angle{0^\circ})\mathbf{ZZ}}{\mathbf{ZZ}+\mathbf{ZZ}+\mathbf{ZZ}}$$

$$= [(V_P\angle{-120^\circ}) + (V_P\angle{120^\circ}) + (V_P\angle{0^\circ})]/3$$

$$\mathbf{V}_{Nn} = 0 \tag{12.3-10}$$

When a three-wire Y-to-Y circuit is balanced, it is not necessary to write and solve a node equation to find $\mathbf{V}_{Nn}$ because $\mathbf{V}_{Nn}$ is known to be zero. Recall that $\mathbf{V}_{Nn} = 0$ in the four-wire Y-to-Y circuit. The balanced three-wire Y-to-Y circuit acts like the balanced four-wire Y-to-Y circuit. In particular, the line currents are given by Eq. 12.3-4, and the average power delivered to the load is given by Eq. 12.3-6.

Ideally, the transmission line connecting the load to the source can be modeled using short circuits. That's what was done in both Figure 12.3-1 and Figure 12.3-2. Sometimes it is appropriate to model the lines connecting the load to the source as impedances. For example, this is done when comparing the power that is delivered to the load to the power that is absorbed by the transmission line. Figure 12.3-3 shows a three-wire Y-to-Y circuit in which the transmission line is modeled by the line impedances $\mathbf{Z}_{aA}$, $\mathbf{Z}_{bB}$, and $\mathbf{Z}_{cC}$. The line impedances do not significantly complicate the analysis of the circuit because each line impedance is connected in series with a load impedance. After replacing series impedances by equivalent impedances, the analysis proceeds as before. If the circuit is not balanced, a node equation is written and solved to determine $\mathbf{V}_{Nn}$. Once $\mathbf{V}_{Nn}$ has been determined, the line currents can be calculated. Both the power delivered to the load and the power absorbed by the line can be calculated from the line currents and the load and line impedances.

Analysis of balanced Y-to-Y circuits is simpler than analysis of unbalanced Y-to-Y circuits in several ways:

1. $\mathbf{V}_{Nn} = 0$. It is not necessary to write and solve a node equation to determine $\mathbf{V}_{Nn}$.
2. The line currents have equal magnitudes and differ in phase by 120°. $\mathbf{I}_{bB}$ and $\mathbf{I}_{cC}$ can be calculated from $\mathbf{I}_{aA}$ by subtracting and adding 120° to the phase angle of $\mathbf{I}_{aA}$.
3. Equal power is absorbed by each impedance of the three-phase load, $\mathbf{Z}_A$, $\mathbf{Z}_B$, and $\mathbf{Z}_C$. It is not necessary to calculate P_A, P_B, and P_C separately. The average power delivered to the load can be determined by calculating P_A and multiplying by 3.

The key to analysis of the balanced Y-to-Y circuit is calculation of the line current, $\mathbf{I}_{aA}$. The *per-phase equivalent circuit* provides the information needed to the line current, $\mathbf{I}_{aA}$. This equivalent circuit consists of the voltage source and impedances in one phase of the three phases of the three-phase circuit. Figure 12.3-4 shows the per-phase equivalent circuit corresponding to the three-phase circuit shown in Figure 12.3-3. The

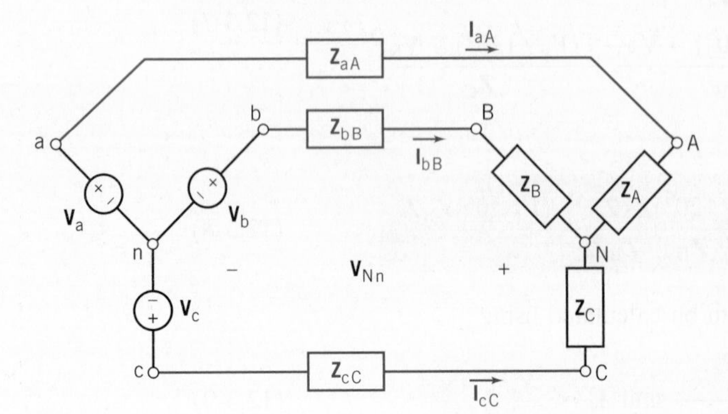

FIGURE 12.3-3 A three-wire Y-to-Y circuit with line impedances.

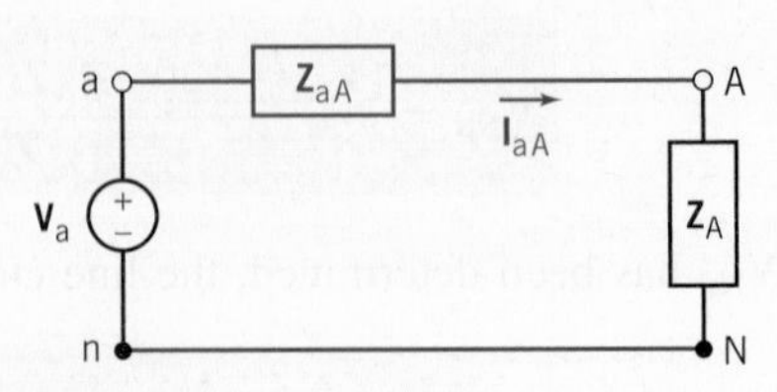

FIGURE 12.3-4 Per-phase equivalent circuit for the three-wire Y-to-Y circuit with line impedances.

Table 12.3-1 The Balanced Y-to-Y Circuit

Phase voltages	$\mathbf{V}_a = V_p\angle 0^\circ$
	$\mathbf{V}_b = V_p\angle -120^\circ$
	$\mathbf{V}_c = V_p\angle -240^\circ$
Line-to-line voltages	$\mathbf{V}_{ab} = \sqrt{3}\, V_p\angle 30^\circ$
	$\mathbf{V}_{bc} = \sqrt{3}\, V_p\angle -90^\circ$
	$\mathbf{V}_{ca} = \sqrt{3}\, V_p\angle -210^\circ$
	$V_L = \sqrt{3}\, V_p$
Currents	$\mathbf{I}_L = \mathbf{I}_p$ (line current = phase current)
	$\mathbf{I}_A = \dfrac{\mathbf{V}_a}{\mathbf{Z}_Y} = I_p\angle -\theta$ with $\mathbf{Z}_p = Z\angle\theta$
	$\mathbf{I}_B = \mathbf{I}_A\angle -120^\circ$
	$\mathbf{I}_C = \mathbf{I}_A\angle -240^\circ$

Note: p = phase, L = line.

neutral nodes, n and N, are connected by a short circuit in the per-phase equivalent circuit to indicate that $\mathbf{V}_{Nn} = 0$ in a balanced Y-to-Y circuit. The per-phase equivalent circuit can be used to analyze either three-wire or four-wire balanced Y-to-Y circuits, but it can be used only for *balanced* circuits.

The behavior of a balanced Y-to-Y circuit is summarized in Table 12.3-1.

EXAMPLE 12.3-1 Four-wire Unbalanced Y-Y Circuit

Determine the complex power delivered to the three-phase load of a four-wire Y-to-Y circuit such as the one shown in Figure 12.3-1. The phase voltages of the Y-connected source are $\mathbf{V}_a = 110\angle 0^\circ$ V rms, $\mathbf{V}_b = 110\angle -120^\circ$ V rms, and $\mathbf{V}_c = 110\angle 120^\circ$ V rms. The load impedances are $\mathbf{Z}_A = 50 + j80\ \Omega$, $\mathbf{Z}_B = j50\ \Omega$, and $\mathbf{Z}_C = 100 + j25\ \Omega$.

Solution

The line currents of an *unbalanced* four-wire Y-to-Y circuit are calculated using Eq. 12.3-1. In this example,

$$\mathbf{I}_{aA} = \frac{\mathbf{V}_a}{\mathbf{Z}_A} = \frac{110\angle 0^\circ}{50 + j80},\quad \mathbf{I}_{bB} = \frac{\mathbf{V}_b}{\mathbf{Z}_B} = \frac{110\angle -120^\circ}{j50},\quad \text{and}\quad \mathbf{I}_{cC} = \frac{\mathbf{V}_c}{\mathbf{Z}_C} = \frac{110\angle 120^\circ}{100 + j25}$$

so

$$\mathbf{I}_{aA} = 1.16\angle -58^\circ \text{ A rms},\quad \mathbf{I}_{bB} = 2.2\angle 150^\circ \text{ A rms},\quad \text{and}\quad \mathbf{I}_{cC} = 1.07\angle 106^\circ \text{ A rms}$$

The complex power delivered to $\mathbf{Z}_A$ is

$$\mathbf{S}_A = \mathbf{I}_{aA}^{*}\,\mathbf{V}_a = (1.16\angle -58^\circ)^{*}\,(110\angle 0^\circ) = (1.16\angle 58^\circ)(110\angle 0^\circ) = 68 + j109 \text{ VA}$$

Similarly, we calculate the complex power delivered to $\mathbf{Z}_B$ and $\mathbf{Z}_C$ as

$$\mathbf{S}_B = (2.2\angle 150^\circ)^{*}\,(110\angle -120^\circ) = j242 \text{ VA}$$

and

$$\mathbf{S}_C = (107\angle 106^\circ)^{*}\,(110\angle 120^\circ) = 114 + j28 \text{ VA}$$

The total complex power delivered to the three-phase load is

$$\mathbf{S}_A + \mathbf{S}_B + \mathbf{S}_C = 182 + j379 \text{ VA}$$

Example 12.3-2 Four-wire Balanced Y-Y Circuit

Determine the complex power delivered to the three-phase load of a four-wire Y-to-Y circuit such as the one shown in Figure 12.3-1. The phase voltages of the Y-connected source are $\mathbf{V}_a = 110\angle 0°$ V rms, $\mathbf{V}_b = 110\angle -120°$ V rms, and $\mathrm{V}_c = 110\angle 120°$ V rms. The load impedances are $\mathbf{Z}_A = \mathbf{Z}_B = \mathbf{Z}_C = 50 + j80\ \Omega$.

Solution

This example is similar to the previous example. The important difference is that this three-phase circuit is balanced. We need to calculate only one line current, $\mathbf{I}_{aA}$, and the complex power, $\mathbf{S}_A$, delivered to only one of the load impedances, $\mathbf{Z}_A$. The power delivered to the three-phase load is $3\mathbf{S}_A$. We begin by calculating $\mathbf{I}_{aA}$ as

$$\mathbf{I}_{aA} = \frac{\mathbf{V}_a}{\mathbf{Z}_A} = \frac{110\angle 0°}{50 + j80} = 1.16\angle -58° \text{ A rms}$$

The complex power delivered to $\mathbf{Z}_A$ is

$$\mathbf{S}_A = \mathbf{I}_{aA}^* \mathbf{V}_a = (1.16\angle -58°)^* (110\angle 0°) = (1.16\angle 58°)(110\angle 0°) = 68 + j109 \text{ VA}$$

The total power delivered to the three-phase load is

$$3\mathbf{S}_A = 204 + j326 \text{ VA}$$

(The currents $\mathbf{I}_{bB}$ and $\mathbf{I}_{cC}$ can also be calculated using Eq. 12.3-1. Verify that $\mathbf{I}_{bB} = 1.16\angle -177°$ A rms and $\mathbf{I}_{cC} = 1.16\angle 62°$ A rms. Notice that $\mathbf{I}_{bB}$ and $\mathbf{I}_{cC}$ can be calculated from $\mathbf{I}_{aA}$ by subtracting and adding 120° to the phase angle of $\mathbf{I}_{aA}$. Also, check that the complex power delivered to $\mathbf{Z}_B$ and to $\mathbf{Z}_C$ is equal to the complex power delivered to $\mathbf{Z}_A$. That is, $\mathbf{S}_B = 68 + j109$ VA and $\mathbf{S}_C = 68 + j109$ VA.)

Example 12.3-3 Three-Wire Unbalanced Y-Y Circuit

Determine the complex power delivered to the three-phase load of a three-wire Y-to-Y circuit such as the one shown in Figure 12.3-2. The phase voltages of the Y-connected source are $\mathbf{V}_a = 110\angle 0°$ V rms, $\mathbf{V}_b = 110\angle -120°$ V rms, and $\mathbf{V}_c = 110\angle 120°$ V rms. The load impedances are $\mathbf{Z}_A = 50 + j80\ \Omega$, $\mathbf{Z}_B = j50\ \Omega$, and $\mathbf{Z}_C = 100 + j25\ \Omega$.

Solution

This example seems similar to Example 12.3-1 but considers a three-wire Y-to-Y circuit instead of the four-wire circuit considered in Example 12.3-1. Because the circuit is unbalanced, $\mathbf{V}_{Nn}$ is not known. We begin by writing and solving a node equation to determine $\mathbf{V}_{Nn}$. The solution of that node equation is given in Eq. 12.3-8 to be

$$\mathbf{V}_{Nn} = \frac{(110\angle -120°)(50 + j80)(100 + j25) + (110\angle 120°)(50 + j80)(j50) + (\angle 1100°)(j50)(100 + j25)}{(50 + j80)(100 + j25) + (50 + j80)(j50) + (j50)(100 + j25)}$$

$$= 56\angle -151° \text{ V rms}$$

Now that $\mathbf{V}_{Nn}$ is known, the line currents are calculated as

$$\mathbf{I}_{aA} = \frac{\mathbf{V}_a - \mathbf{V}_{Nn}}{\mathbf{Z}_A} = \frac{110\angle 0° - 56\angle -151°}{50 + j80} = 1.71\angle -48° \text{ A rms}$$

$$\mathbf{I}_{bB} = \frac{\mathbf{V}_b - \mathbf{V}_{Nn}}{\mathbf{Z}_B} = 2.45\angle 3° \text{ A rms} \quad \text{and} \quad \mathbf{I}_{cC} = \frac{\mathbf{V}_c - \mathbf{V}_{Nn}}{\mathbf{Z}_C} = 1.19\angle 79° \text{ A rms}$$

The complex power delivered to $\mathbf{Z}_A$ is

$$\mathbf{S}_A = \mathbf{I}_{aA}^* \mathbf{V}_a = \mathbf{I}_{aA}^* (\mathbf{I}_{aA} \mathbf{Z}_A) = (1.71 \angle -48°)^* (1.71 \angle -48°)(50 + j80) = 146 + j234 \text{ VA}$$

Similarly, we calculate the complex power delivered to $\mathbf{Z}_B$ and $\mathbf{Z}_C$ as

$$\mathbf{S}_B = \mathbf{I}_{bB}^* (\mathbf{I}_{bB} \mathbf{Z}_B) = j94 \text{ VA} \quad \text{and} \quad \mathbf{S}_C = \mathbf{I}_{cC}^* (\mathbf{I}_{cC} \mathbf{Z}_C) = 141 + j35 \text{ VA}$$

The total complex power delivered to the three-phase load is

$$\mathbf{S}_A + \mathbf{S}_B + \mathbf{S}_C = 287 + j364 \text{ VA}$$

EXAMPLE 12.3-4 Three-Wire Balanced Y-Y Circuit

Determine complex power delivered to the three-phase load of a three-wire Y-to-Y circuit such as the one shown in Figure 12.3-2. The phase voltages of the Y-connected source are $\mathbf{V}_a = 110 \angle 0°$ V rms, $\mathbf{V}_b = 110 \angle -120°$ V rms, and $\mathbf{V}_c = 110 \angle 120°$ V rms. The load impedances are $\mathbf{Z}_A = \mathbf{Z}_B = \mathbf{Z}_C = 50 + j80\ \Omega$.

Solution

This example is similar to Example 12.3-3. The important difference is that this three-phase circuit is balanced, so $\mathbf{V}_{Nn} = 0$. It is not necessary to write and solve a node equation to determine $\mathbf{V}_{Nn}$.

Balanced three-wire Y-to-Y circuits and balanced four-wire Y-to-Y circuits are analyzed in the same way. We need to calculate only one line current, $\mathbf{I}_{aA}$, and the complex power, $\mathbf{S}_a$, delivered to only one of the load impedances, $\mathbf{Z}_A$. The power delivered to the three-phase load is $3\mathbf{S}_a$.

The line current is calculated as

$$\mathbf{I}_{aA} = \frac{\mathbf{V}_a}{\mathbf{Z}_A} = \frac{110 \angle 0°}{50 + j80} = 1.16 \angle -58° \text{ A rms}$$

The total power delivered to the three-phase load is

$$3\mathbf{S}_A = 3\mathbf{I}_{aA}^* \mathbf{V}_a = 204 + j326 \text{ VA}$$

EXAMPLE 12.3-5 Line Losses

Figure 12.3-5*a* shows a balanced three-wire Y-to-Y circuit. Determine average power delivered by the three-phase source, delivered to the three-phase load, and absorbed by the three-phase line.

Solution

The three-wire Y-to-Y circuit in Figure 12.3-5*a* looks different from the three-wire Y-to-Y circuit in Figure 12.3-2. One difference is cosmetic. The circuits are drawn differently, with all circuit elements drawn vertically or horizontally in Figure 12.3-5*a*. A more important difference is that the circuit in Figure 12.3-2 is represented in the frequency domain, using phasors and impedances, whereas the circuit in Figure 12.3-5*a* is represented in the time domain. Because the circuit is represented in the time domain, the magnitude, rather than the effective value, of the source voltage is given.

Because this three-phase circuit is balanced, it can be analyzed using a per-phase equivalent circuit. Figure 12.3-5*b* shows the appropriate per-phase equivalent circuit.

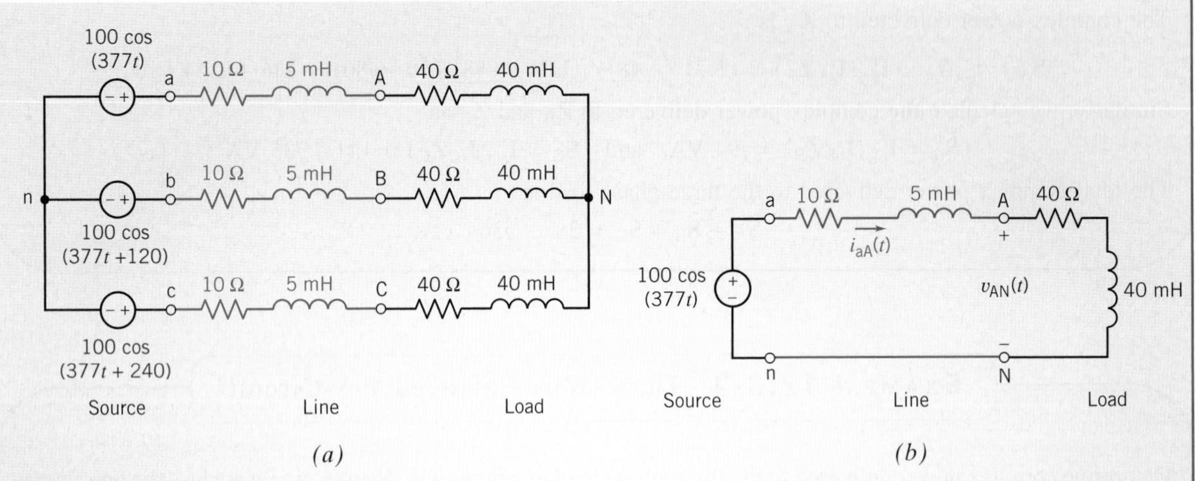

FIGURE 12.3-5 (*a*) A balanced three-wire Y-to-Y circuit and (*b*) the per-phase equivalent circuit.

The line current is calculated as

$$\mathbf{I}_{aA}(\omega) = \frac{100}{50 + j(377)(0.045)} = 1.894\angle{-18.7^\circ}\text{ A}$$

The phase voltage at the load is

$$\mathbf{V}_{AN}(\omega) = (40 + j(377)(0.04))\mathbf{I}_{aA}(\omega) = 81\angle{2^\circ}\text{ V}$$

Because peak values of the sinusoidal voltages and currents have been used instead of effective values, the appropriate formulas for power are those given in the ''peak values'' column of Table 12.1-1. The power delivered by the source is calculated as

$$\mathbf{I}_{aA}(\omega) = 1.894\angle{-18.7^\circ}\text{ A} \quad \text{and} \quad \mathbf{V}_{an}(\omega) = 100\angle{0^\circ}\text{ V}$$

so

$$P_a = \frac{(100)(1.894)}{2}\cos(18.7^\circ) = 89.7\text{ W}$$

The power delivered to the load is calculated as

$$\mathbf{I}_{aA}(\omega) = 1.894\angle{-18.7^\circ}\text{ A} \quad \text{and} \quad R_A = 40\ \Omega, \quad \text{so} \quad P_A = \frac{1.894^2}{2}40 = 71.7\text{ W}$$

The power lost in the line is calculated as

$$\mathbf{I}_{aA}(\omega) = 1.894\angle{-18.7^\circ}\text{ A} \quad \text{and} \quad R_{aA} = 10\ \Omega, \quad \text{so} \quad P_{aA} = \frac{1.894^2}{2}10 = 17.9\text{ W}$$

The three-phase load receives $3P_A = 215.1$ W, and $3P_{aA} = 53.7$ W is lost in the line. A total of 80 percent of the power supplied by the source is delivered to the load. The other 20 percent is lost in the line. The three-phase source delivers $3P_a = 269.1$ W.

EXAMPLE 12.3-6 Reducing Line Losses

As noted in Example 12.3-5, 80 percent of the power supplied by the source is delivered to the load, and the other 20 percent is lost in the line. The loss in the line can be reduced by reducing the current in the line. Reducing the current in the load would reduce the power delivered to the load. Transformers provide a way of reducing the line current without reducing the load current.

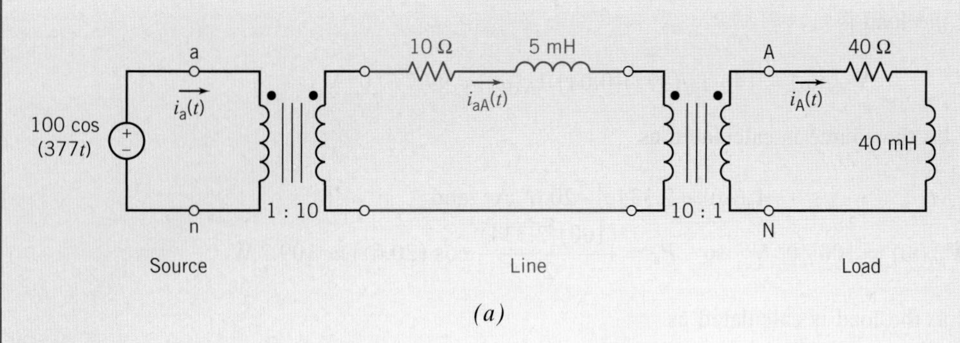

(*a*)

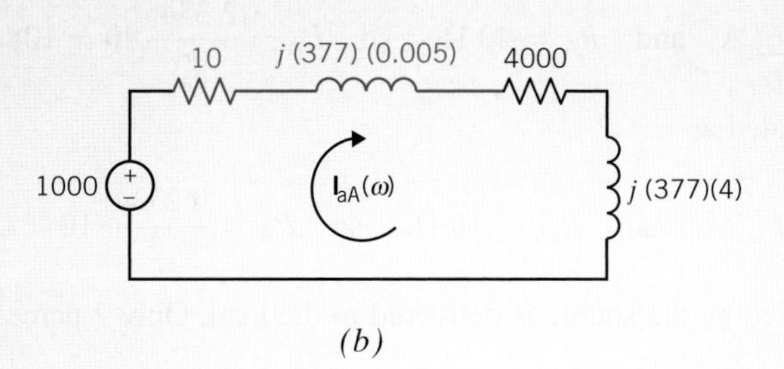

(*b*)

FIGURE 12.3-6 (*a*) A per-phase equivalent circuit for a balanced Y-to-Y circuit with step-up and step-down transformers and (*b*) the corresponding frequency-domain circuit used to calculate the line current.

In this example, two three-phase transformers are added to the three-phase circuit considered in Example 12.3-5. A transformer at the source steps up the voltage and steps down the current. Conversely, a transformer at the load steps down the voltage and steps up the current. Because the turns ratios of these transformers are reciprocals of each other, the voltage and current at the load are unchanged. The current in the line will be reduced to reduce the power lost in line. The line voltage will increase. The higher line voltage will require increased insulation and increased attention to safety.

Figure 12.3-6*a* shows the per-phase equivalent circuit of the balanced three-wire Y-to-Y circuit that includes the two transformers. Determine the average power delivered by the three-phase source, delivered to the three-phase load, and absorbed by the three-phase line.

Solution

To analyze the per-phase equivalent circuit in Figure 12.3-6*a*, notice that

1. The secondary voltage of the left-hand transformer is 10 times the primary voltage, that is, 1000 cos (377*t*).
2. The impedance connected to the secondary of the right-hand transformer can be reflected to the primary of this transformer by multiplying by 100. The result is a 4000-Ω resistor in series with a 4-H inductor.

These observations lead to the one-mesh circuit shown in Figure 12.3-6*b*. The mesh current in this circuit is the line current of the three-phase circuit. This line current is calculated as

$$\mathbf{I}_{\mathrm{aA}}(\omega) = \frac{1000}{4010 + j(377)(4.005)} = 0.2334\,\underline{/-20.6^\circ}\ \mathrm{A}$$

The current into the dotted end of the secondary of the left-hand transformer in Figure 12.3-6*a* is $-\mathbf{I}_{\mathrm{aA}}(\omega)$, so the current into the dotted end of the primary of this transformer is

$$\mathbf{I}_{\mathrm{a}}(\omega) = -10(-\mathbf{I}_{\mathrm{aA}}(\omega)) = 2.334\,\underline{/-20.6^\circ}\ \mathrm{A}$$

The current into the dotted end of the primary of the right-hand transformer is $\mathbf{I}_{\mathrm{aA}}(\omega)$, so the current into the dotted end of the secondary is

$$\mathbf{I}_{\mathrm{A}}(\omega) = -(-10\,\mathbf{I}_{\mathrm{aA}}(\omega)) = 2.334\,\underline{/-20.6^\circ}\ \mathrm{A}$$

The phase voltage at the load is

$$\mathbf{V}_{AN}(\omega) = (40 + j(377)(0.04))\mathbf{I}_A(\omega) = 99.77\angle 0^\circ \text{ V}$$

The power delivered by the source is calculated as

$$\mathbf{I}_a(\omega) = 2.334\angle -20.6^\circ \text{ A} \quad \text{and}$$

$$\mathbf{V}_{an}(\omega) = 100\angle 0^\circ \text{ V} \quad \text{so} \quad P_a = \frac{(100)(2.334)}{2}\cos(20.6^\circ) = 109.2 \text{ W}$$

The power delivered to the load is calculated as

$$\mathbf{I}_A(\omega) = 2.334\angle -20.6^\circ \text{ A} \quad \text{and} \quad R_A = 40\ \Omega, \quad \text{so} \quad P_A = \frac{2.334^2}{2}40 = 108.95 \text{ W}$$

The power lost in the line is calculated as

$$\mathbf{I}_{aA}(\omega) = 0.2334\angle -20.6^\circ \text{ A} \quad \text{and} \quad R_{aA} = 10\ \Omega, \quad \text{so} \quad P_A = \frac{0.2334^2}{2}10 = 0.27 \text{ W}$$

Now 98 percent of the power supplied by the source is delivered to the load. Only 2 percent is lost in the line.

EXERCISE 12.3-1 Determine complex power delivered to the three-phase load of a four-wire Y-to-Y circuit such as the one shown in Figure 12.3-1. The phase voltages of the Y-connected source are $\mathbf{V}_a = 120\angle 0^\circ$V rms, $\mathbf{V}_b = 120\angle -120^\circ$ V rms, and $\mathbf{V}_c = 120\ \angle 120^\circ$V rms. The load impedances are $\mathbf{Z}_A = 80 + j50\ \Omega$, $\mathbf{Z}_B = 80 + j80\ \Omega$, and $\mathbf{Z}_C = 100 - j25\ \Omega$.

Answer: $\mathbf{S}_A = 129 + j81$ VA, $\mathbf{S}_B = 90 + j90$ VA, $\mathbf{S}_C = 136 - j34$ VA, and $\mathbf{S} = 355 + j137$ VA

EXERCISE 12.3-2 Determine complex power delivered to the three-phase load of a four-wire Y-to-Y circuit such as the one shown in Figure 12.3-1. The phase voltages of the Y-connected source are $\mathbf{V}_a = 120\angle 0^\circ$ V rms, $\mathbf{V}_b = 120\angle -120^\circ$ V rms, and $\mathbf{V}_c = 120\ \angle 120^\circ$ V rms. The load impedances are $\mathbf{Z}_A = \mathbf{Z}_B = \mathbf{Z}_C = 40 + j30\ \Omega$.

Answer: $\mathbf{S}_A = \mathbf{S}_B = \mathbf{S}_C = 230 + j173 =$ VA and $\mathbf{S} = 691 + j518$ VA

EXERCISE 12.3-3 Determine complex power delivered to the three-phase load of a three-wire Y-to-Y circuit such as the one shown in Figure 12.3-2. The phase voltages of the Y-connected source are $\mathbf{V}_a = 120\angle 0^\circ$ V rms, $\mathbf{V}_b = 120\angle -120^\circ$ V rms, and $\mathbf{V}_c = 120\ \angle 120^\circ$ V rms. The load impedances are $\mathbf{Z}_A = 80 + j50\ \Omega$, $\mathbf{Z}_B = 80 + j80\ \Omega$, and $\mathbf{Z}_C = 100 - j25\ \Omega$.

Intermediate Answer: $\mathbf{V}_{nN} = 28.89\angle -150.5$ V rms

Answer: $\mathbf{S} = 392 + j142$ VA

EXERCISE 12.3-4 Determine complex power delivered to the three-phase load of a three-wire Y-to-Y circuit such as the one shown in Figure 12.3-2. The phase voltages of the Y-connected source are $\mathbf{V}_a = 120\angle 0^\circ$ V rms, $\mathbf{V}_b = 120\angle -120^\circ$ V rms, and $\mathbf{V}_c = 120\angle 120^\circ$ V rms. The load impedances are $\mathbf{Z}_A = \mathbf{Z}_B = \mathbf{Z}_C = 40 + j30\ \Omega$.

Answer: $\mathbf{S}_A = \mathbf{S}_B = \mathbf{S}_C = 230 + j173$ VA and $\mathbf{S} = 691 + j518$ VA

12.4 THE Δ-CONNECTED SOURCE AND LOAD

The Δ-connected source is shown in Figure 12.2-5*b*. This generator connection, however, is seldom used in practice because any slight imbalance in magnitude or phase of the three-phase voltages will not result in a zero sum. The result will be a large circulating current in the generator coils that will heat the generator and depreciate the efficiency of the generator. For example, consider the condition

$$\begin{aligned} \mathbf{V}_{ab} &= 120\,\underline{/0^\circ} \\ \mathbf{V}_{bc} &= 120.1\,\underline{/-121^\circ} \\ \mathbf{V}_{ca} &= 120.2\,\underline{/121^\circ} \end{aligned} \qquad (12.4\text{-}1)$$

If the total resistance around the loop is 1 Ω, we can calculate the circulating current as

$$\begin{aligned} \mathbf{I} &= (\mathbf{V}_{ab} + \mathbf{V}_{bc} + \mathbf{V}_{ca})/1 \\ &= 120 + 120.1(-0.515 - j0.857) + 120.2(-0.515 + j0.857) \\ &\cong 120 - 1.03(120.15) \\ &\cong -3.75 \text{ A} \end{aligned} \qquad (12.4\text{-}2)$$

which would be unacceptable.

Therefore, we will consider only a Y-connected source as practical at the source side and consider both the Δ-connected load and the Y-connected load at the load side.

The Δ-to-Y and Y-to-Δ transformations convert Δ-connected loads to equivalent Y-connected loads and vice versa. These transformations are summarized in Table 12.4-1. Given the impedances, $\mathbf{Z}_1, \mathbf{Z}_2, \mathbf{Z}_3$ of a Δ-connected load, Table 12.4-1 provides the formulas that are required to determine the impedances, $\mathbf{Z}_A$, $\mathbf{Z}_B$, $\mathbf{Z}_C$, of the equivalent Y-connected load. These three-phase loads are said to be equivalent because replacing the Δ-connected load by the Y-connected load will not change any of the voltages or currents of the three-phase source or three-phase line.

The Δ-to-Y and Y-to-Δ transformations are significantly simpler when the loads are balanced. Suppose the Δ-connected load is balanced, that is, $\mathbf{Z}_1 = \mathbf{Z}_2 = \mathbf{Z}_3 = \mathbf{Z}_\Delta$. The equivalent Y-connected

Table 12.4-1 Y-to-Δ and Δ-to-Y Conversions

DESCRIPTION	CIRCUIT	CONVERSION FORMULAS (UNBALANCED)	CONVERSION FORMULAS (BALANCED)
Y-connected load	A, B, $\mathbf{Z}_A$, $\mathbf{Z}_B$, N, $\mathbf{Z}_C$, C	$\mathbf{Z}_A = \dfrac{\mathbf{Z}_1\mathbf{Z}_3}{\mathbf{Z}_1+\mathbf{Z}_2+\mathbf{Z}_3}$ $\mathbf{Z}_B = \dfrac{\mathbf{Z}_2\mathbf{Z}_3}{\mathbf{Z}_1+\mathbf{Z}_2+\mathbf{Z}_3}$ $\mathbf{Z}_C = \dfrac{\mathbf{Z}_1\mathbf{Z}_2}{\mathbf{Z}_1+\mathbf{Z}_2+\mathbf{Z}_3}$	When $\mathbf{Z}_1 = \mathbf{Z}_2 = \mathbf{Z}_3 = \mathbf{Z}_\Delta$ then $\mathbf{Z}_A = \mathbf{Z}_B = \mathbf{Z}_C = \dfrac{\mathbf{Z}_\Delta}{3}$
Δ-connected load	A, B, $\mathbf{Z}_3$, $\mathbf{Z}_1$, $\mathbf{Z}_2$, C	$\mathbf{Z}_1 = \dfrac{\mathbf{Z}_A\mathbf{Z}_B+\mathbf{Z}_B\mathbf{Z}_C+\mathbf{Z}_A\mathbf{Z}_C}{\mathbf{Z}_B}$ $\mathbf{Z}_2 = \dfrac{\mathbf{Z}_A\mathbf{Z}_B+\mathbf{Z}_B\mathbf{Z}_C+\mathbf{Z}_A\mathbf{Z}_C}{\mathbf{Z}_A}$ $\mathbf{Z}_3 = \dfrac{\mathbf{Z}_A\mathbf{Z}_B+\mathbf{Z}_B\mathbf{Z}_C+\mathbf{Z}_A\mathbf{Z}_C}{\mathbf{Z}_C}$	When $\mathbf{Z}_A = \mathbf{Z}_B = \mathbf{Z}_C = \mathbf{Z}_Y$ then $\mathbf{Z}_1 = \mathbf{Z}_2 = \mathbf{Z}_3 = 3\mathbf{Z}_Y$

load will also be balanced, so $\mathbf{Z}_A = \mathbf{Z}_B = \mathbf{Z}_C = \mathbf{Z}_Y$. Then, we have

$$\mathbf{Z}_Y = \frac{\mathbf{Z}_\Delta}{3} \tag{12.4-3}$$

Therefore, if we have a Y-connected source and a balanced Δ-connected load with $\mathbf{Z}_\Delta$, we convert the Δ load to a Y load with $\mathbf{Z}_Y = \mathbf{Z}_\Delta/3$. Then the line current is

$$\mathbf{I}_A = \frac{\mathbf{V}_a}{\mathbf{Z}_Y} = \frac{3\mathbf{V}_a}{\mathbf{Z}_\Delta} \tag{12.4-4}$$

Thus, we will consider only the Y-to-Y configuration. If the Y-to-Δ configuration is encountered, the Δ-connected load is converted to a Y-connected load equivalent, and the resulting currents and voltages are calculated.

EXAMPLE 12.4-1 Y and Δ Connected Loads

Figure 12.4-1*a* shows a three-phase load that consists of a parallel connection of a Y-connected and Δ-connected load. Convert this load to an equivalent Y-connected load.

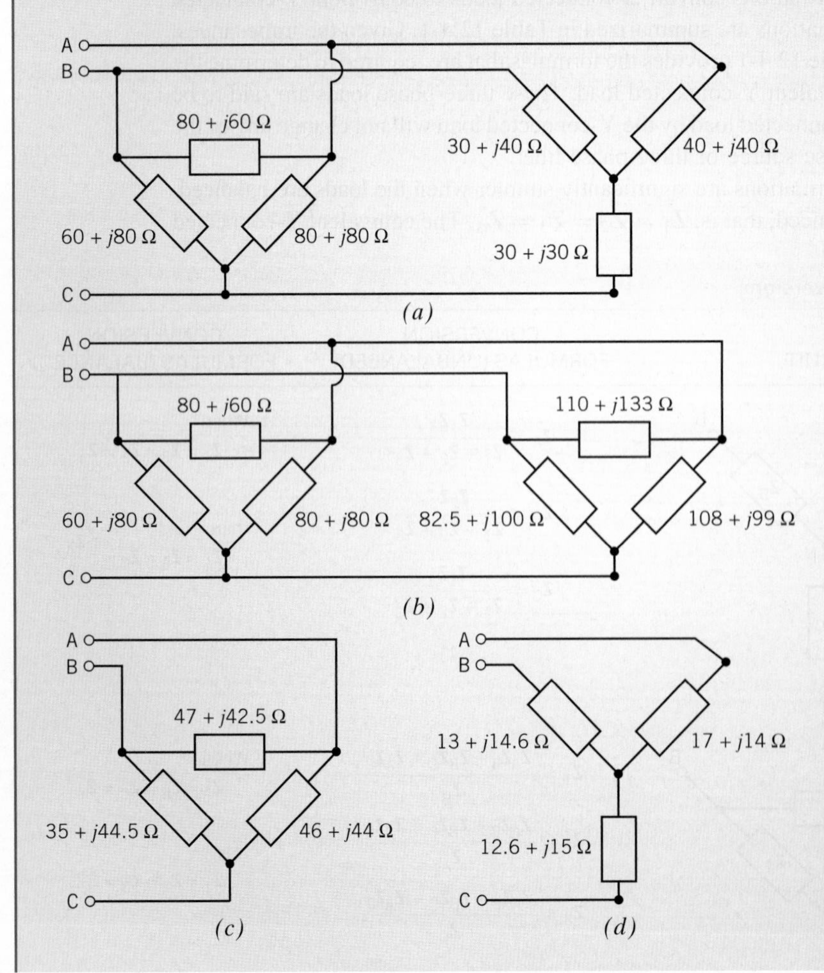

FIGURE 12.4-1 Example of Y-Δ conversions. (*a*) Parallel Y-connected and Δ-connected loads. (*b*) The Y-connected load is converted to a Δ-connected load. (*c*) The parallel Δ-connected loads are replaced by a single equivalent Δ-connected load. (*d*) The Δ-connected load is converted to a Y-connected load.

Solution

First, convert the Y-connected load to a Δ-connected load as shown in Figure 12.4-1*b*. Notice, for example, that both of the Δ-connected loads in Figure 12.4-1*b* have an impedance connected between terminals A and B. These impedances are in parallel and can be replaced by a single equivalent impedance. Replace the parallel Δ-connected loads by a single equivalent Δ-connected load as shown in Figure 12.4-1*c*. Finally, convert the Δ-connected load to a Y-connected load as shown in Figure 12.4-1*d*.

12.5 THE Y-TO-Δ CIRCUIT

Now, let us consider the Y-to-Δ circuit as shown in Figure 12.5-1. Applying KCL at the nodes of the Δ-connected load shows that the relation between the line currents and phase currents is

$$\mathbf{I}_{aA} = \mathbf{I}_{AB} - \mathbf{I}_{CA}$$

$$\mathbf{I}_{bB} = \mathbf{I}_{BC} - \mathbf{I}_{AB}$$

and

$$\mathbf{I}_{cC} = \mathbf{I}_{CA} - \mathbf{I}_{BC} \qquad (12.5\text{-}1)$$

The goal is to calculate the line and phase currents for the load.

The phase currents in the Δ-connected load can be calculated from the line-to-line voltages. These line-to-line voltages appear directly across the impedances of the Δ-connected load. For example, $\mathbf{V}_{AB}$ appears across $\mathbf{Z}_3$, so

$$\mathbf{I}_{AB} = \frac{\mathbf{V}_{AB}}{\mathbf{Z}_3} \qquad (12.5\text{-}2)$$

Similarly,

$$\mathbf{I}_{CA} = \frac{\mathbf{V}_{CA}}{\mathbf{Z}_2} \quad \text{and} \quad \mathbf{I}_{BC} = \frac{\mathbf{V}_{BC}}{\mathbf{Z}_1} \qquad (12.5\text{-}3)$$

When the load is balanced, the phase currents in the load have the same magnitude and have phase angles that differ by 120°. For example, if the three-phase source has the *abc* sequence and

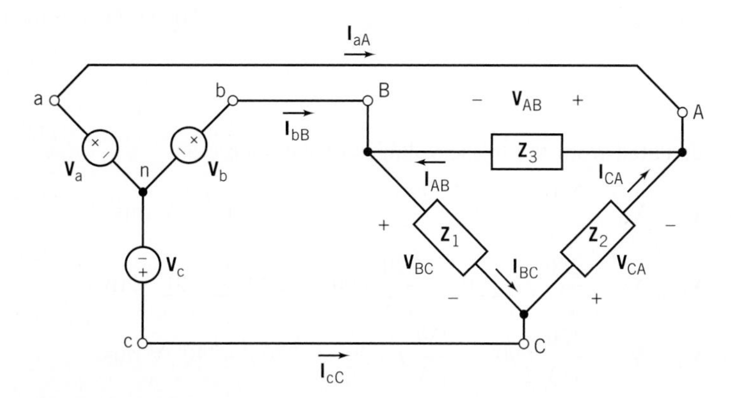

FIGURE 12.5-1 A Y-to-Δ three-phase circuit.

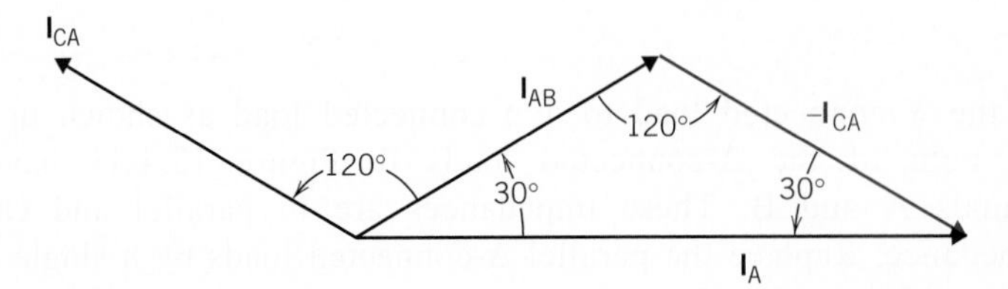

FIGURE 12.5-2 Phasor diagram for currents of a Δ load.

$\mathbf{I}_{AB} = I\,\underline{/\phi}$, then $\mathbf{I}_{CA} = I\,\underline{/\phi + 120^\circ}$. The line current $\mathbf{I}_{aA}$ is calculated as

$$\begin{aligned}
\mathbf{I}_{aA} &= \mathbf{I}_{AB} - \mathbf{I}_{CA} \\
&= I\cos\phi + jI\sin\phi - I\cos(\phi + 120^\circ) - jI\sin(\phi + 120^\circ) \\
&= -2I\sin(\phi + 60^\circ)\sin(-60^\circ) + j2I\cos(\phi + 60^\circ)\sin(-60^\circ) \\
&= \sqrt{3}\,I\,[\sin(\phi + 60^\circ) - j\cos(\phi + 60^\circ)] \\
&= \sqrt{3}\,I\,[\cos(\phi - 30^\circ) - j\sin(\phi - 30^\circ)] \\
&= \sqrt{3}\,I\,\underline{/\phi - 30^\circ}\text{ A}
\end{aligned} \tag{12.5-4}$$

Therefore,

$$|\mathbf{I}_{aA}| = \sqrt{3}|\mathbf{I}| \tag{12.5-5}$$

or

$$I_L = \sqrt{3}I_p$$

and the line current magnitude is $\sqrt{3}$ times the phase current magnitude. This result can also be obtained from the phasor diagram shown in Figure 12.5-2. In a Δ connection, the line current is $\sqrt{3}$ times the phase current and is displaced -30° in phase. The line-to-line voltage is equal to the phase voltage.

EXAMPLE 12.5-1 Balanced Y-Δ Circuit

Consider the three-phase circuit shown in Figure 12.5-1. The voltages of the Y-connected source are

$$\mathbf{V}_a = \frac{220}{\sqrt{3}}\underline{/-30^\circ}\text{ V rms},\quad \mathbf{V}_b = \frac{220}{\sqrt{3}}\underline{/-150^\circ}\text{ V rms},\quad\text{and}\quad \mathbf{V}_c = \frac{220}{\sqrt{3}}\underline{/90^\circ}\text{ V rms}$$

The Δ-connected load is balanced. The impedance of each phase is $\mathbf{Z}_\Delta = 10\,\underline{/-50^\circ}\ \Omega$. Determine the phase and line currents.

Solution

The line-to-line voltages are calculated from the phase voltages of the source as

$$\begin{aligned}
\mathbf{V}_{AB} &= \mathbf{V}_a - \mathbf{V}_b = \frac{220}{\sqrt{3}}\underline{/-30^\circ} - \frac{220}{\sqrt{3}}\underline{/-150^\circ} = 220\,\underline{/0^\circ}\text{ V rms} \\
\mathbf{V}_{BC} &= \mathbf{V}_b - \mathbf{V}_c = \frac{220}{\sqrt{3}}\underline{/-150^\circ} - \frac{220}{\sqrt{3}}\underline{/90^\circ} = 220\,\underline{/-120^\circ}\text{ V rms} \\
\mathbf{V}_{CA} &= \mathbf{V}_c - \mathbf{V}_a = \frac{220}{\sqrt{3}}\underline{/90^\circ} - \frac{220}{\sqrt{3}}\underline{/-30^\circ} = 220\,\underline{/-240^\circ}\text{ V rms}
\end{aligned}$$

The phase voltages of a Δ-connected load are equal to the line-to-line voltages. The phase currents are

$$\mathbf{I}_{AB} = \frac{\mathbf{V}_{AB}}{\mathbf{Z}} = \frac{220\angle 0^\circ}{10\angle -50^\circ} = 22\angle 50^\circ \text{ A rms}$$

$$\mathbf{I}_{BC} = \frac{\mathbf{V}_{BC}}{\mathbf{Z}} = \frac{220\angle -120^\circ}{10\angle -50^\circ} = 22\angle -70^\circ \text{ A rms}$$

$$\mathbf{I}_{CA} = \frac{\mathbf{V}_{CA}}{\mathbf{Z}} = \frac{220\angle -240^\circ}{10\angle -50^\circ} = 22\angle -190^\circ \text{ A rms}$$

The line currents are

$$\mathbf{I}_{aA} = \mathbf{I}_{AB} - \mathbf{I}_{CA} = 22\angle 50^\circ - 22\angle -190^\circ = 22\sqrt{3}\angle 20^\circ \text{ A rms}$$

Then

$$\mathbf{I}_{bB} = 22\sqrt{3}\angle -100^\circ \text{ A rms} \quad \text{and} \quad \mathbf{I}_{cC} = 22\sqrt{3}\angle -220^\circ \text{ A rms}$$

The current and voltage relationships for a Δ load are summarized in Table 12.5-1.

Table 12.5-1 The Current and Voltage for a Δ Load

Phase voltages	$\mathbf{V}_{AB} = V_{AB}\angle 0^\circ$
Line-to-line voltages	$V_{AB} = V_L$ (linear voltage = phase voltage)
Phase currents	$\mathbf{I}_{AB} = \frac{\mathbf{V}_{AB}}{\mathbf{Z}_P} = \frac{\mathbf{V}_L}{\mathbf{Z}_\Delta} = I_p\angle -\theta$ with $\mathbf{Z}_P = \mathbf{Z}/\theta$
	$\mathbf{I}_{BC} = \mathbf{I}_{AB}\angle -120^\circ$
	$\mathbf{I}_{CA} = \mathbf{I}_{AB}\angle -240^\circ$
Line currents	$\mathbf{I}_A = \sqrt{3}I_p\angle -\theta - 30^\circ$
	$\mathbf{I}_B = \sqrt{3}I_p\angle -\theta - 150^\circ$
	$\mathbf{I}_C = \sqrt{3}I_p\angle -\theta + 90^\circ$
	$I_L = \sqrt{3}I_p$

Note: L = line, p = phase.

EXERCISE 12.5-1 Consider the three-phase circuit shown in Figure 12.5-1. The voltages of the Y-connected source are

$$\mathbf{V}_a = \frac{360}{\sqrt{3}}\angle -30^\circ \text{ V rms}, \quad \mathbf{V}_b = \frac{360}{\sqrt{3}}\angle -150^\circ \text{ V rms}, \quad \text{and} \quad \mathbf{V}_c = \frac{360}{\sqrt{3}}\angle 90^\circ \text{ V rms}$$

The Δ-connected load is balanced. The impedance of each phase is $\mathbf{Z}_\Delta = 180\angle 45^\circ\ \Omega$. Determine the phase and line currents when the line-to-line voltage is 360 V rms.

Partial Answer: $\mathbf{I}_{AB} = 2\angle 45^\circ$ A rms and $\mathbf{I}_{aA} = 3.46\angle 15^\circ$ A rms

12.6 BALANCED THREE-PHASE CIRCUITS

We have only two possible practical configurations for three-phase circuits, Y-to-Y and Y-to-Δ, and we can convert the latter to a Y-to-Y form. Thus, a practical three-phase circuit can always be converted to the Y-to-Y circuit.

Balanced circuits are easier to analyze than unbalanced circuits. Earlier, we saw that balanced three-phase Y-to-Y circuits can be analyzed using a per-phase equivalent circuit.

The circuit shown in Figure 12.6-1*a* is a balanced Y-to-Δ circuit. Figure 12.6-1*b* shows the equivalent Y-to-Y circuit in which

$$\mathbf{Z}_Y = \frac{\mathbf{Z}_\Delta}{3}$$

This Y-to-Y circuit can be analyzed using the per-phase equivalent circuit shown in Figure 12.6-1*c*.

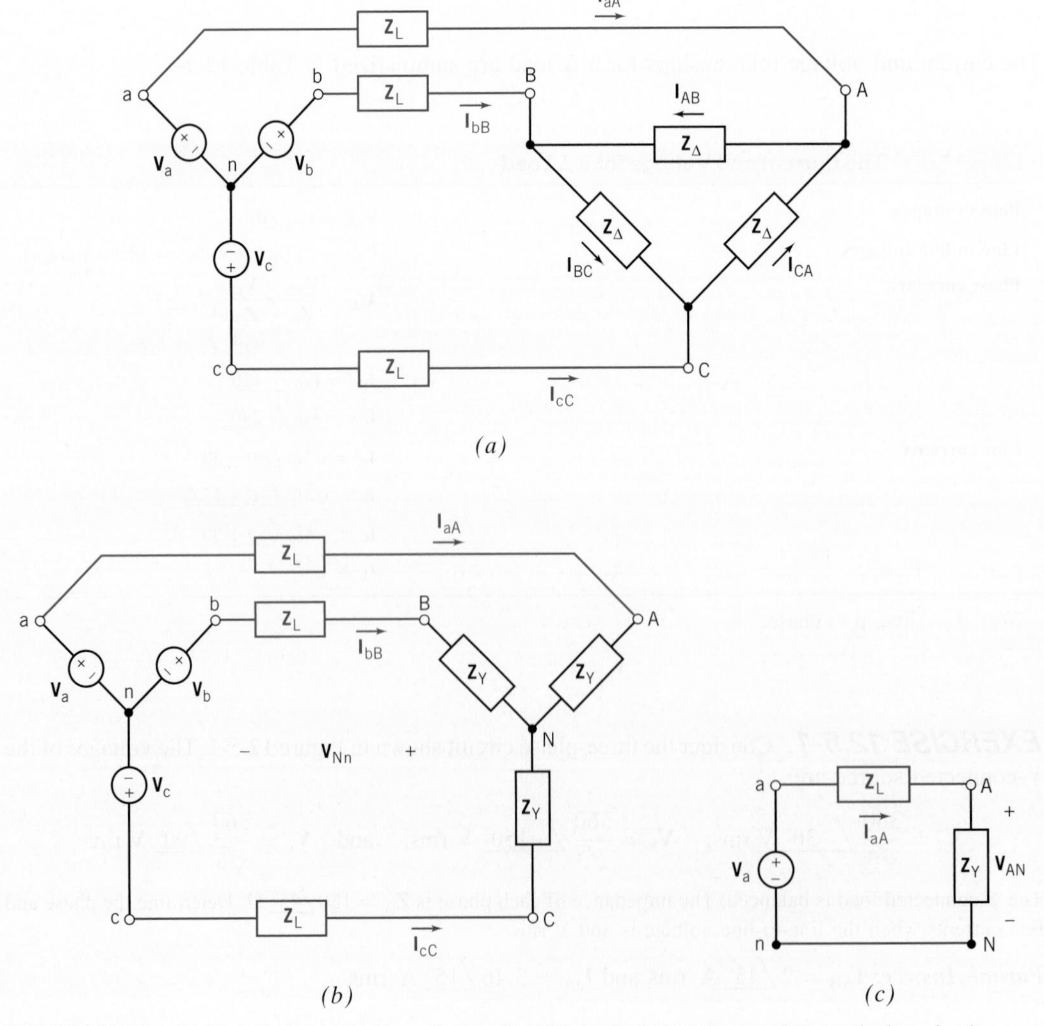

FIGURE 12.6-1 (*a*) A Y-to-Δ circuit, (*b*) the equivalent Y-to-Y circuit, and (*c*) the per-phase equivalent circuit.

EXAMPLE 12.6-1 Per-Phase Equivalent Circuit

Figure 12.6-1*a* shows a balanced Y-to-Δ three-phase circuit. The phase voltages of the Y-connected source are $\mathbf{V}_a = 110\angle 0°$ V rms, $\mathbf{V}_b = 110\angle -120°$ V rms, and $\mathbf{V}_c = 110\angle 120°$ V rms. The line impedances are each $\mathbf{Z}_L = 10 + j5\ \Omega$. The impedances of the Δ-connected load are each $\mathbf{Z}_\Delta = 75 + j225\ \Omega$. Determine the phase currents in the Δ-connected load.

Solution

Convert the Δ-connected load to a Y-connected load using the Δ-to-Y transformation summarized in Table 12.4-1. The impedances of the balanced equivalent Y-connected load are

$$\mathbf{Z}_Y = \frac{75 + j225}{3} = 25 + j75\ \Omega$$

The per-phase equivalent circuit for the Y-to-Y circuit is shown in Figure 12.6-1*c*. The line current is given by

$$\mathbf{I}_{aA} = \frac{\mathbf{V}_a}{\mathbf{Z}_L + \mathbf{Z}_Y} = \frac{110\angle 0°}{(10 + j5) + (25 + j75)} = 1.26\angle -66° \text{ A rms} \tag{12.6-1}$$

The line current, $\mathbf{I}_{aA}$, calculated using the per-phase equivalent circuit, is also the line current, $\mathbf{I}_{aA}$, in the Y-to-Y circuit, as well as the line current, $\mathbf{I}_{aA}$, in the Y-to-Δ circuit. The other line currents in the balanced Y-to-Y circuit have the same magnitude but differ in phase angle by 120°. These line currents are

$$\mathbf{I}_{bB} = 1.26\angle -186° \text{ A rms} \quad \text{and} \quad \mathbf{I}_{cC} = 1.26\angle 54° \text{ A rms}$$

(To check the value of $\mathbf{I}_{bB}$, apply kVL to the loop in the Y-to-Y circuit that starts at node n, passes through nodes b, B, N, and returns to node n. The resulting KVL equation is

$$\mathbf{V}_b = \mathbf{Z}_L\mathbf{I}_{bB} + \mathbf{Z}_Y\mathbf{I}_{bB} + \mathbf{V}_{Nn}$$

Because the circuit is balanced, $\mathbf{V}_{Nn} = 0$. Solving for $\mathbf{I}_{bB}$ gives

$$\mathbf{I}_{bB} = \frac{\mathbf{V}_b}{\mathbf{Z}_L + \mathbf{Z}_Y} = \frac{110\angle -120°}{(10 + j5) + (25 + j75)} = 1.26\angle -186° \text{ A rms} \tag{12.6-2}$$

Comparing Eqs. 12.6-1 and 12.6-2 shows that the line currents in the balanced Y-to-Y circuit have the same magnitude but differ in phase angle by 120°.

The line currents of the Y-to-Δ circuit in Figure 12.6-1*a* are equal to the line currents of the Y-to-Y circuit in Figure 12.6-1*b* because the Y-to-Δ and Y-to-Y circuits are equivalent.

The voltage $\mathbf{V}_{AN}$ in the per-phase equivalent circuit is

$$\mathbf{V}_{AN} = \mathbf{I}_{aA}\mathbf{Z}_Y = (1.26\angle -66°)(25 + j75) = 99.6\angle 5° \text{ V rms}$$

The voltage $\mathbf{V}_{AN}$ calculated using the per-phase equivalent circuit is also the phase voltage, $\mathbf{V}_{AN}$, of the Y-to-Y circuit. The other phase voltages of the balanced Y-to-Y circuit have the same magnitude but differ in phase angle by 120°. These phase voltages are

$$\mathbf{V}_{BN} = 99.6\angle -115° \text{ V rms} \quad \text{and} \quad \mathbf{V}_{CN} = 99.6\angle 125° \text{ V rms}$$

The line-to-line voltages of the Y-to-Y circuit are calculated as

$$\mathbf{V}_{AB} = \mathbf{V}_{AN} - \mathbf{V}_{BN} = 99.5\angle 5^\circ - 99.5\angle -115^\circ = 172\angle 35^\circ \text{ V rms}$$
$$\mathbf{V}_{BC} = \mathbf{V}_{BN} - \mathbf{V}_{CN} = 99.5\angle -115^\circ - 99.5\angle 125^\circ = 172\angle -85^\circ \text{ V rms}$$
$$\mathbf{V}_{CA} = \mathbf{V}_{CN} - \mathbf{V}_{AN} = 99.5\angle 125^\circ - 99.5\angle 5^\circ = 172\angle 155^\circ \text{ V rms}$$

The phase voltages of a Δ-connected load are equal to the line-to-line voltages. The phase currents are

$$\mathbf{I}_{AB} = \frac{\mathbf{V}_{AB}}{\mathbf{Z}_\Delta} = \frac{172\angle 35^\circ}{75 + j225} = 0.727\angle -36^\circ \text{ A rms}$$
$$\mathbf{I}_{BC} = \frac{\mathbf{V}_{BC}}{\mathbf{Z}_\Delta} = \frac{172\angle -85^\circ}{75 + j225} = 0.727\angle -156^\circ \text{ A rms}$$
$$\mathbf{I}_{CA} = \frac{\mathbf{V}_{CA}}{\mathbf{Z}_\Delta} = \frac{172\angle 155^\circ}{75 + j225} = 0.727\angle -84^\circ \text{ A rms}$$

EXERCISE 12.6-1 Figure 12.6-1*a* shows a balanced Y-to-Δ three-phase circuit. The phase voltages of the Y-connected source are $\mathbf{V}_a = 110\angle 0^\circ$ V rms, $\mathbf{V}_b = 110\angle -120^\circ$ V rms, and $\mathbf{V}_c = 110\angle 120^\circ$ V rms. The line impedances are each $\mathbf{Z}_L = 10 + j25\ \Omega$. The impedances of the Δ-connected load are each $\mathbf{Z}_\Delta = 150 + j270\ \Omega$. Determine the phase currents in the Δ-connected load.

Answer: $\mathbf{I}_{AB} = 0.49\angle -32.5^\circ$ A rms, $\mathbf{I}_{BC} = 0.49\angle -152.5^\circ$ A rms, $\mathbf{I}_{CA} = 0.49\angle 87.5^\circ$ A rms

12.7 INSTANTANEOUS AND AVERAGE POWER IN A BALANCED THREE-PHASE LOAD

One advantage of three-phase power is the smooth flow of energy to the load. Consider a balanced load with resistance R. Then the *instantaneous power* is

$$p(t) = \frac{v_{ab}^2}{R} + \frac{v_{bc}^2}{R} + \frac{v_{ca}^2}{R} \tag{12.7-1}$$

where $v_{ab} = V \cos \omega t$, and the other two-phase voltages have a phase of ±120°, respectively. Furthermore,

$$\cos^2 \alpha t = (1 + \cos 2\alpha)/2$$

Therefore,

$$\begin{aligned} p(t) &= \frac{V^2}{2R}[1 + \cos 2\omega t + 1 + \cos 2(\omega t - 120^\circ) + 1 + \cos 2(\omega t - 240^\circ)] \\ &= \frac{3V^2}{2R} + \frac{V^2}{2R}[\cos 2\omega t + \cos (2\omega t - 240^\circ) + \cos (2\omega t - 480^\circ)] \end{aligned} \tag{12.7-2}$$

The bracketed term is equal to zero for all time. Hence,

$$p(t) = \frac{3V^2}{2R}$$

The *instantaneous power* delivered to a balanced three-phase load is a constant.

The total power delivered to a balanced three-phase load can be calculated using the per-phase equivalent circuit. For example, we multiply the complex power delivered to a load in the per-phase equivalent circuit by 3 to obtain the total complex power delivered to the corresponding balanced three-phase load.

Consider, again, Figure 12.6-1. Figure 12.6-1*a* shows a balanced Y-to-Δ circuit. Figure 12.6-1*b* shows the equivalent Y-to-Y circuit, obtained using the Δ-to-Y transformation summarized in Table 12.4-1. Figure 12.6-1*c* shows the per-phase equivalent circuit corresponding to the Y-to-Y circuit. The voltage $\mathbf{V}_{AN} = V_P \angle \theta_{AV}$ and the current $\mathbf{I}_{aA} = I_L \angle \theta_{AI}$ are obtained using per-phase equivalent circuit. The voltage $\mathbf{V}_{AN}$ and the current $\mathbf{I}_{aA}$ are the phase voltage and line current of the Y-connected load in Figure 12.6-1*b*. The total average power delivered to the balanced Y-connected load is given by

$$P_Y = 3\,P_A = 3\,V_P I_L \cos\left(\theta_{AV} - \theta_{AI}\right) = 3\,V_P I_L \cos\left(\theta\right) \tag{12.7-3}$$

where θ is the angle between the phase voltage and the line current, $\cos\theta$ is the power factor, and V_P and I_P are effective values of the phase voltage and line current.

It is easier to measure the line-to-line voltage and the line current of a circuit. Also recall that the line current equals the phase current and that the phase voltage is $V_P = V_L/\sqrt{3}$ for the Y-load configuration. Therefore,

$$P = 3\frac{V_L}{\sqrt{3}} I_L \cos\theta = \sqrt{3}\,V_L I_L \cos\theta \tag{12.7-4}$$

The total average power delivered to the Δ-connected load in Figure 12.6-1*a* is

$$P = 3P_{AB} = 3V_{AB} I_{AB} \cos\theta = 3\left(\sqrt{3}V_P\right)\frac{I_L}{\sqrt{3}} \cos\theta = 3\,V_P I_L \cos\theta \tag{12.7-5}$$

In summary, the total average power delivered to the Δ-connected load in Figure 12.6-1*a* is equal to the total average power delivered to the balanced Y-connected load in Figure 12.6-1*b*. That's appropriate because the two circuits are equivalent. Notice that the information required to calculate the power delivered to a balanced load, Y or Δ, is obtained from the per-phase equivalent circuit.

EXAMPLE 12.7-1 Power Delivered to the Load

Figure 12.6-1*a* shows a balanced Y-to-Δ three-phase circuit. The phase voltages of the Y-connected source are $\mathbf{V}_a = 110\angle 0^\circ$ V rms, $\mathbf{V}_b = 110\angle -120^\circ$ V rms, and $\mathbf{V}_c = 110\angle 120^\circ$ V rms. The line impedances are each $\mathbf{Z}_L = 10 + j5\ \Omega$. The impedances of the Δ-connected load are each $\mathbf{Z}_\Delta = 75 + j225\ \Omega$. Determine the average power delivered to the load.

Solution

This circuit was analyzed in Example 12.6-1. That analysis showed that

$$\mathbf{I}_{aA} = 1.26\angle -66^\circ \text{ A rms}$$

and $$\mathbf{V}_{\mathrm{AN}} = 99.6\angle 5^\circ \text{ V rms}$$

The total average power delivered to the load is given by Eq. 12.7-3 as

$$P = 3(99.6)(1.26)\cos(5^\circ - (-66^\circ)) = 122.6 \text{ W}$$

EXAMPLE 12.7-2 Three-Phase Load

A balanced three-phase load receives 15 kW at a power factor of 0.8 lagging when the line voltage is 480 V rms. Represent this load as a balanced Y-connected load.

Solution

We will represent the load as three Y-connected impedances. Each of these impedances will receive one third of the power delivered to the three-phase load, 5 kW at 0.8 lagging. The complex power received by each impedance will be

$$\mathbf{S} = P + j\frac{P}{pf}\sin\left(\cos^{-1}(pf)\right) = 5 + j\frac{5}{0.8}\sin\left(\cos^{-1}(0.8)\right) = 5 + j3.75 \text{ kVA}$$

The voltage across each impedance of the load will be phase voltage

$$\mathbf{V}_{\mathrm{P}} = \frac{|\mathbf{V}_{\mathrm{L}}|}{\sqrt{3}}\angle\phi = \frac{480}{\sqrt{3}}\angle\phi = 277\angle\phi \text{ V rms}$$

The angle, ϕ, of the phase voltage has not been specified. The voltages across each of the three impedances of the load have the same magnitude but different angles. The current in each of the load impedances is given by

$$\mathbf{I} = \left(\frac{\mathbf{S}}{\mathbf{V}_{\mathrm{P}}}\right)^* = \left(\frac{6250\angle 36.9^\circ}{277\angle\phi}\right)^* = 22.56\angle(\phi - 36.9^\circ) \text{ A rms}$$

Finally, the load impedance is given by

$$\mathbf{Z} = \frac{\mathbf{V}_{\mathrm{P}}}{\mathbf{I}} = \frac{277\angle\phi}{22.56\angle(\phi - 36.9^\circ)} = 12.28\angle 36.9^\circ = 9.82 + j7.37\ \Omega$$

EXAMPLE 12.7-3 Three-Phase Circuit

A balanced three-phase circuit consists of a Y-connected source connected to a balanced load. The line impedances are each $\mathbf{Z}_{\mathrm{L}} = 2 + j0.5\ \Omega$. The balanced three-phase load receives 15 kW at a power factor of 0.8 lagging, and the line voltage at the load is 480 V rms. Determine the required source voltage and the complex power supplied by the three-phase source.

Solution

The three-phase load in this example is the same load encountered in Example 12.7-2. Using the results of Example 12.7-2, we can represent this three-phase circuit, using the per-phase equivalent circuit shown in Figure

12.6-1*c* with $\mathbf{Z}_L = 2 + j0.5\ \Omega$ and $\mathbf{Z}_Y = 9.82 + j7.37\ \Omega$. As in Example 12.7-2, the line current depends on the power received by the load and the line voltage at the load and is given by

$$\mathbf{I}_{aA} = 22.56\angle(\phi - 36.9^\circ)\ \text{A rms}$$

where ϕ has not been specified. Using KVL, the required source voltage can then be expressed as

$$\mathbf{V}_a = (\mathbf{Z}_L + \mathbf{Z}_Y)\mathbf{I}_{aA} = (2 + j0.5 + 9.82 + j7.37)22.56\angle(\phi - 36.9^\circ) = 320.6\angle(\phi - 3.3^\circ)\ \text{V rms}$$

The complex power delivered by the three-phase source is

$$\mathbf{S}_{source} = 3\mathbf{V}_a\mathbf{I}_{aA}^* = 3(320.6\angle(\phi - 3.3^\circ))(22.56\angle(\phi - 36.9^\circ))^* = 21.7\angle 33.6^\circ$$
$$= 18.1 + j12.0\ \text{kVA}$$

It's worth noticing that the power supplied by the three-phase source does not depend on the unspecified angle ϕ. At this point, it may be convenient to specify that $\phi = 3.3^\circ$ so that the Y-connected voltage sources will have phase angles of 0°, 120°, and −120°.

EXERCISE 12.7-1 Figure 12.6-1*a* shows a balanced Y-to-Δ three-phase circuit. The phase voltages of the Y-connected source are $\mathbf{V}_a = 110\angle 0^\circ$ V rms, $\mathbf{V}_b = 110\angle -120^\circ$ V rms, and $\mathbf{V}_c = 110\angle 120^\circ$ V rms. The line impedances are each $\mathbf{Z}_L = 10 + j25\ \Omega$. The impedances of the Δ-connected load are each $\mathbf{Z}_\Delta = 150 + j\,270\ \Omega$. Determine the average power delivered to the Δ-connected load.

Intermediate Answer: $\mathbf{I}_{aA} = 0.848\angle -62.5^\circ$ A rms and $\mathbf{V}_{AN} = 87.3\angle -1.5^\circ$ V rms

Answer: $P = 107.9$ W

12.8 TWO-WATTMETER POWER MEASUREMENT

For many load configurations, for example, a three-phase motor, the phase current or voltage is inaccessible. We may wish to measure power with a wattmeter connected to each phase. However, because the phases are not available, we measure the line currents and the line-to-line voltages. A wattmeter provides a reading of $V_L I_L \cos\theta$ where V_L and I_L are the rms magnitudes and θ is the angle between the line voltage, **V**, and the current, **I**. We choose to measure V_L and I_L, the line voltage and current, respectively. We will show that two wattmeters are sufficient to read the power delivered to the three-phase load, as shown in Figure 12.8-1. We use *cc* to denote current coil and *vc* to denote voltage coil.

Wattmeter 1 reads

$$P_1 = V_{AB} I_A \cos\theta_1 \tag{12.8-1}$$

and wattmeter 2 reads

$$P_2 = V_{CB} I_C \cos\theta_2 \tag{12.8-2}$$

For the *abc* phase sequence for a balanced load,

$$\theta_1 = \theta + 30^\circ$$

and

$$\theta_2 = \theta - 30^\circ \tag{12.8-3}$$

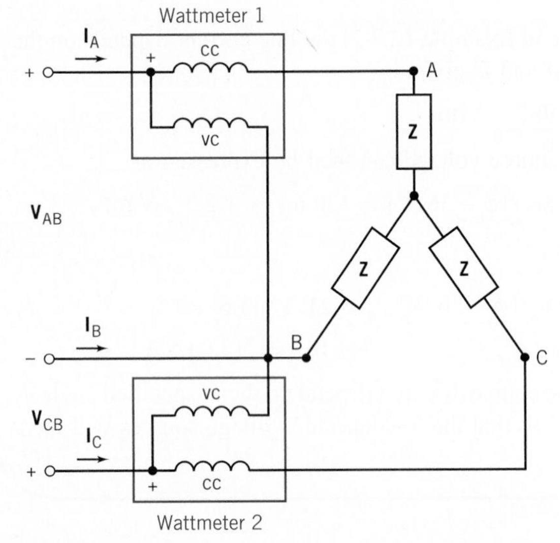

FIGURE 12.8-1 Two-wattmeter connection for a three-phase Y-connected load.

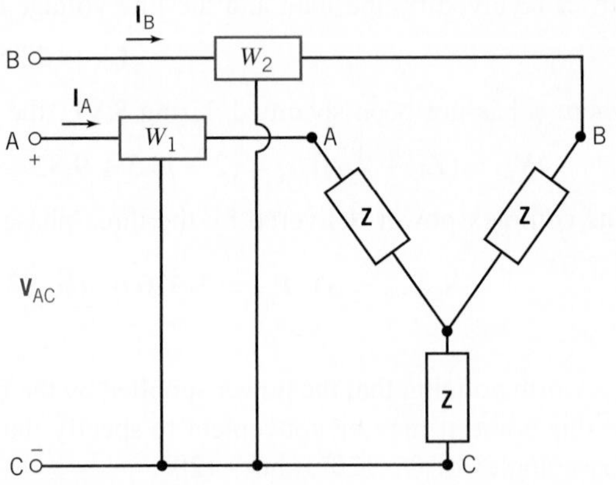

FIGURE 12.8-2 The two-wattmeter connection for Example 12.8-1.

where θ is the angle between the phase current and the phase voltage for phase a of the three-phase source.

Therefore,

$$P = P_1 + P_2 = 2V_L I_L \cos\theta \cos 30^\circ = \sqrt{3} V_L I_L \cos\theta \tag{12.8-4}$$

which is the total average power of the three-phase circuit. The preceding derivation of Eq. 12.8-4 is for a balanced circuit; the result is good for any three-phase, three-wire load, even unbalanced or nonsinusoidal voltages.

The power factor angle, θ, of a balanced three-phase system may be determined from the reading of the two wattmeters shown in Figure 12.8-2.

The total power is obtained from Eqs. 12.8-1 through 12.8-3 as

$$\begin{aligned} P = P_1 + P_2 &= V_L I_L [\cos(\theta + 30^\circ) + \cos(\theta - 30^\circ)] \\ &= V_L I_L \, 2\cos\theta \cos 30^\circ \end{aligned} \tag{12.8-5}$$

Similarly,

$$P_1 - P_2 = V_L I_L (-2 \sin\theta \sin 30^\circ) \tag{12.8-6}$$

Dividing Eq. 12.8-5 by Eq. 12.8-6, we obtain

$$\frac{P_1 + P_2}{P_1 - P_2} = \frac{2\cos\theta \cos 30^\circ}{-2\sin\theta \sin 30^\circ} = \frac{-\sqrt{3}}{\tan\theta}$$

Therefore,

$$\tan\theta = \sqrt{3}\frac{P_2 - P_1}{P_2 + P_1} \tag{12.8-7}$$

where $\theta =$ power factor angle.

EXAMPLE 12.8-1 Two-wattmeter Method

The two-wattmeter method is used, as shown in Figure 12.8-2, to measure the total power delivered to the Y-connected load when $Z = 10\,\angle 45^\circ\ \Omega$ and the supply line-to-line voltage is 220 V rms. Determine the reading of each wattmeter and the total power.

Solution

The phase voltage is

$$\mathbf{V}_\mathrm{A} = \frac{220}{\sqrt{3}}\angle -30^\circ \text{ V rms}$$

Then we obtain the line current as

$$\mathbf{I}_\mathrm{A} = \frac{\mathbf{V}_\mathrm{A}}{\mathbf{Z}} = \frac{220\,\angle -30^\circ}{10\sqrt{3}\,\angle 45^\circ} = 12.7\,\angle -75^\circ \text{ A rms}$$

Then the second line current is

$$\mathbf{I}_\mathrm{B} = 12.7\,\angle -195^\circ \text{ A rms}$$

The voltage $\mathbf{V}_\mathrm{AB} = 220\,\angle 0^\circ$ V rms, $\mathbf{V}_\mathrm{CA} = 220\,\angle +120^\circ$ V rms, and $\mathbf{V}_\mathrm{BC} = 220\,\angle -120^\circ$ V rms. The first wattmeter reads

$$P_1 = I_\mathrm{A} V_\mathrm{AC} \cos\theta_1 = 12.7(220)\cos 15^\circ = 2698 \text{ W}$$

Because $\mathbf{V}_\mathrm{CA} = 220\,\angle +120^\circ$, $\mathbf{V}_\mathrm{AC} = 220\,\angle -60^\circ$. Therefore, the angle θ_1 lies between $\mathbf{V}_\mathrm{AC}$ and $\mathbf{I}_\mathrm{A}$ and is equal to 15°. The reading of the second wattmeter is

$$P_2 = I_\mathrm{B} V_\mathrm{BC} \cos\theta_2 = 12.7(220)\cos 75^\circ = 723 \text{ W}$$

where θ_2 is the angle between $\mathbf{I}_\mathrm{B}$ and $\mathbf{V}_\mathrm{BC}$. Therefore, the total power is

$$P = P_1 + P_2 = 3421 \text{ W}$$

We note that all of the preceding calculations assume that the wattmeter itself absorbs negligible power.

EXAMPLE 12.8-2 Two-Wattmeter Method

The two wattmeters in Figure 12.8-2 read $P_1 = 60$ kW and $P_2 = 180$ W, respectively. Find the power factor of the circuit.

Solution

From Eq. 12.8-7, we have

$$\tan\theta = \sqrt{3}\frac{P_2 - P_1}{P_2 + P_1} = \sqrt{3}\frac{120}{240} = \frac{\sqrt{3}}{2} = 0.866$$

Therefore, we have $\theta = 40.9^\circ$ and the power factor is

$$pf = \cos\theta = 0.756$$

The positive angle, θ, indicates that the power factor is lagging. If θ is negative, then the power factor is leading.

EXERCISE 12.8-1 The line current to a balanced three-phase load is 24 A rms. The line-to-line voltage is 450 V rms, and the power factor of the load is 0.47 lagging. If two wattmeters are connected as shown in Figure 12.8-2, determine the reading of each meter and the total power to the load.

Answers: $P_1 = -371$ W, $P_2 = 9162$ W, and $P = 8791$ W

EXERCISE 12.8-2 The two wattmeters are connected as shown in Figure 12.8-2 with $P_1 = 60$ kW and $P_2 = 40$ kW, respectively. Determine (a) the total power and (b) the power factor.

Answers: **(a)** 100 kW **(b)** 0.945 leading

12.9 HOW CAN WE CHECK . . . ?

Engineers are frequently called upon to check that a solution to a problem is indeed correct. For example, proposed solutions to design problems must be checked to confirm that all of the specifications have been satisfied. In addition, computer output must be reviewed to guard against data-entry errors, and claims made by vendors must be examined critically.

Engineering students are also asked to check the correctness of their work. For example, occasionally just a little time remains at the end of an exam. It is useful to be able to quickly identify those solutions that need more work.

The following examples illustrate techniques useful for checking the solutions of the sort of problem discussed in this chapter.

EXAMPLE 12.9-1 How Can We Check Analysis of Three-Phase Circuits?

Figure 12.9-1*a* shows a balanced three-phase circuit. Computer analysis of this circuit produced the element voltages and currents tabulated in Figure 12.9-1*b*. **How can we check** that this computer analysis is correct?

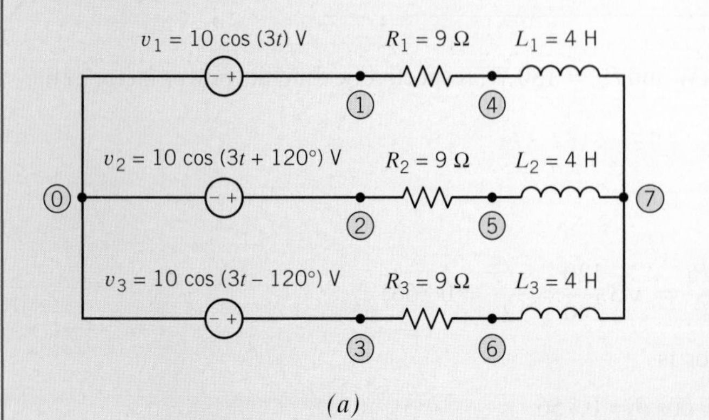

Element	Voltage	Current
V1 1 0 10 ∠0	10 ∠0	0.67 ∠127
V2 2 0 10 ∠120	10 ∠120	0.67 ∠113
V3 3 0 10 ∠–120	10 ∠–120	0.67 ∠7
R1 1 4 9	6 ∠–53	0.67 ∠–53
R2 2 5 9	6 ∠67	0.67 ∠67
R3 3 6 9	6 ∠–173	0.67 ∠–173
L1 4 7 4	8 ∠37	0.67 ∠–53
L2 5 7 4	8 ∠157	0.67 ∠67
L3 6 7 4	8 ∠83	0.67 ∠–173

(*b*)

FIGURE 12.9-1 (*a*) A three-phase circuit. (*b*) The results of computer analysis.

Solution

Because the three-phase circuit is balanced, it can be analyzed by using a per-phase equivalent circuit. The appropriate per-phase equivalent circuit for this example is shown in Figure 12.9-2. This per-phase equivalent circuit can be analyzed by writing a single-mesh equation:

$$10 = (9 + j12)\mathbf{I}_L(\omega)$$

or

$$\mathbf{I}_L(\omega) = 0.67e^{-j53^\circ}\text{ A}$$

where $\mathbf{I}_L(\omega)$ is the phasor corresponding to the inductor current. The voltage across the inductor is given by

$$\mathbf{V}_L(\omega) = j12\ \mathbf{I}_L(\omega) = 8e^{j37^\circ}\text{ V}$$

The voltage across the resistor is given by

$$\mathbf{V}_R(\omega) = 9\ \mathbf{I}_L(\omega) = 6e^{-j53^\circ}\text{ V}$$

These currents and voltages are the same as the values given in the computer analysis for the element currents and voltages of R_1 and L_1. We conclude that the computer analysis of the three-phase circuit is correct.

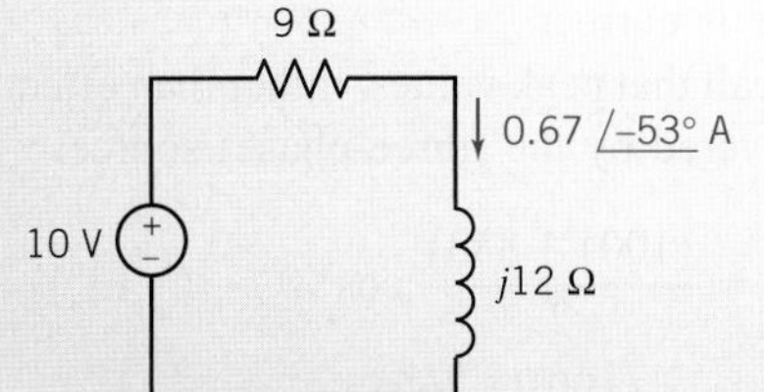

FIGURE 12.9-2 The per-phase equivalent circuit.

EXAMPLE 12.9-2 How Can We Check Unbalanced Three-Phase Circuits?

Computer analysis of the circuit in Figure 12.9-3 shows that $\mathbf{V}_{Nn}(\omega) = 12.67\angle 174.6^\circ$ V. This computer analysis did not use rms values, so 12.67 is the magnitude of the sinusoidal voltage $v_{Nn}(t)$ rather than the effective value. Verify that this voltage is correct.

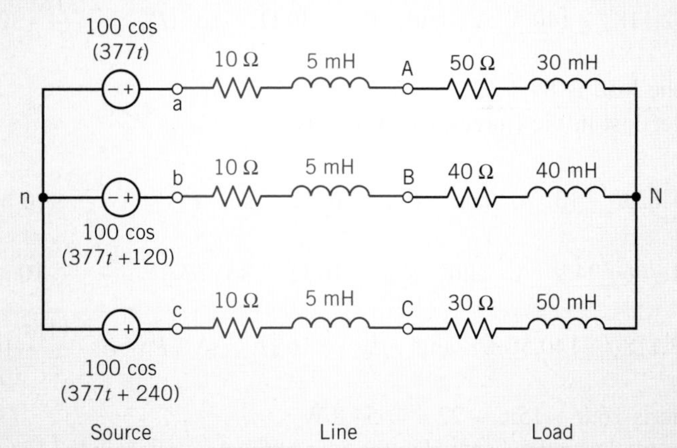

FIGURE 12.9-3 A three-phase circuit.

Solution

This result could be checked by writing and solving a node equation to calculate $\mathbf{V}_{\mathrm{Nn}}(\omega)$, but it is easier to check this result by verifying that KCL is satisfied at node N.

First, calculate the three line currents as

$$\mathbf{I}_{\mathrm{A}}(\omega) = \frac{100 - \mathbf{V}_{\mathrm{Nn}}(\omega)}{60 + j(377)(0.035)} = 1.833\angle{-13^\circ}\ \mathrm{A}$$

$$\mathbf{I}_{\mathrm{B}}(\omega) = \frac{100\angle{120^\circ} - \mathbf{V}_{\mathrm{Nn}}(\omega)}{50 + j(377)(0.045)} = 1.766\angle{94.9^\circ}\ \mathrm{A}$$

$$\mathbf{I}_{\mathrm{C}}(\omega) = \frac{100\angle{-120^\circ} - \mathbf{V}_{\mathrm{Nn}}(\omega)}{40 + j(377)(0.055)} = 2.118\angle{-140.5^\circ}\ \mathrm{A}$$

Next, apply KCL at node N to get

$$1.833\angle{-13^\circ} + 1.766\angle{95.9^\circ} + 2.118\angle{-140.5^\circ} = 0\mathrm{A}$$

Because KCL is satisfied at node N, the given node voltage is correct.

We can also check that average power is conserved. Recall that peak values, rather than effective values, are being used in this example. First, determine the power delivered by the (three-phase) source:

$$\mathbf{I}_{\mathrm{A}}(\omega) = 1.833\angle{-13^\circ}\ \mathrm{A}\ \text{and}\ \mathbf{V}_{\mathrm{an}}(\omega) = 100\angle{0^\circ}\ \mathrm{V},\ \text{so}\ P_{\mathrm{a}} = \frac{(100)(1.833)}{2}\cos\left(0^\circ - (-13^\circ)\right) = 89.3\ \mathrm{W}$$

$$\mathbf{I}_{\mathrm{B}}(\omega) = 1.766\angle{94.9^\circ}\ \mathrm{A}\ \text{and}\ \mathbf{V}_{\mathrm{bn}}(\omega) = 100\angle{120^\circ}\ \mathrm{V},\ \text{so}\ P_{\mathrm{b}} = \frac{(100)(1.766)}{2}\cos\left(120^\circ - (94.9^\circ)\right) = 80\ \mathrm{W}$$

$$\mathbf{I}_{\mathrm{C}}(\omega) = 2.118\angle{-140.5^\circ}\ \mathrm{A}\ \text{and}\ \mathbf{V}_{\mathrm{cn}}(\omega) = 100\angle{240^\circ}\ \mathrm{V},\ \text{so}\ P_{\mathrm{c}} = \frac{(100)(2.118)}{2}\cos\left(0^\circ + 140.5^\circ\right) = 99.2\ \mathrm{W}$$

The power delivered by the source is $89.3 + 80 + 99.2 = 268.5$ W.

Next, determine the power delivered to the (three-phase) load as

$$\mathbf{I}_{\mathrm{A}}(\omega) = 1.833\angle{-13^\circ}\ \mathrm{A}\ \text{and}\ R_{\mathrm{A}} = 50\ \Omega,\ \text{so}\ P_{\mathrm{A}} = \frac{1.833^2}{2}50 = 84.0\ \mathrm{W}$$

$$\mathbf{I}_{\mathrm{B}}(\omega) = 1.766\angle{94.9^\circ}\ \mathrm{A}\ \text{and}\ R_{\mathrm{B}} = 40\ \Omega,\ \text{so}\ P_{\mathrm{B}} = \frac{1.766^2}{2}40 = 62.4\ \mathrm{W}$$

$$\mathbf{I}_{\mathrm{C}}(\omega) = 2.118\angle{-140.5^\circ}\ \mathrm{A}\ \text{and}\ R_{\mathrm{C}} = 30\ \Omega,\ \text{so}\ P_{\mathrm{C}} = \frac{2.118^2}{2}30 = 67.3\ \mathrm{W}$$

The power delivered to the load is $84 + 62.4 + 67.3 = 213.7$ W.

Determine the power lost in the (three-phase) line as

$$\mathbf{I}_{\mathrm{A}}(\omega) = 1.833\angle{-13^\circ}\ \mathrm{A}\ \text{and}\ R_{\mathrm{aA}} = 10\ \Omega,\ \text{so}\ P_{\mathrm{aA}} = \frac{1.833^2}{2}10 = 16.8\ \mathrm{W}$$

$$\mathbf{I}_{\mathrm{B}}(\omega) = 1.766\angle{94.9^\circ}\ \mathrm{A}\ \text{and}\ R_{\mathrm{bB}} = 10\ \Omega,\ \text{so}\ P_{\mathrm{bB}} = \frac{1.766^2}{2}10 = 15.6\ \mathrm{W}$$

$$\mathbf{I}_{\mathrm{C}}(\omega) = 2.118\angle{-140.5^\circ}\ \mathrm{A}\ \text{and}\ R_{\mathrm{cC}} = 10\ \Omega,\ \text{so}\ P_{\mathrm{cC}} = \frac{2.118^2}{2}10 = 22.4\ \mathrm{W}$$

The power lost in the line is $16.8 + 15.6 + 22.4 = 54.8$ W.

The power delivered by the source is equal to the sum of the power lost in the line plus the power delivered to the load. Again, we conclude that the given node voltage is correct.

12.10 DESIGN EXAMPLE

POWER FACTOR CORRECTION

Figure 12.10-1 shows a three-phase circuit. The capacitors are added to improve the power factor of the load. We need to determine the value of the capacitance, C, required to obtain a power factor of 0.9 lagging.

Describe the Situation and the Assumptions

1. The circuit is excited by sinusoidal sources all having the same frequency, 60 Hz or 377 rad/s. The circuit is at steady state. The circuit is a linear circuit. Phasors can be used to analyze this circuit.
2. The circuit is a balanced three-phase circuit. A per-phase equivalent circuit can be used to analyze this circuit.
3. The load consists of two parts. The part comprising resistors and inductors is connected as a Y. The part comprising capacitors is connected as a Δ. A Δ-to-Y transformation can be used to simplify the load.

The per-phase equivalent circuit is shown in Figure 12.10-2.

State the Goal

Determine the value of C required to correct the power factor to 0.9 lagging.

Generate a Plan

Power factor correction was considered in Chapter 11. A formula was provided for calculating the reactance, X_1, needed to correct the power factor of a load

$$X_1 = \frac{R^2 + X^2}{R \tan(\cos^{-1} pfc) - X}$$

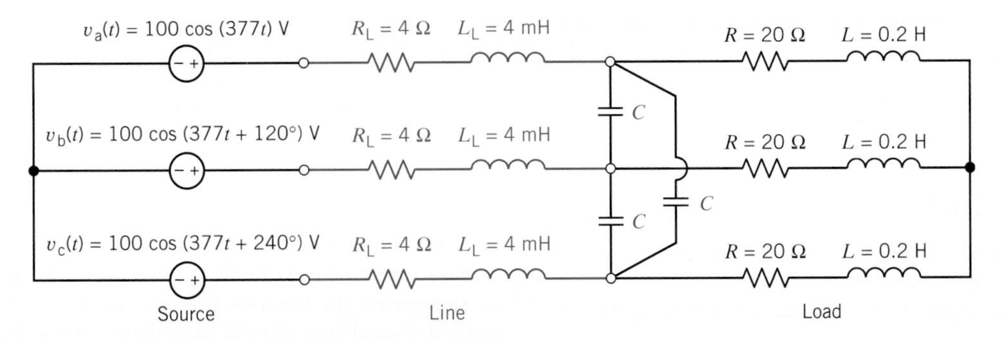

FIGURE 12.10-1 A balanced three-phase circuit.

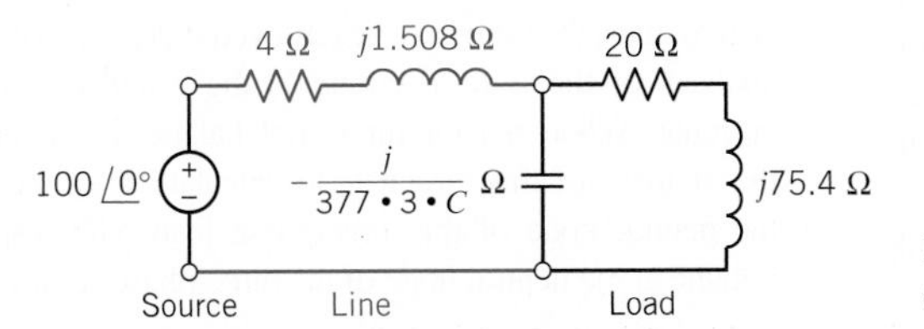

FIGURE 12.10-2 The per-phase equivalent circuit.

where R and X are the real and imaginary parts of the load impedance before the power factor is corrected and *pfc* is the corrected power factor. After this equation is used to calculate X_1, the capacitance, C, can be calculated from X_1. Notice that X_1 will be the reactance of the equivalent Y-connected capacitors. We will need to calculate the Δ-connected capacitor equivalent of the Y-connected capacitor.

Act on the Plan

We note that $\mathbf{Z} = R + jX = 20 + j75.4\ \Omega$. Therefore, the reactance, X_1, needed to correct the power factor is

$$X_1 = \frac{20^2 + 75.4^2}{20 \tan(\cos^{-1} 0.9) - 75.4} = -92.6$$

The Y-connected capacitor equivalent to the Δ-connected capacitor can be calculated from $\mathbf{Z}_Y = \mathbf{Z}_\Delta/3$. Therefore, the capacitance of the equivalent Y-connected capacitor is $3C$.

Finally, because $X_1 = 1/(3C\omega)$, we have

$$C = \frac{1}{\omega \cdot 3 \cdot X_1} = -\frac{1}{377 \cdot 3(-92.6)} = 9.548\ \mu\text{F}$$

Verify the Proposed Solution

When $C = 9.548\ \mu$F, the impedance of one phase of the equivalent Y-connected load will be

$$\mathbf{Z}_Y = \frac{\dfrac{1}{j377 \times 3 \times C}(20 + j75.4)}{\dfrac{1}{j377 \times 3 \times C} + (20 + j75.4)} = 246.45 + j119.4$$

The value of the power factor is

$$pf = \cos\left(\tan^{-1}\left(\frac{119.4}{246.45}\right)\right) = 0.90$$

so the specifications have been satisfied.

12.11 SUMMARY

- The generation and transmission of electrical power are more efficient in three-phase systems employing three voltages of the same magnitude and frequency and differing in phase by 120° from each other.
- The three-phase source consists of either three Y-connected sinusoidal voltage sources or three Δ-connected sinusoidal voltage sources. Similarly, the circuit elements that comprise the load are connected to form either a Y or a Δ. The transmission line connects the source to the load and consists of either three or four wires.
- Analysis of three-phase circuits using phasors and impedances will determine the *steady-state* response of the three-phase circuit. We are particularly interested in the power the three-phase source delivers to the three-phase load. Table 12.1-1 summarizes the formulas that are used to calculate the power delivered to an element when the element voltage and current adhere to the passive convention.
- The current in the neutral wire of a balanced Y-to-Y connection is zero; thus, the wire may be removed if desired. The key to the analysis of the Y-to-Y circuit is the calculation of the line currents. When the circuit is not balanced, the first step in the analysis of this circuit is to calculate $\mathbf{V}_{Nn}$, the voltage at the neutral node of the three-phase load with respect to the voltage at the neutral node of the three-phase source. When the

circuit is balanced, this step isn't needed because $\mathbf{V}_{Nn} = 0$. Once $\mathbf{V}_{Nn}$ is known, the line currents can be calculated. The line current for a balanced Y-to-Y connection is $\mathbf{V}_a/Z$ for phase a, and the other two currents are displaced by $\pm 120°$ from $\mathbf{I}_A$.

- For a Δ load, we converted the Δ load to a Y-connected load by using the relation Δ-to-Y transformation. Then we proceeded with the Y-to-Y analysis.
- The line current for a balanced Δ load is $\sqrt{3}$ times the phase current and is displaced $-30°$ in phase. The line-to-line voltage of a Δ load is equal to the phase voltage.
- The power delivered to a balanced Y-connected load is $P_Y = \sqrt{3}\, V_{AB} I_A \cos\theta$ where V_{AB} is the line-to-line voltage, I_A is the line current, and θ is the angle between the phase voltage and the phase current ($\mathbf{Z}_Y = Z\angle\theta$).
- The two-wattmeter method of measuring three-phase power delivered to a load was described. Also, we considered the usefulness of the two-wattmeter method for determining the power factor angle of a three-phase system.

PROBLEMS

Section 12.2 Three-Phase Voltages

P 12.2-1 A balanced three-phase Y-connected load has one phase voltage:

$$\mathbf{V}_c = 277\angle 45°\ \text{V rms}$$

The phase sequence is *abc*. Find the line-to-line voltages $\mathbf{V}_{AB}$, $\mathbf{V}_{BC}$, and $\mathbf{V}_{CA}$. Draw a phasor diagram showing the phase and line voltages.

P 12.2-2 A three-phase system has a line-to-line voltage

$$\mathbf{V}_{BA} = 12{,}570\angle -35°\ \text{V rms}$$

with a Y load. Find the phase voltages when the phase sequence is *abc*.

P 12.2-3 A three-phase system has a line-to-line voltage

$$\mathbf{V}_{ab} = 1600\angle 30°\ \text{V rms}$$

with a Y load. Determine the phase voltage.

Section 12.3 The Y-to-Y Circuit

P 12.3-1 Consider a three-wire Y-to-Y circuit. The voltages of the Y-connected source are $\mathbf{V}_a = (208/\sqrt{3})\ \angle 0°$ V rms, $\mathbf{V}_b = (208/\sqrt{3})\angle -120°$ V rms, and $\mathbf{V}_c = (208/\sqrt{3})\ \angle 120°$ V rms. The Y-connected load is balanced. The impedance of each phase is $\mathbf{Z} = 12\angle 30°\ \Omega$.

(a) Find the phase voltages.
(b) Find the line currents and phase currents.
(c) Show the line currents and phase currents on a phasor diagram.
(d) Determine the power dissipated in the load.

P 12.3-2 A balanced three-phase Y-connected supply delivers power through a three-wire plus neutral-wire circuit in a large office building to a three-phase Y-connected load. The circuit operates at 60 Hz. The phase voltages of the Y-connected source are $\mathbf{V}_a = 120\angle 0°$ V rms, $\mathbf{V}_b = 120\angle -120°$ V rms, and $\mathbf{V}_c = 120\angle 120°$ V rms. Each transmission wire, including the neutral wire, has a 2-Ω resistance, and the balanced Y load has a 10-Ω resistance in series with 100 mH. Find the line voltage and the phase current at the load.

P 12.3-3 A Y-connected source and load are shown in Figure P 12.3-3. (a) Determine the rms value of the current $i_a(t)$. (b) Determine the average power delivered to the load.

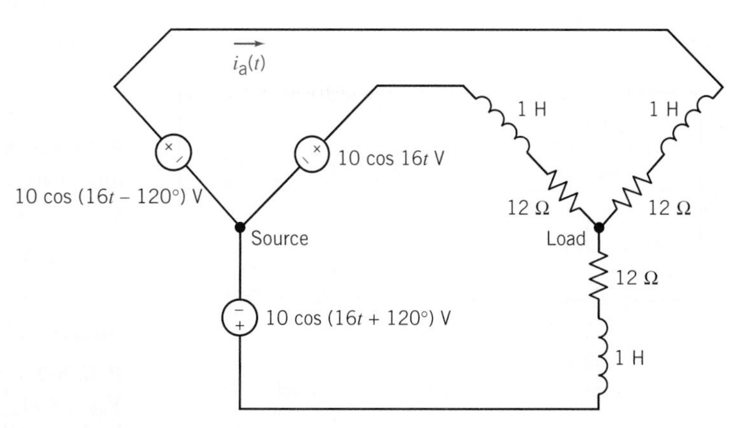

Figure P 12.3-3

P 12.3-4 An unbalanced Y–Y circuit is shown in Figure P 12.3-4. Find the average power delivered to the load.

Hint: $\mathbf{V}_{\mathrm{Nn}}(\omega) = 27.4 \angle -63.6$ V

Answer: 436.4 W

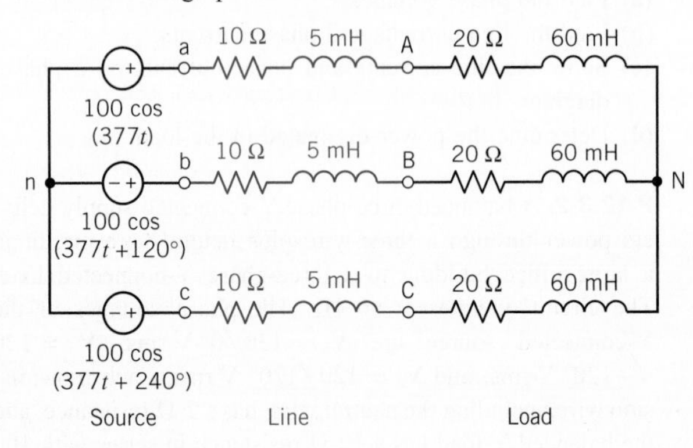

Figure P 12.3-4

P 12.3-5 A balanced Y–Y circuit is shown in Figure P 12.3-5. Find the average power delivered to the load.

Figure P 12.3-5

P 12.3-6 An unbalanced Y–Y circuit is shown in Figure P 12.3-6. Find the average power delivered to the load.

Hint: $\mathbf{V}_{\mathrm{Nn}}(\omega) = 1.755 \angle -29.5$ V

Answer: 436.4 W

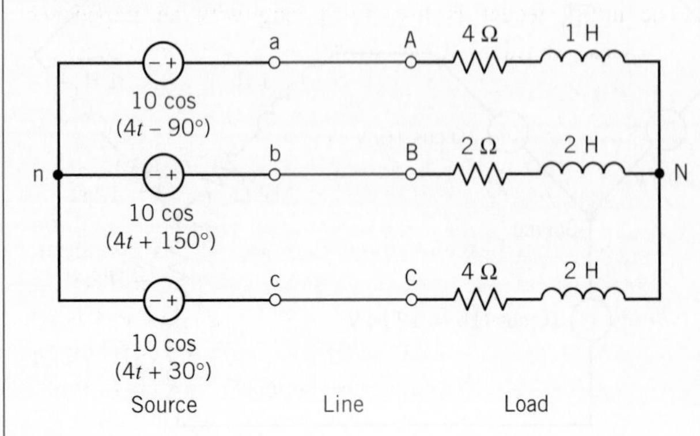

Figure P 12.3-6

P 12.3-7 A balanced Y–Y circuit is shown in Figure P 12.3-7. Find the average power delivered to the load.

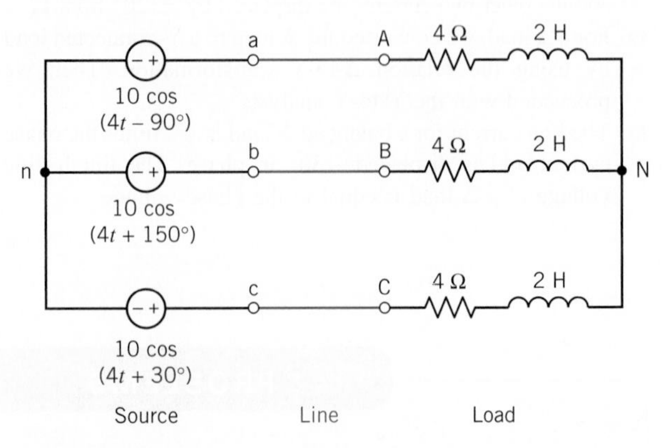

Figure P 12.3-7

Section 12.4 The Δ-Connected Source and Load

P 12.4-1 A balanced three-phase Δ-connected load has one line current:

$$\mathbf{I}_{\mathrm{B}} = 60 \angle -40^\circ \text{ A rms}$$

Find the phase currents $\mathbf{I}_{\mathrm{BC}}$, $\mathbf{I}_{\mathrm{AB}}$, and $\mathbf{I}_{\mathrm{CA}}$. Draw the phasor diagram showing the line and phase currents. The source uses the *abc* phase sequence.

P 12.4-2 A three-phase circuit has two parallel balanced Δ loads, one of 10-Ω resistors and one of 40-Ω resistors. Find the magnitude of the total line current when the line-to-line voltage is 480 V rms.

Section 12.5 The Y-to-Δ Circuit

P 12.5-1 Consider a three-wire Y-to-Δ circuit. The voltages of the Y-connected source are $\mathbf{V}_{\mathrm{a}} = (208/\sqrt{3}) \angle -30^\circ$ V rms, $\mathbf{V}_{\mathrm{b}} = (208/\sqrt{3}) \angle -150^\circ$ V rms, and $\mathbf{V}_{\mathrm{c}} = (208/\sqrt{3}) \angle 90^\circ$ V rms. The Δ-connected load is balanced. The impedance of each phase is $\mathbf{Z} = 12 \angle 30^\circ$ Ω. Determine the line currents and calculate the power dissipated in the load.

Answer: $P = 9360$ W

P 12.5-2 A balanced Δ-connected load is connected by three wires, each with a 4-Ω resistance, to a Y source with $\mathbf{V}_{\mathrm{a}} = (480/\sqrt{3}) \angle -30^\circ$ V rms, $\mathbf{V}_{\mathrm{b}} = (480/\sqrt{3}) \angle -150^\circ$ V rms, and $\mathbf{V}_{\mathrm{c}} = (480/\sqrt{3}) \angle 90^\circ$ V rms. Find the line current $\mathbf{I}_{\mathrm{A}}$ when $\mathbf{Z}_{\Delta} = 45 \angle -40^\circ$ Ω.

Answer: $\mathbf{I}_{\mathrm{A}} = 15.2 \angle 1.9^\circ$ A

P 12.5-3 The balanced circuit shown in Figure P 12.5-3 has $\mathbf{V}_{\mathrm{ab}} = 400 \angle 30^\circ$ V rms. Determine the phase currents in the load when $\mathbf{Z} = 3 + j4$ Ω. Sketch a phasor diagram.

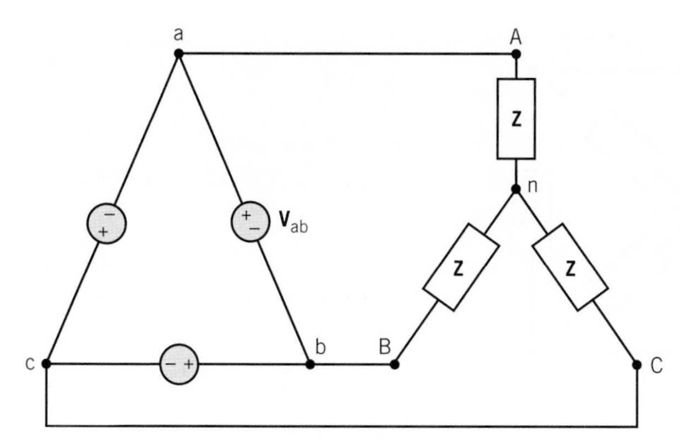

Figure P 12.5-3 A Δ-to-Y circuit.

P 12.5-4 The balanced circuit shown in Figure P 12.5-3 has $\mathbf{V}_{ab} = 400\angle 0°$ V rms. Determine the line and phase currents in the load when $\mathbf{Z} = 9 + j12\ \Omega$.

Section 12.6 Balanced Three-Phase Circuits

P 12.6-1 The English Channel Tunnel rail link is supplied at 25 kV rms from the United Kingdom and French grid systems. When there is a grid supply failure, each end is capable of supplying the whole tunnel but in a reduced operational mode.

The tunnel traction system is a conventional catenary (overhead wire) system similar to the surface mainline electric railway system of the United Kingdom and France. What makes the tunnel traction system different and unique is the high density of traction load and the end-fed supply arrangement. The tunnel traction load is considerable. For each half tunnel, the load is 180 MVA (Barnes and Wong, 1991).

Assume that each line-to-line voltage of the Y-connected source is 25 kV rms and the three-phase system is connected to the traction motor of an electric locomotive. The motor is a Y-connected load with $\mathbf{Z} = 150\angle 25°\ \Omega$. Find the line currents and the power delivered to the traction motor.

P 12.6-2 A three-phase source with a line voltage of 45 kV rms is connected to two balanced loads. The Y-connected load has $\mathbf{Z} = 10 + j20\ \Omega$, and the Δ load has a branch impedance of 60 Ω. The connecting lines have an impedance of 2 Ω. Determine the power delivered to the loads and the power lost in the wires. What percentage of power is lost in the wires?

P 12.6-3 A balanced three-phase source has a Y-connected source with $v_a = 10\cos(2t + 30°)$ connected to a three-phase Y load. Each phase of the Y-connected load consists of a 4-Ω resistor and a 8-H inductor. Each connecting line has a resistance of 2 Ω. Determine the total average power delivered to the load.

Section 12.7 Instantaneous and Average Power in a Balanced Three-Phase Load

P 12.7-1 Find the power absorbed by a balanced three-phase Y-connected load when

$$\mathbf{V}_{CB} = 208\angle 15°\ \text{V rms} \quad \text{and} \quad \mathbf{I}_B = 3\angle 110°\ \text{A rms}$$

The source uses the *abc* phase sequence.

Answer: $P = 620$ W

P 12.7-2 A three-phase motor delivers 20 hp operating from a 480-V rms line voltage. The motor operates at 85 percent efficiency with a power factor equal to 0.8 lagging. Find the magnitude and angle of the line current for phase A.

Hint: 1 hp = 745.7 W

P 12.7-3 A three-phase balanced load is fed by a balanced Y-connected source with a line-to-line voltage of 220 V rms. It absorbs 1500 W at 0.8 power factor lagging. Calculate the phase impedance if it is (a) Δ-connected and (b) Y-connected.

P 12.7-4 A 500-V rms three-phase Y-connected source has two balanced Δ loads connected to the lines. The load impedances are $40\angle 30°\ \Omega$ and $50\angle -60°\ \Omega$, respectively. Determine the line current and the total average power.

P 12.7-5 A three-phase Y-connected source simultaneously supplies power to two separate balanced three-phase loads. The first total load is Δ connected and requires 39 kVA at 0.7 lagging. The second total load is Y connected and requires 15 kW at 0.21 leading. Each line has an impedance $0.038 + j0.072\ \Omega$/phase. Calculate the line-to-line source voltage magnitude required so that the loads are supplied with 208-V rms line-to-line.

P 12.7-6 A building is supplied by a public utility at 4.2 kV rms. The building contains three balanced loads connected to the three-phase lines:

(a) Δ-connected, 500 kVA at 0.85 lagging
(b) Y-connected, 75 kVA at 0.0 leading
(c) Y connected; each phase with a 150-Ω resistor parallel to a 225-Ω inductive reactance

The utility feeder is five miles long with an impedance per phase of $1.69 + j0.78\ \Omega$/mile. At what voltage must the utility supply its feeder so that the building is operating at 4.2 rms?

Hint: 4.2 kV is the line-to-line voltage of the balanced Y-connected source.

P 12.7-7 The diagram shown in P 12.7-7 has two three-phase loads that form part of a manufacturing plant. They are connected in parallel and require 4.16 kV rms. Load 1 is 1.5 MVA, 0.75 lag *pf*, Δ-connected. Load 2 is 2 MW, 0.8 lagging *pf*, Y-connected. The feeder from the power utility's substation transformer has an impedance of $0.4 + j0.8\ \Omega$/phase. Determine the following:

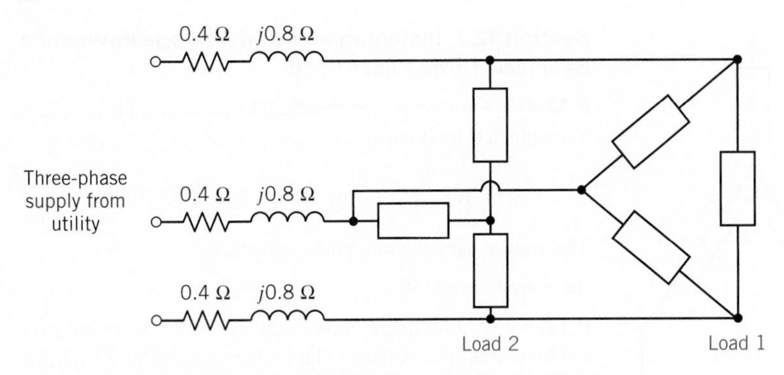

Figure P 12.7-7 A three-phase circuit with a Δ load and a Y load.

(a) The required magnitude of the line voltage at the supply
(b) The real power drawn from the supply
(c) The percentage of the real power drawn from the supply that is consumed by the loads

P 12.7-8 The balanced three-phase load of a large commercial building requires 480 kW at a lagging power factor of 0.8. The load is supplied by a connecting line with an impedance of $5 + j25\ \text{m}\Omega$ for each phase. Each phase of the load has a line-to-line voltage of 500 V rms. Determine the line current and the line voltage at the source. Also, determine the power factor at the source. Use the line-to-neutral voltage as the reference with an angle of 0°.

Section 12.8 Two-Wattmeter Power Measurement

P 12.8-1 The two-wattmeter method is used to determine the power drawn by a three-phase 440-V rms motor that is a Y-connected balanced load. The motor operates at 20 hp at 74.6 percent efficiency. The magnitude of the line current is 52.5 A rms. The wattmeters are connected in the *A* and *C* lines. Find the reading of each wattmeter. The motor has a lagging power factor.

Hint: 1 hp = 745.7 W

P 12.8-2 A three-phase system has a line-to-line voltage of 5000 V rms and a balanced Δ-connected load with $Z = 40 + j30\ \Omega$. The phase sequence is *abc*. Use the two wattmeters connected to lines *A* and *C*, with line *B* as the common line for the voltage measurement. Determine the total power measurement recorded by the wattmeters.

Answer: $P = 1199.5$ kW

P 12.8-3 A three-phase system with a sequence *abc* and a line-to-line voltage of 200 V rms feeds a Y-connected load with $\mathbf{Z} = 70.7\angle 45^\circ\ \Omega$. Find the line currents. Find the total power by using two wattmeters connected to lines *B* and *C*.

Answer: $P = 400$ W

P 12.8-4 A three-phase system with a line-to-line voltage of 210 V rms and phase sequence *abc* is connected to a Y-balanced load with impedance $10\angle -30^\circ\ \Omega$ and a balanced Δ load with impedance $15\angle 30^\circ\ \Omega$. Find the line currents and the total power using two wattmeters.

P 12.8-5 The two-wattmeter method is used. The wattmeter in line *A* reads 920 W, and the wattmeter in line *C* reads 460 W. Find the impedance of the balanced Δ-connected load. The circuit is a three-phase 120-V rms system with an *abc* sequence.

Answer: $\mathbf{Z}_\Delta = 27.1\angle -30^\circ\ \Omega$

P 12.8-6 Using the two-wattmeter method, determine the power reading of each wattmeter and the total power for Problem 12.5-1 when $\mathbf{Z} = 0.868 + j4.924\ \Omega$. Place the current coils in the *A*-to-*a* and *C*-to-*c* lines.

Section 12.9 How Can We Check . . . ?

P 12.9-1 A Y-connected source is connected to a Y-connected load (Figure 12.3-1) with $\mathbf{Z} = 10 + j4\ \Omega$. The line voltage is $V_L = 420$ V rms. A student report states that the line current $\mathbf{I}_A = 39$ A rms and that the power delivered to the load is 16.1 kW. Verify these results.

P 12.9-2 A Δ load with $\mathbf{Z} = 40 + j30\ \Omega$ has a three-phase source with $V_L = 250$ V rms (Figure 12.3-2). A computer analysis program states that one phase current is $5\angle -36.9^\circ$ A. Verify this result.

PSpice Problems

SP 12-1 Use PSpice to determine the power delivered to the load in the circuit shown in Figure SP 12-1.

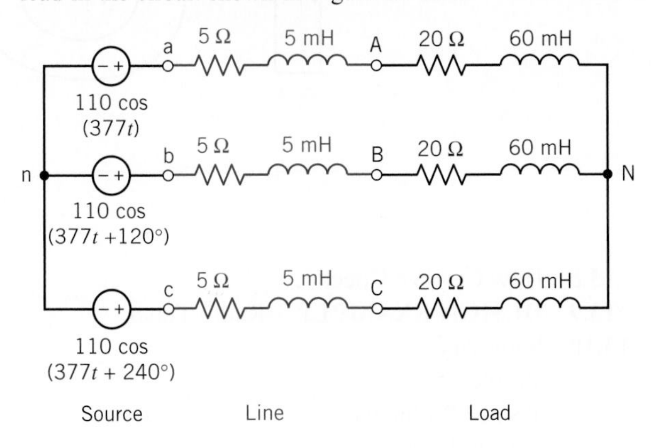

Figure SP 12-1

SP 12-2 Use PSpice to determine the power delivered to the load in the circuit shown in Figure SP 12-2.

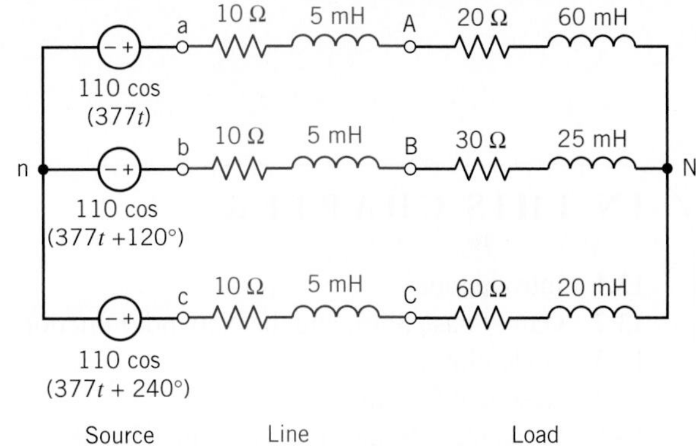

Figure SP 12-2

Design Problems

DP 12-1 A balanced three-phase Y source has a line voltage of 208 V rms. The total power delivered to the balanced Δ load is 1200 W with a power factor of 0.94 lagging. Determine the required load impedance for each phase of the Δ load. Calculate the resulting line current. The source is a 208-V rms *ABC* sequence.

DP 12-2 A three-phase 240-V rms circuit has a balanced Y-load impedance **Z**. Two wattmeters are connected with current coils in lines *A* and *C*. The wattmeter in line *A* reads 1440 W, and the wattmeter in line *C* reads zero. Determine the value of the impedance.

DP 12-3 A three-phase motor delivers 100 hp and operates at 80 percent efficiency with a 0.75 lagging power factor. Determine the required Δ-connected balanced set of three capacitors that will improve the power factor to 0.90 lagging. The motor operates from 480-V rms lines.

DP 12-4 A three-phase system has balanced conditions so that the per-phase circuit representation can be used as shown in Figure DP 12-4. Select the turns ratio of the step-up and step-down transformers so that the system operates with an efficiency greater than 99 percent. The load voltage is specified as 4 kV rms, and the load impedance is $4/3\ \Omega$.

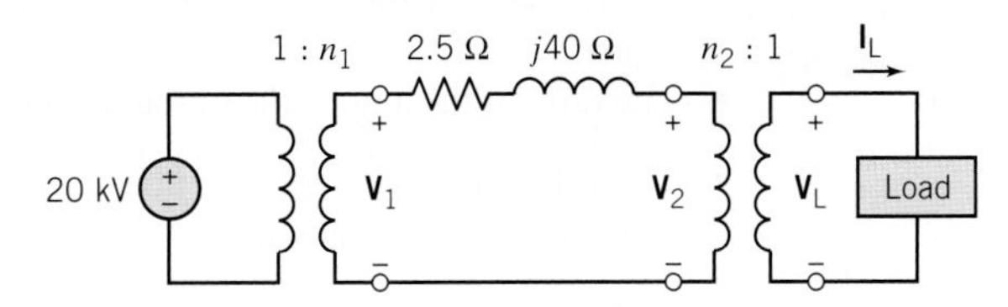

Figure DP 12-4

CHAPTER 

Frequency Response

IN THIS CHAPTER

13.1 Introduction
13.2 Gain, Phase Shift, and the Network Function
13.3 Bode Plots
13.4 Resonant Circuits
13.5 Frequency Response of Op Amp Circuits
13.6 Plotting Bode Plots Using MATLAB
13.7 Using PSpice to Plot a Frequency Response
13.8 How Can We Check . . . ?
13.9 **DESIGN EXAMPLE**—Radio Tuner
13.10 Summary
Problems
PSpice Problems
Design Problems

13.1 INTRODUCTION

Consider the experiment illustrated in Figure 13.1-1. Here a function generator provides the input to a linear circuit and the oscilloscope displays the output, or response, of the linear circuit. The linear circuit itself consists of resistors, capacitors, inductors, and perhaps dependent sources and/or op amps. The function generator allows us to choose from several types of input function.

Suppose we select a sinusoidal input. The function generator permits us to adjust the amplitude, phase angle, and frequency of the input. First, we notice that no matter what adjustments we make, the (steady-state) response is always a sine wave at the same frequency as the input. The amplitude and phase angle of the output differ from the input, but the frequency is always the same as the frequency of the input.

After a little more experimentation, we find that **at any fixed frequency**, the following are true:

- The ratio of the amplitude of the output sinusoid to the amplitude of the input sinusoid is a constant.
- The difference between the phase angle of the output sinusoid and the phase angle of the input sinusoid is also constant.

The situation is not as simple when we vary the frequency of the input. Now the amplitude and phase angle of the output change in a more complicated way.

In this chapter, we will develop analytical tools that enable us to predict how the amplitude and phase angle of the output sinusoid will change as we vary the frequency of the input sinusoid.

13.2 GAIN, PHASE SHIFT, AND THE NETWORK FUNCTION

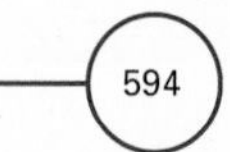

Gain, phase shift, and the network function are properties of linear circuits that describe the effect a circuit has on a sinusoidal input voltage or current. We expect that the behavior of circuits that contain reactive elements, that is, capacitors or inductors, will depend on the frequency of the input sinusoid.

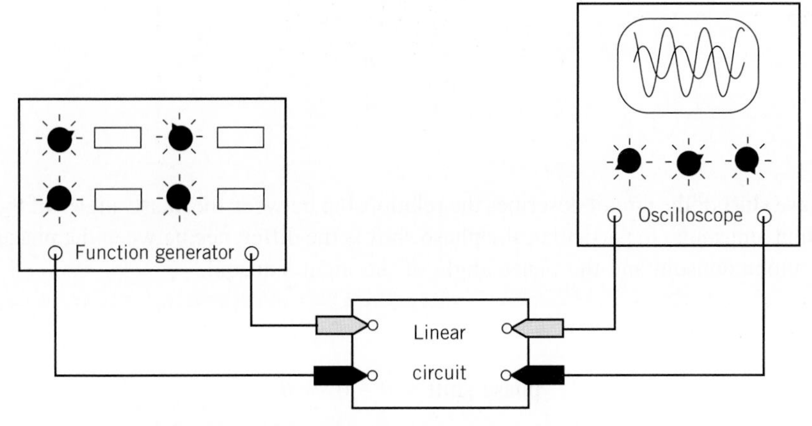

FIGURE 13.1-1 Measuring the input and output of a linear circuit.

Thus, we expect that the gain, phase shift, and network function will all be functions of frequency. Indeed, we will see that this is the case.

We begin by considering the circuit shown in Figure 13.2-1. The input to this circuit is the voltage of the voltage source, and the output, or response, of the circuit is the voltage across the 10-kΩ resistor. When the input is a sinusoidal voltage, the steady-state response will also be sinusoidal and will have the same frequency as the input.

Suppose the voltages $v_{in}(t)$ and $v_{out}(t)$ are measured using an oscilloscope. Figure 13.2-2 shows the waveforms that would be displayed on the screen of the oscilloscope. Notice that the scales are shown, but the axes are not. It is customary to take the angle of the input signal to be 0°, that is,

$$v_{in}(t) = A \cos \omega t$$

Then,

$$v_{out}(t) = B \cos (\omega t + \theta)$$

The **gain** of the circuit describes the relationship between the sizes of the input and output sinusoids. In particular, the gain is the ratio of the amplitude of the output sinusoid to the amplitude of the input sinusoid.

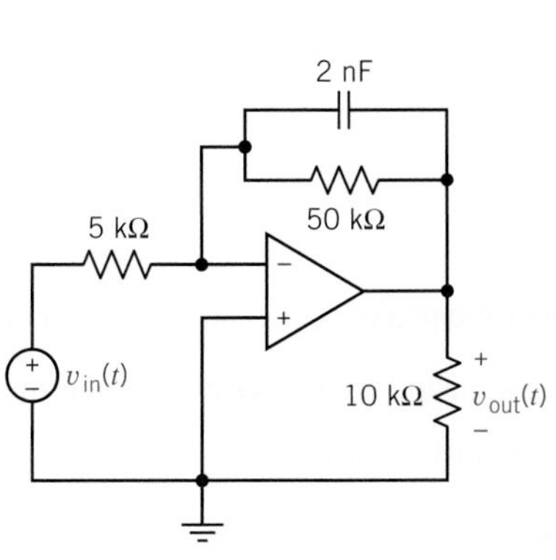

FIGURE 13.2-1 An op amp circuit.

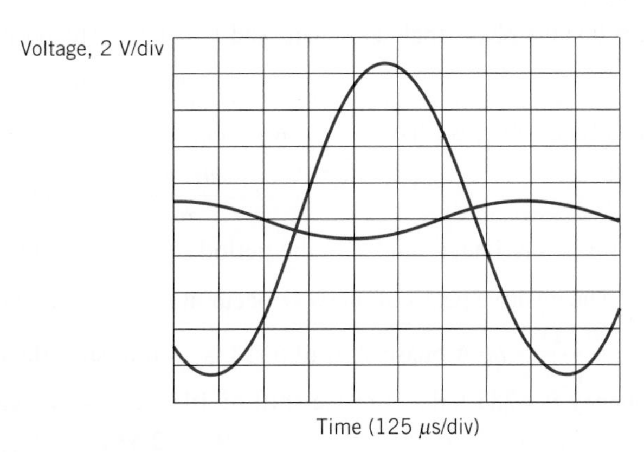

FIGURE 13.2-2 Input and output sinusoids for the op amp circuit of Figure 13.2-1.

That is,

$$\text{gain} = \frac{B}{A}$$

The **phase shift** of the circuit describes the relationship between the phase angles of the input and output sinusoids. In particular, the phase shift is the difference between the phase angle of the output sinusoid and the phase angle of the input sinusoid.

That is,

$$\text{phase shift} = \theta - 0^\circ = \theta$$

To be more specific, we need analytic representations of the sinusoids shown in Figure 13.2-2. The input voltage is the smaller of the two sinusoids and can be represented as

$$v_{in}(t) = 1 \cos 6283t \text{ V}$$

The steady-state response is the larger of the two sinusoids and can be represented as

$$v_{out}(t) = 8.47 \cos (6283t + 148^\circ) \text{ V}$$

The **gain** of this circuit at the frequency $\omega = 6283$ rad/s is

$$\text{gain} = \frac{\text{output amplitude}}{\text{input amplitude}} = \frac{8.47}{1} = 8.47$$

This gain is unitless because both amplitudes have units of volts. Because the gain is greater than 1, the output sinusoid is larger than the input sinusoid. This circuit is said to *amplify* its input. When the gain of a circuit is less than 1, the output sinusoid is smaller than the input sinusoid. This circuit is said to *attenuate* its input.

The *phase shift* of this circuit at the frequency $\omega = 6283$ rad/s is

$$\text{phase shift} = \text{output phase angle} - \text{input phase angle} = 148^\circ - 0^\circ = 148^\circ$$

The phase shift determines the amount of time the output is advanced or delayed with respect to the input. Notice that

$$B \cos (\omega t + \theta) = B \cos \left(\omega \left(t + \frac{\theta}{\omega} \right) \right) = B \cos (\omega (t + t_0))$$

where θ is the phase angle in radians and $t_0 = \theta/\omega$. The positive peaks of $B \cos (\omega t + \theta)$ occur when

$$\omega t + \theta = n(2\pi)$$

and, solving for t, we have

$$t = \frac{n(2\pi)}{\omega} - t_0 = nT - t_0$$

where n is any integer and T is the period of the sinusoid.

The positive peaks of $A \cos \omega t$ occur at $t = \frac{n(2\pi)}{\omega}$ and the positive peaks of $B \cos(\omega t + \theta)$ occur at $t = \frac{n(2\pi)}{\omega} - t_0$. A phase shift of θ rad is seen to shift the output sinusoid by t_0 seconds. When the frequency is 6283 rad/s, a phase shift of 148°or 2.58 rad causes a shift in time equal to

$$t_0 = \frac{\theta}{\omega} = \frac{2.58 \text{ rad}}{6283 \text{ rad/s}} = 410\ \mu\text{s}$$

In Figure 13.2-2, the positive peaks of the input sinusoid occur at 0 ms, 1 ms, 2 ms, 3 ms, ... Positive peaks of the output sinusoid occur at 0.59 ms, 1.59 ms, 2.59 ms, 3.59 ms, ... Peaks of the output sinusoid occur 410 μs before the next peak of the input sinusoid. The output is *advanced* by 410 μs with respect to the input.

Notice that

$$v_{\text{out}}(t) = 8.47 \cos(6283t + 148°) = 8.47 \cos(6283t - 212°)$$

because a phase shift of 360° does not change the sinusoid. A phase shift of −212° or −3.70 rad causes a shift in time of

$$t_0 = \frac{-3.70 \text{ rad}}{6283 \text{ rad/s}} = -590 \ \mu\text{s}$$

Peaks of the output sinusoid occur 590 μs *after* the previous of the input sinusoid. The output is *delayed* by 590 μs with respect to the input.

A phase shift that advances the output is called a **phase lead**. A phase shift that delays the output is called a **phase lag**.

At the frequency $\omega = 6283$ rad/s, this circuit amplifies its input by a factor of 8.47 and advances it by 410 μs or, equivalently, delays it by 590 μs. The circuit of Figure 13.2-1 has a phase lead of 148° or, equivalently, a phase lag of 212°.

Now let us consider this circuit when the frequency of the input is changed. When the input is

$$v_{\text{in}}(t) = 1 \cos 3141.6t \text{ V}$$

the steady-state response of this circuit can be found to be

$$v_{\text{out}}(t) = 9.54 \cos(3141.6t + 163°) \text{ V}$$

The gain and phase shift of this circuit at the frequency $\omega = 3141.6$ rad/s are

$$\text{gain} = \frac{\text{output amplitude}}{\text{input amplitude}} = \frac{9.54}{1} = 9.54$$

and

$$\text{phase shift} = \text{output phase angle} - \text{input phase angle} = 163° - 0° = 163°$$

Changing the frequency of the input has changed the gain and phase shift of this circuit. Apparently, the gain and the phase shift of this circuit are functions of the frequency of the input. Table 13.2-1

Table 13.2-1 Frequency Response Data for a Circuit

f(Hz)	ω (rad/s)	GAIN	PHASE SHIFT
100	628.3	9.98	176°
500	3,141.6	9.54	163°
1,000	6,283	8.47	148°
5,000	31,416	3.03	108°
10,000	62,830	1.57	99°

shows the values of the gain and phase shift corresponding to several choices of the input frequency. As expected, the gain and phase shift changed when the input frequency changed. The **network function** describes the way the behavior of the circuit depends on the frequency of the input. The network function is defined in the frequency domain. It is the ratio of the phasor corresponding to the response sinusoid to the phasor corresponding to the input. Let $\mathbf{X}(\omega)$ be the phasor corresponding to the input to the circuit and $\mathbf{Y}(\omega)$ be the phasor corresponding to the steady-state response of the network. Then,

$$\mathbf{H}(\omega) = \frac{\mathbf{Y}(\omega)}{\mathbf{X}(\omega)} \tag{13.2-1}$$

is the network function. Notice that both $\mathbf{X}(\omega)$ and $\mathbf{Y}(\omega)$ could correspond to either a current or a voltage. Both the gain and the phase shift can be expressed in terms of the network function. The gain is

$$\text{gain} = |\mathbf{H}(\omega)| = \frac{|\mathbf{Y}(\omega)|}{|\mathbf{X}(\omega)|} \tag{13.2-2}$$

and the phase shift is

$$\text{phase shift} = \angle\mathbf{H}(\omega) = \angle\mathbf{Y}(\omega) - \angle\mathbf{X}(\omega) \tag{13.2-3}$$

Consider the problem of finding the network function of a given circuit. To solve such a problem, we do two things. First, we represent the circuit in the frequency domain using impedances and phasors. (We also represented the circuit in the frequency domain when we wanted to find the steady-state response to a sinusoidal input. In that case, the frequency was represented as the value of the frequency of the sinusoidal input, for example, 4 rad/s. When we find the network function, the frequency is represented by a variable, ω). Second, we analyze the circuit to determine the ratio of the phasor corresponding to the circuit output to the phasor corresponding to the circuit input. This analysis might involve mesh equations or node equations or equivalent impedances and voltage or current division. In any case, the analysis is performed in the frequency domain.

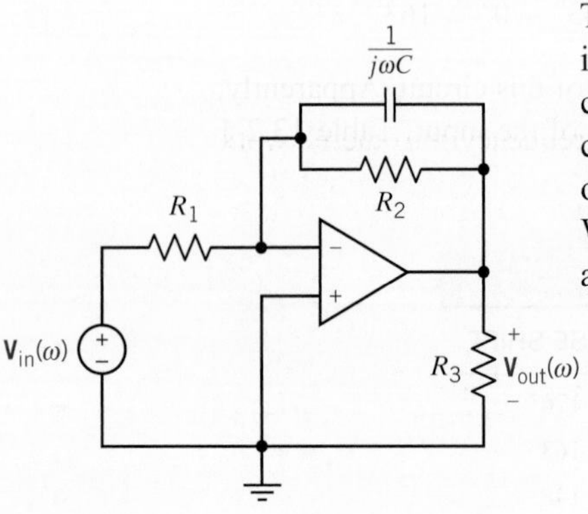

FIGURE 13.2-3 The frequency-domain representation of the op amp circuit of Figure 13.2-1.

Let's find the network function for the circuit shown in Figure 13.2-1. The first step is to represent this circuit in the frequency domain using impedances and phasors. Figure 13.2-3 shows the frequency-domain circuit corresponding to the circuit in Figure 13.2-1. In this example, the phasor corresponding to the input is $\mathbf{V}_{in}(\omega)$, and the phasor corresponding to the output is $\mathbf{V}_{out}(\omega)$. We seek to find the network function $\mathbf{H}(\omega) = \mathbf{V}_{out}/\mathbf{V}_{in}$. Write the node equation at the inverting input node of the op amp and assume an ideal op amp. Then we have

$$\frac{\mathbf{V}_{in}(\omega)}{R_1} + \frac{\mathbf{V}_{out}(\omega)}{R_2} + j\omega C\mathbf{V}_{out}(\omega) = 0$$

This implies

$$\mathbf{H}(\omega) = \frac{\mathbf{V}_{out}(\omega)}{\mathbf{V}_{in}(\omega)} = \frac{-R_2}{R_1 + j\omega C R_1 R_2}$$

The gain of this circuit is

$$\text{gain} = |\mathbf{H}(\omega)| = \mathrm{H} = \frac{R_2/R_1}{\sqrt{1+\omega^2 C^2 R_2^2}}$$

The phase shift of this circuit is

$$\text{phase shift} = \underline{/\mathbf{H}(\omega)} = 180^\circ - \tan^{-1}(\omega C R_2)$$

When $R_1 = 5\,\text{k}\Omega$, $R_2 = 50\,\text{k}\Omega$, and $C = 2\,\text{nF}$,

$$\begin{aligned}
\mathbf{H}(\omega) &= \frac{-10}{1+(j\omega/10,000)} \\
\text{gain} &= |\mathbf{H}(\omega)| = \frac{10}{\sqrt{1+(\omega^2/10^8)}} \\
\text{phase shift} &= \underline{/\mathbf{H}(\omega)} = 180^\circ - \tan^{-1}(\omega/10,000)
\end{aligned}$$

Notice that the frequency of the input has been represented by a variable, ω, rather than by any particular value. As a result, the network function, gain, and phase shift describe the way in which the behavior of the circuit depends on the input frequency. Earlier, we considered the case when $\omega = 6283$ rad/s. Substituting this frequency into the equations for the gain and phase shift gives

$$\text{gain} = \frac{10}{\sqrt{1+\dfrac{6283^2}{10^8}}} = 8.47$$

and

$$\text{phase shift} = 180^\circ - \tan^{-1}(6283/10,000) = 148^\circ$$

These are the same results as were obtained earlier by examining the oscilloscope traces in Figure 13.2-2. Similarly, each line of Table 13.2-1 can be obtained by substituting the appropriate frequency into the equations for the gain and phase shift.

Equations that represent the gain and phase shift as functions of frequency are called the **frequency response** of the circuit. The same information can be represented by a table or by graphs instead of equations. These tables or graphs are also called the frequency response of the circuit.

To see that the network function really does represent the behavior of the circuit, suppose that

$$v_{\text{in}}(t) = 0.4\cos(5000t + 45^\circ)\ \text{V}$$

The frequency of the input sinusoid is $\omega = 5000$ rad/s. Substituting this frequency into the network function gives

$$\mathbf{H}(\omega) = \frac{-10}{1+(j5000/10,000)} = 8.94\,\underline{/153^\circ}$$

Next,

$$\mathbf{V}_{\text{out}}(\omega) = \mathbf{H}(\omega)\mathbf{V}_{\text{in}}(\omega) = (8.94\,\underline{/153^\circ})(0.4\,\underline{/45^\circ}) = 3.58\,\underline{/198^\circ}$$

Back in the time domain, the steady-state response is

$$v_{\text{out}}(t) = 3.58\cos(5000t + 198^\circ)\ \text{V}$$

Notice that the network function contained enough information to enable us to calculate the steady-state response from the input sinusoid. The network function does indeed describe the behavior of the circuit.

EXAMPLE 13.2-1 Network Function of a Circuit

INTERACTIVE EXAMPLE

Consider the circuit shown in Figure 13.2-4*a*. The input to the circuit is the voltage of the voltage source, $v_i(t)$. The output is the voltage, $v_o(t)$, across the series connection of the capacitor and the 16-kΩ resistor. The network function that represents this circuit has the form

$$\mathbf{H}(\omega) = \frac{\mathbf{V}_o(\omega)}{\mathbf{V}_i(\omega)} = \frac{1 + j\dfrac{\omega}{z}}{1 + j\dfrac{\omega}{p}} \tag{13.2-4}$$

The network function depends on two parameters, z and p. The parameter z is called the zero of the circuit and the parameter p is called the pole of the circuit. Determine the values of z and of p for the circuit in Figure 13.2-4*a*.

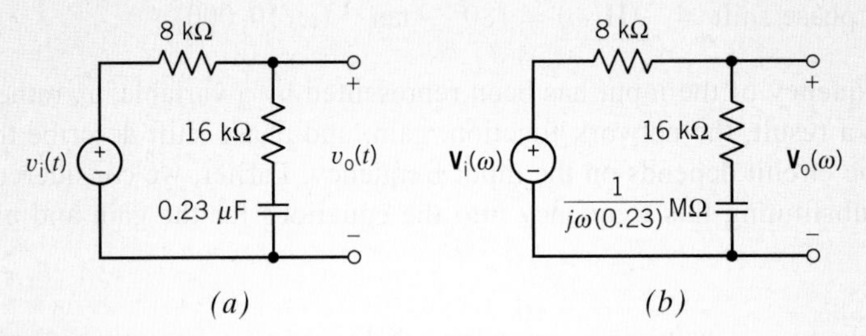

FIGURE 13.2-4 The circuit considered in Example 13.2-1 represented (*a*) in the time domain and (*b*) in the frequency domain.

Solution

We will analyze the circuit to determine its network function and then put the network function into the form given in Eq. 13.2-4. A network function is the ratio of the output phasor to the input phasor. Phasors exist in the frequency domain. Consequently, our first step is to represent the circuit in the frequency domain, using phasors and impedances. Figure 13.2-4*b* shows the frequency-domain representation of the circuit from Figure 13.2-4*a*.

The impedances of the capacitor and the 16-kΩ resistor are connected in series in Figure 13.2-4*b*. The equivalent impedance is

$$\mathbf{Z}_e(\omega) = 16{,}000 + \frac{10^6}{j(0.23)\omega}$$

The equivalent impedance is connected in series with the 8-kΩ resistor. $\mathbf{V}_i(\omega)$ is the voltage across the series impedances, and $\mathbf{V}_o(\omega)$ is the voltage across the equivalent impedance, $\mathbf{Z}_e(\omega)$. Apply the voltage division principle to get

$$\mathbf{V}_o(\omega) = \frac{16{,}000 + \dfrac{10^6}{j(0.23)\omega}}{8000 + 16{,}000 + \dfrac{10^6}{j(0.23)\omega}}\mathbf{V}_i(\omega) = \frac{10^6 + j(0.23)\omega(16{,}000)}{10^6 + j(0.23)\omega(24{,}000)}\mathbf{V}_i(\omega)$$

$$= \frac{\dfrac{10^6 + j(3680)\omega}{10^6}}{\dfrac{10^6 + j(5520)\omega}{10^6}}\mathbf{V}_i(\omega) = \frac{1 + j(0.00368)\omega}{1 + j(0.00552)\omega}\mathbf{V}_i(\omega)$$

Divide both sides of this equation by $\mathbf{V}_i(\omega)$ to obtain the network function of the circuit

$$\mathbf{H}(\omega) = \frac{\mathbf{V}_o(\omega)}{\mathbf{V}_i(\omega)} = \frac{1 + j(0.00368)\omega}{1 + j(0.00552)\omega} \tag{13.2-5}$$

Equating the network functions given by Eq. 13.2-4 and 13.2-5 gives

$$\frac{1+j(0.00368)\omega}{1+j(0.00552)\omega} = \frac{1+j\dfrac{\omega}{z}}{1+j\dfrac{\omega}{p}}$$

Comparing these network functions gives

$$z = \frac{1}{0.00368} = 271.74 \text{ rad/s} \quad \text{and} \quad p = \frac{1}{0.00552} = 181.16 \text{ rad/s}$$

EXAMPLE 13.2-2 Network Function of a Circuit

INTERACTIVE EXAMPLE

Consider the circuit shown in Figure 13.2-5*a*. The input to the circuit is the voltage of the voltage source, $v_i(t)$. The output is the voltage, $v_o(t)$, across the series connection of the inductor and the 2-Ω resistor. The network function that represents this circuit is

$$\mathbf{H}(\omega) = \frac{\mathbf{V}_o(\omega)}{\mathbf{V}_i(\omega)} = 0.2\frac{1+j\dfrac{\omega}{5}}{1+j\dfrac{\omega}{25}} \tag{13.2-6}$$

Determine the value of the inductance, L.

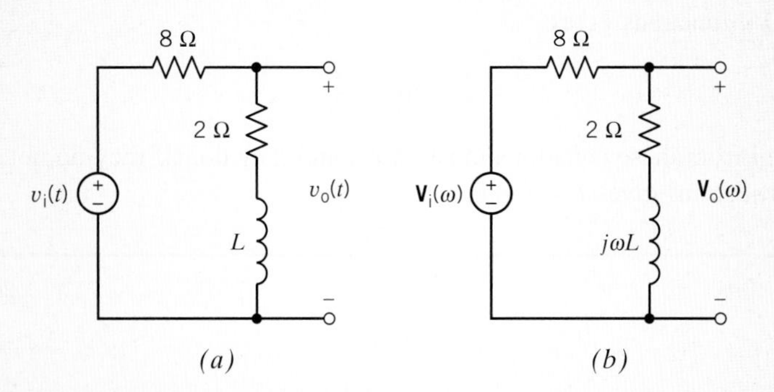

FIGURE 13.2-5 The circuit considered in Example 13.2-2 represented (*a*) in the time domain and (*b*) in the frequency domain.

Solution

The circuit has been represented twice, by a circuit diagram and by a network function. The unknown inductance, L, appears in the circuit diagram but not in the given network function. We can analyze the circuit to determine its network function. This second network function will depend on the unknown inductance. We will determine the value of the inductance by equating the two network functions.

A network function is the ratio of the output phasor to the input phasor. Phasors exist in the frequency domain. Consequently, our first step is to represent the circuit in the frequency domain, using phasors and impedances. Figure 13.2-5*b* shows the frequency-domain representation of the circuit from Figure 13.2-5*a*.

The impedances of the inductor and the 2-Ω resistor are connected in series in Figure 13.2-5*b*. The equivalent impedance is

$$\mathbf{Z}_e(\omega) = 2 + j\omega L$$

The equivalent impedance is connected in series with the 8-Ω resistor. $\mathbf{V}_i(\omega)$ is the voltage across the series impedances, and $\mathbf{V}_o(\omega)$ is the voltage across the equivalent impedance, $\mathbf{Z}_e(\omega)$. Apply the voltage division principle to get

$$\mathbf{V}_o(\omega) = \frac{2 + j\omega L}{8 + 2 + j\omega L}\mathbf{V}_i(\omega) = \frac{2 + j\omega L}{10 + j\omega L}\mathbf{V}_i(\omega)$$

Divide both sides of this equation by $\mathbf{V}_i(\omega)$ to obtain the network function of the circuit:

$$\mathbf{H}(\omega) = \frac{\mathbf{V}_o(\omega)}{\mathbf{V}_i(\omega)} = \frac{2 + j\omega L}{10 + j\omega L}$$

Next, we put the network function into the form specified by Eq. 13.2-6. Factoring 2 out of both terms in the numerator and factoring 10 out of both terms in the denominator, we get

$$\mathbf{H}(\omega) = \frac{2\left(1 + j\omega\frac{L}{2}\right)}{10\left(1 + j\omega\frac{L}{10}\right)} = 0.2\frac{1 + j\omega\frac{L}{2}}{1 + j\omega\frac{L}{10}} \tag{13.2-7}$$

Equating the network functions given by Eqs. 13.2-6 and 13.2-7 gives

$$0.2\frac{1 + j\omega\frac{L}{2}}{1 + j\omega\frac{L}{10}} = 0.2\frac{1 + j\frac{\omega}{5}}{1 + j\frac{\omega}{25}}$$

Comparing these network functions gives

$$\frac{L}{2} = \frac{1}{5} \quad \text{and} \quad \frac{L}{10} = \frac{1}{25}$$

The values of L obtained from these equations must agree, and they do. (If they do not, we've made an error.) Solving each of these equations gives $L = 0.4$ H.

EXAMPLE 13.2-3 Network Function of a Circuit

INTERACTIVE EXAMPLE

Consider the circuit shown in Figure 13.2-6. The input to the circuit is the voltage of the voltage source, $v_i(t)$. The output is the voltage across the capacitor, $v_o(t)$. The network function that represents this circuit is

$$\mathbf{H}(\omega) = \frac{\mathbf{V}_o(\omega)}{\mathbf{V}_i(\omega)} = \frac{3}{\left(1 + j\frac{\omega}{2}\right)\left(1 + j\frac{\omega}{5}\right)} \tag{13.2-8}$$

Determine the value of the inductance, L, and of the gain, A, of the voltage-controlled voltage source (VCVS).

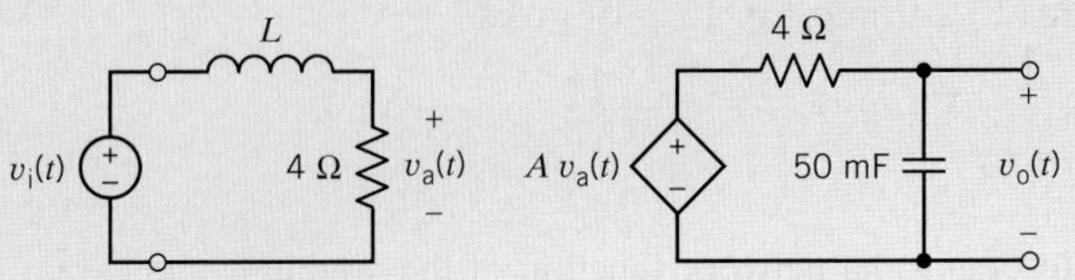

FIGURE 13.2-6 The circuit considered in Example 13.2-3.

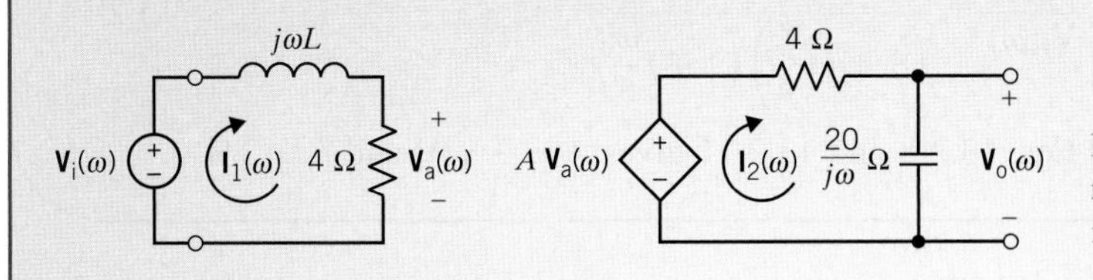

FIGURE 13.2-7 The circuit from Figure 13.2-6, represented in the frequency domain, using impedances and phasors.

Solution

The circuit has been represented twice, by a circuit diagram and by the given network function. The unknown parameters, L and A, appear in the circuit diagram but not in the given network function. We can analyze the circuit to determine its network function. This version of the network function will depend on the unknown parameters. We will determine the value of these parameters by equating the two versions of the network function.

A network function is the ratio of the output phasor to the input phasor. Phasors exist in the frequency domain. Consequently, our first step is to represent the circuit in the frequency domain, using phasors and impedances. Figure 13.2-7 shows the frequency-domain representation of the circuit from Figure 13.2-6.

The circuit in Figure 13.2-7 consists of two meshes. The mesh current of the left-hand mesh is labeled as $\mathbf{I}_1(\omega)$, and the mesh current of the right-hand mesh is labeled as $\mathbf{I}_2(\omega)$. Apply Kirchhoff's voltage law (KVL) to the left-hand mesh to get

$$j\omega L\mathbf{I}_1(\omega) + 4\mathbf{I}_1(\omega) - \mathbf{V}_\mathrm{i}(\omega) = 0$$

Solve for $\mathbf{I}_1(\omega)$ to get

$$\mathbf{I}_1(\omega) = \frac{\mathbf{V}_\mathrm{i}(\omega)}{j\omega L + 4} = \frac{0.25}{1 + j\omega\dfrac{L}{4}}\mathbf{V}_\mathrm{i}(\omega)$$

Next, use Ohm's law to represent $\mathbf{V}_\mathrm{a}(\omega)$ as

$$\mathbf{V}_\mathrm{a}(\omega) = 4\mathbf{I}_1(\omega) = \frac{1}{1 + j\omega\dfrac{L}{4}}\mathbf{V}_\mathrm{i}(\omega) \tag{13.2-9}$$

Apply KVL to the right-hand mesh to get

$$4\mathbf{I}_2(\omega) + \frac{20}{j\omega}\mathbf{I}_2(\omega) - A\mathbf{V}_\mathrm{a}(\omega) = 0$$

Solve for $\mathbf{I}_2(\omega)$ to get

$$\mathbf{I}_2(\omega) = \frac{A}{4 + \dfrac{20}{j\omega}}\mathbf{V}_\mathrm{a}(\omega) = \frac{j\omega A}{j\omega 4 + 20}\mathbf{V}_\mathrm{a}(\omega) = \frac{\dfrac{j\omega}{20}A}{1 + j\dfrac{\omega}{5}}\mathbf{V}_\mathrm{a}(\omega)$$

The output voltage is obtained by multiplying the mesh current $\mathbf{I}_2(\omega)$ by the impedance of the capacitor:

$$\mathbf{V}_\mathrm{o}(\omega) = \frac{20}{j\omega}\mathbf{I}_2(\omega) = \frac{20}{j\omega} \times \frac{\dfrac{j\omega}{20}A}{1 + j\dfrac{\omega}{5}}\mathbf{V}_\mathrm{a}(\omega) = \frac{A}{1 + j\dfrac{\omega}{5}}\mathbf{V}_\mathrm{a}(\omega) \tag{13.2-10}$$

Substituting the expression for $\mathbf{V}_a(\omega)$ from Eq. 13.2-9 into Eq. 13.2-10 gives

$$\mathbf{V}_o(\omega) = \frac{1}{1 + j\omega\frac{L}{4}} \times \frac{A}{1 + j\frac{\omega}{5}}\mathbf{V}_i(\omega) = \frac{A}{\left(1 + j\omega\frac{L}{4}\right)\left(1 + j\frac{\omega}{5}\right)}\mathbf{V}_i(\omega)$$

Divide both sides of this equation by $\mathbf{V}_i(\omega)$ to obtain the network function of the circuit:

$$\mathbf{H}(\omega) = \frac{\mathbf{V}_o(\omega)}{\mathbf{V}_i(\omega)} = \frac{A}{\left(1 + j\omega\frac{L}{4}\right)\left(1 + j\frac{\omega}{5}\right)} \tag{13.2-11}$$

Comparing the network functions given by Eqs. 13.2-8 and 13.2-11 gives $A = 3$ V/V and $L = 2$ H.

The circuit shown in Figure 13.2-1 is an example of a circuit called a first-order low-pass filter. First-order low-pass filters have network functions of the form

$$\mathbf{H}(\omega) = \frac{H_0}{1 + j\frac{\omega}{\omega_0}} \tag{13.2-12}$$

The gain and phase shift of the first-order low-pass filter are

$$\text{gain} = \frac{|H_0|}{\sqrt{1 + \frac{\omega^2}{\omega_0^2}}} \tag{13.2-13}$$

and

$$\text{phase shift} = \angle H_0 - \tan^{-1}(\omega/\omega_0) \tag{13.2-14}$$

The network function of the first-order low-pass filter has two parameters, H_0 and ω_0. At low frequencies, that is, $\omega \ll \omega_0$, the gain is $|H_0|$, so $|H_0|$ is called the dc gain. (When $\omega = 0$, $A\cos\omega t = A$, a constant or dc voltage.)

The other parameter of the network function, ω_0, is called the half-power frequency. To explain this terminology, suppose that the input to the first-order filter in Figure 13.2-1 is

$$v_{in}(t) = A\cos(\omega t)$$

Suppose, for convenience, that $H_0 = 1$. Then the output of the first-order filter in Figure 13.2-1 is

$$v_o(t) = \frac{A}{\sqrt{1 + \frac{\omega^2}{\omega_0^2}}}\cos\left(\omega_0 t - \tan^{-1}\left(\frac{\omega}{\omega_0}\right)\right)$$

In Figure 13.2-1, the output voltage is the voltage across a 10-kΩ resistor. The average power delivered to this resistor is

$$P_{ave} = \frac{A^2}{2(10 \times 10^3)\left(1 + \frac{\omega^2}{\omega_0^2}\right)}$$

At low frequencies, that is, frequencies that satisfy $\omega \ll \omega_0$, the average power is approximately

$$P_1 = \frac{A^2}{2(10 \times 10^3)(1 + 0)} = \frac{A^2}{2(10 \times 10^3)}$$

At the frequency $\omega = \omega_0$, the average power is

$$P_2 = \frac{A^2}{2(10 \times 10^3)(1 + 1)} = \frac{P_1}{2}$$

For this reason, ω_0 is called the half-power frequency.

In words, suppose we hold the input amplitude constant while we vary the frequency, ω, of the input. We find that the value of the output power when $\omega = \omega_0$ is one-half of the value of the output power when $\omega \ll \omega_0$.

Next, consider the problem of designing a first-order low-pass filter. Suppose we are given the following specifications:

$$\text{dc gain} = 2$$
$$\text{phase shift} = 120^\circ \quad \text{when} \quad \omega = 1000 \text{ rad/s}$$

Before designing a circuit to meet these specifications, we need to pay more attention to the phase shift. Consider Eq. 13.2-14. Both ω and ω_0 will be positive, so $\tan^{-1}(\omega/\omega_0)$ will be between 0° and 90°. Also, $\angle H_0$ will be 0° when H_0 is positive and 180° when H_0 is negative. As a result, only phase shifts between $-90°$and 0°or between 90°and 180°can be achieved using a first-order low-pass filter. (Phase shifts that cannot be obtained using a first-order low-pass filter can be obtained using other types of circuit. That's a story for another day.) Table 13.2-2 shows two first-order low-pass filters, one for obtaining phase shifts between 90° and 180° and the other for obtaining phase shifts between $-90°$and 0°. Based on the phase shift, we select the circuit in the first row of Table 13.2-2. The specification on the dc gain gives

$$2 = |H_0| = \frac{R_2}{R_1}$$

The specification on phase shift gives

$$120^\circ = 180^\circ - \tan^{-1}(1000 R_2 C)$$

Table 13.2-2 First-Order Low-Pass Filter Circuits

PHASE SHIFT	FIRST ORDER LOW PASS FILTER CIRCUIT	DESIGN EQUATIONS
$90^\circ \le \text{phase shift} \le 180^\circ$	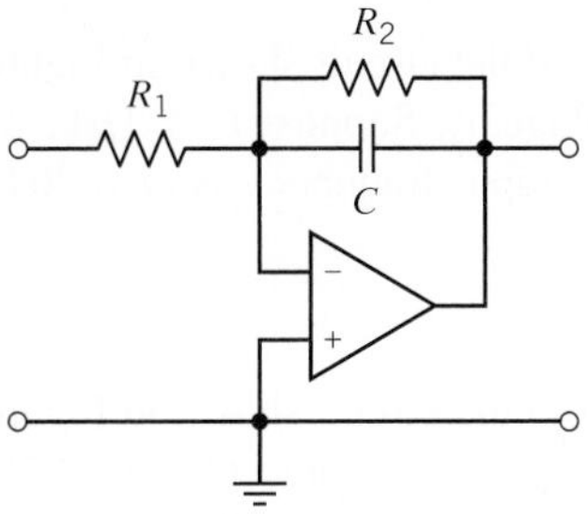	$H_0 = -\frac{R_2}{R_1}$ $\omega_0 = \frac{1}{R_2 C}$
$-90^\circ \le \text{phase shift} \le 0^\circ$	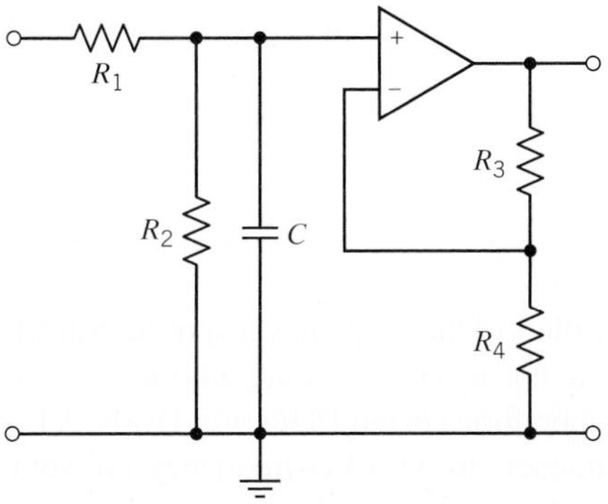	$H_0 = \frac{R_2}{R_1 + R_2}\left(1 + \frac{R_3}{R_4}\right)$ $\omega_0 = \frac{R_1 + R_2}{R_1 R_2 C}$

This is a set of two equations in the three unknowns R_1, R_2, and C. The solution is not unique. We will have to pick a value for one of the unknowns and then solve for values of the other two unknowns. Let's pick a convenient value for the capacitor, $C = 0.1\mu\text{F}$, and calculate the resistances.

$$R_2 = \frac{\tan(60°)}{1000 \times 0.1 \times 10^{-6}} = 17.32\ \text{k}\Omega$$

and

$$R_1 = \frac{R_2}{2} = 8.66\ \text{k}\Omega$$

We conclude that the circuit shown in the first row of Table 13.2-2 will have a dc gain = 2 and a phase shift = 120° at $\omega = 1000$ rad/s when $R_1 = 8.66\ \text{k}\Omega$, $R_2 = 17.32\ \text{k}\Omega$, and $C = 0.1\mu\text{F}$.

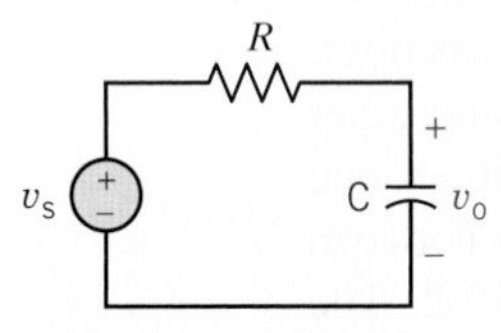

FIGURE E 13.2-1 An *RC* circuit.

EXERCISE 13.2-1 The input to the circuit shown in Figure E 13.2-1 is the source voltage, v_s, and the response is the capacitor voltage, v_o. Suppose $R = 10\ \text{k}\Omega$ and $C = 1\mu\text{F}$. What are the values of the gain and phase shift when the input frequency is $\omega = 100$ rad/s?

Answer: 0.707 and −45°

EXERCISE 13.2-2 The input to the circuit shown in Figure E 13.2-2 is the source voltage, v_s, and the response is the resistor voltage, v_o. $R = 30\ \Omega$ and $L = 2$ H. Suppose the input frequency is adjusted until the gain is equal to 0.6. What is the value of the frequency?

Answer: 20 rad/s

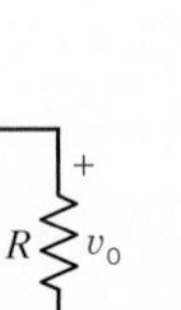

FIGURE E 13.2-2 The *RL* circuit.

EXERCISE 13.2-3 The input to the circuit shown in Figure E 13.2-2 is the source voltage, v_s, and the response is the mesh current, i. $R = 30\ \Omega$ and $L = 2$ H. What are the values of the gain and phase shift when the input frequency is $\omega = 20$ rad/s?

Answer: 0.02 A/V and −53.1°

EXERCISE 13.2-4 The input to the circuit shown in Figure E 13.2-1 is the source voltage, v_s, and the response is the capacitor voltage, v_o. Suppose $C = 1\mu\text{F}$. What value of R is required to cause a phase shift equal to −45° when the input frequency is $\omega = 20$ rad/s?

Answer: $R = 50\ \text{k}\Omega$

EXERCISE 13.2-5 The input to the circuit shown in Figure E 13.2-1 is the source voltage, v_s, and the response is the capacitor voltage, v_o. Suppose $C = 1\mu\text{F}$. What value of R is required to cause a gain equal to 1.5 when the input frequency is $\omega = 20$ rad/s?

Answer: No such value of R exists. The gain of this circuit will never be greater than 1.

13.3 BODE PLOTS

It is common to use logarithmic plots of the frequency response instead of linear plots. The logarithmic plots are called *Bode plots* in honor of H. W. Bode, who used them extensively in his work with amplifiers at Bell Telephone Laboratories in the 1930s and 1940s. A Bode plot is a plot of log-gain and phase angle values versus frequency, using a log-frequency horizontal axis. The use of logarithms expands the range of frequencies portrayed on the horizontal axis.

Table 13.3-1 A Decibel Conversion Table

MAGNITUDE, H	20 log H(dB)
0.1	−20.00
0.2	−13.98
0.4	−7.96
0.6	−4.44
1.0	0.0
1.2	1.58
1.4	2.92
1.6	4.08
2.0	6.02
3.0	9.54
4.0	12.04
5.0	13.98
6.0	15.56
7.0	16.90
10.0	20.00
100.0	40.00

The network function **H** can be written as

$$\mathbf{H} = \mathrm{H}\angle\phi = \mathrm{H}e^{j\phi} \tag{13.3-1}$$

The logarithm of the magnitude is normally expressed in terms of the logarithm to the base 10, so we use

$$\text{logarithmic gain} = 20\log_{10}\mathrm{H} \tag{13.3-2}$$

and the unit is decibel (dB). The logarithmic gain is also called the gain in dB. A decibel conversion table is given in Table 13.3-1.

The unit decibel is derived from the unit bel. Suppose P_1 and P_2 are two values of power. Both P_1/P_2 and log (P_1/P_2) are measures of the relative sizes of P_1 and P_2. The ratio P_1/P_2 is unitless, whereas $\log(P_1/P_2)$ has the bel as its unit. The name *bel* honors Alexander Graham Bell, the inventor of the telephone.

The **Bode plot**, is a chart of gain in decibels and phase in degrees versus the logarithm of frequency.

Let us obtain the Bode plots corresponding to the network function

$$\mathbf{H} = \frac{1}{1 + j\dfrac{\omega}{\omega_0}} = \frac{1}{\sqrt{1 + (\omega/\omega_0)}}\angle \tan^{-1}(\omega/\omega_0) = H\angle\phi \tag{13.3-3}$$

The logarithmic gain is

$$\begin{aligned} 20\log_{10} H &= 20\log_{10}\frac{1}{\sqrt{1+(\omega/\omega_0)^2}} \\ &= 20\log_{10} 1 - 20\log_{10}\sqrt{1+(\omega/\omega_0)^2} = -20\log_{10}\sqrt{1+(\omega/\omega_0)^2} \end{aligned}$$

For small frequencies, that is, $\omega \ll \omega_0$

$$1 + (\omega/\omega_0)^2 \cong 1$$

so the logarithmic gain is approximately

$$20 \log_{10} H = -20 \log_{10} \sqrt{1} = 0 \text{ dB}$$

This is the equation of a horizontal straight line. Because this straight line approximates the logarithmic gain for low frequencies, it is called the low-frequency asymptote of the Bode plot.

For large frequencies, that is, $\omega \gg \omega_0$

$$1 + (\omega/\omega_0)^2 \cong (\omega/\omega_0)^2$$

so the logarithmic gain is approximately

$$\begin{aligned} 20 \log_{10} H &= -20 \log_{10} \sqrt{(\omega/\omega_0)^2} \\ &= -20 \log_{10} \omega/\omega_0 = 20 \log_{10} \omega_0 - 20 \log_{10} \omega \end{aligned}$$

This equation shows one of the advantages of using logarithms. The plot of 20 $\log_{10} H$ versus $\log_{10} \omega$ is a straight line. This straight line is called the high-frequency asymptote of the Bode plot. Figures 13.3-1*a*,*b*

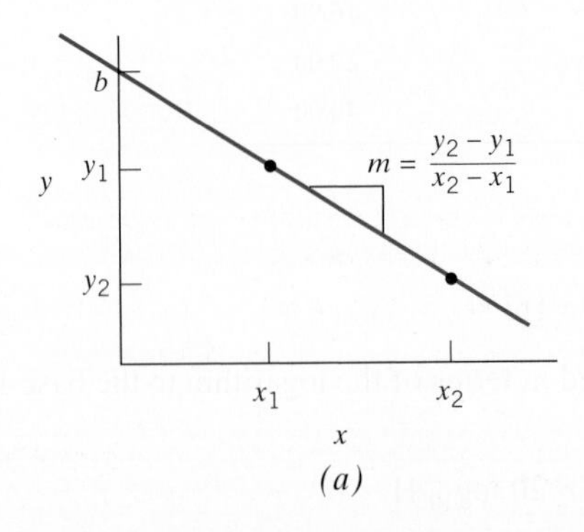

(*a*)

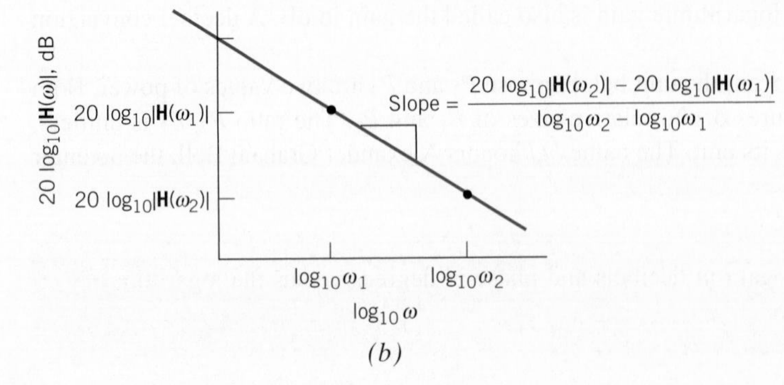

(*b*)

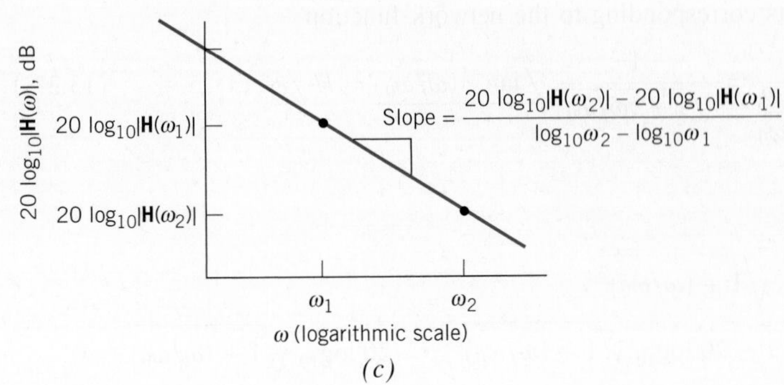

(*c*)

FIGURE 13.3-1 (*a*) Plot of y versus x for the straight line $y = mx + b$. (*b*) Plot of 20 log$|\mathbf{H}(\omega)|$ versus log ω for the straight line 20 log$|\mathbf{H}(\omega)| = 20 \log \omega_0 - 20 \log \omega$. (*c*) Plot of 20 log$|\mathbf{H}(\omega)|$ versus ω for the straight line 20 log$|\mathbf{H}(\omega)| = 20 \log \omega_0 - 20 \log \omega$.

compare the equation of the high-frequency asymptote to the more familiar standard form of the equation of a straight line, $y = mx + b$. The slope of the high-frequency asymptote can be calculated from two points on the straight line. This slope is given using units of dB/decade. In Figure 13.3-1*b* the gain in dB is plotted versus log ω, whereas in Figure 13.3-1*c*, the gain in dB is plotted versus ω using a log scale. It is more convenient to label the frequency axis when a log scale is used for ω. The equation used to calculate the slope from two points on the line is the same in Figure 13.3-1*c* as it is in Figure 13.3-1*b*.

Consider two frequencies, ω_1 and ω_2, with $\omega_2 = 10^x\omega_1$. We say that ω_2 is larger than ω_1 by x decades. Alternately, ω_2 is larger than ω_1 by $x = \log_{10}(\omega_2/\omega_1)$ decades. For example, 1000 rad/s is 2 decades larger than 10 rad/s, and 316 rad/s is 1.5 decades larger than 10 rad/s.

The slope of the high-frequency asymptote is

$$\frac{20\log_{10}|\mathbf{H}(\omega_2)| - 20\log_{10}|\mathbf{H}(\omega_1)|}{\log_{10}\omega_2 - \log_{10}\omega_1} = \frac{20\log_{10}|\mathbf{H}(\omega_2)| - 20\log_{10}|\mathbf{H}(\omega_1)|}{\log_{10}(\omega_2/\omega_1)}$$

The units of this slope are dB/decade. The high-frequency asymptote is characterized by

$$|\mathbf{H}(\omega)| \cong \frac{1}{\dfrac{\omega}{\omega_0}} = \frac{\omega_0}{\omega} \quad \text{when } \omega \gg \omega_0$$

The value of the slope of the high-frequency asymptote is

$$\begin{aligned}\frac{20\log_{10}|\mathbf{H}(\omega_2)| - 20\log_{10}|\mathbf{H}(\omega_1)|}{\log_{10}(\omega_2/\omega_1)} &= \frac{20\log_{10}(\omega_0/\omega_2) - 20\log_{10}(\omega_0/\omega_1)}{\log_{10}(\omega_2/\omega_1)} \\ &= \frac{-20\log_{10}(\omega_2/\omega_1)}{\log_{10}(\omega_2/\omega_1)} = -20 \text{ dB/decade}\end{aligned}$$

The intersection of the low-frequency asymptote with the high-frequency asymptote occurs when

$$0 = 20\log_{10}\omega - 20\log_{10}\omega_0$$

that is, when

$$\omega = \omega_0$$

The low- and high-frequency asymptotes form a corner where they intersect. Because the asymptotes intersect at the frequency $\omega = \omega_0$, ω_0 is sometimes called the **corner frequency**.

Figure 13.3-2 shows the magnitude and phase Bode plots for this network function. The asymptotic curve shown in Figure 13.3-2 is an approximation to the Bode plot. The **asymptotic Bode plot** consists of the low-frequency asymptote for $\omega < \omega_0$ and the high-frequency asymptote for $\omega > \omega_0$. The approximation used to obtain the asymptotic Bode plot is summarized by the following equations:

$$|\mathbf{H}(\omega)| = \frac{1}{\sqrt{1 + (\omega/\omega_0)^2}} \cong \begin{cases} 1 & \omega < \omega_0 \\ \omega_0/\omega & \omega > \omega_0 \end{cases}$$

or

$$20\log_{10}|\mathbf{H}(\omega)| = 20\log_{10}\frac{1}{\sqrt{1 + (\omega/\omega_0)^2}} \cong \begin{cases} 0 & \omega < \omega_0 \\ 20\log_{10}\omega_0 - 20\log_{10}\omega & \omega > \omega_0 \end{cases}$$

The asymptotic Bode plot is a good approximation to the Bode plot when $\omega \ll \omega_0$ or $\omega \gg \omega_0$. Near $\omega = \omega_0$, the asymptotic Bode plot deviates from the exact Bode plot. At $\omega = \omega_0$, the value of the

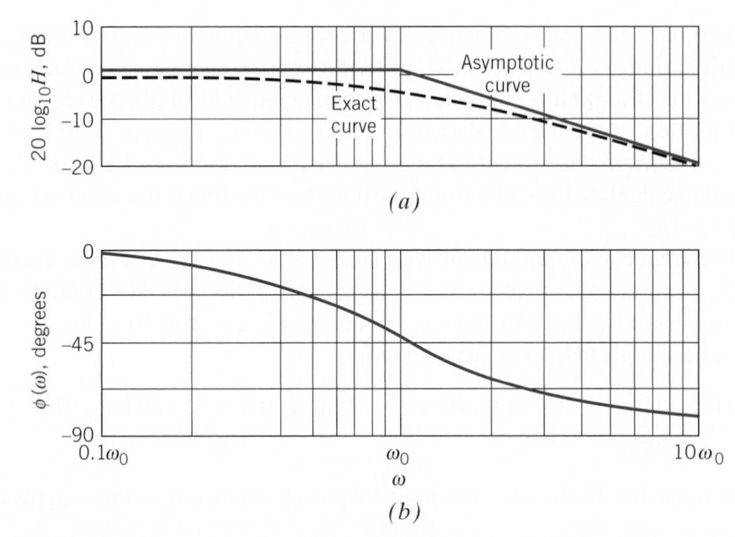

FIGURE 13.3-2 Bode diagram for $\mathbf{H} = (1 + j\omega/\omega_0)^{-1}$. The dashed curve is the exact curve for the magnitude. The solid curve for the magnitude is an asymptotic approximation.

asymptotic Bode plot is 0 dB, whereas the value of the exact Bode plot is

$$20 \log_{10}|\mathbf{H}(\omega_0)| = 20 \log_{10} \frac{1}{\sqrt{1 + (\omega_0/\omega_0)^2}} = 20 \log_{10} \frac{1}{\sqrt{2}} = -3.01 \text{ dB}$$

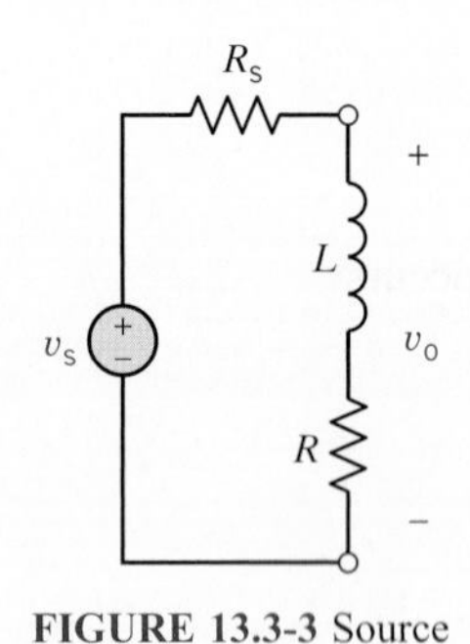

FIGURE 13.3-3 Source voltage delivering power to a load impedance consisting of L and R.

The magnitude characteristic does not exhibit a sharp break. Nevertheless, we designate the frequency at which the magnitude is $1/\sqrt{2}$ times the magnitude at $\omega = 0$ as a special frequency. On the Bode diagram, the magnitude drop of $1/\sqrt{2}$ results in a logarithmic drop of approximately -3 dB at $\omega = \omega_0$. The frequency $\omega = \omega_0$ is often called the *break frequency* or *corner frequency*.

Of course, **H** may take on forms other than that of Eq. 13.3-3. For example, consider the circuit shown in Figure 13.3-3. The network function of this circuit is

$$\mathbf{H} = \frac{\mathbf{V}_o}{\mathbf{V}_s} = \frac{R + j\omega L}{R_s + R + j\omega L}$$

Let's put this network function into the form

$$\mathbf{H} = k \frac{1 + j\dfrac{\omega}{\omega_1}}{1 + j\dfrac{\omega}{\omega_2}} = H \angle \phi$$

This network function has three parameters: k, ω_1, and ω_2. All three parameters have names. The frequencies ω_1 and ω_2 are corner frequencies. Corner frequencies that appear in the numerator of a network function are called zeros, so ω_1 is a zero of the network function. Corner frequencies that appear in the denominator of a network function are called poles, so ω_2 is a pole of the network function. Because

$$k = \lim_{\omega \to 0} H$$

the parameter k is called the low-frequency gain or the dc gain. The network function of this circuit can be expressed as

$$\mathbf{H} = \left(\frac{R}{R + R_s}\right) \frac{1 + j\dfrac{\omega L}{R}}{1 + j\dfrac{\omega L}{R + R_s}}$$

so the dc gain is

$$k = \frac{R}{R + R_s}$$

and the zero and pole frequencies are related by

$$\omega_1 = \frac{R}{L} < \frac{R + R_s}{L} = \omega_2$$

The gain corresponding to a network function of this form is

$$H = k \frac{\sqrt{1 + \left(\frac{\omega}{\omega_1}\right)^2}}{\sqrt{1 + \left(\frac{\omega}{\omega_2}\right)^2}}$$

To obtain the asymptotic Bode plot, we approximate $\sqrt{1 + (\omega/\omega_1)^2}$ by 1 when $\omega < \omega_1$ and by ω/ω_1 when $\omega > \omega_1$. Similarly, we approximate $\sqrt{1 + (\omega/\omega_2)^2}$ by 1 when $\omega < \omega_2$ and by ω/ω_2 when $\omega > \omega_2$. Thus,

$$H \cong \begin{cases} k & \omega < \omega_1 \\ \dfrac{k\omega}{\omega_1} & \omega_1 < \omega < \omega_2 \\ \dfrac{k\omega_2}{\omega_1} & \omega_2 < \omega \end{cases}$$

Next, the logarithmic gain is approximated by

$$20 \log_{10} H \cong \begin{cases} 20 \log_{10} k & \omega < \omega_1 \\ (20 \log_{10} k - 20 \log_{10} \omega_1) + 20 \log_{10} \omega & \omega_1 < \omega < \omega_2 \\ (20 \log_{10} k - 20 \log_{10} \omega_1) + 20 \log_{10} \omega_2 & \omega_2 < \omega \end{cases}$$

These are the equations of the asymptotes of the Bode plot. When $\omega < \omega_1$ and when $\omega > \omega_2$, the asymptotes are horizontal straight lines. The equations for these asymptotes don't include a term involving $\log_{10}\omega$, which means that the slope must be zero. When $\omega_1 < \omega < \omega_2$, the equation of the asymptote does include a term involving $\log_{10} \omega$. The coefficient of $\log_{10}\omega$ is 20, indicating a slope of 20 dB/decade.

The effect of the dc gain k is limited to the term $20 \log_{10} k$, which appears in the equation of each of the three asymptotes. Changing the value of k will shift the Bode plot up (increasing k) or down (decreasing k) but will not change the shape of the Bode plot. For this reason, we sometimes normalize the network function by dividing by the dc gain. The asymptotes of the Bode plot of the normalized network function are given by

$$20 \log_{10}\left(\frac{H}{k}\right) \cong \begin{cases} 0 & \omega < \omega_1 \\ 20 \log_{10} \omega - 20 \log_{10} \omega_1 & \omega_1 < \omega < \omega_2 \\ 20 \log_{10} \omega_2 - 20 \log_{10} \omega_1 & \omega_2 < \omega \end{cases}$$

The phase angle of **H** is

$$\phi = \angle k + \angle\left(1 + j\frac{\omega}{\omega_1}\right) - \angle\left(1 + j\frac{\omega}{\omega_2}\right) = 0 + \tan^{-1}\left(\frac{\omega}{\omega_1}\right) - \tan^{-1}\left(\frac{\omega}{\omega_2}\right)$$

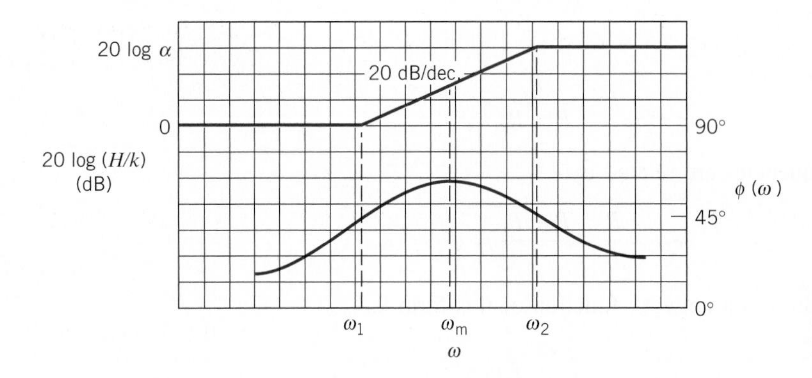

FIGURE 13.3-4 Bode diagram for the circuit of Figure 13.3-3.

The phase Bode plot and the asymptotic magnitude Bode plot are shown in Figure 13.3-4. Notice that the slope of the asymptotic magnitude Bode plot changes as the frequency increases past ω_1 and changes again as the frequency increases past ω_2. Zeros, like ω_1, cause the slope to increase by 20 dB/decade. Poles, like ω_2, cause the slope to decrease by 20 dB/decade. The slope of every asymptote will be an integer multiple of 20 dB/decade.

EXAMPLE 13.3-1 Bode Plot

Find the asymptotic magnitude Bode plot of

$$\mathbf{H}(\omega) = K\frac{j\omega}{1 + j\dfrac{\omega}{p}}$$

Solution

Approximate $\left(1 + j\dfrac{\omega}{p}\right)$ by 1 when $\omega < p$ and by $j\dfrac{\omega}{p}$ when $\omega > p$ to get

$$\mathbf{H}(\omega) \cong \begin{cases} K(j\omega) & \omega < p \\ Kp & \omega > p \end{cases}$$

The logarithmic gain is

$$20\log_{10}|\mathbf{H}(\omega)| \cong \begin{cases} 20\log_{10} K + 20\log_{10}\omega & \omega < p \\ 20\log_{10}(Kp) & \omega > p \end{cases}$$

The asymptotic magnitude Bode plot is shown in Figure 13.3-5. The $j\omega$ factor in the numerator of $\mathbf{H}(\omega)$ causes the low-frequency asymptote to have a slope of 20 dB/decade. The slope of the asymptotic magnitude Bode plot decreases by 20 dB/decade (from 20 dB/decade to zero) as the frequency increases past $\omega = p$.

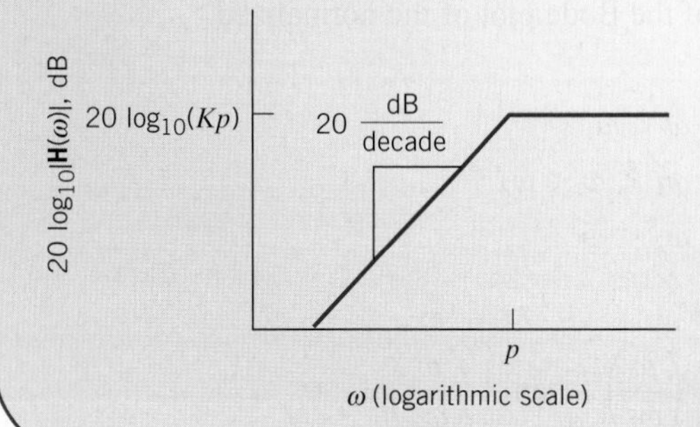

FIGURE 13.3-5 Asymptotic magnitude Bode plot for Example 13.3-1.

EXAMPLE 13.3-2 Bode Plot of a Circuit

INTERACTIVE EXAMPLE

Consider the circuit shown in Figure 13.3-6*a*. The input to the circuit is the voltage of the voltage source, $v_i(t)$. The output is the node voltage at the output terminal of the op amp, $v_o(t)$. The network function that represents this circuit is

$$\mathbf{H}(\omega) = \frac{\mathbf{V}_o(\omega)}{\mathbf{V}_i(\omega)} \tag{13.3-4}$$

The corresponding magnitude Bode plot is shown in Figure 13.3-6*b*. Determine the values of the capacitances, C_1 and C_2.

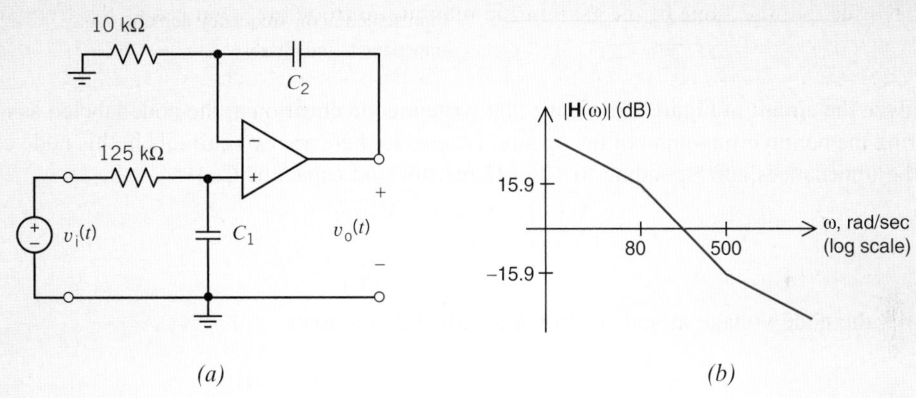

FIGURE 13.3-6 The circuit and Bode plot considered in Example 13.3-2.

Solution

The network function provides a connection between the circuit and the Bode plot. We can determine the network function from the Bode plot, and we can also analyze the circuit to determine its network function. The values of the capacitances are determined by equating the coefficients of these two network functions.

Step 1: Let's make some observations regarding the Bode plot shown in Figure 13.3-6*b*:

1. There are two corner frequencies, at 80 and 500 rad/s. The corner frequency at 80 rad/s is a pole because the slope of the Bode plot decreases at 80 rad/s. The corner frequency at 500 rad/s is a zero because the slope increases at 500 rad/s.
2. The corner frequencies are separated by $\log_{10}\left(\frac{500}{80}\right) = 0.796$ decades. The slope of the Bode plot is $\frac{-15.9 - 15.9}{0.796} = -40$ dB/decade between the corner frequencies.
3. At low frequencies—that is, at frequencies smaller than the smallest corner frequency—the slope is -1×20 dB/decade, so the network function includes a factor $(j\omega)^{-1}$

Consequently, the network function corresponding to the Bode plot is

$$\mathbf{H}(\omega) = k(j\omega)^{-1}\left(\frac{1 + j\dfrac{\omega}{500}}{1 + j\dfrac{\omega}{80}}\right) = k\frac{1 + j\dfrac{\omega}{500}}{j\omega\left(1 + j\dfrac{\omega}{80}\right)} \tag{13.3-5}$$

where k is a constant that is yet to be determined.

Step 2: Next, we analyze the circuit shown in Figure 13.3-6*a* to determine its network function. A network function is the ratio of the output phasor to the input phasor. Phasors exist in the frequency domain. Consequently, our first step is to represent the circuit in the frequency domain, using phasors and impedances. Figure 13.3-7 shows the frequency-domain representation of the circuit from Figure 13.3-6*a*.

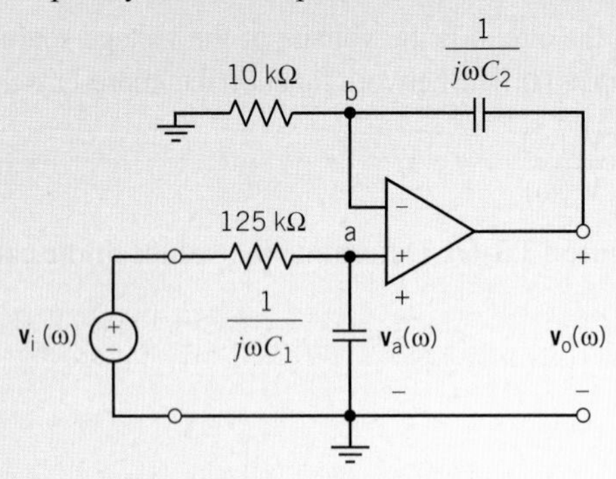

FIGURE 13.3-7 The circuit from Figure 13.3-6*a*, represented in the frequency domain, using impedances and phasors.

To analyze the circuit in Figure 13.3-7, we first write a node equation at the node labeled as node a. (The current entering the noninverting input of the op amp is zero, so there are two currents in this node equation, the currents in the impedances corresponding to 125-kΩ resistor and capacitor C_1.)

$$\frac{\mathbf{V}_{\mathrm{i}}(\omega)-\mathbf{V}_{\mathrm{a}}(\omega)}{125\times 10^3}=\frac{\mathbf{V}_{\mathrm{a}}(\omega)}{\dfrac{1}{j\omega C_1}}$$

where $\mathbf{V}_{\mathrm{a}}(\omega)$ is the node voltage at node a. Doing a little algebra gives

$$\frac{\mathbf{V}_{\mathrm{i}}(\omega)}{125\times 10^3}=\left(\frac{1}{125\times 10^3}+j\omega C_1\right)\mathbf{V}_{\mathrm{a}}(\omega)$$

then
$$\mathbf{V}_{\mathrm{i}}(\omega)=\left(1+j\omega C_1\left(125\times 10^3\right)\right)\mathbf{V}_{\mathrm{a}}(\omega)\;\Rightarrow\;\mathbf{V}_{\mathrm{a}}(\omega)=\frac{\mathbf{V}_{\mathrm{i}}(\omega)}{1+j\omega C_1\left(125\times 10^3\right)}$$

Next, we write a node equation at the node labeled as node b. (The current entering the inverting input of the op amp is zero, so there are two currents in this node equation, the currents in the impedances corresponding to 10-kΩ resistor and capacitor C_2.)

$$\frac{\mathbf{V}_{\mathrm{a}}(\omega)}{10\times 10^3}+\frac{\mathbf{V}_{\mathrm{a}}(\omega)-\mathbf{V}_{\mathrm{o}}(\omega)}{\dfrac{1}{j\omega C_2}}=0$$

Doing some algebra gives

$$\mathbf{V}_{\mathrm{a}}(\omega)+j\omega C_2\left(10\times 10^3\right)\left(\mathbf{V}_{\mathrm{a}}(\omega)-\mathbf{V}_{\mathrm{o}}(\omega)\right)=0$$
$$\left(1+j\omega C_2\left(10\times 10^3\right)\right)\mathbf{V}_{\mathrm{a}}(\omega)=j\omega C_2\left(10\times 10^3\right)\mathbf{V}_{\mathrm{o}}(\omega)$$
$$\left(1+j\omega C_2\left(10\times 10^3\right)\right)\frac{\mathbf{V}_{\mathrm{i}}(\omega)}{1+j\omega C_1\left(125\times 10^3\right)}=j\omega C_2\left(10\times 10^3\right)\mathbf{V}_{\mathrm{o}}(\omega)$$

Finally,
$$\mathbf{H}(\omega)=\frac{\mathbf{V}_{\mathrm{o}}(\omega)}{\mathbf{V}_{\mathrm{i}}(\omega)}=\left(\frac{1}{C_2\left(10\times 10^3\right)}\right)\frac{1+j\omega C_2\left(10\times 10^3\right)}{(j\omega)\left(1+j\omega C_1\left(125\times 10^3\right)\right)}\tag{13.3-6}$$

Step 3: The network functions given in Eqs. 13.3-5 and 13.3-6 must be equal. That is,

$$k\frac{1+j\dfrac{\omega}{500}}{j\omega\left(1+j\dfrac{\omega}{80}\right)}=\mathbf{H}(\omega)=\left(\frac{1}{C_2\left(10\times 10^3\right)}\right)\frac{1+j\omega C_2\left(10\times 10^3\right)}{(j\omega)\left(1+j\omega C_1\left(125\times 10^3\right)\right)}$$

Equating coefficients gives

$$\frac{1}{80} = C_1\left(125 \times 10^3\right), \ \frac{1}{500} = C_2\left(10 \times 10^3\right), \text{ and } k = \frac{1}{C_2\left(10 \times 10^3\right)} = 500$$

so

$$C_1 = \frac{1}{80\left(125 \times 10^3\right)} = 0.1\ \mu\text{F and } C_2 = \frac{1}{500\left(10 \times 10^3\right)} = 0.2\ \mu\text{F}$$

EXAMPLE 13.3-3 Bode Plot of a Circuit

INTERACTIVE EXAMPLE

Consider the circuit shown in Figure 13.3-8*a*. The input to the circuit is the voltage of the voltage source, $v_i(t)$. The output is the node voltage at the output terminal of the op amp, $v_o(t)$. The network function that represents this circuit is

$$\mathbf{H}(\omega) = \frac{\mathbf{V}_o(\omega)}{\mathbf{V}_i(\omega)} \tag{13.3-7}$$

The corresponding magnitude Bode plot is also shown in Figure 13.3-8*b*. Determine the values of the capacitances, C_1 and C_2.

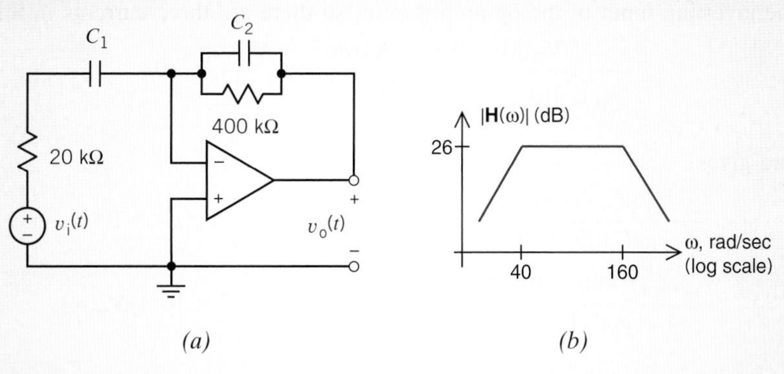

FIGURE 13.3-8 The circuit and Bode plot considered in Example 13.3-3.

Solution

The network function provides a connection between the circuit and the Bode plot. We can determine the network function from the Bode plot, and we can also analyze the circuit to determine its network function. The values of the capacitances are determined by equating the coefficients of these two network functions.

Step 1: First, we make some observations regarding the Bode plot shown in Figure 13.3-8*b*.

1. There are two corner frequencies, at 40 and 160 rad/s. Both corner frequencies are poles because the slope of the Bode plot decreases at both the corner frequencies.
2. Between the corner frequencies, the gain is $|\mathbf{H}(\omega)| = 26 \text{ dB} = 10^{26/20} = 20 \text{ V/V}$.
3. At low frequencies—that is, at frequencies smaller than the smallest corner frequency—the slope is 1×20 dB/decade, so the network function includes a factor $(j\omega)^1$.

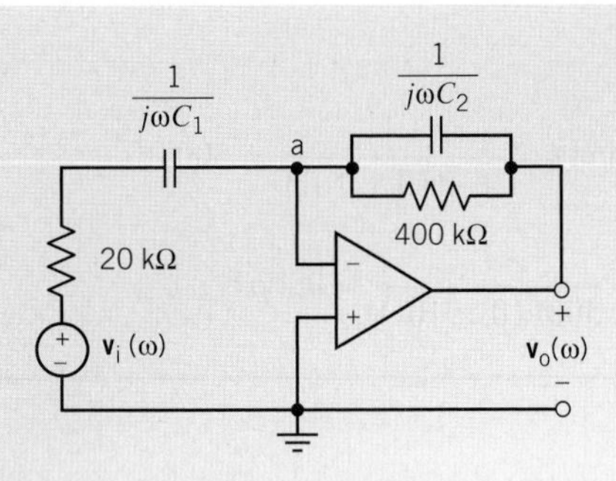

FIGURE 13.3-9 The circuit from Figure 13.3-8a, represented in the frequency domain, using impedances and phasors.

Consequently, the network function corresponding to the Bode plot is

$$\mathbf{H}(\omega) = \frac{k(j\omega)}{\left(1 + j\frac{\omega}{40}\right)\left(1 + j\frac{\omega}{160}\right)} \tag{13.3-8}$$

Step 2: Next, we analyze the circuit shown in Figure 13.3-8a to determine its network function. A network function is the ratio of the output phasor to the input phasor. Phasors exist in the frequency domain. Consequently, our first step is to represent the circuit in the frequency domain, using phasors and impedances. Figure 13.3-9 shows the frequency-domain representation of the circuit from Figure 13.3-8a.

To analyze the circuit in Figure 13.3-9, we write a node equation at the node labeled as node a. In doing so, we will treat the series impedances, 20 kΩ and $\frac{1}{j\omega C_1}$, as a single equivalent impedance equal to $20 \times 10^3 + \frac{1}{j\omega C_1}$. (The node voltage at node a is zero volts because the voltages at the input nodes of an ideal op amp are equal. The current entering the inverting input of the op amp is zero, so there are three currents in this node equation.)

$$\frac{\mathbf{V}_i(\omega)}{20 \times 10^3 + \frac{1}{j\omega C_1}} + \frac{\mathbf{V}_o(\omega)}{400 \times 10^3} + \frac{\mathbf{V}_o(\omega)}{\frac{1}{j\omega C_2}} = 0$$

Doing some algebra gives

$$\frac{(j\omega C_1)\mathbf{V}_i(\omega)}{1 + j\omega C_1(20 \times 10^3)} + \left(\frac{1}{400 \times 10^3} + j\omega C_2\right)\mathbf{V}_o(\omega) = 0$$

$$\frac{(j\omega C_1)(400 \times 10^3)\mathbf{V}_i(\omega)}{1 + j\omega C_1(20 \times 10^3)} = -(1 + j\omega C_2(400 \times 10^3))\mathbf{V}_o(\omega)$$

Finally,

$$\mathbf{H}(\omega) = \frac{\mathbf{V}_o(\omega)}{\mathbf{V}_i(\omega)} = \frac{-j\omega C_1(400 \times 10^3)}{(1 + j\omega C_1(20 \times 10^3))(1 + j\omega C_2(400 \times 10^3))} \tag{13.3-9}$$

Step 3: The network functions given in Eqs. 13.3-8 and 13.3-9 must be equal. That is,

$$\frac{k(j\omega)}{\left(1 + j\frac{\omega}{40}\right)\left(1 + j\frac{\omega}{160}\right)} = \mathbf{H}(\omega) = \frac{-j\omega C_1(400 \times 10^3)}{(1 + j\omega C_1(20 \times 10^3))(1 + j\omega C_2(400 \times 10^3))}$$

Equating coefficients gives

$$\frac{1}{40} = C_1(20 \times 10^3),\ \frac{1}{160} = C_2(400 \times 10^3),\ \text{and } k = -C_1(400 \times 10^3)$$

so

$$C_1 = \frac{1}{40(20 \times 10^3)} = 1.25\ \mu\text{F and } C_2 = \frac{1}{160(400 \times 10^3)} = 15.625\ \text{nF}$$

and

$$k = -C_1(400 \times 10^3) = -(1.25 \times 10^{-6})(400 \times 10^3) = -0.5$$

EXAMPLE 13.3-4 Network Function with Complex Poles

The network function of a second-order low-pass filter has the form

$$\mathbf{H}(\omega) = \frac{k\,\omega_0^2}{(j\omega)^2 + j2\zeta\omega_0\omega + \omega_0^2}$$

This network function depends on three parameters: the dc gain, k; the corner frequency, ω_0; and the damping ratio, ζ. For convenience, we consider the case where $k = 1$. Then, using $j^2 = -1$, we can write the network function as

$$\mathbf{H}(\omega) = \frac{\omega_0^2}{\omega_0^2 - \omega^2 + j2\zeta\omega_0\omega}$$

Determine the asymptotic magnitude Bode plot of the second-order low-pass filter when the dc gain is 1.

Solution

The denominator of $\mathbf{H}(\omega)$ contains a new factor, one that involves ω^2. The asymptotic Bode plot is based on the approximation

$$\left(\omega_0^2 - \omega^2\right) + j2\zeta\omega_0\omega \cong \begin{cases} \omega_0^2 & \omega < \omega_0 \\ -\omega^2 & \omega > \omega_0 \end{cases}$$

Using this approximation, we can express $\mathbf{H}(\omega)$ as

$$\mathbf{H}(\omega) \cong \begin{cases} 1 & \omega < \omega_0 \\ -\dfrac{\omega_0^2}{\omega^2} & \omega > \omega_0 \end{cases}$$

The logarithmic gain is

$$20\log_{10}|\mathbf{H}(\omega)| \cong \begin{cases} 0 & \omega < \omega_0 \\ 40\log_{10}\omega_0 - 40\log_{10}\omega & \omega > \omega_0 \end{cases}$$

The asymptotic magnitude Bode plot is shown in Figure 13.3-10. The actual magnitude Bode plot and the actual phase Bode plot are shown in Figure 13.3-11. The asymptotic Bode plot is a good approximation to the actual Bode plot when $\omega \ll \omega_0$ or $\omega \gg \omega_0$. Near $\omega = \omega_0$, the asymptotic Bode plot deviates from the actual Bode plot. At $\omega = \omega_0$, the value of the asymptotic Bode plot is 0 dB whereas the value of the actual Bode plot is

$$\mathbf{H}(\omega_0) = \frac{1}{2\zeta}$$

As this equation and Figure 13.3-11 both show, the deviation between the actual and asymptotic Bode plot near $\omega = \omega_0$ depends on ζ. The frequency ω_0 is called the *corner frequency*. The slope of the asymptotic Bode plot decreases by 40 dB/decade as the frequency increases past $\omega = \omega_0$. In terms of the asymptotic Bode plot, the denominator of this network function acts like two poles at $p = \omega_0$. If this factor were to appear in the numerator of a network function, it would act like two zeros at $z = \omega_0$. The slope of the asymptotic Bode plot would increase by 40 dB/decade as the frequency increased past $\omega = \omega_0$.

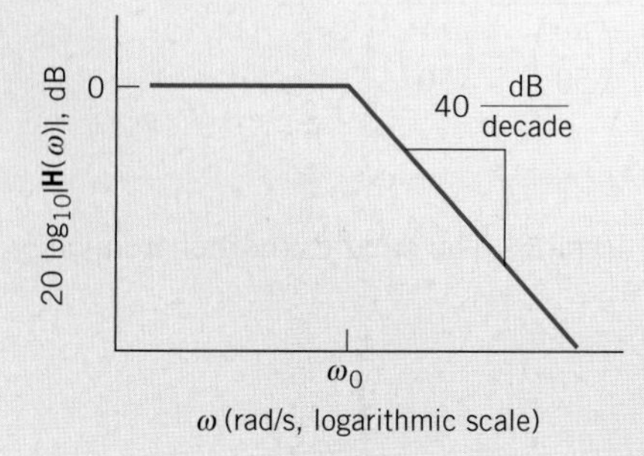

FIGURE 13.3-10 The asymptotic magnitude Bode plot of the second-order low-pass filter when the dc gain is 1.

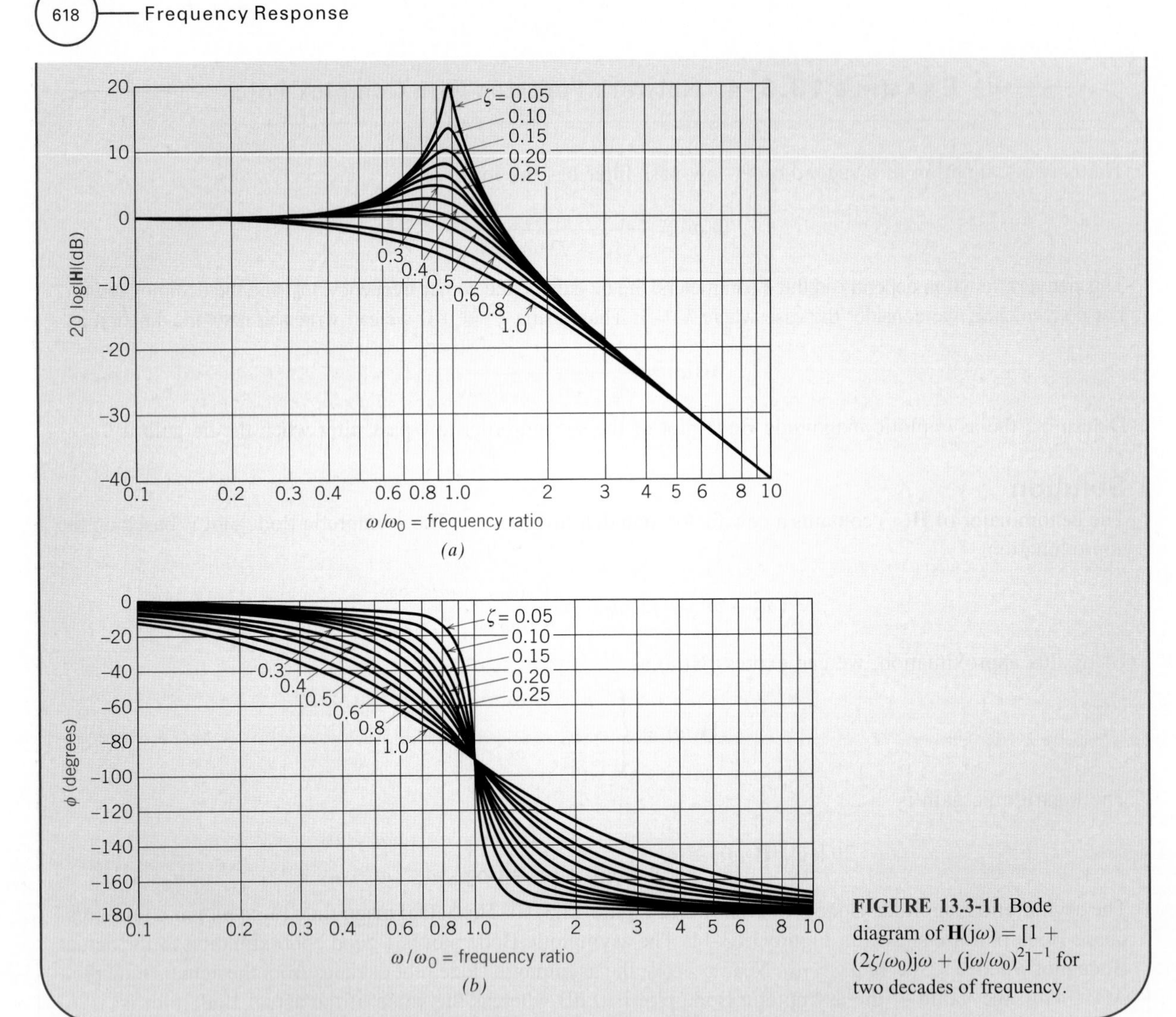

FIGURE 13.3-11 Bode diagram of $\mathbf{H}(j\omega) = [1 + (2\zeta/\omega_0)j\omega + (j\omega/\omega_0)^2]^{-1}$ for two decades of frequency.

EXAMPLE 13.3-5 Magnitude Bode Plot for a Complicated Network Function

Find the asymptotic magnitude Bode plot of

$$\mathbf{H}(\omega) = \frac{5(1 + 0.1j\omega)}{j\omega(1 + 0.5j\omega)\left[1 + 0.6\left(\dfrac{j\omega}{50}\right) - \left(\dfrac{\omega}{50}\right)^2\right]}$$

Solution

The corner frequencies of $\mathbf{H}(\omega)$ are $z = 10$, $p = 2$, and $\omega_0 = 50$ rad/s. The smallest corner frequency is $p = 2$. When $\omega < 2$, $\mathbf{H}(\omega)$ can be approximated as

$$\mathbf{H}(\omega) = \frac{5}{j\omega}$$

so the equation of the low-frequency asymptote is

$$20 \log_{10}|\mathbf{H}| = 20 \log_{10} 5 - 20 \log_{10} \omega$$

The slope of the low-frequency asymptote is −20 dB/decade. Let's find a point on the low-frequency asymptote. When $\omega = 1$,

$$20 \log_{10}|\mathbf{H}| = 20 \log_{10} 5 - 20 \log_{10} 1 = 14 \text{ dB}$$

The low-frequency asymptote is a straight line with a slope of −20 dB/decade passing through the point $\omega = 1$ rad/s, $|\mathbf{H}| = 14$ dB.

The slope of the asymptotic Bode plot will change as ω increases past each corner frequency. The slope decreases by 20 dB/decade at $\omega = p = 2$ rad/s, then increases by 20 dB/decade at $\omega = 10$ rad/s, and finally decreases by 40 dB/decade at $\omega = 50$ rad/s. The asymptotic magnitude Bode plot is shown in Figure 13.3-12.

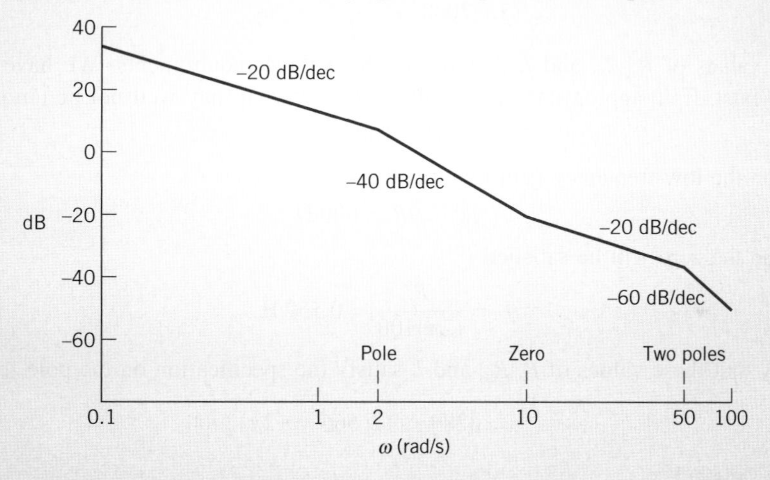

FIGURE 13.3-12 Asymptotic plot for Example 13.3-5.

EXAMPLE 13.3-6 Designing a Circuit to Have a Specified Bode Plot

Let's design the circuit shown in Figure 13.3-3 to satisfy the following specifications.

1. The low-frequency gain is 0.1.
2. The high-frequency gain is 1.
3. The corner frequencies lie in the range of 100 hertz to 2000 hertz.

Solution

We're confronted with two problems. First, can these specifications be satisfied using this circuit? Second, if they can, what values of R, R_s, and L are required?

Our earlier analysis of this circuit showed that the low-frequency gain is less than 1 and that the high-frequency gain is equal to 1. This circuit can be used only to satisfy specifications that are consistent with these facts. Fortunately, the given specifications are consistent with these facts. The first specification requires

$$0.1 = \text{low-frequency gain} = k = \frac{R}{R + R_s}$$

Because this circuit has a high-frequency gain equal to 1, the second specification is satisfied.

Now let's turn our attention to the specifications on the corner frequencies. The specified frequency range is given using units of hertz, whereas the corner frequencies have units of radians/second. Because $\omega_1 > \omega_2$, the third specification requires that

$$(2\pi)100 < \frac{R}{L} = \omega_1$$

and

$$(2\pi)2000 > \frac{R + R_s}{L} = \omega_2$$

Our job is to find values of R, R_s, and L that satisfy these three requirements. We have no guarantee that appropriate values exist. If an appropriate set of values does exist, it may well not be unique. Let's try

$$R = 100\ \Omega$$

The specification on the low-frequency gain requires that

$$R_s = 9R = 900\ \Omega$$

The specification on the zero will be satisfied if

$$L = \frac{R}{(2\pi)100} = 0.159\ \text{H}$$

It remains to verify that these values of R, R_s, and L satisfy the specification on the pole frequency. Because

$$\frac{R + R_s}{L} = 6289 < 12,566 = (2\pi)2000$$

the specification is satisfied.

In summary, when

$$R = 100\ \Omega, \quad R_s = 900\ \Omega, \quad \text{and} \quad L = 0.159\ \text{H}$$

the circuit shown in Figure 13.3-3 satisfies the specifications given above.

This solution is not unique. Indeed, when $R = 100$ and $R_s = 900$, any inductance in the range $0.0796 < L < 0.159$ H can be used to satisfy these specifications.

EXAMPLE 13.3-7 Designing a Circuit to Have a Specified Bode Plot

Design a circuit that has the asymptotic magnitude Bode plot shown in Figure 13.3-13a.

Solution

The slope of this Bode plot is 20 dB/decade for low frequencies, that is, $\omega < 500$ rad/s, so $\mathbf{H}(\omega)$ must have a $j\omega$ factor in its numerator. The slope decreases by 20 dB/decade (from 20 dB/decade to zero) as ω increases past $\omega = 500$ rad/s, so $\mathbf{H}(\omega)$ must have a pole at $\omega = 500$ rad/s. Based on these observations

$$\mathbf{H}(\omega) = \pm k \frac{j\omega}{1 + j\dfrac{\omega}{500}}$$

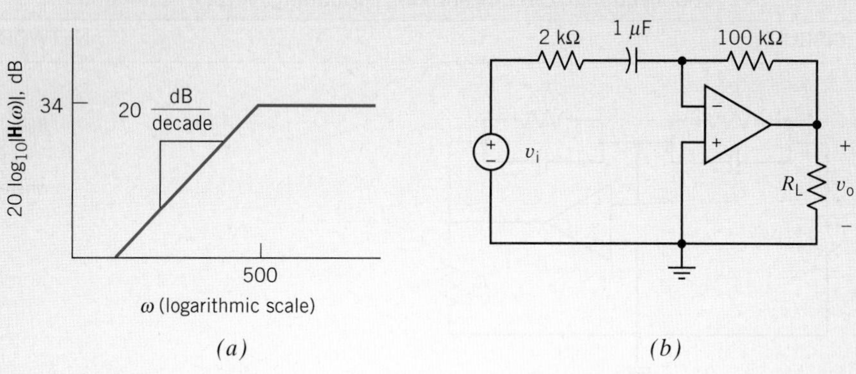

FIGURE 13.3-13 (*a*) An asymptotic magnitude Bode plot and (*b*) a circuit that implements that Bode plot.

The gain of the asymptotic Bode plot is 34 dB = 50 when $\omega > 500$ rad/s, so

$$50 = \pm k \frac{j\omega}{j\dfrac{\omega}{500}} = \pm k \cdot 500$$

Thus, $k = \pm 0.1$ and

$$\mathbf{H}(\omega) = \pm 0.1 \cdot \frac{j\omega}{1 + j\dfrac{\omega}{500}}$$

We need a circuit that has a network function of this form. Table 13.3-2 contains a collection of circuits and corresponding network functions. Row 4 of Table 13.3-2 contains the circuit that we can use. The design equations provided in row 4 of the table indicate that

$$0.1 = R_2 C$$
$$500 = \frac{1}{CR_1}$$

Because there are more unknowns than equations, the solution of these design equations is not unique. Pick $C = 1\,\mu$F. Then

$$R_2 = \frac{0.1}{10^{-6}} = 100\,\text{k}\Omega$$
$$R_1 = \frac{1}{500 \cdot 10^{-6}} = 2\,\text{k}\Omega$$

The finished circuit is shown in Figure 13.3-13*b*.

EXERCISE 13.3-1 (a) Convert the gain $|\mathbf{V}_o/\mathbf{V}_s| = 2$ to decibels. (b) Suppose $|\mathbf{V}_o/\mathbf{V}_s| = -6.02$ dB. What is the value of this gain ‘‘not in dB’’?

Answers: **(a)** 6.02 dB **(b)** 0.5

EXERCISE 13.3-2 In a certain frequency range, the magnitude of the network function can be approximated as $H = 1/\omega^2$. What is the slope of the Bode plot in this range, expressed in decibels per decade?

Answer: −40 dB/decade

Table 13.3-2 A Collection of Circuits and Corresponding Network Functions

CIRCUIT	NETWORK FUNCTION
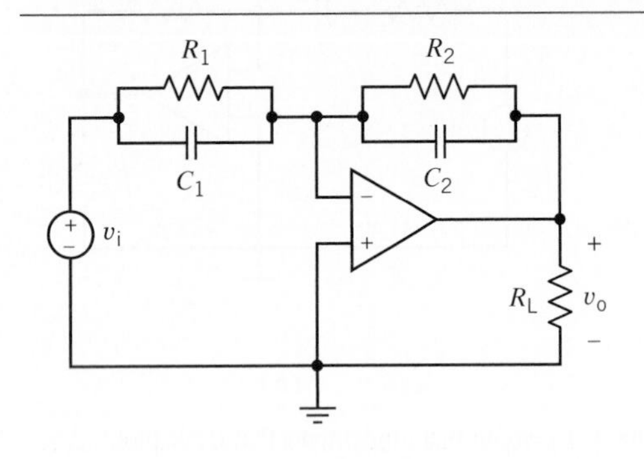	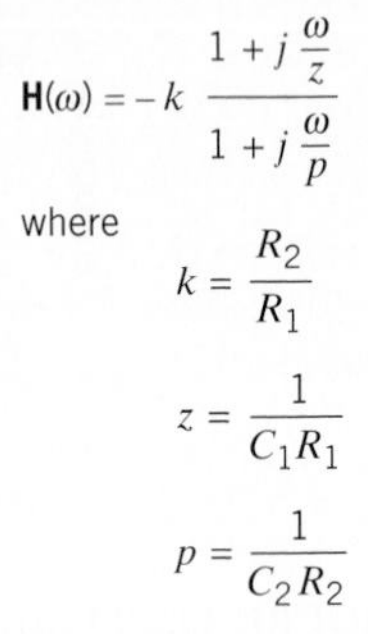 $\mathbf{H}(\omega) = -k \dfrac{1 + j\dfrac{\omega}{z}}{1 + j\dfrac{\omega}{p}}$ where $k = \dfrac{R_2}{R_1}$, $z = \dfrac{1}{C_1 R_1}$, $p = \dfrac{1}{C_2 R_2}$
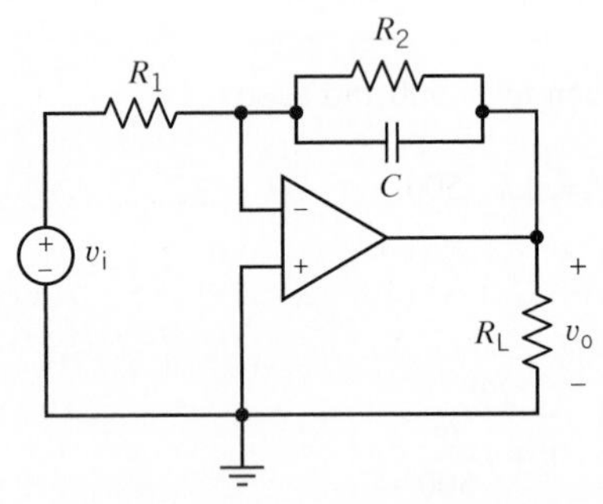	$\mathbf{H}(\omega) = -\dfrac{k}{1 + j\dfrac{\omega}{p}}$ where $k = \dfrac{R_2}{R_1}$, $p = \dfrac{1}{C R_2}$
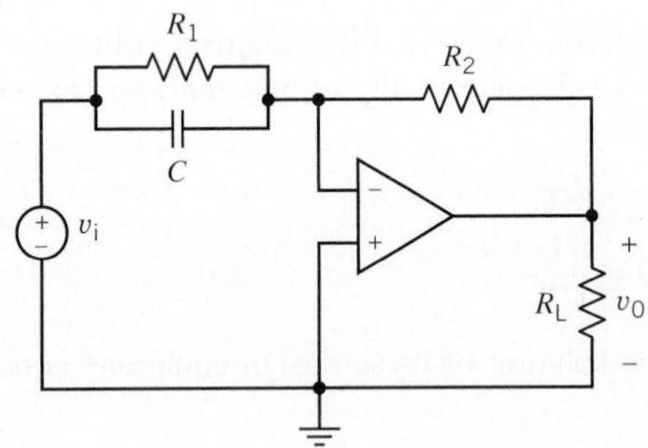	$\mathbf{H}(\omega) = -k\left(1 + j\dfrac{\omega}{z}\right)$ where $k = \dfrac{R_2}{R_1}$, $z = \dfrac{1}{C R_1}$
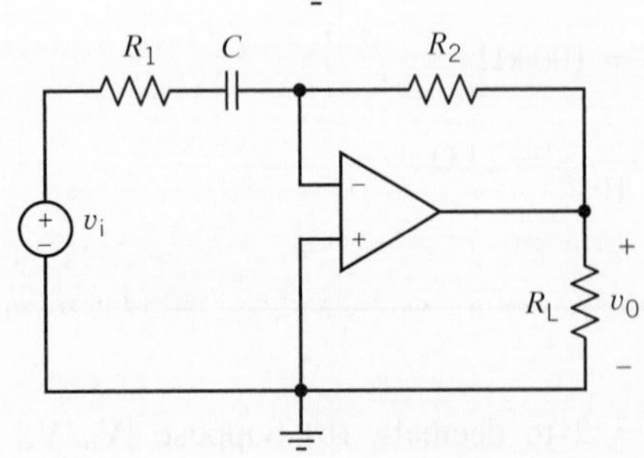	$\mathbf{H}(\omega) = -k \dfrac{j\omega}{1 + j\dfrac{\omega}{p}}$ where $k = R_2 C$, $p = \dfrac{1}{C R_1}$
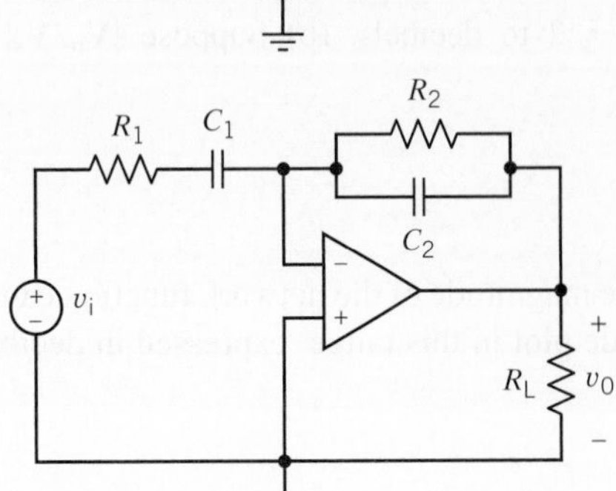	$\mathbf{H}(\omega) = -\dfrac{k(j\omega)}{\left(1 + j\dfrac{\omega}{p_1}\right)\left(1 + j\dfrac{\omega}{p_2}\right)}$ where $k = C_1 R_2$, $p_1 = \dfrac{1}{C_1 R_1}$, $p_2 = \dfrac{1}{C_2 R_2}$

EXERCISE 13.3-3 Consider the network function

$$\mathbf{H}(\omega) = \frac{j\omega A}{B + j\omega C}$$

Find (a) the corner frequency, (b) the slope of the asymptotic magnitude Bode plot for ω above the corner frequency in decibels per decade, (c) the slope of the magnitude Bode plot below the corner frequency, and (d) the gain in decibels for ω above the corner frequency.

Answers: **(a)** $\omega_0 = B/C$ **(b)** zero **(c)** 20 dB/decade **(d)** $20 \log_{10} = \frac{A}{C}$

13.4 RESONANT CIRCUITS

In this section, we will study the behavior of some circuits called *resonant circuits*. We begin with an example.

Consider the situation shown in Figure 13.4-1*a*. The input to this circuit is the current of the current source, and the response is the voltage across the current source. Because the input to the circuit is sinusoidal, we can use phasors to analyze this circuit. We know that the network function of the circuit is the ratio of the response phasor to the input phasor. In this case, that network function will be an impedance

$$\mathbf{Z} = \frac{\mathbf{V}}{\mathbf{I}} = \frac{A\angle\theta}{B\angle 0^\circ}$$

Figure 13.4-1*b* shows some data that were obtained by applying an input with an amplitude of 2 mA and a frequency that was varied. Row 1 of this table describes the performance of this circuit when $\omega = 200$ rad/s. At this frequency, the impedance of the circuit is

$$\mathbf{Z} = \frac{6.6\angle 48^\circ}{0.002} = 3300\angle 48^\circ\ \Omega$$

Let's convert this impedance from polar to rectangular form:

$$\mathbf{Z} = 2208 + j2452\ \Omega$$

This looks like the equivalent impedance of a series resistor and inductor. The resistance would be 2208 Ω. Because the frequency is $\omega = 200$ rad/s, the inductance would be 12.26 H. Recall that in rectangular form impedances are represented as

$$\mathbf{Z} = R + jX$$

where R is called the resistance and X is called the reactance. When $\omega = 200$ rad/s, we say that the reactance of this circuit is inductive because the reactance is positive and therefore could have been caused by a single inductor.

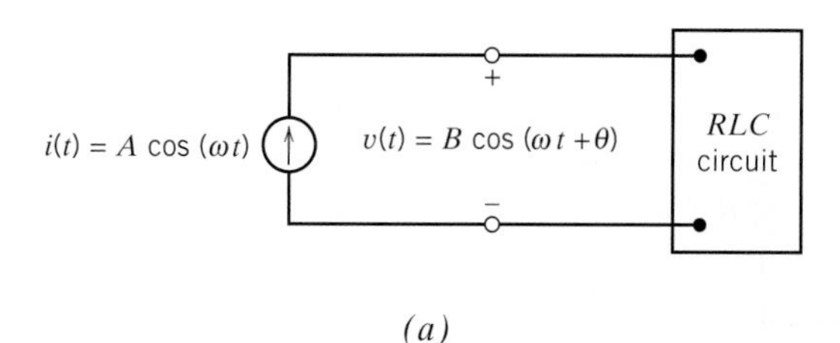

(*a*)

A, A	ω, rad/s	B, V	θ
0.002	200	6.6	48°
0.002	220	8.4	33°
0.002	250	10.0	0°
0.002	270	9.3	–21°
0.002	300	7.4	– 43°

(*b*)

FIGURE 13.4-1 (*a*) An *RLC* circuit with a sinusoidal input and (*b*) some frequency response data.

The last row of the table describes the performance of this circuit when $\omega = 300$ rad/s. Now

$$\mathbf{Z} = \frac{7.4\angle{-43^\circ}}{0.002} = 3700\angle{-43^\circ} = 2706 - j\,2523\ \Omega$$

Because the reactance is negative, it couldn't have been caused by a single inductor. This impedance looks like the equivalent impedance of a single resistor connected in series with a single capacitor:

$$R - j\frac{1}{\omega C} = 2706 - j\,2523\ \Omega$$

Equating the real parts shows that the resistance is 2706 Ω. Equating imaginary parts shows that the capacitance is 1.32 μF.

The reactance of this circuit is inductive at some frequencies and capacitive at other frequencies. We can tell when the reactance will be inductive and when it will be capacitive by looking at the last column of the table. When θ is positive, the reactance is inductive and when θ is negative, the reactance is capacitive. The frequency $\omega = 250$ rad/s is special. When the input frequency is less than 250 rad/s, the reactance is inductive, but when the input frequency is greater than 250 rad/s, the reactance is capacitive. This special frequency is called the **resonant frequency** and is denoted as ω_0. From the third row of the table, we see that when $\omega = \omega_0 = 250$ rad/s

$$\mathbf{Z} = \frac{10\angle{0^\circ}}{0.002} = 5000\angle{0^\circ} = 5000 - j\,0\ \Omega$$

The reactance is zero. At the resonant frequency, the impedance is purely resistive. Indeed, this fact can be used to identify the resonant frequency.

Another observation can be made from Figure 13.4-1. The magnitude of the impedance is maximum when $\omega = \omega_0 = 250$ rad/s. When the frequency is reduced from ω_0 or increased from ω_0, the magnitude of the impedance is decreased.

Next, consider the circuit shown in Figure 13.4-2. This circuit is called the **parallel resonant circuit**. The equivalent impedance of the parallel resistor, inductor, and capacitor is

$$\mathbf{Z} = \frac{1}{\frac{1}{R} + j\omega C + \frac{1}{j\omega L}} = \frac{1}{\sqrt{\left(\frac{1}{R}\right)^2 + \left(\omega C - \frac{1}{\omega L}\right)^2}}\angle{-\tan^{-1} R\left(\omega C - \frac{1}{\omega L}\right)} \tag{13.4-1}$$

This circuit exhibits some familiar behavior. The reactance will be zero when

$$\omega C - \frac{1}{\omega L} = 0$$

The frequency that satisfies this equation is the resonant frequency, ω_0. Solving this equation gives

$$\omega_0 = \frac{1}{\sqrt{LC}}$$

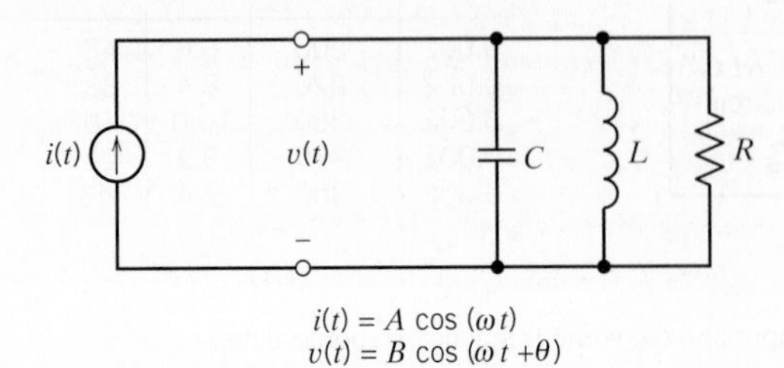

FIGURE 13.4-2 The parallel resonant circuit.

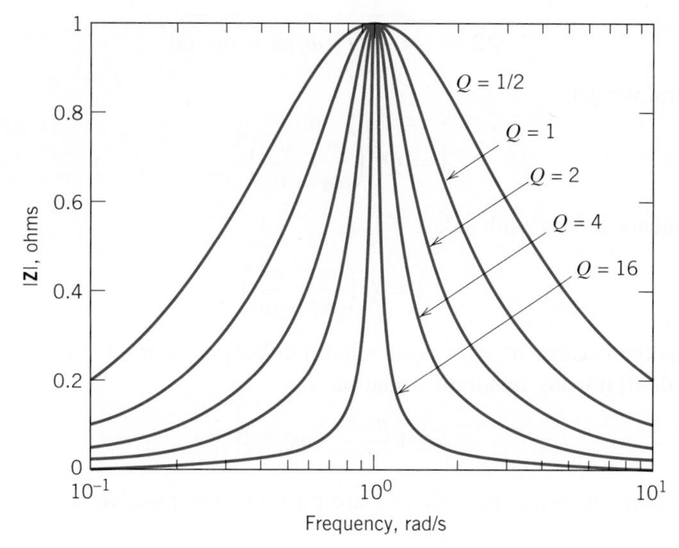

FIGURE 13.4-3 The effect of Q on the frequency response of a resonant circuit.

At $\omega = \omega_0$, $\mathbf{Z} = R$. The magnitude of $\mathbf{Z}$ decreases as ω is either increased or decreased from ω_0. The angle of $\mathbf{Z}$ is positive when $\omega < \omega_0$ and negative when $\omega > \omega_0$, so the reactance is inductive when $\omega < \omega_0$ and capacitive when $\omega > \omega_0$.

The impedance can be put in the form

$$\mathbf{Z} = \frac{k}{1 + jQ\left(\dfrac{\omega}{\omega_0} - \dfrac{\omega_0}{\omega}\right)} \tag{13.4-2}$$

where

$$k = R, \quad Q = R\sqrt{\frac{C}{L}}, \quad \text{and} \quad \omega_0 = \frac{1}{\sqrt{LC}} \tag{13.4-3}$$

The parameters k, Q, and ω_0 characterize the resonant circuit. The resonant frequency, ω_0, is the frequency at which the reactance is zero and where the magnitude of the impedance is maximum. The parameter k is the value of the impedance when $\omega = \omega_0$, so k is the maximum value of the impedance. Q is called the quality factor of the resonant circuit. The magnitude of the impedance will decrease as ω is reduced from ω_0 or increased from ω_0. The **quality factor** controls how rapidly $|\mathbf{Z}|$ decreases. Figure 13.4-3 illustrates the importance of Q. Both k and ω_0 have been set equal to 1 in Figure 13.4-3 to emphasize the relationship between Q and $|\mathbf{Z}|$.

Figure 13.4-3 shows that the larger the value of Q, the more sharply peaked is the frequency response plot. We can quantify this observation by introducing the bandwidth of the resonant circuit. To that end, let ω_1 and ω_2 denote the frequencies where

$$|\mathbf{Z}(\omega)| = \frac{1}{\sqrt{2}}|\mathbf{Z}(\omega_0)| = \frac{k}{\sqrt{2}}$$

There will be two such frequencies, one smaller than ω_0 and the other larger than ω_0. Let $\omega_1 < \omega_0$ and $\omega_2 > \omega_0$. The bandwidth, BW, of the resonant circuit is defined as

$$BW = \omega_2 - \omega_1$$

The frequencies ω_1 and ω_2 are solutions of the equation

$$\frac{k}{\sqrt{2}} = \frac{k}{\sqrt{1 + Q^2(\omega/\omega_0 - \omega_0/\omega)^2}}$$

or

$$\sqrt{2} = \sqrt{1 + Q^2(\omega/\omega_0 - \omega_0/\omega)^2}$$

Squaring both sides, we get

$$1 = Q^2\left(\frac{\omega}{\omega_0} - \frac{\omega_0}{\omega}\right)^2$$

Now, taking the square root of both sides,

$$\pm 1 = Q\left(\frac{\omega}{\omega_0} - \frac{\omega_0}{\omega}\right)$$

(The $\pm$ sign is required because $a^2 = b^2$ is satisfied if either $a = b$ or $-a = b$.) This equation can be rearranged to get the following quadratic equation:

$$\omega^2 \mp \frac{\omega_0 \omega}{Q} - \omega_0^2 = 0$$

This equation has four solutions, but only two are positive. The positive solutions are

$$\omega_1 = -\frac{\omega_0}{2Q} + \sqrt{\left(\frac{\omega_0}{2Q}\right)^2 + \omega_0^2} \quad \text{and} \quad \omega_2 = \frac{\omega_0}{2Q} + \sqrt{\left(\frac{\omega_0}{2Q}\right)^2 + \omega_0^2}$$

Finally, we are ready to calculate the bandwidth

$$BW = \omega_2 - \omega_1 = \frac{\omega_0}{Q} \tag{13.4-4}$$

This equation says that the bandwidth is smaller; that is, the frequency response plot is more sharply peaked; when the value of Q is larger.

EXAMPLE 13.4-1 Series Resonant Circuit

Figure 13.4-4 shows a series resonant circuit. Determine the relationship between parameters k, Q, and ω_0 and the element values R, L, and C for the series resonant circuit.

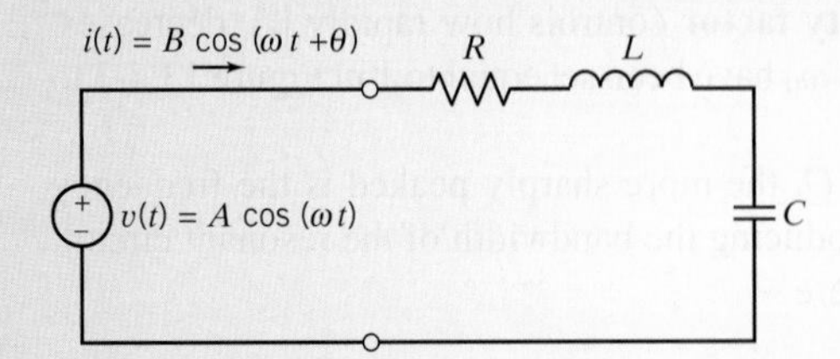

FIGURE 13.4-4 The series resonant circuit.

Solution

The input to this circuit is the voltage source, and the response is the current in the mesh. The network function is the ratio of the response phasor to the input phasor. In this case, the network function is the equivalent admittance of the series resistor, capacitor, and inductor:

$$\mathbf{Y} = \frac{\mathbf{I}}{\mathbf{V}} = \frac{1}{R + j\omega L + \dfrac{1}{j\omega C}} \tag{13.4-5}$$

To identify k, Q, and ω_0, this network function must be rearranged so that it is in the form

$$\mathbf{Y} = \frac{k}{1 + jQ\left(\dfrac{\omega}{\omega_0} - \dfrac{\omega_0}{\omega}\right)} \tag{13.4-6}$$

Rearranging Eq. 13.4-5,

$$\mathbf{Y} = \frac{1}{R + j\left(\omega L - \dfrac{1}{\omega C}\right)} = \frac{1}{R + j\sqrt{\dfrac{L}{C}}\left(\dfrac{\omega}{\dfrac{1}{\sqrt{LC}}} - \dfrac{\dfrac{1}{\sqrt{LC}}}{\omega}\right)} = \frac{\dfrac{1}{R}}{1 + j\dfrac{1}{R}\sqrt{\dfrac{L}{C}}\left(\dfrac{\omega}{\dfrac{1}{\sqrt{LC}}} - \dfrac{\dfrac{1}{\sqrt{LC}}}{\omega}\right)}$$

Comparing this equation to Eq. 13.4-6 gives

$$k = \frac{1}{R}, \quad Q = \frac{1}{R}\sqrt{\frac{L}{C}}, \quad \text{and} \quad \omega_0 = \frac{1}{\sqrt{LC}}$$

EXAMPLE 13.4-2 Frequency Response of a Resonant Circuit

Figure 13.4-5 shows the magnitude frequency response plot of a resonant circuit. What are the values of the parameters k, Q, and ω_0?

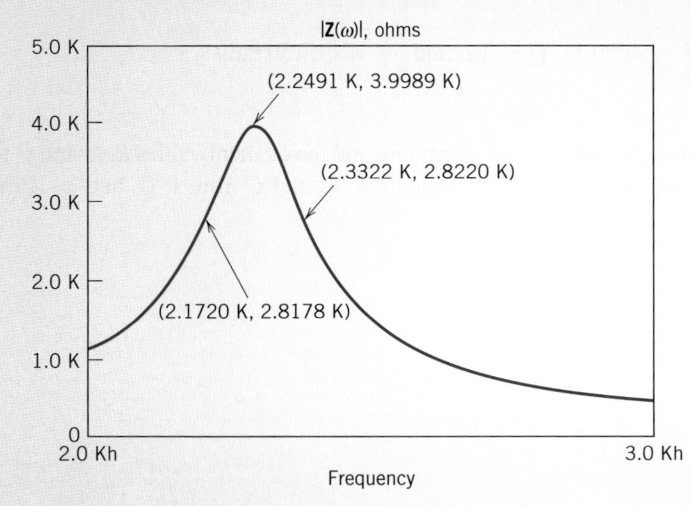

FIGURE 13.4-5 The magnitude frequency response of a resonant circuit.

Solution

The first step is to find the peak of the frequency response and determine the values of the frequency and the impedance corresponding to that point. This frequency is the resonant frequency, ω_0, and the impedance at this frequency is k. This point on the frequency response is labeled in Figure 13.4-5. The frequency is

$$\omega_0 = (2\pi)2249 = 14,130 \text{ rad/s}$$

The impedance is

$$k = 4000\ \Omega$$

Next, the frequencies ω_1 and ω_2 are identified by finding the points on the frequency response where the value of the impedance is $k/\sqrt{2} = 2828\ \Omega$. These points have been labeled in Figure 13.4-5. (The plot shown in Figure 13.4-5 was produced using PSpice and Probe. The cursor function in Probe was used to label points on the frequency response. Each label gives the frequency first, then the impedance. It was not possible to move the cursor to the points where the impedance was exactly 2828 Ω, so the points where the impedance was as close to 2828 Ω as possible were labeled.)

$$\omega_1 = (2\pi)2172 = 13,647 \text{ rad/s} \quad \text{and} \quad \omega_2 = (2\pi)2332 = 14,653 \text{ rad/s}$$

The quality factor, Q, is calculated as

$$Q = \frac{\omega_0}{BW} = \frac{\omega_0}{\omega_2 - \omega_1} = \frac{14,130}{14{,}653 - 13{,}647} = 14$$

Now that the values of the parameters k, Q, and ω_0 are known, the network function can be expressed as

$$\mathbf{Z}(\omega) = \frac{4000}{1 + j14\left(\dfrac{\omega}{14{,}130} - \dfrac{14{,}130}{\omega}\right)}$$

EXAMPLE 13.4-3 Parallel Resonant Circuit

Design a parallel resonant circuit that has $k = 4000\ \Omega$, $Q = 14$, and $\omega_0 = 14,130$ rad/s.

Solution

Table 13.4-1 summarizes the relationship between parameters k, Q, and ω_0 and the element values R, L, and C for the parallel resonant circuit. These relationships can be used to calculate R, L, and C from k, Q, and ω_0. First,

$$R = k = 4000\ \Omega$$

Next,

$$\frac{1}{\sqrt{LC}} = \omega_0 = 14,130$$

and

$$R\sqrt{\frac{C}{L}} = Q = 14$$

Rearranging these last two equations gives

$$\frac{14\sqrt{L}}{4000} = \sqrt{C} = \frac{1}{14,130\sqrt{L}}$$

So,

$$L = \frac{4000}{14{,}130(14)} = 20 \text{ mH} \quad \text{and} \quad C = \frac{1}{14{,}130^2(0.002)} = 0.25\ \mu\text{F}$$

Table 13.4-1 Series and Parallel Resonant Circuits

	SERIES RESONANT CIRCUIT	PARALLEL RESONANT CIRCUIT
Circuit	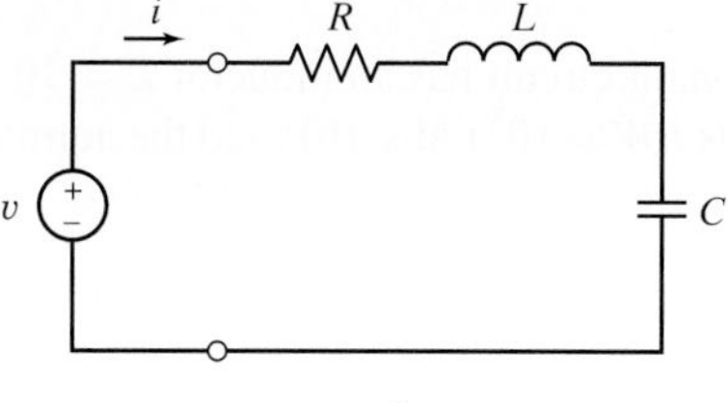	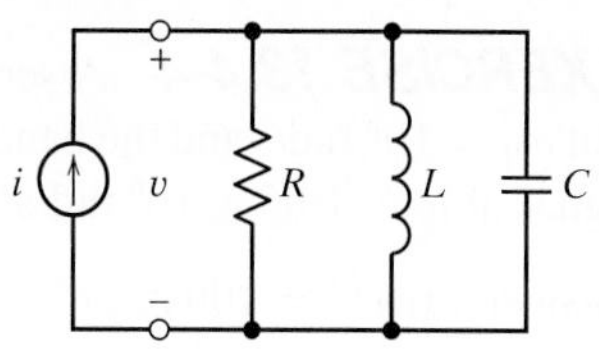
Network function	$\mathbf{Y} = \dfrac{k}{1 + jQ\left(\dfrac{\omega}{\omega_0} - \dfrac{\omega_0}{\omega}\right)}$	$\mathbf{Z} = \dfrac{k}{1 + jQ\left(\dfrac{\omega}{\omega_0} - \dfrac{\omega_0}{\omega}\right)}$
Resonant frequency	$\omega_0 = \dfrac{1}{\sqrt{LC}}$	$\omega_0 = \dfrac{1}{\sqrt{LC}}$
Maximum magnitude	$k = \dfrac{1}{R}$	$k = R$
Quality factor	$Q = \dfrac{1}{R}\sqrt{\dfrac{L}{C}}$	$Q = R\sqrt{\dfrac{C}{L}}$
Bandwidth	$BW = \dfrac{R}{L}$	$BW = \dfrac{1}{RC}$

EXAMPLE 13.4-4 Designing Resonant Circuits

Figure 13.4-5 shows the magnitude frequency response plot of a resonant circuit. Design a circuit that has this frequency response.

Solution

We have already solved this problem. Three things must be done to design the required circuit. First, the parameters k, Q, and ω_0 must be determined from the frequency response. We did that in Example 13.4-2. Second, we notice that the given resonant frequency response is an impedance rather than an admittance, and we choose the parallel resonant circuit from Table 13.4-1. Third, the element values R, L, and C must be calculated from the values of k, Q, and ω_0. We did that in Example 13.4-3.

EXERCISE 13.4-1 For the RLC parallel resonant circuit when $R = 8\ \text{k}\Omega$, $L = 40\ \text{mH}$, and $C = 0.25\ \mu\text{F}$, find (a) Q and (b) bandwidth.

Answers: **(a)** $Q = 20$ **(b)** $BW = 500$ rad/s

EXERCISE 13.4-2 A high-frequency RLC parallel resonant circuit is required to operate at $\omega_0 = 10$ Mrad/s with a bandwidth of 200 krad/s. Determine the required Q and L when $C = 10$ pF.

Answers: $Q = 50$ and $L = 1$ mH

EXERCISE 13.4-3 A series resonant circuit has $L = 1$ mH and $C = 10\ \mu$F. Find the required Q and R when it is desired that the bandwidth be 15.9 Hz.

Answers: $Q = 100$ and $R = 0.1\ \Omega$

EXERCISE 13.4-4 A series resonant circuit has an inductor $L = 10$ mH. **(a)** Select C and R so that $\omega_0 = 10^6$ rad/s and the bandwidth is $BW = 10^3$ rad/s. **(b)** Find the admittance **Y** of this circuit for a signal at $\omega = 1.05 \times 10^6$ rad/s.

Answers: **(a)** $C = 100$ pF, $R = 10\ \Omega$

(b) $\mathbf{Y} = \dfrac{10}{1 + j97.6}$

13.5 FREQUENCY RESPONSE OF OP AMP CIRCUITS

The gain of an op amp is not infinite; rather, it is finite and decreases with frequency. The gain $\mathbf{A}(\omega)$ of the operational amplifier is a function of ω given by

$$\mathbf{A}(\omega) = \frac{A_o}{1 + j\omega/\omega_1}$$

where A_o is the dc gain, and ω_1 is the corner frequency. The dc gain is normally greater than 10^4 and ω_1 is less than 100 rad/s. A circuit model of a frequency-dependent nonideal op amp is shown in Figure 13.5-1. This model is more accurate, but also more complicated, than the ideal op amp model.

Let us consider an example of an op amp circuit incorporating a frequency-dependent op amp.

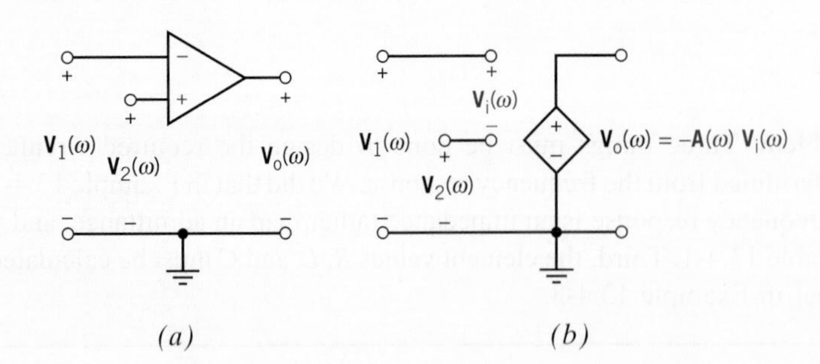

FIGURE 13.5-1 (*a*) An operational amplifier and (*b*) a frequency-dependent model of an operational amplifier.

EXAMPLE 13.5-1 Frequency Response of a Noninverting Amplifier

Consider the noninverting amplifier in Figure 13.5-2*a*. Replacing the op amp with a frequency-dependent op amp gives the circuit shown in Figure 13.5-2*b*. Suppose that $R_2 = 90$ kΩ and $R_1 = 10$ kΩ and that the parameters of the op amp are $A_o = 10^5$ and $\omega_1 = 10$ rad/s. Determine the magnitude Bode plot for both the gain of the op amp, **A** (ω), and the network function of the noninverting amplifier, $\mathbf{V}_o/\mathbf{V}_s$.

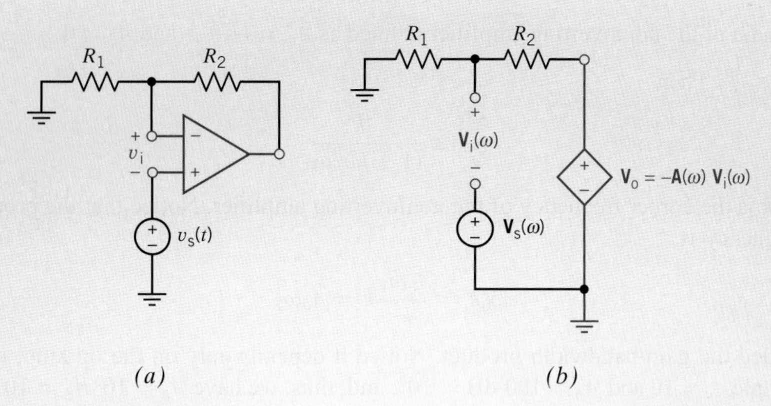

FIGURE 13.5-2 (*a*) A noninverting amplifier and (*b*) an equivalent circuit incorporating the frequency-dependent model of the operational amplifier.

Solution

The Bode plot of 20 log$|\mathbf{A}(\omega)|$ is shown in Figure 13.5-3. Note that the magnitude is equal to 1 (0 dB) at $\omega = 10^6$ rad/s.

Writing a node equation in Figure 13.5-2*b* gives

$$\frac{\mathbf{V}_\mathrm{i} + \mathbf{V}_\mathrm{s}}{R_1} + \frac{\mathbf{V}_\mathrm{i} + \mathbf{V}_\mathrm{s} + \mathbf{A}(\omega)\mathbf{V}_\mathrm{i}}{R_2} = 0$$

The frequency-dependent model of the op amp is described by

$$\mathbf{V}_\mathrm{o} = -\mathbf{A}(\omega)\mathbf{V}_\mathrm{i}$$

Combining these equations gives

$$\frac{\mathbf{V}_\mathrm{o}}{\mathbf{V}_\mathrm{s}} = \frac{\mathbf{A}(\omega)}{1 + \dfrac{\mathbf{A}(\omega)}{k}}$$

where $k = (R_1 + R_2)/R_1$ is the gain of the noninverting amplifier when the op amp is modeled as an ideal op amp. Substituting for $\mathbf{A}(\omega)$, we get

$$\frac{\mathbf{V}_\mathrm{o}}{\mathbf{V}_\mathrm{s}} = \frac{A_\mathrm{o}/(1 + j\omega/\omega_1)}{1 + (A_\mathrm{o}/k)/(1 + j\omega/\omega_1)} = \frac{A_\mathrm{o}}{1 + j\omega/\omega_1 + A_\mathrm{o}/k} = \frac{A_\mathrm{c}}{1 + j\omega/(A_2\omega_1)}$$

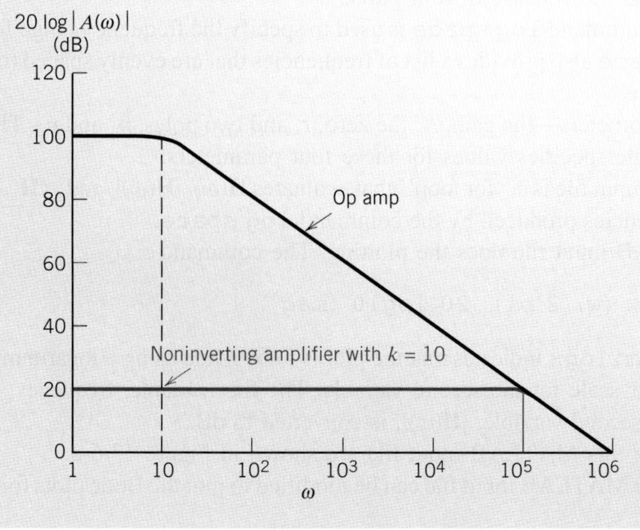

FIGURE 13.5-3 Bode magnitude diagram of the op amp and the noninverting op amp circuit (in color).

where A_c is the dc gain of the noninverting amplifier defined as $A_c = \frac{A_o}{1+\frac{A_o}{k}}$ and $A_2 = 1 + \frac{A_o}{k}$. Usually, $1 \ll \frac{A_o}{k}$, so $A_c \cong k$ and $A_2 \cong \frac{A_o}{k}$. Then

$$\frac{\mathbf{V}_o}{\mathbf{V}_s} \cong \frac{k}{(1 + j\omega/\omega_0)}$$

where $\omega_o = A_o\omega_1/k$ is the corner frequency of the noninverting amplifier. Notice that the product of the dc gain and the corner frequency is

$$\omega_0 k = \frac{A_o\omega_1}{k} k = A_o\omega_1$$

This product is called the gain-bandwidth product. Notice it depends only on the op amp, not on R_1 and R_2.

For this example, $k = 10$ and $A_o = 100\text{ dB} = 10^5$, and, thus, we have $A_c = 10$, $A_2 = 10^4$, and $\omega_1 A_2 = 10^5$. Therefore,

$$\frac{\mathbf{V}_o}{\mathbf{V}_s} = \frac{10}{1 + j10^{-5}\omega}$$

This circuit has a magnitude Bode plot as shown in color in Figure 13.5-3. Note that the noninverting op amp has a low-frequency gain of 20 dB and a break frequency of 10^5 rad/s. The gain-bandwidth product remains 10^6 rad/s.

13.6 PLOTTING BODE PLOTS USING MATLAB

MATLAB can be used to display the Bode plot or frequency response plot corresponding to a network function. As an example, consider the network function

$$\mathbf{H}(\omega) = \frac{K\left(1 + j\frac{\omega}{z}\right)}{\left(1 + j\frac{\omega}{p_1}\right)\left(1 + j\frac{\omega}{p_2}\right)}$$

Figure 13.6-1 shows a MATLAB input file that can be used to obtain the Bode plot corresponding to this network function. This MATLAB file consists of four parts.

In the first part, the MATLAB command `log space` is used to specify the frequency range for the Bode plot. The command `log space` also provides a list of frequencies that are evenly spaced (on a log scale) over this frequency range.

The given network has four parameters—the gain, K; the zero, z; and two poles, p_1 and p_2. The second part of the MATLAB input file specifies values for these four parameters.

The third part of the MATLAB input file is a ''for loop'' that evaluates $\mathbf{H}(\omega)$, $|\mathbf{H}(\omega)|$, and $\angle \mathbf{H}(\omega)$ at each frequency in the list of frequencies produced by the command `log space`.

The fourth part of the MATLAB input file does the plotting. The command

```
semilogx (w/(2*pi), 20*log10 (mag))
```

does several things. The command `semilogx` indicates that the plot is to be made using a logarithmic scale for the first variable and a linear scale for the second variable. The first variable, frequency, is divided by 2π to convert to Hz. The second variable, $|\mathbf{H}(\omega)|$, is converted to dB.

The Bode plots produced using this MATLAB input file are shown in Figure 13.6-2.

The second and third parts of the MATLAB input file can be modified to plot the Bode plots for a different network function.

```
% nf.m - plot the Bode plot of a network function

%------------------------------------------------------------------------
%     Create a list of logarithmically spaced frequencies.
%------------------------------------------------------------------------
wmin=10;                        % starting frequency, rad/s
wmax=100000;                    % ending frequency, rad/s
w = logspace(log10(wmin),log10(wmax));

%------------------------------------------------------------------------
%             Enter values of the parameters that describe the
%                            network function.
%------------------------------------------------------------------------
K= 10;                  % constant
z= 1000;                % zero
p1=100;    p2=10000;    % poles

%------------------------------------------------------------------------
% Calculate the value of the network function at each frequency.
%   Calculate the magnitude and angle of the network function.
%------------------------------------------------------------------------
for k=1:length(w)
    H(k) = K*(1+j*w(k)/z) / ( (1+j*w(k)/p1) * (1+j*w(k)/p2) );
    mag(k) = abs(H(k));
    phase(k) = angle(H(k));
end

%------------------------------------------------------------------------
%                            Plot the Bode plot.
%------------------------------------------------------------------------
subplot(2,1,1), semilogx(w/(2*pi), 20*log10(mag))
xlabel('Frequency, Hz'), ylabel('Gain, dB')
title('Bode plot')
subplot(2,1,2), semilogx(w/(2*pi), phase)
xlabel('Frequency, Hz'), ylabel('Phase, deg')
```

FIGURE 13.6-1 MATLAB input file used to plot the Bode plots corresponding to a network function.

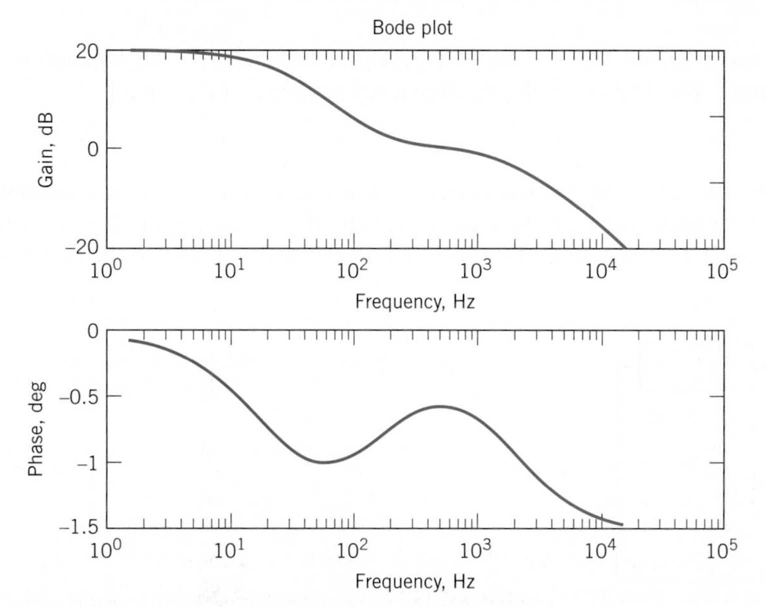

FIGURE 13.6-2 The Bode plots produced using the MATLAB input file given in Figure 13.6-1.

13.7 USING PSPICE TO PLOT A FREQUENCY RESPONSE

To use PSpice to plot the frequency response of a circuit, we do the following:

1. Draw the circuit in the OrCAD Capture workspace.
2. Specify an AC Sweep\Noise simulation.
3. Run the simulation.
4. Plot the simulation results.

The frequency axis of a frequency response plot can be either a linear axis or a logarithmic axis. When a logarithmic axis is used for the frequency variable, the plots are referred to as Bode diagrams or Bode plots. We encounter the terms *octave* and *decade* when working with logarithmic scales. The frequency doubles in an octave and increases by a factor of ten in a decade. (The log of the frequency increases by 1 as the frequency increases by a decade.)

Let $A\angle\theta$ be the phasor of the node voltage at node 2 of a circuit. PSpice uses the notation:

$$\mathrm{V}(2)\angle\mathrm{Vp}(2) = A\angle\theta$$

That is, V(2) denotes the magnitude of the phasor and Vp(2) denotes the angle of the phasor. PSpice gives the angle in degrees. Similarly, V(R2) represents the magnitude of the voltage across resistor R2, whereas Vp(R2) denotes the angle. PSpice indicates that the units are decibels by inserting ''dB'' into the name of a signal just before the parenthesis. For example, VdB(2) denotes the magnitude of the node voltage phasor in dB.

EXAMPLE 13.7-1 Using PSpice to Plot a Frequency Response

The input to the circuit shown in Figure 13.7-1 is the voltage source voltage $v_s(t)$. The response is the voltage, $v_o(t)$, across the 20-kΩ resistor. Use PSpice to plot the frequency response of this circuit.

Solution

We begin by drawing the circuit in the OrCAD workspace as shown in Figure 13.7-2 (see Appendix A). Two nodes of this circuit have been named using a PSpice part called an off-page connector. The particular off-page connector used in Figure 13.7-1 is called an OFFPAGELEFT-R part and is found in the part library named

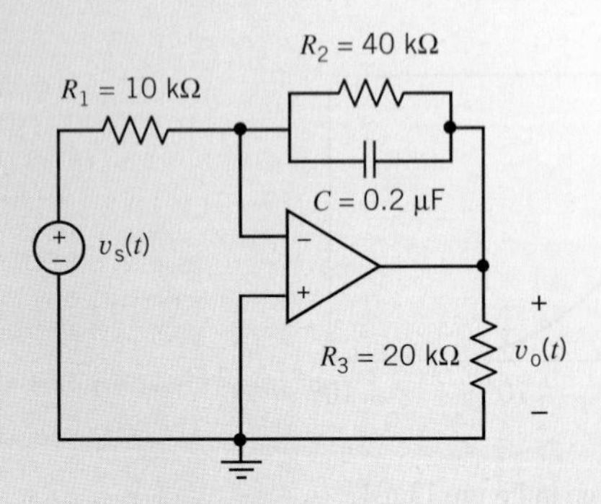

FIGURE 13.7-1 The circuit considered in Example 13.7-1.

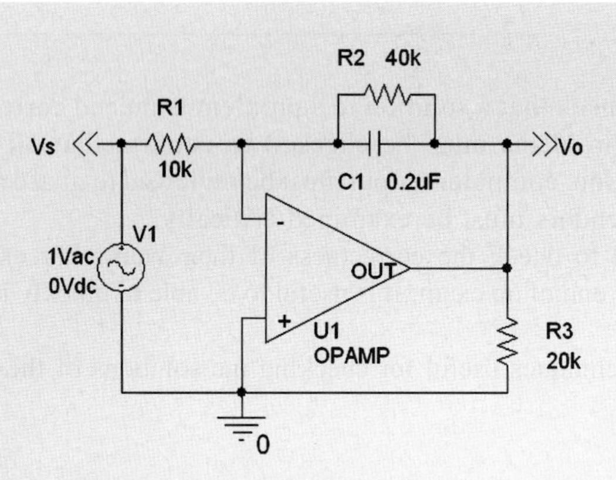

FIGURE 13.7-2 The circuit of Figure 13.7-1 as drawn in the OrCAD workspace.

CAPSYM. To label a node, select Place/Off-Page Connector . . . from the OrCAD capture menus to pop up the Place Off-Page Connector dialog box. Select the library CAPSYM from the list of libraries and then choose OFFPAGELEFT-R. The new connector will be labeled OFFPAGELEFT-R. Use the property editor to change this name to something descriptive, such as Vo. Wire the connector to the appropriate node of the circuit to name that node Vo.

We will perform an AC Sweep\Noise simulation. (Select PSpice\New Simulation Profile from the OrCAD Capture menu bar; then select AC Sweep\Noise from the Analysis Type drop-down list. Set the Start Frequency to 1 and End Frequency to 1000. Select a Logrithmic Sweep and set the Points/Decade to 100.) Select PSpice\Run Simulation Profile from the OrCAD Capture menu bar to run the simulation.

After a successful ACSweep\Noise simulation, OrCAD Capture will automatically open a Schematics window. Select Plot/Add plot from the Schematics menus to add a second plot. Two empty plots will appear, one above the other. Select the top plot by clicking the top plot.

Select Trace/Add Trace from the Schematics menus to pop up the Add Traces dialog box. Select first V(Vo) and then V(Vs) from the list of Simulation Output Variables. The Trace Expression, near the bottom of the dialog box, will be V(Vo)V(Vs). Edit the trace expression to be Vdb(Vo) – Vdb(Vs). Vdb(Vo) – Vdb(Vs) is the gain in decibels. Close the Add Traces dialog box.

Select the bottom plot by clicking the bottom plot. Select Trace/Add Trace to pop up the Add Traces dialog box. Select first V(Vo) and then V(Vs) from the list of Simulation Output Variables. The Trace Expression, near the bottom of the dialog box, will be be V(Vo)V(Vs). Edit the trace expression to be Vp(Vo) – Vp(Vs). Vp(Vo) – Vp(Vs) is the phase shift in degrees. Close the Add Traces dialog box.

Figure 13.7-3 shows resulting plots after labeling some points.

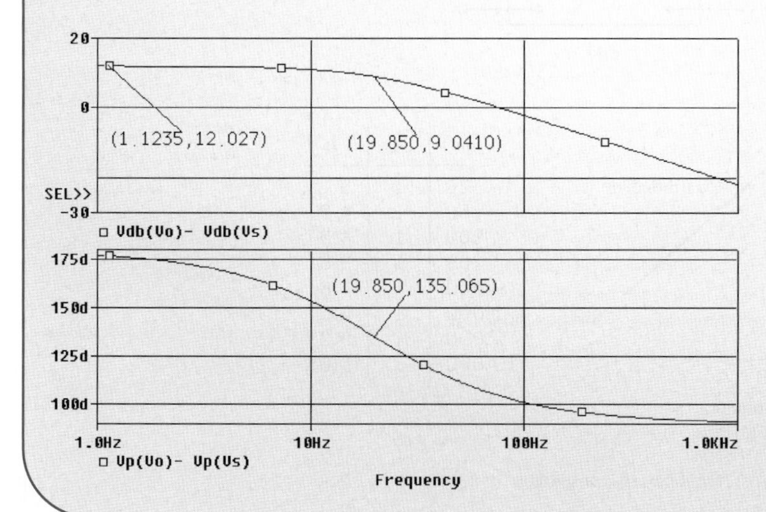

FIGURE 13.7-3 The gain and phase Bode plots.

13.8 HOW CAN WE CHECK . . . ?

Engineers are frequently called upon to check that a solution to a problem is indeed correct. For example, proposed solutions to design problems must be checked to confirm that all of the specifications have been satisfied. In addition, computer output must be reviewed to guard against data-entry errors, and claims made by vendors must be examined critically.

Engineering students are also asked to check the correctness of their work. For example, occasionally just a little time remains at the end of an exam. It is useful to be able to quickly identify those solutions that need more work.

The following examples illustrate techniques useful for checking the solutions of the sort of problem discussed in this chapter.

EXAMPLE 13.8-1 How Can We Check Bode Plots?

Figure 13.8.1*a* shows a laboratory setup for measuring the frequency response of a circuit. A sinusoidal input is connected to the input of a circuit having the network function $\mathbf{H}(\omega)$. An oscilloscope is used to measure the input and output sinusoids. The input voltage is used to trigger the oscilloscope so the phase angle of the input is zero. Frequency response data are collected by varying the input frequency and measuring the amplitude of the input voltage and the amplitude and phase of the output voltage.

In this example, the desired frequency response is specified by the Bode plot shown in Figure 13.8.1*b*. Figure 13.8.1*c* shows frequency response data from laboratory measurements. In this example, the amplitude, but not the phase angle, of the output voltage was measured. **How can we check** that the circuit does indeed have the specified Bode plot?

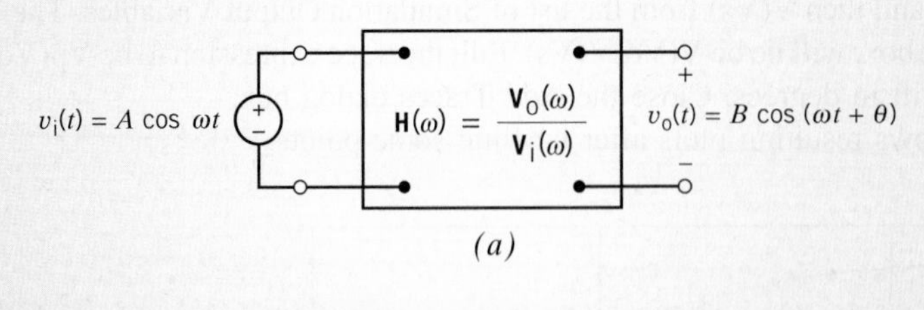

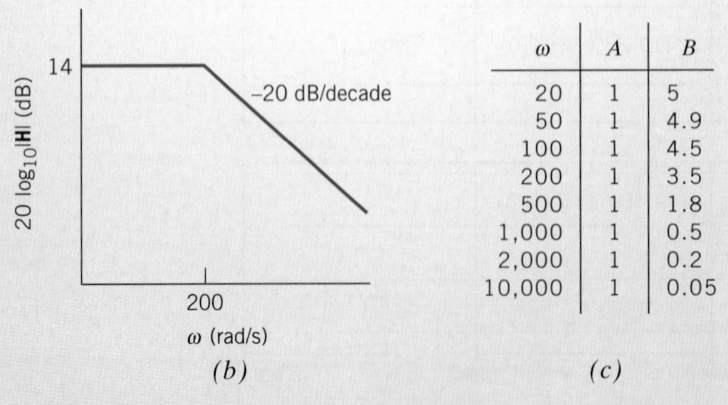

ω	A	B
20	1	5
50	1	4.9
100	1	4.5
200	1	3.5
500	1	1.8
1,000	1	0.5
2,000	1	0.2
10,000	1	0.05

FIGURE 13.8-1 (*a*) A circuit, (*b*) Bode plot, and (*c*) frequency response data.

Solution

The Bode plot has three features that we can look for in the frequency response data.

1. The dc gain is 14 dB.
2. The slope of the Bode plot is -20 dB/decade when $\omega \gg 200$ rad/s.
3. The corner frequency is 200 rad/s.

The lowest frequency at which frequency response data was taken is 20 rad/sec. At this frequency, the gain was measured to be

$$|\mathbf{H}(20)| = \frac{B}{A} = \frac{5}{1} = 14 \text{ dB}$$

which is equal to the dc gain specified by the Bode plot.

To identify the corner frequency from the frequency response data, we look for the frequency at which the gain is

$$\frac{\text{dc gain}}{\sqrt{2}} = \frac{5}{\sqrt{2}} = 3.536$$

The frequency response data indicate that the gain is 3.5 at a frequency of 200 rad/s. That agrees with the corner frequency of 200 rad/s of the specified Bode plot.

The slope of the frequency response at high frequencies is given by

$$\frac{20\log_{10}(0.05) - 20\log_{10}(0.5)}{\log_{10}(10,000) - \log_{10}(1000)} = -20 \text{ dB/decade}$$

which is the same as the slope of the Bode plot.

The frequency response data confirm that the circuit does indeed have the specified Bode plot.

EXAMPLE 13.8-2 How Can We Check Gain and Phase Shift?

Your lab notes indicate that the circuit shown in Figure 13.8.2 was built using $R_1 = 10\,\text{k}\Omega$, $R_2 = 50\,\text{k}\Omega$, and $C = 10\,\text{nF}$. The gain and phase shift of this circuit were measured to be 2.7 and 125°at 500 hertz. **How can we check** whether this information is consistent?

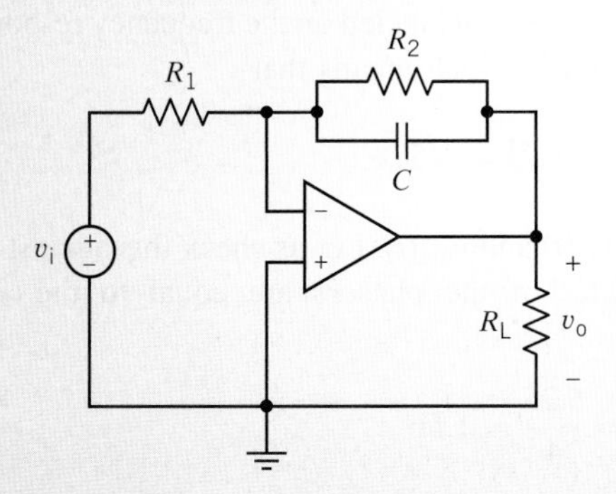

FIGURE 13.8-2 An op amp circuit.

Solution

The network function of this circuit is

$$\mathbf{H}(\omega) = -\frac{\frac{1}{j\omega C} \parallel R_2}{R_1} = \frac{-\frac{R_2}{R_1}}{1 + j\omega R_2 C}$$

$$= \frac{-\frac{50 \cdot 10^3}{10 \cdot 10^3}}{1 + j(2\pi \cdot 500)(50 \cdot 10^3)(10 \cdot 10^{-9})} = 2.685 \angle 122.5^\circ$$

The calculated gain and phase shift agree with the measured gain and phase shift. The lab notes are consistent.

EXAMPLE 13.8-3 How Can We Check Frequency Response?

An old lab report from a couple of years ago includes the following data about a particular circuit:

1. The magnitude and phase frequency responses are as shown in Figure 13.8-3.
2. When the input to the circuit was

   $$v_{in} = 4\cos(2\pi 1200 t)\ \text{V}$$

   the steady-state response was

   $$v_{out} = 6.25\cos(2\pi 1200 t + 110^\circ)\ \text{V}$$

   How can we check whether these data are consistent?

Solution

Three things need to be checked: the frequencies, the amplitudes, and the phase angles. The frequencies of both sinusoids are the same, which is good because the circuit must be linear if it is to be represented by a frequency response, and the steady-state response of a linear circuit to a sinusoidal input is a sinusoid at the same frequency as the input. The frequency of the input and output sinusoids is

$$\omega = 2 \cdot \pi \cdot 1200\ \text{rad/s}$$

or

$$f = 1200\ \text{Hz}$$

Fortunately, the gain and phase shift at 1200 Hz have been labeled on the frequency response plots shown in Figure 13.8-3. The gain at 1200 Hz is labeled as 3.9 dB, which means that

$$\frac{|\mathbf{V}_{out}|}{|\mathbf{V}_{in}|} = 3.9\ \text{dB} = 1.57$$

where $\mathbf{V}_{in}$ and $\mathbf{V}_{out}$ are the phasors corresponding to $v_{in}(t)$ and $v_{out}(t)$. Let us check this against the data about the input and output sinusoids. Because the magnitudes of the phasors are equal to the amplitudes of the corresponding sinusoids,

$$\frac{|\mathbf{V}_{out}|}{|\mathbf{V}_{in}|} = \frac{6.25}{4} = 1.56$$

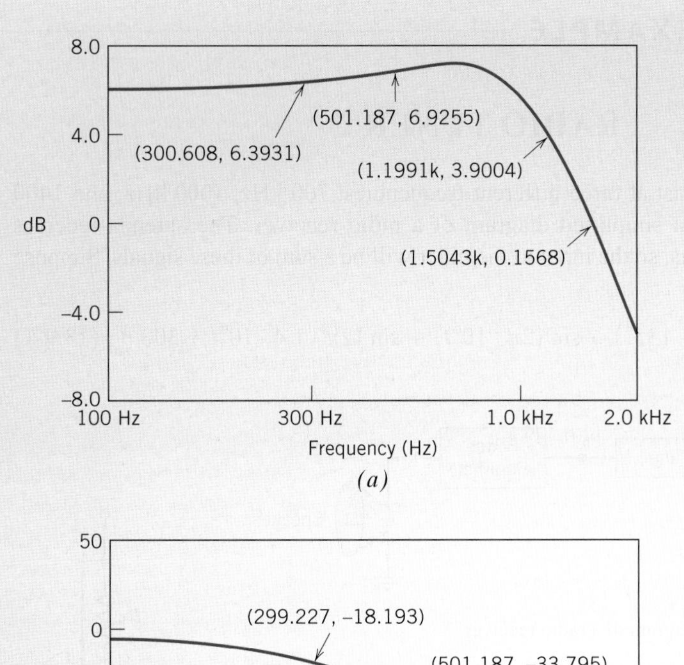

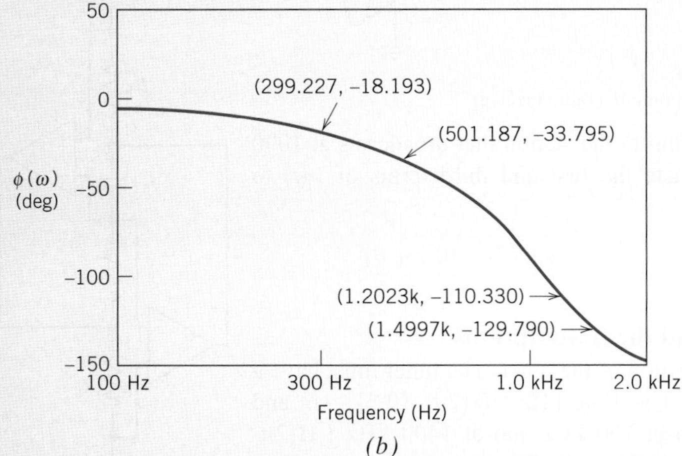

FIGURE 13.8-3 The (*a*) magnitude and (*b*) phase frequency response of the circuit.

This is very good agreement for experimental work.

Next, consider the phase shift. The frequency response indicates that the phase shift at 1200 Hz is $-110°$, which means

$$\angle\mathbf{V}_{\text{out}} - \angle\mathbf{V}_{\text{in}} = -110°$$

Let us check this against the data about the input and output sinusoids. Because the angles of the phasors are equal to the phase angles of the corresponding sinusoids,

$$\angle\mathbf{V}_{\text{out}} - \angle\mathbf{V}_{\text{in}} = 110° - 0° = 110°$$

The signs of the phase angles do not match. At a frequency of 1200 Hz, a phase angle of 110° indicates that the peaks of the output sinusoid will follow the peaks of the input sinusoid by

$$t_0 = \frac{110°}{360°} \cdot \frac{1}{1200} = 0.255 \text{ ms}$$

whereas a phase angle of $-110°$ indicates that the peaks of the output sinusoid will precede the peaks of the input sinusoid by 0.255 ms. It is likely that the angle of the output sinusoid was entered incorrectly in the lab data.

We have found an error in the old lab report and proposed an explanation for the error.

13.9 DESIGN EXAMPLE

RADIO TUNER

Three radio stations broadcast at three different frequencies, 700 kHz, 1000 kHz, and 1400 kHz. Figure 13.9-1 shows a simplified diagram of a radio receiver. The antenna receives signals from all three stations, so the input to the tuner will be a sum of these signals. Suppose this voltage is described by

$$v_i(t) = \sin\left(2\pi \cdot 7 \cdot 10^5 t + 135°\right) + \sin\left(2\pi \cdot 10^6 t\right) + \sin\left(2\pi \cdot 1.4 \cdot 10^6 t + 300°\right) \quad (13.9\text{-}1)$$

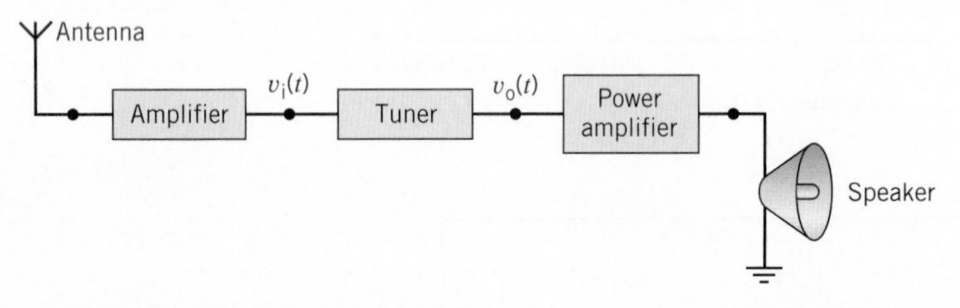

FIGURE 13.9-1 A simplified diagram of a radio receiver.

Consider the problem of tuning to the station that broadcasts at 1000 kHz. The tuner must eliminate the first and third terms of $v_i(t)$ to produce the output signal

$$v_o(t) = \sin\left(2\pi \cdot 10^6 t + \theta\right)$$

Describe the Situation and the Assumptions

Let $\mathbf{H}(\omega)$ be the network function of the tuner. The tuner must have a gain approximately equal to 1 at 1000 kHz ($|\mathbf{H}(2\pi \cdot 10^6)| \cong 1$) and approximately equal to zero at 700 kHz and at 1400 kHz ($|\mathbf{H}(2\pi \cdot 7 \cdot 10^5)| \cong 0$ and $|\mathbf{H}(2\pi \cdot 1.4 \cdot 10^6)| \cong 0$). The tuner output will be

$$\begin{aligned} v_o(t) = & |\mathbf{H}(2\pi \cdot 7 \cdot 10^5)| \sin(2\pi \cdot 7 \cdot 10^5 t + 135° \\ & + \angle\mathbf{H}(2\pi \cdot 7 \cdot 10^5) + |\mathbf{H}(2\pi\, 10^6)| \sin(2\pi\, 10^6 t + \angle\mathbf{H}(2\pi 10^6) \\ & + |\mathbf{H}(2\pi \cdot 1.4 \cdot 10^6)| \sin(2\pi \cdot 1.4 \cdot 10^6 t + 300° \\ & + \angle\mathbf{H}(2\pi \cdot 1.4 \cdot 10^6)) \end{aligned} \quad (13.9\text{-}2)$$

or

$$v_o(t) \cong \sin\left(2\pi \cdot 10^6 t + \theta\right)$$

where

$$\theta = \angle\mathbf{H}(2\pi \cdot 10^6)$$

State the Goal

The goal is to design a circuit consisting of resistors, capacitors, and op amps that has a gain equal to 1 at 1000 kHz and equal to zero at 700 and 1400 kHz.

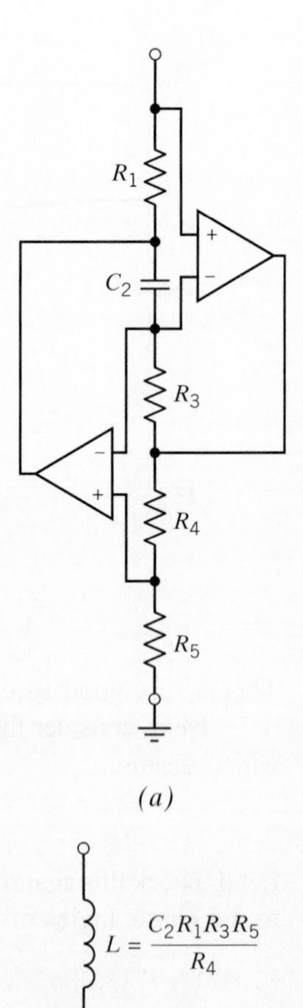

FIGURE 13.9-2 (a) An op amp circuit called a simulated inductor and (b) the equivalent inductor.

Generate a Plan

The tuner will be based on a resonant circuit having $\omega_0 = 2\pi 10^6 = 6.283 \cdot 10^6$ rad/s and $Q = 15$. Figure 13.9-2 shows an op amp circuit

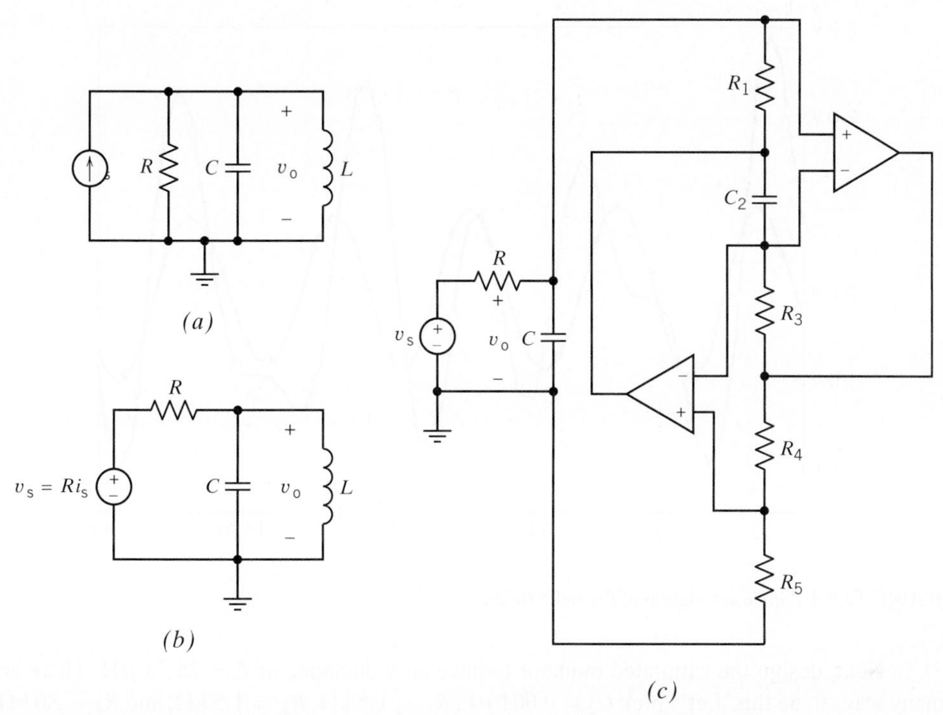

FIGURE 13.9-3 (*a*) A resonant circuit. (*b*) A band-pass filter. (*c*) An RC op amp band-pass filter.

called a simulated inductor. This circuit acts like a grounded inductor having an inductance equal to

$$L = \frac{C_2 R_1 R_3 R_5}{R_4} \tag{13.9-3}$$

Figure 13.9-3 shows how a parallel resonant circuit can be used to design the tuner. A parallel resonant circuit is shown in Figure 13.9-3*a*. The parallel resonant circuit must be modified if it is to be used for the tuner. The input to the tuner is a voltage, but the input to the parallel resonant circuit is a current. A source transformation is used to obtain a circuit that has a voltage input, shown in Figure 13.9-3*b*. Next, the inductor is replaced by the simulated inductor to produce the circuit show in Figure 13.9-3*c*. This is the circuit that will be used as the tuner.

The design will be completed in two steps. First, values of L, R, and C will be calculated so that the parallel resonant circuit has $\omega_0 = 6.283 \cdot 10^6$ rad/s and $Q = 15$. Next, the capacitor and resistors of the simulated inductor will be selected to satisfy Eq. 13.9-3.

Act on the Plan

First, design the resonant circuit to have $\omega_0 = 6.283 \cdot 10^6$ rad/s and $Q = 15$. Pick a convenient value for the capacitance, $C = 0.001 \mu\text{F}$. Then,

$$L = \frac{1}{\omega_0^2 C} = \frac{1}{\left(6.283 \cdot 10^6\right)^2 \cdot 10^{-9}} = 25.33\ \mu H$$

and

$$R = Q\sqrt{\frac{L}{C}} = 15\sqrt{\frac{25.33 \cdot 10^{-6}}{10^{-9}}} = 2387\ \Omega$$

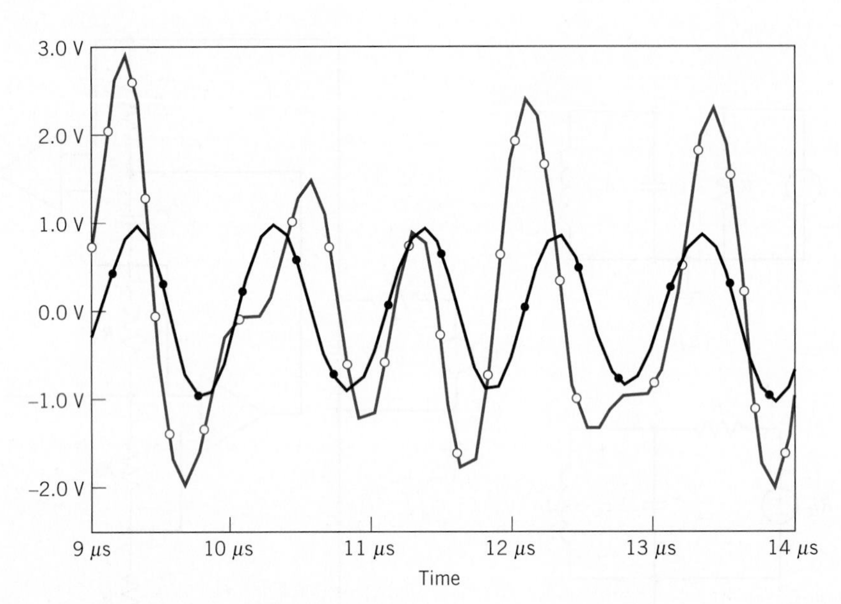

FIGURE 13.9-4 PSpice simulation of the radio tuner.

Next, design the simulated inductor to have an inductance of $L = 25.33\ \mu\text{H}$. There are many ways to do this. Let's pick $C_2 = 0.001\ \mu\text{F}$, $R_1 = 1.5\ \text{k}\Omega$, $R_3 = 1.5\ \text{k}\Omega$, and $R_4 = 80\ \text{k}\Omega$. Then

$$R_5 = \frac{R_4 L}{C_2 R_1 R_3} = \frac{80 \cdot 10^3 \cdot 25.33 \cdot 10^{-6}}{10^{-9} \cdot 1.5 \cdot 10^3 \cdot 1.5 \cdot 10^3} = 900\ \Omega$$

Verify the Proposed Solution

Figure 13.9-4 shows the results of a PSpice simulation of the tuner. The input to the circuit is $v_i(t)$ described by Eq. 13.9-1. This signal is not sinusoidal. The output of the filter is a sinusoid with an amplitude of approximately 1 and a frequency of 1000 kHz, as required by Eq. 13.9-2. Thus, the design specifications are satisfied.

13.10 SUMMARY

- Gain, phase shift, and the network function are properties of linear circuits that describe the effect that a circuit has on a sinusoidal input voltage or current.
- The gain of the circuit describes the relationship between the sizes of the input and output sinusoids. The gain is the ratio of the amplitude of the output sinusoid to the amplitude of the input sinusoid.
- The phase shift of the circuit describes the relationship between the phase angles of the input and output sinusoids. The phase shift is the difference between the phase angle of the output sinusoid and the phase angle of the input sinusoid.
- The network function describes the way the behavior of the circuit depends on the frequency of the input. The network function is defined in the frequency domain. It is the ratio of the phasor corresponding to the response sinusoid to the phasor corresponding to the input.
- Table 13.3-2 tabulates the network functions of several common op amp circuits.
- The frequency response describes the way the gain and phase shift of a circuit depend on frequency. Equations, tables, or plots are each used to express the frequency response.

- Bode plots represent the frequency response as plots of the gain in decibels and the phase using a logarithmic scale for frequency. Asymptotic magnitude Bode plots are approximate Bode plots that are easy to draw. The terms *corner frequency* and *break frequency* are routinely used to describe linear circuits. These terms describe features of the asymptotic Bode plot.
- Some linear circuits exhibit a phenomenon called resonance. These circuits contain reactive elements but act as if they were purely resistive at a particular frequency, called the resonant frequency. Resonant circuits are described using the resonant frequency, quality factor, and bandwidth. Table 13.4-1 summarizes the properties of series and parallel resonant circuits.
- The gain of operational amplifiers depends on the frequency of the input. Using an op amp model that includes a frequency-dependent gain makes our analysis more accurate but also more complicated. We use the more complicated model when we need the additional accuracy, and we use the simpler model when we don't.
- PSpice can be used to analyze a circuit and display its frequency response.
- MATLAB can be used to display the frequency response of a network function.

PROBLEMS

Section 13-2 Gain, Phase Shift, and the Network Function

P 13.2-1 The input to the circuit shown in Figure P 13.2-1 is the voltage of the voltage source, $v_i(t)$. The output is the voltage, $v_o(t)$, across the parallel connection of the capacitor and 15-Ω resistor. Determine the network function, $\mathbf{H}(\omega) = \mathbf{V}_o(\omega)/\mathbf{V}_i(\omega)$, of this circuit.

Answer: $\mathbf{H}(\omega) = \dfrac{0.2}{1 + j12\omega}$

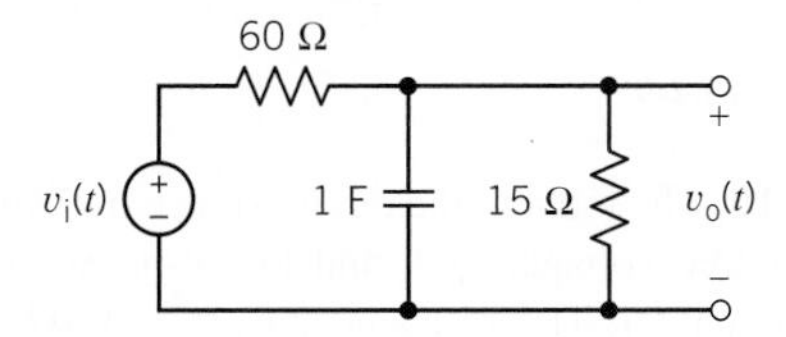

Figure P 13.2-1

P 13.2-2 The input to the circuit shown in Figure P 13.2-2 is the voltage of the voltage source, $v_i(t)$. The output is the voltage, $v_o(t)$, across the series connection of the capacitor and 200-kΩ resistor. Determine the network function, $\mathbf{H}(\omega) = \mathbf{V}_o(\omega)/\mathbf{V}_i(\omega)$, of this circuit.

Answer: $\mathbf{H}(\omega) = \dfrac{1 + j(0.005)\omega}{1 + j(0.00625)\omega}$

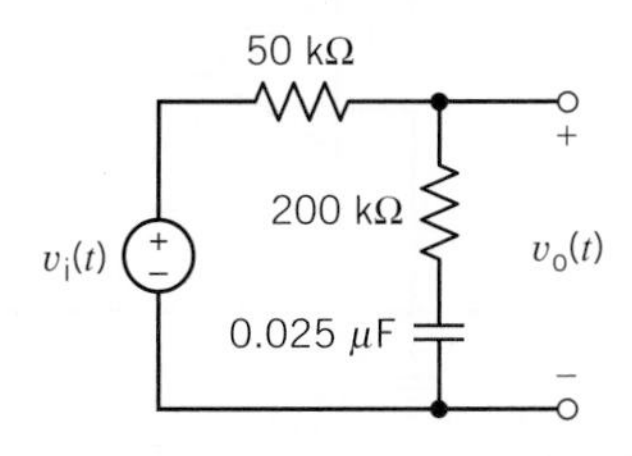

Figure P 13.2-2

P 13.2-3 The input to the circuit shown in Figure P 13.2-3 is the voltage of the voltage source, $v_i(t)$. The output is the voltage, $v_o(t)$, across the 8 Ω resistor. Determine the network function, $\mathbf{H}(\omega) = \mathbf{V}_o(\omega)/\mathbf{V}_i(\omega)$, of this circuit.

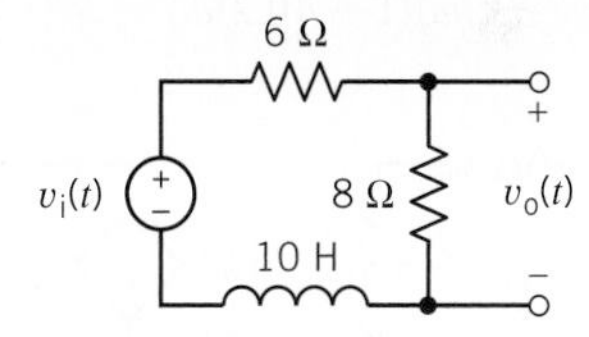

Figure P 13.2-3

P 13.2-4 The input to the circuit shown in Figure P 13.2-4 is the voltage of the voltage source, $v_i(t)$. The output is the voltage, $v_o(t)$, across the series connection of the inductor and 50 Ω resistor. The network function that represents this circuit is

$$\mathbf{H}(\omega) = \frac{\mathbf{V}_o(\omega)}{\mathbf{V}_i(\omega)} = (0.6)\frac{1 + j\dfrac{\omega}{12}}{1 + j\dfrac{\omega}{20}}$$

Determine the values of the inductance, L, and of the resistance, R.

Answers: $L = 4.2$ H and $R = 34$ Ω

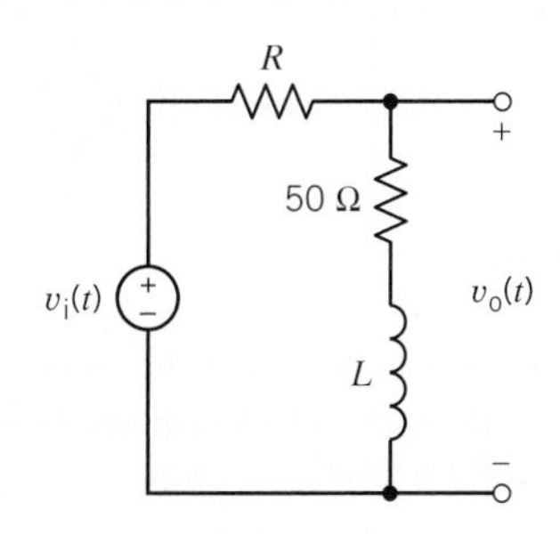

Figure P 13.2-4

P 13.2-5 The input to the circuit shown in Figure P 13.2-5 is the voltage of the voltage source, $v_i(t)$. The output is the voltage, $v_o(t)$, across the paralled connection of the capacitor and 3-Ω resistor. The network function that represents this circuit is

$$\mathbf{H}(\omega) = \frac{\mathbf{V}_o(\omega)}{\mathbf{V}_i(\omega)} = \frac{0.2}{1 + j4\omega}$$

Determine the values of the capacitance, C, and of the resistance, R.

Answers: $C = 1.7$ F and $R = 12\ \Omega$

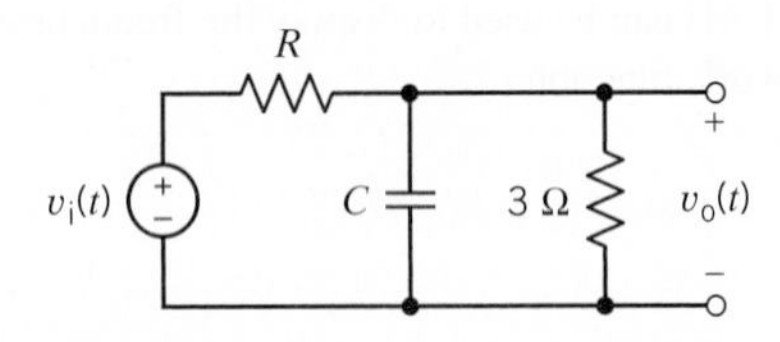

Figure P 13.2-5

P 13.2-6 The input to the circuit shown in Figure P 13.2-6 is the voltage of the voltage source, $v_i(t)$. The output is the voltage, $v_o(t)$, across the capacitor. Determine the network function, $\mathbf{H}(\omega) = \mathbf{V}_o(\omega)/\mathbf{V}_i(\omega)$, of this circuit.

Answer: $\mathbf{H}(\omega) = \dfrac{0.4}{(j\omega)(1 + j(0.2)\omega)}$

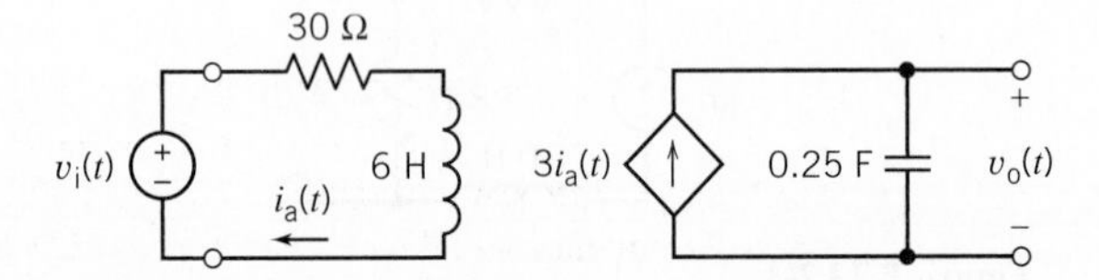

Figure P 13.2-6

P 13.2-7 The input to the circuit shown in Figure P 13.2-7 is the voltage of the voltage source, $v_i(t)$. The output is the voltage, $v_o(t)$, across the 40 kΩ resistor. The network function of this circuit is

$$\mathbf{H}(\omega) = \frac{\mathbf{V}_o(\omega)}{\mathbf{V}_i(\omega)} = \frac{4}{1 + j\dfrac{\omega}{100}}$$

Determine the value of the capacitance, C, and the value of the gain, A, of the VCVS.

Answers: $C = 2.6\mu$F and $A = 6$ V/V

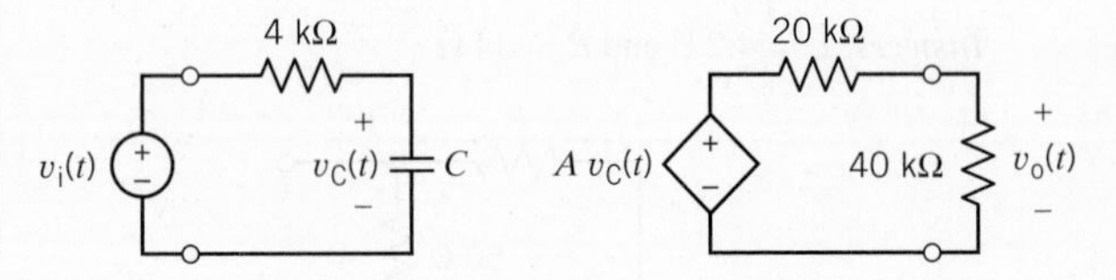

Figure P 13.2-7

P 13.2-8 The input to the circuit shown in Figure P 13.2-8 is the source voltage, $v_i(t)$, and the response is the voltage across R_L, $v_o(t)$. Find the network function.

Answer: $\mathbf{H}(\omega) = -4/(1 + j\omega^3/25)$

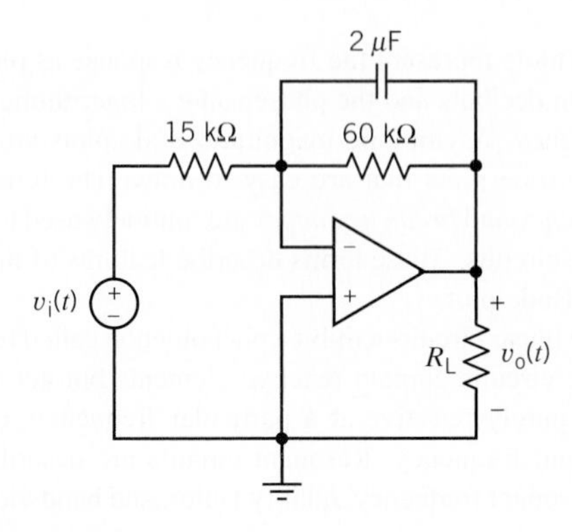

Figure P 13.2-8

P 13.2-9 The input to the circuit shown in Figure P 13.2-9 is the source voltage, $v_i(t)$, and the response is the voltage across R_L, $v_o(t)$. Express the gain and phase shift as functions of the radian frequency, ω.

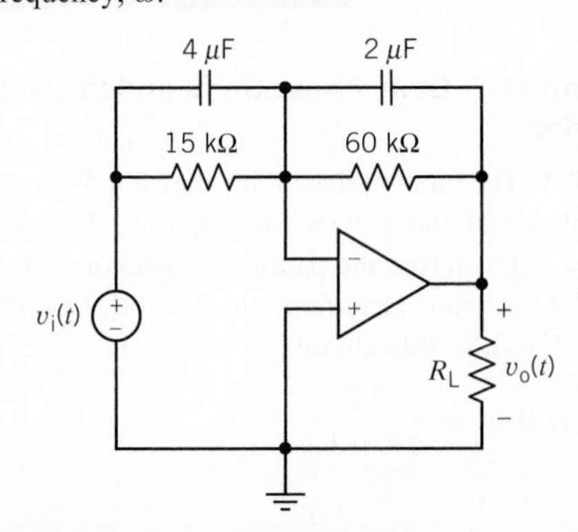

Figure P 13.2-9

P 13.2-10 The input to the circuit shown in Figure P 13.2-10 is the source voltage, $v_i(t)$, and the response is the voltage across R_L, $v_o(t)$. The resistance, R_1, is 20 kΩ. Design this circuit to satisfy the following two specifications:

(a) The gain at low frequencies is 5.
(b) The gain at high frequencies is 2.

Answers: $R_2 = 40$ kΩ and $R_3 = 60$ kΩ

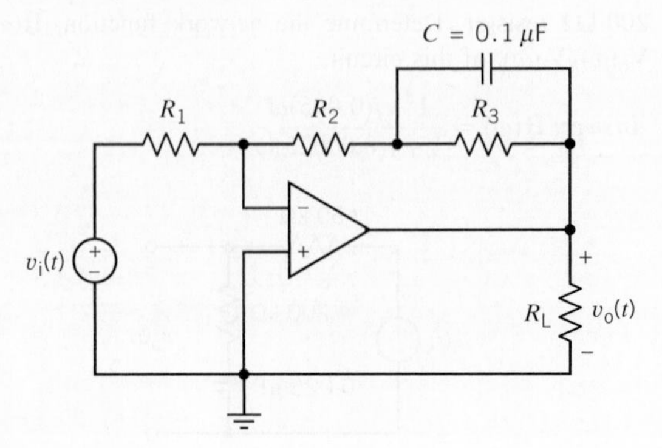

Figure P 13.2-10

P 13.2-11 The input to the circuit shown in Figure P 13.2-11 is the source voltage, $v_i(t)$, and the response is the voltage across R_L, $v_o(t)$. Design this circuit to satisfy the following two specifications:

(a) The phase shift at $\omega = 1000$ rad/s is 135°.
(b) The gain at high frequencies is 10.

Answers: $R_1 = 0.5\ \text{k}\Omega$ and $R_2 = 5\ \text{k}\Omega$

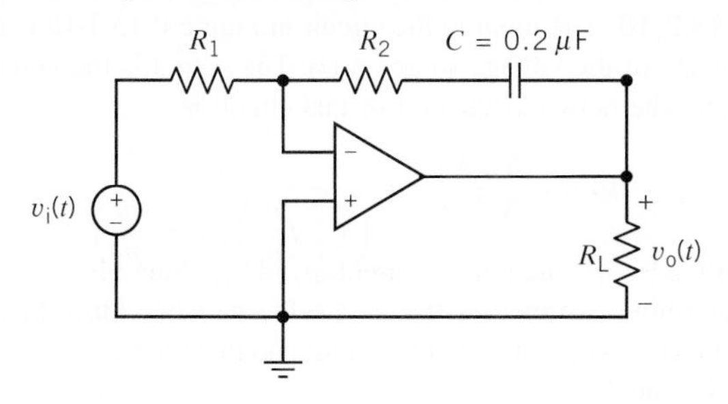

Figure P 13.2-11

P 13.2-12 The input to the circuit shown in Figure P 13.2-12 is the source voltage, $v_i(t)$, and the response is the voltage across R_L, $v_o(t)$. Design this circuit to satisfy the following two specifications:

(a) The phase shift at $\omega = 1000$ rad/s is 225°.
(b) The gain at high frequencies is 10.

Answers: $R_1 = 10\ \text{k}\Omega$ and $R_2 = 100\ \text{k}\Omega$

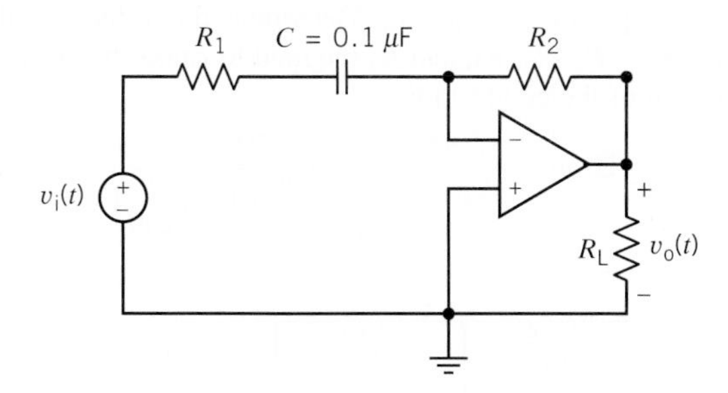

Figure P 13.2-12

P 13.2-13 The input to the circuit of Figure P 13.2-13 is

$$v_s = 50 + 30\cos(500t + 115°) - 20\cos(2500t + 30°)\ \text{mV}$$

Find the steady-state output voltage, v_o, for (a) $C = 0.1\mu\text{F}$ and (b) $C = 0.01\mu\text{F}$. Assume an ideal op amp.

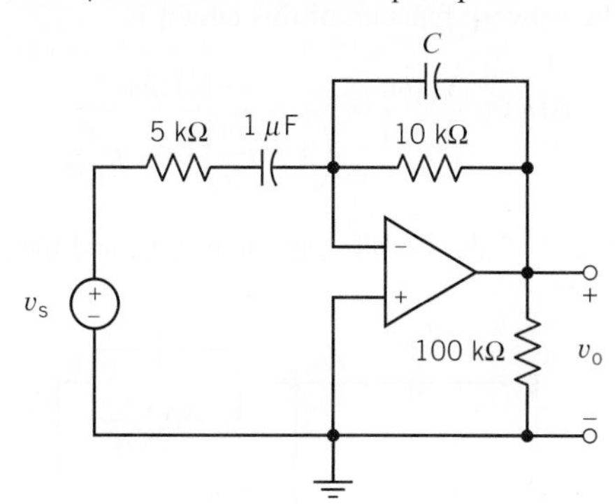

Figure P 13.2-13

P 13.2-14 The source voltage, v_s, shown in the circuit of Figure P 13.2-14*a* is a sinusoid having a frequency of 500 Hz and an amplitude of 8 V. The circuit is in steady state. The oscilloscope traces show the input and output waveforms as shown in Figure P 13.2-14*b*.

(a) Determine the gain and phase shift of the circuit at 500 Hz.
(b) Determine the value of the capacitor.
(c) If the frequency of the input is changed, then the gain and phase shift of the circuit will change. What are the values of the gain and phase shift at the frequency 200 Hz? At 2000 Hz? At what frequency will the phase shift be −45°? At what frequency will the phase shift be −135°?
(d) What value of capacitance would be required to make the phase shift at 500 Hz be −60°? What value of capacitance would be required to make the phase shift at 500 Hz be −300°?
(e) Suppose the phase shift had been −120° at 500 Hz. What would be the value of the capacitor?

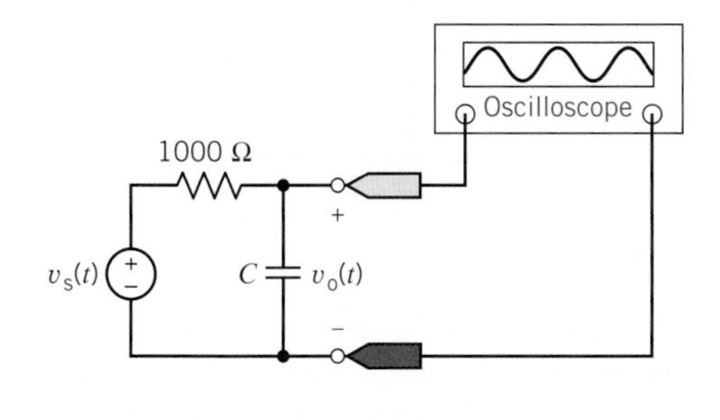

(*a*)

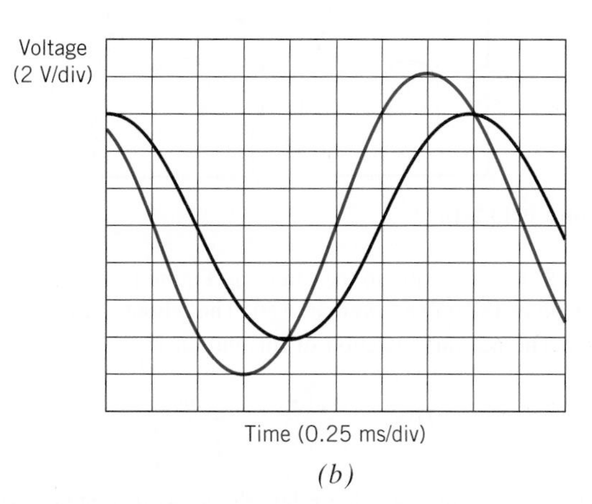

(*b*)

Figure P 13.2-14

Answers: (b) $C = 0.26\mu F$ (e) This circuit can't be designed to produce a phase shift $= -120°$.

P 13.2-15 The input to the circuit in Figure P 13.2-15 is the voltage of the voltage source, $v_i(t)$. The output is the voltage $v_o(t)$. The network function of this circuit is

$$\mathbf{H}(\omega) = \frac{\mathbf{V}_o(\omega)}{\mathbf{V}_i(\omega)} = \frac{(-0.1)j\omega}{\left(1 + j\frac{\omega}{p}\right)\left(1 + j\frac{\omega}{125}\right)}$$

Determine the values of the capacitance, C, and the pole, p.

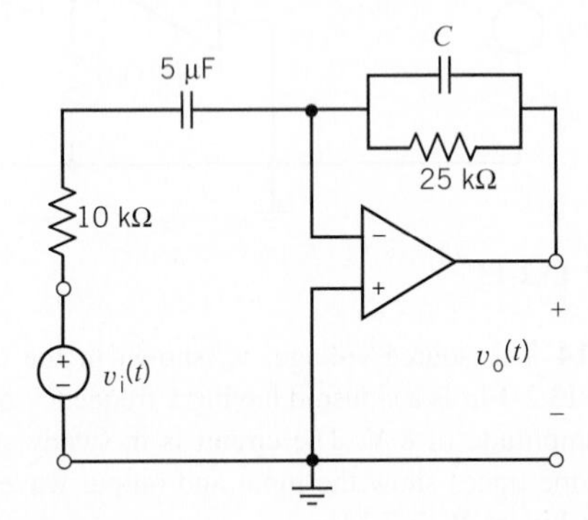

Figure P 13.2-15

P 13.2-16 The input to the circuit in Figure P 13.2-16 is the voltage of the voltage source, $v_s(t)$. The output is the voltage $v_o(t)$. The network function of this circuit is

$$\mathbf{H}(\omega) = \frac{\mathbf{V}_o(\omega)}{\mathbf{V}_s(\omega)} = k\frac{1 + j\frac{\omega}{z}}{1 + j\frac{\omega}{p}}$$

Determine expressions that relate the network function parameters k, z, and p to the circuit parameters R_1, R_2, L, N_1, and N_2.

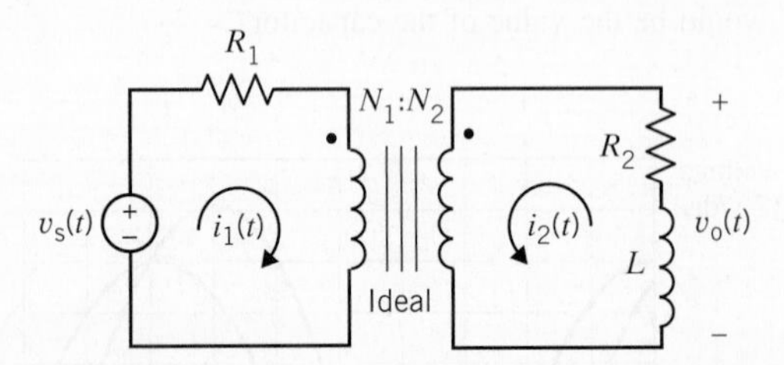

Figure P 13.2-16

P 13.2-17 The input to the circuit in Figure P 13.2-17 is the voltage of the voltage source, $v_s(t)$. The output is the voltage $v_o(t)$. The network function of this circuit is

$$\mathbf{H}(\omega) = \frac{\mathbf{V}_o(\omega)}{\mathbf{V}_s(\omega)} = k\frac{j\omega}{1 + j\frac{\omega}{p}}$$

Determine expressions that relate the network function parameters k and p to the circuit parameters R_1, R_2, M, L_1, and L_2.

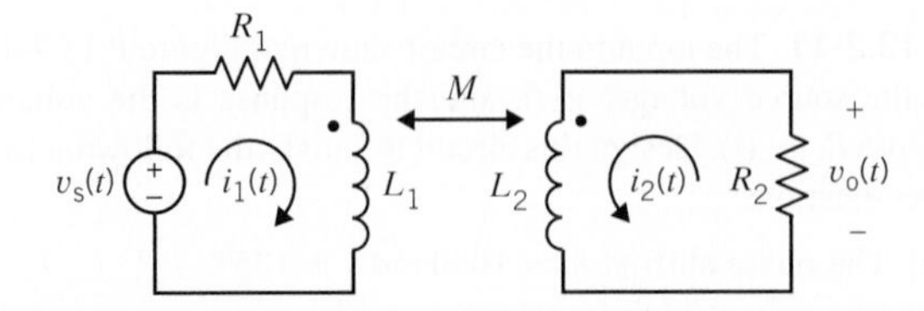

Figure P 13.2-17

P 13.2-18 The input to the circuit in Figure P 13.2-18 is the voltage of the voltage source, $v_i(t)$. The output is the voltage $v_o(t)$. The network function of this circuit is

$$\mathbf{H}(\omega) = \frac{\mathbf{V}_o(\omega)}{\mathbf{V}_i(\omega)} = k\frac{j\omega}{\left(1 + j\frac{\omega}{p_1}\right)\left(1 + j\frac{\omega}{p_2}\right)}$$

Determine expressions that relate the network function parameters k, p_1, and p_2 to the circuit parameters R_1, R_2, R_3, R_4, A, C, and L.

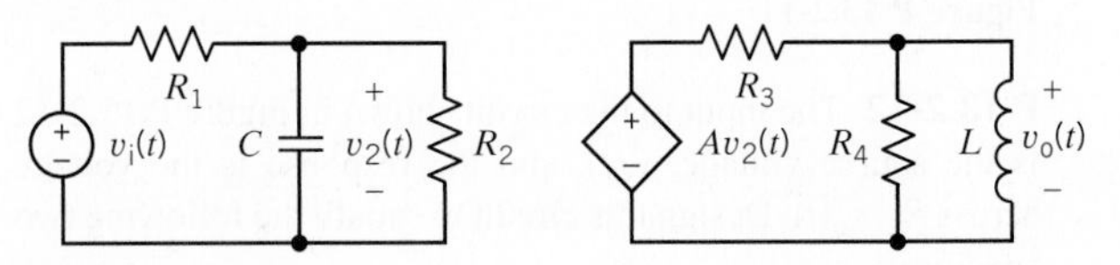

Figure P 13.2-18

P 13.2-19 The input to the circuit shown in Figure P 13.2-19 is the voltage of the voltage source, v_s. The output of the circuit is the capacitor voltage, v_o. Determine the values of the resistances R_1, R_2, R_3, and R_4 required to cause the network function of the circuit to be

$$\mathbf{H}(\omega) = \frac{\mathbf{V}_o(\omega)}{\mathbf{V}_s(\omega)} = \frac{21}{\left(1 + j\frac{\omega}{5}\right)\left(1 + j\frac{\omega}{200}\right)}$$

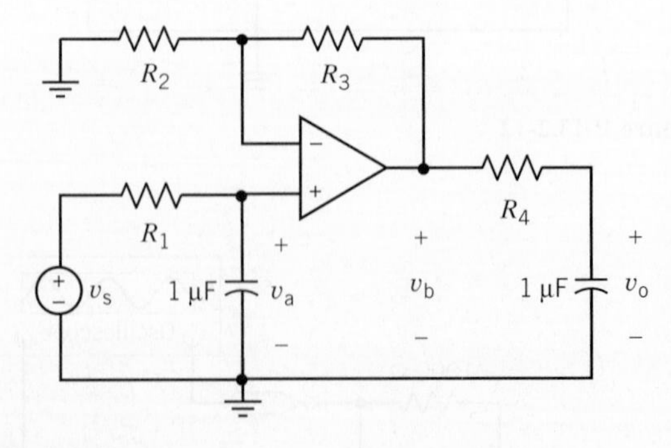

Figure P 13.2-19

P 13.2-20 The input to the circuit shown in Figure P 13.2-20 is the voltage of the voltage source, v_s. The output of the circuit is the voltage v_o. Determine the network function

$$\mathbf{H}(\omega) = \frac{\mathbf{V}_o(\omega)}{\mathbf{V}_s(\omega)}$$

of the circuit.

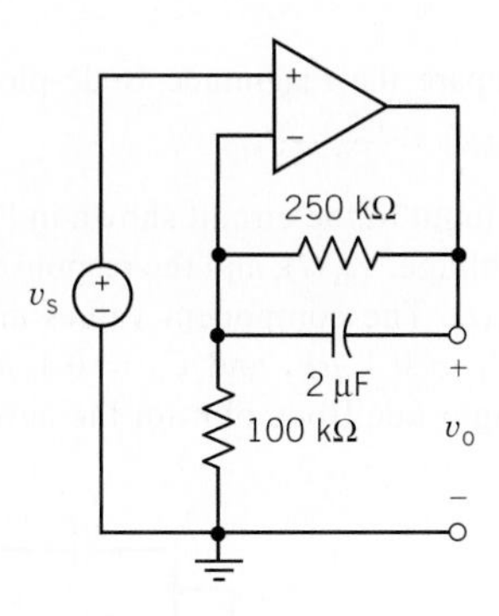

Figure P 13.2-20

P 13.2-21 The input to the circuit shown in Figure P 13.2-21 is the voltage of the voltage source, v_s. The output of the circuit is the capacitor voltage, v_o. Determine the network function

$$\mathbf{H}(\omega) = \frac{\mathbf{V}_o(\omega)}{\mathbf{V}_s(\omega)}$$

of the circuit.

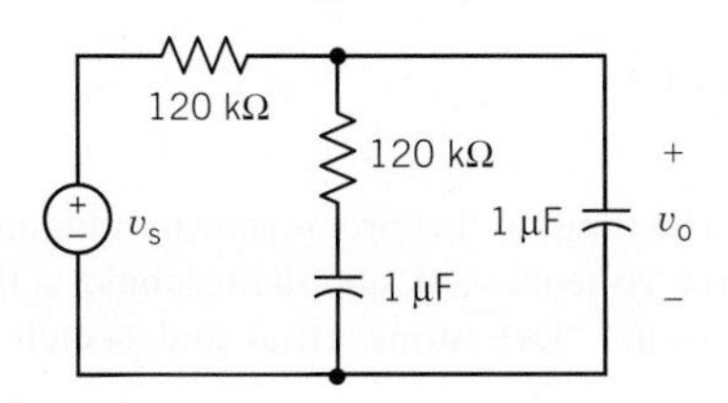

Figure P 13.2-21

P 13.2-22 The input to the circuit shown in Figure P 13.2-22 is the voltage of the voltage source, v_s. The output of the circuit is the capacitor voltage, v_o. The network function of the circuit is

$$\mathbf{H}(\omega) = \frac{\mathbf{V}_o(\omega)}{\mathbf{V}_s(\omega)} = \frac{H_o}{1 + j\dfrac{\omega}{p}}$$

Determine the values of H_o and p.

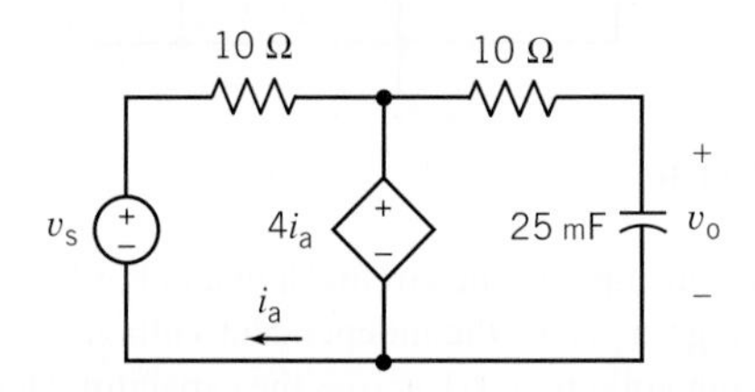

Figure P 13.2-22

P 13.2-23 The input to the circuit shown in Figure P 13.2-23 is the current of the current source, i_s. The output of the circuit is the resistor current, i_o. The network function of the circuit is

$$\mathbf{H}(\omega) = \frac{\mathbf{I}_o(\omega)}{\mathbf{I}_s(\omega)} = \frac{0.8}{1 + j\dfrac{\omega}{40}}$$

Determine the values of the resistances R_1 and R_2.

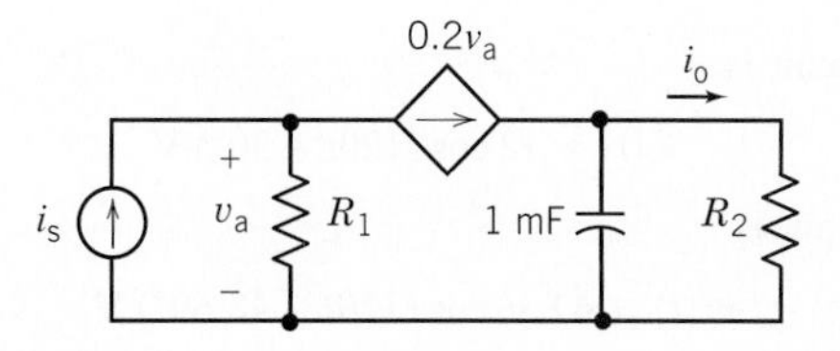

Figure P 13.2-23

P 13.2-24 The input to the circuit shown in Figure P 13.2-24 is the voltage of the voltage source, v_s. The output of the circuit is the resistor voltage, v_o. Specify values for L_1, L_2, R, and K that cause the network function of the circuit to be

$$\mathbf{H}(\omega) = \frac{\mathbf{V}_o(\omega)}{\mathbf{V}_s(\omega)} = \frac{1}{\left(1 + j\dfrac{\omega}{20}\right)\left(1 + j\dfrac{\omega}{50}\right)}$$

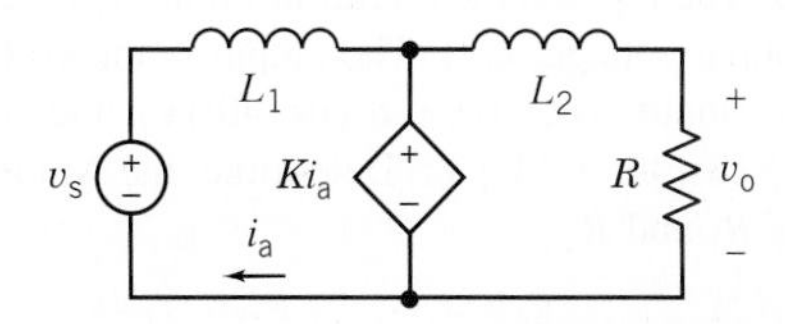

Figure P 13.2-24

P 13.2-25 The input to the circuit shown in Figure P 13.2-25 is the voltage of the voltage source, v_s. The output of the circuit is the resistor voltage, v_o. Specify values for R and C that cause the network function of the circuit to be

$$\mathbf{H}(\omega) = \frac{\mathbf{V}_o(\omega)}{\mathbf{V}_s(\omega)} = \frac{-8}{1 + j\dfrac{\omega}{250}}$$

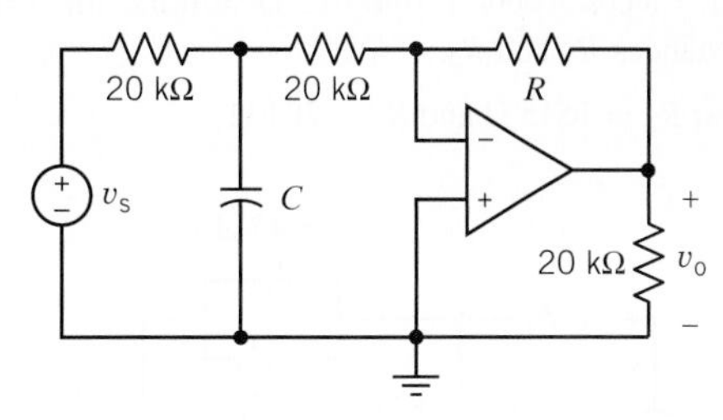

Figure P 13.2-25

P 13.2-26 The network function of a circuit is $\mathbf{H}(\omega) = \frac{\mathbf{V}_o(\omega)}{\mathbf{V}_s(\omega)} = \frac{j40\omega}{120+j20\omega}$. When the input to this circuit is $v_s(t) = 5\cos(5t + 15°)$V, the output is $v_o(t) = A\cos(5t + 65.194°)$ V. On the other hand, when the input to this circuit is $v_s(t) = 5\cos(8t + 15°)$ V, the output is $v_o(t) = 8\cos(8t + \theta)$V. Determine the values of A and θ.

Answers: $A = 6.4018$ V and $\theta = 51.87°$

P 13.2-27 The network function of a circuit is $\mathbf{H}(\omega) = \frac{\mathbf{V}_o(\omega)}{\mathbf{V}_s(\omega)} = \frac{k}{1+j\frac{\omega}{p}}$ where $k > 0$ and $p > 0$. When the input to this circuit is

$$v_s(t) = 12\cos(120t + 30^\circ)\ \text{V}$$

the output is

$$v_o(t) = 42.36\cos(120t - 48.69^\circ)\ \text{V}$$

Determine the values of k and p

Answers: $k = 18$ and $p = 24$ rad/s

P 13.2-28 The network function of a circuit is $\mathbf{H}(\omega) = \frac{20}{8+j\omega}$. When the input to this circuit is sinusoidal, the output is also sinusoidal. Let ω_1 be the frequency at which the output sinusoid is twice as large as the input sinusoid and let ω_2 be the frequency at which output sinusoid is delayed by one tenth period with respect to the input sinusoid. Determine the values of ω_1 and ω_2.

P 13.2-29 The input to the circuit in Figure P 13.2-29 is the voltage source voltage, $v_s(t)$. The output is the voltage $v_o(t)$. When the input is $v_s(t) = 8\cos(40t)$V, the output is $v_o(t) = 2.5\cos(40t + 14^\circ)$ V. Determine the values of the resistances R_1 and R_2.

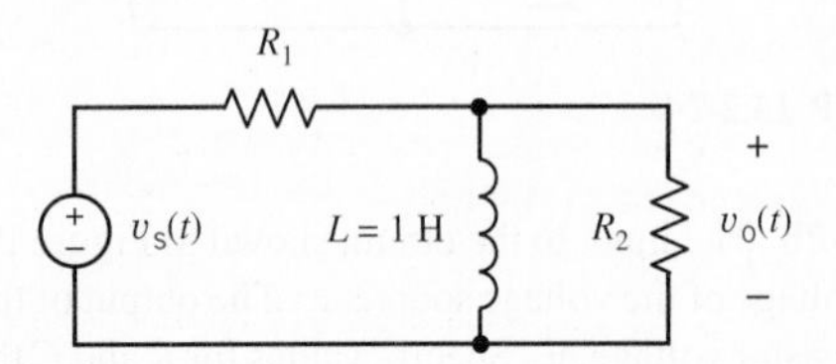

Figure P 13.2-29

P 13.2-30 The input to the circuit shown in Figure P 13.2-30 is the voltage source voltage, $v_s(t)$. The output is the voltage $v_o(t)$. The input $v_s(t) = 2.5\cos(1000t)$ V causes the output to be $v_o(t) = 8\cos(1000t + 104^\circ)$ V. Determine the values of the resistances R_1 and R_2.

Answers: $R_1 = 1515\ \Omega$ and $R_2 = 20\ \text{k}\Omega$

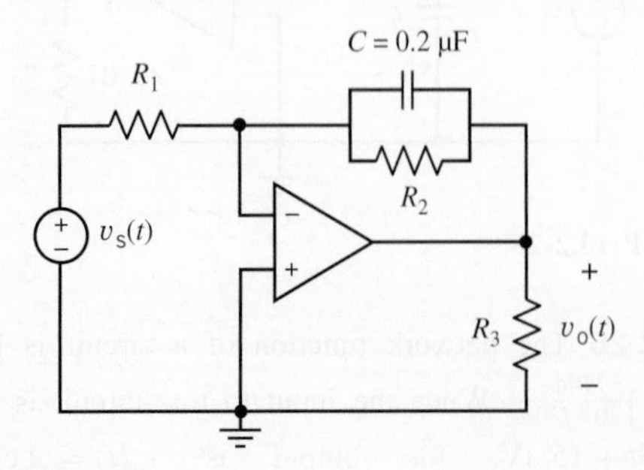

Figure P 13.2-30

Section 13.3 Bode Plots

P 13.3-1 Sketch the magnitude Bode plot of $\mathbf{H}(\omega) = \dfrac{4(5 + j\omega)}{1 + j\dfrac{\omega}{50}}$.

P 13.3-2 Compare the magnitude Bode plots of $\mathbf{H}_1(\omega) = \frac{10(5+j\omega)}{50+j\omega}$ and $\mathbf{H}_2(\omega) = \frac{100(5+j\omega)}{50+j\omega}$.

P 13.3-3 The input to the circuit shown in Figure P 13.3-3 is the source voltage, $v_{in}(t)$, and the response is the voltage across R_3, $v_{out}(t)$. The component values are $R_1 = 10\ \text{k}\Omega$, $R_2 = 20\ \text{k}\Omega$, $C_1 = 0.1\ \mu\text{F}$, and $C_2 = 0.1\ \mu\text{F}$. Sketch the asymptotic magnitude Bode plot for the network function.

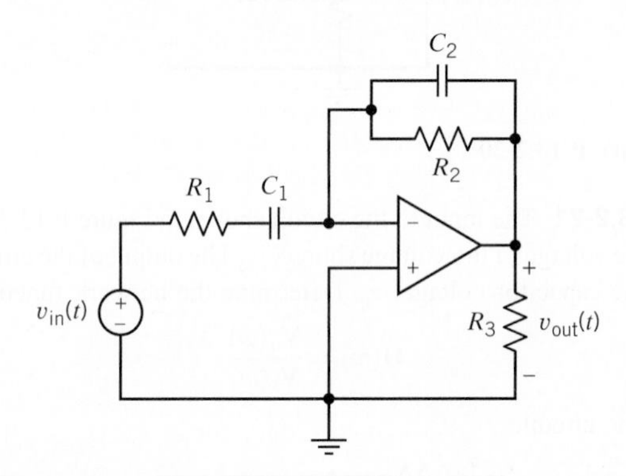

Figure P 13.3-3

P 13.3-4 The input to the circuit shown in Figure P 13.3-4 is the source voltage, $v_s(t)$, and the response is the voltage across R_3, $v_o(t)$. Determine $\mathbf{H}(\omega)$ and sketch the Bode diagram.

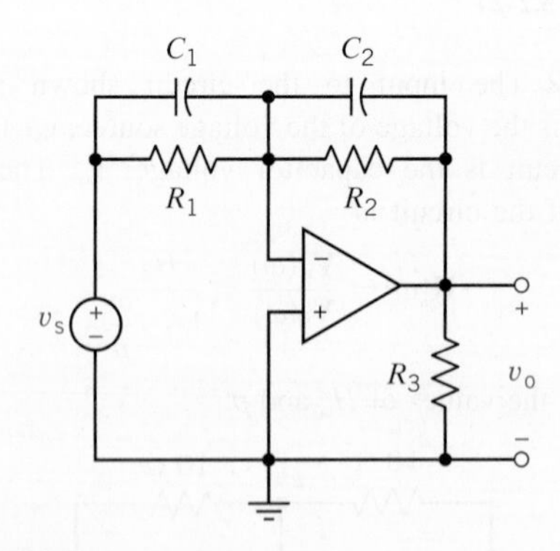

Figure P 13.3-4

P 13.3-5 The input to the circuit shown in Figure P 13.3-5*a* is the voltage, $v_i(t)$, of the independent voltage source. The output is the voltage, $v_o(t)$, across the capacitor. Design this circuit to have the Bode plot shown in Figure P 13.3-5*b*.

Hint: First, show that the network function of the circuit is

$$\mathbf{H}(\omega) = \frac{\mathbf{V}_o(\omega)}{\mathbf{V}_i(\omega)}$$

$$= \frac{j\omega\left(\dfrac{ALR_4}{R_1(R_3 + R_4)}\right)}{\left(1 + j\omega\dfrac{L(R_1 + R_2)}{R_1 R_2}\right)\left(1 + j\omega\dfrac{CR_3R_4}{R_3 + R_4}\right)}$$

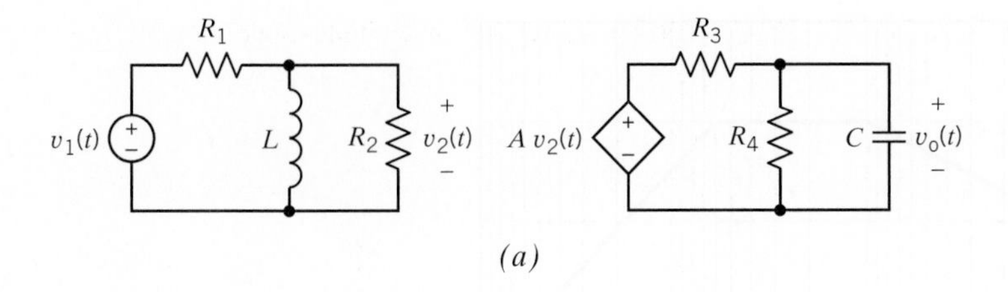

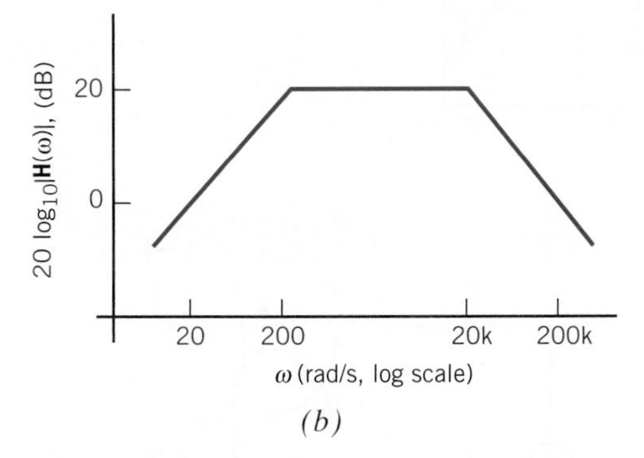

Figure P 13.3-5

P 13.3-6 The input to the circuit shown in Figure P 13.3-6*b* is the voltage of the voltage source, $v_i(t)$. The output is the voltage $v_o(t)$. The network function of this circuit is $\mathbf{H}(\omega) = \mathbf{V}_o(\omega)/\mathbf{V}_i(\omega)$. Determine the values of R_2, C_1, and C_2 that are required to make this circuit have the magnitude Bode plot shown in Figure P 13.3-6*a*.

Answers: $R_2 = 400\ \text{k}\Omega$, $C_1 = 25$ nF, and $C_2 = 6.25$ nF

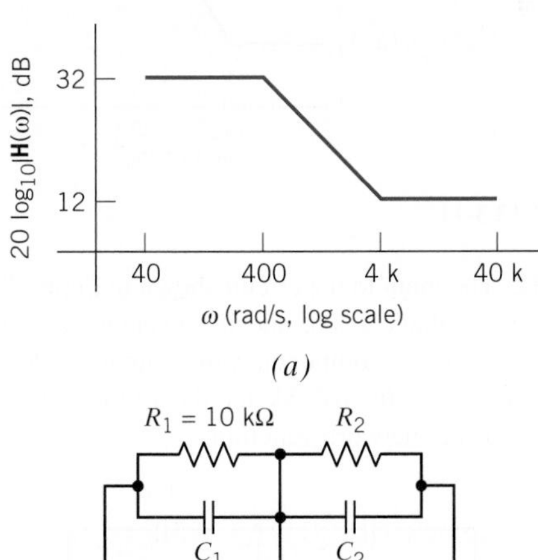

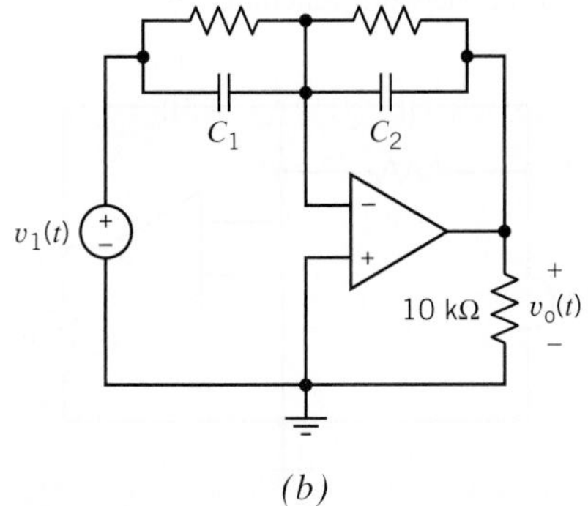

Figure P 13.3-6

P 13.3-7 The input to the circuit shown in Figure P 13.3-7*b* is the voltage of the voltage source, $v_i(t)$. The output is the voltage $v_o(t)$. The network function of this circuit is $\mathbf{H}(\omega) = \mathbf{V}_o(\omega)/\mathbf{V}_i(\omega)$. The magnitude Bode plot is shown in Figure P 13.3-7*a*. Determine values of the corner frequencies, z and p. Determine the value of the low-frequency gain, k.

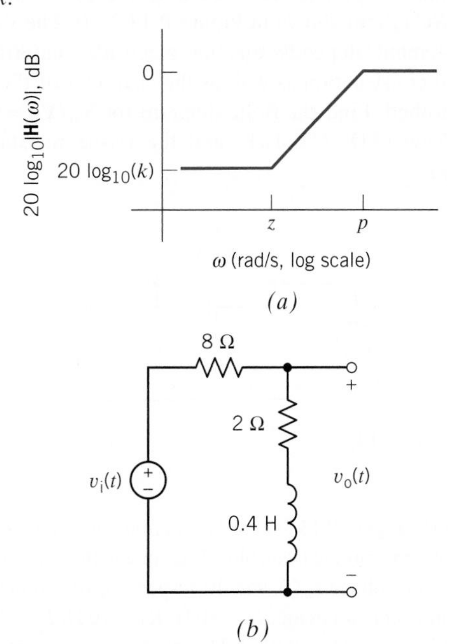

Figure P 13.3-7

P 13.3-8 Determine $\mathbf{H}(j\omega)$ from the asymptotic Bode diagram in Figure P 13.3-8.

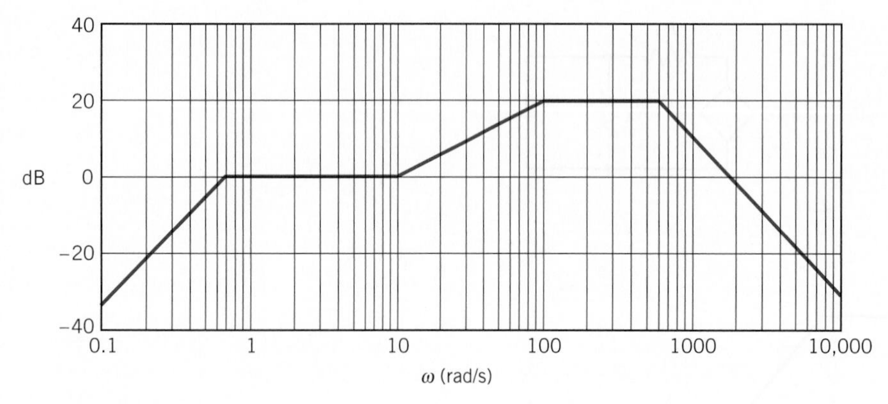

Figure P 13.3-8

P 13.3-9 A circuit has a network function

$$\mathbf{H}(\omega) = \frac{k(1 + j\omega/z)}{j\omega}$$

(a) Find the high- and low-frequency asymptotes of the magnitude Bode plot.

(b) The high- and low-frequency asymptotes comprise the magnitude Bode plot. Over what ranges of frequencies is the asymptotic magnitude Bode plot of $\mathbf{H}(\omega)$ within 1 percent of the actual value of $\mathbf{H}(\omega)$?

P 13.3-10 Physicians use tissue electrodes to form the interface that conducts current to the target tissue of the human body. The electrode in tissue can be modeled by the *RC* circuit shown in Figure P 13.3-10. The value of each element depends on the electrode material and physical construction as well as the character of the tissue being probed. Find the Bode diagram for $\mathbf{V}_o/\mathbf{V}_s = \mathbf{H}(j\omega)$ when $R_1 = 1\ \text{k}\Omega$, $C = 1\mu\text{F}$, and the tissue resistance is $R_t = 5\ \text{k}\Omega$.

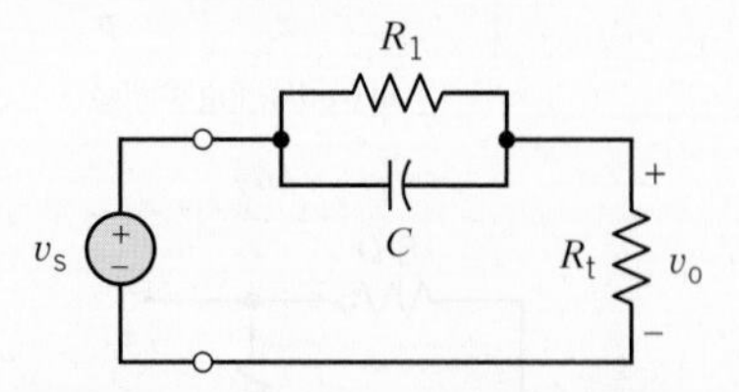

Figure P 13.3-10

P 13.3-11 Figure P 13.3-11 shows a circuit and corresponding asymptotic magnitude Bode plot. The input to this circuit shown is the source voltage $v_{in}(t)$, and the response is the voltage $v_o(t)$. The component values are $R_1 = 80\ \Omega$, $R_2 = 20\ \Omega$, $L_1 = 0.03$ H, $L_2 = 0.07$ H, and $M = 0.01$ H. Determine the values of K_1, K_2, p, and z.

Answers: $K_1 = 0.75$, $K_2 = 0.2$, $z = 333$ rad/s, and $p = 1250$ rad/s

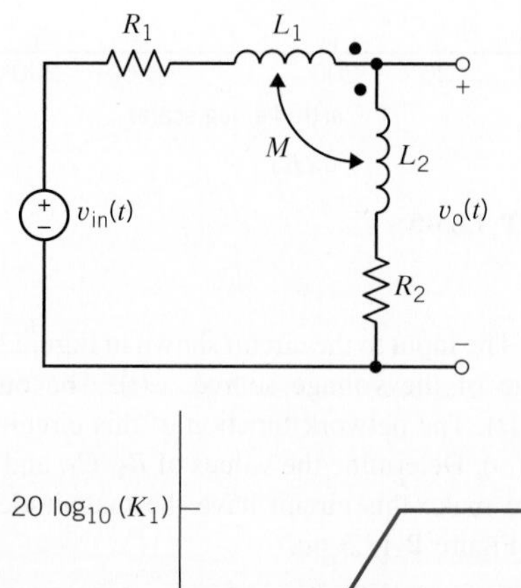

Figure P 13.3-11

P 13.3-12 The input to the circuit shown in Figure P 13.3-12 is the source voltage $v_{in}(t)$, and the response is the voltage across R_3, $v_{out}(t)$. The component values are $R_1 = 20\ \text{k}\Omega$, $C_1 = 0.025\ \mu\text{F}$, and $C_2 = 0.1\ \mu\text{F}$. Sketch the asymptotic magnitude Bode plot for the network function.

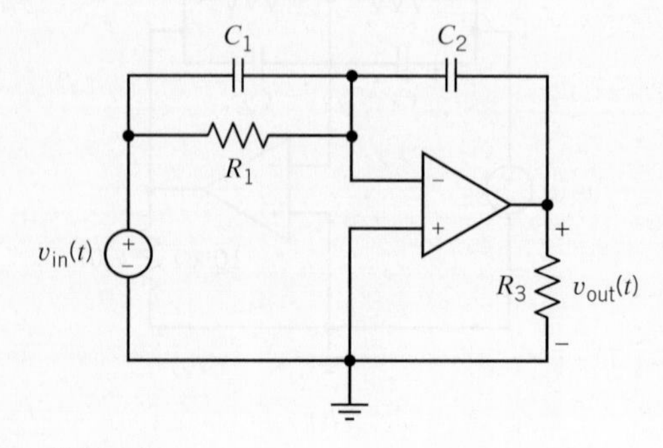

Figure P 13.3-12

P 13.3-13 Design a circuit that has the asymptotic magnitude Bode plot shown in Figure P 13.3-13.

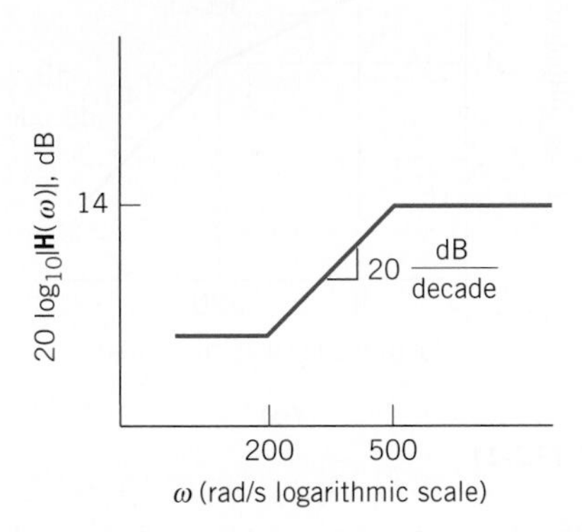

Figure P 13.3-13

P 13.3-14 Design a circuit that has the asymptotic magnitude Bode plot shown in Figure P 13.3-14.

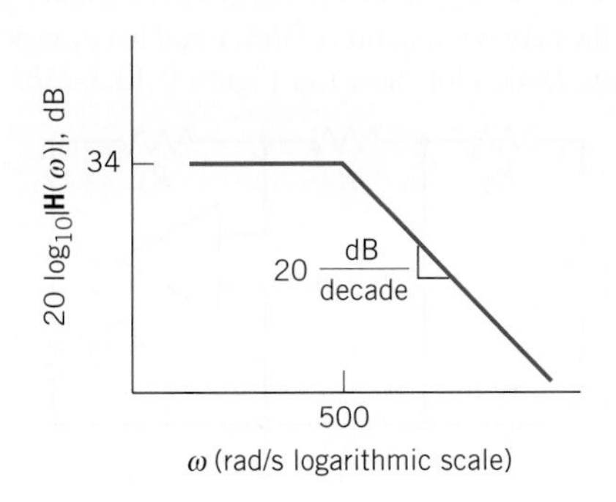

Figure P 13.3-14

P 13.3-15 Design a circuit that has the asymptotic magnitude Bode plot shown in Figure P 13.3-15.

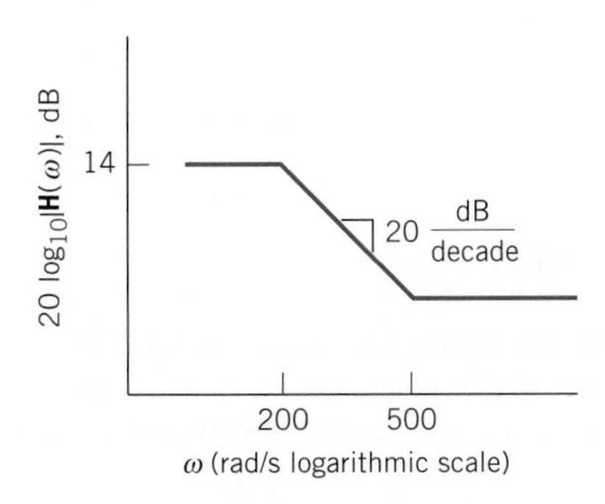

Figure P 13.3-15

P 13.3-16 Design a circuit that has the asymptotic magnitude Bode plot shown in Figure P 13.3-16.

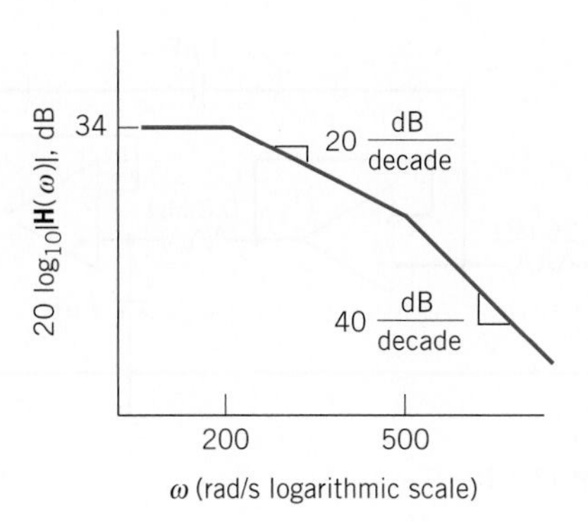

Figure P 13.3-16

P 13.3-17 The cochlear implant is intended for patients with deafness due to malfunction of the sensory cells of the cochlea in the inner ear (Loeb, 1985). These devices use a microphone for picking up sound and a processor for converting it to electrical signals, and they transmit these signals to the nervous system. A cochlear implant relies on the fact that many of the auditory nerve fibers remain intact in patients with this form of hearing loss. The overall transmission from microphone to nerve cells is represented by the gain function

$$\mathbf{H}(j\omega) = \frac{10(j\omega/50 + 1)}{(j\omega/2 + 1)(j\omega/20 + 1)(j\omega/80 + 1)}$$

Plot the magnitude Bode diagram for $\mathbf{H}(j\omega)$ for $1 \leq \omega \leq 100$.

P 13.3-18 An operational amplifier circuit is shown in Figure P 13.3-18, where $R_2 = 10\,\text{k}\Omega$ and $C = 0.04\,\mu\text{F}$.

(a) Find the expression for the network function $\mathbf{H} = \mathbf{V}_o/\mathbf{V}_s$ and sketch the asymptotic Bode diagram.
(b) What is the gain of the circuit, $|\mathbf{V}_o/\mathbf{V}_s|$, for $\omega = 0$?
(c) At what frequency does $|\mathbf{V}_o/\mathbf{V}_s|$ fall to $1/\sqrt{2}$ of its low-frequency value?

Answers: (b) 20 dB and (c) 2500 rad/s

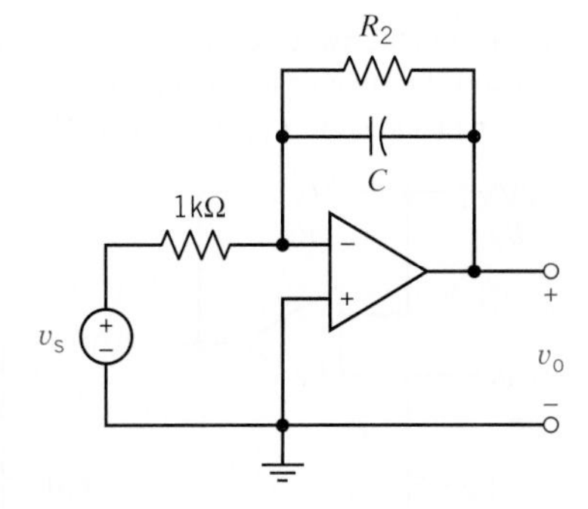

Figure P 13.3-18

P 13.3-19 Determine the network function $\mathbf{H}(\omega)$ for the op amp circuit shown in Figure P 13.3-19 and plot the Bode diagram. Assume ideal op amps.

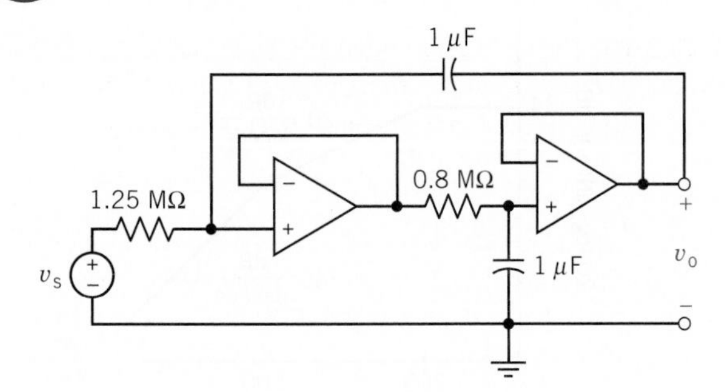

Figure P 13.3-19

P 13.3-20 The network function of a circuit is

$$\mathbf{H}(\omega) = \frac{-3(5 + j\omega)}{j\omega(2 + j\omega)}$$

Sketch the asymptotic magnitude Bode plot corresponding to **H**.

P 13.3-21 The network function of a circuit is

$$\mathbf{H}(\omega) = \frac{(j\omega)^3}{(4 + j2\omega)}$$

Sketch the asymptotic magnitude Bode plot corresponding to **H**.

P 13.3-22 The network function of a circuit is

$$\mathbf{H}(\omega) = \frac{2(j2\omega + 5)}{(4 + j3\omega)(j\omega + 2)}$$

Sketch the asymptotic magnitude Bode plot corresponding to **H**.

P 13.3-23 The network function of a circuit is

$$\mathbf{H}(\omega) = \frac{4(20 + j\omega)(20,000 + j\omega)}{(200 + j\omega)(2000 + j\omega)}$$

Sketch the asymptotic magnitude Bode plot corresponding to **H**.

P 13.3-24 The input to the circuit shown in Figure P 13.3-24*a* is the voltage of the voltage source, v_s. The output of the circuit is the capacitor voltage, v_o. The network function of the circuit is

$$\mathbf{H}(\omega) = \frac{\mathbf{V}_o(\omega)}{\mathbf{V}_s(\omega)}$$

Determine the values of the resistances R_1, R_2, R_3, and R_4 required to cause the network function of the circuit to correspond to the asymptotic Bode plot shown in Figure P 13.3-24*b*.

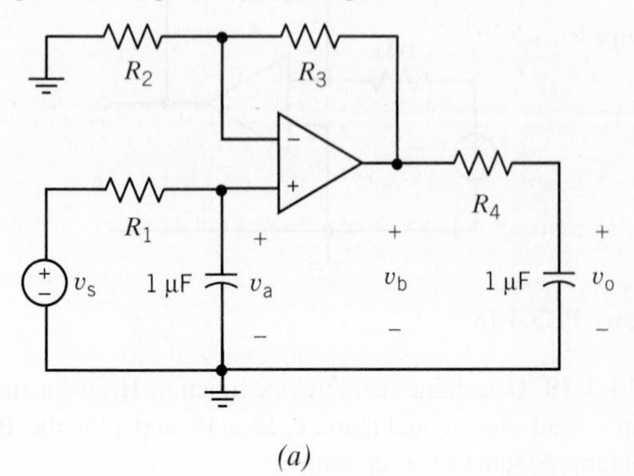

(*a*)

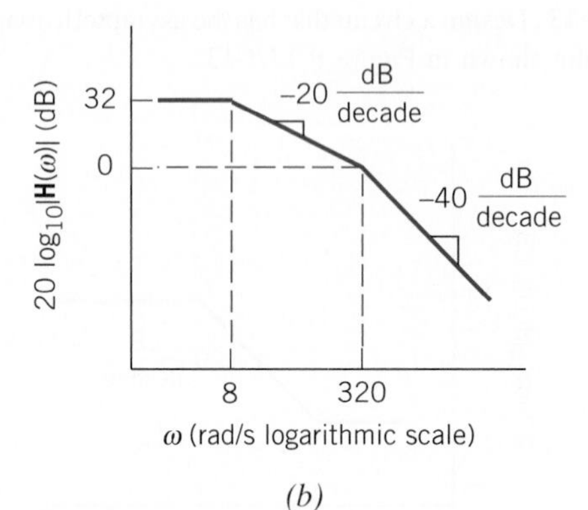

(*b*)

Figure P 13.3-24

P 13.3-25 The input to the circuit shown in Figure P 13.3-25*a* is the voltage of the voltage source, v_s. The output of the circuit is the voltage, v_o. The network function of the circuit is

$$\mathbf{H}(\omega) = \frac{\mathbf{V}_o(\omega)}{\mathbf{V}_s(\omega)}$$

Determine the values of the resistances R_1, R_2, and R_3 required to cause the network function of the circuit to correspond to the asymptotic Bode plot shown in Figure P 13.3-25*b*.

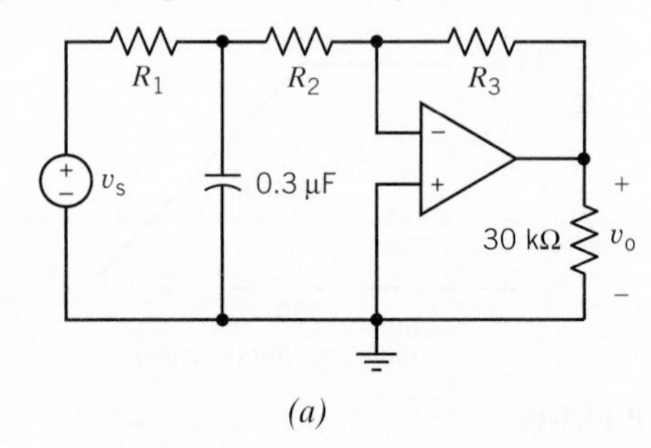

(*a*)

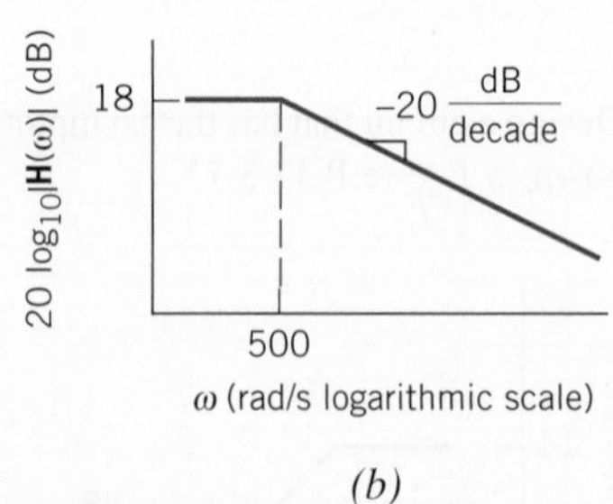

(*b*)

Figure P 13.3-25

P 13.3-26 The input to the circuit shown in Figure P 13.3-26*a* is the voltage of the voltage source, v_s. The output of the circuit is the voltage, v_o. The network function of the circuit is

$$\mathbf{H}(\omega) = \frac{\mathbf{V}_o(\omega)}{\mathbf{V}_s(\omega)}$$

(a) Determine the values of the resistances, R_1 and R_2, required to cause the network function of the circuit to correspond to the asymptotic Bode plot shown in Figure P 13.3-26*b*.

(b) Determine the values of the gains K_1 and K_2 in Figure P 13.3-26*b*.

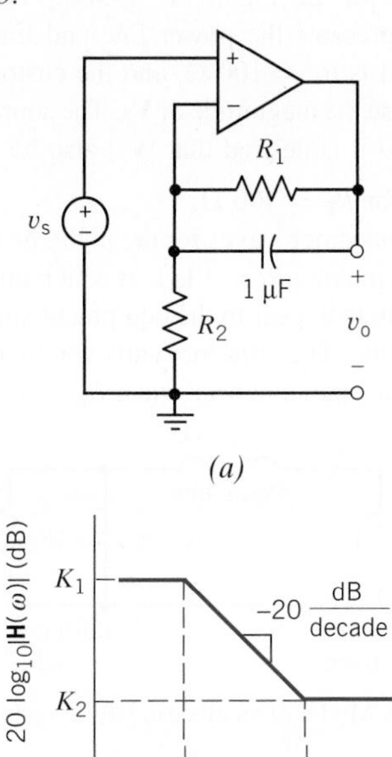

Figure P 13.3-26

P 13.3-27 The input to the circuit shown in Figure P 13.3-27*a* is the voltage of the voltage source, v_s. The output of the circuit is the voltage, v_o. The network function of the circuit is

$$\mathbf{H}(\omega) = \frac{\mathbf{V}_o(\omega)}{\mathbf{V}_s(\omega)}$$

Determine the values of R, C, R_1, and R_2 required to cause the network function of the circuit to correspond to the asymptotic Bode plot shown in Figure P 13.3-27*b*.

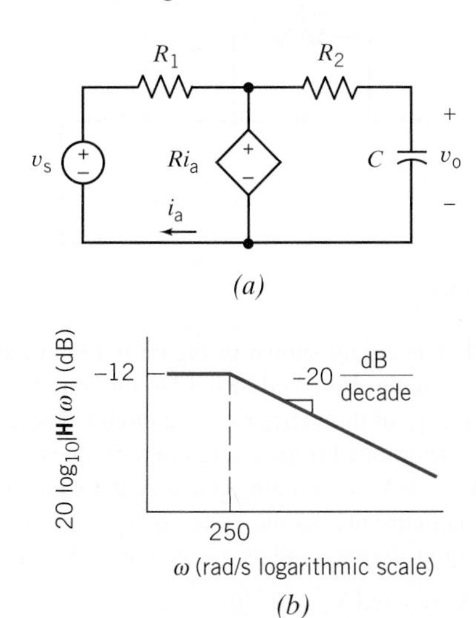

Figure P 13.3-27

P 13.3-28 The input to the circuit shown in Figure P 13.3-28*a* is the current of the current source, i_s. The output of the circuit is the current i_o. The network function of the circuit is

$$\mathbf{H}(\omega) = \frac{\mathbf{I}_o(\omega)}{\mathbf{I}_s(\omega)}$$

Determine the values of G, C, R_1, and R_2 required to cause the network function of the circuit to correspond to the asymptotic Bode plot shown in Figure P 13.3-28*b*.

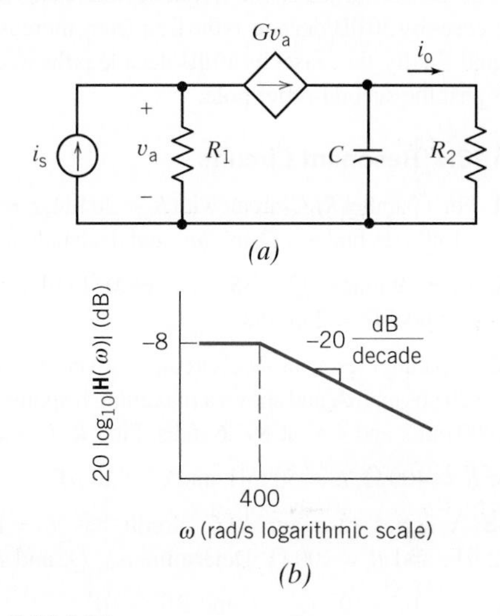

Figure P 13.3-28

P 13.3-29 A first-order circuit is shown in Figure P 13.3-29. Determine the ratio $\mathbf{V}_o/\mathbf{V}_s$ and sketch the Bode diagram when $RC = 0.1$ and $R_1/R_2 = 3$.

Answer: $\mathbf{H} = \left(1 + \frac{R_1}{R_2}\right)\frac{1}{1 + j\omega RC}$

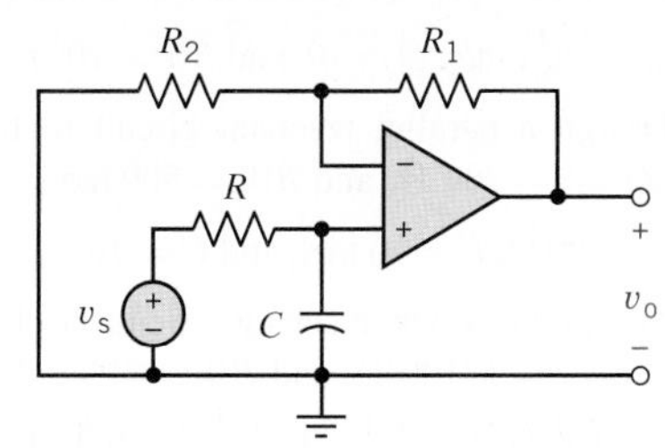

Figure P 13.3-29

P 13.3-30 (a) Draw the Bode diagram of the network function $\mathbf{V}_o/\mathbf{V}_s$ for the circuit of Figure P 13.3-30. (b) Determine $v_o(t)$ when $v_s = 10 \cos 20t$ V.

Answer: (b) $v_o = 4.18 \cos (20t - 24.3°)$ V

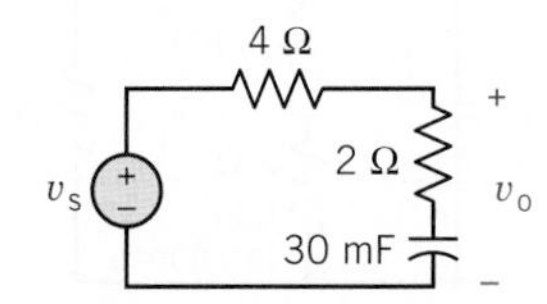

Figure P 13.3-30

P 13.3-31 Draw the asymptotic magnitude Bode diagram for

$$\mathbf{H}(\omega) = \frac{10(1 + j\omega)}{j\omega(1 + j0.5\omega)(1 + j0.6(\omega/50) + (j\omega/50)^2)}$$

Hint: At $\omega = 0.1$ rad/s, the value of the gain is 40 dB and the slope of the asymptotic Bode plot is -20 dB/decade. There is a zero at 1 rad/s, a pole at 2 rad/s, and a second-order pole at 50 rad/s. The slope of the asymptotic magnitude Bode diagram increases by 20 dB/decade as the frequency increases past the zero, decreases by 20 dB/decade as the frequency increases past the pole, and, finally, decreases by 40 dB/decade as the frequency increases past the second-order pole.

Section 13.4 Resonant Circuits

P 13.4-1 For a parallel *RLC* circuit with $R = 20$ kΩ, $L = 1/120$ H, and $C = 1/40$ μF, find ω_0, Q, ω_1, ω_2, and the bandwidth BW.

Answers: $\omega_0 = 69$ krad/s, $Q = 35$, $\omega_1 = 68.02$ krad/s, $\omega_2 = 69.99$ krad/s, and $BW = 2$ krad/s

P 13.4-2 A parallel resonant *RLC* circuit is driven by a current source $i_s = 20 \cos \omega t$ mA and shows a maximum response of 8 V at $\omega = 1000$ rad/s and 4 V at 897.6 rad/s. Find *R*, *L*, and *C*.

Answers: $R = 400$ Ω, $L = 50$ mH, and $C = 20$ μF

P 13.4-3 A series resonant *RLC* circuit has $L = 10$ mH, $C = 0.02$ μF, and $R = 100$ Ω. Determine ω_0, Q, and BW.

Answers: $\omega_0 = 0.7 \times 10^5$, $Q = 7$, and $BW = 10^4$

P 13.4-4 A quartz crystal exhibits the property that when mechanical stress is applied across its faces, a potential difference develops across opposite faces. When an alternating voltage is applied, mechanical vibrations occur and electromechanical resonance is exhibited. A crystal can be represented by a series *RLC* circuit. A specific crystal has a model with $L = 1$ mH, $C = 10\mu$F, and $R = 1$ Ω. Find ω_0, Q, and the bandwidth.

Answers: $\omega_0 = 10^4$ rad/s, $Q = 10$, and $BW = 10^3$ rad/s

P 13.4-5 Design a parallel resonant circuit to have $\omega_0 = 2500$ rad/s, $\mathbf{Z}(\omega_0) = 200$ Ω, and $BW = 500$ rad/s.

Answers: $R = 100$ Ω, $L = 16$ mH, and $C = 10$ μF

P 13.4-6 Design a series resonant circuit to have $\omega_0 = 2500$ rad/s, $\mathbf{Y}(\omega_0) = 1/100$ Ω, and $BW = 600$ rad/s.

Answers: $R = 100$ Ω, $L = 0.17$ H, and $C = 0.9$ μF

P 13.4-7 The circuit shown in Figure P 13.4-7 represents a capacitor, coil, and resistor in parallel. Calculate the resonant frequency, bandwidth, and Q for the circuit.

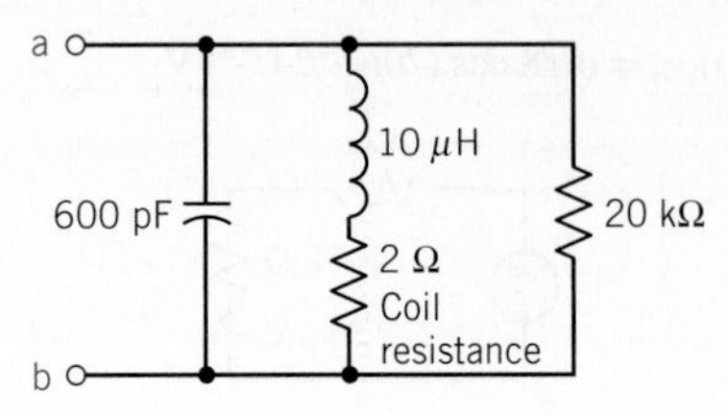

Figure P 13.4-7

P 13.4-8 Consider the simple model of an electric power system as shown in Figure P 13.4-8. The inductance, $L = 0.25$ H, represents the power line and transformer. The customer's load is $R_L = 100$ Ω, and the customer adds $C = 25$ μF to increase the magnitude of $\mathbf{V}_o$. The source is $v_s = 1000 \cos 400t$ V, and it is desired that $|\mathbf{V}_o|$ also be 1000 V.

(a) Find $|\mathbf{V}_o|$ for $R_L = 100$ Ω.

(b) When the customer leaves for the night, he turns off much of his load, making $R_L = 1$ kΩ, at which point, sparks and smoke begin to appear in the equipment still connected to the power line. The customer calls you in as a consultant. Why did the sparks appear when $R_L = 1$ kΩ?

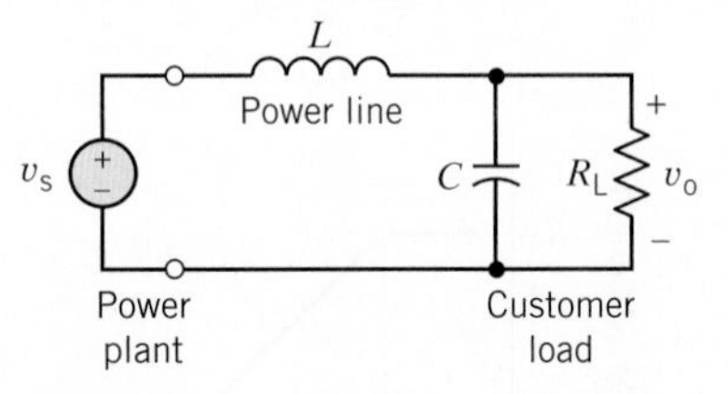

Figure P 13.4-8 Model of an electric power system.

P 13.4-9 Consider the circuit in Figure P 13.4-9. $R_1 = R_2 = 2$ Ω. Select C and L to obtain a resonant frequency of $\omega_0 = 150$ rad/s.

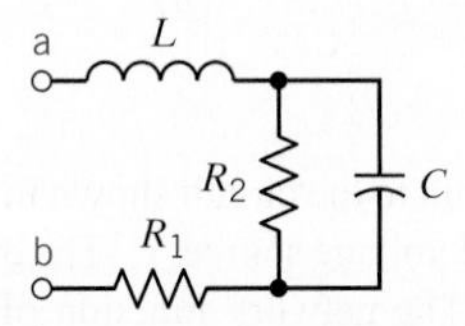

Figure P 13.4-9

P 13.4-10 For the circuit shown in Figure P 13.4-10, (a) derive an expression for the magnitude response $|\mathbf{Z}_{in}|$ versus ω, (b) sketch $|\mathbf{Z}_{in}|$ versus ω, and (c) find $|\mathbf{Z}_{in}|$ at $\omega = 1/\sqrt{LC}$.

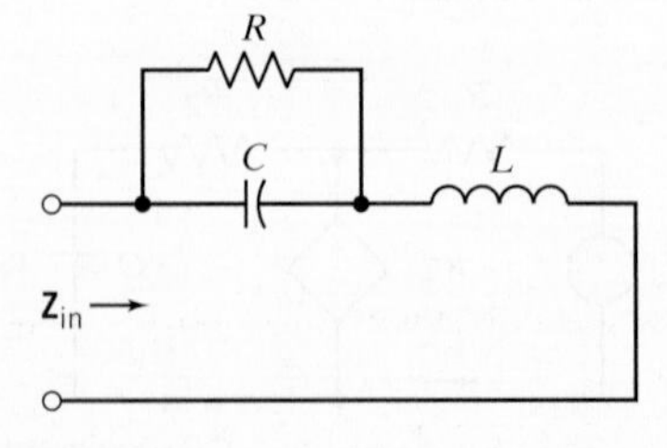

Figure P 13.4-10

P 13.4-11 The circuit shown in Figure P 13.4-11 shows an experimental setup that could be used to measure the parameters k, Q, and ω_0 of this series resonant circuit. These parameters can be determined from a magnitude frequency response plot for $\mathbf{Y} = \mathbf{I}/\mathbf{V}$. It is more convenient to measure node voltages than currents, so the node voltages $\mathbf{V}$ and $\mathbf{V}_2$ have been measured. Express $|\mathbf{Y}|$ as a function of $\mathbf{V}$ and $\mathbf{V}_2$.

Hint: Let $\mathbf{V} = A$ and $\mathbf{V}_2 = B \angle \theta$.

$$\text{Then } \mathbf{I} = \frac{(A - B\cos\theta) - jB\sin\theta}{R}$$

Answer: $|\mathbf{Y}| = \dfrac{\sqrt{(A - B\cos\theta)^2 + (B\sin\theta)^2}}{AR}$

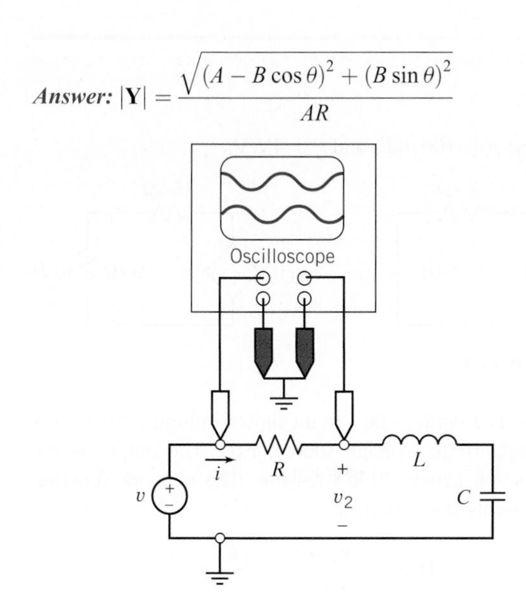

Figure P 13.4-11

Section 13.6 Plotting Bode Plots Using MATLAB

P 13.6-1 The input to the circuit shown in Figure P 13.6-1 is the voltage of the voltage source, v_s. The output of the circuit is the voltage, v_o. Use MATLAB to plot the gain and phase shift of this circuit as a function of frequency for frequencies in the range of $1 < \omega < 1000$ rad/s.

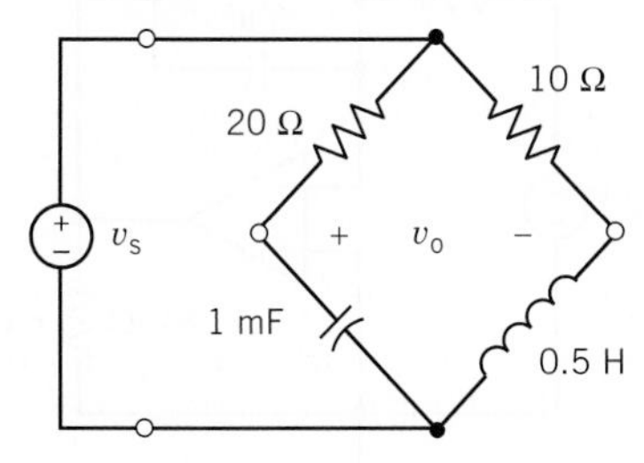

Figure P 13.6-1

P 13.6-2 The input to the circuit shown in Figure P 13.6-2 is the voltage of the voltage source, v_s. The output of the circuit is the voltage, v_o. Use MATLAB to plot the gain and phase shift of this circuit as a function of frequency for frequencies in the range of $1 < \omega < 1000$ rad/s.

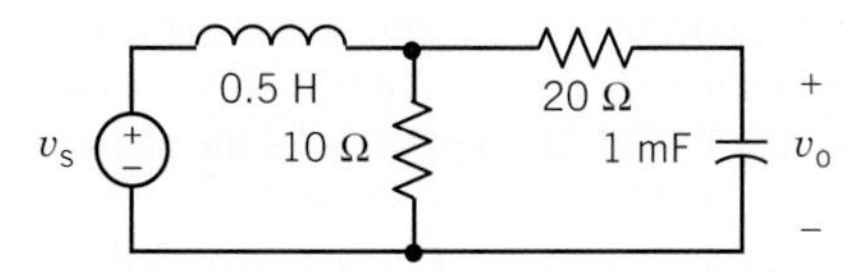

Figure P 13.6-2

P 13.6-3 The input to the circuit shown in Figure P 13.6-3 is the voltage of the voltage source, v_s. The output of the circuit is the voltage, v_o. Use MATLAB to plot the gain and phase shift of this circuit as a function of frequency for frequencies in the range of $1 < \omega < 1000$ rad/s.

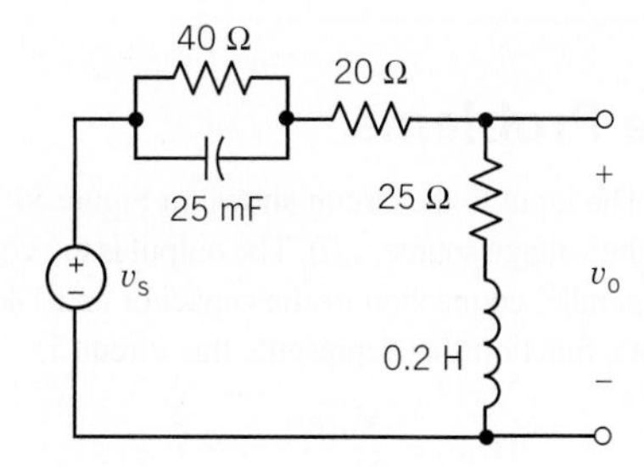

Figure P 13.6-3

Section 13.7 How Can We Check . . . ?

P 13.7-1 Circuit analysis contained in a lab report indicates that the network function of a circuit is

$$\mathbf{H}(\omega) = \frac{1 + j\dfrac{\omega}{630}}{10\left(1 + j\dfrac{\omega}{6300}\right)}$$

This lab report contains the following frequency response data from measurements made on the circuit. Do these data seem reasonable?

ω, rad/s	200	400	795	1585	3162
$\lvert\mathbf{H}(\omega)\rvert$	0.105	0.12	0.16	0.26	0.460
ω, rad/s	6310	12,600	25,100	50,000	100,000
$\lvert\mathbf{H}(\omega)\rvert$	0.71	1.0	1.0	1.0	1.0

P 13.7-2 A parallel resonant circuit (see Figure 13.4-2) has $Q = 70$ and a resonant frequency $\omega_0 = 20,000$ rad/s. A report states that the bandwidth of this circuit is 71.43 rad/s. Verify this result.

P 13.7-3 A series resonant circuit (see Figure P 13.4-4) has $L = 1$ mH, $C = 10\ \mu$F, and $R = 1\ \Omega$. A software program report states that the resonant frequency is $f_0 = 1.59$ kHz and the bandwidth is $BW = 79.6$ Hz. Are these results correct?

P 13.7-4 An old lab report contains the approximate Bode plot shown in Figure P 13.8-4 and concludes that the network function is

$$\mathbf{H}(\omega) = \frac{40\left(1 + j\dfrac{\omega}{200}\right)}{\left(1 + j\dfrac{\omega}{800}\right)}$$

Do you agree?

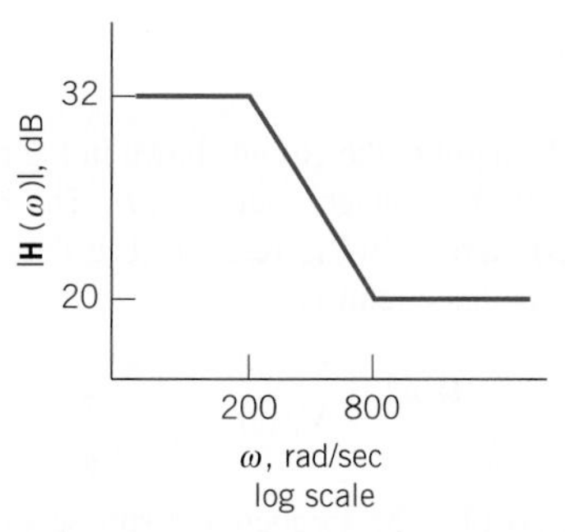

Figure P 13.8-4

PSpice Problems

SP 13-1 The input to the circuit shown in Figure SP 13-1 is the voltage of the voltage source, $v_i(t)$. The output is the voltage, $v_o(t)$, across the parallel connection of the capacitor and 1-kΩ resistor. The network function that represents this circuit is

$$\mathbf{H}(\omega) = \frac{\mathbf{V}_o(\omega)}{\mathbf{V}_i(\omega)} = \frac{k}{1 + j\dfrac{\omega}{p}}$$

Use PSpice to plot the frequency response of this circuit. Determine the values of the pole, p, and of the dc gain, k.

Answers: $p = 250$ rad/s and $k = 0.2$ V/V

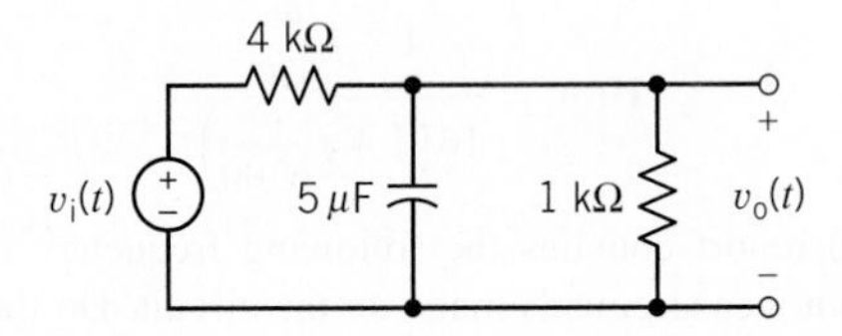

Figure SP 13-1

SP 13-2 The input to the circuit shown in Figure SP 13-2 is the voltage of the voltage source, $v_i(t)$. The output is the voltage, $v_o(t)$, across the series connection of the inductor and 60-Ω resistor. The network function that represents this circuit is

$$\mathbf{H}(\omega) = \frac{\mathbf{V}_o(\omega)}{\mathbf{V}_i(\omega)} = k\frac{1 + j\dfrac{\omega}{z}}{1 + j\dfrac{\omega}{p}}$$

Use PSpice to plot the frequency response of this circuit. Determine the values of the pole, p, of the zero, z, and of the dc gain, k.

Answers: $p = 20$ rad/s, $z = 12$ rad/s, and $k = 0.6$ V/V

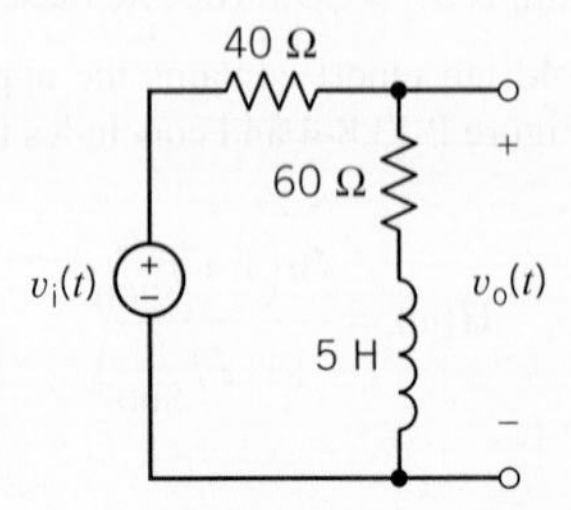

Figure SP 13-2

SP 13-3 The input to the circuit shown in Figure SP 13-3 is the voltage of the voltage source, $v_i(t)$. The output is the voltage, $v_o(t)$, across 30-kΩ resistor. The network function that represents this circuit is

$$\mathbf{H}(\omega) = \frac{\mathbf{V}_o(\omega)}{\mathbf{V}_i(\omega)} = \frac{k}{1 + j\dfrac{\omega}{p}}$$

Use PSpice to plot the frequency response of this circuit. Determine the values of the pole, p, and of the dc gain, k.

Answers: $p = 100$ rad/s and $k = 4$V/V

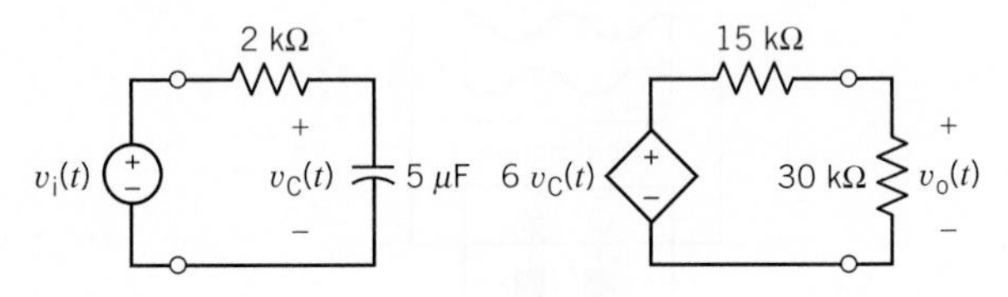

Figure SP 13-3

SP 13-4 The input to the circuit shown in Figure SP 13-4 is the voltage of the voltage source, $v_i(t)$. The output is the voltage, $v_o(t)$, across 20-kΩ resistor. The network function that represents this circuit is

$$\mathbf{H}(\omega) = \frac{\mathbf{V}_o(\omega)}{\mathbf{V}_i(\omega)} = \frac{k}{1 + j\dfrac{\omega}{p}}$$

Use PSpice to plot the frequency response of this circuit. Determine the values of the pole, p, and of the dc gain, k.

Answers: $p = 10$ rad/s and $k = 5$ V/V

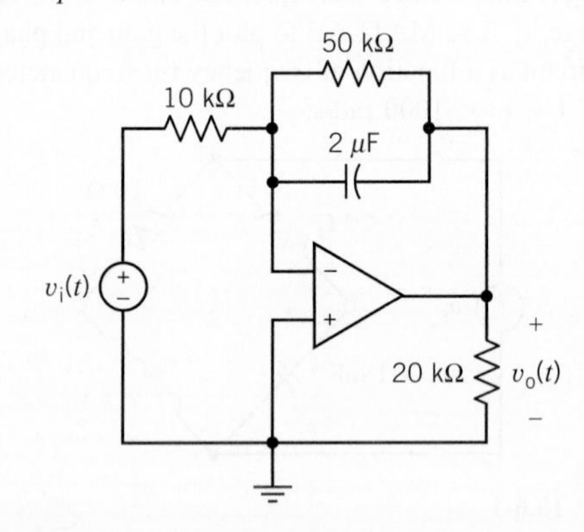

Figure SP 13-4

SP 13-5 Figure SP 13-5 shows a circuit and a frequency response. The frequency response plots were made using PSpice and Probe. V(R3:2) and Vp(R3:2) denote the magnitude and angle of the phasor corresponding to $v_o(t)$. V(V1:+) and Vp(V1:+) denote the magnitude and angle of the phasor corresponding to $v_i(t)$. Hence, V(R3:2)/V(V1:+) is the gain of the circuit and Vp(R3:2) – Vp(V1:+) is the phase shift of the circuit.

Determine values for R and C required to make the circuit correspond to the frequency response.

Hint: PSpice and Probe use m for milli or 10^{-3}. Hence, the label (159.513, 892.827 m) indicates that the gain of the circuit is $892.827 {*} 10^{-3} = 0.892827$ at a frequency of 159.513 Hz ≈ 1000 rad/sec.

Answers: $R = 5$ kΩ and $C = 0.2$ μF

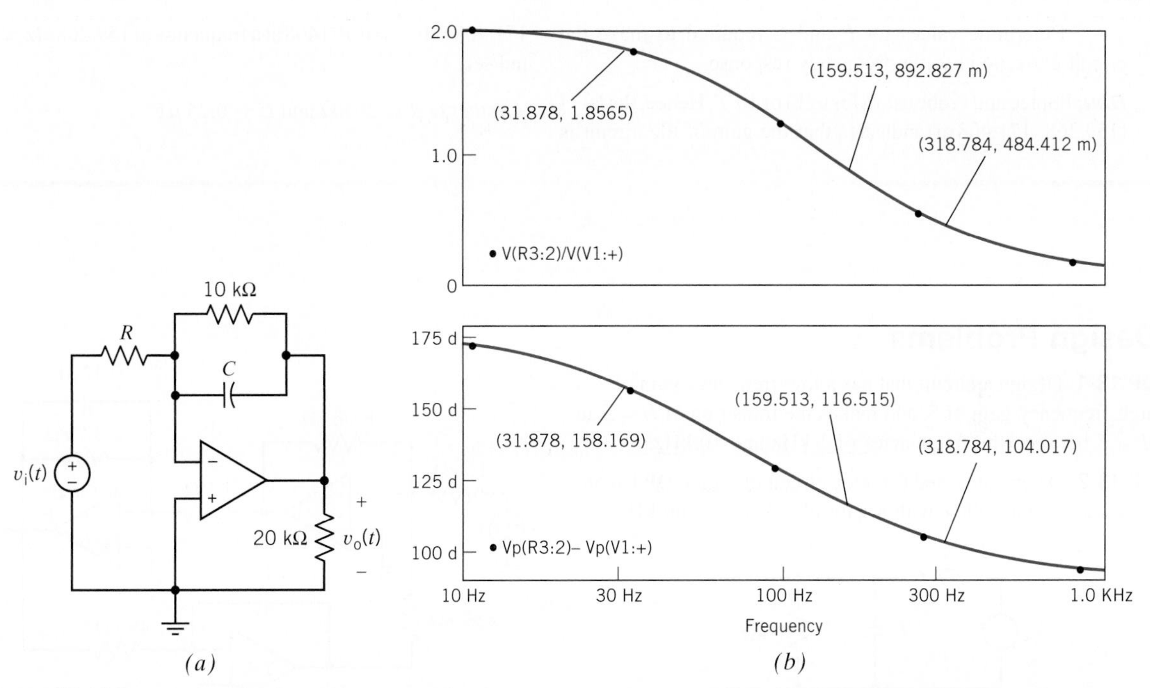

Figure SP 13-5 (*a*) A circuit and (*b*) the corresponding frequency response.

SP 13-6 Figure SP 13-6 shows a circuit and a frequency response. The frequency response plots were made using PSpice and Probe. V(R2:2) and Vp(R2:2) denote the magnitude and angle of the phasor corresponding to $v_o(t)$. V(V1:+) and Vp(V1:+) denote the magnitude and angle of the phasor corresponding to $v_i(t)$. Hence V(R2:2)/V(V1:+) is the gain of the circuit, and Vp(R2:2) – Vp(V1:+) is the phase shift of the circuit.

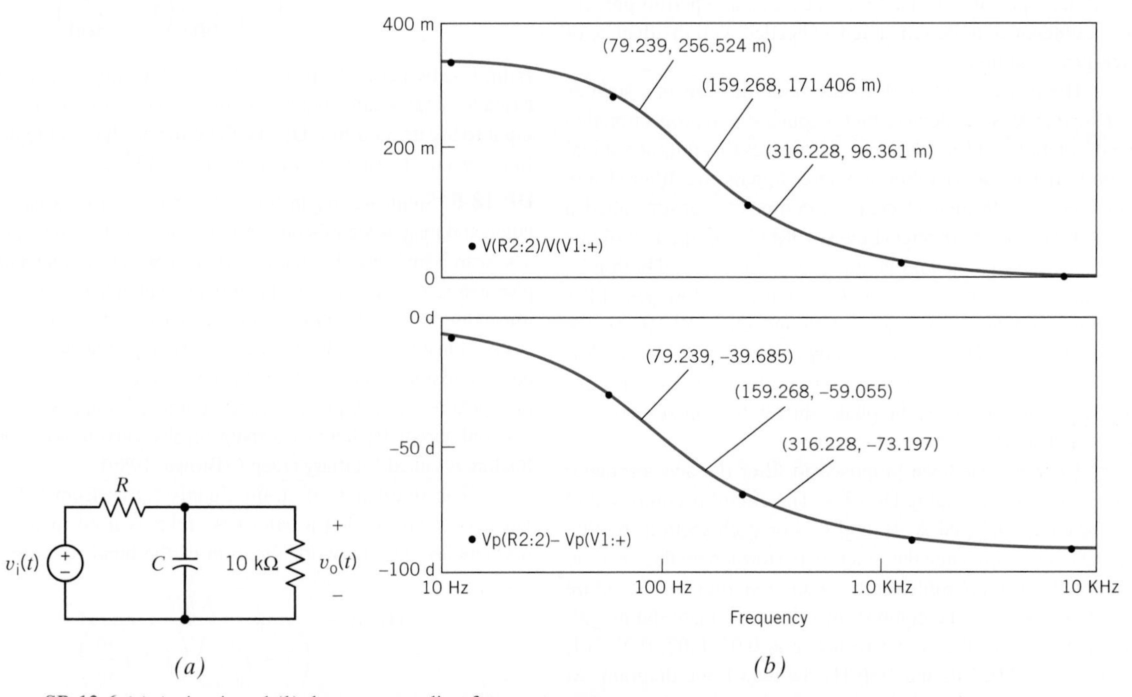

Figure SP 13-6 (*a*) A circuit and (*b*) the corresponding frequency response.

Determine values for R and C required to make the circuit correspond to the frequency response.

Hint: PSpice and Probe use m for milli or 10^{-3}. Hence, the label (159.268, 171.408 m) indicates that the gain of the circuit is $171.408*10^{-3} = 0.171408$ at a frequency of 159.268 Hz ≈ 1000 rad/sec.

Answers: $R = 20\ \text{k}\Omega$ and $C = 0.25\ \mu\text{F}$

Design Problems

DP 13-1 Design a circuit that has a low-frequency gain of 2, a high-frequency gain of 5, and makes the transition of $H = 2$ to $H = 5$ between the frequencies of 1 kHz and 10 kHz.

DP 13-2 Determine L and C for the circuit of Figure DP 13-2 to obtain a low-pass filter with a gain of −3 dB at 100 kHz.

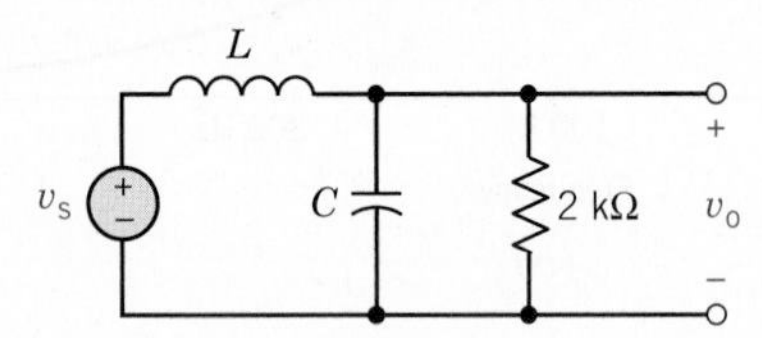

Figure DP 13-2

DP 13-3 British Rail has constructed an instrumented railcar that can be pulled over its tracks at speeds up to 180 km/hr and will measure the track-grade geometry. Using such a railcar, British Rail can monitor and track gradual degradation of the rail grade, especially the banking of curves, and permit preventive maintenance to be scheduled as needed well in advance of track-grade failure.

The instrumented railcar has numerous sensors, such as angular-rate sensors (devices that output a signal proportional to rate of rotation) and accelerometers (devices that output a signal proportional to acceleration), whose signals are filtered and combined in a fashion to create a composite sensor called a compensated accelerometer (Lewis, 1988). A component of this composite sensor signal is obtained by integrating and high-pass filtering an accelerometer signal. A first-order low-pass filter will approximate an integrator at frequencies well above the break frequency. This can be seen by computing the phase shift of the filter-transfer function at various frequencies. At sufficiently high frequencies, the phase shift will approach 90°, the phase characteristic of an integrator.

A circuit has been proposed to filter the accelerometer signal, as shown in Figure DP 13-3. The circuit is composed of three sections, labeled A, B, and C. For each section, find an expression for and name the function performed by that section. Then find an expression for the gain function of the entire circuit, $\mathbf{V}_o/\mathbf{V}_s$. For the component values, evaluate the magnitude and phase of the circuit response at 0.01, 0.02, 0.05, 0.1, 0.2, 0.5, 1.0, 2.0, 5.0, and 10.0 Hz. Draw a Bode diagram. At what frequency is the phase response approximately equal to 0°? What is the significance of this frequency?

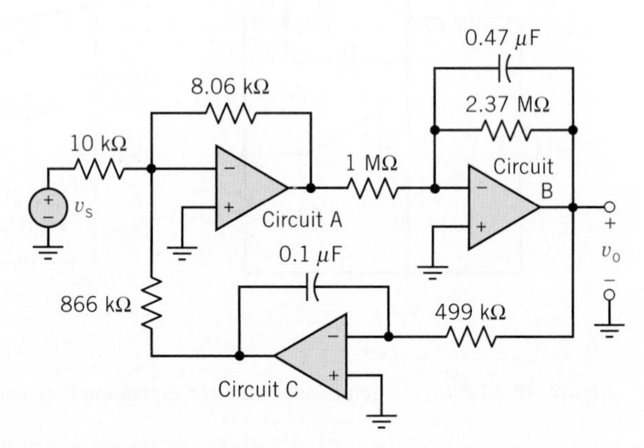

Figure DP 13-3

DP 13-4 Design a circuit that has the network function

$$\mathbf{H}(\omega) = 10\frac{j\omega}{\left(1 + j\frac{\omega}{200}\right)\left(1 + j\frac{\omega}{500}\right)}$$

Hint: Use two circuits from Table 13.4-1. Connect the circuits in cascade. That means that the output of one circuit is used as the input to the next circuit. $\mathbf{H}(\omega)$ will be the product of the network functions of the two circuits from Table 13.3-2.

DP 13-5 Strain-sensing instruments can be used to measure orientation and magnitude of strains running in more than one direction. The search for a way to predict earthquakes focuses on identifying precursors, or changes, in the ground that reliably warn of an impending event. Because so few earthquakes have occurred precisely at instrumented locations, it has been a slow and frustrating quest. Laboratory studies show that before rock actually ruptures—precipitating an earthquake—its rate of internal strain increases. The material starts to fail before it actually breaks. This prelude to outright fracture is called "tertiary creep" (Brown, 1989).

The frequency of strain signals varies from 0.1 to 100 rad/s. A circuit called a band-pass filter is used to pass these frequencies. The network function of the band-pass filter is

$$\mathbf{H}(\omega) = \frac{Kj\omega}{\left(1 + j\frac{\omega}{\omega_1}\right)\left(1 + j\frac{\omega}{\omega_2}\right)}$$

Specify ω_1, ω_2, and K so that the following are the case:

1. The gain is at least 17 dB over the range 0.1 to 100 rad/s.
2. The gain is less than 17 dB outside the range 0.1 to 100 rad/s.
3. The maximum gain is 20 dB.

DP 13-6 Is it possible to design the circuit shown in Figure DP 13-6 to have a phase shift of −45° and a gain of 2 V/V both at a frequency of 1000 radians/second using a 0.1 microfarad capacitor and resistors from the range of 1 k ohm to 200 k ohm?

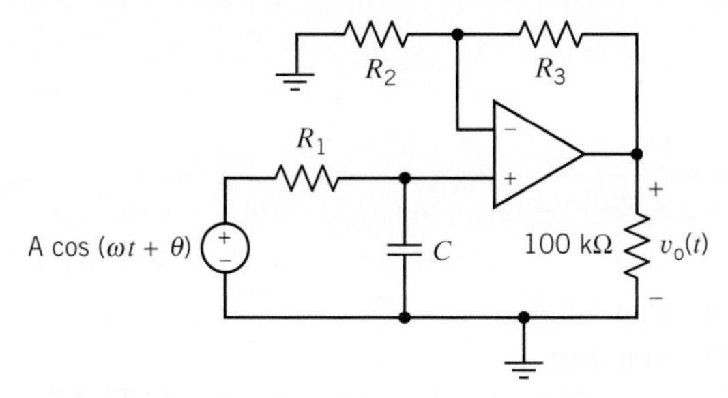

Figure DP 13-6

DP 13-7 Design the circuit shown in Figure DP 13-7*a* to have the asymptotic Bode plot shown in Figure DP 13-7*b*.

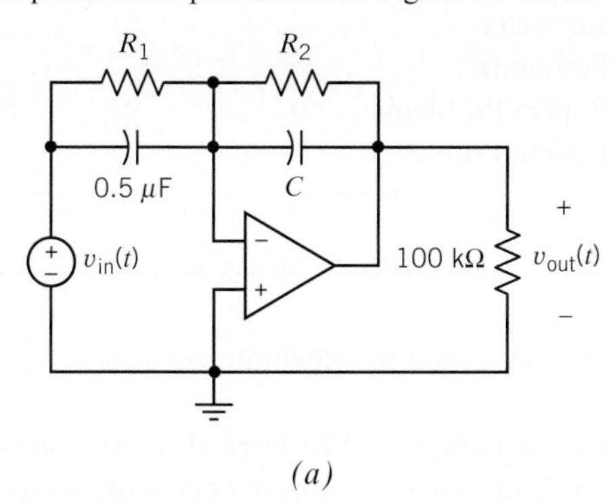

(*a*)

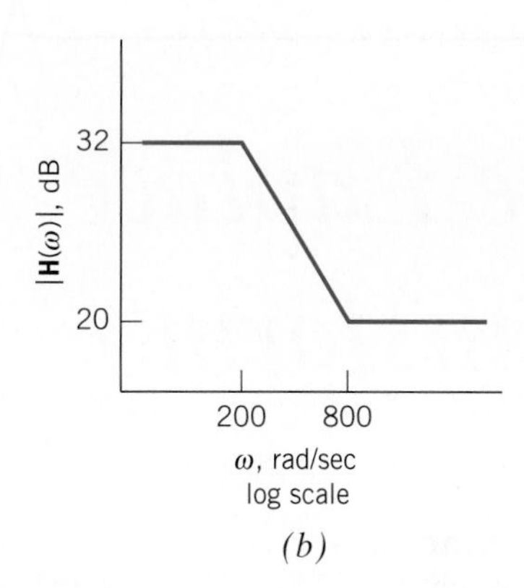

(*b*)

Figure DP 13-7

DP 13-8 For the circuit of Figure DP 13-8, select R_1 and R_2 so that the gain at high frequencies is 10 V/V and the phase shift is 195° at $\omega = 1000$ rad/s. Determine the gain at $\omega = 10$ rad/s.

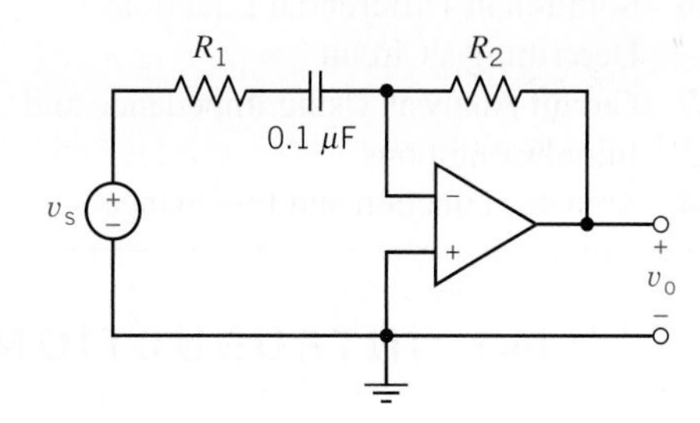

Figure DP 13-8

The Laplace Transform

IN THIS CHAPTER

14.1 Introduction
14.2 Laplace Transform
14.3 Pulse Inputs
14.4 Inverse Laplace Transform
14.5 Initial and Final Value Theorems
14.6 Solution of Differential Equations Describing a Circuit
14.7 Circuit Analysis Using Impedance and Initial Conditions
14.8 Transfer Function and Impedance
14.9 Convolution
14.10 Stability
14.11 Partial Fraction Expansion Using MATLAB
14.12 How Can We Check . . . Transfer Function?
14.13 **DESIGN EXAMPLE**—Space Shuttle Cargo Door
14.14 Summary
Problems
PSpice Problems
Design Problems

14.1 INTRODUCTION

Circuits that have no capacitors or inductors can be represented by algebraic equations.

- Chapters 1–6 described circuits without capacitors or inductors. We learned many things about such circuits, including how to represent them by mesh current equations or node voltage equations.
- Capacitors and inductors are described in Chapter 7.

Circuits that contain capacitors and/or inductors are represented by differential equations. In general, the order of the differential equation is equal to the number of capacitors plus the number of inductors in the circuit. Writing and solving these differential equations can be challenging.

- In Chapter 8, we analyzed first-order circuits.
- In Chapter 9, we analyzed second-order circuits.

The response of a circuit containing capacitors and/or inductors can be separated into two parts: the steady-state response and the transient part of the response.

- In Chapters 10–13, we studied the steady-state response of circuits with sinusoidal inputs. We found that we could analyze such circuits by representing them in the frequency domain. We did not restrict our attention to first- or second-order circuits.

- In this chapter, we find the complete response, transient part plus steady-state part, of circuits with capacitors and/or inductors. We will not restrict our attention to first- or second-order circuits or to circuits with sinusoidal inputs.

In this chapter, we introduce a very powerful tool for the analysis of circuits. The Laplace transform enables the circuit analyst to transform the set of differential equations describing a circuit to the complex frequency domain, where they become a set of linear algebraic equations. Then, using straightforward algebraic manipulation, we solve for the variables of interest. Finally, we use the inverse Laplace transform to go back to the time domain and express the desired response in terms of time. This is a powerful tool indeed!

Next, we learn how to represent the circuit itself in the complex frequency domain. After doing so, we can analyze the circuit by writing and solving a set of algebraic equations, for example, mesh current equations or node voltage equations. In other words, using the complex frequency domain eliminates the need to write the differential equation that represents the circuit.

Finally, we learn how to represent a linear circuit by its transfer function, step response, or impulse response.

14.2 LAPLACE TRANSFORM

As we have seen in earlier chapters, it is useful to *transform* the equations describing a circuit from the time domain into the frequency domain, then perform an analysis and, finally, transform the problem's solution back to the time domain. You will recall that in Chapter 10 we defined the phasor as a mathematical transformation to simplify finding the steady-state response of a circuit to a sinusoidal input. Using the phasor transformation, we solved *algebraic equations* having complex coefficients instead of solving *differential equations*, albeit with real coefficients. The transform method is summarized in Figure 14.2-1.

In this chapter, we will use the Laplace transform, rather than the phasor transformation, to transform differential equations to algebraic equations. This will enable us to determine the complete response to a variety of input functions instead of the steady-state response to sinusoidal inputs. (The complete response consists of the steady-state response together with the transient

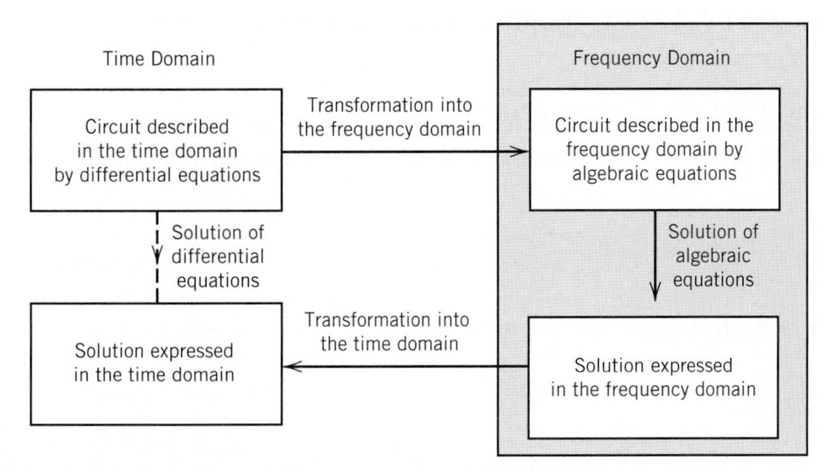

FIGURE 14.2-1 The transform method.

FIGURE 14.2-2 Pierre-Simon Laplace (1749–1827) is credited with the transform that bears his name. Courtesy of Burndy Library.

part of the response. We will have more to say about this later.) Pierre-Simon Laplace, who is shown in Figure 14.2-2, is credited with the transform that bears his name.

The (one-sided or unilateral) **Laplace transform** is defined as

$$F(s) = \mathscr{L}[f(t)] = \int_0^\infty f(t)e^{-st}dt \tag{14.2-1}$$

where s is a complex variable given by

$$s = \sigma + j\omega \tag{14.2-2}$$

The exponent st of e in Eq. 14.2-1 must be dimensionless. Consequently, s has units of frequency. It is customary to refer to s as complex frequency. The lower limit of the integral in Eq. 14.2-1 is 0−, a time just before $t = 0$. As a result, the Laplace transform includes the effects of any discontinuity in $f(t)$ occurring at time $t = 0$. In contrast, the Laplace transform does not include the effect of that part of $f(t)$ occurring for time $t < 0$.

The notation $\mathscr{L}[f(t)]$ indicates taking the Laplace transform of $f(t)$. The result, $F(s)$ is called the Laplace transform of $f(t)$. The function $f(t)$ is said to exist in the time domain whereas the function $F(s)$ is said to exist in the complex-frequency domain or the s-domain. (Occasionally, the complex-frequency domain is referred to casually as the frequency domain when the context makes it clear that frequency domain is short for complex-frequency domain.)

The inverse Laplace transform is defined by the complex inversion integral

$$f(t) = \mathscr{L}^{-1}[F(s)] = \frac{1}{2\pi j}\int_{\alpha-j\infty}^{\alpha+j\infty} F(s)e^{st}ds \tag{14.2-3}$$

The integral in Eq. 14.2-3 is a contour integration in the complex plane. Evaluation of this integral requires complex analysis and is beyond the scope of this book. Instead of evaluating the integral in Eq. 14.2-3, we rely on the fact that the inverse Laplace transform is indeed the inverse of the Laplace transform. That is, if $F(s) = \mathscr{L}[f(t)]$, then also $f(t) = \mathscr{L}^{-1}[F(s)]$. We say that $f(t)$ and $F(s)$ comprise a Laplace transform pair and denote this fact as

$$f(t) \leftrightarrow F(s) \tag{14.2-4}$$

Recalling that the part of $f(t)$ occurring for time $t < 0$ had no effect on $F(s)$, we see that $\mathscr{L}^{-1}[F(s)]$ provides $f(t)$ only for $t > 0$. (Sometimes the uncertainty about $f(t)$ for $t < 0$ is resolved by requiring that $f(t) = 0$ for $t < 0$ for all time domain functions.)

EXAMPLE 14.2-1 Laplace Transform Pairs

(a) Find the Laplace transform of $f(t) = e^{-at}$, where $a > 0$.

(b) Find the Laplace transform of $g(t) = e^{-at}u(t)$, where $a > 0$ and $u(t)$ is the unit step function.

Solution

(a) Using Eq. 14.2-1, we have

$$F(s) = \mathscr{L}[f(t)] = \mathscr{L}[e^{-at}] = \int_{0-}^{\infty} e^{-at}e^{-st}dt = \left.\frac{-e^{-(s+a)t}}{s+a}\right|_{0^-}^{\infty} = \frac{1}{s+a}$$

(b) Again using Eq. 14.2-1, we have

$$G(s) = \mathscr{L}[g(t)] = \mathscr{L}[e^{-at}u(t)] = \int_{0-}^{\infty} e^{-at}u(t)e^{-st}dt = \int_{0-}^{\infty} e^{-at}e^{-st}dt = \left.\frac{-e^{-(s+a)t}}{s+a}\right|_{0-}^{\infty} = \frac{1}{s+a}$$

In this example, $f(t) \neq g(t)$ when $t < 0$, but $f(t) = g(t)$ when $t > 0$. Consequently, $F(s) = G(s)$. The inverse Laplace transform of $F(s) = G(s)$ only provides $f(t)$ or $g(t)$ for $t > 0$. We can summarize the results of this example by the Laplace transform pair:

$$e^{-at} \text{ for } t > 0 \quad \leftrightarrow \quad \frac{1}{s+a}$$

We should stop and ask under what conditions the integral of Eq. 14.2-1 converges to a finite value. It can be shown that the integral converges when

$$\int_{0-}^{\infty} |f(t)| e^{-\sigma_1 t} dt < \infty$$

for some real positive σ_1 If the magnitude of $f(t)$ is $|f(t)| < Me^{at}$ for all positive t, the integral will converge for $\sigma_1 > a$. The region of convergence is therefore given by $\infty > \sigma_1 > a$, and σ_1 is known as the abscissa of absolute convergence. Functions of time, $f(t)$, that are physically possible always have a Laplace transform.

Linearity is an important property of the Laplace transform. Consider

$$f(t) = a_1 f_1(t) + a_2 f_2(t)$$

for arbitrary constants a_1 and a_2. Using Eq. 14.2-1, we have

$$\begin{aligned} F(s) = \mathscr{L}[f(t)] = \mathscr{L}[a_1 f_1(t) + a_2 f_2(t)] &= \int_{0-}^{\infty} (a_1 f_1(t) + a_2 f_2(t)) e^{-st} dt \\ &= a_1 \int_{0^-}^{\infty} f_1(t) e^{-st} dt + a_2 \int_{0^-}^{\infty} f_2(t) e^{-st} dt \\ &= a_1 F_1(s) + a_2 F_2(s) \end{aligned}$$

where $F_1(s)$ and $F_2(s)$ are the Laplace transforms of the time functions $f_1(t)$ and $f_2(t)$, respectively. We can summarize linearity as

$$a_1 f_1(t) + a_2 f_2(t) \quad \leftrightarrow \quad a_1 F_1(s) + a_2 F_2(s) \tag{14.2-5}$$

EXAMPLE 14.2-2 Linearity

Find the Laplace transform of $\sin \omega t$.

Solution

Use Euler's identity to write

$$\sin \omega t = \frac{1}{2j} \left(e^{j\omega t} - e^{-j\omega t} \right)$$

From Example 14.2-1, we have

$$e^{-at} \text{ for } t > 0 \leftrightarrow \frac{1}{s+a}$$

so

$$e^{-j\omega t} \text{ for } t > 0 \leftrightarrow \frac{1}{s+j\omega}$$

and

$$e^{j\omega t} \text{ for } t > 0 \leftrightarrow \frac{1}{s-j\omega}$$

Using superposition, we then have

$$\mathscr{L}[\sin \omega t] = \frac{1}{2j}\mathscr{L}\left[e^{j\omega t} - e^{-j\omega t}\right] = \frac{1}{2j}\left(\frac{1}{s-\omega t} - \frac{1}{s+\omega t}\right) = \frac{(s+j\omega)-(s-j\omega)}{2j(s-\omega)(s+j\omega)} = \frac{\omega}{s^2+\omega^2}$$

We can summarize the results of this example by the Laplace transform pair

$$\sin \omega t \text{ for } t > 0 \leftrightarrow \frac{\omega}{s^2+\omega^2}$$

Let us obtain the transform of the first derivative of $f(t)$. We have

$$\mathscr{L}\left[\frac{df}{dt}\right] = \int_{0^-}^{\infty} \frac{df}{dt} e^{-st} dt$$

In anticipation of integrating by parts, take $u = e^{-st}$ and $dv = \left(\frac{df}{dt}\right)dt = df$. Then $du = -se^{-st}$ and $v = f$. Now integrating by parts gives

$$\mathscr{L}\left[\frac{df}{dt}\right] = s\int_0^{\infty} fe^{-st}dt + fe^{-st}\Big|_0^{\infty} = sF(s) - f(0^-)$$

We can summarize differentiation in the time domain as

$$\frac{df}{dt} \leftrightarrow sF(s) - f(0^-) \qquad (14.2\text{-}6)$$

Thus, the Laplace transform of the derivative of a function is s times the Laplace transform of the function minus the initial condition.

EXAMPLE 14.2-3 Differentiation in the Time Domain

Find the Laplace transform of $\cos \omega t$.

Solution

The cosine is proportional to the derivative of the sine

$$\cos \omega t = \frac{1}{\omega}\frac{d}{dt}\sin \omega t$$

Using linearity

$$\mathscr{L}[\cos \omega t] = \frac{1}{\omega}\mathscr{L}\left[\frac{d}{dt}\sin \omega t\right]$$

Using Eq 14.2-6,

$$\mathscr{L}\left[\frac{d}{dt}\sin \omega t\right] = s\,\mathscr{L}[\sin \omega t] - \sin 0 = s\,\mathscr{L}[\sin \omega t] - 0$$

From Example14.2-2,

$$\mathscr{L}[\sin \omega t] = \frac{\omega}{s^2+\omega^2}$$

Combining these results gives

$$\mathscr{L}[\cos \omega t] = \frac{1}{\omega}(s)\frac{\omega}{s^2+\omega^2} = \frac{s}{s^2+\omega^2}$$

Thus, we use the definition of the Laplace transform given in Eq. 14.2-1 to obtain both Laplace transform pairs and properties of the Laplace transform. Table 14.2-1 provides a collection of important Laplace transform pairs. Table 14.2-2 lists important properties of the Laplace transform.

Table 14.2-1 Laplace Transform Pairs

$f(t)$ for $t > 0$	$F(s) = \mathscr{L}[f(t)u(t)]$
$\delta(t)$	1
$u(t)$	$\frac{1}{s}$
e^{-at}	$\frac{1}{s+a}$
t	$\frac{1}{s^2}$
t^n	$\frac{n!}{s^{n+1}}$
$e^{-at}t^n$	$\frac{n!}{(s+a)^{n+1}}$
$\sin(\omega t)$	$\frac{\omega}{s^2+\omega^2}$
$\cos(\omega t)$	$\frac{s}{s^2+\omega^2}$
$e^{-at}\sin(\omega t)$	$\frac{\omega}{(s+a)^2+\omega^2}$
$e^{-at}\cos(\omega t)$	$\frac{s+a}{(s+a)^2+\omega^2}$

Table 14.2-2 Laplace Transform Properties

PROPERTY	$f(t),\ t > 0$	$F(s) = \mathscr{L}[f(t)u(t)]$
Linearity	$a_1 f_1(t) + a_2 f_2(t)$	$a_1 F_1(s) + a_2 F_2(s)$
Time scaling	$f(at)$, where $a > 0$	$\frac{1}{a}F\left(\frac{s}{a}\right)$
Time integration	$\int_0^t f(\tau)d\tau$	$\frac{1}{s}F(s)$
Time differentiation	$\frac{df(t)}{dt}$	$sF(s) - f(0^-)$
	$\frac{d^2f(t)}{dt^2}$	$s^2F(s) - \left(sf(0^-) + \frac{df(0^-)}{dt}\right)$
	$\frac{d^nf(t)}{dt^n}$	$s^nF(s) - \sum_{k=1}^{n} s^{n-k}\frac{d^{k-1}f(0^-)}{dt^{k-1}}$
Time shift	$f(t-a)u(t-a)$	$e^{-as}F(s)$
Frequency shift	$e^{-at}f(t)$	$F(s+a)$
Time convolution	$f_1(t)^*f_2(t) = \int_0^t f_1(\tau)f_2(t-\tau)d\tau$	$F_1(s)F_2(s)$
Frequency integration	$\frac{f(t)}{t}$	$\int_s^\infty F(\lambda)d\lambda$
Frequency differentiation	$tf(t)$	$-\frac{dF(s)}{ds}$
Initial value	$f(0^+)$	$\lim_{s\to\infty} sF(s)$
Final value	$f(\infty)$	$\lim_{s\to 0} sF(s)$

EXAMPLE 14.2-4 Laplace Transform Pairs and Properties

Find the Laplace transform of $5 - 5e^{-2t}(1 + 2t)$.

Solution

From linearity,
$$\mathscr{L}\left[5 - 5e^{-2t}(1 + 2t)\right] = 5\,\mathscr{L}[1] - 5\,\mathscr{L}\left[e^{-2t}(1 + 2t)\right]$$

Using frequency shift from Table 14.2-2 with $f(t) = 1 + 2t$ gives

$$\mathscr{L}\left[e^{-2t}(1 + 2t)\right] = \mathscr{L}\left[e^{-2t}f(t)\right] = F(s + 2)$$

where
$$F(s) = \mathscr{L}[f(t)] = \mathscr{L}[1 + 2\,t] = \mathscr{L}[1] + 2\,\mathscr{L}[t] = \frac{1}{s} + 2\left(\frac{1}{s^2}\right)$$

Next,
$$F(s + 2) = F(s)|_{s \leftarrow s+2}$$

That is, we must replace each s in $F(s)$ by $s + 2$ to obtain $F(s + 2)$:

$$F(s + 2) = \left(\frac{1}{s} + 2\left(\frac{1}{s^2}\right)\right)\Bigg|_{s \leftarrow s+2} = \frac{1}{s + 2} + 2\left(\frac{1}{(s + 2)^2}\right) = \frac{s + 2 + 2(1)}{(s + 2)^2} + \frac{s + 4}{s^2 + 4s + 4}$$

Putting it all together gives

$$\mathscr{L}\left[5 - 5e^{-2t}(1 + 2t)\right] = 5\left(\frac{1}{s}\right) - 5\left(\frac{s + 4}{s^2 + 4s + 4}\right) = \frac{5(s^2 + 4s + 4) - 5s(s + 4)}{s(s^2 + 4s + 4)} = \frac{20}{s(s^2 + 4s + 4)}$$

EXAMPLE 14.2-5 Laplace Transform Pairs and Properties

Find the Laplace transform of $10\,e^{-4t}\cos(20t + 36.9°)$.

Solution

Table 14.2-1 has entries for $\cos(\omega t)$ and $\sin(\omega t)$ but not for $\cos(\omega t + \theta)$. We can use the trigonometric identity
$$A\cos(\omega t + \theta) = (A\cos\theta)\cos(\omega t) - (A\sin\theta)\sin(\omega t)$$

to write
$$10\cos(20t + 36.9°) = 8\cos(20t) - 6\sin(20t)$$

Now use linearity to write

$$\begin{aligned}\mathscr{L}[10e^{-4t}\cos(20t + 36.9°)] &= \mathscr{L}[e^{-4t}(8\cos(20t) - 6\sin(20t))] \\ &= 8\,\mathscr{L}[e^{-4t}\cos(20t)] - 6\mathscr{L}[e^{-4t}\sin(20t)]\end{aligned}$$

Using frequency shifts from Table 14.2-2 with $f(t) = \cos(20t)$ gives

$$\mathscr{L}\left[e^{-4t}\cos(20t)\right] = \mathscr{L}\left[e^{-4t}f(t)\right] = F(s + 4)$$

where $$F(s) = \mathscr{L}[f(t)] = \mathscr{L}[\cos(20t)] = \frac{s}{s^2+20^2} = \frac{s}{s^2+400}$$

Next, $$F(s+4) = F(s)|_{s \leftarrow s+4}$$

That is, we must replace each s in $F(s)$ by $s+4$ to obtain $F(s+4)$:

$$\mathscr{L}\left[e^{-4t}\cos(20t)\right] = F(s+4) = \left.\frac{s}{s^2+400}\right|_{s\leftarrow s+4} = \frac{s+4}{(s+4)^2+400} = \frac{s+4}{s^2+8s+416}$$

Similarly $$\mathscr{L}\left[e^{-4t}\sin(20t)\right] = \left.\frac{20}{s^2+400}\right|_{s\leftarrow s+4} = \frac{20}{(s+4)^2+400} = \frac{20}{s^2+8s+416}$$

Putting it all together gives

$$\mathscr{L}\left[10e^{-4t}\cos(20t+36.9°)\right] = 8\left(\frac{s+4}{s^2+8s+416}\right) - 6\left(\frac{20}{s^2+8s+416}\right) = \frac{8s-88}{s^2+8s+416}$$

EXAMPLE 14.2-5 Laplace Transform Pairs and Properties

Find the Laplace transform of $2\delta(t) + 3 + 4u(t)$.

Solution

From linearity, $$\mathscr{L}[2\delta(t)+3+4u(t)] = 2\,\mathscr{L}\,[\delta(t)] + 3\,\mathscr{L}\,[1] + 4\mathscr{L}\,[u(t)]$$

Because $1 = u(t)$ for $t \geq 0$, $\mathscr{L}[1] = \mathscr{L}[u(t)]$. Using Table 14.2-1 gives

$$\mathscr{L}[2\delta(t)+3+4u(t)] = 2\,\mathscr{L}\,[\delta(t)] + 3\,\mathscr{L}\,[1] + 4\mathscr{L}\,[u(t)] = 2(1) + 3\left(\frac{1}{s}\right) + 4\left(\frac{1}{s}\right) = 2 + \frac{7}{s}$$

14.3 PULSE INPUTS

The step function, shown in Figure 14.3-1a and represented as

$$u(t) = \begin{cases} 0 & t < 0 \\ 1 & t > 0 \end{cases} \tag{14.3-1}$$

makes an abrupt transition from 0 to 1 at time $t = 0$. Define the impulse function, $\delta(t)$, to be

$$\delta(t) = \frac{d}{dt}u(t) = \begin{cases} 0 & t < 0 \\ \text{undefined} & t = 0 \\ 0 & t > 0 \end{cases} \tag{14.3-2}$$

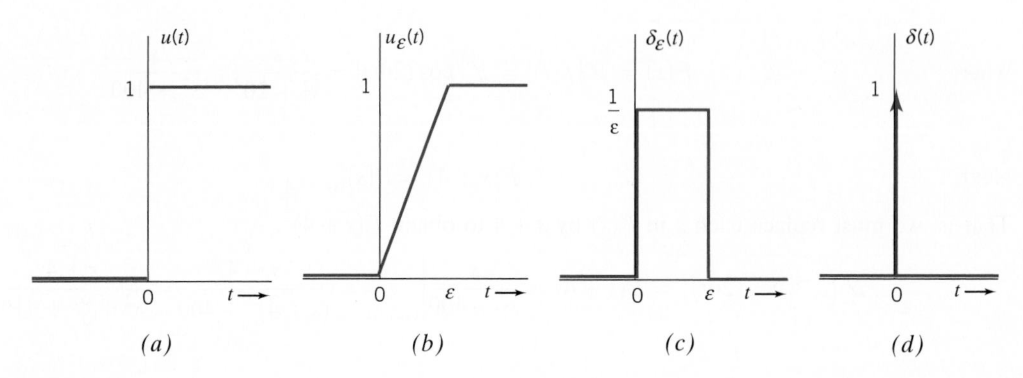

FIGURE 14.3-1 (*a*) A step function, (*b*) an approximation to the step function, (*c*) a pulse function and, (*d*) the impulse function.

Because $\delta(t)$ is undefined at time 0, we consider the function $u_\varepsilon(t)$ shown in Figure 14.3-1*b*. This function makes the transition from 0 to 1 over the time interval from 0 to ε. Notice that

$$\lim_{\varepsilon \to 0} u_\varepsilon(t) = u(t)$$

Let

$$\delta_\varepsilon(t) = \frac{d}{dt} u_\varepsilon(t) = \begin{cases} 0 & t < 0 \\ \frac{1}{\varepsilon} & 0 < t < \varepsilon \\ 0 & t > \varepsilon \end{cases}$$

We see that $\delta_\varepsilon(t)$ is the pulse function shown in Figure 14.3-1*c*. Notice that for any value of ε, the area under the pulse is given by

$$\int_{-\infty}^{+\infty} \delta_\varepsilon(t)dt = \int_0^{\varepsilon} \frac{1}{\varepsilon} dt = 1$$

Now, let

$$\delta(t) = \lim_{\varepsilon \to 0} \delta_\varepsilon(t)$$

This definition of $\delta(t)$ is consistent with the definition given in Eq. 14.3-2. We see that $\delta(t)$ is a pulse having infinite magnitude, infinitesimal duration, and area equal to 1. We can't readily draw such a pulse, so we represent $\delta(t)$ by an arrow as shown in Figure 14.3-1*d*, The height of the arrow is equal to the area of the impulse function. (The area of the impulse function is sometimes called the strength of the impulse. Also, the impulse function is sometimes called the delta function.)

An important property of the impulse function is

$$\int_{-\infty}^{+\infty} f(t)\delta(t)dt = f(0) \tag{14.3-3}$$

Letting $f(t) = 1$ gives

$$\int_{-\infty}^{+\infty} \delta(t)dt = 1$$

showing once again that the area under the impulse function is 1. More interesting, Eq. 14.3-3 can be used to determine the Laplace transform of the impulse function

$$\mathcal{L}[\delta(t)] = \int_{0-}^{\infty} e^{-st}\delta(t)dt = e^0 = 1$$

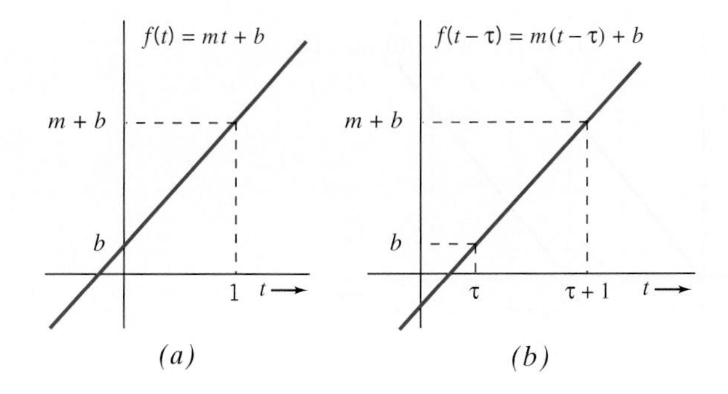

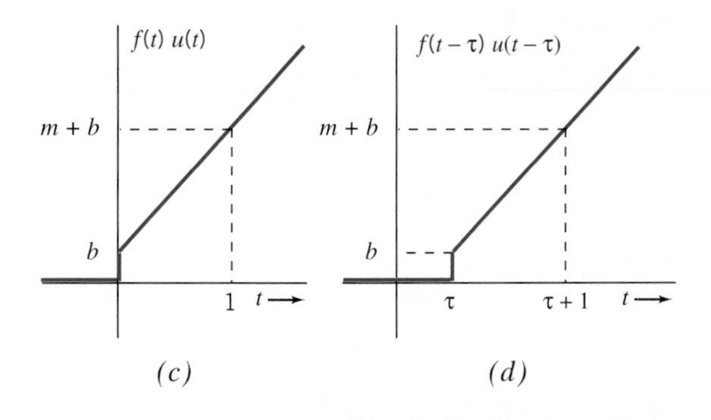

FIGURE 14.3-2 (*a*) A function, (*b*) a delayed copy of the function, (*c*) a new function formed by multiplying $f(t)$ by a step function, and (*d*) a delayed copy of the new function.

Next, we consider some techniques that are useful for finding Laplace transforms of other pulse functions. We can delay a function $f(t)$ by time τ by replacing each occurrence of t by $t - \tau$.

Consider the function

$$f(t) = mt + b$$

shown in Figure 14.3-2*a*. Suppose we wish to shift (delay) it to τ seconds later. This function has a single occurrence of t, so we replace it by $t - \tau$ to obtain

$$f(t - \tau) = m(t - \tau) + b = mt + (b - m\tau)$$

shown in Figure 14.3-2*b*. Next, consider the function

$$g(t) = f(t)u(t) = (mt + b)u(t)$$

This function, shown in Figure 14.3-2*c*, is identical $f(t)$ when $t > 0$ but $g(t) = 0$ when $t < 0$. Suppose we wish to delay $g(t)$ by τ seconds. The function $g(t)$ contains two occurrences of t, and we must replace each occurrence of t by $t - \tau$.

$$g(t - \tau) = (m(t - \tau) + b)u(t - \tau)$$

Shown in Figure 14.3-2*d*, $g(t - \tau)$ is indeed a delayed copy of $g(t)$. Notice that $f(t - \tau)\, u(t - \tau)$ is different than both $f(t - \tau)u(t)$ and $f(t)u(t - \tau)$.

Figure 14.3-3 shows how these techniques can be used to represent pulse functions. Starting with $f(t) = 1.5\, t$, a straight line that passes through the origin in Figure 14.3-3*a*, we multiply by a step function so that the product is 0 for time $t < 0$. The function $f(t)u(t)$ together with a delayed copy, $f(t - 10)u(t - 10)$, are shown in Figure 14.3-3*b*. Subtracting the delayed copy gives

$$g(t) = f(t)\, u(t) - f(t - 10)\, u(t - 10) = 1.5\, t\, u(t) - 1.5(t - 10)\, u(t - 10) \tag{14.3-4}$$

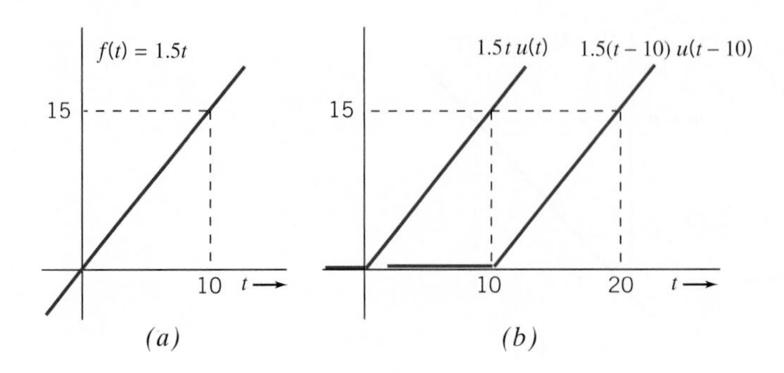

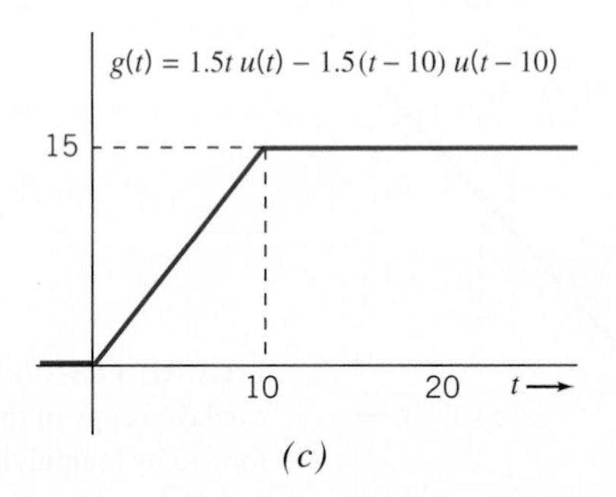

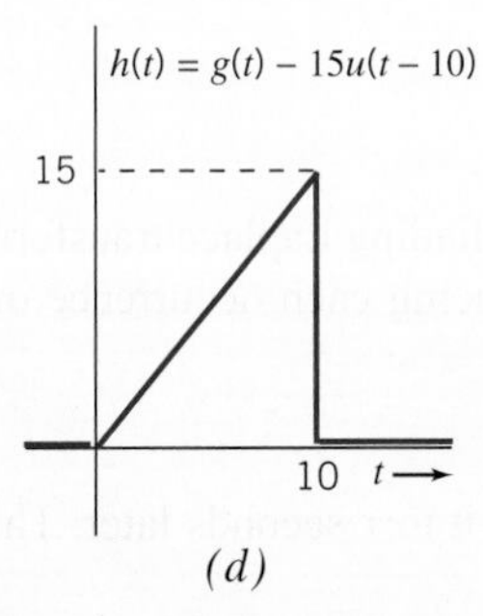

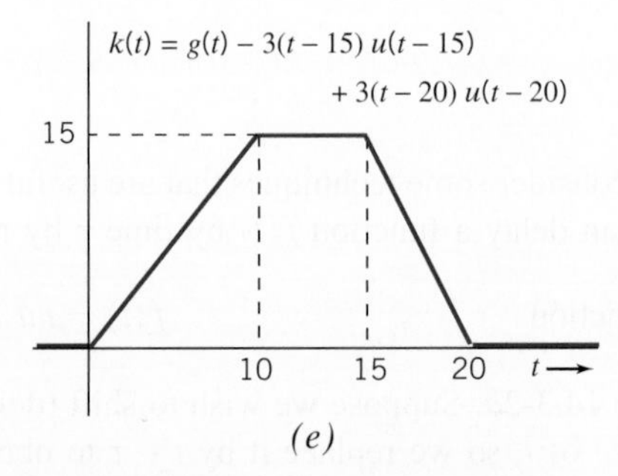

FIGURE 14.3-3 (*a*) A function, (*b*) a ramp function and a delayed copy of the ramp function, (*c*) a new function formed by subtracting the delayed ramp from the ramp, (*d*) a triangular pulse, and (*e*) a trapezoidal pulse.

shown in Figure 14.3-3*c*. Subtracting an appropriately scaled and delayed step function yields the pulse shown in Figure 14.3-3*d*:

$$h(t) = g(t) - 15\,u(t-10) = 1.5\,t\,u(t) - 1.5(t-10)u(t-10) - 15u(t-10) \tag{14.3-5}$$

Alternately, starting with $g(t)$ and then subtracting and adding appropriately scaled and delayed copies of $tu(t)$ yields the pulse shown in Figure 14.3-3*e*:

$$k(t) = g(t) - 3.0(t-15)\,u(t-15) + 3.0(t-20)\,u(t-20) \tag{14.3-6}$$

(Subtracting $3.0(t-15)u(t-15)$ causes $k(t)$ to begin to decrease at $t = 15$ s. Adding $3.0(t-20)\,u\,(t-20)$ causes $k(t)$ to level off at $t = 20$ s. Without this last term, $k(t)$ would continue to decrease.)

To obtain the transform of the time-shifted function, we use the definition of the transform to obtain

$$\mathscr{L}[f(t-\tau)u(t-\tau)] = \int_0^\infty f(t-\tau)u(t-\tau)e^{-st}dt = \int_\tau^\infty f(t-\tau)e^{-st}dt$$

Now let $t - \tau = x$ to obtain

$$\mathscr{L}[f(t-\tau)u(t-\tau)] = \int_0^\infty f(x)e^{-s(\tau+x)}dx = e^{-s\tau}\int_\tau^\infty f(x)e^{-sx}dx = e^{-s\tau}F(s)$$

This result is summarized as

$$f(t-\tau)u(t-\tau) \leftrightarrow e^{-s\tau}F(s) \tag{14.3-7}$$

EXAMPLE 14.3-1 Laplace Transforms of Pulse Functions

Find the Laplace transforms of $g(t)$, $h(t)$ and $k(t)$ shown in Figure 14.3-3.

Solution

After obtaining Eqs. 14.3-4, 14.3-5, and 14.3-6, the required Laplace transforms are easily determined using Eq. 14.3-7:

$$\begin{aligned} G(s) = \mathscr{L}[g(t)] &= \mathscr{L}[1.5t\,u(t)] - \mathscr{L}[1.5(t-10)\,u(t-10)] \\ &= 1.5\left(\frac{1}{s^2}\right) - e^{-10s}\left(1.5\left(\frac{1}{s^2}\right)\right) = \frac{1.5(1-e^{-10s})}{s^2} \end{aligned}$$

$$H(s) = \mathscr{L}[h(t)] = \mathscr{L}[g(t)] - \mathscr{L}[15\,u(t-10)] = \frac{1.5(1-e^{-10s})}{s^2} - e^{-10s}\left(\frac{15}{s}\right)$$

$$\begin{aligned} K(s) &= \mathscr{L}[k(t)] = \mathscr{L}[g(t)] - \mathscr{L}[3.0(t-15)\,u(t-15)] + \mathscr{L}[3.0(t-20)\,u(t-20)] \\ &= \frac{1.5(1-e^{-10s})}{s^2} - e^{-15s}\left(\frac{3.0}{s^2}\right) + e^{-20s}\left(\frac{3.0}{s^2}\right) = \frac{1.5(1-e^{-10s}-2e^{-15s}+2e^{-20s})}{s^2} \end{aligned}$$

14.4 INVERSE LAPLACE TRANSFORM

We will frequently want to find the inverse Laplace transform of a function represented as a ratio of polynomials in s. Consider:

$$F(s) = \frac{N(s)}{D(s)} = \frac{b_m s^m + b_{m-1}s^{m-1} + \cdots + b_1 s + b_0}{s^n + a_{n-1}s^{n-1} + \cdots + a_1 s + a_0} \tag{14.4-1}$$

where the coefficients of the polynomials are real numbers. The function $F(s)$ is said to be a rational function of s because it is the ratio of two polynomials in s. Usually, we have $n > m$, in which case, $F(s)$ is called a proper rational function.

The roots of the denominator polynomial $D(s)$ are the roots of the equation $D(s) = 0$ and are called the **poles** of $F(s)$. Factoring $D(s)$, we obtain

$$D(s) = s^n + a_{n-1}s^{n-1} + \cdots + a_1 s + a_0 = (s-p_1)(s-p_2)\cdots(s-p_n)$$

The poles, p_i, may be either real or complex. Complex poles appear in complex conjugate pairs, that is, if $p_1 = a + jb$ is a pole of $F(s)$, then $F(s)$ will also have a pole, $p_i = p_1^* = a - jb$. A pole p_i of $F(s)$ is said to be a simple pole of $F(s)$ if none of the other poles of $F(s)$ are equal to p_i. In contrast, p_i is a repeated pole of $F(s)$ if at least one of the other poles of $F(s)$ is equal to p_i. The multiplicity of a repeated pole p_i is the number of equal poles, including p_i itself. The roots of the numerator polynomial $N(s)$ are called the **zeros** of $F(s)$.

We will find the inverse Laplace transform of a proper rational function $F(s)$ in three steps. First, we perform a partial fraction expansion to express $F(s)$ as a sum of simpler functions, $F_i(s)$.

$$F(s) = F_1(s) + F_2(s) + \cdots + F_i(s) + \cdots + F_n(s)$$

Next, we use the transform pairs in Table 14.2-1 and properties in Table 14.2-2 to find the inverse Laplace transform of each $F_i(s)$. Finally, using linearity, we sum the inverse transforms of the $F_i(s)$ to obtain the inverse Laplace transform of $F(s)$.

When all of the poles of a proper rational function, $F(s)$, are simple poles, the partial fraction expansion of $F(s)$ is

$$\begin{aligned} F(s) &= \frac{N(s)}{D(s)} = \frac{b_m s^m + b_{m-1}s^{m-1} + \cdots + b_1 s + b_0}{s^n + a_{n-1}s^{n-1} + \cdots + a_1 s + a_0} \\ &= \frac{R_1}{s - p_1} + \frac{R_2}{s - p_2} + \cdots + \frac{R_i}{s - p_i} + \cdots + \frac{R_n}{s - p_n} \end{aligned} \tag{14.4-2}$$

The partial fraction expansion has one term corresponding to each simple pole of $F(s)$. The coefficients, R_i are called residues. Each residue, R_i, corresponds to the pole, p_i, in the same term of Eq. 14.4-2. The residue corresponding to a real pole is a real number. The residues corresponding to complex conjugate poles are themselves complex conjugates. The values of the residues of simple poles are calulated as

$$R_i = (s - p_i)F(s)\big|_{s=p_i} \tag{14.4-3}$$

EXAMPLE 14.4-1 Inverse Laplace Transform: Simple, Real Poles

Find the inverse Laplace transform of $F(s) = \dfrac{s+3}{s^2 + 7s + 10}$.

Solution

The given $F(s)$ is indeed a proper rational function. Factor the denominator and perform a partial fraction expansion.

$$F(s) = \frac{s+3}{s^2 + 7s + 10} = \frac{s+3}{(s+2)(s+5)} = \frac{R_1}{s+2} + \frac{R_2}{s+5}$$

where

$$R_1 = (s+2)\left(\frac{s+3}{(s+2)(s+5)}\right)\bigg|_{s=-2} = \frac{s+3}{s+5}\bigg|_{s=-2} = \frac{-2+3}{-2+5} = \frac{1}{3}$$

and

$$R_2 = (s+5)\left(\frac{s+3}{(s+2)(s+5)}\right)\bigg|_{s=-5} = \frac{s+3}{s+2}\bigg|_{s=-5} = \frac{-5+3}{-5+2} = \frac{2}{3}$$

Then
$$F(s) = \frac{\frac{1}{3}}{s+2} + \frac{\frac{2}{3}}{s+5}$$

Using linearity and taking the inverse Laplace transform of each term gives

$$F(t) = \mathcal{L}^{-1}[F(s)] = \mathcal{L}^{-1}\left[\frac{\frac{1}{3}}{s+2} + \frac{\frac{2}{3}}{s+5}\right] = \frac{1}{3}\mathcal{L}^{-1}\left[\frac{1}{s+2}\right] + \frac{2}{3}\mathcal{L}^{-1}\left[\frac{1}{s+5}\right] = \frac{1}{3}e^{-2t} + \frac{2}{3}e^{-5t} \text{ for } t \geq 0$$

Suppose $F(s)$ has a pair of simple complex conjugate poles $p_1 = -a + jb$ and $p_2 = -a - jb$. The corresponding residues in the partial fraction expansion will also be complex conjugates, say $R_1 = c + jd$ and $R_2 = c - jd$. The partial fraction expansion of $F(s)$ is

$$F(s) = \frac{R_1}{s - p_1} + \frac{R_2}{s - p_2} + F_3(s) = \frac{c + jd}{s - (-a + jb)} + \frac{c - jd}{s - (-a - jb)} + F_3(s) \tag{14.4-4}$$

where $F_3(s)$ is the sum of the terms of the partial fraction expansion due to other poles of $F(s)$. Next, combine the first two terms, using a common denominator, to get

$$\begin{aligned} F(s) &= \frac{c + jd}{s + a - jb} + \frac{c - jd}{s + a + jb} + F_3(s) \\ &= \frac{(c + jd)(s + a - jb) + (c - jd)(s + a - jb)}{(s + a - jb)(s + a + jb)} + F_3(s) \\ &= \frac{2c\, s + 2(a\, c - b\, d)}{s^2 + 2\, a\, s + a^2 + b^2} + F_3(s) \\ &= \frac{2\, c(s + a) - 2\, b\, d}{(s + a)^2 + b^2} + F_3(s) \\ &= 2c\frac{s + a}{(s + a)^2 + b^2} - 2d\frac{b}{(s + a)^2 + b^2} + F_3(s) \end{aligned}$$

Notice that the partial fraction expansion of $F(s)$ can be expressed as

$$F(s) = \frac{K_1 s + K_2}{s^2 + 2\, a\, s + a^2 + b^2} + F_3(s) \tag{14.4-5}$$

where $K_1 = 2\, c$ and $K_2 = 2(a\, c - b\, d)$.

Taking the inverse Laplace transform of the first two terms of the partial fraction expansion gives

$$\mathcal{L}^{-1}\left[2\, c\frac{s + a}{(s + a)^2 + b^2}\right] = 2\, c\, \mathcal{L}^{-1}\left[\frac{s + a}{(s + a)^2 + b^2}\right] = 2\, c\, e^{-at}\, \mathcal{L}^{-1}\left[\frac{s}{s^2 + b^2}\right] = 2\, c\, e^{-at} \cos(bt)$$

and

$$\mathcal{L}^{-1}\left[2\, d\frac{b}{(s + a)^2 + b^2}\right] = 2\, d\, \mathcal{L}^{-1}\left[\frac{b}{(s + a)^2 + b^2}\right] = 2\, d\, e^{-at}\, \mathcal{L}^{-1}\left[\frac{b}{s^2 + b^2}\right] = 2\, d\, e^{-at} \sin(bt)$$

Using linearity, we have

$$\mathcal{L}^{-1}[F(s)] = 2\, c\, e^{-at}\cos(bt) - 2\, d\, e^{-at}\sin(bt) + \mathcal{L}^{-1}[F_3(s)] \tag{14.4-6}$$

EXAMPLE 14.4-2 Inverse Laplace Transform: Simple Complex Poles

Find the inverse Laplace transform of $F(s) = \dfrac{10}{(s^2+6s+10)(s+2)}$.

Solution

The roots of the quadratic $(s^2+6s+10)$ are complex, and we may write $F(s)$ as

$$F(s) = \frac{10}{(s+3-j)(s+3+j)(s+2)}$$

Using a partial fraction expansion, we have

$$F(s) = \frac{10}{(s+3-j)(s+3+j)(s+2)} = \frac{R_1}{s-(-3+j)} + \frac{R_2}{s-(3-j)} + \frac{R_3}{s+2}$$

Using Eq. 14.4-3,

$$\begin{aligned} R_1 &= (s+3-j)\left(\frac{10}{(s+3-j)(s+3+j)(s+2)}\right)\bigg|_{s=-3+j} \\ &= \frac{10}{(s+3+j)(s+2)}\bigg|_{s=-3+j} = \frac{10}{(-3+j+3+j)(-3+j+2)} = -\frac{5}{2}+j\frac{5}{2} \end{aligned}$$

Comparing to Eq. 14.4-4, we see that $a=3$, $b=1$, $c=-2.5$, and $d=2.5$. Next,

$$\begin{aligned} R_2 &= (s+3+j)\left(\frac{10}{(s+3-j)(s+3+j)(s+2)}\right)\bigg|_{s=-3-j} \\ &= \frac{10}{(s+3-j)(s+2)}\bigg|_{s=-3-j} = \frac{10}{(-3-j+3-j)(-3-j+2)} = -\frac{5}{2}-j\frac{5}{2} \end{aligned}$$

and

$$R_3 = (s+2)\left(\frac{10}{(s+3-j)(s+3+j)(s+2)}\right)\bigg|_{s=-2} = \frac{10}{s^2+6s+10}\bigg|_{s=-2} = 5$$

Finally, using Eq. 14.4-6,

$$\begin{aligned} f(t) = \mathcal{L}^{-1}\left[\frac{10}{(s^2+6s+10)(s+2)}\right] &= 2\,c\,e^{-at}\cos(bt) - 2\,d\,e^{-at}\sin(bt) + \mathcal{L}^{-1}\left[\frac{5}{s+2}\right] \\ &= 2(-2.5)e^{-3t}\cos(1t) - 2(2.5)e^{-3t}\sin(1t) + 5e^{-2t} \\ &= -5e^{-3t}\cos(t) - 5e^{-3t}\sin(t) + 5\,e^{-2t} \quad \text{for } t \ge 0 \end{aligned}$$

Alternate Solution

Using Eq. 14.4-5, we can express $F(s)$ as

$$F(s) = \frac{10}{(s^2+6s+10)(s+2)} = \frac{K_1\,s+K_2}{s^2+6s+10} + F_3(s) = \frac{K_1 s+K_2}{s^2+6s+10} + \frac{R_3}{s+2}$$

Using Eq. 14.4-3, we calulate

$$R_3 = (s+2)\left(\frac{10}{(s^2+6s+10)(s+2)}\right)\bigg|_{s=-2} = \frac{10}{s^2+6s+10}\bigg|_{s=-2} = 5$$

Then $$\frac{10}{(s^2+6s+10)(s+2)} = \frac{K_1 s + K_2}{s^2+6s+10} + \frac{5}{s+2} \tag{14.4-7}$$

Multiplying both sides of this equation by the denominator of $F(s)$ gives

$$10 = (K_1+5)s^2 + (2K_1+K_2+30)s + 2K_2 + 50$$

The coefficients of s^2, s^1, and s^0 on the right side of this equation must each be equal to the corresponding coefficients on the left side. (The coefficients of s^2 and s^1 on the left side are zero.) Equating corresponding coefficients gives

$$0 = K_1 + 5, \quad 0 = 2K_1 + K_2 + 30 \quad \text{and} \quad 10 = 2K_2 + 50$$

Solving these equations gives $K_1 = -5$ and $K_2 = -20$. Substituting into Eq 14.4-7 gives

$$\frac{10}{(s^2+6s+10)(s+2)} = \frac{-5s-20}{s^2+6s+10} + \frac{5}{s+2}$$

Next,

$$\frac{-5s-20}{s^2+6s+10} = \frac{-5s-20}{(s^2+6s+9)+1} = \frac{-5s-20}{(s+3)^2+1} = \frac{-5(s+3)-5}{(s+3)^2+1} = -5\left(\frac{s+3}{(s+3)^2+1}\right) - 5\left(\frac{1}{(s+3)^2+1}\right)$$

Then $$\begin{aligned}\mathcal{L}^{-1}\left[\frac{-5s-20}{s^2+6s+10}\right] &= -5\mathcal{L}^{-1}\left[\frac{s+3}{(s+3)^2+1}\right] - 5\mathcal{L}^{-1}\left[\frac{1}{(s+3)^2+1}\right] \\ &= -5e^{-3t}\cos(t) - 5e^{-3t}\sin(t)\end{aligned}$$

Using superposition,

$$f(t) = \mathcal{L}^{-1}\left[\frac{10}{(s^2+6s+10)(s+2)}\right] = -5e^{-3t}\cos(t) - 5e^{-3t}\sin(t) + 5e^{-2t} \text{ for } t \ge 0$$

as before.

Next, suppose $F(s)$ has repeated poles, that is,

$$F(s) = \frac{N(s)}{D(s)} = \frac{b_m s^m + b_{m-1}s^{m-1} + \cdots + b_1 s + b_0}{s^n + a_{n-1}s^{n-1} + \cdots + a_1 s + a_0} = \frac{b_m s^m + b_{m-1}s^{m-1} + \cdots + b_1 s + b_0}{(s-p_1)^q\left(s-p_{q+1}\right)\cdots(s-p_n)}$$

where the integer q is called the multiplicity of the repeated pole, p_1. In this case, the partial fraction expansion of $F(s)$ that includes all powers of the term $(s-p_1)$ up to the multiplicity.

$$F(s) = \frac{R_1}{s-p_1} + \frac{R_2}{(s-p_1)^2} + \cdots + \frac{R_q}{(s-p_1)^q} + \frac{R_{q+1}}{s-p_{q+1}} + \cdots + \frac{R_n}{s-p_n} \tag{14.4-8}$$

The residues corresponding to the repeated poles are given by

$$R_{q-k} = \frac{1}{k!}\left[\frac{d^k}{ds^k}(s-p_1)^q F(s)\right]\Bigg|_{s=p_1} \quad \text{for } k = q-1,\ q-2, \ldots, 2,\ 1,\ 0 \tag{14.4-9}$$

That is,

$$R_1 = \frac{1}{(q-1)!}\left[\frac{d^{q-1}}{ds^{q-1}}(s-p_1)^q F(s)\right]\Bigg|_{s=p_1},$$

$$R_2 = \frac{1}{(q-2)!}\left[\frac{d^{q-2}}{ds^{q-2}}(s-p_1)^q F(s)\right]\Bigg|_{s=p_1}, \ldots$$

$$R_q = \left[(s-p_1)^q F(s)\right]\Big|_{s=p_1}$$

EXAMPLE 14.4-3 Inverse Laplace Transform: Repeated Poles

Find the inverse Laplace transform of $F(s) = \dfrac{4}{(s+1)^2(s+2)}$.

Solution

Using Eq. 14.4-8, we can express $F(s)$ as

$$F(s) = \frac{4}{(s+1)^2(s+2)} = \frac{R_1}{s+1} + \frac{R_2}{(s+1)^2} + \frac{R_3}{s+2}$$

Using Eq. 14.4-3,

$$R_3 = (s+2)\frac{4}{(s+1)^2(s+2)}\bigg|_{s=-2} = \frac{4}{(s+1)^2}\bigg|_{s=-2} = \frac{4}{(-2+1)^2} = 4$$

Using Eq. 14.4-9,

$$R_1 = \frac{d}{ds}\left((s+1)^2\frac{4}{(s+1)^2(s+2)}\right)\bigg|_{s=-1} = \frac{d}{ds}\frac{4}{s+2}\bigg|_{s=-1} = \frac{-4}{(s+2)^2}\bigg|_{s=-1} = -4$$

and

$$R_2 = (s+1)^2\frac{4}{(s+1)^2(s+2)}\bigg|_{s=-1} = \frac{4}{s+2}\bigg|_{s=-1} + \frac{4}{-1+2} = 4$$

Then,

$$F(s) = \frac{4}{(s+1)^2(s+2)} = \frac{-4}{s+1} + \frac{4}{(s+1)^2} + \frac{4}{s+2}$$

Next, using the frequency shift property from Table 14.2-2, we get

$$\mathscr{L}^{-1}\left[\frac{4}{(s+1)^2}\right] = e^{-t}\,\mathscr{L}^{-1}\left[\frac{4}{s^2}\right] = 4\,t\,e^{-t}$$

Finally, using linearity, $\quad f(t) = -4\,e^{-t} + 4\,t\,e^{-t} + 4e^{-2t} \quad \text{for } t \geq 0$

Alternate Solution

Using Eq.14.4-8,

$$F(s) = \frac{4}{(s+1)^2(s+2)} = \frac{R_1}{s+1} + \frac{R_2}{(s+1)^2} + \frac{R_3}{s+2}$$

As before,

$$R_3 = (s+2)\frac{4}{(s+1)^2(s+2)}\bigg|_{s=-2} = \frac{4}{(s+1)^2}\bigg|_{s=-2} = \frac{4}{(-2+1)^2} = 4$$

and

$$R_2 = (s+1)^2\frac{4}{(s+1)^2(s+2)}\bigg|_{s=-1} = \frac{4}{s+2}\bigg|_{s=-1} = \frac{4}{-1+2} = 4$$

so

$$\frac{4}{(s+1)^2(s+2)} = \frac{R_1}{s+1} + \frac{4}{(s+1)^2} + \frac{4}{s+2}$$

Multiplying both sides by $(s+1)^2(s+2)$ gives

$$4 = R_1(s+1)(s+2) + 4(s+2) + 4(s+1)^2 = (R_1+4)s^2 + (3R_1+4+8)s + 2R_1 + 8 + 4$$

The coefficients of s^2, s^1, and s^0 on the right side of this equation must each be equal to the corresponding coefficients on the left side. (The coefficients of s^2 and s^1 on the left side are zero) Equating corresponding coefficients gives

$$0 = R_1 + 4,\ 0 = 3\,R_1 + 4 + 8 \text{ and } 4 = 2\,R_1 + 8 + 4$$

Solving these equations gives $R_1 = -4$. Substituting gives

$$F(s) = \frac{4}{(s+1)^2(s+2)} = \frac{-4}{s+1} + \frac{4}{(s+1)^2} + \frac{4}{s+2}$$

As before $$f(t) = -4e^{-t} + 4t\,e^{-t} + 4\,e^{-2t} \quad \text{for } t \ge 0$$

EXAMPLE 14.4-4 Inverse Laplace Transform: Improper Rational Function

Find the inverse Laplace transform of $F(s) = \dfrac{4\,s^3 + 15\,s^2 + s + 30}{s^2 + 5\,s + 6}$.

Solution

Compare this $F(s)$ with $F(s)$ in Eq. 14.4-1 to see that $m = 3$ and $n = 2$. Because m is not less than n we perform the long division $s^2 + 5s + 6\,\overline{)4s^3 + 15s^2 + s + 30}$ to obtain

$$F(s) = 4s - 5 + \frac{2s}{s^2 + 5s + 6}$$

The last term on the right side is a proper rational function, so we perform partial fraction expansion to get

$$F(s) = 4s - 5 + \frac{2s}{s^2 + 5s + 6} = 4s - 5 + \frac{2s}{(s+3)(s+2)} = 4s - 5 + \frac{6}{s+3} - \frac{4}{s+2}$$

Using the time differentiation property from Table 14.4-4 gives $\mathscr{L}^{-1}[s] = \dfrac{d}{dt}\delta(t)$. Using linearity, we get

$$\mathscr{L}^{-1}\left[\frac{4s^3 + 15s^2 + s + 30}{s^2 + 5s + 6}\right] = 4\frac{d}{dt}\delta(t) - 5\delta(t) + 6e^{-3t} - 4e^{-2t} \quad \text{for } t \ge 0$$

14.5 INITIAL AND FINAL VALUE THEOREMS

The initial value of a function $f(t)$ is the value $t = 0$, provided that $f(t)$ is continuous at $t = 0$. If $f(t)$ is discontinuous at $t = 0$, the initial value is the limit as $t \to 0^+$, where t approaches $t = 0$ from positive time.

A function's **initial value** may be found using

$$f(0+) = \lim_{t \to 0+} f(t) = \lim_{s \to \infty} sF(s) \tag{14.5-1}$$

This equation is called the **initial value theorem**. To prove the initial value theorem, we start with the time differentiation property from Table 14.2-2:

$$sF(s) - f(0-) = \mathscr{L}\left[\frac{df}{dt}\right] = \int_{0-}^{\infty} \frac{df}{dt} e^{-st}\,dt$$

Taking the limit as $s \to \infty$, we get

$$\lim_{s\to\infty}[sF(s) - f(0-)] = \lim_{s\to\infty}\int_{0-}^{0+}\frac{df}{dt}e^{-st}dt + \lim_{s\to\infty}\int_{0+}^{\infty}\frac{df}{dt}e^{-st}dt$$

The first integral on the right is equal to $f(0+) - f(0-)$ because $e^{-st} = 1$ for t between $0-$ and $0+$. The second integral on the right vanishes because $e^{-st} \to 0$ for $s \to \infty$. On the left side, $\lim_{s\to\infty} f(0-) = f(0-)$ because $f(0-)$ is independent of s. Thus,

$$\lim_{s\to\infty} sF(s) - f(0-) = f(0+) - f(0-)$$

Adding $f(0-)$ to each side confirms the initial value theorem given in Eq. 14.5-1.

The **final value** of a function $f(t)$ is $\lim_{t\to\infty} f(t)$ where

$$f(\infty) = \lim_{t\to\infty} f(t) = \lim_{s\to\infty} sF(s) \tag{14.5-2}$$

This equation is called the **final value theorem**. To prove the final value theorem, we again start with the time differentiation property from Table 14.2-2:

$$sF(s) - f(0-) = \mathscr{L}\left[\frac{df}{dt}\right] = \int_0^\infty\left(\frac{df}{dt}\right)e^{-st}dt$$

and we take the limit as $s \to 0$ for both sides to obtain

$$\lim_{s\to 0}[sF(s) - f(0-)] = \lim_{s\to 0}\int_0^\infty\left(\frac{df}{dt}\right)e^{-st}dt = \int_0^\infty\left(\frac{df}{dt}\right)e^{-0t}dt = f(\infty) - f(0-)$$

On the left side, $\lim_{s\to 0} f(0-) = f(0-)$ because $f(0-)$ is independent of s. Thus,

$$\lim_{s\to 0} sF(s) - f(0-) = f(\infty) - f(0-)$$

Adding $f(0-)$ to each side confirms the final value theorem given in Eq. 14.5-2.

EXAMPLE 14.5-1 Initial and Final Value Theorems

Consider the situation in which we build a circuit in the laboratory and analyze the same circuit, using Laplace transfoms. Figure 14.5-1 shows a plot of the circuit output, $v(t)$, obtained by laboratory measurement. Suppose our circuit analysis gives

$$V(s) = \mathscr{L}[v(t)] = \frac{2s^2 + 30s + 136}{s(s^2 + 9s + 34)} \tag{14.5-3}$$

Does the circuit analysis agree with the laboratory measurement?

Solution

Determining the inverse Laplace transform of $V(s)$ requires a partial fraction expansion. Before we do that work, let's use the initial- and final value theorems to see whether it is possible that $V(s)$, given in Eq. 14.5-3, can be the Laplace transform $v(t)$ shown in Figure 14.5-1.

From Figure 14.5-1, we see that the initial and final values are

$$v(0+) = \lim_{t\to 0+} v(t) = 2\text{ V} \quad\text{and}\quad v(\infty) = \lim_{t\to\infty} v(t) = 4\text{ V} \tag{14.5-4}$$

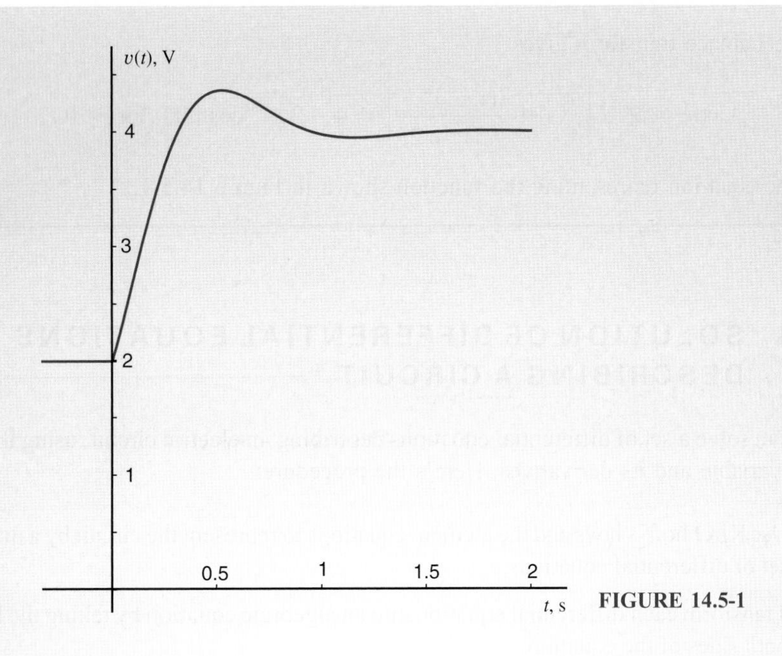

FIGURE 14.5-1

Next, we calculate

$$v(0) = \lim_{s\to\infty} s\left(\frac{2s^2 + 30s + 136}{s(s^2 + 9s + 34)}\right) = \lim_{s\to\infty} \frac{2s^2 + 30s + 136}{s^2 + 9s + 34} = \lim_{s\to\infty} \frac{\dfrac{2s^2}{s^2} + \dfrac{30s}{s^2} + \dfrac{136}{s^2}}{\dfrac{s^2}{s^2} + \dfrac{9s}{s^2} + \dfrac{34}{s^2}} = \frac{2}{1} = 2\text{ V}$$

and

$$v(\infty) = \lim_{s\to 0} s\left(\frac{2s^2 + 30s + 136}{s(s^2 + 9s + 34)}\right) = \lim_{s\to 0} \frac{2s^2 + 30s + 136}{s^2 + 9s + 34} = \frac{136}{24} = 4\text{ V}$$

Because these initial and final values agree, it is possible that $V(s)$, given in Eq. 14.5-3, can be the Laplace transform of $v(t)$ shown in Figure 14.5-1. It is now appropriate to determine the inverse Laplace transform of $V(s)$.

We can express $V(s)$ as

$$V(s) = \frac{2s^2 + 30s + 136}{s(s^2 + 9s + 34)} = \frac{K_1 s + K_2}{s^2 + 9s + 34} + \frac{R_3}{s}$$

where
$$R_3 = s\left(\frac{2s^2 + 30s + 136}{s(s^2 + 9s + 34)}\right)\bigg|_{s=0} = \frac{2s^2 + 30s + 136}{s^2 + 9s + 34}\bigg|_{s=0} = 4$$

Then
$$V(s) = \frac{2s^2 + 30s + 136}{s(s^2 + 9s + 34)} = \frac{K_1 s + K_2}{s^2 + 9s + 34} + \frac{4}{s}$$

Multiplying both sides $s(s^2 + 9s + 34)$ gives

$$2s^2 + 30s + 136 = s(K_1 s + K_2) + 4(s^2 + 9s + 34) = (K_1 + 4)s^2 + (K_2 + 36)s + 136$$

Equating the coefficients of s^2 and s^1 gives $K_1 = -2$ and $K_2 = -6$. Then,

$$V(s) = \frac{2s^2 + 30s + 136}{s(s^2 + 9s + 34)} = \frac{4}{s} - \frac{2s + 6}{s^2 + 9s + 34} = \frac{4}{s} - \frac{2(s + 3)}{(s + 3)^2 + 25}$$

Taking the inverse Laplace transform gives

$$v(t) = \mathscr{L}^{-1}\left[\frac{4}{s} - \frac{2(s+3)}{(s+3)^2+25}\right] = 4 - 2\,e^{-3t}\cos(5t) \ \text{ for } \geq 0$$

which is indeed the equation representing the function shown in Figure 14.5-1.

14.6 SOLUTION OF DIFFERENTIAL EQUATIONS DESCRIBING A CIRCUIT

We can solve a set of differential equations describing an electric circuit, using the Laplace transform of a variable and its derivatives. Here's the procedure:

1. Use Kirchhoff's laws and the element equations to represent the circuit by a differential equation or set of differential equations.
2. Transform each differential equation into an algebraic equation by taking the Laplace transform of both sides of the equation.
3. Solve the algebraic equations to obtain the Laplace transform of the output of the circuit.
4. Take the inverse Laplace transform to obtain the circuit output itself.

The following example illustrates this procedure.

EXAMPLE 14.6-1 Laplace Transforms of Differential Equations

Find $v_C(t)$ for the circuit shown in Figure 14.6-1 when $i_L(0-) = 0.5$ A and $v_C(0-) = 2.5$ V.

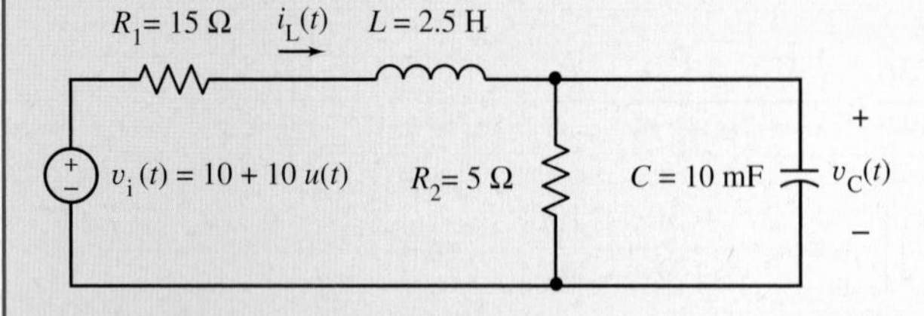

FIGURE 14.6-1 The circuit considered in Example 14.6-1.

Solution

Apply KCL at the top node of R_2 to get

$$i_L(t) = \frac{v_C(t)}{R_2} + C\frac{d\,v_C(t)}{dt} \tag{14.6-1}$$

Apply KVL to the left mesh to get

$$v_1(t) = R_1 i_L(t) + L\frac{di_L(t)}{dt} + v_C(t) \tag{14.6-2}$$

Recall this Laplace transform property from Table 14.2-2:

$$\frac{df}{dt} \leftrightarrow sF(s) - f(0^-)$$

Take the Laplace transform of both sides of Eq. 14.6-1 to get

$$I_L(s) = \frac{V_C(s)}{R_2} + C(V_C(s) - v_C(0-)) \tag{14.6-3}$$

Take the Laplace transform of both sides of Eq. 14.6-2 to get

$$V_i(s) = R_1 I_L(s) + L(I_L(s) - i_L(0-)) + V_C(s) \tag{14.6-4}$$

Substitute the expression for $I_L(s)$ from Eq. 14.6-3 into Eq. 14.6-4 and simplify to get

$$V_i(s) = \left(LCs^2 + \left(\frac{L}{R_2} + R_1 C\right)s + 1 + \frac{R_1}{R_2}\right)V_C(s) - (LCs + R_1 C)v_C(0-) - Li_L(0-) \tag{14.6-5}$$

Noticing that $v_i = 20$ V for $t > 0$, we determine $V_i(s) = \mathscr{L}[20] = \frac{20}{s}$. Then, using the given values of the initial conditions and of the circuit parameters, we obtain

$$\frac{20}{s} = (s^2 + 26s + 160)V_C(s) - (s + 6)(2.5) - 2.5(0.5)$$

Solving for $V_C(s)$ gives

$$V_C(s) = \frac{2.5s^2 + 65s + 800}{s(s^2 + 26s + 160)} = \frac{2.5s^2 + 65s + 800}{s(s + 10)(s + 16)}$$

Performing partial fraction expansion gives

$$V_C(s) = \frac{2.5s^2 + 65s + 800}{s(s + 10)(s + 16)} = \frac{5}{s} + \frac{4.17}{s + 16} - \frac{6.67}{s + 10}$$

Taking the inverse Laplace transform gives

$$v_C(t) = 5 + 4.17e^{-16t} - 6.67e^{-10t} \text{ V for } t > 0$$

14.7 CIRCUIT ANALYSIS USING IMPEDANCE AND INITIAL CONDITIONS

We have seen that we can represent a circuit in the time domain by differential equations and then use the Laplace transform to transform the differential equations into algebraic equations. In this section, we will see that we can represent a circuit in the frequency domain, using the Laplace transform, and then analyze it using algebraic equations. This method will eliminate the need to write differential equations to represent the circuit.

The v–i relationship for the resistor is Ohm's law:

$$v(t) = i(t)R \tag{14.7-1}$$

Therefore, the Laplace transform relationship for a resistor R is

$$V(s) = I(s)R \tag{14.7-2}$$

Figure 14.7.1 shows the representation of the resistor in (*a*) the time domain and (*b*) the frequency domain, using the Laplace transform. As the above equations suggest, the time- and frequency-domain representations of the resistor are very similar.

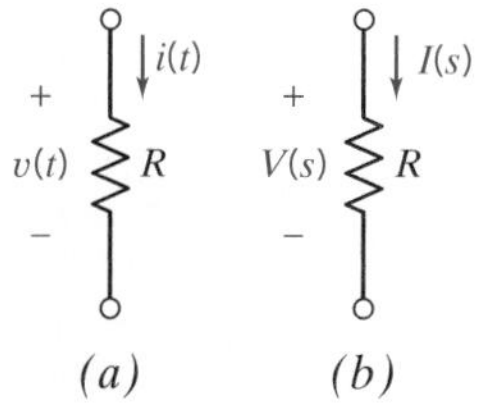

FIGURE 14.7-1 A resistor represented (*a*) in the time domain and (*b*) in the frequency domain using the Laplace transform.

The impedance of an element is defined to be

$$Z(s) = \frac{V(s)}{I(s)} \tag{14.7-3}$$

provided all initial conditions are zero. Notice that the impedance is defined in the frequency domain, not in the time domain.

In the case of the resistor, there is no initial condition to set to zero. Comparing Eqs. 14.7-1 and 14.7-2 shows that the impedance of the resistor is equal to the resistance.

A capacitor is represented by its time-domain equation

$$v(t) = \frac{1}{C}\int_0^t i(\tau)d\tau + v(0) \tag{14.7-4}$$

The Laplace transform of Eq. 14.7-4 is

$$V(s) = \frac{1}{Cs}I(s) + \frac{v(0)}{s} \tag{14.7-5}$$

To determine the impedance of the capacitor, set the initial condition, $v(0)$, to zero. Then, using Eq. 14.7-3, we obtain

$$Z_C(s) = \frac{1}{Cs}$$

as the impedance of the capacitor.

Equation 14.7-5 is used to represent the capacitor in the frequency domain, as shown in Figure 14.7-2*b*. The series connection of elements in Figure 14.7-2*b* corresponds to the sum of voltages in Eq. 14.7-5. The current through the impedance in Figure 14.7-2*b* produces the first voltage on the right side of Eq. 14.7-5, whereas the voltage source in Figure 14.7-2*b* supplies the second voltage on the right side of Eq. 14.7-5.

Solving Eq. 14.7-5 for $I(s)$ gives

$$I(s) = CsV(s) - Cv(0) \tag{14.7-6}$$

Equation 14.7-6 represents the capacitor in the frequency domain, as shown in Figure 14.7-2*c*. The parallel connection of elements in Figure 14.7-2*c* corresponds to the sum of currents in Eq. 14.7-6. The voltage across the impedance in Figure 14.7-2*b* produces the first current on the right

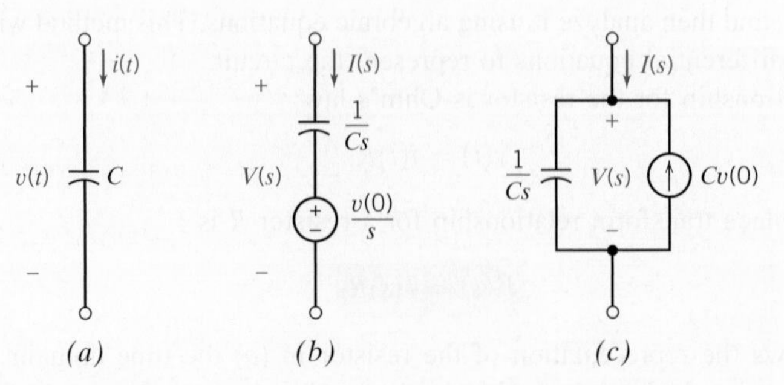

FIGURE 14.7-2 A capacitor represented (*a*) in the time domain and (*b*) in the frequency domain, using the Laplace transform. (*c*) An alternate frequency-domain representation.

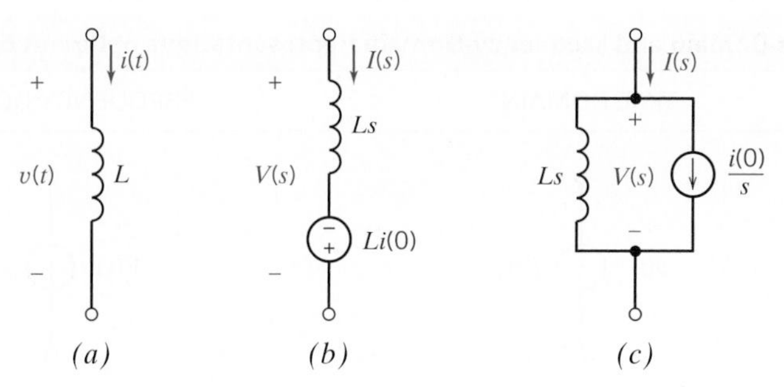

FIGURE 14.7-3 An inductor represented (*a*) in the time domain and (*b*) in the frequency domain, using the Laplace transform. (*c*) An alternate frequency-domain representation.

side of Eq. 14.7-6, whereas the current source in Figure 14.7-2*b* supplies the current on the right side of Eq. 14.7-6. Notice that the reference direction for the current source in Figure 14.7-2*b* was chosen to correspond to the minus sign in Eq. 14.7-6.

An inductor is represented by its time-domain equation,

$$v(t) = L\frac{d}{dt}i(t) \tag{14.7-7}$$

The Laplace transform of Eq. 14.7-7 is

$$V(s) = LsI(s) - Li(0) \tag{14.7-8}$$

To determine the impedance of the inductor, set the initial condition, $i(0)$, to zero. Then, using Eq. 14.7-3, we obtain

$$Z_{\mathrm{L}}(s) = Ls$$

as the impedance of the inductor.

Equation 14.7-8 represents the inductor in the frequency domain, as shown in Figure 14.7-3*b*. The series connection of elements in Figure 14.7-3*b* corresponds to the sum of voltages in Eq. 14.7-8.

Solving Eq. 14.7-8 for $I(s)$ gives

$$I(s) = \frac{1}{Ls}V(s) + \frac{i(0)}{s} \tag{14.7-9}$$

Equation 14.7-9 represents the inductor in the frequency domain, as shown in Figure 14.7-3*c*. The parallel connection of elements in Figure 14.7-3*c* corresponds to the sum of currents in Eq. 14.7-9.

Table 14.7-1 tabulates the time- and frequency-domain representation of circuit elements. In addition to resistors, capacitors, and inductors, Table 14.7-1 shows the frequency-domain representations of independent and dependent sources and of op amps. Independent sources are specified by functions of time, $i(t)$ and $v(t)$, in the time domain and by the corresponding Laplace transforms, $I(s)$ and $V(s)$, in the frequency domain. Dependent sources and op amps operate the same way in the frequency domain as they do in the time domain.

To represent a circuit in the frequency domain, we replace the time-domain representation of each circuit element by its frequency-domain representation.

To find the complete response of a linear circuit, we first represent the circuit in the frequency domain, using the Laplace transform. Next, we analyze the circuit, perhaps by writing mesh or node equations. Finally, we use the inverse Laplace transform to represent the response in the time domain.

Table 14.7-1 Time-Domain and Frequency-Domain Representations of Circuit Elements

NAME	TIME DOMAIN	FREQUENCY DOMIAN
Current source	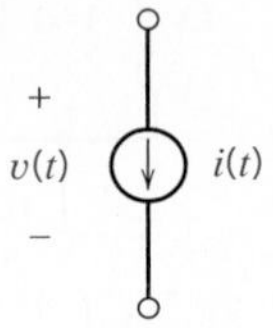 $v(t)$, $i(t)$	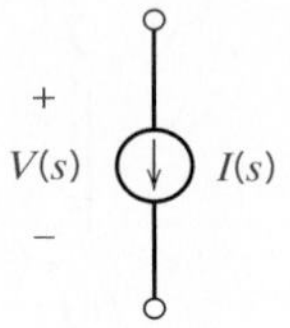 $V(s)$, $I(s)$
Voltage source	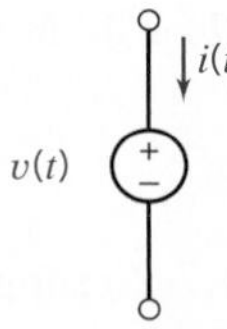 $i(t)$, $v(t)$	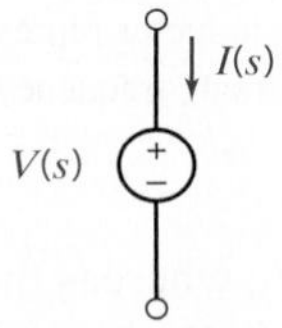 $I(s)$, $V(s)$
Resistor	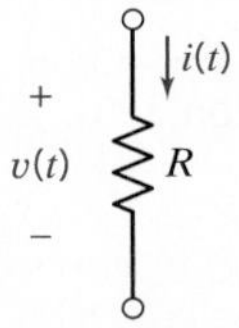 $i(t)$, $v(t)$, R	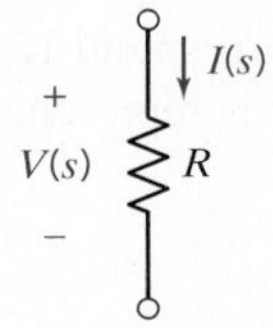 $I(s)$, $V(s)$, R
Capacitor	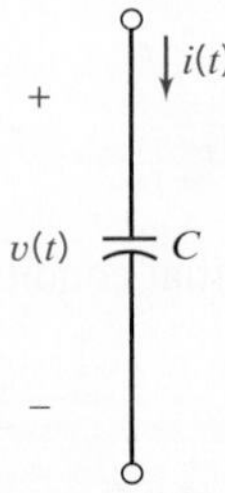 $i(t)$, $v(t)$, C	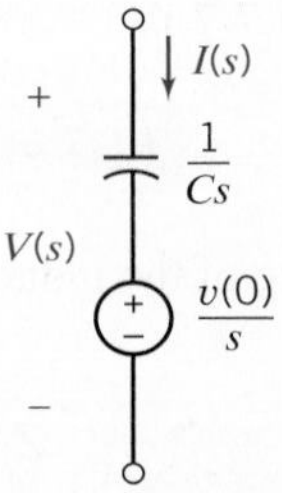 $I(s)$, $V(s)$, $\frac{1}{Cs}$, $\frac{v(0)}{s}$ or $I(s)$, $\frac{1}{Cs}$, $V(s)$, $Cv(0)$ 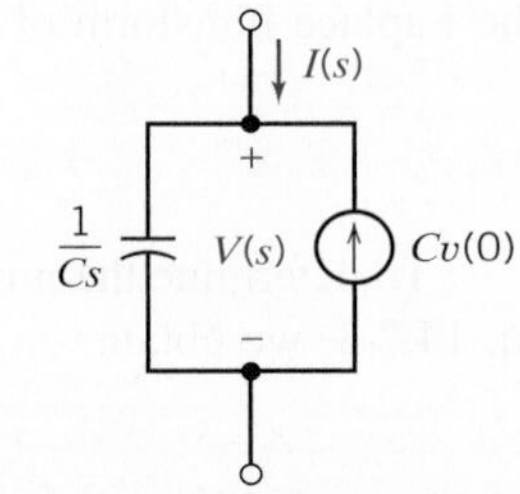
Inductor	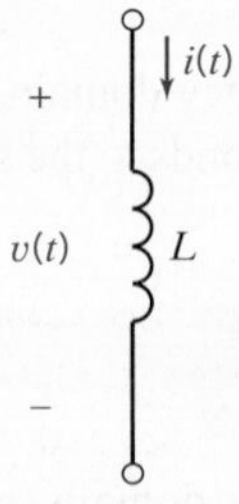 $i(t)$, $v(t)$, L	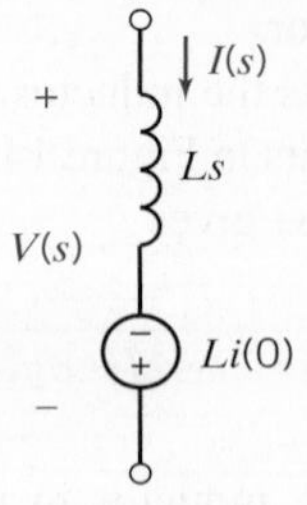 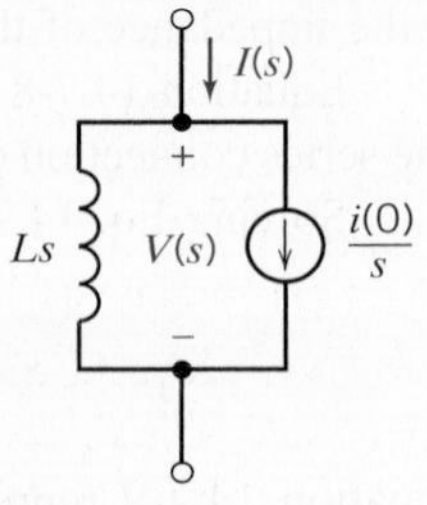 $I(s)$, $V(s)$, Ls, $Li(0)$ or $I(s)$, Ls, $V(s)$, $\frac{i(0)}{s}$
Dependent source	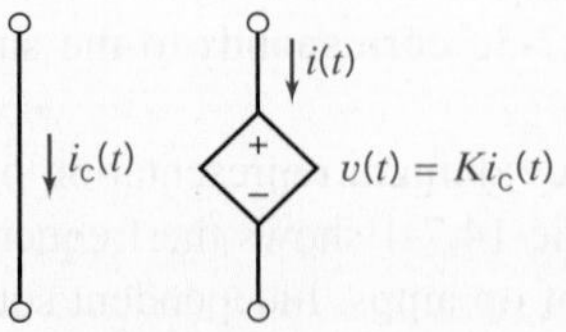 $i_c(t)$, $i(t)$, $v(t) = Ki_c(t)$	$I_c(s)$, $I(s)$, $V(s) = KI_c(s)$
Op amp	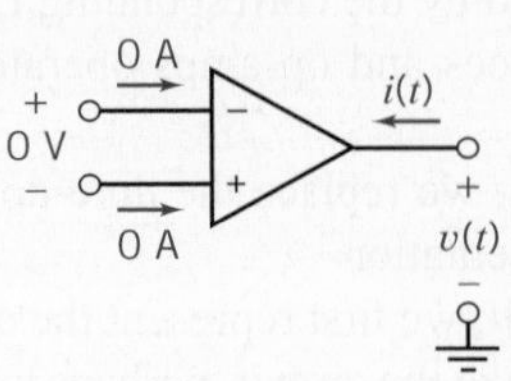 0 A, 0 V, 0 A, $i(t)$, $v(t)$	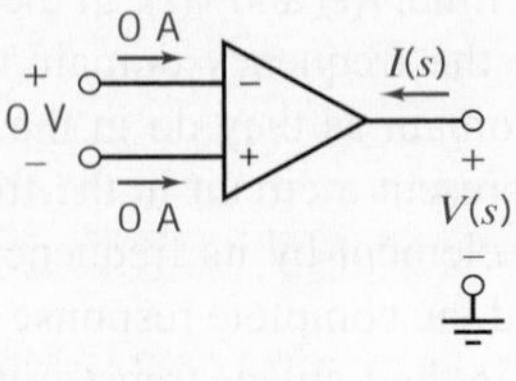 0 A, 0 V, 0 A, $I(s)$, $V(s)$

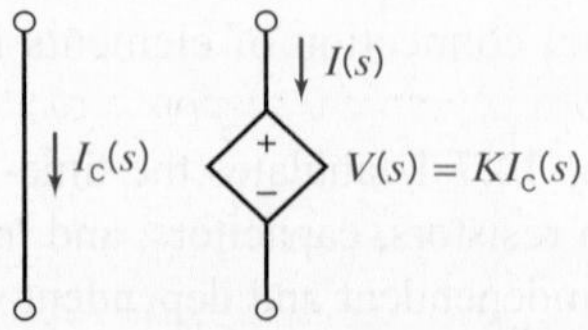

EXAMPLE 14.7-1 Circuit Analysis Using the Laplace Transform

INTERACTIVE EXAMPLE

Consider the circuit shown in Figure 14.7-4. The input to the circuit is the voltage of the voltage source, 24 V. The output of this circuit, the voltage across the capacitor, is given by

$$v_o(t) = 16 - 12e^{-0.6t}\text{ V} \quad \text{when } t > 0 \tag{14.7-10}$$

Determine the value of the capacitance, C.

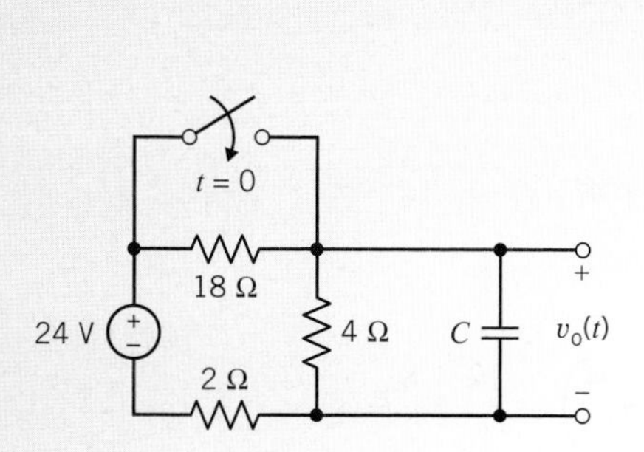

FIGURE 14.7-4 The circuit considered in Example 14.7-1.

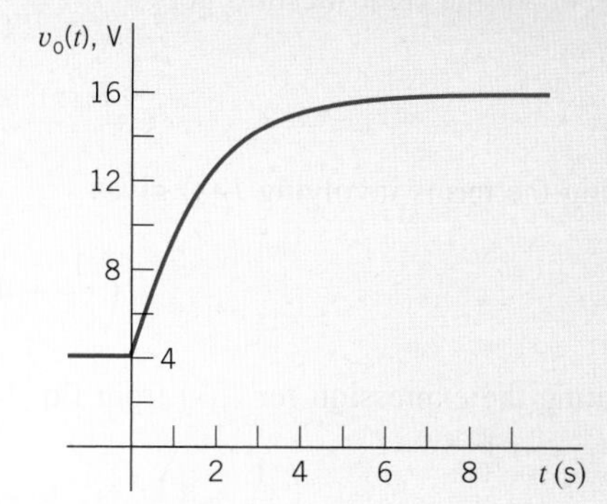

FIGURE 14.7-5 The capacitor voltage, $v_o(t)$, from the circuit shown in Figure 14.7-4.

Solution

Before the switch closes, the circuit will be at steady state. Because the only input to this circuit is the constant voltage of the voltage source, all of the element currents and voltages, including the capacitor voltage, will have constant values. Closing the switch disturbs the circuit by shorting out the 18-Ω resistor. Eventually, the disturbance dies out and the circuit is again at steady state. All the element currents and voltages will again have constant values but, probably, different constant values than they had before the switch closed.

During the disturbance, the element voltages and currents are not constant. For example, Eq. 14.7-10 describes the capacitor voltage after the switch closes. Notice that there are two parts to the capacitor voltage. One part, $12\,e^{-0.6t}$, dies out as the value of t increases. That part is called the transient part of the response, or just the transient response. The other part, 16, does not die out and is the steady-state response. The sum of the transient response and the steady-state response is called the complete response. The output voltage described by Eq. 14.7-10 is a complete response of this circuit.

Figure 14.7-5 shows a plot of the capacitor voltage given by Eq. 14.7-10. Notice that the capacitor voltage is continuous. This is expected because, in the absence of unbounded currents, the voltage of a capacitor must be continuous. In particular, the value of the capacitor voltage immediately after the switch is closed is equal to the value immediately before the switch is closed. From Figure 14.7-5, we see that at time $t = 0$, when the switch closes, the value of the capacitor voltage is $v_o(0) = 4$ V.

How does the value of the capacitance C affect the capacitor voltage? To answer this question, we must analyze the circuit. Because we want to determine the complete response, we will analyze the circuit using Laplace transforms. Figure 14.7-6 shows the frequency-domain representation of the circuit. The closed switch is represented by a short circuit. That short circuit is connected in parallel with the 18-Ω resistor. A short circuit in parallel with a resistor is equivalent to a short circuit, so the closed switch and 18-Ω resistor have been replaced by a single short circuit. The frequency-domain model of the capacitor consists of two parts, an impedance and a voltage source. The voltage of the voltage source depends on the initial condition of the capacitor, that is, $v_o(0) = 4$ V.

We can analyze the circuit in Figure 14.7-6 by writing and solving two mesh equations.

Apply KVL to the left mesh to get

$$4(I_1(s) - I_2(s)) + 2I_1(s) - \frac{24}{s} = 0$$

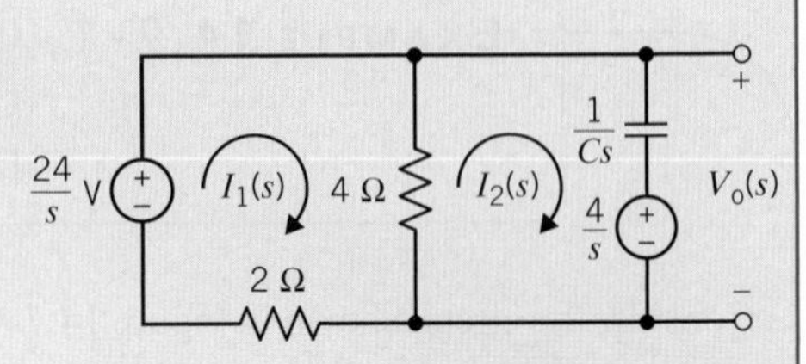

FIGURE 14.7-6 The circuit represented in the frequency domain, using the Laplace transform.

Solving for $I_1(s)$ gives

$$I_1(s) = \frac{2}{3}I_2(s) + \frac{4}{s} \tag{14.7-11}$$

Apply KVL to the right mesh to get

$$\frac{1}{Cs}I_2(s) + \frac{4}{s} - 4(I_1(s) - I_2(s)) = 0$$

Collecting the terms involving $I_2(s)$ gives

$$\left(\frac{1}{Cs} + 4\right)I_2(s) = -\frac{4}{s} + 4I_1(s)$$

Substituting the expression for $I_1(s)$ from Eq. 14.7-11 gives

$$\left(\frac{1}{Cs} + 4\right)I_2(s) = -\frac{4}{s} + 4\left(\frac{2}{3}I_2(s) + \frac{4}{s}\right) = \frac{12}{s} + \frac{8}{3}I_2(s)$$

Collecting the terms involving $I_2(s)$ gives

$$\left(\frac{1}{Cs} + \frac{4}{3}\right)I_2(s) = \frac{12}{s}$$

Multiply both sides of this equation by $\frac{3}{4}s$ to get

$$\left(s + \frac{3}{4C}\right)I_2(s) = 9$$

Solving for $I_2(s)$ gives

$$I_2(s) = \frac{9}{s + \frac{3}{4C}} \tag{14.7-12}$$

Referring to Figure 14.7-6, we see that the capacitor voltage is related to the mesh current of the right mesh by

$$V_o(s) = \frac{1}{Cs}I_2(s) + \frac{4}{s}$$

Substituting the expression for $I_2(s)$ from Eq. 14.7-12 gives

$$V_o(s) = \left(\frac{1}{Cs}\right)\frac{9}{s + \frac{3}{4C}} + \frac{4}{s} = \frac{\frac{9}{C}}{s\left(s + \frac{3}{4C}\right)} + \frac{4}{s}$$

Performing partial fraction expansion gives

$$V_o(s) = \frac{12}{s} - \frac{12}{s + \frac{3}{4C}} + \frac{4}{s} = \frac{16}{s} - \frac{12}{s + \frac{3}{4C}} \tag{14.7-13}$$

Recall that $v_o(t)$ is given in Eq. 14.7-10. Taking the Laplace transform of $v_o(t)$ gives

$$V_o(s) = \mathscr{L}[v_o(t)] = \mathscr{L}\left[\left(16 - 12e^{-0.6t}\right)u(t)\right] = \frac{16}{s} - \frac{12}{s + 0.6} \tag{14.7-14}$$

Comparing Eqs. 14.7-13 and 14.7-14 shows that

$$0.6 = \frac{3}{4C} \quad \Rightarrow \quad C = 1.25\text{ F}$$

EXAMPLE 14.7-2 Circuit Analysis Using the Laplace Transform

INTERACTIVE EXAMPLE

Consider the circuit shown in Figure 14.7-7. The input to the circuit is the voltage of the voltage source, 24 V. The output of this circuit, the voltage across the 6-Ω resistor, is given by

$$v_o(t) = 12 - 6\,e^{-0.35t}\text{ V} \quad \text{when } t > 0 \tag{14.7-15}$$

Determine the value of the inductance, L, and of the resistances, R_1 and R_2.

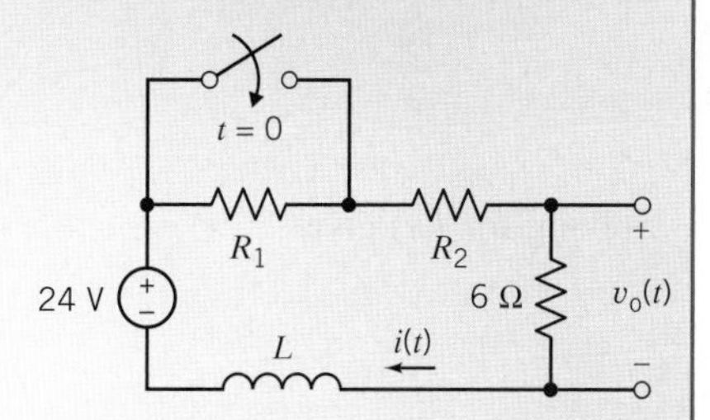

FIGURE 14.7-7 The circuit considered in Example 14.7-2.

Solution

Before the switch closes, the circuit will be at steady state. Because the only input to this circuit is the constant voltage of the voltage source, all of the element currents and voltages, including the inductor current, will have constant values. Closing the switch disturbs the circuit by shorting out the resistor R_1. Eventually, the disturbance dies out and the circuit is again at steady state. All the element currents and voltages will again have constant values but, probably, different constant values than they had before the switch closed.

Equation 14.7-15 describes the output voltage after the switch closes. Notice that there are two parts to this voltage. One part, $-6\,e^{-0.35t}$, dies out as the value of t increases. That part is called the transient part of the response, or just the transient response. The other part, 12, does not die out and is the steady-state response. The sum of the transient response and the steady-state response is called the complete response. The output voltage described by Eq. 14.7-15 is the complete response of this circuit.

How do the values of the circuit parameters L, R_1, and R_2 affect the output voltage? To answer this question, we must analyze the circuit. Because we want to determine the complete response, we will analyze using Laplace transforms. The frequency-domain model of the inductor consists of two parts, an impedance and a voltage or current source. The value of the voltage source voltage or current source current depends on the initial condition of the inductor, that is, the inductor current at time $t = 0$. We need to find the initial inductor current before we can represent the circuit, using Laplace transforms.

Referring to Figure 14.7-7, we see that the inductor current is equal to the current in the 6-Ω resistor. Consequently,

$$i(t) = \frac{v(t)}{6} = \frac{12 - 6e^{-0.35t}}{6} = 2 - e^{-0.35t} \text{ A} \quad \text{when } t > 0 \tag{14.7-16}$$

In the absence of unbounded voltages, the current in any inductor is continuous. Consequently, the value of the inductor current immediately before $t = 0$ is equal to the value immediately after $t = 0$. To find the initial inductor current, we set $t = 0$ in Eq. 14.7-16 to get $i(0) = 1$ A.

Figure 14.7-8 shows the frequency-domain representation of the circuit. We selected the model of the inductor that uses a voltage source to account for the initial condition in anticipation of writing a mesh equation. The voltage of this voltage source is

$$Li(0) = (L)(1) = L$$

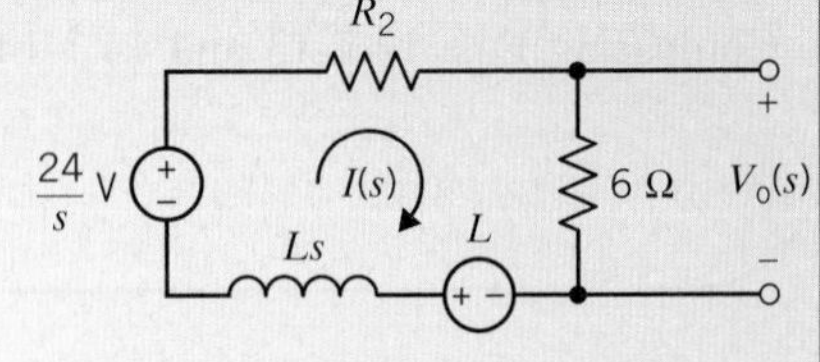

FIGURE 14.7-8 The circuit represented in the frequency domain, using the Laplace transform.

In Figure 14.7-8, the closed switch is represented by a short circuit. That short circuit is connected in parallel with resistor R_1. A short circuit in parallel with a resistor is equivalent to a short circuit, so the closed switch and R_1 have been replaced by a single short circuit.

To analyze the circuit in Figure 14.7-8, we write and solve a single mesh equation. Apply KVL to the mesh to get

$$(R_2 + 6 + Ls)I(s) = L + \frac{24}{s}$$

Solving for $I(s)$ gives

$$I(s) = \frac{L + \frac{24}{s}}{Ls + R_2 + 6} = \frac{s + \frac{24}{L}}{s\left(s + \frac{R_2 + 6}{L}\right)}$$

Using Ohm's law gives

$$V_o(s) = 6I(s) = \frac{6s + \frac{(6)(24)}{L}}{s\left(s + \frac{R_2 + 6}{L}\right)}$$

Partial fraction expansion gives

$$V_o(s) = \frac{\frac{(6)(24)}{R_2 + 6}}{s} - \frac{\frac{6(18 - R_2)}{R_2 + 6}}{s + \frac{R_2 + 6}{L}} \tag{14.7-17}$$

Recall that $v_o(t)$ is given in Eq. 14.7-15. Taking the Laplace transform of $v_o(t)$ gives

$$V_o(s) = \mathscr{L}[v_o(t)] = \mathscr{L}\left[\left(12 - 6\,e^{-0.35t}\right)u(t)\right] = \frac{12}{s} - \frac{6}{s + 0.35} \tag{14.7-18}$$

Comparing Eqs. 14.7-17 and 14.7-18 shows that

$$\frac{(6)(24)}{R_2+6} = 12 \quad \Rightarrow \quad R_2 = 6\,\Omega$$

and

$$0.35 = \frac{R_2+6}{L} = \frac{12}{L} \quad \Rightarrow \quad L = \frac{12}{0.35} = 34.29\,\mathrm{H}$$

How can we find R_1? Resistor R_1 is removed from the circuit by closing the switch, but R_1 was part of the circuit before the switch closed. The initial inductor current depends on the value of the resistance R_1. The only input to the circuit in Figure 14.7-9 is a constant, 24 V. Consequently, when the circuit is at steady state, the inductor will act like a short circuit. Figure 14.7-9 shows the steady-state circuit when the switch is open. The open switch is modeled as an open circuit. The inductor is modeled as short circuit. Writing and solving a mesh equation gives

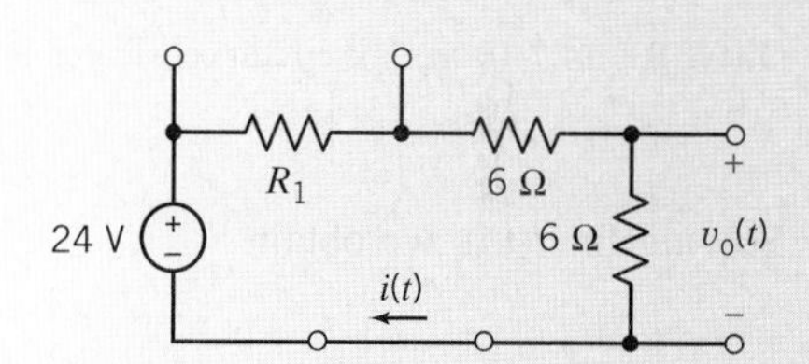

FIGURE 14.7-9 The circuit at steady state before the switch closes.

$$i(t) = \frac{24}{R_1+6+6}$$

Letting $t = 0$ gives

$$\frac{24}{R_1+6+6} = i(0) = 1 \quad \Rightarrow \quad R_1 = 12\,\Omega$$

EXAMPLE 14.7-3 Circuit Analysis Using the Laplace Transform

Consider the circuit shown in Figure 14.7-10*a*. The input to the circuit is the voltage of the voltage source, 12 V. The output of this circuit is the current in the inductor, $i_L(t)$. Determine the current in the inductor, $i_L(t)$, for $t > 0$.

Solution

Let's write and solve mesh equations. The series circuits that represent the capacitor and inductor in the frequency domain contain voltage sources rather than current sources. It's easier to account for voltage sources than current sources when writing mesh equations, so we choose the series representation for both the capacitor and inductor. From Figure 14.7-10*b*, the initial conditions are $v_c(0) = 8$ V and $i_L(0) = 4$ A. Figure 14.7-11*b* shows the frequency-domain representation of the circuit.

The mesh current equations are

$$\left(1+\frac{1}{s}\right) I_1(s) - \frac{1}{s} I_2(s) = \frac{12}{s} - \frac{8}{s}$$

and

$$-\frac{1}{s} I_1(s) + \left(1+s+\frac{1}{s}\right) I_2(s) = 4 + \frac{8}{s}$$

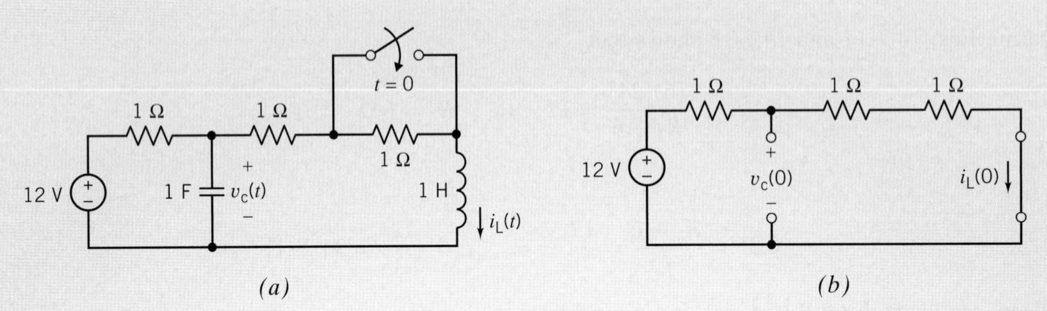

FIGURE 14.7-10 (*a*) The circuit considered in Example 14.7-3. (*b*) The steady-state circuit before the switch closes.

Solving for $I_2(s)$, we obtain

$$I_2(s) = \frac{4(s^2 + 3s + 3)}{s(s^2 + 2s + 2)}$$

The convenient partial fraction expansion is

$$\frac{I_2(s)}{4} = \frac{s^2 + 3s + 3}{s(s^2 + 2s + 2)} = \frac{A}{s} + \frac{Bs + D}{s^2 + 2s + 2}$$

Then, we determine that $A = 1.5$, $B = -0.5$, and $D = 0$. Then, we can state

$$\frac{I_2(s)}{4} = \frac{1.5}{s} + \frac{-0.5s}{(s+1)^2 + 1}$$

Using the Laplace transform Table 14.2-1, we obtain

$$i_L(t) = i_2(t) = \{6 + 2\sqrt{2}e^{-t}\sin(t - 45°)\}\text{ A} \quad \text{for} \quad t > 0$$

Checking the initial value of i_2, we get $i_2(0) = i_L(0) = 4$ A, which verifies the correct initial value. The final value is $i_2(\infty) = 6$ A.

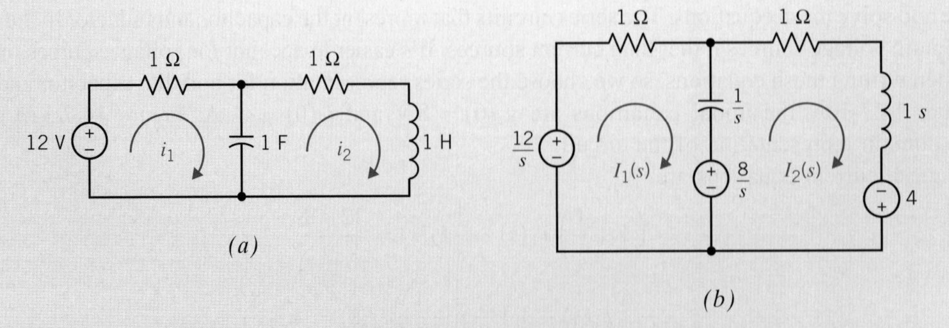

FIGURE 14.7-11 (*a*) Circuit with mesh currents. (*b*) Laplace transform model of circuit.

EXAMPLE 14.7-4 Circuit Analysis Using the Laplace Transform

The switch in the circuit shown in Figure 14.7-12a closes at time $t = 0$. Determine the voltage $v(t)$ after the switch closes.

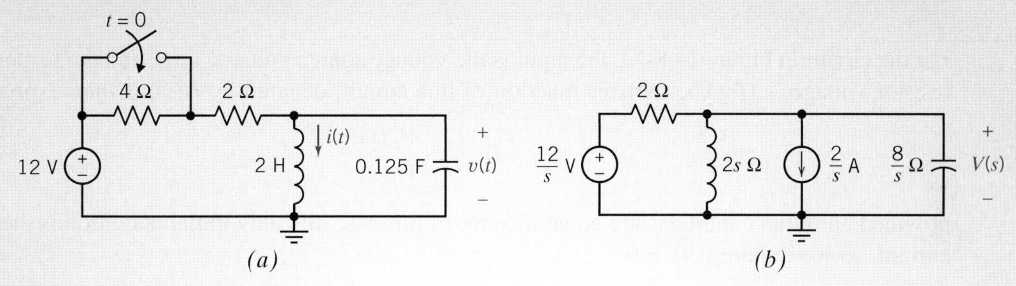

FIGURE 14.7-12 The circuit of Example 14.7-4 represented in the (a) time domain and (b) frequency domain, using Laplace transforms.

Solution

Let's write and solve node equations. In the frequency domain, we will use the parallel model for the capacitor and inductor because the parallel models contain current sources rather than voltage sources. The initial conditions are $i(0) = 2$ A and $v(0) = 0$ V. Because $v(0) = 0$, the current of the current source in the frequency-domain representation of the capacitor is zero. A zero current source is equivalent to an open circuit. Figure 14.7-12b shows the frequency-domain representation of the circuit after the switch has closed.

Apply KCL at the top node of the inductor to get the node equation

$$\frac{V(s) - \dfrac{12}{s}}{2} + \frac{V(s)}{2s} + \frac{2}{s} + \frac{V(s)}{\dfrac{8}{s}} = 0$$

Solving for $V(s)$ gives

$$V(s) = \frac{32}{s^2 + 4s + 4} = \frac{32}{(s+2)^2}$$

Finally, take the inverse Laplace transform to obtain $v(t)$

$$v(t) = \mathscr{L}^{-1}\left[\frac{32}{(s+2)^2}\right] = 32te^{-2t}u(t) \text{ V}$$

EXERCISE 14.7-1 Determine the voltage $v_C(t)$ and the current $i_C(t)$ for $t \geq 0$ for the circuit of Figure E 14.7-1.

Hint: $v_C(0) = 4$ V

Answer: $v_C(t) = (6 - 2e^{-0.67t})u(t)$ V and $i_C(t) = \frac{2}{3}e^{-0.67t}u(t)$ A

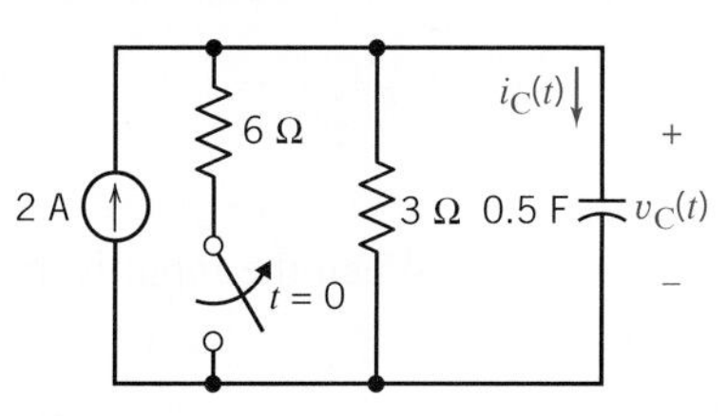

FIGURE E 14.7-1

14.8 TRANSFER FUNCTION AND IMPEDANCE

The **transfer function** of a circuit is defined as the ratio of the Laplace transform of the response of the circuit to the Laplace transform of the input to the circuit when the initial conditions are zero.

For the circuit in Figure 14.8-1*a*, the input is the voltage source voltage, $v_1(t)$, and the response is the resistor voltage, $v_o(t)$. The transfer function of this circuit, denoted by $H(s)$, is then expressed as

$$H(s) = \frac{V_o(s)}{V_1(s)} \tag{14.8-1}$$

provided all initial conditions are equal to zero. In this case, the only initial condition is the inductor current, so we require $i(0) = 0$.

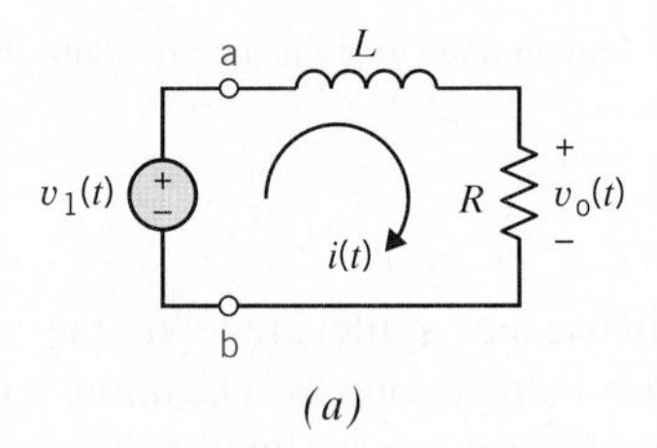

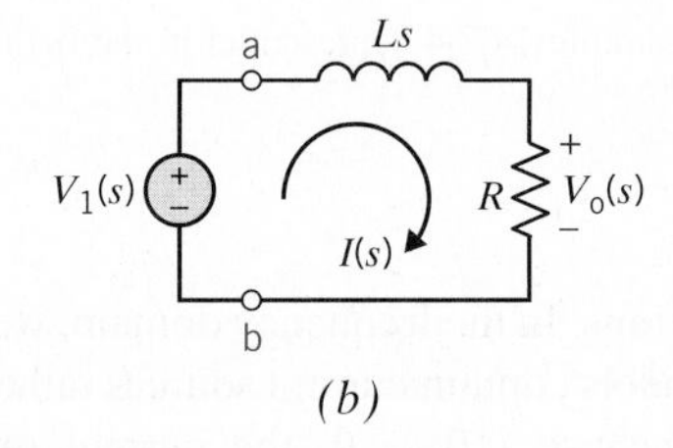

FIGURE 14.8-1 A circuit represented (*a*) in the time domain and (*b*) in the frequency domain, using the Laplace transform.

We can write Eq. 14.8-1 as

$$V_o(s) = H(s)V_1(s) \tag{14.8-2}$$

which says that the Laplace transform of the response is equal to the transfer function times the Laplace transform of the input, provided all initial conditions are equal to zero. We are going to get tired of saying "provided all initial conditions are equal to zero." A response subject to the requirement that all initial conditions be zero is called a zero-state response. With this terminology, we can read Eq. 14.8-1 as "the transfer function is the ratio of the Laplace transform of the zero-state response to the Laplace transform of the input." Similarly, we can read Eq. 14.8-2 as "the Laplace transform of the zero-state response is the product of the transfer function and the Laplace transform of the input."

Two special cases are very significant. When the input is a unit step function, then

$$V_1(s) = \mathcal{L}[u(t)] = \frac{1}{s}$$

and Eq. 14.8-2 becomes

$$V_o(s) = \frac{H(s)}{s}$$

In this case, the zero-state response is called the step response, that is,

$$\text{step response} = \mathcal{L}^{-1}\left[\frac{H(s)}{s}\right] \tag{14.8-3}$$

When the input is an impulse function, then

$$V_1(s) = \mathcal{L}[\delta(t)] = 1$$

and Eq. 14.8-2 becomes

$$V_o(s) = H(s)$$

In this case, the zero-state response is called the impulse response, that is,

$$\text{impulse response} = \mathcal{L}^{-1}[H(s)] \tag{14.8-4}$$

It is important to notice that both the step response and the impulse response are zero-state responses; that is, all initial conditions are set to zero.

Both the input to a circuit and the response of the circuit can be either a current or a voltage. When the input is a current and the response is a voltage, the transfer function is called an impedance. Similarly, when the input is a voltage and the response is a current, the transfer function is called an admittance. This terminology is consistent with our previous use of the term *impedance*. For example, consider the row of Table 14.7-1 corresponding to the capacitor. Consider the frequency-domain representation of the capacitor that contains a voltage source. The restriction that the initial condition be zero, $v(0) = 0$, causes the voltage source to be a zero voltage source, that is, a short circuit. The frequency-domain representation of the capacitor is reduced to a single element. When capacitor current is the input and the capacitor voltage is the response, then the impedance of the capacitor is

$$Z_C(s) = \frac{V(s)}{I(s)} = \frac{1}{Cs} \tag{14.8-5}$$

Next, consider the frequency-domain representation of the capacitor that contains a current source. The restriction that the initial condition be zero, $v(0) = 0$, causes the current source to be a zero current source, that is, an open circuit. The frequency-domain representation of the capacitor is again reduced to a single element. Once again, the impedance of the capacitor is given by Eq. 14.8-5.

A similar argument shows that setting the initial conditions to zero simplifies the frequency-domain representation of the inductor to the single impedance,

$$Z_L(s) = Ls \tag{14.8-6}$$

EXAMPLE 14.8-1 Transfer Function

For the circuit in Figure 14.8-1*a*, the input is the voltage source voltage, $v_1(t)$, and the response is the resistor voltage, $v_o(t)$. Find the transfer function of the circuit.

Solution

Figure 14.8-1*b* shows the frequency-domain representation of the circuit when all of the initial conditions are zero. In this case, the only initial condition is the inductor current, so we require $i(0) = 0$. The requirement that $i(0) = 0$ reduces the frequency-domain representation of the inductor to the impedance of the inductor.

Applying KVL to the mesh of the circuit in Figure 14.8-1*b* gives

$$V_1(s) = LsI(s) + RI(s)$$

Solving for $I(s)$ gives

$$I(s) = \frac{V_1(s)}{Ls + R}$$

The Laplace transform of the response is

$$V_o(s) = RI(s) = \frac{R}{Ls + R} V_1(s)$$

This result could have been obtained using voltage division. Finally, the transfer function is

$$H(s) = \frac{V_o(s)}{V_1(s)} = \frac{R}{Ls + R}$$

EXAMPLE 14.8-2 Step Response

Determine the step response of the circuit shown in Figure 14.8-2*a*.

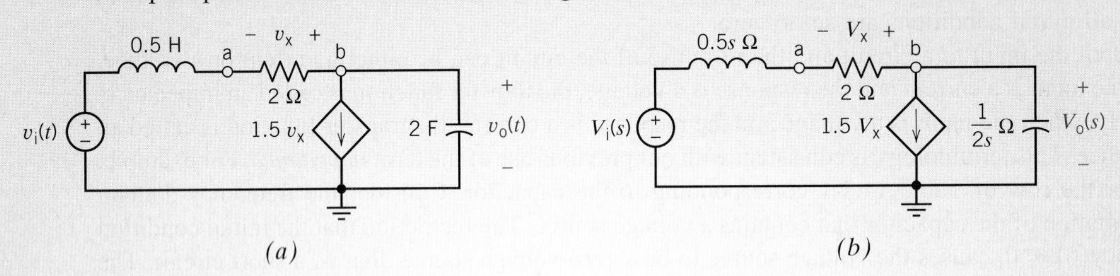

FIGURE 14.8-2 The circuit of Example 14.8-2 represented (*a*) in the time domain, and (*b*) in the frequency domain, using Laplace transforms.

Solution

Figure 14.8-2*b* shows the frequency-domain representation of the circuit when all of the initial conditions are zero.

Denote the node voltages at nodes a and b as V_a and V_b. The node equations are

$$\frac{V_a - V_i}{0.5\ s} - \frac{V_b - V_a}{2} = 0 \quad \Rightarrow \quad (4+s)V_a - sV_b = 4V_i$$

and

$$\frac{V_b - V_a}{2} + 1.5(V_b - V_a) + 2sV_b = 0 \quad \Rightarrow \quad (1+s)V_b = V_a$$

Solving for V_b gives

$$V_b = \frac{4}{(s+2)^2}V_i$$

The response is $V_o = V_b$, so the transfer function is

$$H(s) = \frac{V_o(s)}{V_i(s)} = \frac{V_b(s)}{V_i(s)} = \frac{4}{(s+2)^2}$$

The step response is

$$v_o(t) = \mathscr{L}^{-1}\left[\frac{H(s)}{s}\right] = \mathscr{L}^{-1}\left[\frac{4}{s(s+2)^2}\right] = \left(1 - (1+2t)e^{-2t}\right)u(t)$$

EXAMPLE 14.8-3 Impulse Response

Design the circuit of Figure 14.8-3*a* to have an impulse response equal to

$$h(t) = 2\left(e^{-t} - e^{-2t}\right) \quad t \geq 0$$

Solution

From the given impulse response, we have

$$H(s) = \mathscr{L}^{-1}\left[2\left(e^{-t} - e^{-2t}\right)\right] = 2\left(\frac{1}{s+1} - \frac{1}{s+2}\right) = 2\frac{(s+2)-(s+1)}{(s+1)(s+2)} = \frac{2}{s^2+3s+2} \tag{14.8-7}$$

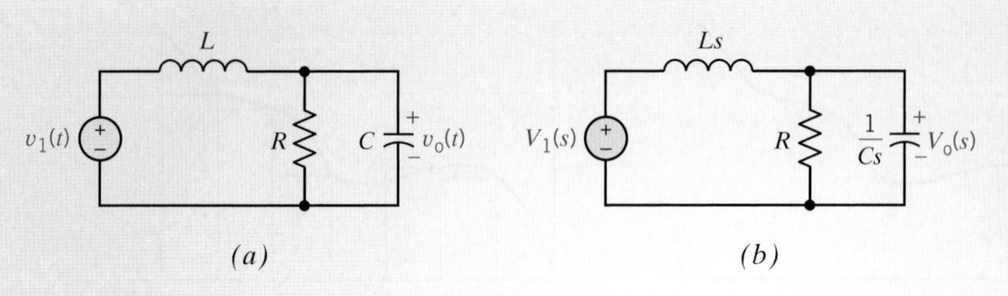

FIGURE 14.8-3 The circuit of Example 14.8-3 represented (*a*) in the time domain and (*b*) in the frequency domain, using the Laplace transform.

Figure 14.8-3*b* shows the circuit represented in the frequency domain, using the Laplace transform. Using the voltage divider principle, we determine the transfer function of this circuit to be

$$H(s) = \frac{V_o(s)}{V_1(s)} = \frac{\dfrac{R\dfrac{1}{Cs}}{R + \dfrac{1}{Cs}}}{\dfrac{R\dfrac{1}{Cs}}{R + \dfrac{1}{Cs}} + Ls} = \frac{\dfrac{1}{LC}}{s^2 + \dfrac{1}{RC}s + \dfrac{1}{LC}} \tag{14.8-8}$$

Comparing Eqs. 14.8-7 and 14.8-8 gives $1/LC = 2$ and $1/RC = 3$. These equations don't have a single unique solution. To obtain one solution, choose $C = 1/12$ F. Then $L = 6$ H and $R = 4\ \Omega$ are the required values. Other solutions can be obtained by changing the value of C and then recalculating L and R.

EXERCISE 14.8-1 The transfer function of a circuit is $H(s) = \dfrac{-5s}{s^2 + 15s + 50}$. Determine the impulse response and step response of this circuit.

Answers: **(a)** impulse response $= \mathscr{L}^{-1}\left[\dfrac{5}{s+5} - \dfrac{10}{s+10}\right] = (5e^{-5t} - 10e^{-10t})u(t)$

(b) step response $= \mathscr{L}^{-1}\left[\dfrac{1}{s+10} - \dfrac{1}{s+5}\right] = (e^{-10t} - e^{-5t})u(t)$

EXERCISE 14.8-2 The impulse response of a circuit is $h(t) = 5e^{-2t}\sin(4t)u(t)$. Determine the step response of this circuit.

Hint: $H(s) = \mathscr{L}[5e^{-2t}\sin(4t)u(t)] = \dfrac{5(4)}{(s+2)^2 + 4^2} = \dfrac{20}{s^2 + 4s + 20}$

Answer: step response $= \mathscr{L}^{-1}\left[\dfrac{H(s)}{s}\right] = \mathscr{L}^{-1}\left[\dfrac{1}{s} - \dfrac{s+4}{s^2 + 4s + 20}\right]$

$$= \left(1 - e^{-2t}\left(\cos 4t + \frac{1}{2}\sin 4t\right)\right)u(t)$$

14.9 CONVOLUTION

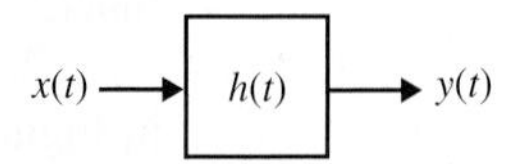

FIGURE 14.9-1 A linear, time-invariant circuit.

In this section, we consider the problem of determining the response of a linear, time-invariant circuit to an arbitrary input, $x(t)$. This situation is illustrated in Figure 14.9-1, in which $x(t)$ is the input to the circuit, $y(t)$ is the output of the circuit, and $h(t)$ is the impulse response of the circuit. We will assume that $x(t) = 0$ when $t < 0$ because $t = 0$ is the time at

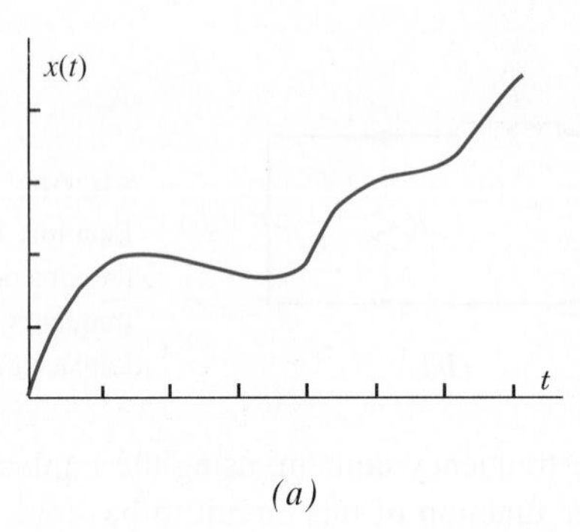

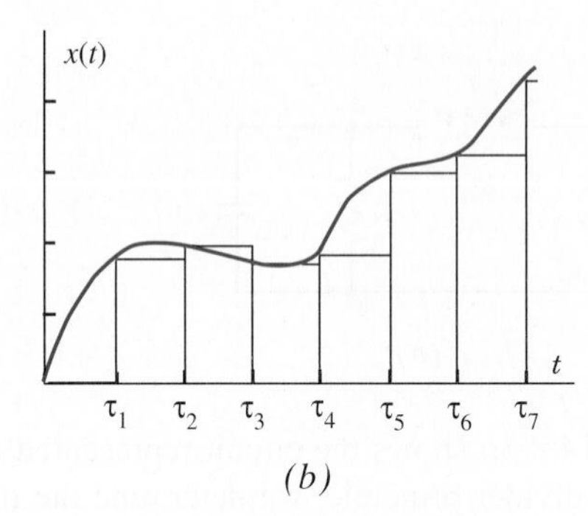

FIGURE 14.9-2 The arbitrary input waveform shown in (*a*) can be approximated, a sequence of pulses as shown in (*b*).

which the input is first applied to the circuit and that $h(t) = 0$ when $t < 0$ because the impulse response cannot precede the impulse that caused it.

It's important to us that the circuit is both linear and time-invariant. To see why, let's use the notation

$$x(t) \rightarrow y(t)$$

to indicate that the input $x(t)$ causes the output $y(t)$. Let k be any constant. Because the circuit is linear,

$$k\,x(t) \rightarrow k\,y(t)$$

(Suppose $k = 2$. The input $2x(t)$ is twice as large as the input $x(t)$, and it causes an output twice as large as the output caused by $x(t)$.) Next, let τ any constant. Because the circuit is time-invariant,

$$x(t-\tau) \rightarrow y(t-\tau)$$

(Suppose $\tau = 4$ s. The input $x(\tau - 4)$ is delayed by 4 s with respect to $x(t)$ and causes an output that is delayed by 4 s with respect to $y(t)$.) Because the circuit is both linear and time-invariant, we have

$$k\,x(t-\tau) \rightarrow k\,y(t-\tau)$$

Next, we use the fact that $h(t)$ is the impulse response of the circuit. Consequently, when the input to the circuit is $x(t) = \delta(t)$, the out put is $y(t) = h(t)$. That is,

$$\delta(t) \rightarrow h(t)$$

Finally,

$$k\,\delta(t-\tau) \rightarrow k\,h(t-\tau) \tag{14.9-1}$$

Consider the arbitrary input waveform $x(t)$ shown in Figure 14.9-2(*a*). This waveform can be approximated by a series of pulses as shown in Figure 14.9-2(*b*). The times, $\tau_1, \tau_2, \tau_3, \ldots$ are uniformly spaced, that is,

$$\tau_{i+1} = \tau_i + \Delta\tau \quad \text{for} \quad i = 1, 2, 3, \ldots$$

where the increment $\Delta\tau$ is independent of the index i. The error in the approximation is small when the increment $\Delta\tau$ is chosen to be small.

Figure 14.9-3(*a*) shows one of the pulses from the approximation of the input waveform. Notice that the area of the pulse is $x(\tau_i)\Delta\tau$. When the time increment $\Delta\tau$ is chosen to be small, this pulse can be approximated by the impulse function having the same area, $x(\tau_i)\Delta\tau\delta\ (t - \tau_i)$. That impulse function is illustrated in Figure 14.9-3(*b*).

The input waveform is represented by the sum of the impulse functions approximating the pulses in Figure 14.9-2(*b*),

$$x(t) = \sum_{i=0}^{\infty} x(\tau_i)\Delta\tau\delta(t-\tau_i) \tag{14.9-2}$$

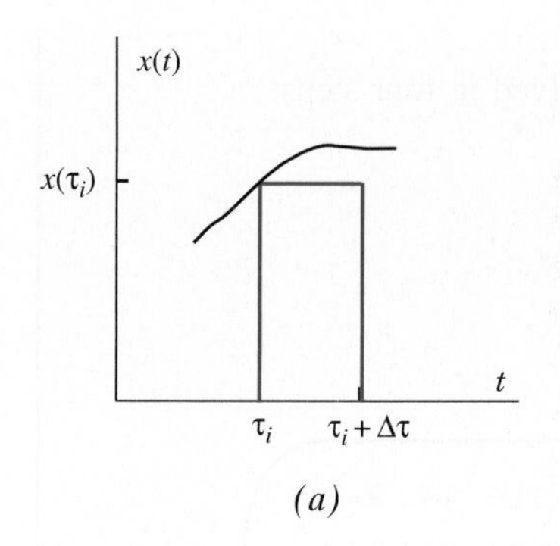

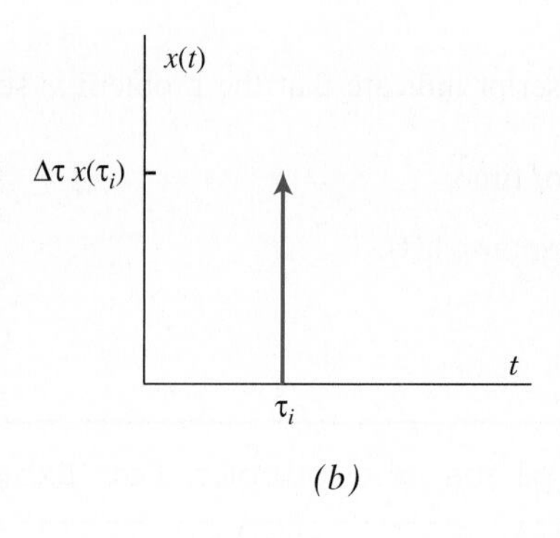

FIGURE 14.9-3 (*a*) A pulse from the approximation of an input waveform and (*b*) the corresponding impulse.

Because the circuit is linear, the response to this sum of impulse inputs is equal to the sum of the responses to the responses to the individual impulse inputs. From Eq. 4.9-1, the responses to the individual impulses inputs are given by

$$(x(\tau_i)\Delta\tau)\delta(t-\tau_i) \rightarrow (x(\tau_i)\Delta\tau)h(t-\tau_i) \quad \text{for } i = 0, 1, 2, 3, \ldots$$

The response of the circuit is

$$y(t) = \sum_{i=0}^{\infty} x(\tau_i)\Delta\tau\, h(t-\tau_i) = \sum_{i=0}^{\infty} x(\tau_i)h(t-\tau_i)\Delta\tau \tag{14.9-3}$$

In the limit as $\Delta\tau$ goes to zero, the summation becomes an integral, and we have

$$y(t) = \int_0^{\infty} x(\tau)h(t-\tau)d\tau \tag{14.9-4}$$

The integral on the right side of Equation 14.9-4 is called the convolution integral and is denoted as $x(t)^*h(t)$. That is,

$$y(t) = x(t)^*h(t) \tag{14.9-5}$$

Equation 14.9-5 indicates that the output of the linear circuit in Figure 14.9-1 can be obtained as the convolution of the input and the impulse response.

MATLAB provides a function called conv that performs convolution. The next example uses this MATLAB function to obtain a plot of the output of a linear, time-invariant circuit.

EXAMPLE 14.9-1 Convolution

Plot the output $y(t)$ for the circuit shown in Figure 14.9-1 when the input $x(t)$ is the triangular waveform shown in Figure 14.9-4 and the impulse response of the circuit is

$$h(t) = \frac{5}{4}(e^{-t} - e^{-5t})u(t)$$

FIGURE 14.9-4 The input for Example 14.9-1.

Solution

Figure 14-9.5 shows a MATLAB script that produces the required plot.

The comments included in the MATLAB script indicate that the problem is solved in four steps:

1. Obtain a list of equally spaced instants of time.
2. Obtain the input x(t) and the impulse response h(t).

```
% convolution.m - plots the output for Example 14.9-1
% -----------------------------------------------------
%  Obtain a list of equally spaced instants of time
% -----------------------------------------------------
t0 = 0; % begin
tf = 12;  % end
N = 5000;  % number of points plotted
dt = (tf-t0)/N; % increment
t = t0:dt:tf; % time in seconds

% -----------------------------------------------------
% Obtain the input x(t) and the impulse response h(t)
% -----------------------------------------------------
for k = 1 : length(t)
    if t(k) < 2
        x(k) = 0;
    elseif t(k) < 5
        x(k) = -8 + 4*t(k); %
    elseif t(k) < 7
        x(k) = 42 - 6*t(k); %
    else
        x(k) = 0;
    end
end
x=x*dt;
h=1.25*exp(-t)-1.25*exp(-5*t);

% -----------------------------------------------------
%            Perform the convolution
% -----------------------------------------------------
y=conv(x,h);

% -----------------------------------------------------
%            Plot the output y(t)
% -----------------------------------------------------
plot(t,y(1:length(t)))
axis([t0, tf, 0, 9])
xlabel('t')
ylabel('y(t)')
```

FIGURE 14.9-5 The MATLAB script for Example 14.9-1.

3. Perform the convolution.
4. Plot the output y(t).

A couple of remarks are helpful for understanding the MATLAB script. First, using the equations of the straight lines that comprise the triangular input waveform, we can write

$$x(t) = \begin{cases} 0 & \text{when} \quad t \le 2 \\ 4t - 8 & \text{when} \quad 2 \le t \le 5 \\ -4t + 42 & \text{when} \quad 5 \le t \le 7 \\ 0 & \text{when} \quad t \ge 7 \end{cases}$$

This equation is implemented by an ''if-then-else'' block in the MATLAB script. For any time, τ_i, this equation produces the corresponding value $x(\tau_i)$. From Eq. 14.9-2, we see that the strengths of the impulse inputs are $x(\tau_i)\Delta t$ rather than $x(\tau_i)$. It is necessary to multiply the values $x(\tau_i)$ by the time increment, and that is accomplished by the line, ''x = x*dt'' in the MATLAB script.

Next, the MATLAB plot function requires two lists of values, t and y, in our case. These lists are required to have the same number of values, but in our case, y is longer than t. The MATLAB expression ''(1:length(t))'' truncates the list y, so that truncated list is the same length as t.

Finally, the plot produced by the MATLAB script is shown in Figure 14.9-6.

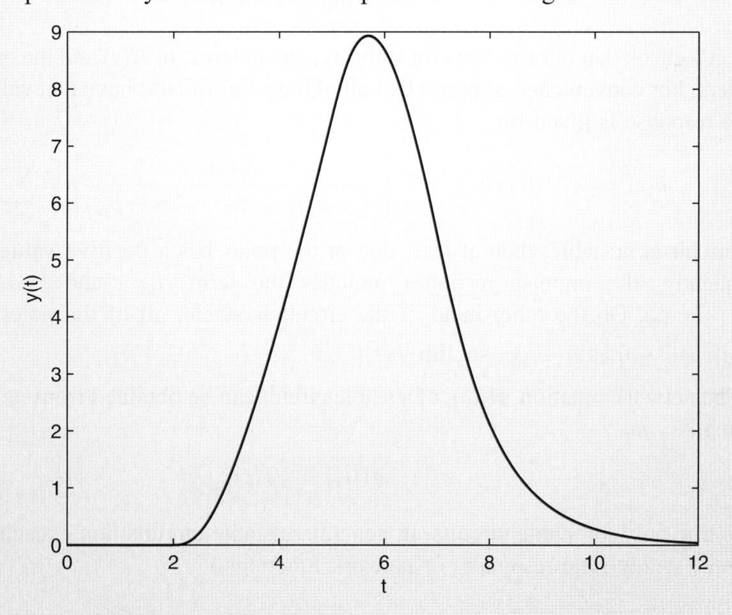

FIGURE 14.9-6 The output for Example 14.9-1.

14.10 STABILITY

A circuit is said to be *stable* when the response to a bounded input signal is a bounded output signal. A circuit that is not stable is said to be *unstable*.

Producing a bounded response to a bounded input is pretty reasonable behavior. As a general rule of thumb, stable circuits are potentially useful, and unstable circuits are potentially dangerous. When we analyze a circuit to see whether it is stable, we are probably trying to do one of two things. First, we may be checking a circuit to see whether it is useful. We will reject the circuit if it is unstable. Second, we may be trying to specify values of the circuit parameters in such a way as to make the circuit stable.

Consider a circuit represented by the transfer function, $H(s)$. Factoring the denominator of the transfer function gives

$$H(s) = \frac{N(s)}{(s-p_1)(s-p_2)\cdots(s-p_N)}$$

The p_i are the poles of the transfer function, also called the poles of the circuit. The poles may have real values or complex values. Complex poles appear in complex conjugate pairs; for example, if $-2+j3$ is a pole, then $-2-j3$ must also be a pole.

A circuit is **stable** if, and only if, all of its poles have negative real parts.

(Real poles must have negative values.) Another way of saying the same thing is that a circuit is stable if, and only if, all of its poles lie in the left half of the *s*-plane.

We can also use the impulse response, $h(t)$, to determine whether a circuit is stable. A circuit is stable if, and only if, its impulse response satisfies

$$\lim_{t\to\infty} |h(t)| = 0$$

Let's check that our two tests for stability, one in terms of $H(s)$ and the other in terms of $h(t)$, are equivalent. For convenience, suppose that all of the poles of $H(s)$ have real values. The corresponding impulse response is given by

$$h(t) = \mathscr{L}^{-1}[H(s)] = \mathscr{L}^{-1}\left[\frac{N(s)}{(s-p_1)(s-p_2)\cdots(s-p_N)}\right] = \sum_{i=1}^{N} A_i e^{p_i t} u(t)$$

If the circuit is unstable, then at least one of the poles has a positive value, for example, $p_4 = 6$. Consequently, the impulse response includes the term $A_4 e^{6t}$ and $|A_4 e^{6t}| \to \infty$ as $t \to \infty$, so $\lim_{t\to\infty} |h(t)| = \infty$. On the other hand, if the circuit is stable, all of the poles have negative values. Each $|A_i e^{p_i t}| \to 0$ as $t \to \infty$, so $\lim_{t\to\infty} |h(t)| = 0$.

The network function, $\mathbf{H}(\omega)$, of a stable circuit can be obtained from its transfer function, $H(s)$, by letting $s = j\omega$.

$$\mathbf{H}(\omega) = |H(s)|_{s=j\omega}$$

(This is true only for stable circuits. In general, unstable circuits don't reach a steady state, so they don't have steady-state responses or network functions.)

EXAMPLE 14.10-1 Stability

The input to the circuit shown in Figure 14.10-1 is the voltage, $v_i(t)$, of the independent voltage source. The output is the voltage, $v_o(t)$, of the dependent voltage source. The transfer function of this circuit is

$$H(s) = \frac{V_o(s)}{V_i(s)} = \frac{\dfrac{k}{RC}s}{s^2 + \dfrac{4-k}{RC}s + \dfrac{2}{R^2C^2}} = \frac{ks}{s^2 + (4-k)s + 2}$$

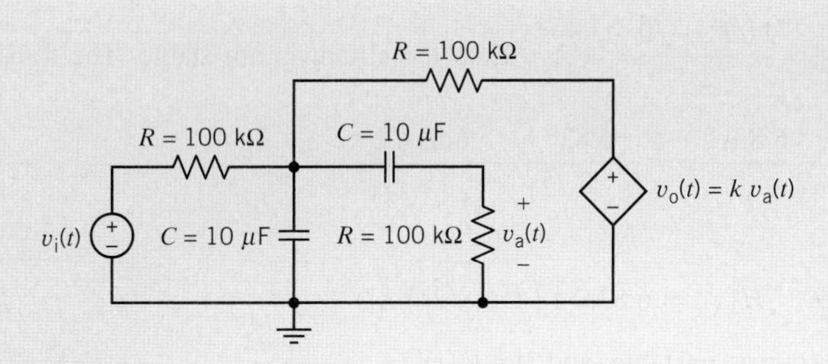

FIGURE 14.10-1 The circuit considered in Example 14.10-1.

Determine the following:

(a) The steady-state response when $v_i(t) = 5\cos 2t$ V and the gain of the VCVS is $k = 3$ V/V

(b) The impulse response when $k = 4 - 2\sqrt{2} = 1.17$ V/V

(c) The impulse response when $k = 4 + 2\sqrt{2} = 6.83$ V/V

Solution

The poles of the transfer function are $p_{1,2} = \dfrac{-(4-k) \pm \sqrt{(4-k)^2 - 8}}{2}$

(a) When $k = 3$ V/V, the poles are $p_{1,2} = \dfrac{-1 \pm \sqrt{-7}}{2} = \dfrac{-1 \pm j7}{2}$, so the circuit is stable. The transfer function is

$$H(s) = \frac{V_o(s)}{V_i(s)} = \frac{3s}{s^2 + s + 2}$$

The circuit is stable when $k = 3$ V/V, so we can determine the network function from the transfer function by letting $s = j\omega$.

$$\frac{\mathbf{V}_o(\omega)}{\mathbf{V}_i(\omega)} = \mathbf{H}(\omega) = H(s)_{s=j\omega} = \left.\frac{3s}{s^2 + s + 2}\right|_{s=j\omega} = \frac{3j\omega}{(2+\omega^2) + j\omega}$$

The input is $v_i(t) = 5\cos 2t$ V. The phasor of the steady-state response is determined by multiplying the network function evaluated at $\omega = 2$ rad/s by the phasor of the input:

$$\mathbf{V}_o(\omega) = \mathbf{H}(\omega)|_{\omega=2} \times \mathbf{V}_i(\omega) = \left(\left.\frac{3j\omega}{(2-\omega^2) + j\omega}\right|_{\omega=2}\right)(5\angle 0^\circ) = \left(\frac{j6}{-2+j2}\right)(5\angle 0^\circ) = 10.61\angle -45^\circ$$

The steady-state response is $v_o(t) = 10.61\cos(2t - 45^\circ)$ V.

(b) When $k = 4 - 2\sqrt{2}$, the poles are $p_{1,2} = \dfrac{-2\sqrt{2} \pm \sqrt{0}}{2} = -\sqrt{2}, -\sqrt{2}$, so the circuit is stable. The transfer function is

$$H(s) = \frac{1.17s}{\left(s+\sqrt{2}\right)^2} = \frac{1.17}{\left(s+\sqrt{2}\right)} - \frac{1.17\sqrt{2}}{\left(s+\sqrt{2}\right)^2}$$

The impulse response is

$$h(t) = \mathscr{L}^{-1}[H(s)] = 1.17e^{-\sqrt{2}t}\left(1 - \sqrt{2}t\right)u(t)$$

We see that when $k = 4 - 2\sqrt{2}$, the circuit is stable, and $\lim_{t\to\infty} h|(t)| = 0$.

(c) When $k = 4 + 2\sqrt{2}$, the poles are $p_{1,2} = \dfrac{2\sqrt{2} \pm \sqrt{0}}{2} = \sqrt{2}, \sqrt{2}$, so the circuit is not stable. The transfer function is

$$H(s) = \frac{6.83s}{\left(s - \sqrt{2}\right)^2} = \frac{6.83}{\left(s - \sqrt{2}\right)} + \frac{6.83\sqrt{2}}{\left(s - \sqrt{2}\right)^2}$$

The impulse response is

$$h(t) = \mathcal{L}^{-1}[H(s)] = 6.83e^{\sqrt{2}t}\left(1 + \sqrt{2}t\right)u(t)$$

We see that when $k = 4 + 2\sqrt{2}$, the circuit is unstable, and $\lim_{t \to \infty} |h(t)| = \infty$.

EXERCISE 14.10-1 The input to a circuit is the voltage $v_i(t)$. The output is the voltage $v_o(t)$. The transfer function of this circuit is

$$H(s) = \frac{V_o(s)}{V_i(s)} = \frac{ks}{s^2 + (3 - k)s + 2}$$

Determine the following:

(a) The steady-state response when $v_i(t) = 5 \cos 2t$ V and the gain of the VCVS is $k = 2$ V/V

(b) The impulse response when $k = 3 - 2\sqrt{2} = 0.17$ V/V

(c) The impulse response when $k = 3 + 2\sqrt{2} = 5.83$ V/V

Answers: **(a)** $v_o(t) = 7.07 \cos(2t - 45°)$ V

(b) $h(t) = 0.17\, e^{-\sqrt{2}t}(1 - \sqrt{2}t)u(t)$

(c) $h(t) = 5.83\, e^{\sqrt{2}t}(1 + \sqrt{2}t)u(t)$

14.11 PARTIAL FRACTION EXPANSION USING MATLAB

MATLAB provides a function called *residue* that performs the partial fraction expansion of a transfer function. Consider a transfer function

$$H(s) = \frac{b_3 s^3 + b_2 s^2 + b_1 s^1 + b_0 s^0}{a_3 s^3 + a_2 s^2 + a_1 s^1 + a_0 s^0} \tag{14.11-1}$$

In Eq. 14.11-1, the transfer function is represented as a ratio of two polynomials in s. In MATLAB, the transfer function given in Eq. 14.11-1 can be represented by two lists. One list specifies the coefficients of the numerator polynomial, and the other list specifies the coefficients of the denominator polynomial. For example,

$$\text{num} = [b_3 \quad b_2 \quad b_1 \quad b_0]$$

and

$$\text{den} = [a_3 \quad a_2 \quad a_1 \quad a_0]$$

(In this case, both polynomials are third-order polynomials, but the order of these polynomials could be changed.)

Partial fraction expansion can represent $H(s)$ as

$$H(s) = \frac{R_1}{s - p_1} + \frac{R_2}{s - p_2} + \frac{R_3}{s - p_3} + k(s) \tag{14.11-2}$$

R_1, R_2, and R_3 are called residues, and p_1, p_2, and p_3 are the poles. In general, both the residues and poles can be complex numbers. The term $k(s)$ will, in general, be a polynomial in s. MATLAB represents this form of the transfer function by three lists:

$$R = [R_1 \quad R_2 \quad R_3]$$

is a list of the residues,

$$p = [p_1 \quad p_2 \quad p_3]$$

is a list of the poles, and

$$k = [c_2 \quad c_1 \quad c_0]$$

is a list of the coefficients of the polynomial $k(s)$.

The MATLAB command

$$[\texttt{R, p, k}] = \texttt{residue (num, den)}$$

performs the partial fraction expansion, calculating the poles and residues from the coefficients of the numerator and denominator polynomials. The MATLAB command

$$[\texttt{n, d}] = \texttt{residue (R, p, k)}$$

performs the reverse operation, calculating the coefficients of the numerator and denominator polynomials from the poles and residues.

Figure 14.11-1 shows a MATLAB screen illustrating this procedure. In this example,

$$H(s) = \frac{s^3 + 2s^2 + 3s + 4}{s^3 + 6s^2 + 11s + 6}$$

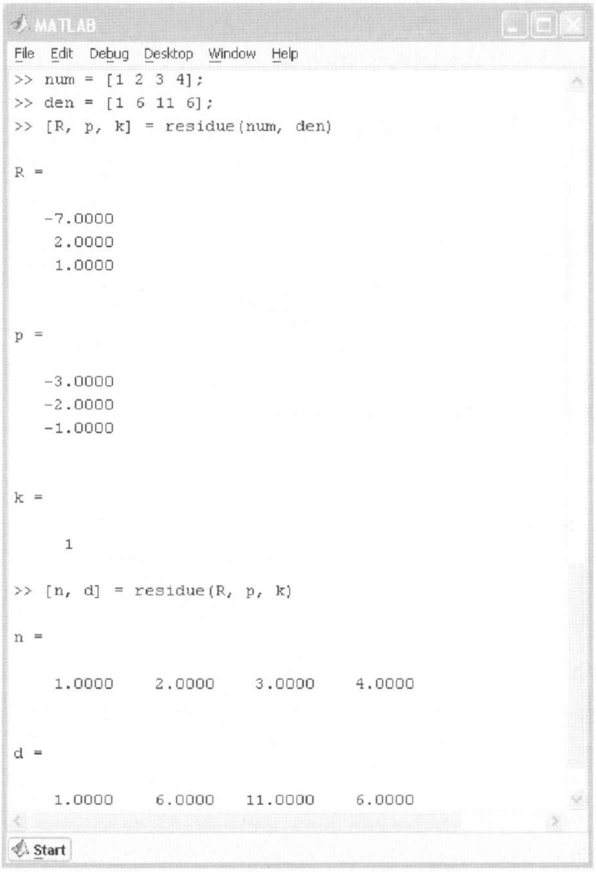

```
MATLAB
File Edit Debug Desktop Window Help
>> num = [1 2 3 4];
>> den = [1 6 11 6];
>> [R, p, k] = residue(num, den)

R =

   -7.0000
    2.0000
    1.0000


p =

   -3.0000
   -2.0000
   -1.0000


k =

     1

>> [n, d] = residue(R, p, k)

n =

    1.0000    2.0000    3.0000    4.0000


d =

    1.0000    6.0000   11.0000    6.0000
Start
```

FIGURE 14.11-1 Using MATLAB to perform partial fraction expansion.

is represented as

$$H(s) = \frac{-7}{s+3} + \frac{2}{s+2} + \frac{1}{s+1} + 1$$

by performing the partial fraction expansion.

The following examples illustrate the use of MATLAB for finding the inverse Laplace transform of functions having complex or repeated poles.

EXAMPLE 14.11-1 Repeated Real Poles

Find the inverse Laplace transform of

$$V(s) = \frac{12}{s(s^2 + 8s + 16)}$$

Solution

First, we will do this problem without using MATLAB. Noticing that $s^2 + 8s + 16 = (s+4)^2$, we begin the partial fraction expansion:

$$V(s) = \frac{12}{s(s^2 + 8s + 16)} = \frac{12}{s(s+4)^2} = \frac{k}{s+4} + \frac{-3}{(s+4)^2} + \frac{\frac{3}{4}}{s}$$

Next, the constant k is evaluated by multiplying both sides of the last equation by $s(s+4)^2$.

$$12 = ks(s+4) - 3s + \frac{3}{4}(s+4)^2 = \left(\frac{3}{4} + k\right)s^2 + (3+4k)s + 12 \;\Rightarrow\; k = -\frac{3}{4}$$

Finally

$$v(t) = \mathscr{L}^{-1}\left[\frac{-\frac{3}{4}}{s+4} + \frac{-3}{(s+4)^2} + \frac{\frac{3}{4}}{s}\right] = \left(\frac{3}{4} - e^{-4t}\left(\frac{3}{4} + 3t\right)\right)u(t)\text{ V}$$

Next, we perform the partial fraction expansion, using the MATLAB function residue:

```
>>num = [12];
>>den = [1 8 16 0];
>>[r, p] = residue(num, den)
```

MATLAB responds

```
r =
    -0.7500
    -3.0000
     0.7500
P =
    -4
    -4
     0
```

A repeated pole of multiplicity m is listed m times corresponding to the m terms

$$\frac{r_1}{s-p}, \frac{r_2}{(s-p)^2}, \ldots \frac{r_m}{(s-p)^m}$$

listed in order of increasing powers of $s-p$. The constants, r_1, r_2 ..., r_m are the corresponding residues, again listed in order of increasing powers of $s-p$. In our present case, the pole $p=-4$ has multiplicity 2, and the first two terms of the partial fraction expansion are

$$\frac{-0.75}{s-(-4)} + \frac{-3}{(s-(-4))^2} = \frac{-0.75}{s+4} + \frac{-3}{(s+4)^2}$$

The entire partial fraction expansion is

$$\frac{-0.75}{s-(-4)} + \frac{-3}{(s-(-4))^2} + \frac{0.75}{s-(0)} = \frac{-0.75}{s+4} + \frac{-3}{(s+4)^2} + \frac{0.75}{s}$$

Finally, as before,

$$v(t) = \mathscr{L}^{-1}\left[\frac{-0.75}{s+4} + \frac{-3}{(s+4)^2} + \frac{0.75}{s}\right] = \left(0.75 - e^{-4t}(0.75+3t)\right)u(t)\text{ V}$$

Example 14.11-2 Complex Poles

Find the inverse Laplace transform of

$$V(s) = \frac{12s+78}{s^2+8s+52}$$

Solution

First, we will do this problem without using MATLAB. Notice that the denominator does not factor any further in the real numbers. Let's complete the square in the denominator

$$V(s) = \frac{12s+78}{s^2+8s+52} = \frac{12s+78}{(s^2+8s+16)+36} = \frac{12s+78}{(s+4)^2+36} = \frac{12(s+4)+30}{(s+4)^2+36} = \frac{12(s+4)}{(s+4)^2+6^2} + \frac{5(6)}{(s+4)^2+6^2}$$

Now, use the property $e^{-at}f(t) \leftrightarrow F(s+a)$ and the Laplace transform pairs

$$\sin \omega t \text{ for } t \geq 0 \quad \leftrightarrow \quad \frac{\omega}{s^2+\omega^2} \quad \text{and} \quad \cos \omega t \quad \text{for } t \geq 0 \leftrightarrow \frac{s}{s^2+\omega^2}$$

to find the inverse Laplace transform:

$$v(t) = e^{-4t}\,\mathscr{L}^{-1}\left[\frac{12s}{s^2+6^2} + \frac{5(6)}{s^2+6^2}\right] = e^{-4t}[12\cos(6t) + 5\sin(6t)] \quad \text{for } t > 0$$

Next, we will use MATLAB to do the partial fraction expansion. First, enter the numerator and denominator polynomials as vectors listing the coefficients in order of decreasing power of s:

```
>>num = [12 78];
>>den = [1 8 52];
```

Now the command

```
>>[r, p] = residue(num, den)
```

tells MATLAB to do the partial fraction expansion return p, a list of the poles of $V(s)$, and r, a list of the corresponding residues. In the present case, MATLAB returns

```
r =
      6.0000 − 2.5000i
      6.0000 + 2.5000i
p =
    −4.0000 + 6.0000i
    −4.0000 − 6.5000i
```

indicating

$$V(s) = \frac{6 - j2.5}{s - (-4 + j6)} + \frac{6 + j2.5}{s - (-4 - j6)}$$

Notice that the first residue corresponds to the first pole and the second residue corresponds to the second pole. (Also, we expect complex poles to occur in pairs of complex conjugates and for the residues corresponding to complex conjugate poles to themselves be complex conjugates.) Taking the inverse Laplace transform, we get

$$v(t) = (6 - j2.5)e^{-(-4+j6)t} + (6 + j2.5)e^{-(-4-j6)t}$$

This expression, containing as it does complex numbers, isn't very convenient. Fortunately, we can use Euler's identity to obtain an equivalent expression that does not contain complex numbers. Because complex poles occur quite frequently, it's worthwhile to consider the general case:

$$V(s) = \frac{a + jb}{s - (c + jd)} + \frac{a - jb}{s - (c - jd)}$$

The inverse Laplace transform is

$$\begin{aligned} v(t) &= (a + jb)e^{(c+jd)t} + (a - jb)e^{(c-jd)t} \\ &= e^{ct}\left[(a + jb)\, e^{j\,dt} + (a + jb)e^{-j\,dt}\right] = e^{ct}\left[2a\left(\frac{e^{j\,dt} + e^{-j\,dt}}{2}\right) - 2b\left(\frac{e^{j\,dt} - e^{-j\,dt}}{2j}\right)\right] \end{aligned}$$

Euler's identity says

$$\frac{e^{j\,dt} + e^{-j\,dt}}{2} = \cos(dt) \quad \text{and} \quad \frac{e^{j\,dt} - e^{-j\,dt}}{2j} = \sin(dt)$$

Consequently,

$$v(t) = e^{ct}[2a\cos(dt) - 2b\sin(dt)]$$

Thus, we have the following Laplace transform pair

$$e^{ct}[2a\cos(dt) - 2b\sin(dt)] \leftrightarrow \frac{a + jb}{s - (c + jd)} + \frac{a - jb}{s - (c - jd)}$$

In the present case, $a = 6$, $b = -2.5$, $c = -4$, and $d = 6$, so we have

$$v(t) = e^{-4t}[12\cos(6t) + 5\sin(6t)] \quad \text{for } t > 0$$

It's sometimes convenient to express this answer in a different form. First, express the sine term as an equivalent cosine:

$$v(t) = e^{-4t}[12\cos(6t) + 5\cos(6t - 90°)] \quad \text{for } t > 0$$

Next, use phasors to combine the cosine terms

$$\mathbf{V}(\omega) = 12\underline{/0°} + 5\underline{/-90°} = 12 - j5 = 13\underline{/-22.62°}$$

Now $v(t)$ is expressed as

$$v(t) = 13e^{-4t}\cos(6t - 22.62°) \quad \text{for } t > 0$$

EXAMPLE 14.11-3 Both Real and Complex Poles

Find the inverse Laplace transform of

$$V(s) = \frac{105s+840}{(s^2+9.5s+17.5)(s^2+8s+80)}$$

Solution

Using MATLAB,

```
>> num=[105 840];
>> den=conv([1 9.5 17.5],[1 8 80]);
>> [r,p] = residue (num, den)
r =
      -0.8087 + 0.2415i
      -0.8087 - 0.2415i
      -0.3196
       1.9371
P =
      -4.0000 + 8.0000i
      -4.0000 - 8.0000i
      -7.0000
      -2.5000
```

Consequently,

$$V(s) = \frac{-0.8087+j0.2415}{s-(-4+j8)} + \frac{-0.8087-j0.2415}{s-(-4-j8)} + \frac{-0.3196}{s-(-7)} + \frac{1.9371}{s-(-2.5)}$$

Using the Laplace transform pair,

$$e^{ct}[2\,a\cos(dt) - 2b\sin(dt)] \quad \leftrightarrow \quad \frac{a+jb}{s-(c+jd)} + \frac{a-jb}{s-(c-jd)}$$

with $a = -0.8087$, $b = 0.2415$, $c = -4$, and $d = 8$, we have

$$\mathcal{L}^{-1}\left[\frac{-0.8087+j0.2415}{s-(-4+j8)} + \frac{-0.8087-j0.2415}{s-(-4-j8)}\right] = e^{-4t}[-1.6174\cos(8t) + 0.483\sin(8t)]$$

Taking the inverse Laplace transform of the remaining terms of $V(s)$, we get

$$v(t) = e^{-4t}[-1.6174\cos(8t) + 0.483\sin(8t)] - 0.3196e^{-7t} + 1.9371e^{-2.5t} \quad \text{for } t > 0$$

14.12 HOW CAN WE CHECK . . . TRANSFER FUNCTIONS?

Engineers are frequently called upon to check that a solution to a problem is indeed correct. For example, proposed solutions to design problems must be checked to confirm that all of the specifications have been satisfied. In addition, computer output must be reviewed to guard against data-entry errors, and claims made by vendors must be examined critically.

Engineering students are also asked to check the correctness of their work. For example, occasionally just a little time remains at the end of an exam. It is useful to be able to quickly identify those solutions that need more work.

The following examples illustrate techniques useful for checking the solutions of the sort of problem discussed in this chapter.

EXAMPLE 14.12-1 How Can We Check Transfer Functions?

A circuit is specified to have a transfer function of

$$H(s) = \frac{V_o(s)}{V_1(s)} = \frac{25}{s^2 + 10s + 125} \tag{14.12-1}$$

and a step response of

$$v_o(t) = 0.1\left(2 - e^{-5t}(3\cos 10t + 2\sin 10t)\right)u(t) \tag{14.12-2}$$

How can we check that these specifications are consistent?

Solution

If the specifications are consistent, then the unit step response and the transfer function will be related by

$$\mathscr{L}[v_o(t)] = H(s)\frac{1}{s} \tag{14.12-3}$$

where $V_1(s) = 1/s$.

This equation can be verified either by calculating the Laplace transform of $v_o(t)$ or by calculating the inverse Laplace transform of $H(s)/s$. Both of these calculations involve a bit of algebra. The final and initial value theorems provide a quicker, though less conclusive, check. (If either the final or initial value theorem is not satisfied, then we know that the step response is not consistent with the transfer function. The step response could be inconsistent with the transfer function even if both the final and initial value theorems are satisfied.) Let us see what the final and initial value theorems tell us.

The final value theorem requires that

$$v_o(\infty) = \lim_{s\to 0} s\left[H(s)\frac{1}{s}\right] \tag{14.12-4}$$

From Eq. 14.12-1, we substitute $H(s)$, obtaining

$$\lim_{s\to 0} s\left[\frac{25}{s^2 + 10s + 125} \cdot \frac{1}{s}\right] = \lim_{s\to 0}\left[\frac{25}{s^2 + 10s + 125}\right] = \frac{25}{125} = 0.2 \tag{14.12-5}$$

From Eq. 14.12-2, we evaluate at $t = \infty$, obtaining

$$v_o(\infty) = 0.1(2 - e^{-\infty}(2\cos\infty + \sin\infty)) = 0.1(2 - 0) = 0.2 \tag{14.12-6}$$

so the final value theorem is satisfied.

Next, the initial value theorem requires that

$$v_o(0) = \lim_{s\to\infty} s\left[H(s)\frac{1}{s}\right] \tag{14.12-7}$$

From Eq. 14.12-1, we substitute $H(s)$, obtaining

$$\lim_{s\to\infty} s\left[\frac{25}{s^2 + 10s + 125} \cdot \frac{1}{s}\right] = \lim_{s\to\infty}\frac{25/s^2}{1 + 10/s + 125/s^2} = \frac{0}{1} = 0 \tag{14.12-8}$$

From Eq. 14.12-1, we evaluate at $t = 0$ to obtain

$$\begin{aligned} v_o(0) &= 0.1(2 - e^{-0}(3\cos 0 + 2\sin 0)) \\ &= 0.1(2 - 1(3 + 0)) \\ &= -0.1 \end{aligned} \tag{14.12-9}$$

The initial value theorem is not satisfied, so the step response is not consistent with the transfer function.

EXAMPLE 14.12-2 How Can We Check Transfer Functions?

A circuit is specified to have a transfer function of

$$H(s) = \frac{V_o(s)}{V_1(s)} = \frac{25}{s^2 + 10s + 125} \tag{14.12-10}$$

and a unit step response of

$$v_o(t) = 0.1\left(2 - e^{-5t}(2\cos 10t + 3\sin 10t)\right)u(t) \tag{14.12-11}$$

How can we check that these specifications are consistent? (This step response is a slightly modified version of the step response considered in Example 14.12-1.)

Solution

The reader is invited to verify that both the final and initial value theorems are satisfied. This suggests, but does not guarantee, that the transfer function and step response are consistent. To guarantee consistency, it is necessary to verify that

$$\mathcal{L}[v_o(t)] = H(s)\frac{1}{s} \tag{14.12-12}$$

either by calculating the Laplace transform of $v_o(t)$ or by calculating the inverse Laplace transform of $H(s)/s$. Recall the input is a unit step, so $V_1(s) = 1/s$. We will calculate the Laplace transform of $v_0(t)$ as follows:

$$\begin{aligned}
\mathcal{L}[0.1(2 - e^{-5t}(2\cos 10t + 3\sin 10t))u(t)] &= 0.1\left[\frac{2}{s} - \frac{2(s+5)}{(s+5)^2 + 10^2} - 3\frac{10}{(s+5)^2 + 10^2}\right] \\
&= 0.1\left[\frac{2}{s} - \frac{2s + 40}{s^2 + 10s + 125}\right] \\
&= \frac{-2s + 25}{s(s^2 + 10s + 125)}
\end{aligned}$$

Because this is not equal to $H(s)/s$, Eq. 14.12-12 is not satisfied. The step response is not consistent with the transfer function even though the initial and final values of $v_o(t)$ are consistent.

EXERCISE 14.12-1 A circuit is specified to have a transfer function of

$$H(s) = \frac{25}{s^2 + 10s + 125}$$

and a unit step response of

$$v_o(t) = 0.1\left(2 - e^{-5t}(2\cos 10t + \sin 10t)\right)u(t)$$

Verify that these specifications are consistent.

14.13 DESIGN EXAMPLE

SPACE SHUTTLE CARGO DOOR

The U.S. space shuttle docked with Russia's *Mir* space station several times. The electromagnet for opening a cargo door on the NASA space shuttle requires 0.1 A before activating. The electromagnetic coil is represented by L, as shown in Figure 14.13-1. The activating current is designated $i_1(t)$. The time period required for i_1 to reach 0.1 A is specified as less than 3 s. Select a suitable value of L.

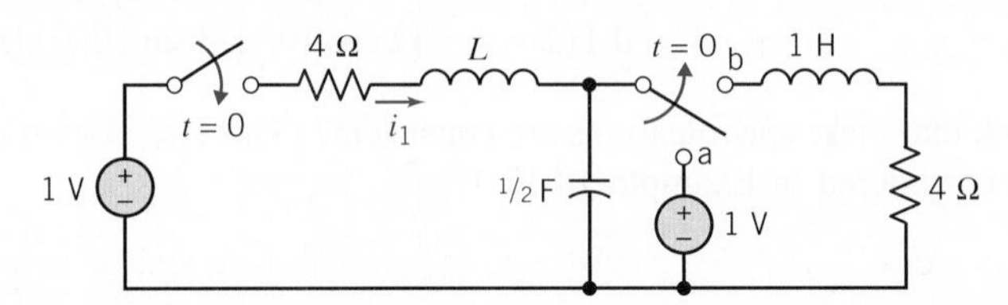

FIGURE 14.13-1 The control circuit for a cargo door on the NASA space shuttle.

Describe the Situation and the Assumptions

1. The two switches are thrown at $t = 0$, and the movement of the second switch from terminal a to terminal b occurs instantaneously.
2. The switches prior to $t = 0$ were in position for a long time.

State the Goal

Determine a value of L so that the time period for the current $i_1(t)$ to attain a value of 0.1 A is less than 3 s.

Generate a Plan

1. Determine the initial conditions for the two inductor currents and the capacitor voltage.
2. Designate two mesh currents and write the two mesh KVL equations, using the Laplace transform of the variables and the impedance of each element.
3. Select a trial value of L and solve for $I_1(s)$.
4. Determine $i_1(t)$.
5. Sketch $i_1(t)$ and determine the time instant t_1 when $i_1(t_1) = 0.1$ A.
6. Check whether $t_1 < 3$ s, and, if not, return to step 3 and select another value of L.

GOAL	EQUATION	NEED	INFORMATION
Determine the initial conditions at $t = 0$	$i(0) = i(0^-)$ $v_c(0) = v_c(0^-)$	Prepare a sketch of the circuit at $t = 0^-$. Find $i_1(0^-)$, $i_2(0^-)$, $v_c(0^-)$.	
Designate two mesh currents and write the mesh KVL equations.		$I_1(s)$, $I_2(s)$; the initial conditions $i_1(0)$, $i_2(0)$	

(continued)

GOAL	EQUATION	NEED	INFORMATION
Solve for $I_1(s)$ and select L.			Cramer's rule
Determine $i_1(t)$.	$i_1(t) = \mathscr{L}^{-1}[I_1(s)]$		Use a partial fraction expansion.
Sketch $i_1(t)$ and find t_1.	$i_1(t_1) = 0.1$ A		

Act on the Plan

First, the circuit with the switches in position at $t = 0^-$ is shown in Figure 14.13-2. Clearly, the inductor currents are $i_1(0^-) = 0$ and $i_2(0^-) = 0$. Furthermore, we have

$$v_c(0) = 1 \text{ V}$$

Second, redraw the circuit for $t > 0$ as shown in Figure 14.13-3 and designate the two mesh currents i_1 and i_2 as shown.

Recall that the impedance is Ls for an inductor and $1/Cs$ for a capacitor. We must account for the initial condition for the capacitor. Recall that the capacitor voltage may be written as

$$v_c(t) = v_c(0) + \frac{1}{C}\int_0^t i_c(\tau)d\tau$$

The Laplace transform of this equation is

$$V_c(s) = \frac{v_c(0)}{s} + \frac{1}{Cs}I_c(s)$$

where $I_c(s) = I_1(s) - I_2(s)$ in this case. We now may write the two KVL equations for the two meshes for $t \geq 0$ with $v_c(0) = 1$ V as

$$\begin{aligned} \text{mesh 1:} &\quad -V_1(s) + (4 + Ls)I_1(s) + V_c(s) = 0 \\ \text{mesh 2:} &\quad (4 + 1s)I_2(s) - V_c(s) = 0 \end{aligned}$$

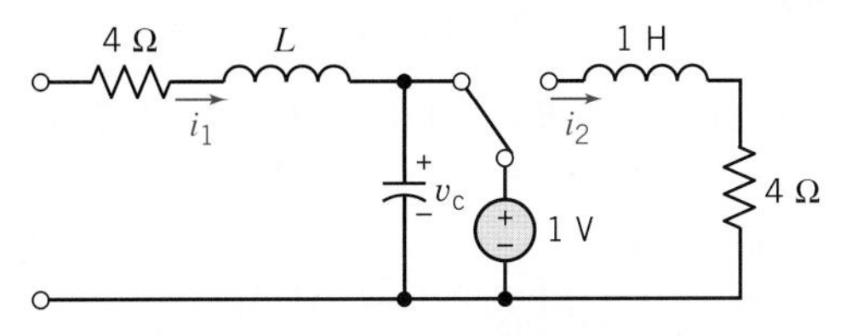

FIGURE 14.13-2 The circuit of Figure14.13-1 at $t = 0^-$.

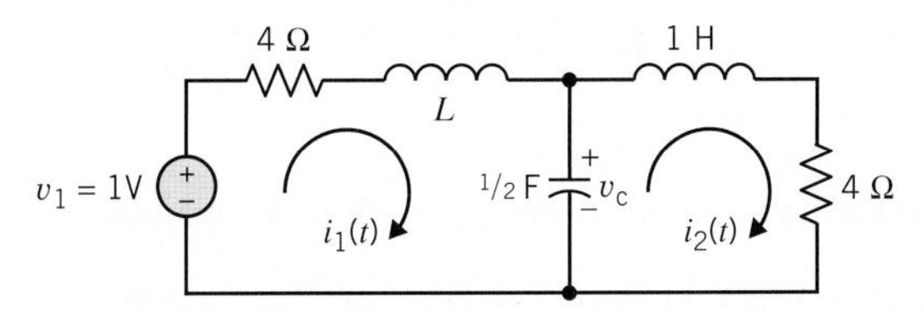

FIGURE 14.13-3 The circuit of Figure 14.13-1 for $t > 0$.

The Laplace transform of the input voltage is

$$V_1(s) = \frac{1}{s}$$

Also, note that for the capacitor we have

$$V_c(s) = \frac{1}{s} + \frac{1}{Cs}(I_1(s) - I_2(s))$$

Substituting V_1 and V_c into the mesh equations, we have (when $C = 1/2\,\text{F}$)

$$\left(4 + Ls + \frac{2}{s}\right)I_1(s) - \left(\frac{2}{s}\right)I_2(s) = 0$$

and

$$-\left(\frac{2}{s}\right)I_1(s) + \left(4 + s + \frac{2}{s}\right)I_2(s) = \frac{1}{s}$$

The third step requires the selection of the value of L and then solving for $I_1(s)$. Examine Figure 14.13-3; the two meshes are symmetric when $L = 1\,\text{H}$. Then, trying this value and using Cramer's rule, we solve for $I_1(s)$, obtaining

$$I_1(s) = \frac{\left(\frac{2}{s}\right)\frac{1}{s}}{\left(4 + s + \frac{2}{s}\right)^2 - \left(\frac{2}{s}\right)^2} = \frac{2}{s(s^3 + 8s^2 + 20s + 16)}$$

Fourth, to determine $i_1(t)$, we will use a partial fraction expansion. Rearranging and factoring the denominator of $I_1(s)$, we determine that

$$I_1(s) = \frac{2}{s(s+4)(s+2)^2}$$

Hence, we have the partial fraction expansion

$$I_1(s) = \frac{A}{s} + \frac{\mathrm{B}}{s+4} + \frac{C}{(s+2)^2} + \frac{D}{s+2}$$

Then, we readily determine that $A = 1/8$, $B = -1/8$, and $C = -1/2$. To find D, we use Eq. 14.4-9 to obtain

$$\begin{aligned} D &= \frac{1}{(2-1)!}\left[\frac{d}{ds}(s+2)^2 I_1(s)\right]_{s=-2} \\ &= \left.\frac{-2(2s+4)}{s^4 + 8s^3 + 16s^2}\right|_{s=-2} \\ &= 0 \end{aligned}$$

Therefore, using the inverse Laplace transform for each term, we obtain

$$i_1(t) = 1/8 - (1/8)e^{-4t} - (1/2)te^{-2t}\ \text{A} \quad t \geq 0$$

Verify the Proposed Solution

The sketch of $i_1(t)$ is shown in Figure 14.13-4. It is clear that $i_1(t)$ has essentially reached a steady-state value of 0.125 A by $t = 4$ seconds.

To find t_1 when

$$i_1(t_1) = 0.1 \text{ A}$$

we estimate that t_1 is approximately 2 seconds. After evaluating $i_1(t)$ for a few selected values of t near 2 seconds, we find that $t_1 = 1.8$ seconds. Therefore, the design requirements are satisfied for $L = 1$ H. Of course, other suitable values of L can be determined that will satisfy the design requirements.

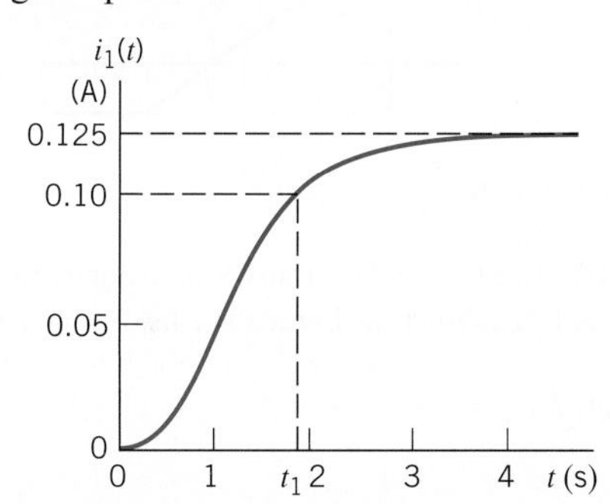

FIGURE 14.13-4 The response of $i_1(t)$.

14.14 SUMMARY

- Pierre-Simon Laplace is credited with a transform that bears his name. The Laplace transform is defined as

$$\mathcal{L}[f(t)] = \int_{0-}^{\infty} f(t)e^{-st}\,dt$$

- The Laplace transform transforms the differential equation describing a circuit in the time domain into an algebraic equation in the complex frequency domain. After solving the algebraic equation, we use the inverse Laplace transform to obtain the circuit response in the time domain. Figure14.2-1 illustrates this process.
- Tables 14.2-1 tabulates frequently used Laplace transform pairs. Table 14.2-2 tabulates some properties of the Laplace transform.
- The inverse Laplace transform is obtained using partial fraction expansion.
- Table 14.7-1 shows that circuits can be represented in the frequency domain in a manner that accounts for the initial conditions of capacitors and inductors.
- To find the complete response of a linear circuit, we first represent the circuit in the frequency domain using the Laplace transform. Next, we analyze the circuit, perhaps by writing mesh or node equations. Finally, we use the inverse Laplace transform to represent the response in the time domain.
- The transfer function, $H(s)$, of a circuit is defined as the ratio of the response $Y(s)$ of the circuit to an excitation $X(s)$ expressed in the complex frequency domain.

$$H(s) = \frac{Y(s)}{X(s)}$$

  This ratio is obtained assuming all initial conditions are equal to zero.
- The step response is the response of a circuit to a step input when all initial conditions are zero. Then step response is related to the transfer function by

$$\text{step response} = \mathcal{L}^{-1}\left[\frac{H(s)}{s}\right]$$

- The impulse response is the response of a circuit to an impulse input when all initial conditions are zero. The impulse response is related to the transfer function by

$$\text{impulse response} = \mathcal{L}^{-1}[H(s)]$$

- A circuit is said to be *stable* when the response to a bounded input signal is a bounded output signal. All the poles of the transfer function of a stable circuit lie in the left-half s-plane.
- MATLAB performs partial fraction expansion.

PROBLEMS

Section 14.2 Laplace Transform

P 14.2-1 Find the Laplace transform, $F(s)$, when $f(t) = A\cos\omega t$, $t \geq 0$.

Answer: $f(s) = \dfrac{As}{s^2 + \omega^2}$

P 14.2-2 Find the Laplace transform, $F(s)$, when $f(t) = t$, $t \geq 0$.

P 14.2-3 Using the linearity property, find the Laplace transform of $f(t) = e^{-5t} + t$, $t \geq 0$.

P 14.2-4 Using the linearity property, find the Laplace transform of $f(t) = A(1 - e^{-bt})u(t)$.

Answer: $F(s) = \dfrac{Ab}{s(s + b)}$

Section 14.3 Pulse Inputs

P 14.3-1 Consider a pulse $f(t)$ defined by

$$\begin{aligned} f(t) &= A \quad 0 \leq t \leq T \\ &= 0 \quad \text{all other } t \end{aligned}$$

Find $F(s)$.

Answer: $F(s) = \dfrac{A(1 - e^{-sT})}{s}$

P 14.3-2 Consider the pulse shown in Figure P 14.3-2, where the time function follows e^{at} for $0 < t < T$. Find $F(s)$ for the pulse.

Answer: $F(s) = \dfrac{1 - e^{-(s-a)T}}{s - a}$

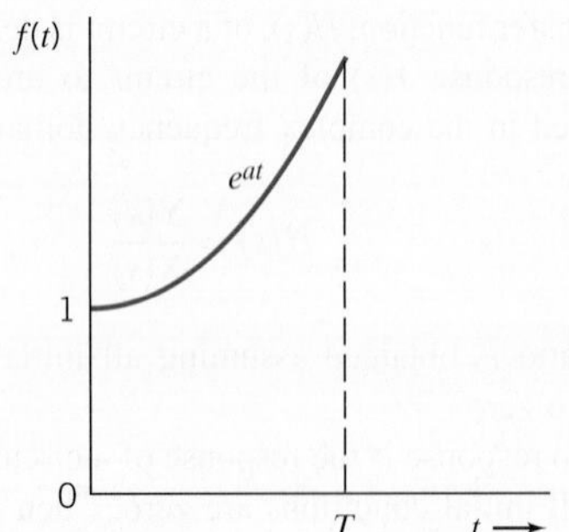

Figure P 14.3-2

P 14.3-3 Find the Laplace transform $F(s)$ for

(a) $f(t) = t^2 e^{-3t}$, $t \geq 0$
(b) $f(t) = \delta(t - T)$, $t \geq 0$
(c) $f(t) = e^{-4t}\sin 6t$, $t \geq 0$

P 14.3-4 Find the Laplace transform for $g(t) = e^{-t}u(t - 0.5)$.

P 14.3-5 Find the Laplace transform for

$$f(t) = \frac{-(t - T)}{T}u(t - T)$$

Answer: $F(s) = \dfrac{-1e^{-sT}}{Ts^2}$

P 14.3-6 Determine the Laplace transform of $f(t)$ shown in Figure P 14.3-6.

Hint: $f(t) = \left(5 - \dfrac{5}{3}t\right)u(t) + \dfrac{5}{3}\left(t - \dfrac{21}{5}\right)u\left(t - \dfrac{21}{5}\right)$

Answer: $F(s) = \dfrac{5e^{-4.2s} + 15s - 5}{3s^2}$

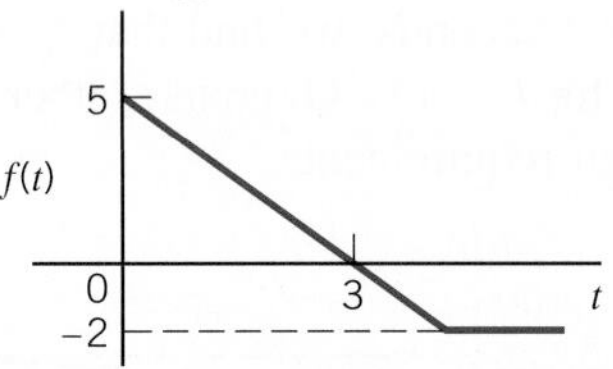

Figure P 14.3-6

P 14.3-7 Use the Laplace transform to obtain the transform of the signal $f(t)$ shown in Figure P 14.3-7.

Answer: $F(s) = \dfrac{3(1 - e^{-2s})}{s}$

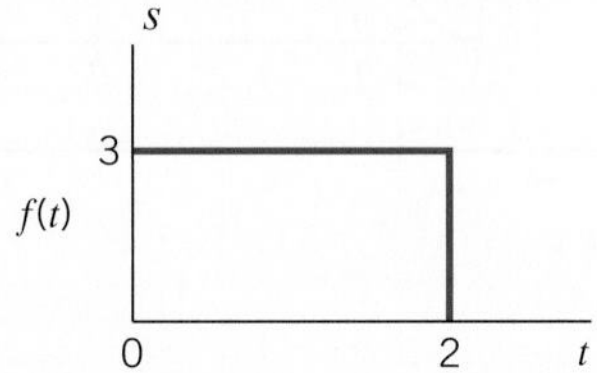

Figure P 14.3-7

P 14.3-8 Determine the Laplace transform of $f(t)$ shown in Figure P 14.3-8.

Answer: $F(s) = \dfrac{5}{2s^2}(1 - e^{-2s} - 2se^{-2s})$

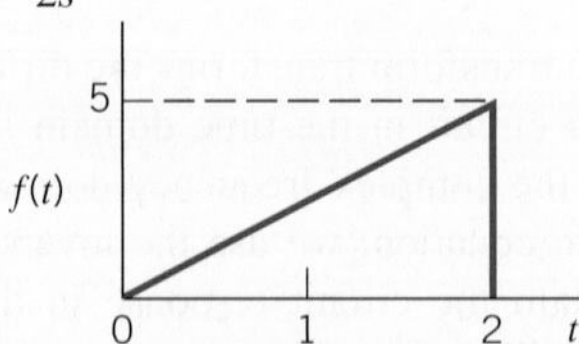

Figure P 14.3-8

Section 14.4 Inverse Laplace Transform

P 14.4-1 Find $f(t)$ when

$$F(s) = \frac{s + 5}{s^3 + 3s^2 + 6s + 4}$$

Answer: $f(t) = \dfrac{4}{3}e^{-t} - \dfrac{4}{3}e^{-t}\cos\sqrt{3}t + \dfrac{1}{\sqrt{3}}e^{-t}\sin\sqrt{3}t$, $t \geq 0$

P 14.4-2 Find $f(t)$ when

$$F(s) = \frac{s^2 - 2s + 2}{s^3 + 3s^2 + 4s + 2}$$

P 14.4-3 Find $f(t)$ when

$$F(s) = \frac{5s - 4}{s^3 - 3s - 2}$$

Answer: $f(t) = e^{-t} + 3te^{-t} + \frac{2}{3}e^{2t}, t \geq 0$

P 14.4-4 Find the inverse transform of

$$Y(s) = \frac{1}{s^3 + 3s^2 + 4s + 2}$$

Answer: $y(t) = e^{-t}(1 - \cos t), t \geq 0$

P 14.4-5 Find the inverse transform of

$$F(s) = \frac{2s + 6}{(s + 1)(s^2 + 2s + 5)}$$

P 14.4-6 Find the inverse transform of

$$F(s) = \frac{2s + 8}{s(s^2 + 3s + 2)}$$

Answer: $f(t) = \left[4 - 6e^{-t} + 2^{e-2t}\right]$

P 14.4-7 Prove that

$$\mathscr{L}^{-1}\left[\frac{cs + (ca - \omega d)}{(s + a)^2 + \omega^2}\right]$$

is $f(t) = me^{-at}\cos(\omega t + \theta)$ where $m = \sqrt{c^2 + d^2}$ and $\theta = \tan^{-1}(d/c)$.

P 14.4-8 Find the inverse transform of $F(s)$, expressing $f(t)$ in cosine and angle forms.

(a) $F(s) = \frac{8s - 3}{s^2 + 4s + 13}$ **(b)** $F(s) = \frac{3e^{-s}}{s^2 + 2s + 17}$

Answers: **(a)** $f(t) = 10.2e^{-2t}\cos(3t + 38.4°), t \geq 0$
(b) $f(t) = \frac{3}{4}e^{-(t-1)}\sin[4(t - 1)], t \geq 1$

P 14.4-9 Find the inverse transform of $F(s)$.

(a) $F(s) = \frac{s^2 - 8}{s(s + 1)^2}$ **(b)** $F(s) = \frac{4s^2}{(s + 3)^3}$

Answers: **(a)** $f(t) = -8 + 9e^{-t} + 7te^{-t}, t \geq 0$
(b) $f(t) = 4e^{-3t} - 24te^{-3t} + 18t^2e^{-3t}, t \geq 0$

Section 14.5 Initial and Final Value Theorems

P 14.5-1 A function of time is represented by

$$F(s) = \frac{2s^2 - 3s + 6}{s^3 + 3s^2 + 2s}$$

(a) Find the initial value of $f(t)$ at $t = 0$.
(b) Find the value of $f(t)$ as t approaches infinity.

P 14.5-2 Find the initial and final values of $v(t)$ when

$$V(s) = \frac{(35 + 16)}{s^2 + 4s + 12}$$

Answer: $v(0) = 3, v(\infty) = 0$ V

P 14.5-3 Find the initial and final values of $v(t)$ when

$$V(s) = \frac{(s + 15)}{(3s^3 + 2s^2 + 1s)}$$

Answers: $v(0) = 0, v(\infty) = 15$ V

P 14.5-4 Find the initial and final values of $f(t)$ when

$$F(s) = \frac{-2(s + 7)}{s^2 - 2s + 10}$$

Answer: initial value = -2; final value does not exist

P 14.5-5 Given that $\mathscr{L}[v(t)] = \frac{a s + b}{s^2 + 8s}$ where $v(t)$ is the voltage shown in Figure P 14.5-5, determine the values of a and b.

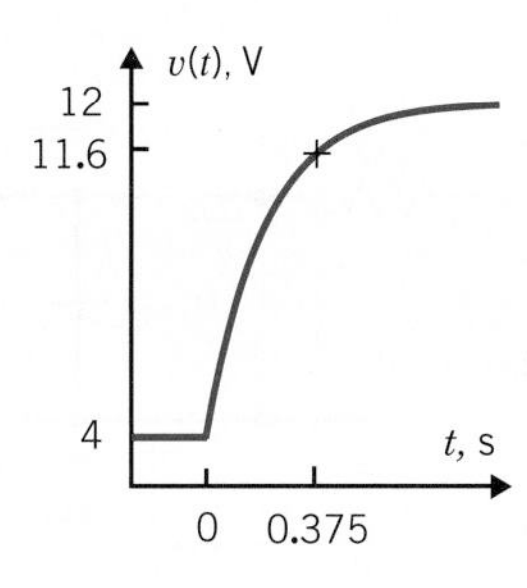

Figure P 14.5-5

P 14.5-6 Given that $\mathscr{L}[v(t)] = \frac{a s + b}{2s^2 + 40s}$ where $v(t)$ is the voltage shown in Figure P 14.5-6, determine the values of a and b.

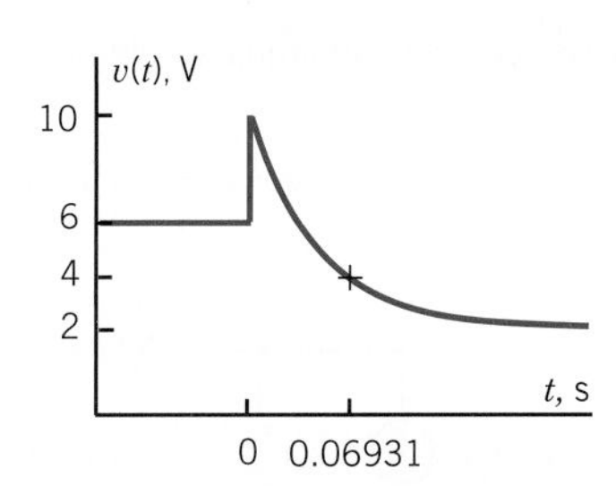

Figure P 14.5-6

Section 14.6 Solution of Differential Equations Describing a Circuit

P 14.6-1 Find $i(t)$ for the circuit of Figure P 14.6-1 when $i(0) = 1$ A, $v(0) = 8$ V, and $v_1 = 2e^{-at}u(t)$ where $a = 2 \times 10^4$.

Answers: $i(t) = \frac{1}{15}\left(-10e^{-bt} + 3e^{-2bt} + 22e^{-4bt}\right)A, t \geq 0$, $b = 10^4$

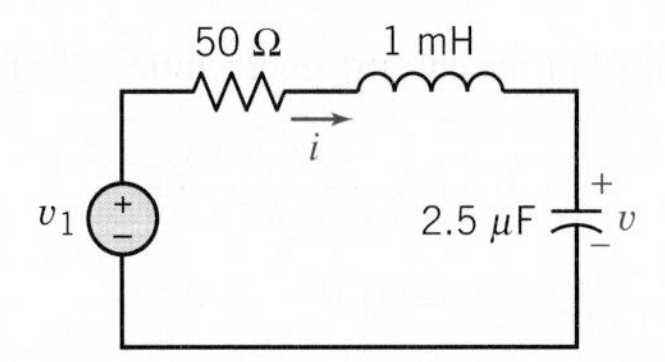

Figure P 14.6-1

P 14.6-2 All new homes are required to install a device called a ground fault circuit interrupter (GFCI) that will provide protection from shock. By monitoring the current going to and returning from a receptacle, a GFCI senses when normal flow is interrupted and switches off the power in 1/40 second. This is particularly important if you are holding an appliance shorted through your body to ground. A circuit model of the GFCI acting to interrupt a short is shown in Figure P 14.6-2. Find the current flowing through the person and the appliance, $i(t)$, for $t \geq 0$ when the short is initiated at $t = 0$. Assume $v = 160 \cos 400t$ and the capacitor is intially uncharged.

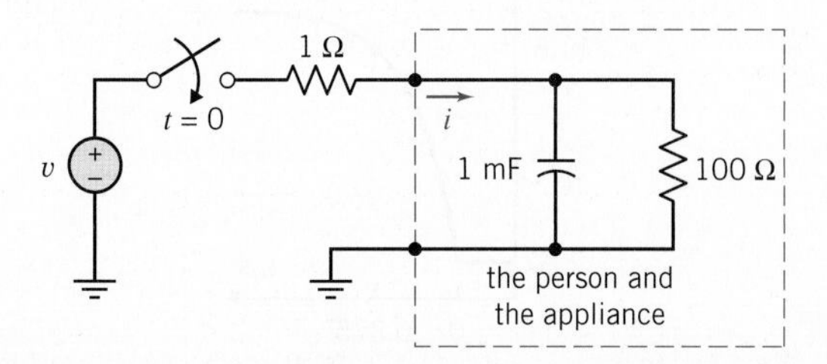

Figure P 14.6-2 Circuit model of person and appliance shorted to ground.

P 14.6-3 Using the Laplace transform, find $v_c(t)$ for $t > 0$ for the circuit shown in Figure P 14.6-3. The initial conditions are zero.

Hint: Use a source transformation to obtain a single mesh circuit.

Answer: $v_c = -6.9e^{-3t} + 7.5\,(\cos 2t + \sin 2t)$ V

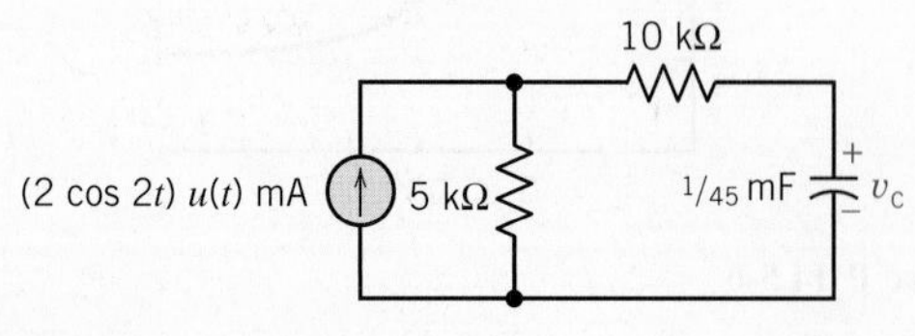

Figure P 14.6-3

P 14.6-4 Using Laplace transforms, find $v_c(t)$ for $t > 0$ for the circuit of Figure P 14.6-4 when (a) $C = 1/18$ F and (b) $C = 1/10$ F.

Answers: (a) $v_c(t) = -8 + 8e^{-3t} + 24te^{-3t}$ V
(b) $v_c(t) = -8 + 10e^{-t} - 2e^{-5t}$ V

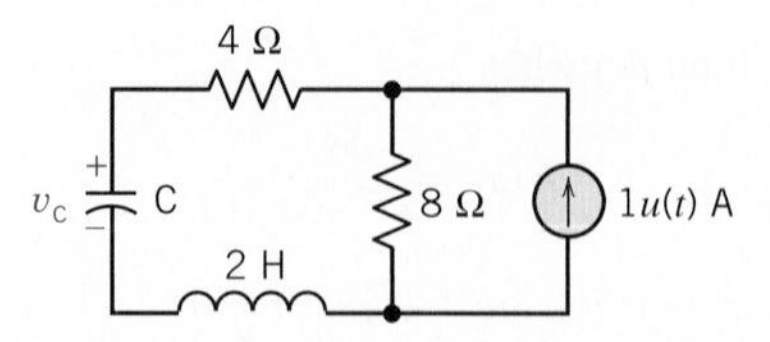

Figure P 14.6-4

P 14.6-5 Find $i(t)$ for the circuit of Figure P 14.6-5. Assume the switch has been open for a long time.

Answer: $i = -0.025e^{-200t} \sin 400t$ A, $t > 0$

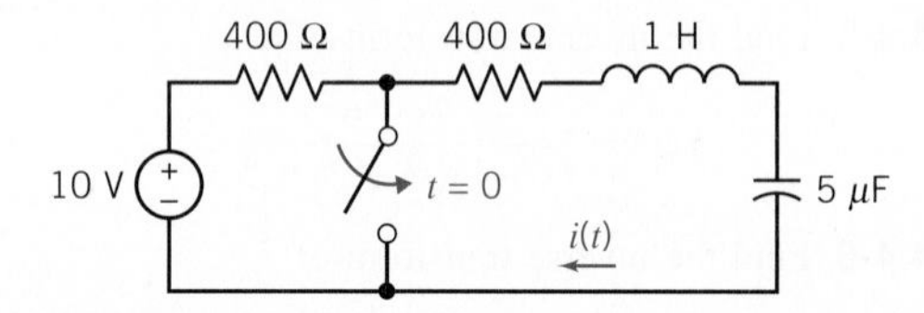

Figure P 14.6-5

P 14.6-6 Determine the inductor current, $i(t)$, in the circuit shown in Figure P 14.6-6.

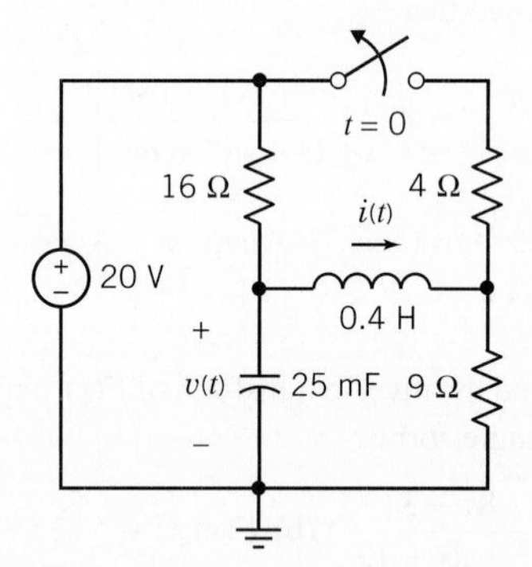

Figure P 14.6-6

P 14.6-7 Find $i(t)$ for the circuit of Figure P 14.6-7 when $i_1(t) = 7e^{-6t}$ A for $t \geq 0$ and $i(0) = 0$.

Answer: $i(t) = -\dfrac{7}{2}\,e^{-6t} + \dfrac{7}{2}\,e^{-3t}$ A $\quad t \geq 0$

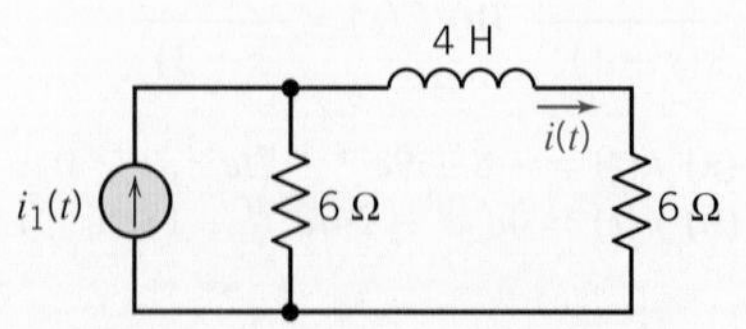

Figure P 14.6-7

P 14.6-8 Find $v_2(t)$ for the circuit of Figure P 14.6-8 for $t \geq 0$.

Hint: Write the node equations at a and b in terms of v_1 and v_2. The initial conditions are $v_1(0) = 10$ V and $v_2(0) = 25$ V. The source is $v_s = 50 \cos 2t\, u(t)$ V.

Answer: $v_2(t) = \frac{23}{3}\,e^{-t} + \frac{16}{3}\,e^{-4t} + 12 \cos 2t + 12 \sin 2t$ V $t \geq 0$

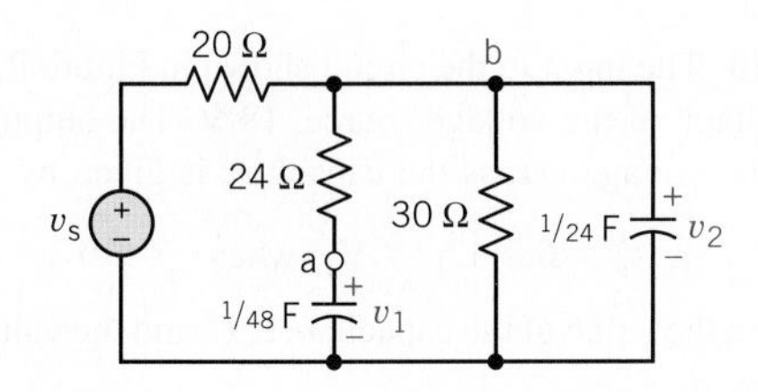

Figure P 14.6-8

P 14.6-9 Using Laplace transforms, find $v(t)$ for $t > 0$ for the circuit shown in Figure P 14.6-9.

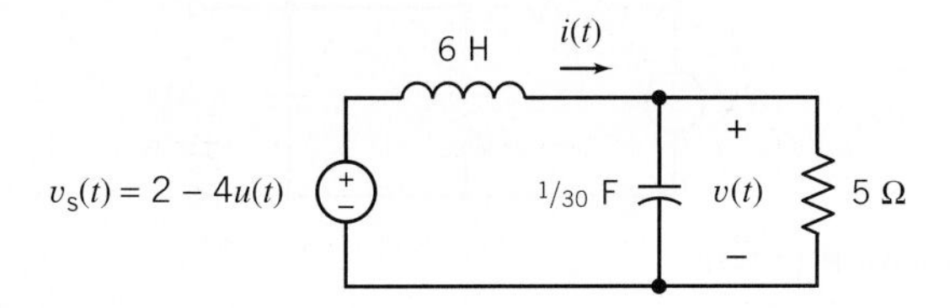

Figure P 14.6-9

P 14.6-10 Using Laplace transforms, find $v_o(t)$ for $t > 0$ for the circuit shown in Figure P 14.6-10.

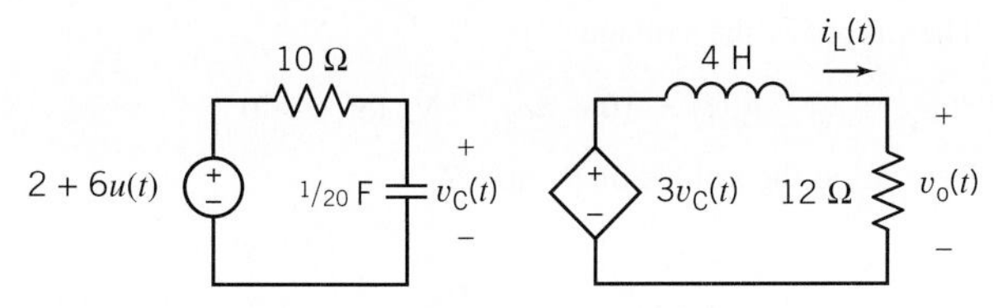

Figure P 14.6-10

Section 14.7 Circuit Analysis Using Impedance and Initial Conditions

P 14.7-1 Using Laplace transforms, find the response $i_L(t)$ for $t > 0$ for the circuit of Figure P 14.7-1.

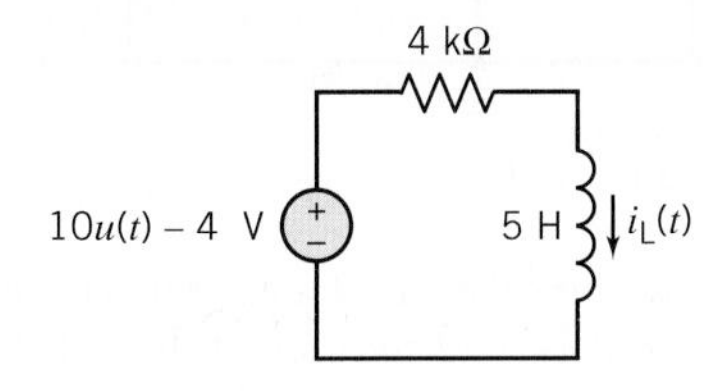

Figure P 14.7-1

P 14.7-2 Using Laplace transforms, find the response $i_L(t)$ for $t > 0$ for the circuit of Figure P 14.7-2.

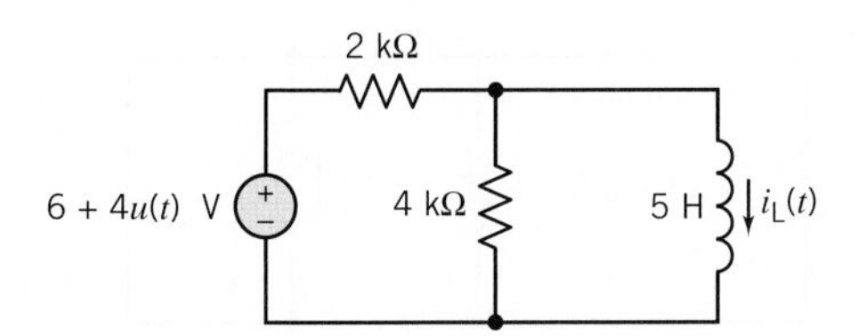

Figure P 14.7-2

P 14.7-3 Using Laplace transforms, find the response $v_c(t)$ for $t > 0$ for the circuit of Figure P 14.7-3.

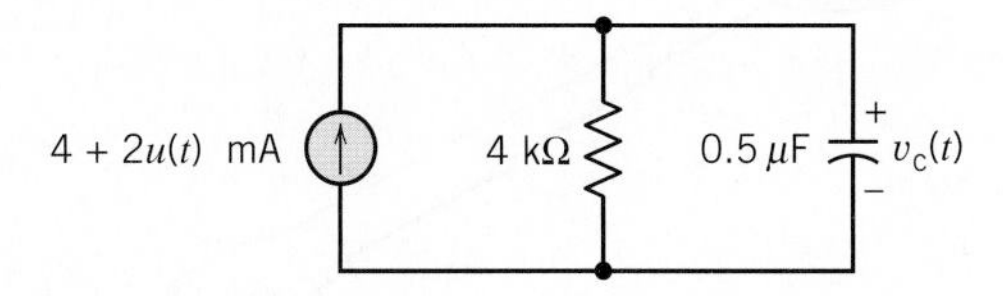

Figure P 14.7-3

P 14.7-4 Using Laplace transforms, find the response $v_c(t)$ for $t > 0$ for the circuit of Figure P 14.7-4.

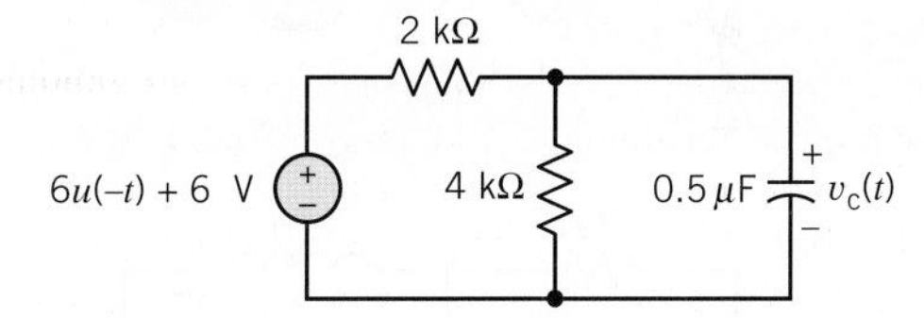

Figure P 14.7-4

P 14.7-5 Using Laplace transforms, find the response $v(t)$ for $t > 0$ for the circuit of Figure P 14.7-5 when $v_s = 6e^{-3t}u(t)$ V.

Answer: $v = \frac{44}{3}e^{-2t} + \frac{1}{3}e^{-5t} - 9e^{-3t}$ V

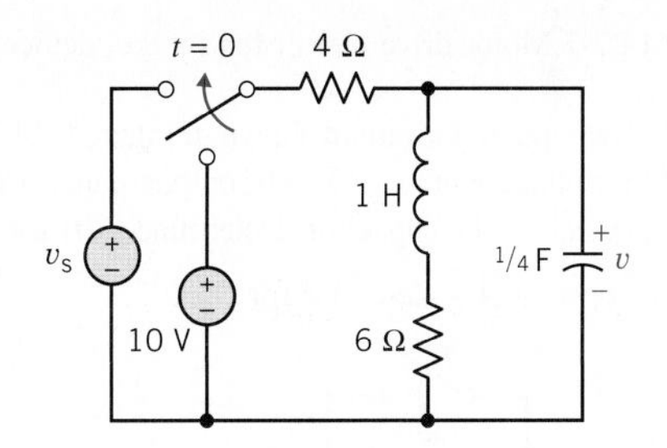

Figure P 14.7-5

P 14.7-6 Determine $v_o(t)$ when the capacitance has an initial voltage $v(0^-) = 5$ V, as shown in Figure P 14.7-6.

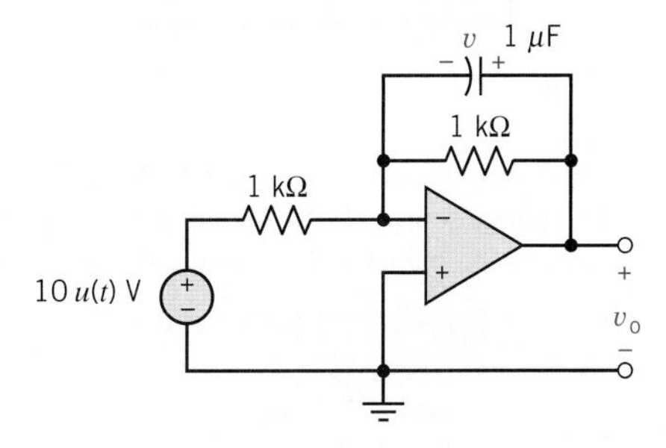

Figure P 14.7-6

P 14.7-7 The motor circuit for driving the snorkel shown in Figure P 14.7-7*a* is shown in Figure P 14.7-7*b*. Find the motor current $I_2(s)$ when the initial conditions are $i_1(0^-) = 2$ A and $i_2(0^-) = 3$ A. Determine $i_2(t)$ and sketch it for 10 s. Does the motor current smoothly drive the snorkel?

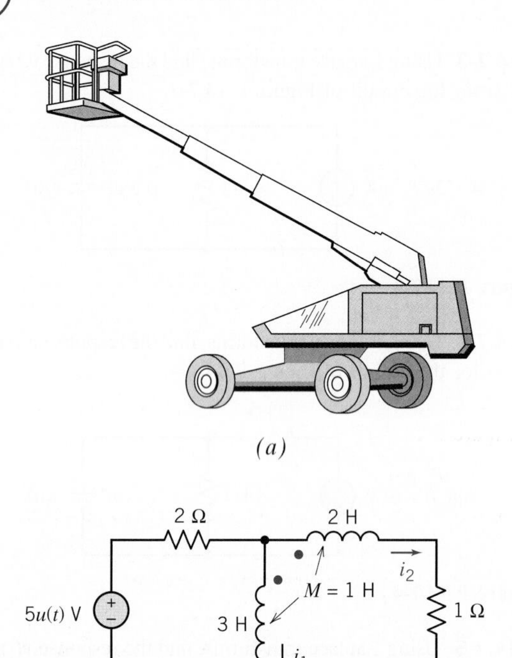

Figure P 14.7-7 Motor drive circuit for snorkel device.

P 14.7-8 The input to the circuit shown in Figure P 14.7-8 is the voltage of the voltage source, 12 V. The output of this circuit is the voltage, $v_o(t)$, across the capacitor. Determine $v_o(t)$ for $t > 0$.

Answer: $v_o(t) = -(4 + 2e^{-t/2})\text{ V}$ for $t > 0$

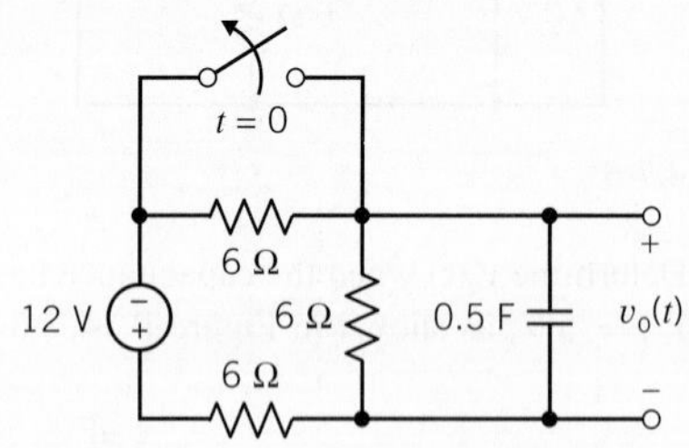

Figure P 14.7-8

P 14.7-9 The input to the circuit shown in Figure P 14.7-9 is the voltage of the voltage source, 12 V. The output of this circuit is the current, $i(t)$, in the inductor. Determine $i(t)$ for $t > 0$.

Answer: $i(t) = -3.5(1 + e^{-0.8t})$ A for $t > 0$

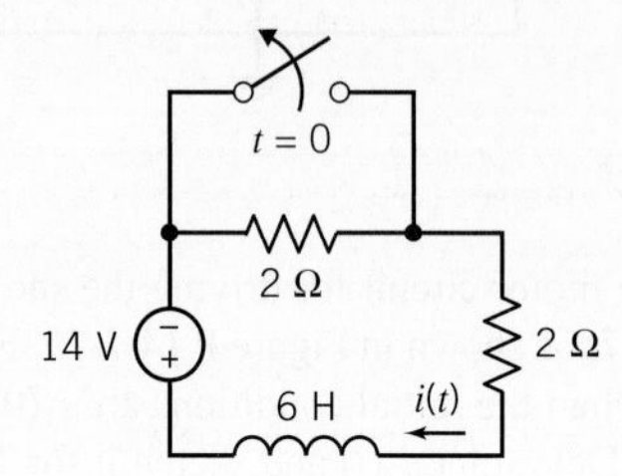

Figure P 14.7-9

P 14.7-10 The input to the circuit shown in Figure P 14.7-10 is the voltage of the voltage source, 18 V. The output of this circuit, the voltage across the capacitor, is given by

$$v_o(t) = 6 + 12e^{-2t}\text{ V} \quad \text{when} \quad t > 0$$

Determine the value of the capacitance, C, and the value of the resistance, R.

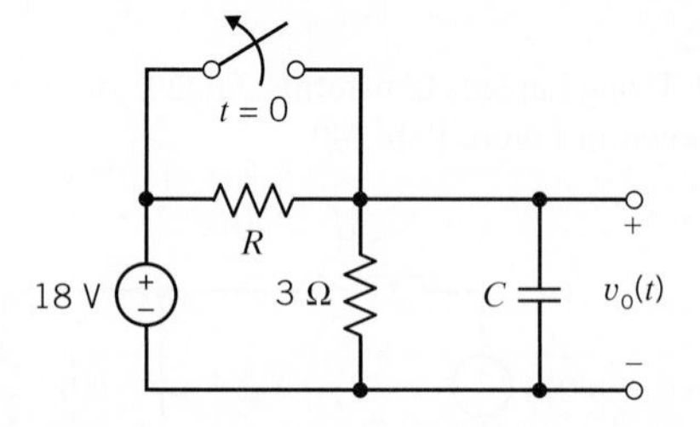

Figure P 14.7-10

P 14.7-11 The input to the circuit shown in Figure P 14.7-11 is the voltage source voltage

$$v_s(t) = 3 - u(t)\text{ V}$$

The output is the voltage

$$v_o(t) = 10 + 5e^{-100t}\text{ V} \quad \text{for} \quad t \geq 0$$

Determine the values of R_1 and R_2.

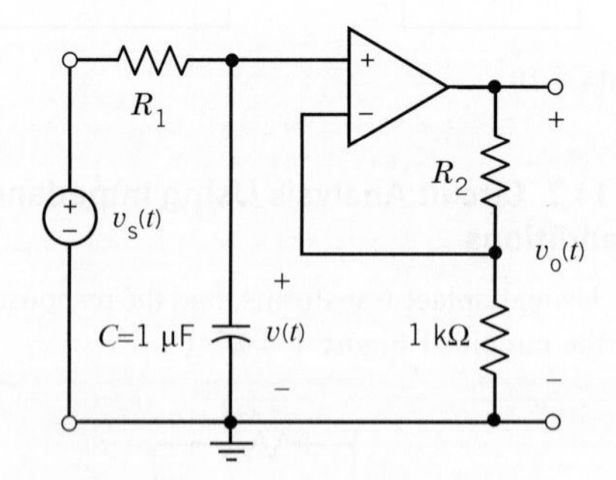

Figure P 14.7-11

P 14.7-12 Determine the inductor current, $i_L(t)$, in the circuit shown in Figure P 14.7-12 for each of the following cases:

(a) $R = 2\,\Omega$, $L = 4.5$ H, $C = 1/9$ F, $A = 5$ mA, $B = -2$ mA
(b) $R = 1\,\Omega$, $L = 0.4$ H, $C = 0.1$ F, $A = 1$ mA, $B = -2$ mA
(c) $R = 1\,\Omega$, $L = 0.08$ H, $C = 0.1$ F, $A = 0.2$ mA, $B = -2$ mA

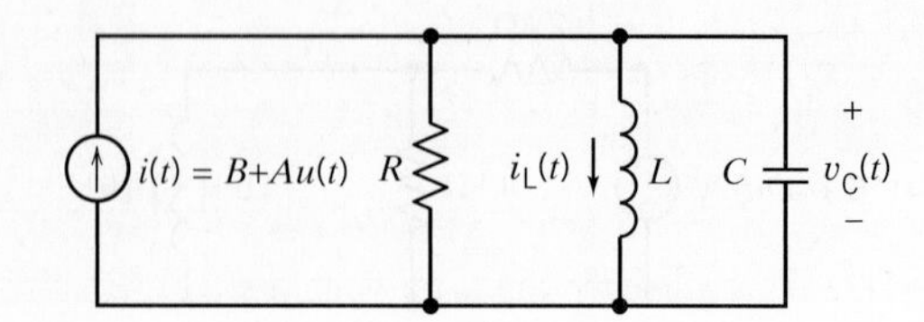

Figure P 14.7-12

P 14.7-13 Determine the capacitor current, $i_c(t)$, in the circuit shown in Figure P 14.7-13 for each of the following cases:

(a) $R = 3\,\Omega,\ L = 2\,\text{H},\ C = 1/24\,\text{F},\ A = 12\,\text{V}$
(b) $R = 2\,\Omega,\ L = 2\,\text{H},\ C = 1/8\,\text{F},\ A = 12\,\text{V}$
(c) $R = 10\,\Omega,\ L = 2\,\text{H},\ C = 1/40\,\text{F},\ A = 12\,\text{V}$

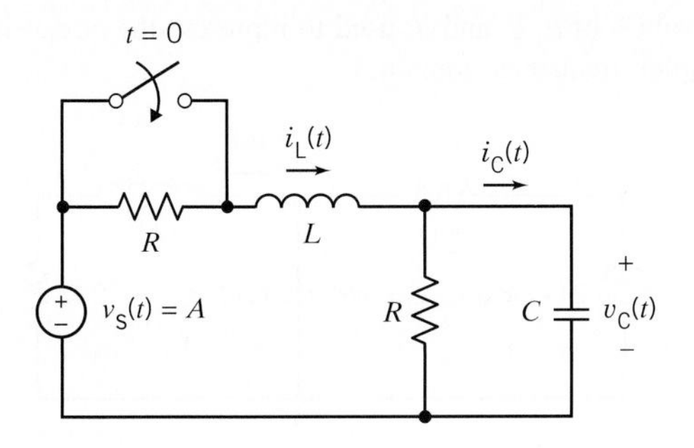

Figure P 14.7-13

P 14.7-14 The voltage source voltage in the circuit shown in Figure P 14.7-14 is

$$v_s(t) = 12 - 6u(t)\ \text{V}$$

Determine $v(t)$ for $t \geq 0$.

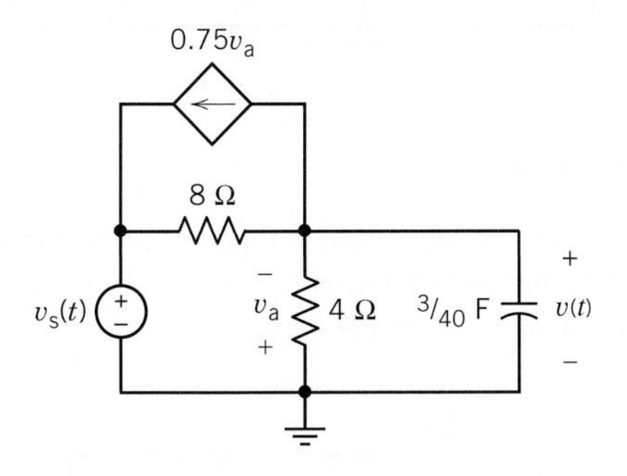

Figure P 14.7-14

P 14.7-15 Determine the output voltage, $v_o(t)$, in the circuit shown in Figure P 14.7-15.

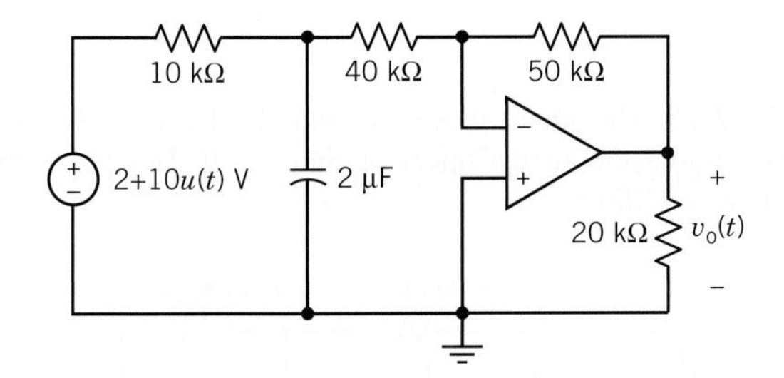

Figure P 14.7-15

P 14.7-16 Determine the capacitor voltage, $v(t)$, in the circuit shown in Figure P 14.7-16.

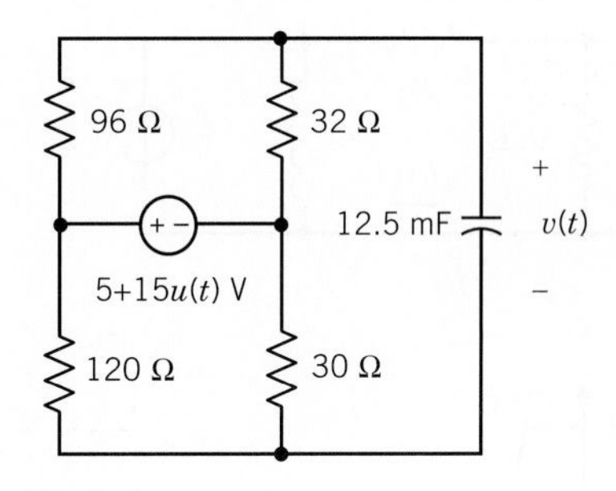

Figure P 14.7-16

P 14.7-17 Determine the voltage $v_o(t)$ for $t \geq 0$ for the circuit of Figure P 14.7-17.

Hint: $v_C(0) = 4\ \text{V}$

Answer: $v_o(t) = 24e^{0.75t}\,u(t)\ \text{V}$ (This circuit is unstable.)

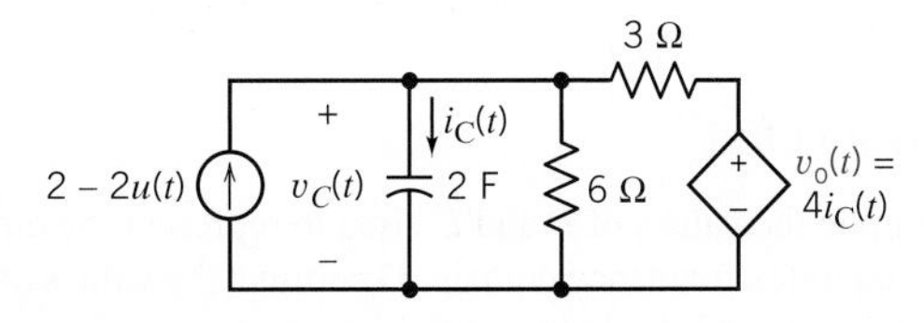

Figure P 14.7-17

P 14.7-18 Determine the current $i_L(t)$ for $t \geq 0$ for the circuit of Figure P 14.7-18.

Hint: $v_C(0) = 8\ \text{V}$ and $i_L(0) = 1\ \text{A}$

Answer: $i_L(t) = \left(e^{-t}\cos 2t + \dfrac{1}{2}e^{-t}\sin 2t\right)u(t)\ \text{A}$

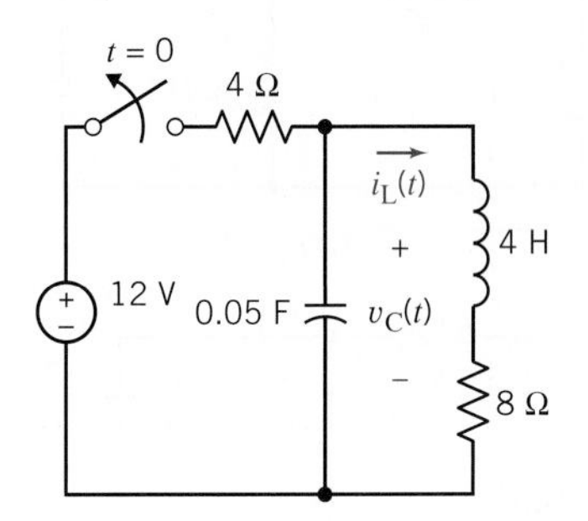

Figure P 14.7-18

P 14.7-19 Figure P 14.7-19*a* shows a circuit represented in the time domain. Figure P 14.7-19*b* shows the same circuit, now represented in the complex frequency domain. Figure P 14.7-19*c* shows a plot of the inductor current.

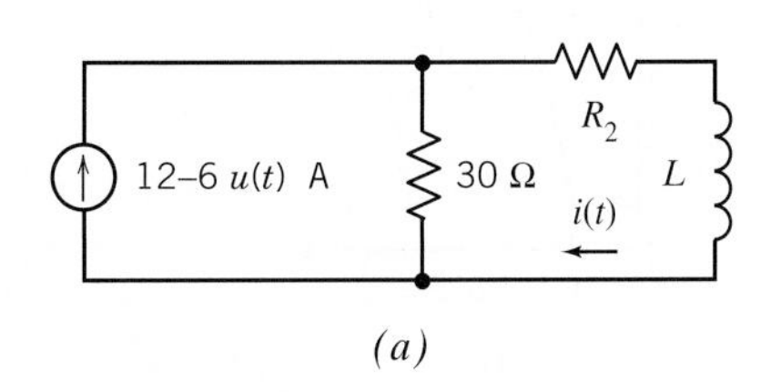

(*a*)

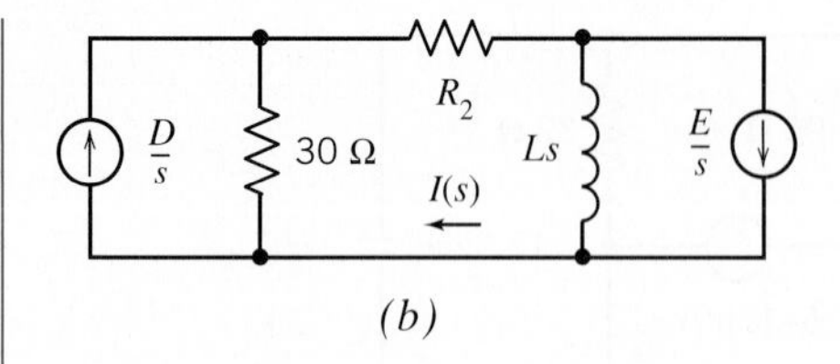

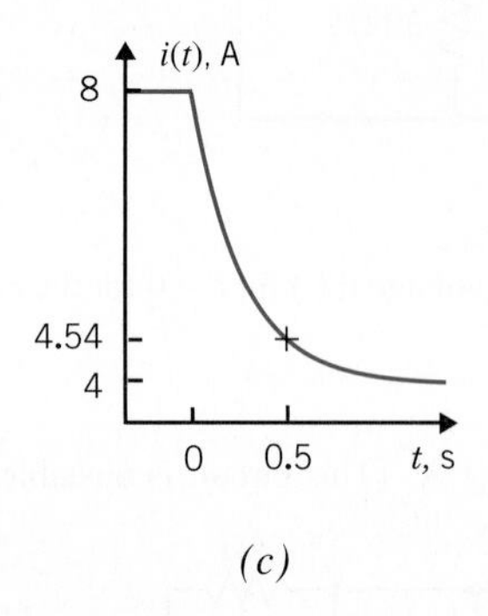

Figure P 14.7-19

Determine the values of D and E, used to represent the circuit in the complex frequency domain. Determine the values of the resistance R_2 and the inductance L.

P 14.7-20 Figure P 14.7-20a shows a circuit represented in the time domain. Figure P 14.7-20b shows the same circuit, now represented in the complex frequency domain. Figure P 14.7-20c shows a plot of the inductor current.

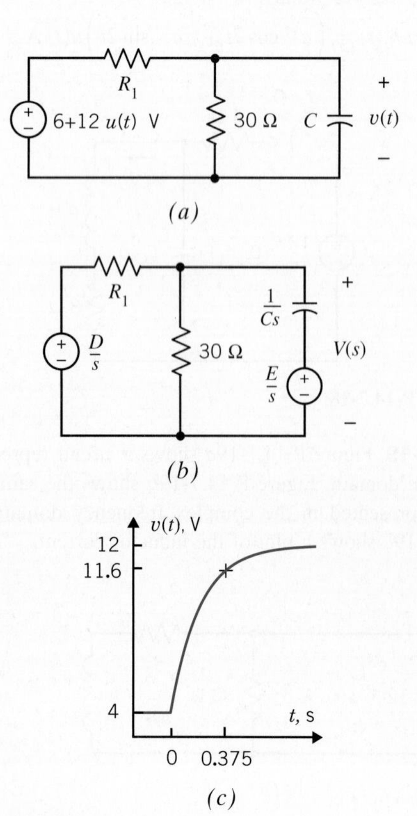

Figure P 14.7-20

Determine the values of D and E, used to represent the circuit in the complex frequency domain. Determine the values of the resistance R_1 and the capacitance C.

P 14.7-21 Figure P 14.7-21a shows a circuit represented in the time domain. Figure P 14.7-21b shows the same circuit, now represented in the complex frequency domain. Determine the values of a, b, and d, used to represent the circuit in the complex frequency domain.

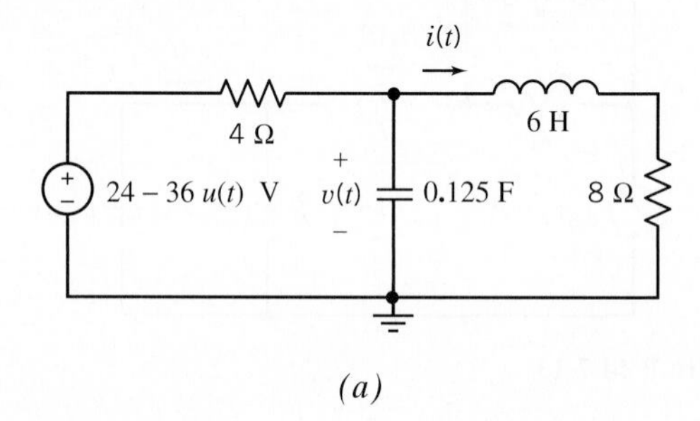

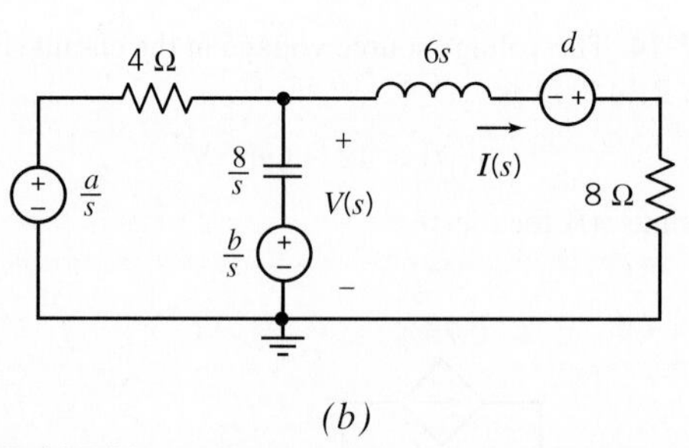

Figure P 14.7-21

P 14.7-22 The circuit shown in Figure P 14.7-22 is at steady state before the switch opens at time $t = 0$. Determine the inductor voltage $v(t)$ for $t > 0$.

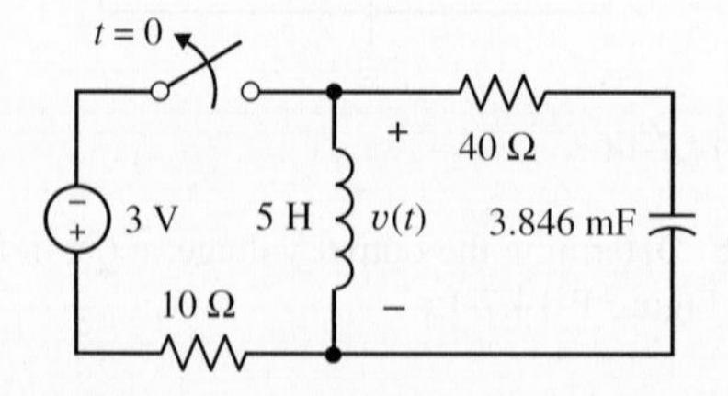

Figure P 14.7-22

P 14.7-23 The circuit shown in Figure P 14.7-23 is at steady state before the switch opens at time $t = 0$. Determine the voltage $v(t)$ for $t > 0$.

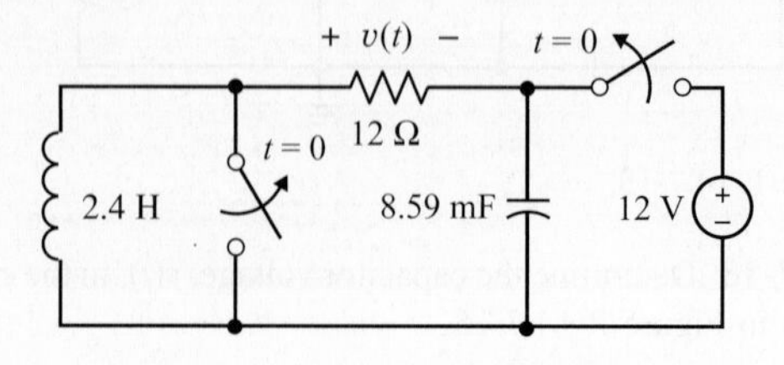

Figure P 14.7-23

P 14.7-24 The circuit shown in Figure P 14.7-24 is at steady state before the switch opens at time $t = 0$. Determine the voltage $v(t)$ for $t > 0$.

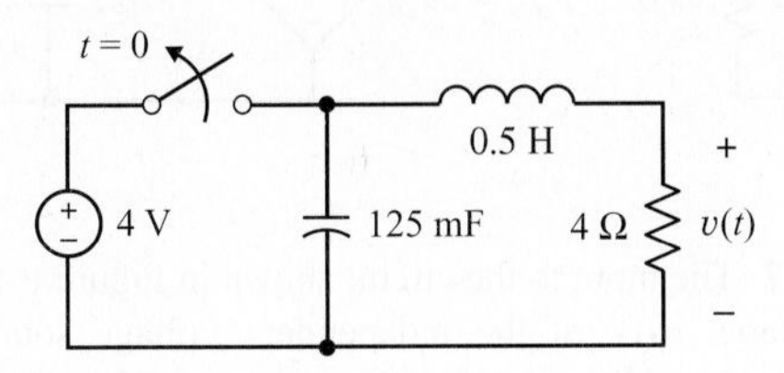

Figure P 14.7-24

P 14.7-25 The circuit shown in Figure P 14.7-25 is at steady state before time $t = 0$. Determine the voltage $v(t)$ for $t > 0$.

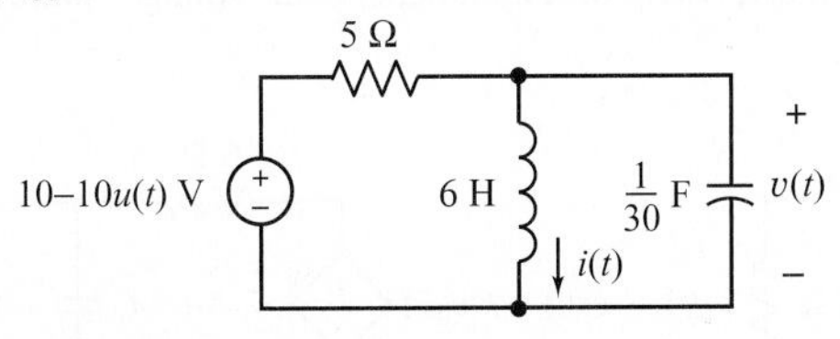

Figure P 14.7-25

P 14.7-26 The input to the circuit shown in Figure P 14.7-26 is the voltage source voltage

$$v_i(t) = 10 + 5u(t) \text{ V} = \begin{cases} 10 \text{ V} & \text{when } t < 0 \\ 15 \text{ V} & \text{when } t > 0 \end{cases}$$

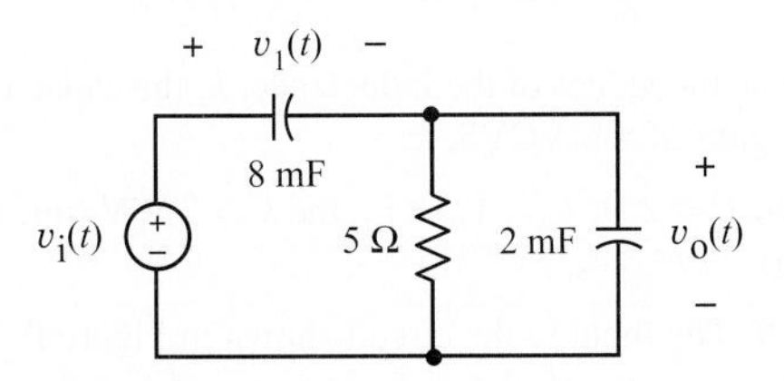

Figure P 14.7-26

Determine the response, $v_o(t)$. Assume that the circuit is at steady state when $t < 0$. Sketch $v_o(t)$ as a function of t.

P 14.7-27 The input to the circuit shown in Figure P 14.7-27 is the current source current

$$i(t) = 25 - 15u(t) \text{ mA} = \begin{cases} 25 \text{ mA} & \text{when } t < 0 \\ 10 \text{ mA} & \text{when } t > 0 \end{cases}$$

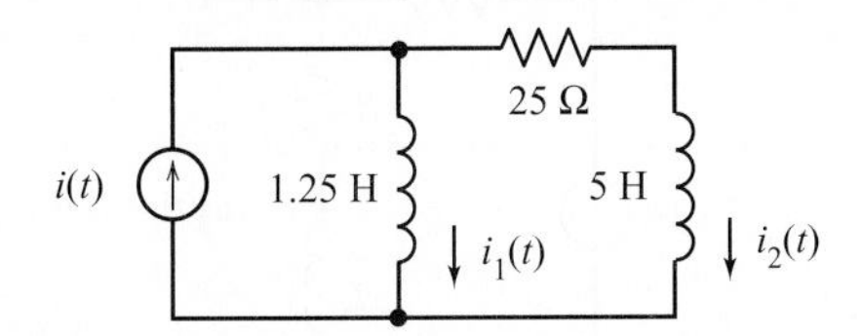

Figure P 14.7-27

Determine the response, $i_2(t)$. Assume that the circuit is at steady state when $t < 0$. Sketch $i_2(t)$ as a function of t.

Section 14.8 Transfer Function and Impedance

P 14.8-1 Consider the circuit of Figure P 14.8-1, where the combination of R_2 and C_2 represents the input of an oscilloscope. The combination of R_1 and C_1 is added to the probe of the oscilloscope to shape the response $v_o(t)$ so that it will equal $v_1(t)$ as closely as possible. Find the necessary relationship for the resistors and capacitors so that $v_o = av_1$ where a is a constant.

Hint: Find the transfer function $V_o(s)/V_1(s)$. Choose R_1 and C_1 so that the transfer function does not depend on s.

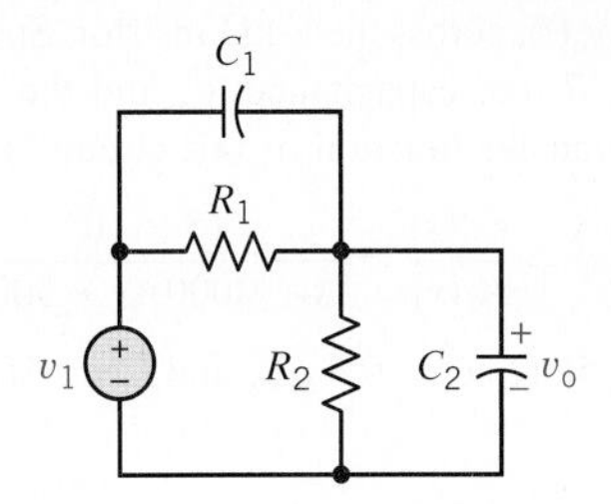

Figure P 14.8-1 Circuit for oscilloscope probe.

P 14.8-2 Consider the circuit shown in Figure P 14.8-2. Show that by proper choice of L, the input impedance $Z = V_1(s)/I_1(s)$ can be made independent of s. What value of L satisfies this condition? What is the value of Z when it is independent of s?

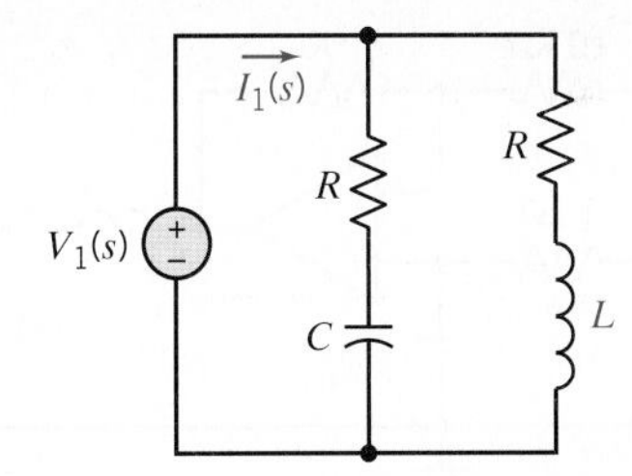

Figure P 14.8-2

P 14.8-3 The input to the circuit shown in Figure P 14.8-3 is the voltage, $v_i(t)$, of the independent voltage source. The output is the voltage, $v_o(t)$, across the capacitor. Determine the transfer function, impulse response, and step response of this circuit.

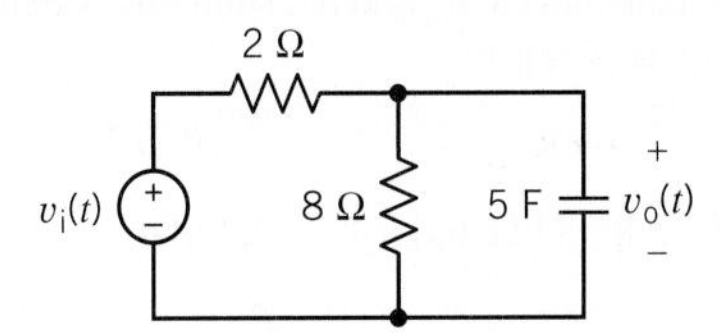

Figure P 14.8-3

P 14.8-4 The input to a linear circuit is the voltage $v_i(t)$ and the response is the voltage $v_o(t)$. The impulse response, $h(t)$, of this circuit is:

$$h(t) = 12te^{-4t}u(t) \text{ V}$$

Determine the step response of the circuit.

Answer: $\left(\frac{3}{4} - e^{-4t}\left(3t + \frac{3}{4}\right)\right)u(t)$ V

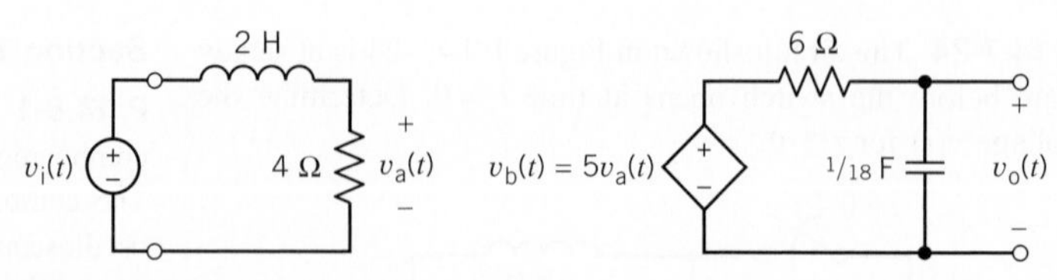

Figure P 14.8-7

P 14.8-5 The input to the circuit shown in Figure P 14.8-5 is the voltage, $v_i(t)$, of the independent voltage source. The output is the voltage, $v_o(t)$, across the 5-kΩ resistor. Specify values of the resistance, R, the capacitance, C, and the inductance, L, such that the transfer function of this circuit is given by

$$H(s) = \frac{V_o(s)}{V_i(s)} = \frac{15 \times 10^6}{(s + 2000)(s + 5000)}$$

Answers: $R = 5\text{k}\,\Omega$, $C = 0.5\,\mu\text{F}$, and $L = 1$ H (one possible solution)

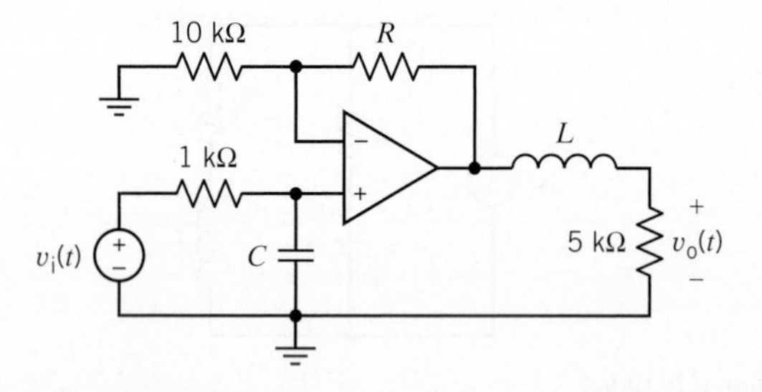

Figure P 14.8-5

P 14.8-6 The input to the circuit shown in Figure P 14.8-6 is the voltage, $v_i(t)$, of the independent voltage source. The output is the voltage, $v_o(t)$, across the 10-kΩ resistor. Specify values of the resistances, R_1 and R_2, such that the step response of this circuit is given by

$$v_o(t) = -4\left(1 - e^{-300t}\right)u(t)\text{ V}$$

Answers: $R_1 = 8.33$ kΩ and $R_2 = 33.3$ kΩ

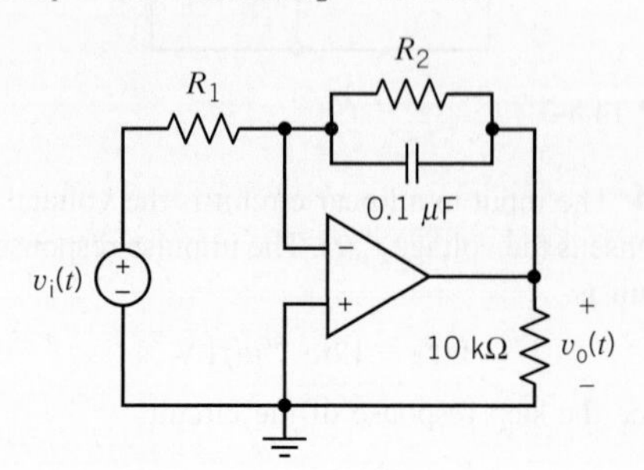

Figure P 14.8-6

P 14.8-7 The input to the circuit shown in Figure P 14.8-7 is the voltage, $v_i(t)$, of the independent voltage source. The output is the voltage, $v_o(t)$, across the capacitor. Determine the step response of this circuit.

Answer: $v_o(t) = [6 - 15e^{-2t} + 10e^{-3t}]u(t)$ V

P 14.8-8 The input to the circuit shown in Figure P 14.8-8 is the voltage, $v_i(t)$, of the independent voltage source. The output is the voltage, $v_o(t)$, across the capacitor. The step response of this circuit is

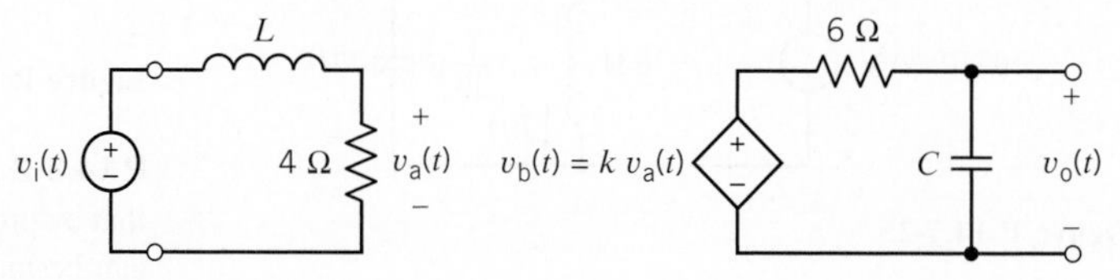

Figure P 14.8-8

$$v_o(t) = \left(2 + 4e^{-3t} - 6e^{-2t}\right)u(t)\text{ V}$$

Determine the values of the inductance, L, the capacitance, C, and the gain of the VCVS, k.

Answers: $L = 2$ H, $C = 1/18$ F, and $k = 2$ V/V (one possible solution)

P 14.8-9 The input to the circuit shown in Figure P 14.8-9 is the voltage, $v_i(t)$, of the independent voltage source. The output is the voltage, $v_o(t)$. The step response of this circuit is

$$v_o(t) = 0.5\left(1 + e^{-4t}\right)u(t)\text{ V}$$

Determine the values of the inductance, L, and the resistance, R.

Answers: $L = 9$ H and $R = 18\ \Omega$

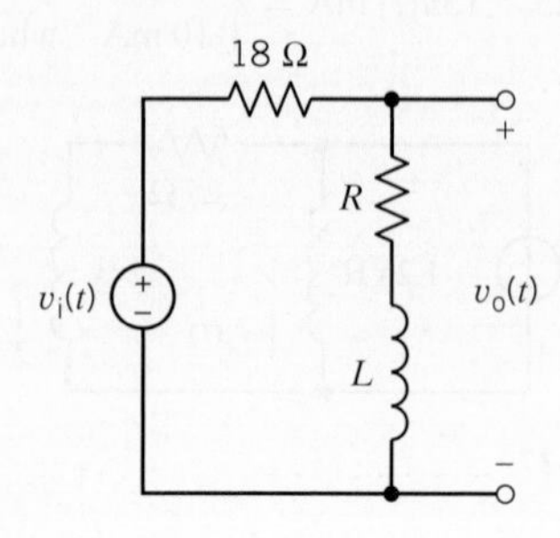

Figure P 14.8-9

P 14.8-10 An electric microphone and its associated circuit can be represented by the circuit shown in Figure P 14.8-10. Determine the transfer function $H(s) = V_o(s)/V(s)$.

Answer: $\dfrac{V_o(s)}{V(s)} = \dfrac{RCs}{(R_1Cs + 2)(2RCs + 1) - 1}$

Figure P 14.8-10 Microphone circuit.

P 14.8-11 Engineers had avoided inductance in long-distance circuits because it slows transmission. Oliver Heaviside proved that the addition of inductance to a circuit could enable it to transmit without distortion. George A. Campbell of the Bell Telephone Company designed the first practical inductance loading coils, in which the induced field of each winding of wire reinforced that of its neighbors so that the coil supplied proportionally more inductance than resistance. Each one of Campbell's 300 test coils added 0.11 H and 12 Ω at regular intervals along 35 miles of telephone wire (Nahin, 1990). The loading coil balanced the effect of the leakage between the telephone wires represented by R and C in Figure P 14.8-11. Determine the transfer function $V_2(s)/V_1(s)$.

Answer: $\dfrac{V_2(s)}{V_1(s)} = \dfrac{R}{RLCs^2 + (L + R_xRC)s + R_x + R}$

Figure P 14.8-11 Telephone and load coil circuit.

P 14.8-12 An op amp circuit for a band-pass filter is shown in Figure P 14.8-12. Determine $V_o(s)/V(s)$. Assume ideal op amps.

Answer: $\dfrac{V_o(s)}{V(s)} = \dfrac{-\dfrac{1}{R_2C_2}s}{s^2 + \dfrac{1}{R_1C_1}s + \dfrac{1}{R_1R_2C_1C_2}}$

Figure P 14.8-12

P 14.8-13 A digital-to-analog converter (DAC) uses an op amp filter circuit shown in Figure P 14.8-13 (Garnett, 1992). The filter receives the pulse output from the DAC and produces the analog voltage, v_o. Determine the transfer function of the filter, $V_o(s)/V(s)$. Assume an ideal op amp.

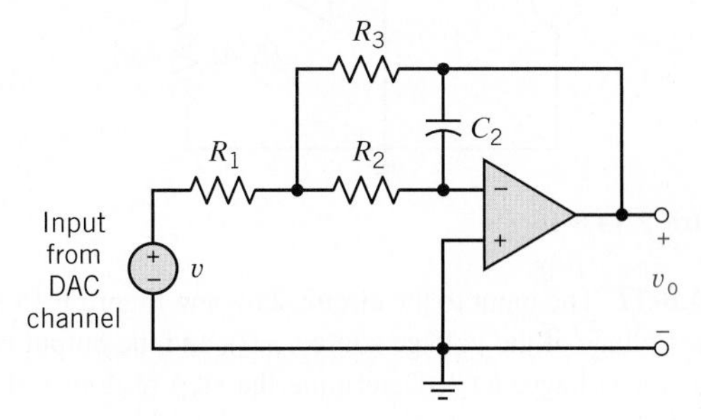

Figure P 14.8-13 Digital-to-analog converter filter.

P 14.8-14 A series RLC circuit is shown in Figure P 14.8-14. Determine (a) the transfer function $H(s)$, (b) the impulse response, and (c) the step response for each set of parameter values given in the table below.

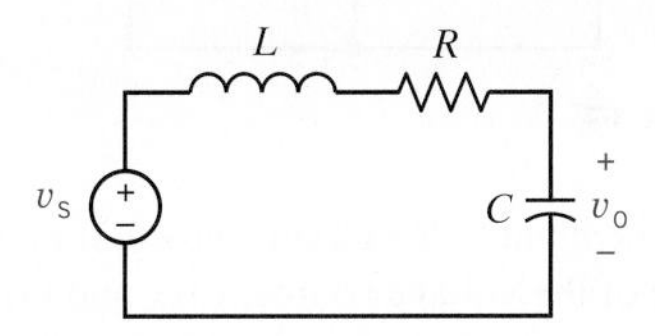

Figure P 14.8-14

	L	C	R
a	4 H	0.025 F	28 Ω
b	2 H	0.025 F	8 Ω
c	1 H	0.391 F	4 Ω
d	2 H	0.125 F	8 Ω

P 14.8-15 A circuit is described by the transfer function

$$\frac{V_o}{V_1} = H(s) = \frac{9s + 27}{3s^3 + 18s^2 + 39s}$$

Find the step response and impulse response of the circuit.

P 14.8-16 The input to the circuit shown in Figure P 14.8-16 is the voltage of the voltage source, $v_i(t)$, and the output is the voltage, $v_o(t)$, across the 15-kΩ resistor.

(a) Determine the steady-state response, $v_o(t)$, of this circuit when the input is $v_i(t) = 1.5$ V.

(b) Determine the steady-state response, $v_o(t)$, of this circuit when the input is $v_i(t) = 4\cos(100t + 30°)$ V.

(c) Determine the step response, $v_o(t)$, of this circuit.

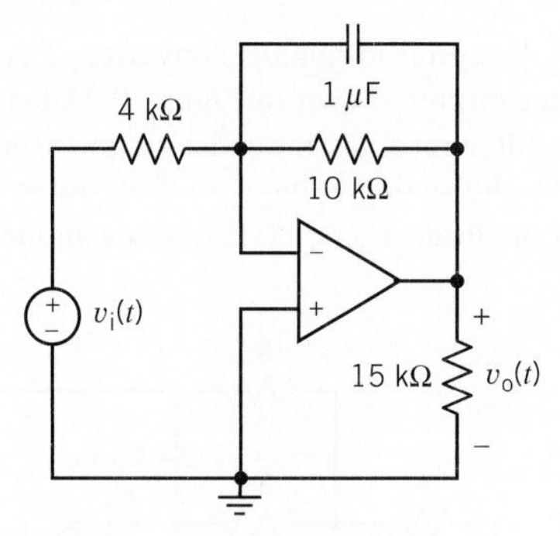

Figure P 14.8-16

P 14.8-17 The input to the circuit shown in Figure P 14.8-17 is the voltage of the voltage source, $v_i(t)$, and the output is the capacitor voltage, $v_o(t)$. Determine the step response of this circuit.

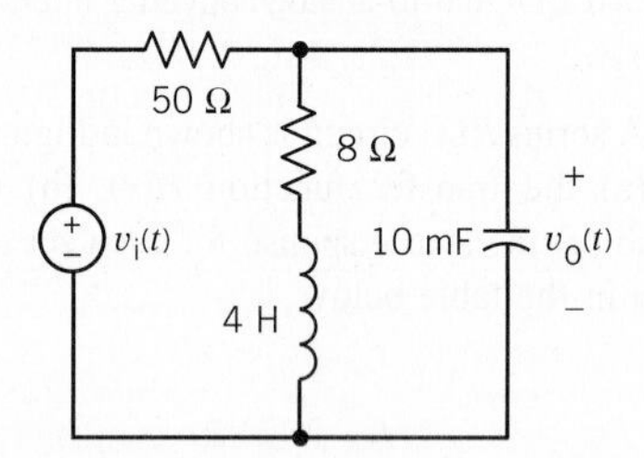

Figure P 14.8-17

P 14.8-18 The input to the circuit shown in Figure P 14.8-18 is the voltage of the voltage source, $v_i(t)$, and the output is the resistor voltage, $v_o(t)$. Specify values for L_1, L_2, R, and K that cause the step response of the circuit to be

$$v_o(t) = \left(1 + 0.667e^{-50t} - 1.667e^{-20t}\right)u(t)\ \mathrm{V}$$

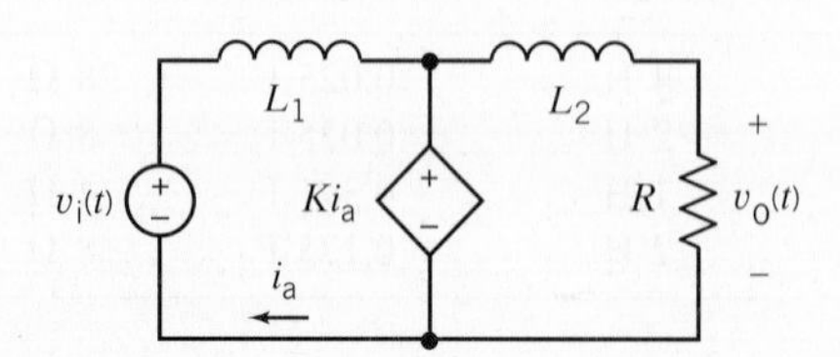

Figure P 14.8-18

P 14.8-19 The input to the circuit shown in Figure P 14.8-19 is the voltage of the voltage source, $v_i(t)$, and the output is the capacitor voltage, $v_o(t)$. Determine the step response of this circuit.

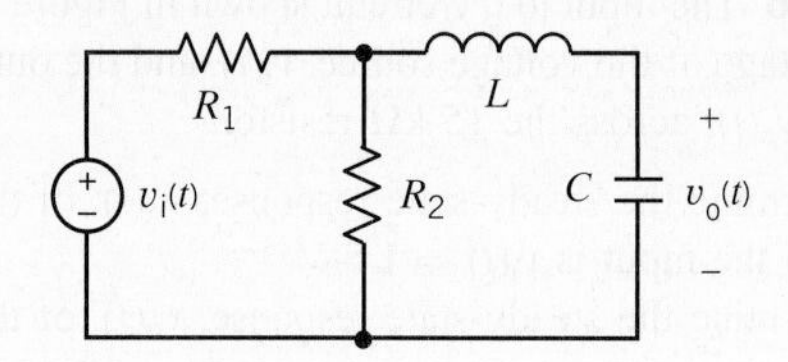

Figure P 14.8-19

P 14.8-20 The input to the circuit shown in Figure P 14.8-20 is the voltage of the voltage source, $v_i(t)$, and the output is the inductor current, $i_o(t)$. Specify values for L, C, and K that cause the step response of the circuit to be

$$v_o(t) = \left(3.2 - \left(3.2e^{-5t} + 16te^{-5t}\right)\right)u(t)\ \mathrm{V}$$

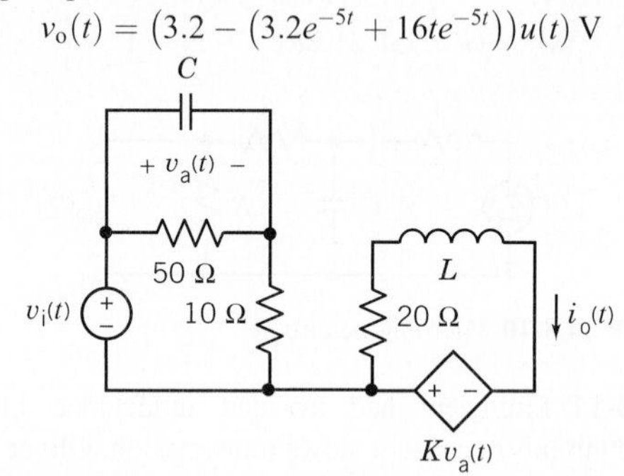

Figure P 14.8-20

P 14.8-21 The input to a circuit is the voltage $v_i(t)$ and the output is the voltage $v_o(t)$. The impulse response of the circuit is

$$v_o(t) = 6.5e^{-2t}\cos\left(2t + 22.6°\right)u(t)\ \mathrm{V}$$

Determine the step response of this circuit.

P 14.8-22 The input to a circuit is the voltage $v_i(t)$ and the output is the voltage $v_o(t)$. The step response of the circuit is

$$v_o(t) = \left[1 - e^{-t}(1 + 3t)\right]u(t)\ \mathrm{V}$$

Determine the impulse response of this circuit.

P 14.8-23 The input to the circuit shown in Figure P 14.8-23 is the voltage of the voltage source, $v_i(t)$, and the output is the voltage, $v_o(t)$. Determine the step response of the circuit.

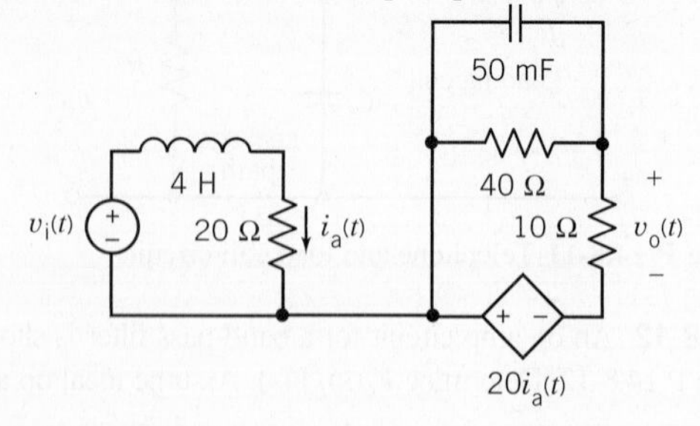

Figure P 14.8-23

P 14.8-24 The transfer function of a circuit is $H(s) = \dfrac{12}{s^2 + 8s + 16}$. Determine the step response of this circuit.

P 14.8-25 The transfer function of a circuit is $H(s) = \dfrac{90s}{s^2 + 8s + 25}$. Determine the step response of this circuit.

Section 14.9 Convolution

P 14.9-1 Let $f(t)$ denote the 1-s pulse given by $f(t) = u(t) - u(t-1)$. Determine the convolution $f(t)^{*}f(t)$, which is the convolution of the pulse with itself.

Answer: $f(t) * f(t) = tu(t) - 2(t-1)u(t-1) + (t-2)u(t-2)$

P 14.9-2 Consider a pulse of amplitude 3 and a duration of 2 s with its starting point at $t = 0$. Find the convolution of this pulse with itself.

P 14.9-3 A circuit is shown in Figure P 14.9-3. Determine (a) the transfer function $V_2(s)/V_1(s)$ and (b) the response $v_2(t)$ when $v_1 = tu(t)$.

Answer: $v_2 = t - \left(1 - e^{-t/RC}\right)/RC,\ t \geq 0$

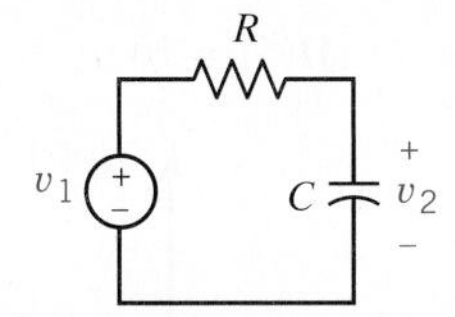

Figure P 14.9-3

P 14.9-4 Find the convolution of $h(t) = t\,u(t)$ and $f(t) = e^{-at}\,u(t)$ for $t > 0$ using the convolution integral and the inverse transform of $H(s)F(s)$.

Answer: $\dfrac{at - 1 + e^{-at}}{a^2},\ t > 0$

Section 14.10 Stability

P 14.10-1 The input to the circuit shown in Figure P 14.10-1 is the voltage, $v_i(t)$, of the independent voltage source. The output is the voltage, $v_o(t)$, across the resistor labeled R. The step response of this circuit is

$$v_o(t) = (3/5)\left(1 - e^{-100t}\right)u(t)\ \text{V}$$

(a) Determine the value of the inductance, L, and the value of the resistance, R.
(b) Determine the impulse response of this circuit.
(c) Determine the steady-state response of the circuit when the input is $v_i(t) = 5\cos 100\,t$ V.

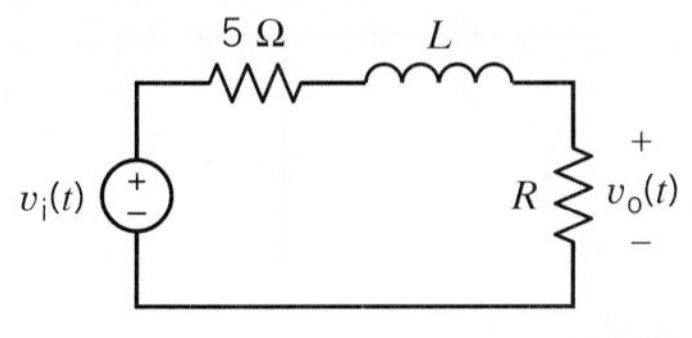

Figure P 14.10-1

P 14.10-2 The input to the circuit shown in Figure P 14.10-2 is the voltage, $v_i(t)$, of the independent voltage source. The output is the voltage, $v_o(t)$, across the capacitor. The step response of this circuit is

$$v_o(t) = \left[5 - 5e^{-2t}(1 + 2t)\right]u(t)\ \text{V}$$

Determine the steady-state response of this circuit when the input is

$$v_i(t) = 5\cos(2t + 45°)\ \text{V}$$

Answer: $v_o(t) = 12.5\cos(2t - 45°)$ V

P 14.10-3 The input to a linear circuit is the voltage $v_i(t)$ and the response is the voltage $v_o(t)$. The impulse response, $h(t)$, of this circuit is

$$h(t) = 30te^{-5t}u(t)\text{V}$$

Determine the steady-state response of this circuit when the input is

$$v_i(t) = 20\cos(3t)\ \text{V}$$

Answer: $v_o(t) = 17.65\cos(3t - 62°)$ V

P 14.10-4 The input to a circuit is the voltage v_s. The output is the voltage v_o. The step response of the circuit is

$$v_o(t) = \left(40 + 1.03e^{-8t} - 41e^{-320t}\right)u(t)$$

Determine the network function

$$\mathbf{H}(\omega) = \frac{\mathbf{V}_o(\omega)}{\mathbf{V}_s(\omega)}$$

of the circuit and sketch the asymptotic magnitude Bode plot.

P 14.10-5 The input to a circuit is the voltage v_s. The output is the voltage v_o. The step response of the circuit is

$$v_o(t) = 60\left(e^{-2t} - e^{-6t}\right)u(t)$$

Determine the network function

$$\mathbf{H}(\omega) = \frac{\mathbf{V}_o(\omega)}{\mathbf{V}_s(\omega)}$$

of the circuit and sketch the asymptotic magnitude Bode plot.

P 14.10-6 The input to a circuit is the voltage v_s. The output is the voltage v_o. The step response of the circuit is

$$v_o(t) = \left(4 + 32e^{-90t}\right)u(t)$$

Determine the network function

$$\mathbf{H}(\omega) = \frac{\mathbf{V}_o(\omega)}{\mathbf{V}_s(\omega)}$$

of the circuit and sketch the asymptotic magnitude Bode plot.

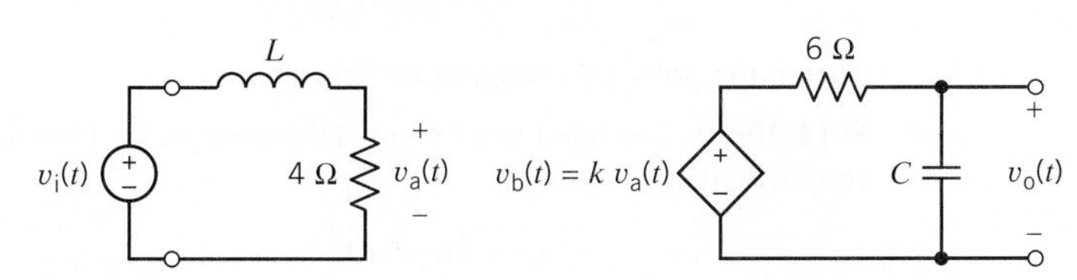

Figure P 14.10-2

P 14.10-7 The input to a circuit is the voltage v_s. The output is the voltage v_o. The step response of the circuit is

$$v_o(t) = \frac{5}{3}\left(e^{-5t} - e^{-20t}\right)u(t) \text{ V}$$

Determine the steady-state response of the circuit when the input is

$$v_s(t) = 12 \cos(30t) \text{ V}$$

P 14.10-8 The input to a circuit is the voltage v_s. The output is the voltage v_o. The impluse response of the circuit is

$$v_o(t) = e^{-5t}(10 - 50t)u(t) \text{ V}$$

Determine the steady-state response of the circuit when the input is

$$v_s(t) = 12 \cos(10t) \text{ V}$$

P 14.10-9 The input to a circuit is the voltage v_s. The output is the voltage v_o. The step response of the circuit is

$$v_o(t) = \left(1 - e^{-20t}(\cos(4t) + 0.5 \sin(4t))\right)u(t) \text{V}$$

Determine the steady-state response of the circuit when the input is

$$v_s(t) = 12 \cos(4t) \text{ V}$$

P 14.10-10 The transfer function of a circuit is if $H(s) = \frac{20}{s+8}$. When the input to this circuit is sinusoidal, the output is also sinusoidal. Let ω_1 be the frequency at which the output sinusoid is twice as large as the input sinusoid and let ω_2 be the frequency at which output sinusoid is delayed by one tenth period with respect to the input sinusoid. Determine the values of ω_1 and ω_2.

P 14.10-11 The input to a linear circuit is the voltage, v_i. The output is the voltage, v_o. The transfer function of the circuit is

$$H(s) = \frac{V_o(s)}{V_i(s)}$$

The poles and zeros of $H(s)$ are shown on the pole-zero diagram in Figure P 14.10-11. (There are no zeros.) The dc gain of the circuit is

$$\mathbf{H}(0) = 5$$

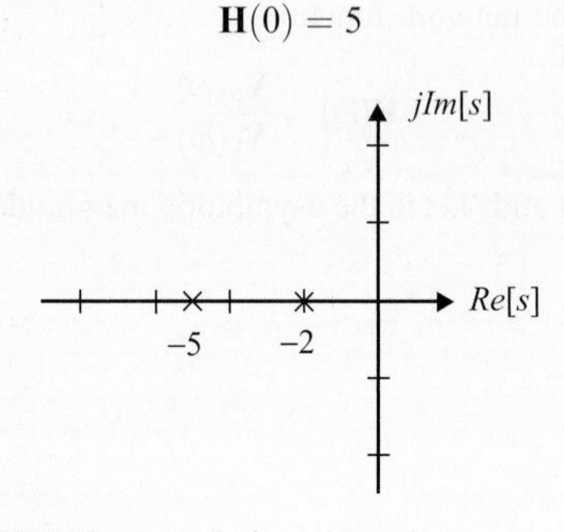

Figure P 14.10-11

Determine the step response of the circuit.

P 14.10-12 The input to a linear circuit is the voltage, v_i. The output is the voltage, v_o. The transfer function of the circuit is

$$H(s) = \frac{V_o(s)}{V_i(s)}$$

The poles and zeros of $H(s)$ are shown on the pole-zero diagram in Figure P 14.10-12. At $\omega = 5$ rad/s, the gain of the circuit is

$$\mathbf{H}(5) = 10$$

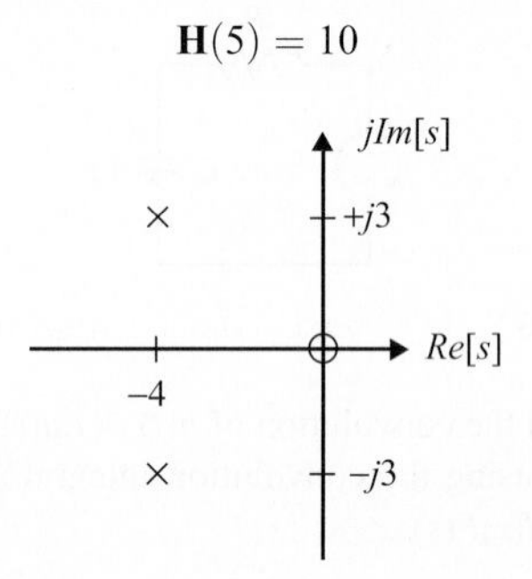

Figure P 14.10-12

Determine the step response of the circuit.

P 14.10-13 The input to a linear circuit is the voltage, v_i. The output is the voltage, v_o. The transfer function of the circuit is

$$H(s) = \frac{V_o(s)}{V_i(s)}$$

The poles and zeros of $H(s)$ are shown on the pole-zero diagram in Figure P 14.10-13. (There is a double pole at $s = -4$.) The dc gain of the circuit is

$$\mathbf{H}(0) = 5$$

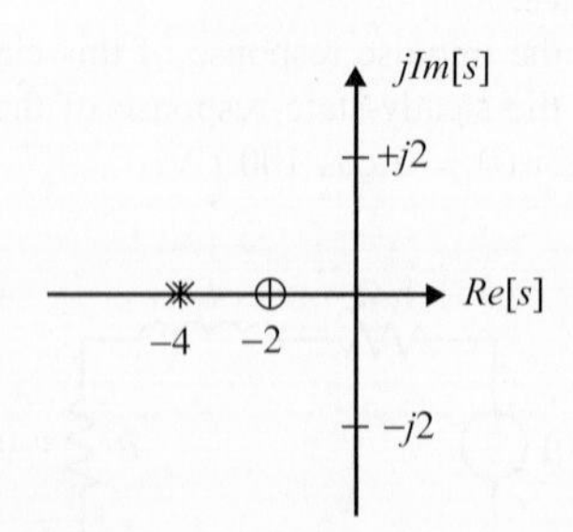

Figure P 14.10-13

Determine the step response of the circuit.

P 14.10-14 The input to a circuit is the voltage, v_i. The step response of the circuit is

$$v_o = 5e^{-4t}\sin(2t)u(t) \text{ V}$$

Sketch the pole-zero diagram for this circuit.

P 14.10-15 The input to a circuit is the voltage, v_i. The step response of the circuit is

$$v_o = 5te^{-5t}u(t) \text{ V}$$

Sketch the pole-zero diagram for this circuit.

Section 14.11 Partial Fraction Expansion Using MATLAB

P 14.11-1 Find the inverse Laplace transform of $V(s) = \dfrac{12s^2 + 91.83s + 186.525}{s^3 + 10.95s^2 + 35.525s + 29.25}$

P 14.11-2 Find the inverse Laplace transform of $V(s) = \dfrac{10s^3 + 139s^2 + 774s + 1471}{s^4 + 12s^3 + 77s^2 + 296s + 464}$

P 14.11-3 Find the inverse Laplace transform of $V(s) = \dfrac{s^2 + 12s + 15}{s^3 + 12s^2 + 48s + 64} = \dfrac{s^2 + 12s + 15}{(s+4)^3}$

P 14.11-4 Find the inverse Laplace transform of $V(s) = \dfrac{-60}{s^2 + 5s + 48.5}$

P 14.11-5 Find the inverse Laplace transform of $V(s) = \dfrac{-30}{s^2 + 25}$

Section 14.12 How Can We Check . . . ?

P 14.12-1 Computer analysis of the circuit of Figure P 14.12-1 indicates that

$$v_C(t) = 6 + 3.3e^{-2.1t} + 2.7e^{-15.9t} \text{ V}$$

and

$$i_L(t) = 2 + 0.96e^{-2.1t} + 0.04e^{-15.9t} \text{ A}$$

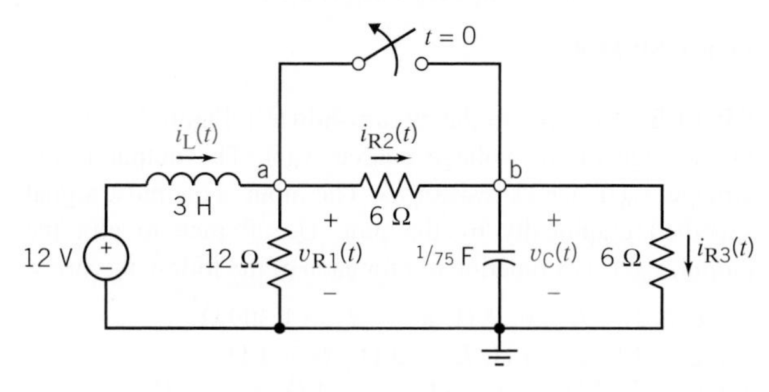

Figure P 14.12-1

after the switch opens at time $t = 0$. Verify that this analysis is correct by checking that (a) KVL is satisfied for the mesh consisting of the voltage source, inductor, and 12-Ω resistor and (b) KCL is satisfied at node b.

Hint: Use the given expressions for $i_L(t)$ and $v_C(t)$ to determine expressions for $v_L(t)$, $i_C(t)$, $v_{R1}(t)$, $i_{R2}(t)$, and $i_{R3}(t)$.

P 14.12-2 Analysis of the circuit of Figure P 14.12-2 when $v_C(0) = -12$ V indicates that

$$i_1(t) = 18e^{0.75t} \text{ A} \quad \text{and} \quad i_2(t) = 20e^{0.75t} \text{ A}$$

after $t = 0$. Verify that this analysis is correct by representing this circuit, including $i_1(t)$ and $i_2(t)$, in the frequency domain, using Laplace transforms. Use $I_1(s)$ and $I_2(s)$ to calculate the element voltages and verify that these voltages satisfy KVL for both meshes.

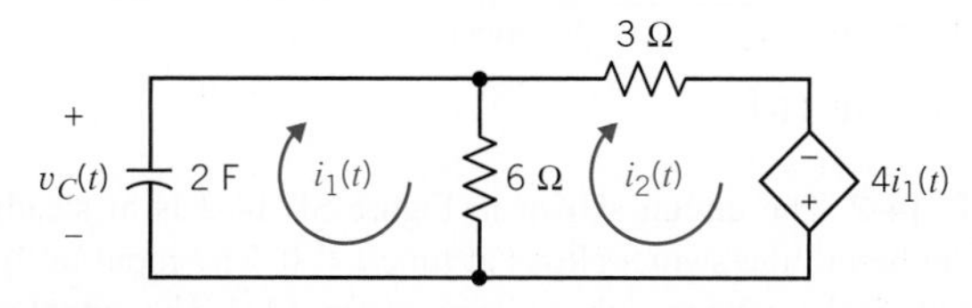

Figure P 14.12-2

P 14.12-3 Figure P 14.12-3 shows a circuit represented in (a) the time domain and (b) the frequency domain, using Laplace transforms. An incorrect analysis of this circuit indicates that

$$I_L(s) = \frac{s+2}{s^2+s+5} \quad \text{and} \quad V_C(s) = \frac{-20(s+2)}{s(s^2+s+5)}$$

(a) Use the initial and final value theorems to identify the error in the analysis. (b) Correct the error.

Hint: Apparently, the error occurred as $V_C(s)$ was calculated from $I_L(s)$.

Answer: $V_C(s) = -\dfrac{20}{s}\left(\dfrac{s+2}{s^2+s+5}\right) + \dfrac{8}{s}$

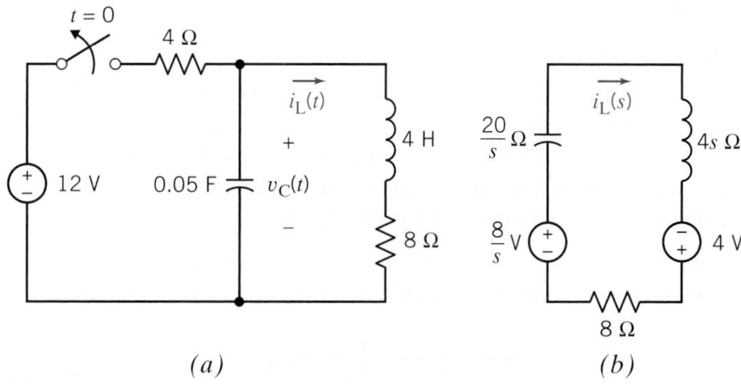

Figure P 14.12-3

PSpice Problems

SP 14-1 The input to the circuit shown in Figure SP 14-1 is the voltage of the voltage source, $v_i(t)$. The output is the voltage across the capacitor, $v_o(t)$. The input is the pulse signal specified graphically by the plot. Use PSpice to plot the output, $v_o(t)$, as a function of t.

Hint: Represent the voltage source using the PSpice part named VPULSE.

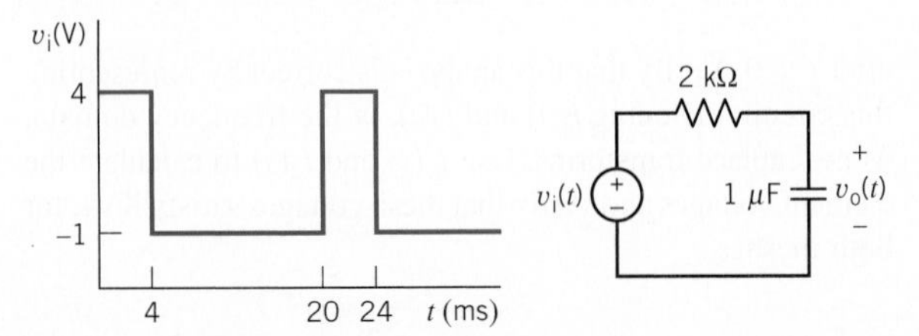

Figure SP 14-1

SP 14-2 The circuit shown in Figure SP 14-2 is at steady state before the switch closes at time $t = 0$. The input to the circuit is the voltage of the voltage source, 12 V. The output of this circuit is the voltage across the capacitor, $v(t)$. Use PSpice to plot the output, $v(t)$, as a function of t. Use the plot to obtain an analytic representation of $v(t)$, for $t > 0$.

Hint: We expect $v(t) = A + \text{B}\, e^{-t/\tau}$ for $t > 0$, where A, B, and τ are constants to be determined.

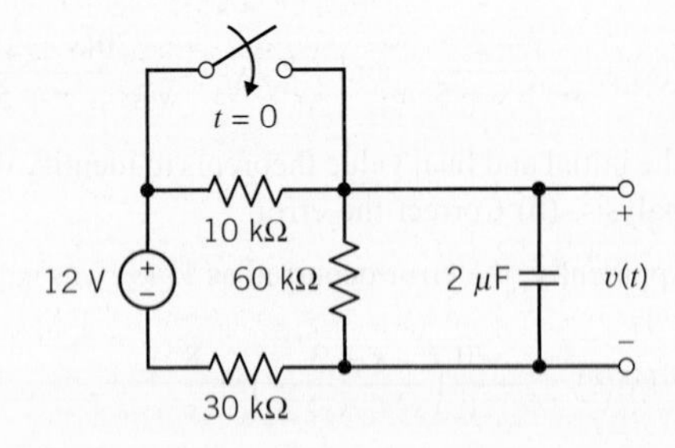

Figure SP 14-2

SP 14-3 The circuit shown in Figure SP 14-3 is at steady state before the switch closes at time $t = 0$. The input to the circuit is the current of the current source, 4 mA. The output of this circuit is the current in the inductor, $i(t)$. Use PSpice to plot the output, $i(t)$, as a function of t. Use the plot to obtain an analytic representation of $i(t)$ for $t > 0$.

Hint: We expect $i(t) = A + \text{B}\, e^{-t/\tau}$ for $t > 0$, where A, B, and τ are constants to be determined.

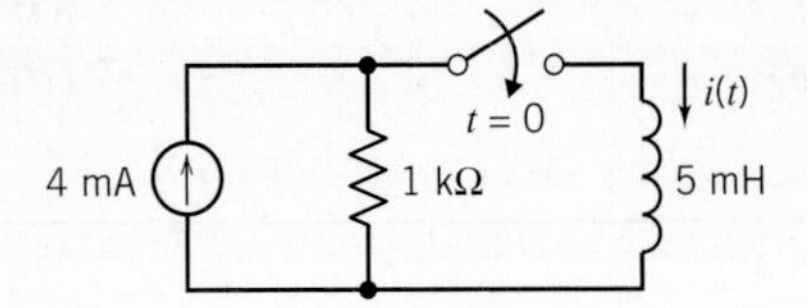

Figure SP 14-3

SP 14-4 The input to the circuit shown in Figure SP 14-4 is the voltage of the voltage source, $v_i(t)$. The output is the voltage across the capacitor, $v_o(t)$. The input is the pulse signal specified graphically by the plot. Use PSpice to plot the output, $v_o(t)$, as a function of t for each of the following cases:

(a) $C = 1\text{ F}$, $L = 0.25\text{ H}$, $R_1 = R_2 = 1.309\ \Omega$
(b) $C = 1\text{ F}$, $L = 1\text{ H}$, $R_1 = 3\ \Omega$, $R_2 = 1\ \Omega$
(c) $C = 0.125\text{ F}$, $L = 0.5\text{ H}$, $R_1 = 1\ \Omega$, $R_2 = 4\ \Omega$

Plot the output for these three cases on the same axis.

Hint: Represent the voltage source, using the PSpice part named VPULSE.

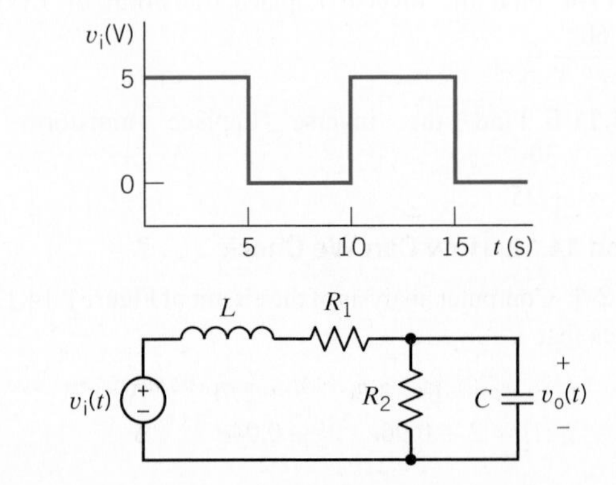

Figure SP 14-4

SP 14-5 The input to the circuit shown in Figure SP 14-5 is the voltage of the voltage source, $v_i(t)$. The output is the voltage, $v_o(t)$, across resistor R_2. The input is the pulse signal specified graphically by the plot. Use PSpice to plot the output, $v_o(t)$, as a function of t for each of the following cases:

(a) $C = 1\text{ F}$, $L = 0.25\text{ H}$, $R_1 = R_2 = 1.309\ \Omega$
(b) $C = 1\text{ F}$, $L = 1\text{ H}$, $R_1 = 3\ \Omega$, $R_2 = 1\ \Omega$
(c) $C = 0.125\text{ F}$, $L = 0.5\text{ H}$, $R_1 = 1\ \Omega$, $R_2 = 4\ \Omega$

Plot the output for these three cases on the same axis.

Hint: Represent the voltage source, using the PSpice part named VPULSE.

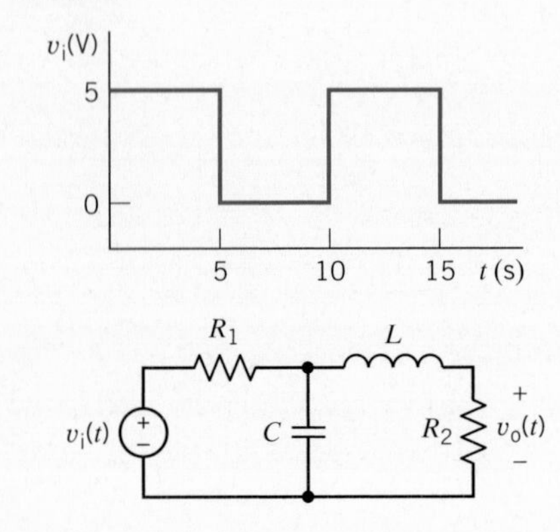

Figure SP 14-5

Design Problems

DP 14-1 Design the circuit in Figure DP 14-1 to have a step response equal to

$$v_o = 5te^{-4t}u(t)\text{ V}$$

Hint: Determine the transfer function of the circuit in Figure DP 14-1 in terms of *k*, *R*, *C*, and *L*. Then determine the Laplace transform of the step response of the circuit in Figure DP 14-1. Next, determine the Laplace transform of the given step response. Finally, determine values of *k*, *R*, *C*, and *L* that cause the two step responses to be equal.

Answer: Pick $L = 1$ H, then $k = 0.625$ V/V, $R = 8\ \Omega$, and $C = 0.0625$ F. (This answer is not unique.)

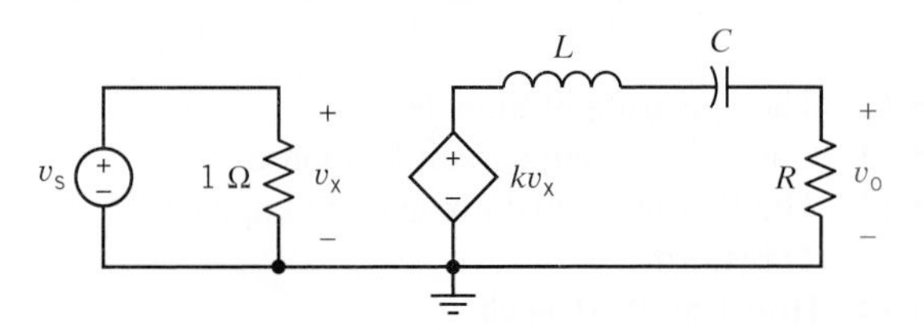

Figure DP 14-1

DP 14-2 Design the circuit in Figure DP 14-1 to have a step response equal to

$$v_o = 5e^{-4t}\sin(2t)u(t)\text{ V}$$

Hint: Determine the transfer function of the circuit in Figure DP 14-1 in terms of *k*, *R*, *C*, and *L*. Then determine the Laplace transform of the step response of the circuit in Figure DP 14-1. Next, determine the Laplace transform of the given step response. Finally, determine values of *k*, *R*, *C*, and *L* that cause the two step responses to be equal.

Answer: Pick $L = 1$ H, then $k = 1.25$ V/V, $R = 8\ \Omega$, and $C = 0.05$ F. (This answer is not unique.)

DP 14-3 Design the circuit in Figure DP 14-1 to have a step response equal to

$$v_o = 10\left(e^{-2t} - e^{-4t}\right)u(t)\text{ V}$$

Hint: Determine the transfer function of the circuit in Figure DP 14-1 in terms of *k*, *R*, *C*, and *L*. Then determine the Laplace transform of the step response of the circuit in Figure DP 14-1. Next, determine the Laplace transform of the given step response. Finally, determine values of *k*, *R*, *C*, and *L* that cause the two step responses to be equal.

Answer: Pick $L = 1$ H, then $k = 3.33$ V/V, $R = 6\ \Omega$, and $C = 0.125$ F. (This answer is not unique.)

DP 14-4 Show that the circuit in Figure DP 14-1 cannot be designed to have a step response equal to

$$v_o = 5\left(e^{-3t} + e^{-4t}\right)u(t)\text{ V}$$

Hint: Determine the transfer function of the circuit in Figure DP 14-1 in terms of *k*, *R*, *C*, and *L*. Then determine the Laplace transform of the step response of the circuit in Figure DP 14-1. Next, determine the Laplace transform of the given step response. Notice that these two functions have different forms and so cannot be made equal by any choice of values of *k*, *R*, *C*, and *L*.

DP 14-5 The circuit shown in Figure DP 14-5 represents an oscilloscope probe connected to an oscilloscope. Components C_2 and R_2 represent the input circuitry of the oscilloscope, and C_1 and R_1 represent the probe. Determine the transfer function $H(s) = V_o(s)/V(s)$. Determine the required relationship so that the natural response of the probe is zero. Determine the required relationship so that the step response is equal to the step input to within a gain constant. Is this achievement physically possible?

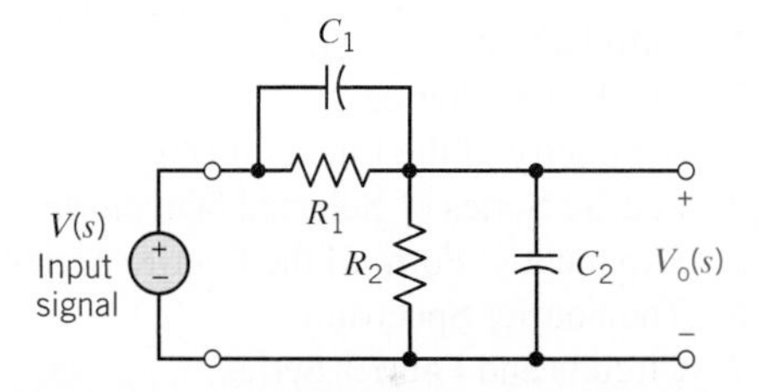

Figure DP 14-5 Oscilloscope probe circuit.

DP 14-6 A bicycle light is a useful accessory if you do a lot of riding at night. By lighting up the road and making you more visible to cars, it reduces the chances of an accident. Although, generator-powered incandescent lights are the most common type used on bikes, there are a number of reasons fluorescent lights are more suitable. For one thing, fluorescent lights shine with a brighter light that fully covers the road, the rider, and the bike and really gets the attention of car drivers. Because they are shaped in narrow tubes that can mount alongside a bike's frame, fluorescent lights offer less wind resistance than a comparable headlight with a flat face. When used with a generator, a fluorescent light offers additional advantages over a conventional incandescent light, first, because it is more efficient—giving more light for the same pedaling effort—and, second, because it cannot be burned out by the overvoltage that a generator can produce when speeding down hills. Fluorescent lights also last longer than do incandescent bulbs, especially on a bicycle, on which vibrations tend to weaken an incandescent bulb's filament.

A model of a fluorescent light for a bike is shown in Figure DP 14-6. Select *L* so that the bulb current rapidly rises to its steady-state value and overshoots its final value only by less than 10 percent.

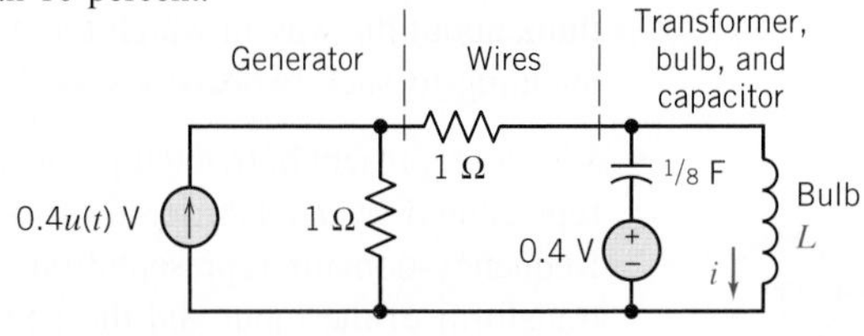

Figure DP 14-6 Fluorescent bicycle light circuit.

CHAPTER

Fourier Series and Fourier Transform

IN THIS CHAPTER

15.1 Introduction
15.2 The Fourier Series
15.3 Symmetry of the Function $f(t)$
15.4 Fourier Series of Selected Waveforms
15.5 Exponential Form of the Fourier Series
15.6 The Fourier Spectrum
15.7 Circuits and Fourier Series
15.8 Using PSpice to Determine the Fourier Series
15.9 The Fourier Transform
15.10 Fourier Transform Properties
15.11 The Spectrum of Signals
15.12 Convolution and Circuit Response
15.13 The Fourier Transform and the Laplace Transform
15.14 How Can We Check . . . ?
15.15 **DESIGN EXAMPLE**—DC Power Supply
15.16 Summary
Problems
PSpice Problems
Design Problems

15.1 INTRODUCTION

This chapter introduces the Fourier series and the Fourier transform. The Fourier series represents a nonsinusoidal periodic waveform as a sum of sinusoidal waveforms. The Fourier series is useful to us in two ways:

- The Fourier series shows that a periodic waveform consists of sinusoidal components at different frequencies. That allows us to think about the way in which the waveform is distributed in frequency. For example, we can give meaning to such expressions as ''the high-frequency part of a square wave.''
- We can use superposition to find the steady-state response of a circuit to an input represented by a Fourier series and, thus, determine the steady-state response of the circuit to the periodic waveform.

We obtain the Fourier transform as a generalization of the Fourier series, taking the limit as the period of a periodic wave becomes infinite. The Fourier transform is useful to us in two ways:

- The Fourier transform represents an aperiodic waveform in the frequency domain. That allows us to think about the way in which the waveform is distributed in frequency. For example, we can give meaning to such expressions as ''the high-frequency part of a pulse.''
- We can represent both the input to a circuit and the circuit itself in the frequency domain: the input represented by its Fourier transform and the circuit represented by its network function. The frequency-domain representation of circuit output is obtained as the product of the Fourier transform of the input and the network function of the circuit.

15.2 THE FOURIER SERIES

Baron Jean-Baptiste-Joseph Fourier proposed in 1807 that any periodic function could be expressed as an infinite sum of simple sinusoids. This surprising claim predicts that even discontinuous periodic waveforms, such as square waves, can be represented using only sinusoids. In 1807, Fourier's claim was controversial. Such famous mathematicians as Pierre Simon de Laplace and Joseph Louis Lagrange doubted the validity of Fourier's representation of periodic functions. In 1828, Johann Peter Gustav Lejeune Dirichlet presented a set of conditions sufficient to guarantee the convergence of Fourier's series. Today, the Fourier series is a standard tool for scientists and engineers.

Let's consider periodic functions. The function $f(t)$ is periodic if there exists a delay τ such that

$$f(t) = f(t - \tau) \tag{15.2-1}$$

for every value of t. This value of τ not unique. In particular, if τ satisfies Eq. 15.2-1, then every integer multiple of τ also satisfies Eq. 15.2-1. In other words, if τ satisfies Eq 15.2-1 and k is any integer, then

$$f(t) = f(t - k\tau)$$

for every value of t. To uniquely define the **period**, T, of the periodic function $f(t)$, we let T be the smallest positive value of τ that satisfies Eq. 15.2-1.

Next, we use the period T to define the **fundamental frequency**, ω_0, of the periodic function $f(t)$,

$$\omega_0 = \frac{2\pi}{T} \tag{15.2-2}$$

The fundamental frequency has units of rad/s. Integer multiples of the fundamental frequency are called harmonic frequencies.

A periodic function $f(t)$ can be represented by an infinite series of harmonically related sinusoids, called the (trigonometric) Fourier series, as follows:

$$f(t) = a_0 + \sum_{n=1}^{\infty} a_n \cos n\,\omega_0 t + \sum_{n=1}^{\infty} b_n \sin n\,\omega_0 t \tag{15.2-3}$$

where ω_0 is the fundamental frequency and the (real) coefficients, and a_0, a_n, and b_n are called the **Fourier trigonometric coefficients**. The Fourier trigonometric coefficients can be calculated using

$$a_0 = \frac{1}{T}\int_{t_0}^{T+t_0} f(t)dt = \text{the average value of } f(t) \tag{15.2-4}$$

$$a_n = \frac{2}{T}\int_{t_0}^{T+t_0} f(t)\cos n\,\omega_0 t\,dt \quad n > 0 \tag{15.2-5}$$

$$b_n = \frac{2}{T}\int_{t_0}^{T+t_0} f(t)\sin n\,\omega_0 t\,dt \quad n > 0 \tag{15.2-6}$$

The conditions presented by Dirichlet are sufficient to guarantee the convergence of the trigonometric Fourier series given in Eq. 15.2-3. The Dirichlet conditions require that the periodic function $f(t)$ satisfies the following mathematical properties:

1. $f(t)$ is a single-valued function except at possibly a finite number of points.
2. $f(t)$ is absolutely integrable, that is, $\int_{t_0}^{t_0+T} |f(t)|dt < \infty$ for any t_0.
3. $f(t)$ has a finite number of discontinuities within the period T.
4. $f(t)$ has a finite number of maxima and minima within the period T.

For our purposes, $f(t)$ will represent a voltage or current waveform, and any voltage or current waveform that we can actually produce will certainly satisfy the Dirichlet conditions. We shall assume that the Dirichlet conditions previously listed are always satisfied for periodic voltage or current waveforms.

A **Fourier series** is an accurate representation of a periodic signal and consists of the sum of sinusoids at the fundamental and harmonic frequencies.

Given a periodic voltage or current waveform, we can obtain the Fourier representation of that voltage or current in four steps:

Step 1 Determine the period T and the fundamental frequency ω_0.

Step 2 Represent the voltage or current waveform as a function of t over one complete period.

Step 3 Use Eqs. 15.2-4, 5 and 6 to determine the Fourier trigonometric coefficients a_0, a_n, and b_n.

Step 4 Substitute the coefficients a_0, a_n, and b_n obtained in Step 3 into Eq. 15.2-3.

The following example illustrates this four-step procedure.

EXAMPLE 15.2-1 Fourier Series of a Full-wave Rectified Cosine

Figure 15.2-1 shows a full-wave rectifier having a cosine input. The output of a full-wave input is the absolute value of its input, shown in Figure 15.2-2. A full-wave rectifier is an electronic circuit often used as a component of such diverse products as power supplies and AM radio receivers. Determine the Fourier series of the periodic waveform shown in Figure 15.2-2.

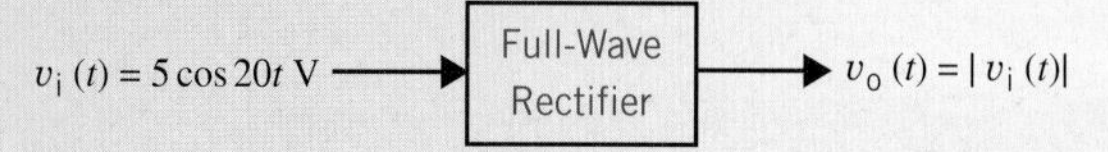

FIGURE 15.2-1 The circuit considered in Example 15.2-1.

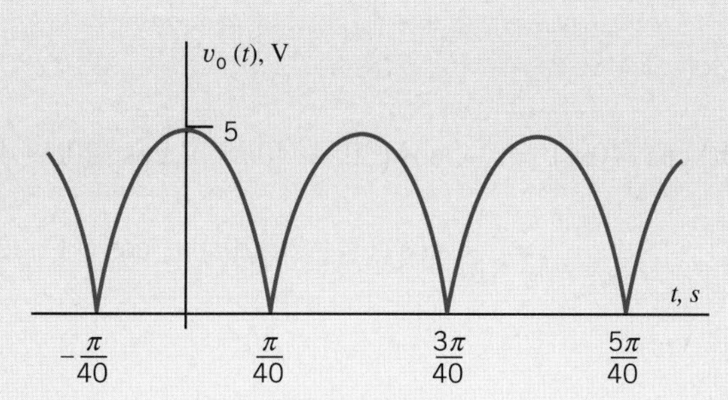

FIGURE 15.2-2 A full-wave rectified cosine.

Solution

Step 1: From Figure 15.2-2, we see that the period of $v_o(t)$ is

$$T = \frac{3\pi}{40} - \frac{\pi}{40} = \frac{\pi}{20} \text{ s}$$

The fundamental radian frequency is

$$\omega_0 = \frac{2\pi}{T} = 40 \text{ rad/s}$$

Step 2: Equations 15.2-4, 5, and 6 require integration over one full period of $v_o(t)$. We are free to choose the starting point of that period, t_o, to make the integration as easy as possible. Often, we choose to integrate either from 0 to T or from $-T/2$ to $T/2$. In this example, the periodic waveform can be represented as

$$v_o(t) = \begin{cases} 5\cos(20t) & \text{when } -\frac{\pi}{40} \leq t \leq \frac{\pi}{40} \\ -5\cos(20t) & \text{when } \frac{\pi}{40} \leq t \leq \frac{3\pi}{40} \end{cases}$$

Consider the calculation of a_0, using Eq. 15.2.4. If we choose to integrate form 0 to T, we have

$$a_0 = \frac{20}{\pi}\int_0^{\pi/20} v_o(t)dt = \frac{20}{\pi}\int_0^{\pi/40} 5\cos(20t)dt + \frac{20}{\pi}\int_{\pi/40}^{\pi/20} -5\cos(20t)dt$$

On the other hand, if we choose to integrate from $-T/2$ to $T/2$, we have

$$a_0 = \frac{20}{\pi}\int_{-\pi/40}^{\pi/40} v_o(t)dt = \frac{20}{\pi}\int_{-\pi/40}^{\pi/40} 5\cos(20t)dt$$

The second equation is simpler, so we choose to integrate from $-T/2$ to $+T/2$ for convenience.

Step 3: Now we will use Eqs. 15.2-4, 5, and 6 to determine the Fourier trigonometric coefficients a_0, a_n, and b_n. First,

$$a_0 = \frac{20}{\pi}\int_{-\pi/40}^{\pi/40} 5\cos(20t)dt = \frac{100}{\pi}\left(\frac{1}{20}\sin(20t)\Big|_{-\pi/40}^{\pi/40}\right) = \frac{5}{\pi}\left(\sin\left(\frac{\pi}{2}\right) - \sin\left(-\frac{\pi}{2}\right)\right) = \frac{10}{\pi}$$

Next,

$$a_n = \frac{40}{\pi}\int_{-\pi/40}^{\pi/40} 5\cos(20t)\cos(n\,\omega_0 t)dt = \frac{40}{\pi}\int_{-\pi/40}^{\pi/40} 5\cos(20t)\cos(40nt)dt$$

Using a trigonometric identity,

$$\begin{aligned}\cos(20t)\cos(40nt) &= \frac{1}{2}(\cos(20t+40nt)+\cos(20t+40nt))\\ &= \frac{1}{2}(\cos((1+2n)20t)+\cos((1-2n)20t))\end{aligned}$$

Then,

$$\begin{aligned}a_n &= \frac{100}{\pi}\int_{-\pi/40}^{\pi/40}(\cos((1+2n)20t)+\cos((1-2n)20t))dt\\ &= \frac{100}{\pi}\left(\frac{\sin((1+2n)20t)}{(1+2n)20}\Bigg|_{-\pi/40}^{\pi/40}+\frac{\sin((1-2n)20t)}{(1-2n)20}\Bigg|_{-\pi/40}^{\pi/40}\right)\\ &= \frac{5}{\pi}\left(\frac{\sin\left((1+2n)\frac{\pi}{2}\right)-\sin\left(-(1+2n)\frac{\pi}{2}\right)}{(1+2n)}+\frac{\sin\left((1-2n)\frac{\pi}{2}\right)-\sin\left(-(1-2n)\frac{\pi}{2}\right)}{(1-2n)}\right)\\ &= \frac{5}{\pi}\left(\frac{2(-1)^n}{(1+2n)}+\frac{2(-1)^n}{(1-2n)}\right)=\frac{20(-1)^n}{\pi(1-4n^2)}\end{aligned}$$

Similarly,

$$\begin{aligned}b_n &= \frac{40}{\pi}\int_{-\pi/40}^{\pi/40}5\cos(20t)\sin(40\,nt)dt\\ &= \frac{100}{\pi}\int_{-\pi/40}^{\pi/40}(\sin((2n+1)20t)+\sin((2n-1)20t))dt\\ &= \frac{100}{\pi}\left(\frac{-\cos((1+2n)20t)}{(1+2n)20}\Bigg|_{-\pi/40}^{\pi/40}+\frac{-\cos((1-2n)20t)}{(1-2n)20}\Bigg|_{-\pi/40}^{\pi/40}\right)=0\end{aligned}$$

In summary,

$$a_0=\frac{10}{\pi},\ a_n=\frac{20(-1)^n}{\pi(1-4n^2)}\text{ and }b_n=0 \tag{15.2-7}$$

Step 4: Substitute the coefficients a_0, a_n, and b_n given in Eq. 15.2-7 into Eq. 15.2-3:

$$v_o(t)=\frac{10}{\pi}+\frac{20}{\pi}\sum_{n=1}^{\infty}\frac{(-1)^n}{1-4n^2}\cos(40\,nt) \tag{15.2-8}$$

Equation 15.2-8 represents the rectified cosine by its Fourier series, but this equation is complicated enough to make us wonder what we have accomplished. How can we be sure that Eq. 15.2-8 actually represents a rectified cosine? Figure 15.2-3 shows a MATLAB script that plots the Fourier series given in Eq. 15.2-8. In particular, notice how the coefficients a_0, a_n, and b_n determined in step 3 are used in the MATLAB script. The plot produced by this MATLAB script is shown in Figure 15.2-4. The wave form in Figure 15.2-4 is indeed a rectified cosine having the correct amplitude, 5 volts, and correct period, $\frac{\pi}{20}\cong 0.16$ seconds. Thus, we see that Eq. 15.2-8 does indeed represent the rectified cosine.

```
% Ex15_2_1.m - full-wave rectified cosine Fourier series
% ----------------------------------------------------------
%         Describe the periodic waveform, v(t)
% ----------------------------------------------------------
T=pi/20;              % period
a0=10/pi;             % average value

% ----------------------------------------------------------
%  Obtain a list of equally spaced instants of time
% ----------------------------------------------------------
w0=2*pi/T;            % fundamental frequency, rad/s
tf=2*T;               % final time
dt=tf/200;            % time increment
t=0:dt:tf;            % time, s

% ----------------------------------------------------------
%  Approximate v(t) using the trig Fourier series.
% ----------------------------------------------------------
v = a0*ones(size(t));  % initialize v(t) as vector
for n=1:100
   an = 20*((-1)^n)/(pi*(1-4*n^2));
   bn = 0;
   v = v + an*cos(n*w0*t) + bn*sin(n*w0*t);
end

% ----------------------------------------------------------
%              Plot the Fourier series
% ----------------------------------------------------------
plot(t, v)
axis([0 tf 0 6])
grid
xlabel('time, s')
ylabel('v(t) V')
title('Full-wave Rectified Cosine')
```

FIGURE 15.2-3 MATLAB script to plot the rectified cosine.

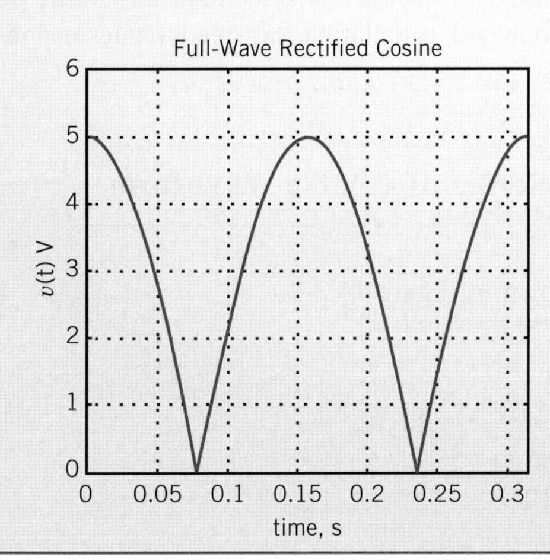

FIGURE 15.2-4 MATLAB plot of the full-wave rectified cosine.

Next, we obtain an alternate representation of the trigonometric Fourier series. The Fourier series, given in Eq 15.2-3, can be written as:

$$f(t) = a_0 + \sum_{n=1}^{\infty} (a_n \cos n\,\omega_0 t + b_n \sin n\,\omega_0 t) \tag{15.2-9}$$

Using a trigonometric identity, the nth term of this series can be written as

$$a_n \cos n\,\omega_0 t + b_n \sin n\,\omega_0 t = a_n \cos n\,\omega_0 t + b_n \cos(n\,\omega_0 t - 90^\circ) \tag{15.2-10}$$

Using phasors, we can represent the right-hand side of Eq 15.2-10 in the frequency domain. Performing a rectangular-to-polar conversion, we obtain

$$a_n \angle 0 + b_n \angle -90^\circ = a_n - jb_n = c_n \angle \theta_n$$

where

$$c_n - \sqrt{a_n^2 + b_n^2} \quad \text{and} \quad \theta_n = \begin{cases} -\tan^{-1}\left(\dfrac{b_n}{a_n}\right) & \text{if } a_n > 0 \\ 180^\circ - \tan^{-1}\left(\dfrac{b_n}{a_n}\right) & \text{if } a_n < 0 \end{cases} \tag{15.2-11}$$

and

$$a_n = c_n \cos\theta_n \quad \text{and} \quad b_n = -c_n \sin\theta_n$$

Back in the time domain, the corresponding sinusoid is

$$c_n \cos(n\,\omega_0 t + \theta_n)$$

After defining c_0 to be

$$c_0 = a_0 = \text{average value of } f(t) \tag{15.2-12}$$

The Fourier series is represented as

$$f(t) = c_0 + \sum_{n=1}^{\infty} c_n \cos(n\,\omega_0 t + \theta_n) \tag{15.2-13}$$

To distinguish between the two forms of the trigonometric Fourier series, we will refer to the series given in Eq. 15.2-3 as the sine-cosine Fourier series and to the series given in Eq. 15.2-13 as the amplitude-phase Fourier series.

In general, it is easier to calculate a_n and b_n than it is to calculate the coefficients c_n and θ_n. We will see in Section 15.3 that this is particularly true when $f(t)$ is symmetric. On the other hand, the Fourier series involving c_n is more convenient for calculating the steady-state response of a linear circuit to a periodic input.

EXAMPLE 15.2-2 Fourier Series of a Pulse Waveform

Determine the Fourier series of the pulse waveform shown in Figure 15.2-5.

Solution

Step 1: From Figure 15.2-5, we see that the period of $v_o(t)$ is

$$T = \frac{\pi}{10}\ \text{s}$$

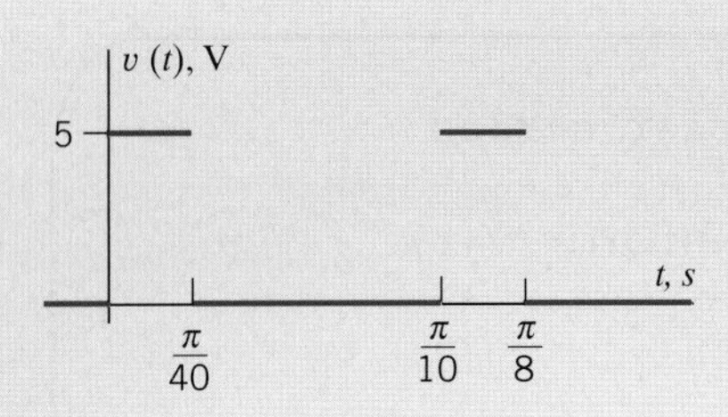

FIGURE 15.2-5 A pulse waveform.

The fundamental frequency is

$$\omega_0 = \frac{2\pi}{T} = 20 \text{ rad/s}$$

Step 2: Over the period from 0 to $\pi/10$, the pulse waveform is given by

$$v(t) = \begin{cases} 5 & \text{when } 0 \le t \le \dfrac{\pi}{40} \\ 0 & \text{when } \dfrac{\pi}{40} \le t \le \dfrac{\pi}{10} \end{cases}$$

Step 3: Next, we will determine the Fourier coefficients a_0, a_n, and b_n. First, we will calculate a_0 as the average value of $v(t)$:

$$a_0 = \frac{\text{area under the curve for the one period}}{\text{one period, } T} = \frac{5\left(\dfrac{\pi}{40}\right) + 0\left(\dfrac{3\pi}{40}\right)}{\dfrac{\pi}{10}} = 1.25 \text{ V}$$

Next,

$$a_n = \frac{20}{\pi}\int_0^{\pi/40} 5\cos(n\,\omega_0 t)dt + \frac{20}{\pi}\int_{\pi/40}^{\pi/10} 0\cos(n\,\omega_0 t)dt = \frac{20}{\pi}\int_0^{\pi/40} 5\cos(20nt)dt$$

$$= \frac{20(5)}{\pi}\left(\left.\frac{\sin(20nt)}{20n}\right|_0^{\pi/40}\right) = \frac{5}{n\pi}\sin\left(\frac{n\pi}{2}\right)$$

Similarly,

$$b_n = \frac{20}{\pi}\int_0^{\pi/40} 5\sin(n\,\omega_0 t)dt = \frac{20}{\pi}\int_0^{\pi/40} 5\sin(20nt)dt = \frac{20(5)}{\pi}\left(\left.\frac{-\cos(20nt)}{20n}\right|_0^{\pi/40}\right) = \frac{5}{n\pi}\left(1 - \cos\left(\frac{n\pi}{2}\right)\right)$$

In summary,

$$a_0 = 1.25,\; a_n = \frac{5}{n\pi}\sin\left(\frac{n\pi}{2}\right) \text{ and } b_n = \frac{5}{n\pi}\left(1 - \cos\left(\frac{n\pi}{2}\right)\right) \tag{15.2-14}$$

Step 4: Substitute the coefficients a_0, a_n, and b_n given in Eq. 15.2-7 into Eq. 15.2-3:

$$v_o(t) = 1.25 + \frac{5}{n\pi}\sum_{n=1}^{\infty}\left(\sin\left(\frac{n\pi}{2}\right)\cos(20nt) + \left(1 - \cos\left(\frac{n\pi}{2}\right)\right)\sin(20nt)\right) \tag{15.2-15}$$

Figure 15.2-6 shows a MATLAB script that plots the Fourier series given in Eq. 15.2-15. In particular, notice how the coefficients a_0, a_n, and b_n given in Eq. 15.2-14 are used in the MATLAB script. The plot produced by this MATLAB script is shown in Figure 15.2-7. The waveform in Figure 15.2-7 is indeed a pulse waveform having the correct amplitude, 5 volts, and correct period, $\frac{\pi}{10} \cong 0.32$ seconds.

```
% Ex15_2_2.m - pulse waveform Fourier series
% -----------------------------------------------------
%       Describe the periodic waveform, v(t)
% -----------------------------------------------------
T=pi/10;             % period
a0=1.25;            % average value

% -----------------------------------------------------
%  Obtain a list of equally spaced instants of time
% -----------------------------------------------------
w0=2*pi/T;           % fundamental frequency, rad/s
tf=2.5*T;            % final time
dt=tf/200;           % time increment
t=0:dt:tf;           % time, s

% -----------------------------------------------------
%  Approximate v(t) using the trig Fourier series.
% -----------------------------------------------------
v = a0*ones(size(t));  % initialize v(t) as vector
for n=1:500
   an = (5/n/pi)*sin(n*pi/2);
   bn = (5/n/pi)*(1-cos(n*pi/2));
   cn = abs(an - j*bn);
   thetan = angle(an - j*bn);
   v = v + cn*cos(n*w0*t + thetan);
end

% -----------------------------------------------------
%            Plot the Fourier series
% -----------------------------------------------------
plot(t, v)
axis([0 tf 0 6])
grid
xlabel('time, s')
ylabel('v(t) V')
title('Pulse Waveform')
```

FIGURE 15.2-6 MATLAB script to plot the pulse waveform.

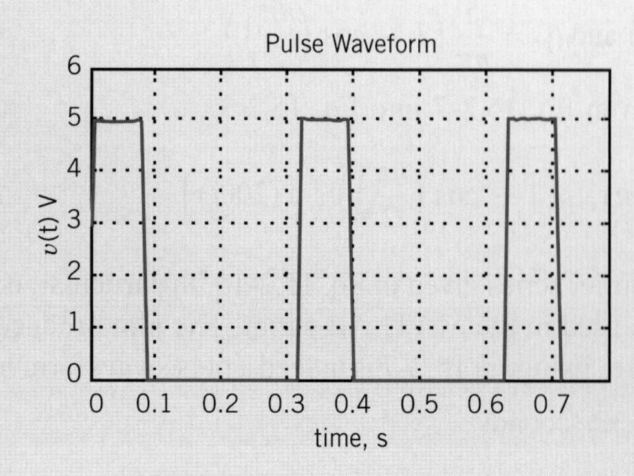

FIGURE 15.2-7 MATLAB plot of the full-wave rectified cosine.

EXERCISE 15.2-1 Suppose $f_1(t)$ and $f_2(t)$ are periodic functions having the same period, T. Then $f_1(t)$ and $f_2(t)$ can be represented by the Fourier series

$$f_1(t) = a_{10} + \sum_{n=1}^{\infty} (a_{1n} \cos(n\omega_0 t) + b_{1n} \sin(n\omega_0 t))$$

and

$$f_2(t) = a_{20} + \sum_{n=1}^{\infty} (a_{2n} \cos(n\omega_0 t) + b_{2n} \sin(n\omega_0 t))$$

Determine the Fourier series of the function

$$f(t) = k_1 f_1(t) + k_2 f_2(t)$$

Answer: $f(t) = (k_1 a_{10} + k_2 a_{20}) + \sum_{n=1}^{\infty} ((k_1 a_{1n} + k_2 a_{2n}) \cos(n\omega_0 t) + (k_1 b_{1n} + k_2 b_{2n}) \sin(n\omega_0 t))$

EXERCISE 15.2-2 Determine the Fourier series when $f(t) = K$, a constant.

Answer: $a_0 = K$ and $a_n = b_n = 0$ for $n \geq 1$

EXERCISE 15.2-3 Determine the Fourier series when $f(t) = A \cos \omega_0 t$.

Answer: $a_0 = 0$, $a_1 = A$, $a_n = 0$ for $n > 1$, and $b_n = 0$

15.3 SYMMETRY OF THE FUNCTION $f(t)$

Four types of symmetry can be readily recognized and then used to simplify the task of calculating the Fourier coefficients. They are the following:

1. Even-function symmetry
2. Odd-function symmetry
3. Half-wave symmetry
4. Quarter-wave symmetry

A function is *even* when $f(t) = f(-t)$, and a function is *odd* when $f(t) = -f(-t)$. The function shown in Figure 15.2-2 is an even function. For even functions, all $b_n = 0$ and

$$a_n = \frac{4}{T} \int_0^{T/2} f(t) \cos n\omega_0 t \, dt$$

For odd functions, all $a_n = 0$ and

$$b_n = \frac{4}{T} \int_0^{T/2} f(t) \sin n\omega_0 t \, dt$$

An example of an odd function is $\sin \omega_0 t$. Another odd function is shown in Figure 15.3-1.

Half-wave symmetry for a function $f(t)$ is obtained when

$$f(t) = -f\left(t - \frac{T}{2}\right) \tag{15.3-1}$$

In these half-wave symmetric waveforms, the second half of each period looks like the first half turned upside down. The function shown in Figure 15.3-2 has half-wave symmetry. If a function has half-wave

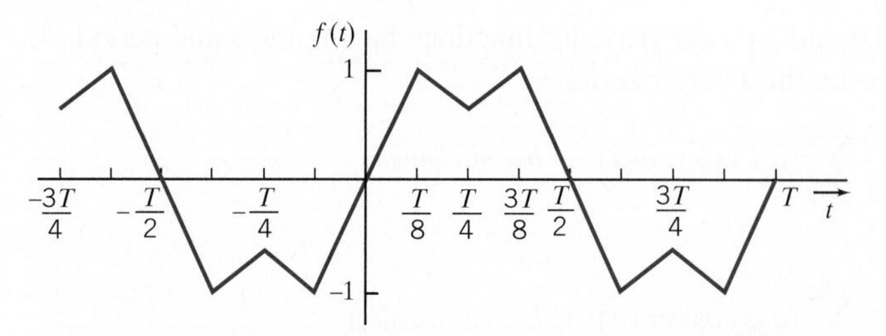

FIGURE 15.3-1 An odd function with quarter-wave symmetry.

symmetry, then both a_n and b_n are zero for even values of n. We see that $a_0 = 0$ for half-wave symmetry because the average value of the function over one period is zero.

Quarter-wave symmetry describes a function that has half-wave symmetry and, in addition, has symmetry about the midpoint of the positive and negative half-cycles. An example of an odd function with quarter-wave symmetry is shown in Figure 15.3-1. If a function is odd and has quarter-wave symmetry, then $a_0 = 0$, $a_n = 0$ for all n, $b_n = 0$ for even n. For odd n, b_n is given by

$$b_n = \frac{8}{T}\int_0^{T/4} f(t)\sin n\omega_0 t\,dt$$

If a function is even and has quarter-wave symmetry, then $a_0 = 0$, $b_n = 0$ for all n, and $a_n = 0$ for even n. For odd n, a_n is given by

$$a_n = \frac{8}{T}\int_0^{T/4} f(t)\cos n\omega_0 t\,dt$$

The calculation of the Fourier coefficients and the associated effects of symmetry of the waveform $f(t)$ are summarized in Table 15.3-1. Often, the calculation of the Fourier series can be simplified by judicious selection of the origin ($t = 0$) because the analyst usually has the choice to select this point arbitrarily.

Table 15.3-1 Fourier Series and Symmetry

SYMMETRY	FOURIER COEFFICIENTS
1. Odd function $f(t) = -f(-t)$	$a_n = 0$ for all n $b_n = \frac{4}{T}\int_0^{T/2} f(t)\sin n\omega_0 t\,dt$
2. Even function $f(t) = f(-t)$	$b_n = 0$ for all n $a_n = \frac{4}{T}\int_0^{T/2} f(t)\cos n\omega_0 t\,dt$
3. Half-wave symmetry $f(t) = -f\left(t - \frac{T}{2}\right)$	$a_0 = 0$ $a_n = 0$ for even n $b_n = 0$ for even n $a_n = \frac{4}{T}\int_0^{T/2} f(t)\cos n\omega_0 t\,dt$ for odd n $b_n = \frac{4}{T}\int_0^{T/2} f(t)\sin n\omega_0 t\,dt$ for odd n
4. Quarter-wave symmetry Half-wave symmetric and symmetric about the midpoints of the positive and negative half-cycles	A. Odd function: $a_0 = 0$, $a_n = 0$ for all n $b_n = 0$ for even n $b_n = \frac{8}{T}\int_0^{T/4} f(t)\sin n\omega_0 t\,dt$ for odd n B. Even function: $a_0 = 0$, $b_n = 0$ for all n $a_n = 0$ for even n $a_n = \frac{8}{T}\int_0^{T/4} f(t)\cos n\omega_0 t\,dt$ for odd n

EXAMPLE 15.3-1 Symmetry and the Fourier Series

Determine the Fourier series for the triangular waveform $v(t)$ shown in Figure 15.3-2.

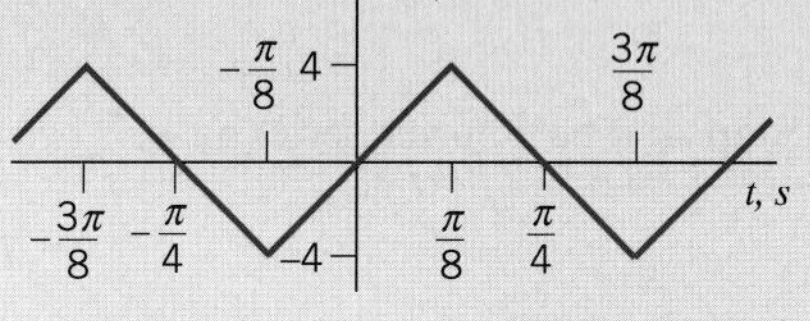

FIGURE 15.3-2 An odd function with half-wave symmetry.

Solution

Step 1: From Figure 15.3-2, we see that the period of $v_o(t)$ is

$$T = \frac{\pi}{4} - \left(-\frac{\pi}{4}\right) = \frac{\pi}{2} \text{ s}$$

The fundamental frequency is

$$\omega_0 = \frac{2\pi}{T} = 4 \text{ rad/s}$$

Step 2: If we don't take advantage of the symmetry of the triangle waveform, determining the Fourier coefficients a_0, a_n, and b_n will require integration over a full period—either from 0 to T or from $-T/2$ to $T/2$. Accordingly, we can represent $v(t)$ from time $-T/2$ to T, that is, from $-\pi/8$ to $\pi/2$ seconds. By writing equations for the various straight-line segments that comprise the triangle waveform, we can represent $v(t)$ as

$$v(t) = \begin{cases} -\dfrac{32}{\pi}t - 8 & \text{when } -\dfrac{3\pi}{8} \leq t \leq \dfrac{\pi}{8} \\ \dfrac{32}{\pi}t & \text{when } -\dfrac{\pi}{8} \leq t \leq \dfrac{\pi}{8} \\ -\dfrac{32}{\pi}t + 8 & \text{when } \dfrac{\pi}{8} \leq t \leq \dfrac{3\pi}{8} \\ \dfrac{32}{\pi}t - 16 & \text{when } \dfrac{3\pi}{8} \leq t \leq \dfrac{5\pi}{8} \end{cases}$$

If we take advantage of symmetry, we will need to integrate only from 0 to $T/2$, that is, from 0 to $\pi/8$ seconds. If we need to represent $v(t)$ only from 0 to $\pi/8$ seconds, we don't have to write equations for so many straight-line segments. In this case, we need to write the equation only for one straight line to represent $v(t)$ as

$$v(t) = \frac{32}{\pi}t \quad \text{when} \quad -\frac{\pi}{8} \leq t \leq \frac{\pi}{8}$$

Step 3: Next, we will determine the Fourier coefficients a_0, a_n, and b_n. First, the average value of the triangle waveform is 0 volt

$$a_0 = \text{the average value of } v(t) = 0$$

The triangle waveform has odd symmetry. From entry 1 of Table 15.3-1, $a_n = 0$ for all n and

$$\begin{aligned} b_n = \frac{4}{T}\int_0^{T/2} v(t)\sin n\omega_0 t\, dt &= \frac{8}{\pi}\int_0^{\pi/4} v(t)\sin 4nt\, dt \\ &= \frac{8}{\pi}\left[\int_0^{\pi/8}\left(\frac{32}{\pi}t\right)\sin 4nt\, dt + \int_{\pi/8}^{\pi/4}\left(-\frac{32}{\pi}t + 8\right)\sin 4nt\, dt\right] \end{aligned}$$

Noticing that the triangle waveform has quarter-wave symmetry provides a simpler equation for determining b_n. Using entry 4A of Table 15.3-1, we see that $b_n = 0$ for even n. For odd n,

$$\begin{aligned} b_n = \frac{8}{T}\int_0^{T/4} v(t)\sin n\omega_0 t\, dt = \frac{512}{\pi^2}\int_0^{\pi/8} t\sin 4nt\, dt &= \frac{512}{\pi^2}\left[\frac{\sin 4nt - 4nt\cos 4nt}{16n^2}\right]_0^{\pi/8} \\ &= \frac{32}{\pi^2 n^2}\left(\sin\left(n\frac{\pi}{2}\right) - 0 - n\frac{\pi}{2}\cos\left(n\frac{\pi}{2}\right) + 0\right) \end{aligned}$$

Because $\cos\left(n\frac{\pi}{2}\right) = 0$ for odd n, we obtain

$$b_n = \frac{32}{\pi^2 n^2}\sin\left(n\frac{\pi}{2}\right) \quad \text{for odd } n$$

In summary,

$$a_0 = 0,\ a_n = 0 \text{ for all } n, \text{ and } b_n = \begin{cases} \dfrac{32}{\pi^2 n^2} \sin\left(n\dfrac{\pi}{2}\right) & \text{for odd } n \\ 0 & \text{for even } n \end{cases}$$

Step 4: The Fourier series is

$$v(t) = \frac{32}{\pi^2} \sum_{odd\ n=1}^{\infty} \frac{1}{n^2} \sin\left(\frac{n\pi}{2}\right) \sin(4nt) \tag{15.3-2}$$

Notice the notation used in Eq. 15.3-2 to indicate that the summation includes only terms corresponding to the odd values of n.

Figure 15.3-3 shows a MATLAB script that plots the Fourier series given in Eq. 15.3-2. The plot produced by this MATLAB script is shown in Figure 15.2-3. The waveform in Figure 15.2-4 is indeed a triangle having the

```
% Ex15_3_1.m - triangle waveform Fourier series
% --------------------------------------------------
%         Describe the periodic waveform, v(t)
% --------------------------------------------------
T=pi/2;                 % period
a0=0;               % average value

% --------------------------------------------------
%  Obtain a list of equally spaced instants of time
% --------------------------------------------------
w0=2*pi/T;              % fundamental frequency, rad/s
tf=1.5*T;               % final time
dt=tf/500;              % time increment
t=0:dt:tf;              % time, s

% --------------------------------------------------
%  Approximate v(t) using the trig Fourier series.
% --------------------------------------------------
v = a0*ones(size(t));  % initialize v(t) as vector
for n=1:2:200
   an = 0;
   bn = (32/n/n/pi/pi)*sin(n*pi/2);
   v = v + bn*sin(n*w0*t);
end

% --------------------------------------------------
%              Plot the Fourier series
% --------------------------------------------------
plot(t, v)
axis([0 tf -5 5])
grid
xlabel('time, s')
ylabel('v(t) V')
title('Triangle Waveform')
```

FIGURE 15.3-3
MATLAB m-file.

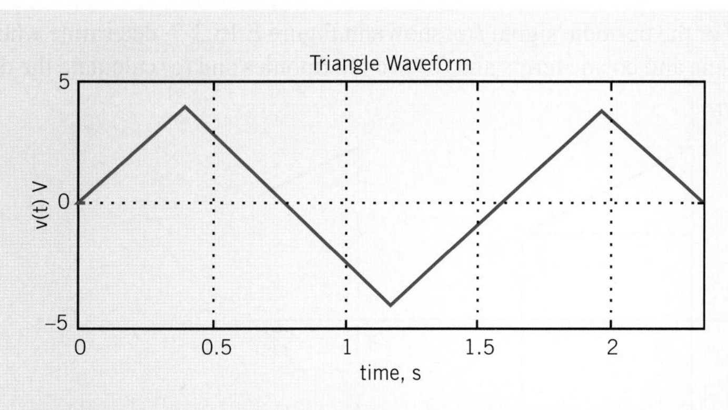

FIGURE 15.3-4 MATLAB output.

correct amplitude, 8 volts peak-to-peak, and correct period, $\frac{\pi}{2} \cong 1.6$ seconds. Thus, we see that Eq. 15.3-2 does indeed represent the triangle waveform.

EXERCISE 15.3-1 Determine the Fourier series for the waveform $f(t)$ shown in Figure E 15.3-1. Each increment of time on the horizontal axis is $\pi/8$ s, and the maximum and minimum are $+1$ and -1, respectively.

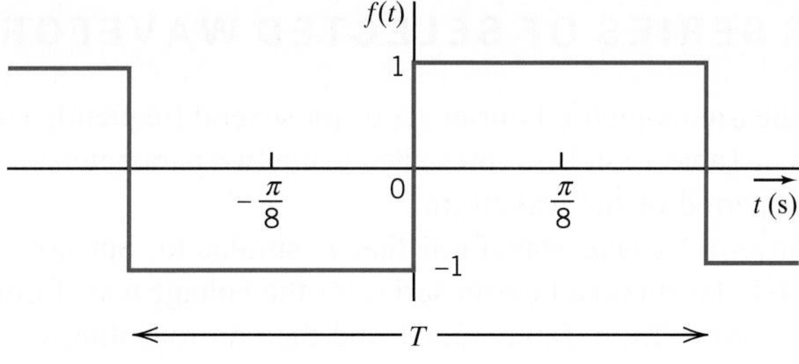

FIGURE E 15.3-1 The period $T = \frac{\pi}{2}$ s.

Answer: $f(t) = \frac{4}{\pi}\sum_{n=1}^{N}\frac{1}{n}\sin n\omega_0 t$ and n odd, $\omega_0 = 4$ rad/s

EXERCISE 15.3-2 Determine the Fourier series for the waveform $f(t)$ shown in Figure E 15.3-2. Each increment of time on the horizontal grid is $\pi/6$ s, and the maximum and minimum values of $f(t)$ are 2 and -2, respectively.

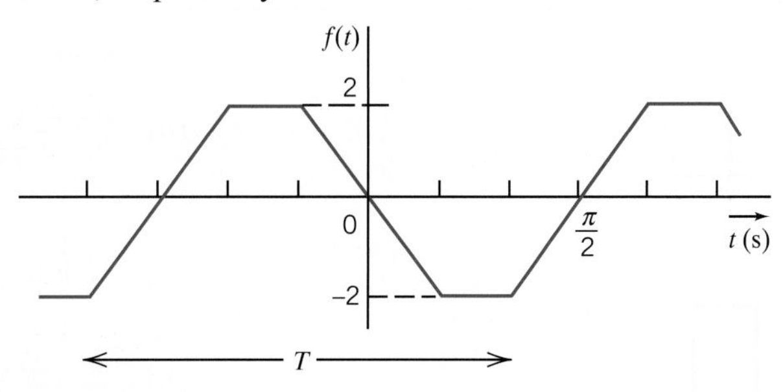

FIGURE E 15.3-2 The period $T = \pi$ s.

Answer: $f(t) = \frac{-24}{\pi^2}\sum_{n=1}^{N}\frac{1}{n^2}\sin(n\pi/3)\sin n\omega_0 t$ and n odd, $\omega_0 = 2$ rad/s

EXERCISE 15.3-3 For the periodic signal $f(t)$ shown in Figure E 15.3-3, determine whether the Fourier series contains (a) sine and cosine terms and (b) even harmonics and (c) calculate the dc value.

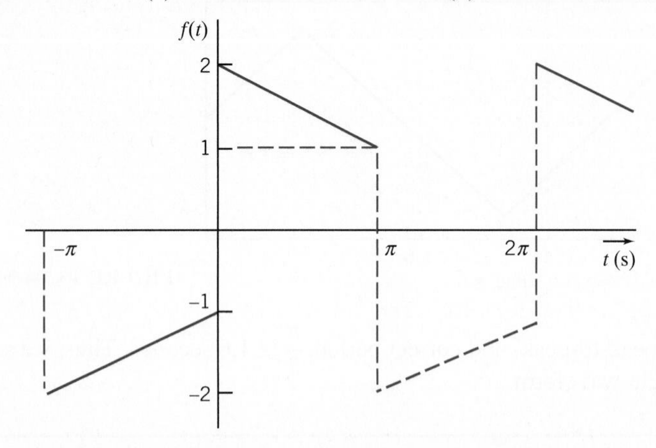

FIGURE E 15.3-3

Answers: **(a)** Yes, both sine and cosine terms; **(b)** no even harmonics; **(c)** $a_0 = 0$

15.4 FOURIER SERIES OF SELECTED WAVEFORMS

Table 15.4-1 provides the trigonometric Fourier series for several frequently encountered waveforms. Each of the waveforms in Table 15.4-1 is represented using two parameters: A is the amplitude of the waveform, and T is the period of the waveform.

Figure 15.4-1 shows a voltage waveform that is similar to, but not exactly the same as, a waveform in Table 15.4-1. To obtain a Fourier series for the voltage waveform, we select the Fourier series of the similar waveform from Table 15.4-1 and then do four things:

1. Set the value of A equal to the amplitude of the voltage waveform.
2. Add a constant to the Fourier series of the voltage waveform to adjust its average value.

Table 15.4-1 The Fourier Series of Selected Waveforms

FUNCTION	TRIGONOMETRIC FOURIER SERIES
$f(t)$, A, 0, $\frac{T}{2}$, T, t	Square wave : $\omega_0 = \frac{2\pi}{T}$ $f(t) = \frac{A}{2} + \frac{2A}{\pi} \sum_{n=1}^{\infty} \frac{\sin((2n-1)\omega_0 t)}{2n-1}$
$f(t)$, A, $-\frac{d}{2}$, $\frac{d}{2}$, T, t	Pulse wave: $\omega_0 = \frac{2\pi}{T}$ $f(t) = \frac{Ad}{T} + \frac{2A}{\pi} \sum_{n=1}^{\infty} \frac{\sin\left(\frac{n\pi d}{T}\right)}{n} \cos(n\omega_0 t)$

(continued)

Table 15.4-1 (*Continued*)

FUNCTION	TRIGONOMETRIC FOURIER SERIES
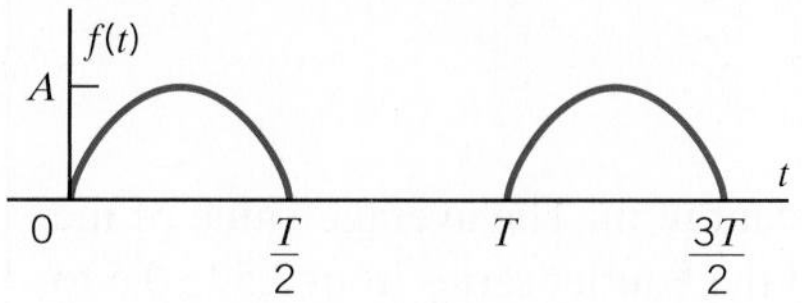	Half-wave rectified sine wave: $\omega_0 = \frac{2\pi}{T}$ $f(t) = \frac{A}{\pi} + \frac{A}{2}\sin\omega_0 t - \frac{2A}{\pi}\sum_{n=1}^{\infty}\frac{\cos(2n\,\omega_0 t)}{4n^2-1}$
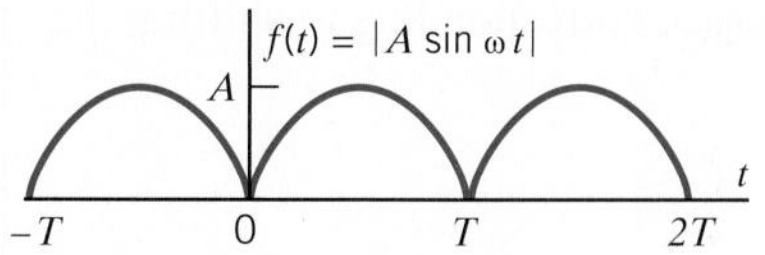	Full-wave rectified sine wave: $\omega_0 = \frac{2\pi}{T}$ $f(t) = \frac{2A}{\pi} - \frac{4A}{\pi}\sum_{n=1}^{\infty}\frac{\cos(n\,\omega_0 t)}{4n^2-1}$
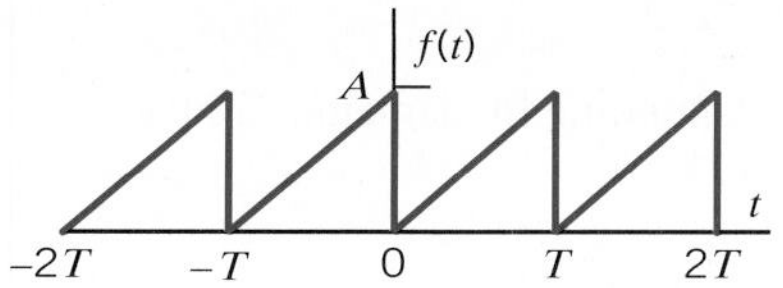	Sawtooth wave: $\omega_0 = \frac{2\pi}{T}$ $f(t) = \frac{A}{2} - \frac{A}{\pi}\sum_{n=1}^{\infty}\frac{\sin(n\,\omega_0 t)}{n}$
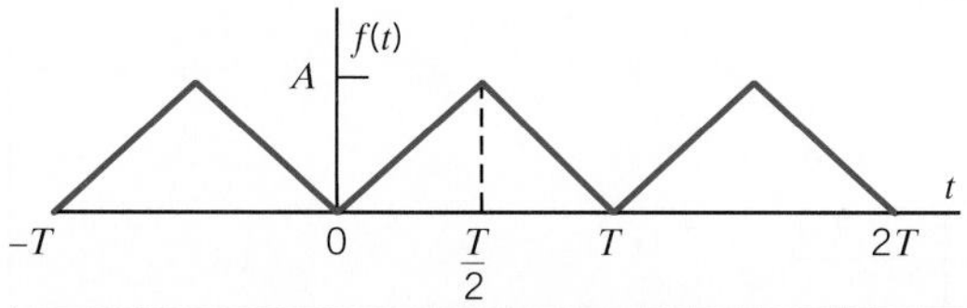	Triangle wave: $\omega_0 = \frac{2\pi}{T}$ $f(t) = \frac{A}{2} - \frac{4A}{\pi^2}\sum_{n=1}^{\infty}\frac{\cos((2n-1)\omega_0 t)}{(2n-1)^2}$

3. Set the value of T equal to the period of the voltage waveform.
4. Replace t by $t - t_o$ when the voltage waveform is delayed by time t_o with respect to the waveform in Table 15.4-1. After some algebra, the delay can be represented as a phase shift in the Fourier series of the voltage waveform.

EXAMPLE 15.4-1

Determine the Fourier series of the voltage waveform shown in Figure 15.4-1.

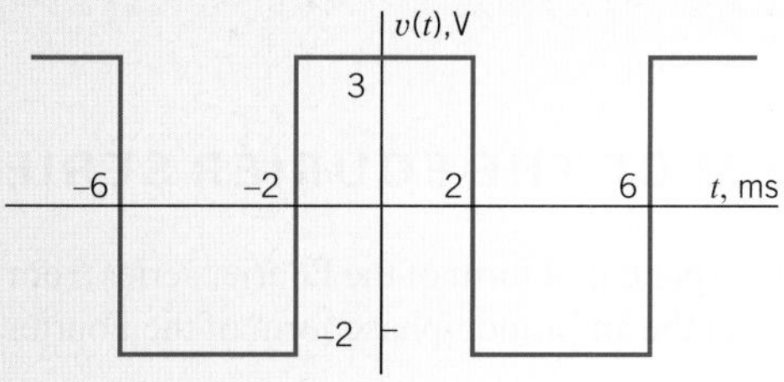

FIGURE 15.4-1 A voltage waveform.

Solution

The voltage waveform is similar to the square wave in Table 15.4-1. The Fourier series of the square is

$$f(t) = \frac{A}{2} + \frac{2A}{\pi}\sum_{n=1}^{\infty}\frac{\sin((2n-1)\omega_0 t)}{2n-1}$$

Step 1: The amplitude of the voltage waveform is $3-(-2)=5$ V. After setting $A=5$, the Fourier series becomes

$$2.5+\frac{10}{\pi}\sum_{n=1}^{\infty}\frac{\sin((2n-1)\omega_0 t)}{2n-1}$$

Step 2: The average value of the Fourier series is 2.5, the value of the constant term. The average value of the voltage waveform is $(3+(-2))/2=0.5$ V. We change the constant term of the Fourier series from 2.5 to 0.5 to adjust its average value. This is equivalent to subtracting 2 from the Fourier series, corresponding to shifting the waveform downward by 2 V:

$$0.5+\frac{10}{\pi}\sum_{n=1}^{\infty}\frac{\sin((2n-1)\omega_0 t)}{2n-1}$$

Step 3: The period of the voltage waveform is $T=6-(-2)=8$ ms. The corresponding fundamental frequency is

$$\omega_0=\frac{2\pi}{0.008}=250\,\pi \text{ rad/s}$$

After setting $\omega_0=250\,\pi$ rad/s, the Fourier series becomes

$$0.5+\frac{10}{\pi}\sum_{n=1}^{\infty}\frac{\sin((2n-1)250\,\pi t)}{2n-1}$$

Step 4: The square wave in Table 15.4-1 has a rising edge at time 0. The corresponding rising edge of the voltage waveform occurs at -2 ms. The voltage waveform is advanced by 2 ms or, equivalently, delayed by -2 ms. Consequently, we replace t by $t-(-0.002)=t+0.002$ in the Fourier series. We notice that

$$\sin((2n-1)250\,\pi(t+0.002))=\sin\left((2n-1)\left(250\,\pi t+\frac{\pi}{2}\right)\right)=\sin((2n-1)(250\,\pi t+90^\circ))$$

After replacing t by $t+0.002$, the Fourier series becomes

$$v(t)=0.5+\frac{10}{\pi}\sum_{n=1}^{\infty}\frac{\sin((2n-1)(250\,\pi t+90^\circ))}{2n-1}$$

15.5 EXPONENTIAL FORM OF THE FOURIER SERIES

Using Euler's identity, we can derive the exponential form of the Fourier series from the trigonometric Fourier series. Recall from Eq. 15.2-13 that the amplitude-phase form of the Fourier series is given by

$$f(t)=c_0+\sum_{n=1}^{\infty}c_n\cos(n\omega_0 t+\theta_n) \tag{15.5-1}$$

Euler's identity is

$$e^{j\theta}=\cos\theta+j\sin\theta \tag{15.5-2}$$

A consequence of Euler's identity is

$$\cos\theta = \frac{1}{2}\left(e^{j\theta} + e^{-j\theta}\right) \tag{15.5-3}$$

Using Euler's identity, the nth term of the Fourier series is written as

$$c_n \cos\left(n\omega_0 t + \theta_n\right) = c_n\left(\frac{e^{j(n\omega_0 t+\theta_n)} + e^{-j(n\omega_0 t+\theta_n)}}{2}\right) = \frac{c_n}{2}\left(e^{j(n\omega_0 t+\theta_n)} + e^{-j(n\omega_0 t+\theta_n)}\right) \tag{15.5-4}$$

Using Eq. 15.5-4 in Eq. 15.5-1 gives

$$f(t) = c_0 + \sum_{n=1}^{\infty}\frac{c_n}{2}\left(e^{j(n\omega_0 t+\theta_n)} + e^{-j(n\omega_0 t+\theta_n)}\right) = c_0 + \sum_{n=1}^{\infty}\left(\frac{c_n}{2}e^{j\theta_n}\right)e^{jn\omega_0 t} + \sum_{n=1}^{\infty}\left(\frac{c_n}{2}e^{-j\theta_n}\right)e^{-jn\omega_0 t} \tag{15.5-5}$$

Define

$$C_0 = c_0, \quad \mathbf{C}_n = \frac{c_n}{2}e^{j\theta_n}, \quad \text{and} \quad \mathbf{C}_{-n} = \frac{c_n}{2}e^{-j\theta_n} \tag{15.5-6}$$

Then $f(t)$ can be expressed as

$$f(t) = C_0 + \sum_{n=1}^{\infty}\mathbf{C}_n e^{jn\omega_0 t} + \sum_{n=1}^{\infty}\mathbf{C}_{-n}e^{-jn\omega_0 t} \tag{15.5-7}$$

Introducing the notation

$$C_0 = C_0 e^{j0} = \mathbf{C}_0$$

we can write Eq 15.5-7 as

$$f(t) = \sum_{n=-\infty}^{\infty}\mathbf{C}_n e^{jn\omega_0 t} \tag{15.5-8}$$

Equation 15.5-8 represents $f(t)$ as an **exponential Fourier series**. The complex coefficients $\mathbf{C}_n$ of the exponential Fourier series can be calculated directly from $f(t)$ using

$$\mathbf{C}_n = \frac{1}{T}\int_{t_0}^{t_0+T} f(t)e^{-jn\omega_0 t}dt \tag{15.5-9}$$

Referring to Eq. 15.5-6, we notice that $\mathbf{C}_{-n}$ is the complex conjugate of $\mathbf{C}_n$, that is, $\mathbf{C}_n = \mathbf{C}^*_{-n}$. Using Eqs. 15.5-6 and 15.2-11, we see that the coefficients of the exponential Fourier series are obtained from the coefficients of the sine-cosine Fourier series, using

$$\mathbf{C}_n = \frac{c_n e^{j\theta_n}}{2} = \frac{a_n - jb_n}{2} \quad \text{and} \quad \mathbf{C}_{-n} = \frac{c_n e^{-j\theta_n}}{2} = \frac{a_n + jb_n}{2} \tag{15.5-10}$$

Equivalently, the coefficients of the sine-cosine Fourier series are obtained from the coefficients of the exponential Fourier series, using

$$a_n = \mathbf{C}_n + \mathbf{C}_{-n} \quad \text{and} \quad b_n = j(\mathbf{C}_n - \mathbf{C}_{-n}) \tag{15.5-11}$$

The coefficients of the exponential Fourier series of selected waveforms are given in Table 15.5-1. Recall that $b_n = 0$ when $f(t)$ is an even function. Consequently, $\mathbf{C}_{-n} = \mathbf{C}_n$ when $f(t)$ is an even function. Similarly, $\mathbf{C}_{-n} = -\mathbf{C}_n$ when $f(t)$ is an odd function.

Table 15.5-1 Complex Fourier Coefficients for Selected Waveform

	WAVEFORM	NAME OF WAVEFORM AND EQUATION	SYMMETRY	C_n
1.	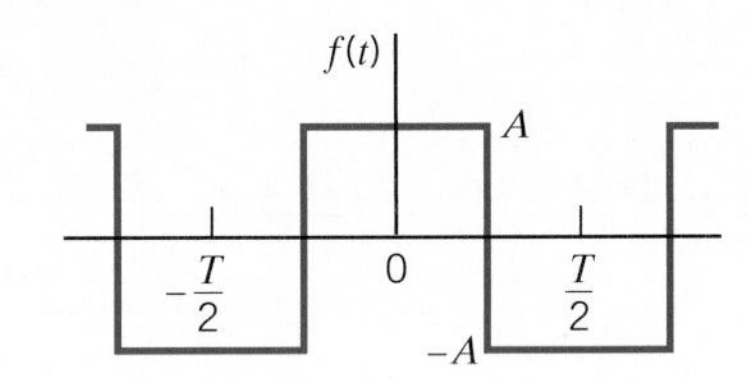	**Square wave** $f(t) = \begin{cases} A, & \frac{-T}{4} < t < \frac{T}{4} \\ -A, & \frac{T}{4} < t < \frac{3T}{4} \end{cases}$	Even	$= A\dfrac{\sin n\pi/2}{n\pi/2}$, n odd $= 0$, $n = 0$ and n even
2.	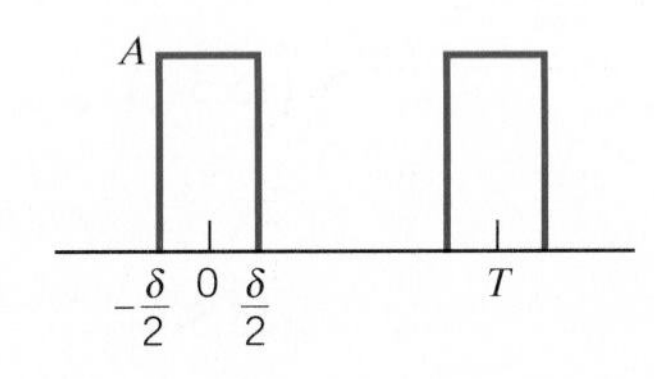	**Rectangular pulse** $f(t) = A, \quad \frac{-\delta}{2} < t < \frac{\delta}{2}$	Even	$= A\dfrac{\delta}{T}\dfrac{\sin(n\pi\delta/T)}{(n\pi\delta/T)}$
3.	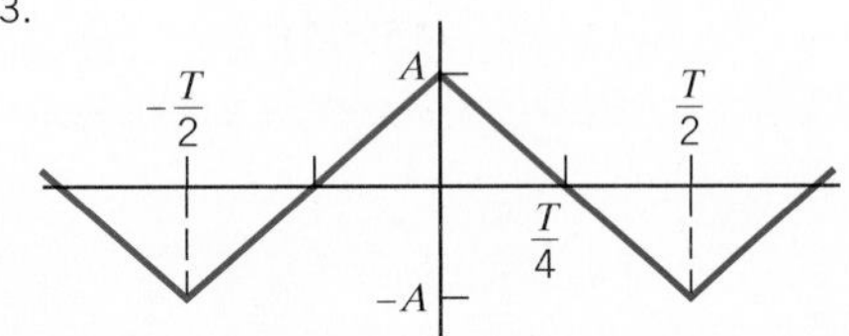	**Triangular wave**	Even	$= A\dfrac{\sin^2(n\pi/2)}{(n\pi/2)^2}, n \neq 0$ $= 0$, $n = 0$
4.	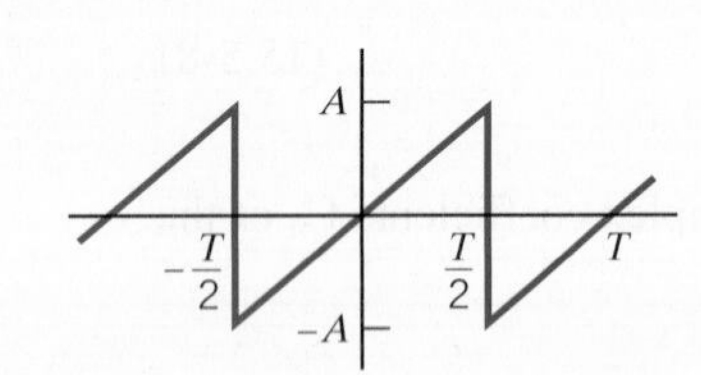	**Sawtooth wave** $f(t) = 2At/T \quad \frac{-T}{2} < t < \frac{T}{2}$	Odd	$= Aj(-1)^n/n\pi$, $n \neq 0$ $= 0$, $n = 0$
5.	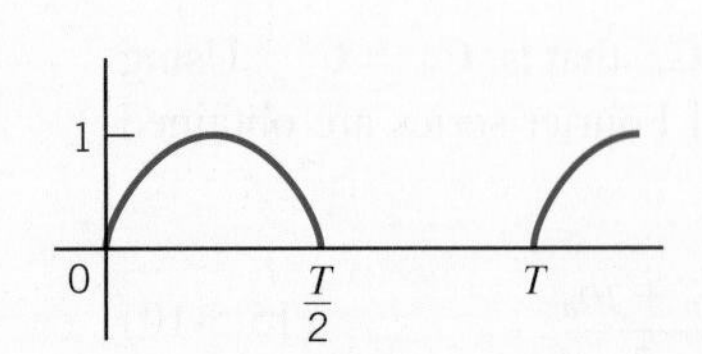	**Half-wave rectified sinusoid** $f(t) = \begin{cases} \sin\omega_0 t, & 0 \le t \le T/2 \\ 0, & -T/2 \le t \le 0 \end{cases}$	None	$= 1/\pi(1-n^2)$, n even $= -j/4$, $n = \pm 1$ $= 0$, otherwise
6.	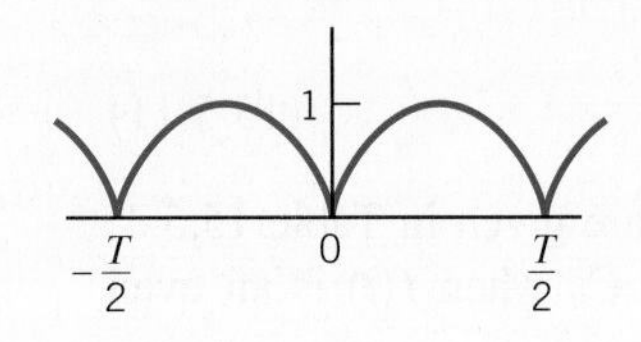	**Full-wave rectified sinusoid** $f(t) = \lvert\sin\omega_0 t\rvert$	Even	$= 2/\pi(1-n^2)$, n even $= 0$, otherwise

EXAMPLE 15.5-1 Exponential Fourier Series

Determine the exponential Fourier series for the function $v(t)$ shown in Figure 15.5-1.

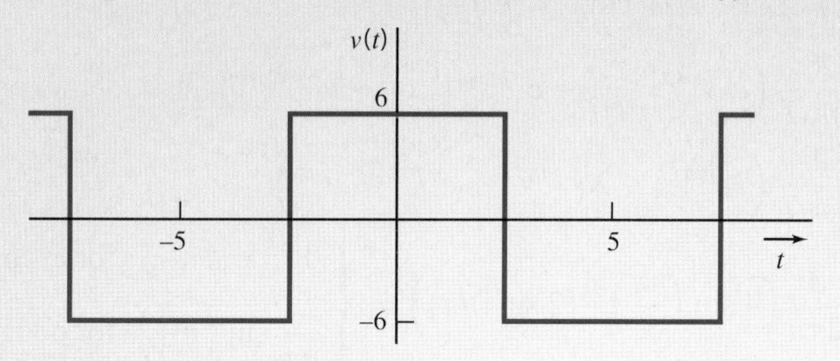

FIGURE 15.5-1 A square wave.

```
% Ex15_5_1.m - Exponential Fourier Series - square wave
% ----------------------------------------------------------
%        Describe the periodic waveform, v(t)
% ----------------------------------------------------------
A=6;
T=10;            % period
c0=0;      % average value

% ----------------------------------------------------------
%  Obtain a list of equally spaced instants of time
% ----------------------------------------------------------
w0=2*pi/T;             % fundamental frequency, rad/s
t0=-T;                 % initial time
tf=1.5*T;              % final time
dt=tf/500;             % time increment
t=-T:dt:tf;            % time, s

% ----------------------------------------------------------
%  Approximate v(t) using the exp Fourier series.
% ----------------------------------------------------------
v = c0*ones(size(t));  % initialize v(t) as vector
for n=1:2:200
   Cn = (2*A/pi/n)*sin(n*pi/2);
   v = v + Cn*exp(j*n*w0*t) + Cn'*exp(-j*n*w0*t);
end

% ----------------------------------------------------------
%              Plot the Fourier series
% ----------------------------------------------------------
plot(t, v)
axis([t0 tf -(A+1) A+1])
grid
xlabel('time, s')
ylabel('v(t) V')
title('Square Wave')
```

FIGURE 15.5-2 MATLAB m-file used in Example 15.5-1.

Solution

The average value of $v(t)$ is zero, so $\mathbf{C}_0 = 0$. Then, using Eq. 15.5-9, with $t_0 = -T/2$, we obtain

$$\begin{aligned}\mathbf{C}_n = \frac{1}{T}\int_{-T/2}^{T/2} v(t)e^{-jn\omega_0 t}dt &= \frac{1}{T}\int_{-T/2}^{-T/4} -Ae^{-jn\omega_0 t}dt + \frac{1}{T}\int_{-T/4}^{T/4} Ae^{-jn\omega_0 t}dt + \frac{1}{T}\int_{T/4}^{T/2} -Ae^{-jn\omega_0 t} \\ &= \frac{A}{jn\omega_0 T}\left(e^{-jn\omega_0 t}\Big|_{-T/2}^{-T/4} - e^{-jn\omega_0 t}\Big|_{-T/4}^{T/4} + e^{-jn\omega_0 t}\Big|_{T/4}^{T/2}\right) \\ &= \frac{A}{jn\omega_0 T}\left(2e^{jn\pi/2} - 2e^{-jn\pi/2} + e^{-jn\pi} - e^{jn\pi}\right) \\ &= \frac{A}{2\pi n}\left(4\sin\left(\frac{n\pi}{2}\right) - 2\sin(n\pi)\right) = \begin{cases} 0 & \text{for even } n \\ \dfrac{2A}{n\pi}\sin\left(n\dfrac{\pi}{2}\right) & \text{for odd } n \end{cases}\end{aligned}$$

Notice that $f(t)$ is an even function, so we expect $\mathbf{C}_{-n} = \mathbf{C}_n$. In particular, we calculate

$$\mathbf{C}_{-1} = \mathbf{C}_1 = \frac{A\sin\pi/2}{\pi/2} = \frac{2A}{\pi}, \quad \mathbf{C}_{-2} = \mathbf{C}_2 = A\frac{\sin\pi}{\pi} = 0 \quad \text{and} \quad \mathbf{C}_{-3} = \mathbf{C}_3 = \frac{A\sin(3\pi/2)}{3\pi/2} = \frac{-2A}{3\pi}$$

Figure 15.5-2 shows a MATLAB script that plots $v(t)$ using its the exponential Fourier series. The plot produced by this MATLAB script is shown in Figure 15.5-3. The waveform in Figure 15.5-3 is indeed a square having the correct amplitude and correct periods.

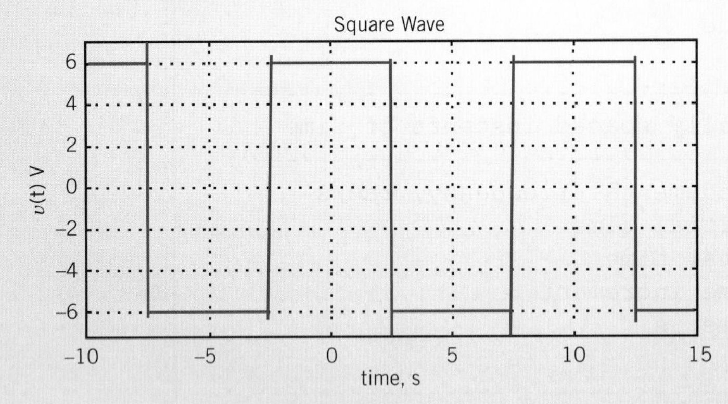

FIGURE 15.5-3 MATLAB output.

MATLAB has a built-in function called FFT (Fast Fourier Transform) that can be used to calculate the coefficients of the exponential Fourier series. Figure 15.5-4 shows a MATLAB function called EFS (for Exponential Fourier Series) that uses FFT to calculate the coefficients of the exponential Fourier series of a periodic function. (EFS follows closely the discussion of Fourier series in Chapter 22 of Hanselman and Littlefield, 2005.) Notice that EFS does not include a description of the periodic function. Instead, EFS calls a MATLAB function named my_periodic_-function. We describe our periodic function, $f(t)$, in the MATLAB function my_periodic_function. As a result, EFS can be used, unchanged, to find the Fourier series coefficients of a variety of periodic functions when we make appropriate changes to my_periodic_function.

The word *function* is being used in two different ways. First, we have the mathematical function, for instance, $f(t)$ as a function of t. Second, we have the MATLAB function, a type of computer program. Although different, these two types of function can be related. In the present case, the MATLAB function my_periodic_function implements the mathematical function $f(t)$ by providing the value of f corresponding to any particular value of t.

The following examples show how to use the MATLAB function EFS to find the exponential Fourier series of periodic functions.

```
function [C0, Cn] = EFS(N, T)
%EFS Exponential Fourier Series
% returns the coefficients of the exponential Fourier
% Series of a periofic function described in the
% MATLAB function named my_periodic_function.m
%
% N = the number of harmonic frequencies
% T = the period of the periodic function
%
% C0=average value
% Cn(1)=C1, Cn(2)=C2, ..., Cn(N)=CN

% -----------------------------------------------------
%  Obtain a list of equally spaced instants of time
% -----------------------------------------------------
n=2*N;
t=linspace(0,T,n+1);
t(end)=[];
% -----------------------------------------------------
%  Obtain values of f(t) at those instants of time
% -----------------------------------------------------
f=my_periodic_function(t,T);
% -----------------------------------------------------
%  Obtain the Fourier coef and do required bookkeeping
% -----------------------------------------------------
Cn=fft(f);
Cn=[conj(Cn(N+1)) Cn(N+2:end) Cn(1:N+1)];
Cn=Cn/n;
C0=Cn(N+1);
Cn=[Cn(N+2:end)];
```

FIGURE 15.5-4 MATLAB function to calculate the coefficients of the exponential Fourier series.

EXAMPLE 15.5-2 Exponential Fourier Series Using MATLAB

Determine the exponential Fourier series for the function $f(t)$ shown in Figure 15.5-1, using MATLAB.

Solution

We need to write the MATLAB function, my_periodic_function, shown in Figure 15.5-5. The inputs to this function are t, a list of times distributed evenly over one period, and T, the period. Let $t(k)$ denote the kth time in the list t and let $f(k)$ denote the value of the periodic function at time $t(k)$. The output of my_periodic_function is a list f of the k values $f(k)$. The for-loop in Figure 15.5-5 indexes through the k times, $t(k)$, and the if-block determines the value of $f(k)$ corresponding to each $t(k)$.

```
function f = my_periodic_function(t, T)
% squarewave with amplitude A and period T

A=6;
for k=1:length(t)
    if (t(k)<T/4 | t(k)>3*T/4) f(k)=A;
    elseif (t(k)>T/4 & t(k)<3*T/4) f(k)=-A;
    else f(k)= 0;
    end
end
```

FIGURE 15.5-5 my_periodic_function for Example 15.5-2.

```
% testEFS.m
% ------------------------------------------------------
%  Obtain a list of equally spaced instants of time
% ------------------------------------------------------
T=10;                   % period
w0=2*pi/T;              % fundamental frequency, rad/s
t0=-T;                  % initial time
tf=1.5*T;               % final time
dt=tf/512;              % time increment
t=-T:dt:tf;             % time, s
% -----------------------------------------------------
%  Call EFS to get exponential Fourier coefficients
% -----------------------------------------------------
N=256; %Number of harmonic frequencies
[C0, Cn] = EFS(N,T);
% -----------------------------------------------------
%  Approximate the function by its Fourier series
% -----------------------------------------------------
v = C0*ones(size(t));  % initialize v(t) as vector
for n=1:N
  v = v + Cn(n)*exp(j*n*w0*t) + Cn(n)'*exp(-j*n*w0*t);
end
% -----------------------------------------------------
%              Plot the Fourier series
% -----------------------------------------------------
plot(t, v)
axis([t0 tf -8 8])
grid
xlabel('time, s')
ylabel('v(t) V')
title('Square Wave')
```

FIGURE 15.5-6 MATLAB script to plot $f(t)$, using the coefficients of the exponential Fourier series.

The values of $f(t)$ at times $T/4$ and $3T/4$ aren't obvious because $f(t)$ is discontinuous at these times. In general, when $f(t)$ is discontinuous at time τ, we will take $f(\tau)$ to be the average of the limits of $f(t)$ as t approaches τ from above and from below. In the present case,

$$f(\tau) = \frac{\lim\limits_{t\to\tau+} f(t) + \lim\limits_{t\to\tau-} f(t)}{2} = \frac{A - A}{2} = 0 \quad \text{when} \quad \tau = \frac{T}{4} \text{ or } \frac{3T}{4}$$

Then, from Figure 15.5-1,

$$f(t) = \begin{cases} A & \text{when } t < T/4 \text{ or } t > 3T/4 \\ -A & \text{when } t > T/4 \text{ and } t < 3T/4 \\ 0 & \text{otherwise} \end{cases}$$

This equation is implemented by the MATLAB function, my_periodic_function, shown in Figure 15.5-5.

Figure 15.5-6 shows a MATLAB script that plots $f(t)$, using the coefficients of the exponential Fourier series. Placing EFS.m, my_periodic_function.m, and testEFS.m in the MATLAB working directory and running testEFS.m produces the same plot obtained in Example 15.5-1 and shown in Figure 15.5-3.

EXAMPLE 15.5-3 Exponential Fourier Series Using MATLAB

Determine the exponential Fourier series for the half-wave rectified sine shown in Figure 15.5-7, using MATLAB.

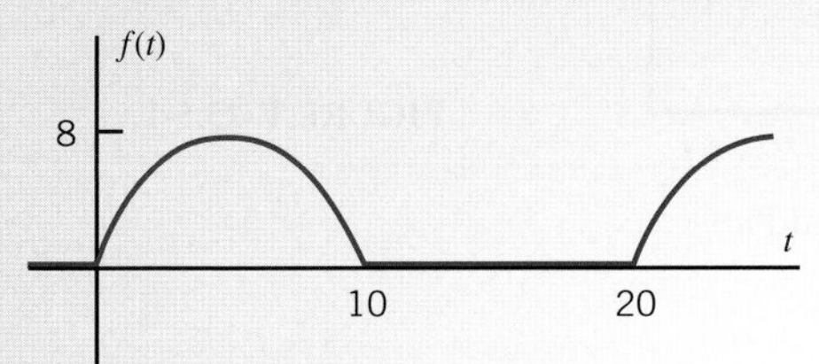

FIGURE 15.5-7 The periodic function for Example 15.5-3.

Solution

We need to do only a couple of things: rewrite the MATLAB function my_periodic_function shown in Figure 15.5-8, change the value of the period T in testEFS.m, and then run testEFS.m to get the plot shown in Figure 15.5-9.

```
function f = my_periodic_function(t, T)
% half-wave rectified sine with amplitude A
% and period T

w=2*pi/T;
A=8;
for k=1:length(t)
    if (t(k)<T/2) f(k)=A*sin(w*t(k));
    else  f(k)=0;
    end
end
```

FIGURE 15.5-8 my_periodic_function for Example 15.5-3.

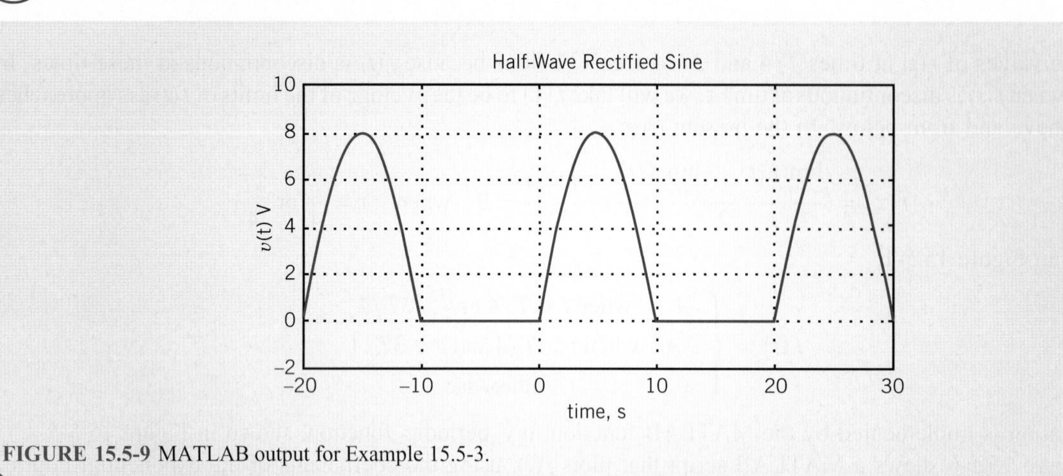

FIGURE 15.5-9 MATLAB output for Example 15.5-3.

EXERCISE 15.5-1 Find the exponential Fourier coefficients for the function shown in Figure E 15.5-1.

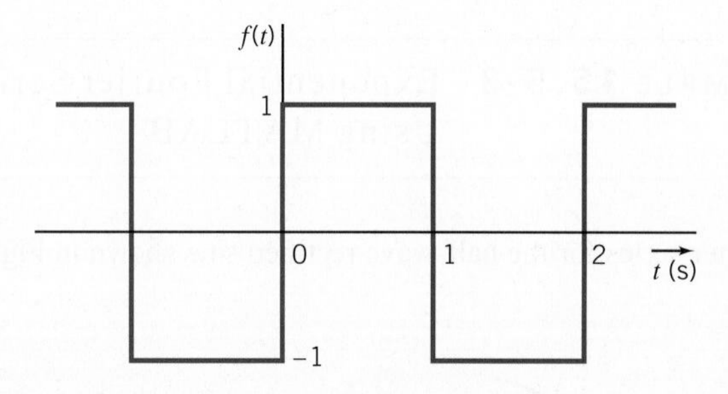

FIGURE E 15.5-1

Answer: $\mathbf{C}_n = 0$ for even n and $\mathbf{C}_n = \dfrac{2}{jn\pi}$ for odd n

EXERCISE 15.5-2 Determine the complex Fourier coefficients for the waveform shown in Figure E 15.5-2.

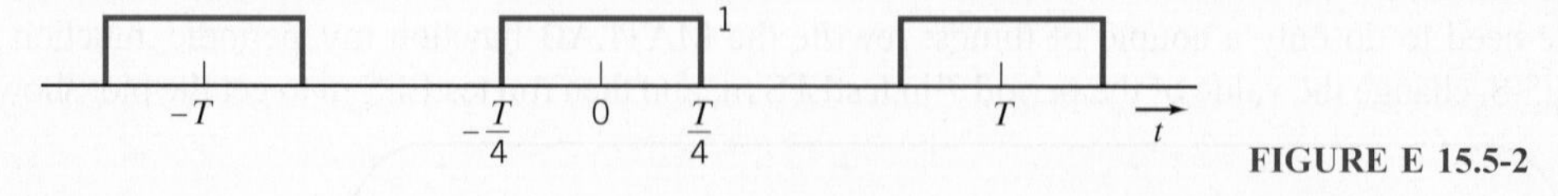

FIGURE E 15.5-2

15.6 THE FOURIER SPECTRUM

If we plot the complex Fourier coefficients $\mathbf{C}_n$ as a function of angular frequency, $\omega = n\omega_0$, we obtain a *Fourier spectrum*. Because $\mathbf{C}_n$ may be complex, we have

$$\mathbf{C}_n = |\mathbf{C}_n| \underline{/\theta_n} \tag{15.6-1}$$

and we plot $|\mathbf{C}_n|$ and $\underline{/\theta_n}$ as the *amplitude spectrum* and the *phase spectrum*, respectively. The Fourier spectrum exists only at the fundamental and harmonic frequencies and, therefore, is called a discrete or line spectrum. The amplitude spectrum appears on a graph as a series of equally spaced vertical lines with heights proportional to the amplitudes of the respective frequency components. Similarly, the phase spectrum appears as a series of equally spaced lines with heights proportional to the value of the phase at the appropriate frequency. The word *spectrum* was introduced into physics by Isaac Newton (1664) to describe the analysis of light by a prism into its different color components or frequency components.

The **Fourier spectrum** is a graphical display of the amplitude and phase of the complex Fourier coefficients at the fundamental and harmonic frequencies.

EXAMPLE 15.6-1 Fourier Spectrum

Determine the Fourier spectrum for the pulse waveform $v(t)$ shown in Figure 15.6-1.

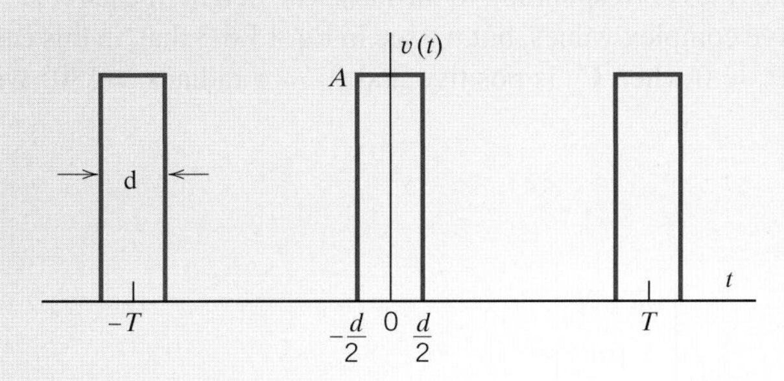

FIGURE 15.6-1 A pulse waveform.

Solution

The Fourier coefficients are

$$\mathbf{C}_n = \frac{1}{T}\int_{-T/2}^{T/2} v(t)e^{-jn\omega_0 t}dt \tag{15.6-2}$$

For $n \neq 0$, we have

$$\begin{aligned}\mathbf{C}_n = \frac{A}{T}\int_{-d/2}^{d/2} e^{-jn\omega_0 t}dt &= \frac{-A}{jn\omega_0 T}\left(e^{-jn\omega_0 d/2} - e^{jn\omega_0 d/2}\right)\\ &= \frac{2A}{n\omega_0 T}\sin\left(\frac{n\omega_0 d}{2}\right) = \frac{A\delta}{T}\sin\left(\frac{n\omega_0 d/2}{n\omega_0 d/2}\right) = \frac{Ad}{T}\frac{\sin x}{x}\end{aligned}$$

where $x = (n\omega_0 d/2)$ and $n \neq 0$. When $n = 0$, we have

$$C_0 = \frac{1}{T}\int_{-d/2}^{d/2} A\,dt = \frac{Ad}{T}$$

One may show that $(\sin x)/x = 1$ for $x = 0$ by using L'Hôpital's rule. In summary,

$$\mathbf{C}_n = \frac{Ad}{T}\frac{\sin(n\omega_0 d/2)}{n\omega_0 d/2} \quad \text{for all } n \tag{15.6-3}$$

The coefficients $\mathbf{C}_n$ correspond to the discrete frequencies $n\omega_0$ where ω_0 is the fundamental frequency, determined from the period T of the periodic function. The amplitude spectrum appears on a graph as a series of equally spaced vertical lines corresponding to the equally spaced frequencies $n\omega_0$. The height of each line represents the amplitude

$$|\mathbf{C}_n| = \left|\frac{Ad}{T}\frac{\sin(n\omega_0 d/2)}{n\omega_0 d/2}\right|$$

The amplitude spectrum, a plot $|\mathbf{C}_n|$ versus $\omega = n\omega_0$, is shown in Figure 15.6-2a for n up to ± 15. Also, $|(\sin x)/x|$ is shown in Figure 15.6-2a in color. Notice that $(\sin x)/x$ is zero whenever x is an integer multiple of π, that is,

$$\frac{\sin(n\pi)}{n\pi} = 0 \quad n = 1, 2, 3, \ldots$$

The phase spectrum, a plot of $\theta_n = \angle\mathbf{C}_n$ versus $\omega = n\omega_0$, is shown in Figure 15.6-2b. The phase spectrum appears on a graph as a series of equally spaced vertical lines corresponding to the $n\omega_0$. The height of each line represents the angle θ_n. In general, the $\mathbf{C}_n$ coefficients have complex values, but we see in Eq. 15.6-3 that, in this case, the $\mathbf{C}_n$ coefficients have real values. Consequently $\theta_n = 0$ when $\mathbf{C}_n$ is positive and $\theta_n = \pi$ radians $= 180°$ when $\mathbf{C}_n$ is negative.

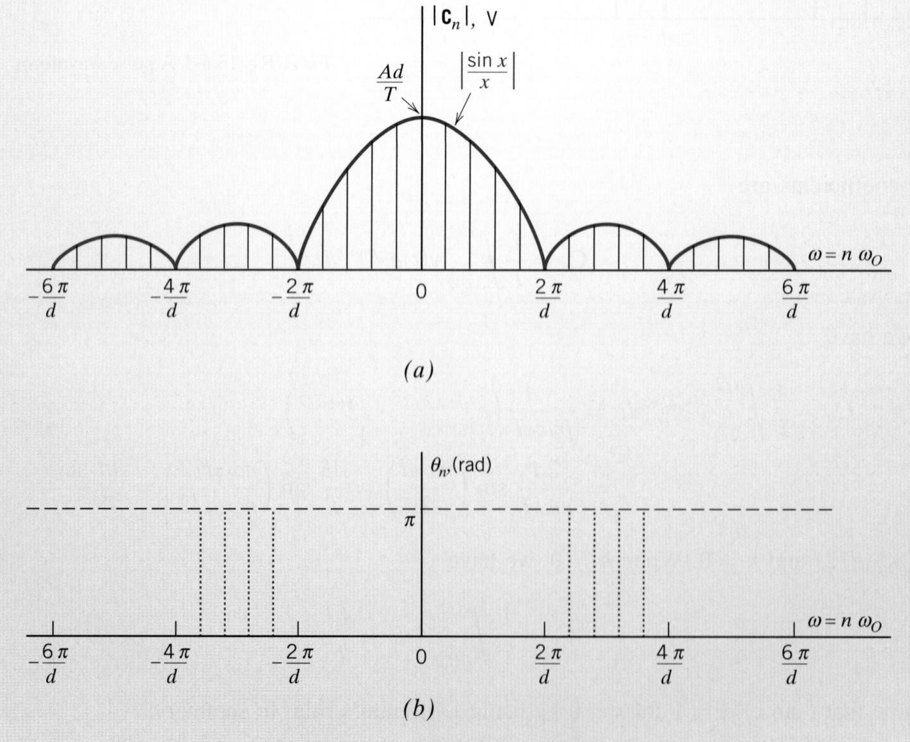

FIGURE 15.6-2 The (a) amplitude and (b) phase Fourier spectra of the waveform.

```
% Spectrum.m
T=20;       % period
N=64;       % Number of harmonic frequencies
% -----------------------------------------------------
%  Obtain a list of equally spaced instants of time
% -----------------------------------------------------
n=2*N;
t=linspace(0,T,n+1);
t(end)=[];
% -----------------------------------------------------
%  Obtain values of f(t) at those instants of time
% -----------------------------------------------------
f=my_periodic_function(t,T);
% -----------------------------------------------------
%  Obtain the Fourier coef and do required bookkeeping
% -----------------------------------------------------
Cn=fft(f);
Cn=[conj(Cn(N+1)) Cn(N+2:end) Cn(1:N+1)];
Cn=Cn/n;

% ----------------------------------------------------
%             Plot the Fourier spectrum
% ----------------------------------------------------
stem(-N:N,abs(Cn))
xlabel('n')
ylabel('|Cn|')
title('Magnitude Spectrum of a Pulse Train')
axis tight
```

FIGURE 15.6-3 MATLAB program to the Fourier spectrum.

Figure 15.6-3 shows a MATLAB program using FFT to plot the Fourier spectrum of a periodic function (Hanselman and Littlefield, 2005).

EXAMPLE 15.6-2 Using MATLAB to Plot the Fourier Spectrum

Use MATLAB to plot the amplitude spectrum for the pulse waveform $v(t)$ in Figure 15.6-1 when $A = 8$ V, $T = 20$ seconds, and $d = T/10$.

Solution

We can use the MATLAB program shown in Figure 15.6-3 to plot the spectrum after doing the following three things:

1. **Specify the values of T and N in the second and third lines.** T is the period in seconds and N determines the number of harmonic frequencies used when plotting the spectrum. The n in $n\omega_0$ varies from −N to N. The values given in Figure 15.6-3 do not need to be changed.

```
function f = my_periodic_function(t, T)
d=T/10;
A=8;
for k=1:length(t)
    if (t(k)<d/2 | t(k)>T-d/2) f(k)=A;
    elseif (t(k)>d/2 & t(k)<T-d/2)  f(k)=0;
    else f(k)= A/2;
    end
end
```

FIGURE 15.6-4 my_periodic_function for Example 15.6-2.

2. **Provide a MATLAB function named my_periodic_function** that describes the pulse train shown in Figure 15.6-1. Figure 15.6-4 provides the required MATLAB function. The inputs to this function are t, a list of time distributed evenly over one period, and T, the period. Let $t(k)$ denote the kth time in the list t and let $f(k)$ denote the value of the periodic function at time $t(k)$. The output of my_periodic_function is a list f of the k values $f(k)$. The for-loop indexes through the k times $t(k)$, and the if-block determines the value of $f(k)$ corresponding to each $t(k)$. (When $f(t)$ is discontinuous at time τ, we will take $f(\tau)$ to be the average of the limits of $f(t)$ as t approaches τ from above and from below.)

3. **Make any desired changes to the plotting statements** at the end of the program. The statement

```
stem (-N : N, abs(Cn))
```

plots the amplitude spectrum. Change `abs(Cn)` to `angle(Cn)` to plot the angle spectrum. Also, the plot labels can be changed as desired. In this case, no changes are required.

Figure 15.6-5 shows the amplitude spectrum plotted using MATLAB.

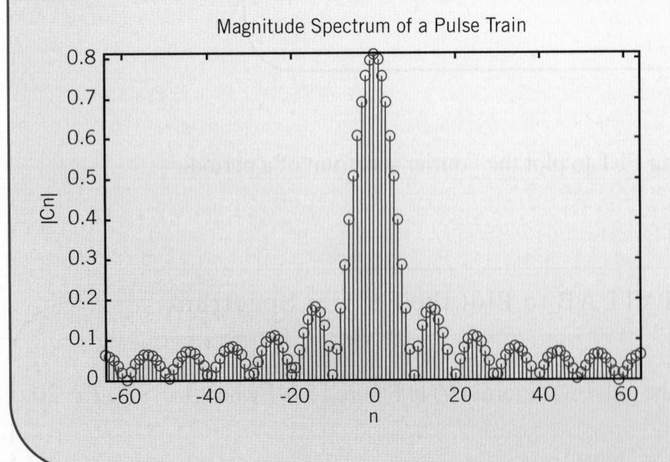

FIGURE 15.6-5 MATLAB output for Example 15.6-2.

15.7 CIRCUITS AND FOURIER SERIES

It is often desired to determine the response of a circuit excited by a periodic input signal $v_s(t)$. We can represent $v_s(t)$ by a Fourier series and then find the response of the circuit to the fundamental and each harmonic. Assuming the circuit is linear and the principle of superposition holds, we can consider that the total response is the sum of the response to the dc term, the fundamental, and each harmonic.

EXAMPLE 15.7-1 Steady-State Response to a Periodic Input

Find the steady-state response, $v_o(t)$, of the *RC* circuit shown in Figure 15.7-1*b*. The input, $v_s(t)$, is the square wave shown in Figure 15.7-1*a*.

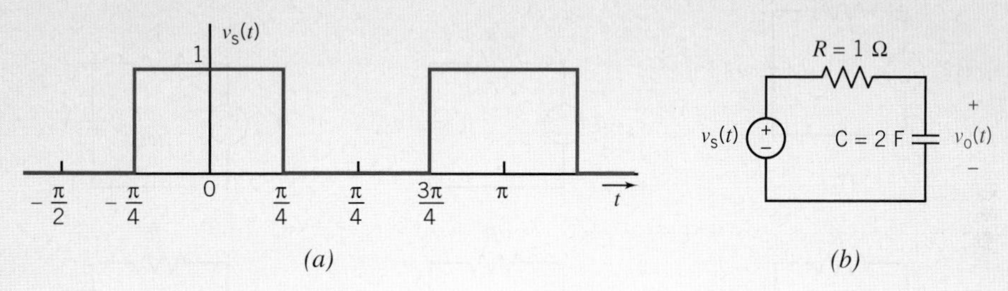

FIGURE 15.7-1 The (*a*) square wave and (*b*) circuit considered in Example 15.7-1.

Solution

Using Table 15.4-1 and proceeding as in Example 15.4-1, we represent $v_s(t)$ by the Fourier series

$$v_s(t) = \frac{1}{2} + \frac{2}{\pi}\sum_{n=1}^{\infty} \frac{\sin((2n-1)(2t+90°))}{2n-1}$$

In this example, we will represent this square wave by the first four terms of its Fourier series

$$v_s(t) = \frac{1}{2} + \frac{2}{\pi}\cos 2t - \frac{2}{3\pi}\cos 6t + \frac{2}{5\pi}\cos 10t$$

We will find the steady-state response, $v_o(t)$, using superposition. It is helpful to let $v_{sn}(t)$ denote the term of $v_s(t)$ corresponding to n. In this example, $v_s(t)$ has four terms, corresponding to $n = 0$, 1, 3, and 5. Then,

$$v_s(t) = v_{s0}(t) + v_{s1}(t) + v_{s3}(t) + v_{s5}(t)$$

where

$$v_{s0}(t) = \frac{1}{2},\ v_{s1}(t) = \frac{2}{\pi}\cos 2t,$$

$$v_{s3}(t) = -\frac{2}{3\pi}\cos 6t, \quad \text{and} \quad v_{s5}(t) = \frac{2}{5\pi}\cos 10t$$

Figure 15.7-2 illustrates the way superposition is used in this example. First, because the series connection of the voltage sources with voltages $v_{s0}(t)$, $v_{s1}(t)$, $v_{s3}(t)$, and $v_{s5}(t)$ is equivalent to a single voltage source having voltage $v_s(t) = v_{s0}(t) + v_{s1}(t) + v_{s3}(t) + v_{s5}(t)$, the circuit shown in Figure 15.7-2*b* is equivalent to the circuit shown in Figure 15.7-2*a*.

Next, the principle of superposition is invoked to break the problem up into four simpler problems, as shown in Figure 15.7-2*c*. Each circuit in Figure 15.7-2*c* is used to calculate the steady-state response to a single one of the voltage sources from Figure 15.7-2*b*. (When calculating the response to one voltage source, the other voltage sources are set to zero; that is, they are replaced by short circuits.) For example, the voltage $v_{o3}(t)$ is the steady-state response to $v_{s3}(t)$ alone. Superposition tells us that the response to all four voltage sources working together is the sum of the responses to the four voltage sources working separately, that is,

$$v_o(t) = v_{o0}(t) + v_{o1}(t) + v_{o3}(t) + v_{o5}(t)$$

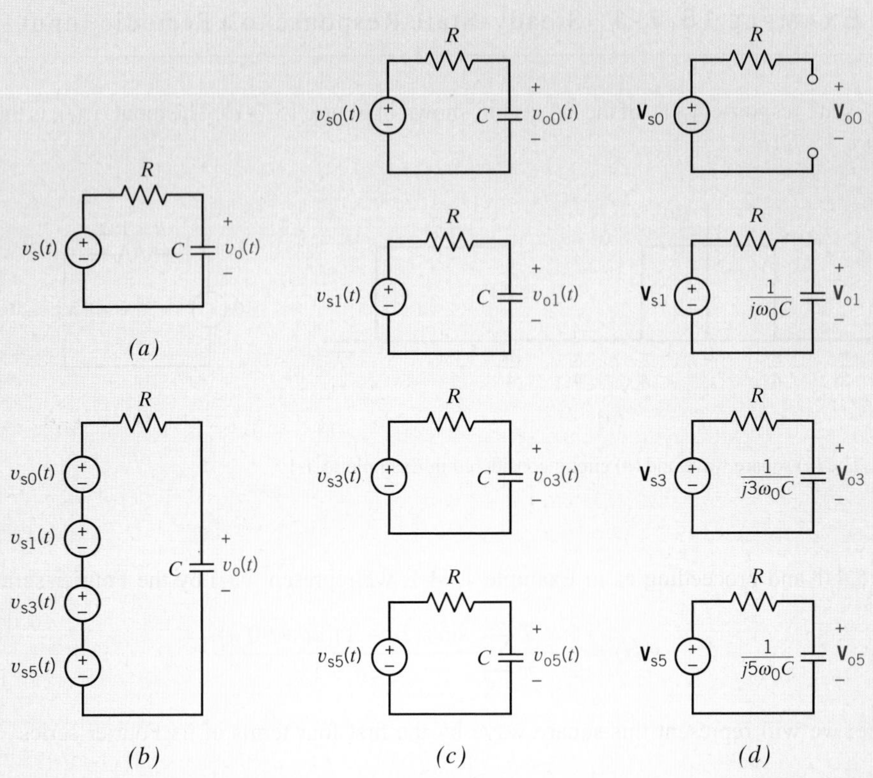

FIGURE 15.7-2 (*a*) An *RC* circuit excited by a periodic voltage $v_s(t)$. (*b*) An equivalent circuit. Each voltage source is a term of the Fourier series of $v_s(t)$. (*c*) Using superposition. Each input is a sinusoid. (*d*) Using phasors to find steady-state responses to the sinusoids.

The advantage of breaking the problem up into four simpler problems is that the input to each of the four circuits in Figure 15.7-2*c* is a sinusoid. The problem of finding the steady-state response to a periodic input has been reduced to the simpler problem of finding the steady-state response to a sinusoidal input. The steady-state response of a linear circuit to a sinusoidal input can be found using phasors. In Figure 15.7-2*d*, the four circuits from Figure 15.7-2*c* have been redrawn using phasors and impedances. The impedance of the capacitor is

$$\mathbf{Z}_c = \frac{1}{jn\omega_0 C} \quad \text{for } n = 0, 1, 3, 5$$

Each of the four circuits corresponds to a different value of n, so the impedance of the capacitor is different in each of the circuits. (The frequency of the input sinusoid is $n\omega_0$, so each of the circuits corresponds to a different frequency.) Notice that when $n = 0$, $Z_c = \infty$ and, therefore, the capacitor acts like an open circuit. The four circuits shown in Figure 15.7-2*d* are very similar. In each case, voltage division can be used to write

$$\mathbf{V}_{on} = \frac{1/(jn\omega_0 C)}{R + 1/(jn\omega_0 C)} \mathbf{V}_{sn} \quad \text{for } n = 0, 1, 3, 5$$

where $\mathbf{V}_{sn}$ is the phasor corresponding to $v_{sn}(t)$ and $\mathbf{V}_{on}$ is the phasor corresponding to $v_{on}(t)$. So,

$$\mathbf{V}_{on} = \frac{\mathbf{V}_{sn}}{1 + jn\omega_0 CR} \quad \text{for} \quad n = 0, 1, 3, 5$$

In this example, $\omega_0 CR = 4$, so

$$\mathbf{V}_{on} = \frac{\mathbf{V}_{sn}}{1 + j4n} \quad \text{for } n = 0, 1, 3, 5$$

Next, the steady-state response can be written as

$$v_{on}(t) = |\mathbf{V}_{on}| \cos(n\omega_0 t + \angle \mathbf{V}_{on}) = \frac{|\mathbf{V}_{sn}|}{\sqrt{1 + 16n^2}} \cos(n\omega_0 t + \angle \mathbf{V}_{sn} - \tan^{-1} 4n)$$

In this example,

$$|\mathbf{V}_{s0}| = \frac{1}{2}$$

$$|\mathbf{V}_{sn}| = \frac{2}{n\pi} \quad \text{for } n = 1, 3, 5$$

$$\angle \mathbf{V}_{sn} = 0 \quad \text{for } n = 0, 1, 5 \quad \text{and} \quad \angle \mathbf{V}_{sn} = 180^\circ \quad \text{for } n = 3$$

Therefore,

$$v_{o0}(t) = \frac{1}{2}$$

$$v_{on}(t) = \frac{2}{n\pi\sqrt{1 + 16n^2}} \cos(n2t + \angle \mathbf{V}_{sn} - \tan^{-1} 4n) \quad \text{for } n = 1, 3, 5$$

Doing the arithmetic yields

$$v_{o0}(t) = \frac{1}{2}$$
$$v_{o1}(t) = 0.154 \cos(2t - 76^\circ)$$
$$v_{o3}(t) = 0.018 \cos(6t + 95^\circ)$$
$$v_{o5}(t) = 0.006 \cos(10t - 87^\circ)$$

Finally, the steady-state response of the original circuit, $v_o(t)$, is found by adding up the partial responses,

$$v_o(t) = \frac{1}{2} + 0.154 \cos(2t - 76^\circ) + 0.018 \cos(6t + 95^\circ) + 0.006 \cos(10t - 87^\circ)$$

It is important to notice that superposition justifies adding the functions of time, $v_{o0}(t)$, $v_{o1}(t)$, $v_{o3}(t)$, and $v_{o5}(t)$ to get $v_o(t)$. The phasors $\mathbf{V}_{o0}$, $\mathbf{V}_{o1}$, $\mathbf{V}_{o3}$, and $\mathbf{V}_{o5}$ each correspond to a different frequency. A sum of these phasors has no meaning.

EXERCISE 15.7-1 Find the response of the circuit of Figure 15.7-2 when $R = 10\,\text{k}\Omega$, $C = 0.4\,\text{mF}$, and v_s is the triangular wave considered in Example 15.3-1 (Figure 15.3-3). Include all terms that exceed 2 percent of the fundamental term.

Answer: $v_o(t) \approx 0.20 \sin(4t - 86^\circ) - 0.008 \sin(12t - 89^\circ)$ V

15.8 USING PSPICE TO DETERMINE THE FOURIER SERIES

The circuit simulation program PSpice (Perry, 1998) provides built-in procedures that make it easy to find the Fourier series of any periodic voltage or current in a simulated circuit. To find a Fourier series using PSpice, we will need to do five things:

Step 1 Represent the circuit and its input in the PSpice workspace.

Step 2 Specify a time domain simulation having a duration that is long enough to include one full period after all transients have died out.

Step 3 Request that the Fourier series coefficients be calculated and printed in the PSpice output file.

Step 4 Simulate the circuit.

Step 5 Interpret the PSpice output.

The following example illustrates this procedure.

EXAMPLE 15.8-1 Fourier Series Using PSpice

Consider the circuit shown in Figure 15.8-1*a*. The input to this circuit is the voltage of the voltage source, $v_i(t)$. The output of the circuit is the voltage, $v_o(t)$, across the 10-kΩ resistor. The input, $v_i(t)$, is the periodic voltage shown in Figure 15.8-1*b*. The output, $v_o(t)$, will also be a periodic voltage. Use PSpice to represent both $v_i(t)$ and $v_o(t)$ by Fourier series.

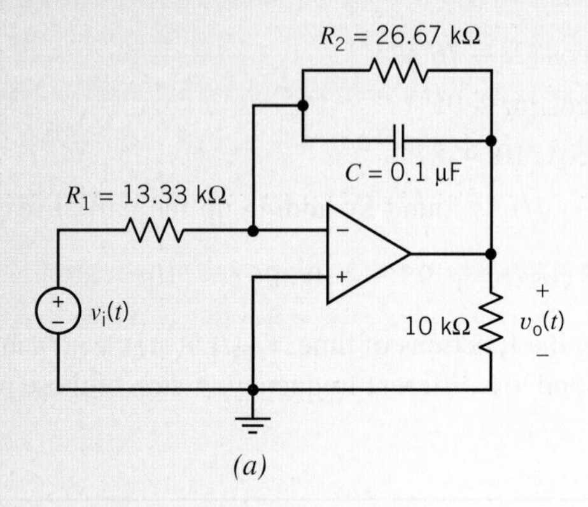

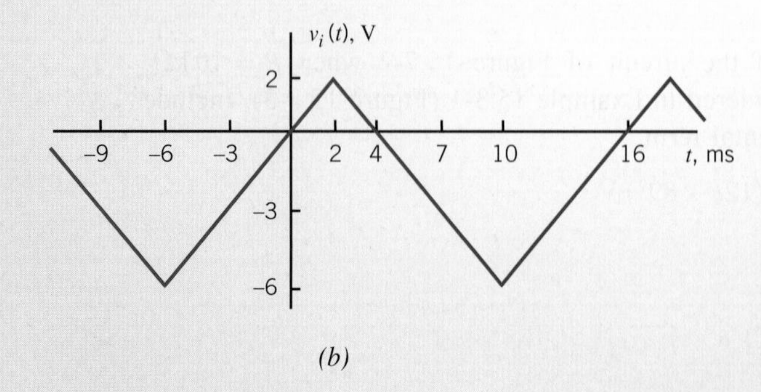

FIGURE 15.8-1 (*a*) A circuit and (*b*) a periodic input voltage.

Solution

Step 1: Represent the circuit and its input in the PSpice workspace.

PSpice refers to circuit elements as parts. Open a new project in PSpice. Place the parts in the PSpice workspace, adjust the resistance and capacitance values, and wire the parts together (Svoboda, 2007). The resulting PSpice circuit is shown in Figure 15.8-2. The voltage source in Figure 15.8-1*a*, corresponds to a PSpice

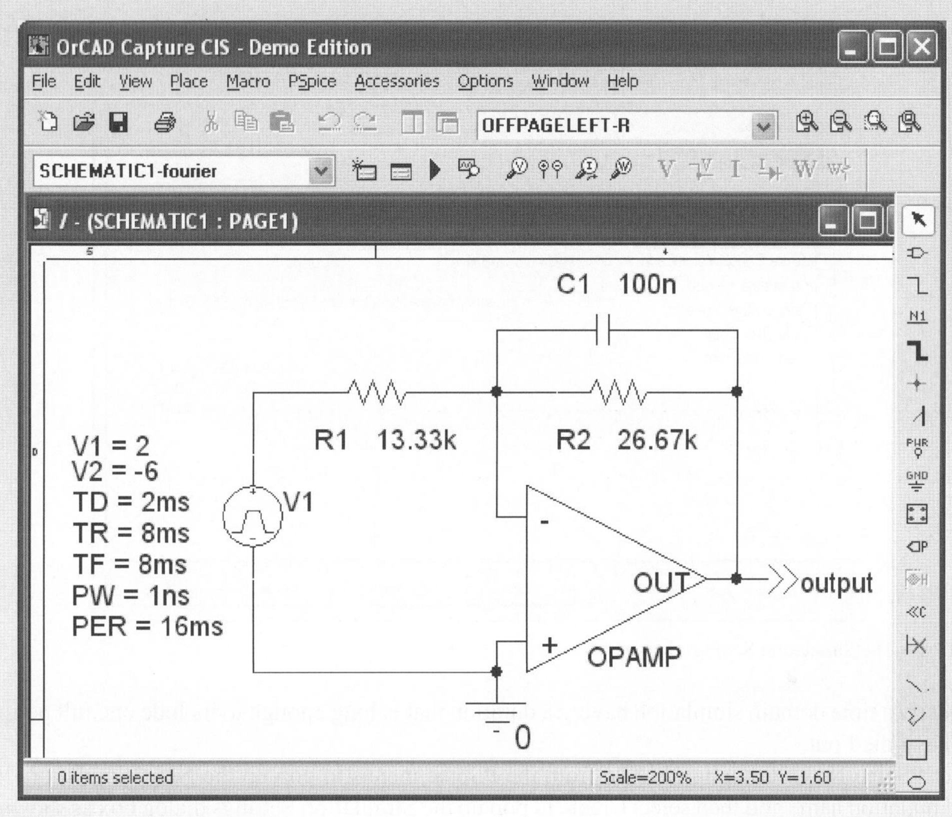

FIGURE 15.8-2 The circuit as described in the PSpice workspace.

part called VPULSE. Figure 15.8-3 shows the symbol for this part together with the voltage waveform that it produces. A VPULSE part is specified by providing values for the parameters $v1$, $v2$, td, tr, tf, pw, and per. The meaning of each parameter is seen by examining Figure 15.8-3*b*. The pulse waveform will simulate the triangle wave when pw is specified to make the time that voltage remains equal to $v2$ negligibly small, and per is specified to make the time that voltage remains equal to $v1$ negligibly small. An appropriate set of parameter values to simulate the input voltage, $v_i(t)$, is

$$v1 = 2\text{ V},\ v2 = -6\text{ V},\ td = 2\text{ ms},\ tr = 8\text{ ms},\ tf = 8\text{ ms},\ pw = 1\text{ ns, and } per = 16\text{ ms}.$$

(PSpice requires $pw > 0$ so we cannot use $pw = 0$. Instead, a value much smaller than both tr and tf is used.)

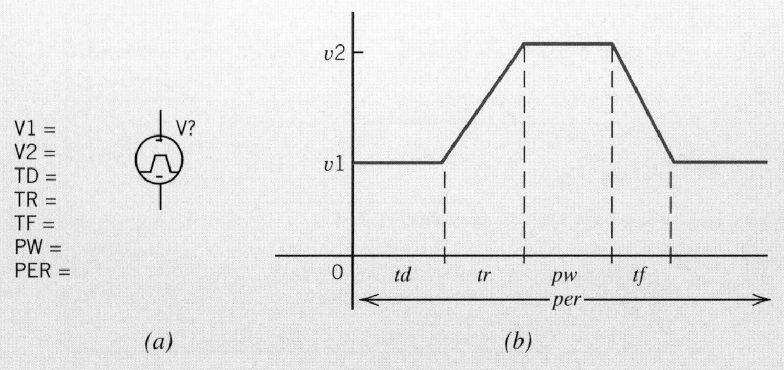

FIGURE 15.8-3 The (*a*) symbol and (*b*) voltage waveform of a VPULSE part.

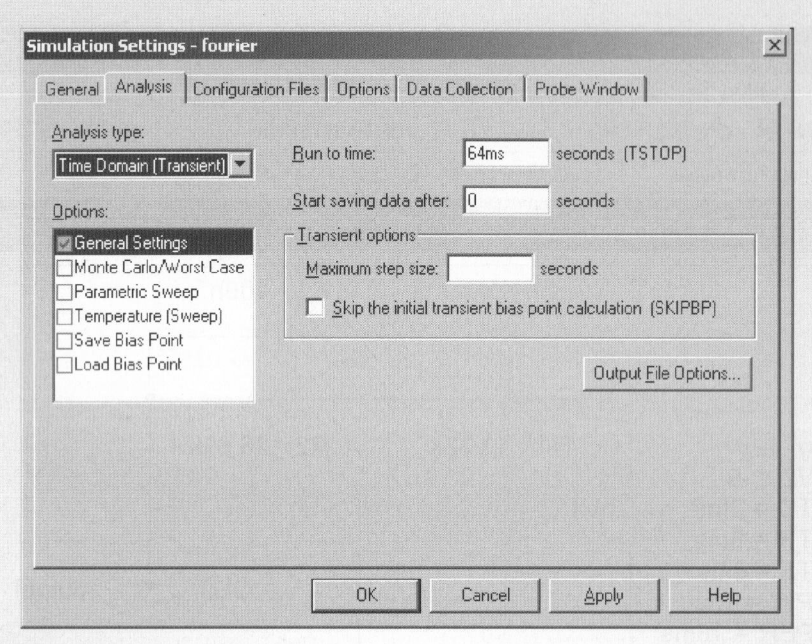

FIGURE 15.8-4 The Simulation Settings dialog box.

Step 2: Specify a time domain simulation having a duration that is long enough to include one full period after all transients have died out.

Select PSpice/New Simulation Profile from the PSpice menus to pop up the New Simulation dialog box. Specify a simulation name and then select Create to pop up the Simulation Settings dialog box as shown in Figure 15.8-4. Select Time Domain(Transient) as the analysis type. Specify the Run To Time as 64 ms to run the simulation for four full periods of the input waveform.

Step 3: Request that the Fourier series coefficients be calculated and printed in the PSpice output file.

Click the Output File Options button to pop up the Transient Output File Options dialog box shown in Figure 15.8-5. Select the Perform Fourier Analysis box. PSpice represents the trigonometric Fourier series using the sine rather than the cosine, that is,

$$v(t) = c_0 + \sum_{n=1}^{N} c_n \sin(n\omega_0 t + \theta_n) \tag{15.8-1}$$

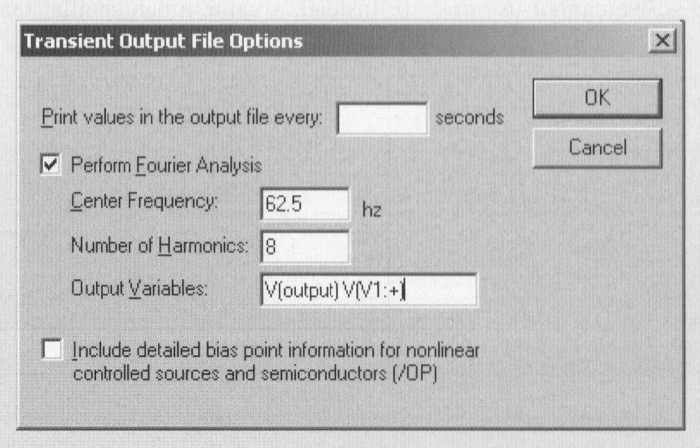

FIGURE 15.8-5 Requesting calculation of the Fourier series coefficients.

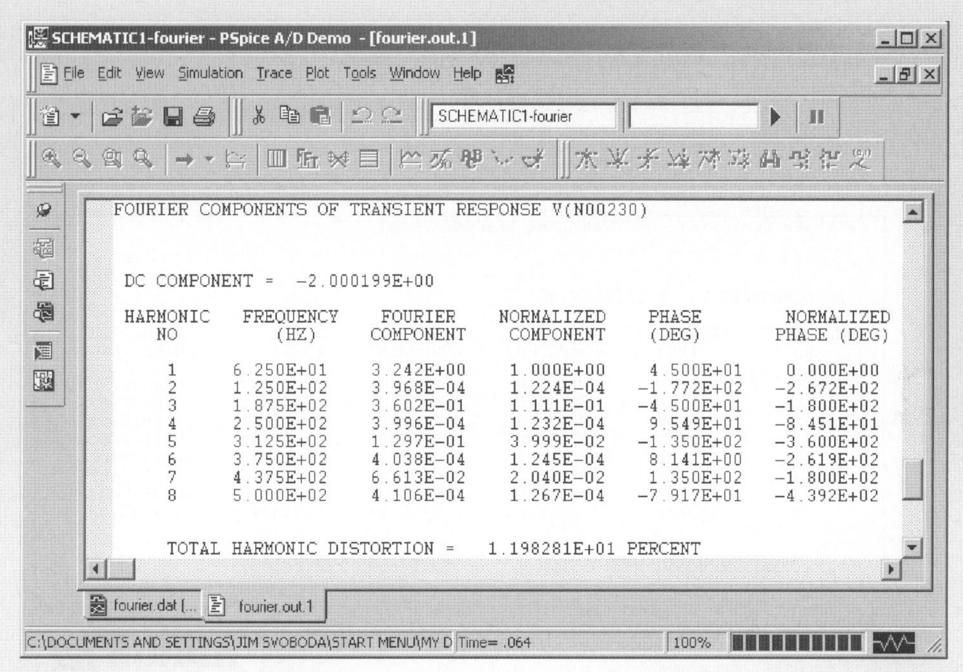

FOURIER COMPONENTS OF TRANSIENT RESPONSE V(N00230)

DC COMPONENT = -2.000199E+00

HARMONIC NO	FREQUENCY (HZ)	FOURIER COMPONENT	NORMALIZED COMPONENT	PHASE (DEG)	NORMALIZED PHASE (DEG)
1	6.250E+01	3.242E+00	1.000E+00	4.500E+01	0.000E+00
2	1.250E+02	3.968E-04	1.224E-04	-1.772E+02	-2.672E+02
3	1.875E+02	3.602E-01	1.111E-01	-4.500E+01	-1.800E+02
4	2.500E+02	3.996E-04	1.232E-04	9.549E+01	-8.451E+01
5	3.125E+02	1.297E-01	3.999E-02	-1.350E+02	-3.600E+02
6	3.750E+02	4.038E-04	1.245E-04	8.141E+00	-2.619E+02
7	4.375E+02	6.613E-02	2.040E-02	1.350E+02	-1.800E+02
8	5.000E+02	4.106E-04	1.267E-04	-7.917E+01	-4.392E+02

TOTAL HARMONIC DISTORTION = 1.198281E+01 PERCENT

FIGURE 15.8-6 The coefficients of the Fourier series of $v_i(t)$.

Enter the fundamental frequency, $f_0 = \omega_0/2\pi$, using units of Hertz, in the Center Frequency text box and N in the Number of Harmonics text box. Enter the PSpice names for voltages or currents that are to be represented by their Fourier series in the Output Variables text box. Click OK to close the Transient Output File Options dialog box and then click OK to close the Simulation Settings dialog box and return to the PSpice workspace.

Step 4: Simulate the circuit.

Select PSpice/Run from the PSpice menus to run the simulation.

Step 5: Interpret the PSpice output.

After a successful Time Domain(Transient) simulation, Probe, the graphical post-processor for PSpice will open automatically in a Schematics window. Select View/Output File from the Schematics menus. Scroll through the output file to find the Fourier coefficients of the input voltage shown in Figure 15.8-6. (PSpice changed the name of the input voltage. We used the name V(V1:+) in the Output Variables text box in the Transient Output File Options dialog box in Figure 15.8-5. Nonetheless, PSpice used the name V(N00230) in Figure 15.8-6.) The table in Figure 15.8-6 has six columns and eight rows. The eight rows correspond to the eight coefficients, c_1, c_2, c_3, . . . c_8. (There are eight rows because $N = 8$ was the number entered in the Number of Harmonics text box in the Transient Output File Options dialog box in Figure 15.8-5.) The first column labels the rows with the subscripts, n, of these coefficients. The second column lists the frequencies, $n\omega_0$, using units of Hertz. The third column lists the coefficients, c_1, c_2, c_3, . . . c_8. The fourth column lists the normalized coefficients $c_1/c_1 = 1$, c_2/c_1, $c_3/c_1, \ldots c_8/c_1$. The fifth column lists the phase angles θ_1, θ_2, $\theta_3, \ldots \theta_8$. The sixth column lists the normalized coefficients, $\theta_1 - \theta_1 = 0$, $\theta_2 - \theta_1$, $\theta_3 - \theta_1, \ldots \theta_8 - \theta_1$.

We expect the even coefficients, c_2, c_4, c_6, . . . c_8, to be zero. They are much smaller than the odd coefficients, so we will interpret them to be zero. The coefficient c_0 is the dc component of the Fourier series and is written above the table in Figure 15.8-5. Finally, PSpice represents the Fourier series, using sine instead of cosine, so the coefficients in Figure 15.8-6 indicate that $v_i(t)$ is represented by the Fourier series

$$v_i(t) = -2.000199 + 3.242 \sin(393t + 45^\circ) + 0.3602 \sin(1178t - 45^\circ) + 0.1297 \sin(1963t - 135^\circ) + 0.06613 \sin(2749t + 135^\circ) + \ldots$$

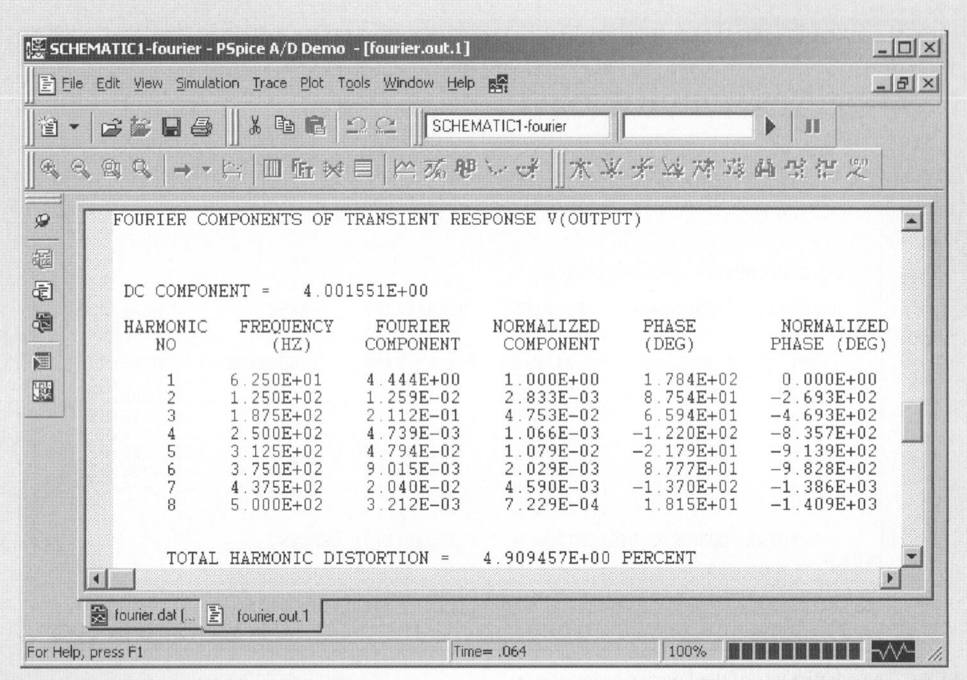

FOURIER COMPONENTS OF TRANSIENT RESPONSE V(OUTPUT)

DC COMPONENT = 4.001551E+00

HARMONIC NO	FREQUENCY (HZ)	FOURIER COMPONENT	NORMALIZED COMPONENT	PHASE (DEG)	NORMALIZED PHASE (DEG)
1	6.250E+01	4.444E+00	1.000E+00	1.784E+02	0.000E+00
2	1.250E+02	1.259E-02	2.833E-03	8.754E+01	-2.693E+02
3	1.875E+02	2.112E-01	4.753E-02	6.594E+01	-4.693E+02
4	2.500E+02	4.739E-03	1.066E-03	-1.220E+02	-8.357E+02
5	3.125E+02	4.794E-02	1.079E-02	-2.179E+01	-9.139E+02
6	3.750E+02	9.015E-03	2.029E-03	8.777E+01	-9.828E+02
7	4.375E+02	2.040E-02	4.590E-03	-1.370E+02	-1.386E+03
8	5.000E+02	3.212E-03	7.229E-04	1.815E+01	-1.409E+03

TOTAL HARMONIC DISTORTION = 4.909457E+00 PERCENT

FIGURE 15.8-7 The coefficients of the Fourier series of $v_o(t)$.

We can represent the series, using cosine, by subtracting 90° from each phase angle. Then,

$$\begin{aligned} v_i(t) &= -2.000199 + 3.242\cos(393t - 45^\circ) + 0.3602\cos(1178t - 135^\circ) \\ &\quad + 0.1297\cos(1963t - 225^\circ) + 0.06613\cos(2749t + 45^\circ) + \ldots \end{aligned} \tag{15.8-2}$$

Scroll through the output file to find the Fourier coefficients of the output voltage shown in Figure 15.8-7. Figure 15.8-7 indicates that the Fourier series of $v_o(t)$ is

$$\begin{aligned} v_o(t) &= 4.001551 + 4.444\cos(393t + 88.4^\circ) + 0.2112\cos(1178t - 24.06^\circ) \\ &\quad + 0.04794\cos(1963t - 118.8^\circ) + 0.02040\cos(2749t - 227^\circ) + \ldots \end{aligned} \tag{15.8-3}$$

15.9 THE FOURIER TRANSFORM

The Fourier transform is closely related to the Fourier series and the Laplace transform. Recall that a periodic waveform $f(t)$ possesses a Fourier series. As we increase the period T, the fundamental frequency ω_0 becomes smaller because

$$\omega_0 = \frac{2\pi}{T}$$

The difference between two consecutive harmonic frequencies is $\Delta\omega = (n+1)\omega_0 - n\omega_0 = \omega_0 = 2\pi/T$. Therefore, as T approaches infinity, $\Delta\omega$ approaches $d\omega$, an infinitesimal frequency increment. Furthermore, the number of frequencies in any given frequency interval increases as $\Delta\omega$ decreases. Thus, in the limit, $n\omega_0$ approaches the continuous variable, ω.

Consider the exponential Fourier series

$$f(t) = \sum_{n=-\infty}^{n=\infty} C_n e^{jn\omega_0 t} \tag{15.9-1}$$

and

$$C_n = \frac{1}{T}\int_{-T/2}^{T/2} f(t)e^{-jn\omega_0 t}\, dt \tag{15.9-2}$$

Multiplying Eq. 15.9-2 by T and letting T approach infinity, we have

$$C_n T = \int_{-\infty}^{\infty} f(t)e^{-j\omega t}\, dt \tag{15.9-3}$$

Let C_nT equal a new frequency function $F(j\omega)$ so that

$$F(j\omega) = \int_{-\infty}^{\infty} f(t)e^{-j\omega t}\, dt \tag{15.9-4}$$

where $F(j\omega)$ is the *Fourier transform* of $f(t)$. The inverse process is found from Eq. 15.9-1, where we let $C_nT = F(j\omega)$ so that

$$f(t) = \lim_{T\to\infty} \sum_{n=-\infty}^{\infty} C_n T e^{jn\omega_0 t}\frac{1}{T} = \lim_{T\to\infty} \sum_{n=-\infty}^{\infty} F(j\omega)e^{jn\omega_0 t}\frac{\omega_0}{2\pi}$$

because $1/T = \omega_0/2\pi$. As $T \to \infty$, the sum becomes an integral, and the increment $\Delta\omega = \omega_0$ becomes $d\omega$. Then, we have

$$f(t) = \frac{1}{2\pi}\int_{-\infty}^{\infty} F(j\omega)e^{j\omega t}\, d\omega \tag{15.9-5}$$

Equation 15.9-5 is called the *inverse Fourier transform*. This pair of equations (Eqs. 15.9-4 and 15.9-5), called the Fourier transform pair, permits us to complete the Fourier transformation to the frequency domain and the inverse process to the time domain.

A given function of time $f(t)$ has a Fourier transform if

$$\int_{-\infty}^{\infty} f(t)\, dt < \infty$$

and if the number of discontinuities in $f(t)$ is finite. From a practical point of view, all pulses of finite duration in which we are interested have Fourier transforms.

The Fourier transform pair is summarized in Table 15.9-1.

Table 15.9-1 The Fourier Transform Pair

EQUATION	NAME	PROCESS
$F(j\omega) = \int_{-\infty}^{\infty} f(t)\, e^{-j\omega t}\, dt$	Transform	Time domain to frequency domain Conversion of $f(t)$ into $F(j\omega)$
$f(t) = \frac{1}{2\pi}\int_{-\infty}^{\infty} F(j\omega)e^{j\omega t}\, d\omega$	Inverse transform	Frequency domain to time domain Conversion of $F(j\omega)$ into $f(t)$

EXAMPLE 15.9-1 Fourier Transform of a Pulse

Derive the Fourier transform of the aperiodic pulse shown in Figure 15.9-1.

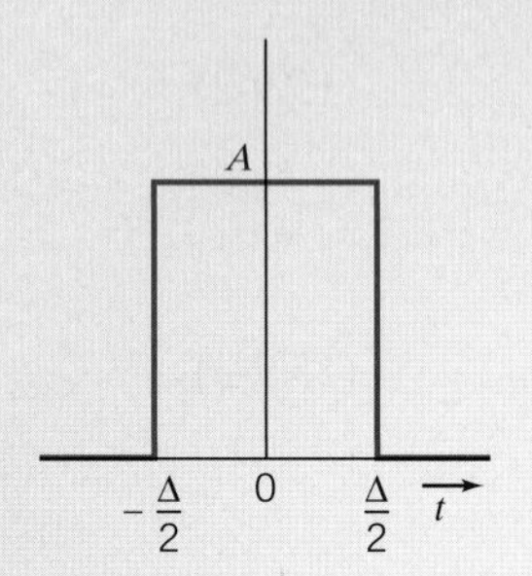

FIGURE 15.9-1 An aperiodic pulse.

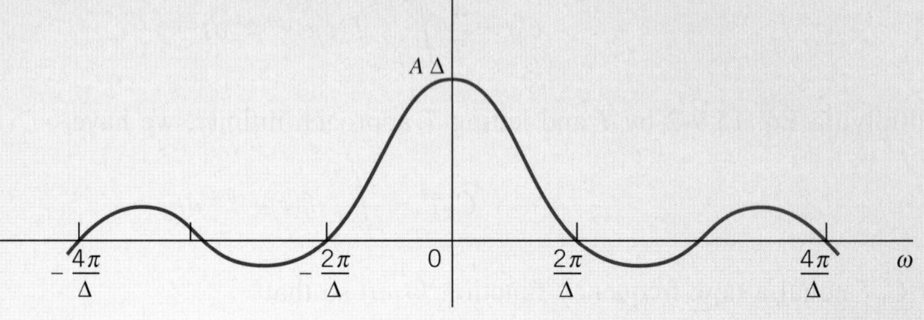

FIGURE 15.9-2 The Fourier transform for the rectangular aperiodic pulse is shown as a function of ω.

Solution

Using the transform, we have

$$\begin{aligned} F(j\omega) &= \int_{-\Delta/2}^{\Delta/2} Ae^{-j\omega t}\,dt = \frac{A}{-j\omega}e^{-j\omega t}\Big|_{-\Delta/2}^{\Delta/2} \\ &= \frac{A}{-j\omega}\left(e^{-j\omega\Delta/2} - e^{j\omega\Delta/2}\right) = A\Delta\frac{\sin(\omega\Delta/2)}{\omega\Delta/2} \end{aligned} \tag{15.9-6}$$

Thus, the Fourier transform is of the form $(\sin x)/x$, where $x = \omega\Delta/2$, as shown in Figure 15.9-2. Note that $(\sin x)/x = 0$ when $x = \omega\Delta/2 = n\pi$ or $\omega = 2n\pi/\Delta$, as shown in Figure 15.9-2. We will denote $(\sin x)/x = Sa(x)$.

Let us consider the shifted version of the rectangular pulse of Figure 15.9-1 where $A = 1/\Delta$ and the width of the pulse approaches zero, $\Delta \to 0$, whereas the area of the rectangle remains equal to 1. Then, we have the *unit impulse* $\delta(t - t_0)$ so that

$$\int_a^b \delta(t - t_0)dt = \begin{cases} 1 & a \le t_0 \le b \\ 0 & \text{otherwise} \end{cases} \tag{15.9-7}$$

We obtain the Fourier transform for a unit impulse at t_0 as

$$F(j\omega) = \int_{t_0-}^{t_0+} \delta(t - t_0)e^{-j\omega t}\,dt = e^{-j\omega t_0} \tag{15.9-8}$$

When $t_0 = 0$, we have the special case,

$$F(j\omega) = 1 \tag{15.9-9}$$

Thus, we note that $F(j\omega) = 1$ of a unit impulse located at the origin is constant and equal to 1 for all frequencies.

EXERCISE 15.9-1 Determine the Fourier transform of $f(t) = e^{-at}u(t)$, where $u(t)$ is the unit step function.

Answer: $F(j\omega) = \dfrac{1}{a + j\omega}$

15.10 FOURIER TRANSFORM PROPERTIES

We can derive some properties of the Fourier transform by writing $F(j\omega)$ in complex form as

$$F(j\omega) = X(\omega) + jY(\omega)$$

Alternatively, we have

$$F(j\omega) = |F(\omega)|e^{j\theta}$$

where $\theta = \tan^{-1}(Y/X)$. Note that we use $F(j\omega) = F(\omega)$ interchangeably. Furthermore,

$$F(-\omega) = F^*(\omega)$$

where $F^*(\omega)$ is the complex conjugate of $F(\omega)$.

If we have the Fourier transform of $f(t)$, we write

$$\mathscr{F}[f(t)] = F(\omega)$$

where the script $\mathscr{F}$ implies the Fourier transform. Then the inverse transform is written as

$$\mathscr{F}^{-1}[F(\omega)] = f(t)$$

Repeating the transformation equation, we have (Table 15.9-1)

$$F(\omega) = \int_{-\infty}^{\infty} f(t)e^{-j\omega t}\,dt \tag{15.10-1}$$

Then, if $\mathscr{F}[af_1(t)] = aF_1(\omega)$ and $\mathscr{F}[bf_2(t)] = bF_2(\omega)$, we have

$$\begin{aligned}\mathscr{F}[af_1 + bf_2] &= \int_{-\infty}^{\infty} [af_1 + bf_2]e^{-j\omega t}\,dt \\ &= \int_{-\infty}^{\infty} af_1 e^{-j\omega t}dt + \int_{-\infty}^{\infty} bf_2 e^{-j\omega t}dt \\ &= aF_1(\omega) + bF_2(\omega)\end{aligned}$$

This is known as the *linearity* property.

We now use the definition of the Fourier transform, Eq. 15.10-1, in the following examples to find several other properties.

EXAMPLE 15.10-1 Fourier Transform Property

Find the Fourier transform of a time-shifted function $f(t - t_0)$.

Solution

$$\mathscr{F}[f(t - t_0)] = \int_{-\infty}^{\infty} f(t - t_0)e^{-j\omega t}\,dt$$

If we let $x = t - t_0$, we have

$$\mathscr{F}[f(t - t_0)] = \int_{-\infty}^{\infty} f(x)e^{-j\omega(x + t_0)}\,dx = e^{-j\omega t_0}F(\omega)$$

where $F(\omega) = \mathscr{F}[f(t)]$.

Selected properties of the Fourier transform are summarized in Table 15.10-1. We can use these properties to derive Fourier transform pairs.

Table 15.10-1 Selected Properties of the Fourier Transform

NAME OF PROPERTY	FUNCTION OF TIME	FOURIER TRANSFORM
1. Definition	$f(t)$	$F(\omega)$
2. Multiplication by constant	$Af(t)$	$AF(\omega)$
3. Linearity	$af_1 + bf_2$	$aF_1(\omega) + bF_2(\omega)$
4. Time shift	$f(t - t_0)$	$e^{-j\omega t_0}F(\omega)$
5. Time scaling	$f(at),\ a > 0$	$\frac{1}{a}F\left(\frac{\omega}{a}\right)$
6. Modulation	$e^{j\omega_0 t}f(t)$	$F(\omega - \omega_0)$
7. Differentiation	$\frac{d^n f(t)}{dt^n}$	$(j\omega)^n F(\omega)$
8. Convolution	$\int_{-\infty}^{\infty} f_1(x)f_2(t - x)dx$	$F_1(\omega)F_2(\omega)$
9. Time multiplication	$t^n f(t)$	$(j)^n \frac{d^n F(\omega)}{d\omega^n}$
10. Time reversal	$f(-t)$	$F(-\omega)$
11. Integration	$\int_{-\infty}^{t} f(\tau)\,d\tau$	$\frac{F(\omega)}{j\omega} + \pi F(0)\delta(\omega)$

With the aid of the properties of the Fourier transform and the original defining equation, we can derive useful transform pairs and develop a table of these relationships. We have already derived the first three entries in Table 15.10-2, and we will add several more by using the properties of Table 15.10-1 and/or the original definition of the transformation.

EXAMPLE 15.10-2 Fourier Transform

Find the Fourier transform of $f(t) = Ae^{-a|t|}$, which is shown in Figure 15.10-1.

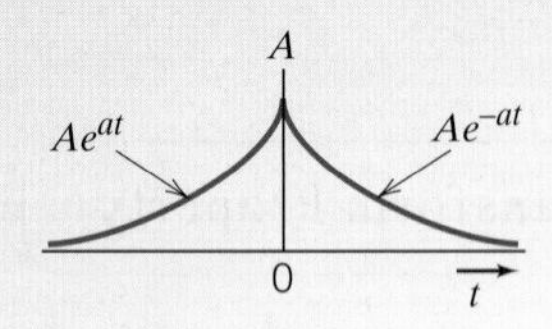

FIGURE 15.10-1 Waveform of Example 15.10-2.

Solution

We will break the function into two symmetric waveforms and use the linearity property. Then,

$$f(t) = f_1(t) + f_2(t) = Ae^{-at}u(t) + Ae^{at}u(-t)$$

We have, from entry 3 of Table 15.10-2,

$$F_1(\omega) = \frac{A}{a + j\omega}$$

From property 10 of Table 15.10-1, we obtain

$$F_2(\omega) = F_1(-\omega) = \frac{A}{a - j\omega}$$

Using the linearity property, we have

$$F(\omega) = F_1(\omega) + F_2(\omega) = \frac{A}{a + j\omega} + \frac{A}{a - j\omega} = \frac{2Aa}{a^2 + \omega^2} \tag{15.10-2}$$

This result is entry 4 in Table 15.10-2. Note that $F(\omega)$ is an even function.

Table 15.10-2 Fourier Transform Pairs

$f(t)$	WAVEFORM	$f(\omega)$
1. Pulse $f_1(t) = Au\left(t + \frac{\Delta}{2}\right) - Au\left(t - \frac{\Delta}{2}\right)$	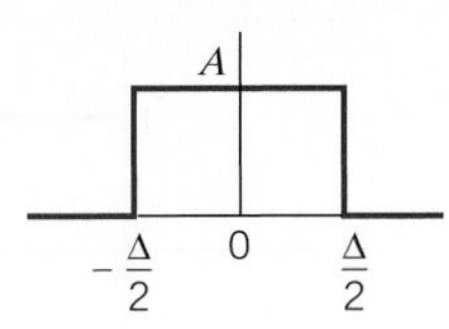	$A\Delta Sa\left(\frac{\omega\Delta}{2}\right)$
2. Impulse $\delta(t - t_0)$	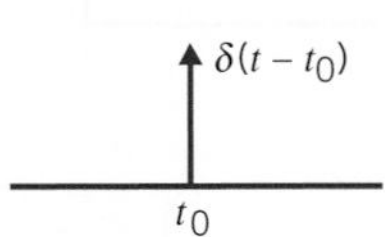	$e^{-j\omega t_0}$
3. Decaying exponential $Ae^{-at}u(t)$ 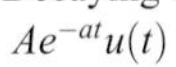	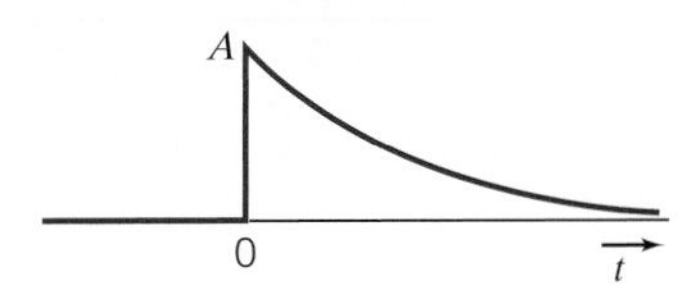	$\frac{A}{a + j\omega}$
4. Symmetric decaying exponential $Ae^{-a\|t\|}$	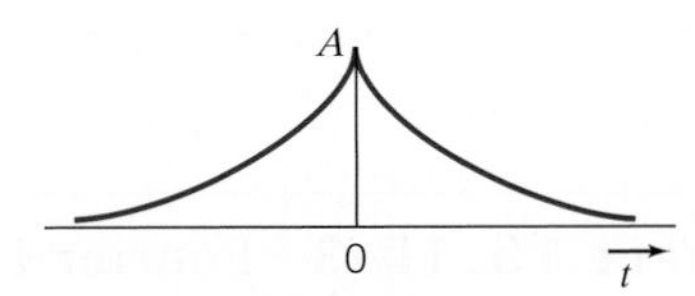	$\frac{2aA}{a^2 + \omega^2}$
5. Tone burst (gated cosine) $Af_1(t) \cos \omega_0 t$	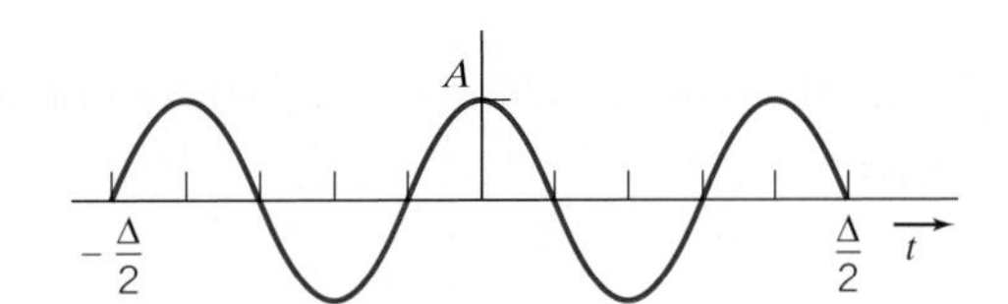	$\frac{A\Delta}{2}[Sa(\omega - \omega_0) + Sa(\omega + \omega_0)]$
6. Triangular pulse	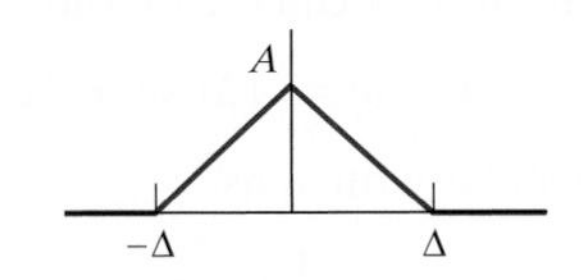	$A\,\Delta Sa^2\left(\frac{\omega\Delta}{2}\right)$
7. $A\,Sa(bt) = A\frac{\sin bt}{bt}$	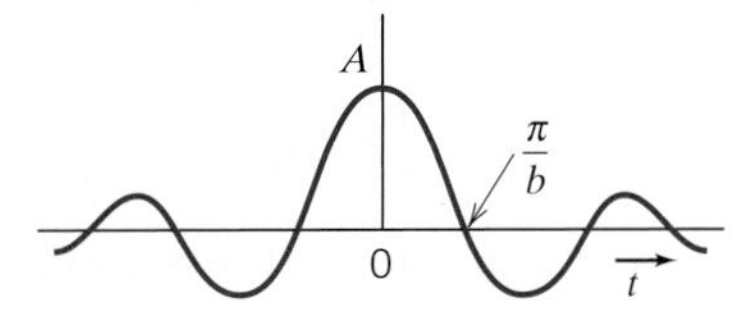	$\begin{cases} \frac{A\pi}{b} & \|\omega\| < b \\ 0 & \|\omega\| > b \end{cases}$
8. Constant dc $f(t) = A$	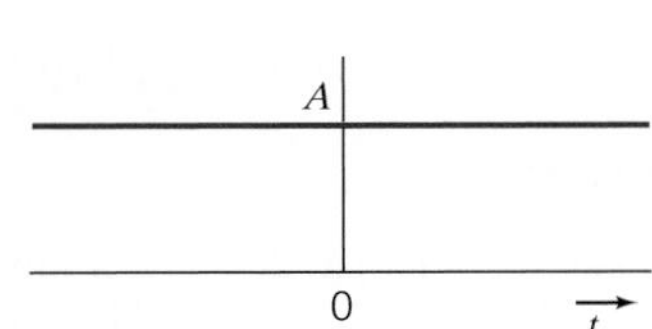	$2\pi A\,\delta(\omega)$
9. Cosine wave $A \cos \omega_0 t$	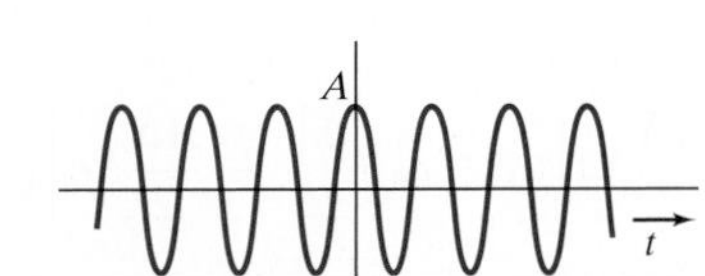	$\pi A[\delta(\omega + \omega_0) + \delta(\omega - \omega_0)]$

(continued)

Table 15.10-2 (*Continued*)

$f(t)$	WAVEFORM	$f(\omega)$
10. Signum $f(t) = \begin{cases} +1 & t > 0 \\ -1 & t < 0 \end{cases}$	1, −1, t	$\frac{2}{j\omega}$
11. Step input $Au(t)$	A, 0, t	$A\left[\pi\delta(\omega) + \frac{1}{j\omega}\right]$

Note: $Sa(x) = (\sin x)/x$.

EXAMPLE 15.10-3 Fourier Transform

Find the Fourier transform of the gated cosine waveform $f(t) = f_1(t) \cos \omega_0 t$, where $f_1(t)$ is the rectangular pulse shown in Figure 15.9-1.

Solution

The Fourier transform of the rectangular pulse is entry 1 in Table 15.10-2 and is written as

$$F_1(\omega) = A\Delta(\sin x)/x$$

where $x = \omega\Delta/2$. The cosine function can be written as

$$\cos \omega_0 t = \frac{1}{2}\left(e^{j\omega_0 t} + e^{-j\omega_0 t}\right)$$

Therefore,

$$f(t) = \frac{1}{2} f_1(t) e^{j\omega_0 t} + \frac{1}{2} f_1(t) e^{-j\omega_0 t}$$

Using the modulation property (entry 6) of Table 15.10-1, we obtain

$$F(\omega) = \frac{1}{2} F_1(\omega - \omega_0) + \frac{1}{2} F_1(\omega + \omega_0)$$

Therefore, using $F_1(\omega)$ from Eq. 15.9-6, we have

$$F(\omega) = \frac{A\Delta}{2} \frac{\sin[(\omega - \omega_0)\Delta/2]}{(\omega - \omega_0)\Delta/2} + \frac{A\Delta}{2} \frac{\sin[(\omega + \omega_0)\Delta/2]}{(\omega + \omega_0)\Delta/2}$$

or, using $Sa(x) = (\sin x)/x$, we have

$$F(\omega) = \frac{A\Delta}{2} Sa\left[(\omega - \omega_0)\frac{\Delta}{2}\right] + \frac{A\Delta}{2} Sa\left[(\omega + \omega_0)\frac{\Delta}{2}\right]$$

EXERCISE 15.10-1 Find the Fourier transform of $f(at)$ for $a > 0$ when $F(\omega) = \mathscr{F}[f(t)]$.

Answer: $\mathscr{F}[f(at)] = \frac{1}{a} F\left(\frac{\omega}{a}\right)$

EXERCISE 15.10-2 Show that the Fourier transform of a constant dc waveform $f(t) = A$ for $-\infty \leq t \leq \infty$ is $F(\omega) = 2\pi A\delta(\omega)$ by obtaining the inverse transform of $F(\omega)$.

15.11 THE SPECTRUM OF SIGNALS

The *spectrum,* also called the *spectral density*, of a signal $f(t)$ is its Fourier transform $F(\omega)$. We can plot $F(\omega)$ as a function of ω to show the spectrum. For example, for a rectangular pulse signal of Figure 15.9-1, we found that

$$F(\omega) = A\Delta Sa(\omega\Delta/2)$$

which is plotted in Figure 15.9-2. The spectrum of the rectangular pulse is real.

The Fourier transform of an impulse $\delta(t)$ is (entry 2 of Table 15.10-2)

$$F(\omega) = 1$$

Thus, the spectrum of an impulse contains all frequencies, and a plot of the spectrum of the impulse is shown in Figure 15.11-1.

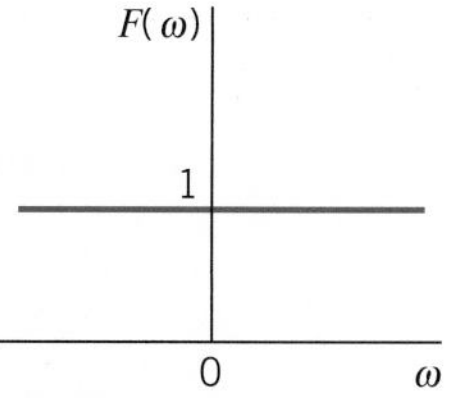

FIGURE 15.11-1 Spectrum of impulse $f(t) = \delta(t)$.

The Fourier transform of a constant dc signal of magnitude A is

$$F(\omega) = 2\pi A\delta(\omega)$$

which has a spectrum as shown in Figure 15.11-2. The integral of the impulse $\delta(\omega)$ has value unity. The symbol for the impulse is a vertical line with an arrowhead.

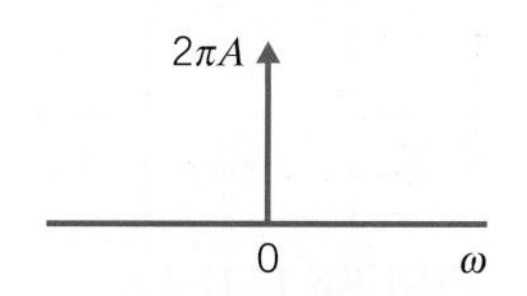

FIGURE 15.11-2 Spectrum of constant dc signal of magnitude A. The symbol for an impulse is a vertical line with an arrowhead.

For completeness, let us examine a function that has a Fourier transform that is complex. When $f(t) = Ae^{-at}u(t)$,

$$F(\omega) = \frac{A}{a + j\omega}$$

To plot the spectrum, we calculate the magnitude and phase of $F(\omega)$ as

$$|F(\omega)| = \frac{A}{(a^2 + \omega^2)^{1/2}}$$

and

$$\phi(\omega) = -\tan^{-1}\omega/a$$

The Fourier spectrum is shown in Figure 15.11-3.

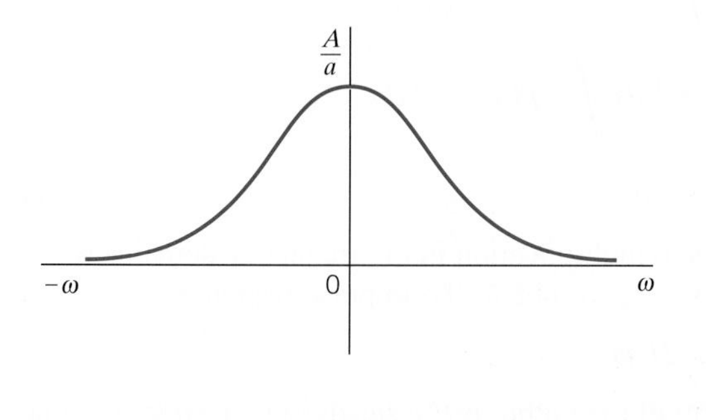

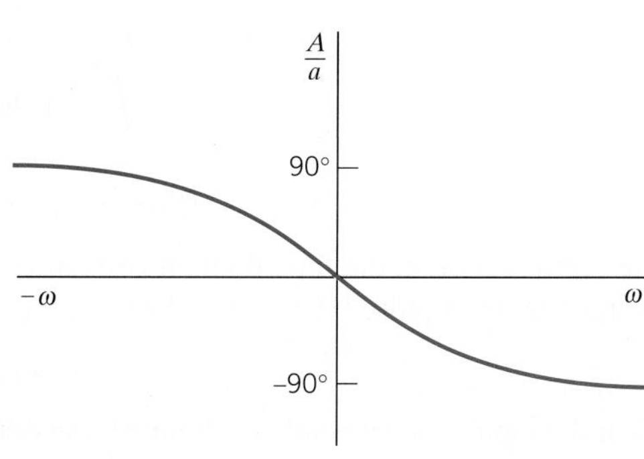

FIGURE 15.11-3 The Fourier spectrum for $f(t) = Ae^{-at}u(t)$.

The **Fourier spectrum** of a signal is a graph of the magnitude and phase of the Fourier transform of the signal.

EXERCISE 15.11-1 Calculate the Fourier transform and draw the Fourier spectrum for $f(t)$ shown in Figure E 15.11-1, where $f(t) = A \cos \omega_0 t$ for all t.

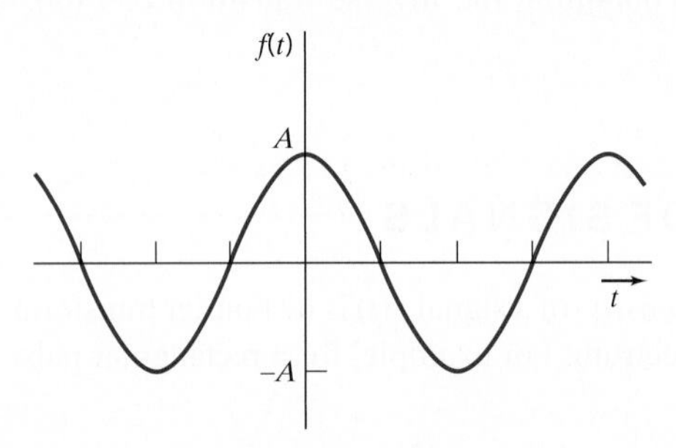

FIGURE E 15.11-1

Answer: $F(\omega) = \pi A \delta(\omega + \omega_0) + \pi A \delta(\omega - \omega_0)$

15.12 CONVOLUTION AND CIRCUIT RESPONSE

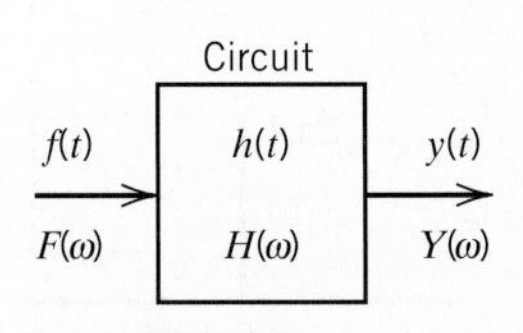

FIGURE 15.12-1
A linear circuit.

A circuit with an impulse response $h(t)$ and an input $f(t)$ has a response $y(t)$ that may be determined from the convolution integral. For the circuit shown in Figure 15.12-1, the convolution integral is

$$y(t) = \int_{-\infty}^{\infty} h(x) f(t - x)\, dx$$

If we use the Fourier transform of the convolution integral, we have

$$\begin{aligned} [y(t)] &= \int_{-\infty}^{\infty} \int_{-\infty}^{\infty} h(x) f(t - x)\, dx\, e^{-j\omega t}\, dt \\ &= \int_{-\infty}^{\infty} h(x) \int_{-\infty}^{\infty} f(t - x)\, e^{-j\omega t}\, dt\, dx \end{aligned}$$

Let $u = t - x$ to obtain

$$\begin{aligned} \mathscr{F}[y(t)] &= \int_{-\infty}^{\infty} h(x) \int_{-\infty}^{\infty} f(u) e^{-j\omega(u + x)}\, du\, dx \\ &= \int_{-\infty}^{\infty} h(x) e^{-j\omega x}\, dx \int_{-\infty}^{\infty} f(u) e^{-j\omega u}\, du \end{aligned}$$

or

$$Y(\omega) = H(\omega) F(\omega) \tag{15.12-1}$$

Thus, convolution in the time domain corresponds to multiplication in the frequency domain. When the input is an impulse, $f(t) = \delta(t)$, because $F(\omega) = 1$, we obtain the impulse response

$$Y(\omega) = H(\omega)$$

When the input is a sinusoid, the Fourier transform of the output is the steady-state response to that sinusoidal driving function.

EXAMPLE 15.12-1 Circuit Analysis Using the Fourier Transform

Find the response, $v_o(t)$, of the RL circuit shown in Figure 15.12-2 when $v(t) = 4e^{-2t}u(t)$V. The initial condition is zero.

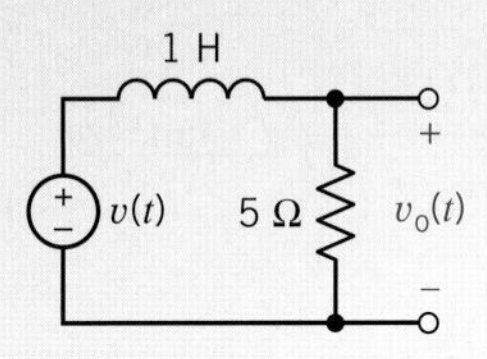

FIGURE 15.12-2 Circuit of Example 15.12-1.

Solution

Because $v(t) = 4e^{-2t}u(t)$, we obtain $V(\omega)$ as

$$V(\omega) = \frac{4}{2 + j\omega}$$

The circuit is represented by $H(\omega)$, and, using the voltage divider principle, we have

$$H(\omega) = \frac{R}{R + j\omega L} = \frac{5}{5 + j\omega}$$

Then, we have

$$V_o(\omega) = H(\omega)V(\omega) = \frac{20}{(5 + j\omega)(2 + j\omega)}$$

Expand, using partial fractions, to obtain[1]

$$V_o(\omega) = \frac{-20/3}{5 + j\omega} + \frac{20/3}{2 + j\omega}$$

Using the inverse transform for each term (entry 3 of Table 15.10-2), we have

$$v_o(t) = \frac{20}{3}(e^{-2t} - e^{-5t})u(t)\text{V}$$

The time-domain responses obtained in this manner are responses of initially relaxed circuits. (No initial energy is stored.)

[1]See Chapter 14, Section 14.4, for a review of partial fraction expansion.

EXAMPLE 15.12-2 Circuit Analysis Using the Fourier Transform

Determine and plot the spectrum of the response $V_o(\omega)$ of the circuit of Figure 15.12-3 when $v = 10e^{-2t}u(t)$ V.

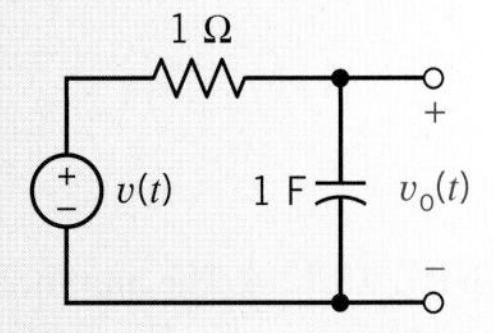

FIGURE 15.12-3 Circuit of Example 15.12-2.

Solution

The input signal $v(t)$ has a Fourier transform

$$V(\omega) = \frac{10}{2 + j\omega} = \frac{10}{(4 + \omega^2)^{1/2}} \angle -\tan^{-1} \omega/2$$

The circuit transfer function is

$$H(j\omega) = \frac{1/(j\omega C)}{R + 1/(j\omega C)} = \frac{1}{1 + j\omega} = \frac{1}{(1 + \omega^2)^{1/2}} \angle -\tan^{-1}\omega$$

Then, the output is

$$V_o(\omega) = H(\omega)V(\omega) = \frac{10}{(2 + j\omega)(1 + j\omega)}$$

Therefore,

$$|V_o| = \frac{10}{[(4 + \omega^2)(1 + \omega^2)]^{1/2}}$$

and

$$\phi(\omega) = V_o(\omega) = -\tan^{-1}\frac{\omega}{2} - \tan^{-1}\omega$$

The calculated magnitude and phase for $V_o(\omega)$ are recorded in Table 15.12-1. For negative ω, $|V_o(\omega)| = |V_o(-\omega)|$ and

$$\phi(-\omega) = -\phi(\omega)$$

Therefore, the Fourier spectrum of $V_o(\omega)$ is represented by the plot shown in Figure 15.12-4.

Table 15.12-1 Fourier Response for Example 15.12-2

ω	0	1	2	3	5	∞
$\lvert V_o \rvert$	5	3.16	1.58	0.88	0.36	0
$\phi(\omega)$	0°	−71.6°	−108.4°	−127.9°	−146.9°	−180°

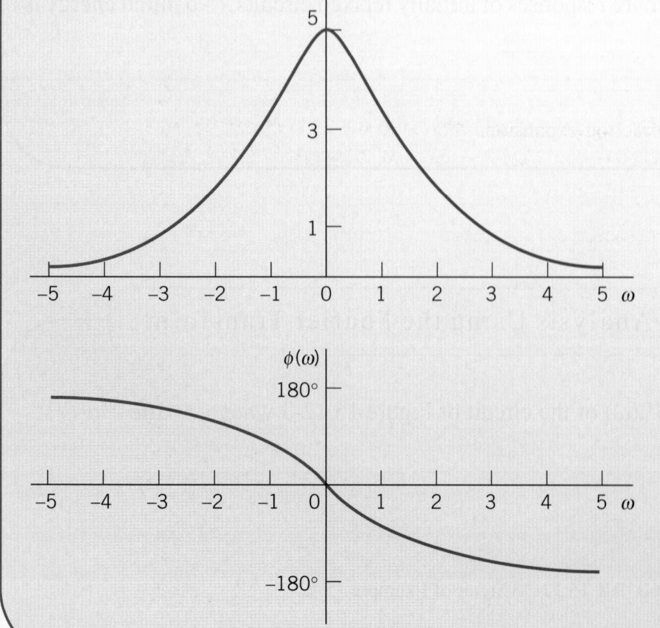

FIGURE 15.12-4 The amplitude and phase versus ω of the output voltage for Example 15.12-2.

EXERCISE 15.12-1 An ideal band-pass filter passes all frequencies between 24 rad/s and 48 rad/s without attenuation and completely rejects all frequencies outside this passband.

(a) Sketch $|V_o|^2$ for the filter output voltage when the input voltage is

$$v(t) = 120e^{-24t}u(t)\text{V}$$

(b) What percentage of the input signal energy is available in the signal at the output of the ideal filter?

Answer: **(b)** 20.5%

15.13 THE FOURIER TRANSFORM AND THE LAPLACE TRANSFORM

The table of Laplace transforms, Table 14.2-1, developed in Chapter 14, can be used to obtain the Fourier transform of a function $f(t)$. Of course, the Fourier transform formally exists only when the Fourier integral, Eq. 15.9-4, converges. The Fourier integral will converge when all the poles of $F(s)$ lie in the left-hand s-plane, not on the $j\omega$-axis or at the origin.

If $f(t)$ is zero for $t \leq 0$ and $\int_0^\infty f(t)\,dt < \infty$, we can obtain the Fourier transform from the Laplace transform of $f(t)$ by replacing s by $j\omega$. Then

$$F(\omega) = F(s)|_{s=j\omega} \tag{15.13-1}$$

where

$$F(s) = \mathscr{L}[f(t)]$$

For example, if (entry 3 of Table 15.10-2)

$$f(t) = Ae^{-at}u(t)$$

then, from Table 14.2-1,

$$F(s) = \frac{A}{s + a}$$

Therefore, with $s = j\omega$ we obtain the Fourier transform:

$$F(\omega) = \frac{A}{a + j\omega}$$

If $f(t)$ is a real function with a nonzero value for negative time only, then we can reflect $f(t)$ to positive time, find the Laplace transform, and then find $F(\omega)$ by setting $s = -j\omega$. Therefore, when $f(t) = 0$ for $t \geq 0$ and $f(t)$ exists only for negative time, we have

$$F(\omega) = \mathscr{L}[f(-t)]|_{s=-j\omega} \tag{15.13-2}$$

For example, consider the exponential function

$$\begin{aligned} f(t) &= 0 \quad t \geq 0 \\ &= e^{at} \quad t < 0 \end{aligned}$$

Then, reversing the time function, we have

$$f(-t) = e^{-at} \quad t > 0$$

and, therefore,

$$F(s) = \frac{1}{s + a}$$

Hence, setting $s = -j\omega$, we obtain

$$F(\omega) = \frac{1}{a - j\omega}$$

Table 15.13-1 Obtaining the Fourier Transform Using the Laplace Transform

CASE	METHOD
A. $f(t)$ nonzero for positive time only and $f(t) = 0,\ t < 0$	Step 1. $F(s) = \mathscr{L}[f(t)]$ 2. $F(\omega) = F(s)\|_{s=j\omega}$
B. $f(t)$ nonzero for negative time only and $f(t) = 0,\ t > 0$	Step 1. $F(s) = \mathscr{L}[f(-t)]$ 2. $F(\omega) = F(s)\|_{s=-j\omega}$
C. $f(t)$ nonzero over all time	Step 1. $f(t) = f^+(t) + f^-(t)$ 2. $F^+(s) = \mathscr{L}[f^+(t)]$ $F^-(s) = \mathscr{L}[f^-(-t)]$ 3. $F(\omega) = F^+(s)\|_{s=j\omega} + F^-(s)\|_{s=-j\omega}$

Note: The poles of $F(s)$ must lie in the left-hand s-plane.

Functions that are nonzero over all time can be divided into positive time and negative time functions. We then use Eqs. 15.13-1 and 15.13-2 to obtain the Fourier transform of each part. The Fourier transform of $f(t)$ is the sum of the Fourier transforms of the two parts.

For example, consider the function $f(t)$ with a nonzero value over all time where

$$f(t) = Ae^{-a|t|}$$

which is entry 4 in Table 15.10-2. The positive time portion of the function will be called $f^+(t)$, and the negative time portion will be called $f^-(t)$. Then,

$$f(t) = f^+(t) + f^-(t)$$

Hence

$$F(\omega) = \mathscr{L}[f^+(t)]_{s=j\omega} + \mathscr{L}[f^-(-t)]_{s=-j\omega}$$

In this case,

$$f^+(t) = Ae^{-at} \qquad t > 0$$

and

$$f^-(t) = Ae^{at} \qquad t < 0$$

Note that $f^-(-t) = Ae^{-at}$. Then,

$$F^+(s) = \frac{A}{s+a} \quad \text{and} \quad F^-(s) = \frac{A}{s+a}$$

We obtain the total $F(\omega)$ as

$$F(\omega) = F^+(s)_{s=j\omega} + F^-(s)_{s=-j\omega} = \frac{A}{a+j\omega} + \frac{A}{a-j\omega} = \frac{2aA}{\omega^2 + a^2}$$

The use of the Laplace transform to find the Fourier transform is summarized in Table 15.13-1. Remember that the method summarized cannot be used for sin ωt, cos ωt, or $u(t)$ because the poles of $F(s)$ lie on the $j\omega$-axis or at the origin.

EXERCISE 15.13-1 Derive the Fourier transform for

$$\begin{aligned} f(t) &= te^{-at} \quad t \geq 0 \\ &= te^{at} \quad t \leq 0 \end{aligned}$$

Answer: $\dfrac{-j4a\omega}{(a^2 + \omega^2)^2}$

15.14 HOW CAN WE CHECK . . . ?

Engineers are frequently called upon to check that a solution to a problem is indeed correct. For example, proposed solutions to design problems must be checked to confirm that all of the specifications have been satisfied. In addition, computer output must be reviewed to guard against data-entry errors, and claims made by vendors must be examined critically.

Engineering students are also asked to check the correctness of their work. For example, occasionally just a little time remains at the end of an exam. It is useful to be able to quickly identify those solutions that need more work.

The following example illustrates techniques useful for checking the solutions of the sort of problem discussed in this chapter.

EXAMPLE 15.14-1 How Can We Check Fourier Series?

Figure 15.14-1 shows the transfer characteristic of the saturation nonlinearity. Suppose that the input to this nonlinearity is

$$v_{\text{in}}(t) = A \sin \omega t$$

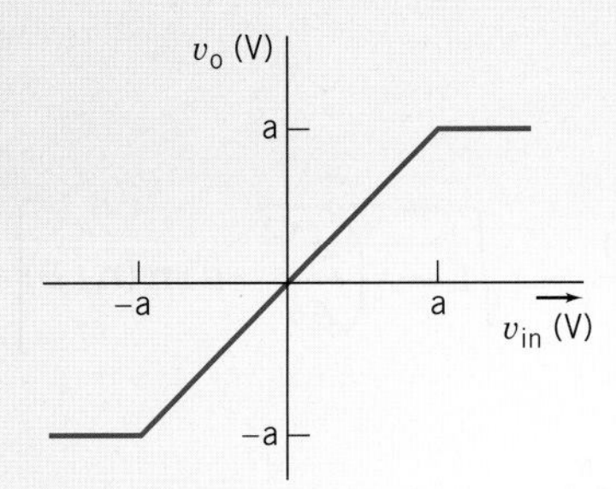

FIGURE 15.14-1 The saturation nonlinearity.

where $A > a$. **How can we check** that the output of the nonlinearity will be a periodic function that can be represented by the Fourier series

$$v_o(t) = b_1 \sin \omega t + \sum_{\substack{n=3 \\ \text{odd}}}^{N} b_n \sin n\omega t \tag{15.14-1}$$

where (Graham, 1971)

$$B = \sin^{-1}\left(\frac{a}{A}\right)$$

$$b_1 = \frac{2}{\pi} A \left[B + \frac{a}{A}\sqrt{1 - \left(\frac{a}{A}\right)^2} \right]$$

and

$$b_n = \frac{4A}{\pi(1 - n^2)} \left[\frac{a}{A} \frac{\cos(nB)}{n} - \sqrt{1 - \left(\frac{a}{A}\right)^2} \sin(nB) \right]$$

Solution

The output voltage, $v_o(t)$, will be a clipped sinusoid. We need to verify that Eq. 15.14-1 does indeed represent a clipped sinusoid. A straightforward, but tedious, way to do this is to plot $v_o(t)$ versus t directly from Eq. 15.14-1. Several computer programs, such as spreadsheets and equation solvers, are available to reduce the work required to produce this plot. Mathcad is one of these programs. In Figure 15.14-2, Mathcad is used to plot $v_o(t)$ versus t. This plot verifies that the Fourier series in Eq. 15.14-1 does indeed represent a clipped sinusoid.

Plot a periodic signal from its coefficients.

Define n, the index for the summation:

$$N := 25 \quad n := 3, 5, \ldots, N$$

Define any parameters that are used to make it easier to enter the coefficient of the Fourier series:

$$A := 12.5 \quad a := 12 \quad B := \operatorname{asin}\left(\frac{a}{A}\right)$$

Enter the fundamental frequency:

$$\omega := 2 \cdot \pi \cdot 1000$$

Define an increment of time. Set up an index to run over two periods of the periodic signal:

$$T := \frac{2 \cdot \pi}{\omega} \quad dt := \frac{T}{200} \quad i := 1, 2, \ldots, 400 \quad t_i := dt \cdot i$$

Enter the formulas for the coefficients of the Fourier series,

$$b_1 := \frac{2}{\pi} \cdot A \left[B + \frac{a}{A} \cdot \sqrt{1 - \left(\frac{a}{A}\right)^2} \right]$$

$$b_n := \frac{4 \cdot A}{\pi \cdot (1 - n^2)} \cdot \left[\frac{a}{A} \cdot \frac{\cos(n \cdot B)}{n} - \sqrt{1 - \left(\frac{a}{A}\right)^2} \cdot \sin(n \cdot B) \right]$$

Enter the Fourier series:

$$v(i) := b_1 \cdot \sin(\omega \cdot t_i) + \sum_{n=3}^{N} b_n \cdot \sin(n \cdot \omega \cdot t_i)$$

Plot the periodic signal:

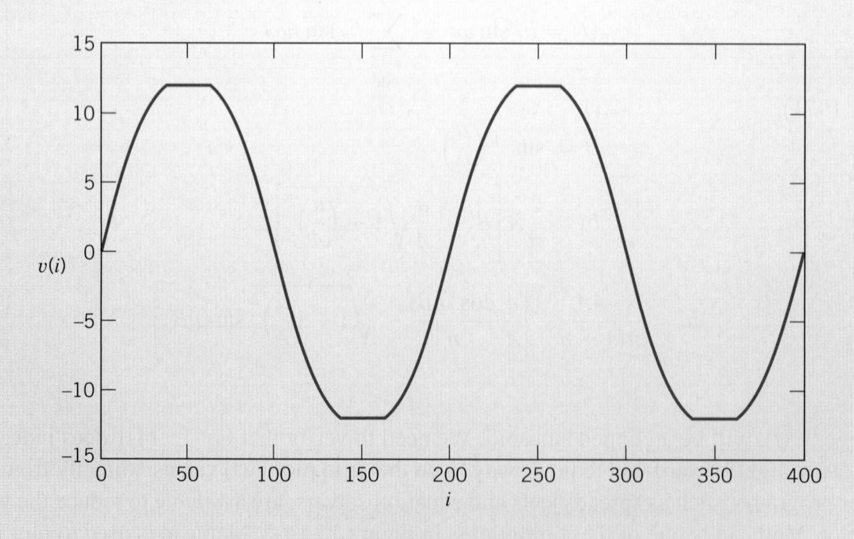

FIGURE 15.14-2 Using Mathcad to verify the Fourier series of a clipped sinusoid.

15.15 DESIGN EXAMPLE

DC POWER SUPPLY

A laboratory power supply uses a nonlinear circuit called a rectifier to convert a sinusoidal voltage input into a dc voltage. The sinusoidal input

$$v_{ac}(t) = A \sin \omega_0 t$$

comes from the wall plug. In this example, $A = 160$ V and $\omega_0 = 377$ rad/s ($f_0 = 60$ Hz). Figure 15.15-1 shows the structure of the power supply. The output of the rectifier is the absolute value of its input, that is,

$$v_s(t) = |A \sin \omega_0 t|$$

The purpose of the rectifier is to convert a signal that has an average value equal to zero into a signal that has an average value that is not zero. The average value of $v_s(t)$ will be used to produce the dc output voltage of the power supply.

The rectifier output is not a sinusoid but a periodic signal with fundamental frequency equal to $2\omega_0$. Periodic signals can be represented by Fourier series. The Fourier series of $v_s(t)$ will contain a constant, or dc, term and some sinusoidal terms. The purpose of the filter shown in Figure 15.15-1 is to pass the dc term and attenuate the sinusoidal terms. The output of the filter, $v_o(t)$, will be a periodic signal and can be represented by a Fourier series. Because we are designing a dc power supply, the sinusoidal terms in the Fourier series of $v_o(t)$ are undesirable. The sum of these undesirable terms is called the ripple of $v_o(t)$.

The challenge is to design a simple filter so that the dc term of $v_o(t)$ is at least 90 V, and the size of the ripple is no larger than 5 percent of the size of the dc term.

Describe the Situation and the Assumptions

1. From Table 15.4-1, the Fourier series of $v_s(t)$ is

$$v_s(t) = \frac{320}{\pi} - \sum_{n=1}^{N} \frac{640}{\pi(4n^2 - 1)} \cos(2 \cdot n \cdot 377 \cdot t)$$

Let $v_{sn}(t)$ denote the term of $v_s(t)$ corresponding to the integer n. Using this notation, we can write the Fourier series of $v_s(t)$ as

$$v_s(t) = v_{s0} + \sum_{n=1}^{N} v_{sn}(t)$$

2. Figure 15.15-2 shows a simple filter. The resistance R_s models the output resistance of the rectifier. We have assumed that the input resistance of the regulator is large enough to be ignored. (The input resistance of the regulator will be in parallel with R and will probably be much larger than R. In this case, the equivalent resistance of the parallel combination will be approximately equal to R.)

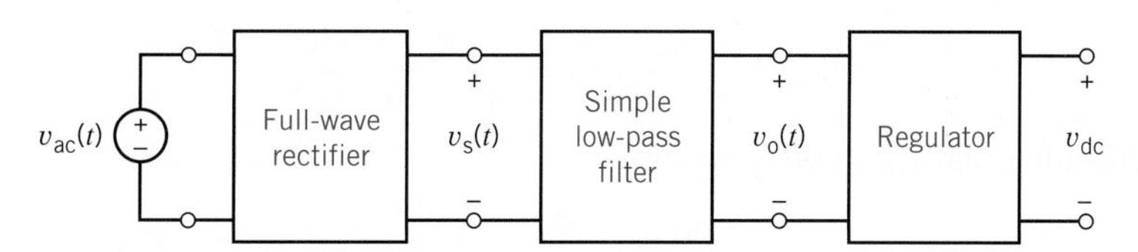

FIGURE 15.15-1 Diagram of a power supply.

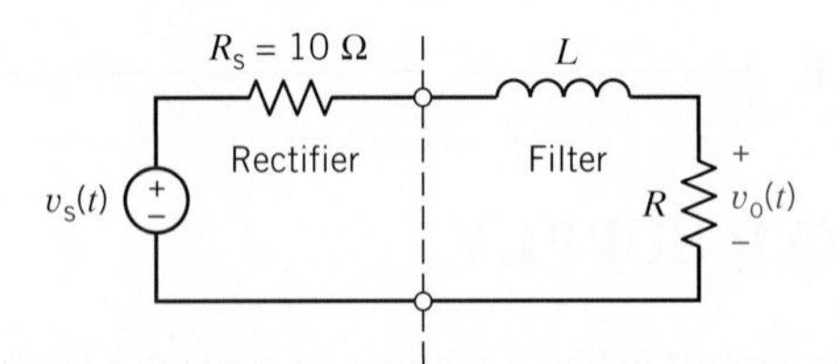

FIGURE 15.15-2 A simple RL low-pass filter connected to the rectifier.

3. The filter output, $v_o(t)$, will also be a periodic signal and will be represented by the Fourier series

$$v_o(t) = v_{o0} + \sum_{n=1}^{N} v_{on}(t)$$

4. Most of the ripple in $v_o(t)$ will be due to $v_{o1}(t)$, the fundamental term of the Fourier series. The specification regarding the allowable ripple can be stated as

$$\text{amplitude of the ripple} \leq 0.05 \cdot \text{dc output}$$

Equivalently, we can state that we require

$$\max\left(\sum_{n=1}^{N} v_{on}(t)\right) \leq 0.05 \cdot v_{o0} \tag{15.15-1}$$

For ease of calculation, we replace Eq. 15.15-1 with the simpler condition

$$v_{o1}(t) \leq 0.04 \cdot v_{o0}$$

That is, the amplitude $v_{o1}(t)$ must be less than 4 percent of the dc term of the output (v_{o0} = dc term of the output).

State the Goal

Specify values of R and L so that

$$\text{dc output} = v_{o0} \geq 90$$

and

$$v_{o1}(t) \leq 0.04 \cdot v_{o0}$$

Generate a Plan

Use superposition to calculate the Fourier series of the filter output. First, the specification

$$\text{dc output} = v_{o0} \geq 90 \text{ V}$$

can be used to determine the required value of R. Next, the specification

$$|v_{o1}(t)| \leq 0.04 \cdot v_{o0}$$

can be used to calculate L.

Act on the Plan

First, we will find the response to the dc term of $v_s(t)$. When the filter input is a constant and the circuit is at steady state, the inductor acts like a short circuit. Using voltage division

$$v_{o0} = \frac{R}{R + R_s} v_{s0} = \frac{R}{R + 10} \cdot \frac{320}{\pi}$$

The specification that $v_{o0} \geq 90$ V requires

$$90 \leq \frac{R}{R + 10} \cdot \frac{320}{\pi}$$

or

$$R \geq 75.9$$

Let us select

$$R = 80\ \Omega$$

When $R = 80\ \Omega$,

$$v_{o0} = 90.54\ \text{V}$$

Next, we find the steady-state response to a sinusoidal term, $v_{sn}(t)$. Phasors and impedances can be used to find this response. By voltage division,

$$\mathbf{V}_{on} = \frac{R}{R + R_s + j2n\omega_0 L}\mathbf{V}_{sn}$$

We are particularly interested in $\mathbf{V}_{o1}$:

$$\mathbf{V}_{o1} = \frac{R}{R + R_s + j2\omega_0 L}\mathbf{V}_{s1} = \frac{80}{90 + j754L} \cdot \frac{640}{\pi \cdot 3}$$

The amplitude of $v_{o1}(t)$ is equal to the magnitude of the phasor $\mathbf{V}_{o1}$. The specification on the amplitude of $v_{o1}(t)$ requires that

$$\frac{80}{\sqrt{90^2 + 754^2 L^2}} \cdot \frac{640}{\pi \cdot 3} \leq 0.04\, v_{o1}$$
$$\leq 0.04 \cdot 90.54$$

That is,

$$L \geq 1.986\ \text{H}$$

Selecting

$$L = 2\ \text{H}$$

completes the design.

Verify the Proposed Solution

Figure 15.15-3*a* displays a plot of $v_s(t)$ and $v_o(t)$, the input and output voltages of the circuit in Figure 15.15-2. Figure 15.15-3*b* shows the details of the output voltage. This plot indicates that the average value of the output voltage is greater than 90 V and that the ripple is no greater than ± 4 V. Therefore, the specifications have been satisfied.

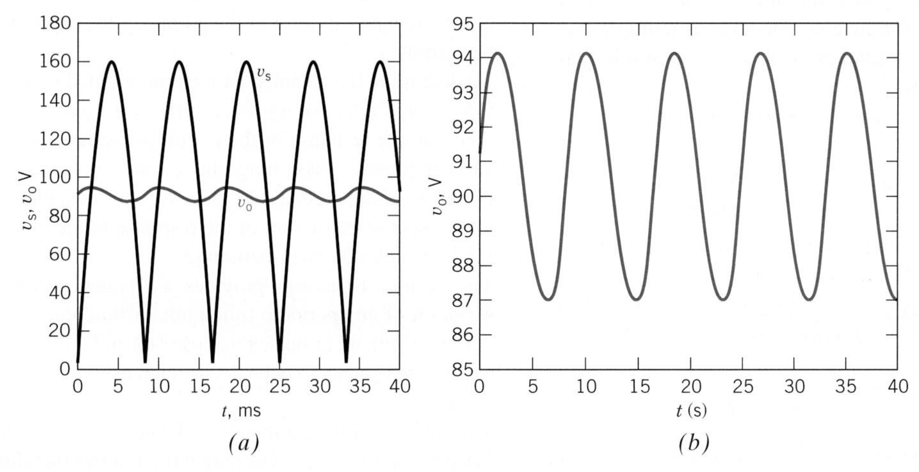

FIGURE 15.15-3 (*a*) Mathcad simulation of the circuit shown in Figure 15.15-2. (*b*) Enlarged plot of the output voltage.

15.16 SUMMARY

- Periodic waveforms arise in many circuits. For example, the form of the load current waveforms for selected loads is shown in Figure15.16-1. Whereas the load current for motors and incandescent lamps is of the same form as that of the source voltage, it is significantly altered for the power supplies, dimmers, and variable-speed drives as shown in Figures 15.16-1*b*,*c*. Electrical engineers have long been interested in developing the tools required to analyze circuits incorporating periodic waveforms.

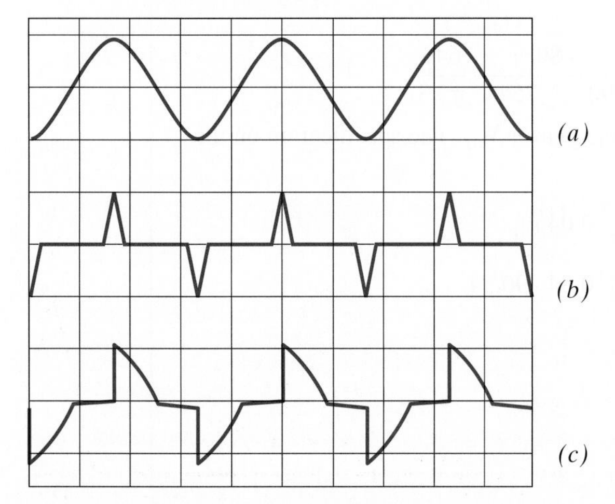

FIGURE 15.16-1 Load current waveforms for (*a*) motors and incandescent lights, (*b*) switch-mode power supplies, and (*c*) dimmers and variable-speed drives. The vertical axis is current, and the horizontal axis is time. *Source:* Lamarre, 1991.

- The brilliant mathematician-engineer Jean-Baptiste-Joseph Fourier proposed in 1807 that a periodic waveform could be represented by a series consisting of cosine and sine terms with the appropriate coefficients. The integer multiple frequencies of the fundamental are called the **harmonic frequencies** (or harmonics).
- The trigonometric form of the Fourier series is

$$f(t) = a_0 + \sum_{n=1}^{N} a_n \cos n\omega_0 t + \sum_{n=1}^{N} b_n \sin n\omega_0 t$$

The coefficients of the trigonometric Fourier series can be obtained from

$$a_0 = \frac{1}{T}\int_{t_0}^{T+t_0} f(t)\, dt$$

$$a_n = \frac{2}{T}\int_{t_0}^{T+t_0} f(t) \cos n\omega_0 t\, dt \quad n > 0$$

$$b_n = \frac{2}{T}\int_{t_0}^{T+t_0} f(t) \sin n\omega_0 t\, dt \quad n > 0$$

- An alternate form of the trigonometric form of the Fourier series is

$$f(t) = c_0 + \sum_{n=1}^{N} c_n \cos(n\omega_0 t + \theta_n)$$

where $c_0 = a_0 =$ average value of $f(t)$ and

$$c_n = \sqrt{a_n^2 + b_n^2} \text{ and } \theta_n = \begin{cases} -\tan^{-1}\left(\dfrac{b_n}{a_n}\right) & \text{if } a_n > 0 \\ 180^\circ - \tan^{-1}\left(\dfrac{b_n}{a_n}\right) & \text{if } a_n < 0 \end{cases}$$

- The Fourier coefficients of some common periodic signals are tabulated in 15.4-1.
- Symmetry can simplify the task of calculating the Fourier coefficients.
- The exponential form of the Fourier series is

$$f(t) = \sum_{-\infty}^{\infty} \mathbf{C}_n e^{jn\omega_0 t}$$

where $\mathbf{C}_n$ is the complex coefficients defined by

$$\mathbf{C}_n = \frac{1}{T}\int_{t_0}^{t_0+T} f(t)\, e^{-jn\omega_0 t}\, dt$$

- The line spectra consisting of the amplitude and phase of the complex coefficients of the Fourier series when plotted against frequency are useful for portraying the frequencies that represent a waveform.
- The practical representation of a periodic waveform consists of a finite number of sinusoidal terms of the Fourier series. The finite Fourier series exhibits the Gibbs phenomenon; that is, although convergence occurs as n grows large, there always remains an error at the points of discontinuity of the waveform.
- To determine the response of a circuit excited by a periodic input signal $v_s(t)$, we represent $v_s(t)$ by a Fourier series and then find the response of the circuit to the fundamental and each harmonic. Assuming the circuit is linear and the principle of superposition holds, we can consider that the total response is the sum of the response to the dc term, the fundamental, and each harmonic.
- The Fourier transform provides a frequency-domain description of an aperiodic time-domain function.
- A circuit with an impulse response $h(t)$ and an input $f(t)$ has a response $y(t)$ that may be determined from the convolution integral.
- The table of Laplace transforms, Table 14.2-1, developed in Chapter 14, can be used to obtain the Fourier transform of a function $f(t)$.

PROBLEMS

Section 15.2 The Fourier Series

P 15.2-1 Find the trigonometric Fourier series for a periodic function $f(t)$ that is equal to t^2 over the period from $t = 0$ to $t = 4$.

P 15.2-2 A ''staircase'' periodic waveform is described by its first cycle as

$$f(t) = \begin{cases} 1 & 0 < t < 0.25 \\ 2 & 0.25 < t < 0.5 \\ 0 & 0.5 < t < 1 \end{cases}$$

Find the Fourier series for this function.

P 15.2-3 Determine the Fourier series for the sawtooth function shown in Figure P 15.2-3.

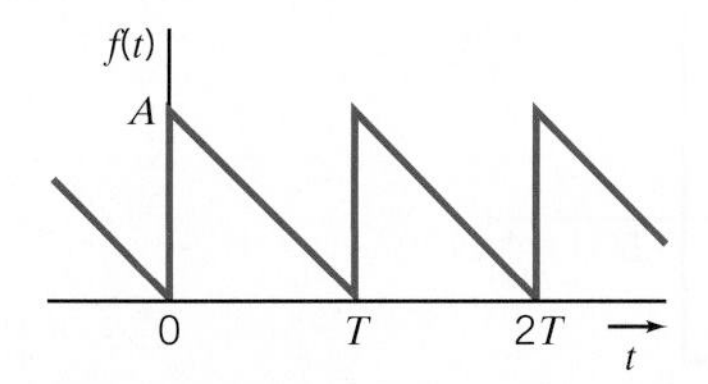

Figure P 15.2-3 Sawtooth wave.

P 15.2-4 Find the Fourier series for the periodic function $f(t)$ that is equal to t over the period from $t = 0$ to $t = 1$ s.

Section 15.3 Symmetry of the Function *f*(*t*)

P 15.3-1 Determine the Fourier series of the voltage waveform shown in Figure P 15.3-1.

Answer: $v_d(t) = \sum_{n=1}^{\infty} \frac{12}{n\pi} \sin(n\pi t)$

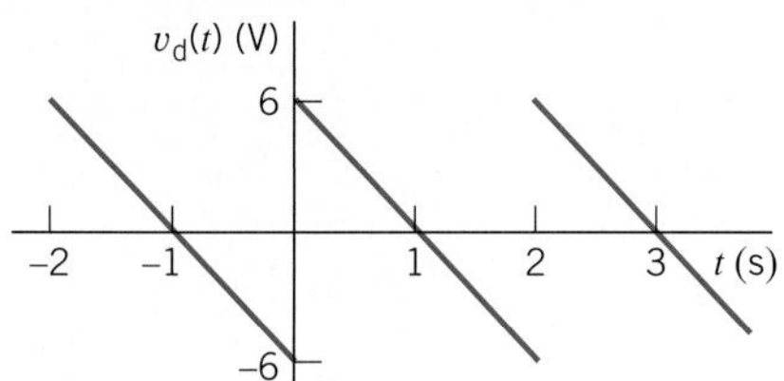

Figure P 15.3-1

P 15.3-2 Determine the Fourier series of the voltage waveform shown in Figure P 15.3-2.

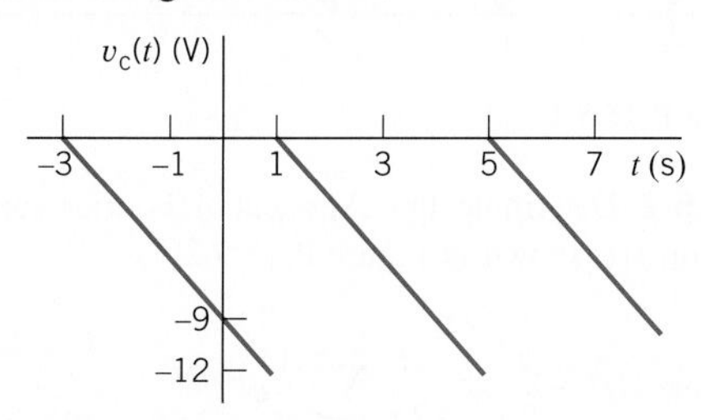

Figure P 15.3-2

Hint: $v_c(t) = v_d(t - 1) - 6$, where $v_d(t)$ is the voltage considered in problem Figure P 15.3-1.

Answer: $v_c(t) = -6 + \sum_{n=1}^{\infty} \frac{12}{n\pi} \sin\left(n\frac{\pi}{2}t - n\frac{\pi}{2}\right)$

P 15.3-3 Determine the Fourier series of the voltage waveform shown in Figure 15.3-3.

Answer:
$$v_a(t) = \frac{1}{2} + \sum_{n=1}^{\infty} \frac{36}{n^e\pi^2}\left(1 - \cos\left(\frac{n\pi}{3}\right)\right)$$
$$\cos\left(n\frac{1000\pi}{3}t\right)$$

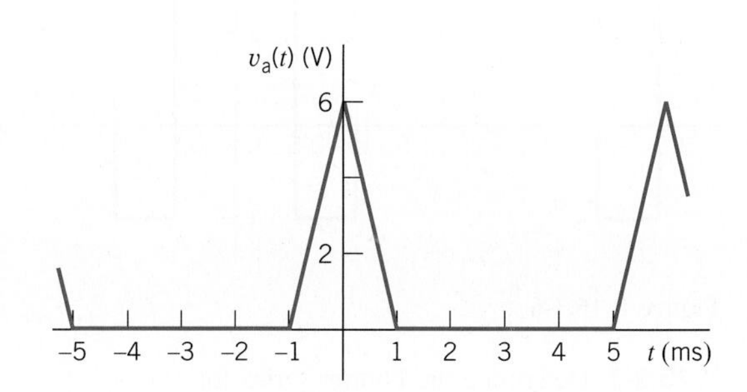

Figure P 15.3-3

P 15.3-4 Determine the Fourier series of the voltage waveform shown in Figure P 15.3-4.

Hint: $v_b(t) = v_a(t - 0.002) - 1$, where $v_a(t)$ is the voltage considered in Problem P 15.4-3.

Answer:
$$v_b(t) = -\frac{1}{2} + \sum_{n=1}^{\infty} \frac{18}{n^2\pi^2}\left(1 - \cos\left(\frac{n\pi}{3}\right)\right)$$
$$\cos\left(n\frac{1000\pi}{3}t - n\frac{2\pi}{3}\right)$$

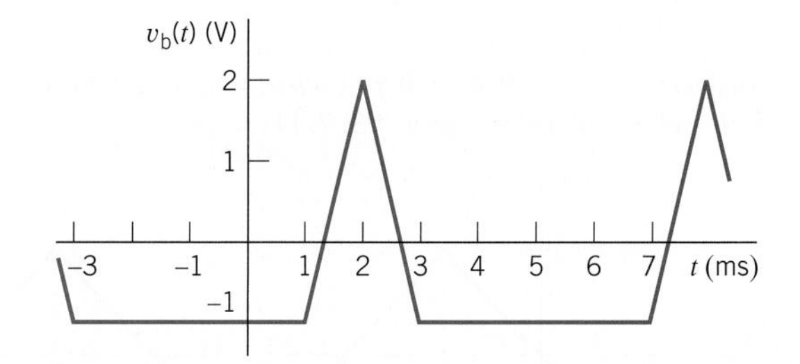

Figure P 15.3-4

P 15.3-5 Find the trigonometric Fourier series of the sawtooth wave, $f(t)$, shown in Figure P 15.3-5.

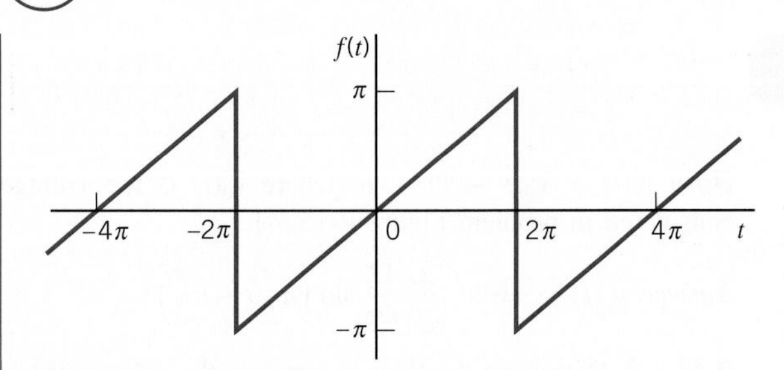

Figure P 15.3-5 Sawtooth wave.

P 15.3-6 Determine the Fourier series for the waveform shown in Figure P 15.3-6. Calculate a_0, a_1, a_2, and a_3.

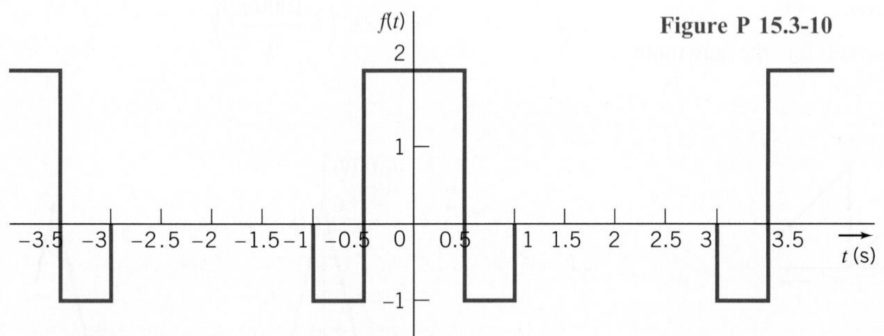

Figure P 15.3-6

P 15.3-7 Determine the Fourier series for

$$f(t) = |A\cos\omega t|$$

P 15.3-8 Find the trigonometric Fourier series for the function of Figure P 15.3-8. The function is the positive portion of a cosine wave.

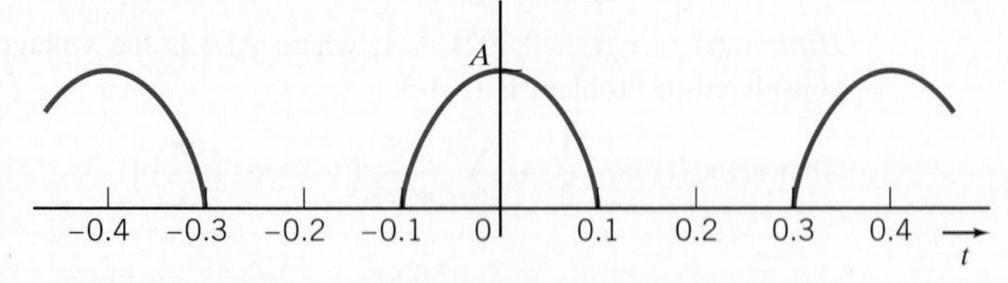

Figure P 15.3-8 Half-wave rectified cosine wave.

P 15.3-9 Determine the Fourier series for $f(t)$ shown in Figure P 15.3-9.

Answer: $a_n = a_0 = 0$; $b_n = 0$ *for even* n, $= 8/(n^2\pi^2)$, *for* $n = 1$, 5, 9, and $= -8/(n^2\pi^2)$ *for* $n = 3, 7, 11$

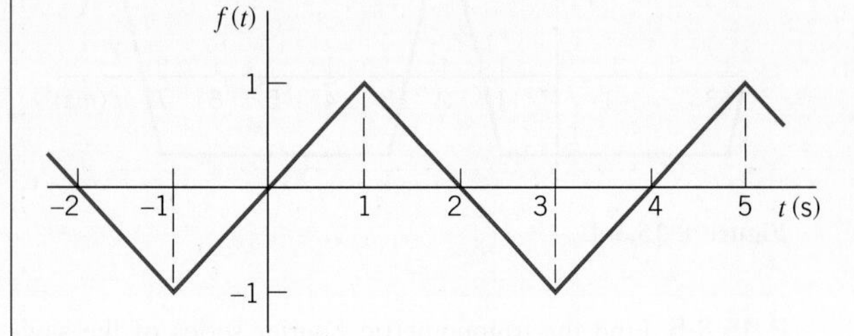

Figure P 15.3-9

P 15.3-10 Determine the Fourier series for the periodic signal shown in Figure P 15.3-10.

Answer:

$$f(t) = \frac{1}{2} + \frac{2}{\pi}\left(\sin t + \frac{1}{3}\sin 3t + \frac{1}{5}\sin 5t + \ldots\right)$$

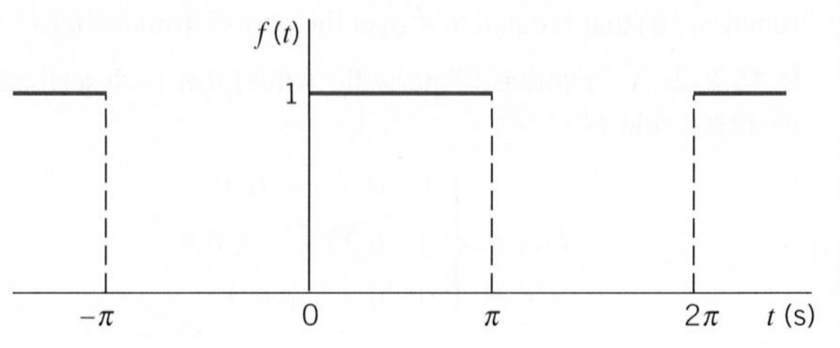

Figure P 15.3-10

Section 15.5 Exponential Form of the Fourier Series

P 15.5-1 Determine the exponential Fourier series of the function

$$f(t) = |A\sin(\pi t)|$$

shown in Figure P 15.5-1.

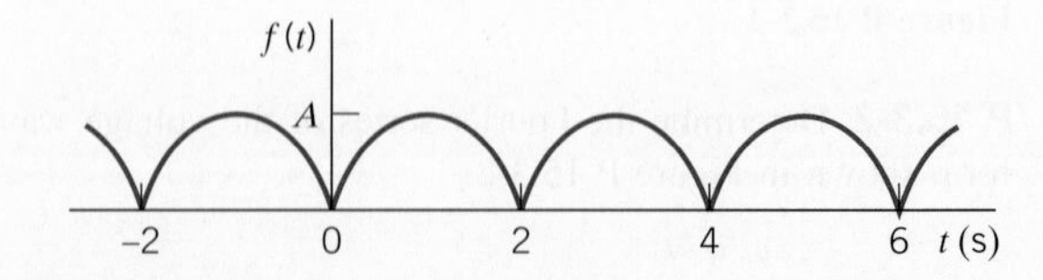

Figure P 15.5-1

P 15.5-2 Determine the exponential Fourier series of the function $f(t)$ shown in Figure P 15.5-2.

Answer: $f(t) = \dfrac{A}{2} + j\dfrac{A}{2\pi}\displaystyle\sum_{\substack{n=-\infty \\ n\neq 0}}^{n=\infty} \frac{1}{n} e^{jn\pi t/T}$

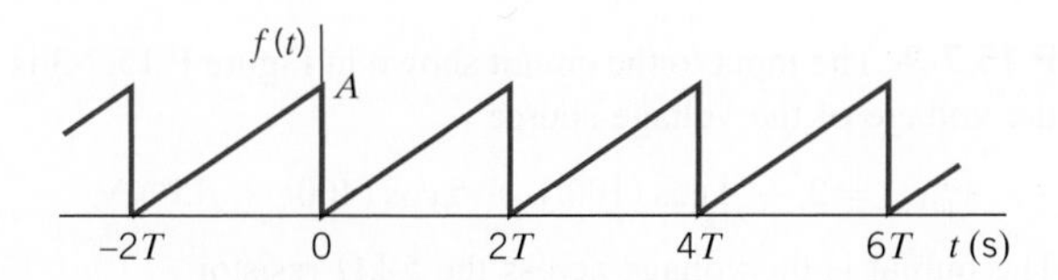

Figure P 15.5-2

P 15.5-3 Determine the exponential Fourier series of the function $f(t)$ shown in Figure P 15.5-3.

Answer: $\mathbf{C}_n = \left(\frac{Ad}{T}\right)\frac{\sin\left(\frac{n\pi d}{T}\right)}{\frac{n\pi d}{T}}$

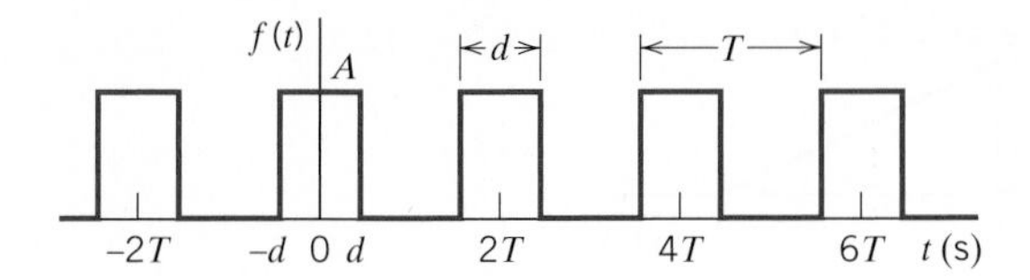

Figure P 15.5-3

P 15.5-4 Consider two periodic functions, $\hat{f}(t)$ and $f(t)$, that have the same period and are related by

$$\hat{f}(t) = af(t - t_d) + b$$

where a, b, and t_d are real constants. Let $\hat{C}_n$ denote the coefficients of the exponential Fourier series of $\hat{f}(t)$ and let C_n denote the coefficients of the exponential Fourier series of $f(t)$. Determine the relationship between $\hat{C}_n$ and C_n.

Answer: $\hat{C}_0 = aC_0 + b$ and $\hat{C}_n = ae^{-jn\omega_0 t_d}C_n \quad n \neq 0$

***P 15.5-5** Determine the exponential form of the Fourier series for the waveform of Figure P 15.3-6.

P 15.5-6 Determine the exponential Fourier series for the waveform of Figure P 15.5-6.

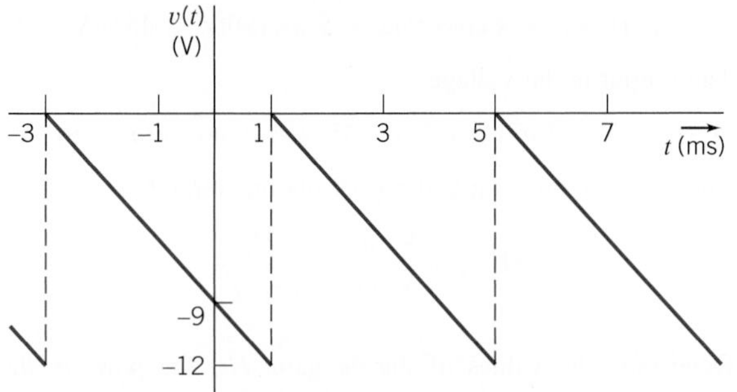

Figure P 15.5-6

***P 15.5-7** A periodic function consists of rising and decaying exponentials of time constants of 0.2 s each and durations of 1 s each as shown in Figure 15.5-7. Determine the exponential Fourier series for this function.

Answer: $C_n = \dfrac{5}{(j\pi n)(5 + j\pi n)}$, $n = 1, 3, 5$

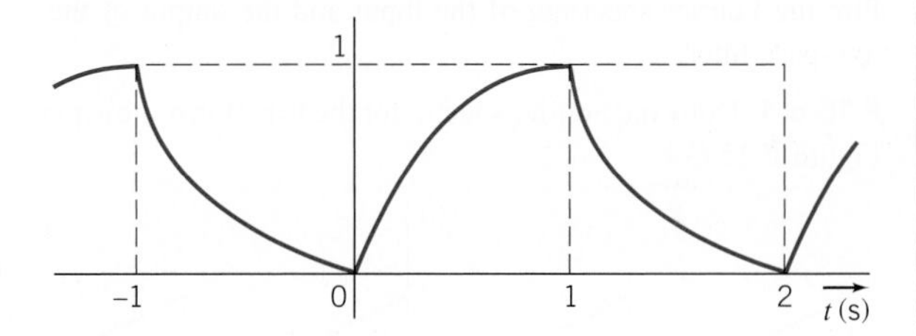

Figure P 15.5-7

Section 15.6 The Fourier Spectrum

P 15.6-1 Determine the cosine-sine Fourier series for the sawtooth waveform shown in Figure P 15.6-1. Draw the Fourier spectra for the first four terms, including magnitude and phase.

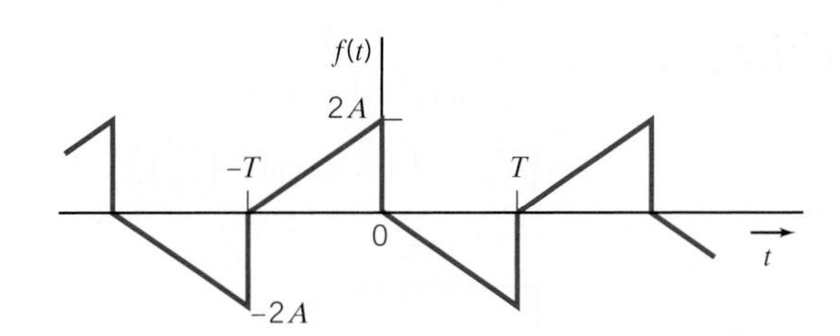

Figure P 15.6-1

P 15.6-2 The load current waveform of the variable-speed motor drive depicted in Figure 15.16-1c is shown in Figure P 15.6-2. The current waveform is a portion of $A \sin \omega_0 t$. Determine the Fourier series of this waveform and draw the line spectra of $|\mathbf{C}_n|$ for the first 10 terms.

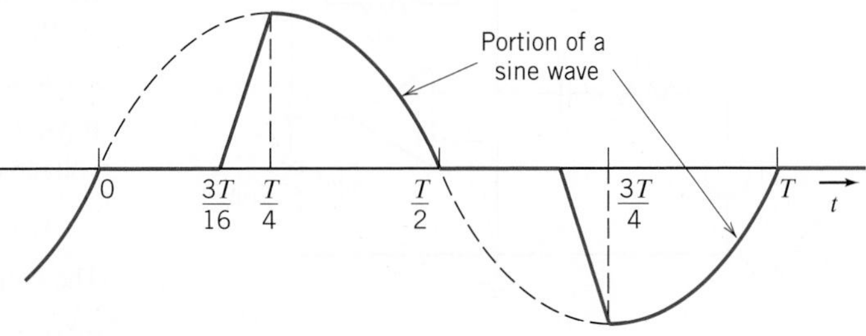

Figure P 15.6-2 The load current of a variable-speed drive.

P 15.6-3 The input to a low-pass filter is

$$v_i(t) = 10\cos t + 10\cos 10t + 10\cos 100t \text{ V}$$

The output of the filter is the voltage $v_o(t)$. The network function of the low-pass filter is

$$\mathbf{H}(\omega) = \frac{\mathbf{V}_o(\omega)}{\mathbf{V}_i(\omega)} = \frac{2}{\left(1 + j\frac{\omega}{5}\right)^2}$$

Plot the Fourier spectrum of the input and the output of the low-pass filter.

P 15.6-4 Draw the Fourier spectra for the waveform shown in Figure P 15.6-4.

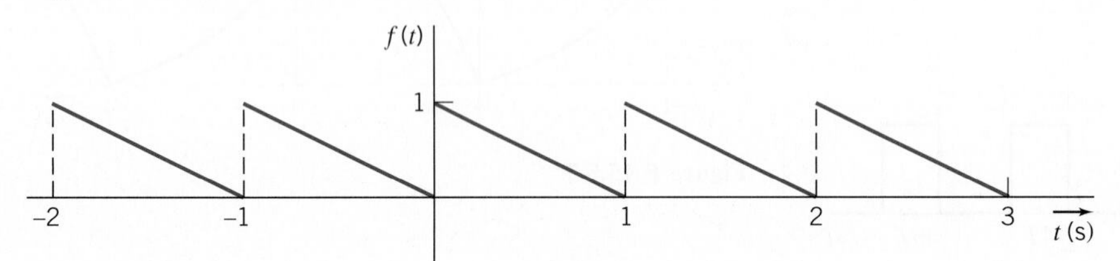

Figure P 15.6-4

Section 15.7 Circuits and Fourier Series

P 15.7-1 Determine the steady-state response, $v_o(t)$, for the circuit shown in Figure P 15.7-1. The input to this circuit is the voltage $v_c(t)$ shown in Figure P 15.3-2.

Answer: $v_o(t) = -6 + \sum_{n=1}^{\infty} \frac{240}{n\pi\sqrt{400 + n^2\pi^2}} \sin\left(n\frac{\pi}{2}t - \left(n\frac{\pi}{2} + \tan^{-1}\left(\frac{n\pi}{20}\right)\right)\right)$

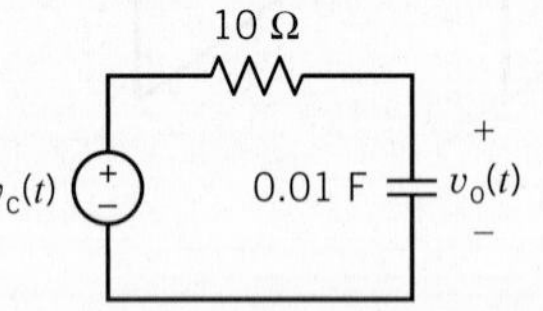

Figure P 15.7-1

P 15.7-2 Determine the steady-state response, $v_o(t)$, for the circuit shown in Figure P 15.7-2. The input to this circuit is the voltage $v_b(t)$, shown in Figure P 15.3-4.

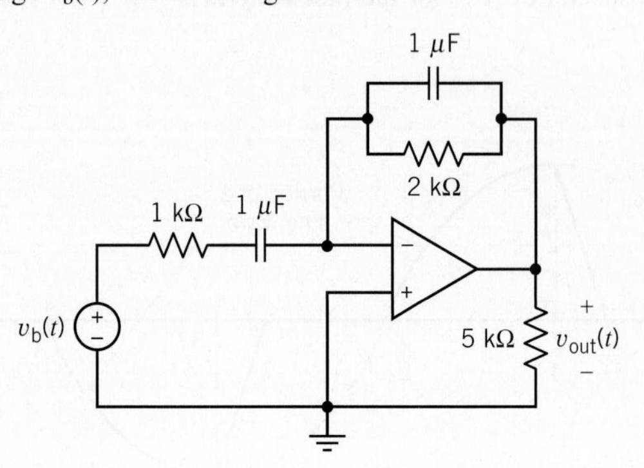

Figure P 15.7-2

P 15.7-3 The input to the circuit shown in Figure P 15.7-3 is the voltage of the voltage source

$$v_{in}(t) = 2 + 4\cos(100t) + 5\cos(400t + 45°) \text{ V}$$

The output is the voltage across the 5-kΩ resistor

$$v_{out}(t) = -5 + 7.071\cos(100t + 135°) + c_4\cos(400t + \theta_4) \text{ V}$$

Determine the values of the resistance, R, the capacitance, C, the coefficient, c_4, and the phase angle, θ_4.

Answers: $R = 25$ kΩ; $C = 0.4$ μF, $c_4 = 3.032$ V, and $\theta_4 = 149°$

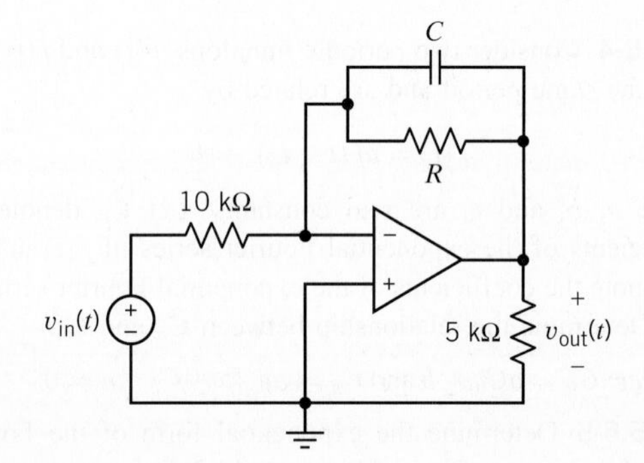

Figure P 15.7-3

P 15.7-4 The input to a circuit is the voltage

$$v_i(t) = 2 + 4\cos(25t) + 5\cos(200t + 45°) \text{ V}$$

The output is the voltage

$$v_o(t) = 5 + 7.071\cos(25t - 45°) + c_4\cos(\omega_4 t + \theta_4) \text{ V}$$

The network function that represents this circuit is

$$\mathbf{H}(\omega) = \frac{\mathbf{V}_o(\omega)}{\mathbf{V}_i(\omega)} = \frac{H_o}{1 + j\frac{\omega}{p}}$$

Determine the values of the dc gain, H_o, the pole, p, the coefficient, c_4, and the phase angle, θ_4.

Answers: $H_o = 2.5$ V/V, $p = 25$ rad/s, $c_4 = 1.55$ V, and $\theta_4 = -38°$

P 15.7-5 The input to the circuit in Figure P 15.7-5 is the voltage of the independent voltage source

$$v_i(t) = 6 + 4\cos(1000t) + 5\cos(3000t + 45°) \text{ V}$$

The output is the voltage across a 500-Ω resistor

$$v_o(t) = 3.75 + 2.34\cos(1000t - 20.5°) + c_3\cos(3000t + \theta_3) \text{ V}$$

Determine the values of the resistance, R_1; the capacitance, C; the coefficient, c_3; and the phase angle, θ_3.

Answer: $R_1 = 300\ \Omega$, $C = 3\ \mu\text{F}$, $c_3 = 0.54$ V, and $\theta_3 = -15.6°$

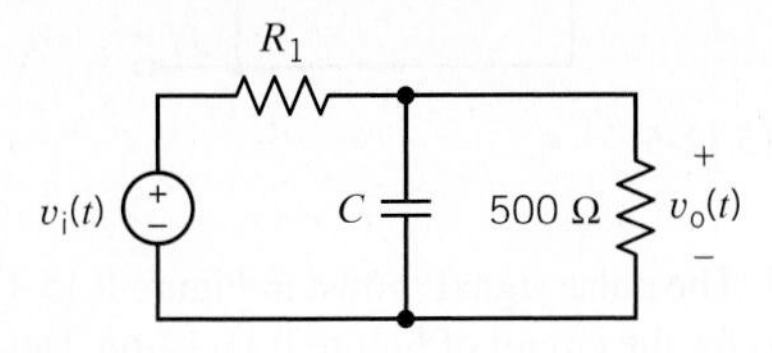

Figure P 15.7-5

P 15.7-6 Find the steady-state response for the output voltage, v_o, for the circuit of Figure P 15.7-6 when $v(t)$ is as described in Figure P 15.5-6.

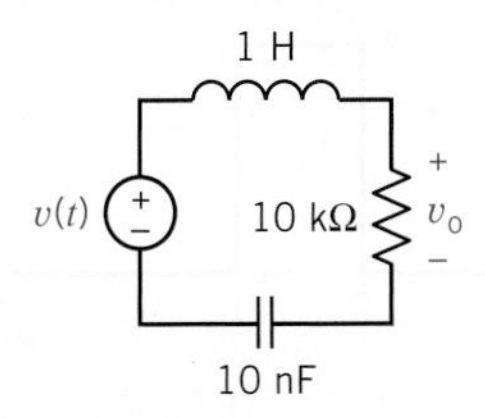

Figure P 15.7-6 An *RLC* circuit.

P 15.7-7 Determine the value of the voltage $v_o(t)$ at $t = 4$ ms when v_{in} is shown in Figure P 15.7-7*a* and the circuit is shown in Figure P 15.7-7*b*.

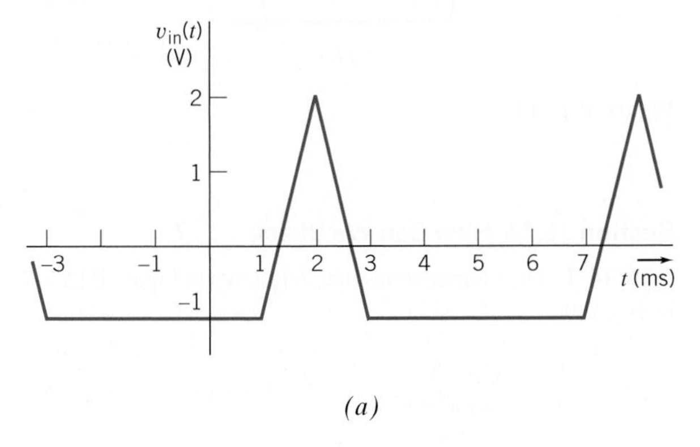

(*a*)

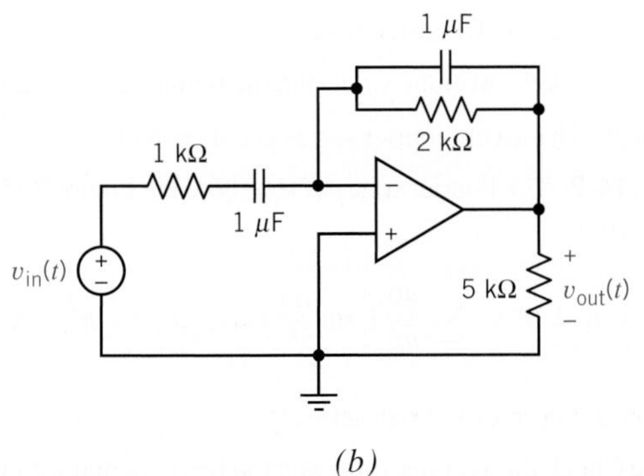

(*b*)

Figure P 15.7-7

Section 15.9 The Fourier Transform

P 15.9-1 Find the Fourier transform of the function

$$f(t) = -u(-t) + u(t)$$

as shown in Figure P 15.9-1. This is called the signum function.

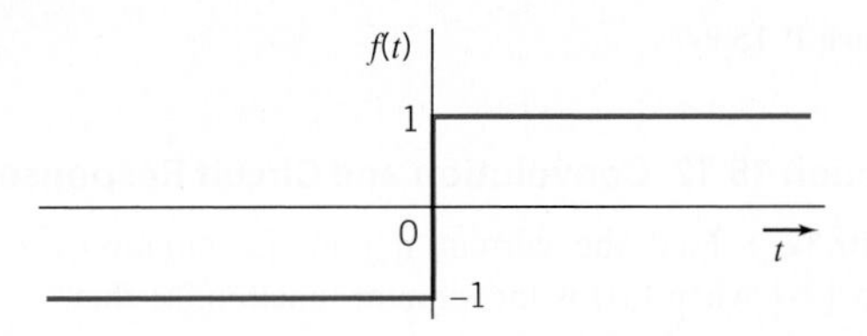

Figure P 15.9-1

P 15.9-2 Find the Fourier transform of $f(t) = Ae^{-at}u(t)$ when $a > 0$.

Answer: $F(\omega) = \dfrac{A}{a + j\omega}$

P 15.9-3 Find the Fourier transform of the waveform shown in Figure P 15.9-3.

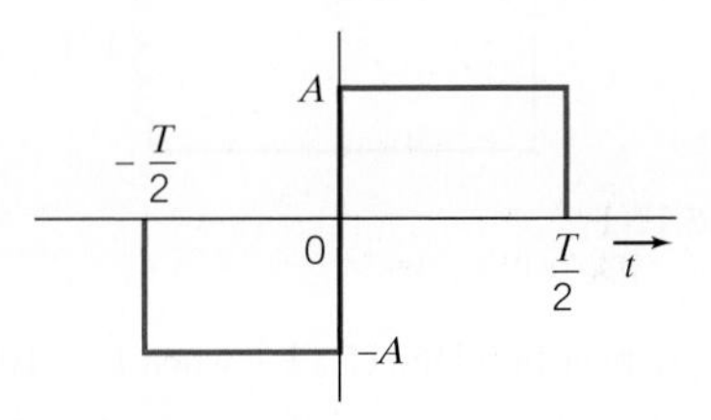

Figure P 15.9-3

P 15.9-4 Determine the Fourier transform of $f(t) = 10 \cos 50\, t$.

Answer: $F(\omega) = 10\pi\delta(\omega - 50) + 10\pi\delta(\omega + 50)$

P 15.9-5 Determine the Fourier transform of the pulse shown in Figure P 15.9-5.

Answer: $F(j\omega) = \frac{2}{\omega}(\sin\omega - \sin 2\omega) + \frac{j2}{\omega}(\cos\omega - \cos 2\omega)$

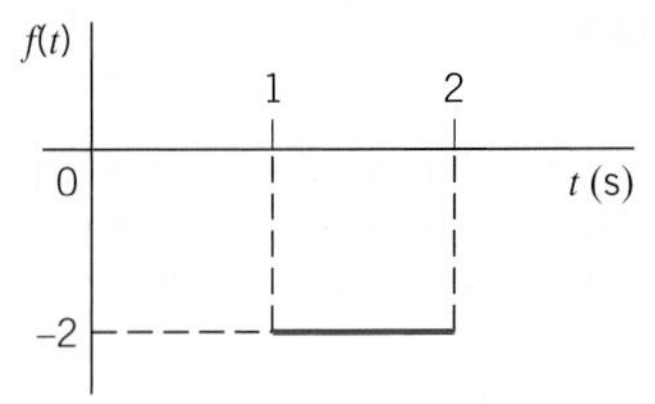

Figure P 15.9-5

P 15.9-6 Determine the Fourier transform of a signal with $f(t) = At/B$ between $t = 0$ and $t = B$ and $f(t) = 0$ elsewhere.

Answer: $F(j\omega) = \dfrac{A}{B}\left[\dfrac{-B}{j\omega}e^{-j\omega B} + \dfrac{1}{\omega^2}e^{-j\omega B} - \dfrac{1}{\omega^2}\right]$

P 15.9-7 Determine the Fourier transform of the waveform $f(t)$ shown in Figure P 15.9-7.

Answer: $F(j\omega) = \dfrac{2}{\omega}(\sin 2\omega - \sin\omega)$

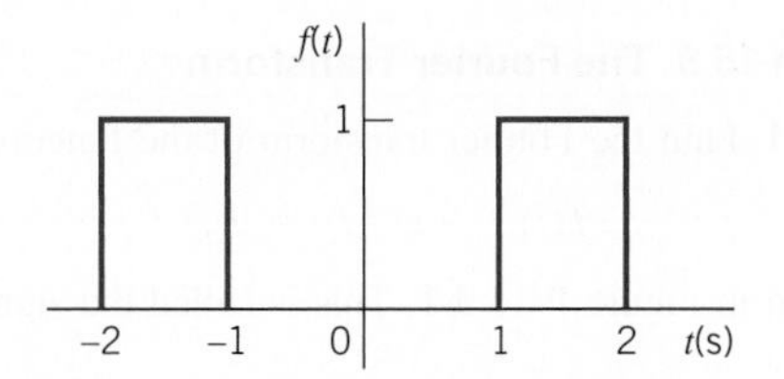

Figure P 15.9-7

Section 15.12 Convolution and Circuit Response

P 15.12-1 Find the current $i(t)$ in the circuit of Figure P 15.12-1 when $i_s(t)$ is the signum function, so that

$$i_s(t) = \begin{cases} +\,40\text{ A} & t > 0 \\ -40\text{ A} & t < 0 \end{cases}$$

Also, sketch $i(t)$.

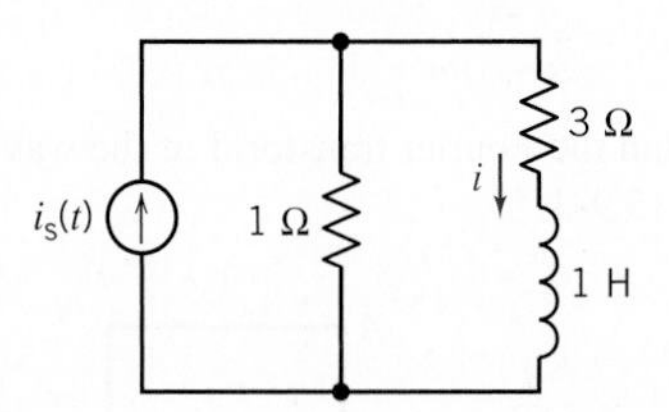

Figure P 15.12-1

P 15.12-2 Repeat Problem 15.12-1 when $i_s = 100 \cos 3t$ A.

P 15.12-3 The voltage source of Figure P 15.12-3 is $v(t) = 10 \cos 2t$ for all t. Calculate $i(t)$ using the Fourier transform.

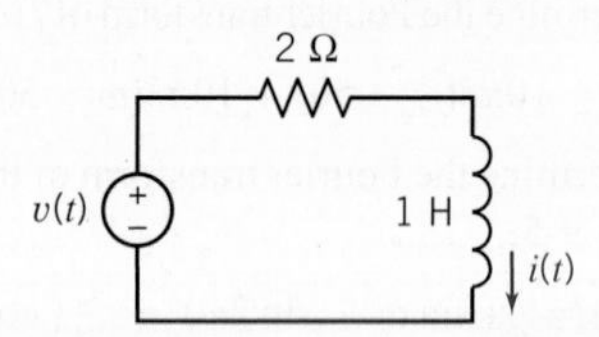

Figure P 15.12-3

P 15.12-4 Find the output voltage $v_o(t)$ using the Fourier transform for the circuit of Figure P 15.12-4 when $v(t) = e^t u(-t) + u(t)$ V.

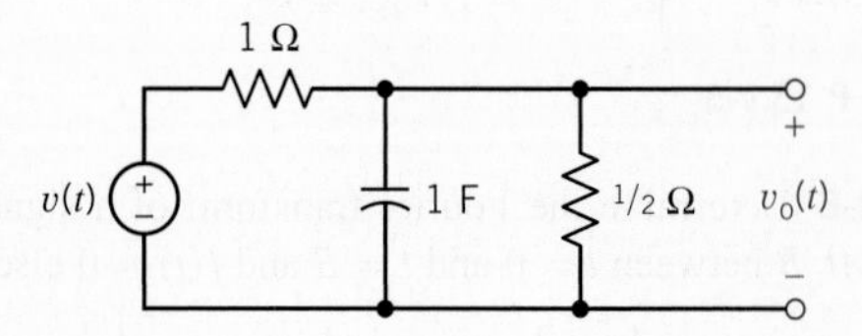

Figure P 15.12-4

P 15.12-5 The voltage source of the circuit of Figure P 15.12-5 is $v(t) = 15e^{-5t}$ V. Find the resistance R when it is known that the energy available in the output signal is two-thirds of the energy of the input signal.

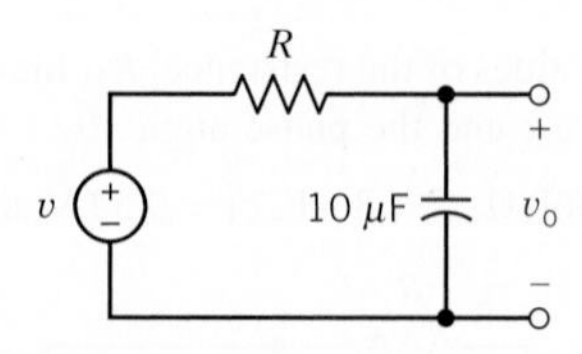

Figure P 15.12-5

P 15.12-6 The pulse signal shown in Figure P 15.12-6*a* is the source $v_s(t)$ for the circuit of Figure P 15.12-6*b*. Determine the output voltage, v_o, using the Fourier transform.

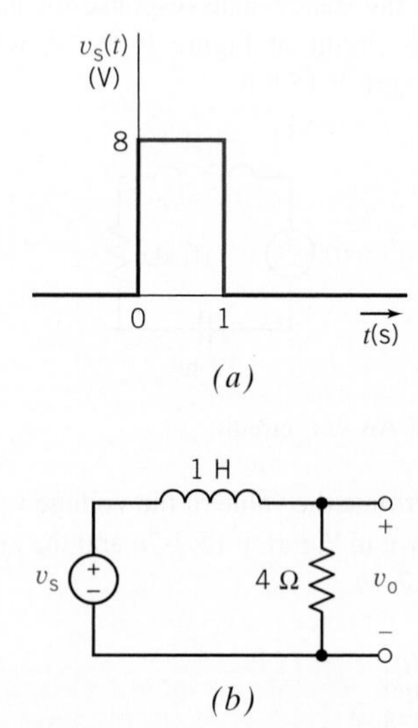

Figure P 15.12-6

Section 15.14 How Can We Check . . . ?

P 15.14-1 The Fourier series of $v_{in}(t)$ shown in Figure P 15.7-7 is given as

$$v_{in}(t) = \frac{1}{2} + \sum_{n=1}^{\infty} \frac{18}{n^2\pi}\left(1 - \cos\frac{n\pi}{3}\right)\cos\left(n\frac{\pi}{3}t - n\frac{2\pi}{3}\right)\text{V}$$

Is this the correct Fourier series?

Hint: Check the average value and the fundamental frequency.

Answer: The given Fourier series is not correct.

P 15.14-2 The Fourier series of $v(t)$ shown in Figure P 15.14-2 is given as

$$v(t) = 9 + \sum_{n=1}^{\infty} \frac{40}{n\pi}\left(\sin\frac{n\pi}{5}\right)\cos\left(n\frac{\pi}{5}t - n\frac{\pi}{5}\right)\text{ V}$$

Is this the correct Fourier series?

Hint: Check the average value and the fundamental frequency.

Answer: The given Fourier series is not correct.

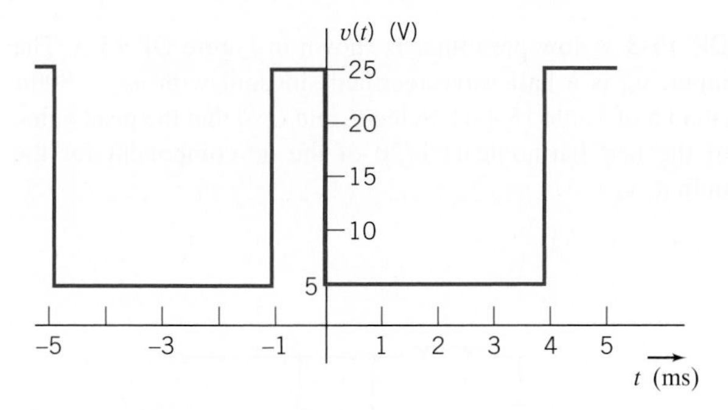

Figure P 15.14-2

P 15.14-3 The Fourier series of $v(t)$ shown in Figure SP 15-2 in the next section is given as

$$v(t) = 2\sum_{n=1}^{\infty} \frac{(-1)^n}{n\pi} \cos(n2\pi t)\ \text{V}$$

Is this the correct Fourier series?

Hint: Check the average value and the fundamental frequency. Check for symmetry.

Answer: The given Fourier series is not correct.

PSpice Problems

SP 15-1 Use PSpice to determine the Fourier coefficients for $v(t)$ shown in Figure SP 15-1.

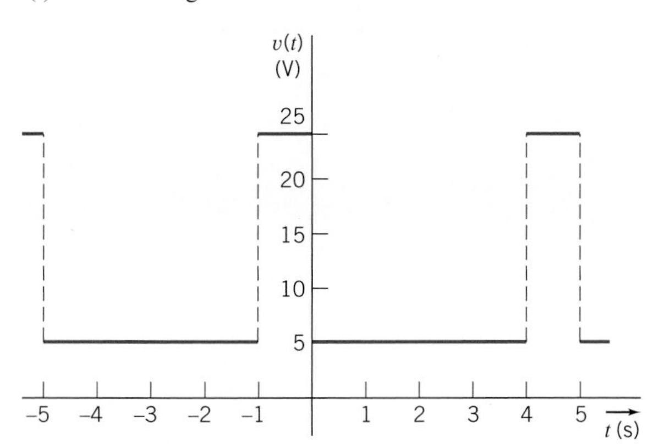

Figure SP 15-1

SP 15-2 Use PSpice to determine the Fourier coei $v(t)$ shown in Figure SP 15-2.

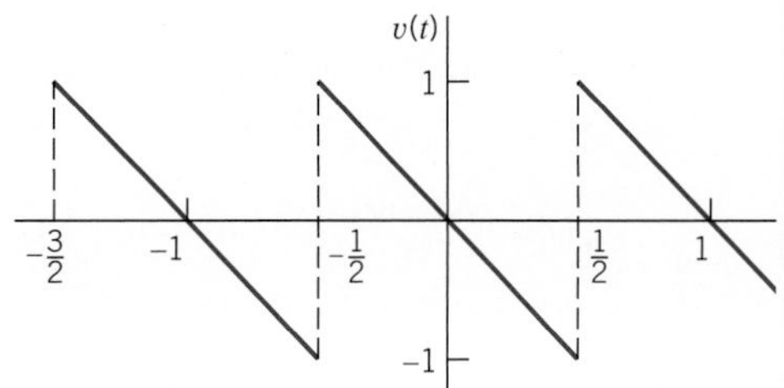

Figure SP 15-2

Design Problems

DP 15-1 A periodic waveform shown in Figure DP 15-1*a* is the input signal of the circuit shown in Figure DP 15-1*b*. Select the capacitance C so that the magnitude of the third harmonic of $v_2(t)$ is less than 1.4 V and greater than 1.3 V. Write the equation describing the third harmonic of $v_2(t)$ for the value of C selected.

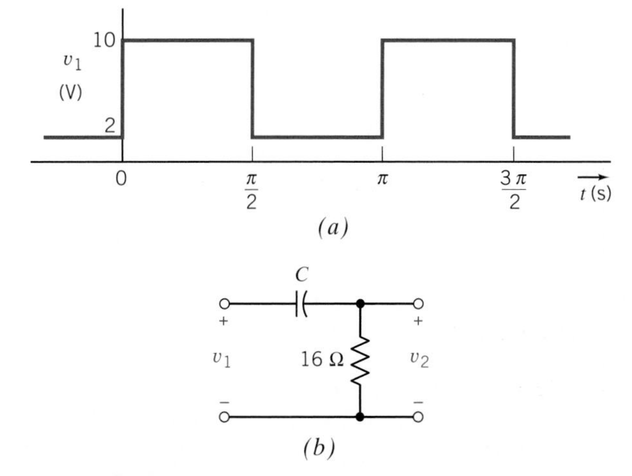

Figure DP 15-1

DP 15-2 A dc laboratory power supply uses a nonlinear circuit to convert a sinusoidal voltage obtained from the wall plug to a constant dc voltage. The wall plug voltage is $A \sin \omega_0 t$, where $f_0 = 60$ Hz and $A = 160$ V. The voltage is then rectified so that $v_s = |A \sin \omega_0 t|$. Using the filter circuit of Figure DP 15-2, determine the required inductance L so that the magnitude of each harmonic (ripple) is less than 4 percent of the dc component of the output voltage.

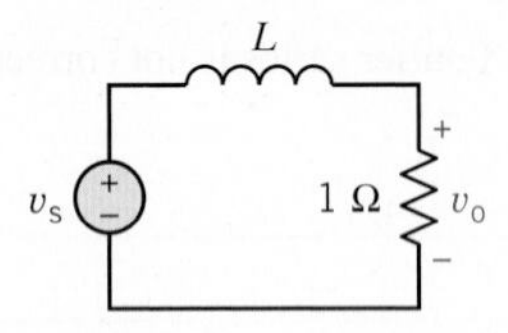

Figure DP 15-2 An *RL* circuit.

DP 15-3 A low-pass filter is shown in Figure DP 15-3. The input, v_s, is a half-wave rectified sinusoid with $\omega_0 = 800\pi$ (item 5 of Table 15.4-1). Select L and C so that the peak value of the first harmonic is 1/20 of the dc component for the output, v_o.

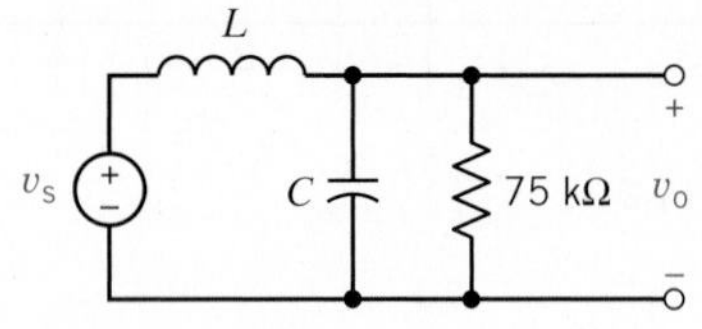

Figure DP 15-3 An *RLC* circuit.

CHAPTER 16

Filter Circuits

IN THIS CHAPTER

16.1 Introduction
16.2 The Electric Filter
16.3 Filters
16.4 Second-Order Filters
16.5 High-Order Filters
16.6 Simulating Filter Circuits Using PSpice
16.7 How Can We Check . . . ?
16.8 **DESIGN EXAMPLE**—Anti-Aliasing Filter
16.9 Summary
Problems
PSpice Problems
Design Problems

16.1 INTRODUCTION

Transfer functions are used to characterize linear circuits. In a previous chapter, we learned how to analyze a circuit so that we could determine its transfer function. In this chapter, we learn how to design a circuit to have a specified transfer function. This design problem does not have a unique solution. There are many ways to obtain a circuit from a specified transfer function. A popular strategy is to design the circuit to be a cascade connection of second-order filter stages. This is the strategy we will use in this chapter.

The problem of designing a circuit that will have a specified transfer function is called filter design. In this chapter we will learn the vocabulary of filter design and describe second-order filter stages. Finally, we will learn how to connect these filter stages to obtain a circuit that has a specified transfer function.

16.2 THE ELECTRIC FILTER

The concept of a filter was conceived early in human history. A paper filter was used to remove dirt and unwanted substances from water and wine. A porous material, such as paper, can serve as a mechanical filter. Mechanical filters are used to remove unwanted constituents, such as suspended particles, from a liquid. In a similar manner, an electric filter can be used to eliminate unwanted constituents, such as electrical noise, from an electrical signal.

The electrical filter was independently invented in 1915 by George Campbell in the United States and K. W. Wagner in Germany. With the rise of radio between 1910 and 1920, a need emerged to reduce the effect of static noise at the radio receiver. As regular radio broadcasting emerged in the 1920s, Campbell and others developed the *RLC* filter, using inductors, capacitors, and resistors. These filters are called *passive filters* because they consist of passive elements. The theory required to design passive filters was developed in the 1930s by S. Darlington, S. Butterworth, and E. A. Guillemin. The Butterworth low-pass filter was reported in *Wireless Engineering* in 1930 (Butterworth, 1930).

When active devices, typically op amps, are incorporated into an electric filter, the filter is called an *active filter*. Because inductors are relatively large and heavy, active filters are usually constructed without inductors—using, for example, only op amps, resistors, and capacitors. The first practical active-*RC* filters were developed during World War II and were documented in a classic paper by R. P. Sallen and E. L. Key (Sallen and Key, 1955).

16.3 FILTERS

We begin by considering an **ideal filter**. For convenience, suppose that both the input and output of this filter are voltages. This ideal filter separates its input voltage into two parts. One part is passed, unchanged, to the output; the other part is eliminated. In other words, the output of an ideal filter is an exact copy of part of the filter input.

This is a familiar use of the word **filter**. For example, we expect an automotive oil filter to separate a mixture of oil and dirt into two parts: oil and dirt. Ideally, the oil filter passes one part of its input, the oil, to its output without changing it in any way. The other part of the input, the dirt, should be completely eliminated. The oil filter stops the dirt from getting to the output.

To understand how an electric filter works, consider an input voltage:

$$v_{\mathrm{i}}(t) = \cos \omega_1 t + \cos \omega_2 t + \cos \omega_3 t$$

This input consists of a sum of sinusoids, each at a different frequency. (For example, periodic voltages can be represented in this way using the Fourier series.) The filter separates the input voltage into two parts, using frequency as the basis for separation. There are several ways of separating this input into two parts and, correspondingly, several types of ideal filter. Table 16.3-1 illustrates the common filter types. Consider the ideal **low-pass filter**, shown in row 1 of the table. The network function of the ideal low-pass filter is

$$\mathbf{H}(\omega) = \begin{cases} 1 \angle 0^\circ & \omega < \omega_{\mathrm{c}} \\ 0 & \omega > \omega_{\mathrm{c}} \end{cases} \tag{16.3-1}$$

The frequency ω_{c} is called the **cutoff frequency**. The cutoff frequency separates the frequency range $\omega < \omega_{\mathrm{c}}$, called the **pass-band**, from the frequency range $\omega > \omega_{\mathrm{c}}$, called the **stop-band**. Those components of the input that have frequencies in the pass-band experience unity gain and zero phase shift. These terms are passed, unchanged, to the output of the filter. Components of the input that have frequencies in the stop-band experience a gain equal to zero. These terms are eliminated or stopped. An ideal filter separates its input into

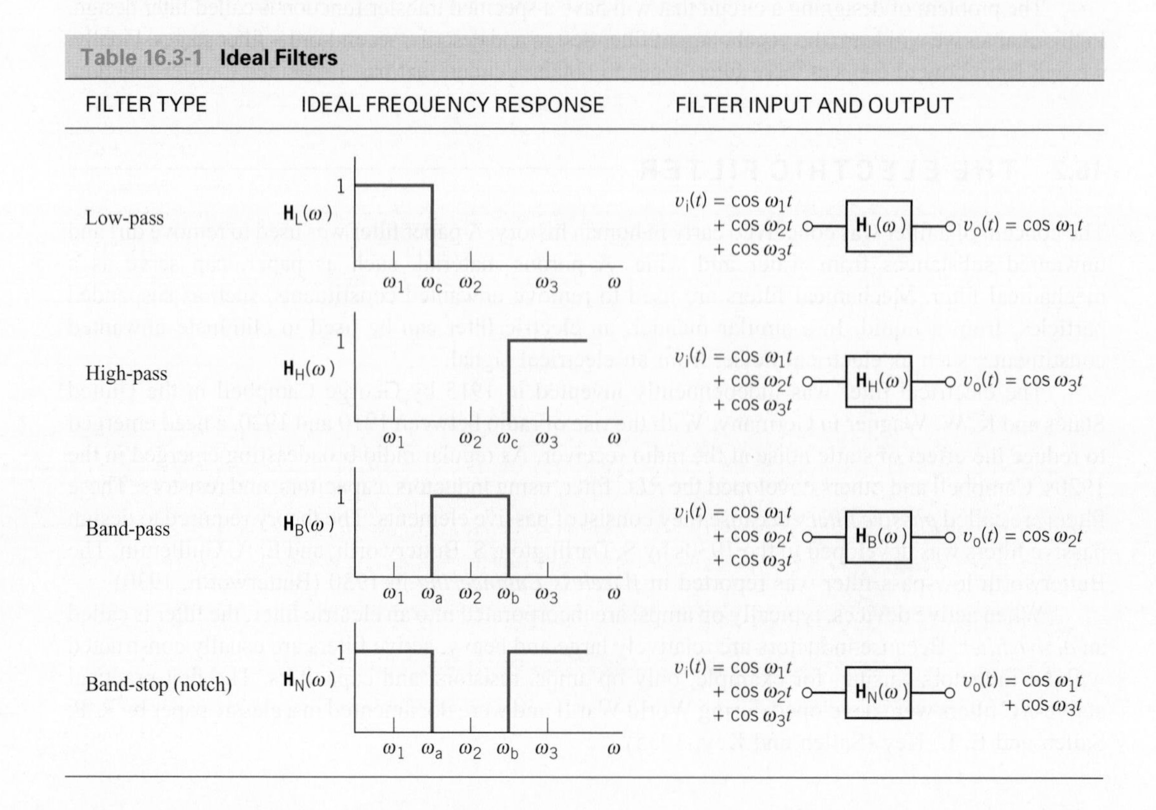

Table 16.3-1 Ideal Filters

FILTER TYPE	IDEAL FREQUENCY RESPONSE	FILTER INPUT AND OUTPUT
Low-pass	$\mathbf{H}_{\mathrm{L}}(\omega)$; 1; ω_1 ω_{c} ω_2 ω_3 ω	$v_{\mathrm{i}}(t) = \cos \omega_1 t + \cos \omega_2 t + \cos \omega_3 t$ → $\mathbf{H}_{\mathrm{L}}(\omega)$ → $v_{\mathrm{o}}(t) = \cos \omega_1 t$
High-pass	$\mathbf{H}_{\mathrm{H}}(\omega)$; 1; ω_1 ω_2 ω_{c} ω_3 ω	$v_{\mathrm{i}}(t) = \cos \omega_1 t + \cos \omega_2 t + \cos \omega_3 t$ → $\mathbf{H}_{\mathrm{H}}(\omega)$ → $v_{\mathrm{o}}(t) = \cos \omega_3 t$
Band-pass	$\mathbf{H}_{\mathrm{B}}(\omega)$; 1; ω_1 ω_{a} ω_2 ω_{b} ω_3 ω	$v_{\mathrm{i}}(t) = \cos \omega_1 t + \cos \omega_2 t + \cos \omega_3 t$ → $\mathbf{H}_{\mathrm{B}}(\omega)$ → $v_{\mathrm{o}}(t) = \cos \omega_2 t$
Band-stop (notch)	$\mathbf{H}_{\mathrm{N}}(\omega)$; 1; ω_1 ω_{a} ω_2 ω_{b} ω_3 ω	$v_{\mathrm{i}}(t) = \cos \omega_1 t + \cos \omega_2 t + \cos \omega_3 t$ → $\mathbf{H}_{\mathrm{N}}(\omega)$ → $v_{\mathrm{o}}(t) = \cos \omega_1 t + \cos \omega_3 t$

Table 16.3-2 **Denominators of Butterworth Low-Pass Filters with a Cutoff Frequency $\omega_c = 1$ rad/s**

ORDER	DENOMINATOR, $D(S)$
1	$s+1$
2	$s^2+1.414s+1$
3	$(s+1)(s^2+s+1)$
4	$(s^2+0.765s+1)(s^2+1.848s+1)$
5	$(s+1)(s^2+0.618s+1)(s^2+1.618s+1)$
6	$(s^2+0.518s+1)(s^2+1.414s+1)(s^2+1.932s+1)$
7	$(s+1)(s^2+0.445s+1)(s^2+1.247s+1)(s^2+1.802s+1)$
8	$(s^2+0.390s+1)(s^2+1.111s+1)(s^2+1.663s+1)(s^2+1.962s+1)$
9	$(s+1)(s^2+0.347s+1)(s^2+s+1)(s^2+1.532s+1)(s^2+1.879s+1)$
10	$(s^2+0.313s+1)(s^2+0.908s+1)(s^2+1.414s+1)(s^2+1.782s+1)(s^2+1.975s+1)$

two parts: those terms that have frequencies in the pass-band and those terms that have frequencies in the stop-band. The output of the filter consists of those terms with frequencies in the pass-band.

Unfortunately, ideal filter circuits don't exist. (This fact can be proved by calculating the impulse response of the ideal filter by taking the inverse Laplace transform of the transfer function. The impulse response of an ideal filter would have to exist before the impulse itself. That is, the response would have to occur before the input that caused the response. Because that can't happen, ideal filter circuits don't exist.) Filters are circuits that approximate ideal filters. Filters divide their inputs into two parts, the terms in the pass-band and the terms in the stop-band. The terms in the pass-band experience a gain that is approximately 1 and experience some phase shift. These terms are passed to the output, but they are changed a little. The terms in the stop-band experience a small gain that isn't quite zero. Because these terms aren't eliminated entirely, some small residue of these terms shows up in the filter output.

Butterworth transfer functions have magnitude frequency responses that approximate the frequency response of an ideal filter. Butterworth low-pass transfer functions are given by

$$H_L(s) = \frac{\pm 1}{D(s)} \tag{16.3-2}$$

We can choose either $+1$ or -1 for the numerator of $H_L(s)$. The polynomial $D(s)$ depends on the cutoff frequency and on the order of the filter. These polynomials, called Butterworth polynomials, are tabulated in Table 16.3-2 for $\omega_c = 1$ rad/s. There is a trade-off involving the order of the filter. The higher the order, the more accurately the filter frequency response approximates the frequency response of an ideal filter; that's good. The higher the filter order, the more complicated the circuit required to build the filter; that's not good.

EXAMPLE 16.3-1 Filter Order

We wish to design a low-pass filter that will approximate an ideal low-pass filter with $\omega_c = 1$ rad/s. Compare the fourth-order Butterworth low-pass filter to the eighth-order Butterworth low-pass filter.

Solution

The fourth row of Table 16.3-2 indicates that the transfer of the fourth-order Butterworth filter is

$$H_4(s) = \frac{1}{(s^2+0.765s+1)(s^2+1.848s+1)} = \frac{1}{(s^2+0.765s+1)} \times \frac{1}{(s^2+1.848s+1)}$$

Similarly, the eighth row of Table 16.3-2 indicates that the transfer function of the eighth-order Butterworth filter is

$$\begin{aligned} H_8(s) &= \frac{1}{(s^2+0.390s+1)(s^2+1.111s+1)(s^2+1.663s+1)(s^2+1.962s+1)} \\ &= \frac{1}{(s^2+0.390s+1)} \times \frac{1}{(s^2+1.111s+1)} \times \frac{1}{(s^2+1.663s+1)} \times \frac{1}{(s^2+1.962s+1)} \end{aligned}$$

Figure 16.3-1 shows the magnitude frequency response plots for these two filters. Both frequency responses show unity gain when $\omega \ll 1$ and a gain of zero when $\omega \gg 1$. Thus, both filters approximate an ideal low-pass filter with $\omega_c = 1$ rad/s. The eighth-order filter makes the transition from the pass-band to the stop-band more quickly, providing a better approximation to the ideal low-pass filter.

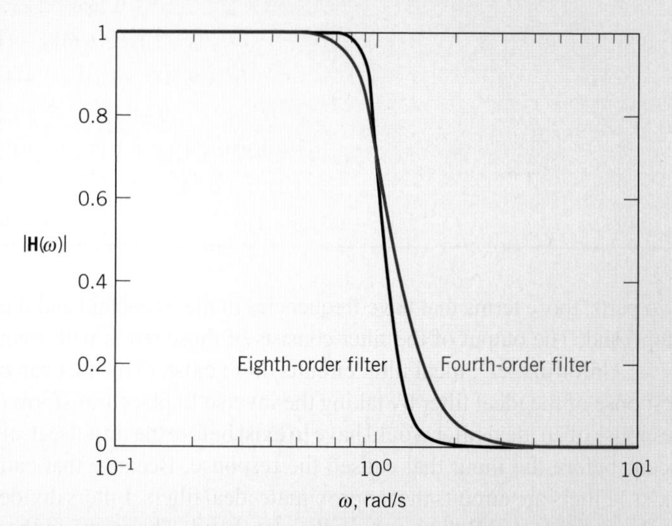

FIGURE 16.3-1 A comparison of the frequency responses of fourth-order and eighth-order Butterworth low-pass filters with $\omega_c = 1$ rad/s.

The transfer function of the fourth-order filter has been expressed as the product of two second-order transfer functions, whereas the transfer function of the eighth-order filter has been expressed as the product of four second-order transfer functions. Each of these second-order transfer functions will be implemented by a second-order circuit. Because all of these second-order circuits will be quite similar, it is reasonable to expect that the eighth-order circuit will be about twice as large as the fourth-order filter. That means twice as many parts, twice the power consumption, twice the assembly cost, twice the space, and so on.

The eighth-order filter performs better, but it costs more. In some applications, the improved performance of the eighth-order filter justifies the additional cost, whereas in other applications, it does not.

EXAMPLE 16.3-2 Frequency Scaling

Determine the transfer function of a third-order Butterworth low-pass filter having a cutoff frequency equal to 500 rad/s.

Solution

Equation 16.3-2 and Table 16.3-2 provide a third-order Butterworth low-pass filter with a cutoff frequency equal to 1 rad/s:

$$H_n(s) = \frac{1}{(s+1)(s^2+s+1)}$$

A technique called **frequency scaling** is used to adjust the cutoff frequency to $\omega_c = 500$ rad/s. Frequency scaling can be accomplished by replacing each s in $H_n(s)$ by s/ω_c. That is,

$$H(s) = \frac{1}{\left(\dfrac{s}{\omega_c}+1\right)\left(\left(\dfrac{s}{\omega_c}\right)^2+\dfrac{s}{\omega_c}+1\right)}$$

In this case, $\omega_c = 500$ rad/s, so

$$
\begin{aligned}
H(s) &= \frac{1}{\left(\frac{s}{500}+1\right)\left(\left(\frac{s}{500}\right)^2+\frac{s}{500}+1\right)} \\
&= \frac{500^3}{(s+500)\left(s^2+500s+500^2\right)} \\
&= \frac{125{,}000{,}000}{(s+500)(s^2+500s+250{,}000)}
\end{aligned}
$$

$H(s)$ is the transfer function of a third-order Butterworth low-pass filter having a cutoff frequency equal to 500 rad/s.

EXERCISE 16.3-1 Find the transfer function of a first-order Butterworth low-pass filter having a cutoff frequency equal to 1250 rad/s.

Answer: $H(s) = \dfrac{1}{\dfrac{s}{1250}+1} = \dfrac{1250}{s+1250}$

16.4 SECOND-ORDER FILTERS

Second-order filters are important for two reasons. First, they provide an inexpensive approximation to ideal filters. Second, they are used as building blocks for more expensive filters that provide more accurate approximations to ideal filters.

The frequency response of second-order filters is characterized by three filter parameters: the gain k, the corner frequency ω_0, and the **quality factor** Q. Filter circuits are designed by choosing the values of the circuit elements in such a way as to obtain the required values of k, ω_0, and Q.

A second-order low-pass filter is a circuit that has a transfer function of the form

$$
H_L(s) = \frac{k{\omega_0}^2}{s^2+\frac{\omega_0}{Q}s+{\omega_0}^2} \tag{16.4-1}
$$

This transfer function is characterized by three parameters: the dc gain k, the corner frequency ω_0, and the quality factor Q. When this circuit is stable, that is, when both $\omega_0 > 0$ and $Q > 0$, the network function can be obtained by letting $s = j\omega$.

$$
\mathbf{H}_L(\omega) = \frac{k{\omega_0}^2}{-\omega^2+j\frac{\omega_0}{Q}\omega+{\omega_0}^2}
$$

The gain of the filter is given by

$$
\begin{aligned}
|\mathbf{H}_L(\omega)| &= \frac{k{\omega_0}^2}{\sqrt{\left(\omega_0^2-\omega^2\right)^2+\left(\frac{\omega_0}{Q}\omega\right)^2}} \\
&\cong \begin{cases} k & \omega \ll \omega_0 \\ 0 & \omega \gg \omega_0 \end{cases}
\end{aligned}
$$

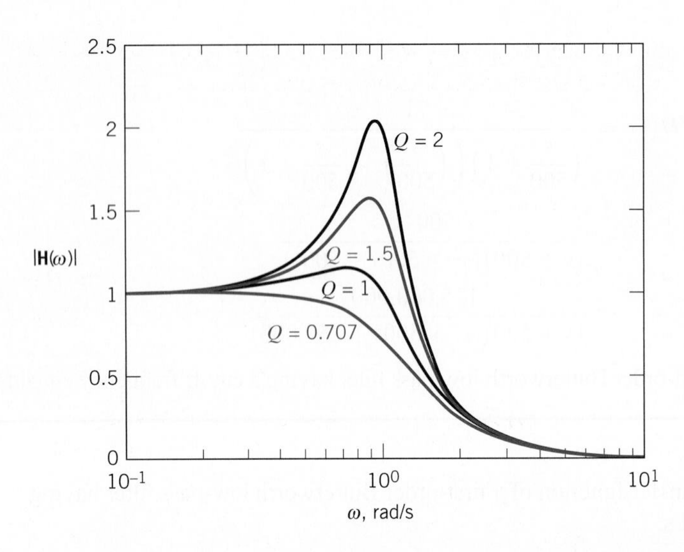

FIGURE 16.4-1 Frequency responses of second-order low-pass filters with four values of Q ($\omega_c = 1$ rad/s).

When $k = 1$, this frequency response approximates the frequency response of an ideal low-pass filter with a cutoff frequency of $\omega_c = \omega_0$. When $k \neq 1$, the low-pass filter approximates an ideal low-pass filter together with an amplifier having gain equal to k. The quality factor, Q, controls the shape of the frequency response during the transition from pass-band to stop-band. Figure 16.4-1 shows the frequency response of a second-order low-pass filter ($k = 1$ and $\omega_c = \omega_0 = 1$) for several choices of Q. A Butterworth approximation to the ideal low-pass filter is obtained by choosing $Q = 0.707$.

Table 16.4-1 Second-Order *RLC* Filters

FILTER TYPE	CIRCUIT	TRANSFER FUNCTION	DESIGN EQUATIONS
Low-pass	R, L, C, $v_i(t)$, $v_o(t)$	$H(s) = \dfrac{\dfrac{1}{LC}}{s^2 + \dfrac{R}{L}s + \dfrac{1}{LC}}$	$\omega_0 = \dfrac{1}{\sqrt{LC}}$ $Q = \dfrac{1}{R}\sqrt{\dfrac{L}{C}}$ $k = 1$
High-pass	C, R, L, $v_i(t)$, $v_o(t)$	$H(s) = \dfrac{s^2}{s^2 + \dfrac{R}{L}s + \dfrac{1}{LC}}$	$\omega_0 = \dfrac{1}{\sqrt{LC}}$ $Q = \dfrac{1}{R}\sqrt{\dfrac{L}{C}}$ $k = 1$
Band-pass	C, L, R, $v_i(t)$, $v_o(t)$	$H(s) = \dfrac{\dfrac{R}{L}s}{s^2 + \dfrac{R}{L}s + \dfrac{1}{LC}}$	$\omega_0 = \dfrac{1}{\sqrt{LC}}$ $Q = \dfrac{1}{R}\sqrt{\dfrac{L}{C}}$ $k = 1$
Band-stop (notch)	R, L, C, $v_i(t)$, $v_o(t)$	$H(s) = \dfrac{s^2 + \dfrac{1}{LC}}{s^2 + \dfrac{R}{L}s + \dfrac{1}{LC}}$	$\omega_0 = \dfrac{1}{\sqrt{LC}}$ $Q = \dfrac{1}{R}\sqrt{\dfrac{L}{C}}$ $k = 1$

Table 16.4-1 provides *RLC* circuits that can be used as second-order filters. Consider the low-pass filter shown in the first row of the table. The transfer function of this circuit is

$$H(s) = \frac{\frac{1}{LC}}{s^2 + \frac{R}{L}s + \frac{1}{LC}} \tag{16.4-2}$$

The relationship between the circuit parameters R, L, and C and the filter parameters k, ω_0, and Q is obtained by comparing Eq. 16.4-2 to Eq. 16.4-1. First, compare the constant terms in the denominators to see that the cutoff frequency of the filter is given by

$$\omega_0 = \frac{1}{\sqrt{LC}}$$

Next, compare the coefficients of s in the denominators to see that

$$\frac{\omega_0}{Q} = \frac{R}{L}$$

Solving these two equations for Q gives

$$Q = \frac{1}{R}\sqrt{\frac{L}{C}}$$

Finally, comparing the numerators gives

$$k{\omega_0}^2 = \frac{1}{LC}$$

So the dc gain is

$$k = 1$$

Notice that ω_0 and Q are determined by the values of R, L, and C but that k is always 1.

Many different circuits are used to build second-order filters. One of the popular filter circuits is called the **Sallen-Key filter**. Table 16.4-2 provides the information required to design Sallen-Key filters.

Table 16.4-2 Sallen-Key Filters

FILTER TYPE	CIRCUIT	DESIGN EQUATIONS
Low-pass	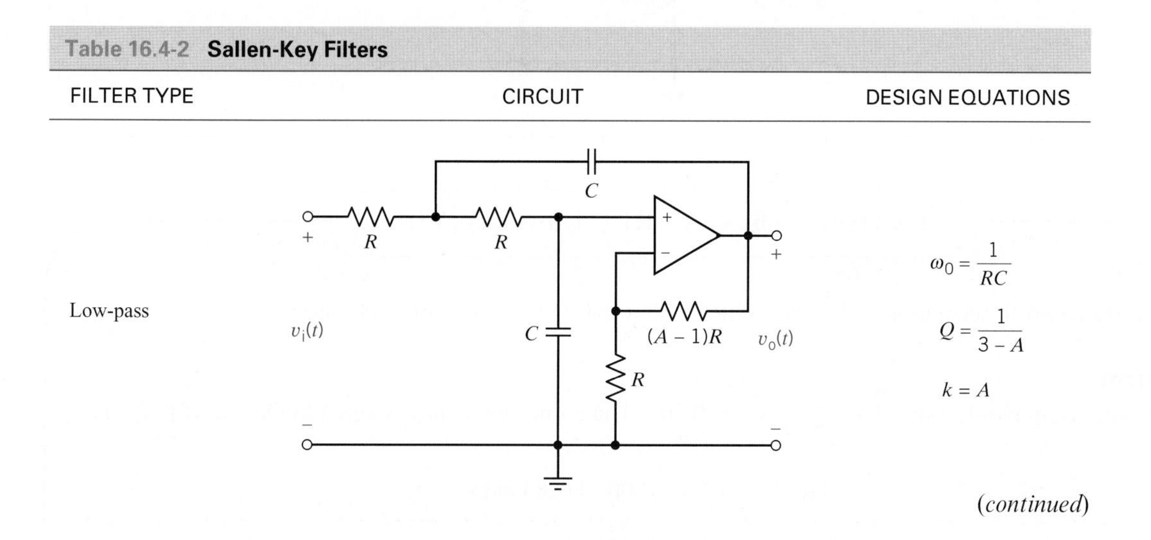	$\omega_0 = \frac{1}{RC}$ $Q = \frac{1}{3 - A}$ $k = A$

(continued)

Table 16.4-2 (*Continued*)

FILTER TYPE	CIRCUIT	DESIGN EQUATIONS
High-pass	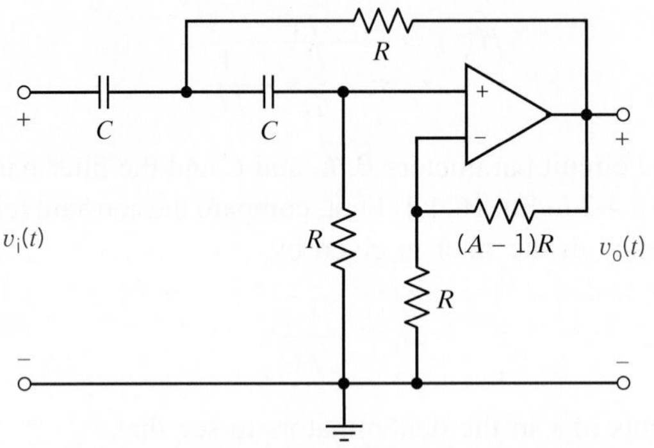	$\omega_0 = \frac{1}{RC}$ $Q = \frac{1}{3 - A}$ $k = A$
Band-pass	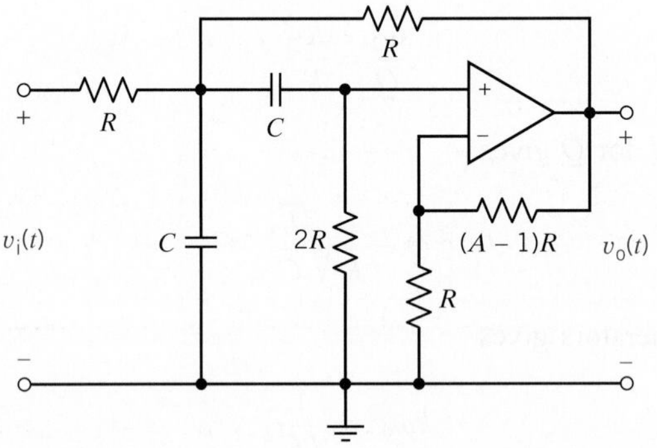	$\omega_0 = \frac{1}{RC}$ $Q = \frac{1}{3 - A}$ $k = AQ$
Band-stop (notch)	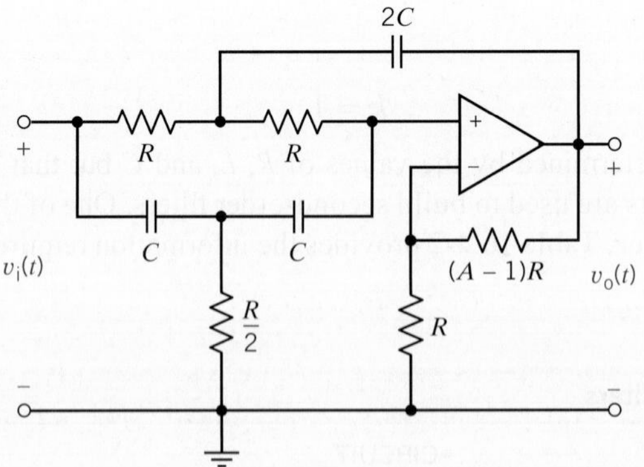	$\omega_0 = \frac{1}{RC}$ $Q = \frac{1}{4 - 2A}$ $k = A$

EXAMPLE 16.4-1 RLC Low-Pass Filter

Design a Butterworth second-order low-pass filter with a cutoff frequency of 1000 hertz.

Solution

Second-order Butterworth filters have $Q = \frac{1}{\sqrt{2}} = 0.707$. The corner frequency is equal to the cutoff frequency, that is,

$$\omega_0 = \omega_c = 2\pi \cdot 1000 = 6283 \text{ rad/s}$$

The *RLC* circuit shown in the first row of Table 16.4-1 can be used to design the required low-pass filter. The design equations are

$$\frac{1}{\sqrt{LC}} = \omega_0 = 6283 \text{ rad/s}$$

and

$$\frac{1}{R}\sqrt{\frac{L}{C}} = Q = \frac{1}{\sqrt{2}}$$

The third design equation indicates that $k = 1$. This last design equation does not constrain the values of R, L, and C. Because we have two equations in three unknowns, the solution is not unique. One way to proceed is to choose a convenient value for one circuit element, say $C = 0.1\ \mu\text{F}$, and then calculate the resulting values of the other circuit elements

$$L = 1/\left(\omega_0{}^2 C\right) = 0.253 \text{ H}$$

and

$$R = \sqrt{\frac{2L}{C}} = 2251\ \Omega$$

If we are satisfied with this solution, the filter design is complete. Otherwise, we adjust our choice of the value of C and recalculate L and R. For example, if the inductance is too large, say $L = 1000$ H, or the resistance is too small, say $R = 0.03\ \Omega$, it will be hard to obtain the parts to build these circuits. Because there is no such problem in this example, we conclude that the circuit shown in the first row of Table 16.4-1 with $C = 0.1\ \mu\text{F}$, $L = 0.253$ H, and $R = 2251\ \Omega$ is the required low-pass filter.

EXAMPLE 16.4-2 Sallen-Key Band-Pass Filter

Design a second-order Sallen-Key band-pass filter with a center frequency of 500 hertz and a bandwidth of 100 hertz.

Solution

The transfer function of the second-order band-pass filter is

$$H(s) = \frac{k\dfrac{\omega_0}{Q}s}{s^2 + \dfrac{\omega_0}{Q}s + \omega_0{}^2}$$

The corresponding network function is

$$\mathbf{H}(\omega) = \frac{jk\dfrac{\omega_0}{Q}\omega}{\omega_0{}^2 - \omega^2 + j\dfrac{\omega_0}{Q}\omega}$$

Dividing numerator and denominator by $j\dfrac{\omega_0}{Q}\omega$ gives

$$\mathbf{H}(\omega) = \frac{k}{1 + jQ\left(\dfrac{\omega}{\omega_0} - \dfrac{\omega_0}{\omega}\right)}$$

We have seen network functions like this one earlier, when we discussed resonant circuits in Chapter 13. The gain, $|\mathbf{H}(\omega)|$, will be maximum at the corner frequency, ω_0. In the case of this band-pass transfer function, ω_0 is also called the center frequency and the resonant frequency. The gain at the center frequency will be

$$|\mathbf{H}(\omega_0)| = k$$

Two frequencies, ω_1 and ω_2, are identified by the property

$$|\mathbf{H}(\omega_1)| = |\mathbf{H}(\omega_2)| = \frac{k}{\sqrt{2}}$$

These frequencies are called the *half-power frequencies* or the 3 *dB frequencies*. The half-power frequencies are given by

$$\omega_1 = -\frac{\omega_0}{2Q} + \sqrt{\left(\frac{\omega_0}{2Q}\right)^2 + \omega_0{}^2} \quad \text{and} \quad \omega_2 = \frac{\omega_0}{2Q} + \sqrt{\left(\frac{\omega_0}{2Q}\right)^2 + \omega_0{}^2}$$

The bandwidth of the filter is calculated from the half-power frequencies

$$BW = \omega_2 - \omega_1 = \frac{\omega_0}{Q}$$

The Sallen-Key band-pass filter is shown in the third row of Table 16.4-2. Our specifications require that

$$\omega_0 = 2\pi \cdot 500 = 3142 \text{ rad/s}$$

and

$$Q = \frac{\omega_0}{BW} = 5$$

From Table 16.4-2, the design equations for the Sallen-Key band-pass filter are

$$\frac{1}{RC} = \omega_0 = 3142$$

and

$$A = 3 - \frac{1}{Q} = 2.8$$

Pick $C = 0.1\ \mu\text{F}$. Then

$$R = \frac{1}{C\omega_0} = 3183\ \Omega$$

Because $k = AQ$, the gain of this band-pass filter at the center frequency is 14. Also, one of the resistances is given by

$$(A - 1)R = 5729\ \Omega$$

The Sallen-Key band-pass filter is shown in Figure 16.4-2.

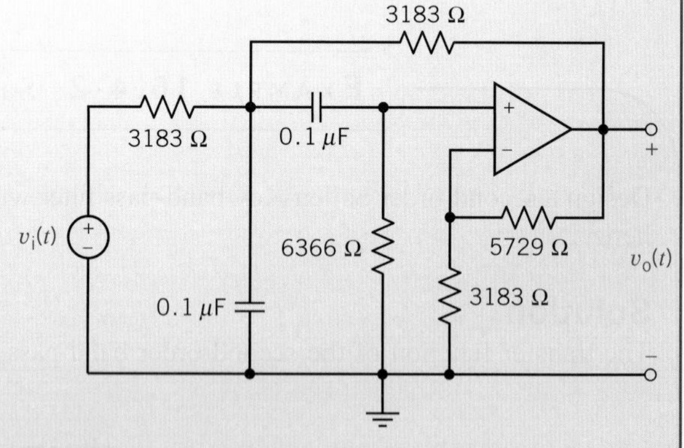

FIGURE 16.4-2 A Sallen-Key band-pass filter.

EXAMPLE 16.4-3 Sallen-Key Band-Stop Filter

Design a second-order band-stop filter with a center frequency of 1000 rad/s and a bandwidth of 100 rad/s.

Solution

The transfer function of the second-order band-stop filter is

$$H(s) = \frac{k(s^2 + \omega_0{}^2)}{s^2 + \frac{\omega_0}{Q}s + \omega_0{}^2}$$

Notice that the transfer functions of the second-order band-pass and band-stop filters are related by

$$\frac{k(s^2+{\omega_0}^2)}{s^2+\dfrac{\omega_0}{Q}s+{\omega_0}^2}=k-\frac{k\dfrac{\omega_0}{Q}s}{s^2+\dfrac{\omega_0}{Q}s+{\omega_0}^2}$$

The network function of the band-stop filter is

$$\mathbf{H}(\omega)=\frac{k({\omega_0}^2-\omega^2)}{{\omega_0}^2-\omega^2+j\dfrac{\omega_0}{Q}\omega}$$

When $\omega \ll \omega_0$ or $\omega \gg \omega_0$, the gain is $|\mathbf{H}(\omega)|=k$. At $\omega=\omega_0$, the gain is zero. The half-power frequencies, ω_1 and ω_2, are identified by the property

$$|\mathbf{H}(\omega_1)|=|\mathbf{H}(\omega_2)|=\frac{k}{\sqrt{2}}$$

The bandwidth of the filter is given by

$$BW=\omega_2-\omega_1=\frac{\omega_0}{Q}$$

The Sallen-Key band-stop filter is shown in the last row of Table 16.4-2. Our specifications require that $\omega_0=$ 1000 rad/s and

$$Q=\frac{\omega_0}{BW}=10$$

Table 16.4-2 indicates that the design equations for the Sallen-Key band-stop filter are

$$\frac{1}{RC}=\omega_0=1000$$

and

$$A=2-\frac{1}{2Q}=1.95$$

Pick $C=0.1\ \mu$F. Then

$$R=\frac{1}{C\omega_0}=10\ \text{k}\Omega$$

The Sallen-Key band-stop filter is shown in Figure 16.4-3.

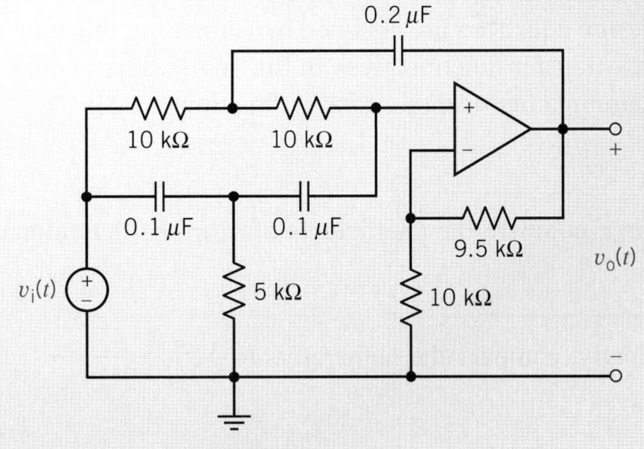

FIGURE 16.4-3 A Sallen-Key band-stop filter.

Example 16.4-4 Tow-Thomas Filter

Figure 16.4-4 shows another circuit that can be used to build a second-order filter. This circuit is called a *Tow-Thomas filter*. This filter can be used as either a band-pass or low-pass filter. When the output is the voltage $v_1(t)$, the transfer function is

$$H_L(s)=\frac{-\dfrac{1}{R_k RC^2}}{s^2+\dfrac{1}{R_Q C}s+\dfrac{1}{R^2C^2}} \tag{16.4-3}$$

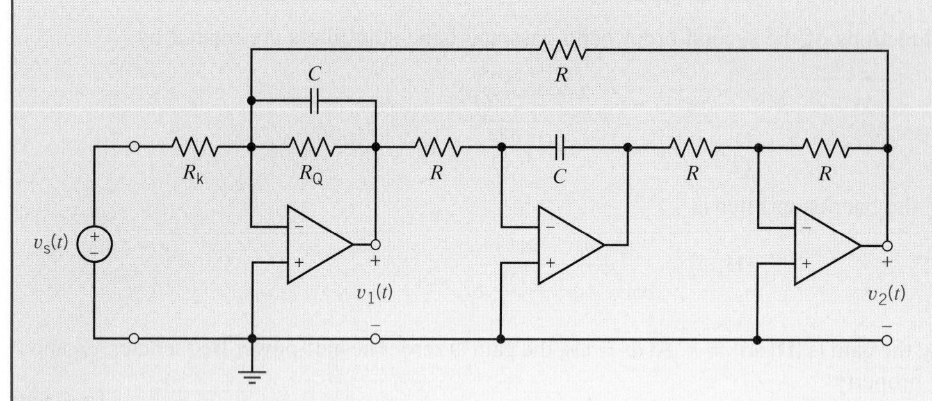

FIGURE 16.4-4 The Tow-Thomas filter.

and the filter is a low-pass filter. If, instead, the voltage $v_2(t)$ is used as the filter output, the network function is

$$H_B(s) = \frac{-\frac{1}{R_k C} s}{s^2 + \frac{1}{R_Q C} s + \frac{1}{R^2 C^2}} \tag{16.4-4}$$

and the Tow-Thomas filter functions as a band-pass filter. Design a Butterworth Tow-Thomas low-pass filter with a dc gain of 5 and a cutoff frequency of 1250 hertz.

Solution

Because the Tow-Thomas filter will be used as a low-pass filter, the transfer function is given by Eq. 16.4-3. Design equations are obtained by comparing this transfer function to the standard form of the second-order low-pass transfer function given in Eq. 16.4-1. First, compare the constant terms (that is, the coefficients of s^0) in the denominators of these transfer functions to get

$$\omega_0 = \frac{1}{RC} \tag{16.4-5}$$

Next, compare the coefficients of s^1 in the denominators of these transfer functions to get

$$Q = \frac{R_Q}{R} \tag{16.4-6}$$

Finally, compare the numerators to get

$$k = \frac{R}{R_k} \tag{16.4-7}$$

Designing the Tow-Thomas filter requires that values be obtained for R, C, R_Q, and R_k. Because there are four unknowns and only three design equations, we begin by choosing a convenient value for one of the unknowns, usually the capacitance. Let $C = 0.01\ \mu$F. Then,

$$R = \frac{1}{\omega_0 C} = \frac{1}{(2\pi)(1250)(0.01)(10^{-6})} = 12{,}732\ \Omega$$

A second-order Butterworth filter requires $Q = 0.707$, so

$$R_Q = QR = (0.707)(12{,}732) = 9003\ \Omega$$

Finally

$$R_k = \frac{R}{k} = 2546\ \Omega$$

and the design is complete.

EXAMPLE 16.4-5 Tow-Thomas High-Pass Filter

Use the Tow-Thomas circuit to design a Butterworth high-pass filter with a high-frequency gain of 5 and a cutoff frequency of 1250 hertz.

Solution

The Tow-Thomas circuit does not implement the high-pass filter, but it does implement the low-pass filter and the band-pass filter. The transfer functions of the second-order high-pass, band-pass, and low-pass filters are related by

$$\begin{aligned} H_{\mathrm{H}}(s) &= \frac{ks^2}{s^2 + \dfrac{1}{R_{\mathrm{Q}}C}s + \dfrac{1}{R^2C^2}} = k + \frac{-\dfrac{1}{R_{\mathrm{k}}C}s}{s^2 + \dfrac{1}{R_{\mathrm{Q}}C}s + \dfrac{1}{R^2C^2}} + \frac{-\dfrac{1}{R_{\mathrm{k}}RC^2}}{s^2 + \dfrac{1}{R_{\mathrm{Q}}C}s + \dfrac{1}{R^2C^2}} \\ &= k + H_{\mathrm{B}}(s) + H_{\mathrm{L}}(s) \end{aligned} \tag{16.4-8}$$

A high-pass filter can be constructed using a Tow-Thomas filter and a summing amplifier. Both the band-pass and low-pass outputs of the Tow-Thomas filter are used. Equation 16.4-8 indicates that the band-pass and low-pass filters must have the same values of k, Q, and ω_0 as the high-pass filter. Thus, we require a Tow-Thomas filter having $k = 5$, $Q = 0.707$, and $\omega_0 = 7854$ rad/s. Such a filter was designed in Example 16.4-4. The high-pass filter is obtained by adding a summing amplifier as shown in Figure 16.4-5.

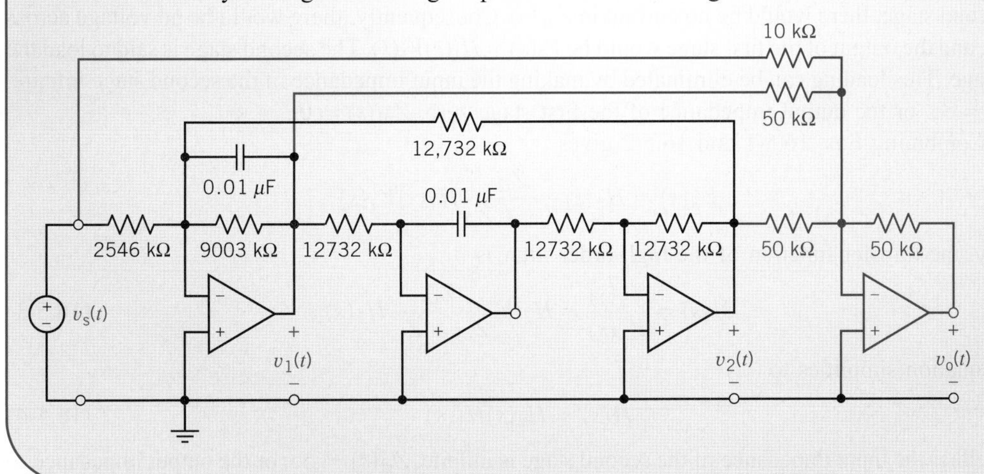

FIGURE 16.4-5 A Tow-Thomas high-pass filter.

16.5 HIGH-ORDER FILTERS

In this section, we turn our attention to filters that have an order greater than 2. These filters are called **high-order filters**. A popular strategy for designing high-order filters uses a cascade connection of second-order filters. The cascade connection is shown in Figure 16.5-1. In this figure, the transfer functions $H_1(s)$, $H_2(s)$, . . . , $H_n(s)$ represent second-order filters that are connected together to build a high-order filter. We refer to the second-order filter as filter stages to distinguish them from the high-order filter. That is, the high-order filter is a cascade connection of second-order filter stages. (When the order of the high-order filter is odd, a first-order filter stage is needed. Nonetheless, we talk about designing high-order filters as a cascade of second-order stages.)

The cascade connection is characterized by the fact that the output of one filter stage is used as the input to the next stage. Unfortunately, the behavior of a stage will sometimes change when another stage is connected to it. We call this phenomenon **loading**, and we say that the second stage loaded the first. Generally, loading is undesirable, and we try to avoid it. Figure 16.5-2 shows a model of a filter

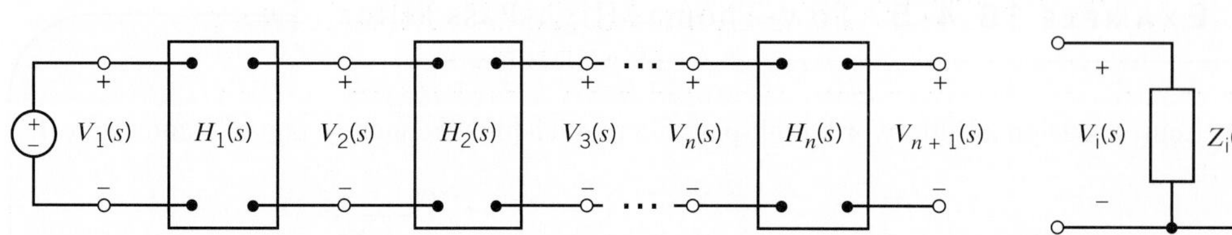

FIGURE 16.5-1 A cascade circuit of n stages.

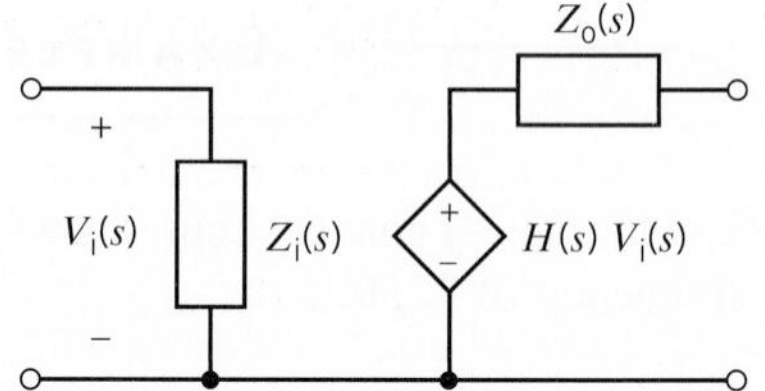

FIGURE 16.5-2 A model of one filter stage.

stage that is appropriate for investigating loading. This model includes the input and output impedance of the filter stage as well as the transfer function.

Figure 16.5-3 shows a high-order filter consisting of the cascade connection of two filter stages. Let's calculate the transfer function of the high-order filter. Starting at the output of the high-order filter, notice that there is no current in the output impedance, $Z_{o2}(s)$, of the second stage. Consequently, there is no voltage across $Z_{o2}(s)$, so

$$V_3(s) = H_2(s)V_2(s) \tag{16.5-1}$$

Next, we use voltage division to find $V_2(s)$.

$$V_2(s) = \frac{Z_{i2}}{Z_{o1} + Z_{i2}} H_1(s)V_1(s) \tag{16.5-2}$$

Connecting the second filter stage to the first stage has changed the output of the first stage. Without the second stage, there would be no current in $Z_{o1}(s)$. Consequently, there would be no voltage across $Z_{o1}(s)$, and the output of the first stage would be $V_2(s) = H_1(s)V_1(s)$. The second stage is said to load the first stage. This loading can be eliminated by making the input impedance of the second stage infinite, $Z_{i2}(s) = \infty$, or the output impedance of the first stage zero, $Z_{o1}(s) = 0$.

Combining Eqs. 16.5-1 and 16.5-2 gives

$$V_3(s) = H_2(s)\frac{Z_{i2}}{Z_{o1} + Z_{i2}} H_1(s)V_1(s)$$

Finally, the transfer function of the high-order filter is

$$H(s) = \frac{V_3(s)}{V_1(s)} = H_2(s)\frac{Z_{i2}}{Z_{o1} + Z_{i2}} H_1(s) \tag{16.5-3}$$

This equation simplifies to

$$H(s) = H_2(s)H_1(s) \tag{16.5-4}$$

when either the input impedance of the second stage is infinite, $Z_{i2}(s) = \infty$, or the output impedance of the first stage is zero, $Z_{o1}(s) = 0$. In other words, Eq. 16.5-4 can be used when the second stage does not load the first stage, but Eq. 16.5-3 must be used when the second stage does load the first stage. We will prove that the Sallen-Key filters have output impedances equal to zero. Therefore, there is no loading when Sallen-Key filter stages are cascaded. The transfer function of the high-order filter is the product of the transfer functions of the individual Sallen-Key filter stages. In contrast, the filters based on the series *RLC* circuit shown in Table 16.4-1 do not have output impedances that are equal to zero or input impedances that are infinite. If these filter stages were cascaded, the transfer function of the high-order filter would not be equal to the product of the transfer functions of the individual filter stages. Thus, we can use cascaded Sallen-Key filter stages to design high-order filters without introducing loading.

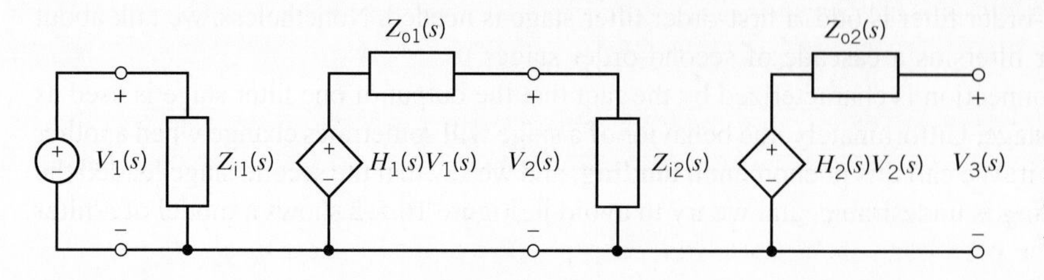

FIGURE 16.5-3 Cascade connection of two filter stages.

Table 16.5-1 Measuring the Parameters of a Filter Stage

PARAMETER	DEFINITION	MEASUREMENTS
Input impedance	$Z_i(s) = \dfrac{V_i(s)}{I_T(s)}$	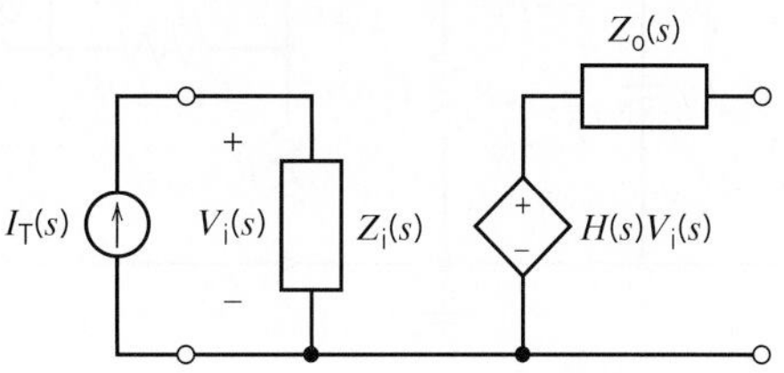
Output impedance	$Z_o(s) = \dfrac{V_T(s)}{I_T(s)}$	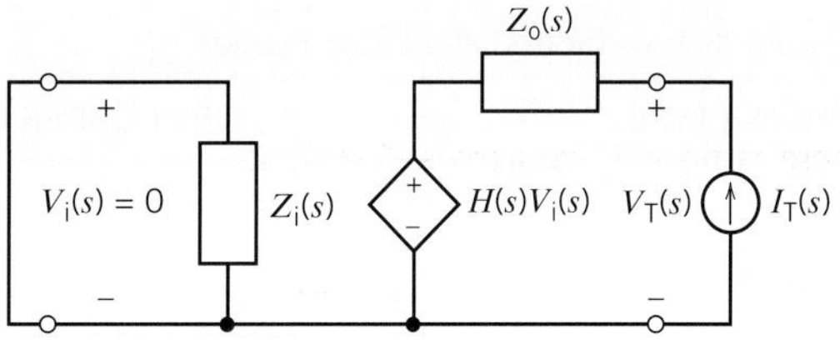
Transfer function	$H(s) = \dfrac{V_o(s)}{V_i(s)}$	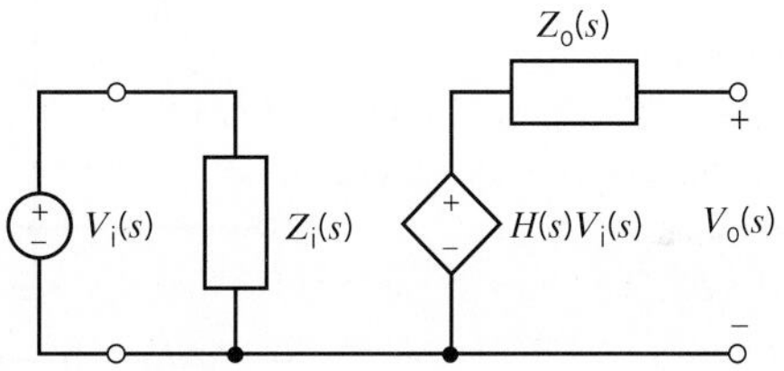

Next, consider calculating the output impedance of a Sallen-Key band-pass filter. Table 16.5-1 shows how the parameters of the model of a filter stage can be calculated or measured. The second row of this table indicates that to calculate the output impedance, a short circuit should be connected to the filter input, and a current source should be connected to the filter output. The voltage across the current source is calculated, and the ratio of this voltage to the current of the current source is the output impedance. Figure 16.5-4 shows a Sallen-Key filter with a short circuit across its input and a current source connected to its output. This circuit can be analyzed by writing node equations at nodes 1, 2, and T:

$$\frac{V_1}{R} + CsV_1 + \frac{V_1 - V_T}{R} + (V_1 - V_2)Cs = 0$$

$$-(V_1 - V_2)Cs + \frac{V_2}{2R} = 0$$

$$\frac{V_2}{R} + \frac{V_2 - V_T}{(A-1)R} = 0$$

Solving these node equations for V_T gives

$$\left[(RCs)^2 + (3 - A)RCs + 1\right] V_T = 0$$

Because the factor in brackets is not zero, this equation indicates that $V_T = 0$. The output impedance of the Sallen-Key band-pass filter is

$$Z_o = \frac{V_T}{I_T} = \frac{0}{I_T} = 0$$

Similarly, each of the Sallen-Key filters shown in Table 16.4-2 has an output impedance equal to zero.

High-order filters can be designed as a cascade connection of second-order filter stages. Filter stages that have an output impedance equal to zero are used so that the transfer function of the high-order filter will be the product of the transfer functions of the cascaded filter stages.

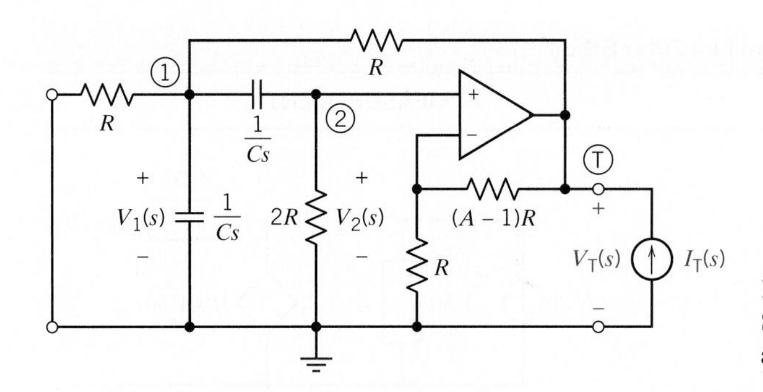

FIGURE 16.5-4 Calculating the output of a Sallen-Key band-pass filter. Circled numbers are node numbers.

Table 16.5-2 First-Order Filter Stages

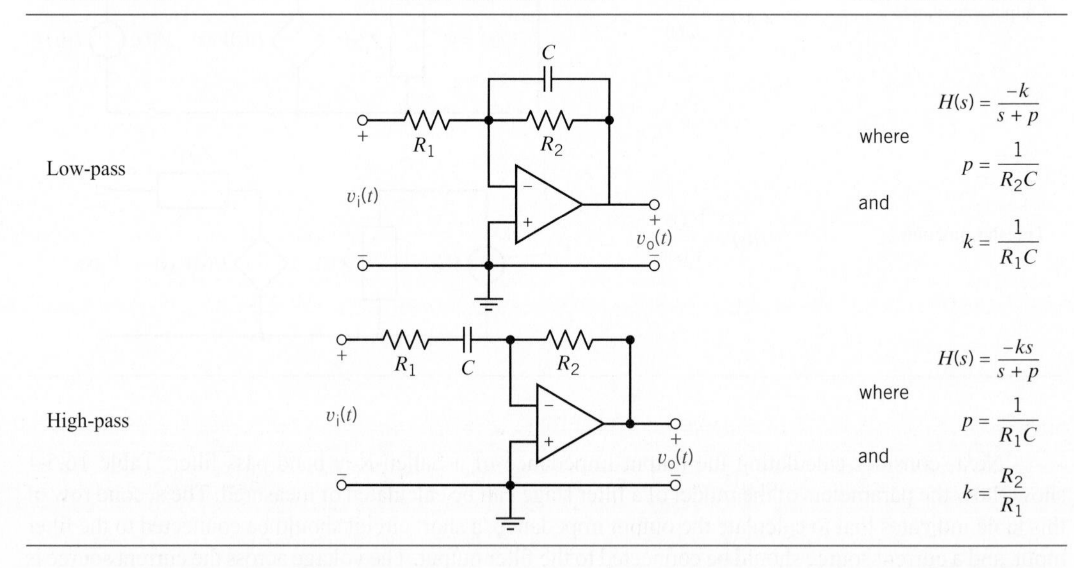

FILTER TYPE	FIRST ORDER CIRCUIT	DESIGN EQUATION
Low-pass		$H(s) = \frac{-k}{s+p}$ where $p = \frac{1}{R_2C}$ and $k = \frac{1}{R_1C}$
High-pass		$H(s) = \frac{-ks}{s+p}$ where $p = \frac{1}{R_1C}$ and $k = \frac{R_2}{R_1}$

Example 16.5-1 Cascade Connection of Filter Stages

Design a third-order Butterworth low-pass filter having a cutoff frequency of $\omega_c = 500$ rad/s and a dc gain equal to 1.

Solution

Equation 16.3-2 and Table 16.3-2 provide a third-order Butterworth low-pass filter having a cutoff frequency equal to 1 rad/s.

$$H_n(s) = \frac{1}{(s+1)(s^2+s+1)}$$

Frequency scaling is used to adjust the cutoff frequency so that $\omega_c = 500$ rad/s.

$$\begin{aligned} H(s) &= \frac{1}{\left(\frac{s}{500}+1\right)\left(\left(\frac{s}{500}\right)^2+\frac{s}{500}+1\right)} \\ &= \frac{500^3}{(s+500)(s^2+500s+500^2)} \end{aligned}$$

$H(s)$ is the transfer function of a third-order Butterworth low-pass filter having a cutoff frequency equal to 500 rad/s. This transfer function can be expressed as

$$H(s) = \frac{-250{,}000}{s^2 + 500s + 250{,}000} \cdot \frac{-500}{s + 500} = H_1(s) \cdot H_2(s) \qquad (16.5\text{-}5)$$

A Sallen-Key low-pass filter can be designed to implement the second-order low-pass transfer function $H_1(s)$. Table 16.5-2 provides circuits and design equations for first-order filter stages. The circuit shown in the first row of this table can be used to implement $H_2(s)$. The first-order filter stages in Table 16.5-2 have output impedances equal to zero. Cascading these filter stages will not cause loading. Cascading the Sallen-Key filter with the first-order filter stage will produce a third-order filter with the transfer function $H(s)$.

First, let's design the Sallen-Key filter with transfer function

$$H_1(s) = \frac{-250{,}000}{s^2 + 500s + 250{,}000}$$

Values of the filter parameters k, ω_0, and Q are determined by comparing $H_1(s)$ with the standard form of the second-order low-pass transfer function given in Eq. 16.4-1. From the constant term in the denominator,

$${\omega_0}^2 = 250{,}000$$

Next, from the coefficient of s in the denominator,

$$\frac{\omega_0}{Q} = 500$$

Finally, from the numerator,

$$k \cdot {\omega_0}^2 = 250{,}000$$

So, $\omega_0 = 500$ rad/s, $Q = 1$, and $k = 1$. The Sallen-Key low-pass filter is shown in row 1 of Table 16.4-2. Designing this filter requires finding values of R, C, and A. The design equations given in row 1 of the table indicate that

$$\omega_0 = \frac{1}{RC} \qquad (16.5\text{-}6)$$

$$Q = \frac{1}{3 - A} \qquad (16.5\text{-}7)$$

$$k = A \qquad (16.5\text{-}8)$$

Equation 16.5-7 gives

$$A = 3 - \frac{1}{Q} = 3 - \frac{1}{1} = 2$$

but Eq. 16.5-8 gives

$$A = k = 1$$

Apparently, we can select A to get the correct value of Q, or we can select A to get the correct value of k, but not both. The dc gain is easy to adjust later, so we pick $A = 2$ to make $Q = 1$ and settle for $k = 2$. Equation 16.5-6 is satisfied by taking $C = 0.1\ \mu$F and

$$R = \frac{1}{C\omega_0} = \frac{1}{(0.1 \times 10^{-6})(500)} = 20\ \text{k}\Omega$$

The Sallen-Key filter stage is shown in Figure 16.5-5a. The transfer function of this stage is

$$H_3(s) = \frac{-500{,}000}{s^2 + 500s + 250{,}000}$$

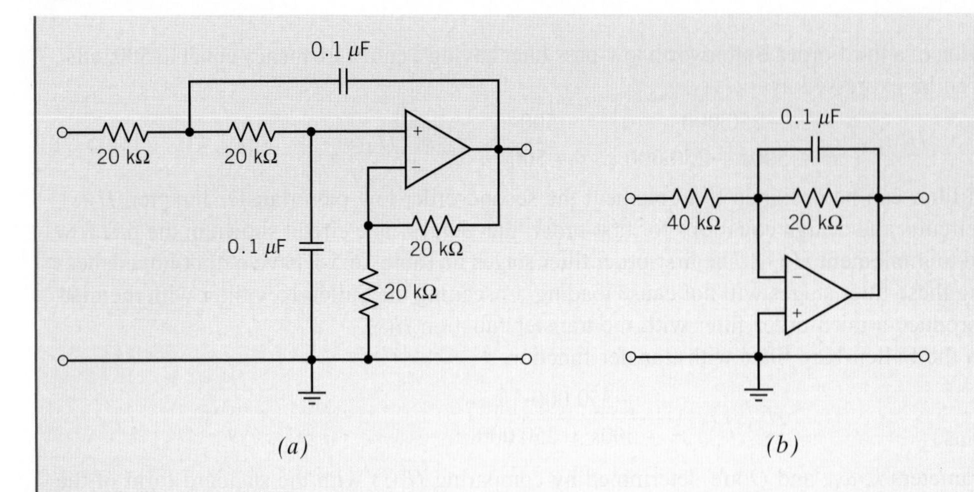

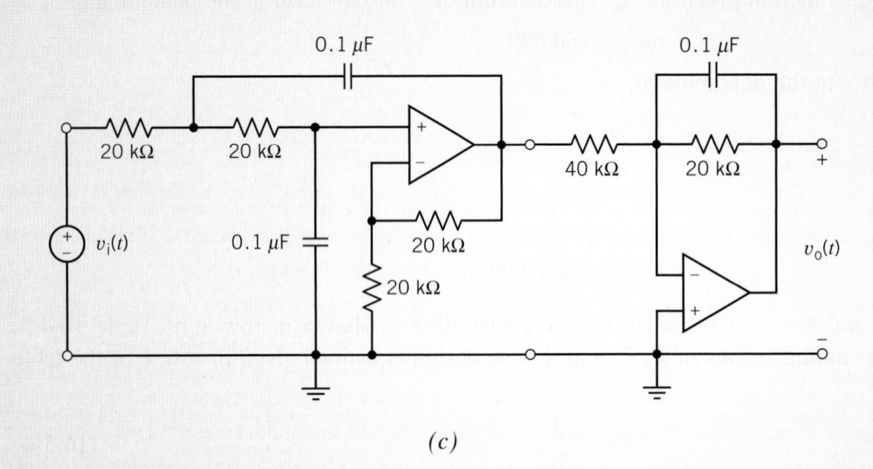

FIGURE 16.5-5 (*a*) A Sallen-Key filter stage, (*b*) a first-order filter stage, and (*c*) a third-order Butterworth filter.

The Sallen-Key filter stage achieved the desired values of ω_0 and $Q = 1$ but not the desired value of the dc gain. To compensate, we will adjust the dc gain of the first-order filter. The desired transfer function of the third-order filter can be expressed as

$$H(s) = \frac{-500{,}000}{s^2 + 500s + 250{,}000} \cdot H_4(s)$$

which requires

$$H_4(s) = \frac{-250}{s + 500}$$

The design equations in row 1 of Table 16.5-2 indicate that

$$500 = \frac{1}{R_2 C}$$

and

$$250 = \frac{1}{R_1 C}$$

Choose $C = 0.1\ \mu\text{F}$. Then

$$R_2 = \frac{1}{500 \cdot C} = \frac{1}{(500)(0.1 \times 10^{-6})} = 20\ \text{k}\Omega$$

and
$$R_1 = \frac{1}{250 \cdot C} = \frac{1}{(250)(0.1 \times 10^{-6})} = 40\ \text{k}\Omega$$

The first-order filter stage is shown in Figure 16.5-5*b*. Cascading the Sallen-Key stage and the first-order stage produces the third-order Butterworth filter shown in Figure 16.5-5*c*.

16.6 SIMULATING FILTER CIRCUITS USING PSPICE

PSpice provides a convenient way to verify that a filter circuit does indeed have the correct transfer function. Figure 16.6-1 illustrates a method of testing a filter design. The filter that is being tested here is a fourth-order notch filter consisting of two Sallen-Key notch filter stages and an inverting amplifier. This filter was designed to have the transfer function

$$H(s) = \frac{4(s^2 + 62{,}500)^2}{(s^2 + 250s + 62{,}500)^2}$$

The voltage source voltage, $v_i(t)$, is used as the input to two separate circuits. One of these circuits is the filter circuit consisting of the Sallen-Key stages and the inverting amplifier. The response of this circuit is the node voltage $v_{o1}(t)$. The other ''circuit'' implements *H*(*s*) directly using a feature of PSpice. The response of this circuit is $v_{o2}(t)$. A single PSpice simulation produces the frequency responses corresponding to the transfer functions of both of these circuits, $V_{o1}(s)/V_i(s)$ and $V_{o2}(s)/V_i(s)$. Next, we use Probe, the graphical post processor included with PSpice, to display both frequency responses on the same axis. If these frequency responses are identical, we know that the filter circuit does indeed implement the transfer function *H*(*s*).

Figure 16.6-2 shows the PSpice input file corresponding to Figure 16.6-1. Two aspects of this file require some explanation. First, notice that parameters are used in the subcircuit that represents the Sallen-Key filter stage. The line

```
.subckt sk_n in out params: C=.1uF w0 = 1 krad/s Q = 0.707
```

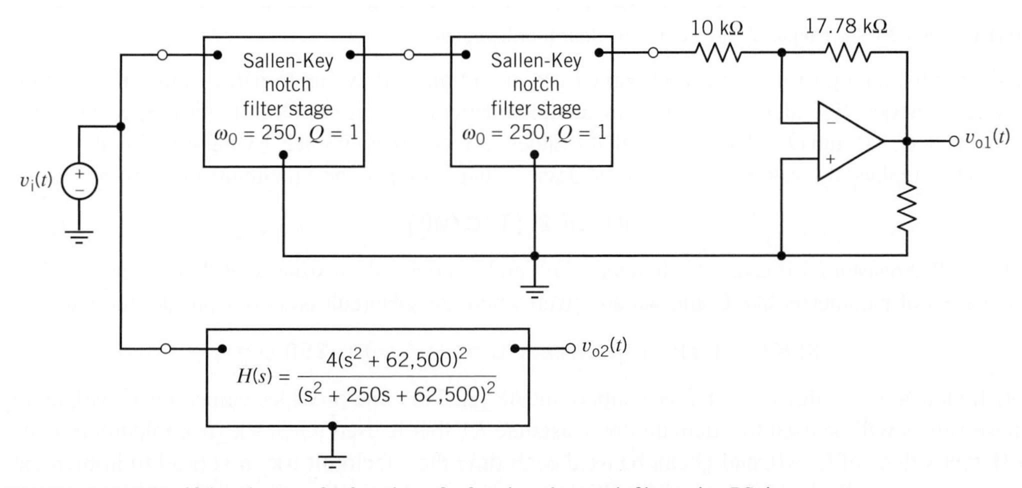

FIGURE 16.6-1 Verifying the transfer function of a fourth-order notch filter using PSpice.

```
Testing a 4th order notch filter

Vin   1   0 ac 1
XSK1 1   2 sk_n   params: C=.1uF w0=250 Q=1
XSK2 2   3 sk_n   params: C=.1uF w0=250 Q=1
R1    3   4 10k
R2    4   5 17.78k
XOA   4   0 5 op_amp
RL    5   0 10G
XLP   1   6 4th_order_notch_filter

.subckt sk_n in out params: C=.1uF w0=1krad/s  Q=0.707
R1   in   2 {1/C/w0}
R2   2    3 {1/C/w0}
C1   in   6 {C}
C2   6    3 {C}
C3   2    out {2*C}
R3   6    0 {1/2/C/w0}
XOA 5    3 out  op_amp
R4   5    0 10kOhm
R5   out 5 {(1-1/Q/2)*10kOhm}
.ends sk_n

.subckt op_amp inv non out
* an ideal op amp
E (out 0) (non inv)  1G
.ends op_amp

.subckt 4th_order_notch_filter in out
R1   in   0 1G
R2   out  0 1G
E1   out  0 LAPLACE {V(in)} = {(4*(s*s+62500)*(s*s+62500)) /
+                    (s*s+250*s+62500) *(s*s+250*s+62500)}
.ends 4th_order_notch_filter

.ac dec 100  1  1000
.probe  V(1)  V(5)  V(6)
.end
```

FIGURE 16.6-2 PSpice input file used to test the fourth-order notch filter.

marks the beginning of the subcircuit named `sk_n`. (PSpice allows us to name, rather than number, nodes. The nodes ''in'' and ''out'' will connect this subcircuit to the rest of the circuit.) Three parameters are defined: C, w0, and Q. All are given default values, as required by PSpice. Expressions involving these parameters replace the values of some of the devices that comprise the subcircuit; for example, the line

```
R1 in 2 {1/C/w0}
```

indicates that resistor $R1$ is connected to nodes ''in'' and 2 and that the resistance of $R1$ is given by 1/C/w0. The values of parameters like C and w0 are given when the subcircuit is used. Consider the line

```
XSK2 2 3 sk_n params : C =.1uf w0 = 250 Q = 1
```

which indicates that device XSK2 is a subcircuit `sk_n`. This line provides values for C, w0, and Q. These values will be used to calculate the resistance $R1$ that is used when `sk_n` implements XSK2. Different values of C, w0, and Q can be used each time the subcircuit `sk_n` is used to implement a different device. Table 16.6-1 provides PSpice subcircuits for the four Sallen-Key filter stages.

Table 16.6-1 PSpice Subcircuits for Sallen-Key Filter Stages

FILTER STAGE	PSPICE SUBCIRCUIT

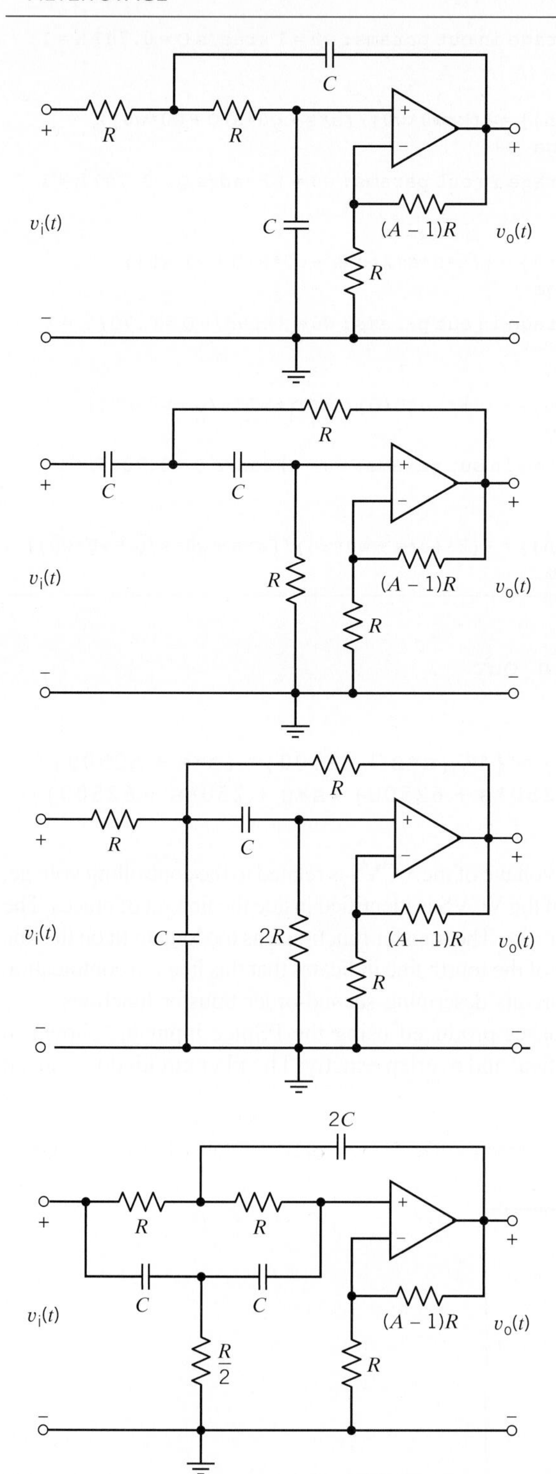


```
.subckt sk_lp in out params: C = .1uF
w0 = 1krad/s Q = 0.707
R1    in    2     {1/C/w0}
R2    2     3     {1/C/w0}
C1    3     0     {C}
C2    2     out   {C}
XOA   5     3     out    op_amp
R3    5     0     10kOhm
R4    out   5     {(2 – 1/Q) *10kOhm}
.ends sk_lp
```

```
.subckt sk_hp in out params: C = .1uF
w0 = 1krad/s Q = 0.707
R1    3     0     {1/C/w0}
R2    2     out   {1/C/w0}
C1    in    2     {C}
C2    2     3     {C}
XOA   5     3     out    op_amp
R3    5     0     10kOhm
R4    out   5     {(2 – 1/Q) *10kOhm}
.ends sk_hp
```

```
.subckt sk_bp in out params: C = .1uF
w0 = 1krad/s Q = 0.707
R1    in    2     {1/C/w0}
R2    2     out   {1/C/w0}
C1    2     3     {C}
C2    2     0     {C}
R3    3     0     {2/C/w0}
XOA   5     3     out    op_amp
R4    5     0     10kOhm
R5    out   5     {(2 – 1/Q)*10kOhm}
.ends sk_bp
```

```
.subckt sk_n in out params: C = .1uF
w0 = 1krad/s Q = 0.707
R1    in    2     {1/C/w0}
R2    2     3     {1/C/w0}
C1    in    6     {C}
C2    6     3     {C}
C3    2     out   {2*C}
R3    6     0     {1/2/C/w0}
XOA   5     3     out    op_amp
R4    5     0     10kOhm
R5    out   5     {(1 – 1/Q/2)*10kOhm}
.ends sk_n
```

Table 16.6-2 PSpice Subcircuits for Second-Order Transfer Functions

TRANSFER FUNCTION	PSPICE SUBCIRCUIT

Low-pass	`.subckt lp_filter_stage in out params: w0 = 1 krad/s Q = 0.707 k = 1`
	`R1 in 0 1G`
	`R2 out 0 1G`
	`E out 0 LAPLACE {V(in)} = {(k*w0*w0)/(s*s + w0*s/Q + w0*w0)}`
	`.ends lp_filter_stage`

High-pass	`.subckt hp_filter_stage in out params: w0 = 1 krad/s Q = 0.707 k = 1`
	`R1 in 0 1G`
	`R2 out 0 1G`
	`E out 0 LAPLACE {V(in)} = {(k*s*s)/(s*s + w0*s/Q + w0*w0)}`
	`.ends hp_filter_stage`

Band-pass	`.subckt bp_filter_stage in out params: w0 = 1krad/s Q = 0.707 k = 1`
	`R1 in 0 1G`
	`R2 out 0 1G`
	`E out 0 LAPLACE {V(in)} = {(k*w0*s/Q)/(s*s + w0*s/Q + w0*w0)}`
	`.ends bp_filter_stage`

Band-stop (notch)	`.subckt n_filter_stage in out params: w0 = 1krad/s Q = 0.707 k = 1`
	`R1 in 0 1G`
	`R2 out 0 1G`
	`E out 0 LAPLACE {V(in)} = {(k*(s*s + w0*w0)/(s*s + w0*s/Q + w0*w0)}`
	`.ends n_filter_stage`

Next, consider the subcircuit

```
.subckt 4 th_order_notch_filter in out
R1     in     0     1G
R2     out     0     1G
E1     out     0     LAPLACE {V(in)} = {4* (s*s + 62500)* (s*s + 62500)/
+                    (s*s + 250*s + 62500) (s*s + 250*s + 62500)}
.ends     4th_order_notch_filter
```

The keyword LAPLACE indicates that controlled voltage of the VCVS is related to the controlling voltage, using a transfer function. The controlling voltage of the VCVS is identified inside the first set of braces. The transfer function is given inside the second set of braces. The transfer function was too long to fit on the line describing the VCVS. The + sign at the beginning of the fourth line indicates that this line is a continuation of the previous line. Table 16.6-2 provides subcircuits describing second-order transfer functions.

Figure 16.6-3 shows the frequency responses produced using the PSpice input file shown in Figure 16.6-2. The frequency responses are identical and overlap exactly. The filter circuit does indeed implement the specified transfer function.

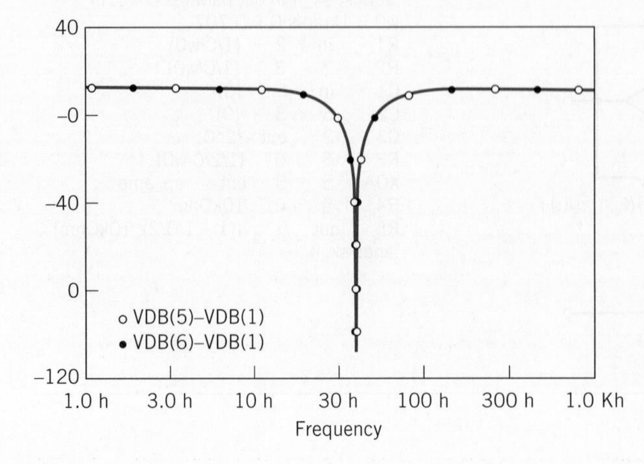

FIGURE 16.6-3 Frequency response plots used to verify the transfer function of the fourth-order notch filter.

16.7 HOW CAN WE CHECK . . . ?

Engineers are frequently called upon to check that a solution to a problem is indeed correct. For example, proposed solutions to design problems must be checked to confirm that all of the specifications have been satisfied. In addition, computer output must be reviewed to guard against data-entry errors, and claims made by vendors must be examined critically.

Engineering students are also asked to check the correctness of their work. For example, occasionally just a little time remains at the end of an exam. It is useful to be able to quickly identify those solutions that need more work.

The following examples illustrate techniques useful for checking the solutions of the sort of problem discussed in this chapter.

EXAMPLE 16.7-1 How Can We Check Filter Frequency Response?

Figure 16.7-1 shows the frequency response of a band-pass filter obtained using PSpice. Such a filter can be represented by

$$\frac{\mathbf{V}_o(\omega)}{\mathbf{V}_{in}(\omega)} = \mathbf{H}(\omega) = \frac{H_0}{1 + jQ\left(\dfrac{\omega}{\omega_0} - \dfrac{\omega_0}{\omega}\right)}$$

where $\mathbf{V}_{in}(\omega)$ and $\mathbf{V}_o(\omega)$ are the input and output of the filter. This filter was designed to satisfy the specifications

$$\omega_0 = 2\pi 1000 \text{ rad/s}, \quad Q = 10, \quad H_0 = 10$$

How can we check that the specifications are satisfied?

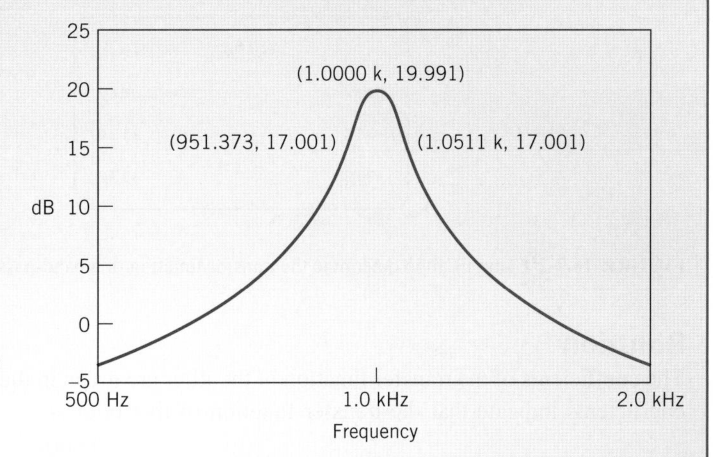

FIGURE 16.7-1 A band-pass frequency response.

Solution

The frequency response was obtained by analyzing the filter using PSpice. The vertical axis of Figure 16.7-1 gives the magnitude of $\mathbf{H}(\omega)$ in decibels. The horizontal axis gives the frequency in hertz. Three points on the frequency response have been labeled, giving the frequency and magnitude at each point. We want to use this information from the frequency response to check the filter to see whether it has the correct values of ω_0, Q, and H_0.

The three labeled points on the frequency response have been carefully selected. One of these labels indicates that the magnitude of $\mathbf{H}(\omega)$ and frequency at the peak of the frequency response are 20 dB and 1000 Hz. This peak occurs at the resonant frequency, so

$$\omega_0 = 2\pi 1000 \text{ rad/s}$$

The magnitude at the resonant frequency is H_0, so

$$20 \log_{10} H_0 = 20$$

or

$$H_0 = 10$$

The other two labeled points were chosen so that the magnitudes are 3 dB less than the magnitude at the peak. The frequencies at these points are 951 Hz and 1051 Hz. The difference of these two frequencies is the bandwidth, BW, of the frequency response. Finally, Q is calculated from the resonant frequency, ω_0, and the bandwidth, BW:

$$Q = \frac{\omega_0}{BW} = \frac{2\pi 1000}{2\pi(1051 - 951)} = 10$$

In this example, three points on the frequency response were used to verify that the band-pass filter satisfied the specifications for its resonant frequency, gain, and quality factor.

EXAMPLE 16.7-2 How Can We Check Filter Transfer Function?

ELab is a circuit analysis program that can be used to calculate the transfer function of a filter circuit (Svoboda, 1997). Figure 16.7-2 shows the result of using ELab to analyze the Sallen-Key band-pass filter shown in Figure 16.4-2. This Sallen-Key filter was designed in Example 16.4-2 to have $\omega_0 = 3142$ rad/s, $Q = 5$, and $k = 14$. **How can we check** that the filter does indeed have the required values of ω_0, Q, and k?

Transfer	Function	Coefficients
s^	numerator	denominator
0	0	9.87e+06
1	8800	629
2	0	1

Transfer	Function Menu
Transfer	Function
Poles	
Zeros	
Frequency	Response
exit to previous menu	

Poles

real	imaginary
-315	3130
-315	-3130

Zeros

real	imaginary
0	0

FIGURE 16.7-2 Using ELab to determine the transfer function of a band-pass filter.

Solution

The coefficients of the transfer function of the filter are given in the upper left-hand portion of Figure 16.7-2. The coefficients indicate that the transfer function of this filter is

$$H(s) = \frac{8800s}{s^2 + 629s + 9.87 \times 10^6} \tag{16.7-1}$$

The general form transfer function of the second-order band-pass filter is

$$H(s) = \frac{k\dfrac{\omega_0}{Q}s}{s^2 + \dfrac{\omega_0}{Q}s + \omega_0^2} \tag{16.7-2}$$

Notice that the coefficient of s^2 in the denominator polynomial is 1 in both of these transfer functions. Values of ω_0, Q, and k are determined by comparing the coefficients of the transfer functions in Eqs. 16.7-1 and 16.7-2.

The square root of the constant term of the denominator polynomial is equal to ω_0. Therefore,

$$\omega_0 = \sqrt{9.87 \times 10^6} = 3142 \text{ rad/s}$$

Next, the coefficient of s in the denominator polynomial is equal to ω_0/Q. Therefore,

$$Q = \frac{\omega_0}{629} = \frac{3142}{629} = 5$$

Finally, the ratio of the coefficient of s in the numerator polynomial to the coefficient of s in the denominator polynomial is equal to k. Therefore,

$$k = \frac{8880}{629} = 14$$

The Sallen-Key band-pass filter shown in Figure 16.4-2 does indeed have the required values of ω_0, Q, and k.

16.8 DESIGN EXAMPLE

ANTI-ALIASING FILTER

Digital signal processing (DSP) frequently involves sampling a voltage and converting the samples to digital signals. After the digital signals are processed, the output signal is converted back into an analog voltage. Unfortunately, a phenomenon called aliasing can cause errors to occur during digital signal processing. Aliasing is a possibility whenever the input voltage contains components at frequencies greater than one-half of the sampling frequency. Aliasing occurs when these components are mistakenly interpreted to be components at a lower frequency. Anti-aliasing filters are used to avoid these errors by eliminating those components of the input voltage that have frequencies greater than one-half of the sampling frequency.

An anti-aliasing filter is needed for a DSP application. The filter is specified to be a fourth-order Butterworth low-pass filter having a cutoff frequency of 500 hertz and a dc gain equal to 1. This filter is to be implemented as an *RC* op amp circuit.

Describe the Situation and the Assumptions

The anti-aliasing filter will be designed as a cascade circuit consisting of two Sallen-Key low-pass filters and perhaps an amplifier. The amplifier will be included if it is necessary to adjust the dc gain of the anti-aliasing filter.

The operational amplifiers in the Sallen-Key filter stages will be modeled as ideal operational amplifiers. Resistances will be restricted to the range of 2 kΩ to 500 kΩ, and capacitances will be restricted to the range of 1 nF to 10 μF.

State the Goal

The transfer function of a fourth-order Butterworth low-pass filter having a cutoff frequency of 500 hertz and a dc gain equal to 1 can be obtained in two steps. First, the transfer function of a fourth-order Butterworth low-pass filter is given by Eq. 16.3-2 and Table 16.3-2 to be

$$H_n(s) = \frac{1}{(s^2 + 0.765s + 1)(s^2 + 1.848s + 1)} \tag{16.8-1}$$

$H_n(s)$ is the transfer function of a filter having a cutoff frequency equal to 1 rad/s. Next, frequency scaling can be used to adjust the cutoff frequency to 500 hertz = 3142 rad/s. Frequency scaling can be accomplished by replacing s by $\dfrac{s}{\omega_c} = \dfrac{s}{3142}$ in $H_n(s)$.

$$\begin{aligned} H(s) &= \frac{1}{\left(\left(\frac{s}{3142}\right)^2 + 0.765\left(\frac{s}{3142}\right) + 1\right)\left(\left(\frac{s}{3142}\right)^2 + 1.848\left(\frac{s}{3142}\right) + 1\right)} \\ &= \frac{3142^4}{\left(s^2 + 2403.6s + 3142^2\right)\left(s^2 + 5806.4s + 3142^2\right)} \end{aligned} \tag{16.8-2}$$

The goal is to design a filter circuit that has this transfer function.

Generate a Plan

We will express $H(s)$ as the product of two second-order low-pass transfer functions. For each of these second-order transfer functions, we will do the following:

1. Determine the values of the filter parameters k, ω_0, and Q.
2. Design a Sallen-Key low-pass filter to have the required values of ω_0 and Q.

It's likely that the Sallen-Key filters won't have the desired values of the dc gain, so an amplifier will be required to adjust the dc gain. The anti-aliasing filter will consist of a cascade connection of the Sallen-Key filter stages and the amplifier.

Act on the Plan

Consider the first factor of the denominator of $H(s)$. From the constant term,

$$\omega_0^2 = 3142^2$$

So $\omega_0 = 3142$ rad/s. Next, from the coefficient of s in the denominator,

$$\frac{\omega_0}{Q} = 2403.6$$

so

$$Q = \frac{3142}{2403.6} = 1.31$$

Next, design a Sallen-Key low-pass filter with $\omega_0 = 3142$ rad/s and $Q = 1.31$. The design equations given in row 1 of Table 16.4-2 indicate that

$$\omega_0 = \frac{1}{RC}$$

and

$$Q = \frac{1}{3 - A}$$

Pick $C = 0.1\ \mu$F. Then,

$$R = \frac{1}{\omega_0 C} = \frac{1}{3142 \cdot 10^{-7}} = 3183\ \Omega$$

Also,

$$A = 3 - \frac{1}{Q} = 3 - \frac{1}{1.31} = 2.24$$

The dc gain of this filter stage is $k = A = 2.24$, so the transfer function of this stage is

$$H_1(s) = \frac{2.24 \cdot 3142^2}{s^2 + 2403.6s + 3142^2}$$

Next, consider the second factor in the denominator of $H(s)$. Once again, the constant term indicates that $\omega_0 = 3142$ rad/s. Now Q can be calculated from the coefficient of s to be

$$Q = \frac{3142}{5806.4} = 0.541$$

We require a Sallen-Key low-pass filter with $\omega_0 = 3142$ rad/s and $Q = 0.541$. Pick $C = 0.1\ \mu$F. Then,

$$R = \frac{1}{\omega_0 C} = \frac{1}{3142 \cdot 10^{-7}} = 3183\ \Omega$$

and

$$A = 3 - \frac{1}{Q} = 3 - \frac{1}{0.541} = 1.15$$

The dc gain of this filter stage is $k = A = 1.15$, so the transfer function of this stage is

$$H_2(s) = \frac{1.15 \cdot 3142^2}{s^2 + 5806.4s + 3142^2}$$

The product of the gains of the filter stages is

$$H_1(s) \cdot H_2(s) = 2.576 \cdot H(s)$$

so

$$H(s) = 0.388 \cdot H_1(s) \cdot H_2(s)$$

The third stage of the anti-aliasing filter is an inverting amplifier having gain equal to 0.388. The anti-aliasing filter is shown in Figure 16.8-1.

Verify the Proposed Solution

Section 16.14 describes a procedure for verifying that a circuit has a specified transfer function. This procedure consists of using PSpice to plot the frequency response of both the circuit and the transfer function. These two frequency responses are compared. If they are the same, the transfer function of the circuit is indeed the specified transfer function.

Figure 16.8-2 shows the PSpice input file used to plot the frequency responses of both the circuit shown in Figure 16.8-1 and the transfer function given in Eq. 16.8-2. These

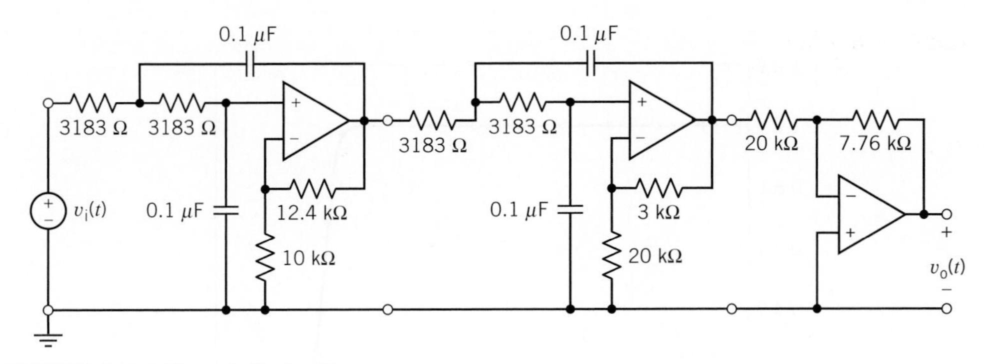

FIGURE 16.8-1 The anti-aliasing filter.

```
Verify the transfer function of the 4th order low-pass filter

Vin  1 0 ac 1
XSK1 1 2 sk_1p params: C={C} w0={w0} Q=1.31
XSK2 2 3 sk_1p params: C={C} w0={w0} Q=0.541
Ri 3 4 20000
Rf 4 5 7760
XOA 4 0 5 op_amp
X1 1 6 H1
X2 6 7 H2

.subckt sk_1p in out params: C=.1uf w0=1krad/s Q=0.707
R1 in  2 {1/C/w0}
R2 2  3 {1/C/w0}
C1 3  0 {C}
C2 2  out  {C}
XOA  5 3  out op_amp
R3  5 0  10kOhm
R4  out  5 {(2-1/Q)*10kOhm}
.ends

.subckt op_amp inv non out
*an ideal op amp
E (out 0) (non inv) 1G
.ends op_amp

.subckt H1 in out
R1 in  0 1G
R2 out 0  1G
E  out 0 LAPLACE {V(in)}={3142*3142/(s*s+2403.6*s+3142*3142)}
.ends H1

.subckt  H2 in out
R1  in 0 1G
R2  out  0 1G
E   out 0 LAPLACE {V(in)}={3142*3142/(s*s+5806.4*s+3142*3142)}
.ends H2

.ac dec 25 .01 5000
.probe  V(7) V(5)
.param: C=0.1uF  w0=3142  Q=2  k=2.5
.end
```

FIGURE 16.8-2 The PSpice input file used to verify that the circuit shown in Figure 16.8-1 has the specified transfer function.

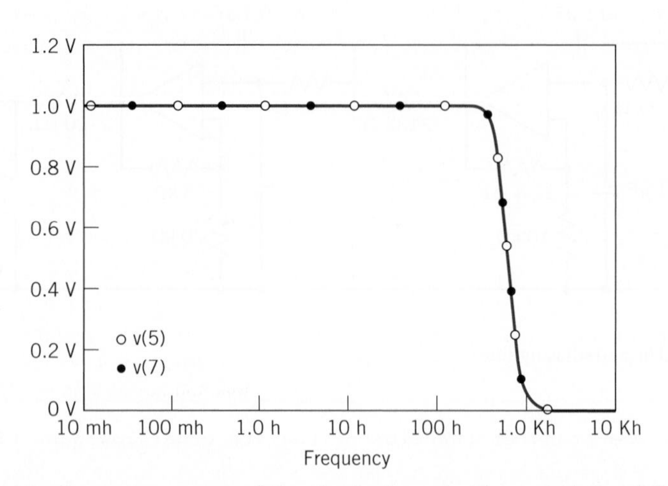

FIGURE 16.8-3 The frequency response of the circuit shown in Figure 16.8-1 and frequency response corresponding to the transfer function given in Eq. 16.8-2 are identical.

frequency responses are shown in Figure 16.8-3. These frequency responses overlap exactly so that the two plots appear to be a single plot. Therefore, the filter does indeed have the required transfer function.

16.9 SUMMARY

- An ideal filter separates its input into two parts. One part is passed, unchanged, to the output; the other part is eliminated. In other words, the output of an ideal filter is an exact copy of part of the filter input.
- There are several ways of separating the filter input into two parts and, correspondingly, several types of ideal filter. Table16.3-1 illustrates the common filter types.
- Unfortunately, ideal filter circuits don't exist. Filters are circuits that approximate ideal filters.
- **Butterworth** transfer functions have magnitude frequency responses that approximate the frequency response of an ideal filter.
- The frequency response of second-order filters is characterized by three filter parameters: a gain k, the corner frequency ω_0, and the **quality factor** Q. Filter circuits are designed by choosing the values of the circuit elements in such a way as to obtain the required values of k, ω_0, and Q.
  1. Table 16.4-1 provides the information required to design second-order RLC filters.
  2. Table 16.4-2 provides the information required to design Sallen-Key filters.
- High-order filters are filters that have an order greater than 2. A popular strategy for designing high-order filters uses a cascade connection of second-order filters.
- PSpice provides a convenient way to verify that a filter circuit does indeed have the correct transfer function.
- PSpice subcircuits reduce the complexity of simulations of high-order filters. Table 16.6-1 provides PSpice subcircuits for the four Sallen-Key filter stages.

PROBLEMS

Section 16.3 Filters

P 16.3-1 Obtain the transfer function of a third-order Butterworth low-pass filter having a cutoff frequency equal to 200 hertz.

Answer: $H_L(s) = \dfrac{1256^3}{(s+1256)(s^2+1256s+1256^2)}$

P 16.3-2 A dc gain can be incorporated into Butterworth low-pass filters by defining the transfer function to be

$$H_L(s) = \frac{\pm k}{D(s)}$$

where $D(s)$ denotes the polynomials tabulated in Table 16.3-2 and k is the dc gain. The dc gain k is also called the pass-band gain. Obtain the transfer function of a third-order Butterworth low-pass filter having a cutoff frequency equal to 200 rad/s and a pass-band gain equal to 5.

P 16.3-3 High-pass Butterworth filters have transfer functions of the form

$$H_{\mathrm{H}}(s)=\frac{\pm k s^{n}}{D_{n}(s)}$$

where n is the order of the filter, $D_n(s)$ denotes the nth order polynomial in Table 16.3-2, and k is the pass-band gain. Obtain the transfer function of a third-order Butterworth high-pass filter having a cutoff frequency equal to 200 rad/s and a pass-band gain equal to 5.

Answer: $H_{\mathrm{H}}(s)=\dfrac{5s^{3}}{(s+200)(s^{2}+200s+40000)}$

P 16.3-4 High-pass Butterworth filters have transfer functions of the form

$$H_{\mathrm{H}}(s)=\frac{\pm k s^{n}}{D_{n}(s)}$$

where n is the order of the filter, $D_n(s)$ denotes the nth order polynomial in Table 16.3-2, and k is the pass-band gain. Obtain the transfer function of a fourth-order Butterworth high-pass filter having a cutoff frequency equal to 400 hertz and a pass-band gain equal to 5.

P 16.3-5 A band-pass filter has two cutoff frequencies, ω_a and ω_b. Suppose that ω_a is quite a bit smaller than ω_b, say $\omega_a < \omega_b/10$. Let $H_L(s)$ be a low-pass transfer function having a cutoff frequency equal to ω_b and $H_H(s)$ be a high-pass transfer function having a cutoff frequency equal to ω_a. A band-pass transfer function can be obtained as a product of low-pass and high-pass transfer functions, $H_B(s) = H_L(s) \cdot H_H(s)$. The order of the band-pass filter is equal to the sum of the orders of the low-pass and high-pass filters. We usually make the orders of the low-pass and high-pass filter equal, in which case the order of the band-pass is even. The pass-band gain of the band-pass filter is the product of pass-band gains of the low-pass and high-pass transfer functions. Obtain the transfer function of a fourth-order band-pass filter having cutoff frequencies equal to 200 rad/s and 1500 rad/s and a pass-band gain equal to 4.

Answer:

$$H_{\mathrm{B}}(s)=\frac{9{,}000{,}000s^{2}}{(s^{2}+282.8s+40{,}000)(s^{2}+2121s+225{,}0000)}$$

P 16.3-6 In some applications, band-pass filters are used to pass only those signals having a specified frequency ω_0. The cutoff frequencies of the band-pass filter are specified to satisfy $\sqrt{\omega_a\omega_b}=\omega_0$. The transfer function of the band-pass filter is given by

$$H_{\mathrm{B}}(s)=k\left(\frac{\frac{\omega_0}{Q}s}{s^{2}+\frac{\omega_0}{Q}s+\omega_0^{2}}\right)^{m}$$

The order of this band-pass transfer function is $n = 2m$. The pass-band gain is k. Transfer functions of the type are readily implemented as the cascade connection of identical second-order filter stages. Q is the quality factor of the second-order filter stage. The frequency ω_0 is called the center frequency of the band-pass filter. Obtain the transfer function of a fourth-order band-pass filter having a center frequency equal to 200 rad/s and a pass-band gain equal to 4. Use $Q = 1$.

Answer: $H_{\mathrm{B}}(s)=\dfrac{250{,}000s^{2}}{(s^{2}+250s+62{,}500)^{2}}$

P 16.3-7 A band-stop filter has two cutoff frequencies, ω_a and ω_b. Suppose that ω_a is quite a bit smaller than ω_b, say $\omega_a < \omega_b/10$. Let $H_L(s)$ be a low-pass transfer function having a cutoff frequency equal to ω_a and $H_H(s)$ be a high-pass transfer function having a cutoff frequency equal to ω_b. A band-stop transfer function can be obtained as a sum of low-pass and high-pass transfer functions, $H_N(s) = H_L(s) + H_H(s)$. The order of the band-pass filter is equal to the sum of the orders of the low-pass and high-pass filters. We usually make the orders of the low-pass and high-pass filter equal, in which case, the order of the band-stop is even. The pass-band gains of both the low-pass and high-pass transfer functions are set equal to the pass-band gain of the band-stop filter. Obtain the transfer function of a fourth-order band-stop filter having cutoff frequencies equal to 100 rad/s and 2000 rad/s and a pass-band gain equal to 2.

Answer:

$$H_{\mathrm{N}}(s)=\frac{2s^{4}+282.8s^{3}+40{,}000s^{2}+56{,}560{,}000s+8\cdot 10^{10}}{(s^{2}+141.4s+10{,}000)(s^{2}+2828s+4{,}000{,}000)}$$

P 16.3-8 In some applications, band-stop filters are used to reject only those signals having a specified frequency ω_0. The cutoff frequencies of the band-stop filter are specified to satisfy $\sqrt{\omega_a\omega_b}=\omega_0$. The transfer function of the band-pass filter is given by

$$H_{\mathrm{N}}(s)=k-H_{\mathrm{B}}(s)=k-k\left(\frac{\frac{\omega_0}{Q}s}{s^{2}+\frac{\omega_0}{Q}s+\omega_0^{2}}\right)^{m}$$

The order of this band-stop transfer function is $n = 2m$. The pass-band gain is k. Transfer functions of the type are readily implemented using a cascade connection of identical second-order filter stages. Q is the quality factor of the second-order filter stage. The frequency ω_0 is called the center frequency of the band-stop filter. Obtain the transfer function of a fourth-order band-stop filter having a center frequency equal to 200 rad/s and a pass-band gain equal to 4. Use $Q = 1$.

Answer: $H_{\mathrm{N}}(s)=\dfrac{4(s^{2}+62{,}500)^{2}}{(s^{2}+250s+62{,}500)^{2}}$

P 16.3-9 Transfer functions of the form

$$H_{\mathrm{L}}(s)=k\left(\frac{\omega_0^{2}}{s^{2}+\frac{\omega_0}{Q}s+\omega_0^{2}}\right)^{m}$$

are low-pass transfer functions. (This is not a Butterworth transfer function.) The order of this low-pass transfer function is $n = 2m$. The pass-band gain is k. Transfer functions of this type are readily implemented using a cascade connection of identical second-order filter stages. Q is the quality factor of the second-order filter stage. The frequency ω_0 is the cutoff frequency, ω_c, of the low-pass filter. Obtain the transfer function of a fourth-order low-pass filter having a cutoff frequency equal to 300 rad/s and a pass-band gain equal to 4. Use $Q = 1$.

P 16.3-10 Transfer functions of the form

$$H_H(s) = k\left(\frac{s^2}{s^2 + \frac{\omega_0}{Q}s + \omega_0{}^2}\right)^m$$

are high-pass transfer functions. (This is not a Butterworth transfer function.) The order of this high-pass transfer function is $n = 2m$. The pass-band gain is k. Transfer functions of the type are readily implemented using a cascade connection of identical second-order filter stages. Q is the quality factor of the second-order filter stage. The frequency ω_0 is the cutoff frequency, ω_c, of the high-pass filter. Obtain the transfer function of a fourth-order high-pass filter having a cutoff frequency equal to 300 rad/s and a pass-band gain equal to 4. Use $Q = 1$.

Section 16.4 Second-Order Filters

P 16.4-1 The circuit shown in Figure P 16.4-1 is a second-order band-pass filter. Design this filter to have $k = 1$, $\omega_0 = 1500$ rad/s, and $Q = 1$.

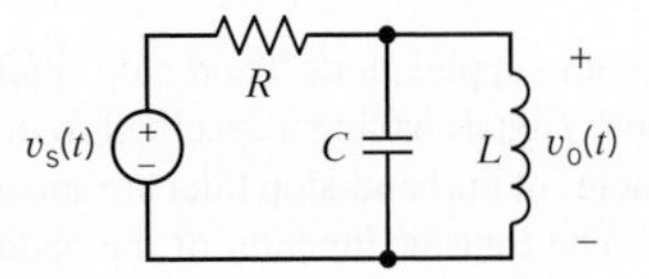

Figure P 16.4-1

P 16.4-2 The circuit shown in Figure P 16.4-2 is a second-order low-pass filter. Design this filter to have $k = 1$, $\omega_0 = 100$ rad/s, and $Q = 0.707$.

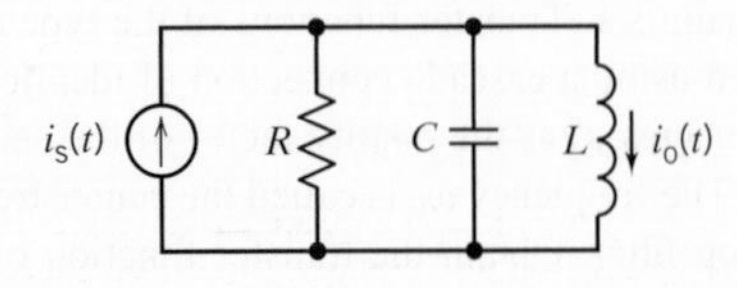

Figure P 16.4-2

P 16.4-3 The circuit shown in Figure P 16.4-3 is a second-order low-pass filter. This filter circuit is called a multiple-loop feedback filter (MFF). The output impedance of this filter is zero, so the MFF low-pass filter is suitable for use as a filter stage in a cascade filter. The transfer function of the low-pass MFF filter is

$$H_L(s) = \frac{-\dfrac{1}{R_1R_3C_1C_2}}{s^2 + \left(\dfrac{1}{R_1C_1} + \dfrac{1}{R_2C_1} + \dfrac{1}{R_3C_1}\right)s + \dfrac{1}{R_2R_3C_1C_2}}$$

Design this filter to have $\omega_0 = 1000$ rad/s and $Q = 8$. What is the value of the dc gain?

Hint: Let $R_2 = R_3 = R$ and $C_1 = C_2 = C$. Pick a convenient value of C and calculate R to obtain $\omega_0 = 2000$ rad/s. Calculate R_1 to obtain $Q = 8$.

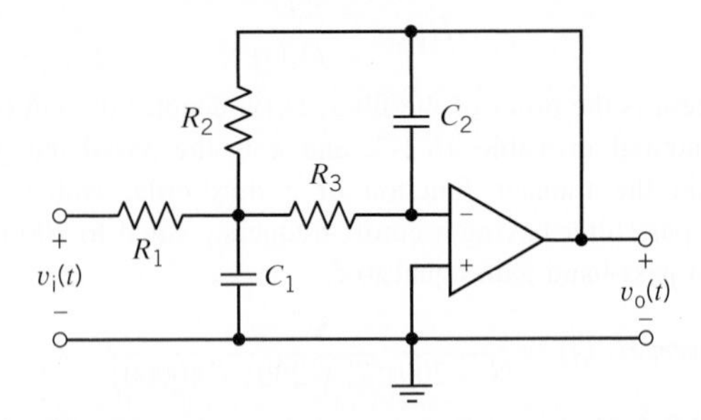

Figure P 16.4-3

P 16.4-4 The circuit shown in Figure P 16.4-4 is a second-order band-pass filter. This filter circuit is called a multiple-loop feedback filter (MFF). The output impedance of this filter is zero, so the MFF band-pass filter is suitable for use as a filter stage in a cascade filter. The transfer function of the band-pass MFF filter is

$$H_B(s) = \frac{-\dfrac{s}{R_1C_2}}{s^2 + \left(\dfrac{1}{R_2C_1} + \dfrac{1}{R_2C_2}\right)s + \dfrac{R_1 + R_3}{R_1R_2R_3C_1C_2}}$$

To design this filter, pick a convenient value of C and then use

$$R_1 = \frac{Q}{k\omega_0 C}, \quad R_2 = \frac{2Q}{\omega_0 C}, \quad \text{and} \quad R_3 = \frac{2Q}{\omega_0 C(2Q^2 - k)}$$

Design this filter to have $k = 5$, $\omega_0 = 1000$ rad/s, and $Q = 8$.

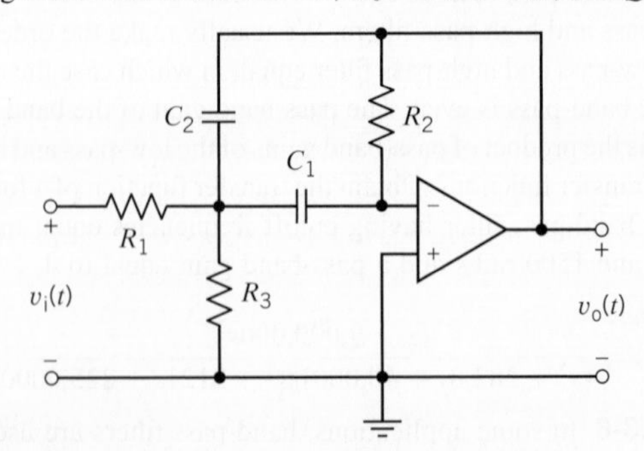

Figure P 16.4-4

P 16.4-5 The circuit shown in Figure P 16.4-5 is a low-pass filter. The transfer function of this filter is

$$H_L(s) = \frac{\dfrac{1}{R_1R_2C_1C_2}}{s^2 + \dfrac{1}{R_1C_1}s + \dfrac{1}{R_1R_2C_1C_2}}$$

Design this filter to have $k = 1$, $\omega_0 = 500$ rad/s, and $Q = 1$.

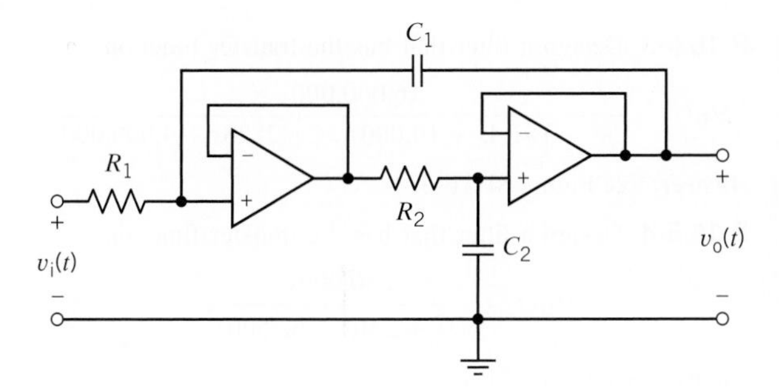

Figure P 16.4-5

P 16.4-6 The *CR:RC* transformation is used to transform low-pass filter circuits into high-pass filter circuits and vice versa. This transformation is applied to *RC* op amp filter circuits. Each capacitor is replaced by a resistor, when each resistor is replaced by a capacitor. Apply the *CR:RC* transformation to the low-pass filter circuit in Figure P 16.4-5 to obtain the high-pass filter circuit shown in Figure P 16.4-6. Design a high-pass filter to have $k = 1$, $\omega_0 = 1000$ rad/s, and $Q = 1$.

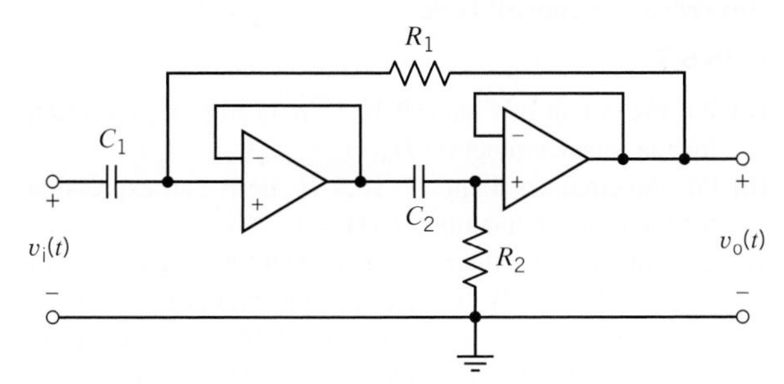

Figure P 16.4-6

P 16.4-7 We have seen that transfer functions can be frequency scaled by replacing s by s/k_f each time that it occurs. Alternately, circuits can also be frequency scaled by dividing each capacitance and each inductance by the frequency scaling factor k_f. Either way, the effect is the same. The frequency response is shifted to the right by k_f. In particular, all cutoff, corner, and resonant frequencies are multiplied by k_f. Suppose that we want to change the cutoff frequency of a filter circuit from ω_{old} to ω_{new}. We set the frequency scale factor to

$$k_f = \frac{\omega_{new}}{\omega_{old}}$$

and then divide each capacitance and each inductance by k_f. Use frequency scaling to change the cutoff frequency of the circuit in Figure P 16.4-7 to 250 rad/s.

Answer: $k_f = 0.05$.

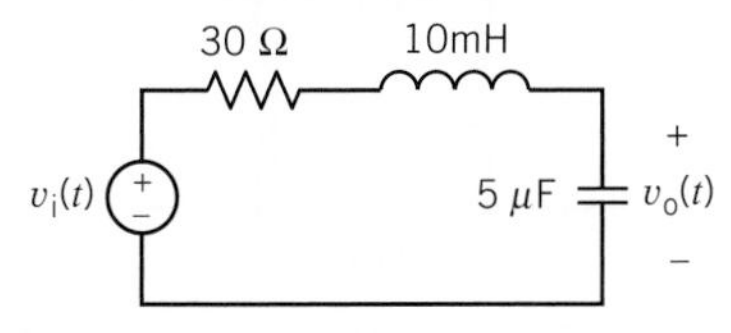

Figure P 16.4-7

P 16.4-8 Impedance scaling is used to adjust the impedances of a circuit. Let k_m denote the impedance scaling factor. Impedance scaling is accomplished by multiplying each impedance by k_m. That means that each resistance and each inductance is multiplied by k_m, but each capacitance is divided by k_m. Transfer functions of the form $H(s) = \frac{V_o(s)}{V_i(s)}$ or $H(s) = \frac{I_o(s)}{I_i(s)}$ are not changed at all by impedance scaling. Transfer functions of the form $H(s) = \frac{V_o(s)}{I_i(s)}$ are multiplied by k_m, whereas transfer functions of the form $H(s) = \frac{I_o(s)}{V_i(s)}$ are divided by k_m. Use impedance scaling to change the values of the capacitances in the filter shown in Figure P 16.4-8 so that the capacitances are in the range of 0.01 μF to 1.0 μF. Calculate the transfer function before and after impedance scaling.

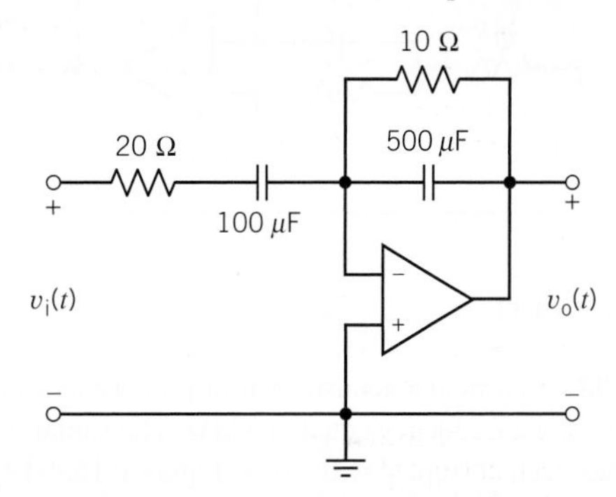

Figure P 16.4-8

P 16.4-9 A band-pass amplifier has the frequency response shown in Figure P 16.4-9. Find the transfer function, $H(s)$.

Hint: $\omega_0 = 2\pi(10 \text{ MHz})$, $k = 10 \text{ dB} = 3.16$, $BW = 0.2$ MHz, $Q = 50$

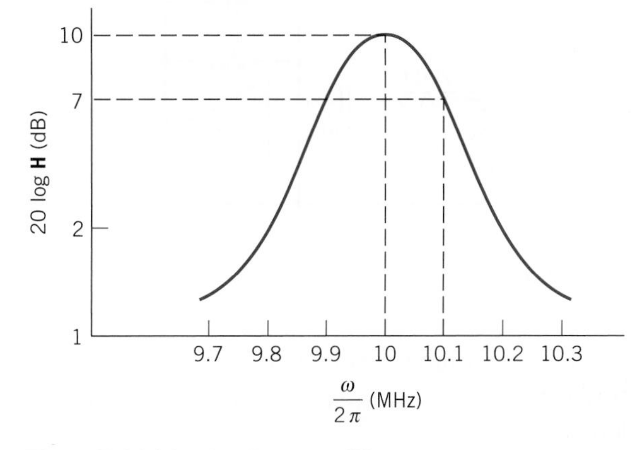

Figure P 16.4-9 A band-pass amplifier.

P 16.4-10 A band-pass filter can be achieved using the circuit of Figure P 16.4-10. Find (a) the magnitude of $\mathbf{H} = \mathbf{V}_o/\mathbf{V}_s$, (b) the low- and high-frequency cutoff frequencies ω_1 and ω_2, and (c) the pass-band gain when $\omega_1 \ll \omega \ll \omega_2$.

Answers:
(b) $\omega_1 = \frac{1}{R_1 C_1}$ and $\omega_2 = \frac{1}{R_2 C_2}$
(c) pass-band gain $= \frac{R_2}{R_1}$

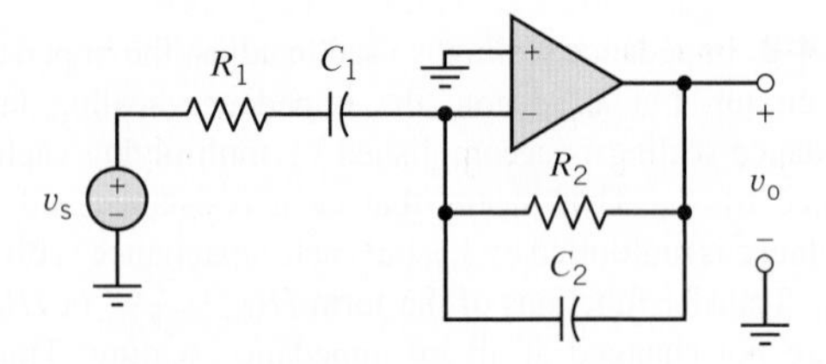

Figure P 16.4-10 A band-pass filter.

P 16.4-11 A unity gain, low-pass filter is obtained from the operational amplifier circuit shown in Figure P 16.4-11. Determine the network function $\mathbf{H}(\omega) = \mathbf{V}_o/\mathbf{V}_s$.

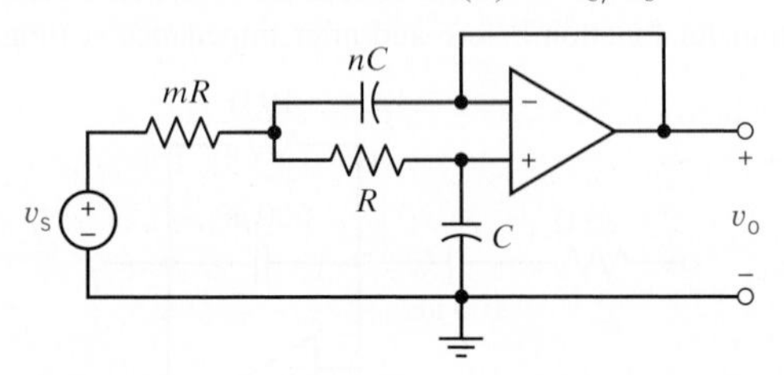

Figure P 16.4-11

P 16.4-12 A particular acoustic sensor produces a sinusoidal output having a frequency equal to 5 kHz. The signal from the sensor has been corrupted with noise. Figure P 16.4-12 shows a band-pass filter that was designed to recover the sensor signal from the noise. The voltage v_s represents the noisy signal from the sensor. The filter output, v_o, should be a less noisy signal. Determine the center frequency and bandwidth of this band-pass filter. Assume that the op amp is ideal.

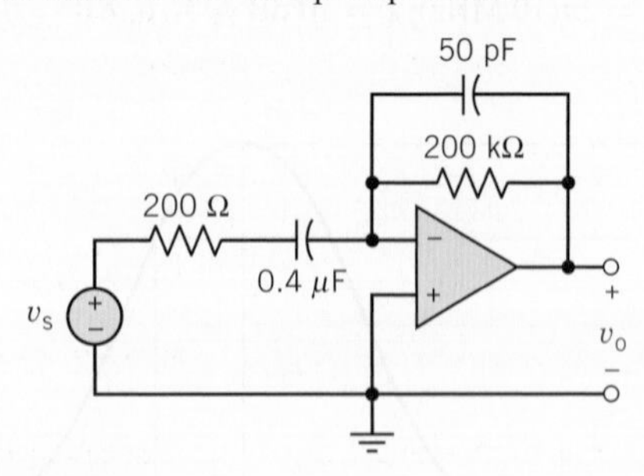

Figure P 16.4-12

Section 16.5 High-Order Filters

P 16.5-1 Design a low-pass filter circuit that has the transfer function

$$H_L(s) = \frac{628^3}{(s+628)(s^2+628s+628^2)}$$

Answer: See Figure SP 16-1.

P 16.5-2 Design a filter that has the transfer function

$$H_H(s) = \frac{5 \cdot s^3}{(s+100)(s^2+100s+10{,}000)}$$

Answer: See Figure SP 16-2.

P 16.5-3 Design a filter that has the transfer function

$$H_B(s) = \frac{16{,}000{,}000 \cdot s^2}{(s^2+141.4s+10{,}000)(s^2+2828s+4{,}000{,}000)}$$

Answer: See Figure SP 16-3.

P 16.5-4 Design a filter that has the transfer function

$$H_B(s) = \frac{250{,}000s^2}{(s^2+250s+62{,}500)^2}$$

Answer: See Figure SP 16-4.

P 16.5-5 Design a filter that has the transfer function

$$H_N(s) = \frac{2s^2}{(s^2+2828s+4{,}000{,}000)} + \frac{20{,}000}{(s^2+141.4s+10{,}000)}$$

Answer: See Figure SP 16-5.

P 16.5-6 Design a filter that has the transfer function

$$H_N(s) = \frac{4(s^2+62{,}500)^2}{(s^2+250s+62{,}500)^2}$$

Answer: See Figure SP 16-6.

P 16.5-7

(a) For the circuit of Figure P 16.5-7*a*, derive an expression for the transfer function $H_a(s) = V_1/V_s$.

(b) For the circuit of Figure P 16.5-7*b*, derive an expression for the transfer function $H_b(s) = V_2/V_1$.

(c) Each of the above filters is a first-order filter. The circuit of Figure P 16.5-7*c* is the cascade connection of the circuits of Figure P 16.5-7*a* and Figure P 16.5-7*b*. Derive an expression for the transfer function $H_c(s) = V_2/V_s$ of the second-order circuit in Figure P 16.5-7*c*.

(d) Why doesn't $H_c(s) = H_a(s)H_b(s)$?

Hint: Consider loading.

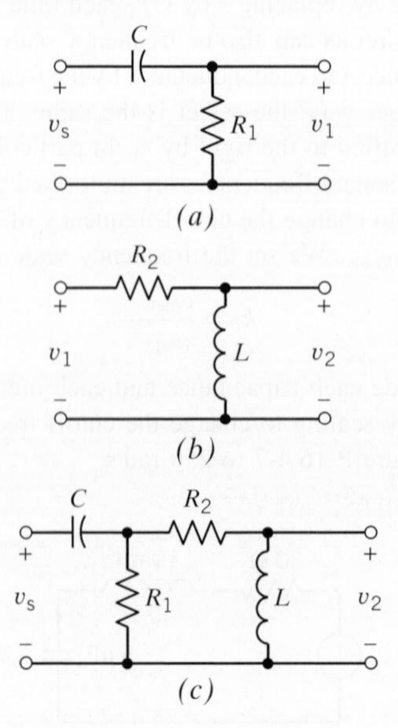

Figure P 16.5-7 (*a*) Circuit for $\mathbf{H}_1$. (*b*) Circuit for $\mathbf{H}_2$. (*c*) Circuit for $\mathbf{H}$.

P 16.5-8 Two filter stages are connected in cascade as shown in Figure P 16.5-8. The transfer function of each filter stage is of the form

$$H(s) = \frac{As}{(1 + s/\omega_L)(1 + s/\omega_H)}$$

Determine the transfer function of the fourth-order filter. (Assume that there is no loading.)

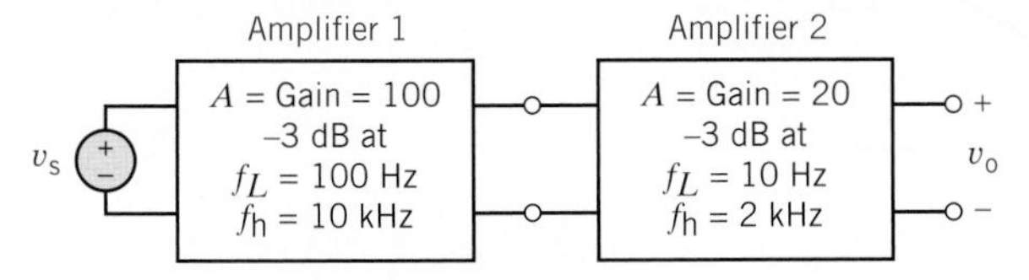

Figure P 16.5-8 Two cascaded amplifiers.

P 16.5-9 A second-order filter uses two identical first-order filter stages as shown in Figure P 16.5-9. Each filter stage is specified to have a cutoff or break frequency at $\omega_c = 1000$ rad/s and a pass-band gain of 0 dB. (a) Find the required R_1, R_2, and C. (b) Find the gain of the second-order filter at $\omega = 10{,}000$ rad/s in decibels.

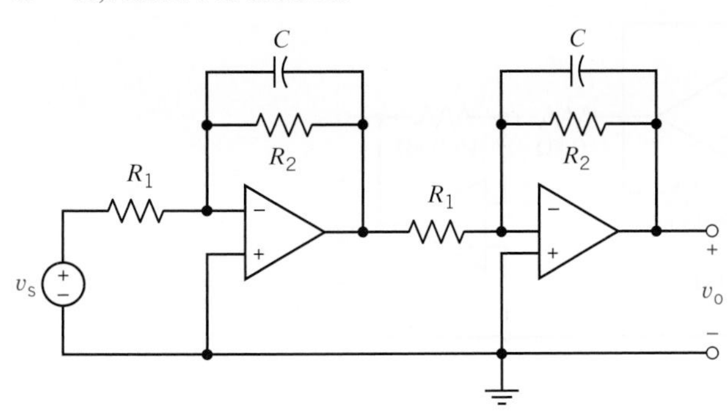

Figure P 16.5-9

Section 16.7 How Can We Check . . . ?

P 16.7-1 The specifications for a band-pass filter require that $\omega_0 = 100$ rad/s, $Q = 4$, and $k = 3$. The transfer function of a filter designed to satisfy these specifications is

$$H(s) = \frac{75s}{s^2 + 25s + 10{,}000}$$

Does this filter satisfy the specifications?

P 16.7-2 The specifications for a band-pass filter require that $\omega_0 = 100$ rad/s, $Q = 4$, and $k = 3$. The transfer function of a filter designed to satisfy these specifications is

$$H(s) = \frac{75s}{s^2 + 25s + 10{,}000}$$

Does this filter satisfy the specifications?

P 16.7-3 The specifications for a low-pass filter require that $\omega_0 = 20$ rad/s, $Q = 0.6$, and $k = 1.5$. The transfer function of a filter designed to satisfy these specifications is

$$H(s) = \frac{600}{s^2 + 25s + 400}$$

Does this filter satisfy the specifications?

P 16.7-4 The specifications for a low-pass filter require that $\omega_0 = 25$ rad/s, $Q = 0.8$, and $k = 1.2$. The transfer function of a filter designed to satisfy these specifications is

$$H(s) = \frac{750}{s^2 + 62.5s + 625}$$

Does this filter satisfy the specifications?

P 16.7-5 The specifications for a high-pass filter require that $\omega_0 = 12$ rad/s, $Q = 5$, and $k = 5$. The transfer function of a filter designed to satisfy these specifications is

$$H(s) = \frac{5s^2}{s^2 + 30s + 144}$$

Does this filter satisfy the specifications?

PSpice Problems

SP 16-1 The filter circuit shown in Figure SP 16-1 was designed to have the transfer function

$$H_L(s) = \frac{628^3}{(s + 628)(s^2 + 628s + 628^2)}$$

Figure SP 16-1

Use PSpice to verify that the filter circuit does indeed implement this transfer function.

SP 16-2 The filter circuit shown in Figure SP 16-2 was designed to have the transfer function

$$H_H(s) = \frac{5 \cdot s^3}{(s + 100)(s^2 + 100s + 10{,}000)}$$

Use PSpice to verify that the filter circuit does indeed implement this transfer function.

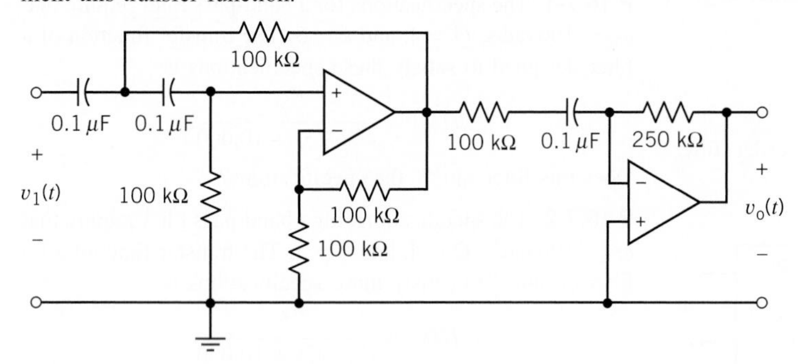

Figure SP 16-2

SP 16-3 The filter circuit shown in Figure SP 16-3 was designed to have the transfer function

$$H_B(s) = \frac{16{,}000{,}000 \cdot s^2}{(s^2 + 141.4s + 10{,}000)(s^2 + 2828s + 4{,}000{,}000)}$$

Use PSpice to verify that the filter circuit does indeed implement this transfer function.

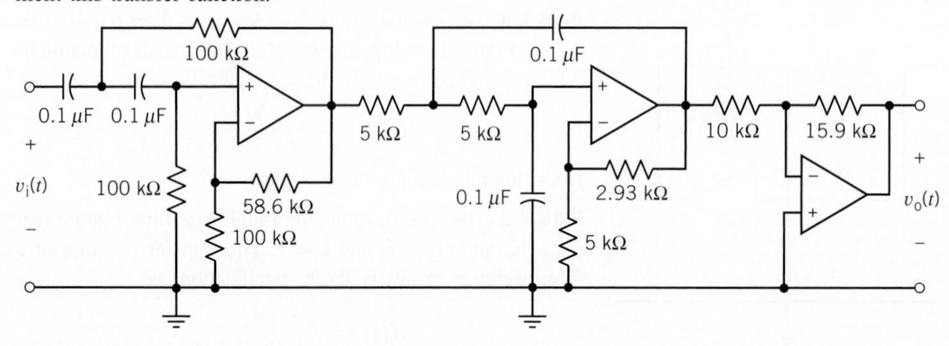

Figure SP 16-3

SP 16-4 The filter circuit shown in Figure SP 16-4 was designed to have the transfer function

$$H_B(s) = \frac{250{,}000s^2}{(s^2 + 250s + 62{,}500)^2}$$

Use PSpice to verify that the filter circuit does indeed implement this transfer function.

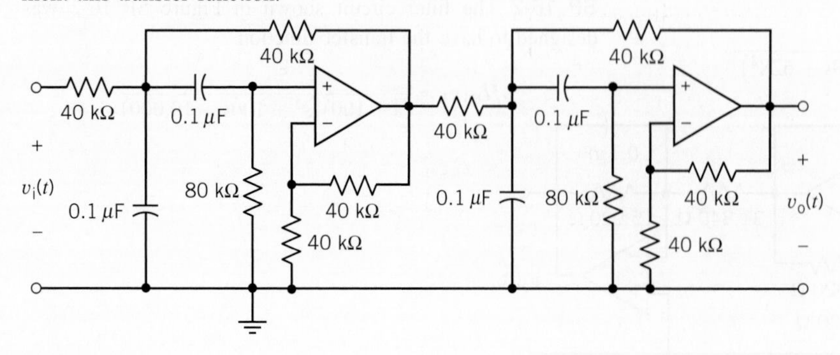

Figure SP 16-4

SP 16-5 The filter circuit shown in Figure SP 16-5 was designed to have the transfer function

$$H_N(s) = \frac{2s^2}{(s^2 + 2828s + 4{,}000{,}000)} + \frac{20{,}000}{(s^2 + 141.4s + 10{,}000)}$$

Use PSpice to verify that the filter circuit does indeed implement this transfer function.

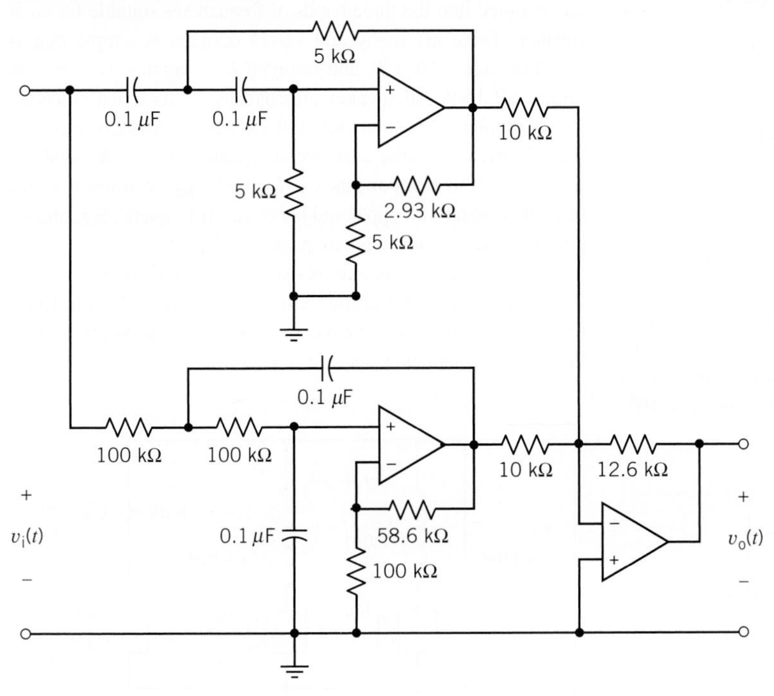

Figure SP 16-5

SP 16-6 The filter circuit shown in Figure SP 16-6 was designed to have the transfer function

$$H_N(s) = \frac{4(s^2 + 62{,}500)^2}{(s^2 + 250s + 62{,}500)^2}$$

Use PSpice to verify that the filter circuit does indeed implement this transfer function.

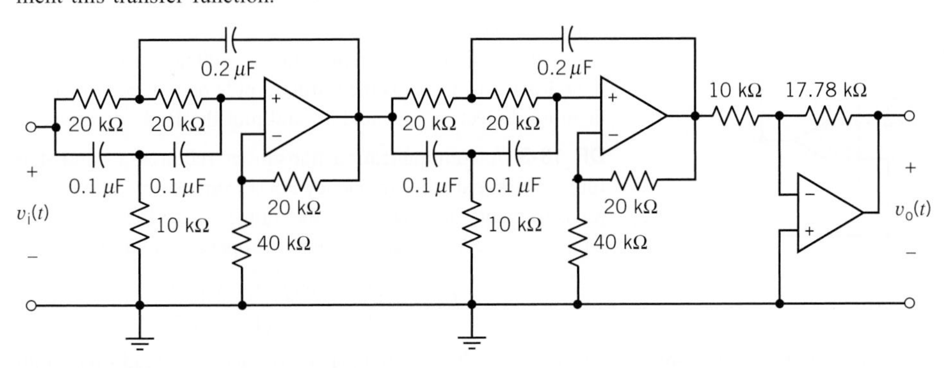

Figure SP 16-6

SP 16-7 A notch filter is shown in Figure SP 16-7. The output of a two-stage filter is v_1, and the output of a three-stage filter is v_2. Plot the Bode diagram of $\mathbf{V}_1/\mathbf{V}_s$ and $\mathbf{V}_2/\mathbf{V}_s$ and compare the results when $L = 10$ mH and $C = 1\mu$F.

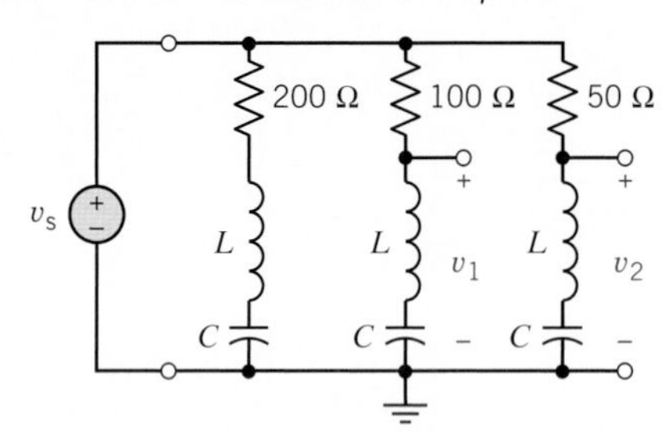

Figure SP 16-7

SP 16-8 An acoustic sensor operates in the range of 5 kHz to 25 kHz and is represented in Figure SP 16-8 by v_s. It is specified that the band-pass filter shown in the figure passes the signal in the frequency range within 3 dB of the center frequency gain. Determine the bandwidth and center frequency of the circuit when the op amp has $R_i = 500$ kΩ, $R_o = 1$ kΩ, and $A = 10^6$.

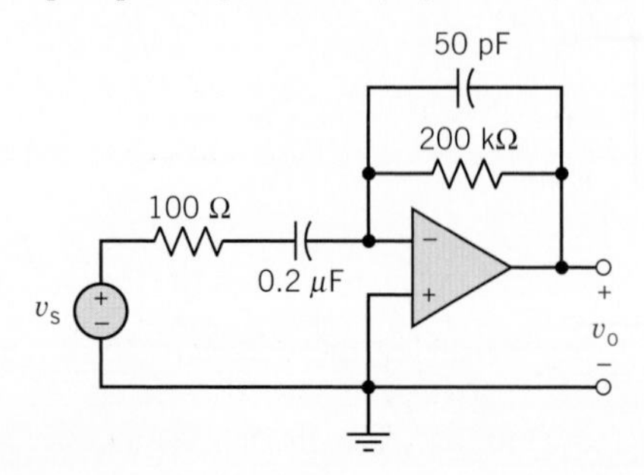

Figure SP 16-8

SP 16-9 Frequently, audio systems contain two or more loudspeakers that are intended to handle different parts of the audio-frequency spectrum. In a three-way setup, one speaker, called a woofer, handles low frequencies. A second, the tweeter, handles high frequencies, and a third, the mid-range, handles the middle range of the audio spectrum.

A three-way filter, called a crossover network, splits the audio signal into the three bands of frequencies suitable for each speaker. There are many and varied designs. A simple one is based on series *LR*, *CR*, and resonant *RLC* circuits as shown in Figure SP 16-9. All speaker impedances are assumed resistive. The conditions are (1) woofer, at the crossover frequency: $X_{L1} = R_W$; (2) tweeter, at the crossover frequency: $X_{C3} = R_T$; and (3) midrange, with components C_2, L_2, and R_{MR} forming a series resonant circuit with upper and lower cutoff frequencies f_u and f_L, respectively. The resonant frequency $= (f_u f_L)^{1/2}$.

When all the speaker resistances are 8 Ω, determine the frequency response and the cutoff frequencies. Plot the Bode diagram for the three speakers. Determine the bandwidth of the midrange speaker section.

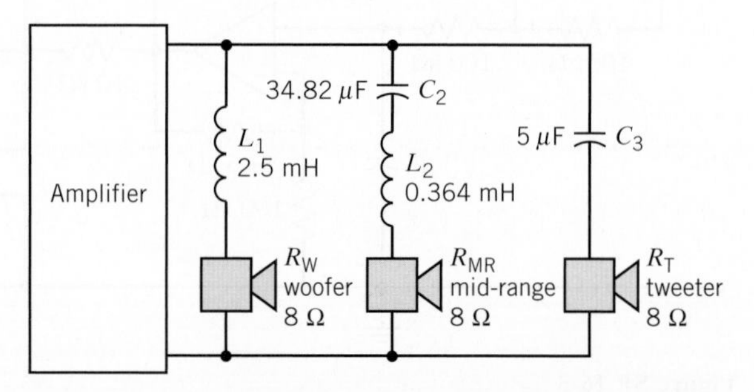

Figure SP 16-9 Three-way filter for a speaker system.

Design Problems

DP 16-1 Design a band-pass filter with a center frequency of 200 kHz and a bandwidth of 10 kHz, using the circuit shown in Figure DP 16-1. Assume that $C = 200$ pF and find R and R_3. Use PSpice to verify the design.

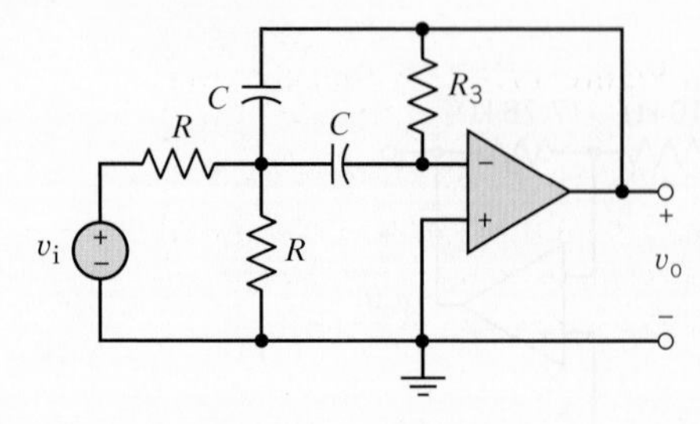

Figure DP 16-1

DP 16-2 A communication transmitter requires a band-pass filter to eliminate low-frequency noise from nearby traffic. Measurements indicate that the range of traffic rumble is $2 < \omega < 12$ rad/s. A designer proposes a filter as

$$H(s) = \frac{(1 + s/\omega_1)^2(1 + s/\omega_3)}{(1 + s/\omega_2)^3}$$

where $s = j\omega$.

It is desired that signals with $\omega > 100$ rad/s pass with less than 3-dB loss, whereas the traffic rumble be reduced by 46 dB or more. Select ω_1, ω_2, and ω_3 and plot the Bode diagram.

DP 16-3 A communication transmitter requires a band-stop filter to eliminate low-frequency noise from nearby auto traffic. Measurements indicate that the range of traffic rumble is 2 rad/s $< \omega <$ 12 rad/s. A designer proposes a filter as

$$H(s) = \frac{(1 + s/\omega_1)^2(1 + s/\omega_3)^2}{(1 + s/\omega_2)^2(1 + s/\omega_4)^2}$$

where $s = j\omega$. It is desired that signals above 130 rad/s pass with less than 4-dB loss, whereas the traffic rumble be reduced by 35 dB or more. Select ω_1, ω_2, ω_3, and ω_4 and plot the Bode diagram.

CHAPTER 17

Two-Port and Three-Port Networks

IN THIS CHAPTER

17.1 Introduction
17.2 T-to-Π Transformation and Two-Port Three-Terminal Networks
17.3 Equations of Two-Port Networks
17.4 *Z* and *Y* Parameters for a Circuit with Dependent Sources
17.5 Hybrid and Transmission Parameters
17.6 Relationships Between Two-Port Parameters
17.7 Interconnection of Two-Port Networks
17.8 How Can We Check . . . ?
17.9 **DESIGN EXAMPLE**—Transistor Amplifier
17.10 Summary
Problems
Design Problems

17.1 INTRODUCTION

Many practical circuits have just two *ports* of access, that is, two places where signals may be input or output. For example, a coaxial cable between Boston and San Francisco has two ports, one at each of those cities. The object here is to analyze such networks in terms of their terminal characteristics without particular regard to the internal composition of the network. To this end, the network will be described by relationships between the port voltages and currents.

We study two-port networks and the parameters that describe them for a number of reasons. Most circuits or systems have at least two ports. We may put an input signal into one port and obtain an output signal from the other. The parameters of the two-port network completely describe its behavior in terms of the voltage and current at each port. Thus, knowing the parameters of a two-port network permits us to describe its operation when it is connected into a larger network. Two-port networks are also important in modeling electronic devices and system components. For example, in electronics, two-port networks are employed to model transistors, op amps, transformers, and transmission lines.

A two-port network is represented by the network shown in Figure 17.1-1. A four-terminal network is called a *two-port network* when the current entering one terminal of a pair exits the other terminal in the pair. For example, I_1 enters terminal a and exits terminal b of the input terminal pair a–b. It will be assumed in our discussion that there are no independent sources or nonzero initial conditions within the linear two-port network. Two-port networks may or may not be purely resistive and can in general be formulated in terms of the *s*-variable or the $j\omega$-variable.

A **two-port network** has two access points appearing as terminal pairs. The current entering one terminal of a pair exits the other terminal in the pair.

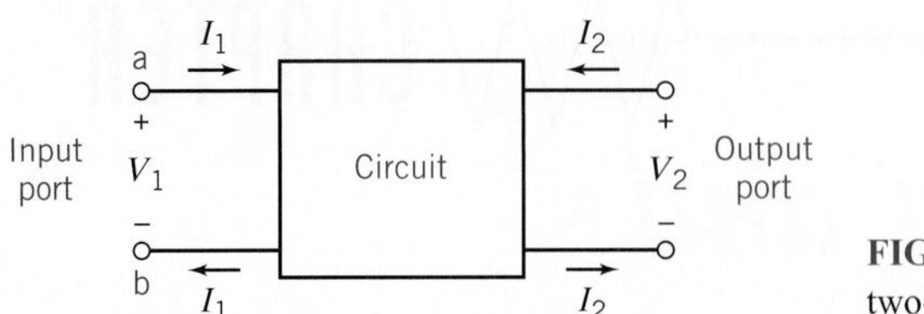

FIGURE 17.1-1 A two-port network.

17.2 T-TO-Π TRANSFORMATION AND TWO-PORT THREE-TERMINAL NETWORKS

Two networks that occur frequently in circuit analysis are the T and Π networks, as shown in Figure 17.2-1. When redrawn, they can appear as the Y or delta (Δ) networks of Figure 17.2-2.

If a network has mirror-image symmetry with respect to some centerline, that is, if a line can be found to divide the network into two symmetrical halves, the network is a *symmetrical network*. The T network is symmetrical when $Z_1 = Z_2$, and the Π network is symmetrical when $Z_A = Z_B$. Furthermore, if all the impedances in either the T or Π network are equal, then the T or Π network is *completely symmetrical*.

Note that the networks shown in Figure 17.2-1 and Figure 17.2-2 have two access ports and three terminals. For example, one port is obtained for the terminal pair a–c and the other port is b–c.

We can obtain equations for direct transformation or conversion from a T network to a Π network, or from a Π network to a T network, by considering that, for equivalence, the two networks must have the same impedance when measured between the same pair of terminals. For example, at port 1 (at a–c) for the two networks of Figure 17.2-2, we require

$$Z_1 + Z_3 = \frac{Z_A(Z_B + Z_C)}{Z_A + Z_B + Z_C}$$

To convert a Π network to a T network, relationships for Z_1, Z_2, and Z_3 must be obtained in terms of the impedances Z_A, Z_B, and Z_C. With some algebraic effort, we can show that

$$Z_1 = \frac{Z_A Z_C}{Z_A + Z_B + Z_C} \tag{17.2-1}$$

$$Z_2 = \frac{Z_B Z_C}{Z_A + Z_B + Z_C} \tag{17.2-2}$$

$$Z_3 = \frac{Z_A Z_B}{Z_A + Z_B + Z_C} \tag{17.2-3}$$

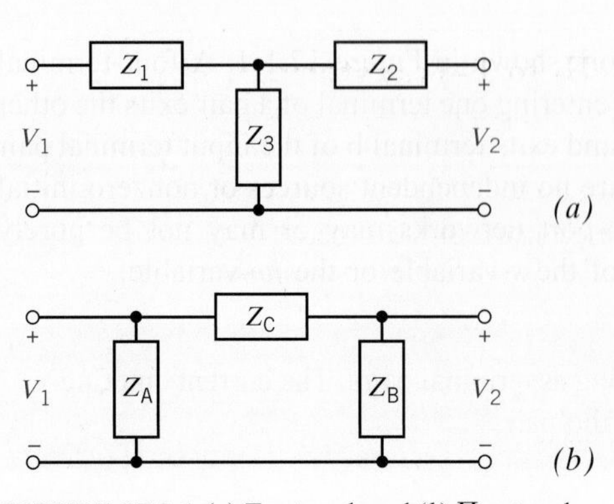

FIGURE 17.2-1 (*a*) T network and (*b*) Π network.

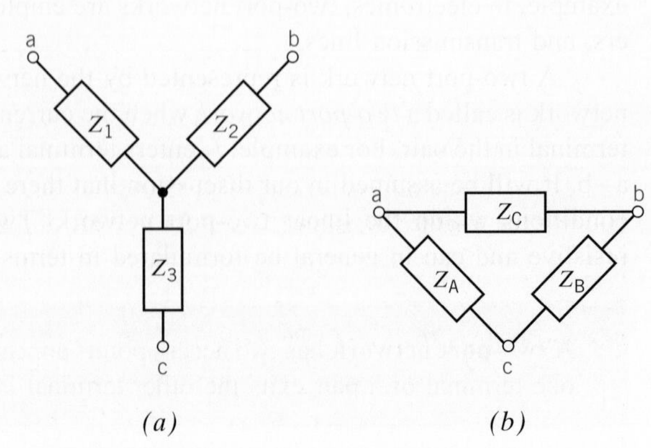

FIGURE 17.2-2 (*a*) Y network and (*b*) Δ network.

Similarly, we can obtain the relationships for Z_A, Z_B, and Z_C as

$$Z_A = \frac{Z_1Z_2 + Z_2Z_3 + Z_3Z_1}{Z_2} \tag{17.2-4}$$

$$Z_B = \frac{Z_1Z_2 + Z_2Z_3 + Z_3Z_1}{Z_1} \tag{17.2-5}$$

$$Z_C = \frac{Z_1Z_2 + Z_2Z_3 + Z_3Z_1}{Z_3} \tag{17.2-6}$$

Each T impedance equals the product of the two adjacent legs of the Π network divided by the sum of the three legs of the Π network. On the other hand, each leg of the Π network equals the sum of the possible products of the T impedances divided by the opposite T impedance.

When a T or a Π network is completely symmetrical, the conversion equations reduce to

$$Z_T = \frac{Z_\Pi}{3} \tag{17.2-7}$$

and

$$Z_\Pi = 3Z_T \tag{17.2-8}$$

where Z_T is the impedance in each leg of the T network and Z_Π is the impedance in each leg of the Π network.

EXAMPLE 17.2-1 T- to Π-Transformation

Find the Π form of the T circuit given in Figure 17.2-3*a*.

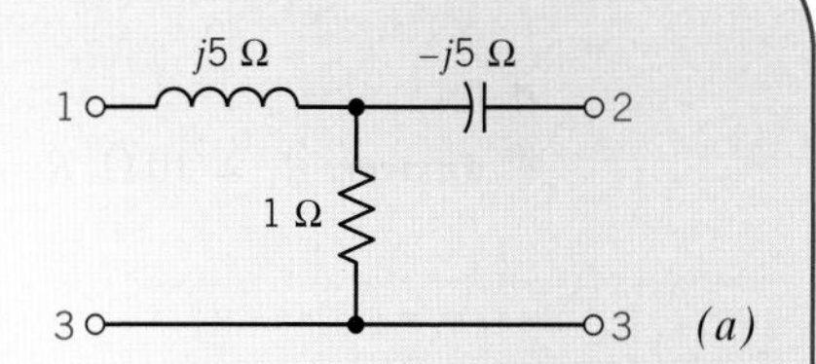

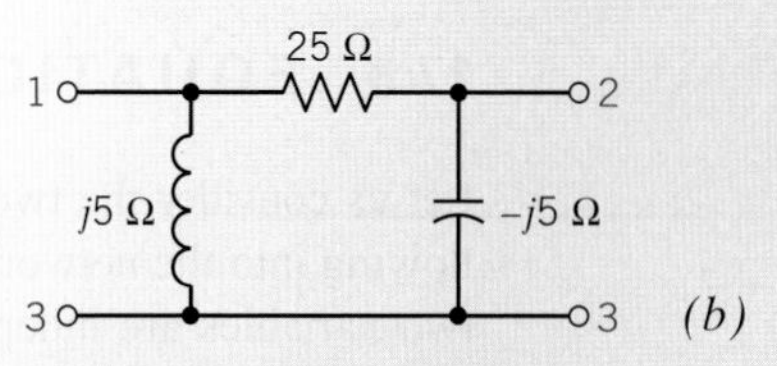

FIGURE 17.2-3 (*a*) T circuit of Example 17.2-1. (*b*) Π equivalent of T circuit.

Solution

The first impedance of the Π network, using Eq. 17.2-4, is

$$Z_A = \frac{Z_1Z_2 + Z_2Z_3 + Z_3Z_1}{Z_2} = \frac{j5(-j5) + (-j5)1 + 1(j5)}{-j5} = j5\ \Omega$$

Similarly, the second impedance, using Eq. 17.2-5, is

$$Z_B = -j5\ \Omega$$

and the third impedance, using Eq. 17.2-6, is

$$Z_C = 25\ \Omega$$

The Π equivalent circuit is shown in Figure 17.2-3*b*.

EXAMPLE 17.2-2 Π- to T-Transformation

Find the T network equivalent to the Π network shown in Figure 17.2-4 in the *s*-domain using the Laplace transform. Then, for $s = j1$, find the elements of the T network.

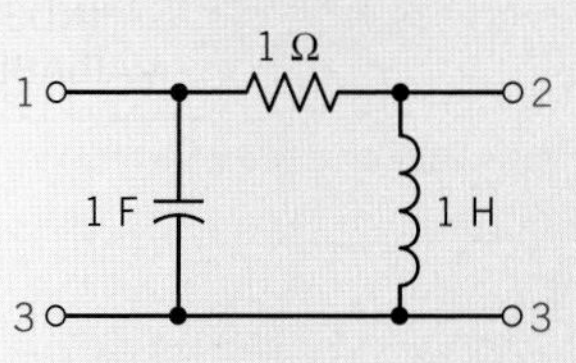

FIGURE 17.2-4 Π circuit of Example 17.2-2.

Solution

First, using Eq. 17.2-1, we have

$$Z_1 = \frac{(1)(1/s)}{s + 1 + 1/s} = \frac{1}{s^2 + s + 1}$$

Then, using Eq. 17.2-2, we have

$$Z_2 = \frac{1(s)}{s+1+1/s} = \frac{s^2}{s^2+s+1}$$

Finally, the third impedance is (Eq. 17.2-3)

$$Z_3 = \frac{s(1/s)}{s+1+1/s} = \frac{s}{s^2+s+1}$$

To find the elements of the T network at $s = j1$, we substitute $s = j1$ and determine each impedance. Then, we have

$$Z_1 = -j, \quad Z_2 = j, \quad Z_3 = 1$$

Therefore, the equivalent T network is as shown in Figure 17.2-5 for the value $s = j1$.

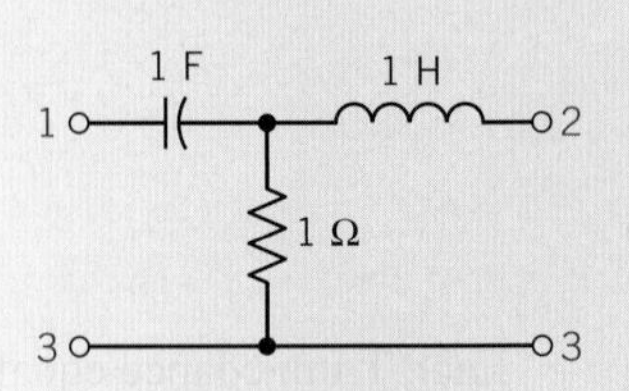

FIGURE 17.2-5 T circuit equivalent of the original Π circuit of Example 17.2-2 for $s = j1$.

EXERCISE 17.2-1 Find the T circuit equivalent to the Π circuit shown in Figure E 17.2-1.

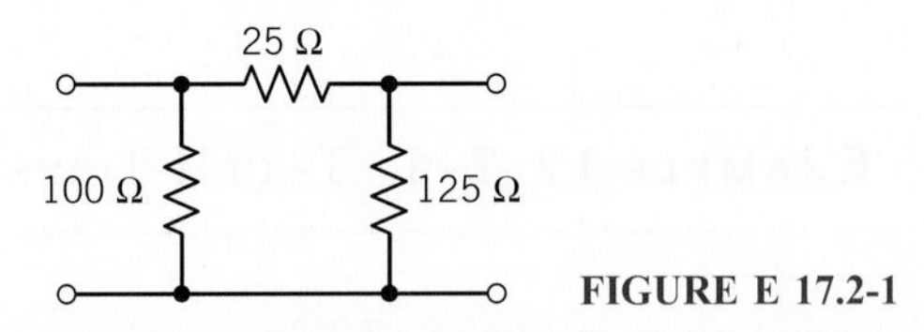

FIGURE E 17.2-1

Answer: $R_1 = 10\ \Omega$, $R_2 = 12.5\ \Omega$, and $R_3 = 50\ \Omega$

17.3 EQUATIONS OF TWO-PORT NETWORKS

Let us consider the two-port network of Figure 17.1-1. By convention, I_1 and I_2 are assumed to be flowing into the network as shown. The variables are V_1, V_2, I_1, and I_2. Within the two-port network, two variables are independent and two are dependent, and we may select a set of two independent variables from the six possible sets: (V_1, V_2), (I_1, I_2), (V_1, I_2), (I_1, V_2), (V_1, I_1), and (V_2, I_2). We will also assume linear elements.

The possibilities for independent (input) variables and the associated dependent variables are summarized in Table 17.3-1. The names of the associated six sets of circuit parameters are also identified in Table 17.3-1. For the case of phasor transforms or Laplace transforms with the circuit of Figure 17.1-1, we

Table 17.3-1 Six Circuit-Parameter Models

INDEPENDENT VARIABLES (INPUTS)	DEPENDENT VARIABLES (OUTPUTS)	CIRCUIT PARAMETERS
I_1, I_2	V_1, V_2	Impedance Z
V_1, V_2	I_1, I_2	Admittance Y
V_1, I_2	I_1, V_2	Inverse hybrid g
I_1, V_2	V_1, I_2	Hybrid h
V_2, I_2	V_1, I_1	Transmission T
V_1, I_1	V_2, I_2	Inverse transmission T'

Table 17.3-2 Equations for the Six Sets of Two-Port Parameters

Impedance Z	$\begin{cases} V_1 = Z_{11}I_1 + Z_{12}I_2 \\ V_2 = Z_{21}I_1 + Z_{22}I_2 \end{cases}$
Admittance Y	$\begin{cases} I_1 = Y_{11}V_1 + Y_{12}V_2 \\ I_2 = Y_{21}V_1 + Y_{22}V_2 \end{cases}$
Hybrid h	$\begin{cases} V_1 = h_{11}I_1 + h_{12}V_2 \\ I_2 = h_{21}I_1 + h_{22}V_2 \end{cases}$
Inverse hybrid g	$\begin{cases} I_1 = g_{11}V_1 + g_{12}I_2 \\ V_2 = g_{21}V_1 + g_{22}I_2 \end{cases}$
Transmission T	$\begin{cases} V_1 = AV_2 - BI_2 \\ I_1 = CV_2 - DI_2 \end{cases}$
Inverse transmission T'	$\begin{cases} V_2 = A'V_1 - B'I_1 \\ I_2 = C'V_1 - D'I_1 \end{cases}$

have the familiar impedance equations in which the output variables are V_1 and V_2, as follows:

$$V_1 = Z_{11}I_1 + Z_{12}I_2 \tag{17.3-1}$$

$$V_2 = Z_{21}I_1 + Z_{22}I_2 \tag{17.3-2}$$

The equations for the admittances are

$$I_1 = Y_{11}V_1 + Y_{12}V_2 \tag{17.3-3}$$

$$I_2 = Y_{21}V_1 + Y_{22}V_2 \tag{17.3-4}$$

It is appropriate, if preferred, to use lowercase letters z and y for the coefficients of Eqs. 17.3-1 through 17.3-4. The equations for the six sets of two-port parameters are summarized in Table 17.3-2.

For linear elements and no dependent sources or op amps within the two-port network, we can show by the theorem of reciprocity that $Z_{12} = Z_{21}$ and $Y_{21} = Y_{12}$. One possible arrangement of a passive circuit as a T circuit is shown in Figure 17.3-1. Writing the two mesh equations for Figure 17.3-1, we can readily obtain Eqs. 17.3-1 and 17.3-2. Therefore, the circuit of Figure 17.3-1 can represent the impedance parameters. A possible arrangement of the admittance parameters as a Π circuit is shown in Figure 17.3-2.

Examining Eq. 17.17, we see that we can measure Z_{11} by obtaining

$$Z_{11} = \left.\frac{V_1}{I_1}\right|_{I_2=0}$$

Of course, $I_2 = 0$ implies that the output terminals are open-circuited. Thus, the Z parameters are often called *open-circuit impedances*.

The Y parameters can be measured by determining

$$Y_{12} = \left.\frac{I_1}{V_2}\right|_{V_1=0}$$

In general, the admittance parameters are called *short-circuit admittance parameters*.

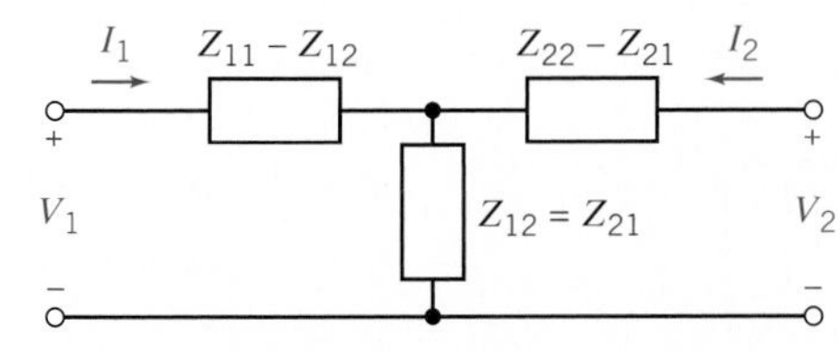

FIGURE 17.3-1 A T circuit representing the impedance parameters.

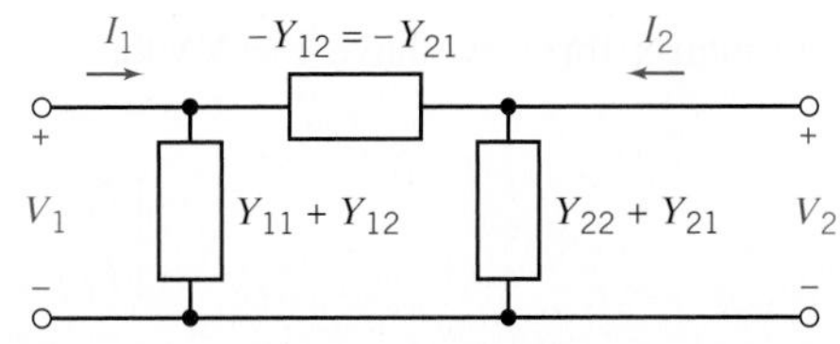

FIGURE 17.3-2 A Π circuit representing the admittance parameters.

Example 17.3-1 Admittance Parameters and Impedance Parameters

Determine the admittance and the impedance parameters of the T network shown in Figure 17.3-3.

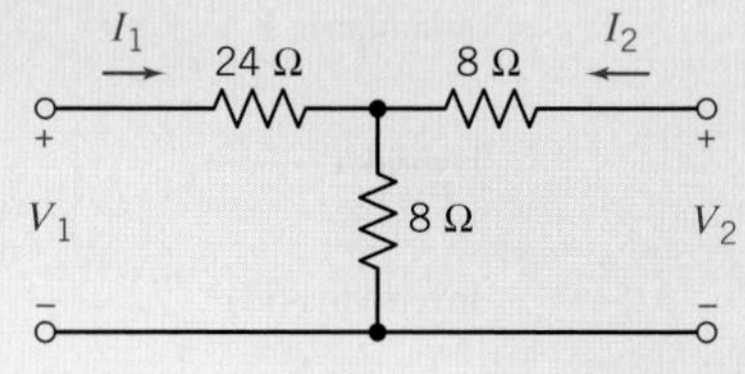

FIGURE 17.3-3 Circuit for Example 17.3-1.

Solution

The admittance parameters use the output terminals shorted and

$$Y_{11} = \left.\frac{I_1}{V_1}\right|_{V_2=0}$$

Then, the two 8-Ω resistors are in parallel and $V_1 = 28I_1$. Therefore, we have

$$Y_{11} = \frac{1}{28}\ \text{S}$$

For Y_{12}, we have

$$Y_{12} = \left.\frac{I_1}{V_2}\right|_{V_1=0}$$

so we short-circuit the input terminals. Then we have the circuit as shown in Figure 17.3-4.

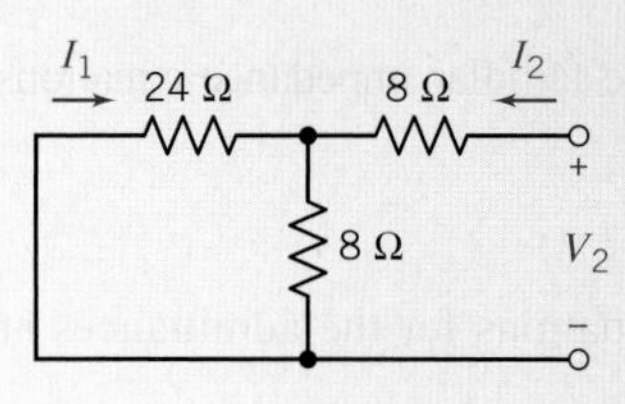

FIGURE 17.3-4 Circuit of Example 17.3-1 with the input terminals shorted.

Employing current division, we have

$$-I_1 = I_2\left(\frac{8}{8+24}\right)$$

and
$$I_2 = \frac{V_2}{8 + [8(24)/(8+24)]} = \frac{V_2}{14}$$

Therefore
$$Y_{12} = \frac{I_1}{V_2} = \frac{-(V_2/14)(1/4)}{V_2} = -\frac{1}{56}\ \text{S}$$

Furthermore,

$$Y_{21} = Y_{12} = -\frac{1}{56}\ \text{S}$$

Finally, Y_{22} is obtained from Figure 17.3-4 as

$$Y_{22} = \left.\frac{I_2}{V_2}\right|_{V_1=0}$$

where
$$I_2 = \frac{V_2}{8 + [8(24)/(8+24)]} = \frac{V_2}{14}$$

Therefore,
$$Y_{22} = \frac{1}{14}\ \text{S}$$

Thus, in matrix form, we have $\mathbf{I} = \mathbf{YV}$ or

$$\begin{bmatrix} I_1 \\ I_2 \end{bmatrix} = \begin{bmatrix} \frac{1}{28} & -\frac{1}{56} \\ -\frac{1}{56} & \frac{1}{14} \end{bmatrix} \begin{bmatrix} V_1 \\ V_2 \end{bmatrix}$$

Now, let us find the impedance parameters. We have

$$Z_{11} = \left.\frac{V_1}{I_1}\right|_{I_2=0}$$

The output terminals are open-circuited, so we have the circuit of Figure 17.3-3. Then,

$$Z_{11} = 24 + 8 = 32\ \Omega$$

Similarly, $Z_{22} = 16\ \Omega$ and $Z_{21} = Z_{12} = 8\ \Omega$. Then, in matrix form, we have $\mathbf{V} = \mathbf{ZI}$ or

$$\begin{bmatrix} V_1 \\ V_2 \end{bmatrix} = \begin{bmatrix} 32 & 8 \\ 8 & 16 \end{bmatrix} \begin{bmatrix} I_1 \\ I_2 \end{bmatrix}$$

The general methods for finding the Z parameters and the Y parameters are summarized in Tables 17.3-3 and 17.3-4, respectively.

Table 17.3-3 Method of Obtaining the Z Parameters of a Circuit

Step IA	To determine Z_{11} and Z_{21}, connect a voltage source V_1 to the input terminals and open-circuit the output terminals.
Step IB	Find I_1 and V_2 and then $Z_{11} = V_1/I_1$ and $Z_{21} = V_2/I_1$.
Step IIA	To determine Z_{22} and Z_{12}, connect a voltage source V_2 to the output terminals and open-circuit the input terminals.
Step IIB	Find I_2 and V_1 and then $Z_{22} = V_2/I_2$ and $Z_{12} = V_1/I_2$.

Note: $Z_{12} = Z_{21}$ only when there are no dependent sources or op amps within the two-port network.

Table 17.3-4 Method for Obtaining the Y Parameters of a Circuit

Step IA	To determine Y_{11} and Y_{21}, connect a current source I_1 to the input terminals and short-circuit the output terminals ($V_2 = 0$).
Step IB	Find V_1 and I_2 and then $Y_{11} = I_1/V_1$ and $Y_{21} = I_2/V_1$.
Step IIA	To determine Y_{22} and Y_{12}, connect a current source I_2 to the output terminals and short-circuit the input terminals ($V_1 = 0$).
Step IIB	Find I_1 and V_2 and then $Y_{22} = I_2/V_2$ and $Y_{12} = I_1/V_2$.

Note: $Y_{12} = Y_{21}$ only when there are no dependent sources or op amps within the two-port network.

EXERCISE 17.3-1 Find the Z and Y parameters of the circuit of E 17.3-1.

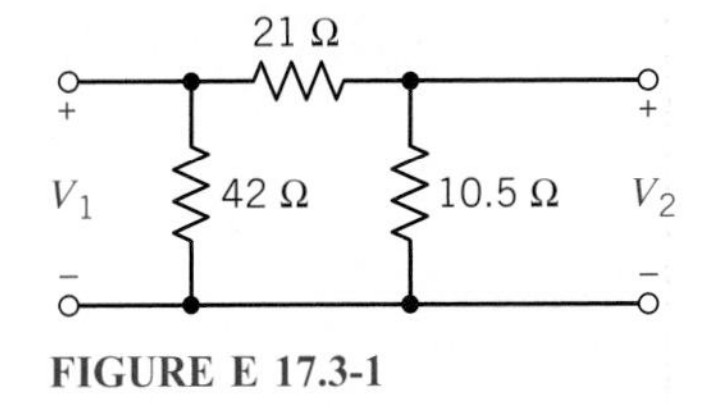

FIGURE E 17.3-1

Answer: $\mathbf{Z} = \begin{bmatrix} 18 & 6 \\ 6 & 9 \end{bmatrix}$, $\mathbf{Y} = \begin{bmatrix} \frac{1}{14} & -\frac{1}{21} \\ -\frac{1}{21} & \frac{1}{7} \end{bmatrix}$

17.4 Z AND Y PARAMETERS FOR A CIRCUIT WITH DEPENDENT SOURCES

When a circuit incorporates a dependent source, it is easy to use the methods of Table 17.3-3 or Table 17.3-4 to determine the Z or Y parameters. When a dependent source is within the circuit, $Z_{21} \neq Z_{12}$ and $Y_{12} \neq Y_{21}$.

EXAMPLE 17.4-1 Impedance Parameters

Determine the Z parameters of the circuit of Figure 17.4-1 when $m = 2/3$.

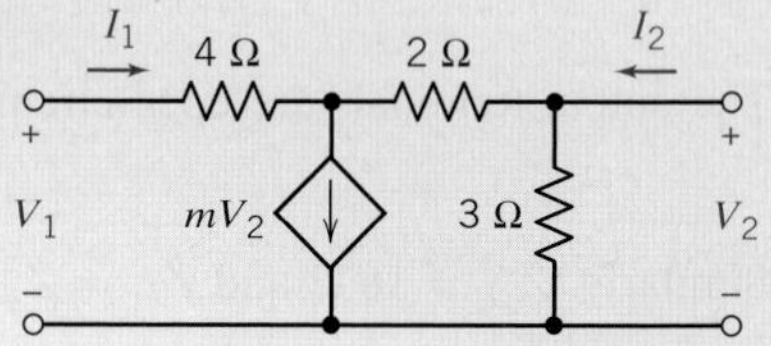

FIGURE 17.4-1 Circuit of Example 17.4-1.

Solution

We determine the Z parameters using the method of Table 17.3-3. Connect a voltage source V_1 and open-circuit the output terminals as shown in Figure 17.4-2*a*.

KCL at node a leads to

$$I_1 - mV_2 - I = 0 \tag{17.4-1}$$

KVL around the outer loop is

$$V_1 = 4I_1 + 5I \tag{17.4-2}$$

Furthermore, $V_2 = 3I$, so $I = V_2/3$. Substituting $I = V_2/3$ into Eq. 17.4-1, we have

$$I_1 = mV_2 + \frac{V_2}{3} = (m + 1/3)V_2 \tag{17.4-3}$$

Therefore,

$$Z_{21} = \frac{V_2}{I_1} = 1\ \Omega$$

Substituting $I = V_2/3$ into Eq. 17.4-2, we obtain

$$V_1 = 4I_1 + \frac{5V_2}{3} = 4I_1 + \frac{5}{3}I_1 \tag{17.4-4}$$

Therefore,

$$Z_{11} = \frac{V_1}{I_1} = \frac{17}{3}\ \Omega$$

To obtain Z_{22} and Z_{12}, we connect a voltage source V_2 to the output terminals and open-circuit the input terminals, as shown in Figure 17.4-2. We can write two mesh equations for the assumed current directions, shown as

$$V_1 + 5I_4 - 3I_2 = 0 \tag{17.4-5}$$

and

$$V_2 + 3I_4 - 3I_2 = 0 \tag{17.4-6}$$

Furthermore, $I_4 = mV_2$, so substituting into Eq. 17.4-6, we have

$$V_2 + 3mV_2 - 3I_2 = 0$$

or

$$V_2 = \frac{3}{3}I_2$$

Therefore,

$$Z_{22} = \frac{V_2}{I_2} = 1\ \Omega$$

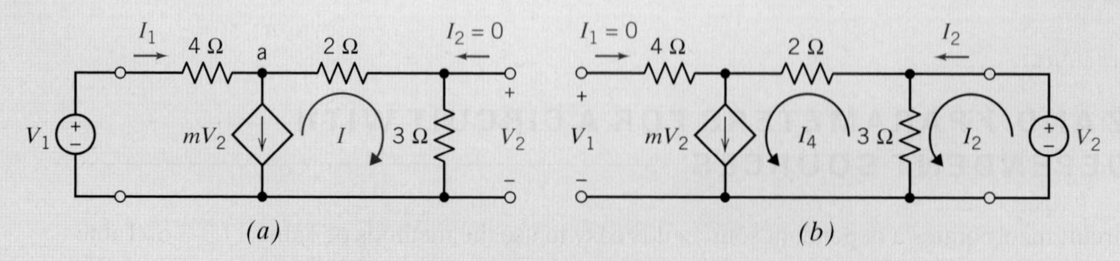

FIGURE 17.4-2 Circuit for determining (*a*) Z_{11} and Z_{21} and (*b*) Z_{22} and Z_{12}.

Substituting $I_4 = mV_2$ into Eq. 17.4-5, we have

$$V_1 + 5mV_2 = 3I_2$$

or

$$V_1 + 5mI_2 = 3I_2$$

Therefore,

$$Z_{12} = \frac{V_1}{I_2} = (3 - 5m) = -\frac{1}{3}\ \Omega$$

Then, in summary, we have

$$\mathbf{Z} = \begin{bmatrix} \frac{17}{3} & -\frac{1}{3} \\ 1 & 1 \end{bmatrix}$$

Note that $Z_{21} \neq Z_{12}$, because a dependent source is present within the circuit.

EXERCISE 17.4-1 Determine the Y parameters of the circuit of Figure 17.4-1.

Answer: $Y = \begin{bmatrix} \frac{1}{6} & \frac{1}{18} \\ -\frac{1}{6} & \frac{17}{18} \end{bmatrix}$

17.5 HYBRID AND TRANSMISSION PARAMETERS

The two-port hybrid parameter equations are based on V_1 and I_2 as the output variables, so that

$$V_1 = h_{11}I_1 + h_{12}V_2 \quad (17.5\text{-}1)$$

$$I_2 = h_{21}I_1 + h_{22}V_2 \quad (17.5\text{-}2)$$

or, in matrix form,

$$\begin{bmatrix} V_1 \\ I_2 \end{bmatrix} = \begin{bmatrix} h_{11} & h_{12} \\ h_{21} & h_{22} \end{bmatrix} \begin{bmatrix} I_1 \\ V_2 \end{bmatrix} = \mathbf{H} \begin{bmatrix} I_1 \\ V_2 \end{bmatrix} \quad (17.5\text{-}3)$$

These parameters are used widely in transistor circuit models. The hybrid circuit model is shown in Figure 17.5-1.

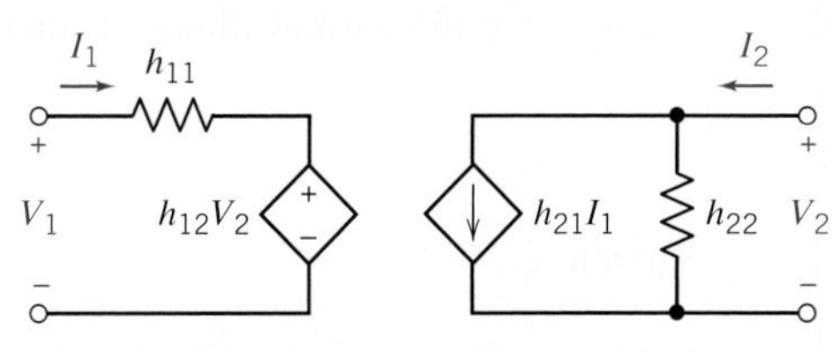

FIGURE 17.5-1 The h-parameter model of a two-port circuit.

The inverse hybrid parameter equations are

$$I_1 = g_{11}V_1 + g_{12}I_2 \quad (17.5\text{-}4)$$

$$V_2 = g_{21}V_1 + g_{22}I_2 \quad (17.5\text{-}5)$$

or, in matrix form,

$$\begin{bmatrix} I_1 \\ V_2 \end{bmatrix} = \begin{bmatrix} g_{11} & g_{12} \\ g_{21} & g_{22} \end{bmatrix} \begin{bmatrix} V_1 \\ I_2 \end{bmatrix} = \mathbf{G} \begin{bmatrix} V_1 \\ I_2 \end{bmatrix} \quad (17.5\text{-}6)$$

The inverse hybrid circuit model is shown in Figure 17.5-2.

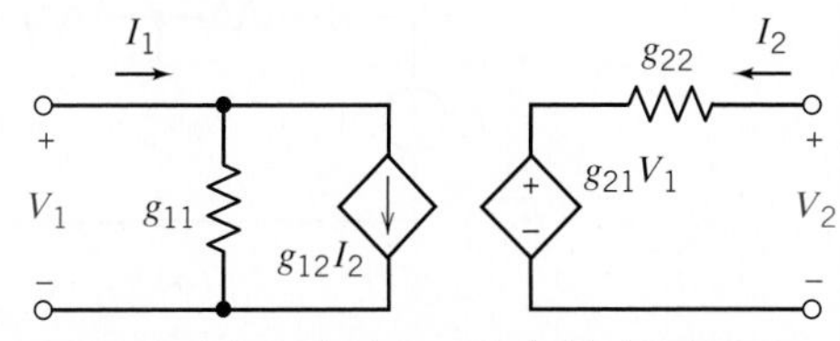

FIGURE 17.5-2 The inverse hybrid circuit (g-parameter) model.

The hybrid and inverse hybrid parameters include both impedance and admittance parameters and are thus called *hybrid*. The parameters h_{11}, h_{12}, h_{21}, and h_{22} represent the short-circuit input impedance, the open-circuit reverse voltage gain, the short-circuit forward current gain, and the open-circuit output admittance, respectively. The parameters g_{11}, g_{12}, g_{21}, and g_{22} represent the open-circuit input admittance, the short-circuit reverse current gain, the open-circuit forward voltage gain, and the short-circuit output impedance, respectively.

The transmission parameters are written as

$$V_1 = AV_2 - BI_2 \tag{17.5-7}$$

$$I_1 = CV_2 - DI_2 \tag{17.5-8}$$

or, in matrix form, as

$$\begin{bmatrix} V_1 \\ I_1 \end{bmatrix} = \begin{bmatrix} A & B \\ C & D \end{bmatrix} \begin{bmatrix} V_2 \\ -I_2 \end{bmatrix} = \mathbf{T} \begin{bmatrix} V_2 \\ -I_2 \end{bmatrix} \tag{17.5-9}$$

Transmission parameters are used to describe cable, fiber, and line transmission. The transmission parameters A, B, C, and D represent the open-circuit reverse voltage gain, the negative short-circuit transfer impedance, the open-circuit transfer admittance, and the negative short-circuit reverse current gain, respectively. The transmission parameters are often referred to as the *ABCD* parameters. We are primarily interested in the hybrid and transmission parameters because they are widely used.

EXAMPLE 17.5-1 Hybrid Parameters and Transmission Parameters

(a) Find the h parameters for the T circuit of Figure 17.5-3 in terms of R_1, R_2, and R_3.

(b) Evaluate the parameters when $R_1 = 1\ \Omega$, $R_2 = 4\ \Omega$, and $R_3 = 6\ \Omega$.

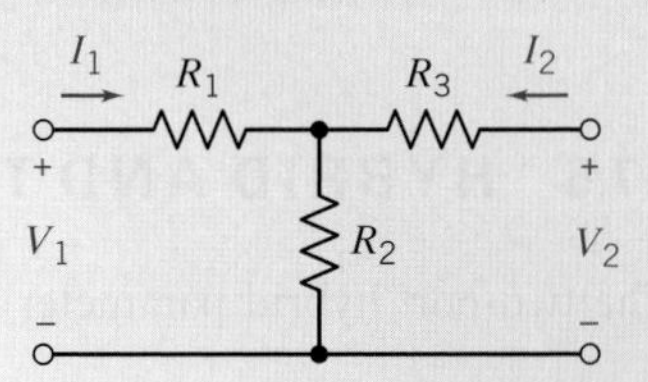

FIGURE 17.5-3 The T circuit of Example 17.5-1.

Solution

(a) First, we find h_{11} and h_{21} by short-circuiting the output terminals and connecting an input current source I_1, as shown in Figure 17.5-4a. Therefore,

$$h_{11} = \left.\frac{V_1}{I_1}\right|_{V_2=0} = R_1 + \frac{R_2R_3}{R_2 + R_3}$$

Then, using the current divider principle, we have

$$I_2 = \frac{-R_2}{R_2 + R_3} I_1$$

Therefore,

$$h_{21} = \left.\frac{I_2}{I_1}\right|_{V_2=0} = \frac{-R_2}{R_2 + R_3}$$

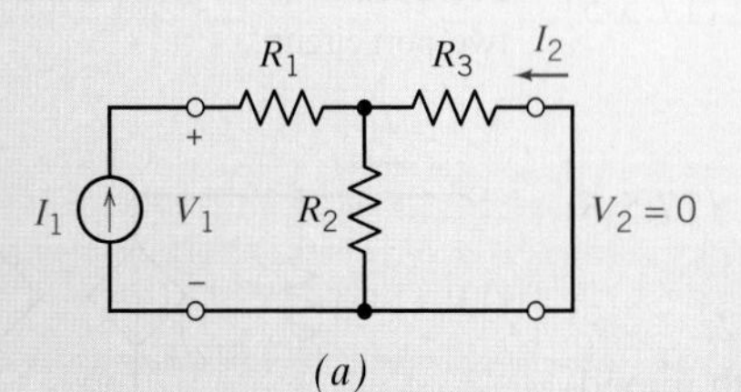

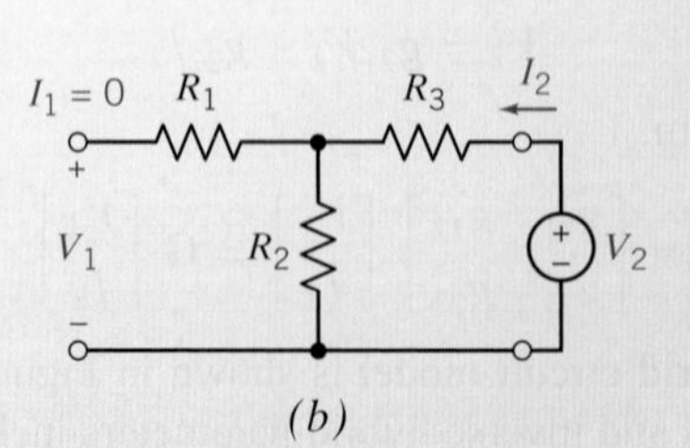

FIGURE 17.5-4 The circuits for determining (a) h_{11} and h_{21} and (b) h_{22} and h_{12}.

The next step is to redraw the circuit with $I_1 = 0$ and to connect the voltage source V_2 as shown in Figure 17.5-4*b*. Then we may determine h_{12} by using the voltage divider principle, as follows:

$$h_{12} = \left.\frac{V_1}{V_2}\right|_{I_1=0} = \frac{R_2}{R_2 + R_3}$$

Finally, we determine h_{22} from Figure 17.5-4*b* as

$$h_{22} = \left.\frac{I_2}{V_2}\right|_{I_1=0} = \frac{1}{R_2 + R_3}$$

It is a property of a passive circuit (no op amps or dependent sources within the two-port network) that $h_{12} = -h_{21}$.

(b) When $R_1 = 1\,\Omega, R_2 = 4\,\Omega$, and $R_3 = 6\,\Omega$, we have

$$h_{11} = R_1 + \frac{R_2 R_3}{R_2 + R_3} = 3.4\,\Omega$$

$$h_{21} = \frac{-R_2}{R_2 + R_3} = -0.4$$

$$h_{12} = \frac{R_2}{R_2 + R_3} = 0.4$$

$$h_{22} = \frac{1}{R_2 + R_3} = 0.1\text{ S}$$

EXERCISE 17.5-1 Find the hybrid parameter model of the circuit shown in Figure E 17.5-1.

Answers: $h_{11} = 0.9\,\Omega$, $h_{12} = 0.1$, $h_{21} = 4.4$, and $h_{22} = 0.6$ S

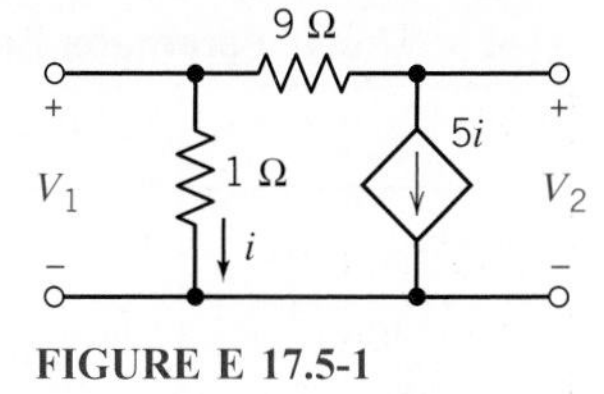

FIGURE E 17.5-1

17.6 RELATIONSHIPS BETWEEN TWO-PORT PARAMETERS

If all the two-port parameters for a circuit exist, it is possible to relate one set of parameters to another because the variables V_1, I_1, V_2, and I_2 are interrelated by the parameters. First, let us consider the relation between the Z parameters and the Y parameters. The matrix equation for the Z parameters is $\mathbf{V} = \mathbf{ZI}$ or

$$\begin{bmatrix} V_1 \\ V_2 \end{bmatrix} = \mathbf{Z} \begin{bmatrix} I_1 \\ I_2 \end{bmatrix} \tag{17.6-1}$$

Similarly, the equation for the Y parameters is $\mathbf{I} = \mathbf{YV}$ or

$$\begin{bmatrix} I_1 \\ I_2 \end{bmatrix} = \mathbf{Y} \begin{bmatrix} V_1 \\ V_2 \end{bmatrix} \tag{17.6-2}$$

Substituting for **I** from Eq. 17.6-2 into Eq. 17.6-1, we obtain

$$\mathbf{V} = \mathbf{ZYV}$$

or

$$\mathbf{Z} = \mathbf{Y}^{-1} \tag{17.6-3}$$

Thus, we can obtain the matrix **Z** by inverting the **Y** matrix. Of course, we can likewise obtain the **Y** matrix if we invert a known **Z** matrix. It is possible that a two-port network has a **Y** matrix or a **Z** matrix but not both. In other words, $\mathbf{Z}^{-1}$ or $\mathbf{Y}^{-1}$ may not exist for some networks.

If we have a known **Y** matrix, we obtain the **Z** matrix by finding the determinant of the **Y** matrix as ΔY and the adjoint of the **Y** matrix as

$$\text{adj}\,\mathbf{Y} = \begin{bmatrix} Y_{22} & -Y_{12} \\ -Y_{21} & Y_{11} \end{bmatrix}$$

Then

$$\mathbf{Z} = \mathbf{Y}^{-1} = \frac{\text{adj}\,\mathbf{Y}}{\Delta Y} \tag{17.6-4}$$

where $\Delta Y = Y_{11}\,Y_{22} - Y_{12}\,Y_{21}$.

The two-port parameter conversion relationships for the Z, Y, h, g, and T parameters are provided in Table 17.6-1.

Table 17.6-1 Parameter Relationships

	Z	Y	h	g	T
Z	$\begin{matrix} Z_{11} & Z_{12} \\ Z_{21} & Z_{22} \end{matrix}$	$\begin{matrix} \frac{Y_{22}}{\Delta Y} & \frac{-Y_{12}}{\Delta Y} \\ \frac{-Y_{21}}{\Delta Y} & \frac{Y_{11}}{\Delta Y} \end{matrix}$	$\begin{matrix} \frac{\Delta h}{h_{22}} & \frac{h_{12}}{h_{22}} \\ \frac{-h_{21}}{h_{22}} & \frac{1}{h_{22}} \end{matrix}$	$\begin{matrix} \frac{1}{g_{11}} & \frac{-g_{12}}{g_{11}} \\ \frac{g_{21}}{g_{11}} & \frac{\Delta g}{g_{11}} \end{matrix}$	$\begin{matrix} \frac{A}{C} & \frac{\Delta T}{C} \\ \frac{1}{C} & \frac{D}{C} \end{matrix}$
Y	$\begin{matrix} \frac{Z_{22}}{\Delta Z} & \frac{-Z_{12}}{\Delta Z} \\ \frac{-Z_{21}}{\Delta Z} & \frac{Z_{11}}{\Delta Z} \end{matrix}$	$\begin{matrix} Y_{11} & Y_{12} \\ Y_{21} & Y_{22} \end{matrix}$	$\begin{matrix} \frac{1}{h_{11}} & \frac{-h_{12}}{h_{11}} \\ \frac{h_{21}}{h_{11}} & \frac{\Delta h}{h_{11}} \end{matrix}$	$\begin{matrix} \frac{\Delta g}{g_{22}} & \frac{g_{12}}{g_{22}} \\ \frac{-g_{21}}{g_{22}} & \frac{1}{g_{22}} \end{matrix}$	$\begin{matrix} \frac{D}{B} & \frac{-\Delta T}{B} \\ \frac{-1}{B} & \frac{A}{B} \end{matrix}$
h	$\begin{matrix} \frac{\Delta Z}{Z_{22}} & \frac{Z_{12}}{Z_{22}} \\ \frac{-Z_{21}}{Z_{22}} & \frac{1}{Z_{22}} \end{matrix}$	$\begin{matrix} \frac{1}{Y_{11}} & \frac{-Y_{12}}{Y_{11}} \\ \frac{Y_{21}}{Y_{11}} & \frac{\Delta Y}{Y_{11}} \end{matrix}$	$\begin{matrix} h_{11} & h_{12} \\ h_{21} & h_{22} \end{matrix}$	$\begin{matrix} \frac{g_{22}}{\Delta g} & \frac{g_{12}}{\Delta g} \\ \frac{-g_{21}}{\Delta g} & \frac{g_{11}}{\Delta g} \end{matrix}$	$\begin{matrix} \frac{B}{D} & \frac{\Delta T}{D} \\ \frac{-1}{D} & \frac{C}{D} \end{matrix}$
g	$\begin{matrix} \frac{1}{Z_{11}} & \frac{-Z_{12}}{Z_{11}} \\ \frac{Z_{21}}{Z_{11}} & \frac{\Delta Z}{Z_{11}} \end{matrix}$	$\begin{matrix} \frac{\Delta Y}{Y_{22}} & \frac{Y_{12}}{Y_{22}} \\ \frac{-Y_{21}}{Y_{22}} & \frac{1}{Y_{22}} \end{matrix}$	$\begin{matrix} \frac{h_{22}}{\Delta h} & \frac{-h_{12}}{\Delta h} \\ \frac{-h_{21}}{\Delta h} & \frac{h_{11}}{\Delta h} \end{matrix}$	$\begin{matrix} g_{11} & g_{12} \\ g_{21} & g_{22} \end{matrix}$	$\begin{matrix} \frac{C}{A} & \frac{-\Delta T}{A} \\ \frac{1}{A} & \frac{B}{A} \end{matrix}$
T	$\begin{matrix} \frac{Z_{11}}{Z_{21}} & \frac{\Delta Z}{Z_{21}} \\ \frac{1}{Z_{21}} & \frac{Z_{22}}{Z_{21}} \end{matrix}$	$\begin{matrix} \frac{-Y_{22}}{Y_{21}} & \frac{-1}{Y_{21}} \\ \frac{-\Delta Y}{Y_{21}} & \frac{-Y_{11}}{Y_{21}} \end{matrix}$	$\begin{matrix} \frac{-\Delta h}{h_{21}} & \frac{-h_{11}}{h_{21}} \\ \frac{-h_{22}}{h_{21}} & \frac{-1}{h_{21}} \end{matrix}$	$\begin{matrix} \frac{1}{g_{21}} & \frac{g_{22}}{g_{21}} \\ \frac{g_{11}}{g_{21}} & \frac{\Delta g}{g_{21}} \end{matrix}$	$\begin{matrix} A & B \\ C & D \end{matrix}$

$\Delta Z = Z_{11}Z_{22} - Z_{12}Z_{21}$, $\Delta Y = Y_{11}Y_{22} - Y_{12}Y_{21}$, $\Delta g = g_{11}g_{22} - g_{12}g_{21}$, $\Delta h = h_{11}h_{22} - h_{12}h_{21}$, $\Delta T = AD - BC$

EXAMPLE 17.6-1 Two-Port Parameter Conversion

Determine the Y and h parameters if

$$\mathbf{Z} = \begin{bmatrix} 18 & 6 \\ 6 & 9 \end{bmatrix}$$

Solution

First, we will determine the Y parameters by calculating the determinant as

$$\Delta Z = Z_{11}Z_{22} - Z_{12}Z_{21} = 18(9) - 6(6) = 126$$

Then, using Table 17.6-1, we obtain

$$Y_{11} = \frac{Z_{22}}{\Delta Z} = \frac{9}{126} = \frac{1}{14}\,\text{S}$$

$$Y_{12} = Y_{21} = \frac{-Z_{12}}{\Delta Z} = \frac{-1}{21}\,\text{S}$$

$$Y_{22} = \frac{Z_{11}}{\Delta Z} = \frac{18}{126} = \frac{1}{7}\,\text{S}$$

$$h_{11} = \frac{\Delta Z}{Z_{22}} = \frac{126}{9} = 14\,\Omega$$

$$h_{12} = \frac{Z_{12}}{Z_{22}} = \frac{6}{9} = \frac{2}{3}$$

$$h_{21} = \frac{-Z_{21}}{Z_{22}} = \frac{-6}{9} = \frac{-2}{3}$$

$$h_{22} = \frac{1}{Z_{22}} = \frac{1}{9}\,\text{S}$$

EXERCISE 17.6-1 Determine the Z parameters if the Y parameters are

$$Y = \begin{bmatrix} \frac{2}{15} & \frac{-1}{5} \\ \frac{-1}{10} & \frac{2}{5} \end{bmatrix}$$

The units are siemens.

Answers: $Z_{11} = 12\,\Omega$, $Z_{12} = 6\,\Omega$, $Z_{21} = 3\,\Omega$, and $Z_{22} = 4\,\Omega$

EXERCISE 17.6-2 Determine the T parameters from the Y parameters of Exercise 17.6-1.

Answer: $A = 4$, $B = 10\,\Omega$, $C = 1/3\,S$, and $D = 4/3$

17.7 INTERCONNECTION OF TWO-PORT NETWORKS

It is common in many circuits to have several two-port networks interconnected in parallel or in cascade. The *parallel* connection of two two-ports shown in Figure 17.7-1 requires that the V_1 of each two-port be equal.

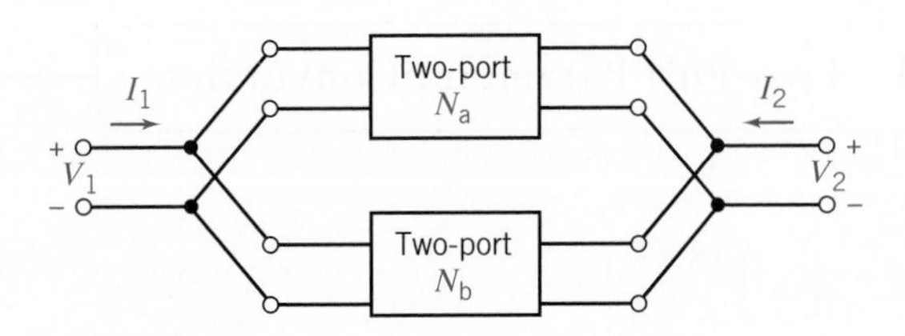

FIGURE 17.7-1 Parallel connection of two two-port networks.

Similarly, at the output port V_2 is the output voltage of both two-port networks. The defining matrix equation for network N_a is

$$\mathbf{I}_a = \mathbf{Y}_a\mathbf{V}_a \tag{17.7-1}$$

and, for network N_b, we have

$$\mathbf{I}_b = \mathbf{Y}_b\mathbf{V}_b \tag{17.7-2}$$

In addition, we have the total current $\mathbf{I}$ as

$$\mathbf{I} = \mathbf{I}_a + \mathbf{I}_b$$

Furthermore, because $\mathbf{V}_a = \mathbf{V}_b = \mathbf{V}$

$$\mathbf{I} = \mathbf{Y}_a\mathbf{V} + \mathbf{Y}_b\mathbf{V} = (\mathbf{Y}_a + \mathbf{Y}_b)\mathbf{Y} = \mathbf{Y}\mathbf{V}$$

Therefore, the Y parameters for the total network of two parallel two-ports are described by the matrix equation

$$\mathbf{Y} = \mathbf{Y}_a + \mathbf{Y}_b \tag{17.7-3}$$

For example,

$$Y_{11} = Y_{11a} + Y_{11b}$$

Hence, to determine the Y parameters for the total network, we add the Y parameters of each network. In general, the Y-parameter matrix of the parallel connection is the sum of the Y-parameter matrices of the individual two-ports connected in parallel.

The series interconnection of two two-port networks is shown in Figure 17.7-2. We will use the Z parameters to describe each two-port and the series combination. The two networks are described by the matrix equations

$$\mathbf{V}_a = \mathbf{Z}_a\mathbf{I}_a \tag{17.7-4}$$

and

$$\mathbf{V}_b = \mathbf{Z}_b\mathbf{I}_b \tag{17.7-5}$$

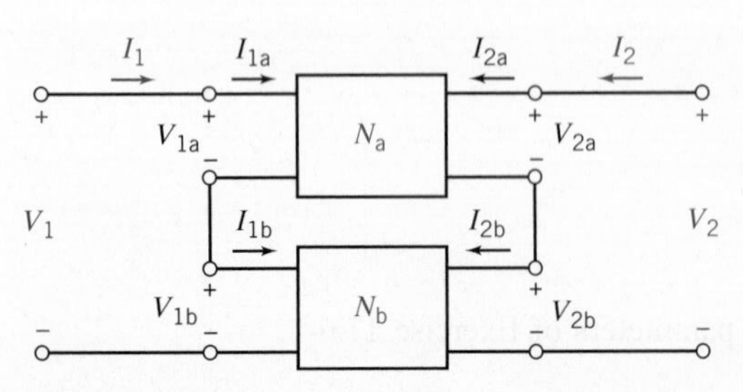

FIGURE 17.7-2 Series connection of two two-port networks.

The terminal currents are

$$\mathbf{I} = \mathbf{I}_a = \mathbf{I}_b$$

Therefore, because $\mathbf{V} = \mathbf{V}_a + \mathbf{V}_b$, we have

$$\begin{aligned}\mathbf{V} &= \mathbf{Z}_a\mathbf{I}_a + \mathbf{Z}_b\mathbf{I}_b \\ &= (\mathbf{Z}_a + \mathbf{Z}_b)\mathbf{I} = \mathbf{Z}\mathbf{I}\end{aligned}$$

or

$$\mathbf{Z} = \mathbf{Z}_a + \mathbf{Z}_b \tag{17.7-6}$$

Therefore, the Z parameters for the total network are equal to the sum of the Z parameters for the networks.

When the output of one network is connected to the input port of the following network, as shown in Figure 17.7-3, the networks are said to be *cascaded*. Because the output variables of the first network become the input variables of the second network, the transmission parameters are used. The first two-port, N_a, is represented by the matrix equation

$$\begin{bmatrix} V_{1a} \\ I_{1a} \end{bmatrix} = \mathbf{T}_a \begin{bmatrix} V_{2a} \\ -I_{2a} \end{bmatrix}$$

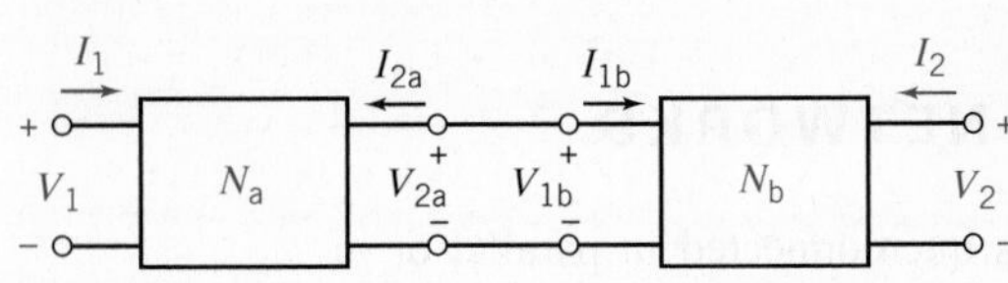

FIGURE 17.7-3 Cascade connection of two two-port networks.

For N_b, we have

$$\begin{bmatrix} V_{1b} \\ I_{1b} \end{bmatrix} = \mathbf{T}_b \begin{bmatrix} V_{2b} \\ -I_{2b} \end{bmatrix}$$

Furthermore, we note that at the input and output, we have

$$\begin{bmatrix} V_1 \\ I_1 \end{bmatrix} = \begin{bmatrix} V_{1a} \\ I_{1a} \end{bmatrix} \quad \text{and} \quad \begin{bmatrix} V_{2b} \\ -I_{2b} \end{bmatrix} = \begin{bmatrix} V_2 \\ -I_2 \end{bmatrix}$$

At the intermediate connection, we have

$$\begin{bmatrix} V_{2a} \\ -I_{2a} \end{bmatrix} = \begin{bmatrix} V_{1b} \\ I_{1b} \end{bmatrix}$$

Therefore,

$$\begin{bmatrix} V_1 \\ I_1 \end{bmatrix} = \mathbf{T}_a\mathbf{T}_b \begin{bmatrix} V_2 \\ -I_2 \end{bmatrix}$$

and

$$\mathbf{T} = \mathbf{T}_a\mathbf{T}_b \tag{17.7-7}$$

Hence, the transmission parameters for the overall network are derived by matrix multiplication, observing the proper order.

All of the preceding calculations for interconnected networks assume that the interconnection does not disturb the two-port nature of the individual subnetworks.

EXAMPLE 17.7-1 Parallel and Cascade Connections of Two-Port Networks

For the T network of Figure 17.7-4, (a) find the Z, Y, and T parameters and (b) determine the resulting parameters after connecting two two-ports in parallel and in cascade. Both two-ports are identical as in Figure 17.7-4.

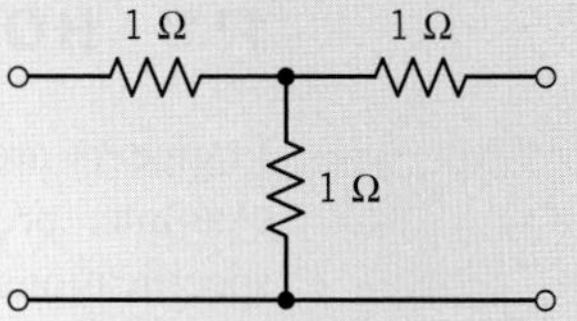

1 Ω 1 Ω 1 Ω

FIGURE 17.7-4 T network of Example 17.7-1.

Solution

First, we find the Z parameters of the T network. Examining the network, we have

$$Z_{12} = Z_{21} = 1\ \Omega$$

$$Z_{22} = Z_{11} = 2\ \Omega$$

Then, using the conversion factors of Table 17.6-1, we find

$$\mathbf{Y} = \begin{bmatrix} \frac{2}{3} & \frac{-1}{3} \\ \frac{-1}{3} & \frac{2}{3} \end{bmatrix}$$

and

$$\mathbf{T} = \begin{bmatrix} 2 & 3 \\ 1 & 2 \end{bmatrix}$$

Two identical networks connected in parallel will have a total $\mathbf{Y}$ matrix of

$$\mathbf{Y} = \mathbf{Y}_a + \mathbf{Y}_b$$

Because $\mathbf{Y}_a = \mathbf{Y}_b$, we have

$$\mathbf{Y} = 2\mathbf{Y}_a = \begin{bmatrix} \frac{4}{3} & \frac{-2}{3} \\ \frac{-2}{3} & \frac{4}{3} \end{bmatrix}$$

Finally, when two identical networks are connected in cascade, we have a total **T** matrix of

$$\mathbf{T} = \mathbf{T}_a\mathbf{T}_b = \begin{bmatrix} 2 & 3 \\ 1 & 2 \end{bmatrix}\begin{bmatrix} 2 & 3 \\ 1 & 2 \end{bmatrix} = \begin{bmatrix} 7 & 12 \\ 4 & 7 \end{bmatrix}$$

EXERCISE 17.7-1 Determine the total transmission parameters of the cascade connection of three two-port networks shown in Figure E 17.7-1.

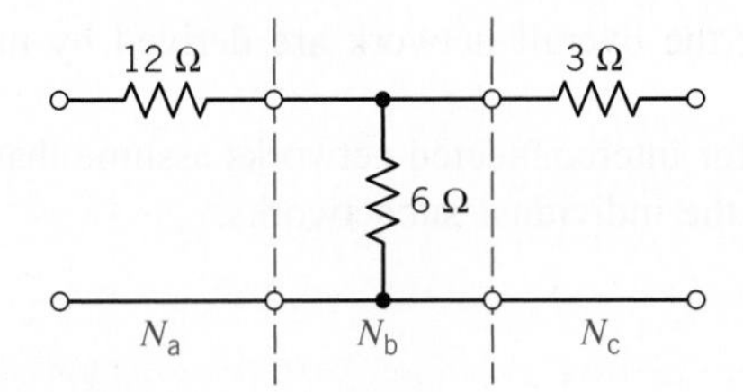

FIGURE E 17.7-1

Answers: $A = 3$, $B = 21\ \Omega$, $C = 1/6$ S, and $D = 3/2$

17.8 HOW CAN WE CHECK . . . ?

Engineers are frequently called upon to check that a solution to a problem is indeed correct. For example, proposed solutions to design problems must be checked to confirm that all of the specifications have been satisfied. In addition, computer output must be reviewed to guard against data-entry errors, and claims made by vendors must be examined critically.

Engineering students are also asked to check the correctness of their work. For example, occasionally just a little time remains at the end of an exam. It is useful to be able to quickly identify those solutions that need more work.

The following example illustrates techniques useful for checking the solutions of the sort of problem discussed in this chapter.

EXAMPLE 17.8-1 How Can We Check Circuits with Two-Port Networks?

The circuit shown in Figure 17.8.1*a* was designed to have a transfer function given by

$$\frac{V_o(s)}{V_{in}(s)} = \frac{2s - 10}{s^2 + 27s + 2}$$

How can we check that the circuit satisfies this specification?

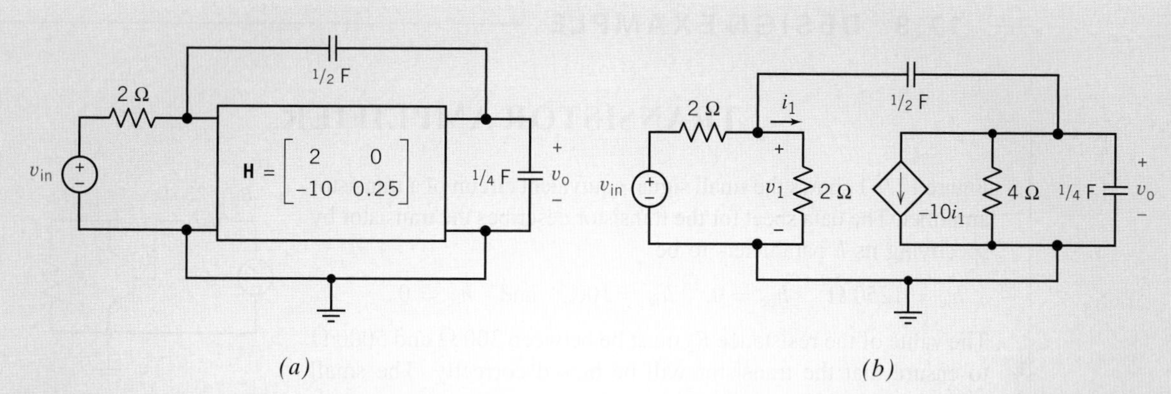

FIGURE 17.8-1 (*a*) A circuit including a two-port network. (*b*) Using the *h*-parameter model to represent the two-port network.

Solution

The *h*-parameter model from Figure 17.5-1 can be used to redraw the circuit as shown in Figure 17.8-1*b*. This circuit can be represented by node equations

$$\begin{bmatrix} \left(1+\frac{s}{2}\right) & -\frac{s}{2} \\ \left(-5-\frac{s}{2}\right) & \left(\frac{3s}{4}+\frac{1}{4}\right) \end{bmatrix} \begin{bmatrix} V_1(s) \\ V_o(s) \end{bmatrix} = \begin{bmatrix} \frac{V_{in}(s)}{2} \\ 0 \end{bmatrix}$$

where $10I_1(s) = 5V_1(s)$ has been used to express the current of the dependent source in terms of the node voltages. Applying Cramer's rule gives

$$\frac{V_o(s)}{V_{in}(s)} = \frac{\frac{1}{2}\left(5+\frac{s}{2}\right)}{\left(1+\frac{s}{2}\right)\left(\frac{3s}{4}+\frac{1}{4}\right)-\frac{s}{2}\left(\frac{s}{2}+5\right)} = \frac{2s+20}{s^2-13s+2}$$

This is not the required transfer function, so the circuit does not satisfy the specification.

EXERCISE 17.8-1 Verify that the circuit shown in Figure E 17.8-1 does indeed have the transfer function

$$\frac{V_o(s)}{V_{in}(s)} = \frac{2s-10}{s^2+27s+2}$$

(The circuits in Figures 17.8-1*a* and E 17.8-1 differ only in the sign of h_{21}.)

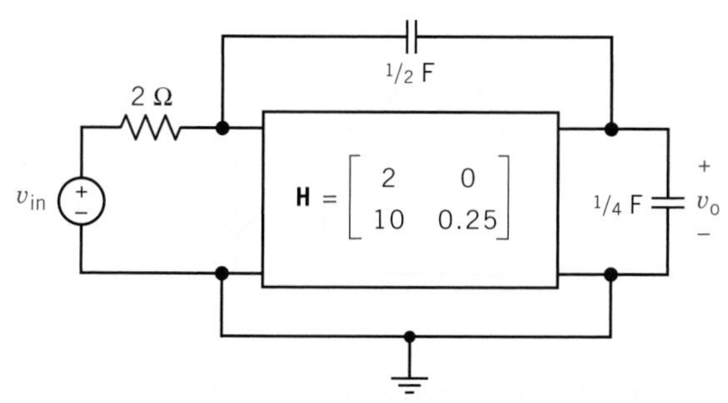

FIGURE E 17.8-1 A modified version of the circuit from Figure 17.8-1.

17.9 DESIGN EXAMPLE

TRANSISTOR AMPLIFIER

Figure 17.9-1 shows the small signal equivalent circuit of a transistor amplifier. The data sheet for the transistor describes the transistor by specifying its h parameters to be

$$h_{\text{ie}} = 1250\ \Omega, \quad h_{\text{oe}} = 0, \quad h_{\text{fe}} = 100, \quad \text{and} \quad h_{\text{re}} = 0$$

The value of the resistance R_{c} must be between 300 Ω and 5000 Ω to ensure that the transistor will be biased correctly. The small signal gain is defined to be

$$A_{\text{v}} = \frac{v_{\text{o}}}{v_{\text{in}}}$$

The challenge is to design the amplifier so that

$$A_{\text{v}} = -20$$

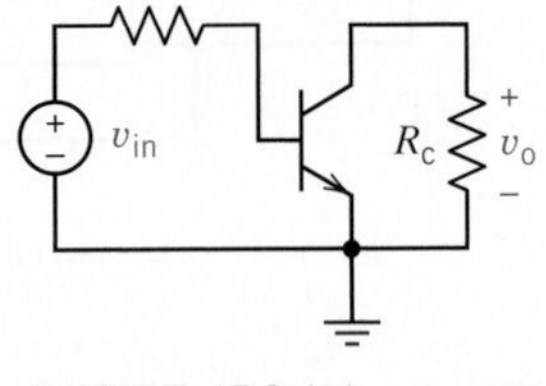

FIGURE 17.9-1 A transistor amplifier.

(There is no guarantee that these specifications can be satisfied. Part of the problem is to decide whether it is possible to design this amplifier so that $A_{\text{v}} = -20$.)

Describe the Situation and the Assumptions

1. R_{c} must be between 300 Ω and 5000 Ω.
2. The transistor is represented by h parameters. Figure 17.9-1a shows that the transistor can be configured to be a two-port network and represented by h parameters. Figure 17.9-2b shows an equivalent circuit for the transistor. This equivalent circuit is based on the h parameters. For this particular transistor, the values of the h parameters are

$$h_{\text{ie}} = 1000\ \Omega, \quad h_{\text{oe}} = 0, \quad h_{\text{fe}} = 100, \quad \text{and} \quad h_{\text{re}} = 0$$

Because

$$\frac{1}{h_{\text{oe}}} = \infty$$

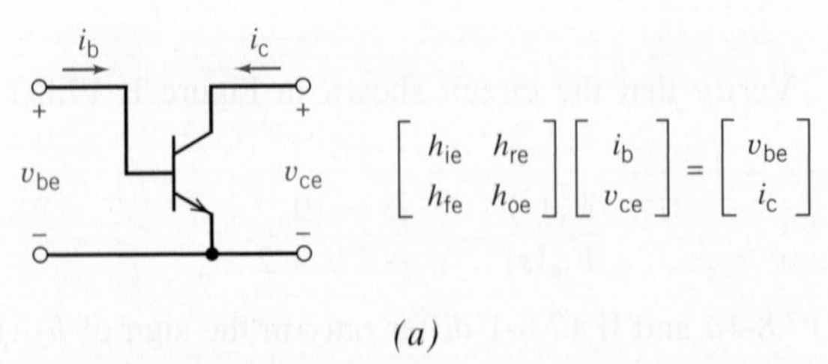

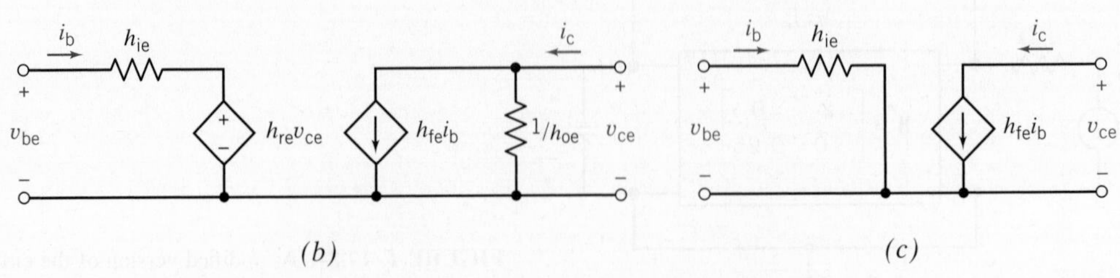

FIGURE 17.9-2 (a) Using h parameters to describe a transistor. (b) An equivalent circuit. (c) A simplified equivalent circuit for $h_{\text{re}} = 0$ and $h_{\text{oe}} = 0$.

the resistor at the right side of the equivalent circuit is an open circuit. Because

$$h_{re} = 0$$

the dependent voltage source is a short circuit. Figure 17.9-2*c* shows the equivalent circuit after these simplifications are made.

3. The voltage gain must be $A_v = -20$.

State the Goal

Select R_c so that $A_v = -20$.

Generate a Plan

Replace the transistor in Figure 17.9.1 by the equivalent circuit in Figure 17.9-2*c*. Analyze the resulting circuit to obtain a formula for the voltage gain, A_v. This formula will involve R_c. Determine the value of R_c that will make $A_v = -20$. If this value of R_c is between 300 Ω and 5000 Ω, the amplifier design is complete. On the other hand, if this value of R_c is not between 300 Ω and 5000 Ω, the specifications cannot be satisfied.

Act on the Plan

Figure 17.9-3 shows the amplifier after the transistor has been replaced by the equivalent circuit. Applying Ohm's law to R_c gives

$$v_o = -R_c 100 i_b$$

where the minus sign is due to reference directions. Next, apply KVL to the left mesh to get

$$v_{in} = 23{,}000 i_b + 1000 i_b$$

Then

$$A_v = \frac{v_o}{v_{in}} = \frac{-100R_c}{24{,}000}$$

Finally, set $A_v = -20$, obtaining

$$-20 = \frac{-100R_c}{24{,}000}$$

Now solve for R_c to determine

$$R_c = 4800\ \Omega$$

Verify the Proposed Solution

First, the resistance $R_c = 4800\ \Omega$ is indeed between 300 Ω and 5000 Ω. Second, the gain of the circuit shown in Figure 17.9-3 is

$$\frac{v_o}{v_{in}} = \frac{-h_{fe}R_c}{R_b + h_{ie}} = -\frac{100 \times 4800}{23{,}000 + 1000} = -20$$

Therefore, both specifications have been satisfied.

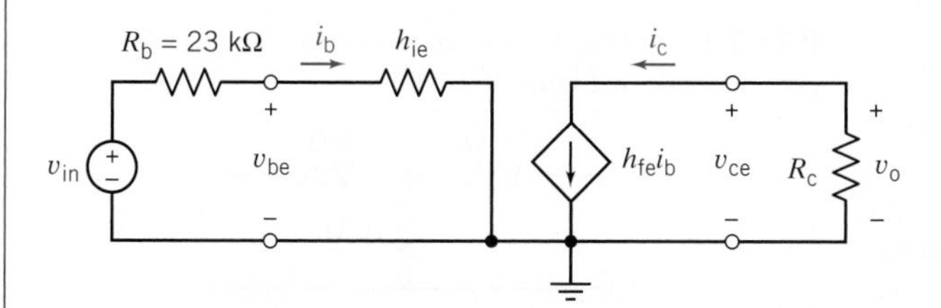

FIGURE 17.9-3 An equivalent circuit for the transistor amplifier.

17.10 SUMMARY

- A port is a pair of terminals together with the restriction that the current directed into one terminal be equal to the current directed out of the other terminal.
- Two-port models of circuits or devices are useful for describing the performance of the circuit or device in terms of the currents and voltages at its ports. The internal details of the circuit or device are not included in the two-port model, so the two-port model of a circuit may be considerably simpler than the circuit itself.
- The two-port model involves four signals—the current and voltage at each port. Two of these signals are treated as inputs, and the other two are treated as outputs. There are six ways of separating the four signals into input and output signals, and so there are six sets of two-port parameters. The six sets of two-port parameters are called the impedance, admittance, hybrid, inverse hybrid, transmission, and inverse transmission parameters. Table 17.3-2 summarizes the six sets of two-port parameters.
- Table 17.6-1 summarizes the equations used to convert one set of two-port parameters into another, for example, to convert impedance parameters into hybrid parameters.
- We may use two-port parameters to describe the performance of the parallel, series, or cascade connection of two or more circuits.

PROBLEMS

Section 17.2 T-to-Π Transformation and Two-Port Three-Terminal Networks

P 17.2-1 Determine the equivalent resistance R_{ab} of the network of Figure P 17.2-1. Use the Π-to-T transformation as one step of the reduction.

Answer: $R_{ab} = 5\ \Omega$

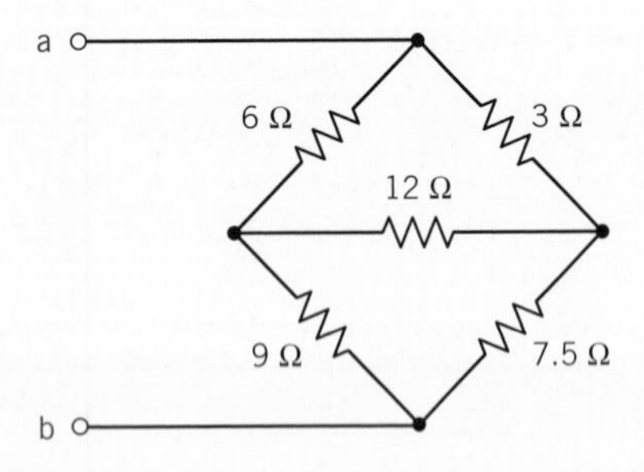

Figure P 17.2-1

P 17.2-2 Repeat problem P 17.2-1 when the 9-Ω resistance is changed to 5 Ω and the 12-Ω resistance is changed to 15 Ω.

P 17.2-3 The two-port network of Figure 17.1-1 has an input source V_s with a source resistance R_s connected to the input terminals so that $V_1 = V_s - I_1 R_s$ and a load resistance connected to the output terminals so that $V_2 = -I_2 R_L = I_L R_L$. Find $R_{in} = V_1/I_1$, $A_v = V_2/V_1$, $A_i = -I_2/I_1$, and $A_p = -V_2 I_2/(V_1 I_1)$ by using the Z-parameter model.

P 17.2-4 Using the Δ-to-Y transformation, determine the current I when $R_1 = 15\ \Omega$ and $R = 20\ \Omega$ for the circuit shown in Figure P 17.2-4.

Answer: $I = 385$ mA

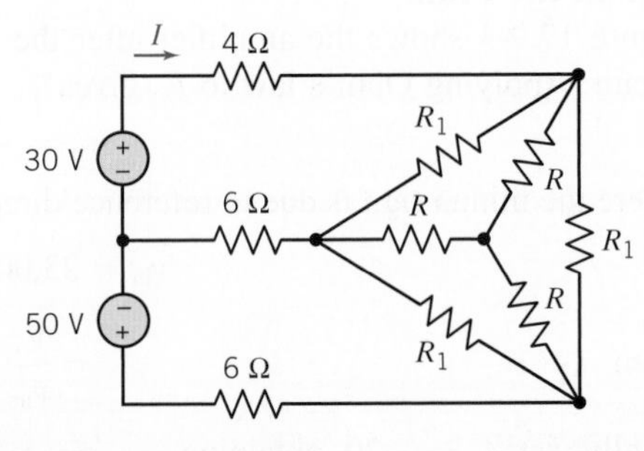

Figure P 17.2-4

P 17.2-5 Use the Y-to-Δ transformation to determine R_{in} of the circuit shown in Figure P 17.2-5.

Answer: $R_{in} = 455.08\ \Omega$

Figure P 17.2-5

Section 17.3 Equations of Two-Port Networks

P 17.3-1 Find the Y parameters and Z parameters for the two-port network of Figure P 17.3-1.

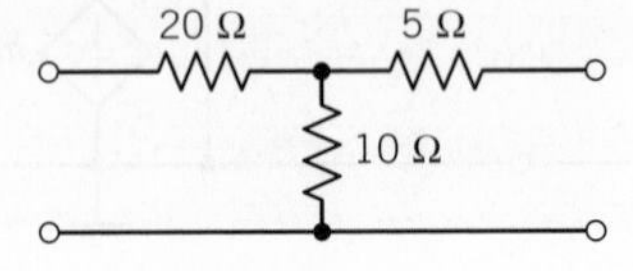

Figure P 17.3-1

P 17.3-2 Determine the Z parameters of the ac circuit shown in Figure P 17.3-2.

Answer: $Z_{11} = 4 - j4\ \Omega$, $Z_{12} = Z_{21} = -j4\ \Omega$, $Z_{22} = -j2\ \Omega$

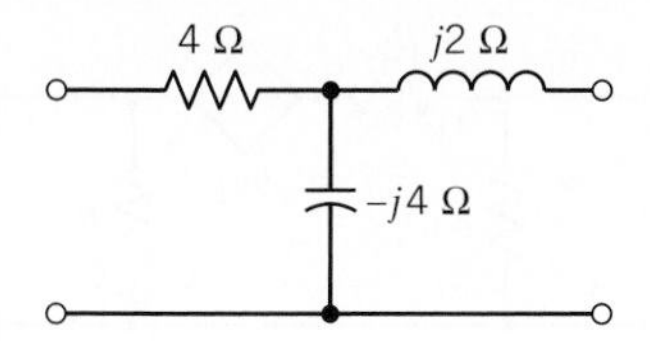

Figure P 17.3-2

P 17.3-3 Find the Y parameters of the circuit of Figure P 17.3-3 when $b = 4$, $G_1 = 2$ S, $G_2 = 1$ S, and $G_3 = 3$ S.

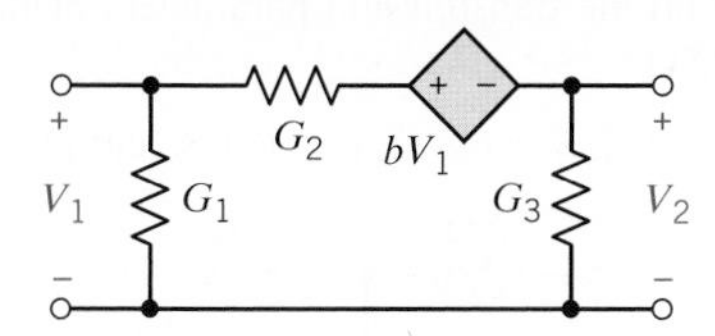

Figure P 17.3-3

P 17.3-4 Find the Y parameters for the circuit of Figure P 17.3-4.

Answers: $Y_{11} = 0.17$ S, $Y_{21} = Y_{12} = -0.07$ S, and $Y_{22} = 0.11$ S

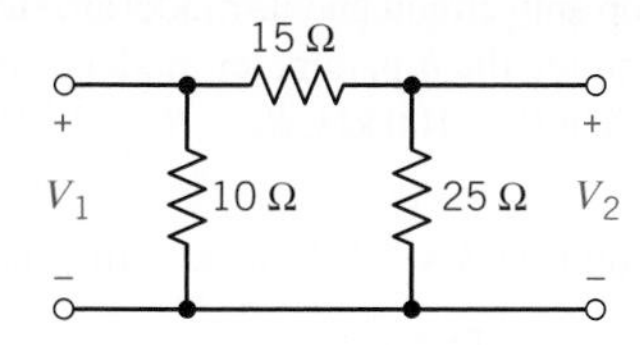

Figure P 17.3-4

P 17.3-5 Find the Y parameters of the circuit shown in Figure P 17.3-5.

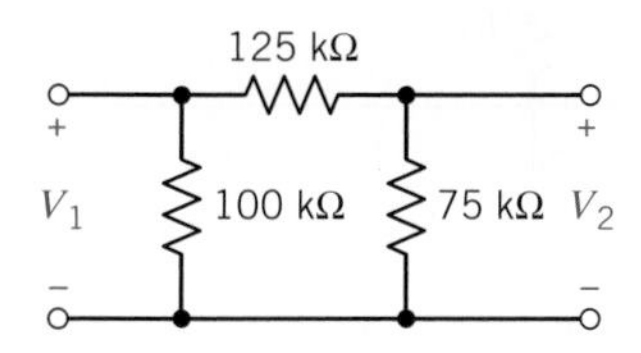

Figure P 17.3-5

P 17.3-6 Find the Z parameters for the circuit shown in Figure P 17.3-6 for sinusoidal steady-state response at $\omega = 3$ rad/s.

Answers: $Z_{11} = 3 + j\ \Omega$, $Z_{12} = Z_{21} = -j2\ \Omega$, and $Z_{22} = -j2\ \Omega$

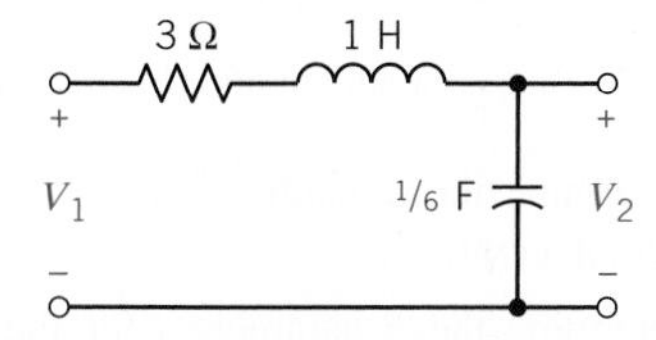

Figure P 17.3-6

P 17.3-7 Determine the impedance parameters in the s-domain (Laplace domain) for the circuit shown in Figure P 17.3-7.

Answers: $Z_{11} = (8s+1)/s$, $Z_{12} = Z_{21} = 1/s$, and $Z_{22} = (2s^2+1)/s$

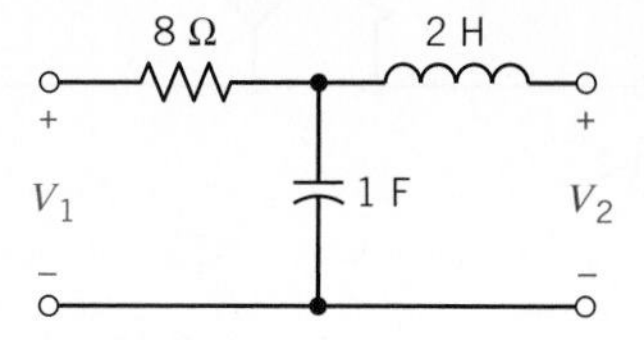

Figure P 17.3-7

P 17.3-8 Determine a two-port network that is represented by the Y parameters:

$$\mathbf{Y} = \begin{bmatrix} \dfrac{2s+1}{s} & -2 \\ -2 & s+2 \end{bmatrix}$$

P 17.3-9 Find a two-port network incorporating one inductor, one capacitor, and two resistors that will give the following impedance parameters:

$$\mathbf{Z} = \frac{1}{\Delta}\begin{bmatrix} (s^2+2s+2) & 1 \\ 1 & (s^2+1) \end{bmatrix}$$

where $\Delta = s^2 + s + 1$.

P 17.3-10 An infinite two-port network is shown in Figure P 17.3-10. When the output terminals are connected to the circuit's characteristic resistance R_o, the resistance looking down the line from each section is the same. Calculate the necessary R_o.

Answer: $R_o = (\sqrt{3} - 1)R$

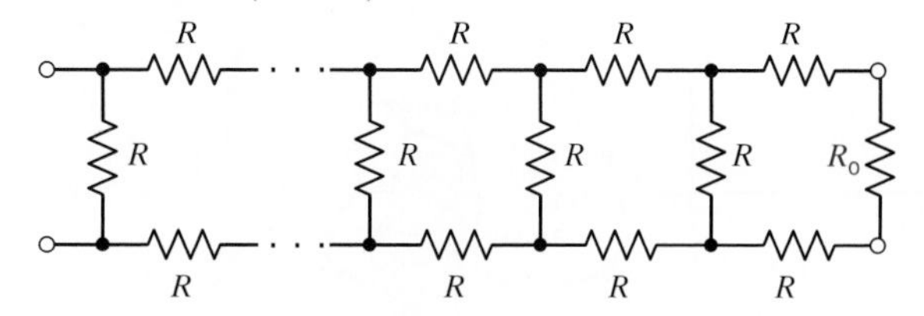

Figure P 17.3-10 Infinite two-port network.

Section 17.4 *Z* and *Y* Parameters for a Circuit with Dependent Sources

P 17.4-1 Determine the Y parameters of the circuit shown in Figure P 17.4-1.

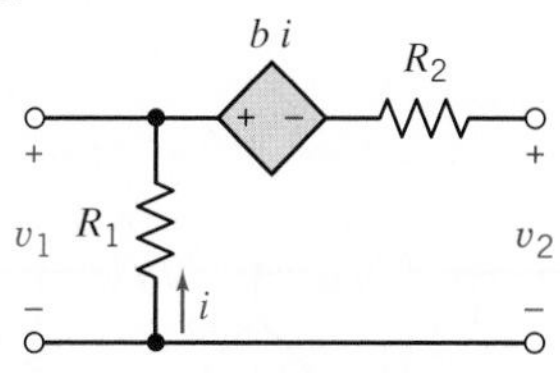

Figure P 17.4-1

P 17.4-2 An electronic amplifier has the circuit shown in Figure P 17.4-2. Determine the impedance parameters for the circuit.

Answers: $Z_{11} = 7$, $Z_{12} = 5(1+\alpha)$, $Z_{21} = 5$, and $Z_{22} = 9 + 5\alpha$

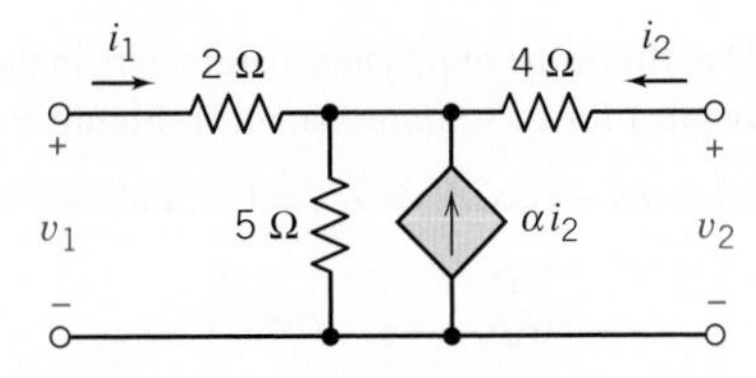

Figure P 17.4-2

P 17.4-3

(a) For the circuit shown in Figure P 17.4-3, determine the two-port Y model using impedances in the s-domain.

(b) Determine the response $v_2(t)$ when a current source $i_1 = 1\ u(t)$ A is connected to the input terminals.

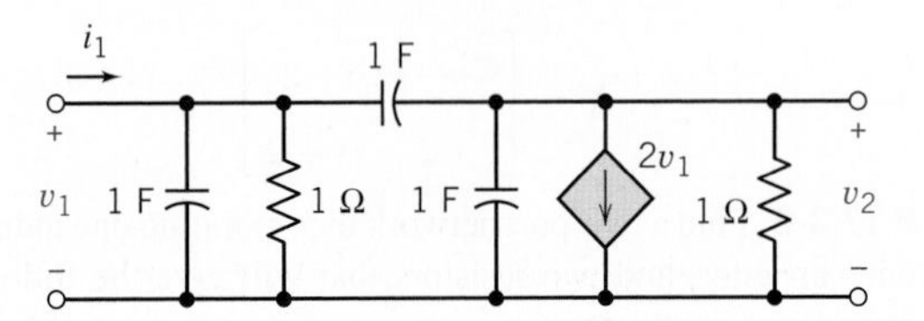

Figure P 17.4-3

P 17.4-4 One form of a heart-assist device is shown in Figure P 17.4-4*a*. The model of the electronic controller and pump/drive unit is shown in P 17.4-4*b*. Determine the impedance parameters of the two-port model.

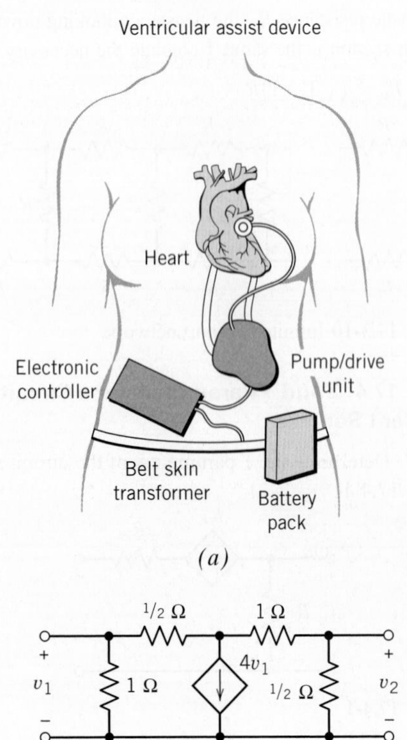

Figure P 17.4-4 (*a*) Heart-assist device and (*b*) model of controller and pump.

P 17.4-5 Determine the Y parameters for the circuit shown in Figure P 17.4-5.

Answer: $Y_{12} = -\frac{1}{R_2}$ and $Y_{21} = \frac{-(1+b)}{R_2}$

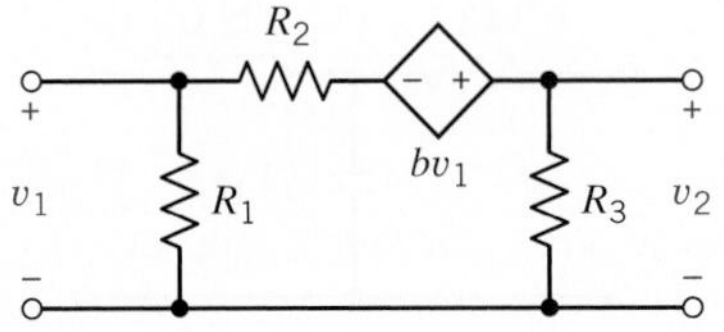

Figure P 17.4-5

Section 17.5 Hybrid and Transmission Parameters

P 17.5-1 Find the transmission parameters of the circuit of Figure P 17.5-1.

Answers: $A = 1.2$, $B = 6.8\ \Omega$, $C = 0.1$ S, and $D = 1.4$

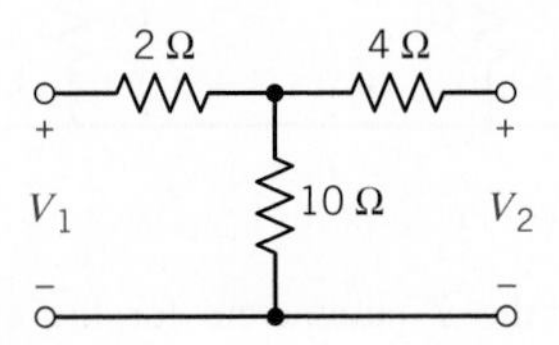

Figure P 17.5-1

P 17.5-2 An op amp circuit and its model are shown in Figure P 17.5-2. Determine the h-parameter model of the circuit and the **H** matrix when $R_i = 100$ kΩ, $R_1 = R_2 = 1$ MΩ, $R_o = 1$ kΩ, and $A = 10^4$.

Answer: $h_{11} = 600$ kΩ, $h_{12} = 1/2$, $h_{21} = -10^6$, and $h_{22} = 10^{-3}$ S

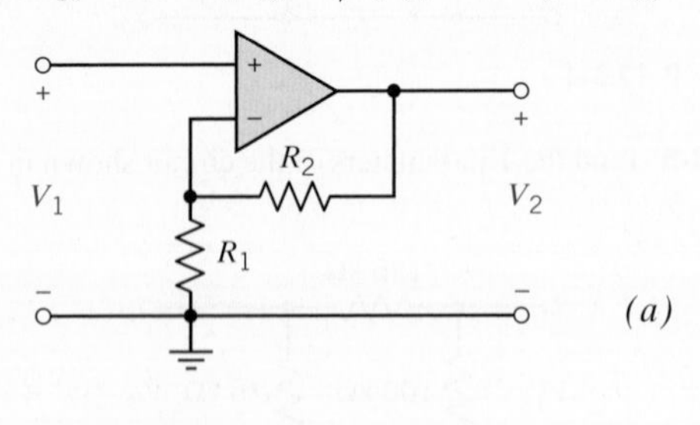

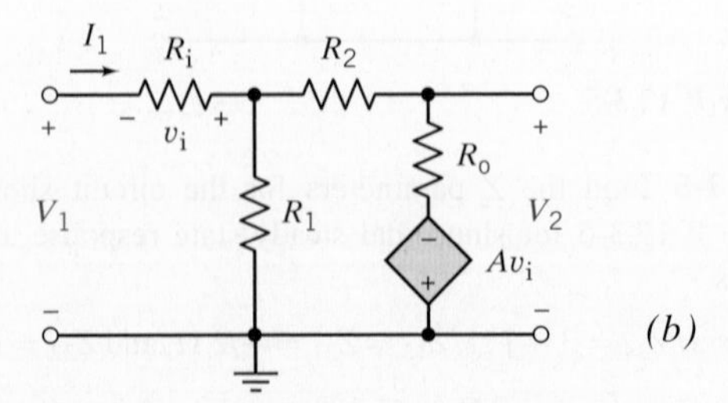

Figure P 17.5-2 (*a*) Op amp circuit and (*b*) circuit model.

P 17.5-3 Determine the h parameters for the ideal transformer of Section 11.11.

P 17.5-4 Determine the h parameters for the T circuit of Figure P 17.5-4.

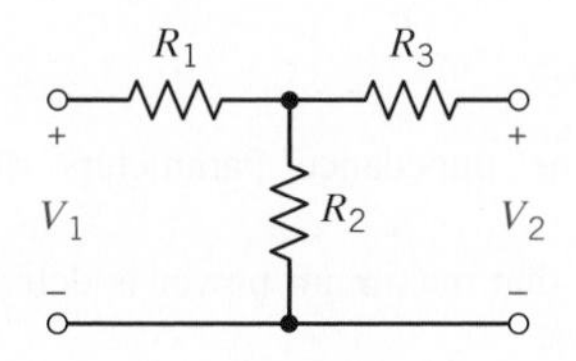

Figure P 17.5-4

P 17.5-5 A simplified model of a bipolar junction transistor is shown in Figure P 17.5-5. Determine the h parameters of this circuit.

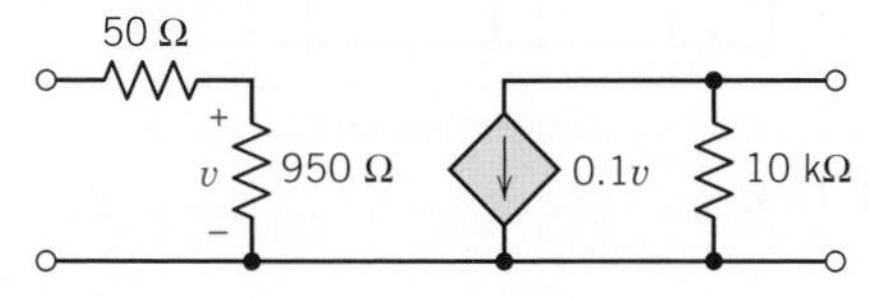

Figure P 17.5-5 Model of bipolar junction transistor.

Section 17.6 Relationships Between Two-Port Parameters

P 17.6-1 Derive the relationships between the Y parameters and the h parameters by using the defining equations for both parameter sets.

P 17.6-2 Determine the Y parameters if the Z parameters are (in ohms):

$$\mathbf{Z} = \begin{bmatrix} 3 & 2 \\ 2 & 6 \end{bmatrix}$$

P 17.6-3 Determine the h parameters when the Y parameters are (in siemens):

$$\mathbf{Y} = \begin{bmatrix} 0.1 & 0.1 \\ 0.4 & 0.5 \end{bmatrix}$$

P 17.6-4 A two-port has the following Y parameters: $Y_{12} = Y_{21} = -0.4$ S, $Y_{11} = 0.5$ S, and $Y_{22} = 0.6$ S. Determine the h parameters.

Answers: $h_{11} = 2\ \Omega$, $h_{21} = -0.8$, $h_{12} = 0.8$, and $h_{22} = 0.28$ S

Section 17.7 Interconnection of Two-Port Networks

P 17.7-1 Connect in parallel the two circuits shown in Figure P 17.7-1 and find the Y parameters of the parallel combination.

Answers: $Y_{11} = 17/6$, $Y_{12} = Y_{21} = -4/3$, and $Y_{22} = 5/3$

3 Ω
a b
1 Ω
c c
(*a*)
1 Ω
a b
2 Ω 3 Ω
c c
(*b*)

Figure P 17.7-1

P 17.7-2 For the T network of Figure P 17.7-2, find the Y and T parameters and determine the resulting parameters after the two two-ports are connected in (a) parallel and (b) cascade. Both two-ports are identical as defined in Figure P 17.7-2

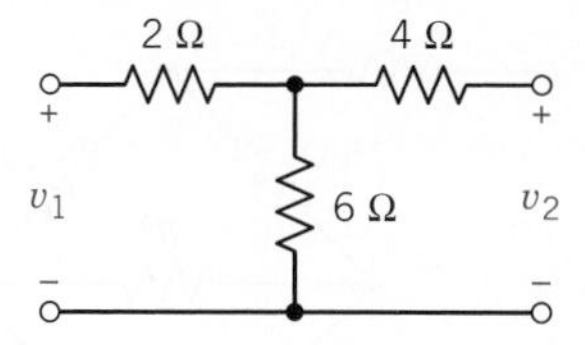

Figure P 17.7-2

P 17.7-3 Determine the Y parameters of the parallel combination of the circuits of Figures P 17.7-3*a*, *b*.

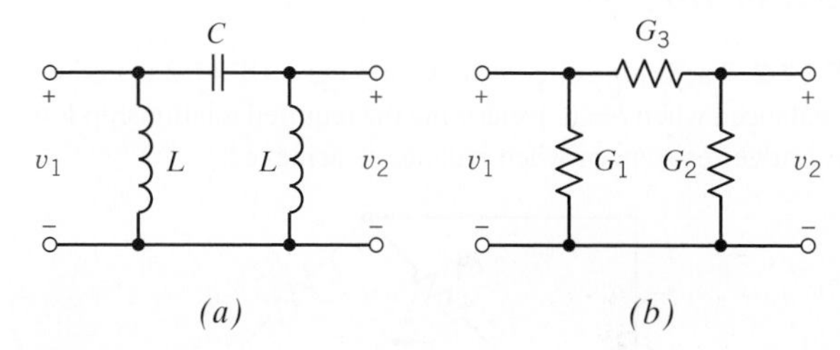

Figure P 17.7-3

Section 17.8 How Can We Check . . . ?

P 17.8-1 A laboratory report concerning the circuit of Figure P 17.8-1 states that $Z_{12} = 15\ \Omega$ and $Y_{11} = 24$ mS. Verify these results.

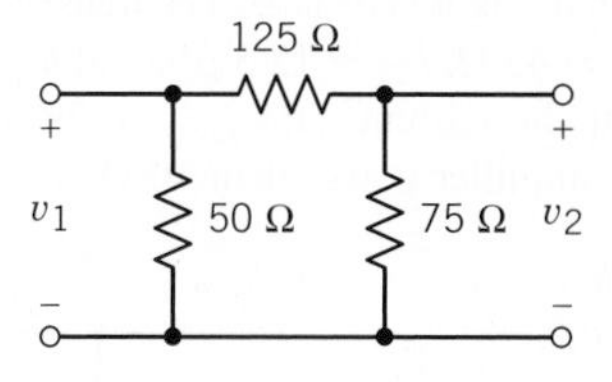

Figure P 17.8-1

P 17.8-2 A student report concerning the circuit of Figure P 17.8-2 has determined the transmission parameters as $A = 2(s + 10)/s$, $D = A$, $C = 10/s$, and $B = (3s^2 + 80s+)400/s^2$. Verify these results when $M = 0.1$ H.

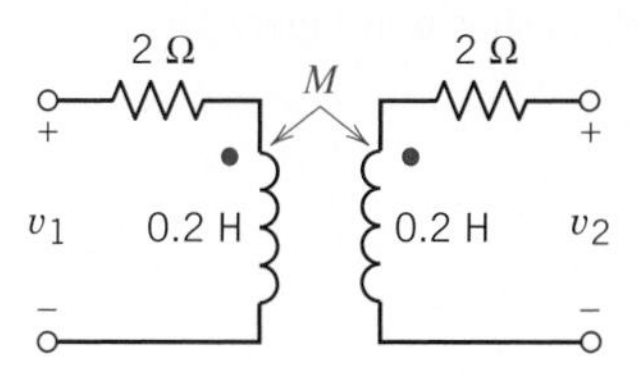

Figure P 17.8-2

Design Problems

DP 17-1 Select R_1 and R so that $R_{in} = 16.6\ \Omega$ for the circuit of Figure DP 17-1. A design constraint requires that both R_1 and R be less than 10 Ω.

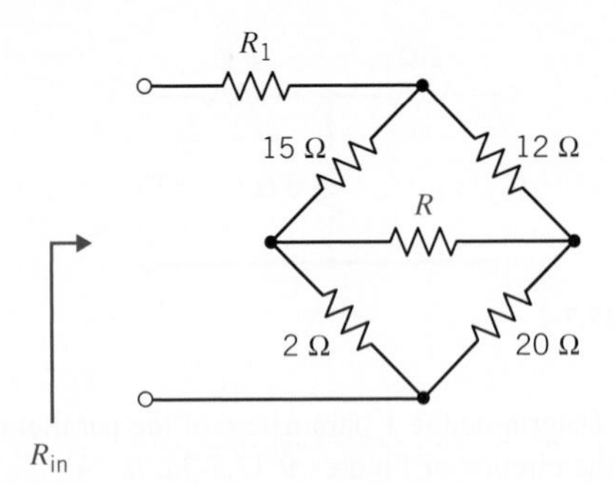

Figure DP 17-1

DP 17-2 The bridge circuit shown in Figure DP 17-2 is said to be balanced when $I = 0$. Determine the required relationship for the bridge resistances when balance is achieved.

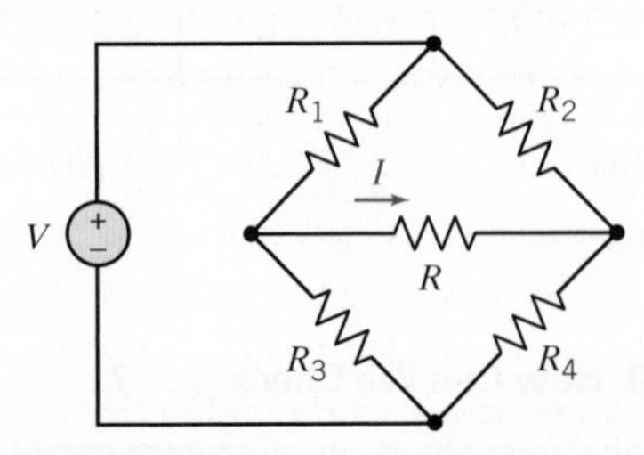

Figure DP 17-2 Bridge circuit.

DP 17-3 A hybrid model of a common-emitter transistor amplifier is shown in Figure DP 17-3. The transistor parameters are $h_{21} = 80$, $h_{11} = 45\ \Omega$, $h_{22} = 12.5\ \mu\text{S}$, and $h_{12} = 5 \times 10^{-4}$. Select R_L so that the current gain $i_2/i_1 = 79$ and the input resistance of the amplifier is less than 10 Ω.

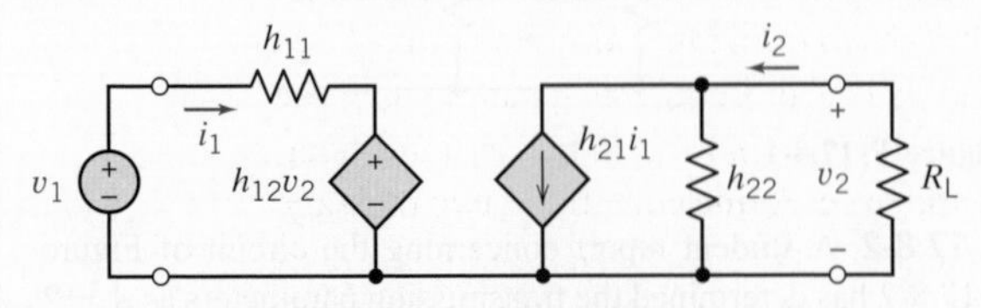

Figure DP 17-3 Model of transistor amplifier.

DP 17-4 A two-port network connected to a source v_s and a load resistance R_L is shown in Figure DP 17-4.

(a) Determine the impedance parameters of the two-port network.

(b) Select R_L so that maximum power is delivered to R_L.

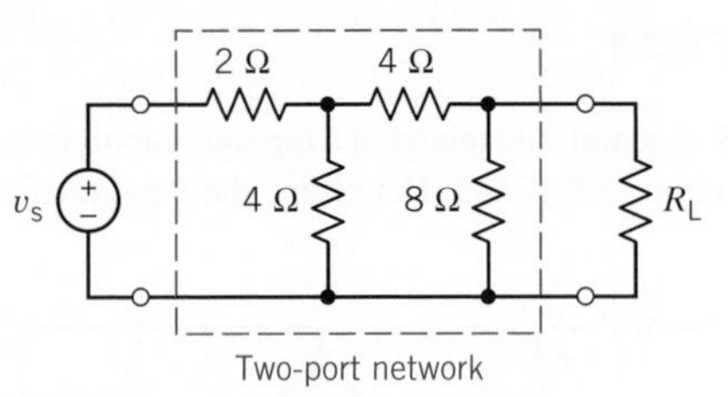

Figure DP 17-4

DP 17-5

(a) Determine the ABCD (transmission matrix) of the two-port networks shown in Figures DP 17-5*a* and DP 17-5*b*.

(b) Using the results of part (a), find the *s*-domain ABCD matrix of the network shown in *c*.

(c) Given $L_1 = (10/\pi)$ mH, $L_2 = (2.5/\pi)$ mH, $C_1 = (0.78/\pi)\ \mu\text{F}$, $C_2 = C_3 = (1/\pi)\ \mu\text{F}$, and $R_L = 100\ \Omega$, find the open-circuit voltage gain V_2/V_1 and the short-circuit current gain I_2/I_1 under sinusoidal-state conditions at the following frequencies: 2.5 kHz, 5.0 kHz, 7.5 kHz, 10 kHz, and 12.5 kHz.

Hint: Use the appropriate entries of the ABCD matrix. Also, note the resonant frequencies of the circuit.

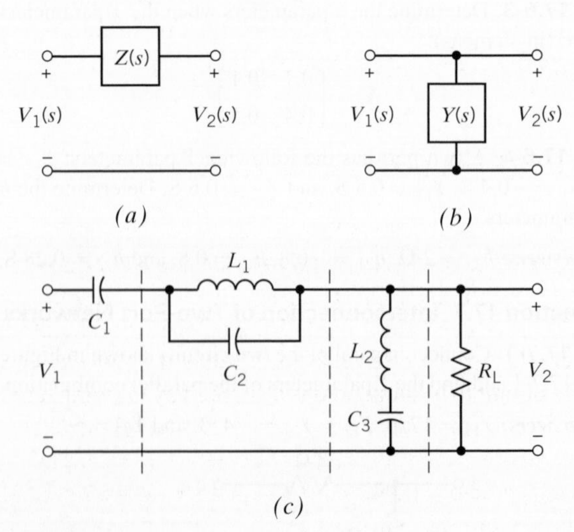

Figure DP 17-5

Getting Started with PSpice

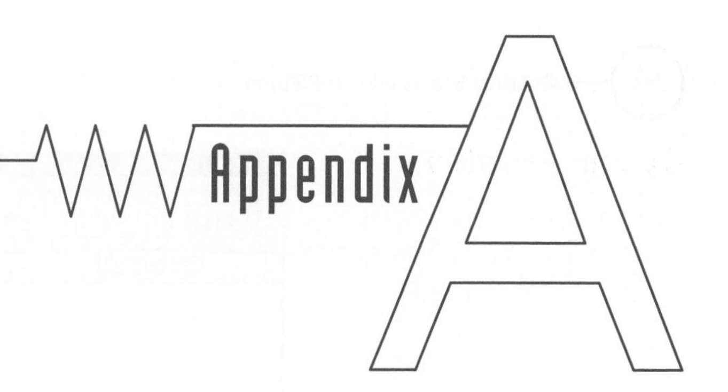

A.1 PSPICE

SPICE, an acronym for Simulation Program with Integrated Circuit Emphasis, is a computer program used for numerical analysis of electric circuits. Developed in the early 1970s at the University of California at Berkeley, it is generally regarded as the most widely used circuit simulation program (Perry, 1998). PSpice is a version of SPICE, designed for personal computers, developed by MicroSim Corporation in 1984 (Tuinenga, 88). SPICE was a text-based program that required the user to describe the circuit using only text, and the simulation results were displayed as text. MicroSim provided a graphical postprocessor, Probe, to plot the results of SPICE simulations. Later, MicroSim also provided a graphical interface called Schematics that allowed users to describe circuits graphically. The name of the simulation program was changed from PSpice to PSpice A/D when it became possible to simulate circuits that contained both analog and digital devices. MicroSim was acquired by ORCAD®, which was in turn acquired by Cadence®. ORCAD improved Schematics and renamed it Capture. ''Using PSpice'' loosely refers to using ORCAD Capture, PSpice A/D, and Probe to analyze an electric circuit numerically.

A.2 GETTING STARTED

Begin by starting the ORCAD Capture program. Figure A.1 shows the opening screen of ORCAD Capture. (If necessary, maximize the Session Log window.) The top line of the screen shows the title of the program, ORCAD Capture CIS – Demo Edition. A menu bar providing menus called File, View, Edit, Options, Window, and Help is located under the title line. A row of buttons is located under the menu bar, and a ruler is located below the row of buttons. A workspace is located beneath the ruler. The circuit to be simulated is described by drawing it in this workspace. A line containing two message fields is located under the workspace. The left message field is of particular interest because it provides information about the Capture screen. For example, move the cursor to one of the buttons. The left message field describes the function of the button. Save Active Document is the function of the third button from the left.

Select File/New/Project from the Capture menus, as shown in Figure A.2. The New Project dialog box, shown in Figure A.3, will pop up. Select Analog Or Mixed A/D, as shown. The New Project dialog box requires a project name and a location. The location is the name of the directory or folder in which Capture should store the project file. The name will be the file name of the project file. ORCAD Capture uses OPJ as a suffix for project files, so choosing Name to be ExampleCircuit and Location to be c:\PSpiceCircuits causes ORCAD to store a file named ExampleCircuit.opj in the c:\PSpiceCircuits folder. Notice that long file names are supported, making it easier to give descriptive names to projects.

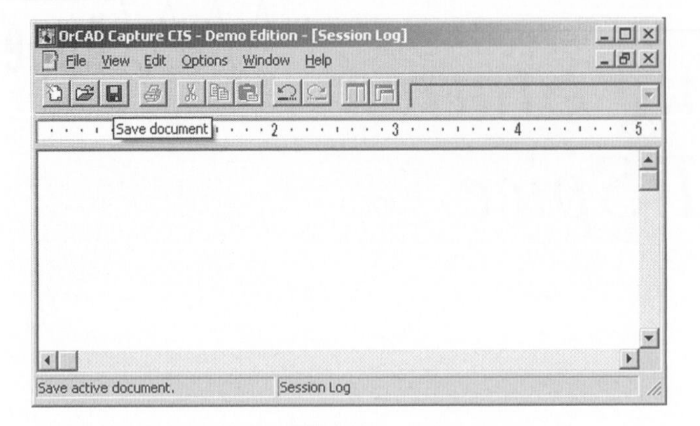

FIGURE A.1 The opening screen of ORCAD Capture CIS demo edition version 15.7.

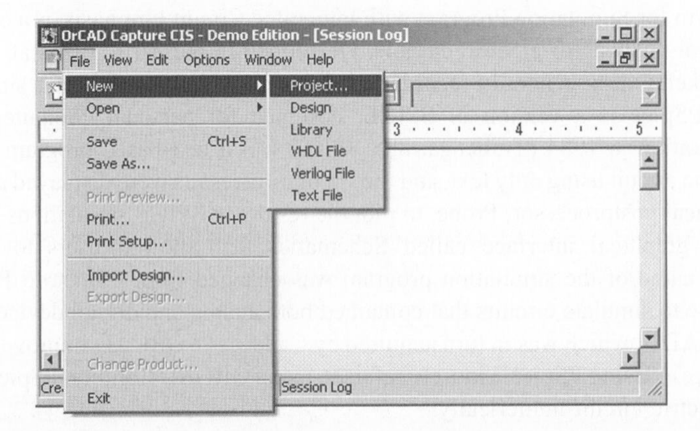

FIGURE A.2 Opening a new project in ORCAD Capture.

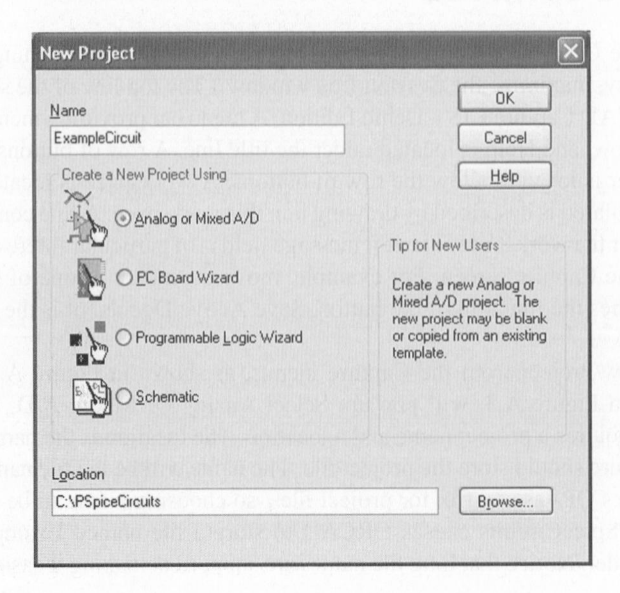

FIGURE A.3 New Project dialog box.

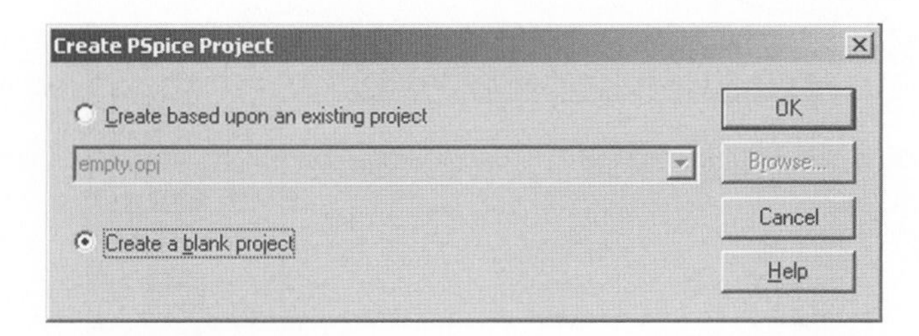

FIGURE A.4 Create PSpice Project dialog box.

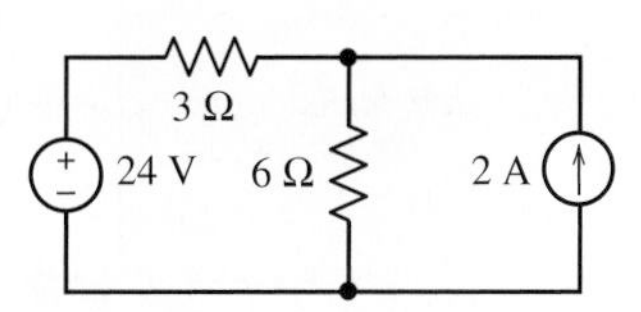

FIGURE A.5 The example circuit.

Click OK in the New Project dialog box to close the New Project dialog box and pop up the Create PSpice Project dialog box shown in Figure A.4. Select Create a blank project and then click OK to return to the ORCAD Capture screen. The Capture screen has changed: Place, Macro, PSpice, and Accessories have been added to the menu bar; there are more buttons; and there is a grid on the workspace.

We are ready to begin our first PSpice simulation. In that first simulation, we will simulate the circuit shown in Figure A.5 to determine its node voltages. We start by drawing the circuit in the ORCAD Capture workspace.

A.3 DRAWING A CIRCUIT IN THE ORCAD CAPTURE WORKSPACE

Drawing a circuit in the ORCAD workspace requires three activites:

1. Placing the circuit elements in the ORCAD Capture workspace
2. Adjusting the values of the circuit element parameters, for example the resistances of the resistors
3. Wiring the circuit to connect the circuit elements

To begin, select Part/Place from the Capture menus to pop up the Place Part dialog box shown in Figure A.6. To obtain a resistor, select ANALOG from the list of libraries and R from the list of parts. Click OK to close the Place Part dialog box and return to the Capture screen. Upon returning to the Capture screen, the cursor will be dragging the symbol for a resistor. Place the resistor, as desired, with a click. The cursor will now be dragging a second resistor symbol. A right-click produces the menu shown in Figure A.7. Selections from this menu will flip or rotate the resistor. Select End Mode to stop placing resistors. (If ANALOG is not listed among the available libraries in the Place Part dialog box, click the Add Library button. ORCAD Capture provides several libraries

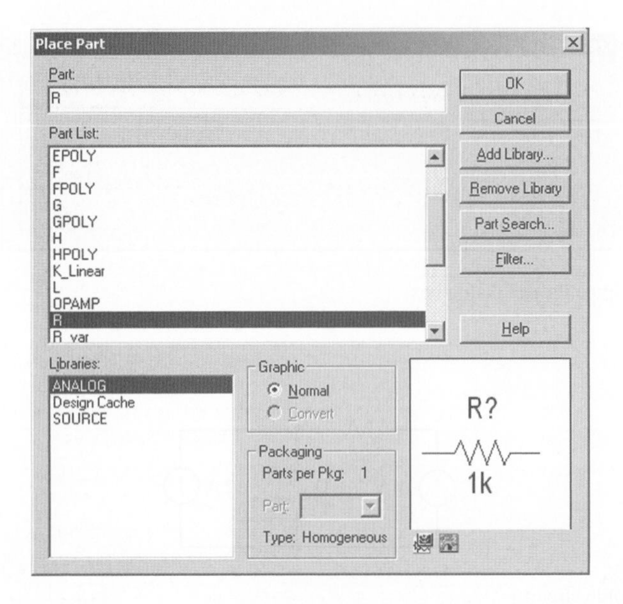

FIGURE A.6 The Place Part dialog box.

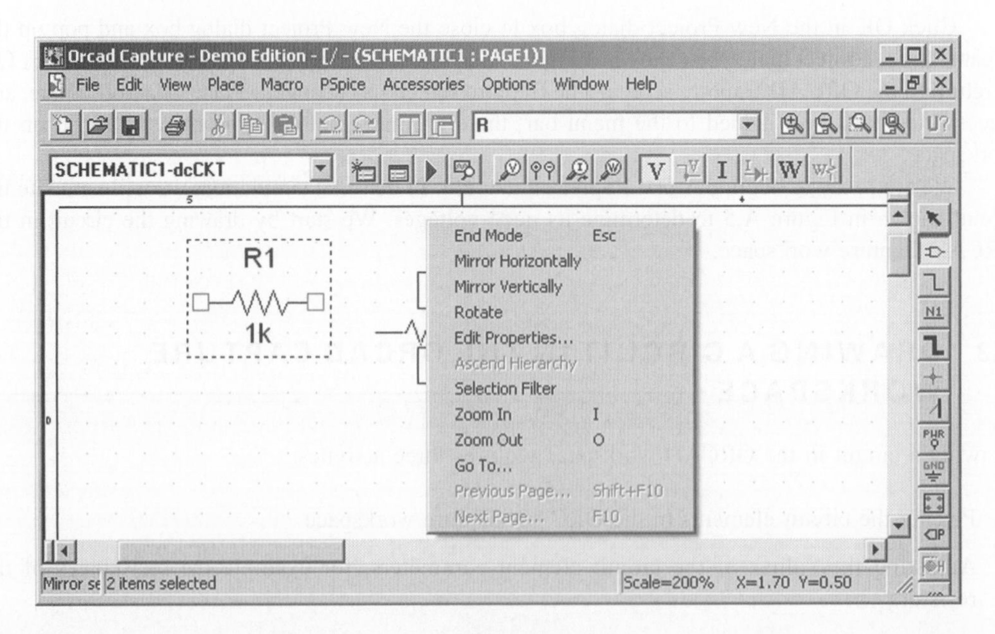

FIGURE A.7 A right-click while placing parts pops up this menu.

containing parts for circuits. File names of parts libraries use the suffix OLB. Select the analog.olb and source.olb libraries.)

SPICE requires every circuit to include a ground node. Select Part/Ground from the Capture menus to pop up the Place Ground dialog box. The ground node is a PSpice part called 0 that is contained in the SOURCE library. (It may be necessary to add this library. Click the Add Library button to pop up a Browse File dialog box. The library file is called source.olb and resides in the PSpice folder. Select the source.olb and library then click Open to make this library available and to return to

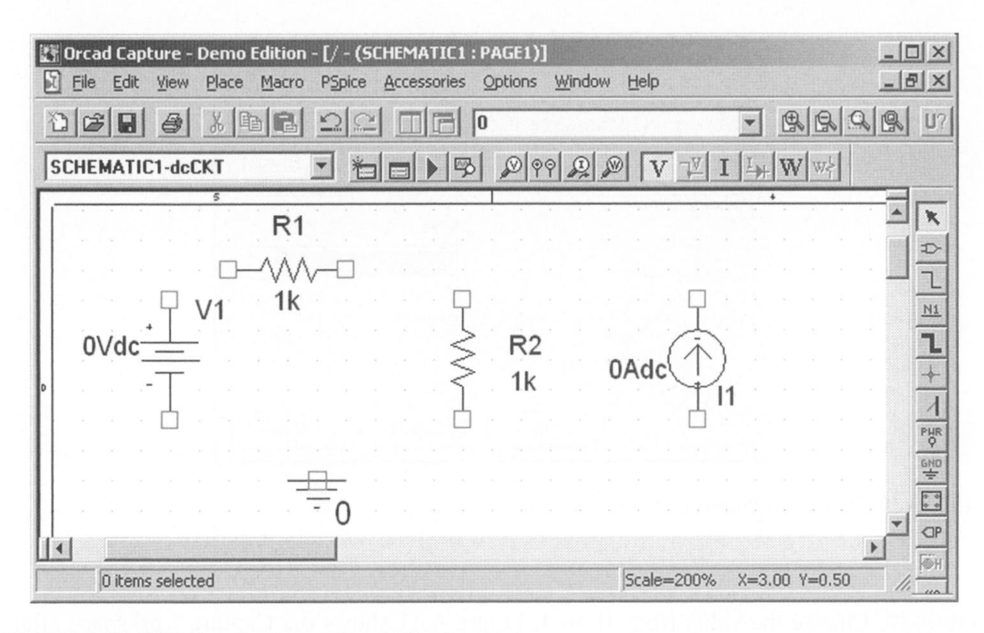

FIGURE A.8 ORCAD Capture screen after placing the parts.

the Place Ground dialog box.) Place the ground node in the Capture workspace. Figure A.8 shows the Capture screen after the parts have been placed.

The resistances of the resistors each has its default value, 1k. Click the 1k of the vertical resistor to select it, then right-click anywhere in the Capture workspace to obtain the menu shown in Figure A.9. Choose Edit Properties to pop up the Display properties dialog box shown in

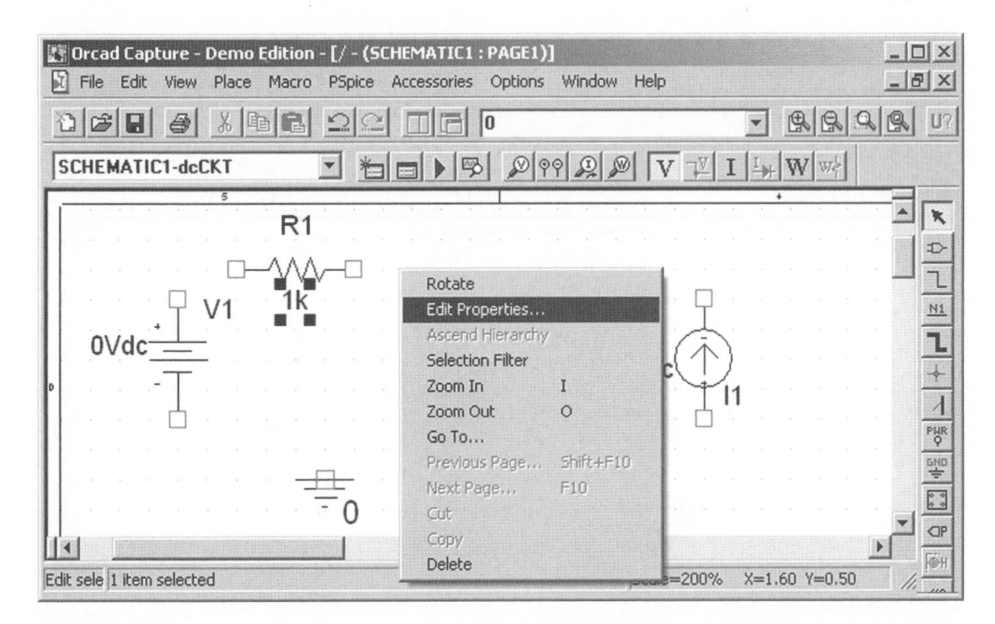

FIGURE A.9 The value, 1k, is shown highlighted. Right-clicking anywhere in the Capture workspace pops up this menu.

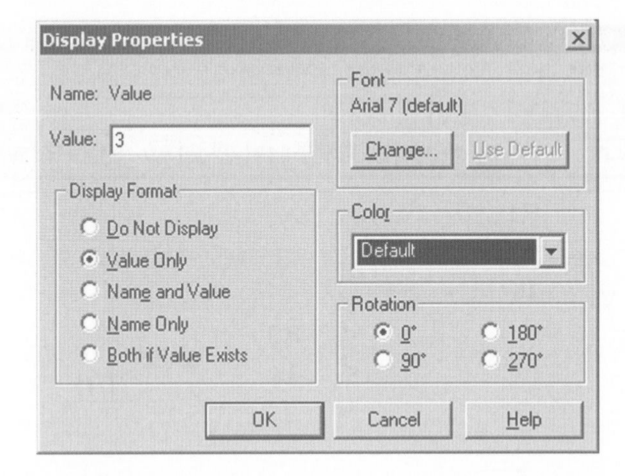

FIGURE A.10 The Display Properties dialog box.

Figure A.10. Change the value from 1k to 3. Figure A.11 shows the Capture workspace after the parameter values of the parts have been adjusted.

Select Parts/Wire to wire the parts together. In Figure A.11, notice that the terminals of each part are marked with small squares. To wire two terminals together, click and hold one terminal, drag the mouse to the other terminal, and then release the mouse. The path of the wire will generally follow the path of the mouse, but wires will be drawn using straight horizontal and vertical lines. Wires can also connect part terminals to wires or connect wires to wires. To stop wiring, right-click and then select End Mode from the menu that appears. Figure A.12 shows the circuit after it has been wired.

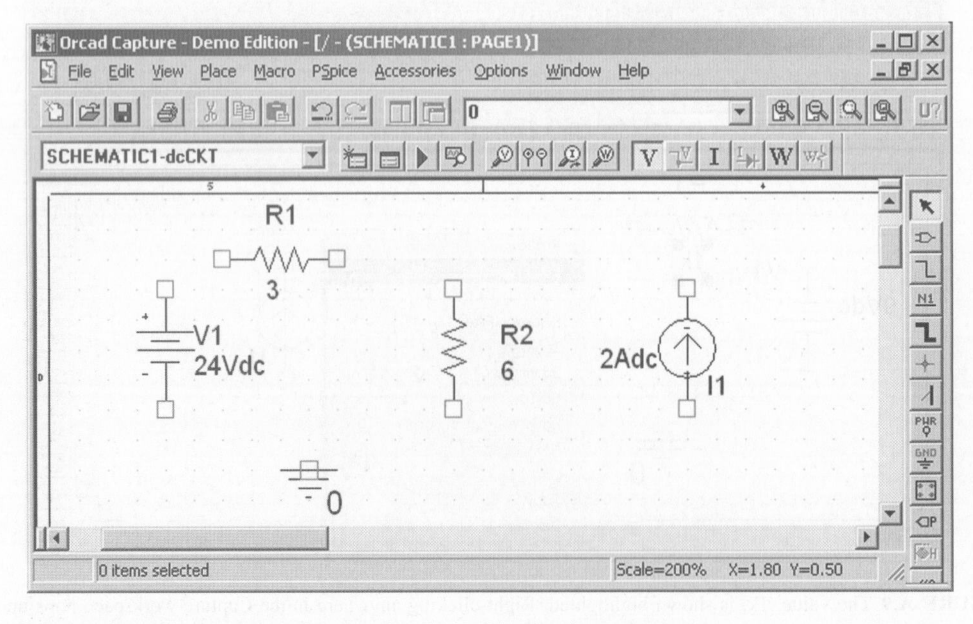

FIGURE A.11 Capture screen after adjusting the values of the circuit parameters.

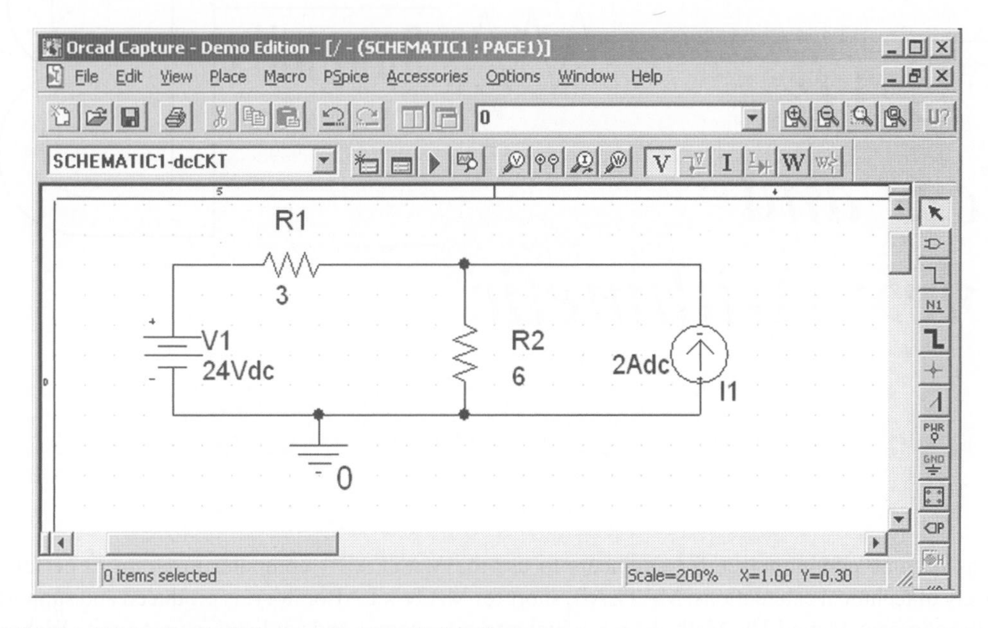

FIGURE A.12 The circuit of Figure A.5 as described in Capture.

A.4 SPECIFYING AND RUNNING THE SIMULATOIN

Select PSpice/New Simulation Profile from the ORCAD Capture menus to pop up the New Simulation dialog box. Provide a name, such as dc analysis, and then click Create. The Simulation Settings dialog box will pop up. Select Bias Point from the Analysis type list and select General Settings under Options. Click OK to close the Simulation Settings dialog box. Select PSpice/Run from the ORCAD menu bar to run the simulation. Figure A.13 shows the simulation results.

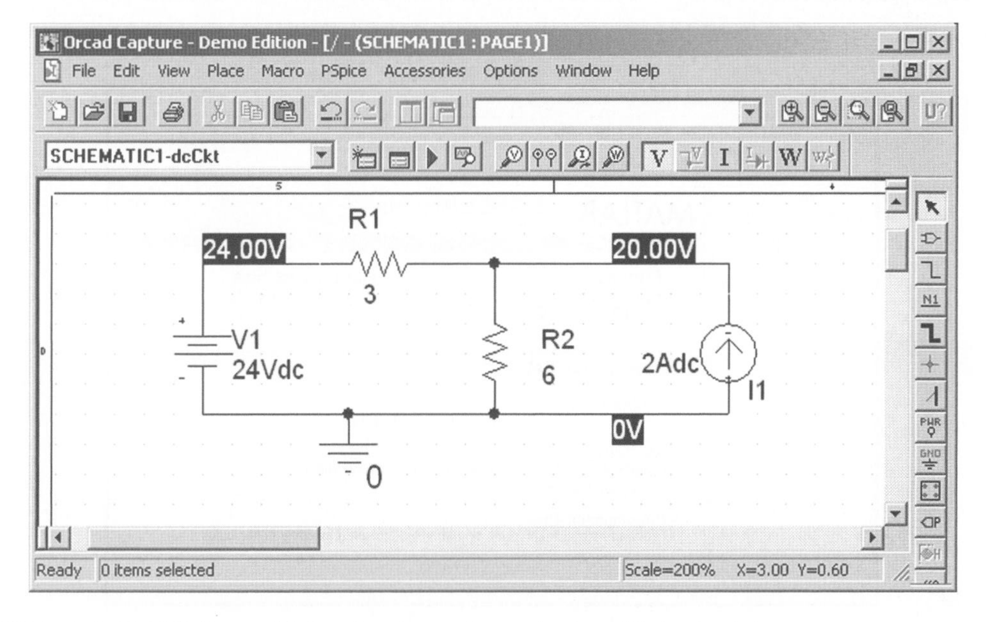

FIGURE A.13 ORCAD Capture labels node voltages after performing a PSpice simulation.

MATLAB, Matrices and Complex Arithmetic

It has become commonplace for engineers to use the MATLAB® computer program to perform a variety of technical calculations. MATLAB, short for MATrix LABoratory, is produced and supported by the company named The Math Works, which provides demos and application notes at its Web site, www.mathworks.com. In addition, MATLAB has extensive built-in help, as shown in Figure B.1.

In this appendix, we will first use MATLAB as a powerful calculator, then use it to solve equations involving matrices or complex numbers and, finally, use it to plot functions.

B.1 USING MATLAB AS A CALCULATOR

Consider the equation

$$C \cdot D = 4A + B \Rightarrow D = \frac{4A + B}{C}$$

Let's use MATLAB to evaluate D when $A = 4$, $B = 7$, and $C = 6$. To do so, we write the equations representing A, B, C, and D in the MATLAB workspace, using the arithmetic operations and functions

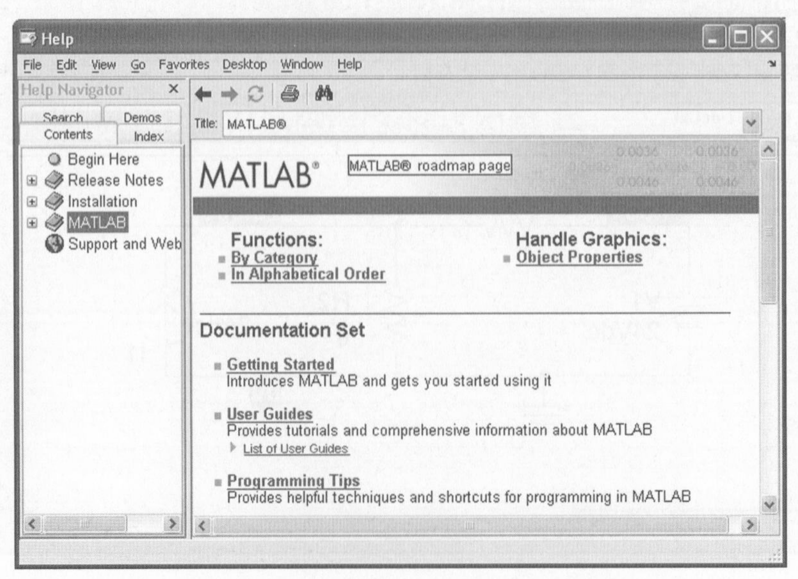

FIGURE B.1 MATLAB Help is accessed by clicking Help on the MATLAB menu bar.

Table B.1 **Arithmetic Operations**

OPERATION	SYMBOL	EQUATION	MATLAB
Addition	+	$4+x$	4 + x
Subtraction	−	$4-x$	4 − x
Multiplication	*	$4\,x$	4 * x
Division	/	$4/x$	4 / x
Power	∧	4^x	4^x

Table B.2 **Built-in Functions**

FUNCTION	EQUATION	MATLAB
sine	$\sin(x)$	sin (x)
cosine	$\cos(x)$	cos (x)
tangent	$\tan(x)$	tan (x)
arc sine	$\sin^{-1}(x)$	asin (x)
arc cosine	$\cos^{-1}(x)$	acos (x)
arc tangent	$\tan^{-1}(x)$	atan (x)
logarithm	$\log_{10}(x)$	log10(x)
natural logarithm	$\ln(x)$	log (x)
exponential	e^x	exp (x)
square root	$\sqrt{x}$	sqrt (x)
absolute value	$\|x\|$	abs (x)

available in MATLAB. Tables B.1 and B.2 list the arithmetic operations and some of the functions available in MATLAB.

Figure B.2 shows the MATLAB workspace. The symbol ≫ is the MATLAB cursor. To indicate that $A = 4$, we type

```
A = 4;<Enter>
```

after the cursor. (<Enter> indicates the Enter key. If we omit the semicolon, MATLAB will tell us the value of A. Because we already know the value of A, we include the semicolon to save space.) MATLAB responds to <Enter> by providing another cursor. We type the equations for B, C, and then D similarly. (MATLAB uses the usual order of precedence for the arithmetic

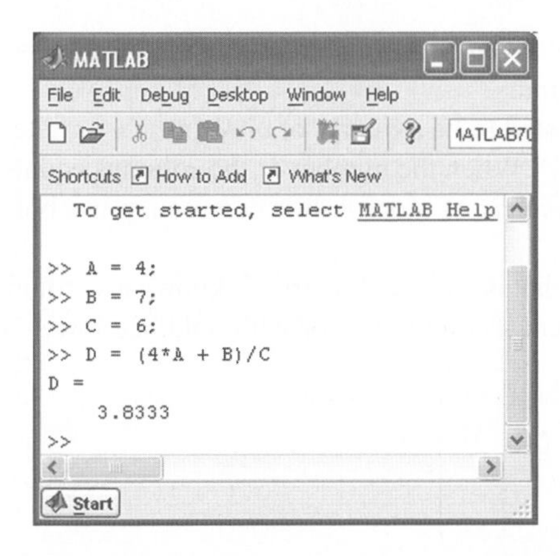

FIGURE B.2 Using MATLAB as a calculator.

operations. Consequently, parentheses are used in the equation representing D to ensure that the addition is performed before the division.) Notice that the semicolon was omitted from the equation representing D, so MATLAB responded to <Enter> by providing the value of D.

EXAMPLE B.1 Trigonometric Functions

Evaluate

$$\theta = \sin^{-1}(\cos(72^\circ))$$

Solution

The trigonometric functions sin, cos, and tan expect an angle in radians, and the inverse trigonometric functions asin, acos, and atan produce an angle in radians. Because we were given an angle in degrees, let's determine the value of θ in degrees. The MATLAB command

```
>> theta = (180/pi) * asin(cos(72*pi/180))
```

produces the result

```
theta = 18.0000
```

The multipliers pi/180 and 180/pi convert units of angles from degrees to radians and vice versa. As a check, the MATLAB command

```
>> phi = (180/pi) * acos(sin(theta*pi/180))
```

produces the result

```
phi = 72
```

B.2 MATRICES, DETERMINANTS, AND SIMULTANEOUS EQUATIONS

There are many situations in circuit analysis in which we have to deal with rectangular arrays of numbers. The rectangular array of numbers

$$\mathbf{A} = \begin{bmatrix} a_{11} & a_{12} & \cdots & a_{1n} \\ a_{21} & a_{22} & \cdots & a_{2n} \\ \vdots & \vdots & & \vdots \\ a_{m1} & a_{m2} & \cdots & a_{mn} \end{bmatrix}$$

is known as a *matrix*. The numbers a_{ij} are called *elements* of the matrix, with the subscript i denoting the row and the subscript j denoting the column.

A matrix with m rows and n columns is said to be a matrix of *order* $m \times n$ or, alternatively, an $m \times n$ matrix. (We read "$m \times n$" as "m by n.") When the number of the columns equals the number of rows, $m = n$, the matrix is called a *square matrix* of order n. It is common to use boldface capital letters to denote an $m \times n$ matrix.

A matrix consisting of only one column, that is, an $m \times 1$ matrix, is known as a column matrix or, more commonly, a *column vector*. We represent a column vector with boldface lowercase letters as

$$\mathbf{x} = \begin{bmatrix} x_1 \\ x_2 \\ \vdots \\ x_m \end{bmatrix}$$

The addition of two matrices is possible for matrices of the same order. The sum of two matrices is obtained by adding the corresponding elements. Thus, if the elements of **A** are a_{ij} and the elements of **B** are b_{ij}, and if

$$\mathbf{C} = \mathbf{A} + \mathbf{B}$$

then the elements of **C** are obtained as

$$c_{ij} = a_{ij} + b_{ij}$$

Matrix addition is commutative, that is,

$$\mathbf{A} + \mathbf{B} = \mathbf{B} + \mathbf{A}$$

Also, the addition operation is associative, so that

$$(\mathbf{A} + \mathbf{B}) + \mathbf{C} = \mathbf{A} + (\mathbf{B} + \mathbf{C})$$

To perform the operation of multiplying matrix **A** by a constant α, every element of the matrix is multiplied by the constant. Therefore, we can write

$$\alpha\mathbf{A} = \begin{bmatrix} \alpha a_{11} & \alpha a_{12} & \cdots & \alpha a_{1n} \\ \alpha a_{21} & \alpha a_{22} & \cdots & \alpha a_{2n} \\ \vdots & \vdots & & \vdots \\ \alpha a_{m1} & \alpha a_{m2} & \cdots & \alpha a_{mn} \end{bmatrix}$$

Matrix multiplication is defined in such a way as to assist in the solution of simultaneous linear equations. The multiplication of two matrices **AB** requires the number of columns of **A** to be equal to the number of rows of **B**. Thus, if **A** is of order $m \times n$ and **B** is of order $n \times q$, the product is a matrix of order $m \times q$. The elements of a product

$$\mathbf{C} = \mathbf{AB}$$

are found by multiplying the ith row of **A** and the jth column of **B** and summing these products to give the element c_{ij}. That is,

$$c_{ij} = a_{i1}b_{1j} + a_{i2}b_{2j} + \cdots + a_{iq}b_{qj} = \sum_{k=1}^{q} a_{ik}b_{kj}$$

Thus we obtain c_{11}, the first element of **C**, by multiplying the first row of **A** by the first column of **B** and summing the products of the elements. We should note that, in general, matrix multiplication is not commutative, that is,

$$\mathbf{AB} \neq \mathbf{BA}$$

EXAMPLE B.2 Matrices in MATLAB

Evaluate

$$\begin{bmatrix} 2 & 1 \\ 4 & 2 \end{bmatrix} + \begin{bmatrix} 6 & 1 \\ 3 & 1 \end{bmatrix}, \begin{bmatrix} 2 & 1 \\ 4 & 2 \end{bmatrix} - \begin{bmatrix} 6 & 1 \\ 3 & 1 \end{bmatrix} \text{ and } \begin{bmatrix} 2 & 1 \\ 4 & 2 \end{bmatrix} * \begin{bmatrix} 6 & 1 \\ 3 & 1 \end{bmatrix}$$

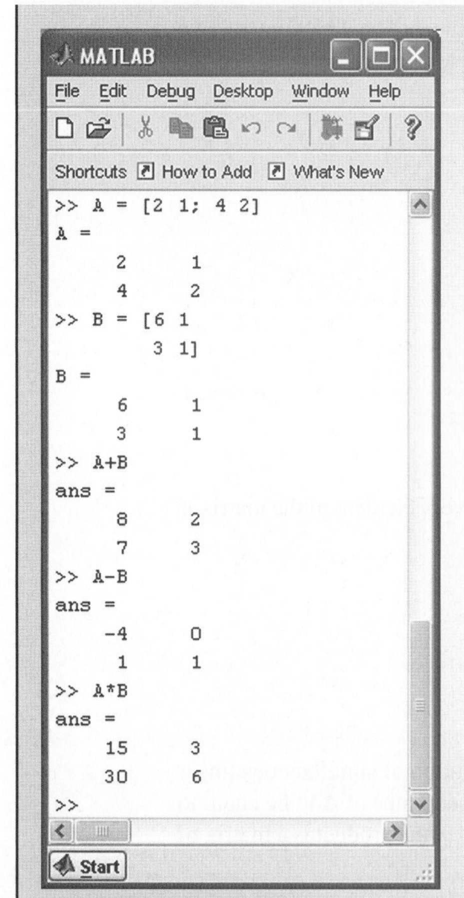

FIGURE B.3 Matrix arithmetic.

Solution

Figure B.3 shows how to do these calculations, using MATLAB. First, two matrix variables

$$\mathbf{A} = \begin{bmatrix} 2 & 1 \\ 4 & 2 \end{bmatrix} \text{ and } \mathbf{B} = \begin{bmatrix} 6 & 1 \\ 3 & 1 \end{bmatrix}$$

are defined. Figure B.3 shows two ways of defining a matrix variable in MATLAB. The command

```
>>A = [2 1; 4 2]
```

uses a space to separate the elements in each row of the matrix and a semicolon to separate the rows of the matrix. The command

```
>> B = [6 1
        2 1]
```

uses a space to separate the elements in a row of the matrix and an <Enter> to separate the rows of the matrix. (After the <Enter>, spaces are used to line up the columns of matrix **B**.) Both commands use the bracket symbols, [and], to indicate the beginning and end of the matrix.

Figure B.3 shows that operations listed in Table B.1 can be used to perform matrix arithmetic. We see that

$$\begin{bmatrix} 2 & 1 \\ 4 & 2 \end{bmatrix} + \begin{bmatrix} 6 & 1 \\ 3 & 1 \end{bmatrix} = \begin{bmatrix} 8 & 2 \\ 7 & 3 \end{bmatrix}, \begin{bmatrix} 2 & 1 \\ 4 & 2 \end{bmatrix} - \begin{bmatrix} 6 & 1 \\ 3 & 1 \end{bmatrix} = \begin{bmatrix} -4 & 0 \\ 1 & 1 \end{bmatrix}$$

$$\text{and } \begin{bmatrix} 2 & 1 \\ 4 & 2 \end{bmatrix} * \begin{bmatrix} 6 & 1 \\ 3 & 1 \end{bmatrix} = \begin{bmatrix} 15 & 3 \\ 30 & 6 \end{bmatrix}$$

A set of simultaneous equations

$$\begin{aligned} a_{11}x_1 + a_{12}x_2 + \cdots + a_{1n}x_n &= b_1 \\ a_{21}x_1 + a_{22}x_2 + \cdots + a_{2n}x_n &= b_2 \\ \vdots \quad\quad \vdots \quad\quad\quad \vdots \quad &\quad \vdots \\ a_{n1}x_1 + a_{n2}x_2 + \cdots + a_{nn}x_n &= b_n \end{aligned} \tag{B-1}$$

can be written in matrix form as

$$\mathbf{Ax} = \mathbf{b} \tag{B-2}$$

where

$$\mathbf{A} = \begin{bmatrix} a_{11} & a_{12} & \cdots & a_{1n} \\ a_{21} & a_{22} & \cdots & a_{2n} \\ \vdots & \vdots & & \vdots \\ a_{n1} & a_{n2} & \cdots & a_{nn} \end{bmatrix}, \mathbf{x} = \begin{bmatrix} x_1 \\ x_2 \\ \vdots \\ x_n \end{bmatrix} \text{ and } \mathbf{b} = \begin{bmatrix} b_1 \\ b_2 \\ \vdots \\ b_n \end{bmatrix}$$

Frequently, we will want to solve a set of simultaneous equations such as Equation B-1. In other words, given the values of the coefficients a_{ij} and b_i, we will want to determine the values of the variables x_i. Using MATLAB, we express the equation in matrix form as shown in Equation B-2, entering matrices **A** and **b** and then giving the MATLAB command

```
>> x = A\b
```

MATLAB will respond with the value of the matrix **x**.

EXAMPLE B.3 Solving Simultaneous Equations Using MATLAB

Solve the simultaneous equations:

$$\begin{aligned} x_1 - 2x_2 + 3x_3 &= 12 \\ 4x_2 - 2x_3 &= -1 \\ 6x_1 - x_2 - x_3 &= 0 \end{aligned}$$

Solution

First, write the simultaneous equations as

$$\mathbf{Ax} = \mathbf{b}$$

where

$$\mathbf{A} = \begin{bmatrix} 1 & -2 & 3 \\ 0 & 4 & -2 \\ 6 & -1 & -1 \end{bmatrix}, \mathbf{b} = \begin{bmatrix} 12 \\ -1 \\ 0 \end{bmatrix} \text{ and } \mathbf{x} = \begin{bmatrix} x_1 \\ x_2 \\ x_3 \end{bmatrix}$$

Next, enter matrices **A** and **b** in the MATLAB command window as shown in Figure B.4. Then, issue the MATLAB command

```
>> x = A\b
```

MATLAB provides the result

$$\mathbf{x} = \begin{bmatrix} 1.2407 \\ 2.3148 \\ 5.1296 \end{bmatrix}$$

indicating that

$$x_1 = 1.2407,\ x_2 = 2.3148, \text{ and } x_1 = 5.1296$$

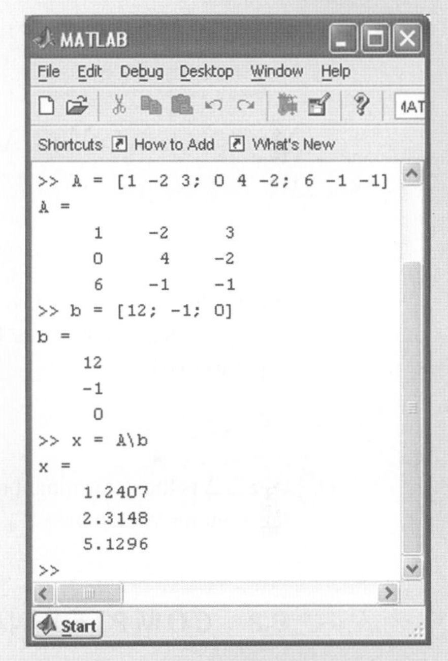

FIGURE B.4 Solving simultaneous equations.

We can also solve simultaneous equations using *Cramer's rule*, which involves determinants, minors, and cofactors. The *determinant* of a matrix is a number associated with a square matrix. We define the determinant of a square matrix **A** as Δ, where

$$\Delta = \begin{vmatrix} a_{11} & a_{12} & \cdots & a_{1n} \\ a_{21} & a_{22} & \cdots & a_{2n} \\ \vdots & \vdots & & \vdots \\ a_{n1} & a_{n2} & \cdots & a_{nn} \end{vmatrix}$$

For example, the determinant of a 2 × 2 matrix

$$\Delta = \begin{vmatrix} a_{11} & a_{12} \\ a_{21} & a_{22} \end{vmatrix} = a_{11}a_{22} - a_{12}a_{21}$$

Similarly, the determinant of a 3 × 3 matrix is

$$\Delta = \begin{vmatrix} a_{11} & a_{12} & a_{13} \\ a_{21} & a_{22} & a_{23} \\ a_{31} & a_{32} & a_{33} \end{vmatrix} = (a_{11}a_{22}a_{33} + a_{12}a_{23}a_{31} + a_{13}a_{32}a_{21}) - (a_{13}a_{22}a_{31} + a_{23}a_{32}a_{11} + a_{33}a_{21}a_{12})$$

In general, we are able to determine the determinant Δ in terms of cofactors and minors. The determinant of a submatrix of **A** obtained by deleting from **A** the ith row and the jth column is called the *minor* of the element a_{ij} and denoted as m_{ij}.

The cofactor c_{ij} is a minor with an associated sign, so that

$$c_{ij} = (-1)^{(i+j)} m_{ij}$$

The rule for evaluating the determinant Δ using the ith row of an $n \times n$ matrix is

$$\Delta = \sum_{j=1}^{n} a_{ij} c_{ij}$$

for a selected value of i. Alternatively, we can obtain Δ by using the jth column and, thus,

$$\Delta = \sum_{j=1}^{n} a_{ij} c_{ij}$$

for a selected value of j.

Cramer's rule states that the solution for the unknown, x_k, of the simultaneous equations of Equation B-1 is

$$x_{\mathrm{k}} = \frac{\Delta_{\mathrm{k}}}{\Delta}$$

where Δ is the determinant of **A** and Δ_k is the determinant formed by replacing the kth column of **A** by the column vector **b**.

B.3 COMPLEX NUMBERS AND COMPLEX ARITHMETIC

We can represent the complex number c as

$$c = a + jb \tag{B-3}$$

where a and b are real numbers and $j = \sqrt{-1}$. It's useful to associate this complex number with a point in the complex plane as shown in Figure B.5a. Figure B.5a shows that the real numbers a and b in Equation B-3 are the projections of the point unto the real and imaginary axes. Consequently, a is called the *real part* of c, and b is called the *imaginary part* of c. We write

$$a = \mathrm{Re}\{c\} \text{ and } b = \mathrm{Im}\{c\}$$

Figure B.5b illustrates an alternate representation of the complex number c, in which a line segment is drawn from the origin of the complex plane to the point representing the complex number. The angle of this line segment, θ, measured counterclockwise from the real axis, is called the *angle* of the complex number. The length of the line segment, r, is called the *magnitude* of the complex number.

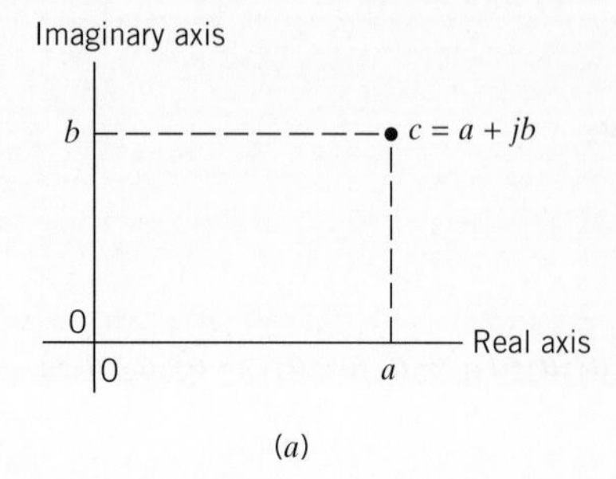

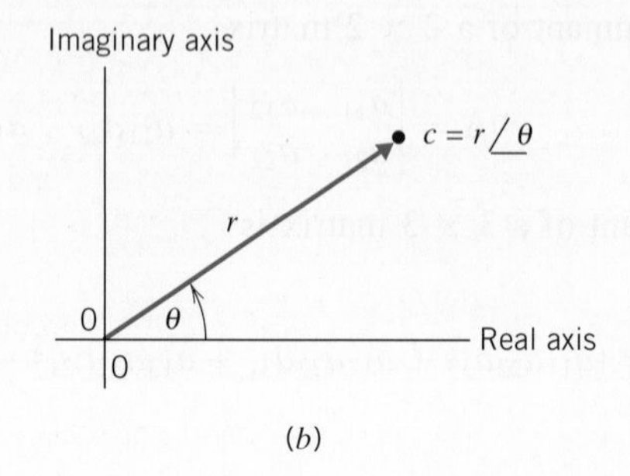

FIGURE B.5 Rectangular (a) and polar (b) forms of a complex number.

The polar form represents the complex number in terms of its magnitude and angle. We write

$$c = r\angle\theta$$

To indicate that r is the magnitude of the complex number c and that θ is the angle of c, we write

$$r = |c| \text{ and } \theta = \angle c$$

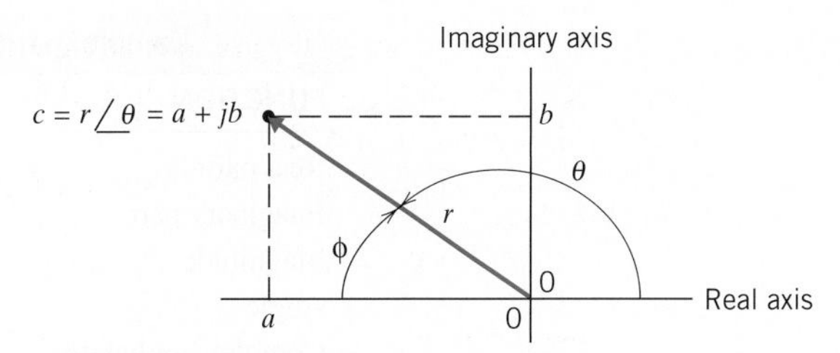

FIGURE B.6 A complex number having a&λτ; 0.

Figure B.6 shows a complex number c with $\text{Re}\{c\} < 0$. Notice that θ, not ϕ, is the angle of c.

Because a complex number can be expressed in both rectangular and polar forms, we write

$$a + jb = c = r\angle\theta$$

The trigonometry of Figure B.4 and Figure B.5 provides the following equations for converting between the rectangular and polar forms of complex numbers.

$$a = r\cos(\theta),\ b = r\sin(\theta),\ r = \sqrt{a^2 + b^2}$$

and

$$\theta = \begin{cases} \tan^{-1}\left(\dfrac{b}{a}\right) & a > 0 \\ 180° - \tan^{-1}\left(\dfrac{b}{-a}\right) & a < 0 \end{cases}$$

Several special cases are worth noticing.

$$1 = 1\angle 0°,\ j = 1\angle 90°,\ -1 = 1\angle \pm 180° \text{ and } -j = 1\angle -90° = 1\angle 270°$$

Next, consider doing arithmetic with complex numbers. We will convert complex numbers to rectangular form before adding or subtracting. Then,

$$(a + jb) + (c + jd) = (a + c) + j(b + d)$$

and

$$(a + jb) - (c + jd) = (a - c) + j(b - d)$$

We will convert complex numbers to polar form before multiplying or dividing. Then,

$$\left(A\angle\theta\right)\left(B\angle\phi\right) = AB\angle(\theta + \phi) \text{ and } \frac{A\angle\theta}{B\angle\phi} = \frac{A}{B}\angle(\theta - \phi)$$

The *conjugate* of the complex number $c = a + jb$ is denoted as c^* and is defined as

$$c^* = a - jb$$

In polar form, we have

$$c^* = r\angle -\theta$$

A third representation of complex numbers, the exponential form, is motivated by Euler's formula. Euler's formula is

$$e^{j\theta} = \cos\theta + j\sin\theta$$

Table B.3 Complex-Arithmetic Functions

FUNCTION	EQUATION	MATLAB
Real part	$\text{Re}\{c\}$	real(c)
Imaginary part	$\text{Im}\{c\}$	imag(c)
Magnitude	$\|c\|$	abs(c)
Angle	$\angle c$	angle(c)
Complex conjugate	c^*	conj(c)

Consequently,

$$r\,e^{j\theta} = r\,\cos\theta + j\,r\,\sin\theta$$

Similarly, when we convert from polar to rectangular form,

$$r\angle\theta = r\,\cos\theta + j\,r\,\sin\theta$$

Noticing that the right-hand sides of the two previous equations are identical establishes the equivalence between the exponential and polar forms of a complex number.

$$r\,e^{j\theta} = r\angle\theta$$

The conversion between the polar and exponential forms is immediate. When using MATLAB, we will represent a polar form complex number by the equivalent exponential form complex number.

It's worth noticing that Euler's formula provides formulas for the sine and cosine.

$$\cos\theta = \frac{1}{2}\left(e^{j\theta} + e^{-j\theta}\right) \text{ and } \sin\theta = \frac{1}{2j}\left(e^{j\theta} - e^{-j\theta}\right)$$

Table B.3 list some of the complex arithmetic functions available in MATLAB.

EXAMPLE B.4 Rectangular and Polar Forms of Complex Numbers

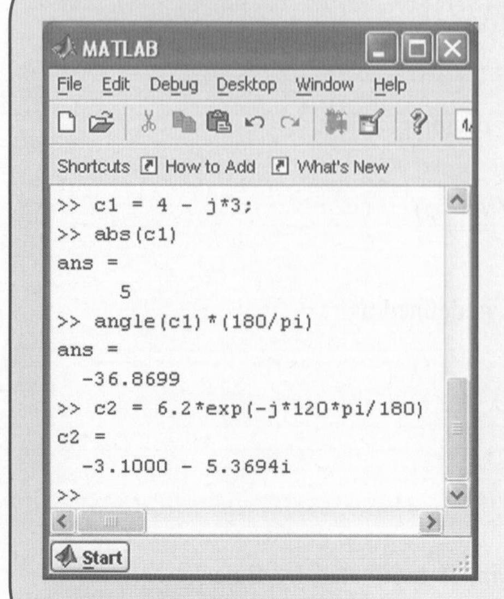

FIGURE B.7 Complex numbers

Express $c_1 = 4 - j3$ in exponential and polar forms. Express $c_2 = 6.2\angle{-120^\circ}$ in rectangular form.

Solution

Doing the conversions by hand yields

$$c_1 = \sqrt{4^2 + (-3)^2}\angle\tan^{-1}\left(\frac{-3}{4}\right) = 5\angle{-36.87^\circ}$$

and

$$c_2 = 6.2\cos(-120^\circ) + j6.2\sin(-120^\circ) = -3.1 - j5.37$$

In Figure B.7, MATLAB does the same conversions with the same results. The factors $180/\pi$ and $\pi/180$ are used to convert radians to degrees and degrees to radians. Notice that the function angle(c1) gives the angle of c1 in radians and the function $\exp(-j^*\theta)$ expects θ to be given in degrees.

EXAMPLE B.5 Arithmetic with Complex Numbers

Find $c + d$, $c - d$, cd, and c/d when $c = 4 - j3$ and $d = 6.2\angle{-120^\circ}$.

Solution

First, let's convert c to polar form and d to rectangular form.

$$c = \sqrt{4^2 + (-3)^2}\angle \tan^{-1}\left(\frac{-3}{4}\right) = 5\angle{-36.87^\circ}$$

and

$$\begin{aligned} d &= 6.2\cos(-120^\circ) + j6.2\sin(-120^\circ) \\ &= -3.1 - j5.37 \end{aligned}$$

Using the rectangular form for addition and subtraction yields

$$\begin{aligned} c + d &= (4 - j3) + (-3.1 - j5.37) \\ &= (4 - 3.1) + j(-3 - 5.37) = 0.9 - j8.37 \end{aligned}$$

and

$$\begin{aligned} c - d &= (4 - j3) - (-3.1 - j5.37) \\ &= (4 + 3.1) + j(-3 + 5.37) = 7.1 + j2.37 \end{aligned}$$

Using the polar form for multiplication and division yields

$$\begin{aligned} cd &= \left(5\angle{-36.87^\circ}\right)\left(6.2\angle{-120^\circ}\right) \\ &= (5 \times 6.2)\angle(-36.87^\circ - 120^\circ) \\ &= 31\angle{-156.87^\circ} \end{aligned}$$

and

$$\begin{aligned} \frac{c}{d} &= \frac{5\angle{-36.87^\circ}}{6.2\angle{-120^\circ}} \\ &= \left(\frac{5}{6.2}\right)\angle(-36.87^\circ + 120^\circ) \\ &= 0.806\angle{83.13^\circ} \end{aligned}$$

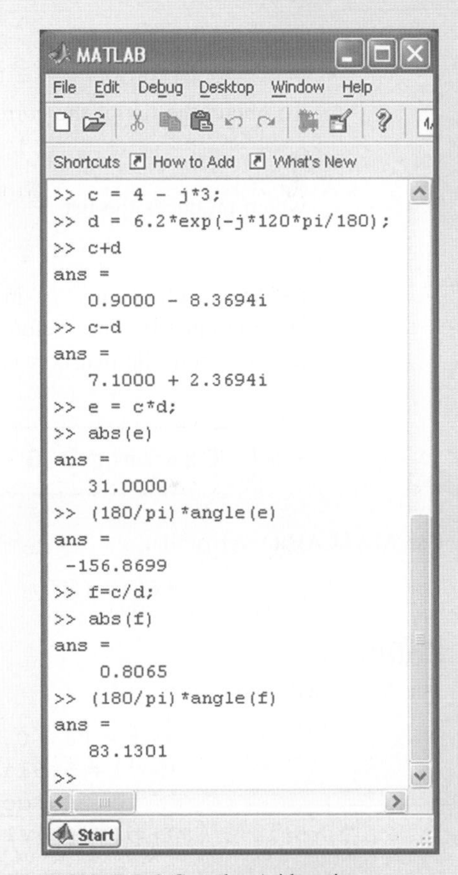

FIGURE B.8 Complex Arithmetic

In Figure B.8, MATLAB does the same arithmetic with the same results.

B.4 PLOTTING FUNCTIONS USING MATLAB

Consider the equation

$$y = 0.2x^2 + 1.6$$

The MATLAB command

```
>> plot(x,y)
```

tells MATLAB to plot **y** as a function of **x**. The command requires **x** to be a row vector, that is, a $1 \times n$ matrix containing a list of equally spaced values of the variable x, and **y** to be a row vector containing a list of the corresponding values of the variable y.

To obtain a list of equally spaced values of the variable x, we issue a MATLAB command of the form

```
>> x = [xs : dx : xf]
```

where xs is the starting value of x, dx is the increment of x, and xf is the final value of x. For example, the MATLAB command

```
>> x = [-5 : 4 : 15]
```

produces the list

```
>> x =   -5   -1   3   7   11   15
```

To obtain the list of the corresponding values of the variable y, we issue the MATLAB command

```
>> y=0.2*x.^2+1.6
```

which produces the list

```
>> y =   0.6   1.4   2.2   3.0   3.8   4.6
```

(Notice the operation ".^" in this command. The operation ^ is the power operation from Table B.1, and x is a matrix. The . before the ^ tells MATLAB to apply the power operation to each element of x rather than to the matrix x itself.)

EXAMPLE B.6 Plotting Functions Using MATLAB

Use MATLAB to verify that

$$5.61\cos(100\,t) - 13.96\sin(100\,t) = 15\,\cos(100\,t + 68.1°)$$

Solution

The MATLAB commands

```
>> t = [0 : 0.001 : 0.12];
>> v1 = 5.61*cos(100*t) - 16.96*sin(100*t);
>> v2 = 15*cos(100*t +68.1*pi/180);
>> plot(t,v1,t,v2)
```

Produce the plot shown in Figure B.9. The MATLAB command

```
>> plot(t,v1,t,v2)
```

tells MATLAB to plot both v1 versus t and v2 versus t on the same axis. Because these plots overlap exactly, we conclude that v1 and v2 are identical functions of t.

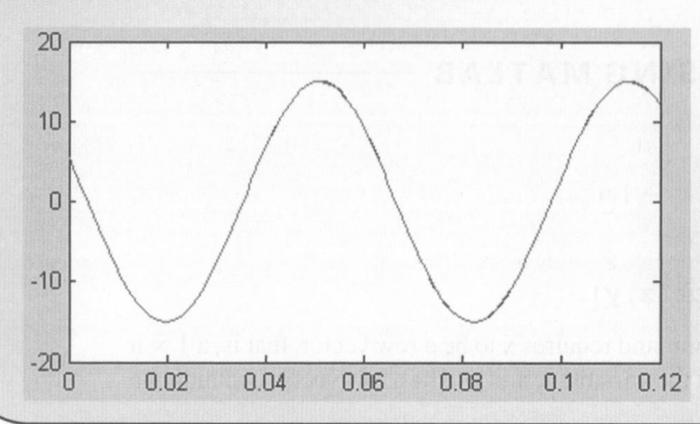

FIGURE B.9 MATLAB plot for Example B.6.

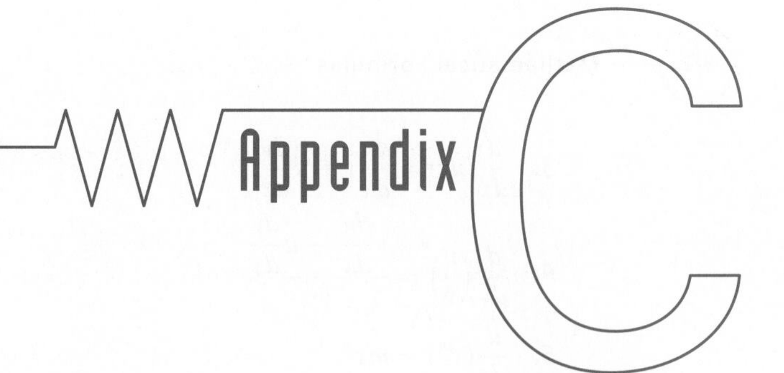

Mathematical Formulas

C.1 TRIGONOMETRIC IDENTITIES

1. $\sin(-\alpha) = -\sin\alpha$
2. $\cos(-\alpha) = \cos\alpha$
3. $\sin\alpha = \cos(\alpha - 90^\circ) = -\cos(\alpha + 90^\circ)$
4. $\cos\alpha = -\sin(\alpha - 90^\circ) = \sin(\alpha + 90^\circ)$
5. $\sin\alpha = -\sin(\alpha \pm 180^\circ)$
6. $\cos\alpha = -\cos(\alpha \pm 180^\circ)$
7. $\sin(\alpha \pm \beta) = \sin\alpha\cos\beta \pm \cos\alpha\sin\beta$
8. $\cos(\alpha \pm \beta) = \cos\alpha\cos\beta \mp \sin\alpha\sin\beta$
9. $\tan(\alpha \pm \beta) = \dfrac{\tan\alpha \pm \tan\beta}{1 \mp \tan\alpha\tan\beta}$
10. $\sin 2\alpha = 2\sin\alpha\cos\alpha$
11. $\cos 2\alpha = \cos^2\alpha - \sin^2\alpha$
12. $2\sin\alpha\sin\beta = \cos(\alpha - \beta) - \cos(\alpha + \beta)$
13. $2\sin\alpha\cos\beta = \sin(\alpha + \beta) + \sin(\alpha - \beta)$
14. $2\cos\alpha\cos\beta = \cos(\alpha + \beta) + \cos(\alpha - \beta)$
15. $2\sin^2\alpha = 1 - \cos 2\alpha$
16. $2\cos^2\alpha = 1 + \cos 2\alpha$
17. $\sin^2\alpha + \cos^2\alpha = 1$

C.2 DERIVATIVES

The letters u and v represent functions of x, whereas a, b, and m are constants.

1. $\dfrac{d}{dx}(au) = a\dfrac{du}{dx}$
2. $\dfrac{d}{dx}(u + v) = \dfrac{du}{dx} + \dfrac{dv}{dx}$

3. $\dfrac{d}{dx}(uv) = \dfrac{du}{dx}v + u\dfrac{dv}{dx}$

4. $\dfrac{d}{dx}\left(\dfrac{u}{v}\right) = \dfrac{v\dfrac{du}{dx} - u\dfrac{dv}{dx}}{v^2}$

5. $\dfrac{d}{dx}(x^m) = mx^{m-1}$

6. $\dfrac{d}{dx}(e^{ax}) = ae^{ax}$

7. $\dfrac{d}{dx}(\ln x) = \dfrac{1}{x}$

8. $\dfrac{d}{dx}\cos(ax+b) = -a\sin(ax+b)$

9. $\dfrac{d}{dx}\sin(ax+b) = a\cos(ax+b)$

C.3 INDEFINITE INTEGRALS

The letters u and v represent functions of x, whereas a and b are constants.

1. $\int au\,dx = a\int u\,dx$

2. $\int (u+v)\,dx = \int u\,dx + \int v\,dx$

3. $\int x^m\,dx = \dfrac{x^{m+1}}{m+1}$ when $m \neq -1$

4. $\int u\dfrac{dv}{dx}\,dx = uv - \int v\dfrac{du}{dx}dx$

5. $\int \dfrac{dx}{x} = \ln|x|$

6. $\int \sin ax\,dx = -\dfrac{1}{a}\cos ax$

7. $\int \cos ax\,dx = \dfrac{1}{a}\sin ax$

8. $\int \sin^2 ax\,dx = \dfrac{x}{2} - \dfrac{\sin 2ax}{4a}$

9. $\int \cos^2 ax\,dx = \dfrac{x}{2} + \dfrac{\sin 2ax}{4a}$

10. $\int \cos ax\sin ax\,dx = \dfrac{\sin^2 ax}{2a}$

11. $\int x\sin ax\,dx = \dfrac{\sin ax - ax\cos ax}{a^2}$

12. $\int x \cos ax \; dx = \dfrac{\cos ax + ax \sin ax}{a^2}$

13. $\int \sin ax \sin bx \; dx = \dfrac{\sin (a-b)x}{2(a-b)} - \dfrac{\sin (a+b)x}{2(a+b)}$ when $b^2 \neq a^2$

14. $\int \cos ax \cos bx \; dx = \dfrac{\sin (a-b)x}{2(a-b)} + \dfrac{\sin (a+b)x}{2(a+b)}$ when $b^2 \neq a^2$

15. $\int \sin ax \cos bx \; dx = -\dfrac{\cos (a-b)x}{2(a-b)} - \dfrac{\cos (a+b)x}{2(a+b)}$ when $b^2 \neq a^2$

16. $\int e^{ax} \, dx = \dfrac{1}{a} e^{ax}$

17. $\int x \, e^{ax} \, dx = \dfrac{ax - 1}{a^2} e^{ax}$

18. $\int e^{ax} \sin bx \; dx = \dfrac{e^{ax}(a \sin bx - b \cos bx)}{a^2 + b^2}$

19. $\int e^{ax} \cos bx \; dx = \dfrac{e^{ax}(a \cos bx + b \sin bx)}{a^2 + b^2}$

Standard Resistor Color Code

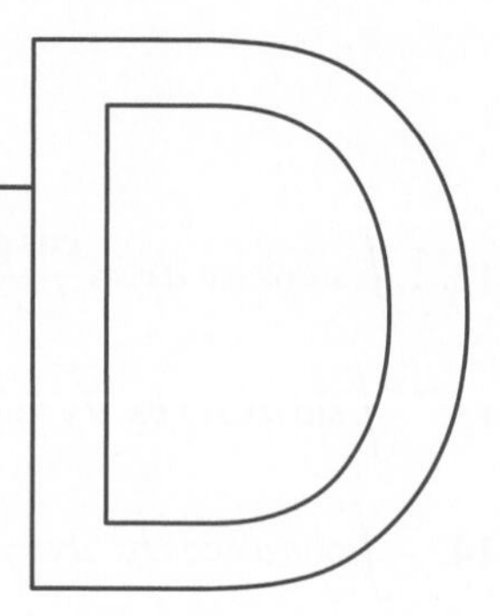

Low-power resistors have a standard set of values. Color-band codes indicate the resistance value as well as a tolerance. The most common types of resistors are the carbon composition and carbon film resistors.

The color code for the resistor value uses two digits and a multiplier digit, in that order, as shown in Figure D.1. A fourth band designates the tolerance. Standard values for the first two digits are listed in Table D.1.

The resistance of a resistor with the four bands of color may be written as

$$R = (a \times 10 + b)m \pm \text{tolerance}$$

where a and b are the values of the first and second bands, respectively, and m is a multiplier. These resistance values are for 2 percent and 5 percent tolerance resistors, as listed in Tables D.1. The color code is listed in Table D.2. The multiplier and tolerance color codes are listed in Tables D.3 and D.4, respectively. Consider a resistor with the four bands, yellow, violet, orange, and gold. We write the resistance as

$$\begin{aligned} R &= (4 \times 10 + 7)\,\text{k}\Omega \pm 5\% \\ &= 47\,\text{k}\Omega \pm 5\% \end{aligned}$$

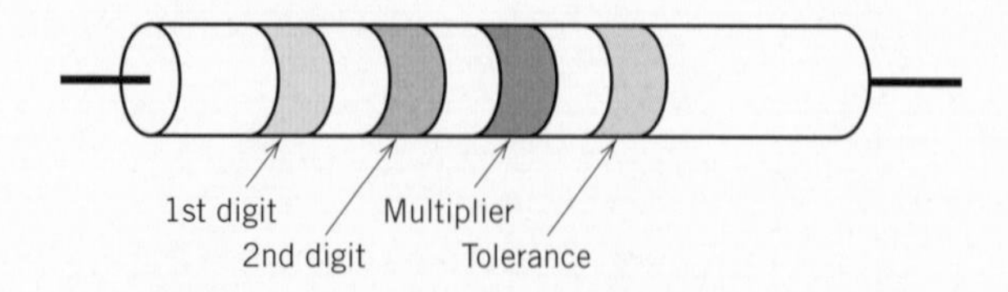

FIGURE D.1 Resistor with four color bands.

Table D.1 Standard Values for First Two Digits for 2 Percent and 5 Percent Tolerance Resistors

10	16	27	43	68
11	18	30	47	75
12	20	33	51	82
13	22	36	56	91
15	24	39	62	100

Table D.2 **Color Code**

0	black
1	brown
2	red
3	orange
4	yellow
5	green
6	blue
7	violet
8	gray
9	white

Table D-3 **Multiplier Color Code**

silver	0.01
gold	0.1
black	1
brown	10
red	100
orange	1 k
yellow	10 k
green	100 k
blue	1 M
violet	10 M
gray	100 M

Table D-4 **Tolerance Band Code**

red	2%
gold	5%
silver	10%
none	20%

References

Adler, Jerry, ''Another Bright Idea,'' *Newsweek,* June 15, 1992, p. 67.

Albean, D. L., ''Single Pot Swings Amplifier Gain Positive or Negative,'' *Electronic Design,* January 1997, p. 153.

Barnes, R., and Wong, K. T., ''Unbalanced and Harmonic Studies for the Channel Tunnel Railway System,'' *IEE Proceedings,* March 1991, pp. 41–50.

Bernstein, Theodore, ''Electrical Shock Hazards,'' *IEEE Transactions on Education,* August 1991, pp. 216–222.

Brown, S. F., ''Predicting Earthquakes,'' *Popular Science,* June 1989, pp. 124–125.

Butterworth, S. ''On the Theory of Filters,'' *Wireless World,* Vol. 7, October 1930, pp. 536–541.

Coltman, John W., ''The Transformer,'' *Scientific American,* January 1988, pp. 86–95.

Doebelin, E. O., *Measurement Systems,* McGraw-Hill, New York, 1966.

Dordick, Herbert S., *Understanding Modern Telecommunications,* McGraw-Hill, New York, 1986.

Dorf, Richard, *The Electrical Engineering Handbook,* CRC Press, 1988.

Dorf, Richard C., *Technology, Society and Man,* Boyd and Fraser, San Francisco, 1974.

Edelson, Edward, ''Solar Cell Update,'' *Popular Science,* June 1992, pp. 95–99.

Gardner, Dana, ''The Walking Piano,'' *Design News,* December 11, 1988, pp. 60–65.

Garnett, G. H., ''A High-Resolution, Multichannel Digital-to-Analog Converter,'' *Hewlett-Packard Journal,* February 1992, pp. 48–52.

Graeme, J., ''Active Potentiometer Tunes Common-Mode Rejection,'' *Electronics,* June 1982, p. 119.

Graham, Dunstan, *Analysis of Nonlinear Control Systems,* Dover Publishing, New York, 1971.

Halliday, D., Resnick, R. and Walker, J., *Fundamentals of Physics*, John Wiley and Sons, New York, 2001.

Hanselman, D., and Littlefield, B., *Mastering MATLAB®*, Prentice Hall, Upper Saddle River, NJ, 2005.

Jurgen, Ronald, ''Electronic Handgun Trigger Proposed,'' *IEEE Institute,* February 1989, p. 5.

Lamarre, Leslie, ''Problems with Power Quality,'' *EPRI Journal,* August 1991, pp. 14–23.

Lenz, James E., ''A Review of Magnetic Sensors,'' *Proceedings of the IEEE,* June 1990, pp. 973–989.

Lewis, Raymond, ''A Compensated Accelerometer,'' *IEEE Transactions on Vehicular Technology,* August 1988, pp. 174–178.

Loeb, Gerald E., ''The Functional Replacement of the Ear,'' *Scientific American,* February 1985, pp. 104–108.

Mackay, Lionel, ''Rural Electrification in Nepal,'' *Power Engineering Journal,* September 1990, pp. 223–231.

Mathcad User's Guide, MathSoft Inc., Cambridge, MA, 1991.

McCarty, Lyle H., ''Catheter Clears Coronary Arteries,'' *Design News*, September 23, 1991, pp. 88–92.

McMahon, A. M., *The Making of a Profession: A Century of Electrical Engineering in America,* IEEE Press, New York, 1984.

Nahin, Paul J., ''Oliver Heaviside,'' *Scientific American,* June 1990, pp. 122–129.

Perry, T. S., ''Donald Pederson: The Father of SPICE,'' *IEEE Spectrum*, June 1998.

Sallen, R. P., and Key, E. L., ''A Practical Method of Designing RC Active Filters,'' *IRE Transactions on Circuit Theory,* Vol. CT-2, March 1955, pp. 74–85.

Smith, E. D., ''Electric Shark Barrier,'' *Power Engineering Journal,* July 1991, pp. 167–177.

Svoboda, J. A., ''Elab, A Circuit Analysis Program for Engineering Education,'' *Computer Applications in Engineering Education,* Vol. 5, No. 2, 1997, pp. 135–149.

Svoboda, J. A., *PSpice for Linear Circuits*, John Wiley and Sons, New York, 2007.

Trotter, D. M., ''Capacitors,'' *Scientific American,* Vol. 259, No. 1, 1988, pp. 86–90.

Tuinenga, P. W., SPICE: A Guide to Circuit Simulation & Analysis Using PSpice, Prentice-Hall, Englewood Cliffs, New Jersey, 1988.

Williams, E. R., ''The Electrification of Thunderstorms,'' *Scientific American,* November 1988, pp. 88–99.

Wright, A., ''Construction and Application of Electric Fuses,'' *Power Engineering Journal,* Vol. 4, No. 3, 1990, pp. 141–148.

Index

2D gel electrophoresis, 18

abc phase sequence, 560
Active element, 25
Admittance, 434, 693
Admittance parameters, 833
Alternating current (ac), 3
Ammeter, 31
Ampere, 5
Amplifier, 143, 189
Amplitude spectrum, 754
Amplitude-phase Fourier series, 736
Analog-to-digital converter (ADC), 239
Anti-aliasing filter, 817
Asymptotic Bode plot, 609
Average power, 499
  three phase-circuit, 579

Balanced three-phase circuits, 576
Balanced three-phase load, 562
Balanced three-phase source, 560
Band-pass filter, 794
Bandwidth, 625
Bell, Alexander Graham, 607
Block diagram, 223, 224, 281
Bode, H.W., 606
Bode plot, 606
  asymptotic, 609
  complex poles, 617
Break frequency, 610
Bridge, 189, 214
Buffer amplifier, 217
Butterworth transfer function, 795

Capacitor, 258, 314, 347
  complex-frequency domain, 682
  dc circuit, 279
  element equation, 293
  two-port networks, 842
Cascade, 805, 808
CCCS, 35
CCVS, 35
Characteristic equation, 374
Characteristic roots, 375
Charge, 2
Circuit drawing, 54
Coaxial cable, 349
Color-code probes, 32, 33
Column vector, 862
Complete response, 313, 337, 386
  first-order circuits, 314
  switched ac circuits, 457
Complex arithmetic, 867
  MATLAB, 869
Complex exponential, 422
Complex frequency, 394, 662
Complex number, 866, 868
  conjugate, 867
  MATLAB, 868
  polar form, 866
  rectangular form, 866
Complex plane, 393
Complex poles, 674
  MATLAB, 705

Complex-frequency domain, 684
Conductance, 26, 436
Conservation of complex power, 507
Controlled source, 33
Convolution, 695, 774
  MATLAB, 697
Corner frequency, 609, 610
Coulomb, 2
Coupled coils, 523
Coupled inductors, 523
  dot convention, 524
  element equation, 541
Coupling coefficient, 526
Cramer's rule, 865
Critically damped, 377, 394
  natural response, 379
Current, 2
Current divider, 67, 87
  design, 70
  frequency domain, 439
Current source, 29
  nonideal, 163
  parallel, 73
Cutoff frequency, 794

Damped resonant frequency, 380, 397
Damping coefficient, 380
Decibel, 607
Delay, 596, 669
Delta-connected three phase source, 561
Delta-Y transformation, 571
Dependent source, 33, 35
  gain, 34
  node equations, 120
  power, 36
Design Example
  ac circuit with op amp, 471
  adjustable voltage source, 84
  airbag igniter, 397
  anti-aliasing filter, 817
  computer and printer, 349
  dc power supply, 781
  integrator and switch, 290
  jet valve controller, 14
  maximum power transfer, 538
  potentiometer angle display, 143
  power factor correction, 587
  radio tuner, 640
  space shuttle cargo door, 710
  strain gauge bridge, 189
  temperature sensor, 42
  transducer interface circuit, 239
  transistor amplifier, 846
Determinant, 865
Device, 2
Dielectric constant, 258
Difference amplifier, 213, 217
Differential equation, 315
  direct method, 369, 370
  first-order circuits, 314
  integrating factor, 335
  Laplace transform, 680
  operator method, 370, 371
  state variable method, 389
Differential operator, 371
Differentiator, 282
Digital signal processing, 817
Dirchlet conditions, 732
Direct current (dc), 3
Dot convention, 524

Effective value, 501
EFS, 750
Electric field, 258, 264
Element, 2
Element equation (constitutive equation), 20
  capacitor, 293
  complex-frequency domain, 684
  coupled inductors, 541
  frequency domain, 433
  ideal transformers, 541
  inductors, 293
Energy, 7
  stored in a capacitor, 264
  stored in an inductor, 274
  stored in coupled coils, 526

Equivalent circuit, 73
coupled inductors, 525
frequency-dependent op amp, 630
hybrid parameters, 837
ideal transformer, 532
inverse hybrid parameters, 837
per-phase, 564, 576
Equivalent resistance, 78, 87
parallel resistors, 67
series resistors, 63
Euler's formula, 868
Even function, 739
Exponential Fourier series, 747
MATLAB, 749

Farad, 258
Faraday, Michael, 258
FFT, 750
Filter, 794
Filter circuits, 793
PSpice, 811
Final value theorem, 678, 708
First-order circuit, 312, 314
network functions, 605
summary, 352
First-order filters, 808
Forced response, 313, 336, 339, 382
Fourier, Jean-Baptise-Joseph, 731
Fourier series, 731
exponential, 746, 748
full-wave rectified cosine, 732
MATLAB, 734
PSpice, 761
triangle waveform, 741
trigonometric, 731, 744
Fourier spectrum, 754, 773
MATLAB, 757
Fourier transform, 766
Laplace transform, 777
properties, 769, 770
table, 771
Franklin, Benjamin, 2
Frequency, 416
Frequency domain, 427
table, 433
Frequency response, 599
PSpice, 634
Frequency scaling, 796
Fundamental frequency, 731

Gain, 220, 594
Gain-bandwidth-product, 632
Ground node, 110, 209
Guidelines for labeling circuit variables, 81

h pararmeters, 845, 846
Harmonics, 731
Heaviside, Oliver, 374
Henry, Joseph, 270
Hertz, 5, 416
Hertz, Heinrich, 416
High-order filters, 805
High-pass filter, 794
Homogeneity, 21
"How can we check . . . ?", 13
ac analysis, 469
AC power, 536
balanced three phase circuits, 584
band-pass filter, 815
capacitor voltage and current, 289
complex arithmetic, 469
first-order circuits, 345
Fourier series, 779
frequency response, 636, 815
gain, 637
hybrid parameters, 844
initial and final values, 708
Kirchhoff's laws, 82
mesh currents, 142
node voltages, 140
Ohm's law, 41

"How can we check . . . ?" (*Continued*)
  operational amplifier, 237
  passive convention, 13
  phase shift, 637
  second-order circuits, 394
  Thévenin equivalent, 188
  unbalanced three phase circuits, 585
Hybrid parameters, 837

Ideal filter, 794
Ideal operational amplifier, 210
Ideal source, 29
Ideal transformers, 531
  element equation, 541
  lossless, 533
Impedance, 434, 693
  complex-frequency domain, 681
Impedance parameters, 833
Impluse function, 667, 693
Independent source, 29
Inductor, 269, 315, 346
  complex-frequency domain, 683
  dc circuit, 278
  element equation, 293
Initial condition, 316, 681
  capacitor, 259
  inductor, 270
  second-order circuits, 377
  switched dc circuits, 277
Initial value theorem, 677, 708
Input and output impedance, 807
Instantaneous power, 498
  three phase circuit, 578
Integrator, 282, 290
Interactive Example, xi
  AC power, 516
  Bode plot, 613, 615
  capacitors, 261, 262
  complete response, 685, 687
  coupled inductors, 528, 529, 530
  equivalent resistance, 75
  inductors, 272
  Kirchhoff's laws, 57
  Kirchhoff's and Ohm's laws, 59, 60
  mesh equations, 132
  network functions, 600, 601, 602
  op amp circuit, 215, 216
  parallel resistors, 69
  RC circuit, 317, 319, 334
  RL circuit, 318, 320, 332
  series resistors, 64
  simple AC circuits, 440, 441, 442
  Thévenin equivalent circuit, 172, 174
  transformers, 534, 535
Inverse Fourier transform, 767
Inverse hybrid parameters, 837
Inverse Laplace transform, 662, 671
Inverting amplifier, 217, 220, 229, 240
  in frequency domain, 456

Joule, 5

Kilo, 6
Kirchhoff, Gustav Robert, 56
Kirchhoff's Current Law (KCL), 56, 83
  in frequency domain, 438
Kirchhoff's Laws, 53
Kirchhoff's Voltage Law (KVL), 57, 83
  in frequency domain, 438

Laplace, Pierre-Simon, 662
Laplace transform, 661
  properties, 665
  table, 665
Line current, 573
Line losses, 567
Linear element, 21
Line-to-line voltage, 561
Loading, 218, 805
Loop, 57, 122
Low-pass filter, 794

Magnetic field, 269, 274
MathCad, 82, 238, 779
  Kirchhoff's Laws, 82
  Fourier series, 779
  simultaneous equations, 238